DATE DUE

JUL 24 '98	
GAYLORD	PRINTED IN U.S.A.

A CATALOG OF THE DIPTERA OF AMERICA NORTH OF MEXICO

Prepared cooperatively by
specialists on the various groups of Diptera
under the direction of

ALAN STONE, CURTIS W. SABROSKY, WILLIS W. WIRTH, RICHARD H. FOOTE, AND JACK R. COULSON

Agricultural Research Service
UNITED STATES DEPARTMENT OF AGRICULTURE

Smithsonian Institution Press
Washington, DC
1983

Copyright © 1983 Smithsonian Institution. All rights reserved.
Reprint, 1983.

Library of Congress Cataloging in Publication Data
Main entry under title:

A Catalog of the Diptera of America north of Mexico.

  Reprint. Originally published: Washington: Supt. of
Docs., U.S. G.P.O., 1965. (Agriculture handbook; no. 276)
  Bibliography: p.
  Includes index.
  Supt. of Docs. no.: SI 1.2:D62
  1. Diptera—North America. 2. Insects—North America.
I. Stone, Alan, 1904-   . II. United States.
Agricultural Research Service. III. Series: Agriculture
handbook (United States. Dept. of Agriculture); no. 276.
QL535.1.A1C37   1983   595.77'0973   83-600004
ISBN 0-87474-890-9

CONTENTS

	Page
Introduction	1
General References	2
Classification	3
Explanatory Information	6
New Names Proposed in This Catalog	11
Acknowledgments	13
Order Diptera	14
Suborder Nematocera	15
Superfamily Tipuloidea	15
Superfamily Psychodoidea	90
Superfamily Culicoidea	98
Superfamily Anisopodoidea	190
Superfamily Bibionoidea	191
Superfamily Mycetophiloidea	196
Suborder Brachycera	296
Superfamily Tabanoidea	296
Superfamily Asiloidea	348
Superfamily Empidoidea	446
Suborder Cyclorrhapha	530
Division Aschiza	530
Superfamily Lonchopteroidea	530
Superfamily Phoroidea	531
Superfamily Syrphoidea	550
Division Schizophora	633
Section Acalyptratae	633
Superfamily Micropezoidea	633
Superfamily Nothyboidea	637
Superfamily Tephritoidea	641
Superfamily Sciomyzoidea	678
Superfamily Lauxanioidea	695
Superfamily Pallopteroidea	710
Superfamily Milichioidea	718
Superfamily Drosophiloidea	734
Superfamily Chloropoidea	773
Unplaced Families of Acalyptratae	793
Acalyptrate Genera of Uncertain Family Position	825
Section Calyptratae	826
Superfamily Muscoidea	826
Superfamily Oestroidea	922

Order Diptera—Continued
  Suborder Cyclorrhapha—Continued
    Division Schizophora—Continued

	Page
Unplaced Genus of Diptera	1113
Unplaced Species of Diptera	1113
Nomina Nuda	1113
Selected Bibliography of North American Diptera	1117
Periodicals Cited, with Explanations	1485
Index	1549

Introduction

The first catalog of the Diptera of North America was published slightly over a century ago by Osten Sacken in 1858, and a second edition by the same author in 1878. Carl Robert Osten Sacken (1828–1906), a Russian diplomat and dipterist, interested himself greatly in the American fauna and arranged for the first major, comprehensive study of that fauna by the German master dipterist Hermann Loew (1807–1878). Loew's ten Centuriae, "Diptera Americae septentrionalis indigena" (1861b–1872a), presented descriptions of 1,000 new species and many new genera. This series, Loew's "Monographs of North American Diptera" (1862c–1873b) and that authored by Osten Sacken (1869b), and the catalogs by the latter, laid a strong foundation for the taxonomy of Diptera on the North American continent. The tradition of comprehensive treatment was continued by two other outstanding dipterists, Samuel Wendell Williston (1852–1918), whose three editions of a "Manual of North American Diptera" (1888a, 1896a, 1908) stimulated and authoritatively influenced several generations of dipterists, and John Merton Aldrich (1866–1934), whose "Catalogue of the Diptera of North America" (1905) is the most recent predecessor of the present work.

Thus the most recent catalog of Nearctic Diptera is now nearly 60 years old. During this period many taxa within the order have been revised extensively and many new names proposed. The need for an up-to-date guide to the literature has long been apparent. In accordance with the interest expressed by many entomologists, the dipterists of the Entomology Research Division, U.S. Department of Agriculture, have prepared, in cooperation with other specialists, this synoptic catalog of the Diptera of America north of Mexico.

In scope, this catalog is a compromise between a checklist on one extreme and a complete work, listing all occurrences of every name in the literature, on the other. The former would be of limited value because of its brevity and the latter too costly, cumbersome, and perhaps noncritical. The most important functions of a catalog are to list all published names with a reference to the original publication of each, to distinguish between valid and synonymous names, to present as sound a classification as possible, and to give an indication of the distribution of the species. A catalog should also serve as a guide to important revisionary works, keys, and significant accessory information. Beyond this limit much could be added, but details can better be left for revisions and monographs of families or other subordinate groups.

The scope of this catalog is as follows: 1. The zoological group treated is the Order Diptera, or two-winged flies, which includes the midges, gnats, and mosquitoes, as well as the insects more commonly regarded as flies. Species known only as fossils are not included. 2. The geographical area treated is North America north of Mexico, including Greenland, Bermuda, and the California Channel Islands, but excluding the Bahama Islands and Commander Islands. 3. It is believed that all pertinent names proposed from 1758 through 1962 are included as well as some 1963 names.

This catalog recognizes 1,971 valid genera and 16,130 valid species, distributed among 105 families. In addition, 374 nontypical subgenera, 236 non-typical subspecies, and 146 varieties are recognized. There are nearly 1,100 generic synonyms and over 3200 specific synonyms, not counting nearly a thousand emendations, errors, and misidentifications. In all, the catalog contains over 25,000 entries.

In the number of recognized taxa, this catalog represents a doubling of genera and a tripling of species since the last previous catalog (Aldrich, 1905). The latter covered a larger area, comprising North America in the widest sense, south to and including Panama, the West Indies, and Trinidad, but a count limited to the area of our present catalog, for direct comparison, showed 952 genera and 5,432 species recognized as of January 1, 1904.

General References

The student of Diptera requires more than a knowledge of species descriptions for an introduction to his subject. To provide a guide to the comprehensive publications that will be of assistance, we have presented for most of the families and some of the lower categories important references to keys, revisions, and works on biology and anatomy. These references are all annotated and unless otherwise stated the annotations pertain to the Nearctic Region only. At the beginning of the catalog proper we have listed the important sources of information pertaining to the Order Diptera as a whole. Most of these sources have bibliographies that will lead the reader to further works.

Certain works that do treat the order wholly or in large part are nevertheless excluded as general references for various reasons. For example the major works of Fabricius, Meigen, Macquart, Say, Osten Sacken, Loew, and Bigot are primarily descriptive in nature and are cited repeatedly under the new species described. Parts of composite series such as Wytsman's "Genera Insectorum," Lindner's "Die

Fliegen der Palaearktischen Region," the "Handbooks of British Insects," the Diptera of Connecticut series, and the California Insect Survey series have been cited at the family level only when relevant or of sufficient value for our region. The present catalog supplants, for the Nearctic Region, the catalogs of Osten Sacken (1858, 1878c), Kertész (1902–1910), Becker et al. (1903–1907), and the many volumes of the "Genera Insectorum," although these are often of historical interest. The information found in Coquillett (1910b) on the type-species of genera is either incorporated or corrected in this catalog. Also excluded are numerous faunal lists, some leading examples being those of Leonard (1928) for New York State, Johnson (1925b) for New England, Brimley (1938) for North Carolina, and Petch and Maltais (1932) for Quebec, but some of these are of doubtful value because the determinations are often questionable or the nomenclature out of date.

Classification

Time has not permitted a thorough study of the suprageneric classification, and this catalog is not a formal contribution to that field. Nevertheless, it was necessary to adopt some arrangement and to make decisions regarding the position of some families and other taxa about which dipterists have differed greatly. After weighing the evidence and opinions of modern authorities, we have adopted the present system. The arrangement of the families within superfamilies follows, with few exceptions, that of Hennig (1954, 1958a).

In the past, two suborders have usually been recognized, either the Nematocera and Brachycera, or the Orthorrhapha and Cyclorrhapha. Neither of these divisions is entirely satisfactory and as a compromise we have recognized three, the Nematocera, Brachycera, and Cyclorrhapha. Since the Suborder Pupipara has long been considered to have no phylogenetic validity, it has been eliminated.

The following list presents the arrangement of the families we have adopted, each with the number of valid genera and valid species in North America north of Mexico.

	Genera	Species
Suborder Nematocera	(417)	(5,014)
Superfamily Tipuloidea	(62)	(1,485)
Family Trichoceridae	3	27
Family Tipulidae	59	1,458
Superfamily Psychodoidea	(17)	(107)
Family Tanyderidae	2	4
Family Psychodidae	12	87
Family Ptychopteridae	3	16
Family Nymphomyiidae	(1 unnamed)	(1 unnamed)

A CATALOG OF DIPTERA OF NORTH AMERICA

	Genera	Species
Superfamily Culicoidea	(121)	(1,300)
Family Blephariceridae	5	22
Family Deuterophlebiidae	1	4
Family Dixidae	1	41
Family Chaoboridae	4	15
Family Culicidae	12	148
Family Thaumaleidae	2	5
Family Ceratopogonidae	27	348
Family Chironomidae	63	601
Family Simuliidae	6	116
Superfamily Anisopodoidea	(3)	(8)
Family Anisopodidae	3	8
Superfamily Bibionoidea	(8)	(80)
Family Bibionidae	6	78
Family Pachyneuridae	2	2
Superfamily Mycetophiloidea	(206)	(2,034)
Family Mycetophilidae	53	612
Family Sciaridae	16	155
Family Hyperoscelididae	1	1
Family Scatopsidae	11	61
Family Cecidomyiidae	125	1,205
Suborder Brachycera	(356)	(4,408)
Superfamily Tabanoidea	(81)	(672)
Family Xylophagidae	7	26
Family Xylomyidae	2	9
Family Stratiomyidae	36	234
Family Pelecorhynchidae	1	1
Family Tabanidae	25	295
Family Rhagionidae	9	104
Family Hilarimorphidae	1	3
Superfamily Asiloidea	(170)	(1,885)
Family Therevidae	13	130
Family Scenopinidae	6	21
Family Apioceridae	2	29
Family Mydidae	6	39
Family Asilidae	82	856
Family Nemestrinidae	3	6
Family Acroceridae	7	59
Family Bombyliidae	51	745
Superfamily Empidoidea	(105)	(1,851)
Family Empididae	61	710
Family Dolichopodidae	44	1,141
Suborder Cyclorrhapha	(1,197)	(6,702)
Division Aschiza	(153)	(1,408)
Superfamily Lonchopteroidea	(1)	(4)
Family Lonchopteridae	1	4
Superfamily Phoroidea	(47)	(293)
Family Phoridae	38	226
Family Platypezidae	9	67
Superfamily Syrphoidea	(105)	(1,111)
Family Pipunculidae	8	105
Family Syrphidae	88	939

INTRODUCTION

	Genera	Species
Family Conopidae	9	67
Division Schizophora	(1,044)	(5,294)
Section Acalyptratae	(423)	(2,351)
Superfamily Micropezoidea	(10)	(33)
Family Micropezidae	8	31
Family Neriidae	2	2
Superfamily Nothyboidea	(6)	(35)
Family Diopsidae	1	1
Family Psilidae	4	32
Family Tanypezidae	1	2
Superfamily Tephritoidea	(102)	(413)
Family Richardiidae	4	7
Family Otitidae	38	127
Family Platystomatidae	4	41
Family Pyrgotidae	3	5
Family Tephritidae	53	233
Superfamily Sciomyzoidea	(35)	(191)
Family Helcomyzidae	2	3
Family Ropalomeridae	1	1
Family Coelopidae	2	5
Family Dryomyzidae	2	8
Family Sepsidae	8	34
Family Sciomyzidae	20	140
Superfamily Lauxanioidea	(31)	(174)
Family Lauxaniidae	23	135
Family Chamaemyiidae	7	36
Family Periscelididae	1	3
Superfamily Pallopteroidea	(9)	(81)
Family Piophilidae	1	31
Family Thyreophoridae	1	2
Family Neottiophilidae	1	1
Family Pallopteridae	1	9
Family Lonchaeidae	5	38
Superfamily Milichioidea	(27)	(197)
Family Sphaeroceridae	3	117
Family Braulidae	1	1
Family Tethinidae	5	22
Family Milichiidae	15	52
Family Canaceidae	3	5
Superfamily Drosophiloidea	(86)	(534)
Family Ephydridae	65	347
Family Curtonotidae	1	1
Family Drosophilidae	17	179
Family Diastatidae	2	6
Family Camillidae	1	1
Superfamily Chloropoidea	(42)	(264)
Family Chloropidae	42	264
Unplaced Families of Acalyptratae	(73)	(426)
Family Odiniidae	3	11
Family Agromyzidae	16	189
Family Clusiidae	5	27
Family Acartophthalmidae	1	2

	Genera	Species
Family Heleomyzidae	23	113
Family Trixoscelididae	5	30
Family Rhinotoridae	1	1
Family Anthomyzidae	5	14
Family Opomyzidae	3	13
Family Chyromyidae	3	9
Family Aulacigastridae	1	1
Family Asteiidae	6	18
Family Cryptochetidae	1	1
Acalyptrate Genera of Uncertain Family Position	2	3
Section Calyptratae	(621)	(2,943)
Superfamily Muscoidea	(131)	(1,220)
Family Anthomyiidae	56	556
Family Muscidae	58	622
Family Gasterophilidae	1	4
Family Hippoboscidae	12	28
Family Streblidae	3	5
Family Nycteribiidae	1	5
Superfamily Oestroidea	(490)	(1,723)
Family Calliphoridae	23	78
Family Sarcophagidae	48	327
Family Tachinidae	414	1,281
Family Cuterebridae	1	26
Family Oestridae	4	11
Unplaced Genus and Species of Diptera	1	6
Totals	1,971	16,130

Explanatory Information

Some remarks on the format, arrangement, and treatment of certain other subjects are presented here. Valid names in the catalog proper (left hand margin) are printed in bold face type; cataloged synonymous names are italicized for both the genus- and species-groups. A separate catalog of names of the family-group is being prepared for publication by Sabrosky; hence, no authorship or references are given for these herein. Throughout the work the references are given by author, date, and page (rarely plate or figure) which will lead the experienced specialist quickly to the original paper. The full reference needed for library use can be found in the bibliography. A complete explanation concerning the format and use of the bibliography is found in its introduction.

Names of the genus-group are usually arranged phylogenetically, although for a few families the authors have preferred an alphabetical arrangement. Citations for the typical subgenus follow the generic rather than the subgeneric centerhead. Synonyms are listed chronologically. Except for errors or emendations, each name, whether valid or synonymous, is followed by the author, date, page, name of type-species, nature of type fixation, and the present valid name of the type-species if there has been a change. For type-species, the original combination is cited. The generic name is omitted if the same as that of the genus for which it is the type. If the type-species is now a junior synonym, the present valid name is given as mentioned above, but its generic name is omitted, it being understood to be the genus under which the species now stands.

Examples (omitted generic names noted in brackets):

Genus VILLA Lioy

Villa Lioy, 1864a: 732. Type-species, *Anthrax concinnus* Meigen (Coquillett, 1910b: 619) = [*Villa*] *abbadon* (Fabricius).

Anthrax, group *Hyalanthrax* Osten Sacken, 1887a: 134. Type-species [*Anthrax*] *faustinus* Osten Sacken (Coquillett, 1910b: 553).

Subgenus HEMIPENTHES Loew

Hemipenthes Loew, 1869b: 28 (Cent. 8, no. 44) (as genus). Type-species, *Musca morio* Linnaeus (Coquillett, 1910b: 550).

Isopenthes Osten Sacken, 1886a: 96. Type-species [*Isopenthes*] *jaennickeana* Osten Sacken (Coquillett, 1910b: 556) = [*Villa*] *sinuosa* (Wiedemann) ssp.

Valid names of the species-group are arranged alphabetically following the generic or subgeneric centerhead, and the synonyms are indented and listed chronologically. Errors and emendations are listed immediately after the name misspelled or emended. Misidentifications are listed last, in alphabetical order. Parentheses, indicating a shift from the original generic combination, are placed around the author's name only for the valid names, not the synonyms. If there has been a generic name change or a change in status, the original generic name and/or status is given in parentheses following the reference. In some instances double references have been used, as explained in the introduction to the bibliography, for certain works that have been published twice, usually with different pagination.

Names below the species level are cited as subspecies or varieties and are arranged alphabetically, each followed by any appropriate synonymy. We recognize that there is great diversity of opinion as to the zoological concepts below the species level but we have accepted each contributor's decision as to which categorical term should be used.

The distributional information is usually presented only to the level of State, Province, or Country. The first item is the type locality (or localities in instances of allopatric syntypes) followed by a semicolon. If the species is confined to a few States or Provinces, these are listed separately. If more widespread, an overall distribution is given by stating the corners of a rectangle or triangle. Distribution is stated almost always in a northwest to northeast and southwest to southeast pattern. Sometimes there may be a combination of these two methods, with an area of general distribution and outlying collection points. Alaska is treated as a separate unit, apart from northern or western United States, and the latter does not include Hawaii. "N.W.T." designates the Canadian province, but "Northwest Territory (U.S.)" means the historical territory of the north-central United States. If questionable records are included, these are preceded by a question mark and are placed after the accepted records. Exotic distribution follows Nearctic distribution.

In the synonymic lists, generic names based on exotic species are not included unless the names have come into significant usage in the Nearctic literature. Generic synonyms based on species described from North America are included, however, even though the names have not since been used in North America. Some spelling errors and emendations are included, but these are confined to those occurring frequently in the literature or those found in standard, frequently consulted works. Citations to these are not given if they can be found in Neave's "Nomenclator Zoologicus," (1939–1950).

Misidentifications are treated in two different ways. If the name of an exotic species has been erroneously applied to only one North American species, it is listed in the synonymy of that species; for example, "*punctatus*, authors, not Fabricius." A reference to the misuse may be given if it has been used only once but in a widely used publication. If the name has been misapplied to several valid North American species, or if the identity of the North American species apparently misnamed is unknown, then the name is cited in the proper alphabetical order in italics without citation, followed by the notation "not Nearctic."

Numerous names of dubious identity occur in the literature. Rather than place these in dubious synonymy preceded by a question mark, we have listed those for which the generic position is reasonably certain in the regular alphabetical order under the genus with the notation "Unrecognized."

Names of uncertain generic position are placed separately at the end of the applicable tribe, subfamily, or family, or if of uncertain subgeneric position, at the end of the genus. *Nomina nuda* are placed in synonymy if the application is known; at the end of the catalog proper if not known.

Certain general matters of nomenclature are briefly noted below:

1. Meigen, 1800. The long controversy over the numerous generic names of Meigen (1800) and those of Meigen (1803) perpetuated conflicting usage of names for some common and widely-studied

genera and families such as *Tendipes* (Tendipedidae) vs. *Chironomus* (Chironomidae). In accordance with an Opinion (I.C.Z.N., 1963b), under Suspension of the Rules by the International Commission on Zoological Nomenclature, we have placed the generic names of Meigen (1800) in synonymy.

2. Type designation. This has been strictly construed, as required by the International Code. On one point, however, our collaborators may not have been consistent. If a new genus is "erected for" or "based upon" a certain species, without use of "type-species," "genotype," or similar term, we construed that technically the author did not designate the type-species in the meaning of the Code. Presumption of intent to do so is not necessarily correct, especially for authors who did not usually designate type-species and who were merely indicating the monobasic nature of the new genus. In most instances, the genera involved are monobasic, and the point is only academic. The term "Gen. et sp. nov." or its equivalent, before 1931, is recorded as original designation, as provided by the Code.

3. Misidentified type-species. A number of genera and subgenera were based on type-species later found to be misidentified. Usually, in the interest of stability and zoological continuity, it is desirable to have the genus based on the zoologically correct species rather than the misidentified one. Under the International Code of Zoological Nomenclature, such cases must be submitted to the International Commission. Our format for such instances shows the originally included nominal species that was named as type, the type-fixation, the abbreviation "misident.," and the actual species involved, here accepted as the type-species. Example: The genus *Corethra*: "Type-species, *Tipula culiciformis* De Geer (mon.; misident.) = *lateralis* Meigen." To conform to nomenclatural requirements, application should be made to the International Commission, but time did not permit securing the necessary decisions for this catalog.

4. Retention of nomenclaturally unacceptable names. For a few taxa, specialists have elected to maintain names that are invalid or unavailable under the International Code. In such instances, authors or the editors have pointed out the name that is correct under the Code. Use of the incorrect name, if judged desirable, would require suspension of the Rules by the International Commission under its plenary powers.

5. Gender. A special effort has been made to use the appropriate gender for each generic name (new Code, Article 30a; Stoll et al., 1961). Following this rule sometimes required a change in the terminations habitually applied to specific names. For example, generic names ending in *-odes* and *-oides* are treated as masculine, and those ending in *-stoma*, *-soma*, *-stigma*, and *-ceras* as neuter. Latinized

forms with changed terminations take the gender appropriate to their endings; hence, names ending in *-stomus, -somus* or *-cerus* are masculine, and those ending in *-cera* are feminine.

Names ending in *-ops*. There has been great divergence of opinion even within the same order, family, or genus, as to whether names of the genus-group ending in *-ops* are masculine or feminine. We have treated them as masculine throughout this catalog. In some instances this may be at variance with the decision of the International Commission in Declaration 36, but we consider our treatment fully justified as promoting uniformity and as a mnemonic aid, in addition to being classically correct. According to Prof. L. W. Grensted, Classical Adviser to the International Commission on Zoological Nomenclature, there is no classical use of the feminine gender for names ending in *-ops*.

The abbreviations used in the catalog and bibliography are as follows:

General

abt.—Abteilung
anat.—anatomy
art.—article
aut.—automatic
beih.—Beiheft
bibl.—bibliography
biol.—biology
cat.—catalog
cent.—central
classif.—classification
comb.—combination
concl.—conclusion
cont.—continued, continuation
des.—designation
descr.—description
div.—division
e.—east, eastern
ed.—edition, editor
emend.—emendation, emended
estab.—established
fasc.—fascicle
fig., figs.—figure, figures
h.—Heft
I., Is.—Island, Islands
I.C.Z.N.—International Commission on Zoological Nomenclature
ident.—identify, identified, identification
introd.—introduced
lfg.—Lieferung
livr.—livraison
misident.—misidentified

mon.—monotypy, monotype, monobasic
monog.—monograph
mt., mts.—mountain, mountains
n.—north, northern; new
ne.—northeast, northeastern
no.—number
nw.—northwest, northwestern
orig.—original
p., pp.—page(s)
pl., pls.—plate, plates
pt.—part
preocc.—preoccupied
rev.—revision
rpt.—report
s.—south, southern
se.—southeast, southeastern
sec.—secretary
sect.—section
ser.—series
sp., spp.—species
ssp., sspp.—subspecies
sub.—subsequent
subg.—subgenus
sw.—southwest, southwestern
syn.—synonym, synonymy
t.—Teil
tax.—taxonomy, taxonomic
var.—variety
vol.—volume
w.—west, western

Geographical Names

Ala.—Alabama
Alta.—Alberta
Amer. (N., S., Cent.)—America
Ariz.—Arizona
Ark.—Arkansas
Baja Calif.—Baja (Lower) California
B.C.—British Columbia
Calif.—California
Chih.—Chihuahua
Coah.—Coahuila
Colo.—Colorado
Conn.—Connecticut
D.C.—District of Columbia
Del.—Delaware
D.F.—Distrito Federal, Mexico
Fla.—Florida
Ga.—Georgia
Ill.—Illinois
Ind.—Indiana
Kans.—Kansas
Ky.—Kentucky
La.—Louisiana
Labr.—Labrador
Man.—Manitoba
Mass.—Massachusetts
Md.—Maryland
Mich.—Michigan
Minn.—Minnesota
Miss.—Mississippi
Mo.—Missouri
Mont.—Montana
N.B.—New Brunswick
N.C.—North Carolina
N. Dak.—North Dakota

Nebr.—Nebraska
Nev.—Nevada
Nfld.—Newfoundland
N.H.—New Hampshire
N.J.—New Jersey
N. Leon—Nuevo Leon
N. Mex.—New Mexico
N.S.—Nova Scotia
N.W.T.—Northwest Territories
N.Y.—New York
Okla.—Oklahoma
Ont.—Ontario
Oreg.—Oregon
Pa.—Pennsylvania
P.E.I.—Prince Edward Island
P.R.—Puerto Rico
Que.—Quebec
R.I.—Rhode Island
Sask.—Saskatchewan
S.C.—South Carolina
S. Dak.—South Dakota
Son.—Sonora
Tam.—Tamaulipas
Tenn.—Tennessee
Tex.—Texas
Va.—Virginia
V.C.—Vera Cruz
Vt.—Vermont
Wash.—Washington
Wis.—Wisconsin
W. Va.—West Virginia
Wyo.—Wyoming
Y.T.—Yukon Territory

New names proposed in this catalog

Anthomyiidae: *Hylemya angustiventralis* Huckett
H. flavidisquama Huckett
Pegomya intersecta arcticola Huckett
P. rubriceps Huckett
Pleurochaetella Vockeroth

Asilidae: *Asilus carolinae* Martin and Wilcox
Laphria asackeni Wilcox

Bombyliidae: *Bombylius lassenensis* Johnson and Johnson
Phthiria loewi Painter
Villa coquilletti Painter

Cecidomyiidae: *Cecidomyia candidipes* Foote
Rhopalomyia truncula Foote

Chironomidae:	*Anatopynia dena* Roback
	Diamesa garretti Sublette and Sublette
	Pentaneura gesta Roback
Dolichopodidae:	*Diaphorus vanduzeei* Robinson
	Medetera wheeleri Foote, Coulson, and Robinson
	Peloropeodes discolor Robinson
	Tachytrechus greenei Foote, Coulson, and Robinson
Empididae:	*Empis barbatoides* Melander
	E. cometes Steyskal
	E. delumbis Melander
	E. desiderata Melander
	E. longeoblita Steyskal
	E. nodipoplitea Steyskal
	Rhamphomyia currani Steyskal
	R. formosula Melander
	R. soleata Melander
Heleomyzidae:	*Suillia thomsoni* Gill
Muscidae:	*Phaonia fuscana* Huckett
	P. pallidosa Huckett
	Pseudophaonia orichalceoides Huckett
	Quadrularia Huckett
	Spilogona bifimbriata Huckett
	S. spinicostalis Huckett
Mycetophilidae:	*Exechia adamsi* Laffoon
	Leia joculator Laffoon
	Mycetophila alea Laffoon
	Platyura willistoni Laffoon
	Sciophila puta Laffoon
Sciomyzidae:	*Limnia loewi* Steyskal
Sphaeroceridae:	*Olinea* Richards (*Copromyza* subg.)
Syrphidae:	*Baccha loewi* Sedman
	Chrysosyrphus Sedman
	Helophilus colei Wirth
	Melangyna coquilletti Sedman
	Melanostoma willistoni Sedman
	Xylota cascadensis Weems
Tachinidae:	*Atactopsis reinhardi* Sabrosky and Arnaud
	Chaetogaedia townsendi Sabrosky and Arnaud
	CHAETOGAEDIINA (subtribe)
	Cylindromyia propusilla Sabrosky and Arnaud
Therevidae:	*Dialineura willistoni* Cole

Tipulidae: *Tipula stonei* Alexander
Trixoscelididae: *Trixoscelis melanderi* Vockeroth

Acknowledgments

This publication would not have been possible without the cooperation of the taxonomists specializing in the various families. The contributors, including the editorial group, and the portion each prepared are here listed.

Alexander, Charles P.—Anisopodidae, Pachyneuridae, Ptychopteridae, Tanyderidae, Tipulidae, Trichoceridae
Arnaud, Paul H., Jr.—Tachinidae (with Sabrosky)
Bequaert, Joseph C.—Hippoboscidae, Nemestrinidae
Beyer, Erwin—Phoridae (with Schmitz)
Camras, Sidney—Conopidae
Cazier, Mont A.—Apioceridae
Chillcott, J. G.—Cuterebridae, Gasterophilidae, Oestridae
Cole, F. R.—Therevidae
Cook, E. F.—Chaoboridae, Hyperoscelididae, Scatopsidae
Coulson, Jack R.—Dolichopodidae (with Foote and Robinson), Bibliography (with Sabrosky and Muller)
Curran, C. H.—Mydidae
Downes, William L., Jr.—Calliphoridae (Melanomyini), Sarcophagidae
Foote, Richard H.—Cecidomyiidae, Cryptochetidae, Dolichopodidae (with Coulson and Robinson), Rhinotoridae, Ropalomeridae, Tephritidae
Frick, Kenneth E.—Agromyzidae
Gill, Gordon D.—Heleomyzidae
Hall, David G.—Calliphoridae (except Melanomyini)
Hardy, D. Elmo—Bibionidae, Pipunculidae, Scenopinidae
Hubert, Alexander A.—Dixidae
Huckett, H. C.—Anthomyiidae (except Scatophaginae), Muscidae
Ide, F. P.—Nymphomyiidae
James, Maurice T.—Hilarimorphidae, Rhagionidae, Stratiomyidae, Xylomyidae, Xylophagidae
Kennedy, Harry D.—Deuterophlebiidae
Kessel, Edward L.—Platypezidae
Laffoon, Jean L.—Mycetophilidae, Sciaridae (with Stone)
Martin, Charles H.—Asilidae (with Wilcox)
McAlpine, J. F.—Camillidae, Chamaemyiidae, Chyromyidae, Diastatidae, Lonchaeidae
Melander, A. L.—Empididae
Muller, Irmgard—Bibliography (with Coulson and Sabrosky)
Painter, Reginald H. and Elizabeth M.—Bombyliidae
Philip, Cornelius B.—Pelecorhynchidae, Tabanidae
Quate, Laurence W.—Psychodidae
Richards, O. W.—Sphaeroceridae
Robinson, Harold—Dolichopodidae (with Foote and Coulson)
Sabrosky, Curtis W.—Anthomyzidae, Asteiidae, Chloropidae, Diopsidae, Milichiidae, Odiniidae, Periscelididae, Tachinidae (with Arnaud), Bibliography (with Coulson and Muller)
Schlinger, Evert I.—Acroceridae

Schmitz, H.—Phoridae (with Beyer)
Sedman, Yale S.—Syrphidae (with Weems and Wirth)
Shewell, G. E.—Lauxaniidae, Pallopteridae, Psilidae
Steyskal, George C.—Acartophthalmidae, Clusiidae, Dryomyzidae, Helcomyzidae, Micropezidae, Neriidae, Otitidae, Piophilidae, Platystomatidae, Pyrgotidae, Richardiidae, Sciomyzidae, Sepsidae, Tanypezidae, Thyreophoridae
Stone, Alan—Blephariceridae, Braulidae, Culicidae, Sciaridae (with Laffoon), Simuliidae, Thaumaleidae
Sublette, James E. and Mary S.—Chironomidae
Vockeroth, J. R.—Anthomyiidae (Scatophaginae), Coelopidae, Neottiophilidae, Opomyzidae, Tethinidae, Trixoscelididae
Weems, Howard V., Jr.—Syrphidae (with Sedman and Wirth)
Wenzel, Rupert L.—Nycteribiidae, Streblidae
Wheeler, Marshall R.—Drosophilidae
Wilcox, J.—Asilidae (with Martin)
Wirth, Willis W.—Aulacigastridae, Canaceidae, Ceratopogonidae, Curtonotidae, Ephydridae, Lonchopteridae, Syrphidae (with Sedman and Weems)

We gratefully acknowledge the valuable information supplied by the following dipterists and bibliographers: Max Beier, David F. Beneway, Thomas Borgmeier, George W. Byers, B. J. Clifton, R. W. Crosskey, P. J. Darlington, Jr., R. Defretin, A. Diakanoff, H. Rodney Dodge, Philip B. Dowden, Claude Dupuis, Paul Freeman, J. G. Franclemont, George Goodwin, L. W. Grensted, B. Herting, Hugh B. Leech, Franklin B. Lewis, B. Mannheims, B. M. McGugan, L. P. Mesnil, C. D. Michener, W. R. Nowell, H. Oldroyd, L. L. Pechuman, H. J. Reinhard, S. S. Roback, H. Sachtleben, W. R. Thompson, J. van der Vecht, and Floyd Werner. We also wish to thank the personnel of the libraries of the Smithsonian Institution and the U.S. Department of Agriculture for their valuable assistance throughout the period of the preparation of this catalog.

Order DIPTERA

REFERENCES: General. Brues, Melander, and Carpenter, 1954; Comstock, 1940; Curran, 1934a, 1942b; Essig, 1958; Hendel, 1928b; Imms, 1957; James, 1948; Lindner, 1925–1929, 1931–1935, 1936a, 1940–1949; Oldroyd, 1949; Roback, 1951; Séguy, 1951; Williston, 1908; Wirth and Stone, 1956.

Biology and immature stages. Brauns, 1954a, 1954b (terrestrial); Chu, 1949; Clausen, 1940 (entomophagous); Hennig, 1948, 1950, 1952; Johannsen, 1934b, 1935 (aquatic); Johannsen and Thomsen, 1937 (aquatic); Malloch, 1917g; Peterson, 1951; Séguy, 1950.

Anatomy and morphology. Cole, 1927 (male terminalia); Comstock, 1918 (wings); Crampton, 1942; Friend, 1942 (wings); Hennig, 1954, 1958a; Peterson, 1916 (head and mouthparts); Snodgrass, 1935.

Suborder NEMATOCERA
Superfamily TIPULOIDEA
Family TRICHOCERIDAE
(Petauristidae, Melusinidae)

By Charles P. Alexander

This is a small family commonly known as the "winter crane-flies," comprising six known genera with fewer than 100 described species. The North American *Trichocera* species need further revisionary studies. The larvae are saprophytic and occur in decaying vegetable matter, such as rotted leaves, roots and tubers, fungi, cattle manure, and other comparable habitats.

REFERENCE: Alexander, 1942b (review).

Genus DIAZOSMA Bergroth

Diazoma Wallengren, 1881: 180 (preocc. Lamarck, 1816). Type-species, *Trichocera hirtipennis* Siebke (mon.).
Diazosma Bergroth, 1913: 583 (n. name for *Diazoma* Wallengren). Type-species, *Trichocera hirtipennis* Siebke (aut.).
subsinuatum (Alexander), 1916d: 124 (*Trichocera*).—Colo.; B.C. to N.B., s. to Calif. and N.Y.

Genus TRICHOCERA Meigen

Petaurista Meigen, 1800: 15 (preocc. Link, 1795). Type-species, *Tipula hiemalis* De Geer (Coquillett, 1910b: 587). Suppressed by I.C.Z.N., 1963b: 339.
Melusina Meigen, 1800: 19. Type-species, *Tipula regelationis* Linnaeus (Hendel, 1908b: 50). Suppressed by I.C.Z.N., 1936b: 339.
Trichocera Meigen, 1803: 262. Type-species, *Tipula hiemalis* De Geer (mon.).
annulata Meigen, 1818: 215—Europe; Alaska, B.C., Oreg., Calif., Nfld.
arctica Lundström, 1915: 28.—Siberia; Alaska.
bimacula Walker, 1848: 84.—N.S.; Wis. to N.S., s. to Kans. and N.C.
  monochroma Harris, 1835: 595 (*Limnobia*). Nomen nudum.
bituberculata Alexander, 1924c: 81.—Alaska; Mass.
borealis Lackschewitz, 1934: 3.—Spitzbergen, U.S.S.R. (Novaya Zemlya, Siberia); N.W.T., Greenland.

brevicornis Alexander, 1952a: 89.—Ga.
brumalis Fitch, 1847: 283.—N.Y.; Man. to Maine and N.J.
    gracilis Walker, 1848: 84.—Ont.
colei Alexander, 1919c: 162.—Oreg.; Alaska, B.C., Wash.
columbiana Alexander, 1927d: 70.—B.C.; Alaska, Wash., Oreg., Calif.
fattigiana Alexander, 1952a: 88.—Ga.; Ill., Tenn.
fernaldi Alexander, 1927d: 70.—Mass.; Mich., e. to Maine and Md.
garretti Alexander, 1927d: 71.—B.C.; B.C. to Maine, s. to Calif. and Md.
hiemalis (De Geer), 1776: 360 (*Tipula*).—Europe; B.C., Maine, Greenland.
hyaloptera Alexander, 1949a: 274.—Oreg.; Alaska, N.W.T., Wash.
longisetosa Alexander, 1927d: 69.—Wash.
lutea Becher, 1886: 64.—Jan Mayen I.; Greenland.
maculipennis Meigen, 1818: 214.—Europe; Alta., B.C., Que., Greenland.
pallens Alexander, 1954: 25.—Oreg.
regelationis (Linnaeus), 1758: 587 (*Tipula*).—Sweden; Canada, Maine, Mass.
salmani Alexander, 1927d: 72.—Mass.; Ont., W. Va., Md.
saltator (Harris).—Not Nearctic.
scutellata Say, 1824a: 360 (1859a: 244).—Minn. Unrecognized.
setosivena Alexander, 1927d: 68.—Alaska; Wash., Oreg., Calif.
tetonensis Alexander, 1945a: 398.—Wyo.; Idaho.
ursamajor Alexander, 1959b: 58.—N.W.T.; Alaska.
venosa Dietz, 1921a: 236.—Pa. ?=*maculipennis* var.

Genus PARACLADURA Brunetti

Paracladura Brunetti, 1911b: 286. Type-species, *gracilis* Brunetti (orig. des.).

trichoptera (Osten Sacken), 1877: 204 (*Trichocera*).—Calif.; B.C., Wash., Oreg.

Family TIPULIDAE
By Charles P. Alexander

    The family Tipulidae is the largest in the Diptera, with about 12,000 described species, arranged in slightly more than 300 genera and subgenera. This is the only family to which the name "crane fly" is properly applied.
    The adult flies commonly frequent wet or humid areas and are particularly numerous near lakes and streams in mountainous sections.

The immature stages are found in a great range of habitats within such wet or moist areas. Many occur in the wet soil at or near the margins of bodies of water; a few are strictly aquatic; many occur in wet mosses or liverworts; several live in the sporophore of fungi, or more commonly in the mycelial growth in wood that is in an advanced state of decay; many species live in and beneath decaying leaves and other organic matter, generally in wooded areas. More specific data are provided under the various genera. Taxonomic papers on the adult flies are limited, in most cases, to certain areas or to particular genera. *Tipula* and *Limonia*, each with far in excess of 1,000 species, are among the largest genera of insects.

REFERENCES: Alexander, 1919h, 1942b (ne. N. Amer.), 1920h (biol.); Rogers, 1933a, 1942 (biol.); Wirth and Stone, 1956 (biol.).

Subfamily TIPULINAE
Tribe TIPULINI

This, the only Nearctic tribe of the subfamily Tipulinae, is divided into three subtribes. These are the Longuriaria, the Ctenophoraria, and the Tipularia.

The immature stages of *Longurio* are aquatic or semi-aquatic. They live in sand or gravel, more rarely in saturated, submerged wood. *Megistocera* is aquatic, occurring in the surface film or neuston fauna in ponds and similar situations (Rogers, 1949). *Brachypremna* lives in saturated black organic soil beneath wet sod (Rogers, 1949). *Holorusia* is found in decaying leaf mold and similar materials in or at the margins of streams and pools. *Prionocera* larvae live in soil where they prey on the early stages of Tabanidae found in the same habitat. The immature stages of *Ctenophora* occur in wood in various stages of decay. The immature stages of *Nephrotoma* frequent wet to moderately damp soil; a few species live in leaf mold in wooded areas. *Dolichopeza* lives in beds or mats of wet to nearly dry mosses and liverworts (Rogers, 1949; Byers, 1961). The early stages of *Tipula* frequent a great range of habitats, as might be expected in a taxon of its magnitude and diversity. The common habitats include the following: Aquatic or semi-aquatic, along streams or marshy areas; semi-aquatic, in wet sand or silt, stream leaf drift, or wet mats or clumps of algae, liverworts, and mosses; bases of grass or sedge tussocks in bogs and marshes; wooded areas in loam beneath decaying leaves; meadows and grasslands (some species of economic importance included); decaying wood.

Genus LONGURIO Loew

Longurio Loew, 1869b: 3 (Cent. 8, no. 2). Type-species, *testaceus* Loew (mon.).
Aeshnasoma Johnson, 1909a: 115. Type-species, *rivertonensis* Johnson (mon.).
Aeschnasoma, error.

Subgenus LONGURIO Loew

minimus Alexander, 1914a: 605.—Ga.; Tenn., N.C., S.C.
pruinosus (Johnson), 1913a: 42 (*Pachyrhina*).—Fla. **N. comb.**
rivertonensis (Johnson), 1909a: 116 (*Aeschnasoma*).—N.J.; Va., N.C.
testaceus Loew, 1869b: 3 (Cent. 8, no. 2).—Mass.; N.Y. to Maine, s. to Tenn. and Fla.

Genus MEGISTOCERA Wiedemann

Megistocera Wiedemann, 1820a: 41 (1821c: 41) (as *Maekistocera*). Type-species, *Tipula filipes* Fabricius (Macquart, 1838a: 63; 1838: 59). [Author rejects the original valid spelling in favor of the later, commonly used spelling, which dates from Wiedemann, 1828: 55.—Eds.].
longipennis (Macquart), 1838a: 61 (1838: 57) (*Tipula*).—Cuba; Tex. to n. Fla., s. to P.R. and Cent. and S. Amer. to Paraguay.

Genus BRACHYPREMNA Osten Sacken

Brachypremna Osten Sacken, 1886b: 161. Type-species, *Tipula dispellens* Walker (Coquillett, 1910b: 545).
dispellens (Walker), 1861: 334 (*Tipula*).—Mexico; Ill. and N.J., s. to Tex. and Fla., Cent. and S. Amer.

Genus HOLORUSIA Loew

Holorusia Loew, 1863b: 277 (Cent. 4, no. 1). Type-species, *rubiginosa* Loew (mon.).

Subgenus HOLORUSIA Loew

rubiginosa Loew, 1863b: 276 (Cent. 4, no. 1).—Calif.; B.C. to Calif. and Ariz.
grandis Bergroth, 1888b: 140 (*Tipula*).—Calif.

Genus PRIONOCERA Loew

Prionocera Loew, 1844a: 170. Type-species, *pubescens* Loew (mon.).
Stygeropis Loew, 1863b: 298 (Cent. 4, no. 42) (unjustified n. name for

TIPULIDAE

Prionocera Loew). Type-species, *Prionocera pubescens* Loew (aut.).

bergrothi (Williston), 1893c: 64 (*Stygeropis*).—Alaska.
broweriana Alexander, 1961b: 79.—Maine.
dimidiata (Loew), 1866a: 129 (Cent. 6, no. 2) (*Stygeropis*).—Hudson Bay Territory; Alaska to Man.
electa Alexander, 1927e: 188.—Labr.; Arctic Canada.
fulvicauda Alexander, 1948b: 121.—Colo.
gracilistyla Alexander, 1956c: 123.—Alaska.
hybrida (Dietz), 1918: 113 (*Pachyrhina*).—Colo.
ominosa (Alexander), 1920i: 199 (*Stygeropis*).—Alaska.
oregonica Alexander, 1943d: 13.—Oreg.; Calif.
oslari (Dietz), 1918: 112 (*Pachyrhina*).—Colo.
parrii (Kirby), 1824: ccxviii (*Ctenophora*).—N.W.T.; Arctic Canada.
primoveris Alexander, 1943b: 723.—Wyo.
rostellata (Doane), 1901: 100 (*Tipula*).—Colo.; Man.
    var. **churchilliana** Alexander, 1942a: 206 (as ssp.).—Man.
    var. **prominens** Alexander, 1942a: 206 (as ssp.).—Man.
sordida (Loew), 1863b: 298 (Cent. 4, no. 42) (*Stygeropis*).—Man.; Alta.
uinticola Alexander, 1948a: 15.—Utah.
unimicra (Alexander), 1915c: 128 (*Stygeropis*).—Colo.

Genus CTENOPHORA Meigen

Ctenophora Meigen, 1803: 263. Type-species, *Tipula pectinicornis* Linnaeus (Westwood, 1840: 128). [In the interest of stability the author rejects the prior type designation of *Tipula atrata* Linnaeus by Latreille, 1810: 442.—Eds.].

Subgenus CTENOPHORA Meigen

apicata Osten Sacken, 1864: 46.—Maine, N.H.; Ont. and Que., s. to N.Y. and R.I.
nubecula Osten Sacken, 1864: 45.—Ill.; Mo. and Tenn., ne. to Que. and N.J.

Subgenus TANYPTERA Latreille

Flabellifera Meigen, 1800: 13. Type-species, *Tipula atrata* Linnaeus (Coquillett, 1910b: 545). Suppressed by I.C.Z.N., 1963b: 339.
Tanyptera Latreille, 1805: 286 (as genus). Type-species, *Tipula ichneumonea* De Geer (mon.) = *atrata* (Linnaeus).
Xiphura Brullé, 1832: 206. Type-species, *villaretiana* Brullé (Guérin-Méneville, 1844: 534) = *atrata* (Linnaeus).

dorsalis Walker, 1848: 76.—Nfld.; Wis. to Nfld., s. to Ill. and N.C.
  succedens Walker, 1856b: 448.—Canada.
  fumipennis Osten Sacken, 1864: 47.—Va.
  frontalis Osten Sacken, 1864: 48.—Mass.
  var. **topazina** Osten Sacken, 1864: 47 (as sp.).—Va.; Wis. to Maine, s. to Ohio and Va. **N. syn.,** G. W. Byers *in litt.*

Subgenus PHOROCTENIA Coquillett

Phoroctenia Coquillett, 1910b: 589 (as genus). Type-species, *Ctenophora angustipennis* Loew (orig. des.)=*vittata* ssp.
Malpighia Enderlein, 1912b: 18. Type-species, *Ctenophora vittata* Meigen (orig. des.).
vittata vittata Meigen.—Not Nearctic.
  ssp. **angustipennis** Loew, 1872a: 51 (Cent. 10, no. 3) (as sp.).—Calif.; B.C., Wash., Oreg.
  similis Williston, 1893c: 63.—Wash.

Genus NEPHROTOMA Meigen

Pales Meigen, 1800: 14. Type-species, *Tipula dorsalis* Fabricius (Hendel, 1908b: 46). Suppressed by I.C.Z.N., 1963b: 339.
Nephrotoma Meigen, 1803: 262. Type-species, *Tipula dorsalis* Fabricius (mon.).
Pachyrhina Macquart, 1834a: 88. Type-species, *Tipula crocata* Linnaeus (Westwood, 1840: 128).
Pachyrrhina, Pachyrina, Pachyrhyna, emends. or errors.

The taxonomy of the genus in our faunal area is in an unsatisfactory state and requires revision following study of the types.

abbreviata (Loew), 1863b: 295 (Cent. 4, no. 36) (*Pachyrhina*).—Miss.
alterna alterna (Walker), 1848: 72 (*Tipula*).—N.S.; Mich. to N.S., s. to Fla.
  incurva Loew, 1863b: 293 (Cent. 4, no. 32) (*Pachyrhina*).—D.C.
  ssp. **nexilis** (Dietz), 1918: 125 (*Pachyrhina;* as sp.).—Colo.; Man., Minn., Wis.
  perdita Dietz, 1918: 116 (*Pachyrhina*).—Man.
altissima (Osten Sacken), 1877: 210 (*Pachyrhina*).—Colo., N. Mex.; B.C. to w. Ont. and Mich., s. to Oreg. and N. Mex.
  erythrophrys Williston, 1893c: 63 (*Pachyrhina*).—Colo.
breviorcornis (Doane), 1908b: 178 (*Pachyrhina*).—Mich.; Iowa to Que., s. to S.C.
  approximata Dietz, 1918: 136 (*Pachyrhina*).—Pa.

stigmatica Dietz, 1918: 137 (*Pachyrhina*).—Pa.
calinota (Dietz), 1918: 121 (*Pachyrhina*).—Mich.; Mich. to N.H., s. to Tenn. and Md.
cingulata (Dietz), 1918: 131 (*Pachyrhina*).—Pa.
clandestina (Dietz), 1921a: 262 (*Pachyrhina*).—Pa.
cornifera (Dietz), 1918: 120 (*Pachyrhina*).—Va.; N.C., Fla.
eucera (Loew), 1863b: 296 (Cent. 4, no. 39) (*Pachyrhina*).—D.C., Ill., N.Y.; Wis. to Que. and Mass., s. to Kans., Tenn., and Va.
euceroides Alexander, 1919c: 172.—N.Y.; Mich. to N.B. and Conn.
evasa (Dietz), 1918: 124 (*Pachyrhina*).—Mich. ?=*calinota.*
excelsior (Bergroth), 1888a: 239 (*Pachyrhina*).—B.C.
ferruginea ferruginea (Fabricius), 1805: 28 (*Tipula*).—N. Amer.; N. Amer. e. from Alta., Colo., and Tex.
    beutenmuelleri Dietz, 1918: 130 (*Pachyrhina*).—Pa. (as N.C., error).
    ssp. **suturalis** (Loew), 1863b: 295 (Cent. 4, no. 37) (*Pachyrhina*; as sp.).—Ga.; coastal, S.C. to Ga.
    costomarginata Dietz, 1918: 129 (*Pachyrhina*).—Fla.
gracilicornis gracilicornis (Loew), 1864a: 66 (Cent. 5, no. 32) (*Pachyrhina*).—N.Y.; Ont., Pa., Md.
    festina Dietz, 1918: 126 (*Pachyrhina;* as *perfida*, p. 128).—Md.
    ssp. **temeraria** (Dietz), 1918: 128 (*Pachyrhina;* as sp.).—Mich.; Wis.
latevittata (Dietz), 1918: 135 (*Pachyrhina*).—Colo.
lineata (Scopoli).—Not Nearctic.
lugens (Loew), 1864a: 63 (Cent. 5, no. 26) (*Pachyrhina*).—N.H.; Minn. to N.S., s. to Ill. and N.C.
lundbecki (Nielsen), 1907: 390 (*Pachyrhina*).—Greenland; N.W.T.
    arcticola Alexander, 1919b: 10.—N.W.T.
macrocera macrocera (Say), 1823: 24 (1859b: 48) (*Tipula*).—Pa.; Wis. to Maine, s. to Kans., Tenn., and Fla.
    gnata Dietz, 1918: 118 (*Pachyrhina;* as var.).—Wis.
    hirsutula Dietz, 1918: 118 (*Pachyrhina;* as *pilosula*, p. 110).—Pa.
    var. **atrocera** (Dietz), 1918: 118 (*Pachyrhina*).—Pa.; Conn.
    ssp. **dietziella** Alexander, 1942b: 230 (as var.; n. name for *virgata* Dietz).—Pa.
    virgata Dietz, 1921a: 260 (*Pachyrhina;* as var.; preocc. Coquillett, 1898).—Pa.
montana (Dietz), 1918: 123 (*Pachyrhina*).—N.C.
navajo Alexander, 1949b: 98.—Ariz.

occidentalis (Doane), 1908b: 177 *(Pachyrhina)*.—Calif.; B.C. and Mont., s. to Calif. and Ariz., also Tex.
occipitalis (Loew), 1864a: 65 (Cent. 5, no. 30) *(Pachyrhina)*.—N.W.T.; N.W.T. to N.B., s. to Colo. and Mich.
okefenoke (Alexander), 1915d: 97 *(Pachyrhina)*.—Ga.; Fla.
opacivittata (Dietz), 1918: 123 *(Pachyrhina)*.—Man.; Wis., Mass.
pedunculata (Loew), 1863b: 293 (Cent. 4, no. 33) *(Pachyrhina)*.— Sask.; Sask. to N.S., s. to Minn. and Pa.
penumbra Alexander, 1915a: 467.—N.H.; N.B., Maine.
perincisa Alexander, 1949b: 99.—Ariz.
polymera (Loew), 1863b: 297 (Cent. 4, no. 40) *(Pachyrhina)*.—Ill.; Wis. to N.H., s. to Kans., Tenn., and S.C.
punctum (Loew), 1863b: 294 (Cent. 4, no. 34) *(Pachyrhina)*.—Ill.; Mich. to Maine, s. to Ill. and N.J.
snowii snowii (Doane), 1908b: 176 *(Pachyrhina)*.—Wyo.
    ssp. **alternata** (Dietz), 1918: 117 *(Pachyrhina;* as var.).—Colo.
sodalis sodalis (Loew), 1864a: 64 (Cent. 5, no. 29) *(Pachyrhina)*.— Conn.; Wis. to Que. and N.H., s. to N.C.
    obliterata Dietz, 1918: 133 *(Pachyrhina)*.—Pa.
    ssp. **nictans** (Dietz), 1918: 129 *(Pachyrhina;* as var.).—Colo.
sphagnicola Alexander, 1920d: 110.—Ill.; Mich.
tenuis tenuis (Loew), 1863b: 297 (Cent. 4, no. 41) *(Pachyrhina)*.— N.Y.; Wis. to Maine, s. to Tenn. and N.C.
    hamata Dietz, 1918: 121 *(Pachyrhina;* as ssp.).—N.Y.
    nigroantennata Dietz, 1921a: 261 *(Pachyrhina;* as var.).— Pa.
    ssp. **fuscostigmosa** Alexander, 1940a: 606.—Tenn.
urocera (Dietz), 1918: 119 *(Pachyrhina)*.—N.C.
virescens (Loew), 1864a: 62 (Cent. 5, no. 25) *(Pachyrhina)*.—D.C.; Mich. to N.H., s. to Ill. and Fla.
vittula (Loew), 1864a: 63 (Cent. 5, no. 27) *(Pachyrhina)*.—N.W.T.; Alaska to Nfld., s. to Alta. and Mass.
wulpiana (Bergroth), 1888a: 200 *(Pachyrhina)*.—Calif.
wyalusingensis (Dietz), 1918: 134 *(Pachyrhina)*.—Pa. Unrecognized.
xanthostigma (Loew), 1864a: 65 (Cent. 5, no. 31) *(Pachyrhina)*.— Ill.; Ont. to Maine, s. to Ill. and S.C.

Genus DOLICHOPEZA Curtis

Dolichopeza Curtis, 1825: pl. 62. Type-species, *sylvicola* Curtis (orig. des.) = *albipes* (Ström).

REFERENCE: Byers, 1961 (rev.).

Subgenus DOLICHOPEZA Curtis

americana Needham, 1908: 211.—N.Y.; Alaska, Alta., S. Dak., Mich. to Labr. and Nfld., s. to Ill. and Ga.

borealis Byers, 1961: 795.—Alaska.

Subgenus OROPEZA Needham

Oropeza Needham, 1908: 211 (as genus). Type-species, *Tipula annulata* Say (orig. des.)=*sayi* (Johnson).

australis Byers, 1961: 796.—Ga.; Ala., Fla.

carolus Alexander, 1942b: 211 (n. name for *albipes* Johnson).—Mass., Vt.; Wis. to Nfld., s. to Tenn. and Fla.

   albipes Johnson, 1909a: 121 (*Oropeza*; preocc. Ström, 1768).—Mass., Vt.

dorsalis (Johnson), 1909a: 119 (*Oropeza*).—Maine; Y.T. to N.S., s. to B.C. and Fla.

   rogersi Alexander, 1922b: 6 (*Oropeza*).—Ind.

   sessilis Alexander, 1941a: 296.—N.C.

johnsonella (Alexander), 1930b: 279 (*Oropeza*).—N.J.; Ill. to Que. and Mass., s. to Ark. and Fla.

obscura (Johnson), 1909a: 122 (*Oropeza*).—Mass.; Alta. to N.S., s. to Ark. and Fla.

polita polita (Johnson), 1909a: 122 (*Oropeza;* as *obscura* var.).—Mass.; Wis. to Que. and Maine, s. to Ga.

  ssp. **pratti** Alexander, 1941d: 192 (as sp.).—Minn.; Wis., Iowa, Kans., Ill., Ark.

  ssp. **cornuta** Byers, 1961: 845.—Ind.; Ill. to Ont., s. to Tenn. and Va.

sayi (Johnson), 1909a: 118 (*Oropeza;* n. name for *annulata* Say).—Pa.; Minn. to N.B. and s. to Fla.

   annulata Say, 1823: 25 (1859b: 49) (*Tipula;* preocc. Linnaeus, 1758).—Pa.

similis (Johnson), 1909a: 119 (*Oropeza*).—Pa., Mass.; Mich. to Que. and Maine, s. to Pa.

subalbipes (Johnson), 1909a: 121 (*Oropeza*).—N.J.; Minn. to N.B., s. to La. and Fla.

subvenosa Alexander, 1940a: 618.—N.C., Tenn.; W. Va. and Md., s. to Tenn. and Ga.

tridenticulata Alexander, 1931b: 177.—Mass.; Man. to Que. and Maine, s. to Mo. and Ga.

venosa (Johnson), 1909a: 120 (*Oropeza*).—Mass.; Y.T. to Nfld., s. to B.C. and S.C.

walleyi (Alexander), 1931c: 139 (*Oropeza*).—Que.; Alta. to N.S., s. to Mo. and Fla.

   dakota Alexander, 1944d: 241.—S. Dak.

Genus TIPULA Linnaeus

Tipula Linnaeus, 1758: 585. Type-species, *oleracea* Linnaeus (Latreille, 1810: 442).

Tipula is divided into a number of subgenera, the largest and most important in the local fauna being *Yamatotipula*, *Oreomyza*, and *Lunatipula*. The two last-named subgenera divide into more or less natural groups of species, for the most part with definite distributional patterns.

REFERENCES: Dietz, 1913, 1914, 1917, 1919, 1921b (revs.); Alexander, 1915a (lectotypes).

Subgenus TRICHOTIPULA Alexander

Tipula, subg. **Trichotipula** Alexander, 1915a: 468. Type-species, *oropezoides* Johnson (orig. des.).
Tipula, subg. *Cinctotipula* Alexander, 1915a: 469. Type-species, *algonquin* Alexander (orig. des.).
Tipula, subg. *Odontotipula* Alexander, 1919h: 939, 943. Type-species, *Pachyrhina unifasciata* Loew (orig. des.) = *stonei* Alexander.
Tipula, subg. *Nitidotipula* Alexander, 1942b: 240. Type-species, *pachyrhinoides* Alexander (orig. des.).

REFERENCE: Alexander, 1946f (key, Calif.).

algonquin Alexander, 1915a: 469.—Ont.; Mich. to Maine, s. to Tenn. and N.C.
apache Alexander, 1916b: 45.—N. Mex.; Ariz.
beatula Osten Sacken, 1877: 209.—Calif.
bituberculata Doane, 1901: 101.—Calif.
cahuilla Alexander, 1920b: 43.—Calif.
capistrano Alexander, 1946f: 4.—Calif.
cazieri Alexander, 1942a: 207.—Nev.; Calif.
cimarronensis Rogers, 1931: 335.—Okla.
desertorum Alexander, 1946f: 6.—Calif.
dis Alexander, 1962: 1.—Nev.
dorsolineata Doane, 1901: 98.—Wash.; B.C., Oreg.
furialis Alexander, 1946f: 7.—Calif.
geronimo Alexander, 1946a: 490.—Ariz.
guasa Alexander, 1916b: 51.—Tex.
hedgesi Alexander, 1961a: 10.—Ariz.
macrophallus (Dietz), 1918: 114 (*Pachyrhina*).—Utah; B.C., Wash., Oreg., Calif.
malkini Alexander, 1955a: 125.—Ariz.
mayedai Alexander, 1946f: 8.—Calif.

megalodonta Alexander, 1946f: 10.—Calif.
mulaiki Alexander, 1948b: 123.—Colo.
oropezoides Johnson, 1909a: 131.—Mass.; Mich. to Nfld., s. to Tenn. and Fla., ?Wash.
pachyrhinoides Alexander, 1915a: 471.—N.H.; Alta., Sask., Nebr., Minn.
prolixa Alexander, 1947b: 68.—Ariz.
puncticollis (Dietz), 1918: 115 (*Pachyrhina*).—Colo.; Nebr.
repulsa Alexander, 1943d: 139.—B.C.; B.C. to Calif.
retinens Alexander, 1946a: 493.—Ariz.
rusticola Doane, 1912: 47.—Wash.; Oreg., Calif.
sayloriana Alexander, 1946f: 12.—Calif.
stonei Alexander (for *unifasciata* Loew).—D.C.; Mich. to Ont., s. to Kans. and Fla. **N. name.**
  unifasciata Loew, 1863b: 294 (Cent. 4, no. 35) (*Pachyrhina*; preocc. Schrank, 1803).—D.C.
subapache Alexander, 1947e: 91.—Calif.
trichophora Alexander, 1920b: 41.—Calif.
unimaculata (Loew), 1864a: 64 (Cent. 5, no. 28) (*Pachyrhina*).—N.Y.; Mich. to Maine, s. to Ill. and N.C.

Subgenus SCHUMMELIA Edwards

Tipula, subg. **Schummelia** Edwards, 1931: 80. Type-species, *variicornis* Schummel (orig. des.).

annulicornis Say, 1829: 151 (1859b: 350).—Ind.; Mich. to Mass., s. to Tenn. and Md.
  jejuna Johnson, 1909a: 132.—Mass.
friendi Alexander, 1941a: 293.—Mass.; Tenn., N.C.
hermannia Alexander, 1915a: 480 (n. name for *fasciata* Loew).—N.Y.; Wis. to Nfld., s. to Kans. and Fla.
  fasciata Loew, 1863b: 279 (Cent. 4, no. 6) (preocc. Linnaeus, 1767).—N.Y.
magnifolia Alexander, 1948a: 18.—Utah; Calif.
stenorhabda Alexander, 1941a: 290.—N.C.
subtenuicornis Doane, 1901: 125.—Wash.; B.C. to Nfld., s. to Oreg., Utah, and Wyo.
  idei Alexander, 1928a: 55.—Ont.
synchroa Alexander, 1927g: 183.—Fla.

Subgenus NOBILOTIPULA Alexander

Tipula, subg. **Nobilotipula** Alexander, 1942b: 239. Type-species, *Pachyrhina nobilis* Loew (orig. des.).

collaris Say, 1823: 25 (1859b: 49).—Pa.; Mich. to Que., s. to S.C.

nobilis (Loew), 1864a: 62 (Cent. 5, no. 24) (*Pachyrhina*).—N.H.; Mich. to N.B., s. to N.C.

Subgenus BELLARDINA Edwards

Tipula, subg. **Bellardina** Edwards, 1931: 82. Type-species, *craverii* Bellardi (orig. des.; as *cravieri*).

abluta Doane, 1901: 122.—Colo. ?=*commiscibilis*.
albimacula Doane, 1912: 51.—Ariz.
aspersa Doane, 1912: 51.—Calif.; Wash., Oreg.
calaveras Alexander, 1944a: 60.—Calif.
catalinensis Alexander, 1950a: 41.—Ariz.
commiscibilis Doane, 1912: 61 (n. name for *contaminata* Doane).— Colo.; Wyo., Utah, N. Mex.
   contaminata Doane, 1901: 121 (preocc. Linnaeus, 1758).—Colo.
faustina Alexander, 1941c: 206.—Colo.
gothicana Alexander, 1943a: 149.—Colo.; B.C., Oreg., Calif., Mont., Utah.
jepsoni Alexander, 1945a: 400.—Wyo.
josephus Alexander, 1954: 26.—Oreg.
pacifica Doane, 1912: 48.—Calif.; B.C. to Calif.
praelauta Alexander, 1949b: 15.—Ariz.
pura Alexander, 1941c: 208.—Colo.; N. Mex.
rastristyla Alexander, 1945d: 129.—Wash.
rupicola Doane, 1912: 50.—Ariz.
sacajawea Alexander, 1945d: 126.—Idaho.
schizomera Alexander, 1940e: 142.—Mexico; Calif., Utah, Ariz., Tex.
shastensis Alexander, 1944a: 58.—Calif.
subcinerea Doane, 1901: 118.—Colo.; B.C. to Oreg., Utah, and Colo.
umbra Alexander, 1959c: 129.—Calif.
warneri Alexander, 1944a: 57.—Idaho.

Subgenus NIPPOTIPULA Matsumura

Nippotipula Matsumura, 1916: 457 (as genus). Type-species, *Tipula nubifera* Coquillett (orig. des.)=*coquilletti* Enderlein.

abdominalis (Say), 1823: 18 (1859b: 45) (*Ctenophora*).—Pa.; Wis. to Nfld. s. to Kans. and Fla.
   albilata Walker, 1848: 65.—?N. Amer.

Subgenus TIPULA Linnaeus

paludosa Meigen, 1830: 289.—Europe; Nfld., N.S. Immigrant.

Subgenus PLATYTIPULA Matsumura

Platytipula Matsumura, 1916: 459 (as genus). Type-species, *moiwana* Matsumura (orig. des.).

carinata Doane, 1901: 103.—Wash.; Oreg.
cunctans Say, 1823: 23 (1859b: 48).—Pa.; Man. to N.B., s. to Colo. and Ala.
    casta Loew, 1863b: 289 (Cent. 4, no. 25).—Pa.
    infuscata Loew, 1863b: 289 (Cent. 4, no. 26).—N.Y.
maritima Alexander, 1930a: 276.—Mass.; Maine.
nebulinervis Alexander, 1940c: 152.—Fla.
paterifera Alexander, 1962: 2.—Tenn.
pendulifera Alexander, 1919c: 166.—Colo.; Alaska to n. Que., s. to Utah and Colo., Nfld.
spenceriana spenceriana Alexander, 1943d: 141.—B.C.; Oreg., Nfld., Maine.
    ssp. **hardyi** Alexander, 1943d: 142 (as sp.).—Utah; Wyo.
tennessa Alexander, 1920d: 226.—Tenn.; Mass.
ultima Alexander, 1915c: 128 (n. name for *flavicans* Fabricius).—N. Amer.; Sask. to N.S., s. to Wyo., Miss., and Fla.
    flavescens Fabricius, 1805: 24 (preocc. Linnaeus, 1758).—N. Amer.
    flavicans Fabricius, 1805: 373 (n. name for *flavescens* Fabricius, but preocc. Müller, 1764).—N. Amer.

Subgenus YAMATOTIPULA Matsumura

Yamatotipula Matsumura, 1916: 461 (as genus). Type-species, *nohirae* Matsumura (orig. des.) = *nova* Walker.

albocaudata Doane, 1901: 123.—Wash.; B.C., Oreg., Wyo., Utah.
alexandriana Dietz, 1917: 146.—Calif.
aprilina Alexander, 1918a: 63.—Va.; Maine, Mass., N.C., S.C.
aspidoptera Alexander, 1916b: 49.—Tex.
brevifurcata Alexander, 1926c: 291.—Tenn.
caloptera Loew, 1863b: 292 (Cent. 4, no. 30). R.I.; Wis. to Nfld., s. to Mo. and Fla.
calopteroides Alexander, 1919c: 168.—N.C.; Pa., S.C.
    antiopa Dietz, 1921a: 266.—Pa.
catawbiana Alexander, 1940a: 609.—Tenn.; N.C.
cayuga Alexander, 1915a: 485.—N.Y.; Mich. to Nfld., s. to Tenn.
cervicula Doane, 1901: 100.—Wash.
cimmeria Speiser, 1909: 57 (n. name for *strigata* Coquillett).—Alaska.
    strigata Coquillett, 1900h: 402 (1904: 16) (preocc. Loew 1866).—Alaska.

cognata Doane, 1901: 123.—Wash.; B.C., Oreg., Utah.
colteri Alexander, 1943b: 725.—Wyo.; B.C., Oreg., Colo.
comanche Alexander, 1916b: 50.—Tex.
concava Alexander, 1926c: 294.—Ind.; Mich. to Maine, s. to Mo. and Tenn.
conspicua Dietz, 1917: 149.—N.C.
dejecta Walker, 1856b: 442.—U.S.; Mich. to Que., s. to Ill. and Va.
    fumosa Doane, 1901: 99. —Ohio.
    brevicollis Loew *in* Alexander, 1915a: 460. Nomen nudum.
edmundsi Alexander, 1948a: 20.—Utah; Oreg.
eluta Loew, 1863b: 290 (Cent. 4, no. 27).—D.C.; Ont. to N.B., s. to Ill. and Fla.
floridensis Alexander, 1926c: 292.—Fla.
footeana Alexander, 1961a: 11.—Idaho; Nfld.
fraterna Loew, 1864a: 56 (Cent. 5, no. 14).—D.C.; N.H. to Fla.
fulvilineata Doane, 1912: 61 (n. name for *graphica* Doane).—Calif.; Oreg.
    graphica Doane, 1901: 124 (preocc. Schiner, 1868).—Calif.
furca Walker, 1848: 70.—N. Amer.; Kans. to Que. and Maine, s. to Tex. and Fla.
    bella Loew, 1863b: 291 (Cent. 4, no. 29).— Conn.
    fuscifer Harris *in* Johnson, 1925c: 61. Nomen nudum.
glendenningi Alexander, 1943d: 140.—B.C.; Oreg.
grenfelli Alexander, 1928b: 96.—Labr.
iroquois Alexander, 1915c: 128 (n. name for *cincta* Loew).—D.C.; Nfld. to Tenn. and N.C.
    cincta Loew, 1863b: 288 (Cent. 4, no. 24) (preocc. Gmelin, 1792).—D.C.
jacintoensis Alexander, 1946e: 65.—Calif.
jacobus Alexander, 1930b: 277.—Fla.; N.S., s. to Tenn. and Fla.
    filipes Walker of Johnson, 1909a: 131.
kennicotti Alexander, 1915a: 480.—N.W.T.; Alta. to N.W.T., s. to Colo. and Ind.
    diluta Doane, 1901: 117 (preocc. Müller, 1764).—Colo.
    parvemarginata Alexander, 1926c: 295.—N. Dak.
    tetra Loew *in* Alexander, 1915a: 482. Nomen nudum.
ludoviciana Alexander, 1919c: 196.—La.; Fla.
maculipleura Alexander, 1927g: 182.—Tenn.
manahatta Alexander, 1919c: 169.—N.Y.; N.Y. to Tenn. and Fla.
meridiana meridiana Doane, 1912: 58.—Ariz.; Calif., Utah.
    ssp. **continentalis** Alexander, 1941c: 85 (as sp.).—Colo.; Alaska to Calif. and Colo.
nephophila Alexander, 1940a: 610.—Tenn.; N.C.

noveboracensis Alexander, 1919c: 167.—N.Y.; Mich. to Nfld., s. to Md.
nuntia Alexander, 1946c: 140.—Wash.
osceola Alexander, 1927g: 181.—Fla.
sackeniana Alexander, 1918a: 62.—Va.; N.Y. and Conn., s. to Tenn. and Ga.
sayi Alexander, 1911c: 194 (n. name for *costalis* Say).—Md., Pa.; Iowa to Nfld., s. to La. and Fla., Bermuda.
    costalis Say, 1823: 23 (1859b: 48) (preocc. Swederus, 1787).—Md., Pa.
spernax spernax Osten Sacken, 1877: 210.—Calif.; B.C., Wash., Oreg., Wyo., Utah.
    ssp. **lanei** Alexander, 1940b: 84 (as sp.).—Oreg.; Wash.
strepens Loew, 1863b: 291 (Cent. 4, no. 28).—N.Y.; Kans. to Nfld., s. to N.J.
subeluta Johnson, 1913a: 42.—Fla.; Mass. to Fla.
succincta Alexander, 1940c: 151.—Ind.
sulphurea sulphurea Doane, 1901: 99.—Mich.; Alta. to Maine, s. to Ind.
    ssp. **jacksonensis** Alexander, 1945a: 402.—Wyo.
tenebrosa Coquillett, 1900h: 403 (1904: 17).—Alaska.
tenuilinea Alexander, 1959a: 69.—Calif.; Wash., Oreg., Utah.
tephrocephala Loew, 1864a: 62 (Cent. 5, no. 23).—N.H.; Wis. to Nfld., s. to Ind. and Pa.
tricolor Fabricius, 1775: 749.—N. Amer.; Wis. to Que. and Maine, s. to Ark. and Fla.
    vitrea Wulp, 1881: 150.—Que.
vicina Dietz, 1917: 148.—Mich.; Mich. to Nfld., s. to Pa.
xanthostigma Dietz, 1917: 150.—Colo.

Subgenus TIPULODINA Enderlein

Tipulodina Enderlein, 1912b: 30 (as genus). Type-species, *magnicornis* Enderlein (orig. des.).

lacteipes Alexander, 1943d: 142.—Calif.

Subgenus ARCTOTIPULA Alexander

Tipula, subg. **Arctotipula** Alexander, 1933b: 410. Type-species, *besselsi* Osten Sacken (orig. des.).

alascaensis Alexander, 1923: 167.—Alaska (Pribilof Is.).
aleutica Alexander, 1923: 164.—Alaska (Pribilof Is.).
bakeriana Alexander, 1954: 28.—Oreg.
besselsi Osten Sacken, 1876b: 42.—Greenland; Alaska.
besselsoides Alexander, 1919b: 15.—N.W.T.

dickinsoni Alexander, 1932: 240.—Wis.; Mich.
kincaidi Alexander, 1949a: 278.—Wash.
loganensis Alexander, 1946b: 67.—Mont.
piliceps Alexander, 1915a: 482.—Hudson Bay Territory.
plutonis Alexander, 1919c: 197.—Calif.; Wash., Oreg., Wyo.
  absaroka Alexander, 1943b: 727.—Wyo.
sacra Alexander, 1946g: 93.—Idaho; Wash., Alta.
semidea Alexander, 1944c: 89.—Oreg.; Wash., Idaho, Mont.
suttoni Alexander, 1934: 5.—N.W.T. (Southampton I.).
thulensis Alexander, 1947c: 245.—N.W.T. (Baffin I.).
tribulator Alexander, 1956a: 179.—Que.
twogwoteeana Alexander, 1945a: 403.—Wyo.; Colo.
williamsiana Alexander, 1940a: 611.—Tenn.; S.C.

Subgenus ANGAROTIPULA Savtshenko

Tipula, subg. **Angarotipula** Savtshenko, 1961: 347. Type-species, *tumidicornis* Lundström (orig. des.).

illustris Doane, 1901: 97.—Idaho; B.C. to Nfld., s. to Colo. and N.J.
  fuscipennis Loew, 1866a: 129 (Cent. 6, no. 3) (*Stygeropis*; preocc. Curtis, 1834).—Ill.
parrioides (Alexander), 1919b: 9 (*Stygeropis*).—Alaska.

Subgenus VESTIPLEX Bezzi

Tipula, subg. **Vestiplex** Bezzi, 1924a: 230. Type-species, *cisalpina* Riedel (orig. des.).

aldrichiana Alexander, 1929d: 16.—Alaska.
arctica Curtis, 1835b: lxxviii.—Arctic Amer.; Greenland, n. Eurasia.
  nodulicornis Zetterstedt, 1838: 841.—Greenland.
  glomerata Walker, 1848: 70.—N. Amer.
balioptera Loew, 1863b: 284 (Cent. 4, no. 15).—Man.; Alta. and Wis. to Que. and Maine.
baliopteroides Alexander, 1945a: 405.—Wyo.; Colo.
bergrothiana Alexander, 1918a: 68.—Alaska.
canadensis Loew, 1864a: 59 (Cent. 5, no. 19).—N.W.T.; N.W.T. to Alta., Ont., and Labr.
caroliniana Alexander, 1916b: 46.—N.C.
centralis Loew, 1864a: 60 (Cent. 5, no. 21).—N.W.T.; N.W.T. to Que. and N.H.
churchillensis Alexander, 1940b: 85.—Man.
fultonensis Alexander, 1918a: 67.—N.Y.; Alaska, se. to Nfld. and N.Y.
  hinei Alexander, 1920i: 200.—Alaska.
leucophaea Doane, 1901: 117.—Colo.; Oreg., Wyo., Utah.

longiventris Loew, 1863b: 278 (Cent. 4, no. 5).—Ill.; Wis. to Maine, s. to Tex. and S.C.
nigrocorporis Doane, 1912: 45.—Colo.; Alta.
  alticola Alexander, 1915c: 141.—Colo.
perretti Alexander, 1928b: 98.—Labr.
platymera Walker, 1856b: 441.—Canada; Ont. to Labr., s. to N.Y. and N.H.
  tessellata Loew, 1863b: 277 (Cent. 4, no. 3) (preocc. Villers, 1789).—Labr.
  septentrionalis Loew, 1863b: 278 (Cent. 4, no. 4).—Labr.
  labradorica Alexander, 1915c: 128 (n. name for *tessellata* Loew).—Labr.
serrulata Loew, 1864a: 58 (Cent. 5, no. 18).—N.W.T.
tacomicola Alexander, 1949a: 280.—Wash.; B.C.

Subgenus ODONATISCA Savtshenko

Tipula, subg. **Odonatisca** Savtshenko, 1956: 130. Type-species, *juncea* Meigen (orig. des.).

breviligula Alexander, 1956a: 178.—Ont.; Alta. to Ont.
optiva Alexander, 1921a: 106.—Wash.
pribilofensis Alexander, 1923: 163.—Alaska (Pribilof Is.).
subarctica Alexander, 1919b: 15.—Alaska.
taenigaster Alexander, 1920i: 201.—Alaska.

Subgenus NESOTIPULA Alexander

Tipula, subg. **Nesotipula** Alexander, 1921b: 183. Type-species, *pribilovia* Alexander (orig. des.).

pribilovia Alexander, 1921b: 183.—Alaska (Pribilof Is.).

Subgenus OREOMYZA Pokorny

Oreomyza Pokorny, 1887b: 50 (as genus). Type-species, *glacialis* Pokorny (Coquillett, 1910b: 580).

accurata Alexander, 1927e: 184.—B.C.; Alta., Oreg., Utah, Colo.
  johannus Alexander, 1945a: 412.—Wyo.
afflicta Dietz, 1915b: 125 (n. name for *suspecta* Dietz).—N.Y.
  suspecta Dietz, 1914: 351 (preocc. Loew, 1863).—N.Y.
albertensis albertensis Alexander, 1927f: 217.—Alta.
  ssp. **fenebris** Alexander, 1955a: 17.—Colo.
alcestis Alexander, 1946e: 175.—Idaho.
alia Doane, 1911: 161 (as *olia*).—Wash.; Oreg., Calif., Wyo.
alta Doane, 1912: 44.—Wyo.

angulata angulata Loew, 1864a: 61 (Cent. 5, no. 22).—Mass.; Wis. to N.S., s. to Ill. and N.Y.
   decora Doane, 1901: 125.—Que.
   ssp. **cherokeana** Alexander, 1940a: 614.—Tenn.; N.C.
appendiculata Loew, 1863b: 287 (Cent. 4, no. 20).—Sask.; Alaska, s. to Utah and e. to Nfld.
   stalactoides Doane, 1901: 102.—Alaska.
   derelicta Dietz, 1914: 358.—Alaska.
   gaspensis Alexander, 1929e: 233.—Que.
athabasca Alexander, 1927f: 217.—Alta.
bakeri Alexander, 1954: 33.—Oreg.
banffiana Alexander, 1946c: 141.—Alta.; Wyo., Colo.
barbata Doane, 1901: 105.—Colo.; Alta., Utah.
borealis Walker, 1848: 66.—N.S.; Wis. to N.S., s. to Kans. and S.C.
   hebes Loew, 1863b: 285 (Cent. 4, no. 18).—Conn.
broweri Alexander, 1940d: 83.—Maine.
cineracea Coquillett, 1900h: 404 (1904: 18).—Alaska.
clathrata Dietz, 1914: 356.—Utah; Colo., N. Mex.
coleana Alexander, 1940a: 615.—Tenn.
coloradensis Doane, 1911: 164.—Colo.; Utah, Wyo., S. Dak.
   fundata Alexander, 1945a: 407.—Wyo.
comstockiana Alexander, 1947g: 36.—Calif.
criddlei Dietz, 1914: 360.—Man.
cylindrata cylindrata Doane, 1912: 46.—Calif.
   ssp. **barda** Alexander, 1959a: 71.—Calif.
diflava Alexander, 1919b: 12.—N.W.T.; Y.T.
doanei doanei Dietz, 1914: 352.—N. Mex.
   ssp. **bifida** Dietz, 1914: 354.—N. Mex.
dorothea Alexander, 1954: 31.—Oreg.
entomophthorae Alexander, 1918a: 385.—N.C.; Alta. to Labr. and Nfld., s. to Ind. and N.C.
   similissima Dietz, 1921a: 263.—Pa.
fallax Loew, 1863b: 281 (Cent. 4, no. 10).—Calif.; Wash., Oreg.
fragilina Alexander, 1919c: 171.—Alaska; B.C., Mont., Colo.
fragilis Loew, 1863b: 279 (Cent. 4, no. 7).—Maine; Alta. to N.B., s. to Ind. and D.C.
   suspecta Loew, 1863b: 280 (Cent. 4, no. 8).—D.C.
gelida Coquillett, 1900h: 404 (1904: 18).—Alaska.
graciae Alexander, 1947e: 93.—Calif.
graminivora Alexander, 1921d: 135.—Calif.
helderbergensis Alexander, 1918a: 64.—N.Y.; Mich. to N.B. and N.Y.

helvocincta Doane, 1901: 101.—Wash.; B.C., Oreg.
hollandi Alexander, 1934: 4.—N.W.T.
huntsmaniana Dietz, 1920: 7.—N.S. Unrecognized.
huron Alexander, 1918a: 66.—Wis.
idahoensis Alexander, 1955a: 16.—Idaho.
ignobilis Loew, 1863b: 280 (Cent. 4, no. 9).—D.C.; Ill. to N.B., s. to Tenn. and N.C.
illinoiensis Alexander, 1915c: 128 (n. name for *versicolor* Loew).—Ill.; Man. to N.H., s. to Ill.
    versicolor Loew, 1863b: 285 (Cent. 4, no. 17) (preocc. Gmelin, 1792).—Ill.
imbellis Alexander, 1927e: 186.—B.C.; Wash.
inclusa Dietz, 1921a: 267.—Pa.
incurva Doane, 1912: 43.—Nebr. ?=*angulata*.
ingrata Dietz, 1914: 355.—Colo.; Alta., Utah.
insignifica Alexander, 1924a: 117.—N.H.; Maine.
inyoensis Alexander, 1946e: 173.—Calif.
kirbyana Alexander, 1918a: 244.—Alaska.
latipennis Loew, 1864a: 60 (Cent. 5, no. 20).—N.H.; Alta. to N.S., s. to Wis. and N.J.
    ottawaensis Dietz, 1914: 349.—Man.
madera Doane, 1911: 162.—Calif.
mandan Alexander, 1915a: 499.—Mont.; Oreg., Wyo., Utah.
    perexigua Alexander, 1924d: 15.—Mont.
margarita Alexander, 1918a: 243.—N.Y.; Ohio.
marina Doane, 1912: 44.—Calif.
nebulipennis Alexander, 1919c: 170.—Labr.; Que., N.H.
neptun Dietz, 1921c: 300.—Colo.; Alaska.
newcomeri Doane, 1911: 163.—Calif.
packardi Alexander, 1928b: 99.—Labr.
paiuta Alexander, 1948a: 23.—Utah.
penobscot Alexander, 1915a: 472.—Maine; Alta. to N.B., s. to Mich. and Pa.
perparvula Alexander, 1926e: 120.—Man.
phoroctenia Alexander, 1919c: 170.—Maine; B.C. to Nfld. and Maine.
productella Alexander, 1928b: 100.—Labr.
pseudotruncorum Alexander, 1920d: 228.—Wash.; B.C., s. to Calif. and Wyo.
    truncorum Meigen of Snodgrass, 1904: 211.
resurgens Walker, 1848: 67.—Nfld.; Alta. to Nfld., s. to Ind. and D.C.
    grata Loew, 1863b: 281 (Cent. 4, no. 11).—D.C.
rohweri Doane, 1911: 165.—Colo.; Wyo., Utah.

senega Alexander, 1915c: 128 (n. name for *pallida* Loew).—Mass.;
   Alta. to Nfld., s. to Iowa and N.J.
   pallida Loew, 1863b: 284 (Cent. 4, no. 16) (preocc. Fabricius,
      1781).—Mass.
sequoicola Alexander, 1947d: 205.—Calif.
serta Loew, 1863b: 283 (Cent. 4, no. 14).—Man.; Alta. to Labr. and
   Nfld., s to Minn. and N.Y., ?Eurasia.
   discolor Loew, 1863b: 282 (Cent. 4, no. 12) (preocc. Gmelin,
      1792).—Mass.
   albonotata Doane, 1901: 120.—Mich.
   ignota Alexander, 1915c: 128 (n. name for *discolor* Loew).—
      Mass.
shoshone Alexander, 1946e: 66.—Wash.; B.C., Oreg.
stylifera Dietz, 1921a: 264.—Pa.
subbarbata Alexander, 1927e: 185.—Sask.; Alta.
subfasciata Loew, 1863b: 282 (Cent. 4, no. 13).—Man.; Wis., N.S.
subserta Alexander, 1928b: 97.—Labr.; Maine.
ternaria Loew, 1864a: 57 (Cent. 5, no. 15).—N.W.T.; Y.T. to Nfld.,
   s. to Alta. and N.H.
tristis Doane, 1901: 102.—Calif.; Oreg.
trivittata trivittata Say, 1823: 26 (1859b: 50).—Pa.; Iowa to Nfld.,
   s. to Tenn. and S.C.
   simulata Walker, 1856b: 441.—Canada.
   ssp. **laetifica** Alexander, 1945a: 411.—Wyo.
variata Alexander, 1927f: 216.—Alta.
whitneyi Alexander, 1923: 161.—Alaska (Pribilof Is.).
yellowstonensis Alexander, 1946b: 65.—Wyo.; Wash., Oreg., Utah.

Subgenus EUMICROTIPULA Alexander

Microtipula, subg. **Eumicrotipula** Alexander, 1922c: 74. Type-
   species, *macrotrichiata* Alexander (orig. des.).

chiricahuensis Alexander, 1946a: 495.—Ariz.
werneri Alexander, 1950a: 43.—Ariz.

Subgenus LUNATIPULA Edwards

Tipula, subg. **Lunatipula** Edwards, 1931: 81. Type-species, *lunata*
   Linnaeus (orig. des.).

acuta Doane, 1901: 116.—Calif.; Oreg., Utah.
acutipleura Doane, 1912: 42.—Calif.
aequalis Doane, 1901: 108.—Wash.; Oreg., Calif.
   reesi Alexander, 1939b: 143.—Calif.
alaska Alexander, 1918a: 412.—Alaska.

albocincta Doane, 1901: 110.—Colo.; Wyo.
albofascia Doane, 1901: 126.—Oreg.; Wash., Calif.
  biarmata Doane, 1912: 55.—Wash.
aperta Alexander, 1918a: 62 (n. name for *imperfecta* Alexander).—
  Labr. ?Subg. *Arctotipula*.
  imperfecta Alexander, 1915a: 484 (preocc. Brunetti, 1913).—
  Labr.
apicalis Loew, 1863b: 277 (Cent. 4, no. 2).—Maine; Mich. to N.S., s.
  to Tenn. and N.C.
armata Doane, 1901: 119.—Wash.; Oreg., Calif.
  varia Doane, 1901: 122.—Wash.
atrisumma Doane, 1912: 42.—Calif.
australis Doane, 1901: 104.—Ga.; Tex. to Ga., n. to Md.
awanichi Alexander, 1947g: 38.—Calif.
bernardinensis Alexander, 1946c: 143.—Calif.
bicornis Forbes, 1890: 78.—Ill.; Wis. to N.B., s. to Kans., Tenn., and
  Va.
  marginalis Harris *in* Johnson, 1925c: 61. Nomen nudum.
bifalcata Doane, 1912: 55.—Calif.
bigeminata Alexander, 1915c: 140.—Nev.
biproducta Alexander, 1947b: 69.—Calif.
bisetosa bisetosa Doane, 1901: 111.—Wash.; Oreg., Calif., Idaho.
  ssp. **percita** Alexander, 1945a: 411.—Wyo.
biuncus Doane, 1912: 58.—Calif.
boregoensis Alexander, 1946b: 45.—Calif.
bucera Alexander, 1927e: 187.—Alta.; Mont.
calcarata Doane, 1901: 107.—Wash.; Oreg., Calif.
carunculata Alexander, 1946c: 186.—Calif.
catawba Alexander, 1915c: 134.—N.C.; S.C.
cladacantha Alexander, 1945d: 130.—Calif.
coconino Alexander, 1946a: 504.—Ariz.
costaloides Alexander, 1915c: 137.—Tex.
degeneri Alexander, 1944a: 166.—Calif.
densursi Alexander, 1943b: 729.—Mont.; Wyo.
diabolica Alexander, 1954: 35.—Calif.; Oreg.
diacanthophora Alexander, 1945e: 33.—Calif.; Oreg.
dido dido Alexander, 1947e: 95.—Calif.
  ssp. **malheurensis** Alexander, 1950a: 156.—Oreg.
dietziana Alexander, 1915a: 501.—D.C.; Kans. to N.Y. and S.C.
dimidiata Dietz, 1921b: 12.—N. Mex.
disjuncta Walker, 1856b: 442.—U.S.; Iowa to Vt., s. to Ill. and Del.
  taughannock Alexander, 1915a: 476.—N.Y.
diversa Dietz, 1921b: 4.—Utah; Wyo., Colo., N. Mex., Ariz.

doaneiana Alexander, 1919c: 195 (n. name for *californica* Doane, 1912).—Calif.
    californica Doane, 1912: 49 (preocc. Doane, 1908).—Calif.
dorsimacula dorsimacula Walker, 1848: 69.—N.S.; B.C. to N.S., s. to Calif. and N.J.
    angustipennis Loew, 1863b: 286 (Cent. 4, no. 19).—Man.
    beaulieui Dietz, 1921c: 301.—Que.
    nubilis Harris, 1835: 595. Nomen nudum.
  ssp. **shasta** Alexander, 1919c: 198 (as sp.).—Calif.; Oreg.
downesi Alexander, 1944a: 171.—Calif.
duplex Walker, 1848: 66.—N.S.; Kans. to N.S. and Fla.
    cinctocornis Doane, 1901: 110.—Pa.
    mingwe Alexander, 1915a: 490.—N.Y.
dupliciformis Alexander, 1940b: 86.—Ill.
evidens Alexander, 1920b: 44.—Calif.
fattigiana Alexander, 1944d: 125.—Ga.
fenderi Alexander, 1954: 36.—Oreg.
filamentosa Alexander, 1959a: 72.—Calif.
flavibasis Alexander, 1918a: 414.—Kans.
flavocauda Doane, 1912: 60.—Ariz. (as Calif., error); N. Mex.
    buenoi Alexander, 1946a: 499.—Ariz.
flavomarginata Doane, 1912: 46.—Calif.
flavoumbrosa Alexander, 1918a: 415.—Kans.; Kans. to Mich., S. C., and Fla.
fuliginosa (Say), 1823: 18 (1859b: 44) (*Ctenophora*).—Mo.; Kans. to Ont. and N.H., s. to N.C.
    speciosa Loew, 1863b: 288 (Cent. 4, no. 22).—Ill.
fulvinodus Doane, 1912: 45.—Wash.
georgiana Alexander, 1915c: 133.—Ga.
    inornata Loew *in* Alexander, 1915c: 134. Nomen nudum.
hastingsae Alexander, 1951: 87.—Calif.
hirsuta Doane, 1901: 113.—Mich.; Wis. to Vt., s. to Pa.
impudica Doane, 1901: 104.—Wash.; Oreg., Colo.
inadusta Alexander, 1946a: 508.—Ariz.
incisa incisa Doane, 1901: 118.—Wash.; B.C., Mont.
    flavicoma Doane, 1912: 57.—Mont.
  ssp. **eriensis** Alexander, 1942b: 285 (as sp.).—Ohio.
  ssp. **kansensis** Alexander, 1918a: 411 (as sp.).—Kans.; Okla.
  ssp. **picturata** Alexander, 1961b: 84.—Ariz.
  ssp. **queres** Alexander, 1946a: 502.—N. Mex.
inusitata Alexander, 1949b: 17.—Calif.
johnsoniana Alexander, 1915a: 505.—Vt.; Mass., Md.
    winnemana Alexander, 1915c: 136.—Md.
kaibabensis Alexander, 1946a: 506.—Ariz.

kirkwoodi Alexander, 1961b: 82.—Ariz.
lamellata Doane, 1901: 105.—Wash.; Oreg., Utah, Mont.
  rangiferina Alexander, 1915a: 498.—Mont.
leechi Alexander, 1938b: 71.—B.C.; Wash., Oreg., Calif.
leiocantha Alexander, 1959c: 132.—Calif.
loewiana Alexander, 1915a: 488.—N.W.T.
  simplex Loew *in* Alexander, 1915a: 490. Nomen nudum.
lucida Doane, 1901: 126.—Idaho; B.C., Wash., Oreg., Calif.
lygropis Alexander, 1920d: 227.—Calif.
lyrifera Dietz, 1921b: 5.—Utah.
macnabi Alexander, 1939a: 92.—Oreg.; Wash.
macracantha Alexander, 1946b: 49.—Ariz.
macrolabis macrolabis Loew, 1864a: 58 (Cent. 5, no. 17).—N.W.T.; Alaska to Labr. and Nfld., s. to Idaho and Mich., Eurasia.
  spectabilis Doane, 1901: 120.—Idaho.
  ssp. **macrolaboides** Alexander, 1918a: 69 (as sp.).—N. Mex.; Alaska, Wyo., Colo.
madina Dietz, 1921b: 6.—Utah.
mainensis Alexander, 1915a: 475.—Maine; Alta. to Nfld., s. to Colo. and Mass.
  laevigata Loew *in* Alexander, 1915a: 476. Nomen nudum.
mallochi Alexander, 1920g: 91.—Ill.; Mo. to Md. and Fla.
mariannae Alexander, 1940c: 153.—Fla.
mariposa Alexander, 1946c: 187.—Calif.
megalabiata megalabiata Alexander, 1915c: 139.—Nev.; Wash., Oreg., Calif.
  ssp. **referta** Alexander, 1947g: 41.—Calif.
megaura Doane, 1901: 112.—Mich.; Minn. to Que., s. to Iowa and Vt.
mesotergata Alexander, 1930b: 277.—Calif.
mitrata Dietz, 1921b: 11.—N. Mex.
miwok Alexander, 1945b: 35.—Calif.
modoc Alexander, 1945e: 35.—Calif.
mohavensis Alexander, 1946b: 47.—Ariz.
mono Alexander, 1945e: 40.—Calif.
monticola Alexander, 1915a: 492.—N.Y.; Ont. to Maine, s. to Pa.
mormon Alexander, 1948a: 27.—Utah.
morrisoni Alexander, 1915a: 507.—Ind.; Kans. to R.I., s. to Miss. and S.C.
occidentalis Doane, 1912: 59.—Calif.
olympia Doane, 1912: 61 (n. name for *concinna* Doane).—Wash.; Oreg., Calif.
  concinna Doane, 1901: 115 (preocc. Philippi, 1865).—Wash.
oxytona Alexander, 1927g: 183.—Fla.

palmarum Alexander, 1947d: 61.—Calif.
parshleyi Alexander, 1915a: 510.—Maine; Alta. to Nfld., s. to Colo.
   and Mass.
   scaphula Loew *in* Alexander, 1915a: 512. Nomen nudum.
pellucida Doane, 1912: 61 (n. name for *clara* Doane).—Wash.;
   Wash. to Calif. and Colo.
   clara Doane, 1901: 107 (preocc. Kirby, 1884).—Wash.
   pyramis Doane, 1912: 53.—Nev.
penicillata Alexander, 1915a: 496.—Hudson Bay Territory.
perfidiosa Alexander, 1945e: 38.—Calif.
perlongipes Johnson, 1909a: 131 (n. name for *filipes* Walker).—Fla.
   longipes (error) Rogers, 1926b: 7.
   filipes Walker, 1848: 65 (preocc. Fabricius, 1805).—Fla.
planicornia Doane, 1912: 52.—Calif.
pleuracicula Alexander, 1915c: 130.—Colo.; Nev., Utah, Ariz.
   arizonica Alexander, 1916b: 53.—Ariz.
   monochroma Dietz, 1919: 88.—Colo.
polingi Alexander, 1950d: 163.—Tex.
polycantha Alexander, 1942a: 209.—N. Mex.; Ariz.
praecisa Loew, 1872a: 51 (Cent. 10, no. 2).—Calif.; Wash., Oreg.
   tingi Alexander, 1939a: 93.—Calif.
pubera Loew, 1864a: 57 (Cent. 5, no. 16).—Calif.; B.C. to Calif.
quaylii Doane, 1909: 18.—Calif.
rabiosa Alexander, 1943a: 154.—Colo.; Wyo., Utah.
rainiericola Alexander, 1946b: 69.—Wash.
ramona Alexander, 1941c: 207.—Colo.
retusa Doane, 1901: 109.—Wash.; B.C., Oreg., Idaho.
rotundiloba Alexander, 1915c: 132.—Tex.
ruidoso Alexander, 1946a: 510.—N. Mex.
sagittifera Alexander, 1948b: 125.—Colo. or Mont.
sanctaeritae Alexander, 1946a: 511.—Ariz.; Utah.
saxemontana Alexander, 1946e: 68.—Wyo.; Alaska, Y.T., B.C., Utah,
   Colo.
saylori Alexander, 1961b: 81.—Calif.; Mexico (Baja Calif.).
seminole Alexander, 1915a: 495.—Ga.; N.C.
sequoiarum Alexander, 1945b: 33.—Calif.
silvestra Doane, 1909: 18.—Calif.
simplex Doane, 1901: 103.—Calif.
sinistra Dietz, 1921b: 8.—Colo.; Wyo.
siskiyouensis Alexander, 1949b: 152.—Oreg.
snoqualmiensis Alexander, 1946c: 190.—Wash.; Oreg.
spaldingi Dietz, 1921b: 7.—Utah.

TIPULIDAE

spatha Doane, 1912: 59.—Ariz.
spernata Dietz, 1921b: 9.—Colo.
sperryana Alexander, 1942a: 208.—Wyo.; Mont.
spinerecta Alexander, 1947b: 70.—Calif.
splendens Doane, 1901: 107.—Wash.; Oreg., Wyo., Ariz.
stalagmites Alexander, 1915c: 130.—N. Mex.
sternata Doane, 1912: 56.—Calif.
    megatergata Alexander, 1920b: 45.—Calif.
submaculata Loew, 1863b: 288 (Cent. 4, no. 23).—Mass.; Wis. to N.S., s. to Tenn. and S.C.
    cuspidata Doane, 1901: 111.—Pa.
    bidens Loew *in* Alexander, 1915a: 464. Nomen nudum.
subtilis Doane, 1901: 106.—Calif.
sylvicola Doane, 1912: 53.—Wash.; Calif.
tanneri Alexander, 1948a: 30.—Utah; Colo.
tenaya Alexander, 1946c: 142.—Calif.
tergata Doane, 1912: 56.—Nev.; Calif.
texensis Alexander, 1916b: 48.—Tex.
timberlakei Alexander, 1947b: 71.—Calif.
translucida Doane, 1901: 109.—Pa.; Ill. to Pa., s. to Okla. and S.C.
    devia Dietz, 1919: 86.—Md.
triplex triplex Walker, 1848: 66.—N.S.; Alta. to Nfld., s. to Wis. and Va.
    inermis Doane, 1901: 112.—Mich.
  ssp. **colei** Alexander, 1942c: 67.—Tenn.
  ssp. **integra** Alexander, 1962: 4.—Ill.
  ssp. **linearis** Alexander, 1940a: 617.—Tenn.
trispinosa trispinosa Lundström.—Not Nearctic.
  ssp. **satyr** Alexander, 1915c: 129 (as sp.).—Colo.; Mont. to Que., s. to Colo.
    claasseni Alexander, 1920d: 111.—Colo.
triton Alexander, 1915a: 487.—Ky.; Ind., D.C., S.C., Ga.
truculenta Alexander, 1943d: 144.—Calif.
tuscarora Alexander, 1915a: 493.—Va.; Mo. to Md., s. to S.C.
    hamata Alexander, 1915a: 495. Nomen nudum.
twightae Alexander, 1959c: 130.—Calif.
umbrosa Loew, 1863b: 292 (Cent. 4, no. 31).—La.; Fla.
ungulata Doane, 1912: 54.—Calif.
unicincta unicincta Doane, 1901: 115.—Idaho; Wash., Oreg., Wyo.
  ssp. **bifila** Alexander, 1954: 41.—Oreg.
usitata usitata Doane, 1901: 124.—Wash.; Oreg.
  ssp. **aurantionota** Alexander, 1946c: 189 (as sp.).—Oreg.
utahicola Alexander, 1948a: 32.—Utah.

valida valida Loew, 1863b: 287 (Cent. 4, no. 21).—Ill.; Minn. to Nfld., s. to Ill. and N.C.
    calva Doane, 1901: 114.—Mich.
    ssp. **atricornis** Alexander, 1940a: 616.—Tenn.; N.C.
vestigipennis Doane, 1908d: 47.—Calif.
vitabilis Alexander, 1947g: 39.—Calif.
vittatipennis Doane, 1912: 61 (n. name for *albovittata* Doane).—Wash.; Oreg., Calif.
    albovittata Doane, 1901: 119 (preocc. Macquart, 1838).—Wash.
williamsii Doane, 1909: 17.—Calif.
willissmithi Alexander, 1945a: 414.—Wyo.
woodi Alexander, 1948a: 34.—Utah.
yosemite Alexander, 1946c: 188.—Calif.
youngi Alexander, 1927f: 218.—Alta.; Alaska to Nfld., s. to Wis. and N.Y.
zelotypa Alexander, 1946e: 177.—Calif.

Subgenus HESPEROTIPULA Alexander

Tipula, subg. **Hesperotipula** Alexander, 1947d: 63. Type-species, *streptocera* Doane (orig. des.).

aitkeniana Alexander, 1944a: 61.—Calif.; Wash.
californica (Doane), 1908b: 176 (*Pachyrhina*).—Calif.; B.C., Wash., Oreg.
    xanthomela Dietz, 1918: 107 (*Pachyrhina;* unjustified n. name for *californica* Doane, 1908).—Calif.
chumash Alexander, 1961a: 13.—Calif.
circularis Alexander, 1947d: 64.—Calif.
contortrix Alexander, 1944a: 169.—Calif.
coronado Alexander, 1946a: 497.—Ariz.
derbyi Doane, 1912: 47.—Calif.
fragmentata Dietz, 1919: 87.—Calif.; Wash., Oreg.
linsdalei linsdalei Alexander, 1951: 85.—Calif.
    ssp. **obispoensis** Alexander, 1962: 4.—Calif.
micheneri Alexander, 1944a: 168.—Calif.
mutica Dietz, 1919: 91.—Calif.
opisthocera Dietz, 1919: 90.—Calif.
ovalis Alexander, 1951: 86.—Calif.
sanctaeluciae Alexander, 1951: 89.—Calif.
streptocera streptocera Doane, 1901: 113.—Idaho; B.C., Wash., Oreg.
    ssp. **pallidocera** Dietz, 1919: 92 (as var.).—Wash. ?Ssp.
supplicata Alexander, 1944a: 170.—Calif.

sweetae Alexander, 1930b: 278.—Calif.
trypetophora Dietz, 1919: 89.—B.C.

Unplaced Species of *Tipula*

frigida Walker, 1848: 68.—N.S. Unrecognized.
hewitti Alexander, 1919b: 14.—N.W.T. ?Subg. *Oreomyza*.
johanseni Alexander, 1919b: 11.—N.W.T. ?Subg. *Oreomyza*.
katmaiensis Alexander, 1920i: 202.—Alaska. ?Subg. *Oreomyza*.
lionota Holmgren, 1883: 188.—n. Siberia; Alaska. ?Subg. *Oreomyza*.
    coracina Alexander, 1918a: 70.—Alaska.
maculatipennis Say, 1824a: 359 (1859a: 243).—Northwest Territory (U.S.). Unrecognized.
    maculipennis, emend.
puncticornis Macquart, 1850: 319 (1850: 15).—N. Amer. Unrecognized.
subpolaris Alexander, 1919b: 14.—N.W.T. ?Subg. *Oreomyza*.
trimaculata (Emmons), 1854: pl. 29, fig. 5 (*Ctenophora;* preocc. Macquart, 1838).—?N.Y. Unrecognized.

Subfamily CYLINDROTOMINAE

The larvae are plant feeders, occurring in wet mosses and similar plants or, in the case of *Cylindrotoma*, on terrestrial angiosperms, including both monocotyledonous and dicotyledonous genera.

REFERENCE: Alexander, 1920h (biol.).

Genus CYLINDROTOMA Macquart

Cylindrotoma Macquart, 1834a: 107. Type-species, *Limnobia distinctissima* Meigen (Westwood, 1840: 128).

americana Osten Sacken, 1865: 236.—N.H.; Mich. to Nfld., s. to Pa.
juncta juncta Coquillett, 1900h: 401 (1904: 15).—Alaska; B.C., Wash., Oreg.
    splendens Doane, 1900: 197.—Alaska.
  ssp. **pallescens** Alexander, 1930b: 280 (as sp.).—Colo.; Wyo.
tarsalis Johnson, 1912a: 2.—N.Y.; N.B., Vt., Conn.
    anomala Johnson, 1912a: 2.—N.Y.

Genus PHALACROCERA Schiner

Phalacrocera Schiner, 1863: 224. Type-species, *Limnobia nudicornis* Schummel (orig. des.)=*replicata* (Linnaeus).

neoxena Alexander, 1914a: 603.—Ont.; Mich. to Que. and Pa.
occidentalis Alexander, 1927c: 10.—Wash.; Oreg. ?*Cylindrotoma*.

tipulina Osten Sacken, 1865: 241.—N.H.; Wis. to Nfld., s. to Pa.
vancouverensis Alexander, 1927e: 189.—B.C.

Genus LIOGMA Osten Sacken

Liogma Osten Sacken, 1869b: 298. Type-species, *Triogma nodicornis* Osten Sacken (Coquillett, 1910b: 561).
nodicornis nodicornis (Osten Sacken), 1865: 239 (*Triogma*).—N.Y., D.C.; Alta. to Nfld., s. to Ill. and D.C.
   ssp. **flaveola** Alexander, 1919c: 195.—Va.; Ind., D.C., Tenn., N.C.

Genus TRIOGMA Schiner

Triogma Schiner, 1863: 223. Type-species, *Limnobia trisulcata* Schummel (orig. des.).
exsculpta Osten Sacken, 1865: 239.—Pa.; Mich. to N.H. and N.J.
   exculpta, error.

Subfamily LIMONIINAE

The subfamily Limoniinae is the largest in the family Tipulidae, and includes small to medium-sized forms. Only a few species in our fauna approach the normal size of the Tipulinae (see *Pedicia*, *Limnophila*, subg. *Eutonia*). Four tribes, the Limoniini, Pediciini, Hexatomini, and Eriopterini, are included.

Tribe LIMONIINI

In our area the tribe Limoniini contains six genera, each of which represents a different subtribe. The immature stages of *Helius* live in saturated, silty sand or organic mud at the margins of marshes (Rogers, 1933a). *Antocha* is strictly aquatic under lotic conditions (Alexander, 1920h). *Elliptera* is semi-aquatic to aquatic, frequenting cliffs or similar hard surfaces that are constantly wet by percolating water. *Dicranoptycha* lives in moderately moist soil beneath a covering of leaf mold (Alexander, 1920h; Rogers, 1933a). The early stages of *Orimarga* occur in rotten wet logs in an advanced stage of decay (Rogers, 1927a, 1933a). The immature stages of *Limonia* are discussed under that genus.

Genus LIMONIA Meigen

Amphinome Meigen, 1800: 15 (preocc. Bruguière, 1792). Type-species, *Tipula tripunctata* Fabricius (Coquillett, 1910b: 505). Suppressed by I.C.Z.N., 1963b: 339.

Limonia Meigen, 1803: 262. Type-species, *Tipula tripunctata* Fabricius (Westwood, 1840: 129).
Limnobia Meigen, 1818: 116 (unjustified n. name for *Limonia* Meigen). Type-species, *Tipula tripunctata* Fabricius (aut.).
Taphrophila Rondani, 1856: 185. Type-species, *Limnobia inusta* Meigen (orig. des.).

The genus *Limonia* is the largest in the Diptera, with nearly 2,000 species distributed in more than a score of subgenera. As might be expected in such a diverse group, the immature stages frequent a range of habitats almost equivalent to that for the entire family. These habitats include semi-aquatic situations such as saturated mats or beds of algae, liverworts, and mosses (subgenera *Limonia*, *Dicranomyia*, *Geranomyia*); decaying vegetable matter; fungi, including rotting wood permeated by mycelia (*Limonia*, *Metalimnobia*); decaying wood (*Rhipidia*, *Discobola*); and mines in leaves (*Dicranomyia*). A few species are marine or virtually so (*Limonia floridana*, *L. marmorata*).

REFERENCE: Doane, 1908a (tax.).

Subgenus LIMONIA Meigen

badia (Walker), 1848: 46 (*Limnobia*).—N.S.
  humidicola, authors, not Osten Sacken.
bistigma (Coquillett), 1905d: 57 (*Limnobia;* as *bestigma*).—B.C.; Wash., Oreg., Calif.
  tributaria Alexander, 1943d: 14.—Calif.
borealis (Doane), 1900: 187 (*Dicranoptycha*).—Alaska. ?=*tripunctata* ssp.
globithorax (Osten Sacken), 1869b: 74 (*Dicranomyia*).—D.C., N.H.; Wis. to Nfld., s. to Tenn. and Fla.
indigena indigena (Osten Sacken), 1859: 215 (*Limnobia*).—D.C., Maine, Wis.; Iowa to Nfld., s. to Tenn. and S.C.
  ssp. **jacksoni** (Alexander), 1917: 199 (*Limnobia*).—Colo.; Wyo., Utah, N. Mex.
  ssp. **loloensis** Alexander, 1958b: 217.—Idaho.
indigenoides (Alexander), 1920i: 193 (*Limnobia*).—Alaska; Maine. ?=*badia*.
macateei (Alexander), 1916b: 42 (*Dicranomyia*).— Md.; Ind. to Maine, s. to Tenn. and Fla.
  varipes Dietz, 1921a: 241 (*Dicranomyia*).—Pa.
maculicosta (Coquillett), 1905d: 57 (*Limnobia*).—B.C.; Alaska to Vt., s. to Calif. and Va., Japan.
nubeculosa nubeculosa (Meigen).—Not Nearctic.
  ssp. **sciophila** (Osten Sacken), 1877: 197 (*Limnobia;* as sp.).— Calif.; Alaska to Calif. and Colo. ?Syn.

parietina (Osten Sacken), 1861: 289 (*Limnobia*).—N.Y.; Mich. to Maine, s. Ill. and N.C.
sociabilis (Osten Sacken), 1869b: 95 (*Limnobia*).—Ill.; Iowa.
tripunctata (Fabricius).—Not Nearctic.
tristigma (Osten Sacken), 1859: 216 (*Limnobia*).—Ill.; Alta. to N.B., s. to Tenn. and N.C.

Subgenus METALIMNOBIA Matsumura

Metalimnobia Matsumura, 1911: 63 (as genus). Type-species, *vittata* Matsumura (mon.) = *quadrimaculata* ssp.

annulus annulus (Meigen).—Not Nearctic.
    ssp. **cinctipes** (Say), 1823: 21 (1859b: 47) (*Limnobia;* as sp.).—Mo.; Alta. to Nfld., s. to Miss. and Fla.
    ssp. **triphaea** Alexander, 1954: 43.—Oreg.
californica (Osten Sacken), 1861: 288 (*Limnobia*).—Calif.; B.C., Wash., Oreg.
dietziana Alexander, 1927f: 219 (n. name for *gracilis* Dietz).—Sask.
    gracilis Dietz, 1915a: 329 (*Limnobia;* preocc. Wiedemann, 1828).—Sask.
fallax (Johnson), 1909a: 125 (*Limnobia*).—N.J., Pa.; Mich. to N.J., s. to Okla. and N.C.
hudsonica (Osten Sacken), 1861: 289 (*Limnobia*).—N.W.T.; Alaska to Nfld., s. to B.C. and N.Y.
immatura (Osten Sacken), 1859: 214 (*Limnobia*).—Wis., Maine, D.C.; B.C. to Maine, s. to Fla.
novaeangliae Alexander, 1929f: 44.—Mass.; Maine, Conn., N.Y.
solitaria (Osten Sacken), 1859: 215 (*Limnobia*).—N.Y.; Alaska to N.S., s. to Minn. and Mass.
triocellata (Osten Sacken), 1859: 216 (*Limnobia*).—Wis., N.Y., D.C.; Alta. to N.S., s. to Tenn. and Ga.

Subgenus DISCOBOLA Osten Sacken

Discobola Osten Sacken, 1865: 226 (as genus). Type-species, *Limnobia argus* Say (Coquillett, 1910b: 534) = *annulata* (Linnaeus).
Trochobola Osten Sacken, 1869b: 97 (unjustified n. name for *Discobola* Osten Sacken). Type-species, *Limnobia argus* Say (aut.) = *annulata* (Linnaeus).

annulata (Linnaeus), 1758: 586 (*Tipula*).—Europe; B.C. to Nfld., s. to Oreg., Tenn., and Va., Eurasia, s. to New Guinea.
    argus Say, 1824a: 358 (1859a: 243) (*Limnobia*).—Minn.

neoelegans Alexander, 1954: 48 (n. name for *elegans* Doane).—Wash.; B.C., Idaho, Oreg., Calif.
  elegans Doane, 1900: 186 (*Trochobola*; preocc. Wiedemann, 1830).—Wash.
nigroclavata Alexander, 1942b: 310.—N.Y.; Maine, Mass.

Subgenus RHIPIDIA Meigen

Rhipidia Meigen, 1818: 153 (as genus). Type-species, *maculata* Meigen (mon.)=*lecontei* Alexander.
Rhipidia, subg. *Arhipidia* Alexander, 1912a: 6. Type-species, *domestica* Osten Sacken (Alexander, 1950b: 195).
Rhipidia, subg. *Monorhipidia* Alexander, 1912a: 6. Type-species, *fidelis* Osten Sacken (Alexander, 1950b: 195).
bryanti (Johnson), 1909a: 123 (*Rhipidia*).—Mass.; Colo. to Maine, s. to Ariz. and Fla.
domestica (Osten Sacken), 1859: 208 (*Rhipidia*).—D.C.; Kans. and Iowa to N.J., s. to Tex. and Fla., Neotropical.
fidelis (Osten Sacken), 1859: 209 (*Rhipidia*).—N.Y.; Alta. to N.S., s. to Oreg., Tenn., and Fla.
gaspicola Alexander, 1941c: 87.—Que.
huachucensis Alexander, 1955a: 127.—Ariz.
lecontei Alexander, 1940a: 624 (as *maculata* ssp.).—Tenn.; Alaska to Nfld., s. to Calif., Tenn., and Va., Eurasia.
  maculata Meigen, 1818: 153 (*Rhipidia*; preocc. Meigen, 1804).—Europe.
schwarzi (Alexander), 1912a: 13 (*Rhipidia*).—Santo Domingo; s. Fla. to Venezuela.
shannoni (Alexander), 1914a: 581 (*Rhipidia*).—Md.; Ind., Tenn., Md. to Fla., Neotropical.

Subgenus DICRANOMYIA Stephens

Furcomyia Meigen, 1818: 133. Type-species, *Limonia lutea* Meigen (Coquillett, 1910b: 546). Unavailable name, cited in specific synonymy.
Dicranomyia Stephens, 1829a: 53 (also 1829b: 243) (as genus). Type-species, *Limnobia modesta* Meigen (Coquillett, 1910b: 533).

acerba Alexander, 1943d: 16.—Wash.; Oreg.
adirondacensis (Alexander), 1922a: 62 (*Dicranomyia*).—N.Y.; Wis. to Que., s. to N.C.
alascaensis (Alexander), 1919b: 4 (*Dicranomyia*).—Alaska.
anteapicalis Alexander, 1946d: 155.—Calif.; Wash.
aquita (Dietz), 1915a: 331 (*Dicranomyia*).—N.W.T. Unrecognized.

athabascae (Alexander), 1927f: 221 (*Dicranomyia*).—Alta.; B.C. to Alta., s. to Calif. and Wyo.
brevivena (Osten Sacken), 1869b: 66 (*Dicranomyia*).—N.Y., D.C.; Oreg. to Nfld., s. to Fla.
brevivenula Alexander, 1929h: 247 (n. name for *flavescens* Dietz).—Pa.
    flavescens Dietz, 1921a: 239 (preocc. Macquart, 1834).—Pa.
brunnea (Doane), 1900: 184 (*Dicranomyia*).—Mass. Unrecognized.
catalinae Alexander, 1944c: 91.—Calif.
chorea (Meigen), 1818: 134 (*Limnobia*).—Europe; B.C.
citrina (Doane), 1900: 183 (*Dicranomyia*).—Wash.; Oreg., Calif., Wyo.
cramptoniana Alexander, 1929h: 247 (n. name for *cramptoni* Alexander).—Mass.
    cramptoni Alexander, 1926c: 47 (*Dicranomyia;* preocc. Alexander, 1912).—Mass.
ctenopyga Alexander, 1943a: 156.—Colo.; Colo. to Mich.
distans (Osten Sacken), 1859: 211 (*Dicranomyia*).—Fla.; Md. to Ala. and Fla., Neotropical, also Calif.
    cervina Doane, 1908a: 8 (*Dicranomyia*).—Calif.
diversoides (Dietz), 1921a: 240 (*Dicranomyia*).—Pa. Unrecognized.
divisa Alexander, 1929h: 247 (n. name for *diversa* Osten Sacken, 1859: 212).—D.C., Md.; Iowa to Mass., s. to Mo. and Fla., Greater Antilles.
    diversa Osten Sacken, 1859: 212 (*Dicranomyia;* preocc. Osten Sacken, 1859: 207).—D.C., Md.
floridana (Osten Sacken), 1869b: 67 (*Dicranomyia*).—Fla.; Md., Va., S.C.
fulva (Doane), 1900: 185 (*Dicranomyia*).—Wash.; Oreg., Calif., Wyo.
    fulvoides Alexander, 1943b: 737 (as ssp.).—Wyo.
geronimo Alexander, 1949b: 101.—Ariz.
geyserensis Alexander, 1943b: 735.—Wyo.
gladiator (Osten Sacken), 1859: 212 (*Dicranomyia*).—D.C.; Alta. to Maine, s. to Colo. and Ga.
haeretica (Osten Sacken), 1869b: 70 (*Dicranomyia*).—N.Y.; Mich. to Nfld. and R.I.
    gibsoni Alexander, 1929b: 17 (*Dicranomyia*).—N.S.
halterata (Osten Sacken), 1869b: 71 (*Dicranomyia*).—Labr.; Alaska to Labr. and Nfld., s. to B.C. and Pa., Eurasia.
    cinereipennis Lundström, 1912: 52 (*Dicranomyia*).—Finland.
halterella (Edwards), 1921a: 201 (*Dicranomyia*).—Scotland; Wash. and Utah to Maine, Europe.
    gracilis Doane, 1900: 184 (*Dicranomyia;* preocc. Wiedemann, 1828).—Idaho.

helva (Doane), 1900: 183 (*Dicranomyia*).—Colo. Unrecognized.
homichlophila Alexander, 1958b: 129.—Calif.
humidicola (Osten Sacken), 1859: 210 (*Dicranomyia*).—N.Y., Conn., D.C.; B.C. to N.S., s. to Calif., Cent. Amer., and Tenn.
  viridicans Doane, 1908a: 7 (*Dicranomyia*).—Calif.
  badia, authors, not Walker.
illustris Alexander, 1944b: 91.—Oreg.; Wash.
immanis Alexander, 1950d: 164.—Mich.
immodesta (Osten Sacken), 1859: 211 (*Dicranomyia*).—Maine, N.Y., D.C.; Alta. to N.S., s. to Iowa and S.C.
immodestoides Alexander, 1919i: 327.—Japan; Oreg. to Nfld., s. to Ind.
  iowensis Rogers, 1926a: 50 (*Dicranomyia*).—Iowa.
inhabilis Alexander, 1949b: 154.—Oreg.
intricata (Alexander), 1927f: 221 (*Dicranomyia*).—Alta.
isabellina (Doane), 1900: 183 (*Dicranomyia*).—Pa. Unrecognized.
lacroixi (Alexander), 1926c: 46 (*Dicranomyia*).—Mass.
liberta (Osten Sacken), 1859: 209 (*Dicranomyia*).—Ala., D.C., Ga., Wis.; Man. to Nfld., s. to Okla. and Fla., Bermuda.
libertoides (Alexander), 1912b: 361 (*Furcomyia*).—Calif.; Oreg.
linsdalei Alexander, 1943f: 253.—Calif.
longipennis (Schummel), 1829: 104 (*Limnobia*).—Europe; Oreg. and Man. to Maine, s. to Colo. and Pa., Eurasia.
  immemor Osten Sacken, 1861: 287 (*Dicranomyia*).—N.Y.
magnicauda magnicauda (Lundström).—Not Nearctic.
  ssp. **broweriana** Alexander, 1941d: 193.—Maine.
marmorata (Osten Sacken), 1861: 288 (*Dicranomyia*).—Calif.; B.C. to Calif. ?Subg. *Idioglochina* Alexander.
  signipennis Coquillett, 1905d: 56 (*Dicranomyia*).—Calif.
  rhipidioides Alexander, 1918a: 381 (*Dicranomyia*).—Calif.
melanderi melanderi Alexander, 1945c: 1.—Idaho; Oreg.
  ssp. **tharpiana** Alexander, 1959a: 74.—Calif.
melleicauda (Alexander), 1917: 22 (*Dicranomyia*).—Colo.; Man., B.C.
michigana Alexander, 1950d: 166.—Mich.
modesta (Meigen), 1818: 134 (*Limnobia*).—Europe; N.W.T., Utah, Greenland, Eurasia.
moniliformis (Doane), 1900: 184 (*Dicranomyia*).—Colo.; B.C. to Ont., s. to Oreg., Colo., and Mich.
  penicillata Alexander, 1927j: 7 (*Dicranomyia*).—N. Dak.
  haeretica, authors, not Osten Sacken.
morioides (Osten Sacken), 1860: 17 (*Dicranomyia*).—N.Y.; Alaska to Nfld., s. to Colo. and N.C.

murina (Zetterstedt), 1851: 3882 (*Limnobia*).—n. Europe; Alta., Eurasia.

    platyrostra Alexander, 1927f: 223 (*Dicranomyia*).—Alta.

neomorio (Alexander), 1927f: 220 (*Dicranomyia;* as *rufiventris* ssp.).—Alta.; Alta. to Maine.

nielseniana Alexander, 1949c: 39.—Wyo.; Oreg.

nycteris (Alexander), 1927f: 220 (*Dicranomyia*).—Alta.; Alta. and Wyo. to Que. and N.B.

particeps (Doane), 1908a: 7 (*Dicranomyia*).—Wash.; Utah.

    uinta Alexander, 1948a: 38.—Utah.

pennsylvanica (Dietz), 1921a: 239 (*Dicranomyia*).—Pa. Unrecognized.

piscataquis Alexander, 1941c: 86.—Maine; Wash., Oreg., Greenland.

    vibei Nielsen, 1951: 185.—Greenland.

profunda (Alexander), 1925a: 173 (*Dicranomyia*).—Mass.; Que. to Nfld., s. to N.Y. and Mass.

pudica (Osten Sacken), 1859: 212 (*Dicranomyia*).—Ill.; Mich. to Maine, s. to Ill. and N.C.

pudicoides Alexander, 1929a: 299.—Tenn.; Ind. and Tenn. to Maine.

rostrifera (Osten Sacken), 1869b: 65 (*Dicranomyia*).—N.Y.; B.C. to N.Y. and Maine.

sera sera (Walker).—Not Nearctic.

    ssp. **erostrata** Alexander, 1930b: 71 (as sp.).—Utah.

    ssp. **forcipula** Meijere, 1918: 128 (*Dicranomyia;* as sp.).—Holland; Man., Europe.

sphagnicola (Alexander), 1925a: 173 (*Dicranomyia*).—Mass.; Colo. to Nfld.

spinifera (Alexander), 1927a: 229 (*Dicranomyia*).—Mass.; Mich. to N.B., s. to N.C.

stigmata (Doane), 1900: 185 (*Dicranomyia*).—Calif.; Wash.

stulta (Osten Sacken), 1859: 210 (*Dicranomyia*).—N.Y., Que., Va.; Wis. to Que. and Maine, s. to Ga.

    monticola Alexander, 1911c: 201 (*Furcomyia*).—Ga.

terraenovae (Alexander), 1920g: 85 (*Dicranomyia*).—Nfld.; Alaska to Nfld., s. to Colo., Europe.

    tenuipes Zetterstedt, 1838: 837 (*Limnobia;* preocc. Say, 1823).—Europe. **N. syn.**

    decora Staeger, 1840: 47 (*Limnobia;* preocc. Haliday, 1833).—Europe. **N. syn.**

uliginosa Alexander, 1929a: 27.—N.Y.; Mich. to Que., s. to Tenn. and S.C.

vulgata (Bergroth), 1888a: 194 (*Dicranomyia*).—Alaska; Alaska, s. to Calif. and Wyo.

    ochracea Doane, 1900: 182 (*Dicranomyia*).—Idaho.

walleyi Alexander, 1942b: 325.—Ont.; Que., N.B., Maine.
willamettensis Alexander, 1949b: 155.—Oreg.; Wyo.
ypsilon Alexander, 1959a: 47.—Calif.; Wash., Oreg., Ariz.

Subgenus ALEXANDRIARIA Garrett

Alexandriaria Garrett, 1922c: 60 (as genus). Type-species, *suffusca* Garrett (orig. des.).
phalangioides Alexander, 1943b: 738.—Wyo.
suffusca (Garrett), 1922c: 60 (*Alexandriaria*).—B.C.
    intermedia Garrett, 1922c: 60 (*Alexandriaria*).—B.C.
    kooteniensis Garrett, 1922c: 61 (*Alexandriaria*).—B.C.
whartoni (Needham), 1908: 211 (*Dicranomyia*).—Mich.

Subgenus GERANOMYIA Haliday

Geranomyia Haliday, 1833: 154 (as genus). Type-species, *unicolor* Haliday (mon.).
canadensis (Westwood), 1835b: 684 (*Limnobiorhynchus*).—Canada; Alaska to Nfld., s. to Calif. and Fla.
communis (Osten Sacken), 1859: 207 (*Geranomyia*).—Wis., D.C. Confused in literature with *canadensis*.
costomaculata (Dietz), 1921a: 237 (*Geranomyia*).—Pa. ?=*diversa*.
distincta (Doane), 1900: 186 (*Geranomyia*).—Mass.; Mich. to Nfld., s. to Tex. and Fla.
diversa (Osten Sacken), 1859: 207 (*Geranomyia*).—N.Y.; Oreg. to Maine, s. to Calif. and S.C.
ibis (Alexander), 1916a: 493 (*Geranomyia*).—Ark.; Calif. to Ark.
parapentheres Alexander, 1948a: 40.—Utah.
perfecta (Alexander), 1928c: 109 (*Geranomyia*).—Mexico; Ariz.
remingtoni Alexander, 1947c: 247.—La.
rostrata (Say), 1823: 22 (1859b: 47) (*Limnobia*).—Md., Pa.; Mich. to Nfld., s. to La. and Fla., Greater Antilles.
valverdensis Alexander, 1946a: 514.—Tex.
vanduzeei (Alexander), 1916a: 488 (*Geranomyia*).—Fla.
virescens (Loew), 1851: 398 (*Aporosa*).—Virgin Is. (St. Thomas I.); s. Fla., Neotropical.
zionana Alexander, 1948a: 41.—Utah.

Unplaced Species of *Limonia*

apicata apicata (Alexander).—Not Nearctic.
    ssp. **subapicata** Alexander, 1931b: 178 (as sp.).—Fla.
argenteceps (Alexander), 1912d: 163 (*Limnobia*).—Ariz.; Colo., N. Mex.

defuncta defuncta (Osten Sacken), 1859: 213 (*Limnobia*).—Que., Maine, N.Y., D.C.; Wis. to Nfld., s. to Ill. and S.C.
  pellucidiguttata Dietz, 1921a: 242 (*Dicranomyia*).—Pa.
  simulans, authors, not Walker.
  ssp. **concinna** (Williston), 1893c: 60 (*Limnobia;* as sp.).—Wash.; B.C. to Wyo., s. to Calif. and N. Mex.
fusca Meigen, 1804: 54.—Europe; Mich. to Nfld., s. to Ga.
  turpis Walker, 1856b: 434 (*Limnobia*).—Canada.
  pubipennis Osten Sacken, 1859: 211 (*Dicranomyia*).—Md.
infuscata (Doane), 1900: 185 (*Dicranomyia*).—Idaho; Alaska to Calif. and Wyo.
  adjecta Doane, 1908a: 8 (*Dicranomyia*).—Calif.
  nitidiuscula Alexander, 1927j: 6.—Oreg.
kuschei Alexander, 1958b: 215.—Ariz.
nelliana (Alexander), 1914a: 579 (*Dicranomyia*).—Colo.
neonebulosa Alexander, 1924e: 555 (n. name for *nebulosa* Alexander).—Japan; Mass., China.
  nebulosa Alexander, 1913b: 203 (*Dicranomyia;* preocc. Zetterstedt, 1838).—Japan.
rara (Osten Sacken), 1869b: 75 (*Dicranomyia*).—N.Y.; Iowa and Wis. to N.Y., s. to Fla.
reticulata (Alexander), 1912b: 334 (*Furcomyia*).—Cuba; s. Fla., Mexico.
rogersiana rogersiana (Alexander), 1926c: 45 (*Dicranomyia*).—Fla.; Ga.
  ssp. **longistylata** Alexander, 1929f: 45.—Fla.
shelfordi Alexander, 1944d: 243.—Ky.; Mich., W. Va.
simulans (Walker), 1848: 45 (*Limnobia*).—Ont.; Que., Maine.
  pemetica Alexander, 1939a: 95.—Maine.
venusta (Bergroth), 1888a: 193 (*Dicranomyia*).—Alaska; Alaska to Calif. and N. Mex.
  duplicata Doane, 1900: 185 (*Dicranomyia*).—Wash.
  negligens Alexander, 1927j: 8 (*Dicranomyia*).—Colo.
yellowstonensis Alexander, 1945d: 155.—Wyo.

Genus HELIUS Lepeletier and Serville

Megarhina Lepeletier and Serville *in* Latreille et al., 1828: 585. Type-species, *Limnobia longirostris* Meigen (mon.).
Helius Lepeletier and Serville *in* Latreille et al., 1828: 831 (unjustified n. name for *Megarhina* Lepeletier and Serville; as Hélius). Type-species, *Limnobia longirostris* Meigen (aut.). [Although Hélius is an invalid name because proposed in the vernacular and was an unjustified new

name for the valid name *Megarhina*, the author prefers to use this name in order not to upset long established usage.—Eds.]

Rhamphidia Meigen, 1830: 281. Type-species, *Limnobia longirostris* Meigen (Westwood, 1840: 129).

Subgenus HELIUS Lepeletier and Serville

flavipes (Macquart), 1855: 37 (1855: 17) (*Rhamphidia*).—Md.; Alta. to N.S., s. to Tex. and Fla.
prominens Walker, 1856b: 435 (*Limnobia*).—N. Amer.
brevirostris Osten Sacken, 1859: 222 (*Rhamphidia*).—Wis.
mainensis (Alexander), 1916a: 498 (*Rhamphidia*).—Maine; Wis. to Maine, s. to Ill. and Md.

Genus ANTOCHA Osten Sacken

Antocha Osten Sacken, 1859: 219. Type-species, *saxicola* Osten Sacken (mon.).

Subgenus ANTOCHA Osten Sacken

biarmata Alexander, 1940a: 620.—Tenn.; N.Y.
capitella Alexander, 1941b: 13.—Tenn.
decurvata Alexander, 1941b: 12.—Tenn.
monticola Alexander, 1917: 23.—Colo.; Y.T., s. to Calif., Colo., and Mexico.
obtusa Alexander, 1925b: 201.—Mich.; Que., N.Y.
opalizans Osten Sacken, 1859: 220.—Que., N.Y., Ga.; Wis. to Que. and Maine, s. to Ga.
saxicola Osten Sacken, 1859: 219.—D.C.; Mich. to Nfld., s. to Mo. and Ga.

Genus THAUMASTOPTERA Mik

Thaumastoptera Mik, 1866: 302. Type-species, *calceata* Mik (mon.).

Unnamed sp.—Calif.

Genus ELLIPTERA Schiner

Elliptera Schiner, 1863: 222. Type-species, *omissa* Schiner (orig. des.).

astigmatica Alexander, 1912d: 164.—B.C.; Alaska to Calif. and Colo.
clausa Osten Sacken, 1877: 198.—Calif.
coloradensis Alexander, 1920d: 109.—Colo.; Wyo.
illini Alexander, 1920g: 86.—Ill.; Tenn.
tennessa Alexander, 1926e: 114.—Tenn.

Genus DICRANOPTYCHA Osten Sacken

Dicranoptycha Osten Sacken, 1859: 217. Type-species, *germana* Osten Sacken (Coquillett, 1910b: 533).

acanthophallus Alexander, 1940a: 621.—Tenn.
australis Alexander, 1926f: 55.—Fla.; Ga.
elsa Alexander, 1929a: 28.—N.Y.; N.Y. to N.C.
germana Osten Sacken, 1859: 217.—N.Y.; Mich. to Que. and Maine, s. to Ill. and N.C.
laevis Alexander, 1948c: 131.—Calif.
megaphallus Alexander, 1926f: 57.—Tenn.; N.C., S.C., Fla.
melampygia Alexander, 1950e: 81.—Calif.; Oreg.
microphallus Alexander, 1947c: 249.—Ga.
minima Alexander, 1919e: 21.—Kans.; Ill.
nigripes Osten Sacken, 1859: 218.—Ga.; N.C.
nigrogenualis Alexander, 1949a: 291.—Wash.; Oreg.
occidentalis Alexander, 1927j: 10.—Calif.; Wash., Oreg.
pallida Alexander, 1926f: 58.—Kans.; Ind.
quadrivittata Alexander, 1919c: 191 (as *sobrina* ssp.).—Colo.; Oreg. to Mont., s. to Ariz.
rogersi Alexander, 1927h: 55.—Fla.
septemtrionis Alexander, 1926f: 56.—Mass.; Mich. to Mass., s. to Ind. and N.C.
sobrina Osten Sacken, 1859: 218.—D.C.; Ind. to N.J., s. to Tenn. and Fla.
sororcula Osten Sacken, 1859: 218.—Ga.
spinosissima Alexander, 1950e: 83.—Calif.; Oreg.
stenophallus Alexander, 1950e: 82.—Calif.; Wash., Oreg.
tennessa Alexander, 1941d: 195.—Tenn.
tigrina Alexander, 1919e: 21.—Kans.; Ind., Ill., Tenn., N.C.
winnemana Alexander, 1916a: 500.—Md.; Mo. to Md., s. to Fla.

Genus ORIMARGA Osten Sacken

Orimarga Osten Sacken, 1869b: 120. Type-species, *Limnobia alpina* Zetterstedt (Coquillett, 1910b: 580) = *attenuata* (Walker).

Subgenus ORIMARGA Osten Sacken

arizonensis Coquillett, 1902g: 83.—Ariz.
sanctaeritae Alexander, 1946a: 515.—Ariz.
wetmorei Alexander, 1920g: 87.—Fla.
zionensis Alexander, 1948a: 43.—Utah.

Subgenus DIOTREPHA Osten Sacken

Diotrepha Osten Sacken, 1878c: 219 (as genus). Type-species, *mirabilis* Osten Sacken (mon.).
Thambeta Williston, 1896a: 32. Type-species, *Diotrepha mirabilis* Osten Sacken (present des.).

mirabilis (Osten Sacken), 1878c: 220 (*Diotrepha*).—Ga.; Ind., Tex., Fla., Greater Antilles.

Tribe PEDICIINI

The tribe Pediciini is a small but very distinct group which is chiefly Holarctic. It has two subtribes, the Ularia, with the single genus *Ula*, and the Pediciaria, including *Pedicia*, *Ornithodes*, *Nasiternella*, and *Dicranota*. The immature stages of *Ula* are fungicolous, and occur in both woody and fleshy fungi, sometimes becoming economically important; *Pedicia* and *Dicranota* frequent wet to saturated, organic soil along streams and in seepage areas.

REFERENCE: Alexander, 1920h (biol.).

Genus ULA Haliday

Ula Haliday, 1833: 153. Type-species, *mollissima* Haliday (mon.)= *sylvatica* (Meigen).

elegans Osten Sacken, 1869b: 276.—N.H.; Alaska to N.S., s. to Wyo. and S.C.
longicornis Dietz, 1921a: 250.—Pa.
paupera Osten Sacken, 1869b: 277.—D.C.; Alaska to Nfld., s. to Calif., Wyo., and N.Y.

Genus PEDICIA Latreille

Pedicia Latreille, 1809: 255. Type-species, *Tipula rivosa* Linnaeus (mon.).
Peditia, error.

Subgenus PEDICIA Latreille

albivitta Walker, 1848: 37.—N.S.; Man. to Nfld., s. to Mo. and S.C.
    goniphora Harris, 1835: 595. Nomen nudum.
contermina Walker, 1848: 38.—N.S.; Mich. to Nfld., s. to N.C.
falcifera Alexander, 1941c: 209.—S. Dak.; Mich.
lewisiana Alexander, 1958b: 131.—Mont.
magnifica Hine, 1903b: 417 (*Peditia*).—B.C.; Oreg., Idaho.
margarita Alexander, 1929a: 300.—Mass.; Nfld. to Tenn.
obtusa Osten Sacken 1877: 205.—Calif.; Oreg.

parvicellula Alexander, 1938b: 72.—B.C.; Wash., Oreg., Colo.
procteriana Alexander, 1939a: 97.—Maine; Ont., Mass.
subobtusa Alexander, 1949b: 19.—Calif.

Subgenus TRICYPHONA Zetterstedt

Tricyphona Zetterstedt, 1837: 65 (as genus). Type-species, *Limonia immaculata* Meigen (mon.).
Amalopis Haliday, 1856: xv. Type-species, *Limnobia occulta* Meigen (orig. des.).

aethiops Alexander, 1955a: 129.—Alta.; B.C., Mont.
ampla ampla (Doane), 1900: 195 (*Amalopis*).—Wash.; B.C., Oreg.
    ssp. **cinereicolor** Alexander, 1958d: 32.—Calif.
    ssp. **euryptera** Alexander, 1949c: 41.—Ariz.
    ssp. **perangusta** Alexander, 1958d: 33.—Oreg.
    ssp. **truncata** Alexander, 1941c: 210.—Calif.
aperta (Coquillett), 1905d: 59 (*Tricyphona*).—B.C.; Alaska to Calif., Colo., and Mont.
aspidoptera aspidoptera (Coquillett), 1905b: 347 (*Limnophila*).—N. Mex.
    ssp. **convexa** Alexander, 1958d: 33.—Colo.
auripennis auripennis (Osten Sacken), 1859: 246 (*Amalopis*).—Mass.; Ont. to N.S., s. to N.Y. and Mass.
    ssp. **attenuata** Alexander, 1941a: 299.—N.C.
    ssp. **breviclava** Alexander, 1941a: 299.—Que.
    ssp. **nephophila** Alexander, 1941a: 300.—Tenn.; N.C.
autumnalis (Alexander), 1917: 30 (*Tricyphona*).—N.Y.; n. Ont. to N.B., s. to Wis. and Pa.
bicomata Alexander, 1943d: 17.—Oreg.; Wash.
bidentifera Alexander, 1950c: 29.—Calif.
brevifurcata (Alexander), 1919b: 6 (*Tricyphona*).—Alaska.
calcar (Osten Sacken), 1859: 247 (*Amalopis*).—Mass., N.H. (Osten Sacken, 1869b: 269); Wis. to Que. and Nfld., s. to S.C.
cascadensis Alexander, 1954: 50.—Oreg.; Wash., Wyo.
cervina (Alexander), 1917: 62 (*Tricyphona*).—Colo.
claggi (Alexander), 1930b: 280 (*Tricyphona*).—Colo.
constans (Doane), 1900: 196 (*Amalopis*).—Wash.; Y.T., B.C., Alta., Oreg.
debilis (Williston), 1893c: 62 (*Rhaphidolabis*).—Calif. Unrecognized. **N. comb.**
degenerata (Alexander), 1917: 209 (*Tricyphona*).—Colo.; Wash., Oreg., Calif., Wyo.
diaphana (Doane), 1900: 195, 198 (*Amalopis;* as *disphana*, p. 195).—Wash.; B.C., Oreg., Calif.

TIPULIDAE

exoloma (Doane), 1900: 194 (*Amalopis*).—Wash.; B.C., Utah.
fenderiana Alexander, 1954: 51.—Oreg.
frigida (Alexander), 1919b: 7 (*Tricyphona*).—Alaska.
fulvicolor Alexander, 1945c: 3.—Idaho; B.C., Oreg.
gigantea Alexander, 1940d: 99.—N.C.
glacialis (Alexander), 1917: 63 (*Tricyphona*).—Alaska; Wash., Oreg.
hannai hannai (Alexander), 1923: 160 (*Tricyphona*).—Alaska (Pribilof Is.).
    ssp. **antennata** Alexander, 1956c: 124.—Alaska.
huffae Alexander, 1940a: 625.—Tenn.
hynesiana Alexander, 1961a: 14.—Calif.
inconstans inconstans (Osten Sacken), 1859: 247 (*Amalopis*).—N.Y., Conn., D.C., Va.; Minn. to Nfld., s. to Mo. and Ga.
    ssp. **calcaroides** Alexander, 1940a: 626.—Tenn.; N.C.
johnsoni (Alexander), 1930a: 277 (*Tricyphona*).—Mass.; coastal, Mass. to Fla.
katahdin (Alexander), 1914a: 598 (*Tricyphona*).—Maine; Mich. to Maine.
macateei (Alexander), 1919c: 166 (*Tricyphona*).—Md.; N.Y. and Mass., s. to N.C.
macrophallus macrophallus Alexander, 1945c: 2.—Oreg.
    ssp. **actaeon** Alexander, 1948c: 132 (as sp.).—Calif.; Oreg.
pahasapa Alexander, 1958d: 31.—S. Dak.
paludicola (Alexander), 1916a: 538 (*Tricyphona*).—N.Y.; Mich. to Nfld. and Mass.
protea (Alexander), 1918a: 242 (*Tricyphona*).—Wash.; Alaska to Calif. and Mont.
pumila Alexander, 1942d: 353.—N.B.; Mich. ?=*paludicola*.
rainieria (Alexander), 1924d: 13 (*Tricyphona*).—Wash.
rubiginosa (Alexander), 1931b: 179 (*Tricyphona*).—Alta.; B.C.
septentrionalis septentrionalis (Bergroth), 1888a: 199 (*Tricyphona*).— Alaska; Alaska to Calif. and N. Mex.
    sparsipuncta Alexander, 1920g: 90 (*Tricyphona*).—Oreg.
    ssp. **vitripennis** (Doane), 1900: 195 (*Amalopis*; as sp.).—Wash.; B.C., Oreg.
shastensis Alexander, 1958d: 35.—Calif.
simplicistyla (Alexander), 1930b: 72 (*Tricyphona*).—Calif.
smithae Alexander, 1941c: 88.—Wash.; Oreg.
steensensis Alexander, 1958d: 33.—Oreg.; Calif.
subaptera (Alexander), 1917: 207 (*Limnophila*).—Calif.
tacoma Alexander, 1949a: 296.—Wash.
townesiana townesiana Alexander, 1942a: 210.—Wash.; Oreg.
    ssp. **majuscula** Alexander, 1954: 54.—Oreg.
unigera Alexander, 1949a: 299.—Oreg.; Wash.

vernalis vernalis (Osten Sacken), 1861: 291 (*Amalopis*).—D.C.;
  Ont. to Nfld., s. to Ga.
  ssp. **catawba** Alexander, 1940a: 628.—Tenn.; N.C.

Genus ORNITHODES Coquillett

Ornithodes Coquillett, 1900h: 400 (1904: 14). Type-species, *harrimani* Coquillett (orig. des.).

brevirostris Alexander, 1955a: 18 (as *harrimani* ssp.).—Nev.; Calif.
harrimani Coquillett, 1900h: 400 (1904: 14).—Alaska; Alaska, s. to Oreg. and Wyo.

Genus NASITERNELLA Wahlgren

Nasiterna Wallengren, 1881: 179 (preocc. Wagler, 1832). Type-species, *Limnobia variinervis* Zetterstedt (mon.).
Nasiternella Wahlgren, 1904: 4 (n. name for *Nasiterna* Wallengren). Type-species, *Limnobia variinervis* Zetterstedt (aut.).

hyperborea (Osten Sacken), 1861: 292 (*Amalopis*).—Labr.; N.W.T. to Labr., s. to N.Y. and N.H. (Hudsonian zone).

Genus DICRANOTA Zetterstedt

Dicranota Zetterstedt, 1838: 851. Type-species, *guerini* Zetterstedt (mon.).

Subgenus EUDICRANOTA Alexander

Dicranota, subg. **Eudicranota** Alexander *in* Curran, 1934a: 46, 52. Type-species, *notabilis* Alexander (orig. des.).

catawbiensis Alexander, 1940a: 628.—Tenn.; N.C.
notabilis Alexander, 1929d: 18.—Tenn.
pallida Alexander, 1914a: 599.—N.H.; Maine.
yonahlossee Alexander, 1941b: 14.—N.C.

Subgenus RHAPHIDOLABINA Alexander

Rhaphidolabis, subg. **Rhaphidolabina** Alexander, 1916a: 540. Type-species, *flaveola* Osten Sacken (orig. des.).

flaveola (Osten Sacken), 1869b: 288 (*Rhaphidolabis*).—Mass., Md.; Mich. to Nfld., s. to N.C.

Subgenus POLYANGAEUS Doane

Polyangaeus Doane, 1900: 196 (as genus). Type-species, *maculatus* Doane (mon.).

maculata (Doane), 1900: 197 (*Polyangaeus*).—Wash.; B.C. to Calif.
megalops Alexander, 1945a: 420.—Wyo.
subapterogyne Alexander, 1943b: 741.—Wyo.; Oreg.

Subgenus DICRANOTA Zetterstedt

argentea Doane, 1900: 196.—Wash.; Alaska to Oreg. and Colo.
  montana Alexander, 1920d: 110.—Colo.
astigma Alexander, 1954: 54.—Oreg.
clementi Alexander, 1956a: 181.—Que.
currani Alexander, 1926c: 50.—Ont.; Minn. to Que.
divaricata Alexander, 1925b: 203.—N.C.
fumipennis Alexander, 1941c: 89.—Minn.
notmani Alexander, 1942b: 358.—N.Y.
noveboracensis Alexander, 1914a: 600.—N.Y.; Que., Mass.
parvella Alexander, 1954: 55.—Oreg.
stainsi Alexander, 1948a: 45.—Utah.
tetonicola Alexander, 1945a: 418.—Wyo.

Subgenus PARADICRANOTA Alexander

Dicranota, subg. **Paradicranota** Alexander *in* Curran, 1934a: 46, 52.
  Type-species, *rivularis* Osten Sacken (orig. des.).

eucera Osten Sacken, 1869b: 281.—D.C.; Minn. to Mass., s. to Va.
iowa Alexander, 1920c: 78.—Iowa; Minn. to Que., s. to Iowa and N.H.
  rogersi Alexander, 1921c: 136.—Mich.
rivularis Osten Sacken, 1859: 249.—D.C.; N.H., Mass., Va.

Subgenus PLECTROMYIA Osten Sacken

Plectromyia Osten Sacken, 1869b: 282 (as genus). Type-species, *modesta* Osten Sacken (mon.).
Astrolabis Osten Sacken, 1865: 225. Nomen nudum.

cascadica Alexander, 1949a: 303.—Wash.; Oreg.
confusa (Alexander), 1924b: 63 (*Rhaphidolabis*).—Mass.; Que. and N.B., s. to S.C.
engelmannia Alexander, 1943a: 164.—Colo.
kulshanensis Alexander, 1949a: 304.—Wash.; Calif.
modesta (Osten Sacken), 1869b: 284 (*Plectromyia*).—N.H.; N.Y., Vt., Nfld.
nooksackiae nooksackiae Alexander, 1949a: 305.—Wash.; Oreg.
  ssp. **latistyla** Alexander, 1958d: 36.—Alta.
  ssp. **subtruncifera** Alexander, 1954: 60.—Oreg.
petiolata (Alexander), 1919c: 194 (*Tricyphona*).—Colo.; Utah to Que. and N.H.

nemoptera Alexander, 1927i: 143 (*Limnophila*).—Utah.
stenoptera Alexander, 1927j: 16 (*Tricyphona*).—Colo.
reducta reducta (Alexander), 1921c: 135 (*Rhaphidolabis*).—Idaho; Wash., Oreg., Calif.
  ssp. **tehamicola** Alexander, 1959a: 75.—Calif.
townesi Alexander, 1940c: 154.—S.C.

Subgenus RHAPHIDOLABIS Osten Sacken

Rhaphidolabis Osten Sacken, 1869b: 284 (as genus). Type-species, *tenuipes* Osten Sacken (orig. des.).

avis (Alexander), 1926c: 50 (*Rhaphidolabis*).—Va.; Maine to N.C.
cayuga (Alexander), 1916a: 543 (*Rhaphidolabis*).—N.Y.; Alaska to Nfld., s. to Oreg., Colo., and N.C.
cazieriana Alexander, 1944c: 93.—Calif.; Oreg., Utah.
fenderi Alexander, 1954: 56.—Oreg.
forceps (Alexander), 1924b: 63 (*Rhaphidolabis*).—N.Y.; Mich. to N.S., s. to Tenn.
hickmanae Alexander, 1940a: 629.—Tenn.
integriloba Alexander, 1943a: 162.—Colo.; Wash. to Calif. and Colo.
major (Alexander), 1917: 210 (*Rhaphidolabis*).—Colo.
neomexicana neomexicana (Alexander), 1912d: 170 (*Rhaphidolabis*).—N. Mex.; Alaska to Calif. and N. Mex.
  ssp. **subtruncata** Alexander, 1949a: 300.—Wash.
nooksackensis nooksackensis Alexander, 1949a: 300.—Wash.; Alaska, Oreg.
  ssp. **brevispinosa** Alexander, 1958d: 36.—Wash.
nuptialis Alexander, 1948c: 134.—Calif.
persimilis (Alexander), 1920c: 79 (*Rhaphidolabis*).—Va.; Md., N.C., S.C.
polymeroides (Alexander), 1914a: 601 (*Rhaphidolabis*).—Calif.; Wash., Oreg., Idaho.
querula Alexander, 1944c: 94 (as *neomexicana* ssp.).—Idaho; Oreg., Calif.
rogersiana (Alexander), 1925b: 204 (*Rhaphidolabis*).—Mich.; Que.
rubescens (Alexander), 1916a: 544 (*Rhaphidolabis*).—N.Y.; N.Y. to N.S. and Mass.
sessilis (Alexander), 1917: 210 (*Rhaphidolabis*).—Colo.
stigma (Alexander), 1924d: 14 (*Rhaphidolabis*).—Wash.; Colo.
subsessilis (Alexander), 1921c: 134 (*Rhaphidolabis*).—Idaho; B.C., Wash., Mont.
tehama Alexander, 1950c: 30.—Calif.

tenuipes (Osten Sacken), 1869b: 287 (*Rhaphidolabis*).—N.Y., Md.;
   Mich. to N.B. and Md.
uniplagia Alexander, 1954: 58.—Oreg.
vanduzeei (Alexander), 1930b: 73 (*Rhaphidolabis*).—Calif.
xanthosoma Alexander, 1944c: 93.—Oreg.; Wash.

Tribe HEXATOMINI

The major tribe Hexatomini includes a somewhat heterogeneous grouping of genera, some of which may prove to be wrongly assigned. There are ten subtribes in our region.

The immature stages of *Paradelphomyia* frequent organic silt along stream margins or in cool seepage areas. *Polymera* inhabits saturated silt or sandy clay soil along water bodies (Rogers, 1933b). *Epiphragma* and *Austrolimnophila* live in wet logs and limbs. *Dactylolabis* is found in algal and other plant associations in dripping water on cliff faces and comparable situations. *Pseudolimnophila* and *Prolimnophila* occur in moist-to-saturated soil along streams or in marshes. *Limnophila*, *Pilaria*, and *Ulomorpha* frequent wet soil ranging from sandy organic silt to black mud. The early stages of *Hexatoma*, with its important subgenus *Eriocera*, are aquatic or semiaquatic, living in or near the sandy or gravelly margins of streams or in saturated organic soil. The early stages of both *Atarba* and *Elephantomyia* occur in wet decaying hardwoods (Rogers, 1927b). Nothing is known about the early stages of *Phyllolabis*.

Genus PARADELPHOMYIA Alexander

Adelphomyia, subg. **Paradelphomyia** Alexander, 1936b: 184. Type-species, *crossospila* Alexander (orig. des.).

Subgenus OXYRHIZA Meijere

Oxydiscus Meijere, 1913b: 350 (preocc. Koken, 1889). Type-species, *nebulosus* Meijere (mon.).
Oxyrhiza Meijere, 1946: 68 (as genus; n. name for *Oxydiscus* Meijere).
   Type-species, *Oxydiscus nebulosus* Meijere (aut.).
americana (Alexander), 1912c: 829 (*Adelphomyia*).—N.Y.; Mich. to
   N.B., s. to S.C.
   senilis Haliday of Alexander, 1911b: 352.
cayuga (Alexander), 1912c: 831 (*Adelphomyia*).—N.Y.; Mich. to
   Maine, s. to Ind. and Pa.
   abnormis Dietz, 1916: 137 (*Ormosia*).—Pa.
   hazletonensis Dietz, 1921a: 252 (*Adelphomyia*).—Pa.

deprivata Alexander, 1954: 60.—Oreg.; Calif.
maddocki (Alexander), 1948a: 46 (*Oxydiscus*).—Utah; Ariz.
minuta (Alexander), 1911a: 286 (*Adelphomyia*).—N.Y.; Mich. to Nfld., s. to N.C.
pacifica (Alexander), 1944b: 92 (*Oxydiscus*).—Oreg.; B.C., Wash., Calif., Utah.
pleuralis (Dietz), 1921a: 251 (*Adelphomyia*).—Pa.; Mich. to N.S., s. to Fla.
sierrensis Alexander, 1958b: 132.—Calif.

Genus POLYMERA Wiedemann

Polymera Wiedemann, 1820a: 40 (1821c: 40). Type-species, *Chironomus hirticornis* Fabricius (mon.).

Subgenus POLYMERA Wiedemann

georgiae Alexander, 1911c: 199.—Ga.; S.C., Fla.
rogersiana Alexander, 1929d: 18.—Fla.

Genus PHYLLOLABIS Osten Sacken

Phyllolabis Osten Sacken, 1877: 202. Type-species, *clavigera* Osten Sacken (Coquillett, 1910b: 590).

bryantiana Alexander, 1931b: 182.—B.C.
clavigera Osten Sacken, 1877: 203.—Calif.
encausta Osten Sacken, 1877: 204.—Calif.
fenderiana Alexander, 1949a: 307.—Oreg.; Wash.
flavida Alexander, 1918b: 287.—Calif.
hirtiloba Alexander, 1947d: 66.—Calif.; Nev.
lagganensis Alexander, 1931b: 183.—Alta.; Alta. and Wash. to Maine, s. to Colo.
latifolia Alexander, 1920g: 90.—Oreg.
meridionalis Alexander, 1945c: 5.—Calif.
myriosticta Alexander, 1945c: 6.—Calif.; Mexico (Baja Calif.).
sequoiensis Alexander, 1945c: 4.—Calif.
zionensis Alexander, 1948a: 47.—Utah.

Genus EPIPHRAGMA Osten Sacken

Limnophila, subg. **Epiphragma** Osten Sacken, 1859: 238. Type-species, *pavonina* Osten Sacken (Coquillett, 1910b: 539)=*fasciapennis* (Say).

Subgenus EPIPHRAGMA Osten Sacken

arizonensis Alexander, 1946a: 517.—Ariz.
fasciapennis (Say), 1823: 19 (1859b: 45) (*Limnobia*).—U.S.; Alta. to Nfld., s. to La. and Fla.
fascipennis, emend.
pavonina Osten Sacken, 1859: 239 (*Limnophila*).—Ill.
ocellaris (Linnaeus), 1761: 433 (*Tipula*).—Sweden; B.C., Eurasia.
solatrix (Osten Sacken), 1859: 238 (*Limnophila*).—D.C.; Mo. to N.Y., s. to La. and Fla., Neotropical.

Genus AUSTROLIMNOPHILA Alexander

Limnophila, subg. **Austrolimnophila** Alexander, 1920a: 4. Type-species, *eutaeniata* Bigot (orig. des.).

Subgenus AUSTROLIMNOPHILA Alexander

badia (Doane), 1900: 191 (*Limnophila*).—Wash.; B.C., Oreg., Calif., Nev., Wyo.
toxoneura (Osten Sacken), 1859: 236 (*Limnophila*).—N.Y.; Wis. to Nfld., s. to Tenn. and Va.

Subgenus ARCHILIMNOPHILA Alexander

Archilimnophila Alexander *in* Curran, 1934a: 47, 52 (as genus). Type-species, *Limnophila unica* Osten Sacken (orig. des.).
harperi (Alexander), 1926e: 23 (*Limnophila*).—Alta.; Alta. to N.Y., s. to Colo.
subunica (Alexander), 1920i: 197 (*Limnophila*).—Alaska; Alaska and Alta., s. to Idaho and S. Dak.
unica (Osten Sacken), 1869b: 205 (*Limnophila*).—N.H.; Wis. to Nfld. and D.C.

Genus DACTYLOLABIS Osten Sacken

Limnophila, subg. **Dactylolabis** Osten Sacken, 1859: 240. Type-species, *montana* Osten Sacken (mon.).

Subgenus DACTYLOLABIS Osten Sacken

adventitia Alexander, 1942a: 211.—Idaho; Wash.
cubitalis (Osten Sacken), 1869b: 229 (*Limnophila*).—Ohio, Va.; Wis., Ind., N.Y., N.C.
hortensia (Alexander), 1914a: 591 (*Limnophila*).—B.C.; Mont.
hudsonica Alexander, 1931b: 181.—Que.; Que. and Nfld., s. to Tenn. and N.C.

imitata Alexander, 1945f: 91.—Calif.
knowltoni Alexander, 1941c: 90.—Utah; Oreg.
luteipyga Alexander, 1955a: 19.—Wyo.; Idaho, Mont.
montana (Osten Sacken), 1859: 240 (*Limnophila*).—Mass., D.C., Ill.; Que. to Nfld., sw. to Ill. and S.C.
nitidithorax (Alexander), 1918b: 288 (*Limnophila*).—Calif.; Wash., Oreg.
parviloba Alexander, 1944c: 92.—Calif.
pemetica Alexander, 1936a: 288.—Maine; Ohio, N.Y., N.H., N.C.
postiana Alexander, 1944b: 94.—Oreg.; Idaho.
pteropoecila (Alexander), 1921a: 104 (*Limnophila*).—Wash.; B.C., Oreg.
rhicnoptiloides (Alexander), 1919b: 6 (*Limnophila*).—N.W.T.; Alaska to Labr.
sparsimacula Alexander, 1942a: 212 (as *hortensia* ssp.).—Wash.
supernumeraria Alexander, 1929f: 46.—N.Y.; Nfld. to N.Y. and Mass.

Subgenus EUDACTYLOLABIS Alexander

Dactylolabis, subg. **Eudactylolabis** Alexander, 1950a: 45. Type-species, *Limnophila damula* Osten Sacken (orig. des.).
damula (Osten Sacken), 1877: 201 (*Limnophila*).—Calif.
vestigipennis Alexander, 1950a: 45.—Ariz.

Genus PSEUDOLIMNOPHILA Alexander

Limnophila, subg. **Pseudolimnophila** Alexander, 1919h: 917. Type-species, *luteipennis* Osten Sacken (Alexander, 1920h: 849).

Subgenus PSEUDOLIMNOPHILA Alexander

australina Alexander, 1927h: 56.—Fla.; Ind. to Md., s. to Ala. and Fla.
contempta (Osten Sacken), 1869b: 218 (*Limnophila*).—Middle States; Mich. to Nfld., s. to Mo. and Fla, ?Wash.
  nigripleura Alexander and Leonard *in* Alexander, 1914a: 592 (*Limnophila*).—N.Y.
inornata inornata (Osten Sacken), 1869b: 219 (*Limnophila*).—Mass.; Mich. to Nfld., s. to Ind. and Md.
  ssp. **vidua** Alexander, 1943b: 743 (as sp.).—Wyo.
luteipennis (Osten Sacken), 1859: 236 (*Limnophila*).—Mass., D.C., Fla.; Maine, Calif.
noveboracensis (Alexander), 1911c: 196 (*Limnophila*).—N.Y.; B.C. to N.S., s. to Utah and S.C.

Genus PROLIMNOPHILA Alexander

Limnophila, subg. **Prolimnophila** Alexander, 1929b: 187. Type-species, *areolata* Osten Sacken (orig. des.).

areolata (Osten Sacken), 1859: 237 (*Limnophila*).—N.Y., Md.; Wis. to Nfld., s. to Tenn. and N.C.

Genus LIMNOPHILA Macquart

Limnophila Macquart, 1834a: 95. Type-species, *Limnobia pictipennis* Meigen (Westwood, 1840: 128).

Subgenus LASIOMASTIX Osten Sacken

Limnophila, subg. **Lasiomastix** Osten Sacken, 1859: 233. Type-species, *Limnobia macrocera* Say (mon.).

macrocera macrocera (Say), 1823: 20 (1859b: 46) (*Limnobia*).—Fla.; Mich. to N.S., s. to Ill. and Fla.
    carbonaria Macquart, 1838a: 70 (1838: 66).—Carolina.
  ssp. **suffusa** Alexander, 1927b: 56.—Fla.; Ga.
subtenuicornis (Alexander), 1918a: 61 (*Lasiomastix*).—N.Y.; Mich., Ont., and N.Y., s. to Tenn.
tenuicornis Osten Sacken, 1869b: 208.—N.H.; Ont. to N.S., s. to S.C.

Subgenus EUTONIA Wulp

Eutonia Wulp, 1874: 147 (as genus). Type-species, *Limonia barbipes* Meigen (mon.).

alleni Johnson, 1909a: 126.—N.H.; Ohio, N.Y., Vt.
marchandi Alexander, 1916d: 118.—Conn.; Mich. to Mass., s. to Fla.
phorophragma Alexander, 1944d: 127.—Ga.; N.C.

Subgenus PRIONOLABIS Osten Sacken

Limnophila, subg. **Prionolabis** Osten Sacken, 1859: 239. Type-species, *rufibasis* Osten Sacken (mon.).

barberi Alexander, 1916d: 122.—Calif.; Wash., Oreg.
boharti Alexander, 1943e: 94.—Oreg.; Wash.
cressoni Alexander, 1917: 208.—Calif.
freeborni Alexander, 1943e: 91.—Calif.
gruiformis Alexander, 1945f: 93.—Idaho.
hepatica Alexander, 1919f: 215.—Calif.; Oreg.
indistincta Doane, 1900: 191.—Idaho; Alaska to Calif. and Idaho.
magdalena Dietz, 1920: 5.—Que. Unrecognized.
munda Osten Sacken, 1869b: 226.—N.H.; Wis. to N.S., s. to N.C.
mundoides Alexander, 1916d: 120.—Md.; Pa., Va., N.C., Fla.

oregonensis Alexander, 1940b: 87.—Oreg.; Wash.
paramunda Alexander, 1949a: 311.—Wash.
politissima Alexander, 1941a: 305.—N.C.; Tenn.
rudimentis Alexander, 1941a: 307.—N.C.; Tenn.
rufibasis rufibasis Osten Sacken, 1859: 239.—N.Y., Mass., D.C.; Wis. to Nfld., s. to S.C.
  ssp. **sedula** Alexander, 1941a: 309.—N.C.; Tenn.
scaria scaria Alexander, 1945f: 92.—Wash.; Oreg.
  ssp. **trifida** Alexander, 1949a: 312.—Wash.; Oreg.
sequoiarum Alexander, 1943e: 89.—Calif.
simplex Alexander, 1911c: 198.—Ga.; N.Y. to N.H., s. to Fla.
terebrans Alexander, 1916d: 121.—Md.; N.J.
vancouverensis Alexander, 1943e: 92.—Oreg.; Alaska, Wash., Calif.
walleyi Alexander, 1929b: 187.—N.Y.; Ill. to Ont., s. to N.C.

Subgenus IDIOLIMNOPHILA Alexander

Limnophila, subg. **Idiolimnophila** Alexander *in* Curran, 1934a: 48, 52. Type-species, *emmelina* Alexander (orig. des.).

emmelina Alexander, 1914a: 597.—Va.; Ont., N.Y., Mass.

Subgenus ELOEOPHILA Rondani

Eloeophila Rondani, 1856: 182 (as genus). Type-species, *Limnobia marmorata* Meigen (orig. des.).
Elaeophila, error.
Ephelia Schiner, 1863: 222. Type-species, *Limnobia marmorata* Meigen (orig. des.).
Limnophila, subg. *Trichephelia* Alexander, 1938b: 73. Type-species, *seticellula* Alexander (orig. des.).

aldrichi aldrichi Alexander, 1927j: 11.—Mont.; Alta., Wash., Oreg.
  ssp. **abrupta** Alexander, 1949a: 309.—Oreg.; Wash.
  ssp. **alticrista** Alexander, 1943a: 165.—Colo.; Wyo.
  ssp. **collata** Alexander, 1948a: 48.—Utah.
aleator Alexander, 1945a: 422.—Wyo.; Oreg., Calif., Utah, N. Mex.
angustior Alexander, 1919c: 163.—Colo.; Oreg., Calif., Mont., Wyo., Utah.
apiculata Alexander, 1919c: 164.—Calif.
aprilina Osten Sacken, 1859: 235.—D.C.; Nfld. to S.C.
bifida Alexander, 1921c: 133.—Idaho; B.C., Oreg.
edentata Alexander, 1919c: 164.—Calif.; Ariz.
irene Alexander, 1927b: 58.—Mich.; Mich. and Ont., s. to S.C.
johnsoni Alexander, 1914a: 591.—N.Y.; Que. and N.B., s. to Tenn. and S.C.
modoc Alexander, 1946g: 95.—Calif.

nupta Alexander, 1947d: 207.—Calif.
sabrina Alexander, 1929b: 189.—Vt.; Que. to N.B., s. to N.Y. and Conn.
serotinella Alexander, 1926a: 110.—Tenn.; S.C., Fla.
seticellula Alexander, 1938b: 74.—S.C.; N.C.
shannoni Alexander, 1921c: 132.—Idaho; Wash., Oreg., Wyo.
solstitialis Alexander, 1926a: 109.—N.Y.; Mich. to Maine, s. to Tenn. and Fla.
superlineata Doane, 1900: 190.—Wash.; Oreg.
vernata Alexander, 1927b: 59.—N.C.
woodgatei Alexander, 1946a: 519.—N. Mex.

Subgenus DICRANOPHRAGMA Osten Sacken

Limnophila, subg. **Dicranophragma** Osten Sacken, 1859: 240. Type-species, *fuscovaria* Osten Sacken (mon.).

angustula Alexander, 1929b: 190.—Mass.; Mich. to Nfld., s. to Tenn. and Fla.
fuscovaria Osten Sacken, 1859: 240.—N.Y., Va.; Wis. to N.S., s. to Mo. and Fla.

Subgenus IDIOPTERA Macquart

Idioptera Macquart, 1834a: 94 (as genus). Type-species, *maculata* Macquart (mon.) = *pulchella* (Meigen).

fasciolata Osten Sacken, 1869b: 206.—Mass.; Mich. to Maine and Conn.
mcclureana Alexander, 1938b: 75.—Man.

Subgenus PHYLIDOREA Bigot

Phylidorea Bigot, 1854: 456 (as genus). Type-species, *Limnobia ferruginea* Meigen (Coquillett, 1910b: 590).

adusta Osten Sacken, 1859: 235.—Wis., Ill., N.Y., Maine; Wis. to Nfld., s. to Ill. and N.C.
adustoides Alexander, 1927b: 63.—Tenn.; Ind. and Tenn., e. to Mass.
aequiatra Alexander, 1949a: 314.—Wash.; Oreg., Calif.
aleutica Alexander, 1920i: 198.—Alaska.
auripennis Alexander, 1926a: 113.—Ill.; Mich. to Nfld., s. to Ill. and N.Y.
biterminata (Walker), 1856b: 437 (*Limnobia*).—U.S. Unrecognized. **N. comb.**, G. W. Byers *in litt.*
brevifilosa Alexander, 1959a: 50.—Calif.
caudifera Alexander, 1927b: 111.—N.Y.; Mich. to Nfld. and Vt.

claggi Alexander, 1930b: 281.—Colo.; B.C. to Mich., s. to Calif. and Colo.
columbiana Alexander, 1927j: 12.—B.C.
consimilis consimilis Dietz, 1921a: 255.—Pa.; Mich. to Que. and Vt., s. to Tenn. and N.C.
    ssp. **griseipleura** Alexander, 1941a: 309.—N.C.
costata Coquillett, 1901g: 149.—N. Mex.
epimicta Alexander, 1927b: 60.—Fla.
flavapila Doane, 1900: 190.—Wash.; B.C., Oreg., Calif.
    strepens Alexander, 1916a: 532.—Calif.
fratria Osten Sacken, 1869b: 220.—Unknown (as N.Y. or N.H.); Maine, Vt., Pa.
frosti Alexander, 1961b: 84.—Fla.
fumidicosta Alexander, 1927b: 115.—N.Y.; Maine, Mass. ?=*consimilis* ssp.
fuscovenosa Alexander, 1927e: 190.—Alta.; B.C.
insularis Johnson, 1913d: 443.—Bermuda.
iowensis Alexander, 1927b: 112.—Iowa; Mich.
lutea Doane, 1900: 191.—Pa.; Nfld., s. to Tenn. and Va.
luteola Alexander, 1927b: 113.—N.Y.; Que., Vt.
microphallus Alexander, 1927j: 13.—Mont.
neadusta Alexander, 1927b: 110.—N.Y.
nevadensis Alexander, 1958b: 218.—Nev.
nigrogeniculata Alexander, 1926a: 114.—Tenn.
novaeangliae Alexander, 1914a: 594.—Maine; Que., N.Y., Mass., Pa.
olympica Alexander, 1949a: 316.—Wash.; Oreg.
osceola Alexander, 1927b: 61.—Fla.
pacalis Alexander, 1949b: 156.—Oreg.
paeneadusta Alexander, 1961b: 85.—Maine.
persimilis Alexander, 1927b: 62.—Ind.
platyphallus Alexander, 1926a: 111.—Mass.; Alta. to Mich. to N.S., s. to Pa. ?=*terraenovae*.
semifacta Alexander, 1948b: 207.—Ariz.
similis Alexander, 1911c: 195.—N.Y.; Ont., Que., Maine, Mass.
siouana Alexander, 1929b: 188.—Iowa.
snoqualmiensis Alexander, 1945f: 94.—Wash.; Oreg.
stupkai Alexander, 1940a: 632.—Tenn.; N.C.
subcostata (Alexander), 1911a: 287 (*Phylidorea*).—N.Y.; Mich. to Nfld., s. to Pa.
subsimilis Alexander, 1927b: 111.—Tenn.
tepida Alexander, 1926e: 119.—Colo.; Alta., Oreg., Utah.
terraenovae Alexander, 1916d: 123.—Nfld.; N.S.
    adjuncta Dietz, 1920: 6.—N.S.

TIPULIDAE 67

Subgenus DENDROLIMNOPHILA Alexander

Limnophila, subg. **Dendrolimnophila** Alexander, 1949a: 314. Typespecies, *Shannonomyia albomanicata* Alexander (orig. des.).

albomanicata (Alexander), 1945f: 96 (*Shannonomyia*).—Wash.; Oreg., Calif.

Unplaced Species of *Limnophila*

albipes Leonard, 1913: 248.—N.J.; Vt. to S.C.
amabilis Alexander, 1950e: 84.—Calif.
antennata Coquillett, 1905d: 58.—B.C.; Wash., Oreg., Calif.
bigladia Alexander, 1945a: 424.—Wyo.; Alta., Wash.
brevifurca Osten Sacken, 1859: 237.—D.C.; Mich. to Nfld., s. to S.C.
bryanti Alexander, 1927f: 224.—Alta.
cherokeensis Alexander, 1940a: 634.—Tenn.; N.C.
euxesta Alexander, 1924d: 11.—Wash.; Oreg.
galactopoda Alexander, 1943b: 741, fig.; 745, descr. (as *Pilaria phaeonota*, error, p. 745).—Wyo.; Mont.
globulifera Alexander, 1941b: 15.—N.C.
irrorata Johnson, 1909a: 127.—N.J.; N.C., Fla.
laricicola Alexander, 1912d: 167.—N.Y.; Mich. to Nfld., s. to Tenn. and Va.
lobifera Alexander, 1955a: 131.—B.C.
mcdunnoughi Alexander, 1926e: 21.—Alta.
nigrofemorata Alexander, 1927j: 14.—Mont.
niveitarsis Osten Sacken, 1869b: 209.—Del., Md.; N.Y. and Mass., s. to Tenn. and N.C.
nycteris Alexander, 1943b: 744.—Wyo.; Oreg., Calif.
occidens Alexander, 1924d: 13.—Wash.; Alaska, s. to Calif. and Colo.
poetica Osten Sacken, 1869b: 207.—Mass.; Alaska to Que., s. to B.C., Ill., and Mass.
rubida Alexander, 1924d: 12.—Oreg.; Calif.
tetonicola Alexander, 1945a: 426.—Wyo.; Wash., Oreg.

Genus SHANNONOMYIA Alexander

Shannonomyia Alexander, 1929c: 142. Type-species, *Limnophila lenta* Osten Sacken (orig. des.).

congenita (Dietz), 1921a: 257 (*Limnophila*).—Pa. ?=*lenta*.
lenta lenta (Osten Sacken), 1859: 241 (*Limnophila*).—Ill., Md., D.C., Va.; Ill. to Nfld., s. to Tenn. and Ga.
    ssp. **gaspeana** Alexander, 1942b: 409.—Que.

oslari (Alexander), 1916d: 123 (*Limnophila*).—Colo.; Wyo., Utah, N. Mex.

Genus PILARIA Sintenis

Pilaria Sintenis, 1889: 398. Type-species, *Limnobia pilicornis* Zetterstedt (Coquillett, 1910b: 591)=*meridiana* (Staeger). *Limnophila*, subg. *Eulimnophila* Alexander, 1919h: 917. Type-species, *Limnobia tenuipes* Say (Ishida, 1959: 8).

arguta Alexander, 1929f: 47.—Fla. ?=*recondita* ssp.
flava (Garrett), 1925b: 4 (*Cladura*).—B.C.; Wash. ?=*recondita*.
harrisoni Alexander, 1936a: 291.—N.H.
imbecilla imbecilla (Osten Sacken), 1859: 237 (*Limnophila*).—Ill.. N.Y., Va.. Ga.; Ill. to Que., s. to Tenn., Ga., and Mass.
　ssp. **illinoiensis** (Alexander), 1920d: 226 (*Limnophila*).—Ill.; Ind.. Ky., Tenn.
meridiana (Staeger), 1840: 41 (*Limnobia*).—Europe; Oreg. to Maine, s. to Colo., Eurasia.
　osborni Alexander, 1914a: 596 (*Limnophila*).—Maine.
microcera Alexander, 1924d: 13.—Oreg.
phaeonota Alexander, 1943b: 746.—Wyo.
quadrata (Osten Sacken), 1859: 241 (*Limnophila*).—Md., Va.; Iowa, to N.S., s. to Fla.
recondita (Osten Sacken), 1869b: 212 (*Limnophila*).—Pa., N.Y., Ga.; Minn. to N.S., s. to La. and Fla.
stanwoodae (Alexander), 1914a: 595 (*Limnophila*).—N.Y.; Maine to S.C.
tenuipes (Say), 1823: 21 (1859b: 46) (*Limnobia*).—Pa.; Wis. to N.B., s. to Kans., Tex., and Fla.
　edwardi Alexander, 1916a: 533 (*Limnophila*).—N.Y.
vermontana Alexander, 1929f: 47.—Vt.; Conn.

Genus ULOMORPHA Osten Sacken

Ulomorpha Osten Sacken, 1869b: 232. Type-species, *Limnophila pilosella* Osten Sacken (mon.).

aridela Alexander, 1927j: 15.—Oreg.
nigrodorsalis Alexander, 1949a: 317.—Wash.; Oreg.
nigronitida Alexander, 1920b: 40.—Calif.; Oreg.
pilosella (Osten Sacken), 1859: 241 (*Limnophila*).—N.Y.; Mich. to N.S., s. to Tenn. and S.C.
quinquecellula Alexander, 1920b: 40.—Calif.
rogersella Alexander, 1929f: 48.—Fla.; N.C., S.C.
sierricola Alexander, 1918a: 163.—Wash.; Oreg.
vanduzeei Alexander, 1920b: 41.—Calif.; Oreg.

TIPULIDAE

Genus Gynoplistia Westwood

annulata Westwood, 1835a: 280.—Australia (as N. Amer., error). Not Nearctic.

Genus HEXATOMA Latreille

Hexatoma Latreille, 1809: 260. Type-species, *nigra* Latreille (mon.).
Anisomera Meigen, 1818: 210. Type-species, *obscura* Meigen (mon.).

Subgenus HEXATOMA Latreille

megacera (Osten Sacken), 1859: 242 (*Anisomera*).—D.C., Md.; Ont. and Que., s. to Ind. and Va.
microcera Alexander, 1926c: 49.—N. Amer.

Subgenus ERIOCERA Macquart

Caloptera Guérin-Méneville, 1831: pl. 20, fig. 2. Type-species, *fasciata* Guérin-Mèneville (orig. des., as gen. n., sp. n.). [In the interest of stability, author rejects this valid prior name.—Eds.].
Eriocera Macquart, 1838a: 78 (1838: 74) (as genus). Type-species, *Limnobia nigra* Wiedemann (mon.)=*wiedemanni* Alexander.
Penthoptera Schiner, 1863: 220. Type-species, *Tipula chirothecata* Scopoli (orig. des.).
Arrhenica Osten Sacken, 1859: 243. Type-species, *spinosa* Osten Sacken (Coquillett, 1910b: 510).

albihirta (Alexander), 1912d: 167 (*Eriocera*).—Calif.
albitarsis (Osten Sacken) 1869b: 257 (*Penthoptera*).—Pa.; Iowa to Pa., s. to Fla.
aurata (Doane), 1900: 194 (*Eriocera*).—N.C.; Tenn., S.C. ,Ga., Fla.
austera (Doane), 1900: 192 (*Eriocera*).—Wash.; Y.T., s. to Calif. and N. Mex.
    obscura Williston, 1893c: 61 (*Eriocera;* preocc. Bigot, 1859).—Wash., Calif.
    parva Doane, 1900: 192 (*Eriocera*).—Calif.
    alberta Alexander, 1930b: 73 (*Eriocera*).—Alta.
azrael Alexander, 1943e: 98.—Calif.
beameri Alexander, 1956b: 212.—Tex.
brachycera (Osten Sacken), 1877: 205 (*Eriocera*).—N.H.; Ont. to Nfld., s. to Tenn. and N.C.
brevioricornis Alexander, 1941a: 311.—Mass.; Que., Conn., N.Y., N.C., Tenn.

brevipila (Alexander), 1918a: 164 (*Eriocera*).—Calif.
californica (Osten Sacken), 1877: 204 (*Eriocera*).—Calif.
cinerea (Alexander), 1912d: 169 (*Eriocera*).—Mass.; Ont., Que., and Maine, s. to Tenn. and S.C.
dayana Alexander, 1959a: 75.—Calif.
dorothea Alexander, 1956b: 215.—Oreg.
eriophora (Williston), 1893c: 61 (*Eriocera*).—Wash.; Oreg., Wyo., Utah, Colo., N. Mex.
fuliginosa (Osten Sacken), 1859: 243 (*Eriocera*).—D.C., Va.; Mich. to N.H., s. to Fla.
fultonensis (Alexander), 1912d: 168 (*Eriocera*).—N.Y.; Mass. to Tenn.
fulvomedia Alexander, 1956b: 123.—Ariz.
gaspensis (Alexander), 1931c: 143 (*Eriocera*).—Que.
gibbosa (Doane), 1900: 193 (*Eriocera*).—Mich.; Mich. to Mass., s. to Ill., Tenn., and Va.
intrita Alexander, 1943e: 97.—Wash.
longicornis (Walker), 1848: 82 (*Anisomera*).—e. N. Amer.; Alaska to Maine, s. to Iowa and N.C.
mariposa Alexander, 1943e: 96.—Calif.
pacifica Alexander, 1956b: 210.—Calif.
palomarensis Alexander 1947g: 43.—Calif.
rubrinota (Alexander), 1918a: 165 (*Eriocera*).—w. U.S.
saturata (Alexander), 1919c: 165 (*Eriocera*).—Calif.
sculleni Alexander, 1943d: 18.—Oreg.; Wash., Calif.
solor Alexander, 1943d: 19.—Oreg.
spinosa (Osten Sacken), 1859: 244 (*Arrhenica*).—N.Y., Mass.; Ill. to Que. and Nfld., s. to Pa.
tristis (Alexander), 1914a: 602 (*Eriocera*).—N.Y.; Ind. to Mass., s. to Va.
velveta velveta (Doane), 1900: 193 (*Eriocera*).—Colo.; Utah, Wyo., S. Dak.
   ssp. **apache** Alexander, 1956b: 215.—N. Mex.
wilsonii (Osten Sacken), 1869b: 255 (*Eriocera*).—Del.; Ohio, Va., N.C.
   antennaria Doane, 1900: 194 (*Eriocera*).—Ohio.

Genus ATARBA Osten Sacken

Atarba Osten Sacken, 1869b: 127. Type-species, *picticornis* Osten Sacken (mon.).

Subgenus ATARBA Osten Sacken

apache Alexander, 1949b: 102.—Ariz.
bellamyi Alexander, 1950d: 168.—Fla.; Ga.
picticornis Osten Sacken, 1869b: 128.—Del.; Mich. to N.H., s. to Mo. and Fla.
werneri Alexander, 1949b: 103.—Ariz.

Genus ELEPHANTOMYIA Osten Sacken

Elephantomyia Osten Sacken, 1859: 220. Type-species, *Limnobiorhynchus canadensis* Westwood (mon.; misident.) = *westwoodi* Osten Sacken.

Subgenus ELEPHANTOMYIA Osten Sacken

curtirostris Alexander, 1947d: 208.—Ariz.
westwoodi westwoodi Osten Sacken, 1869b: 109.—N.Y.; Wis. to Nfld. s. to Ill. and Fla.
  canadensis Westwood of Osten Sacken, 1859: 221.
  ssp. **adirondacensis** Alexander, 1942b: 424.—N.Y.; Maine.

Tribe ERIOPTERINI

The tribe Eriopterini is a vast group of crane flies, in our fauna containing species of small to medium size only. It includes representatives of four subtribes, the Claduraria (*Cladura* to *Neolimnophila*), Gonomyiaria (*Gnophomyia, Gonomyia, Teucholabis*), Eriopteraria (*Cryptolabis* to *Molophilus*), and the Toxorhinaria (*Toxorhina*). *Rhabdomastix, Lipsothrix,* and *Hesperoconopa* are tentatively placed between the Gonomyiaria and Eriopteraria.

The immature stages of *Cladura* and *Chionea* occur in relatively dry woodland soil beneath decaying leaves. *Gonomyia* lives in saturated silt or sand at the margins of streams or ponds. The early stages of both *Gnophomyia* and *Teucholabis* frequent moist rotting hardwoods in various stages of decay. *Rhabdomastix* lives in wet earth near streams. *Lipsothrix* is found in saturated wood along bodies of water, either submerged or constantly moist (Rogers and Byers, 1956). The immature stages of *Erioptera, Ormosia, Tasiocera,* and *Molophilus* are found in soil or in decaying vegetable matter; most species occur in the wet sandy to richly organic soil at stream margins or in bogs and marshes. The immature stages of *Hesperoconopa, Cryptolabis,* and *Toxorhina* remain to be discovered. Habits of the adults of the first two suggest that their larvae may be aquatic; the larvae of *Toxorhina* are presumed to occur in wet-to-saturated soil, but may be found to occupy a specialized habitat within this general category.

Genus CLADURA Osten Sacken

Cladura Osten Sacken, 1859: 229. Type-species, *flavoferruginea* Osten Sacken (mon.).
Pterochionea Alexander, 1916a: 529. Type-species, *bradleyi* Alexander (orig. des.).

bradleyi (Alexander), 1916a: 530 (*Pterochionea*).—B.C.; Oreg., Idaho, Mont.
flavoferruginea Osten Sacken, 1859: 229.—D.C.; Wis. to Que. and Maine, s. to Mo. and Ga.
    indivisa Osten Sacken, 1861: 291.—N.Y., Mass.
macnabi Alexander, 1944b: 95.—Oreg.; Wash.
nigricauda Alexander, 1954: 68.—Oreg.
oregona Alexander, 1919g: 147.—Oreg.; B.C., Wash.

Genus NEOCLADURA Alexander

Cladura, subg. **Neocladura** Alexander, 1920h: 947. Type-species, *delicatula* Alexander (orig. des.).

americana (Alexander), 1917: 29 (*Crypteria*).—Oreg.; Wash.
delicatula (Alexander), 1914a: 589 (*Cladura*).—N.H.; Mich. to Maine, s. to Tenn. and S.C.

Genus CHIONEA Dalman

Chionea Dalman, 1816: 104. Type-species, *araneoides* Dalman (mon.).

albertensis Alexander, 1941c: 211.—Alta.
alexandriana Garrett, 1922c: 62.—B.C.; Alta.
macnabeana Alexander, 1946d: 156.—Oreg.
nivicola Doane, 1900: 188.—Wash.; Mont.
noveboracensis Alexander, 1917: 205.—N.Y.
scita Walker, 1848: 82.—N. Amer.; N.Y., Mass.
    primitiva Alexander, 1917: 204.—N.Y.
stoneana Alexander, 1940d: 100.—Ill.
valga Harris, 1841: 404.—Mass.; Sask. to Labr., s. to Minn. and Mass.
    aspera Walker, 1848: 82.—Ont.
    gracilis Alexander, 1917: 206.—N.Y.
    waughi Curran, 1925a: 24.—Labr.

Genus NEOLIMNOPHILA Alexander

Limnophila, subg. **Neolimnophila** Alexander, 1920b: 37. Type-species, *ultima* Osten Sacken (orig. des.).

appalachicola Alexander, 1941a: 312.—N.C.; Tenn.
brevissima Alexander, 1952b: 233.—Colo.
capnioptera Alexander, 1947c: 251.—Ga.
ultima ultima (Osten Sacken), 1859: 238 (*Limnophila*).—D.C., Maine; Alta. to Que. and Maine, s. to Iowa, Miss., and Ga.
  ssp. **alaskana** (Alexander), 1924d: 10 (*Limnophila*).—Alaska.

Genus GNOPHOMYIA Osten Sacken

Gnophomyia Osten Sacken, 1859: 223. Type-species, *tristissima* Osten Sacken (Coquillett, 1910b: 547).

Subgenus GNOPHOMYIA Osten Sacken

cockerelli Alexander, 1919c: 193 (as *tristissima* ssp.).—Colo.; Colo. to Ind.
tristissima Osten Sacken, 1859: 224.—N.Y., D.C., Va., Wis.; N.W.T. to Que. and Maine, s. to Tex. and Fla.

Subgenus EUGNOPHOMYIA Alexander

Gnophomyia, subg. **Eugnophomyia** Alexander, 1947f: 72. Type-species, *luctuosa* Osten Sacken (orig. des.).

apache Alexander, 1946a: 525.—Ariz.
luctuosa Osten Sacken, 1859: 224.—Fla.; Ill. to D.C., s. to Okla. and Fla.
  nigricola Walker, 1861: 333 (*Limnobia*).—U.S.

Subgenus IDIOGNOPHOMYIA Alexander

Gnophomyia, subg. **Idiognophomyia** Alexander, 1956d: 403. Type-species, *capicola* Alexander (orig. des.).

comstocki Alexander, 1947g: 45.—Calif.

Genus GONOMYIA Meigen

Gonomyia Meigen, 1818: 146. Type-species, *Limnobia tenella* Meigen (mon.). Generic name cited in specific synonymy; needs action by I.C.Z.N.
Goniomyia, emend.

Subgenus PROGONOMYIA Alexander

Gonomyia, subg. **Gonomyella** Alexander, 1916a: 511 (preocc. Meunier, 1899). Type-species, *slossonae* Alexander (mon.).
Gonomyia, subg. **Progonomyia** Alexander, 1920h: 938 (n. name for *Gonomyella* Alexander). Type-species, *slossonae* Alexander (aut.).

hesperia Alexander, 1926g: 78.—Calif.
plumbea Alexander, 1946a: 522.—Ariz.
slossonae Alexander, 1914a: 588.—Fla.; Neotropical.
zionicola Alexander, 1948a: 52.—Utah.

Subgenus IDIOCERA Dale

Limnobia, subg. **Idiocera** Dale, 1842: 431, 433. Type-species, *sexguttata* Dale (mon.).

apicispina Alexander, 1926e: 117.—Ind.
biacus Alexander, 1948b: 209.—Ariz.
blanda Osten Sacken, 1859: 231.—D.C., N.Y.; Alta. to N.S., s. to Mo. and Fla.
brookmani Alexander, 1944b: 96.—Wash.; Oreg.
californica Alexander, 1916c: 324.—Calif.; B.C.
coloradica Alexander, 1920g: 89.—Colo.; Alta., Oreg., Calif., Utah, Ariz.
  icasta Alexander, 1927e: 191.—Alta.
flintiana Alexander, 1961a: 15.—Tex.
gaigei Rogers, 1931: 333.—Okla.
gothicana Alexander, 1943a: 166.—Colo.; Oreg., Calif.
lindseyi Alexander, 1946g: 97.—Calif.; Oreg.
mathesoni Alexander, 1915b: 170.—N.Y.; Iowa and Mich. to Nfld., s. to Mo. and N.C.
multistylata Alexander, 1948a: 54.—Utah.
proserpina Alexander, 1943b: 747.—Wyo.; Calif., Utah.
shannoni Alexander, 1926e: 118.—Wash.; Oreg., Idaho, Calif., Utah, Wyo.
sperryana Alexander, 1948b: 209.—Ariz.

Subgenus EUPTILOSTENA Alexander

Gonomyia, subg. **Euptilostena** Alexander, 1938a: 126. Type-species, *reticulata* Alexander (orig. des.).

knowltoniana Alexander, 1948a: 53.—Utah.
polingi Alexander, 1946a: 523.—Tex.; Calif., Ariz., Mexico.

Subgenus LIPOPHLEPS Bergroth

Leiponeura Skuse, 1890a: 795. Type-species, *gracilis* Skuse (Alexander, 1913a: 503)=*skusei* Alexander. [In the interest of stability, author rejects this valid prior name.—Eds.].
Lipophleps Bergroth, 1915: 55 (as genus; unjustified n. name for *Leiponeura* Skuse). Type-species, *Leiponeura gracilis* Skuse (aut.)=*skusei* Alexander.
burgessi Alexander, 1944d: 245.—La.

manca Osten Sacken, 1869b: 178.—N.J.; Ind. to Mass., s. to Tenn. and Fla.
curvivena Coquillett, 1908: 144 (*Dicranomyia*).—Md.
puer Alexander, 1913a: 506.—Santo Domingo; D.C., S.C., Ga., Fla., Neotropical.
sacandaga Alexander, 1914a: 587.—N.Y.; Mich. to Vt., s. to Mo. and N.C.
sulphurella Osten Sacken, 1859: 230.—N.Y., D.C.; Kans. to Nfld., s. to Tex. and Fla.

Subgenus PARALIPOPHLEPS Alexander

Gonomyia, subg. **Paralipophleps** Alexander, 1947f: 97. Type-species, *Atarba pleuralis* Williston (orig. des.).

pleuralis (Williston), 1896c: 289 (*Atarba*).—St. Vincent I.; Ga., Fla., Bermuda, Neotropical.

Subgenus NEOLIPOPHLEPS Alexander

Gonomyia, subg. **Neolipophleps** Alexander, 1947f: 98. Type-species, *Dicranomyia cinerea* Doane (orig. des.).

alexanderi (Johnson), 1912a: 3 (*Elliptera*).—N.Y.; Ill. to Nfld., s. to Okla. and S.C.
cinerea (Doane), 1900: 182 (*Dicranomyia*).—Wash.; Calif., Utah, Colo.
helophila Alexander, 1916e: 343.—Peru; Okla., Tex., Neotropical.

Subgenus GONOMYIA Meigen

abyssa Alexander, 1946c: 206.—Ariz.
aciculifera Alexander, 1919c: 193.—Calif.; Oreg., Wyo.
armigera Alexander, 1922b: 4.—Ind.; N.Y.
bidentata Alexander, 1922b: 3.—Ind.; Wis. to N.B., s. to Ind. and Conn.
bihamata Alexander, 1943b: 749.—Wyo.; Wash., Oreg., Calif., Utah.
cognatella Osten Sacken, 1859: 230.—D.C.; Wis. to Conn., s. to Mo. and S.C.
currani Alexander, 1926b: 239.—Ont.; N.Y., Mass., Conn., N.C.
extensivena Alexander, 1943a: 167.—Colo.; Wyo.
filicauda filicauda Alexander, 1916c: 320.—Colo.; Alta., Utah.
  ssp. **bidenticulata** Alexander, 1949a: 319.—Wash.
filiformis Alexander, 1948b: 211.—Ariz.
flavibasis Alexander, 1916c: 317.—Calif.; Oreg. to Wyo., s. to Mexico.
florens Alexander, 1916c: 316.—N.Y.; Mich. to Que. and Maine, s. to Ill., Tenn., and N.C.
harmstoni Alexander, 1948a: 56.—Utah; Idaho.

isolata Alexander, 1949a: 319.—Wash.
kansensis Alexander, 1918a: 158.—Kans.; Okla., Mo., Ill., Ind., Mich.
mainensis Alexander, 1919c: 163.—Maine.
noveboracensis Alexander, 1916c: 319.—N.Y.; Mich. to Que. (Gaspé), s. to Ind. and Mass.
paiuta Alexander, 1948a: 59.—Utah.
percomplexa Alexander, 1946c: 206.—Oreg.; Calif.
poliocephala Alexander, 1924d: 8.—Calif.; Colo.
reflexa Alexander, 1927h: 59.—Mich.
sevierensis Alexander, 1948a: 59.—Utah.
spinifer Alexander, 1918a: 384.—Calif.; Utah, Tex.
subcinerea Osten Sacken, 1859: 231.—N.Y., D.C.; B.C. to Nfld., s. to Utah, Kans., and Fla.
    obscuris Doane, 1900: 192 (*Phyllolabis*).—Wash.
taeniata Alexander, 1927h: 59.—Ga.; N.C.
tetonensis Alexander, 1946c: 207.—Wyo.
triformis Alexander, 1946a: 524.—Ariz.; Utah.
vafra Alexander, 1945a: 428.—Wyo.; Utah.
virgata Doane, 1900: 189.—Wash.; Oreg., Calif.

Genus TEUCHOLABIS Osten Sacken

Teucholabis Osten Sacken, 1859: 222. Type-species, *complexa* Osten Sacken (mon.).

Subgenus TEUCHOLABIS Osten Sacken

carolinensis Alexander, 1916b: 44.—S.C.; Fla.
complexa Osten Sacken, 1859: 223.—Ill., N.Y., D.C.; Mich. to Conn., s. to Okla. and Fla.
duncani Alexander, 1946a: 521.—Ariz.
immaculata Alexander, 1922b: 5 (as *complexa* ssp.).—Ind.; Tenn., Ga.
lucida Alexander, 1916b: 43.—D.C.; Mo. to D.C., s. to Fla.
myersi Alexander, 1926d: 228.—Cuba; s. Fla.
rubescens Alexander, 1914a: 582.—N. Mex.; Utah, Ariz.

Genus RHABDOMASTIX Skuse

Rhabdomastix Skuse, 1890a: 828. Type-species, *ostensackeni* Skuse (mon.).

Subgenus RHABDOMASTIX Skuse

nuttingi Alexander, 1950a: 46.—Ariz.

Subgenus SACANDAGA Alexander

Sacandaga Alexander, 1911b: 349 (as genus). Type-species, *flava* Alexander (orig. des.).

borealis Alexander, 1924d: 9.—Alaska.
brachyneura Alexander, 1933a: 94.—S.C.; N.C.
brittoni Alexander, 1933a: 93.—Conn.; N.H., Mass.
californiensis Alexander, 1921a: 103.—Calif.; Utah, Ariz., N. Mex.
caudata (Lundbeck), 1898: 267 (*Gonomyia*).—Greenland; N.W.T. (Baffin I.).
coloradensis Alexander, 1917: 28 (as *flava* ssp.).—Colo.
fascigera Alexander, 1920b: 36.—Calif.
flava (Alexander), 1911b: 351 (*Sacandaga*).—N.Y.; Que. and Nfld., s. to Tenn.
galactoptera (Bergroth), 1888a: 196 (*Gonomyia*).—Alaska.
hansoni Alexander, 1939a: 99.—Mass.
hudsonica Alexander, 1933a: 92.—Que.
ioogoon Alexander, 1948a: 61.—Utah.
leonardi Alexander, 1930b: 75.—Mont.; Wash., Oreg., Wyo., Utah, Colo.
leptodoma Alexander, 1943a: 170.—Colo.; Wyo.
lipophleps Alexander, 1948a: 62.—Utah.
margarita Alexander, 1940a: 639.—Tenn.; N.C.
mediovena Alexander, 1933a: 93.—S.C.
megacantha Alexander, 1959a: 77.—Calif.
monticola Alexander, 1916a: 528.—B.C.; Calif.
neolurida Alexander, 1943a: 171.—Wyo.; Alaska, B.C., Alta.
setigera Alexander, 1943a: 172 (as *neolurida* ssp.).—Colo.
subarctica Alexander, 1933a: 91.—Que.
subcaudata Alexander, 1927e: 191.—Alta.; B.C., Wash.
subfascigera Alexander, 1927e: 192.—Alta.
trichophora Alexander, 1943f: 255.—Wash.; Oreg., Calif.

Genus LIPSOTHRIX Loew

Lipsothrix Loew, 1873c: 67, 69. Type-species, *nobilis* Loew (present des.).

fenderi Alexander, 1946c: 204.—Oreg.; Wash.
nigrilinea (Doane), 1900: 190 (*Limnophila*).—Wash.; Oreg.
shasta Alexander, 1946c: 205.—Calif.
sylvia (Alexander), 1916a: 534. (*Limnophila*).—N.Y.; N.S., s. to Tenn. and N.C.

Genus HESPEROCONOPA Alexander

Erioptera, subg. **Hesperoconopa** Alexander, 1948a: 76. Type-species, *Gnophomyia aperta* Coquillett (orig. des.).

aperta (Coquillett), 1905d: 58 (*Gnophomyia*).—B.C.; Alaska to Oreg., Utah, and Wyo.
    mormon Alexander, 1927i: 144 (*Psiloconopa*).—Utah.
dolichophallus (Alexander), 1948a: 77 (*Erioptera*).—Colo.; Wash., Oreg., Utah.
melanderi (Alexander), 1944a: 217 (*Erioptera*).—Calif.
pilipennis (Alexander), 1918a: 382 (*Erioptera*).—Oreg.
pugilis (Alexander), 1952b: 236 (*Erioptera*).—Colo.

Genus CRYPTOLABIS Osten Sacken

Cryptolabis Osten Sacken, 1859: 224. Type-species, *paradoxa* Osten Sacken (mon.).

Subgenus CRYPTOLABIS Osten Sacken

bidenticulata Alexander, 1949d: 168.—Calif.
bisinuatis Doane, 1900: 189.—Wash.; Oreg., Idaho.
brachyphallus Alexander, 1950a: 159.—Calif.
magnistyla Alexander, 1962: 4.—Calif.
minutula Alexander, 1927h: 62.—Tex.
mixta Alexander, 1949d: 167.—Calif.
molophiloides Alexander, 1943b: 763.—Wyo.
pachyphallus Alexander, 1943f: 257.—Idaho.
paradoxa Osten Sacken, 1859: 225.—Va.; Ont. to N.S., s. to Tenn. and Fla.
retrorsa Alexander, 1950a: 159.—Calif.; Oreg.
sica Alexander, 1946a: 526.—Ariz.; Utah,

Subgenus PHANTOLABIS Alexander

Cryptolabis, subg. **Phantolabis** Alexander, 1956a: 184. Type-species, *Erioptera lacustris* Alexander (orig. des.).

lacustris (Alexander), 1938b: 76 (*Erioptera*).—Mich.

Genus CHEILOTRICHIA Rossi

Cheilotrichia Rossi, 1848: 12. Type-species, *Erioptera imbuta* Meigen (mon.).

Subgenus CHEILOTRICHIA Rossi

alicia (Alexander), 1914a: 585 (*Erioptera*).—Calif.; Wash., Oreg.

Subgenus EMPEDA Osten Sacken

Empeda Osten Sacken, 1869b: 183 (as genus). Type-species, *stigmatica* Osten Sacken (Coquillett, 1910b: 537).

cinereipleura (Alexander), 1917: 200 (*Erioptera*).—Colo.; Utah, N. Mex.
exilistyla (Alexander), 1949c: 42 (*Erioptera*).—Wash.
gloydae (Alexander), 1950a: 157 (*Erioptera*).—Ariz.
noctivagans (Alexander), 1917: 200 (*Erioptera*).—Va.; Fla.
perflavens (Alexander), 1948b: 212 (*Erioptera*).—Ariz.
rectispina (Alexander), 1955b: 13 (*Erioptera*).—Alaska; Y.T., B.C.
stigmatica (Osten Sacken), 1869b: 184 (*Empeda*).--N.Y.; S. Dak. to Nfld., s. to Tenn. and N.C.
subborealis (Alexander), 1955b: 13 (*Erioptera*).—Alaska; Alta.
toklat (Alexander), 1955b: 14 (*Erioptera*).—Alaska.
tristimonia (Alexander), 1943b: 752 (*Erioptera*).—Wyo.; Alaska, Y.T.
umiat (Alexander), 1955b: 15 (*Erioptera*).—Alaska.

Subgenus GONEMPEDA Alexander

Erioptera, subg. **Gonempeda** Alexander, 1924d: 8. Type-species, *Limnobia flava* Schummel (orig. des.).

burra (Alexander), 1924d: 8 (*Erioptera*).—Calif.; B.C., Oreg.
nyctops (Alexander), 1916a: 503 (*Erioptera*).---N.Y.; Mich. to Que. and Maine, s. to Tenn. and N.C.
yellowstonensis (Alexander), 1943b: 751 (*Erioptera*).—Wyo.

Genus GONOMYODES Alexander

Erioptera, subg. **Gonomyodes** Alexander, 1948a: 78. Type-species, *knowltonia* Alexander (orig. des.).

crickmeri (Alexander), 1949d: 167 (*Erioptera*).—Calif.
knowltonia (Alexander,) 1948a: 78 (*Erioptera*).—Utah.
tacoma (Alexander), 1949a: 321 (*Erioptera*).—Wash.; Alaska, Alta.
yohoensis (Alexander), 1952b: 234 (*Erioptera*).—Alta.-B.C. border; Alta.

Genus ARCTOCONOPA Alexander

Erioptera, subg. **Arctoconopa** Alexander, 1955b: 17. Type-species, *Psiloconopa forcipata* Lundström (orig. des.).

aldrichi (Alexander), 1924d: 4 (*Erioptera*).—Alaska; Y.T., B.C., Alta.
carbonipes (Alexander), 1929g: 52 (*Erioptera*).—Wash.
cinctipennis (Alexander), 1918b: 286 (*Erioptera*).—Calif.

forcipata forcipata (Lundström), 1915: 27 (*Psiloconopa*).—Siberia; Alaska, Y.T., B.C., N.W.T.
    angustipennis Alexander, 1919b: 5 (*Erioptera*).—N.W.T.
    alaskensis Alexander, 1924d: 5 (*Erioptera*).—Alaska.
  ssp. **gaspicola** (Alexander), 1929g: 54 (*Psiloconopa;* as sp.).— Que.; Alta. and Colo. to Que.
katmai (Alexander), 1920i: 195 (*Erioptera*).—Alaska; Y.T., B.C.
kluane (Alexander), 1955b: 20 (*Erioptera*).—Y.T.
manitobensis (Alexander), 1929g: 51 (*Erioptera*).—Man.; Alta., Oreg., Wyo., Utah.
pahasapa (Alexander), 1955a: 20 (*Erioptera*).—S. Dak.
painteri (Alexander), 1929d: 19 (*Erioptera*).—Ohio.

Genus ERIOPTERA Meigen

Polymeda Meigen, 1800: 14. Type-species, *Erioptera lutea* Meigen (Coquillett, 1910b: 593). Suppressed by I.C.Z.N., 1963b: 339.
Erioptera Meigen, 1803: 262. Type-species, *lutea* Meigen (Coquillett, 1910b: 540). [In the interest of stability the author rejects the prior designation of *grisea* Meigen, a species of *Molophilus*, designated by Blanchard *in* Audouin et al., 1849: pl. 163, fig. 3.—Eds.].

Subgenus TRIMICRA Osten Sacken

Trimicra Osten Sacken, 1861: 290 (as genus). Type-species, *anomala* Osten Sacken (mon.)=*pilipes* (Fabricius).
pilipes (Fabricius), 1787: 324 (*Tipula*).—Europe; Oreg. to R.I., s. to Tex. and Fla., Mexico, virtually cosmopolitan.
    anomala Osten Sacken, 1861: 290 (*Trimicra*).—D.C.

Subgenus EMPEDOMORPHA Alexander

Empedomorpha Alexander, 1916a: 507 (as genus). Type-species, *Trimicra empedoides* Alexander (orig. des.).
empedoides (Alexander), 1916b: 44 (*Trimicra*).—N. Mex.; N. Dak. to Ariz. and Tex.

Subgenus SYMPLECTA Meigen

Helobia Lepeletier and Serville *in* Latreille et al., 1828: 585 (preocc. Stephens, 1827). Type-species, *Limnobia punctipennis* Meigen (mon.)=*hybrida* (Meigen).
Symplecta Meigen, 1830: 282 (as genus). Type-species, *Limnobia punctipennis* Meigen (Westwood, 1840: 128)=*hybrida* (Meigen).

cana (Walker), 1848: 48 (*Limnobia*).—Ont.; throughout s. Canada and U.S.
    hybrida, authors, not Meigen.
    punctipennis, authors, not Meigen.
hybrida (Meigen), 1804: 57 (*Limonia*).—Europe; Alaska, Eurasia.
sheldoni Alexander, 1955b: 28.—Alaska.
sunwapta Alexander, 1952b: 267.—Alta.; Alaska, Y.T.

Subgenus ERIOPTERA Meigen

bryantiana Alexander, 1929g: 49.—Alta.
chlorophylla Osten Sacken, 1859: 226.—D.C., Ga.; Wis. to N.S., s. to Tenn. and Fla.
chlorophylloides chlorophylloides Alexander, 1919a: 106.—Colo.; Colo. to Que. and Nfld., s. to Tenn.
    ssp. **orthomera** Alexander, 1940c: 155.—Conn.
chrysocoma Osten Sacken, 1859: 226.—D.C.; Que. and N.B., s. to N.C.
chrysocomoides Alexander, 1929g: 50.—Tenn.; Mass.
dyari Alexander, 1924d: 6.—Calif.; Wash., Oreg., Mont.
ebenina Alexander, 1926b: 237.—Ont.; Conn., N.Y.
furcifera Alexander, 1919a: 108.—Md.; Mich., Ont., and Mass., s. to Tenn. and S.C.
gaspeana Alexander, 1929e: 250.—Que.; Mich. to Que.
georgei Alexander, 1956a: 183.—Mich.
hohensis Alexander, 1949c: 44.—Wash.
kluaneana Alexander, 1955b: 31.—Y.T.
leptostyla Alexander, 1940c: 154.—Mich.
megophthalma Alexander, 1918a: 60.—N.Y.; Mich. to Que. and Nfld., s. to Tenn. and N.C.
oregonensis Alexander, 1920g: 87.—Oreg.
osceola Alexander, 1933a: 95.—Fla.
seminole Alexander, 1933a: 96.—Fla.
septemtrionis Osten Sacken, 1859: 226.—Maine, N.Y.; Wash. to Nfld., s. to Calif., Kans., and Fla.
    subseptemtrionis Alexander, 1920d: 109.—Colo.
straminea Osten Sacken, 1869b: 157.—Unknown; Ill. to Que. and Maine, s. to S.C.
subchlorophylla Alexander, 1919a: 107.—N.J.; Mass. to Fla.
subfurcifera Alexander, 1929g: 51.—Mich.; Maine, Conn.
uliginosa Alexander, 1930a: 277.—Conn.; Alaska to Wis. and Maine.
vespertina Osten Sacken, 1859: 226.—D.C., Fla., Wis.; Iowa to N.S., s. to Ala. and Fla.
    holoptica Dietz, 1921a: 245.—Pa.
    fuscoantennata Dietz, 1921a: 246 (as *holoptica* var.).—Pa.

villosa Osten Sacken, 1859: 226.—Middle States; Sask. to Nfld., s. to Utah and Colo.
    dilatata Alexander, 1924d: 7.—Mont.
viridula Alexander, 1929d: 20.—Vt.; Ont. to Nfld., s. to N.Y. and Conn.
yukonensis Alexander, 1955b: 32.—Y.T.

Subgenus MESOCYPHONA Osten Sacken

Erioptera, subg. **Mesocyphona** Osten Sacken, 1869b: 152. Type-species, *caliptera* Say (Brunetti, 1912: 458).

caliptera Say, 1823: 17 (1859b: 44).—Mo.; Calif. to Nfld., s. to Fla., Neotropical.
    caloptera, emend.
distincta Alexander, 1912d: 165.—N. Mex.; Calif. to Wyo. and N. Mex.
dulcis Osten Sacken, 1877: 198.—Calif.; Wash., Oreg., Wyo.
eiseni Alexander, 1913a: 516.—Guatemala; Calif., Utah, Ariz., N. Mex., Neotropical.
evergladea Alexander, 1933a: 97.—Fla.
femoraatra Alexander, 1950d: 170.—Ga.
immaculata Alexander, 1913a: 518.—Nicaragua; Tex., Neotropical.
knabi Alexander, 1913a: 515.—Mexico; Ariz. to Mich., Ind., and Fla.
    hubbelli Rogers, 1931: 332, 333 (as *hubelli*, p. 332).—Okla.
melanderiana Alexander, 1946g: 99.—Wash.; Oreg., Calif., Nev.
    unduligera Alexander, 1946c: 208.—Oreg.
needhami Alexander, 1918a: 383.—N.Y.; Mo. to Ont. and N.S., s. to Fla.
parva Osten Sacken, 1859: 227.—D.C., Ga.; Kans. to Mich. and Conn., s. to Fla.
rubia Alexander, 1914a: 583.—Ariz.; Utah, Colo. ?Not subg. *Mesocyphona*.
serpentina Alexander, 1941a: 318.—N.C.; Mass.
splendida Alexander, 1913a: 514.—Guatemala; Ariz., Neotropical.
tantilla Alexander, 1916a: 502.—Miss.

Subgenus HOPLOLABIS Osten Sacken

Erioptera, subg. **Hoplolabis** Osten Sacken, 1869b: 152. Type-species, *armata* Osten Sacken (mon.)
Haplobasis, error.

armata Osten Sacken, 1859: 227.—D.C., Wis.; Colo. to Nfld., s. to Okla. and Ga.
asiatica Alexander, 1919d: 124.—Japan; Alaska, e. Asia.
bipartita Osten Sacken, 1877: 199.—Calif.
maria Alexander, 1948a: 75.—Utah.

Subgenus PSILOCONOPA Zetterstedt

Psiloconopa Zetterstedt, 1838: 847 (as genus). Type-species, *meigeni* Zetterstedt (mon.).

Ilisia Rondani, 1856: 182. Type-species, *Erioptera maculata* Meigen (orig. des.).

Erioptera, subg. *Acyphona* Osten Sacken, 1869b: 151. Type-species, *venusta* Osten Sacken (Brunetti, 1920: 238).

armillaris Osten Sacken, 1869b: 158.—N.Y., D.C.; Kans. to N.S. and N.C.
bispinigera Alexander, 1930b: 76.—Mont.; Oreg., Utah, Wyo.
bisulca Alexander, 1949d: 166.—Calif.
carsoni Alexander, 1955b: 21.—Alaska.
churchillensis Alexander, 1938b: 77.—Man.
cramptonella (Alexander), 1931c: 144 (*Psiloconopa*).—Que.; P.E.I., Nfld.
denali Alexander, 1955b: 22.—Alaska.
dorothea Alexander, 1914a: 584.—N. Mex.; Colo.
ecalcar Alexander, 1949c: 43.—Utah.
estella Alexander, 1955b: 23.—Alaska.
graphica Osten Sacken, 1859: 227.—D.C.; Nebr. to Ont. and Mass., s. to La. and Fla.
hardyi Alexander, 1948a: 68.—Utah.
hygropetrica Alexander, 1943a: 172.—Colo.; Alaska to Wash. and Colo.
  crassivena Alexander, 1949a: 324.—Wash.
indianensis Alexander, 1922b: 1.—Ind.; Iowa, Mich., Mo., Ill., Ky.
irata Alexander, 1949a: 324.—Oreg.; Wash.
laevis laevis Alexander, 1930b: 77.—Mass.; Maine, Pa.
  ssp. **restricta** Alexander, 1941a: 317.—N.C.
laticeps Alexander, 1916a: 501.—Calif.
lucia Alexander, 1914a: 584.—Colo.; Alta., Utah, Wyo., N. Mex.
mabelana Alexander, 1955b: 25.—Y.T.
margarita Alexander, 1919a: 104.—Colo.; Alaska, Y.T., Wash., Utah, Wyo.
mckinleyana Alexander, 1955b: 26.—Alaska.
megarhabda (Alexander), 1943a: 174 (*Ormosia*).—Colo.; Alaska, Wash., Oreg., Utah, Wyo.
microcellula Alexander, 1914a: 585.—Colo.; Utah.
neomexicana Alexander, 1929g: 53.—N. Mex.; Utah.
peayi Alexander, 1948a: 71.—Utah.
polycantha Alexander, 1945d: 156.—Wash.
rainieria Alexander, 1943f: 46.—Wash.; Oreg., Calif., Mont., Wyo.

recurva Alexander, 1949a: 326.—Wash.
shoshone Alexander, 1945a: 430.—Wyo.; Alaska and Y.T. to Oreg. and Colo.
sinawava Alexander, 1948a: 72.—Utah.
sparsa Alexander, 1919f: 214.—Calif.
stictica stictica (Meigen).—Not Nearctic.
    ssp. **angularis** (Alexander), 1917: 26 (*Trimicra;* as sp.).—Utah.
sweetmani Alexander, 1940d: 102.—Ga.
telfordi Alexander, 1948a: 73.—Utah.
venusta venusta Osten Sacken, 1859: 227.—N.Y., Conn., Va.; Wis. to Que. and Maine, s. to Mo. and Fla.
    ssp. **nubilosa** Alexander, 1956a: 184.—Mich.
zukeli Alexander, 1940b: 89.—Idaho; Oreg.

Genus ORMOSIA Rondani

Ormosia Rondani, 1856: 180. Type-species, *Erioptera nodulosa* Macquart (orig. des.).

Subgenus RHYPHOLOPHUS Kolenati

Rhypholophus Kolenati, 1860: 393, 407 (as genus *Rypholophus*, p. 393). Type-species, *phryganopterus* Kolenati (mon.).

arapaho Alexander, 1958b: 134.—Colo.
bicuspidata Alexander, 1944a: 221.—Wash.
bifidaria Alexander, 1919g: 143.—Colo.; Oreg., Calif., Utah.
fumata (Doane), 1900: 188 (*Rhypholophus*).—Idaho; Wash., Oreg.
hoodiana Alexander, 1944a: 219.—Oreg.
libella Alexander, 1943a: 177.—Colo.
oregonica oregonica Alexander, 1944a: 220.—Oreg.
    ssp. **fugax** Alexander, 1946d: 156 (as sp.).—Oreg.
paradisea Alexander, 1920b: 35.—Wash.; B.C., Alta., Oreg., Wyo.
    garretti Alexander, 1926e: 19.—Alta.
suffumata Alexander, 1943a: 178.—Colo.; Wash., Oreg., Idaho.
wasatchensis Alexander, 1948a: 64.—Utah.

Subgenus SCLEROPROCTA Edwards

Ormosia, subg. **Scleroprocta** Edwards, 1938b: 137. Type-species, *danica* Nielsen (orig. des.).

innocens (Osten Sacken), 1869b: 142 (*Rhypholophus*).—N.J., D.C.; Mich. to N.H., s. to Tenn. and S.C.
tetonica Alexander, 1945a: 434.—Wyo.; Utah.

Subgenus ORMOSIA Rondani

adirondacensis Alexander, 1919g: 145.—N.Y.; N.S. to Tenn.
affinis (Lundbeck), 1898: 266 (*Rhypholophus*).—Greenland; ?B.C.
albertensis Alexander, 1933a: 98.—Alta.; Wash., Oreg., Wyo.
albrighti Alexander, 1954: 75.—Oreg.
apicalis Alexander, 1911c: 200.—Ga.; N.Y. to Ga.
   atriceps Dietz, 1916: 136.—Pa.
arcuata (Doane), 1908c: 201 (*Rhypholophus*).—N.Y.; Alta. to N.B., s. to Tenn.
bilineata Dietz, 1916: 142.—N.Y.; S. Dak. to Nfld., s. to N.C.
   huronis Alexander, 1929d: 20.—Mich.
brachyrhabda Alexander, 1948a: 65.—Utah; Wyo.
brevicalcarata Alexander, 1927h: 61.—N.C.
broweri Alexander, 1939a: 100.—Maine; B.C. to Maine, s. to Colo.
burneyensis Alexander, 1950c: 33.—Calif.
carolinensis Alexander, 1925b: 229.—N.C.
cerrita Alexander, 1949a: 327.—Wash.
cockerelli (Coquillett), 1901g: 149 (*Rhypholophus*).—N. Mex.; Colo.
cornuta (Doane), 1908c: 202 (*Rhypholophus*).—Calif.
curvata Alexander, 1924d: 2.—Alaska; Wash., Oreg.
davisi Alexander, 1954: 77.—Oreg.
decussata Alexander, 1924d: 1.—Alaska; Wash., Oreg.
dedita Alexander, 1943b: 760.—Wyo.
defrenata Alexander, 1948a: 66.—Utah; Oreg., Idaho.
dentifera Alexander, 1919g: 144.—Maine; Que. and Nfld., s. to N.Y. and Conn.
divergens (Coquillett), 1905d: 57 (*Rhypholophus*).—B.C.; Wash., Oreg., Calif.
fascipennis (Zetterstedt), 1838: 831 (*Erioptera*).—N. Europe; Greenland, e. Que.
   cramptoniana Alexander, 1929e: 249.—Que. (Gaspé).
fernaldi Alexander, 1924a: 116.—Mass.
frisoni Alexander, 1920d: 224.—Ill.; Iowa, Mich., Ind.
furibunda Alexander, 1954: 77.—Oreg.
fusiformis fusiformis (Doane), 1900: 187 (*Rhypholophus*).—Idaho; Wash., Oreg., Mont., Utah, Colo.
   divexus Doane, 1908c: 201 (*Rhypholophus*).—Wash.
   ssp. **viduata** Alexander, 1941c: 212 (as sp.).—Colo.
gaspensis Alexander, 1929e: 249.—Que.; N.B.
gerronis Alexander, 1954: 78.—Oreg.
hallahani Alexander, 1943a: 175.—Colo.; Y.T., B.C., Oreg., Wyo., S. Dak.
harrisoniana Alexander, 1940a: 642.—Tenn.
heptacantha Alexander, 1949d: 169.—Calif.

hispa Alexander, 1945d: 158.—Wash.; Oreg.
holotricha (Osten Sacken), 1859: 226 (*Erioptera*).—D.C.; N.H., s. to Tenn. and N.C.
hubbelli Alexander, 1926e: 20.—N. Dak.; Man., Minn.
hynesi Alexander, 1962: 5.—Calif.
ingloria Alexander, 1929d: 21.—Ind.; Ont.
ithacana Alexander, 1929a: 29.—N.Y.
lanuginosa (Doane), 1900: 188 (*Rhypholophus*).—Idaho.
legata Alexander, 1949d: 171.—Calif.
lilliana Alexander, 1940a: 641.—Tenn.; N.C.
longicornis (Doane), 1908c: 201 (*Rhypholophus;* as *longicornus*).—Wash.
luteola Dietz, 1916: 138.—Conn.; Mich. to Maine, s. to Pa.
manicata (Doane), 1900: 187 (*Rhypholophus*).—Idaho; Wash. to Nfld., s. to Calif. and Pa., Japan.
    deviata Dietz, 1916: 143.—Pa.
    fuscopyga Alexander, 1924d: 3.—Calif.
megacera Alexander, 1917: 26.—N.Y.; Que. and Nfld., s. to Pa.
    divergens Dietz, 1916: 144. (preocc. Coquillett, 1905).—Pa.
meigenii (Osten Sacken), 1859: 226 (*Erioptera*).—Middle States; Ont. to N.S., s. to Ill. and S.C.
mesocera Alexander, 1917: 25.—N.Y.; Ont. to N.S., s. to N.Y. and Vt.
mitchellensis Alexander, 1941a: 314.—N.C.
modica Dietz, 1916: 141.—Calif.; Oreg.
    stylifer Alexander, 1919g: 146.—Oreg.
monticola (Osten Sacken), 1869b: 145 (*Rhypholophus*).—N.H.; Mich. to N.B., s. to N.C.
nigripila (Osten Sacken), 1869b: 142 (*Rhypholophus*).—D.C.; Mich. to Nfld., s. to Fla.
nimbipennis Alexander, 1917: 24.—N.Y.; Mich. to Nfld., s. to Conn.
nonacantha Alexander, 1954: 80.—Oreg.
notmani Alexander, 1920d: 225.—N.Y.; Ont., Que., Nfld., Mass., Conn.
onerosa Alexander, 1943b: 761.—Wyo.; Wash., Oreg., Utah., Ariz., N. Mex.
opifex Alexander, 1943b: 753.—Wyo.; Utah.
palpalis Dietz, 1916: 140.—Pa.
parallela (Doane), 1908c: 202 (*Rhypholophus*).—N.Y. Unrecognized.
pernodosa Alexander, 1950c: 34.—Calif.
perplexa Dietz, 1916: 141.—N.Y.
perspectabilis Alexander, 1944a: 218.—Oreg.
pleuracantha Alexander, 1954: 81.—Oreg.; Calif.
profunda Alexander, 1943f: 49.—Calif.; Oreg.

proxima Alexander, 1924d: 3.—Alaska; Oreg.
pugetensis Alexander, 1946g: 100.—Wash.; Oreg.
pygmaea (Alexander), 1912d: 166 (*Trimicra*).—N.Y.; Mich. to N.B., s to N.C.
  pilosa Dietz, 1916: 139.—Pa.
romanovichiana Alexander, 1953b: 327 (n. name for *nubila* Osten Sacken).—D.C.; Ill. to Maine, s. to Tenn. and S.C.
  nubila Osten Sacken, 1859: 227 (*Erioptera;* preocc. Schummel, 1829).—D.C.
rubella (Osten Sacken), 1869b: 144 (*Rhypholophus*).—Del., N.Y.; Wis. to Nfld, s. to Ga.
  enigmatica Dietz, 1921a: 249 (as ?var.).—Pa.
sentis Alexander, 1943b: 757.—Wyo.
serridens Alexander, 1919g: 144.—Va.; Md.
spinifex Alexander, 1943b: 758.—Wyo.; B.C., Utah, Colo.
subcornuta Alexander, 1920g: 88.—Oreg.; B.C., Wash.
subcostata Dietz, 1921a: 249.—Pa. Unrecognized.
subdentifera Alexander, 1941a: 316.—N.C.
subnubila Alexander, 1920i: 196.—Alaska; Alaska to Colo.
taeniocera Dietz, 1916: 145.—Calif.
tahoensis Alexander, 1950c: 32.—Calif.
tennesseensis Alexander, 1940a: 640.—Tenn.
townesi Alexander, 1933a: 99.—N.C.
tricornis Alexander, 1949d: 170.—Calif.
unicornis Alexander, 1954: 83.—Oreg.; Wash.
upsilon Alexander, 1946d: 157.—Oreg.; Wash.

Subgenus OREOPHILA Lackschewitz

Oreophila Lackschewitz, 1935: 7 (as genus). Type-species, *Rhypholophus bergrothi* Strobl (orig. des.).

absaroka Alexander, 1943b: 756.—Wyo.; B.C., Wash., Oreg., Alta.
bucera Alexander, 1954: 76.—Oreg.
flaveola (Coquillett), 1900h: 398 (1904: 12) (*Rhypholophus*).—Alaska; B.C., Oreg.
leptorhabda Alexander, 1943f: 47.—Calif.
sequoiarum Alexander, 1944a: 219.—Calif.
triangularis Alexander, 1949a: 331.—Wash.

Genus TASIOCERA Skuse

Tasiocera Skuse, 1890a: 815. Type-species, *tenuicornis* Skuse (Alexander, 1920e: 52).

Subgenus DASYMOLOPHILUS Goetghebuer

Molophilus, subg. **Dasymolophilus** Goetghebuer *in* Goetghebuer and Tonnoir, 1920: 132. Type-species, *Erioptera murina* Meigen (mon.).

miseranda Alexander, 1950a: 158.—Oreg.
niphadias (Alexander), 1925b: 229 (*Molophilus*).—Mich.; Mich. to Tenn. and Fla.
squiresi Alexander, 1948c: 134.—Oreg.
subnuda (Alexander), 1926g: 77 (*Dasymolophilus*).—Calif.
ursina (Osten Sacken), 1859: 228 (*Erioptera*).—D.C., Md.; Nfld., s. to Tenn. and N.C.

Genus MOLOPHILUS Curtis

Molophilus Curtis, 1833: pl. 444. Type-species, *brevipennis* Curtis (mon.) =*ater* (Meigen). [If the author accepted the nomenclaturally correct type-species for the genus *Erioptera*, *Molophilus* would fall as a synonym of it.—Eds.]

Subgenus MOLOPHILUS Curtis

arapahoensis Alexander, 1958b: 219.—Colo.
arizonicus Alexander, 1946a: 530.—Ariz.
aspersulus Alexander, 1962: 6.—Calif.
bispinosus Alexander, 1919c: 192.—Calif.
colonus Bergroth, 1888a: 195.—Alaska; Alaska to Calif., Utah, and Wyo.
  comatus Doane, 1900: 188 (*Erioptera*).—Wash.
costopunctatus Dietz, 1921a: 248.—Pa.
cramptoni Alexander, 1924b: 61.—N.Y.; Mich. to Maine, s. to Tenn. and S.C.
dirhaphis Alexander, 1958b: 221.—Calif.
falcatus Bergroth, 1888a: 196.—Alaska; B.C., Oreg., Wyo.
fenderi Alexander, 1952b: 269.—Calif.; Oreg.
floridensis Alexander, 1925b: 230.—Fla.
forcipulus forcipulus (Osten Sacken), 1869b: 163 (*Erioptera*).—N.J.; Wis. to N.B., s. to Tenn. and Fla.
  ssp. **heterocerus** Dietz, 1921a: 247.—Pa. ?=color variant only.
fultonensis Alexander, 1916a: 505.—N.Y.; Wis. to Maine, s. to Tenn. and S.C.
gracilipes Alexander, 1959a: 51.—Calif.
harrisoni Alexander, 1945a: 437.—Wyo.; Wash., Oreg., Idaho, Utah.
hirtipennis (Osten Sacken), 1859: 228 (*Erioptera*).—Md.; Ont. to Nfld., s. to Ill., Tenn., and N.C.

huron Alexander, 1929g: 56.—Mich.
kulshanicus Alexander, 1949a: 332.—Wash.; Oreg.
laricicola Alexander, 1929g: 55.—N.Y.; Maine, N.H., Vt.
millardi Alexander, 1944a: 222.—Calif.
nitidulus Alexander, 1946g: 102.—Wash.; Oreg., Idaho.
nitidus Coquillett, 1905d: 58.—Calif.; Wash., Oreg.
novacaesariensis Alexander, 1916a: 506.—N.J.; Ind. to N.J., s. to Fla.
oligacanthus Alexander, 1958b: 135.—Calif.
oregonicola Alexander, 1946d: 158.—Oreg.
palomaricus Alexander, 1947g: 47.—Calif.
paludicola Alexander, 1929g: 57.—Mass.; Maine, Vt., Conn.
paulus Bergroth, 1888a: 196.—Alaska; Wash., Oreg.
perflaveolus Alexander, 1918a: 160.—Calif.; B.C. to Calif., Utah, and Wash., also Tenn., Japan.
    auricomus Alexander, 1926e: 115.—Tenn.
pollex Alexander, 1931c: 146.—Que.; N.B.
pubipennis (Osten Sacken), 1859: 228 (*Erioptera*).—D.C.; Mich. to Que. and Nfld., s. to Fla.
quadrispinosus Alexander, 1924b: 62.—N.Y.; Mich. to Nfld. and Conn.
rainierensis Alexander, 1943f: 50.—Wash.; Oreg.
rostrifer Alexander, 1943b: 762.—Wyo.; Oreg., Calif.
sackenianus Alexander, 1926g: 78.—Calif.; Wash. to Calif.
    distilobatus Alexander, 1945d: 160.—Oreg.
sequoiae Alexander, 1952b: 270.—Calif.
soror Alexander, 1927a: 230.—Maine; Alta. and Mich. to N.B.
sparus Alexander, 1940c: 155.—Fla.
spiculatus spiculatus Alexander, 1918a: 161.—Colo.; Wash., Oreg., Calif.
    ssp. **sigmoideus** Alexander, 1945a: 438.—Wyo.; Alaska, B.C.
squamosus Alexander, 1919c: 191.—Calif.
stolidus Alexander, 1948b: 213.—Ariz.
subnitens Alexander, 1946g: 101.—Wash.
suffalcatus Alexander, 1946d: 159.—Oreg.; B.C.
    neofalcatus Alexander, 1954: 84 (unjustified n. name for *suffalcatus* Alexander).—Oreg.
unispiculatus Alexander, 1959a: 53.—Calif.
ursus Alexander, 1918a: 162.—N. Mex.
xanthus Alexander, 1955a: 21.—Wash.

Genus TOXORHINA Loew

Toxorhina Loew, 1850c: 36 (also 1851: 400). Type-species, *fragilis* Loew (Osten Sacken, 1869b: 113).

Toxorrhina, emend.
magna Osten Sacken, 1865: 232.—N.J.; Mich. to Mass., s. to La. and Fla.
muliebris Osten Sacken, 1865: 233.—Mass.; Wis. to Que. and Maine, s. to Ill. and Va.

Unplaced Species of Tipulidae

gracilis Wiedemann, 1828: 28 (*Limnobia*).—Pa.
humeralis Say, 1823: 22 (1859b: 47) (*Limnobia*).—Pa. ?*Pilaria*.
ignobilis Walker, 1856b: 437 (*Limnobia*).—N. Amer. ?*Hexatoma*.
pratorum Kirby, 1837: 310 (*Tipula*).—Arctic Amer.
retorta Wulp, 1881: 149 (*Tipula*).—Que.

Superfamily PSYCHODOIDEA
Family TANYDERIDAE
By Charles P. Alexander

Two of the ten known genera and four of the approximately 30 species occur in our faunal region. The immature stages are found in wet, sandy soil at margins of major streams, indicating an aquatic or semiaquatic habitat.

REFERENCE: Alexander, 1942b (review).

Genus PROTANYDERUS Handlirsch

Protanyderus Handlirsch, 1909: 270. Type-species, *Protoplasa vipio* Osten Sacken (mon.).

margarita Alexander, 1948a: 13.—Utah; B.C., Oreg., Idaho, Colo.
vanduzeei (Alexander), 1918b: 285 (*Protoplasa*).—Calif.
vipio (Osten Sacken), 1877: 208 (*Protoplasa*).—Calif.; Oreg.

Genus PROTOPLASA Osten Sacken

Protoplasa Osten Sacken, 1859: 251. Type-species, *fitchii* Osten Sacken (mon.).
Protoplasta, emend.
Idioplasta Osten Sacken, 1878c: 222 (unjustified n. name for *Protoplasta* Osten Sacken). Type-species, *Protoplasa fitchii* Osten Sacken (aut.).

fitchii Osten Sacken, 1859: 252.—N.Y. or Vt. (Osten Sacken, 1869b: 319); Que. and N.S., s. to Fla.

Family PSYCHODIDAE

By Laurence W. Quate

Psychodids are small, hairy, moth flies, usually found in protected, moist areas. Adults fly in short, jerky flights, and hold their wings rooflike over the body when at rest. Adult food habits are not known except for the bloodsucking *Phlebotomus* females. In the Tropics, species of this genus are the vectors of several diseases. Immature stages are generally subaquatic, living in mud or moss on stream margins. One group, *Maruina*, is fully aquatic and lives in torrents. *Psychoda* and *Phlebotomus* breed in terrestrial habitats supplied with moist, decaying organic material.

The Nearctic psychodids have been reclassified recently (Quate, 1955) but there are still undescribed species. Little work has been done on the biology of North American species.

REFERENCES: Satchell, 1947 (biol.); Quate, 1955 (rev.); Quate, 1960b (ne. N. Amer.).

Subfamily TRICHOMYIINAE

Genus TRICHOMYIA Curtis

Trichomyia Curtis, 1839: pl. 745.   Type-species, *urbica* Curtis (mon.).

nuda (Dyar), 1926b: 111 (*Maruina*).—Md.; Nebr. and Minn. to N.Y., s. to Kans. and Tenn.
sequoiae Quate, 1955: 120.—Calif.
wirthi Quate, 1955: 119.—Fla.; Tenn., Ga.

Subfamily PHLEBOTOMINAE

Genus PHLEBOTOMUS Rondani and Berté

Phlebotomus Rondani and Berté *in* Rondani, 1840a: 12 (as *Flebotomus*).   Type-species, *Bibio papatasii* Scopoli (mon.). Spelling fixed under suspension of Rules by I.C.Z.N., 1954b: 199.

Subgenus BRUMPTOMYIA França and Parrot

Phlebotomus, subg. **Brumptomyia** França and Parrot, 1921: 281. Type-species, *brumpti* Larrousse (Dyar, 1929a: 112.).
Phlebotomus, subg. *Dampfomyia* Addis, 1945: 120. Type-species, *anthophorus* Addis (mon.).

anthophorus Addis, 1945: 119.—Tex.; Mexico.
aquilonius Fairchild and Harwood, 1962: 244.—Wash.

californicus Fairchild and Hertig, 1957: 328.—Calif.; Wash.
diabolicus Hall, 1936: 28.—Tex.; Mexico.
oppidanus Dampf, 1944: 247.—Mexico; Wash.
shannoni Dyar, 1929a: 121.—Panama; Miss. to N.C. and Fla., Mexico, Cent. and S. Amer.
stewarti Mangabeira and Galindo, 1944: 185.—Calif.; Mexico.
texanus Dampf, 1938: 119.—Tex.; Mexico.
vexator vexator Coquillett, 1907c: 102.—Md.; s. U.S. from Calif. to Md., Mexico.
   ssp. **occidentis** Fairchild and Hertig, 1957: 334.—Calif.; Wash

Subfamily PSYCHODINAE
Tribe PERICOMINI
Genus PERICOMA Walker

Pericoma Walker, 1856a: 256. Type-species, *Trichoptera trifasciata* Meigen (Coquillett, 1910b: 587).

Subgenus PERICOMA Walker

albitarsis (Banks), 1895: 324 (*Psychoda*).—N.Y.; Ohio to Mass., s. to Tenn. and N.C.
ancyla Quate, 1955: 135.—Oreg.; Wash.
aterrima (Banks), 1914a: 128 (*Psychoda*).—N.Y.
bessophila Quate, 1955: 139.—Calif.
bipunctata Kincaid, 1899: 34.—Wash.; Alaska, B.C.
biramus Quate, 1955: 140.—Calif.; Ariz., Baja Calif.
californica Kincaid, 1901: 195.—Calif.; Alaska to Calif. and Nebr.
carolina Banks, 1931: 228.—N.C.
complexa Quate, 1955: 136.—Idaho.
fluviatilis Dyar, 1926a: 106.—Mont.; B.C., Wash., Idaho, Wyo.
kincaidi Quate, 1955: 141.—Alaska; Alaska to Wash., Utah, Wyo., Mich.
lassenica lassenica Quate, 1955: 129.—Calif.; Wash.
   ssp. **hiera** Quate, 1955: 129.—Idaho; B.C., Mont., Wyo.
ludificata Quate, 1955: 150.—Oreg.
marginalis (Banks), 1894: 333 (*Psychoda*).—N.Y.; Minn. to Que., s. to Ga.
melanderi Quate, 1955: 133.—Idaho; B.C., Alta., Mont.
scotiae (Curran), 1924h: 216 (*Psychoda*).—N.S.; Alta. to N.S., s. to Colo. and N.C.
   alberta Curran, 1924h: 219 (*Psychoda*).—Alta.
sicula Quate, 1955: 145.—Mont.; B.C., Wyo.

signata (Banks), 1901: 274 (*Psychoda*).—D.C.; Minn. to Labr., s. to
Tenn. and N.C.
   megantica Curran, 1924h: 217 (*Psychoda*).—Que.
sitchana Kincaid, 1899: 33.—Alaska. Unrecognized.
slossonae (Williston), 1893a: 113 (*Psychoda*; as *slossoni*).—N.Y.;
Man. to Que., s. to Mich. and Pa.
   criddlei Curran, 1924h: 218 (*Psychoda*).—Man.
   augusta Curran, 1926e: 228 (*Psychoda*).—Que.
truncata Kincaid, 1899: 35.—Calif.; Utah, Colo., Nebr.
usingeri Quate, 1955: 142.—Calif.
wirthi Quate, 1955: 154.—Ariz.; Calif., Tex., Fla.

Subgenus CLYTOCERUS Eaton

Phalaenula Meigen, 1800: 18. Type-species, *Trichoptera ocellaris*
Meigen (Coquillett, 1910b: 587). Suppressed by
I.C.Z.N., 1963b: 339.
Clytocerus Eaton, 1904: 59 (as genus). Type-species, *africanus*
Tonnoir (sub. mon., Tonnoir, 1920: 137).

americana Kincaid, 1901: 194 (as *ocellaris* var.).—Maine; Wash. to
Calif. and Colo., Minn. to Maine and s. to Tenn.
   variegata Kincaid, 1899: 33 (preocc. Macquart, 1826).—
Wash.
   interrupta Banks, 1907: 150 (*Psychoda*).—Md.
   satellitia Dyar, 1927: 163.—Md.

Genus BREVISCAPUS Quate

Breviscapus Quate, 1955: 155. Type-species, *Pericoma triloba* Kincaid
(orig. des.).

trilobus (Kincaid), 1899: 33 (*Pericoma*).—Wash.

Genus TELMATOSCOPUS Eaton

Telmatoscopus Eaton, 1904: 58. Type-species, *Pericoma advena*
Eaton (present des.).
Clogmia Enderlein, 1937a: 87. Type-species, *Psychoda albipunctata*
Williston (as *albipennis;* orig. des.).

albipunctatus (Williston), 1893a: 113 (*Psychoda*).—Cuba; N.Y. and
Fla. to Ariz., tropicopolitan.
   albipennis, error.
   snowii Haseman, 1907: 311 (*Psychoda*).—Tex.
arnaudi Quate, 1955: 167.—Mexico (Baja Calif.); Calif.

basalis basalis (Banks), 1907: 150 (*Psychoda*).—Va.; se. Canada, U.S.
   apicalis Banks, 1907: 150 (*Psychoda*).—Va.
   orillia Curran, 1924h: 218 (*Psychoda*).—Ont.
   ssp. **cio** Quate, 1955: 179.—Calif.
   ssp. **dysmicus** Quate, 1955: 177.—Calif.
   ssp. **scalus** (Haseman), 1907: 307 (*Pericoma;* as sp.).—Ariz.; Utah, Calif.
      longiplata Haseman, 1907: 308 (*Pericoma*).—Ariz.
   ssp. **sierra** Quate, 1955: 179.—Calif.; Wash.
furcatus (Kincaid), 1899: 34 (*Pericoma*).—Wash.; Alaska, B.C., U.S.
   trialbawhorla Haseman, 1907: 306 (*Pericoma*).—Mo.
   annulipes Johnson, 1913a: 43 (*Psychoda*).—Fla.
   autumnalis Banks, 1914a: 127 (*Psychoda*).—D.C.
   aldrichanus Dyar, 1926b: 107 (*Pericoma*).—Alaska.
   littoralis Dyar, 1926b: 107 (*Pericoma*).—Calif.
jeanneae Quate, 1955: 183.—Calif.; Baja Calif.
latipenis Quate, 1960a: 144.—Wis.
macdonaldi Quate, 1960a: 146.—Calif.
macneilli Quate, 1955: 164.—Calif.
nebraskensis Quate, 1955: 163.—Nebr.; Wis., Mich.
niger (Banks), 1894: 331 (*Psychoda*).—N.Y.; Minn. to N.Y., s. to Ind. and Va.
   snowhilli Del Rosario, 1936: 140 (*Psychoda*).—Md.
olympia (Kincaid), 1897: 144 (*Psychoda*).—Wash.; Calif.
patibulus Quate, 1955: 167.—Ala.; Tenn., Tex., Miss.
quadripunctatus (Banks), 1907: 149 (*Psychoda*).—Va.; Que., Ont.
sobrinus Quate, 1955: 171.—Tex.; Calif., Ariz.
subtilis Quate, 1960a: 144.—N. Mex.
superbus (Banks), 1894: 332 (*Psychoda*).—N.Y.; Nebr. to Que., s. to Tex. and Ga.
   conspicuus Del Rosario, 1936: 125 (*Psychoda;* as var.).—Md.
varitarsis (Curran), 1924h: 220 (*Psychoda*).—Que.; Ont., Mich., Mass.
   juno Curran, 1926e: 228 (*Psychoda*).—Ont.

Genus BRUNETTIA Annandale

Brunettia Annandale, 1910: 141. Type-species, *Diplonema superstes* Annandale (Brunetti, 1911a: 310).

nitida (Banks), 1901: 275 (*Psychoda*).—D.C.; Nebr. to Mass., s. to Tex. and Fla.
   squamosa Johnson, 1913a: 43 (*Psychoda*).—Fla.
sycophanta Quate, 1955: 190.—Fla. ;Ala.

Tribe PSYCHODINI

REFERENCE: Quate, 1959 (classif.).

Genus TRICHOPSYCHODA Tonnoir

Psychoda, subg. **Trichopsychoda** Tonnoir, 1922: 59. Type-species, *hirtella* Tonnoir (orig. des.).

insulicola (Quate), 1954: 342 (*Psychoda*).—Hawaii; Ga., Ala.

Genus PHILOSEPEDON Eaton

Philosepedon Eaton, 1904: 57. Type-species, *Psychoda humeralis* Meigen (orig. des.).

bishoppi (Del Rosario), 1936: 141 (*Psychoda*).—Md.; Va., Ala.
interdicta (Dyar), 1928a: 88 (*Psychoda*).—N.Y.; Wis. to Mass., s. to Tex. and Fla., Antilles.
opposita (Banks), 1901: 274 (*Psychoda*).—D.C.; Iowa to Md., s. to Tex. and Ga.
tesca (Quate), 1955: 229 (*Psychoda*).—Calif.
tridactila (Kincaid), 1899: 32 (*Pericoma*).—Wash.

Genus THRETICUS Eaton

Threticus Eaton, 1904: 57. Type-species, *Pericoma lucifuga* Walker (Enderlein, 1935b: 249).

bicolor (Banks), 1894: 333 (*Psychoda*).—N.Y.; Mich. to Que. and Va.
jonesi (Quate), 1955: 231 (*Psychoda*).—Wis.; Tenn., Va., Miss.

Genus EURYGARKA Quate

Eurygarka Quate, 1959: 450. Type-species, *Psychoda helicis* Dyar (orig. des.).

helicis (Dyar), 1929c: 64 (*Psychoda*).—Cuba; Fla., Ala.

Genus PSYCHODA Latreille

Psychoda Latreille, 1796: 152. Type-species, *Tipula phalaenoides* Linnaeus (sub. mon., Latreille, 1802: 424).
Trichoptera Meigen, 1803: 261. Type-species, *Tipula phalaenoides* Linnaeus (Coquillett, 1910b: 616).

alternata Say, 1824a: 358 (1859a: 242).—Pa.; U.S., cosmopolitan.
  schizura Kincaid, 1899: 32.—Wash.
  floridica Haseman, 1907: 316.—Fla.
  nocturnala Haseman, 1907: 319.—Mo.
  albimaculata Welch, 1912: 411.—Ill.
  dakotensis Dyar, 1926b: 108.—S. Dak.

alternicula Quate, 1955: 222.—Fla.; Tex., La., Ala., Ga.
cinerea Banks, 1894: 331.—N.Y.; Alta., Ont., U.S., P.R., Chile, Europe, Australia, New Zealand.
    prudens Curran, 1924h: 219.—Alta.
degenera Walker, 1848: 33.—Ont. Unrecognized.
domestica Haseman, 1908: 285.—Mo. Unrecognized.
elegans Kincaid, 1897: 144.—Wash.; Utah.
lativentris Berdén, 1952: 111.—Sweden; Sask., U.S., Mexico, Europe, Africa.
longifringa Haseman, 1907: 318.—Fla. Unrecognized.
minuta Banks, 1894: 331.—N.Y.; Minn. to Maine, s. to Tex. and Md., Europe.
    marylandana Del Rosario, 1936: 111.—Md.
phalaenoides (Linnaeus), 1758: 588 (*Tipula*).—Unknown; Alaska, Canada, U.S., Europe.
    pacifica Kincaid, 1897: 143.—Wash.
    horizontala Haseman, 1907: 313.—Mo.
    tonnoiri Dyar, 1926a: 103.—Unknown.
    angustafona Rapp, 1944: 233.—N.Y.
pusilla Tonnoir, 1922: 83.—Austria; Wash. to N.Y., s. to Calif., Tex., and Ala., Europe.
salicornia Quate, 1954: 350.—Calif.; Hawaii.
satchelli Quate, 1955: 214.—Calif.; Alaska and B.C. to Ont., U.S.
    severini Tonnoir of Del Rosario, 1936: 102.
savaiiensis Edwards, 1928b: 74.—Samoa; Ala., Ga., Fla., Antilles, Cent. Amer., Pacific Is., India.
    rarotongensis Satchell, 1953: 183.—Cook I.
setigera Tonnoir, 1922: 84.—Belgium; Minn. to Ont., s. to Tex.,Ala., and Va., Oreg., Europe, Japan.
    uniformis Del Rosario, 1936: 113.—Md.
sigma Kincaid, 1899: 31.—Wash.; Oreg., Calif.
thrinax Quate, 1955: 199.—Ariz.; Calif., Tex., D.C.
tothastica Quate, 1955: 197.—Tex.; Fla.
trinodulosa Tonnoir, 1922: 86.—Belgium; Utah, Nebr., Minn., Wis., N.Y., Va.
umbracola Quate, 1955: 201.—Calif.; Wash., Minn., N.Y., N.H., Tenn.
uniformata Haseman, 1907: 319.—Mo.; Utah, Nebr., Ont., N.Y.

Tribe MARUININI

Genus MARUINA Müller

Maruina Müller, 1895: 480. Type-species, *pilosella* Müller (Enderlein, 1937a: 110).

    REFERENCE: Quate and Wirth, 1951 (rev.).

lanceolata (Kincaid), 1899: 35 (*Sycorax*).—Calif.; Calif., Nev., Ariz., N. Mex.
californiensis Kellogg, 1901: 46 (*Pericoma*).—Calif.
unipunctata Haseman, 1907: 323 (*Trichomyia*).—Ariz.

Family PTYCHOPTERIDAE
(Liriopeidae)

By Charles P. Alexander

The family includes the subfamily Bittacomorphinae or "phantom crane flies," and the genus *Ptychoptera*. The immature stages are aquatic or semiaquatic, living in organic debris in stagnant or slowly flowing water.

REFERENCE: Alexander, 1920h (immature stages).

Subfamily PTYCHOPTERINAE

Genus PTYCHOPTERA Meigen

Liriope Meigen, 1800: 14. Type-species, *Tipula contaminata* Linnaeus (Coquillett, 1910b: 562). Suppressed by I.C.Z.N., 1963b: 339.
Ptychoptera Meigen, 1803: 262. Type-species, *Tipula contaminata* Linnaeus (Latreille, 1810: 442).
Ctenoceria Rondani, 1856: 187. Type-species, *Ptychoptera pectinata* Macquart (orig. des.).

lenis lenis Osten Sacken, 1877: 206.—Calif.; B.C., Wash., Oreg.
    ssp. **coloradensis** Alexander, 1937: 141.—Colo.; Wyo.
metallica Walker, 1848: 80.—Ont.; Alta., Colo., Minn.
minor Alexander, 1920f: 3.—Calif.; Idaho.
monoensis Alexander, 1947a: 20.—Calif.; Nev.
osceola Alexander, 1959b: 55.—Fla.
pendula Alexander, 1937: 140.—Colo.; B.C., Wyo., Utah, Ariz., N. Mex.
quadrifasciata Say, 1824a: 359 (1859a: 244).—Pa.; Ill. to Que., s. to Fla.
    rufocincta Osten Sacken, 1859: 252.—Pa.
sculleni Alexander, 1943c: 39.—Wash.; Oreg.
townesi Alexander, 1943c: 37.—Wash.; Oreg., Calif.
uta Alexander, 1947a: 19.—Utah.

Subfamily BITTACOMORPHINAE

Genus BITTACOMORPHELLA Alexander

Bittacomorpha, subg. **Bittacomorphella** Alexander, 1916a: 545. Type-species, *jonesi* Johnson (orig. des.).

fenderiana Alexander, 1947a: 22.—Oreg.; B.C., Wash.

jonesi (Johnson), 1905a: 75 (*Bittacomorpha*).—N.C.; Mich. to N.B., s. to S.C.

pacifica Alexander, 1958a: 49.—Wash.; Oreg., Calif.

sackenii (Röder), 1890: 230 (*Bittacomorpha*).—Nev.; Calif.

Genus BITTACOMORPHA Westwood

Bittacomorpha Westwood, 1835a: 281. Type-species, *Tipula clavipes* Fabricius (mon.).

clavipes (Fabricius), 1781: 404 (*Tipula*).—N. Amer.; Man. to Nfld., s. to Ariz. and Fla.

occidentalis Aldrich, 1895: 201.—Wash.; Oreg., Calif.

Family NYMPHOMYIIDAE

By F. P. Ide

This is a family of peculiar, small, delicate flies found in streams. Only a Japanese species (*Nymphomyia alba* Tokunaga) and two unnamed genera and species from Canada and Brazil are known. No larvae have been discovered.

Unnamed Genus

Unnamed sp.—N.B.

Superfamily CULICOIDEA

Family BLEPHARICERIDAE

By Alan Stone

The adults of this worldwide family, known as net-winged midges, are found near rapidly flowing streams. The larvae are attached to rocks in swift water by means of a series of ventral suckers, and the pupae are firmly attached in the same localities. Larval respiration is by means of tracheal gills.

REFERENCES: Kellogg, 1903 (rev.); Walley, 1927b (rev.); Johannsen, 1934b (keys, immatures); Alexander, 1958c (classif., list world spp.).

Genus AGATHON Röder

Agathon Röder, 1890: 230. Type-species, *elegantulus* Röder (mon.).
comstocki (Kellogg), 1903: 192 (*Bibiocephala*).—Calif.; B.C., Oreg., ?Wyo.
canadensis Garrett, 1922a: 89 (*Bibiocephala*).—B.C. N. syn., C. P. Alexander *in litt.*
doanei (Kellogg), 1900b: 39 (*Liponeura*).—Calif.
elegantulus Röder, 1890: 232.—Nev.; Idaho, Colo.

Genus BIBIOCEPHALA Osten Sacken

Bibiocephala Osten Sacken, 1874a: 564. Type-species, *grandis* Osten Sacken (mon.).
Bibionus Curran, 1923g: 266. Type-species, *griseus* Curran (orig. des.).
grandis Osten Sacken, 1874a: 566.—Colo.; B.C., Utah, Idaho, N. Mex.
grisea (Curran), 1923g: 268 (*Bibionus*).—Alta.
kelloggi Garrett, 1922a: 91.—B.C.

Genus BLEPHARICERA Macquart

Blepharicera Macquart, 1843b: 61. Type-species, *limbipennis* Macquart (mon.)=*fasciata* (Westwood).
Blepharocera, emend.
Blepharoptera (error) Loew, 1863b: 298 (Cent. 4, no. 43).

I have not attempted to give the distribution of the eastern species which are in need of revision.

capitata Loew, 1863b: 298 (Cent. 4, no. 43) (as *Blepharoptera*).—D.C.
jordani Kellogg, 1903: 189.—Calif.
micheneri Alexander, 1959b: 41.—Calif.
ostensackeni Kellogg, 1903: 191.—Calif.
separata Alexander, 1963: 54.—Maine; Pa., Tenn.
shastensis Alexander, 1959b: 42.—Calif.
similans Johannsen, 1929b: 123.—Mass.
tenuipes (Walker), 1848: 86 (*Asindulum*).—Ont.
williamsae Alexander, 1953a: 43.—Tenn.
zionensis Alexander, 1953a: 45.—Utah.

Genus DIOPTOPSIS Enderlein

Dioptopsis Enderlein, 1936b: 43. Type-species, *Philorus djordjevici* Komarek (orig. des.).
arizonica Alexander, 1958a: 50.—Ariz.
aylmeri (Garrett), 1923: 244 (*Philorus*).—Alta.; Calif.

cheaini (Garrett), 1925b: 5 *(Philorus).*—B.C.; Calif.
markii (Garrett), 1925b: 5 *(Philorus).*—B.C.; Calif.
sequoiarum (Alexander), 1952a: 91 *(Philorus).*—Calif.

Genus PHILORUS Kellogg

Philorus Kellogg, 1903: 199. Type-species, *Blepharicera yosemite* Osten Sacken (Coquillett, 1910b: 588).
Pelmia Enderlein, 1936b: 42. Type-species, *Blepharicera yosemite* Osten Sacken (orig. des.).

yosemite (Osten Sacken), 1877: 195 *(Blepharicera).*—Calif.
    ancilla Osten Sacken, 1878c: 266 *(Blepharicera).*—Calif.
    N. syn., C. L. Hogue *in litt.*

Family DEUTEROPHLEBIIDAE
By Harry D. Kennedy

This family consists of a few species of relatively rare aquatic insects commonly called mountain midges. The immature stages are found on rocks in mountain streams. There are from one to four generations a year, depending on the species. Overwintering is probably in the egg stage. The adults lack mouthparts.

REFERENCE: Kennedy, 1958 (biol., anat., rev.).

Genus DEUTEROPHLEBIA Edwards

Deuterophlebia Edwards, 1922a: 379. Type-species, *mirabilis* Edwards (orig. des.).

coloradensis Pennak, 1945: 1.—Colo.; Alta, Wyo., Utah.
inyoensis Kennedy, 1960: 192.—Calif. (Mono Co.).
nielsoni Kennedy, 1958: 206.—Calif. (Mono Co.; Tuolumne Co.).
shasta Wirth, 1951c: 51.—Calif. (Siskiyou Co.; Eldorado Co.).

Family DIXIDAE
By Alexander A. Hubert

These small, rather sedentary flies rest head upward on rocks or other substrata along streams or pools. It is not known whether or not they feed as adults. Mating occurs during swarming flights or while resting. The eggs are laid in small gelatinous masses in or just above the water. The larvae rest at the water line either in the shape of an inverted U or with the body perpendicular to the substratum. The pupae are attached by one side to the substratum slightly above the water.

Although the higher classification of the Dixidae has been explored and quite a few species have been described, a great deal of taxonomic work remains to be done.

REFERENCE: Nowell, 1951 (biol., anat., tax., cat.).

Genus DIXA Meigen

Dixa Meigen, 1818: 216. Type-species, *maculata* Meigen (Curtis, 1832: pl. 409).

Subgenus DIXA Meigen

arge Dyar and Shannon, 1924b: 199.—Wash.
blax Dyar and Shannon, 1924b: 199.—Ariz.
brevis Garrett, 1924a: 6.—B.C.
distincta Garrett, 1925a: 12.—B.C.
fraterna Garrett, 1924a: 6.—B.C.
fusca Loew, 1863a: 4 (Cent. 3, no. 5).—N.Y.; Maine, Conn.
garretti Cooper and Rapp, 1944: 247 (n. name for *montanus* Garrett).—B.C.
  montanus Garrett, 1924a: 5 (preocc. Brunetti, 1911).—B.C.
hegemonica Dyar and Shannon, 1924b: 194.—Calif.
inextricata Dyar and Shannon, 1924b: 198.—Va., Md.
johannseni Garrett, 1924a: 7.—B.C.; Alta.
lobata Garrett, 1924a: 6.—B.C.
modesta Johannsen, 1903b: 429.—N.Y.; Maine, Conn., N.C., also Calif.
mystica Dyar and Shannon, 1924b: 197.—Md., Va.
notata Loew, 1863a: 4 (Cent. 3, no. 4).—Md.; Conn., N.J.
parva Garrett, 1924a: 6.—B.C.
plexipus Garrett, 1925b: 5.—B.C.
recens Walker, 1848: 85.—Ont.; N.Y., N.J., Ga.
  terna Loew, 1863a: 3 (Cent. 3, no. 2).—N.Y.
rhathyme Dyar and Shannon, 1924b: 197.—Wash.; Calif.
rudis Garrett, 1924a: 6.—B.C.
similis Johannsen, 1923a: 57.—N.Y.
xavia Dyar and Shannon, 1924b: 195.—Calif.

Subgenus DIXAPUELLA Dyar and Shannon

Dixa, subg. **Dixapuella** Dyar and Shannon, 1924b: 201. Type-species, *marginata* Loew (orig. des.).

marginata Loew, 1863a: 2 (Cent. 3, no. 1).—D.C.

Subgenus DIXELLA Dyar and Shannon

Dixa, subg. **Dixella** Dyar and Shannon, 1924b: 200. Type-species, *lirio* Dyar and Shannon (mon.).

Dixa, subg. *Paradixa* Tonnoir, 1924: 223. Type-species, *neozelandica* Tonnoir (Edwards, 1932a: 9).

aliciae Johannsen, 1924b: 45.—Calif.; Colo.
californica Johannsen, 1923a: 54.—Calif.; B.C., Oreg., Utah, Colo., Ariz.
clavata Loew, 1869b: 2 (Cent. 8, no. 1).—Mass.; Maine, N.Y.
cornuta Johannsen, 1923a: 55.—N.Y.; Ill., Ohio, Ont., N.H., Maine.
dorsalis Garrett, 1924a: 4.—B.C.
dyari Garrett, 1924a: 3.—B.C.
indiana Dyar, 1925: 217.—Ind.; N.Y.
naevia Peus, 1934: 72.—Latvian S.S.R.; Alaska.
neoaliciae Garrett, 1924a: 2.—Mont.; B.C.
nocheles Dyar and Shannon, 1924b: 196.—Fla.
nova Walker, 1848: 85.—Ont.; Alaska, Oreg., Mich. to Maine and N.Y.
    centralis Loew, 1863a: 3 (Cent. 3, no. 3).—N.Y.
occidentalis Garrett, 1924a: 4.—B.C.
serrata Garrett, 1924a: 2.—B.C.
simplex Garrett, 1925a: 12.—B.C.
somnolenta Dyar and Shannon, 1924b: 195.—Idaho; Alaska, B.C., Calif.
    spiralis Garrett, 1924a: 1.—B.C.
thones Dyar and Shannon, 1924b: 193.—Wash.
universitatis Cockerell, 1926: 166.—Colo.
venosa Loew, 1872a: 50 (Cent. 10, no. 1).—Tex.

Subgenus MERINGODIXA Nowell

Meringodixa Nowell, 1951: 229 (as genus). Type-species, *chalonensis* Nowell (orig. des.).

chalonensis (Nowell), 1951: 229 (*Meringodixa*).—Calif.

Family CHAOBORIDAE
By E. F. Cook

This group has usually been considered a subfamily of the Culicidae. The major differences between the Chaoboridae and Culicidae are found in the shortened mouth parts of the adult chaoborids, the prehensile antennae of the larval chaoborids, and the lack of any spiracular apparatus in the larvae of the genus *Chaoborus*. The

larvae are all aquatic and predaceous on small aquatic organisms.
The food habits of the adults are unknown.

REFERENCE: Cook, 1956b (rev.).

Subfamily CHAOBORINAE
Tribe CHAOBORINI
Genus CHAOBORUS Lichtenstein

Chaoborus Lichtenstein, 1800: 174. Type-species, *antisepticus* Lichtenstein (mon.) = *crystallinus* (De Geer).
Corethra Meigen, 1803: 260. Type-species, *Tipula culiciformis* De Geer (mon.; misident.) = *Corethra lateralis* Meigen, ? = *crystallinus* (De Geer).

Subgenus CHAOBORUS Lichtenstein

americanus (Johannsen), 1903b: 395 (*Corethra*; as *plumicornis* var.).—Ill.; Alaska to Que., s. to Utah and Tenn.
hudsoni Felt, 1904: 371 (*Sayomyia*).—N.Y.
borealis Cook, 1956b: 25.—Y.T.; N.W.T., Man.
flavicans (Meigen), 1830: 243 (*Corethra*).—Europe; Alaska to Maine, s. to n. Calif., Mo., and N.J.
albipes Johannsen, 1903b: 398 (*Corethra*).—N.Y.
rotundifolia Felt, 1904: 366 (*Sayomyia*).—N.Y.
eluthera Dyar and Shannon, 1924c: 211.—Idaho.

Subgenus SAYOMYIA Coquillett

Sayomyia Coquillett, 1903c: 190 (as genus). Type-species, *Corethra punctipennis* Say (orig. des.).

albatus Johnson, 1921a: 11.—Mass.; Minn. to Que., s. to La.
annulatus Cook, 1956b: 39.—Fla.; Ga.
astictopus Dyar and Shannon, 1924c: 214.—Calif.; Oreg.
lacustris Freeborn, 1926: 161.—Calif.
punctipennis (Say), 1823: 16 (1859b: 43) (*Corethra*).—Pa.; Sask. to Que., s. to Colo., Tex. and Fla.
appendiculata Herrick, 1884: 11 (*Corethra*).—Minn.

Subgenus SCHADONOPHASMA Dyar and Shannon

Chaoborus, subg. **Schadonophasma** Dyar and Shannon, 1924c: 209. Type-species, *Corethra trivittata* Loew (mon.) = *nyblaei* (Zetterstedt).

nyblaei (Zetterstedt), 1838: 830 (*Erioptera*).—Norway; n. Holarctic, Y.T. to Baffin I., s. to Calif. and Mass.
    trivittata Loew, 1862b: 186 (Cent. 2, no. 1) (*Corethra*).—Maine.
    knabi Dyar, 1905b: 16 (*Sayomyia*).—Mass.

Tribe MOCHLONYCHINI

Genus MOCHLONYX Loew

Mochlonyx Loew, 1844a: 121. Type-species, *Corethra velutina* Ruthe (mon.).

cinctipes (Coquillett), 1903c: 190 (*Corethra*).—N.H.; Wash., Oreg. and Que., s. to Ala.
    obscura Dyar and Shannon, 1924c: 208 (*Corethra*; as var.).—Wash.
fuliginosus (Felt), 1905b: 458 (*Corethra*).—N.Y.; Mass., N.J.
velutinus (Ruthe), 1831: 1205 (*Corethra*).—Europe; Alaska to Que., s. to Utah and N.J.
    culiciformis De Geer, 1776: 372 (*Tipula*; preocc. Linnaeus, 1767).—Europe.
    karnerensis Felt, 1904: 347 (*Corethra*).—N.Y.
    lintneri Felt, 1904: 353 (*Corethra*).—N.Y.

Subfamily EUCORETHRINAE

Genus EUCORETHRA Underwood

Eucorethra Underwood, 1903: 182. Type-species, *underwoodi* Underwood (orig. des., as gen. n., sp. n.).
Pelorempis Johannsen, 1903b: 402. Type-species, *americana* Johannsen (mon.)=*underwoodi* Underwood.

underwoodi Underwood, 1903: 182.—Maine; Alaska to N.B., s. to Calif., Utah, and N.Y.
    americana Johannsen, 1903b: 403 (*Pelorempis*).—N.Y.

Subfamily CORETHRELLINAE

Genus CORETHRELLA Coquillett

Corethrella Coquillett, 1902f: 191. Type-species, *Corethra brakeleyi* Coquillett (orig. des.).

appendiculata Grabham, 1906: 343.—Jamaica; N.C. to Ga., also Tex.
brakeleyi (Coquillett), 1902c: 85 (*Corethra*).—N.J.; s. to Tex. and Fla.
laneana Vargas, 1946: 64.—Mexico; s. Calif.

Family CULICIDAE

By Alan Stone

The family Culicidae, as treated here, includes only the true mosquitoes, the subfamily Culicinae of earlier authors, and not the Chaoboridae and the Dixidae. The adult females of most species suck blood and some species, in addition to being very annoying, carry such diseases as malaria, yellow fever, dengue, and encephalitis. The larvae and pupae live in standing or slowly moving bodies of water that range in size from small accumulations held by plants to vast salt marshes. So much has been written on the biology of mosquitoes that further remarks here are hardly necessary.

REFERENCES: Ross, 1947 (Ill.); Bates, 1949 (biol.); Freeborn and Bohart, 1951 (Calif.); Stage, Gjullin, and Yates, 1952 (nw. U.S.); Carpenter and LaCasse, 1955 (N. Amer.); Horsfall, 1955 (biol.); Owen and Gerhardt, 1957 (Wyo.); Barr, 1958 (Minn.); Stone, Knight, and Starcke, 1959 (world cat.); King, Bradley, Smith, and McDuffie, 1960 (se. U.S.); Christophers, 1960 (biol., anat.); Gjullin, Sailer, Stone, and Travis, 1961 (review, Alaska).

Subfamily ANOPHELINAE

Genus ANOPHELES Meigen

Anopheles Meigen, 1818: 10. Type-species, *maculipennis* Meigen (I.C.Z.N., 1959a: 155).
Coelodiazesis Dyar and Knab, 1906c: 177. Type-species, *Anopheles barberi* Coquillett (orig. des.).

REFERENCES: Aitken, 1945 (w. N. Amer.); Freeborn *in* Boyd, 1949 (Nearctic).

Subgenus ANOPHELES Meigen

atropos Dyar and Knab, 1906b: 160.—Fla.; coastal, N.J. to Tex., Cuba, Jamaica.
barberi Coquillett, 1903e: 310.—Md.; Nebr. to N.Y., s. to Ariz. and Fla., n. Mexico.
bradleyi King, 1939: 468 (as *crucians* var.).—Fla.; coastal, N.Y. to Tex., Mexico.

crucians Wiedemann, 1828: 12.—Pa., La.; Iowa to R.I., s. to N. Mex. and Fla., Mexico to Nicaragua, Cuba, Jamaica, P.R.
earlei Vargas, 1943: 9.—Wis.; Alaska, Canada, Idaho to Maine, s. to Colo. and Conn.
freeborni Aitken, 1939: 192 (as *maculipennis* ssp.).—Calif.; B.C. to Mont., s. to Calif. and Tex., n. Mexico.
georgianus King, 1939: 462 (as *crucians* var.).—Ga.; La. to N.C. and Fla.
occidentalis Dyar and Knab, 1906b: 159.—Calif.; coastal, ?Alaska to s. Calif.
perplexens Ludlow, 1907: 267.—Pa.; e. U.S.
pseudopunctipennis pseudopunctipennis Theobald, 1901b: 305.— Grenada I.; Colo. to Tenn., s. to Argentina, Antilles.
   ssp. **franciscanus** McCracken, 1904: 12 (as sp.).—Calif.; Oreg. to sw. U.S., n. Mexico.
   boydi Vargas, 1939: 361.—Calif.
punctipennis (Say), 1823: 9 (1859b: 39) (*Culex*).—e. U.S.; s. Canada, U.S., Mexico.
   hyemalis Fitch, 1847: 281 (*Culex*).—N.Y.
quadrimaculatus Say, 1824a: 356 (1859a: 241).—s. and w. of Great Lakes; N. Dak. to Que., s. to Tex. and Fla., Mexico.
   annulimanus Wulp, 1867: 129.—Wis.
walkeri Theobald, 1901a: 199.—Ont.; Man. to N.S., s. to Tex. and Fla., also B.C., Mexico.

Subgenus NYSSORHYNCHUS Blanchard

Laverania Theobald, 1902: 183 (preocc. Grassi and Feletti, 1890). Type-species, *Anopheles argyritarsis* Robineau-Desvoidy (orig. des.).

Nyssorhynchus Blanchard, 1902: 795 (as genus; n. name for *Laverania* Theobald). Type-species, *Anopheles argyritarsis* Robineau-Desvoidy (aut.).

albimanus Wiedemann, 1820a: 10 (1821c: 10).—Hispaniola; Fla., Tex., Cent. and S. Amer. to Uruguay, Antilles.

Subfamily TOXORHYNCHITINAE
(Megarhininae)

Genus TOXORHYNCHITES Theobald

Toxorhynchites Theobald, 1901c: 234 (also 1901a: 244). Type-species, *brevipalpis* Theobald (I.C.Z.N., 1959b: 167).

Subgenus LYNCHIELLA Lahille

Megarhinus Robineau-Desvoidy, 1827: 412 (preocc. Rafinesque, 1820). Type-species, *Culex haemorrhoidalis* Fabricius (mon.).

Lynchiella Lahille, 1904: 14 (as genus; n. name for *Megarhinus* Robineau-Desvoidy). Type-species, *Culex haemorrhoidalis* Fabricius (aut.).

rutilus rutilus (Coquillett), 1896a: 44 (*Megarhinus*).—Fla.; Ga., S.C.
  ssp. **septentrionalis** (Dyar and Knab), 1906d: 249 (*Megarhinus;* as sp.).—Va.; Kans. to N.J., s. to Fla. and Tex.
  herrickii Theobald, 1906: 241 (*Megarhinus*).—Miss.

Subfamily CULICINAE

Tribe SABETHINI

Genus WYEOMYIA Theobald

Wyeomyia Theobald, 1901c: 235. Type-species, *grayii* Theobald (Neveu-Lemaire, 1902b: 223).
Wyeomyia, subg. *Phyllozomyia* Dyar, 1924a: 112. Type-species, *Aedes smithii* Coquillett (orig. des.).

haynei Dodge, 1947: 118.—S.C.; N.C., Ala.
mitchellii (Theobald), 1905: 37 (*Dendromyia*).—Jamaica; Fla., Antilles, Mexico to Panama, Venezuela, Trinidad.
  jamaicensis Theobald, 1905: 11 (*Dendromyia*).—Jamaica. Alternate orig. name.
  antoinetta Dyar and Knab, 1909b: 263.—Fla.
smithii (Coquillett), 1901a: 260 (*Aedes*).—N.J.; Man. to N.S., s. to Ill. and Del.
vanduzeei Dyar and Knab, 1906a: 138.—Fla.; Cent. Amer., Antilles.

Tribe CULICINI

Genus MANSONIA Blanchard

Panoplites Theobald, 1900: 5 (also 1901b: 173) (preocc. Gould, 1854). Type-species, *Culex taeniorhynchus* Wiedemann (aut.; misident.) =*titillans* (Walker).
Mansonia Blanchard, 1901: 1046 (n. name for *Panoplites* Theobald). Type-species, *Culex titillans* Walker (Neveu-Lemaire, 1902b: 214).

Panoplites dates from an earlier paper than that cited in I.C.Z.N., 1959c, and *titillans* was not originally included by name. One of the three original species, *"Taeniorhynchus taeniorhynchus* Arri.", was recognized in Theobald, 1901b, as a misidentification of *titillans*. This was accepted by Neveu-Lemaire, who designated *titillans* as type of *Mansonia*.

Subgenus MANSONIA Blanchard

indubitans Dyar and Shannon, 1925a: 41.—Brazil; Ga., Fla., P.R., Jamaica, Cent. and S. Amer. to Bolivia and Uruguay.
titillans (Walker), 1848: 5 *(Culex)*.—Brazil; Fla., Tex., Cent. and S. Amer., Antilles.

Subgenus COQUILLETTIDIA Dyar

Coquillettidia Dyar, 1905c: 47 (as genus). Type-species, *Culex perturbans* Walker (orig. des.).

perturbans (Walker), 1856b: 428 *(Culex)*.—U.S.; B.C. to P.E.I., U.S., Mexico.
    testaceus Wulp, 1867: 128 *(Culex)*.—Wis.
    ochropus Dyar and Knab, 1907a: 100 *(Culex)*.—N.H.

Genus URANOTAENIA Lynch Arribálzaga

Uranotaenia Lynch Arribálzaga, 1891a: 375 (also 1891b: 163). Type-species, *pulcherrima* Lynch Arribálzaga (Neveu-Lemaire, 1902b: 227).

anhydor anhydor Dyar, 1907: 128.—Calif.; Ariz., Nev., Mexico.
    ssp. **syntheta** Dyar and Shannon, 1924a: 189 (as sp.).—Tex.; N. Mex., Okla., Mexico.
lowii Theobald, 1901b: 339.—St. Lucia I.; Ark. to N.C., s. to Tex. and Fla., Neotropical.
    continentalis Dyar and Knab, 1906c: 187.—La.
sapphirina (Osten Sacken), 1868: 47 *(Aedes)*.—D.C., N.Y.; N. Dak. to Que., s. to N. Mex. and Fla., Cuba, P.R., Mexico.
    coquilletti Dyar and Knab, 1906c: 187.—U.S.

Genus ORTHOPODOMYIA Theobald

Orthopodomyia Theobald, 1904: 236. Type-species, *albipes* Leicester (mon.).
Pneumaculex Dyar, 1905c: 46. Type-species, *Culex signifer* Coquillett (orig. des.).

alba Baker, 1936: 1.—N.Y.; N. Mex. to N.Y., s. to Tex. and N.C.
californica Bohart, 1950: 399.—Calif.; Ariz.
kummi Edwards, 1939: 121.—Costa Rica; Ariz. to Panama.

CULICIDAE 109

signifera (Coquillett), 1896a: 43 (*Culex*).—D.C.; N. Dak., Iowa, and Mass., s. to N. Mex. and Fla., Mexico, Jamaica, P.R.

Genus PSOROPHORA Robineau-Desvoidy

Psorophora Robineau-Desvoidy, 1827: 412. Type-species, *Culex ciliatus* Fabricius (Theobald, 1901a: 263).

Subgenus PSOROPHORA Robineau-Desvoidy

ciliata (Fabricius), 1794: 401 (*Culex*).—Carolina; S. Dak. to Que., s. to Tex. and Fla., Neotropical.
  molestus Wiedemann, 1820a: 7 (1821c: 7) (*Culex*).—Ga.
  rubidus Robineau-Desvoidy, 1827: 404 (*Culex*).—Carolina.
  boscii Robineau-Desvoidy, 1827: 413 (*Culex*).—Carolina.
  conterrens Walker, 1856b: 427 (*Culex*).—U.S.
  ctites Dyar, 1918c: 126.—Tex.
howardii Coquillett, 1901a: 258.—S.C.; Nebr. to Md., s. to Nicaragua and Cuba.

Subgenus JANTHINOSOMA Lynch Arribálzaga

Janthinosoma Lynch Arribálzaga, 1891a: 374 (also 1891b: 152) (as genus). Type-species, *Culex discrucians* Walker (Dyar, 1905c: 47).
Conchyliastes Theobald *in* Howard, 1901: 155, 235. Type-species, *Culex musicus* Say (Coquillett, 1910b: 526)=*ferox* (Humboldt).
Lepidosia Coquillett, 1906c: 314. Type-species, *Culex cyanescens* Coquillett (orig. des.).

cyanescens (Coquillett), 1902d: 137 (*Culex*).—Tex.; Nebr. to Va., s. to N. Mex. and Fla., Mexico to Argentina.
ferox (Humboldt), 1819: 340 (*Culex*).—Ecuador; S. Dak. to N.H., s. to Tex. and Fla., S. and Cent. Amer., Greater Antilles.
  musicus Say, 1829: 149 (1859b: 348) (*Culex;* preocc. Leach, 1825).—Ind.
  sayi Dyar and Knab, 1906c: 181 (*Janthinosoma;* n. name for *musicus* Say).—Ind.
  sayi Theobald, 1907: 155 (*Janthinosoma;* n. name for *musicus* Say).—Ind.
horrida (Dyar and Knab), 1908: 56 (*Aedes*).—Miss.; Nebr. to Pa., s. to Tex. and Fla.
johnstonii (Grabham), 1905: 410 (*Janthinosoma*).—Jamaica; Fla., Bahama Is., Cuba, P.R., Virgin Is.

longipalpus Randolph and O'Neill *in* T.S.H.D., 1944: 88.—Tex.; S. Dak. and Mo., s. to Tex. and La. See Stone, 1962.
  longipalpis Roth, 1945: 13.—Ark.
mexicana (Bellardi), 1859: 5 (1861: 205) *(Culex).*—Mexico; Tex.
varipes (Coquillett), 1904a: 10 *(Conchyliastes).*—Mexico; Okla. to N.Y., s. to Tex. and Fla., Cent. and S. Amer. to Argentina.

Subgenus GRABHAMIA Theobald

Grabhamia Theobald, 1903a: 243 (as genus). Type-species, *Culex jamaicensis* Theobald (Felt, 1904: 391*b*)=*confinnis* (Lynch Arribálzaga).
Ceratocystia Dyar and Knab, 1906c: 183. Type-species, *Culex discolor* Coquillett (orig. des.).

confinnis (Lynch Arribálzaga), 1891b: 149 *(Taeniorhynchus).*—Argentina; Calif., S. Dak. and Mass., s. to Tex. and Fla., S. and Cent. Amer., Antilles.
  columbiae Dyar and Knab, 1906a: 135 *(Janthinosoma).*—Va.
  floridense Dyar and Knab, 1906a: 135 *(Janthinosoma).*—Fla.
  texanum Dyar and Knab, 1906a: 135 *(Janthinosoma).*—Tex.
discolor (Coquillett), 1903d: 256 *(Culex).*—N.J.; Nebr. to N.J., s. to N. Mex. and Fla.
pygmaea (Theobald), 1903a: 245 *(Grabhamia).*—Antigua I. and Jamaica; Fla., Antilles, Trinidad.
  nanus Coquillett, 1903d: 256 *(Culex).*—Fla.
signipennis (Coquillett), 1904f: 167 *(Taeniorhynchus).*—Mexico; Sask. to Ky., s. to Ariz. and Tex.

Genus AEDES Meigen

Aedes Meigen, 1818: 13. Type-species, *cinereus* Meigen (mon.).

Subgenus OCHLEROTATUS Lynch Arribálzaga

Ochlerotatus Lynch Arribálzaga, 1891a: 353, 367, 374 (also 1891b: 122, 136, 143) (as genus; as *Ochlerothatus*, p. 374 or p. 143). Type-species, *confirmatus* Lynch Arribálzaga (Coquillett, 1910b: 577)=*scapularis* (Rondani).
Taeniorhynchus Lynch Arribálzaga, 1891a: 374 (also 1891b: 147). Type-species, *Culex taeniorhynchus* Wiedemann (taut.). Suppressed by I.C.Z.N., 1959c: 187.
Culicada Felt, 1904: 391*b*. Type-species, *Culex canadensis* Theobald (orig. des.).

Culicelsa Felt, 1904: 391*b*. Type-species, *Culex taeniorhynchus* Wiedemann (orig. des.).
Pseudoculex Dyar, 1905c: 47. Type-species, *Culex aurifer* Coquillett (orig. des.).
Aedes, subg. *Ochlerotatus*, globus *Hyparctius* Martini, 1930: 264. Type-species, *Culex pullatus* Coquillett (Edwards, 1932a: 135).
Aedes, subg. *Kompia* Aitken, 1941: 81. Type-species, *purpureipes* Aitken (orig. des.).
aboriginis Dyar, 1917b: 99.—Wash.; Alaska to Oreg., Idaho, and Sask.
abserratus (Felt and Young), 1904: 312 (*Culex*).—N.Y.; Minn. to Labr. and N.S., s. to Ill. and N.J.
  centrotus Howard, Dyar, and Knab, 1917: 747.—Ont.
  dysanor Dyar, 1921c: 70.—N.Y.
aloponotum Dyar, 1917b: 98.—Wash.; B.C., Oreg.
atlanticus Dyar and Knab, 1906c: 198.—N.J. and Fla.; Kans. to Tex. and Fla., n. to N.Y.
aurifer (Coquillett), 1903d: 255 (*Culex*).—N.H.; Man. to Que., s. to Iowa and Del.
barri Rueger, 1958: 34.—Minn.; Ont.
bicristatus Thurman and Winkler, 1950: 239.—Calif.
bimaculatus (Coquillett), 1902g: 84 (*Culex*).—Tex.; Mexico, El Salvador.
cacothius Dyar, 1923: 44.—Wyo.
campestris Dyar and Knab, 1907b: 213.—Sask.; Alaska to Que., s. to Oreg. and Tex.
  callithotrys Dyar, 1920a: 16.—Y.T.
canadensis canadensis (Theobald), 1901b: 3 (*Culex*).—Ont.; forest areas of Canada and U.S., Alaska, Mexico.
  nivitarsis Coquillett, 1904f: 168 (*Culex*).—N.J.
  ssp. **mathesoni** Middlekauff, 1944: 42 (as sp.).—Fla.; Ala., Ga., S.C.
cantator (Coquillett), 1903d: 255 (*Culex*).—N.J.; P.E.I. to Va.
cataphylla cataphylla Dyar, 1916b: 86.—Calif.; Alaska to Man., s. to Calif. and Colo., n. and cent. Europe.
  prodotes Dyar, 1917c: 118.—Mont.
  ssp. **pacificensis** Hearle, 1927: 101 (as sp.).—B.C.
communis (De Geer), 1776: 316 (*Culex*).—Europe; Alaska to Labr., s. to Calif. and N.J., Siberia, Europe, Syria.
  lazarensis Felt and Young, 1904: 312 (*Culex*).—N.Y.
  borealis Ludlow, 1911: 178 (*Culex*).—Alaska.
  tahoensis Dyar, 1916b: 82.—Calif.

altiusculus Dyar, 1917b: 100.—Wash.
masamae Dyar, 1920d: 166.—Oreg.
prolixus Dyar, 1922a: 2.—Alaska.
decticus Howard, Dyar, and Knab, 1917: 737.—Ont.; Alaska, Labr., Mich., N.H., N.Y., Mass.
pseudodiantaeus Smith, 1952: 21.—Mass.
impiger, authors, not Walker.
diantaeus Howard, Dyar, and Knab, 1913: pl. 24, fig. 167 (descr. 1917: 758).—N.H.; Alaska to Labr., s. to Wyo. and Mass., n. Europe.
dorsalis (Meigen), 1830: 242 (*Culex*).—Germany; n. Europe and Asia, B.C. to Que., s. to Calif., Miss. and Mass., Formosa, Mexico.
curriei Coquillett, 1901a: 259 (*Culex*).—Idaho.
onondagensis Felt, 1904: 304 (*Culex*).—N.Y.
lativittatus Coquillett, 1906b: 109 (*Culex*).—Calif.
quaylei Dyar and Knab, 1906c: 202.—Calif.
mediolineata Ludlow, 1907: 129 (*Grabhamia*).—N. Dak.
dupreei (Coquillett), 1904a: 10 (*Culex*).—La.; Iowa to Ky. and Md., s. to Tex. and Fla., Mexico.
excrucians (Walker), 1856b: 429 (*Culex*).—N.S.; forest areas of n. Holarctic Region, s. in U.S. to Wash., Utah, Ill., and N.J.
abfitchii Felt, 1904: 381 (*Culex*).—N.Y.
siphonalis Grossbeck, 1904a: 332 (*Culex*).—N.J.
sansoni Dyar and Knab, 1909a: 102.—Alta.
euedes Howard, Dyar, and Knab, 1913: pl. 28, fig. 191 (descr. 1917: 714).—Ont.
fitchii (Felt and Young), 1904: 312 (*Culex*).—N.Y.; Alaska to Labr., s. to Calif., Ill., and N.J.
palustris Dyar, 1916b: 89.—Calif.
pricei Dyar, 1917a: 16 (as *palustris* var.).—Calif.
mimesis Dyar, 1917c: 116.—Mont.
flavescens (Müller), 1764: 87 (*Culex*).—Unknown; Alaska to Labr., s. to Calif., Mo., and N.Y., n. Palaearctic.
fletcheri Coquillett, 1902g: 84 (*Culex*).—Sask.
fulvus fulvus (Wiedemann).—Not Nearctic.
ssp. **pallens** Ross, 1943: 148.—La.; Ill. to Md., s. to Okla., Tex., and Fla., Cuba.
grossbecki Dyar and Knab, 1906c: 201.—N.J.; Mo. to Vt., s. to La. and S.C.
sylvicola Grossbeck, 1906: 129 (*Culex*).—N.J.

hexodontus Dyar, 1916b: 83.—Calif.; N. Amer., arctic tundra and s. in Alpine meadows to Calif. and Colo.
    cyclocerculus Dyar, 1920a: 23.—B.C.
    leuconotips Dyar, 1920a: 24.—B.C.
    labradorensis Dyar and Shannon, 1925b: 78.—Labr.
impiger (Walker), 1848: 6 (*Culex*).—Ont.; treeless arctic areas of Holarctic Region and s. along mt. ranges to Utah and Colo.
    nearcticus Dyar, 1919a: 32.—N.W.T.
implicatus Vockeroth, 1954: 110.—Que.; Alaska to Que., s. to Utah and Mass.
increpitus Dyar, 1916b: 87.—Calif.; B.C. to Sask., s. to Calif. and N. Mex.
    vittata Theobald, 1903c: 313 (*Grabhamia;* preocc. Bigot, 1861).—N. Mex.
    mutatus Dyar, 1919b: 24.—Mont.
    hewitti Hearle, 1923: 5.—B.C.
infirmatus Dyar and Knab, 1906c: 197.—La.; Mo. to N.C., s. to Tex. and Fla., Mexico.
intrudens Dyar, 1919b: 23.—Ont.; Alaska to Labr., s. to Oreg., Colo., and Pa., n. Europe.
melanimon Dyar, 1924b: 126.—Calif.; Wash. to Nebr., s. to Calif. and N. Mex.
    klotsi Matheson, 1933: 69.—Colo.
mitchellae (Dyar), 1905a: 74 (*Culex*).—Fla.; N. Mex. to Ill. and Mass., s. to Fla., Mexico.
monticola Belkin and McDonald, 1957: 179.—Ariz.; Mexico (Baja Calif.).
muelleri Dyar, 1920b: 81.—Mexico; Ariz.
    iridipennis Dyar, 1922b: 92.—Ariz.
nigripes (Zetterstedt), 1838: 807 (*Culex*).—Greenland; treeless arctic areas of Alaska, n. Canada, and n. Eurasia.
    innuitus Dyar and Knab, 1918a: 166.—Greenland.
    alpinus, authors, not Linnaeus.
nigromaculis (Ludlow), 1906c: 83 (1907: 85) (*Grabhamia*).—Mont.; Alta. to Man., s. to Calif., Tex., and Ky.
    grisea Ludlow, 1907: 130 (*Grabhamia*).—Idaho.
niphadopsis Dyar and Knab, 1918a: 166.—Utah; Oreg., Idaho, Nev.
pionips Dyar, 1919b: 19.—Ont.; Alaska to Labr., s. to B.C. and Colo.
pullatus (Coquillett), 1904f: 168 (*Culex*).—B.C.; Alaska to Labr., s. to Calif., Colo., and Mich., n. Eurasia.
    acrophilus Dyar, 1917d: 127.—Alta.
    pearyi Dyar and Shannon, 1925b: 78.—Labr.

punctodes Dyar, 1922a: 1.—Alaska.
punctor (Kirby), 1837: 309 (*Culex*).—N.W.T.; Alaska to Labr., s. to Utah, Mich., and Md., n. Eurasia.
  implacabilis Walker, 1848: 7 (*Culex*).—Ont.
  provocans Walker, 1848: 7 (*Culex*).—N.S.
  auroides Felt, 1905b: 449 (*Culicelsa*).—N.Y.
purpureipes Aitken, 1941: 82.—Mexico (Baja Calif.); Mexico, Ariz.
rempeli Vockeroth, 1954: 112.—Que.; N.W.T.
riparius Dyar and Knab, 1907b: 213.—Man.; Alaska to Ont., s. to Colo., Iowa, and N.Y., n. Eurasia.
scapularis (Rondani), 1848: 109 (*Culex*).—Brazil; Fla., Tex., Greater Antilles, Cent. and S. Amer.
schizopinax Dyar, 1929b: 1.—Mont.; Oreg. to Mont., s. to Calif. and Colo.
sierrensis (Ludlow), 1905a: 231 (*Taeniorhynchus*).—Calif.; B.C., Wash., Idaho, Mont., Oreg.
  varipalpus, authors, not Coquillett.
sollicitans (Walker), 1856b: 427 (*Culex*).—U.S.; Ariz. to Fla., n. to N.S., also N. Dak., Nebr., Ill., Ind., Antilles.
  colon Harris, 1835: 595 (*Culex*). Nomen nudum.
spencerii spencerii (Theobald), 1901b: 99 (*Culex*).—Man.; B.C. to Man., s. to Utah, Kans., and N.Y., Mexico.
  ssp. **idahoensis** (Theobald), 1903a: 250 (*Grabhamia*; as var.).—Idaho; B.C. to N. Dak., s. to Nev. and Colo.
squamiger (Coquillett), 1902g: 85 (*Culex*).—Calif.; Mexico.
  deniedmannii Ludlow, 1904: 234 (*Grabhamia*).—Calif.
sticticus (Meigen), 1838: 1 (*Culex*).—Germany; B.C. to N.B., s. to Calif., Tex., and Fla., n. Eurasia.
  hirsuteron Theobald, 1901b: 98 (*Culex*).—Va.
  pretans Grossbeck, 1904b: 332 (*Culex*).—N.J.
  aestivalis Dyar, 1904b: 245 (*Culex*).—B.C.
  aldrichi Dyar and Knab, 1908: 57.—Idaho.
  gonimus Dyar and Knab, 1918a: 165.—Tex.
  vinnipegensis Dyar, 1919b: 34.—Man.
  lateralis, authors, not Meigen.
stimulans (Walker), 1848: 4 (*Culex*).—N.S.; Alaska to Que., s. to Colo., Miss., and Del.
  subcantans Felt, 1905b: 474 (*Culicada*).—?N.Y.
  mercuator Dyar, 1920a: 13.—Y.T.
  classicus Dyar, 1920c: 113 (as ssp.).—N.J.
  mississippii Dyar, 1920c: 113 (as ssp.).—Miss.
  albertae Dyar, 1920c: 115 (as ssp.).—Alta.

CULICIDAE 115

taeniorhynchus (Wiedemann), 1821a: 43 (1821c: 43) (*Culex*).—
Mexico; coasts and inland saline areas, Calif. to Peru
and Mass. to Brazil, Antilles, Galapagos Is.
damnosus Say, 1823: 11 (1859b: 40) (*Culex*).—Pa.
thelcter Dyar, 1918c: 129.—Tex.; Fla., Okla., Mexico.
keyensis Buren, 1947: 228.—Fla.
thibaulti Dyar and Knab, 1910: 174.—Ark.; Ill. to Ohio and N.C., s.
to Tex. and Fla.
tormentor Dyar and Knab, 1906c: 191.—La.; Okla. to Ohio and N.C.,
s. to Tex. and Fla., Mexico, Guatemala, Honduras.
tortilis (Theobald), 1903d: 281 (*Culex*).—Jamaica; Fla., Bahama Is.,
Virgin Is., Greater Antilles, Mexico, Guatemala.
trichurus (Dyar), 1904a: 170 (*Culex*).—B.C.; B.C. to N.S., s. to
Wash., Mich., and R.I.
cinereoborealis Felt and Young, 1904: 312 (*Culex*).—N.Y.
pagetonotum Dyar and Knab, 1909b: 253.—Ont.
poliochros Dyar, 1919b: 35.—Man.
trivittatus (Coquillett), 1902f: 193 (*Culex*).—N.J.; Idaho to N.S.,
s. to N. Mex. and Ga., Mexico.
inconspicuus Grossbeck, 1904b: 332 (*Culex*).—N.J.
varipalpus (Coquillett), 1902b: 292 (*Culex*).—Ariz.; Utah.
ventrovittis Dyar, 1916b: 84.—Calif.; Wash., Idaho.
fisheri Dyar, 1917a: 19.—Calif.

Subgenus FINLAYA Theobald

Finlaya Theobald, 1903a: 281 (as genus). Type-species, *Culex
kochi* Dönitz (Blanchard, 1905: 415).
Protomacleaya Theobald, 1907: 149 Type-species, *Culex triseriatus*
Say (mon.).
atropalpus (Coquillett), 1902b: 292 (*Culex*).—Md.; Minn. to Labr.
s. to Ariz., Panama, and Ga.
hendersoni Cockerell, 1918: 199 (as *triseriatus* var.).—Wyo.; Mont.,
Colo., Tex.
triseriatus (Say), 1823: 12 (1859b: 40) (*Culex*).—Pa.; B.C. to Que.,
s. to Tex. and Fla., Mexico.
nigra Ludlow, 1905b: 387 (?*Finlaya*).—Ill.
zoosophus Dyar and Knab, 1918a: 165.—Tex.; Kans., Okla., Mexico.
alleni Turner, 1924: 84.—Tex.

Subgenus STEGOMYIA Theobald

Stegomyia Theobald *in* Howard, 1901: 235 (as genus). Type-
species, *Culex fasciatus* Fabricius (Neveu-Lemaire,
1902b: 212) =*aegypti* (Linnaeus).

aegypti (Linnaeus), 1762: 470 (*Culex*).—Egypt; Pantropical, mostly within the 10° C. isotherms. Application for the validation of this name and with its accepted identity has been made to the I.C.Z.N. by Mattingly, Stone, and Knight, 1962: 208.
 taeniatus Wiedemann, 1828: 10 (*Culex*).—Ga.
 excitans Walker, 1848: 4 (*Culex*).—Ga.

Subgenus AEDIMORPHUS Theobald

Aedimorphus Theobald, 1903a: 290 (as genus). Type-species, *Uranotaenia domestica* Theobald (mon.).
Ecculex Felt, 1904: 391c. Type-species, *Culex sylvestris* Theobald (orig. des.) =*vexans* (Meigen).

vexans (Meigen), 1830: 241 (*Culex*).—Germany; Y.T. to N.S., s. to Calif. and Fla., Holarctic, Oriental, Pacific Is., Transvaal.
 sylvestris Theobald, 1901a: 406 (*Culex*; preocc. Ross, 1898).—Ont. and Man.
 montcalmi Blanchard, 1905: 307 (*Culex*; n. name for *sylvestris* Theobald).—Ont., Man.
 euochrus Howard, Dyar, and Knab, 1917: 716.—B.C.

Subgenus AEDES Meigen

cinereus Meigen, 1818: 13.—Europe; Canada, U.S., n. Eurasia.
 fuscus Osten Sacken, 1877: 191.—Mass.
 pallidohirta Grossbeck, 1905: 359 (*Culex*).—N.J.
 hemiteleus Dyar, 1924c: 179 (as race).—Calif.

Genus HAEMAGOGUS Williston

Haemagogus Williston, 1896c: 271. Type-species, *splendens* Williston (mon.).

Subgenus LONGIPALPIFER Levi-Castillo

Haemagogus, subg. **Longipalpifer** Levi-Castillo, 1951: erratas. Type-species, *Haemagogus panarchys* Dyar (orig. des.).

equinus Theobald, 1903d: 282.—Jamaica; Tex., Mexico, Cent. Amer., n. S. Amer.

Genus CULISETA Felt

Theobaldia Neveu-Lemaire, 1902a: 1331 (preocc. Fischer, 1885). Type-species, *Culex annulatus* Schrank (orig. des.).
Culiseta Felt, 1904: 391c. Type-species, *Culex absobrinus* Felt (orig. des.) =*impatiens* (Walker).

CULICIDAE 117

Subgenus CULISETA Felt

alaskaensis (Ludlow), 1906b: 326 (*Theobaldia*).—Alaska; Alaska to Labr., s. in w. to Colo., n. and cent. Palaearctic Region.
impatiens (Walker), 1848: 5 (*Culex*).—Ont.; Alaska to Labr., s. to Calif., Mo., and Mass.
  pinguis Walker *in* Lord, 1866: 337 (*Culex*).—B.C.
  absobrinus Felt, 1904: 318 (*Culex*).—N.Y.
incidens (Thomson), 1869: 443 (*Culex*).—Calif.; Alaska to N.S., s. to Calif., Tex., and Mich.
inornata (Williston), 1893d: 253 (*Culex*).—Calif.; N.W.T. to N.H., s. to Calif., n. Mexico, and Fla.
  magnipennis Felt, 1904: 322 (*Culex*).—N.Y.
particeps (Adams), 1903a: 26 (*Culex*).—Ariz.; Oreg., Calif., Mexico to Costa Rica.
  maccrackenae Dyar and Knab, 1906a: 133.—Calif.

Subgenus CULICELLA Felt

Culicella Felt, 1904: 391*c* (as genus). Type-species, *Culex dyari* Coquillett (orig. des.) =*morsitans* (Theobald).
minnesotae Barr, 1957: 163.—Minn.; Utah, Ont., N.J., Mass.
morsitans (Theobald), 1901b: 8 (*Culex*).—England; Alaska to Labr., s. to Oreg., Colo., and Del., n. Eurasia, Morocco.
  dyari Coquillett, 1902f: 192 (*Culex*).—N.H.
  brittoni Felt, 1905a: 79 (*Culex*).—Conn.
parodites (Dyar), 1928b: 244 (*Culicella*).—Wis.

Subgenus CLIMACURA Howard, Dyar, and Knab

Culex, subg. **Climacura** Howard, Dyar, and Knab, 1915: 452. Type-species, *Culex melanurus* Coquillett (mon.).
melanura (Coquillett), 1902f: 193 (*Culex*).—N.H.; Colo. to Maine, s. to Tex. and Fla.

Genus CULEX Linnaeus

Culex Linnaeus, 1758: 602. Type-species, *pipiens* Linnaeus (Latreille, 1810: 442).
Laiomyia Izquierdo, 1916: 65. Type-species, *Culex stigmatosoma* Dyar (Edwards, 1932a: 200) =*peus* Speiser.
Culex, subg. *Transculicia* Dyar, 1918a: 184. Type-species, *eleuthera* Dyar (mon.) =*bahamensis* Dyar and Knab.

Subgenus NEOCULEX Dyar

Neoculex Dyar, 1905c: 48 (as genus). Type-species, *Culex territans* Walker (orig. des.).

apicalis Adams, 1930a: 26.—Ariz.; Calif., Mexico.
arizonensis Bohart, 1949: 341.—Ariz.; Mexico.
boharti Brookman and Reeves, 1950: 159 (n. name for *reevesi* Bohart).—Calif.; Nev.
  reevesi Bohart, 1949: 342 (preocc. Wirth, 1948).—Calif.
reevesi Wirth *in* Usinger et al., 1948: 230.—Calif.; Mexico (Baja Calif.).
territans Walker, 1856b: 428.—U.S.; Alaska and B.C. to Vt., s. to Calif. and Fla., Europe.
  saxatilis Grossbeck, 1905: 360.—N.J.
  frickii Ludlow, 1906a: 132.—Minn.
  apicalis, authors, not Adams.

Subgenus CULEX Linnaeus

bahamensis Dyar and Knab, 1906c: 210.—Bahama Is.; Fla., Antilles, Trinidad, ?French Guiana.
chidesteri Dyar, 1921d: 117.—Panama; Tex., Cent. and n. S. Amer., Antilles.
coronator Dyar and Knab, 1906c: 215.—Trinidad; Tex. to Argentina.
declarator Dyar and Knab, 1906c: 211.—Trinidad; Tex. to Uruguay and Bolivia.
  virgultus, authors, not Theobald.
erythrothorax Dyar, 1907: 124.—Calif.; Idaho, Utah, Mexico.
  badgeri Dyar, 1924b: 125.—Calif.
interrogator Dyar and Knab, 1906c: 209.—Mexico; Tex. to Panama.
nigripalpus Theobald, 1901b: 322.—St. Lucia I.; Tenn. and N.C., s. to Tex. and Fla., Antilles, Cent. Amer. to Brazil and Ecuador.
peus Speiser, 1904a: 148 (n. name for *affinis* Adams).—Ariz.; Wash. to Okla., s. to Venezuela.
  affinis Adams, 1903a: 25 (preocc. Stephens, 1825).—Ariz.
  stigmatosoma Dyar, 1907: 123.—Calif.
pipiens pipiens Linnaeus, 1758: 602.—Europe, Lapland, America; B.C. to N.S., s. to Calif., Okla., and Ga., Palaearctic, e. and s. Africa, s. S. Amer.
  consobrinus Robineau-Desvoidy, 1827: 408.—Pa.
  var. **pallens** Coquillett, 1898j: 303 (as sp.).—Japan; Calif., across cent. U.S., Mexico, China.
  dipseticus Dyar and Knab, 1909c: 34 (as *quinquefasciatus* ssp.).—Calif.
  comitatus Dyar and Knab, 1909c: 35.—Calif.
  ssp. **quinquefasciatus** Say, 1823: 10 (1859b: 39) (as sp.).—Mississippi River; Utah to Md., s. to Calif. and Fla., cosmotropical.

pungens Wiedemann, 1828: 9.—La.
ferruginosus Wiedemann, 1828: 12 (*Anopheles*).—La.
restuans Theobald, 1901b: 142.—Ont.; B.C. to N.S., s. to Calif. and Fla., Mexico.
brehmei Knab, 1916a: 161.—N.J.
territans, authors, not Walker.
salinarius Coquillett, 1904b: 73.—N.J.; Idaho to N.S., s. to N. Mex. and Fla., Mexico, Bermuda.
tarsalis Coquillett, 1896a: 43.—Calif.; B.C. to Mexico, e. to N.W.T., Mich., S.C., and Fla.
willistoni Giles, 1900: 281.—Calif.
kelloggii Theobald, 1903b: 211.—Calif.
thriambus Dyar, 1921a: 33.—Tex.; Calif., Okla., Mexico.

Subgenus MELANOCONION Theobald

Melanoconion Theobald, 1903a: 238 (as genus). Type-species, *Culex atratus* Theobald (Dyar, 1905c: 49).
Culex, subg. *Choeroporpa* Dyar, 1918b: 103. Type-species, *anips* Dyar (orig. des.).

abominator Dyar and Knab, 1909b: 257.—Tex.
anips Dyar, 1916a: 48.—Calif.; Mexico (Baja Calif.).
atratus Theobald, 1901b: 55.—Jamaica, Trinidad; Fla., Antilles, Brazil, French Guiana, Panama.
erraticus (Dyar and Knab), 1906c: 224 (*Mochlostyrax*).—La.; N. Dak., Mich., and Md., s. to Brazil and Fla.
egberti Dyar and Knab, 1907b: 214.—Fla.
peribleptus Dyar and Kanb, 1918b: 181.—S.C.
pose Dyar and Knab, 1918b: 182.—Tex.
degustator Dyar, 1921b: 39.—Ark.
homoepas Dyar and Ludlow, 1921: 46.—La.
iolambdis Dyar, 1918b: 106.—Panama; Fla., Mexico, Colombia, P.R., Jamaica.
mulrennani Basham, 1948: 1.—Fla.; Cuba.
opisthopus Komp, 1926: 44.—Honduras; Fla., P.R., Mexico, Panama.
peccator Dyar and Knab, 1909b: 256.—Ark.; Kans., Mich., and Md., s. to Mexico and P.R.
incriminator Dyar and Knab, 1909b: 257.—Miss.

Subgenus MOCHLOSTYRAX Dyar and Knab

Mochlostyrax Dyar and Knab, 1906c: 223 (as genus). Type-species, *caudelli* Dyar and Knab (orig. des.).

pilosus (Dyar and Knab), 1906c: 224 (*Mochlostyrax*).—Mexico; Ky. to N.C., s. to Argentina and Antilles.
  floridanus Dyar and Knab, 1906b: 171 (*Mochlostyrax*).—Fla.
  deceptor Dyar and Knab, 1909b: 257.—Fla.

Genus DEINOCERITES THEOBALD

Deinocerites Theobald, 1901c: 235. Type-species, *cancer* Theobald (sub. mon., Theobald, 1901b: 215).
Brachiosoma Theobald, 1901c: 235. Type-species, *Deinocerites cancer* Theobald (Coquillett, 1910b: 515).

cancer Theobald, 1901b: 215.—Jamaica, St. Lucia I.; Fla., Antilles, Mexico, Cent. Amer.
mathesoni Belkin and Hogue, 1959: 426.—Tex.

Family THAUMALEIDAE
(Orphnephilidae)

By Alan Stone

This is a small family of rarely encountered flies. The larvae are found on rocks over which such a thin layer of cold water flows that the back of the larva is exposed. They feed on organic matter on the rocks. Pupation is in wet moss or mud and the adults are found on vegetation usually not far from the larval habitat.

  REFERENCES: Dyar and Shannon, 1924d (rev.); Vaillant, 1959 (tax., biol.).

Genus THAUMALEA Ruthe

Thaumalea Ruthe, 1831: 1211. Type-species, *testacea* Ruthe (mon.).

americana Bezzi, 1913: 250.—N.Y.; Ont. and Que., s. to N.C.
  johannis Dyar and Shannon, 1924d: 434.—Md.
  testacea, authors, not Ruthe.
elnora Dyar and Shannon, 1924d: 434.—Idaho.
fusca (Garrett), 1925b: 10 (*Orphnephila*).—B.C.; Idaho.
thornburghae Vaillant, 1959: 35 (as *thornburghi*).—N.C.

Genus TRICHOTHAUMALEA Edwards

Trichothaumalea Edwards, 1929a: 125. Type-species, *Thaumalea pluvialis* Dyar and Shannon (orig. des.).

pluvialis (Dyar and Shannon), 1924d: 433 (*Thaumalea*).—B.C.
  pilosa Garrett, 1925b: 10 (*Orphnephila*).—B.C.

Family CERATOPOGONIDAE
(Heleidae)

By Willis W. Wirth

The "biting midges" may be found in large numbers in nearly any aquatic or semiaquatic habitat in all regions of the world. Because of their small size (usually 1-4 mm.), they have been little collected and are very poorly known. Adult habits are diverse but most species are adapted to some type of blood-sucking. Species of *Culicoides*, *Lasiohelea*, and *Leptoconops* suck vertebrate blood and some are notorious pests, especially in beach or mountain resort areas. Some *Culicoides* are known vectors of diseases including onchocerciasis of horses and cattle, bluetongue of sheep, horse-sickness, several human filariases, and *Haemoproteus* and *Leucocytozoon* diseases of birds. Some *Atrichopogon* and *Forcipomyia* are ectoparasitic on larger insects, while many genera of Ceratopogoninae are predaceous on other smaller ones. Some genera (e.g. *Dasyhelea*) are exclusively flower-visiting, feeding on nectar, while many others thus supplement their diet or have nonhaematophagous species. The immature stages may be terrestrial under bark or on wet or damp wood (*Forcipomyia*); semiaquatic on wet alga-covered soil, wood, or rocks (*Atrichopogon*, *Dasyhelea*, etc.); in wet, decomposing, plant material such as cactus stems, banana stalks or leaf compost (*Culicoides*, etc.); or aquatic in mud or sand on lake, pond, or stream margins, salt marshes, tree holes, or water-holding plants (most genera).

REFERENCES: Malloch, 1915g (biol., rev. Ill. spp.); Edwards, 1926 (classif., generic rev., British spp.); Thomsen, 1937 (immature stages, biol.); Macfie, 1940a (world rev. of genera); Johannsen, 1943b (synopsis Amer. genera, figs., list N. Amer. spp.); Johannsen, 1952b (keys to genera, spp. of ne. U.S.); Wirth, 1952a (general classif., morph., economics, techniques, biol., immature stages, rev. Calif. spp.).

Subfamily LEPTOCONOPINAE

Genus LEPTOCONOPS Skuse

Leptoconops Skuse, 1889: 288. Type-species, *stygius* Skuse (mon.).
Tersesthes Townsend, 1893f: 370. Type-species, *torrens* Townsend (orig. des.).

REFERENCES: Carter, 1921 (rev.); Smith and Lowe, 1948 (bionomics, Calif.).

Subgenus LEPTOCONOPS Skuse

floridensis Wirth, 1951e: 282.—Fla.
torrens (Townsend), 1893f: 371 (*Tersesthes*).—N. Mex.; Calif. to Colo. and Tex.
   carteri Hoffman, 1926a: 133.—Calif.

Subgenus HOLOCONOPS Kieffer

Holoconops Kieffer, 1918b: 135 (as genus). Type-species, *Leptoconops kerteszi* Kieffer (orig. des.).
bequaerti (Kieffer), 1925b: 405 (*Holoconops;* as *becquaerti*).—Honduras; coastal, Mass., N.C., Miss., Fla., circum-Caribbean.
catawbae (Boesel), 1948: 69 (*Holoconops*).—Ohio; Mich., Ont., Que.
kerteszi Kieffer, 1908a: 576.—Egypt; Wash. to Nebr., s. to Calif. and N. Mex., n. Africa.
   americana Carter, 1921: 22 (as var.).—Utah.

Subgenus STYLOCONOPS Kieffer

Styloconops Kieffer, 1921c: 107 (as genus). Type-species, *Leptoconops albiventris* Meijere (orig. des.).
freeborni Wirth, 1952a: 115.—Calif.

Subfamily FORCIPOMYIINAE

Genus ATRICHOPOGON Kieffer

Ceratopogon, subg. **Atrichopogon** Kieffer, 1906a: 53. Type-species, *exilis* Coquillett (Coquillett, 1910b: 512)=*levis* (Coquillett).

   REFERENCE: Ewen and Saunders, 1958 (rev. based on immature stages).

Subgenus ATRICHOPOGON Kieffer

arcticus (Coquillett), 1900h: 396 (1904: 10) (*Ceratopogon*).—Alaska; Alaska to Colo., s. to Calif. and N. Mex.
corpulentus Ewen *in* Ewen and Saunders, 1958: 681.—B.C.
crinitus Ewen *in* Ewen and Saunders, 1958: 717.—B.C.; Sask.
flavus Ewen *in* Ewen and Saunders, 1958: 697.—Sask.
fusculus (Coquillett), 1901h: 605 (*Ceratopogon*).—N.J.; Sask. to Ont., s. to Calif. and Fla.
fusinervis (Malloch), 1915g: 308 (*Ceratopogon*).—Ill.
gilvus (Coquillett), 1905d: 62 (*Ceratopogon*).—Fla.
humicola Ewen *in* Ewen and Saunders, 1958: 679.—Sask.

inconspicuus Ewen *in* Ewen and Saunders, 1958: 706.—Sask.
levis (Coquillett), 1901h: 604 *(Ceratopogon)*.—Md.; Wash. to N.H., s. to Calif. and Fla.
  exilis Coquillett, 1902g: 86 *(Ceratopogon)*.—D.C.
maculosus Ewen *in* Ewen and Saunders, 1958: 689.—Sask.
minutus (Meigen), 1830: 263 *(Ceratopogon)*.—Europe; Calif., Iowa, Mich., Ind., Md., Va.
occidentalis Wirth, 1952a: 120.—Calif.; Alaska, Mont., Calif.
peregrinus (Johannsen), 1908: 266 *(Ceratopogon)*.—N.Y.
transiens (Walker), 1848: 25 *(Ceratopogon)*.—Hudson Bay.
transversus Wirth, 1952a: 124.—Calif.
warmkei Wirth, 1956c: 243.—P.R.; Fla.
websteri (Coquillett), 1901h: 603 *(Ceratopogon)*.—La.; Calif. to Conn., s. to Fla.

Subgenus MELOEHELEA Wirth

Atrichopogon, subg. **Meloehelea** Wirth, 1956b: 16. Type-species, *meloesugans* Kieffer (orig. des.).

REFERENCE: Wirth, 1956b (rev., habits, parasitic on blister beetles).

epicautae Wirth, 1956b: 21.—Ariz.; B.C., Mont., Calif.
farri Wirth, 1956b: 22.—Mass.; N.H.
oedemerarum Storå, 1939: 16.—Finland; Ont. to Va., Europe.

Genus FORCIPOMYIA Meigen

Forcipomyia Meigen, 1818: 73, 75. Type-species, *Tipula bipunctata* Linnaeus (Application submitted by D. E. Hardy to I.C.Z.N. to validate Meigen's generic name cited in specific synonymy, with *bipunctata* as type-species.)

REFERENCES: Saunders, 1956 (rev. subg., all stages); Wirth, 1956a (classif. spp. parasitic on insects).

Subgenus PTEROBOSCA Macfie

Pterobosca Macfie, 1932: 266 (as genus). Type-species, *Ceratopogon aeschnosuga* Meijere (Macfie, 1940a: 16).

REFERENCE: Wirth, 1956a (rev., checklist, parasitic on Odonata).

fusicornis (Coquillett), 1905d: 63 *(Ceratopogon)*.—Fla.; La., Bermuda.
  fur Johnson, 1913: 444 *(Ceratopogon)*.—Bermuda.
  floridana Johannsen, 1950: 143 *(Pterobosca)*.—Fla.

Subgenus LASIOHELEA Kieffer

Lasiohelea Kieffer, 1921c: 115 (as genus). Type-species, *Atrichopogon pilosipennis* Kieffer (orig. des.)=*velox* (Winnertz).

Females suck vertebrate blood; larvae are found on wet wood.

fairfaxensis Wirth, 1951f: 317.—Va.; Iowa to Mich. and Que., s. to Va.

Subgenus NEOFORCIPOMYIA Tokunaga

Forcipomyia, subg. **Neoforcipomyia** Tokunaga *in* Tokunaga and Murachi, 1959: 200. Type-species, *Ceratopogon pectinunguis* Meijere (orig. des.).

REFERENCE: Wirth, 1956a (rev., habits, ectoparasitic on insects).

baueri Wirth, 1956a: 361.—Ariz.
eques (Johannsen), 1908: 266 (*Ceratopogon*).—N.Y.; Mich. to Que., Europe.
mcateei Wirth, 1956a: 359.—Md.; Mich. to Que., s. to Va.

Subgenus THYRIDOMYIA Saunders

Thyridomyia Saunders, 1925: 268 (as genus). Type-species, *palustris* Saunders (orig. des.).

aspinosa Saunders, 1956: 692.—Sask.
monilicornis (Coquillett), 1905d: 63 (*Ceratopogon*).—B.C.; Alaska to Que., s. to Calif. and Md.
palustris (Saunders), 1925: 269 (*Thyridomyia*).—England; B.C., Tenn., Europe.

Subgenus SYNTHYRIDOMYIA Saunders

Forcipomyia, subg. **Synthyridomyia** Saunders, 1956: 688. Type-species, *Lasiohelea acidicola* Tokunaga (orig. des.).

colemani Wirth, 1952a: 146.—Calif.
johannseni Thomsen, 1935: 286.—N.Y.

Subgenus EUFORCIPOMYIA Malloch

Euforcipomyia Malloch, 1915h: 312 (as genus). Type-species, *hirtipennis* Malloch (orig. des.).

hirtipennis (Malloch), 1915h: 313 (*Euforcipomyia*).—Ill. This may possibly be a synonym of *titillans* (Winnertz); if so then *Euforcipomyia* will replace *Proforcipomyia*.

Subgenus PROFORCIPOMYIA Saunders

Forcipomyia, subg. **Proforcipomyia** Saunders, 1956: 662. Type-species, *wirthi* Saunders (orig. des.).

calcarata (Coquillett), 1905d: 64 (*Ceratopogon*).—Mexico; Miss. to Fla. and Va.

indecora Kieffer, 1914: 269.—s. Africa; Bermuda, pantropical.
  ingrami Carter, 1919: 290.—Gold Coast.
sonora Wirth, 1952a: 145 (as *calcarata* var.).—Calif.; B.C. to Utah, s. to Calif. and N. Mex.
titillans (Winnertz), 1852a: 27 (*Ceratopogon*).—Europe; Ont.
wirthi Saunders, 1956: 663.—Calif.

Subgenus METAFORCIPOMYIA Saunders

Forcipomyia, subg. **Metaforcipomyia** Saunders, 1956: 685. Type-species, *cerifera* Saunders (orig. des.).
pluvialis Malloch, 1923f: 5.—Md.; Ont. to La. and Fla.

Subgenus CALOFORCIPOMYIA Saunders

Forcipomyia, subg. **Caloforcipomyia** Saunders, 1956: 680. Type-species, *caerulea* Saunders (orig. des.).
splendida Wirth, 1951f: 315.—Va.; Mich. to Conn., s. to Fla., also Alta.

Subgenus FORCIPOMYIA Meigen

aurea Malloch, 1915g: 318.—Ill.
bipunctata (Linnaeus), 1767: 978 (*Tipula*).—Europe; entire N. Amer.
brevipennis (Macquart), 1826a: 179 (1826: 121) (*Ceratopogon*).—Europe; B.C. to Que., s. to Calif. and La.
  specularis Coquillett, 1901h: 601 (*Ceratopogon*).—Pa.
christiansoni Wirth and Hubert, 1960a: 640.—Calif.
ciliata (Winnertz), 1852a: 21 (*Ceratopogon*).—Europe; N.Y., Mass.
cilipes (Coquillett), 1900h: 397 (1904: 11) (*Ceratopogon*).—Alaska; Alaska to N.Y., s. to Calif. and Colo.
cinctipes (Coquillett), 1905d: 64 (*Ceratopogon*).—Fla.; Calif., Mexico.
concolor Malloch, 1915g: 319 (as *pergandei* var.).—Ill.
desertensis Wirth and Hubert, 1960a: 639.—Calif.; Mexico (Baja Calif.).
elegantula Malloch, 1915h: 311.—Ill.; D.C., Va.
fimbriata (Coquillett), 1901h: 601 (*Ceratopogon*).—D.C.; Md.
fuliginosa (Meigen), 1818: 86 (*Ceratopogon*).—Europe; entire N. Amer., cosmopolitan.
  erucicida Knab, 1914b: 65.—Fla.
  coquilletti Kieffer, 1917: 297 (*Ceratopogon*).—N.Y.
  brookmani Wirth, 1952a: 140.—Calif.
genualis (Loew), 1866a: 128 (Cent. 6, no. 1) (*Ceratopogon*).—Cuba; La., Ga., Fla.; Bermuda, Neotropical.
  raleighi Macfie, 1938: 160.—Trinidad. **N. syn.**
hurdi Wirth, 1952a: 143.—Calif.

macswaini Wirth, 1952a: 130.—Calif.; Wash., Oreg.
occidentalis Wirth, 1952a: 141.—Calif.; N. Mex.
pallida (Winnertz), 1852a: 15 (*Ceratopogon*).—Europe; N.Y.
pergandei (Coquillett), 1901h: 602 (*Ceratopogon*).—D.C.; Kans., Ill., Md., Va.
pilosa (Coquillett), 1902g: 87 (*Ceratopogon*).—D.C.; Utah, N. Mex.
quatei Wirth, 1952a: 142.—Calif.
solonensis Wirth, 1951f: 315.—Va.
squamipes (Coquillett), 1902g: 88 (*Ceratopogon*).—N. Mex.; U.S.
    brumalis Long, 1902: 3 (*Ceratopogon*).—Tex.
texana texana (Long), 1902: 10 (*Ceratopogon*).—Tex.; Calif. to D.C., s. to Tex.
    ssp. **simulata** Walley, 1932: 165 (as sp.).—Ont.; Alaska to Ont., s. to Calif. and Mass.
    picea, authors, not Winnertz.
townesi Wirth, 1952a: 137.—Calif.; Wash.
varipennis Wirth and Williams, 1957: 8.—Bermuda; Fla., Tex., Guatemala, P.R.

Unplaced Species of *Forcipomyia*

longitarsis (Malloch), 1915h: 314 (*Euforcipomyia*).—Ill.
obscura (Walker), 1848: 26 (*Ceratopogon*).—Hudson Bay.
parva (Walker), 1848: 26 (*Ceratopogon*).—Hudson Bay.
stenammatis (Long), 1902: 10 (*Ceratopogon*).—Conn. Larva only.
wheeleri (Long), 1902: 12 (*Ceratopogon*).—Tex. Larva and pupa only.

Subfamily DASYHELEINAE

Genus DASYHELEA Kieffer

Dasyhelea Kieffer, 1911b: 5. Type-species, *halophila* Kieffer (mon.).
Pseudoculicoides Malloch, 1915g: 309. Type-species, *Ceratopogon mutabilis* Coquillett (orig. des.).

ancora (Coquillett), 1902g: 87 (*Ceratopogon*).—Fla.; Calif., Ariz., N. Mex., Conn.
atlantis Wirth and Williams, 1957: 11.—Bermuda.
atrata Wirth, 1952a: 164.—Calif.; Alaska, La.
bahamensis (Johnson), 1908c: 71 (*Ceratopogon*).—Bahama Is.; Fla.
bermudae Wirth and Williams, 1957: 11.—Bermuda.
bifurcata Wirth, 1952a: 161.—Calif.
brookmani Wirth, 1952a: 152.—Calif.
cactorum Wirth and Hubert, 1960a: 642.—Baja Calif.; Calif., Ariz.

cincta (Coquillett), 1901h: 605 *(Ceratopogon).*—Fla.; Calif. to N.Y., s. to Fla., Bermuda.
festiva Wirth, 1952a: 161.—Calif.; Colo.
grisea (Coquillett), 1901h: 602 *(Ceratopogon).*—D.C.; Calif. to Mich. and Conn., s. to Tex. and Fla.
johannseni (Malloch), 1915g: 311 *(Pseudoculicoides).*—Calif.
luteogrisea Wirth and Williams, 1957: 10.—Bermuda; coastal, Tex. to Fla., n. to Md., Bahama Is.
major (Malloch), 1915g: 311 *(Pseudoculicoides).*—Ill.; Mich. to Conn., s. to Nebr., Ala., and Va.
mutabilis (Coquillett), 1901h: 602 *(Ceratopogon).*—D.C.; entire N. Amer.
oppressa Thomsen, 1935: 285.—N.Y.; Iowa to Que., s. to La. and Fla.
pallens Wirth, 1952a: 152.—Calif.
pentalineata Wirth and Hubert, 1960a: 644.—Mexico (Baja Calif.); Calif.
pollinosa Wirth, 1952a: 156.—Calif.; Wash., Colo.
pritchardi Wirth, 1952a: 163.—Calif.
ryckmani Wirth and Hubert, 1960a: 646.—Calif.; Ariz.
sanctaemariae Wirth, 1952a: 162.—Calif.
scissurae Macfie, 1937: 15.—Trinidad; Bermuda, Neotropical.
scutellata (Meigen), 1830: 262 *(Ceratopogon).*—Europe; Greenland.
subcaerulea Thomsen, 1935: 284.—N.Y.
tenebrosa (Coquillett), 1905d: 64 *(Ceratopogon).*—Calif.; Alaska.
thomsenae Wirth, 1952a: 160.—Calif.; N. Mex., Md., Va.
traverae Thomsen, 1935: 285.—N.Y.; Mich. to N.Y., s. to La. and Va.
tristyla Wirth, 1952a: 165.—Calif.; Calif., Utah., La., Ala., Mass.

Subfamily CERATOPOGONINAE

Tribe CULICOIDINI

Genus CULICOIDES Latreille

Culicoides Latreille, 1809: 251. Type-species, *punctatus* Latreille (mon.).

REFERENCES: Williams, 1951a (bionomics *C. tristriatulus* in Alaska); Williams, 1951b (immature stages); Wirth, 1952e (immature stages); Foote and Pratt, 1954 (rev. spp. e. U.S.); Fox, 1955 (cat. New World spp.); Arnaud, 1956 (world cat.); Wirth and Bottimer, 1956 (population study Tex. spp.); Wirth and Jones, 1957 (systematics *C. variipennis* group); Downes, 1958 (biol.); Jones, 1961a (larval habitats), 1961b (pupae).

Subgenus CULICOIDES Latreille

cockerellii (Coquillett), 1901h: 603 (*Ceratopogon*).—Colo.; Alaska to Calif. and Colo.
luteovenus Root and Hoffman, 1937: 156.—Mexico; Wash. to Utah, s. to Panama.
neopulicaris Wirth, 1955: 355.—Tex.; Mexico.
saltonensis Wirth, 1952a: 173 (as *cockerellii* ssp.).—Calif. **N. status.**
sordidellus (Zetterstedt), 1838: 820 (*Ceratopogon*).—Greenland.
  pulicaris Linnaeus of Fabricius, 1780: 211 (as *Culex pulicans*).
tristriatulus Hoffman, 1925: 294 (as *cockerellii* var.).—Calif.; Alaska.
yukonensis Hoffman, 1925: 291.—Y.T.; Alaska, B.C.

Subgenus HOFFMANIA Fox

Culicoides, subg. **Hoffmania** Fox, 1948: 21. Type-species, *inamollae* Fox and Hoffman (orig. des.)=*insignis* Lutz.

insignis Lutz, 1913b: 51.—Brazil; Fla., Neotropical.
  inamollae Fox and Hoffman, 1944: 110.—P.R.
venustus Hoffman, 1925: 290.—Md.; Wis. to Que., s. to Mo. and Fla.

Subgenus AVARITIA Fox

Culicoides, subg. **Avaritia** Fox, 1955: 218. Type-species, *Ceratopogon obsoletus* Meigen (orig. des.).

alachua Jamnback and Wirth, 1963: 187.—Fla.
chiopterus (Meigen), 1830: 263 (*Ceratopogon*).—Europe; Wash., Ont. to Va.
hirtulus (Coquillett), 1900h: 396 (1904: 10) (*Ceratopogon*).—Alaska; coastal, s. Alaska to n. Calif.
obsoletus (Meigen), 1818: 76 (*Ceratopogon*).—Europe; Alaska to Que., s. to n. Calif., Okla., Tenn., and Va.
pusillus Lutz, 1913b: 52.—Brazil; Fla., Neotropical.
sanguisuga (Coquillett), 1901h: 604 (*Ceratopogon*).—Md.; Alaska to Que., s. to Calif., Kans., and Ga., forest area pest.
  nocivum Harris, 1862: 602 (*Simulium*). Nomen nudum.

Subgenus OECACTA Poey

Oecacta Poey, 1851: 238 (as genus). Type-species, *furens* Poey (mon.).
Haematomyidium Goeldi, 1905: 137. Type-species, *paraense* Goeldi (orig. des.).

alexanderi Wirth and Hubert, 1962: 190.—Mass.; Mich. to Que., s. to Tenn. and Conn.

arboricola Root and Hoffman, 1937: 166.—Md.; Wis. to Md., s. to Tex. and Fla.
arubae Fox and Hoffman, 1944: 109.—Aruba I.; Tex., Neotropical.
barbosai Wirth and Blanton, 1956: 161.—Panama; Fla., Ecuador, Bahama Is., Jamaica, salt marsh pest.
bickleyi Wirth and Hubert, 1962: 188.—Md.; Wis. to Que., s. to Fla.
biguttatus (Coquillett), 1901h: 604 (*Ceratopogon*).—D.C.; Wis. to Ont., s. to La. and Fla.
bottimeri Wirth, 1955: 356.—Tex.
cavaticus Wirth and Jones, 1956: 166.—Calif.; Oreg., Ariz.
daedalus Macfie, 1948: 83.—Mexico; Ariz., Neotropical.
debilipalpis Lutz, 1913b: 60.—Brazil; Md. to Ky. and Fla., Neotropical.
denticulatus Wirth and Hubert, 1962: 193.—Wis.; Minn. to Ont., s. to Md.
  unicolor, authors, not Coquillett.
dickei Jones, 1956: 28.—Wis.; Ont., W. Va.
downesi Wirth and Hubert, 1962: 186.—Ont.; Mich., N.Y., Que.
floridensis Beck, 1951: 135.—Fla.; Bermuda, Bahama Is.
flukei Jones, 1956: 30.—Wis.; N.Y., Va.
furens (Poey), 1851: 238 (*Oecacta*).—Cuba; Gulf and Atlantic Coasts to Mass., Neotropical, salt marsh pest.
  dovei Hall, 1932b: 88.—Ga.
furensoides Williams, 1955: 271.—Mich.
guttipennis (Coquillett), 1901h: 603 (*Ceratopogon*).—Ohio; Minn. to Mass., s. to Okla. and Ga.
jamnbacki Wirth and Hubert, 1962: 192.—N.Y.; Mich., Ont.
khalafi Beck, 1957: 104.—Fla.
luglani Jones and Wirth, 1958: 89.—Tex.; Ariz., Mexico, Panama.
melleus (Coquillett), 1901h: 604 (*Ceratopogon*).—Fla.; Maine to Fla. and Miss., coastal, sandy beach pest.
minutissimus (Zetterstedt), 1855: 4860 (*Ceratopogon*).—Scandinavia; Greenland, Europe.
mohave Wirth, 1952a: 187.—Calif.; Ariz., Mexico.
monoensis Wirth, 1952a: 193.—Calif.
mulrennani Beck, 1957: 103.—Fla.
nanus Root and Hoffman, 1937: 165.—Md.; Wis. to Ont., s. to Tex. and Fla.
niger Root and Hoffman, 1937: 168.—Md.; coastal, Mass. to Fla.
oklahomensis Khalaf, 1952b: 355 (as *villosipennis* ssp.).—Okla.; Ariz., Tex., Mexico.
ousairani Khalaf, 1952b: 354.—Okla.; Tex. and Okla. to Va. and Ala.
palmerae James, 1943a: 151.—Colo.; B.C. to Wyo., s. to Calif. and Nebr.

paraensis (Goeldi), 1905: 137 (*Haematomyidium*).—Brazil; Ohio to Pa., s. to Miss. and Fla., Neotropical.
pecosensis Wirth, 1955: 358.—Tex.
pifanoi Ortiz, 1951: 588.—Venezuela; Fla., Neotropical.
piliferus Root and Hoffman, 1937: 163.—Md.; Wis. to Que., s. to Va.
pseudopiliferus Wirth and Hubert, 1962: 189.—Md.; Wis. to Ont., s. to Fla.
reevesi Wirth, 1952a: 193.—Calif.; Ariz., N. Mex.
riggsi Khalaf, 1957: 198 (as *piliferus* ssp.).—Okla.
salihi Khalaf, 1952b: 351.—Okla.; Calif., N. Mex., Tex.
scanloni Wirth and Hubert, 1962: 187.—Va.; Mass. to Fla.
snowi Wirth and Jones, 1956: 163.—Va.; Ill. to Que., s. to Miss. and Ga.
spinosus Root and Hoffman, 1937: 172.—Md.; Wis. to Ont., s. to Tenn. and Fla.
stellifer (Coquillett), 1901h: 604 (*Ceratopogon*).—D.C.; Mont. to Ont., s. to Calif. and Fla.
stilobezzioides Foote and Pratt, 1954: 33.—N.Y.; Minn. to Que.
stonei James, 1943a: 149.—Colo.; Calif. to S. Dak., s. to Tex.
   weesei Khalaf, 1952a: 65 (also 1952b: 351).—Okla.
testudinalis Wirth and Hubert, 1962: 191.—Pa.; Wis. to Mass., s. to Va.
travisi Vargas, 1949: 233 (n. name for *simulans* Root and Hoffman).—Md.; Wis. to Ont., s. to La. and S.C., also Ariz.
   simulans Root and Hoffman, 1937: 167 (preocc. Vimmer, 1932).—Md.
   horneae Foote and Pratt, 1954: 25.—N.Y. **N. syn.**
unicolor (Coquillett), 1905d: 65 (*Ceratopogon*).—Calif.
usingeri Wirth, 1952a: 192 —Calif.; Utah.
utahensis Fox, 1946: 246.—Utah; Calif., Nev.
villosipennis Root and Hoffman, 1937: 165.—Md.; Wis. to Maine, s. to Okla. and Ala.
wirthi Foote and Pratt, 1954: 36.—Mont.; B.C.

Subgenus DRYMODESMYIA Vargas

Culicoides, subg. **Drymodesmyia** Vargas, 1960: 40. Type-species, *copiosus* Root and Hoffman (orig. des.).

   REFERENCE: Wirth and Hubert, 1960a (rev., larvae in cacti and tree holes).

arizonensis Wirth and Hubert, 1960a: 655.—Ariz.; Calif., Mexico (Baja Calif.).
butleri Wirth and Hubert, 1960a: 650.—Ariz.

cacticola Wirth and Hubert, 1960a: 653.—Calif.; Ariz., Tex.
copiosus Root and Hoffman, 1937: 171.—Mexico; Calif. to Tex.
hinmani Khalaf, 1952b: 353.—Okla.; Ohio to Md., s. to Tex. and Ala.
borinqueni, authors, not Fox and Hoffman.
jamaicensis Edwards, 1922c: 165 (as *loughnani* var.).—Jamaica; s. Tex., Neotropical.
jonesi Wirth and Hubert, 1960a: 650.—Tex.
loughnani Edwards, 1922c: 165.—Jamaica; Tex., Fla., West Indies.
ryckmani Wirth and Hubert, 1960a: 656.—Calif.; Ariz., Tex., Mexico (Baja Calif.).
sitiens Wirth and Hubert, 1960a: 652.—Calif.; Ariz., Mexico (Baja Calif.).

Subgenus DIPHAOMYIA Vargas

Culicoides, subg. **Diphaomyia** Vargas, 1960: 40. Type-species, *baueri* Hoffman (orig. des.).

baueri Hoffman, 1925: 297.—Md.; Wis. to Va., s. to Calif. and Fla.
blantoni Vargas and Wirth, 1955: 33.—Mexico; Tex.
footei Wirth and Jones, 1956: 162.—Va.; Ont., Tenn., Ky.
haematopotus Malloch, 1915g: 302.—Ill.; B.C. to Que., s. to Calif. and Fla.

Subgenus BELTRANMYIA Vargas

Culicoides, subg. **Beltranmyia** Vargas, 1953: 34. Type-species, *crepuscularis* Malloch (orig. des.).

alaskensis Wirth, 1951b: 84.—Alaska.
bermudensis Williams, 1956: 298.—Bermuda; Tex., Fla., Ga.
crepuscularis Malloch, 1915g: 303.—Ill.; B.C. to Que., s. to Calif., Mexico, and Fla., Bermuda.
hollensis (Melander and Brues), 1903: 13 (*Ceratopogon*).—Mass.; coastal, N.S. to Fla., salt marsh pest.
canithorax Hoffman, 1925: 284.—Ga.
knowltoni Beck, 1956: 136.—Fla.
mississippiensis Hoffman, 1926b: 158.—Miss.; La. to Fla., coastal, salt marsh pest.
sphagnumensis Williams, 1955: 269.—Mich.; Alaska, Wis., Ont.
wisconsinensis Jones, 1956: 32.—Wis.; Colo., Mich.

Subgenus MONOCULICOIDES Khalaf

Culicoides, subg. **Monoculicoides** Khalaf, 1954: 39. Type-species, *Ceratopogon nubeculosus* Meigen (orig. des.).

gigas Root and Hoffman, 1937: 172.—Sask.; B.C., Alta., Man.

variipennis variipennis (Coquillett), 1901h: 602 (*Ceratopogon*).—Va.; B.C. to Que., Colo., La., and Fla.
- ssp. **albertensis** Wirth and Jones, 1957: 17.—Alta.; Alta and Mont. to Okla.
- ssp. **australis** Wirth and Jones, 1957: 15.—La.; Kans. to Va., s. to Tex. and S.C.
- ssp. **occidentalis** Wirth and Jones, 1957: 21.—Calif.; B.C. to Calif. and Ariz., Mexico (Baja Calif.).
- ssp. **sonorensis** Wirth and Jones, 1957: 18.—Ariz.; Wash. to Utah, Okla., and Tenn., s. to Calif. and Tex., Mexico.

Subgenus SELFIA Khalaf

Culicoides, subg. **Selfia** Khalaf, 1954: 38. Type-species, *hieroglyphicus* Malloch (orig. des.).

brookmani Wirth, 1952a: 179.—Calif.; Ariz., N. Mex.
denningi Foote and Pratt, 1954: 20.—Sask.; Wash. to Sask., s. to Nev. and Colo.
hieroglyphicus Malloch, 1915g: 297.—Ariz.; Calif. to S. Dak., s. to Mexico.
jamesi Fox, 1946: 244.—Mont.; Wash. to Mont., s. to Calif. and N. Mex.
multipunctatus Malloch, 1915g: 296.—Ill.; Kans., Ill. and Tenn., s. to Tex. and Ala.
tenuistylus Wirth, 1952a: 178.—Calif.; Ariz., N. Mex.

Tribe CERATOPOGONINI

Genus CERATOPOGON Meigen

Helea Meigen, 1800: 18. Type-species, *Ceratopogon communis* Meigen (Coquillett, 1910b: 549). Suppressed by I.C.Z.N., 1963b: 339.

Ceratopogon Meigen, 1803: 261. Type-species, *Tipula barbicornis* Fabricius (mon.; misident.) ?=*communis* Meigen. Edwards (1920: 127) states that Meigen misidentified *barbicornis*, which is an *Orthocladius*, his 1803 material being *Ceratopogon communis*.

Subgenus CERATOPOGON Meigen

borealis (Kieffer), 1919b: 68 (*Psilohelea*).—Norway (Lofoten Is.); Greenland.
culicoidithorax Hoffman, 1926b: 156.—N.Y.; Ont., Que.
lacteipennis Zetterstedt, 1838: 820.—Norway; Greenland.
mallochi (Cole) *in* Cole and Lovett, 1921: 213 (*Hartomyia*).—Oreg.; Calif.

CERATOPOGONIDAE 133

Subgenus NILOHELEA Kieffer

Nilohelea Kieffer, 1921b: 22 (as genus). Type-species, *albipennis* Kieffer (mon.).

longipennis (Wirth), 1952a: 201 (*Helea*).—Calif.
virginianus (Wirth), 1951f: 318 (*Helea*).—Va.; N.J., Md.

Subgenus ISOHELEA Kieffer

Isohelea Kieffer, 1917: 295 (as genus). Type-species, *Ceratopogon lacteipennis* Winnertz (orig. des.)=*sociabilis* Goetghebuer.

pruinosus (Wirth), 1952a: 201 (*Helea*).—Calif.; N. Mex., Colo.
serratus (Lewis), 1956: 46 (*Helea*).—Conn.
stigmalis Coquillett, 1902g: 86.—N. Mex.; Alaska, Calif.

Subgenus BRACHYPOGON Kieffer

Brachypogon Kieffer, 1899: 69 (as genus). Type-species, *Ceratopogon vitiosus* Winnertz (orig. des.).

Unnamed sp.—Fla.

Genus CAMPTOPTEROHELEA Wirth and Hubert

Camptopterohelea Wirth and Hubert, 1960b: 89. Type-species, *hoogstraali* Wirth and Hubert (orig. des.).

Unnamed sp.—s. Calif.

Genus ALLUAUDOMYIA Kieffer

Alluaudomyia Kieffer, 1913d: 12. Type-species, *imparunguis* Kieffer (mon.).
Neoceratopogon Malloch, 1915h: 310. Type-species, *Ceratopogon bellus* Coquillett (orig. des.).
Isoecacta Garrett, 1925b: 9. Type-species, *poeyi* Garrett (orig. des.)= *bella* (Coquillett).

REFERENCES: Wirth, 1952b (rev.); Williams, 1953 (biol., Ga.).

bella (Coquillett), 1902g: 87 (*Ceratopogon*).—D.C.; Alaska to Que., s. to Calif. and Fla.
  poeyi Garrett, 1925b: 9 (*Isoecacta*).—B.C.
  splendida, authors, not Winnertz.
footei Wirth, 1952b: 428.—Fla.; Vt. to Tenn. and Fla.
megaparamera Williams, 1957: 327.—Mich.; N.Y., Md., Va.
needhami Thomsen, 1935: 287.—N.Y.; Sask. to Conn., s. to La. and Fla.

paraspina Wirth, 1952b: 429.—Ga.; coastal, Que. to Fla.
parva Wirth, 1952b: 431.—Fla.; N.Y. and Ont., s. to Fla.
   downesi Wirth, 1952b: 433.—Va.
stictipennis Wirth, 1952a: 197.—Calif.
wirthi Williams, 1957: 328.—Mich.; Sask., Va., Ga.

Tribe STILOBEZZIINI

Genus STILOBEZZIA Kieffer

Stilobezzia Kieffer, 1911a: 118. Type-species, *festiva* Kieffer (orig. des.).
Hartomyia Malloch, 1915g: 339. Type-species, *Ceratopogon pictus* Coquillett (orig. des.)=*coquilletti* Kieffer.

REFERENCE: Wirth, 1953a (rev.).

Subgenus STILOBEZZIA Kieffer

antennalis (Coquillett), 1901h: 606 (*Ceratopogon*).—D.C.; B.C. to Ont., s. to Tex. and Fla.
beckae Wirth, 1953a: 69.—Fla.; Miss., Panama, Peru.
bicolor Lane, 1947: 208.—Brazil; Tex., Neotropical.
bulla Thomsen, 1935: 289.—N.Y.; Ont. and Que., s. to La. and Fla.
coquilletti Kieffer, 1917: 308 (n. name for *pictus* Coquillett).—Va.; Ill. to Md., s. to La. and Ala., Neotropical.
   pictus Coquillett, 1905d: 60 (*Ceratopogon*; preocc. Meigen, 1818).—Va.
diversa (Coquillett), 1901h: 607 (*Ceratopogon*).—N.J.; Va., Ga., Fla.
glauca Macfie, 1939: 204.—Brazil; Va., La., Miss., Fla., Neotropical.
pallidiventris (Malloch), 1915g: 344 (*Hartomyia*).—Ill.; Ind., W. Va., Conn.
pruinosa Wirth, 1952a: 203.—Calif.; Ariz., Va., Fla.
punctipes Wirth, 1953a: 79.—Fla.; Mexico.
rabelloi Lane, 1947: 203.—Brazil; La., Ga., Fla., Neotropical.
sybleae Wirth, 1953a: 82.—Va.; Mich. to Va., Tenn. and Fla., also Calif.
thomsenae Wirth, 1953a: 83.—Fla.
viridis (Coquillett), 1901h: 607 (*Ceratopogon*)—N.J.; N.J. to Fla., also Tex.

Subgenus NEOSTILOBEZZIA Goetghebuer

Stilobezzia, subg. **Neostilobezzia** Goetghebuer *in* Goetghebuer and Lenz, 1934: 53. Type-species, *Ceratopogon ochraceus* Winnertz (Wirth, 1953a: 63).

fuscula Wirth, 1952a: 204.—Calif.; Utah.

lutea (Malloch), 1918c: 18 (*Hartomyia*).—Ill.; Mich. to Ont., s. to Iowa, Tenn., and Fla.
  mallochi Hoffman, 1924: 283.—N.Y., Pa.
stonei Wirth, 1953a: 66.—Va.; N.Y. and Mass. to Tenn. and Fla.

Subgenus EUKRAIOHELEA Ingram and Macfie

Eukraiohelea Ingram and Macfie, 1921: 347 (as genus). Type-species, *africana* Ingram and Macfie (Macfie, 1940a: 22).
elegantula (Johannsen), 1907a: 109 (*Bezzia*).—Kans.; La., Fla., Jamaica.

Genus ECHINOHELEA Macfie

Echinohelea Macfie, 1940b: 187. Type-species, *ornatipennis* Macfie (orig. des.).
lanei Wirth, 1951f: 319.—Va.; Mich. to Mass., s. to Miss. and Fla., Panama, Trinidad.

Genus MONOHELEA Kieffer

Monohelea Kieffer, 1917: 294. Type-species, *hieroglyphica* Kieffer (orig. des.).
  REFERENCE: Wirth, 1953b (rev.).

Subgenus MONOHELEA Kieffer

johannseni Wirth, 1953b: 153.—Va.; Mich. to Md., s. to Iowa and Fla., Panama.
  tessellata, authors, not Zetterstedt.
lanei Wirth, 1953b: 142.—Fla.; Bahama Is.
macfiei Wirth, 1953b: 143.—La.; Mich. to Mass., s. to La. and Va.
maculipennis (Coquillett), 1905d: 64 (*Ceratopogon*).—Fla.; La., Bahama Is., Mexico, and Panama.
nebulosa (Coquillett), 1901h: 606 (*Ceratopogon*). N.J.; Kans. to Mass., s. to Tex. and Fla.
ornata Wirth, 1953b: 144.—Fla.; Panama.
stonei Wirth, 1953b: 148.—La.; Ont., Mass. to La. and Fla., Panama.
texana Wirth, 1953b: 143.—Tex.; Calif., Ariz., N. Mex.

Subgenus SCHIZOHELEA Kieffer

Schizohelea Kieffer, 1917: 295 (as genus). Type-species, *Ceratopogon copiosus* Winnertz (mon.)=*leucopeza* (Meigen).
leucopeza (Meigen), 1804: 29 (*Ceratopogon*).—Europe; B.C. to Que., s. to Calif., Ill., and Mass.
  politus Coquillett, 1901h: 606 (*Ceratopogon*).—Mass.

Genus SERROMYIA Meigen

Serromyia Meigen, 1818: 83. Type-species, *Ceratopogon femoratus* Meigen (mon.). Generic name cited in specific synonymy; needs action by I.C.Z.N.
Ceratolophus Kieffer, 1899: 69 (preocc. Barboza de Bocage, 1873). Type-species, *Ceratopogon femoratus* Meigen (mon.).
Johannseniella Williston, 1907: 1 (n. name for *Ceratolophus* Kieffer). Type-species, *Ceratopogon femoratus* Meigen (aut.).

barberi Wirth, 1952a: 205.—Calif.
crassifemorata Malloch, 1914g: 218.—Ill.; Ill. to Mich. and Md., s. to Tenn., and Va.
femorata (Meigen), 1804: 28 (*Ceratopogon*).—Europe; Alaska to Que., s. to Colo. and Va.

Genus PARABEZZIA Malloch

Parabezzia Malloch, 1915g: 358. Type-species, *petiolata* Malloch (orig. des.).

REFERENCE: Wirth, 1952c (rev.).

inermis (Coquillett), 1902g: 86 (*Ceratopogon*).—Ariz.; N. Mex.
petiolata Malloch, 1915g: 359.—Ill.; S. Dak. to Ont., s. to Ark. and Fla.
uncinata (Johannsen), 1943a: 761 (*Stilobezzia*).—Ala.; Ark., Md., Va., Jamaica.

Tribe HETEROMYIINI

Genus CLINOHELEA Kieffer

Clinohelea Kieffer, 1917: 295. Type-species, *Ceratopogon variegatus* Winnertz (orig. des.).

bimaculata (Loew), 1861b: 311 (Cent. 1, no. 6) (*Ceratopogon*).—D.C.; Mich. to Que., s. to Tex. and Fla.
curriei (Coquillett), 1905d: 62 (*Ceratopogon*).—B.C.; Alaska to Mich. N.Y., and Que.
dimidiata (Adams), 1903a: 27 (*Ceratopogon*).—Ariz.; N. Mex., Utah.
nebulosa (Malloch), 1915g: 322 (*Palpomyia*).—Mich.; Ont., Mich. to Que., s. to W. Va. and Va.
nubifera (Coquillett), 1905d: 61 (*Ceratopogon*).—Fla.
usingeri Wirth, 1952a: 209.—Calif.; Ariz.

Genus HETEROMYIA Say

Heteromyia Say, 1825: pl. 35 (1859a: 79). Type-species, *fasciata* Say (mon.).

Pachyleptus Walker, 1856b: 426. Type-species, *fasciatus* Walker (mon.).

fasciata Say, 1825: pl. 35 (1859a: 80).—N. Amer.; Mich. to Ont., s. to Ala.
  festiva Loew, 1861b: 314 (Cent. 1, no. 13) *(Ceratopogon).*—Pa.
  prattii Coquillett, 1902g: 88.—Va.

Genus NEUROHELEA Kieffer

Neurohelea Kieffer, 1925a: 112. Type-species, *Ceratopogon luteitarsis* Meigen (mon.).

macroneura (Malloch), 1915g: 337 *(Johannsenomyia).*—Kans.; Ill., Tex.
nigra Wirth, 1952a: 208.—Calif.; B.C.

Tribe SPHAEROMIINI

REFERENCE: Wirth, 1962b (rev.).

Genus JOHANNSENOMYIA Malloch

Johannsenomyia Malloch, 1915g: 332. Type-species, *halteralis* Malloch (Wirth, 1952a: 211).
Dicrohelea Kieffer, 1917: 363. Type-species, *Palpomyia filicornis* Kieffer (Macfie, 1940a: 26).

annulicornis Malloch, 1918m: 230.—Ill.; Mich., Ont., Va.
argentata (Loew), 1861b: 310 (Cent. 1, no. 5) *(Ceratopogon).*—D.C.; N. Dak. to Que., s. to Tex. and Fla.
halteralis Malloch, 1915g: 338.—Ill.

Genus JENKINSHELEA Macfie

Jenkinsia Kieffer, 1913e: 161 (preocc. Jordan and Evermann, 1896). Type-species, *setosipennis* Kieffer (orig. des.).
Jenkinshelea Macfie, 1934: 177 (n. name for *Jenkinsia* Kieffer). Type-species, *Jenkinsia setosipennis* Kieffer (aut.).

REFERENCE: Wirth, 1962a (rev.).

albaria (Coquillett), 1895a: 308 *(Ceratopogon).*—Fla.; Ill. to N.Y., s. to Tex. and Fla.
  aequalis Malloch, 1915g: 336 *(Johannsenomyia).*—Ill.
magnipennis (Johannsen), 1908: 268 *(Johannseniella).*—N.Y.; Minn. to Que., s. to Ohio and N.Y.

Genus PROBEZZIA Kieffer

Bezzia, subg. **Probezzia** Kieffer, 1906a: 57. Type-species, *Ceratopogon venustus* Meigen (Coquillett, 1910b: 594).
Dicrobezzia Kieffer, 1919b: 127. Type-species, *Ceratopogon venustus* (orig. des.).

REFERENCE: Wirth, 1951d (rev.).

albiventris (Loew), 1861b: 311 (Cent. 1, no. 7) (*Ceratopogon*).—Ga.; Wis. to N.B., s. to Tex. and Ga.
atriventris Wirth, 1951d: 31.—Mich.
concinna (Meigen), 1818: 77 (*Ceratopogon*).—Europe; Calif., Colo., Ont., Que.
flavonigra (Coquillett), 1905d: 60 (*Ceratopogon*).—B.C.; Alta.
infuscata Malloch, 1915h: 316.—Ill.; Minn. to Ont.
ludoviciana Wirth, 1951d: 28.—La.
pallida Malloch, 1914b: 138.—Ill.; Wis. to Ont., s. to Kans. and Va., also Ariz.
rosewalli Wirth, 1951d: 32.—La.
sabroskyi Wirth, 1951d: 31.—Mich.; Wis. to N.B., s. to Ill. and Pa.
smithii (Coquillett), 1901h: 600 (*Ceratopogon*).—N.J.; Kans. to Que., s. to La. and Va.
  mundus Coquillett *in* Johnson, 1900c: 628 (*Ceratopogon*). Nomen nudum.
unica (Johannsen), 1934a: 345 (*Lasiobezzia*).—N.Y.
xanthogaster (Kieffer), 1917: 329 (*Bezzia;* n. name for *elegans* Coquillett).—N.J.; Wis. to Ont., s. to Ill. and Va.
  elegans Coquillett, 1901h: 599 (*Ceratopogon;* preocc. Winnertz, 1852).—N.J.

Genus SPHAEROMIAS Curtis

Sphaeromias Curtis, 1829: pl. 285. Type-species, *albomarginatus* Curtis (orig. des.)=*fasciatus* (Meigen).

longipennis (Loew), 1861b: 313 (Cent. 1, no. 10) (*Ceratopogon*).—Pa.; Minn. to Ont., s. to Tex. and Ga., also Calif.

Genus NILOBEZZIA Kieffer

Nilobezzia Kieffer, 1921b: 24. Type-species, *armata* Kieffer (mon.).

brevicornis (Wirth), 1952a: 215 (*Sphaeromias*).—Calif.; Calif. to Colo., Kans., and Tex.
mallochi Wirth, 1962b: 284.—Mich.; Sask. to Ont., s. to Okla. and Md.
minor (Wirth), 1952a: 216 (*Sphaeromias*).—Calif.; N.W.T. to Que., s. to Calif. and Tex.

schwarzii (Coquillett), 1901h: 605 *(Ceratopogon)*.—Tex.; La., Fla., S.C., s. through West Indies and Panama to Brazil.
setipes Coquillett, 1905d: 59 *(Ceratopogon)*.—Tex.

Genus MALLOCHOHELEA Wirth

Mallochohelea Wirth, 1962b: 278. Type-species, *Johannsenomyia albibasis* Malloch (orig. des.).

REFERENCE: Wirth, 1962b (rev.).

albibasis (Malloch), 1915h: 315 *(Johannsenomyia)*.—Ill.; N.W.T. to Que., s. to Calif., La., and Ala.
albihalter Wirth, 1962b: 280.—Mich.; Wis. to Que., s. to La. and Md.
atripes Wirth, 1962b: 281.—N.J.; Mich. to Ont., s. to Fla. on e. coast.
caudellii (Coquillett), 1905d: 63 *(Ceratopogon)*.—B.C.; Calif., Ariz., Tex.
flavidula (Malloch), 1914g: 230 *(Johannseniella)*.—Ill.; Wis., La.
pullata (Wirth), 1952a: 213 *(Johannsenomyia)*.—Calif.; Ariz., Tex., Mexico, Panama.
smithi (Lewis), 1956: 47 *(Johannsenomyia)*.—Conn.; Wis. to Que., s. to Tenn. and N.C.
spinipes Wirth, 1962b: 282.—Ga.
sybleae (Wirth), 1952a: 212 *(Johannsenomyia)*.—Calif.; Oreg.
texensis Wirth, 1962b: 283.—Tex.
variegata Wirth, 1962b: 283.—Mich.; Ont.

Tribe STENOXENINI

Genus STENOXENUS Coquillett

Stenoxenus Coquillett, 1899c: 61. Type-species, *johnsoni* Coquillett (mon.).

johnsoni Coquillett, 1899c: 61.—N.J.; Mo., Tex., Mexico, Panama.

Genus PARYPHOCONUS Enderlein

Paryphoconus Enderlein, 1912d: 57. Type-species, *angustipennis* Enderlein (orig. des.).

Unnamed sp.—Tex.

Tribe PALPOMYIINI

Genus PACHYHELEA Wirth

Pachyhelea Wirth, 1959: 50. Type-species, *Ceratopogon magnus* Coquillett (orig. des.) = *pachymera* (Williston).

pachymera (Williston), 1900: 224 *(Ceratopogon).*—Mexico; Tex. to Argentina.
> *magnus* Coquillett, 1905d: 61 *(Ceratopogon).*—Tex.

Genus PALPOMYIA Meigen

Palpomyia Meigen, 1818: 82. Type-species, *Ceratopogon flavipes* Meigen (mon.). Generic name cited in specific synonymy; needs action by I.C.Z.N.

aldrichi (Malloch), 1915g: 326 *(Heteromyia).*—Idaho; Wash. and Idaho to Calif. and Nev.
armatipes Wirth, 1952a: 222.—Calif.; Utah.
cressoni (Malloch), 1915g: 327 *(Heteromyia).*—Pa.
essigi Wirth, 1952a: 225.—Calif.
flavipes (Meigen), 1804: 28 *(Ceratopogon).*—Europe; B.C. to N.H., s. to Calif., Tex., and Va.
> *basalis* Walker, 1848: 27 *(Ceratopogon).*—Ont.
> *flaviceps* Johannsen, 1908: 268 *(Johannseniella).*—N.Y.

hirta (Malloch), 1915g: 330 *(Heteromyia).*—Ill.; Mich.
illinoisensis Malloch, 1914g: 219.—Ill.; N.Y.
kernensis Wirth, 1952a: 223.—Calif.
lineata (Meigen), 1818: 80 *(Ceratopogon).*—Europe; N.J.
linsleyi Wirth, 1952a: 224.—Calif.; Ariz.
nigripes (Meigen), 1830: 265 *(Ceratopogon).*—Europe; Calif.
opacithorax (Malloch), 1915g: 329 *(Heteromyia).*—Ill.; N.Y.
plebeja (Loew), 1861b: 313 (Cent. 1, no. 11) *(Ceratopogon).*—Pa.; Mich. to N.H., s. to Va.
praeusta (Loew), 1869d: 50 *(Ceratopogon).*—Germany; Calif., N. Mex., Mich., Ga.
pruinescens Thomsen, 1935: 290.—N.Y.
rufa (Loew), 1861b: 314 (Cent. 1, no. 12) *(Ceratopogon).*—Pa.; Calif., Mont. to N.Y., s. to La. and Ga.
slossonae (Coquillett), 1905d: 61 *(Ceratopogon).*—N.H.; Mont. to Ont., s. to S. Dak. and Fla.
stonei Wirth, 1951f: 322.—Va.; Ont.
subasper (Coquillett), 1901h: 606 *(Ceratopogon).*—Mexico; Colo. to Ont., s. to Tex. and Fla., Mexico.
tenuicornis (Malloch), 1915g: 328 *(Heteromyia).*—Wis.
tibialis (Meigen), 1818: 82 *(Ceratopogon).*—Europe; Mich. to Ont., s. to La. and Fla.
trifasciata Wirth, 1952a: 222.—Calif.
trivialis (Loew), 1861b: 309 (Cent. 1, no. 4) *(Ceratopogon).*—D.C.; Iowa, Mich., Va.

Genus BEZZIA Kieffer

Bezzia Kieffer, 1899: 69. Type-species, *Ceratopogon ornatus* Meigen (orig. des.).
Lasiobezzia Kieffer, 1925c: 54. Type-species, *Bezzia pilipennis* Lundström (orig. des.).

Subgenus BEZZIA Kieffer

albidorsata Malloch, 1915g: 349.—Ill.; Va.
apicata Malloch, 1914h: 284.—Ill.
atlantica Wirth and Williams, 1957: 13.—Bermuda.
barberi (Coquillett), 1901h: 601 (*Ceratopogon*).—Md.; N.Y.
biannulata Wirth, 1952a: 237.—Calif.
bivittata (Coquillett), 1905d: 60 (*Ceratopogon*).—Calif.; Utah, Mont.
cockerelli Malloch, 1915g: 346.—Colo.; Alaska, Wash., Nev.
coloradensis Wirth, 1952a: 238.—Calif.
copiosa (Thomsen), 1935: 292 (*Probezzia*).—N.Y.; Mich.
dentata Malloch, 1914h: 284.—Ill.; Nebr., Va.
gibbera (Coquillett), 1905d: 60 (*Ceratopogon*).—Cuba; Fla.
glabra (Coquillett), 1902g: 85 (*Ceratopogon*).—Fla.; Calif., Mont. to Ont., s. to La. and Fla.
granulosa Wirth, 1952a: 240.—Calif.
incerta (Malloch), 1915g: 358 (*Probezzia*).—Ill.; Conn., Va.
media (Coquillett), 1904f: 166 (*Ceratopogon*).—N.J.; Va.
modocensis Wirth, 1952a: 233.—Calif.
obscura (Malloch), 1915g: 355 (*Probezzia*).—N.Y.
opaca (Loew), 1861b: 312 (Cent. 1, no. 9) (*Ceratopogon*).—D.C.; Wash. to Ont., s. to Calif. and Fla.
  fulvithorax Malloch, 1915g: 354 (*Probezzia*).—Ill.
pruinosa (Coquillett), 1905d: 59 (*Ceratopogon*).—B.C.
pseudobscura Wirth, 1951f: 324.—Va.
pulverea (Coquillett), 1901h: 600 (*Ceratopogon*).—N.J.; Calif. to Fla., n. to N.J.
punctipennis (Williston), 1896c: 278 (*Ceratopogon*).—St. Vincent I.; Calif., Nev., Fla., Neotropical.
setulosa (Loew), 1861b: 312 (Cent. 1, no. 8) (*Ceratopogon*).—D.C.; Calif., Sask. to Ont., s. to Tex. and Fla.
sordida Wirth, 1952a: 232.—Calif.
varicolor (Coquillett), 1902c: 84 (*Ceratopogon*).—N.Y.; Alaska to Mass., s. to Calif. and Md.

Subgenus PSEUDOBEZZIA Malloch

Pseudobezzia Malloch, 1915g: 351 (as genus). Type-species, *Ceratopogon expolitus* Coquillett (orig. des.).

Allobezzia Kieffer, 1917: 296. Type-species, *Ceratopogon expolitus* Coquillett (orig. des.).
   REFERENCE: Wirth, 1951f (key).
bilineata Wirth, 1952a: 230.—Calif.
expolita (Coquillett), 1901h: 600 (*Ceratopogon*).—N.J.; Mont. to N.J., s. to La. and Va.
   johnsoni Coquillett, 1901h: 600 (*Ceratopogon*).—N.J.
flavitarsis Malloch, 1914h: 283.—Ill.; Calif., Mich. to Conn., s. to Ill. and Ga.
mallochi Wirth, 1951f: 323.—Va.; Mass.

Unplaced Species of Ceratopogonidae

scutellatus Say, 1829: 150 (1859b: 349) (*Ceratopogon*).—Ind.

Family CHIRONOMIDAE
(Tendipedidae)
By James E. Sublette and Mary Smith Sublette

Chironomid midges, with the exception of some terrestrial Orthocladiinae (Metriocnemini, sensu Brundin, 1956), are aquatic. Most species inhabit fresh water, with some Orthocladiinae and a few Tanytarsini being littoral marine. Of the fresh-water forms, perhaps no other aquatic group of insects is more ubiquitous, with species inhabiting practically all ecological niches from mountain torrents to the anaerobic ooze of deep eutrophic lakes. In general, the Tanypodinae are predators, and move freely over the substratum of all types of water bodies. Most other fresh-water forms are tubicolous, as are the marine Orthocladiinae, and feed either on plankton or detritus. Berg (1950b) reviewed some species feeding as leafminers on *Potamogeton* and Brock (1960) recorded some *Cricotopus* symbiotic in the alga *Nostoc*. Diamesine and orthocladiine midges prefer cooler water and are more generally distributed at higher altitudes or latitudes. The Diamesinae are most characteristic of swiftly flowing, clear streams, while the orthocladiines, except for the terrestrial or subaquatic forms, occur in both lakes and streams. Chironominae in general prefer standing or slowly flowing water although some species do occur in fast streams.

A considerable difference of opinion exists between most European entomologists and those of North America as to generic limits within the family, mainly because the German school has emphasized the immature stages in its research. Many European workers have based many genera and subgenera on characters of the immature stages, virtually ignoring an analysis of the corresponding characters of the adults. Adult characters, on the other hand, have been employed extensively by the British and American workers to establish generic limits.

Although Brundin (1956) has shown in his outstanding work on the Orthocladiinae that the two systems can be reconciled, the resulting genera and subgenera are markedly narrowed. This treatment of the Nearctic fauna would cause the genera and subgenera to be so restricted that the types of most Nearctic species would require reexamination. Consequently, the nomenclature used in this catalog is based on the broader generic limits resulting from a study of adult characters alone. It is essentially that presented by Freeman (1955–1958) in his monographic series on African midges, which in turn leans heavily on the earlier monograph of British chironomids by Edwards (1929b).

The principal differences between the American nomenclature of Townes (1945, 1952) and Johannsen (1952a) and that by Freeman (1955–1958), used in this catalog, are outlined as follows:

TOWNES (1945, 1952) JOHANNSEN (1952a)	FREEMAN (1955–1958)
Syndiamesa and *Diamesa*	*Diamesa*
Hydrobaenus (*Diplocladius*)	*Diplocladius*
Hydrobaenus (*Psectrocladius*)	*Psectrocladius*
Hydrobaenus (*Hydrobaenus*)	*Orthocladius* and *Chaetocladius*
Hydrobaenus (*Trichocladius*)	*Trichocladius*
Hydrobaenus (*Eukiefferiella*)	*Nanocladius*
Hydrobaenus (*Smittia*)	*Smittia* (in part)
Hydrobaenus (*Limnophyes*)	*Limnophyes* (in part)
Tendipes (*Tendipes*) *Tendipes* (*Einfeldia*)	*Chironomus* (*Chironomus*)
Tendipes (*Limnochironomus*)	*Chironomus* (*Dicrotendipes*)
Tendipes (*Kiefferulus*)	*Chironomus* (*Kiefferulus*)
Xenochironomus	*Chironomus* (*Xenochironomus*)
Harnischia (*Harnischia*) *Harnischia* (*Cladopelma*) *Cryptochironomus*	*Chironomus* (*Cryptochironomus*)
Kribioxenus	*Nilothauma*
Polypedilum (*Polypedilum*) *Polypedilum* (*Tripodura*)	*Polypedilum* (*Polypedilum*)
Calopsectra (*Micropsectra*)	*Micropsectra*
Calopsectra (*Calopsectra*)	*Tanytarsus* (*Calopsectra*) *Tanytarus* (*Tanytarsus*) *Tanytarsus* (*Rheotanytarsus*) *Tanytarsus* (*Cladotanytarsus*)

Calopsectra (*Stempellina*) .... *Stempellina*
Glyptotendipes (*Demeijerea*) .... } *Chironomus* (*Endochironomus*)
Tanytarsus (*Endochironomus*) ....
Tanytarus (*Tribelos*) .... *Chironomus* (*Tribelos*)
Tanytarsus (*Tanytarsus*) .... *Phaenopsectra*
Tanytarus (*Stictochironomus*) .... *Stictochironomus*

In addition to this large- versus small-genus controversy, further nomenclatural contention exists among taxonomists. Foremost among these arguments is the matter of the Meigen names, discussed in the introduction to this volume. Another is the radical alteration in the use of *Tanytarsus* brought about by Townes' (1945) application of Coquillett's (1910b) inappropriate type designation for the genus. The I.C.Z.N. has set aside all previous type designations and designated *Chironomus signatus* Wulp, 1858, as type (I.C.Z.N., 1961b).

We are using *Tanytarsus* in the sense of *signatus* as type-species, while those species treated by Townes (1945) as *Tanytarsus* are placed in *Phaenopsectra*. Some confusion also exists concerning the type-species of the genera *Tanypus*, *Orthocladius*, and *Metriocnemus*. Pending application to, and ruling by, the I.C.Z.N., in the interest of nomenclatural stability, we also follow Edwards (1929b) in his interpretation of these genera.

We are indebted to S. S. Roback for furnishing two new names in this catalog to replace junior homonyms in *Anatopynia* and *Pentaneura*.

REFERENCES: Johannsen, 1903b, 1905 (N.Y., biol., immature stages); Malloch, 1915g (rev. Ill.); Edwards, 1929b (Britain, classif.); Goetghebuer and Lenz, 1936, 1937–1938, 1939a, b, c, 1940–1944, 1950a, 1950b, 1954–1960 (rev. Palaearctic, classif.); Andersen, 1937, 1946 (Greenland, biol.); Johannsen, 1937a, b (biol., immature stages); Miller, 1941 (biol., ecology, Ont.); Johannsen, 1952a (rev., ne. U.S.); Townes, 1952 (rev. ne. U.S.); Thienemann, 1954 (comprehensive review of biol.); Freeman, 1955–1958 (Ethiopian Region, classif.); Roback, 1957a (immature stages Philadelphia area); Dendy and Sublette, 1959 (Ala.); Bryce, 1960 (larval anat. and generic keys).

Subfamily TANYPODINAE

REFERENCES: Zavřel and Thienemann, 1921 (immature stages); Morrissey, 1950 (Iowa); Fittkau, 1962 (rev.).

Genus ANATOPYNIA Johannsen

Anatopynia Johannsen, 1905: 135. Type-species, *Tanypus plumipes* Fries (orig. des.).
Protanypus, authors, not Kieffer.

CHIRONOMIDAE

Subgenus ANATOPYNIA Johannsen

centralis Malloch, 1934b: 13.—N.W.T.
decolorata (Malloch), 1915g: 370 (*Tanypus*).—Ill.; Wis., Mich., Ohio.
florens (Johannsen), 1908: 272 (*Tanypus*).—Colo., Wash., N.Y.
marginella (Malloch), 1915g: 374 (*Tanypus*).—Ill.; Calif.

Subgenus MACROPELOPIA Thienemann

Macropelopia Thienemann, 1916: 497 (as genus). Type-species, *Pelopia bimaculata* Kieffer (orig. des.)=*nebulosus* (Meigen).

algens (Coquillett), 1902g: 90 (*Tanypus*).—Alaska; Alta., Calif.
decedens (Walker), 1848: 22 (*Tanypus*).—Hudson Bay Territory.
hirtipennis (Loew), 1866b: 5 (Cent. 7, no. 6) (*Tanypus*).—Maine; Man. to Que. and Maine, s. to Colo. and Ill.
miripes (Coquillett), 1905d: 65 (*Tanypus*).—Calif. **N. comb.,** W. W. Wirth *in litt.*
pictipennis (Zetterstedt), 1838: 818 (*Tanypus*).—Lapland; Greenland.

Subgenus PSECTROTANYPUS Kieffer

Psectrotanypus Kieffer, 1909: 42 (as genus). Type-species, *Tipula varia* Fabricius (Fittkau, 1962: 129).
Apsectrotanypus Fittkau, 1962: 141. Type-species, *Tanypus trifascipennis* Zetterstedt (orig. des.). **N. syn.**

dena Roback (for *brunnea* Roback).—Mass.; Pa. **N. name.**
    brunnea Roback, 1955: 2 (preocc. Edwards, 1931).—Mass.
discolor (Coquillett), 1902g: 89 (*Tanypus*).—N.H.
dyari (Coquillett), 1902c: 85 (*Tanypus*).—Mich., N.Y., Mass., D.C.; Man. to Mass., s. to Mo. and Va., also Wash., Idaho.
guttularis (Coquillett), 1902g: 92 (*Tanypus*).—Wash.; n. and cent. Calif.
johnsoni (Coquillett), 1901h: 609 (*Tanypus*).—N.J.; Ill., N.Y., Fla.
venusta (Coquillett), 1902g: 91 (*Tanypus*).—N. Mex.; Calif., Ill.

Unplaced Species of *Anatopynia*

alaskensis (Malloch), 1919f: 35 (*Tanypus*).—Alaska. **N. comb.**
trifascipennis (Zetterstedt).—Not Nearctic.

Genus CLINOTANYPUS Kieffer

Clinotanypus Kieffer, 1913e: 157. Type-species, *Procladius fuscosignatus* Kieffer (present des.).

caliginosus (Johannsen), 1905: 131 (*Procladius*).—N.Y.; Ohio.

flavicinctus (Loew), 1861b: 309 (Cent. 1, no. 2) (*Tanypus*).—Pa.; Ont.
pinguis (Loew), 1861b: 308 (Cent. 1, no. 1) (*Tanypus*).—N.Y.
thoracicus (Loew), 1866b: 4 (Cent. 7, no. 3) (*Tanypus*).—D.C.; Iowa, Mo., Ill., N.J., La., Fla.

Genus COELOTANYPUS Kieffer

Coelotanypus Kieffer, 1913e: 154. Type-species, *Procladius humeralis* Loew (orig. des.).
concinnus (Coquillett), 1895a: 308 (*Tanypus*).—Fla.; Iowa to Ohio, s. to Tex. and Fla.
    flavus Kieffer, 1923a: 296.—Tex. **N. syn.**
scapularis (Loew), 1866b: 2 (Cent. 7, no. 1) (*Tanypus*).—D.C.; Iowa to N.J., s. to Tex. and Fla.
tricolor (Loew), 1861b: 309 (Cent. 1, no. 3) (*Tanypus*).—N.Y.; Ill., Tex., Okla., Fla.

Genus NATARSIA Fittkau

Natarsia Fittkau, 1962: 151. Type-species, *Chironomus punctatus* Fabricius (orig. des.).
fastuosa (Johannsen), 1905: 153 (*Ablabesmyia*).—N.Y.; Wash., Colo., Mich., Pa., N.J., Fla.

Genus PENTANEURA Philippi

Pentaneura Philippi, 1865: 629. Type-species, *grisea* Philippi (mon.) =*cinerea* (Philippi).
Thienemannimyia Fittkau, 1957: 315. Type-species, *Ablabesmyia geijskesi* Goetghebuer (orig. des.). **N. syn.**
Thienemannimyia, subg. *Conchapelopia* Fittkau, 1957: 317. Type-species, *Tanypus pallidulus* Meigen (orig. des.).
Arctopelopia Fittkau, 1962: 194. Type-species, *Tanypus barbitarsis* Zetterstedt (orig. des.). **N. syn.**
Zavrelimyia Fittkau, 1962: 285. Type-species, *Tanypus melanurus* Meigen (orig. des.). **N. syn.**
Labrundinia Fittkau, 1962: 372. Type-species, *Tanypus longipalpis* Goetghebuer (orig. des.). **N. syn.**

    REFERENCES: Walley, 1928b (rev., Canada); Hauber, 1945 (rev., Iowa); Johannsen, 1946 (rev.).

Subgenus PENTANEURA Philippi

americana (Fittkau), 1957: 320 (*Thienemannimyia*).—Conn. **N. comb.**
    vitellina Kieffer of Johannsen, 1946: 280.
apicalis (Walley), 1926a: 64 (*Tanypus*).—Que.; Ont.
aurea (Johannsen), 1907a: 110 (*Ablabesmyia*).—Kans.
barberi (Coquillett), 1902g: 90 (*Tanypus*).—N. Mex.
bifasciata (Coquillett), 1901h: 609 (*Tanypus*).—N.J., Mass.; Pa., N.Y.
carneosa (Fittkau), 1962: 315 (*Zavrelimyia*).—N. Amer.; Iowa to Que. and Maine, s. to Calif. and Fla. **N. comb.**
    carnea, authors, not Fabricius.
cornuticaudata (Walley), 1925: 277 (*Tanypus*).—Que.; Ont., N.Y., Pa.
currani (Walley), 1925: 276 (*Tanypus*).—Que.; Ont., N.Y., Pa.
fimbriata (Walker), 1848: 20 (*Chironomus*).—Hudson Bay.
flavifrons (Johannsen), 1905: 150 (*Ablabesmyia*).—Wash., Idaho, N.Y.; Wash. to Maine, s. to Calif. and Fla.
fragilis (Walley), 1926b: 205 (*Tanypus*).—Ont.
futilis (Wulp), 1867: 130 (*Tanypus*).—Wis.
garretti (Walley), 1925: 275 (*Tanypus*).—B.C.
gesta Roback (for *alba* Roback).—Pa. **N. name.**
    alba Roback, 1957a: 32 (preocc. Tokunaga, 1937).—Pa.
inconspicua (Malloch), 1915g: 371 (*Tanypus*).—Ill.
melanops (Meigen), 1818: 65 (*Tanypus*).—Europe; Nebr. to N.S., s. to N.J.
    unicolor Walker, 1848: 19 (*Chironomus*).—N.S.
melanosoma (Goetghebuer), 1933: 20 (*Ablabesmyia*).—Greenland.
norena Roback, 1957a: 38.—Pa.
okoboji (Walley), 1928b: 582 (*Tanypus*).—Iowa.
ornata (Meigen), 1838: 14 (*Tanypus*).—Europe; Colo., N.Y.
pallens (Coquillett), 1902g: 91 (*Tanypus*).—N. Mex.; N.Y., N.J., Fla.
pilicaudata (Walley), 1925: 277 (*Tanypus*).—Ont.; Que.
pilosella (Loew), 1866b: 5 (Cent. 7, no. 7) (*Tanypus*).—D.C.; N.Y. to Fla., also Iowa.
planensis Johannsen, 1946: 284.—Tex.; Iowa.
rurika Roback, 1957a: 34.—Pa.
senata (Walley), 1925: 276 (*Tanypus*).—Ont.
sinuosa (Coquillett), 1905d: 65 (*Tanypus*).—N.H.; Mich., N.Y.
tetrasticta Kieffer.—Not Nearctic.

Subgenus GUTTIPELOPIA Fittkau

Guttipelopia Fittkau, 1962: 251 (as genus). Type-species, *Tanypus guttipennis* Wulp (orig. des.).

guttipennis (Wulp), 1874: 142 (*Tanypus*).—Netherlands; Iowa, Europe.
multipunctata (Curran), 1930j: 29 (*Tanypus*).—N.Y.

Subgenus NILOTANYPUS Kieffer

Nilotanypus Kieffer, 1923b: 191 (as genus). Type-species, *remotissimus* Kieffer (mon.).

Unnamed sp.—N.Y., Md.
dubia Meigen of Johannsen, 1946: 286.

Genus ABLABESMYIA Johannsen

Ablabesmyia Johannsen, 1905: 125. Type-species, *Tipula monilis* Linnaeus (Coquillett, 1910b: 502).

REFERENCE: Roback, 1959a (rev.).

aequifasciata (Dendy and Sublette), 1959: 507 (*Pentaneura*).—Ala. **N. comb.**
americana Fittkau, 1962: 430.—N. Amer.; Alaska to Que., s. to Colo. and Fla.
monilis, authors, not Linnaeus.
annulata (Say), 1823: 15 (1859b: 43) (*Tanypus*).—Pa.; S. Dak. to Pa., s. to Tex., Fla. **N. comb.**
aspera (Roback), 1959a: 124 (*Pentaneura*).—Maine. **N. comb.**
auriensis (Roback), 1957a: 39 (*Pentaneura*).—Pa.; Idaho, N.J. **N. comb.**
basalis (Walley), 1925: 273 (*Tanypus*).—Ont.; Mich. to Que., s. to Iowa and Fla. **N. comb.**
cinctipes (Johannsen), 1946: 271 (*Pentaneura*).—Fla. **N. comb.**
gera (Roback), 1959a: 126 (*Pentaneura*).—N.H.
idei (Walley), 1925: 272 (*Tanypus*).—Ont. **N. comb.**
illinoensis (Malloch), 1915g: 376 (*Tanypus*).—Ill.; S. Dak. to Ont., s. to La. and Fla. **N. comb.**
janta (Roback), 1959a: 131 (*Pentaneura*).—Md. **N. comb.**
johannseni (Roback), 1959a: 131 (*Pentaneura*).—Ill. **N. comb.**
mallochi (Walley), 1925: 273 (*Tanypus*).—Ont.; S. Dak. to Que. and Maine, s. to Iowa and Ga., also Calif. **N. comb.**
peleensis (Walley), 1926a: 64 (*Tanypus*).—Ont.; Mo., Iowa, N.Y., N.J., Pa., D.C. **N. comb.**
prudens (Walley), 1925: 275 (*Tanypus*).—Man.; Mo., Que. **N. comb.**
pulchripennis (Lundbeck), 1898: 293 (*Tanypus*).—Greenland.

Genus PROCLADIUS Skuse

Procladius Skuse, 1889: 283. Type-species, *paludicola* Skuse (Coquillett, 1910b: 594).

Subgenus PROCLADIUS Skuse

abruptus (Garrett), 1925b: 8 (*Tanypus*).—B.C.
arcuatus (Garrett), 1925b: 7 (*Tanypus*).—B.C.
bifidus (Garrett), 1925b: 8 (*Tanypus*).—B.C.
choreus (Meigen), 1804: 23 (*Tanypus*).—Europe; Greenland, Sask., Wis., Iowa, Tex.
claripennis (Malloch), 1915g: 387 (*Protenthes*).—Mich.; Ont.
crassinervis (Zetterstedt), 1838: 817 (*Tanypus*).—Lapland; Greenland.
culiciformis (Linnaeus), 1767: 978 (*Tipula*).—Sweden; Idaho to Ont. and N.Y., s. to Tex. and Fla., Europe.
lundstromi Goetghebuer *in* Goetghebuer and Lenz, 1936: 11.—Greenland, Scandinavia.
pulcher (Johannsen), 1908: 273 (*Protenthes*).—N.Y.; Ont., Fla.
　　fasciger Curran, 1930j: 29 (*Protenthes*).—N.Y.
riparius (Malloch), 1915g: 389 (*Protenthes*).—Ill.; Pa.
trifolius (Garrett), 1925b: 7 (*Tanypus*).—B.C.
vestitipennis (Kieffer), 1917: 339 (*Trichotanypus*).—N.Y.

Subgenus PSILOTANYPUS Kieffer

Psilotanypus Kieffer, 1906b: 318 (as genus). Type-species, *Tanypus bellus* Loew (Coquillett, 1910b: 597).

adumbratus Johannsen, 1905: 132.—N.Y.; Pa.
bellus (Loew), 1866b: 4 (Cent. 7, no. 4) (*Tanypus*).—D.C.; Mich. to Ont. and N.Y., s. to Tex. and Fla.
　　pusillus Loew, 1866b: 5 (Cent. 7, no. 5) (*Tanypus*).—D.C.
　　flavidus Kieffer, 1923a: 297.—Tex. **N. syn.**
malifero Garrett, 1925b: 10.—B.C.
nubifer (Coquillett), 1905d: 66 (*Tanypus*).—Utah.

Genus TANYPUS Meigen

Pelopia Meigen, 1800: 18. Type-species, *Tipula cincta* Fabricius (Coquillett, 1910b: 586; misident.)=*punctipennis* Meigen. Suppressed by I.C.Z.N., 1963b: 339.
Tanypus Meigen, 1803: 261. Type-species, *Tipula cincta* Fabricius (Latreille, 1810: 442; misident.)=*punctipennis* Meigen. Meigen, 1803, misidentified *cincta* Fabricius, as evidenced in his later works. Recent workers follow

Edwards' (1929b: 299) recognition of *punctipennis* (=*cincta* misident.) as type.
Protenthes Johannsen, 1907b: 400. Type-species, *Tanypus punctipennis* Meigen (orig. des.).

REFERENCE: Walley, 1928b (rev., Canada).

clavatus Beck, 1962: 92.—Fla.
punctipennis Meigen, 1818: 61.—Europe; Wis. to Ont. and N.Y., s. to Tex. and Fla.
americanus Kieffer, 1923a: 297 (*Protenthes*; as var.).—Tex. ?Valid sp.
stellatus Coquillett, 1902g: 89.—Tex.; Minn. to N.Y., s. to Tex. and Fla.

Subfamily PODONOMINAE

REFERENCE: Thienemann, 1937 (world rev.).

Genus BOREOCHLUS Edwards

Boreochlus Edwards, 1938c: 152. Type-species, *thienemanni* Edwards (orig. des.).

persimilis (Johannsen), 1926: 99 (*Trichotanypus*).—N.Y.
posticalis, authors, not Lundbeck.

Genus LASIODIAMESA Kieffer

Syndiamesa, subg. **Lasiodiamesa** Kieffer, 1924c: 48. Type-species, *gracilis* Kieffer (orig. des.).
Linacerus Garrett, 1925b: 9. Type-species, *piloala* Garrett (orig. des.).

piloala (Garrett), 1925b: 9 (*Linacerus*).—B.C.

Genus PODONOMUS Philippi

Podonomus Philippi, 1865: 601. Type-species, *stigmaticus* Philippi (mon.).
Paratanypus Garrett, 1925b: 8. Type-species, *kiefferi* Garrett (orig. des.).

arietinus (Coquillett), 1908: 144 (*Tanypus*).—N.Y.
confusus (Garrett), 1925b: 5 (*Adiamesa*).—B.C.
kiefferi (Garrett), 1925b: 8 (*Paratanypus*).—B.C.; Colo., Maine, N.Y., Greenland.
peregrinus Edwards, 1929b: 296.—England.
tenebrosus (Coquillett), 1905d: 66 (*Tanypus*).—N.H.

Genus TRICHOTANYPUS Kieffer

Trichotanypus Kieffer, 1906b: 319. Type-species, *Tanypus posticalis* Lundbeck (sub. mon., Kieffer, 1906a: 42).

posticalis (Lundbeck), 1898: 295 (*Tanypus*).—Greenland; N.W.T. (Ellesmere I.).

Subfamily DIAMESINAE

REFERENCES: Pagast, 1947 (world rev.); Oliver, 1959 (rev., n. Canada and Alaska).

Genus DIAMESA Waltl

Diamesa Waltl, 1837: 283. Type-species, *cinerella* Waltl (orig. des., as gen. n., sp. n.).
Eutanypus Coquillett, 1899f: 341. Type-species, *borealis* Coquillett (orig. des.) = *nivoriunda* (Fitch).
Diamesa, subg. *Adiamesa* Kieffer, 1906a: 36. Type-species, *tonsa* Haliday (mon.).
Syndiamesa Kieffer, 1918a: 101. Type-species, *Diamesa hygropetrica* Kieffer (Edwards, 1929b: 303).
Psilodiamesa Kieffer, 1918a: 104. Type-species, *spitzbergensis* Kieffer (mon.).
Potthastia Kieffer, 1922c: 362. Type-species, *longimanus* Kieffer orig. des., as gen. n., sp. n.).

aberrata Lundbeck, 1898: 289.—Greenland; N.W.T.
ancysta Roback, 1959b: 1.—Mont.
arctica (Boheman), 1865: 574 (*Chironomus*).—Spitzbergen; N.W.T. (Ellesmere I., D. R. Oliver *in litt.*).
banana Garrett, 1925b: 6.—B.C.
bertrami Edwards, 1935a: 470.—Greenland.
biappendiculata Goetghebuer *in* Remy, 1928: 52.—Greenland.
bohemani Goetghebuer, 1932a: 181.—Spitzbergen; n. Que., Greenland.
caena Roback, 1957b: 9.—Utah.
chorea Lundbeck, 1898: 291.—Greenland.
clavata Edwards, 1933: 615.—N.W.T. (Akpatok I.).
davisi Edwards, 1933: 614.—N.W.T. (Akpatok I.).
fulva Johannsen, 1921: 229.—N.Y.
garretti Sublette and Sublette (for *borealis* Garrett).—B.C. **N. name.**
   borealis Garrett, 1925b: 6 (preocc. Kieffer, 1915).—B.C.
geminata Kieffer, 1926: 79.—Greenland; N.W.T.
   furcata Edwards, 1933: 617.—N.W.T. (Akpatok I.).
gregsoni Edwards, 1933: 618.—N.W.T. (Akpatok I.).

incallida (Walker), 1856a: 183 (*Chironomus*).—England; Utah, Europe.
leona Roback, 1957b: 7.—Utah; Mont.
lindrothi Goetghebuer *in* Goetghebuer and Lindroth, 1931: 281.— Iceland; Greenland, Europe.
longimanus (Kieffer), 1922c: 362 (*Potthastia*).—Germany; Que.
mendotae Muttkowski, 1915: 116.—Wis.
nivoriunda (Fitch), 1847: 282 (*Chironomus*).—N.Y.; Alaska to N.H., s. to N. Mex. and Md. This name has been misapplied to several spp. in *Orthocladius*.
    borealis Coquillett, 1899f: 341 (*Eutanypus*).—U.S.S.R. (Commander Is.).
    heteropus Coquillett, 1905d: 66 (*Tanypus*).—Wash., N. Mex., N.H.
onteona Roback, 1957b: 6.—Utah.
parva Edwards, 1932b: 45.—Scotland; N.W.T., Europe.
pieta Roback, 1957b: 8.—Utah.
polaris Kieffer, 1926: 79.—N.W.T. (Ellesmere I.).
polaris (Kieffer), 1926: 81 (*Syndiamesa;* preocc. in *Diamesa* by Kieffer 1926 but not renamed because of probable synonomy with *parva* Edwards).—N.W.T. (Ellesmere I.).
sequax (Garrett), 1925b: 7 (*Prodiamesa*).—B.C.
simplex Kieffer, 1926: 81.—N.W.T. (Ellesmere I.).
ursa (Kieffer).—Not Nearctic.

Genus HEPTAGYIA Philippi

Heptagyia Philippi, 1865: 635. Type-species, *annulipes* Philippi (mon.).

lurida (Garrett), 1925b: 6 (*Diamesa*).—B.C.; Alaska, Wyo., N.Y.

Genus PAGASTIA Oliver

Pagastia Oliver, 1959: 49. Type-species, *orthogonia* Oliver (orig. des.).

orthogonia Oliver, 1959: 51.—Alaska; Mich.
partica (Roback), 1957b: 4 (*Syndiamesa*).—Utah; Alaska.
    artisia Roback, 1957b: 5 (*Syndiamesa*).—Utah.

Genus PRODIAMESA Kieffer

Prodiamesa Kieffer, 1906a: 37. Type-species, *Diamesa praecox* Kieffer (orig. des.)=*olivacea* (Meigen).

bathyphila Kieffer.—Not Nearctic.

cubita Garrett, 1925b: 7.—B.C.
lutosopra Garrett, 1925b: 7.—B.C.
olivacea (Meigen), 1818: 29 (*Chironomus*).—Europe; Colo., Mich., Greenland, Mass., Pa., N.J.
  turpis Zetterstedt, 1838: 811 (*Chironomus*).—Lapland.
  occidentalis Coquillett, 1902g: 92 (*Tanypus*).—Colo.

Genus Odontomesa Pagast

fulva (Kieffer).—Not Nearctic.

Genus PSEUDODIAMESA Goetghebuer

Syndiamesa, subg. **Pseudodiamesa** Goetghebuer *in* Goetghebuer and Lenz, 1939a: 9. Type-species, *pilosa* Kieffer (orig. des.).

Subgenus PSEUDODIAMESA Goetghebuer

branickii (Nowicki), 1873: 3 (*Diamesa*).—Austria; B.C., Alta., Greenland, Europe, Asia (boreo-alpine).
pertinax (Garrett), 1925b: 6 (*Prodiamesa*).—B.C.; N.W.T., Wyo., Utah.

Subgenus PACHYDIAMESA Oliver

Pseudodiamesa, subg. **Pachydiamesa** Oliver, 1959: 54. Type-species, *Diamesa arctica* Malloch (orig. des.).

arctica (Malloch), 1919f: 37 (*Diamesa*).—N.W.T. (Victoria I.); N.W.T. (Ellesmere I.).

Subfamily ORTHOCLADIINAE

REFERENCES: Thienemann, 1944 (generic key to larvae and pupae of world); Brundin, 1956 (rev., Europe); Wirth, 1949 (rev., biol. intertidal spp.).

Tribe PROTANYPODINI

Genus PROTANYPUS Kieffer

Protanypus Kieffer, 1906b: 318. Type-species, *Tanypus morio* Zetterstedt (Edwards, 1924a: 119).

caudatus Edwards, 1924a: 122.—Norway; N.W.T. (Ellesmere I., D. R. Oliver *in litt.*), Europe.

Tribe ORTHOCLADIINI

Genus BRILLIA Kieffer

Brillia Kieffer, 1913b: 34. Type-species, *bifida* Kieffer (orig. des.)= *modesta* (Meigen).
Eurycnemus, authors, not Wulp.

annuliventris (Malloch), 1915a: 46 (*Metriocnemus*).—Calif. **N. comb.**
flavifrons (Johannsen), 1905: 301 (*Metriocnemus*).—N.Y.
longifurca Kieffer, 1921a: 86.—Germany; Sask.
par (Coquillett), 1901h: 608 (*Orthocladius*).—N.J.; Maine to Fla.
  scitulus Coquillett, 1901h: 608 (*Eurycnemus*).—N.J.
  par Johannsen, 1905: 301 (*Metriocnemus*).—N.Y.
parva Johannsen, 1934a: 351.—N.Y.; Mass.
sera Roback, 1957a: 64.—Pa.

Genus ABISKOMYIA Edwards

Abiskomyia Edwards, 1937: 140. Type-species, *virgo* Edwards (orig. des.).

virgo Edwards, 1937: 141.—Sweden; N.W.T. (Ellesmere I., D. R. Oliver *in litt.*), n. Europe.

Genus CARDIOCLADIUS Kieffer

Cardiocladius Kieffer, 1912c: 22. Type-species, *ceylanicus* Kieffer (orig. des.).

fulvus (Johannsen), 1908: 275 (*Thalassomya*).—N.Y.; Mich., Ill.
  fusca Johannsen, 1903b: 440 (*Thalassomya*). Nomen nudum.
obscurus (Johannsen), 1903b: 437 (*Thalassomya*).—N.Y.; Ill., Pa., Fla.
platypus (Coquillett), 1902g: 93 (*Orthocladius*).—Ariz. **N. comb.**

Genus DIPLOCLADIUS Kieffer

Diplocladius Kieffer, 1908b: 6. Type-species, *cultriger* Kieffer (mon.).

bilobatus Brundin, 1956: 71.—Sweden; N.W.T. (Ellesmere I., D. R. Oliver *in litt.*).
cultriger Kieffer, 1908b: 6.—Germany; N.Y., Conn., Pa., Europe.

Genus NANOCLADIUS Kieffer

Nanocladius Kieffer, 1913d: 31. Type-species, *vitellinus* Kieffer (orig. des.).
Eukiefferiella Thienemann, 1926a: 325. Type-species, *Dactylocladius longicalcar* Kieffer (orig. des.).

alternantherae Dendy and Sublette, 1959: 510.—Ala.; La.
brevinervis (Malloch), 1915g: 526 (*Orthocladius*).—Ill.; Pa.
sordens (Johannsen), 1905: 272 (*Orthocladius*).—N.Y.

Genus HETEROTRISSOCLADIUS Spärck

Heterotrissocladius Spärck, 1923: 94. Type-species, *Metriocnemus cubitalis* Kieffer (Goetghebuer *in* Goetghebuer and Lenz, 1940: 6).
subpilosus (Kieffer) *in* Kieffer and Lundbeck, 1911: 273 (*Dactylocladius*).—Bear I.; N.W.T. (Ellesmere I., D. R. Oliver *in litt.*), Greenland.

Genus TRISSOCLADIUS Kieffer

Trissocladius Kieffer, 1908b: 3. Type-species, *brevipalpis* Kieffer (Edwards, 1929b: 309).
conformis (Holmgren), 1869: 42 (*Chironomus*).—Spitzbergen; Greenland.
   natvigi Goetghebuer, 1933: 25 (*Orthocladius*).—Greenland.
fusistylus (Goetghebuer), 1933: 26 (*Orthocladius*).—Greenland; N.W.T. (Ellesmere I., D. R. Oliver *in litt.*). **N. comb.**, D. R. Oliver *in litt.*
tornetraskensis Edwards *in* Thienemann, 1941: 211 (*Orthocladius*).—Sweden; N.W.T. (Ellesmere I., D. R. Oliver *in litt.*).

Genus ORTHOCLADIUS Wulp

Orthocladius Wulp, 1874: 132. Type-species, *Chironomus sordidellus* Zetterstedt (Kieffer, 1906a: 26; misident.) =*oblidens* (Walker). Edwards' (1929b: 335) examination of type of *sordidellus* Zetterstedt showed it belongs not to the genus as usually defined, but to *Psectrocladius*. He interpreted that Kieffer understood *sordidellus* as a species near *Chironomus oblidens* Walker and named *oblidens* as type of *Orthocladius*. We are following Edwards' interpretation, but the problem should be referred to the I.C.Z.N. for decision.
Orthocladius, subg. *Dactylocladius* Kieffer, 1906c: 356. Type-species, *brevicornis* Kieffer (orig. des.).
Hydrobaenus, authors, not Fries.
Trissocladius, authors, not Kieffer.

albidohalteralis Malloch, 1915g: 528.—Ill. **N. comb.**
anteilis (Roback), 1957b: 14 (*Hydrobaenus*).—Utah; Mont. **N. comb.**

astis (Roback), 1957b: 12 (*Hydrobaenus*).—Utah. **N. comb.**
barbicornis (Linnaeus), 1767: 974 (*Tipula*).—Europe; Wash., Minn.
bifasciatus Malloch, 1918e: 42.—Ill.
carlatus (Roback), 1957a: 77 (*Hydrobaenus*).—Pa. **N. comb.**
claripennis (Lundbeck), 1898: 281 (*Chironomus*).—Greenland.
clepsydrus Coquillett, 1902g: 92.—N. Mex.
consobrinus (Holmgren), 1869: 44 (*Chironomus*).—Spitzbergen, Bear I.; N.W.T. (Ellesmere I., D. R. Oliver *in litt.*), Greenland.
curtistylus Goetghebuer.—Not Nearctic.
decoratus (Holmgren), 1869: 43 (*Chironomus*).—Spitzbergen; n. Canada, Greenland.
difficilis (Lundbeck), 1898: 282 (*Chironomus*).—Greenland.
dorenus (Roback), 1957a: 78 (*Hydrobaenus*).—Pa. **N. comb.**
dubitatus Johannsen, 1942: 72.—N.Y.; Pa.
flavoscutellatus Malloch, 1915g: 523.—Ill.
frigidus (Zetterstedt), 1838: 812 (*Chironomus*).—Greenland, Lapland; n. Que.
gelidus Kieffer.—Not Nearctic.
glacialis (Kieffer), 1926: 86 (*Camptocladius*).—N.W.T. (Ellesmere I.).
graminicola (Lundbeck), 1898: 278 (*Chironomus*).—Greenland.
groenlandensis Goetghebuer, 1933: 26.—Greenland.
groenlandicus, error.
knabeni Goetghebuer, 1933: 26.—Greenland.
mallochi Kieffer, 1919c: 191 (n. name for *lacteipennis* Malloch).—Mich.
lacteipennis Malloch, 1915g: 524 (preocc. Lundström, 1910.—Mich.
minutus (Zetterstedt), 1850: 3522 (*Chironomus*).—Sweden; Greenland.
mixtus (Holmgren), 1869: 45 (*Chironomus*).—Bear I.; N.W.T. (Ellesmere I., D. R. Oliver *in litt.*), Spitzbergen.
nanseni Kieffer, 1926: 84.—Unknown (Arctic Amer.).
nigritus Malloch, 1915g: 525.—Md.; Utah.
nitidoscutellatus Lundström.—Not Nearctic.
oblidens (Walker), 1856a: 180 (*Chironomus*).—England; n. Que.
obumbratus Johannsen, 1905: 281.—N.Y.; Alaska, Colo., Mass.
paradorenus (Roback), 1957a: 79 (*Hydrobaenus*).—Pa. **N. comb.**
pilipes Malloch, 1915g: 522.—Ill.
pleuralis Malloch, 1915g: 527.—Ill.
rivulorum Kieffer.—Not Nearctic.
semivirens Kieffer.—Not Nearctic.
spitzbergensis (Kieffer), 1919a: 116 (*Dactylocladius*).—Spitzbergen; Greenland.

subparallelus Malloch, 1915g: 522.—Ill.
traenis (Roback), 1957b: 12 (*Hydrobaenus*).—Utah. **N. comb.**
trigonolabis Edwards, 1924b: 170.—Spitzbergen; n. Canada, Palaearctic.
xethis (Roback), 1957b: 13 (*Hydrobaenus*).—Utah. **N. comb.**

Genus SYMBIOCLADIUS Kieffer

Symbiocladius Kieffer, 1925d: 565. Type-species, *Phaenocladius rhithrogenae* Zavřel (Goetghebuer *in* Goetghebuer and Lenz, 1950a: 205).

REFERENCE: Roback, 1953 (discussion; larvae ectoparasitic on mayfly nymphs).

equitans (Claassen), 1922: 397 (*Trissocladius*).—Colo.; Calif., Utah Vt.

Genus CRICOTOPUS Wulp

Cricotopus Wulp, 1874: 132. Type-species, *Chironomus tibialis* Meigen (Coquillett, 1910b: 528).

REFERENCE: Walley, 1928a (key).

abanus Curran, 1929a: 1.—Man.
absurdus (Johannsen), 1905: 277 (*Orthocladius*).—N.Y.; Ohio.
alpicola (Zetterstedt), 1850: 3500 (*Chironomus*).—Scandinavia; N.W.T. (Ellesmere I., D. R. Oliver *in litt.*), n. Europe.
aratus Roback, 1957a: 73.—Pa.
basalis (Staeger), 1845b: 351 (*Chironomus*).—Greenland; Wash., N.W.T. (Ellesmere I.), n. Que., n. Eurasia.
belkini Dendy and Sublette, 1959: 510.—Ala.
bicinctus (Meigen), 1818: 41 (*Chironomus*).—Europe; Mich. to Ont. and N.Y., s. to Ill. and Fla.
brunnicans Walley, 1928a: 21.—Sask.
ceris Roback, 1957a: 72.—Pa.
elegans Johannsen, 1943c: 78.—Mich.; Pa.
exilis Johannsen, 1905: 255.—N.Y.; Wis., Ohio.
flavibasis Malloch, 1915g: 502.—Ill.
flavipes Johannsen, 1942: 73.—Mich.
fugax (Johannsen), 1905: 279 (*Orthocladius*).—N.Y.
fuscatus Wirth, 1957: 124.—Calif.
geminatus (Say), 1823: 14 (1859b: 42) (*Chironomus*).—Pa.
glacialis Edwards, 1922b: 209.—Spitzbergen, N. Edinburgh I.; Greenland.
globistylus Roback, 1957b: 10.—Utah.
infuscatus (Malloch), 1915g: 517 (*Orthocladius*).—Ill.
junus Roback, 1957a: 72.—Pa.

nostocicola Wirth, 1957: 122.—Calif.; Oreg.
oceanicus (Packard), 1869b: 42 (*Chironomus*).—Mass.
pavidus (Holmgren), 1869: 42 (*Chironomus*).—Spitzbergen, Bear I.; N.W.T. (Ellesmere I., D. R. Oliver *in litt.*).
pilosellus Brundin, 1956: 114.—Sweden; Greenland, n. Canada, Palaearctic.
polaris Kieffer, 1926: 84.—Greenland.
politus (Coquillett), 1902g: 93 (*Orthocladius*).—D.C.; Colo., Ill., Ohio, N.Y., Md., Fla.
slossonae Malloch, 1915g: 506.—Ill.; N.H., Conn., Pa.
sylvestris (Fabricius), 1794: 252 (*Tipula*).—Europe; Ill.
tremulus (Linnaeus), 1758: 587 (*Tipula*).—Europe; N.Y., N.J.
tricinctus (Meigen), 1818: 41 (*Chironomus*).—Europe; Utah to N.Y., s. to Tex.
trifasciatus (Panzer), 1813: 18 (*Chironomus*).—Germany; Idaho, Ont. to N.Y., s. to Mo. and Fla., Europe.
varipes Coquillett, 1902g: 93.—Md.; Wash., N.Y., Fla.
vitripennis (Meigen), 1818: 32 (*Chironomus*).—Europe; Greenland, Que.
    variabilis Staeger, 1845b: 351 (*Chironomus*).—Greenland.

Genus TRICHOCLADIUS Kieffer

Trichocladius Kieffer, 1906c: 356. Type-species, *Orthocladius fissicornis* Kieffer (orig. des.).

bacilliger Kieffer, 1926: 86 (?*Trichocladius*).—Greenland. Unrecognized.
chapmani (Edwards), 1935a: 471 (*Spaniotoma*).—Greenland. **N. comb.**
distinctus (Malloch), 1915g: 518 (*Orthocladius*).—Ill.; Fla.
    basalis Malloch, 1915g: 519 (*Orthocladius;* as var.).—Ill. ?Valid sp.
    bicolor Malloch, 1915g: 519 (*Orthocladius;* as var.).—Ill. ?Valid sp.
extatus Roback, 1957a: 84.—Pa.
helis Roback, 1957b: 15.—Utah.
lacteipennis Johannsen, 1908: 282.—Pa.
nitidellus (Malloch), 1915g: 515 (*Orthocladius*).—Ill.; N.Y.
nitidus (Malloch), 1915g: 515 (*Orthocladius*).—Ill.
pubitarsis (Zetterstedt), 1838: 811 (*Chironomus*).—Lapland; Greenland, N.W.T.
senex (Johannsen), 1937a: 63 (*Spaniotoma*).—N.Y.; Iowa, Pa.
septris Roback, 1957b: 14.—Utah.
striatus (Malloch). 1915g: 517 (*Orthocladius*).—Ill.

Genus PSECTROCLADIUS Kieffer

Orthocladius, subg. **Psectrocladius** Kieffer, 1906c: 356. Type-species, *psilopterus* Kieffer (orig. des.)=*sordidellus* (Zetterstedt).

aureus Johannsen, 1908: 283.—Kans.
barbatimanus Kieffer, 1926: 82.—N.W.T. (Ellesmere I.).
barbimanus (Edwards), 1929b: 333 (*Spaniotoma*).—England; Greenland, Europe.
    armatus Andersen, 1937: 56.—Greenland.
brevicosta Kieffer, 1917: 362.—Wash.
elatus Roback, 1957a: 89.—N.J.; Pa.
fennicus Storå, 1939: 24.—Finland; N.W.T. (Ellesmere I., D. R. Oliver *in litt.*).
flavus (Johannsen), 1905: 270 (*Orthocladius*).—N.Y.
julia (Curran), 1930j: 34 (*Orthocladius*).—N.Y.
limbatellus (Holmgren), 1869: 44 (*Chironomus*).—Spitzbergen; N.W.T. (Ellesmere I.), Greenland.
nanseni Kieffer, 1926: 82.—N.W.T. (Ellesmere I.).
niger Roback, 1957a: 88.—Pa.
obvius (Walker), 1856a: 174 (*Chironomus*).—England; n. Que.
pilosus Roback, 1957a: 88.—N.J.
polaris Kieffer, 1926: 83.—N.W.T. (Ellesmere I.).
simulans (Johannsen), 1937a: 67 (*Spaniotoma*).—N.Y.
spinifer (Johannsen), 1928a: 34 (*Orthocladius*).—Calif. **N. comb.**
vernalis (Malloch), 1915g: 520 (*Orthocladius*).—Ill.; Ala.

Genus CHASMATONOTUS Loew

Chasmatonotus Loew, 1864a: 51 (Cent. 5, no. 1). Type-species, *unimaculatus* Loew (mon.).

    REFERENCE: Rempel, 1937a (key).

atripes Rempel, 1937a: 253.—N.Y.; Ont., Ill.
bicolor Rempel, 1937a: 254.—Tenn.; N.Y.
bimaculatus Osten Sacken, 1877: 191.—N.Y., Que.
fascipennis Coquillett, 1905d: 66.—B.C.
hyalinus Coquillett, 1905d: 67.—Calif.
maculipennis Rempel, 1937a: 254.—Wash. (quarantine interception from Japan); B.C.
unimaculatus Loew, 1864a: 50 (Cent. 5, no. 1).—N.H.; Ont., Que., N.Y., N.C.
univittatus Coquillett, 1900h: 395 (1904: 9).—Alaska; B.C., Calif.

Tribe METRIOCNEMINI

Genus CHAETOCLADIUS Kieffer

Dactylocladius, subg. **Chaetocladius** Kieffer, 1911d: 182. Type-species, *setiger* Kieffer (Goetghebuer *in* Goetghebuer and Lenz, 1942: 57) =*perennis* (Meigen).

adsimilis (Goetghebuer), 1933: 25 (*Orthocladius*).—Greenland; N.W.T. (Ellesmere I.). **N. comb.**

furcatus (Kieffer) *in* Thienemann, 1916: 535 (*Dactylocladius*).— Sweden; N.Y., Europe. **N. comb.**

grandilobus Brundin, 1956: 126.—Sweden; N.W.T. (Ellesmere I., D.R. Oliver *in litt.*).

holmgreni (Jacobson), 1898: 204 (1898: 34) (*Chironomus*; n. name for *festivus* Holmgren).—Spitzbergen; N.W.T. (Ellesmere I., D. R. Oliver *in litt.*), Bear I.

  festivus Holmgren, 1869: 43 (*Chironomus*; preocc. Say, 1823).—Spitzbergen.

perennis (Meigen), 1830: 249 (*Chironomus*).—Europe; N.W.T. (Ellesmere I., D. R. Oliver *in litt.*).

stamfordi (Johannsen), 1947: 171 (*Hydrobaenus*).—Conn. **N. comb.**

Genus LIMNOPHYES Eaton

Limnophyes Eaton, 1875: 60. Type-species, *pusillus* Eaton (mon.).

atomarius (Zetterstedt), 1850: 3522 (*Chironomus*).—Sweden; Greenland, N.Y.

borealis Goetghebuer, 1933: 29.—Greenland; N.W.T. (Ellesmere I.).
  squamatus Andersen, 1937: 70 (as var.).—Greenland. ?Var.

eltoni (Edwards), 1922b: 203 (*Camptocladius*).—Bear I.; Greenland.
  asquamatus Andersen, 1937: 72 (as var.).—Greenland ?Var.

fumosus (Johannsen), 1905: 261 (*Camptocladius*).—N.Y.

globifer (Lundstrom), 1915: 16 (*Camptocladius*).—Siberia; N.W.T. (Ellesmere I.), Greenland.

groenlandiensis Andersen, 1937: 70.—Greenland.
  groenlandicus, error.

minimus (Meigen), 1818: 47 (*Chironomus*).—Europe; Idaho, Colo., N.Y.

pumilio (Holmgren), 1869: 41 (*Chironomus*).—Spitzbergen; Greenland.

Genus METRIOCNEMUS Wulp

Metriocnemus Wulp, 1874: 136. Type-species, *Chironomus albolineatus* Meigen (Coquillett, 1910b: 569).

abdominoflavatus Picado.—Not Nearctic.

aequalis Johannsen, 1934a: 348.—N.Y., Ill.
atratulus (Zetterstedt), 1850: 3590 (*Chironomus*).—Norway; Colo., N.Y., Greenland, Europe.
debilipennis (Lundbeck), 1898: 286 (*Chironomus*).—Greenland; N.Y.
edwardsi Jones, 1916: 385.—Calif.
exagitans Johannsen, 1905: 303.—N.Y.; Colo., Kans., Wis., Ill.
  brachyneura Malloch, 1915g: 498.—Ill.
fuscipes (Meigen), 1818: 49 (*Chironomus*).—Europe; Greenland, Colo., N.Y.
hamatus Johannsen, 1934a: 349.—N.Y.; Mass.
hygropetricus (Kieffer), 1911d: 181 (*Orthocladius*).—Germany; Greenland, Europe.
  longitarsus Goetghebuer, 1921: 77.—Belgium.
impensus (Walker), 1856a: 184 (*Chironomus*).—England; Greenland.
innocuus Curran, 1930j: 33.—N.Y.
knabi Coquillett, 1904a: 11.—Mass.; Ill., N.C., Fla., Bermuda.
lundbeckii Johannsen, 1905: 302.—Greenland; Mich., Ill., N.Y., Pa., Fla.
  nanus, authors, not Meigen.
mitis Curran, 1930j: 33.—N.Y.
obscuripes (Holmgren), 1869: 38 (*Chironomus*; preocc. Meigen, 1830).—Spitzbergen; N.W.T. (Ellesmere I., D. R. Oliver *in litt.*), Bear I. Not renamed here because of probable synonymy.
perfuscus Malloch, 1934b: 18.—N.W.T.
picipes (Meigen), 1818: 52 (*Chironomus*).—Europe; Greenland.
polaris Kieffer, 1926: 87.—N.W.T. (Ellesmere I.), Greenland.
similis Kieffer, 1922b: 11.—Novaya Zemlya; Greenland.
tristellus Edwards, 1929b: 312.—England; Greenland.
ursinus (Holmgren), 1869: 39 (*Chironomus*).—Spitzbergen; Greenland, n. Que.

Genus PARAKIEFFERIELLA Thienemann

Parakiefferiella Thienemann, 1936: 195. Type-species, *Spaniotoma coronata* Edwards (orig. des.).

nigra Brundin, 1949: 827.—Sweden; N.W.T. (Ellesmere I., D. R. Oliver *in litt.*).

Genus PARAPHAENOCLADIUS Thienemann

Paraphaenocladius Thienemann, 1926b: 9. Type-species, *Metriocnemus ampullaceus* Kieffer (mon.)=*impensus* (Walker).

despectus (Kieffer), 1926: 87 (*Metriocnemus*).—N.W.T. (Ellesmere I.). **N. comb.**, D. R. Oliver *in litt.*

Genus PROSMITTIA Brundin

Prosmittia Brundin, 1956: 165. Type-species, *Pseudosmittia jemtlandica* Brundin (orig. des.).

nanseni (Kieffer), 1926: 86 (*Camptocladius*).—N.W.T. (Ellesmere I.). **N. comb.**, D. R. Oliver *in litt.*

Genus SMITTIA Holmgren

Smittia Holmgren, 1869: 47. Type-species, *Chironomus brevipennis* Holmgren (mon.).
Camptocladius Wulp, 1874: 133. Type-species, *Tipula byssina* Schrank (Coquillett, 1910b: 518)=*stercoraria* (De Geer).

arctica Malloch, 1923i: 175.—Alaska (St. Paul I.).
aterrima (Meigen), 1818: 47 (*Chironomus*).—Europe; Colo. to Ont., s. to Ill. and N.J., also Greenland.
clavicornis (Saunders), 1928: 528 (*Camptocladius*).—B.C.; coastal, B.C. to Calif.
crassicollis (Walker), 1848: 18 (*Chironomus*).—Hudson Bay.
ephemerae Kieffer.—Not Nearctic.
extrema (Holmgren), 1869:40 (*Chironomus*).—Spitzbergen; Greenland.
flavens (Malloch), 1915g: 511 (*Camptocladius*).—Ill.; Mich. **N. comb.**
flavibasis (Malloch), 1915g: 511 (*Camptocladius*).—Ill. **N. comb.**
fulvipluma (Kieffer), 1926: 85 (*Camptocladius*).—N.W.T. (Ellesmere I.).
fumosina (Curran), 1930j: 33 (*Camptocladius*).—N.Y. **N. comb.**
lasiophthalma (Malloch), 1915g: 509 (*Camptocladius*).—Ill. **N. comb.**
lasiops (Malloch), 1915g: 508 (*Camptocladius*).—Ill.; Fla. **N. comb.**
longitibia Goetghebuer, 1933: 29.—Greenland.
marina (Saunders), 1928: 526 (*Camptocladius*).—B.C.; coastal, B.C. to Calif.
neria (Curran), 1930j: 34 (*Camptocladius*).—N.Y. **N. comb.**
oxoniana Edwards.—Not Nearctic.
pacifica (Saunders), 1928: 523 (*Camptocladius*).—B.C.; coastal, B.C. to Calif.
parva (Lundbeck), 1898: 275 (*Chironomus*).—Greenland.
polaris (Kieffer), 1926: 85 (*Camptocladius*).—N.W.T. (Ellesmere I.).
polymorpha Andersen, 1937: 76.—Greenland; N.W.T. (Ellesmere I.).
roena Roback, 1957b: 16.—Utah.
stercoraria (DeGeer), 1776: 388 (*Tipula*).—Europe; Alaska to N.Y., s. to Mo., also Greenland.
  byssina Schrank, 1803: 76 (*Tipula*).—Germany.
subaterrima (Malloch), 1915g: 512 (*Camptocladius*).—Ill. **N. comb.**

subnudipennis Goetghebuer, 1933: 28.—Greenland.
velutina (Lundbeck), 1898: 274 (*Chironomus*).—Greenland.

Genus CLUNIO Haliday

Clunio Haliday, 1855: 62. Type-species, *marinus* Haliday (mon.).
marshalli Stone and Wirth, 1947: 214.—Fla.

Genus ERETMOPTERA Kellogg

Eretmoptera Kellogg, 1900a: 82. Type-species, *browni* Kellogg (orig. des., as gen. n., sp n.).
browni Kellogg, 1900a: 82.—Calif.

Genus TETHYMYIA Wirth

Tethymyia Wirth, 1949: 160. Type-species, *aptena* Wirth (orig. des.).
aptena Wirth, 1949: 161.—Calif.

Tribe TELMATOGETONINI

Genus TELMATOGETON Schiner

Telmatogeton Schiner, 1866b: 931. Type-species, *sanctipauli* Schiner (orig. des.).
japonicus Tokunaga, 1933: 95.—Japan; N.Y., Fla., Hawaii.
macswaini Wirth, 1949: 170.—Calif.

Genus PARACLUNIO Kieffer

Paraclunio Kieffer, 1911c: 103. Type-species, *trilobatus* Kieffer (mon.).
alaskensis (Coquillett), 1900h: 395 (1904: 9) (*Telmatogeton*).—Alaska; B.C., Oreg., Calif.
trilobatus Kieffer, 1911c: 105.—Calif.; Oreg.

Genus THALASSOMYA Schiner

Thalassomya Schiner, 1856: 219. Type-species, *frauenfeldi* Schiner (mon.).
Thalassomyia, emend.
bureni Wirth, 1949: 167.—Fla.

Tribe CORYNONEURINI

Genus CORYNONEURA Winnertz

Corynoneura Winnertz, 1846: 12. Type-species, *scutellata* Winnertz (Coquillett, 1910b: 528).

Subgenus CORYNONEURA Winnertz

celeripes Winnertz, 1852b: 50.—Germany; Greenland, Ill., Ind., Ont., N.Y., Europe.
diara Roback, 1957b: 10.—Utah.
scutellata Winnertz, 1846: 13.—Germany; Ohio, N.Y., N.W.T. (Ellesmere I.), Greenland, Fla.
taris Roback, 1957a: 61.—N.J.; Pa.

Subgenus THIENEMANNIELLA Kieffer

Corynoneura, subg. **Thienemanniella** Kieffer, 1911d: 201. Type-species, *clavicornis* Kieffer (Goetghebuer *in* Goetghebuer and Lenz, 1939c: 7).

cubita Garrett, 1925b: 10.—B.C.
elana Roback, 1957b: 10.—Utah.
similis Malloch, 1915g: 413.—Ill.; Tex., N.Y.
xena Roback, 1957a: 61.—Pa.

Subfamily CHIRONOMINAE
Tribe CHIRONOMINI

REFERENCE: Townes, 1945 (rev.).

Genus CHIRONOMUS Meigen

Tendipes Meigen, 1800: 17. Type-species, *Tipula plumosa* Linnaeus (Coquillett, 1910b: 612). Suppressed by I.C.Z.N., (1963b: 339).
Chironomus Meigen, 1803: 260. Type-species, *Tipula plumosa* Linnaeus (Latreille, 1810: 442).
Einfeldia Kieffer, 1924c: 393. Type-species, *pectoralis* Kieffer (mon.).

Subgenus CHAETOLABIS Townes

Tendipes, subg. **Chaetolabis** Townes, 1945: 114. Type-species, *atroviridis* Townes (orig. des.).

atroviridis (Townes), 1945: 114 (*Tendipes*).—N.Y.; B.C. to Ont. and Mass., s. to Iowa and Va. **N. comb.**
ochreatus (Townes), 1945: 115 (*Tendipes*).—N.J.; Mich. to Maine, s. to Ark. and Fla.

Subgenus CHIRONOMUS Meigen

anthracinus Zetterstedt, 1860: 6499.—Sweden; Calif. to Alta., e. to Mass., Europe.
    meridionalis Johannsen, 1908: 277 (as *hyperboreus* var.).— Unknown.
    rempelii Thienemann, 1941: 234.—Sask.
    hyperboreus, authors, not Staeger.
atrella (Townes), 1945: 124 (*Tendipes*).—Nev.; Alta. to P.E.I., s. to Calif., Colo., and Mass. **N. comb.**
atritibia Malloch, 1934b: 16.—N.W.T. (Southampton I.); B.C., N.Y., Ont., Que.
attenuatus Walker, 1848: 20.—Hudson Bay; B.C. to Que., s. to Calif. and Fla.
    lasiopus Walker, 1848: 19.—Hudson Bay.
    redeuns Walker, 1856b: 422.—U.S.
    similis Johannsen, 1905: 236.—Ill., S. Dak., N.Y.
    decorus Johannsen, 1905: 239.—Wash., Nebr., Kans., Iowa, Ill., Ohio, N.Y.
    maturus Johannsen, 1908: 279.—N.Y.
    cayugae Johannsen *in* Tilbury, 1913: 308.—N.Y.
    distinguendus Kieffer, 1917: 346.—Wash.
    bifimbriatus Kieffer, 1922e: 145.—Tex. **N. syn.**
biseta (Townes), 1945: 127 (*Tendipes*).—Hudson Bay Territory. **N. comb.**
brunneipennis Johannsen, 1905: 205.—N.Y.; Minn. to Mass., s. to Ark. and Fla.
carus (Townes), 1945: 118 (*Tendipes*).—Venezuela; Tex., Fla.
chelonia (Townes), 1945: 114 (*Tendipes*).—N.Y.; N.J., D.C. **N. comb.**
crassicaudatus Malloch, 1915g: 453.—Ill.; Minn. to Ont. and N.J., s. to Tex. and Fla.
decumbens Malloch, 1934b: 16.—N.W.T. (Southampton I.).
dorsalis Meigen, 1818: 25.—Europe; Idaho, Calif., S. Dak. to Mass., s. to Mo. and Fla.
ferrugineus Macquart.—Not Nearctic.
fulvipilus Rempel, 1939: 210.—Brazil; Calif., Tex., Md., Ala., Fla.
hyperboreus Staeger, 1845b: 349.—Greenland; N.W.T., Sask., Colo.
paganus Meigen, 1838: 7.—Belgium; Idaho, S. Dak., Mich., N.Y., Europe.
pilicornis (Fabricius), 1794: 243 (*Tipula*).—Europe; Alaska, Alta., Sask.
    polaris Kirby, 1824: ccxviii.—Amer. within the Arctic Circle.
    conformis Malloch, 1923i: 172 (preocc. Holmgren, 1869).— Alaska (St. Paul I.).

plumosus (Linnaeus), 1758: 587 (*Tipula*).—Europe; B.C. to Que., s. to Calif. and N.C., Greenland, Bermuda.
  annularis De Geer, 1776: 379 (*Tipula*).—Europe.
  cristatus Fabricius, 1805: 39.—N.Y.
  prasinus Meigen, 1818: 22.—Europe.
  ferrugineovittatus Zetterstedt, 1850: 3492.—Sweden.
  imperator Walley, 1926a: 64.—Ont.
pungens (Townes), 1945: 119 (*Tendipes*).—Fla.; D.C.
riparius Meigen, 1804: 13.—Europe; Alaska to Nfld., s. to Calif. and S.C., Fla.
  albistria Walker, 1848: 17.—Hudson Bay, Ont.
  serus Malloch, 1915g: 481.—Ill.
  militaris Johannsen, 1937b: 46.—N.Y.
staegeri Lundbeck, 1898: 271.—Greenland; B.C. to N.W.T. and Labr., s. to Calif. and N.C.
  fasciventris Malloch, 1915g: 438.—Ill.
stigmaterus Say, 1823: 15 (1859b: 42).—U.S.; Wash. to N.Y., s. to Calif. and Fla.
  glaucurus Wiedemann, 1828: 15.—Pa.
tentans Fabricius, 1805: 38.—Europe; B.C. to Que., s. to Utah and Ohio.
  pallidivittatus Malloch, 1915g: 445 (as var.).—Ill.
tuberculatus (Townes), 1945: 128 (*Tendipes*).—Hudson Bay Territory; Alta. **N. comb.**
tuxis Curran, 1930j: 31.—N.Y.; Mich., Mass., N.J., Fla.
utahensis Malloch, 1915g: 438.—Utah; Oreg. to Alta. and Minn., s. to Colo. and Calif.
viridicollis Wulp.—Not Nearctic.

Subgenus CRYPTOCHIRONOMUS Kieffer

Cryptochironomus Kieffer, 1918a: 46 (as genus). Type-species, *Chironomus chlorolobus* Kieffer (orig. des.)=*supplicans* Meigen.
Cladopelma Kieffer, 1921a: 63. Type-species, *laminata* Kieffer (Townes, 1945: 147).
Harnischia Kieffer, 1921a: 69. Type-species, *fuscimana* Kieffer (orig. des., as gen. n., sp. n.).
  REFERENCE: Curry, 1958 (biol., descr. larvae, pupae).

abortivus Malloch, 1915g: 465.—Ill.; S. Dak. to Ont., s. to Okla. and Pa., also Oreg.
alatus Beck, 1962: 91.—Fla.
alboviridis Malloch, 1915g.: 482.—Ill.; Iowa, Okla.
alphaeus (Sublette), 1960: 222 (*Tendipes*).—Calif. **N. comb.**

amachaerus (Townes), 1945: 168 (*Harnischia*).—N.Y.; Calif., Minn., Ill., N.H., Pa., Ala., Fla.
amphitrite (Townes), 1945: 151 (*Harnischia*).—S. Dak. **N. comb.**
argenteus (Townes), 1945: 164 (*Harnischia*).—Ont.; N.J. **N. comb.**
argus (Roback), 1957a: 106 (*Cryptochironomus*).—Pa. **N. comb.**
ariel (Sublette), 1960: 222 (*Tendipes*).—Calif. **N. comb.**
blarina (Townes), 1945: 99 (*Cryptochironomus*).—N.Y.; Mich., Ont., Okla., Tex. **N. comb.**
boydi Beck, 1962: 91.—Fla.
carinatus (Townes), 1945: 158 (*Harnischia*).—Ark.; Ind. to N.Y., s. to Tex. and Fla. **N. comb.**
casuarius (Townes), 1945: 162 (*Harnischia*).—N.Y.; Ont., Fla. **N. comb.**
chaetoala (Sublette), 1960: 220 (*Tendipes*).—Calif. **N. comb.**
claviger (Townes), 1945: 158 (*Harnischia*).—S.C.; Wash., Nebr., Ind. **N. comb.**
collator (Townes), 1945: 169 (*Harnischia*).—N.Y.; Minn., Ohio, Fla. **N. comb.**
cuneatus (Townes), 1945: 163 (*Harnischia*).—N.J.; Iowa to Que. and N.Y., s. to N.J. **N. comb.**
curtilamellatus Malloch, 1915g: 474.—Mich.; Calif., Tex., Ill., Ohio, N.Y., Ont., Va., Fla.
darbyi (Sublette), 1960: 221 (*Tendipes*).—Calif. **N. comb.**
digitatus Malloch, 1915g: 483.—Ill.; S. Dak. to Ont. and Mass., s. to Tex. and Ga., Calif.
directus (Dendy and Sublette), 1959: 514 (*Tendipes*).—Ala.; La. **N. comb.**
doris (Townes), 1945: 151 (*Harnischia*).—Okla.; Iowa. **N. comb.**
edwardsi (Kruseman), 1933: 194 (*Tendipes*).—Netherlands; Calif., Ont., N.Y., Fla., Europe.
elodeae (Townes), 1945: 156 (*Harnischia*).—N.Y. **N. comb.**
emorsus (Townes), 1945: 161 (*Harnischia*).—N.Y.; Iowa, Ohio, N.J., D.C., Fla. **N. comb.**
fastigatus (Townes), 1945: 162 (*Harnischia*).—N.Y.; Ill., Mich., Que. **N. comb.**
forceps (Townes), 1945: 157 (*Harnischia*).—Mass.; Man., Mich. **N. comb.**
frequens Johannsen, 1905: 230.—N.Y.; Mich. to Que. and N.Y., s. to Okla. and Fla., also Oreg. and Calif.
fulvus Johannsen, 1905: 224.—N.Y.; Mont. to Man. and Que., s. to Calif. and Fla.
    abbreviatus Malloch, 1915g: 451 (preocc. Kieffer, 1913).—Ill.
    parvilamellatus Malloch, 1915g: 479.—Ill.
    mallochi Kieffer, 1919c: 191 (n. name for *abbreviatus* Malloch).—Ill.

galapterus (Townes), 1945: 148 (*Harnischia;* n. name for *claripennis* Malloch.).—Mich.; N.Y., Ont., Ill. **N. comb.**
  claripennis Malloch, 1915g: 439 (preocc. Lundbeck, 1898).—Mich.
galeator (Townes), 1945: 170 (*Harnischia*).—Fla.; Iowa. **N. comb.**
griseus Malloch, 1915g: 468.—Mich.; Mont., Ill., Okla., Tex.
incidatus (Townes), 1945: 166 (*Harnischia*).—Okla.; Tex. **N. comb.**
monochromus Wulp, 1874: 129 (n. name for *unicolor* Wulp).—Netherlands; Ill. to Mass., s. to Tex. and Fla., Calif., Europe.
  unicolor Wulp, 1858: 162 (preocc. Walker, 1848).—Netherlands.
nais (Townes), 1945: 149 (*Harnischia*).—N.Y.; Pa. **N. comb.**
nereis (Townes), 1945: 150 (*Harnischia*).—S.C.; Idaho, Ind., N.Y., Pa. **N. comb.**
nigrovittatus Malloch, 1915g: 456.—Ill.; Calif., Tex., Okla., Iowa, N.Y., Fla.
nixe (Townes), 1945: 149 (*Harnischia*).—Colo. **N. comb.**
orbicus (Townes), 1945: 151 (*Harnischia*).—Kans. **N. comb.**
pectinatellae (Dendy and Sublette), 1959: 516 (*Tendipes*).—Ala. **N. comb.**
potamogeti (Townes), 1945: 159 (*Harnischia*).—N.Y.; Man. to Que., s. to Tex. and Fla. **N. comb.**
pseudotener (Goetghebuer), 1922: 38 (*Cryptochironomus*).—Belgium; D.C., Mich., N.Y., Pa., Europe.
psittacinus Meigen, 1830: 247.—Europe; Alaska to N.Y., s. to Oreg. and Ky.
  stylifera Johannsen, 1908: 281.—N.Y.
  obtusilobus Malloch, 1923i: 171.—Alaska (St. George I.).
  farinalis Walley, 1926b: 205.—Alta.
scimitarus (Townes), 1945: 98 (*Cryptochironomus*).—N.Y.; Md., Mich., Ont., Fla. **N. comb.**
sorex (Townes), 1945: 97 (*Cryptochironomus*).—N.Y.; Mich., Okla., Tex., Fla. **N. comb.**
spectabilis (Townes), 1945: 168 (*Harnischia*).—Ont. **N. comb.**
sublettei Beck, 1961: 127.—Fla.
tenuicaudatus Malloch, 1915g: 475.—Ill.; Oreg. to Calif. and N. Mex., Sask. to Mass., s. to Mo. and Ala.
tethys (Townes), 1945: 152 (*Harnischia*).—S. Dak. **N. comb.**
tylus (Townes), 1945: 150 (*Harnischia*).—Ind. **N. comb.**
undine (Townes), 1945: 149 (*Harnischia*).—Oreg.; Ont., N.Y., Pa., Fla. **N. comb.**
varus Goetghebuer, 1921: 162.—Belgium; Man., N.Y., Europe.

viridulus (Linnaeus), 1767: 975 *(Tipula).*—Sweden; Iowa to Ont. and N.Y., s. to Fla., Calif., Europe.

Subgenus DICROTENDIPES Kieffer

Dicrotendipes Kieffer, 1913d: 23 (as genus). Type-species, *pictipennis* Kieffer (mon.).
Limnochironomus Kieffer, 1920: 166. Type-species, *Tendipes falciformis* Kieffer (orig. des.) =*nervosus* Staeger.

aethiops (Townes), 1945: 107 *(Tendipes).*—N. Mex.; Fla.
botaurus (Townes), 1945: 109 *(Tendipes).*—Mont.; Mich. **N. comb.**
californicus Johannsen, 1905: 217.—Calif.; Oreg., Colo., Okla., Tex.
fumidus Johannsen, 1905: 221.—N.Y.; Oreg. to Ont., s. to Utah and W. Va.
  incognitus Malloch, 1915g: 480.—Ill.
leucoscelis (Townes), 1945: 104 *(Tendipes).*—Idaho; Ont., Mich., Mass., R.I., N.Y., N.J., Fla.
lobiger (Kieffer), 1921a: 71 *(Limnochironomus).*—Germany; Mich., N.Y., Europe.
lobus Beck, 1962: 89.—Fla.
milleri (Townes), 1945: 110 *(Tendipes).*—N.Y.; Oreg., Mich., Ont. **N. comb.**
modestus Say, 1823: 13 (1859b: 41).—Pa.; Wash. to Que., s. to Calif. and Va.
neomodestus Malloch, 1915g: 475.—Ill.; S. Dak. to Que., s. to Tex. and Va.
nervosus Staeger, 1839: 567.—?Denmark; Alta. to Que., s. to Calif. and Fla., Greenland, Europe.
  brevitibialis Zetterstedt, 1850: 3537.—Sweden.
  lucifer Johannsen, 1907a: 110.—Kans.
  indistinctus Malloch, 1915g: 477.—Ill.

Subgenus ENDOCHIRONOMUS Kieffer

Endochironomus Kieffer, 1918b: 69 (as genus). Type-species, *Chironomus alismatis* Kieffer (orig. des.) =*tendens* (Fabricius).
Demeijerea Kruseman, 1933: 154. Type-species, *Tipula rufipes* Linnaeus (orig. des.).

abruptus (Townes), 1945: 138 *(Glyptotendipes).*—N.Y.; Mo. **N. comb.**
atrimanus Coquillett, 1902g: 94.—Mo.; Alta. to N.Y., s. to Ala. and Fla.
brachialis Coquillett, 1901h: 607.—N.J.; Oreg. to N.Y., s. to Ala.

nigricans Johannsen, 1905: 219.—N.Y.; B.C. to Que., s. to Calif. and Fla.
>*johnsoni* Kieffer, 1906a: 19.—N.J. Proposed for *Chironomus albipennis* Meigen of Johnson, 1900c: 627. Nomen nudum, see Townes, 1945: 64.
>*albipennis*, authors, not Meigen.

obreptus (Townes), 1945: 139 (*Glyptotendipes*).—S.C.; N.Y. **N. comb.**

oldenbergi (Goetghebuer), 1932b: 130 (*Endochironomus*).—Lapland (Goetghebuer, 1933: 21); Greenland. **N. comb.**

subtendens (Townes), 1945: 65 (*Tanytarsus*).—N.Y.; Alta. to Ont. and Maine., s. to Colo. and Fla. **N. comb.**

Subgenus KIEFFERULUS Goetghebuer

Kiefferulus Goetghebuer, 1922: 40 (as genus). Type-species, *Tanytarsus tendipediformis* Goetghebuer (orig. des.).

dux Johannsen, 1905: 231.—N.Y.; Oreg. to Ont. and R.I., s. to La. and Fla.
>*obscuratus* Malloch, 1915g: 479.—Ill.

Subgenus NILODORUM Kieffer

Nilodorum Kieffer, 1921d: 272 (also 1922f: 45) (as genus). Type-species, *brevibucca* Kieffer (Kieffer, 1922f: 45).

devineyae Beck, 1961: 126.—Fla.

Subgenus TRIBELOS Townes

Tanytarsus, subg. **Tribelos** Townes, 1945: 66. Type-species, *Chironomus dimorphus* Malloch (orig. des.)=*jucundus* Walker.

ater (Townes), 1945: 68 (*Tanytarsus*).—Va.; Minn., Mich., Que., Fla. **N. comb.**

fuscicornis Malloch, 1915g: 466.—Ill.; Minn. to Md., s. to Tex. and Fla.

hesperius (Sublette), 1960: 217 (*Tendipes*).—Calif. **N. comb.**

jucundus Walker, 1848: 16.—Ga.; Alta. to Que., s. to Okla. and Fla.
>*dimorphus* Malloch, 1915g: 464.—Ill. ?Var.
>*fusciventris* Malloch, 1915g: 465.—Wis.

protextus (Townes), 1945: 69 (*Tanytarsus*).—N.Y.; N.J. **N. comb.**

quadripunctatus Malloch, 1915g: 437.—Wis.; Mich., Ind., Ohio, Ont., Que., N.H.

Subgenus WIRTHIELLA Sublette

Tendipes, subg. **Wirthiella** Sublette, 1960: 216. Type-species, *modocensis* Sublette (orig. des.).

modocensis (Sublette), 1960: 216 (*Tendipes*).—Calif. **N. comb.**

Subgenus XENOCHIRONOMUS Kieffer

Xenochironomus Kieffer, 1921a: 69 (as genus). Type-species, *Chironomus xenolabis* Kieffer (Goetghebuer *in* Goetghebuer and Lenz, 1937: 32).
Xenochironomus, subg. *Anceus* Roback, 1963: 237. Type-species, *Chironomus festivus* Say (orig. des.). **N. syn.**

REFERENCE: Roback, 1963 (tax. and immature stages).

dorneri Malloch, 1915g: 471.—Tex.; Fla.
festivus Say, 1823: 13 (1859b: 41).—Ill.; Kans. to Que., s. to Tex.
  lineatus Say, 1823: 14 (1859b: 42).—Pa.
  lineola, emend.
  lasiomerus Walker, 1848: 19.—Hudson Bay.
rogersi (Beck and Beck), 1958: 27 (*Xenochironomus*).—Fla. **N. comb.**
scopula (Townes), 1945: 93 (*Xenochironomus*).—Que.; Alta. to Ont. and N.Y., s. to Kans. and Ohio. **N. comb.**
taenionotus Say, 1829: 149 (1859b: 349).—Ind.; Iowa to Mich. and Pa., s. to Fla.
xenolabis Kieffer *in* Thienemann, 1916: 526.—Sweden; Mich. to Que. and Mass., s. to Okla. and Fla., Oreg., Europe.

Genus GLYPTOTENDIPES Kieffer

Glyptotendipes Kieffer, 1913f: 255. Type-species, *sigillatus* Kieffer (orig. des.).

Subgenus GLYPTOTENDIPES Kieffer

amplus Townes, 1945: 145.—N.J.
dreisbachi Townes, 1945: 145.—Mich.
seminole Townes, 1945: 146.—Fla.
senilis (Johannsen), 1937b: 37 (*Chironomus*).—N.Y.; Iowa, Mich., Ont.
unacus Townes, 1945: 146.—Que.

Subgenus PHYTOTENDIPES Goetghebuer

Glyptotendipes, subg. **Phytotendipes** Goetghebuer, 1934: 394. Type-species, *Chironomus pallens* Meigen (Goetghebuer *in* Goetghebuer and Lenz, 1937: 14).

barbipes (Staeger), 1839: 561 (*Chironomus*).—Denmark; B.C. to N.Y., s. to Ill. and N.J., Europe.
lobiferus (Say), 1823: 12 (1859b: 41) (*Chironomus*).—U.S.; B.C. to Que., s. to Calif. and Fla.
  caliginosus Johannsen, 1905: 205 (*Chironomus*; preocc. Meunier, 1904).—N.Y.
  ithacanensis Johannsen, 1908: 279 (*Chironomus*; n. name for *caliginosus* Johannsen).—N.Y.
  americanus Kieffer, 1917: 355.—Wash.
meridionalis Dendy and Sublette, 1959: 517.—Okla.; Fla., Ala., La.
paripes (Edwards), 1929b: 392 (*Chironomus*).—England; Alta. to Ont. and N.Y., s. to Ala., Europe.
testaceus Townes, 1945: 140.—N.Y.; Okla. to N.Y., s. to La. and Fla.

Genus GRACEUS Goetghebuer

Graceus Goetghebuer, 1928: 15. Type-species, *ambiguus* Goetghebuer (mon.).

Unnamed sp.—Ont. (J. R. Vockeroth *in litt.*).

Genus LAUTERBORNIELLA Bause

Lauterborniella Bause, 1913: 120 (1914: 120). Type-species, *Tanytarsus agrayloides* Kieffer (Kieffer, 1921b: 28).

agrayloides (Kieffer), 1911b: 51 (*Tanytarsus*).—Germany; Calif., N.Y., Ont., Fla., Europe.
perpulchra (Mitchell), 1908: 13 (*Chironomus*).—Md.; Ont. to Maine, s. to Okla. and Fla.
varipennis (Coquillett), 1902g: 94 (*Chironomus*).—N. Mex.; Mich. to Que. and Mass., s. to N. Mex. and Fla.

Genus MICROTENDIPES Kieffer

Microtendipes Kieffer, 1915: 70. Type-species, *Tendipes abbreviatus* Kieffer (orig. des.)=*chloris* (Meigen).

anticus (Walker), 1848: 21 (*Chironomus*).—Ga.; coastal, Va. to Tex.
caducus Townes, 1945: 24 (n. name for *pallidus* Johannsen).—N.Y.; Minn. to Que. and N.Y., s. to Calif. and Ga.
  pallidus Johannsen, 1905: 230 (*Chironomus*; preocc. Fabricius, 1805).—N.Y.
caelum Townes, 1945: 23.—Mass.
pedellus (De Geer), 1776: 378 (*Tipula*).—Sweden; B.C. to Que. and Maine, s. to Calif. and Fla., Europe.
  fascipes Coquillett, 1908: 145 (*Chironomus*).—N.J.

deflexa Walley, 1926b: 206 (*Chironomus*).—Ont.
stygius Townes, 1945: 26 (as var.).—N.Y. ?Var.
var. **aberrans** (Johannsen), 1905: 221 (*Chironomus;* as sp.).—
  Wash., N.Y., Pa., N.J.

Genus NILOTHAUMA Kieffer

Nilothauma Kieffer, 1921b: 27, 37 (also 1921d: 270, key). Type-species, *pictipenne* Kieffer (mon.).
Kribioxenus, authors, not Kieffer.
babiyi (Rempel), 1937b: 274 (*Chironomus*).—N.Y.; Mich., Ont., Ohio, Fla. **N. comb.**
bicorne (Townes), 1945: 35 (*Kribioxenus*).—S.C.; Fla. **N. comb.**
mirabile (Townes), 1945: 35 (*Kribioxenus*).—N.Y.; Fla. **N. comb.**

Genus OMISUS Townes

Omisus Townes, 1945: 27. Type-species, *pica* Townes (orig. des.).
pica Townes, 1945: 27.—N.H.; Idaho, Mass., Conn., Fla.

Genus PARALAUTERBORNIELLA Lenz

Paralauterborniella Lenz, 1941: 48. Type-species, *Lauterborniella brachylabis* Edwards (mon.).
Apedilum Townes, 1945: 32. Type-species, *subcinctum* Townes (as *succinctum;* orig. des.).
elachista (Townes), 1945: 33 (*Apedilum*).—Tex.; Miss., Ala., Fla., Ga., Mass.
nigrohalteralis (Malloch), 1915g: 440 (*Chironomus*).—Ill.; Colo., Ind., Mich., N.Y., Ont., Fla.
subcincta subcincta (Townes), 1945: 33 (*Apedilum*).—Nev.; Calif., Ariz., N. Mex., Fla.
  succinctum, error.
  ssp. **alamedensis** Sublette, 1960: 203.—Calif.

Genus PARATENDIPES Kieffer

Paratendipes Kieffer, 1911b: 41. Type-species, *Chironomus albimanus* Meigen (orig. des.).
albimanus (Meigen), 1818: 40 (*Chironomus;* n. name for *annularis* Meigen).—Europe; Idaho to N.Y., s. to Calif. and N.C.
  annularis Meigen, 1804: 17 (*Chironomus;* preocc. De Geer, 1776).—Europe.
basidens Townes, 1945: 30.—N.J.; Iowa, Ala.

duplicatus (Johannsen), 1937b: 28 (*Chironomus*).—N.Y.; Mo., Ill., Mich., Ohio, R.I., Pa.
    basalis Malloch, 1915g: 441 (*Chironomus*; preocc. Staeger, 1845).—Ill.
fuscitibia Sublette, 1960: 200.—Calif.
nitidulus (Coquillett), 1901h: 608 (*Chironomus*).—N.J.; Mich., Ohio, Pa., Del., Md.
subaequalis (Malloch), 1915g: 440 (*Chironomus*).—Ill.; Calif., Ont., N.Y., N.J., Fla.
thermophilus Townes, 1945: 31.—Ariz.; Calif.

Genus PHAENOPSECTRA Kieffer

Phaenopsectra Kieffer, 1921d: 274. Type-species, *Chironomus connectens* Kieffer (mon.).
Sergentia Kieffer, 1922a: 288. Type-species, *profundorum* Kieffer (orig. des.)=*coracina* (Zetterstedt).

albescens (Townes), 1945: 73 (*Tanytarsus*).—Wash.; B.C. to Alta., s. to Calif. and Colo.
coracina (Zetterstedt), 1850: 3508 (*Chironomus*).—Sweden; Alta., Sask., Mont., Wis., Ont., Greenland, Europe. **N. comb.**
dyari (Townes), 1945: 75 (*Tanytarsus*).—Calif.; Mont., Fla.
flavipes (Meigen), 1818: 50 (*Chironomus*).—Europe; Wash. to Calif., e. to Que. and Fla.
    flavicauda Malloch, 1915g: 493 (*Tanytarsus*).—Ill.
incompta (Zetterstedt), 1838: 816 (*Chironomus*).—Europe; Greenland.
obediens (Johannsen), 1905: 286 (*Tanytarsus*).—Wash., N.Y.; Minn. to Que. and Maine, s. to Mo. and S.C., also Wash., Colo. **N. comb.**
profusa (Townes), 1945: 73 (*Tanytarsus*).—Nev.; Wash. to Mont., s. to Calif. and N. Mex.
punctipes (Wiedemann), 1817: 65 (*Chironomus*).—Germany; Idaho, Que., N.J., N.Y., S.C., Europe.
vittata (Townes), 1945: 77 (*Tanytarsus*).—N.J.; Ont., Que. **N. comb.**

Genus POLYPEDILUM Kieffer

Polypedilum Kieffer, 1913b: 15. Type-species, *emarginatum* Kieffer (orig. des.)=*nubeculosum* (Meigen).
Polypedilum, subg. *Tripodura* Townes, 1945: 36. Type-species, *simulans* Townes (orig. des.).

Subgenus POLYPEDILUM Kieffer

acifer Townes, 1945: 46.—Mich.
albinodus Townes, 1945: 41.—Idaho; Calif., Va.

angustum Townes, 1945: 56.—N.J.; Fla.
apicatum Townes, 1945: 39.—N. Mex.; Calif., Ill.
artifer (Curran), 1930j: 32 (*Chironomus*).—N.Y.; Calif. to Que. and Va.
aviceps Townes, 1945: 61.—N.Y.; Calif., Nev., Idaho, Mont.
braseniae (Leathers), 1922: 8 (*Chironomus*).—N.Y.; Oreg., Minn., Ont., Va., S.C., Fla.
californicum Sublette, 1960: 204.—Calif.
cinctum Townes, 1945: 59.—Nev.
convictum (Walker), 1856a: 161 (*Chironomus*).—England; Wis. to N.Y., s. to Kans. and Fla., Europe.
    flavus Johannsen, 1905: 225 (*Chironomus*).—N.Y.
    flaviventris Johannsen, 1907a: 111 (*Chironomus*).—Kans.
digitifer Townes, 1945: 45.—S. Dak.; S. Dak. to Mich., s. to Tex. and Fla., Calif.
fallax (Johannsen), 1905: 210 (*Chironomus*).—N.Y.; Mich. to Que. and N.H., s. to Tex. and Fla.
    calopterus Mitchell, 1908: 8 (*Chironomus*).—Md. ?Var.
floridense Townes, 1945: 43.—Miss.; Fla.
fuscipenne (Meigen), 1818: 35 (*Chironomus*).—Europe; Wash., Oreg., Mont., Colo.
gomphus Townes, 1945: 42.—Mass.; S.C., Fla.
griseopunctatum (Malloch), 1915g: 428 (*Chironomus*).—Ill.; Mich. to N.Y., s. to Iowa and Va., also Mont., Nev., Fla.
halterale (Coquillett), 1901c: 17 (*Chironomus*).—D.C.; Wis. to N.Y., s. to Okla. and Fla., also Calif., Ariz.
illinoense (Malloch), 1915g: 471 (*Chironomus*).—Ill.; Mont., Calif., Minn. to Que., s. to Tex. and Fla.
    decoloratus Malloch, 1915g: 472 (*Chironomus*; as var.).—Ill. ?Var.
isocerus Townes, 1945: 42.—Calif.; Nev.
labeculosum (Mitchell), 1908: 14 (*Chironomus*).—Ariz.; Calif.
laetum (Meigen), 1818: 38 (*Chironomus*).—Europe; Calif., Nev., N. Mex., Alta., Colo., N.Y., S.C.
nigritum Townes, 1945: 56.—N.Y.; Calif., Idaho, Ont., Fla.
nubeculosum (Meigen), 1804: 18 (*Chironomus*).—Europe; Idaho, Colo., Minn., Mich., N.Y., Pa.
obtusum Townes, 1945: 60.—N.Y.; Mich., Fla., La.
ontario (Walley), 1926b: 206 (*Chironomus*).—Ont.; Tex. to Que., s. to D.C.
    hirtipes Mitchell, 1908: 9 (*Chironomus*; preocc. Macquart, 1834).—Md.
ophioides Townes, 1945: 57.—N.Y.; Calif., Idaho, Mich., Ohio, Ont., Que.

parascalaenum Beck, 1962: 91.—Fla.
pardus Townes, 1945: 41.—Tex.
parvum Townes, 1945: 40.—Miss.; Calif., Tex. to Md., s. to Fla.
pedatum pedatum Townes, 1945: 55.—N.Y.; Mo., La., Mich., Va.
  ssp. **excelsius** Townes, 1945: 55.—Wash.; Calif.
pterospilus Townes, 1945: 40.—Tex.; Ariz., Fla.
scalaenum (Schrank), 1803: 73 (*Tipula*).—Germany; Wash. to Ont. and N.H., s. to Calif. and Fla., Europe.
  needhamii Johannsen, 1908: 278 (*Chironomus*).—Wash., Kans., Ind., N.Y.
simulans Townes, 1945: 43.—N.Y.; Mont. to Ont. and Maine, s. to Calif. and Fla.
subcultellatum Sublette, 1960: 204.—Calif.
sulaceps Townes, 1945: 58.—Idaho; Calif.
trigonus Townes, 1945: 49.—N.Y.; Minn. to N.Y., s. to La. and Fla., Idaho.
vibex Townes, 1945: 52.—Ont.; Wash., Nev., N.J., Fla.
walleyi Townes, 1945: 52.—Que.; Ont.

Subgenus PENTAPEDILUM Kieffer

Pentapedilum Kieffer, 1913b: 25 (as genus). Type-species, *stratiotale* Kieffer (Edwards, 1929b: 376)=*tritum* (Walker).

albulum Townes, 1945: 63.—N.Y.; Iowa, Fla.
sordens (Wulp), 1874: 141 (*Tanytarsus*).—Netherlands; Mich., N.Y., R.I., Va., Europe.
  fulvescens Johannsen, 1905: 293 (*Tanytarsus*).—N.Y.
  americana Kieffer, 1917: 357 (*Calopsectra*).—N.Y.
tritum (Walker), 1856a: 342 (*Chironomus*).—England; Idaho to N.Y., s. to Calif. and Fla., Europe.

Genus PSEUDOCHIRONOMUS Malloch

Pseudochironomus Malloch, 1915g: 500. Type-species, *richardsoni* Malloch (orig. des.).

aix Townes, 1945: 19.—Mich.; Fla.
anas Townes, 1945: 17.—N.Y.
banksi Townes, 1945: 17.—Mass.; N.J., Fla.
chen Townes, 1945: 18.—Md.; Ont.
crassus Townes, 1945: 15.—Hudson Bay Territory; N.W.T.
fulviventris (Johannsen), 1905: 229 (*Chironomus*).—N.Y.; Minn. to Que., s. to Kans. and Fla.
middlekauffi Townes, 1945: 18.—N.Y.; Ill., Fla.
netta Townes, 1945: 19.—N.Y.; Mich., Mass., Ont., N.J.

pseudoviridis (Malloch), 1915g: 450 (*Chironomus*).—Ill.; Colo., Kans., Okla., Tex., Wis., Mich., N.Y.
rex Hauber, 1947: 458.—Iowa; Fla.
richardsoni Malloch, 1915g: 500.—Ill.; Calif. to Ont., and N.Y. to Fla.

Genus STENOCHIRONOMUS Kieffer

Stenochironomus Kieffer, 1919a: 44. Type-species, *Chironomus pulchripennis* Coquillett (Townes, 1945: 84).

aestivalis Townes, 1945: 86.—N.J.; N.Y.
browni Townes, 1945: 86.—Fla.
cinctus Townes, 1945: 87.—Fla.; R.I.
colei (Malloch) *in* Cole and Lovett, 1919: 255 (*Chironomus*).— Oreg.; Wash. to Alta., s. to Calif. and Colo.
hilaris (Walker), 1848: 17 (*Chironomus*).—Unknown; Wis. to Ont. and N.Y., s. to Okla. and Fla.
  nephopterus Mitchell, 1908: 7 (*Chironomus*).—Md.
macateei (Malloch), 1915a: 45 (*Chironomus*).—Md.; Iowa to N.Y., s. to Tex. and Fla.
poecilopterus (Mitchell), 1908: 10 (*Chironomus*).—Md.; Colo. to N.Y., s. to Ark. and S.C., Fla.
pulchripennis (Coquillett), 1902g: 94 (*Chironomus*).—N.H.; Ont., Que., N.Y., N.J., Pa.
taeniapennis (Coquillett), 1901h: 607 (*Chironomus*).—Mass., N.J.; S. Dak. to Que. and Mass., s. to Ill. and N.C., also Calif.
  exquisitus Mitchell, 1908: 11 (*Chironomus*).—Md.
  zonopterus Mitchell, 1908: 12 (*Chironomus*).—N.J.
totifuscus Sublette, 1960: 210.—Calif.
unictus Townes, 1945: 89.—N.J.

Genus STICTOCHIRONOMUS Kieffer

Stictochironomus Kieffer, 1919a: 44. Type-species, *Chironomus pictulus* Meigen (Townes, 1945: 77).

albicrus (Townes), 1945: 79 (*Tanytarsus*).—Kans. **N. comb.**
annulicrus (Townes), 1945: 83 (*Tanytarsus*).—N.Y.; Ont. **N. comb.**
devinctus (Say), 1829: 150 (1859b: 349) (*Chironomus*).—Ind.; Minn. to Maine, s. to Tex. and Fla. **N. comb.**
  compes Coquillett, 1908: 145 (*Chironomus*).—Md.
  ornatipes Kieffer, 1917: 343 (*Chironomus*).—N.Y.
flavicingulus (Walker), 1848: 20 (*Chironomus*).—Hudson Bay. **N. comb.**

lutosus (Townes), 1945: 83 (*Tanytarsus*).—Mass. **N. comb.**
marmoreus (Townes), 1945: 81 (*Tanytarsus*).—Sask.; S. Dak., Mo., N.Y. **N. comb.**
naevus (Mitchell), 1908: 14 (*Chironomus*).—N. Mex.; Calif.
palliatus (Coquillett), 1902g: 95 (*Chironomus*).—D.C.; Wis. to N. Y., s. to Tex. and Ala.
quagga (Townes), 1945: 81 (*Tanytarsus*).—Calif.
rosenscholdi (Zetterstedt), 1838: 811 (*Chironomus*).—Europe; Sask.
unguiculatus (Malloch), 1934b: 16 (*Chironomus*).—N.W.T. (Southampton I.). **N. comb.**
varius (Townes), 1945: 82 (*Tanytarsus*).—N.Y.; Colo., Kans., Iowa, Minn., Mich., Ohio. **N. comb.**
virgatus (Townes), 1945: 84 (*Tanytarsus*).—Mont. **N. comb.**

Tribe TANYTARSINI

Genus MICROPSECTRA Kieffer

Tanytarsus, subg. **Micropsectra** Kieffer, 1909: 50. Type-species, *inermipes* Kieffer (as *inermis*; Kieffer, 1921b: 35) = *brunnipes* (Zetterstedt). Edwards, 1929b, says that the genotype has not been properly described but seems to resemble *Tanytarsus brunnipes* (Zetterstedt) which is therefore to be taken as the type of the genus.

brunnipes (Zetterstedt), 1850: 3518 (*Chironomus*).—Sweden; Mich., Greenland.
  gmundensis Egger, 1863: 1109 (*Chironomus*).—Austria.
connexa (Kieffer), 1906a: 17 (*Chironomus*; n. name for *confinis* Walker).—Hudson Bay. **N. comb.**
  confinis Walker, 1848: 15 (*Chironomus*; preocc. Meigen, 1830).—Hudson Bay.
  brunneus Walker, 1848: 21 (*Chironomus*).—Hudson Bay. **N. syn.**
deflecta (Johannsen), 1905: 288 (*Tanytarsus*).—N.Y. **N. comb.**
dives (Johannsen), 1905: 288 (*Tanytarsus*).—N.Y.; Mont., Colo., Wis., Ky., Mass. **N. comb.**
dubia (Malloch), 1915g: 496 (*Tanytarsus*).—Ill.; Ala.
groenlandica Andersen, 1937: 33, 34.—Greenland.
insignilobus Kieffer, 1924b: 86.—Norway; N.W.T. (Ellesmere I., D. R. Oliver *in litt.*).
logani (Johannsen), 1928a: 33 (*Tanytarsus*).—Utah. **N. comb.**
natvigi Goetghebuer, 1933: 21.—Greenland.
nigripila (Johannsen), 1905: 287 (*Tanytarsus*).—N.Y.; Wash., Utah, Colo., Iowa, Ill., Ohio. **N. comb.**

polita (Malloch), 1915g: 493 (*Tanytarsus*).—Ill.; Utah, Pa., Md. **N. comb.**
recurvata Goetghebuer, 1928: 117.—Belgium; Greenland.
similata (Malloch), 1915g: 494 (*Tanytarsus*).—Wis.; Alaska (St. Paul I.). **N. comb.**
xantha (Roback), 1955: 4 (*Calopsectra*).—Mass. **N. comb.**

Genus STEMPELLINA Bause

Stempellina Bause, 1913: 120 (1914: 120). Type-species, *Tanytarsus bausei* Kieffer (Edwards, 1929b: 491).

johannsenii (Bause), 1913: 67 (1914: 67) (*Tanytarsus*).—N.Y. Larva only; unrecognized.

Genus TANYTARSUS Wulp

Tanytarsus Wulp, 1874: 134. Type-species, *signatus* Wulp (I.C.Z.N., 1961b: 361, under plenary powers).
Holtedahlia Kieffer, 1922b: 5. Type-species, *borealis* Kieffer (orig. des.).

Subgenus TANYTARSUS Wulp

brevipalpis (Kieffer), 1926: 78 (*Holtedahlia*).—N.W.T. (Ellesmere I.).
confusus Malloch, 1915g: 490.—Ill.; N.Y., D.C., Fla., Ala.
dissimilis Johannsen, 1905: 292.—Ont.; Minn., Iowa, Ohio, Okla., Tex.
fatigans Johannsen, 1905: 292.—N.Y.; Colo.
glabrescens Edwards.—Not Nearctic.
gracilentus Holmgren, 1883: 181.—Novaya Zemlya; Alaska, N.W.T. (Ellesmere I.).
    deviatus Malloch, 1923i: 172 (*Chironomus*).—Alaska (St. Paul I.). **N. syn.**
guerlus (Roback), 1957a: 133 (*Calopsectra*).—Pa.; N.J. **N. comb.**
islandicus Malloch, 1934b: 17.—N.W.T.
neoflavellus Malloch, 1915g: 489.—Ill.; Ohio, N.Y., Okla., Tex., Ala.
niger Andersen, 1937: 40.—Greenland; N.W.T. (Ellesmere I.).
varelus (Roback), 1957a: 128 (*Calopsectra*).—Pa. **N. comb.**

Subgenus CALOPSECTRA Kieffer

Tanytarsus, subg. **Calopsectra** Kieffer, 1909: 50. Type-species, *gregarius* Kieffer (Kieffer, 1921b: 36).

gregarius Kieffer.—Not Nearctic.
muticus Johannsen, 1905: 294.—N.Y.; Colo., Wis., Ill., Ohio.

Subgenus CLADOTANYTARSUS Kieffer

Cladotanytarsus Kieffer, 1922d: 100 (as genus). Type-species, *pallidus* Kieffer (mon.).

viridiventris Malloch, 1915g: 491.—Mich.; Ont., N.Y., Ala.

Subgenus LUNDSTROEMIA Kieffer

Lundstroemia Kieffer, 1921b: 36 (as genus). Type-species, *Tanytarsus roseiventris* Kieffer (mon.).
Ditanytarsus Kieffer, 1922d: 98. Type-species, *unicolor* Kieffer (Goetghebuer *in* Goetghebuer and Lenz, 1938: 102).

setosimanus (Goetghebuer), 1933: 21 (*Ditanytarsus*).—Greenland; N.W.T. (Ellesmere I., D. R. Oliver *in litt.*).

Subgenus RHEOTANYTARSUS Bause

Syntanytarsus, subg. **Rheotanytarsus** Bause, 1913: 120 (1914: 120). Type-species, *Tanytarsus pentapoda* Kieffer (Goetghebuer *in* Goetghebuer and Lenz, 1954: 132).

akrina (Roback), 1960: 1 (*Calopsectra*).—Kans. **N. comb.**
exiguus Johannsen, 1905: 294.—N.Y.; Wis. to N.Y., s. to Iowa and Ky., also Idaho.
pellucidus (Walker), 1848: 21 (*Chironomus*).—Hudson Bay.

Unplaced Species of *Tanytarsus*

bausei Kieffer.—Not Nearctic.
flabellatus (Meigen).—Not Nearctic.
flavellus (Zetterstedt).—Probably not Nearctic (type unrecognizable). N. Amer. records unchecked.
junci (Meigen).—Not Nearctic.
tenuis (Meigen).—Not Nearctic.

Genus ZAVRELIA Kieffer

Zavrelia Kieffer *in* Bause, 1913: 73 (1914: 73). Type-species, *pentatoma* Kieffer (mon.).

This genus has been reported from the Nearctic Region from larval identifications (Roback, 1953, 1957a). No species have been definitely identified.

pentatoma Kieffer.—Not Nearctic.

Genus CORYNOCERA Zetterstedt

Corynocera Zetterstedt, 1838: 856. Type-species, *ambigua* Zetterstedt (mon.).

REFERENCE: Hirvenoja, 1961 (tax., biol., immature stages).

ambigua Zetterstedt, 1838: 856.—Sweden (Lapland); N.W.T. (Southampton I.), n. Europe.

Unplaced Species of Chironomidae

baltimoreus Macquart, 1855: 35 (1855: 15) (*Tanypus*).—Md.
bimaculus Walker, 1848: 15 (*Chironomus*).—Hudson Bay.
borealis Curtis, 1835b: lxxvii (*Chironomus*).—Arctic Amer.
excavatus Kieffer, 1917: 348 (*Chironomus*).—N.Y.
excisus Kieffer, 1917: 346 (*Chironomus*).—Pa.
halophilus Packard *in* Verrill, 1873: 539 (*Chironomus*).—Mass.
harti Malloch, 1915g: 457 (*Chironomus*).—Ill.
melanderi Kieffer, 1917: 347 (*Chironomus*).—Wash.
nigritibia Walker, 1848: 16 (*Chironomus*).—Hudson Bay.
tibialis Say, 1823: 15 (1859b: 43) (*Tanypus*).—Pa.
tibialis Staeger, 1845b: 354 (*Tanypus*; preocc. Say, 1823).—Greenland.
trichomerus Walker, 1848: 21 (*Chironomus*).—Hudson Bay.

Family SIMULIIDAE

By Alan Stone

The adults of the family Simuliidae are commonly known as black flies or buffalo gnats. The immature stages are found in flowing water, usually attached to sticks, stones, or living vegetation. Many of the species are bloodsucking and cause a great deal of annoyance; heavy attacks may be fatal to cattle, horses, or poultry. Black flies have been incriminated in the transmission of several diseases of wild and domestic birds. The family is undergoing intensive taxonomic and biological study in the United States and Canada. Recent cytogenetic studies indicate that a number of the named species are complexes of sibling species and in these instances little reliance can be placed on distribution records. Certain Palaearctic species, such as *Simulium latipes* and *S. aureum*, are probably represented in the Nearctic Region by closely related ones.

REFERENCES: Malloch, 1914j (rev.); Dyar and Shannon, 1927 (rev.); Twinn, 1936 (e. Canada); Stone, 1952 (Alaska); Stone and Jamnback, 1955 (N.Y.); Sommerman, Sailer, and Esselbaugh, 1955 (biol.; Alaska); Peterson, 1960b (Utah); Davies, Peterson, and Wood, 1962 (keys, Ont.).

Subfamily PROSIMULIINAE

Tribe GYMNOPAIDINI

Genus GYMNOPAIS Stone

Gymnopais Stone, 1949: 260. Type-species, *dichopticus* Stone (orig. des.).

dichopticus Stone, 1949: 261.—Alaska; N.W.T.
holopticus Stone, 1949: 265.—Alaska; n. Canada.

Genus TWINNIA Stone and Jamnback

Twinnia Stone and Jamnback, 1955: 18. Type-species, *tibblesi* Stone and Jamnback (orig. des.).

nova (Dyar and Shannon), 1927: 5 (*Prosimulium*).—Mont.; B.C. and Mont. to n. Calif. and Wyo.
    biclavata Shewell, 1959: 686.—B.C. **N. syn.**
tibblesi Stone and Jamnback, 1955: 19.—Labr.; Ont., N.Y., N.H., Vt.

Tribe PARASIMULIINI

Genus PARASIMULIUM Malloch

Parasimulium Malloch, 1914j: 24. Type-species, *furcatum* Malloch (orig. des.).

furcatum Malloch, 1914j: 24.—n. Calif.; Oreg.

Tribe PROSIMULIINI

Genus PROSIMULIUM Roubaud

Simulium, subg. **Prosimulium** Roubaud, 1906: 521. Type-species, *hirtipes* Fries (Malloch, 1914j: 16).

Subgenus PROSIMULIUM Roubaud

alpestre Dorogostajskij, Rubzov, and Vlasenko, 1935: 136.—U.S.S.R.; Alaska, Y.T., B.C., e. Siberia. Unrevised species complex.
caudatum Shewell, 1959: 688.—B.C.; Wash.
daviesi Peterson and DeFoliart, 1960: 85.—Utah; Wyo.
decemarticulatum (Twinn), 1936: 110 (*Simulium*).—Ont.; Alaska, Y.T., Man., Wis., Mich., N.H.
dicum Dyar and Shannon, 1927: 7.—Wash.; B.C. to cent. Calif.
doveri Sommerman, 1962a: 226.—Alaska.

exigens Dyar and Shannon, 1927: 10.—Idaho; B.C. to Calif., Ariz., and Colo.
    dicentum Dyar and Shannon, 1927: 7.—Calif.
    hardyi Stains and Knowlton, 1940: 78 (*Simulium*).—Utah.
    tenuicalx (in part) Enderlein, 1925c: 203. Males only.
flaviantennum (Stains and Knowlton), 1940: 79 (*Simulium*).—Utah; Idaho, Mont., Wyo., Colo.
fontanum Syme and Davies, 1958: 708.—Ont.; Que.
formosum Shewell, 1959: 692.—B.C.; Y.T., Alta.
frohnei Sommerman, 1958: 196.—Alaska.
fulvithorax Shewell, 1959: 694.—B.C.
fulvum (Coquillett), 1902g: 96 (*Simulium*).—Alaska, B.C., Mont., Colo.; Alaska and Y.T., s. to Calif. and Colo.
fuscum Syme and Davies, 1958: 702.—Ont.; Wis. to Labr., N.S., and N.Y.
gibsoni (Twinn), 1936: 108 (*Simulium*).—Ont.; n. Man. to Que., s. to Wis.
hirtipes (Fries).—Not Nearctic.
longilobum Peterson and DeFoliart, 1960: 100.—Utah.
magnum Dyar and Shannon, 1927: 6.—Va.; Mich. to Conn., s. to Ill. and Ga.
    frisoni Dyar and Shannon, 1927: 18 (*Eusimulium*).—Ill.
mixtum Syme and Davies, 1958: 706.—Ont.; e. Ont. to Labr., N.S., and N.Y., ?Wis., ?Minn.
multidentatum (Twinn), 1936: 106 (*Simulium*).—Que.; Mich., Ont., Ohio, N.Y.
perspicuum Sommerman, 1958: 199.—Alaska.
pleurale Malloch, 1914j: 17.—B.C.; Alaska, Idaho, Mont.
    tenuicalx Enderlein, 1925c: 203.—Idaho (as U.S.A.; unpublished). **N. syn.** Females only, lectotype to be selected; males=*exigens*.
    pancerastes Dyar and Shannon, 1927: 10.—Idaho.
rhizophorum Stone and Jamnback, 1955: 28.—Pa.; Ill., N. Y., Conn., R.I., Maine.
saltus Stone and Jamnback, 1955: 29.—N.Y.
shewelli Peterson and DeFoliart, 1960: 96.—Wyo.; Utah.
travisi Stone, 1952: 76.—Alaska; Alaska to N.W.T., s. to Calif. and Wyo.
uinta Peterson and DeFoliart, 1960: 91.—Wyo.; Utah.
unicum (Twinn), 1938: 49 (*Simulium*).—Utah.
ursinum (Edwards), 1935b: 535 (*Simulium*).—Bear I. (Greenland Sea); N.W.T., ?Alaska.
    browni Twinn, 1936: 113 (*Simulium*).—N.W.T. (Baffin I.).
vernale Shewell, 1952: 33.—Ont.

Subgenus HELODON Enderlein

Helodon Enderlein, 1921a: 199 (as genus). Type-species, *Simulium ferrugineum* Wahlberg (orig. des.).

onychodactylum Dyar and Shannon, 1927: 4.—Colo.; Alaska, s. to n. Calif. and Colo.

Subfamily SIMULIINAE

Genus CNEPHIA Enderlein

Cnephia Enderlein, 1921a: 199. Type-species, *Simulium pecuarum* Riley (orig. des.).

Subgenus CNEPHIA Enderlein

borealis (Malloch), 1919f: 41 (*Prosimulium*).—N.W.T. (Victoria I.). Unrecognized.
dacotensis (Dyar and Shannon), 1927: 20 (*Eusimulium*).—S. Dak.; Sask. to Nfld., s. to S. Dak., Ill., Pa., and R.I.
    lascivum Twinn, 1936: 127 (*Simulium*).—Ont.
eremites Shewell, 1952: 36.—N.W.T.; Alaska, Man.
freytagi DeFoliart and Peterson, 1960: 216.—Wyo.
jeanae DeFoliart and Peterson, 1960: 218.—Utah; Wyo.
ornithophilia Davies, Peterson, and Wood, 1962: 102.—Ont.
osborni (Stains and Knowlton), 1943: 271 (*Eusimulium*).—Calif.
pecuarum (Riley), 1887: 512 (*Simulium*).—La.; Okla., Ill., and S.C., s. to Tex. and Fla.
saileri Stone, 1952: 82.—Alaska; n. Canada.
saskatchewana Shewell and Fredeen, 1958: 733.—Sask.
sommermanae Stone, 1952: 84.—Alaska.
villosa DeFoliart and Peterson, 1960: 213.—Wyo.; Utah.

Subgenus STEGOPTERNA Enderlein

Stegopterna Enderlein, 1930a: 89 (as genus). Type-species, *richteri* Enderlein (orig. des.).

emergens Stone, 1952: 80.—Alaska; N.W.T., n. Man., Wis., Ont.
minus (Dyar and Shannon), 1927: 21 (*Eusimulium*).—Calif.; Wash., Mont., Wyo.
mutata (Malloch), 1914j: 20 (*Prosimulium*).—N.J.; Alaska to Labr., s. to n. Calif., Utah, and Appalachian Mts. to Ala.
    permutatum Dyar and Shannon, 1927: 17 (*Eusimulium;* as ssp.).—B.C.
stewarti Coleman, 1953: 45.—Calif.

Subgenus ECTEMNIA Enderlein

Ectemnia Enderlein, 1930a: 88 (as genus). Type-species, *Cnetha taeniatifrons* Enderlein (orig. des.).

invenusta (Walker), 1848: 112 (*Simulium*).—Ont.; Que., N.Y.
    loisae Stone and Jamnback, 1955: 35.—N.Y.
taeniatifrons (Enderlein), 1925c: 206 (*Cnetha*).—Ill.; Alta. to Mich. and Ill.

Unplaced Species of Cnephia

abdita Peterson, 1962: 96.—Ont.; N.Y.
denaria Davies, Peterson, and Wood, 1962: 97.—Ont.

Genus SIMULIUM Latreille

Simulium Latreille, 1802: 426. Type-species, *Rhagio colombaschensis* Fabricius (mon.).
Simulia, emend.
Aspathia Enderlein, 1935a: 359. Type-species, *Simulium hunteri* Malloch (orig. des.).

Subgenus EUSIMULIUM Roubaud

Simulium, subg. **Eusimulium** Roubaud, 1906: 521. Type-species, *aureum* Fries (mon.).

aestivum Davies, Peterson, and Wood, 1962: 104.—Ont.
aureum Fries, 1824: 16.—Scandinavia; Holarctic, s. to Ariz., Guatemala, Ark., Ga., Sardinia. Unrevised species complex.
    bracteatum Coquillett, 1898i: 69.—Mass., Calif.
    pilosum Knowlton and Rowe, 1934: 580 (*Eusimulium*).—Utah.
    utahense Knowlton and Rowe, 1934: 582 (*Eusimulium*).—Utah.
baffinense Twinn, 1936: 121.—N.W.T.; Arctic and Transition, Alaska to Utah and Que.
    pallens Twinn, 1936: 123 (as form).—N.W.T.
bicorne Dorogostajskij, Rubzov, and Vlasenko, 1935: 178.—U.S.S.R.; Alaska, N.W.T., B.C., Utah, e. Siberia.
canonicola (Dyar and Shannon), 1927: 22 (*Eusimulium*).—Wyo.; B.C. to Sask., s. to Calif. and Colo.
    quadratus Stains and Knowlton, 1943: 271 (*Eusimulium*).—Utah.
congareenarum (Dyar and Shannon), 1927: 20 (*Eusimulium*).—S.C.; N.Y., Va., Ga., Fla., La., ?Wis.

croxtoni Nicholson and Mickel, 1950: 41.—Minn.; Wis., Ont., N.Y., Maine, Labr.
emarginatum Davies, Peterson, and Wood, 1962: 110.—Ont.
euryadminiculum Davies, 1949: 45.—Ont.; Wis., Que., Labr., Maine.
excisum Davies, Peterson, and Wood, 1962: 113.—Ont.
furculatum (Shewell), 1952: 40 (*Eusimulium*).—Man.; Alaska to Nfld.
gouldingi Stone, 1952: 90.—Pa.; Alaska, Wis., Ont., N.Y., N.H.
impar Davies, Peterson, and Wood, 1962: 116.—Ont.
innocens (Shewell), 1952: 38 (*Eusimulium*).—Ont.
johannseni johannseni Hart *in* Forbes, 1912: 31.—Ill.; Wyo. to N.Y., s. to Miss.
    ssp. **duplex** Shewell and Fredeen, 1958: 734.—Sask.
latipes (Meigen), 1804: 96 (*Atractocera*).—Europe; Alaska to Maine, s. to Calif., Wis., and Pa. Unrevised species complex.
pugetense (Dyar and Shannon), 1927: 23 (*Eusimulium*).—Wash.; Alaska to Que. and Maine, s. to Calif., Utah, and W. Va.
quebecense Twinn, 1936: 117.—Que.; Ont.
rivuli Twinn, 1936: 120.—Ont.; Que., N.H., Conn.
subexcisum Edwards.—Not Nearctic.
wyomingense Stone and DeFoliart, 1959: 395.—Wyo.

Subgenus BYSSODON Enderlein

Byssodon Enderlein, 1925c: 209 (as genus). Type-species, *Simulium forbesi* Malloch (orig. des.)=*meridionale* Riley.

meridionale Riley, 1887: 513.—Miss.; Alaska to Ind., s. to Calif., Mexico, and Fla.
    occidentale Townsend, 1891g: 107.—N. Mex.
    forbesi Malloch, 1914j: 63.—Ill.
rugglesi Nicholson and Mickel, 1950: 60.—Minn.; Alaska, Y.T., Sask. to Labr., s. to Utah, Vt., and N.B.
slossonae Dyar and Shannon, 1927: 34.—Fla.; Va. to Tex. (coastal plain).
transiens Rubzov, 1940a: 361.—e. Siberia; Alta., Sask.

Subgenus PSILOPELMIA Enderlein

Psilopelmia Enderlein, 1934a: 283 (as genus). Type-species, *rufidorsum* Enderlein (orig. des.)=*escomeli* Roubaud.

bivittatum Malloch, 1914j: 31.—N. Mex.; Alta., Sask., and w. U. S., e. to S. Dak., Nebr., Tex.
    clarum Dyar and Shannon, 1927: 21 (*Eusimulium*).—Calif.
    idahoense Twinn, 1938: 50.—Idaho.

griseum Coquillett, 1898i: 69.—Colo.; Sask. and Alta., s. to Calif.
and Tex.
mediovittatum Knab, 1915b: 77.—Tex.
notatum Adams, 1904b: 434.—Ariz.; Calif., Nev.
trivittatum Malloch, 1914j: 30.—Mexico; Calif., N. Mex., Tex.
  distinctum Malloch, 1913p: 133 (preocc. Lutz, 1910).—Tex.
venator Dyar and Shannon, 1927: 36.—Nev.; Calif.
  beameri Stains and Knowlton, 1943: 279.—Calif.

Subgenus PSILOZIA Enderlein

Psilozia Enderlein, 1936c: 113 (as genus). Type-species, *groenlandica* Enderlein (orig. des.) = *vittatum* Zetterstedt.
Simulium, subg. *Neosimulium* Vargas, Martínez Palacios, and Díaz Nájera, 1946: 103, 108. Type-species, *vittatum* Zetterstedt (orig. des.).
Simulium, subg. *Neosimulium* Rubzov, 1940a: 116, 124, 130. Nomen nudum, type-species as above, but no descr.

argus Williston, 1893d: 253.—Calif.; B.C. to Man., s. to s. Calif.
and Tex.
  obtusum Dyar and Shannon, 1927: 15 (*Eusimulium*).—Calif.
  kamloopsi Hearle, 1932: 12.—B.C.
  hearlei Twinn, 1938: 50.—Utah.
encisoi Vargas and Díaz Nájera, 1949: 292.—Mexico; Calif.
vittatum Zetterstedt, 1838: 803.—Greenland; Alaska to Greenland, s. to Calif., Tex., La., and S.C., Iceland.
  tribulatum Lugger, 1897: 179.—Minn.
  glaucum Coquillett, 1902g: 97.—Mo.
  venustoides Hart *in* Forbes, 1912: 42.—Ill.
  groenlandica Enderlein, 1936c: 114 (*Psilozia*; preocc. Enderlein,1935).—Greenland.
  asakakae Smart, 1944: 131 (n. name for *groenlandica* Enderlein).—Greenland.

Subgenus HEMICNETHA Enderlein

Hemicnetha Enderlein, 1934c: 190 (as genus). Type-species, *mexicana* Enderlein (orig. des.) = *paynei* Vargas.

solarii Stone, 1948: 402.—Tex.; Mexico.
virgatum Coquillett, 1902g: 97.—N. Mex.; Calif., e. to S. Dak. and Tex., Mexico.
  cinereum Bellardi, 1859: 13 (1861: 213) (preocc. Macquart, 1834).—Mexico.

Subgenus HEARLEA Vargas, Martínez Palacios, and Díaz Nájera

Simulium, subg. **Hearlea** Vargas, Martínez Palacios, and Díaz Nájera, 1946: 104, 106. Type-species, *canadense* Hearle (orig. des.).

Simulium, subg. *Hearlea* Rubzov, 1940a: 116, 126. Nomen nudum, type-species as above, but no descr.

canadense Hearle, 1932: 14 (as *virgatum* ssp.)—B.C.; B.C. to S. Dak., s. to s. Calif. and N. Mex.

    fraternum Twinn, 1938: 53.—Utah.

Subgenus GNUS Rubzov

Simulium, subg. **Gnus** Rubzov, 1940a: 363. Type-species, *decimatum* Dorogostajskij, Rubzov, and Vlasenko (orig. des.).

arcticum Malloch, 1914j: 37.—B.C.; w.N. Amer., e. to Man., Mont., Colo., and Ariz.

    brevicercum Knowlton and Rowe, 1934: 583.—Utah.

    nigresceum Knowlton and Rowe, 1934: 583.—Utah.

corbis Twinn, 1936: 147.—Que.; Alaska to Nfld., s. to Utah, Wis., and N.H.

defoliarti Stone and Peterson, 1958: 1.—Wyo.; B.C. to Calif., Utah, and Wyo.

malyschevi Dorogostajskij, Rubzov, and Vlasenko, 1935: 142.—Siberia; Alaska, N.W.T., Alta.

nigricoxum Stone, 1952: 94 (n. name for *similis* Malloch).—N.W.T.; Alaska.

    similis Malloch, 1919f: 42 (preocc. Silva, 1917).—N.W.T.

Subgenus HAGENOMYIA Shewell

Simulium, subg. **Hagenomyia** Shewell, 1959: 83. Type-species, *pictipes* Hagen (orig. des.).

longistylatum Shewell, 1959: 84.—Que.; Ont., Labr.

pictipes Hagen, 1880: 306.—N.Y.; Minn. to Que., s. to Ala., also ?Okla., ?Colo.

    innoxium Comstock and Comstock, 1895: 453.—N.Y.

    aldrichiana Enderlein, 1936c: 120 (*Schoenbaueria*).—N.Y.

Subgenus SIMULIUM Latreille

dahlgrueni Enderlein.—Not Nearctic, erroneously ascribed to Greenland by Dyar and Shannon (1927: 29).

decorum Walker, 1848: 112.—Ont.; Alaska to Nfld., s. to Oreg., Colo., and Fla.
    piscicidium Riley *in* McBride, 1870: 367.—N.Y.
    katmai Dyar and Shannon, 1927: 31 (as ssp.).—Alaska.
    ottawaense Twinn, 1936: 146.—Que.
fibrinflatum Twinn, 1936: 141.——Ont.; Ont. to Ala.
hunteri Malloch, 1914j: 59.—Colo.; Alaska and Y.T. to Calif. and N. Mex.
jacumbae Dyar and Shannon, 1927: 44.—Calif.; Nev., Tex.
jenningsi Malloch, 1914j: 41.—Md.; Man. to Maine, s. to Tex. and Fla. Unrevised species complex.
    nigroparvum Twinn, 1936: 142.—Ont.
luggeri Nicholson and Mickel, 1950: 54 (as *jenningsi* ssp.).—Minn.; N.W.T. to Alta., Okla., Miss., and Va.
parnassum Malloch, 1914j: 36.—N.H.; Alleghanian from Ont. and N.S. to Ga.
    hydationis Dyar and Shannon, 1927: 28.—Va.
petersoni Stone and De Foliart, 1959: 394.—Wyo.; Utah, Calif.
piperi Dyar and Shannon, 1927: 38.—Wash.; Calif., Idaho, Utah, Colo.
    sayi Dyar and Shannon, 1927: 40.—Colo.
    knowltoni Twinn, 1938: 53.—Utah.
    stonei Stains and Knowlton, 1943: 277.—Utah.
rubtzovi Smart, 1945: 528 (also 1946: 22) (n. name for *similis* Rubzov).—Siberia; Alaska.
    similis Rubzov, 1940b: 196 (preocc. Silva, 1917).—Siberia.
tuberosum (Lundström), 1911a: 14 (*Melusina*).—Finland; Alaska to Greenland, s. to Calif., Tex., and Fla., Holarctic. Unrevised species complex.
    perissum Dyar and Shannon, 1927: 43.—Va.
    vandalicum Dyar and Shannon, 1927: 44. - Calif.
    turmale Twinn, 1938: 51.—Utah.
    twinni Stains and Knowlton, 1940: 77.—Utah.
venustum Say, 1823: 28 (1859b: 51).—Ohio; Alaska to Greenland, s. to Tex., Miss., and S.C., Holarctic.
    molestum Harris 1841: 405.—?Mass.
    minutum Lugger, 1897: 175.—Minn.
    irritatum Lugger, 1897: 177.—Minn.
    rileyana Enderlein, 1922b: 75 (*Boophthora*).—N.Y.
    groenlandicum Enderlein, 1935a: 363.—Greenland.
verecundum Stone and Jamnback, 1955: 83.—Pa.; Alaska to N.S., s. to Wash., Wyo., Ala., and S.C.

Superfamily ANISOPODOIDEA

Family ANISOPODIDAE
(Rhyphidae, Phryneidae, Sylvicolidae)

By Charles P. Alexander

The adults of this family may be rather common, occurring on tree trunks and similar places, or in small swarms. The immature stages occur in decaying organic matter such as fermenting sap and mammalian manure.

REFERENCES: Edwards, 1928a (rev.); Alexander, 1942b (keys).

Subfamily ANISOPODINAE

Genus SYLVICOLA Harris

Sylvicola Harris, 1776: 100. Type-species, *brevis* Harris (Coquillett, 1910b: 610=*fenestralis* (Scopoli).
Phryne Meigen, 1800: 16. Type-species, *Tipula fuscata* Fabricius (Coquillett, 1910b: 589). Suppressed by I.C.Z.N., 1963b: 339.
Anisopus Meigen, 1803: 264. Type-species, *fuscus* Meigen (Coquillett, 1910b: 507)=*fuscata* (Fabricius).
Rhyphus Latreille, 1804: 188. Type-species, *Tipula fenestralis* Scopoli (as *fenestrarum*; mon.).

alternatus (Say), 1823: 27 (1859b: 51) (*Rhyphus*).—Pa.; Wis. to to Maine, s. to Tex. and Ga.
fenestralis (Scopoli), 1763: 322 (*Tipula*).—Europe; Wash. and Idaho to Calif., Ont. and N.S., s. to Pa. and N.J.
   diversipes Fitch *in* Edwards, 1923: 476. Nomen nudum.
fuscatus (Fabricius), 1775: 755 (*Tipula*).—Sweden; Europe, Alaska, N.H.
marginatus (Say), 1823: 27 (1859b: 50) (*Rhyphus*).—Pa.; Wis. to Maine, s. to Kans. and Va.
   punctatus, authors, not Fabricius.

Genus OLBIOGASTER Osten Sacken

Olbiogaster Osten Sacken, 1886a: 20. Type-species, *Rhyphus taeniatus* Bellardi (Coquillett, 1910b: 579).
REFERENCE: Lane and d'Andretta, 1958 (key).

sackeni Edwards, 1915b: 502.—Mexico; Tex., Fla., S.C.
scalaris (Wiedemann), 1830: 618 (*Rhyphus*).—Brazil (as Ga., error). Not Nearctic.

taeniata (Bellardi), 1862: 5 (1864: 202) (*Rhyphus*).—Mexico; Tex., Fla.
texana Lane and d'Andretta, 1958: 525.—Tex.

Subfamily MYCETOBIINAE

Genus MYCETOBIA Meigen

Mycetobia Meigen, 1818: 229. Type-species, *pallipes* Meigen (Westwood, 1840: 127).

divergens Walker, 1856b: 418.—U.S.; Colo. to Maine, s. to Ariz. and Fla.
    persica Riley, 1867: 397 (*Mycetophila*).—Ill.
    sordida Packard, 1869a: 388.—Unknown.
    marginalis Adams, 1903a: 21.—Mo.

Superfamily BIBIONOIDEA

Family BIBIONIDAE

By D. Elmo Hardy

The biologies of the Bibionidae are not clearly understood. Apparently all species are herbivores. They feed predominantly as scavengers on decaying plant material. Some species are of economic importance, attacking the roots and tubers of a wide variety of crops. The adults of certain species may occur in enormous numbers for a short period, causing much comment and some annoyance.

    REFERENCES: Morris, 1921 (biol.), 1922 (biol.); Hardy, 1945 (rev.); Bollow, 1954 (biol.).

Subfamily HESPERININAE

Genus HESPERINUS Walker

Hesperinus Walker, 1848: 81. Type-species, *brevifrons* Walker (mon.).
Spodius Loew, 1858a: 101. Type-species, *imbecillus* Loew (mon.).

brevifrons Walker, 1848: 81.—Ont.; Alaska, Canada, w. and ne. U.S.
    flagellaria Garrett, 1925a: 11.—B.C.

Subfamily PLECIINAE

Genus PENTHETRIA Meigen

Penthetria Meigen, 1803: 264. Type-species, *funebris* Meigen (sub. mon., Meigen, 1804: 104).
Eupeitenus Macquart, 1838a: 88 (1838: 84). Type-species, *Penthetria atra* Macquart (mon.)=*heteroptera* (Say).
heteroptera (Say), 1823: 77 (1859b: 69) (*Bibio*).—Md.; throughout Nearctic Region.
    atra Macquart, 1834a: 175.—Pa.
    longipes Loew, 1858a: 109 (*Plecia*).—La.
    lugubris Harris, 1835: 596. Nomen nudum.

Genus PLECIA Wiedemann

Plecia Wiedemann, 1828: 72. Type-species, *Hirtea fulvicollis* Fabricius (Blanchard, 1840: 576).
americana Hardy, 1940a: 15.—Fla.; s. U.S., n. Mexico.
nearctica Hardy, 1940a: 20.—Tex.; s. U.S., Mexico, Cent. Amer.

Subfamily BIBIONINAE

Genus BIBIO Geoffroy

Bibio Geoffroy, 1762: 568. Type-species, *Tipula hortulana* Linnaeus (Latreille, 1810: 442). Validated by I.C.Z.N., 1957a: 86.
abbreviatus Loew, 1864a: 54 (Cent. 5, no. 9).—D.C.; Kans., e. to Que. and Ga.
albipennis albipennis Say, 1823: 77 (1859b: 69).—Pa.; Nearctic, s. into Mexcio.
    hirtus Loew, 1864a: 51 (Cent. 5, no. 2).—Calif.
    tenuipes Coquillett, 1902g: 95.—Ariz.
    conjunctivus Hardy, 1937: 200.—Md.
  ssp. **beameri** Hardy, 1945: 451.—Kans.
alexanderi James, 1936b: 1.—Colo.; Okla.
alienus McAtee, 1923: 62.—N.C.
articulatus Say, 1823: 77 (1859b: 69).—Pa. Unrecognized.
atripilosus James, 1936b: 2.—Colo.; Utah.
baltimoricus Macquart, 1855: 37 (1855: 17).—Md. Unrecognized.
brunnipes (Fabricius), 1794: 250 (*Tipula*; n. name for *rufipes* Fabricius).—Nfld. Unrecognized.
    rufipes Fabricius, 1781: 410 (*Tipula*; preocc. Linnaeus, 1761).—Nfld.

carolinus Hardy, 1945: 457 (n. name for *afer* McAtee).—N.C.; se. U.S.
    afer McAtee, 1923: 63 (preocc. Loew, 1854).—N.C.
carri Curran, 1927f: 80.—Alta.
castanipes Jaennicke, 1867: 317 (1868: 9).—Ill. Unrecognized.
cognatus Hardy, 1937: 199.—Calif.
columbiaensis Hardy, 1938a: 207.—B.C.
curtipes James, 1936b: 5.—Colo.; Utah.
femoratus Wiedemann, 1820a: 35 (1821c: 35).—N. Amer.; Canada, U.S. from Colo., e. and s. to Miss.
    fuscipennis Macquart, 1838a: 91 (1838: 87).—N. Amer.
    senilis Wulp, 1869: 81.—Wis.
fluginatus Hardy, 1937: 201.—B.C.
flukei Hardy, 1937: 202.—Colo.
fraternus Loew, 1864a: 54 (Cent. 5, no. 8).—D.C.; e. Canada and U.S., w. to Utah and s. to Miss.
fumipennis Walker, 1848: 122.—Ont.; Commander Is., Alaska, Canada, Labr., Nfld., n. and w. U.S.
    inaequalis Loew, 1864a: 51 (Cent. 5, no. 3).—Alaska.
    simplicis Curran, 1923l: 245.—Alta.
holtii McAtee, 1922: 11.—Ariz.; B.C. to Oreg., Wyo., Colo., Utah.
imparilis Hardy, 1959a: 209.—Calif.
kansensis James, 1936b: 6.—Kans.
knowltoni Hardy, 1937: 202.—Utah; Wash., Ariz.
    paltidus Hardy, 1937: 203 (as var).—Utah.
labradorensis Johnson, 1929a: 133.—Labr.; Alaska, Alta. to Labr.
lobatus Hardy, 1937: 203.—Calif.
longipes Loew, 1864a: 55 (Cent. 5, no. 12).—D.C.; Alta. to N.S., s. to Ariz. and Va.
melanopilosus Hardy, 1936: 195.—Utah; Calif.
    bisepta Hardy, 1937: 204 (as var.).—Utah.
mickeli Hardy, 1937: 204.—Minn.; B.C. and Calif., e. to N.Y.
monstri James, 1936b: 3.—Colo.; Utah.
necotus Hardy, 1937: 205.—Calif.; Wash.
nigrifemoratus Hardy, 1937: 206.—B.C.; w. U.S.
    gilvus Hardy, 1937: 206 (as var.).—Utah.
nigripilus Loew, 1864a: 55 (Cent. 5, no. 10).—Man.; S. Dak. to Que., s. to Mo. and Va.
    lucens Hardy, 1937: 203.—Ont.
nigritus Curran, 1924i: 250 (as *lacteipennis* var.).—B.C.; Labr.
    lacteipennis Curran, 1924i: 250 (preocc. Zetterstedt, 1850).—B.C.
    bryanti Johnson, 1929a: 133.—Labr.

      currani Hardy, 1937: 200 (n. name for *lacteipennis* Curran).—B.C.
painteri James, 1936b: 2.—Kans.; Colo. and Ohio, s. to Ariz. and Okla.
pallipes Say, 1823: 76 (1859b: 68).—Pa. Unrecognized.
pingreensis James, 1936b: 4.—Colo.; Utah.
rufalipes Hardy, 1937: 207.—Tex.
rufithorax Wiedemann, 1828: 78.—Pa.; Pa., s. to Tex. and Fla.
rufitibialis Hardy, 1938a: 209 (as *jacobi* var.).—B.C.
      jacobi Hardy, 1938a: 209 (preocc. Villeneuve, 1924).—B.C.
      neojacobi Hardy, 1945: 476 (n. name for *jacobi* Hardy).—B.C.
sericatus Hardy, 1937: 207.—Wash.; B.C.
sierrae Hardy, 1960a: 255.—Calif.
similis James, 1936b: 5.—Colo.; Utah.
slossonae Cockerell, 1909b: 174 (n. name for *gracilis* Walker).—N.S.; Canada, n. and w. U.S., Ind., Alaska.
      gracilis Walker, 1848: 123 (preocc. Unger, 1841).—N.S.
striatipes Walker, 1848: 122.—Ont.; Man., N.W.T.
      conus Hardy, 1938a: 208.—Man.
teneus Hardy, 1937: 208.—Alaska.
thoracicus Say, 1824a: 368 (1859a: 250).—Fla. Unrecognized.
townesi Hardy, 1945: 487.—R.I.; Ohio, N.Y., N.H., Conn.
tristis Williston *in* Kellogg, 1893: 113.—Kans.; Wash., Utah, Ariz.
utahensis Hardy, 1937: 208.—Utah; Wash., Calif.
velcidus Hardy, 1937: 209.—Ont.; N.B.
velorum McAtee, 1923: 63.—N.C.; ?Colo.
vestitus Walker, 1848: 122.—N.S.; Alaska, Canada, n. and w. U.S.
      nervosus Loew, 1864a: 52 (Cent. 5, no. 4).—Alaska.
      variabilis Loew, 1864a: 53 (Cent. 5, no. 7).—Alaska.
      basalis Loew, 1864a: 55 (Cent. 5, no. 11).—N.H.
xanthopus Wiedemann, 1828: 80.—N.Y.; widespread over U.S. and Canada.
      canadensis Macquart, 1838b: 179 (1839: 295).—Canada.
      humeralis Walker, 1848: 121.—N.S.
      scita Walker, 1848: 122.—N.S.
      obscurus Loew, 1864a: 52 (Cent. 5, no. 5).—Ont.
      lugens Loew, 1864a: 52 (Cent. 5, no. 6).—Man.
      palliatus McAtee, 1922: 16 (as ssp.).—Idaho.
      macateei James, 1936b: 4.—Colo.
      signatus Hardy, 1937: 208.—Utah.

Genus BIBIODES Coquillett

Bibiodes Coquillett, 1904f: 171. Type-species, *halteralis* Coquillett (orig. des.).

aestivus Melander, 1912: 338.—Wash.; B.C. and Mont. to Oreg. and Utah.
femoratus Melander, 1912: 340.—Tex.; Tenn.
halteralis Coquillett, 1904f: 171.—Calif.; Sask., Ariz., N. Mex., Tex., Kans., Mexico (Baja Calif.).

Genus DILOPHUS Meigen

Philia Meigen, 1800: 20. Type-species, *Tipula febrilis* Linnaeus (Coquillett, 1910b: 588). Suppressed by I.C.Z.N., 1963b: 339.
Dilophus Meigen, 1803: 264. Type-species, *Tipula febrilis* Linnaeus (Latreille, 1810: 442).

arizonaensis (Hardy), 1937: 209 (*Philia*).—Ariz.; N. Mex., Nev., Calif.
caurinus McAtee, 1922: 19.—Alaska; Canada, nw. U.S., e. to Mich. and Ohio.
emarginatus McAtee, 1922: 20.—Calif.
fulvicoxa Walker, 1848: 117.—Ont.
longiceps Loew, 1861b: 315 (Cent. 1, no. 14).—Ill. Unrecognized.
obesulus Loew, 1869c: 162 (Cent. 9, no. 60).—D.C.; Canada, Calif., Colo., Pa., Va.
  jamesi Hardy, 1937: 210 (*Philia*).—Colo.
occipitalis Coquillett *in* Baker, 1904: 20.—Calif.; s. U.S., e to La. and N.C.
  oklahomensis Hardy, 1937: 211 (*Philia*).—Okla.
orbatus (Say), 1823: 78 (1859b: 69) (*Bibio*).—Pa.; s. U.S., Calif. to S.C. Current usage is based on probable misidentification by Osten Sacken. A new name may be required.
proximus McAtee, 1922: 22.—Colo.; Wyo.
sectus McAtee, 1922: 22.—N.H.
serotinus Loew, 1861b: 315 (Cent. 1, no. 15).—Ill.; B.C. and Oreg. to Md. and Miss.
serraticollis Walker, 1848: 117.—Ont. Unrecognized.
spinipes Say, 1823: 79 (1859b: 71).—Mo.; throughout U.S.
  thoracicus Say, 1823: 80 (1859b: 71).—Pa.
  bimaculata Walker, 1856b: 422 (*Plecia*).—U.S.
  dimidiatus Loew, 1869b: 4 (Cent. 8, no. 3).—N.Y.

stigmaterus stigmaterus Say, 1823: 78 (1859b: 70).—Mo.; B.C. to N.S., s. to Ariz. and Ind.
    pusillus Wiedemann, 1828: 77.—U.S.
    fraternus Harris, 1835: 596. Nomen nudum.
  ssp. **niger** (Hardy), 1937: 212 (*Philia*).—Minn.; Man.
strigilatus McAtee, 1922: 24.—Calif. (Catalina I.).
stygius Say, 1832: 15 (1859a: 309) (preocc. Say, 1829).—La. Unrecognized.
tibialis tibialis Loew, 1869c: 162 (Cent. 9, no. 61).—Alaska; Canada, U.S.
    atelestes Hardy, 1937: 210 (*Philia*; as *breviceps* var.).—Que.
  ssp. **breviceps** Loew, 1869c: 162 (Cent. 9, no. 59) (as sp.).—N.H.; throughout U.S.
tingi (Hardy), 1942b: 132 (*Philia*).—Calif.; Mexico.

Family PACHYNEURIDAE
By Charles P. Alexander

The members of this family have been placed by some in Anisopodidae, by others in the Bibionidae, but they do not fit satisfactorily in either. The larva of only *Axymyia furcata* is known.

REFERENCE: Edwards, 1928a (rev.); Krogstad, 1959 (biol.).

Genus AXYMYIA McAtee

Axymyia McAtee, 1921a: 49. Type-species, *furcata* McAtee (orig. des.).
furcata McAtee, 1921a: 49.—Pa.; Mass., N.Y., Va.
    fitchii Osten Sacken of Alexander, 1920h: 769 (*Protoplasa*).

Genus CRAMPTONOMYIA Alexander

Cramptonomyia Alexander, 1931a: 7. Type-species, *spenceri* Alexander (orig. des.).
spenceri Alexander, 1931a: 8.—B.C.; Wash., Oreg.

Superfamily MYCETOPHILOIDEA

Family MYCETOPHILIDAE
(Fungivoridae)
By Jean L. Laffoon

The family Mycetophilidae, or fungus gnats, are usually small flies with prominent coxae. The adults are particularly common in dark, humid habitats in wooded areas. The larvae of most species are fungivorous, but those of most Keroplatinae seem to be predominantly predaceous.

Edwards (1925a) provided a generic classification which is generally sound. However, much revisionary work on the Nearctic species is needed. It is presently impossible to make reliable determinations in most genera by means of the literature alone.

Several varieties are recognized in this catalog, but the status of each is considered uncertain. Also several junior homonyms (*Macrocera pilosa* Garrett, *Orfelia nigra* (Cole), *Dziedzickia johannseni* Sherman, *Docosia setosa* Garrett, *Sciophila bicolor* Garrett, *S. fusca* Garrett, and *S. parva* Garrett) are not renamed because some may be junior synonyms.

REFERENCES: Johannsen, 1909 (cat., world), 1910a, 1910b, 1912a, 1912b (revs.); Edwards, 1925a (world genera, key); Landrock, 1926–1927 (rev. Palaearctic); Tonnoir, 1929 (world genera, key); Madwar, 1937 (larvae); Kessel and Kessel, 1939 (biol.); Shaw and Shaw, 1951 (anat. pleuron); Shaw and Fisher, 1952 (key, Conn.); Shaw, 1952b (anat.); Buxton, 1960 (biol.).

Subfamily BOLITOPHILINAE

Most larval records are from Agaricaceae, but larvae have also been found in Boletaceae and Polyporaceae. Most of the hosts grow on the ground.

Genus BOLITOPHILA Meigen

Bolitophila Meigen, 1818: 220. Type-species, *cinerea* Meigen (Westwood, 1840: 127).
Bolitophilella Landrock, 1925a: 179. Type-species, *Bolitophila cinerea* Meigen (Okada, 1939: 292).

REFERENCE: Fisher, 1937 (key); Shaw, 1962 (key).

acuta Garrett, 1925a: 7.—B.C.; N.Y.
alberta Fisher, 1937: 389.—Alta.
atlantica Fisher, 1934: 276.—N.H.
bilobata Garrett, 1925a: 7.—B.C.
bucera Shaw, 1940: 48.—Oreg.
cinerea Meigen, 1818: 221.—Germany, Denmark; B.C., N.Y. to Maine, S.C., Europe, Japan.
clavata Garrett, 1925a: 6.—B.C.
connectans Garrett, 1925a: 6.—B.C.
distus Fisher, 1937: 389.—N.Y.
dubia Siebke, 1863: 185.—Sweden; B.C., Alta.. Idaho, N.Y. to Maine, Europe, Japan.
  disjuncta Loew, 1869a: 19.—Germany.

dubiosa Van Duzee, 1928a: 32.—Calif.; B.C.
dupla Garrett, 1925a: 6.—B.C.
hybrida (Meigen), 1804: 47 (*Macrocera*).—Europe; B.C. to Calif., also Que., s. to Ind. and N.C., Europe.
  fusca Meigen, 1818: 221.—Europe.
montana Coquillett, 1901h: 593.—N.H.; B.C., N.Y., Maine.
patulosa Garrett, 1925a: 7.—Calif.
perlata Garrett, 1925a: 6.—B.C.; Que., Conn.
raca Garrett, 1925a: 6.—B.C.
recurva Garrett, 1925a: 6.—B.C.; Alta.
simplex Garrett, 1925a: 6.—B.C.
subteresa Garrett, 1925a: 7.—B.C.

Subfamily DIADOCIDIINAE

The larvae live in mucus tubes on wood.

REFERENCE: Fisher, 1941 (rev.).

Genus DIADOCIDIA Ruthe

Diadocidia Ruthe, 1831: 1210. Type-species, *flavicans* Ruthe (orig. des., as gen. n., sp. n.)=*ferruginosa* (Meigen).
Macronevra Macquart, 1834a: 146. Type-species, *winthemi* Macquart (mon.)=*ferruginosa* (Meigen).

borealis Coquillett, 1900h: 390 (1904: 4).—B.C.; B.C. to Calif., also N.H., N.C., S.C.
ferruginosa (Meigen), 1830: 294 (*Mycetobia*).—Germany; B.C. to Calif., N.S. to S.C., Europe.
stanfordensis Arnaud and Hoyt, 1956: 87.—Calif.

Subfamily DITOMYIINAE

The larvae of *Ditomyia fasciata* live in hard species of Polyporaceae, those of *Symmerus annulatus* in decaying wood. Those of a Neotropical genus, *Neoditomyia*, live in mucus webs under overhanging rocks and tree trunks; they feed on insects caught in these webs.

REFERENCE: Fisher, 1941 (rev.).

Genus DITOMYIA Winnertz

Ditomyia Winnertz, 1846: 14. Type-species, *trifasciata* Winnertz (mon.)=*fasciata* (Meigen).

euzona Loew, 1869c: 130 (Cent. 9, no. 1).—D.C.; N.H., N.J.
potomaca Fisher, 1941: 278.—Va.; Conn., Md.

Genus SYMMERUS Walker

Symmerus Walker, 1848: 88. Type-species, *ferrugineus* Walker (mon.)=*annulatus* (Meigen).
Plesiastina Winnertz, 1852b: 55. Type-species, *Mycetobia annulata* Meigen (Coquillett, 1910b: 592).

annulatus (Meigen), 1830: 294 (*Mycetobia*).—Europe; ?B.C., ?Oreg.
coqulus Garrett, 1925a: 12.—B.C.; s. Alaska to Maine, s. to Oreg. and N.J.
  annulatus, authors, not Meigen.
dilutus Fisher, 1938a: 196.—W. Va.
lautus (Loew), 1869c: 132 (Cent. 9, no. 3) (*Plesiastina*).—N.Y.; Ind., Vt., Md.
tristis (Loew), 1869c: 131 (Cent. 9, no. 2) (*Plesiastina*).—D.C.; Ind., Maine, Mass., N.C.

Subfamily KEROPLATINAE

The larvae of probably all species spin mucus webs and those of certain species, such as *Orfelia fultoni*, are luminescent. The larvae of *Keroplatus* are found in or on wood or on fungi, and that of one species of *Asindulum* in decaying wood. Those of *Macrocera* and *Orfelia* are found under rocks, in and under logs, or in the soil around roots, and that of *Platyura marginata* is recorded from a cavity in the soil under wood. Some species of *Orfelia* secrete oxalic acid droplets on the webs. The larvae of *Keroplatus* are primarily fungus-spore feeders, of *Orfelia* primarily predatory, and some species of *Macrocera* at least, feed on both fungus spores and small animals. The adults of *Asindulum* have elongate mouthparts and feed on nectar.

REFERENCE: Mansbridge, 1933 (biol.).

Genus ASINDULUM Latreille

Asindulum Latreille, 1805: 290. Type-species, *nigrum* Latreille (mon.).
Asyndulum, error.

coxale Loew, 1869c: 132 (Cent. 9, no. 4).—Canada; Que., N.H., Maine.
montanum Röder, 1887: 116.—N.H.; Alta. to Que., s. to S. Dak. and N.J.
winnertzi (Tarwid), 1936: 6 (*Zelmira;* n. name for *flava* Winnertz).— Europe; N.H., N.C., Europe.
  flava Winnertz, 1846: 17 (*Macrorrhyncha;* preocc. in *Zelmira* by Macquart, 1826, orig. *Platyura*, now *Orfelia*).— Europe.

Genus FENDEROMYIA Shaw

Fenderomyia Shaw, 1948: 94. Type-species, *smithi* Shaw (orig. des.).
smithi Shaw, 1948: 94.—Oreg.

Genus HESPERODES Coquillett

Hesperodes Coquillett, 1900b: 429. Type-species, *johnsoni* Coquillett (orig. des.).
johnsoni Coquillett, 1900b: 429.—N.J.; Mass.

Genus KEROPLATUS Bosc

Keroplatus Bosc, 1792: 42. Type-species, *tipuloides* Bosc (mon.).
Ceroplatus, emend.
> REFERENCES: Edwards, 1929c (subg. of world); Fisher, 1941 (key).

Subgenus CEROTELION Rondani

Cerotelion Rondani, 1856: 191 (as genus). Type-species, *Platyura laticornis* Meigen (orig. des.)=*lineatus* (Fabricius).
johannseni Fisher, 1940: 243.—N.C.; N.H. to Ga., also Iowa.
bellulus, authors, not Williston.

Subgenus EUCEROPLATUS Edwards

Keroplatus, subg. **Euceroplatus** Edwards, 1929c: 174. Type-species, *notaticoxa* Senior-White (orig. des.).
fasciatus (Garrett), 1925a: 12 (*Cerotelion*).—B.C.; Wash., Calif.
fasciolus Coquillett, 1894c: 126.—Wash.; Ark.
fenestralis Fisher, 1938a: 197.—Mich.

Subgenus HETEROPTERNA Skuse

Heteropterna Skuse, 1888: 1129 (as genus). Type-species, *macleayi* Skuse (mon.).
cressoni Fisher, 1941: 293.—Pa.

Subgenus KEROPLATUS Bosc

carbonarius Bosc *in* S.N.A., 1803: 543.—Carolina; Ind., Pa., Tenn.
clausus Coquillett, 1901h: 594.—N.H.; Mich. to Maine, s. to Tex. and S.C.
fernaldi Shaw, 1941d: 20.—Mass.
militaris Johannsen, 1910a: 237.—Pa.; Mich. to Maine, s. to Tenn. and Md., also Wash., Iowa, Fla.
terminalis Coquillett, 1905d: 69.—B.C.; Kans., Tex.

Unplaced Species of *Keroplatus*

apicalis Adams, 1903a: 22.—Kans.

Genus MACROCERA Meigen

Euphrosyne Meigen, 1800: 16. Type-species, *Macrocera lutea* Meigen (Coquillett, 1910b: 542). Suppressed by I.C.Z.N., 1963b: 339.

Macrocera Meigen, 1803: 261. Type-species, *lutea* Meigen (Curtis, 1837a: pl. 637).

beringensis Malloch, 1923i: 177.—Alaska (Pribilof Is.).
bicolor Garrett, 1925a: 7.—B.C.
clara Loew, 1869c: 133 (Cent. 9, no. 6).—D.C.; Wis. to Maine, s. to Tenn. and S.C., also B.C.
clavinervis Van Duzee, 1928a: 35.—Calif.
diluta Adams, 1903a: 22.—Ariz.
distincta Garrett, 1925a: 8.—B.C.
fisherae Shaw, 1935b: 229 (as *fisheri*).—S.C.
floridana Johnson, 1926b: 299.—Fla.; S.C.
formosa Loew, 1866b: 6 (Cent. 7, no. 8).—N.Y.; Maine to N.J., also S.C., Oreg.
    var. **indigena** Johannsen, 1910a: 270.—N.Y.; Conn., S.C. ?Var.
geminata Johannsen, 1910a: 272.—N.Y.
hirsuta Loew, 1869c: 132 (Cent. 9, no. 5).—D.C.; Ind. to Maine, s. to N.C.
hirtipennis Van Duzee, 1928a: 36.—Calif.
hyalipennis Shaw, 1941a: 171.—Tenn., N.C.
immaculata Johnson, 1902a: 240.—N.Y., Pa.; N.C., S.C.
inconcinna Loew, 1869c: 133 (Cent. 9, no. 7).—D.C.; Vt. to Maine, s. to D.C.
nebulosa Coquillett, 1901h: 594.—N.J., N.H.; N.Y. to Maine, s. to N.J.
nobilis Johnson, 1922: 21.—N.H.; Mass.
pilosa Garrett, 1925a: 8 (preocc. Landrock, 1917).—B.C. Unrecognized.
similis Garrett, 1925a: 8.—B.C.
trivittata Johnson, 1922: 21.—Sask.
uniqua Garrett, 1925a: 8.—B.C.
variola Garrett, 1925a: 7.—B.C.
villosa Garrett, 1925a: 8.—B.C.

Genus ORFELIA Costa

Zelmira Meigen, 1800: 16. Type-species, *Platyura fasciata* Meigen (Coquillett, 1910b: 621). Suppressed by I.C.Z.N., 1963b: 339.
Orfelia Costa, 1857: 448. Type-species, *Platyura fasciata* Meigen (Hardy, 1960b: 200).
Proceroplatus Edwards, 1925a: 523. Type-species, *Platyura pictipennis* Williston (orig. des.).
Calliplatyura Malloch, 1928b: 600 (as *Calloplatyura*, p. 601). Type-species, *Platyura elegans* Coquillett (orig. des.). **N. syn.**
Neoplatyura Malloch, 1928b: 601. Type-species, *Platyura setiger* Johannsen (orig. des.). **N. syn.**
Platyura, subg. *Rutylapa* Edwards, 1929c: 171. Type-species, *ruficornis* Zetterstedt (orig. des.). **N. syn.**
Platyura, authors, not Meigen.

A subgeneric classification is available, but most Nearctic species have not yet been assigned to subgenera.

REFERENCES: Edwards, 1929c (tax., as *Platyura*); Fulton, 1941 (biol.)

alexanderi (Shaw), 1941a: 168 (*Platyura*).—Tenn. **N. comb.**
angustata (Van Duzee), 1928a: 34 (*Platyura*).—Calif. **N. comb.**
apicalis (Shaw), 1935b: 229 (*Platyura*).—S.C. **N. comb.**
discoloria (Meigen), 1818: 239 (*Platyura*).—Europe; Wis., Maine to Ga., Europe. **N. comb.**
    diluta Loew, 1869c: 134 (Cent. 9, no. 9) (*Platyura*).—D.C.
divaricata (Loew), 1869c: 134 (Cent. 9, no. 8) (*Platyura*).—Ga.; N.C. **N. comb.**
elegans (Coquillett), 1895a: 307 (*Platyura*).—Ga., Fla.; Iowa to Que., s. to Tenn. and Fla. **N. comb.**
elegantula (Williston), 1900: 218 (*Platyura*).—Mexico; Ariz. **N. comb.**
equalis (Van Duzee), 1928a: 35 (*Platyura*).—Oreg. **N. comb.**
fascipennis (Say), 1824a: 360 (1859a: 244) (*Platyura*).—Northwest Territory (U.S.); Ont. to Que., s. to N.J., also B.C., Alta. **N. comb.**
    var. **sagax** (Johannsen), 1910a: 258 (*Platyura*).—Maine; B.C., Alta. ?Var. **N. comb.**
fultoni (Fisher), 1940: 245 (*Platyura*).—N.C.; Va., S.C. **N. comb.**
genualis (Johannsen), 1910a: 262 (*Platyura*).—Wis., Tenn., N.C.; Wis. to Maine, s. to Tenn. and N.C. **N. comb.**
inops (Coquillett), 1901h: 594 (*Platyura*).—N.J.; Vt. to N.J., also N.C., S.C. **N. comb.**
intermedia (Sherman), 1921: 16 (*Platyura*).—B.C. **N. comb.**

lurida (Coquillett), 1895d: 199 (*Platyura*).—Wash. **N. comb.**
melasoma (Loew), 1869c: 135 (Cent. 9, no. 12) (*Platyura*).—D.C.;
  Que., N.H., Mass., N.J. **N. comb.**
mendica (Loew), 1869c: 135 (Cent. 9, no. 10) (*Platyura*).—N.Y.;
  Maine to N.J., N.C. **N. comb.**
mendosa (Loew), 1869c: 135 (Cent. 9, no. 11) (*Platyura*).—D.C.;
  B.C., Tenn., Maine to S.C. **N. comb.**
mimula (Johannsen), 1910a: 255 (*Platyura*).—Wis., N.H.; N.C.
  N. comb.
miriamae (Shaw), 1941a: 170 (*Platyura*).—N.C. **N. comb.**
mitchellensis (Shaw), 1941a: 168 (*Platyura*).—N.C. **N. comb.**
moerens (Johannsen), 1910a: 262 (*Platyura*).—Wash.; B.C., N.Y.
  N. comb.
moesta (Johannsen), 1910a: 259 (*Platyura*).—Wash. **N. comb.**
nigra (Cole) *in* Cole and Lovett, 1919: 222 (*Platyura*; preocc. Macquart, 1826).—Oreg. **N. comb.** Unrecognized.
nigribarba (Van Duzee), 1928a: 34 (*Platyura*).—Calif. **N. comb.**
nigrita (Johannsen), 1910a: 256 (*Platyura*).—Wash. **N. comb.**
notabilis (Williston), 1893c: 59 (*Platyura*).—Wash. **N. comb.**
palmi (Shaw), 1951b: 277 (*Platyura*).—Wyo. **N. comb.**
pellita (Fisher), 1937: 390 (*Platyura*).—N.S.; N.Y. **N. comb.**
pullata (Coquillett), 1904f: 171 (*Platyura*).—Calif. **N. comb.**
scapularis (Johannsen), 1910a: 263 (*Platyura*).—Wash., Idaho, Calif.
  N. comb.
semirufa (Meigen), 1818: 237 (*Platyura*).—Europe; Mass., N.J.,
  Great Smokies (N.C. or Tenn.), Palaearctic. **N. comb.**
  taeniata Winnertz, 1863: 701 (*Platyura*).—Europe.
setiger (Johannsen), 1910a: 252 (*Platyura*).—Wash. **N. comb.**
subterminalis (Say), 1829: 152 (1859b: 350) (*Platyura*).—Ind.; Wis.,
  Maine to N.J., S.C. **N. comb.**
  var. **nexilis** (Johannsen), 1910a: 261 (*Platyura*).—Wis. ?Var.
  N. comb.
williami (Shaw), 1953a: 66 (*Zelmira*).—Conn. **N. comb.**

Genus PALEOPLATYURA Meunier

Paleoplatyura Meunier, 1899: 164. Type-species, *macrocera* Meunier (mon.).

  REFERENCE: Fisher, 1941 (key).

aldrichii Johannsen, 1909: 10.—Wash.
johnsoni Johannsen, 1910a: 226.—Vt.; N.Y., Mass., Conn., Md., N.C.
melanderi Fisher, 1941: 296.—Wash.

Genus PLATYURA Meigen

Platyura Meigen, 1803: 264. Type-species, *marginata* Meigen (Blanchard *in* Audouin et al., 1849: pl. 164).
Apemon Johannsen, 1909: 20. Type-species, *Platyura pectoralis* Coquillett (orig. des.).

REFERENCE: Fisher, 1941 (key).

manteri (Johnson), 1931: 23 (*Apemon*).—Conn.; N.H., Maine.
maudae Coquillett, 1895d: 199.—Wash.; Oreg.
nigriventris (Johannsen), 1910a: 245 (*Apemon*).—B.C.; Wash., Idaho, Mont.
pectoralis Coquillett, 1895d: 199.—Nev.; Wash., Oreg., Idaho.
    rufa Van Duzee, 1928a: 32 (*Apemon*).—Calif. ?Syn.
pulchra Williston, 1893c: 59.—Wash.
similis Johnson, 1931: 22 (*Apemon*).—N.H.; Que., Maine.
willistoni Laffoon (for *gracilis* Williston).—Wash. **N. name.**
    gracilis Williston, 1893c: 60 (preocc. Skuse, 1890).—Wash.

Subfamily LYGISTORRHININAE
Genus LYGISTORRHINA Skuse

Lygistorrhina Skuse, 1890b: 598. Type-species, *insignis* Skuse (mon.).

Unnamed spp.—se. U.S.

Subfamily MANOTINAE
Genus MANOTA Williston

Manota Williston, 1896c: 260. Type-species, *defecta* Williston (mon.).
Unnamed sp.—B.C. (Sherman, 1920: 15).

Subfamily MYCETOPHILINAE
Tribe EXECHIINI

The larvae of this tribe have been found in Agaricaceae (generally in the stalk), Boletaceae, Polyporaceae, Clavariaceae (one record), and *Peziza*, a genus of Ascomycetes. The host fungi for most species of *Brachypeza* are on wood, for most *Exechia* and *Rymosia* on the ground.

Genus ALLODIA Winnertz

Allodia Winnertz, 1863: 826. Type-species, *Mycetophila ornaticollis* Meigen (Johannsen, 1909: 104).

Brachycampta Winnertz, 1863: 833. Type-species, *Mycetophila alternans* Zetterstedt (Coquillett, 1910b: 515).

actuaria Johannsen, 1912a: 317.—N.Y., Mass.
anglofennica Edwards, 1921b: 122.—Scotland; N.S., n. Europe.
beata Johannsen, 1912a: 319.—N.Y.; Iowa.
bella Johannsen, 1912a: 318.—B.C.; Alta.
bulbosa Johannsen, 1912a: 316.—N.H., N.Y., N.J.
callida Johannsen, 1912a: 319.—Wash., Wyo.; Costa Rica.
cincta Van Duzee, 1928a: 50.—Calif.
crassicornis (Stannius), 1831b: 22 (*Mycetophila*).—Germany; Maine to Md., Europe.
delita Johannsen, 1912a: 320.—Wash., Calif.
elata Johannsen, 1912a: 318.—N.H., Mass.; Maine, Vt.
hirticauda Van Duzee, 1928a: 50.—Calif.
lugens (Wiedemann), 1817: 68 (*Mycetophila*).—Germany; N.S., Europe.
ornaticollis (Meigen), 1818: 269 (*Mycetophila*).—Europe; Alta., N.Y., N.J., Europe.
    falcata Johannsen, 1912a: 317.—N.Y., N.J.
pistillata (Lundström), 1911b: 399 (*Brachycampta*).—Hungary, Rumania; Alta., Europe.
subelata Malloch, 1923i: 178.—Alaska (Pribilof Is.).
truncata Edwards, 1921b: 123.—England; N.S., N.Y., n. Europe.
unicolor (Lundbeck), 1898: 260 (*Brachycampta*).—Greenland.

Genus ANATELLA Winnertz

Anatella Winnertz, 1863: 854. Type-species, *gibba* Winnertz (Johannsen, 1909: 90).

REFERENCE: Fisher, 1938a (key).

affinis Fisher, 1938a: 196.—Md.
ciliata Winnertz, 1863: 856.—Europe; Mo., N.Y., Md., Europe.
difficilis Garrett, 1925a: 5.—B.C.
silvestris Johannsen, 1909: 91.—N.Y.; B.C.
simpatica Dziedzicki, 1923: 6.—Europe; N.S.
    incisurata Edwards, 1925a: 589.—England.

Genus BRACHYPEZA Winnertz

Brachypeza Winnertz, 1863: 806. Type-species, *bisignata* Winnertz (Johannsen, 1909: 101).

brevitibia Van Duzee, 1928a: 47.—Calif.
dentica (Guthrie), 1917: 315 (*Allodia*).—Calif.; Wash., Iowa. **N. comb.**

divergens Johannsen, 1912a: 309 (as *bisignata* var.).—N.H., Vt., Maine; Maine to N.Y., also Iowa, N.C. **N. status.**

Genus EXECHIA Winnertz

Exechia Winnertz, 1863: 879. Type-species, *Mycetophila fungorum* De Geer (Johannsen, 1909: 106; misident.) =*fusca* (Meigen). Unpublished request before I.C.Z.N., according to Melville, 1961: 33, to fix *Mycetophila fusca* Meigen as type.

abrupta Johannsen, 1912b: 68.—N.Y.
absoluta Johannsen, 1912b: 72.—Maine; Man., S. Dak., Que., N.Y., Conn., N.J.
absurda Johannsen, 1912b: 74.—N.Y.; Alta., Maine.
adamsi Laffoon (for *analis* Adams).—Ind. **N. Name.**
    analis Adams, 1907: 37 (*Mycetophila;* preocc. Meigen, 1818).—Ind.
aequalis Van Duzee, 1928a: 54.—Calif.
alexanderi Shaw, 1951a: 65.—Wyo.
angustata Van Duzee, 1928a: 56.—Calif.
assidua Johannsen, 1912b: 71.—Wash.; Alta.
attrita Johannsen, 1912b: 73.—Wis., N.Y., Maine, R.I., N.J.; N.H.
auxiliaria Johannsen, 1912b: 71.—N.Y.; Wis., Maine, S.C.
aviculata Shaw, 1935c: 89.—N.Y.
bella Johannsen, 1912b: 72.—N.Y.; Wis., Maine.
bellula Johannsen, 1912b: 71.—Maine; B.C.
bicincta (Staeger), 1840: 263 (*Mycetophila*).—Denmark; Greenland, Europe.
    interrupta Zetterstedt, 1852: 4240 (*Mycetophila*).—Sweden.
bifurcata Fisher, 1934: 277.—N.Y.
bilobata Shaw, 1951b: 275.—Wyo.
borealis Van Duzee, 1928a: 53.—Alaska.
brevipetiolata Van Duzee, 1928a: 55.—Calif.
canalicula Johannsen, 1912b: 69.—N.C.; N.J.
capillata Johannsen, 1912b: 73.—N.Y.; Calif., Wyo., R.I.
captiva Johannsen, 1912b: 72.—N.J.; N.H., Vt., Mass.
cincinnata Johannsen, 1912b: 69.—Maine; Oreg., Calif., Maine to N.Y., S.C.
clepsydra Fisher, 1937: 397.—N.Y.
contaminata Winnertz, 1863: 891.—Europe; N.Y., Europe.
frigida (Boheman), 1865: 576 (*Mycetophila*).—Spitzbergen; Alaska, Wyo., N.Y., N.W.T., Greenland, n. Palaearctic.
    casta Johannsen, 1912b: 74.—Wyo.
lundstroemi Landrock, 1923: 170.—Finland; N.Y., Europe.

nativa Johannsen, 1912b: 70.—Maine; Ill., N.Y., Vt.
nexa Johannsen, 1912b: 68.—N.Y.; Mass.
nitidicollis Lundström, 1913: 311.—France; Greenland.
noctivaga Van Duzee, 1928a: 54.—Calif.
nugatoria Johannsen, 1912b: 70.—N.Y.; Alta., Wis., R.I.
nugax Johannsen, 1912b: 68.—Que.; B.C., N.Y.
obediens Johannsen, 1912b: 73.—Calif.; Alta., Oreg.
ovata Fisher, 1934: 278.—N.Y.; Iowa.
palmata Johannsen, 1912b: 71.—Wyo.; B.C., Man., Wash., S.C.
perspicua Johannsen, 1912b: 67.—Maine; B.C., Wyo.
plebeia (Walker), 1848: 100 (*Mycetophila*).—Ont.
pollex Shaw, 1935c: 89.—N.Y.
pratti Shaw, 1951a: 66.—Wyo.
quadrata Johannsen, 1912b: 69.—N.Y.; Wis., N.J., N.C.
repanda Johannsen, 1912b: 73.—N.Y.; Mass., Conn., England, Netherlands.
satiata Johannsen, 1912b: 69.—N.Y.
shawi Fisher, 1934: 276.—N.H.
subligulata Shaw, 1952a: 148 (n. name for *ligulata* Shaw).—Wyo.
    ligulata Shaw, 1951a: 66 (preocc. Lundström, 1913).—Wyo.
umbratica (Aldrich), 1897: 186 (*Mycetophila*).—Ind.; B.C., Oreg., N.Y., Pa., S.C.
umbrosa Van Duzee, 1928a: 56.—Calif.
unicincta Van Duzee, 1928a: 57.—Calif.
unicolor Van Duzee, 1928a: 53.—Alaska.

Genus RYMOSIA Winnertz

Rymosia Winnertz, 1863: 810. Type-species, *Mycetophila discoidea* Meigen (Johannsen, 1909: 102)=*fasciata* (Meigen).
Rhymosia, emend.

akeleyi Johannsen, 1912a: 312.—Wis., N.H.; Alta.
beckeri Shaw, 1951a: 68.—Wyo.
    var. **marionae** Shaw, 1951a: 70.—Wyo. ?Var.
brevicornis Sherman, 1921: 19.—B.C.
coheri Shaw, 1951b: 277.—Wyo.
dietrichi Shaw, 1951b: 279.—Wash.
diffissa Johannsen, 1912a: 313.—Calif., Idaho.
domestica (Meigen), 1830: 303 (*Mycetophila*).—Europe; N.Y., Vt., N.H., Mass., Europe, Japan.
    captiosa Johannsen, 1912a: 313.—Vt., N.H., Mass.
faceta Sherman, 1921: 18.—B.C.
filipes Loew, 1869c: 149 (Cent. 9, no. 36).—Conn.; B.C., Ind., N.H.
imitator Johannsen, 1912a: 312.—Wyo., Calif., Tex.; B.C., Pa.

inflata Johannsen, 1912a: 311.—N.Y.
lacki Edwards, 1935a: 468.—Greenland.
parvicauda Van Duzee, 1928a: 48.—Calif.
pectinata Sherman, 1921: 19.—B.C.
pediformis Shaw, 1951a: 68.—Wyo.
plumosa Van Duzee, 1928a: 49.—Calif.
prolixa Sherman, 1921: 18.—B.C.
seminigra Sherman, 1921: 18.—B.C.
sericea (Say), 1824a: 365 (1859a: 248) (*Mycetophila*).—Northwest Territory (U.S.).
serripes Johannsen, 1912a: 311.—N.Y.
spinicauda Van Duzee, 1928a: 47.—Calif.
triangularis Shaw, 1935c: 89.—N.Y.; Pa.

Tribe MYCETOPHILINI

The larvae of most, or possibly all, Mycetophilini are associated with fungi, including Ascomycetes, Myxomycetes, and 11 families of Basidiomycetes. Those of *Cordyla* are in ground-inhabiting fungi, of *Dynatosoma* in wood-inhabiting fungi, and those *Mycetophila* which live in Basidiomycetes ordinarily live in the flesh of the fungi. Species of *Platurocypta* have been reared only from Myxomycetes. Larvae of *Epicypta* carry cases made from their feces and other material; some species of *Phronia* have a protective mucus coating or a firm black excrement case; and certain species of *Trichonta* live in a space between the lower side of a fungus sporophore and a mucus sheet secreted by the larvae.

Genus CORDYLA Meigen

Polyxena Meigen, 1800: 19. Type-species, *Cordyla fusca* Meigen (Stone, 1941: 414). Suppressed I.C.Z.N., 1963b: 339.
Cordyla Meigen, 1803: 263. Type-species, *fusca* Meigen (sub. mon., Meigen, 1804: 93).

confera Garrett, 1925b: 14.—B.C.
gracilis Fisher, 1938a: 198.—Calif.
manca Johannsen, 1912a: 307.—N.Y.; B.C.
neglecta Johannsen, 1912a: 308.—Calif.; B.C.
parva Garrett, 1925b: 14.—B.C.
recens Johannsen, 1912a: 307.—N.Y.
scita Johannsen, 1912a: 307.—Wash.; B.C.
scutellata Garrett, 1925b: 14.—B.C.
verio Garrett, 1925b: 14.—B.C.
volucris Johannsen, 1909: 101.—N.Y.; Iowa, Maine.

Genus DYNATOSOMA Winnertz

Dynatosoma Winnertz, 1863: 947. Type-species, *Mycetophila fuscicornis* Meigen (Johannsen, 1909: 114).
Johannseni Guthrie, 1917: 316. Type-species, *aurei* Guthrie (mon.).

aureum (Guthrie), 1917: 316 (*Johannseni*).—Calif.
bifasciatum (Walker), 1848: 96 (*Mycetophila*).—Ont.; B.C., Alta., Oreg., Ont. to Maine, s. to Mass.
  nigrina Johannsen, 1912b: 75.—Mass.
coquilletti Landrock, 1918: 44 (n. name for *thoracica* Coquillett).—Ill., N.H.
  thoracica Coquillett 1901h: 598 (preocc. Zetterstedt, 1838).—Ill., N.H.
errans Garrett, 1925b: 11.—N.Y.
fulvidum Coquillett, 1895d: 200.—Wash.; B.C., N. Mex., N.Y. to Maine, s. to Conn., S.C.
huliphilum Garrett, 1925b: 12.—B.C.
  var. **grande** Garrett, 1925b: 12.—B.C. ?Var.
montanum Garrett, 1925b: 11.—B.C.
placidum Johannsen, 1912b: 76.—Ont.; B.C., N.Y., N.H., N.C.

Genus EPICYPTA Winnertz

Epicypta Winnertz, 1863: 909. Type-species, *Mycetophila scatophora* Perris (Johannsen, 1909: 110).
Delopsis Skuse, 1890b: 623. Type-species, *flavipennis* Skuse (mon).
REFERENCE: Steenberg, 1938 (biol.).

scatophora (Perris), 1849: 58 (*Mycetophila*).—Europe; Man. to Mass., s. to Iowa and N.C., B.C., Europe.
  pulicaria Loew, 1869c: 151 (Cent. 9, no. 41).—Pa. **N. syn.**
  vitrea Coquillett, 1905d: 68 (*Mycetophila*).—B.C., N.J.
  anomala Johannsen, 1912b: 96 (*Mycetophila*; preocc. Macquart, 1826).—Wis.

Genus MYCETOPHILA Meigen

Fungivora Meigen, 1800: 16. Type-species, *Tipula agarici* Villers (as Olivier; Coquillett, 1910b: 545). Suppressed I.C.Z.N., (1963b: 339.).
Mycetophila Meigen, 1803: 263. Type-species, *Tipula agarici* Villers (Johannsen, 1909: 116). Unpublished request before I.C.Z.N., according to Melville, 1961: 33, to fix *Tipula fungorum* De Geer as type.
Mycothera Winnertz, 1863: 913. Type-species, *Mycetophila dimidiata* Staeger (Johannsen, 1909: 111)=*ocellus* Walker.

Opistholoba Mik, 1891a: 5. Type-species, *Mycetophila caudata* Staeger (orig. des.).
REFERENCE: Laffoon, 1957 (rev.).

alata Guthrie, 1917: 315.—Calif.
   singularis Van Duzee, 1928a: 62.—Calif.
alberta Curran, 1927f: 80.—Alta.; Alaska to Calif.
alea Laffoon (for *guttata* Dziedzicki).—Europe; s. Alaska, s. Canada to cent. Calif. and N.C., Europe. **N. name.**
   guttata Dziedzicki, 1884: 309 (1886: 326) (preocc. Hutton, 1881).—Europe.
alexanderi (Laffoon), 1957: 267 (*Fungivora*).—Iowa; s. Canada, s. to Calif. and Pa. **N. comb.**
analis (Coquillett), 1901h: 598 (*Exechia*).—N.J.; Wis.
arnaudi (Laffoon), 1957: 222 (*Fungivora*).—Calif.; Wash., Oreg. **N. comb.**
attonsa (Laffoon), 1957: 233 (*Fungivora*).—Idaho; Wash. **N. comb.**
bentincki (Laffoon), 1957: 271 (*Fungivora*).—Calif. **N. comb.**
bipunctata Loew, 1869c: 152 (Cent. 9, no. 44).—Wis.; Minn. to se. Canada, s. to Tenn. and N.C.
bohartorum (Laffoon), 1957: 289 (*Fungivora*).—Calif. **N. comb.**
browningi (Laffoon), 1957: 191 (*Fungivora*).—Iowa; Wis. to Mass., s. to Mo. and N.C. **N. comb.**
byersi (Laffoon), 1957: 286 (*Fungivora*).—Iowa; Man. to N.H., s. to Iowa and Mass., Ky. **N. comb.**
capreolata (Laffoon), 1957: 223 (*Fungivora*).—Minn.; N.H. **N. comb.**
carruthi Shaw, 1951b: 276.—Colo.; B.C. to S. Dak., s. to Calif. and N. Mex.
caudata Staeger, 1840: 243.—Denmark; s. Canada, s. to Calif. and Ga., Europe.
   polita Loew, 1869c: 158 (Cent. 9, no. 53).—N.Y.
   ocellata Johannsen, 1909: 126 (*Opistholoba*).—N.Y.
caurina (Laffoon), 1957: 224 (*Fungivora*).—Wyo.; B.C., Wash. **N. comb.**
cavillator (Laffoon), 1957: 212 (*Fungivora*).—Alta.; Calif., Utah. **N. comb.**
celator (Laffoon), 1957: 242 (*Fungivora*).—Iowa; Ga. **N. comb.**
chamberlini (Laffoon), 1957: 260 (*Fungivora*).—Alaska. **N. comb.**
cingulum Meigen, 1830: 299.—Europe; Alaska, N.H., Europe.
clavata Van Duzee, 1928a: 58.—Calif.; B.C. to N.H., s. to Calif. and N.C.
   spinigera Van Duzee, 1928a: 61 (preocc. Tonnoir, 1927).—Calif.
   pacifica Fisher, 1938a: 199.—B.C.
   denningi Shaw, 1951a: 67.—Ariz.

comata (Laffoon), 1957: 196 (*Fungivora*).—Ill.; Iowa, N.Y., Md.,
Del., D.C. **N. comb.**
concinna (Laffoon), 1957: 235 (*Fungivora*).—Iowa; Sask., Mo., Mich.,
Ind., N.Y., D.C. **N. comb.**
consonans (Laffoon), 1957: 223 (*Fungivora*).—Wash.; Oreg., Calif.
N. comb.
contigua Walker, 1848: 96.—N.S.; s. Canada, s. to cent. Calif. and Md.
fallax Loew, 1869c: 156 (Cent. 9, no. 50).—Md.
lassata Johannsen, 1912b: 101.—Calif.
crassiseta (Laffoon), 1957: 209 (*Fungivora*).—N. Mex.; B.C., Wash.
N. comb.
cruciator (Laffoon), 1957: 212 (*Fungivora*).—Calif.; N. Mex., Ont.
N. comb.
dentata Lundström, 1913: 319.—Romania; s. Alaska to cent. Calif.,
Utah, Minn. to N.H., s. to Iowa and Md., Europe.
permata Guthrie, 1917: 314.—Calif.
devia (Laffoon), 1957: 197 (*Fungivora*).—Minn. **N. comb.**
discors (Laffoon), 1957: 265 (*Fungivora*).—Iowa; Minn. to Maine,
s. to Iowa and Pa. **N. comb.**
edura Johannsen, 1912b: 103.—N.Y.; Minn. to N.H., s. to Iowa and
Pa.
exstincta Loew, 1869c: 152 (Cent. 9, no. 43).—Middle States; Wis. to
Que., s. to Iowa and N.C.
faceta (Laffoon), 1957: 203 (*Fungivora*).—Wash.; Calif. **N. comb.**
falcata Johannsen, 1912b: 93.—N.Y.; s. Canada, s. to Calif. and Fla.
fascinator (Laffoon), 1957: 272 (*Fungivora*).—Minn.; Wash., Que.,
N.H., Pa., Calif. **N. comb.**
fatua Johannsen, 1912b: 103.—Idaho; B.C. to Mont., s. to cent.
Calif. and Wyo.
finlandica Edwards, 1913: 377.—Finland, U.S.S.R., Great Britain;
N.Y., Maine, Tenn., Europe.
fisherae (Laffoon), 1957: 180 (*Fungivora*).—Iowa; s. Canada, s. to
Colo., Tex., and Ga. **N. comb.**
foecunda Johannsen, 1912b: 99.—N.Y.; Alaska, Oreg., Idaho, Vt.,
Mass.
frustrator (Laffoon), 1957: 225 (*Fungivora*).—Calif.; Alta., Wash.,
Idaho, Mont., Oreg. **N. comb.**
fungorum (DeGeer), 1776: 361 (*Tipula*).—Scandinavia; Alaska to
Greenland, s. to Calif. and N. Mex., n. Wis., Palaearctic, India, Thailand.
punctata Meigen, 1804: 91.—Europe.
ghanii Shaw, 1951b: 276.—Wyo.; Wash., Oreg.
hiulca (Laffoon), 1957: 223 (*Fungivora*).—N.H.; B.C. **N. comb.**

ichneumonea Say, 1823: 16 (1859b: 43).—Pa.; s. Canada, s. to Mont. and N.C.
  mutica Loew, 1869c: 152 (Cent. 9, no. 45).—Middle States.
illudens (Laffoon), 1957: 188 (*Fungivora*).—Iowa; Wis. to Vt., s. to Mo. and Pa. **N. comb.**
impellans (Johannsen), 1912b: 83 (*Mycothera*).—N.Y.; Alaska to n. Canada, s. to cent. Calif. and N.C.
  edentula Johannsen, 1912b: 105.—B.C.
  pectoralis Van Duzee, 1928a: 62.—Calif.
itascae (Laffoon), 1957: 228 (*Fungivora*).—Minn. **N. comb.**
jucunda Johannsen, 1912b: 90.—N.Y.; Minn., Wis., Que.
jugata Johannsen, 1912b: 104.—Calif.; B.C., Wash., Oreg., Idaho.
laeta Walker, 1848: 97.—N.S.
lenis Johannsen, 1912b: 94.—Maine.
lenta Johannsen, 1912b: 102.—Maine; Alta., Wis., N.Y., Que., N.B., N.S.
limata (Laffoon), 1957: 262 (*Fungivora*).—La.; Iowa, Wis., N.C., S.C. **N. comb.**
luctuosa Meigen, 1830: 299.—Europe; s. Alaska to Que., s. to cent. Calif. and S.C., Europe.
  extenta Johannsen, 1912b: 105.—N.Y.
mitis (Johannsen), 1912b: 82 (*Mycothera*).—Wis.; Minn. to Que., s. to Tex. and S.C., Scotland.
moravica Landrock, 1925b: 38.—Czechoslovakia; s. Alaska to Sask., s. to Ariz., also Minn., N.Y., Tenn., Europe.
napaea (Laffoon), 1957: 292 (*Fungivora*).—Minn.; N.B. **N. comb.**
ocellus Walker, 1848: 95.—England; Hudsonian zone of Alaska and Canada, s. to s. Calif. and Ga., Europe, Japan.
  monochaeta Loew, 1869c: 158 (Cent. 9, no. 54).—D.C.
  fenestrata Coquillett *in* Baker, 1904: 19.—Calif.
  praenubila Johannsen, 1912b: 83 (*Mycothera*; as *fenestrata* var.).—N.J.
  exusta Johannsen, 1912b: 104.—Mass.
  fusca Van Duzee, 1928a: 60.—Calif.
parva Walker, 1848: 97.—Ont. Unrecognized.
parvimaculata Van Duzee, 1928a: 59.—Calif.; B.C., Oreg., Ariz.
  maculosa Guthrie, 1917: 314 (preocc. Meigen, 1818).—Calif.
paula (Loew), 1869c: 151 (Cent. 9, no. 42) (*Mycothera*).—Middle States; s. Alaska to s. Canada, s. to Calif. and Ga.
  trifasciata Coquillett *in* Baker, 1904: 18.—Calif.
paxillata (Laffoon), 1957: 218 (*Fungivora*).—Idaho; Alta., Wash., Oreg. **N. comb.**

pectita Johannsen, 1912b: 101.—B.C.; s. Alaska to s. Canada, s. to cent. Calif. and Pa.
    bispina Van Duzee, 1928a: 58.—Calif.
    ovata Van Duzee, 1928a: 59.—Calif.
percursa (Laffoon), 1957: 245 (*Fungivora*).—Oreg. **N. comb.**
perita Johannsen, 1912b: 90.—N.Y.; s. Canada, s. to Calif. and Gulf of Mexico.
pictula Meigen, 1830: 299.—Europe; Ill. to Que., s. to Va., Europe.
    bimaculata Fabricius, 1805: 59 (*Sciara*; preocc. Meigen, 1804).—Denmark.
    imitator Johannsen, 1912b: 99.—N.Y.
pinguis Loew, 1869c: 153 (Cent. 9, no. 47).—Ont.; Sask. to Maine, s. to Ark. and N.C.
    scalaris Loew, 1869c: 154 (Cent. 9, no. 48).—Middle States.
procera Loew, 1869c: 159 (Cent. 9, no. 55).—N.Y.
propinqua Walker, 1848: 96.—N.S.; Wash. to N.B., s. to Oreg. and N.Y.
    perlonga Johannsen, 1912b: 100.—N.Y.
recta (Johannsen), 1912b: 82 (*Mycothera*).—N.Y.; s. Canada, s. to cent. Calif. and N.C.
    paradoxa Johannsen, 1912b: 82 (*Mycothera*).—N.Y.
recula (Laffoon), 1957: 217 (*Fungivora*).—Calif. **N. comb.**
ruficollis Meigen, 1818: 262.—Austria; Hudsonian zone of Alaska and Canada, s. to Ariz. and N.H., Palaearctic, Ethiopian, Java.
scitula (Laffoon), 1957: 275 (*Fungivora*).—N.Y. **N. comb.**
scotica Edwards, 1941: 80.—Scotland; Alaska, Calif.
seclusa (Laffoon), 1957: 232 (*Fungivora*).—Wash.; Colo. **N. comb.**
sepulta (Laffoon), 1957: 175 (*Fungivora*).—Iowa; s. Alaska to n. Que., s. to Calif., Tex., and Conn. **N. comb.**
sertata (Laffoon), 1957: 225 (*Fungivora*).—Oreg.; Wash., Idaho, Calif., Utah, Wyo., N.H. **N. comb.**
shawi (Laffoon), 1957: 273 (*Fungivora*).—Md.; B.C., Minn. to s. Que., s. to Mo. and Va. **N. comb.**
sierrae (Laffoon), 1957: 234 (*Fungivora*).—Calif. **N. comb.**
sigillata Dziedzicki, 1884: 308 (1886: 265).—Europe; B.C., Oreg., Idaho, Wyo., Que., N.S., Europe.
sigmoides Loew, 1869c: 156 (Cent. 9, no. 51).—Middle States; s. Canada, s. to Wash., Mo., and Ga., U.S.S.R. (Kamchatka).
    fastosa Johannsen, 1912b: 91.—N.Y.
signatoides Dziedzicki, 1884: 310 (1886: 326).—Europe; s. Canada, s. to Oreg. and N.C., Europe.

sordida Wulp, 1874: 125.—Netherlands; s. Alaska and s. Canada, s. to
   n. Calif. and Pa., Europe.
spleniata (Laffoon), 1957: 244 (*Fungivora*).—Mass.; N.B., N.Y.
   N. comb.
stolida Walker, 1856a: 15.—England; s. Alaska, S. Dak. to N.H., s.
   to Kans. and Va., Europe.
   socia Johannsen, 1912b: 106.—N.Y.
stricklandi (Laffoon), 1957: 251 (*Fungivora*).—s. Alaska; Oreg., Alta.,
   Sask., Minn., Wis. **N. comb.**
strigata Staeger, 1840: 242.—Denmark; Wash., Oreg., Iowa, Que. to
   D.C., Europe.
   trichonota Loew, 1869c: 155 (Cent. 9, no. 49).—D.C.
subita (Laffoon), 1957: 283 (*Fungivora*).—Idaho; Wash., Oreg.,
   Calif. **N. comb.**
thioptera Shaw, 1940: 48.—Okla.; Wis. to Mass., s. to Gulf of Mexico.
trinotata Staeger, 1840: 242.—Denmark; s. Alaska, cent. Calif.,
   Sask. to Que., s. to Okla. and S.C., Europe.
   quatuornotata Loew, 1869c: 157 (Cent. 9, no. 52).—Md.
   subquatuornotata Shaw, 1940: 48.—Okla.
uncinata (Laffoon), 1957: 221 (*Fungivora*).—Wyo.; Mont., Utah. **N.
   comb.**
unipunctata Meigen, 1818: 272.—Germany; Minn. to s. Que., s. to
   La. and S.C., Europe.
   discoida Say, 1829: 153 (1859b: 351).—Ind.
   inculta Loew, 1869c: 153 (Cent. 9, no. 46).—D.C.
vegeta (Laffoon), 1957: 192 (*Fungivora*).—Iowa; Calif., Ariz., Ill.,
   Pa., Md., S.C. **N. comb.**
venusta (Laffoon), 1957: 290 (*Fungivora*).—Alaska; s. Alaska to s.
   Que., s. to Tex. and Md. **N. comb.**
verecunda (Laffoon), 1957: 270 (*Fungivora*).—Iowa; Minn., Ill., N.Y.,
   Mass. **N. comb.**
vesca (Laffoon), 1957: 213 (*Fungivora*).—N.Y.; Md., Tenn. **N.
   comb.**
wirthi (Laffoon), 1957: 197 (*Fungivora*).—Va. **N. comb.**

Genus PHRONIA Winnertz

Phronia Winnertz, 1863: 857. Type-species, *rustica* Winnertz
   (Johannsen, 1909: 96)=*exigua* (Zetterstedt).
Macrobrachius Dziedzicki, 1889: 520. Type-species, *kowarzii*
   Dziedzicki (mon.).
Telmaphilus Becker, 1908: 66. Type-species, *biarcuatus* Becker
   (Johannsen, 1909: 126).
   REFERENCES: Steenberg, 1924, 1943 (biol.).

californica Fisher, 1938b: 222 (n. name for *basalis* Van Duzee).—
Calif.
basalis Van Duzee, 1928a: 51 (preocc. Winnertz, 1863).—Calif.
despecta (Walker), 1848: 101 (*Mycetophila*).—Ont.
difficilis Johannsen, 1912b: 61.—N.Y.
exigua (Zetterstedt), 1852: 4246 (*Mycetophila*).—Norway, Sweden;
Greenland, Europe.
rustica Winnertz, 1863: 875.—Europe.
flabellata Van Duzee, 1928a: 51.—Calif.
fusciventris Van Duzee, 1928a: 52.—Calif.
hitchcocki Shaw, 1951a: 67.—Wyo.
incerta (Adams), 1907: 37 (*Mycetophila*).—Ind.
insulsa Johannsen, 1912b: 60.—R.I., N.Y.; B.C., Maine.
nebulosa (Johannsen), 1912b: 64 (*Telmaphilus*).—N.Y.; B.C.
producta Johannsen, 1912b: 60.—Mass.
similis Johannsen, 1912b: 62.—N.Y.
tenebrosa Coquillett, 1904f: 170.—Calif.; B.C., Oreg.
venusta Johannsen, 1912b: 61.—Idaho; B.C., Alta., S. Dak., N.Y.

Genus PLATUROCYPTA Enderlein

Platurocypta Enderlein, 1910a: 76. Type-species, *limbatifemur*
Enderlein (mon.).
Neoepicypta Coher, 1950: 114. Type-species, *Mycetophila punctum*
Stannius (orig. des.). **N. syn.**
Epicypta, authors, not Winnertz.

REFERENCE: Buxton, 1954 (biol.).

punctum (Stannius), 1831b: 16 (*Mycetophila*).—Germany; Mass.,
N.Y., N.J., S.C., Europe.
testata (Edwards), 1925b: 167 (*Epicypta*).—Europe; Que., Mass.,
N.Y., N.J., S.C., Europe.
testacea (error) Shaw and Fisher, 1952: 208.

Genus SCEPTONIA Winnertz

Sceptonia Winnertz, 1863: 907. Type-species, *Mycetophila nigra*
Meigen (Johannsen, 1909: 113).
autumnalis Garrett, 1925b: 16 (as *autumnals*).—B.C.
johannseni Garrett, 1925b: 15 (as *johannsoni*).—B.C.
nigra (Meigen).—Not Nearctic.

Genus TRICHONTA Winnertz

Trichonta Winnertz, 1863: 847. Type-species, *Mycetophila melanura*
Staeger (Johannsen, 1909: 94).

bellula Johannsen, 1912a: 304.—Vt.; N.H., Maine.
chaoi Shaw, 1951b: 279.—Wyo.
cincta Johannsen, 1912a: 303.—Maine.
diffissa Johannsen, 1912a: 305.—N.Y., Mass.; Conn.
foeda Loew, 1869c: 150 (Cent. 9, no. 38).—Middle States.
fusciventris Van Duzee, 1928a: 43.—Calif.
hansoni Shaw, 1940: 50.—Maine.
obesa Winnertz, 1863: 854.—Europe; Greenland, Austria.
patens Johannsen, 1912a: 305.—N.Y.
perspicua Wulp, 1881: 142.—Que.; Maine, Vt., N.J.
sagana Shaw, 1940: 51.—Maine.
triangularis Johannsen, 1912a: 303.—N.Y.
vulgaris Loew, 1869c: 149 (Cent. 9, no. 37).—Md., D.C.

Genus ZYGOMYIA Winnertz

Zygomyia Winnertz, 1863: 901. Type-species, *Mycetophila vara* Staeger (Johannsen, 1909: 112).

bifasciata Garrett, 1925b: 15.—B.C.
christata Garrett, 1925b: 15.—B.C.
christulata Garrett, 1925b: 15.—B.C.
coxalis Garrett, 1925b: 15.—B.C.
ignobilis Loew, 1869c: 150 (Cent. 9, no. 39).—Middle States; N.Y., S.C.
interrupta Malloch, 1914g: 234.—Ill.; Iowa.
ornata Loew, 1869c: 150 (Cent. 9, no. 40).—Pa.; Wis. to N.Y., s. to Iowa and N.C.
pilosa Garrett, 1925b: 14.—B.C.
vara (Staeger), 1840: 266 (*Mycetophila*).—Denmark; Maine, N.Y., Europe.

Subfamily SCIOPHILINAE
Tribe GNORISTINI

Larvae of this tribe have been reported from a variety of habitats such as fungi (*Boletina, Coelosia*), decaying wood (*Boletina, Synapha*), liverwort (*Boletina*), webs in moss (*Gnoriste*), and webs on cave walls (*Speolepta*).

Genus BOLETINA Staeger

Boletina Staeger, 1840: 233. Type-species, *Leia trivittata* Meigen (Johannsen, 1909: 73).

abdominalis Adams, 1903a: 24.—Mo.

akpatokensis Edwards, 1933: 613.—N.W.T. (Akpatok I.).
antica Garrett, 1924c: 165.—B.C.
antoma Garrett, 1924c: 166.—B.C.
arctica Holmgren, 1872: 105.—Greenland.
astacus Garrett, 1924c: 164.—B.C.
atra Cole *in* Cole and Lovett, 1921: 219.—Oreg.
birulai Lundström, 1915: 3.—Siberia; Alaska.
cincta Johannsen, 1912a: 270.—Vt., N.Y.; N.H., Conn., N.C.
crassicauda Van Duzee, 1928a: 44.—Alaska.
delicata Johannsen, 1912a: 276.—Wyo.
differens Garrett, 1924c: 168.—B.C.
gracilis Johannsen, 1912a: 271.—Calif., Wyo.
groenlandica Staeger, 1845b: 356.—Greenland; Alaska, Colo., N.H., Maine, N.J., n. Europe.
hopkinsii (Coquillett), 1895d: 200 (*Mycetophila*).—W. Va.; N.H.
imitator Johannsen, 1912a: 271.—Wash.; B.C., Alta., N.Y.
inops Coquillett, 1900h: 391 (1904: 5).—Alaska; B.C., Idaho.
jucunda Garrett, 1924c: 167.—B.C.
longicornis Johannsen, 1912a: 272.—Idaho.
magna Garrett, 1925a: 5.—B.C.
melancholica Johannsen, 1912a: 271.—Wyo.; B.C., N.C.
montana Garrett, 1924c: 163.—B.C.
nacta Johannsen, 1912a: 277.—Wyo.
notescens Johannsen, 1912a: 272.—N.Y., Mass.; B.C., N.H., Maine, N.C.
obesula Johannsen, 1912a: 276.—Alaska.
obscura Johannsen, 1912a: 270.—N.Y., N.H., Mass., N.J.; Conn., R.I., N.C.
oviducta (Garrett), 1924c: 164 (*Mycomya*).—B.C.
profectus Shaw and Fisher, 1952: 197.—N.Y., N.S.
punctus Garrett, 1925a: 5.—B.C.
sciarina Staeger, 1840: 236.—Denmark; N.Y., N.H., Maine, Conn., Greenland, Europe.
sedula Johannsen, 1912a: 277.—Wash.; Alta.
shermani Garrett, 1924c: 166.—B.C.
sobria Johannsen, 1912a: 274.—Wash.; B.C.
subatra Fisher, 1938b: 222 (n. name for *atra* Van Duzee).—Alaska.
    atra Van Duzee, 1928a: 45 (preocc. Cole, 1921).—Alaska.
tricincta Loew, 1869c: 143 (Cent. 9, no. 25).—Wis., Md.; B.C., Wis., Maine to Md., S.C.
unusa Garrett, 1924c: 168.—N.Y.

Genus COELOSIA Winnertz

Coelosia Winnertz, 1863: 796. Type-species, *Boletina flava* Staeger (Johannsen, 1909: 86).

flava (Staeger).—Not Nearctic.
gracilis Johannsen, 1912a: 294.—Calif., Colo.; B.C.
lepida Johannsen, 1912a: 294.—Calif.; B.C.
modesta Johannsen, 1912a: 294.—Calif.
pygophora Coquillett, 1904f: 170.—Calif.; Oreg.
tenella (Zetterstedt), 1852: 4165 (*Boletina*).—Sweden; Alta., Oreg., Calif., Europe, Mongolia.
    flavicauda Winnertz, 1863: 798.—Europe.
truncata Lundström, 1909: 18.—Finland; Alaska, N.W.T., Sask., Que., Latvia.

Genus DZIEDZICKIA Johannsen

Hertwigia Dziedzicki, 1885: 166 (preocc. Schmidt, 1880). Type-species, *marginata* Dziedzicki (mon.).
Dziedzickia Johannsen, 1909: 44 (n. name for *Hertwigia* Dziedzicki). Type-species, *Hertwigia marginata* Dziedzicki (aut.).
Syntemna, authors, not Winnertz.

columbiana Sherman, 1921: 17.—B.C.
fuscipennis (Coquillett), 1905d: 67 (*Sciophila*).—B.C.
immaculata Cole *in* Cole and Lovett, 1919: 222.—Oreg.
johannseni Sherman, 1921: 17 (preocc. Meunier, 1917).—B.C. Unrecognized.
longicornis (Coquillett), 1901h: 597 (*Docosia*).—N.H.; Mass.
    rejecta Johannsen, 1912a: 296 (*Syntemna*).—Mass.
occidentalis Sherman, 1921: 17.—B.C.
oregona Cole *in* Cole and Lovett, 1919: 223.—Oreg.
polyzona (Loew), 1869c: 142 (Cent. 9, no. 24) (*Syntemna*).—Middle States; Maine, R.I., N.Y., N.J., N.C., S.C.
pullata (Coquillett) *in* Baker, 1904: 19 (*Neoempheria*).—Calif.
rutila Sherman, 1921: 17.—B.C.
separata (Johannsen), 1912a: 297 (*Syntemna*).—Vt.
vernalis Sherman, 1921: 16.—B.C.
vittata (Coquillett), 1901h: 597 (*Docosia*).—N.H.; Maine.
    var. **fasciata** (Johannsen), 1912a: 297 (*Syntemna*).—Maine. ?Var.

Genus GNORISTE Meigen

Gnoriste Meigen, 1818: 243. Type-species, *apicalis* Meigen (mon.).
apicalis Meigen.—Not Nearctic.

groenlandica Lundbeck, 1898: 259.—Greenland; Conn.
macra Johannsen, 1912a: 257.—Wis.; N.H., Maine.
macroides Curran, 1927f: 79.—Ont.; Que.
megarrhina Osten Sacken, 1877: 193.—Calif.; Labr. to N.J., N.C.

Genus HADRONEURA Lundström

Hadroneura Lundström, 1906: 10. Type-species, *palmeni* Lundström (mon.).

kincaidi (Coquillett), 1900h: 391 (1904: 5) (*Neoempheria*).—Alaska (Popof I.).

Genus SPEOLEPTA Edwards

Speolepta Edwards, 1925a: 566. Type-species, *Polylepta leptogaster* Winnertz (orig. des.).

REFERENCE: Schmitz, 1913 (biol.).

leptogaster (Winnertz), 1863: 746 (*Polylepta*).—Europe; Ind., N.H.

Genus SYNAPHA Meigen

Synapha Meigen, 1818: 227. Type-species, *fasciata* Meigen (mon.).
Empalia Winnertz, 1863: 762. Type-species, *Sciophila vitripennis* Meigen (mon.).

bicolor Shaw and Fisher, 1952: 196.—N.Y., R.I.
disjuncta (Garrett), 1925b: 11 (*Empalia*).—B.C. **N. comb.**
tibialis (Coquillett), 1901h: 596 (*Polylepta*).—N.H.; N.Y., N.J.

Tribe MYCOMYINI

The larvae of this tribe have been found under bark, in decaying wood, and in basidiomycete fungi. At least some species of *Mycomya* secrete mucus webs.

REFERENCE: Coher, 1959 (tax.).

Genus MYCOMYA Rondani

Mycomya Rondani, 1856: 194. Type-species, *Sciophila marginata* Meigen (mon.).
Mycomyia, error or emend.
Sciophila, authors, not Meigen.

REFERENCE: Fisher, 1937 (key).

alternata Fisher, 1937: 396.—N.Y.; B.C., Wyo., Iowa, Maine to Tenn.
ampla Garrett, 1924c: 64.—Alta.; B.C.

angulata (Adams), 1903a: 22 (*Sciophila*).—Colo.
armata Garrett, 1924c: 163.—B.C.
ata Garrett, 1924c: 159.—B.C.
autumnalis Garrett, 1924c: 160.—B.C.
biseriata (Loew), 1869c: 140 (Cent. 9, no. 20) (*Sciophila*).—Red River of the North (probably Man.); B.C., Vt., Maine.
brevivitta (Coquillett), 1905d: 67 (*Sciophila*).—B.C.; Wis. to Que., s. to Ill. and N.Y., also B.C.
calcarata (Coquillett) *in* Baker, 1904: 19 (*Sciophila*).—Calif.; B.C.
californica Van Duzee, 1928a: 40.—Calif.
caulfieldi Garrett, 1924c: 62.—B.C.
cranbrooki Garrett, 1924c: 61.—B.C.
curvata Fisher, 1937: 395.—Alta.; N.W.T., B.C., Wash., Mont., Wyo., Maine.
dentata Fisher, 1937: 396.—N.H.; Minn. to Vt., s. to N.C.-Tenn. border.
dichaeta Fisher, 1937: 394.—N.Y.; Iowa, Mass. to Md.
difficilis Garrett, 1924c: 65.—B.C.
dura Garrett, 1924c: 162.—B.C.
echinata Garrett, 1924c: 161.—B.C.; Wash., Idaho, Mont., Utah, Calif.
    abbreviata Van Duzee, 1928a: 42.—Calif.
flavohirta (Coquillett), 1901h: 596 (*Sciophila*).—N.H.; B.C., Wash., Ont., Conn.
fragilis (Loew), 1869c: 138 (Cent. 9, no. 16) (*Polylepta*).—Mass.; Minn., N.H. to Md.
frequens Johannsen, 1910b: 171 (as *littoralis* var.).—Calif.; Alta., N.C.
fulvitibia Van Duzee, 1928a: 38.—Calif.
    longispina Van Duzee, 1928a: 41.—Calif.
fuscipalpis Van Duzee, 1928a: 40.—Calif.
hamata Garrett, 1924c: 160.—B.C.
hirticauda Van Duzee, 1928a: 38.—Calif.; Wash., Oreg.
hirticollis (Say), 1824a: 362 (1859a: 246) (*Sciophila*).—Northwest Territory (U.S.); s. Canada, s. to Oreg. and Tenn.
    appendiculata Loew, 1869c: 139 (Cent. 9, no. 19) (*Sciophila*).—N.Y.
humida Garrett, 1924c: 62.—B.C.; Mont.
imitans Johannsen, 1910b: 177.—B.C., Wis., N.Y., Mass., R.I.; B.C., Alta., Wis. to Que., s. to Iowa and N.C.
incompta Johannsen, 1910b: 186.—B.C., N.Y., Maine.
intermedia Fisher, 1937: 396.—Calif.
kiamichii Shaw, 1940: 50.—Okla.; Iowa.

littoralis (Say), 1824a: 361 (1859a: 245) (*Sciophila*).—Northwest Territory (U.S.); Alta. to Que., s. to Mo. and N.C.
magna Garrett, 1924c: 64.—B.C.
marginalis Johannsen, 1910b: 177.— B.C.; Wash.
maxima Johannsen, 1910b: 179.—Maine; B.C. to Labr., s. to Oreg. and N.Y.
mendax Johannsen, 1910b: 182.—B.C., Idaho, Calif.; Oreg., N.Y., Vt., N.H.
mutabilis Sherman, 1921: 16.—B.C.
nigra Fisher, 1938a: 198.—Md.
nigricauda (Adams), 1903a: 23 (*Sciophila*).—Colo.; N.Y.
nigrihirta Van Duzee, 1928a: 39.—Calif.; Oreg.
nugatoria Johannsen, 1910b: 183.—Wis., N.C.; Iowa, Que., Mass.
obliqua (Say), 1824a: 363 (1859a: 247) (*Sciophila*).—Northwest Territory (U.S.); Man. to Que., s. to Iowa and N.C.
    obtruncata Loew, 1869c: 139 (Cent. 9, no. 18) (*Sciophila*).— D.C.
onusta (Loew), 1869c: 138 (Cent. 9, no. 17) (*Sciophila*).—D.C.; Maine, Mass.
ornata (Meigen).—Not Nearctic.
parascopula Fisher, 1937: 394.—Md.; Iowa, S.C.
polleni Garrett, 1924c: 65.—B.C.
pseudomaxima Fisher, 1937: 393.—N.Y.
recurva Johannsen, 1910b: 185.—Wis.; Maine
    var. **chloratica** Johannsen, 1910b: 185.—Wis. ?Var.
scopula Fisher, 1937: 394.—N.Y.; Que., Vt.
sequax Johannsen, 1910b: 172.—N.Y.; Alta., Maine, S.C.
shermani Garrett, 1924c: 66.—B.C.
sigma Johannsen, 1910b: 180.—N.C.; B.C., Idaho, Wyo., N.Y., Mass.
simplex (Coquillett), 1905d: 67 (*Sciophila*).—B.C.
sphagnicola Shaw, 1941a: 171.—N.C.; B.C., Oreg., N.Y., Nfld.
sublittoralis Shaw, 1941a: 172.—N.C.; s. Canada, s. to Wash. and N.C.
tantilla (Loew), 1869c: 140 (Cent. 9, no. 21) (*Sciophila*).—Nebr.; s. Canada, s. to N. Mex. and N.C., Costa Rica.
tenuis (Walker), 1856a: 37 (*Sciophila*).—England; Greenland, Europe.
terminata Garrett, 1924c: 60.—B.C.; Oreg.
triacantha Shaw, 1941a: 172.—N.C.-Tenn. border; N.H., N.C., Tenn.
turitella Fisher, 1937: 395.—Fla.; N.C.
unicolor (Walker), 1848: 93 (*Leia*).—Ont.
vulgaris Garrett, 1924c: 63.—B.C.; Alta.

Genus NEOEMPHERIA Osten Sacken

Empheria Winnerz, 1863: 738 (preocc. Hagen, 1856). Type-species, *Sciophila striata* Meigen (Coquillett, 1910b: 537).
Neoempheria Osten Sacken, 1878c: 9 (n. name for *Empheria* Winnertz). Type-species, *Sciophila striata* Meigen (aut.).
balioptera (Loew), 1869c: 136 (Cent. 9, no. 13) (*Empheria*).—Ill.; Ont. to Que., s. to Iowa and N.C.
    lutea Tollet, 1948: 1.—Md.
didyma (Loew), 1869c: 136 (Cent. 9, no. 14) (*Empheria*; n. name for *bimaculata* Loew).—Ont.; s. Canada (from B.C. e.), s. to Mich. and N.Y.
    bimaculata Loew, 1866b: 6 (Cent. 7, no. 9) (*Sciophila*; preocc. Roser, 1840).—Ont.
    digitalis Fisher, 1937: 390.—Mich.
illustris Johannsen, 1910b: 163.—N.Y.; Ont., s. to Iowa and Fla.
impatiens Johannsen, 1910b: 161.—R.I.; Que., s. to Tenn. and N.C.
indulgens Johannsen, 1910b: 162.—Que., N.Y., N.C.; Iowa to Que., s. to Tenn. and N.C.
macularis Johannsen, 1910b: 159.—Que., N.Y.; Iowa to Que., s. to Tenn. and N.C.
nepticula (Loew), 1869c: 137 (Cent. 9, no. 15) (*Empheria*).—Ga.; Mass., N.J., N.C.

Tribe SCIOPHILINI

Members of this tribe have been reared from bark fungi (*Sciophila*, *Leptomorphus*), decaying wood (*Monoclona*, *Phthinia*, *Sciophila*), and mosses (*Neuratelia*). The larvae of some are known to make mucus webs.

Genus ACNEMIA Winnertz

Acnemia Winnertz, 1863: 798. Type-species, *Leia nitidicollis* Meigen (Johannsen, 1909: 63).

flaveola Coquillett, 1901h: 598.—N.J.; Que. to N.J., N.C.
psylla Loew, 1869c: 148 (Cent. 9, no. 34).—Md.; B.C., N.Y., Mass.
varipennis Coquillett, 1904f: 169.—Calif.

Genus ALLOCOTOCERA Mik

Eurycera Dziedzicki, 1885: 169 (preocc. de Laporte, 1833). Type-species, *flava* Dziedzicki (mon.)=*pulchella* (Curtis).
Allocotocera Mik, 1886: 102 (n. name for *Eurycera* Dziedzicki). Type-species, *Eurycera flava* Dziedzicki (aut.)=*pulchella* (Curtis).

parvula (Coquillett), 1901h: 597 (*Leptomorphus*).—N.J.; Wis., Maine, Mass.
flavescens Johannsen, 1909: 72.—Wis.

Genus AZANA Walker

Azana Walker, 1856a: 26. Type-species, *scatopsoides* Walker (mon.)=*anomala* (Staeger).
Unnamed sp.—Maine (Johannsen, 1912a: 260), Minn., N.S.

Genus EUDICRANA Loew

Eudicrana Loew, 1869c: 142 (Cent. 9, no. 23). Type-species, *obumbrata* Loew (mon.).

obumbrata Loew, 1869c: 141 (Cent. 9, no. 23).—N.Y.; Maine to Pa.
plexipus Garrett, 1925a: 4.—B.C.

Genus LEPTOMORPHUS Curtis

Leptomorphus Curtis, 1831: pl. 365. Type-species, *walkeri* Curtis (orig. des.).
Diomonus Walker, 1848: 87. Type-species, *nebulosus* Walker (mon.).

bifasciatus (Say), 1824a: 363 (1859a: 246) (*Sciophila*).—Northwest Territory (U.S.); Vt., N.H., Maine, Mass.
hyalinus Coquillett, 1901h: 598.—N.H.; R.I., N.C.
magnificus (Johannsen), 1910b: 155 (*Diomonus*).—Ohio, N.Y., Mass.; Wash.
nebulosus (Walker), 1848: 87 (*Diomonus*).—Ont.; N.H., Maine.
subcaeruleus subcaeruleus (Coquillett), 1901h: 595 (*Sciophila*).—Ont., N.H., Pa.; Alta., Ont. to Maine, s. to N.C.
pulcher Johannsen, 1903a: 14 (*Sciophila*).—N.Y.
ssp. **gurneyi** Shaw, 1947: 155.—Mo. ?Ssp.
walkeri Curtis, 1831: pl. 365.—England; N.J., Europe.
ypsilon Johannsen, 1912a: 265.—N.Y.

Genus MEGALOPELMA Enderlein

Megalopelma Enderlein, 1910b: 165. Type-species, *planiceps* Enderlein (orig. des.).

glabanum (Johannsen), 1910b: 136 (*Sciophila*).—N.Y., Wis.; Alta., Iowa, Vt., Maine, R.I. **N. comb.**
galbana, emend.
var. **germanum** (Johannsen), 1910b: 137 (*Sciophila*).—Wis.; Mass. ?Var. **N. comb.**

var. **socium** (Johannsen), 1910b: 137 (*Sciophila*).—Vt., Mass.; N.Y. ?Var. **N. comb.**

Genus MONOCLONA Mik

Staegeria Wulp, 1876: xlix (preocc. Rondani, 1856). Type-species, *Sciophila halterata* Staeger (mon.)=*rufilatera* (Walker).
Monoclona Mik, 1886: 279 (n. name for *Staegeria* Wulp). Type-species, *Sciophila halterata* Staeger (aut.)=*rufilatera* Walker).

REFERENCE: Fisher, 1946 (key).

elegantula Johannsen, 1910b: 128.—N.Y.; B.C., Que. to N.S., s. to Pa., N.C.
floridensis Fisher, 1946: 2.—Fla.
furcata Johannsen, 1910b: 187.—Maine; B.C., Mass.
idahoensis Fisher, 1946: 1.—Idaho.
simplex Garrett, 1925a: 9.—B.C.

Genus NEURATELIA Rondani

Neuratelia Rondani, 1856: 195. Type-species, *Mycetophila nemoralis* Meigen (orig. des.).
Anaclinia Winnertz, 1863: 770. Type-species, *Mycetophila nemoralis* Meigen (mon.).
Odontopoda Aldrich, 1897: 187. Type-species, *sayi* Aldrich (mon.).

abrevena Garrett, 1925b: 11.—B.C.
coxalis (Coquillett), 1905d: 68 (*Anaclinia*).—B.C.; Oreg.
desidiosa Johannsen, 1912a: 263.—Mass.
distincta (Garrett), 1925a: 4 (*Odontopoda*).—B.C. **N. comb.**
eminens Johannsen, 1912a: 263.—Idaho.
flexa Van Duzee, 1928a: 44.—Calif.
grandis Garrett, 1925a: 5.—B.C.
insignifica Shaw, 1941a: 173.—N.C.-Tenn. border.
nemoralis (Meigen), 1818: 265 (*Mycetophila*).—Europe; Alaska, B.C., Calif., Europe, e. Siberia, Japan.
obscura Garrett, 1925a: 5.—B.C.
sayi (Aldrich), 1897: 187 (*Odontopoda*).—Ind.
scitula Johannsen, 1912a: 263.—Vt., N.J.; Maine, R.I., N.C., S.C.
scituloides Shaw, 1941a: 173.—Tenn., N.C.
silvatica Johannsen, 1912a: 262.—Calif.; S.C.

Genus PARATINIA Mik

Paratinia Mik, 1874: 333. Type-species, *sciarina* Mik (mon.).
recurva Johannsen, 1910b: 144.—N.Y.; Vt., S.C.

Genus PHTHINIA Winnertz

Phthinia Winnertz, 1863: 779. Type-species, *humilis* Winnertz (Johannsen, 1909: 83).
carolina Fisher, 1940: 246.—N.C.
catawbiensis Shaw, 1940: 50.—N.C.
curta Johannsen, 1912a: 291.—N.Y.; B.C., Oreg., Maine.
tanypus Loew, 1869c: 143 (Cent. 9, no. 26).—N.Y.

Genus POLYLEPTA Winnertz

Polylepta Winnertz, 1863: 745. Type-species, *undulata* Winnertz (Johannsen, 1909: 43)=*guttiventris* (Zetterstedt).
modesta Van Duzee, 1928a: 43.—Calif.
nigella Johannsen, 1910b: 148.—Wash.
obediens Johannsen, 1910b: 147.—Wis., N.H., Mass.; B.C., N.Y., Maine, Conn.

Genus SCIOPHILA Meigen

Sciophila Meigen, 1818: 245. Type-species, *hirta* Meigen (Curtis, 1837: pl. 641).
Lasiosoma Winnertz, 1863: 748. Type-species, *Sciophila pilosa* Meigen (Coquillett, 1910b: 558)=*hirta* Meigen.
acuta Garrett, 1925a: 11.—B.C.
agassis Garrett, 1925a: 9.—B.C.
bicolor Garrett, 1925a: 10 (preocc. Dziedzicki, 1885).—B.C. Unrecognized.
bifida Garrett, 1925a: 10.—B.C.
distincta Garrett, 1925a: 10.—B.C.
fasciata Say, 1823: 26 (1859b: 50).—Pa., Md.; Que., Vt., N.J., N. Mex.
fusca Garrett, 1925a: 10 (preocc. Meigen, 1818).—B.C. Unrecognized.
habilis Johannsen, 1910b: 138.—N.Y.; S.C.
hebes Johannsen, 1910b: 139.—Kans., R.I.
hirta Meigen, 1818: 251.—Europe; Oreg., Greenland, Europe.
impar Johannsen, 1910b: 140.—Wash., Wyo.
incallida Johannsen, 1910b: 139.—N.Y.; N.H., Mass., Conn.
longua Garrett, 1925a: 9.—B.C.
neohebes Garrett, 1925a: 9.—B.C.; Alta.
novata Johannsen, 1910b: 140.—N.Y.; Iowa.
nugax Johannsen, 1910b: 137.—N.Y.; Wis.
pallipes Say, 1824a: 361 (1859a: 245).—Northwest Territory (U.S.); Iowa, N.Y., N.H., Maine.

parva Garrett, 1925a: 10 (preocc. Dziedzicki, 1885).—B.C. Unrecognized.
puta Laffoon (for *nitida* Van Duzee).—Calif. **N. name.**
  nitida Van Duzee, 1928a: 37 (preocc. Zetterstedt, 1852).—Calif.
quadratula (Loew), 1869c: 141 (Cent. 9, no. 22) (*Lasiosoma*).—Maine; Minn., N.H.
setosa Garrett, 1925a: 11.—B.C.
severa Johannsen, 1910b: 141.—N.Y.
similis Johannsen, 1910b: 142.—Que.

Tribe TETRAGONEURINI
(Leiini)

Larvae of this tribe are mostly associated with fungi or decaying wood. Species of *Leia* are often found in mucus webs on the lower surface of fungus sporophores although several have been reared from a variety of other places. Species of *Rondaniella* and *Tetragoneura* have been reared from decaying wood and wood-inhabiting fungi. *Docosia fumosa*, a European species, has been bred repeatedly from various bird nests.

Genus DOCOSIA Winnertz

Docosia Winnertz, 1863: 802. Type-species, *Mycetophila sciarina* Meigen (Johannsen, 1909: 92).

aceus Garrett, 1925b: 13.—B.C.
affinis Garrett, 1925b: 12.—B.C.
apicula Garrett, 1925b: 13.—B.C.
defecta Van Duzee, 1928a: 46.—Calif.
dialata Van Duzee, 1928a: 46.—Calif.
dichroa Loew, 1869c: 148 (Cent. 9, no. 35).—D.C.; Wis. to Maine, s. to Kans. and D.C.
  mutor Adams, 1903a: 24 (*Syntemna*).—Mo.
nebulosa Garrett, 1925b: 13.—B.C.
nigella Johannsen, 1912a: 300.—Alaska.
nigrita Garrett, 1925b: 13.—B.C.
nitida Johannsen, 1912a: 300.—S. Dak.; Alta.
obscura Coquillett, 1901h: 597.—N.H.
paradichroa Fisher, 1937: 397.—N.Y.
setosa Garrett, 1925b: 12 (preocc. Landrock, 1916).—B.C. Unrecognized.
similis Garrett, 1925b: 12.—B.C.
vierecki Garrett, 1925b: 13.—B.C.

Genus ECTREPESTHONEURA Enderlein

Ectrepesthoneura Enderlein, 1910b: 155. Type-species, *Tetragoneura hirta* Winnertz (orig. des.).

bicolor (Coquillett), 1901h: 595 (*Tetragoneura*).—N.H.; N.Y., Mass., Conn., N.C.

Genus LEIA Meigen

Leia Meigen, 1818: 253. Type-species, *fascipennis* Meigen (Curtis, 1837a: pl. 645).
Leja, error or emend.
Glaphyroptera Winnertz, 1863: 781 (preocc. Heer, 1852). Type-species, *Leia fascipennis* Meigen (Coquillett, 1910b: 547).
Neoglaphyroptera Osten Sacken, 1878c: 10 (n. name for *Glaphyroptera* Winnertz). Type-species, *Leia fascipennis* Meigen (aut.).

bivittata Say, 1829: 152 (1859b: 351).—Ind.; Minn. to Maine, s. to Kans. and N.C.
    lateralis Wulp, 1867: 131 (*Glaphyroptera*).—Wis.
cephala Garrett, 1925b: 11.—B.C.
cincta (Coquillett), 1895a: 308 (*Neoglaphyroptera*).—Fla.
cuneola (Adams), 1903a: 25 (*Neoglaphyroptera*).—Colo.; Idaho.
decora (Loew), 1869c: 144 (Cent. 9, no 28) (*Glaphyroptera*).—Ga.
dryas Johannsen, 1912a: 287.—Wis.; Iowa, N.C.
hemiata Garrett, 1925a: 11.—B.C.
hyalina (Coquillett), 1905d: 68 (*Lejomya*).—N. Mex.
joculator Laffon (for *nigra* Johannsen).—Wash., Mont. **N. name.**
    nigra Johannsen, 1912a: 281 (preocc. Zetterstedt, 1838).—Wash., Mont.
lineola (Adams), 1903a: 25 (*Neoglaphyroptera*).—Calif.
melaena (Loew), 1869c: 144 (Cent. 9, no. 27) (*Glaphyroptera*).—N.Y.; Ont. to Maine, s. to N.J.
nigricornis Van Duzee, 1928a: 46.—Alaska.
oblectabilis (Loew), 1869c: 146 (Cent. 9, no. 31) (*Glaphyroptera*).—Middle States; Wis. to Maine, s. to Kans. and N.C., Costa Rica.
opima (Loew), 1869c: 145 (Cent. 9, no. 29) (*Glaphyroptera*).—Conn.; Alta., Wyo., Wis. to Maine, s. to Iowa and N.C.
plebeja Johannsen, 1912a: 285.—Wis., Kans.; Iowa.
shermani Garrett, 1925a: 11.—B.C.
striata (Williston), 1893c: 60 (*Neoglaphyroptera*).—Wash.; Alta., N.Y., N.C.

sublunata (Loew), 1869c: 145 (Cent. 9, no. 30) (*Glaphyroptera*).—
N.Y.; B.C., Alta., N.Y. to Maine, s. to N.C.
varia Walker, 1848: 93.—Ont.; Wyo., Iowa, Wis.
ventralis Say, 1824a: 364 (1859a: 247).—Northwest Territory (U.S.); Maine to N.J.
winthemii Lehmann, 1822: 39 (also 1824: 241).—Germany; B.C. to Maine, s. to Oreg. and N.C., Europe, Siberia, Japan, Oriental.
    maculipennis Say, 1824a: 365 (1859a: 248) (*Mycetophila*).— Northwest Territory (U.S.).
    trifasciata Walker, 1848: 93.—Ont.

Genus MEGOPHTHALMIDIA Dziedzicki

Megophthalmidia Dziedzicki, 1889: 404, 525 (as *Megophtalmidia*, pp. 525, 526, 532). Type-species, *zugmayeriae* Dziedzicki (mon.)=*crassicornis* (Curtis).

occidentalis Johannsen, 1909: 89.—Wash.; B.C.

Genus RONDANIELLA Johannsen

Rondaniella Johannsen, 1909: 66. Type-species, *Leia variegata* Winnertz (orig. des.).

sororcula (Loew), 1869c: 147 (Cent. 9, no. 32) (*Leia*).—N.Y.; B.C., Wis. to Maine, s. to Iowa and N.C.
    abbreviata Loew, 1869c: 147 (Cent. 9, no. 33) (*Leia*).— Middle States. **N. syn.**

Genus TETRAGONEURA Winnertz

Tetragoneura Winnertz, 1846: 18. Type-species, *distincta* Winnertz (Johannsen, 1909: 34, 35)=*sylvatica* (Curtis).

arcuata Sherman, 1921: 20.—B.C.
atra Sherman, 1921: 19.—B.C.
fallax Sherman, 1921: 20.—B.C.
longicauda Van Duzee, 1928a: 36.—Calif.
marceda Sherman, 1921: 20.—B.C.
nitida Adams, 1903a: 23.—Mo.; N.C.
pimpla Coquillett, 1901h: 595.—Pa.; B.C., Oreg., N.Y., S.C.
quintana Cole *in* Cole and Lovett, 1921: 218.—Oreg.
robur Garrett, 1925a: 8.—B.C.
similis Garrett, 1925a: 8 (as *similas*, also p. 8).—B.C.

Unplaced Species of Mycetophilidae

grisea Walker, 1848: 92 (*Sciophila*).—Ont.; N.H.
nubila Say, 1829: 153 (1859b: 351) (*Mycetophila*).—Ind.
obscura Walker, 1848: 101 (*Mycetophila*).—Ont.

Family SCIARIDAE

(Lycoriidae)

By Alan Stone and Jean L. Laffoon

Flies of the family Sciaridae are usually small, dark, and rather delicate. They are found in moist places wherever fungus grows. The larvae of some species may be pests in mushroom cellars or greenhouses. Those of certain species may form snakelike masses that move over the ground.

A sound, revisionary study of the genera and species of this family for the Nearctic Region is needed. Frey (1942, 1948) and Tuomikoski (1960) have provided classifications with considerable merit, but based largely on northern European species. The present catalog does little more than list the proposed names, although several changes suggested by Frey and Tuomikoski have been adopted. Most species previously assigned to *Neosciara* are transferred to *Bradysia* to conform with Frey's classification.

"*Sciara ocellaris* Comstock," so cited in the literature, was thought to cause the gall produced by *Cecidomyia ocellaris* Osten Sacken. This was the result of an erroneous association by Comstock (1882) of the pupa and adult of an unknown species of Sciaridae with the gall.

REFERENCES: Johannsen, 1912b (rev.); Pettey, 1918a (rev.); Shaw, 1953b (review).

Genus TRICHOSIA Winnertz

Trichosia Winnertz, 1867: 173. Type-species, *splendens* Winnertz (Coquillett, 1910b: 616)=*hirtipennis* (Zetterstedt).
hebes Loew, 1869c: 161 (Cent. 9, no. 58).—N.Y.; N.J.

Genus METANGELA Rübsaamen

Metangela Rübsaamen, 1894: 24. Type-species, *calliptera* Rübsaamen (Enderlein, 1911a: 126).
toxoneura (Osten Sacken), 1862a: 165 (*Sciara*).—D.C.; Kans., S.C.

Genus PHORODONTA Coquillett

Odontonyx Rübsaamen, 1894: 25 (preocc. Stephens, 1828). Type-species, *Sciara nigra* Wiedemann (Coquillett, 1910b: 578).
Phorodonta Coquillett, 1910b: 589 (n. name for *Odontonyx* Rübsaamen). Type-species, *Sciara nigra* Wiedemann (aut.).

nigra (Wiedemann), 1821a: 44 (1821c: 44) (*Sciara*).—Ga.; N. Mex., Fla., Mexico.

Genus PHYTOSCIARA Frey

Phytosciara Frey, 1942: 27. Type-species, *Sciara halterata* Lengersdorf (orig. des.).

flavipes (Meigen), 1804: 98 (*Sciara*).—Europe; Greenland, Palaearctic.

Genus SCIARA Meigen

Lycoria Meigen, 1800: 17. Type-species, *Tipula thomae* Linnaeus (Coquillett, 1910b: 563). Suppressed by I.C.Z.N., (1963b: 339.).
Sciara Meigen, 1803: 263. Type-species, *Tipula thomae* Linnaeus (mon.).

abdita Johannsen, 1912b: 125.—Ont.
arcuata Garrett, 1925b: 16.—B.C.
cingulata Rübsaamen, 1894: 31.—Ga.
clavata Garrett, 1925b: 16.—B.C.
congregata Johannsen, 1914: 93.—Ark.
cylindrica Pettey, 1918a: 329.—Calif.
dives Johannsen, 1912b: 125.—Calif.
forceps Pettey, 1918a: 328.—Fla.
futilis Johannsen, 1912b: 125.—Wis.
globosa Pettey, 1918a: 330.—Calif.
habilis Johannsen, 1912b: 126.—N.Y.; Man., Ont., Vt., Maine, Mass., N.C.
multisetifera Pettey, 1918a: 328.—Ariz.
ochrolabis Loew, 1869c: 160 (Cent. 9, no. 57).—N.Y.; Wis., Maine to N.C.
polita Say, 1824a: 366 (1859a: 249).—Northwest Territory (U.S.); Ont., N.J.
psittacus Pettey, 1918a: 330.—Maine.
sciophila Loew, 1869c: 160 (Cent. 9, no. 56).—D.C.; Maine to N.C.
  obscura Harris, 1835: 595. Nomen nudum.

townesi Shaw, 1935b: 227.—S.C.
vicina Johannsen, 1912b: 124.—N.Y.; Maine.

Genus EUGNORISTE Coquillett

Eugnoriste Coquillett, 1896c: 321. Type-species, *occidentalis* Coquillett (mon.).
brevirostris Coquillett, 1904f: 169.—Colo.; Ariz., Kans., Mexico.
occidentalis Coquillett, 1896c: 322.—N. Mex.; Alta., Oreg., Idaho, N.H., N.Y., N.J.

Genus RHYNCHOSCIARA Rübsaamen

Rhynchosciara Rübsaamen, 1894: 29. Type-species, *villosa* Rübsaamen (Coquillett, 1910b: 601)=*americana* (Wiedemann).
proboscidea Lengersdorf, 1931: 254.—Wash. (?error). Unrecognized.

Genus ZYGONEURA Meigen

Zygoneura Meigen, 1830: 304. Type-species, *sciarina* Meigen (mon.).
flavicoxa Johannsen, 1912b: 116.—N.Y.; Maine, Mass., N.C., Mexico.
johannseni Shaw, 1941b: 324.—Okla.

Genus LYCORIELLA Frey

Lycoriella Frey, 1942: 36. Type-species, *Sciara vivida* Winnertz (orig. des.; misident)=*auripila* (Winnertz).

Subgenus LYCORIELLA Frey

agraria (Felt), 1898a: 225 (1898: 225) (*Sciara*).—N.Y.; England, New Zealand.
brevipetiolata (Shaw), 1941b: 322 (*Sciara*).—Okla.
caesar (Johannsen), 1929a: 223 (*Sciara*; as *caesar*).—Ont.
multiseta (Felt), 1898a: 223 (1898: 223) (*Sciara*).—N.J.; N.Y., Vt.
pauciseta (Felt), 1898a: 224 (1898: 224) (*Sciara*).—N.J.; B.C., Calif., Ont. and N.H., s. to Pa.
sativae (Johannsen), 1912b: 133 (*Sciara*).—Kans.
similans (Johannsen), 1925: 266 (*Sciara*).—N.Y.
trifolii (Pettey), 1918b: 420 (*Sciara*) (also 1918a: 334 as *Neosciara*).—Idaho.

Subgenus HEMINEURINA Frey

Lycoriella, subg. **Hemineurina** Frey, 1942: 36. Type-species, *Sciara conspicua* Winnertz (orig. des.).

cochleata (Rübsaamen), 1898: 108 (*Sciara*).—Greenland; Europe.
  haemorrhoidalis Lundbeck, 1898: 247 (*Sciara*).—Greenland.
consimilis (Holmgren), 1869: 54 (*Sciara*).—Spitzbergen, Bear I.; Greenland.
permutata (Lundbeck), 1901: 313 (*Sciara*; n. name for *glacialis* Lundbeck).—Greenland.
  glacialis Lundbeck, 1898: 254 (*Sciara*; preocc. Rübsaamen, 1898).—Greenland.
venosa (Staeger), 1840: 285 (*Sciara*).—Denmark; Greenland, Europe.

Genus BRADYSIA Winnertz

Bradysia Winnertz, 1867: 180. Type-species, *angustipennis* Winnertz (Enderlein, 1911a: 127).
Neosciara Pettey, 1918a: 320. Type-species, *Sciara coprophila* Lintner (orig. des.).

Subgenus BRADYSIA Winnertz

actuosa (Johannsen), 1912b: 134 (*Sciara*).—N.Y.; Conn. **N. comb.** ?*Scatopsciara*.
acuta (Johannsen), 1912b: 136 (*Sciara*).—Wash.; B.C., Idaho, S. Dak., Kans., Mexico. ?*Scatopsciara*.
bellingeri Shaw, 1953a: 67.—Conn.
biformis (Lundbeck), 1898: 256 (*Sciara*).—Greenland.
bispina (Fisher), 1938a: 199 (*Sciara*).—Mich. **N. comb.**
bournei (Shaw), 1941c: 175 (*Sciara*).—Man. **N. comb.**
browni (Shaw), 1935b: 229 (*Sciara*).—Ont. **N. comb.**
caldaria (Lintner), 1895: 398 (1896: 398) (*Sciara*).—Idaho; Alta., Utah, N.Y. **N. comb.**
conglomerata (Pettey), 1918a: 335 (*Neosciara*).—Calif. **N. comb.**
coprophila (Lintner), 1895: 394 (1896: 394) (*Sciara*).—N.Y.; Alta. to Maine, s. to Colo. and N.J., N.C.
cucumeris (Johannsen), 1912b: 133 (*Sciara*).—Ill.; Iowa. **N. comb.** ?*Scatopsciara*.
dichaeta (Shaw), 1941b: 321 (*Sciara*).—Okla. **N. comb.**
diluta (Johannsen), 1912b: 135 (*Sciara*).—N.Y.
dolens (Johannsen), 1912b: 134 (*Sciara*).—N.Y. **N. comb.**
dux (Johannsen), 1912b: 127 (*Sciara*).—Wis.; Mich., N.Y., Mass., R.I. **N. comb.**
ericia (Pettey), 1918a: 337 (*Neosciara*).—Mass. **N. comb.**
expolita (Coquillett), 1900h: 392 (1904: 6) (*Sciara*).—Alaska. **N. comb.**
falcata (Pettey), 1918a: 331 (*Neosciara*).—Mass. **N. comb.**
farri (Shaw), 1953a: 67.—Conn.

fatigans (Johannsen), 1912b: 132 (*Sciara*).—N.Y. **N. comb.**
felti (Pettey), 1918a: 339 (*Neosciara*).—N.Y. **N. comb.**
fochi (Pettey), 1918a: 331 (*Neosciara*).—Wyo. **N. comb.**
forcipulata (Lundbeck), 1898: 244 (*Sciara*).—Greenland; Finland.
fulvicauda (Felt), 1898a: 227 (1898: 227) (*Sciara*).—N.J. **N. comb.**
fumida (Johannsen), 1912b: 135 (*Sciara*).—N.Y.; Maine, Mass. **N. comb.**
grandis (Pettey), 1918a: 334 (*Neosciara*).—Ga. **N. comb.**
groenlandica (Holmgren), 1872: 104 (*Sciara*).—Greenland. **N. comb.**
hamata (Pettey), 1918a: 338 (*Neosciara*).—Ga. **N. comb.**
hartii (Johannsen), 1912b: 144 (*Sciara*).—Ill. **N. comb.**
hastata (Johannsen), 1912b: 130 (*Sciara*).—N.Y.; Maine, Mass.
imitans (Johannsen), 1912b: 128 (*Sciara*).—Wash. **N. comb.**
impatiens (Johannsen), 1912b: 136 (*Sciara*).—N.Y.; Mass.
iridipennis (Zetterstedt), 1838: 827 (*Sciara*).—Sweden, Greenland; Alaska, Finland.
joffrei (Pettey), 1918a: 332 (*Neosciara*).—Pa. **N. comb.**
johannseni (Enderlein), 1912a: 282 (*Sciara*; n. name for *nigricans* Johannsen).—R.I.; Kans., N.Y. **N. comb.**
   nigricans Johannsen, 1912b: 134 (*Sciara*; preocc. Enderlein, 1911).—R.I. **N. comb.** *Lycoria nigricans* Enderlein, 1911a: 168, from Brazil, is here transferred to the genus *Bradysia*. **N. comb.**
jucunda (Johannsen), 1912b: 131 (*Sciara*).—R.I.; Wis., N.Y., N.H., Mass., Conn. **N. comb.**
kaiseri (Shaw), 1941b: 320 (*Sciara*).—Okla. **N. comb.**
lobosa (Pettey), 1918a: 333 (*Neosciara*).—B.C. **N. comb.**
longispina (Pettey), 1918a: 325 (*Neosciara*).—Maine; Scotland. **N. comb.**
lugens (Johannsen), 1912b: 132 (*Sciara*).—Maine; Mass. **N. comb.**
luravi (Johannsen), 1929c: 88 (*Sciara*).—Va.; Pa. **N. comb.**
luteola (Pettey), 1918a: 333 (*Neosciara*).—Ga. **N. comb.**
macclurei (Shaw), 1941c: 174 (*Sciara*).—Man. **N. comb.**
macfarlanei (Jones), 1920: 92 (*Neosciara*).—Miss., Ala., S.C., N.C. **N. comb.**
macrodon Frey, 1948: 85 (n. name for *pallidiventris* Holmgren).— Spitzbergen; ?Greenland.
   pallidiventris Holmgren, 1869: 53 (*Sciara*; preocc. Winnertz, 1867).—Spitzbergen.
macroptera (Pettey), 1918a: 339 (*Neosciara*).—Calif. **N. comb.**
mellea (Johannsen), 1912b: 129 (*Sciara*).—Ohio; N.Y., N.H., Mass., Conn. **N. comb.**
mesochra (Shaw), 1941b: 320 (*Sciara*).—N. Mex. **N. comb.**

munda (Johannsen), 1912b: 127 (*Sciara*).—Wash.; B.C., Oreg. **N. comb.**
mutua (Johannsen), 1912b: 131 (*Sciara*).—N.Y. **N. comb.**
neglecta (Johannsen), 1912b: 133 (*Sciara*).—Calif. **N. comb.**
nemoralis (Meigen), 1818: 287 (*Sciara*).—Europe; ?Greenland.
  aprilina Meigen, 1818: 285 (*Sciara*).—Europe.
nigripes (Meigen), 1830: 307 (*Sciara*).—Europe; ?Greenland.
ovata (Pettey), 1918a: 336 (*Neosciara*).—B.C. **N. comb.**
paradichaeta (Shaw), 1941b: 322 (*Sciara*).—Okla. **N. comb.**
parilis (Johannsen), 1912b: 132 (*Sciara*).—Kans.; N.Y., N.H. **N. comb.**
parva (Holmgren), 1869: 52 (*Scaeva*).—Spitzbergen; N.W.T.
penna (Pettey), 1918a: 338 (*Neosciara*).—Calif. **N. comb.**
perfecta (Pettey), 1918a: 341 (*Neosciara*).—Md. **N. comb.**
petaini (Pettey), 1918a: 334 (*Neosciara*).—Md. **N. comb.**
pilata (Pettey), 1918a: 337 (*Neosciara*).—Calif. **N. comb.**
pollicis (Pettey), 1918a: 338 (*Neosciara*).—Ariz. **N. comb.**
polychaeta (Pettey), 1918a: 335 (*Neosciara*).—Ga. **N. comb.**
prolifica (Felt), 1898a: 226 (1898: 226) (*Sciara*).—Mass.; B.C., Idaho, Man., Ont., N.Y., Maine. **N. comb.**
quadrispinosa (Pettey), 1918a: 332 (*Neosciara*).—Mass.; Maine. **N. comb.**
radialis (Shaw), 1934: 233 (*Sciara*).—N.C. **N. comb.**
reynoldsi (Metz), 1938: 177 (*Sciara*).—N.C., Ala. **N. comb.**
scita (Johannsen), 1912b: 135 (*Sciara*).—Oreg.; Wash.
sexdentata (Pettey), 1918a: 340 (*Neosciara*).—Calif. **N. comb.**
silvestrii (Kieffer), 1910: 327 (*Sciara*).—N.Y.; Maine. **N. comb.**
  sylvestris and *sylvestrii*, errors or emends.
spinata (Pettey), 1918a: 324 (*Neosciara*).—R.I. **N. comb.**
subgrandis (Shaw), 1941b: 321 (*Sciara*).—Okla. **N. comb.**
subtrivialis (Pettey), 1918a: 340 (*Neosciara*).—Calif. **N. comb.**
trifurca (Pettey), 1918a: 336 (*Neosciara*).—Calif. **N. comb.**
tritici (Coquillett), 1895i: 408 (*Sciara*).—Unknown (?D.C.); Ohio, England. **N. comb.**
trivialis (Johannsen), 1912b: 136 (*Sciara*).—N.Y. **N. comb.**
unguicauda (Malloch), 1923i: 180 (*Sciara*).—Pribilof Is. **N. comb.**
varians (Johannsen), 1912b: 135 (*Sciara*).—Kans.; N.Y., Maine.

Subgenus SEMNOMYIA Frey

Neosciara, subg. **Semnomyia** Frey, 1942: 32. Type-species, *Sciara psychina* Enderlein (orig. des.).

picea (Rübsaamen), 1894: 32 (*Sciara*).—Ga.; N.C., Fla. **N. comb.**
  americana, authors, not Wiedemann (*Sciara*).

SCIARIDAE

Genus SCATOPSCIARA Edwards

Sciara, subg. **Scatopsciara** Edwards *in* Tonnoir and Edwards, 1927: 798. Type-species, *quinquelineata* Macquart (orig. des.)
Scaptosciara, error.

nacta (Johannsen), 1912b: 132 *(Sciara).*—N.Y.; Mass., Ky., Finland.
vivida (Winnertz), 1867: 156 *(Sciara).*—Europe; Greenland.

Genus PLASTOSCIARA Berg

Pseudosciara Kieffer, 1898b: 194 (preocc. Schiner, 1866). Type-species, *pictiventris* Kieffer (orig. des.).
Plastosciara Berg, 1899: 78 (n. name for *Pseudosciara* Kieffer). Type-species, *Pseudosciara pictiventris* Kieffer (aut.).

Unnamed spp.—Va., Ga.

Genus NIADINA Rapp

Niadina Rapp, 1946a: 126. Type-species, *jauva* Rapp (orig. des.).
jauva Rapp, 1946a: 126.—Ill.

Genus EPIDAPUS Haliday

Epidapus Haliday *in* Walker, 1851a: 7. Type-species, *Chionea venatica* Haliday (sub. mon., Walker, 1856a: 56).
johannseni Shaw, 1953a: 63.—Conn.

Genus PEYERIMHOFFIA Kieffer

Peyerimhoffia Kieffer, 1903: 198. Type-species, *brachyptera* Kieffer (Enderlein, 1911a: 185). This is possibly a syn. of the earlier genus *Aptanogyna* Börner.
johnstoni Shaw, 1935a: 160.—Mass.

Genus PNYXIA Johannsen

Pnyxia Johannsen, 1912b: 114. Type-species, *Epidapus scabiei* Hopkins (orig. des.).
scabiei (Hopkins), 1895: 152 *(Epidapus).*—W. Va.; Mo., Ohio to Mass. and R.I., Europe.

Unplaced Species of Sciaridae

abbreviata Walker, 1848: 109 *(Sciara).*—Ont.; Alaska, N.H., N.J.
atrata Say, 1824a: 366 (1859a: 249) *(Sciara).*—Northwest Territory (U.S.); Ont., N.H.

attenuata Rübsaamen, 1898: 106 (*Sciara*).—Greenland.
    alternata (error) Aldrich, 1905: 149.
    latipennis Lundbeck, 1898: 242 (*Sciara*).—Greenland.
borealis Rübsaamen, 1898: 109 (*Sciara*).—Greenland; Alaska.
delessei Séguy, 1953: 118 (*Lycoria*).—Greenland.
diderma Garrett, 1925b: 16 (*Sciara*).—B.C.
dimidiata Say, 1832: 15 (1859a: 308) (*Sciara*).—La.
diota Garrett, 1925b: 16 (*Sciara*).—B.C.
exigua Say, 1824a: 367 (1859a: 249) (*Sciara*).—Northwest Territory (U.S.); Canada, ?N.H.
exilis Say, 1829: 154 (1859b: 352) (*Sciara*).—Ind.
femorata Say, 1823: 78 (1859b: 70) (*Sciara*).—Pa.; N.Y., N.J.
fraterna Say, 1824a: 367 (1859a: 249) (*Sciara*).—Northwest Territory (U.S.).
fuliginosus Fitch, 1856a: 487 (1856: 255) (*Molobrus*).—N.Y.; N.H., N.J., Fla. Preocc. in *Sciara* by Blanchard, 1852.
fulviventris Wiedemann, 1821a: 44 (1821c: 44) (*Sciara*).—Amer.
fumatella Lundbeck, 1898: 249 (*Sciara*).—Greenland.
glacialis Rübsaamen, 1898: 109 (*Sciara*).—Greenland.
humicola Lundbeck, 1898: 252 (*Sciara*).—Greenland.
inconstans Fitch, 1856a: 487 (1856: 255) (*Molobrus*).—N.Y.; s. Canada, s. to Calif. and Va.
lurida Walker, 1848: 106 (*Sciara*).—N.Y.
mali Fitch, 1856a: 486 (1856: 254) (*Molobrus*).—N.Y.
marginata Rübsaamen, 1898: 107 (*Sciara;* preocc. Skuse, 1891).—Greenland.
nervosa Meigen, 1818: 283 (*Sciara*).—Europe; ?Greenland.
perpusilla Walker, 1848: 106 (*Sciara*).—Ont.
pulicaria Meigen, 1818: 282 (*Sciara*).—Europe; ?Greenland.
punctata Walker, 1848: 106 (*Sciara;* preocc. Fabricius, 1805).—N. Amer.
robusta Walker, 1848: 105 (*Sciara*).—Ont.
rotundipennis Macquart, 1838b: 178 (1839: 294) (*Sciara*).—Carolina; N.C., Mexico.
septemtrionalis Rübsaamen, 1898: 106, 109, 119 (*Sciara;* as *septentrionalis*, p. 119).—Greenland.
tridentata Rübsaamen, 1898: 107 (*Sciara*).—Greenland; B.C., Man., Bear I., Jan Mayer I., Spitzbergen.
    validicornis Lundbeck, 1898: 243 (*Sciara*).—Greenland.
unicorn Garrett, 1925b: 16 (*Sciara*).—B.C.
vulgaris Fitch, 1856a: 487 (1856: 255) (*Molobrus*).—N.Y.; Calif., N.H., N.J.

Family HYPEROSCELIDIDAE

(Corynoscelididae)

By E. F. Cook

This small family, formerly placed as a subfamily of the Scatopsidae, is found in the northern Holarctic Region, Chile, and New Zealand. No life histories are known.

REFERENCES: Tollet, 1959 (tax.); Hardy and Nagatomi, 1960 (key genera).

Genus SYNNEURON Lundström

Synneuron Lundström, 1910: 5. Type-species, *annulipes* Lundström (mon.).

annulipes Lundström, 1910: 6.—Finland; Alaska, Wash., Que.

Family SCATOPSIDAE

By E. F. Cook

The biology of Scatopsidae is practically unknown. The larvae of only four of the Scatopsini and one *Ectaetia* have been described. *Scatopse fuscipes* Meigen and *S. notata* (Linnaeus), both cosmopolitan species, have been found in many types of decaying plant and animal material, including excreta. *Rhexoza similis* (Beekey) has been found under the bark of logs, and *R. incisa* Cook under the bark of elms and cottonwoods. The European species *Ectaetia platyscelis* (Loew) has been found in debris in a tree hole.

Subfamily SCATOPSINAE

Tribe SCATOPSINI

Genus SCATOPSE Geoffroy

Scatopse Geoffroy, 1762: 450. Type-species, *Tipula notata* Linnaeus (I.C.Z.N., 1957a: 86). Conserved as of 1762 under suspension of the Rules (I.C.Z.N., 1957a: 86).
Reichertella Enderlein, 1912a: 268. Type-species, *Scatopse femoralis* Meigen (orig. des.; misident.)=*nigra* Meigen.
Coboldia Melander, 1916: 17. Type-species, *formicarum* Melander (orig. des.)=*fuscipes* Meigen.
Rhaeboza Enderlein, 1936a: 55. Type-species, *Scatopse fuscipes* Meigen (orig. des.).

REFERENCES: Cook, 1957 (rev.); Meade and Cook, 1961 (biol. *S. fuscipes*).

alpestris Cook, 1957: 600.—Colo.; Alaska to Colo.
brevipalpis Cook, 1957: 608.—N.Y.; Maine.
collaris (Melander), 1916: 10 (*Reichertella*).—Wash.; B.C., Idaho, Calif.
fattigi Cook, 1957: 604.—D.C.; Wis. to Ont., s. to Ga.
fuscipes Meigen, 1830: 314.—Europe; N. Amer., cosmopolitan.
    formicarum Melander, 1916: 17 (*Coboldia*).—Wis.
    barrus McAtee, 1921b: 123 (*Rhegmoclema*).—D.C.
    atrata, authors, not Say.
lapponica Duda, 1928a: 27.—Europe; Y.T. to Man. and Labr.
montana Cook, 1957: 608.—Mont.; Wash.
notata (Linnaeus), 1758: 588 (*Tipula*).—Europe; Alaska, Canada, U.S., cosmopolitan.
    femoralis Meigen, 1838: 55.—Europe.
    nitens Walker, 1848: 114.—Ont.
    nitida Harris, 1835: 596.—Nomen nudum.
producta Cook, 1957: 606.—W. Va.; Mass., Tenn.
pulicaria Loew.—Not Nearctic.
uncinata (Melander), 1916: 10 (*Reichertella*).—Alaska; Alaska, s. to Wash. and Wyo.

Genus COLOBOSTEMA Enderlein

Colobostema Enderlein, 1926a: 140. Type-species, *oldenbergi* Enderlein (orig. des.) = *tristis* (Zetterstedt).
    REFERENCE: Cook, 1956a (rev.).
arizonense Cook, 1956a: 331.—Ariz.
variatum Cook, 1956a: 330.—Tex.; La., Fla.
varicorne (Coquillett), 1902g: 96 (*Scatopse*).—D.C.; Wis. to N.Y., s. to Ky. and Va.
    tibialis McAtee, 1921b: 122 (*Scatopse*).—Va.

Tribe RHEGMOCLEMATINI

Genus RHEGMOCLEMA Enderlein

Rhegmoclema Enderlein, 1912a: 276. Type-species, *rufithorax* Enderlein (orig. des.).
Aldrovandiella Enderlein, 1912a: 278. Type-species, *Scatopse halterata* Meigen (orig. des.).
    REFERENCE: Cook, 1955 (rev.).
boreale Cook, 1955: 249.—N.W.T.; Alaska.
halteratum (Meigen).—Not Nearctic.
hubachecki Cook, 1955: 249.—Minn.; Minn. to Maine, s. to W. Va.

majus Cook, 1955: 250.—Iowa; S. Dak. to Que. and N.Y.
minus Cook, 1955: 250.—B.C.; Oreg., Colo.
reticulatum Cook, 1955: 251.—Ont.; Mich., Wis.
truncatum Cook, 1955: 250.—Idaho; Sask. to Labr., s. to Utah and Nebr.

Genus RHEGMOCLEMINA Enderlein

Rhegmoclemina Enderlein, 1936a: 55. Type-species, *Scatopse vaginata* Lundström (orig. des.).

REFERENCE: Cook, 1955 (rev.).

Subgenus RHEGMOCLEMINA Enderlein

bimaculata (Melander), 1916: 14 (*Rhegmoclema*).—Wash.; B.C. to Calif.
melanderi Cook, 1955: 356.—Calif.; Wyo., Colo., Ont.
scrobicollis (Melander), 1916: 15 (*Rhegmoclema*).—Calif.; s. Calif.

Subgenus NEORHEGMOCLEMINA Cook

Rhegmoclemina, subg. **Neorhegmoclemina** Cook, 1955: 358. Type-species, *parvum* Cook (orig. des.).

bisaccata Cook, 1955: 358.—Iowa; N.Y.
parva Cook, 1955: 358.—La.; Va.

Genus PARASCATOPSE Cook

Parascatopse Cook, 1955: 362. Type-species, *wirthi* Cook (orig. des.).
flavida Cook, 1955: 363.—Fla.; Miss.
sonorensis Cook, 1955: 363.—Calif.
wirthi Cook, 1955: 363.—Fla.

Tribe SWAMMERDAMELLINI

Genus SWAMMERDAMELLA Enderlein

Swammerdamella Enderlein, 1912a: 277. Type-species, *Scatopse brevicornis* Meigen (orig. des.).

REFERENCE: Cook, 1956a (rev.).

bispinosa Cook, 1956a: 28.—Calif.; Sask., Mont.
brevicornis (Meigen).—Not Nearctic.
chillcotti Cook, 1956a: 26.—Man.

confusa Cook, 1956a: 22.—Va.
marginata Cook, 1956a: 24.—N. Mex.; Calif., Nev., Mexico.
nevadensis Cook, 1956a: 26.—Nev.
obtusa Cook, 1956a: 26.—Minn.; Sask. to Maine, s. to Kans. and Md.
pusilla (Walker), 1848: 114 (*Scatopse*).—Ont.; Alaska to Wyo. and Maine.
pygmaea (Loew), 1864a: 56 (Cent. 5, no. 13) (*Scatopse*).—D.C.; Md.
   floralis McAtee, 1921b: 123 (*Rhegmoclema*).—Md.
reducta Cook, 1956a: 29.—Ala.; Ind.
sagittata Cook, 1956a: 24.—Tex.; Ill. to Pa. and Fla.

Genus RHEXOZA Enderlein

Rhexoza Enderlein, 1936a: 55. Type-species, *zacheri* Enderlein (orig. des.).

REFERENCE: Cook, 1956a (rev.).

aterrima (Melander), 1916: 14 (*Rhegmoclema*).—Idaho; Calif., Mont.
cryptica Cook, 1956a: 10.—Va.
dampfi (Duda), 1928b: 280 (*Scatopse*).—Mexico; Tex.
grossa Cook, 1956a: 6.—Calif.; Oreg.
incisa Cook, 1956a: 6.—Minn.; Ill., Iowa, Ont.
miniscula Cook, 1956a: 10.—Va.; S.C., Tex.
quatei Cook, 1956a: 8.—Calif.
similis (Beekey), 1938: 151 (*Scatopse*).—Maine; Ont., Que.

Subfamily PSECTROSCIARINAE

Genus PSECTROSCIARA Kieffer

Psectrosciara Kieffer *in* Enderlein, 1911a: 192 (also Kieffer, 1912a: 192). Type-species, *mahensis* Kieffer (orig. des.) = *brunnipes* (Brunetti).

REFERENCE: Cook, 1958 (rev.).

bakeri Cook, 1958: 592.—Calif.
brevipennis Cook, 1958: 594.—Calif.
californica (Cole), 1912b: 151 (*Scatopse*).—Calif.; Mexico.
discata Cook, 1958: 592.—Calif.
elongata Cook, 1958: 594.—Tex.
forcipata Cook, 1958: 591.—Calif.
oregonensis Cook, 1958: 592.—Oreg.
serrata Cook, 1958: 594.—Tex.; Mexico.
stonei Cook, 1958: 593.—Tex.

Genus ANAPAUSIS Enderlein

Anapausis Enderlein, 1912a: 278. Type-species, *Scatopse soluta* Loew (orig. des.).

cismarina McAtee, 1921b: 124.—Va.

Subfamily ECTAETIINAE

Genus ECTAETIA Enderlein

Ectaetia Enderlein, 1912a: 279. Type-species, *Scatopse clavipes* Loew (orig. des.).

clavipes (Loew), 1846b: 333 (*Scatopse*).—Germany; B.C. to Ont., s. to Oreg., Ill., and N.J., Europe.

gracilis (McAtee), 1921b: 123 (*Reichertella*).—D.C.

Subfamily ASPISTINAE

Genus ASPISTES Meigen

Aspistes Meigen, 1818: 319. Type-species, *berolinensis* Meigen (mon.).

Subgenus ASPISTES Meigen

harti Malloch, 1920i: 275.—Ill.

Subgenus ARTHRIA Kirby

Aspistes, subg. **Arthria** Kirby, 1837: 311. Type-species, *analis* Kirby (orig. des.).

analis Kirby, 1837: 311.—N.W.T.

Unplaced Species of Scatopsidae

atrata Say, 1824a: 367 (1859a: 250) (*Scatopse*).—Pa.

obscura Walker, 1848: 114 (*Scatopse*).—Ont.

Family CECIDOMYIIDAE

(Itonididae)

By Richard H. Foote

Although commonly called gall midges, only about two-thirds of the cecidomyiids cause galls on plants. The larvae of most Lestremiinae live in decaying organic matter; a few cause primary damage to fungi. A large number of Cecidomyiinae live as inquilines in the deformations made by other insects, or are predaceous. Paedogenesis is common to many of the genera of Lestremiinae and to a few of the Cecidomyiinae.

The taxonomic status of the family is unsatisfactory. The North American species are little known except in the northeast, the fauna of which E. P. Felt made the basis for his revision published in the bulletins of the New York State Museum. A. Earl Pritchard has revised the Lestremiinae of North America; in so doing he has assigned all the North American genera of the subfamily Heteropezinae to one or the other of the two subfamilies recognized below. This catalog is a synthesis of the conclusions of both these authorities. Many vexing taxonomic problems remain unsolved, and little has been done to integrate the North American fauna with that of other faunal regions. For these reasons the family requires much additional research.

The present International Code of Zoological Nomenclature requires the following usages, which have been adopted here:

(1) Acceptable compounds. A specific name originally proposed in polynomial form is accepted as of the original date and author (and is written as one word) *only* if the two names represent a single entity (e.g., *pini rigidae* from *Pinus rigida* becomes *pinirigidae*), or if originally hyphenated (e.g., *salicis-brassicoides* becomes *salicisbrassicoides*). If not acceptable from the original publication, a polynomial specific name may subsequently be made available by any author who uses a hyphen between the two words, or combines them into one word. In these cases, the author and date of the first subsequent validation are adopted. Sometimes subsequent authors used only one part of the polynomial and validated that part alone as a specific name.

(2) Names based on galls and immature stages. A name based on the "work of an animal" (if proposed before 1931), and names founded on immature stages, are accepted as of their original author and date. I have chosen to give the author, date, and page of a subsequent adult description, often (but not always) the most comprehensive one, in cases in which adults have been *definitely* associated.

The type-localities of many of Felt's species have been supplied from Felt, 1958, or from type-specimens.

REFERENCES: Felt, 1925b (generic key, index to spp.), 1940 (galls, key and figs.); Barnes, 1946a–1956 (biol. economically important spp.); Pritchard, 1953a (Calif.), 1960a (classif. "Heteropezinae").

Subfamily LESTREMIINAE

REFERENCES: Felt, 1913g (rev.); Pritchard, 1958 (rev. ne. spp.).

Tribe CATOTRICHINI

REFERENCES: Pritchard, 1948a (rev.), 1960a (rev.).

Genus CATOTRICHA Edwards

Catotricha Edwards, 1938a: 102. Type-species, *Catocha americana* Felt (orig. des.).

americana (Felt), 1908b: 309 (*Catocha*).—N.H.
subobsoleta (Alexander), 1924c: 82 (*Catocha*).—Wash.; Oreg., Calif.

Tribe CATOCHINI

REFERENCES: Pritchard, 1948a (rev.), 1960a (rev.).

Genus TRITOZYGA Loew

Tritozyga Loew *in* Osten Sacken, 1862b: 179. Type-species, *sackeni* Felt (Felt, 1911e: 32).

sackeni Felt, 1911e: 32.—D.C.; B.C., N.Y., Conn., Md.
    fenestra Felt, 1914b: 117.—N.Y.
    borealis Felt, 1920: 278.—N.Y.
    cranbrooki Felt, 1926c: 266 (*Neocatocha*).—B.C.

Genus NEOCATOCHA Felt

Neocatocha Felt, 1912e: 236. Type-species, *marilandica* Felt (orig. des.).
Konisomyia Felt, 1914b: 118. Type-species, *fusca* Felt (orig. des.)= *marilandica* Felt.

marilandica Felt, 1912e: 236.—Md.; Ont., s. Labr., N.Y.
    fusca Felt, 1914b: 118 (*Konisomyia*).—N.Y.
    borealis Felt, 1920: 279 (*Konisomyia*).—N.Y.

Genus CATOCHA Haliday

Catocha Haliday, 1833: 156. Type-species, *latipes* Haliday (mon.).

slossonae Felt, 1908b: 309.—N.H.; nw. N.W.T., N.Y.
    sylvana Felt, 1920: 280 (*Neocatocha*).—N.Y.

Genus EUCATOCHA Edwards

Eucatocha Edwards, 1938a: 106. Type-species, *Catocha barberi* Felt (orig. des.).

barberi (Felt), 1913g: 131 (*Catocha*).—Wis.
betsyae Pritchard, 1960b: 195.—Calif.

Genus ANOCHA Pritchard

Anocha Pritchard, 1948a: 669. Type-species, *Neocatocha spinosa* Felt (orig. des.).

celesteana Pritchard, 1960b: 197.—Wyo.
spinosa (Felt), 1913g: 152 (*Neocatocha*).—Unknown; Minn., Maine, ?Wis.
    nylanderi Felt, 1926a: 207 (*Neocatocha*).—Maine.

Tribe LESTREMIINI

REFERENCES: Pritchard, 1951 (rev.), 1958 (rev. ne. spp.), 1960a (rev.).

Genus LESTREMIA Macquart

Lestremia Macquart, 1826a: 173 (1826: 115). Type-species, *cinerea* Macquart (mon.).

cinerea Macquart, 1826a: 173 (1826: 115).—France; B.C. to N.H., s. to Calif. and Fla., Europe.
    sylvestris Felt, 1907b: 102 (*Catocha*).—N.C.
    dyari Felt, 1908b: 311.—B.C.
    franconiae Felt, 1908b: 311.—N.H.
    kansensis Felt, 1908b: 311.—Kans.
    fenestrata Malloch, 1914g: 233 (*Zygoneura*).—Ill.
    floridana Felt, 1915a: 226.—Fla.
    garretti Felt, 1926c: 265.—B.C.
leucophaea (Meigen), 1818: 288 (*Sciara*).—Europe; B.C., Wash., Calif., Minn., N.Y.
    sambuci Felt, 1907b: 101 (*Catocha*).—N.Y.
    setosa Felt, 1908b: 311.—N.Y.
    occidentalis Felt, 1926c: 265.—B.C.
solidaginis (Felt), 1907b: 102 (*Catocha*).—N.Y.; Minn., Wis., N.Y.

Genus ANARETELLA Enderlein

Anaretella Enderlein, 1911a: 193. Type-species, *Lestremia defecta* Winnertz (orig. des.).
Neptunimyia Felt, 1912e: 237. Type-species, *tridens* Felt (orig. des.)=*defecta* (Winnertz).

defecta (Winnertz), 1870: 33 (*Lestremia*).—Europe; Wash., Calif., Minn., Wis., N.Y.
    acerifolia Felt, 1907b: 101 (*Campylomyza*).—N.Y.
    pini Felt, 1907b: 103 (*Lestremia*).—N.Y.
    tridens Felt, 1912e: 237 (*Neptunimyia*).—N.Y.
iola Pritchard, 1951: 250.—Wash.
spiraeina (Felt), 1907b: 102 (*Catocha*).—N.Y.; Wash., Calif., Minn., Europe.

Genus ALLARETE Pritchard

Allarete Pritchard, 1951: 250. Type-species, *Lestremia vernalis* Felt (orig. des.).

barberi (Felt), 1908b: 310 (*Lestremia*).—N. Mex.; Calif., Minn.
vernalis (Felt), 1908b: 311 (*Lestremia*).—Kans.; Minn., Tex.

Genus GONGROMASTIX Enderlein

Gongromastix Enderlein, 1936a: 60. Type-species, *andorrana* Enderlein (mon.).

epista Pritchard, 1951: 253.—Wash.; Calif.
schalis Pritchard, 1951: 252.—Minn.

Genus PARARETE Pritchard

Pararete Pritchard, 1951: 253. Type-species, *Lestremia elongata* Felt (orig. des.).

elongata (Felt), 1908b: 310 (*Lestremia*).—Calif.

Genus WASMANNIELLA Kieffer

Wasmanniella Kieffer, 1898a: 49. Type-species, *aptera* Kieffer (mon.).
clauda Pritchard, 1951: 254.—Minn.

Genus CONARETE Pritchard

Conarete Pritchard, 1951: 255. Type-species, *crebra* Pritchard (orig. des.).

crebra Pritchard, 1951: 256.—Minn.
eluta Pritchard, 1951: 257.—Minn.
eschata Pritchard, 1951: 257.—Minn.
texana (Felt), 1913g: 147 (*Microcerata*).—Tex.; Minn.

Genus ANARETE Haliday

Anarete Haliday, 1833: 156. Type-species, *candidata* Haliday (mon.).
Microcerata Felt, 1908b: 309. Type-species, *Micromya corni* Felt (orig. des.).

anepsia Pritchard, 1951: 264.—Calif.; Oreg.
buscki (Felt), 1915f: 198 (*Microcerata*).—Cuba; Ariz., Minn., P.R.
corni (Felt), 1907b: 102 (*Micromya*).—N.Y.; Wash., Oreg., N. Dak., Minn.
    perplexa Felt, 1908b: 310 (*Microcerata*).—N.Y.
    borealis Felt, 1913g: 147 (*Microcerata*).—N.Y.
diervillae (Felt), 1907b: 103 (*Micromya*).—N.Y.
edwardsi Pritchard, 1951: 259.—Minn.
felti Pritchard, 1951: 262.—Minn.
johnsoni (Felt), 1908b: 310 (*Microcerata*).—Pa.; Wash. to Ont., s. to Calif. and Fla.
    cockerelli Felt, 1908b: 310 (*Microcerata*).—N. Mex.
    spinosa Felt, 1913g: 145 (*Microcerata*).—Tex.
    iridis Cockerell, 1914a: 460 (*Microcerata*).—Colo.
    aldrichii Felt, 1915a: 226 (*Microcerata*).—Ind.
lacteipennis Kieffer, 1906b: 345.—Europe; Minn.
rubra Kieffer, 1906b: 343.—Europe; Oreg., Calif., Minn.

Tribe FORBESOMYIINI

REFERENCE: Pritchard, 1960a (rev.).

Genus FORBESOMYIA Malloch

Forbesomyia Malloch, 1914g: 234. Type-species, *atra* Malloch (orig. des.).
atra Malloch, 1914g: 235.—Ill.; B.C., Wash., Mont.

Tribe MOEHNIINI

Genus MOEHNIA Pritchard

Moehnia Pritchard, 1960a: 308. Type-species, *erema* Pritchard (orig. des.).
erema Pritchard, 1960a: 309.—Calif.

Tribe HETEROPEZINI

REFERENCE: Pritchard, 1960a (rev.).

Genus HETEROPEZA Winnertz

Heteropeza Winnertz, 1846: 13. Type-species, *pygmaea* Winnertz (mon.).
Oligarces Meinert, 1865: 237. Type-species, *paradoxus* Meinert (mon.).

pygmaea Winnertz, 1846: 14.—Europe; Calif., N.Y., Pa., N.J., ?Mo.
  noveboracensis Felt, 1908b: 286 (*Oligarces*).—N.Y.
ulmi (Felt), 1911d: 477 (*Oligarces*).—N.Y.

Genus HETEROPEZINA Pritchard

Heteropezina Pritchard, 1960a: 311. Type-species, *cathistes* Pritchard (orig. des.).

cathistes Pritchard, 1960a: 311.—Fla.

Tribe MIASTORINI

REFERENCE: Pritchard, 1960a (rev.).

Genus MIASTOR Meinert

Miastor Meinert, 1864: 42. Type-species, *metraloas* Meinert (mon.)
  REFERENCE: Felt, 1911f (biol.).

metraloas Meinert, 1864: 42.—Denmark; Minn., Ind., N.Y., Conn., N.J., n. Europe.
  americana Felt, 1908b: 286.—N.Y.

Tribe ACOENONIINI

REFERENCE: Pritchard, 1960a (review).

Genus ACOENONIA Pritchard

Acoenonia Pritchard, 1947: 14. Type-species, *perissa* Pritchard (orig. des.).

perissa Pritchard, 1947: 15.—Minn.

Tribe MICROMYINI

REFERENCE: Pritchard, 1947 (rev.).

Genus CAMPYLOMYZA Meigen

Campylomyza Meigen, 1818: 101. Type-species, *flavipes* Meigen (Westwood, 1840: 126).

boulderi Felt, 1908b: 314.—Colo.
 boulderensis Felt, 1913g: 177 (as *Prionellus*), emend.
dilatata Felt, 1907b: 149.—Mass.; N. Dak., Minn., Tex.
 tuckeri Felt, 1908b: 316.—Tex.
 monilis Felt, 1913g: 175 (*Prionellus*).—Tex.
flavipes Meigen, 1818: 102.—Europe; Calif., Minn., Ind., N.Y., Mass.
 graminea Felt, 1907b: 98.—N.Y.
 leguminicola Felt, 1907b: 98.—N.Y.
 balsamicola Felt, 1907b: 99.—N.Y.
 pomiflorae Felt, 1907b: 99.—N.Y.
 karnerensis Felt, 1907b: 101.—N.Y.
 tsugae Felt, 1907b: 101.—N.Y.
 defectiva Felt, 1908b: 314.—N.Y.
 hesperia Felt, 1908b: 315.—N.Y.
 pomifolia Felt, 1908b: 315.—N.Y.
 monilis Felt of Felt, 1915c: 405.
fusca Winnertz, 1870: 12.—Europe; Alta. to Vt., s. to Calif.
 populi Felt, 1907b: 98.—N.Y.
 latipennis Felt, 1908b: 314.—N.Y.
 barlowi Felt, 1908b: 316.—R.I.
 boulderensis (in part) Felt, 1913g: 177 (as *Prionellus*). Oreg. specimen only.
montana (Felt), 1913g: 182 (*Prionellus*).—Colo.
 eremi Felt, 1919: 219 (*Prionellus*).—Colo.
silvana Felt, 1908b: 314.—B.C.
simulator Felt, 1908b: 314.—B.C.

Genus CORDYLOMYIA Felt

Cordylomyia Felt, 1911e: 35. Type-species, *coprophila* Felt (orig. des.)=*texana* (Felt).

denningi Pritchard, 1947: 37.—Minn.
fulva Felt, 1926c: 266.—B.C.; Minn.
 praelonga Felt, 1926c: 266.—B.C.
 scutellata Felt, 1926c: 267.—B.C.
sylvestris Felt, 1907b: 97.—N.C.; N. Dak., Minn.

texana (Felt), 1908b: 316 (*Campylomyza*).—Tex.; Colo., D.C., Ala.
  coprophila Felt, 1911e: 35.—Unknown.
  americana Felt, 1913g: 199.—Colo.
  coloradensis Felt, 1913g: 199.—Colo.
  foliata Felt, 1916e: 195 (*Monardia*).—Colo.
truncata (Felt), 1912c: 102 (*Campylomyza*).—Pa.
xylophila Edwards, 1938a: 200.—England; Minn.

Genus CORINTHOMYIA Felt

Corinthomyia Felt, 1911e: 35. Type-species, *Campylomyza hirsuta* Felt (orig. des.)=*brevicornis* (Felt).
brevicornis (Felt), 1907b: 97 (*Campylomyza*).—N.Y.; B.C., Minn. to Nfld., s. to Fla., also Chile, Europe.
  bryanti Felt, 1908b: 313 (*Campylomyza*).—Nfld.
  luna Felt, 1908b: 313 (*Campylomyza*).—N.Y.
  curreyi Felt, 1908b: 315 (*Campylomyza*).—B.C.
  hirsuta Felt, 1908b: 315 (*Campylomyza*).—N.Y.
  gracilis Felt, 1912c: 102.—Pa.
  tumida Felt, 1913g: 197 (*Cordylomyia*).—N.Y.
  cincinna Felt, 1913g: 200.—N.Y.

Genus XYLOPRIONA Kieffer

Xylopriona Kieffer, 1913a: 291. Type-species, *Campylomyza pulchricornis* Kieffer (orig. des.). Chosen by first reviser, Mani, 1946: 193.
Tetraxyphus Kieffer, 1913a: 290. Type-species, *Campylomyza melanopterus* Kieffer (mon.).
articulosa (Felt), 1908b: 315 (*Campylomyza*).—N.H.
atra (Meigen), 1804: 40 (*Cecidomyia*).—Germany; Calif., also Greenland.
crebra Pritchard, 1947: 45.—Minn.
toxicodendri (Felt), 1907b: 98 (*Campylomyza*).—N.Y.; Calif., Minn., Ill. to N.Y. and Conn.
  toxicodendron, error.
  gilletti Felt, 1908b: 314 (*Campylomyza*).—Pa.
  alexanderi Felt, 1913g: 187 (*Monardia*).—N.Y.
  modesta Felt, 1913h: 142 (*Monardia*).—Conn.
  illinoiensis Felt, 1935d: 47 (*Monardia*).—Ill.

Genus POLYARDIS Pritchard

Polyardis Pritchard, 1947: 47. Type-species, *Campylomyza carpini* Felt (orig. des.).
adela Pritchard, 1947: 49.—Minn.

aporia Pritchard, 1947: 51. —Minn.
carpini (Felt), 1907b: 100 (*Campylomyza*). —N.Y.
    versicolor Felt, 1908b: 314 (*Campylomyza*). —N.Y.
kasloensis (Felt), 1908b: 314 (*Campylomyza*). —B.C.; Calif.
monotheca (Edwards), 1938a: 241 (*Monardia*). —England; B.C., Calif.
vitinea (Felt), 1907b: 98 (*Campylomyza*). —N.Y.

Genus MONARDIA Kieffer

Monardia Kieffer, 1895a: 111. Type-species, *stirpium* Kieffer (mon.).
antennata (Winnertz), 1870: 23 (*Campylomyza*). —Europe; Minn.
canadensis Felt, 1926c: 267. —B.C.
lignivora (Felt), 1907b: 100 (*Campylomyza*). —N.C.
multiarticulata Felt, 1914f: 109. —N.H.

Genus TRICHOPTEROMYIA Williston

Trichopteromyia Williston, 1896c: 255. Type-species, *modesta* Williston (mon.).
modesta Williston, 1896c: 255. —St. Vincent I.; Minn., N.Y., N.H. also England.
    flavoscuta Felt, 1907b: 97 (*Campylomyza*). —N.Y.
    rugosa Felt, 1914f: 110 (*Monardia*). —N.H.

Genus MICROMYA Rondani

Micromya Rondani, 1840b: 23. Type-species, *lucorum* Rondani (mon.).
Micromyia, emend.
Ceratomyia Felt, 1911e: 33. Type-species, *johannseni* Felt (orig. des.).
johannseni (Felt), 1911e: 33 (*Ceratomyia*). —Mexico: Minn.
mana Pritchard, 1947: 63. —Minn.

Genus MYCOPHILA Felt

Mycophila Felt, 1911e: 33. Type-species, *fungicola* Felt (orig. des.).
barnesi Edwards, 1938a: 254. —England; Calif.
fungicola Felt, 1911e: 33. —Calif.; Pa., ?Ind., ?Ohio.
lampra Pritchard, 1947: 65. —Minn.; Wis.
speyeri (Barnes), 1926: 90 (*Pezomyia*). —England; Va.

Genus BRYOMYIA Kieffer

Bryomyia Kieffer, 1895a: 78. Type-species, *bergrothi* Kieffer (mon.).
apsectra Edwards, 1938a: 210. - England; Minn.

cambrica Edwards, 1938a: 210. —England; Minn.
gibbosa (Felt), 1907b: 100 (*Campylomyza*).—N.Y.; Minn.
  cerasi Felt, 1907b: 101 (*Campylomyza*).—N.Y.
  flavida Felt, 1920: 279 (*Neptunimyia*).—N.Y.
producta (Felt), 1908b: 315 (*Campylomyza*).—N.Y.; Minn., Ont.

Genus APRIONUS Kieffer

Aprionus Kieffer, 1894d: 205 (also 1895c: clxxvi, as *Apriona*, error).
  Type-species, *spinigera* Kieffer (Kieffer, 1895a: 93).
asemus Pritchard, 1947: 72.—Minn.
longipennis (Felt), 1908b: 314 (*Campylomyza*).—N.Y.
monticola (Felt), 1920: 281 (*Campylomyza*).—N.Y.
pinicorticis (Felt), 1908b: 315 (*Campylomyza*).—N.J.

Genus PEROMYIA Kieffer

Peromyia Kieffer, 1894d: 205. Type-species, *leveillei* Kieffer (sub.
  mon., Kieffer, 1895c: clxxv).
Joannisia Kieffer, 1894d: 205 (preocc. Monterosato, 1884). Type-
  species, *aurantiaca* Kieffer (Coquillett, 1910b: 556) =
  leveillei Kieffer.
borealis (Felt), 1920: 280 (*Joannisia*).—N.Y.; Minn., also England.
modesta (Felt), 1907b: 99 (*Campylomyza*).—N.Y.
neomexicana (Felt), 1913g: 160 (*Joannisia*).—N. Mex.
ovalis (Edwards), 1938a: 258 (*Joannisia*).—England; Minn.
photophila (Felt), 1907b: 99 (*Campylomyza*).—N.Y.; Minn., Wis., Pa.,
  N.C., also England.
  carolinae Felt, 1907b: 100 (*Campylomyza*).—N.C.
  flavopedalis Felt, 1908b: 313 (*Joannisia*).—N.Y.
  flavoscuta Felt, 1908b: 313 (*Joannisia*).—N.Y.
  pennsylvanica Felt, 1911d: 476 (*Joannisia*).—Pa.

Subfamily CECIDOMYIINAE

REFERENCE: Felt, 1958 (rev., ne. U.S.).

Tribe LEPTOSYNINI

REFERENCE: Pritchard, 1960a (rev.).

Genus HENRIA Wyatt

Henria Wyatt, 1959: 175. Type-species, *psalliotae* Wyatt (orig. des.).
Barnesina Pritchard, 1960a: 314. Type-species, *hylecoites* Pritchard
  (orig. des.).

Leptosyna, authors, not Kieffer.

hylecoites (Pritchard), 1960a: 315 (*Barnesina*).—Calif.
quercivora (Felt), 1912b: 123 (*Leptosyna;* n. name for *quercus* Felt).—N.Y.; Minn.
    quercus Felt, 1911d: 546 (*Leptosyna;* preocc. Kieffer, 1904).—N.Y.
    quercicola Kieffer, 1913c: 54 (*Leptosyna;* n. name for *quercus* Felt).—N.Y.
    americana Felt, 1913g: 216 (*Leptosyna*).—N.Y.

Tribe BRACHINEURINI

Genus EPIMYIA Felt

Epimyia Felt, 1911e: 38. Type-species, *carolina* Felt (orig. des.).
carolina Felt, 1911e: 38.—N.C.

Genus BRACHINEURA Rondani

Brachineura Rondani, 1840b: 16. Type-species, *fuscogrisea* Rondani (mon.).
Brachyneura, emend.

americana Felt, 1908b: 286.—N.Y.
eupatorii Felt, 1908b: 317.—N.Y.
vitis Felt, 1908b: 317.—N.Y.

Tribe PORRICONDYLINI

REFERENCE: Felt, 1915e (rev.).

Genus JOHNSONOMYIA Felt

Johnsonomyia Felt, 1908b: 417. Type-species, *rubra* Felt (orig. des.).
fusca Felt, 1908b: 417.—N.Y.
rubra Felt, 1908b: 417.—Vt.; N.H., Mass.

Genus KRONOMYIA Felt

Kronomyia Felt, 1911d: 476. Type-species, *populi* Felt (orig. des.).
populi Felt, 1911d: 476.—N.Y.

Genus WINNERTZIA Rondani

Asinapta, subg. **Winnertzia** Rondani, 1860: 290. Type-species, *Cecidomyia lugubris* Winnertz (orig. des.).
aceris Felt, 1913f: 213.—N.Y.

CECIDOMYIIDAE 253

ampelophila (Felt), 1907b: 144 (*Porricondyla*).—N.Y.
arizoniensis Felt, 1908b: 421.—Ariz.
calciequina Felt, 1907b: 161.—N.Y.
carpini Felt, 1907b: 148.—N.Y.
fungicola Felt, 1921e: 118.—N.J.
hudsonici Felt, 1908b: 422.—N.Y.
karnerensis Felt, 1908b: 422.—N.Y.
palustris Felt, 1915e: 133.—N.Y.
pectinata Felt, 1911d: 478.—N.Y.
pinicorticis Felt, 1908b: 304.—Va.
rubida Felt, 1908b: 422.—N.Y.
solidaginis Felt, 1907b: 149.—N.Y.

Genus HORMOSOMYIA Felt

Hormosomyia Felt, 1919: 220. Type-species, *oregonensis* Felt (orig. des.).

oregonensis Felt, 1919: 220.—Oreg.

Genus COLPODIA Winnertz

Cecidomyia, subg. **Colpodia** Winnertz, 1853: 185, 293. Type-species, *angustipennis* Winnertz (mon.).

alta Felt, 1908b: 416.—N.Y.
americana Felt, 1914d: 124.—N.Y.
capitata Felt, 1914d: 125.—N.Y.
carolinae (Felt), 1907b: 145 (*Porricondyla*).—N.C.
colei Felt, 1919: 223.—Oreg.
cornuta Felt, 1915e: 153.—Mass.
diervillae (Felt), 1907b: 145 (*Porricondyla*).—N.Y.
graminis (Felt), 1907b: 146 (*Porricondyla*).—N.Y.
maculata Felt, 1908b: 416.—N.Y.
ovata Felt, 1914d: 125.—N.Y.
pectinata (Felt), 1908b: 304 (*Bryocrypta*).—N.Y.
pinea (Felt), 1907b: 145 (*Porricondyla*).—N.C.
porrecta Felt, 1914d: 126.—N.Y.
pratensis Felt, 1908b: 416.—N.Y.
sanguinea Felt, 1908b: 416.—N.Y.; D.C.
sylvestris Felt, 1914d: 126.—N.Y.
temeritatis Felt, 1908b: 416.—N.Y.
terrena Felt, 1908b: 416.—N.Y.
trifolii (Felt), 1907b: 145 (*Porricondyla*).—N.Y.

Genus PARWINNERTZIA Felt

Parwinnertzia Felt, 1920: 281. Type-species, *notmani* Felt (orig. des.).

notmani Felt, 1920: 281.—N.Y.

Genus DIDACTYLOMYIA Felt

Didactylomyia Felt, 1911e: 39. Type-species, *Colpodia longimana* Felt (orig. des.).

capitata Felt, 1913i: 174.—Mass.
flava Felt, 1915e: 145.—N.Y.
longimana (Felt), 1908b: 416 (*Colpodia*).—Mass.; N.Y.
maculata Felt, 1915e: 144.—N.Y.
robusta Felt, 1920: 282.—N.Y.

Genus ASYNAPTA Loew

Cecidomyia, subg. **Asynapta** Loew, 1850b: 21, 39. Type-species, *longicollis* Loew (Coquillett, 1910b: 511).

americana Felt, 1912c: 103.—Pa.
apicalis Felt, 1914d: 127.—N.Y.
borealis Felt, 1920: 285.—N.Y.
canadensis Felt, 1908b: 421.—Canada (unpublished).
caudata Felt, 1908b: 421.—N.Y.
cerasi Felt, 1907b: 147.—N.Y.
dolens Felt, 1920: 285.—N.Y.
flavida Felt, 1908b: 421.—N.Y.
frosti Felt, 1913h: 143.—Mass.
furcata (Felt), 1907b: 148 (*Winnertzia*).—N.Y.
hopkinsi Felt, 1935e: 48.—Fla.
marilandica Felt, 1916c: 412.—Md.
mediana Felt, 1914d: 128.—N.Y.
nobilis Felt, 1913h: 142.—Maine.
saliciperda Felt, 1908b: 421.—Ill.
umbra Felt, 1914d: 128.—N.Y.

Genus HOLONEURUS Kieffer

Holoneura Kieffer, 1894a: 84 (preocc. Tetens, 1891). Type-species, *cincta* Kieffer (Kieffer, 1894b: 316).
Holoneurus Kieffer, 1895b: 115 (n. name for *Holoneura* Kieffer). Type-species, *Holoneura cincta* Kieffer (aut.).

altifilus (Felt), 1907b: 147 (*Porricondyla*).—N.Y.
elongatus Felt, 1908b: 420.—B.C.

humilis (Felt), 1908b: 417 (*Johnsonomyia*).—N.Y.
inflatus Felt, 1920: 287.—N.Y.
multinodus Felt, 1908b: 420.—N.Y.
    multineurus (error) Felt *in* Kieffer, 1913a: 266.
photophilus (Felt), 1907b: 148 (*Asynapta*).—N.Y.
strobilophilus Foote, 1956: 49.—Calif.; Oreg.
tarsalis Felt, 1915e: 192.—N.Y.

Genus RUBSAAMENIA Kieffer

Rubsaamenia Kieffer, 1894a: 83 (also 1894b: 333, as *Ruebsaamenia*). Type-species, *flava* Kieffer (mon.).
keeni Foote, 1956: 48.—Calif.; Oreg., Colo., N. Mex.

Genus DIRHIZA Loew

Cecidomyia, subg. **Dirhiza** Loew, 1850b: 21, 38. Type-species, *lateritia* Loew (mon.).
canadensis Felt, 1908b: 420.—N.Y.
hamata Felt, 1907b: 144.—N.Y.
montana Felt, 1908b: 420.—N.H.
multiarticulata Felt, 1908b: 420.—N.J.; N.Y.
photophila Felt, 1908b: 420.—N.Y.
sylvestris (Felt), 1907b: 146 (*Porricondyla*).—N.Y.

Genus PORRICONDYLA Rondani

Porricondyla Rondani, 1840b: 13. Type-species, *Cecidomyia albitarsis* Meigen (mon.).
antennata Felt, 1915e: 169.—N.Y.
barberi Felt, 1908b: 418.—Ariz.
bidentata Felt, 1920: 283.—N.Y.
borealis Felt, 1907b: 147.—N.Y.
canadensis Felt, 1908b: 418.—Canada (unpublished).
carolina Felt, 1908b: 418.—N.C.
caudata Felt, 1908b: 418.—N.Y.
consobrina Felt, 1919: 221.—Ont.
dietzii Felt, 1912c: 105.—Pa.
dilatata Felt, 1908b: 418.—N.Y.
dorsata Felt, 1912e: 238.—N.Y.
flava Felt, 1907b: 146.—N.Y.
fultonensis Felt, 1919: 222.—N.Y.
hamata Felt, 1907b: 146.—N.C.; Mass.
johnsoni Felt, 1920: 282.—Mass.

juvenalis Felt, 1912e: 239.—N.Y.
karnerensis Felt, 1908b: 418.—N.Y.
minor Felt, 1926c: 268.—B.C.
novaeangliae Felt, 1914f: 110 (as *novae-angliae*).—N.H.
papillata Felt, 1914f: 111.—N.H.
pennulae Felt, 1921d: 142.—Fla.
pini Felt, 1907b: 144.—N.Y.
porrecta Felt, 1912c: 105.—N.C.
quercina Felt, 1907b: 147.—N.Y.
setosa Felt, 1914d: 129.—N.Y.
spinigera Felt, 1920: 284.—N.Y.
sylvestris Felt, 1926a: 207.—Maine.
tuckeri Felt, 1908b: 418.—Kans. (unpublished).
tumidosa Felt, 1920: 284.—N.Y.
vernalis Felt, 1912c: 104.—Pa.
wellsi Felt, 1915a: 227.—N.Y.

Genus CAMPTOMYIA Kieffer

Camptomyia Kieffer, 1894a: 86 (also 1894b: 323). Type-species, *erythromma* Kieffer (orig. des., as gen. n., sp. n.).

aestiva Felt, 1912c: 104.—Pa.; N.Y.
antennata Felt, 1920: 286.—N.Y.
dentata Felt, 1920: 286.—N.Y.
montana Felt, 1915e: 180.—N.Y.
multinoda (Felt), 1908b: 419 (*Porricondyla*).—Ill.
pectinata Felt, 1920: 287.—N.Y.
tsugae Felt, 1913f: 214.—N.Y.

Tribe OLIGOTROPHINI

REFERENCE: Felt, 1915d (rev.).

Genus RHABDOPHAGA Westwood

Cecidomyia, subg. **Rhabdophaga** Westwood, 1847: 588 (as *Rabdophaga*). Type-species, *viminalis* Westwood (mon.).
Epourenia Rapp, 1946a: 128. Type-species, *viva* Rapp (orig. des.).

The identity of *Tipula salicina* Schrank, the type-species of *Phytophaga* Rondani, 1840b: 13, is in doubt. If further study confirms the possibility that it is the same as *viminalis* Westwood, *Phytophaga* replaces the name *Rhabdophaga*.

absobrina Felt, 1907b: 113.—N.Y.
acerifolia Felt, 1907b: 112.—N.Y.

aceris (Shimer), 1868: 281 (*Cecidomyia*).—Ill.; Ont., N.Y.
californica Felt, 1908b: 353.—Calif.
caulicola Felt, 1909a: 290.—Ill.
cephalanthi Felt, 1908b: 355.—N.Y.
consobrina Felt, 1907b: 113.—N.Y.
cornuta (Walsh), 1864b: 625 (*Cecidomyia*).—Ill.
elymi Felt, 1909a: 289.—Calif.
essigi Felt, 1926e: 79.—Calif.
gemmae Felt, 1908b: 354.—N.Y.
globosa Felt, 1908b: 354.—N.Y.; Conn.
hildebrandi Felt, 1928: 58.—Calif.
hirticornis Felt, 1909a: 290.—N.Y.
latebrosa Felt, 1909a: 290.—N.Y.
marginata Felt, 1908b: 352.—N.Y.
normaniana Felt, 1908b: 354.—Man. (unpublished).
occidentalis Felt, 1908b: 353.—Calif.
persimilis Felt, 1908b: 351.—N.Y.
plicata Felt, 1908b: 352.—Unknown.
podagrae Felt, 1908b: 355.—N.Y.
populi Felt, 1907b: 112.—N.Y.
porrecta Felt, 1915d: 101.—N.Y.
    ponecta, error.
pratensis Felt, 1908b: 353.—N.Y.
racemi Felt, 1908b: 352.—Man. (unpublished).
ramuscula Felt, 1908b: 351.—N.Y.
rileyana Felt, 1909a: 289.—Unknown.
rosacea Felt, 1908b: 354.—Man. (unpublished).
salicifolia Felt, 1908b: 293.—N.Y.; Maine, Mass., Conn., R.I.
salicis (Schrank), 1803: 69 (*Tipula*).—Germany; N.Y.
salicisbatatas (Osten Sacken), 1878c: 4 (*Cecidomyia*; orig. as *salicis batatas* Walsh, 1864b: 601).—Ill.
    batatas, emend.
salicisnodulus (Osten Sacken), 1878c: 7 (*Cecidomyia;* orig. as *salicis nodulus* Walsh, 1864b: 599).—Ill.; N.Y., Mass. Gall and larva only; adult descr. Felt, 1915d: 91.
    nodulus, emend.
salicisrhodoides (Osten Sacken), 1878c: 4 (*Cecidomyia;* orig. as *salicis rhodoides* Walsh, 1864b: 587).—Ill.
    rhodoides, emend.
salicistriticoides (Osten Sacken), 1878c: 7 (*Cecidomyia;* orig. as *salicis triticoides* Walsh, 1864b: 598).—Ill.; Ont., Maine, Mass. Gall only; adult descr. Felt, 1915d: 88 (as *triticoides*).
    triticoides, emend.

salicishordoides Osten Sacken, 1878c: 7 (*Cecidomyia*; orig. as *salicis hordeoides* Walsh, 1864b: 599).—Ill. Gall only. *hordoides*, emend.
sodalitatis Felt, 1908b: 351.—N.Y.
strobiloides (Osten Sacken), 1862b: 203 (*Cecidomyia*).—Ill.; Ill., se. N.Y. to Maine.
salicisstrobiloides Osten Sacken, 1878c: 4 (*Cecidomyia;* orig. as *salicis strobiloides* Walsh, 1864b: 580).—Ill.
swainei Felt, 1914e: 77.—Ont.; N.B., P.E.I., Nfld.
viva (Rapp), 1946a: 128 (*Epourenia*).—Ill.

Genus PROCYSTIPHORA Felt

Procystiphora Felt, 1915d: 212. Type-species, *coloradensis* Felt (orig. des.).

coloradensis Felt, 1915d: 212.—Colo.
junci Felt, 1922a: 166.—Ill.

Genus DASINEURA Rondani

Dasineura Rondani, 1840b: 17. Type-species, *luteofusca* Rondani (Coquillett, 1910b: 530).
Dasyneura Agassiz and Loew, 1846: 11, emend. (preocc. Saunders, 1842).
Perrisia Rondani, 1846: 371. Type-species, *Cecidomyia urticae* Perris (orig. des.).
Neocerata Coquillett, 1900e: 47. Type-species, *rhodophaga* Coquillett (mon.).

aberrata Felt, 1908b: 346.—N.Y.
abiesemia Foote, 1956: 52.—Oreg.; Calif.
acerifolia Felt, 1907b: 115.—N.Y.
albohirta Felt, 1908b: 346.—N.Y.
albovittata (Walsh), 1864b: 621 (*Cecidomyia*).—Ill.; N.Y.
albovitta, error.
alopecuri (Reuter), 1895: 3 (*Oligotrophus*).—Finland; Ont., N.B.
americana Felt, 1913h: 136 (n. name for *galii* Felt).—Mass.
galii Felt, 1908b: 348 (preocc. Loew, 1850).—Mass.
galiorum Kieffer, 1913a: 77 (*Perrisia;* n. name for *galii* Felt).—Mass.
ampelophila Felt, 1908b: 343.—N.Y.
anemone Felt, 1908b: 292.—N.Y.
antennata Felt, 1908b: 347.—N.Y.
apicata Felt, 1908b: 345.—N.Y.
aromaticae Felt, 1909a: 289.—Mass.

attenuata Felt, 1908b: 350.—N.Y.
augusta Felt, 1908b: 348.—N.Y.
aurihirta Felt, 1908b: 342.—N.Y.
balsamicola (Lintner), 1888a: 60 (1888: 180) (*Cecidomyia*).—N.Y.;
Ont., e. to Nfld. Gall and larva only. **N. comb.**
bidentata Felt, 1907b: 114.—N.Y.
borealis Felt, 1907b: 117.—N.Y.
californica Felt, 1908b: 347.—Calif.
canadensis Felt, 1907b: 157.—Ont.; Sask.
carbonaria Felt, 1907b: 117.—N.Y.
caricicola (Kieffer), 1913a: 74 (*Perrisia;* n. name for *caricis* Felt).—
N.Y.
caricis Felt, 1907b: 116 (preocc. Kieffer, 1901).—N.Y.
cerocarpi Felt 1913f: 215.—Colo.
cirsioni Felt, 1908b: 346.—N.Y.
clematidis Felt, 1908b: 344.—N.Y.
communis Felt, 1911d: 478.—N.Y.; Mass., Conn., Pa., N.J.
consobrina Felt, 1907b: 118.—N.Y.
corticis Felt, 1909a: 289 (also 1910b: 355).—Mass.
cyanococci Felt, 1908b: 292.—Mass.
denticulata Felt, 1907b: 117.—N.Y.
eugeniae Felt, 1912c: 106.—Fla.
filicis Felt, 1907b: 115.—N.Y.
flavescens Felt, 1908b: 344.—N.Y.
flavicornis Felt, 1908b: 345.—N.Y.
flavoabdominalis Felt, 1908b: 350.—N.Y.
flavoscuta Felt, 1908b: 350.—N.Y.
florida Felt, 1908b: 346.—Fla.
fraxinifolia Felt, 1908b: 293.—N.Y.
fulva Felt, 1908b: 349.—N.Y.
gemmae Felt. 1909a: 288.—Utah.
gentneri Pritchard, 1953b: 131.—Oreg.; Wash., N.Y., also Europe.
gibsoni Felt, 1911: 479.—Ont.; Ind.
glandis Felt, 1908b: 342.—Mo.
gleditchiae (Osten Sacken), 1866: 219 (*Cecidomyia*).—R.I.; Ont.,
N.Y., Vt., Mass., Conn.
gleditschiae, error or emend.
graminis Felt, 1908b: 342.—N.Y.
karnerensis Felt, 1908b: 341.—N.Y.
leguminicola (Lintner), 1879b: 121 (*Cecidomyia;* n. name for *trifolii*
Lintner).—N.Y.; B.C. to N.S., s. to Oreg. and Va.
trifolii Lintner, 1879a: 44 (*Cecidomyia;* preocc. Löw, 1874).—
N.Y.
lepidii Felt, 1908b: 346.—D.C.

lupini Felt, 1916c: 413.—Calif.
lysimachiae (Beutenmüller), 1907b: 305 (*Cecidomyia*).—N.Y.; Mass., N.J.
maculosa Felt, 1908b: 341.—N.Y.
mali (Kieffer), 1904a: 345 (*Perrisia*).—Europe; Mass.
maritima Felt, 1909a: 288.—Mass.
meliloti Felt, 1907b: 116.—N.Y.
modesta Felt, 1908b: 345.—N.Y.
multiannulata Felt, 1908b: 346.—N.Y.
oxycoccana (Johnson) *in* Skinner, 1899: 80 (*Cecidomyia;* n. name for *vaccinii* Smith).—N.J.
  vaccinii Smith, 1890: 31 (*Cecidomyia;* preocc. Osten Sacken, 1862).—N.J.
parthenocissi (Stebbins), 1910: 44 (*Cecidomyia*).—Mass.; Man., Ont., N.Y. Gall only; adult descr. Felt, 1913f: 216.
pedalis Felt, 1908b: 350.—N.Y.
pergandei Felt, 1911d: 480.—Colo.
photophila Felt, 1907b: 114.—N.Y.
piperitae Felt, 1908b: 342.—N.Y.
pseudacaciae (Fitch), 1859: 833 (1859: 53) (*Cecidomyia*).—N.Y.
pudorosa Felt, 1908b: 344.—N.Y.
purpurea Felt, 1908b: 349.—N.Y.; Mass.
pyri (Bouché) 1847: 144 (*Cecidomyia*).—Europe; Conn., N.Y.
quercina Felt, 1907b: 116.—N.Y.
rachiphaga Tripp, 1955: 259.—Ont.; Que.
radifolii Felt, 1909a: 289.—Mass.
rhodophaga (Coquillett), 1900e: 47 (*Neocerata*).—D.C.; Ont. to Mass., s. to Ill. and D.C.
rhois (Coquillett), 1895g: 348 (*Cecidomyia*).—N.Y.; N.Y. to Va.
rosarum (Hardy), 1850: 186 (*Cecidomyia*).—Europe; Mass., N.J.
rubiflorae Felt, 1908b: 343.—Unknown.
rufipedalis Felt, 1908b: 349.—N.Y.
salicifolia Felt, 1908b: 293.—N.Y.; Mass.
sassafras Felt, 1916a: 29.—Ont.
scutata Felt, 1908b: 346.—N.Y.
semenivora (Beutenmüller), 1907a: 390 (*Cecidomyia*).—N.Y.; Ont., N.Y., Mass. Gall and larva only; adult descr. Beutenmüller, 1913a: 414 (as *Dasineura*).
serrulatae (Osten Sacken), 1862b: 198 (*Cecidomyia*).—D.C.; N.Y., Mass.
setosa Felt, 1907b: 115.—N.Y.
simillima (Kieffer), 1913a: 81 (*Perrisia;* n. name for *similis* Felt).— N.Y.
  similis Felt, 1908b: 346 (preocc. Löw, 1888).—N.Y.

simulator Felt, 1908b: 344.—N.Y.
smilacifolia Felt, 1911d: 480.—Mass.
smilacinae Bishop, 1911: 346.—N.Y.
spiraeina Felt, 1908b: 341.—N.Y.
stanleyae (Cockerell), 1914b: 241 (*Perrisia*).—Colo.
torontoensis Felt, 1915c: 405.—Ont.
toweri Felt, 1909a: 289.—Mass.
trifolii (Löw), 1874: 143 (*Cecidomyia*).—Germany; Nebr. to Mass., s. to Ky., Europe.
tumidosae Felt, 1908b: 346.—N.Y.; Mass.
ulmea Felt, 1908b: 349.—Mass.; N.Y.
unguicula Felt, 1908b: 344.—N.Y.
vernalis Felt, 1908b: 341.—N.Y.
vitis Felt, 1908b: 341.—N.Y.
yuccae Felt, 1908b: 343.—Unknown.

Genus PHAENOLAUTHIA Kieffer

Phaenolauthia Kieffer, 1912b: 2 (1912: xi). Type-species, *Ledomyia obscuripennis* Kieffer (orig. des.).
Lasiopteryx Stephens of Felt, 1915d: 189.

arizonensis (Felt), 1915d: 195 (*Lasiopteryx*).—Ariz.
carpini (Felt), 1907b: 119 (*Asphondylia*).—N.Y.
coryli (Felt), 1908b: 292 (*Dasineura*).—N.Y.
crispata (Felt), 1914f: 111 (*Lasiopteryx*).—Mass.
flavotibialis (Felt), 1907b: 157 (*Dasineura*).—N.Y.

Genus NEUROMYIA Felt

Neuromyia Felt, 1911e: 44. Type-species, *Arnoldia minor* Felt (orig. des.).

minor (Felt), 1908b: 290 (*Arnoldia*).—N.Y.

Genus DRYOMYIA Kieffer

Dryomyia Kieffer, 1898a: 17. Type-species, *Cecidomyia cicinans* Giraud (Rübsaamen, 1910: 337).

folliculi (Felt), 1908b: 348 (*Dasineura*).—N.J.

Genus CYSTIPHORA Kieffer

Cystiphora Kieffer, 1892: 212. Type-species, *Cecidomyia hieracii* Löw (Rübsaamen, 1910: 337).

canadensis Felt, 1913b: 417.—Ont.
viburnifolia Felt, 1911d: 480.—Mass.; N.Y., Conn.

Genus ALASSOMYIA Felt

Allomyia Felt, 1918c: 380 (preocc. Banks, 1916). Type-species, *juniperi* Felt (orig. des.).
Alassomyia Felt, 1925b: 147 (n. name for *Allomyia* Felt). Type-species, *Allomyia juniperi* Felt (aut.).
juniperi (Felt), 1918c: 380 (*Allomyia*).—Colo.

Genus RHIZOMYIA Kieffer

Rhizomyia Kieffer, 1898a: 56. Type-species, *perplexa* Kieffer (mon.).
absobrina (Felt), 1908b: 289 (*Arnoldia*).—N.Y.
cerasi (Felt), 1907b: 114 (*Dasineura*).—N.Y.
cincta Felt, 1915d: 204.—N.Y.
fraxinifolia (Felt), 1908b: 289 (*Arnoldia*).—N.J.
hirta Felt, 1911d: 478.—N.Y.
hispida (Felt), 1908b: 290 (*Arnoldia*).—N.Y.
ungulata (Felt), 1908b: 290 (*Arnoldia*).—N.Y.
vitis (Felt), 1908b: 290 (*Arnoldia*).—?D.C.

Genus CTENODACTYLOMYIA Felt

Ctenodactylomyia Felt, 1915f: 199. Type-species, *watsoni* Felt (orig. des.).
watsoni Felt, 1915f: 199.—Fla.

Genus DIARTHRONOMYIA Felt

Diarthronomyia Felt, 1908b: 339. Type-species, *artemisiae* Felt (orig. des.).
artemisiae Felt, 1908b: 339.—Colo.; Utah.
californica Felt, 1912a: 752.—Calif.
chrysanthemi Ahlberg, 1939: 276.—Europe; Oreg. to Que., s. to Calif. and R.I., also New Zealand.
    hypogaea Loew of Felt, 1916d: 51.
floccosa Felt, 1916e: 195.—Calif.
occidentalis Felt, 1916e: 194.—Calif.; Utah.

Genus COCCIDOMYIA Felt

Coccidomyia Felt, 1911e: 45. Type-species, *pennsylvanica* Felt (orig. des.).
erii Felt, 1912d: 147.—Calif.
pennsylvanica Felt, 1911e: 45.—Pa.

Genus FICIOMYIA Felt

Ficiomyia Felt, 1922b: 5. Type-species, *perarticulata* Felt (orig. des.).
birdi Felt, 1934b: 132.—Fla.
perarticulata Felt, 1922b: 5.—Fla.

Genus MAYETIOLA Kieffer

Mayetia Kieffer, 1896b: 5 (preocc. Mulsant and Rey, 1875). Type-species, *Cecidomyia destructor* Say (orig. des.).
Mayetiola Kieffer, 1896c: 89 (n. name for *Mayetia* Kieffer). Type-species, *Cecidomyia destructor* Say (aut.).
Phytophaga Rondani of Felt, 1911e: 45, and authors.

REFERENCES: Barnes, 1927, 1956 (nomenclature), 1958 (cross-breeding of *destructor*).

aceris (Felt), 1907b: 122 (*Oligotrophus*).—N.Y.
americana Felt, 1908b: 370.—Ill.; Conn.
azaleae (Felt), 1907b: 122 (*Oligotrophus*).—N.Y.
balsamifera Felt, 1908b: 369.—N.Y.
californica Felt, 1908b: 370.—Calif.
carpophaga (Tripp), 1955: 261 (*Phytophaga*).—Ont.
caudata (Felt), 1915e: 199 (*Phytophaga*).—N.Y.
caulicola Felt, 1908b: 370.—Ill.
celtiphyllia Felt, 1908b: 371.—Iowa.
destructor (Say), 1817b: 45 (1859b: 4) (*Cecidomyia*).—N. and Middle States; wheat-growing regions of N. Amer.
electra Felt, 1908b: 369.—N.Y.
floridensis (Felt), 1921d: 143 (*Phytophaga*).—Fla.
fraxini (Felt), 1915e: 206 (*Phytophaga*).—N.Y.
latipennis (Felt), 1908b: 353 (*Rhabdophaga*).—Ill.
latipes Felt, 1908b: 370.—N.Y.
occidentalis (Felt), 1926e: 79 (*Phytophaga*).—Calif.
painteri (Felt), 1935c: 7 (*Phytophaga*).—Tex.
perocculta (Cockerell), 1904: 156 (*Cecidomyia*).—Colo.
piceae (Felt), 1926b: 229, 240 (*Phytophaga;* as *picaea*, p. 229).—Que.; N.B.
rigidae (Osten Sacken), 1862b: 189 (*Cecidomyia;* n. name for *salicis* Fitch).—N.Y.; Ill., Ont.
   salicis Fitch, 1845a: 263 (*Cecidomyia;* preocc. in *Cecidomyia* by Schrank, 1803, orig. *Tipula*, now *Rhabdophaga*).— N.Y.

saliciscornu Osten Sacken, 1878c: 4 (*Cecidomyia;* orig. as *salicis cornu* Walsh, 1864b: 590).—Ill. Gall and larva only; adult descr. Walsh, 1866: 224.
cornu, emend.
siliqua Osten Sacken, 1878c: 4 (*Cecidomyia;* orig. as *salicis siliqua* Walsh, 1864b: 591).—Ill.
sabinae (Felt), 1935c: 7 (*Phytophaga*).—Tex.
socialis Felt, 1908b: 369.—N.Y.
tetradymia (Felt), 1925a: 15 (*Phytophaga*).—Calif.
texana (Felt), 1935c: 8 (*Phytophaga*).—Tex.
thalactri (Felt), 1907b: 123 (*Oligotrophus*).—N.Y.
thujae (Hedlin), 1959: 719 (*Phytophaga*).—B.C.
timberlakei (Felt), 1916e: 191 (*Phytophaga*).—Utah.
tsugae (Felt), 1907b: 123 (*Oligotrophus*).—N.Y.; Ont.
tumidosae Felt, 1908b: 370.—Ill.; Ont.
ulmi (Beutenmüller), 1907a: 387 (*Cecidomyia*).—N.Y.; Ont.
violicola (Coquillett), 1900f: 50 (*Diplosis*).—D.C.; N.Y., Va.
virginiana Felt, 1908b: 369.—N.Y.
walshii Felt, 1908b: 371.—Ill.
wellsi (Felt), 1916e: 190 (*Phytophaga*).—Ohio.

Genus JANETIELLA Kieffer

Janetiella Kieffer, 1898a: 23. Type-species, *Cecidomyia thymi* Kieffer (Coquillett, 1910b: 556).

acerifolia (Felt), 1907b: 124 (*Oligotrophus*).—N.Y.
americana Felt, 1908b: 372.—N.Y.
asplenifolia (Felt), 1907b: 159 (*Oligotrophus*).—Mass.
breviaria Felt, 1908b: 372.—N.Y.
brevicauda Felt, 1908b: 372.—N.Y.
brevicornis (Felt), 1907b: 122 (*Oligotrophus*).—N.Y.
coloradensis Felt, 1912d: 148.—Colo.
ligni Felt, 1915e: 220.—N.Y.
nodosa (Felt), 1907b: 123 (*Oligotrophus*).—N.Y.
parma Felt, 1914d: 129.—N.Y.
pini (Felt), 1907b: 124 (*Oligotrophus*).—N.Y.
sanguinea Felt, 1908b: 372.—N.Y.
siskiyou Felt, 1917b: 194.—Oreg.
tiliacea (Felt), 1907b: 121 (*Oligotrophus*).—N.Y.

Genus OLIGOTROPHUS Latreille

Oligotrophus Latreille, 1805: 288. Type-species, *Tipula juniperina* Linnaeus (mon.).

betheli Felt, 1912d: 148.—Colo.; Utah.

betulae (Winnertz), 1853: 234 (*Cecidomyia*).—Europe; N.Y., Conn.
inquilinus Felt, 1908b: 368.—N.Y.
pattersoni White, 1950: 72.—Tex.
salicifolius Felt, 1910b: 354.—N.Y.
vernalis Felt, 1908b: 368.—N.Y.

Genus SACKENOMYIA Felt

Sackenomyia Felt, 1908b: 361. Type-species, *Rhopalomyia acerifolia* Felt (as *Oligotrophus;* orig. des.).
acerifolia (Felt), 1907b: 121 (*Rhopalomyia*).—N.Y.
packardi Felt, 1909a: 290.—Mass.; Ill.
porterae (Cockerell), 1904: 155 (*Rhabdophaga*).—N. Mex.
viburnifolia Felt, 1909a: 290.—Mass.; N.Y.

Genus WALSHOMYIA Felt

Walshomyia Felt, 1908b: 359. Type-species, *juniperina* Felt (orig. des.).
insignis Felt, 1921e: 117.—Tex.
juniperina Felt, 1908b: 360.—Calif.
texana Felt, 1916a: 30.—Tex.

Genus RHOPALOMYIA Rübsaamen

Rhopalomyia Rübsaamen, 1892: 370. Type-species, *Oligotrophus tanaceticola* Karsch (Kieffer, 1896c: 89).
abnormis Felt, 1908b: 365.—N.Y.
albipennis Felt, 1908b: 364.—N.Y.
alticola (Cockerell), 1890a: 281 (*Cecidomyia*).—Colo.
ampullaria Felt, 1916e: 185.—Utah.
antennariae (Wheeler), 1889: 209 (*Cecidomyia*).—Wis.; Maine, N.Y.
anthophila (Osten Sacken), 1869a: 302 (*Cecidomyia*).—N.Y.; Mo., N.C.
apicata Felt, 1908b: 364.—N.Y.
arcuata Felt, 1907b: 158.—N.Y.
astericaulis Felt, 1907b: 159.—Mass.
asteriflorae Felt, 1908b: 298.—N.Y.
audibertiae Felt, 1908b: 299.—N.Y.
baccharis Felt, 1908b: 364.— Unknown; Calif.
betheliana Cockerell, 1909a: 150.—Colo.
bigeloviae (Cockerell), 1889d: 324 (*Cecidomyia*).—Colo. Gall only; adult descr. Cockerell, 1890b: 109.
bigelovioides Felt, 1908b: 366.—Calif.

bulbula Felt, 1908b: 365.—Mass.
californica Felt, 1908b: 364.—Calif.
capitata Felt, 1908b: 363.—N.Y.
carolina Felt, 1908b: 363.—N.C.
castaneae Felt, 1909a: 291.—Mass.
chrysopsidis (Loew) *in* Osten Sacken, 1862b: 203 (*Cecidomyia*).—D.C.
chrysothamni Felt, 1916e: 187.— Utah.
clarkei Felt, 1908b: 299.—N.H.; Ont., Mass., N.C.
cockerelli Felt, 1915e: 237.—Colo.
crassulina (Cockerell), 1908b: 89 (*Cecidomyia*).— Colo.
cruziana Felt, 1908b: 366.—Calif.
enceliae Felt, 1916e: 183.—Unknown.
ericameriae Felt, 1916e: 190.—Calif.
erigerontis Felt, 1916e: 189.—Calif.
fusiformis Felt, 1907b: 120.—N.Y.
glutinosa Felt, 1916e: 188.—Unknown.
gnaphalodis Felt, 1911d: 484.—Colo.
grindeliae Felt, 1916e: 186.—Calif.
grossulariae Felt, 1911c: 347.—Ohio.
gutierreziae (Cockerell), 1901: 23 (*Asphondylia*).—N. Mex.
hirtipes (Osten Sacken), 1862b: 195 (*Cecidomyia*).—D.C.; N.Y., Mass.
inquisitor Felt, 1908b: 364.—N.Y. (unpublished).
lanceolata Felt, 1908b: 367.—Ill.
lateriflorae Felt, 1908b: 364.—N.Y.; Mass.
lobata Felt, 1908b: 366.—N.Y.
lonicera Felt, 1925a: 15.—Calif.
major Felt, 1907b: 121.—N.Y.
palustris Felt, 1908b: 365.—N.Y.
pedicellata Felt, 1908b: 365.—N.Y.
pilosa Felt, 1908b: 366.—N.W.T.
pini Felt, 1907b: 120.—N.Y.
racemicola (Osten Sacken), 1862b: 196 (*Cecidomyia*).—D.C.; N.C. Gall and larva only.
    racemicola Felt, 1907b: 120.—N.C.
sabinae Patterson, 1919: 345, 346.—Tex.; Utah, Colo. Gall only.
    sabinae Felt, 1921e: 115.—Colo. Adult.
salviae Felt, 1916e: 184.—Calif.
solidaginis (Loew) *in* Osten Sacken, 1862b: 194 (*Cecidomyia*).—D.C.
thompsoni Felt, 1907b: 159.—Mass.
    thomi, error.
tridentatae Rübsaamen, 1893: 163.—N. Amer.

truncata (Felt), 1907b: 160 (*Hormomyia*).—Mass. (unpublished).
  uniformis Felt, 1915e: 244 (unjustified n. name for *truncata* Felt, 1907).—Mass.
truncula Foote (for *truncata* Felt, 1908).—Calif. **N. name.**
  truncata Felt, 1908b: 365 (preocc. Felt, 1907).—Calif.
utahensis Felt, 1916e: 186.—Utah.
weldi Felt, 1921e: 116.—Ill.

Tribe ASPHONDYLIINI

REFERENCE: Felt, 1916d (rev.).

Genus SCHIZOMYIA Kieffer

Schizomyia Kieffer, 1889: 183. Type-species, *galiorum* Kieffer (Coquillett, 1910b: 604).

altifila (Felt), 1907b: 119 (*Asphondylia*).—N.Y.
caryaecola Felt, 1908b: 378.—N.Y.
impatientis (Osten Sacken), 1862b: 204 (*Cecidomyia*).—D.C. Gall and larva only. **N. comb.**
macrofila (Felt), 1908b: 297 (*Asphondylia*).—Calif.
petiolicola Felt, 1908b: 379.—N.Y.; Mass.
rivinae Felt, 1908b: 379.—Fla.
rubi (Felt), 1907b: 119 (*Asphondylia*).—N.Y.
speciosa Felt, 1914f: 112.—N.H.
viburni Felt, 1908b: 378.—N.Y.
vitiscoryloides (Packard), 1869a: 377 (*Cecidomyia*; orig. as *vitis coryloides* Walsh and Riley, 1869b: 106).—Ill.; Ont., N.Y., Mass. Gall and larva only; adult descr. Felt, 1916d: 108.
  coryloides, emend.
vitispomum (Osten Sacken), 1878c: 7 (*Cecidomyia*; orig. as *vitis pomum* Walsh and Riley, 1869b: 106).—Va.; N.Y., Mass., N.J. Gall and larva only; adult descr. Felt, 1916d: 109.
  pomum, emend.

Genus ASPHONDYLIA Loew

Cecidomyia, subg. **Asphondylia** Loew, 1850b: 21, 37. Type-species, *sarothamni* Loew (Coquillett, 1910b: 511).

abutilon Felt, 1935c: 1.—Tex.
adenostoma Felt, 1916e: 177.—Calif.
amaranthi Felt, 1935c: 1.—Tex.
antennariae (Wheeler), 1889: 212 (*Asynapta*).—Wis.; Maine.

arizonensis Felt, 1908b: 294.—Ariz.; Calif., Colo.
artemisiae Felt, 1908b: 377.—Ariz.
atriplicicola (Cockerell), 1898a: 326 (*Diplosis*).—N. Mex.
atriplicis (Townsend), 1893j: 1021 (*Cecidomyia*).—N. Mex.; Calif.
Gall and pupa only; adult descr. Cockerell, 1895: 766.
auripila Felt, 1908b: 294.—Ariz.
autumnalis Beutenmüller, 1907a: 386.—N.C.
azaleae Felt, 1908b: 295.—N.Y.
baroni Felt, 1908b: 377.—Ariz. (unpublished).
bea Felt *in* Needham, 1925: 19.—Calif.
betheli Cockerell, 1907: 324.—Colo.
bidens Johannsen, 1945: 9.—Fla.
brevicauda Felt, 1908b: 295.—Ariz.
buddleia Felt, 1935c: 2.—Tex.
bumeliae Felt, 1908b: 296.—Tex.
ceanothi Felt, 1908b: 377.—Calif.
chrysothamni Felt, 1916e: 178.—Utah.
clematidis Felt, 1935c: 3.—Tex.
diervillae Felt, 1907b: 165.—N.Y.
diplaci Felt, 1912d: 151.—Calif.
dondiae Felt, 1918c: 381.—Calif.
enceliae Felt, 1912d: 152.—Calif.
eupatorii Felt, 1911d: 546.—N.Y.
florida Felt, 1908b: 376.—Fla. (unpublished).
fulvopedalis Felt, 1907b: 118.—N.Y.; Vt.
garryae Felt, 1912i: 757 (also 1925b: 54).—Calif. Gall only; adult descr. Felt, 1934c: 34.
helianthiflorae Felt, 1908b: 376.—N.Y.; Que.
helianthiglobulus Osten Sacken, 1878c: 5 (orig. as *helianthi globulus* Osten Sacken, 1869a: 301).—Ill.; N.Y., N.J. Pupa only; adult descr. Felt, 1916d: 131.
globulus, emend.
hydrangeae Felt, 1908b: 296.—Va.; Mo.
ilicicola Foote, 1953: 197.—Va.; W. Va., Md., N.J.
ilicoides Felt, 1908b: 296.—N.Y.
integrifoliae Felt, 1908b: 376.—Calif.; Utah.
johnsoni Felt, 1908b: 377.—Pa.
lacinariae Felt, 1935c: 3.—Tex.
mentzeliae Cockerell, 1900b: 302.—N. Mex.
mimosae Felt, 1934a: 77.—Tex.
monacha Osten Sacken, 1869a: 299.—N.Y.; Ont. to Maine, s. to N.J., also N.C.
patens Beutenmüller, 1907a: 386.—N.C.

neomexicana (Cockerell), 1896: 204 (*Cecidomyia*).—N. Mex.
opuntiae Felt, 1908b: 377.—Tex.; Calif., Ariz., Colo., Mexico.
photiniae Pritchard, 1952: 16.—Calif.
prosopidis Cockerell, 1898a: 329.—N. Mex.; Tex.
ratibidae Felt, 1935c: 5.—Tex.
rudbeckiaeconspicua Osten Sacken, 1878c: 5 (orig. as *rudbeckiae conspicua* Osten Sacken, 1870: 51).—Pa.; Ariz., Mo., Ohio, Mass. to N.C. *conspicua*, emend.
salictaria Felt, 1908b: 297.—Ind.
sambuci Felt, 1908b: 377.—Unknown.
shepherdiae Felt, 1916c: 414.—Colo.
smilacinae Felt, 1908b: 298.—D.C.
thalictri Felt, 1911d: 547.—Mass.; N.Y., N.H.
verbenae Felt, 1935c: 6.—Tex.
vernoniae Felt, 1908b: 377.—Va.
websteri Felt, 1917a: 562.—Ariz.; N. Mex.
xanthii Felt, 1936a: 231.—Tex.

Genus CINCTICORNIA Felt

Cincticornia Felt, 1908b: 379. Type-species, *Asphondylia transversa* Felt (orig. des.).

americana Felt, 1908b: 380.—N.Y.
canadensis Felt, 1908b: 380.—Ont.
caryae Felt, 1908b: 380.—Mass.; Maine.
connecta Felt, 1908b: 381.—Conn.
cornifolia (Felt), 1907b: 124 (*Oligotrophus*).—N.Y.
globosa Felt, 1909a: 291.—Mass.; Conn.
multifila (Felt), 1907b: 118 (*Asphondylia*).—N.Y.
pilulae (Beutenmüller), 1892: 269 (*Cecidomyia*; orig. as *quercus pilulae* Walsh, 1864a: 481).—Ill.; N.Y. to Maine, s. to N.J. Gall only; adult descr. Felt, 1916d: 164.
podagrae Felt, 1909a: 291.—Mass.
pustulata Felt, 1909a: 291.—Mass.
quercifolia Felt, 1908b: 380.—D.C.
rhoina (Felt), 1907b: 123 (*Oligotrophus*).—N.Y.
serrata Felt, 1908b: 380.—Mass.
simpla Felt, 1909a: 291.—Mass.; N.Y.
sobrina (Felt), 1907b: 158 (*Asphondylia*).—Mass.
transversa (Felt), 1907b: 118 (*Asphondylia*).—N.Y

Genus CARYOMYIA Felt

Caryomyia Felt, 1909a: 292. Type-species, *Cecidomyia tubicola* Osten Sacken (orig. des.).

antennata Felt, 1909a: 292.—N.Y.
arcuaria (Felt), 1908b: 388 (*Hormomyia*).—Mass.
consobrina Felt, 1909a: 292.—N.Y.; Mass., Conn.
holotricha (Osten Sacken), 1862b: 193 (*Cecidomyia*).—D.C.; Ont., Ind., Ohio, Mass. to N.C. Gall and larva only; adult descr. Felt, 1921c: 101.
inanis Felt, 1909a: 292.—N.Y.; Mass.
persicoides (Osten Sacken), 1862b: 193 (*Cecidomyia*).—N.Y.; Mass., Conn. Gall only; adult descr. Felt, 1921c: 113.
sanguinolenta (Osten Sacken), 1862b: 192 (*Cecidomyia*).—D.C.; N.Y., Mass., Conn., N.J. Gall and larva only; adult descr. Felt, 1921c: 105.
similis Felt, 1909a: 292.—N.Y.
thompsoni (Felt), 1908b: 388 (*Hormomyia*).—Mass.; N.Y., Conn.
tubicola (Osten Sacken), 1862b: 192 (*Cecidomyia*).—D.C.; N.Y., Mass. Gall and larva only; adult descr. Felt, 1921c: 108.

Unplaced Species of Asphondyliini

caryaecola Osten Sacken, 1862b: 192 (*Cecidomyia*).—D.C.; Ont., Ohio, N.Y., Mass., N.J. Gall only. ?*Caryomyia*.
cynipsea Osten Sacken, 1862b: 193 (*Cecidomyia*).—D.C.; Ohio, N.J. Gall and larva only. ?*Caryomyia*.
glutinosa Osten Sacken, 1862b: 193 (*Cecidomyia*).—D.C.; N.Y. Larva only. ?*Caryomyia*.
nucicola Osten Sacken, 1878c: 6 (*Cecidomyia*; orig. as *caryae nucicola* Osten Sacken, 1870: 53).—N.Y. Gall and larva only. ?*Caryomyia*.

Tribe LASIOPTERINI

REFERENCE: Felt, 1918a (rev.).

Genus LASIOPTERA Meigen

Lasioptera Meigen, 1818: 88. Type-species, *Cecidomyia albipennis* Meigen (Coquillett, 1910b: 558).

abhamata Felt, 1907b: 108.—N.Y.
allioides Pritchard, 1961a: 56.—Calif.
allioniae Felt, 1911d: 482.—Colo.
apocyni Felt *in* Weiss and West, 1921: 149.—N.J.
argentisquama Felt, 1908b: 324.—N.Y.

arizonensis Felt, 1908b: 325 (also 1911d: 482).—Ariz.
  arizonae Kieffer, 1913a: 31 (unjustified n. name for *arizonensis* Felt, 1911d).—Ariz.
asterspinosae White, 1950: 67.—Tex.
basiflava Felt, 1908b: 324.—N.Y.
carbonitens Cockerell, 1902: 183.—N. Mex.
caryae Felt, 1907b: 106.—N.Y.
cassiae Felt, 1909a: 287.—Ariz.
caulicola Felt, 1907b: 162.—?N.Y.
centerensis Felt, 1918a: 170.—N.Y.
cinerea Felt, 1907b: 104.—N.Y.
clarkei Felt, 1909a: 287.—Mass.
clavula (Beutenmüller), 1892: 269 (*Cecidomyia*).—N.Y. Gall only; adult descr. Beutenmüller, 1907a: 396.
colorati Felt, 1918a: 129.—S. Dak.
consobrina Felt, 1907b: 104.—N.Y.
convolvuli Felt, 1907b: 149.—?N.Y.; Ill., Mass.
corni Felt, 1907b: 107.—N.Y.; Mass., Conn.
cylindrigallae Felt, 1907b: 150.—N.Y.; Mass.
danthoniae Felt, 1909a: 287.—N.Y.
desmodii Felt, 1907b: 106.—N.Y.
diplaci Felt, 1912d: 151.—Calif.
echinochloa Felt, 1916e: 181.—S. Dak.
ephedrae Cockerell, 1898a: 327.—N. Mex.
ephedricola Cockerell, 1902: 184.—N. Mex.
excavata Felt, 1908b: 287.—N.Y.; Mass.
farinosa (Osten Sacken), 1862b: 204 (*Cecidomyia*).—D.C.; Ont., N.Y. Mass. Gall and larva only; adult descr. Felt, 1918a: 134.
flavipes Felt, 1908b: 325.—N.Y.
fraxinifolia Felt, 1908b: 327.—N.Y.
fructuaria Felt *in* Woods, 1916: 266.—Maine.
galeopsidis Felt, 1909a: 287.—Mass.
hamata Felt, 1907b: 107.—N.Y.
hecate Felt, 1908b: 326.—N.Y.
howardi Felt, 1921a: 94.—Calif.
humulicaulis Felt, 1907b: 151.—N.Y.
impatientifolia Felt, 1907b: 106.—N.Y.; Mass.
inustorum Felt, 1916e: 182.—S. Dak.
juvenalis Felt, 1908b: 327.—N.Y.
kallstroemia Felt, 1935c: 6.—Tex.
lactucae Felt, 1907b: 151.—Mass.; N.Y., N.H.
linderae Beutenmüller, 1907a: 398.—N.Y. Gall and larva only; adult descr. Beutenmüller, 1913a: 415.

lupini Felt, 1908b: 326.—Calif.
lycopi Felt, 1907b: 152.—N.Y.; Mass.
mitchellae Felt, 1908b: 325.—D.C. (unpublished).
murtfeldtiana Felt, 1909a: 288.—Mo.; Kans.
nassauensis Felt, 1908b: 324.—N.Y.
neofusca Felt, 1908b: 327.—N.Y.
nodulosa Beutenmüller, 1907a: 397.—N.Y.; Mass., R.I., N.J.
palustris Felt, 1907b: 162.—N.Y.
panici Felt, 1908b: 326.—N.Y.
psederae Felt, 1934a: 77.—Ohio.
querciflorae Felt, 1908b: 325.—Ariz.
quercina Felt, 1907b: 108.—N.Y.
querciperda Felt, 1908b: 324.—Va.; Mass., Conn., R.I.
quercirami Felt, 1926e: 80.—Utah.
riparia Felt, 1909a: 287.—N.Y.
rudbeckiae Felt, 1908b: 324.—Pa.
serotina Felt, 1908b: 326.—N.Y.
solidaginis Osten Sacken, 1863: 370.—Unknown; N.Y., N.H., Mass.
    tumifica Beutenmüller, 1907a: 394.—N.Y.
    dorsimaculata Felt, 1908b: 325.—N.Y.
spinulae Felt, 1908b: 325.—Ariz.
spiraeafolia Felt, 1909a: 287.—Mass.; Maine.
tertia Cockerell, 1898a: 328.—N. Mex.
tibialis Felt, 1914b: 119.—Calif.
tripsaci Felt, 1910a: 10.—Tex.
ventralis Say, 1824a: 357 (1859a: 242).—Pa.
verbenae Felt, 1912d: 150.—Calif.
vernoniae (Beutenmüller), 1907a: 389 (*Cecidomyia*).—N.C.; Ind., N.Y., D.C. Gall and larva only; adult descr. Beutenmüller, 1913a: 415 (as *Lasioptera*).
    vernoniflorae Felt, 1908b: 324.—Unknown.
viburni Felt, 1907b: 104.—N.Y.
virginica Felt, 1909a: 287.—Mass.
vitis Osten Sacken, 1862b: 201.—D.C.; N.Y., Mass., Conn.
weldi Felt, 1908b: 326.—Ill. (unpublished).
willistoni Cockerell, 1898a: 327.—N. Mex.
ziziae Felt, 1908b: 327.—Ill. (unpublished).
    zigiae, error.

Genus NEOLASIOPTERA Felt

Neolasioptera Felt, 1908b: 330. Type-species, *Lasioptera vitinea* Felt (Coquillett, 1910b: 575).
agrostis Felt, 1908b: 331 (also 1911d: 483).—Unknown.
    agrostidis Kieffer, 1913a: 22 (unjustified n. name for *agrostis* Felt, 1911d).—Unknown.

albipes Felt, 1918a: 190.—N.J.
albitarsis (Felt), 1907b: 153 (*Choristoneura*).—N.Y.
albolineata Felt, 1908b: 332.—N.Y.
ambrosiae Felt, 1909a: 288.—N.Y.
asclepiae Felt, 1908b: 332.—N.Y.
basalis (Felt), 1907b: 109 (*Choristoneura*).—N.Y.
celastri Felt, 1908b: 330.—N.Y.
cinerea (Felt), 1907b: 111 (*Choristoneura*).—N.Y.
clematidis (Felt), 1908b: 287 (*Choristoneura*).—N.Y.; Mass.
coloradensis (Cockerell), 1908c: 421 (*Hormomyia*).—Colo.
  coloradensis Felt, 1918a: 194.—Colo.
cornicola (Beutenmüller), 1907a: 394 (*Lasioptera*).—N.Y.; R.I., D.C.
erigerontis (Felt), 1907b: 163 (*Choristoneura*).—N.Y.; Mass., D.C.
  podagrae Beutenmüller, 1913a: 414 (*Lasioptera*).—N.Y., N.J.
eupatorii (Felt), 1907b: 154 (*Choristoneura*).—N.Y.
flavomaculata Felt, 1908b: 332.—N.Y.
flavoventris Felt, 1908b: 333.—N.Y.
hamamelidis (Felt), 1907b: 111 (*Choristoneura*).—N.Y.
hamata (Felt), 1907b: 155 (*Choristoneura*).—N.Y.
helianthi (Felt), 1908b: 288 (*Choristoneura*).—N.Y.
hibisci (Felt), 1907b: 155 (*Choristoneura*).—N.Y.
hirsuta Felt, 1908b: 331.—Ariz.
liriodendri (Felt), 1907b: 109 (*Choristoneura*).—N.Y.
major Felt, 1918a: 175.—Colo.
menthae Felt, 1909a: 288.—Ill.
mimuli Felt, 1908b: 332.—Calif.
perfoliata (Felt), 1907b: 156 (*Choristoneura*).—Mass.; N.Y., Conn.
ramuscula (Beutenmüller), 1907a: 392 (*Cecidomyia*).—N.C.; N.Y., Maine, Mass., Conn. Gall and larva only; adult descr. Beutenmüller, 1913a: 416 (as *Rhopalomyia*).
  remuscula, error.
sambuci (Felt), 1906: 131 (*Cecidomyia*).—N.Y.; Mass., Conn. Gall and larva only; adult descr. Beutenmüller, 1907a: 396 (as *Lasioptera*).
sexmaculata Felt, 1908b: 331.—N.Y.
solani (Felt), 1907b: 164 (*Choristoneura*).—D.C.
squamosa Felt, 1908b: 333 (also 1911d: 483).—Mo. (unpublished).
  subsquamosa Kieffer, 1913a: 23 (unjustified n. name for *squamosa* Felt, 1911d).—Mo.
tenuitas Felt, 1908b: 331.—N.Y.
tiliaginea Felt, 1908b: 332.—N.Y.
trimera Felt, 1911d: 484.—Ariz.
tripunctata Felt, 1908b: 331.—N.Y.

viburnicola (Beutenmüller), 1907a: 398 (*Lasioptera*).—N.Y.; Conn.
vitinea (Felt), 1907b: 153 (*Lasioptera*).—N.Y.; Mass.

Genus PROTAPLONYX Felt

Protaplonyx Felt, 1916b: 202. Type-species, *hagani* Felt (orig. des.).
hagani Felt, 1916b: 202.—Utah.

Genus ASTEROMYIA Felt

Asteromyia Felt, 1910b: 348. Type-species, *Cecidomyia carbonifera* Osten Sacken (as *Lasioptera carbonifera* Felt; orig. des.).
Baldratia, authors, not Kieffer.

abnormis (Felt), 1907b: 110 (*Choristoneura*).—N.Y.
agrostis (Osten Sacken), 1862b: 204 (*Cecidomyia*).—N.Y.; Ont., Ill., Ohio, N.H. to D.C. Gall and immature stages only; adult descr. Felt, 1918a: 224.
    muhlenbergiae Marten, 1893: 155 (*Lasioptera*).—Ill.
albomaculata (Felt), 1907b: 111 (*Choristoneura*).—N.Y.; N.J.
asterifoliae (Beutenmüller), 1907a: 395 (*Lasioptera*).—N.C.; N.Y., Mass.
canadensis (Felt), 1907b: 105 (*Lasioptera*).—N.Y.
carbonifera (Osten Sacken), 1862b: 195 (*Cecidomyia*).—Unknown; Que. to Maine, s. to D.C. Gall only.
    carbonifera Felt, 1906: 116 (*Lasioptera*).—Unknown. Adult descr.
chrysothamni Felt, 1918a: 214.—Colo.; Utah.
convoluta (Felt), 1907b: 110 (*Choristoneura*).—N.Y.
divaricata (Felt), 1908b: 330 (*Baldratia*).—N.Y.; N.H., Conn.
dumosae (Felt), 1909a: 286 (*Baldratia*).—Mass.
flavoanulata (Felt), 1908b: 329 (*Baldratia*).—N.Y.
flavolunata (Felt), 1907b: 154 (*Choristoneura*).—N.Y.; Mass.
flavomaculata (Felt), 1908b: 329 (*Baldratia*).—N.Y.
flavoscuta (Felt), 1908b: 328 (*Baldratia*).—N.Y.
grindeliae Felt, 1912d: 149.—Calif.
gutierreziae Felt, 1916e: 179.—Utah.
laeviana (Felt), 1907b: 108 (*Choristoneura*).—N.Y.; Mass., Conn., ?Ont.
modesta (Felt), 1907b: 163 (*Choristoneura*).—N.Y.
nigrina Felt, 1911d: 481.—Mass.; Conn.
nitida (Felt), 1909a: 286 (*Baldratia*).—N.Y.
paniculata (Felt), 1907b: 109 (also 1908b: 329, as *Baldratia*) (*Choristoneura*).—N.Y.
petiolicola (Felt), 1908b: 328 (*Baldratia*).—R.I.; Mo., Va.

phragmites Felt, 1936b: 9.—Mich.
pustulata (Felt), 1908b: 328 (*Baldratia*).—N.Y.
reducta Felt, 1911d: 481.—Mass.; Mo., Conn.
rosea (Felt), 1907b: 152 (*Lasioptera*).—N.Y.; Conn.
rubra (Felt), 1907b: 103 (*Lasioptera*).—N.Y.; Ont., Mo., Ill., Ohio, N.H. to D.C.
socialis (Felt), 1908b: 328 (*Baldratia*).—N.Y.
squarrosae (Felt), 1908b: 329 (*Baldratia*).—N.Y.; Ont.
sylvestris Felt, 1915a: 228.—N.Y.
vesiculosa (Felt), 1909a: 286 (*Baldratia*).—Mass.; Conn.
waldorfi (Felt), 1909a: 286 (*Baldratia*).—N.Y.

Genus APLONYX De Stefani Perez

Aplonyx De Stefani Perez, 1908: 174. Type-species, *chenopodii* De Stefani Perez (orig. des.).

sarcobati Felt, 1914c: 93.—Colo.

Genus CLINORHYNCHA Loew

Lasioptera, subg. **Clinorhyncha** Loew, 1850b: 21. Type-species, *chrysanthemi* Loew (mon.).

eupatoriflorae (Felt), 1908b: 287 (*Lasioptera*).—N.Y.; Mass., Conn.
filicis Felt, 1907b: 108.—N.Y.
karnerensis Felt, 1908b: 333.—N.Y.
millefolii Wachtl, 1884: 161.—Europe; N.Y.

Genus CAMPTONEUROMYIA Felt

Camptoneuromyia Felt, 1908b: 334. Type-species, *Dasineura virginica* Felt (Coquillett, 1910b: 518).

adhesa (Felt), 1908b: 291 (*Dasineura*).—N.Y.
brevicauda (Felt), 1908b: 343 (*Dasineura*).—N.Y.
flavescens (Felt), 1908b: 327 (*Lasioptera*).—N.Y.; N.J.
fulva Felt, 1908b: 334.—N.Y.
hamamelidis (Felt), 1907b: 116 (*Dasineura*).—N.Y.
rubifolia Felt, 1908b: 334.—Mass.; Conn.
virginica (Felt), 1907b: 115 (*Dasineura*).—N.Y.

Genus TROTTERIA Kieffer

Choristoneura Rübsaamen, 1892: 342 (preocc. Lederer, 1859). Type-species, *Cecidomyia obtusa* Loew (mon.).
Trotteria Kieffer, 1901: 561 (n. name for *Choristoneura* Rübsaamen). Type-species, *Cecidomyia obtusa* Loew (aut.).

argenti Felt, 1908b: 335.—N.Y.
caryae (Felt), 1907b: 110 (*Choristoneura*).—N.Y.
caudata Felt, 1908b: 335.—N.Y.
karnerensis Felt, 1908b: 335.—N.Y.
metallica Felt, 1908b: 335.—N.Y.
solidaginis Felt, 1908b: 335.—N.Y.
squamosa Felt, 1908b: 335.—N.Y.
subfuscata Felt, 1908b: 335.—N.Y.
tarsata Felt, 1908b: 335.—N.Y.

Tribe CONTARINIINI

REFERENCE: Felt, 1918b (rev.).

Genus DENTIFIBULA Felt

Dentifibula Felt, 1908b: 385, 389. Type-species, *Cecidomyia viburni* Felt (orig. des.).
caryae (Felt), 1907b: 132 (*Contarinia*).—N.Y.
cocci Felt, 1908b: 389.—Ill.
viburni (Felt), 1907b: 132 (*Contarinia*).—N.Y.

Genus TOXOMYIA Felt

Toxomyia Felt, 1911b: 302. Type-species, *fungicola* Felt (orig. des.).
americana Felt, 1914d: 130.—N.Y.

Genus LOBOPTEROMYIA Felt

Lobopteromyia Felt, 1908b: 385, 389. Type-species, *Contarinia filicis* Felt (orig. des.).
abdominalis Felt, 1908b: 390.—N.Y.
apicalis Felt, 1908b: 390.—N.Y.
caricis Felt, 1908b: 390.—N.Y.
consobrina (Felt), 1907b: 161 (*Contarinia*).—N.Y.
filicis (Felt), 1907b: 131 (*Contarinia*).—N.Y.
foetedi Felt, 1908b: 390.—N.Y.
symplocarpi Felt, 1908b: 390.—N.Y.
tiliae (Felt), 1907b: 161 (*Contarinia*).—N.Y.
venae Felt, 1914b: 120 (as *venitalis*, p. 122).—N.Y. Not ?*Cecidomyia venae* Stebbins.

Genus ENDAPHIS Kieffer

Endaphis Kieffer, 1896d: 383. Type-species, *perfidus* Kieffer (mon.).
americana Felt, 1911a: 129.—Ariz.

Genus CONTARINIA Rondani

Cecidomyia, subg. **Contarinia** Rondani, 1860: 289. Type-species, *Tipula loti* De Geer (orig. des.).

agrimoniae Felt, 1908b: 302.—N.Y.
ampelophila Felt, 1907b: 132.—N.Y.
balsamifera (Felt), 1907b: 131 (*Mycodiplosis*).—N.Y.
canadensis Felt, 1908b: 394.—Ont. (unpublished); N.Y., Maine, Mass., Conn.
clematidis Felt, 1908b: 393.—N.Y. (unpublished); Mass.
coloradensis Felt, 1912e: 240.—Colo.
constricta Condrashoff, 1961: 126.—B.C.; ?nw. U.S.
cuniculator Condrashoff, 1961: 128.—B.C.; ?nw. U.S.
divaricata Felt, 1908b: 392.—N.Y.
flavolinea Felt, 1908b: 392.—N.Y.
gossypii Felt, 1908a: 210.—Barbados Is.; Fla., West Indies.
johnsoni Felt, 1909b: 15.—N.Y.
    johnsoni Slingerland and Johnson, 1904: 72. Unavailable name; orig. genus not given.
juniperina Felt, 1939a: 159.—Mo.; Nebr., Kans., Ky.
maculosa Felt, 1908b: 393.—N.Y.
negundifolia Felt, 1908b: 394.—Ont.
obesa Felt, 1918b: 110.—N.Y.
oregonensis Foote, 1956: 54.—Oreg.; B.C., Wash.
perfoliata Felt, 1908b: 391.—N.Y.
pseudotsugae Condrashoff, 1961: 124.—B.C.; Mont.
pyrivora (Riley), 1886a: 287 (*Cecidomyia*).—Conn.; N.Y., N.J., also Europe.
rumicis (Loew), 1850b: 34 (*Cecidomyia*).—Europe; N.Y.
sambucifolia Felt, 1907b: 131.—N.Y.
setigera (Lintner), 1897: 168 (1897: 168) (*Diplosis*).—Mass.; N.Y.
sorghicola (Coquillett), 1899g: 82 (*Diplosis*).—Tex.; Kans. to Va., s. to N. Mex. and Fla., also Neotropical, Hawaii, Sudan, cent. India, ne. Australia.
spiraeina Felt, 1911d: 547.—Mass.; Maine.
trifolii Felt, 1907b: 131.—N.Y.
tritici (Kirby).—Not Nearctic.
truncata Felt, 1908b: 393.—N.Y.
viatica Felt, 1908b: 393.—N.Y.
virginianiae (Felt), 1906: 130 (*Cecidomyia*).—N.Y.; Man., Ont., Que., Maine to Conn. Larva only; adult descr. Felt, 1908b: 392.
viridiflava Felt, 1908b: 392.—N.Y.
washingtonensis Johnson, 1963: 94.—Wash.; B.C., Oreg.

Genus THECODIPLOSIS Kieffer

Thecodiplosis Kieffer, 1895e: cxciv. Type-species, *Cecidomyia brachyntera* Schwägrichen (orig. des.).

cockerelli Felt, 1918c: 381.—Colo.
cupressiananassa (Osten Sacken), 1878c: 3 (*Cecidomyia*; orig. as
   cupressi ananassa Riley, 1870b: 244).—Tenn.; La., Ala.
   ananassa, emend.
dulichii Felt, 1912e: 241.—Mass.
hudsonici (Felt), 1908b: 393 (*Contarinia*).—N.Y.
liriodendri (Osten Sacken), 1862b: 202 (*Cecidomyia*).—D.C.; N.Y.,
   Mass., Conn. Gall and larva only; adult descr. Felt,
   1918b: 126.
piniradiatae Snow and Mills, 1900: 491 (*Diplosis;* as *pini-radiatae*).—
   Calif.
   radiatae, emend.
piniresinosae Kearby and Benjamin, 1963: 414.—Wis.
quercifolia (Felt), 1908b: 391 (*Contarinia*).—Va.
zauschneriae Felt, 1912d: 152.—Calif.

Genus SITODIPLOSIS Kieffer

Sitodiplosis Kieffer, 1913c: 49. Type-species, *Cecidomyia mosellana*
   Géhin (orig. des.).

mosellana (Géhin), 1857: 21 (*Cecidomyia*).—France; wheat-growing
   regions of N. Amer. and Palaearctic Region.
   amyotii Fitch, 1861: 773 (1865: 31) (*Cecidomyia*).—N.Y.

Genus PECTINODIPLOSIS Felt

Pectinodiplosis Felt, 1918b: 132. Type-species, *Contarinia erratica*
   Felt (orig. des.).

erratica (Felt), 1908b: 391 (*Contarinia*).—Unknown.

Genus ZEUXIDIPLOSIS Kieffer

Zeuxidiplosis Kieffer, 1904a: 349. Type-species, *Thecodiplosis giardiana* Kieffer (orig. des.)=*giardi* (Kieffer).

giardi (Kieffer), 1896d: 383 (*Diplosis*).—France; Calif. Introduced

Genus STENODIPLOSIS Reuter

Stenodiplosis Reuter, 1895: 9. Type-species, *geniculati* Reuter
   Kieffer, 1896c: 92).

geniculati Reuter, 1895: 10.—Finland; Ont.

Unplaced Species of Contariniini

brachynteroides Osten Sacken, 1862b: 198 (*Cecidomyia*).—D.C. Gall and larva only. ?*Thecodiplosis*.
pinirigidae Packard, 1878: 527 (*Diplosis;* as *pini-rigidae*).—Maine. ?*Thecodiplosis*. *rigidae*, emend.

Tribe CECIDOMYIINI

REFERENCES: Felt, 1918b, 1921c (revs.).

Genus YOUNGOMYIA Felt

Youngomyia Felt, 1908b: 398. Type-species, *Dicrodiplosis podophyllae* Felt (orig. des.).
pennsylvanica Felt, 1912c: 106.—Pa.
podophylli (Felt), 1907b: 126 (*Dicrodiplosis*).—N.Y.
producta Felt, 1918b: 150.—N.Y.
quercina Felt, 1911d: 551.—Calif.
rubida Felt, 1908b: 399.—N.Y.
umbellicola (Osten Sacken), 1878c: 7 (*Cecidomyia;* orig. as *sambuci umbellicola* Osten Sacken, 1870: 52).—N.J.; Mo., Ill., N.Y. to R.I. Gall and larva only; adult descr. Felt, 1918b: 149.
vernoniae Felt, 1911d: 552.—Va.

Genus DICRODIPLOSIS Kieffer

Dicrodiplosis Kieffer, 1895e: cxciv. Type-species, *fasciata* Kieffer (orig. des.).
Dichrodiplosis, error or emend.
androgynes Felt, 1908b: 394.—N.Y.
annulata (Felt), 1907b: 113 (*Rhabdophaga*).—N.Y.
antennata Felt, 1912e: 243.—Maine.
borealis (Felt), 1907b: 113 (*Rhabdophaga*).—N.Y.
californica Felt, 1912e: 244.—Calif.
fulva Felt, 1918b: 153.—Mass.
gillettei Felt, 1911d: 549.—Colo.
helena Felt, 1912e: 245.—Mass.
insolens Felt, 1920: 288.—N.Y.
longicornis Felt, 1918b: 156.—N.Y.
populi Felt, 1908b: 394.—N.J.; Mass.
rubida Felt, 1918b: 158.—N.Y.
venitalis Felt, 1914b: 121.—N.Y.

Genus APHIDOLETES Kieffer

Aphidoletes Kieffer, 1904b: 385. Type-species, *abietis* Kieffer (Felt, 1911e: 53).
Phaenobremia Kieffer, 1912b: 1 (1912: x). Type-species, *Aphidoletes urticariae* Kieffer (orig. des.).

basalis Felt, 1908b: 397.—N.Y.
borealis Felt, 1908b: 397.—N.Y.
cucumeris (Lintner), 1888b: 725 (*Cecidomyia*).—Mass.; N.Y.
doutti (Pritchard), 1961b: 100 (*Phaenobremia*).—Calif. **N. comb.**
flavida Felt, 1908b: 397.—N.Y.
fulva Felt, 1908b: 397.—N.Y.; Ont.
hamamelidis (Felt), 1907b: 125 (*Bremia*).—N.Y.
marginata Felt, 1908b: 397.—N.Y.
marina Felt, 1908b: 397.—N.Y.
meridionalis Felt, 1908b: 397.—D.C.; Oreg., Calif., Colo., Wis. to Ont. and N.Y., s. to Iowa and Va.
recurvata Felt, 1908b: 397.—Mass.; N.Y.
rosivora (Coquillett), 1900e: 46 (*Diplosis*).—D.C.; N.Y., Mass., N.J.
thompsoni Möhn, 1954: 462.—Europe; N.B., N.S., Nfld. Introduced.

Genus BREMIA Rondani

Diplosis, subg. **Bremia** Rondani, 1860: 289. Type-species, *Cecidomyia decorata* Loew (orig. des.).

borealis Felt, 1914d: 130.—N.Y.
caricis (Felt), 1907b: 128 (also 1908b: 395) (*Mycodiplosis*).—N.Y.
filicis Felt, 1907b: 125.—N.Y.
montana Felt, 1914d: 131.—N.Y.
podophylli Felt, 1907b: 126.—N.Y.
sylvestris Felt, 1920: 289.—N.Y.
tristis Felt, 1914d: 131.—N.Y.

Genus CLEODIPLOSIS Felt

Cleodiplosis Felt, 1922c: 1. Type-species, *aleyrodici* Felt (orig. des.).

REFERENCE: Koebele, 1893: 38 (biol. *koebelei*, as unnamed sp.).

koebelei (Felt), 1932: 167 (*Silvestrina*).—Calif.

Genus THOMASINIANA Strand

Thomasia Rübsaamen, 1910: 288 (also 1911: 168) (preocc. Poche, 1908). Type-species, *Diplosis oculiperda* Rübsaamen (mon.).

Thomasiniana Strand, 1927: 66 (n. name for *Thomasia* Rübsaamen).
Type-species, *Diplosis oculiperda* Rübsaamen (aut.).
californica (Felt), 1914d: 132 (*Thomasia*).—Calif.

Genus KALODIPLOSIS Felt

Kalodiplosis Felt, 1915a: 229. Type-species, *Dicrodiplosis multifila* Felt (orig. des.).
floridana Felt, 1915a: 230.—Fla.

Genus PERIDIPLOSIS Felt

Peridiplosis Felt, 1918b: 160. Type-species, *Cecidomyia quercina* Felt (orig. des.).
quercina (Felt), 1907b: 137 (*Cecidomyia*).—N.Y.; Ga.
quercina Felt, 1908b: 300 (*Dichrodiplosis*).—Ga.

Genus LOBODIPLOSIS Felt

Lobodiplosis Felt, 1908b: 397. Type-species, *Mycodiplosis acerina* Felt (orig. des.).
acerina (Felt), 1907b: 129 (*Mycodiplosis*).—N.Y.; Mass.
borealis Felt, 1920: 289.—N.Y.
cincta Felt, 1918b: 165.—N.Y.
quercina (Felt), 1907b: 130 (*Mycodiplosis*).—N.Y.; Mass.
speciosa Felt, 1913h: 143.—Mass.
triangularis Felt, 1918b: 163.—Mass.

Genus COQUILLETTOMYIA Felt

Coquillettomyia Felt, 1908b: 398. Type-species, *Mycodiplosis lobata* Felt (orig. des.).
bryanti Felt, 1913h: 144.—Mass.
dentata Felt, 1908b: 398.—N.Y.; Mass.
lobata (Felt), 1907b: 127 (*Mycodiplosis*).—N.Y.
texana Felt, 1908b: 398.—Tex.

Genus FELTIELLA Rübsaamen

Feltiella Rübsaamen, 1910: 285 (also 1911: 280). Type-species, *tetranychi* Rübsaamen (orig. des., as gen. n., sp. n.).
acerifolia (Felt), 1907b: 127 (*Mycodiplosis*).—N.Y.
americana Felt, 1916a: 33.—N.Y.
davisi Felt, 1915c: 406.—Ind.

emarginata (Felt), 1907b: 129 (*Mycodiplosis*).—N.Y.
ithacae Felt, 1926d: 141.—N.Y.
minuta (Felt), 1907b: 127 (*Mycodiplosis*).—N.Y.
pini (Felt), 1907b: 128 (*Mycodiplosis*).—N.Y.
spinosa (Felt), 1911d: 550 (*Mycodiplosis*).—Unknown.
venatoria Felt, 1917b: 195.—Ill.

Genus KARSCHOMYIA Felt

Karschomyia Felt, 1908b: 398 (as *Karshomyia;* emend. Felt, 1911e: 54). Type-species, *Mycodiplosis viburni* Felt (orig. des.).

viburni (Felt), 1907b: 130 (*Mycodiplosis*).—N.Y.

Genus CLINODIPLOSIS Kieffer

Clinodiplosis Kieffer, 1895d: cclxxx. Type-species, *Diplosis cilicrus* Kieffer (orig. des.).

araneosa Felt, 1912d: 154.—Md.; D.C.
caulicola (Coquillett), 1895h: 401 (*Diplosis*).—N.H.
examinis Felt, 1913a: 306.—N.Y.
pucciniae Pritchard, 1948b: 29.—Calif.

Genus MYCODIPLOSIS Rübsaamen

Mycodiplosis Rübsaamen, 1895a: 186. Type-species, *Cecidomyia coniophaga* Winnertz (orig. des.).

acarivora (Felt), 1907a: 242 (*Cecidomyia*).—Calif.
aestiva Felt, 1908b: 402.—N.Y.
alternata Felt, 1907b: 126.—N.Y.; Mass., Conn.
angulata (Felt), 1907b: 135 (*Cecidomyia*).—N.Y.
    urticae Felt, 1907b: 136 (*Cecidomyia*).—N.Y.
aurata Felt, 1908b: 402.—N.Y.
captiva Felt, 1908b: 401.—N.Y.
carolina Felt, 1911d: 549.—N.C.
cerasifolia (Felt), 1908b: 302 (*Cecidomyia*).—N.J.; Mass.
cincta Felt, 1918b: 203.—N.Y.
conicola Foote, 1956: 54.—Calif.; Oreg.
contracta Felt, 1908b: 401.—N.Y.
coryli Felt, 1907b: 128.—N.Y.
corylifolia Felt, 1908b: 301.—N.Y.
coryloides Foote, 1956: 55.—Calif.; Oreg.
cucurbitae Felt, 1911d: 550.—Unknown.

cyanococci Felt, 1907b: 128.—N.Y.
fibulata Felt, 1908b: 401.—N.Y.
fungiperda Felt, 1915c: 407.—D.C. (unpublished).
holotricha Felt, 1908b: 401.—N.Y. (unpublished); Mass.
hudsoni Felt, 1907b: 129.—N.Y.
impatientis Felt, 1908b: 401.—N.Y.
intermedia Felt, 1920: 290.—N.Y.
lenis Felt, 1920: 290.—N.Y.
macgregori Felt, 1915b: 149.—S.C.
modesta Felt, 1908b: 402.—N.Y.
obscura Felt, 1908b: 402.—N.Y.
packardi Felt, 1918c: 382.—N.Y.
perplexa Felt, 1908b: 402.—N.Y.
populifolia Felt, 1908b: 400.—N.Y. (unpublished); N.H.
radicis Felt, 1936a: 232.—Ala.
reducta Felt, 1908b: 400.—N.Y.
robusta Felt, 1908b: 401.—N.Y.
rotundata Felt, 1908b: 401.—N.Y.
silvana Felt, 1908b: 402.—N.Y.
tenuitas Felt, 1908b: 401.—N.Y.
tsugae Felt, 1907b: 130.—N.Y.
variabilis Felt, 1908b: 402.—N.Y.

Genus ARTHROCNODAX Rübsaamen

Arthrocnodax Rübsaamen, 1895b: 189. Type-species, *vitis* Rübsaamen (Coquillett, 1910b: 510.).

acerinus (Felt), 1907b: 136 (*Cecidomyia*).—N.Y.
apiphilus Felt, 1908b: 301.—Calif.; Wash., also N.Y.
carolina Felt, 1913e: 488.—S.C.
cinctus (Felt), 1907b: 143 (*Cecidomyia*).—N.Y.
fenestra Felt, 1908b: 404.—N.Y.
filicis (Felt), 1907b: 136 (*Cecidomyia*).—N.Y.
fraxini (Felt), 1907b: 138 (*Cecidomyia*).—N.Y.
incisus (Felt), 1907b: 140 (*Cecidomyia*).—N.Y.
macrofilus (Felt), 1908b: 302 (*Cecidomyia*).—N. Mex.
obscurus Felt, 1908b: 404.—N.Y.
occidentalis Felt, 1912h: 402.—Calif.; B.C.
rhoinus Felt, 1908b: 404.—N.Y.
rufus Felt, 1908b: 403.—N.Y.
sambucifolius Felt, 1908b: 404.—N.Y.; Mass.
sylvestris (Felt), 1907b: 143 (*Cecidomyia*).—N.C.

Genus PRODIPLOSIS Felt

Prodiplosis Felt, 1908b: 403. Type-species, *Cecidomyia floricola* Felt (orig. des.).

fitchii Felt, 1912f: 288.—N.Y.
floricola (Felt), 1908b: 302 (*Cecidomyia*).—N.Y.

Genus GIARDOMYIA Felt

Giardomyia Felt, 1908b: 405. Type-species, *Cecidomyia photophila* Felt (orig. des.).

emarginata Felt, 1908b: 405.—N.Y.
hudsonica Felt, 1908b: 406.—N.Y.
menthae Felt, 1908b: 405.—N.Y.
montana Felt, 1908b: 406.—N.Y.
noveboracensis Felt, 1908b: 405.—N.Y.
photophila (Felt), 1907b: 134 (*Cecidomyia*).—N.Y.
rhododendri Felt, 1939b: 42.—N.Y.

Genus HYPERDIPLOSIS Felt

Hyperdiplosis Felt, 1908b: 405. Type-species, *Cecidomyia lobata* Felt (orig. des.).

bryanti Felt, 1913h: 146.—Mass.; Conn.
fungicola Felt, 1911d: 552.—D.C.
insolens Felt, 1920: 291.—N.Y.
lobata (Felt), 1907b: 136 (*Cecidomyia*).—N.Y.
meibomiifoliae (Beutenmüller), 1907b: 306 (*Cecidomyia*).—N.J.
phlox Greene, 1941a: 547.—Ohio.

Genus LESTODIPLOSIS Kieffer

Lestodiplosis Kieffer, 1894a: 84. Type-species, *Cecidomyia pictipennis* Perris (Kieffer, 1895d: cclxxx)=*septemguttata* Kieffer. Spelling, from Kieffer, 1895d (as 1894: 280), unnecessarily validated by I.C.Z.N., 1958a: 293.
Leptodiplosis Kieffer, 1894c: xxviii, error or emend.

apocyniflorae Felt, 1908b: 409.—N.Y.
asclepiae Felt, 1908b: 409.—N.Y.
asteris (Felt), 1907b: 142 (*Cecidomyia*).—N.Y.
basalis Felt, 1908b: 408.—N.Y.
carolinae (Felt), 1907b: 139 (*Cecidomyia*).—N.C.
cerasi Felt, 1908b: 407.—N.Y.

cincta Felt, 1908b: 408.—N.Y.
clematiflorae Felt, 1908b: 409. N.Y.
crataegifolia Felt, 1908b: 408.—N.Y. (unpublished); Mass.
eupatorii (Felt), 1907b: 140 (*Cecidomyia*).—N.Y.
flavomarginata (Felt), 1907b: 138 (*Cecidomyia*).—N.Y.
florida Felt, 1908b: 409.—Fla. (unpublished).
floridana Johannsen, 1945: 8.—Fla.
fraxinifolia Felt, 1908b: 408.—N.J.
globosa Felt, 1908b: 409.—N.J.
grassator (Fyles), 1883: 237 (*Diplosis*).—Que.; Ont.
hicoriae (Felt), 1907b: 137 (*Cecidomyia*).—N.Y.
juniperina (Felt), 1907b: 141 (*Cecidomyia*).—N.Y.
maculipennis Greene, 1941a: 550.—Ohio.
novangliae Felt, 1933: 114.—Mass.
platanifolia Felt, 1908b: 410.—N.Y.
populifolia Felt, 1908b: 408.—N.Y.
rudbeckiae (Beutenmüller), 1907a: 388 (*Cecidomyia*).—N.C.
rugosa (Felt), 1907b: 141 (*Cecidomyia*).—N.Y.
rumicis Felt, 1908b: 410.—N.Y.
satiata Felt, 1920: 292.—N.Y.
scrophulariae (Felt), 1908b: 303 (*Cecidomyia*).—N.Y.
septemmaculata (Walsh), 1864b: 630 (*Cecidomyia;* as *septem-maculata*).—Ill.
solidaginis Felt, 1908b: 409.—N.Y.
spiraeafolia Felt, 1908b: 410.—N.Y.
taxiconis Foote, 1956: 56.—Oreg.
triangularis (Felt), 1907b: 138 (*Cecidomyia*).—N.Y.
tsugae (Felt), 1907b: 139 (*Cecidomyia*).—N.Y.
verbenifolia Felt, 1908b: 408.—N.Y.
yuccae Felt, 1908b: 408.—Unknown.

Genus EPIDIPLOSIS Felt

Epidiplosis Felt, 1908b: 406. Type-species, *sayi* Felt (orig. des.).

sayi Felt, 1908b: 406.—N.Y.

Genus METADIPLOSIS Felt

Metadiplosis Felt, 1908b: 406. Type-species, *spinosa* Felt (orig. des.).

spinosa Felt, 1908b: 406.—N.Y.

Genus OBOLODIPLOSIS Felt

Obolodiplosis Felt, 1908b: 410. Type-species, *Cecidomyia orbiculata* Felt (orig. des.)=*robiniae* (Haldeman).

robiniae (Haldeman), 1847: 193 (*Cecidomyia*).—Pa.; N.Y., Conn., Maine.
    orbiculata Felt, 1907b: 133 (*Cecidomyia*).—N.Y.

Genus PARALLELODIPLOSIS Rübsaamen

Parallelodiplosis Rübsaamen, 1910: 287 (also 1911: 120). Type-species, *Diplosis galliperda* Löw (mon.).

acernea (Felt), 1907b: 143 (*Cecidomyia*).—N.Y.
carpinicola (Kieffer), 1913a: 214 (*Cecidomyia;* n. name for *carpini* Felt).—N.Y.
    carpini Felt, 1907b: 135 (*Cecidomyia;* preocc. Löw, 1874).—N.Y.
caryae (Felt), 1907b: 141 (*Cecidomyia*).—N.Y.; Mass.
cattleyae (Molliard), 1903: 165 (*Cecidomyia*).—Brazil; Mass.
    cattleyae Felt, 1908b: 412 (*Clinodiplosis*).—Mass. (unpublished).
cinctipes Felt, 1914f: 113.—Mass.
clarkeae Felt, 1911d: 553.—Mass.
corticis Felt, 1915c: 407.—Unknown.
coryli (Felt), 1907b: 142 (*Cecidomyia*).—N.Y.
extensa (Felt), 1908b: 412 (*Clinodiplosis*).—N.Y.
florida (Felt), 1908b: 411 (*Clinodiplosis*).—Fla.
montana (Felt), 1908b: 412 (*Clinodiplosis*).—N.Y.
pratensis (Felt), 1908b: 412 (*Clinodiplosis*).—N.Y.
rubisolita (Felt), 1908b: 412 (*Clinodiplosis*).—N.Y.; Conn.
rubroscuta (Felt), 1907b: 142 (*Cecidomyia*).—N.Y.
spirae (Felt), 1909a: 293 (*Clinodiplosis*).—Mass.; N.Y.
subtruncata (Felt), 1907b: 140 (*Cecidomyia*).—N.Y.
triangularis (Felt), 1908b: 411 (*Clinodiplosis*).—N.Y.

Genus PARADIPLOSIS Felt

Paradiplosis Felt, 1908b: 410. Type-species, *Cecidomyia obesa* Felt (orig. des.).

obesa (Felt), 1907b: 134 (*Cecidomyia*).—N.Y.
partheniicola (Cockerell), 1900a: 201 (*Diplosis*).—N. Mex.

Genus CECIDOMYIA Meigen

Itonida Meigen, 1800: 19. Type-species, *Tipula pini* De Geer (Coquillett, 1910b: 556). Suppressed by I.C.Z.N., (1963b: 339.).

Cecidomyia Meigen, 1803: 261. Type-species, *Tipula pini* De Geer (mon.).

Cecidomyia, subg. *Diplosis* Loew, 1850b: 20. Type-species, *Tipula pini* De Geer (Rondani, 1860: 289).

Retinodiplosis Kieffer, 1912b: 1 (1912: xi). Type-species, *Diplosis resinicola* Osten Sacken (orig. des.).

REFERENCE: Vockeroth, 1960 (tax.).

accola Vockeroth, 1960: 76.—Ont.
banksianae Vockeroth, 1960: 73.—Man.; Sask., Ont.
candidipes Foote (for *albitarsis* Felt, 1918).—N.Y.; Ont., Que., N.B., Conn. N. name.
   albitarsis Felt, 1918c: 383 (*Retinodiplosis;* preocc. Meigen, 1830.—N.Y.
palustris (Felt), 1915c: 408 (*Retinodiplosis*).—Ala.
piniinopis Osten Sacken, 1862b: 196 (as *pini inopis*).—D.C.; Wash. to Conn., s. to Calif. and Va. Larva only; adult descr. Felt, 1912g: 368.
   inopis, emend.
reeksi Vockeroth, 1960: 70.—Man.; Sask., Ont., Mich.
resinicola (Osten Sacken), 1871: 345 (*Diplosis*).—N.Y.; Mass., Conn.
resinicoloides Williams, 1909: 2.—Calif.

Genus DYODIPLOSIS Rübsaamen

Dyodiplosis Rübsaamen, 1910: 287. Type-species, *Hormomyia arenariae* Rübsaamen (mon.).

davisi Felt, 1921c: 208.—N.Y.

Genus HORMOMYIA Loew

Cecidomyia, subg. **Hormomyia** Loew, 1850b: 20, 31. Type-species, *crassipes* Loew (Coquillett, 1910b: 553).

alexanderi Felt, 1921c: 220.—N.Y.
americana Felt, 1907b: 125.—N.Y.
atlantica Felt, 1908b: 387.—N.J.
caudata Felt, 1916e: 176.—Tenn.
cincta Felt, 1921c: 216.—N.H.
fulva Felt, 1926a: 208.—Mass.

maxima Felt, 1921c: 216.—N.Y.
montana Felt, 1921c: 217.—Colo.
needhami Felt, 1907b: 160.—Ill.
palustris Felt, 1908b: 300.—N.Y.
pudica Felt, 1913h: 146.—N.H.

Genus TRISHORMOMYA Kieffer

Trishormomya Kieffer, 1912b: 2 (1912: xi). Type-species, *Hormomyia strobli* Kieffer (orig. des.).
Trishormomyia, error or emend.

bulla (Felt), 1914a: 286 (*Hormomyia*).—Ill.; Ont.
canadensis (Felt), 1908b: 388 (*Hormomyia*).—Mass.; N.Y., Maine.
clarkei (Felt), 1908b: 388 (*Hormomyia*).—Mass.; N.Y.
consobrina (Felt), 1908b: 299 (*Hormomyia*).—N.Y.
crataegifolia (Felt), 1907b: 160 (*Hormomyia*).—N.Y.; Mich., Ont., Mass.
dilatata Felt, 1921c: 226.—N.Y.
fenestra (Felt), 1915a: 231 (*Hormomyia*).—N.Y.
helianthi (Brodie), 1894: 44 (*Diplosis*).—Ont.
incisa Felt, 1921c: 226.—N.Y.
johnsoni (Felt), 1908b: 299 (*Hormomyia*).—Mass.
modesta (Felt), 1913h: 145 (*Hormomyia*).—Mass.
proteana (Felt), 1914f: 113 (*Hormomyia*).—Mass.
salicisverruca (Osten Sacken), 1878c: 7 (*Cecidomyia;* orig. as *salicis verruca* Walsh, 1864b: 606).—Ill.; Mass. Gall and larva only; adult descr. Felt, 1921c: 230.
  verruca, emend.
saturni (Felt), 1914d: 133 (*Hormomyia*).—N.Y.
shawi (Felt), 1913h: 145 (*Hormomyia*).—N.H.

Genus ODONTODIPLOSIS Felt

Odontodiplosis Felt, 1908b: 404. Type-species, *Cecidomyia karnerensis* Felt (orig. des.).

americana Felt, 1908b: 404.—N.Y.
karnerensis (Felt), 1907b: 141 (*Cecidomyia*).—N.Y.
montana Felt, 1908b: 404.—N.Y.

Genus TRISOPSIS Kieffer

Trisopsis Kieffer, 1912d: 171. Type-species, *oleae* Kieffer (orig. des.).
hibisci Felt, 1935b: 76.—La.

Genus ADIPLOSIS Felt

Adiplosis Felt, 1908b: 405. Type-species, *Cecidomyia toxicodendri* Felt (orig. des.).

toxicodendri (Felt), 1907b: 137 (*Cecidomyia*).—N.Y.; Conn.

Genus MONARTHROPALPUS Rübsaamen

Monarthropalpus Rübsaamen, 1892: 381. Type-species, *Tipula flava* Schrank (mon.) ?=*buxi* (Laboulbène).

buxi (Laboulbène), 1873: 321 (*Cecidomyia*).—France; N.Y. to R.I., s. to Md. and Del., Europe. ?=*flava* (Schrank).

Genus ONODIPLOSIS Felt

Onodiplosis Felt, 1916e: 175. Type-species, *sarcobati* Felt (orig.des.).

sarcobati Felt, 1916e: 176.—Utah.

Genus CYSTODIPLOSIS Kieffer and Jörgensen

Cystodiplosis Kieffer and Jörgensen, 1910: 395. Type-species, *longipennis* Kieffer and Jörgensen (mon.).

eugeniae Felt, 1913c: 175.—Fla.

Unplaced Species of Cecidomyiini

agraria Felt, 1908b: 413 (*Cecidomyia*).—N.Y.
albotarsus Felt, 1907b: 132 (*Cecidomyia*).—N.Y.
americana Felt, 1908b: 413 (*Cecidomyia*).—N.Y.
antennata Felt, 1908b: 414 (*Cecidomyia*).—N.Y.
anthici Felt, 1913d: 278 (*Itonida*).—Ala.; Ark., Miss., Ala.
aphidivora Felt, 1912e: 245 (*Itonida*).—N.Y.
apicis Kieffer, 1913a: 213 (*Cecidomyia*; n. name for *apicalis* Felt).—N.Y.
    apicalis Felt, 1908b: 413 (*Cecidomyia*; preocc. Walker, 1856).—N.Y.
apocyni Felt, 1908b: 414 (*Cecidomyia*).—N.Y.
aprilis Felt, 1912e: 247 (*Itonida*).—N.Y.
canadensis Felt, 1911d: 558 (*Itonida*).—Mass.
caryae Felt, 1907b: 124 (*Oligotrophus*).—N.Y.
    abdominalis Felt, 1921c: 207 (*Itonida*; unjustified n. name for *caryae* Felt).—N.Y.
catalpae Comstock, 1881: 266 (*Diplosis*).—D.C.; N.Y., Conn.

cinctella Kieffer, 1913a: 214 (*Cecidomyia;* n. name for *cincta* Felt).—Unknown.
    cincta Felt, 1911d: 558 (*Itonida;* preocc. in *Cecidomyia* by Felt, 1907, orig. *Cecidomyia,* now *Arthrocnodax*).—Unknown.
citrulli Felt, 1935a: 79 (*Itonida*).—Ariz.
claytoniae Felt, 1907b: 133 (*Cecidomyia*).—N.Y.
cucurbitae Felt, 1911d: 555 (*Itonida*).—Unknown.
cupressi Schweinitz, 1822: 92 (*Merulius,* a fungus).—Ga.; Mo., Ark., Fla.
    taxodii Felt, 1911d: 556 (*Itonida*).—Fla.
    taxodii Felt, 1916c: 415 (*Retinodiplosis*).—Mo.
emarginata Felt, 1907b: 134 (*Cecidomyia*).—N.Y.
excavata Felt, 1907b: 139 (*Cecidomyia*).—N.Y.
    excavationis Felt, 1908b: 415, emend.
explicata Felt, 1908b: 413 (*Cecidomyia*).—N.Y.
flavoscuta Felt, 1907b: 137 (*Cecidomyia*).—N.Y.
foliora Russell and Hooker, 1908: 350 (*Cecidomyia*).—Mass.; Conn.
fragariae Felt, 1907b: 133 (*Cecidomyia*).—N.Y.
hartmaniae Felt, 1921c: 201 (*Itonida*).—N.Y.
hopkinsi Felt, 1911d: 554 (*Cecidomyia*).—Calif.
hudsoni Felt, 1907b: 135 (*Cecidomyia*).—N.Y.
infirma Felt, 1908b: 413 (*Cecidomyia*).—N.Y.
myricae Beutenmüller, 1907b: 306 (*Cecidomyia*).—N.J.
nixoni Felt, 1908b: 414 (*Cecidomyia*).—N.Y.
opuntiae Felt, 1910a: 10 (*Cecidomyia*).—N.Y.
paucifili Felt, 1908b: 413 (*Cecidomyia*).—N.Y.
pinifoliae Felt, 1936b: 7 (*Itonida*).—Conn.; N.Y.
piperitae Felt, 1908b: 303 (*Cecidomyia*).—N.Y.
pugionis Felt, 1911d: 557 (*Itonida*).—N.Y.
putrida Felt, 1912e: 246 (*Itonida*).—N.Y. (unpublished).
quercina Felt, 1907b: 137 (*Cecidomyia*).—N.Y.
ramuli Felt, 1907b: 164 (*Cecidomyia*).—N.Y.
recurvata Felt, 1907b: 134 (*Cecidomyia*).—N.Y.
reflexa Felt, 1913h: 146 (*Itonida*).—N.H.
reginae Felt, 1921c: 196 (*Itonida*).—N.W.T.
ruricola Felt, 1908b: 413 (*Cecidomyia*).—N.Y.
sanguinia Felt, 1908b: 413 (*Cecidomyia*).—N.Y.
setariae Felt, 1908b: 303 (*Cecidomyia*).—N.Y.
spiraeaflorae Felt, 1908b: 304 (*Cecidomyia*).—N.Y.
spiraeina Felt, 1911d: 555 (*Itonida*).—Mass.
tecomae Felt, 1906: 127 (*Bremia*).—N.Y.

terrestris Felt, 1908b: 413 (*Cecidomyia*).—N.Y.
texana Felt, 1921c: 204 (*Itonida*).—Tex.
tolhurstae Felt, 1908b: 414 (*Cecidomyia*).—N.Y.
tritici Felt, 1912f: 289 (*Itonida*).—Unknown.
   triticicola Kieffer, 1913a: 220 (*Cecidomyia;* unjustified n. name for *tritici* Felt).—Unknown.
uliginosa Felt, 1914d: 133 (*Hormomyia*).—N.Y.
verbenae Beutenmüller, 1907b: 306 (*Cecidomyia*).—N.J., N.Y.
   urtifolia Felt, 1908b: 414 (*Cecidomyia*).—Unknown.

Unplaced Genera of Cecidomyiidae

Cecidomyiaceltis Patton, 1897: 247 (as *Cecidomyia-celtis*). Type-species, *deserta* Patton (orig. des., as gen. n., sp. n.). All included species described from galls only, and unrecognized.
Lasiopteryx Stephens, 1829a: 53 (also 1829b: 239). Type-species, *Cecidomyia obfuscata* Meigen (Westwood, 1840: 126). Unrecognized.

Unplaced Species of Cecidomyiidae

Adults Known

annulipes Walsh, 1864b: 629 (*Cecidomyia*).—Ill.
atricornis Walsh, 1864b: 628 (*Cecidomyia*).—Ill.
atrocularis Walsh, 1864b: 626 (*Cecidomyia*).—Ill.
bigeloviae Cockerell, 1889c: 106 (*Cecidomyia*).—Colo.
caliptera Fitch, 1845b: 262 (*Cecidomyia*).—N.Y.
caryae Felt, 1907b: 143 (*Dirhiza*).—N.Y.
caryae Osten Sacken, 1862b: 191 (*Diplosis*).—D.C. *D. caryae* Osten Sacken of Felt, 1921c: 97, is an unnamed sp. of *Caryomyia*.
cerasiphila Felt, 1911d: 554 (*Cecidomyia*).—Unknown.
coloradella Cockerell, 1904: 155 (*Diplosis*).—Colo.
cossae Shimer, 1869: 395 (*Cecidomyia*).—Ill.
culmicola Morris *in* Gambel, 1849: 194 (*Cecidomyia*).—Pa.
decemmaculata Walsh, 1864b: 631 (*Cecidomyia;* as *decem-maculata*).—Ill.
erigeroni Brodie, 1894: 13 (*Diplosis;* as *eregeroni*).—Ont.
frater Cockerell, 1890a: 280 (*Cecidomyia*).—Colo.
fuscoanulata Felt, 1908b: 329 (*Baldratia*).—Unknown. ?=*Asteromyia asterifoliae* (Beutenmüller).

graminis Fitch, 1861: 832 (1865: 90) (*Cecidomyia;* n. name for *cerealis* Fitch).—Mass.
cerealis Fitch, 1845b: 263 (*Cecidomyia;* preocc. in *Cecidomyia* by Sauter, 1817, orig. *Tipula,* now *Porricondyla*).— Mass.
grossulariae Fitch, 1855: 880 (1856: 176) (*Cecidomyia*).—N.Y.
helena Felt, 1908b: 288 (*Choristoneura*).—N.Y. ?=*Asteromyia asterifoliae* (Beutenmüller).
inimica Fitch, 1861: 830 (1865: 88) (*Cecidomyia*).—N.Y.
lateralis Felt, 1914f: 109 (*Monardia*).—Mass.
maccus Loew *in* Osten Sacken, 1862b: 187 (*Diplosis*).—D.C.
meibomiae Beutenmüller, 1907a: 390 (*Cecidomyia*).—N.C., N.Y. Gall and larva only; adult descr. Beutenmüller, 1913a: 415.
monardi Brodie, 1894: 109 (*Diplosis*).—Ont.
negundinis Gillette, 1890: 392 (*Cecidomyia*).—Iowa.
nyssaecola Beutenmüller, 1907a: 387 (*Cecidomyia*).—N.Y.; Ill., Pa., N.J., Ky., Va.
orbitalis Walsh, 1864b: 623 (*Cecidomyia*).—Ill.
ornata Say, 1824a: 357 (1859a: 242) (*Cecidomyia*).—Pa.
parva Walker, 1848: 29 (*Lasioptera*).—Ont.
salicisbrassicoides Packard, 1869a: 377 (*Cecidomyia;* orig. as *salicis brassicoides* Walsh, 1864b: 577).—Ill.
brassicoides, emend.
salicisgnaphaloides Osten Sacken, 1878c: 4 (*Cecidomyia;* orig. as *salicis gnaphaloides* Walsh, 1864b: 583).—Ill.
gnaphaloides, emend.
salicisstrobiliscus Osten Sacken, 1878c: 4 (*Cecidomyia;* orig. as *salicis strobiliscus* Walsh, 1864b: 582).—Ill. Gall and cocoon only; adult descr. Walsh, 1866: 223.
strobiliscus, emend.
scutellata Say, 1823: 17 (1859b: 44) (*Campylomyza*).—Mo.; ?Ohio.
solidaginis Beutenmüller, 1907b: 305 (*Asphondylia*).—N.Y., N.J., N.C. ?=*Asphondylia monacha* Osten Sacken.
spongivora Walker, 1848: 30 (*Cecidomyia*).—Ont.
tergata Fitch, 1845b: 264 (*Cecidomyia*).—N.Y.
togata, error.
thoracica Fitch, 1845b: 264 (*Cecidomyia*).—N.Y.
thurstoni Brodie, 1894: 73 (*Diplosis*).—Ont.
tuberculata Felt, 1907b: 103 (*Lasioptera*).—N.Y. ?=*Asteromyia rubra* (Felt).
unguicula Beutenmüller, 1907a: 388 (*Cecidomyia*).—Ohio.

Adults Unknown

angelicae Beutenmüller, 1908: 74 (*Cecidomyia*).—N.J. Gall and larva only.
bifolia Stebbins, 1910: 49 (*Cecidomyia*).—Mass. Gall only.
bigeloviaebrassicoides Townsend, 1893i: 491 (*Cecidomyia;* as *bigeloviae-brassicoides*).—N. Mex. Gall only.
brassicoides, emend.
bigeloviaestrobiloides Townsend, 1894b: 176 (*Cecidomyia;* as *bigeloviae-strobiloides*).—N. Mex. Gall only.
strobiloides, emend.
boehmeriae Beutenmüller, 1908: 74 (*Cecidomyia*).—N.Y. Gall and larva only.
capsularis Patton, 1897: 248 (*Cecidomyiaceltis*).—Unknown. Gall only.
capsularis Patterson, error.
castaneae Stebbins, 1910: 17 (*Cecidomyia*).—Mass. Gall only.
celastri Stebbins, 1910: 41 (*Cecidomyia*).—Mass. Gall and larva only.
cerasiserotinae Osten Sacken, 1871: 346 (*Cecidomyia;* as *cerasi serotinae*).—N.Y. Gall and larva only.
serotinae, emend.
chinquapin Beutenmüller, 1907a: 389 (*Cecidomyia*).—N.C. Gall and larva only.
citrina Osten Sacken, 1878c: 6 (*Cecidomyia;* orig. as *tiliae citrina* Osten Sacken, 1870: 53).—N.Y. Gall and larva only.
citricola (error) Thompson, 1915: 58.
collinsoniae Beutenmüller, 1908: 73 (*Cecidomyia*).—N.Y. Gall and larva only.
collinsonifolia Beutenmüller, 1908: 74 (*Cecidomyia*).—N.Y. Gall and larva only.
crataegibedeguar Osten Sacken, 1878c: 6 (*Cecidomyia;* orig. as *crataegi bedeguar* Walsh, 1869: 79).—Ill. Gall and larva only.
bedeguar, emend.
crotalariae Stebbins, 1910: 40 (*Cecidomyia*).—Mass. Gall and larva only.
deserta Patton, 1897: 247 (*Cecidomyiaceltis*).—Conn. Gall only.
deserta Patterson, error.
erubescens Osten Sacken, 1862b: 200 (*Cecidomyia*).—D.C. Gall only.
eupatoriflorae Beutenmüller, 1907a: 391 (*Cecidomyia*).—N.C., N.Y. Gall and larva only.
euthamiae Stebbins, 1910: 53 (*Cecidomyia*).—Mass. Gall only.
fulva Beutenmüller, 1908: 75 (*Cecidomyia*).—N.Y. Gall and larva only.

gemmaria Stebbins, 1910: 53 (*Cecidomyia*).—Mass. Gall only.
hageni Aldrich, 1905: 162 (*Cecidomyia*).—Mass. Gall only.
helianthibulla Walsh, 1866: 228 (*Cecidomyia;* as *helianthi-bulla*).— Ill. Gall only.
  bulla, emend.
inaequalis Stebbins, 1910: 48 (*Cecidomyia*).—Mass. Gall only.
irregularis Stebbins, 1910: 9 (*Cecidomyia*).—Mass. Gall only.
lappa Stebbins, 1910: 35 (*Cecidomyia*).—Mass. Gall only.
majalis Osten Sacken, 1878c: 6 (*Cecidomyia;* orig. as *quercus majulis* Osten Sacken, 1870: 53).—N.Y.; Ont., Maine, s. to D.C. Gall and larva only.
muscosa Stebbins, 1910: 35 (*Cecidomyia*).—Mass. Gall only.
niveipila Osten Sacken, 1862b: 199 (*Cecidomyia*).—D.C. Gall and larva only.
ocellaris Osten Sacken, 1862b: 199 (*Cecidomyia*).—D.C. Gall and larva only.
  ocellata (error) Jarvis, 1907: 66.
oviformis Patton, 1897: 248 (*Cecidomyiaceltis*).—Unknown. Gall only.
  oviformis Patterson, error.
palmeri Felt, 1925b:59 (*Cecidomyia*).—N. Mex. Gall only. ?*Lasioptera*.
pellex Osten Sacken, 1862b: 199 (*Cecidomyia*).—D.C. Gall and larva only.
peritomatis Cockerell, 1913a: 280 (*Cecidomyia*).—N. Mex. Gall and larva only.
poculum Osten Sacken, 1862b: 201 (*Cecidomyia*).—D.C. Gall and larva only.
potentillaecaulis Stebbins, 1910: 37 (*Cecidomyia*).—Mass. Gall only.
pubescens Patton, 1897: 248 (*Cecidomyiaceltis*).—Unknown. Gall only.
  pubescens Patterson, error.
pudibunda Osten Sacken, 1862b: 202 (*Cecidomyia*).—D.C. Gall and larva only.
punicei Brodie, 1909: 159 (*Diplosis*).—Ont. Gall and larva only.
pustuloides Beutenmüller, 1907a: 390 (*Cecidomyia*).—N.C., N.J. Gall and larva only. ?*Cincticornia*.
q-oruca Walsh *in* Felt, 1925b: 61 (*Cecidomyia*).—Unknown. Gall and larva only.
  oruca, emend.
racemi Stebbins, 1910: 39 (*Cecidomyia*).—Mass. Gall only.

recondita Osten Sacken, 1875e: 202 (*Asphondylia*).—N.Y. Gall and exuvium only.
reniformis Stebbins, 1910: 36 (*Cecidomyia*).—Mass. Gall only.
salicifoliae Osten Sacken, 1866: 220 (*Cecidomyia*).—Que. Gall and larva only.
saliciscoryloides Osten Sacken, 1878c: 7 (*Cecidomyia*; orig. as *saliciscoryloides* Walsh, 1864b: 588).—Ill. Gall and larva only.
    coryloides, emend.
semenrumicis Patton, 1897: 248 (*Cecidomyiaceltis*).—Unknown. Gall only.
    semenrumicis Patterson, error.
spiniformis Patton, 1897: 248 (*Cecidomyiaceltis*).—Unknown. Gall only.
    spiniformis Patterson, error.
squamulicola Stebbins, 1910: 16 (*Cecidomyia*).—Mass. Gall only.
strobiligemma Stebbins, 1910: 53 (*Cecidomyia*).—Mass. Gall only.
symmetrica Osten Sacken, 1862b: 200 (*Cecidomyia*).—D.C. Gall and larva only. ?*Cincticornia*.
torreyi Felt, 1925b: 63 (*Cecidomyia*).—N. Mex. Gall only.
triadenii Beutenmüller, 1908: 74 (*Cecidomyia*).—N.J. Gall and larva only.
tuba Stebbins, 1910: 46 (*Cecidomyia*).—Mass. Gall and larva only.
tulipiferae Osten Sacken, 1862b: 202 (*Cecidomyia*).—D.C. Gall and larva only.
urnicola Osten Sacken, 1875e: 202 (*Cecidomyia*).—N.Y. Gall and larva only.
vaccinii Osten Sacken, 1862b: 196 (*Cecidomyia*).—D.C.; N.Y., Mass., Maine. Gall and larva only.
    gaylussacii Felt, 1925b: 55 (*Cecidomyia*; unjustified n. name for *vaccinii* Osten Sacken).—D.C.
venae Stebbins, 1910: 39 (?*Cecidomyia*).—Mass. Gall only.
verbesinae Beutenmuller, 1907a: 391 (*Cecidomyia*).—N.C. Gall and larva only.
verrucicola Osten Sacken, 1875e: 201 (*Cecidomyia*).—Mass., N.Y. Gall only.
viticola Osten Sacken, 1862b: 202 (*Cecidomyia*).—D.C. Gall and larva only.
    vitislituus Osten Sacken, 1878c: 7 (*Cecidomyia*; orig. as *vitis lituus* Walsh and Riley, 1870: 28).—Ill. Gall only.
    lituus, emend.

Suborder BRACHYCERA

Superfamily TABANOIDEA

Family XYLOPHAGIDAE

(Coenomyiidae, Erinnidae)

By Maurice T. James

Adults of Xylophagidae frequent forested or wooded areas and feed on sap, nectar of flowers, or other liquid matter. Larvae are predators or scavengers, so far as known; they occur in soil rich in decaying vegetable matter (*Coenomyia*), under bark of trees (*Xylophagus*), or in decaying logs (*Rachicerus*).

REFERENCE: Leonard, 1930 (keys, in Rhagionidae).

Subfamily XYLOPHAGINAE

Genus COENOMYIA Latreille

Coenomyia Latreille, 1796: 159. Type-species, *Musca ferruginea* Scopoli (sub. mon., Latreille, 1802: 439).

ferruginea (Scopoli), 1763: 340 (*Musca*).—Europe; Nebr. to Que., s. to Fla.
  pallida Say, 1824a: 369 (1859a: 251).—Unknown.
  cinereibarbis Bigot, 1879a: 194.—Md.

Genus RACHICERUS Walker

Rachicerus Walker, 1854: 103. Type-species, *fulvicollis* Walker (mon.).
Rhachicerus, error or emend.

fulvicollis Walker, 1854: 104.—Ga. (Walker, 1848: 124, nomen nudum); Mass. to Ga.
  ruficollis, error.
honestus Osten Sacken, 1877: 211.—Calif.
niger Leonard, 1930: 13.—Calif.
nitidus Johnson, 1903b: 22.—Pa.; Maine to Va.
obscuripennis Loew, 1863: 4 (Cent. 3, no. 6).—Ill.; Mich., s. to Kans. and Va.

Genus XYLOPHAGUS Meigen

Erinna Meigen, 1800: 21. Type-species, *Nemotelus cinctus* De Geer (Coquillett, 1910b: 539). Suppressed by I.C.Z.N., 1963b: 339.
Xylophagus Meigen, 1803: 266. Type-species, *Nemotelus cinctus* De Geer (mon.).

REFERENCES: Malloch, Greene, and McAtee, 1931 (keys); Curran, 1933b (keys).

Subgenus ANAXYLOPHAGUS Malloch

Xylophagus, subg. **Anaxylophagus** Malloch *in* Malloch, Greene, and McAtee, 1931: 216. Type-species, *nitidus* Adams (orig. des.).

nitidus Adams, 1904b: 435.—N.H.

Subgenus ARCHIMYIA Enderlein

Archimyia Enderlein, 1920: 281 (as genus). Type-species, *Xylophagus ater* Meigen (sub. mon., Enderlein, 1921c: 175).

decorus Williston, 1885d: 121.—Wash.; B.C. to N.H., s. to Oreg. and Va.
longicornis Loew, 1869c: 163 (Cent. 9, no. 62).—Mass.; Mich., Ohio, N.C.
politus Malloch *in* Malloch, Greene, and McAtee, 1931: 217.—Colo.; B.C.
reflectens Walker, 1848: 129.—N.Y.; Mont. to Que. and Va.
   persequus Walker, 1850: 1.—N. Amer.
rufipes Loew, 1869c: 163 (Cent. 9, no. 63).—Mass. ?=*reflectens*.

Subgenus XYLOPHAGUS Meigen

abdominalis Loew, 1869c: 163 (Cent. 9, no. 64).— Tex.; Oreg. to Que., s. to Tex. and Ga.
   fasciatus Walker, 1848: 128 (preocc. Say, 1829).—Ont.
gracilis Williston, 1885d: 121.—Wash., Oreg.; B.C. to Calif. and Colo.
lugens Loew, 1863a: 6 (Cent. 3, no. 8).—Ill.; Minn. to N. H. and N.C.
   laceyi Curran, 1933b: 2.—N.Y. **N. syn.**
triangularis Say, 1823: 30 (1859b: 52).—Mo.

Subfamily ARTHROCERATINAE

Genus ARTHROCERAS Williston

Arthroceras Williston, 1886b: 107. Type-species, *pollinosum* Williston (Coquillett, 1910b: 510).

leptis (Osten Sacken), 1878c: 223 (*Arthropeas*).—N.H.; Mich., Ont., N.Y.
pollinosum Williston, 1886b: 108.—Wash.; Oreg., Idaho, Colo.
    pruinosus Bigot, 1887d: 115 (*Leptis*).—Oreg.

Genus ARTHROPEAS Loew

Arthropeas Loew, 1850a: 304. Type-species, *siberica* Loew (mon.).

americana Loew, 1861b: 316 (Cent. 1, no. 16).—Wis.; Wis. to Mass. and Va.
jonesi Cresson, 1919: 176.—Calif.
magna Johnson, 1913b: 11.—Man.; Alta., Mont.

Genus GLUTOPS Burgess

Glutops Burgess, 1878: 321. Type-species, *singularis* Burgess (mon.).
    REFERENCE: Wirth, 1954b (key).

punctatus Wirth, 1954b: 138.—Calif.; B.C., Wash., Oreg.
rossi Pechuman, 1945: 134.—B.C.; Calif., Mont.
singularis Burgess, 1878: 322.—Mass.; N.Y., N.H., Conn., Pa.

Genus BOLBOMYIA Loew

Bolbomyia Loew, 1850c: 39. Type-species, *nana* Loew (sub. mon., Loew, 1862b: 188; Cent. 2, no. 5).
Misgomyia Coquillett, 1908: 145. Type-species, *obscura* Coquillett (orig. des.)=*nana* Loew.
    REFERENCE: Chillcott, 1961c (rev.).

macgillisi Chillcott, 1961c: 634.—B.C.; Alaska to Wash. and Idaho.
nana Loew, 1862b: 188 (Cent. 2, no. 5).—D.C.; Mich. to N.S., s. to Va.
    obscura Coquillett, 1908: 146 (*Misgomyia*).—Va.
    mitis Curran, 1931f: 249 (*Ptiolina*).—Que.

Family XYLOMYIDAE

By Maurice T. James

Xylomyid larvae are found under loose bark of trees; they are predaceous or saprophagous.

    REFERENCE: Leonard, 1930 (rev., in Rhagionidae).

Genus SOLVA Walker

Solva Walker, 1859: 98. Type-species, *inamoena* Walker (mon.).
Solva, subg. *Phloophila* Hull, 1944d: 263. Type-species, *Subula pallipes* Loew (orig. des.).
REFERENCE: Steyskal, 1947a (rev.).

crepuscula Hull, 1944d: 263.—Miss.; Conn. to Va., Miss.
pallipes (Loew), 1863a: 6 (Cent. 3, no. 9) (*Subula*).—Ill., Wis.; Mont. to N.B., s. to Calif. and Miss.
   pygmaea Hull, 1944d: 264.—Miss. **N. syn.**

Genus XYLOMYA Rondani

Subula Meigen, 1820: 15 (preocc. Schumacher, 1817). Type-species, *Xylophagus varius* Meigen (Rondani, 1856: 172). A manuscript name cited in specific synonymy.
Xylomya Rondani, 1861a: 11 (n. name for *Subula* Meigen). Type-species, *Xylophagus varius* Meigen (aut.).
Subulaomyia Williston, 1896a: 43 (as *Subula Omyia;* n. name for *Subula* Meigen, but deleted in Corrigenda, p. iv). Type-species, *Xylophagus varius* Meigen (aut.).
REFERENCE: Steyskal, 1947a (rev.).

americana (Wiedemann), 1821b: 1 (1821c: 51) (*Xylophagus*).—N. Amer.; Wyo. to Que., s. to Tex. and Miss., Mexico.
aterrima Johnson, 1903b: 24.—Ill., N.H.; N.B. to Va. and Ill.
fasciata (Say), 1829: 155 (1859b: 353) (*Xylophagus*).—Ind.
pallidifemur Malloch, 1917g: 343.—Ill.; Ill. to Okla., e. to Mass. and Ga.
parens (Williston), 1885d: 122 (*Subula*).—Wash.; B.C. to Calif., Utah.
   farcus (error) Washburn, 1905: 85.
simillima Steyskal, 1947a: 183.—Mich.; Ill. to Pa., s. to Md. and Tenn.
tenthredinoides (Wulp), 1867: 132 (*Subula*).—Wis.; Minn. to Pa.

Family STRATIOMYIDAE

By Maurice T. James

The adults of Stratiomyidae, or "soldier flies," occur chiefly in wooded or forested areas, or in sedge or grass meadows in the neighborhood of water. Many species are attracted to flowers, particularly umbellifers. Larvae are terrestrial or (Stratiomyinae and, in part, Clitellariinae) aquatic.

REFERENCES: Malloch, 1917g (larvae); Johannsen, 1923b (larvae); Curran, 1927l (rev. Canadian spp.); Johannsen, 1935 (larvae).

Subfamily CHIROMYZINAE

Genus ALTERMETOPONIA Miller

Metoponia Macquart, 1847: 44 (1847: 28) (preocc. Duponchel, 1845).
Type-species, *rubriceps* Macquart (mon.).
Altermetoponia Miller, 1945: 72 (n. name for *Metoponia* Macquart).
Type-species, *Metoponia rubriceps* Macquart (aut.).

rubriceps (Macquart), 1847: 44 (1847: 28) (*Metoponia*).—Australia; Calif. A sod pest, introduced by commerce.

Subfamily BERIDINAE

REFERENCE: James, 1939c (rev.).

Genus ALLOGNOSTA Osten Sacken

Allognosta Osten Sacken, 1883a: 297. Type-species, *Beris fuscitarsis* Say (Coquillett, 1910b: 505).

brevicornis Johnson, 1923b: 71.—Vt.; Maine to N.Y.
fuscitarsis (Say), 1823: 29 (1859b: 52) (*Beris*).—Pa.; Nebr. to Que., s. to Okla. and Ga.
  dorsalis Say, 1824a: 377 (1859a: 257) (*Sargus*).—Ky.
  pallipes Wiedemann, 1830: 41 (*Sargus*).—Pa.
  brevis Walker, 1848: 127 (*Beris*).—N.Y.
  lata Walker, 1848: 127 (*Beris*).—N.Y.
obscuriventris (Loew), 1863b: 299 (Cent. 4, no. 45) (*Metoponia*).— D.C.; Minn. to Que., s. to Okla. and N.C.
similis (Loew), 1863b: 299 (Cent. 4, no. 44) (*Metoponia*).—N.Y.

Genus ACTINA Meigen

Actina Meigen, 1804: 116. Type-species, *Beris nitens* Latreille (Rondani, 1863: 87; 1864: 87).
Hemiberis Enderlein, 1921c: 209. Type-species, *Beris quadridentata* Walker (orig. des.) = *viridis* (Say).
Allactina Curran, 1924b: 24. Type-species, *Beris viridis* Say (orig. des.).

viridis (Say), 1824a: 368 (1859a: 251) (*Beris*).—Pa.; Alta. to Nfld., s· to Kans. and Ga., also Ariz.
  quadridentata Walker, 1848: 127 (*Beris*).—Ont.
  obscuripes Johnson, 1926c: 90 (as var.).—Nfld. ?Var.

Genus BERIS Latreille

Beris Latreille, 1802: 447. Type-species, *Stratiomys sexdentata* Fabricius (mon.)=*chalybeata* (Forster).

annulifera annulifera (Bigot), 1887a: 21 (*Oplacantha*).—Ga.; Alaska to Nfld., s. to Calif. and Ga.
    ssp. **brunnipes** Johnson, 1926d: 109 (as sp.).—Labr. ?Full sp.
    ssp. **luteipes** Johnson, 1926d: 109 (as sp.).—Wash. ?Full sp.
californica James, 1939c: 546.—Calif.; Wash. to Calif.
canadensis (Cresson), 1919: 174 (*Actina*).—Man.; Alta.

Genus EXODONTHA Rondani

Exodontha Rondani, 1856: 169. Type-species, *pedemontana* Rondani (orig. des.)=*dubia* (Zetterstedt).
Hexodonta, emend.
Scoliopelta Williston, 1885c: 152, 154. Type-species, *luteipes* Williston (mon.).

grandis (James), 1938c: 156 (*Scoliopelta*).—Oreg.
luteipes (Williston), 1885c: 154 (*Scoliopelta*).—Vt.; Idaho, N.H., Mass., Tenn.

Subfamily SARGINAE

Larvae are terrestrial scavengers in decaying fruits, bulbs, or other vegetable matter, or in excrement of cattle, horses, and other vertebrates.

REFERENCE: James, 1936g (rev.).

Genus SARGUS Fabricius

Sargus Fabricius, 1798: 549, 566. Type-species, *Musca cupraria* Linnaeus (Latreille, 1810: 442). Alleged preocc. by *Sargus* Klein, a pre-Linnaean name.
Geosargus Bezzi, 1907: 53 (unjustified n. name for *Sargus* Fabricius). Type-species, *Musca cupraria* Linnaeus (aut.).

Subgenus SARGUS Fabricius

bipunctatus (Scopoli), 1763: 341 (*Musca*).—Italy; Wash., Oreg., Europe.
    perpulcher James, 1936g: 271 (*Geosargus*).—Wash.
cuprarius (Linnaeus), 1758: 598 (*Musca*).—Europe; B.C. to Que., s. to Calif., N. Mex., and D.C., n. Asia.
    cuprinus, error.
    nubeculosus, authors, not Zetterstedt.

decorus decorus Say, 1824a: 376 (1859a: 257).—Pa., Fla.; Alaska to Que., s. to Calif. and Fla.
    xanthopus Wiedemann, 1830: 40.—Pa.
    debilis Walker, 1851b: 83.—U.S.
    marginatus Wulp, 1867: 134.—Wis.
    picticornis Bigot, 1887a: 27.—Wash.
    puntifer Bigot, 1887a: 27.—Colo.
    punctifer, emend.
    pallipes Bigot, 1887a: 28 (preocc. Bigot, 1879).—Oreg.
    bigoti Brunetti, 1923: 157 (n. name for *pallipes* Bigot).—Oreg.
  ssp. **alaskensis** James, 1951: 343.—Alaska; B.C., Alta.
elegans Loew, 1866b: 7 (Cent. 7, no. 10).—N.Y.; Iowa to Que., s. to Tex. and Fla.
viridis Say, 1823: 87 (1859b:77).—U.S.; Alaska to Que., s. to Baja Calif., Tex., and Md.
    nigribarbis Bigot, 1879a: 224.—Calif.
    caerulea Bigot, 1887a: 29 (*Myochrysa*).—N. Amer. **N. syn.**

Subgenus PEDICELLINA James

Pedicellina James, 1952: 225 (as genus). Type-species, *Sargus notatus* Wiedemann (orig. des.).
lucens Loew, 1866b: 7 (Cent. 7, no. 11).—Cuba; Mich. to N.Y. and Fla., Mexico, Cent. Amer., West Indies.
    clavis Williston, 1885d: 123 (*Macrosargus*).—Va., N.C.

Genus CEPHALOCHRYSA Kertész

Cephalochrysa Kertész, 1912: 99. Type-species, *Sargus hovas* Bigot (orig. des.).
Isosargus James, 1936g: 273. Type-species, *Chrysonotus nigricornis* Loew (orig. des.).
canadensis (Curran), 1927l: 197 (*Chrysochroma*).—Ont.; ?Okla.
nigricornis (Loew), 1866b: 9 (Cent. 7, no. 14) (*Chrysonotus*).—D.C.; Wis. to Que., s. to Kans. and S.C.
    atriventris Graenicher, 1913: 176 (*Chrysochroma*).—Wis.
similis (James), 1936g: 274 (*Isosargus*).—La.; Ga.
texana (Melander), 1904: 19 (*Sargus*).—Tex.; Ga.

Genus PTECTICUS Loew

Ptecticus Loew, 1855a: 142. Type-species, *Sargus testaceus* Fabricius (orig. des.).
Plecticus, error.

Pedicella Bigot, 1856: 85. Type-species, *Sargus tenebrifer* Walker (aut.).
Macrosargus Bigot, 1879a: 187 (unjustified n. name for *Pedicella* Bigot). Type-species, *Sargus tenebrifer* Walker (Aldrich, 1933b: 165).

sackenii Williston, 1885d: 124.—Fla.; Wis. to Ont., s. to Ariz. and Fla., ?B.C.
trivittatus (Say), 1829: 159 (1859b: 355) (*Sargus*).—Ind.; Colo. to Wis. and Mass., s. to Tex. and Ga.
   similis Williston, 1885d: 124.—Va., Ga.
   var. **melanopus** James, 1941a: 106 (as ssp.).—Ohio.

Genus MICROCHRYSA Loew

Microchrysa Loew, 1855a: 146. Type-species, *Musca polita* Linnaeus (orig. des.).

flavicornis (Meigen), 1822: 112 (*Sargus*).—Europe; Wash. to Nfld., s. to Colo. and Pa.
polita (Linnaeus), 1758: 598 (*Musca*).—Europe; B.C. to N.S., s. to Calif. and Ga.

Genus Chloromyia Duncan

Chloromyia Duncan, 1837: 164. Type-species, *Musca formosa* Scopoli (Verrall, 1909: 188).

formosa (Scopoli), 1763: 399 (*Musca*).—Italy; ?N.Y., Europe. Immigrant, ?estab.

Genus MEROSARGUS Loew

Merosargus Loew, 1855a: 144. Type-species, *Sargus obscurus* Wiedemann (orig. des.).

beameri James, 1941a: 107.—Ariz.
caerulifrons (Johnson), 1900b: 325 (*Sargus*).—N.J.; Iowa to Conn., s. to Tex.

Subfamily CYPHOMYIINAE

Larvae, as far as known, are scavengers in decaying plant tissues.
   REFERENCE: James, 1940 (rev. New World spp.).

Genus DICYPHOMA James

Dicyphoma James, 1937a: 151. Type-species, *Cyphomyia schaefferi* Coquillett (orig. des.).

schaefferi (Coquillett), 1904c: 32 *(Cyphomyia)*.—Tex.; Kans. to Tex., Mexico and Panama. Larvae in *Opuntia* cactus.

Genus CYPHOMYIA Wiedemann

Cyphomyia Wiedemann, 1819: 54. Type-species, *auriflamma* Wiedemann (Brauer, 1882: 87; 1882: 31).
Rondania Jaennicke, 1867: 324 (1868: 16) (preocc. Robineau-Desvoidy, 1850; Bigot, 1854). Type-species, *obscura* Jaennicke (mon.).
Neorondania Osten Sacken, 1878c: 50 (n. name for *Rondania* Jaennicke). Type-species, *Rondania obscura* Jaennicke (aut.)
Gyneuryparia Enderlein, 1914b: 604. Type-species, *Cyphomyia pilosissima* Gerstäcker (orig. des.).
bicarinata Williston, 1900: 244.—Mexico; Ariz., N. Mex., Tex., Guatemala. Larvae in *Opuntia*.
marginata Loew, 1866a: 148 (Cent. 6, no. 31).—Cuba; Fla., P.R., Haiti.

Subfamily HERMETIINAE

The larvae are terrestrial scavengers. *Hermetia illucens* (Linnaeus) breeds in excrement, decaying vegetable matter, decaying and waste animal matter, wax in beehives, et cetera; sometimes the species is involved in human enteric myiasis. It has been shown to compete in excrement with the larva of the house fly, to the detriment of the latter.

REFERENCE: James, 1935b (rev.).

Genus HERMETIA Latreille

Hermetia Latreille, 1804: 192. Type-species, *Musca illucens* Linnaeus (mon.).
aurata aurata Bellardi, 1859: 27 (1861: 227).—Mexico; Utah to Kans., s. to Ariz. and Tex.
  chrysopila Loew, 1872a: 56 (Cent. 10, no. 11).—Tex.
  ssp. **eiseni** Townsend, 1895b: 594 (as sp.).—Baja Calif.; Calif., Ariz.
comstocki Williston, 1885d: 125.—Ariz.; N. Mex., Tex., Mexico.
concinna Williston, 1900: 241.—N. Mex., Ariz.; Calif., Mexico.
hunteri Coquillett, 1909: 212.—Tex.; Mexico (Chih., Baja Calif.).
illucens (Linnaeus), 1758: 589 *(Musca)*.—S. Amer.; Calif. to Mass., s. throughout tropical Amer., Hawaii, s. Europe, Old World tropics. Spread by commerce.
  mucens, error.

lativentris Bellardi, 1859: 27 (1861: 227).—Mexico; Ariz., N. Mex., and Tex. to Panama.
reinhardi James, 1935b: 166.—Tex.
sexmaculata Macquart, 1834a: 229.—P.R.; Tenn., Fla., Cuba, Brazil.
  nucis James, 1935b: 166.—Cuba.

Subfamily CHRYSOCHLORINAE
Genus CHRYSOCHLORINA James

Chrysochlorina James, 1939d: 33. Type-species, *Sargus vespertilio* Fabricius (orig. des.).
Chrysochlora, authors, not Latreille.

REFERENCE: James, 1939d (rev.).

quadrilineata (Bigot), 1887a: 26 (*Chrysochlora*).—Cuba; Fla. (Key West).

Subfamily CLITELLARIINAE

The larvae are scavengers in decaying plant tissues (Clitellariini), or are aquatic (Oxycerini, Nemotelini); adults of aquatic genera commonly occur in wet grass or sedge meadows.

Tribe CLITELLARIINI
Genus ADOXOMYIA Bezzi

Adoxomyia Kertész, 1907a: 499. Nomen nudum.
Adoxomyia Bezzi, 1908: 75. Type-species, *Clitellaria dahlii* Meigen (orig. des.).
Euclitellaria Kertész, 1923: 96, 101. Type-species, *Clitellaria heminopla* Wiedemann (orig. des.). **N. syn.**
Clitellaria, authors, not Meigen.

REFERENCE: James, 1943b (rev.).

albopilosa (Cresson), 1919: 173 (*Aochletus*).—N. Mex.
appressa appressa James, 1935a: 63.—N. Mex.; Calif., Ariz.
  ssp. **cibolae** James, 1950b: 71.—N. Mex.
argentata (Williston), 1885d: 127 (*Clitellaria*).—Calif.; Ariz., N. Mex.
claripennis James, 1935a: 62.—Ariz.
lata (Loew), 1872a: 55 (Cent. 10, no. 9) (*Clitellaria*).—Calif.; Wash., Oreg., Ariz.
  nigra Bigot, 1879a: 204 (*Euparyphus*).—Calif.
  nigropilosa Cresson, 1919: 174 (*Aochletus*).—Calif.

micheneri James, 1950b: 71.—Tex.
nigribarba James, 1943b: 168.—Oreg.; Calif.
rustica (Osten Sacken), 1877: 213 (*Clitellaria*).—Calif.; B.C. to Mont., s. to Calif. and Utah.
subulata (Loew), 1866a: 147 (Cent. 6, no. 29) (*Clitellaria*).—Va.; Ind. to N.J., s. to Tex. and N.C.
texana James, 1935a: 63.—Tex.; Okla., Ark.

Genus DIEURYNEURA James

Dieuryneura James, 1937a: 152. Type-species, *callosa* James (orig. des.) = *obscura* (Coquillett).
obscura (Coquillett), 1902g: 98 (*Aochletus*).—Calif., Ariz.; N. Mex. w. Tex., n. Mexico.
  callosa James, 1937a: 153.—Ariz.

Genus BRACHYCARA Thomson

Brachycara Thomson, 1869: 460. Type-species, *ventralis* Thomson (mon.).
Euryneurasoma Johnson, 1913a: 51. Type-species, *slossonae* Johnson (orig. des.).
Neurota Curran, 1931a: 2. Type-species, *Sargus bicolor* Wiedemann (orig. des.; misident.) = *slossonae* (Johnson).
slossonae (Johnson), 1913a: 51 (*Euryneurasoma*).—Fla.; Bermuda, Cuba, P.R., Bahamas.
  bicolor, authors, not Wiedemann.

Genus EURYNEURA Schiner

Euryneura Schiner, 1867a: 308. Type-species, *Stratiomys fascipennis* Fabricius (orig. des.).
propinqua Schiner, 1868: 57.—Colombia; Ariz., Mexico to Colombia.

Tribe OXYCERINI

This section was prepared with the aid of John A. Quist, Ph. D. thesis, Washington State University.

Genus OXYCERA Meigen

Hermione Meigen, 1800: 22. Type-species, *Musca hypoleon* Linnaeus (Coquillett, 1910b: 551) = *trilineata* (Linnaeus). Suppressed by I.C.Z.N., 1963b: 339.

Oxycera Meigen, 1803: 265. Type-species, *Musca trilineata* Linnaeus (Curtis, 1833: pl. 441).
REFERENCE: Malloch, 1917g (rev.).
albovittata Malloch, 1917g: 330.—Ill.; Minn. to Ont., s. to Mo. and Pa.
aldrichi Malloch, 1917g: 329.—Ind.
approximata Malloch, 1917g: 326.—Ill.; Mich., Ind., Ont., N.Y.
centralis Loew, 1863a: 8 (Cent. 3, no. 14).—Man. (Red River); Alaska, N.W.T.
maculata Olivier, 1812: 600.—Carolina; Iowa to Ont. and N.C., ?La.
picta Wulp, 1867: 133.—Wis.
variegata Olivier, 1812: 600.—Carolina; Iowa and Wis., e. to Pa. and Fla.
unifasciata Loew, 1863a: 9 (Cent. 3, no. 15).—Pa.

Genus EUPARYPHUS Gerstäcker

Euparyphus Gerstäcker, 1857: 314. Type-species, *Cyphomyia elegans* Wiedemann (mon.).

Subgenus EUPARYPHUS Gerstäcker

albipilosus Adams, 1903a: 30.—Colo., Ariz.; Wash. to N. Mex. and Ariz.
apicalis Coquillett, 1902g: 99.—Calif.
carbonarius Giglio-Tos, 1891a: 2; also 1892: 12 (1893: 109).—Mexico; N. Mex., Utah, Guatemala.
lagunae Cole, 1912b: 151.—Calif.
limbiventris Curran, 1927l: 217.—Man.; Alta.
limbrocutris Adams, 1903a: 31.—Wash.; Idaho, Utah.
mutabilis mutabilis Adams, 1903a: 29.—Wyo.; Mont. to Iowa, s. to Ariz.
 ssp. **latelimbatus** Curran, 1927l: 219 (as sp.).—Alta; Sask., Man.
 ssp. **vanduzeei** James, 1936f: 89 (as sp.).—Nev.
ornatus Williston, 1885d: 126.—Wash.; B.C., Calif.
pardalinus James, 1936f: 87.—Calif.; Utah.
sabroskyi James, 1936f: 88.—Kans.; Okla.
stigmaticalis Loew, 1866b: 10 (Cent. 7, no. 17).—D.C.; B.C. and Wash., Ont. to Ind. and D.C.
nigrostigma Curran, 1927l: 220.—Ont.
tricolor Osten Sacken, 1886a: 40.—Mexico; Ariz.

Subgenus AOCHLETUS Osten Sacken

Aochletus Osten Sacken, 1886a: 38 (as genus). Type-species, *cinctus* Osten Sacken (mon.).

brevicornis Loew, 1866b: 10 (Cent. 7, no. 16).—D.C.; Idaho to Ont., s. to Calif. and D.C.
  brucensis Steyskal, 1951b: 273.—Ont.
cinctus (Osten Sacken), 1886a: 38 (*Aochletus*).—Mexico; Mont. and S. Dak., s. to Calif. and Ariz.
  quadrimaculatus Cresson, 1919: 172.—Calif.

Genus CALOPARYPHUS James

Euparyphus, subg. **Caloparyphus** James, 1939e: 49. Type-species *Oxycera crotchi* Osten Sacken (orig. des.).

amplus (Coquillett), 1902g: 100 (*Euparyphus*).—Colo.; Calif. to N. Mex. and Colo.
atriventris (Coquillett), 1902g: 100 (*Euparyphus*).—Colo.; B.C. to Minn., s. to Colo. and Kans.
crotchi (Osten Sacken), 1877: 212 (*Oxycera*).—Calif.; B.C. to Colo., s. to Calif. and N. Mex.
  septemmaculatus Adams, 1903a: 31 (*Euparyphus*).—Calif.
crucigerus crucigerus (Coquillett), 1902g: 99 (*Euparyphus*).—Colo.; Wash. to Calif. and Colo.
  ssp. **nicolensis** (Curran), 1927l: 221 (*Euparyphus;* as sp.).—B.C.; Wash., Idaho, Alta.
  ssp. **tahoensis** (Coquillett), 1902g: 98 (*Euparyphus;* as sp.).—Calif.; B.C. to Calif.
    obliquus Hine, 1904b: 87 (*Euparyphus*).—B.C.
currani (James), 1939e: 52 (*Euparyphus*).—Man.
flaviventris (James), 1936f: 89 (*Euparyphus*).—Calif.; B.C., Oreg., Mont.
greylockensis (Johnson), 1912a: 5 (*Euparyphus*).—Mass.; Mich. to Que. and N.Y.
  adaleonora Steyskal, 1941b: 123 (*Euparyphus*).—Mich.
major (Hine), 1901b: 112 (*Euparyphus*).—Colo.; Wash. to Alta., s. to Calif. and Kans.
  octomaculatus Curran, 1927l: 222 (*Euparyphus*).—B.C.
mariposa (James), 1939e: 54 (*Euparyphus*).—Calif.; B.C.
pretiosus (Banks), 1920: 65 (*Euparyphus*).—B.C.
tetraspilus (Loew), 1866b: 9 (Cent. 7, no. 15) (*Euparyphus*).—N.Y.; S. Dak to N.S., s. to Kans. and Pa.
  bellus Loew, 1866b: 11 (Cent. 7, no. 18) (*Euparyphus*).—Mass.

Tribe NEMOTELINI

Genus NEMOTELUS Geoffroy

Nemotelus Geoffroy, 1762: 450, 542. Type-species, *Musca pantherina* Linnaeus (I.C.Z.N., 1957a: 85). Conserved as of 1762 under Suspension of the Rules (I.C.Z.N., 1957a: 85).
Nematotelus, emend.

REFERENCE: James, 1936c (rev.).

Subgenus NEMOTELUS Geoffroy

abdominalis Adams, 1903b: 221.—Kans.; Man., N. Dak., Colo., Fla.
albirostris Macquart, 1850: 359 (1850: 55).—Va.; Ariz. to Fla., n. to Va.
   wheeleri Melander, 1903b: 182.—Tex.
bellulus Melander, 1903b: 183.—Tex.; Fla.
bonnarius Johnson, 1912a: 4.—Sask.; B.C. to Minn.
bruesii Melander, 1903b: 179.—Tex.; Iowa.
immaculatus Johnson, 1895a: 304.—Fla.
kansensis kansensis Adams, 1903b: 221.—Kans.; Alta. to Iowa, s. to Tex.
   plesius Curran, 1927l: 225.—Nebr.
   ssp. **trinotatus** Melander, 1903b: 180 (as sp.).—Tex.; Okla., Mexico. **N. status.**
knowltoni James, 1936c: 89.—Utah; s. Calif. to s. Utah and w. Texas.
montanus James, 1936c: 90.—Colo.; Y.T. to Colo. and Calif.
pallipes Say, 1823: 29 (1859b: 52).—Pa.
quadrinotatus Johnson, 1913a: 50.—Fla.; Tex.
rufoabdominalis Cole, 1923a: 459.—Mexico; Calif., Ariz., N. Mex., Baja Calif.

Subgenus CAMPTOPELTA Williston

Camptopelta Williston, 1917: 23 (as genus). Type-species, *aldrichi* Williston (mon.). **N. syn.**
Nemotelinus Enderlein, 1936a: 79. Type-species, *Nemotelus nigrinus* Fallén (mon.) **N. syn.**
Nematolinus, error.
Nemotelus, subg. *Melanonemotelus* Hanson, 1958: 1356. Type-species, *Nemotelus nigrinus* Fallén (orig. des.). **N. syn.**

REFERENCE: Hanson, 1958 (rev.), 1961 (syn.).

albimarginatus James, 1936c: 86.—Utah; B.C. to S. Dak., s. to Utah and Calif.

aldrichi (Williston), 1917: 24 (*Camptopelta*).—N. Mex. **N. comb.**
   lambda James, 1933a: 70.—N. Mex. **N. syn.**
arator Melander, 1903b: 179.—Calif.; Wash., Oreg.
beameri James, 1933a: 70.—Colo.; Mont. to Ill., s. to Ariz.
canadensis Loew, 1863a: 7 (Cent. 3, no. 12).—N.W.T.; s. Alaska to Minn., s. to n. Calif. and Colo.
centralis Hanson, 1958: 1369.—Minn.; S. Dak., Mich., Ohio., Ont.
communis Hanson, 1958: 1376.—Utah; Oreg. to Man. and Ind., s. to Ariz.
flavicornis Johnson, 1894: 272.—Jamaica; Fla.
glaber Loew, 1872a: 56 (Cent. 10, no. 10).—Tex.; Kans. to N.Y., s. to Miss. and Fla., Panama.
   flavicornis James, 1932a: 7 (preocc. Johnson, 1894).—Kans.
   fulvicornis James, 1936c: 87 (n. name for *flavicornis* James).—Kans.
halophilus Hanson, 1958: 1372.—Tex.; N. Mex., Kans.
jamesi Hanson, 1958: 1367.—Wash.; Wash. to Calif. and Nebr.
melanderi Banks, 1920: 65.—Md. (as Ind., error); Kans. to N.S., s to Tex. and Fla.
nigrinus Fallén, 1817b: 6.—Europe; s. Alaska to Que., s. to Calif., Ill., and R.I., mts. of n. Mexico.
   carneus Walker, 1849: 521.—Ont.
   crassus Loew, 1863a: 7 (Cent. 3, no. 10).—R.I.
   unicolor Loew, 1863a: 7 (Cent. 3, no. 11).—Ill.
   carbonarius Loew, 1869b: 5 (Cent. 8, no. 6).—Mass.
picinus Hanson, 1958: 1378.—Ill.; Minn. and Iowa, e. to Que.
politus Hanson, 1958: 1364.—Utah; Wash. to Mont., s. to Calif. and Colo.
   simplex Snow, 1904: 341.   Nomen nudum.
sabroskyi Hanson, 1958: 1375.—N.C.; N.J. to N.C.
slossonae Johnson, 1895a: 304.—Fla.
tenuistylus Hanson, 1958: 1379.—N. Mex.; Idaho to Kans., s. to Calif. and N. Mex., Mexico (Baja Calif.).
tristis Bigot, 1887a: 30.—Calif.
variabilis Hanson, 1958: 1368.—Fla.; Calif. to Fla., Mexico.

Genus AKRONIA Hine

Akronia Hine, 1901b: 113. Type-species, *frontosa* Hine (orig. des.).
frontosa Hine, 1901b: 113.—Ohio.

Subfamily STRATIOMYINAE

The larvae, so far as known, are aquatic, occurring in a variety of habitats including muddy bottoms, submerged or floating vegetation, along shores both below and just above the waterline, in brackish water, and in hot springs.

REFERENCES: Hart, 1896 (biol.); Kuster, 1934 (biol.).

Tribe STRATIOMYINI

REFERENCE: James and Steyskal, 1952 (rev.).

Genus STRATIOMYS Geoffroy

Stratiomys Geoffroy, 1762: 475. Type-species, *Musca chamaeleon* Linnaeus (I.C.Z.N., 1957b: 123). Conserved as of 1762 under suspension of the Rules (I.C.Z.N., 1957b: 123).
Stratiomyia, emend.

adelpha Steyskal *in* James and Steyskal, 1952: 393.—Mich.; Wash. to Que., s. to Tex.
    discalis, authors, not Loew.
badia Walker, 1849: 529.—N.H. (as New Holland, error; Walker, 1849: 1157); B.C. to Nfld., s. to Colo. and N.Y.
    ischiaca Walker, 1849: 529.—U.S.
    picipes Loew, 1866b: 13 (Cent. 7, no. 21).—Ont.
barbata Loew, 1866a: 133 (Cent. 6, no. 9).—Calif.; Alaska to Man., s. to Calif. and N. Mex.
    calopus Bigot, 1887a: 23.—Colo.
    atra Cole *in* Cole and Lovett, 1919: 223.—Oreg. ?Melanic var.
browni Curran, 1927l: 202.—Okla.; Kans. to Ohio, s. to Tex. and Miss.
    beameri James, 1933a: 66.—Kans.
bruneri Johnson, 1895b: 233.—S. Dak.; N.W.T. to Mich., s. to Utah.
currani currani James, 1932a: 5.—Colo.; Sask. to Colo. and Calif.
  ssp. **boharti** James, 1955c: 47.—Calif.; Oreg. Higher elevations.
discalis Loew, 1866a: 136 (Cent. 6, no. 14).—Ill.; Colo. to Wis. and Que.
    media James, 1933a: 67.—Colo.
discaloides Curran, 1922i: 281.—B.C.; B.C. to Mont., Wyo. and Utah, s. to Calif.
floridensis Steyskal *in* James and Steyskal, 1952: 399.—Fla.
griseata Curran, 1923f: 74 (n. name for *velutina* Curran).—B.C.; B.C. and Alta., s. to Oreg. and Utah.
    velutina Curran, 1922i: 283 (preocc. Bigot, 1877).—B.C.
hirsutissima James, 1932a: 1.—Colo.
hulli Steyskal *in* James and Steyskal, 1952: 394.—Miss.

jamesi Steyskal *in* James and Steyskal, 1952: 398.—Tex.; Kans., Mo., Ill.
laticeps Loew, 1866b: 12 (Cent. 7, no. 20).—Hudson Bay Territory: B.C. to Ont., s. to Calif. and Colo.
   occidentis Banks, 1926: 42.—Utah.
   occidentalis, emend. or error.
lativentris Loew, 1866a: 132 (Cent. 6, no. 8).—Barnston, Lake Superior; Mich., Que., Ont.
maculosa Loew, 1866b: 12 (Cent. 7, no. 19).—Calif.; B.C. to Utah and Calif.
   insignis Loew, 1872a: 54 (Cent. 10, no. 7).—Calif.
   dentata Bigot, 1879a: 210.—Calif.
   lacerata Bigot, 1879a: 211.—Calif.
meigenii Wiedemann, 1830: 61.—Ga.; Wis. to Que., s. to Tex. and Fla.
   robusta Walker, 1854: 37.—N. Amer.
   angularis Loew, 1866a: 138 (Cent. 6, no. 16).—Pa.
   marginalis Loew, 1866a: 138 (Cent. 6, no. 17).—Pa.
   rubra James, 1933a: 68.—Kans.
   quaternaria, authors, not Loew.
melastoma Loew, 1866a: 134 (Cent. 6, no. 10).—Calif.; B.C. to Colo. and Calif.
   melanostoma, emend.
nevadae Bigot, 1887a: 24.—Nev.; B.C. to N. Dak. and Colo., s. to Calif.
nigriventris Loew, 1866a: 137 (Cent. 6, no. 15).—Nebr.; Nebr. to Tex.
norma Wiedemann, 1830: 62.—N. Amer.; S. Dak. to Que., s. to Kans. and Pa.
   quadrigemina Loew, 1866a: 129 (Cent. 6, no. 4).—Conn.
normula normula Loew, 1866a: 130 (Cent. 6, no. 5).—N.Y.; B.C. to Que., s. to Colo. and S.C.
   quaternaria Loew, 1866a: 135 (Cent. 6, no. 12).—Ill.
   apicula Loew, 1866a: 136 (Cent. 6, no. 13).—Ill.
   notata Loew, 1866a: 139 (Cent. 6, no. 18).—Nebr.
   ssp. **angulicincta** James, 1932a: 3 (as sp.).—Colo.; Utah, N. Mex.
   ssp. **senaria** Loew, 1866a: 132 (Cent. 6, no. 7) (as sp.).—Fla.
   ssp. **unilimbata** Loew, 1866a: 131 (Cent. 6, no. 6) (as sp.).—Wis.; Alta. to Colo. and Mich.
      jonesi James, 1932a: 4.—Colo.
   ssp. **wyomingensis** James, 1932a: 3 (as sp.).—Wyo.; Mont., N. Dak.
nymphis Walker, 1849: 530.—Hudson Bay Territory. ?=*laticeps*.

obesa Loew, 1866a: 134 (Cent. 6, no. 11).—Ill.; Sask. to Que., s. to
Nebr. and N.Y.
lativentris, authors, not Loew.
ohioensis Steyskal *in* James and Steyskal, 1952: 400.—Ohio; Mich.,
N.Y.
tularensis James, 1957b: 43.—Calif.

Genus HOPLITIMYIA James

Hoplitimyia James, 1934a: 443. Type-species, *Stratiomys constans* Loew (orig. des.).

Subgenus HOPLITIMYIA James

constans (Loew), 1872a: 55 (Cent. 10, no. 8) (*Stratiomys*).—Tex.;
Ariz. to Kans. and Ark., ?Ind.
fasciata (Fabricius), 1787: 331 (*Stratiomys*).—French Guiana; ?Colo.,
Neotropical.
vespoides James, 1933a: 67 (*Stratiomys*).—?Colo. Colorado
label probably refers to Colorado State University
collection rather than to Colorado as a locality.
mutabilis (Fabricius), 1787: 331 (*Stratiomys*).—French Guiana; Ariz.,
Tex., tropical Amer. to Argentina.

Genus LABOSTIGMINA Enderlein

Labostigmina Enderlein, 1930b: 70. Type-species, *Odontomyia occipitalis* Johnson (orig. des.).

annamariae (Brimley), 1925: 76 (*Odontomyia*). —N.C.; Ala.
defecta James *in* James and Steyskal, 1952: 409.—La.
flavicornis (Olivier), 1812: 433 (*Odontomyia*).—N. Amer.; Kans. to
Ont. and Maine, s. to Tex., and Fla.
flaviceps Macquart, 1834a: 245 (*Stratiomys*).—Pa.
coronata Guérin-Méneville, 1835: pl. 98, fig. 6 (*Stratiomys*).—
Unknown.
pulchella Macquart, 1838a: 184 (1838: 180) (*Stratiomys*).—
Ga.
vicina Macquart, 1838a: 185 (1838: 181) (*Stratiomys*).—Pa.
lasiophthalma Loew, 1866a: 142 (Cent. 6, no. 23) (*Odontomyia*).—N.Y.
fulvicornis (Curran), 1927l: 213 (*Odontomyia*).—Okla.; Tex.
gagatigaster Steyskal, 1949a: 173.—Mich.
hieroglyphica (Olivier), 1812: 434 (*Odontomyia*).—Carolina; Ill. to
Maine and Ga.
fallax Johnson, 1895b: 257 (*Odontomyia*).—Ga.

hypomelas James *in* James and Steyskal, 1952: 407.—Tex.
johnsoni (Curran), 1925e: 255 (*Odontomyia*).—La.; Kans. to Pa., s. to Tex. and Ala.
megantica (Curran), 1925e: 254 (*Odontomyia*).—Que.; Mich., Ont.
micheneri James *in* James and Steyskal, 1952: 408.—Tenn.
novella Steyskal, 1938a: 1.—Tex.; Ariz.
obscura (Olivier), 1812: 433 (*Odontomyia*).—Carolina; Wis. to Va., s. to Tex. and Fla.
    brevipennis Olivier, 1812: 434 (*Odontomyia*).—Carolina.
occipitalis (Johnson), 1895b: 268 (*Odontomyia*).—Pa., Va.; Mich. to N.J., s. to Tenn. and Ga.
rufipennis James *in* James and Steyskal, 1952: 410.—Tex.
similis (Johnson), 1895b: 267 (*Odontomyia*).—Colo.; Alta. to Utah and Colo.
texasiana (Johnson), 1895b: 259 (*Odontomyia*).—Tex.
viridis (Bellardi), 1859: 36 (1861: 236) (*Odontomyia*).—Mexico; S. Dak. to Ind., s. to Tex.
    snowi Hart, 1896: 256 (*Odontomyia*).—Ill.

Tribe ODONTOMYIINI

REFERENCE: James, 1936d (rev.).

Genus ANOPLODONTA James

Anoplodonta James, 1936a: 35. Type-species, *Odontomyia nigrirostris* Loew (orig. des.).

nigrirostris (Loew), 1866a: 140 (Cent. 6, no. 19) (*Odontomyia*).—Wis.; Alta. to Mich., s. to Ariz., Tex., and Mexico.
nuda James, 1932b: 436 (*Odontomyia*).—Colo.

Genus HEDRIODISCUS Enderlein

Hedriodiscus Enderlein, 1914b: 608. Type-species, *Odontomyia brevifacies* Macquart (orig. des.).

bellulus (James), 1936d: 528 (*Odontomyia*).—Tex.
currani (James), 1932a: 6 (*Odontomyia;* as *truquii* var.).—Ariz.; Calif. to Tex., Mexico.
dorsalis (Fabricius), 1805: 82 (*Stratiomys*).—West Indies; Tex., Fla., Mexico, Cent. and S. Amer.
trivittatus trivittatus (Say), 1829: 160 (1859b: 356) (*Stratiomys*).—Mexico; Minn. s. to Calif. and Fla., Cent. Amer.
    ssp. **leucogaster** (James), 1933a: 69 (*Odontomyia;* as sp.).—Tex.; Calif. to Tex., Mexico. ?Valid sp.

truquii (Bellardi), 1859: 34 (1861: 234) (*Odontomyia*).—Mexico; B.C. to Ont. and N.Y., s. to Calif.
    megacephala Loew, 1866a: 140 (Cent. 6, no. 20) (*Odontomyia*).—Calif.
    binotatus Loew, 1866a: 142 (Cent. 6, no. 22) (*Odontomyia*).—Ill.
    bicolor Day, 1882: 78 (*Odontomyia*).—Calif.
    innotatus Curran, 1927l: 210 (*Odontomyia;* as var.).—Man.
varipes (Loew), 1866a: 141 (Cent. 6, no. 21) (*Odontomyia*).— ?Carolina; Alta. to Ont., s. to Calif.
    alberta Curran, 1922i: 279 (*Odontomyia*).—Alta.
vertebratus (Say), 1824a: 369 (1859a: 251) (*Odontomyia*).—Northwest Territory (U.S.); Alta. to Que., s. to Calif. and Ga.
    willistoni Day, 1882: 78 (*Odontomyia*).—N.Y.

Genus ODONTOMYIA Meigen

Eulalia Meigen, 1800: 21. Type-species, *Musca hydroleon* Linnaeus (Coquillett, 1910b: 541). Suppressed by I.C.Z.N., 1963b: 339.

Odontomyia Meigen, 1803: 265. Type-species, *Musca hydroleon* Linnaeus (Westwood, 1840: 130).

Trichacrostylia Enderlein, 1914b: 607. Type-species, *Stratiomys angulata* Panzer (orig. des.).

Subgenus ODONTOMYIA Meigen

alticola James, 1932b: 437.—Colo.; Mont. to Nebr., s. to Calif. and N. Mex.
arcuata Loew, 1872a: 52 (Cent. 10, no. 4).—Calif.; Oreg. to Colo., s. to Calif. and N. Mex.
bermudensis Johnson, 1913d: 445.—Bermuda.
cincta Olivier, 1812: 432.—Carolina; Idaho to N.B., s. to Calif. and Fla.
    extremis Day, 1882: 80.—Conn., Calif.
communis James, 1939b: 220.—Colo.; Wash. to Mich., s. to Calif. and Tex.
    inaequalis Loew of James, 1936d: 544.
discolorata James, 1936d: 548.—Tex.; Mexico, ?Costa Rica.
evansi (James), 1957a: 15 (*Eulalia*).—Fla.
flava Day, 1882: 76.—Wyo.
hirtocculata James, 1936d: 540 (n. name for *pacifica* Curran).—Calif.
    pacifica Curran, 1927l: 215 (preocc. Meigen, 1822).—Calif.
idahoensis James, 1932a: 6.—Idaho.

inaequalis Loew, 1866a: 143 (Cent. 6, no. 24).—Hudson Bay Territory; Oreg. to Man., s. to Calif., high mts. of w. Mexico.
    confusa James, 1936d: 543.—Colo.
microstoma Loew, 1866a: 146 (Cent. 6, no. 28).—Mass.; Maine to Md., near coast.
occidentalis James, 1936d: 545.—Wyo.; Idaho to Colo., s. to Calif.
rufipes Loew, 1866a: 144 (Cent. 6, no. 25).—Cuba; Fla.
tumida Banks, 1926: 42.—Calif. (unpublished); Wash. to Mont., s. to Calif. and Kans.

Subgenus CATATASINA Enderlein

Catatasina Enderlein, 1914b: 608 (as genus). Type-species, *Stratiomys argentata* Fabricius (orig. des.).
Achlyomyia Pleske, 1922: 334. Type-species, *Musca microleon* Linnaeus (orig. des.).

colei James, 1936d: 532.—Calif.; s. Idaho.
hoodiana Bigot, 1887a: 25.—Oreg.; B.C. to Ohio, s. to Calif. and Kans.
interrupta Olivier, 1812: 433.—Carolina; Alta. to Que., s. to Kans. and Fla.
    intermedia Wiedemann, 1830: 64 (*Stratiomys*).—N. Amer.
melantera James, 1939b: 220.—Ont.
    nigerrima Loew of James, 1936d: 533.
nigerrima Loew, 1872a: 53 (Cent. 10, no. 6).—Middle States; Alta., Sask., Ont.
painteri James, 1936d: 530.—Kans.; Nebr., N.J.
pilosa Day, 1882: 76.—Calif.; Wash., Oreg., Utah, Ariz.
    pyrrhostoma Bigot, 1887a: 25.—Oreg.
profuscata Steyskal, 1938a: 3.—Ont.; Mich., Ohio, N.J.
pubescens Day, 1882: 77.—Calif.; B.C. to Que., s. to Calif. and N.C.

Subgenus ODONTOMYIINA Enderlein

Odontomyiina Enderlein, 1930b: 70 (as genus). Type-species, *Stratiomys virgo* Wiedemann (orig. des.).

aldrichi Johnson, 1895b: 262.—Kans.; Nebr. to Tex., e. to Fla.
americana Day, 1882: 77.—Calif.
borealis James, 1936d: 537.—Ont.; Wis. to Que., s. to Kans. and Fla.
hydroleonoides Johnson, 1895b: 261.—Utah, Mich., Ill., Ont.; Oreg. to Nfld., s. to Kans. and Va.
pilimana Loew, 1866a: 146 (Cent. 6, no. 27).—Ill.; Oreg. to Ont., s. to n. Calif. and Va.

plebeja Loew, 1872a: 53 (Cent. 10, no. 5).—Conn.; S. Dak. to Mass., s. to Tex. and Ga.
  plebia, error or emend.
virgo (Wiedemann), 1830: 69 (*Stratiomys*).—Ga.; B.C. to Que., s. to Calif. and Fla.
  paron Walker, 1849: 536.—N.Y.
  nigra Day, 1882: 75.—Kans.

Tribe MYXOSARGINI

REFERENCE: James, 1942b (rev.).

Genus MYXOSARGUS Brauer

Myxosargus Brauer, 1882: 77, 88 (1882: 21, 32). Type-species, *fasciatus* Brauer (orig. des.).
knowltoni Curran, 1929b: 2.—Utah; Idaho and Utah to Calif.
nigricornis Greene, 1918: 71 (as *nigricormis*).—D.C.; Okla. to Ind., e. to Md. and Fla.
pilosus James, 1942b: 59.—Ariz.; Ariz. to Panama.
texensis Curran, 1929b: 4.—Tex.; Okla.

Genus NOTHOMYIA Loew

Nothomyia Loew, 1869b: 4 (Cent. 8, no. 4, note). Type-species, *scutellata* Loew (Brauer, 1882: 88; 1882: 32).
calopus Loew, 1869b: 5 (Cent. 8, no. 5).—Cuba; Fla., P.R.
viridis Hine, 1911b: 301.—Ohio; P.R. Immigrant, ?estab.

Subfamily PACHYGASTRINAE

REFERENCE: Kraft and Cook, 1961 (rev).

Genus NEOPACHYGASTER Austen

Neopachygaster Austen, 1901: 245. Type-species, *Pachygaster meromelaena* Dufour (orig. des.).
maculicornis (Hine), 1902: 228 (*Pachygaster*).—Kans.; Sask., Mont. and Utah to N.Y., s. to Kans. and n. Miss.
occidentalis Kraft and Cook, 1961: 15.—Oreg.; B.C.
reniformis Hull, 1942b: 70.—Miss.; Wash. to Que., s. to Miss. and Ga.
vitrea Hull, 1930b: 103.—Iowa; Ill., Mich., N.Y., Pa.

Genus EUPACHYGASTER Kertész

Eupachygaster Kertész, 1911: 31. Type-species, *Pachygaster tarsalis* Zetterstedt (orig. des.).

fusca Kraft and Cook, 1961: 20.—Md.; Ont., Ill., Pa., N.Y., D.C.
henshawi Malloch, 1917g: 338.—Ill.; N. Mex., Tex., Kans.
punctifera Malloch, 1915f: 316.—Ill.; Wash. to Mass., s. to Tex.

Genus PACHYGASTER Meigen

Pachygaster Meigen, 1803: 266. Type-species, *Nemotelus ater* Panzer (orig. des.).

characta Kraft and Cook, 1961: 23.—Tex.
montana Kraft and Cook, 1961: 23.—Ariz.; N. Mex.
pulchra Loew, 1863a: 10 (Cent. 3, no. 16).—D.C.; Kans. to Wis., e. to Md. and Va.

Genus ZABRACHIA Coquillett

Zabrachia Coquillett, 1901f: 585. Type-species, *polita* Coquillett (orig. des.).

albipila Kraft and Cook, 1961: 11.—Ariz.; Calif., Tex.
beameri Kraft and Cook, 1961: 10.—Calif.; Utah.
cornuta Kraft and Cook, 1961: 9.—Ariz.
hebicornuta Kraft and Cook, 1961: 10.—Calif.
knowltoni Kraft and Cook, 1961: 13.—Idaho.
lopha Kraft and Cook, 1961: 12.—Calif.
microcephala Kraft and Cook, 1961: 10.—Tex.
parva Kraft and Cook, 1961: 11.—Calif.
plicata Kraft and Cook, 1961: 12.—Idaho; B.C., Oreg., Calif., Colo., Ariz.
polita Coquillett, 1901f: 585.—N.Y.; B.C. to Que., s. to Calif. and Fla.

Genus BERKSHIRIA Johnson

Berkshiria Johnson, 1914b: 158. Type-species, *albistylum* Johnson (orig. des.).
Johnsonomyia Malloch, 1915f: 313 (preocc. Felt, 1908). Type-species, *aldrichi* Malloch (orig. des.)=*albistylum* Johnson.

albistylum Johnson, 1914b: 158.—Mass.; Kans. to s. Ont., s. to Ala.
  aldrichi Malloch, 1915f: 313 (*Johnsonomyia*).—Ind.

Unplaced Species of Stratiomyidae

caloceps Bigot, 1879a: 217 (*Exochostoma*).—Colo.
canadensis Walker, 1854: 310 (*Stratiomys*).—Que.
diademata Bigot, 1887a: 23 (*Stratiomys*).—Ga., Colo.
lineolata Macquart, 1850: 352 (1850: 48) (*Stratiomys*).—Va.
magnicornis Cresson, 1919: 171 (*Zabrachia*).—N. Mex.
nigrifrons Walker, 1849: 531 (*Stratiomys*).—Hudson Bay.
pallipes Fabricius, 1781: 417 (*Stratiomys*).—S. Amer. (as N. Amer., error). Not Nearctic.
simplex Bigot, 1887a: 24 (*Stratiomys*).—Tex., Colo.

Family PELECORHYNCHIDAE
By Cornelius B. Philip

Originally described in the Tabanidae, *Bequaertomyia* was transferred to the "Coenomyiidae" by Steyskal (1953a) and is hereby assigned to the subfamily Bequaertomyiinae (**N. status**) of the family Pelecorhynchidae, a primitive Neotropical and Australian group of which it is the only North American representative. Nothing is known about the biology or habits of the American species.

Subfamily BEQUAERTOMYIINAE
Genus BEQUAERTOMYIA Brennan

Bequaertomyia Brennan, 1935: 376. Type-species, *anthracina* Brennan (mon.) =*jonesi* (Cresson).

jonesi (Cresson), 1919: 176 (*Arthropeas*).—Calif.; B.C., s. to Calif.
anthracina Brennan, 1935: 377.—Calif.

Family TABANIDAE
By Cornelius B. Philip

Biological data are available for only a small proportion of species of Tabanidae. Most of these data concern mature larvae collected in semi-aquatic environments and induced to pupate in the laboratory. Several are known, or are thought, to spend their earlier larval life under still or running water, or their entire immature life in moist terrestrial environments. The larvae are often predaceous and cannibalistic. Eggs have been observed in single or multi-layered masses, mostly on vegetation over or near water. Blood is required for egg maturation in most species. Some species are widespread; others are restricted to special habitats such as sea beaches, ricefields, desert springs, brackish-water swamps, stream beds, special bogs, or northern muskeg swamps. Most species probably overwinter as larvae. Observations suggest that in the Southern States there may be two generations a year; in the north, probably one.

The opinions and/or distributional data provided by J. F. McAlpine, L. L. Pechuman, Alan Stone, and many others are gratefully acknowledged.

REFERENCES: Marchand, 1920 (immatures); Bromley, 1926 (anat.); Cameron, 1926 (biol.); Philip, 1947, 1950c (cat.); Bonhag, 1951 (anat.); Mackerras, 1954 (classif.).

Subfamily PANGONIINAE

REFERENCES: Brennan, 1935 (rev.); Philip, 1954b (rev.).

Tribe PANGONIINI

Genus Pangonius Latreille

macroglossus Westwood, 1835a: 449.—Unknown (as Ga., error). Not Nearctic.

Genus APATOLESTES Williston

Apatolestes Williston, 1885c: 12. Type-species, *comastes* Williston (mon.).

REFERENCE: Philip, 1954b (key).

actites Philip and Steffan, 1962: 41.—Calif. coast; w. Calif.
aitkeni Philip, 1941a: 191.—Ariz.; n. Mexico.
albipilosus Brennan, 1935: 371.—cent. Calif.; Oreg.
ater Brennan, 1935: 371.—s. Calif.
colei Philip, 1941a: 192.—s. Calif.
comastes Williston, 1885c: 12.—Calif.; B.C. to Mont., s. to Calif. and Ariz.
    var. **fulvipes** Philip, 1960a: 364.—n. Calif.
    var. **willistoni** Brennan, 1935: 373.—Calif.; B.C. to Mont., s. to Calif. and Ariz.
hinei Brennan, 1935: 374.—s. Calif.
parkeri Philip, 1941a: 193.—Ariz.; sw. Calif.
rossi Philip, 1950a: 451.—Calif.
villosulus (Bigot), 1892b: 684 (*Tabanus*).—Calif.; Ariz.
    similis Brennan, 1935: 374.—Calif.

Genus BRENNANIA Philip

Apatolestes, subg. *Comops* Brennan, 1935: 375 (preocc. Aldrich, 1934). Type-species, *Pangonia hera* Osten Sacken (mon.).

Brennania Philip, 1941a: 196 (n. name for *Comops* Brennan). Type-species, *Pangonia hera* Osten Sacken (aut.).

hera (Osten Sacken), 1877: 214 (*Pangonia*).—Calif.; w. cent. Calif.
    var. **fusca** Philip, 1954b: 19 (as ssp.).—Calif.; w. cent. Calif.

Genus PILIMAS Brennan

Pilimas Brennan *in* Philip, 1941d: 130. Type-species, *Diatomineura californica* Bigot (orig. des.).

abaureus (Philip), 1941d: 114 (*Stonemyia*).—Calif.; n. and cent. Calif.
californicus (Bigot), 1892b: 618 (*Diatomineura*).—Calif.; B.C. to Mont., s. to s. Utah and Calif.
    dives Williston, 1887b: 130 (*Pangonia*; preocc. Macquart, 1857).—Calif.
    jonesi Cresson, 1919: 175 (*Silvius*).—Calif.
    var. **beameri** Philip, 1942a: 61 (as sp.).—Calif.
ruficornis (Bigot), 1892b: 615 (*Corizoneura*).—n. Calif.

Genus STONEMYIA Brennan

Stonemyia Brennan, 1935: 360. Type-species, *Pangonia tranquilla* Osten Sacken (orig. des.).

isabellina (Wiedemann), 1828: 112 (*Silvius*).—N. Amer.; N.Y., s. to Tenn. and Ga.
    pigra Osten Sacken, 1875b: 367 (*Pangonia*).—Ky.
rasa (Loew), 1869b: 5 (Cent. 8, no. 7) (*Pangonia*).—Wis.; Wis. to N.B., s. to Ga.
tranquilla tranquilla (Osten Sacken), 1875b: 367 (*Pangonia*).—N.H.; Wis. to Que., s. to Ga.
    ssp. **fera** (Williston), 1887b: 130 (*Pangonia;* as sp.).—Oreg.; B.C. and Alta. to Calif., also Maine, s. to Tenn.
velutina (Bigot), 1892b: 615 (*Corizoneura*).—Calif.; cent. Calif.
    albomacula Stone, 1940: 60.—Calif.

Genus ASAPHOMYIA Stone

Asaphomyia Stone, 1953: 256. Type-species, *texensis* Stone (orig. des.).

texensis Stone, 1953: 256.—Tex.

Genus ESENBECKIA Rondani

Esenbeckia Rondani, 1863: 83, 95 (as *Esenbekia*, p. 83) (1864: 83). Type-species, *Silvius vulpes* Wiedemann (orig. des.).

delta (Hine), 1920: 313 (*Pangonia*).—Ariz.; Tex., n. Mexico.
incisuralis (Say), 1823: 31 (1859b: 53) (*Pangonia*).—Ark. (Territory); Ariz., e. to Kans., Okla., and Tex., n. Mexico.
  incisa Wiedemann, 1828: 90 (*Pangonia*).—Ark.
  latiflagrum Enderlein, 1925a: 291 (*Ricardoa*).—Tex.
  var. **tinkhami** Philip, 1954b: 50 (as ssp.).—Tex.; Ariz., Colo.
micheneri Philip, 1954b: 48.—n. Mexico; Tex.

Tribe SCIONINI

Genus GONIOPS Aldrich

Goniops Aldrich, 1892b: 236. Type-species, *hippoboscoides* Aldrich (mon.) = *chrysocoma* (Osten Sacken).

chrysocoma (Osten Sacken), 1875b: 368 (*Pangonia*).—N.Y.; Ont. to Ark. and S.C.
  hippoboscoides Aldrich, 1892b: 237.—Pa.

Subfamily CHRYSOPINAE

REFERENCE: Mackerras, 1955 (classif.).

Tribe BOUVIEROMYIINI

Genus MERYCOMYIA Hine

Merycomyia Hine, 1912: 515. Type-species, *geminata* Hine (orig. des.) = *whitneyi* (Johnson).

REFERENCE: Philip, 1954b (key).

brunnea Stone, 1953: 256.—Fla.
mixta Hine, 1912: 516.—Ga.
whitneyi (Johnson), 1904: 15 (*Tabanus*).—N.Y.; Ont. to Ind., Conn., and Fla.
  geminata Hine, 1912: 515.—Conn.

Tribe CHRYSOPINI

Genus SILVIUS Meigen

Silvius Meigen, 1820: 27. Type-species, *Tabanus vituli* Fabricius (mon.).

REFERENCE: Philip, 1954b (key).

Subgenus SILVIUS Meigen

gigantulus (Loew), 1872a: 57 (Cent. 10, no. 12) (*Chrysops*).—Calif., B.C. to Calif., N. Mex., and Nebr.
    trifolium Osten Sacken, 1875b: 395.—B.C. (Vancouver I.)
microcephalus Wehr, 1922a: 109 (=p. 3).—Colo.

Subgenus GRISEOSILVIUS Philip

Silvius, subg. **Griseosilvius** Philip, 1961b: 235. Type-species, *Chrysops quadrivittatus* Say (orig. des.).

abdominalis Philip, 1954b: 55.—sw. Calif.; Nev.
notatus (Bigot), 1892b: 623 (*Diachlorus*).—Calif.; Wash., Oreg.
    laticallus Brennan, 1935: 353.—Oreg.
pollinosus pollinosus Williston, 1880: 244.—Kans.; S. Dak. to Ariz. and Tex.
    ssp. **jeanae** Pechuman, 1960: 793.—Tex.; Calif.
quadrivittatus (Say), 1823: 33 (1859b: 54) (*Chrysops*).—Near the Rocky Mts.; Mont. to Ill., s. to Calif. and Tex.
    var. **texanus** Pechuman, 1938b: 166.—Tex.; Okla., n. Mexico.
sayi Brennan, 1935: 357.—Tex.

Subgenus ZEUXIMYIA Philip

Silvius, subg. **Zeuximyia** Philip, 1941a: 186. Type-species, *philipi* Pechuman (orig. des.).

philipi Pechuman, 1938b: 165.—Oreg.; n. Calif.

Subgenus ASSIPALA Philip

Assipala Philip, 1941b: 4, 9 (as genus). Type-species, *Chrysops tanycerus* Osten Sacken (orig. des.). **N. status.**

REFERENCE: Philip, 1954b (key).

ceras (Townsend), 1897e: 38 (*Chrysops*).—N. Mex.; n. Mexico.

Genus CHRYSOPS Meigen

Chrysops Meigen, 1800: 23. Type-species, *Tabanus caecutiens* Linnaeus (sub. mon., Meigen, 1803: 267). Suppressed by I.C.Z.N., 1963b: 339.
Chrysops Meigen, 1803: 267. Type-species, *Tabanus caecutiens* Linnaeus (mon.).
Chrysops, subg. *Heterochrysops* Kröber, 1920: 50. Type-species, *flavipes* Meigen (Bequaert, 1924: 31).

REFERENCE: Philip, 1955 (keys).

Subgenus CHRYSOPS Meigen

abatus Philip, 1941d: 120.—Fla.; s. Miss., n. to N.C.
aberrans Philip, 1941d: 122.—Minn.; Minn. to N.B., s. to Kans. and N.J.
aestuans Wulp, 1867: 135.—Wis.; Alaska to N.S., s. to Calif., Okla., and Pa.
    moerens Walker, 1848: 201 (preocc. Fabricius, 1787).—N.S.
    var. **abaestuans** Philip, 1941d: 121 (as ssp.).—Kans.; Sask. to Que., s. to Idaho and Nebr.
    var. **pseudoconfusus** Philip, 1959: 200.—S. Dak.
amazon amazon Daecke, 1905: 250.—N.J.; N.H. to N.J.
    ssp. **hubbelli** Philip, 1955: 88.—Fla.
asbestos Philip, 1950a: 455.—Mont.; B.C. to Calif. and Wyo.
atlanticus Pechuman, 1949: 79.—Del.; Mass. to Fla.
beameri Brennan, 1935: 265.—Kans.; Kans. to Mass. and s. to Fla.
bishoppi Brennan, 1935: 266.—Calif.; Oreg.
    var. **gilvus** Philip, 1959: 200.—Calif.
bistellatus Daecke, 1905: 249.—N.J.; N.J. to La. and Fla.
brimleyi Hine, 1904a: 55.—N.C.; N.J. to Miss. and Fla.
brunneus Hine, 1903c: 34.—Ohio; Ariz. to Ont., N.J., and Fla.
callidus Osten Sacken, 1875b: 379.—Ill.; B.C. to Maine, s. to Tex. and Fla.
    callidula Philip, 1941d: 117 (unjustified n. name for *callidus* Osten Sacken, not *calidus* Walker).—Ill.
    var. **confusus** Kröber, 1926: 284 (as *moerens* var.).—B.C.; Wash.
carbonarius Walker, 1848: 203.—N.S.; Alaska to Labr., s. to Calif. and Fla.
    ater Macquart, 1850: 344 (1850: 40).—Nfld.
    fugax Osten Sacken, 1875b: 375.—Maine.
    niger Macquart of Walker, 1848: 202.
    var. **nubiapex** Philip, 1955: 92 (as ssp.).—N.Y.; Ont., s. to Pa.
cincticornis cincticornis Walker, 1848: 201.—Unknown; Sask. to N.S., s. to Okla. and Ga.
    celer Osten Sacken, 1875b: 376.—Mass.
    ssp. **nigropterus** Fairchild, 1937a: 59 (as *celer* var.).—Fla.
clavicornis Brennan, 1935: 277.—Calif.; Nev.
    var. **brennani** Philip, 1955: 94 (as ssp.).—Nev.; sw. Calif.
coloradensis Bigot, 1892b: 605.—Colo.; B.C. to Calif. and Colo.
coquillettii Hine, 1904d: 220.—Calif.; Utah.
    var. **robustus** Brennan, 1935: 333 (as sp.).—Calif.
cuclux Whitney, 1879: 35.—N.H.; Wis. to N.S., s. to Ohio and N.C.
cursim Whitney, 1879: 36.—N.H.; N.H. to Miss. and Fla.

dacne Philip, 1955: 99.—Ga.; Ill. to Conn., s. to La. and Fla.
dawsoni Philip, 1959: 196.—Minn.; Man. and Minn. to Que.
delicatulus Osten Sacken, 1875b: 380.—N.H.; Mich. to Maine and Del.
dimmocki Hine, 1905: 393.—Mass.; Mich. to Mass., s. to La. and Fla.
discalis Williston, 1880: 245.—Wyo.; B.C. to Man., s. to Calif. and Nebr.
dissimilis Brennan, 1935: 288.—N. Mex.; Tex.
divisus Walker, 1848: 204.—Fla.; Ga.
    atropos Osten Sacken, 1875b: 372.—Fla.
dorsovittatus Hine, 1907b: 229.—Ga., Fla.; Md. to Fla.
excitans Walker, 1850: 72.—Cape Breton I.; Alaska to Labr., s. to Calif. and N.J.
    lumbalis Harris *in* Johnson, 1925c: 68. Nomen nudum.
facialis Townsend, 1897e: 39.—N. Mex.; Ariz., n. Mexico.
fascipennis Macquart, 1834a: 216.—Pa. Unrecognized.
flavidus Wiedemann, 1821b: 55 (1821c: 105).—Ga.; Mich. to Mass., s. to Tex. and Fla., Bahama Is., Cuba, Mexico.
    pallida Macquart, 1838a: 166 (1838: 162).—Unknown.
    canifrons Walker, 1848: 197.—Fla.
    pallidus Bellardi, 1859: 73 (1861: 273) (preocc. Macquart, 1838).—Mexico.
  var. **celatus** Pechuman, 1949: 82 (as ssp.).—N.J.; N.Y. to Fla. (coastal).
  var. **reicherti** Fairchild, 1937a: 60 (as sp.).—Fla.; Tenn. and Va., s. to La. and Fla.
frigidus Osten Sacken, 1875b: 384.—N.Y.; B.C. to Labr., s. to Oreg., Colo., and N.J.
    canadensis Kröber, 1926: 277.—Ont.
  var. **xanthas** Philip, 1950a: 453 (as ssp.).—Mont.; Mont. to N.H. and Mass.
fuliginosus Wiedemann, 1821b: 59 (1821c: 109).—N. Amer.; N.S. to Fla.
    plangens Wiedemann, 1828: 210.—Ga.
    confusus Harris, 1835: 597. Nomen nudum.
fulvaster Osten Sacken, 1877: 221.—Colo.; Alta. to Minn., s. to Calif. and Okla.
fulvistigma Hine, 1904a: 55.—N.C.; La. to N.J. and Fla.
  var. **dorsopunctus** Fairchild, 1937a: 59.—Fla.; La. to Fla.
furcatus Walker, 1848: 199.—Ont.; Alaska to Labr., s. to Calif. and Wis.
    lupus Whitney, 1904: 205.—Colo.
  var. **chagnoni** Philip, 1955: 106 (as ssp.).—Que.; Alaska and B.C. to Labr. and N.S.

geminatus Wiedemann, 1828: 205.—Unknown; Wis. to Que., s. to Okla. and Fla.
  fallax Osten Sacken, 1875b: 392.—N.Y.
  var. **impunctus** Kröber, 1926: 301.—Ont.; Mich. to Ont., s. to Tenn. and Ala.
hinei Daecke, 1907: 143.—N.J.; Mass. to Miss. and Fla.
hirsuticallus Philip, 1941d: 126.—Calif.; cent. Calif.
indus Osten Sacken, 1875b: 383.—N.Y.; Colo., e. to Que. and N.C.
  pilumnus Kröber, 1926: 278.—Ont.
lateralis Wiedemann, 1828: 209.—Unknown; Ont. and N.S. to Md.
  hilaris Osten Sacken, 1875b: 391.—N.H.
latifrons Brennan, 1935: 312.—Nev.; sw. Calif.
luteopennis Philip, 1936e: 159.—Minn.; Ohio, Mich.
macquarti Philip, 1961a: 161.—Md.; Nebr. and Minn. to Maine, s. to La. and Fla.
  univittatus, authors, not Macquart.
mitis Osten Sacken, 1875b: 374.—Canada; Alaska to Labr., s. to Calif. and W. Va.
moechus Osten Sacken, 1875b: 387.—D.C.; Minn. to Que., s. to La. and Ga.
montanus Osten Sacken, 1875b: 382.—N.Y.; Man. to N.B., s. to e. Tex. and Fla.
  var. **perplexus** Philip, 1955: 111 (as ssp.).—N.C.; N.C. to Fla.
niger Macquart, 1838a: 165 (1838: 161).—N. Amer.; Mont. to N.S., s. to Okla. and Fla.
  var. **taylori** Philip, 1955: 112 (as ssp.).—Fla.; Ark., N.C.
nigribimbo Whitney, 1879: 36.—N.H.; N.H. to Fla.
nigripes Zetterstedt, 1838: 519.—N. Lapland; Alaska to Labr., s. to Idaho and Maine, n. Europe.
noctifer noctifer Osten Sacken, 1877: 220.—Calif.; Idaho and Oreg., s. to n. Mexico.
  ssp. **pertinax** Williston, 1887b: 132 (as sp.).—Wash. Territory; Yukon, s. to Calif. and Colo.
  nigriventris Bigot, 1892b: 604.—Wash. Territory.
obsoletus Wiedemann, 1821b: 58 (1821c: 108).—Ga.; Mass. to Fla. (coastal).
  trinotata Macquart, 1838a: 165 (1838: 161).—Pa.
  morosus Osten Sacken, 1875b: 389.—Md.
  var. **lugens** Wiedemann, 1821b: 59 (1821c: 109) (as sp.).—Ga.; Conn. to Fla. (coastal).
  ultimus Whitney, 1914: 345.—Fla.

pachycerus Williston, 1887b: 134.—Calif.; Utah and sw. Calif. to Tex.
  dilatus Rowe and Knowlton, 1936: 256.—Utah.
  var. **hungerfordi** Brennan, 1935: 306 (as sp.).—N. Mex.; sw. Calif. to Tex., n. Mexico.
parvulus Daecke, 1907: 142.—N.J.; Ark. to N.J. and Fla.
pechumani Philip, 1941d: 128.—Calif.; n. and cent. Calif.
pikei Whitney, 1904: 205.—Mo.; Nebr. to Ont., s. to Tex. and Fla.
proclivis Osten Sacken, 1877: 222.—Calif.; Y.T. to Alta., s. to Calif. and Colo.
  var. **atricornis** Bigot, 1892b: 603 (as sp.).—Colo.; B.C. to Calif.
  imfurcatus Philip, 1936e: 157 (as *proclivis* var.).—Wash.
provocans Walker, 1850: 73.—N.S. Unrecognized, ?=*carbonarius*.
pudicus Osten Sacken, 1875b: 381.—Mass.; Wis. to Mass., s. to Okla. and Fla.
sackeni Hine, 1903c: 42.—Ohio; Man. to Que., s. to Okla. and Fla.
separatus Hine, 1907b: 228.—N.C.; Okla. and La. to N.C.
sequax Williston, 1887b: 133.—Kans.; Colo. to Ohio., s. to Okla. and Ga.
  var. **tau** Philip, 1955: 113 (as ssp.).—Ill.; Ohio, Ark.
shermani Hine, 1907b: 229.—N.C.; Wis. to N.B. and N.C.
sordidus Osten Sacken, 1875b: 376.—N.H.; Que. and Labr., s. to Mich. and N.Y.
striatus Osten Sacken, 1875b: 391.—?D.C.; N. Dak. to N. B., s. to La. and N.C.
surdus Osten Sacken, 1877: 223.—Calif.; B.C., Nev.
  var. **piceus** Philip, 1936e: 157 (as *proclivis* var.).—Calif.; n. and cent. Calif.
tidwelli Philip and Jones, 1962: 67.—Fla.
univittatus Macquart, 1855: 56 (1855: 36).—Md.; Minn. to N.S., s. to Kans., La., and Fla.
  wiedemanni Kröber, 1926: 267.—Que.
  fraternus Kröber, 1926: 317.—Unknown.
upsilon Philip, 1950a: 458.—Ga.; N.C., La.
venus Philip, 1950a: 457.—Wis.; e. Man. to Que.
virgulatus Bellardi, 1859: 71 (1861: 271).—Mexico; sw. Calif. to Tex.
vittatus Wiedemann, 1821b: 56 (1821c: 106).—N. Amer.; Minn. to N.S., s. to Tex. and Fla.
  areolatus Walker, 1848: 197.—N.Y.
  lineatus Jaennicke, 1867: 334 (1868: 26).—Ill.
  ornatus Kröber, 1926: 328.—Ont.
  var. **floridanus** Johnson, 1913a: 52.—Fla.; Va. to Fla.
wileyae Philip, 1955: 96.—Calif.; Oreg. to Utah, s. to Calif., Mexico (Baja Calif.).

zinzalus Philip, 1942a: 62.—N.B.; N.S.
  lapponicus Loew of Brennan, 1935: 378.

Subgenus LIOCHRYSOPS Philip

Chrysops, subg. **Liochrysops** Philip, 1955: 87. Type-species, *hyalinus* Shannon (orig. des.).
hyalinus Shannon, 1924c: 178 (n. name for *vitripennis* Shannon).—Md.; N.C., n. Fla.
  vitripennis Shannon, 1916b: 69 (preocc. Meigen, 1820).—Md.
  claripennis Kröber, 1926: 230 (n. name for *vitripennis* Shannon).—Md.

Genus NEOCHRYSOPS Walton

Neochrysops Walton *in* McAtee and Walton, 1918: 191. Type-species, *globosus* Walton (mon.).
globosus Walton *in* McAtee and Walton, 1918: 192.—Md.

Subfamily TABANINAE

REFERENCE: Stone, 1938 (rev.).

Tribe DIACHLORINI

REFERENCE: Mackerras, 1954 (classif.).

Genus DIACHLORUS Osten Sacken

Diabasis Macquart, 1834a: 207 (preocc. Hoffmansegg, 1817). Type-species, *Tabanus bicinctus* Fabricius (Coquillett, 1910b: 532).
Diachlorus Osten Sacken, 1876a: 475 (n. name for *Diabasis* Macquart). Type-species, *Tabanus bicinctus* Fabricius (aut.).
badius Kröber, 1928: 50.—Unknown (as Ga., probably in error). Not Nearctic.
ferrugatus (Fabricius), 1805: 111 (*Chrysops*).—Carolina; N.J. to La. and Fla., Mexico, Honduras.
  americanus Palisot de Beauvois, 1819: 222 (*Tabanus;* preocc. Forster, 1771).—U.S.
  ataenia Macquart, 1838a: 156 (1838: 152) (*Diabasis*).—Carolina.
  approximans Walker, 1848: 198 (*Chrysops*).—Fla.

Genus STENOTABANUS Lutz

Stenotabanus Lutz, 1913a: 487. Type-species, *Tabanus taeniotes* Wiedemann (sub. mon., Lutz and Neiva, 1914: 73).

Subgenus STENOTABANUS Lutz

daedalus Stone, 1938: 32.—Fla.; Ga.
flavidus (Hine), 1904d: 236 (*Tabanus*).—Ariz., n. Mexico.
floridensis (Hine), 1912: 513 (*Tabanus*).—Fla.; ?Ga.
guttatulus (Townsend), 1893b: 134 (*Diachlorus*).—N. Mex.; Utah, sw. Calif., nw. Mexico.
   cribellum, authors, not Osten Sacken.

Subgenus AEGIALOMYIA Philip

Aegialomyia Philip, 1941b: 10 (as genus). Type-species, *Tabanus psammophilus* Osten Sacken (orig. des.).

atlanticus (Johnson), 1913d: 445 (*Tabanus*).—Bermuda.
magnicallus (Stone), 1935: 19 (*Tabanus*; n. name for *nanus* Macquart).—Tex.; La., e. Mexico.
   nanus Macquart, 1846: 170 (1846: 42) (*Tabanus*; preocc. Wiedemann, 1821).—Tex.
   maritimus Townsend, 1898a: 167 (*Tabanus*; preocc. Scopoli, 1763).—Tex.
psammophilus (Osten Sacken), 1876a: 445 (*Tabanus*).—Fla.

Genus ANACIMAS Enderlein

Anacimas Enderlein, 1923: 545. Type-species, *limbellatus* Enderlein (orig. des.).

dodgei (Whitney), 1879: 37 (*Tabanus*).—Nebr.; Nebr. to Okla.
geropogon Philip, 1936b: 229.—N.C.; N.C. to Fla.
limbellatus Enderlein, 1923: 545.—U.S.; S.C. to Fla.

Genus MICROTABANUS Fairchild

Tabanus, subg. **Microtabanus** Fairchild, 1937b: 10. Type-species, *Tabanus pygmaeus* Williston (mon.).
pygmaeus (Williston), 1887b: 141 (*Tabanus*).—Fla.; N.Y. to Fla.

Genus CHLOROTABANUS Lutz

Chlorotabanus Lutz, 1909: 30. Type-species, *Tabanus mexicanus* Linnaeus (mon.).

REFERENCE: Philip and Fairchild, 1956 (key).

crepuscularis (Bequaert), 1926b: 234 (*Tabanus;* n. name for *flavus* Macquart).—U.S.; Ariz. to Fla., n. to N.J.
> *flavus* Macquart, 1834a: 200 (*Tabanus;* preocc. Wiedemann, 1828).—U.S.
> *mexicanus,* authors, not Linnaeus.
> *sulphureus,* authors, not Palisot de Beauvois.

Genus BOLBODIMYIA Bigot

Bolbodimyia Bigot, 1892c: 162. Type-species, *bicolor* Bigot (mon.).
Snowiellus Hine, 1904d: 230. Type-species, *atratus* Hine (mon.).
> REFERENCE: Stone, 1954 (key).

atrata (Hine), 1904d: 230 (*Snowiellus*).—Ariz.; Mexico.

Tribe HAEMATOPOTINI

Genus HAEMATOPOTA Meigen

Chrysozona Meigen, 1800: 23. Type-species, *Tabanus pluvialis* Linnaeus (Coquillett, 1910b: 524). Suppressed by I.C.Z.N., 1963b: 339.
Haematopota Meigen, 1803: 267. Type-species, *Tabanus pluvialis* Linnaeus (mon.).

americana Osten Sacken, 1875b: 395.—N.W.T.; Alaska to Labr., s. to Calif. and N. Mex.
champlaini (Philip), 1953: 248 (*Chrysozona*).—Pa.; R.I.
punctulata Macquart, 1838a: 167 (1838: 163).—Carolina; N.J. to Fla.
rara Johnson, 1912d: 182.—Pa.; N.H. to Va.
willistoni (Philip), 1953: 249 (*Chrysozona*).—Calif.

Tribe TABANINI

Genus ATYLOTUS Osten Sacken

Tabanus, subg. **Atylotus** Osten Sacken, 1876a: 426. Type-species, *bicolor* Wiedemann (Hine, 1900: 247).

bicolor (Wiedemann), 1821b: 46 (1821c: 96) (*Tabanus*).—N. Amer.; Alta. to Maine, s. to Wis. and D.C.
> *fulvescens* Walker, 1848: 171 (*Tabanus*).—Mass. and Unknown.
> *ruficeps* Macquart, 1855: 55 (1855: 35) (*Tabanus*).—Md.

duplex (Walker), 1854: 173 (*Tabanus;* n. name for *imitans* Walker, 1848: 173).—Ont.; Alta. to Que.
> *imitans* Walker, 1848: 173 (*Tabanus;* preocc. Walker, 1848: 146).—Ont.

incisuralis (Macquart), 1847: 37 (1847: 21) (*Tabanus*).—Amer.; Alaska to Ont., s. to Calif. and Colo.
  intermedius Walker, 1848: 173 (*Tabanus*).—Ont.
  insuetus Osten Sacken, 1877: 219 (*Tabanus*).—Calif.
  var. **utahensis** (Rowe and Knowlton), 1935: 242 (*Tabanus*; as sp.).—Utah.
ohioensis (Hine), 1901a: 28 (*Tabanus*; n. name for *pruinosus* Hine).— Ohio; Alta. to Pa.
  pruinosus Hine, 1900: 248 (*Tabanus*; preocc. Bigot, 1892).— Ohio.
pemeticus (Johnson), 1921a: 11 (*Tabanus*.)—Maine; Minn. to Nfld., s. to Pa.
thoracicus (Hine), 1900: 248 (*Tabanus*).—N.Y.; ne. Calif. to N.S., s. to Del.
tingaureus (Philip), 1936a: 159 (*Tabanus*; as *insuetus* var.).—Mont.; Alaska to n. Calif.

Genus WHITNEYOMYIA Bequaert

Tabanus, subg. **Whitneyomyia** Bequaert, 1933b: 85. Type-species, *beatificus* Whitney (orig. des.).

beatifica (Whitney), 1914: 344 (*Tabanus*).—Fla.; Tex. to N.C. and Fla.
  ater Palisot de Beauvois, 1811: 101 (*Tabanus*; preocc. Rossi, 1790).—S.C.
  lugubris Macquart, 1838a: 149 (1838: 145) (*Tabanus*; preocc. Linnaeus, 1776).—Carolina.
  stygius Enderlein, 1925a: 353 (*Snowiellus*).—Unknown.
  var. **atricorpus** Philip, 1950b: 122 (as ssp.).—Fla.; Tex.

Genus LEUCOTABANUS Lutz

Leucotabanus Lutz, 1913a: 487. Type-species, *Tabanus leucaspis* Wiedemann (sub. mon., Lutz and Neiva, 1914: 71)= *exaestuans* (Linnaeus).

ambiguus Stone, 1938: 26.—Ariz.; Guatemala.
  albiscutellatus Macquart of Hine, 1925: 34.
annulatus (Say), 1823: 32 (1859b: 53) (*Tabanus*).—Mo.; Kans. to Del., s. to Tex. and Fla.
  argenteus Wiedemann *in* Philip, 1950b: 121 (*Tabanus*). Nomen nudum.

Genus TABANUS Linnaeus

Tabanus Linnaeus, 1758: 601. Type-species, *bovinus* Linnaeus (Latreille, 1810: 443).
Neotabanus Lutz, 1909: 30. Type-species, *Tabanus trilineatus* Latreille (Bequaert, 1924: 29).
Taeniotabanus Kröber, 1931b: 68. Type-species, *Tabanus dorsiger* Wiedemann (mon.).

Subgenus TABANUS Linnaeus

aar Philip, 1941c: 105.—Miss.; La. to Fla.
abactor Philip, 1936d: 153.—Tex.; Kans. to Tex.
abditus Philip, 1941c: 142.—Ariz.
abdominalis Fabricius, 1805: 96.—Carolina; Nebr. to Mass., s. to Tex. and Fla.
    limbatinevris Macquart, 1847: 32 (1847: 16).—Unknown (as Tasmania, error).
acutus (Bigot), 1892b: 660 (*Atylotus*).—La.; La. to Fla.
aegrotus Osten Sacken, 1877: 219.—Calif.; B.C. to Mont. and Calif.
americanus Forster, 1771: 100.—Va., N.Y.; Ont. to Tex. and Fla.
    plumbeus Drury, 1773: 2, of index (1770: 103, unnamed descr.).—N.Y., Va.
    ruficornis Fabricius, 1775: 789.—Amer.
    limbatus Palisot de Beauvois, 1806: 54.—U.S.
aranti Hays, 1961: 127.—Ala.; Ark., Va., Ga.
atratus Fabricius, 1775: 789.—Amer.; Wash. to Que., s. to N. Mex. and Fla.
    americanus Drury, 1773: 2, of index (1770: 104, unnamed descr.) (preocc. Forster, 1771).—N.Y.
    niger Palisot de Beauvois, 1806: 54.—Pa.
    validus Wiedemann, 1828: 113.—Pa.
    var. **fulvopilosus** Johnson, 1919c: 164.—Fla.; Mass. to Fla.
    var. **nantuckensis** Hine, 1917: 271 (as sp.).—Mass.; N.H. to N.C.
birdiei Whitney, 1914: 343.—Fla.
bishoppi Stone, 1933: 77.—Fla.
boharti Philip, 1950b: 115.—Ariz.
    pruinosus, authors, not Bigot.
calens Linnaeus, 1758: 601.—Amer.; Wis. to Conn., s. to Tex. and Fla.
    giganteus DeGeer, 1776: 226.—Pa.
    lineatus Fabricius, 1781: 455.—Amer.
    pallidus Palisot de Beauvois, 1809: 100.—U.S.
    bicolor Macquart, 1847: 37 (1847: 21) (preocc. Wiedemann, 1821).—S.C.
    coesiofasciatus Macquart, 1855: 52 (1855: 32).—Md.

catenatus Walker, 1848: 148.—Mass.; Minn. to N.S., s. to N.C.
  orion Osten Sacken, 1876a: 442.—Canada, Mass., Conn., N.Y.
cayensis Fairchild, 1935: 53.—Fla.
cheliopterus Rondani, 1850a: 192.—Carolina; Tex. to Fla., n. to Va.
  subfronto Philip, 1936c: 100.—N.C.
  var. **fronto** Osten Sacken, 1876a: 431 (as sp.).—Ga.; Va., Fla., Tex.
cingulatus Macquart, 1838a: 148 (1838: 144) (preocc. Thunberg, 1827).—Pa. Unrecognized.
coarctatus Stone, 1935: 13.—Fla.; Ga.
colombensis Macquart, 1846: 165 (1846: 37).—Colombia; s. Tex., Cent. Amer., n. S. Amer., Jamaica.
  amplifrons Kröber, 1933: 354.—Venezuela.
commixtus Walker.—Not Nearctic.
cymatophorus Osten Sacken, 1876a: 444.—Ky.; Ill. to Md., s. to Tex. and Ga.
derivatus Walker, 1848: 151.—Unknown (as N. Amer., error). Not Nearctic.
dietrichi Pechuman, 1956: 39.—Ariz.
dorsifer Walker, 1860a: 273.—Mexico; Ariz. to Tex.
  intensivus Townsend, 1897e: 93.—N. Mex.
  hyalinipennis Hine, 1903a: 244.—Ariz.
dorsonotatus Macquart, 1847: 38 (1847: 22).—Carolina. Unrecognized.
eadsi Philip, 1962: 171.—Tex.
endymion Osten Sacken, 1878a: 556.—Ga.; La. to Fla. and N.C.
equalis Hine, 1923b: 205 (n. name for *uniformis* Hine).—Kans.; Kans. to Ill., s. to Tex. and Fla.
  uniformis Hine, 1917: 270 (preocc. Ricardo, 1911).—Kans.
erythraeus (Bigot), 1892b: 661 (*Atylotus*).—Mexico; Ariz.
eurycerus Philip, 1937b: 66.—Ariz.
exilipalpis Stone, 1938: 54.—S.C.; Miss., Ala.
fairchildi Stone, 1938: 63.—N.Y.; Wis. to N.S., s. to Okla. and Fla.
ferrugineus Palisot de Beauvois, 1819: 221 (preocc. Strøm, 1768).—U.S. Unrecognized.
fulvicallus Philip, 1931: 106.—Minn.; Man. and Minn. to N.Y.
fulvulus Wiedemann, 1828: 153.—Amer.; Ill. to se. N.Y., s. to Okla., La., and Fla.
  fulvofrater Walker, 1848: 181.—Ill.
  mutatus Walker, 1850: 23.—U.S.
fumipennis Wiedemann, 1828: 119.—Ga.; S.C. to Fla.
  rufus Palisot de Beauvois, 1809: 100 (preocc. Scopoli, 1763).—U.S.

fuscicostatus Hine, 1906a: 24.—La.; Kans. to N.C., s. to Tex. and La.
fusconervosus Macquart, 1838a: 151 (1838: 147).—Unknown; Mass. to Fla.
    confusus Walker, 1848: 147.—Ga.
    recedens Walker, 1848: 147.—Fla.
    fur Williston, 1887b: 139.—Fla.
gilanus Townsend, 1897e: 92.—N. Mex.; Ariz., Tex., Colo.
gladiator Stone, 1935: 12.—S.C.; Tenn. to Va., s. to Tex. and Fla.
gracilis Wiedemann, 1828: 156.—Ga.; N.J., s. to Miss. and Fla.
guttatus Wiedemann, 1821b: 23 (1821c: 73) (preocc. Donovan, 1815).—Unknown (as Amer., error; probably Georgian S.S.R.). Unrecognized, not Nearctic.
imitans Walker, 1848: 146.—Ga.; Tex. to N.C. and Fla.
    fuscopunctatus Macquart, 1850: 338 (1850: 34).—Ga.
    var. **excessus** Stone, 1938: 87.—Fla.; S.C., Ga.
    var. **pechumani** Philip, 1960b: 171.—Fla.; Ga.
johnsoni Hine, 1907b: 225.—Fla.
kesseli Philip, 1950b: 117.—Mont.; B.C. and Calif., e. to Wis.
kisliuki Stone, 1940: 59.—Miss.; Miss. to s. Fla.
laticeps Hine, 1904d: 239.—Calif., Wash.; B.C., Oreg.
leucomelas Walker, 1848: 175.—Unknown (as Ga., error). Unrecognized, not Nearctic.
lineola Fabricius, 1794: 369.—N. Amer.; Wis. to Maine, s. to Tex. and Fla., Bahama Is.
    var. **hinellus** Philip, 1960a: 366.—La.; Miss., Tex.
    quinquemaculatus (error for *quinquevittatus*) Wiedemann of Hine, 1904e: 58.
longiusculus Hine, 1907b: 226.—N.C., Ga.; N.C. to Fla.
longus Osten Sacken, 1876a: 447.—Middle Atlantic States; Pa., s. to Tex. and Ga.
marginalis Fabricius, 1805: 99.—N. Amer.; Idaho to Man. and N.S., s. to Del.
    nivosus Osten Sacken, 1876a: 445.—N.J.
melanocerus Wiedemann, 1828: 122.—Ky.; Ark. to Conn., s. to La. and Fla.
    var. **lacustris** Stone, 1935: 13 (as sp.).—Fla.; Ga.
moderator Stone, 1938: 98.—Ga.; Ark., Miss., N.C., S.C.
molestus Say, 1823: 31 (1859b: 53).—Mo.; Wis. to N.J., s. to Tex. and Fla.
    tenessensis Bigot, 1892b: 660 (*Atylotus*).—Tenn.
    var. **mixis** Philip, 1950b: 241 (as ssp.).—Ga.; Ky. to Va., s. to Tex. and Fla.
monoensis Hine *in* Webb and Wells, 1924: 29.—Calif.; Oreg., Idaho.
morbosus Stone, 1938: 89.—Ariz.; n. Mexico.

mularis Stone, 1935: 15.—La.; Ill. and Md., s. to Tex. and Fla.
nebulosus Palisot de Beauvois, 1819: 222 (preocc. De Geer, 1776).—U.S. Unrecognized.
nefarius Hine, 1907b: 224.—La.; Tex.
nigrescens Palisot de Beauvois, 1809: 100.—U.S.; Wis. to Mass., s. to Tex. and Fla.
    var. **atripennis** Stone, 1935: 15 (as ssp.).—Okla.; Kans. to Va., s. to Tex. and Fla.
nigripes Wiedemann, 1821b: 25 (1821c: 75).—Ga.; Ont., s. to Tex. and Fla.
    coffeatus Macquart, 1847: 39 (1847: 23).—Pa.
    winthemi Kröber, 1931a: 295.—Unknown (as S. Amer., error).
nigrovittatus Macquart, 1847: 40 (1847: 24).—N.S.; P.E.I. to Tex. (coastal).
    simulans Walker, 1848: 182.—N.S.
    vicarius Walker, 1848: 187.—N. Amer.
    conterminus Walker, 1850: 24.—U.S.
    allynii Marten, 1883: 110.—N.C.
    floridanus Szilády, 1926a: 24.—Fla.
    divisus Harris, 1833: 593. Nomen nudum.
    var. **fulvilineis** Philip, 1957: 3 (as ssp.).—Miss.; Tex. to Fla., Bahama Is.
novaescotiae Macquart, 1847: 40 (1847: 24) (as *novae-scoeiae*).—N.S.; Minn. to N.S., s. to N.C.
    actaeon Osten Sacken, 1876a: 443.—Mass.
orbicallus Philip, 1936a: 157.—Kans.; S. Dak. to Okla.
pallidescens Philip, 1936d: 150 (as *fulvulus* var.).—Miss.; Mo. to Md., s. to Okla., La., and Fla.
palpinus Palisot de Beauvois, 1819: 221.—U.S. Unrecognized, ?=*wiedemanni*.
petiolatus Hine, 1917: 270.—La.; Tex. to Fla., n. to N.J.
    yulenus Philip, 1950b: 243.—La.
productus Hine, 1904d: 242.—Wyo.; Calif. to Mont. and N. Mex.
proximus Walker, 1848: 147.—Fla.; Ill. to Va., s. to Tex. and Fla.
    benedictus Whitney, 1904: 206.—Mo.
pumilus Macquart, 1838a: 150 (1838: 146).—Carolina; Ont., s. to Tex. and Fla.
punctifer Osten Sacken, 1876a: 453.—Utah; B.C. to Kans., s. to Calif. and Tex.
quaesitus Stone, 1938: 54.—Tex.; La.
quinquevittatus Wiedemann, 1821b: 34 (1821c: 84).—Ga. (as Mexico, error); S. Dak. to N.S. and e. U.S. to Tex.
    costalis Wiedemann, 1828: 173.—Ky.

      manifestus Walker, 1850: 41.—Unknown (as ?S. Amer., error).
      baltimorensis Macquart, 1855: 54 (1855: 34).—Md.
quirinus Philip, 1950b: 120.—Fla.; n. Fla.
reinwardtii Wiedemann, 1828: 130.—Pa.; Alta. to N.S., s. to Tex. and Ga.
      erythrotelus Walker, 1850: 25.—Unknown.
rufofrater Walker, 1850: 26 (as *rufofrator*).—Ga.; Miss. to N.C. and Fla.
      unicolor Macquart, 1847: 38 (1847: 22) (preocc. Wiedemann, 1828).—Carolina.
      lateritius Rondani, 1863: 80 (1864: 80) (n. name for *unicolor* Macquart).—Carolina.
      tener Osten Sacken, 1876a: 440.—Ga., Fla.
sackeni Fairchild, 1934: 141.—Ky.; Kans. to N.H. and Ga.
sagax Osten Sacken, 1876a: 452.—Ill.; Minn. to Maine, s. to La. and S.C.
      baal Townsend, 1895a: 58 (*Atylotus*).—Va.
      dawsoni Philip, 1931: 105.—Minn.
schwardti schwardti Philip, 1942b: 29 (as *vittiger* ssp.).—Tenn.; Ariz. to Mass. and Fla., Mexico.
    ssp. **nippontucki** Philip, 1942b: 32 (as *vittiger* ssp.).—Calif.; Calif. to Tex.
similis Macquart, 1850: 335 (1850: 31).—Unknown (as Tasmania, error); B.C. to N.S., s. to Calif. and Ga.
      scutellaris Walker, 1850: 27.—N. Amer.
sparus Whitney, 1879: 38.—N.H.; Wis. to N.H., s. to Okla. and Fla.
    var. **milleri** Whitney, 1914: 344 (as sp.).—Fla.; Okla. to N.Y. and Fla.
stonei Philip, 1941c: 144.—Mont.; B.C. to Mont., s. to Calif. and Tex.
    var. **jellisoni** Philip, 1941c: 146 (as ssp.).—Mont.; Mont. to Colo.
stygius Say, 1823: 33 (1859b: 54).—Ark. Territory; Minn. to Maine, s. to Tex. and Fla.
sublongus Stone, 1938: 74.—Md.; Okla. to N.Y. and N.C.
subniger Coquillett, 1906a: 48.—Ill.; Ont., N.J.
sulcifrons Macquart, 1855: 53 (1855: 33) (as *fulcifrons*).—Md.; Wis. to Maine, s. to Tex. and Fla.
      variegatus Fabricius, 1805: 95 (preocc. De Geer, 1776).—N. Amer.
      tectus Osten Sacken, 1876a: 436.—Pa.
      exul Osten Sacken, 1878a: 558.—D.C., Md., Pa., N.J.
superjumentarius Whitney, 1879: 37.—N.H.; Ohio to N.H., s. to Ark. and Ga.

tetropsis Bigot, 1892b: 681.—Unknown (as Ga., error). Unrecognized, not Nearctic.
texanus Hine, 1907b: 228.—Tex.; La.
trijunctus Walker, 1854: 182.—Fla.; Ga., Miss., Ala., Bahama Is.
trimaculatus Palisot de Beauvois, 1806: 56.—N. Amer.; Minn. to Mass., s. to Tex. and Fla.
    quinquelineatus Macquart, 1834a: 200.—Ga.
turbidus Wiedemann, 1828: 124.—Ky.; Ark. to N.C., s. to La. and Fla.
venustus Osten Sacken, 1876a: 444.—Tex.; Wis., s. to Kans. and Tex.
vittiger vittiger Thomson.—Not Nearctic.
    ssp. **guatemalanus** Hine, 1906a: 24 (as sp.).—Guatemala; s. Fla., Cent. Amer., Antilles.
    bellardii Szilády, 1926a: 23.—Cuba.
    caymanicus Fairchild, 1942: 181 (as *vittiger* ssp.).—Cuba.
vivax Osten Sacken, 1876a: 446.—N.Y.; Wis. to Labr., s. to N.J.
    arborealis Stone, 1935: 14.—Vt.
wiedemanni Osten Sacken, 1876a: 455.—Fla.; N.C. to Fla.
    ater Palisot de Beauvois of Wiedemann, 1828: 136.
wilsoni Pechuman, 1962: 66.—Ark.; La.
zythicolor Philip, 1936d: 152.—N.C.; N.J. to La. and Fla.

Subgenus GLAUCOPS Szilády

Tabanus, subg. **Glaucops** Szilády, 1923: 17. Type-species, *Tabanus hirsutus* Villers (mon.). **N. status.**

fratellus Williston, 1887b: 140.—Wash. Territory; Alaska, s. to Oreg. and Mont.
    haematopotides Bigot, 1892b: 624 (?*Diachlorus*).—Wash. Territory.

Genus HYBOMITRA Enderlein

Hybomitra Enderlein, 1922c: 347. Type-species, *solox* Enderlein (orig. des.) =*rhombica* (Osten Sacken).
Dasyommia Enderlein, 1922c: 346. Type-species, *Tabanus cinctus* Fabricius (orig. des.).
Tylostypia Enderlein, 1922c: 347. Type-species, *Tabanus astur* Erichson (orig. des.).
Sziladynus Enderlein, 1925b: 181. Type-species, *Tabanus aterrimus* Meigen (orig. des.).
Therioplectes, authors, not Zeller.

aasa Philip, 1954a: 28.—Oreg.; B.C. to Calif.
aatos Philip, 1941c: 148.—Wyo.; Colo., Utah, Ariz.

aequetincta (Becker), 1900: 8 (*Therioplectes*).—Siberia; Labr., Que., n. Eurasia.
  flavipes Wiedemann, 1828: 137 (*Tabanus;* preocc. Gravenhorst, 1807).—Labr.
  nigrotuberculatus Fairchild, 1934: 139 (*Tabanus;* n. name for *flavipes* Wiedemann).—Labr.
affinis (Kirby), 1837: 313 (*Tabanus*).—Canada; Alaska to Labr., s. to Ariz. and N.J.
  triligatus Walker, 1854: 183 (*Tabanus*).—Arctic Amer.
arpadi (Szilády), 1923: 7 (*Tabanus*).—Russian Lapland and Amur District; Alaska to Labr., s. to Minn. and Maine, n. and e. Siberia.
  gracilipalpis Hine, 1923a: 143 (*Tabanus*).—Alaska.
  cristatus Curran, 1927f: 81 (*Tabanus*).—Alta.
astuta (Osten Sacken), 1876a: 471 (*Tabanus*).—N.H.; Alaska to Labr. and Ill.
atrobasis (McDunnough), 1921: 144 (*Tabanus*).—B.C.; B.C. to Oreg. and Wyo.
aurilimbus (Stone), 1938: 130 (*Tabanus*).—Maine; Minn. to Nfld., s. to N.J.
brennani (Stone), 1938: 157 (*Tabanus*).—N.H.; Que.
californica (Marten), 1882: 210 (*Tabanus*).—Calif.; B.C., Oreg., Idaho.
captonis (Marten), 1882: 211 (*Tabanus*).—Calif.; Y.T., s. to Calif. and Colo.
  comastes Williston, 1887b: 137 (*Tabanus*).—Wash. Territory and Oreg.
cincta (Fabricius), 1794: 366 (*Tabanus*).—Va.; Man. to Maine, s. to Fla.
comes (Walker), 1849: 1152 (*Tabanus;* n. name for *inscitus* Walker).—Canada. Unrecognized, ?=*astuta*.
  inscitus Walker, 1848: 172 (*Tabanus;* preocc. Walker, 1848: 161).—Canada.
criddlei (Brooks), 1946b: 234 (*Tabanus*).—Man.; Man. and Colo., e. to Ont.
daeckei (Hine), 1917: 269 (*Tabanus*).—N.J.; Maine to Ga.
difficilis (Wiedemann), 1828: 165 (*Tabanus*).—Unknown; Ohio to Vt., s. to Okla. and Fla.
  carolinensis, authors, not Macquart.
epistates (Osten Sacken), 1878a: 555 (*Tabanus;* n. name for *socius* Osten Sacken).—N.W.T.; Alaska to N.S., s. to Oreg., Colo., and N.J., Manchuria.
  socius Osten Sacken, 1876a: 467 (*Tabanus;* preocc. Walker, 1848).—N.W.T.

frenchii (Marten), 1883: 111 (*Therioplectes*).—Mont. Unrecognized.
frontalis (Walker), 1848: 172 (*Tabanus*).—N.S.; Alaska to Labr., s. to Utah and N.Y.
  incisus Walker, 1850: 26 (*Tabanus*).—N.S.
  septentrionalis Loew, 1858c: 592 (*Tabanus*).—Labr.
  labradorensis Enderlein, 1925a: 363 (*Tylostypia*).—Labr.
  canadensis Curran, 1927f: 82 (*Tabanus*).—Man.
frosti Pechuman, 1960: 794.—Ont.; N.S., Que., N.B., Maine, Pa.
fulvilateralis (Macquart), 1838a: 137 (1838: 133) (*Tabanus*).— Unknown (as Cayenne, error); B.C. to Man., s. to Calif. and N. Mex.
  recedens Walker, 1854: 201 (*Tabanus*; preocc. Walker, 1848).—w. coast Amer.
  haemaphorus Marten, 1882: 210 (*Tabanus*).—Calif.
hearlei (Philip), 1936a: 150 (*Tabanus*).—B.C.; B.C. to Labr.
hinei hinei (Johnson), 1904: 15 (*Tabanus*; n. name for *politus* Johnson).—N.J.; Wis. to Mass., s. to N.C.
  politus Johnson, 1900b: 325 (*Therioplectes*; preocc. Walker, 1871).—N.J.
  ssp. **wrighti** (Whitney), 1915: 380 (*Tabanus*; as sp.).—Fla.; Miss., Ga., N.C.
illota (Osten Sacken), 1876a: 469 (*Tabanus*).—N.W.T.; Alaska to Que., s. to Wash., Iowa, and Pa.
itasca (Philip), 1936a: 149 (*Tabanus*).—Minn.; Alaska and Sask. to Labr.
lanifera (McDunnough), 1922: 239 (*Tabanus*).—Alta.; Alaska to Alta., s. to Oreg. and Colo.
lasiophthalma (Macquart), 1838a: 147 (1838: 143) (*Tabanus*).— Carolina; Alaska to N.S., s. to Tex. and Fla.
  punctipennis Macquart, 1847: 39 (1847: 23) (*Tabanus*); preocc. Macquart, 1838).—Pa.
  notabilis Walker, 1848: 166 (*Tabanus*).—N.Y., ?Ga.
  fretus Stone, 1938: 154 (*Tabanus*).—Conn.
  guttiferus Harris *in* Johnson, 1925c: 70 (*Tabanus*). Nomen nudum.
laticallus (Philip), 1936a: 150 (*Tabanus*).—Minn.; n. Minn.
laticornis (Hine), 1904d: 239 (*Tabanus*).—Ariz.; N. Mex.
liorhina (Philip), 1936a: 151 (*Tabanus*).—B.C.; B.C. to Labr., s. to Minn.
longiglossa (Philip), 1931: 110 (*Tabanus*).—Minn.; Minn. to Nfld., s. to Mass.
melanorhina (Bigot), 1892b: 642 (*Therioplectes*).—Wash. Territory; B.C., s. to Calif. and Colo.

metabola (McDunnough), 1922: 239 (*Tabanus*).—Alta.; Alaska to Labr., s. to Colo. and N.Y.
microcephala (Osten Sacken), 1876a: 470 (*Tabanus*).—N.H.; Man. and Minn. to N.S., s. to N.C.
minuscula (Hine), 1907b: 226 (*Tabanus*).—Maine, N.Y., Mass., Ont.; Minn. to Labr., s. to W. Va.
nigricans (Wiedemann), 1828: 157 (*Tabanus*).—Unknown (Amer. bor. on a syntype); Okla. to N.C. and Ga.
   patulus Walker, 1848: 175 (*Tabanus*).—Ga. **N. syn.**
   oklahomensis Stone, 1933: 76 (*Tabanus*).—Okla. **N. syn.**
nuda (McDunnough), 1921: 143 (*Tabanus*).—Ont.; Alaska to N.B., s. to Wyo. and N.J.
opaca (Coquillett) *in* Baker, 1904: 21 (*Tabanus*).—Nev.; Alta. and Sask., s. to Calif. and Colo.
pediontis (McAlpine), 1961b: 907 (*Tabanus*).—Man.; s. Alta. to s. Man., s. to Utah and S. Dak. **N. comb.**
philipi (Stone), 1938: 133 (*Tabanus*).—Wash.; Calif., B.C.
polaris (Frey), 1915: 7 (*Tabanus*).—n. Siberia; Alaska, n. Eurasia.
   boreus Stone, 1938: 147 (*Tabanus*).—Alaska. **N. syn.**, L. L. Pechuman *in litt.*
procyon (Osten Sacken), 1877: 216 (*Tabanus*).—Calif.; B.C. to Calif. and Wyo.
rhombica (Osten Sacken), 1876a: 472 (*Tabanus*).—Colo.; Alaska to Sask. and Wis., s. to Oreg. and Ariz.
   centron Marten, 1882: 211 (*Tabanus*).—Colo.
   solox Enderlein, 1922c: 347.—Colo.
   var. **osburni** (Hine), 1904d: 241 (*Tabanus;* as sp.).—B.C., Alta., Mont., Wash., Alaska; Alaska, s. to Oreg. and Minn.
rupestris (McDunnough), 1921: 143 (*Tabanus*).—Mont.; B.C. to Oreg. and S. Dak.
sequax (Williston), 1887b: 137 (*Tabanus*).—Oreg.; B.C. to Utah and Mont.
   leucophorus Bigot, 1892b: 640 (*Therioplectes*).—Oreg.
   fuscipalpis Bigot, 1892b: 681 (*Tabanus*).—Wash. Territory.
sexfasciata (Hine), 1923a: 144 (*Tabanus*).—Alaska; Alaska to ne. Man., Europe.
sonomensis (Osten Sacken), 1877: 216 (*Tabanus*).—Calif.; Alaska to Calif. and Wis.
   maculifer Bigot, 1892b: 641 (*Therioplectes*).—Wash. Territory.
   var. **phaenops** (Osten Sacken), 1877: 217 (*Tabanus;* as sp.).—Calif.; B.C. to Calif. and Wyo.
susurra (Marten), 1883: 111 (*Therioplectes*).—Mont. Unrecognized.

tetrica (Marten), 1883: 111 (*Therioplectes*).—Mont.; B.C. to Ont., s. to Calif. and N. Mex.
    var. **hirtula** (Bigot), 1892b: 641 (*Therioplectes;* as sp.).—Wash. Territory; B.C. to Calif. and S. Dak.
    var. **rubrilata** (Philip), 1937b: 64 (*Tabanus*).—Colo.; s. Utah, Ariz., N. Mex.
    laticornis Enderlein, 1925a: 363 (*Tylostypia;* preocc. Hine, 1904).—Colo.
trepida (McDunnough), 1921: 142 (*Tabanus*).—Ont.; Alaska to N.S., s. to Idaho, Ohio, and Pa.
trispila trispila (Wiedemann), 1828: 150 (*Tabanus*).—Ky.; coastal plain, N.Y. to S.C., w. Fla. and s. Miss.
    ssp. **sodalis** (Williston), 1887b: 139 (*Tabanus;* as sp.).—N.H., Conn.; Wis. to Maine, s. to Ga., in mts.
    aestivalis Harris *in* Johnson, 1925c: 70 (*Tabanus*). Nomen nudum.
typhus (Whitney), 1904: 206 (*Tabanus*).—N.H.; Alaska to Labr., s. to Mont. and Ga.
zonalis (Kirby), 1837: 314 (*Tabanus*).—Canada; N.W.T. to Labr., s. to Mont. and Pa.
    tarandi Walker, 1848: 156 (*Tabanus*).—Hudson Bay Territory.
    terraenovae Macquart, 1850: 339 (1850: 35) (*Tabanus*).—New World.
    flavocinctus Bellardi, 1859: 61 (1861: 261) (*Tabanus*).—Mexico (probably error).
zygota (Philip), 1937a: 52 (*Tabanus*).—Oreg.; B.C.

Genus AGKISTROCERUS Philip

Dicladocera, subg. **Agkistrocerus** Philip, 1941b: 13. Type-species, *Tabanus megerlei* Wiedemann (orig. des.).

finitimus (Stone), 1938: 15 (*Dicladocera*).—Fla.; Tex., Miss., Ga.
megerlei (Wiedemann), 1828: 132 (*Tabanus*).—Unknown; Tex. to N.C. and Fla.

Genus HAMATABANUS Philip

Hamatabanus Philip, 1941b: 13. Type-species, *Tabanus scitus* Walker (orig. des.)=*carolinensis* (Macquart).

annularis (Hine), 1917: 269 (*Tabanus*).—Miss.; s. Miss., Fla.
carolinensis (Macquart), 1838a: 149 (1838: 145) (*Tabanus*).—Carolina; Wis. to Md., s. to Iowa and Fla.
    scitus Walker, 1848: 181 (*Tabanus*).—Ga.

hirtioculatus Macquart, 1855: 53 (1855: 33) (*Tabanus*).—Md.
  cerastes Osten Sacken, 1876a: 462 (*Tabanus*).—Ky., Wis.
sexfasciatus (Stone), 1935: 11 (*Dicladocera*).—Fla.

Unplaced Species of Tabaninae

vicinus Macquart, 1838a: 147 (1838: 143) (*Tabanus*).—Carolina. ?*Hybomitra*.

Family RHAGIONIDAE
(Leptidae)
By Maurice T. James

The adults and larvae of Rhagionidae are predaceous. Adults of *Symphoromyia* and *Suragina*, as far as known, suck blood of man and other vertebrates; some are vicious biters. Larvae of *Atherix* are aquatic; other larvae, as far as known, are terrestrial, either excavating ant-lion-like pits in dry sand (*Vermileo*) or living in wet peaty soil, rotting wood, soil under fallen leaves, soil mixed with rotting leaves, and other decaying plant materials, and similar places.

REFERENCE: Leonard, 1930 (rev.).

Subfamily VERMILEONINAE
Genus VERMILEO Macquart

Vermileo Macquart, 1834a: 428. Type-species, *degeerii* Macquart (mon.) =*vermileo* (Fabricius):

REFERENCE: Pechuman, 1938a (key).

comstocki Wheeler, 1918: 84.—Calif.
opacus (Coquillett) *in* Baker, 1904: 21 (*Pheneus*).—Nev.; Calif. to N. Mex. and Colo.

Subfamily RHAGIONINAE
Genus DIALYSIS Walker

Dialysis Walker, 1850: 4. Type-species, *dissimilis* Walker (mon.)= *elongata* (Say).
Triptotricha Loew, 1872a: 59 (Cent. 10, no. 15). Type-species, *lauta* Loew (present des.).
Agnotomyia Williston, 1886b: 106. Type-species, *Stygia elongata* Say (mon.).

Agnotemyia, error.

aldrichi Williston, 1895b: 265.—Idaho; Wash. to Calif.
discolor (Loew), 1874b: 379 (*Triptotricha*).—Calif.; B.C. to Calif.
dispar Bigot, 1879a: 197.—Calif.
disparilis Bergroth, 1889: 296.—B.C.; B.C. to Calif.
elongata (Say), 1823: 41 (1859b: 58) (*Stygia*).—Pa.; Que. to Ga.
    dissimilis Walker, 1850: 4.—N. Amer.
fasciventris (Loew), 1874b: 380 (*Triptotricha*).—Pa.; Ill. to Pa., s. to Va.
kesseli Hardy, 1948b: 129.—Nev.
lauta (Loew), 1872a: 59 (Cent. 10, no. 15) (*Triptotricha*.)—Calif.
rufithorax (Say), 1823: 36 (1859b: 56) (*Leptis*).—Pa.; Mo. to N.J. and Fla.

Genus SYMPHOROMYIA Frauenfeld

Symphoromyia Frauenfeld, 1867: 496. Type-species, *Atherix melaena* Meigen (orig. des.).

    REFERENCES: Aldrich, 1915d (rev.); Sommerman, 1962b (biol., immature stages).

algens Leonard, 1931: 1.—N.H.; Que.
atripes Bigot, 1887d: 111.—Oreg.; Alaska to Alta., s. to Calif. and Colo.
barbata Aldrich, 1915d: 120.—Calif.
cinerea Johnson, 1903b: 25.—N.J.; N.W.T., Calif., D.C.
cruenta Coquillett, 1894d: 55.—Calif.
currani Leonard, 1931: 2.—Conn.
fulvipes Bigot, 1887d: 110.—Oreg.; Alta., s. to Calif. and Colo., also Ky.
hirta Johnson, 1897: 120.—Pa.; Alta. to Pa., s. to N. Mex. and Ala.
    flavipalpis Adams, 1904b: 439.—Colo., Utah.
inquisitor Aldrich, 1915d: 127.—Wash.; Idaho.
inurbana Aldrich, 1915d: 127.—Idaho; Wash. to Mont. and Nev.
johnsoni Coquillett, 1894d: 54.—Wash. and B.C.; Calif.
kincaidi Aldrich, 1915d: 129.—Wash.; B.C.
limata Coquillett, 1894d: 54.—Calif.
montana Aldrich, 1915d: 133.—Mont.; Wash. to n. Que. and N.H.
pachyceras Williston, 1886a: 287.—Calif.; Wash., Oreg., Mont., Ariz.
    comata Bigot, 1887d: 111.—Calif.
pilosa Aldrich, 1915d: 135.—Calif.
plagens Williston, 1886a: 287.—Wash. and Oreg.; B.C. to Calif. and N. Mex.
    latipalpis Bigot, 1887d: 108.—Wash.
    picticornis Bigot, 1887d: 109.—Wash.

pleuralis Curran, 1930j: 40.—N.Y.
plumbea Aldrich, 1915d: 138.—Mont.; Wash., Colo.
pullata Coquillett, 1894d: 56.—Colo.
sackeni Aldrich, 1915d: 139.—Wash.; Wash. to Calif.
securifera Coquillett, 1904f: 171.—Calif.
trivittata Bigot, 1887d: 109.—Colo.; Calif.
    fera Coquillett, 1894d: 56.—Colo.
trucis Coquillett, 1894d: 55.—Calif.
varicornis (Loew), 1872a: 58 (Cent. 10, no. 13) (*Atherix*).—Calif.
    modesta Coquillett, 1894d: 54.—Calif.

Genus ATHERIX Meigen

Atherix Meigen, 1803: 271. Type-species, *Rhagio diadema* Fabricius (Coquillett, 1910b: 511; misident.)=*ibis* (Fabricius). Situation needs clarification by I.C.Z.N. Meigen misidentified *Rhagio diadema* Fabricius; designations by Latreille (1810: 443), of *Atherix maculata* Meigen and by Westwood (1840: 134), of *Leptis ibis* Fabricius both invalid, not originally included species.

A. *ibis* and A. *variegata* females gather together in masses and lay their eggs on their own bodies, which ultimately fall into water; larvae are aquatic and predaceous.

pachypus Bigot, 1887d: 117.—Wash.
variegata Walker, 1848: 218.—Ont.; N.W.T. to Que., s. to Calif. and Ga.

Genus SURAGINA Walker

Suragina Walker, 1859: 110. Type-species, *illucens* Walker (mon.).
    REFERENCE: Malloch, 1932a (key).

concinna (Williston), 1901: 266 (*Atherix*).—Mexico; Tex.

Genus RHAGIO Fabricius

Rhagio Fabricius, 1775: 761. Type-species, *Musca scolopacea* Linnaeus (Latreille, 1810: 443).
Leptis Fabricius, 1805: 69 (unjustified n. name for *Rhagio* Fabricius). Type-species, *Musca scolopacea* Linnaeus (aut.).

albicornis (Say), 1823: 38 (1859b: 56) (*Leptis*).—Pa.; Pa. to Miss. and Fla.
boscii (Macquart), 1840: 30 (1841: 308) (*Leptis*).—Carolina.
brunneipennis Leonard, 1930: 92.—Calif.
californicus Leonard, 1930: 93.—Calif.

costatus (Loew), 1862b: 187 (Cent. 2, no. 4) (*Leptis*).—Calif.; Wash., Oreg.
    limbatus Leonard, 1930: 96 (as var.).—Calif.
dimidiatus dimidiatus (Loew), 1863a: 10 (Cent. 3, no. 17) (*Leptis*).—Alaska; Alaska to Calif. and Nev.
    flavoniger Coquillett *in* Baker, 1904: 20 (*Leptis*).—Nev. **N. syn.**
    pleuralis Adams, 1904b: 441 (*Leptis*).—Wash. **N. syn.**
  ssp. **albibarbis** (Bigot), 1887d: 114 (*Leptis;* as sp.).—Wash.; B.C., Oreg., Nev.
gracilis (Johnson), 1912a: 3 (*Leptis*).—Vt.; Maine to N.Y.
hoodianus (Bigot), 1887d: 115 (*Leptis*).—Oreg. ?=*maculifer*.
incisus (Loew), 1872a: 59 (Cent. 10, no. 16) (*Leptis*).—Calif.; Oreg., ?N.Y.
intermedius Walker, 1848: 212.—Ont.
maculifer maculifer (Bigot), 1887d: 113 (*Leptis*).—Wash.; B.C., Oreg.
  ssp. **concavus** Leonard, 1930: 94 (as sp.).—Idaho; B.C. to Calif.
mystaceus (Macquart), 1840: 30 (1841: 308) (*Leptis*).—N. Amer.; Man. to Nfld., s. to Mo. and Fla.
ochraceus (Loew), 1862b: 187 (Cent. 2, no. 3) (*Leptis*).—N.Y.; Mass., Pa.
palpalis (Adams), 1904b: 442 (*Leptis*).—Wash.
plumbeus (Say), 1823: 39 (1859b: 56) (*Leptis*).—Pa.; Iowa to Ont., s. to N.C.
    griseolus Wulp, 1867: 142 (*Leptis*).—Wis.
pollinosus Leonard, 1930: 116.—Calif.
punctipennis (Say), 1823: 34 (1859b: 55) (*Leptis*).—Pa.; Minn. to Que., s. to Fla.
    filius Walker, 1848: 219 (*Atherix*).—N.Y., N.J.
terminalis (Loew), 1861b: 317 (Cent. 1, no. 20) (*Leptis*).—N.Y.; Wis., Va.
vertebratus (Say), 1823: 38 (1859b: 56) (*Leptis*).—Fla.; B.C. to N.B., s. to Calif. and Fla.
    hirtus Loew, 1861b: 318 (Cent. 1, no. 21) (*Leptis*).—Ill.
    scapularis Loew, 1861b: 318 (Cent. 1, no. 22) (*Leptis*).—Ill.

Genus CHRYSOPILUS Macquart

Chrysopilus Macquart, 1826b: 403 (1826: 82). Type-species, *Rhagio diadema* Fabricius (Westwood, 1840: 134).
Chrysopila, emend.
    REFERENCE: Hardy, 1949 (rev.).
alaskaensis Hardy, 1949: 147.—Alaska.

andersoni Leonard, 1930: 131.—Pa.; Ill., Ind., Ohio.
angustifacies Hardy, 1949: 148.—Ariz.
anthracinus Bigot, 1887d: 105.—Calif.; Wash. to s. Calif.
arctiventris James, 1936e: 343.—Colo.; Wash. to Wyo., s. to Calif. and N. Mex.
basilaris (Say), 1823: 36 (1859b: 55) (*Leptis*).—Pa.; Mich. to N.H., s. to Tex. and Fla., ?Guatemala.
beameri Hardy, 1949: 151.—Ga.; Fla.
connexus Johnson, 1912b: 108.—N.C.; Fla.
davisi Johnson, 1912a: 4.—Ga.; N.C.
dilatus Cresson, 1919: 177.—Calif.
divisus Hardy, 1949: 152.—Ariz.
fasciatus fasciatus (Say), 1823: 37 (1859b: 56) (*Leptis*).—Pa.; N.H. to N.C.
    par Walker, 1848: 215 (*Leptis*).—N. Amer.
  ssp. **infuscatus** Leonard, 1930: 141 (as var.).—N.C.; Ind., La.
flavibarbis flavibarbis Adams, 1904b: 438.—Colo.; B.C. to Man., s. to Calif. and N. Mex., also Mass.
    cameroni Curran, 1926d: 170.—Sask.
  ssp. **aldrichi** James, 1936e: 343 (as sp.).—Colo.; Wash. to Alta., s. to c. Calif. and Colo., at high altitudes.
foedus Loew, 1861b: 317 (Cent. 1, no. 18).—Ill.; S. Dak. to Ind. and Kans.
georgianus Hardy, 1949: 154.—Ga.
griffithi Johnson, 1897: 119.—Va.; Fla.
humilis Loew, 1874b: 379.—Calif.; N. Mex., Colo., ?Ill.
kincaidi Hardy, 1949: 156.—Wash.
longipalpis Hardy, 1949: 157.—Wash.; B.C. to Calif.
lucifer Adams, 1904b: 437.—Wash.; B.C. and Alta., s. to n. Calif. and Utah.
modestus Loew, 1872a: 58 (Cent. 10, no. 14).—Tex.; Nebr. to N.Y., s. to Tex. and Va.
nudus Cresson, 1919: 176.—Calif.; Wash.
ornatus (Say), 1823: 34 (1859b: 54) (*Leptis*).—Pa.; Minn. to Ont., s. to Kans. and Fla.
    servillei Guérin-Méneville, 1835: pl. 96, fig. 3 (*Leptis*).—Unknown.
pilosus Leonard, 1930: 152.—Ill.; Colo. and Nebr., e. to Ohio and Mich.
proximus (Walker), 1848: 214 (*Leptis*).—N.S.; S. Dak. to N.S., s. to Fla., ?Calif., ?Nev.
    propinquus Walker, 1848: 215 (*Leptis*).—N.Y.
    simillimus Walker, 1848: 215 (*Leptis*).—N.Y.

quadratus (Say), 1823: 35 (1859b: 55) (*Leptis*).—Pa., Mo.; B.C. to Nfld., s. to Calif. and Fla.
 fumipennis Say, 1823: 37 (1859b: 56) (*Leptis*).—Pa.
 reflexus Walker, 1848: 216 (*Leptis*).—Ohio, N.S.
 dispar Wulp, 1867: 143.—Wis.
 flavidus Bigot, 1887d: 104.—Canada. ?Valid sp.
 limbipennis Bigot, 1887d: 106 (*Leptipalpus*).—Rocky Mts.
 obscuripennis Bigot, 1887d: 107 (*Leptipalpus*; preocc. Loew, 1873).—Rocky Mts.
rotundipennis Loew, 1861b: 317 (Cent. 1, no. 19).—Ga.; Mass. to Fla. and Ala.
testaceipes Bigot, 1887d: 105.—Wash.; B.C. and Alta. to s. Calif. and N. Mex.
 bellus Adams, 1904b: 438.—Calif.
thoracicus (Fabricius), 1805: 70 (*Leptis*).—Carolina; Ill. to Ont. and N.C.
tomentosus Bigot, 1887d: 104.—Wash., Colo.; Wash. to s. Calif. and Colo.
velutinus Loew, 1861b: 316 (Cent. 1, no. 17).—Ill.; Ky., Md., N.C., Fla.
xanthopus Hardy, 1949: 163.—Ariz.; N. Mex., Tex.

Genus PTIOLINA Zetterstedt

Ptiolina Zetterstedt, 1842: 226. Type-species, *Leptis obscura* Fallén (Frauenfeld, 1867: 497).

REFERENCE: Hardy and McGuire, 1947 (key).

alberta Leonard *in* Curran, 1931f: 250.—Alta.; Alaska.
augusta Curran, 1931f: 249.—Que.
edeta (Walker), 1849: 489 (*Spania*).—Ont.; Alaska, N.H.
fasciata Loew, 1869c: 164 (Cent. 9, no. 65).—Hudson Bay Territory; Alta., Colo.
grisea Curran, 1931f: 251.—N.H.
majuscula Loew, 1869c: 165 (Cent. 9, no. 66).—Hudson Bay Territory; Alaska.
mallochi Hardy and McGuire, 1947: 8 (n. name for *arctica* Malloch).—Alaska.
 arctica Malloch, 1923i: 181 (preocc. Becker, 1921).—Alaska.
nigripilosa Hardy and McGuire, 1947: 9.—N.Y.
nitidifrons Hardy and McGuire, 1947: 10.—Alaska; N.H.
obsoleta Leonard *in* Curran, 1931f: 250.—Mich.
vicina Hardy and McGuire, 1947: 12.—Oreg.
zonata Hardy and McGuire, 1947: 13.—Wash.

Genus SPANIA Meigen

Spania Meigen, 1830: 335. Type-species, *nigra* Meigen (mon.).
nigra nigra Meigen—Not Nearctic.
 ssp. **americana** Johnson, 1923b: 70 (as var.).—Maine.

Unplaced Species of Rhagionidae

vidua Walker, 1849: 1153 (*Atherix*).—Ont. Possibly a *Ptiolina*.

Family HILARIMORPHIDAE

By Maurice T. James

Opinions vary as to the systematic position of *Hilarimorpha*. The genus has been placed variously in the Rhagionidae, Bombyliidae, Empididae, and in a family by itself. The last course, which is adopted here, is favored by the phylogenetic studies of Hennig (1954).

Genus HILARIMORPHA Schiner

Hilarimorpha Schiner, 1860b: 54. Type-species, *singularis* Schiner (orig. des.).
Hilaromorpha, error or emend.
mikii Williston, 1888b: 100.—Ill.; Oreg., Colo.
obscura Bigot, 1887b: cxli (also 1889: 129).—Calif.
pusilla Johnson, 1923b: 70.—N.H.; Vt.

Superfamily ASILOIDEA

Family THEREVIDAE

By F. R. Cole

Adult therevids, or "stiletto flies", are occasionally seen on blossoms of shrubs or on foliage, but are probably more common on dry ground, or sandy beach or dune areas. The flight is often quick but usually of short duration. Because of a resemblance to Asilidae, adult therevids have been called "robber flies" dating back to the time of Schiner and Williston. No authentic observation exists of a true predatory habit, however, and the proboscis has fleshy labella, not at all suited for predation.

The observed larvae are predaceous and even cannibalistic; they may be found in sand or in decaying wood, especially alder branches. Larvae of *Thereva comata* Loew, kept in captivity at the University of

California, devour housefly maggots (MS note by A. Barnes); they are also known to feed on small earthworms. These larvae move like an eel or snake, not being able to creep. The group may have some economic importance, especially where they encounter root-feeding larvae of other insects.

REFERENCES: Collinge, 1909 (biol.); Hyslop, 1910 (biol.); Kröber, 1912 (rev.); Kröber, 1913 (rev. genera of world); Cole, 1923d (rev.).

Genus HENICOMYIA Coquillett

Henicomyia Coquillett, 1898h: 187. Type-species, *hubbardii* Coquillett (orig. des.).

hubbardii Coquillett, 1898h: 187.—Ariz.
varipes Kröber, 1912: 213.—Mexico; Colo.

Genus ATAENOGERA Kröber

Ataenogera Kröber, 1914: 31. Type-species, *abdominalis* Kröber (mon.).

Unnamed sp.—Calif., Costa Rica.

Genus NEBRITUS Coquillett

Nebritus Coquillett, 1894e: 98. Type-species, *pellucidus* Coquillett (mon.).

pellucidus Coquillett, 1894e: 98.—Calif.

Genus ZIONEA Hardy

Zionea Hardy, 1938b: 144. Type-species, *tanneri* Hardy (orig. des.).
tanneri Hardy, 1938b: 144.—Utah; Calif.

Genus OZODICEROMYIA Bigot

Ozodiceromyia Bigot, 1889: 321. Type-species, *mexicana* Bigot (orig. des.).

argentifera (Kröber), 1929: 418 (*Phycus*).—Mexico; Ariz.

Genus PHEROCERA Cole

Pherocera Cole, 1923d: 20. Type-species, *signatifrons* Cole (orig. des.).

albihalteralis Cole, 1923d: 22.—N. Mex.; Calif., Ariz.
flavipes Cole, 1923d: 22.—Ariz.; Calif., Utah.
signatifrons Cole, 1923d: 21.—N. Mex.; Calif., Ariz.

Genus CHROMOLEPIDA Cole

Chromolepida Cole, 1923d: 23. Type-species, *Psilocephala pruinosa* Coquillett (orig. des.).

bella Cole, 1923d: 24.—Calif.; Nev., Idaho.
mexicana Cole, 1923a: 460.—Mexico; Ariz., N. Mex.

Genus FURCIFERA Kröber

Furcifera Kröber, 1911: 524. Type-species, *fascipennis* Kröber (Cole, 1960a: 165).
Epomyia Cole, 1923d: 26. Type-species, *Thereva pictipennis* Wiedemann (orig. des.).

REFERENCE: Cole, 1960a (rev.).

bella (Cole), 1923d: 32 (*Epomyia*).—Tex.
hardyi Cole, 1960a: 167 (n. name for *flavipes* Hardy).—Tex.
  flavipes Hardy, 1943a: 26 (*Epomyia*; preocc. Kröber, 1928).—Tex.
pictipennis (Wiedemann), 1821b: 63 (1821c: 113) (*Thereva*).—Ga.; e. N. Amer. to Mich. and Tex.
  erythrura Loew, 1869c: 172 (Cent. 9, no. 75) (*Psilocephala*).—Middle States.
rufiventris (Loew), 1869b: 12 (Cent. 8, no. 17) (*Psilocephala*).—Nebr.; e. N. Amer. to Nebr., Mont. and Man.
  lacteipennis Kröber, 1914: 53 (*Psilocephala*).—Fla.
scutellaris (Loew), 1869c: 171 (Cent. 9, no. 74) (*Psilocephala*).—D.C.; N.Y., Conn., Va., N.C.

Genus PSILOCEPHALA Zetterstedt

Psilocephala Zetterstedt, 1838: 525. Type-species *Bibio imberbis* Fallén (Coquillett, 1910b: 597).

REFERENCES: Coquillett, 1893g (synopsis); Kröber, 1914 (key).

acuta Adams, 1903b: 222.—Kans.
albertensis Cole, 1925: 86.—Alta.; Canada.
aldrichii Coquillett, 1893g: 227.—Calif., Mont., Wyo., N.J.; B.C. to Man., s. to Calif. and Colo. N.J. syntype=*flavipennis*.
amplifrons Cole, 1925: 85 (n. name for *latifrons* Cole).—N.Y.; Calif., Ont., Que., N.H., Pa.
  latifrons Cole, 1923d: 73 (preocc. Frey, 1921).—N.Y.
argentifrons Cole, 1923d: 56.—Pa.
arizonensis Cole, 1923d: 45.—Ariz.
aurantiaca Coquillett, 1904f: 177.—Calif.; Ariz., N. Mex.
baccata Coquillett, 1893g: 226.—Calif.

brunnea Kröber, 1914: 46.—Ariz.
bussi James *in* James and Huckett, 1952: 265.—Y.T.
canadensis Cole, 1923d: 57.—Ont.; Que.
cinerea Cole, 1923d: 65.—N. Mex.
coloradensis James, 1936e: 341.—Colo.; Ariz., Okla., Kans.
conspicua (Walker), 1848: 223 (*Thereva*).—N.S.
costalis Loew, 1869b: 11 (Cent. 8, no. 16).—Calif.; Wash., Nev., Idaho.
davisi Johnson, 1926b: 300.—N.C.
festina Coquillett, 1893g: 225.—Fla.; Ariz., Mexico.
flavipennis Cole, 1923d: 42.—Md.; R.I. to Va., also Miss.
frontalis Cole, 1923d: 40.—N.Y.; Wyo. to Que., s. to La. and N.J.
fuscipennis Cole, 1923d: 62.—Wash.
germana (Walker), 1848: 222 (*Thereva*).—Fla.
grandis Johnson, 1902a: 241.—Que.
haemorrhoidalis (Macquart), 1840: 26 (1841: 304) (*Thereva*; as *hoemorrhoidalis*).—Carolina; Mich. to N.Y., s. to Tex. and Ala.
johnsoni Coquillett, 1893g: 228.—Fla.
lateralis Adams, 1904b: 444.—Ariz.; Calif., Colo.
levigata Loew, 1876: 319.—Calif.
limata Coquillett, 1894e: 99.—Wash., Colo.; B.C., Calif.
marcida Coquillett, 1893g: 228.—Calif.; Ariz.
melampodia Loew, 1869b: 9 (Cent. 8, no. 12).—Ill.; Mass., N.C.
montiradicis James, 1949: 10.—Colo.
montivaga Coquillett, 1893g: 226.—Calif.
morata Coquillett, 1893g: 225.—N.J., Fla.; N.Y.
munda Loew, 1869b: 9 (Cent. 8, no. 13).—Wis.; Y.T. to Que., s. to Calif., Colo., and Pa.
    melanoprocta Loew, 1869b: 11 (Cent. 8, no. 15).—Maine, Hudson's Bay Territory.
nigrimana Kröber, 1912: 238.—Colo.
nigrina Kröber, 1914: 53.—Colo.
notata (Wiedemann), 1821b: 64 (1821c: 114) (*Thereva*).—Ga.; Fla.
obscura Coquillett, 1893g: 229.—Jamaica; Fla., West Indies.
occipitalis Adams, 1904b: 443.—Ariz.; Calif.
pallida Kröber, 1914: 45.—Tex.
pavida Coquillett, 1893g: 226.—Calif.; Colo.
pilosa Kröber, 1914: 47.—Ariz.
placida Coquillett, 1894e: 99.—Fla.
platancala Loew, 1876: 321.—Tex.; Ariz., Colo., Kans., Pa.
pollinosa Cole, 1923d: 72.—Calif.
senilis (Fabricius), 1805: 68 (*Bibio*).—N. and S. Amer.; Ga. and Brazil.

signatipennis Cole, 1923d: 47.—Oreg.; Calif., Idaho.
slossonae Coquillett, 1893g: 227 (as *slossoni*).—N.H.
squamosa Hardy, 1943a: 24.—Fla.
subnotata Johnson, 1926b: 299.—Fla.
subrufa Cole, 1923d: 68.—Ariz.
tergisa (Say), 1823: 39 (1859b: 57) (*Thereva*).—Fla.
    tergissa, emend.
    corusca Wiedemann, 1828: 232 (*Thereva*).—Fla.
variegata variegata Loew, 1869c: 170 (Cent. 9, no. 73).—Canada; Mich., Ont., N.Y.
    ssp. **flavipilosa** Cole, 1923d: 62.—Calif.
    ssp. **occidentalis** Cole, 1923d: 61.—Oreg.
vicina (Walker), 1848: 222 (*Thereva*).—N.S.

Genus DIALINEURA Rondani

Dialineura Rondani, 1856: 155. Type-species, *Musca anilis* Linnaeus (orig. des.).

melanophleba (Loew), 1876: 317 (*Thereva*).—Calif.; Nev.
willistoni Cole (for *crassicornis* Williston).—Calif.; Wash., Oreg. N. name.
    crassicornis Williston, 1886a: 293 (*Thereva*; preocc. Bellardi, 1861).—Calif.

Genus TABUDA Walker

Tabuda Walker, 1852: 197. Type-species, *fulvipes* Walker (mon.).

borealis Cole, 1923d: 82.—Sask.
fulvipes Walker, 1852: 197.—Unknown; Mass., N.J., N.C., Ga.

Genus METAPHRAGMA Coquillett

Metaphragma Coquillett, 1894e: 97. Type-species, *Xestomyza planiceps* Loew (mon.).

planiceps (Loew), 1872a: 75 (Cent. 10, no. 38) (*Xestomyza*).—Calif.; Wash. to Wyo., s. to n. Calif. and Colo.

Genus THEREVA LATREILLE

Thereva Latreille, 1796: 167. Type-species, *Musca plebeja* Linnaeus (sub. mon., Latreille, 1802: 441).

    REFERENCES: Coquillett, 1893f (synopsis); Kröber, 1914 (key).

albiceps Loew, 1869c: 166 (Cent. 9, no. 69).—Canada; N. Dak., Minn.
albifrons Say, 1829: 156 (1859b: 353).—Ind., Mass.
albopilosa Kröber, 1912: 256.—Colo.

anomala Adams, 1904b: 444.—Ariz.; Mexico.
aurofasciata Kröber, 1912: 263.—Colo.; Wis.
bakeri Cole, 1923d: 124.—Calif.
bella Kröber, 1914: 64.—Mass.; Conn., R.I., N.J.
bimaculata Cole, 1923d: 98.—N.C.; Va.
borealis Cole, 1923d: 126.—Mich.
brunnea Cole, 1923d: 108.—B.C.; Wash., Idaho.
californica Kröber, 1912: 259.—Calif.
candidata Loew, 1869b: 8 (Cent. 8, no. 10).—Wis.; Man. to N.S., s. to Ohio and Va.
cinerascens Cole, 1923d: 97.—Oreg.; B.C.
cingulata Kröber, 1912: 267.—Colo.; Mont., Wyo.
cockerelli Cole, 1923d: 99.—Colo.; Alta., Sask., Man., Mich.
comata Loew, 1869b: 7 (Cent. 8, no. 9).—Calif.; Wash.
concavifrons Kröber, 1914: 70.—N. Mex.
diversa Coquillett, 1894e: 100.—Colo.; Mont., ?Fla.
duplicis Coquillett, 1893f: 199.—S. Dak.; Mont., Sask., Man.
egressa Coquillett, 1894e: 99.—Colo.; Calif.
flavicauda Coquillett *in* Baker, 1904: 23.—Nev.; Calif.
flavicincta Loew, 1869c: 168 (Cent. 9, no. 70.)—Wis. and N.H.; Mich., Ont., Mass.
   gilvipes Loew, 1869c: 168 (Cent. 9, no. 71).—Mass.
flavipilosa Cole, 1923d: 125.—Calif.
flavohirta Kröber, 1914: 70.—Colo.
foxi Cole, 1923d: 112.—Wash.
frontalis Say, 1824a: 370 (1859a: 252).—Northwest Territory (U.S.); B.C. and Wash. to Que. and N.Y., s. to Ill.
fucata Loew, 1872a: 74 (Cent. 10, no. 37).—Calif.; Oreg., Utah, Wyo., Colo.
fucatoides Bromley, 1937b: 99.—Utah.
hirticeps Loew, 1874b: 382.—Calif.
johnsoni Coquillett, 1893f: 200.—Wash.; Oreg., Calif.
macdunnoughi Cole, 1925: 87.—Alta.
melanoneura Loew, 1872a: 74 (Cent. 10, no. 36).—Calif.; N. Mex., Colo.
metallica Kröber, 1914: 68.—N. Mex.
nanella Cole, 1960b: 118 (n. name for *nana* Cole).—Calif.; Colo.
   pygmaea Cole, 1923d: 89 (preocc. Fallén, 1820).—Calif.
   nana Cole, 1959: 148 (n. name for *pygmaea* Cole, but preocc. Fallén, 1820.—Calif.
nebulosa Kröber, 1912: 264.—Calif.; Wash.
neomexicana Cole, 1923d: 117.—N. Mex.
nigrimana Kröber, 1914: 65 (as *bella* var.).—Mass.
nigripilosa Cole, 1923d: 110.—B.C.

nitoris Coquillett, 1894e: 101.—Mo.
nivea Kröber, 1914: 64.—N. Mex.
niveipennis Kröber, 1914: 66.—Calif.
novella Coquillett, 1893f: 200.—Calif.; Colo.
otiosa Coquillett, 1893f: 199.—Calif.
pacifica Cole, 1923d: 103.—Calif.
pseudoculata Cole, 1923d: 121.—Utah; Wash., Idaho, Mont., Colo.
semitaria Coquillett, 1893f: 198.—Calif.; Idaho, Ariz.
senex Walker, 1848: 224.—N.S.
strigipes Loew, 1869c: 169 (Cent. 9, no. 72).—Man.; Colo., N.Y., Vt., N.H.
ustulata Kröber, 1912: 265.—Que.
utahensis Hardy, 1938b: 145.—Utah.
vanduzeei Cole, 1923d: 105.—Calif.
vialis Osten Sacken, 1877: 274.—Calif.; B.C., Oreg., Man.
xanthobasis James, 1949: 12.—Colo.

Unplaced Species of Therevidae

nervosa Walker, 1848: 223 (*Thereva*).—Ga. ?*Tabuda*.
nigra Say, 1823: 40 (1859b: 57) (*Thereva*).—Pa. ?=*Psilocephala haemorrhoidalis*.
ruficornis Macquart, 1840: 25 (1841: 303) (*Thereva*).—Carolina.
varia Walker, 1848: 221 (*Thereva*).—Fla.

Family SCENOPINIDAE

(Omphralidae)

By D. Elmo Hardy

The habits of scenopinids are not definitely known, but apparently the larvae are all predators upon a variety of insects; those of the "window flies" *Scenopinus fenestralis* (Linnaeus) and *glabrifrons* Meigen prey especially on stored food pests, household pests, bark beetles, powder post beetles, et cetera. The North American species have been greatly confused, and are being reclassified by L. P. Kelsey on the basis of genitalic characters.

REFERENCES: Cresson, 1907b (rev.); Kröber, 1937 (rev.); Hardy, 1944b (rev.).

Genus METATRICHIA Coquillett

Metatrichia Coquillett, 1900c: 500. Type-species, *Scenopinus bulbosa* Osten Sacken (orig. des.).

bulbosa (Osten Sacken), 1877: 275 (*Scenopinus*).—Mo.; w. U.S.

Genus BREVITRICHIA Hardy

Brevitrichia Hardy, 1944b: 32. Type-species, *Pseudatrichia griseola* Coquillett (orig. des.).

griseola (Coquillett), 1900c: 501 (*Pseudatrichia*).—N. Mex.; w. U.S.
helenae (James), 1938b: 22 (*Pseudatrichia*).—Colo.; Utah, Wyo., N. Mex.

Genus PSEUDATRICHIA Osten Sacken

Atrichia Schrank of Loew, 1866b: 42 (Cent. 7, no. 76).
Pseudatrichia Osten Sacken, 1877: 276. Type-species, *Atrichia longurio* Loew (mon.).

albocincta Van Duzee, 1926d: 164.—Calif.
parva Hardy, 1944b: 36.—Ariz.
unicolor Coquillett, 1900c: 500.—N. Mex.

Genus BELOSTA Hardy

Belosta Hardy, 1944b: 37. Type-species, *albipilosa* Hardy (orig. des.).

albipilosa Hardy, 1944b: 38.—Idaho; nw. U.S.
flaviceps (Coquillett), 1902g: 102 (*Pseudatrichia*).—Ariz.
pilosa (Coquillett), 1902g: 102 (*Pseudatrichia*).—Ariz.

Genus OMPHRALOSOMA Kröber

Omphralosoma Kröber, 1937: 214, 219. Type-species, *Scenopinus squamosa* Villeneuve (orig. des.).

albifasciatum Hardy, 1944b: 41.—Calif.

Genus SCENOPINUS Latreille

Omphrale Meigen, 1800: 29. Type-species, *Musca senilis* Fabricius (sub. mon., Hendel, 1908b: 58) =*fenestralis* (Linnaeus). Suppressed by I.C.Z.N., 1963b: 339.
Scenopinus Latreille, 1802: 463. Type-species, *Musca fenestralis* Linnaeus (mon.).

Subgenus PAROMPHRALE Kröber

Paromphrale Kröber, 1937: 214, 222 (as genus). Type-species, *Scenopinus glabrifrons* Meigen (orig. des.).

glabrifrons Meigen, 1824: 114.—Europe; entire N. Amer., widespread over much of world by commerce.

Subgenus SCENOPINUS Latreille

beameri beameri (Hardy), 1944b: 43 (*Omphrale*).—Calif.; Nev.
    ssp. **fuscus** (Hardy), 1944b: 43 (*Omphrale;* as var.).—Calif.
cavifrons (Kröber), 1937: 231 (*Omphrale*).—?Ont. (Joinville, Algonquin).
electus Adams, 1904b: 445.—Ariz.
fenestralis (Linnaeus), 1758: 597 (*Musca*).—Europe; entire N. Amer., widespread over much of world by commerce.
    pallipes Say, 1823: 100 (1859b: 86).—Pa.
kuiterti (Hardy), 1944b: 46 (*Omphrale*).—Ariz.
mirabilis Adams, 1904b: 445.—Ariz.; N. Mex.
nubilipes Say, 1829: 170 (1859b: 363).—Ind.
ramaleyi James, 1938b: 22.—Colo.; Wyo.
valgus (Hardy), 1944b: 50 (*Omphrale*).—Calif.
whittakeri (James), 1955c: 47 (*Omphrale*).—Wash.

Family APIOCERIDAE

By Mont A. Cazier

Apiocerid flies apparently are restricted to arid or semiarid regions and are not common in collections. The adults hover and make a loud noise in flight. Partial biologies of only two species are known. English (1947) described portions of the life history and immature stages of *Apiocera maritima* Hardy in Australia, the larvae of which occur in beach sand on the seacoast. The author is publishing the partial biology of a new species of *Apiocera* from Arizona whose eggs are laid in loose sandy soil near the base of various plant species in dry desert environments. Nothing is known about larval habits or the natural food of either the larvae or adults of any of the species. Additional new North American species are in collections and considerable taxonomic work remains to be done, especially on the internal genitalic structures.

    REFERENCE: Cazier, 1941 (rev.).

Subfamily APIOCERINAE

Genus APIOCERA Westwood

Apiocera Westwood, 1835a: 448. Type-species, *fuscicollis* Westwood (Coquillett, 1910b: 508).

aldrichi Painter, 1938: 193.—Ariz.; Calif.
alleni Cazier, 1941: 606.—Calif.
augur Osten Sacken, 1887a: 212.—Mexico; N. Mex., Tex.

APIOCERIDAE

beameri Painter, 1938: 198.—Calif.
bilineata Painter, 1932b: 351.—N. Mex.; Ariz., Tex., Mexico.
caloris Painter, 1938: 194.—Ariz.; Calif.
convergens Painter, 1938: 196.—Calif.
exta Cazier, 1941:609.—Calif.
haruspex Osten Sacken, 1877: 283.—Calif.; B.C. to Wyo., s. to Calif. and Ariz.
hispida Cazier, 1941: 605.—Calif.
infinita Cazier, 1941: 613.—Calif.
interrupta Painter, 1938: 192.—Calif.
intonsa Cazier, 1941: 610.—Ariz.; Calif.
martinorum Painter, 1938: 197.—Oreg.; Idaho.
melanura Cazier, 1941: 611.—Ariz.
mexicana Cazier, 1954: 8.—Mexico; Tex.
notata Painter, 1938: 199.—Calif.
parkeri Cazier, 1941: 607.—Calif.; Ariz., Nev.
pearcei Cazier, 1941: 603.—Calif.; Nev.
trimaculata Painter, 1938: 195.—Calif.

Genus RHAPHIOMIDAS Osten Sacken

Rhaphiomidas Osten Sacken, 1877: 281. Type-species, *episcopus* Osten Sacken (orig. des.).
Apomidas Coquillett, 1892d: 315. Type-species, *trochilus* Coquillett (orig. des.).

REFERENCE: Cazier, 1954 (key).

abdominalis Cazier, 1941: 624.—Calif.
acton Coquillett, 1891a: 85.—Calif.
aitkeni Cazier, 1941: 623.—Calif.
maculatus Cazier, 1941: 628.—Calif.
maehleri Cazier, 1941: 627.—Calif.
painteri Cazier, 1941: 622.—N. Mex.
parkeri Cazier, 1941: 625.—Calif.
terminatus Cazier, 1941: 622.—Calif.
trochilus (Coquillett), 1892d: 315 (*Apomidas*).—Calif.

Family MYDIDAE

(Mydaidae, Mydasidae)

By C. H. Curran

Mydas flies are moderately to very large, elongated flies, usually resembling wasps. Little is known of the biology, though Walsh (1864c) reared *Mydas tibialis* Wiedemann from a hollow in a decaying sycamore tree in Illinois; Malloch (1917g) reared *Mydas clavatus* (Drury) from a rotten stump in Illinois, and Genung (1959) found *Mydas maculiventris* Westwood to be an important larval predator of phytophagous scarab beetle larvae in Florida.

The revision of the North American Mydidae by Johnson (1926a) still stands as the best available for the genus *Mydas*, although Hardy (1943b, 1944a, and 1950) has brought several other genera up to date.

[It has been pointed out to us by L. W. Grensted that the name most likely was derived from King Midas, whose ass-ears would resemble the mydid antennae, and that the correct root of the genus is Myd-, resulting in the formation of the family name Mydidae.— Eds.]

REFERENCE: Séguy, 1928 (generic key).

Genus HETEROMYDAS Hardy

Heteromydas Hardy, 1944a: 227. Type-species, *bicolor* Hardy (orig. des.).

bicolor Hardy, 1944a: 227.—Calif.

Genus MYDAS Fabricius

Mydas Fabricius, 1794: 252. Type-species, *Musca clavata* Drury (Latreille, 1810: 443).
Midas, emend.
Lampromydas Séguy, 1928: 144. Type-species, *belus* Séguy (mon.) = *tricinctus* Bellardi, **n. syn.**

audax Osten Sacken, 1874b: 186.—Ky.; Pa., Va.
boonei Curran, 1953b: 1.—Ariz.
brunneus Johnson, 1926a: 136.—Ark.
carbonifer Osten Sacken, 1874b: 186.—N.Y.; Mich., N.C., Miss., Ga., Fla.
chrysites Osten Sacken, 1886a: 72 (*Midas*).—Mexico; Ariz.
clavatus (Drury), 1773: 2, of index (1770: 103, unnamed descr.) (*Musca*).—N. Amer.; U.S., Ont.
    filata Fabricius, 1775: 757 (*Bibio*).—Amer.
    asiloides De Geer, 1776: 204 (*Nemotelus*).—N. Amer.
cleptes Osten Sacken, 1886a: 72 (*Midas*).—Mexico; Ariz., Tex.
fulvifrons Illiger, 1801: 207 (as 206) (*Midas*).—Ga.; Tex., Okla., Ohio, N.J., Tenn., Fla.
    chrysostoma Osten Sacken, 1874b: 187.—Tex.
lividus Curran, 1953b: 2.—Ariz.; Fla.
luteipennis Loew, 1866b: 14 (Cent. 7, no. 23) (*Midas*).—Tex. (as N. Mex., error); Ariz., N. Mex.

maculiventris Westwood, 1835a: 281 (*Midas*).—Ga.; se. U.S.
  incisus Macquart, 1838b: 11 (1839: 127) (*Midas*).—Carolina.
  pachygaster Westwood, 1841: 53 (*Midas*).—Ga.
  parvulus Westwood, 1841: 53 (*Midas*).—N. Amer.
militaris Gerstäcker, 1868: 99 (n. name for *vittatus* Macquart).—Mexico; Ariz.
  vittatus Macquart, 1850: 364 (1850: 60) (preocc. Wiedemann, 1828).—Mexico.
pertenuis Johnson, 1926a: 137.—Ariz.
simplex Loew, 1866b: 15 (Cent. 7, no. 25) (*Midas*).—Tex. (as N. Mex., error); La.
tibialis Wiedemann, 1831: 42 (*Midas*).—Md.; Mich. to Pa., s. to Ill. and Va.
  fulvipes Walsh, 1864c: 306 (*Midas*).—Ill.
ventralis Gerstäcker, 1868: 102 (*Midas*; n. name for *rufiventris* Loew).—Calif.; Ariz., Mexico.
  rufiventris Loew, 1866b: 14 (Cent. 7, no. 22) (*Midas*; preocc. Macquart, 1850).—Calif.
  abdominalis Adams, 1904b: 434.—Ariz.
xanthopterus Loew, 1866b: 14 (Cent. 7, no. 24) (*Midas*).—Tex. (as N. Mex., error); Ariz., N. Mex., Colo.

Genus NEMOMYDAS Curran

Nemomydas Curran, 1934a: 165. Type-species, *Leptomydas pantherinus* Gerstäcker (orig. des.).
  REFERENCES: Hardy, 1950 (rev.); Steyskal, 1956b (table).
bifidus Hardy, 1950: 22.—Calif.
brachyrhynchus (Osten Sacken), 1886a: 69 (*Leptomydas*).—Mexico; Calif., Tex.
desideratus (Johnson), 1912c: 151 (*Leptomydas*).—Ga.
intonsus intonsus Hardy, 1950: 27.—Calif.
  ssp. **fumosus** Hardy, 1950: 29 (as var.).—Calif.
jonesii (Johnson), 1926a: 143 (*Leptomydas*).—Fla.; Ga.
lara Steyskal, 1956b: 3.—Fla.
melanopogon Steyskal, 1956b: 2.—Fla.
pantherinus (Gerstäcker), 1868: 85 (*Leptomydas*).—Calif.; B.C. to Idaho., s. to Calif. and Colo.
solitarius (Johnson), 1926a: 142 (*Leptomydas*).—Colo.
tenuipes (Loew), 1872a: 61 (Cent. 10, no. 20) (*Midas*).—Calif.
venosus (Loew), 1866b: 15 (Cent. 7, no. 26) (*Midas*).—Tex. (as N. Mex., error); Ariz., N. Mex., Colo., Kans., Fla.

Genus PSEUDONOMONEURA Bequaert

Pseudonomoneura Bequaert, 1961: 13. Type-species, *Leptomydas hirtus* Coquillett (orig. des.).
Nomoneura, authors, not Bezzi.
> REFERENCE: Hardy, 1950 (rev.).

californica (Hardy), 1950: 11 (*Nomoneura*).—Calif. **N. comb.**
hirta (Coquillett) *in* Baker, 1904: 39 (*Leptomydas*).—Calif.
> **concinnus** Coquillett *in* Baker, 1904: 39 (*Leptomydas*).—Calif.

micheneri (James), 1938a: 63 (*Nomoneura*).—Calif. **N. comb.**
tinkhami (Hardy), 1950: 18 (*Nomoneura*).—Calif.; Ariz. **N. comb.**

Genus OPOMYDAS Curran

Opomydas Curran, 1934a: 165. Type-species, *Ectyphus limbatus* Williston (orig. des.).

limbatus (Williston), 1886a: 292 (*Ectyphus*).—Ariz.; Calif., Utah, ?Fla.
townsendi (Williston), 1898: 58 (*Ectyphus*).—N. Mex.

Genus PHYLLOMYDAS Bigot

Phyllomydas Bigot, 1880b: xlvi. Type-species, *phyllocerus* Bigot (mon.).
> REFERENCE: Hardy, 1943b (rev.).

bruesii Johnson, 1926a: 140.—Tex.
currani Hardy, 1943b: 51.—Ariz.; N. Mex.
phyllocerus Bigot, 1880b: xlvii.—Rocky Mts.; Calif., N. Mex., Colo., Tex., Kans.
scitulus (Williston), 1886a: 291 (*Mydas*).—Ariz.

Family ASILIDAE

By Charles H. Martin and J. Wilcox

The adult "robber flies" are predaceous, feeding on other insects and occasionally on their own kind. In some areas they are serious pests of honey bees. Most asilids are found in definite ecological niches and often they are endemic to very restricted areas. Many genera are confined to the West or the extreme Southwest, and others have evolved most of their species there. Special habitat preferences are found in *Laphria* (forests), Leptogastrinae (mostly on grass), *Nannocyrtopogon* (on bare soil, rocks, stumps or logs), *Cophura*, *Eucyrtopogon*, *Heteropogon* and *Nicocles* (on tips of dead branches waiting for prey to fly by).

Oviposition habits vary and are correlated with the ovipositor structure of the females; some species lay eggs in soil, scattered on the ground, or in or on bark of trees or on plants. Larval habits are not well known and most larvae are assumed to be carnivorous like the adults, having been reported from grasshopper eggs, white grubs and other coleopterous larvae and pupae, but Melin (1923) concludes that many groups are vegetarian. Larvae have been found most often in the soil or in stumps and fallen trees. The complete life history probably usually requires from one to three years. Most species emerge as adults in a definite, usually short, period of the season. Adults also usually appear only during a certain time of the day. Some species of *Cyrtopogon* have a characteristic mating dance.

REFERENCES: Harris, 1862, Perris, 1870, Riley, 1878 (larval habits); Back, 1909 (rev. Dasypogoninae and Leptogastrinae); Malloch, 1917g (biol., immature stages); Davis, 1919 (biol. of *Promachus vertebratus*); Melin, 1923 (biol., immature stages Swedish asilids); Bromley, 1934a (key to subfamilies, genera, spp. of Tex.); E. G. Reinhard, 1938 (oviposition); Ritcher, 1940 (larval habits); Bromley, 1946 (key to subfamilies, genera, spp. of ne. U.S.); Martin, 1957 (rev. Leptogastrinae); Hull, 1962 (generic classif. with keys, figs. and spp. check list for Asilidae of the World).

Subfamily LEPTOGASTRINAE

REFERENCE: Martin, 1957 (rev.).

Genus APACHEKOLOS Martin

Apachekolos Martin, 1957: 352. Type-species, *Leptogaster scapularis* Bigot (orig. des.).

confusio Martin, 1957: 354.—Ariz.
crinita Martin, 1957: 354.—N. Mex.; Ariz.
scapularis (Bigot), 1878: 444 (*Leptogaster*).—Calif.; Tex.
tenuipes (Loew), 1862b: 192 (Cent. 2, no. 14) (*Leptogaster*).—D.C.; Okla. to Del., s. to Tex. and Fla.
weslacensis (Bromley), 1951: 3 (*Leptogaster*).—Tex.

Genus BEAMEROMYIA Martin

Beameromyia Martin, 1957: 355. Type-species, *Leptogaster pictipes* Loew (orig. des.).

bifida (Hardy), 1942a: 59 (*Leptogaster*).—Ariz.; Calif.
chrysops Martin, 1957: 356.—Fla.; Ala., Ga.
disfascia Martin, 1957: 357.—N.J.; Ohio, N.Y., Mass., Va.
floridensis (Johnson), 1913a: 60 (*Leptogaster*).—Fla.; Va.
kawiensis Martin, 1957: 358.—Kans.

lacinia Martin, 1957: 358.—Ariz.; Calif.
lunula Martin, 1957: 359.—Ariz.
macula Martin, 1957: 360.—Ariz.; Mexico.
monticola Martin, 1957: 360.—Ariz.
occidentis (Hardy), 1942a: 61 (*Leptogaster*).—Ariz.; N. Mex.
pictipes (Loew), 1862b: 189 (Cent. 2, no. 7) (*Leptogaster*).—Ill.; narrow belt from Kans. to Md.
  varipes Loew, 1862b: 189 (Cent. 2, no. 8) (*Leptogaster*).—D.C.
prairiensis Martin, 1957: 361.—Kans.
punicea Martin, 1957: 362.—Ariz.
silvacola Martin, 1957: 362.—N. Mex.; Ariz.
vulgaris Martin, 1957: 363.—D.C.; Ind. to N.J., s. to Ala. and Fla.

Genus LEPTOGASTER Meigen

Leptogaster Meigen, 1803: 269. Type-species, *Asilus tipuloides* Fabricius (mon.)=*cylindrica* (De Geer).
Gonypes Latreille, 1805: 309. Type-species, *Asilus tipuloides* Fabricius (mon.)=*cylindrica* (De Geer).

aegra Martin, 1957: 381.—N.C.; S.C.
altacola Martin, 1957: 373.—Ariz.
arborcola Martin, 1957: 368.—Ariz.
arenicola James, 1937b: 13.—Colo.; Nebr.
arida Cole *in* Cole and Lovett, 1919: 229.—Oreg.; Wash., Calif.
atridorsalis Back, 1909: 159.—Pa.; Ind. to s. Pa., s. to N.C.
brevicornis Loew, 1872a: 62 (Cent. 10, no. 23).—Tex.; Kans. to Md., s. to Tex. and Fla.
californica Martin, 1957: 373.—Calif.
carolinensis Schiner, 1866a: 696 (n. name for *nitida* Macquart, 1838).—Carolina.
  nitida Macquart, 1838b: 155 (1839: 271) (*Gonypes;* preocc. Macquart, 1826).—Carolina.
coloradensis James, 1937b: 14.—Colo.; S. Dak., Nebr., Kans.
cultaventris Martin, 1957: 375.—Calif.; Wash., Oreg.
eudicrana Loew, 1874a: 353.—Tex.; Ariz., Nev., N. Mex., Colo.
flavipes Loew, 1862b: 193 (Cent. 2, no. 15).—Nebr.; Minn. to Maine, s. to Kans. and Ga.
  favillaceus Loew, 1862b: 191 (Cent. 2, no. 12).—Conn.
  flavicornis Wulp, 1867: 136.—Wis.
  loewi Banks, 1914b: 133.—Va., N.Y.
fornicata Martin, 1957: 377.—Idaho.
hesperis Martin, 1957: 378.—Ariz.; N. Mex.

hirtipes Coquillett, 1904f: 178.—Colo.; Ariz., N. Mex.
incisuralis Loew, 1862b: 190 (Cent. 2, no. 11).—Ill.; Minn. to N.Y., s. to Tex. and Ga.
  ochraceus Schiner, 1867b: 359.—Pa.
lanata Martin, 1957: 382.—Utah; Idaho, Ariz., Tex.
lerneri Curran, 1953a: 1.—Bahama Is.; Fla.
murina Loew, 1862b: 190 (Cent. 2, no. 9).—Nebr.; S. Dak. to Mich., s. to Tex. and Ark.
nitoris Martin, 1957: 379.—Wash.; Oreg., Calif., Nev.
obscuripennis Johnson, 1895a: 304.—Fla.; N.C., S.C.
obscuripes Loew, 1862b: 191 (Cent. 2, no. 13).—Cuba; Tex., Fla.
panda Martin, 1957: 372.—Kans.
parvoclava Martin, 1957: 371.—N. Mex.
patula Martin, 1957: 380.—Ariz.
salvia Martin, 1957: 380.—Idaho; Oreg.
schaefferi Back, 1909: 170.—Tex.
texana Bromley, 1934a: 88.—Tex.
virgata Coquillett, 1904f: 177.—Tex.; Okla., N.Y., Pa., Md., Ga.

Genus LEPTOPTEROMYIA Williston

Leptopteromyia Williston, 1907: 2. Type-species, *gracilis* Williston (sub. mon., Williston, 1908: fig. 77–35).

americana Hardy, 1947a: 74.—Tex.

Genus PSILONYX Aldrich

Psilonyx Aldrich, 1923c: 5. Type-species, *Leptogaster annulatus* Say (orig. des.).

annulatus (Say), 1823: 75 (1859b: 68) (*Leptogaster*).—Pa.; Ohio to Mass., s. to Okla. and Ga.
  histrio Wiedemann, 1828: 535 (*Leptogaster*).—Pa.

Genus TIPULOGASTER Cockerell

Tipulogaster Cockerell, 1913b: 214. Type-species, *Leptogaster badia* Loew (orig. des.)=*glabrata* (Wiedemann).

glabrata (Wiedemann), 1828: 534 (*Leptogaster*).—Unknown; Nebr. to Que., s. to Tex. and Fla.
  badia Loew, 1862b: 188 (Cent. 2, no. 6) (*Leptogaster*).—Ill. **N. syn.**
  testaceus Loew, 1862b: 190 (Cent. 2, no. 10) (*Leptogaster*).—N.Y. **N. syn.**

Subfamily DASYPOGONINAE

REFERENCE: Back, 1909 (rev. N. Amer. genera and spp.).

Genus ABLAUTUS Loew

Ablautus Loew, 1866b: 37 (Cent. 7, no. 63). Type-species, *trifarius* Loew (mon.).
Ablautatus, emend.

REFERENCE: Wilcox, 1935a (key).

californicus Wilcox, 1935a: 226.—Calif.
coquilletti Wilcox, 1935a: 226.—Calif.
flavipes Coquillett, 1904f: 178.—Calif.; Ariz., N. Mex.
mimus Osten Sacken, 1877: 290 (*Ablautatus*).—Calif.; Ariz., N. Mex., Utah.
  squamipes Cole, 1924b: 11.—Ariz. **N. syn.**
nigronotum Wilcox, 1935a: 224.—Oreg.; Calif., Colo.
rubens Coquillett, 1904f: 178.—Wash.
rufotibialis Back, 1909: 182.—Tex.; N. Mex.
trifarius Loew, 1866b: 36 (Cent. 7, no. 63).—Calif.
vanduzeei Wilcox, 1935a: 225.—Calif.

Genus ARCHILESTRIS Loew

Archilestes Schiner, 1866a: 672 (preocc. Selys, 1862). Type-species, *Dasypogon capnopterus* Wiedemann (Schiner, 1868: 168).
Archilestris Loew, 1874a: 377 (n. name for *Archilestes* Schiner). Type-species, *Dasypogon capnopterus* Wiedemann (aut.).

magnifica (Walker), 1854: 427 (*Dasypogon*).—Mexico; Ariz.

Genus BACKOMYIA Wilcox and Martin

Backomyia Wilcox and Martin, 1957b: 1. Type-species, *Eucyrtopogon limpidipennis* Wilcox (orig. des.).

REFERENCE: Wilcox and Martin, 1957b (key).

anomala Wilcox and Martin, 1957b: 3.—Calif.
hannai Wilcox and Martin, 1957b: 4.—Calif.
limpidipennis (Wilcox), 1936d: 204 (*Eucyrtopogon*).—N. Mex.
schlingeri Wilcox and Martin, 1957b: 5.—Calif.

Genus BLEPHAREPIUM Rondani

Blepharepium Rondani, 1848: 89. Type-species, *luridum* Rondani (mon.).

Planetolestes Lynch Arribálzaga, 1879: 147. Type-species, *Laphria coarctata* Perty (mon.).

secabile (Walker), 1860a: 276 (*Dasypogon*).—Mexico; Ariz.

Genus BROMLEYUS Hardy

Bromleyus Hardy, 1944a: 226. Type-species, *flavidorsus* Hardy (orig. des.).

flavidorsus Hardy, 1944a: 226.—Ariz.

Genus CALLINICUS Loew

Callinicus Loew, 1872a: 71 (Cent. 10, no. 32). Type-species, *calcaneus* Loew (mon.).
Chrysoceria Williston, 1907: 1. Type-species, *Laparus pictitarsis* Bigot (mon.).

REFERENCE: Wilcox, 1936c (key).

calcaneus Loew, 1872a: 70 (Cent. 10, no. 32).—Calif.
    bilimbatum Bigot, 1878: 411 (*Dasypogon*).—Calif.
pictitarsis (Bigot), 1878: 417 (?*Laparus*).—Calif.; Ariz., Utah.
pollenius (Cole) *in* Cole and Lovett, 1919: 237 (*Chrysoceria*).—Oreg.; Wash. to Wyo., s. to Calif.
vittatus Wilcox, 1936c: 209.—Calif.

Genus CERATURGUS Wiedemann

Ceraturgus Wiedemann, 1824: 12. Type-species, *Dasypogon aurulentus* Fabricius (mon.).
Ceraturgopsis Johnson, 1903d: 111. Type-species, *Dasypogon cornutus* Wiedemann (orig. des.; misident.)=n. sp. in press.

REFERENCE: Brimley, 1924 (key).

aurulentus (Fabricius), 1805: 166 (*Dasypogon*).—N.Y.; N.J., Pa.
cornutus (Wiedemann), 1828: 382 (*Dasypogon*).—Unknown; Fla.
cruciatus (Say), 1823: 52 (1859b: 66) (*Dasypogon*).—Ark.; S. Dak. to Mass., s. to Ark. and Fla.
    fasciatus Walker, 1849: 367.—N.Y.
elizabethae Brimley, 1924: 8.—N.C.
mabelae Brimley, 1924: 11.—N.C.
mitchelli Brimley, 1924: 9.—N.C.
nigripes Williston, 1886a: 287.—Ga.; N.C.
oklahomensis (Bromley), 1934b: 225 (*Ceraturgopsis*).—Okla.
similis Johnson, 1912c: 152.—Mass.; Vt.

Genus COLEOMYIA Wilcox and Martin

Coleomyia Wilcox and Martin, 1935: 205. Type-species, *Metapogon setiger* Cole (orig. des.).

REFERENCE: Martin, 1953 (key).

alticola James, 1941b: 37.—Colo.; N. Mex.
crumborum Martin, 1953: 25.—Calif.
hinei Wilcox and Martin, 1935: 207.—Oreg.; Idaho.
rainieri Wilcox and Martin, 1935: 211.—Wash.
rubida Martin, 1953: 27.—Oreg.
sculleni Wilcox and Martin, 1935: 209.—Oreg.
setigera (Cole) *in* Cole and Lovett, 1919: 235 (*Metapogon*).—Oreg.; Wash., Calif.

Genus COMANTELLA Curran

Comantella Curran, 1923g: 93. Type-species, *Cyrtopogon maculosus* Coquillett (orig. des.; misident.)=*fallei* (Back). See Melander, 1923c: 207.

REFERENCE: James, 1937c (key).

cristata (Coquillett), 1893b: 33 (*Blacodes*).—Calif.
fallei (Back), 1909: 378 (*Cophura*).—Colo.
pacifica Curran, 1926g: 311.—B.C.; Wash.
rotgeri James, 1937c: 61.—Colo.; Alta., N. Mex.

Genus COPHURA Osten Sacken

Blax Loew, 1872a: 65 (Cent. 10, no. 24) (preocc. Koch, 1840). Type-species, *bellus* Loew (mon.).
Blacodes Loew, 1874a: 377 (n. name for *Blax* Loew, but preocc. Dejean, 1859). Type-species, *Blax bellus* Loew (aut.).
Cophura Osten Sacken, 1887a: 181. Type-species, *sodalis* Osten Sacken (mon.).
Loewiella Williston, 1896a: 57 (n. name for *Blacodes* Loew, but preocc. Meunier, 1894). Type-species, *Blax bellus* Loew (aut.).
Buckellia Curran, 1925d: 156. Type-species, *Cophura albosetosa* Hine (orig. des.).

REFERENCES: Pritchard, 1943 (rev.); Wilcox, 1959 (rev. *clausa* group).

albosetosa Hine, 1908: 202.—B.C.; Wash., Oreg.
  cyrtopogona Cole *in* Cole and Lovett, 1919: 236.—Oreg.
ameles Pritchard, 1943: 298.—N. Mex.
arizonensis (Schaeffer), 1916: 65 (*Lasiopogon*).—Ariz.
  drakei Pritchard, 1935: 6 (*Buckellia*).—Ariz.

bella (Loew), 1872a: 63 (Cent. 10, no. 24) (*Blax*).—Tex.; Ariz., N. Mex.
brevicornis (Williston), 1883a: 22 (*Taracticus*).—Wash.; Wash. to Nebr., s. to Calif. and Colo.
  melanochaeta Melander, 1923c: 210.—Idaho. **N. syn.**
caca Pritchard, 1943: 306.—N. Mex.
clausa (Coquillett), 1893b: 34 (*Blacodes*).—Calif.
daphne Pritchard, 1943: 290.—Mexico; Ariz.
dora Pritchard, 1943: 296.—Nebr.; Ariz.
fur (Williston), 1885a: 53 (*Aphamartania*).—Ariz.
getzendaneri Wilcox, 1959: 123.—Calif.; Ariz., Mexico (Baja Calif.).
hennei Wilcox and Martin, 1945: 11.—Calif.
hesperia (Pritchard), 1935: 7 (*Buckellia*).—Ariz.
  vera Pritchard, 1935: 8 (*Buckellia*).—Ariz. **N. syn.**
painteri Pritchard, 1943: 293.—Ariz.
pollinosa Curran, 1930f: 10.—Ariz.; N. Mex., Colo., Okla., Tex.
  lutzi Curran, 1931b: 7.—Colo. **N. syn.**
  wilcoxi Pritchard, 1935: 8 (*Buckellia*).—Okla. **N. syn.**
pulchella Williston, 1901: 314.—Mexico; Utah.
scitula (Williston), 1883a: 19 (?*Nicocles*).—Wash.; Oreg.
sculleni Wilcox, 1937: 39.—Ariz.; N. Mex.
sodalis Osten Sacken, 1887a: 181.—Mexico; Ariz.
stylosa Curran, 1931b: 7.—Okla.; Colo., Kans., Tex.
texana Bromley, 1934a: 92.—Tex.; Okla.
tolandi Wilcox, 1959: 125.—Calif.; Ariz.
trunca (Coquillett), 1893b: 34 (*Blacodes*).—Calif.
  highlandica Cole, 1916b: 63.—Calif.
vitripennis (Curran), 1927f: 85 (*Buckellia*).—B.C.; Mont.

Genus CYRTOPOGON Loew

Dasypogon, subg. **Cyrtopogon** Loew, 1847c: 516. Type-species, *Asilus ruficornis* Fabricius (Rondani, 1856: 157).
Euarmostus Walker, 1851b: 102. Type-species, *bimacula* Walker (mon.).
Enarmostus, error.
  REFERENCE: Wilcox and Martin, 1936a (rev.).

ablautoides Melander, 1923a: 111.—Wash.
albifacies Johnson, 1942: 1.—Utah.
albifrons Wilcox and Martin, 1936a: 83.—Idaho; Oreg.
albovarians Curran, 1924j: 279.—Alta.
aldrichi Wilcox and Martin, 1936a: 51.—Calif.
alleni Back, 1909: 261.—N.H.; Que. and N.B. to Mass., also N.C.

anomalus Cole *in* Cole and Lovett, 1919: 231.—Oreg.; B.C., Wash., Idaho.
auratus Cole *in* Cole and Lovett, 1919: 230.—Oreg.; Wash., Alta., Idaho, Utah, Wyo.
  albitarsis Curran, 1922i: 278.—Alta.
aurifex Osten Sacken, 1877: 301.—Calif.; B.C., Wash., Oreg.
auripilosus Wilcox and Martin, 1936a: 30.—Wash.; Oreg.
banksi Wilcox and Martin, 1936a: 79.—Wash.; Y.T. and B.C. to Wyo., s. to Calif. and Colo.
basingeri Wilcox and Martin, 1936a: 49.—Calif.
beameri Wilcox and Martin, 1936a: 84.—Ariz.
bigelowi Curran, 1924j: 277.—Ont.; Que.
bimacula (Walker), 1851b: 102 (*Euarmostus*).—N. Amer.; B.C. to Hudson Bay, s. to Wash. and N. Mex., also N.H.
  melanopleurus Loew, 1866b: 35 (Cent. 7, no. 61).—N.H.
caesius Melander, 1923a: 112.—Wash.; Oreg., Wyo.
callipedilus callipedilus Loew, 1874a: 358.—Calif.
  ssp. **nigritarsus** Wilcox and Martin, 1936a: 39.—Calif.
chagnoni Curran, 1939c: 1.—Que.
curtipennis Wilcox and Martin, 1936a: 74.—Calif.
curtistylus Curran, 1923g: 133.—Utah; Idaho to Mont., s. to Calif.
cymbalista Osten Sacken, 1877: 297.—Calif.; Nev.
dasyllis Williston, 1893c: 66.—Colo.; Alaska to Alta., s. to Oreg. and Idaho.
dasylloides Williston, 1883a: 11.—Wash.; Oreg., Calif., Idaho, N.Y.
dubius Williston, 1883a: 13.—Oreg.; Wash.
  tacomae Melander, 1923a: 116.—Wash.
evidens Osten Sacken, 1877: 306.—Calif.; Wash., Oreg.
falto (Walker) ,1849: 355 (*Dasypogon*).—N.S.; Alta. to N.S., s. to Ill. and Fla.
  chrysopogon Loew, 1866b: 34 (Cent. 7, no. 59). Mass.
fumipennis Wilcox and Martin, 1936a: 71.—Wash.; B.C., Idaho.
glarealis Melander, 1923a: 113.—Wash.; Oreg., Calif., Idaho, Mont., Wyo.
idahoensis Wilcox and Martin, 1936a: 82.—Idaho; Utah.
infuscatus Cole *in* Cole and Lovett, 1919: 233.—Oreg.; B.C., Wash., Calif., Idaho.
inversus Curran, 1923g: 172.—B.C.; Wash., Oreg., Wyo., Colo.
jemezi Wilcox and Martin, 1936a: 47.—N. Mex.; Ariz.
laphriformis Curran, 1923a: 59.—N.H.
leptotarsus Curran, 1923g: 186.—Ont.
leucozona Loew, 1874a: 364.—Calif.; B.C. to Colo., s. to Oreg. and N. Mex.

lineotarsus Curran, 1923g: 187.—Alta.; Mont.
longimanus Loew, 1874a: 360.—Calif.
lutatius (Walker), 1849: 357 (*Dasypogon*).—N.S.; Mich. to N.S., s. to Pa.
lyratus Osten Sacken, 1878c: 232.—N.Y.; Maine, N.H., N.C.
maculipennis (Macquart), 1834a: 298 (*Dasypogon*).—Europe; Ont.
marginalis Loew, 1866b: 35 (Cent. 7, no. 60).—Mass.; Ont. to N.H., s. to N.C.
montanus montanus Loew, 1874a: 362.—Calif.; B.C. to Alta., s. to Oreg. and N. Mex.
    latericaudus Curran, 1923g: 170 (as var.).—Alta. ?Var.
  ssp. **wilcoxi** James, 1942a: 124–126.—Colo.
nitidus Cole, 1924b: 10.—Wash.
nugator Osten Sacken, 1877: 307.—Calif.; B.C., Wash., Oreg., Idaho, N. Mex.
perspicax Cole *in* Cole and Lovett, 1919: 233.—Oreg.; Calif.
planitarsus Wilcox and Martin, 1936a: 61.—Oreg.
platycauda Curran, 1924i: 251.—Man.
plausor Osten Sacken, 1877: 297.—N. Mex.; Idaho, Utah, Colo., Nebr.
praepes Williston, 1883a: 12.—Wash.; Oreg., Calif.
predator Curran, 1923g: 188.—B.C.; Wash., Idaho.
princeps Osten Sacken, 1877: 302.—Calif.; Wash., Oreg.
    cretaceus Osten Sacken, 1877: 302.—Calif.
profusus Osten Sacken, 1877: 305.—N. Mex.; Colo., Kans.
pulcher Back, 1909: 274.—Colo.; Utah, Ariz.
rainieri Wilcox and Martin, 1936a: 77.—Wash.; Oreg.
rattus Osten Sacken, 1877: 308.—Calif.; Oreg.
rejectus Osten Sacken, 1877: 307.—Calif.; Wash., Oreg., Idaho.
    positivus Osten Sacken, 1877: 307.—Calif.
rufotarsus Back, 1909: 275.—Mont.; Oreg., Utah, Colo.
sansoni Curran, 1923g: 138.—Alta.
semitarius semitarius Melander, 1923a: 115.—Wash.; Oreg.
  ssp. **californicus** Wilcox and Martin, 1936a: 37.—Calif.
stenofrons Wilcox and Martin, 1936a: 52.—N. Mex.; Ariz.
sudator Osten Sacken, 1877: 307.—Calif.; B.C., Wash., Oreg., Idaho.
swezeyi Wilcox and Martin, 1936a: 64.—Utah; Wash.
tenuis Bromley, 1924: 126.—Maine.
thompsoni Cole *in* Cole and Lovett, 1921: 255.—Oreg.
tibialis Coquillett, 1904f: 183.—Ariz.; N. Mex.
vanduzeei Wilcox and Martin, 1936a: 76.—Calif.; Wash., Oreg.
vandykei Wilcox and Martin, 1936a: 27.—Calif.
varans Curran, 1923g: 141.—Que.; Ont.
vulneratus Melander, 1923a: 118.—Ont.

willistoni Curran, 1922i: 277.—B.C.; B.C. to Alta., s. to Calif. and Wyo.

Genus DICOLONUS Loew

Dicolonus Loew, 1866b: 32 (Cent. 7, no. 56). Type-species, *simplex* Loew (mon.).

simplex Loew, 1866b: 32 (Cent. 7, no. 56).—Calif.
sparsipilosum Back, 1909: 247.—Mont.; Calif., Colo.

Genus DIOCTRIA Meigen

Dioctria Meigen, 1803: 270. Type-species, *Asilus oelandicus* Linnaeus (Latreille, 1810: 443).

REFERENCE: Wilcox and Martin, 1941 (rev.).

Subgenus BOHARTIA Hull

Bohartia Hull, 1958a: 317 (as genus). Type-species, *bromleyi* Hull (orig. des.). **N. status.**

bromleyi (Hull), 1958a: 318 (*Bohartia*).—Nev.

Subgenus DIOCTRIA Meigen

baumhaueri Meigen, 1820: 245.—Europe; Mich., N.Y., Conn., Mass.
henshawi Johnson, 1918: 103 (n. name for *flavipes* Banks).—Wash.
    flavipes Banks, 1917b: 119 (preocc. Meigen, 1804).—Wash.
pleuralis Banks, 1917b: 118.—Calif.
pusio Osten Sacken, 1877: 288.—Calif.; Wash., Oreg., Colo.
seminole Bromley, 1924: 125.—Fla.
vera Back, 1909: 256.—Calif.
vertebrata Cole *in* Cole and Lovett, 1919: 230.—Oreg.; Wash.

Subgenus EUDIOCTRIA Wilcox and Martin

Dioctria, subg. **Eudioctria** Wilcox and Martin, 1941: 8. Type-species, *albius* Walker (orig. des.).

albius Walker, 1849: 301.—N.Y.; Wis. to Que., s. to Pa.
    aurifacies Wilcox and Martin, 1941: 11 (as form).—Wis.
    xanthopennis Wilcox and Martin, 1941: 12 (as form).—Wis.
beameri Wilcox and Martin, 1941: 13.—Calif.
brevis Banks, 1917b: 117.—N.Y.; Ohio to Mass., s. to Tenn. and N.C.
doanei Melander, 1923c: 214.—Calif.
media Banks, 1917b: 118.—Calif.; Wash., Oreg.
monrovia Wilcox and Martin, 1941: 15.—Calif.

nitida Williston, 1883a: 8.—Wash.; B.C., Oreg., Calif.
  denuda Wilcox and Martin, 1941: 17 (as form).—Calif.
parvula Coquillett, 1893c: 80.—Calif.
propinqua Bromley, 1924: 125.—Mass.; N.Y., Que., N.S.
sackeni Williston, 1883a: 8.—Wash.; B.C., Oreg., Idaho.
  rivalis Melander, 1923c: 215 (as form).—Idaho.
tibialis Banks, 1917b: 118 (as *longicornis* var.).—Va.; Pa., N.J., Md.
  longicornis Banks, 1917b: 118 (preocc. Meigen, 1820).—Va.
  banksi Johnson, 1918: 103 (n. name for *longicornis* Banks).—Va.

Subgenus METADIOCTRIA Wilcox and Martin

Dioctria, subg. **Metadioctria** Wilcox and Martin, 1941: 19. Type-species, *rubida* Coquillett (orig. des.).

resplendens Loew, 1872a: 62 (Cent. 10, no. 21).—Calif.
rubida Coquillett, 1893c: 80.—Calif.
  nigripilosa Wilcox and Martin, 1941: 19 (as form).—Calif.
  atripes Wilcox and Martin, 1941: 20 (as form).—Calif.

Subgenus NANNODIOCTRIA Wilcox and Martin

Dioctria, subg. **Neodioctria** Wilcox and Martin, 1941: 7 (preocc. Ricardo, 1912). Type-species, *albicornis* Wilcox and Martin (orig. des.).
Dioctria, subg. **Nannodioctria** Wilcox and Martin, 1942: 35 (n. name for *Neodioctria* Wilcox and Martin). Type-species, *albicornis* Wilcox and Martin (aut.).

albicornis Wilcox and Martin, 1941: 7.—Calif.

Genus DIOGMITES Loew

Diogmites Loew, 1866b: 21 (Cent. 7, no. 36). Type-species, *platypterus* Loew (Coquillett, 1910b: 533).
Deromyia, authors, not Philippi.
  REFERENCE: Bromley, 1936 (key).
angustipennis Loew, 1866b: 23 (Cent. 7, no. 41).—Kans.; Colo., Ill.
basalis (Walker), 1851b: 95 (*Dasypogon*).—Atlantic States; Ill. to N.H., s. to N.J.
  umbrinus Loew, 1866b: 24 (Cent. 7, no. 43).—N.Y.
bilineatus Loew, 1866b: 23 (Cent. 7, no. 40).—Cuba; ?Fla.
coloradensis (James), 1933b: 2 (*Deromyia*).—Colo.; Ariz.
contortus Bromley, 1936: 235.—Ariz.; Calif.
crudelis Bromley, 1936: 233.—N.C.; S.C., Fla.
discolor Loew, 1866b: 21 (Cent. 7, no. 37).—Pa.; Pa. to Mass., s. to Md. and Del.

esuriens Bromley, 1936: 230.—Fla.; S.C., Ga.
fragilis Bromley, 1936: 235.—Tex.
grossus Bromley, 1936: 236.—Colo.; Idaho to Colo., s. to Calif. and N. Mex.
herennius (Walker), 1849: 339 (*Dasypogon*).—Ohio. ?=*basalis*.
misellus Loew, 1866b: 22 (Cent. 7, no. 39).—D.C.; Ill. to Mass., s. to Tex. and Fla.
missouriensis Bromley, 1951: 13.—Mo.; Miss., Ohio.
neoternatus (Bromley), 1931: 433 (*Deromyia*).—Kans.; Colo. to Ind., s. to Tex. and S.C.
perplexus (Back), 1909: 360 (*Deromyia*).—N. Mex.
platypterus Loew, 1866b: 20 (Cent. 7, no. 36).—Ill.; Nebr. to Ill., s. to La. and Miss.
pritchardi Bromley, 1936: 237.—Okla.; Tex.
properans Bromley, 1936: 232.—Ala.; Miss., Fla., Ga.
pulcher (Back), 1909: 361 (*Deromyia*).—Calif.
rufescens (Macquart), 1834a: 295 (*Dasypogon*).—Pa. ?=*discolor*.
sallei (Bellardi).—Not Nearctic.
salutans Bromley, 1936: 233.—N.C.; Miss. to Va., s. to Fla.
symmachus Loew, 1872a: 66 (Cent. 10, no. 26).—Tex.
ternatus Loew, 1866b: 22 (Cent. 7, no. 38).—Cuba; La.
texanus Bromley, 1934a: 91.—Tex.
unicolor Hull, 1958b: 106.—Ariz.

Genus DIZONIAS Loew

Dizonias Loew, 1866b: 29 (Cent. 7, no. 53). Type-species, *phoenicurus* Loew (Coquillett, 1910b: 534)=*tristis* (Walker).

bicinctus Loew, 1866b: 30 (Cent. 7, no. 54).—N. Mex.
lucasi (Bellardi).—Not Nearctic.
pilatei Johnson, 1903d: 112.—Ga.
tristis (Walker), 1851b: 93 (*Dasypogon*).—Amer.; N. Mex. to Fla., s. to Mexico (Tam.).
    albifasciatus Back, 1904: 292 (*Ospriocerus*).—Fla.

Genus ECHTHOPODA Loew

Echthopoda Loew, 1866b: 16 (Cent. 7, no. 27). Type-species, *pubera* Loew (mon.).

carolinensis Bromley, 1951: 9.—N.C.
formosa Loew, 1872a: 62 (Cent. 10, no. 22).—Pa.; Mass., Va., N.C.
pubera Loew, 1866b: 15 (Cent. 7, no. 27).—Nebr.; Wash. to S. Dak., s. to Wyo. and Nebr.

Genus EUCYRTOPOGON Curran

Eucyrtopogon Curran, 1923g: 95. Type-species, *Cyrtopogon nebulo* Osten Sacken (orig. des.).

REFERENCE: Curran, 1923g (key).

albibarbus Curran, 1923g: 117.—Sask.
calcaratus Curran, 1923g: 119.—Alta.; B.C.
comantis Curran, 1923g: 116.—B.C.
diversipilosis Curran, 1923g: 118.—B.C.; Alta.
kelloggi Wilcox, 1936d: 205.—N. Mex.
maculosus (Coquillett), 1904f: 184 (*Cyrtopogon*).—Wash.
nebulo (Osten Sacken), 1877: 309 (*Cyrtopogon*).—Calif.; B.C.
nigripes (Jones), 1907a: 279 (*Heteropogon*).—Nebr.
punctipennis (Melander), 1923a: 114 (*Cyrtopogon*).—Idaho; Wash.
spiniger Curran, 1923g: 117.—B.C.
varipennis (Coquillett), 1904f: 184 (*Cyrtopogon*).—Wash.; B.C.

Genus HADROKOLOS Martin

Hadrokolos Martin, 1959a: 3. Type-species, *Holopogon texanus* Bromley (orig. des.).

cazieri Martin, 1959a: 3.—Tex.
pritchardi Martin, 1959a: 4.—Okla.
texanus (Bromley), 1934a: 89 (*Holopogon*).—Tex.; Okla.

Genus HAPLOPOGON Engel

Haplopogon Engel, 1930: 409. Type-species, *nudus* Engel (orig. des.).

REFERENCE: Martin, 1955b (key).

bullatus (Bromley), 1934a: 89 (*Holcocephala*).—Tex.
erinus Pritchard, 1941: 351.—Ariz.
latus (Coquillett), 1904c: 33 (*Holopogon*).—Tex.
 lautus, error.
triangulatus Martin, 1955b: 316.—Tex.

Genus HETEROPOGON Loew

Dasypogon, subg. **Heteropogon** Loew, 1847c: 488. Type-species, *manicatus* Meigen (Back, 1909: 318).
Anisopogon Loew, 1874a: 377 (unjustified n. name for *Heteropogon* Loew). Type-species, *Dasypogon manicatus* Meigen (aut.).

REFERENCE: Wilcox, 1941 (key).

arizonensis Wilcox, 1941: 55.—Ariz.; Utah.
currani Pritchard, 1935: 3.—Okla.
duncani Wilcox, 1941: 54.—Ariz.
lautus Loew, 1872a: 72 (Cent. 10, no. 34).—Tex.
ludius (Coquillett), 1893a: 20 (*Anisopogon*).—Calif.
macerinus (Walker), 1849: 356 (*Dasypogon*).—N.J.; N.Y., Pa., Ky.
    gibbus Loew, 1866b: 33 (Cent. 7, no. 58).—Pa.
maculinervis James, 1937b: 12.—Colo.; Ariz., Utah, Mont.
patruelis (Coquillett), 1893a: 21 (*Anisopogon*).—Tex.; Ariz.
paurosomus Pritchard, 1935: 4.—Ariz.
phoenicurus Loew, 1872a: 71 (Cent. 10, no. 33).—Tex.
rejectus Williston, 1901: 307.—Mexico; Ariz.
rubidus (Coquillett), 1893a: 21 (*Anisopogon*).—Calif.
rubrifasciatus Bromley, 1931: 432.—N.C.
senilis (Bigot), 1878: 423 (*Anisopogon*).—Calif.; Oreg.
spatulatus Pritchard, 1935: 5.—Ariz.
vespoides (Bigot), 1878: 423 (*Anisopogon*).—Calif.
wilcoxi James, 1934b: 84.—Colo.; Ariz., N. Mex., Wyo., Ill.

Genus HODOPHYLAX James

Hodophylax James, 1933b: 1. Type-species, *aridus* James (orig. des.).

    REFERENCE: Wilcox, 1961 (rev.).

aridus James, 1933b: 2.—Colo.; Ariz., N. Mex., Tex., Kans.
    mcgregori Bromley, 1934a: 88 (*Ablautus*).—Tex.
basingeri Pritchard, 1938c: 130.—Calif.; Ariz.
halli Wilcox, 1961: 115.—Calif.
tolandi Wilcox, 1961: 114.—Ariz.; N. Mex.

Genus HOLCOCEPHALA Jaennicke

Discocephala Macquart, 1838b: 50 (1839: 166) (preocc. Laporte, 1832). Type-species, *rufiventris* Macquart (Coquillett, 1910b: 534) = *abdominalis* (Say).
Holcocephala Jaennicke, 1867: 359 (1868: 51) (n. name for *Discocephala* Macquart). Type-species, *Discocephala rufiventris* Macquart (aut.) = *abdominalis* (Say).

    REFERENCE: Pritchard, 1938a (key).

abdominalis (Say), 1823: 50 (1859b: 64) (*Dasyopogon*).—Pa.; U.S. e. of Rocky Mts.
    rufiventris Macquart, 1838b: 50 (1839: 166) (*Discocephala*).— Carolina.

aeta Walker, 1849: 362 (*Dasypogon*).—Fla.
laticeps Wulp, 1867: 137 (*Dasypogon*).—Wis.
calva (Loew), 1872a: 73 (Cent. 10, no. 35) (*Discocephala*).—Tex.;
   Kans. to N.Y., s. to Tex. and Fla.
fusca Bromley, 1951: 10.—Tex.; Ohio, Tenn.

Genus HOLOPOGON Loew

Dasypogon, subg. **Holopogon** Loew, 1847c: 473. Type-species,
   nigripennis Meigen (Coquillett, 1910b: 552).

REFERENCE: Martin, 1959a (rev.).

Subgenus DASYHOLOPOGON Martin

Holopogon, subg. **Dasyholopogon** Martin, 1959a: 34. Type-species,
   umbrinus Back (orig. des.).

caesariatus Martin, 1959a: 35.—Idaho; Oreg., Utah.
crinitus Martin, 1959a: 37.—Calif.
umbrinus Back, 1909: 317.—Calif.

Subgenus HOLOPOGON Loew

acropennis Martin, 1959a: 13.—N. Mex.; Ariz., Colo.
albipilosus Curran, 1923j: 207.—B.C.; B.C. to Man., s. to Nev. and
   Wyo.
atrifrons Cole, 1924b: 8.—Calif.
atripennis Back, 1909: 312.—N. Mex.; Ariz., Colo.
currani Martin, 1959a: 17.—Ariz.; N. Mex., Nev.
guttulus (Wiedemann), 1821c: 228 (*Dasypogon*).—Ga.; Tex., Pa.,
   and N.J., s. to Fla.
   philadelphicus Schiner, 1867b: 360.—Pa.
mingusae Martin, 1959a: 21.—Ariz.
oriens Martin, 1959a: 22.—N.C.; Minn. to Conn., s. to Tenn. and
   S.C.
phaeonotus Loew, 1874a: 366.—Tex.; Wis. to Mass., s. to Tex. and
   Fla.
   tibialis Curran, 1923j: 207.—Ont.
seniculus Loew, 1866b: 36 (Cent. 7, no. 62).—Nebr.; Nev. to Sask.,
   s. to Utah and Colo.
snowi Back, 1909: 316.—Kans.; Okla., Tex.
stellatus Martin, 1959a: 28.—Calif.; B.C., Wash., Oreg., Nev.
vockerothi Martin, 1959a: 31.—Ont.; Man. to N.Y., s. to Ill.
wilcoxi Martin, 1959a: 33.—Ariz.

Genus ITOLIA Wilcox

Itolia Wilcox, 1936d: 201. Type-species, *maculata* Wilcox (orig. des.).
REFERENCE: Wilcox, 1949 (key).

atripes Wilcox, 1949: 193.—Ariz.; Calif.
maculata Wilcox, 1936d: 202.—Ariz.
timberlakei Wilcox, 1949: 192.—Calif.

Genus LAPHYSTIA Loew

Laphystia Loew, 1847c: 538. Type-species, *sabulicola* Loew (mon.).
Laphyctis Loew, 1858d: 338. Type-species, *Stichopogon gigantellus* Loew (mon.).
Asicya Lynch Arribálzaga, 1880: 224. Type-species, *fasciata* Lynch Arribálzaga (mon.).
REFERENCE: Wilcox, 1960 (rev.).

albiceps (Macquart), 1846: 197 (1846: 69) (*Dasypogon*).—Tex.
annulata Hull, 1957a: 72.—Ariz.
  interrupta Hull, 1957a: 73 (as ssp.).—Ariz.
bromleyi Wilcox, 1960: 331.—Okla.; Ark.
brookmani Wilcox, 1960: 332.—Calif.
canadensis Curran, 1927f: 87.—Man.; Mont. and Man. to Iowa, s. to Wyo. and Kans.
cazieri Wilcox, 1960: 334.—Calif.
confusa Curran, 1927f: 86.—Okla.; Wyo., Tex.
duncani Wilcox, 1960: 334.—Ariz.; Calif.
flavipes Coquillett, 1904f: 180.—Mont.; Mont. to Man., s. to Colo. and N.C.
howlandi Wilcox, 1960: 335.—Calif.
jamesi Wilcox, 1960: 336.—Calif.
laguna Wilcox, 1960: 336.—Tex.
lanhami James, 1941b: 35.—Colo.; N. Mex.
limatula Coquillett, 1904f: 180.—N. Mex.; Ariz.
litoralis Curran, 1931b: 16.—N.Y.; N.Y. to Fla.
martini Wilcox, 1960: 338.—Calif.
notata (Bigot), 1878: 433 (*Triclis*).—N. Amer.; Kans., Miss.
ochreifrons Curran, 1931b: 16.—Ky.; La., Miss., Ga., Fla.
opaca Coquillett, 1904f: 180.—Tex.; La.
rubra Hull, 1957a: 74.—Ariz.
rufiventris Curran, 1931b: 17.—Wyo.; Colo.
rufofasciata Curran, 1931b: 13.—Wyo.; Colo.
sexfasciata (Say), 1823: 50 (1859b: 64) (*Dasypogon*).—Mo.; Nebr., Kans.
  subfasciata, error.

sillersi Hull, 1963: 202.—Ariz.
snowi Wilcox, 1960: 343.—Kans.
texensis Curran, 1931b: 14.—Tex.; La.
tolandi Wilcox, 1960: 344.—Nev.; Utah.
torpida Hull, 1957a: 70.—Calif.
utahensis Wilcox, 1960: 345.—Utah; Ariz.
varipes Curran, 1931b: 15.—Okla.; N. Mex., Tex., Kans.

Genus LASIOPOGON Loew

Dasypogon, subg. **Lasiopogon** Loew, 1847c: 508. Type-species, *pilosellus* Loew (Hull, 1962: 115). Designation by Rondani, 1856: 156, of *Dasypogon hirtellus* Meigen invalid because it was originally included as species inquirenda.
Daulopogon Loew, 1874a: 377 (unjustified n. name for *Lasiopogon* Loew). Type-species, *Dasypogon pilosellus* Loew (aut.).
Alexiopogon Curran, 1934a: 183.—Type-species, *Daulopogon terricola* Johnson (orig. des.).

REFERENCE: Cole and Wilcox, 1938 (key).

actius Melander, 1923b: 138.—Wash.; Oreg.
albidus Cole and Wilcox, 1938: 21.—Wash.
aldrichii Melander, 1923b: 139.—Idaho; Wash., Oreg., Alta., Utah.
arenicola (Osten Sacken), 1877: 310 (*Daulopogon*).—Calif.
aridus Cole and Wilcox, 1938: 25.—N. Mex.
atripennis Cole and Wilcox, 1938: 27.—Oreg.
bivittatus Loew, 1866b: 33 (Cent. 7, no. 57).—Calif.; Wash., Oreg.
californicus Cole and Wilcox, 1938: 31.—Calif.; Nev.
canus Cole and Wilcox, 1938: 32.—Alaska.
carolinensis Cole and Wilcox, 1938: 34.—N.C.; Ind., Ohio, S.C.
chaetosus Cole and Wilcox, 1938: 36.—Wash.; Oreg.
cinereus Cole *in* Cole and Lovett, 1919: 229.—Oreg.; Wash. to Alta., s. to Calif. and Wyo.
currani Cole and Wilcox, 1938: 40.—Mass.; N.Y. to N.H., s. to Ga.
delicatulus Melander, 1923b: 140.—Wash.
dimicki Cole and Wilcox, 1938: 44.—Oreg.
drabicola Cole, 1916b: 65.—Calif.
fumipennis Melander, 1923b: 141.—Wash.; B.C., Oreg., Calif., Colo.
    coloradensis Cole and Wilcox, 1938: 49 (as var.).—Colo.
    olympia Cole and Wilcox, 1938: 49 (as var.).—Wash., B.C.
gabrieli Cole and Wilcox, 1938: 49.—Calif.
hinei Cole and Wilcox, 1938: 51.—Alaska.
littoris Cole, 1924b: 8.—Calif.

martinensis Cole and Wilcox, 1938: 54.—Wash.
monticola Melander, 1923b: 142.—Wash.; B.C., Oreg., Idaho.
oklahomensis Cole and Wilcox, 1938: 57.—Okla.
opaculus (Loew), 1874a: 367 (*Daulopogon*).—Ill.; Ohio, Md., N.C.
pacificus Cole and Wilcox, 1938: 61.—Oreg.; B.C., Wash.
pugeti Cole and Wilcox, 1938: 62.—Wash.; Oreg.
quadrivittatus Jones, 1907a: 278.—Nebr.; Alta., Mont., N. Dak.
ripicola Melander, 1923b: 143.—Wash.; Oreg., Idaho.
shermani Cole and Wilcox, 1938: 67.—S.C.; N.C.
slossonae Cole and Wilcox, 1938: 70.—N.H.; N.J., Va.
terricola (Johnson), 1900b: 326 (*Daulopogon*).—Mass.; Alta. to N.J., s. to N. Dak. and Va.
testaceus Cole and Wilcox, 1938: 72.—Calif.
tetragrammus (Loew), 1874a: 368 (*Daulopogon*).—Canada; Que., N.Y., N.H., Mass., Conn.
trivittatus Melander, 1923b: 144.—Mont.
willametti Cole and Wilcox, 1938: 77.—Oreg.; Wash.
yukonensis Cole and Wilcox, 1938: 79.—Y.T.
zonatus Cole and Wilcox, 1938: 80.—Calif.

Genus LESTOMYIA Williston

Lestomyia Williston, 1883a: 19. Type-species, *Clavator sabulonum* Osten Sacken (orig. des.).
Clavator, authors, not Philippi.

REFERENCE: Curran, 1942a (key).

atripes Wilcox, 1937: 37.—N. Mex.; Ariz.
fraudigera Williston, 1883a: 21.—Calif.
montis Cole, 1916b: 67.—Calif.
sabulona (Osten Sacken), 1877: 292 (*Clavator*).—Calif.
    redlandae Cole, 1916b: 64.—Calif.
strigipes Curran, 1931b: 3.—Wyo.
unicolor Curran, 1942a: 57.—Ariz.

Genus METAPOGON Coquillett

Metapogon Coquillett, 1904f: 181. Type-species, *gilvipes* Coquillett (orig. des.).

REFERENCE: Melander, 1923c (key).

albulus Melander, 1923c: 211.—Wash.; Idaho.
gibber (Williston), 1883a: 14 (?*Cyrtopogon*).—Calif.
gilvipes Coquillett, 1904f: 182.—Calif.
pictus Cole, 1916b: 65.—Calif.
punctipennis Coquillett, 1904f: 182.—N. Mex.; Ariz.

Genus MICROSTYLUM Macquart

Microstylum Macquart, 1838b: 26 (1839: 142). Type-species, *Dasypogon venosus* Wiedemann (Back, 1909: 213).
Megapollyon Walker, 1854: 452. Type-species, *Microstylum acutirostre* Loew (Coquillett, 1910b: 566).
Megapollion, error.

galactodes Loew, 1866b: 25 (Cent. 7, no. 44).—N. Mex.; Ariz. Kans., Tex.
morosum Loew, 1872a: 67 (Cent. 10, no. 27).—Tex.; Ariz.
pollens Osten Sacken, 1878c: 230 (note 100).—Tex.

Genus MYELAPHUS Bigot

Myelaphus Bigot, 1882b: xci. Type-species, *melas* Bigot (orig. des.).

lobicornis (Osten Sacken), 1877: 287 (*Ceraturgus*).—Idaho; Calif., Nev.
melas Bigot, 1882b: xci.—Calif.
rufus Williston, 1883a: 7.—Calif.

Genus NANNOCYRTOPOGON Wilcox and Martin

Nannocyrtopogon Wilcox and Martin, 1936b: 449. Type-species, *Cyrtopogon cerussatus* Osten Sacken (orig. des.).

REFERENCE: Wilcox and Martin, 1957a (key).

antennatus Wilcox and Martin, 1957a: 378.—Calif.
aristatus James, 1942a: 126.—Colo.; Utah.
arnaudi Wilcox and Martin, 1957a: 379.—Calif.
atripes Wilcox and Martin, 1936b: 456.—Idaho; Wash., Oreg., Calif.
bruneri Wilcox and Martin, 1957a: 380.—Calif.
cerussatus (Osten Sacken), 1877: 308 (*Cyrtopogon*).—Calif.
crumbi Wilcox and Martin, 1957a: 380.—Ariz.
deserti Wilcox and Martin, 1957a: 380.—Calif.
howlandi Wilcox and Martin, 1957a: 382.—Calif.
inyoi Wilcox and Martin, 1957a: 384.—Calif.
irvinei Wilcox and Martin, 1957a: 384.—Calif.
j-beameri Wilcox and Martin, 1957a: 385.—Calif.
lestomyiformis Wilcox and Martin, 1936b: 454.—Calif.
mingusi Wilcox and Martin, 1957a: 386.—Ariz.; Utah.
minutus Wilcox and Martin, 1936b: 455.—Calif.
monrovia Wilcox and Martin, 1936b: 453.—Calif.
neoculatus Wilcox and Martin, 1957a: 386.—Calif.
nevadensis Wilcox and Martin, 1957a: 387.—Nev.
nigricolor (Coquillett), 1904f: 183 (*Cyrtopogon*).—Calif.
nitidus Wilcox and Martin, 1957a: 388.—Calif.

oculatus Wilcox and Martin, 1936b: 452.—Calif.
richardsoni Wilcox and Martin, 1957a: 389.—Calif.
sequoia Wilcox and Martin, 1957a: 389.—Calif.
stonei Wilcox and Martin, 1957a: 389.—Calif.
timberlakei Wilcox and Martin, 1957a: 390.—Calif.
tolandi Wilcox and Martin, 1957a: 391.—Calif.
vanduzeei Wilcox and Martin, 1936b: 451.—Calif.
vandykei Wilcox and Martin, 1957a: 391.—Calif.

Genus NICOCLES Jaennicke

Pygostolus Loew, 1866b: 16 (Cent. 7, no. 28) (preocc. Haliday, 1833).
  Type-species, *argentifer* Loew (Coquillett, 1910b: 598)=
  politus (Say).
Nicocles Jaennicke, 1867: 355 (1868: 47). Type-species, *analis*
  Jaennicke (mon.).

  REFERENCE: Wilcox, 1946 (key).

abdominalis Williston, 1883a: 17.—Calif.
aemulator (Loew), 1872a: 66 (Cent. 10, no. 25) (*Pygostolus*).—Calif.
argentatus Coquillett, 1893d: 119.—Calif.
bromleyi Hardy, 1943a: 28.—Ariz.
canadensis Curran, 1923j: 208.—B.C.; Wash., Oreg.
dives (Loew), 1866b: 17 (Cent. 7, no. 29) (*Pygostolus*).—Calif.;
  Wash., Oreg.
engelhardti Wilcox, 1946: 163.—S.C.; N.C.
lomae Cole, 1916b: 67.—Calif.
pictus (Loew), 1866b: 17 (Cent. 7, no. 30) (*Pygostolus*).—D.C.;
  N.J., Ala., Ga., Fla.
  amastris Walker, 1849: 362 (*Dasypogon*).—Ga.
politus (Say), 1823: 52 (1859b: 65) (*Dasypogon*).—Pa.; N.Y. and
  Mass., s. to Fla.
  argentifer Loew, 1866b: 16 (Cent. 7, no. 28) (*Pygostolus*).—
  D.C.
pollinosus Wilcox, 1946: 164.—Mont.; B.C., Calif.
reinhardi Bromley, 1934a: 91.—Tex.
rufus Williston, 1883a: 18.—Wash.; Oreg., Calif.
utahensis Banks, 1920: 66.—Utah; Wash., Oreg., Alta.
  punctipennis Melander, 1923c: 217.—Wash.

Genus OMNIABLAUTUS Pritchard

Omniablautus Pritchard, 1935: 1. Type-species, *arenosus* Pritchard
  (orig. des.).
arenosus Pritchard, 1935: 1.—N. Mex.

Genus ORRHODOPS Hull

Orrhodops Hull, 1958a: 324. Type-species, *Damalis americanus* Curran (orig. des.).

americanus (Curran), 1930h: 5 (*Damalis*).—Ariz.

Genus OSPRIOCERUS Loew

Ospriocerus Loew, 1866b: 29 (Cent. 7, no. 51). Type-species, *Dasypogon aeacus* Wiedemann (Back, 1909: 184) = *abdominalis* (Say).

abdominalis (Say), 1824a: 375 (1859a: 255) (*Asilus*).—Northwest Territory (U.S.); Wash. to Nebr., s. to Calif. and Tex., also Pa.
   aeacus Wiedemann, 1828: 390 (*Dasypogon;* n. name for *abdominalis* Say, when preocc. in *Dasypogon*).—Pa.
   aeacides Loew, 1866b: 29 (Cent. 7, no. 51).—Calif.
   ventralis Coquillett, 1898e: 37.—Ariz. **N. syn.**
aeacidinus (Williston), 1886a: 289 (*Stenopogon*).—Kans.; Mont., Colo., Nebr., Tex. **N. comb.**
arizonensis (Bromley), 1937a: 305 (*Stenopogon*).—N. Mex.; Ariz. **N. comb.**
ebyi (Bromley), 1937a: 305 (*Stenopogon*).—Tex. **N. comb.**
eutrophus Loew, 1874a: 355.—Tex.; Kans.
latipennis (Loew), 1866b: 28 (Cent. 7, no. 49) (*Stenopogon*).—N. Mex.; S. Dak. to N. Mex. and Tex. **N. comb.**
   consanguineus Loew, 1866b: 27 (Cent. 7, no. 48) (*Stenopogon*).—Nebr. **N. syn., n. comb.**
longulus (Loew), 1866b: 28 (Cent. 7, no. 50) (*Stenopogon*).—N. Mex.; Ariz., Tex. **N. comb.**
minos Osten Sacken, 1877: 291.—Colo.; Ariz., N. Mex.
nitens (Coquillett), 1904c: 34 (*Stenopogon*).—Tex.
   monki Bromley, 1934b: 225.—Tex.
parksi Bromley, 1934a: 89.—Tex.
rhadamanthus Loew, 1866b: 29 (Cent. 7, no. 52).—N. Mex.; Kans., Tex.
tenebrosus (Coquillett), 1904c: 33 (*Stenopogon*).—Tex.; N. Mex. **N. comb.**

Genus PARATARACTICUS Cole

Parataracticus Cole, 1924b: 11. Type-species, *rubidus* Cole (orig. des.).

niger Martin, 1955a: 118.—s. Calif.

rubidus Cole, 1924b: 12.—cent. Calif.
wyliei Martin, 1955a: 116.—s. Calif.

Genus PERASIS Hermann

Perasis Hermann, 1905: 37. Type-species, *sareptana* Hermann (orig. des., as gen. n., sp. n.).
Triclis, authors, not Loew.

argentifacies (Williston), 1901: 310 (*Triclis*).—Mexico; Ariz.

Genus PLESIOMMA Macquart

Plesiomma Macquart, 1838b: 54 (1839: 170). Type-species, *testacea* Macquart (Back, 1909: 306).

unicolor Loew, 1866b: 20 (Cent. 7, no. 35).—N. Mex.; Tex.

Genus PSILOCURUS Loew

Psilocurus Loew, 1874a: 373. Type-species, *nudiusculus* Loew (mon.).
Orthoneuromyia Williston, 1893c: 67. Type-species, *modesta* Williston (mon.).
REFERENCE: Curran, 1931b (key).

birdi birdi Curran, 1931b: 9.—Okla.; Miss.
   ssp. **pallustris** Hull, 1961: 104.—Miss.
modestus (Williston), 1893c: 68 (*Orthoneuromyia*).—S. Dak.; Kans., Tex.
nudiusculus Loew, 1874a: 370.—Tex.; La., Miss.
puellus Bromley, 1934a: 92.—Tex.
pygmaeus Hull, 1961: 101.—Tex.
reinhardi Bromley, 1951: 16.—Tex.
tibialis Hull, 1961: 102.—Tex.

Genus PYCNOPOGON Loew

Dasypogon, subg. **Pycnopogon** Loew, 1847c: 526. Type-species, *mixtus* Loew (Rondani, 1856: 157).
REFERENCE: Wilcox, 1941 (key).

cirrhatus Osten Sacken, 1877: 293.—Calif.; Wash., Nebr.
johnsoni (Back), 1904: 293 (*Anisopogon*).—Colo.; Ariz., N. Mex., Tex.

Genus SAROPOGON Loew

Dasypogon, subg. **Saropogon** Loew, 1847c: 439. Type-species, *luctuosus* Wiedemann (Coquillett, 1910b: 603).
REFERENCE: Curran, 1931b (key).

abbreviatus Johnson, 1903d: 113.—Tex.
    bicolor Johnson, 1903d: 113.—Tex.
albifrons Back, 1904: 291.—Ariz.; Calif.
birdi Curran, 1931b: 2.—Okla.
combustus Loew, 1874a: 373.—Tex.; Colo., Nebr., Kans., Okla.
    adustus Loew, 1874a: 375.—Tex.
coquilletti Back, 1909: 348.—N. Mex.; Ariz.
dispar Coquillett, 1902d: 139.—Tex.; Okla.
fletcheri Bromley, 1934a: 91.—Tex.
hyalinus Coquillett, 1904f: 185.—Calif.
hypomelas (Loew), 1866b: 24 (Cent. 7, no. 42) (*Diogmites*).—N. Mex.; Tex.
laparoides Bromley, 1951: 14.—Tex.
luteus Coquillett, 1904f: 185.—Calif.
    rufus Back, 1904: 290.—Calif.
pritchardi Bromley, 1934a: 90.—Tex.; Okla.
purus Curran, 1930h: 3.—Ariz.
semiustus Coquillett, 1904f: 186.—Calif.
senex Osten Sacken, 1887a: 179.—Mexico; Ariz.
    aridus Curran, 1930h: 3.—Ariz. **N. syn.**
solus Bromley, 1951: 15.—Tex.

Genus STENOPOGON Loew

Dasypogon, subg. **Stenopogon** Loew, 1847c: 453. Type-species, *Asilus sabaudus* Fabricius (Coquillett, 1904f: 179).
Gonioscelis Schiner, 1866a: 670. Type-species, *Dasypogon hispidus* Wiedemann (orig. des.).
Scleropogon Loew, 1866b: 26 (Cent. 7, no. 45). Type-species, *picticornis* Loew (mon.).

    REFERENCE: Bromley, 1937a (key).

albibasis Bigot, 1878: 422.—Calif.
andersoni Bromley, 1937a: 302.—Calif.
boharti Bromley, 1951: 6.—Ariz.; Calif.
bradleyi Bromley, 1937a: 309.—Calif.
breviusculoides Bromley, 1937a: 299.—Calif.
breviusculus Loew, 1872a: 68 (Cent. 10, no. 28).—Calif.
californiae (Walker), 1849: 322 (*Dasypogon*).—Calif.
californioides Bromley, 1937a: 304.—Calif.
cazieri Brookman, 1941: 78.—Calif.
cinerascens Back, 1909: 208.—Tex.
coyote Bromley, 1931: 429.—Wyo.; Alta., Colo., Ariz., N. Mex.
dispar Bromley, 1937a: 306.—Ariz.
duncani Bromley, 1937a: 307.—N. Mex.; Calif., Ariz.,

engelhardti Bromley, 1937a: 301.—Calif.
felis Bromley, 1931: 429.—Calif.
floridensis Bromley, 1951: 6.—Fla.
gratus Loew, 1872a: 69 (Cent. 10, no. 31).—Calif.
    univittatus Loew, 1872a: 69 (Cent. 10, no. 29).—Calif.
helvolus (Loew), 1874a: 355 (*Scleropogon*).—Tex.; Nebr., Kans.
huachucanus Hardy, 1942a: 57.—Ariz.
indistinctus Bromley, 1937a: 308.—Ariz.; Okla.
inquinatus Loew, 1866b: 27 (Cent. 7, no. 47).—Nebr.; B.C. to Minn., s. to Calif. and Ariz.
    modestus Loew, 1866b: 26 (Cent. 7, no. 46).—Minn.
    morosus Loew, 1874a: 356.—Minn.
jubatoides Bromley, 1937a: 297.—Calif.
jubatus Coquillett *in* Baker, 1904: 38.—Calif.
kelloggi Wilcox *in* Bromley, 1937a: 307.—N. Mex.; Ariz.
martini Bromley, 1937a: 303.—Idaho; Calif., Mont., Colo., Kans.
mexicanus Cole, 1923a: 463.—Mexico; ?Ariz.
neglectus Bromley, 1931: 430.—Wyo.; Wash. to Wyo., s. to Oreg. and Utah.
neojubatus Wilcox and Martin, 1945: 10.—Calif.
nigritulus Coquillett, 1904f: 179.—Calif.
nigriverticellus Bromley, 1937a: 298.—Calif.
obscuriventris Loew, 1872a: 69 (Cent. 10, no. 30).—Calif.
picticornis (Loew), 1866b: 26 (Cent. 7, no. 45) (*Scleropogon*).—Calif.
propinquus Bromley, 1937a: 298.—Oreg.; Calif.
pumilis Coquillett, 1904c: 33.—Tex.; Kans.
rufibarbis Bromley, 1931: 431.—Calif.; Wash., Oreg., Utah.
rufibarboides Bromley, 1937a: 301.—Calif.
similis (Jones), 1907a: 274 (*Scleropogon*).—Nebr.
subulatus (Wiedemann), 1828: 375 (*Dasypogon*).—Ga.; N.C.
texanus Bromley, 1931: 431.—Tex.
timberlakei Bromley, 1937a: 302.—Calif.
tinkhami Bromley, 1951: 7.—Tex.
uhleri Banks, 1920: 66.—Colo.
utahensis Bromley, 1951: 8.—Utah.
wilcoxi Bromley, 1937a: 300.—Calif.

Genus STICHOPOGON Loew

Dasypogon, subg. **Stichopogon** Loew, 1847c: 499. Type-species, *elegantulus* Wiedemann (Back, 1909: 332).
Neopogon Bezzi, 1910: 147. Type-species, *Dasypogon trifasciatus* Say (orig. des.).

REFERENCE: Wilcox, 1936d (key).

abdominalis Back, 1909: 332.—Fla.
arenicola Wilcox, 1936d: 207.—Ariz.
argenteus (Say), 1823: 51 (1859b: 65) (*Dasypogon*).—Pa.; Colo. to Pa., s. to Md.
catulus Osten Sacken, 1887a: 170.—Mexico; Ariz.
colei Bromley, 1934a: 90.—Tex.
coquilletti (Bezzi), 1910: 151 (*Neopogon*).—Calif.
fragilis Back, 1909: 334.—N. Mex.; Ariz.
pritchardi Bromley, 1951: 8.—Kans.; Colo., Okla., Tex.
salinus (Melander), 1923c: 216 (*Neopogon*).—Utah.
trifasciatus (Say), 1823: 51 (1859b: 64) (*Dasypogon*).—Pa.; U.S.
    plagiata Walker, 1848: 223 (*Thereva*).—Mass.
    gelascens Walker, 1860a: 277 (*Dasypogon*).—Mass.
    snowii Bezzi, 1910: 149.—Kans.

Genus TARACTICUS Loew

Taracticus Loew, 1872a: 64 (Cent. 10, no. 24). Type-species, *Dioctria octopunctata* Say (mon.).
Dioctrodes Coquillett, 1904f: 181. Type-species, *flavipes* Coquillett (orig. des.)=*octopunctatus* (Say).

REFERENCE: Pritchard, 1938b (key).

dimidiatus (Macquart), 1847: 51 (1847: 35) (*Dasypogon*).—Mexico.
    N. comb. Not n. of Mexico; record by Back, 1909: 236 (in *Ceraturgus*) incorrect.
octopunctatus (Say), 1823: 49 (1859b: 63) (*Dioctria*).—U.S.; S. Dak. to N.H., s. to Mo. and Fla.
    flavipes Coquillett, 1904f: 181 (*Dioctrodes*).—Mo.
    rufipes Jones, 1907a: 276 (*Dioctria*).—Nebr.
paulus Pritchard, 1938b: 189.—Calif.
ruficaudus Curran, 1930h: 4.—Ariz.

Genus TOWNSENDIA Williston

Townsendia Williston, 1895d: 107. Type-species, *minuta* Williston (mon.).

minuta Williston, 1895d: 108.—Mexico; N. Mex.
nigra Back, 1909: 175.—N.J.; Tex., La., Ky., Ga.
pulcherrima Back, 1909: 177.—Tex.

Genus WILCOXIA James

Wilcoxia James, 1941b: 38. Type-species, *cinerea* James (orig. des.).
cinerea James, 1941b: 39.—Colo.

Genus WILLISTONINA Back

Willistonina Back, 1909: 337. Type-species, *Habropogon bilineatus* Williston (orig. des.).

bilineata bilineata (Williston), 1883a: 11 (*?Habropogon*).—Calif.; Wash., Oreg.
  ssp. **nigrofemorata** Wilcox, 1935b: 33.—Calif.; B.C., Oreg., Mont.

Genus ZABROPS Hull

Zabrops Hull, 1957b: 90. Type-species, *Triclis tagax* Williston (orig. des.).

flavipilis (Jones), 1907a: 275 (*Triclis*).—Nebr.
tagax (Williston), 1883a: 9 (*Triclis*).—Calif.

Subfamily LAPHRIINAE
Genus ANDRENOSOMA Rondani

Andrenosoma Rondani, 1856: 160. Type-species, *Asilus ater* Linnaeus (orig. des.).
Elaeotoma Costa, 1963: 49. Type-species, *adustiventris* Costa (mon.)=*albibarbe* (Meigen).
Pilica Curran, 1931b: 20. Type-species, *Laphria formidolosa* Walker (orig. des.).
Nusa, authors, in part, not Walker.
  REFERENCE: McAtee, 1919b (key).

cruentum (McAtee), 1919b: 244 (*Nusa*).—Fla.
fulvicauda (Say), 1823: 53 (1859b: 66) (*Laphria*).—Mo.; Calif. to Ont., s. to Tex. and Fla., Neotropical.
  lutea McAtee, 1919b: 246 (*Nusa*; as var.).—Colo.
rubidum (Williston), 1901: 318 (*Nusa*).—Mexico; Tex.
sicarium (McAtee), 1919b: 246 (*Nusa*).—Tex.

Genus ATOMOSIA Macquart

Atomosia Macquart, 1838b: 73 (1839: 189). Type-species, *incisuralis* Macquart (Coquillett, 1910b: 512)=*puella* (Wiedemann).
  REFERENCES: Curran, 1930h, 1935a (keys).

echemon (Walker), 1849: 386 (*Laphria*).—Ohio. ?=*puella*.
glabrata (Say), 1823: 53 (1859b: 66) (*Laphria*).—U.S.; Ohio, Pa., Tex.

melanopogon Hermann, 1912: 144.—Tex.; Ariz., Colo., Iowa.
mucida Osten Sacken, 1887a: 184.—Mexico; Tex.
mucidoides Bromley, 1951: 19.—Tex.
puella (Wiedemann), 1828: 531 (*Laphria*).—Unknown.; Ohio to Pa., s. to Tex. and Fla.
pusilla Macquart, 1838b: 76 (1839: 192).—N. Amer.; Mo., Pa.
rufipes Macquart, 1847: 55 (1847: 39).—Pa.; Tex., Fla.
    soror Bigot, 1878: 236.—Mexico. **N. syn.**
sayii Johnson, 1903d: 113.—Pa.; Tex., Fla.

Genus ATOMOSIELLA Wilcox

Atomosiella Wilcox, 1937: 40. Type-species, *Atomosia antennata* Banks (orig. des.).

antennata (Banks), 1920: 66 (*Atomosia*).—Ariz.; Calif., N. Mex.

Genus ATONIOMYIA Hermann

Atonia Williston, 1889c: 257 (preocc. Gistl, 1847). Type-species, *mikii* Williston (Williston, 1901: 316).
Atoniomyia Hermann, 1912: 81 (n. name for *Atonia* Williston). Type-species, *Atonia mikii* Williston (aut.).
Neatonia Bromley, 1935b: 130. Unavailable name (no type des.).

brevistylata (Williston), 1901: 316 (*Atonia*).—Mexico; Ariz.
duncani (Wilcox), 1937: 42 (*Atonia*).—Ariz.

Genus CEROTAINIA Schiner

Cerotainia Schiner, 1866a: 673. Type-species, *Laphria xanthoptera* Wiedemann (orig. des.).

    REFERENCE: Curran, 1930h, 1934d (keys).

albipilosa Curran, 1930h: 13.—N.C.; N.Y., Va.
atrata Jones, 1907a: 280.—Nebr.
macrocera (Say), 1823: 73 (1859b: 67) (*Laphria*).—Pa.; Ohio, N.J. to Fla.

Genus CEROTAINIOPS Curran

Cerotainiops Curran, 1930f: 11.—Type-species, *rufiventris* Curran (orig. des.)=*abdominalis* (Brown).
Nusa, authors, in part, not Walker.

    REFERENCE: Martin, 1959b (key).

abdominalis (Brown), 1897: 103 (*Nusa*).—N. Mex.; Calif., Ariz., Kans., Okla., Tex.
   similis Brown, 1897: 103 (*Nusa*).—N. Mex.
   atripes McAtee, 1919b: 247 (*Nusa*; as var.).—N. Mex.
   rufiventris Curran, 1930f: 11.—Ariz.
kernae Martin, 1959b: 52.—Calif.
lucyae Martin, 1959b: 50.—Tex.
mcclayi Martin, 1959b: 51.—Calif.
omus Pritchard, 1942: 22.—Calif.
wilcoxi Pritchard, 1942: 23.—Ariz.; Calif.
   abdominalis, authors, not Brown.

Genus DASYLECHIA Williston

Dasylechia Williston, 1907: 1. Type-species, *Hyperechia atrox* Williston (mon.).

atrox (Williston), 1883a: 28 (*Hyperechia*).—Pa.; Utah to Mich., s. to N.J.

Genus LAMPRIA Macquart

Lampria Macquart, 1838b: 60 (1839: 176). Type-species, *Laphria clavipes* Fabricius (Coquillett, 1910b: 557).

bicolor (Wiedemann), 1828: 522 (*Laphria*).—Unknown; Ind. to Conn., s. to Tex. and Fla.
   saniosa Say, 1829: 158 (1859b: 355) (*Laphria*).—Ind.
   megacera Macquart, 1834a: 284 (*Laphria*).—Pa.
   antaea Walker, 1849: 379 (*Laphria*).—Fla.
corallogaster (Bigot), 1878: 226 (*Laphria*).—N. Amer.
rubriventris (Macquart), 1834a: 284 (*Laphria*).—Pa.; Ga., Tex.

Genus LAPHRIA Meigen

Lapria Meigen, 1800: 25. Type-species, *Asilus gibbosus* Linnaeus (Coquillett, 1910b: 557). Suppressed by I.C.Z.N., 1963b: 339.
Laphria Meigen, 1803: 270. Type-species, *Asilus gibbosus* Linnaeus (Latreille, 1810: 443).
Ropalocera Meigen, 1820: 301. Type-species, *Laphria nigripennis* Meigen (mon.).
Bombomima Enderlein, 1914c: 253. Type-species, *Laphria fulvithorax* Fabricius (orig. des.) =*thoracica* Fabricius.
Dasyllis, authors, not Loew.

   REFERENCES: Banks, 1917a (key); McAtee, 1919a (key).

ASILIDAE 389

aeata Walker, 1849: 381.—Hudson Bay; Alta.
affinis Macquart, 1855: 74 (1855: 54).—Md.; N.J., Ala., Ga.
aimatis McAtee, 1919a: 160.—Calif.; Colo.
aktis McAtee, 1919a: 152.—Pa.; Ohio, Va., N.C.
altitudinum Bromley, 1924: 126.—Maine; N.H., N.Y.
apila (Bromley), 1951: 22 (*Bombomima*).—Ala.; Ga., Tenn.
asackeni Wilcox (for *sackeni* Wilcox).—Calif.; Alaska, B.C., Oreg., Idaho, Mont. **N. name.**
    sackeni Wilcox, 1936b: 8 (preocc. Banks, 1917).—Calif.
astur Osten Sacken, 1877: 285.—Calif.; Wash., Oreg.
    californica Banks, 1917a: 54 (*Dasyllis*).—Calif.
asturina (Bromley), 1951: 22 (*Bombomima*).—B.C.; Alaska, Wash., Oreg., Calif.
    astur, authors, not Osten Sacken.
canis Williston, 1883a: 31.—Conn.; Ohio, Pa., N.Y., Md., Va.
    dispar Banks, 1911a: 130 (preocc. Coquillett, 1898).—N.Y.
    disparella Banks, 1913: 52 (n. name for *dispar* Banks).—N.Y.
carbonaria Snow, 1896: 181 (n. name for *anthrax* Williston).—Calif.; N. Mex.
    anthrax Williston, 1883a: 29 (preocc. Meigen, 1804).—Calif.
carolinensis Schiner, 1867b: 380.—Carolina.
champlainii (Walton), 1910: 243 (*Dasyllis*).—Pa.; Conn.
cinerea (Back), 1904: 289 (*Dasyllis*).—N.C.; N.Y.
columbica Walker *in* Lord, 1866: 338.—B.C.; Wash., Oreg.
coquillettii McAtee, 1919a: 257.—Calif.
divisor (Banks), 1917a: 54 (*Dasyllis*).—N.C.; Wis., Ill., Maine, Pa.
engelhardti (Bromley), 1931: 434 (*Bombomima*).—Colo.; Ariz., N. Mex.
fattigi (Bromley), 1951: 23 (*Bombomima*).—Ga.
felis (Osten Sacken), 1877: 286 (*Lampria*).—Calif.; B.C., Wash., Oreg., Utah, Wyo., Colo.
    xanthippe Williston, 1883a: 31.—Oreg.
    atripes McAtee, 1919a: 161 (as var.).—Colo.
    crocea McAtee, 1919a: 162 (as var.).—Wash.
    varipes McAtee, 1919a: 162 (as var.).—Colo.
fernaldi (Back), 1904: 290 (*Dasyllis*).—Colo.; Wash., Oreg. Utah.
ferox Williston, 1883a: 29.—Wash.; B.C., Mont.
flavescens Macquart, 1838b: 69 (1839: 185).—Carolina and Pyrenees.
flavicollis Say, 1824a: 374 (1859a: 255).—Northwest Territory (U.S.); Iowa to Que., s. to Tex. and Fla.
    melanopogon Wiedemann, 1828: 520.—Ky.
flavipila Macquart, 1834a: 282.—U.S.

franciscana Bigot, 1878: 225.—Calif.; B.C., Wash.
georgina Wiedemann, 1821c: 235.—Ga.
gilva (Linnaeus), 1758: 605 (*Asilus*).—Europe; Wash. and Oreg. to Mass.
    bilineata Walker, 1849: 1156.—Ont.
grossa (Fabricius), 1775: 791 (*Asilus*).—Amer.; Pa., Ga., Fla.
    tergissa Say, 1823: 74 (1859b: 67).—Pa.
    analis Macquart, 1838b: 68 (1839: 184).—N. Amer.
    flavibarbis Harris, 1862: 604.—Unknown.
huron (Bromley), 1929: 159 (*Bombomima*).—Ont.; N.B., N.Y.
index McAtee, 1919a: 164.—Pa.; e. Canada, N.H., N.Y., N.J., Va.
insignis (Banks), 1917a: 54 (*Dasyllis*).—Labr.; Calif.
ithypyga McAtee, 1919a: 165.—Pa.; Md.
janus McAtee, 1919a: 153.—N.H.; B.C. and Wash. to Maine, s. to Colo. and N.Y.
lasipes Wiedemann, 1828: 502.—Ky.
lata Macquart, 1850: 379 (1850: 75) (n. name for *analis* Macquart, 1846).—Tex.; La.
    analis Macquart, 1846: 206 (1846: 78) (*Mallophora*; preocc. Macquart, 1838).—Tex.
macquarti (Banks), 1917a: 54 (*Dasyllis*).—Tex.
melanogaster Wiedemann, 1821c: 236.—Mexico; Tex., Ga.
milvina Bromley, 1929: 160.—B.C.; Oreg.
nigella (Bromley), 1934a: 93 (*Bombomima*).—Tex.
partitor (Banks), 1917a: 54 (*Dasyllis*).—B.C.; Wash., Oreg.
posticata Say, 1824a: 374 (1859a: 255).—Northwest Territory (U.S.); Wis., Ont., Que., N.Y.
    brunnea Bromley, 1929: 159 (*Bombomima*; as var.).—Ont. ?Var.
    scutellaris Bromley, 1929: 159 (*Bombomima*; as var.).—Ont.
rapax Osten Sacken, 1877: 286.—Calif.; Wash., Oreg.
royalensis (Bromley), 1950b: 2 (*Bombomima*).—Mich.
sackeni (Banks), 1917a: 54 (*Dasyllis*).—Calif.; Oreg.
sacrator Walker, 1849: 382.—N.S.; Wis. to Que., s. to Conn.
sadales Walker, 1849: 378.—N.Y.; B.C. to N.H., s. to Calif.
    pubescens Williston, 1883a: 32.—Wash.
saffrana Fabricius, 1805: 160.—Carolina; Va, N.C., Ga., Fla.
scorpio McAtee, 1919a: 163.—N.H.; N.Y., Vt.
semitecta (Coquillett), 1910a: 124 (*Dasyllis*).—Man.
sericea Say, 1823: 74 (1859b: 67).—U.S.; Ill., Atlantic Seaboard States.
sicula McAtee, 1919a: 165.—Md.; Ill., Ohio, Pa., Va.
terraenovae Macquart, 1838b: 69 (1839: 185).—Nfld.

thoracica Fabricius, 1805: 158.—N. Amer.; e. of the Rocky Mts.
   fulvithorax Fabricius, 1805: 373 (unjustified n. name for *thoracica* Fabricius).—N. Amer.
   alcanor Walker, 1849: 383.—Mass.
trux McAtee, 1919a: 158.—Calif.
   audax McAtee, 1919a: 158 (as var.).—Calif.
unicolor Williston, 1883a: 26 (*Dasyllis*).—Wash.
ventralis Williston, 1885a: 55.—Calif.; Wash.
virginica (Banks), 1917a: 53(*Dasyllis*).—Va.
vivax Williston, 1883a: 30.—Wash.; B.C., Colo.
   anthemon McAtee, 1919a: 156 (as ssp.).—N. Mex.
vorax (Bromley), 1929: 158 (*Bombomima*).—Kans.; Nebr.
vultur Osten Sacken, 1877: 286.—Calif.; Wash., Oreg.
winnemana McAtee, 1919a: 168.—Md.; Canada, Pa., Va.

Genus ORTHOGONIS Hermann

Orthogonis Hermann, 1914: 132. Type-species, *Laphria scapularis* Wiedemann (Hull, 1962: 330.).

stygia (Bromley), 1931: 433 (*Laphria*).—N.C.; Miss., Fla.

Genus POGONOSOMA Rondani

Pogonosoma Rondani, 1856: 160. Type-species, *Asilus maroccanus* Fabricius (orig. des.).

dorsatum (Say), 1824b: pl. 6 (1859a: 13) (*Laphria*).—Pa.; Wash., Idaho, Conn., N.J., Va., S.C., and Fla.
   melanoptera Wiedemann, 1828: 514 (*Laphria*).—Carolina (unpublished).
ridingsi Cresson, 1920c: 214.—Colo.; Calif.

Subfamily ASILINAE

Genus ASILUS Linnaeus

Asilus Linnaeus, 1758: 605. Type-species, *crabroniformis* Linnaeus (Latreille, 1810: 443). The only North American species that is known to be congeneric with the genotype of *Asilus* is *sericeus* Say. Many species formerly assigned to *Asilus* have been moved to other genera, leaving most of the following species tentatively awaiting generic reassignment.

REFERENCE: Hine, 1909 (key to spp. in *Asilus* and allied genera).

albicomus Hine, 1909: 150.—Mont.
angustipennis Hine, 1909: 152.—N.C.; Md., Va.
astutus Williston, 1893c: 70.—Calif.
auriannulatus (Hine), 1906b: 29 (*Stilpnogaster*).—B.C.; Wash., Calif.
auricomus Hine, 1909: 148.—Ohio; Ill.; Conn., N.J., Pa.
autumnalis Banks, 1914b: 131.—Va.
blantoni Bromley, 1940: 19.—Fla.
californicus Hine, 1909: 164.—Calif.
carolinae Martin and Wilcox (for *annulipes* Macquart).—Carolina. **N. name.**
      annulipes Macquart, 1838b: 149 (1839: 265) (preocc. Brullé, 1832).—Carolina.
citus Hine, 1918b: 321.—Ariz.
comosus Hine, 1918b: 319.—Calif.
compositus Hine, 1918b: 321.—Calif.
delicatulus Hine, 1918b: 320.—N. Mex.
erythocnemius Hine, 1909: 163.—Mass.; Mont. to Mass., s. to Wyo. and Md.
fattigi Bromley, 1940: 19.—Ga.
floridensis Bromley, 1940: 18.—Fla.
formosus Hine, 1918b: 321.—Kans.
frosti Bromley, 1950a: 238.—Fla.; S.C.
fulviventris Schaeffer, 1916: 69.—Ariz.
gilvipes Hine, 1918b: 319.—Colo.; N. Mex.
gracilis Wiedemann, 1828: 445.—Ga.; Tex., N.C., Fla.
      auratus Johnson, 1895a: 305.—Fla.
hubbelli Bromley, 1950b: 3.—Fla.
hypopygialis Schaeffer, 1916: 68.—Utah.
knulli Bromley, 1940: 17.—Ariz.
latipennis Hine, 1909: 152.—N.Y.; Mass.
lecythus Walker, 1849: 451.—N.S.; Ohio to N.S., s. to N.C.
      femoralis Macquart, 1847: 61 (1847: 45) (preocc. Zeller, 1840).—Pa.
lepidus Hine, 1909: 150.—Colo.; N. Mex.
mesae (Tucker), 1907: 92 (*Tolmerus*).—Colo.; Wyo., Kans.
montanus Hine, 1909: 149.—Calif.; B.C.
mydas Brauer, 1885: 387.—Mexico; Ariz., N. Mex.
      midas, emend.
piceus Hine, 1909: 149.—Mass.
platyceras Hine, 1922a: 7.—Wash.
      schuhi Bromley, 1940: 17.—Oreg. **N. syn.**
rubicundus Hine, 1909: 162.—Kans.; Ill.
sackeni Banks, 1920: 67.—Calif.; Oreg.

sericeus Say, 1823: 48 (1859b: 63).—Pa.; Kans., Ind., Conn., Mass., N.J.
    herminius Walker, 1849: 410.—Mass.
vescus Hine, 1918b: 320.—Calif.
virginicus Banks *in* McAtee and Banks, 1920: 31.—Va.; Md.
willistoni Hine, 1909: 150 (n. name for *angustifrons* Williston).— Wash.; B.C.
    angustifrons Williston, 1893c: 71 (preocc. Loew, 1849).— Wash.

Genus ECCRITOSIA Schiner

Eccritosia Schiner, 1866a: 674. Type-species, *Asilus barbatus* Fabricius (orig. des.).
    REFERENCE: Curran, 1934d (key).
amphinome (Walker), 1849: 387 (*Asilus*).—Mexico; Tex.
zamon (Townsend), 1895b: 600 (*Proctacanthus*).—Mexico (Baja Calif.); Ariz.

Genus EFFERIA Coquillett

Efferia Coquillett, 1893e: 175. Type-species, *candida* Coquillett (Coquillett, 1910b: 536).
Nerax Hull, 1962: 476. Type-species, *Asilus aestuans* Linnaeus (orig. des.).
Erax, authors, not Scopoli.
    REFERENCES: Hine, 1919 (keys to groups and spp. under generic name *Erax*); Martin, 1961, 1962b (syn.).
aestuans (Linnaeus), 1763: 413 (*Asilus*).—Pa.; entire e. N. Amer. to Wyo. and Tex. **N. comb.**
    niger Wiedemann, 1821c: 196 (*Asilus*).—Ga. **N. comb**
    bastardi Macquart, 1838b: 117 (1839: 233) (*Erax*).—N. Amer. **N. comb.**
    incisuralis Macquart, 1838b: 117 (1839: 233) (*Erax*).—Pa. **N. comb.**
    tibialis Macquart, 1838b: 118 (1839: 234) (*Erax*).—Pa. **N. comb.**
affinis (Bellardi).—Not Nearctic. **N. comb.**
albibarbis (Macquart), 1838b: 118 (1839: 234) (*Erax*).—N. Amer.; U.S., coast to coast, s. to Guatemala.
    furax Williston, 1885a: 67 (*Erax*).—Wash.
    barbatus, authors, not Fabricius (*Erax*).
anacapai (Wilcox and Martin), 1945: 13 (*Erax*).—Calif. **N. comb.**
anomala (Bellardi).—Not Nearctic. **N. comb.**

apicalis (Wiedemann), 1821c: 191 (*Asilus*).—N. Amer.; Tex., N.C., Ga., Fla. **N. comb.**
    vicinus Macquart, 1846: 213 (1846: 85) (*Erax*).—Tex. **N. comb.**
argentifrons (Hine), 1911c: 308 (*Erax*).—Kans. **N. comb.**
argyrosoma (Hine), 1911c: 310 (*Erax*).—N. Mex. **N. comb.**
arida (Williston), 1893d: 254 (*Erax*).—Calif. **N. comb.**
armata (Hine), 1918a: 4 (*Erax*).—Tex. **N. comb.**
aurimystacea (Hine), 1919: 122 (*Erax*).—Kans. **N. comb.**
auripila (Hine), 1916: 22 (*Erax*).—Tex. **N. comb.**
belfragei (Hine), 1919: 121 (*Erax*).—Tex. **N. comb.**
benedicti (Bromley), 1940: 15 (*Erax*).—Ariz. **N. comb.**
bexarensis (Bromley), 1934a: 94 (*Erax*).—Tex. **N. comb.**
bicaudata (Hine), 1919: 138 (*Erax*).—Colo. **N. comb.**
bicolor (Bellardi), 1861: 47 (1864: 147) (*Erax*).—Mexico; Ariz., N. Mex., Tex., Miss. **N. comb.**
californica (Schaeffer), 1916: 67 (*Erax*).—Calif.; Wash., Oreg., Nev., Mont. **N. comb.**
cana (Hine), 1916: 22 (*Erax*).—Calif. **N. comb.**
candida Coquillett, 1893e: 176.—Calif.; Ariz., N. Mex., Kans.
cannella (Bromley), 1934a: 95 (*Erax*).—Ariz. **N. comb.**
clementi (Wilcox and Martin), 1945: 14 (*Erax*).—Calif. **N. comb.**
coquillettii (Hine), 1919: 149 (*Erax*).—Calif.; Ariz. **N. comb.**
costalis (Williston), 1885a: 64 (*Erax*).—Unknown; Mont., Wyo., Colo. **N. comb.**
cressoni (Hine), 1919: 134 (*Erax*).—N. Mex.; Tex. **N. comb.**
cuervana (Hardy), 1943a: 27 (*Erax*).—N. Mex. **N. comb.**
dubia (Williston), 1885a: 64 (*Erax*).—Wash.; Ariz., Wyo. Name in key; "*Erax* n. sp.," p. 68, may be the descr. **N. comb.**
femorata (Macquart), 1838b: 115 (1839: 231) (*Erax*).—Carolina; La., Fla. **N. comb.**
grandis (Hine), 1919: 111 (*Erax*).—Tex. **N. comb.**
harveyi (Hine), 1919: 115 (*Erax*).—B.C. **N. comb.**
helenae (Bromley), 1951: 30 (*Erax*).—Colo. **N. comb.**
inflata (Hine), 1911c: 310 (*Erax*).—Calif. **N. comb.**
interrupta (Macquart), 1834a: 310 (*Asilus*).—Ga.; entire U.S., also Mexico. **N. comb.**
    lateralis Macquart, 1838b: 116 (1839: 232) (*Erax*).—Pa. **N. comb.**
    ambiguus Macquart, 1846: 212 (1846: 84) (*Erax*).—Tex., Mexico. **N. comb.**
jubata (Williston), 1885a: 66 (*Erax*).—N. Mex.; Ariz., Colo. **N. comb.**
kansensis (Hine), 1919: 122 (*Erax*).—Kans. **N. comb.**

knowltoni (Bromley), 1937b: 104 (*Erax*).—Utah. **N. comb.**
latruncula (Williston), 1885a: 67 (*Erax*).—Ariz.; Mont. **N. comb.**
leucocoma (Williston), 1885a: 69 (*Erax*).—Kans.; Tex. **N. comb.**
macrolabis (Wiedemann), 1828: 458 (*Asilus*).—Ky. **N. comb.**
mesquite (Bromley), 1951: 27 (*Erax*).—Tex. **N. comb.**
mexicana (Hine), 1919: 123 (*Erax*).—Mexico; Tex. **N. comb.**
monki (Bromley), 1951: 28 (*Erax*).—Tex. **N. comb.**
nemoralis (Hine), 1911c: 311 (*Erax*).—La.; Tex., Va. **N. comb.**
pallidula (Hine), 1911c: 309 (*Erax*).—N. Mex.; Colo. **N. comb.**
pernicis Coquillett, 1893e: 175.—Calif.; Ariz.
pilosa (Hine), 1919: 150 (*Erax*).—Tex. **N. comb.**
plena (Hine), 1916: 21 (*Erax*).—Kans.; Okla. **N. comb.**
pogonias (Wiedemann), 1821c: 198 (*Asilus*; n. name for *barbatus* Fabricius when preocc. in *Asilus*).—N. Amer.; s. U.S. e. of Rocky Mts.
    barbatus Fabricius, 1805: 169 (*Dasypogon;* preocc. Fabricius, 1787).—N. Amer.
    rufibarbis Macquart, 1838b: 116 (1839: 232) (*Erax*).—N. Amer.
    completus Macquart, 1838b: 117 (1839: 233) (*Erax*).—N. Amer.
    dascyllus Walker, 1849: 401 (*Asilus*).—Mass.
    virginianus Wulp, 1882b: 109 (*Proctacanthus*).—Va.
    rava Coquillett, 1893e: 176.—Tex.
    aestuans, authors, not Linnaeus.
prairiensis (Bromley), 1934a: 95 (*Erax*).—Tex. **N. comb.**
prattii (Hine), 1919: 117 (*Erax*).—Tex. **N. comb.**
producta (Hine), 1919: 136 (*Erax*).—Calif. **N. comb.**
rapax (Osten Sacken), 1887a: 201 (*Erax*).—Mexico; Ariz., N. Mex. **N. comb.**
slossonae (Hine), 1919: 121 (*Erax*).—Fla. **N. comb.**
snowi (Hine), 1919: 116 (*Erax*).—Kans.; Colo., N. Mex., Tex. **N. comb.**
spiniventris (Hine), 1919: 135 (*Erax*).—Ariz. **N. comb.**
staminea (Williston), 1885a: 68 (*Erax*).—Mont.; Wyo. **N. comb.**
    stramineus, emend.
subarida (Bromley), 1940: 14 (*Erax*).—Ariz. **N. comb.**
subcuprea (Schaeffer), 1916: 66 (*Erax*).—Ariz.; Nev., Mont., Colo., N. Mex. **N. comb.**
subpilosa (Schaeffer), 1916: 67 (*Erax*).—Utah; Calif., Nev., N. Mex. **N. comb.**
tabescens (Banks) *in* Hine 1919: 126 (*Erax*).—Fla. **N. comb.**
tagax (Williston), 1885a: 65 (*Erax*).—Ariz. **N. comb.**
    similis Williston, 1885a: 68 (*Erax*).—Ariz. **N. comb.**

tanneri (Bromley), 1937b: 105 (*Erax*).—Colo. **N. comb.**
texana (Banks) *in* Hine 1919: 151 (*Erax*).—Tex. **N. comb.**
tricella (Bromley), 1951: 32 (*Erax*).—Ariz.; N. Mex. **N. comb.**
truncata (Hine), 1911c: 309 (*Erax*).—Ariz. **N. comb.**
tuberculata (Coquillett), 1904c: 34 (*Erax*).—Calif.; Tex. **N. comb.**
utahensis (Bromley), 1937b: 103 (*Erax*).—Utah. **N. comb.**
varipes (Williston), 1885a: 71 (*Erax*).—Ariz.; N. Mex., Kans. **N. comb.**
vertebrata (Bromley), 1940: 14 (*Erax*).—Calif. **N. comb.**
wilcoxi (Bromley), 1940: 16 (*Erax*).—Tex. **N. comb.**
willistoni (Hine), 1919: 110 (*Erax*).—Ariz.; Colo. **N. comb.**
zonata (Hine), 1919: 112 (*Erax*).—Ariz.; Calif., N. Mex. **N. comb.**

Genus MACHIMUS Loew

Machimus Loew, 1849: 1. Type-species, *Asilus chrysitis* Meigen (Coquillett, 1910b: 564).

avidus (Wulp), 1869: 82 (*Asilus*).—Wis.
griseus Hine, 1906b: 29.—Colo.
occidentalis (Hine), 1909: 147 (*Asilus*).—B.C.; Wash., Oreg., Nev., Calif.
tenebrosus (Williston).—Not Nearctic.

Genus MALLOPHORA Macquart

Mallophora Macquart, 1834a: 300. Type-species, *Asilus bomboides* Wiedemann (Coquillett, 1910b: 565).
Megaphorus Bigot, 1857b: 542. Type-species, *Mallophora heteroptera* Macquart (mon.). ?=*Mallophorina*.

REFERENCES: Curran, 1930f, 1931b (keys).

ardens Macquart, 1834a: 302.—N. Amer.
belzebul Schiner, 1867b: 385.—Brazil; Tex.
bomboides (Wiedemann), 1821c: 203 (*Asilus*).—Ga.; Fla.
bromleyi Curran, 1930f: 13.—Tex.; Ariz.
chrysomela Bromley, 1925: 193.—Miss.; Ga.
fautricoides Curran, 1930f: 13.—Calif.; Ariz., Colo.
fulva Banks, 1911a: 130.—Ariz.
fulviventris Macquart, 1850: 381 (1850: 77).—Mexico; Tex.
heteroptera Macquart, 1838b: 90 (1839: 206).—Pa.; Ky. ?=*Mallophorina laphroides*.
minuta Macquart, 1834a: 302.—Pa.; Fla.
nigra Williston, 1885a: 58.—Minn.; Fla.
orcina (Wiedemann), 1828: 477 (*Asilus*).—Ga.; Ariz., D.C., Va., Fla.

perpusilla (Walker), 1851b: 123 (*Trupanea*).—U.S.
rex Bromley, 1925: 192.—Miss.; N.C., Fla.

Genus MALLOPHORINA Curran

Mallophorina Curran, 1934a: 183. Type-species, *Mallophora guildiana* Williston (orig. des.).
REFERENCES: Curran, 1930f, 1931b (keys, in *Mallophora*).
acra (Curran), 1931b: 21 (*Mallophora*).—Okla.
clausicella (Macquart), 1850: 383 (1850: 79) (*Mallophora*).—Va.; Colo., N. Mex., Pa., N.J., Fla.
  intermedia Tucker, 1907: 92 (*Mallophora*; as var.).—Colo.
frustra Pritchard, 1935: 11.—Ariz.
guildiana (Williston), 1885a: 60 (*Mallophora*).—Kans.; Mont., N.C.
laphroides (Wiedemann), 1828: 483 (*Asilus*).—Ky.; Fla.
megachile (Coquillett), 1893d: 118 (*Mallophora*).—Calif.
pallida (Johnson), 1958: 41 (*Mallophora*).—Utah.
prudens Pritchard, 1935: 10.—Ariz.
pulchra Pritchard, 1935: 11.—N. Mex.

Genus NEGASILUS Curran

Negasilus Curran, 1934a: 184. Type-species, *belli* Curran (orig. des.).
belli Curran, 1934a: 184.—Nev.; Oreg., Colo.

Genus NEOITAMUS Osten Sacken

Itamus Loew, 1849: 84 (preocc. Schmidt-Goebel, 1846).—Type-species, *Asilus cyanurus* Loew (Coquillett, 1910b: 556).
Neoitamus Osten Sacken, 1878c: 82 (n. name for *Itamus* Loew). Type-species, *Asilus cyanurus* Loew (aut.).
affinis (Williston), 1893c: 73 (*Asilus*).—Wash.; Calif.
auceps (Wulp), 1869: 84 (*Asilus*).—Wis. Unrecognized.
brevicomus (Hine), 1909: 155 (*Asilus*).—B.C.; Calif.
coquillettii (Hine), 1909: 154 (*Asilus*).—Calif.
flavofemoratus (Hine), 1909: 153 (*Asilus*; n. name for *flavipes* Williston).—Pa.; Canada, Ill. to Mass., s. to N.C.
  flavipes Williston, 1893c: 72 (*Asilus*; preocc. Meigen, 1820).—Pa.
hardyi Bromley, 1938: 61.—Utah.
orphne (Walker), 1849: 456 (*Asilus*).—N.Y.; Colo., Ill., Wis., Maine, N.H., Conn., N.C.
  distinctus Williston, 1893c: 73 (*Asilus*).—N.H., Conn.
terminalis (Hine), 1909: 155 (*Asilus*).—Calif.

Genus NIGRASILUS Hine

Nigrasilus Hine, 1908: 203. Type-species, *nitidifacies* Hine (mon.).
nitidifacies Hine, 1908: 204.—B.C.; Wash., Oreg.

Genus OMMATIUS Wiedemann

Ommatius Wiedemann, 1821c: 213. Type-species, *Asilus marginellus* Fabricius (Coquillett, 1910b: 579).
REFERENCE: Wilcox, 1936a (key).
baboquivari Wilcox, 1936a: 174.—Ariz.
beameri Wilcox, 1936a: 174.—Ariz.
bromleyi Pritchard, 1935: 9.—Ariz.
gemma Brimley, 1928: 205.—N.C.; Okla., Ark., Miss., Va., Fla.
maculatus Banks, 1911a: 128.—Ariz.
parvulus Schaeffer, 1916: 69.—Ariz.
pretiosus Banks, 1911a: 129.—Ariz.
tibialis Say, 1823: 49 (1859b: 63).—Pa.; Kans. to Conn., s. to Tex. and Fla.

Genus PHILONICUS Loew

Philonicus Loew, 1849: 144. Type-species, *Asilus albiceps* Meigen (mon.).
arizonensis (Williston), 1893c: 76 (*Stenoprosopus*).—Ariz.; N. Mex.
fuscatus (Hine), 1909: 168 (*Asilus*; n. name for *obscurus* Hine).—Ky., N.J., D.C., N.C.; Pa. to Mass., s. to Ky. and N.C.
  obscurus Hine, 1907a: 117 (preocc. Meigen, 1820).—Ky., N.J., D.C., N.C.
limpidipennis (Hine), 1909: 167 (*Asilus*).—Colo.; Ariz.
  arizonensis Williston of Hine, 1909: 167 (*Asilus*).
persimilis (Banks), 1920: 67 (*Asilus*).—Calif.
rufipennis Hine, 1907a: 117.—Kans.; Ariz., N. Mex., Okla., Ill.
truquii (Bellardi), 1861: 52 (1864: 152) (*Asilus*).—Mexico; Ariz.

Genus PROCTACANTHELLA Bromley

Proctacanthella Bromley, 1934a: 96. Type-species, *Asilus cacopilogus* Hine (orig. des.).
cacopiloga (Hine), 1909: 165 (*Asilus*).—Tex.; Nebr., Kans., Okla., Ill., N.J.
exquisita (Osten Sacken), 1887a: 206 (*Proctacanthus*).—Mexico; Ariz., Colo., Okla. **N. comb.**
jamesi Pritchard, 1935: 13.—Okla.

leucopogon (Williston), 1893c: 75 (*Asilus*).—S. Dak.; Ariz., Kans., Nebr.
robusta Bromley, 1951: 26.—Tex.
wilcoxi Bromley, 1935a: 5.—Tex.

Genus PROCTACANTHUS Macquart

Proctacanthus Macquart, 1838b: 120 (1839: 236). Type-species, *philadelphicus* Macquart (Coquillett, 1910b: 595).
Proctacantha, emend.

REFERENCES: Hine, 1911a, Curran, 1934d (keys).

brevipennis (Wiedemann), 1828: 431 (*Asilus*).—Ky. Unrecognized.
coquillettii Hine, 1911a: 160.—Calif.
distinctus (Wiedemann), 1828: 432 (*Asilus*).—Ga.
duryi Hine, 1911a: 160.—Ohio; Kans., Ky.
fulviventris Macquart, 1850: 392 (1850: 88).—Fla.
gracilis Bromley, 1928: 15.—Ga.; Fla.
heros (Wiedemann), 1828: 427 (*Asilus*).—Ky.; Miss. to Ky.
hinei Bromley, 1928: 13.—N. Mex.; Ohio, Ky., Fla.
 rufus Williston of Hine, 1911a: 157.
longus (Wiedemann), 1821c: 183 (*Asilus*).—Ga.; Fla., Tex.
micans Schiner, 1867b: 397.—N. Amer.; Ariz., N. Mex., Colo.
milbertii Macquart, 1838b: 124 (1839: 240).—N. Amer.; U.S., coast to coast.
 agrion Jaennicke, 1867: 365 (1868: 57) (*Asilus*).—Ill. ?Valid sp.
 missouriensis Riley, 1870c: 122 (*Asilus*).—Mo.
nearno Martin, 1962a: 187.—Ariz.; Calif., Utah, N. Mex., Tex.
 arno Townsend of Hine, 1911a: 161.
nigriventris Macquart, 1838b: 124 (1839: 240).—Pa., Carolina; N.J.
nigrofemoratus Hine, 1911a: 161.—Mexico; ?Tex.
occidentalis Hine, 1911a: 159.—Calif.; Idaho.
philadelphicus Macquart, 1838b: 123 (1839: 239).—Pa.; Conn., Mass., N.J., Md., Va.
rodecki James, 1933b: 2.—Colo.; Kans.
rufus Williston, 1885a: 74.—N.C.; Ohio, Mass., N.J., Ky., Fla.
 rufiventris Macquart of Hine, 1911a: 158.

Genus PROMACHINA Bromley

Promachina Bromley, 1934a: 96. Type-species, *Promachus trapezoidalis* Bellardi (orig. des.).

pilosa Wilcox, 1937: 43.—Ariz.
trapezoidalis (Bellardi), 1861: 28 (1864: 128) (*Promachus*).—Mexico; Tex.

Genus PROMACHUS Loew

Trupanea Macquart, 1838b: 91 (1839: 207) (preocc. Guettard, 1762).
Type-species, *Asilus maculatus* Fabricius (orig. des.).
Promachus Loew, 1848: 390. Type-species, *Asilus maculatus* Fabricius (Coquillett, 1910b: 595).
REFERENCE: Hine, 1911a (key).

albifacies Williston, 1885a: 63.—Ariz.; Calif., Colo., N. Mex.
aldrichii Hine, 1911a: 171.—Utah; Idaho.
atrox Bromley, 1940: 13.—Ariz.
bastardii (Macquart), 1838b: 104 (1839: 220) (*Trupanea*).—N. Amer.; Kans. to Mass., s. to Tex. and Ga.
    ultimus Walker, 1851b: 136 (*Asilus*).—N. Amer.
    laevinus Walker, 1849: 392 (*Asilus*).—Mass.
    rubiginis Walker, 1852: 123 (*Trupanea*).—N. Amer.
    philadelphicus Schiner, 1867b: 389.—Pa.
dimidiatus Curran, 1927f: 87.—Man.
fitchii Osten Sacken, 1878c: 234 (n. name for *apivora* Fitch).—Nebr.; Kans., Mo., Tex., Conn., Fla.
    apivora Fitch, 1864: 63 (*Trupanea*; preocc. Walker, 1860).—Nebr.
giganteus Hine, 1911a: 172.—Tex.
hinei Bromley, 1931: 435.—Ohio; Tex., Iowa, Ind.
    rufipes Fabricius of Hine, 1911a: 166.
magnus Bellardi, 1861: 26 (1864: 126).—Mexico; Tex.
minusculus Hine, 1911a: 171.—N. Mex.; Tex.
nigripes Hine, 1911a: 170.—N. Mex.
nigropilosus Schaeffer, 1916: 68.—Calif.
oklahomensis Pritchard, 1935: 12.—Okla.; Nebr., Kans., Ind.
painteri Bromley, 1934a: 93.—Tex.
princeps Williston, 1885a: 62.—Wash.; Oreg., Calif.
quadratus (Wiedemann), 1821c: 201 (*Asilus*).—Ga.; La., Fla.
rufipes (Fabricius), 1775: 794 (*Asilus*).—Amer.; Iowa to D.C., s. to Miss. and Fla.
sackeni Hine, 1911a: 166.—Ariz.
texanus Bromley, 1934a: 94.—Tex.
truquii Bellardi, 1861: 30 (1864: 130).—Mexico; Ariz.
vertebratus (Say), 1823: 47 (1859b: 62) (*Asilus*).—Mo.; Colo., Kans., Wis., Ill., Ohio.

Genus REGASILUS Curran

Regasilus Curran, 1931b: 24. Type-species, *strigarius* Curran (orig. des.).

blantoni Bromley, 1951: 35.—Nev.

Genus TOLMERUS Loew

Tolmerus Loew, 1849: 94. Type-species, *Asilus pyragrus* Zeller (Coquillett, 1910b: 615).

antimachus (Walker), 1849: 454 (*Asilus*).—N.Y.; Kans., Mo., Ind., Ohio, Va.
callidus Williston, 1893c: 75.—Wash.; B.C., Oreg.
delusus Tucker, 1907: 92 (as *annulipes* var.).—Kans., Colo.; Mont.
johnsoni (Hine), 1909: 159 (*Asilus*).—Pa.
maneei (Hine), 1909: 158 (*Asilus*).—N.C.; Fla.
notatus (Wiedemann), 1828: 451 (*Asilus*).—Ga.; Kans. to Maine, s. to D.C., also Ga.
    alethes Walker, 1849: 454 (*Asilus*).—N.Y.
novaescotiae (Macquart), 1847: 62 (1847: 46) (*Asilus*).—N.S.; N.S. to N.C.
paropus (Walker), 1849: 455 (*Asilus*).—N.Y.; Canada, Wyo. to N.H., s. to N. Mex. and Conn.
prospectus Tucker, 1907: 94.—Colo.
prairiensis Tucker, 1907: 93.—Kans.; Colo., N. Mex., Tex.
sadyates (Walker), 1849: 453 (*Asilus*).—Ohio; Ind. to Mass., s. to N.C.
    sadytes, error.
    tibialis Macquart, 1834a: 313 (*Asilus*; preocc. Fabricius, 1793).—Pa.
snowii (Hine), 1909: 160 (*Asilus*; n. name for *annulatus* Williston).— S. Dak., Kans., N.H., Mass., Conn.; e. N. Amer.
    annulatus Williston, 1893c: 70 (*Asilus*; preocc. Fabricius, 1775).—S. Dak., Kans., N.H., Mass., Conn.

Unplaced Species of Asilidae

apicalis Walker, 1854: 497 (*Discocephala*).—w. coast of Amer.
appendiculatus Bigot, 1878: 438 (?*Holopogon*).—Calif.
longicella Macquart, 1850: 399 (1850: 95) (*Asilus*).— N. Amer.
niger Macquart, 1838b: 25 (1839: 141) (*Ceraturgus*).—N. Amer.
nitidiventris Bigot, 1878: 437 (*Holopogon*).—Calif.
ochraceus Wulp, 1870: 213 (*Stenopogon*).—N. Amer.
quadrinotatus Bigot, 1878: 412 (*Dasypogon*).—Calif.
rubiginosus Bigot, 1878: 419 (*Seilopogon*).—N. Amer.

Family NEMESTRINIDAE

By Joseph C. Bequaert

Nemestrinids are medium to large size "tanglewing flies," not commonly seen, and then usually hovering and darting with characteristic loud buzz in open fields. Two species, *Neorhynchocephalus sackenii* (Williston) and *Trichopsidea clausa* (Osten Sacken), are sometimes very effective internal parasites of grasshoppers in the Western States (Prescott, 1960, biol., immature stages).

REFERENCE: Bequaert, 1934 (rev.), 1947 (cat.).

Subfamily HIRMONEURINAE

Genus HIRMONEURA Meigen

Hirmoneura Meigen, 1820: 132. Type-species, *obscura* Wiedemann (mon.).
Hermoneura, emend.

Subgenus NEOHIRMONEURA Bequaert

Hirmoneura, subg. **Neohirmoneura** Bequaert, 1920: 306. Type-species, *flavipes* Williston (orig. des.).

bradleyi Bequaert, 1920: 301.—Tex.
flavipes Williston, 1886a: 292.—U.S.; Ariz.

Subgenus HYRMOPHLAEBA Rondani

Hyrmophlaeba Rondani, 1863: 51 (1864: 51) (as genus). Type-species, *Hirmoneura brevirostris* Macquart (orig. des.).
Hirmophloeba, error.
Hyrmophloeba, error.

texana texana Cockerell, 1908a: 253.—Tex.; Ariz., Mexico, Panama.
ssp. **arizonensis** Bequaert, 1934: 180 (as var.).—Ariz.

Subfamily NEMESTRININAE

Genus NEORHYNCHOCEPHALUS Lichtwardt

Neorhynchocephalus Lichtwardt, 1909: 512. Type-species, *Rhynchocephalus volaticus* Williston (Bequaert, 1930: 287).
Rhynchocephalus, subg. *Nemestrinopsis* Cockerell, 1910b: 285. Type-species, *volaticus* Williston (orig. des.).

sackenii (Williston), 1880: 243 (*Rhynchocephalus*).—Wash.; Wash. to Iowa, s. to Calif. and Ga., also Mich., Mexico.
   subnitens Cockerell, 1908a: 250 (*Rhynchocephalus*).—Kans.
volaticus (Williston), 1883b: 71 (*Rhynchocephalus*).—Fla.; Ariz. to Mo. and Fla., Mexico, Cent. Amer.
   maculatus Curran, 1931d: 69 (*Rhynchocephalus*).—Kans.
   flavus Curran, 1931d: 70 (*Rhynchocephalus*).—Kans.

Subfamily TRICHOPSIDEINAE

Genus TRICHOPSIDEA Westwood

Trichopsidea Westwood, 1839: 151. Type-species, *oestracea* Westwood (mon.).

Subgenus PARASYMMICTUS Bigot

Parasymmictus Bigot, 1879c: lxvii (as genus). Type-species, *Hirmoneura clausa* Osten Sacken (orig. des.).
Parasymmyctus, error.

clausa (Osten Sacken), 1877: 225 (*Hirmoneura*).—Tex.; B.C. to Man. and Kans., s. to Calif. and Fla.

Family ACROCERIDAE
(Cyrtidae)

By Evert I. Schlinger

Adult acrocerids are commonly known as small-headed flies. All known species are solitary internal parasites of true spiders (Araneae). Eggs are deposited in large numbers (up to 5,000 per female) away from the host, and the first-instar larva is a planidium. The mature larva pupates free from its host. Adult longevity varies from 1 to 6 weeks, and mating takes place in flight.

REFERENCES: Cole, 1919 (rev.); Curran, 1934a (key to genera); Clausen, 1940 (biol.); Sabrosky, 1948a (rev.).

Subfamily PANOPINAE

Genus OCNAEA Erichson

Ocnaea Erichson, 1840: 155. Type-species, *micans* Erichson (Coquillett, 1910b: 577).
Pialeoidea Westwood, 1876: 514. Type-species, *Cyrtus magnus* Walker (mon.).
Pialoidea, error.

REFERENCES: Aldrich, 1932b (key); Jenks, 1938 (biol.).

auripilosa Johnson, 1923c: 50.—Ariz.
coerulea Cole, 1919: 26.—Tex.; Ariz.
gloriosa (Sabrosky), 1943b: 176 (*Pialeoidea*).—Tex.
helluo Osten Sacken, 1877: 278.—Tex.; N.C.
loewi Cole, 1919: 26.—Tex.
magna (Walker), 1849: 511 (*Cyrtus*).—Ga.

sequoia Sabrosky, 1948a: 386.—Calif.
smithi Sabrosky, 1948a: 385.—s. Calif.
   smithi Jenks (ex Cole), 1938: 820–828. Nomen nudum.
xuthogaster Schlinger, 1961: 8.—n. Calif.

Genus LASIA Wiedemann

Lasia Wiedemann, 1824: 11. Type-species, *splendens* Wiedemann (mon.).
   REFERENCE: Bequaert, 1931 (rev.).
klettii Osten Sacken, 1875a: 805.—Ariz.; N. Mex., Mexico (Son., Chih.).
purpurata Bequaert, 1933a: 1.—Okla.; Ark., Tex.

Genus EULONCHUS Gerstäcker

Eulonchus Gerstäcker, 1856: 359. Type-species, *smaragdinus* Gerstäcker (mon.).

A number of species and subspecies now being described show the western distribution of the genus to be B.C. to Mont., s. to Baja Calif. and Ariz.

   REFERENCES: Sabrosky, 1948a (key); Schlinger, 1960b (partial key, biol.).
halli Schlinger, 1960b: 418.—s. Calif.; Mexico (Baja Calif.).
marginatus Osten Sacken, 1877: 277.—n. Calif.
marialiciae Brimley, 1925: 77.—N.C.
sapphirinus Osten Sacken, 1877: 276.—n. Calif.
smaragdinus smaragdinus Gerstäcker, 1856: 360.—Calif.; n. Calif. to Mexico (Baja Calif.).
   ssp. **pilosus** Schlinger, 1960b: 418.—s. Calif.
tristis Loew, 1872a: 60 (Cent. 10, no. 19).—Calif.; n. Calif.

Genus PTERODONTIA Gray

Pterodontia Gray *in* Griffith and Pidgeon, 1832: 779. Type-species, *flavipes* Gray (mon.).
Nothra Westwood, 1876: 514. Type-species, *bicolor* Westwood (mon.).
   REFERENCES: King, 1916 (biol.); Sabrosky, 1948a (key).
flavipes Gray *in* Griffith and Pidgeon, 1832: 779.—Ga.; Oreg. to Que. and N.B., s. to Calif. and Md.
   flavoscutellata Steyskal, 1941c: 140.—Mich.
   obesus Harris, 1835: 596 (*Cyrtus*). Nomen nudum.
johnsoni Cole, 1919: 42.—Wash.; Oreg., Calif., Idaho.

misella Osten Sacken, 1877: 277.—Oreg.; B.C. to Mont., s. to Calif. and Utah.
　　americana Bigot, 1889: 320 (*Nothra*).—Wash.
notomaculata Sabrosky, 1948a: 393.—Calif.; Oreg., Idaho.
vix Townsend, 1895b: 607.—s. Calif.
westwoodi Sabrosky, 1948a: 392 (n. name for *analis* Westwood).—Ga.; N.Y., Mass., Conn., also Costa Rica.
　　analis Westwood, 1848: 97 (preocc. Macquart, 1846).—Ga.

Subfamily ACROCERINAE

Genus OPSEBIUS Costa

Opsebius Costa, 1856: 20. Type-species, *perspicillatus* Costa (mon.).

REFERENCES: Sabrosky, 1948a (key); Schlinger, 1952 (biol.).

diligens Osten Sacken, 1877: 278.—B.C.; B.C. to Calif., Ariz., and Mexico (Durango).
　　paucus Osten Sacken, 1877: 279.—Calif.
　　hyalinus Cole, 1919: 47 (as var.).—Calif.
gagatinus Loew, 1866a: 150 (Cent. 6, no. 34).—Pa.
sulphuripes Loew, 1869c: 166 (Cent. 9, no. 68).—N.Y.; Minn. to Que., s. to Tex. and Fla.
　　pterodontinus Osten Sacken, 1883b: 299.—Tex. **N. syn.**
　　agelenae Melander, 1902b: 180.—Tex.

Genus ACROCERA Meigen

Acrocera Meigen, 1803: 266. Type-species, *Syrphus globulus* Panzer (mon.) = *orbiculus* (Fabricius).
Paracrocera Mik, 1886: 276. Type-species, *Acrocera tumida* Erichson (Coquillett, 1910b: 583) = *orbicula* (Fabricius).

REFERENCE: Sabrosky, 1948a (rev.).

arizonensis Cole, 1919: 51 (as *bakeri* var.).—Ariz.; Mexico (Son.).
bakeri Coquillett *in* Baker, 1904: 23.—Nev.; Calif., Ariz.
bimaculata Loew, 1866a: 149 (Cent. 6, no. 33).—D.C.; Mich. to Que., s. to Nebr. and N.C.
bulla Westwood, 1848: 98.—N.Y.; B.C. to Calif., and Wis. to Mass., also Tex. and Ga.
convexa Cole, 1919: 53.—Calif.; B.C. to Mich., s. to Calif. and Colo.
fasciata Wiedemann, 1830: 16.—Ga.; Alta. to Que., s. to Iowa and N.C.
　　hungerfordi (in part) Sabrosky, 1944: 406. Pa. males only.
flaveola Sabrosky, 1944: 411.—N.Y.; Alta., Iowa. Mich., Ont.
fumipennis Westwood, 1848: 98.—Ga.

melanderi Cole, 1919: 55 (as *bulla* var.).—Mont.; Wash. and Alta. to Maine, s. to Calif., Tex., and Ga.
    steyskali Sabrosky, 1944: 401.—Mich. **N. syn.**
melanogaster Schlinger, 1961: 10.—Ariz.
nigrina Westwood, 1848: 97.—Ga.
obsoleta Wulp, 1867: 139.—Wis.
orbicula (Fabricius), 1787: 340 (*Syrphus*).—Europe; N.W.T. to Ont. and Pa., s. to Calif. and Ark.
    hubbardi Cole, 1919: 58.—Ariz. **N. syn.**
    hungerfordi Sabrosky, 1944: 406.—Mich. Females only. **N. syn.**
stansburyi Johnson, 1923c: 49.—Utah.
subfasciata Westwood, 1848: 98.—N.Y.; Wash. to Calif., S. Dak., and Ill. to Mass.
    liturata Williston, 1886a: 294.—Wash. **N. syn.**
unguiculata Westwood, 1848: 98.—Ga.; Iowa, Ill., Mich., ?Calif.

Genus OGCODES Latreille

Ogcodes Latreille, 1796: 154. Type-species, *Musca gibbosa* Linnaeus (sub. mon., Latreille, 1802: 432).
Oncodes, emend.
Henops Meigen, 1803: 266 (preocc. Illiger, 1798). Type-species, *Musca gibbosa* Linnaeus (mon.).

    REFERENCES: Kaston, 1937 (biol.); Schlinger, 1960a (rev., biol.); Lamore, 1960 (biol.; immature stages).

Subgenus OGCODES Latreille

adaptatus Schlinger, 1960a: 297.—Calif.; Alaska to N.W.T., s. to Calif. and Colo.
boharti Schlinger, 1960a: 305.—Ariz.
borealis Cole, 1919: 68.—Que.; B.C. to Que., s. to Calif., Mich., and N.J.
canadensis Schlinger, 1960a: 307.—Ont.
colei Sabrosky, 1948a: 423.—Ariz.; Calif.
dispar (Macquart), 1855: 87 (1855: 67) (*Henops*).—Md.; Minn. to Que., s. to Tex. and N.C., also Costa Rica.
    vittatus Johnson, 1923c: 50.—N.J.
eugonatus Loew, 1872a: 60 (Cent. 10, no. 18) (*Oncodes*).—Tex.; B.C. to Que. and Maine, s. to Calif., Tex., and N.C., also s. Mexico.
    marginatus Cole, 1919: 67 (preocc. Meigen, 1822).—Wyo.
    albicinctus Cole, 1923c: 47 (n. name for *marginatus* Cole).—Wyo.

floridensis Sabrosky, 1948a: 421.—Fla.
hennigi Schlinger, 1960a: 305.—N.Y.
melampus Loew, 1872a: 60 (Cent. 10, no. 17) (*Oncodes*).—Calif.; Alaska to Calif. and Nev., also Minn.
niger Cole, 1919: 65.—Utah.
pallidipennis Loew, 1866a: 149 (Cent. 6, no. 32) (*Oncodes*).—Pa.; Alta. to N.B., s. to Mexico, Miss., and Va., also Costa Rica.
  costatus Loew, 1869c:165 (Cent. 9, no. 67) (*Oncodes*).—Mass.
  incultus Osten Sacken, 1877: 279 (*Oncodes*).—N.H.
  humeralis Osten Sacken, 1887a: 164 (*Oncodes*).—Mexico (Son.).
  aedon Townsend, 1895b: 608 (*Oncodes*).—Mexico (Baja Calif.).
  fasciatus Harris, 1835: 596 (*Cyrtus*). Nomen nudum.
rufoabdominalis Cole, 1919: 68.—Utah.
sabroskyi Schlinger, 1960a: 307.—Ga.
shewelli Sabrosky, 1948a: 422.—Ont.; N.Y.
vittisternum Sabrosky, 1948a: 420.—Oreg.; Wash.

Subgenus NEOGCODES Schlinger

Ogcodes, subg. **Neogcodes** Schlinger, 1960a: 309. Type-species, *albiventris* Johnson (orig. des.).

albiventris Johnson, 1904: 18 (*Oncodes*).—Ont.; B.C., Calif., Mich.

Family BOMBYLIIDAE
By Reginald H. and Elizabeth M. Painter

Adult Bombyliidae, or bee flies, are nectar and pollen feeders, a few apparently being specific to certain flowers. The species of one Nearctic genus (*Oestranthrax*) have vestigial mouth parts. Bee flies frequently rest on bare spaces on the ground, on rocks, or on fragments of dead wood or leaves. The species are far more common in sandy areas and in the southwestern United States than elsewhere.

In most recorded life histories, bombyliid larvae are endo- or ectoparasites of larvae of other Endopterygota or predaceous on the egg pods of Acrididae.

Distribution records are primarily from the author's collection and identifications, the latter often based on a comparison with types wherever available. Distribution records of older authors have been accepted only where they agree with recent records. Say's species and those of other earlier writers whose types were not available have generally been accepted on the basis of identification by Osten Sacken and Coquillett. Acknowledgment is made to Jack Hall, Mr. and Mrs. D. E. Johnson, and Norman Marston for the use of unpublished distribution records, and to the Bache Fund grants Nos. 334 and 453, and National Science grant 10664 for financial assistance.

The limits of variation and distribution of some species and species groups appear open to question on the basis of available information. Such species are found in *Anthrax, Aphoebantus,* and *Villa.* In *Villa,* this is particularly true of the *alternata-agrippina-molitor-sabina, catulina-webberi-eumenes-morio,* and *syrtis-consul* groups and the species related to these groups. Some names in these groups which have been considered synonyms have been left as valid species in this catalog, with only the type locality recorded, until further studies on adequate materials can be made. The new names proposed herein should be attributed to the senior author.

REFERENCES: Coquillett, 1894a, Becker, 1913, Cockerell, 1914c, Bezzi, 1924b, Engel, 1932–1937, Maughan, 1935, Hesse, 1938–1956 (all tax.); Graenicher, 1910c, Malloch, 1915h, 1917g, Brooks, 1952 (all biol.); Painter and Painter, 1962 (redescr. types).

BOMBYLIIDAE HOMOEOPHTHALMAE

Subfamily BOMBYLIINAE

Genus BOMBYLIUS Linnaeus

Bombylius Linnaeus, 1758: 606. Type-species, *major* Linnaeus (Latreille, 1810: 443).

Larvae are found in nests of solitary bees.

REFERENCES: Johnson, 1907c (rev.); Painter, 1940 (notes, key e. spp.).

albicapillus albicapillus Loew, 1872a: 78 (Cent. 10, no. 42).—Calif.; Wash. to Alta. and Mont., s. to Calif. and Utah, also N.W.T.

atricapillus, error.

ssp. **diegoensis** Painter, 1933: 16 (as var.).—Calif.; Ariz.

atriceps atriceps Loew, 1863b: 301 (Cent. 4, no. 49).—Fla., Va.; Minn. to Conn., s. to Okla. and Ga.

ssp. **fulvibasoides** Painter *in* Painter and Painter, 1962: 8.— N.Y.; Wis. to Vt., s. to Okla. and N.J.

fulvibasis, authors, not Macquart.

aurifer aurifer Osten Sacken, 1877: 249.—Calif.; Wash. to Sask. and Mont., s. to Calif. and Colo.

ssp. **pendens** Cole *in* Cole and Lovett, 1919: 226.—Oreg.; Wash., Idaho, Utah, Colo.

austini Painter, 1933: 16.—Tex.
cachinnans Osten Sacken, 1877: 250.—Calif.
cinerivus Painter *in* Painter and Painter, 1962: 5 (n. name for *cinereus* Bigot).—Calif.
    cinereus Bigot, 1892a: 364 (preocc. Olivier, 1789.)—Calif.
comanche Painter *in* Painter and Painter, 1962: 11.—Tex.; S. Dak., s. to Colo., Tex., and Ark.
    io, authors, not Williston.
duncani Painter, 1940: 275.—Ariz.; Calif.
eboreus Painter, 1940: 270.—N. Mex..
facialis Cresson, 1919: 187.—Ariz.; Calif., Utah, Colo., N. Mex.
flavipilosus Cole, 1923a: 308.—Mexico (Baja Calif.); Ariz., N. Mex.
fraudulentus Johnson, 1907c: 99.—Mass.; Conn., N.Y., N.J., Ga.
incanus Johnson, 1907c: 97.—Mass.; N.Y., N.J.
    philadelphicus, authors, not Macquart.
lancifer Osten Sacken, 1877: 251.—Calif.; B.C., s. to Calif. and N. Mex.
lassenensis Johnson and Johnson (for *pallescens* Johnson and Maughan).—Calif. **N. name**.
    pallescens Johnson and Maughan, 1953: 20 (preocc. Hesse, 1938).—Calif.
maculifer Walker, 1852: 200.—Unknown; Md., N.C., S.C., Ga.
    azaleae Shannon, 1916b: 71.—Md.
major Linnaeus, 1758: 606.—Europe; B.C. to Maine, s. to Calif. and Ga.
    aequalis Fabricius, 1781: 473.—N. Amer.
    fratellus Wiedemann, 1828: 583.—Ga.
    vicinus Macquart, 1840: 98 (1841: 376).—Pa.
    albipectus Macquart, 1855: 102 (1855: 82).—Md.
medorae Painter, 1940: 273.—Kans.; Tex.
metopium Osten Sacken, 1877: 249.—Calif.; Oreg., Idaho, Utah.
mexicanus Wiedemann, 1821c: 166.—Ga. (Painter and Painter, 1962: 15); Minn. to N.Y., s. to Miss. and Fla.
    philadelphicus Macquart, 1840: 99 (1841: 377).—Pa.
    fulvibasis Macquart, 1855: 102 (1855: 82).—Md.
    flavibasis, error.
pulchellus Loew, 1863b: 300 (Cent. 4, no. 47).—Ill.; Alta. to Maine, s. to Kans. and N.J.
pygmaeus pygmaeus Fabricius, 1781: 474.—N. Amer.; Alta. to Maine, s. to Ill. and Ga.
    ssp. **canadensis** Curran, 1933b: 2 (as sp.).—Que.; Mont., Colo., Wis., Mich., N.Y.
ravus Loew, 1863b: 301 (Cent. 4, no. 50).—Mexico; Colo., Tex.
silvus Cole *in* Cole and Lovett, 1919: 225.—Oreg.

subvarius Johnson, 1907c: 98.—N.J., Pa.
texanus Painter, 1933: 17.—Tex.; Kans., Okla.
validus Loew, 1863b: 300 (Cent. 4, no. 48).—Ill., Va.; Minn., Mich.
varius Fabricius, 1805: 132.—N. Amer.; Mo. to N.H., s. to Ga.

Genus PARABOMBYLIUS Williston

Parabombylius Williston, 1907: 1. Type-species, *Thlipsogaster ater* Coquillett (Coquillett, 1910b: 583)=*coquilletti* (Williston).

REFERENCES: Painter, 1926a (key and notes), 1940 (notes); Curran, 1930b (key).

albopenicillatus (Bigot), 1892a: 363 (*Bombylius*).—Mexico; Tex., Ark.
coquilletti (Williston), 1899: 331 (*Bombylius*; n. name for *ater* Coquillett).—La.; Kans., Tex., Ark.
   ater Coquillett, 1894a: 108 (*Thlipsogaster*; preocc. in *Bombylius* by Scopoli, 1763).—La.
dolorosus (Williston), 1901: 286 (*Bombylius*).—Mexico; Ariz., Tex.
maculosus Painter, 1926a: 78.—Ariz.
nigrofemoratus Painter, 1940: 278 (n. name for *vittatus* Curran).—Ariz.
   vittatus Curran, 1930b: 7 (preocc. Painter, 1926).—Ariz.
pulcher Painter, 1926a: 76.—Tex.
subflavus Painter, 1926a: 76.—Ariz.
syndesmus (Coquillett), 1894a: 108 (*Thlipsogaster*).—Calif.; Ariz., Tex.
vittatus Painter, 1926a: 77.—N. Mex.

Genus SYSTOECHUS Loew

Systoechus Loew, 1855b: 34. Type-species, *Bombylius sulphureus* Mikan (Coquillett, 1910b: 611).

REFERENCES: Berg, 1940, Parker and Wakeland, 1957 (biol., predator on grasshopper egg pods); Painter, 1962 (rev., biol.).

candidulus Loew, 1863b: 302 (Cent. 4, no. 51).—Wis.; Oreg. to Mich., s. to Ariz. and Ala.
fumipennis Painter, 1962: 257.—Wyo.; Wash., Utah, Colo.
oreas Osten Sacken, 1877: 254.—Calif.; B.C. to Mont., s. to Calif. and N. Nex.
solitus (Walker), 1849: 288 (*Bombylius*).—Fla.; N.J., S.C., Ga., Ala., Miss., La.
vulgaris Loew, 1863b: 302 (Cent. 4, no. 52).—Nebr.; Oreg. to Alta. and Mass., s. to Ariz., Tex., and N.J.

Genus ANASTOECHUS Osten Sacken

Anastoechus Osten Sacken, 1877: 251. Type-species, *barbatus* Osten Sacken (mon.).

REFERENCES: Parker and Wakeland, 1957 (biol., predator on grasshopper egg pods); Painter, 1962 (rev., biol.).

barbatus Osten Sacken, 1877: 252.—Calif., Wyo., Colo., Mass.; Y.T., N.W.T., Alta. to Mass., s. to Oreg., Calif., Tex., and Md.
  nitidulus, authors, not Fabricius.
hessei Hall, 1958: 195 (n. name for *deserticola* Hall).—Ariz.; Calif., Tex.
  deserticola Hall, 1956: 200 (preocc. Hesse, 1938).—Ariz.
melanohalteralis Tucker, 1907: 91.—Colo.; Alta. to S. Dak., s. to Calif. and Tex.
  fulvipennis Tucker, 1907: 91 (as var.).—Colo.

Genus HETEROSTYLUM Macquart

Heterostylum Macquart, 1848a: 195 (1848: 35). Type-species, *flavum* Macquart (orig. des.).
Comastes Osten Sacken, 1877: 256. Type-species, *robustus* Osten Sacken (mon.).

The larvae are parasitic on Megachilidae.

REFERENCES: Painter, 1930a (rev.), 1940 (notes); Bohart, Stephen, and Eppley, 1960 (biol.).

Subgenus HETEROSTYLUM Macquart

croceum Painter, 1930a: 6.—Kans.; Colo., N. Mex., Tex.
deani Painter, 1930a: 5.—Kans.; Colo.
engelhardti Painter, 1930a: 7.—Tex.; Ariz.
laticeps (Bigot), 1892a: 363 (*Bombylius*).—Calif.
robustum (Osten Sacken), 1877: 257 (*Comastes*).—Tex.; Oreg. and Calif. to Ga.
sackeni (Williston), 1893d: 255 (*Comastes*).—Calif.; Ariz.

Subgenus TRIPLOECHUS Edwards

Triploechus Edwards, 1936: 31 (as genus). Type-species, *Bombylius heteroneurus* Macquart (orig. des.).
Triplasius, authors, not Loew.

REFERENCE: Paramonov, 1947 (rev.).

novum (Williston), 1893d: 254 (*Triplasius*).—Calif.; Ariz.
  recurvus Coquillett, 1920g: 100 (*Bombylius*).—Calif.
vierecki Cresson, 1919: 186.—N. Mex.; Ariz.

Genus LORDOTUS Loew

Lordotus Loew, 1863b: 303 (Cent. 4, no. 53). Type-species, *gibbus* Loew (mon.).

REFERENCES: Painter, 1940 (key); Hall, 1954a (rev.); Johnson and Johnson, 1959b (notes).

abdominalis Johnson and Johnson, 1959b: 10.—Ariz.
albidus Hall, 1954a: 19.—Calif.; Nev., Ariz.
apiculus Coquillett, 1887c: 116.—Colo.; Oreg. to Wyo., s. to Calif. and Tex.
arizonensis Johnson and Johnson, 1959b: 19.—Ariz.; N. Mex.
bipartitus Painter, 1940: 284.—N. Mex.; Oreg., Calif., Ariz., Tex.
bucerus Coquillett, 1894a: 110.—Calif.; Ariz.
cingulatus cingulatus Johnson and Johnson, 1959b: 21.—Calif.; Ariz.
   ssp. **rufotibialis** Johnson and Johnson, 1959b: 23.—Ariz.
diversus diversus Coquillett, 1891b: 198.—Calif.; Oreg., Ariz., N. Mex., Tex., Mexico (Baja Calif.).
   ssp. **diplasus** Hall, 1954a: 17.—Calif.; Ariz.
divisus Cresson, 1919: 186.—N. Mex.; Calif., Ariz., Tex., Mexico (Baja Calif.).
   niger Cresson, 1923: 367.—N. Mex.
ermae Hall, 1952: 49.—Calif.
gibbus gibbus Loew, 1863b: 303 (Cent. 4, no. 53).—Mexico; Oreg. to Alta., s. to Calif. and Mexico.
   flavus Jaennicke, 1867: 346 (1868: 38) (*Adelidea*).—Mexico.
   ssp. **striatus** Painter, 1940: 287.—N. Mex.; Oreg. to Idaho, s. to Calif. and Tex.
hurdi Hall, 1957: 145.—Calif.
junceus Coquillett, 1891b: 198.—Calif.; Ariz., N. Mex., Mexico.
lutescens Johnson and Johnson, 1959b: 15.—N. Mex.
miscellus miscellus Coquillett, 1887c: 116.—Calif.; Nev., Utah, Ariz.
   ssp. **melanosus** Johnson and Johnson, 1959b: 24.—Utah; Nev.
perplexus Johnson and Johnson, 1959b: 16.—Ariz.; Calif.
planus Osten Sacken, 1877: 258.—Calif.
puellus Williston, 1893c: 64.—Calif.
pulchrissimus pulchrissimus Williston, 1893c: 64.—Nev.; Wash. to Wyo. and Kans., s. to Mexico (Baja Calif.).
   pulcherrimus, emend.
   carus Cresson, 1923: 366.—Calif.
   ssp. **luteolus** Hall, 1954a: 26.—Calif.; Nev., Ariz., Tex.
sororculus sororculus Williston, 1893d: 255.—Calif.; Nev., Ariz., Tex.
   ssp. **nigriventris** Johnson and Johnson, 1959b: 17.—Utah.
zonus Coquillett, 1887c: 116.—Calif.; Idaho, Nev., Ariz.

Genus GEMINARIA Coquillett

Geminaria Coquillett, 1894a: 109. Type-species, *Lordotus canalis* Coquillett (orig. des.).

REFERENCE: Painter, 1940 (key).

canalis (Coquillett), 1887c: 115 (*Lordotus*).—Calif.; Ariz.
pellucida Coquillett, 1894a: 109.—Calif.

Genus SPARNOPOLIUS Loew

Sparnopolius Loew, 1855b: 43. Type-species, *Bombylius fulvus* Wiedemann (Coquillett, 1910b: 606)=*lherminierii* (Macquart).

The larvae are parasites of Scarabaeidae.

REFERENCE: Painter, 1940 (key).

anomalus Painter, 1940: 281.—Colo.
brevicornis Loew, 1872a: 79 (Cent. 10, no. 43).—Tex.; N. Mex.
coloradensis Grote, 1867: 445.—Colo.; Utah, N. Mex.
cumatilis Grote, 1867: 445.—Colo.
lherminierii (Macquart), 1840: 103 (1841: 381) (*Bombylius*).—Carolina; Mont. to Maine, s. to Tex. and Ga.
  fulvus Wiedemann, 1821c: 172 (*Bombylius;* preocc. Meigen, 1820).—N. Amer.
  brevirostris Macquart, 1840: 103 (1841: 381) (*Bombylius;* preocc. Olivier, 1789).—Carolina.
  fuscipes Bigot, 1892a: 369 (*Dischistus*).—N. Amer.

Genus CONOPHORUS Meigen

Conophorus Meigen, 1803: 268. Type-species, *Bombylius maurus* Mikan (mon.)=*virescens* (Fabricius).
Ploas Latreille, 1804: 190. Type-species, *hirticornis* Latreille (sub. mon., Latreille, 1805: 300)=*virescens* (Fabricius).
Calopelta Greene, 1921b: 23. Type-species, *fallax* Greene (mon.).

REFERENCE: Priddy, 1958 (rev.).

amabilis (Osten Sacken), 1877: 261 (*Ploas*).—Calif.; Wash., s. to Calif. and Colo.
atratulus (Loew), 1872a: 79 (Cent. 10, no. 44) (*Ploas*).—Calif.; Oreg., Utah, Mexico (Baja Calif.).
auratus Priddy, 1954: 55.—Calif. (as Wyo., error); Oreg.
chinooki Priddy, 1954: 53.—Wash.; Oreg., Idaho.
collini Priddy, 1958: 31.—Calif.

columbiensis Priddy, 1954: 54.—B.C.; Wash., Mont.
cristatus Painter, 1940: 294.—Calif.
fallax (Greene), 1921b: 23 (*Calopelta*).—Colo.; B.C. to Sask., s. to Calif. and N. Mex.
fenestratus (Osten Sacken), 1877: 260 (*Ploas*).—Calif.; Oreg., Idaho, Utah.
hiltoni Priddy, 1958: 24.—Calif.
limbatus (Loew), 1869b: 31 (Cent. 8, no. 51) (*Ploas*).—N. Mex.; Tex.
melanoceratus Bigot, 1892a: 361.—Calif.; Oreg., Utah.
nigripennis (Loew), 1872a: 80 (Cent. 10, no. 45) (*Ploas*).—Calif.; B.C. to Calif. and N. Mex.
obesulus (Loew), 1872a: 80 (Cent. 10, no. 46) (*Ploas*).—Calif.; B.C. to Calif. and Colo.
painteri Priddy, 1958: 12.—Utah; Idaho, Colo., Calif., N. Mex.
pictipennis (Macquart), 1840: 107 (1841: 385) (*Ploas*).—Carolina.
rufulus (Osten Sacken), 1877: 261 (*Ploas*).—Calif.; Wash., Oreg.
sackenii Johnson and Maughan, 1953: 22.—Calif.; B.C. to Sask., s. to Calif. and Colo.
serratus (Coquillett), 1894a: 102 (*Ploas*).—Calif.; Wash.

Genus ALDRICHIA Coquillett

Aldrichia Coquillett, 1894a: 93. Type-species, *ehrmanii* [Coquillett (mon.).
auripuncta Painter, 1940: 294.—Ohio.
ehrmanii Coquillett, 1894a: 94.—Pa.; Mich. to N.Y., s. to Kans. and Ohio.
ehrmanni, error or emend.

Subfamily CYTHEREINAE

The Palaearctic species are predators in egg pods of Acrididae.

Genus PANTARBES Osten Sacken

Pantarbes Osten Sacken, 1877: 254. Type-species, *capito* Osten Sacken (mon.).
REFERENCE: Painter, 1940 (key).

capito Osten Sacken, 1877: 256.—Calif.; Wyo.
pusio Osten Sacken, 1887a: 153.—Mexico; Oreg. to Wyo., s. to Calif. and Mexico.
willistoni Osten Sacken, 1887a: 153.—Ariz.; Calif., Tex.

Subfamily PLATYPYGINAE
(Cyrtosiinae)

Cyrtomorpha sp. is predaceous on egg pods of Acrididae in Australia.

REFERENCE: Melander, 1950b (key).

Genus PLATYPYGUS Loew

Platypygus Loew, 1844a: 127. Type-species, *chrysanthemi* Loew (mon.).

americanus Melander, 1950b: 140.—Calif.

Subfamily MYTHICOMYIINAE
(Glabellulinae)

REFERENCE: Melander, 1950b (key).

Genus GLABELLULA Bezzi

Platygaster Zetterstedt, 1838: 574 (preocc. Latreille, 1809). Type-species, *arctica* Zetterstedt (mon.).
Glabellula Bezzi, 1902: 191 (n. name for *Platygaster* Zetterstedt). Type-species, *Platygaster arctica* Zetterstedt (aut.).
Pachyneres Greene, 1924b: 62. Type-species, *crassicornis* Greene (mon.).
Reared from decaying wood.

REFERENCES: Melander, 1946b, 1950b (keys).

crassicornis (Greene), 1924b: 62 (*Pachyneres*).—D.C.; Idaho, Man., Pa.
fasciata Melander, 1950b: 143.—Wash.
metatarsalis Melander, 1950b: 143.—Calif.
nanella Melander, 1950b: 144.—Calif.
pumila Melander, 1950b: 144.—Calif.
rotundipennis Melander, 1950b: 145.—Calif.

Genus EMPIDIDEICUS Becker

Empidideicus Becker, 1907d: 97. Type-species, *carthaginensis* Becker (mon.).

REFERENCES: Melander, 1946b, 1950b (revs.).

flavifrons Melander, 1950b: 141.—Calif.
humeralis Melander, 1946b: 455.—Calif.; Ariz.

propleuralis Melander, 1946b: 458.—Calif.
scutellaris Melander, 1946b: 456.—Calif.

Genus MYTHICOMYIA Coquillett

Mythicomyia Coquillett, 1893h: 208. Type-species, *rileyi* Coquillett (mon.).

REFERENCES: Cresson, 1915c (key); Melander, 1902a, 1961 (revs.).

acuta Melander, 1961: 179.—Calif.
agilis Melander, 1961: 180.—s. Calif.
angusta Melander, 1961: 180.—s. Calif.
annulata Melander, 1961: 181.—s. Calif.; Mexico (Baja Calif.), Ariz.
anomala Melander, 1961: 182.—s. Calif.
antecessor Melander, 1961: 182.—s. Calif.
apricata Melander, 1961: 183.—s. Calif.; Wash., Utah, Ariz.
armata Cresson, 1915c: 455.—N. Mex.; Wash., Idaho, Calif., Ariz., Mexico.
armipes Cresson, 1915c: 454.—N. Mex.
atra Cresson, 1915c: 456.—N. Mex.; Calif., Idaho, Utah, Ariz., Tex., Mexico.
atrita Melander, 1961: 186.—N. Mex.
aureola Melander, 1961: 187.—Calif.; Ariz.
    var. **flavida** Melander, 1961: 187.—Ariz.; s. Calif.
aurifera Melander, 1961: 187.—s. Calif.
bibosa Melander, 1961: 188.—s. Calif.
bilychnis Melander, 1961: 189.—s. Calif.
bivulneris Melander, 1961: 190.—s. Calif.
bucinator Melander, 1961: 190.—s. Calif.
cala Melander, 1961: 190.—s. Calif.
    var. **bella** Melander, 1961: 191.—s. Calif.
californica Greene, 1924b: 61.—Calif.; Oreg., Nev.
caligula Melander, 1961: 192.—s. Calif.
callima Melander, 1961: 193.—s. Calif.
calva Melander, 1961: 194.—s. Calif.; Wash.
carptura Melander, 1961: 194.—s. Calif.; Wash., Oreg.
collina Melander, 1961: 195.—s. Calif.
    var. **actites** Melander, 1961: 196.—Calif.
comma Melander, 1961: 196.—N. Mex.
comparata Melander, 1961: 197.—s. Calif.
compta Melander, 1961: 197.—s. Calif.
concinna Melander, 1961: 198.—s. Calif.
concrescens Melander, 1961: 198.—s. Calif.

cressoni Melander, 1961: 199.—s. Calif.; Utah, N. Mex.
  rileyi, authors, not Coquillett.
cristata Melander, 1961: 199.—Wash.
crocina Melander, 1961: 200.—Calif.
  var. **aspilota** Melander, 1961: 200.—Calif.
cylla Melander, 1961: 201.—s. Calif.
diadela Melander, 1961: 201.—Calif.
dipura Melander, 1961: 202.—s. Calif.
diropeda Melander, 1961: 202.—s. Calif.
enoria Melander, 1961: 203.—s. Calif.
eremica Melander, 1961: 204.—s. Calif.
fasciolata Melander, 1961: 204.—s. Calif.
flavipes Cresson, 1915c: 452.—Tex.; Calif., Utah, Ariz., N. Mex., Mexico.
flaviventris Melander, 1961: 205.—s. Calif.; Wash.
formosa Melander, 1961: 206.—Ariz.; Calif.
frontalia Melander, 1961: 206.—s. Calif.; Ariz.
fulgida Melander, 1961: 207.—s. Calif.; Ariz.
galbea Melander, 1961: 208.—s. Calif.
gausa Melander, 1961: 208.—s. Calif.; Mexico (Baja Calif.).
gibba Melander, 1961: 209.—Ariz.; s. Calif., Mexico.
gibbera Melander, 1961: 210.—s. Calif.
gracilis Melander, 1961: 210.—Calif.
habra Melander, 1961: 211.—Ariz.
hamata Melander, 1961: 211.—Calif.
hiata Melander, 1961: 212.—s. Calif.
hoplites Melander, 1961: 212.—s. Calif.
hormatha Melander, 1961: 213.—s. Calif.; Utah.
hybos Melander, 1961: 214.—s. Calif.
illustris Melander, 1961: 215.—Ariz.; Mexico.
imbellis Melander, 1961: 215.—s. Calif.
indicata Melander, 1961: 216.—Calif.
insignis Melander, 1961: 216.—s. Calif.
intermedia Melander, 1961: 217.—s. Calif.; Utah, Ariz.
introrsa Melander, 1961: 218.—s. Calif.; N. Mex.
irrupta Melander, 1961: 218.—s. Calif.; Wash., Ariz., N. Mex., Mexico.
laticlavia Melander, 1961: 220.—s. Calif.; Ariz.
lenticularis Melander, 1961: 220.—Calif.
levigata Melander, 1961: 221.—s. Calif.
liticen Melander, 1961: 221.—s. Calif.
longimana Melander, 1961: 222.—s. Calif.
marginata Melander, 1961: 222.—s. Calif.; Ariz.

minima Melander, 1961: 223.—s. Calif.
    var. **lucens** Melander, 1961: 224.—s. Calif.
ministra Melander, 1961: 224.—s. Calif.
minor Melander, 1961: 224.—Mexico; s. Calif.
minuscula Melander, 1961: 225.—s. Calif.
minuta Greene, 1924b: 62.—N. Mex.
mira Melander, 1961: 226.—Calif.
mirifica Melander, 1961: 226.—s. Calif.
mitrata Melander, 1961: 227.—Calif.
modesta Melander, 1961: 227.—Mexico (Son.); s. Calif.
mulsea Melander, 1961: 228.—s. Calif.; N. Mex.
murina Melander, 1961: 229.—s. Calif.
mutabilis Melander, 1961: 229.—s. Calif.; Nev., Ariz.
napaea Melander, 1961: 230.—Calif.
nigricans Melander, 1961: 231.—s. Calif.
nitida Melander, 1961: 231.—s. Calif.
nitidula Melander, 1961: 231.—s. Calif.
ocreata Melander, 1961: 232.—sw. Ariz.; s. Calif.
oporina Melander, 1961: 232.—s. Calif.
optata Melander, 1961: 233.—s. Calif.
    var. **diloga** Melander, 1961: 234.—s. Calif.
orchestes Melander, 1961: 234.—s. Calif.
ornata Melander, 1961: 234.—s. Calif.
ornatula Melander, 1961: 235.—Oreg.; Wash., Calif., Ariz., Mexico.
ostenta Melander, 1961: 236.—s. Calif.
parma Melander, 1961: 236.—s. Calif.
penicula Melander, 1961: 237.—Ariz.
petena Melander, 1961: 237.—s. Calif.
petes Melander, 1961: 238.—Ariz.; Calif., Utah.
petiolata Melander, 1961: 238.—Calif.
phacodes Melander, 1961: 239.—Calif.
phalerata Melander, 1961: 239.—Wash.
pharetra Melander, 1961: 239.—Calif.
picta Melander, 1961: 240.—s. Calif.; Oreg., Idaho.
pictipes Coquillett, 1902g: 102.—Ariz.; N. Mex.
platycheira Melander, 1961: 241.—Oreg.
polygena Melander, 1961: 242.—Utah; s. Calif.
potrix Melander, 1961: 242.—s. Calif.; Wash., Oreg., Ariz.
pravipes Melander, 1961: 243.—Calif.; Wash., Ariz., Mexico.
pruinosa Melander, 1961: 244.—s. Calif.; Mexico.
pulla Melander, 1961: 245.—Nev.
pusilla Melander, 1961: 245.—s. Calif.
rhaeba Melander, 1961: 246.—Calif.; Idaho, Mont., Ariz., Mexico.

rileyi Coquillett, 1893h: 209.—Calif.; Wash. to Idaho, s. to Calif. and N. Mex., Mexico.
robiginosa Melander, 1961: 249.—Calif.
salpinx Melander, 1961: 249.—s. Calif.
scapulata Melander, 1961: 250.—s. Calif.; N. Mex.
scutellata Coquillett, 1902g: 102.—Ariz.; Calif., Utah, N. Mex., Mexico (Baja Calif.)
sorbens Melander, 1961: 252.—s. Calif.
sugens Melander, 1961: 252.—Oreg.; Calif.
tagax Melander, 1961: 253.—s. Calif.
tenthes Melander, 1961: 253.—s. Calif.; Mexico.
tibialis Coquillett, 1895m: 409.—Calif.; N. Mex.
trifaria Melander, 1961: 254.—s. Calif.; Nev., Utah, Ariz.
triformis Melander, 1961: 255.—s. Calif.; Idaho, Mexico.
  var. **chitona** Melander, 1961: 256.—s. Calif.
  var. **monacha** Melander, 1961: 256.—s. Calif.
tristis Melander, 1961: 257.—Calif.
tubicen Melander, 1961: 257.—Mexico (Baja Calif.); Ariz.
tumescens Melander, 1961: 258.—s. Calif.
uncata Melander, 1961: 258.—Calif.
vestis Melander, 1961: 259.—Utah.
vilis Melander, 1961: 259.—s. Calif.
virgata Melander, 1961: 260.—s. Calif.
virgo Melander, 1961: 260.—Utah; s. Calif.
vulnerata Melander, 1961: 261.—Calif.

Subfamily PHTHIRIINAE

Genus PHTHIRIA Meigen

Phthiria Meigen, 1803: 268. Type-species, *Bombylius pulicarius* Mikan (mon.).
Neacreotrichus Cockerell, 1917: 377. Type-species, *Acreotrichus atratus* Coquillett (orig. des.).

REFERENCES: Coquillett, 1894a (key); Painter and Painter, 1962 (partial keys).

Subgenus PHTHIRIA Meigen

americana (Coquillett), 1895l: 273 (*Acreotrichus*).—Wash.; Oreg., Calif.
cingulata Loew, 1846b: 383.—Mexico; Ariz., Calif.
diversa Coquillett, 1894a: 103.—Calif.
egerminans Loew, 1872a: 80 (Cent. 10, no. 47).—Calif.
floralis Coquillett, 1894a: 103.—Calif.; Utah.

humilis Osten Sacken, 1877: 264.—Calif.; Ariz., Tex., Utah.
maculipennis (Cole), 1923b: 24 (*Acreotrichus*).—Calif.
melanoscuta Coquillett, 1904f: 172.—N. Mex.; Ariz.
picturata Coquillett, 1904f: 175.—N. Mex.; Ariz.
similis Coquillett, 1894a: 103.—Calif.; Nev., Utah.

Subgenus POECILOGNATHUS Jaennicke

Poecilognathus Jaennicke, 1867: 350 (1868: 42) (as genus). Type-species, *thlipsomyzoides* Jaennicke (mon.).

aldrichi Johnson, 1903e: 184.—Idaho; S. Dak., N. Mex., Tex.
alterans Williston, 1901: 291.—Mexico; Ariz.
amplicella Coquillett, 1904f: 175.—Tex.
badia Coquillett, 1904f: 174.—Tex.; Honduras.
bicolor Coquillett, 1904f: 176.—N. Mex.
borealis Johnson, 1910a: 229.—Maine.
coquilletti Johnson, 1902a: 240.—N.J.; Ill., Mich., Mass.
cyanoceps Johnson, 1903e: 184.—Mass.; Minn., Ill., Mich., N.Y.
flaveola Coquillett, 1904f: 175.—N. Mex., Calif.
fulvida Coquillett, 1904f: 172.—Mexico; Ariz.
inornata Coquillett, 1904f: 174.—Tex.
loewi Painter (for *notata* Loew).—Calif.; Calif. to Utah, s. to Tex.
    N. name.
        notata Loew, 1863a: 11 (Cent. 3, no. 19) (preocc. Bigot, 1862).—Calif.
marginata Coquillett, 1904f: 173.—N. Mex.
nubeculosa Coquillett, 1904f: 173.—N. Mex.
psi Cresson, 1919: 186.—Calif.
punctipennis Walker, 1849: 294.—Ga.; S. Dak. to Ill., s. to Ariz. and Ga.
scolopax Osten Sacken, 1877: 263.—Colo.; Calif., Ariz., N. Mex.
sulphurea Loew, 1863a: 11 (Cent. 3, no. 18).—N.J.; Oreg. to Mass., s. to Ariz. and Fla.
thlipsomyzoides (Jaennicke), 1867: 351 (1868: 43) (*Poecilognathus*).—Mexico; Ariz., N. Mex., Colo., Tex.
unimaculata Coquillett, 1904c: 32.—N. Mex., Tex.; Ariz.
vittiventris Coquillett, 1904f: 173.—N. Mex.; Ariz., Utah.

Genus DESMATOMYIA Williston

Desmatomyia Williston, 1895c: 268. Type-species, *anomala* Williston (mon.).

anomala Williston, 1895c: 268.—Colo.; Ariz.

Genus APOLYSIS Loew

Apolysis Loew, 1860c: 86 (also 1860d: 269; 1860: 197). Type-species, *humilis* Loew (mon.).

REFERENCE: Melander, 1946b (key).

aperta Melander, 1946b: 459.—Calif.
disjuncta Melander, 1946b: 459.—Calif.
druias Melander, 1946b: 460.—Calif.; Ariz.
glauca Melander, 1946b: 461.—Calif.
minutissima Melander, 1946b: 461.—Calif.
mohavea Melander, 1946b: 462.—Calif.
petiolata Melander, 1946b: 462.—Calif.; Ariz.
timberlakei Melander, 1946b: 463.—Calif.

Genus OLIGODRANES Loew

Oligodranes Loew, 1844a: 160. Type-species, *obscuripennis* Loew (Becker, 1913: 484).
Rhabdopselaphus Bigot, 1886b: ciii. Type-species, *mus* Bigot (mon.).
Pseudogeron Cresson, 1915b: 201. Type-species, *mitis* Cresson (orig. des.).

REFERENCE: Melander, 1946b (key).

acrostichalis acrostichalis Melander, 1946b: 471.—Calif.
    ssp. **matutinus** Melander, 1946b: 471 (as var.).—Calif.
analis Melander, 1946b: 471.—Calif.
anthonomus Melander, 1946b: 472.—Calif.
ater (Cresson), 1915b: 204 (*Pseudogeron*).—N. Mex.
bicolor Melander, 1946b: 473.—Calif.
bifarius Melander, 1946b: 473.—Calif.
bilineatus Melander, 1946b: 474.—Calif.
bivittatus (Cresson), 1915b: 206 (*Pseudogeron*).—Tex.
capax (Coquillett), 1892c: 126 (*Geron*).—Calif.
chalybeus Melander, 1946b: 474.—Oreg.
cincturus (Coquillett), 1894a: 111 (*Geron*).—Calif.; Oreg., Utah.
cinereus Melander, 1946b: 475.—Calif.
cockerelli Melander, 1946b: 476.—N. Mex.
colei Melander, 1946b: 476.—Calif.
comosus Melander, 1946b: 477.—Calif.
dissimilis Melander, 1946b: 478.—Calif.
distinctus Melander, 1946b: 478.—Calif.
divisus Melander, 1946b: 479.—Calif.
dolorosus Melander, 1946b: 480.—Calif.
eremitis Melander, 1946b: 480.—Calif.
fasciolus (Coquillett), 1892c: 125 (*Geron*).—Calif.; Wyo.

formosus (Cresson), 1915b: 205 (*Pseudogeron*).—N. Mex.
instabilis Melander, 1946b: 481.—Calif.
knabi (Cresson), 1915b: 206 (*Pseudogeron*).—N. Mex.; Utah.
lasius Melander, 1946b: 482.—Calif.
longirostris Melander, 1946b: 483.—Calif.
loricatus Melander, 1946b: 483.—Calif.
lugens Melander, 1946b: 484.—Calif.
maculatus Melander, 1946b: 484.—Calif.
marginalis (Cresson), 1915b: 207 (*Pseudogeron*).—N. Mex.
mitis (Cresson), 1915b: 205 (*Pseudogeron*).—N. Mex.; Utah.
montanus Melander, 1946b: 485.—Calif.
mus (Bigot), 1886b: civ (*Rhabdopselaphus*).—Calif.
neuter Melander, 1946b: 486.—Calif.
obscurus (Cresson), 1915b: 204 (*Pseudogeron*).—N. Mex.; Oreg., Utah, Tex.
palpalis Melander, 1946b: 486.—Wash.; Calif., Mont.
panneus Melander, 1946b: 487.—Calif.
parkeri Melander, 1946b: 487.—Ariz.
polius Melander, 1946b: 488.—Calif.; Ariz.
pulcher Melander, 1946b: 489.—Calif.
pullatus Melander, 1946b: 490.—Calif.
quinquenotatus (Johnson), 1903e: 185 (*Phthiria*).—Colo.; Idaho.
retrorsus Melander, 1946b: 490.—Calif.
scapularis Melander, 1946b: 491.—Calif.
scapulatus Melander, 1946b: 492.—Calif.
setosus (Cresson), 1915b: 203 (*Pseudogeron*).—Calif.; N. Mex.
sigma (Coquillett), 1902g: 101 (*Geron*).—Colo., N.C., Ala.; Oreg., Utah, Calif., N.Y.
sipho Melander, 1946b: 492.—Calif.
speculifer Melander, 1946b: 493.—Calif.
togatus Melander, 1946b: 494.—Calif.
trifidus Melander, 1946b: 495.—Calif.
trochilus (Coquillett), 1894a: 111 (*Geron*).—Calif.; Ariz.

Genus GERON Meigen

Geron Meigen, 1820: 223. Type-species, *gibbosus* Meigen (Rondani, 1856: 165) =*gibbosus* (Olivier).

The larvae are parasites in various Pyralidae in India and Europe.

REFERENCE: Painter, 1932a (monog.).

Subgenus GERON Meigen

albarius Painter, 1932a: 159.—Tex.; Wyo., Colo., Kans., Ariz.
albidipennis Loew, 1869c: 174 (Cent. 9, no. 78).—Calif.

arenicola Painter, 1932a: 160.—Kans.; Minn., Nebr., N.Y., Mass.
argutus Painter, 1932a: 158.—Kans.; Oreg. to Wyo. and Kans., s. to Calif. and Tex.
digitarius Cresson, 1919: 184.—N. Mex.; Ariz., Colo.
grandis Painter, 1932a: 150.—Tex.; Ariz., N. Mex.
holosericeus Walker, 1849: 295.—Ga.; Kans. to N.J., s. to Ariz. and Ga.
    versicolor Painter, 1932a: 156.—Kans.
johnsoni Painter, 1932a: 155.—Mass.; N.Y.
litoralis Painter, 1932a: 146.—Honduras; Tex.
nigripes Painter, 1932a: 152.—Calif.; Oreg.
niveus Cresson, 1919: 185.—Calif.; N. Mex.
parvidus Painter, 1932a: 155.—Calif.
robustus Cresson, 1919: 185 (as *digitarius* var.).—Md.
subauratus Loew, 1863b: 304 (Cent. 4, no. 55).—Pa.; Ohio to Mass., s. to N.C.
vitripennis Loew, 1869c: 173 (Cent. 9, no. 77).—Middle States; Ill. to N.J., s. to Tex. and Ga.
winburni Painter, 1932a: 157.—Calif.

Subgenus EMPIDIGERON Painter

Geron, subg. **Empidigeron** Painter, 1932a: 143. Type-species, *snowi* Painter (orig. des.).

aequalis Painter, 1932a: 162.—Calif.
aridus Painter, 1932a: 163.—Ariz.; Colo., N. Mex.
calvus Loew, 1863b: 303 (Cent. 4, no. 54).—N.Y.; N.H., Mass., Pa., N.J.
    macropterus Loew, 1869c: 172 (Cent. 9, no. 76).—N.Y.
hybus Coquillett, 1894a: 112.—Calif.
nudus Painter, 1932a: 163.—N. Mex.; Colo.
snowi Painter, 1932a: 164.—N. Mex.; Ariz., Colo.

Subfamily HETEROTROPINAE

REFERENCE: Melander, 1950b (key).

Genus HETEROTROPUS Loew

Heterotropus Loew, 1873c: 182. Type-species, *albidipennis* Loew (mon.).

senex Melander, 1950b: 150.—Ariz.

Genus CAENOTUS Cole

Caenotus Cole, 1923d: 14. Type-species, *inornatus* Cole (orig. des.).
REFERENCE: Melander, 1950b (key).
canus Melander, 1950b: 149.—Calif.
hospes Melander, 1950b: 149.—Ariz.
inornatus Cole, 1923d: 16.—N. Mex.
minutus Cole, 1923d: 15.—N. Mex.

Genus PRORATES Melander

Prorates Melander, 1906: 372. Type-species, *claripennis* Melander (mon.).
claripennis Melander, 1906: 373.—N. Mex.; Calif.

Genus APYSTOMYIA Melander

Apystomyia Melander, 1950b: 146. Type-species, *elinguis* Melander (orig. des.).
elinguis Melander, 1950b: 147.—Calif.

Subfamily SYSTROPODINAE

REFERENCE: Enderlein, 1926b (key).

Genus SYSTROPUS Wiedemann

Systropus Wiedemann, 1820b: 18. Type-species, *macilentus* Wiedemann (mon.).
Cephenus Berthold, 1827: 506 (unjustified n. name for *Systropus* Wiedemann). Type-species, *Systropus macilentus* Wiedemann (aut.).

The larvae are parasitic on Eucleidae (Limacodidae), Lepidoptera.

REFERENCE: Townsend, 1901 (key).
ammophiloides Townsend, 1901: 159.—N. Mex.; Ariz.
angulatus (Karsch), 1880: 657 (*Cephenus*).—Tex.; Va., N.C., S.C., Ga., Ala.
arizonicus Banks, 1909: 18.—Ariz.
macer Loew, 1863b: 305 (Cent. 4, no. 56).—Wis., Pa.; Wis. to Mass., s. to Tex. and Ga.
  infuscatus Karsch, 1880: 657 (*Cephenus*).—Tex.
  imbecillus Karsch, 1880: 658 (*Cephenus*).—Ga.

Genus DOLICHOMYIA Wiedemann

Dolichomyia Wiedemann, 1830: 642. Type-species, *nigra* Wiedemann (mon.).

gracilis Williston, 1894: 41.—Colo.; Ariz.

Subfamily TOXOPHORINAE

REFERENCE: d'Andretta and Carrera, 1950 (key).

Genus TOXOPHORA Meigen

Toxophora Meigen, 1803: 270. Type-species, *Asilus maculatus* Rossi (sub. mon., Meigen, 1804: 273).

The larvae are parasitic on Aculeate Hymenoptera, especially *Eumenes* and *Odynerus*.

REFERENCE: Coquillett, 1891b (key).

americana Guérin-Méneville, 1835: pl. 95, fig. 1.—N. Amer. Unrecognized.

amphitea Walker, 1849: 298.—Fla.; Mo. to Md., s. to Tex. and Fla.

leucopyga Wiedemann, 1828: 361.—Unknown; La., Ala., Fla., Ga., N.C.

 fulva Gray *in* Griffith and Pidgeon, 1832: 779.—Unknown.

maxima Coquillett, 1886c: 222.—Calif.; Oreg., Ariz., N. Mex., Tex., Okla., Kans.

pellucida Coquillett, 1886c: 222.—Calif.; Oreg. to Alta., s. to Calif., Tex., and Kans.

vasta Coquillett, 1891b: 199.—Calif.; Ariz., Colo.

virgata Osten Sacken, 1877: 266.—Tex., Ga.; Calif., Ariz., N. Mex., Utah, Colo., Okla.

Genus LEPIDOPHORA Westwood

Lepidophora Westwood, 1835a: 447. Type-species, *Ploas aegeriiformis* Gray (mon.)=*lepidocera* (Wiedemann).

REFERENCES: Painter, 1925, Paramonov, 1949 (revs.).

lepidocera (Wiedemann), 1828: 360 (*Toxophora*).—Unknown; Iowa, s. to Tex. and La., Ohio, S.C. to Fla.

 aegeriiformis Gray *in* Griffith and Pidgeon, 1832: 779, pl. 126, fig. 6 (*Ploas*; as *aegeriformis*, p. 779).—Ga. (Westwood, 1835a: 447).

 appendiculata Macquart, 1846: 246 (1846: 118) (*Toxophora*).—Tex.

lutea Painter *in* Painter and Painter, 1962: 51.—N.J.; Minn. to Ont. and Maine, s. to Mo., La., and Fla.

lepidocera, authors, not Wiedemann.

vetusta Walker, 1857: 145.—Mexico (as Valley of the Amazon, error; Painter and Painter, 1962: 52); Tex., Mexico, Cent. Amer.

Subfamily CYLLENIINAE

REFERENCE: Hall, 1957 (partial key).

Genus ECLIMUS Loew

Eclimus Loew, 1844a: 154. Type-species, *perspicillaris* Loew (Coquillett, 1910b: 536).

Thevenemyia Bigot, 1875b: clxxiv. Type-species, *californica* Bigot (mon.).

Thevenetimyia, emend.

Epibates Osten Sacken, 1877: 268. Type-species, *funestus* Osten Sacken (Coquillett, 1910b: 538).

REFERENCE: Osten Sacken, 1877 (key).

californicus (Bigot), 1875b: clxxv (*Thevenemyia*).—Calif.

celer Cole *in* Cole and Lovett, 1919: 224.—Oreg.; Wash., Calif.

funestus (Osten Sacken), 1877: 271 (*Epibates*).—N.H.; Mont., Utah, Que., Tex., Ill.

harrisi (Osten Sacken), 1877: 273 (*Epibates*).—Unknown (probably n. U.S.); N.H.

laniger Cresson, 1919: 182.—Calif.; Wash., Utah.

leechi Hall, 1954b: 147.—Calif.

lotus Williston, 1893c: 66.—Calif.; Wash., Oreg.

auratus Williston, 1893c: 66.—Wash.

luctifer (Osten Sacken), 1877: 271 (*Epibates*).—B.C.; Wash., Oreg., Calif.

lucifer, error.

auripilus Bigot, 1892a: 372 (*Amictus*; preocc. Osten Sacken, 1887).—Wash. Territory.

magnus (Osten Sacken), 1877: 272 (*Epibates*).—B.C.; B.C. to Mont., s. to Calif. and Utah.

melanopogon Bigot, 1892a: 370 (*Thevenetimyia*).—Wash. Territory.

marginatus (Osten Sacken), 1877: 272 (*Epibates*).—Calif.; Oreg.

melanosus Williston, 1893c: 65.—Calif.

muricatus (Osten Sacken), 1877: 272 (*Epibates*).—Calif.; Oreg., Idaho, Utah, Colo.
niger (Macquart), 1834a: 390 (*Apatomyza*).—Ga.; Miss.
  aegiale Walker, 1849: 296 (*Cyllenia*).—Ga.
ostensackenii (Burgess), 1878: 323 (*Epibates*).—Colo.; Calif., Ariz., Utah.
sodalis Williston, 1893c: 65.—Wash.; Oreg., Calif., Ariz., Wyo.
yosemite Cresson, 1919: 183.—Calif.; Oreg.

Genus AMPHICOSMUS Coquillett

Amphicosmus Coquillett, 1891c: 219. Type-species, *elegans* Coquillett (mon.).

REFERENCE: Hall, 1957 (key).

arizonensis Johnson and Johnson, 1960: 67.—Ariz.
elegans Coquillett, 1891c: 220.—Calif.
vanduzeei Cole, 1923b: 22.—Calif.

Genus METACOSMUS Coquillett

Metacosmus Coquillett, 1891c: 220. Type-species, *exilis* Coquillet (mon.).

REFERENCE: Melander, 1950b (key).

exilis Coquillett, 1891c: 221.—Calif.
mancipennis Coquillett, 1910c: 41.—Pa.; Mass., N.J.
nitidus Cole, 1923b: 23.—Calif.

Genus PARACOSMUS Osten Sacken

Allocotus Loew, 1872a: 82 (Cent. 10, no. 48) (preocc. Mayr, 1864). Type-species, *edwardsii* Loew (mon.).
Paracosmus Osten Sacken, 1877: 262 (n. name for *Allocotus* Loew). Type-species, *Allocotus edwardsii* Loew (aut.).

REFERENCE: Melander, 1950b (key).

edwardsii (Loew), 1872a: 81 (Cent. 10, no. 48) (*Allocotus*).—Calif.; Utah.
insolens Coquillett, 1891c: 221.—Calif.
morrisoni Osten Sacken, 1887a: 155.—Mexico; Calif., Ariz., Tex.
rubicundus Melander, 1950b: 155.—Mexico; Calif., Ariz.
similis Hall, 1957: 143.—Calif.

BOMBYLIIDAE TOMOPHTHALMAE

Subfamily LOMATIINAE

REFERENCES: Röder, 1886, Melander, 1950a (keys).

Genus OGCODOCERA Macquart

Ogcodocera Macquart, 1840: 83 (1841: 361). Type-species, *dimidiata* Macquart (mon.)=*leucoprocta* (Wiedemann).
Oncodocera, emend.
Anisotamia, authors, not Macquart.

REFERENCES: Röder, 1886, Paramonov, 1930 (keys).

analis Williston, 1901: 283 (*Oncodocera*).—Mexico; s. Ariz.
leucoprocta (Wiedemann), 1828: 330 (*Mulio*).—Unknown; Iowa and Mich. to Que. and Mass., s. to Kans., Ark., and Ga.
    dimidiata Macquart, 1840: 84 (1841: 362).—N. Amer.
    leucotelus Walker, 1852: 175 (*Anthrax*).—Unknown.
valida (Wiedemann), 1830: 636 (*Anthrax*).—Mexico; Ariz.
    eximia Macquart, 1850: 419 (1850: 115) (*Anisotamia*).—Mexico.

Genus EPACMUS Osten Sacken

Leptochilus Loew, 1872a: 78 (Cent. 10, no. 40) (preocc. Saussure, 1852). Type-species, *modestus* Loew (mon.).
Epacmus Osten Sacken, 1887a: 142 (n. name for *Leptochilus* Loew). Type-species, *Leptochilus modestus* Loew (aut.).

REFERENCE: Melander, 1950a (key).

cirratus Melander, 1950a: 6.—N. Mex.
clunalis Melander, 1950a: 8.—Wash.
connectens Melander, 1950a: 10.—Ariz.; Calif.
labiosus Melander, 1950a: 11.—Utah.
litus (Coquillett), 1886a: 84 (*Aphoebantus*).—Calif.; Ariz.
modestus (Loew), 1872a: 77 (Cent. 10, no. 40) (*Leptochilus*).—Tex.; N. Mex.
morsicans Melander, 1950a: 11.—Ariz.
nebritus Coquillett, 1894a: 104.—Calif.; Ariz.
nitidus Cole *in* Cole and Lovett, 1921: 249.—Oreg.
pallidus Cresson, 1919: 181.—Tex.; Utah.
ponderosus Melander, 1950a: 13.—Ariz.
pulvereus Melander, 1950a: 13.—Ariz.
tomentosus Melander, 1950a: 14.—Ariz.; Calif.

Genus APHOEBANTUS Loew

Aphoebantus Loew, 1872a: 77 (Cent. 10, no. 39). Type-species, *cervinus* Loew (mon.).
Triodites Osten Sacken, 1877: 245. Type-species, *mus* Osten Sacken (mon.).
Predators on egg pods of Acrididae; parasites of *Tiphia* in India.
REFERENCES: Coquillett, 1894a, Melander, 1950a (keys).

abnormis Coquillett, 1891d: 14 (1891: 262).—Calif.
    inermis, error.
altercinctus Melander, 1950a: 21.—Ariz.
arenicola Melander, 1950a: 22.—Tex.; Ariz., N. Mex.
balteatus Melander, 1950a: 23.—Ariz.
bisulcus Osten Sacken, 1887a: 148.—Mexico; Ariz.
borealis Cole *in* Cole and Lovett, 1921: 251. Oreg.; Calif.
brevistylus Coquillett, 1891d: 16 (1891: 264).—Calif.
capax Coquillett, 1891d: 13(1891: 261).—Calif.
carbonarius Osten Sacken, 1887a: 149.—Mexico; Wash., Calif., Ariz., Kans.
catenarius Melander, 1950a: 24.—Ariz.
catulus Coquillett, 1894a: 107.—Calif.
cervinus Loew, 1872a: 76 (Cent. 10, no. 39).—Tex.; Kans., Okla.
concinnus (Coquillett), 1892b: 10 (*Epacmus*).—Calif.
contiguus contiguus Melander, 1950a: 25.—Utah.
    ssp. **separatus** Melander, 1950a: 25 (as var.).—Ariz.; Calif.
conurus Osten Sacken, 1887a: 148.—Calif.; Oreg., Ariz.
cyclops Osten Sacken, 1887a: 146.—Mexico; Ariz.
denudatus Melander, 1950a: 27.—Ariz.
desertus Coquillett, 1891d: 13 (1891: 261).—Calif.
eremicola Melander, 1950a: 28.—Calif.; Ariz.
fumidus Coquillett, 1891d: 15 (1891: 263).—Calif.
fumosus (Coquillett), 1892b: 11 (*Epacmus*).—Calif.; Ariz.
    fucatus Coquillett, 1894a: 108.—Calif.
gluteatus Melander, 1950a: 29.—Ariz.
halteratus Melander, 1950a: 30.—Calif.
hians Melander, 1950a: 30.—Ariz.
hirsutus Coquillett, 1886a: 85.—Calif.; Oreg., Ariz.
interruptus Coquillett, 1891d: 11 (1891: 259).—Calif.; Oreg., Tex.
inversus Melander, 1950a: 32.—Utah.
leviculus Coquillett, 1894a: 107.—Calif.; Wyo., Colo., N. Mex., Tex.
maculatus Melander, 1950a: 33.—Ariz.
marcidus Coquillett, 1891d: 10 (1891: 258).—Calif.
    squamosus (in part) Coquillett, 1891d: 15 (1891: 263). Male only.

marginatus Cole, 1923a: 310.—Mexico (Baja Calif.); Ariz., Tex.
micropyga Melander, 1950a: 34.—Ariz.; Mexico.
mixtus Coquillett, 1891d: 11 (1891: 259).—Calif.
mormon Melander, 1950a: 35.—Utah; Ariz.
mus mus (Osten Sacken), 1877: 246 (*Triodites*).—Utah; Calif.
    ssp. **barbatus** Melander, 1950a: 36 (as var.).—Ariz.
obtectus Melander, 1950a: 36.—Mexico; Ariz., Calif.
parkeri Melander, 1950a: 37.—Ariz.; Utah.
pavidus Coquillett, 1886a: 87 (as *cervinus* var.).—Calif.
pellucidus (Coquillett), 1892b: 10 (*Epacmus*).—Calif.
peodes Osten Sacken, 1887a: 149.—Mexico; Oreg., Calif., Ariz.
rattus Osten Sacken, 1887a: 147.—Tex.
scalaris Melander, 1950a: 39.—Calif.; Ariz.
schlingeri Hall, 1957: 147.—Calif.
scriptus Coquillett, 1891d: 12 (1891: 260).—Calif. (unpublished).
sperryorum Melander, 1950a: 40.—Calif.
squamosus Coquillett, 1891d: 15 (1891: 263).—Calif. Female only; male=*marcidus*.
tardus Coquillett, 1891d: 10 (1891: 258).—Calif.
timberlakei Melander, 1950a: 42.—Calif.
transitus (Coquillett), 1886a: 83 (*Leptochilus*).—Calif.
ursula Melander, 1950a: 42.—Calif.
varius Coquillett, 1891d: 8 (1891: 256).—Calif.
vasatus Melander, 1950a: 44.—Calif.
vittatus Coquillett, 1886a: 86.—Calif.
vulpecula Coquillett, 1894a: 107.—Calif.; Ariz., Utah.

Genus DESMATONEURA Williston

Desmatoneura Williston, 1895c: 267. Type-species, *argentifrons* Williston (mon.).

argentifrons Williston, 1895c: 267.—N. Mex.; Calif., Utah, Ariz., Tex., Kans.

Genus EXEPACMUS Coquillett

Exepacmus Coquillett, 1894a: 101. Type-species, *johnsoni* Coquillett (mon.).

johnsoni Coquillett, 1894a: 101.—Calif.
    nasalis Melander, 1950b: 152.—Calif.

Genus EUCESSIA Coquillett

Eucessia Coquillett, 1886a: 82. Type-species, *rubens* Coquillett (mon.).

rubens Coquillett, 1886a: 82.—Calif.

Subfamily ANTHRACINAE

Genus ANTHRAX Scopoli

Anthrax Scopoli, 1763: 358. Type-species, *Musca morio* Linnaeus (mon.; misident.)=*anthrax* (Schrank). See Aldrich, 1926e: 12.

Spogostylum Macquart, 1840: 53 (1841: 331). Type-species, *mystaceum* Macquart (mon.).

Spongostylum, emend.

Argyromoeba Schiner, 1860b: 51. Type-species, *Anthrax tripunctata* Wiedemann (Coquillett, 1910b: 510).

Argyramoeba, emend.

The larvae are parasites of Aculeate Hymenoptera and *Cicindela*, or predators on egg pods of Acrididae (in Palaearctic Region.)

REFERENCE: Curran, 1927f (key).

albofasciatus Macquart, 1840: 67 (1841: 345) (n. name for *analis* Macquart).—Ga.; Alta., Que., Tex., Fla.
    analis Macquart, 1834a: 407 (preocc. Say, 1823).—Ga.

albosparsus (Bigot), 1892a: 348 (*Argyromoeba*).—Colo.

analis Say, 1823: 45 (1859b: 60).—Ga.; Mont., Ont., Ariz., Fla.

antecedens Walker, 1852: 193.—U.S.

argentatus (Cole) *in* Cole and Lovett, 1919: 227 (*Spogostylum*).—Oreg.; Wash., Idaho, Calif., Utah.

argyropygus Wiedemann, 1828: 313, 591 (as *argyropya*, p. 313).—Unknown; Ont., Va., Tenn., Fla., Ariz.
    contigua Loew, 1869b: 30 (Cent. 8, no. 50) (*Argyromoeba*).—Va.

aterrimus (Bigot), 1892a: 349 (*Argyromoeba*).—Md. (unpublished); Ind., Ky., La., Va., S.C., Fla.
    slossonae Johnson, 1913a: 55 (*Spogostylum*).—Ky. **N. syn.**

cedens Walker, 1852: 190.—U.S.

cephus Fabricius.—Not Nearctic.

cintalapa Cole, 1957: 200.—Mexico; Calif., Utah, Nebr., Tex.

cybele (Coquillett), 1894a: 96 (*Argyromoeba*).—Calif.; Ariz., Tex., Okla.

daphne (Osten Sacken), 1886a: 104 (*Argyromoeba*).—Mexico; Oreg., Calif., N. Mex., Colo.

delila (Loew), 1869b: 28 (Cent. 8, no. 45) (*Argyromoeba*).—Calif.
fur (Osten Sacken), 1877: 244 (*Argyromoeba*).—Tex.; Nebr., Kans., Okla., Ark.
georgicus Macquart, 1834a: 406.—Ga.
grossbecki (Johnson), 1913a: 55 (*Spogostylum*).—Fla.; Ga.
irroratus Say, 1823: 46 (1859b: 61).—Rocky Mts.; Alaska to Que., s. to Calif. and Fla.
    oedipus, authors, not Fabricius.
latelimbatus (Bigot), 1892a: 351 (*Hemipenthes*).—Carolina.
limatulus Say, 1829: 157 (1859b: 354).—Ind.; Del., Mo., Ark., Tex., Fla.
    obsoleta Loew, 1869b: 29 (Cent. 8, no. 47) (*Argyromoeba*).—Mo.
melanopogon (Bigot), 1892a: 348 (*Argyromoeba*).—Wash. (as N. Amer.).
nidicola Cole, 1952: 126.—Calif.
occidentalis (Johnson), 1913a: 56 (*Spogostylum*).—Wash., Colo.
pauper (Loew), 1869b: 29 (Cent. 8, no. 48) (*Argyromoeba*).—Ill.; Nebr., Ont., Tex., Va.
plesius Curran, 1927f: 84.—B.C.; Calif., Wyo., Colo.
pluricellus Williston, 1901: 277.—Mexico; Ariz.
    pleuricellus, error.
    capucina Fabricius of Rau, 1940: 594.
pluto Wiedemann, 1828: 261.—Ky.
seriepunctatus (Osten Sacken), 1886a: 103 (*Argyromoeba*).—Mexico; Ariz.
stellans (Loew), 1869b: 28 (Cent. 8, no. 46) (*Argyromoeba*).—Oreg.
tigrinus (De Geer), 1776: 206 (*Nemotelus*).—Pa.; Calif. to Ont., s. to Kans. and Fla., Neotropical.
    simson Fabricius, 1805: 119.—S. Amer.
    scriptus Say, 1823: 43 (1859b: 59).—Pa.
varicolor (Bigot), 1892a: 347 (*Argyromoeba*).—N. Amer. (as Columbia).
varius Fabricius.—Not Nearctic.
viereki (Cresson), 1919: 180 (*Spogostylum*).—N. Mex.; Calif., Ariz., Tex.

Genus DICRANOCLISTA Bezzi

Coquillettia Williston, 1896a: 65 (preocc. Uhler, 1890). Type-species, *Spogostylum vandykei* Coquillett (sub. mon., Aldrich, 1905: 221).

Dicranoclista Bezzi, 1924b: 178. Type-species, *simpsoni* Bezzi (orig. des.).

fasciata Johnson and Johnson, 1960: 70.—Ariz.; Utah.
vandykei (Coquillett), 1894a: 94 (*Spogostylum*).—Calif.; Wash., Oreg., Utah, Tex.

Subfamily EXOPROSOPINAE

REFERENCE: Painter and Hall, 1960 (key).

Genus VILLA Lioy

Villa Lioy, 1864a: 732. Type-species, *Anthrax concinnus* Meigen (Coquillett, 1910b: 619) = *abbadon* (Fabricius).
Anthrax, group *Hyalanthrax* Osten Sacken, 1887a: 134. Type-species, *faustinus* Osten Sacken (Coquillett, 1910b: 553).
Anthrax, authors, not Scopoli.

The larvae are external or internal parasites of various Diptera, Lepidoptera, and Hymenoptera, especially Ichneumonidae, Tiphiidae, Tenthredinidae, Eumenidae, *Bembex, Anthophora, Megachile,* and *Andrena.* Also parasites of Noctuidae and Tenebrionidae and predators in egg pods of grasshoppers.

Species have been assigned to subgenera wherever possible; those left unplaced include (1) species intermediate between two subgenera, (2) species unknown to us, and (3) species that may require new subgenera for their reception. The latter should be named only after thorough restudy of most of the species in this large group. Some of these subgenera have been considered genera by some authors, especially for Palaearctic and Ethiopian species (Bezzi, 1924b; Hesse, 1938–1956; Engel, 1933–1937).

REFERENCES: Coquillett, 1887a, 1892a, 1894a (keys to spp.); Painter and Hall, 1960 (key to subg.).

Subgenus VILLA Lioy

adusta (Loew), 1869b: 26 (Cent. 8, no. 41) (*Anthrax*).—Cuba; Fla.
aenea (Coquillett), 1887a: 165 (*Anthrax*).—Calif.; Ariz., N. Mex., Utah, Tex., Kans.
agrippina (Osten Sacken), 1887a: 139 (*Anthrax*).—Mexico; Wash., Calif., Utah, Colo., Ariz., N. Mex.
alternata alternata (Say), 1823: 45 (1859b: 61) (*Anthrax*).—Mo., Pa.
  bastardi Macquart, 1840: 60 (1841: 338) (*Anthrax*).—N. Amer.
  consanguinea Macquart, 1840: 69 (1841: 347) (*Anthrax*).—Pa.
  albipectus Macquart, 1848a: 194 (1848: 34) (*Anthrax*).—N. Amer.

ssp. **nigropecta** Cresson, 1916c: 443 (as var.).—N. Mex.; Utah, Kans., Minn.
chromolepida Cole, 1923b: 21.—Calif.
compressa Painter, 1926b: 211.—Tex.
connexa (Macquart), 1855: 96 (1855: 76) (*Anthrax*).—Md.
consessor (Coquillett), 1887a: 165 (*Anthrax*).—Calif.; B.C., Utah, Colo., Kans., Ariz., N. Mex.
faustina (Osten Sacken), 1887a: 136 (*Anthrax*).—Mexico; Oreg., Calif., Colo., Ariz.
flavocostalis Painter, 1926b: 210.—Tex.; Colo.
fulviana fulviana (Say), 1824a: 372 (1859a: 253) (*Anthrax*).— Northwest Territory (U.S.); Wash. to N.S., s. to Calif., N. Mex., and Iowa.
  ssp. **nigricauda** (Loew), 1869b: 24 (Cent. 8, no. 38) (*Anthrax;* as sp.).—Mass.; Minn. to Maine, also Colo.
fumicosta Painter *in* Painter and Painter, 1962: 106.—Fla.; Ohio, Tex., Ga., Fla., Mexico.
  fauna, authors, not Fabricius.
gracilis (Macquart), 1840: 76 (1841: 354) (*Anthrax*).—Pa. Unrecognized.
handfordi Curran, 1935b: 1.—Man.
harveyi (Hine), 1904b: 88 (*Anthrax*).—B.C.; Wash., Wyo., Colo.
hypomelas (Macquart), 1840: 76 (1841: 354) (*Anthrax*).—N. Amer.; Alta. to Maine, s. to S. Dak. and N.C.
lateralis lateralis (Say), 1823: 42 (1859b: 59) (*Anthrax*).—Md., Pa.; B.C. to Maine, s. to Calif. and Fla., Mexico.
  ssp. **arenicola** (Johnson), 1908a: 15 (*Anthrax*).—N.J., Mass.; Wash., Nev., Utah, Wyo.
  ssp. **atra** Painter, 1926b: 209 (as var.).—N.J.; Minn. to Ont., s. to N. Mex. and Ga., also Oreg.
  ssp. **johnsoni** Painter, 1926b: 207 (as var.).—N.Y., Pa., N.J., N.C.; N.Y. to N.C.
    gracilis Macquart of Johnson, 1908a: 15 (*Anthrax*).
  ssp. **nigra** Cresson, 1916c: 442 (as var.).—N.Y.; Minn. to Mass., s. to Colo. and Ga.
  ssp. **semifulvipes** Painter *in* Painter and Painter, 1962: 108 (n. name for *fulvipes* Coquillett).—Ariz.; Wash. to Mich., s. to Calif., Tex., and Va.
    fulvipes Coquillett, 1887a: 166 (*Anthrax;* as *alternata* var.; preocc. Loew, 1860).—Ariz.
livia (Osten Sacken), 1887a: 139 (*Anthrax*).—Mexico; Ariz., Tex.
molitor (Loew), 1869b: 26 (Cent. 8, no. 42) (*Anthrax*).—Calif.
moneta (Osten Sacken), 1887a: 138 (*Anthrax*).—Mexico (n. Son.); s. Ariz.

mucorea (Loew), 1869b: 27 (Cent. 8, no. 43) (*Anthrax*).—Nebr.; Alta., Wyo., Colo., Ariz.
muscaria (Coquillett), 1892a: 178 (*Anthrax*).—Calif.; Alta., Colo., Kans.
nebulo (Coquillett), 1887a: 165 (*Anthrax*).—Wash. Territory.
pretiosa (Coquillett), 1887a: 168 (*Anthrax*; as *molitor* var.).—Calif.; Alta., Utah.
sabina (Osten Sacken), 1887a: 137 (*Anthrax*).—Mexico (n. Son.); s. Ariz.
salebrosa Painter, 1926b: 211.—Tex.; Wash., Alta., Colo., Iowa.
scrobiculata (Loew), 1869b: 24 (Cent. 8, no. 39) (*Anthrax*).—Ill.
shawii (Johnson), 1908a: 14 (*Anthrax*).—N.H.; Mass., N.Y., N.J., Del.
squamigera (Coquillett), 1892a: 181 (*Anthrax*).—Calif.
stenozona (Loew), 1869b: 25 (Cent. 8, no. 40) (*Anthrax*).—Ill.
supina (Coquillett), 1887a: 169 (*Anthrax*).—Calif.; Ariz., Utah.
vacans (Coquillett), 1887a: 168 (*Anthrax*; as *molitor* var.).—Wash. Territory.
vestita (Walker), 1849: 258 (*Anthrax*).—N.S.

Subgenus HEMIPENTHES Loew

Hemipenthes Loew, 1869b: 28 (Cent. 8, no. 44) (as genus). Type-species, *Musca morio* Linnaeus (Coquillett, 1910b: 550).
Isopenthes Osten Sacken, 1886a: 96. Type-species, *jaennickeana* Osten Sacken (Coquillett, 1910b: 556)=*sinuosa* (Wiedemann) ssp.

REFERENCE: Coquillett, 1894a (key); Painter and Painter, 1962 (key, *celer* group).

bigradata (Loew), 1869b: 23 (Cent. 8, no. 37) (*Anthrax*).—Cuba; Calif., N. Mex.
castanipes (Bigot), 1892a: 350 (*Hemipenthes*).—N. Amer.
catulina (Coquillett), 1894a: 100 (*Anthrax*).—Wash., Calif.; Wash. to Mont., Calif. to N. Mex., Minn. and Ill. to N.Y. and N.J.
celer (Wiedemann), 1828: 310 (*Anthrax*).—Ky.
chimaera (Osten Sacken), 1887a: 131 (*Anthrax*).—Mexico; s. Ariz.
comanche Painter *in* Painter and Painter, 1962: 96.—N. Mex.; Calif. to Nebr. and N. Mex.
    chimaera, authors, not Osten Sacken.
curta (Loew), 1869b: 22 (Cent. 8, no. 35) (*Anthrax*).—Calif.; Ariz., N. Mex., Tex., Mexico.
edwardsii (Coquillett), 1894f: 102 (*Anthrax*).—B.C., Calif.
eumenes (Osten Sacken), 1887a: 131 (*Anthrax*).—Mexico; Wash. to Alta. and Mont., s. to Calif. and N. Mex.

floridana (Macquart), 1850: 416 (1850: 112) *(Anthrax).*—Fla.; Ga.
incisiva Painter *in* Painter and Painter, 1962: 112 (n. name for *incisa* Walker).—N. Amer.; Mexico. Possibly not Nearctic.
    incisa Walker, 1852: 187 *(Anthrax;* preocc. Macquart, 1847).—N. Amer.
inops (Coquillett), 1887a: 169 *(Anthrax).*—Calif.; Wash., Oreg., Utah, Ariz., Colo.
lepidota (Osten Sacken), 1887a: 130 *(Anthrax).*—Mexico; Alta. to Calif., Mexico, and La.
mobile (Coquillett), 1894a: 100 *(Anthrax).*—Calif.
morio (Linnaeus), 1758: 590 *(Musca).*—Europe; Wash. to Maine, s. to Calif., N. Mex., and Mo., Holarctic.
    morioides Say, 1823: 42 (1859b: 58) *(Anthrax).*—Mo.
pima Painter *in* Painter and Painter, 1962: 92.—Ariz.; N. Mex., Tex., Mexico.
pullata (Coquillett), 1894a: 98 *(Anthrax).*—Calif.; Ariz.
sagata (Loew), 1869b: 21 (Cent. 8, no. 34) *(Anthrax).*—Mexico; Ariz., Tex., Mo., Miss.
    orbitalis Williston, 1901: 281 *(Anthrax).*—Mexico.
scylla (Osten Sacken), 1887a: 132 *(Anthrax).*—Mexico; Ariz., Tex.
    succincta Coquillett, 1894a: 96 *(Argyromoeba).*—Ariz. **N. syn.**
seminigra (Loew), 1869b: 27 (Cent. 8, no. 44) *(Hemipenthes).*—Sask.; Utah.
sinuosa sinuosa (Wiedemann), 1821c: 244 *(Anthrax).*—Ga.; S. Dak. to Vt., s. to Tex. and Ga.
    nycthemerus Macquart, 1840: 67 (1841: 345) *(Anthrax).*—Ga.
    concisa Macquart, 1840: 68 (1841: 346) *(Anthrax).*—Carolina.
    assimilis Macquart, 1846: 242 (1846: 114) *(Anthrax).*—Tex.
    gideon, authors, not Fabricius.
    ssp. **blanchardiana** (Jaennicke), 1867: 341 (1868: 33) *(Exoprosopa;* as sp.).—Mexico; Calif., Ariz., Tex.
    ssp. **jaennickeana** (Osten Sacken), 1886a: 97 *(Isopenthes;* as sp.).—Mexico; Oreg. to Mont., s. to Calif., Mexico, and w. Tex.
webberi Johnson, 1919b: 11.—Mass.; Wyo., Ky., Ont., Que., Vt., Conn.
wilcoxi Painter, 1933: 15.—Wash.; Calif.
yaqui Painter *in* Painter and Painter, 1962: 94.—Ariz.; Mexico (Son.).

Subgenus THYRIDANTHRAX Osten Sacken

Anthrax, group **Thyridanthrax** Osten Sacken, 1886a: 123. Type-species, *selene* Osten Sacken (Coquillett, 1910b: 615).

The larvae are predators on egg pods of Acrididae in Palaearctic Region.

atrata (Coquillett), 1887a: 171 (*Anthrax*).—Calif.; Wash., Oreg., Nev.
bifenestrata (Bigot), 1892a: 356 (*Anthrax*).—Calif. ?=*nugator*.
fenestratoides (Coquillett), 1892a: 185 (*Anthrax*).—Calif.; Oreg. to Kans., s. to Calif., also Mass.
  macula Cole *in* Cole and Lovett, 1919: 226 (*Anthrax*).— Oreg. N. syn.
melasoma (Wulp), 1882a: 74 (*Anthrax*).—Ariz.; Oreg., Utah.
nugator (Coquillett), 1887a: 178 (*Anthrax*).—Calif.; Oreg., Utah.
pallida (Coquillett), 1887a: 179 (*Anthrax*; as *nugator* var.).—Ariz., Calif.; Utah.
pertusa (Loew), 1869b: 18 (Cent. 8, no. 28) (*Anthrax*).—N. Mex.; Calif., Ariz., Colo.
selene (Osten Sacken), 1886a: 122 (*Anthrax*).—Mexico; s. Calif., Ariz.
  otiosa Coquillett, 1887a: 182 (*Anthrax*).—Ariz.
utahensis Maughan, 1935: 51.—Utah; Tex.

Subgenus CHRYSANTHRAX Osten Sacken

Anthrax, group **Chrysanthrax** Osten Sacken, 1886a: 121. Type-species, *fulvohirta* Wiedemann (Coquillett, 1910b: 523)=*cypris* (Meigen).

adumbrata (Coquillett), 1887a: 176 (*Anthrax*).—Calif.
alta (Tucker), 1907: 89 (*Anthrax*).—Colo.; N. Mex.
arenosa (Coquillett), 1892a: 187 (*Anthrax*).—N. Mex.
arizonensis (Coquillett), 1887a: 182 (*Anthrax*).—Ariz.
cinefacta (Coquillett), 1892a: 180 (*Anthrax*).—Calif.
crocina (Coquillett), 1892a: 183 (*Anthrax*).—Calif.; Nev.
cypris (Meigen), 1820: 158 (*Anthrax*).—Unknown (as Europe, error); Iowa to Mass., s. to Tex. and Fla.
  fulvohirta Wiedemann, 1821b: 99 (1821c: 149) (*Anthrax*).— Ga.
  conifacies Macquart, 1850: 416 (1850: 112) (*Anthrax*).—Va.
dispar (Coquillett), 1887a: 177 (*Anthrax*).—Fla.; Minn. to Ont., s. to Ariz. and Fla.
edititia (Say), 1829: 156 (1859b: 353) (*Anthrax*).—Unknown; Oreg. to Mich., s. to Calif., Tex., and Ohio.
  impiger Coquillett, 1887a: 177 (*Anthrax*).—Ariz.

sabulosa Coquillett, 1892a: 186 (*Anthrax*).—Calif., N. Mex. **N. syn.**
eudora (Coquillett), 1887a: 169 (*Anthrax*).—Calif.; Ariz., N. Mex., Utah.
hircina (Coquillett), 1892a: 182 (*Anthrax*).—Calif.
junctura (Coquillett), 1887a: 163 (*Anthrax*).—Calif.; Ariz., N. Mex.
lepidotoides Johnson, 1919b: 12.—N.J.
lepidota, authors, not Osten Sacken.
mira (Coquillett), 1887a: 179 (*Anthrax*).—Calif.; Calif. to Colo., Okla., and Tex., also Fla.
nebulosa (Coquillett), 1894a: 99 (*Anthrax*).—Calif.; Oreg., Utah.
pallidula (Coquillett), 1894a: 99 (*Anthrax*).—Calif.
scitula (Coquillett), 1887a: 172 (*Anthrax*).—Calif.
tantilla (Coquillett), 1892a: 184 (*Anthrax*).—Calif.
turbata (Coquillett), 1887a: 168 (*Anthrax*).—Calif.; Colo., Nebr., Kans., N. Mex., Tex.
comparata Tucker, 1907: 90 (*Anthrax*).—Colo. **N. syn.**
vana (Coquillett), 1887a: 173 (*Anthrax*).—Calif.; Oreg., Utah.
variata (Coquillett), 1892a: 184 (*Anthrax*).—Calif.
vulpina (Coquillett), 1892a: 183 (*Anthrax*).—Calif.

Subgenus RHYNCHANTHRAX Painter

Villa, subg. **Rhynchanthrax** Painter, 1933: 6. Type-species, *Anthrax parvicornis* Loew (orig. des.).

REFERENCE: Painter, 1933 (key).

parvicornis (Loew), 1869b: 23 (Cent. 8, no. 36) (*Anthrax*).—Ill.; Colo. to Ill., s. to Tex. and Miss.
quivera Painter, 1933: 9.—N. Mex.
rex (Osten Sacken), 1886a: 127 (*Anthrax*).—Mexico; Ariz., Tex.
plagosa Coquillett, 1887a: 178 (*Anthrax*).—Ariz.
texana Painter, 1933: 8.—Tex.; N. Mex., Kans., Mexico.

Subgenus PARAVILLA Painter

Villa, subg. **Paravilla** Painter, 1933: 10. Type-species, *edititoides* Painter (orig. des.).

apicola Cole, 1952: 128.—Calif.
castanea (Jaennicke), 1867: 338 (1868: 30) (*Anthrax*).—Mexico; Calif.
cinerea Cole, 1923a: 299.—Mexico (Baja Calif.); Ariz., Tex.
consul (Osten Sacken), 1886a: 125 (*Anthrax*).—Guatemala; Ariz.
coquilletti Painter (for *obscura* Coquillett).—Calif.; Oreg. **N. name.**
obscura Coquillett, 1894a: 99 (*Anthrax*; preocc. Macquart, 1846).—Calif.

cunicula (Osten Sacken), 1886a: 125 (*Anthrax*).—Mexico; Ariz.
diagonalis (Loew), 1869b: 21 (Cent. 8, no. 33) (*Anthrax*).—Calif.; Tex., Kans.
edititoides Painter, 1933: 11.—Colo.; Oreg. to Iowa, s. to Calif. and Tex.
emulata Painter *in* Painter and Painter, 1962: 103.—N. Mex.; Colo., Calif.
    eurhinata, authors, not Bigot.
epheba (Osten Sacken), 1886a: 124 (*Anthrax*).—Mexico; N. Mex., Utah.
extremitis (Coquillett), 1902d: 138 (*Anthrax*).—Mexico; Colo.
flavipilosa Cole, 1923a: 303.—Mexico (Baja Calif.: San José I.); Ariz.
fulvicoma (Coquillett), 1887a: 176 (*Anthrax;* as *edititia* var.).—Calif., Kans.
fumida (Coquillett), 1887a: 177 (*Anthrax*).—Calif.; Oreg.
lacunaris (Coquillett), 1892a: 185 (*Anthrax*).—Calif.
mercedis (Coquillett), 1887a: 166 (*Anthrax*).—Calif.; Ariz., Colo.
nigronasica Painter, 1933: 14.—Nebr.; Idaho, Utah, Colo., Kans., Tex.
palliata (Loew), 1869b: 20 (Cent. 8, no. 32) (*Anthrax*).—Ill.; Colo., S. Dak., Minn., Nebr., Kans., Okla.
perplexa (Coquillett), 1887a: 176 (*Anthrax*).—Calif.
separata (Walker), 1852: 177 (*Anthrax*).—N. Amer.; Oreg., S. Dak., Minn., Wis., Mich., Ohio, Ga.
    nemakagonensis Graenicher, 1910a: 26 (*Anthrax*).—Wis.
spaldingi Painter, 1933: 13.—Utah; Calif., Nev.
syrtis (Coquillett), 1887a: 173 (*Anthrax*).—Calif.; Nev., Utah, Colo., Ariz., N. Mex., Okla.
tricellula Cole, 1952: 129.—Calif.
vasta (Coquillett), 1892a: 184 (*Anthrax*).—Calif.
vigilans (Coquillett), 1887a: 176 (*Anthrax*).—Calif.
xanthina Painter, 1933: 13.—Kans.; Wyo. and S. Dak., s. to Ariz. and Tex.

Unplaced Species of *Villa*

albicincta Cole, 1923a: 297.—Mexico; Ariz.
albovittata (Macquart), 1850: 417 (1850: 113) (*Anthrax*).—?N. Amer.
anna (Coquillett), 1887a: 169 (*Anthrax*).—Calif.; Ariz.
caprea (Coquillett), 1887a: 170 (*Anthrax*).—Calif.; Calif. to Nebr. and N. Mex.
cautor (Coquillett), 1887a: 175 (*Anthrax*).—Calif.; Ariz., Utah.
costata (Say), 1824a: 373 (1859a: 254) (*Anthrax*).—Northwest Territory (U.S.). Unrecognized.

fissa (Bigot), 1892a: 354 (*Anthrax*).—N. Amer. ?=*inops*.
gemella (Coquillett), 1892a: 182 (*Anthrax*).—Calif.
inculta (Coquillett), 1892a: 181 (*Anthrax*).—Calif.; Ariz., Utah, Colo.
levicula (Coquillett), 1894a: 99 (*Anthrax*).—Calif.
meridionalis Cole, 1923a: 292.—Mexico (Baja Calif.: Ceralbo I.); Calif.
miscella (Coquillett), 1887a: 171 (*Anthrax*).—Wash. Territory, Calif.; Oreg.
psammina Cole, 1960b: 118 (n. name for *arenicola* Cole).—Mexico (Baja Calif.); Utah, Mexico.
    arenicola Cole, 1923a: 294 (preocc. Johnson, 1908).—Mexico (Baja Calif.).
sini Cole, 1923a: 303.—Mexico (Baja Calif.: Angel de la Guardia I.); Calif.
sodom (Williston), 1893d: 254 (*Anthrax*).—Calif.
telluris (Coquillett), 1892a: 182 (*Anthrax*).—Calif.
terrena (Coquillett), 1892a: 181 (*Anthrax*).—Calif.
vanduzeei Cole, 1923a: 306.—Mexico; Utah, Colo., Tex.

Genus STONYX Osten Sacken

Stonyx Osten Sacken, 1886a: 94. Type-species, *clelia* Osten Sacken (Coquillett, 1910b: 609).

clelia Osten Sacken, 1886a: 95.—Mexico; Ariz.
    keenii Coquillett, 1887a: 164 (*Anthrax*).—Ariz.

Genus LEPIDANTHRAX Osten Sacken

Lepidanthrax Osten Sacken, 1886a: 107. Type-species, *Anthrax disjunctus* Wiedemann (Coquillett, 1910b: 559).

REFERENCE: Curran, 1930b (key).

agrestis (Coquillett), 1887a: 171 (*Anthrax*).—Calif.; Nev., Utah.
    agrestris, error.
angulus Osten Sacken, 1886a: 111.—Mexico; Calif., Ariz.
campestris (Coquillett), 1887a: 171 (*Anthrax*; as *camprestris*, corrected in Errata at end of vol.).—Calif.
disjunctus (Wiedemann), 1830: 639 (*Anthrax*; as *disiuncta*).—Mexico; Ariz., N. Mex.
hyalinipennis Cole, 1923a: 307.—Mexico (Baja Calif.: Espiritu Santo I.); Calif., Nev., Utah, Ariz.
inauratus (Coquillett), 1887a: 170 (*Anthrax*).—Wash. Territory; Oreg., Calif., Nev., Ariz.
lautus (Coquillett), 1887a: 171 (*Anthrax*).—Calif.

lutzi Curran, 1930b: 2.—Ariz.; N. Mex.
morna Curran, 1930b: 3.—Ariz.; N. Mex., Tex.
proboscideus (Loew), 1869b: 17 (Cent. 8, no. 27) (*Anthrax*).—Mexico; Ariz., N. Mex.
rufolimbatus (Bigot), 1892a: 359 (*Epacmus*).—Calif.

Genus DIPALTA Osten Sacken

Dipalta Osten Sacken, 1877: 236. Type-species, *serpentina* Osten Sacken (mon.).

Parasitic on larvae of Myrmeleontidae (Smith, 1934).

banksi Johnson, 1921a: 12.—Va.; Ohio.
serpentina Osten Sacken, 1877: 237.—Calif., Colo.; B.C. to Mich., s. to Calif. and Fla., Mexico, Guatemala.

Genus NEODIPLOCAMPTA Curran

Neodiplocampta Curran, 1934a: 200. Type-species, *Diplocampta roederi* Curran (orig. des.).

paradoxa (Jaennicke), 1867: 339 (1868: 31) (*Anthrax*).—Mexico; Tex.

Genus ASTROPHANES Osten Sacken

Astrophanes Osten Sacken, 1886a: 106. Type-species, *adonis* Osten Sacken (mon.).

adonis Osten Sacken, 1886a: 107.—Mexico; Ariz., Utah, N. Mex., Tex., Okla., Kans.

Genus MANCIA Coquillett

Mancia Coquillett, 1886b: 159. Type-species, *nana* Coquillett (mon.).
nana Coquillett, 1886b: 159.—Calif.

Genus POECILANTHRAX Osten Sacken

Anthrax, group **Poecilanthrax** Osten Sacken, 1886a: 119. Type-species, *alcyon* Say (Coquillett, 1910b: 593).

The larvae are mostly parasites on larvae of Noctuidae (cutworms).

REFERENCES: Johnson and Johnson, 1957 (key); Painter and Hall, 1960 (monog.).

alcyon (Say), 1824a: 371 (1859a: 252) (*Anthrax*).—Northwest Territory (U.S.); Alta. to Ont., s. to Calif. and Va.
halcyon, emend.

alpha (Osten Sacken), 1877: 239 (*Anthrax*).—Wyo.; Oreg. and Alta. to N. Dak., s. to Calif. and N. Mex.
apache Painter and Hall, 1960: 31.—Ariz.; Calif. to Wyo.
arethusa (Osten Sacken) 1886a: 116 (*Anthrax*).—Mexico; Calif. to Wyo. and Nebr., s. to Mexico (Baja Calif.) and Panama.
autumnalis (Cole), 1917: 71 (*Anthrax;* as *arethusa* var.).—Calif.; Idaho, Colo., Ariz., Nev., Utah.
californicus (Cole), 1917: 69 (*Anthrax*).—Calif.; Oreg. to Wyo., s. to Calif. and N. Mex., Mexico (Baja Calif.).
colei Johnson and Johnson, 1957: 10.—Calif.
demogorgon (Walker), 1849: 265 (*Anthrax*).—Fla.; Minn. to N.Y., s. to Kans. and Fla.
    demorgon, error.
    ceyx Loew, 1869b: 19 (Cent. 8, no. 30) (*Anthrax*).—Va.
effrenus (Coquillett), 1887a: 182 (*Anthrax*).—Ariz.; Calif., N. Mex., Okla., Tex., Mexico.
eremicus Painter and Hall, 1960: 51.—Calif.; Ariz.
fasciatus Johnson and Johnson, 1957: 11.—Colo.; Kans., Tex.
flaviceps flaviceps (Loew), 1869b: 18 (Cent. 8, no. 29) (*Anthrax*).— Mexico; N. Mex., Tex., Kans.
    ssp. **fuliginosus** (Loew), 1869b: 20 (Cent. 8, no. 31) (*Anthrax;* as sp.).—Calif.; Ariz., N. Mex., Colo., Mexico.
    butleri Johnson and Johnson, 1957: 7.—Ariz.
hyalinipennis Painter and Hall, 1960: 64.—Ariz.; Calif., Nev., Utah.
ingens Johnson and Johnson, 1957: 13.—Ariz.; Mexico.
johnsonorum Painter and Hall, 1960: 69.—Utah.
litoralis Painter and Hall, 1960: 71.—Calif.
lucifer (Fabricius), 1775: 759 (*Bibio*).—Amer.; e. Colo. to N.C., s. to Panama and West Indies.
marmoreus Johnson and Johnson, 1957: 16.—Ariz.; Calif., Utah, Colo., Tex.
moffitti Painter and Hall, 1960: 78.—Calif.
montanus Painter and Hall, 1960: 82.—Calif.; Wash., Oreg., Idaho, Wyo., N. Mex.
nigripennis (Cole), 1917: 70 (*Anthrax*).—Md.; Mich., N.Y., Va., Tenn., N.C., Ga.
painteri Maughan, 1935: 56.—Utah; Oreg., Calif., Idaho, Mont., Wyo., Colo.
pilosus (Cole), 1917: 76 (*Anthrax*).—Calif.
poecilogaster poecilogaster (Osten Sacken), 1886a: 118 (*Anthrax*).— Mexico; Oreg. to Idaho, s. to Mexico.
    ssp. **interruptus** Painter and Hall, 1960: 97.—Calif.
robustus Johnson and Johnson, 1957: 20.—Calif.; Mexico (Baja Calif.).

sackenii sackenii (Coquillett), 1887a: 180 (*Anthrax;* as *tegminipennis* var.).—Ariz.; B.C. to Man., s. to Calif., Ariz., and Colo.
  ssp. **monticola** Johnson and Johnson, 1957: 23.—Utah; B.C. to Wyo., s. to Utah and Nebr.
signatipennis (Cole), 1917: 74 (*Anthrax*).—Nev., Wyo., Mont.; Wash. to Alta., s.to Calif. and Colo.
  yellowstonei Cole, 1917: 80 (*Anthrax*).—Nev., Wyo., Mont. Mislabeling of pl.
  marginatus Johnson and Johnson, 1957: 14.—Utah.
tanbarkensis Painter and Hall, 1960: 21, 109, 110 (also as *tanbarkenis*, p. 109).—Calif.
tegminipennis (Say), 1824a: 371 (1859a: 253) (*Anthrax*).—Northwest Territory (U.S.); Alta. to Que., s. to Calif., Nebr., and N.C.
  fuscipennis Macquart, 1834a: 410 (*Anthrax*).—N. Amer.
varius Painter and Hall, 1960: 114.—Calif.; Idaho and Mont., s. to Calif. and Colo.
vexativus Painter and Hall, 1960: 119.—N. Mex.; Ariz., Colo., Mexico.
willistonii (Coquillett), 1887a: 181 (*Anthrax*).—Calif., N. Mex.; B.C. to Man., s. to Calif. and Okla., Mexico.
zionensis Johnson and Johnson, 1957: 6 (as *alpha* ssp.).—Utah.

Genus EXOPROSOPA Macquart

Exoprosopa Macquart, 1840: 35 (1841: 313). Type-species, *Anthrax pandora* Fabricius (Coquillett, 1910b: 544).
Exoptata Coquillett, 1887b: 13. Type-species, *divisa* Coquillett (mon.).
Litorhynchus, authors, not Macquart.

The larvae are parasitic on Hymenoptera and Diptera (Asilidae).

A number of subgenera have been used in Europe (Engel, 1932–1937) and Africa (Bezzi, 1924b, Hesse, 1956), but the relationships of these subgenera to Nearctic species have not been studied sufficiently to permit assignment of species where this is possible. Rondani's (1856: 162) type designation of *Bibio capucina* Fabricius is invalid, as the species was not originally included; Coquillett's designation of *pandora* was accomplished by indicating this species as a synonym of *capucina* in citing Rondani's designation.

REFERENCES: Osten Sacken, 1877, 1886a, Williston, 1901, Coquillett, 1892c, Curran, 1930f (partial keys).

agassizii Loew, 1869b: 16 (Cent. 8, no. 24).—Calif.; Utah, Colo., Ariz., N. Mex., Tex.

albifrons Curran, 1930f: 8.—Ariz.
anomala Painter, 1934: 68.—Tex.; Ariz., N. Mex.
arenicola Johnson and Johnson, 1959a: 71.—Utah.
bifurca Loew, 1869b: 15 (Cent. 8, no. 23).—Calif.; Ariz., N. Mex.
brevistylata Williston, 1901: 272.—Mexico; Tex.
butleri Johnson and Johnson, 1959a: 74.—Ariz.
californiae (Walker), 1852: 172 (*Anthrax*).—Calif. ?=*caliptera*.
caliptera (Say), 1823: 46 (1859b: 62) (*Anthrax*).—Colo. (as Arkansas River); B.C. to Calif., e. to e. face of Rocky Mts. and Black Hills, S. Dak.
capucina (Fabricius).—Not Nearctic.
    nigrita Fabricius, 1775: 757 (*Bibio*).—Amer. (error).
    nigratus, error.
celer Cole, 1916a: 463.—Calif.
clarki Curran, 1930f: 4.—Ariz.
decora Loew, 1869b: 13 (Cent. 8, no. 19).—Wis.; Alta. to Maine, s. to Ariz. and Ga.
divisa (Coquillett), 1887b: 13 (*Exoptata*).—Calif., Ariz.; Oreg., Kans., s. to Calif. and Ala.
dodrans Osten Sacken, 1877: 234.—Colo.; Ariz., N. Mex., Kans.
dodrina Curran, 1930f: 5.—Colo.; Ariz., Utah, N. Mex.
dorcadion Osten Sacken, 1877: 231.—Wash., Calif., Colo., N.H., Maine; B.C. to N.B., s. to Calif., Tex., and Mo.
doris Osten Sacken, 1877: 235.—Nev.; Wash. to Idaho, s. to Calif. and Tex.
    pallens Bigot, 1892a: 345.—Calif.
eremita Osten Sacken, 1877: 236.—Calif.; Wash. to S. Dak., s. to s. Calif.
    grata Coquillett, 1892c: 124.—Calif., Wash.
fasciata Macquart, 1840: 51 (1841: 329).—U.S.; Wyo. to N.H., s. to N. Mex. and Fla.
    rubiginosa Macquart, 1840: 51 (1841: 329).—Pa.
    longirostris Macquart, 1850: 412 (1850: 108).—Va.
    americana Wulp, 1867: 141 (*Mulio*).—Wis.
fascipennis fascipennis (Say), 1824a: 373 (1859a: 254) (*Anthrax*).—Man.; Man. to Mass., s. to Tex. and Ga.
    philadelphica Macquart, 1840: 52 (1841: 330).—Pa.
    coniceps Macquart, 1850: 412 (1850: 108).—Va.
    melanura Bigot, 1892a: 344.—N. Amer.
  ssp. **albicollaris** Painter *in* Painter and Painter, 1962: 145.—Kans.; Wyo. to Minn. and Ohio, s. to N. Mex. and Tex.
  ssp. **noctula** (Wiedemann), 1830: 635 (*Anthrax*; as sp.).—N. Amer.; Ala., S.C. to Fla.
fumosa Cresson, 1919: 177.—Mexico; s. Ariz.

hulli Painter, 1930b: 798.—Miss.; Kans., Mo., Ark., Tex., Mexico, Honduras.
hyalipennis Cole, 1923a: 290.—Mexico (Baja Calif.); Ariz.
ingens Cresson, 1919: 178.—Ariz.; N. Mex., Tex.
iota Osten Sacken, 1886a: 82.—Mexico; Calif., Tex., Kans., Ark.
jonesi Cresson, 1919: 178.—Calif.
junta Curran, 1930f: 9.—Colo.
lutzi Curran, 1930f: 5.—Ariz.
meigenii (Wiedemann), 1828: 278 (*Anthrax*).—Unknown; Kans. to N.J., s. to Tex. and Fla.
    emarginata Macquart, 1840: 51 (1841: 329).—Pa.
painterorum Johnson and Johnson, 1960: 74 (n. name for *cingulata* Johnson and Johnson).—Calif.
    cingulata Johnson and Johnson, 1959a: 76 (preocc. Wulp, 1885).—Calif.
pardus Osten Sacken, 1886a: 88.—Mexico; Ariz.
pueblensis Jaennicke, 1867: 342 (1868: 38).—Mexico; Nebr. to N.C., s. to Tex. and Fla.
rhea Osten Sacken, 1886a: 83.—Mexico; Ariz., N. Mex., Tex.
rostrifera Jaennicke, 1867: 341 (1868: 33).—Mexico; Ariz., N. Mex.
sharonae Johnson and Johnson, 1959a: 78.—Utah.
sima Osten Sacken, 1877: 231.—Nev.
socia Osten Sacken, 1886a: 87.—Mexico; Ariz., N. Mex.
sordida Loew, 1869b: 14 (Cent. 8, no. 21).—Mexico; Colo., Kans., Okla., N. Mex., Tex.
texana Curran, 1930f: 6.—Tex.; Ariz.
tiburonensis Cole, 1923a: 291.—Mexico (Baja Calif.: Tiburon I.); Ariz.
titubans Osten Sacken, 1877: 233.—Colo.; Utah, N. Mex., Kans.
utahensis Johnson and Johnson, 1959a: 81.—Utah.
xanthina Painter, 1934: 69.—Calif.; Nev., Utah.

Genus LIGYRA Newman

Ligyra Newman, 1841: 220. Type-species, *Anthrax bombyliformis* MacLeay (mon.) =*sylvanus* (Fabricius).
Velocia Coquillett, 1886b: 158 (preocc. Robineau-Desvoidy, 1863). Type-species, *Anthrax cerberus* Fabricius (mon.).
Hyperalonia, authors, not Rondani.

The larvae are parasites of Tiphiidae in India.

gazophylax (Loew), 1869b: 12 (Cent. 8, no. 18) (*Exoprosopa*).—Calif.; Ariz.
pilatei (Macquart), 1846: 238 (1846: 110) (*Exoprosopa*).—Mexico; s. Ariz.

Genus OESTRANTHRAX Bezzi

Oestranthrax Bezzi, 1921: 130. Type-species, *Anthrax obesus* Loew (mon.).

farinosus Johnson and Maughan, 1953: 18.—Utah.

Superfamily EMPIDOIDEA

Family EMPIDIDAE

(Empidae, Hybotidae)

By A. L. Melander

The Empididae, or dance-flies, exhibit more diversity in wing venation and genitalic structure than almost any other fly family. The adults are from 1 to about 15 mm. long and in many genera have holoptic eyes. Probably all adult empidids are predatory, mostly on small Diptera. One or more pairs of legs of some species are enlarged and spined. Many species are anthophilous. In temperate regions Empididae are especially abundant in spring and early summer, and many, if not most, species occur in rather humid, hilly or mountainous environments. Empidids are noted for their swarming activities and for their unique and varied epigamic behavior, which includes offering prey to the female, or the construction by the male of a frothy "balloon" or a ball of silk.

The larvae live in damp earth, decaying wood, or vegetation, under the bark of trees, or are aquatic, and probably all are predators. Terestrial larvae are slender-muscoid in appearance, and aquatic forms have tapered posterior processes and well developed ventral pseudopods.

The present catalog is arranged exactly as my Genera Insectorum volume (Melander, 1927) which served, to that date, as a revision of the North American species of many genera.

[Unfortunately, Dr. Melander was unable to make a final review of this manuscript due to failing health and his death on Aug. 14, 1962. Since that time several changes have been made to bring the taxonomy of the family in line with current developments.—Eds.]

REFERENCES: Melander, 1902a (rev.); Lundbeck, 1910 (rev., Denmark); Melander, 1927 (cat., world); Engel, 1938–1954, Frey, 1955–1956 (rev., Palaearctic); Kessel, 1952c, 1955, 1959b (biol.); Collin, 1961 (rev., Britain).

Subfamily BRACHYSTOMATINAE

Genus BRACHYSTOMA Meigen

Brachystoma Meigen, 1822: 12. Type-species, *Syrphus vesiculosus* Fabricius (Blanchard, 1840: 582).

Subgenus BLEPHAROPROCTA Loew

Blepharoprocta Loew, 1862b: 194 (Cent. 2, no. 17) (as genus). Type-species, *Brachystoma nigrimana* Loew (Coquillett, 1903a: 246).

nigrimanum Loew, 1862b: 194 (Cent. 2, no. 17).—Ill.
serrulatum Loew, 1861b: 319 (Cent. 1, no. 23).—Ga.; Ohio, Pa., N.Y., D.C.
    serratula, error or emend.
    binummus Loew, 1862b: 193 (Cent. 2, no. 16).—D.C.

Subgenus BRACHYSTOMA Meigen

occidentale Melander, 1902a: 260.—Wash., Idaho; B.C. to Nev.
robertsonii Coquillett, 1895m: 393.—Ill.; Ohio, Tenn.

Genus ANOMALEMPIS Melander

Anomalempis Melander, 1927: 14. Type-species, *tacomae* Melander (orig. des.).

archon Melander, 1945: 79.—Alaska.
tacomae Melander, 1927: 15.—Wash.

Subfamily HYBOTINAE

Genus HYBOS Meigen

Noeza Meigen, 1800: 27. Type-species, *Musca grossipes* Linnaeus (Coquillett, 1910b: 576). Suppressed by I.C.Z.N., 1963b: 339.
Hybos Meigen, 1803: 269. Type-species, *funebris* Meigen (Curtis, 1837a: pl. 661)=*grossipes* (Linnaeus).

reversus Walker, 1849: 487.—N.Y.; Wis. to Que. and Maine, s. to Ga.
    var. **slossonae** Coquillett, 1895m: 437 (as sp.).—N.H.; Wis. to Maine, s. to N.C.

Genus EUHYBUS Coquillett

Euhybus Coquillett, 1895m: 437. Type-species, *Hybos purpureus* Walker (Coquillett, 1903a: 250).

Euhybos, emend.

Adults of *Euhybus* are abundant in meadows and other flat, vegetated areas. The females of many species are so similar they are difficult to identify.

baropeodes Melander, 1927: 32.—N.H.; Ont.
coquilletti Melander, 1927: 27.—N.C.
cuspidatus Melander, 1927: 32.—Mass.
duplex (Walker), 1849: 486 (*Hybos*).—N.Y.; Ont. to Ga.
genitivus Melander, 1927: 28.—N.H.
metatarsalis Melander, 1927: 28.—Ala., Ga., S.C.
nigripes Melander, 1927: 28.—Mo., Ill., N.Y., Vt., Mass., Pa., N.J., Va.; N.C.
purpureus (Walker), 1849: 486 (*Hybos*).—Ga.
sordipes Melander, 1927: 28.—Ont.; Que., N.B., P.E.I.
strumaticus Melander, 1927: 26.—Md.; Md. to Tex. and Ga.
subjectus (Walker), 1849: 487 (*Hybos*).—Ont.; Ont. to N.H., s. to Fla., P.R.
triplex (Walker), 1849: 486 (*Hybos*).—N.Y.; Nev. to Maine, s. to Tex. and Fla., Cent. and S. Amer.
  var. **simplex** Melander, 1927: 30.—Mass.; Vt. to Tex.

Genus SYNDYAS Loew

Syndyas Loew, 1857b: 369. Type-species, *opaca* Loew (Coquillett, 1903a: 257).

dorsalis Loew, 1861b: 320 (Cent. 1, no. 26).—N.Y.; Ill. to Maine, s. to N.J.
polita Loew, 1861b: 321 (Cent. 1, no. 27).—Carolina; Minn. to Que., s. to Kans. and Ala.

Genus SYNECHES Walker

Syneches Walker, 1852: 165. Type-species, *simplex* Walker (mon.) = *phthia* (Walker).

Syneches is a worldwide genus, the species of which occur in nonmountainous areas.

albonotatus Loew, 1862b: 195 (Cent. 2, no. 18.).—D.C.
ater Melander, 1927: 40.—Pa.
debilis Coquillett, 1895m: 436.—D.C., Md.; Ga.
hyalinus Coquillett, 1895m: 437.—Md.; N.J.
longipennis Melander, 1902a: 346.—N.C.

phthia (Walker), 1849: 492 (*Gloma*).—N.Y.; Wis. to Que., s. to Fla.
simplex Walker, 1852: 165.—U.S.
punctipennis Wulp, 1867: 139.—Wis.
pusillus Loew, 1861b: 320 (Cent. 1, no. 25).—N.Y., Ill.; S. Dak., Ont., and N.J., s. to Fla., Panama, West Indies.
rufus Loew, 1861b: 320 (Cent. 1, no. 24).—Ill., N.Y.; Wis., Ind., Ohio.
testaceus Melander, 1927: 44.—Mo.
thoracicus (Say), 1823: 76 (1859b: 68) (*Hybos*).—Pa.; Minn. to Que. and Mass., s. to Kans. and Tenn.

Genus MEGHYPERUS Loew

Meghyperus Loew, 1850a: 303. Type-species, *sudeticus* Loew (mon.).
nitidus Melander, 1902a: 255.—Idaho.
occidens Coquillett, 1895m: 435.—s. Calif.

Subfamily OCYDROMIINAE

Genus EUTHYNEURA Macquart

Euthyneura Macquart, 1836: 518. Type-species, *myrtilli* Macquart (mon.).
aperta Melander, 1902a: 348.—N. Mex.
argyria Melander, 1927: 53.—Oreg., Idaho, Colo.
bucinator Melander, 1902a: 348.—Pa.; Wis. to N.H., s. to Mo., also Idaho.
crocata (Coquillett), 1900h: 413 (1904: 27) (*Microphorus*).—Alaska.
crocota, error.
matura Melander, 1927: 53.—Wash.
spinipes Melander, 1927: 54.—B.C., Wash., Idaho.

Genus TRICHINA Meigen

Trichina Meigen, 1830: 335. Type-species, *clavipes* Meigen (Rondani, 1856: 152).
atripes (Melander), 1902a: 349 (*Euthyneura*).—s. Calif.
basalis Melander, 1927: 55.—Calif.
clavipes Meigen, 1830: 336.—Europe; Maine, N.H., Mass.
flavipes Meigen, 1830: 336.—Europe; Maine.
nitida Melander, 1927: 56.—Mo.
nura (Melander), 1902a: 349 (*Euthyneura*).—Mass.; Maine, N.Y.
pullata Melander, 1927: 57.—Wash.; Mont.

Genus ANTHALIA Zetterstedt

Anthalia Zetterstedt, 1838: 538. Type-species, *schoenherri* Zetterstedt (Melander, 1927: 57). Coquillett's type designation (1903a: 246), of *gyllenhalli* Zetterstedt makes this name a synonym of *Euthyneura* Macquart, 1836. Action by the I.C.Z.N. is needed to preserve this name.

bulbosa (Melander), 1902a: 349 (*Euthyneura*).—Pa.; Ont., N.Y., N.H., Maine.
femorata Melander, 1927: 59.—Wash., Idaho.
flava Coquillett, 1903a: 268.—N.H.; Ont., N.Y. Maine.
gilvihirta (Coquillett), 1903a: 268 (*Microphorus*).—N.H.; Ont.
inornata Melander, 1927: 61.—Wash.
interrupta Melander, 1927: 60.—Wash.
lacteipennis Melander, 1927: 59.—Idaho; B.C., Wash., Calif.
mandalota Melander, 1927: 58.—Wash., Calif.
schoenherri Zetterstedt, 1838: 539.—Lapland; Mo., Maine, Mass., D.C., Europe.
scutellaris Melander, 1927: 60.—Wash.
stigmalis Coquillett, 1903a: 268.—B.C.; Alaska to Calif.
   var. **petiolata** Melander, 1927: 60.—Idaho.

Genus ALLANTHALIA Melander

Allanthalia Melander, 1927: 61. Type-species, *Anthalia pallida* Zetterstedt (orig. des.).

pallida (Zetterstedt), 1838: 539 (*Anthalia*).—n. Europe; Mo., Que., Md., Va.

Genus OEDALEA Meigen

Oedalea Meigen, 1820: 355. Type-species, *Empis hybotina* Fallén (Westwood, 1840: 133).

The adults of *Oedalea* are yellow legged and black bodied, and are very sluggish in flight.

astylata Melander, 1927: 63.—Ga.
lanceolata Melander, 1927: 63.—Wash.; Oreg., Idaho, Mont.
   var. **testacea** Melander, 1927: 64.—Wash.
ohioensis Melander, 1902a: 256.—Ohio; Maine, Pa., Md.
   stigmatella Zetterstedt of Coquillett *in* Johnson, 1900c: 654.
pruinosa Coquillett, 1903a: 267.—N.H.; N.Y., N.J.

Genus LEPTOPEZA Macquart

Leptopeza Macquart, 1827b: 143 (as *Lemtopeza;* emend. Macquart, 1834a: 320). Type-species, *Ocydromia flavipes* Meigen (mon.) = *ruficollis* (Meigen).

antennalis Melander, 1927: 66.—D.C.; N.C.
borealis Zetterstedt, 1842: 243.—n. Europe; Ont. and Que. to S.C., Europe.
compta Coquillett, 1895m: 435.—N.H., Mass.; Wis. to Que., s. to Tenn., also Idaho.
disparilis Melander, 1902a: 258.—Idaho, Calif.; B.C. to Calif., ?N.Y.
ruficollis (Meigen), 1820: 353 (*Ocydromia*).—Europe; Alaska to Que., s. to N.C., Europe.
flavipes Meigen, 1820: 353 (*Ocydromia*).—Europe.

Genus OCYDROMIA Meigen

Ocydromia Meigen, 1820: 351. Type-species, *Empis glabricula* Fallén (Westwood, 1840: 133).

glabricula (Fallén), 1816a: 33 (*Empis*).—Europe; Alaska, B.C. to Que., s. to N. Mex. and N.C.
peregrinata Walker, 1849: 488.—N.Y.

Genus BICELLARIA Macquart

Bicellaria Macquart, 1823: 155. Type-species, *nigra* Macquart (mon.) = *spuria* (Fallén).
Cyrtoma Meigen, 1824: 1. Type-species, *atra* Meigen (Westwood, 1840: 133 = *spuria* (Fallén).

angustifurca Melander, 1927: 75.—Wash.; B.C., Oreg., Idaho.
brevifurca Melander, 1927: 75.—Wash.; B.C., Alta.
furcifer Melander, 1927: 75.—Mont.; B.C., Alta., Sask., Idaho, Wyo.
halteralis (Loew), 1862b: 206 (Cent. 2, no. 46) (*Cyrtoma*).—D.C.; Wis. to Que. and N.S., s. to N.C.
longipes (Loew), 1862b: 206 (Cent. 2, no. 47) (*Cyrtoma*).—Ill.; B.C. to Que., s. to N. Mex. and Ga.
lugubris Melander, 1927: 76.—Wyo.; Alaska, Y.T., B.C., Alta., Sask., Mont.
pectinata Melander, 1927: 76.—Wash.; B.C.
pilipes (Loew), 1862b: 207 (Cent. 2, no. 48) (*Cyrtoma*).—Ill.; Alaska, Ill., Ont. to N.B., s. to N.Y.
spuria (Fallén), 1816a: 33 (*Empis*).—Europe; Ont., Mass.
uvens Melander, 1927: 74.—Que.; Alaska to Labr.

Genus HOPLOCYRTOMA Melander

Hoplocyrtoma Melander, 1927: 78. Type-species, *Cyrtoma procera* Loew (orig. des.).

femorata (Loew), 1864a: 84 (Cent. 5, no. 69) (*Cyrtoma*).—N.H.; Ont. N.Y., Mass., N.C.
procera (Loew), 1864a: 85 (Cent. 5, no. 70) (*Cyrtoma*).—Alaska; B.C., Wash., Oreg.

Subfamily EMPIDINAE

Genus PARATHALASSIUS Mik

Parathalassius Mik, 1891c: 217. Type-species, *blasigii* Mik (orig. des.).
Adults frequent the sands of the seashore.

aldrichi Melander, 1906: 374.—Calif.; Wash.
candidatus Melander, 1906: 375.—Wash.
melanderi Cole, 1912b: 154.—Calif.

Genus MICROPHORELLA Becker

Microphorella Becker, 1909a: 28. Type-species, *Microphorus praecox* Loew (orig. des.).

acroptera Melander, 1927: 88.—Wash.; Calif.
chiragra Melander, 1927: 89.—Wash.
longitarsis Melander, 1927: 89.—Idaho.
ornatipes Melander, 1927: 88.—Idaho.
tubifera Melander, 1927: 88.—Calif.

Genus MICROPHORUS Macquart

Microphorus Macquart, 1827b: 139 (as *Microphor;* emend. Macquart, 1834a: 345). Type-species, *velutinus* Macquart (Rondani, 1856: 151).
Holoclera Schiner of Melander, 1902a: 333.

Subgenus MICROPHORUS Macquart

REFERENCE: Melander, 1940a (rev.).

armipes Melander, 1927: 91.—Wash.
atratus Coquillett, 1900h: 412 (1904: 26).—Alaska; Wash., Idaho. Maine, N.H.
bilineatus (Melander), 1902a: 334 (*Holoclera*).—La.
cirripes Melander, 1940a: 64.—Oreg.; B.C. to Calif.

discalis Melander, 1940a: 64.—Calif.; Wash.
drapetoides Walker, 1849: 489.—Ont. Unrecognized.
evisceratus Melander, 1940a: 65.—Idaho; Mont.
isommatus Melander, 1927: 92.—B.C.; Wash.
obscurus Coquillett, 1903a: 268.—N.H.; N.Y. to N.H., s. to Va.
ravidus Coquillett, 1895m: 409.—Calif.
ravus Melander, 1940a: 67.—s. Calif.; Wash.
robustus Melander, 1927: 91.—Pa.
strigilifer Melander, 1940a: 68.—Calif.
sycophantor (Melander), 1902a: 334 (*Holoclera*).—Idaho; Wash. to Alta., s. to Calif. and Wyo.
tacomae Melander, 1940a: 69.—Wash.

Subgenus SCHISTOSTOMA Becker

Schistostoma Becker, 1902: 46 (as genus). Type-species, *eremita* Becker (mon.) =*truncatus* Loew, n. syn.

yakimensis Melander, 1927: 94.—Wash.; Alta.

Genus HORMOPEZA Zetterstedt

Hormopeza Zetterstedt, 1838: 540. Type-species, *obliterata* Zetterstedt (mon.).

REFERENCE: Kessel, 1958 (biol. *copulifera*).

brevicornis Loew, 1864a: 83 (Cent. 5, no. 65).—Alaska; Alaska to Calif. and S. Dak., N.W.T., N.H.
bullata Melander, 1902a: 274.—Wyo.; Ont.
copulifera Melander, 1927: 96.—Idaho; Alaska to Wash. and Idaho.
nigricans Loew, 1864a: 83 (Cent. 5, no. 66).—Alaska; Y.T., Alta., Idaho.
senator Melander, 1927: 95.—D.C.
virgator Melander, 1927: 96.—Wash.; Idaho.

Genus OREOGETON Schiner

Oreogeton Schiner, 1860b: 53. Type-species, *Gloma basalis* Loew (orig. des.).

capnopterus Melander, 1927: 99.—Wash.; Alta.
cymballista Melander, 1927: 99.—B.C.
heterogamus Melander, 1927: 100.—Oreg.
mitrephorus Melander, 1927: 99.—Idaho.
obscurus (Loew), 1864a: 84 (Cent. 5, no. 68) (*Gloma*).—N.H.; Alaska, Idaho, Maine to N.Y. and Mass.

rufus (Loew), 1864a: 84 (Cent. 5, no. 67) (*Gloma*).—N.H.; Maine, Mass.
scopifer (Coquillett), 1900h: 412 (1904: 26) (*Gloma*).—Alaska.
xanthus Melander, 1945: 85.—Wash.

Genus APALOCNEMIS Philippi

Apalocnemis Philippi, 1865: 752. Type-species, *obscura* Philippi (mon.).

REFERENCE: Melander, 1946a (key).

hirsuta Melander, 1946a: 40.—Oreg.
oreas Melander, 1946a: 39.—Calif.

Genus GLOMA Meigen

Gloma Meigen, 1822: 14. Type-species, *fuscipennis* Meigen (mon.).

REFERENCE: Melander, 1945 (key).

fuscipes Melander, 1945: 84.—Wash.; Oreg., Idaho.
luctuosa Melander, 1927: 102.—Wash.
pectinipes Melander, 1945: 84.—Alaska.

Genus ITEAPHILA Zetterstedt

Iteaphila Zetterstedt, 1838: 540.—Type-species, *macquarti* Zetterstedt (Coquillett, 1903a: 251).

REFERENCE: Melander, 1946a (key)

cana Melander, 1946a: 33.—Alta.; Sask.
conjuncta (Coquillett), 1900h: 411 (1904: 25) (*Empis*).—Alaska; B.C., Idaho.
cormus (Walker), 1849: 496 (*Empis*).—Ont.; N.H., Maine.
fuliginosa Melander, 1946a: 34.—Wash.; Oreg.
luctuosa (Kirby), 1837: 311 (*Empis*).—n. Canada; B.C., N.H.
    geniculata Kirby, 1837: 312 (*Empis*).—n. Canada.
macquarti Zetterstedt, 1838: 541.—n. Europe; Alaska to Labr., s. to Oreg., Colo., and Maine.
    curva Curran, 1925a: 24.—Labr. [See Tuomikoski, 1958: 130.]
napaea Melander, 1946a: 36.—Idaho; B.C., Wash., Calif.
nitidula Zetterstedt, 1838: 541.—Lapland; B.C. to Que., s. to Oreg., N. Mex., and Mo., Europe.
    americana Melander, 1946a: 32.—Wyo. [See Tuomikoski, 1958: 130.].
orchestris Melander, 1902a: 354.—N. Mex.; B.C., Alta., Wash., Colo.

testacea Melander, 1946a: 37.—Idaho; B.C.
triangula (Coquillett), 1900h: 410 (1904: 24) (*Empis*).—Alaska; B.C., Wash., Mont., N. Mex.
vetula Melander, 1946a: 37.—Wash.; Idaho.

Genus ANTHEPISCOPUS Becker

Anthepiscopus Becker, 1891: 281. Type-species, *ribesii* Becker (Coquillett, 1903a: 246).
flavicoxa Melander, 1927: 107.—Wash.
flavipilosus (Coquillett), 1900h: 413 (1904: 27) (*Microphorus*).—B.C.
hirsutus Melander, 1927: 106.—Wash.
longipalpis Melander, 1927: 106.—Wash.
nuptus Melander, 1927: 106.—Wash.
polygynus Melander, 1927: 105.—Wash.
stentor (Melander), 1902a: 348 (*Euthyneura*).—N. Mex.

Genus BROCHELLA Melander

Brochella Melander, 1927: 108. Type-species, *monticola* Melander (orig. des.).
monticola Melander, 1927: 109.—Wash.

Genus RAGAS Walker

Ragas Walker, 1837: 229. Type-species, *unica* Walker (mon.).
primigenia Melander, 1945: 83.—Calif.

Genus PHILETUS Melander

Philetus Melander, 1927: 110. Type-species, *memorandus* Melander (orig. des.).
memorandus Melander, 1927: 110.—Wash.; Wyo.
schizophorus Melander, 1927: 110.—Wash.

Genus HESPEREMPIS Melander

Hesperempis Melander, 1906: 377. Type-species, *Ragas mabelae* Melander (mon.).
mabelae (Melander), 1902a: 277 (*Ragas*).—Idaho; B.C. and N.W.T. to Colo.
sanduca Melander, 1927: 111.—Calif.; Oreg.

Genus HILARA Meigen

Hilara Meigen, 1822: 1. Type-species, *Empis maura* Fabricius (Curtis, 1826: pl. 130).

Hilara adults are common along small streams, often performing an aerial dance. The males of some species envelop their prey in a web spun from the mouth.

REFERENCE: Melander, 1940b (biol.).

argyrata Curran, 1930j: 45.—N.Y.
atra Loew, 1862b: 205 (Cent. 2, no. 42).—Ill.; Oreg., Colo., N. Mex., N.H. to Fla.
aurata Coquillett, 1900h: 411 (1904: 25).—Alaska; Maine.
auripila Curran, 1926f: 247.—Alta.
baculifer Melander, 1902a: 271.—Ga.
basalis Loew, 1862b: 206 (Cent. 2, no. 45).—Ill.
bella Melander, 1902a: 271.—Mass.
brevipila Loew, 1862b: 204 (Cent. 2, no. 41).—Ill.; D.C.
cana Coquillett, 1895m: 395.—s. Calif.
carbonaria Melander, 1902a: 272.—Mass.
cavernicola Melander, 1945: 86.—Wash.
congregaria Melander, 1902a: 272.—Calif.
crickmayi Curran, 1926f: 246.—N.W.T.
femorata Loew, 1862b: 202 (Cent. 2, no. 35).—Md.; Wis., Ohio, Maine to N.J.
garretti Curran, 1926f: 248.—Alta.
gracilis Loew, 1862b: 205 (Cent. 2, no. 44).—Pa.; Tex., Maine to N.J.
granditarsis Curran, 1926f: 248.—Alta.
johnsoni Coquillett, 1895m: 395.—Ala.
juno Curran, 1930j: 44.—N.Y.
leucoptera Loew, 1862b: 205 (Cent. 2, no. 43).—Fla.; Ill., Que. to N.J.
lutea Loew, 1863a: 18 (Cent. 3, no. 33).—D.C.; N.Y. and Mass. to Va.
macroptera Loew, 1863a: 18 (Cent. 3, no. 32).—D.C.; Minn., N.H. to N.C.
migrata Walker, 1849: 491.—Ont.
mutabilis Loew, 1862b: 204 (Cent. 2, no. 40).—Ill.; Que. to Va.
nearctica Shewell, 1955: 45.—N.W.T.
nigriventris Loew, 1862b: 203 (Cent. 2, no. 38).—Pa.; Maine.
nugax Melander, 1902a: 273.—Calif.
plebeia Walker, 1857: 148.—U.S.
quadrivittata Meigen, 1822: 7.—Europe; Alaska.
rufopuncta Curran, 1926f: 245.—Alta.
seriata Loew, 1864a: 82 (Cent. 5, no. 63).—N.H.; Mass., N.Y., N.J.
testacea Loew, 1864a: 82 (Cent. 5, no. 64).—N.H.; N.Y., Mass., R.I., Pa., N.J.
transfuga Walker, 1849: 491.—Ont.; Alaska.

tristis Loew, 1864a: 82 (Cent. 5, no. 62).—N.H.; Que. to Nfld., s. to N.J.
trivittata Loew, 1862b: 204 (Cent. 2, no. 39).—Ill.; N.H., Conn., N.J., Tex.
umbrosa Loew, 1862b: 202 (Cent. 2, no. 34).—Ill.; Que. to N.J.
 brachystoma Coquillett *in* Johnson, 1900c: 652 (*Empis*). Nomen nudum.
unicolor Loew, 1862b: 203 (Cent. 2, no. 37).—Md.; Labr., Mass., N.Y.
varipennis (Curran), 1930j: 45 (*Empis*).—N.Y.
velutina Loew, 1862b: 203 (Cent. 2, no. 36).—D.C.; Que. to Md.
wheeleri Melander, 1901: 214.—Wyo.; Calif.

Genus TOREUS Melander

Toreus Melander, 1906: 376. Type-species, *Empis neomexicanus* Melander (orig. des.).

neomexicanus (Melander), 1902a: 352 (*Empis*).—N. Mex.

Genus EMPIS Linnaeus

Empis Linnaeus, 1758: 603. Type-species, *pennipes* Linnaeus (Latreille, 1810: 443).
Platyptera Meigen, 1803: 269. Type-species, *Empis borealis* Linnaeus (Melander, 1927: 144). By tautonymy, *Empis platyptera* Panzer is the proper type-species, but acceptance of that fixation would require that *Platyptera* replace *Rhamphomyia* Meigen, 1822. Because of the size of the latter genus and extensive usage of the name, suspension of the rules should be requested.
Pachymeria Stephens, 1829a: 55 (also 1829b: 262). Type-species, *Empis ruralis* Meigen (mon.)=*femorata* Fabricius.
Eriogaster Macquart, 1838b: 162 (1839: 278) (preocc. Germar, 1811). Type-species, *Empis laniventris* Eschscholtz (orig. des.).
Enoplempis Bigot, 1880b: xlvii. Type-species, *mira* Bigot (mon.).
Empimorpha Coquillett, 1895m: 396. Type-species, *comantis* Coquillett (orig. des.).
Empis, subg. *Pterempis* Bezzi, 1909: 87. Type-species, *pennipes* Linnaeus (Engel, 1945: 341).
Empis, subg. *Xanthempis* Bezzi, 1909: 88. Type-species, *stercorea* Linnaeus (orig. des.).
Empis, subg. *Anacrostichus* Bezzi, 1909: 93. Type-species, *nitida* Meigen (orig. des.).

Empis, subg. *Polyblepharis* Bezzi, 1909: 95. Type-species, *albicans* Meigen (orig. des.).

Empis, subg. *Argyrandrus* Bezzi, 1909: 100. Type-species, *dispar* Scholz (orig. des.).

Empis, subg. *Coptophlebia* Bezzi, 1909: 100. Type-species, *hyalipennis* Fallén (orig. des.).

Empis, subg. *Acallomyia* Melander, 1927: 140. Type-species, *Iteaphila peregrina* Melander (orig. des.).

Empis, subg. *Pyrrempis* Melander, 1927: 144. Type-species, *rufescens* Loew (orig. des.).

Adults of *Empis* are noted for their interesting and sometimes intricate mating habits. The larvae are usually found in rich, moist soil.

[Recent work by Collin (1961) indicates that a revision of the subgeneric classification of the North American species of *Empis* is advisable. A preliminary examination of our species in the light of Collin's work, and a comparison of the type-species of the subgenera, has strengthened that idea. Our list is, therefore, not divided among the subgenera, which are, however, listed above. It seems highly unlikely that any of our species can actually be referred to the subgenera *Empis*, *Pachymeria*, *Platyptera*, or *Argyrandrus*, while on the other hand it appears likely that some new subgenera will be needed.— George C. Steyskal.]

REFERENCES: Melander, 1946c (keys, descr., N. Amer., Oriental *Coptophlebia*); Kessel and Karabinos, 1947 (biol. *geneatis*); Kessel and Kessel, 1951 (biol. *bullifera*); Tilden *in* Skinner, 1962 (biol.).

abcirus Walker, 1849: 494.—Ga.
aeripes Melander, 1902a: 328.—Idaho.
aerobatica Melander, 1902a: 323.—Calif., Idaho.
agasthus Walker, 1849: 496.—Ont.
aldrichii Melander, 1902a: 309.—Oreg., Idaho.
amytis Walker, 1849: 493.—N.Y.
    amystis, error.
anthophila Melander, 1946c: 109.—N. Mex.; Tex.
armipes Loew, 1861b: 323 (Cent. 1, no. 32).—N.Y.
arthritica Melander, 1902a: 318.—Pa.
asema Melander, 1902a: 294.—Tex.; Colo.
barbatoides Melander (for *barbata* Loew).—Calif.; Wash., Oreg. **N. name.**
    barbata Loew, 1862b: 195 (Cent. 2, no. 19) (preocc. Macquart, 1823).—Calif.
bigoti Melander, 1902a: 319 (n. name for *cinerea* Bigot).—Calif.
    cinerea Bigot, 1882b: xci (*Enoplempis;* preocc. Fabricius, 1775).—Calif.

brachysoma Coquillett, 1900h: 409 (1904: 23).—Alaska.
brevis (Loew), 1862b: 196 (Cent. 2, no. 22) (*Pachymeria*).—D.C.; Mo.
browni Curran, 1931e: 93.—Que.
brunnea Coquillett, 1903a: 270.—Calif.
bullifera Kessel and Kessel, 1951: 137.—Calif.
cacuminifer Melander, 1902a: 304.—Ohio, Ala.
caeligena Melander, 1902a: 314.—Ala.
canaster Melander, 1902a: 326.—Oreg., Idaho.
captus Coquillett, 1895m: 405.—N.C., Ga.
clausa Coquillett, 1895m: 401.—Ill.; S. Dak. to Ohio, s. to Tex.
colonica Walker, 1849: 498.—N.S.
comantis (Coquillett), 1895m: 396 (*Empimorpha*).—n. Calif.; Oreg.
   comata (error) Kessel, 1955: 98.
cometes Steyskal (for *comantis* Coquillett, 1895m: 402).—n. Calif. **N. name.**
   comantis Coquillett, 1895m: 402 (preocc. Coquillett, 1895m: 396).—n. Calif.
compta Coquillett, 1895m: 405.—Ill.; La.
ctenocnema Melander, 1945: 87.—N.Y.
dactylica Melander, 1946c: 110.—Ont.; Alta.
delumbis Melander (for *clauda* Coquillett).—Alaska. **N. name.**
   clauda Coquillett, 1900h: 407 (1904: 21) (preocc. Schrank, 1803).—Alaska.
desiderata Melander (for *avida* Coquillett).—Ill. **N. name.**
   avida Coquillett, 1895m: 405 (preocc. Harris, 1776).—Ill.
deterra Walley, 1927a: 96.—Ont.; N.Y.
distans Loew, 1869b: 32 (Cent. 8, no. 54).—Ga.; N.H., Mass., Conn., Tex.
dolabraria dolabraria Melander, 1902a: 325.—Calif.
   ssp. **disconvenita** Melander, 1902a: 326.—Calif.
enodis Melander, 1902a: 303.—Ill.
eudamides Walker, 1849: 493.—N. Amer.
exilis Coquillett, 1903a: 269.—Mo.
falcata Melander, 1902a: 326.—Idaho, Calif.
frontalis Coquillett, 1903a: 271.—Alaska (Pribilof Is.).
fumida Coquillett, 1900h: 409 (1904: 23).—Alaska.
geneatis (Melander), 1902a: 329 (*Empimorpha*).—Calif.
gladiator Melander, 1902a: 316.—Kans.
gulosa Coquillett, 1895m: 408.—Ill.; Mich.
hirticrus Melander, 1927: 141, 158 (n. name for *hirtipes* Coquillett).—N. Mex.; Colo.
   hirtipes Coquillett, 1903a: 270 (preocc. Wiedemann, 1824).—N. Mex.

tenebrosa Coquillett, 1903a: 270 (preocc. Coquillett, 1895).—
N. Mex.
humilis Coquillett, 1895m: 403.—Ill.; N.Y., N.J.
infumata Coquillett, 1900h: 409 (1904: 23).—Alaska.
johnsoni Melander, 1902a: 303.—Pa.
labiata Loew, 1861b: 323 (Cent. 1, no. 33).—D.C.
laevigata Loew, 1864a: 75 (Cent. 5, no. 49).—N.H.
laniventris Eschscholtz, 1823: 113.—Alaska; B.C., Wash.
latrappensis Ouellet, 1942: 78.—Que.
leptogastra Loew, 1863a: 17 (Cent. 3, no. 30).—D.C.; N.C.
levicula Coquillett, 1895m: 406.—Ill.
longeoblita Steyskal (for *longipes* Loew).—N.Y.; Maine, Vt., N.J.
N. name.
longipes Loew, 1864a: 76 (Cent. 5, no. 51) (preocc. Meigen, 1804).—N.Y.
loripedis Coquillett, 1895m: 400.—Ill.; Ohio, N.J.
manca Coquillett, 1895m: 406.—Calif.
mira (Bigot), 1880b: xlvii *(Enoplempis)*.—Calif.
mixopolia Melander, 1902a: 327.—Idaho.
montiradicis James, 1942c: 163.—Colo.
nodipoplitea Steyskal (for *nodipes* Melander).—N. Mex. **N. name.**
nodipes Melander, 1902a: 324 (preocc. Fallén, 1816).—
N. Mex.
nuda Loew, 1862b: 195 (Cent. 2, no. 20).—Ill.; Minn. to Que., s. to
Iowa and Ill.
obesa Loew, 1861b: 321 (Cent. 1, no. 28).—Mass.; Labr., N.H.
ravida Coquillett, 1895m: 403.—N.H.
ollius Walker, 1849: 493.—N.S.
otiosa Coquillett, 1895m: 407.—Ill., Conn.; Minn. to Que., s. to La.
pallida Loew, 1861b: 322 (Cent. 1, no. 30).—N.Y.; Ont., N.H.,
Vt., Mass.
pellucida Coquillett, 1900h: 408 (1904: 22).—Alaska.
peregrina (Melander), 1902a: 331, 361 (as *perigrina*, p. 331) *(Iteaphila)*.—N. Mex. (as Calif., error; Melander, 1927: 168).
plectrum Melander, 1946c: 113.—Tex.
podagra Melander, 1902a: 318.—Idaho.
poeciloptera Loew, 1861b: 322 (Cent. 1, no. 31).—N.Y.; Ont., Vt., Mass., Conn.
poplitea Loew, 1863a: 16 (Cent 3, no. 29).—Alaska; Alaska to N. Mex.
serperastrorum Melander, 1902a: 324.—Idaho, Colo.
pudica (Loew), 1861b: 324 (Cent. 1, no. 35) *(Pachymeria)*.—D.C.;
N. Mex., Ill., N.Y. to Va.
reciproca Walker, 1857: 147.—U.S.

rufescens Loew, 1864a: 76 (Cent. 5, no. 52).—N.H.; N.Y., Vt., Maine, Mass.
scatophagina Melander, 1902a: 351.—Alaska.
scoparia Coquillett, 1903a: 269.—N.H.
sordida Loew, 1861b: 321 (Cent. 1, no. 29).—D.C.; N.Y., N.H., Conn.
spectabilis Loew, 1862b: 196 (Cent. 2, no. 21).—Md.; Conn. to N.C.
stenoptera Loew, 1864a: 75 (Cent. 5, no. 50).—N.H.; N.Y., Vt.
subinfumata Malloch, 1923i: 184.—Alaska (Pribilof Is.).
tenebrosa Coquillett, 1895m: 404.—Tex.; Colo.
teres Melander, 1902a: 315.—Idaho.
tersa Coquillett, 1895m: 404.—N.C., N.Y., Mass.
tridentata Coquillett, 1901h: 609.—Pa.; N.H., Conn., N.J.
vaginifer Melander, 1902a: 352.—D.C.; Md., Tenn.
valentis Coquillett, 1895m: 402.—n. Calif.
varipes Loew, 1861b: 324 (Cent. 1, no. 34).—Pa.; Ont. and Que. to Mass., s. to Pa.
virgata Coquillett, 1895m: 408.—Wash.; Alaska, B.C.

Genus RHAMPHOMYIA Meigen

Rhamphomyia Meigen, 1822: 42. Type-species, *Empis sulcata* Meigen (Curtis, 1834: pl. 517).
Megacyttarus Bigot, 1880b: xlvii. Type-species, *argenteus* Bigot (mon.) = *limbata* Loew.
Neocota Coquillett, 1895m: 434. Type-species, *weedii* Coquillett (orig. des.).
Rhamphomyia, subg. *Pararhamphomyia* Frey, 1922: 33. Type-species, *Empis plumipes* Meigen (as Fallén; orig. des.).
Rhamphomyia, subg. *Dasyrhamphomyia* Frey, 1922: 65. Type-species, *Empis vesiculosa* Fallén (orig. des.).

Rhamphomyia contains a great many undescribed species. The genus in North America requires a thorough revision. [A classification of the genus, with several new subgenera, is being prepared from a manuscript left by Dr. Melander. Meanwhile, the species are listed without reference to subgenera.—George C. Steyskal.]

REFERENCES: Steyskal, 1941a, 1942c, 1950 (biol., *longicauda*); Frohne, 1952, 1959 (biol.); Hubert, 1953 (biol.); Chillcott, 1959 (tax., *basalis* group); Crane, 1961 (biol., *scutellaris*).

adversa Coquillett, 1900h: 418 (1904: 32).—Alaska.
agasicles Walker, 1849: 499.—Ont.
albata Coquillett, 1902g: 103.—Ariz.
albopilosa Coquillett, 1900h: 418 (1904: 32).—Alaska; Y.T., N.W.T.
ambocnema Chillcott, 1959: 267.—Man.; Ont., Que., N.Y.
americana Wiedemann, 1830: 8.—N. Amer.; Colo., N.H., Maine.

amplicella Coquillett, 1895m: 431.—s. Calif.; Oreg.
amplipedis Coquillett, 1895m: 422.—Mass.; Que. to Maine, s. to N.J.
    ethellia Curran, 1933b: 4.—Que.
anaxo Walker, 1849: 500.—Ont.
angustipennis Loew, 1861b: 336 (Cent. 1, no. 55).—N.Y.; Ill. to N.H., s. to N.J.
anthracodes Coquillett, 1900h: 420 (1904: 34).—Alaska.
aperta Loew, 1862b: 199 (Cent. 2, no. 27).—Ill.; Maine to N.J.
arctotibia Chillcott, 1959: 271.—Que.; Ont.
arcuata Coquillett, 1895m: 421.—Mass.; Maine, N.H.
atrata Coquillett, 1900h: 420 (1904: 34).—Alaska.
avida Coquillett, 1895m: 425.—Mass.; Maine to N.J.
barypoda Coquillett, 1900h: 417 (1904: 31).—Alaska.
basalis Loew, 1864a: 77 (Cent. 5, no. 54).—N.H.; Mich. to N.S., s. to N.C.
bifilata Coquillett, 1895m: 424.—s. Calif.
bigelowi Walley, 1927a: 97.—Ont.
birdi Curran, 1929a: 7.—Man.
borealis (Fabricius), 1780: 211 (*Empis*).—Greenland.
brevis Loew, 1861b: 334 (Cent. 1, no. 52).—D.C.; Alaska, B.C. to Que., s. to Calif. and Ga.
    var. **corvina** Loew, 1861b: 307, 334 (as *cervina*, 334) (Cent. 1, no. 51) (as sp.).—N.Y.; Alaska, B.C., Oreg., N.Y. to N.C.
californica Coquillett, 1895m: 420.—Calif.
calvimontis Cockerell, 1916: 123.—Colo.
candicans Loew, 1864a: 81 (Cent. 5, no. 61).—N.H.; Maine to Md.
ciliata Coquillett, 1895m: 428.—N.H.
cilipes (Say), 1823: 95 (1859b: 83) (*Empis*).—Ohio.
cinefacta Coquillett, 1900h: 419 (1904: 33).—Alaska.
cineracea Coquillett, 1900h: 416 (1904: 30).—Alaska.
clauda Coquillett, 1901h: 610.—N.H., N.J.; Ont. to N.S., s. to Md.
clavator Coquillett, 1901h: 611 (n. name for *macrura* Coquillett).—Alaska.
    macrura Coquillett, 1900h: 421 (1904: 35) (preocc. Loew, 1871).—Alaska.
clavigera Loew, 1861b: 335 (Cent. 1, no. 53).—N.Y.; Md., N.C.
colorata Coquillett, 1895m: 420.—Tex.
compta Coquillett, 1895m: 423.—U.S.; Nev., Mo., N.Y. and N.H. to N.J.
conjuncta Loew, 1861b: 336 (Cent. 1, no. 56).—D.C.
conservativa Malloch, 1919f: 48.—N.W.T.; Y.T., Man.
cophas Walker, 1849: 499.—N.Y.

currani Steyskal (for *rufipes* Curran).—Okla. N. name.
  rufipes Curran, 1933b: 5 (*Neocota*; preocc. Zetterstedt, 1838).
    —Okla.
curvipes Coquillett *in* Baker, 1904: 24.—Calif.; Oreg., Nev.
dana Walker, 1849: 502.—Ont.
daria Walker, 1849: 503.—N.Y.
debilis Loew, 1861b: 330 (Cent. 1, no. 45).—Sask.; Minn., N.H., Maine.
dimidiata Loew, 1861b: 325 (Cent. 1, no. 36).—Md.; Ill., N.Y., N.H., Mass.
disconcerta Curran, 1930j: 46.—N.Y.; Md., Va.
disparilis Coquillett, 1900h: 415 (1904: 29).—Alaska.
diversa Coquillett, 1901h: 611.—N.J.
duplicis Coquillett, 1895m: 424.—s. Calif.
ecetra Walker, 1849: 500.—Ga.
effera Coquillett, 1895m: 427.—Colo. (unpublished); Mass., R.I.
erinacioides Malloch, 1919f: 45.—Alaska.
exigua Loew, 1862b: 201 (Cent. 2, no. 32).—Ill.; Mass. to Va.
expulsa Walker, 1857: 148.—U.S.; N.H.
falcipedia Chillcott, 1959: 268.—Que.; Mich.
ficana Walker, 1849: 501.—Ont.
filicauda Henriksen and Lundbeck, 1917: 608.—Greenland.
fimbriata Coquillett, 1895m: 429.—Calif.
flavirostris Walker, 1849: 501.—Ont.; Alaska.
flexuosa Coquillett, 1895m: 433.—Colo.; Alta., N. Mex.
formosula Melander (for *pulchra* Loew).—N.Y.; N.Y. to Ga. N. name.
  pulchra Loew, 1861b: 327 (Cent. 1, no. 40) (preocc. Egger. 1860).—N.Y.
frontalis Loew, 1862b: 199 (Cent. 2, no. 28).—Ill.
fumosa Loew, 1861b: 327 (Cent. 1, no. 39).—N.Y.; Minn. to Conn., s. to Ky.
gilvipes Loew, 1861b: 332 (Cent. 1, no. 48).—N.Y.; Ill., Ont. and Que., s. to N.J.
gilvipilosa Coquillett, 1895m: 434.—Ill.; Que.
glabra Loew, 1861b: 328 (Cent. 1, no. 41).—Ill., Va.; Mo. to Vt., s. to N.C.
glauca Coquillett, 1900h: 416 (1904: 30).—Alaska.
gracilis Loew, 1861b: 329 (Cent. 1, no. 43).—Pa.; Que. to N.S., s. to Pa. and N.J.
  bipunctata Curran, 1930j: 47.—N.Y.
herschelli Malloch, 1919f: 47.—Y.T.
  herscheli, emend.
hirticula Collin, 1937a: 407.—Greenland.

hirtipes Loew, 1864a: 80 (Cent. 5, no. 59).—N.H.; Ont., Que., Mass.
hirtula Zetterstedt, 1842: 421.—Greenland.
hoeli Frey, 1950: 100 (as *hoelsi*; emend. Frey, 1955: 482).—Greenland.
hovgaardii Holmgren, 1880: 21.—U.S.S.R. (Novaya Zemlya); Y.T., n. Asia.
impedita Loew, 1862b: 201 (Cent. 2, no. 31).—Ill.; R.I., D.C.
incompleta Loew, 1863a: 17 (Cent. 3, no. 31).—D.C.; Labr., Md.
insecta Coquillett, 1895m: 426.—Tex.
irregularis Loew, 1864a: 81 (Cent. 5, no. 60).—N.H.; Alaska, Minn. to Que., s. to Colo. and N.C.
jubata Chillcott, 1959: 266.—Que.
laevigata Loew, 1861b: 325 (Cent. 1, no. 37).—Nebr.; Mont. to Man. to Que., s. to Va.
leucoptera Loew, 1861b: 340 (Cent. 1, no. 62).—D.C.; N.Y. to D.C.
limata Coquillett, 1900h: 417 (1904: 31).—Alaska.
limbata Loew, 1861b: 338 (Cent. 1, no. 60).—D.C.; Alaska, Mont. to N.H., s. to Colo. and D.C.
    argentea Bigot, 1880b: xlvii (*Megacyttarus*).—Colo.
liturata Loew, 1861b: 339 (Cent. 1, no. 61).—Md.; Maine to D.C.
longicauda Loew, 1861b: 326 (Cent. 1, no. 38).—D.C.; Mich. to Que., s. to Mo. and N.C.
    fumosa, authors, not Loew.
longicornis Loew, 1861b: 332 (Cent. 1, no. 47).—D.C.; Que., Pa.
longipennis Loew, 1861b: 331 (Cent. 1, no. 46).—D.C.; N.H., Mass.
loripedis Coquillett, 1895m: 419.—Calif.
luctifera Loew, 1861b: 333 (Cent. 1, no. 50).—N.Y.; N.H., N.C.
luctuosa Loew, 1872a: 114 (Cent. 10, appendix) (n. name for *lugens* Loew).—Calif.
    lugens Loew, 1862b: 200 (Cent. 2, no. 30) (preocc. Zetterstedt, 1859).—Calif.
luteiventer Curran, 1929a: 7.—Man.
luteiventris Loew, 1864a: 79 (Cent. 5, no. 57).—N.H.; Ont. to N.B., s. to Va.
macilenta Loew, 1864a: 78 (Cent. 5, no. 55).—N.H.; Maine to Va.
mallos Walker, 1849: 502.—Hudson Bay Territory.
manca Coquillett, 1895m: 427.—N.C.; Maine to Fla.
minytus Walker, 1849: 502.—Ont.; Ont. and N.H., s. to Pa. and R.I.
    minutus (error) Johnson, 1910b: 760.
mutabilis Loew, 1862b: 198 (Cent. 2, no. 26).—Ill.; Ill. to N.B., s. to N.C.
nana Loew, 1861b: 341 (Cent. 1, no. 64).—Md.; Ariz., Mo., N.Y. and N.H. to Va.
nasoni Coquillett, 1895m: 423.—Ill.; Ariz., Kans., Mo.
    masoni, error.

nigricans Loew, 1864a: 80 (Cent. 5, no. 58).—N.H.; Ont., N.Y., Maine, ?Oreg.
nigrita Zetterstedt, 1838: 567.—Greenland; n. Europe, n. Asia. ?=*borealis*.
nitidivittata Macquart, 1846: 225 (1846: 97).—Tex.
novecarolina Beutenmüller, 1913b: 130.—N.C.
opacithorax Malloch, 1923i: 185.—Alaska (Pribilof Is.).
otiosa Coquillett, 1895m: 425.—Colo.; Que., N.J.
pachymera Bigot, 1887b: cxlii (also 1889: 133).—Calif.
parva Coquillett, 1895m: 433.—Mass.
pectinata Loew, 1861b: 333 (Cent. 1, no. 49).—D.C.; Md., Va.
pectoris Coquillett, 1895m: 420.—Ga.; N.C.
phemius Walker, 1849: 500.—Ont.; Alaska, B.C., Maine, Mass.
piligeronis Coquillett, 1895m: 432.—Ill.; Ont.
polita Loew, 1862b: 200 (Cent. 2, no. 29).—Ill.; N.Y. to Ga.
prava Chillcott, 1959: 266.—Que.; N.H.
priapulus Loew, 1861b: 335 (Cent. 1, no. 54).—Md.; Maine to Va.
pulla Loew, 1861b: 330 (Cent. 1, no. 44).—Conn.; Minn. to Que., s. to N.J.
pusio Loew, 1861b: 340 (Cent. 1, no. 63).—Md.; Maine to Md.
quinquelineata (Say), 1823: 95 (1859b: 82) (*Empis*).—Mo.; N.C.
rava Loew, 1862b: 198 (Cent. 2, no. 25).—Ill.; Nev., N.C.
    morissoni Bigot, 1887b: cxli (also 1889: 132).—Nev.
ravida Coquillett, 1895m: 418.—N. Mex., Tex., Ill.; N. Mex. to Ind. and N.C.
rufirostris Say, 1829: 159 (1859b: 355) (as *rufirostra*, error).—Ind.
rustica Loew, 1864a: 79 (Cent. 5, no. 56).—N.H.; Que. to N.C.
scaurissima Wheeler, 1896d: 189.—Calif.
scolopacea (Say), 1823: 96 (1859b: 83) (*Empis*).—Pa.; Conn., R.I., N.J., Md.
scutellaris Coquillett, 1895m: 429.—n. Calif.
    geniculata Bigot, 1887b: cxlii (also 1889: 134) (preocc. Meigen, 1830).—Calif.
sellata Loew, 1861b: 328 (Cent. 1, no. 42).—D.C.; Md.
setosa Coquillett, 1895m: 426.—N.H.; Alaska, Man. to Labr., s. to N.J.
similata Malloch, 1919f: 46.—N.W.T.
soccata Loew, 1861b: 342 (Cent. 1, no. 67).—Miss.; Mass.
sociabilis (Williston), 1893c: 76 (*Empis*).—Wash.; Wash. to N. Mex.
    abdita Coquillett, 1895m: 430.—Wash.
soleata Melander (for *argentea* Curran).—N.Y. **N. name.**
    argentea Curran, 1930j: 47 (preocc. Bigot, 1880).—N.Y.

sordida Loew, 1861b: 337 (Cent. 1, no. 58).—D.C.; N.Y. to R.I., s. to D.C.
    crassinervis Loew, 1861b: 338 (Cent. 1, no. 59).—N.Y.
stylata Coquillett, 1895m: 432.—Calif.
sudigeronis Coquillett, 1895m: 431.—Calif.; B.C., Oreg.
    nigrita Bigot, 1887b: cxlii (also 1889: 133) (preocc. Zetterstedt, 1838).—Calif.
tersa Coquillett, 1895m: 422.—N.H.; Ont. to Que., s. to Md.
testacea Loew, 1862b: 197 (Cent. 2, no. 24).—Ill.; Tex., N.Y. to D.C.
tristis Walker, 1857: 148.—U.S.
umbilicata Loew, 1861b: 342 (Cent. 1, no. 65).—Maine (as Mexico, error), Pa.; N.Y. to Maine, s. to D.C.
    ungulata Loew, 1861b: 342 (Cent. 1, no. 66).—Maine (as Mexico, error).
umbrosa Loew, 1864a: 77 (Cent. 5, no. 53).—N.H.; Que. to N.S., s. to N.J.
unimaculata Loew, 1862b: 201 (Cent. 2, no. 33).—Ill.; Md., D.C.
ursinella Melander, 1927: 209 (n. name for *ursina* Malloch).—N.W.T.
    ursina Malloch, 1919f: 46 (preocc. Oldenberg, 1915).—N.W.T.
valga Coquillett, 1895m: 428.—N.H.; Que. to Nfld., s. to N.Y.
vara Loew, 1861b: 337 (Cent. 1, no. 57).—Nebr.; Nebr. to Maine, s. to N.J.
versicolor Chillcott, 1959: 269.—Que.; Ont., N.Y., N.H.
villipes Coquillett, 1900h: 414 (1904: 28).—Alaska; N.W.T.
virgata Coquillett, 1895m: 430.—Mass.; Ont. to Nfld., s. to Pa.
vittata Loew, 1862b: 197 (Cent. 2, no. 23).—Ill.; S. Dak. to Mass., s. to Mo.
weedii (Coquillett), 1895m: 434 (*Neocota*).—Miss.

Subfamily CLINOCERINAE

Adults of this subfamily tend to inhabit situations very close to water, and some of them have been seen to enter the water for their prey. All the known larvae are aquatic.

Genus BOREODROMIA Coquillett

Boreodromia Coquillett, 1903a: 247, 260. Type-species, *Synamphotera bicolor* Loew (orig. des.).
Boreomyia (error) Aldrich, 1905: 316.
Synamphotera Loew of Melander, 1902a: 231.

    bicolor (Loew), 1863a: 18 (Cent. 3, no. 34) (*Synamphotera*).—Alaska; Wash.

Genus NIPHOGENIA Melander

Niphogenia Melander, 1927: 217. Type-species, *eucera* Melander (orig. des.).

eucera Melander, 1927: 217.—Wash.

Genus CERATEMPIS Melander

Ceratempis Melander, 1927: 218. Type-species, *longicornis* Melander (orig. des.).

longicornis Melander, 1927: 218.—Wash.

Genus PROCLINOPYGA Melander

Proclinopyga Melander, 1927: 220. Type-species, *amplectens* Melander (orig. des.).

amplectens Melander, 1927: 221.—Calif.
exporrecta Melander, 1927: 222.—Mont.
fistulator Melander, 1927: 222.—Que.
monogramma Melander, 1927: 222.—Mont; Wash., Idaho, Calif.
  var. **flavicoxa** Melander, 1927: 223.—Idaho; Wash.
solivaga Melander, 1927: 223.—Wash.

Genus OREOTHALIA Melander

Oreothalia Melander, 1902a: 232. Type-species, *pelops* Melander (mon.).

pelops Melander, 1902a: 233.—Idaho.
rupestris Vaillant, 1960: 118.—N.C.; Tenn.

Genus HELEODROMIA Haliday

Heleodromia Haliday, 1833: 159. Type-species, *immaculata* Haliday (Curtis, 1834: pl. 519).
Sciodromia Westwood, 1840: 132. Type-species, *Heleodromia immaculata* Haliday (orig. des.).

pullata (Melander), 1902a: 345 (*Sciodromia*).—N. Mex.

Genus ROEDERIODES Coquillett

Roederiodes Coquillett, 1901f: 585. Type-species, *juncta* Coquillett (orig. des.).
Roederioides (error) Curran, 1934a: 214.

REFERENCE: Chillcott, 1961b (rev.).

distinctus Chillcott, 1961b: 425.—Colo.
junctus Coquillett, 1901f: 585.—N.Y.; Ont., Que., N.H.
recurvatus Chillcott, 1961b: 424.—Que.; Ont., Maine.
retroversus Chillcott, 1961b: 427.—Calif.
vockerothi Chillcott, 1961b: 424.—Fla.
wirthi Chillcott, 1961b: 426.—N. Mex.

Genus CLINOCERA Meigen

Atalanta Meigen, 1800: 31. Type-species, *Clinocera nigra* Meigen (Coquillett, 1910b: 511). Suppressed by I.C.Z.N., 1963b: 339.
Clinocera Meigen, 1803: 271. Type-species, *nigra* Meigen (sub. mon., Meigen, 1804: 292).

Subgenus BERGENSTAMMIA Mik

Bergenstammia Mik, 1881: 326 (as genus). Type-species, *Clinocera nudipes* Loew (mon.).

brunnipennis Melander, 1927: 232.—Calif.
dolicheretma Melander, 1902a: 241.—Idaho.

Subgenus CLINOCERA Meigen

fuscipennis Loew, 1876: 324.—N.H.
lineata Loew, 1862b: 207 (Cent. 2, no. 50).—Pa.; Maine to Pa., ?Wash.
olivacea Melander, 1927: 233.—Alaska.
prasinata Melander, 1927: 232.—Calif.
trunca Melander, 1927: 230.—Wash.; ?Labr.

Subgenus HYDRODROMIA Macquart

Hydrodromia Macquart, 1835: 658 (as genus). Type-species, *Heleodromia stagnalis* Haliday (Coquillett, 1903a: 251).

appendiculata (Zetterstedt), 1838: 559 (*Wiedemannia*).—Lapland; N.W.T., Que.
    aucta Zetterstedt, 1849: 3019 (*Brachystoma*).—Lapland.
binotata Loew, 1876: 325.—N.Y.; Ont.
    bicincta (error) Tucker, 1907: 97.
conjuncta Loew, 1860b: 80.—D.C.; Labr. to D.C.
genualis Coquillett, 1910a: 124.—Alta.
longifurca Melander, 1927: 231.—N.H.; Labr.
longipes (Walker), 1849: 504 (*Heleodromia*).—Ont.
maculata Loew, 1860b: 79.—D.C.; Maine to D.C.

stagnalis (Haliday), 1833: 159 (*Heleodromia*).—Europe; Greenland, Iceland, Asia, Africa.
taos Melander, 1902a: 242.—N.H.; N.Y.
undulata Melander, 1927: 230.—Idaho; Wash.

Subgenus PHAEOBALIA Mik

Phaeobalia Mik, 1881: 326 (as genus). Type-species, *trinotata* Mik (Coquillett, 1903a: 255).
brevitibia Melander, 1927: 235.—Wash.
lecta Melander, 1902a: 243.—Idaho.

Genus WIEDEMANNIA Zetterstedt

Wiedemannia Zetterstedt, 1838: 559. Type-species, *borealis* Zetterstedt (Coquillett, 1903a: 258)=*bistigma* (Curtis). *Wiedemannia* of Zetterstedt, 1833: 207 and 1837: 32 are nomina nuda.

Subgenus PHILOLUTRA Mik

Philolutra Mik, 1881: 327 (as genus). Type-species, *Clinocera phantasma* Mik (Coquillett, 1903a: 255).
simplex (Loew), 1862b: 207 (Cent. 2, no. 49) (*Clinocera*).—Ont.; Alta., Wash., Mont.

Subgenus CHAMAEDIPSIA Mik

Chamaedipsia Mik, 1881: 326 (as genus). Type-species, *Clinocera hastata* Mik (mon.).
comata Melander, 1927: 232.—Wash.; Wyo.
ctenistes Melander, 1927: 234.—N.H.; Maine.
gubernans Melander, 1927: 234.—B.C.
hamifera Melander, 1927: 233.—Maine, N.Y., Conn.
lepida (Melander), 1902a: 241 (*Clinocera*).—Idaho.
minor Melander, 1927: 233.—N.Y.

Subgenus ROEDERIA Mik

Roederia Mik, 1881: 326. Type-species, *Clinocera longipennis* Mik (mon.).
Atalanta, subg. *Roederella* Engel, 1918: 79. Type-species, *Roederia czernyi* Bezzi (orig. des.; misident.)=*longipennis* (Mik).
fumosa Vaillant, 1960: 119.—N.Y.
saltans Vaillant, 1960: 122.—N.Y.

Genus DOLICHOCEPHALA Macquart

Dolichocephala Macquart, 1823: 147. Type-species, *maculata* Macquart (mon.)=*irrorata* (Fallén).
Ardoptera Macquart, 1827b: 105. Type-species, *Tachydromia irrorata* Fallén (mon.).

argus Melander, 1927: 246.—Wash.
irrorata (Fallén), 1815: 13 (*Tachydromia*).—Europe; Wash., Idaho, N.H., Vt.

Subfamily HEMERODROMIINAE
(Phyllodromiinae)

The oldest name for this subfamily is Phyllodromiinae, but we elect to retain Hemerodromiinae because of its long established usage, following Art. 23d of the new International Code (Stoll et al., 1961: 23).

Adults of this subfamily are found in shady foliage overhanging streams. All the known larvae are aquatic.

REFERENCE: Melander, 1947 (rev.).

Genus HEMERODROMIA Meigen

Hemerodromia Meigen, 1822: 61. Type-species, *Tachydromia oratoria* Fallén (Rondani, 1856: 148). Acceptance of Westwood's (1840: 132) type-designation of *mantispa* Panzer, which equals *Phyllodromia melanocephala* (Fabricius), would require that *Hemerodromia* replace *Phyllodromia* Zetterstedt, 1837. Action by the I.C.Z.N. is needed to preserve the present widespread usage of these two names. See Melander, 1927: 252.

brevifrons Melander, 1947: 248 (as *empiformis* var.).—Calif.
brunnea Melander, 1927: 255.—Ga.
captus Coquillett, 1895m: 391.—N.Y.; Minn. to Que., s. to Va.
coleophora Melander, 1927: 256 (as *empiformis* ssp.).—Mont.; B.C., Wash.
empiformis (Say), 1823: 99 (1859b: 85) (*Ochthera*).—Ill.; Minn. to Maine, s. to Fla.
  empidiformis, error.
exhibitor Melander, 1947: 248 (as *empiformis* var.).—Ga.
haruspex Melander, 1947: 249.—Fla.
jugulator Melander, 1927: 256.—N.Y.; Wis. and Ont. to Mass., s. to Md.

melanosoma Melander, 1947: 250.—Ont.; N.Y., Conn.
rogatoris Coquillett, 1895m: 392.—N.C.; B.C. to Que., s. to Calif. and Ga.
stellaris Melander, 1947: 251.—Tex.
sufflexa Melander, 1947: 248 (as *empiformis* var.).—Idaho; Wash.
superstitiosa Say, 1824a: 376 (1859a: 256).—Northwest Territory (U.S.); Ont. to N.H., s. to Tex. and Fla.
vates Melander, 1947: 252.—Conn.
vittata Loew, 1862b: 210 (Cent. 2, no. 56).—D.C.; Conn.

Genus CHELIFERA Macquart

Chelifera Macquart, 1823: 150. Type-species, *raptor* Macquart (mon.) = *precatoria* (Fallén).
Mantipeza Rondani, 1856: 148. Type-species, *Hemerodromia monostigma* Meigen (orig. des.).

banksi Melander, 1947: 253.—Conn., N.C.
cirrata Melander, 1947: 254.—Wyo.
ensifera Melander, 1947: 254.—Wash.
lovetti Melander, 1947: 255.—Oreg.; Wash., Idaho.
notata (Loew), 1862b: 209 (Cent. 2, no. 53) (*Hemerodromia*).—Ill.; Ill. to Maine.
obsoleta (Loew), 1862b: 208 (Cent. 2, no. 52) (*Hemerodromia*).—Ill.; N.Y. and Mass. to Fla.
palloris (Coquillett), 1895m: 392 (*Mantipeza*).—N.H.; Wash. to Que., s. to Pa.
precatoria (Fallén), 1815: 10 (*Tachydromia*).—Sweden; Ont., n. Europe.
rastrifera Melander, 1947: 257.—N.C.
scrotifera Melander, 1947: 257.—Alaska.
valida (Loew), 1862b: 208 (Cent. 2, no. 51) (*Hemerodromia*).—Ont.; Alaska, N.Y. ?=*precatoria*.
varix Melander, 1947: 258.—Idaho, Mont.

Genus METACHELA Coquillett

Metachela Coquillett, 1903a: 253, 263. Type-species, *Hemerodromia collusor* Melander (orig. des.).

albipes (Walker), 1849: 505 (*Hemerodromia*).—Ont.; Labr., N.H.
collusor (Melander), 1902a: 235 (*Hemerodromia*).—Wyo.; Wash. to Mont., s. to Calif. and Colo.

Genus NEOPLASTA Coquillett

Neoplasta Coquillett, 1895m: 392. Type-species, *Hemerodromia scapularis* Loew (orig. des.).

hebes Melander, 1947: 261.—Calif.; Wash., Oreg., Idaho.
scapularis (Loew), 1862b: 209 (Cent. 2, no. 54) (*Hemerodromia*).— Md.; B.C. to Maine, s. to Calif. and Ga.
  maculipes Bigot, 1887d: 118 (*Clinocera*).—Calif.
  var. **alleghani** Melander, 1947: 263.—N.Y.; N.Y. to Ga.
  var. **megorchis** Melander, 1947: 263.—Wash.; Wash. to Mont., s. to Calif. and Colo.
  var. **radialis** Melander, 1947: 264.—Vt.

Genus THANATEGIA Melander

Chelifera, subg. **Thanategia** Melander, 1927: 263. Type-species, *Hemerodromia defecta* Loew (orig. des.).

defecta (Loew), 1862b: 210 (Cent. 2, no. 55) (*Hemerodromia*).— D.C.; Mass., Conn., N.J.
  deflecta, error.
recurvata Melander, 1947: 260.—B.C.; Wash.
stuprator Melander, 1947: 259.—Wash.

Genus CHELIPODA Macquart

Chelipoda Macquart, 1823: 148. Type-species, *Tachydromia mantispa* Panzer (orig. des.; misident.) = *vocatoria* (Fallén).
Chiromantis Rondani, 1856: 205 (also as *Chyromantis*, p. 148; *Chiromantis* preocc. Peters, 1854). Type-species, *Tachydromia vocatoria* Fallén (orig. des.).
Litanomyia Melander, 1902a: 231. Type-species, *Sciodromia mexicana* Wheeler and Melander (Coquillett, 1903a: 252).
Phyllodromia, authors, not Zetterstedt.

albiseta (Zetterstedt), 1838: 544 (*Hemerodromia*).—Sweden; N.Y., Mass., n. and cent. Europe.
contracta Melander, 1947: 265.—Mass.; B.C., Wash., Mont., Ont. to Maine, s. to Conn.
elongata (Melander), 1902a: 232 (*Litanomyia*).—S. Dak. (Melander, 1947: 267); S. Dak., Wis. to R.I., s. to Fla.
praestans Melander, 1947: 267.—Conn.; Ont. to Maine, s. to Va.
sicaria Melander, 1947: 268.—Tenn., N.C.
vocatoria (Fallén), 1815: 12 (*Tachydromia*).—Sweden; N.Y., N.J.

Genus PHYLLODROMIA Zetterstedt

Hemerodromia, subg. **Phyllodromia** Zetterstedt, 1837: 31. Type-species, *Empis melanocephala* Fabricius (Rondani, 1856: 148).

americana Melander, 1947: 269.—N.H.; N.Y., R.I., Va., Ga.

Subfamily TACHYDROMIINAE

Genus TACHYPEZA Meigen

Tachypeza Meigen, 1830: 341. Type-species, *Tachydromia nervosa* Meigen (Rondani, 1856: 147)=*nubila* (Meigen).

Tachypeza adults are often found running on the smooth bark of trees with their wings folded. They will not fly easily.

annularis Melander, 1927: 274.—Wash., Calif.
binotata Melander, 1927: 275.—Wash., Calif.
brachialis (Melander), 1902a: 343 (*Tachydromia*).—N.J.; Ont., N.H., Va.
clavipes Loew, 1864a: 86 (Cent. 5, no. 73).—Ill.; Pa., N.J., Md.
corticalis (Melander), 1902a: 343 (*Tachydromia*).—N. Mex.; B.C. to Calif. ?=*portaecola*.
discifera Melander, 1927: 273.—Idaho, Mont.; B.C., Alta.
distans Melander, 1927: 273.—Wash.
dolorosa Melander, 1927: 275.—N. Mex.
excisa Melander, 1927: 274.—N.Y.; Que.
fenestrata (Say), 1823: 95 (1859b: 82) (*Sicus*).—Middle States; Wis. to Maine, s. to Mo. and Va.
    similis Walker, 1849: 506 (*Tachydromia*).—Hudson Bay Territory.
    rapax Loew, 1864a: 85 (Cent. 5, no. 71).—Ill.
humeralis Melander, 1927: 276.—D.C.; Ont., Que.
inusta (Melander), 1902a: 226 (*Tachydromia*).—Idaho, N. Mex.; Wash., Oreg., Calif., Colo.
portaecola (Walker), 1849: 506 (*Tachydromia*).—Hudson Bay Territory.
postica (Walker), 1857: 149 (*Tachydromia*).—U.S.; N. Mex., Kans., D.C.
pruinosa Coquillett, 1903a: 267.—Mo.
rostrata Loew, 1864a: 86 (Cent. 5, no. 72).—N.H.; Maine to Va.
vittipennis (Walker), 1857: 149 (*Tachydromia*).—U.S.
winthemi Zetterstedt, 1838: 548.—Europe; N.W.T., B.C., Alta., Man., Mont., Que., Labr., N.H., Mass., n. Europe, n. Asia.

Genus TACHYDROMIA Meigen

Sicus Latreille, 1796: 158 (preocc. Scopoli, 1763). Type-species, *Musca cimicoides* Fabricius (sub. mon., Latreille, 1805: 312)=*arrogans* (Linnaeus).
Coryneta Meigen, 1800: 27. Type-species, *Tachydromia connexa* Meigen (Coquillett, 1910b: 528). Suppressed by I.C.Z.N., 1963b: 339.
Tachydromia Meigen, 1803: 269. Type-species, *Musca cimicoides* Fabricius (Curtis, 1833: pl. 477; misident.)=*connexa* Meigen.
Sicodus Rafinesque, 1815: 131 (n. name for *Sicus* Latreille). Type-species, *Musca cimicoides* Fabricius (aut.)=*arrogans* (Linnaeus).
Phoneutisca Loew, 1863a: 19 (Cent. 3, no. 35). Type-species, *bimaculata* Loew (mon.).
Tachista Loew, 1864g: 15 (as *Tachysta*, p. 19). Type-species, *Musca cimicoides* Fabricius (as Meigen; Coquillett, 1903a: 258; misident.)=*connexa* Meigen.

Tachydromia adults run rapidly over the ground or on tree trunks and leaves and rarely take flight.

bimaculata (Loew), 1863a: 19 (Cent. 3, no. 35) (*Phoneutisca*).—Alaska; B.C., Idaho, Mont.
    maculipennis Walker of Coquillett, 1903a: 266.
chelana Melander, 1927: 282.—Wash.
diversipes Melander, 1910: 55 (as *schwarzii* var.).—Tex.
enecator Melander, 1902a: 226.—Wyo., Que.; Alaska, Maine.
harti Malloch, 1919i: 248.—Ill.; Ind.
hirtipes Melander, 1927: 282.—Oreg.
maculipennis Walker, 1849: 507.—Ont.; Alta. to Vt., s. to Mexico and Fla.
    pusilla Loew, 1864a: 87 (Cent. 5, no. 74).—Ill.
    bimaculata Loew of Melander, 1902a: 204.
monaca Melander, 1927: 283.—Wash.
phengites Melander, 1927: 282.—Ont., Va.
pseliophora Melander, 1927: 281.—Calif.
schwarzii Coquillett, 1895m: 440.—Calif., Utah; Wash. to Calif. and Tex., Mexico.
tacoma Melander, 1927: 281.—Wash., Idaho.
varipennis Coquillett, 1903a: 266.—N.H.; Wash., Oreg., Md.

Genus TACHYEMPIS Melander

Tachyempis Melander, 1927: 288. Type-species, *Tachydromia agens* Melander (orig. des.).

agens (Melander), 1910: 59 (*Tachydromia*).—Wash.
calva (Melander), 1910: 58 (*Tachydromia*).—Ga.
cinerea Melander, 1927: 290.—N. Mex.
longipennis Melander, 1958: 296.—Ind.
nervosa Melander, 1927: 290.—Calif.
universalis (Melander), 1910: 60 (*Tachydromia*).—Tex., Ill., Pa.

Genus CHARADRODROMIA Melander

Charadrodromia Melander, 1927: 292. Type-species, *microphona* Melander (orig. des.).

arnaudi Melander, 1960: 129.—Calif.
microphona Melander, 1927: 293.—Wash.
syletor Melander, 1927: 293.—Wash.

Genus CHERSODROMIA Walker

Chersodromia Walker, 1851a: 137. Type-species, *Tachypeza brevipennis* Zetterstedt (Rondani, 1856: 147) = *arenaria* (Haliday).
Coloboneura Melander, 1902a: 229. Type-species, *inusitata* Melander (mon.).
Thinodromia Melander, 1906: 370. Type-species, *inchoata* Melander (mon.).

Chersodromia adults occur on the wet or dry beaches of both fresh and salt water.

REFERENCE: Melander, 1945 (key).

cana Melander, 1945: 82.—Calif.
houghii (Melander), 1902a: 206 (*Stilpon*).—Mass.; N.Y., Conn.
inchoata (Melander), 1906: 370 (*Thinodromia*).—Calif.
insignita Melander, 1945: 81.—Calif.
inusitata (Melander), 1902a: 230 (*Coloboneura*).—Mass., Fla.; Maine to Fla.
magacetes Melander, 1945: 80.—Calif.
nana (Coquillett), 1903a: 267 (*Coloboneura*).—Fla.
parallela (Melander), 1927: 297 (*Thinodromia*).—Wash.

Genus MICREMPIS Melander

Micrempis Melander, 1927: 298. Type-species, *nana* Melander (orig. des.).

minuta (Melander), 1902a: 339 (*Stilpon*).—N. Mex.
nana Melander, 1927: 299.—Tex.
obliqua Melander, 1927: 299.—Iowa.
testacea Melander, 1927: 299.—Va.; Md.

Genus SYMBALLOPHTHALMUS Becker

Macroptera Becker, 1889a: 80 (preocc. Lioy, 1864). Type-species, *pictipes* Becker (mon.)=*dissimilis* (Fallén).
Symballophthalmus Becker, 1889b: 285 (n. name for *Macroptera* Becker). Type-species, *Macroptera pictipes* Becker (aut.)=*dissimilis* (Fallén).
masoni Chillcott, 1958b: 647.—Ont.; Mich.

Genus MEGAGRAPHA Melander

Megagrapha Melander, 1927: 301. Type-species, *Drapetis pubescens* Loew (orig. des.).
REFERENCE: Chillcott, 1958a (rev.).
pubescens (Loew), 1862b: 210 (Cent. 2, no. 57) (*Drapetis*).—N.Y.; Alta., Ont. to Mass., s. to Ill. and Ga.
  exquisita Malloch, 1923g: 5 (*Coloboneura*).—Md.

Genus STILPON Loew

Agatachys Meigen, 1830: 343. Unavailable, cited in specific synonymy.
Drapetis, subg. **Stilpon** Loew, 1859a: 34. Type-species, *Tachydromia graminum* Fallén (Coquillett, 1903a: 257).

Subgenus STILPON Loew

curvipes Melander, 1927: 303.—Wis.; Ont., Que., N.Y.
pauciseta Melander, 1927: 304.—Mass., Md.
pectiniger Melander, 1902a: 205.—Wis., Mass.
spinipes Melander, 1927: 303.—Ga.
varipes Loew, 1862b: 211 (Cent. 2, no. 58).—Pa.

Subgenus TETRANEURELLA Dahl

Tetraneurella Dahl, 1909: 362 (as genus). Type-species, *beckeri* Dahl (orig. des., as gen. n., sp. n.)=*graminum* (Fallén) var.
pleuriticus Melander, 1927: 302.—N.H.; Que., Pa., N.C., Fla.

Genus DRAPETIS Meigen

Drapetis Meigen, 1822: 91. Type-species, *exilis* Meigen (mon.).

Subgenus DRAPETIS Meigen

aliternigra Melander, 1918: 192.—B.C., Wash., S. Dak., Tex., Que., N.Y., Mass., Pa.; Ill.
    nigra, authors, not Meigen.
assimilis (Fallén), 1815: 8 (*Tachydromia*).—Europe; Canada, S. Dak., Ill.
    nigra Meigen, 1830: 344.—Europe.
bispina Melander, 1918: 192.—Ind.
deceptor Curran, 1929a: 7.—Man.
divergens Loew, 1872a: 90 (Cent. 10, no. 62).—Tex.; Utah, N. Mex. to Ga. to D.C., Cent. Amer., West Indies.
diversa Melander, 1918: 193.—N. Mex.
dividua Melander, 1902a: 208.—Idaho; B.C. to Oreg. and N. Mex.
    nigripes Melander, 1902a: 339 (*Stilpon*).—N. Mex.
infumata Melander, 1918: 194.—B.C.; Idaho.
latipennis Melander, 1902a: 209.—Kans.; Wis.
mariae Steyskal, 1953b: 257.—Mich.
micropyga Melander, 1918: 195.—Wash.; B.C. to Calif., ?D.C.
naica Melander, 1918: 195.—Wash., Idaho, N. Mex., Tex.
pilosa Melander, 1918: 196.—Ind.; Ont., Alta., Ill.
populi Steyskal, 1953b: 256.—Mich.
setulosa Melander, 1918: 196.—Wash.; B.C., Calif.
trichura Melander, 1918: 197.—Tex.; Calif.

Subgenus CROSSOPALPUS Bigot

Crossopalpus Bigot, 1857b: 557, 563 (as genus). Type-species, *Platypalpus ambiguus* Macquart (mon.) =*flexuosa* Loew.
Drapetis, subg. *Eudrapetis* Melander, 1918: 187. Type-species, *spectabilis* Melander (orig. des.).

armata Melander, 1918: 197.—Wash., Idaho, Maine, Mass.; Alaska to N.S., Mont., Va.
discalis Melander, 1918: 198.—Wash., Calif.
diversipes Melander, 1918: 198.—Calif.
facialis Melander, 1918: 200.—Alta., Ga.
gilvipes Loew, 1872a: 89 (Cent. 10, no. 61).—Tex.; Maine, Mass., Conn., West Indies.
inculta (Coquillett), 1895m: 439 (*Platypalpus*).—Calif.; Tex.
lata (Coquillett), 1903a: 266 (*Tachydromia*).—Fla.
medetera Melander, 1902a: 208.—Wyo., Colo., Ariz.; Wash., Mont., Man., Utah.

parvicornis Melander, 1918: 202.—Wash., D.C.; Calif., Maine to D.C.
plumipes Melander, 1918: 203.—Tex.; Ala., Mexico.
scissa Melander, 1918: 204.—Wash., Idaho, Wyo.; B.C. and Wash. to Man. and Wyo.
>*medetera* (in part) Melander, 1902a: 208. Idaho specimen only.

septentrionalis Melander, 1902a: 211.—Mich.; Wash. to Vt., s. to Va.
spectabilis Melander, 1902a: 212.—Mass.; Maine to N.Y. and R.I.
unipila Loew, 1872a: 88 (Cent. 10, no. 60).—Tex.; Wash. to Iowa, s. to Calif. and Tex.
>var. **nitida** Melander, 1902a: 207 (as sp.).—Calif., Tex.; Wash.

xanthopoda Williston, 1896c: 308.—West Indies (St. Vincent I.); Tex., Ga.

Subgenus ELAPHROPEZA Macquart

Elaphropeza Macquart, 1827b: 86 (as genus). Type-species, *Tachydromia ephippiata* Fallén (mon.).

vittata Melander, 1918: 214.—Fla.; Cuba.

Genus PLATYPALPUS Macquart

Platypalpus Macquart, 1827b: 92. Type-species, *Musca cursitans* Fabricius (Westwood, 1840: 132).
Tachydromia, authors, not Meigen.

achlytarsis Chillcott, 1962: 124.—N.W.T.; Y.T., Man., Que., Labr.
aequalis Loew, 1864a: 88 (Cent. 5, no. 75).—Ill.; Oreg. to Que., s. to Calif., La., and N.C., Mexico.
alexippus Walker, 1849: 510.—Ont.
alumnus Melander, 1927: 330.—Wyo.
anatolicus Chillcott, 1962: 132.—Que.; Labr., N.H.
apicalis Loew, 1864a: 90 (Cent. 5, no. 79).—Pa.; N.Y., N.H., Mass., R.I.
arcticus Melander, 1927: 344.—Labr.
armillatus Melander, 1927: 344.—Wash.; Idaho, Mont.
ballistrarius Melander, 1927: 338.—Va.
ballucatus Melander, 1927: 336.—Wash., Ill.
bicornis Melander, 1927: 329.—Wash.
callithrix Melander, 1927: 325.—Ill.; N.Y.
canus Melander, 1902a: 220.—Calif.
carectorum Chillcott, 1962: 134.—N.W.T.; Alaska, Y.T.
cellarius Melander, 1927: 337.—B.C.; Wyo.
churchillensis Chillcott, 1962: 136.—Man.
continguus Melander, 1927: 320.—B.C.

coquilletti Melander, 1924a: 83.—Mass.; N.Y.
    trivialis Loew of Melander, 1902a: 216.
crassifemoris (Fitch), 1856a: 533 (1856: 301) (*Oscinis*).—N.Y.;
    B.C. and Wash. to Que., s. to Oreg. and Fla.
    var. **debilis** Loew, 1863a: 20 (Cent. 3, no. 37) (as sp.).—D.C.;
        Mass.
    var. **melanocerus** Melander, 1927: 337.—Wash.
    var. **mollis** Melander, 1927: 337.—Wash., Idaho, Wis., Pa.
crepidarius Melander, 1927: 340.—Wash.
cuneipennis Melander, 1924a: 83.—Vt.; Wis.
decolor Melander, 1927: 321.—Mont.
direptor Melander, 1927: 319.—B.C.
discifer Loew, 1863a: 20 (Cent. 3, no. 36).—D.C.; N.Y., ? Utah.
dissimilipes Melander, 1927: 318.—Idaho; B.C., Wash., Oreg., Colo.
diversipes Coquillett, 1900h: 422 (1904: 36).—Alaska.
enervatus Melander, 1927: 333.—Calif.
fasciventris Melander, 1927: 343.—Wash.
flammifer Melander, 1924a: 84.—N.Y.; B.C., Mont., Ont., Que.,
    N.H., Vt., Mass.
flavirostris Loew, 1864a: 90 (Cent. 5, no. 80).—N.H.; Alaska, Wash.
    to Mont., Ont. to Maine, s. to N.Y. and Mass.
    var. **dilutior** Melander, 1927: 325.—Wash.
    var. **microcerus** Melander, 1927: 325.—Idaho.
    var. **vittiger** Melander, 1927: 323.—Alaska; Mont.
gesticulor Melander, 1927: 328.—Mont.
glacialis Melander, 1927: 331.—Mont.
gravidus Melander, 1902a: 221.—Calif.
harpestylis Chillcott, 1962: 135.—N.W.T.; Alaska, Y.T., Man.
harpiger Melander, 1924a: 84.—Mass.
hastatus Melander, 1902a: 222.—Kans.; Idaho, Colo., Que., N.Y.
hians Melander, 1902a: 220.—Colo.; Maine.
    var. **fuscohalteratus** Melander, 1924a: 85.—N.Y.; Colo., Ont. and
        N.Y. to N.B.
holosericus Melander, 1924a: 85.—Mass.; Iowa, Ont., Que., N.B.,
    N.S.
hyaenoides Melander, 1927: 332 (as *juvenis* var.).—Wash.; B.C.,
    Alta. [See Chillcott, 1962: 126.]
impexus Melander, 1902a: 219.—S. Dak.; Mich.
incurvus Melander, 1902a: 221.—Calif.
inferialis Melander, 1927: 321.—Wash.; Idaho.
inops Melander, 1902a: 220.—Wyo.; B.C., Mont.
    var. **aequicornis** Melander, 1927: 330.—Mont.
juvenis Melander, 1927: 332.—B.C. (lectotype Chillcott, 1962:
    128); B.C. to Calif. and Wyo., Maine.

lacertosus Melander, 1927: 319.—Idaho.
laetabilis Melander, 1927: 324.—N.Y.
laetus Loew, 1864a: 91 (Cent. 5, no. 81).—N.H.
lateralis Loew, 1864a: 89 (Cent. 5, no. 78).—N.H.; Alaska, Maine to N.J.
    collateralis Melander, 1927: 327.—Alaska. [See Chillcott, 1962: 128.]
luctator Melander, 1927: 335.—Idaho.
lupatus Melander, 1902a: 340.—N. Mex.
lyristes Melander, 1927: 338.—Ariz.
manni Chillcott, 1962: 142.—B.C.
masoni Chillcott, 1962: 121.—B.C.
melanogaster Melander, 1927: 323.—Labr.
melleus Melander, 1927: 321.—N.Y.
mesogrammus Loew, 1863a: 21 (Cent. 3, no. 38).—D.C.; N.Y. to Tenn. and N.C.
mimus Melander, 1927: 324.—N.Y.
monticola Melander, 1902a: 217.—Colo.
murphyi Chillcott, 1962: 133.—N.W.T.; Alaska.
nitidipleura Melander, 1927: 329.—Mont.
ochricollis Melander, 1927: 323.—Wash.; Utah.
oculeus Melander, 1927: 342.—Pa.
pachycnemus Loew, 1864a: 89 (Cent. 5, no. 77).—D.C.; N.Y., N.J.
paluster Chillcott, 1962: 137.—Y.T.; Alaska, N.W.T.
pectinator Melander, 1924a: 85.—Idaho; Alaska, B.C., Wash. and Alta. to Mass.
peneprorsus Chillcott, 1962: 130.—Alaska; Y.T., B.C.
pilatus Melander, 1927: 340.—Wash.
pluto Melander, 1902a: 217.—Calif.; Idaho.
politellus Melander, 1927: 382 (n. name for *politus* Melander).—Calif.; Mo., Md., D.C.
    politus Melander, 1927: 333 (preocc. Collin, 1926).—Calif.
    var. **nitens** Melander, 1927: 334.—Calif., Mo., D.C.; Md.
porrectus Melander, 1924a: 86.—Wash.; Calif., Colo., Maine.
    var. **suffasciatus** Melander, 1927: 336.—Calif., Colo.
postpositus Melander, 1927: 343.—Wash.
prorsus Melander, 1927: 328.—Que. (as Labr., error; Chillcott, 1962: 124); Man., Labr., N.H.
pubescens Melander, 1927: 331.—Wash.
pudens Melander, 1927: 329.—Wash.; Idaho.
puerinus Melander, 1927: 332 (as *juvenis* var.).—B.C. (lectotype Chillcott, 1962: 120); Wash., ?Maine. [See Chillcott, 1962: 120.]

pulverulentus Melander, 1927: 341.—Calif.
recurvus Melander, 1927: 339.—Calif.; Oreg.
richardsi Chillcott, 1962: 139.—Que.; Labr., N.H.
rubefactus Melander, 1927: 322.—Ill.; Md.
rufiventris Melander, 1902a: 341.—N. Mex.
satyriacus Melander, 1927: 327.—Que. (as Labr., error; Chillcott, 1962: 141); Y.T., B.C., Sask., Mich.
sericatus Melander, 1927: 320.—B.C.
simplicipes Melander, 1927: 331.—Wash.
soccatus Melander, 1927: 341.—Idaho.
spinosus Melander, 1927: 319.—Calif.
splendens Melander, 1927: 365 (n. name for *montanus* Melander).—Colo.
    montanus Melander, 1902a: 213 (*Elaphropeza;* preocc. Becker, 1887).—Colo.
sutor Melander, 1924a: 87.—Wash.; B.C., Maine, N.Y., Mass.
tachistiformis Melander, 1927: 320.—Mont.
talaris Melander, 1927: 342.—Wash.; Wis., Ont., Que.
tenax Melander, 1927: 339.—Wash.
tenellus Melander, 1902a: 223.—S. Dak., Ill.; Ariz.
tenuis Melander, 1927: 326.—Wash.; B.C.
tersus Coquillett, 1895m: 439.—N.C., Ga.; Iowa, N.Y., La.
trivialis Loew, 1864a: 88 (Cent. 5, no. 76).—Maine, D.C.; Minn. to Que., s. to N.J.
truncatus Chillcott, 1962: 122.—B.C.
valgus Melander, 1927: 338.—Wash.
velox Melander, 1927: 336.—Wash.
venaticus Melander, 1927: 335.—Wash.; B.C., Idaho.
verpus Melander, 1927: 333.—Wash.
versipes Melander, 1927: 318.—Va.
versutus Melander, 1924a: 87.—Va.; Wis., N.Y., Vt.
vicarius Walker, 1857: 148.—U.S.; Maine.
vierecki Melander, 1902a: 340.—N. Mex.
vittatus Melander, 1927: 334.—Wash., Wyo.; Alta.
    var. **perimerus** Melander, 1927: 334.—Alta.
vulnificus Melander, 1927: 339.—Mont.
xanthochiton Melander, 1927: 322.—Wash.
xanthopodus Melander, 1927: 367 (n. name for *gilvipes* Coquillett).—Alaska; B.C., Y.T., Ont., Que., Labr.
    xanthopus, error.
    gilvipes Coquillett, 1900h: 422 (1904: 36) (preocc. Meigen, 1822).—Alaska.

Family DOLICHOPODIDAE
(Dolichopidae)

By Richard H. Foote, Jack R. Coulson and Harold Robinson

Flies of the family Dolichopodidae are ubiquitous, although many species have restricted and localized habitats near water or in moist places. The adults are predaceous and are found on foliage, tree trunks, or damp earth, usually in swamps or along lightly shaded streams where they prefer small areas of sunlight. The males of some species, especially those of the genus *Dolichopus*, have characteristic mating dances. As far as known, most dolichopodid larvae are predaceous, and, except for a few species which mine stems, occur under the bark of trees, or are found in decaying vegetation, they are aquatic. Dolichopodids typically pupate within a cocoon made by the larva from available materials such as sand, mud, or bits of wood.

Because the relationships of the Nearctic and Old World faunas are poorly understood, many undetected synonyms probably exist in the present listing of Nearctic species. Further, much taxonomic work is yet needed to determine the best generic and subfamily placements of some of the species listed below. Hence, the present catalog is to be considered provisional; there is no up-to-date revision of the family in the Nearctic Region.

REFERENCES: Lundbeck, 1912 (rev., Denmark, biol.); Becker, 1922 (rev., Nearctic and Neotropical); Stackelberg, 1930–1941 (rev., Palearctic); Parent, 1938 (rev., France); Smith, 1952b (biol., immature stages); Wirth and Stone, 1956 (keys, biol. aquatic forms).

Subfamily SCIAPODINAE
(Psilopodinae, Chrysosomatinae, Agonosomatinae)

Genus MESORHAGA Schiner

Mesorhaga Schiner, 1868: 217. Type-species, *tristis* Schiner (orig. des.).
Aptorthus Aldrich, 1893b: 48. Type-species, *albiciliatus* Aldrich (Coquillett, 1910b: 509).

REFERENCE: Parent, 1929b (key).

albiciliata (Aldrich), 1893b: 48 (*Aptorthus*).—N.J.; N.C., Ga.
borealis (Aldrich), 1893b: 49 (*Aptorthus*).—Minn.
caerulea Van Duzee, 1930a: 1.—Conn.
caudata Van Duzee, 1915d: 94.—Ga.
clavicauda Van Duzee, 1925c: 154.—Mich.; Ill.
flavipes Van Duzee, 1932b: 9.—N.Y.

jucunda Becker, 1922: 377.—Ga., Paraguay.   Neotropical specimen probably different sp.
nigripes (Aldrich), 1893b: 49 (*Aptorthus*).—Calif.
pallidicornis Van Duzee, 1925e: 178.—Ohio; Man. to Ont., s. to Tex.
pallicornis, error.
townsendii (Aldrich), 1893b: 50 (*Aptorthus*).—Ariz.; Utah. ?Mass., ?N.J.
tristis Schiner.—Not Nearctic.
varipes Van Duzee, 1917a: 123.—Mass.; N.J.

Genus CONDYLOSTYLUS Bigot

Condylostylus Bigot, 1859b: 215, 223.   Type-species, *Psilopus bituberculatus* Macquart (mon.).
Laxina Curran, 1934a: 230.   Type-species, *Dolichopus patibulatus* Fabricius (orig. des.).
Psilopodinus, authors, not Bigot.
Psilopus, authors, not Meigen.
Sciapus, authors, not Zeller.

Condylostylus is a complex genus with wide distribution. In the Tropics, this genus and closely related ones are the predominant dolichopodids. *Laxina* is not a natural group as previously recognized; its relationship to *Condylostylus* and *Sciapus* needs further study.

REFERENCES: Van Duzee, 1915b (key, as *Sciapus*); Parent, 1929a (key); Curran, 1942a (rev., *Laxina*).

albicoxa (Walker), 1849: 651 (*Psilopus*).—Ohio, Mass., N.S. Unrecognized.
banksii (Van Duzee), 1915b: 23 (*Sciapus*).—Va.; N.Y.
calcaratus (Loew), 1861a: 93 (*Psilopus*).—Carolina; Iowa to Que., s. to Carolina.
caudatus (Wiedemann), 1830: 224 (*Psilopus*).—Ga.; Calif. to Que., s. to Fla.
virgo Wiedemann, 1830: 224 (*Psilopus*).—N.Y.
caudatulus Loew, 1861a: 93 (*Psilopus*).—Miss.
chalybeus (Van Duzee), 1914a: 390 (*Sciapus*).—Pa.
chrysoprasi (Walker), 1849: 646 (*Psilopus*).—West Indies; N.C., Fla., Mexico, Cent. Amer.
chrysoprasius, emend.
ciliipes Aldrich, 1901: 355 (*Psilopus*).—Mexico.
clavatus (Van Duzee), 1929: 15 (*Psilopus*).—Panama; ?Mass.
coloradensis Van Duzee, 1932b: 4.—Colo.; Nebr.

comatus (Loew), 1861a: 89 (*Psilopus*).—Middle States; Kans., Iowa, N.J., Fla.
    comatus Schiner (error) of authors.
connectans (Curran), 1942a: 61 (*Laxina*).—Conn.; Mass., N.C.
crinitus (Aldrich), 1904: 283 (*Psilopodinus*).—Kans., Fla.; Ga.
    albiapicatus Parent, 1929a: 206.—Ga.
debilis Becker, 1922: 356.—Ga.
delicatus (Walker), 1849: 645 (*Psilopus*).—N.Y.; N.C.
    pallescens Bigot, 1888c: xxix (also 1890: 289) (*Psilopodinus*).—N.C.
diffusus (Wiedemann), 1830: 221 (*Psilopus*).—Brazil (as Savannah, error; Loew, 1864e: 234). Not Nearctic.
dimidiatus (Loew), 1862b: 216 (Cent. 2, no. 70) (*Psilopus*).—Mexico; Ariz.
erectus Becker, 1922: 296.—N.Y., Paraguay, Argentina. Neotropical specimens probably different sp.
femoratus (Say), 1823: 86 (1859b: 76) (*Dolichopus*).—Pa. Unrecognized.
flavipes (Aldrich), 1904: 284 (*Psilopodinus*).—S. Dak., Mass.; Man. to Maine, s. to S. Dak. and Fla.
    femoratus Say of Say, 1829: 168 (1859b: 361).
furcatus (Van Duzee), 1915b: 21, key (descr., 1915d: 90) (*Sciapus*).—Ga., Fla.
fusitarsis Van Duzee, 1933a: 2.—N.Y.
graenicheri (Van Duzee), 1927c: 73 (*Psilopus*).—Fla.; Mexico to Colombia, West Indies.
guttulus (Wiedemann), 1830: 222 (*Psilopus*).—Brazil (as Savannah, error; Loew, 1864e: 237). Not Nearctic.
hirsutus Becker.—Not Nearctic.
inermis (Loew), 1861a: 93 (*Psilopus*).—Pa.; Wis. to Que., s. to N.C.
inornatus (Aldrich), 1901: 356 (*Psilopus*).—Mexico; Ariz.
jucundus (Loew), 1861a: 87 (*Psilopus*).—Cuba; ?Mass., ?Pa.
    sipho, authors, not Say.
leonardi (Van Duzee), 1915b: 19, key (descr., 1915d: 89) (*Sciapus*).—Ga.; Fla. **N. comb.**
longitalus (Van Duzee), 1923d: 72 (*Psilopus*).—La.
melampus (Loew), 1862b: 215 (Cent. 2, no. 69) (*Psilopus*).—Mexico; Nev. to Colo., s. to Mexico.
mundus (Wiedemann), 1830: 227 (*Psilopus*).—Ga.; N.C., Fla., West Indies, Brazil.
    ciliatus Loew, 1861a: 88 (*Psilopus*).—Fla.
nigrofemoratus (Walker), 1849: 650 (*Psilopus*).—N.S.; Colo. to Que. and N.S., s. to Ga.
    nigrifemoratus, error.

scobinator Loew, 1861a: 91 (*Psilopus*).—N.Y., Ill.
scrobinator, error.
cockerelli Van Duzee, 1927c: 73 (*Psilopus*).—Colo.
scutellatus Harris, 1835: 597 (*Psilopus*). Nomen nudum.
noveboracensis (Van Duzee), 1915b: 25 (*Sciapus*).—N.Y.; Ind.
occidentalis (Bigot), 1888c: xxix (also 1890: 290) (*Psilopodinus*).—
Calif.; Mexico.
ogilvii (Malloch), 1932a: 124 (*Sciapus*).—Bermuda.
parvicauda (Van Duzee), 1927c: 72 (*Psilopus*).—Ont.
patibulatus (Say), 1823: 87 (1859b: 76) (*Dolichopus*).—Fla.; Nebr. to Que., s. to Fla., ?Alta.
amatus Walker, 1849: 648 (*Psilopus*).—N.Y.
carolinensis Bigot, 1888c: xxix (also 1890: 291) (*Psilopodinus*).—Carolina.
pilicornis (Aldrich), 1904: 282 (*Psilopodinus*).—Wash., Idaho, Calif.; B.C.
filicornis, error.
portoricensis (Macquart), 1834a: 450 (*Psilopus*).—P.R.; Fla.
radians (Macquart), 1834a: 450 (*Psilopus*).—N. Amer.; Fla., Cuba.
semicomatus (Van Duzee), 1929: 4 (*Psilopus*).—Guatemala; Tex., British Guiana. **N. comb.**
similis (Aldrich), 1901: 359 (*Psilopus*).—Mexico, Brazil; Tex.
sipho (Say), 1823: 84 (1859b: 75) (*Dolichopus*).—U.S.; se. Canada to Tex. and Fla.
gemmifer Walker, 1849: 646 (*Psilopus*).—N.Y.
scaber Loew, 1861a: 85 (*Psilopus*).—Pa.
superbus (Wiedemann).—Not Nearctic.
tonsus (Aldrich), 1901: 364 (*Psilopus*).—Mexico; N.C., Fla.
viridicoxa (Aldrich), 1904: 284 (*Psilopodinus*).—La.; Mo., Ind., N.J.
viridis Parent, 1929a: 238.—Pa.

Genus SCIAPUS Zeller

Psilopus Meigen, 1824: 35 (preocc. Poli, 1795). Type-species, *Dolichopus platypterus* Fabricius (Westwood, 1840: 134).
Sciapus Zeller, 1842: 831 (n. name for *Psilopus* Meigen). Type-species, *Dolichopus platypterus* Fabricius (aut.).
Sciopus, error.
Psilopodinus Bigot, 1888b: xxiv (also 1888c: xxix). Type-species, *Dolichopus platypterus* Fabricius (orig. des.).
Gnamptopsilopus Aldrich, 1893b: 48. Type-species, *Psilopus scintillans* Loew (Coquillett, 1910b: 547).

Agonosoma, authors, not Guérin-Méneville.

REFERENCE: Van Duzee, 1915b (rev.).

amabilis Parent, 1929a: 239.—Ga.
bicolor (Loew), 1861a: 96 (*Psilopus*).—Middle States; N.C.
bradleii Van Duzee, 1915b: 24.—Ga.; N.C., Fla.
costalis (Aldrich), 1904: 286 (*Agonosoma*).—Ga.; Fla.
divergens Van Duzee, 1933a: 3.—Wash.
dorsalis (Loew), 1866a: 180 (Cent. 6, no. 85) (*Psilopus*).—Cuba; Mo., D.C., S.C., Fla.
    viridivittatus Robinson, 1960: 272 (*Condylostylus*).—D.C.
filipes (Loew), 1861a: 99 (*Psilopus*).—Middle States; Minn., Conn.
fuscinervis (Van Duzee), 1926g: 56 (*Psilopus*).—Ont.
infumatus (Aldrich), 1901: 365 (*Gnamptopsilopus*).—Mexico; Ariz.
lectus Becker, 1922: 366.—Amer. Probably not Nearctic.
pallens (Wiedemann), 1830: 219 (*Psilopus*).—N.Y.; Mich., Mass. to N.C.
pollinosus Van Duzee, 1915b: 22, key (descr., 1915d: 93).—Ga.
pressipes Parent, 1929a: 244.—Carolina.
pruinosus Coquillett, 1904f: 186.—Fla.
psittacinus (Loew), 1861a: 96 (*Psilopus*).—Fla.; N.J., Ga., West Indies.
rotundiceps (Aldrich), 1904: 286 (*Agonosoma*).—Fla.
scintillans (Loew), 1861a: 94 (*Psilopus*).—Middle States; Minn. and Iowa to N.H.
    fulgidus Parent, 1929a: 242.—N.H.
    dubiosus Van Duzee, 1932b: 8.—N.Y.
tener (Loew), 1862b: 217 (Cent. 2, no. 71) (*Psilopus*).—Pa.; Mich., Que. and Maine to Fla.
trisetosus Van Duzee, 1932a: 2.—Fla.
unifasciatus (Say), 1823: 85 (1859b: 75) (*Dolichopus*).—Pa.; Ont. and Mich. to R.I., s. to La. and Fla.
    sayi Wiedemann, 1830: 219 (*Psilopus*).—Pa.
    ungulivena Walker, 1857: 149 (*Psilopus*).—U.S.
    plumosa Van Duzee, 1932a: 1 (*Chrysosoma*).—Ill. **N. syn.**
variegatus (Loew), 1861a: 95 (*Psilopus*).—Fla.; N.J., Tex., Ga., Neotropical.

Genus PSILOPIELLA Van Duzee

Psilopiella Van Duzee, 1914c: 438. Type-species, *rutila* Van Duzee (orig. des.).
rutila Van Duzee, 1914c: 439.—Fla.; Nebr.

Subfamily DOLICHOPODINAE
Genus DOLICHOPUS Latreille

Dolichopus Latreille, 1796: 159. Type-species, *Musca ungulata* Linnaeus (Latreille, 1810: 443).
Hygroceleuthus Loew, 1857a: 10. Type-species, *Dolichopus latipennis* Fallén (Coquillett, 1910b: 554).
Spathichira Bigot, 1888b: xxiv (also 1888c: xxx). Type-species, *Dolichopus funditor* Loew (orig. des.).
Spatichira Bigot, 1888c: xxx, emend. or error.

The genus *Dolichopus* contains some of the largest and most common members of the family, and is by far the largest North American dolichopodid genus. Aside from the observations by Steyskal, little is known about the biology of the adults or immature stages in North America.

REFERENCES: Van Duzee, Cole, and Aldrich, 1921 (rev.); Van Duzee and Curran, 1934a (key, males), 1934b (key, females); Steyskal, 1938b, 1942b, 1947c, 1959b (biol.).

abbreviatus Van Duzee, 1921c: 144.—Labr.; Maine.
aboriginis Harmston and Knowlton, 1943a: 102.—Utah.
abrasus Van Duzee, 1921c: 162.—N.Y.
abruptus Aldrich, 1922b: 14.—N.Y.; Mich.
absonus Van Duzee, 1921c: 294.—N.Y.; Iowa, Mo., Mich., Ind., Ont., Que.
    absconus, error.
accidentalis Harmston and Knowlton, 1941b: 93.—Colo.
acricola Van Duzee, 1921c: 58.—Calif.
acuminatus Loew, 1861a: 12.—D.C.; Man. to Que., s. to Colo. and D.C.
acutus Van Duzee, 1921c: 142.—Mass.
adaequatus Van Duzee, 1921c: 53.—Nev.; B.C. to Minn., s. to N. Mex.
adultus Van Duzee, 1921c: 98.—N.Y.; Mich., Maine, Mass., N.J.
aequalis Van Duzee, 1921c: 81.—N.Y.; Que.
aeratus Van Duzee, 1921c: 211.—Colo.; B.C., Alta., Wyo.
aethiops Van Duzee, 1921c: 296.—Ind.
affinis Walker, 1849: 659.—N.S.; Wis. and Ill. to N.S.
    splendidulus Loew, 1864a: 91 (Cent. 5, no. 82).—N.H.
afflictus (Osten Sacken), 1877: 313 (*Hygroceleuthus*).—Calif.; Wash. to Wyo., s. to Calif. and N. Mex.
affluens Van Duzee, 1921c: 114.—Wash.; Colo.
agronomus Melander and Brues, 1900: 140.—Mass.; N.H.
ainsliei Van Duzee, 1921c: 272.—Minn.; Nebr., Que., N.Y.

alacer Van Duzee, 1921c: 86.—Ind.; Ind. to Que., s. to La. and Fla., ?Man.
albertensis Curran, 1922i: 286.—Alta.; Colo.
albiciliatus Loew, 1862b:211 (Cent. 2, no. 59).—Ill.; Iowa to Que., s. to N.J.
    atrovirens Harris *in* Johnson, 1925c: 77. Nomen nudum.
albicoxa Aldrich, 1893a: 10.—Mass., Conn.; Wis. to Que., s. to N.J., ?Alta.
aldrichii (Wheeler), 1899: 3 (*Hygroceleuthus*).—Idaho, Wyo.; B.C., Wash., Utah, Colo.
amnicola (Melander and Brues), 1900: 130 (*Hygroceleuthus*).—Colo.; Idaho to Man., s. to Calif. and Colo.
    var. **robertsoni** Curran, 1923h: 191 (as sp.).—Man.; Colo.
amphericus Melander and Brues, 1900: 146.—Wis.; Alaska, Alta., Colo.
amplipennis Van Duzee, 1921c: 197.—Colo.; Mich.
andersoni Curran, 1924m: 304.—B.C.
angustatus Aldrich, 1893a: 15.—Mass.; Minn., Vt., N.H., Maine.
angusticornis Van Duzee, 1921c: 79.—Ind.
apheles Melander and Brues, 1900: 144.—Wis.; Sask., Mich., Que.
appendiculatus Van Duzee, 1921c: 72.—Nev.; Utah.
argentipes Van Duzee, 1921c: 61.—Wash.
arizonicus Harmston, 1951b: 103.—Ariz.; Tex.
aurifacies Aldrich, 1893a: 20.—Tenn., Kans.
aurifex Van Duzee, 1921c: 224.—Oreg.; N. Mex.
bakeri Cole, 1912a: 839.—Calif.; B.C. to Sask., s. to N. Mex.
barbaricus Van Duzee, 1921c: 41.—Colo.
barbicauda Van Duzee, 1921c: 76.—Ont.; B.C. to P.E.I., s. to Iowa and N.Y.
barbipes Van Duzee, 1921c: 129.—Nev.; Colo.
barycnemus Coquillett, 1900h: 424 (1904: 38).—Alaska.
    braycnemus, error.
beameri Harmston and Knowlton, 1941b: 92.—Ariz.
beatus Van Duzee, 1921c: 63.—Idaho.
bifractus Loew, 1861a: 19.—Ill., Nebr.; Alta. to Que., s. to Utah, Mo., and N.Y., ?Mexico.
bisetosus Van Duzee, 1921c: 77.—Idaho.
blandus Van Duzee, 1921c: 176.—Colo.; Mont., Utah, ?Ont.
bolsteri Van Duzee, 1921c: 249.—Nfld.
brevicauda Van Duzee, 1921c: 108.—N.H.
breviciliatus Van Duzee, 1930c: 71.—Alta.
brevimanus Loew, 1861c: 14.—D.C.; N.W.T., Man. to Que., s. to Wis. and D.C.

brevipennis Meigen, 1824: 89.—Sweden; Alaska to N.W.T., s. to B.C. and P.E.I., n. Europe.
brevipilosus Van Duzee, 1933b: 20 (n. name for *breviciliatus* Van Duzee).—Que.
   breviciliatus Van Duzee, 1933a: 13 (preocc. Van Duzee, 1930).—Que.
bruesi Van Duzee, 1921c: 223 (n. name for *propinquus* Melander and Brues).—B.C.; Wash., Calif., Wyo.
   propinquus Melander and Brues, 1900: 132 (*Hygroceleuthus*; as *consanguineus* var.; preocc. Zetterstedt, 1852).—B.C.
bruneifacies Van Duzee, 1933a: 14.—Colo.
brunneus Aldrich, 1893a: 14.—S. Dak.; Colo.
   brunneus Van Duzee (error) Van Duzee and Curran, 1934a: 4.
bryanti Van Duzee, 1921c: 104.—Labr.; Que.
burnesi Van Duzee, 1921c: 64.—N.Y.; Mich., Que., Mass., N.J.
calainus Melander and Brues, 1900: 138.—Ill.; Maine.
calcaratus Aldrich, 1893a: 8.—N.J.; Ont. to Que., s. to Iowa and N.J.
californicus Van Duzee, 1921c: 99.—Calif.; Wash.
calvimontis James, 1939a: 221.—Colo.
canadensis Van Duzee, 1921c: 141.—Ont.; Sask. to Que., s. to Wyo. and Mass.
canaliculatus Thomson, 1869: 512.—Calif.; Nev.
carolinensis Van Duzee, 1921c: 239.—N.C.; Fla.
cavatus Van Duzee, 1921c: 226. —Oreg.; Calif.
celeripes Van Duzee, 1921c: 243. —Oreg.; Utah, Colo., Ont.
chrysostomus Loew, 1861a: 23.—D.C.; Ont. to N.S., s. to Ga.
coercens Walker, 1849: 661.—N.Y.; S. Dak. and Iowa to N.S.
   conterminus Walker, 1849: 664.—N.Y.
   batillifer Loew, 1861a: 15.—Middle States.
coloradensis Aldrich, 1893a: 26 (n. name for *agilis* Aldrich).—Colo.; Alta., N. Mex., Ont., Que.
   agilis Aldrich, 1893a: 16 (preocc. Meigen, 1824).—Colo.
comatus Loew, 1861a: 23.—Pa., Md.; Wis. to Que., s. to Tex. and N.C.
   cornutus (error) Johnson, 1910b: 756.
compactus Van Duzee, 1921c: 206.—Oreg.; B.C.
completus Van Duzee, 1921c: 210.—Calif.; B.C.
comptus Van Duzee, 1921c: 160.—Calif.
confinis Walker, 1849: 664.—Ont. Unrecognized.
consanguineus (Wheeler), 1899: 5 (*Hygroceleuthus*).—Calif.; Idaho, Utah, Colo.
conspectus Van Duzee, 1921c: 30, 65 (as *conspicuus*, p. 13).—Idaho; B.C. to Man., s. to Colo.

contiguus Walker, 1849: 663.—N.Y.; N. Dak. to Que., s. to S. Dak., Ill., and Va.
  splendidus Loew, 1861a: 14.—Ill.
convergens Aldrich, 1893a: 9.—Wash., Oreg.
coquilletti Aldrich, 1893a: 19.—Calif.; Wash. to P.E.I., s. to Calif. and Colo.
corax Osten Sacken, 1877: 314.—Calif.; ?B.C.
correus Steyskal, 1959b: 1.—Mich.; Minn. to N.Y., s. to Tenn. and N.C.
crassicornis Aldrich, 1922b: 10.—Alaska.
crenatus (Osten Sacken), 1877: 312 (*Hygroceleuthus*).—Calif.; B.C. to P.E.I., s. to Calif. and Ariz.
cuprinus Wiedemann, 1830: 230 (n. name for *cupreus* Say).—Md.; N.W.T., B.C. to N.S., s. to Nev. and Va.
  cupreus Say, 1823: 86 (1859b: 76) (preocc. Fallén, 1823).—Md.
czekanowskii Stackelberg, 1928: 263.—n. Siberia; n. Alaska.
dakotensis Aldrich, 1893a: 11.—S. Dak.; Mont. and Man. to Que., s. to Ill.
dasyops Malloch, 1919f: 49.—N.W.T.
dasypodus Coquillett, 1910c: 42.—N.H.; Alaska, Mich., Que., Maine, N.Y.
decorus Van Duzee, 1921c: 153.—Ill.; Mich. to N.Y.
defectus Van Duzee, 1921c: 143.—Ont.; Man. to Que., s. to Conn., ?N. Mex.
  deflectus, error.
delicatus Aldrich, 1922b: 12.—Labr.
demissus Van Duzee, 1921c: 63.—N.Y.; N.J.
detersus Loew, 1866b: 44 (Cent. 7, no. 79).—N.Y.; Alaska, Man. to Que., s. to S. Dak., Iowa, and N.Y.
digitus Van Duzee, 1921c: 283.—La.
discessus Walker, 1849: 662.—Mass. Unrecognized.
discolor Van Duzee, 1921c: 116.—N.Y.; N. Dak. to Vt., s. to Ill.
distractus Walker, 1849: 662.—N.Y. Unrecognized.
diversipennis Curran, 1922i: 286.—Alta.
divigatus Harmston, 1952: 284.—Oreg.
dolosus Parent, 1934b: 267.—Wis.
domesticus Van Duzee, 1921c: 246.—N.Y.; Mich. to N.Y., N.J., Va.
dorsalis Van Duzee, 1921c: 161.—N. Mex.; Ariz.
dorycerus Loew, 1864a: 93 (Cent. 5, no. 85).—N.H.; N.H. to Ga.
duplicatus Aldrich, 1893a: 18.—Wash.; Wash. to Mont., s. to Calif. and Utah.
elegans Aldrich, 1922b: 12.—Colo.

enigma Melander and Brues, 1900: 139.—Colo.; Idaho, Wyo., Utah, Mich.
eudactylus Loew, 1861a: 16.—N.Y.; Kans. and Iowa to N.S., s. to N.C.
    edactylus, error.
evolvens Parent, 1929a: 173.—N.H.
exclusus Walker, 1849: 663.—Ont. Unrecognized.
facirecedens Harmston and Knowlton, 1939c: 83.—S. Dak.; Iowa.
fallax Van Duzee, 1933a: 14.—Colo.
finitus Walker, 1849: 662.—N.Y.; Kans., Wis. to N.S., s. to Pa.
    scoparius Loew, 1864e: 70.—Maine, Mass.
    atricornis Harris, 1835: 597. Nomen nudum.
flagellitenens Wheeler, 1890: 339.—Wis.; S. Dak. to N.H., s. to Colo.
flaviciliatus Van Duzee, 1921c: 152.—Ont.
flavicoxa Van Duzee, 1921c: 188.—Wis.; B.C. to N.Y., s. to W. Va.
flavifacies Van Duzee, 1933a: 15.—Que.
flavilacertus Van Duzee, 1921c: 110.—Md.; Ont. to Mass., s. to Tenn. and N.C.
formosus Van Duzee, 1921c: 42.—Calif.
fortis Aldrich, 1922b: 8.—Alaska.
frontalis Van Duzee, 1928b: 40.—Sask.; Alaska.
fucatus Van Duzee, 1921c: 113.—Wash.
fulvipes Loew, 1862b: 212 (Cent. 2, no. 61).—Ill.; Alta. to N.S., s. to N.J.
fumosus Van Duzee, 1921c: 74.—Sask.
funditor Loew, 1861a: 22.—Middle States; Ind. to Ont., s. to La. and Ga.
    frauditor, error.
    var. **distinctus** Van Duzee, 1921c: 274.—La.; Ont., N.J.
genualis Van Duzee, 1921c: 119.—Maine; Ont., Que., Nfld.
gladius Van Duzee, 1921c: 136 (as *socius* var.).—Ont.; Ont. to N.S., s. to Mich. and N.Y., also Alaska.
grandis Aldrich, 1893a: 21.—Calif.; Wash., Oreg., Nev.
gratus Loew, 1861a: 11.—N.Y.; Wis. to Que., s. to Ga.
    mercieri Parent, 1929a: 175.—Ga.
groenlandicus Zetterstedt, 1843: 528.—Greenland; Alaska, Colo., Labr.
harbecki Van Duzee, 1921c: 233.—Pa.; Que. and Maine to Ga.
hardyi Harmston, 1951b: 105.—Ariz.; N. Mex.
hastatus Loew, 1864e: 59.—Alaska; Alaska to Calif.
helenae James, 1939a: 220.—Colo.
hirsutitarsis Harmston, 1952: 281.—Calif.
humilis Van Duzee, 1921c: 108.—Alaska; N.W.T., Greenland.

idahoensis (Aldrich), 1894: 154 (*Hygroceleuthus*).—Idaho; Oreg. to S. Dak., s. to Calif. and Colo.
idoneus Van Duzee, 1921c: 140.—N.Y.; Vt.
imperfectus Van Duzee, 1921c: 240.—Ill.
incisuralis Loew, 1861a: 25.—N.Y.; N.W.T., Ont. to N.H., s. to Ga.
    platyprosopus Loew, 1866b: 44 (Cent. 7, no. 80).—N.W.T.
incongruus Wheeler, 1890: 338.—Wis.; Wis., N.H., s. to Tenn. and N.C.
indianus Harmston and Knowlton, 1946a: 672.—Ind.
indigenus Van Duzee, 1921c: 139.—Idaho; Wash. and Man. to N.S., s. to Iowa.
inflatus Aldrich, 1922b: 11.—Alaska.
integripes Parent, 1929a: 174.—B.C.
intentus Melander and Brues, 1900: 137.—Ill.
interjectus Van Duzee, 1923d: 70.—Mont.
iowaensis Harmston and Knowlton, 1939c: 84.—Iowa.
jaquesi Harmston and Knowlton, 1939d: 87.—Iowa.
johnsoni Aldrich, 1893a: 7.—N.J.; Ind. to N.H., s. to N.C.
jugalis Tucker, 1911: 106.—Colo.; Wash. to Man., s. to Utah.
    procerus Van Duzee, 1921c: 209.—Colo.
kansensis Aldrich, 1893a: 8.—Kans.
kleini Curran *in* Van Duzee and Curran, 1934a: 25.—N.J.
laciniatus Coquillett, 1910c: 42.—Pa.; Mich. and Ont. to N.C.
lamellipes Walker, 1849: 660.—Ont.; Alaska, Man. to Labr., Europe.
    boreus Van Duzee, 1921c: 204.—Labr.
laticornis Loew, 1861a: 12.—Conn. (Loew, 1864e: 29); Wyo. to Que., s. to Fla.
latipes (Loew), 1861a: 5 (*Hygroceleuthus*).—Ill.; B.C. to Que., s. to Wyo., Iowa, and Mass.
    var. **cognatus** (Melander and Brues), 1900: 129 (*Hygroceleuthus*).— Mass., Ill.
latronis Van Duzee, 1921c: 232.—Labr.
leucacra James, 1939a: 225.—Colo.
litoralis Van Duzee, 1921c: 82.—Wash.; Alaska.
lobatus Loew, 1861a: 24.—Ont.; B.C. and Idaho to N.Y., s. to Iowa.
longicornis Stannius, 1831a: 53.—Germany; ?Alaska.
    acuticornis, authors, not Wiedemann.
longimanus Loew, 1861a: 14.—Ont.; Alaska, B.C. to N.H., s. to Colo. and Va.
longipennis Loew, 1861a: 21.—Ill.; B.C. to Que., s. to Fla.
longus Aldrich, 1922b: 13.—Colo.
lundbecki Curran, 1923k: 236.—Ont.

luteipennis Loew, 1861a: 18.—D.C.; B.C. to Mich. and Que., s. to Va.
    greenei Van Duzee, 1921c: 192.—Va. **N. syn.**
maculitarsis Van Duzee, 1925e: 184.—Man.; N.W.T., Alta. to Man., Colo.
magnantenna James, 1939a: 222.—Colo.
manicula Van Duzee, 1921c: 56.—Colo.; Alaska, B.C. and Alta. to N. Mex.
mannerheimi Zetterstedt, 1838: 707.—Lapland; Alaska, n. Europe.
marginatus Aldrich, 1893a: 17.—Conn.; Man. to N.S., s. to Fla.
melanderi Van Duzee, 1921c: 70.—Wash.; B.C.
melanocerus Loew, 1864a: 93 (Cent. 5, no. 86).—Canada; Mich. to Nfld. and Conn.
micropygus Wahlberg, 1851: 216.—Sweden; ?Colo.
monticola Van Duzee, 1921c: 40.—Wash.; ?B.C.
multisetosus Van Duzee, 1921c: 49.—Calif.; Utah, Colo., ?Maine.
myosotus Osten Sacken, 1887a: 213.—Mexico; B.C. to Iowa, s. to Mexico, ?N.J.
neomexicanus Harmston, 1951b: 104.—N. Mex.
nigricauda Van Duzee, 1921c: 46.—Colo.; N.W.T., B.C. to Man., s. to Calif. and Colo.
nigricornis Meigen, 1824: 82.—Europe; Alaska, B.C. to N.S., s. to Colo. and N.Y., Europe.
    discifer Stannius, 1831a: 57.—Germany.
    tanypus Loew, 1861a: 24.—Man., Ont.
nigricoxa Van Duzee, 1926h: 230.—B.C.; Utah, Colo.
nigrilineatus Van Duzee, 1924c: 248.—Alta.; Man., Colo.
nigrimanus Van Duzee, 1921c: 45.—Idaho; B.C. and Idaho to Ont.
nigroapicalis Van Duzee, 1930d: 125.—Colo.
    nigriapicalis, error.
nodipennis Van Duzee, 1921c: 102.—N.Y.; Que to Ont. and Conn.
nomadus Harmston and Knowlton, 1942c: 18.—Alta.
nubifer Van Duzee, 1921c: 75.—Nev.; Man., Idaho, Utah, Nebr., Mich.
nudus Loew, 1864e: 41.—N.W.T.; Alaska to N.W.T., s. to Alta. and Que.
obcordatus Aldrich, 1893a: 14.—Wyo., Colo.; Alaska to Alta., s. to Calif. and N. Mex., Mich.
obsoletus Van Duzee, 1921c: 121.—N.H.; Maine.
occidentalis Aldrich, 1893a: 19.—Wash.; B.C. and Idaho to Calif., ?P.E.I.
omnivagus Van Duzee, 1921c: 216.—N.Y.; Alaska, Alta. to Maine, s. to Mont. and Iowa.

opportunus Van Duzee, 1921c: 107.—Idaho; Utah.
oregonensis Van Duzee, 1927e: 148.—Oreg.
ornatipennis Van Duzee, 1921c: 132.—Mass.; N.Y. to N.S., s. to N.C.
ovatus Loew, 1861a: 13.—Middle States; B.C. to Que., s. to Idaho and N.C., ?Calif.
pachycnemus Loew, 1861a: 13.—Middle States; Man. to Que., s. to Mont. and Iowa.
packardi Van Duzee, 1921c: 83.—Labr.; Alta., Que.
  refulgens Harmston and Knowlton, 1939b: 349.—Alta. N. syn., F. C. Harmston *in litt.*
palaestricus Loew, 1864a: 92 (Cent. 5, no. 84).—N.H.; Ont. and N.H., s. to Ill. and N.J.
paluster Melander and Brues, 1900: 136.—Calif.; Alta., Oreg.
pantomimus Melander and Brues, 1900: 142.—Mass.; Ont. to Mass., s. to Tenn. and N.C.
partitus Melander and Brues, 1900: 135.—Colo.
parvicornis Van Duzee, 1921c: 231.—Ont.; Minn.
parvimanus Van Duzee, 1933a: 16.—Colo.
penicillatus Van Duzee, 1921c: 227 (n. name for *ciliatus* Aldrich).— S. Dak.; B.C. to Maine, s. to Calif. and N.Y.
  ciliatus Aldrich, 1893a: 25 (*Hygroceleuthus*; preocc. Walker, 1849).—S. Dak.
pensus Aldrich, 1922b: 8.—Alaska.
pernix Melander and Brues, 1900: 141.—B.C.; Wash., Alta., Sask.
perplexus Van Duzee, 1923d: 71 (n. name for *melanderi* Becker).— Wyo.
  misellus Melander, 1900: 136 (preocc. Boheman, 1851).— Wyo.
  melanderi Becker, 1922: 15 (as *melandri;* n. name for *misellus* Melander, but preocc. Van Duzee, 1921).—Wyo.
phyllocerus Vockeroth, 1962: 502.—N.C.
pilatus Van Duzee, 1921c: 167.—Labr.; Alta.
pingreensis James, 1939a: 223.—Colo.
plumipes (Scopoli), 1763: 334 (*Musca*).—Yugoslavia; Alaska to N.W.T., s. to Calif., Ark., and Mich., Greenland, Mexico, Europe.
  ciliatus Walker, 1849: 661.—Ont.
  sequax Walker, 1849: 666.—Ont.
plumitarsis Fallén, 1823c: 10.—Sweden; Alaska, Ont., n. Europe.
  pulmitarsis, error.
plumosus Aldrich, 1893a: 18.—Wash.; B.C., Alta., Nev.
pollex Osten Sacken, 1877: 314.—Calif.; Alta., Oreg., Idaho, Utah, Nev.

porphyrops Van Duzee, 1921c: 169.—N.H.; Ont. to N.S., s. to Mich., N.Y., and Mass.
praeustus Loew, 1862b: 212 (Cent. 2, no. 62).—Ill.; Mo., Ind., Conn.
puberiseta Parent, 1934b: 268.—Va.
pugil Loew, 1866b: 43 (Cent. 7, no. 77).—Canada; Ont. to N.S., s. to N.J.
    henshawi Wheeler, 1890: 340.—Mass.
pulcher Walker, 1852: 215.—U.S. Unrecognized.
pulchrimanus (Bigot), 1888c: xxx (also 1890: 292) (*Spathichira*).—Rocky Mts.; Wis. to N.Y., s. to Tex. and La.
    willistonii Aldrich, 1893a: 22.—Kans.
quadrilamellatus Loew, 1864a: 92 (Cent. 5, no. 83).—N.J. (as N.Y., error); Ont. to Conn., s. to Tenn. and N.C.
ramifer Loew, 1861a: 19.—Man., Nebr.; Alaska, Oreg. to Que., s. to Calif., Tex., and Va.
recticosta Aldrich, 1922b: 13.—Fla.
reflectus Aldrich, 1893a: 12.—Pa.; Ont. and Que., s. to Kans. and Fla.
remipes Wahlberg, 1839: 13.—Sweden; Wash. to Man. and Maine, s. to Wis., n. Europe.
remotus Walker, 1849: 666.—N. Amer.; N.Y.
    cuniculus Van Duzee, 1921c: 145.—N.Y.
remus Van Duzee, 1921c: 96.—N.Y.; Ill., Ohio, Ont., Vt.
renidescens Melander and Brues, 1900: 143.—Colo.; Alaska, B.C. to Man., s. to S. Dak. and Iowa, ?Maine.
reticulus Van Duzee, 1926h: 231.—Oreg.; B.C.
retinens Van Duzee, 1921c: 90.—Ind.; S. Dak. and Nebr. to Ont., s. to Ga.
ruficornis Loew, 1861a: 21.—Middle States; Kans. to N.Y., s. to Ark. and Va.
rupestris Haliday, 1833: 164.—England; Alaska, n. Europe.
    festinans Zetterstedt, 1838: 708.—Lapland.
sarotes Loew, 1866b: 44 (Cent. 7, no. 81).—Ill.; Kans. to Ont. and Mass.
    sorotes, error.
scapularis Loew, 1861a: 22.—Ill.; Wis. to N.H., s. to Tex. and Ga.
scopifer James, 1939a: 224.—Colo.
sedulus Van Duzee, 1921c: 93.—Idaho; N.W.T., Ind.
separatus Walker, 1849: 665.—Ont. Unrecognized.
serratus Van Duzee, 1921c: 155.—Maine; Wis., Mich., N.Y., Labr.
setifer Loew, 1861a: 12.—N.Y.; Ont. and Que., s. to N.C.
    michiganus Harmston and Knowlton, 1945: 78.—Mich. **N. syn.**, F. C. Harmston *in litt.*
setosus Loew, 1862b: 213 (Cent. 2, no. 63).—Mass.; Ont. to N.S., s. to N.C., ?B.C.

sexarticulatus Loew, 1864e: 62.—D.C.; Ill. and Mich. to D.C., s. to Tex. and S.C.
shelfordi Curran *in* Van Duzee and Curran, 1934a: 26.—Man.
sicarius Van Duzee, 1921c: 261.—Ont.; Mich., Que., N.S.
silvicola Harmston, 1951b: 106.—N. Mex.
simplicipes Aldrich, 1922b: 9.—Alaska.
simulans Van Duzee, 1926h: 231.—Man.
sincerus Melander, 1900: 136.—Wis.; Oreg., Idaho, Wis. to N.S., s. to N.C.
    var. **subdirectus** Van Duzee, 1921c: 118.—Mass.; N.Y., N.H., Maine.
slossonae Van Duzee, 1921c: 235.—N.H.; Que. to Tenn. and N.C.
soccatus Walker, 1849: 666.—Ont. Unrecognized.
socius Loew, 1862b: 211 (Cent. 2, no. 60).—Ill.; Oreg. to Que. and N.S., s. to Iowa and N.J.
solidus Van Duzee, 1921c: 104.—Alaska; Colo.
sordidatus Van Duzee, 1921c: 41.—Idaho.
speciosus Van Duzee, 1921c: 208.—N. Mex.; Alta., Utah, Colo.
sphaeristes Brues, 1901a: 44.—Tex.; Tenn.
sporadicus Harmston and Knowlton, 1942c: 17.—Utah.
squamosus Van Duzee, 1921c: 43.—Idaho; Calif., Nev., Wyo., Utah.
stenhammari Zetterstedt, 1843: 521 (n. name for *annulipes* Zetterstedt).—Lapland; Alaska, Ont. to Labr., s. to N.Y., n. Europe.
    annulipes Zetterstedt, 1838: 710 (preocc. in *Dolichopus* by Meigen, 1824, orig. *Porphyrops*, now *Sympycnus*).—Lapland.
stricklandi Harmston and Knowlton, 1939b: 350.—Alta.
subciliatus Loew, 1864e: 42.—N.W.T.; Ont., N.S.
subcostatus Van Duzee, 1930d: 124.—Calif.
subspina Van Duzee, 1928b: 41.—Sask.
sufflavus Van Duzee, 1921c: 213.—Idaho; Alta., Wash., Mont., Utah, Colo.
    subflavus, error.
superbus Van Duzee, 1921c: 287.—Calif.
talus Van Duzee, 1921c: 268.—Calif.
tener Loew, 1861a: 17.—Ill.; Wis., Ont., Que., N.Y.
tenuimanus Van Duzee, 1932a: 12.—N.C.
tenuipes Aldrich, 1894: 155.—Idaho; B.C. and Idaho to Wash., Calif.
terminalis Loew, 1866b: 43 (Cent. 7, no. 78).—N.Y.; Wyo. to Que., s. to Iowa, ?Wash., ?D.C.
    germanus Wheeler, 1890: 341 (preocc. Wiedemann, 1817).—Wis.
terminatus Walker, 1849: 665.—N. Amer. Unrecognized.

tetricus Loew, 1864e: 33.—N.W.T.; Que.
tonsus Loew, 1861a: 16.—D.C.; Mich. and Ind. to Mass., s. to N.C.
townsendi Aldrich, 1922b: 15.—N. Mex.; Ariz.
trisetosus Van Duzee, 1921c: 122.—Mass.; Ont. to N.B., s. to N.Y.
uliginosus Van Duzee, 1923d: 69.—B.C.
umbrosus Van Duzee, 1921c: 100.—Wis.; Ont.
ungulatus (Linnaeus), 1758: 598 (*Musca*).—Europe; Wis.
 aeneus De Geer, 1776: 194 (*Nemotelus*).—Europe.
utahensis Harmston and Knowlton, 1943a: 101.—Utah.
uxorcula Van Duzee, 1921c: 186.—Alaska; Alaska, Alta. to Man., s. to Calif.
vanduzeei Curran, 1922i: 285.—Alta.
variabilis Loew, 1861a: 17.—N.Y.; B.C. to N.S., s. to Colo. and N.C.
 var. **gracilis** Aldrich, 1893a: 15 (as sp.).—Pa.; Ont. to Maine, s. to Tenn.
varipes Coquillett, 1900h: 425 (1904: 39).—Alaska; Alaska to N. Mex.
vegetus Harmston, 1952: 283.—Alaska.
vernaae Harmston and Knowlton, 1940b: 129.—Utah; Wyo.
versutus Van Duzee, 1921c: 253.—Ont.; Ont. and Que., s. to Kans. and Va.
vigilans Aldrich, 1893a: 13.—Kans.; Ont. to N.B., s. to Kans. and Ohio.
virga Coquillett, 1910c: 41.—N.J.; Ont. to N.S., s. to N.J.
virginiensis Van Duzee, 1921c: 236.—Va.; N.H. to Ga.
 lobipennis Van Duzee, 1930a: 3.—Conn.
viridis Van Duzee, 1921c: 44.—Idaho; B.C., Mont., Utah, Colo.
vittatus Loew, 1861a: 20.—Ill.; S. Dak. to Que., s. to Kans. and N.J.
walkeri Van Duzee, 1921c: 207.—Colo.; Man., N. Mex.
wheeleri (Melander and Brues), 1900: 126 (*Hygroceleuthus*).—Mass.; Man. to N.S., s. to Mass.
xanthocnemus Loew, 1864e: 31.—Alaska; B.C., Colo., Ont., Que., N.H.

Genus HERCOSTOMUS Loew

Hercostomus Loew, 1857a: 9. Type-species, *Sybistroma longiventris* Loew (orig. des.).

albipodus Harmston and Knowlton, 1941a: 131.—Utah.
aurifer (Thomson), 1869: 512 (*Dolichopus*).—Calif.
cacheae Harmston and Knowlton, 1941a: 131.—Utah; Calif., Wyo.
chaetilamellus Harmston and Knowlton, 1941a: 127.—Calif.
coloradensis Harmston, 1952: 286.—Colo.

convergens (Van Duzee), 1920: 49 (*Gymnopternus*).—Nev.; S. Dak.
costalis Van Duzee, 1923d: 65.—Ont.; Iowa, Mich., Ohio.
cryptus Harmston and Knowlton, 1941a: 130.—Utah.
dorsalis (Van Duzee), 1914c: 434 (*Neurigona*).—N.Y.; N.H.
flavicornis (Van Duzee), 1918: 48 (*Paraclius*).—Ariz.
flutatus Harmston and Wheeler, 1945: 79.—Mich.; Tenn.
impudicus Wheeler, 1899: 10.—Calif.
indianus Harmston, 1952: 285.—Ind.; Iowa.
longilamellus Harmston and Knowlton, 1940f: 127.—Utah.
neocryptus Harmston and Knowlton, 1941a: 130.—Alta.; Wash.
nigrifacies Van Duzee, 1933a: 19.—Oreg.
occidentalis Cole, 1912a: 839.—Calif.
orbicularis Harmston, 1952: 286.—Calif.
ornatus (Van Duzee), 1921b: 128 (*Paraclius*).—N.Y.; Mich. to Que., s. to Tenn. and N.C.
    ornatipes (error) Curran, 1925j: 100.
    dreisbachi Harmston and Knowlton, 1945: 80.—Mich.
setosus (Van Duzee), 1913b: 54 (*Neurigona*).—N. Mex.; Utah, Colo.
    torridus Harmston and Knowlton, 1941a: 129.—Utah.
stanfordi Harmston and Knowlton, 1940f: 126.—Utah.
tibialis (Van Duzee), 1913b: 55 (*Neurigona*).—N.Y.; Tenn., N.C., S.C.
truncatus Harmston and Knowlton, 1940f: 127.—Utah.
unicolor Loew, 1864e: 117.—N.W.T.; Alaska, Wash. and Alta. to Que., s. to n. Calif., Mich., and Maine, also Ariz.
    poenitens Wheeler, 1890: 355 (*Gymnopternus*).—Wis.
    ornaticauda Van Duzee, 1933a: 16.—Wash.
utahensis Harmston and Knowlton, 1940f: 125.—Utah.
wasatchensis Harmston and Knowlton, 1943a: 103.—Utah.

Genus GYMNOPTERNUS Loew

Gymnopternus Loew, 1857a: 10. Type-species, *Dolichopus cupreus* Fallén (Coquillett, 1910b: 548).

albiceps Loew, 1861a: 30.—Middle States (Loew, 1864e: 85); Maine to Fla.
anarmostus (Melander), 1900: 139 (*Hercostomus*).—Ill.
    anormostus, error.
annulatus Van Duzee, 1926g: 58.—N.S.; Mich., Conn., Va., N.C.
barbatulus Loew, 1861a: 29.—Middle States; Iowa to Que., s. to Tenn. and N.C.
brevipes (Van Duzee), 1933a: 17 (*Hercostomus*).—Iowa; Colo.
brunneifacies (Robinson), 1960: 273 (*Hercostomus*).—N.Y.; S.C. **N. comb.**

californicus Van Duzee, 1920: 48.—Calif.; Oreg., Colo.
consanguineus (Harmston), 1952: 287 (*Hercostomus*).—Wash.
coxalis Loew, 1864a: 94 (Cent. 5, no. 87).—N.Y.; N.H.
crassicauda Loew, 1861a: 35.—N.Y.; Iowa to Que. and N.S., s. to Kans. and Ga.
currani (Van Duzee), 1930a: 4 (*Hercostomus*).—Conn.; N.Y., N.H.
debilis Loew, 1861a: 35.—Pa.; Iowa, Mich., Maine to Fla.
despicatus Loew, 1861a: 33.—Middle States; Que. and Maine to N.Y.
difficilis Loew, 1861a: 33.—N.Y.; Ont. to N.S., s. to Fla.
exilis Loew, 1861a: 30.—Pa.; N.H. and Mass. to Fla.
fimbriatus Loew, 1861a: 32.—Md.; Que. to Ga.
flaviciliatus Van Duzee, 1914b: 404.—Ga.
flavitarsis (Van Duzee), 1925e: 187 (*Hercostomus*).—N.Y.
flavus Loew, 1861a: 28.—Pa.; Ont. and Que. to Ga.
frequens Loew, 1861a: 32.—N.Y.; Ont. and Que., s. to Iowa and Ga.
humilis Loew, 1864e: 336.—Ill., N.Y.; Kans. and Mich. to Que., s. to N.C.
 humeralis (error) Criddle, 1921: 85.
laevigatus Loew, 1861a: 31.—Middle States; Maine to Ga.
lividifrons Van Duzee, 1926g: 58.—Ont.; Mich., Que.
lunifer Loew, 1861a: 32.—N.Y.; Ill. to Que., s. to Ga.
 exiguus Loew, 1864e: 337.—Ill.
maculiventris (Van Duzee), 1925e: 188 (*Hercostomus*).—N.Y.
meniscus Loew, 1864a: 94 (Cent. 5, no. 88).—D.C.; Iowa to N.Y., s. to S.C.
metatarsalis (Thomson), 1869: 512 (*Dolichopus*).—Calif.; B.C.
 procerus Wheeler, 1899: 8 (*Hercostomus*).—Calif.
minutus Loew, 1861a: 35.—Middle States (Loew, 1864e: 96); Iowa.
mirificus Melander, 1900: 137.—Mass.
nigribarbus Loew, 1861a: 33.—Pa.; Mich. to N.S., s. to N.J.
 phyllophorus Loew, 1866b: 45 (Cent. 7, no. 82).—N.Y.
nigricoxa Van Duzee, 1924a: 103.—N.Y.; Que., Mass.
 nigricera (error) Leonard, 1928: 782.
obscurus (Say), 1823: 85 (1859b: 75) (*Dolichopus*).—Pa.
obtusicauda Van Duzee, 1924a: 103.—Maine; N.Y. to N.S.
opacus Loew, 1861a: 34.—N.Y.; Maine to Mo. and Ga.
ovaticornis (Van Duzee), 1933a: 20 (*Hercostomus*).—N.J.
pallidiciliatus (Van Duzee), 1930b: 85 (*Hercostomus*).—Mass.
parvicornis Loew, 1861a: 34.—Middle States; N.Y. and Mass. to Ga.
politus Loew. 1861a: 34.—N.Y.; Mich., N.Y. to Tenn. and S.C.
purpuratus (Van Duzee), 1925e: 185 (*Hercostomus*).—Man.; Mich.
pusillus Loew, 1864e: 334.—Ill.; Iowa, Tenn.
robustus (Van Duzee), 1925e: 189 (*Hercostomus*).—N.Y.

scotias Loew, 1861a: 29.—Ont.; Man. to N.S., s. to Iowa and N.J.
  browni Van Duzee, 1933a: 18 (*Hercostomus*).—Que.
singularis Van Duzee, 1924a: 102.—R.I.
spectabilis Loew, 1861a: 30.—N.Y.; Ont. and Que. to Tenn. and Ga.
  chalcochrus Loew, 1864e: 335.—N.Y., D.C.
subdilatatus Loew, 1861a: 31.—Middle States; Maine to Tenn. and Ga.
subulatus Loew, 1861a: 29.—N.Y.; Ont. to N.S., s. to N.C.
tenuicauda Van Duzee, 1928d: 88.—N.C.
tibialis Van Duzee, 1928d: 88.—N.C.
tristis Loew, 1864e: 83.—Alaska; B.C., Que., Vt., Mass.
vanduzeei Curran, 1930l: 287.—Oreg.
ventralis Loew, 1861a: 36.—N.Y.; Ont. and Que. to N.C.
vernaculus Van Duzee, 1924a: 104.—N.Y.; Maine.
vetitus (Melander), 1900: 138 (*Hercostomus*).—N.J.; N.Y.
  vestitus and *vetius*, errors.
violaceus (Van Duzee), 1921b: 123 (*Proarchus*).—N.Y.; Mich. to Mass., s. to Tenn. and S.C.

Genus PARACLEIUS Bigot

Paracleius Bigot, 1859b: 215, 227. Type-species, *Dolichopus heteroneurus* Macquart (mon.).
Paraclius Loew, 1864e: 97, emend. Since *heteroneurus* Macquart is unrecognized, Loew's concept may not be congeneric. Coquillett (1910b: 583) has even designated a type-species for this emendation.

The species are found on moist soil or sand or on low herbaceous foliage near water. Most of the species are maritime.

REFERENCE: Aldrich, 1904 (key).

alternans (Loew), 1864a: 95 (Cent. 5, no. 91) (*Pelastoneurus*).—N.Y.; Maine to Fla.
  vicinus Aldrich, 1904: 277.—Mass.
claviculatus Loew, 1866b: 45 (Cent. 7, no. 83).—N.Y.; Ont. to N.S., s. to N.C.
  fraternus Van Duzee, 1933a: 21.—N.Y.
consors (Walker), 1852: 213 (*Dolichopus*).—U.S. Unrecognized.
filifer Aldrich, 1896b: 314.—St. Vincent I.; Fla., Mexico, West Indies.
flagellatus (Harmston), 1952: 293 (*Hercostomus*).—Ala. (immigrant, found in hold of ship). **N. comb.**
heteroneurus (Macquart), 1850: 432 (1850: 128) (*Dolichopus*).—N. Amer. Unrecognized.
  heteropterus, error.
hybridus Melander, 1900: 141.—Mass.; Maine, N.Y., Conn., N.C.

magnicornis Van Duzee, 1927e: 146.—Idaho.
minutus Van Duzee, 1921b: 127.—Fla.
nigrocaudatus Van Duzee, 1918: 47.—Mont.
ovatus Van Duzee, 1914c: 436.—Ga.; Mich. to N.Y., s. to Fla., ?Panama.
propinquus Wheeler, 1899: 18.—Fla.; Mass.
pumilio Loew, 1872a: 90 (Cent. 10, no. 63).—Tex.; Iowa, Mich., Ind., Md., Va., Tenn., Mexico.
quadrinotatus Aldrich, 1902: 81.—Grenada I.; Tex., Fla., West Indies.
utahensis Harmston and Knowlton, 1946b: 23.—Utah.

Genus PELASTONEURUS Loew

Pelastoneurus Loew, 1861a: 36. Type-species, *vagans* Loew (Coquillett, 1910b: 586).
Metapelastoneurus Aldrich, 1894: 152. Type-species, *kansensis* Aldrich (mon.).

REFERENCE: Van Duzee, 1923b (rev.).

abbreviatus Loew, 1864a: 94 (Cent. 5, no. 89).—N.Y.; Ont. to N.S., s. to Ill. and N.C., ?Mont.
aldrichi Van Duzee, 1923b: 36.—Utah.
asciaeformis Becker, 1922: 63.—Ga.
aurifacies Van Duzee, 1923b: 35.—Fla.; Ala.
    minutus Van Duzee, 1933a: 27.—Ala.
bifrons (Walker), 1852: 212 (*Dolichopus*).—U.S.
cognatus Loew, 1861a: 40.—Middle States; Ill., Tex., Fla., Mexico.
cyaneus Wheeler, 1899: 17.—Calif.; Oreg. to S. Dak., s. to Calif., Mexico.
dissimilipes Wheeler, 1899: 16.—Calif.; Ariz.
    nigrescens Wheeler, 1899: 78, 79 (pls. 1, 2, figs. 21, 27) (alternate orig. spelling).—Calif.
floridanus Wheeler, 1899: 13.—Fla.; Tex., N.C.
furcifer Loew, 1872a: 91 (Cent. 10, no. 64).—Tex.
hebes (Walker), 1852: 213 (*Dolichopus*).—U.S. Unrecognized.
ineptus (Walker), 1852: 214 (*Dolichopus*).—U.S. Unrecognized.
irrasus (Walker), 1849: 667 (*Dolichopus*).—Fla. Unrecognized.
kansensis (Aldrich), 1894: 153 (*Metapelastoneurus*).—Kans.; S. Dak., Mich., Ill., Ind.
laetus Loew, 1861a: 38.—Ga.; Mich. to Que., s. to La. and Fla.
    falcatus Aldrich, 1904: 277.—Que.
    ramosus Van Duzee, 1923b: 33.—Ind., Va.
lamellatus Loew, 1864a: 95 (Cent. 5, no. 90).—N.Y.; Iowa to Maine, s. to Tex. and Fla.
    cristatus Van Duzee, 1924a: 105.—Mass.

latifacies Van Duzee, 1930d: 123.—Calif.
longicauda Loew, 1861a: 37.—N.Y.; N.C., Fla., ?Calif.
    quadricincta Van Duzee, 1928d: 89.—N.C.
lugubris Loew, 1861a: 38.—N.Y.; Mich., Vt. to Tex. and Fla., Mexico.
maculipes (Walker), 1852: 214 (*Dolichopus*).—U.S. Unrecognized.
neglectus Wheeler, 1899: 12.—Wis., Ill.; Wis. to N.H., s. to La. and Ga.
nigricornis Van Duzee, 1923b: 39.—Mo.; Tex., La., Tenn.
    penicillatus Parent, 1929a: 181.—Tex.
occidentalis Wheeler, 1899: 13.—Calif.; Oreg., ?Va.
parvus Aldrich, 1904: 276.—La.; Ga., Fla.
proximus Aldrich, 1904: 278.—La.; Mich., Ind., Tenn., N.C.
    arboreus Van Duzee, 1923b: 34.—Ind.
scutatus Aldrich, 1904: 276.—Fla.
semiplumatus Becker, 1922: 69.—Amer. Possibly not Nearctic.
seticauda Van Duzee, 1930b: 85.—Mo.; Tenn.
tibialis Van Duzee, 1923b: 41.—N. Mex.
umbripictus Becker, 1922: 73.—Carolina, Colombia; Mich., N.C.
vagans Loew, 1861a: 39.—Middle States; Alaska, Oreg. to Que., s. to N. Mex. and Ga.
    vagrans, error.
    longilamellatus Parent, 1929a: 180.—Tex., Ga.
varius (Walker), 1852: 215 (*Dolichopus*).—U.S.; Ga., Fla.
    pictipennis Wheeler, 1899: 14.—Fla.
    punctipennis, error (not Say).
wheeleri Melander, 1900: 140.—Tex.; Calif., Mich., Tenn., N.C., S.C., Ga.

Genus TACHYTRECHUS Haliday

Ammobates Stannius, 1831a: 33, 268 (preocc. Latreille, 1809). Type-species, *notatus* Stannius (Rondani, 1856: 143).
Hammobates, error or emend.
Tachytrechus Haliday, 1851: 173 (n. name for *Ammobates* Stannius).
    Type-species, *Ammobates notatus* Stannius (aut.).
    Tachytrechus of Stannius, 1831a: 261, is a nomen nudum.
Tetrechus, error.
Macellocerus Mik, 1878b: ?3 (1878: 5). Type-species, *Tachytrechus moechus* Loew (orig. des.).

    REFERENCES: Greene, 1922 (key); Harmston and Knowlton, 1940a (key, males).

albonotatus (Loew), 1864e: 102 (*Paraclius*).—La.; Utah to Vt., s. to La. and Fla., Cent. Amer.
angulatus (Van Duzee), 1914c: 436 (*Paraclius*). N.Y.; Tenn., N.C.

angustipennis Loew, 1862b: 213 (Cent. 2, no. 64).—D.C.; Calif. to D.C., s. to Fla., Mexico.
argyropus Becker, 1922: 79.—Amer. Possibly not Nearctic.
auratus (Aldrich), 1896a: 83 (*Macellocerus*).—Idaho; Calif., Utah, Colo., Ariz.
binodatus Loew, 1866b: 46 (Cent. 7, no. 84).—N.Y.; Alaska, Ont. and Que., s. to Tenn.
floridensis Aldrich, 1896a: 82.—Fla.
granditarsis Greene, 1922: 17.—Calif.; Utah, Ariz.
greenei Foote, Coulson, and Robinson (for *bipunctatus* Greene).— Idaho; Alta., Utah. **N. name.**
    bipunctatus Greene, 1922: 9 (preocc. Macquart, 1842).— Idaho.
indianus (Harmston and Knowlton), 1946b: 24 (*Paraclius*).—Ind.; Tenn., N.C., S.C., Fla.
laticrus Van Duzee, 1918: 46.—N.J.
moechus Loew, 1861a: 40.—N.Y.; Ont. and Que., s. to Iowa and S.C.
mysticus (Becker), 1922: 52 (*Paraclius*).—Ga.
olympiae (Aldrich), 1896a: 83 (*Macellocerus*).—Wash.; Wash. to Ill., s. to Calif.
protervus Melander, 1900: 143.—N.J.; N.H. to Fla.
    junctus Coquillett, 1910a: 125.—N.J.
rotundipennis Greene, 1922: 5.—Del.; N.J.
sanus Osten Sacken, 1877: 316.—Calif.; Wash. to Alta., s. to Calif. and Colo.
simulatus Greene, 1922: 4.—Idaho; B.C., Utah.
spinitarsis (Van Duzee), 1924d: 43 (*Tetrechus*).—Calif.
tahoensis Harmston and Knowlton, 1940a: 114.—Calif.
tenuiseta Greene, 1922: 10.—Oreg.
utahensis Harmston and Knowlton, 1940a: 112.—Utah.
volitans Melander, 1900: 143.—Wyo.; Tex.
vorax Loew, 1861a: 41.—D.C.; Calif. to Que., s. to N.C.

Genus POLYMEDON Osten Sacken

Polymedon Osten Sacken, 1877: 317. Type-species, *flabellifer* Osten Sacken (mon.).

    REFERENCE: Van Duzee, 1927d (key).

californicus Harmston and Knowlton, 1943a: 105.—Calif.
castus Wheeler, 1899: 6.—Ariz; Utah.
    castor, error.
dilaticosta Van Duzee, 1927d: 124.—Ariz.
flabellifer Osten Sacken, 1877: 317.—Calif.; Idaho.
flavitibialis Van Duzee, 1930c: 71.—Ariz.

nigrifemoratus Van Duzee, 1927d: 125.—Ariz.
nimius Aldrich, 1901: 334.—Mexico; Idaho, Utah, Ariz.
nitidus Van Duzee, 1927d: 125.—Ariz.

Unplaced Species of Dolichopodinae

contingens Walker, 1852: 213 (*Dolichopus*).—U.S.

Subfamily PLAGIONEURINAE

Genus PLAGIONEURUS Loew

Plagioneurus Loew, 1857c: 43. Type-species, *univittatus* Loew (mon.).

univittatus Loew, 1857c: 43.—Cuba; S. Dak. to Mass., s. to Fla., Neotropical.

Subfamily HYDROPHORINAE

Genus HYDROPHORUS Fallén

Hydrophorus Fallén, 1823c: 2. Type-species, *jaculus* Fallén (Macquart, 1827a: 249; 1827b: 37).
Parhydrophorus Wheeler, 1896c: 185. Type-species, *canescens* Wheeler (mon.).
Millardia Curran, 1934a: 231 (preocc. Thomas, 1911). Type-species, *Medeterus viridiflos* Walker (orig. des.).

Flies of the genus *Hydrophorus* skate over the quiet surface of large bodies of water. The adults of *H. pacificus* Van Duzee, a Hawaiian species, have been seen pulling the larvae of Chironomidae from the mud of wet shores.

REFERENCES: Aldrich, 1911a (rev.); Van Duzee, 1923c, 1926f (keys).

agalma Wheeler, 1899: 66.—Mich.; Man., N.Y.
alboflorens (Walker), 1849: 656 (*Medeterus*).—N.S. Unrecognized.
albomaculatus Van Duzee, 1926c: 47.—Man.
algens Wheeler, 1899: 63.—Wyo.; Alta., Man., Colo.
alter Parent, 1934a: 19.—Amer. Unrecognized.
   glaber, authors, not Walker.
altivagus Aldrich, 1911a: 67.—Colo.; Alaska, Alta., Idaho, Man.
amplectens Aldrich, 1911a: 67.—S. Dak.; Mich., Que., N.Y.
ampullaceus Van Duzee, 1924c: 247.—Alta.
aquatilis Aldrich, 1922b: 17.—Alaska.
argentatus Van Duzee, 1918: 50.—Calif.
argentifacies Van Duzee, 1926c: 51.—Man.

brevicauda Van Duzee, 1923c: 259.—Alaska.
breviseta (Thomson), 1869: 510 (*Medeterus*).—Calif.; Alaska, Wash.
browni Curran, 1930i: 73.—Que.
canescens (Wheeler), 1896c: 187 (*Parhydrophorus*).—Wyo.; Calif., Utah., S. Dak., Kans.
canities Van Duzee, 1923c: 253.—Alaska.
cerutias Loew, 1872a: 91 (Cent. 10, no. 65).—Tex.; Alta., S. Dak., Colo., Kans., N. Mex.
chrysologus (Walker), 1849: 655 (*Medeterus*).—Ont.; Alaska, Man. to N.S., s. to Colo. and N.Y.
  chrysolygus, error.
claripennis Van Duzee, 1924c: 246.—Man.; Sask.
criddlei Van Duzee, 1925e: 181.—Man.
extrarius Aldrich, 1911a: 65.—S. Dak.; Ont., Que.
flavihirtus Van Duzee, 1923c: 259.—Alaska; Colo.
flavipennis Van Duzee, 1926c: 48.—Ont.; Mich., N.Y., ?Calif.
fulvidorsum Van Duzee, 1925e: 182.—Alta.
fumipennis Van Duzee, 1921a: 167.—Alaska; Alaska (Pribilof Is.), N.W.T. (Baffin I.).
glaber (Walker), 1849: 655 (*Medeterus*).—Ont.; Alaska, N.W.T.
gratiosus Aldrich, 1911a: 49.—Idaho; Alaska to Calif. and Colo.
hirtipes Van Duzee, 1933b: 17.—Que.
innotatus Loew, 1864e: 212.—Alaska; Alaska (Pribilof Is.) to n. N.W.T., s. to Oreg.
lividipes Van Duzee, 1926c: 50.—Man.
maculipennis Van Duzee, 1926c: 45.—Oreg.
magdalenae Wheeler, 1899: 67.—N. Mex.; Wash. to Nev. and N. Mex.
manicatus Collin, 1935: 372.—Que.
minimus Van Duzee, 1924f: 15.—Alaska.
nigribarbus Van Duzee, 1923c: 257.—Alaska.
nigrinervis Van Duzee, 1926c: 46.—B.C.
oregonensis Van Duzee, 1933b: 19.—Oreg.
parvus Loew, 1862b: 214 (Cent. 2, no. 67).—Ill.; Mich., Pa., N.H., Mass., Conn.
pectinipes Van Duzee, 1923c: 254.—Alaska.
pensus Aldrich, 1911a: 68.—Idaho; Wash., Oreg., Utah.
philombrius Wheeler, 1890: 378.—Wis.; Wash. to Que., s. to Calif. and Tex.
phoca Aldrich, 1911a: 63.—B.C.
pilitarsis Malloch, 1919f: 51.—Alaska.
pirata Loew, 1861a: 71.—Pa., D.C. (Loew, 1864e: 214); Mich., Que. to Pa. and D.C.

plumbeus Aldrich, 1911a: 50.—Wash.
praecox (Lehmann), 1822: 42 (also 1824: 243) (*Dolichopus*).—Germany; s. Canada, entire U.S., Europe.
    cinerea Perris, 1847: 492 (*Aphrozeta*).—France.
    aestuum Loew, 1869b: 36 (Cent. 8, no. 60).—R.I.
    aestuans, error.
    eldoradensis Wheeler, 1899: 65.—Calif., Wyo., N. Mex., Tex., Kans.
propinquus Van Duzee, 1923c: 256.—Alaska.
purus Curran, 1924g: 193.—Man.; Alta.
signifer Coquillett, 1899f: 344.—U.S.S.R. (Bering I.); Alaska.
sodalis Wheeler, 1899: 68.—Wyo.; Nev., Utah, Colo., Mich., N. Mex.
spinosus Van Duzee, 1933b: 16.—Oreg.
vandykei Van Duzee, 1926f: 4.—Oreg.
viridifacies Van Duzee, 1923c: 255.—Alaska; Colo.
viridiflos (Walker), 1852: 212 (*Medeterus*).—N. Amer.; Maine to N.J.
    intentus Aldrich, 1911a: 51.—Mass.
    indentus, error.

Genus SCELLUS Loew

Scellus Loew, 1857a: 22. Type-species, *Hydrophorus spinimanus* Zetterstedt (Coquillett, 1910b: 603).

REFERENCE: Greene, 1924a (rev.).

amplus Curran, 1923e: 73.—B.C.; Oreg., Idaho, Mont.
avidus Loew, 1864e: 207.—N.W.T.; Alta. to N. Mex.
coloradensis Harmston and James *in* Harmston and Knowlton, 1942b: 82.—Colo.
crinipes Van Duzee, 1925b: 177.—Utah; Idaho, Wyo.
exustus (Walker), 1852: 211 (*Medeterus*).—N. Amer.; Wash. and Alta. to Maine, s. to Colo. and N.Y.
filifer Loew, 1864e: 209.—N.W.T.; Alaska to N.W.T., s. to Colo.
knowltoni Harmston, 1939: 71.—Utah; Mont.
monstrosus Osten Sacken, 1877: 319.—B.C.; B.C. to Minn., s. to Utah.
spinimanus (Zetterstedt), 1843: 445 (*Hydrophorus*).—Scandinavia; N.W.T.
varipennis Van Duzee, 1925b: 180.—Calif.
vigil Osten Sacken, 1877: 318.—Calif.; Wash. to Alta., s. to Calif. and Utah.
virago Aldrich, 1907a: 133.—Calif.; Sask. to Calif. and Utah.

Genus MELANDERIA Aldrich

Melanderia Aldrich, 1922c: 146. Type-species, *mandibulata* Aldrich (orig. des.).

A genus of intertidal dolichopodids, the larvae of which have highly developed labellae which are used like mandibles. The biology of these flies is similar to that of *Hydrophorus*.

REFERENCE: Arnaud, 1958b (rev.).

Subgenus MELANDERIA Aldrich

crepuscula Arnaud, 1958b: 181.—Calif.
mandibulata Aldrich, 1922c: 146.—Wash.; Calif.

Subgenus WIRTHIA Arnaud

Melanderia, subg. **Wirthia** Arnaud, 1958b: 184. Type-species, *Hydrophorus curvipes* Van Duzee (orig. des.).

curvipes (Van Duzee), 1918: 48 (*Hydrophorus*).—Calif.

Genus LIANCALUS Loew

Anoplomerus Rondani, 1856: 141 (preocc. Dejean, 1835). Type-species, *Musca regius* Fabricius (orig. des.)=*virens* (Scopoli).
Liancalus Loew, 1857a: 22 (n. name for *Anoplomerus* Rondani). Type-species, *Musca regius* Fabricius (aut.)=*virens* (Scopoli).

REFERENCE: Van Duzee, 1917a (key).

genualis Loew, 1861a: 70.—Middle States (Loew, 1864e: 199); Minn. to Que., s. to Ga.
hydrophilus Aldrich, 1893c: 569.—S. Dak.; Alta., Utah, Colo.
limbatus Van Duzee, 1917a: 127.—Calif.; B.C. to Calif.
querulus Osten Sacken, 1877: 318.—Calif.; Idaho, Wyo., Utah.
similis Aldrich, 1893c: 571.—Wash.; Idaho.

Genus THINOPHILUS Wahlberg

Thinophilus Wahlberg, 1844a: 37 (also Schiödte, 1844: 44). Type-species, *Rhaphium flavipalpe* Zetterstedt (mon.).

REFERENCE: Van Duzee, 1926b (rev.).

armiger Van Duzee, 1926b: 43.—Fla.; Tex., N.C.
bimaculatus Johnson, 1921a: 13.—Fla.; Md., N.C.
brevipes Van Duzee, 1932a: 10.—Fla.

canities Van Duzee, 1926b: 44 (as *cantities*, p. 36).—Kans.; Nebr., Iowa, N. Mex.
delicatus Van Duzee, 1926b: 40.—Mo.
depressus Van Duzee, 1926b: 39.—Mexico (Baja Calif.); Utah.
frontalis Van Duzee, 1914b: 406.—Fla.; La., N.C.
insulanus Van Duzee, 1926b: 49.—Calif.
latimanus Van Duzee, 1926b: 37.—Colo.
magnipalpus Van Duzee, 1926b: 45.—N. Mex.
neglectus Wheeler, 1899: 70.—N.J.; Fla.
ochrifacies Van Duzee, 1924a: 101.—N.Y.; N.S. to Mo. and Fla.
pectinifer Wheeler, 1896b: 155.—Wyo.; Calif., Utah, Colo.
prasinus Johnson, 1921a: 13.—Mass.; N.Y., N.H., N.C., Fla.
pruinosus Van Duzee, 1930b: 84.—Colo.
rufibarbis Van Duzee, 1926b: 48.—Utah.
russelli Curran, 1927f: 89.—N.W.T.
scopiventris Harmston and Knowlton, 1940e: 108.—Calif.
spinipes Van Duzee, 1926b: 46.—Man.; Nev., Utah, Colo., Nebr.
thallasinus Van Duzee, 1926b: 42.—Fla.
viridifacies Van Duzee, 1924a: 102.—Mass.; Tex., N.C.
    quadratus Van Duzee, 1924b: 28 (*Chrysotus*).—Mass.
    nigripilosus Van Duzee, 1926b: 44.—Tex.

Genus DIOSTRACUS Loew

Diostracus Loew, 1861a: 43. Type-species, *prasinus* Loew (mon.).
olga Aldrich, 1911b: 71.—Wash.
prasinus Loew, 1861a: 44.—N.Y.; Vt., Mass., Va., Tenn., N.C.

Genus HYPOCHARASSUS Mik

Hypocharassus Mik, 1878a: 627. Type-species, *gladiator* Mik (orig. des.).
Drepanomyia Wheeler, 1898: 217. Type-species, *pruinosus* Wheeler (Coquillett, 1910b: 535).

Hypocharassus larvae are intertidal and form cocoons of sand in which the insect pupates. Adults are occasionally found in immense numbers on floating seaweed.

    REFERENCE: Smith, 1952b (biol.).

gladiator Mik, 1878a: 629.—Ga.; N.C., Fla.
    johnsonii Wheeler, 1898: 219 (*Drepanomyia*).—Fla.
pruinosus (Wheeler), 1898: 218 (*Drepanomyia*).—Fla.; Mass., R.I., Va., N.C., Ga.

Subfamily APHROSYLINAE

Genus APHROSYLUS Haliday

Aphrosylus Haliday, 1851: 220. Type-species, *raptor* Haliday (Coquillett, 1910b: 508).
Aphrosylus, subg. *Paraphrosylus* Becker, 1922: 127. Type-species, *praedator* Wheeler (present des.).

The immature stages of *Aphrosylus* species breed in intertidal seaweeds. The adults are found flying over exposed plants at low tides.

REFERENCES: Wheeler, 1897 (rev., larvae); Saunders, 1928 (biol.); Wirth and Stone, 1956 (biol.).

californicus Harmston, 1952: 292.—Calif.
direptor Wheeler, 1897: 148.—Calif.; B.C.
grassator Wheeler, 1897: 149.—Calif.
nigripennis Van Duzee, 1924e: 75.—Alaska.
praedator Wheeler, 1897: 146.—Calif.; B.C.
wirthi Harmston, 1951a: 13.—Calif.

Subfamily MEDETERINAE

Genus MEDETERA Fischer von Waldheim

Medetera Fischer von Waldheim, 1819: 7 and pl. Type-species, *carnivora* Fischer von Waldheim (mon.)=*diadema* (Linnaeus).
Medeterus, error or emend.

Adults of *Medetera* are usually found on tree trunks, logs, or old wooden fences. The larvae apparently feed on the immature stages of wood-boring beetles.

REFERENCES: Van Duzee, 1928e (key); De Leon, 1935 (biol.).

aberrans Wheeler, 1899: 22.—N.J.; Iowa, Ont., N.Y., Tenn., N.C.
    lobatus Van Duzee, 1914c: 441.—N.J.
aeneiventris Van Duzee, 1933c: 152 (n. name for *aeneus* Van Duzee).—Calif.
    aeneus Van Duzee, 1919a: 263 (preocc. Meigen, 1838).—Calif.
aequalis Van Duzee, 1919a: 265.—Calif.
albiciliata Van Duzee, 1933b: 13.—Ont.
albosetosa Van Duzee, 1928e: 36.—Tex.
aldrichii Wheeler, 1899: 24.—Idaho.
alpina Harmston and Knowlton, 1941b: 95.—Calif.

arctica Van Duzee, 1933c: 152 (n. name for *bicolor* Van Duzee).—
Alaska.
bicolor Van Duzee, 1923c: 249 (preocc. Meigen, 1838).—
Alaska.
arnaudi Harmston, 1951a: 12.—Calif.
aurivittata Wheeler, 1899: 29.—Idaho.
bistriata Parent, 1929a: 183.—Ga.
caerulescens Malloch, 1919m: 8.—Ill.; Alaska, N.Y., N.J.
distinctus Van Duzee, 1919a: 266.—N.Y.
californiensis Wheeler, 1899: 27.—Calif.
longimana Van Duzee, 1933b: 12.—Calif.
ciliata Van Duzee, 1928e: 37.—Ont.; N.Y.
crassivenis Curran, 1928e: 199.—N.Y.
cuneiformis Van Duzee, 1919a: 263.—Calif.
cyanogaster Wheeler, 1899: 27.—Wash.
emarginata Van Duzee, 1914c: 439.—Ont.; N.Y.
excipiens Becker, 1922: 133.—S.C., Paraguay. Neotropical specimen
probably different sp.
falcata Van Duzee, 1919a: 261.—Calif.
flavicosta Van Duzee, 1932a: 11.—N.Y.
frontalis Van Duzee, 1919a: 265.—N.Y.; Ont., Que.
furcata Curran, 1928e: 200.—N.Y.
halteralis Van Duzee, 1919a: 267.—Ont.; Que.
idahoensis Harmston and Knowlton, 1943a: 106.—Idaho.
intermedia Van Duzee, 1928b: 40.—Sask.
longinervis Van Duzee, 1928e: 36.—Oreg.
longinqua Van Duzee, 1919a: 262.—Calif.
maura Wheeler, 1899: 23.—N.H.; Vt., Mass., N.C.
minima Van Duzee, 1925e: 180.—Calif.; Alaska.
modesta Van Duzee, 1914c: 440.—N.J.; Maine to N.J.
nigripes Loew, 1861a: 73.—Middle States; Iowa, Ont. to Mass., s.
to Fla.
nitidiventris Van Duzee, 1919a: 264.—Calif.
nova Van Duzee, 1919a: 262.—Va.; Md., Tenn., N.C., S.C.
obscuripennis Van Duzee, 1919a: 266.—Calif.
orbiculata Van Duzee, 1932a: 12.—Calif.
oregonensis Van Duzee, 1919a: 268.—Oreg.
parva Van Duzee, 1923c: 249.—Alaska.
petulca Wheeler, 1899: 21.—Wash.
princeps Wheeler, 1899: 25.—N.J.; N.H., Mass., R.I.
similis Van Duzee, 1919a: 261.—Calif.
simplicipes Curran, 1928e: 202.—N.Y.; Tenn., N.C.
trisetosa Van Duzee, 1924c: 246.—Alta.; Ont.

univittata Van Duzee, 1933c: 151 (n. name for *obesus* Van Duzee).—N.Y.; N.J.
   obesus Van Duzee, 1919a: 264 (preocc. Kowarz, 1877).—N.Y.
vanduzeei Curran, 1928e: 203.—Que.
veles Loew, 1861a: 73.—Fla.; Idaho to Man. to N.H., s. to Utah and Fla.
veneta Curran, 1928e: 201.—N.Y.
vidua Wheeler, 1899: 24.—Wash.; Alaska, Oreg., Idaho, ?Ont.
viridifacies Van Duzee, 1923c: 248.—Alaska.
vittata Van Duzee, 1919a: 268.—Ont.; Iowa, Mich., Ill., Que., N.Y., Tenn.
wheeleri Foote, Coulson, and Robinson (for *appendiculatus* Wheeler).—Wyo. **N. name.**
   appendiculatus Wheeler, 1899: 29 (preocc. Macquart, 1827).—Wyo.
xerophila Wheeler, 1899: 28.—Calif.

Genus THRYPTICUS Gerstäcker

Thrypticus Gerstäcker, 1864: 43. Type-species, *smaragdinus* Gerstäcker (mon.).

Aphantotimus Wheeler, 1890: 375. Type-species, *willistoni* Wheeler (Coquillett, 1910b: 508).

In contrast to those of *Medetera*, the larvae of *Thrypticus* are said to be phytophagous.

   REFERENCE: Van Duzee, 1921b (key).

abdominalis (Say), 1829: 169 (1859b: 362) (*Chrysotus*).—Ind.; Kans. to Mich. and Pa., s. to La. and N.C.
aurinotatus Van Duzee, 1915c: 87.—Ga.
comosus Van Duzee, 1915c: 86.—N.Y., Ont.; Mich.
fraterculus (Wheeler), 1890: 376 (*Aphantotimus*).—Wis.; Alaska, Minn. to Ont., s. to Calif. and Tenn., Mexico.
longicauda Van Duzee, 1921b: 126.—Calif.
minutus Parent, 1929a: 186.—Tex.; Pa.
muhlenbergiae Johannsen and Crosby, 1913: 164.—N.Y.; Mo., Ohio, Md., N.C.
nigripes Van Duzee, 1921b: 125.—Nev.; Utah.
tectus Van Duzee, 1915c: 87.—N.Y.; Iowa to Conn., s. to N.C.
vietus Van Duzee, 1915c: 86.—Fla.
willistoni (Wheeler), 1890: 376 (*Aphantotimus*).—Wis.; Wyo., Iowa and Mo. to Mass., s. to Tenn. and N.C.

Subfamily RHAPHIINAE

Genus RHAPHIUM Meigen

Rhaphium Meigen, 1803: 272. Type-species, *macrocerum* Meigen (Curtis, 1835a: pl. 568).
Xiphandrium Loew, 1857a: 36. Type-species, *Rhaphium quadrifilatum* Loew (as *quadrifilum;* Coquillett, 1910b: 621).
Porphyrops, authors, not Meigen.

Adults of *Rhaphium* are usually found on foliage in moist places.

REFERENCES: Curran, 1926j, 1927k (rev.).

aequale Van Duzee, 1927e: 147.—Idaho.
albibarbum (Van Duzee), 1924f: 10 (*Porphyrops*).—Alaska.
aldrichi (Van Duzee), 1922a: 86 (*Xiphandrium*).—Alaska.
arboreum Curran, 1924f: 140.—Que.; Ind.
armatum Curran, 1924f: 136.—Que.; N.Y., N.C.
atkinsoni Curran, 1926j: 257, key (descr. 1927k: 156).—Sask.; Mich.
banksi Van Duzee *in* Curran, 1926j: 257, key (descr. 1927k: 109).— Va.; N.C.
barbipes (Van Duzee), 1923a: 239 (*Porphyrops*).—Maine.
boreale (Van Duzee), 1923c: 246 (also 1924f: 9) (*Porphyrops*).— Alaska.
brevicorne (Van Duzee), 1923a: 241 (*Porphyrops*).—Oreg.
brevilamellatum Van Duzee *in* Curran, 1926j: 254, key (descr. 1927k: 119).—Va.
browni Curran, 1931c: 5.—Que.
calcaratum Van Duzee, 1928c: 167.—N.Y.
campestre Curran, 1924f: 133.—Man.; B.C. to Man.
canadense Curran, 1924f: 137.—Que.; Ohio to Que., s. to Ga.
caudatum Van Duzee *in* Curran, 1926j: 254, key (descr. 1927k: 117).—N.Y., Ont.
ciliatum Curran, 1929c: 30.—B.C.
coloradense Curran, 1926j: 256, key (descr. 1927k: 162) (n. name for *longicorne* Van Duzee).—Colo.
  longicorne Van Duzee, 1922a: 85 (*Xiphandrium;* preocc. Meigen, 1824).—Colo.
colute Harmston and James *in* Harmston and Knowlton, 1942b: 83.—Colo.; Utah.
crassipes (Meigen), 1824: 50 (*Porphyrops*).—Europe; Alaska, B.C. to Que., Europe, Asia.
discolor Zetterstedt, 1838: 704.—Lapland; Alaska, Scandinavia.
  consobrinum Zetterstedt, 1843: 471.—Scandinavia.

dubium (Van Duzee), 1922a: 85 (*Xiphandrium*; as *femineum* var.).—
N.Y.; Ont., Que.
effilatum (Wheeler), 1899: 34 (*Porphyrops*).—Wis., Wyo.; B.C. to
N.S., s. to Utah and Colo.
elegantulum (Meigen), 1824: 51 (*Porphyrops*).—Germany; Alaska,
Que.
elongatum Van Duzee, 1933a: 4.—Que.
exile Curran, 1926j: 257, key (descr. 1927k: 165).—Ind.; Tenn.
fascipes (Meigen), 1824: 54 (*Porphyrops*).—Germany; Alaska, Kans.,
Mich., Que., N.Y.
femineum (Van Duzee), 1922a: 84 (*Xiphandrium*).—N.Y.; Mich., Ind.
femoratum (Van Duzee), 1922a: 81 (*Xiphandrium*).—Nev.; Alaska,
Alta., Utah.
flavicoxa (Van Duzee), 1922a: 84 (*Xiphandrium*).—Ga.; Tenn.
foliatum Curran, 1926j: 256, key (descr. 1927k: 150).—Ohio; N.Y.,
Tenn.
fuscicosta Van Duzee, 1928c: 167.—N.Y.
furcifer Curran, 1931c: 2.—Que.
gibsoni Curran, 1926d: 171.—N.S.
gracile Curran, 1924o: 228.—Mass.
grande (Curran), 1923j: 210 (*Porphyrops*).—B.C.
hirtimanum Van Duzee, 1933a: 6.—Oreg.
impetuum Curran, 1926j: 254, key (descr. 1927k: 172).—Md.
insolitum Curran, 1926j: 256, key (descr. 1927k: 151).—Fla.; Md.,
N.C., ?Ont.
johnsoni (Van Duzee), 1923a: 240 (*Porphyrops*).—N.J.; N.H., R.I.,
Va.
latifacies Van Duzee, 1930c: 53.—Alta.
longibara Van Duzee, 1930c: 53.—Alta.
longipalpe Curran, 1926j: 253, key (descr. 1927k: 148).—Colo.;
Idaho, Utah.
longipes (Loew), 1864a: 95 (Cent. 5, no. 92) (*Porphyrops*).—N.H.;
N.B. to N.Y. and Conn.
lugubre Loew, 1861a: 49.—Carolina; N.Y., Conn., Pa., N.J., Md.
melampus (Loew), 1861a: 50 (*Porphyrops*).—Middle States; Mich. to
Que., s. to Tenn., and S.C.
americanus Van Duzee, 1914b: 404 (*Systenus*).—N.C.
montanum (Van Duzee), 1920: 47 (*Porphyrops*).—Calif.
nigricoxa (Loew), 1861a: 51 (*Porphyrops*).—Md.; Maine to Md.
nigrociliatum Curran, 1926j: 252, key (descr. 1927k: 99).—Colo.
nigrovittatum Curran, 1926j: 254, key (descr. 1927k: 121).—Idaho.
nigrum (Van Duzee), 1923c: 245 (*Porphyrops*).—Alaska.

nudum (Van Duzee), 1924f: 9 (*Porphyrops*).—Alaska; Wash., Ont., Que.
obtusum Van Duzee, 1928c: 166.—Nev.
occipitale Curran, 1927k: 171.—La.
orientale Curran, 1926j: 254, key (descr. 1927k: 113).—N.B.; Mass.
ornatum (Van Duzee), 1923a: 242 (*Porphyrops*).—N.Y.; Ont., Que., Vt.
    petchi Curran, 1924f: 138.—Que.
palpale Curran, 1926j: 256, key (descr. 1927k: 149).—N.J.
pollex (Van Duzee), 1922a: 82 (*Xiphandrium;* as *femoratum* var.).— Nev.; Alaska, Oreg., Idaho, Utah.
punctitarse Curran, 1924f: 135.—Que.; Man., Mich.
robustum Curran, 1926j: 256, key (descr. 1927k: 145).—Idaho.
rossi Harmston and Knowlton, 1940d: 60.—Ill.
rotundiceps (Loew), 1861a: 51 (*Porphyrops*).—D.C.; Mich., Ind., Que. to Fla.
    fumipennis Loew, 1861a: 51 (*Porphyrops*).—Middle States.
septentrionale Curran, 1931c: 4.—Que.
shannoni Curran, 1926j: 257, key (descr. 1927k: 155).—Md.; Mass. to Pa. and Va.
signiferum (Osten Sacken), 1878c: 242 (*Porphyrops*).—N.Y.; Ont., Mich., N.S., Mass., Tenn., Va., N.C.
simplicipes Curran, 1926j: 256, key (descr. 1927k: 169).—Colo.
slossonae (Johnson), 1906: 59 (*Leucostola*).—N.H.; Oreg., Idaho, Colo., Que. to N.S., s. to N.Y.
spinitarse Curran, 1924f: 139.—Man.; Alta., Calif., Nev., Utah, Colo.
subarmatum Curran, 1924o: 228.—N.B.; N.H., N.C.
subfurcatum Van Duzee, 1932a: 8.—Wyo.
temerarium (Becker), 1922: 151 (*Xiphandrium*).—Colo.; Y.T., Man., Utah.
terminale (Van Duzee), 1924f: 7 (*Porphyrops*).—Alaska; N.W.T.
triangulatum (Van Duzee), 1922a: 83 (*Xiphandrium*).—Ariz.; Utah, Colo.
tricaudatum (Van Duzee), 1923c: 243 (*Porphyrops*).—Alaska.
tripartitum (Frey) *in* Lundstrom and Frey, 1913: 13 (*Porphyrops*).— n. U.S.S.R.; N.W.T.
unistylatum Vockeroth, 1952: 276.—N.W.T. (Mackenzie Delta); Utah, Colo.
vanduzeei Curran, 1926j: 253, key (descr. 1927k: 104).—Ind.; Nebr., Ont., Que., N.Y., Tenn., N.C.
wheeleri Van Duzee, 1932a: 9.—Wis.
xipheres (Wheeler), 1899: 34 (*Porphyrops*).—Pa.

DOLICHOPODIDAE 515

Genus NEMATOPROCTUS Loew

Nematoproctus Loew, 1857a: 40. Type-species, *Porphyrops annulata* Macquart (Coquillett, 1910b: 574)=*distendens* (Meigen).

The adults of *Nematoproctus* are usually found on herbaceous foliage.

REFERENCE: Van Duzee, 1930f (rev.).

cylindricus (Van Duzee), 1924c: 248 (*Diaphorus*).—Que.; Ont.
flavicoxa Van Duzee, 1930f: 168.—Conn.; Va.
jucundus Van Duzee, 1927b: 53.—Pa.; N.C.
    junctus (error) Van Duzee, 1930f: 168.
metallicus Van Duzee, 1930f: 169.—Pa.; Mich., N.C., ?Ind., ?Tenn.
terminalis (Van Duzee), 1914b: 405 (*Leucostola*).—Va.; Ohio, N.Y.
varicoxa Van Duzee, 1930f: 170.—Conn.
venustus Melander, 1900: 142.—N.J.; ?Iowa.
    venustus Malloch (error) Johannsen, 1928b: 774 (*Leucostola*).

Genus KEIROSOMA Van Duzee

Keirosoma Van Duzee, 1929: 24. Type-species, *albicinctum* Van Duzee (orig. des.).

slossonae Van Duzee, 1933b: 2.—Fla.
    albicinctum (in part) Van Duzee, 1929: 24. Fla. specimens only.

Genus SYNTORMON Loew

Syntormon Loew, 1857a: 35. Type-species, *Rhaphium metathesis* Loew (Coquillett, 1910b: 611).
Synarthrus Loew, 1857a: 35. Type-species, *Musca pallipes* Fabricius, (once as *tarsatum*, error) (mon.).

REFERENCE: Van Duzee, 1925a (rev.).

affine (Wheeler), 1899: 38 (*Synarthrus*).—Calif.; Nev., Utah.
bisinuatum Van Duzee, 1925a: 282.—Oreg.; Calif.
californicum Harmston, 1951a: 15.—Calif.
cinereiventre (Loew), 1861a: 48 (*Synarthrus*).—Middle States; Alta. and Idaho to Que., s. to Tex. and Tenn.
clavatum Van Duzee, 1925a: 281.—Nev.; Utah.
dissimilipes Van Duzee, 1925a: 283.—Nev.; Idaho, Utah.
femoratum Van Duzee, 1925a: 280.—Ind.
kennedyi Harmston and Knowlton, 1942d: 23.—Colo.; Utah.
nubilum Van Duzee, 1933b: 3.—Que.
oregonense Harmston and Knowlton, 1942d: 25.—Oreg.

ornatipes Van Duzee, 1925a: 277.—Colo.; Utah.
palmare (Loew), 1864e: 135 (*Synarthrus*).—Alaska; Wash., Idaho, Calif., Utah, Colo., Mich.
simplicitarse Van Duzee, 1925a: 286.—Ind.
strataegum (Wheeler), 1899: 39 (*Synarthrus*).—Wyo., Calif.; Oreg., Idaho, Utah, Colo.
tricoloripes Curran, 1923j: 209.—B.C.; Oreg., Idaho, Ont., Mich., Que.
uintaense Harmston and Knowlton, 1940b: 130.—Utah.
utahense Harmston and Knowlton, 1942d: 24.—Utah.
vanduzeei Curran, 1931c: 1.—Ont.
variegatum Harmston, 1952: 291.—Calif.

Genus PARASYNTORMON Wheeler

Parasyntormon Wheeler, 1899: 41. Type-species, *asellus* Wheeler (Coquillett, 1910b: 585).
Neosyntormon Curran, 1934a: 230. Type-species, *Parasyntormon montivagum* Wheeler (orig. des.).

appendiculatum Harmston and Knowlton, 1943b: 63.—Utah.
asellus Wheeler, 1899: 42.—Calif.; Colo.
classicum Harmston and Knowlton, 1943b: 64.—Utah.
emarginatum Wheeler, 1899: 45.—Calif.
emarginicorne Curran, 1923h: 192.—Alta.; B.C.
flavicoxa Van Duzee, 1922b: 89.—Calif.
fraterculus Van Duzee, 1922b: 88.—Calif.
hendersoni Harmston and Knowlton, 1939a: 256.—Utah.
hinnulus Wheeler, 1899: 44.—Wyo.
lagotis Wheeler, 1899: 43.—Calif.
lepus Van Duzee, 1918: 45.—Calif.
longicorne Van Duzee, 1933b: 4.—Oreg.
montivagum Wheeler, 1899:46.—Wyo.; Calif.
mulinum Van Duzee, 1922b: 89.—Calif.
nigripes Harmston and Knowlton, 1943b: 64.—Utah.
occidentale (Aldrich), 1894:153 (*Sympycnus*).—Wyo.; Utah, Nev.
petiolatum Van Duzee, 1933b: 5.—Calif.
rotundicorne Van Duzee, 1926g: 57.—N.S.
utahnum Van Duzee, 1933d: 64.—Utah.
virens Harmston and Knowlton, 1943b: 65.—Utah.

Genus PELOROPEODES Wheeler

Peloropeodes Wheeler, 1890: 373. Type-species, *salax* Wheeler (mon.).

Kophosoma Van Duzee, 1926a: 39. Type-species, *Sympycnus falco* Aldrich (orig. des.).

REFERENCE: Van Duzee, 1926a (as *Kophosoma*).

acuticornis (Van Duzee), 1926a: 44 (*Kophosoma*).—Ind.; Utah to Mich. to D.C., s. to Ga., Mexico.
apicales Harmston and Knowlton, 1946c: 139.—Idaho.
bicolor (Van Duzee), 1926a: 41 (*Kophosoma*).—Pa.; Iowa to Pa., s. to S.C.
brevis (Van Duzee), 1926a: 45 (*Kophosoma*).—N.Y.; Ont. to Conn., s. to Ga.
crassitibia Van Duzee, 1930a: 1 (*Campsicnemus*).—Conn.
cornutus (Van Duzee), 1926a: 42 (*Kophosoma*).—Calif.; Oreg., Idaho, Mich., Mexico.
discolor Robinson (for *frater* Robinson).—N.C.
[Name published by Robinson, 1964, Ent. Soc. Amer. Misc. Pub. 4: 126.—EDS.]
frater Robinson, 1960: 271 (preocc. Aldrich, 1902).—N.C.
fuscipes (Van Duzee), 1926a: 40 (*Kophosoma*).—N. Mex.; Idaho, Mexico.
salax Wheeler, 1890: 374.—Wis.; Wis. to Va., s. to Tenn.
flavipes Van Duzee, 1914c: 437.—Va.

Genus SYSTENUS Loew

Systenus Loew, 1857a: 34. Type-species, *Rhaphium adpropinquans* Loew (present des.) = *pallipes* (Roser).

All species of which the larval stages are known, have been reared from moist tree holes and the ulcerated or decaying wood of trees.

REFERENCE: Wirth, 1952d (rev.).

albimanus Wirth, 1952d: 240.—Va.; Tex., Tenn., Ala.
apicalis Wirth, 1952d: 237.—Va.; Md., Ala.
minutus (Van Duzee), 1913b: 60 (*Neurigona*).—Pa.; Ky., Tenn., Ala.
shannoni Wirth, 1952d: 240.—Md.; N.H.

Subfamily NEURIGONINAE

Genus NEURIGONA Rondani

Neurigona Rondani, 1856: 142. Type-species, *Musca quadrifasciata* Fabricius (orig. des.).
Neurogona, Neurogonia, errors.

Saucropus Loew, 1857a: 41 (unjustified n. name for *Neurigona* Rondani). Type-species, *Musca quadrifasciata* Fabricius (aut.).
Dactylomyia Aldrich, 1894: 151. Type-species, *gracilipes* Aldrich (mon.) = *lateralis* (Say).

The adults of most *Neurigona* species are found on tree trunks and low foliage. The larval habitats are unknown.

REFERENCE: Van Duzee, 1913b (rev.).

aestiva Van Duzee, 1913b: 50.—N.Y., Vt., Md.; Iowa, Ont., Mich.
albospinosa Van Duzee, 1913b: 59.—Wash., Idaho, Calif.; B.C.
aldrichii Van Duzee, 1913b: 40.—Kans.; Iowa, Mich., N.Y.
arcuata Van Duzee, 1913b: 45.—Ont.; Mich., N.Y., Tenn.
australis Van Duzee, 1913b: 59.—N. Mex.
bivittata Van Duzee, 1913b: 51.—Colo.; B.C.
carbonifer (Loew), 1869c: 177 (Cent. 9, no. 84) (*Saucropus*).—N.Y., Minn. to Que., s. to Va.
ciliata Van Duzee, 1913b: 56.—Wash.
cilimanus Van Duzee, 1924c: 245.—Man.
deformis Van Duzee, 1913b: 46.—N.Y., Ont.; N.H., Maine.
dimidiata (Loew), 1861a: 75 (*Saucropus*).—Fla.; N.H. and Vt., s. to Fla.
disjuncta Van Duzee, 1913b: 42.—Canada, Vt., N.Y.; Mich. to se. Canada, s. to Ga.
flava Van Duzee, 1913b: 40.—Idaho.
floridula Wheeler, 1899: 72.—N.J., Md.; Wis. to Maine, s. to N.C.
  var. **infuscata** Van Duzee, 1913b: 39.—N.Y.; Wis. to Ont. and Mass., s. to N.J.
kesseli Hendrickson, 1961: 278.—Calif.
lateralis (Say), 1829: 169 (1859b: 362) (*Medeterus*).—Ind.; S. Dak. to Que., s. to Fla.
  superbiens Loew, 1861a: 76 (*Saucropus*).—Fla.
  gracilipes Aldrich, 1894: 151 (*Dactylomyia*).—S. Dak.
lienosa Wheeler, 1899: 73.—Calif.
maculata Van Duzee, 1913b: 36.—Canada, Wis., Mich., N.Y., N.H., Mass., Pa., N.C.; Wis. to Que., s. to N.C.
nigricornis Van Duzee, 1914c: 433.—N.Y.
nigrimanus Van Duzee, 1930c: 70.—Alta.
nitida Van Duzee, 1913b: 33.—Wis.; Maine, N.H., Vt.
ornata Van Duzee, 1930c: 55.—Alta.
pectoralis Van Duzee, 1913b: 49.—N. Mex.; ?Ont.
perbrevis Van Duzee, 1913b: 57.—N. Mex.
perplexa Van Duzee, 1913b: 29.—Pa.; Mich., Maine, N.H.
planipes Van Duzee, 1924c: 245.—Alta.; B.C.

rubella (Loew), 1861a: 76 (*Saucropus*).—Va.; Iowa to Mass., s. to Kans. and Va.
sombrea Harmston and Knowlton, 1945: 77.—Mich.
tarsalis Van Duzee, 1913b: 51.—Mich., N.Y., Pa.; Mich. to Vt., s. to N.C.
tenuis (Loew), 1864e: 228 (*Saucropus*).—Middle States; Mo., Mich., Mass. to N.C., ?N. Mex.
terminalis Van Duzee, 1924c: 244.—B.C.
torrida Harmston, 1951a: 14.—Calif.
transversa Van Duzee, 1913b: 41.—Calif.
tridens Van Duzee, 1913b: 34.—Wash. (as Idaho, error); B.C.
uinta Harmston and Knowlton, 1942b: 80.—Utah.
viridis Van Duzee, 1913b: 43.—N.Y., N.H., Va.; Tenn.
zionensis Harmston and Knowlton, 1942b: 81.—Utah.

Subfamily DIAPHORINAE
Genus DIAPHORUS Meigen

Diaphorus Meigen, 1824: 32. Type-species, *flavocinctus* Meigen (Westwood, 1840: 134) = *oculatus* (Fallén).
 REFERENCE: Van Duzee, 1915a (rev.).
aldrichi Van Duzee, 1915a: 185.—Idaho; Mont., Nev., Utah, Colo., Mich.
 albifacies Parent, 1929a: 191.—Mont.
alienus Van Duzee, 1915a: 176.—Oreg.
argentifacies Van Duzee, 1932a: 5.—Nev.
australis Van Duzee, 1915a: 194, on printed slip pasted in separates (n. name for *femoratus* Van Duzee).—La. No further reference to this name has been found in the literature. If *australis* is not considered published in the sense of the International Code, its use in this catalog is to be construed as a formal proposal of the replacement name, credited to Van Duzee.
 femoratus Van Duzee, 1915a: 194 (preocc. Walker, 1852).—La.
basalis Van Duzee, 1915a: 168.—Va.; Iowa, N.C.
brevinervis Van Duzee, 1924f: 2.—Alaska.
californicus Van Duzee, 1917c: 33.—Calif.
funeralis Parent, 1929a: 188.—Ga.; Fla.
fuscus Van Duzee, 1921b: 122 (n. name for *adustus* Van Duzee).—Idaho, Nev.; Alaska, N.W.T., Utah.
 adustus Van Duzee, 1915a: 172 (preocc. Wiedemann, 1830, orig. *Dolichopus*, now *Lyroneurus*).—Idaho, Nev.

gibbosus Van Duzee, 1915a: 173.—N.Y., Ont., Maine, Vt., Mass., Pa.; Alaska, Calif., Ont. and Wis. to Maine, s. to N.C.
hirsutus Van Duzee, 1921b: 120.—Calif.
inornatus Van Duzee, 1917c: 38.—Calif.
junctus Van Duzee, 1917c: 35.—Calif.
lamellatus Loew, 1864e: 165.—Middle States; N.Y. to S.C.
latifacies Van Duzee, 1932a: 6.—Okla.
leucostoma Loew, 1861a: 58.—Md.; S. Dak. to Que., s. to La. and Fla.
    vittatas Van Duzee, 1915a: 179.—Va.
    var. **infuscatus** Van Duzee, 1915a: 179.—D.C.; N.Y., Mass., Conn., N.J.
mundus Loew, 1861a: 57.—Pa.; Idaho, Que. to Ga.
nigricans Meigen, 1824: 33.—Germany; Utah to Maine, s. to Tenn. and Fla., Europe, Mexico, Cent. Amer.
    opacus Loew, 1861a: 56.—N.Y.
nudus Van Duzee, 1917c: 34.—Va.; N.Y., Pa., N.C.
occidentalis Van Duzee, 1915a: 180.—Oreg.; Alaska, Calif., Colo., N. Mex.
parmatus Van Duzee, 1915a: 181.—Calif.
quadratus Van Duzee, 1915a: 179.—Ont.
remulus Van Duzee, 1915a: 182.—S. Dak.
repandus Van Duzee, 1915a: 183.—Calif.
similis Van Duzee, 1915a: 186.—Pa.; S.C., Ga.
slossonae Van Duzee, 1932a: 5.—Fla.
snowii Van Duzee, 1917c: 36.—Ariz.; Alta., Colo.
sodalis Loew, 1861a: 58.—N.Y.; Iowa to N.H., s. to Ga.
sparsus Van Duzee, 1917c: 37.—Va.; N.Y.
spectabilis Loew, 1861a: 57.—D.C.; S. Dak. to Conn., s. to La. and Ga., Mexico, Cent. Amer.
spinitalus Van Duzee, 1923d: 67.—Calif.
texanus Van Duzee, 1932a: 7.—Tex.
triangulatus Van Duzee, 1915a: 189.—Idaho; Utah.
trivittatus Van Duzee, 1915a: 187.—Fla.; Mexico.
usitatus Van Duzee, 1915a: 184.—Oreg., Idaho.
vanduzeei Robinson (for *communis* Van Duzee).—Okla. [Name published by Robinson, 1964. Ent. Soc. Amer. Misc. Pub. 4: 138.—Eds.]
    communis Van Duzee, 1932a: 4 (preocc. White, 1917).—Okla.
variabilis Van Duzee, 1915a: 191.—Fla., Ga.; Minn., Iowa, Ill., N.Y. to Fla.
versicolor Van Duzee, 1932a: 6.—Tex.
vulsus Van Duzee, 1917c: 38.—Ariz.

Genus CHRYSOTUS Meigen

Chrysotus Meigen, 1824: 40. Type-species, *Musca nigripes* Fabricius (Westwood, 1840: 134). Further study to determine the true type-species is necessary. Fabricius' *nigripes* is at present unrecognized; Meigen's usage of it was probably a mixed species. See Kowarz, 1874: 471.

Chrysotus is a widespread genus, and its species are found in a wide variety of habitats. Females can seldom be determined to species, and as a group are hard to distinguish from those of *Diaphorus*.

REFERENCE: Van Duzee, 1924b (key, males).

affinis Loew, 1861a: 64.—Middle States; Ont., Que., Maine, Conn., N.Y.
albohirtus Van Duzee, 1924b: 41.—S. Dak.
aldrichi Van Duzee, 1924b: 47.—Ga.
annulatus Van Duzee, 1924b: 27.—Ga.; Mich., Conn., Va. to Fla.
anomalus Malloch, 1914g: 238.—La.; N.Y., Ga.
arcuatus Van Duzee, 1924b: 48.—Calif.; Oreg., Idaho, Mexico.
argentatus Van Duzee, 1924b: 41.—Calif.; Idaho, Mich.
arkansensis Van Duzee, 1930b: 84.—Ark.
atratus Van Duzee, 1930a: 2.—Conn.
auratus Loew, 1861a: 65.—N.Y. Unrecognized.
badius Van Duzee, 1932b: 15.—Que.
barbatus (Loew), 1861a: 48 (*Synarthrus*).—Middle States (Loew, 1864e: 138); Wis. to R.I., s. to Iowa and Fla., Mexico, West Indies.
    validus Loew, 1861a: 63.—Middle States.
    americanum Wheeler, 1896b: 154 (*Xiphandrium*).—Wis., Ill., Ind.
barbipes Van Duzee, 1932b: 14.—Colo.
bellulus Van Duzee, 1932b: 14.—Que.
bellus Van Duzee, 1924b: 39.—D.C.; Mich., N.Y. to Maine, s. to N.C.
bracteatus Van Duzee, 1924b: 36.—Ga.
caerulus Van Duzee, 1924b: 21.—Wash.; B.C.
californicus Van Duzee, 1924b: 16.—Calif.
canadensis Van Duzee, 1924b: 19.—Ont.; Mich.
caudatus Van Duzee, 1924b: 29.—N.Y.
chlanoflavus Harmston and Knowlton, 1940d: 59.—N. Dak.
choricus Wheeler, 1890: 357.—Wis.; Ont., Iowa, Ill., ?Mexico, ?West Indies.
    chloricus, error.
    ciliatus Malloch, 1914g: 236.—Ill.
coloradensis Van Duzee, 1924b: 14.—B.C.

convergens Van Duzee, 1924b: 17.—N.Y.; Mich., Que.; N.C.
cornutus Loew, 1862b: 214 (Cent. 2, no. 65) (also 1864e: 174).—Ill.;
 Iowa, N.Y.
costalis Loew, 1861a: 64.—Md., Fla.; Mich., N.Y., Ga.
  costatus, error.
cressoni Van Duzee, 1924b: 17.—N.Y.; N.C.
crosbyi Van Duzee, 1924b: 43.—Mo.; N.C.
currani Van Duzee, 1924b: 15.—Que.; Ont., N.Y., Del., Tenn., N.C.
dakotensis Harmston, 1952: 290.—S. Dak.
discolor Loew, 1861a: 65.—Middle States; Oreg. to Man. and Que.,
 s. to Iowa and Fla.
  cobaltinus Van Duzee, 1924b: 38.—Fla.
disjunctus Van Duzee, 1924b: 33.—Va.; Mass.
distinctus Van Duzee, 1924b: 23.—Ga.; Mich., Tenn.
dividuus Van Duzee, 1924b: 12.—N.Y.; Mich., Iowa, Ga.
dorsalis Van Duzee, 1924b: 44.—Ga.; La. to S. C.
emarginatus Van Duzee, 1928d: 87.—Iowa.
excisus Aldrich, 1896b: 325.—St. Vincent I.; ?N.Y.
exilis Van Duzee, 1924b: 32.—Ga.; N.C.
flavicauda Van Duzee, 1928d: 87.—Iowa.
flavisetus Malloch, 1914g: 239.—Ill.; Iowa, Tenn.
frontalis Van Duzee, 1924b: 24.—Ga.
fulvohirtus Van Duzee, 1915d: 95.—Ga.; Mass.
gilvipes Van Duzee, 1924b: 25.—Ga.
halteralis Van Duzee, 1924b: 40.—N.Y.; Que., N.C.
hastatus Van Duzee, 1924b: 46.—N.Y.
hirtipes Van Duzee, 1924b: 22.—N.Y.; Alta., S. Dak., Iowa, Ont.,
 Que.
  var. **dubius** Van Duzee, 1924b: 22.—S. Dak.; Que., Ont., N.Y.
humilis Parent, 1928: 164.—Costa Rica; Tex.
idahoensis Van Duzee, 1924b: 18.—Wash.
incertus Walker, 1849: 651.—U.S.
intrudus Harmston, 1951b: 107.—Okla.; Kans.
johnsoni Van Duzee, 1924b: 21.—Maine; Mich., Que., N.C., S.C.
junctus Van Duzee, 1924b: 13.—Fla.; Ga.
kansensis Harmston, 1952: 289.—Kans.
longihirtus Van Duzee, 1932b: 13.—Que.
longimanus Loew, 1861a: 62.—Middle States; Oreg., Calif., Colo.,
 Kans.
magnicornis Van Duzee, 1924b: 13.—N.Y.; Iowa.
major Van Duzee, 1924b: 34.—N.Y.
nigriciliatus Van Duzee, 1933b: 2.—N.H.
nigrifrons Parent, 1929a: 187.—Tex.
nigripalpis Van Duzee, 1924b: 32.—Ga.

nudipes Van Duzee, 1924b: 33.—Calif.
obliquus Loew, 1861a: 63.—N.Y.; Iowa to Ont., Que., and N.S., s. to Kans. and Fla., ?S. Amer.
pallipes Loew, 1861a: 66.—Middle States; Iowa to Conn., s. to Mo. and Ga., West Indies, S. Amer.
palpiger (Wheeler), 1890: 360 (*Diaphorus*).—Wis.; Alaska, Wash. to Ont. and Mass., s. to Calif, Ill., and N.Y.
    spinifer Malloch, 1914g: 238.—Ill.
parvicornis Van Duzee, 1924b: 12.—N.Y., Va.; Iowa, Ohio, Del., Tenn., N.C., S.C.
    exiguus Van Duzee, 1924b: 23.—N.Y.
parvulus Van Duzee, 1924b: 25.—N. Mex. (as N.H., error).
pectoralis Van Duzee, 1924b: 24.—Ga.; Fla.
philtrum Melander, 1903a: 72.—Pa., Tex., La.; Iowa, Mo., Mich., Ind.
picticornis Loew, 1862b: 214 (Cent. 2, no. 66) (also 1864e: 184).—Ill.; Calif. to Mass., s. to Fla., Mexico, S. Amer.
pomeroyi Parent, 1934b: 262.—Va.; Conn., N.C.
rauterbergi (Wheeler), 1890: 360 (*Diaphorus*).—Nebr.
sagittarius Van Duzee, 1924b: 42.—N.Y.
silvicola Harmston. 1951a: 11.—Calif.
simulans Van Duzee, 1924b: 31.—Va.; Fla.
subcostatus Loew, 1864e: 181.—Ill.; Nebr. and Mich. to Mass., s. to S.C., ?Calif.
    pratincola Wheeler, 1890: 357.—Nebr.
subjectus Van Duzee, 1924b: 40.—Va.
tarsalis Van Duzee, 1924b: 30.—Ga.; Tenn. and N.C., s. to La. and Fla.
teapanus Aldrich, 1901: 347.—Mexico; Ga., Fla.
terminalis Van Duzee, 1924b: 28.—Md.
tibialis Van Duzee, 1924b: 16.—Idaho; Man., Mich., N.Y.
triangularis Van Duzee, 1924b: 46.—Md.
varipes Van Duzee, 1924b: 20.—N.Y.
vividus Loew, 1864e: 178.—Ill.; Kans., N.Y., D.C., Fla., Mexico.
vulgaris Van Duzee, 1924b: 15.—N.Y.; Mich., Conn., N.C.
wisconsinensis Wheeler, 1890: 356.—Wis.; Que., Vt., N.Y., N.C.,
    wisconensis, error.
xanthocal Harmston and Knowlton, 1940e: 110.—Calif.

Genus ASYNDETUS Loew

Asyndetus Loew, 1869b: 34 (Cent. 8, no. 58). Type-species, *ammophilus* Loew (Coquillett, 1910b: 511).

REFERENCE: Van Duzee, 1919b (key, males).

ammophilus Loew, 1869b: 34 (Cent. 8, no. 58).—R.I.; Iowa, Mass. to S.C.
appendiculatus Loew, 1869b: 36 (Cent. 8, no. 59).—R.I.; Iowa, Mich., Ind., Fla.
caudatus Van Duzee, 1916b: 92.—Ariz.
cornutus Van Duzee, 1916b: 89.—Calif., Utah.
flavipalpus Van Duzee, 1932b: 10.—Nev.
harbeckii Van Duzee, 1914c: 442.—N.J.
interruptus (Loew), 1861c: 37 (*Diaphorus*).—Cuba; Fla.
johnsoni Van Duzee, 1916b: 93.—Conn.; N.H., N.C.
latus Van Duzee, 1916b: 91.—Ariz.; Calif., ?Mich.
nigripes Van Duzee, 1916b: 91.—Calif.
occidentalis Van Duzee, 1919b: 249.—Calif.
parvicornis Van Duzee, 1932b: 11.—Wyo.
scopifer Harmston, 1952: 288.—Oreg.
severini Harmston and Knowlton, 1939b: 351.—S. Dak.
spinitarsis Harmston, 1951b: 108.—Calif.
syntormoides Wheeler, 1899: 32.—Mass., N.J.; Kans., Mass. to Fla.
texanus Van Duzee, 1916b: 90.—Ariz., Tex.
utahensis Harmston and Knowlton, 1942b: 85.—Utah.

Genus ARGYRA Macquart

Porphyrops Meigen, 1824: 45. Type-species, *Musca diaphana* Fabricius (Curtis, 1835a: pl. 541). See discussion below.
Argyra Macquart, 1834a: 456. Type-species, *Musca diaphana* Fabricius (Rondani, 1856: 141). See discussion below.
Leucostola Loew, 1857a: 39. Type-species, *Argyra vestita* Wiedemann (mon.).

The name *Porphyrops* has priority over *Argyra*. However, the presently established use of *Argyra* and the former usage of *Porphyrops* by many authors in Rhaphiinae, make a suspension of the latter name in favor of *Argyra* advisable to avoid confusion. Application has been made to the I.C.Z.N.

Adults of *Argyra* are frequently encountered on moist soil and herbaceous foliage near water.

REFERENCE: Van Duzee, 1925d (rev.).

abdominalis (Say), 1829: 170 (1859b: 362) (*Dolichopus*).—Ind.
albicans Loew, 1861a: 45.—D.C.; Minn. to Que., s. to Kans. and S.C.
  quadriplagiatus Harris, 1835: 597 (*Porphyrops*). Nomen nudum.
albicoxa Van Duzee, 1925d: 30.—Que.; N.Y.
albiventris Loew, 1864e: 128.—Alaska; Alaska, s. to Oreg. and Utah.

aldrichi Johnson, 1904: 18.—N.J.; Mass. to S.C.
angustata Van Duzee, 1925d: 10.—Maine; Wis. to Maine and Mass.
argentiventris Van Duzee, 1925d: 15.—Calif.
barbipes Van Duzee, 1925d: 11.—Calif.
basalis Van Duzee, 1932c: 186.—N.Y.; S.C.
bimaculata Van Duzee, 1925d: 19.—Que.; Oreg. to Maine, s. to Ill. and Mass.
brevipes Van Duzee, 1925d: 11.—La.
calceata Loew, 1861a: 47.—Middle States; Ont. to Que., s. to Ind. and Pa.
calcitrans Loew, 1861a: 46.—N.Y.; Ont. to Maine, s. to Iowa and S.C.
californica Van Duzee, 1925d: 26.—Calif.
ciliata Van Duzee, 1924f: 5.—Alaska.
cingulata (Loew), 1861: 53 (*Leucostola*).—D.C.; Kans. to Maine, s. to La. and Fla.
condomina Harmston and Knowlton, 1946c: 137.—Utah; Colo., S. Dak.
currani Van Duzee, 1925d: 24.—Ont.; Mich., Que.
cylindrica Loew, 1864e: 132.—Alaska; Wash., Calif.
dakotensis Harmston and Knowlton, 1939c: 85.—S. Dak.; Utah.
fasciventris Van Duzee, 1930a: 2.—Conn.
femoralis Van Duzee, 1925d: 16.—Calif.
flavicornis Van Duzee, 1925d: 35.—Kans.; Mo., La.
flavicoxa Van Duzee, 1925d: 39.—Fla.
flavipes Van Duzee, 1925d: 34.—Md.; Ind., N.Y., Tenn.
hirta Van Duzee, 1925d: 7.—Colo.
idahona Harmston and Knowlton, 1946c: 138.—Idaho.
inaequalis Van Duzee, 1925d: 40.—Ind.; Tenn.
involuta Van Duzee, 1925d: 38.—Ind.; S. Dak.
johnsoni Van Duzee, 1925d: 37.—N.J.; Iowa, Ill., Pa.
minuta Loew, 1861a: 46.—D.C.; Iowa, Ind. to Ont. and Vt., s. to D.C.
nigricoxa Van Duzee, 1925d: 26.—Ohio; Ont., Va., S.C.
nigripes Loew, 1864e: 127.—Alaska; Wash. and Idaho, s. to Calif.
nigriventris Van Duzee, 1925d: 14.—Wash.
obscura Van Duzee, 1925d: 36.—N.H.
robusta Johnson, 1906: 59.—Que.; Idaho and Utah to Maine.
scutellaris Van Duzee, 1925d: 12.—Oreg.; Idaho.
sericata Van Duzee, 1925d: 28.—Maine; Iowa, Ind., Que., Mass.
setipes Van Duzee, 1925d: 31.—N.Y.; Ont. to Mass., s. to N.C.
similis Harmston and Knowlton, 1940d: 58.—Ill.; S.C.
spina Van Duzee, 1925d: 41.—Ind.

splendens Van Duzee, 1933c: 151 (n. name for *splendida* Van Duzee).—
Calif.
splendida Van Duzee, 1925d: 21 (preocc. Meijere, 1916).—
Calif.
thoracica Van Duzee, 1925d: 23.—Ont.; Ind., Tenn., N.Y.
utahna Harmston, 1951a: 16.—Utah.
velutina Van Duzee, 1925d: 20.—Que.; Mich.

Subfamily SYMPYCNINAE

(Xanthochlorinae, Campsicneminae)

Genus CAMPSICNEMUS Haliday

Medeterus, subg. *Camptosceles* Haliday, 1832: 357. Type-species, *Dolichopus scambus* Fallén (Coquillett, 1910b: 518). Suppressed by I.C.Z.N., 1958b: 349.
Campsicnemus Haliday 1851: 187. Type-species, *Dolichopus curvipes* Fallén (Coquillett, 1910b: 518).

The species of *Campsicnemus* are small and are found on moist sand or soil and on low herbaceous foliage near the margins of streams and lakes. Adults sometimes run about on the surface of still water.

REFERENCES: Curran, 1933d (key); Harmston and Knowlton, 1942a (rev.).

americanus Van Duzee, 1924f: 3.—Alaska; Colo., Ont., Mich., Maine, N.Y., Mass.
arcuatus Van Duzee, 1917a: 125.—Colo.
bryanti Malloch, 1932a: 121.—Man.
claudicans Loew, 1864e: 194.—Alaska; Alaska to Calif. and Idaho, also Colo.
clandicans, error.
curvispina Van Duzee, 1930a: 2 (n. name for *calcaratus* Van Duzee).—
Alaska.
calcaratus Van Duzee, 1924f: 3 (preocc. Grimshaw, 1901).—
Alaska.
degener Wheeler, 1899: 58.—Calif.; Alaska, Wash. to Calif. and Colo., also Ont., Que., ne. U.S.
hirtipes Loew, 1861a: 68.—Pa.; Kans to Mich. and Que., s. to Fla.
melanus Harmston and Knowlton, 1942a: 11.—Utah.
montanus Harmston and Knowlton, 1942a: 14.—Mont.; Wyo.
nigripes Van Duzee, 1917a: 126.—Calif.; Mont. to Calif. and Colo.
oedipus Wheeler, 1899: 60.—Wyo.; Oreg., Utah, Colo.
philoctetes Wheeler, 1899: 59.—Wyo., S. Dak.; Wash., Oreg., Idaho, Utah.

thersites Wheeler, 1899: 61.—Wyo.; Mont., Idaho, Utah.
utahensis Harmston and Knowlton, 1942a: 13.—Utah.
vanduzeei Curran, 1933d: 6.—Man.
wheeleri Van Duzee, 1923d: 64.—Maine; Que.
variabilis Van Duzee, 1932a: 9 (*Thinophilus*).—Que.

Genus SYMPYCNUS Loew

Sympycnus Loew, 1857a: 42. Type-species, *Porphyrops annulipes* Meigen (Coquillett, 1910b: 610).

Most species of *Sympycnus* are found in deep shade and often hover over moist rocks and falling water.

REFERENCE :Van Duzee, 1930e (rev.).

aldrichi Van Duzee, 1930e: 39.—Ind.
aurifacies Van Duzee, 1923c: 248.—Alaska.
basistylatus Parent, 1929a: 196.—B.C.
binodatus Harmston and Knowlton, 1940c: 397.—Utah.
brevicauda Van Duzee, 1932b: 19.—Que.
brevipes Van Duzee, 1933b: 6.—N.Y.
breviventris Van Duzee, 1930e: 40.—Calif.; Utah.
calcaratus Van Duzee, 1930e: 41.—Colo.; Utah, N. Mex.
caudatus Van Duzee, 1917b: 338.—Calif.
clavatus Van Duzee, 1913a: 271.—N. Mex.; Utah, Colo., Ariz., Mexico.
cuprinus Wheeler, 1899: 50.—Calif.; Alaska, Alta., Utah.
fasciventris Van Duzee, 1917b: 337.—Calif.; Oreg., Colo.
globulicauda Van Duzee, 1930e: 43.—Colo.
hardyi Harmston and Knowlton, 1940c: 397.—Utah.
inaequalis Van Duzee, 1930e: 44.—Calif.
isoaristus Harmston and Knowlton, 1940c: 402.—Utah.
laevigatus Van Duzee, 1930e: 45.—Calif.
latitarsis Van Duzee, 1930e: 46.—Utah.
jamesi Harmston and Knowlton, 1939a: 257 (*Peloropeodes*).— Utah. **N. syn.**, F. C. Harmston *in litt.*
lineatus Loew, 1861a: 67.—N.Y., Va.; Minn. to Que., s. to Kans. and Va.
longinervis Van Duzee, 1932b: 18.—N.Y.
marcidus Wheeler, 1899: 48.—Wyo.; Alta., Calif., Utah.
minuticornis Van Duzee, 1932b: 17.—Colo.
montanus Van Duzee, 1930e: 47.—Idaho.
pectoralis Van Duzee, 1933b: 7.—N.Y.
pennarista Harmston and Knowlton, 1940c: 399.—Utah.
pictipes Harmston and Knowlton, 1940c: 401.—Utah.

pugil Wheeler, 1899: 51.—Wash.; Oreg., Calif., Utah.
pulvillus Van Duzee, 1930e: 49.—Colo.; Calif.
setosus Van Duzee, 1930e: 50.—Colo.
tertianus Loew, 1864e: 187.—Alaska.
tripilus Van Duzee, 1930e: 52.—Calif.; Utah.
utahensis Harmston and Knowlton, 1939a: 258.—Utah.

Genus CALYXOCHAETUS Bigot

Calyxochaetus Bigot, 1888b: xxiv. Type-species, *Sympycnus nodatus* Loew (as *notatus*; orig. des.).
Nothosympycnus Wheeler, 1899: 51. Type-species, *vegetus* Wheeler (Coquillett, 1910b: 576).
Notosympycnus, error or emend.

REFERENCE: Van Duzee, 1930e (rev.).

abbreviatus (Van Duzee), 1917b: 341 (*Nothosympycnus*).—Kans.
cilifemoratus (Van Duzee), 1924f: 12 (*Nothosympycnus*).—Alaska; Utah, Colo.
    fuscitibialis Harmston and Knowlton, 1940c: 400 (*Sympycnus*).—Utah. **N. syn.**, F. C. Harmston *in litt.*
clavicornis (Van Duzee), 1930e: 55 (*Sympycnus*).—Calif.
distortus (Van Duzee), 1930e: 55 (*Sympycnus*).—Utah.
fortunatus (Wheeler), 1899: 52 (*Nothosympycnus*).—Pa.; Ont. to Maine, s. to N.C.
frontalis (Loew), 1861a: 67 (*Sympycnus*).—Pa.; Iowa to Maine, s. to Tenn. and N.C.
hastatus (Van Duzee), 1930e: 57 (*Sympycnus*).—Calif.; Mexico.
inornatus (Van Duzee), 1917b: 340 (*Nothosympycnus*).—Calif.
insolitus (Van Duzee), 1932b: 20 (*Sympycnus*).—Colo.; Utah.
luteipes (Van Duzee), 1923d: 63 (*Nothosympycnus*).—Maine; Ont., N.Y.
    tarsalis Curran, 1924e: 108 (*Nothosympycnus*).—Ont.
monticola (Van Duzee), 1932b: 20 (*Sympycnus*).—Colo.
nodatus (Loew), 1862b: 215 (Cent. 2, no. 68) (*Sympycnus*).—Ill.; Man. to N.Y., s. to Tex.
    notatus, error.
oreas (Wheeler), 1899: 55 (*Nothosympycnus*).—Wyo.; Calif., Utah.
    spatulatus Harmston and Knowlton, 1940c: 398 (*Sympycnus*).—Utah.
sobrinus (Wheeler), 1899: 54 (*Nothosympycnus*).—Idaho; Calif.
vegetus (Wheeler), 1899: 53 (*Nothosympycnus*).—Calif.; Oreg., Idaho, Utah, ?N.S.

Genus CHRYSOTIMUS Loew

Chrysotimus Loew, 1857a: 48. Type-species, *pusio* Loew (Coquillett, 1910b: 524).

REFERENCE: Curran, 1923h (key).

delicatus Loew, 1861a: 74.—N.Y.; Ont. to Maine, s. to Ga.
flavicornis Van Duzee, 1916a: 24.—N.Y.; Utah, Wis., S.C.
luteopalpus Curran, 1923h: 190.—Man.
luteus Curran, 1930j: 52.—N.Y.; Md., S.C., Ga.
occidentalis Harmston, 1951b: 109.—Ariz.
pusio Loew, 1861a: 74.—N.Y.; N.C.

Genus XANTHOCHLORUS Loew

Xanthochlorus Loew, 1857a: 42.—Type-species, *Medeterus ornatus* Haliday (Coquillett, 1910b: 620).

helvinus Loew, 1861a: 75.—Ill.; Wis. to Vt., s. to Iowa and Ga.

Genus TEUCHOPHORUS Loew

Teuchophorus Loew, 1857a: 44. Type-species, *Dolichopus spinigerellus* Zetterstedt (Coquillett, 1910b: 613).

REFERENCE: Harmston and Knowlton, 1946a (key).

clavigerellus Wheeler, 1899: 57.—S. Dak.; Mich.
condylus Harmston and Knowlton, 1946a: 671.—Ind.; Conn.
diminucosta Harmston and Knowlton, 1942c: 21.—Utah.
utahensis Harmston and Knowlton, 1942c: 20.—Utah.

Genus TELMATURGUS Mik

Telmaturgus Mik, 1874: 349. Type-species, *Sympycnus tumidulus* Raddatz (orig. des.).

parvus (Van Duzee), 1924b: 45 (*Chrysotus*).—N.Y.; Mich., Pa., Tenn.
 perparvus (?error) Johannsen, 1928b: 773.

Genus ENLINIA Aldrich

Collinellula Aldrich, 1932b: 4 (preocc. Strand, 1928). Type-species, *magistri* Aldrich (orig. des.).
Enlinia Aldrich, 1933b: 168 (n. name for *Collinellula* Aldrich). Type-species, *Collinellula magistri* Aldrich (aut.).

 E. magistri is one of the smallest species of Dolichopodidae. The adults fly very rapidly close to the ground.

magistri (Aldrich), 1932b: 5 (*Collinellula*).—N.Y.; D.C.

Genus MICROMORPHUS Mik

Micromorphus Mik, 1878b: ?4 (1878: 6). Type-species, *Hydrophorus albipes* Zetterstedt (orig. des.).

albipes (Zetterstedt), 1843: 454 (*Hyrdophorus*).—Sweden; Md., S.C., Fla., Mexico, Costa Rica, Europe, ?New Zealand.
fulvosetosus Parent, 1929a: 193.—Tex.; Md., S.C., Fla.
minimus (Van Duzee), 1925e: 179 (*Neurigona*).—N.Y.; Que.

Genus LAMPROCHROMUS Mik

Lamprochromus Mik, 1878b: ?5 (1878: 7). Type-species, *Chrysotus elegans* Meigen (orig. des.).

canadensis (Van Duzee), 1917b: 339 (*Sympycnus*).—Ont.; N.Y., N.C., ?Utah.
    brevicornis Robinson, 1960: 271 (*Telmaturgus*).—N.C.
satrapa (Wheeler), 1890: 359 (*Diaphorus*).—Nebr. N. comb.

Unplaced Species of Dolichopodidae

adjacens Walker, 1849: 661 (*Dolichopus*).—Ont.
deremptus Walker, 1849: 667 (*Orthochile*).—N. Amer.
lamellicornis Thomson, 1869: 511 (*Dolichopus*).—Calif.
nubilus Say, 1829: 168 (1859b: 361) (*Chrysotus*).—Ind.
pilosicornis Walker, 1849: 653 (*Porphyrops*).—Ont.
viridifemora Macquart, 1850: 428 (1850: 124) (*Chrysotus*).—N. Amer.

Suborder CYCLORRHAPHA

Division ASCHIZA

Superfamily LONCHOPTEROIDEA

Family LONCHOPTERIDAE

(Musidoridae)

By Willis W. Wirth

The "pointed-wing" flies are a small family of slender, yellowish to brownish flies which are found in moist, shady places. The larvae live under leaves and decaying vegetation. *Lonchoptera furcata* (Fallén) is a worldwide, parthenogenetic species, which has been studied genetically by Stalker (1956).

REFERENCES: Curran, 1934b (rev.); Czerny, 1934a (anat., biol., immature stages, rev. Palaearctic spp.); Hardy, 1952 (syn.).

Genus LONCHOPTERA Meigen

Musidora Meigen, 1800: 30. Type-species, *Lonchoptera lutea* Panzer (Coquillett, 1910d: 377). Suppressed by I.C.Z.N., 1963b: 339.
Lonchoptera Meigen, 1803: 272. Type-species, *lutea* Panzer (sub. mon., Panzer 1809c: 20).

borealis Curran, 1934b: 4.—Que.; Alaska, Colo., Ont., N.Y.
furcata (Fallén), 1823b: 1 (*Dipsa*).—Europe; N. Amer., worldwide.
  dubia Curran, 1934b: 5.—N.Y.
occidentalis Curran, 1934b: 4.—Calif.; Alaska, Idaho.
uniseta Curran, 1934b: 2.—Colo.; B.C., Que.

Superfamily PHOROIDEA

Family PHORIDAE

By H. Schmitz[1] and Erwin Beyer

Flies of the family Phoridae are black, brown, or yellow and are small, humpbacked, and usually inconspicuous. They are common around decaying vegetation or animal matter, and sometimes in and around nests of ants, termites, and bees. The adults move about with a characteristic quick, jerky movement. Larval habits are diverse (see under genera). Some species are parasitic, others are scavengers, and many have been reared from fleshy and woody fungi. Parasitic forms are often found in nests of ants, termites, bees, and wasps, and on beetles, caterpillars, land mollusks, and diplopods.

The extensive work on North American phorids early in the century by Aldrich, Brues, and Malloch is now badly out of date and needs to be revised in the light of modern revisionary studies by Schmitz and Borgmeier. While this catalog was in preparation, and after the death of Father Schmitz on September 1, 1960, Father T. Borgmeier made an extensive study of types and other material in the Brues Collection of the Museum of Comparative Zoology in Cambridge, Mass., and the Aldrich and Malloch collections at the U.S. National Museum in Washington, D.C. Father Borgmeier has gen-

---
[1] Father Schmitz died September 1, 1960.

erously allowed us to incorporate the resulting new synonymies, new generic assignments, and additional distributions in the present catalog.

REFERENCES: Brues, 1903 (monog.), 1906a (cat., keys to genera for world), 1915 (world cat.), 1950 (keys genera and spp. ne. U.S.); Malloch, 1912b (rev.); Lundbeck, 1922 (rev., Denmark); Schmitz, 1929b (world rev. genera).

Subfamily PHORINAE

Genus ANEVRINA Lioy

Anevrina Lioy, 1864b: 77. Type-species, *Phora urbana* Meigen (Coquillett, 1910b: 506, by inference as "the first species").
Aneurina, emend.
Chaetoneura Malloch, 1909: 26 (preocc. Felder, 1862). Type-species, *Trineura thoracica* Meigen (Malloch, 1910: 19).
Stenophora Malloch, 1909: 27 (preocc. Labbé, 1899). Type-species, *Trineura unispinosa* Zetterstedt (Malloch, 1910: 89).
Pseudostenophora Malloch, 1912b: 412 (n. name for *Stenophora* Malloch). Type-species, *Trineura unispinosa* Zetterstedt (aut.).
Chaetoneurophora Malloch, 1912b: 422 (n. name for *Chaetoneura* Malloch). Type-species, *Trineura thoracica* Meigen (aut.).

aureiventris (Brues), 1913: 90 (*Chaetoneurophora*).—R.I.; N.H., Mass.
californica (Van Duzee), 1933d: 63 (*Hypocera*).—Calif. **N. comb.**, T. Borgmeier *in litt.*
curvinervis (Becker), 1901: 33 (*Phora*).—Germany; Wash., N.H., Maine, Europe.
luggeri (Aldrich), 1892a: 145 (*Phora*).—Minn.; Minn. to N.H., s. to Kans. and Md.
macateei (Malloch), 1913b: 273 (*Chaetoneurophora*).—Md.; Mich.
olympiae (Aldrich) *in* Brues, 1903: 344 (*Phora*).—Wash.; N.Y.
spinipes (Coquillett), 1895b: 105 (*Phora*).—Conn.; Wash., Oreg., Idaho, Ohio.
sulcatifemur Borgmeier, 1962a: 65.—Calif.; Oreg.
thoracica (Meigen), 1804: 313 (*Trineura*).—Europe; Alaska, Mont., Mich., N.H.
urbana (Meigen), 1830: 215 (*Phora*).—Europe; Utah.
variabilis (Brues), 1908: 199 (*Phora*).—Wash.; Oreg., Colo.

Genus CHAETOPLEUROPHORA Schmitz

Chaetopleurophora Schmitz, 1922: 131. Type-species, *Phora erythronota* Strobl (Schmitz, 1927c: 31).

Larval stages have been found in dead snails.

REFERENCE: Borgmeier, 1961 (key).

atra Borgmeier, 1962a: 66.—Calif.
erythronota (Strobl), 1892: 195 (*Phora*).—Austria; se. Canada, Tenn., Md., Va.
multiseriata (Aldrich) *in* Brues, 1903: 345 (*Phora*).—Kans.; Nebr. to Conn., s. to Miss. and Va.
pennsylvanica (Malloch), 1914e: 175 (*Paraspiniphora*).—Pa.; Ohio, Conn., Md.
   spinosissima Strobl of Malloch, 1912b: 426.
rubricornis Borgmeier, 1962a: 67.—N.Y.
rufithorax Brues, 1943: 50.—N.Y.
spinosissima (Strobl).—Not Nearctic.

Genus SPINIPHORA Malloch

Spiniphora Malloch, 1909: 26. Type-species, *Phora maculata* Meigen (Malloch, 1910: 20).
Paraspiniphora Malloch, 1912b: 425 (unjustified n. name for *Spiniphora* Malloch). Type-species, *Phora maculata* Meigen (aut.).

Larvae are found in dead snails.

bergenstammii (Mik), 1864: 793 (*Phora*).—Austria; Calif., Europe, n. Africa, Uruguay.
excisa (Becker), 1901: 28 (*Phora*).—Europe; Tex., N.Y., Md.
   comstocki (Aldrich) *in* Brues, 1903: 346 (*Phora*).—N.Y.
slossonae (Malloch), 1912b: 428 (*Paraspiniphora*).—N.H.; N.Y.
spinulosa (Malloch), 1912b: 429 (*Paraspiniphora*).—N.Y. **N. comb.**
trispinosa (Malloch), 1912b: 427 (*Paraspiniphora*).—B.C.; Idaho.

Genus TRIPHLEBA Rondani

Triphleba Rondani, 1856: 136. Type-species, *hyemalis* Rondani (orig. des.) = *hyalinata* (Meigen).
Trupheoneura Malloch, 1909: 27. Type-species, *Phora perennis* Meigen (Malloch, 1910: 88).
Woodia Malloch, 1909: 28 (preocc. Deshayes, 1860). Type-species, *Phora gracilis* Wood (mon.).

Parastenophora Malloch, 1910: 17, 90 (n. name for *Woodia* Malloch). Type-species, *Phora gracilis* Wood (aut.).

The larvae of most species are necrophagous.

bispinosa (Malloch), 1914e: 173 (*Pseudostenophora*).—N.J.
carbonaria Borgmeier, 1962a: 70.—Alaska.
forcipata Borgmeier, 1962a: 69.—Wash.
labida Borgmeier, 1962a: 69.—Wash.; Alaska, Calif.
laticosta Borgmeier, 1962a: 68.—B.C.
lugubris (Meigen), 1830: 217 (*Phora*).—Europe; Wyo., N.H., Md., Tenn.
    fratercula Brues, 1903: 341 (*Phora*).—Wyo. **N. syn.**
microcephala (Loew), 1866b: 51 (Cent. 7, no. 96) (*Phora*).—D.C.; Mass.
occidentalis (Brues), 1908: 200 (*Phora*).—Idaho; Wash.
pachyneura (Loew), 1866b: 52 (Cent. 7, no. 97) (*Phora*).—Alaska; Alaska to Que. and Maine, s. to Idaho and Md.
subfusca (Malloch), 1912b: 422 (*Trupheoneura*).—Mass.; Mich., Mass. to Md.
suspecta (Malloch), 1912b: 420 (*Trupheoneura*).—N. Dak.
trinervis (Becker), 1901: 19 (*Phora*).—Germany; B.C., Europe.
    gilsoni Schmitz, 1915: 497 (*Trupheoneura*).—B.C. **N. syn.**
varipes (Malloch), 1912b: 419 (*Trupheoneura*).—Kans.; Ill., Md.
vitrinervis (Malloch), 1912b: 419 (*Trupheoneura*).—N.H.

Genus HYPOCERA Lioy

Hypocera Lioy, 1864b: 78. Type-species, *Trineura mordellaria* Fallén (Brues, 1906a: 6).

americana Borgmeier, 1962a: 71.—Md.; Mass.
ehrmanni Aldrich *in* Brues, 1903: 353.—Pa.; Ont. to N.C.
mordellaria (Fallén).—Probably not Nearctic.
    subsultans, authors, not Linnaeus.

Genus DIPLONEVRA Lioy

Diplonevra Lioy, 1864b: 77. Type-species, *Bibio florea* Fabricius (Enderlein, 1924: 272).
Diploneura, emend.

aberrans Borgmeier, 1962a: 73.—Iowa.
funebris (Meigen), 1830: 221 (*Phora*).—Europe; Idaho to Mass., s. to Calif. and Tex.
    cimbicis Aldrich, 1892a: 143 (*Phora*).—S. Dak.
gaudialis (Cockerell), 1915: 351 (*Dohrniphora*).—Calif.

hamata Borgmeier, 1962a: 72.—Va.; Md.
  alleni (Brues) of Brues, 1919b: 499, 501. Va. specimens only.
nitidula (Meigen), 1830: 221 (*Phora*).—Europe; Minn. to Maine, s. to Ind. and Va.
  nitidifrons Brues, 1903: 347 (*Phora*).—N.Y., Mass., Pa.
  concinna, authors, not Meigen.

Genus DOHRNIPHORA Dahl

Dohrniphora Dahl, 1898: 188. Type-species, *dohrni* Dahl (orig. des.).
castaneicoxa Borgmeier, 1960: 272.—Brazil; Okla.
cornuta (Bigot), 1857a: 348 (1857: 827) (*Phora*).—Cuba; entire N. Amer., cosmopolitan.
  venusta Coquillett, 1895b: 107 (*Phora*).—Mass.
divaricata (Aldrich), 1896b: 437 (*Phora*).—St. Vincent I.; Bermuda.
incisuralis (Loew), 1866b: 52 (Cent. 7, no. 98) (*Phora*).—D.C.; Mass., s. to La. and Fla., also Cent. Amer.
perplexa (Brues), 1903: 350 (*Phora;* as *divaricata* var.).—Fla., Ga.

Genus BOROPHAGA Enderlein

Borophaga Enderlein, 1924: 277. Type-species, *Phora flavimana* Meigen (orig. des.)=*femorata* (Meigen).

REFERENCE: Schmitz, 1952b (partial key).

clavata (Loew), 1866b: 51 (Cent. 7, no. 95) (*Phora*).—D.C.; Kans. to Maine and Ga.
femorata (Meigen).—Not Nearctic.
fuscipalpis Schmitz, 1952b: 104.—Va.; Mass., Md., D.C.
okellyi Schmitz, 1937: 91.—Ireland; Alaska to Ont. and Maine, s. to Va., Europe.
verticalis Borgmeier, 1962a: 74.—Mo.; Alta. to Que., s. to Ariz. and D.C.
  clavata, authors, not Loew.

Genus ABARISTOPHORA Schmitz

Abaristophora Schmitz, 1927b: 21. Type-species, *arctophila* Schmitz (Schmitz, 1927c: 64).
diversipennis Borgmeier, 1962a: 75.—Mont.

Genus STICHILLUS Enderlein

Stichillus Enderlein, 1924: 279. Type-species, *acutivertex* Enderlein (orig. des.)=*insperatus* (Brues).

Tressinus Schmitz, 1924b: 149. Type-species, *Hypocera johnsoni* Brues (orig. des.).

adamsi (Brues), 1924b: 156 (*Hypocera*).—Mo.; N.Y., N.J.
johnsoni (Brues), 1903: 352 (*Hypocera*).—N.J.; Va.
laeticornis Borgmeier, 1962a: 76.—Miss.
planipes Borgmeier, 1962a: 77.—N. Mex.
rectilineatus Schmitz, 1952b: 102.—Ga.

Genus CONICERA Meigen

Conicera Meigen, 1830: 226. Type-species, *atra* Meigen (mon.) = *dauci* (Meigen).

Subgenus CONICERA Meigen

dauci (Meigen), 1830: 223 (*Phora*).—Europe; Oreg. to Mass., s. to Calif. and Tenn., Japan.
 atra Meigen, 1830: 226.—Europe.

Subgenus HYPOCERINA Malloch

Hypocerina Malloch, 1913i: 24 (as genus). Type-species, *barberi* Malloch (orig. des.).

aldrichii Brues, 1903: 379.—Idaho; Oreg.
barberi (Malloch), 1913i: 25 (*Hypocerina*).—Va.

Genus CONICEROMYIA Borgmeier

Coniceromyia Borgmeier, 1923: 338. Type-species, *epicantha* Borgmeier (orig. des.).

arizonensis Borgmeier, 1962a: 78.—Ariz.
pilipleura Borgmeier, 1962a: 77.—Fla.; Tex.

Genus PHORA Latreille

Phora Latreille, 1796: 169. Type-species, *Musca aterrima* Fabricius (sub. mon., Latreille, 1802: 464).
Trineura Meigen, 1803: 276. Type-species, *atra* Meigen (Brues, 1906a: 8) = *aterrima* (Fabricius).

REFERENCE: Schmitz and Wirth, 1954 (rev.).

aerea Schmitz, 1930: 60.—N.S.; Kans. to N.S., s. to Ind., Tenn., and Md.
americana Schmitz and Wirth, 1954: 124.—Mexico; Alaska, Oreg., Calif., Neotropical.

aterrima (Fabricius), 1794: 334 (*Musca*).—Europe; N.Y.
carlina Schmitz, 1930: 59.—N.S.
coangustata Schmitz, 1927a: 154 (n. name for *frontalis* Schmitz).—
Calif.; Wash. to Mont., s. to Calif.
frontalis Schmitz, 1920b: 224 (preocc. Wood, 1909).—Calif.
cristipes Schmitz and Wirth, 1954: 120.—Calif.
holosericea Schmitz, 1920a: 121.—Europe; Canada.
occidentata Malloch, 1912b: 438.—Alaska, Wyo., N.H.; Calif., Colo., Maine.
stictica Meigen, 1830: 225.—Europe; Alaska to N.H., s. to N. Mex.
montana Brues, 1903: 378 (*Trineura*).—N. Mex.
tripliciseta Schmitz and Wirth, 1954: 116.—Alaska.
velutina Meigen.—Not Nearctic.
viridinota Brues, 1916a: 394.—Man.; Iowa, Mich.

Subfamily AENIGMATIINAE

Genus AENIGMATIAS Meinert

Platyphora Verrall, 1877: 259 (preocc. Gistel, 1857). Type-species, *lubbockii* Verrall (mon.).
Aenigmatias Meinert, 1890b: 213. Type-species, *blattoides* Meinert (orig. des., as gen. n., sp. n.) =*lubbockii* (Verrall).

The larvae are associated with ants and termites.

coloradensis (Brues), 1914: 79 (*Platyphora*).—Colo.; Mich.
curvinervis Borgmeier, 1962b: 458.—Calif.
eurynotus (Brues), 1914: 77 (*Platyphora*).—Mass.; Mich., Ont., Pa.
flavofemoratus (Malloch), 1915h: 353 (*Platyphora*).—Ill.; Calif.
schwarzii Coquillett, 1903b: 21.—Ariz.; Mont., Wyo.

Subfamily METOPININAE

Tribe BECKERININI

Genus BECKERINA Malloch

Beckerina Malloch, 1910: 17, 90. Type-species, *Phora umbrimargo* Becker (mon.).

REFERENCE: Malloch, 1923b (key).

aliena Malloch, 1924f: 356.—Md.; Mich.
luteola Malloch, 1919j: 256.—Ill.
orphnephiloides Malloch, 1912b: 441.—Md.
similata Malloch, 1923b: 32.—Md.

Tribe METOPININI

Genus WOODIPHORA Schmitz

Woodiphora Schmitz, 1925b: 73. Type-species, *Phora retroversa* Wood (orig. des.).

velutinipes (Brues), 1924b: 158 (*Aphiochaeta*).—Mo.

Genus MEGASELIA Rondani

Megaselia Rondani, 1856: 137. Type-species, *crassineura* Rondani (orig. des.)=*costalis* (Roser).
Megaselida, emend.

Subgenus MEGASELIA Rondani

agarici (Lintner), 1895: 399 (1896: 399) (*Phora*).—N.Y.; Mich. to Mass., s. to Kans. and S.C.
    setacea Aldrich of Malloch, 1912b: 495.
aletiae (Comstock), 1880a: 208 (*Phora*).—Ala.; Calif. to N.Y., s. to Calif. and Fla.
    fungorum Malloch, 1912b: 473 (*Aphiochaeta*).—Md.
aristalis (Malloch), 1914c: 57 (*Aphiochaeta*).—Ill.
aurea (Aldrich), 1896b: 437 (*Phora*).—St.Vincent I.; Fla., Neotropical.
barberi (Malloch), 1912b: 450 (*Aphiochaeta*).—N. Mex.
bisetulata (Malloch), 1915b: 65 (*Aphiochaeta*).—Ill.
borealis (Malloch), 1912b: 488 (*Aphiochaeta*).—B.C.; Oreg., N.Y.
brunnipes (Malloch), 1912b: 475 (*Aphiochaeta*).—Md.
cata (Melander and Brues), 1903: 16 (*Phora*).—Mass.
cavernicola (Brues), 1906b: 101 (*Aphiochaeta*).—Ind.; Mo. to Mass., s. to N. Mex. and Ala., common in caves.
cayuga (Malloch), 1912b: 474 (*Aphiochaeta*).—N.Y.; Mass., Md.
centralis (Brues), 1919a: 190 (*Aphiochaeta*).—Mich.
chaetoneura (Malloch), 1912b: 490 (*Aphiochaeta*).—Ga.; Man., Ohio, Tenn.
    catana Curran, 1929a: 8 (*Aphiochaeta*).—Man. **N. syn.**, T. Borgmeier *in litt.*
conspicualis (Malloch), 1912b: 487 (*Aphiochaeta*).—Calif.
erecta (Wood), 1910: 202 (*Phora*).—England; N.S., Europe.
evarthae (Malloch), 1912b: 472 (*Aphiochaeta*).—Unknown; Mich.
fisheri (Malloch), 1912b: 463 (*Aphiochaeta*).—Md.
flava (Fallén), 1823b: 7 (*Trineura*).—Sweden; B.C., Mich., Europe.
flavinervis (Malloch), 1912b: 493 (*Aphiochaeta*).—Md.
fungicola (Coquillett), 1895b: 106 (*Phora*).—N. Mex.; N.Y.

fuscopedunculata (Malloch), 1912b: 498 (*Aphiochaeta*).—B.C.
giraudii (Egger), 1862: 1233 (*Phora*).—Austria; B.C., N.S., R.I., Europe.
    dyari Malloch, 1912b: 493 (*Aphiochaeta*).—B.C.
incrassata (Schmitz), 1920a: 145 (*Aphiochaeta*).—Europe; Mich.
infumata (Malloch), 1912b: 490 (*Aphiochaeta*).—Calif.
inornata (Malloch), 1912b: 488 (*Aphiochaeta*).—N.Y.
iroquoiana (Malloch), 1912b: 476 (*Aphiochaeta*).—N.Y., Mich.; Mass., Md.
johannseni (Malloch), 1912b: 474 (*Aphiochaeta*).—N.Y.; Maine.
longipennis (Malloch), 1912b: 473 (*Aphiochaeta*).—La.; Ill.
lutea (Meigen), 1830: 220 (*Phora*).—Europe; Wis., Mich., N.S., La.
minuta (Aldrich), 1892a: 146 (*Phora*).—S. Dak.; Idaho to Maine, s. to Calif. and Fla.
    minor, authors, not Zetterstedt.
modesta (Brues), 1919a: 193 (*Aphiochaeta*).—Wash.
nigra (Meigen), 1830: 218 (*Phora*).—Europe; B.C., Ohio, N.Y., N.J., Md., D.C.
    albidihalteris Felt, 1898b: 228 (1898: 228) (*Phora*).—N.Y.
    smithii Brues, 1909: 106 (*Aphiochaeta*).—N.J.
nubilipennis Schmitz, 1952a: 186.—Ind.
peregrina (Malloch), 1912b: 492 (*Aphiochaeta*).—D.C.; Md.
perplexa (Malloch), 1912b: 489 (*Aphiochaeta*).—B.C.
picta (Lehmann), 1822: 42 (also 1824: 244) (*Phora*).—Germany; Utah to N.S., s. to Va., Europe.
    atlantica Brues, 1903: 362 (*Aphiochaeta*).—N.S., Pa., Mass.
platychira (Malloch), 1919f: 52 (*Aphiochaeta*).—Alaska.
pulicaria (Fallén), 1823b: 7 (*Trineura*).—Europe; Kans. to Maine, s. to Md.
ruficornis (Meigen), 1830: 218 (*Phora*).—Europe; N.H. to Md.
rufipes (Meigen), 1804: 313 (*Trineura*).—Europe; entire N. Amer., cosmopolitan.
rusticata (Malloch), 1912b: 489 (*Aphiochaeta*).—Oreg.; N.Y.
scalaris (Loew), 1866b: 53 (Cent. 7, no. 100) (*Phora*).—Cuba; Ind. to Mass., s. to Tex. and Fla., subtropic and tropic regions of world.
schwarzi (Malloch), 1912b: 517 (*Aphiochaeta*).—Md.
setacea (Aldrich), 1892a: 144 (*Phora*).—S. Dak.
straminea (Malloch), 1912b: 472 (*Aphiochaeta*).—N.Y.
straminipes (Malloch), 1912b: 474 (*Aphiochaeta*).—N.Y.
subpicta (Malloch), 1912b: 452 (*Aphiochaeta*).—Fla.; Brazil.

tertia (Brues), 1915: 134 (*Aphiochaeta;* n. name for *inaequalis* Malloch).—N.Y.; N.H., Mass.
  inaequalis Malloch, 1912b: 464 (*Aphiochaeta;* preocc. Brunetti, 1912).—N.Y.
ursina (Malloch), 1912b: 476 (*Aphiochaeta*).—B.C.; Oreg.

Subgenus APHIOCHAETA Brues

Aphiochaeta Brues, 1903: 337 (as genus). Type-species, *Phora nigriceps* Loew (Brues, 1906a: 9).

aculeata (Schmitz), 1919a: 142 (*Aphiochaeta*).—Netherlands; Mich., Europe.
aequalis (Wood), 1909: 26 (*Phora*).—England; Ont., D.C., Va., Europe.
  subciliata Malloch, 1913i: 26 (*Aphiochaeta*).—D.C., Va.
alaskensis (Malloch), 1919f: 52 (*Aphiochaeta*).—Alaska.
annulipes (Schmitz), 1921: 325 (*Aphiochaeta*).—Romania; Mass., Europe.
anomala (Malloch), 1912b: 484 (*Aphiochaeta*).—N.H.
arizonensis (Malloch), 1912b: 478 (*Aphiochaeta*).—Ariz.
atomella (Malloch), 1912b: 481 (*Aphiochaeta*).—Alaska, Sask.
beckeri (Wood), 1909: 115 (*Aphiochaeta*).—England; Idaho, Europe.
californiensis (Malloch), 1912b: 447 (*Aphiochaeta*).—Calif.
carlynensis (Malloch), 1912b: 468 (*Aphiochaeta*).—Va.; Mass., Md., Brazil.
cirriventris Schmitz, 1929a: 86.—Greenland; high mts. of cent. and n. Europe.
clara (Schmitz), 1921: 320 (*Aphiochaeta*).—Sweden; Greenland, n. Europe.
coaequalis (Schmitz), 1919b: 153 (*Aphiochaeta*).—Netherlands; Mich., Europe.
  ciliata, authors, not Zetterstedt.
conglomerata (Malloch), 1912b: 445 (*Aphiochaeta*).—B.C.; Idaho.
crassipes (Wood), 1909: 59 (*Phora*).—England; N.S., Europe.
difficilis (Malloch), 1912b: 484 (*Aphiochaeta*).—Mass.; N.Y., N.C.
dilatata (Brues), 1919a: 183 (*Aphiochaeta*).—Idaho.
divergens (Malloch), 1912b: 480 (*Aphiochaeta*).—Md., D.C.; Ill.
fenestrata (Malloch), 1912b: 517 (*Aphiochaeta*).—D.C.
franconiensis (Malloch), 1912b: 479 (*Aphiochaeta*).—N.H.
groenlandica (Lundbeck), 1901: 307 (*Phora*).—Greenland; Alaska, B.C., Europe.
  dubitata Malloch, 1912b: 480 (*Aphiochaeta*).—B.C.
johnsoni (Brues), 1916b: 175 (*Aphiochaeta*).—Maine.

marginalis (Malloch), 1912b: 457 (*Aphiochaeta*).—Mo.; Va.
  capillaris Brues, 1919a: 189 (*Aphiochaeta*).—Va.
monticola (Malloch), 1912b: 479 (*Aphiochaeta*).—B.C.
nasoni (Malloch), 1914c: 58 (*Aphiochaeta*).—Ill.; Wash., Mass.
nigriceps (Loew), 1866b: 53 (Cent. 7, no. 99) (*Phora*).—D.C.; entire U.S., N.S., Europe, Japan.
  projecta Becker, 1901: 56 (*Phora*).—Romania.
  flavipalpis Malloch, 1913i: 25 (*Aphiochaeta*).—D.C.
  submanicata Malloch, 1914e: 175 (*Aphiochaeta*).—Pa.
pallidiventris (Malloch), 1919d: 47 (*Aphiochaeta*).—Ill.
palpata (Brues), 1919a: 186 (*Aphiochaeta*).—N.J.; Ill., Va.
perdita (Malloch), 1912b: 459 (*Aphiochaeta*).—Miss.; Md., N.Y.
plebeia (Malloch), 1914c: 58 (*Aphiochaeta*).—Ill.
pleuralis (Wood), 1909: 117 (*Phora*).—England; Alaska to N.S., s. to Wash., Ill., and Mass., Palaearctic.
  approximata Malloch, 1912b: 483 (*Aphiochaeta;* preocc. Brunetti, 1912).—Ill.
  vulgata Malloch, 1912b: 483 (*Aphiochaeta*).—Mo., Maine.
  secunda Brues, 1915: 132 (*Aphiochaeta;* n. name for *approximata* Malloch).—Ill.
pulla (Brues), 1919a: 187 (*Aphiochaeta*).—Idaho.
pusilla (Meigen), 1830: 218 (*Phora*).—Europe; N. Amer.
quadripunctata (Malloch), 1918l: 147 (*Aphiochaeta*).—Ill.
retardata (Malloch), 1912b: 482 (*Aphiochaeta*).—N. Mex.
scopalis (Brues), 1919a: 184 (*Aphiochaeta*).—Wash.
subatomella (Malloch), 1912b: 481 (*Aphiochaeta*).—N.Y.; ?B.C.
sublutea (Malloch), 1912b: 468 (*Aphiochaeta*).—N.H.; N.J.
submarginalis (Malloch), 1912b: 458 (*Aphiochaeta*).—Md.
subobscurata (Malloch), 1912b: 485 (*Aphiochaeta*).—N.H.
subpleuralis (Wood), 1909: 118 (*Phora*).—England; Kans., Idaho, Europe.
ventralis Borgmeier, 1963: 133.—Ariz.

Genus PHYSOPTERA Borgmeier

Physoptera Borgmeier, 1958: 356. Type-species, *Aphiochaeta vesiculata* Borgmeier (orig. des.).

apicinebula (Malloch), 1924f: 355 (*Aphiochaeta*).—Md.; Va.

Genus GYMNOPHORA Macquart

Gymnophora Macquart, 1835: 631. Type-species, *Phora arcuata* Meigen (mon.).

fastigiorum Schmitz, 1952a: 180.—Colo.

luteiventris Schmitz, 1952a: 178.—Tex.; Wis. to N.H., s. to Tex. and Ga.
  quartomollis, authors, not Schmitz.
subarcuata Schmitz, 1952a: 182.—Tenn.; Alaska, Minn. to Maine, s. to Tenn.
  arcuata, authors, not Meigen.

Genus PLASTOPHORA Brues

Plastophora Brues, 1905: 551. Type-species, *beirne* Brues (orig. des.).
  REFERENCE: Colyer, 1957 (habits; key spp. of world).

Some species are parasitic on diplopods.

arcuata (Malloch), 1912b: 460 (*Aphiochaeta*).—Md.; Mich.
juli Brues, 1908: 201.—Wis.; Mich., Pa., Md., Va., Mexico, Brazil.
  xantippe Banks, 1911b: 212 (*Aphiochaeta*).—Va.
winnemana (Malloch), 1912b: 461 (*Aphiochaeta*).—Md.; ?Mass.

Genus PHALACROTOPHORA Enderlein

Phalacrotophora Enderlein, 1912c: 21. Type-species, *bruesiana* Enderlein (orig. des.).

The larvae are parasitic on beetles, bees, *Tremex*, and spiders.

Subgenus PHALACROTOPHORA Enderlein

epeirae (Brues), 1902b: 351 (*Phora*).—Tex.; Mich. to Que., s. to Mexico and Fla.
  fasciata, authors, not Fallén.
halictorum (Melander and Brues), 1903: 14 (*Phora*).—Mass.; Wash., Ind.

Subgenus OMAPANTA Schmitz

Phalacrotophora, subg. **Omapanta** Schmitz, 1932: 117. Type-species, *appendicigera* Borgmeier (orig. des.).

longifrons (Brues), 1906b: 100 (*Aphiochaeta*).—Wis.; Kans., N.Y., Conn., Pa., Md.

Genus SYNEURA Brues

Syneura Brues, 1903: 383. Type-species, *Phora cocciphila* Coquillett (mon.).

The larvae are parasitic in scales of the genus *Icerya* in tropical and subtropical America.

cocciphila (Coquillett), 1895b: 106 (*Phora*).—Mexico; Ariz., Tex.

Genus RHYNCOPHOROMYIA Malloch

Rhyncophoromyia Malloch, 1923c: 143. Type-species, *trivittata* Malloch (orig. des.).

conica (Malloch), 1912b: 462 (*Aphiochaeta*).—D.C.; Md., Va.
    proboscidea Malloch, 1912b: 477 (*Aphiochaeta*).—Md. N. syn., T. Borgmeier *in litt.*

Genus TROPHITHAUMA Schmitz

Trophithauma Schmitz, 1925a: 40. Type-species, *portentum* Schmitz (orig. des.).

rostratum (Melander and Brues), 1903: 15 (*Phora*).—Mass.; Mich. to N.H., s. to Va.
    bicolorata Malloch, 1912b: 486 (*Aphiochaeta*).—Md.

Genus APOCEPHALUS Coquillett

Apocephalus Coquillett *in* Pergande, 1901: 501. Type-species, *pergandei* Coquillett (orig. des.).

REFERENCE: Brues, 1924a (key).

Parasitic on ants.

Subgenus APOCEPHALUS Coquillett

analis Borgmeier, 1958: 318.—Brazil; Tex., Neotropical.
aridus Malloch, 1912b: 444.—Mexico (V.C.); Tex.
borealis Brues, 1924a: 41.—Maine.
coquilletti Malloch, 1912b: 443.—Tenn.; Tex., Pa.
disparicauda Borgmeier, 1962a: 79.—Va.
pergandei Coquillett *in* Pergande, 1901: 501.—Md.; Wash., Wis., D.C.
pictus Malloch, 1918l: 146.—Ill.
similis Malloch, 1912b: 444.—Ariz.

Subgenus MESOPHORA Borgmeier

Apocephalus, subg. **Mesophora** Borgmeier, 1937: 209. Type-species, *mortifer* Borgmeier (orig. des.).

The type-species is parasitic on *Chauliognathus fallax* Germar (Coleoptera).

antennatus Malloch, 1913b: 274.—Md.; Mich., Tenn.
wheeleri Brues, 1903: 373.—Wis.; Va.

Genus CREMERSIA Schmitz

Cremersia Schmitz, 1924a: 33. Type-species, *zikani* Schmitz (orig. des.).

REFERENCE: Borgmeier, 1961 (key).

Parasitic on ants.

adunca Borgmeier, 1961: 32.—Tex.; Mexico, Panama.
pilipes Borgmeier, 1961: 33.—Tex.
spinicosta (Malloch), 1912b: 442 (*Apocephalus*).—Tex.
coeca Greene, 1938: 184 (*Apocephalus*).—Tex.

Genus MYRMOSICARIUS Borgmeier

Myrmosicarius Borgmeier, 1928: 122. Type-species, *gracilipes* Borgmeier (orig. des.).
Attamyia Greene, 1938: 181. Type-species, *texana* Greene (orig. des.).

REFERENCE: Borgmeier, 1931 (rev.).

Parasitic on ants.

grandicornis Borgmeier, 1928: 124.—Brazil; Tex.
texanus (Greene), 1938: 182 (*Attamyia*).—Tex.

Genus AUXANOMMATIDIA Borgmeier

Auxanommatidia Borgmeier, 1924: 175. Type-species, *variegata* Borgmeier (orig. des.).

Unnamed sp.—Mich.

Genus PSEUDACTEON Coquillett

Pseudacteon Coquillett, 1907b: 208. Type-species, *crawfordi* Coquillett (orig. des.).

Parasitic on ants.

crawfordi Coquillett, 1907b: 208.—Tex.; La., ?Miss.
curriei (Malloch), 1912b: 501 (*Plastophora*).—B.C.
grandis Greene, 1941b: 183.—Jamaica; Miss.
onyx Steyskal, 1944b: 1.—Mich.
spatulatus (Malloch), 1912b: 502 (*Plastophora*).—Tex.

Genus METOPINA Macquart

Metopina Macquart, 1835: 666. Type-species, *Phora galeata* Haliday (mon.).

fenyesi Malloch, 1912b: 503.—Mexico; Ill.

Genus CATACLINUSA Schmitz

Cataclinusa Schmitz, 1927c: 73. Type-species, *bucki* Schmitz (orig. des.).

pachycondylae (Brues), 1903: 384 (*Metopina*).—Tex.

Genus PULICIPHORA Dahl

Puliciphora Dahl, 1897: 410. Type-species, *lucifera* Dahl (mon.).
Pachyneurella Brues, 1903: 382. Type-species, *Phora venata* Aldrich (mon.).

borinquenensis Wheeler, 1906: 269.—P.R.; Md., Tenn., Fla., Neotropical.
glacialis Malloch, 1912b: 507.—Mass.; B.C., Mich.
nudipalpis Malloch, 1912b: 504.—N. Mex.
occidentalis (Melander and Brues), 1903: 17 (*Stethopathus*).—Mass.; Md., Tenn.
palposa Malloch, 1912b: 505.—Calif.
sylvatica Brues, 1909: 107.—Wash. ?*Puliciphora*.
virginiensis Malloch, 1912b: 519.—Va.

Genus ACONTISTOPTERA Brues

Acontistoptera Brues, 1902a: 373. Type-species, *melanderi* Brues (mon.).

Associated with ants.

melanderi Brues, 1902a: 374.—Tex.

Genus XANIONOTUM Brues

Xanionotum Brues, 1902a: 376. Type-species, *hystrix* Brues (mon.).

hystrix Brues, 1902a: 377.—Tex.
mexicanum Borgmeier, 1932: 378.—Mexico; Ariz., Tex.
smithii Brues, 1936: 69.—Miss.; Ala.

Genus ECITOMYIA Brues

Ecitomyia Brues, 1901b: 347. Type-species, *wheeleri* Brues (orig. des.).

wheeleri Brues, 1901b: 347.—Tex.

Genus COMMOPTERA Brues

Commoptera Brues, 1901b: 344. Type-species, *solenopsidis* Brues (orig. des.).

solenopsidis Brues, 1901b: 344.—Tex.

Genus TROPHODEINUS Borgmeier

Trophodeinus Borgmeier, 1960: 312. Type-species, *analis* Borgmeier (orig. des.).

barberi Borgmeier, 1962a: 80.—Md.
pygmaeus Borgmeier, 1962a: 81.—Mich.

Genus LECANOCERUS Borgmeier

Lecanocerus Borgmeier, 1962a: 79. Type-species, *compressiceps* Borgmeier (orig. des.).

compressiceps Borgmeier, 1962a: 80.—Va.

Family PLATYPEZIDAE
(Clythiidae)

By Edward L. Kessel

Platypezids or "flat-footed flies" are encountered most frequently in damp woods where filtered sunshine reaches the low-growing vegetation. Here they run about on broad leaves in a characteristic zigzag, stop-and-go fashion. Others will be found hovering motionless in the air or running about on the damp sand of stream beds. The males sometimes dance in swarms, which the females enter to select their mates. Oviposition occurs on fungi, the female forcing her abdomen down into the pore or between the gills to lay her eggs. While only four of the world's twelve genera have been reared from fungi, it is presumed that all the species of the family are fungus-feeders. Late summer and autumn are the seasons of greatest abundance and the adults of some species are found only at these times.

REFERENCES: Meijere, 1901, 1911a (immature stages); Johnson, 1923a (rev., e. N.Amer.); Lundbeck, 1927 (biol., immature stages); Czerny, 1930a (key to genera, biol., immature stages); Hennig, 1952 (immature stages); Kessel, 1952b (key to genera); Kessel, 1961a (biol.).

Genus MELANDEROMYIA Kessel

Melanderomyia Kessel, 1960a: 93. Type-species, *kahli* Kessel (orig. des.).

kahli Kessel, 1960a: 93.—Pa.; Kans., Iowa, Ill., Ohio.

Genus MICROSANIA Zetterstedt

Cyrtoma, subg. **Microsania** Zetterstedt, 1837: 30. Type-species, *stigmaticalis* Zetterstedt (mon.).

Called smoke flies because of positive response to smoke, which was studied in *M. occidentalis* Malloch by Kessel (1947, 1960b).

imperfecta (Loew), 1866a: 179 (Cent. 6, no. 82) (*Platycnema*).— D.C.; Man., Mich., N.Y., Fla.
occidentalis Malloch, 1935c: 66.—Idaho; Alaska to Idaho and Calif.

Genus CALLOMYIA Meigen

Cleona Meigen, 1800: 30. Type-species, *Callomyia elegans* Meigen (Coquillett, 1910b: 525). Suppressed by I.C.Z.N., 1963b: 339.
Callomyia Meigen, 1804: 311. Type-species, *elegans* Meigen (mon.).
Callimyia, emend.
Calomyia, emend.

bertae Kessel, 1961c: 193.—N. Mex.
calla Kessel, 1949a: 141.—Calif.
clara Kessel, 1949a: 146.—Calif.; Wash., Oreg.
cleta Kessel, 1949a: 144.—Calif.
corvina Kessel, 1949a: 140.—Calif.
gilloglyorum Kessel, 1961d: 4.—Calif.
proxima Johnson, 1916a: 32.—N.H.
velutina Johnson, 1916a: 32.—N.Y.; Ont., Mass.
venusta Snow, 1894b: 151.—N. Mex.; Alaska, B.C., Nev., Maine, N.H.

Genus AGATHOMYIA Verrall

Agathomyia Verrall, 1901: 30. Type-species, *Callomyia antennata* Zetterstedt (Coquillett, 1910b: 504).

REFERENCES: Hennig, 1952 (immature stages); Kessel, 1957 (biol.).

aestiva Kessel, 1949b: 217.—Calif.
alaskensis Kessel, 1961c: 217.—Alaska; B.C., Alta.
aquilonia Kessel, 1961c: 199.—Alta.; Y.T., B.C., Mont.

arossi Kessel, 1961c: 213.—Calif.
brooksi Johnson, 1923a: 57.—Mass.
canadensis Johnson, 1923a: 58.—Ont.
colei Kessel, 1961c: 195.—Md.; Mich., Pa.
cushmani Johnson, 1916a: 30.—N.H.; N.Y.
decolor Kessel, 1961c: 202.—Pa.
divergens (Loew), 1866a: 177 (Cent. 6, no. 77) (*Callomyia*).—Pa.;D.C.
dubia Johnson, 1916a: 28.—Mass.
fenderi Kessel, 1949b: 215.—Oreg.
fulva (Johnson), 1908b: 59 (*Callomyia*).—Maine; N.H.
laffooni Kessel, 1961c: 220.—B.C.; Iowa.
leechi Kessel, 1961c: 225.—Calif.
lucifuga Kessel, 1961c: 209.—Calif.; Wash.
lutea Cole *in* Cole and Lovett, 1919: 238.—Oreg.; Calif.
macneilli Kessel, 1961c: 222.—Calif.
monticola Johnson, 1923a: 57.—N.H.
nemophila Kessel, 1961c: 207.—Calif.
notata (Loew), 1866a: 177 (Cent. 6, no. 76) (*Callomyia*).—Pa.; Oreg. to Que., s. to Calif. and Va., also Brazil.
　　　tenera Loew, 1869c: 176 (Cent. 9, no. 82) (*Callomyia*).—N.Y.
　　　aldrichii Snow, 1894b: 152 (*Callomyia*).—Kans.
obscura Johnson, 1916a: 30 (as *vanduzeei* var.).—Mass.
perplexa Johnson, 1916a: 29.—N.Y.
pulchella (Johnson), 1908b: 58 (*Callomyia*).—Vt.; Mich., N.H., Mass., R.I.
stonei Kessel, 1961c: 215.—Alaska.
sylvania Kessel, 1961c: 204.—Calif.
talpula (Loew), 1869c: 175 (Cent. 9, no. 81) (*Callomyia*).—N.H.; Wash., Mass., N.Y.
vanduzeei Johnson, 1916a: 29.—N.Y.; Maine.

Genus PLATYPEZINA Wahlgren

Platypezina Wahlgren, 1910: 30. Type-species, *Platypeza connexa* Boheman (orig. des.).
Platypezoides Johnson, 1923a: 56. Type-species, *diversa* Johnson (orig. des.).
　　REFERENCE: Kessel, 1948 (review).

diversa (Johnson), 1923a: 56 (*Platypezoides*).—N.H.; Mich., N.Y., Maine, Conn., Pa.
johnsoni Kessel, 1961b: 188.—N.H.; Pa., N.Y.
pacifica Kessel, 1948: 54.—Calif.; B.C., Wash., Oreg.

Genus PROTOCLYTHIA Kessel

Protoclythia Kessel, 1950a: 257. Type-species, *californica* Kessel (orig. des.).

REFERENCE: Kessel, 1950a (review).

californica Kessel, 1950a: 259.—Calif.
carbonaria Kessel, 1950a: 263.—N.Y.; Kans., N.C.
pulchra (Snow), 1894b: 149 (*Platypeza*).—N. Mex.; Ariz.
umbrosa (Snow), 1894b: 148 (*Platypeza*).—N. Mex.

Genus PLATYPEZA Meigen

Clythia Meigen, 1800: 30. Type-species, *Platypeza fasciata* Meigen (Coquillett, 1910b: 525). Suppressed by I.C.Z.N., 1963b: 339.

Platypeza Meigen, 1803: 272. Type-species, *fasciata* Meigen (Blanchard *in* Audouin et al., 1849: pl. 170, fig. 7).

REFERENCE: Czerny, 1930a (immature stages; hosts include fungi of Polyporaceae and Agaricaceae).

abscondita Snow, 1895c: 205.—Idaho; Oreg.
agarici Willard, 1914: 167.—Calif.
anthrax Loew, 1869c: 176 (Cent. 9, no. 83).—N.Y.; Mich. to Que., s. to Ind. and N.C.
    elongata Banks, 1915a: 215.—Va.
banksi Johnson, 1923a: 53.—Va.; Vt., Mass., Md.
cinerea Snow, 1894b: 150.—N. Mex.; Alaska to Que., s. to Calif., N. Mex., and N.Y.
coraxa (Kessel), 1950b: 77 (*Clythia*).—Calif.; B.C.
dymka (Kessel), 1961c: 191 (*Clythia*).—Calif.; B.C.
egregia Snow, 1894b: 150.—N. Mex.
flavicornis Loew, 1866a: 178 (Cent. 6, no. 79).—Pa.; Ont. to Mass., s. to Ark. and Ga.
    minorata Banks, 1915a: 214.—Va.
    submacula Banks, 1915a: 214.—Va.
    mediana Banks, 1915a: 215.—Va.
    nitida Banks, 1915a: 215.—Va.
hunteri (Kessel), 1959a: 19 (*Clythia*).—Alaska; B.C.
obscura Loew, 1866a: 178 (Cent. 6, no. 80).—Pa.; Kans., Mich., N.H.
polypori Willard, 1914: 167.—Calif.; Alaska to Que., s. to Calif. and N.Y.
    infumata, authors, not Haliday.
pulla Snow, 1895c: 206.—N. Mex.; Mass.
taeniata Snow, 1894b: 149.—Ill.; Mich. to Maine, s. to Va.

unicolor Snow, 1895c: 206.—Idaho.
velutina Loew, 1866a: 178 (Cent. 6, no. 78).—Pa.; Mich. to Que., s. to Kans. and N.C.

Genus METACLYTHIA Kessel

Metaclythia Kessel, 1952a: 348. Type-species, *currani* Kessel (orig. des.).
currani Kessel, 1952a: 349.—Ont.; Wis.

Genus CALOTARSA Townsend

Calotarsa Townsend, 1894a: 50. Type-species, *ornatipes* Townsend (mon.) = *pallipes* (Loew).
REFERENCE: Kessel, 1963 (review).
calceata (Snow), 1894b: 146 (*Platypeza*).—N. Mex.
insignis Aldrich, 1906: 126.—Calif.; B.C., Oreg., w. Mont.
pallipes (Loew), 1866a: 179 (Cent. 6, no. 81) (*Platypeza*).—D.C.; S. Dak. to Que., s. to Ill. and N.C.
 ornatipes Townsend, 1894a: 52.—Ill.

Superfamily SYRPHOIDEA

Family PIPUNCULIDAE
(Dorilaidae)

By D. Elmo Hardy

Pipunculids ("big-headed flies" or "big-eyed flies") are parasitic on Cicadellidae, Delphacidae, and possibly Cercopidae and some other families of Homoptera. They are evidently of considerable importance as an aid in control of some pest leafhopper species.

REFERENCES: Perkins, 1905 (biol.); Hardy, 1943c (rev.); Aczél, 1948 (monog.).

Subfamily CHALARINAE

Genus CHALARUS Walker

Chalarus Walker, 1834: 269. Type-species, *Cephalops spurius* Fallén (Westwood, 1840: 135).
latifrons Hardy, 1943c: 33.—Ariz.; B.C. to Mass., s. to Calif., N. Mex., and Va.
spurius (Fallén), 1816b: 16 (*Cephalops*).—Europe; Sask. to N.B., s. to Calif., Tex., and Va.

Genus VERRALLIA Mik

Verrallia Mik, 1899: 137. Type-species, *Cephalops aucta* Fallén (orig. des.).
Prothechus, authors, not Rondani. *Prothechus* Rondani (1856: 139) has been suppressed by I.C.Z.N. (1961a: 230), and remains a nomen dubium.
aucta (Fallén), 1817a: 61 (*Cephalops*).—Europe; S. Dak. to Mass. and Va.
    virginica Banks, 1915b: 169.—Va.
csikii Aczél, 1940: 152 (n. name for *opacus* Williston).—Wash.
    opacus Williston, 1886a: 295 (*Pipunculus*; preocc. Fallén, 1817).—Wash.

Genus JASSIDOPHAGA Aczél

Jassidophaga Aczél, 1939: 20. Type-species, *Pipunculus pilosus* Zetterstedt (orig. des.). Enderlein's earlier proposal (1936a: 129) invalid under Art. 25c of the Rules.
fasciata (Hardy), 1939: 16 (*Verrallia*).—Colo.; N. Mex.
pilosa (Zetterstedt), 1838: 579 (*Pipunculus*).—Europe: Que. to N.J.

Subfamily PIPUNCULINAE

Tribe NEPHROCERINI

Genus NEPHROCERUS Zetterstedt

Nephrocerus Zetterstedt, 1838: 578. Type-species, *lapponicus* Zetterstedt (mon.).
daeckei Johnson, 1903c: 107.—N.Y.; N.H. to Pa. and Md.
slossonae Johnson, 1915a: 55.—N.H.

Tribe PIPUNCULINI

Genus PIPUNCULUS Latreille

Dorilas Meigen, 1800: 31. Type-species, *Pipunculus campestris* Latreille (Coquillett, 1910b: 535). Suppressed by I.C.Z.N., 1963b: 339.
Dorylas, emend.
Pipunculus Latreille, 1802: 463. Type-species, *campestris* Latreille (mon.).
Cephalops Fallén, 1810a: 10. Type-species, *aeneus* Fallén (mon.) = *pratorum* (Fallén).

Subgenus PIPUNCULUS Latreille

alpinus Cresson, 1911: 306.—Vt.; Que., Maine, N.H., also Utah.
angus Cresson, 1911: 305.—N. Mex. ?=*fuscus* Loew.
apicarinus Hardy and Knowlton, 1939b: 114 (as *femoratus* var.)—Utah.
ater ater Meigen, 1824: 23.—Europe; N. Dak. to Que., s. to Calif., Tex., and Ga.
    cingulatus Loew, 1866a: 176 (Cent. 6, no. 73).—D.C.
    horvathi Kertész, 1907b: 579.—N.Y.
    townsendi Malloch, 1912a: 292.—N. Mex.
    hertzogi Rapp, 1943a: 118.—N.J.
    fuscus, authors, not Loew.
  ssp. **velutinus** Cresson, 1911: 300 (as *cingulatus* var.).—Pa.; Alta. to N.S., s. to Ariz., Iowa, and Pa.
banksi (Aczél), 1940: 152 (*Dorilas*; n. name for *terminalis* Banks).—Va.; Idaho to N.S., s. to Calif., Ohio, and Va.
    terminalis Banks, 1915b: 168 (preocc. Thomson, 1869).—Va.
fuscus fuscus Loew, 1866a: 175 (Cent. 6, no. 71).—Md.; B.C. to Ont., s. to Calif., Kans., and Fla.
  ssp. **nitidiventris** Loew, 1866a: 175 (Cent. 6, no. 72) (as sp.).—D.C.; B.C. to Que., s. to Ariz., Kans., and Va.
    viduus Cresson, 1911: 301.—Vt.
    sororius Cresson, 1911: 305.—Vt.
    luteicornis Cresson, 1911: 307.—Maine. **N. syn.**
houghi houghi Kertész, 1900a: 244 (n. name for *lateralis* Walker).—N. Amer.; Kans. to Ont., s. to Fla.
    lateralis Walker, 1852: 216 (preocc. Macquart, 1833).—N. Amer.
    femoratus Cresson, 1911: 302.—Vt.
  ssp. **curvitibiae** Hardy, 1939: 19 (as *femoratus* var.).—Ariz.
pallipes Johnson, 1903c: 107.—N.J.; B.C. to Que., s. to N. Mex. and Ga.
trichaetus Malloch, 1912a: 296.—N.H.; Utah, Wyo., Colo., S. Dak.
varius varius Cresson, 1911: 309.—Pa.; B.C. to N.B., s. to Utah, Mo., and Fla.
    mainensis Cresson, 1911: 298.—Maine.
  ssp. **phaethus** Hardy and Knowlton, 1939b: 123 (as var.).—Utah.

Subgenus EUDORYLAS Aczél

Eudorylas Aczél, 1940: 151 (as genus). Type-species, *Pipunculus opacus* Fallén (orig. des.).

aberratus Hardy and Knowlton, 1939a: 87.—Utah; Nev.
aequus aequus Cresson, 1911: 292.—Mass.; Utah to Mich. and Que., s. to Tex. and N.C.
    longipes Hardy and Knowlton, 1939a: 88 (as var.).—Utah.
  ssp. **argryofrons** Hardy and Knowlton, 1939a: 87 (as var.).—Utah.
    oleous Rapp, 1943b: 223 (*Allomethus*).—Calif.
affinis Cresson, 1910: 283.—Que.; B.C. to N.S., s. to Calif., Okla., and N.C.
    globosus Cresson, 1912: 453.—R.I.
alternatus Cresson, 1910: 286.—N. Mex.; Ariz.
apicalis Hardy and Knowlton, 1939a: 88.—Utah.
aquavicinus (Hardy), 1943c: 72 (*Dorilas*).—N. Mex.
arundani (Hardy), 1954: 122 (*Dorilas*).—Ga.
atlanticus Hough, 1899d: 80.—Mass.; Alta. to Que., s. to Utah, Calif., and Fla.
bidactylus (Hardy), 1943c: 80 (*Dorilas*).—Ariz.; Mich., N.Y., Mass.
bilobus (Hardy), 1947b: 148 (*Dorilas*).—N.Y.
caudatus Cresson, 1910: 289.—N.Y.; B.C., Ohio to Que., s. to Va.
    discolor Banks, 1910: 290 (as var.).—N.Y.
cinctus cinctus Banks, 1915b: 169.—Va.; coastal, Va. to Fla.
  ssp. **subtilis** (Hardy), 1943c: 84 (*Dorilas*).—Ariz.; N. Mex., Colo., Kans.
curtus (Hardy), 1943c: 85 (*Dorilas*).—Ariz.
dives (Hardy), 1947b: 149 (*Dorilas*).—Ind.
dreisbachi (Hardy), 1948a: 89 (*Dorilas*).—Mich.
fuscitarsis Adams, 1903a: 36.—N. Mex.; Calif., Utah.
grandis (Hardy), 1943c: 90 (*Dorilas*).—Ariz.
harmstori Hardy and Knowlton, 1939b: 115.—Utah.
huachucanus (Hardy), 1943c: 95 (*Dorilas*).—Ariz.; Kans.
kansensis (Hardy), 1940b: 102 (*Dorilas*).—Kans.; B.C., Ariz., and N. Mex. to Colo. and Iowa.
lasiofemoratus Hardy and Knowlton, 1939b: 116.—Utah.
latipennis Banks, 1915b: 168.—Va.
lautus (Hardy), 1943c: 99 (*Dorilas*).—N. Mex.
loewii Kertész, 1900b: 270 (n. name for *fasciatus* Loew).—Tex.; Minn. to N.H., s. to Ariz., Kans., and Va.
    fasciatus Loew, 1872a: 88 (Cent. 10, no. 59) (preocc. Roser, 1840).—Tex.
    nigricornis Adams, 1903a: 36.—Mo.
    semifasciatus Cresson, 1910: 288.—Mass.
michiganensis (Hardy), 1948a: 90 (*Dorilas*).—Mich.

minor Cresson, 1911: 293.—Conn.; Wash. to Que., s. to Calif., Tex., and N.C., Mexico.
    cressoni Johnson of Hardy, 1943c: 104.
montivagus (Hardy), 1943c: 104 (*Dorilas*).—Colo.
nevadaensis (Hardy), 1943c: 106 (*Dorilas*).—Nev.
nigripes Loew, 1866a: 176 (Cent. 6, no. 75).—Pa.; Alaska, Idaho to Maine, s. to Calif. and Ga.
    dubius Cresson, 1910: 284.—Maine.
    winnemannae Malloch, 1912c: 655.—Md.
reipublicae Walker, 1849: 639.—N.Y.; Alta. to Ont., s. to Okla. and Fla.
    albofasciatus Hough, 1899d: 85.—La.
sabroskyi (Hardy), 1943c: 112 (*Dorilas*).—Mich.
stainsi (Hardy), 1943c: 113 (*Dorilas*).—Utah; Calif.
stigmaticus stigmaticus Malloch, 1912a: 294.—B.C.; Que., Mich., Va.
    ssp. **brachystigmaticus** Hardy and Knowlton, 1939a: 90 (as sp.).—Utah.
subopacus subopacus Loew, 1866a: 176 (Cent. 6, no. 74).—Wash.; Alta. and Wash. to N.H., s. to N. Mex., Kans., and N.C.
    confraternus Banks *in* Cresson, 1910: 285.—N.Y.
    occidentalis Malloch, 1912a: 291.—Alta.
    ssp. **industrius** Knab, 1915a: 83 (as sp.).—Calif.; B.C. to Que., s. to Calif., Tex., and Ohio.
    melanis Hardy and Knowlton, 1939b: 113 (as *confraternus* var.).—Utah.
tarsalis Banks in Cresson, 1911: 309.—N.Y.; Minn. to N.Y., s. to Kans. and Va., also Ariz.
vierecki Malloch, 1912c: 654.—Md.; Tex., Mass., N.C.

Subgenus CEPHALOSPHAERA Enderlein

Cephalosphaera Enderlein, 1936a: 129 (as genus). Type-species, *Pipunculus furcatus* Egger (mon.).

acuminatus Cresson, 1911: 297.—N. Mex.; B.C., Ariz., Kans.
appendiculatus Cresson, 1911: 296.—Vt.; Iowa, Vt. and N.H. to Ga.
biscaynei Cresson, 1912: 453.—Fla.; Kans., Mich., Va., Ala.
brevis Cresson, 1911: 303.—Vt.; Minn. to Que., s. to Kans. and Va.
    eronis Curran, 1927i: 291.—B.C.
constrictus Banks *in* Cresson, 1911: 306.—N.C.; Man. to N.B., s. to Tex. and Fla.
maximus (Hardy), 1943c: 50 (*Cephalosphaera*).—Ariz.
stricklandi Curran, 1927i: 291.—Alta.; Mich., Colo., N. Mex.
tibialis (Hardy), 1943c: 53 (*Cephalosphaera*).—N. Mex.; Nev.

Genus ALLOMETHUS Hardy

Allomethus Hardy, 1943c: 128. Type-species, *brimleyi* Hardy (orig. des.).

brimleyi Hardy, 1943c: 128.—N.C.
mysticus Rapp, 1943b: 223.—Que. Unrecognized.

Genus TOMOSVARYELLA Aczél

Tomosvaryella Aczél, 1939: 22. Type-species, *Pipunculus sylvaticus* Meigen (orig. des.).

agnesea Hardy, 1940b: 103.—Kans.; Wash. to Idaho, s. to Calif., N. Mex., and Kans.
aliena Hardy, 1947b: 147 (n. name for *propinqua* Hardy).—B.C.; B.C. to Man., s. to Calif. and Tex.
  propinqua Hardy, 1943c: 169 (preocc. Becker, 1913).—B.C.
appendipes (Cresson), 1911: 319 (*Pipunculus*).—S.C.; Calif. to Iowa and Vt., s. to Fla.
armata Hardy, 1940b: 106.—Ga.
beameri Hardy, 1940b: 107.—Kans.
bidens (Cresson), 1911: 320 (*Pipunculus*).—Calif.; Calif. to Kans. and Fla., s. to Mexico.
brevijuncta Hardy, 1943c: 155.—Calif.
columbiana (Kertész), 1915: 386 (*Pipunculus*; n. name for *trochanteratus* Malloch).—B.C.
  trochanteratus Malloch, 1912a: 297 (*Pipunculus*; preocc. Becker, 1900).—B.C.
contorta (Hardy), 1939: 18 (*Pipunculus*).—Kans.; Calif. to Man., s. to Kans. and Ga.
coquilletti coquilletti (Kertész), 1907b: 582 (*Pipunculus*).—N.Y.; Alaska to Que., s. to Calif. and Fla.
  proximus Cresson, 1911: 318 (*Pipunculus*).—Idaho.
  nudus Rapp, 1943b: 223 (*Pipunculus*).—Que.
  ssp. **flaviantenna** (Hardy and Knowlton), 1939b: 118 (*Pipunculus proximus* var.).—Utah; Wash., Idaho.
deformis Hardy, 1947b: 151.—Fla.
dissimilis Hardy, 1943c: 161.—Ariz.; N. Mex.
exilidens Hardy, 1943c: 162.—Ariz.; Calif., N. Mex.
floridensis Hardy, 1940b: 109.—Fla.
inconspicua (Malloch), 1912a: 295 (*Pipunculus*).—Alta.
lepidipes Hardy, 1943c: 166.—Nev.; B.C. to Sask. and Pa., s. to Calif. and Tex., S. Amer.
minacis Hardy, 1940b: 110.—Fla.; Keys only.
pauca Hardy, 1943c: 168.—Fla.; Ga.

quadradentis Hardy, 1943c: 172.—Calif.; Nev.
sachtlebeni (Aczél), 1940: 152 (*Pipunculus*; n. name for *unguiculatus* Cresson).—Va.; Sask. to Mich. and N.J., s. to Kans. and Fla.
  unguiculatus Cresson, 1911: 319 (*Pipunculus*; preocc. Zeller, 1860).—Va.
similis (Hough), 1899d: 84 (*Pipunculus*).—Ga.; Utah to Pa., s. to Ariz., Okla., and Ga.
subnitens (Cresson), 1911: 316 (*Pipunculus*).—N. Mex.; Wyo., Calif.
subvirescens (Loew), 1872a: 87 (Cent. 10, no. 58) (*Pipunculus*).—Tex.; entire U.S. and Canada, Bermuda, Neotropical, Palaearctic.
  aridus Williston, 1893d: 255 (*Pipunculus*).—Calif.
  insularis Cresson, 1911: 317 (*Pipunculus*).—Bermuda.
  albiseta Cresson, 1911: 318 (*Pipunculus*).—Bermuda.
  knowltoni Hardy, 1939: 20 (*Pipunculus*).—Utah.
sylvatica (Meigen), 1824: 20 (*Pipunculus*).—Europe; B.C. to Que., s. to Calif. and Ga.
  scoparius Cresson, 1911: 317 (*Pipunculus*).—Maine.
  tangomus Rapp, 1943b: 224 (*Pipunculus*; as *nudus* var.).—Que.
toxodentis (Hardy and Knowlton), 1939b: 118 (*Pipunculus*).—Utah; Calif. and Wyo., s. to Tex.
translata (Walker), 1857: 150 (*Pipunculus*).—U.S. ?=*sachtlebeni*.
tumida Hardy, 1940b: 112.—Fla.; Calif. to Kans., Ga. and Fla.
turgida Hardy, 1940b: 113.—Ga.; Calif. to Sask. and Ohio, s. to Tex. and Ga.
utahensis (Hardy and Knowlton), 1939b: 122 (*Pipunculus*).—Utah; Sask. to Ont., s. to Calif. and N. Mex.
vagabunda (Knab), 1915a: 84 (*Pipunculus*).—Calif.; Wash. to Sask., Mich. and Va., s. to Calif. and Fla.
  tenellus Hardy and Knowlton, 1939b: 121 (*Pipunculus*; as *trochanteratus* var.).—Utah.
wilburi (Hardy), 1939: 22 (*Pipunculus*).—Kans.; N. Mex., Wyo., Colo., Iowa, Mo.
xerophila Hardy, 1943c: 188.—N. Mex.; Calif., Ariz., Tex.

Genus DORYLOMORPHA Aczél

Tomosvaryella, subg. **Dorylomorpha** Aczél, 1939: 22. Type-species, *Pipunculus rufipes* Meigen (orig. des.).

atramontensis (Banks) *in* Cresson, 1911: 312 (*Pipunculus*).—N.C.
canadensis Hardy, 1943c: 133.—Sask.
caudelli (Malloch), 1912a: 298 (*Pipunculus*).—B.C.; Ont.

exilis (Malloch), 1912a: 295 (*Pipunculus*).—Alta.; Alta. to Mich., s. to Calif. and N. Mex.
flavomaculata (Hough), 1899d: 85 (*Pipunculus*).—Mass.; B.C., Calif., Ohio to N.B. and Pa.
occidens (Hardy), 1939: 17 (*Pipunculus;* as *atramontensis* var.).— Idaho; Alaska.
ornata Hardy, 1943c: 139.—B.C.; Alaska, s. to Wyo. and S. Dak.
tridentata Hardy, 1943c: 141.—Calif.; Alaska.
uncinata Hardy, 1943c: 142.—Nev.; Idaho, Colo.

Family SYRPHIDAE

By Willis W. Wirth, Yale S. Sedman, and Howard V. Weems, Jr.

The "flower flies" or "hover flies" of the family Syrphidae are typically medium-sized to large flies which are frequently strikingly marked with black or yellow bands or lunules; many mimic bees or wasps. The adults are habitual visitors of flowers and are considered of importance second only to some species of bees in the cross-pollination of many economic plants. The males are often seen hovering, almost motionless in the air, and dart swiftly aside when disturbed.

The eggs are oval, chalky white in color, with a reticulated pattern on the shell. They are deposited near the intended food of the larvae, those of the aphidophagous species singly, the others in masses of more than a hundred. The larvae present a diversity of forms, the classification of which has been studied mainly by Metcalf (1913, 1916b) and Heiss (1938), and recently in very fine revisions by Dixon (1960) and Hartley (1961). There are two principal types: (1) the primitive, usually saprophagous type in the subfamily Milesiinae, and (2) the carnivorous, usually aphidophagous type in the subfamily Syrphinae (Hartley, 1961). From the short-tailed saprophagous type have evolved various modifications such as the *Microdon*-type, sluglike larvae of the Microdontini, a short-tailed type found in bee or wasp nests or in cacti in the Volucellini, a pointed-tailed type fitted for piercing roots in the Chrysogastrini, the phytophagous larvae of the Cheilosiini, and the long-tailed or rat-tailed aquatic larvae of the Eristalini. Various special habits are discussed under the respective taxa. The puparium is usually pear-shaped or teardrop-shaped in the Syrphinae and barrel-shaped in the Milesiinae.

The classification and nomenclature of higher taxa in this family are in great need of critical review. An attempt has been made in this catalog to work out a new compromise system giving more weight to biology and immature stages, as proposed by Goffe (1952) and Hartley (1961), while at the same time preserving as much as possible the modern classification of adults used by Hull (1949b) and

Coe (1953). This procedure has resulted mainly in the downgrading of the numerous aberrant subfamilies to tribal status.

Probably more problems of nomenclature have been left undecided in the Syrphidae than in any other family of Diptera. Most of these go back to the series of works by Meigen (1800, 1803, 1804) in which erratic and conflicting proposals of genera were finally in large part discarded as preliminary groping by Meigen himself in his definitive classification in the 1822 "Systematische Beschreibung." Nearly all syrphid taxonomists have followed Meigen (1822) in spite of the rising tide of priority adherents who have proposed from time to time to go back to earlier and nomenclaturally correct names. In this catalog we have attempted to determine which names are nomenclaturally correct, but if usage or practicality is overwhelmingly in favor of retention of incorrect names, we have adopted the latter with the statement that a proposal should be made to the I.C.Z.N. for their formal preservation.

REFERENCES: Williston, 1887c (basic rev. N. Amer. spp.); Verrall, 1901 (basic rev., Britain); Metcalf, 1913 (tax., biol., Ohio), 1916b (biol., immature stages, Maine), 1921 (genitalia); Lundbeck, 1916 (basic rev., Denmark); Shannon, 1921, 1922a, 1922b, 1923a (higher classif.); Jones, 1922 (tax., biol., Colo.); Campbell and Davidson, 1924 (biol., economic spp., Calif.); Sack, 1928–1932 (rev. Palaearctic Region); Fluke, 1929 (biol. spp. on pea aphid, keys, Wis.); Brues and Melander, 1932 (higher classif.); Johannsen, 1935 (immature stages of aquatic forms); Heiss, 1938 (tax. immature stages); Telford, 1939 (tax. and biol., Minn.); Shiraki, 1949 (higher classif.); Hull, 1949b (rev. genera of world); Goffe, 1952 (higher classif.); Coe, 1953 (general tax., keys, Britain); Dixon, 1960 (tax. immature stages, Britain); Hartley, 1961 (tax. immature stages, Britain); Glumac, 1961 (higher classif.).

Subfamily SYRPHINAE

Nearly all the larvae of the Syrphinae are aphidophagous or predaceous on small, soft-bodied Homoptera or on Thysanoptera. Syrphine larvae share with coccinellid beetles the role of the most important predators of our more destructive plant lice (Metcalf, 1913, 1916b; Jones, 1922; Fluke, 1929; Wadley, 1931; Telford, 1939).

Tribe SYRPHINI

REFERENCE: Fluke, 1950 (rev., based on male genitalia).

Genus SYRPHUS Fabricius

Syrphus Fabricius, 1775: 762. Type-species, *Musca ribesii* Linnaeus (Rondani, 1844b: 459). To preserve the long and almost universal usage of *Syrphus* in the sense of *Musca* *ribesii* Linnaeus as type-species, the I.C.Z.N. should be asked to suspend the rules and suppress Curtis' (1839: pl. 753) designation of *Musca lucorum* Linnaeus and to place *Syrphus* Fabricius on the Official List of Generic Names with *Musca ribesii* Linnaeus as type-species.

Syrphidis Goffe, 1933: 78. Type-species, *Musca ribesii* Linnaeus (orig. des.).

REFERENCES: Fluke, 1933 (rev.), 1954 (key).

attenuatus Hine, 1922b: 144.—Alaska; Alaska to Man., s. to B.C. and Colo., also Wis., Maine.
autumnalis Fluke, 1954: 3.—Ont.; Alaska to Labr., s. to Wash., Utah, and Conn.
bigelowi Curran, 1924k: 288.—Ont.; Alaska to Que., s. to B.C. and Wis.
concavus (Say), 1823: 89 (1859b: 78) (*Scaeva*).—Pa. Unrecognized.
currani Fluke, 1939: 365.—Utah; Alaska to Calif. and Colo.
dimidiatus Macquart, 1834a: 537 (preocc. Fabricius, 1781).—Ga. Unrecognized, and therefore not renamed here because of possible synonymy with older species.
hinei Fluke, 1933: 73.—Alaska; Alta., Wis., Ont.
jonesi Fluke, 1949: 41 (as *ribesii* var.; n. name for *similis* Jones).— Colo.; Alaska to Labr., s. to B.C. and Colo.
   similis Jones, 1917: 224 (preocc. Blanchard, 1852).—Colo.
knabi Shannon, 1916a: 200.—Md.; Wis. to Ont. and Maine, s. to Kans., Miss., and Fla.
opinator Osten Sacken, 1877: 327.—Calif.; B.C. to Alta. and Wyo., s. to Calif. and Tex.
philadelphicus Macquart, 1842: 153 (1842: 93).—Pa. Unrecognized, ?=*ribesii*.
rectus Osten Sacken, 1875c: 140.—N.Y., N.H.; Man. to Que. and Maine, s. to Colo., Mo., and N.C.
ribesii (Linnaeus), 1758: 593 (*Musca*).—Europe; Alaska to N.B., s. to Cent. Amer. and N.C., Eurasia.
torvus Osten Sacken, 1875c: 139.—Canada, Colo., N.H., Mass., R.I.; Alaska to Greenland, s. to Calif., N. Mex., and S.C., Eurasia.
transversalis Curran, 1921b: 155.—Ont.; Ont. and Que., s. to Wis. and D.C., also B.C.
vitripennis Meigen, 1822: 308.—Europe; Alaska to N.S., s. to Calif., Colo., and N.C., Europe, ?Mexico.
vittafrons Shannon, 1916a: 202 (as *ribesii* var.).—Md.; Man. to N.B., s. to Utah, Ind., and Fla.

Genus METASYRPHUS Matsumura

Metasyrphus Matsumura *in* Matsumura and Adachi, 1917a: 147.
Type-species, *Syrphus corollae* Fabricius (orig. des.).
Posthosyrphus Enderlein, 1938: 204. Type-species, *Syrphus americanus* Wiedemann (orig. des.). Enderlein's description suggests that he misidentified *Eupeodes volucris* Osten Sacken as *americanus* Wiedemann, but we are accepting Fluke's (1950) interpretation as nomenclaturally correct.

REFERENCES: Fluke, 1933, 1952 (revs.).

aberrantis (Curran), 1925k: 90 (*Syrphus*).—Wash.; B.C., Alta., Oreg., Idaho.
americanus (Wiedemann), 1830: 129 (*Syrphus*).—N. Amer.; B.C. to Que., s. to Calif., Mexico, and Fla. Alleged preoccupation by *Musca americana* Swederus erroneous.
    wiedemanni Johnson, 1919a: 32 (*Syrphus;* unjustified n. name for *americanus* Wiedemann).—N. Amer.
astutus Fluke, 1952: 15 (as *luniger* ssp.).—Colo.; Wash. to Mont., s. to Utah and Colo., also Ont.
canadensis (Curran), 1926d: 172 (*Syrphus*).—Man.; Alaska to Maine, s. to Wash. and Wis.
chillcotti Fluke, 1952: 20.—Man.; N.W.T., Que.
confertus Fluke, 1952: 7 (as *nitens* ssp.).—Wis.; Vt., Maine, Mass.
    nitens, authors, not Zetterstedt.
curtus (Hine), 1922b: 145 (*Syrphus*).—Alaska; B.C.
daphne (Hull), 1925a: 280 (*Didea*).—B.C.
depressus Fluke, 1933: 97.—Alaska; Alaska and N.W.T. to Oreg., also Ont. and Que. to Mass.
flukei (Jones), 1917: 222 (*Syrphus*).—Colo.; Wash., Idaho, Mont., Wyo.
fumipennis (Thomson), 1869: 499 (*Syrphus*).—Calif.; Wash., Oreg., Nev.
gentneri Fluke, 1952: 19.—Oreg.; Wash., Calif., Utah.
lapponicus (Zetterstedt), 1838: 598 (*Scaeva*).—Sweden; Alaska to Greenland, s. to Calif., Tex., and Va., Europe.
    agnon Walker, 1849: 579 (*Syrphus*).—Ont., N.S.
    alcidice Walker, 1849: 579 (*Syrphus*).—Ont.
    arcucinctus Walker, 1849: 580 (*Syrphus*).—Ont.
    arcuatus, authors, not Fallén.
latifasciatus (Macquart), 1829: 242 (1829: 94) (*Syrphus*).—France; Alaska to Nfld., s. to Calif., Tex., and Pa., Europe.
    abbreviatus Zetterstedt, 1849: 3136 (*Scaeva*).—Europe.
    pallifrons Curran, 1925k: 172 (*Syrphus*).—Wis.

lebanoensis (Fluke), 1930: 139 *(Syrphus).*—B.C.; B.C. to Alta., s. to Wash., Ariz., and N. Mex.
luniger (Meigen), 1822: 300 *(Syrphus).*—Europe; Greenland, Eurasia.
marginatus (Jones), 1917: 222 *(Syrphus).*—Colo.
meadii (Jones), 1917: 223 *(Syrphus).*—Colo; B.C. to Ont., s. to Calif. and N. Mex.
medius (Jones), 1917: 224 *(Syrphus).*—Colo.; Oreg. to P.E.I., s. to Calif. and La.
montanus (Curran), 1925k: 174 *(Syrphus).*—Mont.; Calif., Utah, Colo.
montivagus (Snow), 1895d: 236 *(Syrphus).*—Colo.; Wash. to Alta., s. to Utah and N. Mex.
neoperplexus (Curran), 1925k: 93 *(Syrphus).*—Man.; Alaska, N. Dak., Ont.
nigrocomus Hull, 1943e: 48.—Wash.
nigroventris Fluke, 1933: 97.—Greenland.
ochrostomus (Zetterstedt), 1849: 3133 *(Scaeva).*—Europe; Alta., Ont., N.Y.
palliventris (Curran), 1925k: 173 *(Syrphus).*—Alta.; B.C. to Ont., s. to Calif.
perplexus (Osburn), 1910: 55 *(Syrphus).*—N. Amer. (12 states and provinces); Alaska to Labr., s. to Oreg., N. Mex., and N.C.
  arcuatus, authors, not Fallén.
  lundbecki, authors, not Soot-Ryen.
pingreensis (Fluke), 1930: 137 *(Syrphus).*—Colo.; Y.T. to Man., s. to Calif. and Colo., ?Wis., ?Md.
pomus (Curran), 1921c: 172 *(Syrphus;* as *americanus* var.).—Ont. (unpublished); Man. to Maine, s. to Okla. and Va.
rufipunctatus (Curran), 1925k: 180 *(Syrphus).*—B.C.; Wash., Colo.
sculleni Fluke, 1952: 12.—Oreg.; B.C. to Idaho, s. to Calif. and N. Mex.
snowi (Wehr), 1922b: 137 (=p. 19) *(Syrphus;* n. name for *ruficauda* Snow).—Colo.; B.C. to Idaho and Nebr., s. to Oreg., Ariz., and N. Mex.
  ruficauda Snow, 1892: 36 *(Syrphus;* preocc. Bigot, 1883).—Colo.
  snowi Curran, 1925k: 173 *(Syrphus;* n. name for *ruficauda* Snow).—Colo.
subsimus Fluke, 1952: 11.—Colo.; Alaska to Calif., also Colo. and N. Mex.
talus Fluke, 1933: 95.—Oreg.
venablesi (Curran), 1929d: 45 *(Syrphus).*—B.C.; Alaska to Alta., s. to Calif. and N. Mex.

vinelandi (Curran), 1921c: 172 (*Syrphus*; as *americanus* var.).—Ont., Wis.; Minn. to Ont. and Conn., s. to Kans. and Va.

vockerothi Fluke, 1952: 17 (as *luniger* ssp.).—N.W.T.; Alaska to Que., s. to Colo. and Md.

Genus EUPEODES Osten Sacken

Eupeodes Osten Sacken, 1877: 328. Type-species, *volucris* Osten Sacken (mon.).

volucris Osten Sacken, 1877: 329.—Calif., Nev., Utah, Colo.; B.C. to Ont., s. to Calif., Mexico, and La.
  perpallidus Bigot, 1884a: 90 (*Syrphus*).—N. Amer.
  braggii Jones, 1917: 221.—Colo. **N. syn.**
  weldoni Jones, 1917: 221.—Colo. **N. syn.**

Genus SCAEVA Fabricius

Scaeva Fabricius, 1805: 248. Type-species, *Musca pyrastri* Linnaeus (Curtis, 1834: pl. 509).

Lasiopthicus Rondani, 1844b: 459. Type-species, *Musca pyrastri* Linnaeus (Rondani, 1856: 51).

Lasiophthicus, Lasiophticus, Lasiopticus, errors or emends.

Catabomba Osten Sacken, 1877: 326. Type-species, *Musca pyrastri* Linnaeus (mon.).

  REFERENCE: Goffe, 1933 (syn.).

pyrastri (Linnaeus), 1758: 594 (*Musca*).—Europe; Alaska to Alta., s. to Calif., N. Mex., and Ark., Eurasia, n. and w. Africa.
  affinis Say, 1823: 93 (1859b: 81).—Ark.
    var. **unicolor** Curtis, 1834: pl. 509 (as sp.).—Britain; Oreg., Idaho, Utah.

selenitica (Meigen), 1822: 304 (*Syrphus*).—Europe; ?N.C. Introduced.

Genus DIDEA Macquart

Didea Macquart, 1834a: 508. Type-species, *fasciata* Macquart (mon.).

  REFERENCE: Hull, 1925a (key).

alneti (Fallén), 1817a: 38 (*Scaeva*).—Europe; Alaska to Labr., s. to Colo. and N.S., Eurasia.

catalina (Curran), 1930f: 14 (*Syrphus*).—Ariz. Unrecognized.

fasciata Macquart, 1834a: 508.—France; B.C. to N.S., s. to Oreg., N. Mex., and N.C., Eurasia.
  fuscipes Loew, 1863b: 318 (Cent. 4, no. 82).—Pa.
laxa Osten Sacken, 1875d: 66.—Mich., N.H., Maine; Alaska to Nfld., s. to Calif., Mexico, and N.C.
  syrphoides Hull, 1925a: 278 (as var.).—Mich., N.Y. **N. syn.**
pacifica Lovett *in* Cole and Lovett, 1919: 246.—Oreg. Unrecognized.

Genus LEUCOZONA Schiner

Leucozona Schiner, 1860c: 214. Type-species, *Musca lucorum* Linnaeus (orig. des.). *Musca lucorum* Linnaeus was also the type-species validly designated by Curtis, 1839: pl. 753, for *Syrphus* Fabricius, but to preserve usage and prevent confusion the I.C.Z.N. should be asked to suspend the rules and designate *Musca ribesii* Linnaeus as type-species of *Syrphus* Fabricius (which see), and to place *Leucozona* Schiner on the Official List of Generic Names.

lucorum (Linnaeus), 1758: 592 (*Musca*).—Europe; Alaska to N.S., s. to Oreg., Colo., and N.Y., Eurasia.
  americana Curran, 1923c: 38 (as var.).—Que. **N. syn.**

Genus DASYSYRPHUS Enderlein

Dasysyrphus Enderlein, 1938: 208. Type-species, *Scaeva albostriata* Fallén (orig. des.).

REFERENCE: Fluke, 1933 (key, as *amalopis* group).

amalopis (Osten Sacken), 1875c: 148 (*Syrphus*).—N.H.; Alaska to Que., s. to Calif., N. Mex., and N.C.
  intrudens Osten Sacken, 1877: 326 (*Syrphus*).—Calif.
arcuatus (Fallén), 1817a: 42 (*Scaeva*).—Europe; Alaska to Que. and Mass., s. to Oreg., Europe.
  venustus Meigen, 1822: 299 (*Syrphus*).—Europe.
  lapponicus, authors, not Zetterstedt.
creper (Snow), 1895d: 234 (*Syrphus*).—Colo., N. Mex.; Wash. to Alta. and Nebr., s. to Calif. and N. Mex.
disgregus (Snow), 1895d: 233 (*Syrphus*).—N. Mex.; Idaho, Colo.
laticaudatus (Curran), 1925k: 176 (*Syrphus*).—B.C.
laticaudus (Curran), 1925k: 175 (*Syrphus*).—Ont.; Oreg., Utah, Mich., N.H., Que., N.B.
limatus (Hine), 1922b: 146 (*Syrphus*).—Alaska; Alaska to Que., s. to B.C. and Colo.

lotus (Williston), 1887c: 75 (*Syrphus*).—Ariz.; B.C. to Idaho and Colo., s. to Calif. and Mexico.
lunulatus (Meigen), 1822: 299 (*Syrphus*).—Europe; Greenland.
osburni (Curran), 1925k: 177 (*Syrphus*).—Ont.; Alaska to Labr., s. to Mass.
pacificus (Lovett) *in* Cole and Lovett, 1919: 245 (*Syrphus*).—Oreg.; Alaska to Alta., s. to Calif. and Utah.
pauxillus (Williston), 1887c: 74 (*Syrphus*).—N. Mex.; Y.T. to Alta., s. to Calif. and N. Mex., also Que.
reflectipennis (Curran), 1921b: 157 (*Syrphus*).—Ont.

Genus EPISTROPHE Walker

Epistrophe Walker, 1852: 242. Type-species, *conjungens* Walker (mon.)=*grossulariae* (Meigen).

REFERENCES: Metcalf, 1917b (biol., immature stages of *divisa*); Fluke, 1931 (biol., immature stages), 1933 (partial key, as *emarginatus* group), 1935 (partial key, combined with *Stenosyrphus* spp.); Goffe, 1944e (syn.).

divisa (Williston), 1882a: 311 (*Xanthogramma*).—Wash. Territory; Y.T. to Que. and Maine, s. to Wash., Colo., and Va.
  disjunctus Williston, 1882a: 314 (*Syrphus*; preocc. Macquart, 1842).—Wash. Territory.
  disjectus Williston, 1887c: 73 (*Syrphus*; n. name for *disjunctus* Williston).—Wash. Territory.
  fragila Fluke, 1922: 237 (*Xanthogramma*).—Wis.
emarginata (Say), 1823: 91 (1859b: 79) (*Scaeva*).—Fla.; Man. to Que., s. to Nebr., Tex., and Fla.
  aenea Jones, 1907b: 93 (*Xanthogramma*).—Nebr. **N. syn.**
felix (Osten Sacken), 1875d: 67 (*Xanthogramma*).—Ill., N.Y., Pa.; Minn. to Maine, s. to Nebr. and Va.
  infuscatus Fluke, 1931: 297 (*Syrphus*).—Wis.
grossulariae (Meigen), 1822: 306 (*Syrphus*).—Europe; Alaska to Que., s. to Calif., Colo., and S.C., Eurasia.
  lesueurii Macquart, 1842: 152 (1842: 92) (*Syrphus*).—Pa.
  conjungens Walker, 1852: 242.—U.S.
  melanis Curran, 1922f: 96 (*Syrphus*; as var.).—Ont. **N. syn.**
hunteri (Curran), 1925k: 171 (*Stenosyrphus*).—Man.; Alaska, Calif., Wis., Maine, Pa.
invigora (Curran), 1921c: 171 (*Syrphus*).—Ont.; Man. to Que. and Mass., s. to Miss. and Ga.
metcalfi (Fluke), 1933: 119 (*Metasyrphus*).—N.C.; Wis.

nitidicollis (Meigen), 1822: 308 (*Syrphus*).—Europe; Alaska to N.B., s. to Calif. and S.C., Europe.
    protritus Osten Sacken, 1877: 328 (*Syrphus*).—Calif.
submarginalis (Curran), 1925k: 101 (*Stenosyrphus*).—Ont.; Alaska to Maine, s. to Calif. and Pa.
terminalis (Curran), 1925k: 98 (*Stenosyrphus*).—Ont.; Maine.
weborgi (Fluke), 1931: 299 (*Syrphus*).—Wis.; Mich., N.Y., Pa.
xanthostoma (Williston), 1887c: 86 (*Syrphus*).—Pa.; Alaska to N.S., s. to Idaho, Kans., and N.C.

Genus ISCHYROSYRPHUS Bigot

Ischyrosyrphus Bigot, 1882a: lxviii. Type-species, *Musca glaucia* Linnaeus (Verrall, 1901: 321).

velutinus (Williston), 1882a: 314 (*Syrphus*). —Oreg.; Alaska to Idaho and Calif.
    tricolor Bigot, 1884a: 73.—Calif.
xylotoides (Johnson), 1916b: 80 (*Syrphus*).—Mass.; Vt., N.H., Conn., Va.

Genus MELANGYNA Verrall

Melangyna Verrall, 1901: 313. Type-species, *Melanostoma quadrimaculatum* Verrall (mon.).
Stenosyrphus Matsumura and Adachi, 1917b: 14. Type-species, *Scaeva lasiophthalma* Zetterstedt (orig. des.). **N. syn.** J.R. Vockeroth *in litt.*
Phalacrodira Enderlein, 1938: 205. Type-species, *Scaeva tarsata* Zetterstedt (orig. des.). **N. syn.**, J. R. Vockeroth *in litt.*
Petersina Enderlein, 1938: 205. Type-species, *lanata* Enderlein (orig. des.)=*groenlandica* (Nielsen). **N. syn.**, J. R. Vockeroth *in litt.*
Epistrophe, subg. Meligramma Frey, 1946a: 165. Type-species, *Scaeva guttata* Fallén (orig. des.). **N. syn.**, J. R. Vockeroth *in litt.*

    REFERENCES: Metcalf, 1917b (biol., immature stages of *triangulifer*, as *oronoensis*); Curran, 1925k (key); Fluke, 1935 (partial key, in *Epistrophe*); Goffe, 1944e (syn.).

abrupta (Curran), 1924p: 80 (*Epistrophe*).—N.H. **N. comb.**
albipunctata (Curran), 1925k: 104 (*Stenosyrphus*).—B.C., Wash., Idaho; Alaska to Nfld., s. to Wash., Colo., and Mich. **N. comb.**

arctica (Zetterstedt), 1838: 604 (*Scaeva*).—Europe; Alaska to Labr., s. to B.C. and Colo. **N. comb.**
  glacialis Johnson, 1898: 18 (*Melanostoma*).—Alaska.
bimaculata (Lovett) *in* Cole and Lovett, 1919: 244 (*Syrphus*).—Oreg.; B.C., Wash. **N. comb.**
bulbosa (Fluke), 1954: 8 (*Stenosyrphus*).—N.W.T. **N. comb.**
cherokeenensis (Jones), 1917: 219 (*Melanostoma*).—Colo.; Alaska to Idaho, s. to Oreg. and Colo., also Ont. **N. comb.**
  remotus Curran, 1925k: 108 (*Stenosyrphus*).—Oreg.
cincta (Fallén), 1817a: 45 (*Scaeva*).—Europe; Alaska, Alta., Oreg., Wis. to Que., s. to N.C., Europe. **N. comb.**
columbiae (Curran), 1925k: 110 (*Stenosyrphus*).—B.C.; Alaska to Wash. and Alta. **N. comb.**
  flavosignatus Hull, 1930a: 139 (*Syrphus*).—B.C.
compositarum (Verrall), 1873: 254 (*Syrphus*).—Europe; Alaska to N. Mex. in Rocky Mts., also N.H., Nfld., Europe. **N. comb.**
conjuncta (Osburn), 1908: 7 (*Syrphus*).—B.C.; B.C. to Oreg. and Alta. **N. comb.**
  melanderi Curran, 1925k: 103 (*Stenosyrphus*).—Wash.
coquilletti Sedman (for *gracilis* Coquillett).—Alaska. **N. name.**
  gracilis Coquillett, 1900h: 432 (1904: 46) (*Syrphus;* preocc. Meigen, 1822).—Alaska.
currani (Fluke), 1935: 29 (*Epistrophe*).—Alaska; B.C.
diversipunctata (Curran), 1925k: 106 (*Stenosyrphus*).—Ont.; Wis., N.H., Que., Maine. **N. comb.**
fisherii (Walton), 1911: 319 (*Syrphus*).—Pa.; Wis. to Ont. and Maine, s. to N.C. **N. comb.**
garretti (Curran), 1925k: 109 (*Stenosyrphus*).—B.C.; Alaska to Man. and Calif. **N. comb.**
geniculata (Macquart), 1842: 161 (1842: 101) (*Syrphus*).—Nfld.; Alaska to Labr., s. to B.C., Colo., and N.Y. **N. comb.**
genualis (Williston), 1887c: 86 (*Syrphus*).—N.H.; Y.T. to N.S., s. to B.C., N. Mex., and Conn. **N. comb.**
groenlandica (Nielsen), 1910: 61 (*Catabomba*).—Greenland. **N. comb.**, J. R. Vockeroth *in litt.*
  nigropilosa Curran, 1927b: 12 (*Epistrophe*).—Greenland. **N. syn.**, J. R. Vockeroth *in litt.*
  monachus Hull, 1930a: 140 (*Syrphus*).—Greenland. **N. syn.**, J. R. Vockeroth *in litt.*
  lanata Enderlein, 1938: 206 (*Petersina*).—Greenland. **N. syn.**, J. R. Vockeroth *in litt.*
  evanescens Enderlein, 1938: 207 (*Petersina;* as *lanata* var.).—Greenland. **N. syn.**, J. R. Vockeroth *in litt.*

extrema Enderlein, 1938: 207 (*Petersina;* as *lanata* var.).—Greenland. **N. syn.**, J. R. Vockeroth *in litt.*
flavifacies Enderlein, 1938: 207 (*Petersina;* as *lanata* var.).—Greenland. **N. syn.**, J. R. Vockeroth *in litt.*
violaceiventris Enderlein, 1938: 207 (*Petersina;* as *lanata* var.).—Greenland. **N. syn.**, J. R. Vockeroth *in litt.*

guttata (Fallén), 1817a: 44 (*Scaeva*).—Europe; Alaska to Ariz. and N. Mex. **N. comb.**
  habilis Snow, 1895d: 238 (*Xanthogramma*).—N. Mex.
imperialis (Curran), 1925k: 100 (*Stenosyrphus*).—Que.; Alaska to Que., s. to Wash. and Idaho. **N. comb.**
insolita (Osburn), 1908: 5 (*Syrphus*).—B.C.; Alaska to Labr., s. to Oreg. and Mont. **N. comb.**
johnsoni (Curran), 1924p: 79 (*Syrphus*).—N.H.; B.C. to N.S., s. to Colo. and N.H. **N. comb.**
lineola (Zetterstedt), 1843: 714 (*Scaeva*).—Europe; Alaska to Que., s. to Oreg., Colo., and N. Mex., Europe. **N. comb.**
macularis (Zetterstedt), 1843: 730 (*Scaeva*).—Europe: Alaska to Oreg. and Alta., Europe. **N. comb.**
maculifrons (Bigot), 1884a: 89 (*Syrphus*).—Oreg. Unrecognized, **N. comb.**
mallochi (Curran), 1923f: 74 (*Syrphus*; n. name for *interruptus* Malloch).—Alaska; N.W.T., Wash., Oreg., Que. **N. comb.**
  interruptus Malloch, 1919f: 55 (*Syrphus*; as *sodalis* var.; preocc. Philippi, 1865).—Alaska.
mediaconstricta (Fluke), 1930: 135 (*Epistrophe*).—Colo. **N. comb.**
mentalis (Williston), 1887c: 72 (*Syrphus*).—Wash. Territory; Alaska to N.B., s. to Oreg., Colo., and Conn. **N. comb.**
nigrifacies (Curran), 1923d: 62 (*Stenosyrphus*).—Alta.; Alaska to Alta. and Oreg. **N. comb.**
nudifrons (Curran), 1925k: 104 (*Stenosyrphus*).—N.Y.; Wis., Labr. **N. comb.**
pullula (Snow), 1895d: 237 (*Syrphus*).—N. Mex.; Alaska to Alta., s. to N. Mex., also Maine. **N. comb.**
quinquelimbata (Bigot), 1884a: 91 (*Syrphus*).—Calif.; B.C. to Idaho, s. to Calif. and Colo. **N. comb.**
rectoides (Curran), 1921b: 159 (*Syrphus*).—B.C.; Alaska to Labr., s. to Calif. and Colo. **N. comb.**
semiinterrupta (Fluke), 1935: 21 (*Epistrophe*).—Ont.; Ont. to Labr., s. to N.C. **N. comb.**
sexquadrata (Walker), 1849: 586 (*Syrphus*).—Ont., N.S.; Alta. to Nfld. **N. comb.**

sodalis (Williston), 1887c: 74 *(Syrphus).*—Colo.; Alaska to Que., s. to Oreg. and Colo. **N. comb.**

subfasciata (Curran), 1925k: 111 *(Stenosyrphus).*—B.C., Alta., Wash., Idaho; Oreg. **N. comb.**

tarsata (Zetterstedt), 1838: 601 *(Scaeva).*—Europe; Alaska to Greenland, s. to Wash., Colo., and N.H., n. Europe.

    adolescens Walker, 1849: 584 *(Syrphus).*—Ont., N.S.

    dryadis Holmgren, 1869: 26 *(Scaeva).*—Spitzbergen.

    contumax Osten Sacken, 1875c: 147 *(Syrphus).*—N.H.

    bryantii Johnson, 1898: 17 *(Syrphus).*—Alaska.

tenuis (Osburn), 1908: 8 *(Xanthogramma).*—B.C.; Alaska to Que., s. to B.C., Colo., and N.J. **N. comb.**

triangulifera (Zetterstedt), 1843: 737 *(Scaeva).*—Europe; Minn. to N.S. and Mass., also Y.T., Alta., Colo., Europe.

    oronoensis Metcalf, 1917b: 162 *(Syrphus).*—Maine.

umbellatarum (Fabricius), 1794: 307 *(Syrphus).*—Europe; Alaska to Nfld., s. to Ariz. and N.C., Europe. **N. comb.**

vittifacies (Curran), 1923b: 66 *(Stenosyrphus).*—N.H.; Pa., Norway. **N. comb.**

Genus MELISCAEVA Frey

Epistrophe, subg. **Meliscaeva** Frey, 1946a: 164. Type-species, *Scaeva cintella* Zetterstedt (orig. des.).

Episyrphus, authors, not Matsumura.

cinctella (Zetterstedt), 1843: 742 *(Scaeva).*—Europe; Alaska to Labr., s. to Calif., Colo., and N.C., Europe.

    diversipes Macquart, 1850: 459 (1850: 155) *(Syrphus).*—Nfld.

diversifasciata (Knab), 1914b: 151 *(Syrphus).*—Calif., Ariz.; B.C. to Utah, s. to Calif. and Ariz., Mexico, Chile.

    rubripleuralis Curran, 1921c: 172 *(Syrphus).*—Calif.

Genus MERCURYMYIA Fluke

Stenosyrphus, subg. **Mercurymyia** Fluke, 1950: 140. Type-species, *Syrphus caldus* Walker (orig. des.).

jactator (Loew), 1861c: 40 *(Syrphus).*—Cuba; Fla. **N. comb.**, J. R. Vockeroth *in litt.*

Genus ALLOGRAPTA Osten Sacken

Allograpta Osten Sacken, 1875d: 49. Type-species, *Scaeva obliqua* Say (mon.).

REFERENCES: Curran, 1932b (key); Fluke, 1942 (key).

cubana Curran, 1932b: 3.—Cuba; Fla.
exotica (Wiedemann), 1830: 136 (*Syrphus*).—Brazil; Oreg. to Nebr.,
s. to Calif. and S.C., Neotropical, Hawaii.
fracta Osten Sacken, 1877: 331.—Calif.
micrura (Osten Sacken), 1877: 330 (*Sphaerophoria*).—Calif.; Wash.,
Tex., Mexico, Ecuador.
picticauda Bigot, 1884a: 102 (*Sphaerophoria*).—Mexico.
obliqua (Say), 1823: 89 (1859b: 78) (*Scaeva*).—U.S.; Wash. to Que.,
s. to Calif. and Fla., also Hawaii, Bermuda, Neotropical.
securiferus Macquart, 1842: 160 (1842: 100) (*Syrphus*).—N.
Amer.
bacchides Walker, 1849: 594 (*Sphaerophoria*).—Fla.
dimensus Walker, 1852: 235 (*Syrphus*).—U.S.
signatus Wulp, 1867: 144 (*Syrphus*).—Wis.
transversa (Hull), 1943c: 32 (*Sphaerophoria*).—Oreg.; Calif., Ariz.,
Tex., Mexico.
venusta Curran, 1927a: 5.—Virgin Is.; Fla.

Genus DOROS Meigen

Doros Meigen, 1803: 274. Type-species, *Syrphus conopseus* Fabricius
(Westwood, 1840: 136).

aequalis Loew, 1863b: 318 (Cent. 4, no. 84).—Pa.; Wis. to Que., s.
to N.C.

Genus XANTHOGRAMMA Schiner

Xanthogramma Schiner, 1860c: 215; 1861a: 318. Type-species, *Syrphus ornatus* Meigen (Williston, 1887c: 91)=*pedissequum* (Harris).
Philhelius Coquillett, 1910d: 378. Type-species, *Musca citrofasciata*
De Geer (orig. des.). Coquillett attributed the name
Philhelius to Stephens, 1841: 201, where it appeared
only in a list of species in the vague combination "*Philhelius ornatus.*" Neither was *Philhelius* validated in
Verrall, 1901: 448, who merely mentioned it in a discussion of the synonymy of the genus *Xanthogramma*.
flavipes (Loew), 1863b: 318 (Cent. 4, no. 83) (*Doros*).—Pa.; Minn to
Que., s. to Nebr. and N.C.

Genus SPHAEROPHORIA Lepeletier and Serville

Sphaerophoria Lepeletier and Serville *in* Latreille et al., 1828: 513.
Type-species, *Musca scripta* Linnaeus (Rondani, 1844b:
458).

Melithreptus Loew, 1840: 577 (unjustified n. name for *Sphaerophoria* Lepeletier and Serville). Type-species, *Musca scripta* Linnaeus (aut.).

REFERENCE: Curran, 1930j (partial key, e. N. Amer.).

cleoae Metcalf, 1917a: 210.—Maine.
cranbrookensis Curran, 1921c: 173.—B.C.
cylindrica (Say), 1824b: pl. 11 (1859a: 22) (*Syrphus*).—Pa.; B.C. to N.B., s. to Calif. and Fla., Japan.
    contigua Macquart, 1847: 78 (1847: 62).—Pa.
dubia Bigot, 1884a: 101.—Calif. Unrecognized.
guttulata Hull, 1942c: 20.—Idaho; Calif., Utah, Mexico.
infumata (Thomson), 1869: 501 (*Syrphus*).—Calif. Unrecognized.
interrupta Jones, 1917: 225.—Colo.; Utah, Wyo.
melanosa Williston, 1887c: 106.—Calif.; B.C. to Minn., s. to Calif. and N. Mex.
menthastri (Linnaeus), 1758: 594 (*Musca*).—Europe; Alaska to N.S., s. to Calif., N. Mex., and Va., Eurasia.
    hieroglyphicus Meigen, 1822: 327 (*Syrphus*).—Europe.
nigratarsi Fluke, 1930: 143.—Colo.; Alaska to B.C., Alta. to N. Mex.
nigricauda Metcalf, 1913: 88.—Ohio.
novaeangliae Johnson, 1916b: 76.—Maine; Alaska, B.C., Wis. to Que., Maine, and Mass.
pyrrhina Bigot, 1884a: 101.— Calif. Unrecognized.
robusta Curran, 1930j: 62.—Maine; Alaska to N.S., s. to Calif., Colo., and Fla.
    scripta, authors, not Linnaeus.
strigata Staeger, 1845b: 362.—Greenland; N.W.T., B.C., Alta.
sulphuripes (Thomson), 1869: 500 (*Syrphus*).—Calif.; Alaska to Calif., N. Mex., and Nebr., also Hawaii.

Genus MESOGRAPTA Loew

Mesogramma Loew, 1866a: 157 (Cent. 6, no. 47) (preocc. Stephens, 1850). Type-species, *parvula* Loew (Williston, 1887c: 98).

Mesograpta Loew, 1872a: 114 (Cent. 10, appendix) (n. name for *Mesogramma* Loew). Type-species, *Mesogramma parvula* Loew (aut.).

Although it is believed that most species of *Mesograpta* have the primitive aphidophagous habit of the Syrphinae, the only species whose feeding habits are definitely known, *polita* (Say), is phytophagous, feeding on corn pollen.

REFERENCES: Curran, 1930c, 1934e (keys); Hull, 1943a (rev.); Fluke, 1953b (syn.).

arethusa (Hull), 1945a: 185 (*Mesogramma*).—Fla.
basilaris (Wiedemann), 1830: 143 (*Syrphus*).—Brazil; Tex., Neotropical.
boscii (Macquart), 1842: 160 (1842: 100) (*Syrphus*).—Carolina; Tex. to Tenn. and N.C., s. to Fla., Antilles, Mexico.
    gurges Walker, 1852: 236 (*Syrphus*).—U.S.
coalescens (Walker), 1852: 237 (*Syrphus*).—U.S. Unrecognized.
corbis (Walker), 1852: 236 (*Syrphus*).—U.S. Unrecognized.
floralis (Fabricius), 1798: 563 (*Syrphus*).—French Guiana; Fla., Neotropical.
    duplicata, authors, not Wiedemann.
marginata (Say), 1823: 92 (1859b: 80) (*Scaeva*).—U.S.; B.C. to Que., s. to Calif., Cent. Amer., and Fla.
    quintius Walker, 1852: 239 (*Syrphus*).—N. Amer.
    limbiventris Thomson, 1869: 495 (*Syrphus*).—Calif.
parvula (Loew), 1866a: 157 (Cent. 6, no. 47) (*Mesogramma*).—Fla.; Wis., Tex.
planiventris (Loew), 1866a: 158 (Cent. 6, no. 49) (*Mesogramma*).—Fla.; N.C., Ga.
polita (Say), 1823: 88 (1859b: 77) (*Scaeva*).—U.S.; Minn. to Que., s. to Ariz. and Fla., Neotropical.
    cingulatulus Macquart, 1850: 459 (1850: 155) (*Syrphus*).—Fla.
    cingulatus, error.
    hecticus Jaennicke, 1867: 398 (1868: 90) (*Syrphus*).—Ill.
slossonae (Curran), 1930c: 8 (*Mesogramma*).—Fla; Ecuador.
subannulata (Loew), 1866a: 157 (Cent. 6, no. 48) (*Mesogramma*).—Cuba; Fla., Antilles, Mexico and Cent. Amer.
teligera Fluke, 1953a: 125.—Ariz.
triradiata (Hull), 1942e: 48 (*Mesogramma*).—Panama; Tex.

Genus TOXOMERUS Macquart

Toxomerus Macquart, 1855: 112 (1855: 92). Type-species, *notatus* Macquart (orig. des.) = *geminatus* (Say).

The larvae are thought to be aphidophagous but no detailed life history studies have been made.

geminatus (Say), 1823: 92 (1859b: 80) (*Scaeva*).—U.S.; Minn. to Que., s. to Colo., Tex., and Fla.
    privernus Walker, 1852: 225 (*Eumerus*).—U.S.
    interrogans Walker, 1852: 238 (*Syrphus*).—N. Amer.
    notatus Macquart, 1855: 113 (1855: 93).—Md.
jussiaeae Vigé, 1939: 66.—La.; Mo., Ill., Tenn., Miss.

occidentalis Curran, 1922a: 258.—B.C.; B.C. to Colo., s. to Calif. and Tex.

Tribe BACCHINI

Genus BACCHA Fabricius

Baccha Fabricius, 1805: 199. Type-species, *Syrphus elongatus* Fabricius (Curtis, 1839: pl. 737).
Bacchina Williston, 1896a: 86. Type-species, *Syrphus elongatus* Fabricius (present des.). **N. syn.**

The larvae of *Baccha* are effective predators of aphids, scale insects, and mealy bugs, including many of economic importance.

REFERENCES: Osten Sacken, 1862c (biol.); Campbell and Davidson, 1924 (biol.); Curran, 1930a, 1941 (keys); Heiss, 1938 (immature stages); Hull, 1943b, 1949a (keys).

Subgenus BACCHA Fabricius

calypso Hull, 1944b: 60.—Fla.
costata Say, 1829: 161 (1859b: 357).—Ind.; Ont. to N.H., s. to La. and Fla.
    costalis Wiedemann, 1830: 97.—Unknown.
    tarchetius Walker, 1849: 549.—Ga.
cubana Hull, 1943b: 62, key (also 1944e: 30, descr.).—Cuba; Fla.
elongata (Fabricius), 1775: 768 (*Syrphus*).—Europe; Alaska to Que., s. to Calif., Ohio, and Va., Europe.
    obscuricornis Loew, 1863a: 15 (Cent. 3, no. 26).—Alaska. **N. syn.**
    cognata Loew, 1863a: 15 (Cent. 3, no. 27).—Wis. **N. syn.**
    angusta Osten Sacken, 1877: 332.—Calif. **N. syn.**
    tricincta Bigot, 1883a: 333.—Wash. Territory. **N. syn.**
lineata Macquart, 1846: 267 (1846: 139).—Yucatan or Tex.; Tex., Fla., Neotropical.
    notata Loew, 1866b: 37 (Cent. 7, no. 65).—Cuba. **N. syn.**
    tropicalis Townsend, 1897c: 172.—Tex.
parvicornis Loew, 1861c: 41.—Cuba; Fla., P.R.

Subgenus DIOPROSOPA Hull

Baccha, subg. **Dioprosopa** Hull, 1949a: 99 (also 1949b: 296). Type-species, *Syrphus clavatus* Fabricius (mon.).

clavata (Fabricius), 1794: 298 (*Syrphus*).—West Indies; Calif. to Nebr., Wis., and N.J., s. to Fla., Neotropical.

babista Walker, 1849: 549.—Ga.
quadrimaculata Ashmead, 1880: 70 (?*Conops*).—Fla.
bacchoides Bigot, 1883a: 326 (*Spazigaster*).—Rocky Mts.

Subgenus OCYPTAMUS Macquart

Ocyptamus Macquart, 1834a: 554 (as genus). Type-species, *fascipennis* Macquart (Coquillett, 1910b: 577) =*fuscipennis* Say.

fascipennis Wiedeman, 1830: 96.—Unknown; Man. to Que., s. to Tex. and Fla.
aurinota Walker, 1849: 548.—Mass. (as U.S.).
funebris (Macquart), 1834a: 554 (*Ocyptamus*).—Brazil; Tex., Neotropical.
fuscipennis Say, 1823: 100 (1859b: 86).—Pa.; Colo. to Wis. and Que., s. to Tex. and Fla., Cent. Amer., Antilles.
fascipennis Macquart, 1834a: 554 (*Ocyptamus;* preocc. Wiedemann, 1830).—Pa.
amissas Walker, 1849: 589 (*Syrphus*).—Ga.
peas Walker, 1849: 590 (*Syrphus*).—Unknown.
radaca Walker, 1849: 590 (*Syrphus*).—Fla.
lugens Loew, 1863a: 14 (Cent. 3, no. 24).—Wis.
longiventris Loew, 1866b: 38 (Cent. 7, no. 66) (*Ocyptamus*).—D.C.
fenestrata Hull, 1949a: 280, fig. 381 (as var.).—Unknown.
cylindrica, authors, not Fabricius.
gastrostacta (Wiedemann), 1830: 123 (*Syrphus*).—Brazil; Tex., Fla., Neotropical.
torva Williston, 1887c: 124.—Tex.
latiuscula (Loew), 1866b: 39 (Cent. 7, no. 68) (*Ocyptamus*).—Cuba; Fla., West Indies.
lemur Osten Sacken, 1877: 331.—Calif., N. Mex., Wyo.; B.C. to Wis., s. to Calif. and Tex.
loewi Sedman (for *scutellata* Loew).—Cuba; Fla., Neotropical. **N. name.**
scutellata Loew, 1866b: 39 (Cent. 7, no. 69) (*Ocyptamus;* preocc. Meigen, 1822).—Cuba.

Subgenus ORPHNABACCHA Hull

Orphnabaccha Hull, 1949a: 93 (as genus). Type-species, *Baccha coerula* Williston (orig. des.).

coerula Williston, 1891: 38.—Mexico (Guerrero); Tex.

Subgenus XESTOPROSOPA Hull

Xestoprosopa Hull, 1949a: 94 (as genus). Type-species, *Baccha delicatula* Hull (orig. des.).

marmorata Bigot, 1883a: 333.—Mexico; Ariz.

Subgenus MIMOCALLA Hull

Baccha, subg. **Mimocalla** Hull, 1943b: 46. Type-species, *capitata* Loew (orig. des.).

nepenthe Hull, 1943d: 40.—Fla.

Genus SALPINGOGASTER Schiner

Salpingogaster Schiner, 1868: 344. Type-species, *pygophora* Schiner (orig. des.).

REFERENCE: Curran, 1932b, 1941 (keys).

punctifrons Curran, 1929f: 493.—Cuba; Fla.
texana Curran, 1932b: 6.—Tex.

Tribe MELANOSTOMATINI

The larvae of the Melanostomatini with known habits are transitional forms between the saprophagous and purely entomophagous types. Larvae of *Melanostoma* and *Platycheirus* have been observed feeding on aphids, small caterpillars, and other insects, as well as on decomposing plant material.

REFERENCES: Curran, 1930j (keys); Fluke, 1958 (rev. by male genitalia).

Genus XANTHANDRUS Verrall

Xanthandrus Verrall, 1901: 316. Type-species, *Musca comtus* Harris (Coquillett, 1910b: 620).

mexicanus Curran, 1930g: 9.—Mexico (Yucatan); Tex., Costa Rica.

Genus MELANOSTOMA Schiner

Melanostoma Schiner, 1860c: 213. Type-species, *Musca mellina* Linnaeus (orig. des.).
Pachysphyria Enderlein, 1938: 196. Type-species, *Scaeva ambigua* Fallén (orig. des.).

REFERENCES: Davidson, 1922 (biol.); Curran, 1930j (keys).

ambiguum (Fallén), 1817a: 47 (*Scaeva*).—Europe; B.C., Oreg.

SYRPHIDAE 575

angustatum Williston, 1887c: 50.—Wash. Territory; Alaska to Maine, s. to Oreg. and N.C.
concinnum Snow, 1895d: 229.—N. Mex., Colo.; Oreg. to Alta., s. to N. Mex.
dubium (Zetterstedt), 1838: 609 (*Scaeva*).—Europe; Utah, Colo., Mexico.
fallax Curran, 1923n: 271.—Alta.; B.C., Wis., Mich., Que.
latum Curran, 1922c: 276.—Y.T.; Y.T. and Alta., s. to Calif. and Colo.
lundbecki Collin, 1931: 68.—Greenland.
    ambiguum, authors, not Fallén.
melanderi Curran, 1930j: 64.—Wash. (unpublished); B.C., Alta.
mellinum (Linnaeus), 1758: 594 (*Musca*).—Europe; Alaska to Que., s. to Wash. and Va., Europe.
montivagum Johnson, 1916b: 78.—N.H.; Ont. to Nfld., s. to N.Y.
pachytarse Bigot, 1884a: 80 (?*Melanostoma*).—Calif. Unrecognized.
pallitarse Curran, 1926b: 83.—Wis.; Man. to Ont., s. to Tenn.
parvum (Williston), 1882a: 307 (*Cheilosia*).—Oreg.; Colo.
    ochripes Bigot, 1883a: 555 (?*Melanogaster*).—Oreg.
pictipes Bigot, 1884a: 78 (?*Melanostoma*).—Calif.; B.C. to N.S., s. to Calif. and Tenn.
    pruinosa Bigot, 1884a: 79 (?*Melanostoma*).—Calif. **N. syn.**
    mellinum, authors, not Linnaeus.
rufipes (Bigot), 1883a: 554 (?*Melanogaster*; preocc. Williston, 1882).—N. Amer. Unrecognized, and therefore not renamed here because of possible synonymy with older species.
willistoni Sedman (for *rufipes* Williston).—Wash. Territory; B.C. and Alta., s. to Calif. and Colo. **N. name.**
    rufipes Williston, 1882a: 306 (*Cheilosia*; preocc. Macquart, 1828).—Wash. Territory.

Genus CARPOSCALIS Enderlein

Carposcalis Enderlein, 1938: 199. Type-species, *Syrphus stegnus* Say (orig. des.).

    REFERENCE: Curran, 1930j (key, as *Melanostoma*, in part).

agens (Curran), 1931f: 253 (*Melanostoma*).—Man.; Sask., Colo., Que.
atra (Curran), 1925k: 114 (*Melanostoma*).—Colo. **N. comb.**
carinata (Curran), 1927b: 11 (*Melanostoma*).—Greenland; Alaska, Wash., Colo.
chaetopoda (Davidson), 1922: 35 (*Melanostoma*).—Calif.; Colo., Mexico.

chilosia (Curran), 1922c: 275 (*Melanostoma*).—Alta.; Wash. **N. comb.**
coerulescens (Williston), 1887c: 49 (*Melanostoma*).—Colo.; B.C. to Que., s. to Oreg., N. Mex., and Tenn.
confusa (Curran), 1925k: 112 (*Melanostoma*).—Ont.; Ont. to N.S., s. to N.J.
johnsoni (Jones), 1917: 220 (*Melanostoma*).—Colo. **N. comb.**
kelloggi (Snow), 1895d: 230 (*Melanostoma*).—Colo.; B.C. **N. comb.**
luteipennis (Curran), 1925k: 114 (*Melanostoma*).—Wash.; Utah, Colo. **N. comb.**
monticola (Jones), 1917: 220 (*Melanostoma*).—Colo.
nitidiventris (Curran), 1931f: 252 (*Melanostoma*).—Ont. **N. comb.**
obscura (Say), 1824b: pl. 11 (1859a: 23) (*Syrphus*).—Pa., Va.; Wash. to Labr., s. to Calif. and N.C.
ontario (Davidson), 1922: 37 (*Melanostoma*).—Ont. **N. comb.**
rostrata (Bigot), 1884a: 80 (?*Melanostoma*).—Calif.; Wash., N. Mex., Colo. **N. comb.**
squamulae (Curran), 1922c: 275 (*Melanostoma*).—B.C.; Alaska, Wash., Oreg.
stegna (Say), 1829: 163 (1859b: 358) (*Syrphus*).—Mexico; B.C. to Mont., s. to Calif., Tex., and Mexico.
  tigrinum Osten Sacken, 1877: 323 (*Melanostoma*).—Calif.
trichopus (Thomson), 1869: 502 (*Syrphus*).—Calif.; Alaska and N.W.T., s. to Calif. and Colo.

Genus PLATYCHEIRUS Lepeletier and Serville

Platycheirus Lepeletier and Serville *in* Latreille et al., 1828: 513. Type-species, *Syrphus scutatus* Meigen (Westwood, 1840: 137).
Platychirus, emend.

  REFERENCES: Curran, 1927b (key, males), 1930j (key, females).

aeratus Coquillett, 1900h: 430 (1904: 44).—Alaska; Oreg.
albimanus (Fabricius), 1781: 434 (*Syrphus*).—Europe; Alaska to Greenland, s. to Calif., Colo., and N.H., Europe.
  ciliatus Bigot, 1884a: 74.—Calif.
amplus Curran, 1927b: 4.—Ont.; Alaska, Utah, Labr., Nfld.
angustatus (Zetterstedt), 1843: 762 (*Scaeva*).—Europe; Alaska to Que. and Maine, s. to Wash., Wis., and Md., Europe.
bigelowi Curran, 1927b: 5.—Ont.; Alaska.
chirosphena Hull, 1944c: 76.—B.C.
clypeatus (Meigen), 1822: 335 (*Syrphus*).—Europe; Alaska to Ont. and Mass., s. to Calif., Colo., and D.C., Eurasia.

discimanus Loew, 1871: 227.—Europe; B.C. to Que., s. to Oreg. and Maine.
erraticus Curran, 1927b: 7.—Ont.; B.C. to Labr., s. to Calif. and Fla. *hyperboreus*, authors, not Staeger.
felix Curran, 1931e: 94.—Que.
femineus Curran, 1931f: 251.—Que.
flabellus Hull, 1944c: 75.—Wash.
frontosus Lovett *in* Cole and Lovett, 1919: 247.—Oreg.
groenlandicus Curran, 1927b: 10.—Greenland.
hyperboreus (Staeger), 1845b: 362 (*Syrphus*).—Greenland; Labr.
immarginatus (Zetterstedt), 1849: 3149 (*Scaeva*).—Europe; Sask. to Que., s. to Utah and N.J., Europe.
inversus Ide, 1926: 156.—Que.; Alaska, B.C., Colo., Labr., N.S.
modestus Ide, 1926: 155.—Que.; Alaska, Alta., Wis. to N.B., s. to Mass.
naso (Walker), 1849: 587 (*Syrphus*).—Ont.
nodosus Curran, 1923n: 272.—Alta.; Alaska to Que.
normae Fluke, 1939: 366.—Wis.; Ont., Que., P.E.I.
occidentalis Curran, 1927b: 9.—Wyo.; B.C., Utah, Colo.
pacilus (Walker), 1852: 240 (*Syrphus*).—N. Amer. Unrecognized.
palmulosus Snow, 1895d: 231.—Colo.; N.Mex.
pauper Hull, 1944c: 77.—Colo.
peltatoides Curran, 1923n: 274.—B.C.; Alaska to Wash. and Colo.
peltatus (Meigen), 1822: 334 (*Syrphus*).—Europe; Alaska to N.S., s. to Oreg., Colo., and N.J., Europe.
perpallidus Verrall, 1901: 290.—England; Alaska to N.B., s. to Utah and Mass., Europe.
podagratus (Zetterstedt), 1838: 606 (*Scaeva*).—Europe; Alta., Colo., Ont.
quadratus (Say), 1823: 90 (1859b: 79) (*Scaeva*).—U.S.; Alaska to Labr., s. to Calif. and S.C.
  fuscanipennis Macquart, 1855: 115 (1855: 95) (*Syrphus*).—Md.
scamboides Curran, 1927b: 6.—Mich.; Wis. to Ont. and Va.
scambus (Staeger), 1843: 325 (*Syrphus*).—Europe; Alaska to Que., s. to Calif., Colo., and S.C., Europe.
  chaetopodus Williston, 1887c: 59.—Wash. Territory.
scutatus (Meigen), 1822: 333 (*Syrphus*).—Europe; Alaska to Ont. and Maine, s. to B.C., Colo., and Mich., Europe.
tarsalis (Schummel) *in* Gravenhorst, 1837: 84 (*Syrphus*).—Europe; Alaska, B.C., Colo.
tenebrosus Coquillett, 1900h: 428 (1904: 42).—Alaska; B.C.
thylax Hull, 1944c: 78.—Que.
varipes Curran, 1923b: 65.—Maine; Ont., Que., Labr.

Genus PYROPHAENA Schiner

Cheilosia Panzer, 1809c, 108: 14. Type-species, *Syrphus rosarum* Fabricius (mon.).
Pyrophaena Schiner, 1860c: 213. Type-species, *Syrphus rosarum* Fabricius (orig. des.).

To prevent confusion, the I.C.Z.N. should be asked to suppress *Cheilosia* Panzer to preserve usage of *Cheilosia* Meigen 1822 (which see) and to place *Pyrophaena* Schiner on the Official List of Generic Names.

apicauda Curran, 1925k: 115 (as *granditarsis* var.).—Calif.; Idaho, Utah.
    apicula (error) Curran, 1930j: 67.
digitalis Fluke, 1939: 367.—Colo.; Alta., Utah.
granditarsa (Forster), 1771: 99 (*Musca*).—Europe; Alaska to Que., s. to Oreg., Colo., and Pa., Europe.
    ocymi Fabricius, 1794: 309 (*Syrphus*).—Europe.
rosarum (Fabricius), 1787: 341 (*Syrphus*).—Europe; Alaska, Wis. to N.S., s. to N.J., Europe.
    var. **duplicata** Fluke, 1922: 228.—Wis.; Mich.

Tribe PARAGINI

Genus PARAGUS Latreille

Paragus Latreille, 1804: 194. Type-species, *Syrphus bicolor* Fabricius (mon.).

The larvae of *Paragus* are common predators of aphids, mostly of noneconomic species.

REFERENCES: Metcalf, 1911, Campbell and Davidson, 1924, Heiss, 1938 (biol., immature stages).

Subgenus PARAGUS Latreille

bicolor (Fabricius), 1794: 297 (*Syrphus*).—n. Africa; B.C. and Y.T. to Que., s. to Calif. and Va., Europe.
    angustifrons Loew, 1863b: 309 (Cent. 4, no. 64).—Va.
transatlanticus Walker, 1849: 544.—N.Y. Unrecognized.

Subgenus PANDASYOPTHALMUS Stuckenberg

Paragus, subg. **Pandasyopthalmus** Stuckenberg, 1954: 100. Type-species, *longiventris* Loew (orig. des.).
tibialis (Fallén), 1817a: 60 (*Pipiza*).—Europe; B.C. to N.S., s. to Calif., Cent. Amer., and Fla., Eurasia, Africa, Australia.
    dimidiatus Loew, 1863b: 308 (Cent. 4, no. 63).—D.C.
    auricaudatus Bigot, 1883a: 540.—Calif.

Tribe PIPIZINI

The larvae of Pipizini are aphidophagous, preferring woolly or root aphids with waxy secretions, and attacking them both above and below ground. The larvae of *Neocnemodon vitripennis* (Meigen) have been found feeding on coccids in Europe. Several European species of *Neocnemodon* have been liberated in North America as balsam woolly aphid predators.

REFERENCES: Metcalf, 1916b (biol., immature stages); Curran, 1921a (rev.); Heiss, 1938 (biol., immature stages); Collin, 1952b (generic classif.); Delucchi and Pschorn-Walcher, 1955 (tax., biol.).

Genus PIPIZA Fallén

Pipiza Fallén, 1810a: 11. Type-species, *Musca noctiluca* Linnaeus (Curtis, 1837a: pl. 669).

atrata Curran, 1922i: 283.—B.C.
crassipes Bigot, 1883a: 557.—N. Amer. Unrecognized.
davidsoni Curran, 1921a: 377.—Calif.; Wash., N. Mex.
distincta Curran, 1921a: 375.—Calif.
femoralis Loew, 1866a: 152 (Cent. 6, no. 38).—Ill.; B.C. to Que., s. to Oreg., Nebr., and Va.
    albipilosa Williston, 1887c: 28.—Pa.
    festiva, authors, not Meigen.
grandifemoralis Curran, 1921a: 383.—Calif.; Colo., N. Mex.
latifrons Curran, 1921a: 378.—Calif.
macrofemoralis Curran, 1921a: 383.—Oreg.; Alaska to Alta., s. to Calif.
    nigripilosa Williston of Fluke, 1949: 54.
nigripilosa Williston, 1887c: 28.—Pa.; Ont. to Que., s. to Ohio and N.C.
nigrotibiata Curran, 1924p: 81.—N.B.; N.Y., N.H., Mass.
oregona Lovett *in* Cole and Lovett, 1919: 246.—Oreg.; B.C. and Alta., s. to Calif. and Colo.
puella Williston, 1887c: 27.—N.H.; Colo., Ont.
quadrimaculata (Panzer), 1804: 19 (*Syrphus*).—Europe; Wash. and Alta. to Que. and Maine, s. to Colo. and N.C., Europe.
severnensis Curran, 1921a: 381.—Ont.
tricolor Curran, 1921a: 379.—Ont.; Alta., Que.
vanduzeei Curran, 1921a: 375.—Calif.; B.C.

Genus HERINGIA Rondani

Heringia Rondani, 1856: 53. Type-species, *Pipiza heringi* Zetterstedt (orig. des.).

Heryngia, error or emend.

californica (Davidson), 1917: 417 (*Pipiza*).—Calif.; B.C., Oreg., Utah.
canadensis Curran, 1921a: 354.—Ont.; Que.
comutata Curran, 1921a: 356.—Oreg.; B.C., Calif., Utah.
intensica Curran, 1921a: 355.—Ont.
salax (Loew), 1866a: 152 (Cent. 6, no. 39) (*Pipiza*).—Pa.; Alaska to Que., s. to Calif. and N.C.
    radicum Walsh and Riley, 1869a: 83, footnote (*Pipiza*).—Ill.
    pistica Williston, 1887c: 29 (*Pipiza*).—Conn.

Genus PARAPENIUM Collin

Parapenium Collin, 1952b: 85. Type-species, *Pipiza carbonaria* Zetterstedt (orig. des.).
Pipizella, authors, not Rondani.
Triglyphus, authors, not Loew.

REFERENCE: Curran, 1924a (rev.).

apisaon (Walker), 1849: 572 (*Chrysogaster*).—N.Y.; Alta. to Que., s. to Tex. and Va. **N. comb.**
    modestus Loew, 1863b: 308 (Cent. 4, no. 62) (*Triglyphus*).—N.Y. **N. comb.**
    nigribarba Loew, 1866a: 153 (Cent. 6, no. 40) (*Pipiza*).—N.Y. **N. comb.**
    pulchella Williston, 1887c: 29 (*Pipiza*).—Conn., Mass. **N. comb.**
australe (Johnson), 1907b: 77 (*Pipiza*).—Fla.; Tex. **N. comb.**
banksi (Curran), 1921a: 349 (*Pipizella;* as *pulchella* ssp.).—Va.; Kans. to Wis. and N.J., s. to Tex. and Fla. **N. comb.**
nigritarse (Curran), 1924a: 341 (*Pipizella*).—Idaho; Wyo., Colo. **N. comb.**
occidentale (Townsend), 1897a: 140 (*Pipiza*).—N. Mex.; Ariz., Utah, S. Dak. **N. comb.**
pubescens (Loew), 1863b: 307 (Cent. 4, no. 61) (*Triglyphus*).—Wis.; Wis. to Que. and Maine, s. to N.C. **N. comb.**
recedens (Walker), 1852: 228 (*Chrysogaster*).—U.S.; Alta. to Maine, s. to Calif., Colo., and N.Y. **N. comb.**
    fraudulenta Loew, 1866a: 153 (Cent. 6, no. 41) (*Pipiza*).—Ill. **N. comb.**
rufithoracicum (Curran), 1921a: 351 (*Pipizella*).—Calif.; Alta., Utah. **N. comb.**
simile (Curran), 1924a: 342 (*Pipizella*).—Mont.; Wis. **N. comb.**

Genus NEOCNEMODON Goffe

Cnemodon Egger, 1865: 573 (preocc. Schoenherr, 1823). Type-species, *latitarsis* Egger (Goffe, 1944c: 128).

Neocnemodon Goffe, 1944c: 128 (n. name for *Cnemodon* Egger). Type-species, *Cnemodon latitarsis* Egger (aut.).

albipleura (Curran), 1921a: 370 (*Cnemodon*).—Calif.
auripleura (Curran), 1921a: 369 (*Cnemodon*).—Oreg.; B.C. to Sask., s. to Calif. and Colo.
calcaratus (Loew), 1866a: 154 (Cent. 6, no. 42) (*Pipiza*).—N.Y.; B.C. to Que., s. to Kans. and Va.
carinatus (Curran), 1921a: 370 (*Cnemodon*).—Ont.; B.C. to Ont., s. to Wis.
cevelatus (Curran), 1921a: 362 (*Cnemodon*).—Ont.
corvallis (Curran), 1921a: 364 (*Cnemodon*).—Oreg.
coxalis (Curran), 1921a: 366 (*Cnemodon*).—Ont.; B.C. to N.S., s. to Wis. and Md.
elongatus (Curran), 1921a: 362 (*Cnemodon*).—Ont.; B.C. to Que., s. to Tex. and Md.
intermedius (Curran), 1921a: 360 (*Cnemodon*).—Ont.
latitarsis (Egger), 1865: 573 (*Cnemodon*).—Europe; ?N.B. Introduced.
longiseta (Curran), 1921a: 361 (*Cnemodon*).—Ont.; Que., N.Y.
lovetti (Curran), 1921a: 365 (*Cnemodon*).—Oreg.
myermus (Curran), 1921a: 371 (*Cnemodon*).—Ont.; Va.
nigricornis (Curran), 1922i: 284 (*Cnemodon*).—Alta.; N.W.T.
nudifrons (Curran), 1921a: 367 (*Cnemodon*).—Oreg.
ontarioensis (Curran), 1921a: 372 (*Cnemodon*).—Ont.; Wis.
pisticoides (Williston), 1887c: 29 (*Pipiza*).—N.H.; Alaska to Ont. and Maine, s. to N. Mex. and S.C.
placidus (Curran), 1921a: 364 (*Cnemodon*).—Oreg.
pubescens (Delucchi and Pschorn-Walcher), 1955: 504 (*Cnemodon*).— Germany; ?N.B. Introduced.
rita (Curran), 1921a: 366 (*Cnemodon*).—Oreg.; B.C. to Alta., s. to Calif. and Colo.
sinuosus (Curran), 1921a: 368 (*Cnemodon*).—Oreg.; Alta., Utah, Nebr.
squamulae (Curran), 1921a: 361 (*Cnemodon*).—Ont.; Que., Wis.
trochanteratus (Malloch), 1918o: 127 (*Cnemodon*).—Ill.
unicolor (Curran), 1921a: 359 (*Cnemodon*).—Ont.
venteris (Curran), 1921a: 371 (*Cnemodon*).—Ont.

Tribe CHRYSOTOXINI

Genus CHRYSOTOXUM Meigen

Antiopa Meigen, 1800: 32. Type-species, *Musca bicincta* Linnaeus (Coquillett, 1910b: 508). Suppressed by I.C.Z.N., 1963b: 339.

Chrysotoxum Meigen, 1803: 275. Type-species, *Musca bicincta* Linnaeus (Latreille, 1810: 443).

The biology of *Chrysotoxum* is very poorly known. The typically syrphine larvae have been taken from ant nests, debris in hollow trees, compost heaps, and under stones. Their structure indicates that they are probably carnivorous.

REFERENCES: Curran, 1924c (key); Shannon, 1926b (rev.); Curran, 1927g (syn.).

Subgenus PRIMOCHRYSOTOXUM Shannon

Chrysotoxum, subg. **Primochrysotoxum** Shannon, 1926b: 5. Type-species, *ypsilon* Williston (orig. des.).

chinook Shannon, 1926b: 6.—Wash.; Alaska to Mont., s. to Calif. and Colo.
> *villosulum* Bigot of Curran, 1924c: 40.

coloradense Greene, 1918: 70.—Colo.; B.C. to Atla., s. to Oreg. and Colo.

derivatum Walker, 1849: 542.—Ont.; Alta., Mich.

fasciolatum (De Geer), 1776: 124 (*Musca*).—Europe; Alaska to N.B., s. to Wash., Wis., and Mass.

flavifrons Macquart, 1842: 77 (1842: 17).—Nfld. Unrecognized.

integre Williston, 1887c: 16.—Ariz.; B.C. to Idaho, s. to Calif., Mexico, and Tex.
> *columbianum* Curran, 1927g: 206.—B.C. **N. syn.**, J. R. Vockeroth *in litt.*

laterale Loew, 1864a: 72 (Cent. 5, no. 42).—Nebr.; Alaska to Ont. and Maine, s. to Nev., Kans., and Ohio.

minor Curran, 1927g: 206.—Man.; B.C., Sask.

occidentale Curran, 1924c: 37.—B.C.; B.C. to Alta., s. to Calif. and Colo., also Wis.

perplexum Johnson, 1924: 99.—N.H.; Man. to N.S., s. to Colo. and N.C.

plumeum Johnson, 1924: 99.—N.J.; Man. to Que. and Mass., s. to Ga.

pubescens Loew, 1860b: 84.—Ill.; Alta. to Que., s. to Kans. and N.C.
> *currani* Wehr, 1922b: 127 (=p. 9).—Nebr.
> *cuneatum* Wehr, 1922b: 128 (=p. 10).—Nebr.
> *luteopilosum* Curran, 1924c: 36.—Man.

radiosum Shannon, 1926b: 10.—Ind.; B.C. to Conn., s. to N. Mex. and N.J.
pubescens Loew of Curran, 1924c: 39.
ventricosum Loew, 1864a: 72 (Cent. 5, no. 44).—Wash.; Alaska to Ont., s. to Ariz. and N. Mex.
villosulum Bigot, 1883a: 323.—Wash. Territory. Unrecognized.
willistoni Curran, 1924c: 59 (as *integre* var.).—N. Mex.
ypsilon Williston, 1887c: 14.—N. Mex.; B.C. to Alta., s. to Oreg. and N. Mex.

Subfamily MILESIINAE
Tribe CHEILOSIINI
Genus CHEILOSIA Meigen

Cheilosia Meigen, 1822: 296 (preocc. Panzer, 1809). Type-species, *Eristalis scutellatus* Fallén (as Fabricius; Rondani, 1856: 51). In order to preserve usage for this economically important genus of phytophagous syrphids, the I.C.Z.N. should be asked to suppress *Cheilosia* Panzer, and to place *Cheilosia* Meigen on the Official List of Generic Names. See *Pyrophaena*.
Chilosia, emend.
Chilosia, subg. *Chilomyia* Shannon, 1922c: 127. Type-species, *occidentalis* Williston (orig. des.).

The larvae of *Cheilosia* are phytophagous and have been found breeding in the stems and roots of a variety of plants, including fungi. The larvae of some American species known as "bark maggots" cause a blemish called "black check" in timber of Western hemlock and other coniferous trees.

REFERENCES: Burke, 1905 (biol., economics); Shannon, 1922c (rev.); Goffe, 1944d (syn.); Fluke and Hull, 1946 (rev., as *Chilomyia*), 1947 (rev., *Cartosyrphus*); Hull and Fluke, 1950 (rev.).

Subgenus CHEILOSIA Meigen

alaskensis Hunter, 1897: 124.—Alaska; B.C. to Oreg., also Ariz.
aldrichi Hunter, 1896b: 229.—Idaho; Oreg., Colo.
angelica Telford, 1939: 35.—Minn.; Alta., Mich., N.H.
atrocapilla Hull and Fluke, 1950: 361.—Wash.
baroni Williston, 1887c: 40.—Wash. Territory, Calif.; Wash., Oreg., Colo.
bicolorata Shannon, 1922c: 139.—Idaho; Alta.

bigelowi Curran, 1926d: 212.—Ont.; Alaska, Wash., Calif.
  var. **callichroma** Hull and Fluke, 1950: 345 (as ssp.).—Calif.
borealis borealis Coquillett, 1900h: 426 (1904: 40).—Alaska; Alta., Oreg.
  ssp. **nigroseta** Hull and Fluke, 1950: 334.—Colo.
browni Curran, 1931e: 96.—Que.
burkei Shannon, 1922c: 141.—Wash.; B.C., Oreg., Idaho, Colo.
canada Hull and Fluke, 1950: 338.—B.C.
catalina Shannon, 1922c: 137.—Ariz.
chalybescens Williston, 1893c: 76.—Calif.; Wash., Oreg.
chintimini Lovett *in* Cole and Lovett, 1921: 277.—Oreg.
chrysochlamys Williston, 1891: 8.—Mexico (Guerrero); Calif., ?Alaska, ?Colo.
cineralis Hull and Fluke, 1950: 317.—Ohio; Kans., Wis., Ill.
coerula Fluke and Hull, 1946: 343.—Wash.
columbiae Curran, 1922e: 69.—B.C.; Idaho.
consentiens Curran, 1926d: 211.—Ont.
cottrelli Telford, 1939: 36.—Minn.; Alta., Wash.
cratorhina Hull and Fluke, 1950: 332.—Oreg.
cynoprosopa Hull and Fluke, 1950: 350.—Wis.
dakota Hull and Fluke, 1950: 316.—S. Dak.; Minn.
decumbens Hull and Fluke, 1950: 345.—Idaho; Colo.
ferruginea Lovett *in* Cole and Lovett, 1919: 238.—Oreg.; B.C., Alta., Wash.
flavipennis Hull and Fluke, 1950: 316 (as *alaskensis* ssp.).—Oreg.
flavosericea Hull and Fluke, 1950: 353.—B.C.
florella Shannon, 1922c: 138.—Idaho; Alaska.
fuma Fluke and Hull, 1946: 341.—Oreg.
hermiona Hull and Fluke, 1950: 360.—Oreg.; Utah.
hesperia (Shannon), 1922c: 132 (*Cartosyrphus*).—Alaska; Alta.
hiantha Hull and Fluke, 1950: 356.—Colo.
hiawatha Shannon, 1922c: 138.—Mass.
hirsuta Hull and Fluke, 1950: 365.—Oreg.; Wash.
hoodiana (Bigot), 1883a: 552 (*Cartosyrphus*).—Oreg.; B.C. to Calif. and N. Mex.
  petulca Williston, 1887c: 39.—Wash. Territory.
hunteri hunteri Curran, 1922d: 17.—Man.; Alta., Wash.
  ssp. **nuda** Hull and Fluke, 1950: 330.—Alta.
  ssp. **truncata** Hull and Fluke, 1950: 330.—Man.
julietta Shannon, 1922c: 140.—Idaho.
lasiophthalma Williston. 1882a: 306.—Colo.; Alaska to Oreg. and Colo., ?N.H.

livida Wehr, 1922b: 142 (=p. 24).—Colo.; Alta., Wash., Oreg., N. Mex.
    var. **lintea** Fluke and Hull, 1946: 347.—Colo.
luna Hull and Fluke, 1950: 340.—Calif.
margarita Hull and Fluke, 1950: 318.—Wash.
meganosa Hull and Fluke, 1950: 325.—Wyo.
minnesotensis Telford, 1939: 36.—Minn.
montanipes Hull and Fluke, 1950: 334.—Colo.
nannomorpha Hull and Fluke, 1950: 338.—Colo.
nasica Hull and Fluke, 1950: 376.—Oreg.
nigrescens Hull and Fluke, 1950: 342.—Oreg.
nigroapicata Curran, 1926d: 173.—Ont.; Alta., N.B., N.H., N.C.
nigrobarba Hull and Fluke, 1950: 350.—Colo.
nigrofasciata Curran, 1926d: 174.—Ont.; Alta.
nigrovittata Lovett *in* Cole and Lovett, 1919: 239.—Oreg.; Wash.
nokomis Hull and Fluke, 1950: 318.—N.H.
obesa Hull and Fluke, 1950: 324.—Wash.; Alta., Wyo.
occidentalis Williston, 1882a: 305.—Calif.; Alaska to S. Dak., s. to Calif. and N. Mex.
orilliaensis Curran, 1922e: 67.—Ont.; Alta. to N.B., s. to Ill. and N.H.
pacifica Hunter, 1897: 127.—Calif. Unrecognized.
pikei Shannon, 1922c: 130 (as *aldrichi* var.).—Colo.; Utah., Wyo.
pilosipes Hull and Fluke, 1950: 337.—Colo.
pluto Hull and Fluke, 1950: 352.—B.C.
pontiaca Shannon, 1922c: 142.—N.H.; Que.
porcina Hull and Fluke, 1950: 326.—Oreg.
primoveris Shannon, 1915b: 168.—Md.; N.Y. and Mass. to Ohio and Va., ?Utah.
promethea Hull and Fluke, 1950: 321.—Calif.
punctulata Hunter, 1897: 128.—Nebr.
rhinoprosopa Hull and Fluke, 1950: 362.—N.H.
robusta Hine, 1922b: 144.—Alaska; Wash.
scilla Hull and Fluke, 1950: 320.—Calif.
seripila Hull and Fluke, 1950: 355.—Wash.; Oreg.
skinneri Johnson, 1903a: 101.—N. Mex.
sonoriana Shannon, 1922c: 136.—N. Mex.; Idaho.
speculum Hull and Fluke, 1950: 327.—Colo.
subchalybea Curran, 1923n: 276.—B.C.; Que.
swannanoa Brimley, 1925: 73.—N.C.
tantalus Hull and Fluke, 1950: 319.—Mich.
varipila Fluke and Hull, 1946: 338.—Alta.
variseta Fluke and Hull, 1946: 336.—Wash.; Mont., Calif., Colo.
yukonensis Shannon, 1922c: 129.—Alaska; B.C., Alta., Colo.

Subgenus CARTOSYRPHUS Bigot

Cartosyrphus Bigot, 1883a: 230 (as genus). Type-species, *Syrphus paganus* Meigen (Coquillett, 1910d: 376).

REFERENCE: Fluke and Hull, 1947 (rev.).

aesyctes (Walker), 1849: 591 (*Syrphus*).—Ont.
    aescytes, error.
brevichaeta (Shannon), 1922c: 133 (*Cartosyrphus*).—Colo.; Wyo.
caltha (Shannon), 1922c: 133 (*Cartosyrphus*).—Ind.; Wis. to Mich., s. to Kans. and Ind.
capillata Loew, 1863b: 309 (Cent. 4, no. 65).—D.C.; Ill. to Ont., s. to Ga.
    lamprurus Bigot, 1883a: 552 (*Cartosyrphus*).—N. Amer.
comosa Loew, 1863b: 309 (Cent. 4, no. 66.)—Man., Ont.; Wash. to Que., s. to Oreg., Colo., and N.J.
laevifrons Jones, 1907b: 90.—Nebr.
laevis (Bigot), 1883a: 553 (*Cartosyrphus*).—Wash. Territory; Wash. to Alta., s. to Oreg. and Colo.
latrans (Walker), 1849: 575 (*Syrphus*).—Ont.
leucoparea Loew, 1863b: 311 (Cent. 4, no. 69).—Carolina; Tenn., N.C., Ga.
lucta Snow, 1895d: 228.—Colo; Alta., N. Mex.
megatarsa Fluke and Hull, 1947: 246.—Colo.
ontario Curran, 1922g: 96 (also 1922h: 191) (error corrected, n. name for *rita* Curran, 1922e: 70).—Ont.; Ont. to N.B., s. to Wis. and N.Y.
    rita Curran, 1922e: 70 (error, preocc. Curran, 1922e: 71).—Ont.
    slossonae Shannon, 1922c: 144 (*Cartosyrphus*).—N.H.
pallipes Loew, 1863b: 311 (Cent. 4, no. 70).—D.C.; D.C. to Que., s. to Calif., Colo., and N.C.
platycera Hine, 1922b: 143.—Alaska.
    kincaidia Shannon, 1922c: 142 (*Cartosyrphus*).—Alaska.
prima Hunter, 1896a: 92.—Pa.; Md., Miss., Ga., Fla.
pulchripes Loew, 1857d: 597.—Europe; B.C., Alta.
sensua Curran, 1922e: 19.—Ont.; N.J.
shannoni (Curran), 1923i: 198 (*Cartosyrphus*; n. name for *similis* Shannon).—Md.; Wis. to Maine, s. to N.C.
    similis Shannon, 1916a: 196 (preocc. Michl, 1911).—Md.
sialia Shannon, 1922c: 132.—N.B.; Alta. to Que., s. to Oreg., Colo., and N.Y.
    rita Curran, 1922e: 71.—Ont.
    var. **alpinensis** Fluke and Hull, 1947: 228.—Colo.; Wash., Oreg.
    var. **argentipila** Fluke and Hull, 1947: 230.—Wis.; Minn., N.H.

sororcula Williston, 1891: 9.—Mexico (Guerrero); Oreg. to Idaho, s. to Ariz., N. Mex., and Mexico.
tarda Snow, 1895d: 228.—Colo.
tristis Loew, 1863b: 312 (Cent. 4, no. 71).—Man.; Alaska to N.S., s, to Oreg., N. Mex., and Va.
    longipilosus Wehr, 1922b: 143 (=p. 25) (*Cartosyrphus*).— Colo.
wisconsinensis Fluke and Hull, 1947: 232.—Wis.

Subgenus HIATOMYIA Shannon

Cartosyrphus, subg. **Hiatomyia** Shannon, 1922c: 127. Type-species, *Chilosia willistoni* Snow (orig. des.).

canadensis (Shannon), 1922c: 133 (*Cartosyrphus*).—B.C.; Alta., Wash., Utah.
chionthrix Hull and Fluke, 1950: 379.—Calif.
chrysothrix Hull and Fluke, 1950: 381.—Oreg.
coriacea Hull and Fluke, 1950: 384.—Oreg.
cyanea Hunter, 1896b: 228.—Idaho; Alaska, Wash., Oreg.
cyanescens Loew, 1863b: 310 (Cent. 4, no. 67).—Ill.; Wis. to Que., s. to Ill. and Va.
    plumata Loew, 1863b: 310 (Cent. 4, no. 68).—Va.
gemini (Shannon), 1922c: 133 (*Cartosyrphus*).—Colo.; Wash.
hecate Hull and Fluke, 1950: 378.—Oreg.
hyacintha Hull and Fluke, 1950: 374.—Oreg.
idahoa Shannon, 1922c: 132.—Idaho.
    var. **eunicea** Hull and Fluke, 1950: 383 (as ssp.).—Wash.
nigrocoerulea Lovett *in* Cole and Lovett, 1921: 279 (n. name for *pacifica* Lovett).—Oreg. ?=*cyanea*.
    pacifica Lovett *in* Cole and Lovett, 1919: 240 (preocc. Hunter, 1897).—Oreg.
nigrocyanea Hull and Fluke, 1950: 385.—Calif.
niveifrons Hull and Fluke, 1950: 373.—Calif.
nyctichroma Hull and Fluke, 1950: 380.—Calif.
olivia Hull and Fluke, 1950: 385.—Oreg.
plumosa Coquillett *in* Baker, 1904: 25.—Nev.; B.C., Sask.
plutonia Hunter, 1897: 125.—Alaska; Alaska to Alta., s. to Wash., Nev., and Colo.
    gracilis Hunter, 1897: 126.—Alaska.
rubroflava Hull and Fluke, 1950: 378.—Calif.
signatiseta Hunter, 1896b: 227.—Idaho; Wash., Oreg.
tessa Hull and Fluke, 1950: 381.—Calif.
townsendi Hunter, 1896a: 94.—Calif.; B.C. to Calif.

willistoni Snow, 1895d: 227 (n. name for *lugubris* Williston).—Calif.; B.C. to Calif. and Colo.
  lugubris Williston, 1887c: 45 (preocc. Zetterstedt, 1838).— Calif.

Genus RHINGIA Scopoli

Rhingia Scopoli, 1763: 358. Type-species, *Conops rostrata* Linnaeus (mon.).

The larva of the European *Rhingia campestris* Meigen breeds in cow dung.

nasica Say, 1823: 94 (1859b: 81).—U.S.; Man. to N.B., s. to Colo. and Ga.

Genus FERDINANDEA Rondani

Ferdinandea Rondani, 1844a: 196. Type-species, *Conops cuprea* Scopoli (Rondani, 1856: 51).
Chrysoclamis Walker, 1851a: 279 (unjustified n. name for *Ferdinandea* Rondani). Type-species, *Conops cuprea* Scopoli (aut.).
Chrysochlamys, emend.

The larvae of the European species of *Ferdinandea* live in exuding sap of trees or in decaying wood.

REFERENCES: Shannon, 1924d (key); Hull, 1942d (rev.).

aenicolor Shannon, 1924d: 214.—Oreg.; N. Mex.
buccata (Loew), 1863b: 313 (Cent. 4, no. 72) (*Chrysochlamys*).—Va.; Wis., N.Y., Pa., N.J., Ga.
croesus (Osten Sacken), 1877: 341 (*Chrysochlamys*).—Utah; B.C. to Minn., s. to Calif. and N. Mex.
  var. **midas** Hull, 1942d: 240.—Ariz.
dives (Osten Sacken), 1877: 340 (*Chrysochlamys*).—Ky.; B.C. to Que., s. to Mo. and Ga.
nigripes (Osten Sacken), 1877: 341 (*Chrysochlamys*).—Mass.; Alaska, Oreg., Alta., Wis., Que. to N.C.

Genus CYNORHINELLA Curran

Cynorhinella Curran, 1922d: 14. Type-species, *canadensis* Curran (orig. des.).
Apicomyia Shannon, 1922c: 122. Type-species, *Myolepta bella* Williston (orig. des.).

REFERENCES: Shannon, 1924a, Curran, 1925k (tax.).

bella (Williston), 1882a: 308 (*Myolepta*).—Wash., Oreg.; Alaska, B.C.
  carbicolor Lovett, 1920: 51 (*Myolepta*).—Wash.

canadensis Curran, 1922d: 15.—B.C.
longinasus Shannon, 1924a: 124.—N.H.; Que., N.C.

Tribe MYOLEPTINI

The larvae of *Myolepta* breed in decaying wood, and the larvae of *Lejota aerea* (Loew) have been found under bark.

REFERENCES: Shannon, 1925d (key to genera); Fluke and Weems, 1956 (rev.).

Genus MYOLEPTA Newman

Myolepta Newman, 1838: 373. Type-species, *Musca luteola* Gmelin (mon.).
Myiolepta, emend.

Subgenus MYOLEPTA Newman

camillae Weems *in* Fluke and Weems, 1956: 5.—Fla.
lunulata Bigot, 1883a: 537.—Oreg.; Wash. to Calif. and Ariz.
    californica Shannon, 1923a: 21.—Calif.
nigra Loew, 1872a: 84 (Cent. 10, no. 52).—Pa.; Minn. to Que., s. to Tex. and Ga.
    tuberans Williston, 1887c: 225 (?*Xylota*).—Tex.
varipes Loew, 1869c: 174 (Cent. 9, no. 79).—Va.; Nebr. to Ont. and N.H., s. to Tex. and Ga.
    pretiosa Hull, 1923: 295.—Miss.

Subgenus EUMYIOLEPTA Shannon

Eumyiolepta Shannon, 1921: 71 (as genus). Type-species, *Myolepta strigilata* Loew (orig. des.).

auricaudata Williston, 1891: 40.—Mexico (Guerrero); Ariz.
aurinota Hine, 1903a: 245.—Ariz.; Calif., N. Mex., Tex.
    cornellia Shannon, 1923a: 20 (*Eumyiolepta*).—Ariz.
strigilata Loew, 1872a: 85 (Cent. 10, no. 54).—Tex.; Kans. to Wis. and Ont., s. to Ariz. and N.C.

Genus CHALCOSYRPHUS Curran

Chalcomyia, subg. **Chalcosyrphus** Curran, 1925k: 122. Type-species, *atra* Curran (orig. des.) = *depressus* (Shannon).

anomalus (Shannon), 1925a: 153 (*Chalcomyia*).—N.J. **N. comb.**, J. R. Vockeroth *in litt.*
aristatus (Johnson), 1929b: 374 (*Xylota*).—N.H. **N. comb.**, J. R. Vockeroth *in litt.*

depressus (Shannon), 1925a: 153 (*Chalcomyia*).—Idaho; Wash., Mo.
   atra Curran, 1925k: 122 (*Chalcomyia*).—Mo.

Genus LEJOTA Rondani

Lejota Rondani, 1857: 176. Type-species, *Psilota ruficornis* Zetterstedt (Goffe, 1944a: 29).
Leiota, error or emend.
Chalcomyia Williston, 1885b: 133. Type-species, *Myolepta aerea* Loew (as *aera*; orig. des.).
   REFERENCES: Shannon, 1925a (rev., as *Chalcomyia*), 1926c (key).

aerea (Loew), 1872a: 84 (Cent. 10, no. 53) (*Myolepta*).—Ill.; Nebr. to N.H., s. to Tex. and N.C.
cyanea (Smith), 1912: 119 (*Chalcomyia*).—N.H.; Alaska, B.C., Mich. Ont., Que.
   calcitrans Curran, 1922b: 260 (*Chalcomyia*).—Ont.
   bigelowi Curran, 1924g: 195 (*Cynorhina*).—Ont. **N. syn.**, J. R. Vockeroth *in litt.*

Genus LEPIDOMYIA Loew

Lepidomyia Loew, 1864a: 69 (Cent. 5, no. 38). Type-species, *calopus* Loew (mon.). Alleged preoccupation by *Lepidomya* Bigot, 1857, erroneous.
Lepidostola Mik, 1886: 278 (unjustified n. name for *Lepidomyia* Loew). Type-species, *Lepidomyia calopus* Loew (aut.).
Lepromyia Williston, 1887c: 31 (unjustified n. name for *Lepidomyia* Loew). Type-species, *Lepidomyia calopus* Loew (aut.).
   REFERENCE: Hull, 1946 (rev., as *Lepidostola*).

micheneri (Fluke), 1953a: 126 (*Lepidostola*).—Tex. **N. comb.**

Tribe CHRYSOGASTRINI

Genus CHRYSOGASTER Meigen

Chrysogaster Meigen, 1800: 32. Type-species, *Eristalis solstitialis* Fallén (Coquillett, 1910b: 523). Suppressed by I.C.Z.N., (1963b: 339.).
Chrysogaster Meigen, 1803: 274. Type-species, *Musca coemiteriorum* Linnaeus (as *Syrphus coemiteriorum* Fabricius; Coquillett, 1910b: 523) ?=*solstitialis* (Fallén).

The larvae of *Chrysogaster* are aquatic. Those of the European *C. hirtella* Loew live among the roots of the aquatic grass *Glycera* and pierce the roots with the spinelike posterior spiracles to obtain oxygen from the plant's intercellular spaces.

REFERENCES: Shannon, 1916c (rev.); Sedman, 1959 (male genitalia).

Subgenus CHRYSOGASTER Meigen

inflatifrons Shannon, 1916c: 107.—N.C.; Okla. to Conn. and Ga.
minuta Hull, 1945a: 215.—Wash.
nigripes Loew, 1863b: 307 (Cent. 4, no. 60).—N.Y.; Wis. to Que., s. to Nebr. and Ga.
    antitheus Walker, 1849: 572.—N.Y. ?Synonym.
    ustulata Loew, 1869c: 175 (Cent. 9, no. 80) (*Orthonevra*).—N.J.
nigrovittata (Loew), 1876: 323 (*Orthonevra*).—Caiif.; Idaho.
    pacifica Shannon, 1916c: 105.—Unknown (?Calif.) **N. syn.**
parva Shannon, 1916c: 104.—Colo.; B.C. to Alta., s. to Calif. and Colo.
robusta Shannon, 1916c: 103.—Calif.; ?Nebr.
sinuosa (Bigot), 1883a: 556 (*Orthonevra*).—Wash. Territory; B.C. to Idaho, s. to Calif. and Colo.
texana Shannon, 1916c: 108.—Tex.; Mich. to Ont., s. to Tex. and Ga.
    greenei Shannon, 1916a: 195.—Va. **N. syn.**
    ontario Curran, 1925k: 117.—Ont. **N. syn.**

Subgenus ORTHONEVRA Macquar

Orthonevra Macquart, 1829: 188 (1829: 40) (as genus). Type-species, *Chrysogaster elegans* Meigen (mon.).
Orthoneura, emend.
Cryptineura Bigot, 1859a: 308. Type-species, *hieroglyphica* Bigot (mon.) = *nitida* Wiedemann.

bellula Williston, 1882a: 304.—Wash. Territory, Calif.; Wash, to Idaho and S. Dak., s. to Calif., N. Mex., and Mexico.
nitida Wiedemann, 1830: 116.—N. Amer.; Wis. to Ont. and Mass., s. to Nebr., Ariz., and Fla., Neotropical.
    aeneus Walker, 1849: 545 (*Paragus*).—Ohio.
    hieroglyphica Bigot, 1859a: 308 (*Cryptineura*).—La.
nitidula Curran, 1925k: 116.—Ariz.
pictipennis (Loew), 1863b: 306 (Cent. 4, no. 58) (*Orthonevra*).—N.Y.; Colo. to Wis. and Que., s. to Kans. and S.C.
pulchella Williston, 1887c: 35.—Canada, N.H., Conn.; Sask. to Que., s. to Colo. and N.C.
stigmata Williston, 1882a: 303.—Calif.; B.C. to Idaho, s. to Calif. and Utah.
unicolor Shannon, 1916c: 106.—Nev.; Oreg.

Genus CHRYSOSYRPHUS Sedman

Chrysogaster, subg. *Barberiella* Shannon, 1922c: 122 (preocc. Poppius, 1914). Type-species, *chilosioides* Shannon (orig. des.).
Chrysosyrphus Sedman (for *Barberiella* Shannon). Type-species, *Chrysogaster chilosioides* Shannon (aut.). **N. name.**

REFERENCE: Fluke, 1949 (key, as *Barberiella*).

alaskensis (Shannon), 1922c: 124 (*Chrysogaster*).—Alaska. **N. comb.**
bigelowi (Curran), 1926b: 82 (*Chrysogaster*).—Ont. **N. comb.**
canadensis (Curran), 1933b: 6 (*Chrysogaster*).—Que. **N. comb.**
chilosioides (Shannon), 1922c: 122 (*Chrysogaster*).—Calif. **N. comb.**
ithaca (Shannon), 1925d: 107 (*Chrysogaster*).—N.Y. **N. comb.**
latus (Loew), 1863b: 306 (Cent. 4, no. 59) (*Chrysogaster*).—Ont.; Oreg., Idaho. Utah, Colo., Nebr. **N. comb.**
nigripennis (Williston), 1882a: 307 (*Cheilosia*).—Oreg.; B.C., Wash. **N. comb.**
  infumatus Bigot, 1883a: 553 (*Cartosyrphus*).—Oreg.
versipellis (Williston), 1887c: 44 (*Cheilosia*).—Wash. Territory; B.C., Oreg. **N. comb.**

Genus BRACHYOPA Meigen

Brachyopa Meigen, 1822: 260. Type-species, *Musca conica* Panzer (Westwood, 1840: 137) =*panzeri* Goffe.

The larvae of *Brachyopa* live in sap exuding from trees.

REFERENCES: Lundbeck, 1916 (biol.); Curran, 1922h (rev.); Goffe, 1945b (type-sp.); Sedman, 1961 (male genitalia).

Subgenus BRACHYOPA Meigen

cinereovittata Bigot, 1883a: 537.—Calif.; B.C., Wash., Idaho.
  basilaris Curran, 1922j: 255 (also 1925k: 120).—Wash. **N. syn.**
cynops Snow, 1892: 37.—Colo.
daeckei Johnson, 1917b: 360.—Pa.; Va.
diversa Johnson, 1917b: 361.—N.H.; Ont., Que., Maine, N.Y.
flavescens Shannon, 1915a: 144.—Va.; Ont. and Que. to Va.
gigas Lovett *in* Cole and Lovett, 1919: 243.—Wash.; B.C., Oreg., Calif., Idaho.
media Williston, 1882a: 308.—Calif.; Alaska to Que., s. to Calif., Man., and Va.
nigricauda Curran, 1922j: 255 (also 1925k: 121).—Conn. (unpublished). ?=*daeckei*.
notata Osten Sacken, 1875d: 68.—N.H.; Wis. to N.S., s. to Mo. and Ga., also B.C.

perplexa Curran, 1922f: 117.—Ont.; Ont. and Que., s. to Wis. and N.C.
punctipennis Curran, 1925k: 181.—Oreg.; Alaska to Que., s. to Calif.
rufiabdominalis Jones, 1917: 226.—Colo.; B.C., Wash., Man.
vacua Osten Sacken, 1875d: 51, 67(as *racua*, p. 67).—Que.; Ont. and Que., s. to Kans. and N.C., also B.C.

Subgenus HAMMERSCHMIDTIA Schummel

Hammerschmidtia Schummel *in* Gravenhorst, 1834: 739 (as genus). Type-species, *vittata* Schummel (mon.)=*ferruginea* (Fallén).
Eugeniamyia Williston, 1882b: 80. Type-species, *rufa* Williston (mon.)=*ferruginea* (Fallén).

ferruginea (Fallén), 1817a: 34 (*Rhingia*).—Europe; Alaska to Que., s. to Ariz., N. Mex., and Mass., Europe.
rufa Williston, 1882b: 80 (*Eugeniamyia*).—Wash. Territory.

Genus PSILOTA Meigen

Psilota Meigen, 1822: 256. Type-species, *anthracina* Meigen (mon.).
The early stages of *Psilota* are unknown.

buccata (Macquart), 1842: 167 (1842: 107) (*Pipiza*).—Carolina; Kans. to N.C:, s. to Tex. and Fla.
flavidipennis Macquart, 1855: 117 (1855: 97).—Pa.; Okla., Tex., Fla.
thatuna Shannon, 1922b: 38.—Idaho; Calif.

Genus SPHEGINA Meigen

Sphegina Meigen, 1822: 193. Type-species, *Milesia clunipes* Fallén (Westwood, 1840: 136).

The larvae of the European *Sphegina clunipes* (Fallén) and *S. kimakowiczi* Strobl have been found in moist places under the bark of trees.

REFERENCES: Malloch, 1922f (key); Cole, 1924a (rev.); Hull, 1935a (key).

armatipes Malloch, 1922a: 141.—Calif.; Wash., Idaho, Oreg.
    var. **rufa** Malloch, 1922a: 142.—Calif.; Oreg.
biannulata Malloch, 1922a: 143.—Va.
brachygaster Hull, 1935a: 376.—N.Y.; Mich.
bridwelli Cole, 1924a: 42.—Oreg.
brimleyi Shannon, 1940: 118.—N.C.; N.H.

californica Malloch, 1922a: 144.—Calif.; Alta., Wash., Oreg.
campanulata Robertson, 1901: 284.—Ill.; Wis. to N.B., s. to S. Dak. and Va.
cressoni Hull, 1935a: 375.—Wash.; B.C.
flavimana Malloch, 1922a: 143.—Md.; Minn. to N.S., s. to Ind. and Ga., also Colo.
flavomaculata Malloch, 1922a: 141.—Va.; Pa., Md.
infuscata Loew, 1863a: 13 (Cent. 3, no. 23).—Alaska; Alaska to Man., s. to Ariz. and Nebr.
keeniana Williston, 1887c: 113.—Pa.; Wis. to N.B., s. to Ind. and N.C., ?Wash.
lobata Loew, 1863a: 12 (Cent. 3, no. 21).—Middle States; Wis. to Que., s. to Nebr., Ohio, and Ga., ?B.C., ?Oreg., ?Idaho.
lobulifera Malloch, 1922f: 269.—Md.; Vt.
melanderi Cole, 1924a: 43.—Wash.
monticola Malloch, 1922a: 142.—N.H.; Maine.
nigrimana Cole, 1924a: 41.—Calif.
notata Hull, 1935a: 377.—Wash.
occidentalis Malloch, 1922a: 142.—Wash.; Oreg., Calif.
perplexa Hull, 1935a: 377.—Que.
petiolata Coquillett, 1910a: 125.—N.H.; Ont. and Que., s. to Wis. and Mass.
pluto Hull, 1935a: 374.—B.C.; Calif.
punctata Cole *in* Cole and Lovett, 1921: 285.—Oreg.; B.C. to Idaho and Calif.
rufiventris Loew, 1863a: 13 (Cent. 3, no. 22).—N.Y.; Wis. to N.B., s. to N.C., ?Wash., ?Oreg.
vittata Cole, 1924a: 43.—Wash.

Genus NEOASCIA Williston

Ascia Meigen, 1822: 185 (preocc. Scopoli, 1777). Type-species, *Syrphus podagricus* Fabricius (Westwood, 1840: 136).
Neoascia Williston, 1887c: 111 (n. name for *Ascia* Meigen). Type-species, *Syrphus podagricus* Fabricius (aut.).

The available information indicates that the larvae of *Neoascia* are aquatic or semiaquatic scavengers.

REFERENCE: Curran, 1925i (rev.).

albipes (Bigot), 1883a: 328 (*Ascia*).—N. Amer.; Mich., Ohio, Mass. to Md.
conica Curran, 1925i: 56.—Alta.; Alaska, Man., Calif., Nev., N. Mex.
distincta Williston, 1887c: 112.—Mass.; Man. to Que., s. to Colo. and N.C.

globosa (Walker), 1849: 546 (*Ascia*).—N.Y.; Alaska to Que., s. to Calif. and Va.
macrofemoralis Curran, 1925i: 58.—Alaska; Alta., N. Dak.
metallica (Williston), 1882a: 315 (*Ascia*).—Oreg.; Alaska to Ont., s. to Oreg., Nev., and Colo.
  quadrinotata Bigot, 1883a: 327 (*Ascia*).—Oreg.
  nasuta Bigot, 1883a: 327 (*Ascia*).—Oreg.
minuta Curran, 1925i: 53.—Colo.; Nev.
sphaerophoria Curran, 1925i: 57.—Alta.; Alaska.
subchalybea Curran, 1925i: 53.—Que.
unifasciata Curran, 1925i: 55.—Man.; Alta., Utah, Wis.

Tribe CALLICERINI

Genus CALLICERA Panzer

Callicera Panzer, 1809a, 104: 17. Type-species, *Bibio aenea* Fabricius (mon.).

The larvae of the European *Callicera rufa* (Schummel) were found in water in a cavity in a pine tree.

  REFERENCES: Curran, 1935b (key); Coe, 1938 (biol., immature stages).

auripila Metcalf, 1916a: 112 (as *johnsoni* var.).—N.C.; Miss.
duncani Curran, 1935b: 5.—Ariz.
johnsoni Hunter, 1896a: 87.—Pa.; Mich., N.Y. to N.C.
montensis Snow, 1892: 34.—Colo.; N. Mex.

Tribe PELECOCERINI

Genus PELECOCERA Meigen

Pelecocera Meigen, 1822: 340. Type-species, *tricincta* Meigen (mon.).
Euceratomyia Williston, 1884b: 185. Type-species, *pergandei* Williston (mon.).

pergandei (Williston), 1884b: 186 (*Euceratomyia*).—D.C.; N.Y. to Miss. and Ga.

Genus CHAMAESYRPHUS Mik

Chamaesyrphus Mik, 1895: 133. Type-species, *Rhingia scaevoides* Fallén (mon.).

  REFERENCE: Curran, 1923a (key).

apichaetus Curran, 1923a: 60.—Calif.
willistonii (Snow), 1895b: 187 (*Pelecocera*).—N. Mex.; Ariz., Colo.

Tribe NAUSIGASTRINI

Genus NAUSIGASTER Williston

Nausigaster Williston, 1883a: 33. Type-species, *punctulata* Williston (mon.).

The larvae of *Nausigaster* have been reared in Brazil from a decaying stem of papaya holding rainwater.

REFERENCES: Curran, 1941 (key); Carrera, Lopes, and Lane, 1947 (biol., immature stages).

clara Curran, 1941: 255.—Tex.; Ariz.
curvinervis Curran, 1941: 256.—Ariz.; Tex.
geminata Townsend, 1897b: 25.—Tex.; Ariz., Mexico.
nova Curran, 1941: 256.—Ariz.
punctulata Williston, 1883a: 34.—N. Mex.; Ariz., Tex.
scutellaris Adams, 1904b: 446.—Ariz.; Calif.
texana Curran, 1941: 257.—Tex.
unimaculata Townsend, 1897b: 24.—Calif.; Ariz., Tex.

Tribe EUMERINI

Genus EUMERUS Meigen

Eumerus Meigen, 1822: 202. Type-species, *Syrphus tricolor* Fabricius (Curtis, 1839: pl. 749). Because of the great amount of economic and taxonomic literature which has appeared for this genus under this name, an application should be prepared for submission to the I.C.Z.N. to suppress the genus *Eumeros* Meigen, 1803 (changed to *Eumerus* by Meigen, 1804) and to place *Eumerus* Meigen, 1822, on the Official List of Generic Names. See *Xylota*.

Pumilio Schembi *in* Kertész, 1910: 313. Nomen nudum. Not Scudder, 1882.

Paragopsis Matsumura, 1916: 250. Type-species, *griseofasciatus* Matsumura (orig. des.)=*strigatus* (Fallén).

Microxylota Jones, 1917: 230. Type-species, *robii* Jones (orig. des.)=*strigatus* (Fallén).

Citabaena, authors, not Walker. *Citabaena* Walker, 1857, is a poorly understood Asian genus.

There is considerable literature on the biology, taxonomy, and economic importance of the larvae of *Eumerus*, known as "lesser bulb flies". The species are probably all associated with decaying plant tissue, and their feeding on healthy plants is questionable. The work of Creager and Spruijt (1935) showed that it is necessary for larvae of *tuberculatus* Rondani to feed on the fungus of basal-rot in order to develop normally. The three North American species were introduced in commerce from Europe in onion, narcissus, and related bulbs.

REFERENCES: Collin, 1920 (tax., biol.); Hodson, 1927 (biol.); Smith, 1928 (rev.); Latta and Cole, 1933 (rev.); Martin, 1934 (biol.).

narcissi Smith, 1928: 139.—Calif.; Oreg., Ont., N.Y., Europe. Immigrant.
strigatus (Fallén), 1817a: 61 (*Pipiza*).—Europe; U.S. and Canada. Immigrant.
     robii Jones, 1917: 231 (*Microxylota*).—Colo.
tuberculatus Rondani, 1857: 93.—Europe; Canada and n. U.S. Immigrant.

Tribe MICRODONTINI

REFERENCES: Curran, 1941 (key to genera); Greene, 1955 (immature stages).

Genus MICRODON Meigen

Microdon Meigen, 1803: 275. Type-species, *Musca mutabilis* Linnaeus (as *Mulio mutabilis* Fabricius; mon.).
Aphritis Latreille, 1804: 193. Type-species, *auropubescens* Latreille (sub. mon, Latreille, 1805: 358)=*mutabilis* (Linnaeus).
Dimeraspis Newman, 1838: 372. Type-species, *podagra* Newman (mon.)=*globosus* (Fabricius).
Mesophila Walker, 1849: 1157. Type-species, *Ceratophya fuscipennis* Macquart (mon.).
Microdon, subg. *Serichlamys* Curran, 1925k: 50. Type-species, *rufipes* Macquart (mon.). **N. syn.**

Due to their very peculiar, hemispherical shape, the larvae of *Microdon* were first described as molluscs, then as coccids. They usually have been found associated with nests of ants where they are believed to act as scavengers.

REFERENCE: Curran, 1925k (rev.).

Subgenus MICRODON Meigen

albipilis Curran, 1925k: 54.—Man.; Alta.
aurulentus (Fabricius), 1805: 185 (*Mulio*).—Carolina; Pa. to N.C., ?Neotropical.
basicornis Curran, 1925k: 79.—N.B.; Mich., N.Y., N.S.
champlaini Curran, 1925k: 71.—Pa.; Ont. to N.S., s. to Ohio and Va.

conflictus Curran, 1924o: 226.—Va.; Kans. to Mich. and Maine, s. to Tex. and Fla.
cothurnatus Bigot, 1883a: 320.—Wash. Territory; B.C. to N.B., s. to Calif. and N.C.
    similis Jones, 1917: 219.—Colo.
    cockerelli Jones, 1922: 17 (as *tristis* var.).—Colo.
    tristis, authors, not Loew.
craigheadii Walton, 1912: 463.—Pa.; Kans., W. Va., Md., Ga.
    laetus, authors, not Loew.
diversipilosus Curran, 1925k: 76.—Kans.; Wis., Ohio, La., N.C.
fuscipennis (Macquart), 1834a: 488 (*Ceratophya*).—Pa.; Minn. to Que., s. to Tex. and Fla.
    agapenor Walker, 1849: 539.—Ga.
    pachystylum Williston, 1887c: 8.—Ga.
globosus (Fabricius), 1805: 185 (*Mulio*).—Carolina; Man. to Que., s. to Fla.
    podagra Newman, 1838: 373 (*Dimeraspis*).—Ill.
hutchingsi Curran, 1927f: 89.—Que.; Ont.
laetoides Curran, 1935b: 3.—Ariz.
laetus Loew, 1864a: 74 (Cent. 5, no. 46).—Cuba; Tex. to Md. and Fla., Antilles.
    scitulus Williston, 1887c: 10.—Fla. **N. syn.**
lanccolatus Adams, 1903b: 222.—Kans.; Alta. to Mont., s. to N. Mex. and Kans.
    coloradensis Cockerell and Andrews, 1916: 53.—Colo.
manitobensis Curran, 1924o: 227.—Man.; B.C. to N.S. and Maine.
marmoratus Bigot, 1883a: 320.—Calif.; B.C., Oreg., Utah.
megalogaster Snow, 1892: 34.—Unknown; Wis. and Ill. to Que., s. to Ga., ?Colo.
    bombiformis Townsend, 1895a: 33.—Va.
modestus Knab, 1917: 139.—Nev.
ocellaris Curran, 1924o: 227.—Pa.; N.Y. to Maine, s. to S.C.
piperi Knab, 1917: 136.—Wash.; B.C. and Alta., s. to Oreg. and Mont.
pseudoglobosus Curran, 1924o: 226.—N.Y.; Man. to Vt., s. to Fla.
robustus Telford, 1939: 14.—Minn.
ruficrus Williston, 1887c: 7 (as *tristis* var.).—Conn.; Mich. to Maine, s. to Ga.
rufipes (Macquart), 1842: 71 (1842: 11) (*Aphritis*).—Pa.; Pa. to La. and Fla.
    limbus Williston, 1887c: 8.—Fla.
scutifer Knab, 1917: 141.—Tex.; La., Ala., Fla.
senilis Knab, 1917: 139.—Calif.
tristis Loew, 1864a: 73 (Cent. 5, no. 45).—Va.; B.C. to Que. and Maine, s. to Fla.

viridis Townsend, 1895b: 610.—Mexico (Baja Calif.); B.C., Calif., ?Tenn., ?Fla.
xanthopilis Townsend, 1895b: 611.—Calif.; Wash., Colo.

Subgenus CHYMOPHILA Macquart

Chymophila Macquart, 1834a: 485 (as genus). Type-species, *splendens* Macquart (mon.) =*fulgens* Wiedemann.
Microdon, subg. *Eumicrodon* Curran, 1925k: 50. Type-species, *fulgens* Wiedemann (mon.). **N. syn.**

fulgens Wiedemann, 1830: 82.—Ga.; Pa., N.J., Fla.
  splendens Macquart, 1834a: 486 (*Chymophila*).—Pa.
  aurifex, authors, not Wiedemann.

Subgenus OMEGASYRPHUS Giglio-Tos

Omegasyrphus Giglio-Tos, 1891b: 4 (as genus). Type-species, *Microdon coarctatus* Loew (sub. mon., Giglio-Tos, 1892b: 3).

baliopterus Loew, 1872a: 86 (Cent. 10, no. 56).—Tex.; Calif. to Kans. and Va., s. to Mexico and Fla.
coarctatus Loew, 1864a: 74 (Cent. 5, no. 47).—D.C.; Nebr. to D.C., s. to Mexico and Fla.
painteri Hull, 1922: 370.—Miss.; Tex., Fla.
pallipennis Curran, 1925k: 89.—Colo.; N. Mex., Tex.

Genus MIXOGASTER Macquart

Mixogaster Macquart, 1842: 74 (1842: 14). Type-species, *conopsoides* Macquart (orig. des.).

REFERENCE: Hull, 1954 (rev.).

breviventris Kahl, 1897: 137.—Kans.; Kans., Va. to Fla.
delongi Johnson, 1926b: 301.—Fla.
johnsoni Hull, 1941b: 162.—Mass.; Conn., N.J.

Genus RHOPALOSYRPHUS Giglio-Tos

Rhopalosyrphus Giglio-Tos, 1891b: 3. Type-species, *Holmbergia guentherii* Lynch Arribálzaga (sub. mon., Giglio-Tos, 1892b: 1).

REFERENCE: Capelle, 1956 (rev.).

carolae Capelle, 1956: 174.—Ariz.; Mexico.

Tribe VOLUCELLINI

REFERENCE: Curran, 1930e (key to genera).

Genus ORNIDIA Lepeletier and Serville

Ornidia Lepeletier and Serville, *in* Latreille et al., 1828: 786. Type-species, *Syrphus obesus* Fabricius (orig. des.).
REFERENCE: Curran, 1930e (key).

obesa (Fabricius), 1775: 763 (*Syrphus*).—Amer.; N. Mex., Tex., Fla., Neotropical. Immigrant throughout world tropics.

Genus VOLUCELLA Geoffroy

Volucella Geoffroy, 1762: 540. Type-species, *Musca pellucens* Linnaeus (Curtis, 1833: pl. 452). Validated by I.C.Z.N., 1957a: 85.
Temnocera Lepeletier and Serville *in* Latreille et al., 1828: 786. Type-species, *violacea* Lepeletier and Serville (orig. des.).
Atemnocera Bigot, 1882c: cxiv. Type-species, *Volucella scutellata* Macquart (mon.).
Camerania Giglio-Tos, 1892a: 45 (1893: 141). Type-species, *macrocephala* Giglio-Tos (Coquillett, 1910b: 517).

The larvae of most Temperate Zone species live as scavengers in the nests of bees or wasps, while the tropical American species live mostly in decomposing cacti.

REFERENCES: Johnson, 1916c (tax., *bombylans*); Gabritschevsky, 1924, 1926 (mimicry, *bombylans*); Curran, 1926h, 1930e, 1939b (keys).

Subgenus VOLUCELLA Geoffroy

abdominalis Wiedemann, 1830: 196.—Cuba; Fla., Bahama Is., Jamaica.
anastasia Hull, 1946: 270.—Miss.; Fla.
anna Williston, 1887c: 138.—Ariz.; N. Mex.
apicalis Loew, 1866a: 151 (Cent. 6, no. 36).—Cuba; Md., Neotropical.
apicifera Townsend, 1895a: 40.—N. Mex.; Calif., Ariz., Mexico.
  clarki Curran, 1930e: 19.—Ariz. **N. syn.**
avida Osten Sacken, 1877: 333.—Calif.; B.C. to Calif. and Tex., Mexico.
barei Curran, 1925e: 255.—Tex.; Kans. to N.Y., s. to Tex. and Fla.
bombylans (Linnaeus), 1758: 591 (*Musca*).—Europe; Alaska to Nfld., s. to Calif. and Ga. The typical variety is not Nearctic.

var. **americana** Johnson, 1916c: 162 (as form).—U.S. (5 states, Maine to Pa.); Man. to N.B., s. to Tenn. and Ga.
var. **arctica** Johnson, 1916c: 163 (as form).—Labr.; Alaska to Nfld.
var. **evecta** Walker, 1852: 251 (as sp.).—U.S.; Alaska to Que., s. to Colo. and Va.
    sanguinea Williston, 1887c: 136 (as *evecta* var.).—Unknown.
var. **plumata** (De Geer), 1776: 134 (*Musca;* as sp.).—Europe; Alaska to Nfld., s. to Calif., Mont., and N.Y., Europe.
    facialis Williston, 1882a: 316.—Calif.
    lateralis Johnson, 1916c: 162 (as *bombylans* form).—N.H., Maine, Nfld.
var. **rufomaculata** Jones, 1917: 227 (as sp.).—Colo.; B.C. to Alta. and Mont., s. to Oreg. and Colo.
comstocki Williston, 1887c: 138.—Ariz., N. Mex.; Utah to Wyo., s. to Ariz., Mexico, and Tex.
eugenia Williston, 1887c: 139.—Bahama Is., Fla.; Antilles.
fasciata Macquart, 1842: 82 (1842: 22).—Carolina; Alta. to N.Y., s. to Tex. and Fla.
florida Hull, 1941c: 278.—Fla. ?=*pusilla*.
fraudulenta Williston, 1891: 48.—Mexico (Guerrero); Calif. to Tex., s. to Cent. Amer.
haagii Jaennicke, 1867: 397 (1868: 89).—Mexico; Calif. to Tex. and Mexico.
    setigera Osten Sacken, 1877: 334 (*Temnocera*).—N. Mex.
isabellina Williston, 1887c: 140.—Ariz.; Calif. to Colo., s. to Mexico.
lutzi Curran, 1930e: 18.—Ariz.; Calif., Tex.
macrocephala (Giglio-Tos), 1892a: 45 (1893: 141) (*Camerania*).—Mexico; Ariz.
    lata Wiedemann of Williston, 1891: 45.
    megacephala Loew of Williston, 1887c: 146.
megacephala (Loew), 1863b: 305 (Cent. 4, no. 57) (*Temnocera*).—Calif.; Mexico.
mexicana Macquart, 1842: 85 (1842: 25).—Mexico; s. Calif. to Tex., s. to Cent. Amer.
    violacea Say, 1829: 166 (1859b: 360) (preocc. Lepeletier and Serville, 1828).—Mexico.
    esuriens, authors, in part, not Fabricius.
nigra Greene, 1923: 165.—Fla.; B.C., s. Calif. to Ga., s. to Mexico and Fla.
    esuriens, authors, in part, not Fabricius.
opalescens Townsend, 1901: 160.—N. Mex. Unrecognized.

pallens Wiedemann, 1830: 204.—Brazil; Ariz. to Fla., N.C., Neotropical.
    sexpunctata Loew, 1861c: 38.—Cuba.
postica Say, 1829: 166 (1859b: 360).—Mexico; Calif. to Colo. and N. Mex., Mexico.
pusilla Macquart, 1842: 81 (1842: 21).—Cuba; Tex. to Fla.
    vacua, authors, not Fabricius.
quadrata Williston, 1891: 46.—Mexico (Guerrero); Ariz.
satura Osten Sacken, 1877: 333.—Colo., Utah; Alta. to Wyo., s. to Calif. and Tex.
sternalis Curran, 1930e: 20.—Mexico (Baja Calif.); Ariz.
tamaulipana Townsend, 1898b: 51.—Tex.; La., Mexico, Cent. Amer., Hawaii (immigrant).
    timberlakei Curran, 1926h: 63.—Tex. **N. syn.**
    pusilla, authors, not Macquart.
tau Bigot, 1883a: 84.—Mexico; Oreg. to Colo., s. to Calif., Mexico, and Tex.
tricincta Bigot, 1875a: 477.—Mexico; Tex., Cent. Amer., Hawaii (immigrant).
unipunctata Curran, 1926h: 63.—West Indies (Desecheot I.); Tex., Neotropical.
vesiculana Curran, 1947b: 5.—Ariz.; Tex.
vesicularia Curran, 1947b: 4.—N.Y.; Wis. to Ont., s. to Okla. and Fla.
    vesiculosa, authors, not Fabricius.
victoria Williston, 1887c: 145.—N. Mex.; s. Calif. to Colo. and La.

Subgenus COPESTYLUM Macquart

Copestylum Macquart, 1846: 252 (1846: 124) (as genus). Type-species, *flaviventris* Macquart (mon.)=*marginata* Say.

The larvae of the subgenus *Copestylum* live in cacti.

    REFERENCES: Curran, 1927j, 1930e, 1935b (keys).

caudata (Curran), 1927j: 45 (*Copestylum*).—Alta.; B.C. to Sask., s. to Calif. and Tex.
lenta (Williston), 1887c: 152 (*Copestylum*; as *marginata* var.).— N. Mex. (unpublished); Idaho, Colo.
limbipennis (Williston), 1887c: 152 (*Copestylum*).—Mexico; Tex.
marginata Say, 1829: 167 (1859b: 360).—Mexico; Calif. to Colo., s. to Tex. and Mexico.
    fax Townsend, 1895a: 42.—Colo.
    inops Townsend, 1895a: 43.—Colo.
similis (Giglio-Tos), 1892c: 2 (*Copestylum*).—Mexico; N. Mex., Tex.

Subgenus VOLUCELLOSIA Curran

Volucellosia Curran, 1930e: 5 (as genus). Type-species, *Volucella fornax* Townsend (orig. des.).

fornax Townsend, 1895b: 613.—Mexico (Baja Calif.); Calif., Ariz.

Tribe SERICOMYIINI

Genus SERICOMYIA Meigen

Cinxia Meigen, 1800: 35. Type-species, *Musca lappona* Linnaeus (Coquillett, 1910b: 524). Suppressed by I.C.Z.N., 1963b: 339.

Sericomyia Meigen, 1803: 274. Type-species, *Musca lappona* Linnaeus (Latreille, 1810: 443).

Condidea Coquillett, 1907a: 75. Type-species, *lata* Coquillett (orig. des.).

The "rat-tailed" larvae of the European *Sericomyia silentis* (Harris) have been found in wet decomposing turf sods in a peat bog.

REFERENCES: Curran, 1923o, 1934c (keys).

arctica Schirmer, 1913: 221.—Lapland; N.W.T., Arctic Eurasia.
bifasciata Williston, 1887c: 154.—N.H.; Mich. to Nfld., s. to Pa.
carolinensis (Metcalf), 1917a: 209 (*Cinxia*).—N.C.; Miss., Ga.
chalcopyga Loew, 1863a: 12 (Cent. 3, no. 20).—Alaska; Alaska to Alta., s. to Oreg., Utah, and Mont.
chrysotoxoides Macquart, 1842: 79 (1842: 19).—Pa.; Wis. to Nfld., s. to Tenn. and S.C.
  limbipennis Macquart, 1847: 74 (1947: 58).—N.S.
  filia Walker, 1849: 596.—Ont., N.S.
cynocephala Hine, 1922b: 147.—Alaska,
lappona (Linnaeus).—Not Nearctic.
lata (Coquillett), 1907a: 75 (*Condidea*).—Mass.; B.C. to N.B., s. to Nebr. and W. Va.
militaris Walker, 1849: 595.—Ont., N.S.; B.C. to Nfld., s. to Oreg., N. Mex., and Pa.
  calcarata Curran, 1923o: 141 (also 1925k: 182).—Mont. **N. syn.**, J. R. Vockeroth *in litt.*
slossonae Curran, 1934c: 4.—N.H.; N.Y.
  sexfasciata Walker of Curran, 1923o: 136.
sexfasciata Walker, 1849: 596.—Ont.; Alaska to Nfld. and Maine.
transversa (Osburn), 1926: 51 (*Condidea*).—N. Dak.; N. Dak. to Labr. and Nfld.

Genus ARCTOPHILA Schiner

Arctophila Schiner, 1860c: 215. Type-species, *Syrphus bombiformis* Fallén (Williston, 1887c: 158).

flagrans Osten Sacken, 1875d: 69.—Colo.; Alaska to Alta., s. to Calif. and N. Mex.
harveyi Osburn, 1908: 9.—B.C.; Wash., Oreg.

Genus PYRITIS Hunter

Pyritis Hunter, 1897: 131. Type-species, *montigena* Hunter (orig. des.).

kincaidii (Coquillett), 1895e: 131 (*Volucella*).—Wash.; B.C. to Sask., s. to Oreg. and Mont.
montigena Hunter, 1897: 132.—Idaho; B.C.

Tribe MILESIINI

The usual larval habitat of the Milesinii is in rotten wood or under bark of trees, but some genera breed in exuding sap or rot holes of trees (*Blera*), or in decaying organic matter (*Syritta, Tropidia*). *Syritta pipiens* (Linnaeus) is one of the commonest and most widespread scavengers in semiliquid, rotting vegetable and animal material. Mimicry of aculeate Hymenoptera has become developed to a considerable extent in the genera placed toward the end of this tribe.

REFERENCE: Shannon, 1926a (partial rev.).

Genus XYLOTA Meigen

Zelima Meigen, 1800: 34. Type-species, *Musca segnis* Linnaeus (Coquillett, 1910b: 621). Suppressed by I.C.Z.N., (1963b: 339.).
Heliophilus Meigen, 1803: 273. Type-species, *Musca sylvarum* Linnaeus (as *Syrphus sylvarum* Fabricius; mon.).
Eumeros Meigen, 1803: 273. Type-species, *Musca segnis* Linnaeus (Coquillett, 1910b: 541).
Eumerus Meigen, 1804: xx, error or emend.
Eumenos Leach *in* Brewster, 1817: 160, error or emend.
Xylota Meigen, 1822: 211 (unjustified n. name for *Heliophilus* Meigen). Type-species, *Musca sylvarum* Linnaeus (aut.).

An application should be prepared for submission to the I.C.Z.N. to suppress *Heliophilus* Meigen, 1803, because of possible confusion with the valid and widely used genus *Helophilus* Meigen, 1822, and to suppress *Eumeros* Meigen, 1803, because of possible confusion with the valid and economically important genus *Eumerus* Meigen, 1822, and to place *Xylota* Meigen, 1822, which has had long and almost universal usage, on the Official List of Generic Names.

REFERENCES: Shannon, 1926a (rev.); Curran, 1941 (key as *Heliophilus*).

Subgenus XYLOTA Meigen

analis Williston, 1887c: 226.—Calif., N. Mex.; B.C. to Colo., s. to Calif., N. Mex., and Mexico.

angustiventris Loew, 1866a: 164 (Cent. 6, no. 58).—Ill., Minn. to Que., s. to Kans., Miss., and Ga.

elongata Williston, 1887c: 234.—N.H., Pa.

annulifera Bigot, 1883a: 545.—N. Amer.; Alaska to Que., s. to Oreg., Tenn., and N.J.

argoi Shannon, 1926a: 38.—Wash.; B.C., Idaho, Colo., N. Mex.

artemita Hull, 1943e: 49.—Miss.; ?Ont.

atlantica Shannon, 1926a: 34 (as *naknek* var.).—Maine; B.C. to Que., s. to Wis. and Maine.

hesperia Shannon *in* Johnson, 1925b: 176 (as *hesperia atlantica*). Nomen nudum.

obscura, authors, not Loew.

barbata Loew, 1864a: 70 (Cent. 5, no. 40).—Alaska; Alaska to Alta., s. to Calif. and Colo.

bicolor Loew, 1864a: 70 (Cent. 5, no. 39).—Ill.; Nebr. to Ont. and Que., s. to Miss. and Fla.

bigelowi (Curran), 1941: 298 (*Heliophilus*).—Ont.; Alta. to Que., s. to Wis. and Mich.

florum, authors, not Fabricius.

confusa Shannon, 1926a: 30.—Maine; Minn., Mich.

ejuncida Say, 1824b: pl. 8 (1859a: 15).—Fla.; Minn. to Va., s. to Miss. and Fla.

viridaenea Shannon, 1926a: 33.—Ga.

flavifrons Walker, 1849: 557.—Ont.; Alaska to Labr., s. to B.C. and Pa.

communis Walker, 1849: 557.—Ont.

obscura Loew, 1866a: 162 (Cent. 6, no. 55).—Man.

flavitibia Bigot, 1883a: 546.—Calif.; Alaska to N.W.T., s. to Calif., N. Mex., and Nebr.

flukei (Curran), 1941: 294 (*Heliophilus*).—Wis.; Ont.

hinei (Curran), 1941: 297 (*Heliophilus*).—Que.; B.C. to N.B., s. to Oreg. and N.C.

lovetti Curran, 1925b: 44 (n. name for *bivittata* Lovett).—Calif.; B.C. to Idaho and Calif.
   bivittata Lovett, 1920: 52 (preocc. Bigot, 1883).—Calif.
   oregona Curran, 1925b: 44.—Oreg.
micrura (Curran), 1941: 295 (*Heliophilus*).—Colo.; Alta.
mixta (Curran), 1941: 295 (*Heliophilus*).—N.H.
naknek Shannon, 1926a: 33.—Alaska; Alaska to Ont., s. to Oreg., Ariz., and S. Dak.
nebulosa Johnson, 1921c. 58.—Tex.; N. Mex., S. Dak.
notha Williston, 1887c: 228.—Colo.; Nev., Utah, N. Mex. ?= *subfasciata*.
ouelleti (Curran), 1941: 296 (*Heliophilus*).—Que.; Sask.
quadrimaculata Loew, 1866a: 163 (Cent. 6, no. 56).—Ill.; B.C. to Que., s. to Miss. and Fla.
   ejuncida, authors, not Say.
rainieri Shannon, 1926a: 34.—Wash.; Oreg., Calif.
scutellarmata Lovett *in* Cole and Lovett, 1919: 241.—Oreg.; Colo.
segnis (Linnaeus), 1758: 595 (*Musca*).—Europe; Que., N.S., N.Y., Va.
stigmatipennis Lovett *in* Cole and Lovett, 1919: 242.—Oreg.
subfasciata Loew, 1866a: 164 (Cent. 6, no. 57).—Man.; Alaska to Que., s. to Oreg. and Va.
tuberculata (Curran), 1941: 294 (*Heliophilus*).—Ont.; Alta., Man., S. Dak., Que.

Subgenus XYLOTOMIMA Shannon

Xylotomima Shannon, 1926a: 15 (as genus). Type-species, *Xylota vecors* Osten Sacken (orig. des.).

althaea Hull, 1943c: 30.—S.C.
anthreas Walker, 1849: 556.—N.Y.; Man. to Que., s. to Nebr. and Ga.
   facialis Coquillett, 1910a: 126.—Mich.
arctica (Curran), 1941: 293 (*Heliophilus*).—Man.
astarte Hull, 1943e: 50.—Miss.
azurea (Fluke), 1953a: 126 (*Heliophilus*).—N. Mex.; Mexico.
chalybea Wiedemann, 1830: 98.—Unknown; Minn. to Que., s. to Kans., Miss., and Ga.
   purpurea Walker, 1849: 560.—Unknown.
curvaria (Curran), 1941: 290 (*Heliophilus*).—Que.; Alaska to N.S., s. to Calif., Colo., and N.Y.
   curvipes, authors, not Loew.
dubia (Shannon), 1926a: 22 (*Xylotomima*).—Idaho; B.C.

libo Walker, 1849: 556.—N.S.; B.C. to N.S., s. to S. Dak. and N.C.
  marginalis Williston, 1887c: 226.—N.H.
metallica Wiedemann, 1830: 102.—Ga.; Ill. to Md., s. to Tex. and
    Fla.
  subtropica Curran, 1925b: 44.—Tenn.
nemorum (Fabricius), 1805: 192 (*Milesia*).—Europe; Alaska to N.S.,
    s. to Calif. and Fla., Europe.
  baton Walker, 1849: 554.—Fla., N.S. **N. syn.**
  fraudulosa Loew, 1864a: 71 (Cent. 5, no. 41).—Wis., Ill.
    N. syn.
  americana Shannon, 1926a: 21 (*Xylotomima*; as var.).—
    Calif. **N. syn.**
nigromaculata Jones, 1917: 229.—Colo. Unrecognized.
pigra (Fabricius), 1794: 295 (*Syrphus*).—Germany; B.C. to Que.,
    s. to Calif., Fla., and Mexico, Europe.
  haematodes Fabricius, 1805: 193 (*Milesia*).—Carolina.
  rubbiginigaster Bigot, 1883a: 543.—Colo.
  rubiginigaster, error.
plesia Curran, 1925b: 45.—N.B.; Ont. to N.S., s. to W. Va.
primavera Hull, 1944a: 37.—Miss.
satanica Bigot, 1883a: 546.—Calif.; B.C., Alta., Wash.
vecors Osten Sacken, 1875d: 69.—N.H.; Alaska to Nfld., s. to Wash.,
    Colo., and N.C.

Subgenus XYLOTODES Shannon

Xylotodes Shannon, 1926a: 22 (as genus). Type-species, *Brachypalpus inarmatus* Hunter (orig. des.).

brevipilosa (Shannon), 1926a: 23 (*Xylotodes*).—Wash.; Idaho, Wyo.,
    N. Mex.
carri (Curran), 1941: 291 (*Heliophilus*).—Alta.; Ont., ?Calif.
cascadensis Weems (for *pigra* Lovett).—Oreg.; B.C. to Alta., s. to
    Oreg. and Utah. **N. name.**
  pigra Lovett *in* Cole and Lovett, 1919: 241 (*Brachypalpus*;
    preocc. Fabricius, 1794).—Oreg.
flexa (Curran), 1941: 292 (*Heliophilus*).—B.C.; Nebr., Md.
inarmata (Hunter), 1897: 142 (*Brachypalpus*).—Idaho; B.C. to
    Que., s. to Idaho and N.C.
  apicaudus Curran, 1922f: 119 (*Brachypalpus*).—B.C.
metallifera Bigot, 1883a: 545.—Colo.; Alta. to Que., s. to Colo.,
    Miss., and Ga.
  rileyi Williston, 1887c: 222 (*Brachypalpus*).—N.C.
ontario (Curran), 1941: 292 (*Heliophilus*).—Ont.; Alaska, Alta., Man.

parva (Williston), 1887c: 222 (*Brachypalpus*).—Colo.; B.C. and N.W.T. to Oreg. and Colo.
sacajaweae (Shannon), 1926a: 24 (*Xylotodes*).—Wash.; Utah.

Genus BRACHYPALPUS Macquart

Brachypalpus Macquart, 1834a: 523. Type-species, *tuberculatus* Macquart (Rondani, 1844b: 456) =*valgus* (Panzer).
margaritus Hull, 1946: 271.—Miss.
oarus (Walker), 1849: 558 (*Xylota*).—N.Y.; Man. to N.B., s. to Kans. and N.C.
    frontosus Loew, 1872a: 83 (Cent. 10, no. 50).—D.C.
trifasciatus Hull, 1944a: 39.—Wash.

Genus CALIPROBOLA Rondani

Caliprobola Rondani, 1844b: 455. Type-species, *Syrphus speciosus* Rossi (as *Milesia speciosa* Fabricius; Rondani, 1856: 47).
Calliprobola, emend.
Chrysosomidia Curran, 1934a: 261. Type-species, *Caliprobola crawfordi* Shannon (orig. des.). **N. syn.**
    REFERENCE: Shannon, 1916c (rev.).
aepalia (Walker), 1849: 557 (*Xylota*).—Ga.; Que., N.J., Tenn., N.C.
    sorosis Williston, 1887c: 223 (*Brachypalpus*).—Ga.
aldrichi Shannon, 1916c: 111.—Wash.; B.C., Oreg., Idaho, Calif.
crawfordi Shannon, 1916c: 112.—Wash.; B.C. to Idaho and Calif.
opaca Shannon, 1916c: 110.—Wash.; Alaska, B.C.
pulchra (Williston), 1882b: 79 (?*Brachypalpus*).—Wash., Oreg.; B.C. to Idaho and Oreg.
    aerea Bigot, 1883a: 352.—Wash. Territory.

Genus TEUCHOCNEMIS Osten Sacken

Teuchocnemis Osten Sacken, 1875d: 58. Type-species, *Pterallastes lituratus* Loew (Williston, 1887c: 199).
bacuntius (Walker), 1849: 563 (*Milesia*).—Ga.; N.Y. to Tex. and Fla.
lituratus (Loew), 1863b: 317 (Cent. 4, no. 81) (*Pterallastes*).—Pa.; Wis. to Que., s. to Mo. and Ga.

Genus SYRITTA Lepeletier and Serville

Syritta Lepeletier and Serville *in* Latreille et al., 1828: 808. Type-species, *Musca pipiens* Linnaeus (as *Xylota pipiens* Meigen; mon.).

pipiens (Linnaeus), 1758: 594 (*Musca*).—Europe; B.C. to Nfld., s. to Calif. and Fla., Eurasia, Africa.
proxima Say, 1824b: pl. 8 (1859a: 16) (*Xylota*).—Pa, Va.

Genus Acrochordonodes Bigot

comstocki (Williston), 1882a: 326 (*Senogaster*).—Unknown (as N.Y., error).=*dentipes* (Fabricius), not Nearctic.

Genus TROPIDIA Meigen

Tropidia Meigen, 1822: 346. Type-species, *Eristalis milesiformis* Fallén (Curtis, 1832: pl. 401)=*scita* (Harris).

REFERENCE: Shannon, 1926a (rev.).

albistylum Macquart, 1847: 76 (1847: 60).—N. Amer.; Mich. to N.J., s. to Okla., Tex., and Fla.
calcarata Williston, 1887c: 208.—Mich.; Wis. to Que., s. to Ind. and S.C.
coloradensis (Bigot), 1883a: 544 (*Xylota*).—Colo.
incana Townsend, 1895a: 52.—Colo.
mamillata Loew, 1861b: 343 (Cent. 1, no. 68).—Ill.; Nebr., Kans., N.C.
montana Hunter, 1896d: 305 (n. name for *nigricornis* Hunter).—Idaho; Oreg.
nigricornis Hunter, 1896c: 215 (preocc. Philippi, 1865).—Idaho.
pygmaea Shannon, 1926a: 10.—Utah.
quadrata (Say), 1824b: pl. 8 (1859a: 14) (*Xylota*).—Pa.; B.C. to Nfld., s. to Calif., La., and N.C.

Genus PTERALLASTES Loew

Pterallastes Loew, 1863b: 317 (Cent. 4, no. 80). Type-species, *thoracicus* Loew (Osten Sacken, 1878c: 250).

thoracicus Loew, 1863b: 317 (Cent. 4, no. 80).—Pa.; Wis. to N.Y., s. to Nebr., Miss., and Ga.

Genus HADROMYIA Williston

Hadromyia Williston, 1882b: 78. Type-species, *grandis* Williston (mon.).

grandis Williston, 1882b: 79.—Wash. Territory; B.C. to Oreg. and Idaho.

morissoni Bigot, 1883a: 355 (*Brachypalpus*).—Wash. Territory.
morrisoni, error.

Genus CRIOPRORA Osten Sacken

Crioprora Osten Sacken, 1878c: 251. Type-species, *Pocota alopex* Osten Sacken (Williston, 1887c: 217).

REFERENCE: Williston, 1887c (key).

alopex (Osten Sacken), 1877: 338 (*Pocota*).—Calif.; B.C. to Calif.
amithaon (Walker), 1849: 567 (*Milesia*).—N.C. Unrecognized.
cyanella (Osten Sacken), 1877: 339 (*Pocota*).—Calif.; B.C., Oreg.
cyanogaster (Loew), 1872a: 83 (Cent. 10, no. 51) (*Brachypalpus*).—Pa.; Colo., Wis., Que., N.J.
femorata Williston, 1882a: 329.—Wash. Territory; B.C. to Mont., s. to Calif. and Colo.

Genus POCOTA Lepeletier and Serville

Pocota Lepeletier and Serville *in* Latreille et al., 1828: 518. Type-species, *Milesia apicata* Meigen (mon.)=*apiformis* (Schrank).

bomboides Hunter, 1897: 141.—Calif.; B.C. to Idaho and Calif.

Genus BLERA Billberg

Blera Billberg, 1820: 118. Type-species, *Musca fallax* Linnaeus (Johnson, 1911: 73).
Criorhina, subg. *Cynorhina* Williston, 1887c: 209. Type-species, *Milesia analis* Macquart (orig. des.).

REFERENCES: Curran, 1925k, 1953b (keys as *Cynorhina*).

analis (Macquart), 1842: 139 (1842: 79) (*Milesia*).—N. Amer.; S. Dak. to Que. and Maine, s. to N.J.
armillata armillata (Osten Sacken), 1875d: 68 (*Criorhina*).—Que.; Alaska to Que., s. to Calif. and N.H.
  ssp. **hunteri** (Curran), 1925k: 136 (*Cynorhina*; as var.).—Man.; Alaska, Utah.
  ssp. **pacifica** (Curran), 1953b: 14 (*Cynorhina*).—Wash.; B.C., Oreg., Idaho.
badia (Walker), 1849: 559 (*Xylota*).—N.Y.; Minn. to Que., s. to Ga.
  intersistens Walker, 1849: 615 (*Eristalis*).—N.Y.
banksi (Hull), 1945b: 72 (*Cynorhina*).—Maine. **N. comb.**
confusa Johnson, 1913c: 294.—Maine; B.C. to N.S., s. to Colo. and N.C.

flukei (Curran), 1953b: 13 (*Cynorhina*).—Wash.; Alaska to Oreg. and Idaho. **N. comb.**
garretti (Curran), 1924g: 194 (*Cynorhina*).—B.C.; Alta., Mont. **N. comb.**
humeralis (Williston), 1882a: 330 (*Criorhina*).—Wash. Territory; Wash. and Idaho to Calif.
johnsoni (Coquillett), 1894c: 125 (*Criorhina*).—Wash.; Oreg.
metcalfi (Curran), 1925k: 132 (*Cynorhina*).—N.C. **N. comb.**
nigra (Williston), 1887c: 214 (*Criorhina*).—N.H.; Wash. to Nfld., s. to N.C.
nigripes (Curran), 1925k: 128 (*Cynorhina*).—B.C.; Wash., Wis. **N. comb.**
notata (Wiedemann), 1830: 109 (*Milesia*).—Ga.; Pa. and N.J. to Fla.
  profusus Walker, 1849: 578 (*Syrphus*).—Ga.
pictipes (Bigot), 1883a: 354 (*Caliprobola*).—Carolina; Mich. to N.Y. s. to Tex. and N.C.
robusta (Curran), 1922d: 14 (*Cynorhina*).—B.C.; Wash., Mont. **N. comb.**
scitula (Williston), 1882a: 331 (*Criorhina*).—Wash. Territory; Alaska to Calif. and Idaho.
umbratilis (Williston), 1887c: 212 (*Criorhina*).—Conn.; Nebr. to N.Y. and Nfld., s. to Tex. and Ga.

Genus CRIORHINA Meigen

Penthesilea Meigen, 1800: 35. Type-species, *Musca ruficauda* De Geer (sub. mon., Bezzi *in* Hendel, 1908b: 67; misident.) = *ranunculi* (Panzer). Suppressed by I.C.Z.N., 1963b: 339.
Criorhina Meigen, 1822: 236. Type-species, *Syrphus asilicus* Fallén (Westwood, 1840: 136).
Criorrhina, emend.
Brachymyia Williston, 1882b: 77. Type-species, *lupina* Williston (Coquillett, 1910b: 516).

REFERENCE: Curran, 1925k (rev.).

aurea Lovett *in* Cole and Lovett, 1919: 248.—Idaho; B.C., Oreg.
caudata Curran, 1925k: 151.—B.C.; Wash., Oreg., Idaho.
coquilletti Williston, 1892a: 145.—Calif.; B.C., Wash.
grandis Lovett *in* Cole and Lovett, 1921: 291.—Oreg.; Wash.
kincaidi Coquillett, 1901h: 611.—Wash.; Alaska to Oreg. and Idaho.
latipilosa Curran, 1925k: 149.—Mont.; Oreg., Ont.
luna Lovett *in* Cole and Lovett, 1919: 249.—Oreg.; Alaska to Oreg.
lupina (Williston), 1882b: 77 (*Brachymyia*).—Calif.; Oreg.
mystaceae Curran, 1925k: 144.—N.S.

nigripes (Williston), 1882b: 78 (*Brachymyia*).—Calif.; B.C. to Calif. and Idaho.
nigriventris Walton, 1911: 318.—Pa.; B.C., Man. to N.S., s. to Va.
    intermedia Johnson, 1917c: 153.—N.H., Mass. **N. syn.**, J. R. Vockeroth *in litt.*
    maritima Curran, 1924i: 252.—N.S. **N. syn.**, J. R. Vockeroth *in litt.*
    aurata Curran, 1925k: 146 (as *verbosa* var.).—Pa. **N. syn.**, J. R. Vockeroth *in litt.*
quadriboscis Lovett *in* Cole and Lovett, 1919: 250.—Oreg.
tricolor Coquillett, 1900h: 436 (1904: 50).—Alaska; Alaska to Calif.
verbosa (Walker), 1849: 568 (*Milesia*).—N. Amer.; Man. to N.S., s. to Ohio and N.C.

Genus SOMULA Macquart

Somula Macquart, 1847: 73 (1847: 57). Type-species, *decora* Macquart (orig. des.).

    REFERENCE: Curran, 1925k (rev.).

decora Macquart, 1847: 73 (1847: 57).—Pa.; Minn. to N.B., s. to Tex. and Ga., also Calif.
mississippiensis Hull, 1922: 371.—Miss.; N.C.
    marivirginiae Brimley, 1923: 278.—N.C.

Genus MERAPIOIDUS Bigot

Merapioidus Bigot, 1879b: 1 (=50). Type-species, *villosus* Bigot (mon.).

villosus Bigot, 1879b: 1 (=50).—Ga.; Ohio and Pa. to Ga.

Genus SPHECOMYIA Latreille

Sphecomyia Latreille *in* Bory, 1829: 545 (also Latreille, 1829: 495). Type-species, *Chrysotoxum vittatum* Wiedemann (Coquillett, 1910b: 607).

    REFERENCES: Shannon, 1925c, Curran, 1932b (keys).

brevicornis Osten Sacken, 1877: 341.—Calif.; B.C. to Idaho, s. to Calif.
    vespiformis, authors, not Gorski.
dyari Shannon, 1925c: 43.—Calif.
nasica Osburn, 1908: 13.—B.C.; Wash., Oreg.
occidentalis Osburn, 1908: 12.—B.C., Wash.; Idaho.

pattonii Williston, 1882a: 328.—Wash. Territory; B.C. to Idaho and Calif.

calorhina Bigot, 1883a: 353 (*Caliprobola*).—Wash. Territory.
vittata (Wiedemann), 1830: 87 (*Chrysotoxum*).—Unknown; Man. to Que., s. to N. Mex. and Fla.

ornatus Wiedemann, 1830: 91 (*Psarus*).—Ga.

Genus SPILOMYIA Meigen

Tritonia Meigen, 1800: 33 (preocc. Cuvier, 1798). Type-species, *Musca vespiformis* Linnaeus (Coquillett, 1910b: 617). Suppressed by I.C.Z.N., 1963b: 339.
Spilomyia Meigen, 1803: 273. Type-species, *Musca diophthalma* Linnaeus (as *Syrphus diophthalmus* Fabricius; Coquillett, 1910b: 607; misident.) = *saltuum* (Fabricius).
Mixtemyia Macquart, 1834a: 491. Type-species, *Paragus quadrifasciatus* Say (mon.).

REFERENCES: Curran, 1951 (key); Vockeroth, 1958 (partial key).

citima Vockeroth, 1958: 287.—B.C.; Wash. Territory.
crandalli Curran, 1951: 7.—Ariz.
foxleei Vockeroth, 1958: 284.—B.C.
fusca Loew, 1864a: 67 (Cent. 5, no. 34).—Pa.; Minn. to N.S., s. to Ga.
hamifera Loew, 1864a: 66 (Cent. 5, no. 33).—Pa.; Wis. to Nfld., s. to Miss. and Fla.
interrupta Williston, 1882a: 327.—Wash. Territory; B.C. to Calif. and Colo.
kahli Snow, 1895d: 245.—N. Mex.; Ariz., Tex., Mexico.

xanthocauda Curran, 1935b: 6.—Ariz.
liturata Williston, 1887c: 245.—N. Mex.; Wash. to Idaho, s. to Ariz. and N. Mex.
longicornis Loew, 1872a: 82 (Cent. 10, no. 49).—Mass., Pa., Tex.; Minn. to Que., s. to Tex., Mexico, and Fla.
quadrifasciata (Say), 1824a: 377 (1859a: 257) (*Paragus*).—Northwest Territory (U.S.); Alta. to N.B., s. to Kans. and N.J.
texana Johnson, 1921c: 57.—Tex.; La., Ala., Ga.

Genus TEMNOSTOMA Lepeletier and Serville

Temnostoma Lepeletier and Serville *in* Latreille et al., 1828: 518. Type-species, *Milesia bombylans* Fabricius (Coquillett, 1910b: 612).

REFERENCES: Curran, 1930j (key), 1939d (rev.); Shannon, 1939 (rev.); Krivosheina and Mamajev, 1962 (immature stages).

acrum Curran, 1939d: 2.—Conn.; Man. to N.S., s. to Kans. and Va.
bombylans, authors, in part, not Fabricius.
alternans Loew, 1864a: 68 (Cent. 5, no. 37).—Pa.; Sask. to N.S., s. to Ohio and Ga.
balyras (Walker), 1849: 577 (*Syrphus*).—N.Y.; Man. to N.B., s. to Miss. and Ga.
bombylans, authors, in part, not Fabricius.
barberi Shannon, 1939: 219.—N.C.; Minn. to Maine, s. to Ill. and Fla.
excentricum (Harris), 1862: 609 (*Milesia*).—Mass.; Wis. to Que., s. to Tenn. and N.C.
obscurum Loew, 1864a: 67 (Cent. 5, no. 35).—Sask.; B.C., Mich., Ont., Que.
pictulum Williston, 1887c: 251.—Pa.; Ohio to N.Y., s. to Miss. and Fla.
greenei Shannon, 1939: 221.—Md. **N. syn.**
trifasciatum Robertson, 1901: 285.—Ill.; Minn. to Que., s. to Tenn. and Fla.
venustum Williston, 1887c: 253.—N.H.; Mich. to Que., s. to N.C.
nipigonense Curran, 1923m: 269.—Ont. **N. syn.**, J. R. Vockeroth *in litt.*
vespiforme (Linnaeus), 1758: 593 (*Musca*).—Europe; Alaska to Que., s. to N. Mex. and Tenn.
aequale Loew, 1864a: 68 (Cent. 5, no. 36).—Ont.
apiforme, authors, not Fabricius.

Genus MILESIA Latreille

Milesia Latreille, 1804: 194. Type-species, *Musca diophthalma* Linnaeus (Rondani, 1844b: 455).
Sphixea Rondani, 1844b: 455. Type-species, *Eristalis fulminans* Fabricius (mon.) = *semiluctifera* (Villers).
Sphyxea, error.

REFERENCE: Hull, 1924 (rev.).

bella Townsend, 1897a: 142.—N. Mex.; Ariz.
mida Moodie, 1905: 138.—Ariz.
scutellata Hull, 1924: 280.—Miss.; La. to N.C. and Fla.
virginiensis (Drury), 1773: 71, pl. 37, fig. 6 (index to vol. 2, p. [2]) (*Musca*).—Va.; Minn. to Ont., s. to Tex. and Fla.
trifasciatus Hausmann, 1799: 37 (*Syrphus*).—N. Amer.
ornata Fabricius, 1805: 188.—Carolina.
fulvifrons Bigot, 1883a: 341 (*Sphixea*).—Ga.

Tribe CERIOIDINI

In the tribe Cerioidini we find some of the most remarkable examples of the mimicry of Hymenoptera in the Diptera, bringing to a climax a tendency, which has been expressed in different degrees in various tribes and genera of Syrphidae. The larvae of *Ceriana* have been found in the exuding sap of ulcerated trees.

REFERENCE: Shannon, 1925b (rev.).

Genus CERIANA Rafinesque

Ceria Fabricius, 1794: 277 (preocc. Scopoli, 1763). Type-species, *clavicornis* Fabricius (Latreille, 1810: 443) = *conopsoides* (Linnaeus).
Cina Fabricius, 1798: 557. Unavailable name cited in generic synonymy.
Ceriana Rafinesque, 1815: 131 (n. name for *Ceria* Fabricius). Type-species, *Ceria clavicornis* Fabricius (aut.) = *conopsoides* (Linnaeus).
Cerioides Rondani, 1850b: 211. Unavailable name cited in synonymy of *Sphiximorpha* Rondani.
Ceriodes, error.
Sphiximorpha Rondani, 1850b: 212. Type-species, *Ceria subsessilis* Illiger (orig. des.).
Sphyximorpha, error.

Because there is conflicting usage between *Cerioides* and *Sphiximorpha*, it is preferable to change to the correct name under the rules, *Ceriana*, hitherto unused.

REFERENCES: Curran, 1925k (rev.), 1941 (key); Goffe, 1945a (syn.).

abdominalis (Curran), 1925k: 26 (*Cerioides*).—Ariz. **N. comb.**
aldrichi (Hull), 1935b: 99 (*Cerioides*).—Calif. **N. comb.**
cylindrica (Curran), 1921c: 175 (*Ceria*).—Calif. **N. comb.**
durani (Shannon), 1925b: 62 (*Cerioides*).—Calif. (unpublished). **N. comb.**
    durani Davidson, 1926: 40 (*Cerioides*).—Calif. **N. syn., n. comb.**
loewii (Williston), 1887c: 260 (*Ceria*).—Ariz. **N. comb.**
ontarioensis (Curran), 1921c: 174 (*Ceria*).—Ont.; Wis., Md. **N. comb.**
schnablei (Williston), 1892b: 76 (*Ceria*).—Mexico (V.C.); Tex. **N. comb.**

signifera (Loew), 1853: 18 (*Ceria*).—Mexico; Minn. to N.Y. and Mass., s. to Tex., Mexico, and Fla. **N. comb.**
willistoni Kahl, 1897: 141 (*Ceria*).—Kans. **N. comb.**

Genus TENTHREDOMYIA Shannon

Tenthredomyia Shannon, 1925b: 50. Type-species, *Ceria abbreviata* Loew (orig. des.).

abbreviata (Loew), 1864a: 75 (Cent. 5, no. 48) (*Ceria*).—Pa.; B.C. and N.W.T. to Que., s. to Colo. and Fla.
proxima Curran, 1924o: 228 (*Cerioides*).—Ont.
mime Hull, 1935b: 100.—La.
pictula (Loew), 1853: 17 (*Ceria*).—Southern States. Unrecognized.
snowi (Adams), 1904b: 447 (*Sphiximorpha*).—Ariz.; Calif.
tridens (Loew), 1872a: 86 (Cent. 10, no. 57) (*Ceria*).—Calif.; Wash. and Idaho to Calif. and Tex.
ancoralis Coquillett, 1902a: 196 (*Sphiximorpha*).—N. Mex. **N. syn.**

Genus POLYBIOMYIA Shannon

Polybiomyia Shannon, 1925b: 56. Type-species, *schwarzi* Shannon (orig. des.).

REFERENCE: Shannon, 1925b (key).

bellardii Shannon, 1925b: 59.—Tex.
capitis (Curran), 1925k: 27 (*Cerioides*).—Mexico; Tex.
engelhardti Shannon, 1925b: 60.—Ariz.
festiva Hull, 1930c: 178.—Tex.
macquarti Shannon, 1925b: 59 (n. name for *scutellata* Williston).—Mexico; Tex.
scutellata Williston, 1887c: 265 (*Ceria*; preocc. Macquart, 1842).—Mexico.
pedicellata (Williston), 1887c: 264 (*Ceria*).—Mexico; Ariz., Tex., Fla.
reinhardi Hull, 1930c: 180.—Tex.
sayi Shannon, 1925b: 61.—Ariz.
townsendi (Snow), 1895d: 246 (*Ceria*).—N. Mex.; Calif., Ariz., Tex.

Genus MONOCEROMYIA Shannon

Cerioides, subg. **Monoceromyia** Shannon, 1922b: 41. Type-species, *Ceria tricolor* Loew (mon.).

floridensis (Shannon), 1922b: 41 (*Cerioides*; as *tricolor* var.).—Fla.

Tribe MERODONTINI

Genus MERODON Meigen

Lampetia Meigen, 1800: 34. Type-species, *Syrphus clavipes* Fabricius (Coquillett, 1910b: 557). Suppressed by I.C.Z.N., 1963b: 339.

Merodon Meigen, 1803: 274. Type-species, *Syrphus clavipes* Fabricius, as *curvipes* (error by Meigen) (Westwood, 1840: 137).

The phytophagous habits and short-tailed structure of the larvae of *Merodon* warrant the exclusion of this genus from the Eristalini and recognition of a separate tribe along the lines proposed by Goffe (1952: 114). The larvae of the narcissus bulb fly, *Merodon equestris* (Fabricius), a species widely spread by commerce, are important primary pests of Narcissus and other bulbs.

REFERENCES: Hodson, 1932, Doucette et al., 1942 (biol., economics).

equestris (Fabricius), 1794: 292 (*Syrphus*).—Europe; B.C. to Calif., Wis. to N.B. and Ga. Immigrant from Europe.
  narcissi Fabricius, 1805: 239 (*Eristalis*).—Europe.
  transversalis Meigen, 1822: 354.—Europe.
  validus Meigen, 1822: 365.—Europe.

Tribe ERISTALINI

The larvae of Eristalini are aquatic and of the long-tailed type. Those of *Eristalis* and *Helophilus* are very commonly found in large numbers breeding in putrid or stagnant water or in moist excrement, and are called "rat-tailed maggots" or "mousies." Other genera breed in wet rot holes in trees, bromeliads, and other accumulations of water.

Genus HELOPHILUS Meigen

Helophilus Meigen, 1822: 368. Type-species, *Musca pendula* Linnaeus (Curtis, 1832: pl. 429). See *Eristalis* and *Xylota* for discussion of confused nomenclature.

REFERENCE: Curran and Fluke, 1926 (rev.).

Subgenus HELOPHILUS Meigen

alaskensis Fluke, 1949: 50.—Alaska.
borealis Staeger, 1845b: 359.—Greenland; Alaska to Greenland, s. to Oreg., Colo., and N.Y., n. Europe.
  glacialis Loew, 1846a: 121.—Labr.

dychei Williston *in* Hunter, 1897: 136.—Alaska.
bruesi Graenicher, 1910b: 40.—Wis.
fasciatus Walker, 1849: 605.—Ont.; B.C. to Que., s. to Calif., Mexico, and Fla.
similis Macquart, 1842: 124 (1842: 64) (preocc. Curtis, 1832).—Ga.
decisus Walker, 1849: 614 (*Eristalis*).—Ont.
sussurans Jaennicke, 1867: 402 (1868: 94).—Ill.
frater (Walker), 1849: 613 (*Eristalis*).—Ont. Unrecognized.
groenlandicus (Fabricius), 1780: 208 (*Tabanus*).—Greenland; Alaska to Greenland, s. to B.C., Colo., and N.H.
bilineatus Curtis, 1835b: lxxviii.—Arctic Amer.
latro Walker, 1849: 607.—Ont., N.S.
androclus Walker, 1849: 612 (*Eristalis*).—Ont.
chalepus Walker, 1852: 247 (*Eristalis*).—Canada.
arcticus, authors, not Zetterstedt.
hybridus Loew, 1846a: 141.—Europe; Alaska to N.S., s. to B.C., Utah, N. Dak., and N.H., Europe.
novaescotiae Macquart, 1847: 76 (1847: 60).—N.S.
latitarsis Hunter, 1897: 134.—Minn.
intentus Curran and Fluke, 1926: 223.—Sask.; Alaska to B.C. and Sask.
pendulus, authors, not Linnaeus.
latifrons Loew, 1863b: 313 (Cent. 4, no. 73).—Nebr.; Alaska to Labr., s. to Calif., Mexico, and Fla.
trivittatus, authors, not Fabricius.
neoaffinis Fluke, 1949: 49.—Alaska.
obscurus Loew, 1863b: 314 (Cent. 4, no. 74).—Colo.; Alaska to Que., s. to B.C., Colo., and N.Y.
pilosus Hunter, 1897: 137.—B.C. Unrecognized.
stricklandi Curran, 1927f: 90.—Alta.; N.W.T., Man.

Subgenus ANASIMYIA Schiner

Helophilus, subg. **Anasimyia** Schiner, 1864: 108. Type-species, *Musca transfuga* Linnaeus (Coquillett, 1910b: 506).
Lejops, authors, not Rondani.

According to A. A. Stackelberg, *in litt.*, the European *Mallota vittata* Meigen, type-species of *Lejops* Rondani, is subgenerically distinct from any other species formerly placed in this subgenus.

bilinearis Williston, 1887c: 295.—Colo.; Alta. to Que., s. to Colo. and N.J., also Wash.
chrysostomus (Wiedemann), 1830: 174 (*Eristalis*).—Ga.; Minn. to Ont. and Mass., s. to Colo. and Fla.

colei Wirth (for *borealis* Cole).—Alaska (St. Paul I.). **N. name.**
   borealis Cole, 1921: 170 (*Pterallastes;* preocc. Staeger, 1845).—
      Alaska (St. Paul I.). **N. comb.**
distinctus Williston, 1887c: 192.—Conn., Pa., Va.; Wis. to Mass., s.
      to Va.
grisescens (Hull), 1943c: 31 (*Lejops*).—N.J. **N. comb.**
lunulatus Meigen, 1822: 370.—Europe; Alaska to Que., s. to Wash.,
      Colo., and N.J., Europe.
   hamatus Loew, 1863b: 316 (Cent. 4, no. 79).—Hudson Bay
      Territory.
orion (Hull), 1943e: 51 (*Lejops*).—Miss.; Ga. **N. comb.**
perfidiosus (Hunter), 1897: 139 (*Pterallastes*).—B.C.; Alaska to Man.,
      s. to Calif. and Colo.
relictus (Curran and Fluke), 1926: 256 (*Lejops*).—Ont.; Wis. to N.B.,
      s. to Colo., Miss., and N.C. **N. comb.**

Subgenus PARHELOPHILUS Girschner

Helophilus, subg. **Parhelophilus** Girschner, 1897: 604. Type-species,
   Syrphus frutetorum Fabricius (Curran and Fluke, 1926:
   230).

anniae Brimley, 1923: 278.—N.C.; Que. to Ga.
currani (Fluke), 1953a: 128 (*Parhelophilus*).—La. **N. comb.**
divisus Loew, 1863b: 316 (Cent. 4, no. 78).—D.C.; Mich. to Ont., s.
      to Ind. and Fla.
flavifacies Bigot, 1883a: 344.—Md. ?=*divisus.*
integer Loew, 1863b: 314 (Cent. 4, no. 76).—N.Y.; Ont. and Que. to
      N.C.
laetus Loew, 1863b: 315 (Cent. 4, no. 77).—N.Y., Wis.; B.C. to Que.,
      s. to N. Mex. and N.C.
   aureopilis Townsend, 1895a: 51.—Mich.
obsoletus Loew, 1863b: 314 (Cent. 4, no. 75).—Hudson Bay Territory; Alaska to Ont. and Maine, s. to B.C. and Wis.
porcus (Walker), 1849: 551 (*Eumerus*).—Ont.; B.C. to N.B., s. to
      Wis. and Md.
rex (Curran and Fluke), 1926: 234 (*Parhelophilus*).—Ont.; B.C. to
      Ont., s. to Colo. and N.Y. **N. comb.**

Subgenus EURIMYIA Bigot

Eurimyia Bigot, 1883b: xx (as genus). Type-species, *rhingioidis*
   Bigot (mon.)=*lineatus* (Fabricius).
Eurhimyia, emend. or error.

stipatus Walker, 1849: 602.—N.Y.; Alaska to Que., s. to Wash., Iowa, and N.J.
   anausis Walker, 1849: 603.—Ont.
   conostomus Williston, 1887c: 193.—Conn., Mass., Ill., Canada.
   lineatus, authors, not Fabricius.

Subgenus LUNOMYIA Curran and Fluke

Lunomyia Curran and Fluke, 1926: 252 (as genus). Type-species, *Tropidia cooleyi* Seamans (orig. des.).

brooksi (Curran), 1927f: 90 (*Parhelophilus*).—Man.; Alta., Minn., Wis. **N. comb.**
   pollinaria Fluke, 1939: 373 (*Lunomyia*).—Wis. **N. comb.**
cooleyi (Seamans), 1917: 342 (*Tropidia*).—Mont.; Oreg. to Sask., s. to Calif., Colo., and S. Dak.
   modestus Williston, 1887c: 192 (preocc. Wiedemann, 1818).—Wyo.

Genus ASEMOSYRPHUS Bigot

Asemosyrphus Bigot, 1882c: cxxviii. Type-species, *oculiferus* Bigot (Aldrich, 1933b: 168) = *mexicanus* (Macquart).

REFERENCES: Curran and Fluke, 1926 (rev.); Curran, 1939c (rev.).

polygrammus (Loew), 1872a: 85 (Cent. 10, no. 55) (*Helophilus*). Calif.; B.C. to Mont., s. to Calif., Mexico, and Colo.
   mexicanus, authors, not Macquart.

Genus ARCTOSYRPHUS Frey

Arctosyrphus Frey, 1918b: 15. Type-species, *nitidulus* Frey (orig. des.).

willingii (Smith), 1912: 118 (*Helophilus*).—Sask.; B.C. and N.W.T. to Man. and N. Dak. **N. comb.**
   canadensis Curran, 1922f: 94 (*Asemosyrphus*).—Sask. **N. comb.**

Genus MALLOTA Meigen

Mallota Meigen, 1822: 377. Type-species, *Syrphus fuciformis* Fabricius (Rondani, 1844b: 452).
Imatisma Macquart, 1842: 127 (1842: 67). Type-species, *Eristalis posticatus* Fabricius (orig. des.).

REFERENCES: Curran, 1930j, 1940 (keys), 1953b (syn.).

albipilis Snow, 1895d: 244.—N. Mex.; Kans., Tex.
    diversipennis Curran, 1922d: 16.—Unknown (?Canada). **N. syn.**, J. R. Vockeroth *in litt.*
bautias (Walker), 1849: 600 (*Merodon*).—Ga.; Wis. to Que., s. to Colo., Tex., and Fla.
    facialis Hunter, 1896a: 100.—Nebr. **N. syn.**, J. R. Vockeroth *in litt.*
    flavoterminata Jones, 1917: 228.—Colo.
    cimbiciformis, authors, not Fallén.
    dentipes Williston *in* Lintner, 1883: 211 (as *cimbiciformis* var.). Nomen nudum.
bequaerti Hull, 1956: 24.—Tex.; Ariz., Mexico.
bipartita (Walker), 1849: 599 (*Merodon*).—Ga. Unrecognized.
illinoensis Robertson, 1901: 284.—Ill.; Nebr., Kans., Wis., ?N.C.
mississippensis Hull, 1946: 268.—Miss.
palmerae Jones, 1917: 229.—Colo.; N. Mex.
posticata (Fabricius), 1805: 237 (*Eristalis*).—Carolina; Minn. to Que., s. to Kans. and Fla.
    coactus Wiedemann, 1830: 165 (*Eristalis*).—Unknown.
    balanus Walker, 1849: 600 (*Merodon*).—N.Y.
    barda (in part) Say, 1829: 163 (1859b: 357) (*Milesia*). Male only.
sackeni Williston, 1882a: 324.—Wash. Territory; B.C., Wash., Oreg., Idaho, Utah.
    columbiae Curran, 1922d: 15.—B.C.
separata Hull, 1945a: 184.—Miss.

Genus POLYDONTOMYIA Williston

Polydonta Macquart, 1850: 448 (1850: 144) (preocc. Fischer, 1807). Type-species, *bicolor* Macquart (orig. des.)=*curvipes* (Wiedemann).
Triodonta Williston, 1885b: 136 (n. name for *Polydonta* Macquart, but preocc. Bory, 1827). Type-species, *Polydonta bicolor* Macquart (aut.) =*curvipes* (Wiedemann).
Polydontomyia Williston, 1896a: 89. Type-species, *Merodon curvipes* Wiedemann (Coquillett, 1910b: 593).

curvipes (Wiedemann), 1830: 149 (*Merodon*).—N. Amer.; Alaska to Man., s. to Calif., Mexico, and Nebr., also e. Que. and N.S., s. to N.J.
    albiceps Macquart, 1846: 260 (1846: 132) (*Helophilus*).—N.S.
    morosus Walker, 1849: 599 (*Merodon*).—N.S.
    bicolor Macquart, 1850: 448 (1850: 144) (*Polydonta*).—N.S.

Genus MEROMACRUS Rondani

Plagiocera Macquart, 1842: 119 (1842: 59) (preocc. Klug, 1834). Type-species, *Milesia crucigera* Wiedemann (Coquillett, 1910b: 591) =*acutus* (Fabricius).
Meromacrus Rondani, 1848: 70. Type-species, *ghiliani* Rondani (mon.).

REFERENCE: Hull, 1942a (rev.).

acutus (Fabricius), 1805: 189 (*Milesia*).—Carolina; Va. to Fla. and Tex., Neotropical.
  crucigera Wiedemann, 1830: 105 (*Milesia*).—Ga.
draco Hull, 1942a: 3.—Guatemala; Ariz., Tex., Cent. Amer.
gloriosus Hull, 1941a: 166.—N. Mex.; Tex.
ruficrus (Wiedemann), 1830: 105 (*Milesia*).—Cuba; Fla., Jamaica.

Genus ERISTALIS Latreille

Tubifera Meigen, 1800: 34. Type-species, *Musca tenax* Linnaeus (Coquillett, 1910b: 618). Suppressed by I.C.Z.N., 1963b: 339.
Elophilus Meigen, 1803: 274. Type-species, *Musca tenax* Linnaeus (as *Eristalis tenax* Fabricius; Latreille, 1810: 443).
Elophila Meigen, 1804: xx, error.
Helophila Fabricius, 1805: 232, error. In specific synonymy of *Eristalis flavicinctus*.
Helophilus Fabricius, 1805: 233, 234, 238, error. In specific synonymies of *Eristalis pendulus, nemorum,* and *tenax*.
Helophilus Leach *in* Brewster, 1817: 159, emend.
Eristalis Latreille, 1804: 194. Type-species, *Musca tenax* Linnaeus (Curtis, 1832: pl. 432).
Eristaloides Rondani, 1844b: 453. Type-species, *Musca tenax* Linnaeus (Coquillett, 1910b: 540).
Eristalomya Rondani, 1857: 38. Type-species, *Musca tenax* Linnaeus (orig. des.).
Eristalomyia, error or emend.

An application should be prepared for submission to the I.C.Z.N. to suppress *Elophilus* Meigen, 1803, and the emendation *Helophilus* Leach, 1817, because of possible confusion with the valid genus *Helophilus* Meigen, 1822, and to preserve the long and almost universal usage of *Eristalis* Latreille in syrphid literature by placing this name on the Official List of Generic Names.

REFERENCES: Hull, 1925b (rev.); Curran, 1930d (key); Goffe, 1944b (nomenclature); Telford, 1949 (key); Bean, 1949 (rev., male genitalia).

Subgenus ERISTALIS Latreille

agrorum (Fabricius), 1787: 335 *(Syrphus).*—West Indies; La., Fla., Neotropical.
    cubensis Macquart, 1842: 102 (1842: 42).—Cuba.
albifrons Wiedemann, 1830: 189.—Brazil; La. to Fla. and N.C., Neotropical.
    albiceps Macquart, 1842: 116 (1842: 56).—Carolina.
alhambra Hull, 1925b: 39.—Calif.; Ariz.
anthophorinus (Fallén), 1817a: 28 *(Syrphus).*—Europe; Alaska to Maine, s. to Calif., N. Mex., and Mass., Europe.
    anthorinus, error.
    montanus Williston, 1882a: 322.—Wyo.
    perplexus Hull, 1925b: 22 (as var.).—Unknown.
arbustorum (Linnaeus), 1758: 591 *(Musca).*—Europe; Wis. to Labr., s. to Kans. and S.C. Immigrant.
bardus (Say), 1829: 163 (1859b: 357) *(Milesia).*—Ind.; Alaska to N.B., s. to Calif., Colo., and N.C. ?=*oestraceus.*
    flavipes Walker, 1849: 633.—Ont., N.S.
    melanostomus Loew, 1866a: 173 (Cent. 6, no. 69).—Oreg., Minn.
    var. **rufipilis** Hull, 1925b: 23 (as *flavipes* var.).—Alaska, Oreg., Idaho, Wis.
basilaris Macquart, 1834a: 502.—Ga. Unrecognized.
bastardii Macquart, 1842: 95 (1842: 35).—N. Amer.; Alta. to Nebr., s. to Ill. and Va.
    nebulosus Walker, 1849: 616.—Ont., N.S., N.Y.
    semimetallicus Macquart, 1850: 444 (1850: 140).—N.S.
bellardii Jaennicke 1867: 400 (1868: 92).—Mexico; Ariz., Neotropical.
beltrami Telford, 1949: 40.—Unknown. Unrecognized.
brousii Williston, 1882a: 323.—Mass.; Alaska to N. S., s. to B.C., N. Mex., and Va.
    androclus, authors, not Walker.
    meigenii, authors, not Wiedemann.
compactus Walker, 1849: 619.—Ont.; Alaska to N.S., s. to Colo. and N.H.
    atriceps Loew, 1866a: 169 (Cent. 6, no. 64).—N.H.
dimidiatus Wiedemann, 1830: 180.—N. Amer.; Alta. to N.S., s. to Kans. and N.C.
    niger Macquart, 1834a: 505.—N. Amer.
    chalybeus Macquart, 1842: 115 (1842: 55).—Carolina.
    lherminierii Macquart, 1842: 115 (1842: 55).—Carolina.
    inflexus Walker, 1849: 617.—Ont., N.S.
    incisuralis Macquart, 1850: 443 (1850: 139).—N. Amer.

duncani Curran, 1935b: 7.—Ariz.; Calif., Nev.
everes Walker, 1852: 246.—N. Amer. Unrecognized.
hirtus Loew, 1866a: 170 (Cent. 6, no. 66).—Calif.; Alaska to Wis., s. to Calif., N. Mex., and Kans.
    temporalis Thomson, 1869: 490.—Calif.
    alpha Hull, 1925b: 299 (as *temporalis* var.).—Alta.
    beta Hull, 1925b: 299 (as *temporalis* var.).—Alta.
inflatus Macquart, 1834a: 507.—N. Amer. Unrecognized.
inornatus Loew, 1866a: 172 (Cent. 6, no. 68).—Man.; B.C. to Nfld., s. to Nebr. and Maine.
latifrons Loew, 1866a: 169 (Cent. 6, no. 65).—Mexico; B.C. to N.S., s. to Calif., Mexico, and Fla.
    stipator Osten Sacken, 1877: 336.—Calif., N. Mex., Colo.
    maculipennis Townsend, 1897e: 93 (as var.).—N. Mex.
mellissoides Hull, 1925b: 19.—B.C.; Oreg.
nemorum (Linnaeus), 1758: 591 (*Musca*).—Europe; Wash. and Oreg. to Que., s. to Colo. and Conn., Eurasia.
    neomorum, error.
nitidus Wehr, 1922b: 151 (=p. 33) (preocc. Wulp, 1884).—Colo. Unrecognized, and therefore not renamed here because of possible synonymy with older species.
obscurus Loew, 1866a: 171 (Cent. 6, no. 67).—Man.; Alaska to Que., s. to Wash., N. Mex., and Mich.
occidentalis Williston, 1882a: 322.—Wash. Territory; Alaska to Minn., s. to Calif.
oestraceus (Linnaeus), 1758: 592 (*Musca*).—Europe; Colo., Ont.
    oestriformis Walker, 1849: 573 (*Syrphus*).—Ont.
parens Bigot, 1880a: 216.—N. Amer. Unrecognized.
pilosus Loew, 1866a: 174 (Cent. 6, no. 70).—Greenland; Alaska, N.W.T., Man., Labr.
pusillus Macquart, 1842: 114 (1842: 54).—Mexico; Tex., Neotropical.
    tricolor Jaennicke, 1867: 400 (1868: 92).—Mexico.
rufiventris Macquart, 1846: 257 (1846: 129).—Colombia; Tex., Neotropical.
rupium Fabricius, 1805: 241.—Europe; B.C. to N.Y., s. to Oreg. and Colo.
saxorum Wiedemann, 1830: 158.—Ga.; Wyo. to N.Y., s. to Colo. and Fla.
    pervagus Walker, 1849: 618.—U.S.
scutellaris (Fabricius), 1805: 190 (*Milesia*).—S. Amer.; N. Mex., Neotropical.
    rileyi Williston, 1887c: 178 (*Doliosyrphus*).–N. Mex.
tenax (Linnaeus), 1758: 591 (*Musca*).—Europe; Alaska to Labr., s. to Calif. and Fla., cosmopolitan. Immigrant.

testaceicornis Macquart, 1850: 442 (1850: 138).—Mexico; Calif., Ariz., Neotropical.
  obsoletus, authors, not Wiedemann.
texanus Hull, 1925b: 28.—Tex.; Calif., N. Mex., Fla.
transversus Wiedemann, 1830: 188.—N. Amer.; Mont. to N.S., s. to Utah, Tex., and Fla.
  vittatus Macquart, 1834a: 307.—N. Amer.
  philadelphicus Macquart, 1842: 94 (1842: 34).—N. Amer.
  pumilus Macquart, 1842: 117 (1842: 57).—N. Amer.
  zonatus Bigot, 1880a: 217.—N. Amer.
triangularis Giglio-Tos, 1892c: 6.—Mexico; Calif., Neotropical.
vinetorum (Fabricius), 1798: 562 (*Syrphus*).—West Indies; Wis. to Pa., s. to Tex. and Fla., Neotropical.

Subgenus LATHYROPHTHALMUS Mik

Lathyrophthalmus Mik, 1897: 114 (as genus). Type-species, *Conops aeneus* Scopoli (orig. des.).

aeneus (Scopoli), 1763: 356 (*Conops*).—Europe; Calif. to Minn. and Ont., s. to Tex. and Va., Eurasia, Africa, Bermuda, Hawaii.
  cuprovittatus Wiedemann, 1830: 190.—N. Amer.
  sincerus Walker, 1849: 611.—U.S.

Unplaced Species of Syrphidae

americana Swederus, 1787: 288 (*Musca*; preocc. Fabricius, 1775).—N. Amer. ?=*Eristalis bardus*.
  tomentosa Osten Sacken, 1878c: 136 (*Musca*). Nomen nudum, error for *Musca* (*Syrphus*) *americana* Swederus in catalog citation.
bicruciatum Bigot, 1884a: 79 (?*Melanostoma*).—Calif.
monoculus Swederus, 1787: 287 (*Musca*).—N. Amer.
sexmaculatus Palisot de Beauvois, 1819: 224 (*Syrphus*).—Hispaniola, Southern States.
vivax Fabricius, 1780: 206 (*Musca*).—Greenland.

Family CONOPIDAE

By Sidney Camras

Conopids are solitary internal parasites, mainly of Hymenoptera. The adults, often called thick-headed flies, are common on flowers.

REFERENCES: Kröber, 1919a, b, (biol., cat., keys); Parsons, 1948 (synopsis, N. Amer.); Camras and Hurd, 1957 (keys, Calif.).

Subfamily CONOPINAE

Tribe CONOPINI

Genus CONOPS Linnaeus

Conops Linnaeus, 1758: 604. Type-species, *flavipes* Linnaeus (Curtis, 1831: pl. 377).
REFERENCE: Camras, 1955 (rev.).

Subgenus ASICONOPS Chen

Conops, subg. **Asiconops** Chen, 1939: 170. Type-species, *aureomaculatus* Kröber (orig. des.).

bermudensis Parsons, 1940: 28.—Bermuda.

Genus PHYSOCONOPS Szilády

Physoconops Szilády, 1926b: 588. Type-species, *Conops brachyrhynchus* Macquart (orig. des.; misident.) =*obscuripennis* (Williston).
REFERENCE: Camras, 1955 (rev.).

Subgenus PACHYCONOPS Camras

Physoconops, subg. **Pachyconops** Camras, 1955: 161. Type-species, *Conops bulbirostris* Loew (orig. des.).

brachyrhynchus (Macquart), 1843a: 172 (1843: 15) (*Conops*).—N. Amer.; Wyo. to Mass., s. to N. Mex., Mexico, and Fla.
  xanthopareus Williston, 1882c: 332 (*Conops*).—Tex., Mass., Conn.
  xanthostomus (error) Banks, 1916: 200.
  fenestratus Kröber, 1915b: 134 (*Conops*).—Nebr.
bulbirostris (Loew), 1853: 30 (*Conops*).—N. Amer. (Osten Sacken, 1878c: 140); Ind. to N.J., s. to Tex. and Fla., mainland s. to Paraguay.
excisus (Wiedemann), 1830: 234 (*Conops*).—Ga.; Colo. to N.Y., s. to Okla. and Fla.
  sugens Wiedemann, 1830: 236 (*Conops*).—Unknown.
floridanus Camras, 1955: 163.—Fla.; N.C.
gracilis (Williston), 1885e: 377 (*Conops*).—Ariz.; Calif. and Colo. s. to Mexico and Tex., also Colombia.
townsendi Camras, 1955: 170 (n. name for *auratus* Townsend).—N. Mex.; Calif., Ariz., Tex.
  auratus Townsend, 1901: 161 (*Conops;* preocc. Walker, 1871). N. Mex.

Subgenus GYROCONOPS Camras

Physoconops, subg. **Gyroconops** Camras, 1955: 174. Type-species, *Conops sylvosus* Williston (orig. des.).

sylvosus (Williston), 1882c: 329 (*Conops*).—Mass., Conn.; Colo. to N.H., s. to Calif., Mexico, and Fla.
arizonicus Banks, 1916: 191 (*Conops*).—Ariz.

Subgenus PHYSOCONOPS Szilády

analis (Fabricius), 1805: 175 (*Conops*).—S. Amer.; Tex. to Paraguay.
angustifrons Williston, 1892c: 44 (*Conops*).—Brazil.
discalis (Williston), 1892b: 80 (*Conops*).—Mexico; Utah, Ariz., N. Mex., Kans., Tex., s. to Argentina.
semifuscus Banks, 1916: 192 (*Conops*; as *brachyrhynchus* var.).—N. Mex.
fronto (Williston), 1885e: 378 (*Conops*).—Kans.; Wash. and Alta. to Mass., s. to Calif., Mexico, and Fla.
pulchellus Kröber, 1915b: 134 (*Conops*).—S.C.
argentifacies Van Duzee, 1927a: 574 (*Conops*).—Idaho.
fraterculus Van Duzee, 1927a: 575 (*Conops*).—Idaho.
rubicundulus Van Duzee, 1927a: 576 (*Conops*).—Idaho.
nigrimanus (Bigot), 1887a: 38 (*Conops*).—Ga.; Nebr. to N.J., s. to Tex. and Fla.
striatifrons Kröber, 1915b: 132 (*Conops*).—Ga., Nebr.
limuva Brimley, 1927: 235 (*Conops*).—N.C.
obscuripennis (Williston), 1882c: 328 (*Conops*).—S.C.; B.C. to Mass., s. to Tex. and Fla.
foxi Van Duzee, 1927a: 574 (*Conops*).—Wash.
brachyrhynchus, authors, not Macquart.

Tribe PHYSOCEPHALINI

Genus PHYSOCEPHALA Schiner

Physocephala Schiner, 1861b: 137. Type-species, *Conops rufipes* Fabricius (orig. des.).

REFERENCE: Camras, 1957 (rev.).

burgessi (Williston), 1882c: 337 (*Conops*).—Calif.; B.C. and Alta., s. to Calif. and Tex.
brevirostris Van Duzee, 1927a: 579.—Calif.
floridana Camras, 1957: 214.—Fla.; La., Ga.
analis, authors, not Fabricius.

furcillata (Williston), 1882c: 336 *(Conops).*—N.H.; Alta. to N.S., s. to Wis. and Pa., also Mexico, ?Calif.
  lucida Van Duzee, 1931: 284.—Ont.
marginata (Say), 1823: 82 (1859b: 73) *(Conops).*—Mo.; B.C. to N.H., s. to Calif. and Fla.
  dakotensis Van Duzee, 1934: 317.—N. Dak.
  stylifer Van Duzee, 1934: 318.—N. Dak.
sagittaria (Say), 1823: 83 (1859b: 73) *(Conops).*—Pa.; N. Dak. to Mass., s. to N. Mex. and Fla.
  aethiops Walker, 1849: 671 *(Conops).*—N. Amer.
  dimidiata Walker, 1852: 254 *(Conops).*—Unknown.
  genualis Loew, 1853: 32 *(Conops).*—Ky.
  castanoptera Loew, 1853: 33 *(Conops).*—Ga., Carolina.
  ruficornis Van Duzee, 1934: 316.—N. Dak.
texana (Williston), 1882c: 338 *(Conops).*—Tex.; B.C. to Que., s. to Mexico, also ?Ga.
  affinis Williston, 1882c: 339 *(Conops).*—Wash.
  ochreiceps Bigot, 1887a: 39 *(Conops).*—Ga. (?error).
  humeralis Van Duzee, 1927a: 580.—Idaho.
  simulans Van Duzee, 1927a: 581 (as *humeralis* ssp.).—Calif.
  aurifacies Van Duzee, 1927a: 581.—Calif.
  buccalis Van Duzee, 1927a: 582.—Utah.
  rubida Van Duzee, 1934: 315.—Oreg.
tibialis (Say), 1829: 171 (1859b: 363) *(Conops).*—Ind.; Wis. to Mass., s. to Tex. and Fla.
  nigricornis Wiedemann, 1830: 236 *(Conops).*—Pa.
  fulvipennis Macquart, 1843a: 170 (1843: 13) *(Conops).*—Ga.
  lugubris Macquart, 1843a: 173 (1843: 16) *(Conops).*—Unknown.

Subfamily MYOPINAE

Genus ZODION Latreille

Zodion Latreille, 1796: 162. Type-species, *Myopa cinerea* Fabricius (sub. mon., Latreille, 1802: 444).

REFERENCES: Camras, 1943 and 1944 (rev.).

abitus Adams, 1903a: 33.—Kans., Mass.; Idaho to Que., s. to Ariz. and Fla.
  bicolor Adams, 1903a: 35.—Kans.
albonotatum Townsend, 1897c: 175.—Tex.; Ariz., N. Mex., Mexico, Guatemala.

americanum Wiedemann, 1830: 242.—Uruguay; Wash. and Alta. to N.S., s. to Uruguay, also West Indies.
  nanellum Loew, 1866b: 42 (Cent. 7, no. 75).—D.C.
  pygmaeum Williston, 1885e: 381.—Calif.
  albifacies Van Duzee, 1927a: 588.—Ariz.
anale Kröber, 1915a: 113.—Chile; Calif. to Ill. to Ga., s. to Chile.
californicum Camras, 1954: 165.—Calif.
cinereiventre Van Duzee, 1927a: 585.—Calif.; B.C. to P.E.I., s. to Calif. and N.C.
cyanescens Camras, 1943: 188.—N.C.; Ill. to N.J., s. to Miss. and Fla.
fulvifrons Say, 1823: 83 (1859b: 74).—Md., Pa.; Canada, U.S., Mexico.
  abdominalis Say, 1823: 84 (1859b: 74).—Nebr.
  rubrifrons Robineau-Desvoidy, 1830: 247 (*Myopa*).—Pa.
  bistria Walker, 1849: 679 (*Myopa*).—N. Amer.
  lativentre Graenicher, 1910a: 26.—Wis.
  obscurum Banks, 1916: 194.—Calif.
  sayi Banks, 1916: 194.—Va.
  reclusum Banks, 1916: 195.—Calif.
  bilineata Van Duzee, 1927a: 586.—Oreg.
intermedium Banks, 1916: 193.—Pa.; B.C. to N.S., U.S., Mexico.
  occidentale Banks, 1916: 194.—Oreg.
nigrifrons Kröber, 1915a: 97.—Calif.; Oreg.
  hirtipes Van Duzee, 1927a: 587.—Calif.
obliquefasciatum (Macquart), 1846: 269 (1846: 141) (*Myopa*).— Tex.; Wash. and Alta. to Wis., s. to Calif. and La., Mexico, Guatemala.
  leucostoma Williston, 1885e: 380.—Kans., Ariz., Mont.
  tectura Adams, 1903a: 35 (*Myopa*).—Unknown.
perlongum Coquillett, 1902a: 199.—Colo., N. Mex., Mexico; Idaho to Maine, s. to Calif., Mexico, and N.C.
pictulum Williston, 1885e: 379.—N. Mex.; Ariz., Colo.
triste Bigot, 1887a: 203.—Calif.; Wash., Idaho, Utah.

Genus ROBERTSONOMYIA Malloch

Robertsonomyia Malloch, 1919n: 205. Type-species, *Zodion palpalis* Robertson (orig. des.).

  REFERENCE: Camras, 1944 (review, in *Zodion*); Parsons, 1948 (synopsis).

palpalis (Robertson), 1901: 284 (*Zodion*).—Ill.; Wyo. to Va., s. to Ariz., Mexico, and Ga.
  scapularis Adams, 1903a: 34 (*Zodion*).—Ariz.

parva (Adams), 1903a: 33, 34 (*Zodion*; as *parvis*, p. 34).—Ariz.; Nev., Wyo., and Nebr., s. to Ariz. and N. Mex.
   lovetti Van Duzee, 1934: 323 (*Zodion*).—Colo.

Genus MYOPA Fabricius

Myopa Fabricius, 1775: 798. Type-species, *Conops buccatus* Linnaeus (Curtis, 1838: pl. 677). Latreille (1810: 444) designated *Conops ferrugineus* Linnaeus (as "*Myopa ferruginea*, F."), which is the type of *Sicus* Scopoli, 1763, a different genus. Long-standing usage is maintained here, and the case has been submitted to the International Commission on Zoological Nomenclature.

Gonirhyncus Rondani, 1856: 58. Type-species, *dispar* Rondani (orig. des.) = *occulta* Wiedemann.

REFERENCE: Camras, 1953 (rev.).

bohartorum Camras, 1953: 107.—Calif.
castanea (Bigot), 1887a: 207 (*Gonirhyncus*).—Nev.; Calif.
clausa Loew, 1866b: 41 (Cent. 7, no. 72).—Maine; B.C. to Maine, s. to Calif. and Va.
curticornis Kröber, 1916: 32.—Colo., Calif.; Wash. to Wyo., s. to Calif. and Ariz., also Mich.
flavopilosa Kröber, 1916: 30.—Colo.; Alta., Wis. and Ill., sw. to Calif.
   varians Banks, 1916: 196 (as *vesiculosa* var.).—Nebr.
longipilis Banks, 1916: 197.—Wash.; B.C. and Utah, sw. to Calif.
melanderi Banks, 1916: 197.—Wash.; Idaho, Oreg., Calif.
perplexa Camras, 1953: 103.—Calif.; Wash. and Idaho, s. to Calif. and Ariz.
plebeia Williston, 1885e: 384.—Ariz.
rubida (Bigot), 1887a: 206 (*Glossigona*).—Colo.; B.C. to Wyo., s. to Calif. and Colo.
   aperta Röder, 1889: 5 (as *clausa* var.).—B.C., Nev.
   seminuda Banks, 1916: 198.—Oreg.
vesiculosa Say, 1823: 80 (1859b: 72).—Pa.; B.C. to Que., s. to Calif. and Fla.
   apicalis Walker, 1849: 679.—N. Amer.
   conjuncta Thomson, 1869: 515.—Calif.
   maculifrons Bigot, 1887a: 206 (*Glossigona*).—Nev.
   utahensis Stains and Knowlton, 1939: 51.—Utah.
vicaria Walker, 1849: 679.—N.S.; Alaska to N.S., s. to Calif. and Ga.
   pilosa Williston, 1885e: 383.—Calif.
virginica Banks, 1916: 198.—Va.; S. Dak. to Ont. and Vt., s. to Ark. and N.C.

willistoni Banks, 1916: 197 (n. name for *pictipennis* Williston).—Calif.; Wash., Oreg., Nev., Ariz., Mexico.
    pictipennis Williston, 1885e: 382 (preocc. Robineau-Desvoidy, 1830).—Calif.

Genus THECOPHORA Rondani

Thecophora Rondani, 1845a: 15. Type-species, *Myopa atra* Fabricius (Rondani, 1856: 58).
Thecomyia, error.
Occemya Robineau-Desvoidy, 1853a: 92, 130. Type-species, *Myopa atra* Fabricius (orig. des.).
Occemyia and *Oncomyia*, emend.
Eccemyia, error.
    REFERENCE: Camras, 1945 (rev.).

abbreviata (Loew), 1866b: 41 (Cent. 7, no. 73) (*Oncomyia*).—D.C.; Utah, Mont. to N.B., s. to Ill. and N.C. **N. comb.**
    melanopoda Williston, 1885e: 393 (*Oncomyia*; as *modesta* var.).—N.H.
longicornis (Say), 1823: 81 (1859b: 72) (*Myopa*).—Mo.; Alaska and Y.T. to N.B., s. to Calif. and Ga. **N. comb.**
    infuscipes Van Duzee, 1927a: 592 (*Oncomyia*).—Ont.
luteipes (Camras), 1945: 220 (*Occemya*).—Wash.; B.C. to Colo., sw. to Calif. **N. comb.**
modesta (Williston), 1883c: 96 (*Oncomyia*).—Wash.; B.C. to Sask., s. to Calif. and N. Mex. **N. comb.**
    basalis Van Duzee, 1927a: 586 (*Zodion*).—Idaho. **N. syn., n. comb.**
nigra (Van Duzee), 1927a: 596 (*Oncomyia*).—Oreg.; Wash., Idaho, Calif., N. Mex., Maine. **N. comb.**
nigripes (Camras), 1945: 218 (*Occemya*).—Ont.; B.C. to N.S., s. to Mexico and Ga., Guatemala. **N. comb.**
occidensis (Walker), 1849: 676 (*Zodion*).—Ohio; Y.T., B.C. to Que., s. to Calif., Mexico, and Ga. **N. comb.**
    loraria Loew, 1866b: 41 (Cent. 7, no. 74) (*Oncomyia*).—N.H.
    baroni Williston, 1883c: 97 (*Oncomyia*).—Calif.
    brevirostris Van Duzee, 1927a: 593 (*Oncomyia*).—Wash.
    aequalis Van Duzee, 1927a: 594 (*Oncomyia*).—Calif.
    terminalis Van Duzee, 1927a: 594 (*Oncomyia*).—Oreg.
    bimacula Curran, 1933b: 7 (*Zodion*).—Ont.
    frontalis Van Duzee, 1934: 322 (*Oncomyia*).—Wash.

propinqua (Adams), 1903a: 32 (*Oncomyia*).—Unknown; B.C. to N.S., s. to Calif. and Ala. **N. comb.**
  angusticornis Van Duzee, 1927a: 589 (*Zodion*).—Calif.
  angusticornis Van Duzee, 1927a: 595 (*Oncomyia*).—Idaho.
  longipalpis Van Duzee, 1934: 321 (*Oncomyia*).—Wash.

Subfamily DALMANNIINAE

Genus DALMANNIA Robineau-Desvoidy

Dalmannia Robineau-Desvoidy, 1830: 248. Type-species, *Myopa punctata* Fabricius (Rondani, 1856: 59).

REFERENCE: Bohart, 1938 (rev.).

blaisdelli Cresson, 1919: 190.—Colo.; Oreg., Calif., Idaho, Wyo.
heterotricha Bohart, 1938: 134.—Calif.
nigriceps Loew, 1866b: 40 (Cent. 7, no. 71).—Va.; Colo. to Que., s. to Ga.
pacifica Banks, 1916: 199.—Oreg.; Calif., Utah.
  hirsuta Van Duzee, 1927a: 591.—Oreg.
picta Williston, 1883c: 94.—N. Mex.; B.C. to Colo., s. to Calif. and N. Mex.
vitiosa Coquillett, 1892e: 150.—Calif.; Idaho to N.H., s. to Calif. and Ga.

Subfamily STYLOGASTRINAE

Genus STYLOGASTER Macquart

Stylogaster Macquart, 1835: 38. Type-species, *Conops stylatus* Fabricius (mon.).
Stylomyia Westwood, 1852: 268. Type-species, *leonum* Westwood (Coquillett, 1910b: 610).

REFERENCES: Aldrich, 1930b (rev.); Curran, 1942a (rev.).

biannulata (Say), 1823: 81 (1859b: 72) (*Myopa*).—Pa.; Nebr. to R.I., s. to Tex. and Fla.
  confusa Westwood, 1852: 269 (*Stylomyia*).—Unknown.
  stylata, authors, not Fabricius.
neglecta Williston, 1883c: 91.—Conn.; Nebr. to Mass., s. to Ariz. and Ga.

Unplaced Species of Conopidae

flaviceps Macquart, 1843a: 172 (1843: 15) (*Conops*).—N. Amer.

Division SCHIZOPHORA
Section ACALYPTRATAE
Superfamily MICROPEZOIDEA

Family MICROPEZIDAE
(Tylidae, Calobatidae)

By George C. Steyskal

The family Micropezidae is of cosmopolitan distribution, but largely in the Tropics. The biology of the few species for which it is known indicates saprophagous habits.

Cardiacephala triluminata Cresson, *Grallipeza nigrinotata* Hennig, and *G. ornatithorax* (Enderlein), all described from South America, have not been included. The known distribution of these species indicates that the single records of the first from Pennsylvania and of the others from California are based on mislabeled specimens.

REFERENCES: Hennig, 1934, 1935a,b, 1936 (rev., biol.); Cresson, 1938a (rev., biol.); Aczél, 1951 (anat.).

Subfamily CALOBATINAE

Genus CNODACOPHORA Czerny

Trepidaria, subg. **Cnodacophora** Czerny, 1930b: 4. Type-species, *Calobata adusta* Loew (Cresson, 1938a: 343).

nasoni (Cresson), 1914c: 459 (*Calobata*).—Ill.; Alaska, Wash. and Alta. to Que., s. to Colo. and N.Y.

Genus COMPSOBATA Czerny

Trepidaria, subg. **Compsobata** Czerny, 1930b: 5. Type-species, *Musca cibaria* Linnaeus (Cresson, 1938a: 332).
Paracalobata, authors, not Hendel.

Subgenus COMPSOBATA Czerny

univitta (Walker), 1849: 1049 (*Calobata*).—N.Y., Ont.; B.C. to Que., s. to Calif., N. Mex., Ill., Md. **N. comb.**
  univittata, error or emend.
  albiceps Wulp, 1883: 50 (*Calobata*).—Que.
  agilis Harris, 1835: 600 (*Calobata*). Nomen nudum.

Subgenus TRILOPHYROBATA Hennig

Trepidaria, subg. **Trilophyrobata** Hennig, 1938a: 9. Type-species, *commutata* Czerny (orig. des.).

kennicotti (Banks), 1926: 43 (*Calobata*).—Man.; Mich. **N. comb.**
microfulcrum (James), 1946: 129 (*Paracalobata*).—Colo.; Ariz. **N. comb.**
mima (Hennig), 1936: 134 (*Trepidaria*).—Colo., Alta., Wash.: B.C. to Sask., s. to Calif. and Colo. **N. comb.**
pallipes (Say), 1823: 97 (1859b: 84) (*Calobata*).—Mo.; Alaska, Alta. to N.B., s. to Nev., Mo., and N.J. **N. comb.**
  alesia Walker, 1849: 1048 (*Calobata*).—Ont. **N. comb.**

Subfamily MICROPEZINAE

Genus MICROPEZA Meigen

Tylos Meigen, 1800: 31. Type-species, *Musca corrigiolata* Linnaeus (Coquillett, 1910b: 618). Suppressed by I.C.Z.N., 1955b: 267.
Micropeza Meigen, 1803: 276. Type-species, *Musca corrigiolata* Linnaeus (mon.).

Subgenus MICROPEZA Meigen

abnormis Cresson, 1938b: 72.—Ariz.
ambigua Cresson, 1908: 3 (as *turcana* var.).—N. Mex.; Calif., Ariz.
atra Cresson, 1938b: 74.—Ariz.; N. Mex.
compar Cresson, 1938b: 73.—Ariz.; N. Mex.
lineata Van Duzee, 1926e: 2.—Utah; Alta. to Man., s. to Calif., N. Mex., and Nebr.
nitidor Cresson, 1938a: 319 (n. name for *nitidus* Hennig).—Ariz.
  nitidus Hennig, 1936: 212 (*Tylos*; preocc. Robineau-Desvoidy, 1830).—Ariz.
setaventris Cresson, 1938b: 74.—Utah; Utah and N. Dak., s. to Calif. and N. Mex.
turcana Townsend, 1893c: 136.—Ariz.; Alta. to Man., s. to Ariz. and Kans.
  jamesi Cresson, 1935b: 229.—Colo.
ventralis Cresson, 1930c: 356.—Mexico (D.F.); Ariz.

Subgenus NERIOCEPHALUS Enderlein

Neriocephalus Enderlein, 1922a: 160 (as genus). Type-species, *Micropeza appendiculata* Schiner (orig. des.).
Metopobrachia Enderlein, 1922a: 161. Type-species, *Micropeza obscura* Bigot (orig. des.).

Cliopeza Enderlein, 1922a: 162. Type-species, *Calobata pectoralis* Wiedemann (orig. des.).
Tylus, subg. *Neotylus* Hendel, 1932b: 121. Type-species, *Micropeza brasiliensis* Schiner (orig. des.).
Tylos, subg. *Protylos* Aczél, 1949: 238. Type-species, *Micropeza stigmatica* Wulp (orig. des.).
bisetosa Coquillett, 1902e: 177.—Ariz.; N. Mex.
californica Van Duzee, 1926e: 1.—Calif.
producta Walker, 1849: 1056.—Ga.; Fla., Cuba, Panama.
ruficeps Wulp, 1897: 365.—Mexico (Son.); Calif., Ariz.
flaviventris Cole, 1923a: 477.—Baja Calif.
stigmatica Wulp, 1897: 366.—Mexico; Ariz., Kans., Tex., s. to Argentina.
texana Cresson, 1938b: 75.—Tex.

Subfamily TAENIAPTERINAE

Genus CALOBATINA Enderlein

Calobatina Enderlein, 1922a: 194.—Type-species, *texana* Enderlein (orig. des.).
Meganeria Cresson, 1926: 271. Type-species, *geometroides* Cresson (orig. des.).
geometra (Robineau-Desvoidy), 1830: 736 (*Neria*).—Carolina. Unrecognized.
geometroides (Cresson), 1926: 271 (*Meganeria;* n. name for *varipes* Johnson).—Fla.; Mo. and Ky., s. to Tex. and Fla.
varipes Johnson, 1895a: 306 (*Calobata;* preocc. Walker, 1856).—Fla.
texana Enderlein, 1922a: 194.—Tex.; Pa.
daeckei Cresson, 1926: 272 (*Meganeria*).—Pa.

Genus GRALLIPEZA Rondani

Grallipeza Rondani, 1850a: 180. Type-species, *Calobata unimaculata* Macquart (orig. des.).
Systellapha Enderlein, 1922a: 189. Type-species, *ornatithorax* Enderlein (orig. des.).
nebulosa (Loew), 1866b: 48 (Cent. 7, no. 89) (*Calobata*).—Fla.; Ga., La., Haiti, Cuba, Costa Rica.
baracoa Cresson, 1926: 265 (*Systellapha*).—Cuba.

Genus HOPLOCHEILOMA Cresson

Hoplocheiloma Cresson, 1926: 272. Type-species, *Musca fasciata* Fabricius (orig. des.).
Gymnosphen Frey, 1927: 71. Type-species, *macropyga* Frey (orig. des.).

fasciatum (Fabricius), 1775: 781 (*Musca*).—West Indies; Fla. to Brazil.

Genus RAINIERIA Rondani

Rainieria Rondani, 1843: 40. Type-species, *Calobata calceata* Fallén (orig. des.).
Tanipoda Rondani, 1856: 116. Type-species, *Calobata calceata* Fallén (orig. des.).
Tanypoda, emend.

antennaepes (Say), 1823: 97 (1859b: 83) (*Calobata*).—Pa.; Alta. to N.S., s. to Tex. and Fla.
  antennipes, error or emend.
  var. **brunneipes** (Cresson), 1938b: 76 (*Taeniaptera;* as sp.).—D.C.; Ont. to Que., s. to Mich. and Va.

Genus TAENIAPTERA Macquart

Taeniaptera Macquart, 1835: 491. Type-species, *trivittata* Macquart (mon.).
Grallopoda Rondani, 1850a: 178, 180. Type-species, *Calobata albimana* Macquart (orig. des.)=*trivittata* Macquart.
Grallomya Rondani, 1850a: 180. Type-species, *Calobata tarsata* Wiedemann (orig. des.).

lasciva (Fabricius), 1798: 564 (*Musca*).—French Guiana; Calif., Tex., Ga., Fla., Neotropical.
trivittata Macquart, 1835: 491.—N. Amer.; N.Y. to Fla. and Tex., also Mich.
  albimana Macquart, 1843a: 402 (1843: 245) (*Calobata*).—Pa. (here restricted; Java, Australia, Cuba also in the mixed series).
  valida Walker, 1852: 390 (*Calobata*).—U.S.
  divaricata Cresson, 1914c: 459.—Ga.

Unplaced Species of Micropezidae

atripes Robineau-Desvoidy, 1830: 738 (*Neria*).—Carolina.
carolinensis Robineau-Desvoidy, 1830: 738 (*Neria*).—Carolina.

Family NERIIDAE

By George C. Steyskal

The family Neriidae, largely tropical, is closely related to the Micropezidae. *Odontoloxozus* has been reared in Texas from decaying roots of papaya.

REFERENCES: Hennig, 1937 (rev.).; Cresson, 1938a (rev.); Aczél, 1951 (anat.).

Genus ODONTOLOXOZUS Enderlein

Odontoloxozus Enderlein, 1922a: 158. Type-species, *punctulatus* Enderlein (orig. des.)=*longicornis* (Coquillett).

longicornis (Coquillett), 1904f: 188 (*Nerius*).—Tex.; Calif., Ariz., Mexico, Costa Rica.
punctulatus Enderlein, 1922a: 158.—Mexico.

Genus ONCOPSIA Enderlein

Oncopsia Enderlein, 1922a: 152. Type-species, *mexicana* Enderlein (orig. des.)=*flavifrons* (Bigot).
Cerantichir Enderlein, 1922a: 155. Type-species, *Nerius flavifrons* Bigot (orig. des.).
Brachycrotaphus Czerny, 1932: 301. Type-species, *Nerius flavifrons* Bigot (orig. des.).

flavifrons (Bigot), 1886a: 372 (*Nerius*).—Mexico; Ariz., Neotropical.
mexicana Enderlein, 1922a: 153.—Mexico.

Superfamily NOTHYBOIDEA

Family DIOPSIDAE

By Curtis W. Sabrosky

The small family Diopsidae, including the curious stalk-eyed flies, is confined to the Old World Tropics except for one relict species in eastern North America. Little is known of the habits, but apparently the larvae breed in decaying plant material. Adults of our species are usually collected near water, sitting on rocks or on skunk cabbage or other vegetation.

REFERENCE: Lavigne, 1962 (biol., immature stages).

Genus SPHYRACEPHALA Say

Sphyracephala Say, 1828: pl. 52 (1859a: 116). Type-species, *Diopsis brevicornis* Say (mon.).

brevicornis (Say), 1817a: 23 (1859b: 3) (*Diopsis*).—Pa.; Minn. to Que., s. to Colo. and N.C.
    bicornis (error) Peterson, 1916: 183.
    subbifasciata Fitch, 1855: 774 (1856: 70).—Ill., N.Y.

Family PSILIDAE
By G. E. Shewell

The family Psilidae occurs in the Holarctic, Ethiopian, and Oriental Regions. It is poorly represented in South America, and is unknown in the Australian Region except for the introduced pest *Psila rosae* (Fabricius), the carrot rust fly, in New Zealand and southern Australia.

The family is probably most closely related to the Micropezidae-Tanypezidae-Megamerinidae complex, the genus *Megamerina* Rondani having in several respects a striking resemblance to *Chyliza* Fallén.

The larvae are phytophagous, living in the roots and stems of plants and under the bark of trees to which they gain access through wounds. Known wild hosts in North America are *Lupinus*, *Pinus*, and *Ulmus*, and in Europe, *Carex*, *Juncus*, *Neottia*, *Orobanche*, *Scabiosa*, and *Spiraea*. The adults rest on foliage and herbage and are mostly shade-loving. The most injurious species is the carrot rust fly ("carrot fly" of European literature).

REFERENCE: Melander, 1920b (rev.).

Subfamily PSILINAE
Genus LOXOCERA Meigen

Loxocera Meigen, 1803: 275. Type-species, *Musca ichneumonea* Linnaeus (mon.).

REFERENCE: Capelle, 1953 (rev.).

californica Capelle, 1953: 106.—Calif.; ?Tenn.
collaris Loew, 1869c: 184 (Cent. 9, no. 97).—D.C.; Wis. to N.S., s. to N.C., also Wash., Oreg., Idaho, Mont.
cylindrica cylindrica Say, 1823: 98 (1859b: 84).—Pa.; s. Man. to N.S., s. to Kans. and N.C.
    pleuritica Loew, 1869b: 38 (Cent. 8, no. 65).—Conn.
    atricornis Harris, 1835: 600. Nomen nudum.
    ssp. **obsoleta** Johnson, 1920: 16 (as var.).—e. U.S. (7 States); Iowa to Vt., s. to Tex. and S.C.

ssp. **pectoralis** Loew, 1869b: 38 (Cent. 8, no. 64) (as sp.).—D.C.;
Nebr. to s. Que., s. to Tex. and N.C.
fumipennis Coquillett, 1901h: 617.—Kans.; s. Sask. to Iowa, s. to
Ariz. and Tex.
microps Melander, 1920b: 92.—Wash.; Alaska, s. B.C.

Genus PSILA Meigen

Psila Meigen, 1803: 278. Type-species, *Musca fimetaria* Linnaeus
(aut.).
Psilomyia Latreille, 1829: 525 (unjustified n. name for *Psila* Meigen).
Type-species, *Musca fimetaria* Linnaeus (Westwood,
1840: 146).

Subgenus CHAMAEPSILA Hendel

Chamaepsila Hendel, 1917: 37 (as genus). Type-species, *Musca
rosae* Fabricius (orig. des.).
atrata Melander, 1920b: 96.—Idaho; Alta., Oreg., Mont., Wis.
?=*washingtona*.
dimidiata Loew, 1869b: 40 (Cent. 8, no. 69).—Red River of the North;
Alaska to Nfld., s. to Ariz. and Maine.
bicolor, authors, not Meigen.
levis Loew, 1869b: 40 (Cent. 8, no. 71).—N.H.; Alaska to n. Que.,
s. to B.C. and Maine, also N.C.
microcera Melander, 1920b: 95.—Oreg.; Wash. to Calif., s. Alta.,
Sask., Utah.
nigricornis Meigen, 1826: 359.—Germany; B.C. Immigrant.
rosae (Fabricius), 1794: 356 (*Musca*).—Germany (Kiel); n. B.C. and
s. Alta., s. to Utah and Colo., also Wis. to Nfld., s. to
Ill. and Md., Europe, s. Australia, and New Zealand.
Immigrant, N. Amer. and Australian Region.
sternalis Loew, 1869b: 40 (Cent. 8, no. 70).—Middle States; Wis. to
s. Que., s. to Iowa, Ill., and N.C.
washingtona Melander, 1920b: 96.—Wash.; Alaska to Wash.

Subgenus PSEUDOPSILA Johnson

Pseudopsila Johnson, 1920: 17 (as genus). Type-species, *Loxocera
fallax* Loew (orig. des.). **N. status.**
angustata Cresson, 1919: 193.—N.Y.; s. Man., Mich., and Ill. to Vt.
and Conn., also Tenn.
bivittata Loew, 1869b: 39 (Cent. 8, no. 67).—Conn.; Mich. to se.
Canada and Maine, s. to Ga.
dorsalis Harris *in* Johnson, 1925c: 98 (*Loxocera*). Nomen
nudum.

collaris Loew, 1869b: 39 (Cent. 8, no. 68).—Conn.; Wis. to N.S., s. to Ga.
fallax (Loew), 1869c: 185 (Cent. 9, no. 98) (*Loxocera*).—Canada; s. Ont. to N.S., s. to N.Y., also Ind. **N. comb.**
lateralis Loew, 1860b: 81.—D.C.; S. Dak. to Mass., s. to Tex. and Ga.
  colorata Melander, 1920b: 95.—Tex. **N. syn.**
perpolita (Johnson), 1920: 18 (*Pseudopsila*).—Maine, N.H.; s. B.C. to s. Que., s. to Minn. and Mich. **N. comb.**

Subgenus TETRAPSILA Frey

Tetrapsila Frey, 1925: 48 (as genus). Type-species, *Psila obscuritarsis* Loew (orig. des.).

frontalis Coquillett, 1901h: 617.—N.H.; s. Ont. and Que., Maine, Vt., Mass., Mich.

Subfamily CHYLIZINAE

Genus CHYLIZA Fallén

Chyliza Fallén, 1820d: 6.—Type-species, *Musca leptogaster* Panzer (Westwood, 1840: 146).
Tetradiscus Bigot, 1886a: 370. Type-species, *pictus* Bigot (Coquillett, 1910b: 613)=*apicalis* Loew.

annulipes Macquart, 1835: 380.—France; Mass., Va., Europe.
apicalis Loew, 1860b: 82.—D.C.; Colo. to se. Canada, s. to La. and Ga., also Mexico.
  pictus Bigot, 1886a: 370 (*Tetradiscus*).—Rocky Mts.
erudita Melander, 1920b: 99.—Mass.; s. B.C., Alta., se. Canada, N.Y., Vt., N.C.
leguminicola Melander, 1920b: 99.—Oreg.; s. B.C.
nigroviridis Walker, 1861: 330.—U.S. Unrecognized.
notata Loew, 1869c: 185 (Cent. 9, no. 99).—D.C.; s. B.C., Wash., Idaho, Mich., se. Canada, s. to Tenn. and N.C.
robusta Coquillett *in* Baker, 1904: 28.—Nev.; Calif.
scrobiculata Melander, 1920b: 98.—Wash.; s. B.C., Oreg.
similis Johnson, 1913a: 85.—Fla.

Subfamily STRONGYLOPHTHALMYIINAE

Genus STRONGYLOPHTHALMYIA Heller

Strongylophthalmus Hendel, 1902a: 179 (preocc. Faust, 1894). Type-species, *Chyliza ustulata* Zetterstedt (orig. des.).

Strongylophthalmyia Heller, 1902: 226 (n. name for *Strongylophthalmus* Hendel). Type-species, *Chyliza ustulata* Zetterstedt (aut.).
angustipennis Melander, 1920b: 100.—Wash.; B.C. to N.S., s. to Wyo., Mich., and Mass.
longula Johnson, 1921a: 14 (*Psila*).—Mass. **N. syn., n. comb.**

Unplaced Species of Psilidae

metallica Walker, 1849: 1045 (*Chyliza*).—n. Ont.
quadrilinea Walker, 1861: 329 (?*Loxocera*).—U.S.

Family TANYPEZIDAE

By George C. Steyskal

The family Tanypezidae is principally Neotropical in distribution but is also represented in Europe. The larvae are unknown.

Genus TANYPEZA Fallén

Tanypeza Fallén, 1820d: 4. Type-species, *longimana* Fallén (mon.).
longimana Fallén.—Not Nearctic.
luteipennis Knab and Shannon, 1916: 34.—Ill.; Alta. to Maine, s. to Ill. and N.Y.
picticornis Knab and Shannon, 1916: 35.—Va.; Mich., N.Y., Md.

Superfamily TEPHRITOIDEA

Family RICHARDIIDAE

By George C. Steyskal

The habits of all stages are poorly known. Larvae seem to be associated with decaying vegetable matter, and adults have been caught in fruit-fly traps. The family is principally Neotropical in distribution.

The genus *Epiplatea* Loew is not Nearctic; for *dohaniani* Johnson, see *Automola* Loew, and for *scutellaris*, see *Curranops* Harriot (Otitidae).

REFERENCE: Steyskal, 1958c (rev.).

Genus AUTOMOLA Loew

Automola Loew, 1873b: 118. Type-species, *Ortalis atomaria* Wiedemann (Coquillett, 1910b: 512).

rufa Cresson, 1906: 282.—N. Mex.; Ariz., Tex., Mexico.
dohaniani Johnson, 1921c: 58 (*Epiplatea*).—Tex.

Genus CONICEPS Loew

Coniceps Loew, 1873b: 177. Type-species, *niger* Loew (mon.).

niger Loew, 1873b: 178.—Tex.; S. Dak., Kans.

Genus ODONTOMERA Macquart

Odontomera Macquart, 1843a: 372 (1843: 215). Type-species, *ferruginea* Macquart (mon.).

ferruginea Macquart, 1843a: 372 (1843: 215).—Unknown; N.J., s. to Fla., Cent. and S. Amer.
dorotheae Brimley, 1925: 74.—N.C.
limbata Steyskal, 1958c: 304.—Fla.; S.C.

Genus SEPSISOMA Johnson

Sepsisoma Johnson, 1900b: 327. Type-species, *flavescens* Johnson (mon.).

REFERENCE: Steyskal, 1961a (key).

flavescens Johnson, 1900b: 327.—N.J.; Kans. and N.Y., s. to N.C.
minimum Steyskal, 1961a: 85.—Kans.
sabroskyi Steyskal, 1961a: 83.—Kans.

Family OTITIDAE

(Ortalidae, Ulidiidae, Pterocallidae)

By George C. Steyskal

The principal distribution of the family Otitidae is in the Holarctic and Neotropical Regions. The larvae of most species are saprophagic, although a few, such as *Tritoxa flexa* and *Tetanops myopaeformis*, attack living plant tissue.

REFERENCES: Hennig, 1939, 1940 (monog. Palaearctic spp.); Steyskal, 1961b (generic rev.).

Subfamily OTITINAE

Genus CALLOPISTROMYIA Hendel

Callopistria Loew, 1873b: 140 (preocc. Hübner, 1816). Type-species, *Platystoma annulipes* Macquart (mon.).
Callopistromyia Hendel, 1907a: 98 (n. name for *Callopistria* Loew). Type-species, *Platystoma annulipes* Macquart (aut.).
annulipes (Macquart), 1855: 141 (1855: 121) (*Platystoma*).—Md.; Utah to Maine, s. to Kans., Ohio, and Ga.

Genus CEPHALIA Meigen

Cephalia Meigen, 1826: 293. Type-species, *nigripes* Meigen (Rondani, 1856: 115)=*rufipes* Meigen.
Myrmecomya Robineau-Desvoidy, 1830: 721. Type-species, *formicaria* Robineau-Desvoidy (Coquillett, 1910b: 573)=*rufipes* Meigen.
Myrmecomyia, error or emend.

flavoscutellata Becker, 1900: 60.—Siberia; N.W.T.
fulvicornis Bigot, 1886a: 386.—Calif. Unrecognized.
maculipennis Bigot, 1886a: 385.—Rocky Mts. Unrecognized.
rufipes Meigen, 1826: 294.—Europe; Ariz., N. Mex.

Genus CEROXYS Macquart

Meckelia Robineau-Desvoidy, 1830: 714 (preocc. Leuckart, 1828). Type-species, *Oscinis elegans* Robineau-Desvoidy (Coquillett, 1910b: 564)=*hortulanus* (Rossi).
Ceroxys Macquart, 1835: 437. Type-species, *Musca urticae* Linnaeus (Westwood, 1840: 149).
Ceratoxys Rondani, 1861a: 10 (n. name for *Meckelia* Robineau-Desvoidy). Type-species, *Oscinis elegans* Robineau-Desvoidy (aut.)=*hortulanus* (Rossi).
Anacampta Loew, 1868b: 7. Type-species. *Musca urticae* Linnaeus (Loew, 1873b: 58).
Ortalis, authors, not Fallén.

latiusculus (Loew), 1873b: 130 (*Anacampta*).—Calif.; B.C. and Mont., s. to Calif. and Tex., also N.C. and Mexico, immigrant to Samoa and Hawaii.

Genus CURRANOPS Harriot

Curranops Harriot, 1942b: 249. Type-species, *Tetanops apicalis* Cole (orig. des.).

REFERENCE: Steyskal, 1962b (rev.).

apicalis (Cole) *in* Cole and Lovett, 1921: 328 (*Tetanops*).—Oreg.; Wash. to Calif.

scutellaris (Coquillett), 1900d: 25 (*Epiplatea*).—Calif.; Oreg.

Genus DELPHINIA Robineau-Desvoidy

Delphinia Robineau-Desvoidy, 1830: 719. Type-species, *thoracica* Robineau-Desvoidy (mon.) = *picta* (Fabricius).

Camptoneura Macquart, 1843a: 357 (1843: 200). Type-species, *Musca picta* Fabricius (orig. des.).

picta (Fabricius), 1781: 452 (*Musca*).—N. Amer.; Minn. to Maine, s. to Kans. and Fla.
  conica Fabricius, 1805: 318 (*Tephritis*).—N. Amer.
  thoracica Robineau-Desvoidy, 1830: 720.—Carolina.
  nigriventris Macquart, 1855: 144 (1855: 124) (*Urophora*).—Md.

Genus DIACRITA Gerstäcker

Diacrita Gerstäcker, 1860: 195. Type-species, *costalis* Gerstäcker (mon.).

REFERENCE: Steyskal, 1947b (rev.).

costalis Gerstäcker, 1860: 197.—Mexico; s. Calif. to Tex.
  aemula Loew, 1873b: 114.—Calif.
plana Steyskal, 1947b: 151.—N. Mex.; Colo.

Genus DYSCRASIS Aldrich

Dyscrasis Aldrich, 1932b: 7. Type-species, *hendeli* Aldrich (orig. des.).

hendeli Aldrich, 1932b: 8.—Tex.; Ariz., Mexico.

Genus HAIGIA Steyskal

Haigia Steyskal, 1961b: 406. Type-species, *nevadana* Steyskal (orig. des.).

nevadana Steyskal, 1961b: 406.—Nev.; Wash.

Genus HERINA Robineau-Desvoidy

Herina Robineau-Desvoidy, 1830: 724. Type-species, *liturata* Robineau-Desvoidy (Rondani, 1856: 109) =*germinationis* (Rossi).
Tephronota Loew, 1868b: 6. Type-species, *Ortalis gyrans* Loew (Loew, 1873b: 57) =*tristis gyrans* (Loew).

REFERENCE: McAlpine, 1951 (rev.).

canadensis (Johnson), 1902b: 144 (*Tephronota*).—Que.; B.C. to N.B., s. to Nebr., Ill., and Mass.
narytia (Walker), 1849: 1020 (*Trypeta*).—Fla.; N.H. to Fla., Bahama Is.
nigribasis McAlpine, 1951: 312.—Man.; B.C., Wyo.
ruficeps Wulp, 1867: 156.—Wis.; Ont. to N.S., s. to Tex. and Va.
   humilis Loew, 1873b: 123 (*Tephronota*).—N.Y., Tex., Va.

Genus HIATUS Cresson

Hiatus Cresson, 1906: 280. Type-species, *fulvipes* Cresson (mon.).
fulvipes Cresson, 1906: 281.—N. Mex.

Genus IDANA Loew

Idana Loew, 1873b: 115. Type-species, *Ortalis marginata* Say (mon.).
marginata (Say), 1830: 183 (1859b: 368) (*Ortalis*).—Ind.; Mich. and Ind. to Mass., s. to Va.

Genus MELIERIA Robineau-Desvoidy

Melieria Robineau-Desvoidy, 1830: 715. Type-species, *gangrenosa* Robineau-Desvoidy (Rondani, 1856: 108) =*crassipennis* (Fabricius).

REFERENCE: Steyskal, 1962a (rev.).

cana (Loew), 1858b: 374 (*Ortalis*).—Europe; Alaska to P.E.I., s. to Ariz., Kans., and Ind., Palaearctic.
   obscuricornis Loew, 1873b: 126 (*Ceroxys*).—Nebr.
occidentalis Coquillett *in* Baker, 1904: 29.—Nev.; B.C. to Minn., s. to Calif. and Tex.
ochricornis (Loew), 1873b: 126 (*Ceroxys*).—Wis.; Alaska and Calif. to N.B.
picta (Meigen), 1826: 276 (*Ortalis*).—England; Alaska, Europe.
sabuleti Steyskal, 1962a: 253.—Mich.
similis (Loew), 1873b: 127 (*Ceroxys*).—Conn.; B.C. to N.S., s. to Ohio and Conn.

Genus MYIOMYRMICA Steyskal

Myiomyrmica Steyskal, 1961b: 404. Type-species, *Cephalia fenestrata* Coquillett (orig. des.).

fenestrata (Coquillett), 1900d: 24 (*Cephalia*).—Kans.; Iowa, Wis., Ill.

Genus MYRMECOTHEA Hendel

Myrmecothea Hendel, 1910b: 310. Type-species, *Cephalia myrmecoides* Loew (sub. mon., Hendel, 1914f: 16).

myrmecoides (Loew), 1860b: 83 (*Cephalia*).—D.C.; N.Y. to Vt., s. to S.C.

Genus NOTOGRAMMA Loew

Notogramma Loew, 1868a: 289. Type-species, *cimiciforme* Loew (mon.).

cimiciforme Loew, 1868a: 289.—Cuba; Tex., Cent. and S. Amer., Hawaii, Mariana Is., Caroline Is.
purpuratum Cole, 1923a: 474.—Mexico (Baja Calif.); Calif. to Colo. and Tex., s. to Guatemala, Jamaica, Cuba.
stigma (Fabricius).—Not Nearctic.

Genus OTITES Latreille

Otites Latreille, 1804: 196. Type-species, *porcus* Latreille (as *Musca porca* Bosc; mon.) =*formosa* (Panzer).
Ortalimyia Curran, 1934a: 284. Type-species, *Ortalis snowi* Cresson (orig. des.).

REFERENCE: Hendel, 1911 (key, as *Ortalis*).

bimaculata (Hendel), 1911: 22 (*Ortalis*).—Colo.; N.W.T., s. to n. Calif. and Colo. **N. comb.**
    longicauda Hendel, 1913: 635 (*Ortalis*).—Nev. **N. syn.**
    carbona Cresson, 1914c: 458 (*Tetanops*).—Wyo. **N. syn.**
erythrocephala (Hendel), 1911: 23 (*Ortalis*).—Calif. **N. comb.**
pyrrhocephala (Loew), 1876: 335 (*Anacampta*).—Calif.; Nev. **N. comb.**
snowi Cresson, 1924a: 234 (*Ortalis*).—Kans.; Calif., Utah, N. Mex. **N. comb.**
stigma (Hendel), 1911: 23 (*Ortalis*).—Colo., S. Dak.; B C. to Man., s. to Colo. and Nebr. **N. comb.**

Genus PSEUDOTEPHRITIS Johnson

Stictocephala Loew, 1873b: 61 (preocc. Stål, 1869). Type-species, *Ortalis vau* Say (orig. des.).

Pseudotephritis Johnson, 1902b: 144 (n. name for *Stictocephala* Loew). Type-species, *Ortalis vau* Say (aut.).

REFERENCE: Steyskal, 1961b, 1962a (revs.).

Subgenus PSEUDOTEPHRITIS Johnson

approximata Banks, 1914c: 138, 185 (as *appoximata*, p. 138).—Va.; Iowa, Mich., and Pa., s. to Miss and Va.

corticalis (Loew), 1873b: 136 (*Stictocephala*).—N.Y.; Alta. to Conn., s. to Ohio and Va., also U.S.S.R.

vau (Say), 1830: 184 (1859b: 369) (*Ortalis*).—Ohio; B.C. to Que., s. to Tex. and Va.

   conjuncta Johnson, 1921a: 15 (as var.).—Maine.

   var. **californica** Steyskal, 1962a: 260 (as form).—Calif.

   var. **idahoana** Steyskal, 1962a: 260 (as form).—Wash., Idaho.

   var. **metzi** Johnson, 1915c: 49 (as sp.).—N.Y.; Minn. to Maine, s. to Va.

Subgenus PSEUDOTEPHRITINA Malloch

Pseudotephritina Malloch, 1931c: 5 (as genus). Type-species, *Stictocephala cribellum* Loew (orig. des.).

cribellum (Loew), 1873b: 134 (*Stictocephala*).—Nebr.; Mont. to Pa., s. to N. Mex. and Miss.

   cribrum Loew, 1873b: 135 (*Stictocephala*).—Middle States.

inaequalis Malloch, 1931c: 5.—Colo.; Wash. to Ind., s. to Calif. and Tex.

   var. **davisensis** Steyskal, 1962a: 262 (as form).—Wash., Idaho, Calif., Nev., Kans., Tex.

Genus PTEROCALLA Rondani

Pterocalla Rondani, 1848: 83. Type-species, *Dictya ocellata* Fabricius (mon.).

strigula Loew, 1873b: 133.—Ga.; Man. to Maine, s. to N. Mex. and Ga.

Genus SEIOPTERA Kirby

Ortalis Fallén, 1810a: 17 (preocc. Merrem, 1786). Type-species, *Musca vibrans* Linnaeus (Westwood, 1840: 149).

Seioptera Kirby *in* Kirby and Spence, 1817: [305]. Type-species, *Musca vibrans* Linnaeus (mon.).

Seoptera, emend.

Myodina Robineau-Desvoidy, 1830: 727. Type-species, *Musca urticae*
Fabricius (mon.) =*vibrans* (Linnaeus).

REFERENCE: Steyskal, 1956a (rev.).

albipes Cresson, 1919: 192.—Pa.; Mich. to Que., s. to Kans. and Ga.
currani Harriot, 1942a: 196.—N.Y.
colon Loew, 1868a: 296.—Ill., N.Y.; Ill. to Vt., s. to Miss.
costalis (Walker), 1849: 995 (*Ortalis*).—Ont.
dubiosa Johnson, 1921a: 15.—Maine; Man.
vibrans (Linnaeus), 1758: 599 (*Musca*).—Europe; B.C. to N.S., s.
to Oreg., Colo., and D.C., Palaearctic.

Genus TETANOPS Fallén

Tetanops Fallén, 1820c: 2. Type-species, *myopinus* Fallén (mon.).
Terelliosoma Rondani, 1856: 109. Type-species, *heryngii* Rondani
(orig. des.) =*flavescens* (Macquart).

The North American records of *rufifrons* may refer to an undescribed species.

REFERENCE: Harriot, 1942b (key).

Subgenus TETANOPS Fallén

cazieri Harriot, 1942b: 250.—n. Calif.; Oreg.
integer Loew, 1873b: 121.—Ill.; N. Dak. to Ont., s. to N. Mex.
and Tex.
luridipennis Loew, 1873b: 119.—Nebr.; N. Dak. to Mass., s. to
Ariz. and Fla.
magdalenae Cresson, 1924a: 233.—N. Mex.; Ariz.
rufifrons Wulp, 1899: 391.—Mexico; N. Mex., ?Kans.

Subgenus EURYCEPHALOMYIA Hendel

Eurycephala Röder, 1881: 211 (preocc. Laporte, 1833). Type-species,
myopaeformis Röder (mon.).
Eurycephalomyia Hendel, 1907a: 98 (as genus; n. name for *Eurycephala* Röder). Type-species, *myopaeformis* Röder
(aut.).

REFERENCE: Gojmerac, 1956 (descr., biol. *myopaeformis*).

myopaeformis (Röder), 1881: 212 (*Eurycephala*).—Calif.; B.C. to
Man., s. to Calif. and N. Mex.
polita Coquillett, 1900d: 22.—Colo.
aldrichi Hendel, 1911: 20.—Idaho.

Subgenus PSAEROPTERELLA Hendel

Psaeropterella Hendel, 1914b: 158 (as genus). Type-species, *macrocephala* Hendel (orig. des.). **N. status.**

macrocephala (Hendel), 1914b: 159 (*Psaeropterella*).—Calif. **N. comb.**

punctifrons (Hendel), 1914b: 160 (*Psaeropterella*).—B.C. **N. comb.**

Genus TETROPISMENUS Loew

Tetropismenus Loew, 1876: 333, 334. Type-species, *hirtus* Loew (mon.).

Califortalis Curran, 1934a: 284. Type-species, *hirsutifrons* Curran (orig. des.)=*hirtus* Loew. **N. syn.**

hirtus Loew, 1876: 333.—Calif.
  hirsutifrons Curran, 1934a: 284 (*Califortalis*).—Calif. **N. syn.**

Genus TRITOXA Loew

Tritoxa Loew, 1873b: 102. Type-species, *Trypeta flexa* Wiedemann (Coquillett, 1910b: 617).

REFERENCES: Chittenden, 1927 (bionomics, *flexa*); Manis, 1941 (bionomics, morphology *flexa*); Harriot, 1942c (key).

cuneata Loew, 1873b: 107.—Nebr.; Oreg., Alta., and Nebr., s. to Calif. and N. Mex.

flexa (Wiedemann), 1830: 483 (*Trypeta*).—Ga.; Man. and N. Dak. to Conn., s. to Ill. and Ga.
  arcuata Walker, 1852: 383 (*Trypeta*).—U.S.

incurva Loew, 1873b: 104.—Ill.; Wis. to Conn., s. to Colo., Tex., and Ga.

pollinosa Cole *in* Cole and Lovett, 1919: 252.—Oreg.; Wash., Colo.

ra Harriot, 1942c: 23.—Calif.

Genus TUJUNGA Steyskal

Tujunga Steyskal, 1961b: 405. Type-species, *mackenziei* Steyskal (orig. des.).

mackenziei Steyskal, 1961b: 405.—Calif.

Genus ULIDIOTITES Steyskal

Ulidiotites Steyskal, 1961b: 406. Type-species, *dakotana* Steyskal (orig. des.).

dakotana Steyskal, 1961b: 406.—S. Dak.; Nebr.

Subfamily ULIDIINAE

Genus ACROSTICTA Loew

Acrosticta Loew, 1868a: 293. Type-species, *scrobiculata* Loew (Coquillett, 1910b: 503).

apicalis (Williston), 1896c: 375 (*Euxesta*).—St. Vincent I.; D.C. to Fla., Neotropical, Pacific Is., Ghana (cent. Africa).
compta (Cole), 1912b: 158 (*Euxesta*).—Calif. **N. comb.**
    similis Johnson, 1919d: 165 (*Ulidia*).—Calif.
dichroa Loew, 1874b: 384.—Calif.
fulvipes Coquillett, 1900d: 24.—Calif.
rubida (Loew), 1876: 337 (*Ulidia*).—Calif.; Mexico (Baja Calif.). **N. comb.**
rufiventris Hendel, 1910a: 52.—Tex.; Calif., Fla., Cent. Amer., West Indies.

Genus CHAETOPSIS Loew

Chaetopsis Loew, 1868a: 315. Type-species, *Ortalis aenea* Wiedemann (Coquillett 1910b: 521).

Most of the published locality records of *aenea*, *fulvifrons*, and *massyla* are unreliable.

    REFERENCE: Curran, 1928d (rev.).

aenea (Wiedemann), 1830: 462 (*Ortalis*).—La.; Mich., R.I., s. to La. and Fla.
    trifasciata Say, 1830: 184 (1859b: 368) (*Ortalis*).—U.S.
apicalis apicalis Johnson, 1900b: 326.—N.J., Fla.; Mass. to Tenn. and Fla.
    ssp. **duplicata** Johnson, 1921a: 16 (as sp.).—Maine; Maine to Mass. **N. status.**
debilis Loew, 1868a: 318.—Cuba; Kans., Tex. and Mexico to Fla., Cuba, ?Bermuda.
    trifasciata Say, 1831: 19 (*Trypeta*; preocc. Say, 1830).—La.
fulvifrons (Macquart), 1855: 145 (1855: 125) (*Urophora*).—Md.; Wis. to R.I., s. to Calif. and Fla.
    aenea Wulp, 1867: 157 (*Trypeta*; preocc. Wiedemann, 1830).—Wis.
magna Cresson, 1924a: 241.—Kans.
massyla (Walker), 1849: 992 (*Ortalis*).—N. Amer.; Alta. to Maine and Mass., s. to N. Mex.
quadrifasciata Curran, 1928d: 80.—P.R.; Fla.

Genus EUMETOPIELLA Hendel

Eumetopia Macquart, 1847: 103 (1847: 87) (preocc. Westwood, 1837). Type-species, *rufipes* Macquart (orig. des.).
Eumetopiella Hendel, 1907a: 98 (n. name for *Eumetopia* Macquart). Type-species, *Eumetopia rufipes* Macquart (aut.).
rufipes (Macquart), 1847: 104 (1847: 88) (*Eumetopia*).—Pa.; Wis. to Mass., s. to Kans. and Fla.
varipes (Loew), 1866a: 181 (Cent. 6, no. 87) (*Eumetopia*).—Cuba; Mexico to Fla.

Genus EUXESTA Loew

Euxesta Loew, 1868a: 297. Type-species, *Ortalis notata* Wiedemann (Coquillett, 1910b: 543).
Aloceuxesta Hendel, 1936b: 78. Type-species, *Euxesta spoliata* Loew (orig. des.). **N. syn.**
Euxestina Enderlein, 1937b: 438 (preocc. Curran, 1934). Type-species, *Musca costalis* Fabricius (orig. des.).

The genus *Euxesta* is widespread and among the most commonly encountered of the ulidiine genera. Many published records of *notata* may refer to *anna*, *pechumani*, or undescribed species.

REFERENCE: Curran, 1935b (key).

abana Curran, 1935b: 13.—Utah; N. Mex.
abdominalis Loew, 1868a: 307.--Cuba; Cent. Amer., Fla., Bermuda, West Indies.
alternans Loew, 1868a: 308.—?Cuba, ?Brazil; Bermuda, Neotropical.
anna Harriot, 1942c: 25.—Calif.
annonae (Fabricius), 1794: 358 (*Musca*).—West Indies; Fla., Bermuda, Neotropical.
 quadrivittata Macquart, 1835: 456 (*Urophora*).—Cuba.
basalis (Walker), 1852: 373 (*Ortalis*).—U.S.; Ga., Fla., Bahama Is.
bicolor (Cresson), 1906: 285 (*Acrosticta*).—Tex.
brookmani Harriot, 1942c: 24.—Calif.
contorta Curran, 1935b: 16.—Utah.
eluta Loew, 1868a: 312.—Cuba; Tex., Fla , Mexico, West Indies, S. Amer.
fervida Curran, 1935b: 14.—Utah; Idaho, Nev., N. Mex.
 knowltoni Curran, 1935b: 15.—Utah. **N. syn.**
juncta Coquillett, 1904e: 95.—Nicaragua; Tex., Bahama Is., P.R.
lutzi Curran, 1935b: 16.—Colo.; Nev. to Kans. and Tex
magdalenae Cresson, 1924a: 240.—N. Mex.; Wyo.
minor Cresson, 1906: 286.—N. Mex.
nigriceps Curran, 1935b: 19.—Utah.

nitidiventris Loew, 1873b: 157.—Tex.; Calif., Nebr., and Ga., s. through Neotropical Region, Galapagos Is., also Italy.
notata (Wiedemann), 1830: 462 (*Ortalis*).—N.Y., Ga.; Calif. and Mexico (Baja Calif.) to Wis. and Vt., s. to Fla., also Bahama Is., Galapagos Is.
pechumani Curran, 1938: 4.—N.Y.; Mich. to N.Y., s. to Ariz. and Fla.
pulchella Cresson, 1906: 287.—N. Mex.; Utah.
pusio Loew, 1868a: 299.—Cuba; Bermuda, Jamaica.
quaternaria Loew, 1868a: 302.—Cuba; Tex. to Fla., s. to Cent. Amer., West Indies, and Bahama Is.
rubida Curran, 1935b: 13.—Utah; Idaho.
sanguinea Hendel, 1913: 635.—Utah; N. Mex.
scoriacea Loew, 1876: 336.—Tex.; Mass., s. to Tex. and Fla., Bahama Is.
scutellaris Curran, 1935b: 17.—Utah.
spoliata Loew, 1868a: 298.—Cuba; Tex., Fla., Cent. and S. Amer.
stigmatias Loew, 1868a: 310.—Cuba; Fla., Mexico, West Indies, S. Amer.
thomae Loew, 1868a: 306.—St. Thomas I.; Tex., Fla., Cent. Amer., West Indies.
willistoni Coquillett, 1900d: 24 (n. name for *spoliata* Williston).—Calif.; Ariz.
  spoliata Williston, 1893d: 257 (preocc. Loew, 1867).—Calif.
xeres Curran, 1935b: 17.—Utah; N. Mex.

Genus HOMALOCEPHALA Zetterstedt

Homalocephala Zetterstedt, 1838: 749. Type-species, *albitarsis* Zetterstedt (mon.).
Psairoptera Wahlberg, 1839: 18. Type-species, *biumbrata* Wahlberg (Zetterstedt, 1847: 2264) = *albitarsis* Zetterstedt.
Psaeroptera, error or emend.

The larvae live under the bark of various kinds of trees. *H. apicalis* has been reared from the sap of *Pinus monticola* in Montana.

REFERENCE: Hennig, 1940 (key).

albitarsis Zetterstedt, 1838: 749.—Europe; Alaska to N.S., s. in mts. to Ariz. and N. Mex.
  nubecula Johnson, 1921a: 14 (*Chyliza*).—Maine. **N. syn.**
apicalis (Wahlberg), 1839: 21 (*Psairoptera*).—Europe; Alaska, B.C., Mont.

bipunctata (Loew), 1854: 22 (?*Psairoptera*).—Europe; Oreg., Ont., N.H.
    diopsides Walker, 1849: 995 (*Ortalis*).—Ont. **N. syn.**
similis (Cresson), 1924a: 236 (*Psairoptera*).—Sask.; Alaska, Alta. to Que., s. in mts. to Ariz.

Genus OEDOPA Loew

Oedopa Loew, 1868a: 287. Type-species, *capito* Loew (mon.).
ascriptiva Hendel, 1909: 267.—Colo.; Utah.
capito Loew, 1868a: 287.—Nebr.; Utah to Nebr., s. to Ariz. and Kans.

Genus PAROEDOPA Coquillett

Paroedopa Coquillett, 1900d: 22. Type-species, *punctigera* Coquillett (orig. des.).
punctigera Coquillett, 1900d: 23.—Ariz., N. Mex.; Tex.

Genus PHYSIPHORA Fallén

Physiphora Fallén, 1810a: 11. Type-species, *Chrysomyza splendida* Fallén (aut.) = *demandata* (Fabricius).
Chrysomyza Fallén, 1817c: 3 (unjustified n. name for *Physiphora* Fallén. Type-species, *splendida* Fallén (mon.) = *demandata* (Fabricius).
Cliochloria Enderlein, 1927: 103. Type-species, *Musca aenea* Fabricius (orig. des.).

    REFERENCES: Knab, 1916b (distribution, biol., *aenea*); Hennig, 1940 (rev.); Séguy, 1941 (key).

aenea (Fabricius), 1794: 335 (*Musca*).—Asia; Calif., Mich., La., Ala., Ga., Mexico, S. Amer., Australia.
demandata (Fabricius), 1798: 564 (*Musca*).—Europe; B.C. to Mass., s. to Calif. and S.C., Mexico, S. Amer., Asia, Africa.

Genus STENERETMA Loew

Steneretma Loew, 1873b: 186. Type-species, *laticauda* Loew (mon.).
laticauda Loew, 1873b: 187.—Tex.; Kans., N.C.

Genus STENOMYIA Loew

Stenomyia Loew, 1868a: 320. Type-species, *tenuis* Loew (mon.).
    REFERENCE: Cresson, 1913 (key).

hendeli (Johnson), 1913a: 83 (*Chaetopsis*).—N.J., Ga.; N.Y. (Long I.) to Ga.
nasoni Cresson, 1913: 320.—Ill.; Ariz. to Mass., s. to Fla.
tenuis Loew, 1868a: 321.—Tex., Ga.; Kans. to Mass., s. to Tex. and Fla.
tenuissima (Hendel), 1910a: 28 (*Euxesta*).—Ga.; Man. to N.J., s. to Utah and Fla., Bahama Is., Cuba.
  fasciapennis Cresson, 1913: 320.—Minn. **N. syn.**
  brooksi Johnson, 1926b: 301 (*Chaetopsis*).—Fla. **N. syn., n. comb.**

Genus STICTOMYIA Bigot

Stictomyia Bigot, 1885d: clxvi. Type-species, *longicornis* Bigot (mon.).

REFERENCE: Hunter, Pratt, and Mitchell, 1913 (habits, *longicornis*).

longicornis Bigot, 1885d: clxvi.—Mexico; Calif. to Tex.
punctata Coquillett, 1900d: 23.—N. Mex.; Tex.

Genus TEXASA Steyskal

Texasa Steyskal, 1961b: 407. Type-species. *chaetifrons* Steyskal (orig. des.).

chaetifrons Steyskal, 1961b: 407.—Tex.

Genus ZACOMPSIA Coquillett

Zacompsia Coquillett, 1901b: 15. Type-species, *fulva* Coquillett (orig. des.).

fulva Coquillett, 1901b: 15.—Tex., La.; Tex. to S.C., s. to Fla.

Unplaced Genus and Species of Otitidae

Prionella Robineau-Desvoidy, 1830: 759. Type-species, *beauvoisii* Robineau-Desvoidy (Coquillett, 1910b: 594).

beauvoisii Robineau-Desvoidy, 1830: 760.—?U.S. Unrecognized.
villosa Robineau-Desvoidy, 1830: 760.—?U.S. Unrecognized.

Unplaced Species of Otitidae

philadelphica Robineau-Desvoidy, 1830: 715 (*Meckelia*).—Pa.

Family PLATYSTOMATIDAE

(Platystomidae)

By George C. Steyskal

Most of the platystomatids are found in the Old World Tropics. The species are small to large, and usually handsome. The biology of very few of the species is known. They seem to be saprophagous.

REFERENCES: Hendel, 1914a (cat.), 1914f (rev.); Steyskal, 1961b (rev.).

Genus AMPHICNEPHES Loew

Amphicnephes Loew, 1873b: 83. Type-species, *pertusus* Loew (mon.) =*pullus* (Wiedemann).

fasciola Coquillett, 1900d: 21.—Kans.; Ariz.
pullus (Wiedemann), 1830: 506 (*Trypeta*).—Unknown; Mich. to Mass., s. to Tex. and Fla.
pertusus Loew, 1873b: 84.—Carolina, D.C., Conn.

Genus POGONORTALIS Hendel

Pogonortalis Hendel *in* Meijere, 1911b: 370. Type-species, *uncinata* Meijere (mon.).

doclea (Walker), 1849: 1035 (*Trypeta*).—Australia; Calif., also Java.

Genus RIVELLIA Robineau-Desvoidy

Rivellia Robineau-Desvoidy, 1830: 729. Type-species, *herbarum* Robineau-Desvoidy (Rondani, 1869: 8, 28; 1869: 8, 28)=*syngenesiae* (Fabricius).

REFERENCE: Namba, 1956 (rev.).

atriventris Hendel, 1914f: 182.—Ga.
australis Namba, 1956: 32.—Ariz.
bipars (Walker), 1861: 326 (*Ortalis*).—Tasmania (as U.S., error). Not Nearctic.
boscii Robineau-Desvoidy, 1830: 730.—Carolina; Ark., N.C. to Fla.
brevifasciata Johnson, 1900b: 326.—N.J.; Colo. to Mass., s. to Miss. and Ga.
cognata Cresson, 1919: 191.—Pa.; S. Dak. to Mass., s. to Kans. and Fla.
colei Namba, 1956: 71.—Minn.; Minn. to Conn., s. to La. and Tenn.
conjuncta Loew, 1873b: 88.—Md.; Kans. to Mass., s. to Tex. and La., also Mexico (V.C.).

coquilletti Hendel, 1914f: 155 (n. name for *basilaris* Coquillett).—
Colo.; Mont., Man., and Minn., s. to Colo. and Iowa.
basilaris Coquillett, 1900d: 21 (preocc. Wiedemann, 1830).—
Colo.
flavimana Loew, 1873b: 92.—Nebr.; Man. to Que., s. to Nebr. and N.Y.
floridana Johnson, 1900a: 247.—Fla.
imitabilis Namba, 1956: 43.—Ga.; Mich.
inaequata Namba, 1956: 62.—Fla.; N.J., also Tex. to Fla.
interrupta (Macquart), 1835: 459 (*Urophora*).—N. Amer. Unrecognized.
maculosa Namba, 1956: 35.—Ga.; Alta., also Ala., N.C. to Fla.
melliginis (Fitch), 1855: 769 (1856: 65) (*Tephritis*).—N.Y.; Minn., Ont., and Mass., s. to Iowa and S.C.
metallica (Wulp), 1867: 154 (*Herina*).—Wis.; Minn. to Ont., s. to Mo. and Va.
micans Loew, 1873b: 94.—Tex.; Mont. and S. Dak., s. to Calif. and Tex.
michiganensis Namba, 1956: 58.—Mich.
munda Namba, 1956: 40.—Kans.; S. Dak. and Minn., s. to Tex.
occulta Wulp, 1898: 382.—Mexico; Ariz., N. Mex., also Costa Rica.
otroeda (Walker), 1849: 992 (*Ortalis*).—N. Amer. Unrecognized.
pallida Loew, 1873b: 95.—D.C.; Minn. to R.I., s. to D.C.
quadrifasciata (Macquart), 1835: 433 (*Herina*).—N. Amer.; Mont. to Que., s. to Tex. and Fla.
rufitarsis (Macquart), 1855: 143 (1855: 123) (*Herina*).—Md. Unrecognized.
severini Blanton, 1937: 139.—S. Dak.; Idaho to Ind., s. to Utah, Kans., and Iowa.
socialis Namba, 1956: 68.—Iowa; Ill.
steyskali Namba, 1956: 64.—Mich.; Minn. to N.Y., s. to Tex. and Fla.
succinata (Wiedemann), 1830: 526 (*Dacus*).—Unknown; Fla., ?S. Africa.
tersa Namba, 1956: 31.—N. Mex.; Ariz., also Mexico (V.C.).
texana Namba, 1956: 52.—Tex.; Ark.
vaga Namba, 1956: 50.—Fla.; Alta., Iowa to N.C., s. to Tex. and Fla.
variabilis Loew, 1873b: 91.—D.C.; Ont. and Que., s. to Ill. and Fla.
viridulans Robineau-Desvoidy, 1830: 729.—N. Amer.; Iowa to Mass., s. to Miss. and Ga.
winifredae Namba, 1956: 44.—Minn.; Utah to Que., s. to Kans. and Ga.

Genus SENOPTERINA Macquart

Senopterina Macquart, 1835: 453. Type-species, *Dacus brevipes* Fabricius (mon.).
Stenopterina, emend.
Bricinnia Walker, 1861: 324. Type-species, *flexivitta* Walker (mon.).
Briciniella Giglio-Tos, 1893b: 13. Type-species, *cyanea* Giglio-Tos (mon.).

REFERENCE: Shewell, 1962 (rev.).

brevipes (Fabricius), 1805: 272 (*Dacus*).—Cent. Amer. Probably not Nearctic.
caerulescens Loew, 1873b: 97.—Tex.
foxleei Shewell, 1962: 197.—B.C.; Idaho, Colo., Kans., Ill., La.
mexicana (Macquart), 1843a: 365 (1843: 208) (*Herina*).—Mexico; Ariz., Tex.
varia Coquillett, 1900d: 25.—Fla.
bicolor Johnson, 1900a: 246.—Fla.

Family PYRGOTIDAE

By George C. Steyskal

The family Pyrgotidae is worldwide in distribution. Its members are largely nocturnal in habit and, as far as known, parasitic on scarabaeid beetles. No modern revision of the North American species is available. However, Aczél's revision (1956a-c) of the Central and South American forms contains a wealth of general information and and excellent bibliography.

Genus PYRGOTA Wiedemann

Pyrgota Wiedemann, 1830: 580. Type-species, *undata* Wiedemann (mon.).
Oxycephala Macquart, 1843a: 354 (1843: 197). Type-species, *fuscipennis* Macquart (mon.)=*undata* Wiedemann.

fenestrata (Macquart), 1851: 254 (1851: 281) (*Oxycephala*).—Amer.; Carolina to Fla.
vespertilio Gerstäcker, 1860: 189.—Carolina. **N. syn.**
undata Wiedemann, 1830: 581.—N. Amer.; Man. to Que., s. to Tex. and Fla.
nigripennis Gray *in* Griffith and Pidgeon, 1832: 779 (pl. 125, fig. 5) (*Myopa*).—Unknown.
fuscipennis Macquart, 1843a: 355 (1843: 198) (*Oxycephala*).— N. Amer.
pterophorina Gerstäcker, 1860: 190.—Carolina. **N. syn.**

Genus PYRGOTELLA Curran

Pyrgotella Curran, 1934a: 270. Type-species, *Pyrgota chagnoni* Johnson (orig. des.).

chagnoni (Johnson) 1900a: 246 (*Pyrgota*).—Que.; Minn., Ont., and Que., s. to N.C.

Genus SPHECOMYIELLA Hendel

Sphecomyiella Hendel, 1933a: 1. Type-species, *Sphecomyia valida* Harris (mon.).

maculipennis (Macquart) 1846: 338 (1846: 210) (*Oxycephala*).—Tex.; Ky. to Md., s. to Tex. and Fla. **N. comb.**
  filiola Loew, 1876: 332 (*Pyrgota*).—Tex. **N. syn.**
  debilis Osten Sacken, 1877: 343 (*Pyrgota*).—Ky.
valida (Harris), 1841: 410 (*Sphecomyia*).—Mass.; Minn. to Ont. and Mass., s. to Fla.
  millepunctata Loew, 1854: 22 (*Pyrgota*).—Carolina, D.C., N.Y., Ill.

Family TEPHRITIDAE

(Trypetidae, Trupaneidae)

By Richard H. Foote

Females of most species of Tephritidae oviposit in living, healthy plant tissue and their larvae live and feed in various parts of the plant. Some form stem and root galls, a very few mine leaves, while others develop in fleshy fruits and in the seeds and ovaries of various flowers, principally composites. The larvae of fruit flies of the greatest economic importance develop within fleshy fruits and vegetables, their eggs being inserted directly beneath the epidermis. They take a heavy toll of these plant products in commercial plantings over the entire world and constitute one of the most economically important families of the Diptera. A supergeneric classification of the family was proposed by Hering; it has not been adopted in this catalog because the Nearctic genera have not been studied adequately.

REFERENCES: Snodgrass, 1924 (larval morphology); Greene, 1929 (larval tax.); Benjamin, 1934 (adult tax., principally Fla.); Curran, 1934a (key to genera); Phillips, 1946 (larval tax.); Hering, 1947 (subfamilies and tribes); Peterson, 1951 (larval tax.); Christenson and Foote, 1960 (biol.).

Genus TOXOTRYPANA Gerstäcker

Toxotrypana Gerstäcker, 1860: 191. Type-species, *curvicauda* Gerstäcker (mon.).
Mikimyia Bigot, 1884c: xxix. Type-species, *furcifera* Bigot (mon.) = *curvicauda* Gerstäcker.

The single species is the well-known papaya fruit fly of the New World. It is commonly found in both wild and cultivated papayas and has also been recorded from mango, the only other known host.

REFERENCES: Knab and Yothers, 1914 (biol.); Benjamin, 1934 (descr.); Baker et al., 1944 (biol., descr., distribution).

curvicauda Gerstäcker, 1860: 194.—Virgin Is. (St. John); s. Tex. and Fla., s. to Brazil, West Indies, Bahama Is.
furcifera Bigot, 1884c: xxix (*Mikimyia*).—Brazil.

Genus RHYNENCINA Johnson

Rhynencina Johnson, 1922: 24. Type-species, *longirostris* Johnson (orig. des.).
Aleomyia Phillips, 1923: 123. Type-species, *alpha* Phillips (orig. des.) = *longirostris* Johnson.

longirostris Johnson, 1922: 24.—Pa.; Md., Ga.
alpha Phillips, 1923: 124 (*Aleomyia*).—Md. **N. syn.**

Genus UROPHORA Robineau-Desvoidy

Euribia Meigen, 1800: 36. Type-species, *Musca cardui* Linnaeus (Hendel, 1927a: 41). Suppressed by I.C.Z.N., 1963b: 339.
Urophora Robineau-Desvoidy, 1830: 769. Type-species, *sonchi* Robineau-Desvoidy (Westwood, 1840: 149) = *cardui* (Linnaeus).

REFERENCE: Bezzi, 1923 (rev.).

formosa (Coquillett), 1894b: 71 (*Trypeta*).—Calif.; Oreg., Idaho, Ariz.
caurina Doane, 1899: 182 (*Rhagoletis*).—Oreg. **N. syn.**
grindeliae (Coquillett), 1908: 146 (*Rhagoletis*).—Tex.; Calif., Colo.
jaceana (Hering), 1935: 169 (*Euribia*).—Poland; Nfld., N.S.
rufipes (Curran), 1932c: 2 (*Aleomyia*).—Ariz.
timberlakei Blanc and Foote, 1961: 81.—Calif.

Genus CALLACHNA Aldrich

Callachna Aldrich, 1929a: 11. Type-species, *Trypeta gibba* Loew (orig. des.).

gibba (Loew), 1873b: 260 (*Trypeta*).—Tex.; Mich. to Pa., s. to Tex. and La.

Genus PERONYMA Loew

Peronyma Loew, 1873b: 250. Type-species, *Trypeta sarcinata* Loew (orig. des.).
Tomoplagina Curran, 1932e: 14. Type-species, *maculata* Curran (orig. des.) = *sarcinata* (Loew).

REFERENCE: Benjamin, 1934 (descr.).

sarcinata (Loew), 1862b: 218 (Cent. 2, no. 73) (*Trypeta*).—Carolina; N.C., s. to Ala. and Fla.
maculata Curran, 1932e: 14 (*Tomoplagina*).—Fla.

Genus PROCECIDOCHARES Hendel

Oedaspissolidago Patton, 1897: 247. Type-species, *Trypeta atra* Loew (mon.). [In the interest of stability, the author has applied to the I.C.Z.N. for suppression of this valid name.]
Procecidochares Hendel, 1914e: 91 (also 1914c: 42). Type-species, *Trypeta atra* Loew (orig. des.).
Oedaspis, authors, not Loew.

REFERENCE: Aldrich, 1929a (rev.).

anthracina (Doane), 1899: 180 (*Oedaspis*).—Idaho; Oreg. to Mich., s. to Calif. and Ariz.
atra (Loew), 1862b: 219 (Cent. 2, no. 74) (*Trypeta*).—N.Y.; Kans. to Maine, s. to Fla.
setigera Coquillett, 1899e: 262 (*Oedaspis*).—Va.
australis Aldrich, 1929a: 9 (as *atra* var.).—Tex.; S.C., Fla.
blantoni Hering, 1940d: 12.—Oreg.
flavipes Aldrich, 1929a: 5.—Mexico (Socorro I. in Baja Calif.); Calif.
grindeliae Aldrich, 1929a: 6.—Calif.; Colo.
minuta (Snow), 1894a: 164 (*Oedaspis*).—Mont.; Wash. to Mich., s. to Calif. and Tex.
montana (Snow), 1894a: 163 (*Oedaspis*).—Mont.
pleuralis Aldrich, 1929a: 9.—Ariz.

polita (Loew), 1862c: 77 (*Trypeta*).—Miss., Wash.; Wash., Kans. to Mass., s. to Miss. and Fla.
stonei Blanc and Foote, 1961: 77.—Calif.

Genus PROCECIDOCHAROIDES Foote

Procecidocharoides Foote, 1960c: 671. Type-species, *Trypeta penelope* Osten Sacken (orig. des.).
caliginosus Foote, 1960c: 674.—N. Mex.
flavissimus Foote, 1960c: 672.—Calif.
penelope (Osten Sacken), 1877: 346 (*Trypeta*).—N.Y.; Mich., Ill., Ohio, Pa., Mass., N.J.
pullatus Foote, 1960c: 673.—Ariz.; N. Mex.

Genus EUTRETA Loew

Icaria Schiner, 1868: 276 (preocc. Saussure, 1853). Type-species, *Trypeta sparsa* Wiedemann (orig. des.).
Eutreta Loew, 1873b: 276. Type-species, *Trypeta sparsa* Wiedemann (Coquillett, 1910b: 543).

REFERENCE: Curran, 1932e (key, descrs.).

angusta Banks, 1926: 44.—Tex.
baccharis (Coquillett), 1894b: 73 (*Trypeta*).—Calif.; s. Calif. to Kans., s. to n. Mexico.
    fasciata Adams, 1904b: 449 (*Icterica*).—Ariz.
diana (Osten Sacken), 1877: 347 (*Trypeta*).—Mo.; Wash. to Mont., s. to Calif. and Mo.
    var. **tricolor** Snow, 1894a: 168.—Mont.
facialis Curran, 1932e: 17.—Mont.
frontalis Curran, 1932e: 16.—N.C.; Nev. to N.Y., s. to Tex. and N.C.
hespera Banks, 1926: 44.—Calif., Colo.; Nev., N. Mex., Tex.
jonesi Curran, 1932e: 19.—Oreg.; Oreg. to Wyo., s. to Calif. and Nev.
longicornis Snow, 1894a: 168.—Mont.; Idaho.
oregona Curran, 1932e: 18.—Oreg.; Calif., Nev.
pacifica Curran, 1932e: 17.—Calif.; Colo., Ariz., Tex.
pollinosa Curran, 1932e: 18.—Oreg.; Wash. and Idaho, s. to Calif. and Utah.
rotundipennis (Loew), 1862c: 79 (*Trypeta*).—Middle States; Kans. to Md., s. to N.C.
simplex Thomas, 1914: 425.—Colo.
sparsa (Wiedemann), 1830: 492 (*Trypeta*).—Unknown; Colo. to Maine, s. to La. and N.C.
    caliptera Say, 1830: 187 (1859b: 370) (*Trypeta*).—Ind.
    calyptera, error or emend.

latipennis Macquart, 1843a: 357 (1843: 200) (*Platystoma*).—Unknown.
novaeboracensis Fitch, 1855: 771 (1856: 67) (*Acinia*).—N.Y.
cribripennis Harris *in* Johnson, 1925c: 97 (*Trypeta*). Nomen nudum.

Genus CRYPTOTRETA Blanc and Foote

Cryptotreta Blanc and Foote, 1961: 82. Type-species, *Eurosta pallida* Cole (orig. des.).

pallida (Cole), 1923a: 472 (*Eurosta*).—Mexico (Baja Calif.); Calif.

Genus PARACANTHA Coquillett

Paracantha Coquillett, 1899e: 264. Type-species, *Trypeta culta* Wiedemann (orig. des.).
Carphotricha, authors, not Loew.
Scriptotricha (error) Cockerell, 1889a: 1.
REFERENCE: Malloch, 1941a (rev.).

culta (Wiedemann), 1830: 486, 680 (*Trypeta;* as *cutta*, p. 486).—Ga.; Wash. to Del., s. to Calif. and Fla.
fimbriata Macquart, 1843a: 385 (1843: 228) (*Acinia*).—Carolina.
sarnia Walker, 1849: 1029 (*Trypeta*).—Unknown.
cultaris (Coquillett), 1894b: 72 (*Trypeta*).—s. Calif.; Wash. to Nebr., s. to Calif. and Tex.
forficula Benjamin, 1934: 31.—Fla.
genalis Malloch, 1941a: 40.—Calif.; Idaho and Mont., s. to Calif. and Tex.
gentilis Hering, 1940a: 54.—Wyo.; Oreg. to Mont., s. to Calif. and Tex.
mimetica Malloch, 1941a: 41.—Oreg.
elongata Malloch, 1941a: 41 (as *mimetica* var.).—N. Mex.
marginepunctata (Macquart), 1835: 464 (*Carphotricha*).—Pa. Unrecognized.

Genus EUROSTA Loew

Eurosta Loew, 1873b: 280. Type-species, *Trypeta solidaginis* Fitch (Coquillett, 1910b: 543) ?=*asteris* (Harris).
Eurostina Curran, 1932e: 4. Type-species, *Trypeta latifrons* Loew (orig. des.).

The larvae form galls in roots and stems of a wide variety of plants. The species appear to be rather host-specific, and each forms a gall of characteristic appearance on its host-plant group.

REFERENCES: Phillips, 1923 (key); Benjamin, 1934 (descrs. Fla. spp.).

asteris (Harris), 1841: 417 (*Tephritis;* preocc. Haliday, 1838).—Mass. Unrecognized.
comma (Wiedemann), 1830: 478 (*Trypeta*).—Ky.; Colo. to Maine, s. to n. Fla.
    alvea Walker, 1849: 1027 (*Trypeta*).—N. Amer. (as Australia, error; Hardy, 1959c: 208).
    dertona Walker, 1849: 1028 (*Trypeta*).—N. Amer. (Hardy, 1959c: 213).
conspurcata Doane, 1899: 186.—Wash.; N.Y., N.H., Mass., N.J.
donysa (Walker), 1849: 1007 (*Trypeta*).—Unknown; Fla.
    nicholsoni Benjamin, 1934: 27.—Fla.
elsa Daecke, 1910: 342.—N.Y., Md.; Maine to Md.
fenestrata Snow, 1894a: 169.—Ariz.; Fla.
latifrons (Loew), 1862c: 89 (*Trypeta*).—Carolina; Wis. to Maine, s. to N.C.
    cribrata Wulp, 1867: 158 (*Trypeta*).—Wis.
    confusa Curran, 1934a: 293 (*Eurostina*). Nomen nudum.
reticulata Snow, 1894a: 170.—Mont.; Mont. to N.H., s. to Colo. and Pa.
solidaginis (Fitch), 1855: 771 (1856: 67) (*Acinia*).—N.Y.; s. N.W.T. to Maine, s. to B.C. and Kans. ?=*asteris*.
    var. **fascipennis** Curran, 1923p: 302.—Ont.; Sask., Pa.
    var. **subfasciata** Curran, 1923p: 302.—B.C.

Genus ACIDOGONA Loew

Acidogona Loew, 1873b: 285. Type-species, *Trypeta melanura* Loew (orig. des.).
    REFERENCE: Benjamin, 1934 (descr.).

melanura (Loew), 1873b: 283 (*Trypeta*).—D.C.; Mass. to Fla.

Genus JAMESOMYIA Quisenberry

Jamesomyia Quisenberry, 1949a: 49. Type-species, *Trypeta geminata* Loew (orig. des.).

geminata (Loew), 1862b: 220 (Cent. 2, no. 75) (*Trypeta*).—Pa.; Mich. to N.H., s. to Mo. and W. Va.

Genus XANTHOMYIA Phillips

Xanthomyia Phillips, 1923: 140. Type-species, *Trypeta platyptera* Loew (orig. des.).

nora (Doane), 1899: 184 (*Eutreta*).—Idaho; Colo.

platyptera (Loew), 1873b: 306 (*Trypeta*).—Conn.; Mich. to Vt., s. to Ohio and Va., ?Colo.

Genus XENOCHAETA Snow

Xenochaeta Snow, 1894a: 166. Type-species, *dichromata* Snow (orig. des.).

REFERENCE: Foote, 1960h (rev.).

aurantiaca (Doane), 1899: 185 (*Eutreta*).—Wash.
dichromata Snow, 1894a: 166.—Oreg.; Wash., Calif.

Genus ACINIA Robineau-Desvoidy

Acinia Robineau-Desvoidy, 1830: 775. Type-species, *jaceae* Robineau-Desvoidy (Rondani, 1871b: 4; 1871: 4) = *corniculata* (Zetterstedt).

REFERENCE: Benjamin, 1934 (descr.).

picturata (Snow), 1894a: 173 (*Tephritis*).—Fla.; s. Calif. to Fla., N.Y. to Ga., Mexico, West Indies.
fucata, N. Amer. authors, not Fabricius.

Genus ICTERICA Loew

Icterica Loew, 1873b: 287. Type-species, *Trypeta seriata* Loew (Coquillett, 1910b: 555).

REFERENCES: Phillips, 1923 (key, descrs.); Steyskal and Foote, 1962 (notes).

circinata (Loew), 1873b: 288 (*Trypeta*).—N.Y.; Wis., N.J.
seriata (Loew), 1862c: 84 (*Trypeta*).—Middle States; Nebr. to Maine, s. to Ariz. and N.J.
sericata, error.

Genus ACROTAENIA Loew

Acrotaenia Loew, 1873b: 274. Type-species, *Trypeta testudinea* Loew (orig. des.).

REFERENCE: Foote, 1960h (distribution, descr.).

testudinea (Loew), 1873b: 272 (*Trypeta*).—Cuba; Fla.

Genus OXYNA Robineau-Desvoidy

Oxyna Robineau-Desvoidy, 1830: 755. Type-species, *Musca parietina* Linnaeus (Rondani, 1856: 110).

REFERENCE: Quisenberry, 1949c (rev.).

aterrima (Doane), 1899: 187 (*Eurosta*).—Colo.; Oreg., Utah, Calif.
palpalis (Coquillett), *in* Baker, 1904: 30 (*Tephritis*).—Nev.; Wash. and Idaho, s. to Calif. and Utah.
utahensis Quisenberry, 1949c: 73.—Utah; Wash., Calif.

Genus EUARESTA Loew

Euaresta Loew, 1873b: 296. Type-species, *Trypeta festiva* Loew (Coquillett, 1910b: 540).
Camaromyia Hendel, 1914e: 95. Type-species, *Trypeta bullans* Wiedemann (orig. des.).
Euaresta, subg. *Setigeresta* Benjamin, 1934: 50. Type-species, *Trypeta aequalis* Loew (orig. des.).
REFERENCE: Quisenberry, 1950 (rev.).

aequalis (Loew), 1862c: 86 (*Trypeta*).—Pa.; s. Canada, throughout U.S.
    gemella Coquillett, 1902e: 181 (*Tephritis*).—N. Mex.
bella (Loew), 1862c: 88 (*Trypeta*).—Wash., N.Y.; s. Canada, throughout U.S.
bellula Snow, 1894a: 172.—Ariz.; Wash. and Idaho, s. to s. Calif. and Tex.
    stelligera Coquillett, 1894b: 74 (*Trypeta*).—s. Calif.
bullans (Wiedemann), 1830: 506 (*Trypeta*).—Uruguay; Calif., Chile, Brazil, Uruguay, s. Africa, Australia.
    adspersa Coquillett *in* Baker, 1904: 30.—Calif.
    wolffi Cresson, 1931a: 5 (*Tephritis*).—s. Calif.
festiva (Loew), 1862c: 86 (*Trypeta*).—Pa.; Idaho to Conn., s. to Colo. and N.C.
jonesi Curran, 1932c: 9.—Oreg.; Wash.
stigmatica Coquillett, 1902e: 180.—Ariz.; Mont. to Mich., s. to Calif. and Tex.
tapetis (Coquillett), 1894b: 75 (*Trypeta*).—N. Mex.; Wash. to Kans., s. to Calif. and N. Mex.

Genus PAROXYNA Hendel

Paroxyna Hendel, 1927a: 146. Type-species, *Trypeta tessellata* Loew (orig. des.).
Ensina, authors, not Robineau-Desvoidy.

albiceps (Loew), 1873b: 302 (*Trypeta*).—Canada, Maine; Mich. ne. U.S.
    euryptera Loew, 1873b: 304 (*Trypeta*).—N.Y.
americana Hering, 1944: 11 (as *difficilis* ssp.).—Colo; Calif.
    franciscana Hering, 1947: 5.—Calif.

clathrata (Loew), 1862c: 80 (*Trypeta*).—Middle States; throughout U.S.
coloradensis Quisenberry, 1949b: 85.—Colo.
corpulenta (Cresson), 1907a: 103 (*Tephritis*).—N. Mex.; Calif.
distincta Quisenberry, 1949b: 85.—Colo.
dupla (Cresson), 1907a: 102 (*Tephritis*).—N. Mex.
fibulata (Wulp), 1900: 421 (*Tephritis*).—Mexico; N. Mex., ?Utah.
genalis (Thomson), 1869: 585 (*Trypeta*).—Calif.
maculifemorata Hering, 1947: 6.—Wash.
murina (Doane), 1899: 189 (*Tephritis*).—Wash.; Oreg., Calif., N. Mex.
pallidipennis (Cresson), 1907a: 104 (*Tephritis*).—Colo.; Calif.
snowi Hering, 1944: 8 (n. name for *obscuripennis* Snow).—Oreg.; Calif.
    obscuripennis Snow, 1894a: 174 (*Tephritis;* preocc. Loew, 1850).—Oreg. (Foote, 1962: 176).
tenebrosa (Coquillett), 1899e: 264 (*Tephritis*).—Colo.
variabilis (Doane), 1899: 188 (*Tephritis*).—Wash.; Oreg., Calif.

Genus DIOXYNA Frey

Dioxyna Frey, 1945: 62. Type-species, *Trypeta sororcula* Wiedemann (orig. des.).

picciola (Bigot), 1857a: 347 (1857: 824) (*Acinia*).—Cuba; Calif. to Fla., s. to Costa Rica and Cuba.
    humilis Loew, 1862c: 81 (*Trypeta*).—Cuba.
    aurifera Thomson, 1869: 585 (*Trypeta*).—Calif.
    sororcula, authors, not Wiedemann.
thomae (Curran), 1928d: 70 (*Ensina*).—Virgin Is. (St. Thomas I.); Fla., West Indies.

Genus TRUPANEA Schrank

Trupanea Guettard, 1762: 170–173. Unavailable name (author not binominal).
Trupanea Schrank, 1795: 147. Type-species, *radiata* Schrank (mon.) = *stellata* (Fuessley).
Trypanea, emend.
Urellia Robineau-Desvoidy, 1830: 774. Type-species, *calcitrapae* Robineau-Desvoidy (Coquillett, 1910b: 618) = *stellata* (Fuessley).

    Larvae develop in the ovaries of various flowers, principally composites.

    REFERENCES: Malloch, 1942 (rev.); Foote, 1960g (rev.).

actinobola (Loew), 1873b: 326 (*Trypeta*).—Tex.; Idaho to Mass., s. to Calif., n. Mexico, and Fla.
ageratae Benjamin, 1934: 56.—Fla.
arizonensis Malloch, 1942: 15.—Ariz.; Calif., Tex., Mexico (Baja Calif.).
bisetosa (Coquillett), 1899e: 266 (*Urellia*).—N. Mex.; Idaho and Wyo., s. to Calif. and Tex.
californica Malloch, 1942: 17.—Calif.; Wash. to Colo., s. to Calif. and Tex.
   microsetulosa Malloch, 1942: 17.—Calif.
conjuncta (Adams), 1904b: 451 (*Urellia*).—Ariz.; Calif.`
dacetoptera Phillips, 1923: 148.—N.Y.; Mich., Maine to Fla.
eclipta Benjamin, 1934: 57.—Fla.
femoralis (Thomson), 1869: 582 (*Trypeta*).—Calif.; Wyo., Kans.
   occidentalis Adams, 1904b: 452 (*Urellia*).—Calif.
imperfecta (Coquillett), 1902e: 181 (*Urellia*).—Ariz.; Calif., Nev.
jonesi Curran, 1932e: 6.—Oreg.; Oreg. to Mont., s. to Calif. and Tex.
   microstigma Curran, 1932e: 7.—Oreg.
maculigera Foote, 1960g: 13.—Calif.
mevarna (Walker), 1849: 1023 (*Trypeta*).—Fla.; Kans., R.I. to Ala. and Fla.
   maverna, error.
   solaris Loew, 1862c: 84 (*Trypeta*).—Ga.
   daphne, authors, not Wiedemann.
nigricornis (Coquillett), 1899e: 266 (*Urellia*).—Colo.; Oreg. to Wyo., s. to Calif. and N. Mex.
pseudovicina Hering, 1947: 12 (n. name for *texana* Hering).—Tex.; Calif., Ariz., N. Mex.
   texana Hering, 1942: 29 (preocc. Malloch, 1942).—Tex.
radifera (Coquillett), 1899e: 267 (*Urellia*).—Ariz.; Alta. and Mont., s. to Calif. and Tex.
   hebes Curran, 1932e: 9.—Wyo.
signata Foote, 1960g: 22.—Calif.; Oreg., Nebr., Ariz., Tex.
texana Malloch, 1942: 13.—Tex.; Ariz. to Okla. and La.
vicina (Wulp), 1900: 427 (*Urellia*).—Mexico; Calif. to La., s. to Mexico (D.F.).
wheeleri Curran, 1932e: 7.—Calif.; Utah to Mexico (Baja Calif.) and N. Mex.

Genus TEPHRITIS Latreille

Tephritis Latreille, 1804: 196. Type-species, *Musca arnicae* Linnaeus (Cresson, 1914b: 278).

Trupanea, subg. *Tephritoides* Benjamin, 1934: 58. Type-species, *Euaresta subpura* Johnson (orig. des.).

REFERENCES: Quisenberry, 1951 (rev.); Foote, 1960e (key, distribution).

angustipennis (Loew), 1844b: 382 (*Trypeta*).—Scandinavia; Alaska, sw. Canada, N.Y. to Maine, Holarctic.
araneosa (Coquillett), 1894b: 74 (*Trypeta*).—Calif.; Wash. to Mich., s. to Calif. and Ga.
  aldrichii Doane, 1899: 192 (*Urellia*).—S. Dak.
  pacifica Doane, 1899: 192 (*Urellia*).—Oreg.
arizonaensis Quisenberry, 1951: 62.—Ariz.; Calif.
  arizonensis, emend. or error.
californica Doane, 1899: 190.—Calif.; n. Mexico.
candidipennis Foote, 1960e: 82.—Calif.; Oreg., Mich.
labecula Foote, 1959a: 13.—Utah; Calif., Nev., Wyo.
michiganensis Quisenberry, 1951: 68.—Man.; Mich.
opacipennis Foote, 1960e: 74.—Calif.
ovatipennis Foote, 1960e: 81.—Calif.
pura (Loew), 1873b: 320 (*Trypeta*).—Mass.; Man., Minn., Mich., Maine to Fla.
  tribulis Harris, 1835: 600 (*Trypeta*). Nomen nudum.
rufipennis Doane, 1899: 190.—Calif.
signatipennis Foote, 1960e: 77.—Calif.; Nev.
stigmatica (Coquillett), 1899e: 266 (*Urellia*).—Colo.; Wash. to Mont., s. to Calif. and Tex.
subpura (Johnson), 1909b: 114 (*Euaresta*).—N.J.; N.Y. to Fla.
webbii Doane, 1899: 189.—Idaho; Idaho to Man. and Minn., s. to Calif. and Nev.

Genus DYSEUARESTA Hendel

Dyseuaresta Hendel, 1928a: 368. Type-species, *Euaresta adelphica* Hendel (orig. des.).

REFERENCE: Benjamin, 1934 (descr.).

mexicana (Wiedemann), 1830: 511 (*Trypeta*).—Mexico; Ariz., Tex., Fla., West Indies.

Genus NEOTEPHRITIS Hendel

Neotephritis Hendel, 1935b: 54. Type-species, *Trypeta finalis* Loew (orig. des.).

REFERENCE: Foote, 1960f (rev.).

finalis (Loew), 1862b: 222 (Cent. 2, no. 78) (*Trypeta*).—Calif.; Wash. and Alta. to Va., s. to Calif., n. Mexico, and Ga.
affinis Snow, 1894a: 172 (*Tephritis*).—Wash., Mont., Calif.
inornata (Coquillett), 1902e: 181 (*Tephritis*).—N. Mex.; Utah and Colo., s. to Calif. and Tex.
rava Foote, 1960f: 150.—Ariz.

Genus EUARESTOIDES Benjamin

Trupanea, subg. **Euarestoides** Benjamin, 1934: 57. Type-species, *Trypeta abstersa* Loew (orig. des.).

REFERENCE: Foote, 1958 (rev.).

abstersus (Loew), 1862b: 221 (Cent. 2, no. 77) (*Trypeta*).—N. Amer.; N.Y. to Fla.
acutangulus (Thomson), 1869: 583 (*Trypeta*).—Calif.; Alta. to Mich., s. to Calif. and n. Mexico.
arnaudi Foote, 1958: 291.—Calif.; Ariz.
flavus (Adams), 1904b: 450 (*Urellia*).—Ariz.; N. Mex.

Genus MYLOGYMNOCARENA Foote

Mylogymnocarena Foote, 1960h: 111. Type-species, *Urellia apicata* Thomas (orig. des.).

apicata (Thomas), 1914: 428 (*Urellia*).—Colo.

Genus METATEPHRITIS Foote

Metatephritis Foote, 1960h: 110. Type-species, *fenestrata* Foote (orig. des.).
fenestrata Foote, 1960h: 110.—Wyo.

Genus MYOLEJA Rondani

Myoleja Rondani, 1856: 112. Type-species, *Tephritis lucida* Fallén (orig. des.).
Myiolia, error or emend.
Pseudeuleia Hering, 1940c: 16. Type-species, *Aciura limata* Coquillett (orig. des.).
Aciura, in part, authors, not Robineau-Desvoidy.

REFERENCE: Blanc and Foote, 1961 (key).

limata (Coquillett), 1899e: 263 (*Aciura*).—Mass.; Mich., Mass. to Fla.
nigricornis (Doane), 1899: 183 (*Aciura*).—Pa.; Mich. to Maine, s. to Fla.
unifasciata Blanc and Foote, 1961: 73.—Calif.

Genus ACIURINA Curran

Aciurina Curran, 1932e: 9. Type-species, *trixa* Curran (orig. des.).
Aciura, in part, authors, not Robineau-Desvoidy.
Tephrella, authors, not Bezzi.

REFERENCES: Curran, 1932e (key); Bates, 1935 (rev., as *Tephrella*).

aplopappi (Coquillett), 1894b: 72 (*Trypeta*).—s. Calif.; Ariz.
bigeloviae (Cockerell), 1890c: 324 (*Trypeta*).—Colo.; Utah, Ariz., N. Mex.
  bigeloviae Townsend, 1893k: 49 (*Eurosta*).—Colo.
  var. **disrupta** (Cockerell), 1890c: 324 (*Trypeta*).—Colo.
ferruginea (Doane), 1899: 182 (*Aciura*).—Wash.; Wash. to Wyo., s. to Calif. and N. Mex.
lutea (Coquillett), 1899e: 264 (*Aciura*).—Utah; Nev.
maculata (Cole) *in* Cole and Lovett, 1919: 252 (*Aciura*).—Oreg.; Wash., Calif.
  pacifica Curran, 1932e: 10.—Wash.
notata (Coquillett), 1899e: 262 (*Trypeta*).—N. Mex.
opaca (Coquillett), 1899e: 263 (*Aciura*).—Nev.; Colo.
  johnsoni Thomas, 1914: 426 (*Acidia*).—Colo.
semilucida (Bates), 1935: 111 (*Tephrella*).—Wash.
thoracica Curran, 1932e: 11.—Calif.; Ariz.
trilitura Blanc and Foote, 1961: 75.—Calif.
trixa Curran, 1932e: 11.—Utah; Idaho and Mont., s. to Calif. and N. Mex.

Genus VALENTIBULLA Foote and Blanc

Valentibulla Foote and Blanc, 1959: 149. Type-species, *Trypeta californica* Coquillett (orig. des.).

californica (Coquillett), 1894b: 73 (*Trypeta*).—s. Calif.; Nev., Utah.
  mundula Coquillett, 1899e: 265 (*Euaresta*).—Utah.
munda (Coquillett), 1899e: 265 (*Euaresta*).—Nev.; Idaho, Mont., Colo.
  euarestoides Bates, 1935: 106 (*Tephrella*).—Colo.
thurmanae Foote *in* Foote and Blanc, 1959: 154.—Calif.

Genus XANTHACIURA Hendel

Xanthaciura Hendel, 1914e: 86. Type-species, *Trypeta chrysura* Thomson (orig. des.).
Tetraciura Hendel, 1914e: 90. Type-species, *quadrisetosa* Hendel (orig. des.).

Aciura, subg. *Eucosmoptera* Phillips, 1923: 131.   Type-species, *tetraspina* Phillips (Bates, 1933: 55).
Aciura, in part, authors, not Robineau-Desvoidy.

REFERENCE: Benjamin, 1934 (descrs.).

connexionis Benjamin, 1934: 45.—Fla.
  brevinervis Malloch, 1933c: 269.   Nomen nudum.
insecta (Loew), 1862c: 72 (*Trypeta*).—Cuba; Tex. to N.C., s. to Fla., n. Mexico, West Indies, Bahama Is.
tetraspina (Phillips), 1923: 132 (*Aciura*).—Mo.; Utah to Ind., s. to n. Mexico and Fla.

Genus STENOPA Loew

Stenopa Loew, 1873b: 234.   Type-species, *Trypeta vulnerata* Loew (orig. des.).

REFERENCE: Quisenberry, 1949b (descr.).

affinis Quisenberry, 1949b: 87.—Colo.; Ariz.
vulnerata (Loew), 1873b: 232 (*Trypeta*).—Mass.; throughout U.S.

Genus ORELLIA Robineau-Desvoidy

Orellia Robineau-Desvoidy, 1830: 765.   Type-species, *flavicans* Robineau-Desvoidy (mon.) = *punctata* (Schrank).

REFERENCE: McFadden and Foote, 1961 (rev.).

occidentalis (Snow), 1894a: 163 (*Trypeta*).—Colo.; B.C. and Mont., s. to Calif. and Tex.
  straminea Doane, 1899: 179 (*Trypeta*).—Wash.
palposa (Loew), 1862c: 74 (*Trypeta*).—Wis.; Oreg. to Nfld., s. to Calif. and Ga.
ruficauda (Fabricius), 1794: 353 (*Musca*).—France; B.C. to Nfld., s. to Calif.
  florescentiae, authors, not Linnaeus.
undosa (Coquillett), 1899e: 262 (*Trypeta*).—Colo.; Idaho, Wyo., Calif.

Genus PARATERELLIA Foote

Paraterellia Foote, 1960a: 121.   Type-species, *Trypeta varipennis* Coquillett (orig. des.).

superba Foote, 1960a: 123.—Calif.; Utah, Colo., Ariz.
varipennis (Coquillett), 1902e: 180 (*Trypeta*).—Ariz.; Oreg., Calif., Nev., N. Mex.
  versatilis Curran, 1932e: 13 (*Trypeta*).—Oreg.
ypsilon Foote, 1960a: 124.—Calif.

Genus NEASPILOTA Osten Sacken

Aspilota Loew, 1873b: 286 (preocc. Förster, 1862). Type-species, *Trypeta alba* Loew (Coquillett, 1910b: 511).

Neaspilota Osten Sacken, 1878c: 192 (n. name for *Aspilota* Loew). Type-species, *Trypeta alba* Loew (aut.).

Aspilomyia Hendel, 1907a: 98 (n. name for *Aspilota* Loew). Type-species, *Trypeta alba* Loew (aut.).

> REFERENCE: Quisenberry, 1949b (key).

achilleae Johnson, 1900b: 328.—N.J., Pa., Ga.; Maine to Fla.
alba (Loew), 1861b: 345 (Cent. 1, no. 72) (*Trypeta*).—Pa.; Mont. to R.I., s. to N. Mex. and N.C.
albidipennis (Loew), 1861b: 345 (Cent. 1, no. 73) (*Trypeta*).—Pa.; Mich., Mass., Conn., N.J.
  albipennis, error.
brunneostigmata Doane, 1899: 187.—Wash.; Oreg., Calif., N. Mex.
  brunneistigma, error.
dolosa Benjamin, 1934: 39.—Fla.
punctistigma Benjamin, 1934: 38.—Fla.
signifera Coquillett, 1894b: 73.—s. Calif.; Ariz.
vernoniae (Loew), 1861b: 346 (Cent. 1, no. 74) (*Trypeta*).—Pa.; Mich. to Mass., s. to Kans. and Fla.
viridescens Quisenberry, 1949b: 82.—Colo.
wilsoni Blanc and Foote, 1961: 78.—Calif.

Genus CERATITIS Macleay

Ceratitis Macleay, 1829: 482. Type-species, *citriperda* Macleay (mon.)=*capitata* (Wiedemann).

The single species treated below is not now known to occur on the continent north of Nicaragua. On three previous occasions, however, outbreaks of this fly have threatened the commercial citrus growing region of Florida.

> REFERENCES: Newell, 1930 (summary, 1928 Fla. infestation); Steiner et al., 1961 (summary, 1956 Fla. infestation).

capitata (Wiedemann), 1824: 55 (*Trypeta*).—East Indies (probably error); Bermuda, Nicaragua, Costa Rica, Brazil, Argentina, Uruguay, Peru, Azores, Canary Is., Mediterranean, Africa, Madagascar, Mauritius, Zanzibar, Australia, Hawaii.

Genus ANASTREPHA Schiner

Anastrepha Schiner, 1868: 263. Type-species, *Dacus serfentinus* Wiedemann (orig. des.).

Several *Anastrepha* species, especially *ludens*, the Mexican fruit fly, are inimical to commercial fruit growers throughout the tropical and subtropical Americas. The genus is of primary economic importance.

REFERENCES: Stone, 1942 (rev.); Baker et al., 1944 (biol., *ludens* and others, descrs.).

chiclayae Greene, 1934b: 167.—Peru; s. Tex., n. Mexico to Panama.
distincta Greene, 1934b: 149.—Peru; s. Tex. to Peru and Brazil.
edentata Stone, 1942: 48.—Fla.; P.R.
fraterculus (Wiedemann), 1830: 524 (*Dacus*).—Brazil; s. Tex. to Chile and Argentina, also Trinidad, British Guiana, Tobago.
    unicolor Loew, 1862c: 70 (*Trypeta*).—Colombia.
interrupta Stone, 1942: 62.—Fla.
lathana Stone, 1942: 105.—Mexico; s. Tex.
limae Stone, 1942: 67.—Panama; s. Tex.
ludens (Loew), 1873b: 223 (*Trypeta*).—Mexico; s. Tex., n. Mexico to Costa Rica.
mombinpraeoptans Seín, 1933: 187 (as *fraterculus* var.).—P.R.; s. Tex., Fla., Neotropical.
    acidusa, authors, not Walker.
nigrifascia Stone, 1942: 91.—Fla.
ocresia (Walker), 1849: 1016 (*Trypeta*).—Jamaica; Fla., Cuba, Hispaniola.
    tricincta Loew, 1873b: 225 (*Trypeta*).—Haiti.
serpentina (Wiedemann), 1830: 521 (*Dacus*).—Brazil; s. Tex. to Peru, also Trinidad, British Guiana.
spatulata Stone, 1942: 61.—Mexico; s. Tex. to Panama.
striata Schiner, 1868: 264.—S. Amer.; s. Tex., Neotropical.
suspensa (Loew), 1862c: 69 (*Trypeta*).—Cuba; s. Fla., Greater Antilles.
zuelaniae Stone, 1942: 82.—Panama Canal Zone; s. Tex. to Panama.

Genus LUCUMAPHILA Stone

Lucumaphila Stone, 1939b: 340. Type-species, *sagittata* Stone (orig. des.).

dentata Stone, 1939b: 343.—Mexico; s. Tex.
sagittata Stone, 1939b: 347.—Mexico; s. Tex.

Genus PSEUDODACUS Hendel

Anastrepha, subg. **Pseudodacus** Hendel, 1914e: 97. Type-species, *daciformis* Bezzi (orig. des.).

REFERENCE: Stone, 1939a (rev.).

bicolor Stone, 1939a: 288.—Tex.; Mexico (Morelos).
pallens (Coquillett), 1904c: 35 (*Anastrepha*).—Tex.; s. Tex. to Honduras.

Genus RHAGOLETIS Loew

Rhagoletis Loew, 1862a: 44. Type-species, *Musca cerasi* Linnaeus (mon.).
Zonosema Loew, 1862a: 43. Type-species, *Trypeta meigenii* Loew (Coquillett, 1910b: 622).

The larvae of *Rhagoletis* species attack commercially grown cherries, apples, walnuts, and roses, as well as many wild fruits in the United States and Canada.

REFERENCES: Illingworth, 1912 (biol., *pomonella*); Cresson, 1929 (rev.); Curran, 1932c (key); Boyce, 1935 (biol., *completa*); Pickett, 1937 (tax., males); Balduf, 1959 (biol., *basiola*).

basiola (Osten Sacken), 1877: 348 (*Trypeta*).—Mass.; Alaska to Maine, s. to Calif. and N.C.
    setosa Doane, 1899: 178 (*Spilographa*).—Idaho.
    flavonotata Macquart of Loew, 1873b: 244.
berberis Curran, 1932c: 8.—Oreg.; Wash. to Calif.
boycei Cresson, 1929: 413.—Ariz.
cingulata cingulata (Loew), 1862c: 76 (*Trypeta*).—Middle States; Mich. to N.H., s. to Fla.
    ssp. **indifferens** Curran, 1932c: 8 (as sp.).—Oreg.; Wash. and Idaho to Calif.
completa Cresson, 1929: 412 (as *suavis* ssp.).—Calif.; Minn. to Calif. and Tex.
fausta (Osten Sacken), 1877: 346 (*Trypeta*).—N.H.; B.C. to Que., s. to Calif. and N.Y.
    intrudens Aldrich, 1909: 70.—B.C.
juglandis Cresson, 1920b: 65.—Ariz.; Calif., N. Mex.
mendax Curran, 1932c: 7.—Maine; N.S., N.J.
pomonella (Walsh), 1867: 343 (*Trypeta*).—Ill.; e. N. Dak. to N.S., s. to e. Tex. and Fla.
    albiscutellata Harris, 1835: 600 (*Trypeta*). Nomen nudum.
ribicola Doane, 1898: 69.—Unknown; Wash., Idaho, Mont., Calif.
striatella Wulp, 1899: 408.—Mexico; N. Mex., ?Wis., ?Ill.

suavis (Loew), 1862c: 75 (*Trypeta*).—Middle States; Minn. to Mass., s. to Ala. and N.C.
tabellaria (Fitch), 1855: 770 (1856: 66) (*Tephritis*).—N.Y.; B.C. to Maine, s. to Calif. and Ariz.
  juniperinus Marcovitch, 1915: 171.—N.Y.
zephyria Snow, 1894a: 164.—s. Calif.; B.C., Wash., Oreg.
  symphoricarpi Curran, 1924d: 63.—B.C.

Genus RHAGOLETOIDES Foote

Rhagoletoides Foote, 1960d: 145. Type-species, *Spilographa latifrons* Wulp (orig. des.).

latifrons (Wulp), 1899: 407 (*Spilographa*).—Mexico; Ariz. to Mexico (V.C.).

Genus ZONOSEMATA Benjamin

Zonosemata Benjamin, 1934: 17. Type-species, *Trypeta electa* Say (orig. des.).

REFERENCE: Foote, 1960h (rev.).

electa (Say), 1830: 185 (1859b: 369) (*Trypeta*).—Ind.; Okla. to Ont. and s. to e. Tex. and Fla.
  flavonotata Macquart, 1855: 145 (1855: 125) (*Tephritis*).—Md.
vittigera (Coquillett), 1899e: 261 (*Zonosema*).—Tex.; Ariz. to e. Tex., s. to Mexico, ?Calif.

Genus RHAGOLETOTRYPETA Aczél

Rhagoletotrypeta Aczél, 1950: 313. Type-species, *xanthogastra* Aczél (orig. des.).

Unnamed sp.—N.J.

Genus OEDICARENA Loew

Oedicarena Loew, 1873b: 247. Type-species, *Trypeta tetanops* Loew (orig. des.).

REFERENCE: Foote, 1960h (rev.).

persuasa (Osten Sacken), 1877: 344 (*Trypeta*).—Colo.; Kans., Tex.

Genus GYMNOCARENA Hering

Gymnocarena Hering, 1940b: 4. Type-species, *Oedicarena diffusa* Snow (orig. des.).

REFERENCE: Foote, 1960h (rev.).

bicolor Foote, 1960h: 113.—Ariz.
diffusa (Snow), 1894a: 161 (*Oedicarena*).—Kans.; Wash. to N. Dak., s. to Ariz. and N. Mex.
tricolor (Doane), 1899: 191 (*Euaresta*).—S. Dak.; Wis.

Genus STRAUZIA Robineau-Desvoidy

Strauzia Robineau-Desvoidy, 1830: 718. Type-species, *inermis* Robineau-Desvoidy (Coquillett, 1910b: 609)=*longipennis* (Wiedemann).
Straussia, emend.

Studies are now underway to determine the exact status of the seven varieties described by Loew.

REFERENCE: Phillips, 1923 (descrs.).

longipennis (Wiedemann), 1830: 483 (*Trypeta*).—N. Amer.; Mont. to Maine, s. to Calif. and N.C.
  inermis Robineau-Desvoidy, 1830: 718.—Pa.
  armata Robineau-Desvoidy, 1830: 719.—Pa.
  trimaculata Macquart, 1843a: 383 (1843: 226) (*Tephritis*).—N. Amer.
  cornigera Walker, 1849: 1010 (*Trypeta*).—N. Amer.
  cornifera Walker, 1849: 1011 (*Trypeta*).—U.S.
  sepentaria Harris, 1835: 600 (*Trypeta*). Nomen nudum.
  septenaria, serpentaria, errors.
  var. **arculata** (Loew), 1873b: 242 (*Trypeta*).—Ill.; Mont., Colo.
  var. **confluens** (Loew), 1873b: 241 (*Trypeta*).—Conn.
  var. **intermedia** (Loew), 1873b: 241 (*Trypeta*).—Unknown; N.Y.
  var. **longitudinalis** (Loew), 1873b: 240 (*Trypeta*).—N.Y.; Colo., Pa., Conn., N.C.
  var. **perfecta** (Loew), 1873b: 239 (*Trypeta*).—Unknown; Conn., Kans.
  var. **typica** (Loew), 1873b: 239 (*Trypeta*).—Pa.; Calif., Kans., Conn.
  var. **vittigera** (Loew), 1873b: 241 (*Trypeta*).—Nebr.; Mont., Calif., Kans., Ill.

Genus TRYPETA Meigen

Trypeta Meigen, 1803: 277. Type-species, *Musca artemisiae* Fabricius (Coquillett, 1910b: 618).
Spilographa Loew, 1862a: 39. Type-species, *Trypeta hamifera* Loew (Coquillett, 1910b: 607).

REFERENCE: Foote, 1960b (rev.).

angustigena Foote, 1960b: 258.—Calif.

flaveola Coquillett, 1899f: 345.—Commander Is.; Alaska.
fractura (Coquillett), 1902g: 125 (*Spilographa*).—N. Mex.; Calif., Ariz.
inaequalis (Coquillett) *in* Baker, 1904: 29 (*Spilographa*).—Nev.; Wyo., Colo.
maculosa (Coquillett), 1899e: 261 (*Spilographa*).—Colo.; N. Mex.
  dubia Johnson, 1903a: 102 (?*Zonosema*).—N. Mex.
sigma (Phillips), 1923: 129 (*Acidia*).—Md.; Pa.
tortilis Coquillett, 1894b: 71.—Wash.; N. Mex., Mich.

Genus EULEIA Walker

Euleia Walker, 1836: 81. Type-species, *Musca onopordinis* Fabricius (mon.) = *heraclei* (Linnaeus).
Acidia, authors, not Robineau-Desvoidy.

Euleia fratria, the parsnip leaf miner, is of considerable economic importance in the United States. A Palaearctic representative, *heraclei* (Linnaeus), does great damage by mining celery leaves and stalks.

  REFERENCE: R. H. Foote, 1959b (rev.).

fratria (Loew), 1862c: 67 (*Trypeta*).—U.S.; Ont., Que., throughout U.S.
  liogaster Thomson, 1869: 578 (*Trypeta*).—Calif.
uncinata (Coquillett), 1899e: 260 (*Acidia*).—Alaska (Pribilof Is.); Alaska.

Genus EPOCHRA Loew

Epochra Loew, 1873b: 238. Type-species, *Trypeta canadensis* Loew (mon.).
Epochroa, error.

  REFERENCE: Foote, 1959c (tax.).

canadensis (Loew), 1873b: 235 (*Trypeta*).—Canada; B.C. to Maine, s. to Calif. and N.Y.
  lunifera Hering, 1940d: 5.—Wash.

Genus CHETOSTOMA Rondani

Chetostoma Rondani, 1856: 112. Type-species, *curvinerve* Rondani (orig. des.).
Chaetostoma, emend.

  REFERENCE: Blanc, 1959 (tax.).

californicum Blanc, 1959: 202.—Calif.; Man., Colo.

rubidum (Coquillett), 1899e: 260 (*Epochra*).—Colo.; Calif.
elizabethae Quisenberry, 1949b: 81.—Colo.

Genus TOMOPLAGIA Coquillett

Plagiotoma Loew, 1873b: 252 (preocc. Dujardin, 1841). Type-species, *Trypeta obliqua* Say (Coquillett, 1910b: 591).
Tomoplagia Coquillett, 1910b: 591, 615 (n. name for *Plagiotoma* Loew). Type-species, *Trypeta obliqua* Say (aut.).

REFERENCE: Aczél, 1955 (rev.).

cressoni Aczél, 1955: 347.—Calif.; Ariz., N. Mex., Tex.
obliqua (Say), 1830: 186 (1859b: 370) (*Trypeta*).—Ind.; Nebr. to N.Y., s. to Ariz., n. Mexico, and Fla., also Cuba.

Superfamily SCIOMYZOIDEA

Family HELCOMYZIDAE

By George C. Steyskal

The few species of Helcomyzidae are found on seashores in northern Eurasia, northern North America, and the Antarctic, developing in rotting seaweed.

REFERENCE: Steyskal, 1958a (rev.).

Genus HELCOMYZA Curtis

Helcomyza Curtis, 1825: pl. 66. Type-species, *ustulata* Curtis (orig. des.).
Actora Meigen, 1826: 403. Type-species, *aestuum* Meigen (mon.)= *ustulata* Curtis.
Macromelanderia Curran, 1934a: 382. Type-species, *Helcomyza mirabilis* Melander (mon.).

mirabilis Melander, 1920a: 309.—Wash.; Oreg.

Genus HETEROCHEILA Rondani

Heterostoma Rondani, 1856: 104 (preocc. Hartmann, 1843). Type-species, *Heteromyza buccata* Fallén (orig. des.).
Heterocheila Rondani, 1857: 13 (n. name for *Heterostoma* Rondani). Type-species, *Heteromyza buccata* Fallén (aut.).
Oedoparea Loew, 1862e: 10. Type-species, *Heteromyza buccata* Fallén (mon.).
Pseudosciomyza Malloch, 1923i: 214. Type-species, *Dryomyza hannai* Cole (orig. des.).

buccata (Fallén), 1820e: 2 (*Heteromyza*).—Sweden; ?N.S., Europe.
hannai (Cole), 1921: 174 (*Dryomyza*).—Alaska; Wash., Oreg.
nudiseta Curran, 1933b: 8.—Oreg.

Family ROPALOMERIDAE

(Rhopalomeridae)

By Richard H. Foote

The family Ropalomeridae is distributed throughout the Tropics and Subtropics of the New World. Adults of the single North American species are attracted to fresh exudates of various kinds of palm trees. Nothing is known about larvae or larval habitats.

REFERENCE: Malloch, 1941b (key to genera).

Genus RHYTIDOPS Lindner

Rhytidops Lindner, 1930: 135. Type-species, *chacoensis* Lindner (mon.).

floridensis (Aldrich), 1932b: 10 (*Kroeberia*).—Fla.
floridana (error) Malloch, 1941b: 51.

Family COELOPIDAE

(Phycodromidae)

By J. R. Vockeroth

All species of Coelopidae, or seaweed flies, are restricted to seacoasts. The known larvae live in decaying seaweed. *Coelopa frigida* (Fabricius) in Europe consists of at least two morphologically indistinguishable species (Remmert, 1959), and the same could be true of Nearctic *frigida*.

REFERENCES: Aldrich, 1929b (rev.); Malloch, 1933a (key to genera); Egglishaw, 1960 (biol., immature stages).

Genus COELOPA Meigen

Coelopa Meigen, 1830: 8. Type-species, *Musca frigida* Fabricius (mon.; misident.)=*pilipes* Haliday.

Subgenus COELOPA Meigen

stejnegeri Aldrich, 1929b: 5.—St. Paul I. (Bering Sea); is. of Bering Sea, s. Alaska, Vancouver I.

Subgenus FUCOMYIA Haliday

Coelopa, subg. **Fucomyia** Haliday, 1838: 186. Type-species, *Musca frigida* Fabricius (Westwood, 1840: 144).

frigida (Fabricius), 1805: 307 (*Musca*).—Norway; n. Man., Southampton I., cent. Labr., s. to R.I., is. of N. Atlantic, Europe, and China.
  gravis Haliday, 1833: 167.—Ireland.
  parvula Haliday, 1833: 167.—Ireland.
  nitidula Zetterstedt, 1847: 2473.—Scandinavia.
  eximia Stenhammar, 1854: 318 (1855: 60).—Norway.
nebularum Aldrich, 1929b: 5.—St. Paul I. (Bering Sea); is. of Bering Sea, s. Alaska, Vancouver I.

Subgenus NEOCOELOPA Malloch

Coelopa, subg. **Neocoelopa** Malloch, 1933a: 345. Type-species, *vanduzeei* Cresson (orig. des.).

vanduzeei Cresson, 1914c: 457.—Calif.; s. Alaska to s. Calif.

Genus COELOPINA Malloch

Coelopina Malloch, 1933a: 350. Type-species, *Coelopa anomala* Cole (mon.).

anomala (Cole), 1923a: 470 (*Coelopa*).—Mexico (Baja Calif.); Calif.

Family DRYOMYZIDAE

By George C. Steyskal

The small family Dryomyzidae contains species often abundant in moist woods. The immature stages are unknown.

REFERENCES: Steyskal, 1957 (rev.), 1958a (male terminalia).

Genus DRYOMYZA Fallén

Dryomyza Fallén, 1820b: 15. Type-species, *anilis* Fallén (Zetterstedt, 1846: 2082).
Dryope Robineau-Desvoidy, 1830: 618. Type-species, *communis* Robineau-Desvoidy (Coquillett, 1910b: 536)=*flaveola* (Fabricius).
Neuroctena Rondani, 1868b: 56 (1868: 254). Type-species, *Dryomyza anilis* Fallén (mon.).

anilis Fallén, 1820b: 16.—Europe; Alaska to Labr., s. to Oreg., Ill., and W. Va.
  pallida Day, 1881: 89.—Conn.

bergi Steyskal, 1957: 60.—Alaska.
ferruginea Melander, 1920a: 311.—Vt.; Maine, N.H. ?=*flaveola*.
flaveola (Fabricius), 1794: 343 (*Musca*).—Europe; Alaska to N.S., Kans., N.Y., Pa., Maine.
  dayi Cresson, 1920a: 34.—N.Y.
melanderi Steyskal, 1957: 63 (n. name for *maculipennis* Melander).— Alaska, Idaho; B.C., Que.
  maculipennis Melander, 1920a: 311 (preocc. Macquart, 1850).—Alaska, Idaho.
setosa (Bigot), 1886a: 386 (?*Odontomera*).—Wash.; Alaska and Y.T. to N. Mex., also n. Que.
  fumida Coquillett, 1901h: 616 (*Neuroctena*).—N. Mex.
simplex Loew, 1862c: 128.—Middle States; Mich. to Maine, s. to e. Tenn. and Ga.
  ruthae Brimley, 1925: 75 (*Neuroctena*).—N.C.

Genus OEDOPARENA Curran

Oedoparena Curran, 1934a: 383. Type-species, *Oedoparea glauca* Coquillett (mon.).

glauca (Coquillett), 1900h: 458 (1904: 72) (*Oedoparea*).—Alaska; Alaska to Calif.

Family SEPSIDAE
By George C. Steyskal

The members of the small family Sepsidae are found most often upon excrement and carrion, but also upon many other kinds of decaying animal and vegetable matter. Owing perhaps to their habits, many species have a very broad distribution.

REFERENCES: Melander and Spuler, 1917 (rev.); Steyskal, 1943b (key to genera); Hennig, 1949 (monog. Palaearctic spp., including many Holarctic spp.).

Genus DECACHAETOPHORA Duda

Decachaetophora Duda, 1926: 27. Type-species, *Sepsis aeneipes* Meijere (mon.).

aeneipes (Meijere), 1913a: 119 (*Sepsis*).—Formosa; B.C. to s. Calif., n. India to Japan.

Genus MEROPLIUS Rondani

Meroplius Rondani, 1874: 170 (1874: 4). Type-species, *Nemopoda stercoraria* Robineau-Desvoidy (orig. des.).

stercorarius (Robineau-Desvoidy), 1830: 745 (*Nemopoda*).—Europe; B.C. to Que., s. to Wash., Utah, and Fla., also Asia.
  minuta Wiedemann, 1830: 468 (*Sepsis*).—N.Y.

Genus NEMOPODA Robineau-Desvoidy

Nemopoda Robineau-Desvoidy, 1830: 743. Type-species, *putris* Robineau-Desvoidy (Rondani, 1874: 170, 178; 1874: 4, 12)=*nitidula* (Fallén).

aterrima Bigot, 1886a: 390.—Calif. Unrecognized.
coeruleifrons Macquart, 1847: 110 (1847: 94).—Pa. Unrecognized.
  caeruleiformis (error) Aldrich, 1905: 620.
nitidula (Fallén), 1820c: 21 (*Sepsis*).—Europe; Alta. to Que., s. to Mich. and Va.
  cylindrica Fabricius, 1794: 336 (*Musca*; preocc. De Geer, 1776).—Europe.
obscuripennis Bigot, 1886a: 392.—Calif. Unrecognized.

Genus ORYGMA Meigen

Orygma Meigen, 1830: 6. Type-species, *luctuosum* Meigen (mon.).
Eugenacephala Johnson, 1922: 22. Type-species, *salsa* Johnson (orig. des.)=*luctuosum* Meigen.

luctuosum Meigen, 1830: 6.—Europe; Greenland, Labr. to Mass.
  salsa Johnson, 1922: 22 (*Eugenacephala*).—Mass.
  ruficeps Curran, 1925a: 25 (*Eugenacephala*).—Labr.

Genus PALAEOSEPSIS Duda

Sepsis, subg. **Palaeosepsis** Duda, 1926: 43. Type-species, *haemorrhoidalis* Schiner (Hennig, 1949: 61).

pleuralis (Coquillett), 1904c: 35 (*Sepsis*).—Tex.
pusio (Schiner), 1868: 262 (*Sepsis*).—S. Amer.; Tex., Fla., West Indies, S. and Cent. Amer.
  insularis Williston, 1896c: 431 (*Sepsis*).—St. Vincent I.

Genus SALTELLA Robineau-Desvoidy

Saltella Robineau-Desvoidy, 1830: 746. Type-species, *nigripes* Robineau-Desvoidy (Westwood, 1840: 147).
Pandora Haliday, 1833: 169 (preocc. Brugière, 1797, perhaps also 1792). Type-species, *Piophila scutellaris* Fallén (Hennig, 1949: 27)=*sphondylii* (Schrank).

Pseudopandora Rapp, 1946b: 500 (n. name for *Pandora* Haliday).
Type-species, *Piophila scutellaris* Fallén (aut.)=*sphondylii* (Schrank).

sphondylii (Schrank), 1803: 149 (*Trupanea*).—Germany; Wash. to Que., s. to Calif. and Md., Europe.
   scutellaris Fallén, 1820e: 10 (*Piophila*).—Europe.
   ruficoxa Macquart, 1835: 481 (*Nemopoda*).—France.
   nigerrima Rondani, 1874: 179 (1874: 13).—Italy.
   parmensis Rondani, 1874: 179 (1874: 13).—Italy.

Genus SEPSIS Fallén

Sepsis Fallén, 1810a: 17. Type-species, *Musca cynipsea* Linnaeus (Curtis, 1829: pl. 245).
Sepsidimorpha Frey, 1908: 584. Type-species, *Sepsis loewi* Hendel (mon.)=*pilipes* Wulp.

biflexuosa Strobl, 1893a: 225.—Europe; B.C. to Que., s. to N. Mex., Ill., and N.J., also Mexico, Hawaii, Canary Is.
   signifera Melander and Spuler, 1917: 26.—Pa.
   curvitibia Melander and Spuler, 1917: 28 (as *signifera* var.).—Wash.
brunnipes (Melander and Spuler), 1917: 33 (*Sepsidimorpha*; as *secunda* var.).—Pa.; Kans. **N. comb., n. status.**
cynipsea (Linnaeus),—Probably not Nearctic.
ecalcarata Thomson, 1869: 588.—Calif. Unrecognized.
flavimana Meigen, 1826: 288.—Europe; Alaska, Mich.
   simplex Goetghebuer and Bastin, 1925: 129.—Europe.
fulvicoxalis (Bigot), 1886a: 390 (*Nemopoda*).—Calif., Cuba. Unrecognized.
luteipes Melander and Spuler, 1917: 29.—Europe, N. Amer.
neocynipsea Melander and Spuler, 1917: 28.—N. Amer. (14 States and provinces); Alaska to Que., s. to Calif. and Ala., also Europe, Asia.
   melanderi Duda, 1926: 61, 135 (as var.).—N. Amer.
   pectoralis Macquart of Melander and Spuler, 1917: 23.
piceipes (Melander and Spuler), 1917: 33 (*Sepsidimorpha*; as *secunda* var.).—Idaho. **N. comb., n. status.**
punctum (Fabricius), 1794: 351 (*Musca*).—Europe; Alaska to Que., all U.S., Bermuda, Mexico, n. Africa.
   similis Macquart, 1851: 269 (1851: 296).—N. Amer.
   hecate Melander and Spuler, 1917: 22 (as *violacea* var.).—Colo.
   violacea, authors, not Meigen.
referens Walker, 1849: 999.—N. Amer. Unrecognized.

secunda (Melander and Spuler), 1917: 32 (*Sepsidimorpha*).—Wash. **N. comb.**
vicaria Walker, 1849: 998.—Fla.; Alaska to Que., all U.S., Mexico, also Bermuda.
    pyrrhosoma Melander and Spuler, 1917: 25.—Ind., Pa. **N. syn.**
violacea Meigen, 1826: 289.—Europe. Questionably distinct from *punctum*; if distinct, probably not Nearctic.

Genus THEMIRA Robineau-Desvoidy

Themira Robineau-Desvoidy, 1830: 745. Type-species, *pilosa* Robineau-Desvoidy (Rondani, 1874: 170, 178; 1874: 4, 12) = *putris* (Linnaeus).
Cheligaster Macquart, 1835: 479. Type-species, *Musca putris* Linnaeus (Coquillett, 1910b: 522).

REFERENCE: Steyskal, 1946 (key).

Subgenus THEMIRA Robineau-Desvoidy

arctica (Becker), 1915: 67 (*Cheligaster*).—U.S.S.R. (Polar Urals); N.W.T., Greenland, Labr.
flavicoxa Melander and Spuler, 1917: 46.—N.Y.; Ill. to Conn.
latitarsata Melander and Spuler, 1917: 45 (as *incisurata* var.).—Wash.; Wash. to Vt., s. to Oreg. and Pa.
maculitarsis Curran, 1929a: 10.—Man.; Que.
malformans Melander and Spuler, 1917: 46.—Hudson Bay; also n. Europe.
nigricornis (Meigen), 1826: 291 (*Sepsis*).—Europe; Ill., Mass., Conn., N.H.
notmani Curran, 1927c: 2.—N.Y.
pusilla (Zetterstedt), 1847: 2295 (*Sepsis*).—Sweden; Wash.
    incisurata Melander and Spuler, 1917: 44.—Wash.
putris (Linnaeus), 1758: 597 (*Musca*).—Sweden; B.C. to Que., s. to Calif., Ill., and N.Y., also Asia.

Subgenus ENICITA Westwood

Enicopus Walker, 1833: 253 (preocc. Stephens, 1830). Type-species, *Sepsis annulipes* Meigen (mon.).
Sepsis, subg. **Enicita** Westwood, 1840: 148 (n. name for *Enicopus* Walker). Type-species, *annulipes* Meigen (aut.).

annulipes (Meigen), 1826: 292 (*Sepsis*).—Europe; Que., Maine, Mich., also Asia.
    elegantipes Ouellet, 1940: 226 (*Enicita*).—Que.
bispinosa (Melander and Spuler), 1917: 40 (*Enicita*).—Tex.

Subgenus ENICOMIRA Duda

Enicomira Duda, 1926: 27 (as genus). Type-species, *Sepsis minor* Haliday (mon.).

minor (Haliday), 1833: 170 (*Sepsis*).—Ireland; Idaho to Que., s. to Mich. and N.Y., also Asia, Africa.

Family SCIOMYZIDAE
(Tetanoceridae)

By George C. Steyskal

Flies of the worldwide family Sciomyzidae are, as far as known, parasitic or predaceous on fresh-water snails. The most recent revisions of the family (1920) are superseded in part by modern revisions of genera.

REFERENCES: Cresson, 1920a (rev.); Melander, 1920a (rev.); Berg, 1953, 1961 (biol.); Foote, 1961a (keys and distribution, nw. spp.); Steyskal, 1961c (key, genera of Sciomyzinae).

Subfamily SCIOMYZINAE

Genus ATRICHOMELINA Cresson

Atrichomelina Cresson, 1920a: 40 (as *Achaetomelina*, p. 30). Type-species, *Sciomyza pubera* Loew (orig. des.).

pubera (Loew), 1862c: 106 (*Sciomyza*).—Middle States; Wash. and Alta. to Ont., s. to Calif., Mexico (D.F.), and Fla.

Genus COLOBAEA Zetterstedt

Colobaea Zetterstedt, 1837: 53 (also 1838: 762). Type-species, *Opomyza bifasciella* Fallén (as *bifasciata*, in 1837; mon.).

americana Steyskal, 1954a: 61.—Que.; Man., N.Y.

Genus OIDEMATOPS Cresson

Oidematops Cresson, 1920a: 36. Type-species, *ferrugineus* Cresson (orig. des.).

ferrugineus Cresson, 1920a: 36.—Vt.; Mich. to New England.

Genus PHERBELLIA Robineau-Desvoidy

Pherbellia Robineau-Desvoidy, 1830: 695. Type-species, *vernalis* Robineau-Desvoidy (mon.) = *schoenherri* (Fallén).

Melina Robineau-Desvoidy, 1830: 695 (preocc. Retzius, 1799). Type-species, *riparia* Robineau-Desvoidy (mon.)=*dubia* (Fallén).

This genus is still in part a residuum. When a thoroughgoing study of the genitalia of the species of the Holarctic Region is made, doubtless many of the species will have to be segregated under some of the many generic names that have been proposed by Hendel, Sack, and Enderlein.

REFERENCE: Steyskal, 1961c (key, *fuscipes* group).

albocostata (Fallén), 1820b: 12 (*Sciomyza*).—Europe; Alaska to Maine, s. to B.C., Colo., and Conn.
albovaria (Coquillett), 1901h: 616 (*Sciomyza*).—N.Y., N.H., N.C.; Mich., Ind., Maine.
annulipes (Zetterstedt), 1846: 2113 (*Sciomyza*).—Europe; Que., N.Y.
beatricis Steyskal, 1949a: 178.—Mich.; Ind., Ohio, Miss.
brunnipes (Meigen), 1838: 364 (*Sciomyza*).—Europe; Idaho, Wyo.
footei Steyskal, 1961c: 409.—Idaho.
fusca (Cresson), 1920a: 43 (*Melina*).—N. Dak.
    spadix Cresson, 1920a: 42 (*Melina*; alternate name for *fusca*).—N. Dak.
fuscipes (Macquart).—Not Nearctic.
griseola (Fallén), 1820b: 14 (*Sciomyza*).—Europe; Wash. to Mont., s. to Utah, and Mich. to Que. and N.Y., Europe.
grisescens (Meigen), 1830: 20 (*Sciomyza*).—Europe; throughout s. Canada and U.S.
    humilis Loew, 1876: 330 (*Sciomyza*).—Tex.
guttata (Coquillett), 1901h: 615 (*Sciomyza*).—Tex.; Colombia.
idahoensis Steyskal, 1961c: 411.—Idaho; Wash., Oreg.
luctifera (Loew), 1861b: 345 (Cent. 1, no. 71) (*Sciomyza*).—Pa.; Ind., Ohio, Ont.
nana (Fallén), 1820b: 15 (*Sciomyza*).—Europe; Alaska to Que., s. to Mexico (D.F.) and Fla.
    transducta Walker, 1861: 320 (*Sciomyza*).—N. Amer.
obtusa (Fallén), 1820b: 13 (*Sciomyza*).—Europe; Calif., Nev., Alta., Ill., Que., N.J.
oregona Steyskal, 1961c: 411.—Oreg.
quadrata Steyskal, 1961c: 412.—Mich.; Idaho, N. Dak., N.Y.
schoenherri schoenherri (Fallén).—Not Nearctic.
    ssp. **maculata** (Cresson), 1920a: 48 (*Melina;* as sp.).—Ill.; Alaska to Nfld., s. to Colo., Ill., and N.Y.
seticoxa Steyskal, 1961c: 414.—Mich.; Nebr., N.Y.

similis (Cresson), 1920a: 44 (*Melina;* as *vitalis* var.).—Wis.; D.C.
tenuipes (Loew), 1872a: 99 (Cent. 10, no. 80) (*Sciomyza*).—Middle States; Alaska, B.C., S. Dak., Mich., Maine, Va.
trabeculata (Loew), 1872a: 100 (Cent. 10, no. 81) (*Sciomyza*).—Tex.; N. Mex., Mexico (V.C., D.F.).
trivittata (Cresson), 1920a: 50 (*Melina*).—Nebr.; S. Dak., Wis., Ill., Ind., Tex.
ventralis (Fallén), 1820b: 14 (*Sciomyza*).—Europe; Wash., Oreg., Idaho.
vitalis (Cresson), 1920a: 43 (*Melina*).—Calif.; B.C., Idaho, Maine.
  palustris Melander, 1920a: 316 (*Melina*).—Idaho. **N. syn.**

Genus PTEROMICRA Lioy

Pteromicra Lioy, 1864a: 1012. Type-species, *Sciomyza glabricula* Fallén (mon.).
Dichrochira Hendel, 1901b: 199. Type-species, *Sciomyza glabricula* Fallén (present des.).

REFERENCES: Steyskal, 1954d (rev.); B.A. Foote, 1959a (key).

albicalceata (Cresson), 1920a: 39 (*Dichrochira*).—Mass.; N.H., Maine, Conn.
anopla Steyskal, 1954d: 260.—Mich.; Sask.
apicata (Loew), 1876: 331 (*Sciomyza*).—N.W.T.
glabricula (Fallén), 1820b: 15 (*Sciomyza*).—Europe; Alaska (Popof I.).
  angustipennis Staeger, 1845a: 40 (*Sciomyza*).—Unknown.
inermis Steyskal, 1956c: 76.—N.Y.; S. Dak.
leucopeza (Meigen), 1838: 380 (*Opomyza*).—Europe; ?S. Dak., ?N.H., ?Mass. Amer. records on females only, which may be *steyskali*.
leucothrix Melander, 1920a: 313.—Wash.; Idaho.
melanothrix Melander, 1920a: 313.—Wyo.
nigrimana (Meigen), 1830: 14 (*Sciomyza*).—Europe; Man. to Que., s. to Wash. and Pa.
  pleuralis Cresson, 1920a: 39 (*Dichrochira*).—Pa.
pectorosa (Hendel), 1902c: 61 (*Dichrochira*).—Europe; Calif., Man., S. Dak., Mich., N.Y.
perissa Steyskal, 1958b: 271.—Colo.
similis Steyskal, 1954d: 266.—N.Y.; Mich., Que.
sphenura Steyskal, 1954d: 268.—Mich.; S. Dak., D.C.
steyskali B.A. Foote, 1959a: 14.—N.Y.; Iowa.

Genus SCIOMYZA Fallén

Sciomyza Fallén, 1820b: 11. Type-species, *simplex* Fallén (Westwood, 1840: 145).

Bischofia Hendel, 1902c: 52. Type-species, *Sciomyza simplex* Fallén (Coquillett, 1910b: 514).

REFERENCES: Steyskal, 1954c (key); B. A. Foote, 1959b (key, biol.).

aristalis (Coquillett), 1901h: 617 (*Dryomyza*).—Ont.; Ont. to Maine, s. to Mich. and Mass.
dryomyzina Zetterstedt, 1846: 2094.—Europe; Alaska, Idaho, Nfld.
simplex Fallén, 1820b: 12.—Europe; Alaska to Man., s. to Oreg. and Mich.
varia (Coquillett), 1904a: 12 (*Bischofia*).—Que.; Mich., Ont.

Subfamily TETANOCERINAE

Genus ANTICHAETA Haliday

Antichaeta Haliday, 1838: 187. Type-species, *Tetanocera vittata* Haliday (mon.)=*analis* (Meigen).
Lioya Enderlein, 1939: 205. Type-species, *Sciomyza atriseta* Loew (orig. des.).

REFERENCES: Steyskal, 1960a (rev.); Foote, 1961b (partial key).

Subgenus ANTICHAETA Haliday

analis (Meigen).—Not Nearctic.
borealis Foote, 1961b: 161.—Idaho; N.Y.
canadensis (Curran), 1923n: 277 (*Pteromicra*).—Alta.; Alta. to Que., also Idaho, Wis., Mich.
fulva Steyskal, 1960a: 20.—N.Y.; Idaho.
melanosoma Melander, 1920a: 318.—Wis.; Idaho, N. Dak. to Que., N.Y., and Ohio, also Utah.
robiginosa Melander, 1920a: 317.—Mont.; Wash., Oreg., Calif., N.S.
testacea Melander, 1920a: 318.—Mont.; Idaho, Utah, S. Dak., N. Mex.

Subgenus PARANTICHAETA Enderlein

Parantichaeta Enderlein, 1936a: 145 (as genus). Type-species, *Antichaeta bisetosa* Hendel (mon.).
johnsoni (Cresson), 1920a: 51 (*Hemitelopteryx*).—N.H.; Ont., N.Y., Mass.

Genus DICTYA Meigen

Dictya Meigen, 1803: 277. Type-species, *Musca umbrarum* Linnaeus (I.C.Z.N. designation will be applied for, to avoid confusion in the literature by preserving *Dictya* in its universally accepted meaning).
Monochaetophora Hendel, 1900: 355. Type-species, *Musca umbrarum* Linnaeus (mon.).

American specimens of the genus were long identified as *umbrarum* (Linnaeus), but none identical with European material has been examined from North America.

REFERENCE: Steyskal, 1954b (rev.).

atlantica Steyskal, 1954b: 524.—Va.; Mich. to Que., s. to Mo. and Va.
borealis Curran, 1932a: 6.—Man.; Sask. to Ont., s. to Tex. and N.C.
brimleyi Steyskal, 1954b: 526.—N.C.; N.C. to Fla.
  confusa Malloch *in* Wray, 1950: 32. Nomen nudum.
expansa Steyskal, 1938a: 9.—Mich.; B.C. to Que., s. to Tex. and Ga.
floridensis Steyskal, 1954b: 528.—Fla.; N.C. to Fla.
gaigei Steyskal, 1938a: 9.—Mich.
hudsonica Steyskal, 1954b: 530.—n. Que.
incisa Curran, 1932a: 6.—Ariz.; Wash., Utah, N. Mex.
iron Steyskal, 1960b: 38.—Miss.
laurentiana Steyskal, 1954b: 532.—Que.; Ont. to Que., s. to Mich. and Va.
lobifera Curran, 1932a: 6.—Fla.; Mich. to N.J., s. to Fla., also Cuba.
montana Steyskal, 1954b: 534.—Calif.; B.C. to Sask., s. to Calif. and Ariz.
oxybeles Steyskal, 1960b: 40.—S.C.; N.S. to Miss., near coast.
pictipes (Loew), 1859b: 292 (*Tetanocera*).—D.C.; Sask. to Ont., s. to Colo., Ill., and N.C.
ptyarion Steyskal, 1954b: 522.—N.C.; N.C. to Fla.
sabroskyi Steyskal, 1938a: 8.—Kans.; S. Dak. to Mich., s. to Tex.
stricta Steyskal, 1938a: 10.—La.; B.C. to N.B., s. to Kans., Tex., Miss., N.C.
texensis Curran, 1932a: 6.—Tex.; Idaho to Mass., s. to Ariz. and S.C.
umbrarum (Linnaeus).—Not Nearctic.
umbroides Curran, 1932a: 4.—Alta.; Alaska to Nfld., s. to Colo. and Mich.
  currani Steyskal, 1939: 78.—Que.

Genus DICTYACIUM Steyskal

Dictyomyia Cresson, 1920a: 82 (preocc. Tavares, 1919). Type-species, *Tetanocera ambigua* Loew (orig. des.).

Dictyacium Steyskal, 1956c: 78 (n. name for *Dictyomyia* Cresson). Type-species, *Tetanocera ambigua* Loew (aut.).

ambiguum (Loew), 1864a: 97 (Cent. 5, no. 95) (*Tetanocera*).—Maine; Ont. to N.B., s. to S. Dak. and Maine.

firmum Steyskal, 1956c: 79.—Sask.; Sask. to Ont., s. to Colo., Mich., W. Va.

Genus ELGIVA Meigen

Elgiva Meigen, 1838: 365. Type-species, *Musca cucularia* Linnaeus (Rondani, 1856: 106).

Ilione Haliday *in* Curtis, 1837b: 280, 288. Nomen nudum.

Ilione Hendel (ex Haliday), 1901a: 141. Type-species, *Musca cucularia* Linnaeus (present des.).

Hedroneura Hendel, 1902b: 265. Type-species, *Musca cucularia* Linnaeus (mon.).

REFERENCE: Steyskal, 1954c (key).

connexa (Steyskal), 1954c: 60 (*Hedroneura*).—Alaska; Alaska to w. Ont., s. to Oreg. and N. Dak. **N. comb.**

sundewalli Kloet and Hincks, 1945: 391 (n. name for *rufa* Panzer).— Europe; Alaska to N.S., s. to Oreg., Nev., and N.Y. Previous uses of *sundewalli* are unavailable, as citations in synonymy.

   rufa Panzer, 1798: 17 (*Musca*; preocc. Scopoli, 1763).— Europe.

   lineata Day, 1881: 88 (*Tetanocera*; preocc. Fallén, 1820).— Conn.

Genus EUTHYCERA Latreille

Euthycera Latreille, 1829: 529. Type-species, *Musca chaerophylli* Fabricius (mon.).

arcuata (Loew), 1859b: 292 (*Tetanocera*).—Middle States; S. Dak. to to Ont., s. to Tex. and N.C.

   flavescens Loew, 1847a: 123 (*Tetanocera*; preocc. Robineau-Desvoidy, 1830).—Carolina.

   flavipes (error) Gibson, 1917: 157 (*Tetanocera*).

   uniformis Cresson, 1920a: 74 (as var.; n. name for *flavescens* Loew).—Carolina.

   borealis Cresson, 1920a: 74.—N.C.

   cribraria Harris, 1835: 600 (*Tetanocera*). Nomen nudum.

SCIOMYZIDAE 691

Genus HEDRIA Steyskal

Hedria Steyskal, 1954a: 62. Type-species, *mixta* Steyskal (orig. des.).

mixta Steyskal, 1954a: 63.—N.W.T.; Alta., Sask., Minn., Wis., Mich., Maine.

Genus HOPLODICTYA Cresson

Hoplodictya Cresson, 1920a: 67. Type-species, *Tetanocera setosa* Coquillett (orig. des.).

REFERENCE: Neff and Berg, 1962 (biol., immature stages).

kincaidi (Johnson), 1913d: 449 (*Tetanocera*).—Bermuda.
setosa (Coquillett), 1901h: 615 (*Tetanocera*).—Mass., Ga.; Atlantic coast, Maine to Ga.
spinicornis (Loew), 1866a: 181 (Cent. 6, no. 86) (*Tetanocera*).— Cuba; Wash. to Guatemala, and N.J. to Fla. and Cuba, also Ill., Tex., and Miss.
    acuticornis Wulp, 1897: 358 (*Tetanocera*).—Mexico. **N. syn.**

Genus LIMNIA Robineau-Desvoidy

Limnia Robineau-Desvoidy, 1830: 684. Type-species, *limbata* Robineau-Desvoidy (Cresson, 1920a: 75)=*unguicornis* (Scopoli).

This genus is in need of revision. Preliminary work on male genitalia has shown that more species exist than available names, and, that type examination will be necessary to establish identities. Except for *boscii* and *shannoni*, which are distinctive, only type localities are cited.

armipes Melander, 1920a: 324 (as *saratogensis* var.).—Wash.
boscii (Robineau-Desvoidy), 1830: 690 (*Pherbina*).—Carolina; Alaska to Nfld., s. to Colo. and N.C.
    combinata Loew, 1859b: 295 (*Tetanocera*).—Middle States.
    sparsa Loew, 1862c: 117 (*Tetanocera*).—Middle States.
    inopa Adams, 1904b: 448 (*Tetanocera*).—Wash.
brevicostalis Melander, 1920a: 323 (as *costalis* var.).—Idaho.
georgiae Melander, 1920a: 324.—Ga.
loewi Steyskal (for *costalis* Loew).—Ill. **N. name.**
    costalis Loew, 1862c: 118 (*Tetanocera*; preocc. Walker. 1837).—Ill.
louisianae Melander, 1920a: 325.—La.
ottawensis Melander, 1920a: 324 (as *saratogensis* var.).—Ont.

pubescens (Day), 1881: 86 (*Tetanocera*).—Wash.
saratogensis (Fitch), 1855: 772 (1856: 68) (*Tetanocera*).—N.Y.
septentrionalis Melander, 1920a: 325 (as *louisianae* var.).—D.C.
severa Cresson, 1920a: 80 (as *saratogensis* var.).—Calif.
shannoni Cresson, 1920a: 78.—Md.
vittata Melander, 1920a: 323 (as *costalis* var.).—Wash., Idaho, Mont.

Genus PHERBECTA Steyskal

Pherbecta Steyskal, 1956c: 82. Type-species, *limenitis* Steyskal (orig. des.).
limenitis Steyskal, 1956c: 83.—N.Y.; Ont.

Genus POECILOGRAPHA Melander

Poecilomyia Melander, 1913b: 58 (preocc. Hendel, 1911). Type-species, *Sapromyza decora* Loew (mon.).
Poecilographa Melander, 1913d: 205 (n. name for *Poecilomyia* Melander). Type-species, *Sapromyza decora* Loew (aut.).
decora (Loew), 1864a: 97 (Cent. 5, no. 96) (*Sapromyza*).—N.Y.; S. Dak. to Que., s. to Colo. and N.J.

Genus RENOCERA Hendel

Renocera Hendel, 1900: 333. Type-species, *stroblii* Hendel (Cresson, 1920a: 51) =*fuscinervis* (Zetterstedt).
  REFERENCE: Curran, 1933d (key).
amanda Cresson, 1920a: 53.—Maine; Ont. and Que., s. to Mich. and Maine.
brevis (Cresson), 1920a: 58 (*Chaetomacera*).—N.Y.
cyathiformis Melander, 1920a: 319.—Wash.; Alaska, Idaho, Colo., Mich.
  bergi Steyskal, 1954c: 64.—Alaska. **N. syn.**
johnsoni Cresson, 1920a: 53.—Maine; Alaska to Alta., s. to Idaho and Colo.
longipes (Loew), 1876: 328 (*Sciomyza*).—N.H.; Que., New England, N.Y., Pa., W. Va.
pacifica Curran, 1933d: 9.—Oreg.
quadrilineata Melander, 1920a: 320.—Mont.

Genus SEPEDON Latreille

Sepedon Latreille, 1804: 196. Type-species, *Syrphus sphegeus* Fabricius (mon.).

REFERENCE: Steyskal, 1951c (rev.).

anchista Steyskal, 1956c: 87.—Mont.; Sask., Idaho.
armipes Loew, 1859b: 298.—Middle States; B.C. to Nfld., U.S., except s. Calif. and Fla.
bifida Steyskal, 1951c: 280.—Calif.
borealis Steyskal, 1951c: 283.—Idaho; Alaska, B.C. to Nfld., s. to Calif., Ill., and Pa.
fuscipennis fuscipennis Loew, 1859b: 299.—Middle States; Alaska to Nfld., s. to Oreg., Colo., Tex., and Ga.
    ssp. **floridensis** Steyskal, 1951c: 290.—Fla.; La.
lignator Steyskal, 1951c: 286.—Mich.; B.C. to Nfld., s. to Mont., Mich., and Maine.
macropus Walker, 1849: 1078.—Jamaica; Tex., Mexico, Cent. and S. Amer.
    nigriventris Wulp, 1897: 359.—Mexico.
melanderi Steyskal, 1951c: 282.—Wash.
neili Steyskal, 1951c: 286.—Mich.; Idaho and Man. to Ont., s. to Utah and N.C.
praemiosa Giglio-Tos, 1893b: 8.—Mexico; B.C. to Alta., s. to Mexico, N. Mex., and Kans.
    pacifica Cresson, 1914c: 457.—Calif.
pusilla Loew, 1859b: 299.—Middle States; Ind. to Md., s. to Ky. and Ga.
spinipes spinipes (Scopoli).—Not Nearctic.
    ssp. **americana** Steyskal, 1951c: 277.—Mich.; Alaska, B.C. to Que., s. to Oreg., Utah, Ohio, and N.Y.
tenuicornis Cresson, 1920a: 84.—D.C.; Ont. and Que., s. to Tex. and Fla.

Genus TETANOCERA Duméril

Tetanocera Duméril, 1800: 439 (as "Tétanocère"). Type-species, *Musca elata* Fabricius (I.C.Z.N. designation, and validation of the generic name from 1800, will be applied for, to avoid confusion in the literature by preserving the long accepted meaning of the genus).
Chaetomacera Cresson, 1920a: 54.—Type-species, *Musca elata* Fabricius (orig. des.).

REFERENCE: Steyskal, 1959a (rev.)

annae Steyskal, 1938a: 5.—Mich.; B.C., Ont., Minn. to N.H.

bergi Steyskal, 1954c: 66.—Alaska; B.C. to Man.
brevihirta Steyskal, 1959a: 77.—Alaska.
clara Loew, 1862c: 109.—N.Y.; Wis. to Que. and N.H., s. to Ill. and Ga.
elata (Fabricius).—Probably not Nearctic.
ferruginea Fallén, 1820b: 9.—Europe; Alaska to Nfld., s. to Oreg., Colo., Ill., and N.J.
    triangularis Loew, 1861b: 344 (Cent. 1, no. 69).—Ont.
    huronensis Steyskal, 1938a: 6.—Mich.
iowensis Steyskal, 1938a: 4.—Iowa; Mo., Ohio.
latifibula Frey, 1924: 51.—Europe; Alaska to Alta., s. to Wash., Utah, S. Dak.
    hespera Steyskal, 1959a: 71.—Utah.
loewi Steyskal, 1959a: 68.—Mich.; Wis. and Ind. to Ont. and N.Y.
mallochi Steyskal, 1959a: 75.—Man.; Alaska to Labr., s. to Alta. and N.H.
melanostigma Steyskal, 1959a: 80.—N.Y.; Ont., Que., Mich.; N.H.
mesopora Steyskal, 1959a: 70.—Colo.; Sask. to Nfld., s. to Colo. and N.Y.
    parvipyga Malloch *in* Jaques, 1939: 283. Nomen nudum.
montana Day, 1881: 87.—Wyo.; Alaska and Y.T. to Que., s. to Idaho and N.Y., also Europe.
    borealis Frey, 1924: 51.—Europe.
    ornatifrons Frey, 1924: 51.—Europe. **N. syn.**
nanciae Brimley, 1925: 75 (as *vicina* ssp.).—N.C.; Alaska to Nfld., s. to Nev., N. Mex., and N.C.
nigricosta Rondani, 1868b: 25 (1868: 223).—Europe; Alaska to N.S., s. to N. Mex., Mich., and N.H.
    phyllophora Melander, 1920a: 330.—Wash.
obtusifibula Melander, 1920a: 328.—Wash.; B.C. to Que., s. to Calif. and Idaho.
oxia Steyskal, 1959a: 80.—Mich.; Alta., Sask., Mont., Nfld.
plebeja Loew, 1862c: 120.—Middle States; Alaska to Labr. and Nfld., s. to Calif., N. Mex., Ill., N.C.
plumosa Loew, 1847b: 201.—Alaska. Unrecognized.
robusta Loew, 1847b: 197.—Europe; Alaska to Nfld., s. to N. Mex., Colo., and Mich.
    papillifera Melander, 1920a: 329.—Wash.
rotundicornis Loew, 1861b: 344 (Cent. 1, no. 70).—Ont.; Alaska to Nfld., s. to Oreg., Ill., and N.Y.
silvatica Meigen, 1830: 41.—Europe; Alaska to Labr., s. to B.C. and S. Dak., also Ariz.
soror Melander, 1920a: 328.—Wash.; Wash. to Mont., s. to Nev.

spirifera Melander, 1920a: 330.—Mont.; Y.T., B.C. to Man., s. to Wyo. and S. Dak.
stricklandi Steyskal, 1959a: 90.—Alta.
struthio Walker, 1849: 1086.—Ont. Unrecognized.
unicolor Loew, 1847b: 199.—Europe; Alaska to Nfld., s. to Ariz., Iowa, and Maine.
valida Loew, 1862c: 110.—Unknown; B.C. to Nfld., s. to N. Mex. and Pa.
  reticulata Harris *in* Johnson, 1925c: 95. Nomen nudum.
vicina Macquart, 1843a: 337 (1843: 180).—Europe; B.C. to Nfld., s. to Ariz., Colo., and N.C.

Genus TRYPETOPTERA Hendel

Trypetoptera Hendel, 1900: 352. Type-species, *Musca punctulata* Scopoli (Cresson, 1920a: 66).

canadensis (Macquart), 1843a: 338 (1843: 181) (*Tetanocera*).—Canada; B.C. to Que., s. to Utah, Ill., and N.C.
  pallida Loew, 1859b: 294 (*Tetanocera*).—Middle States.

Unplaced Species of Sciomyzidae

antica Walker, 1852: 400 (*Sciomyza*).—U.S.
parallela Walker, 1852: 401 (*Sciomyza*).—U.S.

Superfamily LAUXANIOIDEA

Family LAUXANIIDAE

(Sapromyzidae)

By G. E. Shewell

The family Lauxaniidae is worldwide in distribution, with a rich tropical element. The classification is at present somewhat artificial. No suitable subfamily or tribal divisions have yet been proposed, and a natural arrangement of genera is not attempted.

The larvae are saprophagous, living under fallen leaves, in straw and other vegetable trash, in bird nests, and in similar situations. In Europe, larvae of *Calliopum* have been found in large numbers in stems of overwintered clover where they had evidently eaten out the root collars. Adults are normally shade-loving, and are found in woodlands and in dense vegetation by water, sometimes on flowers. They are often attracted to light and occasionally to traps baited for fruit flies or flesh flies.

REFERENCES: Hendel, 1908a (monog., cat.), 1925 (key, world genera); Meijere, 1909 (biol., European spp.); Melander, 1913b (rev.); Malloch and McAtee, 1924b (rev., e. U.S.); McAtee, 1927 (biol.); Shewell, 1938 (rev., e. Canada).

Genus CALLIOPUM Strand

Lauxania, subg. *Calliope* Westwood, 1840: 151 (preocc. Gould, 1836). Type-species, *Lauxania scutellata* Meigen (orig. des.)= *aeneum* (Fallén).
Halidayella Hendel, 1925: 107 (n. name for *Calliope* Westwood, but preocc. Dalla Torre, 1897). Type-species, *Lauxania scutellata* Meigen (aut.)=*aeneum* (Fallén).
Calliopum Strand, 1928: 48 (n. name for *Calliope* Westwood). Type-species, *Lauxania scutellata* Meigen (aut.)=*aeneum* (Fallén).

blaisdelli (Cresson), 1920b: 66 (*Sapromyza*).—Calif. **N. comb.**
elisae (Meigen), 1826: 297 (*Lauxania*).—Europe; ?Wash., ?N.S.
livingstoni (Coquillett), 1898d: 278 (*Sapromyza*).—s. B.C.; Calif., Nev.
　　vanduzeei Cresson, 1920b: 67 (*Sapromyza*).—Calif. **N. comb.**
nigerrimum (Melander), 1913b: 71 (*Calliope*).—Calif.
pacificum (Cole), 1912b: 158 (*Lauxania*).—Calif. **N. comb.**
quadrisetosum (Thomson), 1869: 569 (*Lauxania*).—Calif.; B.C., Wash., Idaho.

Genus CAMPTOPROSOPELLA Hendel

Camptoprosopella Hendel, 1907b: 223. Type-species, *melanoptera* Hendel (orig. des.)=*dolorosa* (Williston).

REFERENCE: Shewell, 1939a (key).

acuticornis Shewell, 1939a: 150.—S. Dak.
angulata Shewell, 1939a: 147.—Md.
antennalis (Fitch), 1856a: 532 (1856: 300) (*Chlorops*).—N.Y. Unrecognized.
borealis Shewell, 1939a: 143.—s. Ont.; Canadian and Transition Zones from latitude 60° to Colo. and N.Y.
cincta (Loew), 1861b: 349 (Cent. 1, no. 81) (*Sapromyza*).—Cuba; ?s. Fla.
confusa Shewell, 1939a: 148.—s. Ont.; s. Man. to N.S., s. to Tex. and N.C.
decolor Shewell, 1939a: 149.—Tex.; Tenn.

diversa Curran, 1926a: 13.—P.R.; Fla.
dolorosa (Williston) *in* Adams, 1903a: 37 (*Pachycerina*).—Colo.; Ariz., N. Mex., Mexico.
gracilis Shewell, 1939a: 145.—Va.; Md., D.C., Miss.
imitatrix Shewell, 1939a: 147.—Mexico; Tex.
maculipennis Malloch, 1923j: 47.—Ariz.; Nicaragua.
mallochi Shewell, 1939a: 146.—Md.; D.C., Va.
media Shewell, 1939a: 150.—Mexico; Tex.
ocellaris (Townsend), 1892g: 303 (*Sapromyza*).—N. Mex.; Kans. to s. Ont. and N.J., s. to N. Mex. and Ga.
  inaequalis Shewell, 1939a: 148.—Tex.
plumata (Wulp), 1867: 159 (*Sapromyza*).—Wis. Unrecognized.
resinosa (Wiedemann), 1830: 456 (*Sapromyza*).—Ga. **N. comb.**
setipalpis Shewell, 1939a: 145.—D.C.; Tenn., Va.
slossonae Shewell, 1939a: 146.—N.H.; Mo., N.C.
texana Shewell, 1939a: 151.—Tex.; La.
verticalis (Loew), 1861b: 349 (Cent. 1, no. 82) (*Pachycerina*).—Fla.; Ga.
  claripennis Coquillett, 1898d: 280 (*Pachycerina;* as *clavipennis*).—Fla.
vulgaris (Fitch), 1856a: 532 (1856: 300) (*Chlorops*).—N.Y. Unrecognized.

Genus DEUTOMINETTIA Hendel

Deutominettia Hendel, 1925: 106, key (1926: 125, descr.). Type-species, *pulchrifrons* Hendel (orig. des.).

REFERENCE: Malloch, 1928a (key).

geniseta (Malloch), 1926: 16 (*Minettia*).—Costa Rica; Tex.

Genus HOMONEURA Wulp

Homoneura Wulp, 1891a: 213. Type-species, *picea* Wulp (mon.).
Sapromyza, subg. *Sapromyzosoma* Malloch, 1920d: 127 (preocc. Lioy, 1864). Type-species, *citreifrons* Malloch (mon.).
Sapromyzama, error.
Mallochomyza Hendel, 1925: 112 (n. name for *Sapromyzosoma* Malloch). Type-species, *Sapromyza citreifrons* Malloch (aut.).
Homoneura, subg. *Tarsohomoneura* Hendel, 1933b: 76. Type-species, *Sapromyza americana* Wiedemann (orig. des.).

REFERENCES: Malloch and McAtee, 1924b (key, e. U.S.); Shewell, 1938 (key, e. Canada).

aequalis (Malloch), 1914a: 36 (*Sapromyza*).—Ill.; Sask., S. Dak., Nebr., Vt., Va.

americana (Wiedemann), 1830: 453 (*Sapromyza*).—N. Amer. (as Brazil, error); S. Dak. to N.S., s. to Tex. and Ga.

    compedita Loew, 1861b: 347 (Cent. 1, no. 76) (*Sapromyza*).—Pa.

bispina (Loew), 1861b: 348 (Cent. 1, no. 79) (*Sapromyza*).—Nebr.; s. Alta. to N.S., s. to Kans. and N.C.

    bispinosa, error.

cilifera (Malloch), 1914a: 33 (*Sapromyza*).—Ill.; Md.

citreifrons (Malloch), 1920d: 127 (*Sapromyza*).—Ill.; Ont. to Maine, s. to Ill. and n. Ga.

conjuncta (Johnson), 1914a: 22 (*Sapromyza*).—R.I.; s. Ont. and Que., s. to Ill. and S.C.

disjuncta (Johnson), 1914a: 22 (*Sapromyza*).—R.I.; s. Ont. to N.S., s. Mo. and Fla.

    similata Malloch, 1914a: 30 (*Sapromyza*).—Ill.

fraterculus (Malloch), 1920d: 128 (*Sapromyza*).—Mont.; s. Alta. to s. Ont., s. to Nebr. and Ga.

fraterna (Loew), 1861b: 347 (Cent. 1, no. 77) (*Sapromyza*).—Pa.; s. Man. to s. Que., s. to Kans. and N.C.

fuscibasis (Malloch), 1920d: 126 (*Sapromyza*).—Ill.; Ill. to sw. Ont. and N.J., s. to Miss. and Fla.

harti (Malloch), 1914a: 32 (*Sapromyza*).—Ill.; s. Alta. to s. Ont., also Wyo., Kans., Okla.

houghii (Coquillett), 1898d: 277 (*Sapromyza*).—Mass.; s. Sask. to P.E.I., s. to S. Dak. and Ga.

imitatrix (Malloch), 1920d: 128 (*Sapromyza*).—N.J.; Conn., s. to La. and Fla.

inaequalis (Malloch), 1914a: 35 (*Sapromyza*).—Ill.; s. parts of B.C., Alta., and Sask.

incerta (Malloch), 1914a: 36 (*Sapromyza*).—Md.; s. Man. to se. Canada and N.H., s. to Mo. and n. Ga.

lamellata (Becker), 1895: 204 (*Sapromyza*).—Russia; Y.T. to e. Que., s. to Mass., Holarctic.

    deceptor Malloch *in* Malloch and McAtee, 1924b: 24 (*Sapromyzosoma*).—N.H. **N. syn.**

    armata Shewell, 1939b: 264 (preocc. Malloch, 1925).—s. Alta. **N. syn.**

    cactifera Shewell, 1940: 86 (n. name for *armata* Shewell).—s. Alta. **N. syn.**

littoralis (Malloch), 1915a: 47 (*Sapromyza*).—Mich.; n. Alta. to P.E.I., s. to s. N.Y.

melanderi (Johnson), 1914a: 21 (*Sapromyza*).—Mass.; Alaska to Que., s. to S. Dak. and N.J.
notata (Fallén).—Not Nearctic. N. Amer. records refer to spp. of *fraterna* group.
nubila (Melander), 1913b: 74 (*Minettia*).—Ill.; sw. Ont., Kans., Ark., N.C., Va.
nubilifera (Malloch), 1920d: 126 (*Sapromyza*).—Ill.; Ind., Tenn., Md.
occidentalis (Malloch), 1920d: 127 (*Sapromyza*).—Calif.; s. B.C., Utah, Ariz., N. Mex.
   var. **nudifemur** (Malloch) *in* Malloch and McAtee, 1924b: 24 (*Sapromyzosoma;* as sp.).—s. B.C.
ornatipes (Johnson), 1914a: 20 (*Sapromyza*).—Mass.; se. Que., N.Y., N.C., n. Ga.
pernotata (Malloch), 1920d: 128 (*Sapromyza*).—Ill.; s. Sask. to N.S., s. to Wis. and N.Y.
philadelphica (Macquart), 1843a: 348 (1843: 191) (*Sapromyza*).—Pa.; S. Dak. and s. Man. to se. Canada and Maine, s. to La. and Ga.
seticauda (Malloch), 1914a: 34 (*Sapromyza*).—Ill.; s. Alta., se. Canada, Iowa, Mo., Md., Va.
setitibia Shewell, 1940: 86 (n. name for *praeapicalis* Shewell).— s. Alta.; S. Dak., Kans.
   praeapicalis Shewell, 1939b: 264 (preocc. Malloch, 1925).— s. Alta.
severini Shewell, 1939b: 265.—N.W.T.; s. Alaska and Y.T. to n. Man., s. to Utah, Minn., and s. Ont.
sheldoni (Coquillett), 1898d: 277 (*Sapromyza*).—N.Y.; s. Ont., s. Que., Maine, Mass.
tenuispina (Loew), 1861b: 349 (Cent. 1, no. 80) (*Sapromyza*).— Nebr.; Nebr. to N.H., s. to Tex. and N.C.

Genus LAUXANIA Latreille

Lauxania Latreille, 1804: 196. Type-species, *Musca cylindricornis* Fabricius (mon.).

   REFERENCE: Malloch and McAtee, 1924b (key).

albiseta Coquillett, 1898d: 280.—Calif.
cylindricornis (Fabricius), 1794: 332 (*Musca*).—France; Alaska and Canada at treeline, s. to Ariz. and Fla., Holarctic.
nigrimana Coquillett, 1902e: 179.—Ariz.; Y.T. to Ariz.

Genus LAUXANIELLA Malloch

Lauxaniella Malloch *in* Malloch and McAtee, 1924b: 9. Type-species *Lauxania femoralis* Loew (orig. des.).

femoralis (Loew), 1861b: 353 (Cent. 1, no. 89) (*Lauxania*).—Ga.; Kans., Tenn., Pa., N.J. to N.C.
manuleata (Loew), 1861b: 352 (Cent. 1, no. 88) (*Lauxania*).—Pa.; Ariz., N.C., Ga.
opaca (Loew), 1861b: 350 (Cent. 1, no. 84) (*Lauxania*).—Fla.; Tenn., N.J., N.C., S.C.
signatifrons (Coquillett), 1904f: 189 (*Lauxania*).—Tex.; Ariz.
trivittata (Loew), 1861b: 353 (Cent. 1, no. 90) (*Lauxania*).—Ga.; Tenn. and Md., s. to La. and Fla.
    facialis Coquillett, 1898d: 280 (*Lauxania*).—Fla., Ga.

Genus LYCIELLA Collin

Lycia Robineau-Desvoidy, 1830: 637 (preocc. Hübner, 1823). Type-species, *flava* Robineau-Desvoidy (Coquillett, 1910b: 563) = *rorida* (Fallén).
Lyciella Collin, 1948: 237 (n. name for *Lycia* Robineau-Desvoidy). Type-species, *Lycia flava* Robineau-Desvoidy (aut.) = *rorida* (Fallén).

REFERENCE: Shewell, 1938 (key, *Sapromyza* part, e. Canada).

annulata (Melander), 1913b: 73 (*Minettia*).—Wash. (here restricted); Alaska to Que., s. to Oreg. **N. comb.**
    annularis Melander, 1913d: 205 (*Minettia;* unjustified n. name for *annulata* Melander).—Wash. **N. comb.**
    ouelleti Shewell, 1938: 115 (*Sapromyza*).—s. Que. **N. syn., n. comb.**
aspinosa (Shewell), 1938: 116 (*Sapromyza*).—s. Que.; s. Ont. **N. comb.**
browni (Curran), 1933b: 8 (*Sapromyza*).—s. Ont.; Wis. to N.B., s. to Ga. **N. comb.**
currani (Shewell), 1938: 118 (*Sapromyza*).—s. Que.; s. Ont., N.Y. **N. comb.**
discolor (Cresson), 1920b: 67 (*Sapromyza*).—Calif.; Oreg. **N. comb.**
    fusciventris Malloch, 1923j: 53 (*Sapromyza*).—Calif. **N. syn.**
novaescotiae (Shewell), 1938: 118 (*Sapromyza*).—N.S.; s. Ont., s. Que., N.H., N.C., Ga. **N. comb.**
obtusilamellata (Malloch), 1923j: 52 (*Sapromyza*).—Md.; Maine, N.H., Vt., Mass., Pa. **N. comb.**
pictiventris (Malloch), 1923j: 52 (*Sapromyza*).—Va.; Ill., Tenn., Md., N.C., Ga. **N. comb.**

quadrilineata (Loew), 1861b: 348 (Cent. 1, no. 78) (*Sapromyza*).— Pa.; s. Que. to Pa. Published records are unreliable. **N. comb.**
serrata (Malloch), 1923j: 52 (*Sapromyza*).—D.C.; s. Ont. and Que., s. to S.C. **N. comb.**
spatulata (Shewell), 1938: 116 (*Sapromyza*).—Que.; s. B.C., s. Ont. **N. comb.**
subserrata (Shewell), 1938: 117 (*Sapromyza*).—s. Que.; s. Ont., N.Y., Ga. **N. comb.**

Genus MINETTIA Robineau-Desvoidy

Minettia Robineau-Desvoidy, 1830: 646. Type-species, *nemorosa* Robineau-Desvoidy (Westwood, 1840: 150–151) =*rivosa* (Meigen).

REFERENCES: Malloch and McAtee, 1924b (key, e. U.S.); Shewell, 1938 (key, e. Canada).

americana Malloch, 1923j: 53.—Md.; s. Ont. and Que., s. to Kans. and Va.
americanella Shewell, 1938: 108.—s. Que.; s. B.C. to s. Que., N.Y., Pa., N.C.
buchanani Malloch *in* Malloch and McAtee, 1924b: 16.—Md.; Tenn., N.C., Ga.
caesia (Coquillett) *in* Baker, 1904: 31 (*Sapromyza*).—Nev.; Wash., Oreg., Calif.
cana Melander, 1913b: 72.—Mass.; s. Que.
  cena, error.
flaveola (Coquillett), 1898d: 279 (*Sapromyza*).—Wash., Oreg., Calif.; s. B.C., Utah.
fumipennis Melander, 1913b: 75.—Calif.
glauca (Coquillett), 1902e: 177 (*Sapromyza*).—Md.; s. Ont., Pa., Va., N.C., Ga.
hubbardii (Coquillett), 1898d: 277 (*Sapromyza*).—Ariz.; N. Mex. **N. comb.**
lobata Shewell, 1938: 108.—s. Que.; Ill. to s. Que. and Mass., s. to Ga.
longipennis (Fabricius).—Not Nearctic. N. Amer. records refer to spp. of *obscura* group.
lupulina (Fabricius), 1787: 344 (*Musca*).—Denmark; Alaska and Canada at treeline, s. to Ariz. and N.C., Holarctic.
lyraformis Shewell, 1938: 109.—s. Que.; s. Man. to s. Que., N.Y., Tenn., N.C., Ga.
magna (Coquillett), 1898d: 279 (*Sapromyza*).—D.C.; Mich. to N.J., s. to Tex. and Ga.

obscura (Loew), 1861b: 351 (Cent. 1, no. 86) (*Lauxania*).—Ont., Pa. Published records are unreliable.
rivosa (Meigen), 1826: 265 (*Sapromyza*).—Germany; s. B.C., Europe.
univittata (Coquillett) *in* Baker, 1904: 32 (*Sapromyza*).—Calif.; Wash., Oreg.

Genus NEODECEIA Malloch

Deceia, subg. **Neodeceia** Malloch *in* Malloch and McAtee, 1924b: 12. Type-species, *Lauxania cineracea* Coquillett (orig. des.). **N. status.**

cineracea (Coquillett), 1902e: 179 (*Lauxania*).—Fla.; Cuba. **N. comb.**
  cinerea, error.

Genus NEOGRIPHONEURA Malloch

Neogriphoneura Malloch *in* Malloch and McAtee, 1924b: 11. Type-species, *Sapromyza sordida* Wiedemann (orig. des.).

sordida (Wiedemann), 1830: 456 (*Sapromyza*).—West Indies; Va., s. to Tex. and Fla.

Genus PACHYCERINA Macquart

Pachycerina Macquart, 1835: 511. Type-species, *Lauxania seticornis* Fallén (mon.).

flavida (Wiedemann), 1824: 57 (*Lauxania*).—S. Amer.; Fla.

Genus PHYSEGENUA Macquart

Physegenua Macquart, 1848a: 220 (1848: 60). Type-species, *vittata* Macquart (orig. des.).
Physogenia, error or emend.

nasalis (Thomson), 1869: 568 (*Lauxania*).—Calif. (?error).

Genus PHYSOCLYPEUS Hendel

Physoclypeus Hendel, 1907b: 226. Type-species, *Chlorops flavus* Wiedemann (orig. des.).

coquilletti (Hendel), 1908a: 30 (*Lauxania*; n. name for *lutea* Coquillett).—Fla.; N.C., S.C., Ga. **N. comb.**
  lutea Coquillett, 1902e: 179 (*Lauxania*; preocc. Wiedemann, 1830).—Fla.

Genus POECILOMINETTIA Hendel

Minettia, subg. **Poecilominettia** Hendel, 1932b: 103. Type-species, *Sapromyza picticornis* Coquillett (orig. des.).

REFERENCE: Malloch and McAtee, 1924b (key, *Minettia* part, e. U.S.).

ordinaria (Melander), 1913b: 74 (*Minettia*).—Kans., Ill.; Kans. to s. Ont., s. to Tenn. and Va., also Fla. **N. comb.**
picticornis (Coquillett), 1904f: 189 (*Sapromyza*).—Nicaragua; ?Tex.
puncticeps (Coquillett), 1902e: 178 (*Sapromyza*).—N.H.; s. Ont., s. Que., Md. **N. comb.**
punctifer (Malloch), 1920d: 126 (*Minettia*).—Fla. **N. comb.**
valida (Walker), 1858: 232 (*Drosophila*).—U.S.; Kans. to s. Ont. and N.Y., s. to Tex. and Ga.
   macula Loew, 1872a: 101 (Cent. 10, no. 82) (*Sapromyza*).—Tex.

Genus PSEUDOCALLIOPE Malloch

Pseudocalliope Malloch, 1928a: 10. Type-species, *Lauxania flaviceps* Loew (orig. des.).

eucephala (Loew), 1872a: 101 (Cent. 10, no. 83) (*Lauxania*).—Tex.; Ariz., Miss., Ga. **N. comb.**
flaviceps (Loew), 1866b: 49 (Cent. 7, no. 91) (*Lauxania*).—D.C.; s. Ont., N.Y., Mass., Md., Va., Tex.
longicornis (Coquillett), 1902e: 178 (*Lauxania*).—Ariz.; N. Mex., Tex. **N. comb.**
variceps (Coquillett), 1902e: 178 (*Lauxania*).—Ariz.; Tex. **N. comb.**

Genus PSEUDOGRIPHONEURA Hendel

Pseudogriphoneura Hendel, 1907b: 225. Type-species, *cinerella* Hendel (orig. des.).
Deceia Malloch, 1923j: 49. Type-species, *Sapromyza crevecoeuri* Coquillett (orig. des.).
Deceia, subg. *Melanomyza* Malloch, 1923j: 50. Type-species, *Lauxania gracilipes* Loew (orig. des.).

REFERENCE: Malloch and McAtee, 1924b (key); Curran, 1942a (key).

crevecoeuri (Coquillett), 1898d: 278 (*Sapromyza*).—Kans.; Ill., Mo., Md., Tex., Fla.
floridensis Curran, 1942a: 72.—Fla.
gracilipes (Loew), 1861b: 351 (Cent. 1, no. 85) (*Lauxania*).—Pa.; Kans. to s. Que. and N.H., s. to Miss. and Fla.

incongrua (Malloch), 1923j: 50 (*Deceia*).—Fla.
intermedia (Malloch), 1923j: 50 (*Deceia*).—Md.; Ill. to Mass., s. to Ga.
scutellata (Malloch), 1923j: 51 (*Deceia*).—Md.; Kans., Ill., D.C.
wetmorei (Malloch), 1923j: 49 (*Deceia*).—Fla.

Genus SAPROMYZA Fallén

Sapromyza Fallén, 1810a: 18. Type-species, *Musca flava* Linnaeus (mon.; misident.)=*obsoleta* Fallén.

REFERENCES: Malloch and McAtee, 1924b (key); Shewell, 1938 (key, e. Canada).

brachysoma Coquillett, 1898d: 278.—N.H.; Alaska and n. Y.T. to Que., s. to s. B.C. and Va.
    brachystoma, error.
    fusca Shewell, 1938: 113.—Que.
    hemingfordensis Shewell, 1946: 166. Nomen nudum.
cyclops Melander, 1913b: 75.—Wyo.; Alaska, B.C., s. Alta., s. Sask., n. Que.
hyalinata (Meigen), 1826: 300 (*Lauxania*).—Europe; Alaska to n. Que., s. to Wash. and Ga.
    nigrans Melander, 1913b: 72 (*Minettia*).—Wash.
monticola Melander, 1913b: 75.—Wash., Idaho; Alaska, n. Y.T., and N.W.T., s. to Wash. and s. Alta., also n. Que. and Maine.
nigripalpus (Walker), 1849: 1068 (*Sciomyza*).—n. Ont.; Alaska to s. B.C. **N. comb.**
rotundicornis Loew, 1863a: 30 (Cent. 3, no. 56).—Alaska; Alaska and Y.T., s. to Wash., Idaho, and Alta., also Minn. to N.S., s. to N.C.
slossonae Coquillett, 1898d: 278.—Fla.
sororia Williston, 1896c: 385.—St. Vincent I.; Bermuda.
    saroria, error.
umbrosa Loew, 1863a: 30 (Cent. 3, no. 57).—D.C.; Mass., s. to La. and Fla.
vittigera Coquillett, 1902e: 178.—Ga.; s. N.Y. to Tenn. and Fla., also Mexico.

Genus STEGANOPSIS Meijere

Steganopsis Meijere, 1910: 145. Type-species, *pupicola* Meijere (mon.).

Subgenus STEGANOLAUXANIA Frey

Steganolauxania Frey, 1918a: 2 (as genus). Type-species, *Lauxania latipennis* Coquillett (orig. des.).

latipennis (Coquillett), 1898d: 279 (*Lauxania*).—Fla., Ga.; N.J., N.C.

Genus TRIGONOMETOPUS Macquart

Trigonometopus Macquart, 1835: 419. Type-species, *Tetanocera frontalis* Meigen (mon.).

REFERENCES: Knab, 1914c (key); Malloch, 1923j (key).

punctipennis Coquillett, 1898d: 280.—Colo.

vittatus Loew, 1869b: 51 (Cent. 8, no. 98).—Ga.; Va. to Fla., also Tex.

Genus TRIVIALIA Malloch

Deceia, subg. **Trivialia** Malloch, 1923j: 53. Type-species, *fuscocapitata* Malloch (orig. des.). **N. status.**

fuscocapitata (Malloch), 1923j: 53 (*Deceia*).—Va.; N.C., Ga. **N. comb.**

Genus TRYPETISOMA Malloch

Trypetisoma Malloch *in* Malloch and McAtee, 1924b: 25. Type-species, *Sapromyza stictica* Loew (orig. des.).

sticticum (Loew), 1863a: 30 (Cent. 3, no. 58) (*Sapromyza*).—D.C.; N.C., Tex.

Genus XENOCHAETINA Malloch

Xenochaetina Malloch, 1923j: 49. Type-species, *Lauxania muscaria* Loew (orig. des.).

REFERENCE: Malloch, 1923j (key).

muscaria (Loew), 1861b: 352 (Cent. 1, no. 87) (*Lauxania*).—Cuba; Tenn. and N.J., s. to La. and Fla.

Genus XENOPTERELLA Malloch

Xenopterella Malloch, 1926: 2. Type-species, *obliqua* Malloch (orig. des.).

costalis Curran, 1942a: 77.—Colo.; Mexico.

Unplaced Species of Lauxaniidae

amida Walker, 1849: 988 (*Sapromyza*).—Ga. ?=*Neogriphoneura sordida*.
connexa Say, 1829: 177 (1859b: 367) (*Sapromyza*).—Ind. ?*Homoneura*.
planiscuta Thomson, 1869: 568 (*Lauxania*).—Calif. ?*Physegenua*.

Family CHAMAEMYIIDAE
(Ochtiphilidae)
By J. F. McAlpine

Members of the genera *Cremifania, Leucopis, Lipoleucopis,* and *Pseudoleucopis* are larval predators of aphids, scale insects, and mealybugs. The one Nearctic species of *Paraleucopis* has been reared from birds' nests. Little is known about the habits of the remaining genera. Many undescribed species are known in North America, particularly in the genus *Leucopis*, and a revision of the family is needed.

REFERENCES: Malloch, 1940 (rev.); Brown and Clark, 1956 (field ident., aphid predators); McAlpine, 1960a (key, world genera), 1963 (subfamilies and tribes).

Subfamily CREMIFANIINAE

Genus CREMIFANIA Czerny

Cremifania Czerny, 1904a: 169. Type-species, *nigrocellulata* Czerny (mon.).

REFERENCE: McAlpine, 1963 (rev.).

nearctica McAlpine, 1963: 252.—N. Mex.
nigrocellulata Czerny, 1904a: 170.—Europe; Oreg., N.B., Nfld. Introduced.

Subfamily CHAMAEMYIINAE

Tribe CHAMAEMYIINI

Genus PLUNOMIA Curran

Plunomia Curran, 1934a: 365. Type-species, *elegans* Curran (mon.).
elegans Curran, 1934a: 365.—Ill.; B.C., Wash., Man. to Que., s. to Ill. and N.H. Specific name credited to Curran be-

cause of intentional use of misidentification (new International Code of Zoological Nomenclature, Article 70b).
flavicornis Malloch, 1940: 272.—Man.
elegans Panzer of Malloch, 1921j: 347 (*Chamaemyia*).
obtusa Malloch, 1940: 271.—Alaska; Ont.
tibialis Malloch, 1940: 271.—Ind.; Ont., Mich.
transversa Malloch, 1940: 270.—Va.; s. Alaska and B.C. to Que., s. to Colo. and Va.

Genus ACROMETOPIA Schiner

Acrometopia Schiner, 1862b: 434. Type-species, *Oxyrhina wahlbergi* Zetterstedt (Coquillett, 1910b: 503).
Acrometopa, error.
punctata Coquillett, 1902e: 185.—Ga.; Ala., Fla.
reticulata Johnson, 1913a: 81 (*Trigonometopus*).—Fla.

Genus CHAMAEMYIA Meigen

Chamaemyia Meigen, 1803: 278. Type-species, *elegans* Panzer (sub. mon., Panzer, 1809b: 12).
Ochtiphila Fallén, 1823b: 9. Type-species, *aridella* Fallén (Westwood, 1840: 151, or Blanchard, 1840: 625).
Ochthiphila, emend.

Illiger (1807: 481) associated *Musca flava* Linnaeus with *Chamaemyia* 2 years before Panzer's action, but was not followed. That species is the type of *Chyromya*, type-genus of Chyromyidae. It would be undesirable to involve two family names in confusing changes at this late date. The case has been submitted to the International Commission on Zoological Nomenclature.

REFERENCES: Coe, 1942, 1943 (rev. British spp.).

aridella (Fallén), 1823b: 10 (*Ochtiphila*).—Sweden; Alta., Europe.
elegans Panzer.—Probably not Nearctic.
flavipalpis (Haliday), 1838: 187 (*Ochtiphila*).—England; ?Ont., Europe.
maritima Zetterstedt, 1846: 1946 (*Ochtiphila*).—Sweden.
fumicosta Malloch, 1940: 267.—Ariz.
geniculata (Zetterstedt), 1838: 720 (*Ochtiphila*).—Lapland; Alaska, B.C., Europe.
herbarum (Robineau-Desvoidy), 1830: 635 (*Estelia*).—Europe; B.C. to P.E.I.
juncorum (Fallén), 1823b: 9 (*Ochtiphila*).—Sweden; B.C. to N.S., s. to Minn., Ind., and Mass.
polystigma Meigen, 1830: 92 (*Ochtiphila*).—Europe.

Genus PARALEUCOPIS Malloch

Paraleucopis Malloch, 1913n: 148. Type-species, *corvina* Malloch (orig. des.).

corvina Malloch, 1913n: 149.—N. Mex.; Tex.

Genus PSEUDODINIA Coquillett

Pseudodinia Coquillett, 1902e: 187. Type-species, *varipes* Coquillett (orig. des.).

antennalis Malloch, 1940: 269.—Va.; Man. to Ont., s. to Utah and Ga.

nitens (Melander and Spuler), 1917: 70 (*Piophila;* n. name for *nitida* Wulp).—Wis.; B.C. to Man., s. to Idaho and Ill. **N. comb.**, C. W. Sabrosky *in litt.*

    nitida Wulp, 1867: 160 (*Piophila;* preocc. Brullé, 1832).—Wis. **N. comb.**, C. W. Sabrosky *in litt.*

    nitida Melander, 1913a: 295 (preocc. Wulp, 1867).—Idaho.

polita Malloch, 1915j: 152.—Ill.; Md., N.C.

varipes Coquillett, 1902e: 187.—N. Mex.; B.C. to Man., s. to Idaho, N. Mex., and Tex.

    pruinosa Melander, 1913a: 295.—Tex.

Genus Pseudoleucopis Malloch

Pseudoleucopis Malloch, 1925c: 93. Type-species, *magnicornis* Malloch (orig. des.).

benefica Malloch, 1930: 490.—Australia; ?N. Amer. Introduced.

Tribe LEUCOPINI

Genus Lipoleucopis Meijere

Lipoleucopis Meijere, 1928: 76. Type-species, *praecox* Meijere (mon.).

praecox Meijere, 1928: 76.—Holland; ?N.B. Introduced.

Genus LEUCOPIS Meigen

Leucopis Meigen, 1830: 133. Type-species, *Anthomyza griseola* Fallén (Blanchard, 1840: 627).

Leucopis, subg. *Leucopina* Malloch, 1927c: 575. Type-species, *bella* Loew (orig. des.).

Subgenus NEOLEUCOPIS Malloch

Leucopis, subg. **Neoleucopis** Malloch, 1921j: 357. Type-species, *pinicola* Malloch (orig. des.).

obscura Haliday, 1833: 173.—Ireland; Nfld. to Vt. and ?N.Y., also Oreg., ?Wash. Introduced.
pinicola Malloch, 1921j: 357.—Ill.; Ont., N.B., Nfld.

Subgenus LEUCOPOMYIA Malloch

Leucopis, subg. **Leucopomyia** Malloch, 1921j: 355. Type-species, *pulvinariae* Malloch (orig. des.).

pulvinariae Malloch, 1921j: 356.—Ill.; Wash., Sask. to N.S., N.Y.

Subgenus XENOLEUCOPIS Malloch

Leucopis, subg. **Xenoleucopis** Malloch, 1933c: 384. Type-species, *cilifemur* Malloch (orig. des.).

Unnamed spp.—B.C., Tex. Otherwise, the subgenus is known only from Chile.

Subgenus LEUCOPIS Meigen

americana Malloch, 1921j: 354.—Ill.; Man. to N.S., s. to Ill. and Fla.
atrifacies Aldrich, 1925a: 152.—Calif.
bella Loew, 1866a: 186 (Cent. 6, no. 99).—Cuba. Records on continent probably misidentified.
bellula Williston *in* Riley, 1889: 258.—Tex.; Alta., Sask.
bivittata Malloch, 1940: 273.—Calif.
flavicornis Aldrich, 1914a: 404.—Tex.
griseola (Fallén), 1823a: 8 (*Anthomyza*).—Scandinavia. N. Amer. records probably misidentified.
maculata Thompson, 1910: 241.—Nebr.; Alta., Man., Ont.
major Malloch, 1921j: 352.—Ill.
minor Malloch, 1921j: 354.—Ill.; ?Wash.
nigricornis Egger, 1862: 782.—Europe. N. Amer. records probably misidentified.
ocellaris Malloch, 1940: 272.—Alta.; Man.
orbitalis Malloch, 1921j: 352.—Ill.; Ont., N.B., Nfld.
parallela Malloch, 1921j: 353.—Ill.
pemphigae Malloch, 1921j: 350.—Ill.; Alta.
piniperda Malloch, 1921j: 351.—Ill.; Colo., Que., N.B.
simplex Loew, 1869b: 51 (Cent. 8, no. 96).—N.Y.; Ont., ?Wash.
    phylloxerae Riley, 1883: 39.—Unknown.
verticalis Malloch, 1940: 273.—Calif.

Family PERISCELIDIDAE

By Curtis W. Sabrosky

The species of Periscelididae are seldom collected. They are especially attracted to bleeding wounds on trees. *Periscelis wheeleri* has been reared from fermenting oak sap, and this may be the normal breeding place for the family.

REFERENCE: Sturtevant, 1954 (rev.).

Genus PERISCELIS Loew

Periscelis Loew, 1858a: 113. Type-species, *annulipes* Loew (Sturtevant, 1923: 1). Oldenberg (1914: 39) was formerly construed to have fixed the type by elimination, in the spirit of Opinion 6, I.C.Z.N.
Microperiscelis Oldenberg, 1914: 39, 42. Type-species, *Notiphila annulata* Fallén (Sturtevant, 1923: 1).
Phorticoides Malloch, 1915c: 86. Type-species, *flinti* Malloch (orig. des.) = *annulata* (Fallén).
Sphyroperiscelis Sturtevant, 1923: 1. Type-species, *wheeleri* Sturtevant (orig. des.).

annulata (Fallén), 1813: 250 (*Notiphila*).—Sweden; Wash., Alta., S. Dak. to Que., s. to Ariz. and Fla., Europe.
   flinti Malloch, 1915c: 87 (*Phorticoides*).—Ill.
occidentalis Sturtevant, 1954: 554.—Calif.; Wash., Ariz., N. Mex., Tex.
wheeleri (Sturtevant), 1923: 2 (*Sphyroperiscelis*).—Mass.; B.C., Oreg., Calif., Pa., N.J., Fla.

Superfamily PALLOPTEROIDEA

Family PIOPHILIDAE

By George C. Steyskal

The family Piophilidae is a small group containing several species of cosmopolitan or Holarctic distribution. They are scavengers in animal products and fungi; one species, the well-known "cheese fly" or "cheese-skipper," *Piophila casei* (Linnaeus), has been recorded as an important cause of human intestinal myiasis.

REFERENCES: Melander and Spuler, 1917 (rev.); Melander, 1924b (rev.); Hennig, 1943 (monog., Palaearctic spp., including several Holarctic and cosmopolitan spp.).

Genus PIOPHILA Fallén

Piophila Fallén, 1810a: 20. Type-species, *Musca casei* Linnaeus (mon.).
Tyrophaga Kirby *in* Kirby and Spence, 1826: 78. Type-species, *Musca casei* Linnaeus (mon.).

Subgenus PIOPHILA Fallén

casei (Linnaeus), 1758: 597 (*Musca;* as *putris* var.).—Europe; cosmopolitan.
  pusilla Meigen, 1838: 360.—Germany.

Subgenus ALLOPIOPHILA Hendel

Allopiophila Hendel, 1917: 37 (as genus). Type-species, *luteata* Haliday (orig. des.).
anomala Malloch, 1923i: 219.—Alaska (Pribilof Is.).
arctica Holmgren, 1883: 177.—U.S.S.R. (Vaigach I.); Alaska, N.W.T., n. Que.
  aterrima Becker, 1897: 402.—Novaya Zemlya.
atrifrons Melander and Spuler, 1917: 66.—Wash., Idaho; Maine.
fulviceps Holmgren, 1883: 177.—n. U.S.S.R.; Alaska, Man.
  borealis Malloch, 1919f: 84.—Alaska.
liturata Melander and Spuler, 1917: 60.—Idaho, Wash.
nitidissima Melander and Spuler, 1917: 66.—Wash., Idaho, Calif.; Colo.
setosa Melander and Spuler, 1917: 64.—Alaska.
vulgaris Fallén, 1820e: 9.—Europe; Alaska to Que., s. to N.Y. and Mass., Greenland.
  oriens Melander and Spuler, 1917: 63.—N.Y.
xanthopoda Melander and Spuler, 1917: 59 (in key), [102].—Mont., Idaho.
  flavipes Melander and Spuler, 1917: 63 (in descr.).—Mont., Idaho.
  xanthostoma (error in correcting *flavipes*) Melander and Spuler, 1917: [103].

Subgenus AMPHIPOGON Wahlberg

Macrochira Zetterstedt, 1838: 784 (preocc. Meigen, 1803). Type-species, *flava* Zetterstedt (mon.).
Amphipogon Wahlberg, 1845: 217 (as genus). Type-species, *spectrum* Wahlberg (mon.)=*flava* (Zetterstedt).
Ambopogon Greene, 1919: 126. Type-species, *hyperboreus* Greene (orig. des.).

hyperborea (Greene), 1919: 127 (*Ambopogon*).—Y.T.; Idaho.

Subgenus ARCTOPIOPHILA Duda

Piophila, subg. **Arctopiophila** Duda, 1924d: 109. Type-species, *nigerrima* Lundbeck (mon.).

nigerrima Lundbeck, 1901: 301.—Greenland.

Subgenus LASIOPIOPHILA Duda

Piophila, subg. **Lasiopiophila** Duda, 1924d: 109. Type-species, *pilosa* Staeger (mon.).

pilosa Staeger, 1845b: 368.—Greenland; n. Alaska, N.W.T.

Subgenus LIOPIOPHILA Duda

Piophila, subg. **Liopiophila** Duda, 1924d: 109. Type-species, *varipes* Meigen (Hennig, 1943: 28).

halterata Melander and Spuler, 1917: 62 (as *affinis* var.).—Wash. ?=*varipes* var.
morator Melander, 1924b: 85.—Wash.
nigricoxa Melander and Spuler, 1917: 64.—Wash., Idaho, Mont.
nigrimana Meigen, 1826: 396.—Europe; B.C. to Mass., s. to Oreg., Tex., and Ill., also s. Chile.
  affinis Meigen, 1830: 383.—Belgium.
  privigna Melander, 1924b: 86.—B.C., Wash., Idaho, Wyo., Ill., N.Y., Mass.
  pusilla Meigen of Melander and Spuler, 1917: 61.
occipitalis Melander and Spuler, 1917: 65.—Ill. ?=*nigricornis* Meigen.
varipes Meigen, 1830: 384.—Germany; Wash. to Ont. and Mass., Europe and Asia.
  affinis Meigen of Melander and Spuler, 1917: 61.

Subgenus MYCETAULUS Loew

Mycetaulus Loew, 1845: 37 (as genus). Type-species, *hoffmeisteri* Loew (mon.)=*bipunctata* (Fallén).

bipunctata (Fallén), 1823e: 3 (*Geomyza*).—Sweden; Wash. to Maine, s. to Calif., Mich., and Ga.
  pulchellus Banks, 1915c: 145 (*Mycetaulus*).—Va.
costalis (Melander), 1924b: 79 (*Mycetaulus*).—Oreg.; Calif.
longipennis (Loew), 1869c: 186 (Cent. 9, no. 100) (*Mycetaulus*).—Hudson Bay Territory; Mich. to Maine and Va.
nigritella (Melander), 1924b: 79 (*Mycetaulus*).—Wash., Idaho, Mont.
polypori (Melander), 1924b: 80 (*Mycetaulus*).—Wash.

subdola (Johnson), 1922: 25 (*Geomyza*).—Mass.; Ont. to Mass. **N. comb.**
testacea (Melander), 1924b: 81 (*Mycetaulus*).—Wash., Idaho.

Subgenus PROCHYLIZA Walker

Prochyliza Walker, 1849: 1045 (as genus). Type-species, *xanthostoma* Walker (mon.).

brevicornis (Melander), 1924b: 81 (*Prochyliza*).—Mont., Wyo., Ill.; Mont. and Wyo. to Mich. **N. comb.**

xanthostoma (Walker), 1849: 1045 (*Prochyliza*).—Ont.; Alaska to Que., s. to Calif., Tex., and Ala.

Subgenus PROTOPIOPHILA Duda

Piophila, subg. **Protopiophila** Duda, 1924d: 109. Type-species, *latipes* Meigen (mon.).

latipes Meigen, 1838: 360.—Germany; Wis., Mich., Pa., widespread in Palaearctic, Oriental, and Australian Regions.

hornigi Cresson, 1919: 193 (*Mycetaulus*).—Pa.

Subgenus STEARIBIA Lioy

Stearibia Lioy, 1864a: 1105 (as genus).—Type-species, *Piophila foveolata* Meigen (mon.).

foveolata Meigen, 1826: 396.—Europe; B.C. to Que. and Maine, s. to Wash., Ill., and Ala., also s. Chile.

nigriceps Meigen, 1826: 397.—Europe.

Unplaced Species of *Piophila*

senescens Melander and Spuler, 1917: 72 (n. name for *nigriceps* Macquart).—N. Amer.

nigriceps Macquart, 1851: 276 (1851: 303) (preocc. Meigen, 1826).—N. Amer.

Family THYREOPHORIDAE

By George C. Steyskal

Little is known of the rarely collected family Thyreophoridae, here recorded from North America for the first time. The genus *Omomyia* was described in the Coelopidae and later transferred to the Pallopteridae, but its relationships are still somewhat obscure. The habits of the type-species were described by Barber (1908).

Genus OMOMYIA Coquillett

Omomyia Coquillett, 1907a: 76. Type-species, *hirsuta* Coquillett (orig. des.).

hirsuta Coquillett, 1907a: 76.—Calif.
regularis Curran, 1935b: 21.—Ariz.

Family NEOTTIOPHILIDAE
By J. R. Vockeroth

The family Neottiophilidae has not been recorded previously from North America. The habits of the species listed below are unknown; the larva of the only other species referred to the family is an ectoparasitic blood-sucker on nestling birds in Europe.

Genus ACTENOPTERA Czerny

Gymnomyza Strobl, 1894b: 85 (preocc. Fallén, 1810). Type-species, *Heteromyza hilarella* Zetterstedt (mon.).
Actenoptera Czerny, 1904b: 202 (n. name for *Gymnomyza* Strobl). Type-species, *Heteromyza hilarella* Zetterstedt (aut.).

hilarella (Zetterstedt), 1847: 2467 (*Heteromyza*).—Norway; n. Que., n. and cent. Europe.

Family PALLOPTERIDAE
By G. E. Shewell

The family Pallopteridae occurs in the north temperate regions, temperate South America, and New Zealand. With the exception of *Palloptera* Fallén, the genera are little known, and their relationships imperfectly understood. The family is usually placed close to Lonchaeidae and Otitidae.

No biological information on North American species has been published, but records from Europe indicate that the larvae are phytophagous or carnivorous. They have been found in flower buds and stems of Compositae and Umbelliferae; also under bark of felled coniferous and deciduous trees infested with larvae of Cerambycidae and Scolytidae on which they prey.

REFERENCES: Melander, 1913b (rev.); Johnson, 1921b (key); Malloch, and McAtee, 1924b (rev.); Meijere, 1944, and Morge, 1956 (biol. European spp., figs.).

Genus PALLOPTERA Fallén

Palloptera Fallén, 1820c: 23. Type-species, *Musca umbellatarum* Fabricius (Westwood, 1840: 150).

albertensis Johnson, 1921b: 22.—Alta.; s. B.C., Idaho, ?Calif.
arcuata (Fabricius).—Not Nearctic. Records may refer to *jucunda*.
claripennis Malloch *in* Malloch and McAtee, 1924b: 7.—Calif.; s. B.C.
jucunda Loew, 1863a: 29 (Cent. 3, no. 55).—Alaska; Alaska and Y.T., s. to Calif. and Colo., also e. Que. and Labr.
setosa Melander, 1913b: 81.—Mich.; s. Ont., s. Que.
similis Johnson, 1910a: 233.—Maine; Que., N.H.
subarcuata Johnson, 1921b: 21.—Maine; s. Alta. to s. Que., N.H.
subusta Malloch *in* Malloch and McAtee, 1924b: 7.—Calif.; s. B.C.
superba Loew, 1861b: 346 (Cent. 1, no. 75).—Pa.; s. Sask. to Maine, s. to S.C.
terminalis Loew, 1863a: 29 (Cent. 3, no. 54).—Alaska; B.C., Wash., Oreg.
ustulata Fallén.—Not Nearctic. Records may refer to *claripennis* or an undescribed sp.

Family LONCHAEIDAE

By J. F. McAlpine

Numerous undescribed species of lonchaeid flies are known in the Nearctic Region (McAlpine, rev. in progress), and the latest key (Malloch and McAtee, 1924b) cannot be depended upon.

The immature stages of most species are secondary invaders in diseased or injured plant material. Many species, particularly of the genera *Dasiops* and *Lonchaea*, live under the bark of dead and dying trees or in injured, overripe, or decaying fruits and vegetables. Some species are associated with primary insect invaders; e.g., larvae of *Lonchaea corticis* Taylor live in close association with the larvae of *Pissodes strobi* (Peck), larvae of the *L. watsoni* complex are associated with various bark beetles, and a large number of *Silba* species live in conjunction with members of the family Tephritidae. A few species are primary invaders of plant tissue (e.g., cone-infesting species of the genus *Earomyia*, see McAlpine, 1956). Some species of the genera *Dasiops*, *Lonchaea*, and *Silba* apparently attack certain fruits and vegetables in a similar manner.

REFERENCES: Morge, 1959 (monog., Palaearctic spp.); McAlpine, 1960c (key, world genera).

Genus DASIOPS Rondani

Dasiops Rondani, 1856: 120. Type-species, *Lonchaea latifrons* Meigen (I.C.Z.N., 1963a: 114, under plenary powers).
Arctobiella Coquillett, 1902e: 188. Type-species, *obscura* Coquillett (orig. des.).
Acucula Townsend, 1913a: 264. Type-species, *saltans* Townsend (orig. des.).

albiceps (Malloch), 1914a: 39 (*Lonchaea*).—D.C.; Alaska, B.C., Que.
alveofrons McAlpine, 1961a: 539.—Calif.; Ariz., Tex., Mexico.
arkansensis (Malloch), 1923j: 45 (*Lonchaea*).—Ark. **N. comb.**
latiterebrus (Czerny), 1934b: 10 (*Psilolonchaea*).—e. Siberia; Alaska, n. Eurasia.
obscurus (Coquillett), 1902e: 188 (*Arctobiella*).—B.C.
parvicornis (Meigen).—Not Nearctic.
quadrisetosus (Malloch), 1914a: 40 (*Lonchaea*).—Md.; Ill., D.C., Fla. **N. comb.**
vibrissatus (Malloch), 1914a: 37 (*Lonchaea*).—Ill.; Calif., Minn., Ont., Que., Mass. to Va.

Genus CHAETOLONCHAEA Czerny

Earomyia, subg. **Chaetolonchaea** Czerny, 1934b: 26. Type-species, *Lonchaea dasyops* Meigen (orig. des.).

pallipennis (Zetterstedt), 1855: 4786 (*Lonchaea*).—Sweden; B.C., Alta., Mont., Colo., Europe. Nearctic material is possibly not *pallipennis*.

Genus EAROMYIA Zetterstedt

Earomyia Zetterstedt, 1842: 78. Type species, *lonchaeoides* Zetterstedt (sub. mon., Zetterstedt, 1848: 2690).

REFERENCE: McAlpine, 1956 (partial key).

aberrans (Malloch), 1920e: 131 (*Lonchaea*).—Ill.; Ont. to Nfld., also Fla.
abietum McAlpine, 1956: 180.—B.C.; B.C. to Calif., also Colo.
aquilonia McAlpine, 1956: 184.—B.C.; Alta.
aterrima (Malloch), 1920e: 129 (*Lonchaea*).—Maine; Ont., Que., N.H.
barbara McAlpine, 1956: 187.—B.C.; Oreg., Calif., Alta., Colo.
brevistylata McAlpine, 1956: 190.—Calif.; Oreg., Colo.
hirtithorax (Aldrich), 1925c: 8 (*Lonchaea*).—Oreg.; B.C.
longistylata McAlpine, 1956: 192.—Calif.; Oreg.
viridana (Meigen).—Not Nearctic.

Genus LONCHAEA Fallén

Lonchaea Fallén, 1820c: 25. Type-species, *Musca chorea* Fabricius (Westwood, 1840: 150).

affinis Malloch, 1920e: 130.—N.H.; Alaska, B.C., Ont. to N.B., N.H., N.C.
    subpolita Malloch, 1923j: 45.—N.C. **N. syn.**
albitarsis Zetterstedt.—Not Nearctic.
angustitarsis Malloch, 1920e: 131.—Maine.
atritarsis Malloch, 1923j: 47.—B.C.
caerulea Walker, 1849: 1004.—Ga.; s. Ont. to Fla., also Tex., La.
chorea (Fabricius), 1781: 444 (*Musca*).—Europe; B.C. to Oreg., e. to Que. and Mass.
    vaginalis Fallén, 1820c: 26.—Sweden.
coloradensis Malloch, 1923j: 46.—Colo.
corticis Taylor, 1928: 192.—Mass.; Ont. to N.B., s. to Ohio and Pa.
deutschi Zetterstedt.—Not Nearctic.
flavipennis Morge, 1959: 938.—Unknown; B.C., Oreg., Calif., Europe.
    flavidipennis, authors, not Zetterstedt.
hirta Malloch, 1920e: 129.—Mass.; Ont., Que., N.B., Mich., N.H.
hyalipennis Zetterstedt.—Not Nearctic.
    hyalinipennis, error.
laticornis Meigen.—Not Nearctic.
laxa Collin, 1953b: 204.—Scotland; Alaska and B.C. to Nfld., s. to Mich. and N.Y.
    affinis, authors, not Malloch.
longicornis Williston, 1896c: 378.—St. Vincent I.; Fla.
marylandica Malloch, 1923j: 46.—Md.; Ont. to N.B., s. to Ohio and Md.
nigrociliata Malloch, 1920e: 131.—Maine.
nudifemorata Malloch, 1914a: 38.—Md.; s. Ont. to Mass., s. to Ill. and Va.
pleuriseta Malloch, 1920e: 132.—Minn.; B.C., Sask.
polita Say, 1830: 188 (1859b: 371).—Ind.; s. Canada, widespread U.S.
    rufitarsis Macquart, 1851: 273 (1851: 300).—N. Amer.
ruficornis Malloch, 1920e: 130.—Ill.; Mich., Mass.
striatifrons Malloch, 1920f: 246.—Calif.; Ariz., N. Mex., Tex., Mexico.
    occidentalis Malloch, 1923j: 46.—Ariz. **N. syn.**
tarsata Fallén.—Not Nearctic.
ursina Malloch, 1920e: 132.—Alaska; B.C.
watsoni Curran, 1926d: 213.—Ont.; Que.
wiedemanni Townsend.—Not Nearctic.
winnemanae Malloch, 1914a: 38.—Va.; Minn., Tex., N.Y., Pa., N.J., Md.

Genus SILBA Macquart

Silba Macquart, 1851: 277 (1851: 304). Type-species, *virescens* Macquart (orig. des.).
Lonchaea, subg. *Carpolonchaea* Bezzi, 1920: 199. Type-species, *plumosissima* Bezzi (orig. des.).

glaberrima (Wiedemann), 1830: 475 (*Lonchaea*).—West Indies; Fla.

Superfamily MILICHIOIDEA

Family SPHAEROCERIDAE
(Borboridae)
By O. W. Richards

The American species of this family need to be thoroughly revised from their types. The following list is largely a compilation.

Larvae live as scavengers in excrement and decaying vegetable matter.

REFERENCES: Spuler, 1924d (keys, N. Amer. genera and subg.); Richards, 1930 (biol., figs. of puparia); Goddard, 1938 (figs. of puparia); Hammer, 1941 (biol.); Laurence, 1955 (biol., figs. of puparia); Richards, 1961 (nomenclatural problems).

Genus SPHAEROCERA Latreille

Sphaerocera Latreille, 1804: 197. Type-species, *curvipes* Latreille (sub. mon., Latreille, 1805: 394).
Sphaerocera, subg. *Parasphaerocera* Spuler, 1924b: 67. Type-species, *bimaculata* Williston (orig. des.).

REFERENCE: Malloch, 1925d (rev.).

Subgenus SPHAEROCERA Latreille

annulicornis Malloch, 1913j: 363.—Mass.; Mass. to Va., also Tenn., Mo.
bimaculata Williston, 1896c: 435.—St. Vincent I.; Fla., Mexico, Costa Rica.
curvipes Latreille, 1805: 394.—Europe; widespread in Alaska, Canada, U.S., cosmopolitan.
    planipes Harris, 1835: 600 (*Borborus*). Nomen nudum.
    subsultans, authors, not Linnaeus.
varipes Malloch, 1925d: 121.—Costa Rica; Ga., Cent. Amer.

Subgenus LOTOBIA Lioy

Lotobia Lioy, 1864a: 1114 (as genus). Type-species, *Borborus pallidiventris* Meigen (mon.).
Allosphaerocera Hendel, 1920b: 54. Type-species, *Borborus hyalipennis* Meigen (orig. des.)=*pallidiventris* (Meigen).

nigrifemur Malloch, 1925d: 122.—Md.
pusilla (Fallén), 1820e: 8 (*Copromyza*).—Sweden; B.C., Wash., Idaho, Ill., Mich., Mass., D.C., Europe.
scabra Spuler, 1924b: 68.—Wash.; Mich.
striata Malloch, 1925d: 122.—Fla.
vaporariorum Haliday, 1836: 319.—United Kingdom; Mich., Europe.

Genus COPROMYZA Fallén

Copromyza Fallén, 1810a: 19. Type-species, *equina* Fallén (Zetterstedt, 1847: 2475).
Borborus, subg. *Trichiaspis* Duda, 1923a: 55. Type-species, *Copromyza equina* Fallén (Richards, 1930: 267).
Borborus, authors, not Meigen.
Cypsela, authors, not Meigen.

REFERENCE: Spuler, 1925a (rev.).

Subgenus COPROMYZA Fallén

calcitrans (Spuler), 1925a: 141 (*Borborus*).—Wash.
equina Fallén, 1820e: 6.—Sweden; widespread in Alaska, Canada, U.S., Europe.
neglecta (Malloch), 1913j: 364 (*Borborus*).—Mass.; Wash., Mich.
stercoraria (Meigen), 1830: 202 (*Borborus*).—Europe; Oreg.
nigrifemoratus Macquart, 1835: 567 (*Borborus*).—France.
subaptera (Malloch), 1923i: 211 (*Borborus*).—Pribilof Is.

Subgenus OLINEA Richards

Copromyza, subg. **Olinea** Richards (for *Olina*, authors, as descr. by Duda, 1923a: 55, 99). Type-species, *Borborus ater* Meigen (present des.). **N. subg.**
Copromyza, subg. *Olinea* Richards, 1961: 561. Nomen nudum (type des., but no descr. of subg.).
Olina, authors, not Robineau-Desvoidy.
Scatophora, authors, not Robineau-Desvoidy.

atra (Meigen), 1830: 203 (*Borborus*).—Europe; B.C. to Mass., s. to Calif. and Ga., Europe.
carolinensis, authors, not Robineau-Desvoidy (*Scatophora*).
geniculatus, authors, probably not Macquart (*Borborus*).

Subgenus FUNGOBIA Lioy

Fungobia Lioy, 1864a: 1114 (as genus). Type-species, *Borborus nitidus* Meigen (mon.).
Borborus, subg. *Stratioborborus* Duda, 1923a: 54. Type-species, *nitidus* Meigen (Richards, 1930: 267).

maculipennis (Spuler), 1925a: 6 (*Borborus*).—Wash.; Idaho.
setitibialis (Spuler), 1925a: 6 (*Borborus*).—N. H.

Subgenus CRUMOMYIA Macquart

Crumomyia Macquart, 1835: 569 (as genus). Type-species, *Borborus glacialis* Meigen (orig. des.).

annulus (Walker), 1849: 1129 (*Borborus*).—Hudson Bay; Pribilof Is.
immensa (Spuler), 1925a: 4 (*Borborus*).—N.H.
nigra (Meigen), 1830: 201 (*Borborus*).—Europe; ?Ga.

Subgenus BORBORILLUS Duda

Borborus, subg. **Borborillus** Duda, 1923a: 54. Type-species, *uncinatus* Duda (Richards, 1930: 267).

arctica (Malloch), 1913j: 367 (*Borborus;* as *articus*).—Labr.; ?Wash.
frigipennis (Spuler), 1925a: 9 (*Borborus*).—Fla.
peltastes (Spuler), 1925a: 10 (*Borborus*).—Tex.
scripta (Malloch), 1915b: 64 (*Borborus*).—Ill.
singularis (Spuler), 1925a: 12 (*Borborus*).—Wash.
sordida Zetterstedt, 1847: 2484.—Sweden; B.C. to Mass., s. to Calif., N. Mex., and D.C., also Bermuda, Hawaii, Palaearctic, e. India, s. Africa, Canary Is.
    brevisetus Malloch, 1913j: 365 (*Borborus*).—D.C., N. Mex.
    minutus Johnson, 1913d: 449 (*Borborus*).—Bermuda.
    marmoratus Becker of Spuler, 1925a: 8, 11, and authors (*Borborus*).

Genus LEPTOCERA Olivier

Leptocera Olivier, 1813: 489. Type-species, *nigra* Olivier (mon.).
Limosina, subg. *Paracollinella* Duda, 1924b: 166. Type-species, *Copromyza fontinalis* Fallén (Richards, 1930: 267).

REFERENCES: Spuler, 1924a, b, c, 1925b, c (rev.); Sabrosky, 1949b (rev., *Rachispoda* part); Richards, 1960 (key, *Coproica*).

Subgenus LEPTOCERA Olivier

caenosa (Rondani), 1880: 38 (1880: 36) (*Limosina*).—Italy; Sask. to N.S., Mich., Europe.

downesi Richards, 1944: 137.—Scotland (ship from Argentina); Ill., N.Y., Hawaii, Micronesia.
fontinalis (Fallén), 1826: 16 (*Copromyza*).—Sweden; widespread, Alaska, Canada, U.S., Bermuda, P.R., Europe.
hoplites Spuler, 1924a: 115.—Wash.; Oreg., Mich., Vt.
pararoralis Duda, 1925a: 51.—Algonquin (presumably Ont.).

Subgenus RACHISPODA Lioy

Rachispoda Lioy, 1864a: 1116 (as genus). Type-species, *Copromyza limosa* Fallén (mon.).
Limosina, subg. *Collinella* Duda, 1918: 27 (preocc. Schmidt, 1879). Type-species, *Copromyza limosa* Fallén (Richards, 1930: 266).
Leptocera, subg. *Collinellula* Strand, 1928: 49 (n. name for *Collinella* Duda). Type-species, *Copromyza limosa* Fallén (aut.).

atra (Adams), 1903b: 223 (*Limosina*).—Kans.; Wash. to Md., s. to Mexico and Fla.
barbata Sabrosky, 1949b: 16.—Mich.; Alaska, N.H., Utah.
cryptica Sabrosky, 1949b: 12.—Idaho; Wash. to Man., s. to Calif. and Colo.
echinaspis Spuler, 1924a: 108.—Calif.
forceps Sabrosky, 1949b: 10.—Kans.; Wash., Oreg., Nev., Utah, Tex.
frosti Johnson, 1915b: 21.—Mass.
fumipennis Spuler, 1924a: 110.—B.C.; Wash., Idaho, Calif., Ill., Tex.
fuscipennis (Haliday), 1833: 178 (*Borborus*).—Ireland; Mich., Ill., N.Y., Md., Europe.
latiforceps Sabrosky, 1949b: 21.—B.C.; Calif.
limosa (Fallén), 1820e: 8 (*Copromyza*).—Sweden; Wash. to Ont. and Pa., s. to Calif. and Ga.
lutosa (Stenhammar), 1854: 380 (1855: 122) (*Limosina*).—Sweden; Alaska, Man., Europe.
melanderi Sabrosky, 1949b: 20.—Wash.
michigana Sabrosky, 1949b: 14.—Mich.; Alaska, Wash. to Maine, s. to Calif. and Pa.
omega Sabrosky, 1949b: 21.—Colo.
richardsi Sabrosky, 1949b: 18.—Kans.; Calif. to Mich., s. to La. and Md.
spuleri Sabrosky, 1949b: 8.—La.; Kans. to Mass., s. to Mexico, Tex., and Fla.
suberecta Sabrosky, 1949b: 20.—Colo.; Wash., Calif.
subpiligera (Malloch), 1914e: 176 (*Limosina*).—Pa.
tenaculata Sabrosky, 1949b: 16.—Calif.; Wash.

trochanterata (Malloch), 1913k: 462 (*Limosina*).—Va.; Mich., Ind., Tex., Fla.
urodela Sabrosky, 1949b: 18.—Idaho; Wash. to Nebr., s. to Calif. and Tex.
weemsi Sabrosky, 1956: 74.—Fla.

Subgenus OPACIFRONS Duda

Limosina, subg. **Opacifrons** Duda, 1918: 28. Type-species, *coxata* Stenhammar (Spuler, 1924c: 121).

caelobata Spuler, 1924c: 126.—Wash.
convexa Spuler, 1924c: 130.—Idaho; Wash.
coxata (Stenhammar), 1854: 396 (1855: 138) (*Limosina*).—Sweden; B.C. and Idaho to Calif., also S. Dak., Europe, Congo (Belgian).
grandis Spuler, 1924c: 123.—Oreg.; Wash.
pellucida Spuler, 1924c: 127.—Wash.; B.C.
sciaspidis Spuler, 1924c: 124.—Wash.; Idaho.
wheeleri Spuler, 1924c: 128.—Tex.; Kans., Ill., Wis., Mich., Ont.

Subgenus PTEREMIS Rondani

Pteremis Rondani, 1856: 124 (as genus). Type-species, *Borborus nivalis* Haliday (orig. des.)=*fenestralis* (Fallén).
Limosina, subg. **Stenhammaria** Duda, 1918: 28. Type-species, *Copromyza fenestralis* Fallén (mon.).

flavifrons Spuler, 1924c: 133.—Va.
parvipennis Spuler, 1924c: 132.—Alaska; B.C.
unica Spuler, 1924c: 134.—Wyo.

Subgenus THORACOCHAETA Duda

Limosina, subg. **Thoracochaeta** Duda, 1918: 32. Type-species, *Borborus zosterae* Haliday (Spuler, 1925b: 120).

brachystoma (Stenhammar), 1854: 393 (1855: 135) (*Limosina*).—Sweden; N.Y., Bermuda, S. Amer., Europe, Seychelles.
johnsoni Spuler, 1925b: 121.—Wash.
rufa Spuler, 1925b: 122.—Fla.; Mass.
seticosta Spuler, 1925b: 120.—Wash.

Subgenus POECILOSOMELLA Duda

Leptocera, subg. **Poecilosomella** Duda, 1925a: 78. Type-species, *Copromyza punctipennis* Wiedemann (Richards, 1930: 268).

SPHAEROCERIDAE

angulata (Thomson), 1869: 602 (*Limosina*).—Brazil; Tex. to Fla., Bermuda, West Indies, Cent. and S. Amer., Hawaii.
  venalicia Osten Sacken, 1878c: 263 (*Borborus*).—Cuba.

Subgenus PTEROGRAMMA Spuler

Leptocera, subg. **Pterogramma** Spuler, 1924d: 376. Type-species, *Limosina sublugubrina* Malloch (orig. des.; misident.) = *substituta* Richards.
intrudens Malloch, 1922d: 87.—Md.
palliceps Johnson, 1915b: 22.—N.J.; Md.
  sublugubrina Malloch of Spuler (in part), 1923: 376 (also 1925b: 101).
  substituta (in part) Richards, 1961: 563. Md. records only.

Subgenus BROMELOECIA Spuler

Leptocera, subg. **Bromeloecia** Spuler, 1924d: 375. Type-species, *Limosina bromeliarum* Knab and Malloch (orig. des.).
winnemana Malloch, 1925a: 97.—Md.; Minn., Tenn.

Subgenus LIMOSINA Macquart

Limosina Macquart, 1835: 571 (as genus). Type-species, *Borborus silvaticus* Meigen (Westwood, 1840: 145).
Limosina, subg. *Scotophilella* Duda, 1918: 34, 104. Type-species, *Borborus silvaticus* Meigen (Duda, 1924c: 6).
Leptocera, subg. *Spelobia* Spuler, 1924d: 376. Type-species, *Limosina tenebrarum* Aldrich (orig. des.). **N. syn.**
Leptocera, subg. *Americaptilotus* Richards, 1951: 845. Type-species, *Aptilotus borealis* Malloch (orig. des.). **N. syn.**
Aptilotus, Amer. authors, not Mik.
abundans Spuler, 1925c: 74, 151.—Idaho; B.C. to Mont., s. to Calif. and N. Mex., also Pa.
albifrons Spuler, 1925c: 73, 147.—Idaho; Wash., Calif., Ill., Mich.
aldrichi (Williston), 1893d: 259 (*Limosina*).—Calif.
borealis (Malloch), 1913j: 361 (*Aptilotus*).—Alaska; Wash., Oreg., Mont., Wyo.
  politus Williston of Coquillett, 1900h: 464 (1904: 78) (*Aptilotus*).
carinata Spuler, 1925c: 75, 153.—Ill.; S. Dak., Mich., Mass.
cellularis Spuler, 1925c: 79.—Tex.; Mich.
concava Spuler, 1925c: 83.—Calif.; Wash.
crassimana (Haliday), 1836: 328 (*Limosina*).—United Kingdom; Alaska, B.C. to Mont., s. to Calif., also Kans., Ill., Mich., Vt., Europe.

curtipennis Spuler, 1925c: 74, 150.—Wash.
dissimilicosta Spuler, 1925c: 73, 148.—Wash.; Idaho.
elegans Spuler, 1925c: 73, 149.—Bermuda; Oreg., Iowa, Ill., Mich., D.C.
empirica (Hutton), 1901: 94 (*Limosina*).—New Zealand; Sask., Europe, Juan Fernandez I.
gracilipennis Spuler, 1925c: 78.—Wash.; Oreg., Idaho, Ill., Mich., Mass.
heteroneura (Haliday), 1836: 331 (*Limosina*).—United Kingdom; Iowa, Europe.
levifrons Spuler, 1925c: 77.—Idaho; Wash.
levigena Spuler, 1925c: 74, 152.—N.Y.; Wash. to Mont., s. to Calif.
longicosta Spuler, 1925c: 75, 155.—Wash.; B.C. to Calif., Kans. to N.H.
lucifuga Spuler, 1925b: 117.—Oreg.
luctuosa Spuler, 1925c: 76, 157.—Wash.; Idaho.
maculipennis Spuler, 1925c: 80.—Wash.; Alaska, B.C., Calif.
mirabilis (Collin), 1902: 59 (*Limosina*).—England; Wash., Oreg., Mont., Kans. to Mass., also Va.
nasuta Spuler, 1925c: 84.—Wash.; Idaho.
nigrifrons Spuler, 1925c: 76, 158.—Wash.; Oreg., Idaho, Calif.
obfuscata (Tucker), 1907: 103 (*Limosina*).—Colo.
occidentalis (Adams), 1904b: 455 (*Limosina*).—Calif.
ochripes (Meigen), 1830: 209 (*Borborus*).—Germany; Mich., Europe.
opacella Richards, 1961: 564 (n. name for *opaca* Aldrich).—Colo.
    opaca Aldrich, 1932b: 6 (preocc. Duda, 1925).—Colo.
ordinaria Spuler, 1925c: 76, 159.—Idaho; B.C. to Mont., s. to Calif. and N. Mex., also Ill., N.H.
parva (Malloch), 1913j: 371 (*Limosina*).—D.C.; B.C., Ill.
polita (Williston), 1893d: 259 (*Apterina*).—Calif.
rara Spuler, 1925c: 155.—Calif.
schmitzi (Duda), 1918: 111 (*Limosina*).—Europe; Wash., Oreg.
    curtiventris Stenhammar of Spuler, 1925c: 148.
setigera (Adams), 1903b: 223 (*Limosina*).—Kans., N. Mex.
silvatica (Meigen), 1830: 207 (*Borborus*).—Europe; Va.
sordipes (Adams), 1904b: 455 (*Limosina*).—S. Dak.; Wash. to N.Y. and Pa., s. to Kans.
    evanescens Tucker, 1907: 102- (*Limosina*).—S. Dak. **N. syn.**, C. W. Sabrosky *in litt.*
tenebrarum (Aldrich), 1897: 190 (*Limosina*).—Ind.; Ky., Tenn.
    stygia Coquillett *in* Call, 1897: 384 (*Limosina*).—Ky.
varicosta (Malloch), 1914l: 14 (*Limosina*).—Costa Rica; Md., Mexico, P.R.

SPHAEROCERIDAE 725

Subgenus TRACHYOPELLA Duda

Limosina, subg. **Trachyopella** Duda, 1918: 34, 195. Type-species, *melania* Haliday (Spuler, 1925b: 103).
lineafrons Spuler, 1925b: 103.—Wash.; Mich.

Subgenus HALIDAYINA Duda

Limosina, subg. **Halidayina** Duda, 1918: 32. Type-species, *spinipennis* Haliday (mon.).
spinipennis (Haliday), 1836: 331 (*Limosina*).—United Kingdom; B.C. to Calif., also S. Dak., Mich., N.H., N.J., Europe.

Subgenus ELACHISOMA Rondani

Elachisoma Rondani, 1880: 5 (1880: 3) (as genus). Type-species, *Limosina nigerrima* Haliday, 1836 (orig. des.)=*aterrima* (Haliday, 1833). [*Borborus nigerrimus* Haliday, 1833, is a nomen nudum.]
approximata Malloch, 1913q: 135.—Tex.; Ariz., Kans., Tenn.

Subgenus COPROICA Rondani

Heteroptera Macquart, 1835: 570 (preocc. Rafinesque, 1814). Type-species, *Limosina pusilla* Meigen (mon.; misident.)= *acutangula* (Zetterstedt).
Coproica Rondani, 1861a: 10 (as genus; n. name for *Heteroptera* Macquart). Type-species, *Limosina pusilla* Meigen (aut.; misident.)=*acutangula* (Zetterstedt).
Limosina, subg. *Coprophila* Duda, 1918: 45. Type-species, *Borborus vagans* Haliday (Spuler, 1925b: 122).
acutangula (Zetterstedt), 1847: 2499 (*Limosina*).—Sweden; Man. to Que. and Conn., s. to Utah, Tex., and Ga., also Europe, Madeira Is., Congo (Belgian).
cacti Richards, 1960: 203.—Ariz.; Calif., Mexico.
ferruginata (Stenhammar), 1854: 397 (1855: 139) (*Limosina*).— Sweden; B.C., widespread U.S., Bermuda, cosmopolitan.
  illotus Williston, 1896c: 434 (*Borborus*).—St. Vincent I.
hirtula (Rondani), 1880: 40 (1880: 38) (*Limosina*).—Italy; widespread U.S., Bermuda, cosmopolitan.
  exigua Adams, 1904b: 454 (*Limosina;* preocc. Rondani, 1880).—N. Mex.
  exiguella Spuler, 1925b: 123 (n. name for *exigua* Adams).— N. Mex.
mitchelli Malloch, 1913q: 135.—Tex.; Ariz., Costa Rica.
urbana Richards, 1960: 204.—Ill.; Minn.

vagans (Haliday), 1833: 178 (*Borborus*).—Ireland; Oreg., Utah, Wyo., Colo., Va., cosmopolitan.
    albipennis Rondani, 1880: 41 (1880: 39) (*Limosina*).—Italy.

Family BRAULIDAE
By Alan Stone

This family consists of probably only one small, wingless, highly specialized species of Diptera living in the hives of honey bees, usually attached to the bees. The adults apparently eat nectar and pollen at the bee's mouth. The species has been an immigrant to the United States a number of times but has rarely become established in apiaries. Phillips (1925) has discussed *Braula coeca* in considerable detail.

Genus BRAULA Nitzsch

Braula Nitzsch, 1818: 314. Type-species, *coeca* Nitzsch (mon.).

coeca Nitzsch, 1818: 315.—?Europe; Africa, N. and S. Amer.

Family TETHINIDAE
By J. R. Vockeroth

The genera *Tethina*, *Phycomyza*, and *Neopelomyia* are apparently confined to seashores. *Pelomyia* and *Pelomyiella* are coastal or inland; if inland, they are usually, but not invariably, in alkaline areas. The larval habits and larvae are unknown.

    REFERENCE: Melander, 1952a (rev.).

Genus PELOMYIA Williston

Pelomyia Williston, 1893d: 258. Type-species, *occidentalis* Williston (mon.).

coronata (Loew), 1866a: 185 (Cent. 6, no. 98) (*Rhicnoessa*).—Ga.
    This species, as usually identified, is an unworked complex.
cruciata Hendel, 1934b: 52.—Mo.
nubila Melander, 1952a: 195.—Calif.
occidentalis Williston, 1893d: 258.—Calif.

Genus PELOMYIELLA Hendel

Pelomyiella Hendel, 1934b: 39. Type-species, *Pelomyia hungarica* Czerny (orig. des.).

mallochi (Sturtevant), 1923: 7 (*Pelomyia*).—Mass.; s. B.C. to s. Man., s. to s. Calif. and Colo. (inland only), n. Man. and cent. Baffin I. to s. Greenland, s. to N.Y. (coastal only), Europe.

maritima (Melander), 1913a: 297 (*Tethina*).—Tex.

melanderi (Sturtevant), 1923: 7 (*Pelomyia*).—Calif.; cent. B.C. to s. Calif. and Ariz., Mexico (coastal and inland).

Genus NEOPELOMYIA Hendel

Neopelomyia Hendel, 1917: 46. Type-species, *Tethina rostrata* Hendel (orig. des.).

rostrata (Hendel), 1911: 41 (*Tethina*).—B.C. (as Idaho, error), Wash.; s. B.C. to Calif.

Genus PHYCOMYZA Melander

Phycomyza Melander, 1952a: 198. Type-species, *Rhicnoessa milichioides* Melander (orig. des.).

milichioides (Melander), 1913a: 299 (*Rhicnoessa*).—Wash.; Wash. to Calif.

Genus TETHINA Haliday

Opomyza, subg. **Tethina** Haliday, 1838: 188. Type-species, *illota* Haliday (mon.).

Rhicnoessa Loew, 1862d: 174. Type-species, *cinerea* Loew (mon.) = *grisea* (Fallén).

albula (Loew), 1869b: 44 (Cent. 8, no. 80) (*Rhicnoessa*).—R.I.; Mass. to Fla. and Tex.

angustifrons Melander, 1952a: 199.—Calif.

angustipennis (Melander), 1952a: 203 (*Rhicnoessa*).—Calif. **N. comb.**

bermudaensis (Melander), 1952a: 203 (*Rhicnoessa*).—Bermuda. **N. comb.**

denudata (Melander), 1952a: 204 (*Rhicnoessa*).—Calif. **N. comb.**

horripilans (Melander), 1952a: 204 (*Rhicnoessa*).—Wash.; Oreg., Calif. **N. comb.**

lavendula (Melander), 1952a: 205 (*Rhicnoessa*).—Calif. **N. comb.**

parvula (Loew), 1869b: 45 (Cent. 8, no. 81) (*Rhicnoessa*).—R.I.; Que. to Md.

whitmani Melander, 1913a: 298 (*Rhicnoessa*).—Mass.

prognatha (Melander), 1952a: 206 (*Rhicnoessa*).—Calif. **N. comb.**
seriata (Melander), 1952a: 206 (*Rhicnoessa*).—Fla. **N. comb.**
spinulosa Cole, 1923a: 478.—Mexico (Baja Calif.); s. Calif.
texana (Malloch), 1913n: 148 (*Rhicnoessa*).—Tex.
variseta (Melander), 1952a: 209 (*Rhicnoessa*).—Calif. **N. comb.**

Family MILICHIIDAE

(Phyllomyzidae, Carnidae)

By Curtis W. Sabrosky

Flies of the family Milichiidae are small, usually black, and inconspicuous. In some species, the dorsum of the male abdomen is brilliantly silvered, flashing in the sun. The larvae are saprophagous or coprophagous, and the family is often associated with carrion, excrement, or decaying plant material. Some species have been reared from detritus deep in the nests of leaf-cutting ants (*Atta*). Others are reared from birds' nests. *Carnus hemapterus* is a bloodsucking ectoparasite on birds. Some appear to have a commensal relationship with predatory insects, riding on them, and sucking exudates from the victims.

REFERENCES: Melander, 1913a (rev.); Malloch, 1913n (partial rev.).

Subfamily CARNINAE

Genus HEMEROMYIA Coquillett

Hemeromyia Coquillett, 1902e: 190. Type-species, *obscura* Coquillett (orig. des.).
Paramadiza Melander, 1913a: 245. Type-species, *washingtona* Melander (orig. des.).

obscura Coquillett, 1902e: 190.—N. Mex.; Calif., Ariz., Tex., n. Mexico.
washingtona (Melander), 1913a: 246 (*Paramadiza*).—Wash.; Idaho, Calif., Colo., N. Mex.
  nitida Malloch, 1913n: 146.—Colo.
  nitens Malloch *in* Melander, 1913c: 169. Nomen nudum.

Genus MEONEURA Rondani

Meoneura Rondani, 1856: 128. Type-species, *Agromyza obscurella* Fallén (orig. des.).

REFERENCE: Sabrosky, 1959a (rev.).

californica Sabrosky, 1961b: 229.—Calif.
digitata Sabrosky, 1959a: 19.—Nebr.
flavifacies Collin, 1930b: 85.—England; Alaska and N.W.T. to Labr., s. to Oreg., Utah, and Mich., cent. Europe.
  lacteipennis, authors, not Fallén.
forcipata Sabrosky, 1959a: 24.—B.C.
lamellata Collin, 1930b: 83, 86.—England; Alaska, Europe.
nigrifrons Malloch, 1915a: 47.—Ill.; Nebr., Ont.
obscurella (Fallén), 1823a: 6 (*Agromyza*).—Sweden; Minn., Ont., N.Y., Mass., Europe.
  tritici Fitch, 1856a: 535 (1856: 303) (*Agromyza*).—N.Y.
polita Sabrosky, 1959a: 20.—Tex.; Wash. and Idaho, s. to Calif., Mexico, and Tex.
pteropleuralis Sabrosky, 1959a: 19.—Nebr.
seducta Collin, 1937b: 250.—England; Wash., Idaho, Calif., Tex., Mich., Ont., Md.
triangularis Collin, 1930b: 88.—England; Alaska to Man., s. to Oreg. and Colo., cent. Europe.
vagans (Fallén), 1823a: 5 (*Agromyza*).—Sweden; Mich. *Vagans* of Amer. authors equals several spp.
wirthi Sabrosky, 1959a: 25.—Calif.; Wash.

Genus CARNUS Nitzsch

Carnus Nitzsch, 1818: 284, 305. Type-species, *hemapterus* Nitzsch (mon.).

  REFERENCE: Bequaert, 1942b (biol., figs.).

hemapterus Nitzsch, 1818: 306.—Germany; Calif., Ariz., N.Y., N.B., Fla., Mexico (Baja Calif.), Europe.

Subfamily MADIZINAE
Genus PHYLLOMYZA Fallén

Phyllomyza Fallén, 1810a: 20. Type-species, *securicornis* Fallén (sub. mon., Fallén, 1823b: 8).

hirtipalpis Malloch, 1913n: 137.—Md.; Tenn., Va., Tex.
milnei Steyskal, 1942a: 84.—Va.; Ill., Md., N.C.
securicornis Fallén, 1823b: 8.—Sweden; Wis. to Que. and Maine, s. to N.C., also B.C. and Alta.

Genus NEOPHYLLOMYZA Melander

Neophyllomyza Melander, 1913a: 243. Type-species, *quadricornis* Melander (orig. des.).

nitens Melander, 1913a: 244.—Idaho.
quadricornis Melander, 1913a: 243.—Wash., Idaho, N. Mex., La.; Mich. to N.H., s. to N. Mex. and Fla., also Alaska, Wash., Idaho.
  approximata Malloch, 1913n: 138 (*Phyllomyza*).—D.C.

Genus PARAMYIA Williston

Paramyia Williston, 1897: 1. Type-species, *nigra* Williston (mon.). ?=*nitens* (Loew).

nitens (Loew), 1869b: 45 (Cent. 8, no. 82) (*Phyllomyza*).—Pa.; S. Dak. to Que., s. to Fla., also B.C., Alta., Ariz., ?to Chile and s. Brazil.

Genus STOMOSIS Melander

Stomosis Melander, 1913a: 242. Type-species, *Desmometopa luteola* Coquillett (orig. des.)=*innominata* (Williston).
  REFERENCE: Sabrosky, 1958 (key).

flava Sabrosky, 1958: 172.—Ind.; Mich. to Va., s. to Ala. and Fla.
innominata (Williston), 1896c: 443 (*Agromyza*).—St. Vincent I.; Ariz., Honduras, Costa Rica, Panama.
  luteola Coquillett, 1902e: 188 (*Desmometopa*).—Ariz.

Genus ALDRICHIOMYZA Hendel

Aldrichiella Hendel, 1911: 35 (preocc. Vaughan, 1903). Type-species, *agromyzina* Hendel (orig. des.).
Aldrichiomyza Hendel, 1914d: 73 (n. name for *Aldrichiella* Hendel). Type-species, *Aldrichiella agromyzina* Hendel (aut.).

agromyzina (Hendel), 1911: 37 (*Aldrichiella*).—S. Dak.; S. Dak. to Ont., s. to Kans. and N.C.

Genus MADIZA Fallén

Madiza Fallén, 1810a: 19. Type-species, *glabra* Fallén (Hendel; 1903: 251).
Desmomyza Curran, 1934a: 338. Type-species, *confusa* Curran (orig. des.)=*glabra* Fallén.

  The earliest type-designation for *Madiza* is actually *oscinina* Fallén by Rondani (1856: 128). However, this species had correctly been referred to Chloropidae by earlier revisers. Later workers in general have followed those revisers, adopted *Siphonella* Macquart, 1835, for *oscinina,* and used *Madiza* in Milichiidae. This usage is adopted here in the interests of stability and universality, but suspension of the rules is needed.

glabra Fallén, 1820a: 9.—Sweden; B.C. to N.S., s. to Nev. and Va., Holarctic.
    confusa Curran, 1934a: 338 (*Desmomyza*).—N.Y.

Genus LEPTOMETOPA Becker

Leptometopa Becker, 1903: 188. Type-species, *rufifrons* Becker (mon.).
Hypaspistomyia Hendel, 1907b: 240. Type-species, *coquilletti* Hendel (mon.).
Paramadiza Malloch *in* Melander, 1913c: 169 (preocc. Melander, 1913). Type-species, *Desmometopa halteralis* Coquillett (mon.). **N. syn.**
Mallochiella Melander, 1913c: 169 (n. name for *Paramadiza* Malloch). Type-species, *Desmometopa halteralis* Coquillett (aut.). **N. syn.**
Desmometopina Curran, 1930j: 81. Type-species, *Agromyza latipes* Meigen (orig. des.).

halteralis (Coquillett), 1900g: 267 (*Desmometopa*).—P.R.; B.C. to N.B., s. to Mexico and Fla., also Bahama Is., P.R. **N. comb.**
latipes (Meigen), 1830: 177 (*Agromyza*).—Germany; Alaska to N.S., U.S., Cuba.
    orillia Curran, 1927e: 50 (*Mallochiella*).—Ont.

Genus DESMOMETOPA Loew

Desmometopa Loew, 1866a: 184 (Cent. 6, no. 96). Type-species, *Agromyza m-atrum* Meigen (Hendel, 1903: 251)= *sordida* (Fallén).

m-nigrum (Zetterstedt), 1848: 2743 (*Agromyza*).—Sweden; Iowa and Mich. to N.H. and Md., thence coastal States to Fla. and Tex., also Ariz. and Calif.
sordida (Fallén), 1820a: 10 (*Madiza*).—Sweden; B.C. to N.S., s. to Oreg., Utah, Tex., and Ga.
tarsalis Loew, 1866a: 184 (Cent. 6, no. 96).—Cuba; Nebr. and Mich. to D.C., s. to Calif., Panama, and Fla., West Indies.

Subfamily MILICHIINAE

Genus PHOLEOMYIA Bilimek

Pholeomyia Bilimek, 1867: 903. Type-species, *leucozona* Bilimek (mon.).
Rhynchomilichia Hendel, 1903: 250. Type-species, *Lobioptera argyrophenga* Schiner (orig. des.; misident.)=*schineri* Hendel.

REFERENCES: Sabrosky, 1959b (rev.), 1961b (partial key).

comans Sabrosky, 1959b: 330.—La.
decorior Steyskal, 1943a: 100.—N.C.; N.J. to Fla.
dispar (Becker), 1907a: 524 (*Rhynchomilichia*; as *leucogastra* var.).— Ga.; Tex., Ala., Fla., Va.
expansa Aldrich, 1925c: 1.—Calif.
indecora (Loew), 1869b: 50 (Cent. 8, no. 94) (*Lobioptera*).—Nebr.; B.C. to N.S., s. to Calif., Kans., and Ga.
leucogastra (Loew).—Neotropical.
myopa Melander, 1913a: 238.—Haiti; Fla., West Indies, Mexico, Cent. Amer.
nitidula Sabrosky, 1959b: 327.—Ga.
obscura Sabrosky, 1959b: 330.—Tex.
pseudodecora (Becker), 1907a: 524 (*Rhynchomilichia*).—Ga.; Md., S.C., Fla.
robertsoni (Coquillett), 1902e: 187 (*Milichia*).—Fla.
texensis Sabrosky, 1959b: 329.—Tex.
vockerothi Sabrosky, 1961b: 231.—N.C.

Genus EUSIPHONA Coquillett

Eusiphona Coquillett, 1897: 49. Type-species, *mira* Coquillett (orig. des.).

cooperi Sabrosky, 1955a: 170.—N.Y.; Ohio, Md.
flava Sabrosky, 1953b: 38.—Utah.
mira Coquillett, 1897: 49.—Colo.; Wash. and Alta. to P.E.I., s. to Calif., Mexico (V.C.), and Fla.

Genus MILICHIA Meigen

Milichia Meigen, 1830: 131. Type-species, *speciosa* Meigen (Westwood, 1840: 151).
Lobioptera Wahlberg, 1847: 259. Type-species, *ludens* Wahlberg (mon.).

aethiops Malloch, 1913n: 133.—Tex.; Mexico.

Genus MILICHIELLA Giglio-Tos

Milichiella Giglio-Tos, 1895b: 367. Type-species, *Tephritis argentea* Fabricius (mon.; misident.)=*tosi* Becker.
Ophthalmomyia Williston, 1896c: 426. Type-species, *Lobioptera lacteipennis* Loew (mon.).

arcuata (Loew), 1876: 339 (*Lobioptera*).—N.Y.; Mich. to Que. and Maine, s. to Kans. and Fla., also Ariz.
bisignata Melander, 1913a: 239.—N.J.; Mich. to N.J., s. to Kans. and N.C., also N. Mex., Mexico.
lacteipennis (Loew), 1866a: 185 (Cent. 6, no. 97) (*Lobioptera*).—Cuba; Calif. to Ont. and N.H., s. to Chile and Argentina, also Bermuda, West Indies, Canary and Madeira Is., Africa, Oriental and Australian Regions.
  nigrella Cole, 1912b: 162.—Calif. **N. syn.**
nitida Hendel, 1911: 39.—Calif.
urbana Malloch, 1913c: 284.—D.C.; Mich., Pa., Del., Tenn., Tex.
  populi Steyskal, 1951a: 130.—Mich. **N. syn.**
  cinerea, authors, not Coquillett.

Family CANACEIDAE
By Willis W. Wirth

The "beach flies" are closely similar in appearance and habits to the Ephydridae, although their true relationship is closer to the Milichiidae and Sphaeroceridae. In habit, canaceids are almost all intertidal. The larvae feed on algae on wave-splashed rocks or sandy beaches.

REFERENCES: Williams, 1939 (biol.); Wirth, 1951a (rev.).

Genus CANACE Haliday

Ephydra, subg. **Canace** Haliday, 1839: 411. Type-species, *nasica* Haliday (mon.).
Canacea, error.

aldrichi Cresson, 1936: 264.—Calif.
snodgrassii Coquillett, 1901i: 378.—Galapagos Is.; Atlantic Coast, P.E.I. to Fla., Panama.
  macateei Malloch, 1924i: 52.—Ga.

Genus CANACEOIDES Cresson

Canaceoides Cresson, 1934a: 221. Type-species, *Canace nudata* Cresson (orig. des.).

nudatus (Cresson), 1926: 257 (*Canace*).—Calif.; Hawaii.

Genus NOCTICANACE Malloch

Nocticanace Malloch, 1933b: 4. Type-species, *peculiaris* Malloch (orig. des.).

arnaudi Wirth, 1954a: 59.—Calif.
texensis (Wheeler), 1952a: 92 (*Canaceoides*).—Tex.; Fla., P.R.

Superfamily DROSOPHILOIDEA

Family EPHYDRIDAE

By Willis W. Wirth

Ephydrids, or "shore flies," are mainly beneficial, providing important food items for wildlife, especially waterfowl inhabiting the marshes where these flies so commonly breed. The larvae are aquatic or semiaquatic and the adults are usually found close by on the surface of the mud or water, or on low, emergent vegetation. The "brine flies" become so abundant in the salt and alkaline lakes of our Western States that in times past the Indians often took advantage of their numbers to gather puparia from the windrows where they were stranded on the beaches and prepare them for use as food. In a few areas the leaf mines of *Hydrellia* may become extensive enough to damage crops of watercress, or rice, barley, and other cereals.

REFERENCES: Cresson, 1942a (Psilopinae), 1944b (Notiphilinae), 1946 (Notiphilini), 1949 (Parydrinae) (revs. N. Amer. spp.); Sturtevant and Wheeler, 1954 (key to genera, rev. N. Amer. genera not treated by Cresson); Wirth and Stone, 1956 (key to genera, review biol., immature stages, keys to Calif. spp.); Dahl, 1959 (comprehensive study ecology, etc., in Scandinavia).

Subfamily PSILOPINAE

Tribe GYMNOPINI

Genus MOSILLUS Latreille

Mosillus Latreille, 1804: 196. Type-species, *arcuatus* Latreille (sub. mon., Latreille, 1805: 390) = *subsultans* (Fabricius).
Gymnopa Fallén, 1820a: 10. Type-species, *aenea* Fallén (mon.) = *subsultans* (Fabricius).

bidentatus (Cresson), 1926: 249 (*Gymnopa*).—Utah; B.C. to Man., s. to Calif. and N. Mex.
tibialis Cresson, 1916b: 149.—N.J.; B.C. to Que., s. to Calif. and Fla., West Indies, s. to Ecuador.

Genus ATHYROGLOSSA Loew

Athyroglossa Loew, 1860a: 12. Type-species, *Notiphila glabra* Meigen (mon.).

REFERENCES: Hendel, 1931b (subg. discussion); Sturtevant and Wheeler, 1954 (key).

Subgenus ATHYROGLOSSA Loew

glabra (Meigen), 1830: 69 (*Notiphila*).—Europe; Y.T. to Calif., e. to Mich., N.H., and Que.
glaphyropus Loew, 1878: 197.—Tex.; Ill. to Va., s. to Tex. and Fla., Neotropical.
transversa Sturtevant and Wheeler, 1954: 246.—N. Mex.; Calif., Ariz., Utah.

Subgenus PARATHYROGLOSSA Hendel

Parathyroglossa Hendel, 1931b: 68 (as genus). Type-species, *Athyroglossa ordinata* Becker (orig. des.).

ordinata Becker, 1896: 135.—Romania; B.C. to Que., s. to Ariz. and Md., Europe.

Subgenus OCHTHEROIDEA Williston

Ochtheroidea Williston, 1896c: 401 (as genus). Type-species, *atra* Williston (mon.).

granulosa (Cresson), 1922a: 341 (*Ochtheroidea*).—Pa.; Mich. to Que., s. to Ind. and N.C.
laevis (Cresson), 1918a: 61 (*Ochtheroidea*).—Costa Rica; Ariz., Neotropical.
melanderi (Cresson), 1922a: 339 (*Ochtheroidea*).—Calif.; Ariz., N. Mex.

Tribe ATISSINI

Genus ATISSA Haliday

Atissa Haliday *in* Curtis, 1837b: 281. Type-species, *Ephydra pygmaea* Haliday (mon.).
Parephydra Coquillett, 1902e: 183. Type-species, *humilis* Coquillett (orig. des.) = *pygmaea* (Haliday).

litoralis (Cole), 1912b: 160 (*Allotrichoma*).—Calif.; Wash. to Man. and Calif., s. to Cent. and S. Amer.
pygmaea (Haliday), 1833: 174 (*Ephydra*).—Ireland; Alaska to Wyo., s. to Calif. and Fla., Cent. and S. Amer., Europe.
  humilis Coquillett, 1902e: 183 (*Parephydra*).—Ariz.
  atlantica Cresson, 1926: 253.—N.J.

Genus PTILOMYIA Coquillett

Ptilomyia Coquillett, 1900g: 261. Type-species, *enigma* Coquillett (orig. des.).
Atissiella Cresson, 1918a: 55. Type-species, *setulosa* Cresson (orig. des.).
  REFERENCE: Sturtevant and Wheeler, 1954 (rev.).

alkalinella (Cresson), 1942a: 112 (*Atissiella*).—Nev.; Wash., Calif., Ariz.
enigma Coquillett, 1900g: 262.—P.R.; Calif. to Va. and Fla., Neotropical.
  falsisetae Cresson, 1942a: 110 (*Atissiella*).—Pa.
lobiochaeta Sturtevant and Wheeler, 1954: 254.—Tex.; Calif., Ariz., El Salvador.
mabelae (Cresson), 1926: 253 (*Atissiella*).—Pa.; Tex. to Pa. and Fla., s. to Ecuador.
occidentalis Sturtevant and Wheeler, 1954: 255.—Calif.; N. Mex.
pleuriseta (Cresson), 1942a: 111 (*Atissiella*).—N. Mex.; Calif. to Tex., Baja Calif.

Genus ALLOTRICHOMA Becker

Allotrichoma Becker, 1896: 121. Type-species, *Hecamede lateralis* Loew (orig. des.).

The species are poorly understood and distribution records need revision after a study of male genitalia.

atrilabre Cresson, 1926: 252.—Pa.; Mont. to Mich. and Md., s. to Tex. and Ala.
lacteum Cresson, 1926: 252.—Ariz.; Calif., N. Mex.
lasiocercum Cresson, 1926: 251.—Calif.; Nev., Utah.
laterale (Loew), 1860a: 13 (*Hecamede*).—Europe; Oreg. to Sask., s. to Calif. and Tex.
simplex (Loew), 1861b: 354 (Cent. 1, no. 92) (*Discocerina*).—Md.; general in U.S.
trispinum Becker, 1896: 124.—Germany; general in U.S. and Canada, Europe.
yosemite Cresson, 1926: 252.—Calif.; B.C. to Mont., s. to Calif. and Ariz.

Genus PSEUDOHECAMEDE Hendel

Pseudohecamede Hendel, 1936b: 104. Type-species, *nasalis* Hendel (orig. des.) = *salubris* (Cresson).

abdominalis (Williston), 1896c: 398 (*Hecamede*).—St. Vincent I.; Atlantic and Gulf States, Calif., Neotropical.
facialis Hendel, 1936b: 105.—Brazil; Fla., Neotropical.
slossonae (Cresson), 1942a: 108 (*Allotrichoma*).—Fla. **N. comb.**

Genus HECAMEDE Haliday

Notiphila, subg. **Hecamede** Haliday *in* Curtis, 1837b: 281. Type-species, *albicans* Meigen (mon.).
albicans (Meigen), 1830: 65 (*Notiphila*).—Europe; coastal, Mass. to Md.

Genus GLENANTHE Haliday

Hydrellia, subg. **Glenanthe** Haliday, 1839: 404. Type-species, *ripicola* Haliday (mon.).
fascipennis Sturtevant and Wheeler, 1954: 250.—Tex.
litorea Cresson, 1925: 166.—N.J.; coastal, Alaska to Calif., N.B. to Fla. and Tex., West Indies, Cent. Amer.

Genus PELIGNUS Cresson

Pelignus Cresson, 1926: 254. Type-species, *Atissa durrenbergensis* Loew (orig. des.).
salinus Cresson, 1942a: 109.—Utah.

Genus DIPHUIA Cresson

Diphuia Cresson, 1944a: 3. Type-species, *anomala* Cresson (orig. des.).
   REFERENCE: Wirth, 1956d (tax.).
nitida Sturtevant and Wheeler, 1954: 248.—N.Y.; Fla.

Genus PELIGNELLUS Sturtevant and Wheeler

Pelignellus Sturtevant and Wheeler, 1954: 252. Type-species, *subnudus* Sturtevant and Wheeler (orig. des.).
subnudus Sturtevant and Wheeler, 1954: 252.—Calif.

Tribe DISCOCERININI

Genus DISCOCERINA Macquart

Discocerina Macquart, 1835: 527. Type-species, *Notiphila pusilla* Meigen (Coquillett, 1910b: 534)=*obscurella* (Fallén).

Coquillett's synonymy of *obscurella* with *pusilla* is only implied, i.e., Macquart's first species.

Psilopa, sectio *Clasiopa* Stenhammar, 1844: 159. Type-species, *Notiphila obscurella* Fallén (orig. des.).

Subgenus DISCOCERINA Macquart

flavipes Cresson, 1941a: 35.—Calif.; Ariz., Neotropical.
obscura Williston, 1896c: 397.—St. Vincent I.; N.Y. to Fla., Tex., La., Neotropical.
obscurella (Fallén), 1813: 251 (*Notiphila*).—Sweden; entire U.S. and Canada, Neotropical, Europe.
    parva Loew, 1862c: 146.—D.C.
    nigriventris Cresson, 1916b: 148 (as *parva* var.).—Calif.
trochanterata Cresson, 1940: 3.—Calif.

Subgenus BASILA Cresson

Basila Cresson, 1942a: 116 (as genus). Type-species, *Ditrichophora nadineae* Cresson (orig. des.).

nadineae (Cresson), 1925: 166 (*Ditrichophora*).—Calif.; Oreg., Mexico.

Subgenus LAMPROCLASIOPA Hendel

Discocerina, subg. **Lamproclasiopa** Hendel, 1933b: 79. Type-species, *facialis* Hendel (orig. des.).

brunneonitens Cresson, 1940: 4.—Miss.; Ill., Md., La., S.C., Fla.
turgidula Cresson, 1940: 3.—Calif.; Nev.

Genus HECAMEDOIDES Hendel

Hecamedoides Hendel, 1917: 41. Type-species, *Psilopa glaucella* Stenhammar (orig. des.).

glaucellus (Stenhammar), 1844: 253 (*Psilopa*).—Sweden; Wash. to Ont., s. to Calif. and N.C., Europe.

Genus HYDROCHASMA Hendel

Hydrochasma Hendel, 1936b: 101. Type-species, *zernyi* Hendel (mon.).

buccatum (Cresson), 1930b: 78 (*Hecamedoides*).—N.J.; Alta. to Que., s. to N. Mex. and Fla.
capax Cresson, 1938c: 26.—Guatemala; Calif., Ariz., and Tex., s. to S. Amer.
incisum (Coquillett), 1902e: 182 (*Discocerina*).—P.R.; Fla. to S. Amer.

leucoproctum (Loew), 1861b: 355 (Cent. 1, no. 93) (*Discocerina*).—
Md.; Minn. to Maine, s. to Tex. and Fla.

Genus DICLASIOPA Hendel

Diclasiopa Hendel, 1917: 42. Type-species, *Hecamede xanthocera* Loew (orig. des.)=*lacteipennis* (Loew).

lacteipennis (Loew), 1862c: 145 (*Discocerina*).—D.C.; Wash. to Ont., s. to Ariz., Tex., and Md.
    xanthocera Loew, 1869d: 58 (*Hecamede*).—Germany.

Genus DITRICHOPHORA Cresson

Ditrichophora Cresson, 1924b: 159. Type-species, *exigua* Cresson (orig. des.).
Gymnoclasiopa Hendel, 1930a: 136. Type-species, *Notiphila plumosa* Fallén (orig. des.).

argyrostoma (Cresson), 1916b: 149 (*Discocerina*).—Calif.; Wash. to Sask., s. to Calif. and N. Mex.
    aliena Cresson, 1922b: 137 (*Discocerina*).—Calif.
atrata Cresson, 1940: 7.—Calif.; Wash. to Man. and Que., s. to Calif. and N.Y.
cana Cresson, 1940: 7.—Wash.; Wash. to Calif.
canifrons Cresson, 1926: 250.—Pa.; Que., N.Y., Tenn.
exigua Cresson, 1924b: 159.—Pa.; Mich. to Que. and Fla.
lenis Cresson, 1940: 5.—Wash.; Calif., Tex.
lugubris Cresson, 1940: 7.—Oreg.; Calif.
montana Cresson, 1942a: 120.—Mont.; Alaska to Que., s. to Utah.
occidentalis Cresson, 1942a: 118.—Wash.; Alaska to Calif. and Idaho.
parilis Cresson, 1924b: 160.—Maine; Y.T., N.W.T., Que., New England States.
pulchella (Meigen), 1830: 70 (*Notiphila*).—Europe; Alta. to Ont., s. to Tex.
simiaceps Cresson, 1940: 6.—Wash.; Ind.
subnubila Cresson, 1940: 6.—Wash.; Alaska to Wash., se. to Colo. and Que.
tacoma Cresson, 1924b: 160.—Wash.; Alaska to Wash., e. to Que. and Maine.
valens Cresson, 1942a: 118.—Wash.; Y.T., Idaho, Calif., Tex.

Genus POLYTRICHOPHORA Cresson

Polytrichophora Cresson, 1924b: 161. Type-species, *agens* Cresson (orig. des.).

agens Cresson, 1924b: 161.—Tex.; coastal, Maine to Tex., Bahamas.

conciliata Cresson, 1924b: 161.—N.J.; Mich. to Que., s. to Calif. and Fla.

orbitalis (Loew), 1861b: 354 (Cent. 1, no. 91) (*Discocerina*).—D.C.; Wash. to Que., s. to Calif. and Fla.

setigera (Cresson), 1916b: 148 (*Discocerina*).—Calif.; Wash. to Idaho, s. to Calif. and Tex.

Genus PARATISSA Coquillett

Paratissa Coquillett, 1900a: 36. Type-species, *Drosophila pollinosa* Williston (orig. des.)=*semilutea* (Loew).

semilutea (Loew), 1869b: 51 (Cent. 8, no. 97) (*Cacoxenus*).—Cuba; Fla., Bermuda, West Indies, Panama.

   pollinosa Williston, 1896c: 414 (*Drosophila*).—St. Vincent I.

Tribe DISCOMYZINI

Genus DISCOMYZA Meigen

Discomyza Meigen, 1830: 76. Type-species, *Psilopa incurva* Fallén (mon.).

   REFERENCE: Cresson, 1939 (rev.).

maculipennis (Wiedemann), 1824: 57 (*Notiphila*).—East Indies; Calif., Neotropical, Oriental, Australasian.

u-signata Cresson, 1926: 250.—Tex.

Genus CLANONEURUM Becker

Clanoneurum Becker, 1903: 165. Type-species, *Discomyza cimiciformis* Haliday (Becker, 1926: 24).

   Miners in beets, Chenopodiaceae.

americanum Cresson, 1940: 1.—Calif.; w. N. Amer. and Atlantic and Gulf States in saline habitats.

Tribe PSILOPINI

Genus PSILOPA Fallén

Psilopa Fallén, 1823d: 6. Type-species, *Notiphila nitidula* Fallén (Rondani, 1856: 132).

compta (Meigen), 1830: 68 (*Notiphila*).—Europe; Alaska to Que., s. to Calif. and Va.

   gaudens Cresson *in* Strickland, 1946: 167. Nomen nudum.

dupla Cresson, 1940: 2.—N.J.; Ill. to Md., s. to Tex., Fla., and Bahamas, Ariz.
flavida Coquillett, 1900a: 33.—Mass.; coastal, Maine to Tex.
leucostoma (Meigen), 1830: 68 (*Notiphila*).—Europe; B.C. to Ont., s. to Calif. and Fla.
olga Cresson, 1922b: 137.—Wash.; Alaska to Que., s. to Calif., Colo., and Tex.
    dimidiata Cresson, 1922b: 137.—Idaho.
pulchripes Loew, 1878: 197.—Tex.; s. U.S., Neotropical.

Genus HELAEOMYIA Cresson

Helaeomyia Cresson, 1941a: 35. Type-species, *Psilopa petrolei* Coquillett (orig. des.).

REFERENCES: Thorpe, 1930, 1931b (biol. of *petrolei*, "petroleum fly," breeding in pools of petroleum).

nigra (Williston), 1896c: 393 (*Psilopa*).—St. Vincent I.; Gulf States, Neotropical.
petrolei (Coquillett), 1899a: 8 (*Psilopa*).—Calif.; West Indies.

Genus LEPTOPSILOPA Cresson

Leptopsilopa Cresson, 1922b: 136. Type-species, *Psilopa similis* Coquillett (orig. des.).

atrimana (Loew), 1878: 197 (*Psilopa*).—D.C., Tex.; Man. to Que., s. to Ariz., Fla., Mexico, and Guatemala.
nigrimana (Williston), 1896c: 393 (*Psilopa*).—St. Vincent I.; Tex., Fla., Neotropical.
    willistoni Cresson, 1918a: 53 (*Psilopa*; unjustified n. name for *nigrimana* Williston).—St. Vincent I.
similis (Coquillett), 1900a: 33 (*Psilopa*).—Fla.; Tex. to Fla., Neotropical.
varipes (Coquillett), 1900a: 33 (*Psilopa*).—B.C.; B.C. to Idaho, s. to Calif.
    californica Cresson, 1941a: 35 (*Helaeomyia*).—Calif.

Genus CEROPSILOPA Cresson

Ceropsilopa Cresson, 1917b: 340. Type-species, *nasuta* Cresson (orig. des.).
Batula Cresson, 1940: 2. Type-species, *Psilopa mellipes* Coquillett (orig. des.).

adjuncta Cresson, 1925: 165.—P.R.; Fla.

coquilletti Cresson, 1922b: 136.—Calif.; Calif., e. to Md. and Fla.,
Neotropical.
costalis Wirth, 1956d: 12.—Va.; Fla., Bahamas.
dispar Cresson, 1922b: 135.—Calif.
mellipes (Coquillett), 1900g: 260 (*Psilopa*).—P.R.; Fla., Tex.,
Neotropical.
nasuta Cresson, 1917b: 341.—Trinidad; Fla., Neotropical.
staffordi Cresson, 1925: 166.—La.; coastal, Calif., Tex. to Ga.,
Bahamas.

Genus CRESSONOMYIA Arnaud

Plagiops Cresson, 1918a: 53 (preocc. Townsend, 1911). Type-species,
nitidifrons Cresson (orig. des.).
Plagiopsis Cresson, 1934a: 201 (n. name for *Plagiops* Cresson;
preocc. Brauer and Bergenstamm, 1890). Type-species,
Plagiops nitidifrons Cresson (aut.).
Cressonomyia Arnaud, 1958a: 24 (n. name for *Plagiopsis* Cresson).
Type-species, *Plagiops nitidifrons* Cresson (aut.).

aciculata (Loew), 1862c: 142 (*Psilopa*).—Cuba; Tex., La., Ala., Ga.,
Neotropical. **N. comb.**
aeneonigra (Loew), 1878: 196 (*Psilopa*).—Tex.; coastal, Mass. to
Ga., La.
fulvipennis Hine, 1904c: 64 (*Psilopa*).—La. **N. syn.**
hinei (Cresson), 1922b: 135 (*Plagiops*).—Guatemala; coastal, N.Y.
to Tex., Neotropical. **N. comb.**
skinneri (Cresson), 1922b: 136 (*Psilopa*).—Cuba; Fla. **N. comb.**

Genus CLASIOPELLA Hendel

Clasiopella Hendel, 1914g: 109. Type-species, *uncinata* Hendel (orig.
des.).

uncinata Hendel, 1914g: 110.—Formosa; Fla., Australasian, Ethiopian.

Genus TRIMERINA Macquart

Trimerina Macquart, 1835: 528. Type-species, *Psilopa madizans*
Fallén (Westwood, 1840: 153).

madizans (Fallén), 1813: 252 (*Notiphila*).—Sweden; Sask. to Ont.,
s. to Calif., Colo., and N.Y., Europe.

Genus TRIMERINOIDES Cresson

Trimerinoides Cresson, 1925: 165. Type-species, *Trimerina adfinis* Cresson (orig. des.).

adfinis (Cresson), 1922b: 137 (*Trimerina*).—B.C.; Idaho.

Genus RHYSOPHORA Cresson

Rhysophora Cresson, 1924b: 159. Type-species, *robusta* Cresson (mon.).

robusta Cresson, 1924b: 159.—Va.; Mich. to Que., s. to Fla.
    magna Coquillett *in* Johnson, 1910b: 806 (*Discocerina*). Nomen nudum, **N. syn.**

Subfamily NOTIPHILINAE

Tribe TYPOPSILOPINI

Genus TYPOPSILOPA Cresson

Typopsilopa Cresson, 1916b: 147. Type-species, *flavitarsis* Cresson (orig. des.).

atra (Loew), 1862c: 143 (*Psilopa*).—Middle States; Oreg. to Ont., s. to Calif., Fla., and Cent. Amer.
    scoriacea Loew, 1862c: 142 (*Psilopa*).—N.Y. **N. syn.**
flavitarsis Cresson, 1916b: 147.—Ariz.; s. U.S., Neotropical.

Tribe HYDRELLIINI

Genus HYDRELLIA Robineau-Desvoidy

Hydrellia Robineau-Desvoidy, 1830: 790. Type-species, *aurifacies* Robineau-Desvoidy (Westwood, 1840: 153) = *flaviceps* (Meigen).

Leaf miners in aquatic and semiaquatic plants. The species are poorly known and need revision based on biology and characters of male genitalia.

    REFERENCES: Cresson, 1944b (rev.); Berg, 1950a (biol., spp. in *Potamogeton*); Grigarick, 1959 (biol. *griseola*).

advenae Cresson, 1934b: 236.—Maine.
americana Cresson, 1931b: 106.—Md.; Calif., N.Y., Maine.
ascita Cresson, 1942b: 78.—Mich.

atroglauca Coquillett, 1910a: 131.—Fla.
bergi Cresson, 1941a: 37.—Mich.
bilobifera Cresson, 1936: 262.—Mo.
borealis Cresson, 1944b: 164.—Ont.; N.H.
caliginosa Cresson, 1936: 257.—Maine; Alaska, Mich.
conformis Loew, 1869b: 41 (Cent. 8, no. 73).—R.I.; Mass.
crassipes Cresson, 1931b: 107.—Ohio.
cruralis Coquillett, 1910a: 131.—N.J.; n. U.S.
decens Cresson, 1931b: 107.—Md.
definita Cresson, 1944b: 165.—Utah; S. Dak., Mich.
flavicoxalis Cresson, 1944b: 167.—Colo.
formosa Loew, 1861b: 355 (Cent. 1, no. 94).—Pa.; Ont., e. U.S.
griseola (Fallén), 1813: 250 (*Notiphila*).—Sweden; widespread in U.S. and Canada, Europe. Common leaf-mining pest of irrigated cereals.
    hypoleuca Loew, 1862c: 151.—Middle States.
    scapularis Loew, 1862c: 153.—U.S.
harti Cresson, 1936: 262.—Ill.
ischiaca Loew, 1862c: 150.—Middle States; N. Amer. from Alaska to N.H., s. to Ind. and Va.
lata Cresson, 1944b: 165.—Wash.
luctuosa Cresson, 1942b: 78.—Mich.; Minn.
morrisoni Cresson, 1924b: 162.—N.H.; N.Y.
nobilis (Loew), 1862b: 229 (Cent. 2, no. 92) (*Psilopa*).—D.C.; e. U.S.
notiphiloides Cresson, 1924b: 162.—Ohio; Mass.
obscuriceps Loew, 1862c: 152.—Middle States; Ind. to N.H. and Va.
penicilli Cresson, 1944b: 168.—Ont.; Wash.
platygastra Cresson, 1931b: 105.—Oreg.; Wash. to Calif.
proclinata Cresson, 1915a: 69.—Calif.; Wash. to Idaho, s. to Calif. and N. Mex.
procteri Cresson, 1934b: 235 (as *proctori*).—Maine; N.J.
prudens Curran, 1930j: 78.—N.Y.; Idaho, Maine, e. Canada.
    johnsoni Cresson, 1941a: 37.—Maine.
pulla Cresson, 1931b: 108.—N.Y.; n. U.S.
serena Cresson, 1931b: 104.—Wash.; Alaska to Oreg. and Colo.
subnitens Cresson, 1931b: 106.—Wash.; Oreg.
suspecta Cresson, 1936: 258.—Maine.
tibialis Cresson, 1917b: 341.—Idaho; Wash. to N.Y. and Que., s. to Calif., Fla., Mexico, and Bolivia.
trichaeta Cresson, 1944a: 7.—Conn.; Ill., Mich., Mass., R.I.
valida Loew, 1862c: 153.—Middle States; ne. U.S.
wilburi Cresson, 1944b: 168.—Colo.; Calif.

Tribe PHILYGRIINI

Genus PHILYGRIA Stenhammar

Notiphila, sectio **Philygria** Stenhammar, 1844: 154. Type-species, *flavipes* Fallén (Coquillett, 1910b: 588).
Hydrina, authors (in part), not Robineau-Desvoidy.

REFERENCE: Sturtevant and Wheeler, 1954 (rev., as *Hydrina*).

debilis Loew, 1861b: 357 (Cent. 1, no. 96).—Pa.; B.C. to Que., s. to Calif. and Fla.
fuscicornis Loew, 1862c: 155.—Middle States.
dimidiata (Sturtevant and Wheeler), 1954: 238 (*Hydrina*).—W. Va.; Va., N.C. **N. comb.**
nigrescens (Cresson), 1930b: 80 (*Hydrina*).—B.C.; Alaska to Que., s. to Calif. and Colo.
opposita Loew, 1861b: 356 (Cent. 1, no. 95).—Pa.; B.C. to Que., s. to Calif., Ill., and Md.

Genus NOSTIMA Coquillett

Nostima Coquillett, 1900a: 35. Type-species, *slossonae* Coquillett (orig. des.).
Philygriola Hendel, 1917: 42. Type-species, *Notiphila picta* Fallén (orig. des.).

REFERENCES: Cresson, 1941b; Sturtevant and Wheeler, 1954 (revs.).

approximata Sturtevant and Wheeler, 1954: 240.—Okla.
gilvipes (Coquillett), 1900g: 261 (*Hydrellia*).—P.R.; Fla., Neotropical.
niveivenosa Cresson, 1930b: 80.—P.R.; Fla., Neotropical.
picta (Fallén), 1813: 254 (*Notiphila*).—Sweden; B.C. to Ont., s. to Calif., Europe.
pulchra (Williston), 1896c: 399 (*Hydrellia*).—St. Vincent I.; Fla., Neotropical.
quinquenotata Cresson, 1930b: 79.—Md.; Okla. and Tex. to Md., s. to Fla.
scutellaris scutellaris Cresson, 1933a: 68.—Ind.; Alta. to Ont., s. to Calif. and Ala.
  spp. **occidentalis** Sturtevant and Wheeler, 1954: 242.—Calif.
slossonae Coquillett, 1900a: 35.—Fla.; Neotropical.

Genus LEMNAPHILA Cresson

Lemnaphila Cresson, 1933b: 229. Type-species, *scotlandae* Cresson (mon.).

REFERENCES: Scotlana, 1934, 1940 (biol., leaf miners in *Lemna*).

scotlandae Cresson, 1933b: 229.—N.Y.; Mich.

Tribe ILYTHEINI

Genus ILYTHEA Haliday

Ephydra, subg. **Ilythea** Haliday, 1839: 408. Type-species, *spilota* Curtis (mon.).

caniceps Cresson, 1918a: 50.—Costa Rica; Wash. to Calif., Neotropical.
flaviceps Cresson, 1916b: 147.—Ariz.; Calif.
spilota (Curtis), 1832: pl. 413 (*Ephydra*).—Europe; Alaska to Labr., s. to Calif. and N.C.

Genus ZEROS Cresson

Zeros Cresson, 1943: 10. Type-species, *Ilythea obscura* Cresson (orig. des.).

calverti (Cresson), 1918a: 51 (*Ilythea*).—Costa Rica; Fla., Neotropical.
fenestralis (Cresson), 1918a: 51 (*Ilythea*).—Costa Rica; Fla., Neotropical.
flavipes (Williston), 1896c: 403 (?*Ilythea*).—St. Vincent I.; Ont., s. to Tex. and Fla.; Neotropical.
obscurus (Cresson), 1918a: 52 (*Ilythea*).—Costa Rica; Ariz., N. Mex., Neotropical.
vicinus Cresson, 1943: 12.—Fla.

Tribe NOTIPHILINI

Genus NOTIPHILA Fallén

Notiphila Fallén, 1810a: 22. Type-species, *cinerea* Fallén (Westwood, 1840: 153).

Subgenus NOTIPHILA Fallén

avia Loew, 1878: 193.—Hudson Bay Territory; Alaska to Mich. and Que.
bella Loew, 1862c: 135.—Middle States; Colo., Ala.
carinata Loew, 1862c: 137.—Middle States; Ind. to Ont. and N.C.
  bicolor Cresson, 1917a: 35 (preocc. Waltl, 1837).—Mo.
erythrocera Loew, 1878: 194.—Cuba; Calif. to Ill. and N.J., s. to Fla., Neotropical.
  varia Jones, 1906: 153.—Calif.

floridensis Cresson, 1917a: 46.—Fla.; N.J.
    cognata Cresson, 1917a: 46.—N.J.
loewi Cresson, 1917a: 44.—Ohio; Mich. to Que. and Fla.
    unicolor Loew, 1862c: 137 (preocc. Walker, 1860).—U.S.
    var. **biseriata** Cresson, 1917a: 46 (as sp.).—Ohio.
nudipes Cresson, 1917a: 43.—N.J.; Mich., Ohio, N.Y., Ont.
    latelimbata Curran, 1930j: 77.—N.Y.
riparia Meigen, 1830: 65.—Europe; Alaska to N.S., s. to Calif. and Fla., Mexico, Europe.
vittata Loew, 1862c: 136.—Middle States; Nebr. to Que., s. to Fla.

Subgenus AGROLIMNA Cresson

Notiphila, subg. **Agrolimna** Cresson, 1917a: 48. Type-species, *scalaris* Loew (orig. des.).

aenigma Cresson, 1917a: 54 (as *olivacea* var.).—Wash. **N. status.**
atripes Cresson, 1917a: 50.—Mass.; Wash. to N.S., s. to Calif. and Md.
atrisetis Cresson, 1917a: 52.—Calif.; Wash. to Man., s. to Calif. and N. Mex.
bispinosa Cresson, 1917a: 58.—N.J.; coastal, N.S. to Fla.
decoris Williston, 1893d: 258.—Calif.; Oreg.
frontalis Coquillett, 1904e: 97.—Nicaragua; Tex., Neotropical.
furcata (Coquillett), 1902e: 182 (*Dichaeta*).—Fla.; coastal, Va. to Fla. and La., Neotropical.
hamifera Wheeler, 1961: 87.—Ariz.; Colo.
macrochaeta Loew, 1878: 192.—Tex.; Oreg. to Iowa, s. to Calif. and Ga., Mexico.
    brachychaeta Cresson, 1946: 232 (as ssp.).—Calif.
minima Cresson, 1917a: 52 (as *occidentalis* var.).—N. Mex.; Nev. **N. status.**
occidentalis Cresson, 1917a: 51.—Calif.; Alaska to Man., s. to Colo. and Calif.
olivacea Cresson, 1917a: 52.—Ohio; Alaska to N.S., s. to Calif. and N.Y.
pallidipalpis Cresson, 1940: 8.—Mich.; Sask. to N.S., s. to Iowa and Va.
pulchrifrons Loew, 1872a: 102 (Cent. 10, no. 84).—Tex.; Calif. to Tex. and Mexico.
quadrisetosa Thomson, 1869: 594.—Calif. ?=*occidentalis*.
scalaris Loew, 1862c: 134.—Middle States; Man. to N.S., s. to Ill. and N.C.

sicca Cresson, 1940: 8.—Nev.; Wyo. to Mich., s. to Calif. and N. Mex.
signata Cresson, 1917a: 57 (as *frontalis* var.).—Ga.
solita Walker, 1852: 406.—U.S. ?=*furcata*.
transversa Walker, 1852: 407.—U.S.

Genus DICHAETA Meigen

Dichaeta Meigen, 1830: 61. Type-species, *Notiphila caudata* Fallén (mon.).

atriventris Cresson, 1915a: 68.—Colo.; Oreg. to Colo. and N. Mex., Mexico (Morelos).
caudata (Fallén), 1813: 249 (*Notiphila*).—Sweden; Alaska to Que., s. to Wash., Kans., and Fla., Europe.
   brevicauda Loew, 1860a: 5.—Germany.

Genus PARALIMNA Loew

Paralimna Loew, 1862c: 138. Type-species, *appendiculata* Loew (mon.)=*punctipennis* (Wiedemann).

REFERENCES: Cresson, 1916a, 1946 (revs.).

Subgenus PARALIMNA Loew

multipunctata Williston, 1896c: 390.—St. Vincent I.; Calif. to Tex. and Fla., Neotropical.
punctipennis (Wiedemann), 1830: 590 (*Notiphila*).—Unknown; S. Dak. to Minn. and N.Y., s. to Calif. and Fla.
   appendiculata Loew, 1862c: 138.—Middle States and Ga.
texana Cresson, 1915a: 69.—Tex.; Kans., Mo., Okla.

Subgenus PHAIOSTERNA Cresson

Phaiosterna Cresson, 1916a: 104 (as genus). Type-species, *Paralimna decipiens* Loew (orig. des.).

decipiens Loew, 1878: 195.—Tex.; Calif. to Mich. and Fla., Neotropical.
obscura Williston, 1896c: 391.—St. Vincent I.; s. Fla., Bermuda, Neotropical.

Genus OEDENOPS Becker

Oedenops Becker, 1903: 178. Type-species, *isis* Becker (mon.).
nudus (Coquillett), 1902e: 182 (*Paralimna*).—Mexico; Calif. to Miss.

Subfamily PARYDRINAE

REFERENCE: Sturtevant and Wheeler, 1954 (rev.).

Tribe PARYDRINI

Genus PARYDRA Stenhammar

Napaea Robineau-Desvoidy, 1830: 799 (preocc. Hübner, 1819). Type-species, *stagnicola* Robineau-Desvoidy (Westwood, 1840: 153) =*coarctata* (Fallén).

Ephydra, sectio **Parydra** Stenhammar, 1844: 144. Type-species, *aquila* Fallén (Coquillett, 1910b: 585).

Subgenus PARYDRA Stenhammar

bituberculata Loew, 1862c: 165.—Middle States; Man. to Que., s. to Iowa and Ala.
incommoda Cresson, 1930b: 81.—Idaho; Alaska to Alta., s. to Calif. and Colo.
metallica Cole, 1921: 176.—Alaska; Y.T., N.W.T.
nitida Cresson, 1915a: 70.—Idaho; Alaska to Sask., s. to Calif. and N. Mex.
papulata Cresson, 1949: 247.—Wash.; Alaska to Ont., s. to Calif., Colo., and N.H.
quadrituberculata Loew, 1862c: 165.—Middle States; Alta. to Que., s. to Tex. and Fla.
tibialis Cresson, 1916b: 150.—Ariz.; B.C. to Mich., s. to Calif. and Tex.

Subgenus CHAETOAPNAEA Hendel

Napaea, subg. **Chaetoapnaea** Hendel, 1930a: 150. Type-species, *Ephydra pusilla* Meigen (orig. des.).

abbreviata Loew, 1861b: 357 (Cent. 1, no. 97).—Pa.; Ont. to W. Va. and Md.
alpina (Cresson), 1924b: 163 (*Napaea*).—Wash.; Wash. and Oreg. to Man., Mich., and Que.
appendiculata Loew, 1878: 202.—Tex.; Wash. to Mich., s. to Calif. and Tex., Mexico.
aurata Jones, 1906: 154.—Calif.; B.C. to Idaho, s. to Calif. and N. Mex., Mexico (Baja Calif.).
borealis (Cresson), 1949: 235 (*Napaea*).—Idaho; Alaska to Labr., s. to N.Y. **N. comb.**

breviceps breviceps Loew, 1862c: 167.—Middle States; Ariz. and Colo., e. to Iowa and Ont., s. to Fla.
  limpidipennis Loew, 1878: 201.—D.C.
 ssp. **vicina** Cresson, 1940: 9 (as sp.).—Calif.; Oreg., Nev., Ariz.
halteralis (Cresson), 1930b: 81 (*Napaea*).—Wash.; Idaho to Calif. and N. Mex.
hulli (Cresson), 1934a: 212 (*Napaea*).—Miss.; Fla.
humilis Williston, 1897: 7.—Brazil; Calif., Neotropical.
imitans Loew, 1878: 201.—Mass.; coastal, N.S. to Ga.
parva Cresson, 1949: 241 (n. name for *undulata* Cresson).—D.C.; Mass., Md.
  undulata Cresson, 1934a: 212 (*Napaea*; preocc. Becker, 1926).—D.C.
paullula Loew, 1862c: 167.—U.S.; Alaska and B.C., e. to N.H. and Colo.
pinguis (Walker), 1852: 409 (*Ephydra*).—U.S.; Iowa to Mass., s. to Tex. and S.C.
socia (Cresson), 1934a: 213 (*Napaea*).—Calif.; Wash., Oreg., N. Mex.
transversa Cresson, 1940: 10.—Fla.
trituberculata (Sturtevant and Wheeler), 1954: 225 (*Napaea*).—Ala.; Miss., Ga. **N. comb.**
unituberculata Loew, 1878: 200.—D.C.; Kans. to Mass., s. to Tex. and S.C.
vanduzeei (Cresson), 1933a: 68 (*Napaea*).—N.Y.; Ga.
varia Loew, 1863b: 326 (Cent. 4, no. 100).—Alaska; Alaska to Que. and N.H., s. to Calif. and Colo.
vulgaris (Cresson), 1949: 233 (*Napaea*).—Wyo.; Alaska to Que., s. to Calif. and Colo.
yukonensis (Cresson), 1949: 234 (*Napaea*).—Alaska; Que. **N. comb.**

Subgenus CALLINAPAEA Sturtevant and Wheeler

Napaea, subg. **Callinapaea** Sturtevant and Wheeler, 1954: 220. Type-species, *aldrichi* Sturtevant and Wheeler (orig. des.).

aldrichi (Sturtevant and Wheeler), 1954: 220 (*Napaea*).—Calif.; Alaska, Wash., Alta., Iowa, Que. **N. comb.**

Tribe LIPOCHAETINI

Genus LIPOCHAETA Coquillett

Lipochaeta Coquillett, 1896b: 220. Type-species, *slossonae* Coquillett (orig. des.).

slossonae Coquillett, 1896b: 221.—Fla.; coasts and saline habitats, Calif., Okla., Tex., Mass., Md., Fla., Mexico.
texensis Townsend, 1898a: 168.—Tex.

Genus ASMERINGA Becker

Asmeringa Becker, 1903: 174. Type-species, *inermis* Becker (mon.).

lindsleyi Sturtevant and Wheeler, 1954: 242.—Calif.

Tribe HYADININI

Genus HYADINA Haliday

Hydrina Robineau-Desvoidy, 1830: 794 (preocc. Rafinesque, 1815). Type-species, *vernalis* Robineau-Desvoidy (Coquillett, 1910b: 553)=*guttata* (Fallén). Cresson (1930a: 93) attempted to set aside Coquillett's type des. as invalid (but without proper reason), substituting *maculipennis* Robineau-Desvoidy as type.

Ephydra, subg. **Hyadina** Haliday, 1839: 406. Type-species, *Notiphila guttata* Fallén (Westwood, 1840: 153).

albovenosa Coquillett, 1900a: 34.—Ga. and La.; Sask. to Que., s. to Tex. and Ga.

binotata (Cresson), 1926: 256 (*Hydrina*).—Calif.; Alaska to Ont., s. to Calif. and Ga.
  macquarti Cresson, 1930b: 80.—Alaska.

corona (Cresson), 1926: 256 (*Hydrina*).—Pa.; Iowa to Mich., N.Y. and N.J.

longicornis Sturtevant and Wheeler, 1954: 214.—Ohio.

neglecta Sturtevant and Wheeler, 1954: 214.—N.Y.; R.I.

octonotata (Walker), 1849: 1106 (*Ephydra*).—Ont. **N. comb.**

producta (Walker), 1849: 1099 (*Notiphila*).—Ont. **N. comb.**

pruinosa (Cresson), 1926: 256 (*Hydrina*).—Calif.; Wash. to S. Dak., s. to Calif. and N. Mex.

subnitida Sturtevant and Wheeler, 1954: 215.—Minn.; B.C. to Que., s. to Ind.

Genus AXYSTA Haliday

Ephydra, subg. **Axysta** Haliday, 1839: 406. Type-species, *Hydrina viridula* Robineau-Desvoidy (mon.; misident.)=*cesta* (Haliday).

bradleyi Cresson, 1930b: 79.—Ga.; Mich. to Fla.

cesta (Haliday), 1833: 177 (*Ephydra*).—Ireland; N.W.T. to Que. and N.H., s. to Wash. and Mich., Europe.

extera (Cresson), 1942b: 162 (also 1926: 257) (*Lytogaster*).—N.J.; Mich. to Que., s. to Ill. and N.J.

Genus LYTOGASTER Becker

Lytogaster Becker, 1896: 202. Type-species, *Philygria abdominalis* Stenhammar (orig. des.).

abdominalis (Stenhammar), 1844: 238 (*Notiphila*).—Sweden; Sask. to Mich. and Que., Europe.
angustata Cresson, 1934a: 206.—Cuba; Ga., Fla., Neotropical.
excavata (Sturtevant and Wheeler), 1954: 213 (*Hyadina;* as *gravida* ssp.).—N.Y.; Wis. to Que., s. to Tex. and Fla. **N. comb.**
flavipes (Sturtevant and Wheeler), 1954: 212 (*Hyadina*).—Calif.; Calif. to Tex., Mexico. **N. comb.**
furva Cresson, 1926: 257.—N.Y.; Ariz., Tex. to Ind., N.Y. and Fla., Bermuda.
gravida (Loew), 1863b: 325 (Cent. 4, no. 98) (*Hyadina*).—Alaska; Alaska to Alta., s. to Calif. and N. Mex., Mexico.
  willistoni Cresson, 1916b: 150.—Calif.
pallipes Cresson, 1914a: 248.—Costa Rica; Ga. and Fla. to Cuba and Costa Rica.

Genus PELINA Haliday

Ephydra, subg. **Pelina** Haliday, 1839: 407. Type-species, *Notiphila aenea* Fallén (mon.).

aenescens (Stenhammar), 1844: 210 (*Notiphila*).—Sweden; Y.T. to Mich., s. to B.C. and Colo., Europe.
canadensis Cresson, 1934a: 208.—Man.; Alaska to Minn., s. to Calif.
compar Cresson, 1934a: 207.—Wash.; B.C. to Man., s. to Calif. and Colo.
rudis Cresson, 1940: 9.—Mich.; B.C. to Que., s. to Calif. and N.J.
truncatula Loew, 1878: 198.—Tex.; B.C. to Que., s. to Calif. and Tex., Mexico.

Genus GASTROPS Williston

Gastrops Williston, 1897: 3. Type-species, *niger* Williston (mon.).
Ventrops Williston *in* Grossbeck, 1912: 378. Nomen nudum.

REFERENCE: Wirth, 1958 (rev.).

nebulosus Coquillett, 1900a: 34.—N.C. and Ga.; Mich. to Mass., s. to Tex. and Fla.
niger Williston, 1897: 3.—Brazil; Tex., Neotropical.

Genus OCHTHERA Latreille

Ochthera Latreille, 1802: 462. Type-species, *Musca manicata* Fabricius (Latreille, 1810: 444) =*mantis* (De Geer).

baia Cresson, 1931b: 169.—Mexico (Baja Calif.); Calif., Ariz.
cuprilineata Wheeler, 1896a: 123.—St. Vincent I.; Calif. to Okla. and Tex., Neotropical.
  cuprilineata Williston, 1896c: 402.—St. Vincent I.
  melanderi Cresson, 1944a: 8.—Calif. **N. syn.**, A. H. Sturtevant *in litt.*
exsculpta Loew, 1862c: 160.—Cuba; N.C. to Fla., West Indies.
lauta Wheeler, 1896a: 121.—Wis.; Iowa, Mo., Okla., Tex.
loreta Cresson, 1931b: 168.—Mexico (Baja Calif.); Fla. (Keys), Bahama Is., Cuba, Mexico.
mantis (De Geer), 1776: 143 (*Musca*).—Europe; Alaska to N.S., s. to Calif., Tex., and N.C.
  rapax Loew, 1862c: 162.—Carolina.
pilosa Cresson, 1926: 255.—N. Mex.; Ariz.
tuberculata Loew, 1862c: 161.—Ill.; Ill. to Mass., s. to Tex. and Fla.

Genus BRACHYDEUTERA Loew

Brachydeutera Loew, 1862c: 162. Type-species, *dimidiata* Loew (mon.) =*argentata* (Walker).

argentata (Walker), 1852: 407 (*Notiphila*).—U.S.; Calif. to Ont. and Fla.
  dimidiata Loew, 1862c: 163.—D.C.

Subfamily EPHYDRINAE

REFERENCE: Sturtevant and Wheeler, 1954 (rev.).

Tribe EPHYDRINI

Genus CIRRULA Cresson

Cirrula Cresson, 1915a: 70. Type-species, *gigantea* Cresson (mon.).
Pogonephydra Hendel, 1917: 42. Type-species, *chalybea* Hendel (mon.) =*gigantea* Cresson.

gigantea Cresson, 1915a: 71.—Mass.; coastal, Nfld. to Conn.
  chalybea Hendel, 1917: 42 (*Pogonephydra*).—N.H.

Genus HYDROPYRUS Cresson

Hydropyrus Cresson, 1934a: 216. Type-species, *Ephydra hians* Say (orig. des.).

REFERENCE: Aldrich, 1912a (biol. of *hians*).

hians (Say), 1830: 188 (1859b: 371) (*Ephydra*).—Mexico; B.C. to Minn., s. to Calif. and Okla., Mexico.
    crassimana Loew, 1866a: 182 (Cent. 6, no. 88) (*Ephydra*).—Mexico.
    californica Packard, 1871: 103 (*Ephydra*).—Calif.
    tarsata Williston, 1893d: 257 (*Ephydra*).—Calif.
    salina Curran, 1931c: 7 (*Ephydra*; preocc. von Heyden, 1843).—Okla.

Genus EPHYDRA Fallén

Ephydra Fallén, 1810a: 22. Type-species, *riparia* Fallén (Curtis, 1832: pl. 413).

REFERENCE: Aldrich, 1912a (biol. spp. w. N. Amer.).

auripes Aldrich, 1912b: 100.—Utah; Wash. to Calif., Nev., Utah.
bruesi Cresson, 1934a: 215.—Wyo.
cinerea Jones, 1906: 159.—Calif.; Calif. to Utah and Tex., Hawaii, Mexico, West Indies.
    gracilis Packard, 1871: 105.—Utah.
macellaria Egger, 1862: 779.—Europe; Calif. to Mass., s. to Va., Mexico, Europe.
    subopaca Loew, 1864a: 98 (Cent. 5, no. 99).—Conn.
niveiceps Cresson, 1916b: 151.—Wash.; Alaska to Que., s. to Calif. and Kans.
obscuripes Loew, 1866b: 50 (Cent. 7, no. 92).—Mass.; Alaska, Que., Labr., Nfld.
pectinulata Cresson, 1916b: 151.—Wyo.; N.W.T. to Wash., e. to Minn. and Nebr.
riparia Fallén, 1813: 246.—Sweden; entire N. Amer., Europe, Asia, Africa.
    halophila Packard, 1869b: 49.—Ill.
    millbrae Jones, 1906: 155.—Calif.
thermophila Cresson, 1934a: 215.—Wyo.

Genus SETACERA Cresson

Setacera Cresson, 1930a: 116. Type-species, *Ephydra pacifica* Cresson (orig. des.).

aldrichi Cresson, 1935a: 348.—Idaho; Calif., Wyo., Man.

atrovirens (Loew), 1862c: 169 *(Ephydra).*—Middle States; Y.T. to N.S., s. to Kans. and Pa.
durani Cresson, 1935a: 348.—Calif.; Wash. to Calif., Ariz., Colo., Tex.
needhami Johannsen, 1935: 53.—Calif.
    needhami Cresson, 1935a: 347.—Calif.
pacifica (Cresson), 1925: 167 *(Ephydra).*—B.C.; B.C. to Man., s. to Calif., N. Mex., and Nebr.
pilicornis (Coquillett), 1902e: 184 *(Ephydra).*—Fla.
    knabi Cresson, 1935a: 346.—Fla.

Genus DIMECOENIA Cresson

Dimecoenia Cresson, 1916b: 152. Type-species, *Coenia spinosa* Loew (orig. des).
austrina (Coquillett), 1900a: 36 *(Ephydra).*—Fla.; coastal, Calif., Tex. to Va., Mexico, Bermuda.
    virida Hine, 1904c: 65 *(Coenia).*—Tex.
spinosa (Loew), 1864a: 99 (Cent. 5, no. 100) *(Coenia).*—Mass.; coastal, N.B. to Fla., Tex., Calif., Mexico (Baja Calif.).

Tribe SCATELLINI

Genus THIOMYIA Wirth

Thiomyia Wirth, 1954c: 196. Type-species, *quatei* Wirth (orig. des.).
quatei Wirth, 1954c: 196.—Calif.

Genus COENIA Robineau-Desvoidy

Coenia Robineau-Desvoidy, 1830: 800. Type-species, *caricicola* Robineau-Desvoidy (mon.) = *palustris* (Fallén).
Caenia, emend.
curvicauda (Meigen), 1830: 116 *(Ephydra).*—Europe; Alaska to N.H., s. to Wyo., Wis., and Mass.

Genus PARACOENIA Cresson

Paracoenia Cresson, 1935a: 356. Type-species, *Coenia bisetosa* Coquillett (orig. des.).
bisetosa (Coquillett), 1902e: 183 *(Coenia).*—Utah; B.C. to Ont., s. to Calif., Tex., and Mass., Mexico (Baja Calif.).
fumosalis Cresson, 1935a: 356.—Mass.; Alaska to Ont., s. to Calif. and Mass.

paurosoma (Sturtevant and Wheeler), 1954: 165 (*Coenia*).—Wyo.; Colo. **N. comb.**
platypelta Cresson, 1935a: 356.—Calif.; Wash. to Alta., s. to Calif., Colo., and Ariz.
turbida (Curran), 1927f: 91 (*Coenia*).—Wyo.; Alaska to Man., s. to Mexico (Baja Calif.) and N. Mex. **N. comb.**

Genus PHILOTELMA Becker

Philotelma Becker, 1896: 163. Type-species, *anomalum* Becker (mon.) =*nigripenne* (Meigen).
alaskense Cresson, 1935a: 357.—Alaska; Alaska to Ont., s. to Nev.

Genus LAMPROSCATELLA Hendel

Lamproscatella Hendel, 1917: 42. Type-species, *Ephydra sibilans* Haliday (orig. des.).
brunnipennis (Malloch), 1923i: 221 (*Scatella*).—Alaska; Alaska and Arctic Canada to Greenland.
cephalotes Cresson, 1935a: 360.—Utah; Wash. to Mont., s. to Calif. and Utah.
dichaeta (Loew), 1860a: 40 (*Scatella*).—Europe; Wash. and Alta. to N.S., s. to Calif., Miss., and N.C., Asia, North Africa.
nivosa Cresson, 1935a: 361.—Wyo.; Wash. to Man., s. to Calif. and N. Dak.
quadrisetosa (Becker), 1896: 229 (*Scatella*).—Norway; Alaska to Greenland, s. to Calif. and N.B., Europe.
salinaria (Sturtevant and Wheeler), 1954: 176 (*Scatella*).—Calif.; Oreg., Utah. **N. comb.**
sibilans (Haliday), 1833: 175 (*Ephydra*).—Ireland; Alaska to Que., s. to Calif. and Colo., Europe.

Genus NEOSCATELLA Malloch

Neoscatella Malloch, 1933b: 9. Type-species, *atra* Malloch (orig. des.).
crassicosta (Becker), 1896: 234 (*Scatella*).—Germany; Alaska to Labr., Europe.
obscuriceps (Cresson), 1915a: 72 (*Scatella;* as *intermedia* var.).—Pa.; Mass. to Fla.
setosa (Coquillett), 1900h: 462 (1904: 76) (*Scatella*).—Alaska; Alaska to Nfld., s. to Calif. and Ala.
    intermedia Cresson, 1915a: 72 (*Scatella*).—Calif.

Genus PARASCATELLA Cresson

Parascatella Cresson, 1935a: 357. Type-species, *Scatella pilifera* Cresson (orig. des.).

marinensis Cresson, 1935a: 358.—Calif.; Wash., Ariz.
melanderi Cresson, 1935a: 358.—Wash.; Oreg., Calif.
triseta (Coquillett), 1902e: 184 (*Scatella*).—Ariz.; Oreg. to Alta. and S. Dak., s. to Calif. and Tex.

Genus SCATELLA Robineau-Desvoidy

Scatella Robineau-Desvoidy, 1830: 801. Type-species, *buccata* Robineau-Desvoidy (Coquillett, 1910b: 603) = *stagnalis* (Fallén).

arizonensis Cresson, 1935a: 365.—Ariz.; Calif., N. Mex., Mexico (Son.).
favillacea Loew, 1862c: 170.—Middle States; Sask. to Que., s. to Tex. and Fla., Neotropical.
  nitidifrons Cresson, 1918a: 67.—Costa Rica.
laxa Cresson, 1933a: 70.—Calif.; Alaska to Que., s. to Calif. and Tex.
obscura Williston, 1896c: 403.—St. Vincent I.; Fla., Neotropical.
obsoleta Loew, 1861b: 358 (Cent. 1, no. 98).—D.C.; Alaska to Que., s. to Calif. and Fla.
paludum (Meigen), 1830: 118 (*Ephydra*).—Europe; B.C. to Ont., s. to Calif. and Tex., Africa.
pentastigma (Thomson), 1869: 591 (*Ephydra*).—Calif.; Wash. to Calif. and N. Mex.
picea (Walker), 1849: 1105 (*Ephydra*).—Ont.; N.W.T. to Labr., s. to Wash., La., and Fla.
  lugens Loew, 1862c: 171.—Middle States.
quadrinotata Cresson, 1933a: 70.—N.Y.; Iowa to Que., s. to Tex. and N.C.
stagnalis (Fallén), 1813: 248 (*Ephydra*).—Sweden; entire N. Amer., cosmopolitan.
thermarum Collin, 1930c: 138.—Iceland; Greenland.
troi Cresson, 1933a: 69.—Idaho; Alaska, Wash.

Genus LIMNELLIA Malloch

Limnellia Malloch, 1925c: 331. Type-species, *maculipennis* Malloch (orig. des.).
Eustigoptera Cresson, 1930a: 126. Type-species, *Notiphila quadrata* Fallén (orig. des.).
Stictoscatella Collin, 1930c: 133. Type-species, *Notiphila quadrata* Fallén (orig. des.).

anna Cresson, 1935a: 363.—Mass.; S. Dak. to Que., s. to Mo. and N.C.
oscitans (Walker), 1849: 1106 (*Ephydra*).—Ont. **N. comb.**
quadrata (Fallén), 1813: 255 (*Notiphila*).—Sweden; Alaska to Idaho, s. to Calif. and Ariz., Europe.
sejuncta (Loew), 1863b: 326 (Cent. 4, no. 99) (*Scatella*).—Alaska; Wash., Calif.
stenhammari (Zetterstedt), 1846: 1842 (*Ephydra*).—Sweden; Alaska to Que., s. to Wash., Iowa and S.C., Europe.

Genus SCATOPHILA Becker

Scatophila Becker, 1896: 237. Type-species, *Ephydra caviceps* Stenhammar (orig. des.).

REFERENCES: Bolwig, 1940 (biol.); Sturtevant and Wheeler, 1954 (rev.).

adamsi Cresson, 1935a: 369.—Mo.; Nebr. and Iowa to Tex., Md.
arenaria Cresson, 1935a: 372.—Wash.; Wash. to Calif., Nebr., Minn.
bipiliaris Sturtevant and Wheeler, 1954: 187.—Calif.
bisignata Cresson, 1935a: 370.—Wash.; Calif.
carinata Sturtevant and Wheeler, 1954: 188.—Iowa; Alta.
conifera Sturtevant and Wheeler, 1954: 189.—Calif.
cribrata (Stenhammar), 1844: 269 (*Ephydra*).—Sweden; Alaska to Greenland, s. to B.C. and Idaho, Europe.
despecta (Haliday), 1839: 409 (*Ephydra*).—Ireland; B.C. to Maine, s. to Calif. and Fla., Europe.
  striata Walker, 1849: 1107 (*Ephydra*).—Ont. ?Valid sp.
  variabilis Cresson, 1917b: 341.—Calif.
disjuncta Cresson, 1935a: 371.—Calif.; Wash. to Calif., Ariz., and N. Mex.
exilis Cresson, 1935a: 366.—Colo.; Wash. to Man., s. to Calif., N. Mex., and Okla., Mexico (Baja Calif.).
hesperia Sturtevant and Wheeler, 1954: 192.—Calif.; Oreg., Utah.
hirsuta Sturtevant and Wheeler, 1954: 193.—Ariz.
hirtirostris Sturtevant and Wheeler, 1954: 194.—Alaska; N.W.T.
iowana Wheeler, 1961: 86.—Iowa.
mesogramma (Loew), 1869b: 42 (Cent. 8, no. 74) (*Scatella*).—R.I.; Alaska, N.W.T., Labr. to Md.
ordinaria Sturtevant and Wheeler, 1954: 195.—Tex.; Wash., Calif. to La.
picta Sturtevant and Wheeler, 1954: 196.—Alaska.
pulchra Sturtevant and Wheeler, 1954: 196.—Calif.; Ariz., N. Mex.

rubribrunnea Sturtevant and Wheeler, 1954: 197.—Calif.; Alaska to
Wyo., s. to Calif. and N. Mex., Mexico.
unicornis Czerny, 1900: 205.—Austria; Calif., Mo., Ont., Europe.
Probably introduced into N. Amer. by commerce.
tuberculosa Cresson, 1935a: 368.—Mo.
variofacialis Sturtevant and Wheeler, 1954: 199.—Wash.; Alaska to
N.W.T., s. to Calif. and Colo.
viridella Sturtevant and Wheeler, 1954: 200.—Calif.; B.C. to Calif.,
also Ariz., Wyo., Nebr.

Unplaced Species of Ephydridae

brevis Walker, 1858: 233 (*Ephydra*).—U.S.
frontalis Curran, 1933d: 7 (*Diclasiopa*).—Que.
lata Walker, 1858: 233 (*Ephydra*).—U.S.
nana Walker, 1858: 234 (*Ephydra*).—U.S.
nigroaenea Walker, 1852: 413 (*Gymnopa*).—U.S.
tarsalis Walker, 1852: 413 (*Gymnopa*).—U.S.

Family CURTONOTIDAE
By Willis W. Wirth

Curtonotids are a small family of small, grayish, hump-backed, robust flies closely related to the Drosophilidae; following Hennig (1958a) they are accorded family status. The breeding habits have only recently been discovered by Greathead (1958) in Eritrea, where the larvae were scavengers in egg pod beds of the desert locust (*Schistocerca*). We have only one species of *Curtonotum*, a genus better represented in tropical America and Africa.

REFERENCES: Sturtevant, 1921 (discussion); Greathead, 1958 (immature stages, biol.).

Genus CURTONOTUM Macquart

Curtonotum Macquart, 1843a: 350 (1843: 193). Type-species, *Musca gibba* Fabricius (mon.).
Cyrtonotum, emend.
Diplocentra Loew, 1862e: 13 (unjustified n. name for *Curtonotum* Macquart). Type-species, *Musca gibba* Fabricius (aut.).

helvum (Loew), 1862b: 228 (Cent. 2, no. 91) (*Diplocentra*).—Minn.;
Man. to Ont., s. to Ill. and Ga.

Family DROSOPHILIDAE

By Marshall R. Wheeler

Drosophilid flies are found most often around spoiled fruit, slime fluxes, rotting cacti, and in fungi, the larvae feeding primarily on the microorganisms living in such situations. Some species occur in flowers, on fresh sap from tree wounds, or in decaying organic matter. Some *Scaptomyza* larvae are leaf miners, some *Clastopteromyia* larvae are ectoparasitic on nymphal Cercopidae, and *Pseudiastata* larvae are mostly predaceous on mealybugs.

Economic pests are rare, though several domestic *Drosophila* must be controlled around fruit and vegetable canneries and are annoying in markets. Some species are involved in the transmission of fungus diseases of plants, and many could serve as mechanical vectors of disease-producing microorganisms. Many *Drosophila* can be reared easily in the laboratory on agar-based media, and have been used extensively in research in genetics, cytology, and insect physiology.

REFERENCES: Sturtevant, 1921; Malloch and McAtee, 1924a; Patterson, 1943; Wheeler, 1952b (keys to genera); Patterson and Stone, 1952 (biol., distribution, genetics); Wheeler, 1949b (keys, distribution), 1954a, 1957 (distribution); Patterson and Mainland, 1944 (distribution); Patterson, 1942 (*Drosophila* hybridization).

Subfamily STEGANINAE

Genus STEGANA Meigen

Stegana Meigen, 1830: 79. Type-species, *nigra* Meigen (Zetterstedt, 1847: 2577) = *curvipennis* (Fallén).

Subgenus STEGANA Meigen

curvipennis (Fallén).—Not Nearctic.
vittata (Coquillett), 1901h: 618 (*Phortica*).—N.J.; Maine to Pa., Va., N.C. and Tenn.

Subgenus STEGANINA Wheeler

Stegana, subg. **Steganina** Wheeler, 1960a: 110. Type-species, *Musca coleoptrata* Scopoli (orig. des.).

antigua Wheeler, 1960a: 110.—Va.; Md., N.Y.
coleoptrata (Scopoli), 1763: 338 (*Musca*).—Europe; N.W.T., B.C., Wash., Mich., Tenn., Que. to Va.

Genus AMIOTA Loew

Amiota Loew, 1862b: 229 (Cent. 2, no. 93). Type-species, *leucostoma* Loew (Coquillett, 1910b: 505).

Subgenus AMIOTA Loew

alboguttata (Wahlberg).—Not Nearctic.
buccata Wheeler, 1952b: 171.—N. Mex.; Ariz., Utah.
humeralis Loew, 1862b: 229 (Cent. 2, no. 93).—D.C. Unrecognized.
leucostoma Loew, 1862b: 230 (Cent. 2, no. 94).—Pa.; Que. to Ind., Tenn., and N.C.
minor (Malloch), 1921o: 312 (*Phortica*).—Ill.; Nebr. to Maine, s. to Tex. and Ga., also Wash., Ariz., Mexico.
nigrescens Wheeler, 1952b: 170.—Ariz.; N. Mex.
setigera Malloch, 1924c: 51.—Ill.; Mo., Tenn., Md.
steganoptera Malloch, 1926: 31.—Costa Rica; Miss., Ala., Va., Fla., Neotropical.
subtusradiata Duda, 1934b: 32 (as *alboguttata* var.).—Europe; Sask. and Ind. to N.H. and Va., Eurasia.

Subgenus PHORTICA Schiner

Phortica Schiner, 1862b: 433 (as genus). Type-species, *Drosophila variegata* Fallén (orig. des.).
huachucae Wheeler, 1960c: 90.—Ariz.
albavictoria, authors, not Patterson and Mainland.

Subgenus SINOPHTHALMUS Coquillett

Sinophthalmus Coquillett, 1904f: 190 (as genus). Type-species, *pictus* Coquillett (orig. des.).
picta (Coquillett), 1904f: 191 (*Sinophthalmus*).—Calif.; Nev., Ariz., N. Mex., Mexico. **N. comb.**

Genus LEUCOPHENGA Mik

Leucophenga Mik, 1886: 317. Type-species, *Drosophila maculata* Dufour (orig. des.).

Subgenus LEUCOPHENGA Mik

maculosa (Coquillett), 1895a: 317 (*Drosophila*).—Fla.; Minn. to Mass., s. to Tex. and Fla., Nev., Neotropical.
neovaria Wheeler, 1960c: 89.—P.R.; Fla., Neotropical.
varia (Walker), 1849: 1109 (*Drosophila*).—Ga.; Nebr. to Mass., s. to Tex. and Fla., Neotropical.
quadrimaculata Walker, 1852: 410 (*Drosophila*).—U.S.
signicosta Walker, 1861: 330 (*Opomyza*).—U.S.

Subgenus NEOLEUCOPHENGA Oldenberg

Paraleucophenga Oldenberg, 1914: 18 (preocc. Hendel, 1914). Type-species, *Leucophenga quinquemaculata* Strobl (orig. des.).

Neoleucophenga Oldenberg, 1915: 93 (as genus; n. name for *Paraleucophenga* Oldenberg). Type-species, *Leucophenga quinquemaculata* Strobl (aut.).

guttata Wheeler, 1952b: 188.—Ariz.; N. Mex.
montana Wheeler, 1952b: 188.—Oreg.; Nev., B.C., Calif., Utah.
pulcherrima Patterson and Mainland, 1944: 14.—Mexico; Ariz., N. Mex., Neotropical.
trisphenata Wheeler, 1952b: 187.—Ariz.; N. Mex.

Genus PSEUDIASTATA Coquillett

Pseudiastata Coquillett, 1908: 148. Type-species, *nebulosa* Coquillett (orig. des.).

REFERENCES: Sabrosky, 1951c, Hardy, 1959b (revs.).

Subgenus PSEUDIASTATA Coquillett

nebulosa Coquillett, 1908: 148.—Md.; Va., Ga.

Subgenus HYALISTATA Wheeler

Pseudiastata, subg. **Hyalistata** Wheeler, 1960b: 67. Type-species, *pictiventris* Wheeler (orig. des.).

pictiventris Wheeler, 1960b: 68.—Mexico; Fla.

Genus GITONA Meigen

Gitona Meigen, 1830: 129. Type-species, *distigma* Meigen (mon.).

americana Patterson, 1943: 33.—Tex.; s. Calif. to Okla., Mexico.
bivisualis Patterson, 1943: 35.—Tex.; Ariz. to Fla., Neotropical.
sonoita Wheeler, 1949b: 158.—Ariz.; Tex.

Genus PARACACOXENUS Hardy

Paracacoxenus Hardy *in* Hardy and Wheeler, 1960: 358. Type-species, *guttatus* Hardy and Wheeler (orig. des.).

guttatus Hardy and Wheeler, 1960: 358.—Wash.; Oreg.

Genus RHINOLEUCOPHENGA Hendel

Rhinoleucophenga Hendel, 1917: 44. Type-species, *pallida* Hendel (orig. des.).

Pseudophortica Sturtevant, 1918b: 37.   Type-species, *Drosophila obesa* Loew (orig. des.).

obesa (Loew), 1872a: 102 (Cent. 10, no. 85) (*Drosophila*).—Tex.; Tex. to N.J. and Tenn., s. to Fla., Neotropical.
hirtifrons Johnson, 1913a: 88 (*Phortica*).—Fla.

Genus TRACHYLEUCOPHENGA Hendel

Trachyleucophenga Hendel, 1917: 44.   Type-species, *flavocostata* Hendel (orig. des.).

Unnamed sp.—La. (Wheeler, 1954a: 63).

Subfamily DROSOPHILINAE

Genus DROSOPHILA Fallén

Drosophila Fallén, 1823e: 4.   Type-species, *Musca funebris* Fabricius (Zetterstedt, 1847: 2542).
Drosophila, subg. *Acrodrosophila* Duda, 1924a: 203.   Type-species, *testacea* Roser (mon.).

REFERENCES: Mainland, 1941 (rev. *macrospina* group); Sturtevant, 1942 (key N. Amer. spp.); Streisinger, 1946 (rev. *cardini* group); Stalker, 1953 (rev. *cardini* group); Carson, 1958 (population genetics of *D. robusta*); Miller, 1958 (rev. *affinis* group); Wheeler, 1959 (world cat.).

Subgenus DROSOPHILA Fallén

acutilabella Stalker, 1953: 345.—Fla.; Antilles.
aldrichi Patterson and Crow, 1940: 251.—Tex.; Neotropical.
americana americana Spencer, 1938: 169 (as *virilis* ssp.).—Ohio; Mont. to Maine, s. to Tex., N.C., and Tenn.
　　ssp. **texana** Patterson *in* Patterson, Stone, and Griffen, 1940: 219 (as *virilis* ssp.).—Tex.; Va. and Tenn., s. to Tex., and Fla.
arizonensis Patterson and Wheeler, 1942: 96.—Ariz.; N. Mex., Mexico.
bifurca Patterson and Wheeler, 1942: 85.—Tex.; Ariz., Neotropical.
borealis Patterson, 1952: 20.—Minn.; Idaho to Man., s. to Colo. and Wis.
bromeliae Sturtevant, 1921: 72.—Cuba; Fla., Neotropical.
californica Sturtevant, 1923: 9.—Calif.; Oreg. to Idaho, s. to Calif. and N. Mex., Mexico.
　　fuliginea Patterson and Wheeler, 1942: 80.—N. Mex.   **N. syn.**
carbonaria Patterson and Wheeler, 1942: 103.—Tex.; Ariz., Mexico.
cardini Sturtevant, 1916: 336.—Cuba; Tex., Fla., Neotropical.

carsoni Wheeler, 1957: 95.—Wis.; S. Dak. to Maine, also N. Mex.
colorata Walker, 1849: 1110.—N.Y.; Wash. to Maine, s. to Mo., Miss., and Va.
    sulcata Sturtevant, 1916: 330.—Md.
deflecta Malloch *in* Malloch and McAtee, 1924a: 36.—D.C.; Mich., Fla.
euronotus Patterson and Ward, 1952: 158.—La.; N.C. to La. and Fla.
falleni Wheeler, 1960d: 161.—Mo.; B.C. to Que., s. to Colo. and Tex.
flavomontana Patterson, 1952: 21.—Idaho; Wyo., Utah, Colo., N. Mex.
flavopinicola Wheeler, 1954a: 47.—Calif.; s. B.C.
fulvalineata Patterson and Wheeler, 1942: 106.—Ariz.; N. Mex., Neotropical.
funebris (Fabricius), 1787: 345 (*Musca*).—Europe; entire U.S., cosmopolitan.
guttifera Walker, 1849: 1110.—Fla.; Nebr. to Mass., s. to Tex. and Fla.
    multipunctata Loew, 1866b: 50 (Cent. 7, no. 93).—D.C.
hamatofila Patterson and Wheeler, 1942: 91.—Tex.; Ariz. to Tex. and Okla., Neotropical.
hydei Sturtevant, 1921: 101.—Fla.; entire U.S., cosmopolitan.
immigrans Sturtevant, 1921: 83.—N.Y.; entire U.S., Alaska, cosmopolitan.
innubila Spencer *in* Patterson, 1943: 94.—Ariz., N. Mex., Mexico (unpublished).
lacicola Patterson, 1944: 102.—Minn.; Wis., Ont., Que.
longicornis Patterson and Wheeler, 1942: 90.—Tex.; Ariz., Neotropical.
macroptera Patterson and Wheeler, 1942: 105.—Colo.; N. Mex. to Tex. and Colo., Mexico.
macrospina macrospina Stalker and Spencer, 1939: 110.—Tex.; Mont. to N.Y., s. to Ariz. and Fla.
    ssp. **limpiensis** Mainland, 1941: 160.—Tex.; Ariz. to Tex., Mexico.
    ssp. **ohioensis** Spencer, 1940: 304.—Ohio; Mich., N.Y. ?Valid sp.
magnafumosa Stalker and Spencer, 1939: 112.—Tenn.; N.C. ?= *ordinaria*.
magnaquinaria Wheeler, 1954a: 48.—Wash.; B.C. to Oreg.
mainlandi Patterson, 1943: 147.—Calif.
melanderi Sturtevant, 1916: 337.—Wash.; Alaska, Mont., Minn.
melanica Sturtevant, 1916: 332.—Ala.; Colo. to Mich. and N.Y., s. to Ariz. and Fla., Mexico.
melanissima Sturtevant, 1916: 333.—Ala.; Tex. to N.C. and Fla.

melanopalpa Patterson and Wheeler, 1942: 77.—Tex.; Ariz., Neotropical.
melanura Miller, 1944: 86.—N.Y.; Wash. and Oreg. to Que., s. to Va.
mercatorum Patterson and Wheeler, 1942: 93.—Calif.; Ariz., La., Neotropical, Hawaii.
meridiana meridiana Patterson and Wheeler, 1942: 99.—Tex.; Nebr., Neotropical.
    ssp. **rioensis** Patterson, 1943: 152.—Tex.; Mexico.
micromelanica Patterson *in* Sturtevant and Novitski, 1941: 394.—Tex.; Ariz. to Va. and Fla., Mexico.
mojavensis Patterson and Crow, 1940: 251 (as *mulleri* ssp.).—Calif.; Mexico (Baja Calif.).
montana Stone, Griffen, and Patterson, 1941: 172.—Wyo.; Alaska to Wash., Idaho, Calif. to N. Mex.
    montana Patterson and Wheeler, 1942: 75.—Wyo.
mulleri Sturtevant, 1921: 101.—Tex.; Okla., Nebr., Ark., La., Fla., Mexico, Jamaica.
munda Spencer, 1942: 58.—N. Mex., Ariz., Mexico (unpublished).
nigrohydei Patterson and Wheeler, 1942: 84.—Tex.; Ariz., Mexico.
nigromelanica Patterson and Wheeler, 1942: 100.—Tex.; Minn. to N.Y. and Conn., s. to Tex. and Fla.
nigrospiracula Patterson and Wheeler, 1942: 81.—Ariz.; Mexico.
novamexicana Patterson, 1941: 535 (as *virilis* ssp.).—N Mex.; Ariz., Utah, Colo.
occidentalis Spencer, 1942: 60.—Calif. (unpublished); Y.T., Wash. to Mont., s. to Calif. and Colo.
ordinaria Coquillett, 1904f: 190.—N.H.; Que., Mass.
palustris Spencer, 1942: 63 (n. name for *lativittata* Malloch).—Va.; Nebr., Minn., Ohio, Ill., N.Y., N.J.
    lativittata Malloch *in* Malloch and McAtee, 1924a: 36 (preocc. Malloch, 1923).—Va.
    mallochi Frota-Pessoa, 1946: 155 (n. name for *lativittata* Malloch).—Va.
paramelanica Patterson, 1942: 12 (as *melanica* ssp.).—Conn.; Minn. to Que., s. to Nebr. and Ga.
peninsularis Patterson and Wheeler, 1942: 92.—Fla.; Neotropical.
phalerata Meigen.—Not Nearctic.
pinicola Sturtevant, 1942: 40.—Calif.; Wash.
polychaeta Patterson and Wheeler, 1942: 102.—Tex.; Panama, Brazil, Hawaii, and Micronesia.
putrida Sturtevant, 1916: 339.—Mass.; Minn. to Ont., s. to Tex. and Fla.
    var. *pseudomelanica* Sturtevant, 1916: 333 (as sp.).—Va. Melanistic winter phase. Note: The name *pseudome-*

lanica has page precedence, but *putrida* is hereby selected as the name long used in genetics and cytology. **N. syn., N. status.**

quinaria Loew, 1866a: 182 (Cent. 6, no. 90).—N.Y.; Minn. to Que., s. to Mo. and Va.

recens Wheeler, 1960d: 162.—Mich.; N. Dak., Ont., Que., Maine.

rellima Wheeler, 1960d: 162.—Nebr.; Oreg., Calif.

repleta Wollaston, 1858: 117.—Madeira; entire U.S., cosmopolitan.

ritae Patterson and Wheeler, 1942: 87.—Tex.; Ariz., Mexico.

robusta Sturtevant, 1916: 331.—Ala.; Mont., Nebr., Tex., Maine to Fla.

rubrifrons Patterson and Wheeler, 1942: 107.—Ariz.

stalkeri Wheeler, 1954a: 52.—Fla.

subfunebris Stalker and Spencer, 1939: 108.—Calif.; Wash.

suboccidentalis Spencer, 1942: 61.—Idaho, Utah, Colo., S. Dak. (unpublished); Mont., Nebr.

subpalustris Spencer, 1942: 64.—Ohio (unpublished); S.C.

subquinaria Spencer, 1942: 59.—Utah, Wyo., Colo. (unpublished); Alaska to Ont., s. to Oreg. and N. Mex. ?=*transversa*.

suffusca Spencer *in* Patterson, 1943: 92.—Ariz. (unpublished); N. Mex.

tenebrosa Spencer *in* Patterson, 1943: 93.—N. Mex., Mexico (unpublished); Ariz.

tenuipes (Walker), 1849: 1112 (*Diastata*).—Ont.

testacea Roser, 1840: 62.—Europe; Wash. to Ont., s. to Calif. and Tenn., also Alaska.

transversa Fallén.—Not Nearctic.

tripunctata Loew, 1862b: 231 (Cent. 2, no. 97).—D.C.; Iowa to N.J., s. to Tex. and Fla.

modesta Sturtevant, 1916: 338.—Ala.

trispina Wheeler, 1949b: 180.—Calif.; Ariz.

virilis Sturtevant, 1916: 330.—N.Y.; Calif. and Utah to Ind., Tenn., and Fla., Mexico, Argentina, Brazil, China, Japan.

wheeleri Patterson and Alexander, 1952: 129.—Calif.

Subgenus DORSILOPHA Sturtevant

Drosophila, subg. **Dorsilopha** Sturtevant, 1942: 28. Type-species, *busckii* Coquillett (orig. des.).

busckii Coquillett, 1901c: 18 (as *buskii*).—D.C.; entire U.S., cosmopolitan.

Subgenus HIRTODROSOPHILA Duda

Drosophila, subg. **Hirtodrosophila** Duda, 1923b: 41. Type-species, *carinata* Duda (Frota-Pessoa, 1945: 470).

REFERENCE: Frota-Pessoa, 1945 (rev.).

alabamensis Sturtevant, 1918b: 38.—Ala.; Mich. to Ont., s. to Nebr., Tex., and Ala.
chagrinensis Stalker and Spencer, 1939: 111.—Ohio; Wis., Iowa.
cinerea Patterson and Wheeler, 1942: 71.—Tex.; Nebr.
duncani Sturtevant, 1918a: 446.—Ill.; Minn. to R.I., s. to Tex. and Fla.
grisea Patterson and Wheeler, 1942: 72.—Ariz.
longala Patterson and Wheeler, 1942: 71.—N. Mex.; Mexico.
nigrohalterata Duda, 1925b: 195.—Costa Rica; Tex., Neotropical.
orbospiracula Patterson and Wheeler, 1942: 70.—Tex.; Ariz., Mexico.
pictiventris Duda, 1925b: 211.—Costa Rica; Fla., Neotropical.
thoracis Williston, 1896c: 411.—St. Vincent I.; Fla., Neotropical.

Subgenus LORDIPHOSA Basden

Drosophila, subg. **Lordiphosa** Basden, 1961: 186. Type-species, *fenestrarum* Fallén (orig. des.).

basdeni Wheeler, 1957: 94.—Mich.; Ohio.

Subgenus PHLORIDOSA Sturtevant

Drosophila, subg. **Phloridosa** Sturtevant, 1942: 28. Type-species, *floricola* Sturtevant (orig. des.).

floricola Sturtevant, 1942: 42.—Calif.; Ariz., Neotropical.
lutzii Sturtevant, 1916: 340.—Cuba; Fla., Neotropical.

Subgenus PHOLADORIS Sturtevant

Drosophila, subg. **Pholadoris** Sturtevant, 1942: 28. Type-species, *victoria* Sturtevant (orig. des.).

REFERENCE: Wheeler, 1949a (rev.).

brooksi Pipkin, 1961: 146.—Ariz.
latifasciaeformis Duda, 1940: 22.—Uganda; S.C. to Fla., circumtropical.
    baeomyia Wheeler, 1949a: 145.—Mexico.
lebanonensis lebanonensis Wheeler.—Not Nearctic.
   ssp. **casteeli** Pipkin, 1961: 149.—Ariz.; Utah, N. Mex.
victoria Sturtevant, 1942: 33.—Calif.; Oreg. and Calif. to Nebr. and Tex., ?N.Y.

Subgenus SIPHLODORA Patterson and Mainland

Drosophila, subg. **Siphlodora** Patterson and Mainland, 1944: 25. Type-species, *sigmoides* Loew (orig. des.).

sigmoides Loew, 1872a: 103 (Cent. 10, no. 86).—Tex.; Ill. to N.Y., s. to Tex. and Ala.

Subgenus SOPHOPHORA Sturtevant

Drosophila, subg. **Sophophora** Sturtevant, 1939: 139. Type-species, *melanogaster* Meigen (orig. des.).

affinis affinis Sturtevant, 1916: 334.—Ala.; Mich. to Ont., s. to Nebr., Tex., and Fla.
   ssp. **iroquois** Sturtevant and Dobzhansky, 1936: 576.—Mass.; N.Y., N.J., Miss. ?Valid sp.
algonquin Sturtevant and Dobzhansky, 1936: 575.—Mass.; N. Dak. to Que., s. to Tex. and Miss.
ananassae Doleschall, 1858: 128.—Molucca Is. (Amboina I.); Tex., Fla., circumtropical.
   caribea Sturtevant, 1916: 335.—Cuba.
athabasca athabasca Sturtevant and Dobzhansky, 1936: 576.— Alaska; Alaska to Que., s. to Oreg., N. Mex., and Tenn.
   ssp. **mahican** Sturtevant and Dobzhansky, 1936: 576.—N.H.; Mass., N.J. ?Valid sp.
azteca Sturtevant and Dobzhansky, 1936: 577.—Mexico; Calif. to Tex., Neotropical.
melanogaster Meigen, 1830: 85.—Europe; entire U.S., cosmopolitan.
   ampelophila Loew, 1862b: 231 (Cent. 2, no. 99).—Cuba.
miranda Dobzhansky, 1935: 378.—Wash.; B.C. to Idaho, s. to Calif.
narragansett Sturtevant and Dobzhansky, 1936: 577.—Mass.; Mich. to Maine, s. to Nebr., Tex., and Fla.
nebulosa Sturtevant, 1916: 327 (n. name for *limbata* Williston).— St. Vincent I.; Tex., Fla., ?Nebr., Neotropical.
   limbata Williston, 1896c: 414 (preocc. Roser, 1840).—St. Vincent I.
obscura Fallén.—Not Nearctic.
persimilis Dobzhansky and Epling, 1944: 7.—Oreg.; B.C. to Idaho and Calif.
populi Wheeler and Throckmorton, 1961: 138.—Alaska.
pseudoobscura Frolova *in* Frolova and Astaurov, 1929: 212.— N. Amer.; B.C. to Mont., s. to Calif., Mo. and Tex., Neotropical.
seminole Sturtevant and Dobzhansky, 1936: 577.—Ala.
simulans Sturtevant, 1919: 153.—Fla.; entire U.S., cosmopolitan.

willistoni Sturtevant, 1916: 327 (n. name for *pallida* Williston).—
St. Vincent I.; Tex., Fla., Neotropical.
pallida Williston, 1896c: 415 (preocc. Zetterstedt, 1847).—
St. Vincent I.

Genus SCAPTOMYZA Hardy

Scaptomyza Hardy, 1849: 361. Type-species, *Drosophila graminum* Fallén (Coquillett, 1910b: 603).

REFERENCES: Stalker, 1945 (biol., genetics); Hackman, 1955, 1959 (revs.).

Subgenus SCAPTOMYZA Hardy

flaveola flaveola (Meigen), 1830: 66 (*Notiphila*).—Europe; Calif., Ind., Conn. to Va.
    nigrocella Wheeler, 1949b: 167.—N.Y.
  ssp. **montana** Wheeler, 1949b: 166 (as sp.).—Mont.; Alaska, Wash. to Calif., Idaho, Mich., Europe.
flaviventris Hackman, 1959: 63.—Wash.; Calif.
graminum (Fallén), 1823e: 8 (*Drosophila*).—Sweden; Wash. to Calif., Mich., Nfld., Maine to Va., Europe.
    borealis Wheeler, 1952b: 204.—N.H.
nigrita Wheeler, 1952b: 205.—Calif.; Alaska, Alta., Idaho, Wyo.
teinoptera Hackman, 1955: 82.—Finland; Alaska, Europe.

Subgenus HEMISCAPTOMYZA Hackman

Scaptomyza, subg. **Hemiscaptomyza** Hackman, 1959: 19. Type-species, *Geomyza unipunctum* Zetterstedt (orig. des.).

apicata (Thomson), 1869: 597 (*Drosophila*).—Calif.; Wash. to Mexico (Baja Calif.).
bipunctipennis Wheeler, 1952b: 206.—Calif.; Wash., Idaho.
hsui Hackman, 1955: 88.—Calif.; Wash., Oreg., Idaho, Nev.
parapicata Hackman, 1959: 54.—Ecuador; Ariz. to Tex., Neotropical.
terminalis (Loew), 1863a: 32 (Cent. 3, no. 60) (*Drosophila*).—Alaska.
trochanterata Collin, 1953a: 150.—England; Alaska, N.W.T., Man., Nfld., N.Y., Europe.
unipunctum (Zetterstedt), 1847: 2533 (*Geomyza*).—Europe; Alaska.

Subgenus MESOSCAPTOMYZA Hackman

Scaptomyza, subg. **Mesoscaptomyza** Hackman, 1959: 17. Type-species, *wheeleri* Hackman (orig. des.).

adusta (Loew), 1862b: 231 (Cent. 2, no. 98) (*Drosophila*).—D.C., B.C., Wash., Idaho, Ariz., Ont. and Maine to Fla.; Bermuda, Neotropical.

hirsuta Wheeler, 1949b: 166.—Mexico; Ariz., N. Mex.
paradusta Wheeler, 1952b: 198.—Calif.; Wash., Oreg., Ariz.
paravittata Wheeler, 1952b: 200.—Calif.; Neotropical.
vittata (Coquillett), 1895a: 318 (*Drosophila*).—Fla.; Miss. to Fla., Neotropical.
wheeleri Hackman, 1959: 49.—Va.; Tenn. and Va. to Fla., Neotropical.

Subgenus PARASCAPTOMYZA Duda

Drosophila, subg. **Parascaptomyza** Duda, 1924a: 203. Type-species, *Drosophila graminum* Fallén (mon.; misident.) = *pallida* (Zetterstedt).

pallida (Zetterstedt), 1847: 2571 (*Drosophila*).—Europe; entire U.S., cosmopolitan.
graminum, authors, not Fallén.

Genus CHYMOMYZA Czerny

Chymomyza Czerny, 1903b: 199. Type-species, *Drosophila fuscimana* Zetterstedt (Sturtevant, 1921: 61).

REFERENCE: Wheeler, 1952b (key).

aldrichii Sturtevant, 1916: 325.—Idaho; Alaska, Wash. to Calif. and Ariz., e. to Minn., Colo., Maine.
tetonensis Wheeler, 1949b: 163.—Wyo.
amoena (Loew), 1862b: 230 (Cent. 2, no. 96) (*Drosophila*).—D.C.; Minn. to Ont., s. to Nebr., Tex., and Fla., also Ariz., Utah, Mexico.
subfasciata Harris, 1835: 600. Nomen nudum.
caudatula Oldenberg, 1914: 14.—Hungary; Alaska, Wash. to Wyo., s. to Calif. and Ariz., also Mich., Maine, Europe.
coxata Wheeler, 1952b: 177.—Colo.; Alaska, Wyo.
olympia Wheeler, 1960c: 89.—Wash.
procnemis (Williston), 1896c: 412 (*Drosophila*).—St. Vincent I.; Tex. to Fla., Neotropical, Hawaii.
procnemoides Wheeler, 1952b: 175.—N. Mex.; Mich. to N.Y., s. to Ariz. and Fla.
wirthi Wheeler, 1954a: 62.—Va.; Alaska, Ill., Ont.

Genus MYCODROSOPHILA Oldenberg

Mycodrosophila Oldenberg, 1914: 4. Type-species, *Amiota poecilogastra* Loew (mon.).

REFERENCE: Wheeler and Takada, 1963 (rev.).

claytonae Wheeler and Takada, 1963: 394.—Ark.; Mont. to Que., s. to Tex. and Fla.
dimidiata (Loew), 1862b: 230 (Cent. 2, no. 95) (*Drosophila*).—Ill.; Idaho to Maine, s. to Tex. and Fla.
stalkeri Wheeler and Takada, 1963: 393.—Tex.; Ill. to Ont., s. to Tex. and Fla.

Genus CLADOCHAETA Coquillett

Cladochaeta Coquillett, 1900g: 263. Type-species, *nebulosa* Coquillett (orig. des.).

nebulosa Coquillett, 1900g: 263.—P.R.; Fla., Neotropical.

Genus CLASTOPTEROMYIA Malloch

Clastopteromyia Malloch *in* Malloch and McAtee, 1924a: 31. Type-species, *Drosophila inversa* Walker (orig. des.).

floridana Malloch, 1924k: 10.—Fla.
inversa (Walker), 1861: 331 (*Drosophila*).—U.S.; B.C. to Que., s. to Calif. and Va.
opaca (Williston), 1896c: 411 (*Drosophila*).—St. Vincent I.; Fla., Neotropical.

Genus DETTOPSOMYIA Lamb

Dettopsomyia Lamb, 1914: 349. Type-species, *formosa* Lamb (orig. des.).

nigrovittata (Malloch), 1924l: 352 (*Drosophila*).—Australia; Calif., Hawaii, Japan, Africa.

Genus PARAMYCODROSOPHILA Duda

Paramycodrosophila Duda, 1924a: 191. Type-species, *Drosophila pictula* Meijere (mon.).

anomala Wheeler, 1954a: 60.—Tex.; Miss.
centralis Wheeler, 1954a: 58.—Tex.; Fla., Neotropical.

Genus MICRODROSOPHILA Malloch

Microdrosophila Malloch, 1921o: 312. Type-species, *Drosophila quadrata* Sturtevant (orig. des.).

quadrata (Sturtevant), 1916: 341 (*Drosophila*).—Ala.; Ill. to Ind. and Va., s. to Tex. and Fla., Neotropical.

Unplaced Species of Drosophilidae

albipes Walker, 1852: 410 (*Drosophila*).—U.S.
brevis Walker, 1852: 411 (*Drosophila*).—U.S.
fronto Walker, 1852: 410 (*Drosophila*).—U.S.
linearis Walker, 1852: 411 (*Drosophila*).—U.S.
minuta Walker, 1852: 412 (*Drosophila*).—U.S.

Family DIASTATIDAE
By J. F. McAlpine

The most recent treatment of the Nearctic Diastatidae (Melander, 1913a) includes all the described species in the region, but a number of undescribed species are now known. Virtually nothing is known about their biology.

REFERENCE: Melander, 1913a (key).

Genus DIASTATA Meigen

Diastata Meigen, 1830: 94. Type-species, *Geomyza obscurella* Fallén (Westwood, 1840: 152; misident.) =*vagans* Loew.
Trichoptera Lioy, 1864a: 1109 (preocc. Meigen, 1803). Type-species, *Diastata adusta* Meigen (Coquillett, 1910b: 616).
Calopterella Coquillett, 1910b: 517 (n. name for *Trichoptera* Lioy). Type-species, *Diastata adusta* Meigen (aut.), despite Coquillett's designation of *Diastata vagans* Loew.

eluta Loew, 1863a: 31 (Cent. 3, no. 59).—Alaska; B.C., Wash., Oreg., Idaho.
modesta Melander, 1913a: 290.—Wash.; B.C. ?=*inornata* Loew.
nebulosa (Fallén).—Probably not Nearctic.
repleta (Walker), 1849: 1099 (*Notiphila*).—Ont.; Man., Que., Pa.
  decemguttata Walker, 1852: 411 (*Drosophila*).—U.S.
  pulchra Loew, 1861b: 359 (Cent. 1, no. 100).—Pa.
vagans Loew, 1864c: 362.—Europe; B.C., Wash., Alta., Ont., N.H., Eurasia.

Genus CAMPICHOETA Macquart

Campichoeta Macquart, 1835: 547. Type-species, *rufipes* Macquart (mon.) =*obscuripennis* (Meigen).
Campichaeta, emend.
Thryptocheta Rondani, 1856: 134. Type-species, *Diastata punctum* Meigen (orig. des.).
Thryptochaeta and *Tryptochaeta*, emends.

REFERENCE: McAlpine, 1962 (rev., world).

griseola (Zetterstedt), 1855: 4799 (*Geomyza*).—Sweden; Alaska, B.C. to Que., s. to Calif., N. Mex., and Mich., also Va., Europe.

micans Hendel, 1911: 44 (*Thryptocheta*).—Wash.

latigena McAlpine, 1962: 4.—Calif.; Ariz.

Family CAMILLIDAE
By J. F. McAlpine

This family is monogeneric for *Camilla* Haliday, with less than a dozen known species, all from the Northern Hemisphere. Nothing is known of the biology of *C. glabra*, but a few species of the genus have been reared from or taken near rodents' nests.

REFERENCES: Duda, 1934a (rev. Palaearctic spp.); Collin, 1956 (key, British spp.); McAlpine, 1960b (tax.).

Genus CAMILLA Haliday

Diastata, subg. **Camilla** Haliday, 1838: 188. Type-species, *Drosophila glabra* Fallén (mon.).

glabra (Fallén), 1823e: 8 (*Drosophila*).—Sweden; Ont., Europe.

Superfamily CHLOROPOIDEA

Family CHLOROPIDAE
By Curtis W. Sabrosky

The family Chloropidae contains small flies that are commonly collected by sweeping grass and other low herbage. Some frequent flowers, and some (eye gnats or *Hippelates* flies) are annoying to man and animals. Most species of the subfamily Chloropinae are yellow and black striped, and this is the common conception of the family; the more abundant species of the Oscinellinae are chiefly dark gray to black, and less conspicuous.

The larvae of many species are saprophagous. Others are phytophagous and feed especially on grasses; among these are the destructive frit flies and wheat stem maggots. A few species are gall formers. *Pseudogaurax* larvae devour masses of eggs, such as those of spiders, mantids, and some Lepidoptera. The larvae of *Thaumatomyia glabra* are predaceous on root aphids, and in the western United States are an important factor in the control of the sugarbeet root aphid.

The only monograph of the Nearctic species (Becker, 1912a) is greatly out of date. Keys to genera by Malloch (1914d) and Curran (1934a) are useful except for the modern groups split from the broad genera *Oscinella* and *Siphonella*, some of which can be identified by keys to Palaearctic Chloropidae. Generic revisions partially replace Becker's monograph.

REFERENCES: Duda, 1932–1933 (monog., Palaearctic spp.); Sabrosky, 1935a (keys to Kans. spp., useful for e. U.S.), 1941b (type-spp., genera of world); Collin, 1946 (key, British genera of Oscinellinae).

Subfamily OSCINELLINAE

Genus LASIOPLEURA Becker

Lasiopleura Becker, 1910a: 130. Type-species, *Oscinis longepilosa* Strobl (mon.).
Pseudohippelates Malloch, 1913o: 261. Type-species, *Hippelates capax* Coquillett (orig. des.).

REFERENCE: Sabrosky, 1951b (rev.).

barberi Sabrosky, 1951b: 339.—Fla.; Miss., Ala., Ga.
capax (Coquillett), 1898f: 48 (*Hippelates*).—Ill.; S. Dak. to Mich., s. to Ill. and Ohio, also Miss., N.C.
grisea Malloch, 1934a: 417.—Ariz.; Wash. and Idaho, s. to Guatemala.
hirta (Loew), 1863a: 39 (Cent. 3, no. 75) (*Oscinis*).—Ill.; S.C.
hirtoides Sabrosky, 1951b: 341.—Md.; Mich. to Md., s. to Kans. and N.C., also Calif.
itascae Sabrosky, 1951b: 342.—Minn.; Mich., N.J., N.C., Ga.
longula (Becker), 1912a: 89 (*Hippelates*).—Grenada (as Canada, error). Not Nearctic.
shewelli Sabrosky, 1951b: 342.—Ont.; N.W.T., Mich.
willistoni Sabrosky, 1951b: 340.—N. Mex.

Genus HIPPELATES Loew

Hippelates Loew, 1863a: 35, 36 (Cent. 3, no. 67, 68). Type-species, *plebejus* Loew (Coquillett, 1910b: 552).
Hippelates, subg. *Siphomyia* Williston, 1896c: 418. Type-species, *proboscideus* Williston (Coquillett, 1910b: 606).
Liohippelates Duda, 1929: 165, 169. Type-species, *Hippelates pusio* Loew (Sabrosky, 1941b: 755).

Some of the species of this genus are annoying eye gnats; others feed at the body openings or on serous exudates of sores or wounds.

REFERENCE: Sabrosky, 1941a (key).

apicatus Malloch, 1930o: 248.—P.R.; Fla.
bicolor Coquillett, 1898f: 48.—Fla.; coastal, N.Y. to Tex., Bahama Is.
bishoppi Sabrosky, 1941a: 26.—Fla.; Sask. to Que. and N.H., s. to Colo., Tex., and Fla.
collusor (Townsend), 1895b: 619 (*Oscinis*).—Mexico (Baja Calif.); Calif., Nev., Ariz., n. Mexico.
convexus Loew, 1866a: 184 (Cent. 6, no. 94).—Cuba; Fla., Bermuda, Neotropical.
dissidens (Tucker), 1908: 274 (*Oscinis*).—Tex.; Iowa to Mich. and N.Y., s. to Ariz., n. Mexico, and Fla.
    texanus Malloch, 1930o: 251.—Tex.
dorsalis Loew, 1869b: 42 (Cent. 8, no. 75).—Cuba; Calif. to Tex., Fla., Neotropical.
flaviceps (Loew), 1863a: 38 (Cent. 3, no. 71) (*Oscinis*).—Cuba; Tex., Ga., Fla., Neotropical.
genalis Thomson, 1869: 608.—Calif. (?error). Probably not Nearctic.
hermsi Sabrosky, 1941a: 27.—Calif.; Calif. and Nev. to n. Mexico and Tex.
impressus Becker, 1912a: 92.—Tex.; Colo., Ariz., N. Mex., Mexico.
microcentrus Coquillett *in* Baker, 1904: 28.—Calif.; Oreg. to Colo., s. to Calif. and N. Mex.
montanus Sabrosky, 1941a: 27.—Colo.; Calif. to Colo. and N. Mex.
nobilis Loew, 1863a: 35 (Cent. 3, no. 67).—Ill.; N. Dak. to Ont. and Mass., s. to Colo., Tex. and Fla.
pallipes (Loew), 1863a: 37 (Cent. 3, no. 69) (*Oscinis*).—Cuba; N. Dak. to Que. and Maine, s. to Utah, Tex., and Fla., also Calif., Cuba.
    partitus Becker, 1912a: 89.—Ala.
    nitidifrons Malloch, 1930o: 243.—Md.
    flavipes, authors, not Loew.
particeps (Becker), 1912a: 115 (*Oscinella*).—Pa., Tex., Ala.; N. Dak. to Que., s. to Utah, Mexico, and Fla., also Oreg.
    subvittatus Malloch, 1930o: 251.—Tex.
peruanus Becker, 1912a: 170.—Peru, Argentina, Paraguay; s. Fla.
plebejus Loew, 1863a: 36 (Cent. 3, no. 68).—D.C.; S. Dak. to Que., s. to N. Mex., Mexico, and Fla.
proboscideus Williston, 1896c: 418.—St. Vincent I.; Tex., Ga., Fla., Neotropical.
pusio Loew, 1872a: 103 (Cent. 10, no. 87).—Tex.; Wash. to N. Dak. to Pa., s. to Calif., Mexico, and Fla., Bermuda.
    splendens Adams, 1904b: 453.—Wyo.
    lituratus Becker, 1912a: 87 (in key).—Tex. (unpublished, from type).

robertsoni Sabrosky, 1941a: 26.—Calif.; Oreg., Calif. to Okla., s. to Mexico, ?Guatemala.
saundersi Kumm, 1936: 326.—Jamaica; Tex., Miss., Ga., Fla.
tener Coquillett, 1900g: 265.—P.R.; Tex., s. Fla., Neotropical.

Genus CADREMA Walker

Cadrema Walker, 1859: 117. Type-species, *lonchopteroides* Walker (mon.).
Prohippelates Malloch, 1913o: 260. Type-species, *Hippelates pallidus* Loew (orig. des.).

pallida (Loew), 1866a: 184 (Cent. 6, no. 93) (*Hippelates*).—Cuba; s. Fla., nearly tropicopolitan.

Genus CERATOBARYS Coquillett

Ceratobarys Coquillett, 1898f: 45. Type-species, *Hippelates eulophus* Loew (orig. des.).

eulophus (Loew), 1872a: 104 (Cent. 10, no. 88) (*Hippelates*).—Tex.; Kans. to Del., s. to Tex. and Fla.

Genus ELACHIPTERA Macquart

Elachiptera Macquart, 1835: 621. Type-species, *Chlorops brevipennis* Meigen (orig. des.).
Crassiseta Roser, 1840: 63. Type-species, *Oscinis cornuta* Fallén (Corti, 1909: 121).
Melanochaeta, subg. *Doliomyia* Johannsen, 1924a: 89. Type-species, *longiventris* Johannsen (orig. des.).

REFERENCE: Sabrosky, 1948b (synopsis).

Subgenus MELANOCHAETA Bezzi

Crassiseta, subg. *Pachychoeta* Bezzi, 1895: 72 (preocc. Bigot, 1857). Type-species, *Elachiptera aterrima* Strobl (orig. des.)=*capreola* (Curtis).
Melanochaeta Bezzi, 1906: 50 (as genus; n. name for *Pachychoeta* Bezzi). Type-species, *Elachiptera aterrima* Strobl (aut.)=*capreola* (Curtis).

eunota (Loew), 1872a: 104 (Cent. 10, no. 89) (*Crassiseta*).—Tex.; Alta. to Que., s. to Oreg., Tex., and Va.
kaw (Sabrosky), 1948b: 371 (*Melanochaeta*).—Kans.; Kans. to Mich. and Md., s. to Tex. and Fla.
melampus (Becker), 1912a: 84 (*Melanochaeta*).—Idaho; Calif., Mich.

Subgenus ELACHIPTERA Macquart

angusta Sabrosky, 1948b: 376.—Mass.; Ill., N.J., Va. to Fla.
angustifrons Sabrosky, 1948b: 378.—Fla.; coastal states, Md. to Tex.
angustistylum Sabrosky, 1948b: 378.—Utah; Oreg.
californica Sabrosky, 1948b: 380.—Calif.
costata (Loew), 1863a: 33 (Cent. 3, no. 62) (*Crassiseta*).—D.C.; Wash. and Alta. to Que. and Maine, s. to Ariz., Tex., and S.C.
decipiens (Loew), 1863a: 40 (Cent. 3, no. 76) (*Oscinis*).—Alaska; Alaska and N.W.T. to Que., s. to Calif. and Colo.
erythropleura Sabrosky, 1948b: 375.—Mich.; N. Dak. to Mass., s. to Colo. and Ga.
flaviceps Sabrosky, 1948b: 382.—Idaho; Oreg., Calif.
formosa (Loew), 1863a: 32 (Cent. 3, no. 61) (*Crassiseta*).—D.C.; Ill., Ind., Del. to Va., Fla.
knowltoni Sabrosky, 1948b: 381.—Utah; Idaho, Mont., Nev., Colo.
longiventris (Johannsen), 1924a: 89 (*Melanochaeta*).—N.Y.; Mich., Tex.
nigriceps (Loew), 1863a: 33 (Cent. 3, no. 63) (*Crassiseta*).—Pa.; Mont. to Que. and Maine, s. to Tex. and Fla., also Wash. to Calif.
pechumani Sabrosky, 1948b: 377.—N.Y.; Ill., Mich.
penita (Adams), 1908: 152 (*Crassiseta*).—Wis.; ?Mass.
punctulata Becker, 1912b: 645 (n. name for *nigroscutellata* Becker, 1912).—N. Amer. (?error).
    nigroscutellata Becker, 1912a: 80 (preocc. Becker, 1911). N. Amer. (?error).
tau Sabrosky, 1948b: 374.—Ky.
vittata Sabrosky, 1948b: 380 (n. name for *bilineata* Adams).—Ariz.; Wash. and Alta. to Ont. and N.H., s. to Calif., Ariz., Minn., and Md.
    bilineata Adams, 1904b: 453 (preocc. Bigot, 1891).—Ariz.
willistoni Sabrosky, 1948b: 373.—Fla.; Tex.
    flavida, authors, not Williston.

Genus ERIBOLUS Becker

Eribolus Becker, 1910a: 127. Type-species, *sudeticus* Becker (Enderlein, 1911b: 206)=*nanus* (Zetterstedt).
    REFERENCE: Sabrosky, 1950b (key).

californicus Sabrosky, 1950b: 91.—Calif.; Mexico.

longulus (Loew), 1863a: 34 (Cent. 3, no. 64) (*Crassiseta*).—D.C.; Alta. to Que. and Maine, s. to Colo., Tex., and Fla., ?Idaho.
    intermedia Becker, 1912a: 83 (*Melanochaeta*).—Pa.
nanus (Zetterstedt), 1848: 2650 (*Oscinis*).—Sweden; Alaska and N.W.T. to Que. and Maine, s. to Calif., Colo., S. Dak., and Mass., Europe.
    sudeticus Becker, 1910a: 127.—Silesia, Bohemia.
    planicollis Becker, 1912a: 114 (*Oscinella*).—Idaho. **N. syn.**
nearcticus Sabrosky, 1948b: 370.—N. Dak.; Alaska, Wash., Man. to Ont., s. to Ill.

Genus OSCINISOMA Lioy

Oscinisoma Lioy, 1864a: 1125. Type-species, *Chlorops vitripennis* Meigen (Coquillett, 1910b: 582) = *cognatum* (Meigen). *Oscinosoma*, emend. or error.

alienum (Becker), 1912a: 81 (*Elachiptera*).—Mass.; S. Dak. to Ont. and Mass., s. to Mo. and S.C. **N. comb.**

Genus HAPLEGINELLA Duda

Hapleginella Duda, 1933: 77. Type-species, *Oscinis laevifrons* Loew (orig. des.).

conicola (Greene), 1918: 69 (*Madiza*).—Oreg.; Calif., Utah, Wis., Ont., Que., ?Ark. **N. comb.**

Genus MONOCHAETOSCINELLA Duda

Monochaetoscinella Duda, 1930: 107. Type-species, *Oscinis anonyma* Williston (Duda, 1931: 166).

anonyma (Williston), 1896c: 423 (*Oscinis*).—St. Vincent I.; Ariz., Tex., Fla., Bermuda, Neotropical.
nigricornis (Loew), 1863a: 34 (Cent. 3, no. 65) (*Crassiseta*).—La.; Kans. to Va., s. to Tex. and Fla.

Genus STENOSCINIS Malloch

Stenoscinis Malloch, 1918d: 21. Type-species, *Oscinis longipes* Loew (orig. des.).
    REFERENCE: Sabrosky, 1961a (key).

adachiae Sabrosky, 1961a: 22.—Ariz.
atriceps (Loew), 1863a: 39 (Cent. 3, no. 74) (*Oscinis*).—Pa.; Wyo. to Ont. and Vt., s. to Okla. and N.C.

longipes (Loew), 1863a: 41 (Cent. 3, no. 77) (*Oscinis*).—D.C.; S. Dak. to Conn., s. to Tex. and Ga.

Genus RHOPALOPTERUM Duda

Oscinella, subg. **Rhopalopterum** Duda, 1929: 167. Type-species, *limitata* Becker (mon.).
Rhopalopternum, emend. or error.

REFERENCE: Sabrosky, 1961a (key, with *Stenoscinis*).

limitatum (Becker), 1912a: 211 (*Oscinella*).—Haiti; Fla., Bermuda, Neotropical.

Genus OSCINELLA Becker

Oscinella Becker, 1909b: 120. Type-species, *deficiens* Becker (orig. des., as gen. n., sp. n.).
Botanobia, authors, not Lioy.
Oscinis, authors, not Latreille.

The following list contains four or five distinct species groups, but complications of type designations, needed comparisons with Palaearctic and Neotropical species, and lack of available names for some of the groups have made division impracticable at this time.

REFERENCES: Sabrosky, 1936 (key), 1939a (key, *O. frit* and relatives), 1940 (partial key).

beameri Sabrosky, 1940: 221.—Kans.
bispina (Malloch), 1918g: 109 (*Botanobia*).—Ill.; Man. to Maine, s. to Colo. and W. Va.
carbonaria (Loew), 1869b: 42 (Cent. 8, no. 76) (*Oscinis*).—D.C.; Ont. to N.S., entire U.S., Mexico.
    salinaria Curran, 1927i: 292 (*Botanobia*).—N.S.
coxendix (Fitch), 1856a: 533 (1856: 301) (*Oscinis*).—N.Y.; Alaska to Que., entire U.S., Mexico.
    pullicornis Sabrosky, 1936: 721 (as var.).—17 States and D.C.
criddlei (Aldrich), 1918c: 341 (*Oscinis*).—Man.; Man. to Mich., s. to Kans. and Ind., also Utah.
dorsata (Loew), 1872a: 115 (Cent. 10, appendix) (*Oscinis;* n. name for *dorsalis* Loew, 1869).—R.I.; S. Dak. to Maine, s. to Kans. and R.I.
    dorsalis Loew, 1869b: 43 (Cent. 8, no. 77) (*Oscinis;* preocc. Loew, 1863).—R.I.
    spiniger Malloch, 1918g: 109 (*Botanobia*).—Ill.

frit (Linnaeus), 1758: 598 (*Musca*).—Sweden; Canada, U.S. except Fla., Mexico, Palaearctic.
> *nigra* Tucker, 1908: 272 (*Oscinis*).—Colo.

frontoorbitalis Sabrosky, 1940: 221.—Utah; Calif. to Utah and N. Mex., s. to Mexico.

fuscipalpis Becker, 1912a: 111.—Wash.

gigas Sabrosky, 1940: 224.—N.Y.; S. Dak., Minn., Iowa, Ill., W. Va.

grandissima Sabrosky, 1940: 226.—Colo.; N. Mex.

hesperia Sabrosky, 1940: 219.—Utah; B.C. to S. Dak., s. to Mexico.

incerta Becker, 1912a: 116.—Idaho; N.W.T. to Labr., s. to Ariz., Kans., and Mass., also Tenn.
> *opacifrons* Aldrich *in* Sabrosky, 1936: 724. Nomen nudum.

infesta Becker, 1912a: 108.—Mass.; coastal states, Maine to Tex.

insularis Malloch, 1914i: 26 (*Botanobia*).—Va.; Mass., Md., Ga.

lugubria Sabrosky, 1940: 225.—Maine; Ont. and Que. to Conn., also Wis.

luteiceps Sabrosky, 1940: 222.—Ill.; S. Dak. to Ont. and Maine, s. to Kans. and Tenn.

mallochi Sabrosky, 1938: 417 (n. name for *halterata* Malloch).—D.C.; S. Dak. to Mass., s. to Okla. and Md.
> *halterata* Malloch, 1913f: 47 (*Botanobia*; preocc. Lamb, 1912).—D.C.

minor (Adams), 1905a: 110 (*Oscinis*).—La.; Sask. to Ont. and N.H., s. to Ariz., Mexico, and Fla.

neocoxendix Sabrosky, 1940: 215.—Ga.; Man. to Que. and Maine, s. to Tex. and Fla., Mexico.

nitidissima (Meigen), 1838: 388 (*Chlorops*).—Bavaria; Alaska to Ont. and Vt., s. to Calif., N. Mex., Kans., Tenn., and Md., Palaearctic.

nudiuscula (Loew), 1863a: 37 (Cent. 3, no. 70) (*Oscinis*).—Ga.; Wyo. to Ont., s. to Miss. and Ga.

ochripes Sabrosky, 1940: 218.—Calif.; Utah.

ovalis (Adams), 1905a: 110 (*Oscinis*).—Ga.; Man. to Mass., s. to Tex. and Fla., also Ariz.
> *marginalis* Malloch, 1914i: 25 (*Botanobia*).—Fla.
> *proxima* Malloch, 1914i: 25 (*Botanobia*).—Va.

painteri Sabrosky, 1940: 223.—Md.; Man. to Conn., s. to Tex. and Fla.

pectoralis (Coquillett), 1898f: 49 (*Oscinis;* preocc. Becker, 1910).—N.H. Not now renamed because it may not remain in the genus.

soror (Macquart), 1851: 279 (1851: 306) (*Chlorops*).—N. Amer.; Man. to Que., s. to Utah, Tex., and Fla.
> *tibialis* Fitch, 1856a: 532 (1856: 300) (*Oscinis*).—N.Y.

variabilis Loew, 1863a: 42 (Cent. 3, no. 79) (*Oscinis*).—D.C.
ainsliei Curran, 1927i: 292 (*Botanobia*).—Minn.
umbrosa (Loew), 1863a: 39 (Cent. 3, no. 73) (*Oscinis*).—Pa.; Man. to Que. and Maine, s. to Tex. and Fla., also Utah.
virgata (Coquillett), 1898f: 49 (*Oscinis*).—Colo.; Wash. to Man. to N.H., s. to Utah and Kans.

Genus OSCINOIDES Malloch

Oscinoides Malloch, 1916b: 86. Type-species, *arpidia* Malloch (orig. des.).

> REFERENCE: Malloch, 1918d (key to vars.).

arpidia Malloch, 1916b: 87.—Ill.; S. Dak. to Maine, s. to Tex. and Ga.
    var. **ater** Malloch, 1918d: 19.—Ill.; Minn., Iowa, Maine, Tex.
    dorri Johnson, 1925a: 47 (*Gaurax*).—Maine.
    var. **elegans** Malloch, 1918d: 19.—Ill.
    var. **humeralis** Malloch, 1918d: 19.—Ill.; S. Dak., Mich., N.Y., N.C., Ga.

Genus RHODESIELLA Adams

Rhodesiella Adams, 1905b: 197. Type-species, *tarsalis* Adams (orig. des.).

> REFERENCE: Sabrosky, 1947 (tax.).

brimleyi Sabrosky, 1947: 174.—N.C.; Tex., Miss., Fla., Md.

Genus GAURAX Loew

Gaurax Loew, 1863a: 35 (Cent. 3, no. 66). Type-species, *festivus* Loew (mon.).
Botanobia Lioy, 1864a: 1125. Type-species, *Oscinis dubia* Macquart (mon.).
Neogaurax Malloch, 1914d: 119. Type-species, *Gaurax montanus* Coquillett (orig. des.).

> REFERENCE: Sabrosky, 1951a (rev.).

apicalis Malloch, 1915k: 160.—Ill.; Wis. to N.H., s. to Ill. and Va.
atrilinea Sabrosky, 1951a: 415.—N.Y.
atripalpus Sabrosky, 1951a: 427.—Mich.; Ont.
basitarsalis Sabrosky, 1951a: 415.—Mass.
dorsalis (Loew), 1863a: 38 (Cent. 3, no. 72) (*Oscinis*).—Pa.; Alta. to Que., s. to Tenn. and Va., also N. Mex., Tex.

dubius (Macquart), 1835: 604 (*Oscinis*).—France; Mich., N.H.
    ephippium Zetterstedt, 1848: 2664 (*Oscinis*).—Sweden.
festivus Loew, 1863a: 35 (Cent. 3, no. 66).—Pa.; Iowa, Mass., Md., Va.
fumipennis (Malloch), 1915e: 108 (*Neogaurax*).—Ill.; ?D.C.
maculicornis Sabrosky, 1951a: 422.—Mass.; Ill., Ont., N.H., Va., Ga.
maculipes Sabrosky, 1951a: 424.—Mich.; Ill.
melanotum Sabrosky, 1951a: 429.—Ind.; Ill. and Mich. to Ont. and Maine.
mesopedalis Sabrosky, 1951a: 427.—Ky.; Tenn.
montanus Coquillett, 1898f: 48.—N.H.; Man., Wis., Ill., Que. to Va.
obscuripennis Johnson, 1913e: 35.—Mass.
ocellaris Sabrosky, 1951a: 418.—Mich.; Ill.
pallidipes Malloch, 1915h: 362.—Ill.; Ohio, Vt., Mass., N.J., Md.
pilosulus (Becker), 1912a: 111 (*Oscinella;* as *pilosella,* p. 106).—La.
pseudostigma Johnson, 1913e: 35.—Ont.; Iowa to Maine, s. to Ala. and N.C.
    flavidulus Malloch, 1915h: 361.—Ill.
    interruptus Malloch, 1915h: 363 (preocc. Becker, 1912).—Ill.
semivittatus Sabrosky, 1951a: 419.—Tenn.; Mass.
shannoni Sabrosky, 1951a: 425.—Md.; Minn. to Que., s. to Ga.
splendidus Malloch, 1915k: 161.—Ill.; Iowa to Que., s. to Tenn. and Va.
tripus Sabrosky, 1951a: 417.—Tenn.; Mich., Mass., N.C.
varihalteratus (Malloch), 1913b: 274 (*Botanobia*).—D.C.; Ill.
vittipes Sabrosky, 1951a: 419.—Maine.

Genus PSEUDOGAURAX Malloch

Pseudogaurax Malloch, 1915k: 159. Type-species, *Gaurax anchora* Loew (orig. des.).

anchora (Loew), 1866b: 51 (Cent. 7, no. 94) (*Gaurax*).—N.Y.; Mich. to Mass., s. to Tex. and Fla.
    dispar Williston *in* Forbush and Fernald, 1896: 390 (*Elachiptera*).—Mass.
floridensis Sabrosky, 1950a: 33.—Fla.; Jamaica.
signatus (Loew), 1876: 338 (*Gaurax*).—Tex.; Mo. to Md., s. to Mexico and Ga., also Ariz., Calif., Cuba.
    araneae Coquillett *in* Davidson, 1896: 320 (*Gaurax*).—Calif.
    mallochi Duda, 1930: 87 (*Gaurax;* as var.).—Calif.

Genus DICRAEUS Loew

Dicraeus Loew, 1873a: 51. Type-species, *obscurus* Loew (mon.)= *raptus* (Haliday).

REFERENCE: Sabrosky, 1950e (rev.).

aberrans Sabrosky, 1950e: 56.—N. Mex.
elongatus Sabrosky, 1950e: 55.—Ariz.; ?N. Mex.
incongruus Aldrich, 1918c: 339.—Idaho, Mont., Man., Utah; Alaska to Que., s. to Calif., Okla., and Tenn.
ingratus (Loew), 1866c: 26 (*Eutropha*).—Germany; Idaho, Mich., Europe.
  ruficeps, authors, not Meigen.
tibialis (Macquart), 1835: 602 (*Oscinis*).—France; Alaska, Sask., N. Dak., S. Dak., Mich., Pa., Europe.
wilburi Sabrosky, 1950e: 61.—Kans.; Minn., Mich., Ont., Mo.

Genus OPETIOPHORA Loew

Opetiophora Loew, 1872a: 105 (Cent. 10, no. 90). Type-species, *straminea* Loew (mon.).

straminea Loew, 1872a: 105 (Cent. 10, no. 90).—Tex.; Utah to Minn. and Mich., s. to Mexico.

Genus EUGAURAX Malloch

Eugaurax Malloch, 1913f: 46. Type-species, *floridensis* Malloch (orig. des.).

REFERENCE: Sabrosky, 1950c (rev.).

floridensis floridensis Malloch, 1913f: 46.—Fla.
  spp. **vittatus** Sabrosky, 1950c: 187.—Ind.; Ill., Que., N.J., Va.

Genus CONIOSCINELLA Duda

Oscinella, subg. **Conioscinella** Duda, 1929: 166. Type-species, *soluta* Becker (Sabrosky, 1941b: 751).

REFERENCE: Sabrosky, 1936 (key).

aequa (Becker), 1912a: 102 (*Siphonella*).—Idaho; B.C., Oreg., Colo., Minn. to Mich., Que. **N. comb.**
difficilis (Becker), 1912a: 107 (*Oscinella*).—Wash.; B.C. **N. comb.**
  apparens Becker, 1912a: 109 (*Oscinella*).—Wash. **N. comb.**
finalis (Becker), 1912a: 99 (*Siphonella*).—Alaska; Calif. **N. comb.**
flavescens (Tucker), 1908: 272 (*Oscinis*).—Colo.; S. Dak. to N.Y., s. to Calif. and Fla. **N. comb.**

grisescens (Sabrosky), 1940: 227 (*Oscinella*).—Ga.; Ala., Fla. **N. comb.**

hinkleyi (Malloch), 1915d: 12 (*Botanobia*).—Ill.; Kans. to Pa., s. to La. and Ga. **N. comb.**

melancholica (Becker), 1912a: 109 (*Oscinella*).—Mich.; Alaska to Labr., entire U.S., Mexico. **N. comb.**

nuda (Adams), 1905a: 110 (*Oscinis*).—Ga.; Ariz. to Kans. and Md., s. to Mexico and Fla. **N. comb.**

> *frontalis* Tucker, 1908: 273 (*Oscinis*).—Tex. **N. comb.**

triangulata (Becker), 1912a: 102 (*Siphonella*).—Wash.; Man. **N. comb.**

triorbiculata (Sabrosky), 1940: 229 (*Oscinella*).—Ga.; Minn. to N.Y., s. to Tex. and Fla. **N. comb.**

Genus DASYOPA Malloch

Dasyopa Malloch, 1918d: 20. Type-species, *pleuralis* Malloch (orig. des.) =*latifrons* (Loew).

latifrons (Loew), 1872a: 106 (Cent. 10, no. 91) (*Siphonella*).—Tex.; S. Dak. to Pa., s. to Ariz. and Ga.

> *pleuralis* Malloch, 1918d: 20.—Ill.

Genus OLCELLA Enderlein

Olcella Enderlein, 1911b: 231. Type-species, *mendozana* Enderlein (orig. des.).

REFERENCE: Sabrosky, 1936 (key).

cinerea (Loew), 1863a: 43 (Cent. 3, no. 81) (*Siphonella*).—Fla.; Mich. to Mass., s. to N. Mex., Mexico, and Fla. **N. comb.**

> *trivittata* Sabrosky, 1935c: 109 (*Madiza*).—Kans. **N. comb.**

parva (Adams), 1904c: 104 (*Siphonella*).—Colo.; B.C. to Ont. and Mass., s. to Mexico and Fla. **N. comb.**

projecta (Malloch), 1913c: 283 (*Madiza*).—N. Mex.; Calif. to S. Dak., s. to Mexico and Ark. **N. comb.**

provocans (Becker), 1912a: 98 (*Siphonella*).—Wis.; Alaska, Sask. to Que., s. to Utah and N.J., also Calif. **N. comb.**

punctifrons (Becker), 1912a: 101 (*Siphonella*).—Utah; Oreg. to S. Dak., s. to Mexico. **N. comb.**

pygmaea (Becker), 1912a: 97 (*Siphonella*).—Wyo. **N. comb.**

quadrivittata (Sabrosky), 1935c: 108 (*Madiza*).—Kans.; Minn. to Maine, s. to Kans. and Md., also Utah, N. Mex., Fla. **N. comb.**

submarginalis (Sabrosky), 1935c: 111 (*Madiza*).—Kans.; S. Dak. to Ariz. and Fla. **N. comb.**
trigramma (Loew), 1863a: 42 (Cent. 3, no. 80) (*Oscinis*).—D.C.; S. Dak. to Mass., s. to Tex. and Fla. **N. comb.**
quinquelineata Adams, 1904c: 104 (*Siphonella*).—Ga. **N. comb.**

Genus TRICIMBA Lioy

Tricimba Lioy, 1846a: 1125. Type-species, *Oscinis lineella* Fallén (Enderlein, 1911b: 207).
Notonaulax Becker, 1903: 153. Type-species, *Oscinis lineella* Fallén (Enderlein, 1911b: 207).
Hammaspis Duda, 1930: 76. Type-species, *Tricimba spinigera* Malloch (mon.).

REFERENCE: Sabrosky, 1938 (key).

brunnicollis (Becker), 1912a: 103 (*Notonaulax*).—Wash.; Oreg., Calif.
cincta (Meigen), 1830: 162 (*Chlorops*).—Germany; B.C., Calif., Europe.
lineella (Fallén), 1820a: 8 (*Oscinis*).—Sweden; Nebr. to Mich. and Tenn., also Que., Mass., N.J., Europe.
occidentalis Sabrosky, 1938: 431.—Calif.; Wash., Oreg., Idaho, Mont.
spinigera Malloch, 1913g: 60.—D.C.; Iowa to Maine, s. to Tex. and Ga.
trisulcata (Adams), 1905a: 111 (*Oscinis*).—La.; S. Dak. to Ont. and N.H., s. to Tex. and S.C.

Genus APHANOTRIGONUM Duda

Aphanotrigonum Duda, 1932: 35. Type-species, *Chlorops trilineata* Meigen (orig. des.).

darlingtoniae (Jones), 1916: 389 (*Botanobia*).—Calif. **N. comb.**
scabrum (Aldrich), 1918c: 342 (*Oscinis*).—Man.; Alaska and N.W.T. to Maine, s. to Utah, Ind., and Conn. **N. comb.**

Genus CHAETOCHLOROPS Malloch

Chaetochlorops Malloch, 1914d: 120 (as *Chactochlorops*, p. 116). Type-species, *Siphonella inquilina* Coquillett (orig. des.).

REFERENCE: Sabrosky, 1950c (synopsis, biol.).

inquilinus (Coquillett), 1898f: 48 (*Siphonella*).—Mo., Va.; S. Dak. to Ont. and Mass., s. to Tex. and Fla.

Genus GONIOPSITA Duda

Goniopsita Duda, 1930: 69, 72. Type-species, *Goniopsis verrucosa* Duda (Sabrosky, 1941b: 753).

catalpae (Malloch), 1913g: 63 (*Botanobia*).—Pa.; Kans. to Mich. and Mass., s. to Tex. and Ga. **N. comb.**

magnipalpis (Becker), 1912a: 111 (*Oscinella*).—Idaho. **N. comb.**

Genus LIPARA Meigen

Lipara Meigen, 1830: 1. Type-species, *lucens* Meigen (mon.).

Lipara lucens Meigen causes a conspicuous gall on the giant reed, *Phragmites*.

lucens Meigen, 1830: 1.—Germany; Conn., Europe. Immigrant.

Genus CALAMONCOSIS Enderlein

Calamoncosis Enderlein, 1911b: 235. Type-species, *Lipara rufitarsis* Loew (orig. des.; misident.)=*minima* (Strobl).

Unnamed sp.—Mich.

Genus SIPHONELLA Macquart

Siphonella Macquart, 1835: 584. Type-species, *Madiza oscinina* Fallén (orig. des.).

Madiza, authors, not Fallén, as recognized in Milichiidae.

Rondani's (1856) designation of *oscinina* as type-species of *Madiza* Fallén unfortunately made the latter the senior synonym of *Siphonella*. However, widespread usage confined *Madiza* to the family Milichiidae, where it is the type of a subfamily, and that usage has been maintained in order to minimize confusion. Application to the I.C.Z.N. is required.

Of the following species, only *oscinina* properly belongs to *Siphonella*. The other species are left in the list pending study of their relationships.

REFERENCE: Sabrosky, 1936 (key).

abdominalis Becker, 1912a: 99.—Kans.; S. Dak. to Ont., s. to Kans. and Ohio.

bifurca Becker, 1912a: 109 (*Oscinella*).—Kans.

diabolus Becker, 1912a: 99.—Idaho; Oreg., Calif.

excipiens Becker, 1912a: 101.—Idaho; Mont.

extrema Becker, 1912a: 100.—Calif.

neglecta Becker, 1912a: 100.—Calif.; B.C. to Man. and Mich., s. to Mexico.

nigripalpis (Malloch), 1913c: 282 (*Madiza*).—Md.; Sask. to Que., s. to Ariz. and Fla.
oscinina (Fallén), 1820a: 9 (*Madiza*).—Sweden; Alaska to Labr. and Nfld., s. to Calif., N. Mex., Mo., Ala., and N.C.' Palaearctic.
setulosa (Malloch), 1918g: 110 (*Madiza*).—Ill.; Mont. to Ont. and Vt., s. to Utah and N.C.

Subfamily CHLOROPINAE

Genus THAUMATOMYIA Zenker

Thaumatomyia Zenker, 1833: 344. Type-species, *prodigiosa* Zenker (mon.) =*notata* (Meigen).
Chloropisca Loew, 1866c: 79. Type-species, *Chlorops glaber* Meigen (Coquillett, 1910b: 522).
Pseudochlorops Malloch, 1914d: 119. Type-species, *Chlorops unicolor* Loew (orig. des.; misident.) =*pulla* (Adams).

REFERENCE: Sabrosky, 1943a (key).

annulata (Walker), 1849: 1119 (*Chlorops*).—Ont.; Alaska and Y.T. to Que., s. to Calif., N. Mex., Nebr., and N.J.
  variceps Loew, 1863a: 46 (Cent. 3, no. 86) (*Chlorops*).—Pa.
  prolifica Osten Sacken *in* Lintner, 1888a: 70 (1888: 190) (*Chloropisca*).—N.Y.
  marianapolitana Ouellet, 1934: 320 (*Chloropisca*).—Que.
apache Sabrosky, 1943a: 111.—Ariz.; N. Mex., Tex., n. Mexico.
  sulfurifrons, authors, not Duda.
appropinqua (Adams), 1903a: 39 (*Chlorops*).—Colo., Wyo., Kans.; Wash. to Sask. and N. Dak., s. to Calif., n. Mexico, and Tex.
bistriata (Walker), 1849: 1119 (*Chlorops*).—Ont.; N. Dak. to Ont. and Maine, s. to Utah and Fla.
  clypeata Malloch, 1914d: 119 (*Chloropisca;* as *glabra* var.).—Ill.
glabra (Meigen), 1830: 149 (*Chlorops*).—Europe; Alaska, Canada, U.S., n. Mexico.
  assimilis Macquart, 1851: 279 (1851: 306) (*Chlorops*).—N. Amer.
  obesa Fitch, 1856a: 531 (1856: 299) (*Siphonella*).—N.Y.
  trivialis Loew, 1863a: 47 (Cent. 3, no. 87) (*Chlorops*).—D.C.
  hortensis Fitch, 1872: 363 (*Chlorops*).—N.Y.
  halteralis Adams, 1903a: 41 (*Chlorops*).—Ariz.
grata (Loew), 1863a: 50 (Cent. 3, no. 92) (*Chlorops*).—Pa.; B.C. to Que. and Maine, s. to Oreg., Colo., Ohio, and N.C.

obtusa (Malloch), 1914d: 118 (*Chloropisca*).—Ill.; Mich., Ont., Mass., R.I.

parviceps (Malloch), 1915k: 158 (*Chloropisca*).—Ill.; Minn. to Vt., s. to Ill. and Va., also Ark.

pulla (Adams), 1904a: 303 (*Chlorops*).—Colo.; B.C. to Que., entire U.S., n. Mexico.

    monticola Becker, 1912a: 30 (*Chloropisca*).—Colo.

    punctum Becker, 1912a: 35 (*Chloropisca*).—Tex.

    integra Becker, 1912a: 66 (*Chlorops*).—Mass.

pullipes (Coquillett), 1898f: 47 (*Chlorops*).—Colo., N. Mex.; Alaska, Oreg. to Mont., s. to Ariz. and N. Mex.

rubida (Coquillett), 1898f: 46 (*Chlorops*).—Calif., Colo.; Oreg. to Mont., s. to Calif. and N. Mex.

rubrivittata Sabrosky, 1943a: 110.—Calif.

trifasciata (Zetterstedt), 1848: 2609 (*Oscinis*).—Scandinavia; Alaska to Calif. and Sask., also Labr.

Genus TRIGONOMMA Enderlein

Trigonomma Enderlein, 1911c: 124. Type-species, *lippulum* Enderlein (orig. des.).

fossulatum (Loew), 1863a: 43 (Cent. 3, no. 82) (*Chlorops*).—Cuba; s. Tex., Neotropical. **N. comb.**

    atra Curran, 1926a: 3 (*Chloropisca*).—P.R. **N. comb.**

Genus CETEMA Hendel

Centor Loew, 1866c: 7 (preocc. Schönherr, 1848). Type-species, *Oscinis cereris* Fallén (Coquillett, 1910b: 519).

Cetema Hendel, 1907a: 98 (n. name for *Centor* Loew). Type-species, *Oscinis cereris* Fallén (aut.).

procera (Loew), 1872a: 106 (Cent. 10, no. 92) (*Chlorops*).—Conn.; New England, also Md.

subvittata (Loew), 1863a: 41 (Cent. 3, no. 78) (*Oscinis*).—D.C.; Man. to Que. and Maine, s. to Iowa and Va., also Utah, Colo., and N. Mex.

    hypocera Becker, 1912a: 27.—Wis.

Genus ANTHRACOPHAGA Loew

Anthracophaga Loew, 1866c: 15. Type-species, *Musca strigula* Fabricius (Coquillett, 1910b: 508).

declinata Becker, 1912a: 45.—Mich.; Colo. to Mich. and N.Y., s. to Tex. and N.C.

pulvera Harbeck *in* Fox, 1905: 62 (*Chlorops*). Nomen nudum.

Genus CHLOROPS Meigen

Chlorops Meigen, 1803: 278. Type-species, *Musca pumilionis* Bjerkander (I.C.Z.N., 1955a: 423).
Oscinis Latreille, 1804: 196. Type-species, *Musca lineata* Fabricius (Zetterstedt, 1838: 778) = *pumilionis* (Bjerkander). Zetterstedt actually designated *Musca nasuta* Schrank, but showed the originally included *M. lineata* as its synonym.

The suggestion that *Titania* Meigen (1800: 35) was the same as *Chlorops* is patently erroneous, and the former is here regarded as a genus dubium (also suppressed by I.C.Z.N., 1955a: 423).

The genus *Chlorops* needs revision, and some of the species listed here will undoubtedly have to be removed.

abdominalis Coquillett, 1895a: 318.—Fla.; Ill. and Mich. to Mass., s. in coastal states to Fla. and Ala.
adamsi Sabrosky, 1935b: 81 (n. name for *annulata* Adams).—La.
    annulata Adams, 1904a: 304 (preocc. Walker, 1849).—La.
brunnipennis Becker, 1912a: 58.—N.J., Tex.
certimus Adams, 1904a: 304.—Mass.; Alaska to Que., s. to Utah, Tex., and N.C.
cinerapennis Adams, 1903a: 40.—Kans.; S. Dak. to N.Y., s. to Okla., Ohio, and N.C., also Miss.
    albifascies Adams, 1903a: 42.—Mo.
constrictus Becker, 1912a: 57.—Mich.; S. Dak., Kans., Ill., Ind.
crocotus Loew, 1863a: 48 (Cent. 3, no. 89).—Pa.; Minn. to Mass., s. to Colo., Pa., and Va.
distichliae (Malloch), 1918k: 386 (*Anthracophaga*).—Calif. **N. comb.**
egregius Becker, 1912a: 55.—Oreg.; Wash.
genarum Becker, 1912a: 62.—S. Dak.; Idaho, Kans.
gramineus Coquillett, 1898f: 47.—Calif.
ingratus Williston, 1893b: 156.—Ohio; Kans., Mich., Ind.
    interrupta Becker, 1912a: 44 (*Anthracophaga*).—Kans. **N. comb.**
    fossae Becker, 1912a: 52.—Mich., Kans. **N. syn.**
laevis Becker, 1912a: 63.—Wash.
languidus Becker, 1912a: 58.—Pa.; Mass.
lascivus Adams, 1904a: 303.—La.
lituratus Adams, 1903a: 40.—Wyo.; Oreg. to Mont., s. to Calif. and Colo.

melanocerus Loew, 1863a: 49 (Cent. 3, no. 91).—D.C.; Minn. to Que. and Mass., s. to La. and N.C.
melleus Loew, 1872a: 111 (Cent. 10, no. 100).—Tex.; Colo.
oblitus Becker, 1912a: 65.—Mont.
obscuricornis Loew, 1863a: 49 (Cent. 3, no. 90).—D.C.; Oreg. to Man., s. to Calif., n. Mexico, and Va., also Fla.
    obscuripennis (error) Cole and Lovett, 1921: 336.
palpalis Adams, 1903a: 42.—Mo.; S. Dak. to Mich., s. to Kans. and Mo.
perflavus Walker, 1849: 1120.—Ont.; Colo., Minn., Maine, N.S.
pilosulus Becker, 1912b: 645 (n. name for *horrida* Becker).—Tex.
    horrida Becker, 1912a: 66 (preocc. Becker, 1910).—Tex.
productus Loew, 1863a: 52 (Cent. 3, no. 96).—Alaska.
proximus Say, 1830: 187 (1859b: 370).—Ind. Unrecognized.
pubescens Loew, 1863a: 47 (Cent. 3, no. 88).—Fla.
quinquepunctatus Loew, 1863a: 51 (Cent. 3, no. 94).—Nebr.; S. Dak., s. to Ariz., n. Mexico, and Tex., also Ill.
rectinervis Becker, 1912a: 58.—Wash.
rubicundus Adams, 1903a: 43.—Wyo.; Oreg. and Alta. to Man., s. to Calif. and Colo.
rubrivittatus Adams, 1904a: 304.—La.; Kans.
rufescens Coquillett, 1910c: 45.—Pa., N.J., D.C.; Ont. to Maine, s. to N. Mex., Ark., Pa., and N.C.
ruginosus Becker, 1912a: 62.—Colo., Kans., Mich.; S. Dak. to Pa., s. to Colo. and Mo., also Va.
sabulonus Becker, 1912a: 67.—Calif.; Wash., Oreg.
sahlbergii Loew, 1863a: 51 (Cent. 3, no. 95).—Alaska.
scaber Coquillett, 1898f: 46.—N.Y.
seminiger Becker, 1912a: 66.—Que.
sordidellus Becker, 1912a: 54.—Colo., Nev.; Oreg. to Nebr., s. to Calif. and N. Mex.
stigmaticalis Becker, 1912b: 645 (n. name for *quadrimaculatus* Becker).—Calif.
    quadrimaculatus Becker, 1912a: 53 (preocc. Czerny, 1909).—Calif.
stigmatus Becker, 1912a: 60.—B.C., Wash.; Idaho, Mont.
subniger Coquillett, 1910c: 45.—N.J.; Wyo., Nebr., Mich., Que., Nfld., Maine.
sulphureus Loew, 1863a: 44 (Cent. 3, no. 83).—Ont.; Kans., Mich.
surdus Curran, 1930j: 79.—N.Y.; Minn., Wis., Mich., Mass.
tarsalis Becker, 1912a: 62.—Mich.
testaceus Macquart, 1851: 279 (1851: 306).—N. Amer. Unrecognized.

CHLOROPIDAE 791

Genus EPICHLOROPS Becker

Epichlorops Becker, 1910a: 77. Type-species, *Oscinis puncticollis* Zetterstedt (orig. des.).

exilis (Coquillett), 1898f: 45 (*Eurina*).—Colo., Mass.; Wash. to Sask., s. to Nev. and Colo., also Iowa, Mich., Ind., Mass.
puncticollis (Zetterstedt), 1848: 2636 (*Oscinis*).—Scandinavia; Alaska, S. Dak. to Ont., s. to Utah and Ind., Europe.

Genus ECTECEPHALA Macquart

Ectecephala Macquart, 1851: 253 (1851: 280). Type-species, *albistylum* Macquart (orig. des.).

REFERENCE: Sabrosky, 1941c (key).

albistylum Macquart, 1851: 253 (1851: 280).—N. Amer.; Utah to Minn. to N.J., s. to n. Mexico and Fla.
  laevifrons Becker (in part), 1912a: 153. Kans. and Tex. syntypes only.
laticornis Coquillett, 1910c: 46.—Colo., N.C., Ga.; S. Dak. to N.C., s. to Ariz. and Ga., also N.Y.
sulcata Sabrosky, 1941c: 78.—Ind.; Iowa to Mich. to Md., s. to Tex. and N.C.
sulcifrons Coquillett, 1910c: 46.—Kans.; ?Ariz.
unicolor (Loew), 1863a: 51 (Cent. 3, no. 93) (*Chlorops*).—Miss.; Okla. to N.Y., s. to Tex. and Fla. **N. comb.**
  similis Becker, 1912a: 72.—N. Amer.
  capillata, authors, not Coquillett.

Genus LASIOSINA Becker

Lasiosina Becker, 1910a: 73. Type-species, *Chlorops cinctipes* Meigen (orig. des.).
Euchlorops Malloch, 1913n: 139. Type-species, *vittatus* Malloch (orig. des.).

approximatonervis (Zetterstedt), 1848: 2622 (*Oscinis*).—Scandinavia; Alaska and Y.T. to Que., U.S., Europe.
  microcera Loew, 1872a: 108 (Cent. 10, no. 95) (*Chlorops*).—Tex. **N. comb., n. syn.**
  parva Adams, 1903a: 42 (*Chlorops*).—Kans. **N. comb., n. syn.**
  var. **transversalis** Malloch, 1919f: 53 (*Leptocera*; as sp.).—Alaska; N.W.T., n. Man. **N. comb., n. syn.**
canadensis Aldrich, 1918c: 337.—Sask.; N.W.T. to N.Y., s. to Wash. and Wyo.

similis (Malloch), 1913n: 140 (*Euchlorops*).—N. Mex.; Colo., Kans.
vittatus (Malloch), 1913n: 139 (*Euchlorops*).—Kans.; S. Dak., Iowa.

Genus PARECTECEPHALA Becker

Parectecephala Becker, 1910a: 105. Type-species, *Oscinis longicornis* Fallén (Duda, 1933: 208).

aristalis (Coquillett), 1898f: 46 (*Chlorops*).—N.C., Ga.; Ill., Ala., Fla.
dissimilis Malloch, 1914i: 24.—Pa.; Minn. to Maine, s. to Iowa and Va.
eucera (Loew), 1863a: 45 (Cent. 3, no. 85) (*Chlorops*).—D.C.; Minn. to N.S., s. to Kans., Ala., and N.C.
maculiceps Becker, 1912a: 70.—Tex., La.; Ill., Tenn., Ga., Fla.
maculosa (Loew), 1872a: 111 (Cent. 10, no. 99) (*Chlorops*).—Tex.; Ariz., N. Mex., Kans. **N. comb.**
sanguinolenta (Loew), 1863a: 44 (Cent. 3, no. 84) (*Chlorops*).— Carolina; Minn. to Maine, s. to Ill. and Carolina. **N. comb.**

Genus DIPLOTOXA Loew

Diplotoxa Loew, 1863a: 54 (Cent. 3, no. 97). Type-species, *Chlorops versicolor* Loew (mon.).

alternata (Loew), 1872a: 109 (Cent. 10, no. 97) (*Chlorops*).—Tex.; Kans. to Tex., and Va. to Fla.
inclinata Becker, 1912a: 43.—Tex.; Calif. to Que., N.J., and Tex.
  confluens Loew, 1872a: 107 (Cent. 10, no. 94) (*Chlorops*; preocc. Meigen, 1830).—Tex.
messoria (Fallén), 1820a: 5 (*Oscinis*).—Sweden; B.C. to Que., s. to Oreg., Utah, and Mass., Europe.
  bilineata Adams, 1903a: 40 (*Chlorops*).—Colo.
nigripes (Coquillett), 1910c: 44 (*Chlorops*).—N.J.; S. Dak. to Que. and Maine, s. to Kans. and D.C.
  major Becker, 1912a: 40.—Ohio.
recurva (Adams), 1903a: 41 (*Chlorops*).—Wyo.; Alta. to Man., s. to Calif., Ariz., and Iowa, also Mich., N.Y.
unicolor Becker, 1912a: 41.—Calif.; Oreg., Ariz., n. Mexico.
versicolor (Loew), 1863a: 53 (Cent. 3, no. 97) (*Chlorops*).—D.C.; B.C. to N.S., U.S.

Genus ELLIPONEURA Loew

Elliponeura Loew, 1869b: 44 (Cent. 8, no. 79). Type-species, *debilis* Loew (mon.)

debilis Loew, 1869b: 44 (Cent. 8, no. 79).—D.C.; Ill. and Mich. to N.Y., s. to N. Mex. and Fla.

diplotoxoides Becker, 1912a: 26.—Idaho; Wash. to Que. and Maine, s. to Calif., Tex., and Pa.

Genus NEODIPLOTOXA Malloch

Neodiplotoxa Malloch, 1914d: 116. Type-species, *Chlorops nigricans* Loew (orig. des.).

nigricans (Loew), 1872a: 110 (Cent. 10, no. 98) (*Chlorops*).—Tex.

pulchripes (Loew), 1872a: 108 (Cent. 10, no. 96) (*Chlorops*).—Tex.; Man., s. to Ariz. and Tex. **N. comb.**

Genus MEROMYZA Meigen

Meromyza Meigen, 1830: 163. Type-species, *Musca saltatrix* Linnaeus (Macquart, 1835: 589).

> REFERENCE: Rockwood et al., 1947 (biol., seasonal forms, key).

americana Fitch, 1856a: 531 (1856: 299).—N.Y.; S. Dak. to Maine, s. to n. Mexico and S.C.

> *flavipalpis* Malloch, 1914d: 117.—Ill.

pratorum Meigen, 1830: 165.—Germany; Alaska and N.W.T. to Maine, s. to Calif., N. Mex., N. Dak., and N.Y., also n. Mexico.

> *lineola* Curran, 1923n: 278.—N.W.T.

saltatrix (Linnaeus), 1761: 555 (*Musca*).—Sweden; Wash. and Alta., s. to Calif. and Utah, Palaearctic.

> *nigriventris* Macquart, 1835: 590.—France.
> *punctifer* Becker, 1912a: 24.—Idaho, Wash.
> *marginata* Becker, 1912a: 25.—Idaho, Oreg.

Unplaced Species of Chloropinae

ater Macquart, 1851: 280 (1851: 307) (*Chlorops*).—N. Amer.

Unplaced Families of Acalyptratae

Family ODINIIDAE

By Curtis W. Sabrosky

Odiniid flies are seldom collected. They have been reared in association with wood-boring beetles and moths, possibly as scavengers. Adults have been collected at bleeding wounds on trees, on

Polyporus fungi, tree trunks, and rotting stumps, and in fruit-fly traps.

REFERENCE: Shewell, 1960 (key, world genera).

Genus ODINIA Robineau-Desvoidy

Odinia Robineau-Desvoidy, 1830: 648. Type-species, *trinotata* Robineau-Desvoidy (Rondani, 1875a: 167; 1875: 2) = *maculata* (Meigen).

REFERENCE: Sabrosky, 1959c (rev.).

betulae Sabrosky, 1959c: 230.—Maine; B.C., Que., N.H.
biguttata Sabrosky, 1959c: 226.—St. Croix I.; Fla., P.R., Virgin Is.
boletina (Zetterstedt), 1848: 2721 (*Milichia*).—Sweden; Alta., Ont., Que., Europe.
conspicua Sabrosky, 1959c: 228.—Md.; Mass., Ga.
coronata Sabrosky, 1959c: 234.—El Salvador; Ariz., Tex., Panama.
meijerei Collin, 1952a: 115.—England; Mich. to N.Y., s. to Va., also Tex., ?Calif., ?Ariz., Europe.
picta (Loew), 1861b: 358 (Cent. 1, no. 99) (*Milichia*).—Ga.; Pa.
xanthocera Collin, 1952a: 112.—Holland, Germany; B.C., Europe.

Genus NEOALTICOMERUS Hendel

Neoalticomerus Hendel, 1903: 252. Type-species, *Odinia formosa* Loew (orig. des.).

seamansi Shewell, 1960: 629.—Alta.; Ariz., Nebr., N.B.

Genus TRAGINOPS Coquillett

Traginops Coquillett, 1900b: 429. Type-species, *irroratus* Coquillett (orig. des.).

REFERENCE: Steyskal, 1963 (rev.).

irroratus Coquillett, 1900b: 430.—Ga.; Mich. to Que., s. to N. Mex. and Fla.
purpurops Steyskal, 1963: 53.—Mich.; Mich. to Md., s. to Tex. and Va.

Family AGROMYZIDAE
By Kenneth E. Frick

The larvae of the majority of species in the family Agromyzidae mine the leaves of a very large number of plants. Very little is known about larval taxonomy. Several North American species, especially those belonging to *Liriomyza* and *Phytomyza*, are economically important, attacking various crop plants. Although centered in the Palaearctic Region, the family has been found on every continent.

REFERENCES: Frick, 1952, 1953 (rev., cat.), 1957b (nomenclature), 1959 (synopsis).

Subfamily AGROMYZINAE

Genus AGROMYZA Fallén

Agromyza Fallén, 1810a: 21. Type-species, *reptans* Fallén (Rondani, 1875a: 177; 1875: 12). In order to conserve the present usage of this name, Rondani's earlier designation (1856: 121) of *aeneoventris* Fallén is disregarded. Westwood's designation (1840: 151) of *nigripes* Meigen is invalid. Application should be made to the I.C.Z.N. concerning the type-species of this genus.

albitarsis Meigen, 1830: 171.—Europe; Wash., Calif., Pa.
ambigua Fallén, 1823a: 4 (as *reptans* var.).—Sweden; Alaska, Palaearctic.
    kincaidi Malloch, 1913e: 285.—Alaska.
aristata Malloch, 1915d: 13.—Ill.; Iowa to N.Y., s. to Va.
    ulmi Frost, 1924: 54.—Pa.
barberi Frick, 1952: 372 (n. name for *abbreviata* Malloch).—N. Mex.
    abbreviata Malloch, 1913e: 285 (preocc. Fallén, 1823).—N. Mex.
canadensis Malloch, 1913e: 299.—Ont.
diversa Johnson, 1922: 26.—Mass.; Iowa, Ill., Ont., Vt.
invaria Walker, 1858: 232.—U.S. Unrecognized.
isolata Malloch, 1913e: 306.—Calif.; D.C.
nigripes Meigen, 1830: 170.—Europe; Iowa to Que., s. to Mass.
    dubitata Malloch, 1913e: 311.—Mass.
niveipennis Zetterstedt, 1848: 2741.—Sweden; Wash., Utah, Kans., Europe.
pallidiseta Malloch, 1924d: 192.—D.C.
parvicornis Loew, 1869b: 49 (Cent. 8, no. 92).—D.C.; Ont., throughout U.S.
reptans Fallén, 1823a: 3.—Sweden; Wash., Calif., Europe.
rubi Brischke, 1881: 250.—Danzig; Wash., Idaho, Calif., Europe.
    sulfuriceps Strobl, 1898a: 270.—Europe.
rutiliceps Melander, 1913a: 261.—Mont.
spiraeae Kaltenbach, 1867: 104.—Europe; Calif., N.Y., Pa., N.J.
    fragariae Malloch, 1913e: 307.—Calif.

subnigripes Malloch, 1913e: 334.—N.H.; Man., Iowa, Ill., Ind.
  aprilina Malloch, 1915h: 359.—Ill.
varifrons Coquillett, 1902e: 189.—D.C.; Iowa, Ill., Pa.

Genus JAPANAGROMYZA Sasakawa

Japanagromyza Sasakawa, 1958: 138. Type-species, *Agromyza duchesneae* Sasakawa (orig. des.).

viridula (Coquillett), 1902e: 190 (*Agromyza*).—D.C.; Mass. to Ga., also Ind. **N. comb.**

Genus MELANAGROMYZA Hendel

Melanagromyza Hendel, 1920a: 114, 126. Type-species, *Agromyza aeneoventris* Fallén (orig. des.).
Limnoagromyza Malloch, 1921f: 147. Type-species, *diantherae* Malloch (orig. des.).

angelicae (Frost), 1934: 40 (*Agromyza*).—N.Y.; Wash., Idaho, Calif.
burgessi (Malloch), 1913e: 323 (*Agromyza*).—Mass.; N. Dak. to Mass., s. to Colo. and Kans.
diantherae (Malloch), 1921f: 147 (*Limnoagromyza*).—Ill.; Ind.
  dianthereae, error.
gibsoni (Malloch), 1915i: 106 (*Agromyza*).—Ariz.; s. Calif., Colo., Tex.
minima (Malloch), 1913e: 328 (*Agromyza*).—Trinidad; Miss., Mexico, Guatemala, Canal Zone, Puerto Rico.
riparella (Hendel), 1923: 145 (*Agromyza;* n. name for *riparia* Malloch).—Ill.
  riparia Malloch, 1915i: 105 (*Agromyza;* preocc. Wulp, 1871).—Ill.
salicis (Malloch), 1913e: 314 (*Agromyza*).—Mass.; Ill., Ohio, N.Y.
schineri (Giraud), 1861: 484 (*Agromyza*).—Europe; Wash., Colo., Ont., Mass.
setifrons (Melander), 1913a: 260 (*Agromyza;* as *maura* var.).—Idaho.
similata (Malloch), 1918i: 178 (*Agromyza*).—Ill.
simplex (Loew), 1869b: 46 (Cent. 8, no. 84) (*Agromyza*).—Pa.; throughout U.S., also Europe.
subvirens (Malloch), 1915i: 105 (*Agromyza*).—Ill.; Iowa, Pa., Va.
tamia (Melander), 1913a: 258 (*Domomyza*).—Wash.
tiliae (Couden), 1908: 35 (*Agromyza*).—Mo.; Ont., Ill.
virens (Loew), 1869b: 46 (Cent. 8, no. 85) (*Agromyza*).—Pa.; Ill. to Mass., s. to D.C.
viridis (Frost), 1931: 277 (*Agromyza*).—s. Calif.
winnemanae (Malloch), 1913e: 314 (*Agromyza*).—Md.

Genus OPHIOMYIA Braschnikov

Agromyza, subg. **Ophiomyia** Braschnikov, 1897: 40. Type-species, *Agromyza pulicaria* Meigen (mon.; misident.)=*maura* (Meigen).

congregata (Malloch), 1913e: 328 (*Agromyza*).—Ariz.; Idaho, N. Dak., Colo.
coniceps (Malloch), 1915i: 107 (*Agromyza*).—Utah; Man. to Que., s. to Calif. and La.
lantanae (Froggatt), 1919: 665 (*Agromyza*).—Australia; s. Calif., s. Tex., Fla. Introduced.
major (Strobl), 1898b: 587 (also 1900: 642) (*Agromyza;* as *curvipalpis* var.).—Europe; Ga.
  vibrissata Malloch, 1913e: 316 (*Agromyza*).—Ga.
maura (Meigen), 1838: 399 (*Agromyza*).—Europe; Mich. to N.Y., s. to Ga., also Calif.
  affinis Malloch, 1913e: 317 (*Agromyza*).—Md.
proboscidea (Strobl), 1900: 641 (*Agromyza*).—Europe; Mich., Ind., N.Y., N.J., La.
texana (Malloch), 1913e: 319 (*Agromyza*).—Tex.; Wash. to Mich., s. to N. Mex. and Tex.

Genus TYLOMYZA Hendel

Ophiomyia, subg. **Tylomyza** Hendel, 1931a: 181. Type-species, *Madiza pinguis* Fallén (orig. des.).

nasuta (Melander), 1913a: 260 (*Agromyza;* as *maura* var.).—Wash. Ont., Que., n. half of U.S., Europe.
  youngi Malloch, 1914f: 312 (*Agromyza*).—N.Y.
  madizina Hendel, 1920a: 130 (*Ophiomyia*).—Austria, Germany.

Subfamily PHYTOMYZINAE

Genus PHYTOBIA Lioy

Phytobia Lioy, 1864a: 1313. Type-species, *Agromyza errans* Meigen (mon.).
Dizygomyza, subg. **Dendromyza** Hendel, 1931a: 22. Type-species, *Agromyza carbonaria* Zetterstedt (orig. des.).

Subgenus PHYTOBIA Lioy

amelanchieris (Greene), 1917: 316 (*Agromyza*).—W. Va.; Wash., Mich. to Mass., s. to Tenn. and N.C.
indecora (Malloch), 1918h: 132 (*Agromyza*).—Ill.

picta (Coquillett), 1902e: 188 (*Agromyza*).—Mexico; N. Mex. to Panama Canal Zone.
pruinosa (Coquillett), 1902e: 189 (*Agromyza*).—Colo.; Ill., N.Y., D.C.
pruni (Grossenbacher), 1915: 235 (*Agromyza*).—N.Y.
setosa (Loew), 1869b: 45 (Cent. 8, no. 83) (*Agromyza*).—D.C.; Iowa to Que., s. to Va.
  aceris Greene, 1917: 313 (*Agromyza*).—Va.
waltoni (Malloch), 1913e: 303 (*Agromyza*).—N.Y.; Tenn.

Subgenus NEMORIMYZA Frey

Dizygomyza, subg. **Nemorimyza** Frey, 1946b: 42. Type-species, *Agromyza posticata* Meigen (orig. des.).

posticata (Meigen), 1830: 172 (*Agromyza*).—Europe; Ont., Que., throughout U.S.
  terminalis Coquillett, 1895a: 318 (*Agromyza*).—Pa.
  taeniola Coquillett, 1904f: 191 (*Agromyza*).—Calif.

Subgenus AMAUROMYZA Hendel

Dizygomyza, subg. **Amauromyza** Hendel, 1931a: 59. Type-species, *Agromyza lamii* Kaltenbach (orig. des.).

abnormalis (Malloch), 1913e: 320 (*Agromyza*).—D.C.; Kans., Iowa, cent. Europe.
maculosa (Malloch), 1913e: 302 (*Agromyza*).—N.Y.; throughout U.S., Hawaii, Neotropical.

Subgenus CEPHALOMYZA Hendel

Dizygomyza, subg. **Cephalomyza** Hendel, 1931a: 32. Type-species, *luteiceps* Hendel (orig. des.).

albidohalterata (Malloch), 1916d: 52 (*Agromyza*).—Ill.; Iowa.
auriceps (Melander), 1913a: 262 (*Agromyza*).—Idaho; Colo.
indecisa (Malloch), 1913e: 292 (*Agromyza*).—N. Mex.

Subgenus POEMYZA Hendel

Dizygomyza, subg. **Poemyza** Hendel, 1931a: 35. Type-species, *Agromyza pygmaea* Meigen (orig. des.).

angulata (Loew), 1869b: 47 (Cent. 8, no. 87) (*Agromyza*).—Pa.; e. Canada, throughout U.S.
  neptis Loew, 1869b: 50 (Cent. 8, no. 93) (*Agromyza*).—D.C.
  cinereifrons Frost, 1931: 276 (*Agromyza*).—N.Y.
incisa (Meigen), 1830: 182 (*Agromyza*).—Europe; Ont., n. half of U.S.
inconspicua (Malloch), 1913e: 310 (*Agromyza*).—Colo.

lateralis (Macquart), 1835: 609 (*Agromyza*).—Europe; Man., n. half of U.S.
 coquilletti Malloch, 1913e: 295 (*Agromyza*).—Colo.
muscina (Meigen), 1830: 177 (*Agromyza*).—Europe; Wash. to Mass., s. to Calif. and D.C., also N.W.T.
 marginata Loew, 1869b: 49 (Cent. 8, no. 91) (*Agromyza*).—D.C.
subangulata (Malloch), 1916d: 51 (*Agromyza*).—Ill.

Subgenus DIZYGOMYZA Hendel

Dizygomyza Hendel, 1920a: 114, 130 (as genus). Type-species, *Agromyza morosa* Meigen (orig. des.).

iraeos (Robineau-Desvoidy), 1851a: 393 (*Agromyza*).—Europe; Alaska, Mich., Ind., La.
iridis (Hendel), 1927b: 253 (*Dizygomyza*).—Europe; Calif., Mich.
luctuosa (Meigen), 1830: 182 (*Agromyza*).—Europe; Wash., Oreg., Calif., N.Y., N.S.
magnicornis (Loew), 1869b: 46 (Cent. 8, no. 86) (*Agromyza*).—D.C.; Mich. to N.H., s. to Ill. and D.C., also B.C.
morosa (Meigen), 1830: 170 (*Agromyza*).—Europe; S. Dak., Ill., Ind., Md.
thompsoni Frick, 1952: 396.—Mass.; Ill., Mich., N.Y., Pa.
 laterella, authors, not Zetterstedt.
 magnicornis Loew of Thompson, 1907: 74.

Subgenus ICTEROMYZA Hendel

Dizygomyza, subg. **Icteromyza** Hendel, 1931a: 51. Type-species, *Agromyza geniculata* Fallén (orig. des.).

capitata (Zetterstedt), 1848: 2750 (*Agromyza*).—Europe; Wash. to Calif. and Colo., e. to Maine.
 genualis Melander, 1913a: 261 (*Agromyza*).—Wash.
 coloradensis Malloch, 1913e: 297 (*Agromyza*).—Colo.
longipennis (Loew), 1869b: 48 (Cent. 8, no. 90) (*Agromyza*).—D.C.; Man., Que., throughout U.S.
pollinosa (Melander), 1913a: 263 (*Agromyza*).—Alaska.

Subgenus CALYCOMYZA Hendel

Dizygomyza, subg. **Calycomyza** Hendel, 1931a: 65. Type-species, *Agromyza artemisiae* Kaltenbach (orig. des.).

allecta (Melander), 1913a: 257 (*Agromyza;* as *platyptera* var.; n. name for *lateralis* Williston).—St. Vincent I.; U.S. e. of Rocky Mts., West Indies.

lateralis Williston, 1896c: 428 (*Agromyza;* preocc. Macquart, 1835).—St. Vincent I.

ambrosiae Frick, 1956: 299.—Tenn.; Ind., Fla.

artemisiae (Kaltenbach), 1856: 236 (*Phytomyza*).—Europe; B.C., Ont., throughout U.S., also Guatemala.

cynoglossi Frick, 1956: 295.—Ind.; s. Canada, Kans., Maine to Va.

flavinotum Frick, 1956: 297.—N.Y.; Wis., Ont., Pa., Mass.

gigantea Frick, 1956: 296.—Md.; Ont., Que., Ill., Va.

humeralis (Roser), 1840: 63 (*Agromyza*).—Europe; throughout Canada and U.S., Hawaii, Japan, Africa.

jucunda (Wulp), 1867: 161 (*Agromyza*).—Wis.; throughout s. Canada and U.S., Hawaii.

platyptera Thomson, 1869: 608 (*Agromyza*).—Calif.

coronata Loew, 1869b: 48 (Cent. 8, no. 89) (*Agromyza*).—Pa.

lantanae Frick, 1956: 287.—s. Tex.; Mexico, P.R., Trinidad.

majuscula Frick, 1956: 295.—Calif.; Wash., Idaho, Sask., Ont.

malvae (Burgess) *in* Comstock, 1880b: 202 (*Oscinis*).—D.C.; Calif., Ariz., N. Mex., Ind., N.Y. to Fla., Panama Canal Zone.

promissa Frick, 1956: 287.—Calif.; Man., Ont.

solidaginis (Kaltenbach), 1869: 196 (*Agromyza*).—Europe; e. Canada, throughout U.S.

verbenae (Hering), 1951: 42 (*Dizygomyza*).—N. Mex.; Ariz. to N.Y., s. to Miss.

Subgenus TRILOBOMYZA Hendel

Dizygomyza, subg. **Trilobomyza** Hendel, 1931a: 71. Type-species, *Agromyza flavifrons* Meigen (orig. des.).

calyptrata (Hendel), 1923: 145 (*Agromyza;* n. name for *nigrisquama* Malloch, 1916).—Ill.

nigrisquama Malloch, 1916d: 53 (*Agromyza;* preocc. Malloch, 1914).—Ill.

pleuralis (Malloch), 1914f: 311 (*Agromyza*).—Ill.; Ohio.

varia (Melander), 1913a: 264 (*Agromyza*).—Idaho.

Subgenus PRASPEDOMYZA Hendel

Dizygomyza, subg. **Praspedomyza** Hendel, 1931a: 77. Type-species, *approximata* Hendel (orig. des.).

clara (Melander), 1913a: 265 (*Agromyza*).—Wash.; Calif., Mich., Ont., Maine, Tenn.

citreifrons Malloch, 1913e: 290 (*Agromyza*).—Calif.

morio (Brischke), 1881: 258 (*Agromyza*).—Europe; Calif., Md.

subinfumata (Malloch), 1915i: 108 (*Agromyza;* n. name for *infumata* Malloch).—Ill.

infumata Malloch, 1915d: 15 (*Agromyza;* preocc. Czerny and Strobl, 1909).—Ill.

Genus CERODONTHA Rondani

Odontocera Macquart, 1835: 614 (preocc. Serville, 1833). Type-species, *Chlorops denticornis* Panzer (orig. des.).
Cerodontha Rondani, 1861a: 10 (n. name for *Odontocera* Macquart). Type-species, *Chlorops denticornis* Panzer (aut.).

Subgenus CERODONTHA Rondani

denticornis (Panzer).—Not Nearctic.
dorsalis (Loew), 1863a: 54 (Cent. 3, no. 98) (*Odontocera*).—D.C.; throughout Canada and U.S.
femoralis, authors (N. Amer.), not Meigen.

Subgenus XENOPHYTOMYZA Frey

Cerodontha, subg. **Xenophytomyza** Frey, 1946b: 51. Type-species, *Haplomyza atronitens* Hendel (orig. des.).
illinoensis (Malloch), 1934a: 483 (*Agromyza*).—Ill.; Va.

Genus LIRIOMYZA Mik

Liriomyza Mik, 1894b: 289. Type-species, *urophorina* Mik (mon.).
allia (Frost), 1943: 257 (*Agromyza*).—Kans.
alliovora Frick, 1955: 88.—Iowa.
angulicornis (Malloch), 1918h: 79 (*Agromyza*).—Ill.; Man.
archboldi Frost, 1962: 51.—Fla.
assimilis (Malloch), 1918h: 80 (*Agromyza*).—Ill.; Pa.
baptisiae (Frost), 1931: 275 (*Agromyza*).—Pa.
borealis (Malloch), 1913e: 280 (*Agromyza*).—B.C.
brassicae (Riley), 1885: 322 (*Oscinis*).—Mo.; Man., throughout U.S., Europe.
  mitis Curran, 1931e: 97 (*Phytomyza*).—Man.
chlamydata (Melander), 1913a: 250 (*Antineura*).—Wash.
deceptiva (Malloch), 1918h: 78 (*Agromyza*).—Ill.; Va.
dianthi Frick, 1958: 1.—Calif.
discalis (Malloch): 1913e: 277 (*Agromyza*).—Ariz.
eupatorii (Kaltenbach), 1874: 320 (*Agromyza*).—Europe; Wash.
felti (Malloch), 1914f: 310 (*Agromyza*).—N.Y.; Ill.
flaveola (Fallén), 1823a: 6 (*Agromyza*).—Sweden; cent. Calif., Europe.
flavonigra (Coquillett), 1902e: 189 (*Agromyza*).—N. Mex.
fumicosta (Malloch), 1914f: 310 (*Agromyza*).—Ill.

guytona Freeman, 1958: 344.—Ala.; Tex., Fla., Jamaica.
langei Frick, 1951: 81.—Calif.; Wash., Oreg.
lima (Melander), 1913a: 265 (*Agromyza*).—Idaho; S. Dak.
    holti Malloch, 1924d: 191 (*Agromyza*).—S. Dak.
lutea (Meigen), 1830: 177 (*Agromyza*).—Europe; Alaska.
marginalis (Malloch), 1913e: 283 (*Agromyza;* as *melampyga* var.).—
    S.C.; Ill., Tex.
melampyga (Loew), 1869b: 48 (Cent. 8, no. 88) (*Agromyza*).—D.C.;
    Mich., N.Y., Mass., N.J., Md.
    flaviventris Johnson, 1902a: 242 (*Agromyza;* preocc. Strobl,
    1898).—N.Y.
munda Frick, 1957a: 61.—Calif.
pacifica (Melander), 1913a: 264 (*Agromyza*).—Wash.; Alaska, B.C.,
    Man. to Que., s. to Ill. and Ind.
    longispinosa Malloch, 1913e: 276 (*Agromyza*).—B.C.
phaseolunata (Frost), 1943: 256 (*Agromyza*).—N.J.
pictella (Thomson), 1869: 609 (*Agromyza*).—Calif.; Ariz.
propepusilla Frost, 1954: 73 (n. name for *subpusilla* Frost).—Kans.
    subpusilla Frost, 1943: 255 (*Agromyza;* preocc. Malloch,
    1914).—Kans.
quadrisetosa (Malloch), 1913e: 332 (*Agromyza*).—Tex.; Ind., N.Y.
reverberata (Malloch), 1924d: 191 (*Agromyza*).—Md.; Ont.
sorosis (Williston), 1896c: 429 (*Agromyza*).—St. Vincent I.; U.S. e. of
    Rocky Mts.
trifolii (Burgess) *in* Comstock, 1880b: 201 (*Oscinis*).—D.C.; Wash. to
    Calif., also Ind., Fla.
variata (Malloch), 1913e: 277 (*Agromyza*).—Maine.
verbenicola Hering, 1951: 43.—N. Mex.; Utah.

Genus METOPOMYZA Enderlein

Metopomyza Enderlein, 1936a: 180. Type-species *Agromyza*
    flavonotata Haliday (mon.).
interfrontalis (Melander), 1913a: 263 (*Agromyza*).—Wash.; n. Calif.,
    Man., Kans., Tex., Ill., Mich., Va.

Genus HAPLOMYZA Hendel

Antineura Melander, 1913a: 249 (preocc. Osten Sacken, 1881).
    Type-species, *togata* Melander (orig. des.).
Haplomyza Hendel, 1914d: 73 (n. name for *Antineura* Melander).
    Type-species, *Antineura togata* Melander (aut.).
minuta (Frost), 1924: 86 (*Phytomyza*).—N. Dak.; Wash.
palliata (Coquillett), 1902e: 191 (*Phytomyza*).—N. Mex.

togata (Melander), 1913a: 250 (*Antineura*).—Wash.; Calif. to Tex., also N. Dak., Kans.

Genus PHYTOLIRIOMYZA Hendel

Liriomyza, subg. **Phytoliriomyza** Hendel, 1931a: 203. Type-species, *Agromyza perpusilla* Meigen (mon.).

arctica (Lundbeck), 1901: 304 (*Agromyza*).—Greenland; B.C.
immaculata (Coquillett), 1902e: 185 (*Odinia*).—N.H.; B.C., Man., throughout U.S.
perpusilla (Meigen), 1830: 181 (*Agromyza*).—Europe; Wash., Mont., Calif., N. Mex.

Genus XYRAEOMYIA Frick

Xyraeomyia Frick, 1952: 412. Type-species, *conjunctimontis* Frick (orig. des.).

conjunctimontis Frick, 1952: 413.—Calif.

Genus PHYTAGROMYZA Hendel

Phytagromyza Hendel, 1920a: 115, 145. Type-species, *Domomyza flavocingulata* Strobl (orig. des.).

lonicerae (Robineau-Desvoidy), 1851a: 396 (*Phytomyza*).—Europe; Calif.
nitida (Malloch), 1913e: 288 (*Agromyza*).—Md.; Iowa, Ill., Va., N.C.
orbitalis (Melander), 1913a: 271 (*Phytomyza*).—Wash.; Idaho, Calif.
plagiata (Melander), 1913a: 273 (*Napomyza*).—Idaho; Mont.
  brevicostalis Malloch, 1913e: 283 (*Agromyza*).—Mont.
populicola (Walker), 1853: 247 (*Phytomyza*).—England; Ont., Europe.

Genus PSEUDONAPOMYZA Hendel

Pseudonapomyza Hendel, 1920a: 115. Type-species, *Phytomyza atra* Meigen (orig. des.).

atra (Meigen), 1830: 191 (*Phytomyza*).—Europe; Wash., Calif.
  nitidula Malloch, 1913n: 151 (*Phytomyza*).—Calif.
lacteipennis (Malloch), 1913n: 152 (*Phytomyza*).—N. Mex.; Wash., N. Dak., Kans., Mich.

Genus NAPOMYZA Westwood

Phytomyza, subg. **Napomyza** Westwood, 1840: 152. Type-species, *Phytomyza festiva* Meigen (orig. des.)=*elegans* (Meigen).

davisii (Walton), 1912: 463 *(Agromyza).*—Ind.; Ont., Wis., Mich., Mo.
lateralis (Fallén), 1823b: 3 *(Phytomyza).*—Sweden; Alaska, N.W.T. to Calif., also Ont., Que.
parvicella (Coquillett), 1902e: 189 *(Agromyza).*—Alaska.

Genus PHYTOMYZA Fallén

Phytomyza Fallén, 1810a: 21, 26. Type-species, *flaveola* Fallén (mon.).

affinalis Frost, 1924: 84.—Sask.
agromyzina Meigen, 1830: 191.—Europe; Wash., Calif.
albiceps Meigen, 1830: 194.—Europe; Calif.
angelicella Frost, 1927: 218.—N.Y.
aquilegiana Frost, 1930: 459.—Pa.; Wash., Idaho, Calif., N.Y.
atricornis Meigen, 1838: 404.—Europe; B.C., s. to Calif., also N.Y., Mass., Asia, Africa.
    chrysanthemi Kowarz *in* Lintner, 1892: 243.—Europe.
    affinis, authors, not Fallén.
atripalpis Aldrich, 1929e: 89.—B.C.
auricornis Frost, 1927: 217.—N.Y.
bicolor Coquillett, 1902e: 191.—N.Y.
clemativora Coquillett, 1910a: 131.—s. Tex.
crassiseta Zetterstedt, 1860: 6469.—Europe; s. Calif., Wash., Idaho.
delphiniae Frost, 1928: 77.—N.Y.; Ohio, Pa.
diminuta Walker, 1858: 233.—U.S.
    dimidiata, error.
dura Curran, 1931c: 10.—Que.
flavicornis Fallén, 1823b: 4.—Sweden; Mich., Ind., Europe.
flavinervis Frost, 1924: 85.—Tex.
genalis Melander, 1913a: 272.—Ill.
gregaria Frick, 1954: 371.—Calif.; Wash., Idaho.
ilicicola Loew, 1872a: 114 (Cent. 10, appendix) (n. name for *illicis* Loew).—D.C.; Tex., Mass. to Ala.
    ilicis Loew, 1863a: 54 (Cent. 3, no. 99) (preocc. Curtis, 1846).—D.C.
ilicis Curtis, 1846: 444.—England; B.C., Wash., Oreg., Calif., Europe.
lactuca Frost, 1924: 85.—Pa.; Mich., N.Y.
loewii Hendel, 1923: 145 (n. name for *clematidis* Loew).—D.C.; Wash., Idaho, Ind., La.
    clematidis Loew, 1863a: 55 (Cent. 3, no. 100) (preocc. Kaltenbach, 1859).—D.C.

major Malloch, 1913n: 150.—Labr.; Man.
marginalis Frost, 1927: 219.—N.Y.
melanella Frost, 1924: 86.—Calif.
minuscula Goureau, 1851: 153.—Europe; Wash. to Calif., Ill. to Conn., s. to D.C.
    nitida Melander, 1913a: 271.—Idaho.
nervosa Loew, 1869b: 52 (Cent. 8, no. 99).—D.C.; Iowa, Kans.
nigra Meigen, 1830: 191.—Europe; Wash., Oreg.
nigrinervis Frost, 1924: 87.—Colo.
nigripennis Fallén, 1823b: 2.—Sweden; Wis., Ont., Europe.
periclymeni Hendel, 1922: 71.—Europe; Wash., Calif.
persicae Frick, 1954: 369.—Ohio; Ont., Conn., Va.
plantaginis Robineau-Desvoidy, 1851a: 404.—Europe; throughout U.S.
    genualis Loew, 1869b: 52 (Cent. 8, no. 100).—D.C.
plumiseta Frost, 1924: 87.—Pa.
ranunculi ranunculi (Schrank).—Not Nearctic.
    ssp. **albipes** Meigen, 1830: 195 (as sp.).—Europe; Wash.
    ssp. **flavoscutellata** Fallén, 1823b: 4 (as sp.).—Sweden; Oreg., Idaho.
    ssp. **praecox** Meigen, 1830: 194 (as sp.).—Europe; Idaho.
rufipes Meigen, 1830: 192.—Europe; Oreg., Nfld., N.B.
solita Walker, 1858: 232.—U.S. ?=*atricornis*.
spondylii Robineau-Desvoidy, 1851a: 400.—Europe; Calif.
    sphondylii, emend.
subtenella Frost, 1924: 89.—Wash.; Wis.
trivittata Frost, 1924: 89.—Nev.; Calif.

Family CLUSIIDAE

(Clusiodidae)

By George C. Steyskal

The family Clusiidae contains about 150 species from all parts of the world except Africa. The species are usually considered to be rare, but adults may sometimes be found in numbers on rotten wood. Very little is known of the habits or immature stages.

    REFERENCES: Melander and Argo, 1924 (rev.); Czerny, 1928b (monog., Palaearctic spp., figs.); Hennig, 1938b (male postabdomen); Frey, 1960 (world cat.).

Subfamily CLUSIINAE

Genus CHAETOCLUSIA Coquillett

Chaetoclusia Coquillett, 1904e: 93. Type-species, *bakeri* Coquillett (orig. des.).

affinis Johnson, 1913f: 101.—N.J.

Genus CLUSIA Haliday

Heteroneura, subg. **Clusia** Haliday, 1838: 188. Type-species, *Heteromyza flava* Meigen (mon.).

czernyi Johnson, 1913f: 100.—Maine; Ill. to Ont. and Que., s. to Va.
  flava (Meigen).—Not Nearctic.
lateralis (Walker), 1849: 1095 (?*Helomyza*).—N. Amer.; Ill. to Ont. and Que., s. to N.C.
  spectabilis Loew, 1860b: 82 (*Heteroneura*).—D.C.
  pallida Harris, 1835: 600 (*Tetanura*). Nomen nudum.
occidentalis Malloch, 1918n: 4.—Wash.; B.C. to Calif.

Genus HETEROMERINGIA Czerny

Heteromeringia Czerny, 1903a: 72. Type-species, *Heteroneura nigrimana* Loew (mon.).

annulipes Johnson, 1913f: 99.—N.C.
flavipes (Williston), 1896c: 387 (*Heteroneura*).—St. Vincent I.; Fla., West Indies, Nicaragua.
nitida Johnson, 1913f: 99.—N.J.; Mich. to Mass., s. to Ill. and N.C., also Ariz., N. Mex.
  var. **nigripes** Melander and Argo, 1924: 31.—N. Mex.; Ariz., N.J., Va.

Genus SOBAROCEPHALA Czerny

Sobarocephala Czerny, 1903a: 85. Type-species, *ruebsaameni* Czerny (mon.).

flava Melander and Argo, 1924: 40.—Va.; Mich., Md., La.
flaviseta (Johnson), 1913f: 99 (*Heteromeringia*).—N.J.; Mich. to N.H., s. to Mo. and Ga.
  convergens Malloch, 1922b: 50 (*Heteromeringia*).—Ill.
lachnosternum Melander and Argo, 1924: 42.—Va.; Mich.
latifrons (Loew), 1860b: 83 (*Heteroneura*).—D.C.; Mich. to Mass., s. to Ill. and Va.
populi Steyskal, 1951a: 129.—Mich.
setipes Melander and Argo, 1924: 47.—Md.

Subfamily CLUSIODINAE

Genus CLUSIODES Coquillett

Heteroneura Fallén, 1823a: 2 (preocc. Fallén, 1810). Type-species, *geomyzina* Fallén (Zetterstedt, 1848: 2787).
Clusiodes Coquillett, 1904e: 93 (n. name for *Heteroneura* Fallén, 1823). Type-species, *Heteroneura geomyzina* Fallén (aut.).
Clusiodes, subg. *Clusiaria* Malloch, 1922b: 47. Type-species, *duplicata* Malloch (orig. des.)=*melanostoma* (Loew).

Subgenus CLUSIODES Coquillett

apicalis (Zetterstedt), 1848: 2789 (*Heteroneura;* as *geomyzina* var.).—Denmark; Idaho.
ater Melander and Argo, 1924: 18.—Mass.; N.Y.
geomyzinus (Fallén), 1823a: 2 (*Heteroneura*).—Europe; Ariz., Maine, N.H.
melanostoma (Loew), 1864a: 98 (Cent. 5, no. 97) (*Heteroneura*).—N.H.; Wash. and Alta. to Que., s. to Ill. and Va.
  duplicata Malloch, 1922b: 49.—Md.
  flavifascies Coquillett *in* Slosson, 1897: 239 (*Heteroneura*). Nomen nudum.
nitidus Melander and Argo, 1924: 20.—N. Mex.; Colo.
  var. **scutellatus** Melander and Argo, 1924: 20.—Colo.
orbitalis Malloch, 1922b: 50.—Maine.
pictipes (Zetterstedt), 1855: 4816 (*Heteroneura*).—Sweden; ?N. Mex., ?N.H.
ruficollis (Meigen), 1830: 128 (*Heteroneura*).—Europe; ?Ont., ?Que.
terminalis Melander and Argo, 1924: 22.—N.H.; ?Ill.

Subgenus COLUMBIELLA Malloch

Clusiodes, subg. **Columbiella** Malloch, 1922b: 47. Type-species, *apiculata* Malloch (orig. des.).

albimanus (Meigen), 1830: 128 (*Heteroneura*).—Europe; Wash. to Que., s. to N.Y. and N.J.
americanus Malloch, 1922b: 48.—Md.; Maine to Md.
apiculatus Malloch, 1922b: 49.—N.H.; Vt., N.C.
johnsoni Malloch, 1922b: 49.—Maine; Mich. to Maine, s. to Va.
  nigripalpis Malloch, 1922b: 49.—Md.
niger Melander and Argo, 1924: 17.—N. Mex.

Family ACARTOPHTHALMIDAE

By George C. Steyskal

This family contains only a single genus formerly classed as a subfamily of the Clusiidae. Nothing is known about its biology except that adults have been taken on rotten fungi and carrion.

REFERENCES: Melander and Argo, 1924 (rev.); Czerny, 1928b (rev., fig.); Frey, 1946c (key).

Genus ACARTOPHTHALMUS Czerny

Acartophthalmus Czerny, 1902: 256. Type-species, *Anthophilina nigrina* Zetterstedt (mon.).

bicolor Oldenberg, 1910: 284.—Germany; Mass., Europe.
nigrinus (Zetterstedt), 1848: 2697 (*Anthophilina*).—Sweden; Alaska, Wash., Oreg., and Colo. to Mass., Europe.

Family HELEOMYZIDAE

(Helomyzidae)

By Gordon D. Gill

As far as known, species of the subfamily Suilliinae breed in fungi, and the adults are commonly collected in shaded areas. The Heleomyzinae breed in various kinds of decaying plant and animal matter, and are commonly taken from birds' nests and mammal burrows.

REFERENCE: Gill, 1962 (rev.).

Subfamily SUILLIINAE

Genus PORSENUS Darlington

Porsenus Darlington *in* Aldrich and Darlington, 1908: 69. Type-species, *johnsoni* Darlington (mon.).

johnsoni Darlington *in* Aldrich and Darlington, 1908: 70.—Mass.

Genus SUILLIA Robineau-Desvoidy

Suillia Robineau-Desvoidy, 1830: 642. Type-species, *fungorum* Robineau-Desvoidy (Coquillett, 1910b: 610), probably =*rufa* (Fallén).
Heleomyza, authors, not Fallén.

REFERENCE: Steyskal, 1944a (key).

apicalis (Loew), 1862b: 226 (Cent. 2, no. 86) (*Helomyza*).—D.C.; Alaska, Oreg. to N.S., s. to Tenn. and N.C.
barberi (Darlington) *in* Aldrich and Darlington, 1908:93 (*Helomyza*).— N. Mex.; Alta. to N.S., s. to Calif. and Tex.
convergens (Walker), 1849: 983 (*Dryomyza*).—N.S.; Alaska to Nfld., s. to Calif. and N.C. **N. comb.**
    loewi Garrett, 1925a: 3.—Mass. **N. syn.**
    bicolor, authors, not Zetterstedt.
    zetterstedti, authors, not Loew.
longipennis (Loew), 1862b: 228 (Cent. 2, no. 90) (*Helomyza*).—N.Y.; Alaska and Y.T. to N.S., s. to Idaho, Minn., Tenn., and N.C.
nemorum (Meigen), 1830: 52 (*Helomyza*).—Germany; Alaska to Que. and Vt., in w., s. to Ariz. and N. Mex., Europe.
    assimilis Loew, 1862b: 226 (Cent. 2, no. 87) (*Helomyza*).— Hudson Bay Territory.
plumata (Loew), 1862b: 227 (Cent. 2, no. 88) (*Helomyza*).—N.Y.; B.C. to N.S., s. to Calif., Kans., and Ga.
    chaetomera Czerny, 1933: 236.—Calif.
quinquepunctata (Say), 1823: 101 (1859b: 86) (*Helomyza*).—Mo.; Minn. to Ont. and Conn., s. to Tex. and Fla.
    latericia Loew, 1862b: 227 (Cent. 2, no. 89) (*Helomyza*).— Conn.
sororcula Czerny, 1926: 53.—Calif.; cent. Calif.
thomsoni Gill (for *limbata* Thomson).—Calif.; Alaska, B.C. to Sask., s. to Calif. and N. Mex. **N. name.**
    limbata Thomson, 1869: 569 (*Helomyza*; preocc. Walker, 1857).—Calif.

Genus ALLOPHYLA Loew

Allophyla Loew, 1862e: 43. Type-species, *Helomyza atricornis* Meigen (mon.).
atricornis (Meigen), 1830: 54 (*Helomyza*).—Europe; ?Alaska.
laevis Loew, 1862b: 225 (Cent. 2, no. 85).—Ont.; B.C. to N.B., s. to Calif. in w., and to Ga. in e.

Subfamily HELEOMYZINAE
Genus BORBOROPSIS Czerny

Borboropsis Czerny, 1902: 256. Type-species, *Anthomyza fulviceps* Strobl (mon.).
fulviceps (Strobl), 1898a: 269 (*Anthomyza*).—Austria; Alaska, N.W.T., B.C. to Que., Europe.

Genus OLDENBERGIELLA Czerny

Oldenbergiella Czerny, 1924: 69. Type-species, *callosa* Czerny (Aldrich, 1926f: 102).

brumalis Czerny, 1924: 70.—Germany; Man., Europe.

Genus ORBELLIA Robineau-Desvoidy

Orbellia Robineau-Desvoidy, 1830: 656. Type-species, *myopiformis* Robineau-Desvoidy (Coquillett, 1910b: 580).
Crymobia Loew, 1862e: 45. Type-species, *hiemalis* Loew (mon.).
Anarostomoides Malloch, 1916a: 15. Type-species, *petersoni* Malloch (orig. des.).
Anorostomoides, error.
Barbastoma Garrett, 1921: 123. Type-species, *barbatum* Garrett (orig. des.).

barbata (Garrett), 1921: 123 (*Barbastoma*).—B.C.; Wash., Alta.
hiemalis (Loew), 1862e: 46 (*Crymobia*).—Germany; ?Colo., ?Maine.
petersoni (Malloch), 1916a: 15 (*Anarostomoides*).—Ill.; Wash. to Maine, Europe.

Genus HETEROMYZA Fallén

Heteromyza Fallén, 1820e: 1. Type-species, *oculata* Fallén (Westwood, 1840: 145).

oculata Fallén, 1820e: 2.—Sweden; Alaska, Ont., Que., Labr., N.H., Europe.
 eriphides Walker, 1849: 1088.—Ont.
 flavipes Walker, 1849: 1089.—Ont.

Genus TEPHROCHLAMYS Loew

Tephrochlamys Loew, 1862e: 72. Type-species, *Helomyza rufiventris* Meigen (Coquillett, 1910b: 613).

flavitarsis Darlington *in* Aldrich and Darlington, 1908: 71.—N.H.; B.C., Ont. to N.S., also N.C.
rufiventris (Meigen), 1830: 58 (*Helomyza*).—Europe; Alaska to Nfld., s. to Calif. and N.C.
 canescens Meigen, 1830: 57 (*Helomyza*).—Europe.
 bicolor Walker, 1849: 974 (*Cordylura*).—Ont. **N. syn.,** C. W. Sabrosky *in litt.*

Genus AECOTHEA Haliday

Aecothea Haliday, 1838: 187. Type-species, *Helomyza fenestralis* Fallén (mon.).
Oecothea, emend.

REFERENCE: Czerny, 1928a (key).

aristata Malloch, 1919f: 82.—N.W.T.; Alaska.
fenestralis (Fallén), 1820e: 5 (*Helomyza*).—Sweden; B.C., Minn., Iowa to Mass., Europe.
specus (Aldrich), 1897: 189 (*Blepharoptera*).—Ind.; Alaska, B.C. to Ont., s. to Calif. and Ga.
  canadensis Garrett, 1921: 129.—B.C.

Genus ECCOPTOMERA Loew

Eccoptomera Loew, 1862e: 47. Type-species, *Helomyza longiseta* Meigen (Coquillett, 1910b: 536).
Viatica Garrett, 1924b: 32. Type-species, *spinosa* Garrett (orig. des.).

aldrichi Czerny, 1928a: 53.—Wyo.; Sask.
callipus Garrett, 1925a: 2.—Oreg.; Calif.
crypta Gill, 1962: 526.—Idaho; Wash., Calif.
garretti Gill, 1962: 527.—Calif.
melanderi (Garrett), 1925a: 2 (*Viatica*).—Idaho; Alta., Calif., Colo.
simplex Coquillett *in* Baker 1904: 32.—Nev.; B.C. to Alta. and Iowa, s. to Calif. and Colo.
  americana Darlington *in* Aldrich and Darlington, 1908: 74.—Idaho.
spinosa (Garrett), 1924b: 32 (*Viatica*).—Calif.; Wash.

Genus LUTOMYIA Aldrich

Lutomyia Aldrich, 1922a: 108. Type-species, *spurca* Aldrich (mon.).
Criddleria Curran, 1929c: 31. Type-species, *hemiptera* Curran (orig. des.).

REFERENCE: Sabrosky, 1949a (rev.).

aldrichi Sabrosky, 1949a: 4.—Wash.
distincta Garrett, 1924b: 30.—B.C.; Wash.
hemiptera (Curran), 1929c: 32 (*Criddleria*).—Man.
spurca Aldrich, 1922a: 109.—N.Y.; Wis., Va.

Genus PSEUDOLERIA Garrett

Pseudoleria Garrett, 1921: 128. Type-species, *Blepharoptera pectinata* Loew (orig. des.).

crassata Garrett, 1925b: 3.—Ohio; Oreg. to Ont., s. to Calif. and N.C.
dubia Garrett, 1925b: 3.—Tex.
intermedia Garrett, 1925b: 3.—N. Mex.; Calif. to Wyo. and N. Mex., also Man., Kans., Tenn.
longigena Garrett, 1925b: 2.—Calif.; Oreg., Ariz.
longigenoidea Gill, 1962: 534.—Wash.; Calif., N. Mex., Wyo.
media Garrett, 1925b: 3.—Kans.; Calif., Ariz., Tex.
parvitarsus Garrett, 1925b: 2.—B.C.; B.C. to Man., s. to Calif. and N. Mex., also Ind.
pectinata (Loew), 1872a: 99 (Cent. 10, no. 79) (*Blepharoptera*).—Tex.; B.C. to D.C., s. to Calif. and Ga.
> *pectinerata* Garrett, 1921: 128.—B.C.

robusta Garrett, 1925b: 2.—B.C.; Wash., Mont., Calif., Utah.
similis Garrett, 1925b: 2.—Idaho; Calif.
subrobusta Gill, 1962: 540.—Calif.; Tex.
vulgaris Garrett, 1925b: 2.—Wash.; B.C. to Nebr., s. to Oreg. and N. Mex.

Genus ANOROSTOMA Loew

Anorostoma Loew, 1862e: 47. Type-species, *marginatum* Loew (sub. mon., Loew, 1862b: 223; Cent. 2, no. 81).

> REFERENCE: Curran, 1933c (rev.).

alternans Garrett, 1925b: 4.—Wash.; Oreg., Calif.
carbona Curran, 1933c: 7.—Wyo.; Colo., Nebr.
cinereum Curran, 1932c: 10.—Oreg.; Wash.
coloradense Garrett, 1924b: 28.—Colo.; N.W.T. and B.C. to Que., s. to Calif. and N.J.
currani Garrett, 1922b: 176.—Man.; N.W.T. and B.C. to Man., s. to Wash. and Colo.
fumipenne Gill, 1962: 552.—Wash.; Oreg.
grande Darlington *in* Aldrich and Darlington, 1908: 75.—Calif.; Wash.
hinei Garrett, 1925b: 4.—Alaska.
jamesi Gill, 1962: 553.—Calif.
jersei Garrett, 1924b: 29.—N.J.; Alaska to Que., s. to Calif., N. Mex., and Nebr., ?Mich.
longipile Gill, 1962: 552.—N. Mex.
lutescens Curran, 1933c: 8.—Oreg.; B.C., Wash.
maculatum Darlington *in* Aldrich and Darlington, 1908: 76.—Calif.
marginatum Loew, 1862b: 223 (Cent. 2, no. 81).—Ont.; Alta. to Que., s. to Mich. and N.Y.
> *raca* Garrett, 1923: 244.—Man.

opacum Coquillett, 1901h: 614.—s. Calif.
wilcoxi Curran, 1933c: 4.—Oreg.; Wash., Calif.

Genus NEOLERIA Malloch

Neoleria Malloch, 1919f: 83. Type-species, *rotundicornis* Malloch (orig. des.) = *prominens* (Becker).
Postleria Garrett, 1921: 124. Type-species, *fuscolinea* Garrett (orig. des.).

czernyi (Garrett), 1925a: 2 (*Postleria;* as *czerni*).—Alaska (St. Paul I.).
diversa (Garrett), 1925b: 4 (*Postleria*).—B.C.
fuscolinea (Garrett), 1921: 124 (*Postleria*).—B.C.; Alaska, Man., Mont.
inscripta (Meigen), 1830: 59 (*Helomyza*).—Germany; Alaska, B.C. to Oreg., e. to Maine, Europe.
    leucostoma Loew, 1863a: 28 (Cent. 3, no. 53) (*Blepharoptera*).—Alaska.
lutea (Loew), 1863a: 28 (Cent. 3, no. 52) (*Blepharoptera*).—Alaska; Oreg., Calif., Mont., Que., N.H.
prominens (Becker), 1897: 402 (*Tephrochlamys*).—Novaya Zemlya; Alaska, N.W.T., Man., Greenland, n. Europe.
    rotundicornis Malloch, 1919f: 83.—Alaska.
    septentrionalis Collin, 1923: 121 (*Leria*).—Bear I. (n. of Norway).
    crassipes, authors, not Loew.
    tibialis, Czerny and authors, not Zetterstedt.
ruficauda (Zetterstedt), 1847: 2456 (*Helomyza*).—Scandinavia; ?Labr., Europe.
tibialis (Zetterstedt), 1838: 767 (*Helomyza*).—Lapland; Alaska, Europe.

Genus SPANOPAREA Czerny

Spanoparea Czerny, 1924: 103. Type-species, *Helomyza ruficornis* Meigen (Aldrich, 1926f: 102).

discolor (Loew), 1872a: 99 (Cent. 10, no. 78) (*Blepharoptera*).—N.H.
laffooni Gill, 1962: 563.—Minn.
walkeri Garrett, 1925b: 3.—B.C.; Wyo. and Colo. to Que. and Maine.

Genus MORPHOLERIA Garrett

Morpholeria Garrett, 1921: 127. Type-species, *melaneura* Garrett (mon.) = *tristis* (Loew).

tristis (Loew), 1862b: 225 (Cent. 2, no. 84) (*Blepharoptera*).—Man.; Alaska, B.C. to Ont.
  melaneura Garrett, 1921: 127.—B.C.

Genus ACANTHOLERIA Garrett

Acantholeria Garrett, 1921: 130. Type-species, *Blepharoptera cineraria* Loew (orig. des.).

abnormalis Garrett, 1921: 131.—B.C.; Alaska, Mont., Wyo., Colo.
armipes (Loew), 1862b: 224 (Cent. 2, no. 83) (*Blepharoptera*).—Hudson Bay Territory; Y.T. and N.W.T., s. to Ariz., Colo., and S. Dak.
  oediemus Garrett, 1921: 131.—B.C.
  oedicnema, emend.
  cineraria, authors, not Loew.
moscowa Garrett, 1925a: 3.—Idaho.

Genus SCHROEDERELLA Enderlein

Schroederia Enderlein, 1914a: 314 (preocc. Schmidt, 1911). Type-species, *Helomyza iners* Meigen (mon.).
Schroederella Enderlein, 1921b: 231 (n. name for *Schroederia* Enderlein). Type-species, *Helomyza iners* Meigen (aut.).
Amoebaleria, subg. *Edioamoeba* Garrett, 1925b: 3. Type-species, *luteoala* Garrett (orig. des.)=*iners* (Meigen). **N. syn.**

fuscopicea Gill, 1962: 570.—Wash.; Idaho, Md., Va.
iners (Meigen), 1830: 57 (*Helomyza*).—Germany; Minn. and Iowa to Ont. and Mass., Europe.
  luteoala Garrett, 1925b: 3 (*Amoebaleria*).—Ill.

Genus SCOLIOCENTRA Loew

Scoliocentra Loew, 1862e: 43. Type-species, *Helomyza villosa* Meigen (mon.).
Achaetomus Coquillett, 1907a: 75. Type-species, *pilosus* Coquillett (orig. des.)=*tincta* (Walker).

fraterna Loew, 1863a: 27 (Cent. 3, no. 51).—Alaska; Alaska to Greenland, s. to Colo., Minn., and N.Y.
  hyalina Garrett, 1925a: 4 (*Amoebaleria*; as var.).—B.C.
thoracica Collin, 1935: 381.—N.W.T. (Akpatok I.).
tincta (Walker), 1849: 1092 (*Helomyza*).—N.S.; Alaska, B.C. to N.S., s. to Calif. and Pa.
  ferruginea Walker, 1849: 1066 (*Actora*).—N.S.

pubescens Loew, 1862b: 224 (Cent. 2, no. 82) (*Blepharoptera*).—Mass.
pilosus Coquillett, 1907a: 75 (*Achaetomus*).—Mass.
gigas Garrett, 1921: 126 (*Amoebaleria*).—B.C.

Genus ANYPOTACTA Czerny

Anypotacta Czerny, 1924: 143. Type-species, *setulosa* Czerny (Aldrich, 1926f: 101).
aldrichi (Garrett), 1921: 121 (*Leria*).—B.C.
czernyi Garrett *in* Gill, 1962: 575.—N. Mex.

Genus AMOEBALERIA Garrett

Amoebaleria Garrett, 1921: 125. Type-species, *scutellata* Garrett (Aldrich, 1926f: 101).
Chaetomus Czerny, 1924: 128. Type-species, *Helomyza flavotestacea* Zetterstedt (Aldrich, 1926f: 101).
caesia (Meigen), 1830: 56 (*Helomyza*).—Europe; ?N. Amer.
confusa (Wahlgren), 1918: 6 (*Helomyza*).—Sweden; ?Calif., Europe.
defessa (Osten Sacken) *in* Packard, 1877: 168 (*Blepharoptera*).—Ky.; Mich. to Pa., s. to Mo. and Ala.
flavotestacea (Zetterstedt), 1838: 765 (*Helomyza*).—Lapland; Alaska to Labr., s. to Calif. and N.C., Europe.
glauca (Aldrich) *in* Aldrich and Darlington, 1908: 87 (*Leria*).—Wash.; Calif.
gonea Garrett, 1925a: 3.—B.C.; Utah, Ill.
helvola (Loew), 1862b: 223 (Cent. 2, no. 80) (*Scoliocentra*).—Ill.; S. Dak. to Que., s. to Ark. and Ga.
    angustifrons Banks, 1926: 43 (*Leria;* as var.).—N.Y., D.C., Va., N.C.
infuscata Gill, 1962: 586.—Idaho; Calif., Colo.
perplexa Garrett, 1924b: 27.—B.C.; Alaska, Wash.
sabroskyi Gill, 1962: 587.—Iowa; Idaho, Nev.
sackeni Garrett, 1925a: 3.—N.Y.; Minn. to Que., s. to Mo. and Tenn.
scutellata Garrett, 1921: 125.—B.C.
spectabilis (Loew), 1862e: 58 (*Blepharoptera*).—Europe; Calif.
triangulata Garrett, 1925a: 4.—B.C.; Wash., Idaho, Wyo., Colo.
tularensis Gill, 1962: 588.—Calif.

Genus HELEOMYZA Fallén

Heleomyza Fallén, 1810a: 19. Type-species, *Musca serrata* Linnaeus (mon.).
Helomyza, emend.

Leria Robineau-Desvoidy, 1830: 653. Type-species, *domestica* Robineau-Desvoidy (Rondani, 1866b: 5, 42; 1867: 89, 126) =*serrata* (Linnaeus).

Blephariptera Macquart, 1835: 412. Type-species, *Musca serrata* Linnaeus (Westwood, 1840: 145).

Blepharoptera, emend.

bisetata (Garrett), 1922b: 175 (*Amoebaleria*).—Man.; Alaska, B.C., Mich., Md., Va.

brachypterna (Loew), 1873a: 49 (*Blepharoptera*).—Hungary; Alaska, B.C. to N.S., s. to Colo., Ark., and Va., Europe.
    latens Aldrich, 1897: 188 (*Blepharoptera*).—Ind.
    caccabata Tucker, 1909: 301 (*Leria*).—Kans.
    serrataria Garrett, 1924b: 26 (*Leria*).—B.C.

carolinensis (Robineau-Desvoidy), 1830: 629 (*Scatophaga*).—Carolina. Unrecognized.

czernyi Collart, 1933: 402.—Belgium; Greenland, France.

difficilis Gill, 1962: 594.—Alaska; N.W.T., Que., Labr.

fusca (Macquart), 1843a: 420 (1843: 263) (*Heteromyza*).—N. Amer. Unrecognized.

genalis (Coquillett), 1910a: 130 (*Leria*).—B.C.

modesta Meigen, 1838: 369.—Bavaria; Greenland, Europe.

nebulosa (Coquillett), 1910a: 129 (*Leria*).—Alta.

pleuralis (Becker), 1907b: 1 (*Blepharoptera*).—Siberia; Alaska, B.C., N. Mex., Labr., N.H.
    pleuralis Coquillett, 1910a: 130 (*Leria*).—Alaska.

serrata (Linnaeus), 1758: 597 (*Musca*).—Sweden; Alaska to Que., s. to Calif. and Va., Europe.
    nigricana Garrett, 1922b: 176 (*Leria;* as var.).—B.C.
    vinus Garrett, 1922b: 177 (*Leria;* as var.).—B.C.
    americana Garrett, 1925a: 3 (as form).—U.S., Canada.

tristissima (Garrett), 1921: 122 (*Leria*).—Nfld.; Alaska.

Genus TRICHOCHLAMYS Czerny

Trichochlamys Czerny, 1924: 155. Type-species, *borealis* Czerny (mon.).

borealis Czerny, 1924: 155.—Alaska; Bering Strait, Copper I., Commander Is.

Unplaced Species of Heleomyzidae

fasciata Walker, 1849: 1094 (*Helomyza*).—N.S. Type lost.

Family TRIXOSCELIDIDAE

By J. R. Vockeroth

Most species of Trixoscelididae occur in sandy or arid areas. *Neossos marylandicus* Malloch has been reared from birds' nests; the larval habits of the other species are unknown.

REFERENCES: Melander, 1952b (rev.); Wheeler, 1955 (key to genera).

Genus NEOSSOS Malloch

Neossos Malloch, 1927b: 90. Type-species, *marylandicus* Malloch (orig. des.).
Neossus (error) Melander, 1952b: 38.

californicus Melander, 1952b: 38.—Calif.
marylandicus Malloch, 1927b: 90.—Md.; B.C., e. Que.

Genus PARANEOSSOS Wheeler

Paraneossos Wheeler, 1955: 107. Type-species, *arizonicus* Wheeler (orig. des.).

arizonicus Wheeler, 1955: 108.—Ariz.

Genus SPILOCHROA Williston

Spilochroa Williston, 1907: 2. Type-species, *Heterochroa ornata* Johnson (mon.).

REFERENCE: Wheeler, 1955 (key).

albibasis (Malloch), 1931b: 30 (*Diastata*).—N. Mex.
geminata Sabrosky, 1961b: 233.—Ariz.; Mexico (Son.).
ornata (Johnson), 1895a: 306 (*Heterochroa*).—Fla.; Tex., West Indies.
polita (Malloch), 1931b: 30 (*Diastata*).—N. Mex.; Tex.
punctipennis Melander, 1913c: 167.—N. Mex.

Genus TRIXOSCELIS Rondani

Trixoscelis Rondani, 1856: 134. Type-species, *Geomyza obscurella* Fallén (orig. des.).
Trichoscelis, emend.
Parodinia Coquillett, 1902e: 186. Type-species, *cinerea* Coquillett (orig. des.).

buccata Melander, 1952b: 43.—Ariz.; Calif., Colo.
cinerea (Coquillett), 1902e: 186 (*Parodinia*).—Calif.; Colo.
claripennis (Malloch), 1913d: 276 (*Parodinia*).—Ariz.

deserta Melander, 1952b: 44.—Calif.
flavens Melander, 1952b: 45.—Wash.
flavida Melander, 1952b: 45.—Wash.; Oreg.
frontalis (Fallén), 1823a: 7 (*Anthomyza*).—Sweden; B.C. to Mont., s. to Calif. and Ariz., Europe.
fumipennis Melander, 1913c: 168.—Man.; Wash., s. Alta. to s. Man., Nebr.
litorea (Aldrich) *in* Aldrich and Darlington, 1908: 100 (*Siligo*).—Calif.
melanderi Vockeroth (for *dimidiata* Melander).—Calif. **N. name.**
  dimidiata Melander, 1952b: 45 (preocc. Hendel, 1913).—Calif.
mohavea Melander, 1952b: 47.—Calif., Wash.
nitidiventris Melander, 1952b: 47.—Calif.
nuda (Coquillett), 1910a: 130 (*Leria*).—Calif.; Wash., Ariz.
  prima Hendel, 1911: 43.—Calif. (as N.H., error).
plebs Melander, 1952b: 48.—Calif.
pygochroa Melander, 1952b: 49.—Calif.
sagulata Melander, 1952b: 49.—Calif.; Utah, Ariz.
signifera Melander, 1952b: 49.—Calif.
suffusa Melander, 1952b: 50.—Calif.; Wash.
triplex Melander, 1952b: 50.—Ariz.; Calif., Mexico.
tumida Melander, 1952b: 51.—Wash.; Oreg., Calif.

Genus ZAGONIA Coquillett

Zagonia Coquillett *in* Baker, 1904: 27. Type-species, *flava* Coquillett (orig. des.).
Siligo Aldrich *in* Aldrich and Darlington, 1908: 98. Type-species, *oregona* Aldrich (orig. des.)=*flava* Coquillett.

flava Coquillett *in* Baker, 1904: 27.—Calif.; Wash., Idaho, Oreg.
  oregona Aldrich *in* Aldrich and Darlington, 1908: 99 (*Siligo*).—Oreg.
flavicornis Melander, 1952b: 39.—Wash.; B.C., Calif.

Family RHINOTORIDAE
By Richard H. Foote

The family Rhinotoridae is confined to the New World Tropics and Subtropics. The larvae and larval habitats are unknown. Adults of the single North American species were attracted to banana bait traps during late twilight hours.

REFERENCE: Wheeler, 1954b (descr.).

Genus RHINOTORA Schiner

Rhinotora Schiner, 1868: 233. Type-species, *pluricellata* Schiner (orig. des.).

diversa Giglio-Tos, 1893b: 13.—Mexico; Ariz., N. Mex.

Family ANTHOMYZIDAE
By Curtis W. Sabrosky

Anthomyzids are small, slender flies, commonly swept from grass and low vegetation, especially in marshy areas. Larvae have been found associated with *Juncus*, *Typha*, *Elymus*, and so forth, but it is not clear whether they are phytophagous or saprophagous.

REFERENCES: Melander, 1913a (rev.); Sturtevant, 1954 (key to genera).

Genus ANTHOMYZA Fallén

Anthomyza Fallén, 1810a: 20. Type-species, *gracilis* Fallén (Westwood, 1840: 152).
Anthophilina Zetterstedt, 1837: 55. Type-species, *Anthomyza gracilis* Fallén (mon.).

The genus needs revision; the only available key is probably unreliable for all the species.

concolor (Thomson), 1869: 596 (*Piophila*).—Calif. **N. comb.**
gracilis Fallén, 1823a: 8.—Sweden; Alaska to Maine, s. to Calif., N. Mex., and W. Va., Europe.
pallida (Zetterstedt), 1848: 2702 (*Anthophilina*).—Scandinavia; Alaska to S. Dak., s. to Calif. and Utah, also Mich., Ont., Europe.
tenuis (Loew), 1863b: 324 (Cent. 4, no. 95) (*Anthophilina*).—Alaska (Sitka); Alaska to Wash., also Minn. to Que., s. to Va.
variegata (Loew), 1863b: 324 (Cent. 4, no. 96) (*Anthophilina*).— D.C.; Wash. to Maine, s. to Calif. and Va., also Ga.

Genus ISCHNOMYIA Loew

Ischnomyia Loew, 1863b: 325 (Cent. 4, no. 97). Type-species, *vittula* Loew (mon.)=*albicosta* (Walker).

albicosta (Walker), 1849: 1113 (?*Diastata*).—Unknown; Wis. to N.Y., s. to N.C.
vittula Loew, 1863b: 325 (Cent. 4, no. 97).—Pa.
vittata (error) Curran, 1934a: 330.
spinosa Hendel, 1911: 45.—Mich.; Wis. to N.H., s. to N.C.

Genus MUMETOPIA Melander

Mumetopia Melander, 1913a: 293. Type-species, *occipitalis* Melander (orig. des.).

nigrimana (Coquillett), 1900g: 264 (*Anthomyza*).—P.R.; Fla., Neotropical.
occipitalis Melander, 1913a: 294.—Tex. and La.; S. Dak. to Mass., s. to ne. Mexico and Fla., also Bermuda.
terminalis (Loew), 1863b: 324 (Cent. 4, no. 94) (*Anthophilina*).— N.H. (as Carolina, error); Mich. to Que., s. to Va.
nitens Melander, 1913a: 294.—Mass.

Genus STENOMICRA Coquillett

Stenomicra Coquillett, 1900g: 262. Type-species, *angustata* Coquillett (orig. des.).

angustata Coquillett, 1900g: 262.—P.R.; Wis. to Que., s. to Tex. and Fla., Neotropical.

Genus CYAMOPS Melander

Cyamops Melander, 1913a: 291. Type-species, *nebulosus* Melander (orig. des.).

This genus has usually been placed in the Anthomyzidae, but it may go elsewhere when family assignments in the Acalyptratae are more thoroughly studied.

REFERENCE: Sabrosky, 1958 (key).

halteratus Sabrosky, 1958: 170.—Wis.; Mich., Ont., N.H., Mass.
imitatus Sturtevant, 1954: 559.—Ind.
nebulosus Melander, 1913a: 292.—Mass.; Ind. to N.S., s. to Fla.

Family OPOMYZIDAE

(Geomyzidae)

By J. R. Vockeroth

The known larvae of *Opomyza* and *Geomyza* feed in stems of a wide variety of Gramineae.

REFERENCES: Melander, 1913a (key); Vockeroth, 1961a (rev.).

Genus OPOMYZA Fallén

Opomyza Fallén, 1820d: 10. Type-species, *Musca germinationis* Linnaeus (Westwood, 1840: 152).

germinationis (Linnaeus), 1758: 600 (*Musca*).—Sweden; N.S., Europe. Immigrant.
petrei Mesnil, 1934: 202.—France; sw. B.C., Europe. Immigrant.

Genus ANOMALOCHAETA Frey

Anomalochaeta Frey, 1921: 25. Type-species, *Opomyza guttipennis* Zetterstedt (orig. des.).

guttipennis (Zetterstedt), 1838: 762 (*Opomyza*).—Sweden; s. Alaska to n. Ont. (Hudsonian and Canadian zones), n. Europe, Siberia.

Genus GEOMYZA Fallén

Geomyza Fallén, 1810a: 18. Type-species, *Musca combinata* Linnaeus (mon.).
Balioptera Loew, 1864b: 347. Type-species, *Musca combinata* Linnaeus (Coquillett, 1910b: 513).
Mutiloptera Coquillett, 1908: 147. Type-species, *apicalis* Coquillett (orig. des.) =*coquilletti* (Hendel).

apicalis (Meigen), 1830: 109 (*Opomyza*).—Europe; s. Ont., N.S.
balachowskyi Mesnil, 1934: 197.—France; s. B.C., Wash., Oreg., N.S., Conn. Immigrant.
combinata (Linnaeus).—Not Nearctic.
coquilletti (Hendel), 1917: 39 (*Mutiloptera*; n. name for *apicalis* Coquillett).—N. Dak.; s. Alta., s. Sask., S. Dak., Mich.
   apicalis Coquillett, 1908: 148 (*Mutiloptera*; preocc. Meigen, 1830).—N. Dak.
dolomata Vockeroth, 1961a: 518.—Colo.; Utah.
lurida (Loew), 1864a: 98 (Cent. 5, no. 98) (*Opomyza*).—Alaska; s. Alaska, s. to s. Calif., w. of Coast Ranges.
monostigma Melander, 1913a: 288.—Calif.; Wash.
monticola Vockeroth, 1961a: 516.—s. Alta.; s. Sask.
parvistigma Vockeroth, 1961a: 518.—s. B.C.
velata Vockeroth, 1961a: 519.—Sask.; Alta.
venusta (Meigen).—Not Nearctic.
vespertina Vockeroth, 1961a: 517.—Calif.; Utah.

Family CHYROMYIDAE

By J. F. McAlpine

Nine described species of chyromyid flies occur in North America, but apparently undescribed species are known in at least the genera *Aphaniosoma* Becker and *Gymnochiromyia* Hendel.

Several species have been reared from birds' nests and mammals' burrows, and a few from rotting wood. Adults are often taken on windows and on vegetation. Collections of the family made at the same time and place frequently contain a mixture of species.

REFERENCE: Wheeler, 1961 (key).

Genus APHANIOSOMA Becker

Aphaniosoma Becker, 1903: 186. Type-species, *approximatum* Becker (mon.).

REFERENCE: Wheeler, 1961 (key).

aldrichi Wheeler, 1961: 90.—Tex.
frontatum Wheeler, 1961: 89.—Calif.; Utah, Ariz., Mexico.
quadrivittatum Malloch, 1915h: 357.—Ill.; Ont.

Genus CHYROMYA Robineau-Desvoidy

Chyromya Robineau-Desvoidy, 1830: 620. Type-species, *fenestratum* Robineau-Desvoidy (mon.)=*flava* (Linnaeus).
Chyromyia and *Chiromyia*, emends.

REFERENCE: Malloch, 1914k (rev.).

femorella (Fallén), 1820c: 34 (*Sapromyza*).—Sweden; temperate N. Amer., Europe.
flava (Linnaeus), 1758: 600 (*Musca*).—Europe; e. N. Amer.
  ochrapesus Rathvon *in* Reigert, 1855: 9 (*Musca*).—Pa.
oppidana (Scopoli), 1763: 349 (*Musca*).—Yugoslavia; Que., D.C., Europe.

Genus GYMNOCHIROMYIA Hendel

Gymnochiromyia Hendel, 1933c: 43. Type-species, *Peletophila minima* Becker (orig. des.).

concolor (Malloch), 1914k: 181 (*Chyromya*).—Ill.; Man. to N.B., Tenn. **N. comb.**
minima (Becker), 1904b: 133 (*Peletophila*).—Estonia; Wash., Calif., Europe.
nigrimana (Malloch), 1914k: 181 (*Chyromya*).—Ill.; Calif. **N. comb.**

Family AULACIGASTRIDAE
By Willis W. Wirth

The family status follows Hennig (1958a). There is only one North American species, which is found on, and breeds in, sap fluxes from wounds on tree trunks.

REFERENCES: Malloch and McAtee, 1924a (biol., immature stages); Robinson, 1953 (morphology and life history).

Genus AULACIGASTER Macquart

Aulacigaster Macquart, 1835: 579. Type-species, *rufitarsis* Macquart (mon.) = *leucopeza* (Meigen).
Aulacogaster, emend.

leucopeza (Meigen), 1830: 100 (*Diastata*).—Europe; Sask. to Que., s. to Fla. and Mexico.

Family ASTEIIDAE
By Curtis W. Sabrosky

Adult asteiids are small, delicate, and rarely collected. Almost nothing is known of their biology.

REFERENCE: Sabrosky, 1957 (rev., W. Hemisphere).

Subfamily SIGALOESSINAE
Genus LEIOMYZA Macquart

Leiomyza Macquart, 1835: 605. Type-species, *Agromyza glabricula* Meigen (Blanchard, 1840: 629) = *scatophagina* (Fallén).
Liomyza, emend.

curvinervis (Zetterstedt), 1838: 785 (*Anthophilina*).—Lapland; Alaska to Mont., s. to Calif. and N. Mex., also Ont., Que., N.Y., N.H., Maine, Europe.
   melanderi Aldrich, 1919a: 140, 141.—Idaho.
   slossonae Aldrich, 1919a: 140.—N.H.
laevigata (Meigen), 1830: 179 (*Agromyza*).—Europe; Mich., Ont.
scatophagina (Fallén), 1823a: 3 (*Heteroneura*).—Sweden; Mich. to N.H., s. to Tenn., Europe.
wheeleri Sabrosky, 1957: 49.—N. Mex.; Utah.

Genus PHLEBOSOTERA Duda

Phlebosotera Duda, 1927: 119, 125. Type-species, *mollis* Duda (orig. des.).

angustigena Sabrosky, 1957: 50.—Tex.
setipalpis Sabrosky, 1943c: 511.—Utah; N. Mex., Ill.
shewelli Sabrosky, 1957: 51.—Ont.; ?Mich.

Genus ASTIOSOMA Duda

Astiosoma Duda, 1927: 119, 127. Type-species, *rufifrons* Duda (orig. des.).

aridum Sabrosky, 1957: 60.—Calif.; s. Calif.
flaveolum (Coquillett), 1898f: 49 (*Sigaloessa*).—D.C.; Kans., Ill., Ohio, N.H. to Fla.
hirtum (Aldrich), 1915c: 97 (*Sigaloessa*).—Idaho; Calif., ?Tenn., ?N.C.
lineatum (Aldrich), 1915c: 96 (*Sigaloessa*).—Idaho; Wash., Calif., N. Mex.

Genus SIGALOESSA Loew

Sigaloessa Loew, 1866a: 186 (Cent. 6, no. 100). Type-species, *bicolor* Loew (mon.).

bicolor Loew, 1866a: 186 (Cent. 6, no. 100).—Cuba; Ill., Tex., La., Bermuda, Neotropical.
flavifrons Sabrosky, 1957: 56.—Fla.; Bahama Is.
obscura Sabrosky, 1957: 57.—Md.; Tex., Fla.
semiglabra Sabrosky, 1957: 56.—Fla.

Subfamily ASTEIINAE

Genus ASTEIA Meigen

Asteia Meigen, 1830: 88. Type-species, *amoena* Meigen (Westwood, 1840: 152).

beata Aldrich, 1915c: 95.—Mass.; B.C., Wash., Oreg., Mont., Ill. to Maine.
expansa Sabrosky, 1957: 46.—Costa Rica; Fla., Haiti, Ecuador.
multipunctata Sabrosky, 1939b: 165.—Idaho; B.C. to Sask., s. to N. Mex.

Genus LOEWIMYIA Sabrosky

Loewimyia Sabrosky, 1943c: 503. Type-species, *bifurcata* Sabrosky (orig. des.).

Unnamed sp. (Sabrosky, 1957: 48).—Tex.

Family CRYPTOCHETIDAE
By Richard H. Foote

The population center of the Cryptochetidae is tropical Africa. As far as known, the larvae of all species parasitize monophlebine coccids, and on account of their great differences in form are much more easily identified than are adults. The single North American species was introduced from Australia into California in an attempt to control the cottony-cushion scale, *Icerya purchasi*. This introduction has been a very successful one.

REFERENCE: Thorpe, 1931a (review).

Genus CRYPTOCHETUM Rondani

Cryptochetum Rondani, 1875a: 167 (as *Cryptochoetum*, p. 172) (1875: 2, 7). Type-species, *grandicorne* Rondani (orig. des.).
Cryptochaetum, emend.
Lestophonus Williston, 1888c: 21. Type-species, *iceryae* Williston (mon.).

iceryae (Williston), 1888c: 21 (*Lestophonus*).—Australia; Calif. Introduced.
  monophlebi, authors, not Skuse.

Acalyptrate Genera of Uncertain Family Position

Genus CINDERELLA Steyskal

Cinderella Steyskal, 1949b: 135. Type-species, *lampra* Steyskal (mon.).

lampra Steyskal, 1949b: 134.—Okla.; Oreg., Ariz., Tex., Tenn., Ga., Fla.

Genus LATHETICOMYIA Wheeler

Latheticomyia Wheeler, 1956: 306. Type-species, *tricolor* Wheeler (orig. des.).

lineata Wheeler, 1956: 310.—Ariz.
tricolor Wheeler, 1956: 307.—Ariz.; Utah.

Section CALYPTRATAE

Superfamily MUSCOIDEA

Family ANTHOMYIIDAE

(Muscidae, in part)

By H. C. Huckett (subfamilies Fucelliinae and Anthomyiinae) and
J. R. Vockeroth (subfamily Scatophaginae)

Adults of Anthomyiidae resemble many species of Muscidae in being dark-bodied, rather slender flies, with wings of moderate length. The larvae, especially of the subfamily Anthomyiinae, are commonly known as root maggots, but many species inhabit the aerial portions of plants as well; they are usually primary invaders of plant tissue. The larvae of several species of Scatophaginae feed in dung. Some are aquatic, and sometimes they are predaceous.

The family has often been treated taxonomically as a part of the Muscidae, which it very closely resembles in morphology. A supposed representative of the subfamily Eginiinae, *Lutzomyia*, is herein referred to the tachinid subfamily Rhinophorinae.

REFERENCES: Schnabl and Dziedzicki, 1911 (Palaearctic); Malloch, 1917f (key, subfamilies); Stein, 1920 (rev.); Séguy, 1937 (cat., key genera of world); Ringdahl, 1959 (Sweden).

Subfamily SCATOPHAGINAE

(Scatomyzidae, Scopeumatidae, Cordiluridae)

By J. R. Vockeroth

The catalog of this subfamily is based on the examination of almost all available material, including most of the existing types. A revision, in which new synonymies and new combinations will be substantiated and many new species described, is approaching completion.

The biology of most species is unknown, but available records indicate that habitats are extremely varied. The Delinini are probably all leaf miners in plants of the families Orchidaceae, Liliaceae, and (in Japan) Commelinaceae. Scatophagini of the genera *Norellia*, *Cordilura*, *Nanna*, *Gonarcticus*, *Gimnomera*, and *Hydromyza* feed in a variety of families of plants, a few *Scatophaga* in rotten seaweed, and many *Scatophaga* in dung; *Spaziphora* live in water.

REFERENCES: Curran, 1934a (key to genera); Séguy, 1952 (cat.; key, world genera; summary of hosts and larval habitats).

Tribe SCATOPHAGINI

Genus NORELLIA Robineau-Desvoidy

Norellia Robineau-Desvoidy, 1830: 673. Type-species, *pseudonarcissi* Robineau-Desvoidy (mon.)=*tipularia* (Fabricius), **n. syn.**

Subgenus NORELLISOMA Wahlgren

Norellisoma Wahlgren, 1917: 148 (as genus). Type-species, *Cordilura spinimana* Fallén (present des.). **N. status.**
Norellisoma Hendel, 1910b: 308. Nomen nudum.
Norelliosoma Hendel, 1930b: 2, error.

spinimana (Fallén), 1819: 7 (*Cordilura*).—Sweden; Alaska, Oreg., Colo., s. Que., Europe, U.S.S.R. (Kamchatka).
    occidentalis Malloch, 1919e: 311.—Oreg. **N. syn.**
    septentrionalis Hendel, 1930b: 2 (*Norelliosoma*).—U.S.S.R. (Kamchatka). **N. syn.**

Genus CORDILURA Fallén

Cordilura Fallén, 1810a: 15. Type-species, *Musca pubera* Linnaeus (mon.; misident.)=*rufipes* Meigen.
Cordylura, error.
Paratidia Malloch, 1931a: 432. Type-species, *Acicephala intermedia* Curran (orig. des.). **N. syn.**
Cordilura, subg. *Snyderia* James, 1955b: 98. Type-species, *praeusta* Loew (orig. des.). **N. syn.**

Subgenus CORDILURA Fallén

REFERENCE: James, 1955b (key).

adrogans Cresson, 1918b: 135.—Calif.; Wash., Oreg.
angustifrons Loew, 1863a: 24 (Cent. 3, no. 45).—Wis.; Minn. to N.B., s. to Mass.
atrata Zetterstedt, 1846: 2002.—Norway, Sweden; Alaska to Labr., s. to Calif. and cent. Que., n. Europe, U.S.S.R. (Kamchatka).
    beringensis Malloch, 1923i: 198.—Bering Sea (St. George I.). **N. syn.**
atripennis James, 1955b: 100.—Utah.
carbonaria (Walker), 1849: 1047 (*Lissa*).–N.Y.; cent. Ont. to N.C.
    terminalis Loew, 1863a: 21 (Cent. 3, no. 39).—Pa.
ciliatipes James, 1955b: 93.—Calif.

confusa Loew, 1863a: 23 (Cent. 3, no. 43).—N.W.T.; cent. Alaska to Nfld., s. to N. Mex., Minn., and N.Y.
criddlei Curran, 1929e: 131.—B.C.; B.C. to Labr., s. to Calif. and Colo.
masconina Curran, 1931c: 11.—Que.
gagatina Loew, 1869c: 182 (Cent. 9, no. 93).—Canada; n. Man. to s. Labr., s. to Mich. and N.J.
intermedia (Curran), 1927h: 259 (*Acicephala*).—Man.; s. Alta. to s. Man., s. to Calif. and Nev.
latifrons Loew, 1869c: 181 (Cent. 9, no. 92).—Middle States; cent. Alaska to Nfld., s. to Calif., N. Mex., and Va.
inversa Curran, 1929e: 131.—B.C.
loewi James, 1955b: 88 (n. name for *flavipes* Loew).—Wis.; s. Man. to Mich., s. to S. Dak. and Ind.
flavipes Loew, 1863a: 25 (Cent. 3, no. 46) (preocc. Fallén, 1819).—Wis.
ontario Curran, 1929e: 132.—Ont.; cent. Alaska to s. Labr., s. to Calif., Colo., Minn., and Mass.
passiva Curran, 1929e: 130.—Colo.; s. Alta. and Man., s. to Colo.
picticornis Loew, 1864f: 22.—Siberia; cent. Alaska to n. Que. and s. Labr., s. to Calif. and N. Mex., n. Europe, Siberia.
pictipennis (error) Osten Sacken, 1878c: 173.
vierecki Cresson, 1918b: 134.—N. Mex. **N. syn.**
praeusta Loew, 1864a: 96 (Cent. 5, no. 93).—Canada; cent. Ont. to N.S., s. to Iowa and N.C.
proboscidea Zetterstedt, 1838: 728.—Lapland; cent. Alaska, n. Man., n. Europe.
pudica Meigen, 1826: 231.—Europe; cent. Alaska to s. Que., s. to Colo., Minn., and Maine.
alberta Curran, 1929e: 132.—Alta. **N. syn.**
rufimana Meigen, 1826: 232.—Europe; cent. Alaska, n. Man.
setosa Loew, 1860b: 81.—D.C.; cent. Ont. to N.B., s. to Wis. and n. Ga.
variabilis Loew, 1876: 326.—Mass.; Mich. to s. Que. and Maine, s. to N.J.
varicornis Curran, 1929e: 130.—Alta.

Subgenus ACICEPHALA Coquillett

Acicephala Coquillett, 1898g: 163 (as genus). Type-species, *polita* Coquillett (orig. des.). **N. status.**

polita (Coquillett), 1898g: 163 (*Acicephala*).—Colo.; cent. Sask. to s. Que., s. to Colo. and Mich. **N. comb.**

Subgenus ACHAETELLA Malloch

Achaetella Malloch, 1923h: 140 (as genus). Type-species, *Lissa varipes* Walker (orig. des.). **N. status.**

deceptiva Malloch, 1923h: 180.—Mich.; Wis. to s. Que., s. to W. Va.
varipes (Walker), 1849: 1046 (*Lissa*).—Ohio; cent. Alta. to Maine, s. to Colo. and n. Ga. **N. comb.**
    bimaculata Loew, 1860b: 80.—D.C.
    maculipennis Wulp, 1867: 152.—Wis.
    polita Harris, 1835: 600 (*Lissa*). Nomen nudum.

Subgenus CORDILURINA James

Parallelomma Becker, 1894: 94 (preocc. Becker, 1894). Type-species, *Cordilura albipes* Fallén (orig. des.). Becker's formal proposal, and the usage of authors was unfortunately antedated by the use of *Parallelomma* Becker *in* Strobl, 1894b, now recognized as a distinct genus.
Cordilura, subg. **Cordilurina** James, 1955b: 96. Type-species, *vittipes* Loew (orig. des.)=*fuscipes* Zetterstedt.

    REFERENCE: Malloch, 1923h (key, as *Parallelomma*).

albicoxa James, 1955b: 97.—Calif.
banksi (Malloch), 1923h: 180 (*Parallelomma*).—N.Y.; cent. Ont. to N.Y., s. to Tenn. **N. comb.**
dimidiata (Cresson), 1918b: 135 (*Parallelomma*).—Pa.; cent. Ont. to s. Que., s. to Mo., Tenn., and Va. **N. comb.**
emarginata (Malloch), 1923h: 179 (*Parallelomma*).—Mass.; Pa. to Mass., s. to N.C. **N. comb.**
    dorsalis Malloch, 1923h: 180 (*Parallelomma*; as var.).—Va. **N. syn.**
fasciventris Curran, 1927h: 258.—B.C.
fuscipes Zetterstedt, 1838: 726.—Sweden; cent. Alaska to n. Labr., s. in mts. to N. Mex. and N.H., n. Europe, U.S.S.R. (Kamchatka).
    pallida Walker, 1849: 981 (*Scatophaga*).—Ont. **N. syn.**
    vittipes Loew, 1872a: 96 (Cent. 10, no. 74).—Alaska. **N. syn.**
    lutea Loew, 1872a: 96 (Cent. 10, no. 75).—Alaska. **N. syn.**
    fulvithorax Curran, 1929e: 131 (as *fasciventris* var.).—Alta. **N. syn.**
    browni Curran, 1931c: 12.—Que. **N. syn.**
glabra Loew, 1869c: 180 (Cent. 9, no. 90).—N.H.; cent. Ont. to s. Que., s. to Wis., Tenn., and N.J.
    similata Malloch, 1923h: 178 (*Parallelomma*).—Ont. **N. syn.**

gracilipes Loew, 1869c: 178 (Cent. 9, no. 87).—N.H.; cent. B.C. to s. Labr., s. to Mich. and N.C.
luteola Malloch, 1924g: 14.—Oreg.; cent. B.C. to Idaho, s. to Calif.
munda Loew, 1869c: 180 (Cent. 9, no. 91).—N.W.T.; cent. Alaska to cent. Que., s. to Kans. and Va.
pleuritica Loew, 1863a: 23 (Cent. 3, no. 42).—Ont.; cent. B.C. to s. Labr., s. to Colo. and N.C.
  slossonae Coquillett, 1898g: 164.—N.H., Mass.
  vicina Cresson, 1918b: 136 (*Parallelomma*).—Pa.
scapularis Loew, 1869c: 179 (Cent. 9, no. 89).—Ont.; cent. Sask. to w. Ont., s. Colo. and Minn. L
tarsalis (Malloch), 1923h: 177 (*Parallelomma*).—Va.; cent. Ont., Iowa. **N. comb.**

Subgenus PSEUDACICEPHALA Malloch

Pseudacicephala Malloch, 1931a: 431 (as genus). Type-species, *Acicephala pilosella* Coquillett (orig. des.). **N. status.**
REFERENCE: Malloch, 1931a (key).

alberta (Curran), 1927h: 259 (*Acicephala*).—Alta.; cent. Alta. to cent. Sask., s. to Mont. **N. comb.**
marginata (Malloch), 1931a: 432 (*Pseudacicephala*).—Nev.; Oreg., Calif., Utah. **N. comb.**
pilosella (Coquillett), 1898g: 163 (*Acicephala*).—Nev., Colo.; Utah, N. Mex. **N. comb.**

Subgenus SCOLIAPHLEPS Becker

Scoliaphleps Becker, 1894: 98 (as genus). Type-species, *Cordilura ustulata* Zetterstedt (orig. des.). **N. status.**

ustulata Zetterstedt, 1838: 727.—Sweden; Alta., Sask., n. Europe.

Genus GONATHERUS Rondani

Gonatherus Rondani, 1856: 99. Type-species, *Cordilura planiceps* Fallén (orig. des.).

planiceps (Fallén), 1826: 12 (*Cordilura*).—Sweden; s. Alaska, Europe, U.S.S.R. (Kamchatka).

Genus NANNA Becker

Clidogastra, subg. **Nanna** Becker *in* Strobl, 1894b: 78. Type-species, *Cordilura flavipes* Fallén (present des.). **N. status.**
Amaurosoma Becker, 1894: 109. Type-species, *Cordilura flavipes* Fallén (orig. des.). **N. syn.**

Pselaphephila Becker, 1894: 122. Type-species, *loewi* Becker (orig. des.). **N. syn.**

REFERENCE: Curran, 1927i (key).

atripes (Malloch), 1931a: 436 (*Amaurosoma*).—Y.T.; n. Y.T. to n. Que., s. to s. Alaska, s. Ont., and s. Labr. **N. comb.**
bispinosa (Malloch), 1920j: 285 (*Amaurosoma*).—Alaska. **N. comb.**
brunneicosta (Johnson), 1927: 100 (*Amaurosoma*).—Maine; cent. Ont. to Maine, s. to W. Va. **N. comb.**
katmaiensis (Malloch), 1920j: 284 (*Amaurosoma*).—Alaska; nw. N.W.T. to s. Labr., s. to s. Alaska, s. Alta., and cent. Que., U.S.S.R. (Kamchatka). **N. comb.**
  kamtschatkense Hendel, 1930b: 9 (*Amaurosoma*).—U.S.S.R. (Kamchatka). **N. syn.**
leucostoma (Zetterstedt), 1846:2063 (*Cordilura*).—Lapland; s. Alaska, n. Que., n. Europe. **N. comb.**
  acuticornis Loew, 1869c: 182 (Cent. 9, no. 94) (*Cordilura*).— Hudson Bay Territory. **N. syn.**
loewi (Becker), 1894: 123 (*Pselaphephila*).—Unknown; cent. Alta. to N.B., s. to Minn., U.S.S.R. (Kamchatka). **N. comb.**
  argyriceps Curran, 1927h: 254 (*Pselaphephila*).—Ont. **N. syn.**
  carbonaria Hendel, 1930b: 11 (*Amaurosoma*).—U.S.S.R. (Kamchatka). **N. syn.**
pallidipes (Malloch), 1922c: 77 (*Amaurosoma*).—N.H.; n. B.C. to Maine, s. to Mich. and in mts. to Tenn. **N. comb.**
similis (Coquillett), 1902g: 124 (*Pselaphephila*).—Mass.; s. N.W.T. to cent. Que., s. to B.C. and Va. **N. comb.**
unispinosa (Malloch), 1920j: 285 (*Amaurosoma*).—Alaska. **N. comb.**

Genus ORTHACHETA Becker

Orthacheta Becker, 1894: 101. Type-species, *Cordilura pilosa* Zetterstedt (orig. des.).
Orthochaeta Aldrich, 1905: 567, emend.

cornuta (Loew), 1863a: 26 (Cent. 3, no. 48) (*Cordilura*).—Ont.; n. Y.T. to s. Labr., s. to s. Alaska, cent. Alta., and Pa., also Colo., U.S.S.R. (Kamchatka).
  amoena Cresson, 1918b: 133.—Pa. **N. syn.**
  brunneipennis Johnson, 1927: 102.—N.H. **N. syn.**
  fuscipennis Hendel, 1930b: 8.—U.S.S.R. (Kamchatka). **N. syn.**
dissimilis Malloch, 1924h: 194.—Ill.; Minn. to cent. Que., s. to Ill. and Va.

hirtipes Johnson, 1927: 103.—N.H.; s. B.C. to cent. Que., s. to Utah, Iowa, and Mass., also Va.
pilosa (Zetterstedt).—Not Nearctic.
strigipes Johnson, 1927: 102.—Colo.; cent. Alta. to s. Man., s. to N. Mex.

Genus GONARCTICUS Becker

Gonarcticus Becker, 1894: 103. Type-species, *Scatomyza antennata* Zetterstedt (orig. des.).

alberta (Curran), 1927i: 293 (*Amaurosoma*).—Alta.; s. Alaska to cent. Que., s. to Wash. and sw. Alta., also Colo. **N. comb.**
arcticus (Becker), 1907c: 412 (*Pselaphephila*).—Greenland; n. Alaska to n. Greenland, s. to n. B.C. and n. Que. **N. comb.**
  atricornis Malloch, 1919f: 77 (*Gonatherus*).—N.W.T. **N. syn.**

Genus BUCEPHALINA Malloch

Bucephalina Malloch, 1919f: 76, 77. Type-species, *Cordilura megacephala* Loew (orig. des.).

Subgenus BUCEPHALINA Malloch

megacephala (Loew), 1869c: 183 (Cent. 9, no. 95) (*Cordilura*).—D.C.; Pa., Md.
setipes (Coquillett), 1910c: 43 (*Parallelomma*).—Pa.; Iowa to N.Y., s. to Va. **N. comb.**
  flavovaria Coquillett, 1910c: 44 (*Parallelomma*).—Pa. **N. syn.**

Subgenus NEOGIMNOMERA Malloch

Neogimnomera Malloch, 1920c: 36 (as genus). Type-species, *Cordilura amans* Cresson (orig. des.). **N. status.**

amans (Cresson), 1918b: 134 (*Cordilura*).—Calif. **N. comb.**

Genus MEGAPHTHALMA Becker

Megaphthalma Becker, 1894: 105. Type-species, *Cordilura pallida* Fallén (orig. des.).
Megophthalma, emend.
Megophthalmum Hendel, 1910b: 308, emend.

pallida pallida (Fallén).—Not Nearctic.
   ssp. **americana** Malloch, 1924g: 14 (as sp.).—Oreg.; s. B.C. to cent. Que., s. to Calif., Mich., and S.C. **N. status.**

Genus MEGAPHTHALMOIDES Ringdahl

Megaphthalmoides Ringdahl, 1936: 179. Type-species, *Cordilura unilineata* Zetterstedt (orig. des.).

unilineatus (Zetterstedt), 1838: 727 (*Cordilura*).—Sweden; cent. Alaska to n. Que. and N.S., s. to s. B.C. and N.C., Europe.

Genus POGONOTA Zetterstedt

Cordilura, subg. **Pogonota** Zetterstedt, 1860: 6333. Type-species, *hircus* Zetterstedt (Becker, 1894: 138)=*barbata* (Zetterstedt).

Pogonota, subg. **Okenina** Malloch, 1931a: 427. Type-species, *Cordilura fulvibarba* Loew (orig. des.)=*gilvipes* (Loew). **N. syn.**

Subgenus POGONOTA Zetterstedt

barbata (Zetterstedt), 1838: 734 (*Cordilura*).—Sweden; Alaska, n. Man., n. Europe.
   hircus Zetterstedt, 1838: 735 (*Cordilura*).—Sweden.
gilvipes (Loew), 1863a: 26 (Cent. 3, no. 49) (*Cordilura*).—Ont.; cent. Alaska to s. Labr., s. to Colo. and Mich. **N. comb.**
   fulvibarba Loew, 1872a: 97 (Cent. 10, no. 76) (*Cordilura*).—N.W.T. **N. syn.**
pallida Malloch, 1931a: 429.—Ont.; cent. Alta. to cent. Que., s. to Mich. and Maine.

Subgenus LASIOSCELUS Becker

Lasioscelus Becker, 1894: 143 (as genus). Type-species, *Cordilura clavata* Zetterstedt (orig. des.)=*immunda* (Zetterstedt). **N. status.**

immunda (Zetterstedt), 1838: 733 (*Cordilura*).—Norway; n. Y.T. to n. Labr., s. to sw. Alaska and n. Man., n. Europe. **N. comb.**
sahlbergi (Becker), 1900: 51 (*Lasioscelus*).—nw. Siberia; nw. N.W.T. to s. Baffin I., s. to se. N.W.T. **N. comb.**

Genus OKENIELLA Hendel

Okenia Zetterstedt, 1838: 734 (preocc. Menke, 1830). Type-species, *Cordilura caudata* Zetterstedt (Becker, 1894: 141).

Okeniella Hendel, 1907a: 98 (n. name for *Okenia* Zetterstedt). Type-species, *Cordilura caudata* Zetterstedt (aut.).

dasyprocta (Loew), 1864f: 25 (*Cordilura*).—Sweden; n. Alaska to N.W.T. (s. Baffin I.)., s. in mts. to cent. Alta. and cent. Que., also Colo., n. Europe.
    kincaidi Coquillett, 1900h: 455 (1904: 69) (*Pogonota*).—Alaska. **N. syn.**

Genus COSMETOPUS Becker

Cosmetopus Becker, 1894: 146. Type-species, *Cordilura dentimana* Zetterstedt (orig. des.).

longus (Walker), 1849: 976 (*Cordilura*).—Ont.; n. N.W.T. to n. Que., s. to s. Alaska and cent. Ont., n. Europe, n. Siberia. **N. comb.**
    bergrothi Becker, 1900: 48.—nw. Siberia. **N. syn.**
    bryanti Malloch, 1932b: 303.—N.W.T. **N. syn.**

Genus PLEUROCHAETELLA Vockeroth

Pleurochaeta Becker, 1915: 63 (preocc. Beddard, 1883). Type-species, *fulvisetis* Becker (mon.) = *simplicipes* (Becker).

Pleurochaetella Vockeroth (for *Pleurochaeta* Becker). Type-species, *Pleurochaeta fulvisetis* Becker (aut.) = *simplicipes* (Becker). **N. name.**

simplicipes (Becker), 1900: 50 (*Cosmetopus*).—nw. Siberia; n. Alaska to n. Labr., s. to sw. Alaska and n. Man. **N. comb.**
    dissimilis Malloch, 1920j: 286 (*Microprosopa*).—Alaska. **N. syn.**

Genus ALLOMYELLA Malloch

Allomyia Malloch, 1919f: 80 (preocc. Banks, 1916). Type-species, *unguiculata* Malloch (mon.).

Allomyella Malloch, 1923i: 199 (n. name for *Allomyia* Malloch). Type-species, *Allomyia unguiculata* Malloch (aut.).

borealis Curran, 1927h: 260.—Alaska.

brevipennis Malloch, 1923i: 199.—Alaska (Pribilof Is.).

crinipes (Ringdahl), 1928a: 21 (*Microprosopa*).—Sweden; n. Y.T., n. N.W.T., n. Que. **N. comb.**

frigida (Holmgren), 1883: 176 (*Cordilura*).—U.S.S.R. (Novaya Zemlya); n. Alaska, n. Siberia. **N. comb.**
portenkoi (Stackelberg), 1952: 406 (*Microprosopa*).—ne. U.S.S.R. (Wrangel I.); n. Alaska and n. N.W.T., s. to cent. Alaska and cent. Y.T. **N. comb.**
unguiculata (Malloch), 1919f: 80 (*Allomyia*).—N.W.T.; n. Y.T.

Genus MICROPROSOPA Becker

Microprosopa Becker, 1894: 147. Type-species, *Cordilura haemorrhoidalis* Meigen (orig. des.).

diversipes Curran, 1927h: 256.—Alta.; nw. B.C., Wyo., Colo.
flavinervis Malloch, 1924h: 193.—Mass.; N.Y., N.H., Conn., N.J., Va., N.C.
haemorrhoidalis (Meigen), 1826: 237 (*Cordilura*).—Sweden; n. Alaska to s. Greenland, s. to s. Alaska and s. Labr., and in mts. to cent. Alta., N.H., and Maine, n. Europe.
    volucricaput Walker, 1849: 977 (*Cordilura*).—Ont. **N. syn.**
    triseta Malloch, 1920j: 286.—Alaska. **N. syn.**
heteromyzina (Zetterstedt), 1838: 723 (*Scatomyza*).—Sweden; s. Alaska and B.C. to Nfld., also Colo., n. Europe.
lineata (Zetterstedt), 1838: 732 (*Cordilura*).—Sweden; cent. Alaska, N.W.T., n. Que., n. Europe.
    ciliatus Van Duzee, 1927a: 599 (*Sicus*).—Alaska (St. Paul I.). **N. syn.**
pallidicauda (Zetterstedt), 1838: 733 (*Cordilura*).—Sweden; Alaska, Y.T., w. N.W.T., n. Europe
    pallicauda, error.
    arctica Malloch, 1920j: 285.—Alaska. **N. syn.**
varicornis Curran, 1927h: 257.—B.C.; n. B.C. to se. N.W.T., s. to s. B.C. and s. Alta.

Genus ACANTHOCNEMA Becker

Acanthocnema Becker, 1894: 136 (preocc. Costa, 1859). Type-species, *Cordilura nigrimana* Zetterstedt (orig. des.). Name conserved I.C.Z.N., 1954c: 91.
Clinoceroides Hendel, 1917: 36. Type-species, *Cordilura glaucescens* Loew (orig. des.).

albibarba (Loew), 1869c: 183 (Cent. 9, no. 96) (*Cordilura*).—N.H.; N.Y.
capillata (Loew), 1872a: 98 (Cent. 10, no. 77) (*Cordilura*).—N.H.; s. Que.

nigrimana (Zetterstedt).—Not Nearctic.
ruficauda Curran, 1929e: 133.—Colo.; Utah.

Genus STAEGERIA Rondani

Staegeria Rondani, 1856: 99. Type-species, *Cordilura kunzei* Zetterstedt (orig. des.).

Unnamed sp.—cent. Alta. to cent. Que., s. to Minn.

Genus SPAZIPHORA Rondani

Spaziphora Rondani, 1856: 99. Type-species, *Cordilura hydromyzina* Fallén (orig. des.).
Spathiophora Mik, 1884: 254, emend.
Spathiphora, emend.

cincta (Loew), 1863a: 25 (Cent. 3, no. 47) (*Cordilura*).—D.C.; cent. Alaska to cent. Que., s. to s. Alta., Iowa, and Va.
   litoralis Curran, 1927h: 256 (as var.).—Ont. **N. syn.**
   fascipes, authors, not Becker.

Genus CORDYLURELLA Malloch

Cordylurella Malloch, 1919f: 78. Type-species, *Cordilura nebulosa* Coquillett (orig. des.).

nana (Loew), 1864a: 96 (Cent. 5, no. 94) (*Cordilura*).—Canada; cent. Ont. to s. Labr., s. to N.J. **N. comb.**
   costalis Curran, 1929e: 133.—Que.
nebulosa (Coquillett), 1898g: 164 (*Cordilura*).—Ill.; cent. Ont. to N.S., s. to Ill. and Maine.
rufula Curran, 1927h: 257.—Que.; N.Y.

Genus TRICHOPALPUS Rondani

Trichopalpus Rondani, 1856: 100 (as *Tricopalpus*, p. 225). Type-species, *Cordilura fraterna* Meigen (orig. des.).
Opsiomyia Coquillett, 1898g: 162. Type-species, *palpalis* Coquillett (orig. des.). **N. syn.**
Microprosopa, subg. *Paramicroprosopa* Ringdahl, 1936: 169, 177. Type-species, *subarctica* Ringdahl (mon.)=*obscurella* (Zetterstedt).

nigribasis Curran, 1927h: 255.—Alta.; s. Alaska to cent. N.W.T., s. to s. B.C., Colo., and s. Man., n. Europe.
   pilirostris Ringdahl, 1936: 178 (*Chaetosa*).—Norway. **N. syn.**

obscurella (Zetterstedt), 1846: 2043 (*Cordilura*).—Sweden; cent. Alaska, N.W.T., n. Que., n. Europe.
palpalis (Coquillett), 1898g: 162 (*Opsiomyia*).—N.H.; s. B. C. to cent. Ont. and Maine, s. to Minn. and Ind. **N. comb.**

Genus CHAETOSA Coquillett

Chaetosa Coquillett, 1898g: 163. Type-species, *Cordilura punctipes* Meigen (orig. des.).

churchilli Malloch, 1931a: 434.—Man.; n. B.C., N.W.T., n. Man., n. Que.

punctipes (Meigen), 1826: 239 (*Cordilura*).—Europe; cent. Alaska to s. Labr., s. to Calif., Colo., Minn., and Maine.

Genus ERNONEURA Becker

Ernoneura Becker, 1894: 135. Type-species, *Scatomyza argus* Zetterstedt (orig. des.).

argus (Zetterstedt), 1838: 724 (*Scatomyza*).—Sweden; s. Alaska, n. Y.T., N.W.T., n. Que., n. Europe.

Genus SCATOPHAGA Meigen

Scopeuma Meigen, 1800: 36. Type-species, *Musca merdaria* Fabricius (Coquillett, 1910b: 604) = *sterocoraria* (Linnaeus). Suppressed by I.C.Z.N., 1963b: 339.

Scatophaga Meigen, 1803: 277 (as *Scathophaga*). Type-species, *Musca merdaria* Fabricius (mon.) = *stercoraria* (Linnaeus). The spelling *Scatophaga* has been used by all later authors, and by Meigen himself in his later works, and the I.C.Z.N. has been asked to conserve it (Vockeroth, 1961b: 296).

Pyropa Illiger, 1807: 475. Type-species, *Musca stercoraria* Linnaeus (present des.).

Scatomyza Fallén, 1810a: 15. Type-species, *Musca scybalaria* Linnaeus (Lucas *in* d'Orbigny, 1848: 411).

Conisternum Becker *in* Strobl, 1894b: 79. Type-species, *Cordilura obscura* Fallén (mon.).

Coniosternum Becker, 1894: 176 (as *Koniosternum*, p. 85). Type-species, *Cordilura obscura* Fallén (orig. des.). Formal proposal by Becker, unexpectedly antedated by Becker *in* Strobl.

Pseudopogonota Malloch, 1920c: 35. Type-species, *aldrichi* Malloch (orig. des.). **N. syn.**

REFERENCES: Malloch, 1935a (key); James, 1950a (key, w. N. Amer.).

aldrichi (Malloch), 1920c: 35 (*Pseudopogonota*).—Idaho; s. Alaska, B.C., w. Alta., Colo. **N. comb.**
  pallida Malloch, 1920c: 36 (*Pseudopogonota*; as var.).—Idaho.
  palpalis Malloch, 1931a: 443.—Alta.
apicalis Curtis, 1835b: lxxx.—Arctic N. Amer.; n. Alaska to n. Greenland, s. to n. Que., n. Europe, n. Siberia.
  lanata Lundbeck, 1901: 294.—Greenland. **N. syn.**
  nigrolanata Cresson, 1918b: 136.—Greenland. **N. syn.**
  rubicunda Malloch, 1919f: 81.—Alaska (Pribilof Is.). **N. syn.**
crinita Coquillett, 1901h: 612.—Commander Is. (Bering I.); is. of Bering Sea.
dasythrix Becker, 1894: 173.—Bering Strait; w. Alaska to Calif., littoral.
frigida Coquillett, 1900h: 454 (1904: 68).—Alaska; sw. Alaska to cent. Calif., littoral.
furcata (Say), 1823: 98 (1859b: 85) (*Pyropa*).—Mo.; cent. Alaska to s. Greenland, s. to s. Calif., Mexico (D.F.), Miss., and Ga., Palearctic.
  squalida Meigen, 1826: 252.—Germany.
  bicolor Walker, 1849: 982.—Ont.
  pubescens Walker, 1849: 982.—Ont. **N. syn.**
  canadensis Walker, 1858: 218.—Canada. **N. syn.**
  suisterci Townsend, 1891j: 153 (*Cleigastra*).—D.C.
  postilena Harris, 1835: 600. Nomen nudum.
grisea Malloch, 1920c: 34.—Utah; Oreg. to cent. Alta., s. to Calif. and Colo.
hiemalis (James), 1950a: 348 (*Scopeuma*).—Colo.
incola Becker, 1900: 54.—nw. Siberia; Alaska, N.W.T., n. Labr., n. Europe.
intermedia Walker, 1849: 980.—N.S.; sw. Alaska to s. B.C., coast of Hudson Bay, Atlantic coast from N.W.T. (s. Baffin I.) to Mass., littoral.
  impudica, authors, not Reiche.
  litorea, authors, not Fallén.
islandica Becker, 1894: 175.—Labr., Iceland. Perhaps a mixture of *litorea* and *intermedia*.
lapponica (Ringdahl), 1920: 39 (*Coniosternum*).—Sweden; sw. Alaska. **N. comb.**

litorea (Fallén), 1819: 4 (*Scatomyza*).—Sweden; s. Greenland, Iceland, Europe, n. Africa, littoral.
    impudica Reiche, 1857: ix (*Anthomyia*).—Greenland. **N. syn.**
lutaria (Fabricius).—Not Nearctic.
mollis Becker, 1894: 171.—Siberia; sw. Alaska to n. Y.T. and n. Man., s. to Utah and Colo.
monticola Malloch, 1924h: 195.—N.H.; s. Alaska, B.C., cent. Que.
multisetosa (Holmgren), 1883: 174 (*Scatomyza*).—U.S.S.R. (Novaya Zemlya and Vaigach I.); n. Alaska to N.W.T. (n. Ellesmere I., Baffin I.), n. Siberia.
    vulpina Coquillett, 1898g: 162.—Alaska. **N. syn.**
nigripalpis Becker, 1907c: 413.—Greenland; cent. Alaska to n. Greenland, s. to se. N.W.T. and s. Baffin I.
    picipes Malloch, 1935a: 263.—Y.T.
nigrolimbata Cresson, 1918b: 137.—N.Y.; Minn. to s. Labr., s. to Tenn. and N.C.
    pallida, authors, not Walker.
obscura (Fallén), 1819: 9 (*Cordilura*).—Sweden; sw. Alaska, nw. N.W.T., Europe.
    minuta Malloch, 1935a: 255.—N.W.T. **N. syn.**
obscurinervis Becker, 1900: 55.—nw. Siberia; Alaska to n. Que., n. Europe, n. Siberia.
    futilis Malloch, 1935a: 264.—N.W.T. **N. syn.**
orcasae Malloch, 1935a: 265.—Wash.; B.C., Oreg.
pictipennis Oldenberg, 1923: 307.—Austria; cent. B.C., cent. Ont., n. and cent. Europe.
reses Giglio-Tos ,1893b: 7.—Mexico; Ariz. and N. Mex., s. to Guatemala and British Honduras.
robusta (Curran), 1927h: 260 (*Allomyella*).—Alta. **N. comb.**
stercoraria (Linnaeus), 1758: 599 (*Musca*).—Sweden; s. Alaska to Nfld., s. to s. Calif., Mexico, and Ga., also Haiti, Europe, Asia.
    merdaria Fabricius, 1794: 344 (*Musca*).—Germany.
subpolita Malloch, 1935a: 262.—Y.T. Unrecognized.
suilla (Fabricius), 1794: 343 (*Musca*).—Germany; cent. Alaska to n. Que., s. to Oreg., Colo., s. Man., and N.C.
varipes (Holmgren), 1883: 175 (*Scatomyza*).—U.S.S.R. (Novaya Zemlya); is. of Bering Sea, n. Alaska, n. N.W.T., Hudson Strait.
    nubifera Coquillett, 1901h: 612.—Alaska. **N. syn.**

Genus CERATINOSTOMA Meade

Ceratinostoma Meade, 1885: 152. Type-species, *maritimum* Meade (Sack, 1937: 64) =*ostiorum* (Curtis).

ostiorum (Curtis), 1832: pl. 405 (*Scatophaga*).—Ireland; se. Que. to Mass., Europe, littoral.

Genus GIMNOMERA Rondani

Gimnomera Rondani, 1866b: 21 (as *Gymnomera*, p. 51) (1867: 105, 135). Type-species, *Cordilura tarsea* Fallén (orig. des.).
Dasypleuron Malloch, 1919f: 79. Type-species, *tibialis* Malloch (orig. des.). **N. syn.**

cerea (Coquillett), 1908: 146 (*Scatophaga*).—N.J.; Oreg. to n. Que., s. to Minn. and N.J., and in mts. to N. Mex. and N.C. **N. comb.**
    atrifrons Malloch, 1920c: 37.—Minn. **N. syn.**
    wilcoxi Curran, 1933d: 11 (*Megaphthalma*).—Oreg. **N. syn.**
fasciventris Malloch, 1920c: 38.—Ill.; Sask., Tex.
incisurata Malloch, 1920c: 37.—Ill.; Ont.
subvittata (Malloch), 1919f: 78 (*Cordylurella*).—N.W.T.; Alaska, Y.T., n. Que. **N. comb.**
tibialis (Malloch), 1919f: 79 (*Dasypleuron*).—Alaska; Y.T., n. B.C., N.W.T. **N. comb.**

Genus HYDROMYZA Fallén

Hydromyza Fallén, 1813: 243. Type-species, *Musca livens* Fabricius (mon.).

confluens Loew, 1863a: 27 (Cent. 3, no. 50).—Ont.; s. N.W.T. to Maine, s. to s. Man. and N.Y.

Tribe DELININI

Genus PARALLELOMMA Becker

Clidogastra, subg. **Parallelomma** Becker *in* Strobl, 1894b: 78. Type-species, *Cordilura vittata* Meigen (mon.).
Chylizosoma Hendel, 1924: 83. Type-species, *Parallelomma medium* Becker (orig. des.). **N. syn.**

vittatum (Meigen), 1826: 236 (*Cordilura*).—Europe; s. B.C. to N.S., s. to Calif., Mich., and N.C.
    inermis Loew, 1869c: 178 (Cent. 9, no. 88) (*Cordilura*).—N.H. **N. syn.**
    nudicornis Cresson, 1918b: 135.—Calif. **N. syn.**

Genus AMERICINA Malloch

Americina Malloch, 1923h: 139. Type-species, *Cordilura adusta* Loew (orig. des.).

adusta (Loew), 1863a: 22 (Cent. 3, no. 41) (*Cordilura*).—N.J.; cent. Ont., s. to Kans. and Ga.

Genus DELINA Robineau-Desvoidy

Delina Robineau-Desvoidy, 1830: 669. Type-species, *dejeanii* Robineau-Desvoidy (Séguy, 1952: 56) =*nigrita* (Fallén).

nigrita (Fallén), 1819: 10 (*Cordilura*).—Sweden; s. Alaska to n. N.W.T. and n. Que., s. to cent. Sask. and cent. Ont., also Colo., Europe.
 dejeanii Robineau-Desvoidy, 1830: 670.—France. **N. syn.**
 cornuta Walker, 1849: 1047 (*Lissa*).—Ont. **N. syn.**

Genus HEXAMITOCERA Becker

Hexamitocera Becker, 1894: 107. Type-species, *Cordilura loxocerata* Fallén (orig. des.).

loxocerata (Fallén), 1826: 12 (*Cordilura*).—Sweden; n. B.C., Europe.
tricincta (Loew), 1869c: 177 (Cent. 9, no. 85) (*Coenosia*).—N.H.; cent. Alta. to cent. Que., s. to Calif., Colo., and Conn.
 flavida Coquillett, 1901h: 612.—N.H.
vittata Coquillett, 1898g: 165.—Colo.; cent. B.C. to cent. Sask., s. to Utah and Colo.

Genus PLETHOCHAETA Coquillett

Plethochaeta Coquillett, 1901h: 613. Type-species, *varicolor* Coquillett (orig. des.).
Mesamyia Malloch, 1931a: 437. Type-species, *testacea* Malloch (orig. des.). **N. syn.**

testacea (Malloch), 1931a: 437 (*Mesamyia*).—Colo.; cent. Alta., cent. Ont., s. Que., Utah, Mich. **N. comb.**
varicolor Coquillett, 1901h: 614.—Pa.; Wis. to s. Que., s. to Ohio and Md.

Genus NEOCHIROSIA Malloch

Neochirosia Malloch, 1917a: 36. Type-species, *setigera* Malloch (mon.).

This genus and its type-species were originally described in Coenosiinae, and are here transferred to the Scatophaginae.

atrifrons (Coquillett), 1910c: 44 (*Plethochaeta*).—N.H.; Tenn. **N. comb.**
nuda (Malloch), 1922c: 78 (*Amaurosoma*).—Mass.; cent. Alta. to s. Labr., s. to s. Man. and Mass. **N. comb.**
setiger Malloch, 1917a: 36.—Ill.

Unplaced Species of Scatophaginae

cloacaris Fabricius, 1780: 204 (*Musca*).—Greenland.
estotilandica Rondani, 1863: 35 (1864: 35) (*Scatina*).—Labr.
exotica Wiedemann, 1830: 448 (*Scatophaga*).—La.

Subfamily FUCELLIINAE
By H. C. Huckett

The larvae of Fucelliinae are aquatic or semi-aquatic, and zoophagous.

Genus FUCELLIA Robineau-Desvoidy

Scatophaga, subg. *Halithea* Haliday, 1838: 185 (preocc. Savigny, 1817). Type-species, *maritima* Haliday (mon.).
Fucellia Robineau-Desvoidy, 1842: 269. Type-species, *arenaria* Robineau-Desvoidy (mon.)=*maritima* (Haliday).
Fucellina Schnabl and Dziedzicki, 1911: 123 (1911: 71). Type-species, *Scatomyza griseola* Fallén (Séguy, 1937: 41).

REFERENCE: Aldrich, 1918a (rev.).

aestuum Aldrich, 1918a: 178.—Wash.; B.C., Oreg.
albeola Huckett, 1927: 163.—N.Y.
antennata Stein, 1910: 23.—Alaska; N.W.T., Wash., Oreg., Man., Que.
ariciiformis (Holmgren), 1872: 103 (*Scatomyza*).—n. Greenland; Alaska, N.W.T., Man., Que., Nfld.
    hinei Aldrich, 1918a: 178.—Alaska.
assimilis Malloch, 1918a: 317.—Calif.
costalis Stein, 1910: 21.—Calif.
fucorum (Fallén), 1819: 5 (*Scatomyza*).—Sweden; Alaska to Oreg., also Greenland, Europe.
hypopygialis Ringdahl, 1930: 7.—U.S.S.R. (Kamchatka); Alaska, U.S.S.R. (Commander Is.).
intermedia Lundbeck, 1901: 291.—Greenland; Ill. to Labr. and Nfld., s. to Fla., Bermuda.
    marina, authors, not Macquart.
    maritima, authors, not Haliday.

kamtchatica Ringdahl, 1930: 7.—U.S.S.R. (Kamchatka); Alaska, U.S.S.R. (Commander Is.).
pacifica Malloch, 1923d: 427 (as *maritima* ssp.).—Mexico (Baja Calif.); Wash., Calif.
   maritima Haliday of Malloch, 1918a: 318.
pictipennis Becker, 1907c: 411.—e. Greenland; Alaska, Y.T., N.W.T.
   punctipennis (error) Malloch, 1919f: 75.
rejecta Aldrich, 1918a: 171.—Calif.; Oreg.
rufitibia Stein, 1910: 25.—Calif.; B.C. to Calif.
separata Stein, 1910: 24.—Wash., Calif.; B.C., Oreg.
thinobia (Thomson), 1869: 563 (*Scatophaga*).—Calif.; Alaska, Oreg., U.S.S.R. (Commander Is.).
   bicruciata Stein, 1910: 20.—Bering Straits (?Miednaja).
   evermanni Aldrich, 1918a: 173.—Calif.
vibei Collin, 1951: 187.—N.W.T. (Ellesmere I.); N.W.T. (Baffin I.).

Genus CIRCIA Malloch

Circia Malloch, 1929: 100. Type-species, *tinctipennis* Malloch (orig. des.).

tinctipennis Malloch, 1929: 100.—Alta.; Alaska.

Subfamily ANTHOMYIINAE
By H. C. Huckett

The larval habitats of Anthomyiinae are widely varied. Some of the species of the genera *Hylemya* and *Pegomya* are of primary economic importance.

REFERENCE: Ringdahl, 1959 (rev.).

Tribe MYOPININI

Genus MYOPINA Robineau-Desvoidy

Myopina Robineau-Desvoidy, 1830: 675. Type-species, *reflexa* Robineau-Desvoidy (mon.)=*myopina* (Fallén).
Gonarcticoides Curran, 1927h: 253. Type-species, *argenticeps* Curran (orig. des.)=*myopina* (Fallén).

crassipalpis Ringdahl, 1937: 124.—Sweden; N.W.T. (Baffin I.).
myopina (Fallén), 1824: 65 (*Musca*).—Sweden; Alaska, N.W.T., Ont. to Nfld., n. Europe, U.S.S.R. (Kamchatka).
   reflexa Robineau-Desvoidy, 1830: 676.—France.
   argenticeps Curran, 1927h: 254 (*Gonarcticoides*).—Ont.

scoparia (Zetterstedt), 1845: 1504 *(Aricia)*.—Sweden; Alaska, Y.T., N.W.T. (Baffin I.), Que., Lapland.

Genus PSEUDOCHIROSIA Ringdahl

Pseudochirosia Ringdahl, 1928b: 22, 23. Type-species, *Chirosia fractiseta* Stein (orig. des.).

albipennis Ringdahl, 1928b: 23.—Sweden; Alaska, N.W.T. (Victoria and Baffin Is.).

fractiseta (Stein), 1908: 8 *(Chirosia)*.—Sweden; Alaska, N.W.T., Man., Que., Labr.

Genus CHIROSIA Rondani

Chirosia Rondani, 1856: 102. Type-species, *Aricia albitarsis* Zetterstedt (orig. des.).

hirtipes Stein, 1907: 368.—Tibet; Alaska to w. N.W.T., s. to Calif. and Colo., also Labr., Europe, Asia.
  subarctica Ringdahl, 1937: 125.—Sweden.
idahensis Stein, 1898: 251.—Idaho; B.C. to Calif.

Genus CHIASTOCHETA Pokorny

Chiastocheta Pokorny, 1889: 568. Type-species, *Aricia trollii* Zetterstedt (orig. des.)=*inermella* (Zetterstedt).
Chiastochaeta, emend. or error.

REFERENCES: Hennig, 1953 (rev., Palaearctic); Collin, 1954 (rev., British Isles).

glauca (Coquillett), 1900h: 452 (1904: 66) *(Chirosia)*.—Alaska; Calif.

Tribe CHELISIINI

Genus CHELISIA Rondani

Chelisia Rondani, 1856: 101. Type-species, *Coenosia monilis* Meigen (orig. des.).

elegans Stein, 1920: 62.—Calif.

Tribe ANTHOMYIINI

REFERENCE: Huckett, 1924 (keys, rev., N.Y.).

Genus HYLEMYA Robineau-Desvoidy

Hylemya Robineau-Desvoidy, 1830: 550. Type-species, *strenua* Robineau-Desvoidy (Rondani, 1866a: 74, 184; 1866: 7,117) =*strigosa* (Fabricius).
Hylemyia, emend.

REFERENCE: Collin, 1927 (tax.).

Subgenus HYLEMYA Robineau-Desvoidy

alcathoe (Walker), 1849: 937 (*Anthomyia*).—N.S.; Alaska, Wash. and Alta. to N.S., s. to Colo. and Ga.
 flavicaudata Bigot, 1885a: 299.—Wash. Territory.
 strigata Stein, 1898: 211.—Idaho, Wash.
 tenax Johannsen, 1916: 388.—N.Y.
variata (Fallén), 1823f: 59 (*Musca*).—Sweden; Alaska and B.C. to Que., s. to Ariz., La., and Va., Europe.

Subgenus HYLEMYZA Schnabl and Dziedzicki

Hylemyia, subg. **Hylemyza** Schnabl and Dziedzicki, 1911: 94 (1911: 42). Type-species, *Anthomyza lasciva* Zetterstedt (Séguy, 1937: 70).
lasciva (Zetterstedt), 1838: 666 (*Anthomyza*).—Sweden; Alaska, Alta. to Que., s. to Utah and Pa.
 marginata Stein of Coquillett, 1900h: 448 (1904: 62).

Subgenus PYCNOGLOSSA Coquillett

Pycnoglossa Coquillett, 1901h: 613 (as genus). Type-species, *flavipennis* Coquillett (orig. des.)=*flavipennis* (Fallén).
Pogonomyza Schnabl, 1911: 78. Type-species, *Musca flavipennis* Fallén (Karl, 1928: 177).

REFERENCE: Huckett, 1949 (rev.).

cinerosa (Zetterstedt), 1845: 1450 (*Aricia*).—Denmark; Mich., Ont., Que., N.Y., Mass., Europe.
delicata Huckett, 1949: 55.—N.Y.; N.C.
filicis Huckett, 1949: 56.—N.Y.; Mich., Maine
flavipennis (Fallén), 1823f: 59 (*Musca*).—Sweden; Wash., Wis. to Que. and N.S., s. to Pa.
 flavipennis Coquillett, 1901h: 613 (*Pycnoglossa*).—Wash.
gleniensis (Huckett), 1924: 40 (*Pogonomyza*).—N.Y.; Wis., Que.
proboscidalis (Malloch), 1920a: 185 (*Pogonomyza*).—Pa.; Mich., Que., and Maine, s. to Tenn. and N.C.
 campestris Huckett, 1924: 40 (*Pogonomyza*).—N.Y.
pusillans Huckett, 1949: 57.—N.Y.; Ont., Tenn.

spinosissima Malloch, 1919g: 95.—Ont.; Wis., Mich., Que.
stratifrons Huckett, 1949: 58.—Que.; Mich. to Que., s. to e. Tenn. and R.I.

Subgenus PHORBIA Robineau-Desvoidy

Phorbia Robineau-Desvoidy, 1830: 559 (as genus). Type-species, *musca* Robineau-Desvoidy (Coquillett, 1910b: 589).
Chortophila Macquart, 1835: 323. Type-species, *Anthomyia sepia* Meigen (Westwood, 1840: 142).

REFERENCE: Huckett, 1948 (rev.).

barbicula Huckett, 1948: 114.—Alta.; Sask.
genitalis Schnabl *in* Schnabl and Dziedzicki, 1911: 248 (1911: 196). Poland; Alaska, N.W.T., Sask. to Labr., s. to Colo., Europe.
impula Huckett, 1948: 117.—B.C.
lobata Huckett, 1929: 137.—Alta.; Alaska to Ariz. and Colo.
masculans Huckett, 1948: 117.—N. Mex.
nitidula (Coquillett) *in* Johnson, 1903a: 103 (*Pegomyia*).—N. Mex.; Alaska, Ariz., Colo., Que.
conicans Huckett, 1948: 116.—Colo.
penicillaris (Stein), 1916: 193 (*Chortophila*).—Germany; Y.T., Alta., Sask., Ont.
portensis Huckett, 1948: 118.—Oreg.
sepia (Meigen), 1826: 152 (*Anthomyia*).—Germany; Alta., Ont., Que.
curvicauda Zetterstedt, 1845: 1618 (*Aricia*).—Sweden.
sinuata Malloch, 1924h: 196.—Mass.; Mich., N.Y., N.H.
sepia Meigen of Leonard, 1928: 839.

Subgenus DELIA Robineau-Desvoidy

Delia Robineau-Desvoidy, 1830: 571 (as genus). Type-species, *floricola* Robineau-Desvoidy (Coquillett, 1910b: 531)= *cardui* (Meigen).
Hylemyia, subg. *Crinura* Schnabl, 1911: 71. Type-species, *Chortophila cilicrura* Rondani (mon.)=*platura* (Meigen)

The subgenus *Delia* contains most of the economically important species of *Hylemya*. *H. antiqua*, the onion maggot, *brassicae*, the radish maggot, and *platura*, the seed-corn maggot (heretofore commonly known as *cilicrura*), all cause primary damage to plant tissue above or below ground.

REFERENCES: Huckett, 1951b (rev. *setiventris* complex), 1952, 1953a (revs. *Delia*, in part); Brooks, 1951 (ident. cruciferous root maggots).

aemene (Walker), 1849: 937 (*Anthomyia*).—N.S.; B.C. to N.S., s. to Utah and Nebr.
    testacea Stein, 1898: 208.—S. Dak.
alaba (Walker), 1849: 948 (*Anthomyia*).—N. Amer.; Alaska to N. Mex., Wis. to Nfld., s. to N.Y.
    relata Stein, 1901: 206.—U.S.
    lateralis Stein, 1920: 77.—Wash., Que.
    innocua Malloch, 1920a: 186.—Mass.
albula (Fallén), 1825: 74 (*Musca*).—Sweden; Alaska to Utah, also Ill., Mich., Que., Europe.
    biciliata Coquillett, 1900h: 451 (1904: 65) (*Phorbia*).—Alaska.
angustifrons (Meigen), 1826: 146 (*Anthomyia*).—Germany; Alaska to Greenland, Europe.
    scatophagina Zetterstedt, 1838: 677 (*Anthomyza*).—Sweden.
    brevitarsata Malloch, 1919f: 73.—Alaska.
angustitarsis Malloch, 1920j: 277.—Alaska; Y.T., N.W.T., Man., Mich., Labr., U.S.S.R. (Kamchatka).
angustiventralis Huckett (for *angustiventris* Malloch).—N. Mex.; Calif., Utah, Colo. **N. name.**
    angustiventris Malloch, 1918a: 315 (preocc. Zetterstedt, 1845).—N. Mex.
antiqua (Meigen), 1826: 166 (*Anthomyia*).—Germany; B.C. to Que., s. to Kans. and N.J., Europe.
    ceparum Meigen, 1830: 376 (*Anthomyia*).—Germany.
    cepetorum Meade, 1883: 218 (*Phorbia*).—England.
aquitima Huckett, 1929: 138.—Alta.
armata (Stein), 1920: 86 (*chortophila*).—Wash.; Alaska and Y.T., s. to Calif. and Colo.
arnolitra Huckett, 1924: 22.—N.Y.; Mich., Conn.
atrovittata Malloch, 1920j: 280.—Alaska.
attenuata Malloch, 1920a: 188.—Calif.; Wash., Alta.
brassicae (Bouché), 1833: 131 (also 1834: 74) (*Anthomyia*).—Germany; Alaska to Calif. and Colo., Man. to Nfld., s. to Ill. and N.C., Europe.
    floccosa Macquart, 1835: 326 (*Chortophila*).—n. France.
    raphani Harris, 1841: 415 (*Anthomyia*).—Mass.
brunnescens (Zetterstedt), 1845: 1455 (*Aricia*).—Sweden; B.C., Europe.
bucculenta (Coquillett), 1904f: 188 (*Pegomya*).—Calif.; Alaska and N.W.T. to Labr., s. to Calif. and Conn.
    occidentalis Malloch, 1920a: 191.—Wash.
canadensis Huckett, 1929: 162.—Alta.; Alaska, Y.T., N.W.T., B.C., Que.

capito (Coquillett), 1902g: 123 *(Chirosia)*.—N.J.; Mich., N.Y. to Maine, s. to N.J.
cerealis (Gillette), 1904: 14 *(Pegomyia)*.—Colo.; Alta. to Calif. and N. Mex.
cilifera Malloch, 1918a: 311.—Mont.; N.W.T., Wash., Alta., Ont. to Labr.
  ventralis Stein, 1920: 81.—Wash.
coenosiaeformis Stein, 1904: 477.—N.H.; Alaska to Nfld., s. to Nev. and N.Y.
  varipes Curran, 1927h: 255 *(Pycnoglossa)*.—Ont.
curvipes Malloch, 1918a: 316.—Ill.; B.C. to Que., s. to Kans. and Ga.
depressa Stein, 1898: 214.—Pa.; Alaska to Labr., s. to Ariz. and Conn.
dorsilinea (Stein), 1920: 87 *(Chortophila)*.—Colo.; N.W.T., Que.
echinata (Séguy), 1923: 360 *(Chortophila)*.—France; Alaska to Greenland and Nfld., s. to Oreg. and Conn., Europe.
  abdena Hall, 1937a: 201.—Oreg.
  florilega, authors, not Zetterstedt.
egleformis Huckett, 1929: 142.—Ont.; Alaska, Que. and Labr., s. to N.Y. and Maine.
equifrons Huckett, 1929: 163.—Alta. Unrecognized, a gynandromorph.
exigua (Meade), 1883: 220 *(Phorbia)*.—England; Alaska, Europe.
extenuata Huckett, 1952: 118.—Wash.; Oreg.
extremitata Malloch, 1919e: 309.—Mont.; Alaska to N.W.T., s. to Ariz. and Colo.
  megatricha Kertész of Stein, 1920: 83.
fabricii (Holmgren), 1872: 101 *(Aricia)*.—w. Greenland; Alaska to Greenland, s. to Colo. and Labr., also U.S.S.R. (Kamchatka).
fasciventris Ringdahl, 1933a: 12.—Sweden; Alaska, N.W.T. (Baffin I.), Colo.
fennica (Karl), 1930: 174 *(Chortophila)*.—Europe; Alaska to N.W.T., s. to Ariz. and Que., Europe.
  setigera Stein, 1920: 70 *(Pegomyia*; preocc. Johannsen, 1916).—Wash. **N. syn.**
floralis (Fallén), 1824: 71 *(Musca)*.—Sweden; Alaska to Labr., s. to B.C. and Que., Europe.
  crucifera Huckett, 1929: 93.—B.C.
florilega (Zetterstedt), 1845: 1555 *(Aricia)*.—Sweden; Alaska to N.W.T. to Que., s. to N. Mex. and N.Y., Europe.
  liturata Meigen, 1838: 329 *(Anthomyia*; preocc. Robineau-Desvoidy, 1830).—Germany.

trichodactyla Rondani, 1866a: 164 (1866: 97) (*Chortophila*).—
Italy.
fracta Malloch, 1918a: 305.—N. Mex.; Alaska to N.W.T., also Ariz.
frontulenta Huckett, 1929: 161.—Alta.; B.C., Colo., N. Mex.
garretti Huckett, 1929: 117.—B.C.; Alaska to Labr., s. to Calif. and Que.
gracilipes Malloch, 1920a: 187.—Mont.; Alaska to Oreg. and Colo.
grandivillosa Huckett, 1924: 26.—N.Y.
hinei Malloch, 1920j: 278.—Alaska; Alaska to se. N.W.T., s. to Oreg. and N.H.
ithacensis Huckett, 1924: 28.—N.Y.
hirtitibia (Stein), 1916: 172 (*Chortophila*).—U.S.S.R. (Kola Peninsula); Alaska, Man., Que.
inaequalis Malloch, 1920a: 190.—Ill.
inconspicua Huckett, 1924: 27.—N.Y.; N.W.T., B.C. to Que., s. to Iowa and Pa.
incrassata (Stein), 1920: 88 (*Chortophila*).—Wash.; B.C.
pilicauda Huckett, 1929: 141.—B.C. **N. syn.**
ineptifrons Huckett, 1951b: 258.—Wash.; Idaho, Nev., Utah.
laevis (Stein), 1898: 231 (*Chortophila*).—Mass.; Ill. to Que., s. to e. Tenn.
lamellicauda Huckett, 1952: 119.—Ariz.; Wash. to Alta., s. to Ariz. and Colo.
linearis Stein, 1898: 219.—Minn.; Alaska to sw. N.W.T., s. to Colo. and N.Y.
lineariventris (Zetterstedt), 1845: 1541 (*Aricia*).—Norway, Sweden; Alaska and N.W.T. to Labr., s. to Colo. and N.Y.,. Europe, U.S.S.R. (Kamchatka).
uniseriata Stein, 1914: 51 (*Chortophila*).—Germany.
longicauda (Strobl), 1898a: 245 (*Anthomyia*).—Austria; Alaska to N.W.T., s. to Calif., N. Mex., and Conn., Europe.
seamansi Huckett, 1929: 143.—Alta.
longula (Fallén), 1824: 72 (*Musca*).—Sweden; Alaska, Y.T., Colo., Man. to Labr., Europe.
lupini (Coquillett), 1901d: 206 (*Phorbia*).—Calif.; Oreg., Ariz., Fla.
megacephala Malloch, 1920a: 188.—Pa.; Mich., N.Y., Mass., Conn.
montana Malloch, 1919b: 134.—Colo.; Y.T., Alta., Wash. to Calif. and N. Mex.
coarctata Fallén of Stein, 1920: 82.
montivagans Huckett, 1952: 121.—Colo.; Mont., Utah.
mutans Huckett, 1929: 113.—Alta.
neomexicana Malloch, 1918a: 310.—N. Mex.; Alaska to sw. N.W.T., s. to Ariz. and N. Mex.

nigricaudata Huckett, 1929: 116.—Alta.; Alaska, Y.T., B.C., Wash., Mont.
nigrithorax Stein, 1920: 79.—Wash.
normalis Malloch, 1919e: 309.—Mont.; Alta. to Nev. and N. Mex.
  similis Stein, 1920: 80.—Colo.
oppidans Huckett, 1929: 165.—B.C.
pilifemur Ringdahl, 1933a: 14.—Sweden; Alaska and N.W.T. to Que., s. to Man., Mich., and Conn.
  fusciceps Zetterstedt of Malloch, 1920j: 273, 274.
pilitarsis Stein, 1920: 76.—Colo.; Alaska to N.W.T., s. to N. Mex., also Que., Labr.
  angusta Stein, 1898: 218 (preocc. Macquart, 1835).—Colo.
  pilimana Stein, 1920: 80.—Colo.
  tenuis Drew, 1963: 248 (n. name for *angusta* Stein).—Colo.
planipalpis (Stein), 1898: 234 (*Chortophila*).—Idaho; Alaska to N.W.T, s. to Calif. and Colo.
  anthracina Malloch, 1918a: 314 (preocc. Czerny, 1906).—Oreg.
  anthracodes Malloch, 1920a: 194 (n. name for *anthracina* Malloch).—Oreg.
  vilis Stein, 1920: 91 (*Chortophila*).—Wash., Idaho.
platura (Meigen), 1826: 171 (*Anthomyia*).—Germany; Alaska to Greenland, s. to Calif. and Fla., cosmopolitan.
  cana Macquart, 1835: 340 (*Anthomyia*).—n. France.
  ruficeps Meigen of Staeger, 1845b: 366 (*Aricia*).
  fusciceps Zetterstedt, 1845: 1552 (*Aricia*).—Denmark.
  tinia Walker, 1849: 949 (*Anthomyia*).—Ont.
  perrima Walker, 1849: 950 (*Anthomyia*).—Ont.
  viana Walker, 1849: 951 (*Anthomyia*).—N.S.
  marginata Walker, 1849: 964 (*Eriphia*).—Ont.
  cupreifrons Walker, 1849: 966 (*Dialyta*).—Ont.
  deceptiva Fitch, 1856a: 533 (1856: 301).—N.Y.
  similis Fitch, 1856a: 533 (1856: 301) (*Hylemyia*).—N.Y.
  cilicrura Rondani, 1866a: 165 (1866: 98) (*Chortophila*).—Italy.
  cilioraca (error) Hagen, 1881a: 50.
  zeas Riley, 1869: 154 (*Anthomyia*).—Mo.
  zeae, error.
  calopteni Riley, 1877: 92 (*Anthomyia*; as *radicum* var.).—Minn.
  rupecula Bigot, 1885a: 285 (*Homalomyia*).—Rocky Mts.
pluvialis Malloch, 1918j: 310.—Ont.; Alaska to Nfld., s. to Colo. and N.Y.
  aniseta Stein, 1920: 77.—Colo.

pratensis (Meigen), 1826: 158 (*Anthomyia*).—Germany; Oreg., Idaho, Calif., Colo.
propinquina Huckett, 1929: 118.—Alta.; Alaska to Oreg. and Idaho, also Que. and Labr.
recurva Malloch, 1919e: 308.—Calif.; Oreg., Colo.
repleta Huckett, 1929: 112.—Alta.; Alaska, Y.T., N.W.T., Man.
rondanii (Ringdahl), 1918: 191 (*Chortophila*).—Sweden; Alaska to N.W.T., s. to Man. and Labr., Lapland.
seriata Stein, 1920: 80.—Colo.; Alaska to N.W.T., s. to Calif. and Colo.
    setifer Malloch, 1920a: 192.—Colo.
    pentaformis Huckett, 1929: 110.—Alta.
setifirma Huckett, 1951b: 259.—Oreg.; Mont. to Nev. and Colo.
setiseriata Huckett, 1952: 116, key; 1953a: 10, descr.—Wash.; Utah.
setisissima Huckett, 1929: 164.—B.C.; Wash., Mont., Calif.
setitarsata Huckett, 1924: 32.—N.Y.; Wis. to N.B., s. to Ga.
setiventris setiventris Stein, 1898: 216.—Idaho; Alaska to Calif. and Colo.
    ssp. **alternata** Huckett, 1951b: 255.—Oreg.; Alaska to Utah and Colo.
    ssp. **extensa** Huckett, 1951b: 257.—Alaska; Alaska, Wash. to Mont., s. to Calif.
    ssp. **rainieri** Huckett, 1951b: 256.—Wash.; Alaska and w. N.W.T., s. to Calif. and Wyo.
    ssp. **reliquens** Huckett, 1951b: 256.—Idaho.
    ssp. **sobrians** Huckett, 1951b: 256.—Idaho; Y.T., Wash., Utah, Colo.
simpla Coquillett, 1900h: 450 (1904: 64).—Alaska.
simulata Huckett, 1952: 117, key; 1953a: 11, descr.—Mont.; Alaska, Alta., Idaho, N. Mex., Que., Labr.
spinipes (Bigot), 1885a: 279 (*Chortophila*).—Rocky Mts. Unrecognized.
tarsata (Ringdahl), 1918: 194 (*Chortophila*).—Sweden; Alaska to Labr., s. to Wyo. and N.H.
    gemina Huckett, 1929: 95.—Alta.
triseriata Malloch, 1920j: 276.—Alaska; Calif.
vesicata Huckett, 1952: 115, key; 1953a: 12, descr.—Calif.; Oreg.
winnemana Malloch, 1919c: 209.—Md.; N.Y.
    unidorsalis Huckett, 1924: 35.—N.Y.

Subgenus BOTANOPHILA Lioy

Botanophila Lioy, 1864a: 990 (as genus). Type-species, *Anthomyia varicolor* Meigen (mon.).

Euryparia Ringdahl, 1929b: 269 (preocc. Becker, 1911). Type-species, *Anthomyia varicolor* Meigen (orig. des.).

REFERENCE: Huckett, 1947 (rev.)

acuticauda Huckett, 1947: 17.—Wash.; B.C. to Sask., s. to Calif. and N. Mex.
anane (Walker), 1849: 927 (*Anthomyia*).—Ont.
brevipalpis Huckett, 1929: 180.—Alta.; B.C., Wyo.
fibulans Huckett, 1947: 20.—Idaho; Alta., Oreg., Wyo.
formiceps Huckett, 1947: 27.—Calif.
hedleya Huckett, 1947: 22.—B.C.; Alta.
inornata Stein, 1898: 220.—Mass.; Alaska, N.W.T., Alta., Wis., Ohio, Maine to Ga.
marginata Stein, 1898: 221.—Colo.; Alaska to Calif. and Colo.
marginella Malloch, 1918a: 311.—Colo.; Y.T. and Alta., s. to Calif. and N. Mex.
    nigribasis Stein (in part), 1920: 78.—Colo. Also partly *spinidens*.
piloseta Malloch, 1918a: 313.—Oreg.; Calif., Wyo., Colo.
robusta Stein, 1920: 76.—Idaho; Wash.
setigera (Johannsen), 1916: 387 (*Hammomyia*).—N.Y.; Alta. to Que., s. to Wyo. and N.H.
spinidens Malloch, 1920a: 194 (n. name for *spinilamellata* Malloch).—Utah; Alaska to N.W.T., s. to Ariz. and N. Mex., U.S.S.R. (Kamchatka).
    spinilamellata Malloch, 1918a: 312 (preocc. Stein, 1904).—Utah.
    nigribasis Stein (in part), 1920: 78.—Colo. Also partly *marginella*.
spiniventris Coquillett, 1900h: 449 (1904: 63).—Alaska; Alaska to N.W.T., s. to Utah and N. Mex.
    spinilamellata Stein, 1904: 476.—Alaska.
subspinata Huckett, 1947: 10.—Calif.; Mont., Utah.
trifurcata Huckett, 1947: 14.—Wash.; Wash. to Alta., s. to Calif. and Utah.
varicolor (Meigen), 1826: 167 (*Anthomyia*).—Germany; Alaska and Y.T. to Utah, also N.C., Europe.

Subgenus PEGOHYLEMYIA Schnabl

Hylemyia, subg. **Pegohylemyia** Schnabl, 1911: 75. Type-species *Musca cinerea* Fallén (present des.).

abitibiensis Huckett, 1929: 167.—Ont.; Y.T. and N.W.T., s. to B.C. and Que.

ANTHOMYIIDAE

aliena Malloch, 1920j: 282.—Alaska; Y.T., N.W.T., Que.
apiciseta Ringdahl, 1933a: 16.—Sweden; Que.
appendiculata Malloch, 1920j: 281.—Alaska; Alaska to Oreg. and Wyo., also Ont., Que., Labr., N.H.
betarum (Lintner), 1883: 208 (*Chortophila*).—N.Y.; Alaska to N.W.T. (Baffin I.), s. to N. Mex. and Nfld.
    substriata Stein, 1898: 233 (*Chortophila*).—Idaho.
bidens Ringdahl, 1933a: 14.—Sweden; Alaska, N.W.T.
coronata (Ringdahl), 1951: 35 (*Pegohylemyia*).—Sweden; Y.T., N.W.T.
discreta (Meigen), 1826: 172 (*Anthomyia*).—Germany; Wash., Calif., Colo., Mich., Ind., Europe, U.S.S.R. (Kamchatka).
flavidisquama Huckett (for *flavisquama* Malloch).—Alaska (Pribilof Is.); Alaska. **N. name.**
    flavisquama Malloch, 1923i: 193 (preocc. Stein, 1916).— Alaska (Pribilof Is.).
fugax (Meigen), 1826: 174 (*Anthomyia*).—Germany; Alaska to Calif., Wis. to Que., s. to N.Y. and N.S., Europe, U.S.S.R. (Kamchatka).
    striolata Zetterstedt, 1838: 684 (*Anthomyza*).—Sweden.
    denticauda Malloch, 1920j: 281.—Wash.
hucketti Ringdahl, 1935: 26.—Sweden; Alaska to N.W.T., s. to B.C., Mich., and Labr., Lapland.
impersonata Huckett, 1929: 114.—Man.; Alta., Wash., Oreg., Utah, S. Dak.
incursa Malloch, 1920j: 282.—Alaska; Y.T., N.W.T., B.C.
latifrontalis Huckett, 1924: 30.—N.Y.; Ariz., N. Mex., Mich., Mass. to Ga.
moriens (Zetterstedt), 1845: 1505 (*Aricia*).—Sweden; Alaska to Greenland, s. to B.C. and Labr., n. Europe.
parvicornis Malloch, 1920j: 283.—Alaska.
    parvaicornis (error) Drew, 1963: 244.
profuga Stein, 1916: 141.—Sweden, Finland; Alaska to Greenland, s. to Colo. and Labr., n. Europe.
    acrostichalis Malloch, 1919f: 72.—Alaska.
    tridens Malloch, 1920j: 284.—Alaska.
ringdahli Drew, 1963: 247 (as *ringdahl;* n. name for *appendiculata* Malloch).—Alaska; Alaska to Oreg. and Wyo., also Ont., Que., Labr., N.H.
    appendiculata Malloch, 1920j: 281 (preocc. Bigot, 1885).— Alaska.
sericea Malloch, 1920j: 280. Alaska; Alaska to N.W.T., s. to Que. and Labr., also Colo.
sobrina (Collin), 1931: 86 (*Pegohylemyia*).—Greenland.

subnitida Malloch, 1920j: 283.—Alaska.
trilineata (Stein), 1898: 245 (*Pegomyia*).—S. Dak.; Alta., Colo., Mich.
trivittata (Stein), 1898: 246 (*Pegomyia*).—Wash., Mass.; B.C. to Utah, Ont., and Que., s. to Va.

Subgenus EGLE Robineau-Desvoidy

Egle Robineau-Desvoidy, 1830: 584 (as genus). Type-species, *parva* Robineau-Desvoidy (Coquillett, 1910b: 536).
Xenophorbia Malloch, 1920a: 175. Type-species, *Stomoxys muscaria* Fabricius (orig. des.).

REFERENCE: Huckett, 1928 (rev.).

atomaria (Zetterstedt), 1845: 1624 (*Aricia*).—Sweden; Alaska to Labr., s. to Calif. and N.Y.
  fuscohalterata Malloch, 1920j: 279.—Alaska.
  tantalisa Huckett, 1928: 76.—N.Y.
  parva, authors, not Robineau-Desvoidy.
bicaudata Malloch, 1920a: 193.—Ill.; Alta., Oreg., Wis. to Que., s. to Ill. and N.Y.
brevicornis (Zetterstedt), 1838: 683 (*Anthomyza*).—Sweden; Alaska, Oreg., Calif., Man.
longipalpis Malloch, 1924h: 197.—Maine; Alaska to sw. N.W.T., s. to Alta., Mich., and Maine, Sweden.
  parva Robineau-Desvoidy of Huckett, 1924: 10.
muscaria (Fabricius), 1794: 395 (*Stomoxys*).—Denmark; Alaska and N.W.T. to Labr., s. to Calif. and N.S., Europe.
  determinata Walker, 1849: 954 (*Anthomyia*).—N.S.
  ciliata Walker, 1849: 961 (*Eriphia*).—Ont.
parvaeformis (Schnabl) *in* Schnabl and Dziedzicki, 1911: 105 (1911: 53) (*Egle*; as *steini* var.).—Europe; Alaska to N.W.T., also Man., Mich., Labr., Europe.
pilitibia (Ringdahl), 1918: 193 (*Chortophila*).—Sweden; Alaska, Man., Que.
salicola Huckett, 1928: 79.—N.Y.; Alaska to Man., Iowa, Mich., Que.
steini (Schnabl) *in* Schnabl and Dziedzicki, 1911: 105 (1911: 53) (*Egle*).—Europe; Alaska, Y.T., Man.

Subgenus PAREGLE Schnabl

Hylemyia, subg. **Paregle** Schnabl, 1911: 71. Type-species, *Musca radicum* Linnaeus (Huckett, 1924: 39).

aestiva (Meigen), 1826: 169 (*Anthomyia*).—Germany; Alaska to N.W.T. (Baffin I.), s. to Wash., Colo., and N.S., Europe.

cinerella (Fallén), 1825: 77 (*Musca*).—Sweden; sw. N.W.T. to Greenland, s. to Ariz. and Ga., Europe, U.S.S.R., Africa.
mystacea (Coquillett), 1900h: 447 (1904: 61) (*Anthomyia*).—Alaska; Alaska to n. Man., s. to N. Mex., also Que., Labr.
    hirta Malloch, 1918a: 299 (*Egle*).—N. Mex.
radicum (Linnaeus), 1761: 454 (*Musca*).—Scandinavia; Alaska to Greenland, s. to Calif. and N.J., Europe.
    ruficeps Meigen, 1826: 177 (*Anthomyia*).—Germany.
    campestris Robineau-Desvoidy, 1830: 585 (*Egle*).—Pa., France.
    tristicula Holmgren, 1872: 101 (*Aricia*).—n. Greenland.

Subgenus CRASPEDOCHAETA Macquart

Craspedochaeta Macquart, 1851: 241 (1851: 268) (as *Craspedochoeta*, as genus). Type-species, *Anthomyia punctipennis* Wiedemann (mon.).
Melinia Ringdahl, 1929b: 270. Type-species, *Aricia pullula* Zetterstedt (orig. des.).

REFERENCE: Huckett, 1946 (rev.).

facialis Malloch, 1918a: 306.—Calif.; B.C. to Nev. and Ariz., also Tex., Wis.
mimetica Malloch, 1918a: 313.—N. Mex.; Alaska to Alta., Utah, and Colo., Man., Mich. to Mass., Sweden.
nigriceps Huckett, 1946: 116.—Calif.
ochripes (Thomson), 1869: 553 (*Anthomyia*).—Calif.; Ariz.
pullula (Zetterstedt), 1845: 1449 (*Aricia*).—Denmark, Sweden; Alaska to Que., s. to Calif., N. Mex., and N.Y., Europe.

Subgenus LASIOMMA Stein

Chortophila, subg. **Lasiomma** Stein, 1916: 168. Type-species, *Lasiops ctenocnema* Kowarz (Séguy, 1937: 123).

abietis Huckett, 1953b: 107.—Oreg.; Calif.
anthracina (Czerny), 1906: 251 (*Chortophila*).—Austria; Alaska, Oreg., Man., Ont. to Maine, s. to N.Y., Europe.
    eriophthalma (Zetterstedt).—Not Nearctic.
morionella (Zetterstedt), 1838: 687 (*Anthomyza*).—Sweden; Alaska to N.W.T., Lapland.
nidicola Aldrich, 1919c: 380.—Wash.; Calif., Utah, Wis., N.Y.
octoguttata octoguttata (Zetterstedt), 1845: 1570 (*Aricia*).—Sweden; B.C. and Alta. to Utah, also Wis. to N.H., s. to N.C., Europe.
    spizellae Huckett, 1924: 33.—N.Y.

ssp. **moesta** (Holmgren), 1872: 102 (*Aricia;* as sp.)—w. Greenland; Alaska to N.W.T., s. to Man. and Labr.

Subgenus ACROSTILPNA Ringdahl

Hylemyia, subg. **Acrostilpna** Ringdahl, 1929b: 269. Type-species, *Anthomyza latipennis* Zetterstedt (orig. des.).

REFERENCE: Huckett, 1946 (rev.).

atricauda (Zetterstedt), 1845: 1529 (*Aricia*).—n. Sweden; Alaska to Nfld., s. to Colo. and Tenn., Europe.
collini Ringdahl, 1929b: 271.—Sweden; Alaska to Labr., s. to Colo.
consobrina Huckett, 1929: 187.—N.B.; N.Y., Que., Maine.
latipennis (Zetterstedt), 1838: 676 (*Anthomyza*).—Norway, Sweden; Alaska to Wash. and Wyo., Alta. to Que., s. to N.Y.
replicata Huckett, 1929: 136.—Alta.; Alaska to Que., s. to B.C., Wis., and N.H., Sweden.
restorata Huckett, 1946: 123.—Ohio, Mass., Conn., N.Y., N.J., N.C., Ga.; Mich. to Maine, s. to Ga.
  latipennis, authors, not Zetterstedt.
sedula Huckett, 1929: 183.—Man.; Mich., N.J.

Subgenus CRINURINA Karl

Kingia Malloch, 1921g: 53 (preocc. Theobald, 1910). Type-species, *Hylemyia quintilis* Malloch (orig. des.)=*cuneicornis* (Zetterstedt).
Crinurina Karl, 1928: 185 (as genus). Type-species, *Aricia pictiventris* Zetterstedt (orig. des.)=*cuneicornis* (Zetterstedt).
Kingiella Séguy, 1937: 143 (n. name for *Kingia* Malloch). Type-species, *Hylemyia quintilis* Malloch (aut.)=*cuneicornis* (Zetterstedt).
Kingiella Malloch *in* Curran, 1934a: 391. No type desig., hence unavailable. Nomen nudum.

cuneicornis (Zetterstedt), 1838: 682 (*Anthomyza*).—Sweden; Alaska to Labr., s. to B.C. and N.J.
  quintilis Malloch, 1919k: 275.—Que.
quinquelineata Ringdahl, 1926: 114.—Sweden; Alaska, Man.

Genus MACROPHORBIA Malloch

Macrophorbia Malloch, 1920a: 173. Type-species, *houghi* Malloch (orig. des.).

houghi Malloch, 1920a: 173.—Mass.; N.Y., N.H., N.J., Ga.

ANTHOMYIIDAE

Genus ALLIOPSIS Schnabl and Dziedzicki

Alliopsis Schnabl and Dziedzicki, 1911: 92 (1911: 40). Type-species, *Aricia glacialis* Zetterstedt (mon.).

Alliops Malloch, 1934b: 26. Type-species, *Alliopsis obesa* Malloch (orig. des.).

glacialis (Zetterstedt), 1845: 1521 (*Aricia*).—Sweden; arctic N.W.T. and Greenland.
obesa Malloch, 1919f: 70.—N.W.T.; Alaska, N.W.T. (Victoria and Baffin Is.), Que.

Genus EREMOMYIA Stein

Eremomyia Stein, 1898: 223. Type-species, *humeralis* Stein (Coquillett, 1901e: 137).

REFERENCE: Huckett, 1951a (rev.).

albidosa Huckett, 1951a: 83.—Kans.; Colo., Iowa.
fumipennis Huckett, 1951a: 87.—Wash.; Oreg.
humeralis Stein, 1898: 224.—Idaho; B.C., Wash., Calif., Utah.
impolita Huckett, 1951a: 81.—Oreg.; B.C., Calif., Utah, N. Mex.
lucescens Huckett, 1951a: 88.—Alta.; B.C., Wash.
medicaginis Huckett, 1951a: 82.—Alta.
obversa Huckett, 1951a: 85.—N.Y.; Ont., Que., Conn.
  humeralis Stein of Huckett, 1924: 10.
parafacialis Huckett, 1951a: 86.—Alaska; N.W.T., B.C., Alta.
pilimana (Ringdahl), 1918: 190 (*Chortophila*).—Sweden; Wis., Ont. to Labr., s. to Ga.
  vernalis Huckett, 1924: 11.—N.Y.
turbida Huckett, 1951a: 89.—N.W.T.; Alaska, Y.T., Man.

Genus EREMOMYIOIDES Malloch

Eremomyioides Malloch, 1918b: 67. Type-species, *Eremomyia cylindrica* Stein (orig. des.).

REFERENCE: Huckett, 1944b (rev.).

conscriptus Huckett, 1944b: 366.—B.C.
cylindricus (Stein), 1898: 226 (*Eremomyia*).—Minn., Ind., Mass.; Alta., Kans., Que. to Fla.
fuscipes Malloch, 1920a: 182.—Ill.; Alta., Wis., Mich., Ohio, Que. to Ga.
parkeri Malloch, 1918b: 67.—Mont.; Alta., Sask.
setosus (Stein), 1898: 247 (*Pegomyia*).—Idaho; Alaska, Wash., Oreg.
similis Malloch, 1920a: 183.—Ill.; Alta., Colo., S. Dak., Wis.

Genus EMMESOMYIA Malloch

Emmesomyia Malloch, 1917c: 114. Type-species, *unica* Malloch (orig. des.) =*socialis* (Stein).

apicalis Malloch, 1917c: 115.—Ill.; S. Dak., Nebr., Okla., Ont. to Fla.
socialis (Stein), 1898: 193 (*Spilogaster*).—N.C., Ga.; Okla., La., Ill., N.Y., N.J.
    unica Malloch, 1917c: 114.—Ill.

Genus PEGOMYA Robineau-Desvoidy

Pegomya Robineau-Desvoidy, 1830: 598. Type-species, *Musca hyoscyami* Panzer (Coquillett, 1901e: 140).
Pegomyia, emend.
    REFERENCE: Huckett, 1941 (rev.).

Subgenus PEGOMYA Robineau-Desvoidy

affinis Stein, 1898: 286.—Va.; Wash. to Que., s. to Ariz. and Va.
    vicina Lintner of Stein, 1898: 239.
alticola Huckett, 1939: 25.—Utah; Alaska, Wash.
aninotata Huckett, 1939: 15.—Alta.
anorufa Stein, 1920: 67.—Colo.; Alta., Mont., Wyo.
    subgrisea Malloch, 1920a: 180.—Mont.
apicalis (Stein), 1898: 227 (*Eremomyia*).—Idaho; B.C. and Alta., s. to Calif. and N. Mex.
assimilis Huckett, 1939: 22.—Calif.
atlanis Huckett, 1939: 4.—N.Y.
balteata (Holmgren), 1883: 172 (*Anthomyza*).—U.S.S.R. (Novaya Zemlya); arctic Alaska and N.W.T.
banffi Huckett, 1939: 33.—Alta.; Wash., Wyo., Colo.
bicolor (Wiedemann), 1817: 77 (*Anthomyia*).—Germany; Alaska to Calif. and Wyo., Tex., Wis. to Pa. and Labr., Europe.
caduca Huckett, 1939: 35.—Wash.; Y.T., Wyo.
caesia Stein, 1906: 78.—Germany; Alaska, Y.T., B.C., Alta., Calif., Europe.
carduorum Huckett, 1939: 1.—Wis.; Wash., Alta., S. Dak., N.Y., N.C.
cedrica Huckett, 1939: 21.—Ohio.
chrysida Huckett, 1939: 27.—Calif.; Oreg., Wyo., Utah, Ariz.
cognata Stein, 1920: 67.—Calif.; Oreg., Mexico (Guadalupe I.).
connexa Stein, 1920: 68.—Pa.; Wash., Alta., Minn. to Ill. to Ohio, Que. to Ga.
    emmesia Malloch, 1920a: 179.—Ill.

convergens Huckett, 1939: 5.—Sask.; Kans.
corrupta Huckett, 1939: 13.—Alaska; Alaska to N.W.T. (Baffin I.), s. to Que. and Labr.
    albimargo, authors, not Pandellé.
costalis Stein, 1898: 243.—S. Dak.
cresca Huckett, 1939: 30.—B.C.; Alaska to Alta., s. to Colo.
debilis Stein, 1901: 194.—U.S. Unrecognized.
duplicata (Malloch), 1918a: 308 (*Hylemyia*).—Calif.; B.C. and Alta., s. to Calif. and N. Mex., also N. Dak., S. Dak.
    fuscinervis Stein, 1920: 69.—Calif.
finitima Stein, 1898: 241.—S. Dak.; S. Dak., Mich., s. to Kans. and Ohio, also N.H., Mass.
flavicans Stein, 1898: 213.—S. Dak.; Alta.
flavifrons (Walker), 1849: 966 (*Eriphia*).—Ont.; Alaska, Alta. to Que., s. to Oreg. and N.C.
    fringilla Malloch, 1920a: 181.—Pa.
flavipalpis (Zetterstedt), 1845: 1707 (*Anthomyza*).—Sweden; Alaska, N.W.T. to Labr. and N.S., s. to B.C. and Colo., Europe.
frigida (Zetterstedt), 1845: 1685 (*Anthomyza*).—Norway, Sweden; Alaska to Que., s. to Oreg. and e. Tenn., Europe, U.S.S.R. (Kamchatka).
    bivittata Stein, 1906: 98.—Germany.
    lativittata Malloch, 1920j: 272.—Alaska.
fumipennis Huckett, 1939: 20.—Wash.; B.C.
fuscicauda Huckett, 1939: 31.—Calif.; B.C. and Alta. to Calif., also Colo.
geniculata (Bouché), 1834: 81 (*Anthomyia*).—Germany; Alaska to Oreg., Man. to N.B., s. to S. Dak. and e. Tenn., Europe.
gilva (Zetterstedt), 1846: 1789 (*Anthomyza*).—Norway, Sweden; Alaska to Oreg., Alta. to Que., s. to Wis. and e. Tenn., Europe.
    pallida Stein, 1906: 69.—Germany.
    luteola Malloch, 1920a: 175.—Maine.
glabra (Stein), 1920: 88 (*Chortophila*).—Wash.; B.C. to Calif., also Colo., N. Mex., Mich., Conn.
gopheri Johnson, 1913a: 77.—Fla.; Kans. to Tex., Wis. to N.Y., s. to Ill. and Fla.
haemorrhoa (Zetterstedt), 1838: 692 (*Anthomyza*).—Sweden; Man., Greenland.
holmgreni (Boheman), 1858: 55 (*Anthomyza*).—Sweden; Alaska, n. Europe.
hyoscyami hyoscyami (Panzer), 1809c: 13 (*Musca*).—Germany; Alaska to Calif. and N. Mex., B.C. to Que., s. to Ind. and N.J., Europe.

vicina Lintner, 1883: 209.—N.Y.
> ssp. **betae** (Curtis), 1847: 412 (*Anthomyia*; as sp.).—England; Alta., Calif., Ind., N.Y., Europe.

icterica (Holmgren), 1872: 102 (*Aricia*).—Greenland; N.W.T. (Baffin I.).
incisiva Stein, 1906: 73.—Germany, Czechoslovakia; Alaska and Y.T., s. to Colo. and Vt., Europe.
incompleta (Stein), 1898: 228 (*Eremomyia*).—Minn.; Wash., Alta., Idaho, Utah.
indicta Huckett, 1939: 3.—Alaska.
intersecta intersecta (Meigen), 1826: 175 (*Anthomyia*).—Germany; B.C. and Alta., s. to Utah and Nebr., also Mich., Maine, Europe.
> ssp. **arcticola** Huckett (for *arctica* Ringdahl).—Greenland; Alaska to Greenland, s. to Calif. and Labr. **N. name.**
> *arctica* Ringdahl, 1933b: 16 (*Hylemyia;* as *intersecta* var.; preocc. Schnabl, 1915).—Greenland.

jacobi Malloch, 1920j: 272.—Alaska.
juvenilis (Stein), 1898: 211 (*Hylemyia*).—Pa.; S. Dak., Tex., Wis. to Que., s. to Ala. and Ga.
labradorensis Malloch, 1920a: 176.—Labr.; Alaska, Alta., Man., Que.
lipsia (Walker), 1849: 928 (*Anthomyia*).—Ont.; Alta., Ariz., Kans. to N.B., s. to Ala. and Ga.
> *substituta* Walker, 1849: 971 (*Coenosia*).—Mass.

lividiventris Huckett, 1939: 29.—Calif.; B.C. to Nev., also Mont.
longicornis Huckett, 1939: 14.—B.C.; Y.T., Alta., Mich.
longimana (Pokorny), 1887a: 405 (*Chortophila*).—Austria; Y.T. to Ariz. and N. Mex., also S. Dak., Que., Europe.
lunatifrons (Zetterstedt), 1845: 1708 (*Anthomyza*).—Sweden; Alaska to N.W.T., s. to Oreg. and Maine, Europe, U.S.S.R. (Kamchatka).
lundbeckii Ringdahl, 1918: 188.—Sweden; Que., Lapland.
major (Malloch), 1919e: 310 (*Eremomyia*).—Calif.; B.C. and Alta., s. to Calif. and Mont.
mallochi Huckett, 1939: 15.—N.Y.; Okla., Tex., Mich., Ohio, N.H. to Ga.
> *winthemi* Meigen of Malloch, 1920b: 122.

marginata Huckett, 1939: 7.—Wash.; Alaska and Y.T., s. to Idaho.
minuta Malloch, 1918a: 302.—N. Mex.; Calif., Utah.
nigritarsis (Zetterstedt), 1838: 696 (*Anthomyza*).—Sweden; ?N.Y.
palpata Stein, 1906: 101.—Germany; Alta., Vt., Europe.

palposa (Stein), 1897: 320 (*Hydrophoria*).—Germany; Mich., Ohio, N.Y., N.C., Ga., Europe.
  orientalis Huckett, 1924: 16 (*Hydrophoria*).—N.Y.
  palpata Stein of Schnabl and Dziedzicki, 1911: 109 (1911: 57).
partita Huckett, 1939: 33.—Alta.; B.C., Wash., Idaho.
pertusa Huckett, 1939: 8.—Alta.
pilosa Stein, 1900: 332.—Lapland; Alaska to Greenland, s. to Mich. and N.S., Lapland.
  flavipes, authors, not Fallén.
pollinosa Ringdahl, 1938: 209.—Sweden; Alaska, Alta., S. Dak., Mich. to Que. and N.S.
quadrispinosa Malloch, 1920a: 181.—Mont.; Alta., Wash. to Ariz. and Colo.
rubivora (Coquillett) *in* Slingerland, 1897: 162 (*Phorbia*).—N.Y.; Oreg., Man., Mich. to Que. and Mass.
  dentiens Pandellé, 1900: 268 (*Anthomyia*).—France.
rubriceps Huckett (for *ruficeps* Stein).—La.; Wash. to Ariz., Mich to. Que., s. to Tex. and Fla. **N. name.**
  ruficeps Stein, 1898: 286 (preocc. Zetterstedt, 1838).—La.
rufescens Stein, 1898: 238.—Ont.; Wis., Ill., Mich.
rufina (Fallén), 1825: 92 (*Musca*).—Sweden; Colo., Europe.
  squamifera Stein, 1906: 63.—Germany.
rufipes (Fallén), 1825: 85 (*Musca*).—Sweden; Alaska to Greenland, s. to Mich. and Labr., Europe.
  pedestris Malloch, 1919k: 274 (*Hylemya*).—Que.
rumicifoliae Huckett, 1941: 81.—N.Y.; Oreg., Wis. to Que. and R.I.
  calyptrata Zetterstedt of Frost, 1919: 229.
setaria (Meigen), 1826: 178 (*Anthomyia*).—Germany; B.C. to Mass.
  polygoni Seamans, 1923: 221.—Man.
setiformis Huckett, 1939: 32.—Alta.; B.C., Wash., Idaho, Wyo.
sitiens Huckett, 1939: 9.—Alta.; Alaska to Idaho and Wyo., Sask. to Wis. and Maine, also N.C.
socculata (Zetterstedt), 1845: 1683 (*Anthomyza*).—Sweden; Alaska, Y.T., Europe.
solitaria Stein, 1906: 80.—Germany; Alaska to N.W.T., s. to Idaho and Que., Europe, U.S.S.R. (Kamchatka).
spinigerella Malloch, 1920a: 178.—Ill.; S. Dak., Ariz., Tex., Fla.
spinosissima Stein, 1898: 242.—Kans.; N. Dak., Okla., Tex.
striata Stein, 1920: 71.—La.
substriatella (Malloch), 1918a: 309 (*Hylemyia*).—Va.; Kans., Ohio, Md., D.C.
tacta Huckett, 1939: 16.—Alta.; Alaska, Y.T., Man., Wyo., Mich.

tarsata (Wulp), 1867: 151 (*Anthomyia*).—Wis.; Alaska, Y.T., Alta., S. Dak. to N.B.
>*costalis* Stein of Coquillett, 1900h: 451 (1904: 65).

tenera (Zetterstedt), 1838: 697 (*Anthomyza*).—Sweden; Alaska to Greenland, s. to Wyo. and Labr., Europe.
>var. **obscurior** Collin, 1931: 84.—w. Greenland; Alaska, Labr.
>*conformis* Fallén of Lundbeck 1901: 287.
>*hyoscyami* Panzer of Lundbeck *in* Henriksen and Lundbeck, 1917: 640.

thrixia Huckett, 1939: 23.—Iowa; Alta., Minn.
tinctisquama Huckett, 1939: 26.—B.C.; Y.T., Alta., Wash., Calif.
triseta Malloch, 1920a: 177.—Que.; B.C. to Que., s. to Calif. and R.I.
>*calyptrata* Zetterstedt of Stein, 1898: 237.

tunicata (Zetterstedt), 1846: 1787 (*Anthomyza*).—Sweden; Alaska to Greenland, s. to Que., arctic U.S.S.R.
>*analis* Schnabl, 1915: 18.—U.S.S.R. (Karskaja Tundra).
>*alpina* Ringdahl, 1916: 237 (*Chortophila*).—Sweden.

unguiculata Malloch, 1920a: 176.—Alta.; Alaska, N.W.T., B.C.
unicolor Stein, 1898: 236.—Pa.; Wis. and Ill. to Que., also Okla., Ga.
univittata (Roser), 1840: 59 (*Anthomyia*).—Germany; Alaska and B.C. to Que., s. to Calif. and e. Tenn., Europe.
vanduzeei Malloch, 1919e: 307.—Calif.; Alaska to Mont., also Man., Mich., Ohio, N.Y.
>*affinis* Stein of Frost, 1919: 240.

variegata Huckett, 1939: 11.—Md.; N.Y., N.C.
vicaria Huckett, 1939: 6.—Wash.; Nev., Colo.
vittigera (Zetterstedt), 1838: 697 (*Anthomyza*).—Sweden; Alaska to Alta., Mich. to Que. and N.S., Europe, U.S.S.R. (Kamchatka).
winthemi (Meigen), 1826: 186 (*Anthomyia*).—Germany; Alaska to Wash., Man. to Que., s. to Okla. and Ga., Europe.
>*latitarsis* Zetterstedt, 1846: 1754 (*Anthomyza*).—Sweden, Denmark.
>*fuscofasciata* Malloch, 1920a: 178.—Mass.

Subgenus NUPEDIA Karl

Chortophila, subg. *Nudaria* Karl, 1928: 171 (preocc. Haworth, 1909). Type-species, *Anthomyia dissecta* Meigen (orig. des.).
Chortophila. subg. **Nupedia** Karl, 1930: 174 (n. name for *Nudaria* Karl). Type-species, *Anthomyia dissecta* Meigen (aut.).

abnormis Stein, 1920: 65.—Idaho.
acutipennis Malloch, 1918a: 301.—N. Mex.; B.C. and Alta., s. to Ariz. and Tex., also Ill., Wis., N.Y.

anabnormis Huckett, 1939: 20.—B.C.; B.C. and Alta., s. to Calif., N. Mex., and S. Dak.
cuticornis Huckett, 1939: 18.—B.C.; Alaska, Y.T., Alta., Colo.
dissecta (Meigen), 1826: 176 (*Anthomyia*).—Germany; Alaska, Alta., Mich., Ont. and N.Y. to Labr. and Nfld., Europe.
latipalpis (Stein), 1920: 89 (*Chortophila*).—Wash.; B.C., Alta., Mich.
nigroscutellata (Stein), 1920: 90 (*Chortophila*).—Wash.; Alaska and B.C. to Nfld., s. to S. Dak. and N.H.
    slossonae Malloch, 1920b: 127.—N.H.
patellans (Pandellé), 1900: 222 (*Anthomyia*).—France; Alaska, U.S.S.R. (Commander Is.).
pseudodissecta (Ringdahl), 1926: 111 (*Hylemyia*).—Sweden; Alaska to Greenland, s. to Nfld.
rectifrons Huckett, 1939: 19.—Idaho; Calif.

Genus HYDROPHORIA Robineau-Desvoidy

Hydrophoria Robineau-Desvoidy, 1830: 503. Type-species, *littoralis* Robineau-Desvoidy (Coquillett, 1910b: 554). The identity of *littoralis* is in doubt, as is the identity of an earlier-suggested senior synonym, *nigrita* Fallén, which has been variously referred to such widely different genera as *Hebecnema* and *Musca*. Common usage of *Hydrophoria* has followed the interpretation of Rondani, 1866a: 72 (1866: 5) who designated as type-species *Anthomyia conica* Wiedemann, a species not originally included. The author prefers to maintain this established usage of *Hydrophoria*, either by arbitrarily considering *littoralis* as a true *Hydrophoria*, or if necessary by applying to the I.C.Z.N. to fix the type-species as *conica* Wiedemann. (See Huckett, 1944a: 195.)
Acroptena Pokorny, 1893a: 60. Type-species, *simonyi* Pokorny (mon.).

    REFERENCES: Malloch, 1920g (key); Huckett, 1944a (rev.).

alaskensis Malloch, 1920g: 257.—Alaska (Pribilof Is.); Alaska to N.W.T. (Baffin I.), s. to Que. and Labr.
alpina Huckett, 1944a: 272.—N.H.; Alaska to N.W.T. (Baffin I.), s. to N.H. and Labr.
altilega Huckett, 1944a: 275.—B.C.; Wyo., Colo., Man., Que.
ambigua (Fallén), 1823f: 56 (*Musca*).—Sweden; Alaska to N.W.T., s. to Colo., Maine, and Labr., Europe, U.S.S.R. (Kamchatka).
    coloradensis Malloch, 1920a: 172.—Colo.

borealis Malloch, 1920a: 172.—Alaska; Alaska to Calif., also Wis., Minn., Ont., N.H.
brunneifrons (Zetterstedt), 1838: 690 (*Anthomyza*).—Sweden; Alaska to Greenland, s. in w. to Calif. and Colo., in e. to Ont., Lapland.
   occidentalis Malloch, 1920a: 170.—Nev.
conica (Wiedemann), 1817: 79 (*Anthomyia*).—Germany; Wash., Mich. to Que., Conn., Europe, U.S.S.R. (Kamchatka).
divisa (Meigen), 1826: 99 (*Anthomyia*).—Germany; Alaska to Greenland, s. to Calif. and N.C., Europe.
fasciculata (Schnabl), 1915: 15 (*Acroptena*).—U.S.S.R. (Karskaja Tundra); Alaska to N.W.T., s. to Alta. and Man., Sweden, arctic U.S.S.R.
flavohalterata Malloch, 1920a: 171.—Idaho; Alaska, Wash., Utah.
frontata (Zetterstedt), 1838: 669 (*Anthomyza*).—Lapland; Y.T. to Greenland, s. to Que. and Labr., Europe.
galeata Malloch, 1920j: 270.—Alaska; Alaska to N.W.T., s. to Calif. and S. Dak.
ignobilis (Zetterstedt), 1845: 1448 (*Aricia*).—n. Sweden; Alaska to N.W.T., also Alta., Wash., Idaho, n. Europe, U.S.S.R. (Kamchatka).
implicata Huckett, 1944a: 287.—Ill., Mass.; N.W.T., B.C. to Que., s. to Ill. and D.C.
   ambigua Fallén of Stein, 1898: 208.
laticornis (Ringdahl), 1916: 236 (*Acroptena*).—Sweden; Alaska to N.W.T., s. to n. B.C., Minn. and Labr., n. Europe.
   congrua Malloch, 1920j: 271.—Alaska.
lucidiventris (Zetterstedt), 1845: 1554 (*Aricia*).—Sweden; Alaska.
nigerrima Malloch, 1920a: 169.—Wash.; Mont., Calif.
nuda (Schnabl) *in* Schnabl and Dziedzicki, 1911: 255 (1911: 203) (*Acroptena*).—U.S.S.R.; Alaska, Y.T., Europe.
packardi Malloch, 1924b: 514.—Labr.
polita Malloch, 1920a: 170.—Mont.; Y.T. and N.W.T., s. to Man. and Que., Nev., Colo.
proxima Malloch, 1920a: 171.—Maine; Alaska, Wash., Ont. to Labr., n. Europe.
seticauda Malloch, 1919e: 306.—Calif.; Alaska to N.W.T., s. to Calif. and Nev.
subpellucida Malloch, 1918a: 296.—N. Mex.; Alta., S. Dak., Ariz., Ohio to N.Y., s. to N.C.
   subpellucens, error.
   ruralis, authors, not Meigen.

teate (Walker), 1849: 931 (*Anthomyia*).—Ont.; Alaska to Greenland, s. to Ont. and Que., U.S.S.R. (Kamchatka).
    katmaiensis Malloch, 1920a: 170.—Alaska.
    kamtchatica Ringdahl, 1930: 8 (*Acroptena*).—U.S.S.R. (Kamchatka).
tristis (Ringdahl), 1926: 107 (*Acroptena*).—Sweden; Alaska to N.W.T., s. to Man. and Que.
uniformis Malloch, 1918a: 297.—Ill.; Idaho, Ont. and Que., s. to Ill. and Ga.
verticina (Zetterstedt), 1838: 665 (*Anthomyza*).—Sweden; Alaska to Greenland, s. to Que., also Colo., n. Europe.
    arctica Malloch, 1919f: 69.—N.W.T.
wierzejskii (Mik), 1867: 420 (*Spilogaster*).—Poland; w. N.W.T. and B.C. to N.Y., s. to Calif., Europe.
    elongata Malloch, 1920a: 172.—Wash.
zetterstedtii (Ringdahl), 1918: 185 (*Acroptena*).—Sweden; Alaska to N.W.T., s. to Colo. and Ont., Sweden, U.S.S.R. (Kamchatka).

Genus ANTHOMYIA Meigen

Anthomyia Meigen, 1803: 281. Type-species, *Musca pluvialis* Linnaeus (Westwood, 1840: 143).
Anthomyza, error. Not *Anthomyza* Fallén.

oculifera Bigot, 1885a: 299.—Md.; Wash. to Calif., Que to Md.
pluvialis (Linnaeus), 1758: 597 (*Musca*).—Scandinavia; Alaska to Ariz. to Ga., also Alta., S. Dak., Que., Europe.

Genus ANTHOMYIELLA Malloch

Anthomyiella Malloch, 1920a: 174. Type-species, *Anthomyia pratincola* Panzer (orig. des.).

pratincola (Panzer), 1809c: 12 (*Anthomyia*).—Germany; Wash. to Que., s. to Colo., Okla., and N.C., Europe.

Genus CALYTHEA Schnabl and Dziedzicki

Pegomyia, subg. **Calythea** Schnabl and Dziedzicki, 1911: 111 (1911: 59). Type-species, *Musca albicincta* Fallén (mon.).

albicincta (Fallén).—Not Nearctic.
bidentata (Malloch), 1913m: 606 (*Anthomyia*).—B.C.; Wash.
micropteryx (Thomson), 1869: 555 (*Anthomyia*).—Calif.; sw. N.W.T., Wash. to Mich., s. to Ariz. and Tex., Mass. to Va.
monticola (Bigot), 1885a: 297 (*Anthomyia*).—Rocky Mts.

separata Malloch, 1924h: 198.—R.I.; Wash. to N. Mex. and Tex., Alta. to N.H., s. to Va., also Bermuda.

Genus PARAPROSALPIA Villeneuve

Prosalpia Pokorny, 1893a: 55 (preocc. Koch, 1872). Type-species, *styriaca* Pokorny (Coquillett, 1901e: 140) = *billbergi* (Zetterstedt).

Prosalpia, subg. **Paraprosalpia** Villeneuve, 1922a: 511. Type-species, *rambolitensis* Villeneuve (Séguy, 1937: 127).

REFERENCES: Collin, 1943 (rev. British spp.); Huckett, 1950 (rev.).

angustitarsis (Malloch), 1920a: 184 (*Prosalpia*).—Maine; Alaska to Labr., s. to Alta. and N.Y.
arelate (Walker), 1849: 961 (*Eriphia*).—Ont.; Alaska, N.W.T., Alta., Man.
    pretiosa Walker, 1849: 965 (*Eriphia*).—Ont.
badia (Walker), 1849: 950 (*Anthomyia*).—Ont. Unrecognized.
benanderi (Ringdahl), 1926: 114 (*Hylemyia*).—Sweden; Alaska, Y.T.
brevitarsis (Malloch), 1918a: 309 (*Hylemyia*).—Calif.
brunetta (Huckett), 1929: 139 (*Hylemyia*).—B.C.; Que.
conifrons (Zetterstedt) 1845: 1569 (*Aricia*).—Sweden; Alaska to Labr., s. to Utah in w. and s. to N.H. in e., Europe.
constrictor (Malloch), 1920j: 277 (*Hylemyia*).—Alaska; Y.T., N.W.T., Alta., Man., Labr.
denticauda (Zetterstedt), 1838: 675 (*Anthomyza*).—Sweden; Ont., Lapland.
dentiventris (Ringdahl), 1918: 189 (*Prosalpia*).—Sweden; Alaska, Lapland.
genalis Huckett, 1950: 140.—Calif.; Utah.
gentilis Huckett, 1950: 142.—Colo.
incisa (Ringdahl), 1926: 101 (*Prosalpia*).—Sweden; Ont., Que., Labr., Lapland.
littoralis (Malloch), 1920b: 127 (*Pegomyia*).—Maine; Y.T., Que., N.Y., N.H.
longipennis (Ringdahl), 1918: 189 (*Chortophila*).—Sweden; Alaska, N.W.T., Alta., Idaho, Man. to Labr., Lapland.
    bicruciata Malloch, 1920a: 190 (*Hylemyia*).—Labr.
moerens (Zetterstedt), 1838: 681 (*Anthomyza*).—Sweden; Alaska, N.W.T., Alta., Man., Que., Labr., Lapland, Sweden.
pilitarsis (Stein), 1900: 313 (*Prosalpia*).—Hungary; Idaho, Mont., Vt., Europe.
sepiella (Zetterstedt), 1845: 1541 (*Aricia*).—Sweden; Alaska, Y.T., Alta., Europe.
    pedicillaris Huckett, 1929: 181 (*Hylemyia*).—Alta.

silvestris (Fallén), 1824: 70 (*Musca*).—Sweden; Alaska, N.W.T. to Nfld., s. to S. Dak. and N.C., Europe, U.S.S.R. (Kamchatka).
    apina Walker, 1849: 927 (*Anthomyia*).—N.S.
    donuca Walker, 1849: 946 (*Anthomyia*).—N.S.
    grisea Walker, 1849: 962 (*Eriphia*).—Ont.
sitiens (Collin), 1943: 85 (*Prosalpia*).—Scotland; Vt.

Genus NEOHYLEMYIA Malloch

Neohylemyia Malloch, 1917b: 37. Type-species, *proboscidalis* Malloch (orig. des.).
Pergandea Aldrich, 1919b: 106 (also as *Pergandia*, p. 106; preocc. Ashmead, 1905). Type-species, *apivora* Aldrich (orig. des.).
Ganperdea Aldrich, 1921b: 98 (n. name for *Pergandea* Aldrich). Type-species, *Pergandea apivora* Aldrich (aut.).

apivora (Aldrich), 1919b: 106 (*Pergandea*).—Mo.; N. Mex.
mallochii Huckett, 1924: 37.—N.Y.
proboscidalis Malloch, 1917b: 38.—Ill.

Genus LEUCOPHORA Robineau-Desvoidy

Leucophora Robineau-Desvoidy, 1830: 562. Type-species, *cinerea* Robineau-Desvoidy (Coquillett, 1901e: 138) = *albiseta* (Roser).
Hylephila Rondani, 1877: 13 (preocc. Billberg, 1820). Type-species, *Musca buccata* Fallén (orig. des.).
Hammomyia Rondani, 1877: 13, 236. Type-species, *Aricia albescens* Zetterstedt (orig. des.) = *albiseta* (Roser).

The larvae of *Leucophora* species live as inquilines or parasites in the nests of solitary bees and wasps.

    REFERENCE: Huckett, 1940 (rev.).

albiseta (Roser), 1840: 59 (*Anthomyia*).—Germany; Alaska to Calif. and Colo., also N.S., Europe.
annexa Huckett, 1940: 355.—Wash.; Idaho.
fusca Huckett, 1940: 353.—Idaho; Y.T., Alta., Wash. to Calif., also Ind.
johnsoni (Stein), 1898: 215 (*Hylemyia*).—Pa.; Idaho, Tex., La., Que. to Nfld., s. to e. Tenn. and N.C.
maculata (Stein), 1898: 229 (*Hammomyia*).—Idaho; N.W.T., B.C. to Ariz. and Colo., Man. to S. Dak.
marylandica (Malloch), 1920a: 185 (*Hammomyia*).—Md.; B.C. and Alta., s. to Colo., Mich. to Labr., s. to Ga., also Tex.

obtusa (Zetterstedt), 1838: 682 (*Anthomyza*).—Sweden; Alaska to Calif. and Tex., Wis. to Que., s. to Va., Europe.
sociata (Meigen), 1826: 98 (*Anthomyia*).—Germany; Alaska to Calif. and Colo., also Man., Wis., N.H., Md., Europe.
  depressa Malloch, 1918a: 304 (*Eremomyia*).—Idaho.
unilineata (Zetterstedt), 1838: 675 (*Anthomyza*).—Sweden; Alta., s. to Calif. and e. to Mich., Mass. to D.C., Europe.
unistriata (Zetterstedt), 1838: 677 (*Anthomyza*).—Sweden; Wash., Alta. to Ind., also Que., N.Y., Mass., N.S., Europe.
  paludis Johannsen, 1917: 323 (*Hammomyia*).—Mass.

Genus PROBOSCIMYIA Bigot

Proboscimyia Bigot, 1883c: xxx. Type-species, *siphonina* Bigot (mon.).
Proboscidomyia, emend.
Dolichoglossa Stein, 1898: 230. Type-species, *americana* Stein (orig. des.) = *siphonina* Bigot.
Dolichogaster (error) Coquillett, 1901e: 136.

  REFERENCE: Huckett, 1940 (rev.).

brevis Huckett, 1940: 357.—N.Y.; N.S., D.C., N.C., S.C.
siphonina Bigot, 1883c: xxx.—Rocky Mts.; S. Dak. to N. Mex. and Tex., also Ill.
  americana Stein, 1898: 230, 286 (*Dolichoglossa*).—S. Dak.

Genus MACATEEIA Malloch

Macateeia Malloch, 1919a: 1. Type-species, *protuberans* Malloch (orig. des.).
atra Malloch, 1919e: 305.—Calif.
protuberans Malloch, 1919a: 1.—Md.; N.Y.

Genus EUSTALOMYIA Kowarz

Dendrophila Lioy, 1864a: 909 (preocc. Swainson, 1837). Type-species, *Musca hilaris* Fallén (mon.).
Eustalomyia Kowarz, 1873: 461. Type-species, *Musca hilaris* Fallén (mon.).

festiva (Zetterstedt), 1845: 1424 (*Aricia*).—Sweden; Wash., Nev., Iowa to Que. and Conn., Europe.
  brixia Walker of Johnson, 1925b: 233.
histrio (Zetterstedt), 1838: 676 (*Anthomyza*).—Lapland, Sweden; Mich., Que., N.S., N.H.
  brixia Walker, 1849: 946 (*Anthomyia*).—N.S.

vittipes (Zetterstedt), 1845: 1649 (*Anthomyza*).—Sweden; Alaska, Oreg., Colo., Mich. to Que., s. to Tenn. and N.C., Europe.

Unplaced Species of Anthomyiini

cupricrus Walker, 1849: 974 (*Cordylura*).—Ont.
flavipennis Walker, 1849: 975 (*Cordylura*).—Ont.

Unplaced Species of Anthomyiinae

anthracina Bigot, 1885a: 298 (*Anthomyia*).—Rocky Mts.
lamnia Walker, 1849: 964 (*Eriphia*).—Ont.
nigricauda Bigot, 1885a: 276 (*Hydrophoria*).—Rocky Mts. ?*Neohylemyia*. Possibly not Nearctic.
stygia Meigen, 1826: 155 (*Anthomyia*).—Europe; Greenland.

Family MUSCIDAE
By H. C. Huckett

Adult Muscidae vary from 2 to 12 mm. in body length. Some are yellowish or black, but most species are varying shades of gray and brown. The wings are usually unmarked. Adults of most genera are predaceous or anthophilous. The larvae are ubiquitous and may be phytophagous, saprophagous, scatophagous, or zoophagous. They may also live as commensals, or actually feed on the blood of living animals. For additional information, see the discussions of the various subfamilies.

Many authors have included the Scatophaginae, Anthomyiinae, and Fucelliinae (all Anthomyiidae) with the Muscidae in previous taxonomic works.

REFERENCES: Séguy, 1937 (cat., key gen. world); Ringdahl, 1954–1959 (Sweden); Hennig, 1955–1957, 1958b, 1959–1963 (Palaearctic).

Subfamily COENOSIINAE

REFERENCES: Malloch, 1917a, 19211 (keys, genera).

Genus SCHOENOMYZA Haliday

Schoenomyza Haliday, 1833: 166. Type-species, *Sciomyza fasciata* Meigen (Westwood, 1840: 143)=*litorella* (Fallén).
Litorella Rondani, 1856: 101. Type-species, *Ochtiphila litorella* Fallén (orig. des.).

REFERENCE: Huckett, 1934 (rev.).

chrysostoma Loew, 1869c: 177 (Cent. 9, no. 86).—N.H.; Alaska, N.W.T. to Labr., s. to Oreg., N. Mex., and Fla.
  leucostoma, error.
dorsalis Loew, 1872a: 95 (Cent. 10, no. 73).—D.C.; Alaska to Labr., s. to Calif. and Fla.
  sulfuriceps Malloch, 1918a: 288 (as var.).—Calif.
lispina (Thomson), 1869: 599 (*Ochtiphila*).—Calif.; B.C. to Calif. **N. comb.**
  convexifrons Malloch, 1918a: 287.—Calif. **N. syn.**
  flaviceps Stein, 1920: 105.—Wash., Calif. **N. syn.**
litorella (Fallén), 1823b: 10 (*Ochtiphila*).—Sweden; Alaska to Labr., s. to Calif. and N.Y., also Tenn., Europe, U.S.S.R. (Kamchatka).
  partita Malloch, 1918a: 289 (*as dorsalis* var.).—Calif.

Genus COENOSIA Meigen

Coenosia Meigen, 1826: 210. Type-species, *Musca tigrina* Fabricius (Westwood, 1840: 143).
Caricea Robineau-Desvoidy, 1830: 530. Type-species, *communis* Robineau-Desvoidy (Hennig, 1961: 518)=*tigrina* (Fabricius).

REFERENCE: Huckett, 1934 (rev.).

Subgenus COENOSIA Meigen

humilis Meigen, 1826: 220.—Germany; Wash. to Calif. and Colo., Wis. and Ill. to Que. and N.J., Europe.
  nana Zetterstedt, 1845: 1716 (*Anthomyza*).—Sweden.
tigrina (Fabricius), 1775: 779 (*Musca*).—England; B.C. and Alta. to Calif., nw. Ont. and Que. to Mich. and Maine, Europe.
  sexmaculata Walker, 1849: 970.—Ont.

Subgenus LIMOSIA Robineau-Desvoidy

Limosia Robineau-Desvoidy, 1830: 535 (as genus). Type-species, *campestris* Robineau-Desvoidy (Coquillett, 1901e: 138).
Macrocoenosia Malloch, 1920a: 162 (as *Macorcoenosia*, error). Type-species, *Coenosia triseta* Stein (orig. des.).
Macrocoenia, error.

albibasis Stein, 1920: 95.—Wash.; B.C. to Calif.
aliena Malloch, 1921m: 134.—Mont.; Alaska to N.W.T., s. to Calif. and Colo.
alticola Malloch, 1919e: 303.—Calif.; Oreg., Utah.
antennalis Stein, 1898: 272.—Ga.; N.Y., N.C., S.C., Fla.

anthracina Malloch, 1921m: 134.—Mont.; Alaska to w. N.W.T., s. to Ariz., Labr.
argentata Coquillett *in* Baker, 1904: 33.—Calif.; Wash.
    argenticolor Stein, 1920: 96.—Calif.
argenticeps Malloch, 1920a: 166.—Mont.; Alta., Wash. to Calif. and Colo.
armiger Huckett, 1934: 181.—Que.; Mich.
atrata Walker, 1852: 369.—U.S.; Wash. and Alta. to Que., s. to Ariz. and Fla.
    canescens Stein, 1898: 265.—Ill., Pa., Va.
atritibia Ringdahl, 1926: 107.—Sweden; Alaska, Y.T., N.W.T., Man., Que.
bonita Huckett, 1934: 179.—Calif.; Wash., Oreg., Idaho, Utah.
canadensis Curran, 1933b: 9.—Que.; N.W.T., Man., Ont., Labr.
candida Huckett, 1934: 185.—Wash.
cilicauda Malloch, 1920h: 103.—Mont.; sw. N.W.T., B.C., and Man., s. to Utah and Colo., Que.
comita (Huckett), 1936: 214 (*Macrorchis*).—Alaska; Alaska to N.W.T., s. to Mont.
compressa Stein, 1904: 489.—N.H.; Que. to N.S., s. to Minn. and Pa.
conforma Huckett, 1934: 190 (as *compressa* var.).—Man.; Alaska to N.W.T., s. to Colo. and Nfld., also e. Tenn.
dorsovittata Malloch, 1920a: 167.—Colo.; Sask.
effulgens Huckett, 1934: 151.—Calif.; Wash., Oreg.
errans Malloch, 1920a: 165 (n. name for *steini* Johnson).—Ga., Mass.; Iowa, Ohio, N.H. to Fla.
    flavipes Stein, 1898: 268 (preocc. Williston, 1896).—Ga., Mass.
        steini Johnson, 1913a: 78 (n. name for *flavipes* Stein, but preocc. Verrall, 1912).—Ga., Mass.
    geniculata, authors, not Fallén.
flavifrons Stein, 1898: 261.—Ga.
flavipes Williston.— Not Nearctic.
fraterna Malloch, 1918a: 282.—Calif.; Sask. to Calif. and Tex.
frisoni Malloch, 1920a: 165.—Ill.; B.C. to N.W.T. and Que., s. to Utah and Ill.
furtiva Huckett, 1934: 178.—N.J.; Ill., Ind., N.Y. to Fla.
fuscifrons Malloch, 1919h: 96.—Ont.; Wash., Oreg., Wyo., Minn. to Nfld.
impunctata Malloch, 1920a: 165.—Alaska; Wash.
incisurata Wulp, 1869: 86.—Wis.; B.C., Alta., Idaho, Man. to Labr., s. to Ill. and D.C.
    flavicoxa Stein, 1898: 271.—Ill.

johnsoni Malloch, 1920a: 167.—Ont.; Alaska, B.C., Alta., N. Mex., Mich. to Que.
laricata Malloch, 1920a: 166.—Ill.
lata Walker, 1852: 368.—U.S.; Wash., Alta. to Que., s. to La. and Fla.
    hypopygialis Stein, 1898: 268.—Ga., Mass., Ill.
longimaculata Stein, 1920: 97.—Wash.; Alaska to N.W.T., s. to B.C. and Mich.
maculiventris Huckett, 1934: 187.—Wash.
modesta Loew, 1872a: 95 (Cent. 10, no. 72).—D.C.; Man. to Que., s. to Nebr. and Ill.
    pallida Stein, 1920: 100.—S. Dak., Wis.
nigrescens Stein, 1920: 98.—Colo., Que.; Alaska to Nfld., s. to Calif. and Pa.
nigricoxa Stein, 1920: 99.—Colo.; Alaska, Alta., Man.
nivea Loew, 1872a: 94 (Cent. 10, no. 70).—Pa.; Wash. to Calif., Iowa to N.H., s. to Fla.
    var. **brunnescens** Malloch, 1920a: 165.—Pa.; Calif., Ohio, N.Y., D.C.
nudipes Stein, 1920: 99.—Wash.
nudiseta Stein, 1898: 273.—Mass.; Que. to N.J.
oregonensis Malloch *in* Cole and Lovett, 1919: 254.—Oreg.; Alaska to N.W.T., s. to Calif.
pallipes Stein, 1898: 270.—Ill., Ont.; Alaska, B.C. to Que., s. to Ill. and Vt.
    denticornis Malloch, 1920a: 164.—Sask.
pedella (Fallén), 1825: 88 (*Musca*).—Sweden; Alaska to w. N.W.T., s. to Colo., also Que., Europe.
    dichaeta Malloch, 1920a: 163.—Colo.
perspicua Huckett, 1934: 176.—Calif.; Mont.
pilosissima Stein, 1920: 100.—Wash., Idaho, Calif. (as Colo., error); Alaska to Calif. and Okla.
    longispinosa Malloch, 1920a: 166.—Colo.
pulicaria (Zetterstedt), 1845: 1733 (*Anthomyza*).—Sweden; Alaska and Y.T. to Colo.
    pygmaea Zetterstedt of Huckett, 1934: 152.
pumila (Fallén), 1825: 88 (*Musca*).—Sweden; Alaska, N.W.T., Man., Que., Europe.
rubrina Huckett, 1934: 194.—Calif.
rufibasis Stein, 1920: 101.—Wash.; B.C. and Idaho to Calif.
setigera Malloch, 1918a: 284.—N. Mex.; Colo.
sexnotata Meigen, 1826: 213.—Germany; Alaska to Wash, and Alta., Europe.
strigifemur Stein, 1920: 103.—Wash.

tausa Huckett, 1934: 156.—Wash.; Calif.
toshua Huckett, 1934: 192.—Man.; Alaska, N.W.T., N. Dak., Ont., Que.
triseta Stein, 1898: 262.—Mass.; Alaska to N.W.T., s. to Colo. and Nfld., also N.C.
verralli Collin, 1953c: 171 (n. name for *steini* Verrall).—Wales; Alaska to N.W.T., s. to Man. and Labr., England.
  steini Verrall, 1912: 196 (preocc. Strobl, 1908).—Wales.

Subgenus HOPLOGASTER Rondani

Hoplogaster Rondani, 1856: 98 (as genus *Oplogaster;* emend. Rondani, 1871a: 331). Type-species, *Musca mollicula* Fallén (orig. des.).

albifacies (Johnson), 1922: 23 (?*Micropselapha*).—Maine; Y.T., N.W.T. **N. comb.**, C. W. Sabrosky *in litt.*
californica (Malloch), 1920a: 194 (*Hoplogaster*; n. name for *parvisquama* Malloch).—Calif.; Oreg., Colo., Ariz., N. Mex.
  californiensis (error) Malloch, 1921n: 205.
  parvisquama Malloch, 1919e: 304 (preocc. Stein, 1900).—Calif.
flavibasis Huckett, 1934: 90.—Mont.; Ariz., Colo., N. Mex.
flavidipalpis Huckett, 1934: 95.—B.C.; Alaska, Y.T., Alta., Oreg., Mont.
fumipennis Huckett, 1934: 104.—N.Y.; Mich.
intacta Walker, 1852: 369.—U.S.; Mich. to Que. and Pa., also Tenn.
laeta Huckett, 1934: 98.—Ont.
mollicula (Fallén), 1825: 90 (*Musca*).—Sweden; Alaska to Nfld., s. to N. Mex. and N.Y., Europe.
morrisoni (Malloch), 1924e: 172 (*Hoplogaster*).—N.H.; Alaska to N.H. and Labr.
nigritarsis (Stein), 1898: 252 (*Hoplogaster*).—N.J. (as N.Y., error); Que. to N.S., s. to Mich., Tenn., and S.C.
nigritella Huckett, 1934: 93.—Oreg.; Wash.
octopunctata (Zetterstedt), 1838: 693 (*Anthomyza*).—Sweden; Alaska to Labr., s. to Colo., U.S.S.R. (Kamchatka).
  octomaculata (error) Malloch, 1919f: 68.

Subgenus NEODEXIOPSIS Malloch

Neodexiopsis Malloch, 1920a: 162 (as genus). Type-species, *Dexiopsis basalis* Stein (orig. des.).
Xenocoenosia Malloch, 1920a: 162. Type-species, *Coenosia calopyga* Coquillett (orig. des.).

REFERENCE: Snyder, 1958 (rev., *ovata* group, New World).

arizona (Snyder), 1958: 25 (*Neodexiopsis*).—Ariz.
basalis (Stein), 1898: 259 (*Dexiopsis*).—Ill.; Calif. to Mass. and Ga.
borea (Snyder), 1958: 26 (*Neodexiopsis*).—Wis.
calopyga Loew, 1872a: 94 (Cent. 10, no. 71).—Pa.; Wis. to Que., s. to Miss. and Ga.
    calophaga (error) Johnson, 1925b: 237.
floridensis (Malloch), 1920a: 163 (*Xenocoenosia*).—Fla.; N.Y., Pa., Ohio, e. Tenn., N.C.
major (Malloch), 1920a: 163 (*Xenocoenosia*).—Fla.; Okla. to N.Y., s. to Fla.
ovata Stein, 1898: 263.—Ill., Ga.; B.C. to Que., s. to s. Calif. and Fla.
pectoralis Huckett, 1934: 76.—Ariz.; B.C., Utah, Colo., N. Mex.
peninsula (Snyder), 1958: 23 (*Neodexiopsis*).—Fla.; Bahama Is.
rufitibia Stein, 1919: 161 (n. name for *tibialis* Stein).—Ill., Pa., Ga.; s. Calif. to Mich. and N.Y., s. to Fla.
    tibialis Stein, 1898: 275 (preocc. Macquart, 1842).—Ill., Pa., Ga.
simplex Stein, 1920: 102.—Oreg.; Utah, Calif.

Genus BITHORACOCHAETA Stein

Bithoracochaeta Stein, 1911: 177. Type-species, *Anthomyia despecta* Walker (orig. des.) = *leucoprocta* (Wiedemann).
Cariciella Malloch, 1920a: 162. Nomen nudum; syn. by J. R. Malloch *in litt.*

REFERENCE: Huckett, 1934 (rev.).

leucoprocta (Wiedemann), 1830: 433 (*Anthomyia*).—West Indies; Okla. to N.C., s. to Tex. and Fla., Cent. Amer., Antilles.
    despecta Walker, 1852: 364 (*Anthomyia*).—Brazil.
    antica Walker, 1852: 367 (*Coenosia*).—U.S.
    insignis Stein, 1898: 257 (*Caricea*).—Fla.

Genus ALLOGNOTA Pokorny

Allognota Pokorny, 1893a: 64. Type-species, *Coenosia agromyzella* Rondani (mon.) = *agromyzina* (Fallén).

REFERENCE: Huckett, 1934 (rev.).

semivitta Malloch, 1918a: 282.—Ill.; S. Dak. to Kans., Mich., Ont., N.Y.

Genus ATHERIGONA Rondani

Atherigona Rondani, 1856: 97. Type-species, *Coenosia varia* Meigen (orig. des.).

REFERENCE: Huckett, 1936 (rev., N. Amer.).

Subgenus ACRITOCHAETA Grimshaw

Acritochaeta Grimshaw, 1901: 41 (as genus). Type-species, *pulvinata* Grimshaw (mon.) = *orientalis* Schiner.

orientalis Schiner, 1868: 295.—Nicobar Is.; Calif. to Ga. and Fla., cosmopolitan.
    excisa Thomson, 1869: 560 (*Coenosia*).—Burma (Ross I.).
    pulvinata Grimshaw, 1901: 42 (*Acritochaeta*).—Hawaii.
    varia Meigen of Malloch, 1921l: 107.

Genus MACRORCHIS Rondani

Macrorchis Rondani, 1877: 17, 280. Type-species, *Musca meditata* Fallén (orig. des.).

REFERENCE: Huckett, 1936 (rev.).

alone (Walker), 1849: 941 (*Anthomyia*).—Ont.; Alaska, B.C. to Labr.
ausoba (Walker), 1849: 938 (*Anthomyia*).—N.S.; B.C. to Nfld., s. to Utah and Ga.
    aurifrons Stein, 1898: 260 (*Coenosia*).—Ont., Mass., Ill., Pa.
majuscula (Coquillett) *in* Baker, 1904: 34 (*Coenosia*).—Calif.; Oreg.

Genus PSEUDOCOENOSIA Stein

Pseudocoenosia Stein, 1916: 113. Type-species, *Aricia longicauda* Zetterstedt (Karl, 1928: 207) = *solitaria* (Zetterstedt).

REFERENCE: Huckett, 1936 (rev.).

brevicauda Huckett, 1936: 207.—Alta.; Alaska to N.W.T., s. to Utah and Colo., also Labr.
fletcheri (Malloch), 1919k: 274 (*Helina*).—Sask.; Alaska to N.W.T., s. to Colo., also Que.
    uralica Stein, 1920: 60.—Colo.
nigriventris Huckett, 1936: 206.—Calif.; Wash.
solitaria (Zetterstedt), 1838: 677 (*Anthomyza*).—Sweden; Alaska and Y.T., s. to Utah and Colo., also Labr., Europe.
    longicauda Zetterstedt, 1860: 6230 (*Aricia*).—Sweden.

Genus PHYLLOGASTER Stein

Tetrachaeta Stein, 1898: 254 (preocc. Ehrenberg, 1844). Type-species, *unica* Stein (mon.).
Phyllogaster Stein, 1898: 256. Type-species, *cordyluroides* Stein (mon.).
Tetramerinx Berg, 1898: 17 (n. name for *Tetrachaeta* Stein). Type-species, *Tetrachaeta unica* Stein (aut.).
Parasteinia Cockerell, 1905: 361 (n. name for *Tetrachaeta* Stein). Type-species, *Tetrachaeta unica* Stein (aut.).

REFERENCE: Huckett, 1936 (rev.).

cordyluroides Stein, 1898: 256.—Mass., Fla.; Calif. to Maine and Fla., Bermuda, Bahama Is.
inermis Stein, 1920: 58.—Wash., Calif.; Oreg.
littoralis Malloch, 1917e: 228.—Ill., Alta. to S. Dak., s. to Calif. and Tex., Ill. to Que. and N.J.
mallitosa Huckett, 1936: 199.—Md.; Conn., Va., N.C., Fla.
parvimaculata (Stein), 1920: 60 (*Tetramerinx*).—Tex.
robusta Johnson, 1917a: 148.—Mass.; R.I., N.Y., Fla.
  maximus Stein, 1920: 59.—Mass.
rufitibia (Stein), 1911: 144 (*Tetramerinx*).—Peru, Chile; Oreg., Calif., Fla., Bahama Is.
  californiensis Malloch, 1918a: 274 (*Tetramerinx*).—Calif.
unica (Stein), 1898: 254 (*Tetrachaeta*).—Mass.; B.C. and Alta., s. to Ariz. and Tex., also Mich. to Que., s. to Va.

Genus LIMNOSPILA Schnabl

Limnospila Schnabl, 1902: 111. Type-species, *Aricia albifrons* Zetterstedt (orig. des.).

REFERENCE: Huckett, 1936 (rev.).

albifrons (Zetterstedt), 1849: 3301 (*Aricia*).—Denmark; Alaska to Man., s. to Nev., Colo., Que. to N.J., Europe.

Genus LISPOCEPHALA Pokorny

Lispocephala Pokorny, 1893b: 532. Type-species, *Anthomyia alma* Meigen (orig. des.).

REFERENCE: Malloch, 1935b (rev.).

alma (Meigen), 1826: 188 (*Anthomyia*).—Germany; Alaska to Nfld., s. to Utah and Tenn., U.S.S.R. (Kamchatka), Europe.
  pallipalpis Zetterstedt, 1845: 1678 (*Anthomyza*).—Sweden.
brevitarsis Malloch, 1935b: 570.—N.W.T.; Alaska to Labr., also Colo.

erythrocera (Robineau-Desvoidy), 1830: 534 (*Caricea*).—France; Alaska to Labr., s. to Calif. and Fla., Europe, U.S.S.R. (Kamchatka).
  lacteipennis Zetterstedt, 1845: 1722 (*Anthomyza*).—Sweden.
  intacta Walker, 1861: 318 (*Coenosia;* preocc. Walker, 1852).—U.S.
rubricornis (Zetterstedt).—Not Nearctic.
setipes Malloch, 1935b: 567.—Calif.; Oreg., Nev.
tinctinervis Malloch, 1935b: 565.—Que.; Alta., Mont., Mich.
varians Malloch, 1935b: 568.—B.C.; Alaska and Y.T. to Alta., Ont. to Labr.
verna (Fabricius), 1794: 330 (*Musca*).—France; N.H., Europe.

Genus PENTACRICIA Stein

Pentacricia Stein, 1898: 249. Type-species, *aldrichii* Stein (mon.)
aldrichii Stein, 1898: 249.—Kans., Ill., Ga.; Alta. to n. Ont., s. to Calif. and Fla.

Unplaced Species of Coenosiinae

albicornis Meigen, 1826: 220 (*Coenosia*).—Europe; ?N. Amer.
fuscopunctata Macquart, 1851: 243 (1851: 270) (*Coenosia*).—N. Amer.
imperator Walker, 1849: 975 (*Cordylura*).—Ont.
solita Walker, 1852: 368 (*Coenosia*).—U.S.
spinosa Walker, 1849: 967 (*Coenosia*).—Ont.
tenuior Walker, 1849: 977 (*Cordylura*).—Ont.

Subfamily LISPINAE

Genus LISPE Latreille

Lispe Latreille, 1796: 169. Type-species, *Musca tentaculata* De Geer (sub. mon., Latreille, 1802: 462).
Lispa, error or emend.

Adults are usually found near the sunny margins of standing or running water, or occasionally near carrion or feces. The larvae of a few species are aquatic.

  REFERENCE: Snyder, 1954 (rev.).

albitarsis Stein, 1898: 277.—Kans., Ill., Pa., Mass., Ga.; Kans. to Que., s. to La. and Fla.
antennata Aldrich, 1913: 144.—Nev. (as Utah, error); Alta., Utah.
argentea Snyder, 1954: 27.—Calif.

brevipes Aldrich, 1913: 137.—Idaho; Wash. to Calif. and Colo., also Man., Wis., Ind.
canadensis Snyder, 1954: 29.—N.W.T.; Man., Labr.
cinifera Becker.—Not Nearctic.
consanguinea Loew, 1858e: 8.—Germany, Sweden; ?N. Amer.
cotidiana Snyder, 1954: 22.—Alta.; Alaska to Labr., s. to N. Mex. and N.Y., cent. and e. Asia.
flavicincta Loew.—Not Nearctic.
frigida Erichson *in* Ménétries, 1851: 67.—U.S.S.R. (Boganida); N.W.T., Man., Que., Europe, n. Asia.
    bohemica Becker, 1904a: 53.—Poland.
hispida Walker, 1849: 971.—Ont.; N.H., N.J.
jamesi Snyder, 1954: 33.—Oreg.; Alaska, Wash., Calif.
johnsoni Aldrich, 1913: 138.—Mass.; Alta. to Maine and Mass.
nasoni Stein, 1898: 280.—S. Dak., Ill., Ga.; Alaska to Que., s. to Calif. and Fla., Bahama Is.
neouliginosa Snyder, 1954: 24.—Calif.; Wash. to Calif. and N. Mex., also Sask., Ont.
nudifacies Snyder, 1954: 18.—Wis.; Alta., Man., Wis., Mich.
palposa (Walker), 1849: 926 (*Anthomyia*).—Ont.; B.C. to Que., s. to Nev. and Md.
    simillima Walker, 1849: 972.—Ont.
    nigromaculata Stein, 1898: 278.—Idaho, S. Dak., Kans., Pa.
patellata Aldrich, 1913: 140.—Idaho; Wash. to Mont., s. to Calif. and N. Mex.
polita Coquillett *in* Baker, 1904: 34.—Nev.; Alta., Oreg. to Calif. and Colo.
probohemica Speiser, 1914: 93.—w. Poland; Wash. to Kans., s. to Calif. and Miss., Europe.
    spinipes Aldrich, 1913: 136 (preocc. Bigot, 1885).—Idaho, Calif.
salina Aldrich, 1913: 134.—Calif.; N.W.T. to Calif. and Colo.
sociabilis Loew, 1862b: 217 (Cent. 2, no. 7).—D.C.; Wis. to Que., s. to Tex. and Ga.
sordida Aldrich, 1913: 132.—Utah; Wash. to Wyo., s. to Calif. and N. Mex.
tentaculata (De Geer), 1776: 86 (*Musca*).—Sweden; Alaska to Labr., s. to Colo., Tex., and S.C., Europe, n. Asia.
    acela Walker, 1849: 962 (*Eriphia*).—Ont.
    consanguinea Loew of Osten Sacken, 1878c: 171
uliginosa Fallén, 1825: 93.—Sweden; Wash. to Man., also N.Y., Md., Greenland, Europe, n. Asia.

Subfamily LIMNOPHORINAE

REFERENCES: Huckett, 1932 (rev.); Hennig, 1959–1960 (rev., Palaearctic).

Genus LISPOIDES Malloch

Lispoides Malloch, 1920a: 146. Type-species, *Limnophora aequifrons* Stein (orig. des.).

aequifrons (Stein), 1898: 205 (*Limnophora*).—Idaho, S. Dak.; Alaska, Alta. to Que., s. to Ariz. and Ga.
    aequalis (error) Malloch, 1921c: 61.

Genus SPILOGONA Schnabl

Limnophora, subg. **Spilogona** Schnabl, 1911: 92. Type-species, *Aricia carbonella* Zetterstedt (mon.).
Paralimnophora Malloch, 1913m: 604. Type-species, *brunnesquama* Malloch (orig. des.)=*narina* (Walker).
Parlimnophora, error.
Sphenomyia Aldrich, 1919b: 108. Type-species, *kincaidi* Aldrich (orig. des.)=*leucogaster* (Zetterstedt).
Melanochelia Rondani of Malloch, 1921c: 61.

REFERENCES: Malloch, 1921c (synopsis); Collin, 1930a (rev., Greenland).

acuticornis (Malloch), 1920a: 147 (*Limnophora*).—Pa.; Alaska to Calif. and Colo., B.C. to Man., Mich., and Que.
aerea (Fallén), 1825: 76 (*Musca*).—s. Sweden; Alaska to Labr., s. to Oreg. and Mass., Europe.
alberta (Huckett), 1932: 143 (*Limnophora*).—Alta.; Que.
albisquama (Ringdahl), 1932: 156 (*Limnophora*).—Sweden; N.W.T., Man.
almquistii (Holmgren), 1880: 17 (also 1883: 167) (*Aricia*).—U.S.S.R. (Novaya Zemlya); N.W.T., Greenland, Colo.
    angulata Malloch, 1920a: 151 (*Limnophora*).—Greenland.
alpica (Zetterstedt), 1845: 1624 (*Aricia*).—Sweden; Que., Labr., Greenland, Europe.
alticola (Malloch), 1920a: 153 (*Limnophora*).—N.H.; Alaska to Labr., s. to B.C. and Conn.
anthrax (Bigot), 1885a: 274 (*Limnophora*).—Mexico; Alaska to Colo., also N.W.T., N.Y.
    squamosa Stein, 1920: 53 (*Limnophora*).—Wash.
arctica (Zetterstedt), 1838: 669 (*Anthomyza*).—Sweden; Alaska to Greenland, s. to N.H. and Maine, n. Europe.
    fumipennis Zetterstedt, 1845: 1465 (*Aricia*).—Sweden.

arenosa (Ringdahl), 1918: 155 (*Limnophora*).—Sweden; Alaska, Wash. to Man. and Que., s. to Wyo., Greenland, n. Europe.
argenticeps Malloch, 1924h: 200.—N.H.; N.Y.
argentiventris (Malloch), 1920a: 149 (*Limnophora*).—Mont.; Alaska to N.W.T., s. to Ariz. and Iowa.
    occidentalis Huckett, 1932: 150 (*Limnophora*; as var.).—Wash.
atrisquamula Hennig, 1959: 277 (n. name for *atrisquama* Ringdahl).—Sweden; N.W.T., Man.
    atrisquama Ringdahl, 1932: 156 (*Limnophora*; preocc. Stein, 1904).—Sweden.
baltica (Ringdahl), 1918: 165 (*Limnophora*).—Germany; Alaska to N.W.T., also Idaho, Europe.
    fumipennis Zetterstedt of Stein, 1916: 101.
bifimbriata Huckett (for *fimbriata* Huckett).—B.C.; Alaska to N.W.T., s. to Alta. and n. Man. **N. name.**
    fimbriata Huckett, 1932: 305 (*Limnophora*; preocc. Schnabl, 1915).—B.C.
bisetosa (Huckett), 1932: 300 (*Limnophora*).—Oreg.; Wash. to Calif. and Colo.
    litorea Fallén of Stein, 1920: 56.
brevicornis (Malloch), 1917d: 226 (*Tetramerinx*).—Ill.; Oreg., N.W.T. to Colo. and S. Dak,. N.Y., Que., Maine.
cana (Huckett), 1932: 299 (*Limnophora*).—Wash.; Oreg., Wyo.
caroli (Malloch), 1920a: 154 (*Limnophora*).—Vt.; N.Y., N.H., Maine, N.C.
clarans (Huckett), 1932: 298 (*Limnophora*).—Ont.
comata (Huckett), 1932: 146 (*Limnophora*).—Alaska.
concolor (Stein), 1920: 55 (*Limnophora*).—Wyo., Colo.; Alaska to Utah and Colo., also N.W.T., Man.
contractifrons (Zetterstedt), 1838: 683 (*Anthomyza*).—Sweden; Alaska to Labr., s. to B.C., Europe, U.S.S.R. (Kamchatka).
crepusculenta (Huckett), 1932: 144 (*Limnophora*).—Alaska.
cretans (Huckett), 1932: 151 (*Limnophora*).—Alta.; Wyo., Colo.
deflorata (Holmgren), 1872: 102 (*Aricia*).—w. Greenland; Alaska to Labr., s. to n. Man.
    addicta Huckett, 1932: 133 (*Limnophora*).—Alaska.
denudata (Holmgren), 1869: 30 (*Aricia*).—Spitzbergen; Alaska to Greenland, s. to Que., Spitzbergen.
    ranunculi Holmgren, 1869: 34 (*Aricia*).—Spitzbergen.
dorsata (Zetterstedt), 1845: 1472 (*Aricia*).—Sweden; Alaska to Greenland, s. to n. Man. and Que., n. Europe. See Ringdahl, 1931: 174.

hyperborea Boheman, 1865: 571 (*Aricia*).—Spitzbergen.
labiosa Boheman, 1865: 571 (*Aricia*).—Spitzbergen.
pearyi Malloch, 1920a: 151 (*Limnophora*).—w. Greenland.
extensa (Malloch), 1920a: 150 (*Limnophora*).—w. Greenland; N.W.T. (Victoria, Ellesmere, and Baffin Is.).
fatima (Huckett), 1932: 127 (*Limnophora*).—Que.; N.W.T., Man. to Que., s. to Mich. and N.Y.
fimbriata (Schnabl), 1915: 42 (*Limnaricia*).—U.S.S.R. (Karskaja Tundra); Alaska (Point Barrow).
gibsoni (Malloch), 1920a: 152 (*Limnophora*).—N.B.; Alta., Mich., Que., and N.B. to N.Y.
glauca (Stein), 1916: 90 (*Limnophora*).—Germany, Poland; Alaska, N.W.T., Europe.
imitatrix (Malloch), 1921c: 64 (*Melanochelia*).—Labr.; Alaska, Alta. to Utah and Colo., n. Man. to Que.
incauta (Huckett), 1932: 288 (*Limnophora*).—Alaska; N.W.T.
instans (Huckett), 1932: 116 (*Limnophora*).—Alaska; N.W.T., n. Man., Labr.
latilamina (Collin), 1930a: 266 (*Limnophora*).—Greenland; Alaska to N.W.T., s. to Man. and Labr.
leucogaster (Zetterstedt), 1838: 674 (*Anthomyza*).—Sweden; Alaska to Labr., s. to Calif., Mich., and Que., n. Europe.
   biquadrata Walker, 1849: 963 (*Eriphia*).—Ont.
   kincaidi Aldrich, 1919b: 108 (*Sphenomyia*).—Alaska.
   nitidifrons Stein, 1920: 50 (*Limnophora*).—Wash.
   banffi Seamans, 1926: 175 (*Sphenomyia*).—Alta.
limnophorina (Stein), 1898: 200 (*Spilogaster*).—Pa.; Alaska, Oreg., Idaho, Ohio, Que. to Pa.
   rufitibia Stein, 1920: 52 (*Limnophora*).—Idaho.
magnipunctata (Malloch), 1919e: 301 (*Limnophora*).—Calif.; Alaska to Calif. and Colo., also Labr., Que., N.H.
   fumosa Stein, 1920: 48 (*Limnophora*).—Wash., Colo.
malaisei (Ringdahl), 1920: 29 (*Limnophora*).—Sweden; Alaska to Greenland, s. to B.C. and Que., Lapland.
megastoma (Boheman), 1865: 572 (*Aricia*).—Spitzbergen; Alaska to Greenland, s. to Que., Norway.
   pauxilla Holmgren, 1869: 32 (*Aricia*).—Spitzbergen.
melanosoma (Huckett), 1932: 110 (*Limnophora*).—Alaska; Alaska to N.W.T., s. to n. Man. and Que.
   dorsata Zetterstedt of Malloch, 1934b: 30.
micans (Ringdahl), 1918: 156 (*Limnophora*).—Sweden; Alaska to Greenland, s. to n. Man., Que., and Labr.
monacantha (Collin), 1930a: 271 (*Limnophora*).—Greenland; Alaska to Greenland, s. to n. Man. and Que.

narina (Walker), 1849: 933 (*Anthomyia*).—N.S.; Alaska to Labr., s. to Oreg., N.Y., and N.S.
    brunnesquama Malloch, 1913m: 605 (*Paralimnophora*).—N.H.
    nigrifrons Stein, 1920: 49 (*Limnophora*).—N.Y.
    velutina Malloch, 1920a: 147 (*Limnophora*; unjustified n. name for *brunnesquama* Malloch, as *brunneisquama*).—N.H.
nigriventris (Zetterstedt), 1845: 1442 (*Aricia*).—Sweden; Alaska, Man., Que., N.H., Labr., Europe.
    monticola Malloch, 1920a: 152 (*Limnophora*).—N.H.
nobilis (Stein), 1898: 207 (*Limnophora*).—Alaska; B.C. to Calif.
norvegica (Ringdahl), 1932: 158 (*Limnophora*).—Norway; Alaska to N.W.T., s. to Man. and Que., Norway, Sweden.
    carbonella Zetterstedt of Huckett, 1932: 140.
novemmaculata (Zetterstedt), 1860: 6219 (*Aricia*).—Sweden; Alaska, N.W.T., Man., Que., Labr., Lapland.
obscuripennis (Stein), 1916: 93 (*Limnophora*).—Lapland; Alaska and Y.T., s. to Colo., also Que., Labr., n. Europe, U.S.S.R. (Kamchatka).
obsoleta (Malloch), 1920a: 149 (*Limnophora*).—w. Greenland; Alaska to w. Greenland, s. to Que.
    hirticauda Malloch, 1921p: 181 (*Melanochelia*).—Alaska (Pribilof Is.).
opaca (Schnabl), 1915: 28 (*Limnophora*).—U.S.S.R. (Ural Mts.); Alaska to Greenland, s. to Colo. and Labr., n. Europe.
    freyii Ringdahl, 1918: 150 (*Limnophora*).—Sweden.
    alliterata Huckett, 1932: 280 (*Limnophora*).—Alaska.
parvimaculata (Stein), 1920: 50 (*Limnophora*).—N.Y.; Ill. to Ont., Maine to Tenn.
    clivicola Malloch, 1920a: 155 (*Limnophora*).—Ill.
placida (Huckett), 1932: 154 (*Limnophora*).—Alaska; Alaska and Y.T., s. to Colo., also Que., N.H., Sweden.
pruinella (Huckett), 1932: 302 (*Limnophora*; as *bisetosa* var.).—Calif.; Wash. to Calif., also Wyo., Colo.
pseudodispar (Frey), 1915: 24 (*Limnophora*).—U.S.S.R. (Lena Estuary); Alaska, N.W.T. (Baffin I.), n. B.C. to n. Man., also Que., n. Europe.
    spinitibia Ringdahl, 1918: 151 (*Limnophora*).—Sweden.
puberula (Ringdahl), 1918: 151 (*Limnophora*).—Sweden; Alaska.
pulchra (Huckett), 1932: 313 (*Limnophora*).—Alaska.
pulvicrura (Huckett), 1932: 138 (*Limnophora*).—Alaska; N.W.T.
pusilla (Huckett), 1932: 134 (*Limnophora*).—Colo.; Alaska and Y.T. to Colo., B.C. to Labr.

quinquelineata (Zetterstedt), 1838: 671 (*Anthomyza*).—Sweden; Alaska, n. Man., Que.
reflecta (Huckett), 1932: 316 (*Limnophora*).—Mont.; Alaska to N.W.T., Mont.
rostrata (Ringdahl), 1920: 26 (*Limnophora*).—Sweden; Que., Lapland.
rufitarsis (Stein), 1920: 51 (*Limnophora*).—Wash.; Alaska, Alta., Oreg.
sanctipauli (Malloch), 1921p: 180 (*Melanochelia*).—Alaska (Pribilof Is.); Alaska to Greenland, s. to Que. and Labr., n. U.S.S.R.
sectata (Huckett), 1932: 112 (*Limnophora*).—B.C.; Wash., Alta.
    armipes Stein, 1916, of Stein, 1920: 56.
semiglobosa (Ringdahl), 1916: 236 (*Limnophora*).—Sweden; Alaska to Greenland, s. to Wash. and N.H.
setilamellata (Huckett), 1932: 118 (*Limnophora*).—N.H.; Maine, Labr.
setinervis (Huckett), 1932: 155 (*Limnophora*).—Alaska; Alaska to Nfld., s. to n. Man. and Que.
sospita (Huckett), 1932: 115 (*Limnophora*).—Alaska; N.W.T., U.S.S.R.
spinicostalis Huckett (for *spinicosta* Malloch).—Alaska (Pribilof Is.); Y.T. **N. name.**
    spinicosta Malloch, 1921p: 181 (*Melanochelia;* preocc. Stein, 1907).—Alaska (Pribilof Is.).
subrostrata (Stein), 1920: 54 (*Limnophora*).—B.C.; Alaska to Wash. and Colo., N.W.T. to Man., Mich., and N.H.
surda (Zetterstedt), 1845: 1476 (*Aricia*).—Denmark, Sweden; Alaska to Calif. and Wyo., N.W.T. to Man., N.Y., and Que., Europe, U.S.S.R. (Kamchatka).
suspecta (Malloch), 1920a: 154 (*Limnophora*).—Maine; Alaska to Labr., s. to Mont. and N.Y.
tendipes (Malloch), 1920j: 270 (*Limnophora*).—Alaska.
tetrachaeta (Malloch), 1920a: 153 (*Limnophora*).—Oreg.; Calif., Alta. to Colo.
tornensis (Ringdahl), 1926: 102 (*Limnophora*; n. name for *seticosta* Ringdahl).—Sweden; Alaska to Greenland, s. to Wash. and Que., Lapland.
    seticosta Ringdahl, 1920: 28 (*Limnophora;* preocc. Schnabl, 1915).—Sweden.
    pluvialis Huckett, 1932: 306 (*Limnophora*).—Wash.
torreyae (Johannsen), 1916: 391 (*Limnophora*).—N.Y.; N.H.
    umbrina Stein, 1920: 54 (*Limnophora*).—N.Y.

triangulifera (Zetterstedt), 1838: 680 (*Anthomyza*).—Sweden; Alaska to N.W.T., s. to Que., Greenland, Lapland.
  tristiola (in part) Zetterstedt, 1838: 675 (*Anthomyza*). Female only.
trigonata (Zetterstedt), 1838: 680 (*Anthomyza*).—Sweden; Alaska to N.W.T., s. to n. Man. and Labr., Lapland.
trigonifera (Zetterstedt), 1838: 669 (*Anthomyza*).—Greenland, Lapland; Alaska to Utah, also Ont. and Labr. to Mass., Greenland, Lapland.
  novaeangliae Malloch, 1920a: 151 (*Limnophora*).—N.H.
trilineata (Huckett), 1932: 283 (*Limnophora*).—Alaska; Alaska to N.W.T., s. to n. Man. and Labr., also Utah.
tundrae (Schnabl), 1915: 39 (*Limnaricia*).—U.S.S.R. (Karskaja Tundra); Alaska to Greenland, s. to Que. and Labr.
  macropyga Frey, 1915: 23 (*Limnophora*).—U.S.S.R. (Taimur Peninsula).
vana (Zetterstedt), 1845: 1465 (*Aricia*).—Sweden; Alaska, s. to Wash. and Idaho, and e. to Ont. and Labr., Europe.
  nupta Zetterstedt, 1860: 6217 (*Aricia*).—Sweden.
  crassiventris Huckett, 1932: 292 (*Limnophora*).—Alaska.
zetterstedtii Ringdahl, 1918: 173 (*Limnophora*).—Sweden; Alaska to N.W.T., s. to Wash., Colo., and Que.
  fuscomarginata Huckett, 1932: 290 (*Limnophora*).—Alta.
  trianguligera Zetterstedt of Stein, 1916: 111.

Genus MYDAEINA Malloch

Mydaeina Malloch, 1919f: 62. Type-species, *obscura* Malloch (mon.).
obscura Malloch, 1919f: 62.—N.W.T.

Genus GYMNODIA Robineau-Desvoidy

Gymnodia Robineau-Desvoidy, 1863b: 635. Type-species, *pratensis* Robineau-Desvoidy (mon.)=*polystigma* (Meigen).
Eulimnophora Malloch, 1920a: 145. Type-species, *Limnophora arcuata* Stein (orig. des.).

arcuata (Stein), 1898: 201 (*Limnophora*).—Ga.; N. Dak. to Ont., s. to Calif. and Fla.
cilifera (Malloch), 1920a: 145 (*Eulimnophora*).—Ill.; Ariz. to Pa., s. to Fla.
debilis (Williston), 1896c: 369 (*Limnophora*).—St. Vincent I.; Tex., Fla., West Indies.
  dorsovittata Malloch, 1920a: 146 (*Eulimnophora*).—Jamaica.

humilis (Zetterstedt), 1860: 6221 (*Aricia*).—Sweden; Mich., Ont., Europe.

Genus LIMNOPHORA Robineau-Desvoidy

Limnophora Robineau-Desvoidy, 1830: 517. Type-species, *palustris* Robineau-Desvoidy (Coquillett, 1910b: 561).

discreta Stein, 1898: 204.—Ill.; Alaska to Ariz. and N. Mex., B.C. to Que., s. to Ill. and Fla.
groenlandica Malloch, 1920a: 147.—w. Greenland.
incrassata Malloch, 1919e: 299.—Calif.; Alaska, Wash., Mont., Oreg., Wyo.
narona (Walker), 1849: 945 (*Anthomyia*).—Fla.; Wash. and Alta. to Que., s. to Ariz. and Fla., Bermuda, Bahama Is.
    dentata Bigot, 1885a: 284 (*Homalomyia*).—Rocky Mts.
    cyrtoneurina Stein, 1898: 203.—Wash., N. Dak., S. Dak., Kans., Minn., Ill., Mass., Ga.
sinuata Collin, 1930a: 256.—e. Greenland.
uniseta Stein, 1916: 94.—Sweden; Alaska to Greenland, also B.C., Wash., n. Europe.

Genus PSEUDOLIMNOPHORA Strobl

Pseudolimnophora Strobl, 1893b: 272. Type-species, *Musca triangula* Fallén (Coquillett, 1901e: 140).

nigripes (Robineau-Desvoidy), 1830: 541 (*Limosia*).—France; Alaska to Greenland, s. to Wash., Que., and Labr., also S. Dak., Europe.
rotundata Collin, 1930a: 256.—w. Greenland.
triangula (Fallén).—Not Nearctic.

Genus BUCEPHALOMYIA Malloch

Bucephalomyia Malloch, 1918a: 273. Type-species, *Tetramerinx femorata* Malloch (mon.).

femorata (Malloch), 1913m: 603 (*Tetramerinx*).—Calif.; Ariz.

Subfamily MYDAEINAE

REFERENCES: Hennig, 1955–1957, 1958b (rev., Palaearctic).

Genus HELINA Robineau-Desvoidy

Helina Robineau-Desvoidy, 1830: 493. Type-species, *euphemioidea* Robineau-Desvoidy (Coquillett, 1901e: 137) = *pertusa* (Meigen).

Aricia Robineau-Desvoidy, 1830: 486 (preocc. Savigny, 1822). Type-species, *impunctata* Robineau-Desvoidy (Coquillett, 1901e: 135) =*impuncta* (Fallén).
Spilogaster Macquart, 1835: 293. Type-species, *Musca quadrum* Fabricius (Westwood, 1840: 142).
Yetodesia Rondani, 1861a: 9 (n. name for *Aricia* Robineau-Desvoidy). Type-species, *Aricia impunctata* Robineau-Desvoidy (aut.) =*impuncta* (Fallén).
Hyetodesia, error or emend.
Mydaea, subg. *Spilaria* Schnabl, 1911: 96. Type-species, *Spilogaster pubescens* Stein (Séguy, 1937: 292).
Caricea, authors, not Robineau-Desvoidy.

> REFERENCE: Snyder, 1949a (rev.).

abiens (Stein), 1898: 193 (*Spilogaster*).—Ont., Mass.; Ariz., N. Dak. to s. Que., s. to N.C.
algonquina Malloch, 1922e: 96.—Ill.; N.C., Fla.
barpana (Walker), 1849: 933 (*Anthomyia*).—N.S.
basalis (Zetterstedt), 1838: 663 (*Anthomyza*).—Sweden; Alaska to N.W.T., s. to Calif. and Labr., Europe, U.S.S.R. (Kamchatka).
> *flavisquama* Zetterstedt, 1849: 3287 (*Aricia*).—Sweden
> *flavocalyptrata* Stein, 1920: 31 (*Mydaea*).—B.C.
> *hylemyioides* Malloch, 1920a: 137.—B.C.

bicolorata (Malloch) *in* Cole and Lovett, 1919: 253 (*Aricia*).—Oreg.; B.C., Wash., Calif.
> *aperta* Stein, 1920: 28 (*Mydaea*).—Wash., Calif.

bispinosa Malloch, 1920a: 142.—Ill.; B.C. and Alta., s. to Calif. and N. Mex.
bohemani (Ringdahl), 1916: 235 (*Mydaea*).—Sweden; B.C. to Man., s. to Calif. and N. Mex., also Ill., Lapland.
canadensis Snyder, 1949a: 136.—Ont.; Que., N.Y.
cinerella (Wulp), 1867: 150 (*Aricia*).—Wis.; Alaska to Labr., s. to Calif. and Mich., Netherlands, U.S.S.R. (Kamchatka).
> *vanderwulpi* Schnabl, 1888a: 387 (*Aricia*).—Netherlands.
> *brevis* Stein, 1898: 180 (*Aricia*).—Ill. Male only.
> *aldrichi* Snyder, 1949a: 122.—Alaska.

consimilata Malloch, 1920a: 144.—Mass.; Wash., Colo., Wis. to N.H. and Mass.
cothurnata (Rondani), 1866a: 116 (1866: 49) (*Spilogaster*).—Italy; Alaska to N.W.T., s. to Calif. and N.H., Europe, U.S.S.R. (Kamchatka).
> *obscuripes* Zetterstedt of Stein, 1916: 65.

cruciata Snyder, 1941: 9.—Ariz.

depuncta (Fallén).—Not Nearctic.
duplicata (Meigen), 1826: 92 (*Anthomyia*).—Germany; Alaska to Labr., s. to Calif. and N.Y., Europe.
exilis (Stein), 1920: 30 (*Mydaea*).—Mass.; Wis., Mich., N.Y., Que., Conn.
floridensis Snyder, 1949a: 126.—Fla.
fulvisquama (Zetterstedt), 1845: 1491 (*Aricia*).—Sweden; Alaska to Labr., s. to Calif. and N.H., Lapland, U.S.S.R. (Kamchatka).
    tuberculata Malloch, 1919l: 277.—Labr.
garretti Snyder, 1949a: 140.—B.C.
griseogaster Snyder, 1949a: 138.—Colo.
humilis (Stein), 1920: 32 (*Mydaea*).—Wash.; B.C. to Calif., Ariz., and Colo.
johnsoni Malloch, 1920a: 141.—Mass.; Wis., Que. and N.S., s. to Ga.
keremeosa Snyder, 1949a: 156.—B.C.
lasiosterna Snyder, 1941: 9.—Mexico; Ariz.
laxifrons (Zetterstedt), 1860: 6200 (*Aricia*).—Sweden; Alaska to Que., s. to Utah and Ga., Europe.
    nigripennis Schnabl, 1888a: 378 (*Aricia;* preocc. Walker, 1849).—Poland.
    nigricans Stein, 1898: 198 (*Spilogaster*).—Colo.
    tinctipennis Stein, 1916: 69 (*Mydaea;* n. name for *nigripennis* Schnabl).—Poland.
linearis Malloch, 1920a: 139.—Mont.; Wash., Alta., Mont., Minn. to Que., s. to Ill.
longicornis (Zetterstedt), 1838: 678 (*Anthomyza*).—Sweden; Y.T., N.W.T., Man., n. Europe.
luteisquama (Zetterstedt), 1845: 1492 (*Aricia*).—Sweden; Alaska to N.W.T., s. to Man. and Labr., Lapland.
maculipennis (Zetterstedt), 1845: 1475 (*Aricia*).—Sweden; Alaska, B.C. and Alta., s. to Calif. and N. Mex., also Man. to Labr., s. to Wis. and Mass., n. Europe, U.S.S.R. (Kamchatka).
    poeciloptera Malloch, 1918a: 271 (*Aricia;* preocc. Schiner, 1868).—N. Mex.
    neopoeciloptera Malloch, 1920a: 139 (n. name for *poeciloptera* Malloch).—N. Mex.
marguerita Snyder, 1949a: 137.—Wash.; Y.T., N.W.T., Ont., Labr.
mulcata (Giglio-Tos), 1893a: 7 (*Hyetodesia*).—Mexico; ?B.C., ?Kans.
multiseriata Malloch, 1922e: 95.—Wash.; B.C. and Alta. to Utah and Colo.

nigribasis Malloch, 1920a: 143.—Ill.; Alaska to Ariz., Wis. and Ill. to Md., s. to Fla.
    anceps Zetterstedt of Stein, 1904: 449.
    copiosa Wulp of Stein, 1920: 41.
nigripennis (Walker), 1849: 929 (*Anthomyia*).—Ont.; Alaska to Calif. and Colo., also Man. to Nfld., s. to Mich.
    nitida Stein, 1898: 185 (*Aricia;* preocc. Macquart, 1851).—Ont.
    crepuscularis Stein, 1898: 201 (*Spilogaster*).—Colo.
nigrita Malloch, 1920a: 139.—Mont.; Alta. to Man., s. to Calif. and N. Mex.
nudibasis Snyder, 1949a: 147.—N. Mex.
obscurata (Meigen), 1826: 89 (*Anthomyia*).—Germany; Alaska to Labr., s. to Utah, Ill., and N.Y., Europe, U.S.S.R. (Kamchatka).
    nasoni Malloch, 1920a: 138.—Ill.
obscuratoides (Schnabl), 1887: 347 (*Aricia*).—Poland; Alta., Man., S. Dak., Europe.
obscurinervis (Stein), 1898: 199 (*Spilogaster*).—Pa., Ga.; Oreg., Calif., Kans. to Que., s. to Tex. and Fla.
orbitaseta (Stein), 1898: 186 (*Aricia*).—Idaho; Wash. to Calif. and Ariz., also Mo.
oregonensis (Malloch) *in* Cole and Lovett, 1919: 254 (*Aricia*).—Oreg.; B.C., Idaho, Utah.
pectinata (Johannsen), 1916: 392 (*Mydaea*).—N.Y.; B.C. to Man., s. to Ariz. and Ark., also Que., N.Y., N.H. ?=*mulcata*.
    biseriata Stein, 1920: 29 (*Mydaea*).—N.Y.
    mimetica Malloch, 1920a: 142.—N.H.
    brevis (in part) Stein, 1898: 180 (*Aricia*). Female only (Ark.).
procedens (Walker), 1861: 315 (*Aricia*).—Mexico; B.C. and Alta. to Mexico, Minn. to Ill., Que. to N.C.
    uniseta Stein, 1898: 192 (*Spilogaster*).—Ill., Ont., Mass., Conn.
punctata (Robineau-Desvoidy), 1830: 492 (*Rohrella*).—France; S. Dak. and Minn., s. to Colo. and Mo., Mass. to Ga., Europe.
    uliginosa Fallén, 1825: 81 (*Musca;* preocc. Linnaeus, 1767).—Sweden.
refusa (Giglio-Tos), 1893a: 8 (*Spilogaster*).—Mexico. Probably not Nearctic.
rothi Ringdahl, 1939: 150.—Europe; Alaska to Labr., s. to Calif. and N.Y., Europe, U.S.S.R. (Kamchatka).
    marmorata, authors, not Zetterstedt.

rufitibia (Stein), 1898: 181 (*Aricia*).—Kans., Ill. Pa., Ga.; Alta. to Que., s. to Ariz. and Ga.
spinicosta (Zetterstedt), 1845: 1641 (*Anthomyza*).—Sweden; Alaska, N.W.T., n. Europe, U.S.S.R. (Kamchatka).
spinilamellata Malloch, 1920a: 140.—Mont.
spinosa (Walker), 1849: 926 (*Anthomyia*).—Ont.; Alaska to Labr., s. to Calif. and N. Mex.
 latifrontata Malloch, 1918a: 270 (*Aricia*).—N. Mex.
spuria Malloch, 1920a: 144.—Calif.; Oreg., Colo.
squalens (Zetterstedt), 1838: 669 (*Anthomyza*).—Sweden; Alaska to Labr., s. to Man., Lapland.
toga Snyder, 1949a: 158.—Mich.
troene (Walker), 1849: 936 (*Anthomyia*).—N.S.; B.C. to Nfld., s. to Calif. and Ga.
 lysinoe Walker, 1849: 938 (*Anthomyia*).—N.S.
 amoeba Stein, 1898: 190 (*Spilogaster*).—Idaho, Ill., Mass.
 pubiceps Stein, 1898: 194 (*Spilogaster*).—Idaho.
 var. **fulviventris** Bigot, 1885a: 291 (*Spilogaster*; as sp.).—Calif.; Wash., Idaho.
 varia Stein, 1920: 36 (*Mydaea*).—Wash., Calif.
ute Snyder, 1949a: 155.—Utah.
villihumilis Snyder, 1949a: 156.—Calif.; B.C.

Genus QUADRULARIA Huckett

Aricia, subg. *Quadrula* Pandellé, 1898: 51 (preocc. Rafinesque, 1820). Type-species, *Anthomyza annosa* Zetterstedt (Coquillett, 1901e: 141).
Quadrularia Huckett (for *Quadrula* Pandellé). Type-species, *Anthomyza annosa* Zetterstedt (aut.). **N. name.**
Spilaria, authors, not Schnabl.

 REFERENCE: Malloch, 1921d (rev.).

annosa (Zetterstedt), 1838: 663 (*Anthomyza*).—Sweden; Alaska to Labr., s. to Ariz. and N.Y., Europe. **N. comb.**
 multisetosa Strobl, 1898a: 238 (*Aricia*).—Austria.
laetifica (Robineau-Desvoidy), 1830: 500 (*Mydina*).—France; Alaska to Labr., s. to Calif. and Tenn., Europe, U.S.S.R. (Kamchatka). **N. comb.**
 lucorum Fallén, 1823f: 55 (*Musca*; preocc. Linnaeus, 1758).—s. Sweden.
 pylone Walker, 1849: 928 (*Anthomyia*).—N. Amer.
 incerta Walker, 1852: 354 (*Anthomyia*).—U.S.
 solita Walker, 1852: 354 (*Anthomyia*).—U.S.
 abacta Giglio-Tos, 1893a: 8 (*Hyetodesia*).—Mexico.

var. **nivalis** (Zetterstedt), 1838: 663 (*Anthomyza*; as sp.).—
Sweden; Alta., Mich. to Que., e. Tenn., Lapland.
N. comb.

punctata (Stein), 1898: 182 (*Aricia*).—S. Dak., Colo.; B.C. to Alta. and N. Dak., s. to Ariz. and N. Mex. **N. comb.**

Genus ARICIELLA Malloch

Ariciella Malloch, 1918b: 66. Type-species, *flavicornis* Malloch (orig. des.)=*rubripalpis* (Wulp).

rubripalpis (Wulp), 1896: 320 (*Spilogaster*).—Mexico; Tex.
flavicornis Malloch, 1918b: 66.—Tex.

Genus HEBECNEMA Schnabl

Hebecnema Schnabl, 1889: 331. Type-species, *Anthomyia umbratica* Meigen (Coquillett, 1901e: 137).
Xenaricia Malloch, 1918a: 272. Type-species, *Spilogaster fulvus* Bigot (orig. des.).

REFERENCE: Malloch, 1921e (key).

affinis Malloch, 1921e: 214.—Vt.; Alaska to Calif., Mich. to Que. and Vt., Europe.

fulva (Bigot), 1885a: 289 (*Spilogaster*).—Wash. Territory; Alaska, Calif.

nigricolor (Fallén), 1825: 81 (*Musca*).—Sweden; Alaska to Oreg., also N. Dak., Mich., Mass., Europe.
pallipes Malloch, 1920j: 269.—Alaska.

umbratica (Meigen), 1826: 88 (*Anthomyia*).—Germany; Alaska to Que., s. to Colo. and Ga., Europe.

vespertina (Fallén), 1823f: 58 (*Musca*).—Sweden; Alaska to Labr., s. to Ariz. and Ga., Europe.

Genus MYOSPILA Rondani

Myospila Rondani, 1856: 91. Type-species, *Musca meditabunda* Fabricius (orig. des.).
Myiospila, error or emend.

meditabunda (Fabricius), 1781: 444 (*Musca*).—Denmark, Italy; Alaska to N.W.T., s. to Calif. and Ga., Europe, U.S.S.R. (Kamchatka).
quadrisetosa Thomson, 1869: 549 (*Cyrtoneura*).—Calif.
quadrisignata (error) Coquillett, 1900h: 441 (1904: 55).
anthomydea Bigot, 1887c: clxxxii (also 1887d: 614) (*Curtonevra*).—U.S. (Rocky Mts.).

Genus MYDAEA Robineau-Desvoidy

Mydaea Robineau-Desvoidy, 1830: 479. Type-species, *scutellaris* Robineau-Desvoidy (Coquillett, 1901e: 139) = *pagana* (Fabricius).

Opsolasia Coquillett, 1910b: 580. Type-species, *Lasiops calvicrura* Coquillett (orig. des.) = *orichalcea* (Zetterstedt).

REFERENCES: Malloch, 1923e (synopsis); Snyder, 1949b (rev.).

brevipilosa Malloch, 1920a: 135.—Ill.; Alta. to Calif., Wis. to Que., s. to Ga.

consors (Zetterstedt).—Not Nearctic.

discimana Malloch, 1920a: 136.—Mass.; Alaska to Labr., s. to B.C., Colo., and N.C.

discimanoides Snyder, 1949b: 19.—Ariz.

electa (Zetterstedt), 1860: 6271 (*Anthomyza*).—Sweden; Alaska, B.C., N.Y., N.H., N.C., Europe.

flavicornis Coquillett, 1902g: 123.—Que.; Que. to Okla. and N.C.

furtiva Stein, 1920: 32.—Wash.; Alaska to Labr., s. to Calif. and Mich.
  persimilis Malloch, 1920a: 134.—Alta.

impedita Stein, 1920: 33.—Mass.; Mass. to Ga.
  flavidipalpis Malloch, 1923e: 221.—Md.

narona Snyder, 1949b: 23.—Md.

neglecta Malloch, 1920a: 136.—N.J.; Ont. and Que., s. to Okla. and Ga.

neobscura Snyder, 1949b: 26.—N.C.; Ind., Wis., N.H. to N.C.
  obscura, authors, not Stein.

nubila Stein, 1916: 65 (n. name for *obscura* Stein).—Pa.; B.C. and Alta., s. to Oreg., N. Mex., Okla., Mich. to Que., s. to Md.
  obscura Stein, 1898: 197 (*Spilogaster;* preocc. Wulp, 1896).—Pa.
  hirtiventris Malloch of Johnson, 1925b: 227.

obscurella Malloch, 1921b: 10.—Unknown (?Idaho); Alaska, N.W.T., Mich. to Que., Tenn.
  anicula Zetterstedt of Strickland, 1946: 170.

occidentalis Malloch, 1920a: 134.—N.H.; Alaska to N.W.T., s. to Utah and N. Mex., Minn., Mich., Que. to Tenn.
  humeralis (in part) Zetterstedt, 1845: 1697 (also 1860: 6284) (*Anthomyza*). N.Y. specimen only.
  pagana, authors, not Fabricius.
  tincta, authors, not Zetterstedt.

orichalcea (Zetterstedt), 1849: 3285 (*Aricia*).—Sweden; Alaska to Labr., s. to Oreg. and Colo., Europe, U.S.S.R. (Kamchatka).

calvicrura Coquillett, 1900h: 444 (1904: 58) (*Lasiops*).—Alaska (Popof I.).

astuta Stein, 1920: 29.—Colo.

palpalis Stein, 1916: 56.—U.S.S.R. (Kanin Peninsula); Alaska to Labr., s. to N. Mex., n. Europe.

hannai Malloch, 1921d: 109 (*Helina*).—Alaska (Pribilof Is.).

occimons Snyder, 1949b: 6.—N. Mex.

rugia Walker of Malloch, 1921p: 178.

sootryeni Ringdahl, 1928c: 45.—Norway; Y.T., Que.

urbana (Meigen), 1826: 118 (*Anthomyia*).—Germany; Mich., Que. to N.C., Europe.

winnemana Malloch, 1919b: 133.—Md.; N.H. to Fla.

Genus XENOMYDAEA Malloch

Xenomydaea Malloch, 1920a: 144. Type-species, *buccata* Malloch (orig. des.) = *otiosa* (Stein).

REFERENCE: Snyder, 1949b (rev.).

armatipes (Malloch), 1921b: 10 (*Mydaea*; n. name for *armata* Malloch).—Mont.; Y.T., Oreg. and Mont., s. to Ariz. and N. Mex.

armata Malloch, 1920a: 135 (*Mydaea*; preocc. Stein, 1918).—Mont.

armipes (error for *armata*) Malloch, 1921b: 10.

fuscomarginata (Malloch), 1919e: 298 (*Helina*).—Calif.

pulla Stein, 1920: 35 (*Mydaea*).—Calif.

hirtiventris (Malloch), 1920j: 269 (*Mydaea*).—Alaska.

nudiseta (Stein), 1920: 33 (*Mydaea*).—Idaho; Wash.

otiosa (Stein), 1920: 34 (*Mydaea*).—Calif., Idaho; Alaska to Calif. and Colo., also N.H.

buccata Malloch, 1920a: 144.—Mont.

pogonoides Snyder, 1949b: 34.—Wash.

rufinervis (Pokorny), 1889: 554 (*Spilogaster*).—Italy-Switzerland (Stelvio Pass); Alaska.

Subfamily FANNIINAE

REFERENCES: Hennig, 1955, 1956 (rev., Palaearctic); Chillcott, 1961a (rev.).

Genus FANNIA Robineau-Desvoidy

Fannia Robineau-Desvoidy, 1830: 567. Type-species, *saltatrix* Robineau-Desvoidy (mon.) = *scalaris* (Fabricius).

Anthomyia, subg. *Homalomyia* Bouché, 1834: 89. Type-species, *Musca canicularis* Linnaeus (Westwood, 1840: 143).
Steinomyia Malloch, 1912c: 656. Type-species, *steini* Malloch (orig. des.) = *scalaris* (Fabricius).

Some species of *Fannia* are closely associated with man. In many subarctic regions, *canicularis* (L.) completely replaces *Musca domestica* L. as the common housefly. The most usual larval habitat is in mushrooms and related fungi, but the larvae of many widely distributed species are found in decaying plant and animal matter, including excrement on the ground and occasionally in birds' nests.

REFERENCES: Malloch, 1924a, 1924b (key, notes).

abrupta Malloch, 1924a: 423.—N.H.; Alaska, Labr. to N.Y., also N.C.
aethiops Malloch, 1913l: 628.—N.H.; Alaska to Labr., s. to Calif. and N.C.
alaskensis Chillcott, 1961a: 198.—Alaska.
americana Malloch, 1927a: 176.—Va.; Minn. to Que., s. to Mo. and Ga.
arizonensis Chillcott, 1961a: 160.—Ariz.
armata (Meigen), 1826: 139 (*Anthomyia*).—Germany; ?Greenland, Europe.
atra (Stein), 1895: 125 (*Homalomyia*).—Austria; B.C., Alta., Que., N.Y.
atripes Stein, 1916: 79.—Europe; Y.T. and N.W.T., s. to Alta. and Labr., Europe.
benjamini Malloch, 1913l: 625.—Calif.; Idaho and Wyo., s. to Calif. and Ark., Mexico.
bifimbriata Collin *in* Carpenter, 1938: 550.—w. Greenland; Alaska, Y.T., B.C., Wis. to Que., s. to Ga.
    enotahensis Seago, 1953: 141.—Ga.
bigelowi Chillcott, 1961a: 115.—N.W.T.; Alaska.
binotata Chillcott, 1961a: 203.—Mexico; Oreg., Calif.
borealis Chillcott, 1961a: 127.—Alta.
bradorei Chillcott, 1961a: 173.—Que.
brevicauda Chillcott, 1961a: 106.—Calif.; B.C. and Alta., s. to Calif. and Wyo.
brevipalpis Chillcott, 1961a: 133.—N.Y.; Wis. to N.S., s. to Kans. and Ga.
brooksi Chillcott, 1961a: 117.—Ont.; Mich. to Maine, s. to Mo. and Va.
canadensis Malloch, 1924a: 423.—Ont.; Alta. to N.S., s. to S. Dak., Ill., and Conn.
    nigra Stein, 1920: 43 (preocc. Malloch, 1910).—Que.

  immaculata Malloch, 1924a: 424 (n. name for *nigra* Stein).—Que.
canicularis (Linnaeus), 1761: 454 (*Musca*; unjustified n. name for *lateralis* Linnaeus).—Europe; Alaska to Greenland, s. to Calif. and Fla., cosmopolitan.
  lateralis Linnaeus, 1758: 597 (*Musca*).—Europe. [Cited as an error by Linneaus, 1761: 454, in synonymy with *canicularis*. Application should be made to the I.C.Z.N. to suppress this name for purposes of homonymy and synonymy in order to preserve the well-known name *canicularis*.—Eds.]
  isura Walker, 1849: 952 (*Anthomyia*).—N.S.
  prunivora Walsh, 1870: 138 (*Homalomyia*).—Ill. Larva only.
carbonaria (Meigen), 1826: 154 (*Anthomyia*).—Germany; B.C., Wash., Europe.
ceringogaster Chillcott, 1961a: 202.—Que.; Maine, N.S.
ciliatissima Chillcott, 1961a: 65.—Que.; Alaska, s. to Colo., also B.C. to Labr. and Nfld., s. to Pa.
columbiana Chillcott, 1961a: 67.—B.C.
conspicua Malloch, 1913l: 624.—Ariz.
coracina (Loew), 1873a: 47 (*Homalomyia*).—Hungary; Calif., Minn. to Ont., s. to Conn.
corvina (Verrall), 1892: 149 (*Homalomyia*).—England; Alaska to Que., s. to Kans., Ill., and Ga., Europe.
  carbonaria, authors, not Meigen.
curvipes Malloch, 1924a: 421.—Md.; Ind., Ohio, N.Y. to Ga.
  fasciculata, authors, not Loew.
depressa (Stein), 1898: 173 (*Homalomyia*).—Mass.; Alaska, Man., Que. to Mo. and Ga.
difficilis (Stein), 1895: 58 (*Homalomyia*).—Austria; Alaska, N.W.T., Colo., Minn. to Mass., s. to Ga., Europe.
elongata Chillcott, 1961a: 149.—Calif.
enigmata Chillcott, 1961a: 107.—Alaska; B.C.
falcata Chillcott, 1961a: 91.—Man.; N.W.T., s. to Colo., B.C. to Labr.
femoralis (Stein), 1898: 282 (*Homalomyia*).—La.; Mont., Calif. to, W. Va., s. to La. and Ga.
flavibasis (Stein), 1898: 171 (*Homalomyia*).—Ill.; Mont., Wis., Mich.
flavipalpis Stein, 1911: 103.—Chile. Probably not Nearctic.
flavitibia Stein, 1920: 42.—Idaho; se. Alaska to Calif. and Utah.
  minuta Stein, 1920: 44.—Wash.
fuscitibia Stein, 1920: 42.—Idaho; s. Alaska to Idaho and Alta., Calif., N. Mex.

fuscula (Fallén), 1825: 86 (*Musca*).—Sweden; Alaska to N.W.T. and Que., s. to Calif. and Ga., n. Europe.
   tetracantha Loew, 1872a: 93 (Cent. 10, no. 69) (*Homalomyia*).—Middle States.
garretti Chillcott, 1961a: 128.—B.C.; ?Conn.
genualis (Stein), 1895: 126 (*Homalomyia*).—Germany; Alaska and Y.T. to Colo., Labr., Que., N.H., Europe.
glaucescens (Zetterstedt), 1845: 1586 (*Aricia*).—Sweden; Alaska to N.S., s. to Iowa, Europe.
hinei Chillcott, 1961a: 93.—Alaska.
hirticeps (Stein), 1892: 70 (*Homalomyia*).—Germany; Alaska.
howardi Malloch, 1913l: 626.—D.C.; Kans. to Iowa and Conn., s. to La. and Ga.
   brevis Rondani of Stein, 1898: 176.
   trimaculata, authors, not Stein.
immutica Collin, 1939: 142.—Scotland; Alaska, Wash., Oreg., Que.
incisurata (Zetterstedt), 1838: 679 (*Anthomyza*).—Sweden; B.C. to Que., s. to Colo., Miss., and Va., Europe.
intermedia Chillcott, 1961a: 168.—Alaska; B.C., Alta.
kowarzi (Verrall), 1892: 149 (*Homalomyia*).—England; Alaska to Labr., s. to Colo. and N.C., Europe.
laevis (Stein), 1898: 174 (*Homalomyia*).—Mass.; B.C. and Alta., s. to Ariz. and Tex., Que. to Mass. and Pa.
lasiops Malloch, 1920a: 168.—Ill.; B.C., N. Mex., Kans., Mich., Ont.
latifrons Malloch, 1914g: 240.—Ill.; Minn., Iowa, Ont.
leidyi (Walsh), 1870: 138 (*Homalomyia*).—Pa. Larva only. Unrecognized.
leucogaster Chillcott, 1961a: 148.—Utah; Oreg., Idaho.
leucosticta (Meigen), 1838: 328 (*Anthomyia*).—Germany; Wash., Oreg., Kans. to Wis., Ont. and N.Y., s. to Ga., Europe.
   brevis Rondani, 1866a: 132 (1866: 65) (*Homalomyia*).—Malta, Italy.
lucida Chillcott, 1961a: 129.—Sask.; Alaska to Wash. and Nebr., Alta. to Que.
lugubrina (Zetterstedt), 1838: 687 (*Anthomyza*).—Sweden; Alaska to Labr., s. to Colo., n. Europe.
macalpinei Chillcott, 1961a: 175.—Ont.; Que.
manicata (Meigen), 1826: 140 (*Anthomyia*).—Germany; Alaska to Greenland, s. to Colo. and Ga., Europe.
   uxama Walker, 1849: 948 (*Anthomyia*).—Ont.
   acra Walker, 1849: 951 (*Anthomyia*).—Ont.
melanura Chillcott, 1961a: 125.—Alaska; B.C. and Wash. to Que.
meridionalis Chillcott, 1961a: 171.—Ind.; N. Mex., Minn. to Que., s. to Mo. and Tenn.

minutipalpis (Stein), 1895: 106 (*Homalomyia*).—Germany; Alaska, Y.T., n. B.C., Europe.
morrisoni Malloch, 1913l: 627.—N.H.; Que., Mont., Colo., N. Mex.
  plebia Malloch, 1918a: 293.—N. Mex.
multisetosa Chillcott, 1961a: 108.—Ariz.
mutica (Zetterstedt), 1845: 1580 (*Aricia*).—Sweden; Alaska to Oreg. and Colo., Mich., Que., N.Y.
neomexicana Chillcott, 1961a: 147.—N. Mex.
neopolychaeta Chillcott, 1961a: 136.—Sask.; B.C., s. to Ariz. and Tex. and e. to Minn.
neotomaria Chillcott, 1961a: 159.—N. Mex.
nidicola Malloch, 1927b: 92.—Md.; Alaska to N.W.T., s. to Utah and Ga., Europe.
  subpubescens Collin, 1958: 89.—England.
  pubescens, authors, not Stein.
ochrogaster (Thomson), 1869: 557 (*Anthomyia*).—Calif.; B.C. to Oreg. and Colo.
  splendida Stein, 1898: 170 (*Homalomyia*).—Idaho.
operta Chillcott, 1961a: 160.—Calif.
oregonensis Chillcott, 1961a: 83.—Oreg.; Y.T. to Alta. and Ariz.
ornata (Meigen), 1826: 191 (*Anthomyia*).—Germany; ?se. Alaska.
pallidiventris Malloch, 1924a: 422.—Mass.; Alta., Minn. to Que., s. to Ill. and Mass., also Ga.
pellucida (Stein), 1898: 283 (*Homalomyia*).—Ga.; Wash. to Calif. and N. Mex., N.Y. to Ga.
penepretiosa Chillcott, 1961a: 144.—Que.; Alaska and B.C. to Que. and N.Y., also N.C.
  pretiosa, authors, not Schiner.
polychaeta (Stein).—Not Nearctic.
postica (Stein), 1895: 89 (*Homalomyia*).—England; Alaska to Calif. and Colo., N.W.T. to Nfld. to N.C.
presignis Chillcott, 1961a: 78.—Ont.; Labr.
pusio (Wiedemann), 1830: 437 (*Anthomyia*).—S. Amer.; Ariz. to Fla., N.Y. to D.C., Cuba, Bermuda, S. Amer.
  femorata Loew, 1861c: 42 (*Homalomyia*).—Cuba.
robusta Chillcott, 1961a: 220.—B.C.; Alaska to Wash., B.C. to Que.
rondanii (Strobl), 1893b: 241 (*Homalomyia;* n. name for *carbonaria* Rondani).—Italy; Alaska to Idaho and Wyo., Mich. to Que., s. to Tenn. and N.C., Europe. [The identity of the type specimen of *rondanii* is in doubt; further study may show this species to require a new name. The name *aerea* (Zetterstedt) may not be used; Article 49 of the International Rules of Zoological Nomenclature prohibits the reuse of a name for a misidentification.]

carbonaria Rondani, 1871a: 324 (*Homalomyia;* preocc. Meigen, 1826).—Italy.
aerea Fallén of Meigen, 1826: 157, and authors.
aerea Meigen of Zetterstedt, 1845: 1605, and authors.
aerea Zetterstedt of Chillcott, 1961a: 181.
scalaris (Fabricius), 1794: 332 (*Musca*).—Denmark; all states and provinces n. of Mexico, Europe.
propinqua Macquart, 1851: 241 (1851: 268) (*Anthomyia*).— N.Y. (as Amer.; Séguy, 1938: 120).
scyphocerca Chillcott, 1961a: 118.—Que.; Alaska, Man.
sequoiae Chillcott, 1961a: 112.—Calif.; B.C.
serena (Fallén), 1825: 76 (*Musca*).—Sweden; Alaska to Calif. and Colo., Mich. to Labr., s. to N.C., Europe.
serrata Chillcott, 1961a: 184.—N.Y.
setifer Chillcott, 1961a: 162.—Tex.
snyderi Seago, 1954: 2.—Md.; Conn., W. Va., Tenn., La., Ga.
sociella (Zetterstedt), 1845: 1564 (*Aricia*).—Denmark; Alaska and N.W.T., s. to N. Mex. and Colo., also Wis. to Labr., s. to Ga., n. Europe.
socialis (error) Malloch, 1924a: 419.
parallela Johannsen, 1916: 390.—N.Y.
spathiophora Malloch, 1918a: 294.—Ont.; Alaska to w. N.W.T., s. to Ariz. and N. Mex., Minn. to Labr., s. to N.C., n. Europe.
tescorum Chillcott, 1961a: 161.—N. Mex.; Ariz.
tibialis Malloch, 1913k: 461.—B.C.; Alaska to sw. N.W.T., s. to Calif. and Wyo.
trianguligera Malloch, 1918a: 292.—N. Mex.; Ariz., Tex.
trigonifera Chillcott, 1961a: 137.—Calif.; Wash. to s. Calif.
tuberculata (Zetterstedt), 1849: 3296 (*Aricia*).—Denmark; Alaska to N.W.T., Denmark.
tuberosa (Curtis), 1845: 817 (*Anthomyia*).—Great Britain; ?N. Amer. ?=*canicularis*.
tundrarum Chillcott, 1961a: 116.—N.W.T.
ungulata Chillcott, 1961a: 98.—Kans.; Alaska to Wyo., Alta. to Ont., s. to Kans. and N.C.
variabilis Chillcott, 1961a: 196.—Idaho; B.C., Ariz., Ont., N.J., Ga.
vesparia (Meade), 1891: 42 (*Homalomyia*).—England; Calif., Europe.
wilsoni (Walsh), 1870: 138 (*Homalomyia*).—Mass. Larva only. Unrecognized.

Genus COELOMYIA Haliday

Coelomyia Haliday *in* Westwood, 1840: 143. Type-species, *mollissima* Haliday (mon.).

mollissima Haliday *in* Westwood, 1840: 143.—Unknown (?n. Ireland); Alaska, Y.T., N.W.T., Europe.
   spathulata Zetterstedt, 1845: 1543 (*Aricia*).—Sweden.
subpellucens (Zetterstedt), 1845: 1561 (*Aricia*).—Sweden; Alaska to Labr., s. to Calif. and N.C., n. Europe, U.S.S.R. (Kamchatka).
   flavivaria Coquillett, 1900h: 446 (1904: 60) (*Homalomyia*).—se. Alaska.

Genus EURYOMMA Stein

Euryomma Stein, 1899: 19. Type-species, *hispaniense* Stein (orig. des.) = *peregrinum* (Meigen).
americanum Chillcott, 1961a: 226.—Calif.
peregrinum (Meigen), 1826: 187 (*Anthomyia*).—Germany; B.C., Calif., Ill. to Que., s. to Ala. and Va., cosmopolitan.
   communis Walker, 1852: 366 (*Anthomyia*).—U.S.

Genus PIEZURA Rondani

Piezura Rondani, 1866a: 71, 122 (1866: 4, 55). Type-species, *pardalina* Rondani (orig. des.) = *graminicola* (Zetterstedt).
graminicola (Zetterstedt), 1846: 1747 (*Anthomyza*).—Sweden; Mich., N.H., Europe.
nearctica Chillcott, 1961a: 230.—Maine.

Genus PLATYCOENOSIA Strobl

Platycoenosia Strobl, 1894a: 72. Type-species, *mikii* Strobl (mon.).
Choristomma Stein, 1895: 138. Type-species, *pokornyi* Stein (mon.) = *mikii* Strobl.
mikii Strobl, 1894a: 72.—Germany; Wash., Oreg., Wis. to Maine, s. to La. and Tenn., Europe.
   pokornyi Stein, 1895: 138 (*Choristomma*).—Germany.

Genus AZELIA Robineau-Desvoidy

Azelia Robineau-Desvoidy, 1830: 592. Type-species, *florea* Robineau-Desvoidy (Rondani, 1866a: 72, 135; 1866: 5, 68) ?= *triquetra* (Wiedemann).
Atomogaster Macquart, 1835: 329. Type-species, *Anthomyia triquetra* Wiedemann (orig. des.).

   REFERENCE: Hennig, 1956 (rev., Palaearctic).

aequa Stein, 1920: 46.—Wash.
cilipes (Haliday), 1838: 185 *(Anthomyia)*.—Unknown (?n. Ireland); Alaska to Ariz. and Colo., Ill. to Que., s. to Ga., Europe.
  staegeri Zetterstedt, 1845: 1592 *(Aricia)*.—Sweden.
  maculiventris Malloch, 1918a: 276 *(Trichopticus)*.—Idaho.
  pretiosina Curran, 1930j: 86 *(Fannia)*.—N.Y.
gibbera (Meigen), 1826: 152 *(Anthomyia)*.—Germany; Alaska to Oreg., Ill., Que. to Tenn., Europe.
triquetra (Wiedemann), 1817: 85 *(Anthomyia)*.—Germany; Alaska to N.W.T. and Que., s. to Colo. and Mass., Europe.
zetterstedtii Rondani, 1866a: 135 (1866: 68).—Denmark; Alaska, Europe.

Subfamily PHAONIINAE

Genus HYDROTAEA Robineau-Desvoidy

Hydrotaea Robineau-Desvoidy, 1830: 509. Type-species, *Musca meteorica* Linnaeus (Curtis, 1839: pl. 768).
Achaetina Malloch, 1918b: 67. Type-species, *Musca ciliata* Fabricius (orig. des.) =*bimaculata* (Meigen).

Adults of some species of *Hydrotaea* annoy man by feeding on his skin secretions. The larvae are commonly found in association with decaying vegetable and animal matter; some larvae feed on excrement, and several species have been reared from birds' nests.

  REFERENCE: Huckett, 1954 (rev.).

acuta Stein, 1898: 167.—Ga.; Calif., Mo., Ariz., Fla.
  dissimilis Aldrich, 1926d: 5.—Calif.
armipes (Fallén) 1825: 75 *(Musca)*.—Sweden; B.C. to Labr., s. to Ariz. and N.C., Europe.
basdeni Collin, 1939: 135.—England; Iowa to Que., s. to Md.
bimaculata (Meigen), 1826: 160 *(Anthomyia)*.—Germany; Alaska, N.W.T., Greenland, n. Europe.
  ciliata Fabricius, 1794: 333 *(Musca;* preocc. Müller, 1776).—Germany.
  spinipes Fallén, 1823f: 61 *(Musca;* preocc. Scopoli, 1763).—Sweden.
bispinosa (Zetterstedt), 1845: 1428 *(Aricia)*.—Greenland, Sweden; Alaska to Greenland, B.C., Oreg., boreal Europe.
cressoni Malloch, 1914e: 172.—N. Mex.; B.C. to N. Mex.
cristata Malloch, 1918f: 93.—Mass.; Alaska to Calif. and N. Mex., N.W.T. to Nfld.

dentipes (Fabricius), 1805: 303 (*Musca*).—Denmark; Alaska to N.W.T. and Que., Europe.
depressa Huckett, 1954: 327.—Calif.; Alta., Idaho.
houghi Malloch, 1916c: 110.—Ill.; Alaska to Nfld., s. to Calif. and Fla.
    caerulescens Stein, 1920: 41 (as *dentipes* var.).—Wash., Calif.
irritans (Fallén), 1823f: 62 (*Musca*).—Sweden. Probably not Nearctic.
lasiophthalma Malloch, 1919e: 297.—Calif.; Alta., Colo.
meteorica (Linnaeus), 1758: 597 (*Musca*).—Europe; s. B.C., s. to Ariz. and N. Mex., e. to Maine.
militaris (Meigen), 1826: 136 (*Anthomyia*).—Germany; s. Alaska to Nfld., s. to N. Mex. and Ga., Europe.
    impexa Loew, 1873c: 243.—Germany.
nidicola Malloch, 1925b: 184.—N.Y.; Que., Europe.
occulta (Meigen), 1826: 133 (*Anthomyia*).—Germany; Alaska to Labr., s. to Calif. and Ga., Europe.
    idyla Walker, 1849: 948 (*Anthomyia*).—Ont.
    lata Walker, 1849: 963 (*Eriphia*).—Ont.
palaestrica (Meigen), 1826: 135 (*Anthomyia*).—Germany; B.C., s. to Ariz. and N. Mex. and e. to N.S., Europe.
pilipes Stein, 1903: 312.—nw. U.S.S.R.; Alaska, s. to Idaho, Ind., and N.H., U.S.S.R. (Kamchatka).
    orbitalis Aldrich, 1918b: 311.—Ind.
pilitibia Stein, 1916: 73.—Sweden; Alaska to N.W.T., s. to Calif. and Colo., Mich. to Que., s. to Tenn. and N.C.
    abdominalis Aldrich, 1926d: 6.—B.C.
ringdahli Stein, 1916: 74.—Sweden; Alaska to Que., s. to Wash. and N.Y.
    comata Aldrich, 1918b: 312.—Wash.
scambus (Zetterstedt), 1838: 668 (*Anthomyza*).—Sweden; Alaska to Calif. and N. Mex., N.W.T. to Que., s. to Mich., n. Europe.
spinifemorata Huckett, 1954: 328.—Alaska; B.C., Wash., Alta., Mich.
succedens Stein, 1901: 212.—s. Ga.
tuberculata Rondani, 1866a: 79 (1866: 12).—Italy; B.C. to Man., s. to Idaho and S. Dak., Ill. to Que., s. to W. Va., Europe.
    metatarsata Stein, 1898: 166.—Pa., Mass.
unispinosa Stein, 1898: 165.—Colo., Ont.; Mont. to Colo., Minn to Que., s. to Ky.

Genus NEOHYDROTAEA Malloch

Neohydrotaea Malloch, 1924h: 202. Type-species, *hirtipes* Malloch (orig. des.).

hirtipes Malloch, 1924h: 202.—Mass.; N.Y.

Genus TRICHOPTICOIDES Ringdahl

Trichopticoides Ringdahl, 1931: 173. Type-species, *Musca decolor* Fallén (mon.).

Unnamed spp.—Alaska and Y.T. to Colo., Alta. to N.B., s. to Mich. and N.Y.

Genus OPHYRA Robineau-Desvoidy

Ophyra Robineau-Desvoidy, 1830: 516. Type-species, *nitida* Robineau-Desvoidy (Rondani, 1866a: 70, 84; 1866: 3, 17)=*leucostoma* (Wiedemann).

The larvae of *Ophyra* are scavengers in decaying vegetation, animal matter, and excrement.

REFERENCE: Sabrosky, 1949c (rev., Pacific).

aenescens (Wiedemann), 1830: 435 (*Anthomyia*).—La.; Oreg. to Ariz., Ill. to N.Y., s. to La. and Fla., ?Bermuda, Neotropical, Galapagos Is., Hawaii, Nauru and Ocean Is. (w. of Gilbert Is.).
  argentina Bigot of Johnson, 1913a: 76.
capensis (Wiedemann), 1818: 46 (*Anthomyia*).—S. Africa; N.Y., Europe, Asia, Africa. Immigrant.
chalcogaster (Wiedemann), 1824: 52 (*Anthomyia*).—Java; Bermuda, Hawaii, widespread in Ethiopian, Oriental, and Australian Regions.
leucostoma (Wiedemann), 1817: 82 (*Anthomyia*).—Germany; B.C. to Nfld., s. to Ariz. and Fla., Europe, China, Japan.
  opalia Walker, 1849: 956 (*Anthomyia*).—N.S.

Genus POGONOMYIA Rondani

Pogonomyia Rondani, 1871a: 336. Type-species, *alpicola* Rondani (mon.).
Eriphia, subg. *Neoeriphia* Schnabl and Dziedzicki, 1911: 195 (1911: 143). Type-species, *metatarsata* Stein (mon.).

REFERENCE: Malloch, 1918a (key, descr.).

aldrichi Malloch, 1918a: 281.—Idaho; Alta., Wash. to Calif., Wyo.
  unicolor Stein, 1920: 22.—Wash., Idaho.

alpicola Rondani, 1871a: 337.—Italy; Alta. to N. Mex., also Labr., cent. Europe.

aterrima Wulp.—Not Nearctic.

flavinervis Malloch, 1915h: 356.—Ill.; Alta., Sask., S. Dak., Wis. to s. Que.

    nitens Stein, 1898: 199 (*Spilogaster;* preocc. Macquart, 1855).—Ont. (as Mass., error).

    flavipennis Stein, 1920: 21.—Wis., Ill., Ont.

latifrons Malloch, 1918a: 281.—Colo.

meadei Pokorny.—Not Nearctic.

metatarsata (Stein), 1907: 326 (*Eriphia*).—Tibet. Probably not Nearctic.

minor Malloch, 1918a: 280.—N. Mex.; Alta. and Sask., s. to Calif. and N. Mex.

quadrisetosa Malloch, 1919f: 66.—N.W.T. (Victoria I.); cent. Alaska to Labr.

similis Malloch, 1918a: 279.—N. Mex.; Alaska to N.W.T. and Que., s. to N. Mex.

spinitarsis Aldrich, 1918d: 184.—Colo.

Genus EUPOGONOMYIA Malloch

Eupogonomyia Malloch, 1921p: 178. Type-species, *pribilofensis* Malloch (orig. des.).

    REFERENCE: Malloch, 1921p (key).

groenlandica (Lundbeck), 1901: 281 (*Ophyra*).—Greenland; Y.T., N.W.T. (Baffin and Ellesmere Is.).

neoborealis Snyder, 1949a: 122 (*Helina;* n. name for *borealis* Malloch).—N.W.T.; Alaska, N.W.T. (Victoria and Ellesmere Is.).

    borealis Malloch, 1919f: 64 (*Aricia;* preocc. Zetterstedt, 1838).—N.W.T.

pribilofensis Malloch, 1921p: 179.—Alaska (Pribilof Is.); Alaska, Y.T., N.W.T. (Prince Patrick, Victoria, and Baffin Is.).

Genus POGONOMYIOIDES Malloch

Pogonomyioides Malloch, 1919f: 67. Type-species, *atratus* Malloch (orig. des.)=*segnis* (Holmgren).

segnis (Holmgren), 1883: 169 (*Aricia*).—U.S.S.R. (Novaya Zemlya); Alaska to Greenland, s. to n. Man. and Que., Holarctic.

    atratus Malloch, 1919f: 67.—N.W.T.

    decolor Fallén of Nielsen, 1907: 397.

Genus BEBRYX Gistel

Eriphia Meigen, 1826: 206 (preocc. Latreille, 1817). Type-species, *cinerea* Meigen (mon.).
Bebryx Gistel, 1848: ix (n. name for *Eriphia* Meigen). Type-species, *Eriphia cinerea* Meigen (aut.).

Unnamed sp.—Alaska (Pribilof Is.), Y.T., n. N.W.T.
cinerea Meigen of Malloch, 1921p: 178.

Genus LASIOPS Meigen

Lasiops Meigen, 1838: 323. Type-species, *Anthomyia semicinerea* Wiedemann (Karl, 1928: 32).
Thricops Rondani, 1856: 96. Type-species, *Anthomyza hirtula* Zetterstedt (orig. des.).
Trichops (error) Rondani, 1866a: 217 (1866: 151).
Tricophthicus Rondani, 1861a: 9 (unjustified n. name for *Thricops* Rondani). Type-species, *Anthomyza hirtula* Zetterstedt (aut.).
Trichophticus (error) Rondani, 1871a: 335.
Trichopticus, emend.

REFERENCES: Malloch, 1921a (synopsis); Ringdahl, 1947 (rev., Palaearctic).

Subgenus LASIOPS Meigen

albibasalis (Zetterstedt), 1849: 3317 (*Anthomyza*).—Norway; Alaska to Calif. and Wyo., Mich., N.Y. to Labr. and Nfld., U.S.S.R. (Kamchatka), n. Europe.
  conformis Malloch, 1920a: 157 (*Trichopticus*).—Alta.
coquilletti (Malloch), 1920a: 156 (*Trichopticus*).—Alaska (Popof I.); Y.T., sw. N.W.T., Alta., Que., n. Sweden.
  hirsutulus Zetterstedt of Coquillett, 1900h: 444 (1904: 58).
fimbriatus (Coquillett) *in* Baker, 1904: 35 (*Phaonia*).—Nev.; Calif.
furcatus (Stein), 1916: 40 (*Trichopticus*).—Austria, Italy; Alaska to Calif. and Colo., Ont., N.H., U.S.S.R. (Kamchatka), Europe.
  melanderi Malloch, 1920a: 159 (*Trichopticus*).—Wash.
hirtulus (Zetterstedt), 1838: 673 (*Anthomyza*).—Lapland; Alaska to N.W.T. (Baffin I.), s. to N.H. and Maine, Europe.
  subrostratus Zetterstedt, 1845: 1496 (*Aricia*).—Sweden.
innocuus (Zetterstedt), 1838: 674 (*Anthomyza*).—Sweden; Alaska to Ariz. and Colo., N.W.T. to Nfld., s. to Mich. and N.Y., Europe.
  cunctans Meigen of Slosson, 1895: 320.

lividiventris (Zetterstedt), 1845: 1444 (*Aricia*).—Sweden; Alaska, Wash., Wyo., Lapland.
medius (Stein), 1920: 19 (*Trichopticus*).—Colo.; Alaska to Alta. and Wash., Wyo., Nebr.
  brevitarsis Malloch, 1920a: 161 (*Trichopticus*).—Wash. **N. syn.**
rufisquama (Schnabl), 1915: 45 (*Mydaea*).—U.S.S.R. (n. Ural Mts.); Alaska to sw. N.W.T., s. to Wash. and Alta., Mich. to Que., s. to Mass., Tenn., and N.H.
  johnsoni Malloch, 1920a: 160 (*Trichopticus*).—Mass.
septentrionalis (Stein), 1898: 184 (*Aricia*).—se. Alaska; Alaska to Nfld., s. to Ariz. and N.Y., U.S.S.R. (Kamchatka, Commander Is.).
spiniger (Stein), 1904: 428 (*Trichopticus*).—N.H.; Alaska to Nfld., s. to Utah and N.H.
  nigrifrons Walker, 1849: 932 (*Anthomyia*; preocc. Robineau-Desvoidy, 1830).—Ont.
  diffinis Malloch, 1920a: 159 (*Trichopticus*).—Utah. **N. syn.**
tarsalis (Walker), 1852: 355 (*Anthomyia*).—U.S.; Alta., S. Dak., Mich. to Que. and Vt.
  infestus Stein, 1920: 9 (*Phaonia*).—S. Dak., N.Y. **N. syn.**
  latipennis Malloch, 1920a: 158 (*Trichopticus*).—Vt.
villicrura (Coquillett), 1900h: 443 (1904: 57) (*Hyetodesia*).—se. Alaska; Alaska (Aleutian Is.), Mont., Colo.
  aldrichi Ringdahl, 1947: 95.—Alaska. **N. syn.**

Subgenus ALLOEOSTYLUS Schnabl

Alloeostylus Schnabl, 1888b: 49 (as genus). Type-species, *sudeticus* Schnabl (mon.).
Pterocanthus Malloch, 1921i: 418. Type-species, *Anthomyza sundewalli* Zetterstedt (orig. des.).

diaphanus (Wiedemann), 1817: 81 (*Anthomyia*).—Germany; Alaska to Nfld., s. to Calif. and S.C., Europe.
  signius Walker, 1849: 939 (*Anthomyia*).—N.S.
  geldrius Walker, 1849: 940 (*Anthomyia*).—N.S.
ineptus (Stein), 1920: 10 (*Phaonia*).—B.C.

Genus DIALYTA Meigen

Dialyta Meigen, 1826: 208. Type-species, *Musca erinacea* Fallén (mon.).

flavitibia Johannsen, 1916: 394.—N.Y.; Wis. to Que., s. to N.Y.
  rufitibia Stein, 1920: 23.—Wis.
nigropolita Malloch, 1919e: 302.—Calif.

Genus DENDROPHAONIA Malloch

Dendrophaonia Malloch, 1923a: 237. Type-species, *Spilogaster hilariformis* Stein (orig. des.) = *scabra* (Giglio-Tos).

REFERENCE: Snyder, 1956 (rev.).

marguerita Snyder, 1956: 450.—Minn.; Tenn.
querceti (Bouché), 1834: 82 (*Anthomyia*).—Germany; Alta. to s. Man., s. to Colo. and Ill., Que. to Ga., Europe.
scabra (Giglio-Tos), 1893a: 9 (*Spilogaster*).—Mexico; Wash. to Conn., s. to Calif., Mexico, and Fla.
    hilariformis Stein, 1898: 196 (*Spilogaster*).—Pa., Va.

Genus COENOSOPSIA Malloch

Coenosopsia Malloch, 1924j: 74. Type-species, *prima* Malloch (orig. des.).

prima Malloch, 1924j: 74.—Costa Rica; Tex., Mexico, British Honduras, Panama, ?Brazil.

Genus PHAONIA Robineau-Desvoidy

Phaonia Robineau-Desvoidy, 1830: 482. Type-species, *viarum* Robineau-Desvoidy (Coquillett, 1901e: 140) = *erratica* (Fallén).

REFERENCE: Malloch, 1923a (rev.).

Subgenus PHAONIA Robineau-Desvoidy

aberrans Malloch, 1919c: 208.—Md.; N.J., Tenn., N.C.
albocalyptrata Malloch, 1920j: 267.—s. Alaska.
alpicola (Zetterstedt), 1845: 1401 (*Aricia*).—Norway, Sweden; Alaska, N.W.T., Man., Que., Europe, U.S.S.R. (Kamchatka).
alticola Malloch, 1923a: 260.—N. Mex.; Alaska, N.W.T.
apicalis Stein, 1914: 46.—Germany; Alaska, Mich., Europe.
apicata Johannsen, 1916: 396.—N.Y.; Alta., Iowa, Wis., Mich., Que. to Ga.
    var. **solitaria** Stein, 1920: 15 (as sp.).—Wis., N.Y.
atlanis Malloch, 1923a: 279.—N.J.; Mich. to N.Y., s. to Miss. and Fla.
atrocitrea Malloch, 1923a: 262.—s. Alaska; Y.T.
atrocyanea Ringdahl, 1916: 234.—Sweden; Alaska, Y.T., Que., Labr.
    citreibasis Malloch, 1920j: 268.—Alaska.
aurea Malloch, 1923a: 256.—Wash.
azygos Malloch, 1923a: 261.—N.Y.

basiseta Malloch, 1920a: 133.—Mont.; Alta. to N.W.T., s. to Alta., S. Dak. and Minn., U.S.S.R. (Kamchatka).
bidentata Ringdahl, 1933b: 17.—ne. Greenland; N.W.T., Que., Labr.
brevispina Malloch, 1923a: 269.—Ill.; Wash., Oreg., Idaho, Ill. to N.H. and Va.
bysia (Walker), 1849: 936 (*Anthomyia*).—N.S.; Alaska to Oreg., Alta., and Idaho, Mich. to Que., s. to Ga.
 pallicornis Stein, 1920: 12.—Wash., Idaho.
caerulescens (Stein), 1898: 187 (*Aricia*).—Idaho; Wash. to Alta., s. to Calif. and Utah.
consobrina (Zetterstedt), 1838: 665 (*Anthomyza*).—Sweden; Alaska to Greenland, s. to Que., Europe, U.S.S.R. (Commander Is., Kamchatka).
 serva Meigen of Lundbeck, 1898: 309.
curvinervis Malloch, 1923a: 275.—Conn.; N.S.
curvipes (Stein), 1920: 19 (*Trichopticus*).—Mass.; Mich. to Que., s. to N.C.
deleta (Stein), 1898: 178 (*Aricia*).—Ill., Pa.; N.W.T., Tex., Iowa to Que., s. to Pa.
diruta (Stein), 1898: 188 (*Spilogaster*).—Pa.; S. Dak., Kans., Mich. to Conn., s. to Ga.
dissimilis Malloch, 1923a: 263.—s. Alaska.
errans (Meigen), 1826: 112 (*Anthomyia*).—Germany; Alaska to Labr., s. to Calif. and N.J., Europe, U.S.S.R. (Kamchatka).
 var. **completa** Malloch, 1923a: 258.—N.H.
 var. **luteva** (Walker), 1849: 934 (*Anthomyia*; as sp.).—N.S.; Alaska to Calif., Mich. to Que. to N.S.
  varipes Coquillett, 1900h: 441 (1904: 55) (*Hyetodesia*).—Alaska (Popof I.).
flava Stein, 1920: 6.—Wash.; B.C., Oreg.
flavibasis Malloch, 1919c: 208.—Mass.; N.Y., N.H.
fraterna Malloch, 1923a: 251.—Mass.
fuscana Huckett (for *fusca* Stein).—Kans., Va., Mass.; Wash., Oreg., Wis. to Que., s. to Tex. and Ga. **N. name.**
 fusca Stein, 1898: 189 (*Spilogaster*; preocc. Meade, 1897).—Kans., Va., Mass.
fuscicauda Malloch, 1918a: 269.—Calif.; Wash., Oreg.
 fuscinervis Stein, 1920: 7.—Wash.
harti Malloch, 1923a: 266.—Ill.; Alta., Idaho, Ariz., Kans. to Mich. and Mass., s. to Va.
hybrida (Schnabl), 1888a: 396 (*Aricia*).—Germany; Alaska, Y.T., Europe, U.S.S.R. (Commander Is., Kamchatka).

imitatrix Malloch, 1919f: 61.—N.W.T.
laeta (Fallén), 1823f: 56 (*Musca*).—Sweden; Mich., Ont., N.H., Ohio, Europe.
laticornis Malloch, 1923a: 279.—N.H.; Wis., Mich., Ill., N.H. to N.C.
limbinervis Stein, 1918: 208.—Mexico; Oreg., Calif., Ariz., N. Mex.
marylandica Malloch, 1923a: 265.—Md.; Maine.
monticola Malloch, 1918a: 266.—N. Mex.; Alaska to Nfld., s. to Ariz. and Colo.
    morio Zetterstedt of Stein, 1920: 17.
morio (Zetterstedt), 1845: 1399 (*Aricia*).—Sweden; Alaska, N.W.T., Greenland, Europe.
    plumbea Zetterstedt of Lundbeck, 1898: 308.
nigricans Johannsen, 1916: 395.—N.Y.; Alta. to S. Dak., Utah, Wis. to Que., s. to D.C.
    cayugae Johannsen, 1917: 327 (unjustified n. name for *nigricans* Johannsen).—N.Y.
    nervosa Stein, 1920: 12.—N.Y.
nigricauda Malloch, 1918a: 268.—Calif.
pallidisquama (Zetterstedt), 1849: 3288 (*Aricia*).—Sweden; N.W.T. (Baffin I.), Greenland, Que., Lapland.
pallidosa Huckett (for *pallida* Stein).—Idaho; B.C. and Idaho to Calif. **N. name.**
    pallida Stein, 1920: 22 (*Dialyta*; preocc. Fabricius, 1794).—Idaho.
pallidula Coquillett, 1902g: 122.—Ga.; Tex. to s. Ga. to Mass.
    dulcis Stein, 1920: 5.—Tex., N.J.
parviceps Malloch, 1918a: 267.—Calif.; Oreg.
    caesia Stein, 1920: 4.—Calif.
perfida Stein, 1920: 13.—Unknown; Calif., Utah.
prisca Stein, 1920: 14.—N.Y.
protuberans Malloch, 1923a: 247.—N.H.; Wash. to Colo., Alta. to Labr., s. to N.Y.
proxima (Wulp), 1869: 85 (*Aricia*).—Wis.; Alaska, Y.T., Wash., Mich.
pruinosa (Macquart), 1846: 329 (1846: 201) (*Aricia*).—Tex. Unrecognized.
pudoa Hall, 1937a: 215.—Idaho; Wash., Mont.
pulvillata (Stein), 1904: 422 (*Aricia*).—Carolina; N.Y. to Ga.
quieta Stein, 1920: 14.—Wash.; Oreg., Calif., Utah.
rufibasis Malloch, 1919c: 207.—Mass.; Alta., Utah, Que. to Md.
rugia (Walker), 1849: 923 (*Anthomyia*).—Ont.; Alaska to Labr., s. to Utah and Colo.
    brunneinervis Stein, 1898: 183 (*Aricia*).—Idaho.

incerta Malloch, 1923a: 250.—Alaska.
  consobrina Zetterstedt of Stein, 1920: 17.
savonoskii Malloch, 1923a: 248.—s. Alaska; Wash., Alta., Man.
serva (Meigen), 1826: 86 (*Anthomyia*).—Germany; N.W.T. to Nfld., s. to Wyo. and Ga., Europe, U.S.S.R. (Kamchatka).
soccata (Walker), 1849: 941 (*Anthomyia*).—Ont.; Alta., Mich. to Vt.
striata (Stein), 1898: 179 (*Aricia*).—Idaho; Wash., Calif.
subfusca Malloch, 1923a: 273.—Ill.; Ark., Ohio, N.Y.
subfuscinervis (Zetterstedt), 1838: 673 (*Anthomyza*).—Sweden; Alaska to Greenland, s. to Man. and Que., Sweden.
  consobrina Zetterstedt of Henriksen and Lundbeck, 1917: 628.
texensis texensis Malloch, 1923a: 271.—Tex.
  ssp. **flavofemorata** Malloch, 1923a: 271 (as var.).—Fla.; Ariz., Tex.
tipulivora Malloch, 1923a: 252.—Md.; N.Y., N.H., Mass., Conn.
trivialis Malloch, 1923a: 278.—Alta.
uniseriata Malloch, 1923a: 268.—Wash.
versicolor Stein, 1920: 16.—Colo.
winnemanae Malloch, 1919a: 3.—Md.; Wis., Mich., N.Y., Mass., Maine.
  apta Stein, 1920: 4.—Wis., Mass.

Subgenus LOPHOSCELES Ringdahl

Lophosceles Ringdahl, 1922: 3 (as genus). Type-species, *Musca mutata* Fallén (orig. des.).

alaskensis Malloch, 1923a: 272.—Alaska; N.W.T.
cinereiventris (Zetterstedt), 1845: 1500 (*Aricia*).—Sweden; Alaska to Que., s. to N.Y. and N.H., Europe, U.S.S.R. (Kamchatka).
  morrisoni Malloch, 1923a: 264.—N.H.
frenata (Holmgren), 1872: 103 (*Aricia*).—w. Greenland; Alaska to Greenland, s. to Wash., Alta., and N.H.
hians (Zetterstedt), 1838: 698 (*Anthomyza*).—Sweden; Alaska, Lapland.
minima Malloch, 1919f: 61.—Alaska; Alaska (Pribilof Is.) to Que.

Subgenus WAHLGRENIA Ringdahl

Wahlgrenia Ringdahl, 1929a: 12 (as genus). Type-species, *Anthomyza magnicornis* Zetterstedt (orig. des.).

magnicornis (Zetterstedt), 1845: 1666 (*Anthomyza*).—Sweden; Alaska, Y.T., boreal Europe.

Genus BIGOTOMYIA Malloch

Bigotomyia Malloch, 1921h: 173. Type-species, *Spilogaster trispila* Bigot (orig. des.).

californiensis Malloch, 1923a: 236.—Calif.
houghii (Stein), 1898: 177 (*Aricia*).—Ont.; B.C. to Que., s. to Calif., Okla., and N.J.
  inculta Stein, 1920: 8 (*Phaonia*).—S. Dak.

Genus PSEUDOPHAONIA Malloch

Pseudophaonia Malloch, 1918b: 66. Type-species, *Aricia orichalcea* Stein (orig. des.) = *orichalceoides* Huckett.
Polietella Ringdahl, 1922:2. Type-species, *Trichopticus steini* Ringdahl (orig. des.).

griseocaerulea Malloch, 1923a: 235.—N.H.; Alaska, Wash.
orichalceoides Huckett (for *orichalcea* Stein).—Idaho; Alta. to Calif. and Colo., Mich. to Que., s. to N.Y. and Maine. **N. name.**
  orichalcea Stein, 1898: 183 (*Aricia;* preocc. Zetterstedt, 1849).—Idaho.

Genus MUSCINA Robineau-Desvoidy

Muscina Robineau-Desvoidy, 1830: 406. Type-species, *Musca stabulans* Fallén (Coquillett, 1910b: 571).

REFERENCE: Snyder, 1956 (rev.).

assimilis (Fallén), 1823f: 56 (*Musca*).—s. Sweden; Alaska to Labr., s. to Ariz. and Ga., Europe, U.S.S.R. (Kamchatka).
  omole Walker, 1849: 930 (*Anthomyia*).—Ont.
  similis Walker, 1849: 930 (*Anthomyia*).—Ont.
  nigra Walker, 1849: 931 (*Anthomyia*).—Ont.
dorsilinea (Wulp), 1896: 308 (*Clinopera*).—Mexico; Wyo., S. Dak., s. Calif. to Fla.
  aurantiaca Hough, 1899b: 25.—Ga.
flukei Snyder, 1956: 446.—Colo.; Oreg., Calif., Nev.
fulvacrura Snyder, 1956: 449.—Wash.; Wyo., Colo.
stabulans (Fallén), 1817d: 252 (*Musca*).—s. Sweden; Alaska to Nfld., s. to Ariz. and Ga., Europe, U.S.S.R. (Kamchatka).

Subfamily MUSCINAE

Flies of the subfamily Muscinae are ubiquitous. Adults are chiefly anthophilous or coprophilous, and the larvae mostly scavengers.

Species of the genus *Musca*, notably the house fly, *Musca domestica* Linnaeus, are important as pests and disseminators of disease.

REFERENCES: Zimin, 1951 (monog., U.S.S.R., figs.); West, 1951 (house fly); Eldridge and James, 1957 (Calif.).

Genus PARARICIA Brauer and Bergenstamm

Pararicia Brauer and Bergenstamm, 1891: 391 (1891: 87). Type-species, *Musca pascuorum* Meigen (Brauer, 1893: 508).

REFERENCE: Snyder, 1956 (rev.).

pabulorum (Fallén), 1817d: 252 (*Musca*).—s. Sweden; Mont., Ariz., Ohio, N.Y., Mass., Europe.

pascuorum (Meigen), 1826: 74 (*Musca*).—Germany; S. Dak., Minn., Mich., Que. to Ga., Europe.

Genus MESEMBRINA Meigen

Mesembrina Meigen, 1826: 10. Type-species, *Musca meridiana* Linnaeus (Westwood, 1840: 141).

Eumesembrina Townsend, 1908: 124. Type-species, *Mesembrina latreillii* Robineau-Desvoidy (orig. des.).

latreillii Robineau-Desvoidy, 1830: 401.—N.S.; Alaska to Labr., s. to Colo., N. Dak., Mich., Tenn., and W. Va., n. Europe, U.S.S.R. (Kamchatka).
resplendens Wahlberg, 1844b: 66.—n. Sweden.
alascensis Townsend, 1908: 124 (*Eumesembrina*).—Alaska.

Genus HYPODERMODES Townsend

Hypodermodes Townsend, 1912c: 46. Type-species, *Musca mystacea* Linnaeus (orig. des.).

solitarius Knab, 1914a: 325.—Alta.; Alaska to Ont., s. to Idaho, Colo., Minn., and Mich.
mystacea Linnaeus of Hine, 1907c: 98.

Genus GRAPHOMYA Robineau-Desvoidy

Graphomya Robineau-Desvoidy, 1830: 403. Type-species, *Musca maculata* Scopoli (Rondani, 1856: 91).
Graphomyia, emend.

maculata (Scopoli), 1763: 326 (*Musca*).—Carniola; Alaska to Greenland, s. to Calif. and Ga., Europe, Asia.
americana Robineau-Desvoidy, 1830: 404.—N. Amer.

idessa Walker, 1849: 908 (*Musca*).—Ont. **N. syn.,** C. W. Sabrosky *in litt.*
picta Zetterstedt, 1855: 4715 (*Cyrtoneura*).—Sweden.

Genus SYNTHESIOMYIA Brauer and Bergenstamm

Synthesiomyia Brauer and Bergenstamm, 1893: 96 (1893: 8). Type-species, *brasiliana* Brauer and Bergenstamm (orig. des.) =*nudiseta* (Wulp).

nudiseta (Wulp), 1883: 42 (*Cyrtoneura*).—Argentina; Calif. to Tex., and N.C. to Fla., Bermuda, Neotropical, also Hawaii, S. Pacific islands, Queensland, India, Seychelles Is., s. Africa, Canary and Madeira Is.
brasiliana Brauer and Bergenstamm, 1893: 96, 110 (1893: 8, 22).—Brazil.

Genus NEOMUSCINA Townsend

Neomuscina Townsend, 1919b: 541. Type-species, *cavicola* Townsend (orig. des.) =*tripunctata* (Wulp).

Subgenus NEOMUSCINA Townsend

tripunctata (Wulp), 1896: 305 (*Muscina*).—Mexico; Ariz., N. Mex., Tex.
texana Hough, 1899b: 25 (*Muscina*).—Tex.
cavicola Townsend, 1919b: 541.—Ariz.

Subgenus SPILOPTEROMYIA Malloch

Spilopteromyia Malloch, 1921i 422 (as genus). Type-species, *Spilogaster apicata* Stein (orig. des.).
rufoscutella Dodge, 1955a: 150.—Fla. (Keys); Cuba.

Genus PHILORNIS Meinert

Philornis Meinert, 1890a: 304. Type-species, *molesta* Meinert (mon.).
Neomusca Malloch, 1921k: 41. Type-species, *Mydaea obscura* Wulp (mon.).

obscurus (Wulp), 1896: 317 (*Mydaea*).—Mexico; Tex. to Cent. Amer.
porteri Dodge, 1955a: 147.—Fla.

Genus MORELLIA Robineau-Desvoidy

Morellia Robineau-Desvoidy, 1830: 405. Type-species, *agilis* Robineau-Desvoidy (Townsend, 1916d: 8)=*hortorum* Fallén).

micans (Macquart), 1855: 136 (1855: 116) (*Cyrtoneura*).—Md.; Alaska to Nfld., s. to Calif., Okla., Ill., and Ga.
  recurva Thomson, 1869: 548 (*Cyrtoneura*).—Calif.
podagrica (Loew), 1857c: 45 (*Cyrtoneura*).—Austria; Alaska to Que., s. to B.C., Mont., and N.Y., Europe.
  pulchra Curran, 1926d: 213.—Ont.
scapulata (Bigot), 1878: 35 (*Pyrellia*).—Mexico; s. Tex., s. Fla., Neotropical.

Genus PYRELLIA Robineau-Desvoidy

Pyrellia Robineau-Desvoidy, 1830: 462. Type-species, *vivida* Robineau-Desvoidy (Townsend, 1916d: 8)=*cadaverina* (Linnaeus).

cadaverina (Linnaeus), 1758: 595 (*Musca*).—Sweden; ?N. Amer., Palaearctic.
cyanicolor Zetterstedt, 1845: 1323.—Sweden; Alaska to Que., s. to Calif., S. Dak., Mich., and Ga., Europe.
  occidentis Walker, 1852: 347 (*Musca;* preocc. Walker, 1852: 332).—U.S.
  setosa Loew, 1869b: 37 (Cent. 8, no. 63).—Ill.
  serena, authors, not Meigen.

Genus ORTHELLIA Robineau-Desvoidy

Orthellia Robineau-Desvoidy, 1863b: 837. Type-species, *rectinervis* Robineau-Desvoidy (Townsend, 1916d: 8)=*cornicina* (Fabricius).
Euphoria Robineau-Desvoidy, 1863b: 799 (preocc. Burmeister, 1842). Type-species, *nitidula* Robineau-Desvoidy (Townsend, 1916d: 7)=*caesarion* (Meigen).
Cryptolucilia Brauer and Bergenstamm, 1893: 206 (1893: 118). Type-species, *asiatica* Brauer and Bergenstamm (mon.)=*caesarion* (Meigen).
Pseudopyrellia Girschner, 1893: 306. Type-species, *Musca cornicina* Fabricius (orig. des.).

caesarion (Meigen), 1826: 57 *(Musca).*—Germany; N.W.T. to Que., s. to Calif. and Ga., Europe, Asia. Texas
carolinensis Robineau-Desvoidy, 1830: 457 *(Lucilia).*—Carolina.
compar Robineau-Desvoidy, 1830: 457 *(Lucilia).*—Pa.
viridis Wiedemann, 1830: 354 *(Idia).*—N. Amer.
heraea Walker, 1849: 881 *(Musca).*—?N. Amer.
frontalis Thomson, 1869: 545 *(Pyrellia).*—Calif.
morrilli Townsend, 1908: 120 *(Lucilia).*—Tex.
amoena Harris, 1835: 599 *(Musca).* Nomen nudum.
cornicina, authors (N. Amer.), not Fabricius.

Genus MUSCA Linnaeus

Musca Linnaeus, 1758: 589. Type-species, *domestica* Linnaeus (I.C.Z.N. 1925: 1).
Byomya Robineau-Desvoidy, 1830: 392. Type-species, *violacea* Robineau-Desvoidy (Townsend, 1915k: 434).
Biomyia, emend.
Eumusca Townsend, 1911b: 170 (as subgenus). Type-species, *Musca corvina* Fabricius (orig. des.)=*autumnalis* De Geer.
Promusca Townsend, 1915k: 434. Type-species, *Musca domestica* Linnaeus (orig. des.).
Emusca Malloch, 1925e: 372. Type-species, *Musca autumnalis* De Geer (orig. des.).

autumnalis De Geer, 1776: 83.—Scandinavia; Colo. and N. Dak. to N.S., s. to Kans., Tenn., and Ga., Palaearctic. Immigrant, spreading rapidly since first record in 1952.
domestica Linnaeus, 1758: 596.—Scandinavia; widespread in N. Amer., including Greenland, cosmopolitan.
vicina Macquart, 1851: 226 (1851: 253).—Amer.
contigua Walker, 1852: 344.—U.S. **N. syn.,** C. W. Sabrosky *in litt.*
harpyia Harris, 1869: 335.—Unknown (?Mass.).
flavipennis Bigot, 1887c: clxxxi (also 1887d: 605).—Rocky Mts.
corvina, authors (N. Amer.), not Fabricius.

Subfamily STOMOXYINAE

Flies of the subfamily Stomoxyinae have a slender, strongly sclerotized proboscis. They are vicious biters and suck the blood of warm-blooded vertebrates. Adults are common in fields, woods, and along shorelines, and they frequent buildings occupied by animals. Larvae live in excrement, manure, or piles of decaying plant material. Stable flies (*Stomoxys*) and horn flies (*Haematobia*) are especially important pests.

Genus HAEMATOBIA Lepeletier and Serville

Haematobia Lepeletier and Serville *in* Latreille et al., 1828: 499. Type-species, *Conops irritans* Linnaeus (Westwood, 1840: 140).
Lyperosia Rondani, 1856: 93. Type-species, *Conops irritans* Linnaeus (orig. des.).
Siphona, authors, not Meigen.

irritans (Linnaeus), 1758: 604 (*Conops*).—Sweden; B.C. to Que., s. to Calif. and Fla., West Indies, Europe.
  serrata Robineau-Desvoidy, 1830: 389.—s. France.
  cornicola Williston, 1889b: 180.—e. N. Amer.

Genus LYPEROSIOPS Townsend

Lyperosiops Townsend, 1912c: 47. Type-species, *Stomoxys stimulans* Meigen (orig. des.).

alcis (Snow), 1891: 88 (*Haematobia*).—n. Minn.; Alaska, Ont., Nfld., Mich.

Genus STOMOXYS Geoffroy

Stomoxys Geoffroy, 1762: 449, 538. Type-species, *Conops calcitrans* Linnaeus (I.C.Z.N., 1957a: 85). Conserved I.C.Z.N., 1957a: 85.

calcitrans (Linnaeus), 1758: 604 (*Conops*).—Sweden; Alaska, B.C. to N.S., throughout U.S., s. to Argentina and Chile, cosmopolitan.
  parasita Fabricius, 1781: 467.—N. Amer. (Wiedemann, 1830: 253).
  dira Robineau-Desvoidy, 1830: 387.—N. Amer.
  inimica Robineau-Desvoidy, 1830: 387.—N. Amer.
  cybira Walker, 1849: 1159.—N.S.
  occidentis Walker, 1852: 332 (*Musca*).—U.S.

Unplaced Genus and Species of Muscidae

Genus ACRIDOMYIA Stackelberg

Acridomyia Stackelberg, 1929: 121, 126. Type-species, *sacharovi* Stackelberg (orig. des.).

REFERENCE: Snyder, 1940 (rev.).

canadensis Snyder, 1940: 1.—Man.; Sask.
fumisquama Snyder, 1940: 2.—N.C.

Unplaced Species of Muscidae

aea Walker, 1849: 978 (*Cordylura*).—Ont.
cadaverum Kirby, 1837: 316 (*Musca*).—N. Amer.
dejeanii Robineau-Desvoidy, 1830: 558 (*Nerina*).—Pa.
dubia Curtis, 1835b: lxxix (*Anthomyia*).—N.W.T. (Boothia Peninsula).
nigriceps Bigot, 1887c: clxxxii (also 1887d: 615) (*Curtonevra*).—U.S. (Rocky Mts.).
pallida Say, 1829: 175 (1859b: 366) (*Mesembrina*).—Ind.
terminalis Walker, 1852: 356 (*Anthomyia*).—U.S.

Family GASTEROPHILIDAE

By J. G. Chillcott

The larvae are endoparasitic in mammals.

REFERENCES: Wells and Knipling, 1938 (keys to adults, keys and figs. for 1st and 3d stage larvae; biol.); James, 1948 (synopsis, keys to adults and 1st and 3d stage larvae); Zumpt and Paterson, 1953 (keys to adults and larvae).

Subfamily GASTEROPHILINAE

The larvae develop in the gut of Equidae, and are known as horse bots.

Genus GASTEROPHILUS Leach

Gasterophilus Leach, 1817: 2 (Leach *in* Brewster, 1817: 162). Type-species, *Oestrus equi* Clark (Curtis, 1826: pl. 146) =*intestinalis* (De Geer).
Gastrophilus, emend.
Gastrus, emend.
Enteromyza Rondani, 1857: 20 (unjustified n. name for *Gasterophilus* Leach). Type-species, *Oestrus equi* Clark (aut.) = *intestinalis* (De Geer).

Rhinogastrophilus Townsend, 1918c: 152. Type-species, *Oestrus nasalis* Linnaeus (orig. des.).

Haemorrhoestrus Townsend, 1934c: 406. Type-species, *Oestrus haemorrhoidalis* Linnaeus (orig. des.).

Enteromyia Enderlein, 1934b: 425. Type-species, *Oestrus haemorrhoidalis* Linnaeus (orig. des.).

haemorrhoidalis (Linnaeus), 1758: 584 (*Oestrus*).—Europe; s. Canada and n. U.S., s. to Calif., Mo., and Va., also cosmopolitan.

inermis Brauer, 1858: 464 (*Gastrus*).—Austria; Ill., Europe.

intestinalis (De Geer), 1776: 292 (*Oestrus*).—Europe; B.C. to N.B., throughout U.S., also cosmopolitan.

    equi Clark, 1797: 326 (*Oestrus*).—Europe.

nasalis (Linnaeus), 1758: 584 (*Oestrus*).—Europe; N.W.T. to N.S., s. to Calif., Miss., and Va., cosmopolitan.

    veterinus Clark, 1797: 328 (*Oestrus*).—Europe.

    subjacens Walker, 1849: 687 (*Gastrus*).—N.S.

Family HIPPOBOSCIDAE

By Joseph C. Bequaert

The adults of both sexes are hematophagous and obligate ectoparasites of birds or mammals. Most species are restricted to a narrow range of hosts. Reproduction is by integral viviparity (pupiparity).

REFERENCES: Bequaert, 1942a (rev. of Melophaginae); Bequaert, 1953 and 1954–1957 (rev.).

Subfamily ORNITHOICINAE

Genus ORNITHOICA Rondani

Ornithoica Rondani, 1878: 159. Type-species, *beccariina* Rondani (orig. des.) = *confluenta* (Say).

Anthoica, error.

Ornithoeca, emend.

confluenta (Say), 1823: 103 (1859b: 88) (*Ornithomyia*).—U.S.; Fla., Antilles, S. Amer., Old World.

    confluens, emend.

vicina (Walker), 1849: 1144 (*Ornithomyia*).—Jamaica; B.C. to Maine, s. to Calif., La., and S.C., Mexico, Cent. and S. Amer., Antilles.

    promiscua Ferris and Cole, 1922: 203.—Calif.

    confluens, authors, not Say.

    confluenta, authors, not Say.

Subfamily ORNITHOMYIINAE

Tribe ORNITHOMYIINI

Genus ORNITHOMYIA Latreille

Ornithomyia Latreille, 1802: 466 (as *Ornithomya*). Type-species, *Hippobosca avicularia* Linnaeus (mon.). [Because of long usage the author prefers to use the emended spelling of this name—Eds.]

fringillina Curtis, 1836: pl. 585.—England; Alaska to St. Pierre and Miquelon, s. to Calif., Tex., and Miss.
    pallida Say, 1823: 103 (1859b: 87) (preocc. Latreille, 1812).— U.S.
    chloropus Bergroth, 1901: 146.—Finland.
    anchineuria Speiser, 1905: 348 (n. name for *pallida* Say).— U.S.
    avicularia, authors, New World, not Linnaeus.

Genus MYIOPHTHIRIA Rondani

Myiophthiria Rondani, 1875b: 464 (as *Myophthiria*). Type-species, *reduvioides* Rondani (mon.). [Because of long usage the author prefers to use the emended spelling of this name—Eds.]

Subgenus BRACHYPTEROMYIA Williston

Brachypteromyia Williston, 1896b: 184 (as genus). Type-species, *femorata* Williston (mon.) = *fimbriata* (Waterhouse).

fimbriata (Waterhouse), 1887: 164 (*Anapera*).—N. Mex.; Wyo. and Nebr. to Ariz. and N. Mex.
    femorata Williston, 1896b: 185 (*Brachypteromyia*).—Wyo.

Genus ORNITHOCTONA Speiser

Ornithoctona Speiser, 1902a: 328. Type-species, *Ornithomyia erythrocephala* Leach (orig. des.).

erythrocephala (Leach), 1817: 13 (1818: 559) (*Ornithomyia*).—Brazil; B.C. to N.S., s. to Kans., Miss., and Fla., Mexico, Cent. and S. Amer.
    nebulosa Say, 1823: 102 (1859b: 87) (*Ornithomyia*).—U.S.
    buteonis Swenk, 1916: 133 (*Ornithomyia*).—Nebr.

fusciventris (Wiedemann), 1830: 611 (*Ornithomyia*).—Ky.; Wash., Calif., and Minn. to Que., s. to Nebr. and Va., Mexico, Cent. and S. Amer., Antilles.
  strigilecula Ferris, 1923: 57 (*Ornithomyia*).—Peru.

Genus STILBOMETOPA Coquillett

Stilbometopa Coquillett, 1899b: 336. Type-species, *Ornithomyia fulvifrons* Walker (mon.).

fulvifrons (Walker), 1849: 1145 (*Ornithomyia*).—Jamaica; N.J., Antilles, Costa Rica.
impressa (Bigot), 1885b: 237 (*Olfersia*).—Calif.; Nev., Ariz., N. Mex., Mexico.
podopostyla Speiser, 1904b: 394.—Brazil; Nebr., Iowa, Okla., Tex., Mexico, Cent. and S. Amer.

Genus LYNCHIA Weyenbergh

Lynchia Weyenbergh, 1881: 195. Type-species, *penelopes* Weyenbergh (mon.).
Ornithoponus Aldrich, 1923b: 77. Type-species, *Feronia americana* Leach (orig. des.).

albipennis (Say), 1823: 101 (1859b: 87) (*Olfersia*).—U.S.; B.C. to Que. and s. to Wash., Tex., and Fla., Mexico, Cent. and S. Amer., Antilles, Old World.
  ardeae Macquart, 1835: 640 (*Olfersia*).—Sicily.
  botaurinorum Swenk, 1916: 128 (*Olfersia*).—Nebr.
  scutellaris Swenk, 1916: 129 (*Olfersia*).—Mich.
americana (Leach), 1817: 11 (1818: 557) (*Feronia*).—Ga.; Wash. to N.S., s. to Calif., Tex., and Fla., Mexico, Cent. and S. Amer.
  fusca Macquart, 1846: 346 (1846: 218) (*Olfersia*).—Colombia.
  bubonis Packard, 1869a: 417 (*Hippobosca*).—Mass.
  falconis Harris, 1833: 594 (*Ornithomyia*). Nomen nudum.
angustifrons (Wulp), 1903: 430 (*Olfersia*).—Mexico, Costa Rica; Minn. to N.H., s. to Tex. and Va., Cent. and S. Amer.
hirsuta Ferris, 1927: 249.—Calif.; Idaho, Nev., Utah.
holoptera (Lutz, Neiva, and Lima), 1915: 184 (*Olfersia*).—Brazil; Wis., Ohio, Pa., Mass., S.C., Colombia.
nigra (Perty), 1833: 190 (*Hippobosca*).—Brazil; B.C. to Ariz. and Tex., Minn., Wis., Que., N.Y., Mexico, Cent. and S. Amer.

intertropica Walker, 1849: 1144 (*Ornithomyia*).—Galapagos Is.
pallidilabris Rondani, 1878: 161 (*Olfersia*).—Mexico.
wolcotti (Swenk), 1916: 132 (*Olfersia*).—Mich.; Nebr., N.H., Mexico, Cent. and S. Amer.

Genus MICROLYNCHIA Lutz, Neiva, and Lima

Microlynchia Lutz, Neiva, and Lima, 1915: 185. Type-species *Lynchia pusilla* Speiser (mon.).
pusilla (Speiser), 1902b: 157 (*Lynchia*).—Cuba; Idaho, N. Dak., to Iowa, s. to Calif. and Tex., D.C., Fla., Mexico, Cent. and S. Amer., Antilles.

Genus PSEUDOLYNCHIA Bequaert

Pseudolynchia Bequaert, 1926a: 271. Type-species, *Olfersia maura* Bigot (orig. des.)=*canariensis* (Macquart).
brunnea (Latreille) *in* Olivier, 1812: 544 (*Ornithomyia*).—Carolina; Minn. to Mass., s. to N. Mex., Tex., and Fla., Mexico, Cent. and S. Amer., Antilles.
canariensis (Macquart), 1839: 119 (*Olfersia*).—Canary Is.; Old World, immigrant to New World, Calif. to Wis. and Mass., s. to Tex. and Fla., Mexico, Cent. and S. Amer., Antilles.
maura Bigot, 1885b: 237 (*Olfersia*).—Algeria.

Tribe OLFERSIINI

Genus OLFERSIA Leach

Feronia Leach, 1817: 11 (1818: 557) (preocc. Latreille, 1817). Type-species, *spinifera* Leach (aut.).
Olfersia Leach *in* Brewster, 1817: 162 (n. name for *Feronia* Leach). Type-species, *Feronia spinifera* Leach (Speiser, 1899: 202).
Pseudolfersia Coquillett, 1899b: 336. Type-species, *maculata* Coquillett (mon.)=*fumipennis* (Sahlberg).

bisulcata Macquart, 1847: 111 (1847: 95).—Chile; Tex., ?La.
fumipennis (Sahlberg), 1886: 150 (*Lynchia*).—Finland; B.C., Oreg., Ariz., Minn. to N.H. and s. to Tex. and Fla., Mexico, Cent. and S. Amer., Antilles, Old World.
maculata Coquillett, 1899b: 336 (*Pseudolfersia*).—Wis.

sordida Bigot, 1885b: 239.—Guatemala; Oreg., Utah, Tex., La., Fla., Mexico, Cent. and S. Amer., Antilles.
spinifera (Leach), 1817: 11 (1818: 557) (*Feronia*).—Unknown; La., Fla., Mexico, Cent. and S. Amer., Antilles, Old World.

Subfamily MELOPHAGINAE
Tribe LIPOPTENINI
Genus NEOLIPOPTENA Bequaert

Neolipoptena Bequaert, 1942a: 47. Type-species, *Lipoptena ferrisi* Bequaert (orig. des.).

ferrisi (Bequaert), 1935: 170 (*Lipoptena*).—Calif.; B.C. to Calif. and e. to S. Dak., Mexico.
 subulata Coquillett of Ferris and Cole, 1922: 187.

Genus LIPOPTENA Nitzsch

Lipoptena Nitzsch, 1818: 310. Type-species, *Hippobosca cervina* Nitzsch (mon.)=*cervi* (Linnaeus).

Subgenus LIPOPTENA Nitzsch

cervi (Linnaeus), 1758: 611 (*Pediculus*).—Europe; Palaearctic Region, immigrant to N.H., Mass., N.Y., Pa.
 subulata Coquillett, 1907d: 290.—N.H.

Subgenus LIPOPTENELLA Bequaert

Lipoptena, subg. **Lipoptenella** Bequaert, 1942a: 55. Type-species, *Melophagus depressus* Say (orig. des.).

depressa (Say), 1823: 104 (1859b: 88) (*Melophagus*).—U.S.; B.C. to Calif. and S. Dak.
mazamae Rondani, 1878: 153.—Tropical Amer.; Tex. to S.C. and Fla., Mexico, Cent. and S. Amer.

Tribe MELOPHAGINI
Genus MELOPHAGUS Latreille

Melophagus Latreille, 1802: 466. Type-species, *Hippobosca ovina* Linnaeus (mon.).

ovinus (Linnaeus), 1758: 607 (*Hippobosca*).—Europe; B.C. to Maine and s. to Calif., La., and N.C., Mexico, Cent. and S. Amer. Immigrant.

montanus Ferris and Cole, 1922: 192 (as ssp.).—Alaska-Yukon border.

Family STREBLIDAE

By Rupert L. Wenzel

Members of this family are pupiparous, obligate, blood-sucking ectoparasites of bats. The females of one Old World genus burrow into the skin and become endoparasites. A few species are subapterous and flightless. The soft white pupae are deposited on the walls of caves, trunks of trees, and so forth, in the immediate vicinity of the host bat roosts, and quickly harden and color. The species are chiefly tropical and subtropical in distribution in both the Old and New Worlds.

REFERENCES: Kessel, 1925 (world rev.); Jobling, 1949 (host relationships, key to New World genera).

Subfamily TRICHOBIINAE

Genus NYCTEROPHILIA Ferris

Nycterophilia Ferris, 1916: 436. Type-species, *coxata* Ferris (orig. des.).

coxata Ferris, 1916: 437.—Calif.; Ariz. to Brazil, Antilles.

Genus TRICHOBIUS Gervais

Trichobius Gervais, 1844: 14. Type-species, *parasiticus* Gervais (mon.).

REFERENCES: Jobling, 1938 (rev.); Ross, 1961 (biol.).

adamsi Augustson, 1943: 52.—Ariz.; s. Calif. to Mexico.
corynorhini Cockerell, 1910a: 59.—Colo.; w. U.S. from Wash., s. to Mexico, Okla. to W. Va.
  quadrisetosus Kessel, 1925: 15 (as *major* ssp.).—?Mexico.
major Coquillett, 1899b: 334.—Fla.; Calif. to Kans., s. to Guatemala, Tex. and Fla.
sphaeronotus Jobling, 1939: 494.—Mexico; Ariz., Tex.

Genus PARATRICHOBIUS Lima

Paratrichobius Lima, 1921: 20. Type-species, *Trichobius longicrus* Ribeiro (orig. des.).

Unnamed sp., not *longicrus* (Ribeiro).—Ariz.

Family NYCTERIBIIDAE

By Rupert L. Wenzel

Members of this family are wingless, spiderlike flies, closely related to the Streblidae and very similar to them in biology and distribution.

REFERENCE: Guimarães and d'Andretta, 1956 (rev., New World).

Genus BASILIA Ribeiro

Basilia Ribeiro, 1903: 177. Type-species, *ferruginea* Ribeiro (orig. des., as gen. n., sp. n.).

REFERENCE: Peterson, 1960a (key).

antrozoi (Townsend), 1893d: 79 (*Nycteribia*).—N. Mex.; Oreg. to Kans., s. to Calif. and La., Mexico.
boardmani Rozeboom, 1934: 315.—Fla.; s. Ill., Ga.
corynorhini (Ferris), 1916: 435 (*Penicillidia*).—Calif.; Utah, Okla., Tex.
forcipata Ferris, 1924: 196.—Calif.; B.C. to Mont., s. to Mexico and La.
   calverti Fox and Stabler, 1953: 22.—Colo.
rondanii Guimarães and d'Andretta, 1956: 50.—Guatemala; Tex. to Honduras.

Superfamily OESTROIDEA

Family CALLIPHORIDAE

By David G. Hall

The Calliphoridae are the well-known, large, metallic blue, green, or black blow flies that occur in great numbers in almost every region of the world. The adults feed at flowers and on various kinds of organic matter. The larvae are omnivorous, carnivorous, or parasitic. Because of their breeding and feeding habits, blow flies are of considerable importance in the transmission of human disease and in myiasis. Several of the species are responsible for almost all subcutaneous myiasis in the region covered by this catalog.

The section on the tribe Melanomyini has been contributed by William L. Downes, Jr.

REFERENCES: Hall, 1948 (monog.); James, 1955a (Calif.).

Subfamily RHINIINAE
Genus STOMORHINA Rondani

Idia Wiedemann, 1820b: 21 (preocc. Hübner, 1813). Type-species, *Musca lunata* Fabricius (Brauer and Bergenstamm, 1889: 154 (1889: 86); also Brauer, 1893: 507).
Stomorhina Rondani, 1861a: 9 (n. name for *Idia* Wiedemann). Type-species, *Musca lunata* Fabricius (aut.).
Stomatorrhina, Stomorrhina, errors or emends.

lunata (Fabricius), 1805: 292 (*Musca*).—Madeira; Bermuda, s. Europe, n. Africa, s. Asia.

Subfamily CHRYSOMYINAE
Tribe CHRYSOMYINI
Genus CHLOROPROCTA Wulp

Chloroprocta Wulp, 1896: 296. Type-species, *semiviridis* Wulp (mon.)=*fuscanipennis* (Macquart).

fuscanipennis (Macquart), 1851: 223 (1851: 250) (*Lucilia*).—Brazil; s. Tex. through Cent. Amer. to Brazil.
  semiviridis Wulp, 1896: 296.—Mexico.

Genus COCHLIOMYIA Townsend

Callitroga Brauer, 1883: 74 (in specific synonymy) (also Brauer and Bergenstamm, 1893: 194 (1893: 106), in generic synonymy). Unavailable.
Cochliomyia Townsend, 1915i: 646. Type-species, *Musca macellaria* Fabricius (orig. des.).
Callitroga Hall, 1948: 120. Type-species, *Musca macellaria* Fabricius (orig. des.).
Chrysomya, Chrysomyia, authors, not Robineau-Desvoidy.
Compsomyia, authors, not Brauer and Bergenstamm.

In recent years, the generic name *Callitroga* has been used by North American authors for the New World screw-worm flies. However, the new International Code of Zoological Nomenclature rejects availability by citation in synonymy, and *Cochliomyia* is again the valid name.

aldrichi Del Ponte, 1938: 274.—Bahama Is.; S. Fla., Bermuda, West Indies.
  laniaria, authors, not Wiedemann.

hominivorax (Coquerel), 1858: 173 (*Lucilia*).—French Guiana; Calif. to Mont., Minn., Ind., and N.J., s. to Chile and Argentina.
  americana Cushing and Patton, 1933: 540.—Tex.
macellaria (Fabricius), 1775: 776 (*Musca*).—West Indies; Oreg. to Que. and Maine, s. to Chile and Argentina.
  lherminieri Robineau-Desvoidy, 1830: 446 (*Chrysomya*).—Carolina.
  coerulescens Robineau-Desvoidy, 1830: 447 (*Chrysomya*).—Carolina.
  caerulescens, error.
  vittata Macquart, 1843a: 298 (1843: 141) (*Lucilia*).—Unknown (as Australia, error).
  certima Walker, 1849: 873 (*Musca*).—Fla.
minima Shannon, 1926d: 124.—Santo Domingo; Fla. (Keys), West Indies.

Genus PARALUCILIA Brauer and Bergenstamm

Paralucilia Brauer and Bergenstamm, 1891: 391 (1891: 87). Type-species, *Calliphora fulvipes* Macquart (mon.)=*fulvicrura* (Robineau-Desvoidy).
Compsomyiops Townsend, 1918c: 153. Type-species, *Calliphora fulvipes* Macquart (orig. des.)=*fulvicrura* (Robineau-Desvoidy).
wheeleri (Hough), 1899a: 284 (*Chrysomyia*).—Calif.; Wash. to Colo., s. to s. Mexico.

Tribe PHORMIINI

Genus PHORMIA Robineau-Desvoidy

Phormia Robineau-Desvoidy, 1830: 465. Type species, *Musca regina* Meigen (Robineau-Desvoidy, 1849: v).
Euphormia Townsend, 1919b: 542. Type-species, *Musca regina* Meigen (orig. des.).
regina (Meigen), 1826: 58 (*Musca*).—Germany; Alaska and N.W.T. to Labr., s. to Mexico and Ga., n. Eurasia, Hawaii.
  philadelphica Robineau-Desvoidy, 1830: 466.—Pa.
  mollis Walker, 1849: 892 (*Musca*).—Ont.
  proxima Walker, 1852: 341 (*Musca*).—Calif.
  rufipalpis Jaennicke, 1867: 375 (1868: 67) (*Lucilia*).—Ill.
  stigmaticalis Thomson, 1869: 544 (*Lucilia*).—Calif.
  nigrina Bigot, 1877: 247 (*Somomyia*).—Ill.

rupicola Bigot, 1887c: clxxx (also 1887d: 603) (*Somomyia*).—Rocky Mts.
rufigena Bigot, 1887c: clxxxi (also 1887d: 598) (*Somomyia*).—Rocky Mts.

Genus BOREELLUS Aldrich and Shannon

Boreellus Aldrich and Shannon *in* Shannon, 1923b: 107. Type-species, *aristatus* Aldrich and Shannon (mon.) = *atriceps* (Zetterstedt).
Mallochomyia Townsend, 1926: 25. Type-species, *johanseni* Townsend (orig. des.) = *atriceps* (Zetterstedt).

atriceps (Zetterstedt), 1845: 1311 (*Sarcophaga*).—Sweden; n. of Arctic Circle, Alaska to Greenland, s. to n. Que., n. Eurasia.
   caerulea Malloch, 1919f: 59 (*Phormia*; preocc. Robineau-Desvoidy, 1830).—N.W.T.
   aristatus Aldrich and Shannon *in* Shannon, 1923b: 107 (n. name for *caerulea* Malloch).—N.W.T.
   johanseni Townsend, 1926: 25 (*Mallochomyia*; n. name for *caerulea* Malloch).—N.W.T.

Genus PROTOPHORMIA Townsend

Protophormia Townsend, 1908: 123. Type-species, *Phormia terraenovae* Robineau-Desvoidy (mon.).

terraenovae (Robineau-Desvoidy), 1830: 467 (*Phormia*).—Nfld.; Alaska to Greenland, s. to Calif., Ariz., and n. Ga., n. and montane areas of Europe, Siberia.
   groenlandica Zetterstedt, 1838: 657 (*Musca*).—Greenland, Sweden.
   terraenovae Macquart, 1851: 224 (1851: 251) (*Lucilia*).—Nfld.

Genus PROTOCALLIPHORA Hough

Protocalliphora Hough, 1899c: 66. Type-species, *Musca azurea* Fallén (orig. des.).
Apaulina Hall, 1948: 179. Type-species, *Protocalliphora avium* Shannon and Dobroscky (orig. des.).

The larvae are obligatory, blood-sucking parasites of nestling birds. There are a number of undescribed species. A revision is being completed by Curtis W. Sabrosky and Gordon F. Bennett.

aenea Shannon and Dobroscky, 1924: 251 (as *splendida* var.).—N.H. B.C. to Que., s. in mts. to Colo. and Va.

asiovora Shannon and Dobroscky, 1924: 250 (as *avium* var.).—Wash.; B.C. to Mont., s. to Calif., N. Mex., and Colo.
  basingeri Hall, 1948: 190 (*Apaulina*).—Calif. **N. syn.,** C. W. Sabrosky and G. F. Bennett *in litt.*
avium Shannon and Dobroscky, 1924: 250.—N.Y.; Wis., Ont., N.Y., Pa.
azurea (Fallén).—Not Nearctic.
chrysorrhoea (Meigen).—Not Nearctic.
cuprina (Hall), 1948: 191 (*Apaulina*).—Calif.; Wash., Oreg., Idaho, Mont.
hesperia Shannon and Dobroscky, 1924: 251 (as *splendida* var.).—B.C.; sw. Canada, nw. U.S.
hirudo Shannon and Dobroscky, 1924: 252.—Colo.; B.C. to Ont., in west s. to Oreg. and N. Mex., in east s. to n. Ga.
  cuprea Shannon and Dobroscky, 1924: 253 (as var.).—Wash.
  parva Shannon and Dobroscky, 1924: 253 (as ssp. or var.).—Colo. (as Kans., error).
hirundo Shannon and Dobroscky, 1924: 251 (as *splendida* ssp.).—Wash.; B.C. to Mont., s. to Calif. and Colo., also Ont., N.Y., Mass.
metallica (Townsend), 1919a: 379 (*Phormia*).—N.H.; B.C. to Que., s. to Calif., Colo., and Va.
  splendida, authors, not Macquart.
sapphira (Hall), 1948: 200 (*Apaulina*).—Alaska; N.W.T.
sialia Shannon and Dobroscky, 1924: 251 (as *splendida* var.).—N. Amer.; B.C. to Que., s. to Calif., S. Dak., and n. Ga.

Subfamily CALLIPHORINAE
Tribe LUCILIINI
Genus FRANCILIA Shannon

Francilia Shannon, 1924b: 74. Type-species, *alaskensis* Shannon (mon.) =*fuscipalpis* (Zetterstedt).

fuscipalpis (Zetterstedt), 1845: 1306 (*Sarcophaga*).—Scandinavia; Alaska to Labr., n. Scandinavia.
  alaskensis Shannon, 1924b: 74.—Alaska.

Genus BUFOLUCILIA Townsend

Bufolucilia Townsend, 1919b: 542. Type-species, *Lucilia bufonivora* Moniez (orig. des.).
  REFERENCE: James, 1953 (key).

elongata (Shannon), 1924b: 76 (*Lucilia*).—Wash.; B.C., Oreg., Calif., ?Colo.
silvarum (Meigen), 1826: 53 (*Musca*).—Austria; s. Canada, s. to Calif., Okla., and Va., Eurasia, n. Africa.
    sylvarum, emend.
    brunicosa Robineau-Desvoidy, 1830: 459 (*Lucilia*).—N. Amer.
    nigripalpis Townsend, 1908: 120 (*Lucilia*).—Ohio.
thatuna (Shannon), 1926d: 132 (*Lucilia*).—Idaho; B.C. to Mont., s. to s. Calif. and Colo.

Genus LUCILIA Robineau-Desvoidy

Lucilia Robineau-Desvoidy, 1830: 452. Type-species, *Musca caesar* Linnaeus (Macquart, 1834b: 162; 1834: 26).

illustris (Meigen), 1826: 54 (*Musca*).—Germany; Alaska and N.W.T. to s. Labr., s. to n. Mexico.
    consobrina Macquart, 1848a: 217 (1848: 57).—N. Amer. ?Syn.
    fraterna Macquart, 1848a: 217 (1848: 57).—N. Amer. ?Syn.
    muralis Walker, 1849: 888 (*Musca*).—Ont.
    purpurea Townsend, 1908: 122.—Alaska. ?Syn.
    caesar, authors (N. Amer.), not Linnaeus.

Genus PHAENICIA Robineau-Desvoidy

Phaenicia Robineau-Desvoidy, 1863b: 750. Type-species, *concinna* Robineau-Desvoidy (Townsend, 1916d: 8)=*sericata* (Meigen).

The species of *Phaenicia* generally are included in *Lucilia* by European authors.

cluvia (Walker), 1849: 885 (*Musca*).—West Indies; Miss. to N.C. and Fla.
    pilatei Hough, 1899a: 287 (*Lucilia*).—Ga.
coeruleiviridis (Macquart), 1855: 133 (1855: 113) (*Lucilia*).—Md.; Mich. to Md., s. to Tex. and Fla.
    caeruleiviridis, error or emend.
    australis Townsend, 1908: 122 (*Lucilia*).—Tenn.
    oculata Townsend, 1908: 123 (*Lucilia*).—Ky.
eximia (Wiedemann), 1819: 53 (*Musca*).—Brazil; s. Tex., La., Cent. and S. Amer., P.R.
    sylphida Bigot, 1877: 45 (*Somomyia*).—La.

mexicana (Macquart), 1843a: 300 (1843: 143) (*Lucilia*).—Mexico; cent. Calif. to Utah and Tex., s. to Brazil.
  unicolor Townsend, 1908: 121 (*Lucilia*).—N. Mex.
  infuscata Townsend, 1908: 123 (*Lucilia*).—N. Mex.
pallescens (Shannon), 1924b: 78 (*Lucilia*).—N.C.; Calif., Tenn., and D.C., s. to Mexico and Fla. Syn. of *P. cuprina cuprina* (Wiedemann) by some recent authors, notably Waterhouse and Paramonov (1950) and James (1953).
problematica (Johnson), 1913d: 448 (*Lucilia*).—Bermuda.
sericata (Meigen), 1826: 53 (*Musca*).—Austria; s. Canada to Mexico, cosmopolitan.
  sayi Jaennicke, 1867: 375 (1868: 67) (*Lucilia*).—Ill.
  barberi Townsend, 1908: 121 (*Lucilia*).—Ariz.
  giraulti Townsend, 1908: 121 (*Lucilia*).—Tex.

Unplaced Species of Luciliini

rectinervis Bigot, 1887c: clxxxi (also 1887d: 600) (*Somomyia*).—Rocky Mts.

Tribe CALLIPHORINI

Genus ACRONESIA Hall

Acronesia Hall, 1948: 272. Type-species, *Steringomyia aldrichia* Shannon (orig. des.).
Steringomyia, authors, not Pokorny.

abina Hall, 1948: 274.—Colo.
alaskensis (Shannon), 1923b: 112 (*Steringomyia*).—Alaska; Wyo., Colo.
aldrichia (Shannon), 1923b: 112 (*Steringomyia*).—Colo.; Alaska to Mont., s. to Wash. and Colo., also Que.
anana Hall, 1948: 278.—Alaska; N.W.T., s. Baffin I., n. Que., n. Labr.
collini Hall, 1948: 279.—n. Que.; Y.T. to s. Baffin I., s. to n. Man. and Labr.
montana (Shannon), 1926d: 135 (*Steringomyia*).—Alta.; s. Alaska, n. Man., Ont., Labr.
popoffana (Townsend), 1908: 117 (*Calliphora*).—Alaska; n. Canada.

Genus EUCALLIPHORA Townsend

Eucalliphora Townsend, 1908: 118. Type-species, *Calliphora latifrons* Hough (mon.)=*lilaea* (Walker).

lilaea (Walker), 1849: 894 *(Musca)*.—Ont.; Alaska to Ont., s. to n. Mexico and Colo.
    ilerda Walker, 1849: 895 *(Musca)*.—Ont.
    latifrons Hough, 1899a: 286 *(Calliphora)*.—Idaho.

Genus ALDRICHINA Townsend

Aldrichiella Rohdendorf, 1931: 177 (preocc. Vaughan, 1903). Type-species, *Calliphora grahami* Aldrich (orig. des.).
Aldrichina Townsend, 1934b: 111 (n. name for *Aldrichiella* Rohdendorf). Type-species, *Calliphora grahami* Aldrich (aut.).
grahami (Aldrich), 1930a: 1 *(Calliphora)*.—China (Szechuan Province); Wash. to Colo., s. to Calif. and N. Mex., Siberia to China.

Genus CALLIPHORA Robineau-Desvoidy

Calliphora Robineau-Desvoidy, 1830: 433. Type-species, *Musca vomitoria* Linnaeus (orig. des.).
    REFERENCE: James, 1953 (key).

alpina (Zetterstedt), 1845: 1304 *(Sarcophaga)*.—Sweden; Greenland, n. Que., n. Europe.
coloradensis Hough, 1899a: 286.—Colo.; Alaska to Ont. and Ind., s. to Mexico.
livida Hall, 1948: 296.—Ga.; Alaska and Y.T. to Ont., s. to Calif. and Ga.
    viridescens authors (N. Amer.), not Robineau-Desvoidy.
morticia Shannon, 1923b: 116.—Alaska.
rufipalpis Macquart, 1851: 216 (1851: 243) (preocc. Macquart, 1834).—Amer. Unrecognized.
stygia (Fabricius).—Not Nearctic.
terraenovae Macquart, 1851: 217 (1851: 244).—Nfld.; Alaska to Greenland, s. to Calif., Colo., and n. Fla.
    nigribucca Hough, 1899c: 66. Nomen nudum.
    nigribarba Shannon, 1923b: 116 (as *vomitoria* var.).—N.Y.
uralensis Villeneuve, 1922b: 515.—U.S.S.R.; Greenland, n. Europe, cent. Europe in mts.
vicina Robineau-Desvoidy, 1830: 435.—La.; Alaska to Que., s. to Mexico, Holarctic, s. S. Amer., s. Africa, s. Australia.
    erythrocephala Meigen, 1826: 62 *(Musca*; preocc. De Geer, 1776).—Europe.
    rufifacies Macquart, 1851: 216 (1851: 243).—N.Y.
viridescens Robineau-Desvoidy, 1830: 437.—Carolina. Unrecognized.

vomitoria (Linnaeus), 1758: 595 *(Musca)*.—Sweden; Alaska to Greenland, s. to Mexico and Va., Hawaii, s. Africa.
  obscoena Eschscholz, 1823: 113 *(Musca)*.—Alaska.
  rubrifrons Townsend, 1908: 116.—B.C.

Genus ONESIA Robineau-Desvoidy

Onesia Robineau-Desvoidy, 1830: 365. Type-species, *floralis* Robineau-Desvoidy (Townsend, 1916d: 8) = *sepulchralis* (Meigen).
bisetosa Hall, 1948: 322.—N.J.
  agilis Meigen of Shannon, 1923b: 108.
townsendi Hall, 1948: 324.—Mass.
  aculeata Pandellé of Shannon, 1923b: 108.

Genus CYNOMYOPSIS Townsend

Cynomyopsis Townsend, 1915m: 118. Type-species, *Cynomya cadaverina* Robineau-Desvoidy (orig. des.).
cadaverina (Robineau-Desvoidy), 1830: 365 *(Cynomya)*.—Carolina; Alaska to Greenland, s. to Calif., Tex., and Ga.
  myoidea Robineau-Desvoidy, 1830: 436 *(Calliphora)*.—Pa.
  aurulans Robineau-Desvoidy, 1830: 437 *(Calliphora)*.—Carolina.
  compressa Robineau-Desvoidy, 1830: 438 *(Calliphora)*.—Carolina.
  mortisequa Kirby, 1837: 316 *(Musca)*.—Alaska.
  americana Hough, 1898a: 105 *(Cynomyia)*.—La.
  texensis Townsend, 1908: 116 *(Calliphora)*.—Tex.

Genus CYANUS Hall

Cyanus Hall, 1948: 331. Type-species, *Cynomyia elongata* Hough (orig. des.).
elongatus (Hough), 1898a: 106 *(Cynomyia)*.—Wyo.; Oreg. and Alta. to N. Dak., s. to Calif. and Colo.

Genus CYNOMYA Robineau-Desvoidy

Cynomya Robineau-Desvoidy, 1830: 363. Type-species, *Musca mortuorum* Linnaeus (Macquart, 1834b: 176; 1834: 40).
Cyanomyia (error) Wilson, 1932: 89.
Cynomyia, error or emend.

Carcinomyia Townsend, 1915a: 21. Type-species, *Cynomyia hirta* Hough (orig. des.).

hirta Hough, 1898b: 166.—Alaska.
mortuorum (Linnaeus), 1761: 452 (*Musca*).—Sweden; Alaska, Greenland, n. Eurasia.

Unplaced species of Calliphorini

flavipalpis Macquart, 1851: 209 (1851: 236) (*Cynomyia*).—Nfld.

Subfamily POLLENIINAE
Tribe POLLENIINI
Genus POLLENIA Robineau-Desvoidy

Pollenia Robineau-Desvoidy, 1830: 412. Type-species, *Musca rudis* Fabricius (orig. des.).

rudis (Fabricius), 1794: 314 (*Musca*).—Germany; B.C. to N.S., U.S., Europe, n. Africa.
    familiaris Harris, 1869: 336 (*Musca*; preocc. Panzer, 1804).—Mass.
    obscura Bigot, 1887c: clxxxi (also 1887d: 597).—N. Amer.
    hirticollis Harris, 1835: 599 (*Musca*). Nomen nudum.
vagabunda (Meigen), 1826: 72 (*Musca*).—Europe; B.C., P.E.I., N.S.
vespillo (Fabricius).—Not Nearctic.

Tribe MELANODEXIINI
Genus MELANODEXIA Williston

Melanodexia Williston, 1893d: 256. Type-species, *tristis* Williston (mon.).
Melanodexiopsis Hall, 1948: 356. Type-species, *tristina* Hall (orig. des.).

californica Hall, 1948: 354.—Calif.
glabricula (Bigot), 1887c: clxxxii (also 1887d: 594) (*Nitellia*).—Calif.
grandis Shannon, 1926d: 138.—Calif.
    pacifica Hall, 1948: 359 (*Melanodexiopsis*).—Calif.
idahoensis (Hall), 1948: 357 (*Melanodexiopsis*).—Idaho.
nox (Hall), 1948: 358 (*Melanodexiopsis*).—Oreg.; Wash., Calif.
satanica Shannon, 1926d: 138.—Calif.; Wash.

tristina (Hall), 1948: 359 (*Melanodexiopsis*).—Calif.
tristis Williston, 1893d: 257.—Calif.

Tribe MELANOMYINI
By William L. Downes, Jr.

Although members of this group have also been placed in the Sarcophagidae or Tachinidae, a better case can be made for assigning them to the Calliphoridae. Four of the species have been reared from snails, including, in North America, *Melanomya obscura* and *M. ordinaria*.

Genus MELANOMYA Rondani

Melanomya Rondani, 1856: 88. Type-species, *Dexia nana* Meigen (orig. des.).

Subgenus ANGIONEURA Brauer and Bergenstamm

Angioneura Brauer and Bergenstamm, 1893: 187 (1893: 99) (as genus). Type-species, *Calobataemyia vetusta* Brauer and Bergenstamm (as "*Myobia vetusta* Stein"; mon.) = *acerba* (Meigen). **N. syn.**
Opelousia Townsend, 1919b: 547. Type-species, *obscura* Townsend (orig. des.). **N. syn.** (genus and subg.).
Opsodexiopsis Townsend, 1935b: 69. Type-species, *Opsodexia abdominalis* Reinhard (orig. des.). **N. syn.** (genus and subg.).

abdominalis (Reinhard), 1929: 6 (*Opsodexia*).—N.H.; Mich. **N. comb.**
flavescens (Reinhard), 1929: 3 (*Opelousia*).—Tex.; ?N.C. **N. comb.**
mitis (Reinhard), 1945b: 76 (*Opelousia*).—Man. **N. comb.**
obscura (Townsend), 1919b: 547 (*Opelousia*).—La.; Y.T. to Que., s. to Tex. and Va. **N. comb.**
 addenda West, 1925: 129 (*Chaetona*).—N.Y. **N. syn., n. comb.**
ordinaria (West), 1925: 129 (*Chaetona*).—N.Y.; Alaska to Que., s. to Mich. and N.Y. **N. comb.**

Subgenus OPSODEXIA Townsend

Opsodexia Townsend, 1915a: 20 (as genus). Type-species, *Chaetona bicolor* Coquillett (orig. des.). **N. syn.**
Phalacrodexia Townsend, 1915a: 21. Type-species, *Chaetona flavipennis* Coquillett (orig. des.). **N. syn.** (genus and subg.).

Opelodexia Reinhard, 1945b: 74. Type-species, *artata* Reinhard (orig. des.)=*grisea* (Coquillett). **N. syn.** (genus and subg.).

bicolor (Coquillett), 1899d: 221 (*Chaetona*).—N.H.; Que. and Maine, s. to Tenn. and Ga. **N. comb.**
flavipennis (Coquillett), 1902g: 121 (*Chaetona*).—Ohio; N.C. **N. comb.**
grisea (Coquillett), 1899d: 222 (*Chaetona*).—Ga.; Ill., N.C., S.C. **N. comb.**
    artata Reinhard, 1945b: 75 (*Opelodexia*).—N.C. **N. syn.**

Unplaced Species of Calliphoridae

splendida Macquart, 1846: 324 (1846: 196) (*Calliphora*).—Tex.

Family SARCOPHAGIDAE
(Stephanostomatidae, Metopiidae)

By William L. Downes, Jr.

Most adult sarcophagids, or flesh flies, are strong fliers. Apparently they need sugars for survival: Nectar is used, perhaps exclusively by some flies (*Senotainia* spp.), and homopterous honeydew attracts others. The Sarcophaginae will use tree sap and juices from damaged fruits, as well as nectar and honeydew. As far as known, the males of this family take "stations" on various surfaces, and sally out after passing females (or objects mistaken for them). The larval food is exceedingly diverse but almost never consists of plant materials.

Dr. H. R. Dodge read the manuscript and kindly furnished additional distribution data, as well as records of some species not previously known from the Nearctic Region.

REFERENCES: Aldrich, 1916a (rev., biol., except Miltogrammini), 1930c (syn.); Greene, 1925 (immature stages); Roback, 1954 (male terminalia, phylogeny); Sanjean, 1957 (larvae, N.Y.).

Subfamily MILTOGRAMMINAE
Tribe MILTOGRAMMINI

With rare exceptions, the larvae of this tribe feed on provisions in the nests of wasps or bees, sometimes first destroying the egg of the host. In any event the host larva does not survive. Some females of *Senotainia* locate the larval food by trailing host wasps carrying prey to their nests. This habit is probably common to all members of that genus and others, such as *Amobia*, in which the females have greatly enlarged facets in the front part of their eyes. Nontrailing females of species known to search out the nests of their hosts have typical eyes and, in addition, somewhat flattened fore tarsi with a mat of fine yellow hairs on the under surface. Almost nothing is known about the biology of the miltogrammines, in which these modifications are absent.

REFERENCE: Allen, 1926b (rev., biol.).

Genus AMOBIA Robineau-Desvoidy

Amobia Robineau-Desvoidy, 1830: 96. Type-species, *conica* Robineau-Desvoidy (mon.) = *signata* (Meigen).
Ammobia Bezzi and Stein, 1907: 519 (preocc. Billberg, 1820), emend.
Pachyophthalmus Brauer and Bergenstamm, 1889: 117 (1889: 49). Type-species, *Tachina signata* Meigen (mon.).

Larvae of this genus develop on the provisions in nests of sphecid wasps such as *Sceliphron* and *Trypoxylon*, and vespids such as *Symmorphus*, *Eumenes*, and *Stenodynerus*.

Subgenus AMOBIA Robineau-Desvoidy

aurifrons (Townsend), 1891c: 354 (*Pachyophthalmus*.—Ill.; Oreg. to Maine, s. to Ariz. and Va. **N. comb.**
  signata, authors, not Meigen.
distorta (Allen), 1926b: 15 (*Pachyophthalmus*).—Pa.; B.C. to Maine, s. to Ariz. and Ga. ?also Palaearctic.

Subgenus SARCOMACRONYCHIA Townsend

Sarcomacronychia Townsend, 1892a: 100 (as genus). Type-species *unica* Townsend (orig. des.) = *erythrura* (Wulp).

aberrans Reinhard, 1955a: 128.—Utah.
erythrura (Wulp), 1890: 89 (*Miltogramma*).—Mexico; Wash. to N.Y., s. to s. Calif., Mexico, and Fla. **N. comb.**
  unica Townsend, 1892a: 101 (*Sarcomacronychia*).—N. Mex. **N. syn.**
  sarcophagoides Townsend, 1892e: 165 (*Sarcomacronychia*).—N. Mex. **N. syn.**
floridensis (Townsend), 1892h: 80 (*Pachyophthalmus*).—Fla.; Wash. to N.J., s. to s. Calif. and Fla.
  trypoxylonis Townsend, 1893e: 165 (*Sarcomacronychia*).—Ohio.
hinei (Allen), 1926b: 12 (*Pachyophthalmus*).—La.

Genus EUMACRONYCHIA Townsend

Eumacronychia Townsend, 1892a: 98. Type-species, *decens* Townsend (orig. des.).

REFERENCE: Reinhard, 1939b (rev.).

agnella Reinhard, 1939b: 64.—Tex.
alternata Reinhard, 1939b: 61.—Tex.
crassipalpis Reinhard, 1939b: 60.—Calif. ?=*rohweri*.
decens Townsend, 1892a: 99.—N. Mex.; Calif.
duplicata Reinhard, 1944b: 57.—N. Mex.
elita Townsend, 1892a: 100.—N. Mex.; Calif.
    fasciata Coquillett, 1897: 81 (*Senotainia*).—N. Mex.
elongata Allen, 1926b: 93.—Man.
montana Allen, 1926b: 88.—N. Mex.; Calif., Ariz.
nigricornis Allen, 1926b: 90.—Ohio; Mo., N.J., Ga.
prolixa Reinhard, 1939b: 67.—Tex.; Mexico.
rohweri Allen, 1926b: 92.—Colo.; Oreg., Calif.
scitula Reinhard, 1953a: 50.—Calif.
sternalis Allen, 1926b: 89.—Tex.
tortilis Reinhard, 1939b: 65.—Tex.
tricosa Reinhard, 1939b: 63.—Tex.

Genus EUPHYTO Townsend

Euphyto Townsend, 1908: 63. Type-species, *Leucostoma subopaca* Coquillett (orig. des.).
Tetropsis Coquillett, 1910a: 128. Type-species, *modesta* Coquillett (orig. des.)=*subopaca* (Coquillett).
Euphytomima James, 1955e: 283. Type-species, *nomiivora* James (orig. des.). **N. syn.**

E. nomiivora has been reared from nests of *Nomia* bees; *pollinaris* feeds on the stores of *Perdita* bees.

caesia (Reinhard), 1963a: 78 (*Euphytomima*).—Ariz. **N. comb.**
nomiivora (James), 1955e: 284 (*Euphytomima*).—Utah; Calif., Ariz. **N. comb.**
pollinaris Reinhard, 1945b: 69.—Tex.; N. Dak., Kans.
rixosa Reinhard, 1956c: 103.—Calif.
ruficeps Reinhard, 1953a: 49.—Calif.
subopaca (Coquillett), 1897: 69 (*Leucostoma*).—N.J.; Md.
    modesta Coquillett, 1910a: 128 (*Tetropsis*).—N.J.

Genus GYMNOPROSOPA Townsend

Gymnoprosopa Townsend, 1892a: 108. Type-species, *polita* Townsend (orig. des.).

argentifrons Townsend, 1892a: 109.—Fla.; Iowa to Md., s. to Tex. and Fla.
filipalpus Allen, 1926b: 100.—Miss.; Tex., Ala., Ga., Fla.
inflaticornis Allen, 1926b: 102.—N. Mex.
latifasciata Reinhard, 1945b: 72.—Tex.
milanoensis Reinhard, 1945b: 74.—Tex.
pallida Allen, 1926b: 101.—Idaho.
polita Townsend, 1892a: 109.—Fla.; Ill. to Conn., s. to Fla.
    clarifrons Townsend, 1892a: 109.—Ill.

Genus HILARELLA Rondani

Hilarella Rondani, 1856: 70. Type-species, *Miltogramma hilarella* Zetterstedt (taut.).

H. hilarella is generally associated with fossorial sphecids, but there is one authentic record of it as a parasite of a camel cricket.

hilarella (Zetterstedt), 1844: 1212 (*Miltogramma*).—Scandinavia; B.C. to Mass., s. to s. Calif. and Fla., Mexico, Palaearctic.
    siphonina, authors, not Zetterstedt.

Genus MACRONYCHIA Rondani

Macronychia Rondani, 1859: 229, 239 (as *Macronichia*, p. 229). Type-species, *Tachina agrestis* Fallén (mon.).
Amobiopsis Townsend, 1915a: 20. Type-species, *Amobia aurata* Coquillett (orig. des.).

aurata (Coquillett), 1902g: 119 (*Amobia*).—N.H.; B.C. to Maine, s. to Calif. and Fla.
confundens (Townsend), 1915a: 20 (*Amobiopsis*).—N.H.; B.C. to N.S., s. to Mexico and Ga.
    utahensis Smith, 1916: 95 (*Amobia*).—Utah. **N. syn.**

Genus METOPIA Meigen

Metopia Meigen, 1803: 280. Type-species, *Musca leucocephala* Rossi (as Panzer; mon.) = *argyrocephala* (Meigen).

M. argyrocephala has been reared from nests of *Lasioglossum* (Halictidae) and *Sphex*, and from nests of other bees and wasps in Europe.

Subgenus METOPIA Meigen

argyrocephala (Meigen), 1824: 372 (*Tachina*).—Europe; Y.T. to Nfld., s. to Mexico and Fla., Palaearctic.
  leucocephala Rossi, 1790b: 306 (*Musca*; preocc. Villers, 1789).—Italy.
  luggeri Townsend, 1892e: 69.—Minn.
campestris (Fallén), 1820g: 8 (*Tachina*).—Sweden; Alaska to N.S., s. to Calif. and Ga., Palaearctic.
inermis Allen, 1926b: 54.—Utah; s. Man.
lucipeda Reinhard, 1961: 213.—Calif.
opaca Allen, 1926b: 50.—Oreg.; B.C. to Oreg., e. to Mass., also N. Mex.
tessellata Allen, 1926b: 55.—Ohio; Iowa to Maine, s. to Mo. and N.C., also Ariz., ?S. Amer.

Subgenus ALLENANICIA Townsend

Parametopia Townsend, 1916h: 619 (preocc. Reitter, 1884). Type-species, *morrisoni* Townsend (orig. des.)=*lateralis* (Macquart).
Allenanicia Townsend, 1935b: 68 (as genus). Type-species, *Metopia sinipalpis* Allen (orig. des.). **N. status.**

krombeini Sabrosky, 1953a: 50.—Tex.; D.C., Va., N.C.
lateralis (Macquart), 1848a: 208 (1848: 48) (*Degeeria*).—N. Amer.; Wis. to N.H., s. to Kans. and Fla., also Ariz. and ?B.C.
  morrisoni Townsend, 1916h: 619 (*Parametopia*).—N.H.
lateropili Allen, 1926b: 57.—Cuba; Calif. to Iowa, Ohio, and Va., s. to Tex. and Fla.
sinipalpis Allen, 1926b: 60.—Va.; Iowa to Mass., s. to Miss.

Genus OPSIDIA Coquillett

Opsidia Coquillett, 1895k: 102. Type-species, *gonioides* Coquillett (orig. des.).
Opsidiopsis Townsend, 1919b: 544. Type-species, *oblata* Townsend (orig. des.). **N. syn.**

gonioides Coquillett, 1895k: 102.—N.J.; Wis. to Que. and Mass., s. to Ind. and Md.
oblata (Townsend), 1919b: 544 (*Opsidiopsis*).—N. Mex.; Wash., Calif., Tex. **N. comb.**
  modesta Reinhard, 1922: 334 (*Chaetoplagia*).—Tex.

Genus OPSIDIOTROPHUS Reinhard

Opsidiotrophus Reinhard, 1963a: 80. Type-species, *micidus* Reinhard (orig. des.).

micidus Reinhard, 1963a: 80.—Calif.; Ariz., N. Mex.

Genus PHROSINELLA Robineau-Desvoidy

Phrosinella Robineau-Desvoidy, 1863b: 82, 101 (as *Phrosina*, p. 101, the latter preocc. Risso, 1822). Type-species, *Phrosina argyrina* Robineau-Desvoidy (mon.)=*nasuta* (Meigen).

Euhilarella Townsend, 1915a: 22. Type-species, *Gymnoprosopa fulvicornis* Coquillett (orig. des.)

aldrichi Allen, 1926b: 75.—Idaho; B.C., Alta., Wash., Iowa.
fulvicornis (Coquillett), 1895k: 106 (*Gymnoprosopa*).—N.J.; Y.T. to Labr., s. to Tex. and Fla.
fumosa Allen, 1926b: 74.—Va.; Del., Md.
pilosifrons Allen, 1926b: 76.—Wash.; N.W.T., B.C., Oreg., Calif.
talpina Reinhard, 1961: 212.—Calif.

Genus PTYCHONEURA Brauer and Bergenstamm

Ptychoneura Brauer and Bergenstamm, 1889: 104 (1889: 36). Type-species, *Tachina rufitarsis* Meigen (mon.).

Oestrohilarella Townsend, 1919c: 162. Type-species, *Hilarella aristalis* Coquillett (orig. des.). **N. syn.**

Ouelletia Curran, 1934a: 410. Type-species, *aristalis* Curran (orig. des.)=*aristalis* (Coquillett). **N. syn.**

aristalis (Coquillett), 1897: 129 (*Hilarella*).—Ill.; B.C. to Que., s. to Calif., Ill., and N.Y.
  aristalis Curran, 1934a: 410 (*Ouelletia*).—Que. **N. syn.**

Genus SENOTAINIA Macquart

Senotainia Macquart, 1846: 295 (1846: 167) (as *Senetainia*, pp. 295 and 167; as *Senotaina*, pp. 364 and 236). Type-species, *rubriventris* Macquart (orig. des.).

Eusenotainia Townsend, 1915a: 22. Type-species, *Hilarella rufiventris* Coquillett (orig. des.).

Microsenotainia Townsend, 1916h: 618. Type-species, *Senotainia nana* Coquillett (orig. des.).

The adults of this genus frequent sandy areas and are probably all "parasitic" in the nests of fossorial wasps such as *Bicyrtes*.

arenicola Reinhard, 1963a: 77.—Kans.

flavicornis (Townsend), 1891c: 355 (*Miltogramma*).—Ill.; Nebr., Iowa, Mo., Ga., Fla.
  similis Townsend, 1891c: 357 (*Miltogramma*).—Ill.
inyoensis Reinhard, 1955a: 131.—Calif.
kansensis (Townsend), 1892e: 68 (*Miltogramma*).—Kans.; Wash. to Iowa, s. to s. Calif. and Fla.
litoralis Allen, 1924: 90.—Miss.; Mass., N.C., Fla.
nana Coquillett, 1897: 81.—N. Mex.; Calif., Ariz., Mexico.
opiparis Reinhard, 1955a: 129.—Tex.
rubriventris Macquart, 1846: 295 (1846: 167).—Tex.; Wash. to Ont. and Conn., s. to Calif. and Fla.
  decisa Townsend, 1892h: 81 (*Miltogramma*).—Fla.
rufiventris (Coquillett), 1897: 129 (*Hilarella*).—Miss.; Calif.
setulicosta Allen, 1926b: 31.—Calif.
sinopsis Reinhard, 1955a: 130.—Tex.
trilineata (Wulp), 1890: 89 (*Miltogramma*).—Mexico; B.C. to N.S., s. to Calif. and Fla., ?S. Amer.
  americanus Brauer and Bergenstamm, 1891: 361 (1891: 57) (*Arrenopus*).—Ga.
  argentifrons Townsend, 1891c: 357 (*Miltogramma*).—Ill.
  cinerascens Townsend, 1891c: 358 (*Miltogramma*).—Ill.
vigilans Allen, 1924: 89.—Miss.; Minn. to Mass., s. to Iowa and N.J., also Miss.

Genus SPHENOMETOPA Townsend

Sphenometopa Townsend, 1908: 64. Type-species, *Araba nebulosa* Coquillett (orig. des.).
Euaraba Townsend, 1915a: 20. Type-species, *Araba tergata* Coquillett (orig. des.).
Arabiopsis Townsend, 1915c: 285. Type-species, *cocklei* Townsend (orig. des.).

cocklei (Townsend), 1915c: 286 (*Arabiopsis*).—B.C.; Mont.
nebulosa (Coquillett), 1902a: 200 (*Araba*).—Mexico; Calif., Ariz.
planitarsis Reinhard, 1945b: 72.—Tex.
tergata (Coquillett), 1895k: 103 (*Araba*).—Ill.; B.C. to N.H., s. to Calif., Mexico, and Ga.
violae Reinhard, 1945b: 71.—Tex.; Ga.

Genus TAXIGRAMMA Perris

Taxigramma Perris, 1852: 209. Type-species, *pipiens* Perris (orig. des., as genus n., sp. n.) = *heteroneura* (Meigen). As "Taxigramme" in Macquart, 1849: 359, a vernacular name.

Heteropterina Macquart, 1854: 426. Type-species, *Miltogramma heteroneura* Meigen (mon.).
Nasonimyia Townsend, 1916h: 619. Type-species, *Heteropterina nasoni* Coquillett (orig. des.) = *heteroneura* (Meigen).

heteroneura (Meigen), 1830: 367 (*Miltogramma*).—Germany; B.C. to Man., s. to Calif. and Tex., also Ill., Mich., Ont., and Conn., Palaearctic.
  nasoni Coquillett, 1895f: 207 (*Heteropterina*).—Ill.

Tribe PARAMACRONYCHIINI

The larvae of *Brachicoma* feed on bumble bee larvae, and those of *Pseudosarcophaga* parasitize larvae of *Choristoneura fumiferanae* and other Lepidoptera. The larvae of *Wohlfahrtia vigil* develop in pustules in the skin of many young mammals, occasionally including man.

Genus BRACHICOMA Rondani

Brachicoma Rondani, 1856: 69. Type-species, *Tachina nitidula* Meigen (orig. des.; misident.) = *devia* (Fallén).
Brachycoma, emend. or error.
Laccoprosopa Townsend, 1891c: 365. Type-species, *sarcophagina* Townsend (orig. des.).
Bombobrachycoma Townsend, 1919c: 157. Type-species, *Brachycoma davidsoni* Coquillett (orig. des.).
Kennesawmyia Dodge, 1956a: 182. Type-species, *truncatipennis* Dodge (orig. des.) = *sarcophagina* (Townsend). **N. syn.**
Lacchoprosopa Curran *in* Dobroscky, 1925: 280 (also *in* Johannsen, 1928b: 825). Nomen nudum.

davidsoni Coquillett in Davidson, 1894: 172 (*Brachycoma*).—Calif.; Wyo., Colo.
devia (Fallén), 1820g: 6 (*Tachina*).—Sweden; Alaska to e. Ont., ?Wash., ?Idaho, Europe.
sarcophagina (Townsend), 1891c: 366 (*Laccoprosopa*).—Ill.; Iowa to Mass., s. to Mo. and Ga.
  truncatipennis Dodge, 1956a: 182 (*Kennesawmyia*).—Ga. **N. syn.**
  avium Curran *in* Dobroscky, 1925: 280 (also *in* Johannsen, 1928b: 825) (*Lacchoprosopa*). Nomen nudum.
setosa Coquillett, 1902g: 117 (*Brachycoma*).—N. Mex.; B.C. to Mont., s. to Calif. and N. Mex., also Maine.

Genus CATTASOMA Reinhard

Cattasoma Reinhard, 1947b: 97. Type-species, *mediocre* Reinhard (orig. des.).
festinans Reinhard, 1947b: 98.—Tex.; Mexico.
mediocre Reinhard, 1947b: 97.—Tex.

Genus ERYTHRANDRA Brauer and Bergenstamm

Erythrandra Brauer and Bergenstamm, 1891: 368 (1891: 64). Type-species, *picipes* Brauer and Bergenstamm (mon.).
Trixoclista Townsend, 1892a: 102. Type-species, *distincta* Townsend (orig. des.). **N. syn.**
Sarcoclista Townsend, 1892a: 122. Type-species, *dakotensis* Townsend (orig. des.) =*distincta* (Townsend). **N. syn.**
Eubrachycoma Townsend, 1916e: 19. Type-species, *Brachycoma apicalis* Coquillett (orig. des) =*picipes* Brauer and Bergenstamm.
Rabunmyia Dodge, 1956a: 184. Type-species, *compressa* Dodge (orig. des.) =*distincta* (Townsend). **N. syn.**

distincta (Townsend), 1892a: 103 (*Trixoclista*).—Ill.; Ariz. to S. Dak. and Minn.; se. to Ga. **N. comb.**
   dakotensis Townsend, 1892a: 123 (*Sarcoclista*).—S. Dak. **N. syn.**
   compressa Dodge, 1956a: 184 (*Rabunmyia*).—Ga. **N. syn.**
   fattigi Dodge, 1956a: 186 (*Rabunmyia*).—Ga. **N. syn.**
picipes Brauer and Bergenstamm, 1891: 368 (1891: 64).—Ga.; Sask. to Mass., s. to Tex. and Fla.
   apicalis Coquillett, 1897: 131 (*Brachycoma*).—Conn.
   atlantica Parker, 1919b: 203 (*Sarcofahrtia*).—Conn.

Genus PSEUDOSARCOPHAGA Kramer

Pseudosarcophaga Kramer, 1908b: 201. Type-species, *Musca affinis* Fallén (Enderlein, 1928: 53).
Agria, authors, not Robineau-Desvoidy.

Unnamed sp.—B.C. to N.B., s. to Ariz., Minn., and Pa.
   affinis, authors, not Fallén.

Genus SARCOFAHRTIA Parker

Sarcofahrtia Parker, 1916d: 131. Type-species, *ravinia* Parker (orig. des.).
Thelodiscus Aldrich, 1916a: 63.—Type-species, *indivisus* Aldrich (orig. des.) =*ravinia* Parker.

montanensis Parker, 1919b: 201.—Mont.; B.C., Wash., Idaho, Oreg.
    madisoni Parker, 1919b: 201.—Mont. **N. syn.**
    femoralis Reinhard, 1937: 63.—Wash.
ravinia Parker, 1916d: 133.—Mass.; N. Dak. to N.B., s. to Ohio and N.J., also Ga.
    indivisus Aldrich, 1916a: 64 (*Thelodiscus*).—Mass.

Genus WOHLFAHRTIA Brauer and Bergenstamm

Wohlfahrtia Brauer and Bergenstamm, 1889: 123 (1889: 55). Type-species, *Sarcophila magnifica* Schiner (Brauer, 1893: 501).
Paraphyto Coquillett, 1895k: 105. Type-species, *chittendeni* Coquillett (orig. des.) = *vigil* (Walker).

vigil (Walker), 1849: 831 (*Sarcophaga*).—N.S.; Alaska to N.S., s. to s. Calif., N. Mex., Iowa, and N.J., Europe.
    meigenii Schiner, 1862a: 567 (*Sarcophila*).—Austria.
    chittendeni Coquillett, 1895k: 105 (*Paraphyto*).—N.Y.
    opaca Coquillett, 1897: 123 (*Paraphyto*).—Colo., N. Mex.

Subfamily SARCOPHAGINAE

Much of the synonymy and other details of the following generic classification depends on unpublished information, which is not included because of present space limitations. The subgeneric classifications have been adopted for convenience, but available evidence suggests that not all of these groupings are natural.

Genus ANOLISIMYIA Dodge

Anolisimyia Dodge, 1955b: 183. Type-species, *blakeae* Dodge (orig. des.).

The type-species of this genus was reared from a subcutaneous lesion in the chameleon, *Anolis carolinensis*.

blakeae Dodge, 1955b: 183.—N.C.; Fla.

Genus ARCHIMIMUS Reinhard

Archimimus Reinhard, 1952a: 140. Type-species, *camatus* Reinhard (orig. des.).

camatus Reinhard, 1952a: 140.—Ariz.
sternalis (Reinhard), 1939a: 62 (*Emblemasoma*).—Tex.

Genus ARGORAVINIA Townsend

Argoravinia Townsend, 1917b: 190, 195. Type-species, *Sarcophaga argentea* Townsend (orig. des.) = *modesta* (Wiedemann).

modesta (Wiedemann), 1830: 363 (*Sarcophaga*).—Brazil; Tex., La., Mexico.

Genus BLAESOXIPHA Loew

Blaesoxipha Loew, 1861d: 386. Type-species, *grylloctona* Loew (mon.) ?=*laticornis* (Meigen).
Protodexia Townsend, 1912a: 117. Type-species, *synthetica* Townsend (orig. des.) = *hunteri* (Hough). **N. syn.**
Opsophyto Townsend, 1915a: 23. Type-species, *Sarcophaga opifera* Coquillett (orig. des.).
Tephromyiella Townsend, 1918b: 164 (as *Tephomyiella*, also p. 164). Type-species, *frankliniana* Townsend (orig. des.) = *atlanis* (Aldrich).
Stenaulacotheca Townsend, 1919c: 162. Type-species, *Sarcophaga spatulata* Aldrich (orig. des.). **N. syn.**
Stenolaucotheca, error.

Subgenus ACANTHODOTHECA Townsend

Acanthodotheca Townsend, 1918b: 159 (as genus). Type-species, *Sarcophaga prohibita* Aldrich (orig. des.). **N. syn.**
Acandotheca (error) Roback, 1954: 87.
Eleodiomyia Townsend, 1918b: 160. Type-species, *Sarcophaga eleodis* Aldrich (orig. des.). **N. syn.** (genus).
Sarcophodexia Townsend, 1918b: 161. Type-species, *Sarcophaga hamata* Aldrich (orig. des.). **N. syn.** (genus).
Notochaetopsis Townsend, 1918b: 162. Type-species, *Sarcophaga masculina* Aldrich (orig. des.). **N. syn.** (genus).

Members of this subgenus have been recorded as parasites of the adult scarabaeids *Canthon*, *Copris*, *Phyllophaga*, *Ligyrus*, and *Thyce*, adult tenebrionids, and the cockroach, *Arenivaga bolliana*.

acridiophagoides (Lopes and Downs), 1951: 588 (*Acanthodotheca*).— Brazil; Mich., N.Y. **N. comb.** ?=*alcedo*.
alcedo (Aldrich), 1916a: 132 (*Sarcophaga*).—Ind.; Kans. and Iowa to N.Y., s. to La. and N.C. **N. comb.**
apertella (Parker), 1920b: 106 (*Sarcophaga*).—B.C. **N. comb.**
beameri (Hall), 1931d: 52 (*Sarcophaga*).—Calif. **N. comb.**
complosa (Reinhard), 1947b: 123 (*Sarcophaga*).—Calif.; Alta., Ariz. **N. comb.**

compressa (Reinhard), 1947b: 126 (*Sarcophaga*).—Tex.; Calif. **N. comb.**

eleodis (Aldrich), 1916a: 128 (*Sarcophaga*).—N. Mex.; B.C. to Kans., s. to Calif. and N. Mex. **N. comb.**

excisa (Aldrich), 1916a: 127 (*Sarcophaga*).—Mass.; Minn., Kans. **N. comb.**

hamata (Aldrich), 1916a: 272 (*Sarcophaga*).—La.; Ala., Ga., ?N.C. **N. comb.**

magna (Aldrich), 1916a: 117 (*Sarcophaga*).—Fla.; Ga. **N. comb.**

masculina (Aldrich), 1916a: 130 (*Sarcophaga*).—Ga.; N.C. **N. comb.**

omani (Hall), 1931d: 53 (*Sarcophaga*).—N. Mex.; Calif., Tex. **N. comb.**

prohibita (Aldrich), 1916a: 133 (*Sarcophaga*).—Kans.; Utah, N. Mex., Tex. **N. comb.**

reperta (Reinhard), 1947b: 122 (*Sarcophaga*).—Oreg.; Wash., Calif. **N. comb.**

rudis (Aldrich), 1916a: 125 (*Sarcophaga*).—Ind.; Kans., Mo., Ark., Del., La. **N. comb.**

savoryi (Parker), 1920b: 105 (*Sarcophaga*).—B.C.; Wash., Ariz., N. Mex. **N. comb.**

spretor (Reinhard), 1947b: 121 (*Sarcophaga*).—Utah. **N. comb.**

thyceae (Reinhard), 1945c: 11 (*Sarcophaga*).—Calif. **N. comb.**

Subgenus ACRIDIOPHAGA Townsend

Acridiophaga Townsend, 1917a: 46 (as genus). Type-species, *Sarcophaga aculeata* Aldrich (orig. des.). **N. syn.**

Acridophaga, error.

The members of this subgenus parasitize adult and nymphal grasshoppers.

aculeata (Aldrich), 1916a: 143 (*Sarcophaga*).—Kans.; B.C. to Mass., s. to Utah and Fla. **N. comb.**

angustifrons (Aldrich), 1916a: 142 (*Sarcophaga*).—N. Mex.; B.C., Kans., N.Y., Tenn., N.C., Ga. **N. comb.**

caridei, authors, not Brèthes.

caridei (Brèthes), 1906: 299 (*Sarcophaga*).—Argentina; introd. Ont., not estab.

gavia (Aldrich), 1916a: 145 (*Sarcophaga*; as *aculeata* var.).—N. Mex.; B.C., Kans. **N. comb.**

potanini Rohdendorf, 1928: 46.—Mongolia; Wash., Oreg.

reversa (Aldrich), 1916a: 135 (*Sarcophaga*).—Ind.; B.C. to Que., s. to Calif., Mo., and Mass. **N. comb.**

taediosa (Aldrich), 1916a: 145 (*Sarcophaga*; as *aculeata* var.).— S. Dak.; B.C., Wyo., Kans., N. Mex. **N. comb.**

Subgenus BLAESOXIPHA Loew

As far as known the members of this subgenus normally parasitize grasshoppers.

atlanis (Aldrich), 1916a: 100 (*Sarcophaga*).—N.H.; B.C. to Que., s. to Calif. and Va. **N. comb.**

frankliniana Townsend, 1918b: 164 (*Tephromyiella*).—N.H.

australis (Blanchard), 1942: 344 (*Protodexia*).—Argentina; introd. B.C., Alta., Sask., Ont., not estab.

devulsa (Reinhard), 1947b: 101 (*Sarcophaga*).—Ohio. **N. comb.**

hunteri (Hough) *in* Hunter, 1898: 207 (*Sarcophaga*).—Kans.; B.C. to Mass., s. to Calif. and Fla., Mexico. **N. comb.**

synthetica Townsend, 1912a: 117 (*Protodexia*).—Mass.

aenigma Reinhard, 1947b: 103 (*Sarcophaga*; as ssp.).—N.Y.

neuquenensis (Blanchard), 1942: 356 (*Tephromyiella*).—Argentina; introd. Alta., Ont., not estab.

opifera (Coquillett), 1892f: 22 (*Sarcophaga*).—Calif.; B.C. to Iowa, s. to Calif., Mexico, and Tex. **N. comb.**

spatulata (Aldrich), 1916a: 105 (*Sarcophaga*).—Ohio; S. Dak. **N. comb.**

Subgenus FLETCHERIMYIA Townsend

Fletcherimyia Townsend, 1917b: 191 (as genus). Type-species, *Sarcophaga fletcheri* Aldrich (orig. des.). **N. syn.**

Peltopyga Townsend, 1917b: 191. Type-species, *Sarcophaga celarata* Aldrich (orig. des.). **N. syn.** (genus).

The larvae of this subgenus live in the water in pitcher plant leaves, and feed on other insects drowned therein.

celarata (Aldrich), 1916a: 242 (*Sarcophaga*).—Ala.; N.C. **N. comb.**

fletcheri (Aldrich), 1916a: 96 (*Sarcophaga*).—N.J.; Man., N.Y., N.C. **N. comb.**

jonesi (Aldrich), 1916a: 241 (*Sarcophaga*).—S.C.; Ga. **N. comb.**

rileyi (Aldrich), 1916a: 239 (*Sarcophaga*).—S.C.; N.C., Ala. **N. comb.**

Subgenus KELLYMYIA Townsend

Kellymyia Townsend, 1917b: 191 (as genus). Type-species, *Sarcophaga kellyi* Aldrich (orig. des.). **N. syn.**

Hystricocnema Townsend, 1919c: 160. Type-species, *Sarcophaga robusta* Aldrich (orig. des.)=*plinthopyga* (Wiedemann). **N. syn.** (genus and subg.).

Imparia Roback, 1954: 57. Type-species, *Sarcophaga impar* Aldrich (orig. des.). **N. syn.** (genus and subg.).

Kurtomyia Roback, 1954: 84. Type-species, *Sarcophaga postilla* Reinhard (orig. des.)=*californica* (Parker). **N. syn.** (genus and subg.).

The members of this subgenus are not so strictly parasitic as other members of the genus. Rearing records include adult beetles, grasshoppers, carrion, and old wounds in mammals.

californica (Parker), 1918b: 32 *(Sarcophaga).*—Calif. **N. comb.**
    postilla Reinhard, 1947b: 111 *(Sarcophaga).*—Calif.
cessator (Aldrich), 1916a: 84 *(Sarcophaga).*—Calif.; Oreg., Idaho, Ariz., N. Mex., Mexico. **N. comb.**
impar (Aldrich), 1916a: 220 *(Sarcophaga).*—Tex.; Iowa, s. to N. Mex. and Fla. **N. comb.**
kellyi (Aldrich), 1914b: 443 *(Sarcophaga).*—Kans.; B.C. to Ont., s. to N. Mex. **N. comb.**
    vericauda Coquillett *in* Hungerford, 1915: 135 *(Sarcophaga).* Nomen nudum.
plinthopyga (Wiedemann), 1830:360 *(Sarcophaga).*—Virgin Is.; Wash., s. to Mexico and Tex., also Fla., West Indies. **N. comb.**
    robusta Aldrich, 1916a: 268 *(Sarcophaga).*—Tex.

Subgenus SERVAISIA Robineau-Desvoidy

Servaisia Robineau-Desvoidy, 1863b: 429 (as genus). Type-species, *Sarcophaga erythrura* Meigen (as *erythrocera;* orig. des.).
Blaesoxiphotheca Townsend, 1918b: 159. Type-species, *caudata* Townsend (orig. des.) ?=*coloradensis* (Aldrich). **N. syn.** (genus and subg.).
Amblycoryphenes Townsend, 1918b: 161. Type-species, *Sarcophag amblycoryphae* Coquillett (orig. des.). **N. syn.** (genus and subg.).
Mantidophaga Townsend, 1919c: 160. Type-species, *stagmomantidis* Townsend (orig. des.). **N. syn.** (genus and subg.).
Servaisia, subg. *Sarpedia* Roback, 1954: 87. Type-species, *Sarcophaga setigera* Aldrich (orig. des.). **N. syn.** (genus).

Species of this subgenus have been reared from *Amblycorypha* (Tettigoniidae), *Melanoplus* spp. (Acrididae), and *Stagmomantis* (Mantidae).

alopecis (Reinhard), 1947b: 106 *(Sarcophaga).*—Iowa. **N. comb.**
amblycoryphae (Coquillett), 1904f: 187 *(Sarcophaga).*—Mass.; S. Dak., Iowa, Mo., N.Y. **N. comb.**
    lorena Roback, 1952: 45 *(Sarcophaga).*—N.Y.

coloradensis (Aldrich), 1916a: 139 (*Sarcophaga*).—Colo.; Calif. **N. comb.**
  caudata Townsend, 1918b: 159 (*Blaesoxiphotheca*).—Colo. ?Syn.
convena (Reinhard), 1947b: 103 (*Sarcophaga*).—Tex. **N. comb.**
falciformis (Aldrich), 1916a: 137 (*Sarcophaga*).—Calif.; Alta. to Man., s. to Calif. and N.C. **N. comb.**
flavipes (Aldrich), 1916a: 146 (*Sarcophaga*).—Ga. **N. comb.**
ignipes (Reinhard), 1947b: 105 (*Sarcophaga*).—Tex. **N. comb.** Probably=*amblycoryphae*.
pagella (Reinhard), 1947b: 100 (*Sarcophaga*).—Tex. **N. comb.**
prolepsis (Reinhard), 1947b: 120 (*Sarcophaga*).—Ariz. **N. comb.**
putilla (Reinhard), 1947b: 104 (*Sarcophaga*).—Ohio. **N. comb.**
setigera (Aldrich), 1916a: 138 (*Sarcophaga*).—Mass.; Iowa, Conn. **N. comb.**
stagmomantidis (Townsend), 1919c: 160 (*Mantidophaga*).—D.C.; Ill., Tex., N.C. **N. comb.** Probably=*flavipes*.
  austinana Reinhard, 1942d: 17 (*Sarcophaga*).—Tex.
uncata (Wulp), 1895: 268, key (descr. 1896: 277) (*Sarcophaga*).—Mexico; Idaho to Que., s. to Utah, Okla., and N.C., also Mexico. **N. comb.**
  marginata Aldrich, 1916a: 136 (*Sarcophaga*).—Ind.
websteri (Aldrich), 1916a: 141 (*Sarcophaga*).—Va.; Ohio. **N. comb.**

Subgenus SPECIOSIA Roback

Servaisia, subg. **Speciosia** Roback, 1954: 87. Type-species, *Fletcherimyia speciosa* Lopes (orig. des.). **N. syn.**

speciosa (Lopes), 1946: 121 (*Fletcherimyia*).—Mexico; Ariz., La. **N. comb.**

Subgenus SPIROBOLOMYIA Townsend

Spirobolomyia Townsend, 1917a: 43 (as genus). Type-species, *Sarcophaga singularis* Aldrich (orig. des.). **N. syn.**

  B. *flavipalpis* and *singularis* have been reared from millipeds of the genus *Spirobolus*.

basalis (Walker), 1852: 328 (*Sarcophaga*).—U.S.; Iowa to W. Va., s. to Ala. and Ga. **N. comb.**
  deceptiva Aldrich, 1916a: 186 (*Sarcophaga*).—Ala.
flavipalpis (Aldrich), 1916a: 256 (*Sarcophaga*).—Va.; Ind., N.Y., N.C. **N. comb.**
ohioensis (Hall), 1927: 176 (*Sarcophaga*).—Ohio. **N. comb.**
singularis (Aldrich), 1916a: 184 (*Sarcophaga*).—Va.; Ala., Ga. **N. comb.**

Unplaced Species of *Blaesoxipha*

hetaera (Reinhard), 1952a: 146(*Sarcophaga*).—Tex.; Ariz. **N. comb.**

Genus BOETTCHERIA Parker

Boettcheria Parker, 1914a: 65. Type-species, *latisterna* Parker (orig. des.).

The larvae of this genus are scavengers in dead insects, but sometimes may be parasitic. *B. litorosa* has been recorded as a parasite of bumblebees.

bisetosa Parker, 1914a: 69.—N.Y., Vt., Mass.; Alta. to N.H., s. to Fla.
cimbicis (Townsend), 1892f: 126 (*Sarcophaga*).—S. Dak.; S. Dak. to Maine, s. to Miss. and Fla.
  fernaldi Parker, 1914a: 72.—Mass.
latisterna Parker, 1914a: 67.—Vt.; Minn. to Que., s. to Tex. and Fla.
  irrisoris Reinhard, 1952a: 142 (*Sarcophaga;* as ssp.).—Tex.
litorosa (Reinhard), 1947b: 115 (*Sarcophaga*).—Calif.; Wash., Oreg., ?S. Dak.
  carata Roback, 1952: 48.—Calif. **N. syn.**
praevolans (Wulp), 1895: 268, key (descr. 1896: 275) (*Sarcophaga*).—Mexico; Ariz.
siccana Reinhard, 1947b: 117 (*Sarcophaga*).—Ariz.

Genus CAMPTOPS Aldrich

Camptops Aldrich, 1916a: 34. Type-species, *unicolor* Aldrich (orig. des.).

unicolor Aldrich, 1916a: 34.—La.; Iowa, Tex., Ala.

Genus CISTUDINOMYIA Townsend

Cistudinomyia Townsend, 1917a: 48. Type-species, *Sarcophaga cistudinis* Aldrich (orig. des.).

The sole species of this genus parasitizes tortoises.

  REFERENCE: Knipling, 1937 (biol.).

cistudinis (Aldrich), 1916a: 278 (*Sarcophaga*).—N.J.; Colo. to N.J., s. to Tex. and Fla.

Genus COLCONDAMYIA Reinhard

Colcondamyia Reinhard, 1963a: 82. Type-species, *falcifera* Reinhard (orig. des.).

Golcondamyia Reinhard, 1963b: 152, emend.
falcifera Reinhard, 1963a: 82.—Nev.

Genus COMASARCOPHAGA Hall

Comasarcophaga Hall, 1931b: 280. Type-species, *texana* Hall (orig. des.).
Tejasomyia Reinhard, 1945b: 68. Type-species, *nexilis* Reinhard (orig. des.) **N. syn.**

nexilis (Reinhard), 1945b: 69 (*Tejasomyia*).—Tex. **N. comb.**
texana Hall, 1931b: 280.—Tex.; Calif.

Genus EMBLEMASOMA Aldrich

Emblemasoma Aldrich, 1916a: 56. Type-species, *erro* Aldrich (orig. des.).

An undescribed species of this genus parasitizes cicadas.

albicoma Reinhard, 1939a: 63.—S. Dak.
erro Aldrich, 1916a: 56.—N.J.; Minn. to N.J., s. to Kans. and Ga., also ?B.C.
faciale Aldrich, 1916a: 58.—Ga.; Fla. Probably=*erro*.

Genus ERUCOPHAGA Reinhard

Erucophaga Reinhard, 1963a: 75. Type-species, *triloris* Reinhard (orig. des.).

triloris Reinhard, 1963a: 75.—Ariz.; Tex.

Genus HELICOBIA Coquillett

Helicobia Coquillett, 1895a: 317. Type-species, *Sarcophaga helicis* Townsend (orig. des.)=*rapax* (Walker).

The Nearctic species of this genus are scavengers or perhaps occasionally parasitic in insects and snails.

morionella (Aldrich), 1930c: 31 (*Sarcophaga*).—Cuba; N.C. to s. Calif., s. to Mexico and Fla., West Indies.
rapax (Walker), 1849: 818 (*Sarcophaga*).—?N. Amer.; Alta. to Que., s. to s. Calif., Mexico, and Fla.
helicis Townsend, 1892k: 220 (*Sarcophaga*).—Ohio.

Genus JOHNSONIA Coquillett

Johnsonia Coquillett, 1895a: 316. Type-species, *elegans* Coquillett (orig. des.).

Subgenus JOHNSONIA Coquillett

J. elegans is reported to parasitize the snail, *Succinia brevis*.

REFERENCE: Hallock, 1938 (key).

bivittata Curran, 1928d: 95.—P.R.; s. Tex.
borealis Reinhard, 1937: 62.—Mich.; Minn., Ohio, Pa., Conn.
elegans Coquillett, 1895a: 316.—Fla.
frontalis Aldrich, 1929c: 1.—Fla.

Subgenus STHENOPYGA Aldrich

Sthenopyga Aldrich, 1916a: 59 (as genus). Type-species, *globosa* Aldrich (orig. des.) = *rufitibia* (Wulp).
Camptopsis Townsend, 1918b: 162. Type-species, *miamensis* Townsend (orig. des.). **N. syn.** (genus and subg.).

miamensis (Townsend), 1918b: 163 (*Camptopsis*).—Fla.; N.C. **N. comb.**
rufitibia (Wulp), 1895: 270 (*Sarcophaga*).—Mexico; Tex., La., N.C.
  globosa Aldrich, 1916a: 59 (*Sthenopyga*).—La.
strigosa (Reinhard), 1945b: 67 (*Camptopsis*).—N.Y. **N. comb.**

Genus MECYNOCORPUS Roback

Acandotheca, subg. **Mecynocorpus** Roback, 1954: 89. Type-species, *Sarcophaga salva* Aldrich (orig. des.).

salvum (Aldrich), 1916a: 237 (*Sarcophaga*).—Ind.; Nebr. to Pa., s. to Tex. and Ga.

Genus METOPOSARCOPHAGA Townsend

Metoposarcophaga Townsend, 1917a: 46. Type-species, *Sarcophaga pachyprocta* Parker (orig. des.) = *importuna* (Walker).
Thelylepticocnema Townsend, 1917a: 43. Type-species, *Sarcophaga incurva* Aldrich (orig. des.).
Zygastropyga Townsend, 1917b: 191. Type-species, *aurea* Townsend (orig. des.).
Sabinata Parker, 1921: 112. Type-species, *catalina* Parker (orig. des.) = *villipes* (Wulp).
Cacotrophus Reinhard, 1947b: 99. Type-species, *beameri* Reinhard (orig. des.). **N. syn.**

The larvae of this genus occur in turtle eggs, carrion, mammal wounds, and caterpillars of *Megathymus* butterflies.

aldrichi (Parker), 1921: 114 (*Sabinata*).—Calif.

arizonica (Parker), 1921: 113 (*Sabinata*).—Ariz.; N. Mex.
    cantenea Roback, 1952: 48 (*Zygastropyga*).—Ariz. **N. syn.**
aurea (Townsend), 1917b: 195 (*Zygastropyga*).—Ariz.; Calif.
beameri (Reinhard), 1947b: 99 (*Cacotrophus*).—Ariz.; ?Mexico. **N. comb.**
conabilis (Reinhard), 1947b: 124 (*Sarcophaga*).—Tex. **N. comb.**
importuna (Walker), 1849: 819 (*Sarcophaga*).—?N. Amer.; Minn., Iowa, N.Y., Mass., Conn.
    pachyprocta Hagen, 1881b: 149 (*Sarcophaga*).—N.Y., Mass. **N. syn.**
    pachyprocta Parker, 1916c: 171 (*Sarcophaga*).—Mass.
    larga Aldrich, 1916a: 147 (*Sarcophaga*).—N.Y.
incurva (Aldrich), 1916a: 260 (*Sarcophaga*).—N. Mex.; B.C., Oreg., Utah.
    insurgens Aldrich, 1916a: 262 (*Sarcophaga*).—N. Mex.
    tothilli Parker, 1919a: 155.—B.C. **N. syn.**
pachyproctosa Parker, 1919a: 157.—B.C.; Calif.
sulculata (Aldrich), 1916a: 223 (*Sarcophaga*).—Tex.; Iowa, Kans., Okla.
villipes (Wulp), 1895: 269 (*Sarcophaga*).—Mexico; Ariz.
    catalina Parker, 1921: 112 (*Sabinata*).—Ariz.

Genus MICROCERELLA Macquart

Microcerella Macquart, 1851: 209 (1851: 236). Type-species, *rufomaculata* Macquart (orig. des.).
Xenoppia Townsend, 1915a: 20. Type-species, *hypopygialis* Townsend (orig. des.). **N. syn.**
Camptopyga Aldrich, 1916a: 41. Type-species, *aristata* Aldrich (orig. des.) = *hypopygialis* (Townsend). **N. syn.**
Hypopelta Aldrich, 1916a: 49. Type-species, *scrofa* Aldrich (orig. des.). **N. syn.**
Gymnopsoa Townsend, 1919c: 161. Type-species, *texana* Townsend (orig. des.). **N. syn.**

An undescribed species of this genus parasitizes snails.

hypopygialis (Townsend), 1915a: 20 (*Xenoppia*).—Ga.; N.J., Tex., La., Fla. **N. comb.**
    aristata Aldrich, 1916a: 42 (*Camptopyga*).—Tex.
scrofa (Aldrich), 1916a: 50 (*Hypopelta*).—Ind.; Wis. to N.J., s. to Kans., La., and Fla. **N. comb.**

texana (Townsend), 1919c: 161 (*Gymnopsoa*).—Tex.; N.Y., Conn., N.C., S.C., Fla. **N. comb.**
  nox Hall, 1931a: 217 (*Sarcophaga*).—S.C. **N. syn.**
  monela Reinhard, 1947b: 95 (*Xenoppia*).—Fla. **N. syn.**
valgata (Reinhard), 1947b: 96 (*Xenoppia*).—Tex. **N. comb.**

Genus NEOPHYTO Townsend

Neophyto Townsend, 1908: 55. Type-species, *Phyto setosa* Coquillett (orig. des.).
Phytodes Coquillett, 1910a: 127. Type-species, *hirculus* Coquillett (orig. des.). **N. syn.**
Oppiopsis Townsend, 1915a: 20. Type-species, *Brachycoma sheldoni* Coquillett (orig. des.). **N. syn.**
Harbeckia Aldrich, 1916a: 46. Type-species, *tessellata* Aldrich (orig. des.)=*sheldoni* (Coquillett). **N. syn.**

hirculus (Coquillett), 1910a: 127 (*Phytodes*).—Tex. **N. comb.**
nocturnalis Walton, 1915b: 162.—Md.; Ga.
olmaba Brimley, 1927: 235.—N.C.
setosa (Coquillett), 1895k: 99 (*Phyto*).—Ill.; Minn., Mo., Ga.
  anomala Townsend, 1908: 55.—Mo. **N. syn.**
sheldoni (Coquillett), 1898c: 236 (*Brachycoma*).—N.Y.: Conn., Pa., N.J., W. Va., Md. **N. comb.**
  tessellata Aldrich, 1916a: 47 (*Harbeckia*).—Pa.

Genus OXYSARCODEXIA Townsend

Oxysarcodexia Townsend, 1917b: 191. Type-species, *Sarcophaga peltata* Aldrich (orig. des.).

Subgenus OXYSARCODEXIA Townsend

As far as known, Nearctic members of this subgenus normally breed in dung.

bakeri (Aldrich), 1916a: 270 (*Sarcophaga*).—Cuba; Tex., Mexico, Brazil.
cingarus (Aldrich),1916a: 288 (*Sarcophaga*).—Pa.; Iowa to Que., s. to Kans. and Fla.
comparilis (Reinhard), 1939a: 66 (*Sarcophaga*).—Tex.
galeata (Aldrich), 1916a: 280 (*Sarcophaga*).—Ind.; Iowa to N.Y., s. to Kans., Ark., and Fla.
ochripyga (Wulp), 1895: 269, key (descr. 1896: 285) (*Sarcophaga*).—Mexico; Calif., Tex., La., S. Amer.
peltata (Aldrich), 1916a: 216 (*Sarcophaga*).—P.R.; Fla.
plebeja Lopes, 1946: 142.—Mexico; Tex.

ramosa (Reinhard), 1939a: 64 (*Sarcophaga*).—Tex.
trivialis (Wulp), 1895: 268, key (descr. 1896: 277) (*Sarcophaga*).—
Mexico; Ariz., N. Mex.
ventricosa (Wulp), 1895: 268, key (descr. 1896: 274) (*Sarcophaga*).—
Mexico; Ont. to Que., s. to Mexico and Fla.
assidua, authors, not Walker.

Subgenus DEXOSARCOPHAGA Townsend

Dexosarcophaga Townsend, 1917c: 221 (as genus). Type-species,
transita Townsend (orig. des.). **N. syn.**
Sarcomyia Roback, 1954: 64. Type-species, *Sarcophaga scelesta* Hall
(orig. des.) = *transita* (Townsend). **N. syn.** (genus and
subg.).
transita (Townsend), 1917c: 221 (*Dexosarcophaga*).—Brazil; Tex.
N. comb.
scelesta Hall, 1931b: 285 (*Sarcophaga*).—Tex. **N. syn.**

Genus PACHYGRAPHIA Brauer and Bergenstamm

Pachygraphia Brauer and Bergenstamm, 1891: 372, 379 (1891: 68,
75). Type-species, *Dexia virgata* Wiedemann (Townsend, 1916d: 8).
Unnamed sp.—Ga.

Genus PECKIA Robineau-Desvoidy

Peckia Robineau-Desvoidy, 1830: 335. Type-species, *imperialis*
Robineau-Desvoidy (Coquillett, 1910b: 585) = *praeceps*
(Wiedemann).
hillifera (Aldrich), 1916a: 210 (*Sarcophaga*).—Fla.
spectabilis (in part) Aldrich, 1916a: 212 (*Sarcophaga*).
U.S. records only.
volucris (Wulp), 1895: 269, key (descr. 1896: 285) (*Sarcophaga*).—
Mexico; Tex.

Genus RAVINIA Robineau-Desvoidy

Ravinia Robineau-Desvoidy, 1863b: 434. Type-species, *Sarcophaga
haematodes* Meigen (orig. des.) = *pernix* (Harris).
Euravinia Townsend, 1917b: 191. Type-species, *Ravinia communis*
Parker (orig. des.) = *querula* (Walker).
Miltoravinia Townsend, 1917b: 191. Type-species, *Sarcophaga planifrons* Aldrich (orig. des.).

The larvae of members of this genus normally occur in dung. *R. (Chaetoravinia) anandra* seems to be strictly parthenogenetic.

Subgenus CHAETORAVINIA Townsend

Chaetoravinia Townsend, 1917b: 190 (as genus). Type-species, *Helicobia quadrisetosa* Coquillett (orig. des.)=*derelicta* (Walker).

anandra (Dodge), 1956a: 187 (*Chaetoravinia*).—Ga.; Wis. to N.Y., s. to Kans., La., and Fla., also Hawaii. **N. comb.**

assidua (Walker), 1852: 328 (*Sarcophaga*).—U.S.; Ga., Fla.

coachellensis (Hall), 1931c: 182 (*Sarcophaga*).—Calif.

derelicta (Walker), 1852: 322 (*Sarcophaga*).—U.S.; Man. to N.Y., s. to Mexico and Fla.

    quadrisetosa Coquillett, 1901c: 17 (*Helicobia*).—D.C.

effrenata (Walker), 1861: 309 (*Sarcophaga*).—Mexico; Ariz., Tex., S. Amer.

errabunda (Wulp), 1895: 268, key (descr. 1896: 278) (*Sarcophaga*).— Mexico; Calif. to Kans., s. to Mexico.

    reinhardii Hall, 1928: 346 (*Sarcophaga*).—Tex.

laakei (Hall), 1931c: 181 (*Sarcophaga*).—Tex.; Calif. to N.Y. and Fla.

latisetosa Parker, 1914a: 63.—?Mass.; B.C. to Que., s. to Calif. and Fla. Probably=*assidua*.

Subgenus RAVINIA Robineau-Desvoidy

acerba (Walker), 1849: 824 (*Sarcophaga*).—N.S.; Alaska to N.S., s. to Oreg. and Ohio.

    peniculata Parker, 1914a: 58.—Maine.

floridensis (Aldrich), 1916a: 249 (*Sarcophaga*).—Fla.; Ga. Probably=*ochracea*.

lherminieri (Robineau-Desvoidy), 1830: 339 (*Myophora*).—Carolina; Wash. to Ont. and N.Y., s. to Mexico and Ga., also Hawaii.

    anxia Walker, 1849: 818 (*Sarcophaga*).—N. Amer.

    rediviva Walker, 1849: 823 (*Sarcophaga*).—Ont.

    comes Walker, 1852: 323 (*Sarcophaga*).—U.S.

    pallinervis Thomson, 1869: 535 (*Sarcophaga*).—Hawaii.

ochracea (Aldrich), 1916a: 255 (*Sarcophaga*; as *communis* var.)— Miss.; Ariz. to N.Y. and Fla.

pectinata (Aldrich), 1916a: 251 (*Sarcophaga*).—Ind.; Utah. to N.Y., s. to Ariz. and Ga.

    orbitalis Hall, 1928: 337 (*Sarcophaga*).—Colo. (as Colombia, error).

planifrons (Aldrich), 1916a: 249 (*Sarcophaga*).—N. Mex.; B.C. to
Sask. and Iowa, s. to s. Calif. and Tex.
duplicata Hall, 1928: 338 (*Sarcophaga*).—N. Mex.
pusiola (Wulp), 1895: 268, key (descr. 1896: 278) (*Sarcophaga*).—
Mexico; Minn. to s. Ont., s. to Ariz. and Ohio.
querula (Walker), 1849: 821 (*Sarcophaga*).—?N. Amer.; B.C. to
N. S., s. to n. Calif., Mo., and n. Ga.
avida Walker, 1849: 822 (*Sarcophaga*).—N.S.
rabida Walker, 1849: 823 (*Sarcophaga*).—N.S.
aspera Walker, 1849: 825 (*Sarcophaga*).—?N. Amer.
communis Parker, 1914a: 55.—?Mass.
sueta (Wulp), 1895: 268, key (descr. 1896: 281) (*Sarcophaga*).—
Mexico; Calif.

Genus SARCODEXIA Townsend

Sarcodexia Townsend, 1892j: 105. Type-species, *sternodontis* Townsend (orig. des.).

Subgenus RAFAELIA Townsend

Rafaelia Townsend, 1917a: 45 (as genus). Type-species, *rufiventris* Townsend (orig. des.). **N. syn.**
Tylomyia Roback, 1954: 83. Type-species, *Sarcophaga texana* Aldrich (orig. des.). **N. syn.** (genus and subg.).

S. texana (Aldrich) has been reared from carrion and old wounds on mammals.

ampulla (Aldrich), 1916a: 151 (*Sarcophaga*).—Tex. **N. comb.**
rufiventris (Townsend), 1917a: 45 (*Rafaelia*).—Mexico; Okla., Tex. **N. comb.**
texana (Aldrich), 1916a: 294 (*Sarcophaga*).—Tex. **N. comb.**
vernilis (Reinhard), 1947b: 117 (*Sarcophaga*).—Tex. **N. comb.**

Subgenus SARCODEXIA Townsend

The larvae of *S. sternodontis* are scavengers in dead insects and other arthropods, and have also been reported as parasitic on various insects.

sternodontis Townsend, 1892j: 106.—Jamaica; Tex., Ga., Fla., West Indies, S. Amer.
lambens, authors, not Wiedemann.

Subgenus SARCODEXIOPSIS Townsend

Sarcodexiopsis Townsend, 1917b: 190 (as genus). Type-species, *Sarcophaga biseriata* Aldrich (orig. des.). **N. syn.**

Aphelomyia Roback, 1954: 81. Type-species, *Sarcophaga welchi* Hall (orig. des.). **N. syn.** (genus and subg.).

S. welchi has been reared from dead crabs.

servilis (Aldrich), 1916a: 243 (*Sarcophaga*).—Ind.; Kans., Ill., N.Y., Ga. **N. comb.**

welchi (Hall), 1930: 2 (*Sarcophaga*).—Cuba; Fla. **N. comb.**

Subgenus TITANOGRYPA Townsend

Titanogrypa Townsend, 1917a: 44 (as genus). Type-species, *Sarcophaga alata* Aldrich (orig. des.). **N. syn.**

Cucullomyia Roback, 1954: 83. Type-species, *Sarcophaga pedunculata* Hall (orig. des.). **N. syn.** (genus and subg.).

Acandotheca, subg. *Lepyria* Roback, 1954: 89. Type-species, *Sarcophaga melampyga* Aldrich (orig. des.). **N. syn.** (genus and subg.).

The larvae of *S. alata* have been reported from mammal wounds.

alata (Aldrich), 1916a: 109 (*Sarcophaga*; as *melampyga* var.).—Fla. **N. comb.**

melampyga (Aldrich), 1916a: 107 (*Sarcophaga*).—Tex.; Ont., s. to Tex. and Fla. **N. comb.**

pedunculata (Hall), 1931b: 284 (*Sarcophaga*).—Tex.; Mexico. **N. comb.**

   placida (in part) Aldrich, 1925c: 24 (*Sarcophaga*). Tex. records only.

Genus SARCOFAHRTIAMYIA Hall

Sarcofahrtiamyia Hall, 1937b: 353. Type-species, *tenta* Hall (orig. des.).

tenta Hall, 1937b: 353.—Argentina; Fla.

Genus SARCOFAHRTIOPSIS Hall

Sarcofahrtiopsis Hall, 1933b: 261. Type-species, *Sarcofahrtia capitata* Curran (mon.).

Unnamed sp.—Fla.

Genus SARCOPHAGA Meigen

Sarcophaga Meigen, 1826: 14. Type-species, *Musca carnaria* Linnaeus (Westwood, 1840: 140).

Sarcotachinella Townsend, 1892a: 110. Type-species, *intermedia* Townsend (orig. des.) = *sinuata* Meigen.

Wohlfahrtiopsis Townsend, 1917a: 45. Type-species, *Sarcophaga johnsoni* Aldrich (orig. des.).
Bercaeopsis Townsend, 1917b: 192, 193, 195. Type-species, *Sarcophaga tetra* Aldrich (orig. des.).
Sarraceniomyia Townsend, 1917b: 192. Type-species, *Sarcophaga sarraceniae* Riley (orig. des.).
Scarabaeophaga Townsend, 1918b: 160. Type-species, *Sarcophaga utilis* Aldrich (orig. des.).
Petrosarcophaga Townsend, 1919b: 543. Type-species, *arizonica* Townsend (orig. des.).
Arachnidomyia Townsend, 1934b: 111. Type-species, *Sarcophaga davidsonii* Coquillett (orig. des.).
Sapromyia Roback, 1954: 64. Type-species, *Sarcophaga bullata* Parker (orig. des.). **N. syn.**
Idoneamima Dodge, 1956b: 242. Type-species, *sabroskyi* Dodge (orig. des.). **N. syn.**

The larvae of this genus feed on a wide variety of foods. Some species feed on carrion in the broadest sense, including dead insects, dead snails, and dead crabs and *Limulus* tossed up on sea beaches, and on dung and other decaying materials. One species has caused ulcerative myiasis in ground squirrels, a kitten, and man. Some species are parasitic in larval or pupal Lepidoptera, grasshoppers, adult scarabaeids (*Phyllophaga*), a snail (*Ventridens*), spiders of several genera, and larvae (presumably) of *Polistes* wasps. Larvae of one species devour eggs in the egg sacs of certain spiders.

acadiana Reinhard, 1947b: 109.—La.
acrophila (Dodge), 1956b: 247 (*Idoneamima*).—Ga.; N.C. **N. comb.**
aldrichi Parker, 1916b: 438.—Mass.; B.C. to se. Canada, s. to Ill. and Conn.
aratrix Pandellé, 1896: 191.—France; Man.
argyrostoma (Robineau-Desvoidy), 1830: 340 (*Myophora*).—Cape Province; Calif. to Que., s. to Tex. and N.C., cosmopolitan.
    barbata Thomson, 1869: 533.—Hawaii.
    falculata Pandellé, 1896: 185.—France.
arizonica (Townsend), 1919b: 544 (*Petrosarcophaga*).—Ariz.
beameri (Dodge), 1956b: 247 (*Idoneamima*).—Kans. **N. comb.**
bicolor (Dodge), 1956b: 247 (*Idoneamima*).—Ga. **N. comb.** ?= *tarsata*.
bishoppi Aldrich, 1916a: 258.—Tex.; Ariz. to Fla.
bullata Parker, 1916a: 359.—N.Y.; B.C. to Que., s. to Calif. and Fla.
    bison Aldrich, 1916a: fig. 110.—Unknown.

canadensis Hall, 1929: 322.—Alta.
carinata (Dodge), 1956b: 248 (*Idoneamima*).—Ohio. **N. comb.**
citellivora Shewell, 1950: 245.—Alta.; B.C.
cockerellae Aldrich, 1916a: 70.—N. Mex.; B.C., Colo.
cooleyi Parker, 1914b: 417.—Mont.; B.C. to Sask., s. to Ariz. and Kans., also Ont., N.Y.
crassipalpis Macquart, 1839: 112.—Canary Is.; Calif. to Que., s. to N.C., nearly cosmopolitan.
    securifera Villeneuve *in* Becker, 1908: 123.—Canary Is.
davidsonii Coquillett, 1892f: 24.—Calif.; Idaho, Tex., Mexico, West Indies.
elanis Reinhard, 1947b: 108.—N.Y.
elongata Aldrich, 1916a: 198.—Colo.; N. Mex.
epitheca Reinhard, 1952a: 143.—N.C.; Ga.
fattigi (Dodge), 1956b: 250 (*Idoneamima*).—Ga. **N. comb.**
fortisa Reinhard, 1947b: 110.—Ohio; Pa.
gracilis Aldrich, 1916a: 202.—Colo.; Utah, Wyo.
haemorrhoidalis (Fallén), 1817d: 237 (*Musca*).—Europe; Oreg. to Que., s. to s. Calif. and Fla., nearly cosmopolitan.
    georgina Wiedemann, 1830: 357.—Ga.
    aegra Walker, 1849: 821.—Mass.
harpax Pandellé, 1896: 189.—Europe; B.C. to Que., s. to Ariz., Kans., and N.Y.
helicivora (Dodge), 1956b: 250 (*Idoneamina*).—Mich. **N. comb.**
hesterna Reinhard, 1947b: 108.—N.Y.; Iowa, Ont., Vt., Ga.
hinei Aldrich, 1916a: 71.—Ohio; Wash., Alta., Ont., ?N.C.
houghi Aldrich, 1916a: 170.—Mass.; Wash., Iowa., Ill., N.Y., W. Va., Ga.
idonea Aldrich, 1916a: 90.—Mass.; N.Y., W. Va.
    montanensis Hallock, 1938: 98.—N.Y. **N. syn.**
johnsoni Aldrich, 1916a: 162.—Va.; Mass. to Tex. along seacoast.
juliaetta Aldrich, 1916a: 200.—Idaho.
libera Aldrich, 1916a: 235.—Mont.; B.C. to Ont., s. to Tenn. and Fla.
louisianensis (Dodge), 1956b: 253 (*Idoneamima*).—La. **N. comb.**
melanura Meigen, 1826: 23.—Europe; Que., N.Y., Mass., W. Va.
mimoris Reinhard, 1947b: 110.—Ohio; Mich. to Pa., s. to Ind. and D.C.
    neali Steyskal *in* Dodge, 1956b: 253. Nomen nudum.
minutissima Hall, 1929: 320.—D.C.; Pa., Va.
monticola (Dodge), 1956b: 254 (*Idoneamima*).—Ga.; Iowa, N.C. **N. comb.** ?=*elanis*.
morosa Aldrich, 1925c: 26.—Ont.; Mich., Ohio, N.Y.

nearctica Parker, 1916a: 422 (as *scoparia* ssp.).—?Mass.; B.C. to Maine, s. to Calif. and Ga. Probably=*scoparia*.
occidentalis Aldrich, 1916a: 198.—Idaho; B.C. to Calif., Mont.
ontariensis Hall, 1932a: 102.—Ont.; Mo., N.Y., Ga.
parallela Aldrich, 1916a: 123.—N.J.; Mo. and Ark. to N.Y., s. to Ga.
paulina Hall, 1937a: 212.—Ohio; Md., Ga.
perissa Reinhard, 1952a: 145.—Utah.
perspicax Aldrich, 1916a: 201.—Colo.; Utah, N. Mex.
pleomenda Reinhard, 1953b: 243 (n. name for *mendax* Reinhard).— Tenn.; Ga.
    mendax Reinhard, 1947b: 107 (preocc. Walker, 1859).— Tenn.
polistensis Hall, 1933a: 110.—Tex.; Ariz., Okla. ?=*libera*.
pratti (Dodge), 1956b: 258 (*Idoneamima*).—Ga.; W. Va. **N. comb.**
pubicornis (Coquillett), 1902g: 116 (*Brachycoma*).—Idaho. **N. comb.**
pulla Aldrich, 1916a: 72.—Pa.; Ohio, Ga.
rabunensis (Dodge), 1956b: 258 (*Idoneamima*).—Ga. **N. comb.**
sabroskyi (Dodge), 1956b: 245 (*Idoneamima*).—Ga.; Iowa to Mass., s. to Ga. **N. comb.**
sarraceniae Riley, 1874: 238.—S.C.; N.Y., N.C., Ala., Ga.
sarracenioides Aldrich, 1916a: 227 (as *tuberosa* var., but as sp. in key, p. 177).—Ind.; B.C. to Que., s. to Calif., Tex., and N.C.
scoparia Pandellé, 1896: 189.—East Prussia; Alaska.
seagoi (Dodge), 1956b: 259 (*Idoneamima*).—Ga.; N.C. **N. comb.**
semimarginalis Hall, 1931b: 283.—Tex.
shermani Parker, 1923: 124.—B.C.; B.C. to Que., s. to Calif., Kans., and N.Y.
    exuberans, authors, not Pandellé.
sigilla Reinhard, 1947b: 118.—Ariz.
sima Aldrich, 1916a: 91.—Md.; Mo., Tenn., W. Va., Ga.
sinuata Meigen, 1826: 22.—Europe; Alaska to Maine, s. to Utah and N.C., Palaearctic.
    intermedia Townsend, 1892a: 111 (*Sarcotachinella*).—Ill.
smithi (Dodge), 1956b: 260 (*Idoneamima*).—W. Va. **N. comb.**
snyderi (Dodge), 1956b: 260 (*Idoneamima*).—La. **N. comb.**
statuta Reinhard, 1952a: 144.—Mont.
subaenescens Aldrich, 1925c: 27.—N.J.
subdiscalis Aldrich, 1916a: 219.—N.J.; Iowa.
sudiai (Dodge), 1956b: 261 (*Idoneamima*).—Ga. **N. comb.**
sutilis Reinhard, 1952a: 144.—Tex.; Md., D.C., Ga., Ala. ?=*tarsata*.
tarsata Aldrich, 1916a: 75.—Fla.; La., Ga.

tetona Reinhard, 1952a: 141.—Wyo.
tetra Aldrich, 1916a: 89.—Ohio; Ont., Que., Ga.
thatuna Aldrich, 1916a: 196.—Idaho; B.C., Oreg.
triplasia Wulp, 1895: 269, key (descr. 1896: 283).—Mexico; Iowa, N.Y., La., Fla.
   dissidia Parker, 1917: 157 (as *fulvipes* ssp.).—N.Y.
   fulvipes, authors, not Macquart.
uliginosa Kramer, 1908a: 153.—Germany; Ont., Mass., Palaearctic.
utilis Aldrich, 1915a: 151.—Ind.; Utah to Que., s. to Ala. and Ga.
vancouverensis Parker, 1918a: 122.—B.C.; Wash., Colo.
wrangeliensis Parker, 1920b: 107.—B.C.; Alaska, Wash., Calif.
   cosmeta Reinhard, 1947b: 112.—Wash.
yorkii Parker, 1920a: 265.—N.Y. Probably=*subdiscalis*.

Genus TRICHARAEA Thomson

Tricharaea Thomson, 1869: 540. Type-species, *scatophagina* Thomson (mon.).
Sarcophagula Wulp, 1887: 173. Type-species, *Musca occidua* Fabricius (Coquillett, 1910b: 602). **N. syn.**
Sarothromyia Brauer and Bergenstamm, 1891: 365 (1891: 61). Type-species, *Sarcophila femoralis* Schiner (mon.). **N. syn.**

T. occidua has been reared from dung, decaying vegetable matter, and other wastes; *T. simplex*, from dead snails.

canuta (Wulp), 1896: 289 (*Sarcophagula*).—Mexico; Tex., Fla., S. Amer. **N. comb.**
occidua (Fabricius), 1794: 315 (*Musca*).—West Indies; Tex., Fla., S. Amer. **N. comb.**
simplex (Aldrich), 1916a: 39 (*Sarothromyia*; as *femoralis* var.).—Fla.; S.C. to Tex. along sea coast. **N. comb.**
   femoralis, authors, not Schiner.

Genus UDAMOPYGA Hall

Udamopyga Hall, 1938: 255. Type-species, *smagra* Hall (orig. des.).
niagarana (Parker), 1918c: 28 (*Sarcophaga*).—N.Y.; Iowa, Ga.

Unplaced Genus and Species of Sarcophagidae

Genus TEPHROMYIOPSIS Townsend

Tephromyiopsis Townsend, 1919b: 544. Type-species, *Megerlea rufocaudatata* Bigot (orig. des.).

rufocaudatata (Bigot), 1888a: 269 (*Megerlea*).—Rocky Mts.

Unplaced Species of Sarcophagidae

anaces Walker, 1849: 833 (*Sarcophaga*).—?N. Amer.
argyrocephala Macquart, 1846: 320 (1846: 192) (*Sarcophaga*).—Tex.
aterrima Robineau-Desvoidy, 1830: 336 (*Peckia*).—Carolina.
consobrina Robineau-Desvoidy, 1830: 344 (*Myophora*).—Pa.
epilepsalis French, 1900: 263 (*Gastrophilus*).—Ill. First stage larva only.
fulvipes Walker, 1852: 328 (*Sarcophaga*; preocc. Macquart, 1843).—U.S.
grisea Robineau-Desvoidy, 1830: 131 (*Araba*).—N. Amer.
incerta Walker, 1852: 324 (*Sarcophaga*).—Jamaica. Name applied to various spp. in U.S.
leucaniae Townsend, 1893h: 468 (*Sarcophaga*).—Ill.
mantivora Riley, 1875: 180 (*Sarcophaga*; as *carnaria* var.).—Ill. Probably *Blaesoxipha* (*Servaisia*) sp.
pallipes Walker, 1852: 329 (*Sarcophaga*).—U.S. Probably *Blaesoxipha* (*Spirobolomyia* sp.).
stimulans Walker, 1849: 817 (*Sarcophaga*).—?N. Amer. Probably *Ravinia latisetosa* Parker.
viridescens Robineau-Desvoidy, 1830: 342 (*Myophora*).—N.S.

Family TACHINIDAE

(Larvaevoridae)

By Curtis W. Sabrosky and Paul H. Arnaud, Jr.[1]

The larvae of the family Tachinidae are internal parasites in many kinds of insects but especially in the Lepidoptera, to a lesser extent in Coleoptera, Hemiptera, Orthoptera, and Hymenoptera (sawflies), and occasionally in a few others. The parasites usually attack immature stages, but some species regularly parasitize adults. A very few species attack other arthropods, such as terrestrial isopods. The larval feeding usually results in the death of the host. Depending on the species, a parasite may issue from larval, pupal, or adult stage of the host.

Most adult tachinids, which are usually heavily bristled and stout bodied, are strong fliers. Some adults visit flowers, and others may be observed walking rapidly over vegetation or ground, presumably

---

[1] The authors are especially indebted for invaluable assistance and advice to H. J. Reinhard, J. F. McAlpine, L. P. Mesnil, R. W. Crosskey, and D. F. Beneway, to the first two for critically reading the manuscript during its development, and to Dr. Mesnil for numerous helpful discussions and material. Grateful acknowledgement is also made to the National Science Foundation for Grant G-14359 to the junior author for assistance in the preliminary stages of the catalog.

in search of hosts or the plants upon which the hosts feed. Reproduction may be by minute microtype eggs, laid on plants and ingested by the hosts when they consume the plant tissue, by macrotype eggs laid on the host itself, or by larvae deposited in or near the host.

Many species have been utilized in biological control projects against insect pests, such as the gypsy and brown-tail moths and the European corn borer.

The taxonomy of the family is difficult and confused. The widely divergent points of view represented by the restricted genera of Townsend and the inclusive genera of Aldrich and others are in strong contrast. In general, our approach has been conservative, though each problem has been approached on its individual merits. In some tribes, the present classification can be regarded as satisfactory, or reasonably so, but for many others this catalog is necessarily only a preliminary organization. Most of the tribes recognized by Townsend (1934a–1942) in his "Manual of Myiology" have been adopted for present convenience, in the absence of any other published arrangement of the Nearctic genera, though with some combinations and generic transfers, notably where we agree with the recent work of Mesnil and coworkers in Europe. This is especially true in the Goniinae. Prior to and during the preparation of this catalog, there appeared a number of fascicles of Mesnil's section on the tachinids in Lindner's "Die Fliegen der Palaearktischen Region" (Mesnil, 1944–1963). Wherever possible, we have used and evaluated this work in relation to the Nearctic fauna. It should be noted that many of the Townsend tribes appear in the Mesnil classification as subtribes, a change that sometimes alters the relationship of the groups but not necessarily their composition.

No available generic key is entirely satisfactory, but any may prove helpful in troublesome cases. However, great differences in the classifications and names adopted will cause considerable difficulty.

Type designations are accepted from Brauer and von Bergenstamm (1889–1893) for some genera but not for others, although some workers have regarded the format of that work as indicating type designation. The International Commission on Zoological Nomenclature ruled in Opinion 98 (I.C.Z.N., 1928: 1) that Brauer and von Bergenstamm did not fix types for older generic names "except in the cases where they distinctly state that the species mentioned is the type of the genus." For new genera, where a new generic and a new specific name depend on the same description, the format is essentially "gen.n., sp.n." and is accepted as original designation (new International Code, Article 68a, i, Stoll et al., 1961: 67).

Distribution is difficult to determine in a family in which there has been so much misunderstanding and misidentification. For maximum reliability and uniformity of treatment, we have based the distributions primarily on the identifications of H. J. Reinhard, the leading living American authority on the family. These records have been supplemented from other reliable modern revisions, from material verified in the collections before us, and in certain genera by records considered unquestionably acceptable.

REFERENCES: Coquillett, 1897 (rev.); Adams *in* Williston, 1908; Curran, 1934a (keys to genera); Pantel, 1910, 1913 (methods of reproduction, immature stages); Greene, 1921a (puparia); Townsend, 1936a, 1936b (keys, tribes and genera of world); Clausen, 1940 (biol., immature stages), 1956 (biol. control, continental U.S.); Thompson, 1943a–1958 (cat., hosts and parasites); Mesnil, 1944–1963 (monog., Palaearctic Region, in progress); Emden, 1954 (keys, British genera, spp.); Herting, 1960 (biol., especially host lists); McLeod, McGugan, and Coppel, 1962 (biol. control, Canada).

Subfamily RHINOPHORINAE

(Melanophorinae)

The subfamily Rhinophorinae has been placed in the Sarcophagidae or the Tachinidae, or treated as a separate family. The postscutellum (infrascutellum) is distinctly though narrowly developed, and accordingly we believe that the group is best placed as a primitive subfamily of the Tachinidae. *Melanophora roralis*, probably the commonest species, is a parasite of terrestrial isopods.

Genus BEZZIMYIA Townsend

Bezzimyia Townsend, 1919b: 591. Type-species, *busckii* Townsend (orig. des.).
Lutzomyia Curran, 1934a: 387 (preocc. França, 1927). Type-species, *americana* Curran (mon.).
Pseudolutzomyia Rapp, 1945: 278 (n. name for *Lutzomyia* Curran). Type-species, *Lutzomyia americana* Curran (aut.).

americana (Curran), 1934a: 387 (*Lutzomyia*).—Ariz.
    latifrons Curran, 1934a: 396, 398 (figs. 52, 59) (*Lutzomyia*).—Ariz. Rejected alternate original name.
busckii Townsend, 1919b: 592.—Panama; ?s.Tex.

Genus MELANOPHORA Meigen

Melanophora Meigen, 1803: 279. Type-species, *Musca grossificationis* Linnaeus (mon.)=*roralis* (Linnaeus).

roralis (Linnaeus), 1758: 597 (*Musca*).—Europe; Kans. to Mich. and N.H., s. to La. and Fla., Palaearctic. ?Immigrant.
    stygia Harris, 1835: 599. Nomen nudum.

Genus STYLONEURIA Brauer and Bergenstamm

Styloneuria Brauer and Bergenstamm, 1891: 365 (1891: 61). Type-species, *manni* Brauer and Bergenstamm (mon.).

discrepans (Pandellé), 1896: 132 (*Phyto*).—France; Nfld. (W. L. Downes *in litt.*). ?Immigrant.

Subfamily PHASIINAE

(Gymnosomatinae)

The species of Phasiinae are chiefly parasites of adult Hemiptera-Heteroptera, with a few attacking adult Coleoptera.

REFERENCE: Dupuis, 1963 (monog.).

Tribe GYMNOSOMATINI

REFERENCE: Brooks, 1946a (rev.).

Genus CISTOGASTER Latreille

Cistogaster Latreille, 1829: 511. Type-species, *Musca globosa* Fabricius (Blanchard, 1840: 612).
Pallasia Robineau-Desvoidy, 1830: 239. Type-species, *Musca globosa* Fabricius (Coquillett, 1910b: 582).

amplifrons (Brooks), 1946a: 225 (*Pallasia*).—Alta.

Genus GYMNOCLYTIA Brauer and Bergenstamm

Gymnoclytia Brauer and Bergenstamm, 1893: 157 (1893: 69). Type-species, *Cistogaster divisa* Loew (mon.)=*occidua* (Walker).
Procistogaster Townsend, 1934c: 208. Type-species, *ferruginea* Townsend (orig. des.) =*immaculata* (Macquart).
Siphopallasia Brooks, 1946a: 225. Type-species, *Gymnosoma dubia* West (orig. des.).

atrota (Reinhard), 1935b: 173 (*Cistogaster*).—Ohio. **N. comb.**
dubia (West), 1925: 121 (*Gymnosoma*).—N.Y.; Man. to N.S., s. to Va., also ?Calif.
immaculata (Macquart), 1843a: 233 (1843: 76) (*Cistogaster*).—Carolina; B.C. to Que., entire U.S.; Mexico.
 pallasii Townsend, 1891b: 142 (*Cistogaster*).—S. Dak.
 ferruginea Townsend, 1934c: 208 (*Procistogaster*).—Amer.
minuta Brooks, 1946a: 227.—N.J.; D.C., Va., Tex.

occidentalis Townsend, 1908: 128.—N. Mex.; B.C., Idaho, Calif., Utah, Colo.
occidua (Walker), 1849: 692 (*Gymnosoma*).—N.S.; Mich. to N.S., sw. to Ariz., Mexico, and Ga.
    divisa Loew, 1863b: 321 (Cent. 4, no. 88) (*Cistogaster*).—Conn.
    americana Brauer and Bergenstamm, 1891: 388 (1891: 84) (*Eliozeta*).—Ga.
unicolor (Brooks), 1946a: 229 (*Procistogaster*).—Tex.; Calif., Utah, Mich., Ark., N.C., Fla.

Genus GYMNOSOMA Meigen

Rhodogyne Meigen, 1800: 39. Type-species, *Musca rotundata* Linnaeus (sub. mon., Hendel, 1908b: 66). Suppressed by I.C.Z.N., 1963b: 339.
Gymnosoma Meigen, 1803: 278. Type-species, *Musca rotundata* Linnaeus (mon.).

canadense (Brooks), 1946a: 220 (*Rhodogyne*).—Que.; Mich. to Que. and N.S., s. to N.C., also B.C. to Nev.
filiola Loew, 1872a: 92 (Cent. 10, no. 66).—Tex.; B.C. to Man., s. to Calif. and La., also Mich.
fuliginosum Robineau-Desvoidy, 1830: 237.—Carolina; B.C. to N.H., s. to Calif., Mexico, and S.C.
    latreillii Robineau-Desvoidy, 1830: 237.—Unknown.
occidentale Curran, 1927m: 144.—B.C.; B.C. to Ont., s. to Calif., Tex., and Va.
par Walker, 1849: 692.—N.S; Idaho, Sask. to N.S., s. to Va., also Ga.

Tribe TRICHOPODINI

The species of the tribe Trichopodini are parasites in adult Hemiptera-Heteroptera. The commonest and best-known species is *Trichopoda pennipes*, a parasite of the squash bug.

REFERENCE: Sabrosky, 1950d (key to genera).

Genus Eutrichopoda Townsend

Eutrichopoda Townsend, 1908: 134. Type-species, *nigra* Townsend (orig. des.).

Unnamed sp. (Sabrosky, 1950d: 366).—Panama (as Fla., error).

Genus TRICHOPODA Berthold

Trichopoda Berthold, 1827: 508. Type-species, *Thereva plumipes* Fabricius (Coquillett, 1910b: 616). As "Trichopode" (vernacular) by Latreille, 1825: 498.
Trichiopoda, emend.
Polistomyia Townsend, 1908: 132. Type-species, *Trichopoda trifasciata* Loew (orig. des.) = *plumipes* (Fabricius).

Subgenus GALACTOMYIA Townsend

Galactomyia Townsend, 1908: 135 (as genus). Type-species, *Trichopoda radiata* Loew (orig. des.) = *lanipes* (Fabricius).
Trichopodopsis Townsend, 1913b: 148, 313. Type-species, *Musca pennipes* Fabricius (orig. des.).

REFERENCE: Townsend, 1897b (key).

aurantiaca Townsend, 1891b: 140.—Va. Unrecognized.
lanipes (Fabricius), 1805: 220 (*Thereva*).—Carolina; Kans. to Conn., s. to Mexico and Fla.
   formosa Wiedemann, 1830: 268.—Ga.
   radiata Loew, 1863b: 321 (Cent. 4, no. 89).—D.C.
pennipes (Fabricius), 1781: 450 (*Musca*).—N. Amer.; Wash. (introd.), Calif. to Ont. and Mass., s. to Mexico and Fla.
   hirtipes Fabricius, 1805: 219 (*Thereva*).—Carolina.
   pennipes Fabricius, 1805: 219 (*Thereva*; preocc. Fabricius, 1781).—Carolina.
   jugatoria Say, 1829: 172 (1859b: 364) (*Phasia*).—Ind.
   flavicornis Robineau-Desvoidy, 1830: 284.—Carolina.
   cilipes Wiedemann, 1830: 276 (n. name for *pennipes* Fabricius, 1805).—Carolina.

Subgenus TRICHOPODA Berthold

indivisa Townsend, 1897b: 281 (as *histrio* var.).—Mexico; Calif., Ariz., N. Mex., Tex.
plumipes (Fabricius), 1805: 220 (*Thereva*).—Carolina; Kans. to Conn., s. to Tex. and Fla.
   histrio Walker, 1849: 697.—Unknown.
   trifasciata Loew, 1863b: 322 (Cent. 4, no. 90).—Conn.
subdivisa (Townsend), 1908: 133 (*Polistomyia*).—Calif.

Genus XANTHOMELANODES Townsend

Xanthomelana Wulp, 1892: 188 (preocc. Waterhouse, 1889). Type-species, *gracilenta* Wulp (Coquillett, 1910b: 620).

Xanthomelanodes Townsend, 1893l: 167 (n. name for *Xanthomelana* Wulp). Type-species, *Xanthomelana gracilenta* Wulp (aut.).
Erythrophasia Townsend, 1917d: 127. Type-species, *atripennis* Townsend (orig. des.)=*atripennis* (Say).
  REFERENCE: Sabrosky, 1950d (key).
arcuatus (Say), 1829: 173 (1859b: 365) (*Ocyptera*).—Ind.; Wash., Calif. to Wis. to N.H., s. to Mexico and Fla.
atripennis (Say), 1829: 172 (1859b: 363) (*Phasia*).—Ind.; Mich. to Vt., s. to Ark. and Fla., also? Mexico.
  corythus Walker, 1849: 797 (*Tachina*).—Ga.
  atripennis Townsend, 1891b: 145 (*Wahlbergia*).—Va.
  atripennis Townsend, 1917d: 127 (*Erythrophasia*).—Fla.
californicus Townsend, 1908: 129.—Calif.; B.C. to Calif., Mexico, and Tex.
flavipes (Coquillett), 1897: 72 (*Xanthomelana*).—Mass.; Que., N.Y., Vt., Va.
pictipes (Bigot), 1888a: 254 (*Stevenia*).—Wash. Unrecognized.

Tribe PHASIINI

Members of the tribe Phasiini parasitize adult Hemiptera-Heteroptera. The widely used name *Alophora* Robineau-Desvoidy (*Allophora*, emend.) does not apply to Nearctic species.
  REFERENCE: Brooks, 1945b (rev.).

Genus ALOPHORELLA Townsend

Alophorella Townsend, 1912c: 45. Type-species, *Thereva obesa* Fabricius (orig. des.).
Alophoropsis Townsend, 1915a: 20. Type-species, *Alophora phasioides* Coquillett (orig. des.).
Euphorantha Townsend, 1915a: 20. Type-species, *Alophora diversa* Coquillett (orig. des.).
Oedematopteryx Townsend, 1916h: 633. Type-species, *Alophora pulverea* Coquillett (orig. des.).
aeneoventris (Williston), 1886a: 296 (*Hyalomyia*).—Wash.; B.C. to N.S., s. to Calif., Tex., and N.C.
  robertsonii Townsend, 1891b: 136 (*Hyalomyia*).—Ill.
  brevineura West, 1925: 122 (*Phasia*).—N.Y.
  furva West, 1925: 123 (*Phasia*).—N.S.
alaskensis (Brooks), 1945b: 663 (*Alophoropsis*).—Alaska. **N. comb.**

diversa (Coquillett), 1897: 45 (*Alophora*).—Mass.; Kans. to Ont. and Que., s. to Tex. and Ga. **N. comb.**
  divisa (error) Johnson, 1925b: 184.
fumosa (Coquillett), 1897: 46 (*Alophora*).—N.J.; S. Dak. to N.H., s. to Ariz. and Va. **N. comb.**
nitida (Coquillett), 1897: 45 (*Alophora*).—Va. **N. comb.**
occidentalis (Brooks), 1945b: 664 (*Alophoropsis*).—N. Mex.; B.C., Alta., Mont., Ariz. **N. comb.**
opaca (Coquillett), 1897: 44 (*Alophora*).—Wash.; B.C. **N. comb.**
phasioides (Coquillett), 1897: 46 (*Alophora*).—N.H.; Alaska, B.C., Calif., Ariz., Colo., also Mich. to Que. and N.J. **N. comb.**
polita Brooks, 1945b: 669.—N. Mex.; B.C. to Mont., s. to Ariz. and N. Mex.
pulverea (Coquillett), 1897: 46 (*Alophora*).—Ont.; Wis. to N.B., s. to Miss. and Va., also Wash. **N. comb.**
subopaca (Coquillett), 1897: 47 (*Alophora*).—N.J.; B.C. to Calif. and Tex., also Wis. to N.B., s. to Ill. and S.C. **N. comb.**

Genus EUCLYTIA Townsend

Euclytia Townsend, 1908: 60. Type-species, *Clytia flava* Townsend (orig. des.).

flava (Townsend), 1891c: 372 (*Clytia*).—Ill.; B.C., Calif. to Maine, s. to Tex. and Va.

Genus HYALOMYA Robineau-Desvoidy

Hyalomya Robineau-Desvoidy, 1830: 298. Type-species, *Phasia semicinerea* Meigen (Westwood, 1840: 140) = *pusilla* (Meigen).
Hyalomyia, emend.
Alophorellopsis Townsend, 1927: 209. Type-species, *capitata* Townsend (orig. des.).
Hyalomyiopsis Brooks, 1945b: 676. Type-species, *Hyalomyia aldrichii* Townsend (orig. des.).

aldrichii Townsend, 1891b: 136.—S. Dak.; Alaska, B.C. to Vt., s. to Calif., Mexico, and Ala.
  celer Townsend, 1895a: 65.—N. Mex.
  pruinosa Robertson, 1901: 285 (*Phorantha*).—Ill.
  cara West, 1925: 123 (*Phasia*).—N.Y.
argentifrons (Brooks), 1945b: 675 (*Alophorellopsis*).—B.C.; ?Calif., ?Mexico. **N. comb.**

purpurascens Townsend, 1891b: 137.—Ill.; Wash., Idaho, Calif. to Que., s. to Mexico and Ga.
  calyptrata Coquillett, 1897: 44 (*Phorantha*).—Ill., Ky., D.C., Va.
  humeralis Robertson, 1901: 286 (*Phorantha*).—Ill.
robusta (Brooks), 1945b: 678 (*Hyalomyiopsis*).—Oreg.; B.C. to Calif. and Kans.

Genus PARAPHORANTHA Townsend

Paraphorantha Townsend, 1915a: 20. Type-species, *Alophora grandis* Coquillett (orig. des.).
auricaudata Brooks, 1945b: 661.—Oreg.; Wash. to Ark., s. to Calif., Mexico, and Tex.
grandis (Coquillett), 1897: 45 (*Alophora*).—Tex.; Ill., Mass., Va., N.C., Miss., Ga.
pollinosa Brooks, 1945b: 660.—Md.; Calif., Ariz., Tex., Miss., Ga.

Genus PHASIA Latreille

Phasia Latreille, 1804: 195. Type-species, *Conops subcoleoptratus* Linnaeus (sub. mon., Latreille, 1805: 379; also desig. Latreille, 1810: 444; as *Thereva subcoleoptrata* Fabricius).
Phorantha Rondani, 1861b: 209. Type-species, *musciformis* Rondani (mon.).
Paraphasia Townsend, 1915a: 20. Type-species, *Alophora fenestrata* Bigot (orig. des.).
albipennis (Brooks), 1945b: 657 (*Paraphasia*).—Sask. **N. comb.**
fenestrata (Bigot), 1888a: 255 (*Alophora*).—Nev.; Wash. to Calif. and Utah, also Ont. and Que., s. to Kans. and N.C.
  bridwelli Hine, 1902: 229 (*Phorantha*).—Kans.
  bidwelli, error.
  magnapennis Johnson, 1904: 19 (*Alophora*).—Que.
  phasiatrata Smith, 1915b: 98.—Mass.
nigra (Brooks), 1945b: 658 (*Paraphasia*).—Calif.; Wash. to Calif. and Utah. **N. comb.**
  nigrens, authors, not Wulp.

Genus PHASIOMYIA Townsend

Phasiomyia Townsend, 1915a: 20. Type-species, *Alophora splendida* Coquillett (orig. des.).

splendida (Coquillett), 1902g: 105 (*Alophora*).—N.H.; B.C., Alta., Wash., also Mich. to N.B., s. to N.C.
    meliceris Reinhard, 1955b: 234.—B.C. **N. syn.**

Genus PHORANTHELLA Brooks

Phoranthella Townsend, 1915a: 23. Nomen nudum (Opinion 205, I.C.Z.N., 1954a: 311).
Phoranthella Townsend, 1936a: 58 (also 1938b: 68). Nomen nudum.
Phoranthella Brooks, 1945b: 672. Type-species, *morrisoni* Brooks (mon.)=*punctigera* (Townsend). The generic name *Phoranthella* appears to date from Brooks (1945b) by a peculiar set of circumstances. Townsend's original publication did not make the name available in nomenclature, and his later works appeared when the "gen.n., sp.n." rule was no longer in effect.

punctigera (Townsend), 1891b: 135 (*Hyalomyia*).—Va.; Idaho to Ariz. to Ga. to Mass., also Ill., Ind.
    morrisoni Townsend, 1915a: 23 (also 1936a: 58, 1938b: 68). Nomen nudum.
    morrisoni Brooks, 1945b: 673.—Ga.

Genus TRICHOCLYTIA Townsend

Trichoclytia Townsend, 1916h: 633. Type-species, *Clytiomyia atrata* Coquillett (orig. des.).

atrata (Coquillett), 1895j: 53 (*Clytiomyia*).—Wash.; B.C. to Mont., s. to Calif., also Maine to Mass.

Unplaced Species of Phasiini

luctuosa Bigot, 1888a: 255 (*Alophora*).—Rocky Mts. Unrecognized.
occidentis Walker, 1852: 260 (*Hyalomya*).—U.S. Unrecognized.

Tribe FRERAEINI

The known hosts of the tribe Freraeini are adult carabid beetles.

Genus EUGYMNOGASTER Townsend

Gymnogaster Townsend, 1926: 25 (preocc. Gronovius, 1763). Type-species, *Gymnophania montana* Coquillett (orig. des.).
Eugymnogaster Townsend, 1931b: 328 (n. name for *Gymnogaster* Townsend). Type-species, *Gymnophania montana* Coquillett (aut.).

Gymnophania, authors, not Brauer and Bergenstamm.
montana (Coquillett), 1897: 50 (*Gymnophania*).—N.H.; Wash. and Alta. to Minn., s. to Calif. and Colo., also Ohio, N.H., and Maine.

Tribe CATHAROSIINI

Species of the tribe Catharosiini are parasites of adult Lygaeidae.

Genus CATHAROSIA Rondani

Catharosia Rondani, 1868a: 46 (1868: 86). Type-species, *Thereva pygmaea* Fallén (orig. des.).
Sciasma Coquillett, 1897: 69. Type-species, *nebulosa* Coquillett (orig. des.).
Microsciasma Townsend, 1915f: 234. Type-species, *minuta* Townsend (orig. des.). **N. syn.**

lustrans (Reinhard), 1944b: 58 (*Sciasma*).—Calif.; B.C.
minuta (Townsend), 1915f: 234 (*Microsciasma*).—Ariz. **N. comb.**
nebulosa (Coquillett), 1897: 69 (*Sciasma*).—Mass.; Mass. to Tex. and Fla.

Genus PROCATHAROSIA Villeneuve

Petia Coquillett, 1910a: 126 (preocc. Gray, 1839). Type-species, *calva* Coquillett (orig. des.).
Procatharosia Villeneuve, 1924: 31. Type-species, *Leucostoma flavicornis* Zetterstedt (mon.).

calva Coquillett, 1910a: 127.—Idaho; B.C. to Calif. and Tex.
frontalis (Smith), 1917a: 56 (*Sciasma*).—Mass. **N. comb.**

Tribe CYLINDROMYIINI

Members of the tribe Cylindromyiini are chiefly parasites of adult Hemiptera-Heteroptera, although a few parasitize adult Coleoptera.

Genus APINOPS Coquillett

Apinops Coquillett, 1897: 67. Type-species, *ater* Coquillett (orig. des.).

ater Coquillett, 1897: 68.—Ill.; Que. and Maine to Kans. and Ga. also Tex.

Genus APOSTROPHUS Loew

Apostrophus Loew, 1871: 310, 311. Type-species, *suspectus* Loew (Coquillett, 1910b: 509). Coquillett's designation, which referred to a publication by Loew, 1870, that is now known to be 1872, is nevertheless accepted under Article 67g of the International Code of Zoological Nomenclature (Stoll et al., 1961: 65).

anthophilus Loew, 1871: 310.—Germany; N.W.T., Alta.
incompletus (Curran), 1926b: 84 (*Besseria*).—Ont. **N. comb.**

Genus BESKIA Brauer and Bergenstamm

Beskia Brauer and Bergenstamm, 1889: 139 (1889:-71). Type-species, *cornuta* Brauer and Bergenstamm (orig. des., as gen. n., sp. n.) ?=*aelops* (Walker).

aelops (Walker), 1849: 796 (*Tachina*).—Ga.; Ariz. to Va., s. to Mexico and Fla., ?Neotropical.

Genus BESSERIA Robineau-Desvoidy

Besseria Robineau-Desvoidy, 1830: 232. Type-species, reflexa Robineau-Desvoidy (mon.)=*melanura* (Meigen).
Oedemasoma Townsend, 1908: 80. Type-species, *nuda* Townsend (orig. des.)=*brevipennis* (Loew).

brevipennis (Loew), 1863b: 322 (Cent. 4, no. 91) (*Wahlbergia*).— Nebr.; B.C. to N. Dak., s. to Calif., Ariz., and Nebr.
nuda Townsend, 1908: 80 (*Oedemasoma*).—Nev.

Genus CYLINDROMYIA Meigen

Cylindromyia Meigen, 1803: 279. Type-species, *Musca brassicaria* Fabricius (mon.).
Ocypterodes Townsend, 1916h: 631. Type-species, *Ocyptera euchenor* Walker (orig. des.; misident.)=*fumipennis* (Bigot).
Aldrichocyptera Townsend, 1936c: 488. Type-species, *Cylindromyia alticola* Aldrich (mon.).
Ocyptera, authors, not Latreille.

In the Eastern Hemisphere and in older American literature, *Cylindromyia* has usually been known as *Ocyptera* Latreille (1804), but the type-species of the latter, by designation of Latreille (1810), is *Musca lateralis* Fabricius, now a synonym of *Eriothrix rufomaculata* De Geer in the subfamily Dexiinae.

REFERENCE: Aldrich, 1926b (rev.).

Subgenus APINOCYPTERA Townsend

Apinocyptera Townsend, 1915k: 94 (as genus). Type-species, *signata* Townsend (orig. des.).
Odontocyptera Townsend, 1915f: 233. Type-species, *nana* Townsend (orig. des.).

nana (Townsend), 1915f: 233 (*Odontocyptera*).—Mexico (Chih.); Wash. to Kans., s. to Calif., Mexico, and Miss., also N.C., Fla.
signatipennis (Wulp), 1892: 187 (*Ocyptera*).—Mexico; Calif., Tex., Guatemala.
  limbata Aldrich, 1926b: 25.—Calif. **N. syn.**

Subgenus CYLINDROMYIA Meigen

alticola Aldrich, 1926b: 16.—Colo.; Mont., S. Dak., Man.
armata Aldrich, 1926b: 22.—Md.; Mont., Calif. to Kans. and N. Mex., Mich., Mass. to Ga.
atra (Röder), 1885: 344 (*Ocyptera*).—P.R.; s. Tex., Mexico.
binotata (Bigot), 1878: 44 (*Ocyptera*).—Md.; Man., Colo. and S. Dak. to N.Y., s. to Tex. and Fla.
  argentea Townsend, 1891b: 144 (*Ocyptera*).—D.C., Iowa.
californica (Bigot), 1878: 42 (*Ocyptera*).—Calif.; B.C. to N. Dak., s. to Calif. and Tex., also Mich. and Ind. to N.Y.
  intermedia Meigen of Aldrich, 1926b: 12.
decora Aldrich, 1926b: 21.—N.H.; B.C. to N.H., s. to Calif., Tex., and Md.
euchenor (Walker), 1849: 696 (*Ocyptera*).—Mass., Nfld.; Minn. to Nfld., s. to N. Mex. and Fla., also Calif.
  lateralis Harris, 1835: 599 (*Ocyptera*). Nomen nudum.
fumipennis (Bigot), 1878: 43 (*Ocyptera*).—Calif.; B.C. to Conn., s. to Calif. and Fla.
  vulgaris Aldrich, 1926b: 20.—Va.
propusilla Sabrosky and Arnaud (for *pusilla* Aldrich).—Va.; Wis. to Mass., s. to Va., also Mexico, Tex., Fla. **N. name.**
  nigra Aldrich, 1926b: 11 (preocc. Villeneuve, 1917).—Va.
  pusilla Aldrich, 1927: 18 (n. name for *nigra* Aldrich but preocc. Meigen, 1824).—Va.
  argentea, authors, not Townsend.
uniformis Aldrich, 1926b: 24.—Mexico (D.F.); Calif., Ariz., El Salvador, Trinidad, Dominica.

Subgenus NEOCYPTERA Townsend

Neocyptera Townsend, 1916e: 32 (as genus). Type-species, *Ocyptera dosiades* Walker (orig. des.).

Aubaeina Enderlein, 1937b: 442. Type-species, *Ocyptera dosiades* Walker (orig. des.).

compressa Aldrich, 1926b: 10.—Alta.; Alta. to Mich., s. to Nev. and Colo.
dosiades (Walker), 1849: 695 (*Ocyptera*).—N.S.; Alaska to Calif., also N. Dak. to N.S., s. to Colo. and N.J.

Unplaced Species of *Cylindromyia*

carolinae (Robineau-Desvoidy), 1830: 232 (*Parthenia*).—Carolina. Unrecognized.
epytus (Walker), 1849: 694 (*Ocyptera*).—Ga. Unrecognized.

Genus EPIGRIMYIA Townsend

Epigrimyia Townsend, 1891c: 375. Type-species, *polita* Townsend (orig. des.).

illinoensis Robertson, 1901: 286.—Ill.; Mich., Ohio, Md., Va., and Tex. to Ga.
polita Townsend, 1891c: 376.—Va.; Kans. to Mich. to N.H., s. to Tex. and Fla., also Ariz.

Genus EUSCOPOLIA Townsend

Euscopolia Townsend, 1892a: 123. Type-species, *dakotensis* Townsend (orig. des.).
Politomyia Reinhard, 1935b: 165. Type-species, *angulineura* Reinhard (orig. des.)=*dakotensis* Townsend.

dakotensis Townsend, 1892a: 124.—S. Dak.; B.C. to Oreg., also Nebr., Mich., Ohio, Va.
    angulineura Reinhard, 1935b: 166 (*Politomyia*).—Wash. **N. syn.**

Genus HEMYDA Robineau-Desvoidy

Hemyda Robineau-Desvoidy, 1830: 226. Type-species, *aurata* Robineau-Desvoidy (mon.).

aurata Robineau-Desvoidy, 1830: 226.—Pa.; B.C. to N.H., s. to Calif., Mexico, and Ga.
    latipennis Curran, 1924i: 252.—Man.

Genus ICHNEUMONOPS Townsend

Ichneumonops Townsend, 1908: 82. Type-species, *mirabilis* Townsend (orig. des.).

mirabilis Townsend, 1908: 84.—N. Mex.; Ariz., Colo.

Tribe IMITOMYIINI

Genus IMITOMYIA Townsend

Himantostoma Loew, 1863b: 320 (Cent. 4, no. 87) (preocc. Agassiz, 1862). Type-species, *sugens* Loew (mon.).
Imitomyia Townsend, 1912c: 49 (n. name for *Himantostoma* Loew). Type-species, *Himantostoma sugens* Loew (aut.).
Saskatchewania Smith, 1915a: 153. Type-species, *canadensis* Smith (orig. des.)=*sugens* (Loew).

sugens (Loew), 1863b: 320 (Cent. 4, no. 87) (*Himantostoma*).—Ill.; Oreg. to Sask. to Mich., s. to Calif. and Colo.
  canadensis Smith, 1915a: 153 (*Saskatchewania*).—Sask.

Tribe CINOCHIRINI

The European *Cinochira atra* Zetterstedt has been reared from a species of Lygaeidae.

Genus CINOCHIRA Zetterstedt

Cinochira Zetterstedt, 1845: 1358. Type-species, *atra* Zetterstedt (mon.).
Baromyia Reinhard, 1957: 100. Type-species, *mitis* Reinhard (orig. des.). **N. syn.**

mitis (Reinhard), 1957: 101 (*Baromyia*).—Tex. **N. comb.**

Tribe LEUCOSTOMATINI

The species of Leucostomatini are parasites of adult Hemiptera and Coleoptera.

Genus CALYPTROSOMUS Reinhard

Calyptrosomus Reinhard, 1956c: 105. Type-species, *dapsilis* Reinhard (orig. des.).

dapsilis Reinhard, 1956c: 105.—Calif.; Ariz., Kans., Okla., Tex.

Genus CLAIRVILLIA Robineau-Desvoidy

Clairvillia Robineau-Desvoidy, 1830: 234. Type-species, *Ocyptera pusilla* Meigen (mon.; misident.)=*biguttata* (Meigen).
Neodionaea Townsend, 1916h: 631. Type-species, *Dionaea nitoris* Coquillett (orig. des.).

amicta Reinhard, 1962a: 169.—Ariz.

nitoris (Coquillett), 1898c: 235 (*Dionaea*).—Oreg.; B.C. to Calif. and Nev.
timberlakei (Walton), 1914c: 91 (*Dionaea*).—Utah; B.C. to S. Dak., s. to Calif., Tex., and Mo.

Genus LEUCOSTOMA Meigen

Leucostoma Meigen, 1803: 279. Type-species, *Ocyptera simplex* Fallén (sub. mon., Meigen, 1824: 234).
Cyclodionaea Townsend, 1915f: 233. Type-species, *acuminata* Townsend (orig. des.)=*aterrimum* (Villers).
Paradionaea Townsend, 1916h: 631. Type-species, *Leucostoma atra* Townsend (orig. des.)=*simplex* (Fallén).
Neopsalida Townsend, 1916h: 632. Type-species, *Leucostoma neomexicana* Townsend (orig. des.)=*aterrimum* (Villers).
Parapsalida Townsend, 1916h: 632. Type-species, *Phyto nigricornis* Townsend (orig. des.)=*gravipes* Wulp.

REFERENCE: Reinhard, 1956b (rev.).

acirostre Reinhard, 1956b: 162.—Tex.; Wash. to Calif. and Tex., also Ind. to N.J. and Va., Miss., Fla., ?Kans., ?Tenn.
aterrimum (Villers), 1789: 548 (*Musca*).—Europe; Calif. to Wyo. and Kans., s. to Mexico and Tex., also Wis., Mich., Ind., Europe.
   neomexicana Townsend, 1892e: 169.—N. Mex.
   acuminata Townsend, 1915f: 234 (*Cyclodionaea*).—Calif.
effrenatum Reinhard, 1956b: 165.—Utah.
gravipes Wulp, 1890: 207.—Mexico; Wash. to Nebr., s. to Calif., Mexico, and Tex., also Mich. to N.Y., s. to Va.
   senilis Townsend, 1892e: 81 (*Phyto*).—N.Y.
   nigricornis Townsend, 1892e: 170 (*Phyto*).—N. Mex.
perrarum Reinhard, 1956b: 164.—Utah; Oreg., Calif., Nev., Ariz.
simplex (Fallén), 1820f: 8 (*Ocyptera*).—Sweden; B.C. to N.H., s. to Calif., Tex., and Va.
   atra Townsend, 1891c: 380.—Ill.
vapulare Reinhard, 1956b: 165.—Calif.

Tribe STRONGYGASTRINI

The species of Strongygastrini are parasites of adult Coleoptera.

Genus Campogaster Rondani

Campogaster Rondani, 1856: 80. Type-species, *parvula* Rondani (orig. des.)=*exigua* (Meigen).

Syntomogaster Schiner, 1861b: 140. Type-species, *Tachina singularis* Egger (Townsend, 1916b: 9)=*exigua* (Meigen).

exigua (Meigen), 1824: 367 (*Tachina*).—Europe; introd. Wash., Man., N. Dak., Minn., Nebr., N.J., ?estab.

Genus CLISTOMORPHA Townsend

Clistomorpha Townsend, 1892e: 79. Type-species, *hyalomoides* Townsend (orig. des.)=*didyma* (Loew).

REFERENCE: Brooks, 1942 (rev.).

didyma (Loew), 1863b: 320 (Cent. 4, no. 86) (*Xysta*).—Ill.; Calif., Alta., S. Dak. to N.Y., s. to Colo. and Va.

hyalomoides Townsend, 1892e: 80.—N.Y.

Genus HYALOMYODES Townsend

Hyalomyodes Townsend, 1893g: 429. Type-species, *weedii* Townsend (mon.)=*triangulifer* (Loew).

REFERENCE: Brooks, 1942 (key); Thompson, 1954 (biol., immature stages).

californicus Townsend, 1908: 126.—Calif.; ?B.C. ?=*triangulifer*.

robustus Townsend, 1908: 125.—N. Mex.; B.C., Wash., Idaho, Ariz., N.S. to Pa.

triangulifer (Loew), 1863b: 319 (Cent. 4, no. 85) (*Hyalomyia*).—N.Y.; B.C. to N.S., s. to Calif., Mexico, and Ga.

weedii Townsend, 1893g: 430.—N.H.

Tribe HESPEROPHASIINI

REFERENCE: Curran, 1927i (rev., as *Hesperophasia, sens. lat.*).

Genus COLEOPHASIA Townsend

Coleophasia Townsend, 1931b: 327. Type-species, *Hesperophasia pacifica* Curran (orig. des.).

pacifica (Curran), 1927i: 301 (*Hesperophasia*).—Oreg.; Calif.

Genus HESPEROPHASIA Townsend

Hesperophasia Townsend, 1915f: 220. Type-species, *setosa* Townsend (orig. des.).

setosa Townsend, 1915f: 221.—N. Mex.; Ariz., Mexico, Nebr. to N.Y., s. to Kans. and Md.

Genus HESPEROPHASIOPSIS Townsend

Hesperophasiopsis Townsend, 1915f: 221. Type-species, *californica* Townsend (orig. des.).

aldrichi (Curran), 1927i: 301 (*Hesperophasia*).—Sask.; Alta., N. Dak.
californica Townsend, 1915f: 222.—Calif.
nigripennis (Curran), 1927i: 303 (*Hesperophasia*).—Calif.; Oreg. **N. comb.**

Tribe EUTHERINI

The few known hosts of Eutherini are adult Pentatomidae.

Genus EUTHERA Loew

Euthera Loew, 1866b: 46 (Cent. 7, no. 85). Type-species, *tentatrix* Loew (mon.).

REFERENCE: Brooks, 1945a (key).

bicolor Coquillett, 1902g: 114.—Tex.; Ark., Mexico.
setifacies Brooks, 1945a: 79.—B.C.; Wash., Maine.
tentatrix Loew, 1866b: 46 (Cent. 7, no. 85).—N.Y.; Ill. to Mass., s. to Tex. and Fla., also Calif., N. Mex.

Tribe CHIRICAHUIINI

Genus CHIRICAHUIA Townsend

Chiricahuia Townsend, 1918a: 177. Type-species, *cavicola* Townsend (mon.).

cavicola Townsend, 1918a: 178.—Ariz.

Tribe PALPOSTOMATINI

As far as known, the species of Palpostomatini parasitize adult scarabaeid beetles.

Genus Eutrixopsis Townsend

Eutrixopsis Townsend, 1919c: 166. Type-species, *javana* Townsend (orig. des.).

javana Townsend, 1919c: 166.—Java; Japan to Java, introd. N.J., not estab. Townsend, 1938b: 216, maintains that the introduced form, from Japan, was a different species.

Genus Hamaxia Walker

Hamaxia Walker, 1860b: 153. Type-species, *incongrua* Walker (orig. des., as gen. n., sp. n.).
Ochromeigenia Townsend, 1919b: 578. Type-species, *ormioides* Townsend (orig. des.) =*incongrua* Walker.

incongrua Walker, 1860b: 153.—Molucca Is. (Amboina); Japan and Korea to East Indies, introd. Pa. and N.J., not estab.
  ormioides Townsend, 1919b: 578 (*Ochromeigenia*).—Java.

Subfamily PROSENINAE
(Dexiinae of authors)

Most of the species of Proseninae parasitize larvae of Coleoptera, but some attack certain lepidopterous larvae and nocturnal Orthoptera.

Tribe GLAUROCARINI

Genus CENOSOMA Wulp

Cenosoma Wulp, 1890: 44, key. Type-species, *signiferum* Wulp (sub. mon., Wulp, 1890: 167).

signiferum Wulp, 1890: 167.—Mexico; Oreg., Calif. to La., also Ont. and Que., s. to Fla.

Genus OESTROPHASIA Brauer and Bergenstamm

Oestrophasia Brauer and Bergenstamm, 1889: 145 (1889: 77). Type-species, *clausa* Brauer and Bergenstamm (Townsend, 1892b: 133).

calva Coquillett, 1920g: 109.—Ariz., Ont.; Calif. to Tex. and Mexico, also Mich. to Mass., s. to Fla.
clausa Brauer and Bergenstamm, 1889: 146 (1889: 78).—Colo.; Idaho, Calif., Utah, Ariz., N. Mex.
  setosa Coquillett, 1902g: 110.—Colo.

Tribe ORMIINI

The species of the tribe Ormiini are parasites of nocturnal Orthoptera.

REFERENCES: Sabrosky, 1953c (rev.); Nutting, 1953 (biol.).

Genus EUPHASIOPTERYX Townsend

Euphasiopteryx Townsend, 1915a: 23. Type-species, *Phasiopteryx australis* Townsend (orig. des.).

brevicornis brevicornis (Townsend), 1919b: 548 (*Ormia*).—Tex.
  ssp. **nuttingi** Sabrosky, 1953c: 293.—Mass.; N.Y. and Mass., s. to Va., also Kans.
dominicana (Townsend), 1919b: 548 (*Ormia*).—Dominican Republic; Fla., Cuba, P.R., Guatemala, Panama.
ochracea (Bigot), 1888a: 268 (*Pyrrosia*).—Mexico; Colo. to Mo., s. to Ariz. and Mexico, also Mich., N.C., Fla.
  montana Townsend, 1912a: 114 (*Phasiopteryx*).—Colo.
reinhardi Sabrosky, 1953c: 291.—Ohio; N.J., Tex.

Genus ORMIA Robineau-Desvoidy

Ormia Robineau-Desvoidy, 1830: 428. Type-species, *punctata* Robineau-Desvoidy (mon.).
Phasiopteryx Brauer and Bergenstamm, 1889: 147 (1889: 79). Type-species, *bilimekii* Brauer and Bergenstamm (Coquillett, 1910b: 588).

bilimekii (Brauer and Bergenstamm), 1889: 147 (1889: 79) (*Phasiopteryx*).—Mexico; Tex.
lineifrons Sabrosky, 1953c: 175.—P.R.; Mexico (Baja Calif.), Tex. to Fla.
punctata Robineau-Desvoidy, 1830: 428.—Antilles; Fla., Cuba, Haiti, P.R.
  punctata Coquillett, 1895j: 52 (*Clytiomyia*).—Fla.

Tribe TRIXODINI

Genus TRIXODES Coquillett

Trixodes Coquillett, 1902a: 201. Type-species, *obesus* Coquillett (orig. des.).

obesus Coquillett, 1902a: 202.—Mexico (Chih.); Ariz., N. Mex.

Tribe TRIXINI

Some members of the tribe Trixini parasitize scarabaeid grubs in the soil. Species of *Phasiops* have been reared from *Crambus* larvae webbing maize roots, and from *Tabanus* larvae.

Genus EUMEGAPARIA Townsend

Eumegaparia Townsend, 1908: 58. Type-species, *Megaparia flaveola* Coquillett (orig. des.).

flaveola (Coquillett), 1902g: 121 (*Megaparia*).—Colo.; B.C. to Mont., s. to Ariz. and Colo.

Genus PHASIOPS Coquillett

Phasiops Coquillett, 1899d: 219. Type-species, *flavus* Coquillett (orig. des., as gen. n., sp. n.).

flavus Coquillett, 1899d: 219.—N.J.; Ill. to N.Y. and N.J., also La., Fla.

Tribe AULACEPHALINI

Flies of the genus *Eutrixa* are nocturnal and larviposit in adult *Phyllophaga*.

Genus EUTRIXA Coquillett

Eutrixa Coquillett, 1897: 72. Type-species, *Tachina masuria* Walker (orig. des.; misident.)=*exilis* (Coquillett).

exilis (Coquillett), 1895j: 53 (*Clytiomyia*).—N.H.; Minn. to N.H., s. to Tex. and Ga., also Wash., Calif., Ariz.
laxifrons Reinhard, 1962b: 216.—Ariz.; Calif.

Genus ISIDOTUS Reinhard

Isidotus Reinhard, 1962b: 215. Type-species, *incanus* Reinhard (orig. des.).

incanus Reinhard, 1962b: 216.—Ariz.

Tribe DEXILLINI

(Dexiini of authors)

The species of the tribe Dexillini are parasites of white grubs such as *Phyllophaga*, *Popillia*, and *Melolontha*. For an explanation of the use of Dexiini, see under *Dexia*.

Genus DEXILLA Westwood

Dexilla Westwood, 1840: 140. Type-species, *Musca rustica* Fabricius (orig. des.).
Dexia, authors, not Meigen.

rustica (Fabricius), 1775: 777 (*Musca*).—Denmark; introd. N.Y., ?estab.
vacua (Fallén), 1817d: 240 (*Musca*).—Sweden; introd. N.Y., ?estab.
ventralis (Aldrich), 1925c: 33 (*Dexia*).—Korea; introd. N.J., estab.

Genus MEGAPARIA Wulp

Megaparia Wulp, 1891b: 212, key. Type-species, *venosa* Wulp (sub. mon., Wulp, 1891b: 240.).

venosa Wulp, 1891b: 240.—Mexico; Ariz.

Genus MEGAPARIOPSIS Townsend

Megapariopsis Townsend, 1915a: 22. Type-species, *Megaparia opaca* Coquillett (orig. des.).

opaca (Coquillett), 1899d: 218 (*Megaparia*).—Fla.; Nebr. to Tex., also Va. to Fla.

Genus MICROPHTHALMA Macquart

Microphthalma Macquart, 1843a: 241 (1843: 84). Type-species, *nigra* Macquart (orig. des.)=*disjuncta* (Wiedemann).
Microphthalmia (error) Adams *in* Williston, 1908: 376.
Eumicrophthalma Townsend, 1915l: 97. Type-species, *shannoni* Townsend (orig. des.).

REFERENCE: Aldrich, 1926c (rev.).

disjuncta (Wiedemann), 1824: 45 (*Tachina*; as *disiuncta*).—N. Amer.; B.C. to Ont., all U.S., Mexico, Guatemala.
    distincta (error) Aldrich, 1926c: 1, 5.
    trifasciata Say, 1829: 174 (1859b: 365) (*Miltogramma*).—Ind.
    nigra Macquart, 1843a: 242 (1843: 85).—Pa.
    apicalis Walker, 1849: 699 (*Trixa*).—Unknown.
    trixoides Walker, 1849: 760 (*Tachina*).—Ga.
europaea Egger, 1860: 801.—Europe; introd. N.Y., not estab.
michiganensis (Townsend), 1892a: 111 (*Megaprosopus*).—Mich.; Alta., S. Dak. to Que., s. to Iowa and Md., also N. Mex., Tex.
    phyllophagae Curran, 1925g: 16.—Que.
obsoleta (Wulp), 1890: 87 (*Trixa*).—Mexico; Calif. to N. Mex., Guatemala.
    sordida Giglio-Tos, 1893a: 3.—Mexico.
    pruinosa Coquillett, 1902a: 200.—N. Mex.
ruficeps Aldrich, 1926c: 6.—N. Mex.; Ariz., Kans., Tex.
shannoni (Townsend), 1915l: 98 (*Eumicrophthalma*).—D.C.

Tribe MYIOPHASIINI

Members of the tribe Myiophasiini are parasites of weevil grubs, and the females larviposit at weevil punctures in plant tissues.

Genus MYIOPHASIA Brauer and Bergenstamm

Myiophasia Brauer and Bergenstamm, 1891: 362 (1891: 58). Type-species, *Tachina aenea* Wiedemann (mon.)=*australis* Townsend.
Phasioclista Townsend, 1891c: 369. Type-species, *metallica* Townsend (orig. des.).
Ennyomma Townsend, 1891c: 371. Type-species, *clistoides* Townsend (orig. des.).
Pseudoclista Brauer and Bergenstamm, 1893: 192 (1893: 104). Type-species, *atra* Brauer and Bergenstamm (orig. des.).
Ennyommopsis Townsend, 1915d: 109. Type-species, *Loewia nigrifrons* Townsend (orig. des.).
Euloewia Townsend, 1915d: 109. Type-species, *Loewia globosa* Townsend (orig. des.).
Megaeuloewia Townsend, 1919b: 545. Type-species, *morinioides* Townsend (orig. des.).

The species of *Myiophasia* are difficult to distinguish and a revision is badly needed. Although the name *aenea* has been applied to several Nearctic species, on the theory that the stated type locality "Montevideo" was an error, it is now known that *Myiophasia* occurs in Uruguay, and we accept the locality as stated.

REFERENCE: Townsend, 1915d (key).

australis Townsend, 1916d: 11 (n. name for *aenea* Wiedemann).—Uruguay. Apparently not Nearctic.
  aenea Wiedemann, 1830: 298 (*Tachina;* preocc. Meigen, 1824).—Uruguay.
clistoides (Townsend), 1891c: 371 (*Ennyomma*).—Ill.; Calif., Kans., N. Mex., Tex.
globosa (Townsend), 1892h: 129 (*Loewia*).—Fla.; Ariz., Tex.
lasia Reinhard, 1959b: 225.—Wyo.; Wash., Oreg., Utah.
mesensis (Townsend), 1915d: 110, 112 (*Ennyomma;* as *clistoides* ssp.).—N. Mex.; Kans., Tex.
metallica (Townsend), 1891c: 370 (*Phasioclista*).—Ill.; Kans. to Mich. to N.H., s. to Tex. and Fla.
morinioides (Townsend), 1919b: 546 (*Megaeuloewia*).—N. Mex.; Colo.
neomexicana (Townsend), 1915d: 110, 112 (*Ennyomma;* as *robusta* ssp.).—N. Mex., Mexico.

nigrifrons (Townsend), 1892e: 77 (*Loewia*).—Ill.; Mich.
  americana Townsend, 1892e: 78 (*Clista*).—Ill.
oregonensis Townsend, 1915d: 111, 112 (as *setigera* ssp.).—Oreg.; B.C. to Calif. and Utah.
robusta Coquillett, 1897: 51.—Calif.; B.C.
ruficornis (Townsend), 1892e: 77 (*Loewia*).—Mich.; Colo.
setigera Townsend, 1908: 56.—N. Mex.; Oreg. to Colo., s. to Calif. and Tex.
sigilla Reinhard, 1959b: 226.—Wyo.

Tribe MELISONEURINI
Genus MICROCHAETINA Wulp

Microchaetina Wulp, 1891b: 212. Type-species, *cinerea* Wulp (sub. mon., Wulp, 1891b: 241).

  REFERENCE: Reinhard, 1942b (key).

cinerea Wulp, 1891b: 241.—Mexico; ?Fla.
mexicana (Townsend), 1892e: 168 (*Rhinophora*).—N. Mex.; Utah to Mo., s. to N. Mex. and Tex., also Fla.
  setifacies Reinhard, 1942b: 88.—Kans. **N. syn.**
teleta Reinhard, 1962b: 217.—Calif.; Utah.
valida (Townsend), 1892e: 167 (*Rhinophora*).—N. Mex.; Idaho to Calif., Mexico, and Tex.

Genus ORESTILLA Reinhard

Orestilla Reinhard, 1944b: 62. Type-species, *primoris* Reinhard (orig. des.).

primoris Reinhard, 1944b: 63.—N. Mex.

Genus ORTHOSIMYIA Reinhard

Orthosia Reinhard, 1944b: 60 (preocc. Ochsenheimer, 1816). Type-species, *montana* Reinhard (orig. des.).
Orthosimyia Reinhard, 1944c: 159 (n. name for *Orthosia* Reinhard). Type-species, *Orthosia montana* Reinhard (aut.).

montana (Reinhard), 1944b: 61 (*Orthosia*).—Calif.
palaga (Reinhard), 1944b: 62 (*Orthosia*).—Calif.

Genus PHALACROPHYTO Townsend

Phalacrophyto Townsend, 1915a: 23. Type-species, *Paraphyto sarcophagina* Coquillett (orig. des.).

sarcophagina (Coquillett), 1902g: 118 (*Paraphyto*).—Ill.; Colo. to Ill., s. to Okla.

Genus REINHARDIANA Arnaud

Hypenomyia Townsend, 1919b: 545 (preocc. Grimshaw, 1901). Type-species, *petiolata* Townsend (orig. des.).
Reinhardiana Arnaud, 1952: 58 (n. name for *Hypenomyia* Townsend). Type-species, *Hypenomyia petiolata* Townsend (aut.).

REFERENCE: Reinhard, 1942b (key).

petiolata (Townsend), 1919b: 545 (*Hypenomyia*).—Ariz.; B. C., Oreg. to Colo., s. to Mexico.

Genus STEVENIOPSIS Townsend

Steveniopsis Townsend, 1919b: 546. Type-species, *sinuata* Towsend (orig. des.).

REFERENCE: Reinhard, 1942b (key).

rubidiapex (Reinhard), 1942b: 89 (*Hypenomyia*).—Utah; Wash. to Wyo., s. to Calif. and N. Mex. **N. comb.**
sinuata Townsend, 1919b: 547.—N. Mex.; Wash., Calif., Colo.
subnitens (Reinhard), 1942b: 91 (*Hypenomyia*).—Utah. **N. comb.**

Genus WEBSTERIANA Walton

Websteriana Walton, 1914a: 180. Type-species, *Tricogena costalis* Coquillett (orig. des.).

costalis (Coquillett), 1897: 130 (*Tricogena*).—Colo.; Idaho, Calif. to Kans., N. Mex.

Tribe PROSENINI

The species of Prosenini are parasites of larvae of *Popillia*, *Phyllophaga*, and other Coleoptera.

Genus ARCTOPHYTO Townsend

Arctophyto Townsend, 1915a: 22. Type-species, *Paraphyto borealis* Coquillett (orig. des.).
Oreophyto Townsend, 1916i: 302. Type-species, *ochreicornis* Townsend (orig. des.).

REFERENCE: Curran, 1924l (partial key).

algens Curran, 1926d: 215.—Ont.; Colo.
borealis (Coquillett), 1900h: 439 (1904: 53) (*Paraphyto*).—Alaska; B.C., Idaho.

erythrocera (Thomson), 1869: 523 (*Miltogramma*).—Calif.
gillettei (Townsend), 1892e: 68 (*Trixa*).—Colo.
   gilletti, error.
johnsoni West, 1924: 187.—Mass.
marginalis Curran, 1924l: 302.—B.C.; B.C. to Mont., s. to Oreg. and Colo. ?=*borealis.*
ochreicornis (Townsend), 1916i: 302 (*Oreophyto*).—Oreg.; B.C., Calif. ?=*erythrocera.*
   ochricornis, error.
regina West, 1924: 187.—Mass.
wickhami Townsend, 1915a: 22.—Alta.; Alaska.
   gillettei Townsend of Coquillett, 1897: 122.

Genus ATELOGLOSSA Coquillett

Ateloglossa Coquillett, 1899d: 19 (as *Atelogossa;* correct in Index to Volume, p. 269). Type-species, *cinerea* Coquillett (orig. des., as gen. n., sp. n.).

REFERENCE: Curran, 1930j (key).

cinerea Coquillett, 1899d: 219.—Maine; Iowa to Maine and Conn., also S.C.
   wheeleri West, 1924: 186.—Conn. **N. syn.**
   calyptrata West, 1925: 128.—N.Y. **N. syn.**
glabra West, 1925: 128.—N.Y.
trivittata Curran, 1930j: 92.—N.Y.

Genus DINERA Robineau-Desvoidy

Dinera Robineau-Desvoidy, 1830: 307. Type-species, *grisea* Robineau-Desvoidy (Townsend, 1916d: 6)=*grisescens* (Fallén).

grisescens (Fallén), 1817d: 243 (*Musca*).—Sweden; B.C. to Maine, s. to Ariz., N. Mex., Kans., and N.J., Europe.
   punctata Robineau-Desvoidy, 1830: 308.—Pa.
   futilis West, 1924: 191.—N.J.
   cremides, authors, not Walker.

Genus DOLICHOCODIA Townsend

Dolichocodia Townsend, 1908: 59. Type-species, *Myocera bivittata* Coquillett (orig. des.).

bivittata (Coquillett), 1902g: 121 (*Myocera*).—N. Mex.; Ariz.
furacis Reinhard, 1958d: 277.—Tex.; Ariz., N. Mex.

Genus HESPERODINERA Townsend

Hesperodinera Townsend, 1919b: 551. Type-species, *cinerea* Townsend (orig. des.).

cinerea Townsend, 1919b: 551.—N. Mex.; B.C. to Man., s. to N. Mex.
   rava, authors, in part, not Wulp.

Genus MOCHLOSOMA Brauer and Bergenstamm

Mochlosoma Brauer and Bergenstamm, 1889: 126 (1889: 58). Type-species, *validum* Brauer and Bergenstamm (orig. des., as gen. n., sp. n.).
   REFERENCE: Reinhard, 1958a (rev.).

duplare Reinhard, 1958a: 100.—Mexico; Ariz.
illocale Reinhard, 1958a: 102.—Oreg.; B.C. to Mont., s. to Calif. and N. Mex., also Ohio.
indutile Reinhard, 1958a: 104.—Calif.; Ariz.
opipare Reinhard, 1958a: 105.—Ariz.; N. Mex., Mexico.
validum Brauer and Bergenstamm, 1889: 126, 168 (1889: 58, 100).—Pa.; B.C. to Man. to Mass., s. to Calif., Mexico, Colo., and Pa.

Genus NIMIOGLOSSA Reinhard

Nimioglossa Reinhard, 1945a: 35. Type-species, *ravida* Reinhard (orig. des.).

planicosta Reinhard, 1945a: 36.—Utah; Calif., Ariz.
ravida Reinhard, 1945a: 36.—N. Mex.; Calif., Utah, Ariz., Mexico.

Genus PROSENA Lepeletier and Serville

Calirrhoe Meigen, 1800: 39. Type-species, *Stomoxys siberita* Fabricius (sub. mon., Hendel, 1908b: 68). Suppressed by I.C.Z.N., 1963b: 339.

Prosena Lepeletier and Serville *in* Latreille et al., 1828: 499, 500. Type-species, *Stomoxys siberita* Fabricius (orig. des.).

siberita (Fabricius), 1775: 798 (*Stomoxys*).—Denmark; Europe, introd. ne. U.S., estab. N.J.
   sibirita, emend. or error.

Genus PROSENOIDES Brauer and Bergenstamm

Prosenoides Brauer and Bergenstamm, 1891: 370 (1891: 66). Type-species, *papilio* Brauer and Bergenstamm (mon.) =*curvirostris* (Bigot).

REFERENCE: Reinhard, 1954 (rev.).

assimilis Reinhard, 1954: 412.—Tex.; N. Mex. to Kans. and Ark., s. to Mexico, also Ohio.
curvirostris, authors, not Bigot.
flavipes Coquillett, 1895a: 314.—Fla.; Ga., Cuba.
grandis Reinhard, 1954: 413.—Ariz.; Mexico (Chih.).

Genus PTILODEXIA Brauer and Bergenstamm

Ptilodexia Brauer and Bergenstamm, 1889: 119 (1889: 51). Type-species, *carolinensis* Brauer and Bergenstamm (orig. des., as gen. n., sp. n.).
Myoceropsis Townsend, 1915a: 23. Type-species, *Rhynchiodexia flavotessellata* Walton (orig. des.).
Rhamphinina, authors, not Bigot.
Rhynchiodexia, authors, not Bigot.
Rhynchodexia, emend.

REFERENCE: Curran, 1930j (key).

A revision of this common but exceedingly difficult group is badly needed. No attempt has been made here to suggest synonymy, or, in most cases, to give distribution. The generic position of most species has been confirmed from the types.

agilis Reinhard, 1943c: 22.—Tex.; N.Y.
albifrons (Walker), 1852: 317 (*Dexia*).—U.S.
arida (Curran), 1930j: 93 (*Rhynchiodexia*).—Ariz.
canescens (Walker), 1852: 310 (*Dexia*).—U.S.
carolinensis Brauer and Bergenstamm, 1889: 119 (1889: 51).—S.C.
cerata (Walker), 1849: 847 (*Dexia*).—N. Amer.
confusa (West), 1924: 185 (*Rhynchodexia*).—N.Y.; Man. to N.H., s. to Tex. and Va.
dubia (Curran), 1930j: 93 (*Rhynchiodexia*).—N.Y.
elevata (West), 1925: 135 (*Rhynchodexia*).—N.Y.; N.J.
levata (error) Curran, 1930j: 93.
flavotessellata (Walton), 1914a: 176 (*Rhynchiodexia*).—N. Mex.
halone (Walker), 1849: 837 (*Dexia*).—Ga.
harpasa (Walker), 1849: 840 (*Dexia*).—N. Amer.
hucketti West, 1925: 131.—N.Y.; N.J.

incerta West, 1925: 131.—N.Y.; Mich., Tex.
leucoptera West, 1925: 132.—N.Y.
major (Bigot), 1888a: 265 (*Rhamphinina*).—Wash., Mexico.
mathesoni (Curran), 1930j: 93 (*Rhynchiodexia*).—Mich.
minor West, 1925: 132.—N.Y.; Ind.
neotibialis West, 1924: 184.—Conn.; N. Dak., Ohio to Mass., s. to Va., also Utah, Colo., Ariz., Tex.
obscura West, 1925: 133.—N.Y.; N.J.
ponderosa (Curran), 1930j: 93 (*Rhynchiodexia*).—Fla.
prexaspes (Walker), 1849: 837 (*Dexia*).—Ga.
proxima West, 1925: 133.—N.Y.
robusta (Curran), 1930j: 93 (*Dinera*).—N.Y.
rufipennis (Macquart), 1843a: 244 (1843: 87) (*Dexia*).—N.S.
simplex (Bigot), 1888a: 266 (*Myocera*).—Mexico; scattered U.S. records (?error).
translucipennis (West), 1925: 135 (*Rhynchodexia*).—N.Y. ?=*leucoptera*.

Unplaced Species of Prosenini

abdominalis Robineau-Desvoidy, 1830: 306 (*Estheria*).—N.S.
tibialis Robineau-Desvoidy, 1830: 306 (*Estheria*).—N.S.

Tribe THERESIINI

Species of Theresiini are parasites in wood-boring coleopterous larvae such as *Monochamus*, in soil-living grubs, and in lepidopterous stalk-borers such as *Diatraea*. A species of *Carinosillus* is a parasite of horse fly larvae.

Genus CARINOSILLUS Reinhard

Carinosillus Reinhard, 1943b: 84. Type-species, *pravus* Reinhard (orig. des.).
Myocera, authors, in part, not Robineau-Desvoidy.
Myiocera, emend. or error.

amicabilis (West), 1925: 130 (*Myiocera*).—N.Y. **N. comb.**, H. J. Reinhard *in litt*.
isolatus (West), 1924: 188 (*Myiocera*).—N.H. **N. comb.**
novaeangliae (West), 1924: 189 (*Myiocera*).—Conn. **N. comb.**
pravus Reinhard, 1943b: 85.—N.Y.; Ariz., Mich., Ohio, N.C.
tabanivorus (Hall), 1937a: 206 (*Myocera*).—Minn.; Wis., Mich. **N. comb.**

Genus EUCHAETOGYNE Townsend

Euchaetogyne Townsend, 1908: 59. Type-species, *Hystrichodexia roederi* Williston (orig. des.).

roederi (Williston), 1893c: 77 (*Hystrichodexia*).—Ariz.; N. Mex., Mexico.

Genus EUTHERESIA Townsend

Eutheresia Townsend, 1911a: 149. Nomen nudum.
Eutheresia Townsend, 1912a: 117. Type-species, *monohammi* Townsend (orig. des., as gen. n., sp. n.).

interrupta Curran, 1929c: 33.—Ill.; Iowa, Pa., N.C., Tex.
monohammi Townsend, 1912a: 117.—Ga.; Maine to Ga., also Minn.
montana West, 1924: 188.—Vt.
nipigonensis (Curran), 1926b: 89 (*Billaea*).—Ont. **N. comb.**, J. F. McAlpine *in litt.*
satisfacta West, 1925: 129.—N.Y.; Calif., Colo., Mich., D.C., Ga.
sibleyi West, 1925: 130.—N.Y.
trivittata Curran, 1929c: 33.—Ont.

Genus NICEPHORUS Reinhard

Nicephorus Reinhard, 1944b: 64. Type-species, *floridensis* Reinhard (orig. des.).

floridensis Reinhard, 1944b: 64.—Fla.; Ga.

Genus OPSOTHERESIA Townsend

Opsotheresia Townsend, 1919b: 552. Type-species, *obesa* Townsend (orig. des.).
Paraprosena, authors, not Brauer and Bergenstamm.

bigelowi (Curran), 1926b: 88 (*Gymnodexia*).—Iowa; Wis. to Ont., s. to Kans. and Pa. **N. comb.**, H. J. Reinhard *in litt.*
    nigricornis Reinhard, 1939a: 67.—Wis. **N. syn.**, H. J. Reinhard *in litt.*
    apicalis, authors (as *Paraprosena*), not Robineau-Desvoidy.
obesa Townsend, 1919b: 552.—Md.; N.Y. and Mass., s. to Ga.
    protrudens West, 1924: 190 (*Myiocera*).—N.Y. **N. syn.**, H. J. Reinhard *in litt.*
    compacta West, 1925: 130 (*Myiocera*).—N.Y. **N. syn.**, H. J. Reinhard *in litt.*

Genus PARATHERESIA Townsend

Paratheresia Townsend, 1915e: 65. Type-species, *signifera* Townsend (orig. des.) =*claripalpis* (Wulp).

The genus *Paratheresia* is predominantly Neotropical, and parasitizes various species of *Diatraea*. There is considerable difference of opinion as to whether proposed names represent environmentally adapted species or races of one variable species.

REFERENCE: Emden, 1949 (nomenclature).

claripalpis (Wulp), 1895: 268, key (1896: 280, descr.) (*Sarcophaga*).—
  Mexico; introd. La. and Fla., estab. Fla., Neotropical.
  signifera Townsend, 1915e: 66.—Peru.

Genus THERESIA Robineau-Desvoidy

Theresia Robineau-Desvoidy, 1830: 325. Type-species, *tandrec* Robineau-Desvoidy (mon.) =*rutilans* (Fabricius).
Sardiocera Brauer and Bergenstamm, 1889: 119 (1889: 51). Type-species, *valida* Brauer and Bergenstamm (orig. des., as gen. n., sp. n.) =*rutilans* (Fabricius).

rutilans (Fabricius), 1781: 436 (*Musca*).—"Americae meridionalis
  insulis" (possibly is. of S.C. and Ga. coast); Nebr. and
  Kans. to Maine, s. to Fla., also N. Mex.
  valida Wiedemann, 1830: 387 (*Musca*).—S.C. (as Unknown).
  tandrec Robineau-Desvoidy, 1830: 326.—Carolina.
  ruficornis Bigot, 1888a: 266 (*Myocera*).—Md.
  melanogaster Bigot, 1888a: 269 (*Phorostoma*).—N.Y.
  valida Brauer and Bergenstamm, 1889: 167 (1889: 99)
    (*Sardiocera*).—S.C.

Genus URSOPHYTO Aldrich

Ursophyto Aldrich, 1926d: 14. Type-species, *rufigena* Aldrich (orig. des.) =*nigriceps* (Bigot).

nigriceps (Bigot), 1888a: 267 (*Myostoma*).—Wash.; B.C. to Mont.,
  s. to Calif. and Colo.
  rufigena Aldrich, 1926d: 14.—Wash. **N. syn.**

Genus VIBRISSOTHERESIA Reinhard

Vibrissotheresia Reinhard, 1943b: 86. Type-species, *pechumani* Reinhard (orig. des.).

pechumani Reinhard, 1943b: 87.—N.Y.

Subfamily TACHININAE
(Larvaevorinae)

Species of the subfamily Tachininae parasitize many kinds of lepidopterous larvae, especially cutworms, armyworms, and stalk borers such as *Diatraea*.

Tribe MACROMYINI

Genus BOMBYLIOMYIA Brauer and Bergenstamm

Bombyliomyia Brauer and Bergenstamm, 1889: 131 (1889: 63). Type-species, *Hystricia flavipalpis* Macquart (mon.).
Tachinalia Curran, 1934a: 442, 466. Type-species, *hispida* Curran (orig. des.) =*soror* (Williston).

soror (Williston), 1886a: 298 (*Hystricia*).—Ariz.; Oreg., Calif., Mexico.
    hispida Curran, 1934a: 466 (*Tachinalia*).—Calif. **N. syn.**
    flavipalpis, authors, not Macquart.

Genus BOMBYLIOPSIS Townsend

Bombyliopsis Townsend, 1915a: 23. Type-species, *Tachina abrupta* Wiedemann (orig. des.).
Hystricia, authors, not Macquart.

REFERENCE: Curran, 1942a (key, as *Hystricia*).

abrupta (Wiedemann), 1830: 293 (*Tachina*).—N. Amer.; B.C. to N.S., s. to Calif., Mexico, Tex., Minn., and Ga.
    vivida Harris, 1841: 411 (*Tachina*).—Mass.
    finitima Walker, 1849: 707 (*Tachina*).—N.S., U.S.
    fulvida Bigot, 1887b: cxxxix (*Hystricia*).—N. Amer.

Genus CHLOROHYSTRICIA Townsend

Chlorohystricia Townsend, 1927: 244. Type-species, *purpurea* Townsend (orig. des.) =*reinwardtii* (Wiedemann).

cyaneiventris (Wulp), 1884: ccxci (*Hystricia*).—Mexico; Ariz.

Genus MACROMYA Robineau-Desvoidy

Macromya Robineau-Desvoidy, 1830: 322. Type-species, *depressa* Robineau-Desvoidy (Townsend, 1916d: 7).

Unnamed sp.—Ariz., Tex.

TACHINIDAE 993

Tribe TACHININI

(Larvaevorini, Echinomyini)

The members of the tribe Tachinini parasitize a great variety of lepidopterous larvae.

Genus METOPOTACHINA Townsend

Metopotachina Townsend, 1915a: 21. Type-species, *Echinomyia palpalis* Coquillett (orig. des.).
Fabriciella, authors, in part, not Bezzi.

intermedia (Reinhard), 1942a: 27 (*Fabriciella*).—Calif. **N. comb.**
margella (Reinhard), 1942a: 28 (*Fabriciella*).—N. Mex. **N. comb.**
orbitalis (Reinhard), 1942a: 26 (*Fabriciella*).—Utah; Calif. to Wyo. and Nebr. **N. comb.**
palpalis (Coquillett), 1902g: 120 (*Echinomyia*).—Calif.; N. Mex.
pictilis (Reinhard), 1942a: 25 (*Fabriciella*).—Oreg. **N. comb.**

Genus NOWICKIA Wachtl

Nowickia Wachtl, 1894: 142. Type-species, *Echinomyia regalis* Rondani (orig. des.)=*marklini* (Zetterstedt).
Fabriciodes Townsend, 1916e: 26. Type-species, *montana* Townsend (orig. des.).
Eularvaevora Townsend, 1916e: 27. Type-species, *Tachina algens* Wiedemann (orig. des.; misident.)=*hispida* (Tothill).
Fabriciella, authors, in part, not Bezzi.

REFERENCES: Tothill, 1924b (rev.); Rowe, 1931 (rev., males); both as *Fabriciella*.

Subgenus ECHINOMYODES Townsend

Echinomyodes Townsend, 1916e: 25 (as genus). Type-species, *piceifrons* Townsend (orig. des.).

actinosa (Reinhard), 1938: 8 (*Fabriciella*).—Colo.; Oreg. **N. comb.**
brevirostris (Tothill), 1924b: 264 (*Fabriciella*).—B.C.; B.C. to Calif. and Colo. **N. comb.**
canadensis (Tothill), 1924b: 264 (*Fabriciella*).—Que.; Alta., Colo., Mich. **N. comb.**
compressa (Tothill), 1924b: 262 (*Fabriciella*).—Colo.; Wash., Oreg., Calif. **N. comb.**
egula (Reinhard), 1938: 9 (*Fabriciella*).—Colo.; Oreg. to Wyo., s. to Calif. and N. Mex. **N. comb.**
evanida (Reinhard), 1953a: 53 (*Fabriciella*).—Colo.; Calif., Utah, Ariz. **N. comb.**

invelata (Reinhard), 1953a:, 51 (*Fabriciella*).—Calif.; Oreg. **N. comb.**
latifacies (Tothill), 1924b: 262 (*Fabriciella*).—Nev.; B.C. to Wyo., s. to Nev. and N. Mex. **N. comb.**
latigena (Tothill), 1924b: 262 (*Fabriciella*).—B.C.; B.C. to Mont., s. to Calif., N. Mex., and Kans., also Mich.—**N. comb.**
longiunguis (Tothill), 1924b: 265 (*Fabriciella*).—B.C. **N. comb.**
lutzi (Curran), 1925e: 256 (*Fabriciella*).—Colo.; Wash. to Wyo., s. to Calif. and N. Mex. **N. comb.**
nigella (Reinhard), 1938: 10 (*Fabriciella*).—Mont.; Calif., Nev., Utah. **N. comb.**
picea (Robineau-Desvoidy), 1830: 44 (*Echinomya*).—N.S. Unrecognized. **N. comb.**
piceifrons (Townsend), 1916e: 25 (*Echinomyodes*).—Vt.; N.W.T. to N.S., s. to Calif., N. Mex., and Mass., also N.C. ?= *picea*. **N. comb.**
plumasana (Reinhard), 1953a: 53 (*Fabriciella*).—Calif.; Ariz. **N. comb.**
spinosa (Tothill), 1924b: 263 (*Fabriciella*).—B.C.; B.C. to Colo., s. to Calif. and Ariz.—**N. comb.**
tahoensis (Reinhard), 1938: 11 (*Fabriciella*).—Nev.; B.C., Calif. Ariz. **N. comb.**

Subgenus FABRICIELLA Bezzi

Fabricia Latreille, 1829: 510 (preocc. de Blainville, 1828). Type-species, *Musca fera* Linnaeus (mon.; misident.)=*ferox* (Panzer). **N. syn.**
Fabricia Robineau-Desvoidy, 1830: 42 (preocc. de Blainville, 1828). Type-species, *Musca ferox* Panzer (mon.). **N. syn.**
Fabriciella Bezzi, 1906: 49 (as genus; n. name for *Fabricia* Robineau-Desvoidy). Type-species, *Musca ferox* Panzer (aut.). **N. syn.**
Larvaevoropsis Townsend, 1916e: 24. Type-species, *Echinomyia dakotensis* Townsend (orig. des.)=*florum* (Walker). **N. syn.**

acuminata (Tothill), 1924b: 260 (*Fabriciella*).—Colo.; Calif., Ariz., N. Mex., Mexico. **N. comb.**
cordiforceps (Rowe), 1931: 649 (*Fabriciella*).—Utah; Wash. to Colo., s. to Calif. and N. Mex. **N. comb.**
florum (Walker), 1849: 722 (*Tachina*).—Ont., N.S.; Wash. to Man. to N.S., s. to Mexico (Chih.), S. Dak., Mich., and Ga. **N. comb.**

dakotensis Townsend, 1892a: 94 (*Echinomyia*).=S. Dak.
N. comb.
orientalis Townsend, 1916e: 25 (*Larvaevoropsis*).—N.H.
N. comb.
planiforceps (Tothill), 1924b: 261 (*Fabriciella*).—N.C. N. comb.

Subgenus NOWICKIA Wachtl

emarginata (Tothill), 1924b: 260 (*Fabriciella*).—Colo.; Alaska, Alta.
N. comb.
hispida (Tothill), 1924b: 265 (*Fabriciella*).—Ont.; Alaska, B.C. to
N.S., s. to Calif., N. Mex., S. Dak., and N.Y.
ampliforceps Rowe, 1931: 673 (*Fabriciella*).—Alta. **N. syn.**,
N. comb.
latiforceps (Tothill), 1924b: 266 (*Fabriciella*).—Que.; B.C., Alta.,
Mont., Ariz., Maine. N. comb.
argentea Rowe, 1931: 654 (*Fabriciella*).—Alta. **N. syn.**,
n. comb.
montana (Townsend), 1916e: 26 (*Fabriciodes*).—N.H.; Oreg., Colo.,
N.Y., Labr., Maine. N. comb.
nivalis (Tothill), 1924b: 264 (*Fabriciella*).—Alaska; Alta. N. comb.
pilosa (Tothill), 1924b: 263 (*Fabriciella*).—B.C.; Alaska to Mont., s.
to Oreg. and N. Mex., also Mich. to N.B. N. comb.

Subgenus RHACHOGASTER Townsend

Rhachogaster Townsend, 1915c: 291 (as genus). Type-species,
kermodei Townsend (orig. des.)=*algens* (Wiedemann).
N. syn.
Upodemocera Townsend, 1915f: 228. Type-species, *robinsoni* Townsend (orig. des.)=*nitida* (Wulp). **N. syn.**
Eupodermocera (error) Tothill, 1924b: 258.
algens (Wiedemann), 1830: 285 (*Tachina*).—N. Amer.; Alaska and
N.W.T. to Nfld., s. to Calif., Ariz., and Mass. N. comb.
lapilaei Robineau-Desvoidy, 1830: 44 (*Echinomya*).—Nfld.
N. comb.
lugubris Wulp, 1883: 20 (*Echinomyia*).—Que. N. comb.
kermodei Townsend, 1915c: 291 (*Rhachogaster*).—B.C. N.
comb.
eurekana (Reinhard), 1942a: 28 (*Fabriciella*).—Calif.; B.C. N. comb.
latianulum (Tothill), 1924b: 266 (*Fabriciella*).—B.C.; B.C. to S. Dak.,
s. to Calif. and N. Mex., also Ont., N.B. N. comb.
latifrons (Tothill), 1924b: 260, 269 (*Fabriciella*).—N.W.T. (Walley,
1933: 168); N.W.T. to Idaho, Colo., and Kans., also
Mich. N. comb.

nitida (Wulp), 1882a: 82 (*Jurinia*).—Ariz.; B.C. to Calif., Ariz., and Kans, also Ont. and Ohio to Mass. **N. comb.**
  robinsoni Townsend, 1915f: 229 (*Upodemocera*).—Wyo. **N. comb.**
rostrata (Tothill), 1924b: 267 (*Fabriciella*).—B.C.; Alaska to Ont., s. to Calif. and N. Mex. **N. comb.**
spineiventer (Tothill), 1924b: 266 (*Fabriciella*).—B.C.; B.C. to Mont., s. to Calif. **N. comb.**

Genus PELETERIA Robineau-Desvoidy

Peleteria Robineau-Desvoidy, 1830: 39. Type-species, *abdominalis* Robineau-Desvoidy (Coquillett, 1910b: 586).
Peletieria, emend.
Echinogaster Lioy, 1864a: 1335. Type-species, *Echinomyia argentifrons* Macquart (mon.) =*prompta* (Meigen).
Chaetopeleteria Mik, 1894a: 100. Type-species, *Echinomyia popelii* Portschinsky (orig. des.).
Peleteriopsis Townsend, 1916h: 630. Type-species, *Echinomyia flaviventris* Wulp (orig. des.).

REFERENCE: Curran, 1925h (rev.).

Subgenus OXYDOSPHYRIA Townsend

Oxydosphyria Townsend, 1926: 40 (as genus). Type-species, *infernalis* Townsend (orig. des.) =*iterans* (Walker).

aclista Reinhard, 1956c: 109.—Mexico; Utah, Colo., N. Mex.
iterans (Walker), 1849: 727 (*Tachina*).—N.S.; B.C. to N.S., s. to Calif., n. Mexico, N. Mex., S. Dak., and Mass.
  punctifera Walker, 1849: 728 (*Tachina*).—Mass.
  infernalis Townsend, 1926: 41 (*Oxydosphyria*).—N. Mex.

Subgenus PANZERIOPSIS Townsend

Panzeriopsis Townsend, 1915c: 290 (as genus). Type-species, *curriei* Townsend (orig. des.).
Chaetopeleteria, authors, not Mik.

alberta Curran, 1925h: 234.—Alta.; Wash., Calif.
cornigera Curran, 1925h: 232.—B.C.; B.C. and Alta., s. to Calif. and Colo.
cornuta Curran, 1925h: 232.—Alta.; Mont., Wyo., Colo.
cornuticaudata Curran, 1925h: 234.—Alta.; B.C., Wyo., Colo.
curriei (Townsend), 1915c: 291 (*Panzeriopsis*).—B.C.
  phairi Curran, 1925h: 233.—B.C.

Subgenus PELETERIA Robineau-Desvoidy

aenea (Staeger) *in* Zetterstedt, 1849: 3217 (*Echinomyia*).—Greenland; N.W.T. (n. Ellesmere I., Baffin I.).
aldrichi Curran, 1925h: 240.—Colo.; Utah., Wyo., N. Mex.
angulata Curran, 1925h: 236.—B.C.; B.C. and Alta., s. to Calif., also Colo.
arctica Malloch, 1919f: 57.—N.W.T.
blanda Curran, 1925h: 241.—Calif.
clara Curran, 1925h: 239.—Alta.; B.C. to Mont., s. to Calif. and N. Mex.
conjuncta Curran, 1925h: 235.—B.C.; Mont., Wyo., Colo., N. Mex.
flaviventris (Wulp), 1888: 32 (*Echinomyia*).—Mexico (Durango); N. Mex.
neglecta (Townsend), 1897a: 148 (*Echinomyia*).—N. Mex.; B.C. to Man., s. to Calif. and N. Mex.
 eronis Curran, 1925h: 238.—B.C.
posticata Curran, 1925h: 241.—Calif.; B.C. to Calif. and Ariz., also Colo.
regalis Curran, 1925h: 235.—Colo.; Calif., Wyo., Ariz.
tessellata (Fabricius).—Not Nearctic.
trifasciata Curran, 1925h: 242.=Calif.

Subgenus SPHYROMYIA Bigot

Sphyromyia Bigot, 1884b: cviii (as genus). Type-species, *malleola* Bigot (mon.).
Aphriosphyria Townsend, 1927: 238. Type-species, *communis* Townsend (orig. des.) =*robusta* (Wiedemann).

anaxias (Walker), 1849: 726 (*Tachina*).—N.S.; Sask. to P.E.I., s. to Tenn., also B.C., Nev.
biangulata Curran, 1925h: 255.—Ariz.; B.C. to Wyo., s. to Calif. and N. Mex.
bryanti Curran, 1925h: 250.—Alta.; B.C. to Man., s. to Calif. and Tex.
compascua (Wulp), 1892: 192 (*Echinomyia*).—Mexico (Guerrero); Colo., Ariz., N. Mex.
convexa Curran, 1925h: 51.—N. Mex.; Wash., Colo., Calif., Mexico (Chih.).
haemorrhoa (Wulp), 1867: 145 (*Echinomyia*).—Wis.; B.C. to N.B. and Md., also Calif., Colo., N. Mex.
 apicalis Walker, 1852: 275 (*Tachina*; preocc. Meigen, 1824.—Calif. (as Columbia, error; Austen, 1907: 333).
 confusa Curran, 1925h: 253.—Ont.

incongrua (Reinhard), 1934e: 68 (*Cuphocera*).—Ariz.; Calif., Colo., Tex. **N. comb.**

incontesta Curran, 1926k: 171 (n. name for *neglecta* Curran).—N. Mex.; Utah, Colo., Ariz., Mexico.
  neglecta Curran, 1925h: 243 (preocc. Townsend, 1897).—N. Mex.

malleola (Bigot), 1884b: cix (*Sphyromyia*).—Calif.; B.C. to Man. and Mich., s. to Calif., Mexico, and Tex.
  campestre Curran, 1925h: 247.—Man.

mediana Reinhard, 1944b: 71.—Oreg.; Wash., Calif.

neotexensis Brooks, 1949: 23.—Tex.; Colo.

obsoleta Curran, 1925h: 251.—N. Mex.; B.C. to Colo., s. to Costa Rica.

setosa Curran, 1925h: 252.—Mexico (Chih.); Utah, Colo., Ariz., N. Mex.

texensis Curran, 1925h: 246.—Tex.; Calif., Ariz., Okla., Mexico, Costa Rica.
  robusta, authors, not Wiedemann.

thomsoni (Williston), 1886a: 302 (*Echinomyia*; n. name for *filipalpis* Thomson).—Calif.; ?Ariz.
  filipalpis Thomson, 1869: 517 (*Echinomyia*; preocc. Rondani, 1863).—Calif.

torta Reinhard, 1943c: 21.—Ariz.; Oreg., Calif.

townsendi Curran, 1925h: 252.—Mexico (Chih.); N. Mex., Tex.

valida Curran, 1925h: 255.—S.C.

Unplaced Species of *Peleteria*

hirta (Curtis), 1835b: lxxix (*Tachina*; preocc. Macquart, 1834).—Arctic N. Amer. ?=*aenea*.

rubrifrons (Bigot), 1887b: cxl (*Echinomyia*).—N. Amer.

Genus Tachina Meigen

Larvaevora Meigen, 1800: 38. Type-species, *Musca grossa* Linnaeus (Coquillett, 1910b: 557). Suppressed by I.C.Z.N., 1963b: 339.

Tachina Meigen, 1803: 280. Type-species, *Musca grossa* Linnaeus (Brauer, 1893: 489).

Echinomya Latreille, 1805: 377. Type-species, *Musca grossa* Linnaeus (Westwood, 1840: 138).

vernalis (Robineau-Desvoidy), 1830: 48 (*Echinomya*).—France; Europe, introd. New England, not estab.
  magnicornis Zetterstedt, 1844: 996 (*Echinomya*).—Sweden.

Unplaced Species of Tachinini

victoria Townsend, 1897a: 148 (*Echinomyia*).—N. Mex.

Tribe DEJEANIINI

Members of the tribe Dejeaniini parasitize lepidopterous larvae such as *Catocala*, *Malacosoma*, armyworms, etc.

Genus ADEJEANIA Townsend

Adejeania Townsend, 1913c: 103, 104. Type-species, *Tachina armata* Wiedemann (orig. des.).
Trichodejeania Townsend, 1913c: 104. Type-species, *Dejeania vexatrix* Osten Sacken (orig. des.).
Dejeania, authors, not Robineau-Desvoidy.

REFERENCE: Curran, 1947a (rev.).

vexatrix (Osten Sacken), 1877: 343 (*Dejeania*).—Colo.; B.C. to Mont., s. to Calif. and N. Mex., Mexico.
rufipalpis, authors, not Macquart.

Genus ARCHYTAS Jaennicke

Archytas Jaennicke, 1867: 392 (1868: 84). Type-species, *bicolor* Jaennicke (mon.)=*diaphanus* (Fabricius).
Pseudoarchytas Townsend, 1915j: 185. Type-species, *marmoratus* Townsend (orig. des.).

REFERENCES: Curran, 1928c (rev.); Sabrosky, 1955b (*piliventris* complex).

Subgenus ARCHYTAS Jaennicke

amethystinus (Macquart), 1843a: 199 (1843: 42) (*Jurinia*).—Ga. Unrecognized.
apicifer (Walker), 1849: 718 (*Tachina*).—N. Amer.; B.C., Calif. to Ont. and Maine, s. to Mexico and Fla.
vulgaris Curran, 1928c: 204, 276.—Kans.
analis, authors, in part, not Fabricius.
basifulvus, authors, not Fabricius.
californiae (Walker), 1852: 270 (*Tachina*).—Calif.; B.C. to N.S., entire U.S.
amethystinus, authors, not Macquart.
analis, authors, in part, not Fabricius.
apicifera Walker of Curran, 1928c: 203, 253.

incertus (Macquart), 1851: 152 (1851: 179) (*Gonia*).—Argentina; introd. Fla., not estab.
lobulatus Curran, 1928c: 204, 279.—Brazil; s. Tex., Mexico, Neotropical.
marmoratus (Townsend), 1915j: 186 (*Pseudoarchytas*).—Peru; Ariz. to Fla. to N.C., also Kans., s. to Peru.
  incertus, N. Amer. authors, not Macquart.
  piliventris, authors, not Wulp.
nivalis Curran, 1928c: 203, 254.—D.C.; Mich. to Mass., s. to Ga.
plangens Curran, 1928c: 204, 255.—Ecuador; s. Tex., Mexico, Cent. Amer.
rufiventris Curran, 1928c: 202, 280.—Fla.; N.C., Ga.

Subgenus NEMOCHAETA Wulp

Nemochaeta Wulp, 1888: 38 (as genus). Type-species, *dissimilis* Wulp (mon.).

aterrimus (Robineau-Desvoidy), 1830: 35 (*Jurinia*).—Carolina; Man. to N.B., s. to Calif., Mexico, and Fla.
  leucostoma Robineau-Desvoidy, 1830: 37 (*Jurinia*).—Carolina.
  metallifera Walker, 1849: 717 (*Tachina*).—Unknown.
  carbonifera Walker, 1849: 721 (*Tachina*).—Unknown.
  virginiensis Macquart, 1851: 144 (1851: 171) (*Jurinia*).—Va.
  atra Walker, 1852: 273 (*Tachina*).—Ga.
  errabunda Harris, 1835: 599 (*Tachina*). Nomen nudum.
convexiforceps Brooks, 1949: 23.—Fla.
instabilis Curran, 1928c: 203, 224.—N.J.; Ont. to Conn., s. to Va.
lateralis (Macquart), 1843a: 199 (1843: 42) (*Jurinia*).—Mexico; Calif. to Iowa, s. to Mexico, also B.C., Mont., Ga., Fla.
metallicus (Robineau-Desvoidy), 1830: 35 (*Jurinia*).—Carolina; Calif. to Ont. and N.H., s. to Mexico and Fla.
  boscii Robineau-Desvoidy, 1830: 36 (*Jurinia*).—Carolina.
  georgica Macquart, 1835: 79 (*Echinomyia*).—Ga.
  fuscipennis Jaennicke, 1867: 391 (1868: 83) (*Jurinea*).—N. Amer.
  hystricoides Williston, 1886a: 300 (*Jurinia*).—Ariz., N. Mex., Conn., D.C., Va., Ga.
  hystrix, authors, not Fabricius.
  pilosa Drury of Curran, 1928c: 206.

Genus JURINELLA Brauer and Bergenstamm

Jurinella Brauer and Bergenstamm, 1889: 132 (1889: 64). Type-species, *Jurinia caeruleonigra* Macquart (mon.).
Pseudohystricia Brauer and Bergenstamm, 1889: 132 (1889: 64). Type-species, *Hystricia ambigua* Macquart (mon.).

REFERENCE: Curran, 1947a (rev.).

lutzi Curran, 1947a: 66.—N. Mex.; Colo., Ariz., Mexico.
  ambigua, authors, not Macquart.
  obesa, authors, not Wiedemann.

Genus JURINIOPSIS Townsend

Juriniopsis Townsend, 1916g: 73. Type-species, *floridensis* Townsend (orig. des.).

REFERENCE: Curran, 1960 (key).

adusta (Wulp).—Apparently not n. of Mexico.
aurifrons Brooks, 1949: 21.—N. Mex.; Utah, Colo., Ariz., Mexico.
  metallica, authors, in part, not Robineau-Desvoidy.
floridensis Townsend, 1916g: 73.—Fla.; Mo. to Mass., s. to Tex. and Fla. ?=*myrrhea*.
  metallica, authors, in part, not Robineau-Desvoidy.
lampuris Reinhard, 1953d: 93.—Mexico (Morelos); Ariz.
myrrhea (Brauer and Bergenstamm), 1889: fig. 234 (1889: fig. 234) (*Jurinea*).—Pa. (Brauer and Bergenstamm, 1891: 409; 1891: 105).

Genus MICROTRICHOMMA Giglio-Tos

Microtrichomma Giglio-Tos, 1893b: 1. Type-species, *Nemoraea forreri* Wulp (Coquillett, 1910b: 570).
Exopalpus, authors, not Macquart.

pompale (Reinhard), 1941: 58 (*Exopalpus*).—Mich.; Minn. to Que., s. to N.C. **N. comb.**
  currani Ouellet, 1942: 81 (*Archytas*).—Que. **N. comb.**
smithi (Wulp), 1890: 50 (*Nemoraea*).—Mexico (V.C.); Fla. **N. comb.**

Genus PARADEJEANIA Brauer and Bergenstamm

Jurinia, subg. **Paradejeania** Brauer and Bergenstamm, 1893: 147, 184 (1893: 59, 96). Type-species, *Dejeania rutilioides* Jaennicke (Coquillett, 1910b: 584).

REFERENCE: Arnaud, 1951 (rev.).

rutilioides rutilioides (Jaennicke), 1867: 394 (1868: 86) (*Dejeania*).—
Mexico; Ariz. and Colo., s. to Cent. Amer.
 ssp. **nigrescens** Arnaud, 1951: 322.—Calif.; B.C., Oreg.

Genus PARARCHYTAS Brauer and Bergenstamm

Pararchytas Brauer and Bergenstamm, 1894: 76 (1895: 612). Type-species, *Tachina decisa* Walker (mon.).

decisus (Walker), 1849: 715 (*Tachina*).—Ont.; B.C. to P.E.I. and Pa., s. to Calif., Mexico, and N. Mex. in west, also Tenn., Ga.

hammondi Brooks, 1945a: 80.—Ont.; S. Dak. to N.B., s. to Kans. and Ga.

Genus PROTODEJEANIA Townsend

Protodejeania Townsend, 1915a: 21. Type-species, *Dejeania hystricosa* Williston (orig. des.).

REFERENCE: Curran, 1947a (key).

echinata (Thomson), 1869: 516 (*Jurinea*).—Calif.; B.C. to Calif.
hystricosa (Williston), 1886a: 297 (*Dejeania*).—N. Mex. (Townsend, 1931a: 163); Colo., Ariz., Mexico.
 willistoni Curran, 1947a: 51.—Ariz. **N. syn.**

Tribe JURINIINI

Species of the tribe Juriniini parasitize lepidopterous larvae.

Genus EPALPUS Rondani

Epalpus Rondani, 1850a: 169. Type-species, *Micropalpus rufipennis* Macquart (Coquillett, 1910b: 538).
Argentoepalpus Townsend, 1919c: 178. Type-species, *Epalpus niveus* Townsend (orig. des.).

albomaculatus (Jaennicke), 1867: 388 (1868: 80) (*Micropalpus*).—
Mexico; Ariz., N. Mex., Guatemala.
 maculata Williston, 1886a: 304 (*Saundersia*).—N. Mex. **N. syn.**

rufipes (Brooks), 1949: 22 (*Argentoepalpus*).—B.C.; B.C. and Idaho to Calif.

signifer (Walker), 1849: 708 (*Tachina*).—N.S. and unknown; B.C. to N.S., s. to Calif., N. Mex., and N.C.
 plagiata Harris, 1835: 599 (*Tachina*). Nomen nudum.

Genus PAREPALPUS Coquillett

Parepalpus Coquillett, 1902g: 120. Type-species, *flavidus* Coquillett (mon.)

REFERENCE: Curran, 1947a (key).

flavidus Coquillett, 1902g: 120.—Colo.; Utah, Ariz., N. Mex.
  nigropilosa Wulp of Coquillett, 1897: 145.

Genus RHACHOEPALPUS Townsend

Rhachoepalpus Townsend, 1908: 114. Type-species, *Saundersia testacea* Wulp (Coquillett, 1910b: 599).

REFERENCE: Curran, 1947a (key).

Unnamed sp.—Ariz.

Genus XANTHOEPALPUS Townsend

Xanthoepalpus Townsend, 1914: 157. Type-species, *Saundersia bipartita* Wulp (orig. des.)=*bicolor* (Williston).

bicolor (Williston), 1886a: 304 (*Saundersia*).—N. Mex.; B.C. to Calif., Mexico, and Colo.
  melanopygatus Bigot, 1887b: cxli (*Cryptopalpus*).—Wash.

Tribe CUPHOCERINI

Members of the tribe Cuphocerini parasitize lepidopterous larvae, especially cutworms and armyworms.

Genus COPECRYPTA Townsend

Copecrypta Townsend, 1908: 109. Type-species, *Schineria ruficauda* Wulp (mon.).

ruficauda (Wulp), 1867: 146 (*Schineria*).—Wis.; S. Dak. to Mass., s. to Ariz., Mexico, and Fla.
  nitens, authors, not Wiedemann.
  undulata Say *in* Brauer and Bergenstamm, 1891: 406 (1891: 102) (*Elachipalpus*). Nomen nudum.

Genus DEOPALPUS Townsend

Deopalpus Townsend, 1908: 110. Type-species, *hirsutus* Townsend (orig. des.).
Spanipalpus Townsend, 1908: 110. Type-species, *Trichophora miscelli* Coquillett (orig. des.).
Cuphocera, authors, not Macquart.

Cyphocera, error or emend.

REFERENCE: Reinhard, 1934e (rev.).

beameri (Reinhard), 1934e: 66 (*Cuphocera*).—Calif. **N. comb.**
californiensis (Macquart), 1851: 148 (1851: 175) (*Micropalpus*).— Calif. **N. comb.** Unrecognized.
conformis (Reinhard), 1934e: 54 (*Cuphocera*).—Ariz. **N. comb.**
contiguus (Reinhard), 1934e: 61 (*Cuphocera*).—Calif.; B.C. to S. Dak. s. to Calif. and Mexico, also N.Y. **N. comb.**
    stricklandi Curran *in* West, 1928: 819 (*Cuphocera*). Nomen nudum.
flavicornis (Reinhard), 1934e: 58 (*Cuphocera*).—Ariz.; Calif. **N. comb.**
geminatus (Reinhard), 1934e: 55 (*Cuphocera*).—Calif. **N. comb.**
hirsutus Townsend, 1908: 110.—Mexico (Chih.); Oreg. to Maine, s. to Calif., Mexico, and Fla.
    aurifrons Reinhard, 1924a: 54 (*Cuphocera*).—Tex.
    fucata, authors, not Wulp.
    furcata, error.
miscelli (Coquillett), 1897: 139 (*Trichophora*).—Calif. **N. comb.**
parksi (Reinhard), 1934e: 50 (*Cuphocera*).—Tex.; Ariz. **N. comb.**
scutellaris (Reinhard), 1934e: 53 (*Cuphocera*).—Ariz.; Calif. **N. comb.**
torosus (Reinhard), 1934e: 67 (*Cuphocera*).—Oreg.; Calif. **N. comb.**

Tribe LINNAEMYINI

The species of Linnaemyini parasitize a large variety of lepidopterous larvae.

Genus BONNETIA Robineau-Desvoidy

Bonnetia Robineau-Desvoidy, 1830: 55. Type-species, *oenanthis* Robineau-Desvoidy (Townsend, 1916d: 6) =*comta* (Fallén).
Marshamia Robineau-Desvoidy, 1830: 57. Type-species, *analis* Robineau-Desvoidy (Townsend, 1916d: 7) =*comta* (Fallén).

REFERENCE: Brooks, 1944b (rev.).

comta (Fallén), 1810b: 277 (*Tachina*).—Sweden; B.C. to Que. and Maine, entire U.S., Mexico, Palaearctic.
    distincta Robineau-Desvoidy, 1830: 54 (*Linnaemya*).—Pa.
    analis Robineau-Desvoidy, 1830: 58 (*Marshamia*; preocc. Robineau-Desvoidy, 1830: 54).—Carolina.
    nigripes Robineau-Desvoidy, 1830: 58 (*Marshamia*).—?Carolina.
    piceus Macquart, 1835: 84 (*Micropalpus*; n. name for *analis* Robineau-Desvoidy, 1830: 58).—Carolina.

TACHINIDAE 1005

Genus CHRYSOTACHINA Brauer and Bergenstamm

Chrysotachina Brauer and Bergenstamm, 1889: 161 (1889: 93). Type-species, *Tachina reinwardtii* Wiedemann (mon.; misident.) = *braueri* Townsend.
Eugymnochaeta Townsend, 1912b: 314. Type-species, *Gymnochaeta alcedo* Loew (orig. des.).

REFERENCE: Curran, 1939a (key).

alcedo (Loew), 1869b: 36 (Cent. 8, no. 61) (*Gymnochaeta*).—Mass.; B.C. to Calif., Mexico, and Colo., also Wis. to N.H., s. to Fla.
equatorialis (Townsend), 1912b: 314 (*Eugymnochaeta*).—Peru; Fla., Brazil.

Genus GYMNOCHETA Robineau-Desvoidy

Gymnocheta Robineau-Desvoidy, 1830: 371. Type-species, *Tachina viridis* Fallén (as Meigen; mon.).
Gymnochaeta, emend.
Chlorometaphyto Townsend, 1919c: 180. Type-species, *Gymnochaeta vivida* Williston (orig. des.).

REFERENCE: Brooks, 1945a (key).

frontalis Brooks, 1945a: 88.—B.C.; Alta.
ruficornis Williston, 1886a: 302.—Pa.; Iowa and Kans. to Mass. and N.J., also Ariz.
rufipalpis Brooks, 1945a: 87.—B.C.
vivida Williston, 1886a: 302.—Pa.; B.C. to Calif. and Colo., also Maine, N.C.
viridis, authors, not Fallén.

Genus HINEOMYIA Townsend

Hinea Townsend, 1916h: 629 (preocc. Gray, 1847). Type-species, *Nemoraea setigera* Coquillett (orig. des.).
Hineomyia Townsend, 1916d: 12 (n. name for *Hinea* Townsend). Type-species, *Nemoraea setigera* Coquillett (aut.).

setigera (Coquillett), 1902g: 111 (*Nemoraea*).—Ohio; Wyo., N.Y.', Vt.

Genus LINNAEMYA Robineau-Desvoidy

Linnaemya Robineau-Desvoidy, 1830: 52. Type-species, *silvestris* Robineau-Desvoidy (Robineau-Desvoidy, 1863a: 131) = *vulpina* (Fallén).
Linnaemyia, emend. or error.

Bonellia Robineau-Desvoidy, 1830: 56 (preocc. Rolando, 1822). Type-species, *tessellans* Robineau-Desvoidy (Townsend, 1916d: 6)=*haemorrhoidalis* (Fallén).
Bonellimyia Townsend, 1919c: 177 (n. name for *Bonellia* Robineau-Desvoidy). Type-species, *Bonellia tessellans* Robineau-Desvoidy (aut.)=*haemorrhoidalis* (Fallén).

REFERENCE: Brooks, 1944b (rev., as *Bonellimyia*).

glauca (Brooks), 1944b: 200 (*Bonellimyia*).—Que.; B.C. and Wash. to Colo., also Minn. to N.B. and N.S., s. to Ohio and Conn. **N. comb.**
haemorrhoidalis (Fallén).—Not Nearctic.
picta (Meigen).—Not Nearctic.
speculifera (Walker), 1849: 731 (*Tachina*).—N. Amer. Unrecognized.
subpolita (Brooks), 1944b: 200 (*Bonellimyia*).—Ont.; Utah to N.S., also B.C. **N. comb.**
tessellata (Brooks), 1944b: 198 (*Bonellimyia*).—N.S.; B.C. to Sask., s. to Calif. and Colo., also Mich. to N.S., s. to Mass., also Tenn. **N. comb.**

Genus NIGROBONELLIA Brooks

Nigrobonellia Brooks, 1944b: 202. Type-species, *Linnaemyia varia* Curran (orig. des.).

nigrescens (Curran), 1925g: 15 (*Linnaemyia*).—B. C.; Man.
varia (Curran), 1925g: 14 (*Linnaemyia*).—Labr.; N.W.T. (Baffin I.).

Genus SPILOCHAETOSOMA Smith

Spilochaetosoma Smith, 1917c: 125. Type-species, *californicum* Smith (orig. des.).

californicum Smith, 1917c: 126.—Calif.; Idaho, Utah, Ariz.

Genus TENUIROSTRA Ringdahl

Tenuirostra Ringdahl, 1933b: 17. Type-species, *Lypha arctica* Sack (mon.).

arctica (Sack), 1923: 7 (*Lypha*).—U.S.S.R. (Novaya Zemlya); N.W.T. (n. Ellesmere I., Akpatok I. in Ungava Bay), Greenland.

Genus THOMPSONOMYIA Brooks

Thompsonomyia Brooks, 1944b: 204. Type-species, *Linnaemyia anthracina* Thompson (orig. des.).

anthracina (Thompson), 1911: 266 (*Linnaemyia*).—Ont.; N.W.T. (Baffin I.).

Genus TRAFOIA Brauer and Bergenstamm

Trafoia Brauer and Bergenstamm, 1893: 142 (1893: 54). Type species, *monticola* Brauer and Bergenstamm (orig. des., as gen. n., sp. n.).
rufipalpis (Bigot), 1888a: 256 (*Exorista*).—Mexico; Colo., N. Mex., El Salvador.
 hispida Wulp, 1890: 65 (*Exorista*).—Mexico. **N. syn.**

Tribe ERNESTIINI

The species of Ernestiini are parasites of lepidopterous larvae, especially those of the family Noctuidae.

Genus APPENDICIA Stein

Appendicia Stein, 1924: 54. Type-species, *Tachina truncata* Zetterstedt (mon.).
frontalis (Tothill), 1921: 204, 228 (*Ernestia*).—Alaska, B.C.; Alaska and B.C. to Ont. and N.Y., also Ariz.
 solita Reinhard, 1937: 66 (*Ernestia*).—Mich.

Genus MELINOCERA Townsend

Melinocera Townsend, 1915a: 22. Type-species, *Meriania chalybaea* Coquillett (orig. des.).
 REFERENCE: Brooks, 1943 (key).
chalybaea (Coquillett), 1902g: 119 (*Meriania*).—Idaho; B.C., Mont., Calif.
flavicornis (Brauer), 1898: 532 (*Panzeria*).—N.H.; B.C. to N.B., s. to Calif., Tex., and N.C.

Genus MERICIA Robineau-Desvoidy

Mericia Robineau-Desvoidy, 1830: 64. Type-species, *erigonea* Robineau-Desvoidy (mon.).
Ernestia, authors, not Robineau-Desvoidy.
 REFERENCES: Tothill, 1921 (rev.); Brooks, 1943 (rev., Canadian spp.).
alberta Curran, 1924n: 248.—Alta.; Calif., Nev., Mich.
aldrichi (Townsend), 1892a: 91 (*Hystricia*).—S. Dak.; B.C., Alta., Sask., Mich., Ariz.

ampelus (Walker), 1849: 732 (*Tachina*).—N.S.; B.C. to N.S., s. to Calif. and Md., also Ga.
  nigricornis Townsend, 1893h: 467 (*Nemoraea*). Nomen nudum.
  nigricornis Williston, 1908: 371, fig. 101 (*Nemoraea*).—Ill. (unpublished). **N. syn.**
arcuata (Tothill), 1921: 205, 248 (*Ernestia*).—Va.; Mich. to N.S., s. to N.C.
bicarina (Tothill), 1921: 204, 272 (*Ernestia*).—B.C.; B.C. to S. Dak., s. to Ariz. and N. Mex., also Ont., N.B.
campestris Curran, 1924n: 249.—Man.; Alta. to N.Y., s. to Ohio.
cobala Reinhard, 1953a: 55.—N. Mex.
degenera (Walker), 1849: 732 (*Tachina*).—Ont. Unrecognized.
fissicarina (Tothill), 1921: 204, 274 (*Ernestia*).—Calif.
hamilla Reinhard, 1953a: 54.—Wash.; Idaho.
incisa (Tothill), 1921: 204, 249 (*Ernestia*).—Pa.; B.C., Mont., Mich., Ohio.
johnsoni (Tothill), 1921: 204, 229 (*Ernestia*).—Mass.; B.C., Mich.
longicarina (Tothill), 1921: 205, 251 (*Ernestia*).—Calif.; Alta., Wash., Oreg.
manitoba Brooks, 1943: 77.—Man.; Mich., N.Y., Vt.
nigropalpis (Tothill), 1921: 204, 247 (*Ernestia*).—B.C.; B.C. to Calif., also Mich. to N.S., s. to Ohio.
occidentalis Brooks, 1943: 77.—B.C.
platycarina (Tothill), 1921: 204, 270 (*Ernestia*).—Va.; B.C. and Wash. to Colo., also Minn. to P.E.I., s. to Va.
radicum (Fabricius).—Not Nearctic.
ruficauda (Brauer), 1898: 539 (*Erigone*).—N. Amer. Unrecognized, probably=*ampelus*.
sulcocarina (Tothill), 1921: 205, 271 (*Ernestia*).—B.C.; Alaska and B.C. to N.Y., also Calif., N. Mex., Colo.
triangularis Curran, 1924n: 247.—Man.; Calif., Ariz., Mich., Ont., N.Y.

Genus METAPHYTO Coquillett

Metaphyto Coquillett, 1897: 89. Type-species, *genalis* Coquillett (orig. des.).
Okanagania Townsend, 1915c: 289. Type-species, *hirta* Townsend (orig. des.).

genalis Coquillett, 1897: 90.—Colo.; B.C. to Sask. and Mont., also Calif. to Utah.
hirta (Townsend), 1915c: 290 (*Okanagania*).—B.C.
setifrons (Brooks), 1943: 71 (*Okanagania*).—B.C.; Idaho, Sask.

Genus OSTRACOPHYTO Townsend

Ostracophyto Townsend, 1915f: 228. Type-species, *aristalis* Townsend (orig. des.).
aristalis Townsend, 1915f: 228.—Calif.; B.C., Wash.

Genus PROMERICIA Brooks

Promericia Brooks, 1943: 69. Type-species, *Mericia fasciventris* Curran (orig. des.).
fasciventris (Curran), 1924n: 248 (*Mericia*).—Que.; B.C., Calif. to Colo., also Man. to Que. and N.Y.

Genus PSEUDOMERIANIA Brooks

Pseudomeriania Brooks, 1943: 69. Type-species, *Ernestia nigrocornea* Tothill (orig. des.).
nigrocornea (Tothill), 1921: 204, 227 (*Ernestia*).—Calif.; Alaska to Sask., s. to Calif. and Colo., also Ont.
fasciata Curran, 1924n: 246 (*Ernestia*).—B.C.
septentrionalis Curran, 1930i: 75 (*Meriania*).—B.C.

Genus XANTHOPHYTO Townsend

Xanthophyto Townsend, 1916h: 627. Type-species, *Nemoraea labis* Coquillett (orig. des.).
Xanthoernestia Townsend, 1926: 39. Type-species, *antennalis* Townsend (orig. des.).
antennalis (Townsend), 1926: 40 (*Xanthoernestia*).—N.H.; B.C., Calif., Utah, N.Y., N.C., Ga.
labis (Coquillett), 1895k: 104 (*Nemoraea*).—Wash.; B.C. to Calif. and N. Mex., also Ont., Vt.

Tribe MELANOPHRYINI

(Melanophryctini)

Genus ATROPHARISTA Townsend

Atropharista Townsend, 1892a: 92. Type-species, *jurinoides* Townsend (orig. des.) = *insolita* (Walker).
insolita (Walker), 1852: 277 (*Tachina*).—U.S.; B.C., Calif. to Colo., S. Dak. to Maine, s. to Ill. and Conn.
jurinoides Townsend, 1892a: 92.—S. Dak.

Genus MELANOPHRYS Williston

Melanophrys Williston, 1886a: 305. Type-species, *flavipennis* Williston (mon.).

flavipennis Williston, 1886a: 306.—Wyo.; Wash. and Alta., s. to Calif. and Wyo.

Tribe APHRIINI

The species of the tribe Aphriini parasitize lepidopterous larvae, including some wood- and stalk-borers.

Genus APACHEMYIA Townsend

Apachemyia Townsend, 1908: 75. Type-species, *Demoticus pallidus* Coquillett (orig. des.).

pallida (Coquillett), 1897: 121 (*Demoticus*).—Colo.; Oreg. to Mont., s. to Calif. and N. Mex.

Genus APHRIA Robineau-Desvoidy

Aphria Robineau-Desvoidy, 1830: 89. Type-species, *abdominalis* Robineau-Desvoidy (Robineau-Desvoidy, 1863a: 767) = *longirostris* (Meigen).

georgiana Townsend, 1908: 68.—Ga.

ocypterata Townsend, 1891c: 361.—S. Dak., Minn.; B.C., Wash. to Que. and Maine, s. to Calif., N. Mex., and Va.

occidentalis Townsend, 1908: 68.—Colo. **N. syn.**

Genus NEOFISCHERIA Townsend

Neofischeria Townsend, 1908: 74. Type-species, *flava* Townsend (orig. des.).

flava Townsend, 1908: 75.—Pa.; Ill. to Conn., s. to Va.

Genus NEOSOLIERIA Townsend

Neosolieria Townsend, 1927: 211, 336. Type-species, *nasuta* Townsend (orig. des.).

Parcipromus Reinhard, 1958c: 227. Type-species, *silus* Reinhard (orig. des.). **N. syn.**

sila (Reinhard), 1958c: 228 (*Parcipromus*).—Mexico (Baja Calif.); Calif., Utah. **N. comb.**

Genus PARADEMOTICUS Townsend

Parademoticus Townsend, 1916h: 629. Type-species, *Demoticus piperi* Coquillett (orig. des.).

piperi (Coquillett), 1897: 122 (*Demoticus*).—Wash.; B.C. to Calif., also Colo.

Genus PARAFISCHERIA Townsend

Parafischeria Townsend, 1908: 74. Type-species, *Drepanoglossa venatoris* Coquillett (orig. des.)=*eucerata* (Bigot).

eucerata (Bigot), 1888a: 263 (*Masicera*).—Calif.; Wash. to Calif. and Colo.
venatoris Coquillett, 1895c: 127 (*Drepanoglossa*).—Wash.

Genus PYRAUSTOMYIA Townsend

Pyraustomyia Townsend, 1916h: 627. Type-species, *Panzeria penitalis* Coquillett (orig. des.).

penitalis (Coquillett), 1897: 89 (*Panzeria*).—Mo.; Wis. to Mass., s. to Kans. and Ga.

Tribe LYPHINI

The hosts of the species of Lyphini are chrysomelid grubs.

Genus LYPHA Robineau-Desvoidy

Lypha Robineau-Desvoidy, 1830: 141. Type-species, *Tachina dubia* Fallén (Robineau-Desvoidy, 1863a: 196).

REFERENCE: Brooks, 1945a (key).

dubia (Fallén), 1810b: 284 (*Tachina*).—Sweden; Palaearctic, introd. Mich., e. Canada, N.Y., Mass., Conn., ?estab.
frontalis Brooks, 1945a: 85. —Ont.
fumipennis Brooks, 1945a: 85.—B.C.; Mich., Ohio.
intermedia Brooks, 1945a: 86.—Ont.
maculipennis Aldrich, 1926d: 24.—Wash.; B.C., Que.
melobosis (Walker), 1849: 743 (*Tachina*).—Fla.
    addita Walker, 1852: 265, 290 (*Tachina*).—U.S.
parva Brooks, 1945a: 86.—Ont.
setifacies (West), 1925: 124 (*Didyma*).—N.Y.; B.C. to N.B., s. to Ill. and Mass.

Tribe GERMARIINI

Members of the tribe Germariini parasitize lepidopterous larvae, except for one genus that contains parasites of earwigs (Dermaptera), and a few species reared from *Gryllus*, beetles, a centipede, and pupae of Tephritidae.

Genus ASSECLAMYIA Reinhard

Asseclamyia Reinhard, 1956c: 108. Type-species, *sphenofrons* Reinhard (orig. des.).

sphenofrons Reinhard, 1956c: 109. Calif.

Genus BIGONICHETA Rondani

Bigonicheta Rondani, 1845b: 34. Type-species, *mariettii* Rondani (mon.)=*setipennis* (Fallén).
Bigonichaeta, Bigonochaeta, emends. or errors.
Digonichaeta, Digonicheta, Digonochaeta, emends. or errors.

Parasites of earwigs.

spinipennis (Meigen), 1824: 350 (*Tachina*).—France; introd. and estab. B.C. to Oreg. and Utah, also Nfld., not estab. Ont.
   setipennis, authors, not Fallén.

Genus CACOZELUS Reinhard

Cacozelus Reinhard, 1943a: 168. Type-species, *riederi* Reinhard (orig. des.).

riederi Reinhard, 1943a: 168.—Oreg.

Genus CHROMATOCERA Townsend

Chromatocera Townsend, 1915a: 21. Type-species, *Eulasiona setigena* Coquillett (orig. des.).

fumator Reinhard, 1962b: 218.—Ariz.
setigena (Coquillett), 1897: 53 (*Eulasiona*).—La.; S. Dak., Kans., Ohio, Md., Tex.

Genus COCKERELLIANA Townsend

Cockerelliana Townsend, 1915f: 216. Type-species, *capitata* Townsend (orig. des.).

capitata Townsend, 1915f: 218.—N. Mex.; Calif., Utah, Ariz.

Genus DICHOCERA Williston

Dichocera Williston, 1895a: 31. Type-species, *lyrata* Williston (mon.).

lyrata Williston, 1895a: 32.—Idaho; B.C. to S. Dak., s. to Calif. and N. Mex., also N.Y., N.H.
robusta Brooks, 1945a: 80.—Alta.; Wash. to Sask., s. to Oreg. and Utah.

Genus EXORISTOIDES Coquillett

Exoristoides Coquillett, 1897: 90. Type-species, *johnsoni* Coquillett (orig. des.).

REFERENCE: Aldrich, 1932b (key).

harrisi Reinhard, 1935b: 160.—Iowa.
johnsoni Coquillett, 1897: 91.—N.C.; B.C. to Calif. and Ariz., also S. Dak., Va., N.C., Tex., and La.

Genus HELIOPLAGIA Townsend

Helioplagia Townsend, 1934c: 211. Type-species, *amazonica* Townsend (orig. des.).

slossonae (Coquillett), 1897: 91 (*Exoristoides*).—Maine, N.H., N.J.; Mich. to Que. and Maine, s. to Miss. and Fla.
spinipennis Coquillett, 1897: 95 (*Exorista*).—Ga.

Genus HOMALACTIA Townsend

Homalactia Townsend, 1915a: 21. Type-species, *Exoristoides harringtoni* Coquillett (orig. des.).

REFERENCE: Curran, 1926b (key).

brimleyi Curran, 1926b: 86.—Ont.
facula Reinhard, 1959a: 161.—Calif.
harringtoni (Coquillett), 1902g: 110 (*Exoristoides*).—Ont.; Man. to N.H., s. to Wis. and Va.

Genus IMPECCANTIA Reinhard

Impeccantia Reinhard, 1961: 204. Type-species, *claterna* Reinhard (orig. des.).

claterna Reinhard, 1961: 205.—Ariz.; N. Mex.

Genus LYDINA Robineau-Desvoidy

Lydina Robineau-Desvoidy, 1830: 124. Type-species, *nitida* Robineau-Desvoidy (Robineau-Desvoidy, 1863a: 111)= *aenea* (Meigen).
Polidea Macquart, 1848b: 92. Type-species, *Tachina aenea* Meigen (Coquillett, 1910b: 593).
Polidaria Curran, 1934a: 429, 444, 464. Type-species, *Tachina areos* (Walker (mon.).

americana (Townsend), 1892e: 78 (*Tryphera*).—Ill.; Minn. and Ill. to Mass. and Conn., also Md., Ga.
areos (Walker), 1849: 766 (*Tachina*).—N. Amer.; Oreg., Idaho, Calif., Ind.
polidoides (Townsend), 1892e: 79 (*Tryphera*).—N.Y.; Idaho, Minn. to N.S., s. to Va.
    americana Townsend, 1892e: 82 (*Polidea*; preocc. Townsend, 1892e: 78).—Mich.

Genus MACTOMYIA Reinhard

Mactomyia Reinhard, 1958c: 228. Type-species, *fracida* Reinhard (orig. des.).

fracida Reinhard, 1958c: 229.—Ariz.; Calif., Utah, N. Mex.

Genus MAUROMYIA Coquillett

Mauromyia Coquillett, 1897: 51. Type-species, *pulla* Coquillett (orig. des.).

pulla Coquillett, 1897: 52.—N.H.; Oreg., Minn. to N.H. and Pa., also N. Mex., Ga.

Genus METATACHINA Townsend

Metatachina Townsend, 1919b: 588. Type-species, *mellifrons* Townsend (orig. des.).

mellifrons Townsend, 1919b: 588.—Maine.

Genus NEODICHOCERA Walton

Neodichocera Walton, 1914a: 184. Type-species, *tridens* Walton (orig. des.)=*orientalis* (Coquillett).
Dichoceropsis Townsend, 1916b: 178. Type-species, *Dichocera orientalis* Coquillett (orig. des.).

orientalis (Coquillett), 1897: 138 (*Dichocera*).—Mass.; Mont., Utah, Ariz., N. Mex.
    tridens Walton, 1914a: 185.—N. Mex.

Genus PARADMONTIA Coquillett

Paradmontia Coquillett, 1902g: 106. Type-species, *brevis* Coquillett (orig. des.).

brevis Coquillett, 1902g: 106.—Fla.; Mont., Iowa to Md. and Va., also La.

Genus PETINARCTIA Villeneuve

Petinarctia Villeneuve, 1928: 306. Type-species, *Peteina stylata* Brauer and Bergenstamm (mon.).
Rhynchopeteina Townsend, 1931b: 459. Type-species, *Peteina stylata* Brauer and Bergenstamm (orig. des.).
Pseudopetina Ringdahl, 1933b: 17. Type-species, *Peteina stylata* Brauer and Bergenstamm (mon.). **N. syn.**

stylata (Brauer and Bergenstamm), 1891: 387 (1891: 83) (*Peteina*).— Greenland; N.W.T., Sweden.

Genus PLAGIOSIPPUS Reinhard

Plagiosippus Reinhard, 1962a: 172. Type-species, *invasor* Reinhard (orig. des.).

invasor Reinhard, 1962a: 173.—Ariz.; Calif.

Genus TROCHILODES Coquillett

Trochilodes Coquillett *in* Johnson, 1903a: 102. Type-species, *skinneri* Coquillett (orig. des.).

leonardi (West), 1925: 134 (*Rhamphina*).—N.Y. **N. comb.**
skinneri Coquillett *in* Johnson, 1903a: 103.—N. Mex.; B.C. to Wyo., s. to Calif. and N. Mex.

Tribe CAMPYLOCHETINI

The species of Campylochetini are parasites of lepidopterous larvae.

Genus CHAETOPHLEPSIS Townsend

Chaetophlepsis Townsend, 1915g: 422. Type-species, *tarsalis* Townsend (orig. des.).
Coloradalia Curran, 1934a: 444. Type-species, *ocellaris* Curran (orig. des.)=*eudryae* (Smith). **N. syn.**

REFERENCE: Reinhard, 1952c (rev.).

atriceps Reinhard, 1952c: 21.—Ohio.
canora Reinhard, 1952c: 18.—N.Y.

eudryae (Smith), 1916: 94 (*Hypochaeta*).—Mass.; Colo., Ark., Conn., Miss.
  ocellaris Curran, 1934a: 467 (*Coloradalia*).—Colo. **N. syn.**
  tarsalis, authors, not Townsend.
nasellensis Reinhard, 1952c: 18.—Wash.
orbitalis Webber, 1931: 2.—Mass.; B.C., Ont. to N.H. and R.I., also Md.
polita Brooks, 1945a: 83.—Miss.; Oreg., Calif., Wis., Ont.
  ucayali Townsend of Reinhard, 1952c: 17.
rindgei Reinhard, 1952c: 20.—Calif.
semiothisae Brooks, 1945a: 82.—Ont.; Minn. to N.S., s. to Ohio and Md., also B.C.
teliosis Reinhard, 1952c: 18.—N.C.; B.C.
townsendi (Smith), 1916: 94 (*Hypochaeta*).—Fla.

Genus PARAHYPOCHAETA Brauer and Bergenstamm

Parahypochaeta Brauer and Bergenstamm, 1891: 337 (1891: 33). Type-species, *heteroneura* Brauer and Bergenstamm (mon.).

heteroneura Brauer and Bergenstamm, 1891: 337 (1891: 33).—N. Amer.

Tribe VORIINI

The species of the tribe Voriini parasitize many kinds of lepidopterous larvae.

Genus ANZAMYIA Reinhard

Agathomyia Reinhard, 1959b: 228 (preocc. Verrall, 1901). Type-species, *cordata* Reinhard (orig. des.).
Anzamyia Reinhard, 1960: 103 (n. name for *Agathomyia* Reinhard). Type-species, *Agathomyia cordata* Reinhard (aut.).

cordata (Reinhard), 1959b: 229 (*Agathomyia*).—Calif.

Genus ATHRYCIA Robineau-Desvoidy

Athrycia Robineau-Desvoidy, 1830: 111. Type-species, *erythrocera* Robineau-Desvoidy (Robineau-Desvoidy, 1863a: 830).
Blepharigena Rondani, 1856: 69. Type-species, *Tachina trepida* Meigen (orig. des.; misident.)=*curvinervis* (Zetterstedt).
Paraplagia Brauer and Bergenstamm, 1891: 354 (1891: 50). Type-species, *Tachina trepida* Meigen (mon.).

cinerea (Coquillett), 1895k: 101 (*Paraplagia*).—Ill., D.C.; B.C. to Colo. to Mass., s. to Mexico (Chih.) and Va. **N. comb.**
  spinosula, authors, not Bigot.

Genus CATALINOVORIA Townsend

Catalinovoria Townsend, 1926: 37. Type-species, *cauta* Townsend (orig. des.).
Sthenopleura Aldrich, 1926a: 18. Type-species, *latifrons* Aldrich (orig. des.)=*cauta* Townsend.

The genus *Catalinovoria* is closely related to the Palaearctic *Hypovoria* Villeneuve, of which it may prove to be a synonym or subgenus.

cauta Townsend, 1926: 38.—Ariz.; B.C. to N.Y., s. to Calif., Mexico, and Va.
    latifrons Aldrich, 1926a: 18 (*Sthenopleura*).—N. Mex.
    rigidirostris, authors, not Wulp.
discalis Brooks, 1945a: 81.—N.B.; Alta., Sask., Idaho, Calif., Ohio.

Genus CHAETOPLAGIA Coquillett

Chaetoplagia Coquillett, 1895k: 98. Type-species, *atripennis* Coquillett (orig. des.).

atripennis Coquillett, 1895k: 98.—D.C.; Calif. to N.Y., s. to Mexico and Fla.
    aterrima (error) Williston, 1908: 368.

Genus CYRTOPHLEBA Rondani

Cyrtophleba Rondani, 1856: 68. Type-species, *Tachina ruricola* Meigen (orig. des.).
Cyrtophlebia, Cyrtophloeba, emends.

coquilletti Aldrich, 1926a: 12 (n. name for *horrida* Coquillett).—Ill., N.Y., Mass.; B.C. to Calif. and N. Mex., also Mich. to Mass., s. to Ill. and N.J.
    horrida Coquillett, 1895k: 101 (preocc. Giglio-Tos, 1893).—Ill., N.Y., Mass.
nitida Curran, 1930i: 74.—Man.; B.C., Alta., Wash., Colo., also Man. to Que. and Va.

Genus EUCYRTOPHLOEBA Townsend

Eucyrtophloeba Townsend, 1916i: 316. Type-species, *rhois* Townsend (orig. des.).

horrida (Giglio-Tos), 1893b: 6 (*Cyrtophloeba*).—Mexico; Ariz., Tex.

Genus EUPTILOPAREIA Townsend

Euptilopareia Townsend, 1961i: 319. Type-species, *Paraplagia erucicola* Coquillett (orig. des.).

erucicola (Coquillett), 1897: 78 (*Paraplagia*).—Mo.; Idaho.
vicinalis Reinhard, 1956a: 125.—Calif.

Genus GINGLYCHAETA Aldrich

Ginglychaeta Aldrich, 1926a: 21. Type-species, *seriata* Aldrich (orig. des.).

seriata Aldrich, 1926a: 21.—Colo.; Wash.

Genus GONIOCHAETA Townsend

Goniochaeta Townsend, 1891c: 351. Type-species, *plagioides* Townsend (orig. des.).

fuscibasis Aldrich, 1926a: 24.—Calif.
plagioides Townsend, 1891c: 352.—N. Mex.; Calif., Utah, Ariz., Tex.

Genus MELETERUS Aldrich

Meleterus Aldrich, 1926a: 20. Type-species, *montanus* Aldrich (orig. des.).

montanus Aldrich, 1926a: 20.—Mexico (Chih.); N. Mex.
nuperus Reinhard, 1956a: 123.—Ga.

Genus MENETUS Aldrich

Menetus Aldrich, 1926a: 23. Type-species, *Brachicoma macropogon* Bigot (orig. des.).

macropogon (Bigot), 1888a: 259 (*Brachicoma*).—Calif.

Genus METAPLAGIA Coquillett

Metaplagia Coquillett, 1895k: 102. Type-species, *occidentalis* Coquillett (orig. des.).

occidentalis Coquillett, 1895k: 103.—Calif.; Tex.

Genus METAVORIA Townsend

Metavoria Townsend, 1915l: 101. Type-species, *orientalis* Townsend (orig. des.).

brevicornis (Brooks), 1945a: 81 (*Metaplagia*).—Man.; Sask., Ont., P.E.I., N.H. **N. comb.**

facialis Reinhard, 1956a: 123.—Utah; Calif.
latifrons Reinhard, 1956a: 121.—Tex.
orientalis Townsend, 1915l: 101.—Va.; Mich. to Mass., s. to Tex. and Ga.

Genus PLAGIOMIMA Brauer and Bergenstamm

Plagiomima Brauer and Bergenstamm, 1891: 384 (1891: 80). Type-species, *disparata* Brauer and Bergenstamm (mon.).
Siphoplagia Townsend, 1891c: 349. Type-species, *anomala* Townsend (orig. des.)=*spinosula* (Bigot).
  REFERENCE: Aldrich, 1926a (rev.).

Subgenus PLAGIOMIMA Brauer and Bergenstamm

alternata Aldrich, 1926a: 27.—Fla.; Ohio.
auriceps Aldrich, 1926a: 28.—Va.; Ohio to Md., s. to Ga.
brevirostris Reinhard, 1962b: 219.—Ariz.
cognata Aldrich, 1926a: 27.—Md.; Iowa to N.J., s. to Tex. and N.C.
euethes Reinhard, 1957: 104.—Ariz.
haustellata Reinhard, 1944b: 59.—Kans.; Utah, Nebr., Tex.
spinosula (Bigot), 1888a: 262 (*Heteropterina*).—N. Amer.; Calif., Ariz., Colo., N. Mex., Tex.
  anomala Townsend, 1891c: 350 (*Siphoplagia*).—N. Mex.

Subgenus SIPHOPLAGIOPSIS Townsend

Siphoplagiopsis Townsend, 1917d: 123 (as genus). Type-species, *similis* Townsend (orig. des.). **N. status.**

abdominalis Aldrich, 1926a: 27.—Colo.
faceta Reinhard, 1957: 105.—Tex.; Mexico.
similis (Townsend), 1917d: 124 (*Siphoplagiopsis*).—Md.; N.Y. and N.J. to S.C., also Colo., Ill., Tex.

Genus PSILOPLEURA Reinhard

Psilopleura Reinhard, 1943a: 165. Type-species, *arida* Reinhard (orig. des.).
arida Reinhard, 1943a: 166.—Utah; Calif., Ariz., N. Mex.

Genus UCLESIA Girschner

Uclesia Girschner, 1901: 69. Type-species, *fumipennis* Girschner (mon.).
Euclesia (error) Aldrich, 1926a: 24.

Atractouclesia Townsend, 1931b: 460. Type-species, *Uclesia retracta* Aldrich (orig. des.). **N. syn.**

Uclesiopsis Townsend, 1931b: 461. Type-species, *Uclesia varicornis* Curran (orig. des.). **N. syn.**

REFERENCE: Curran, 1927i (key).

retracta Aldrich, 1926a: 24.—Wash.; Alta.
varicornis Curran, 1927i: 300.—N. Dak.
zonalis Curran, 1927i: 299.—B.C.; Wash., Utah, Wyo.

Genus VORIA Robineau-Desvoidy

Voria Robineau-Desvoidy, 1830: 195. Type-species, *latifrons* Robineau-Desvoidy (mon.)=*ruralis* (Fallén).

Plagia Meigen, 1838: 201. Type-species, *Tachina verticalis* Meigen (Rondani, 1856: 69)=*ruralis* (Fallén).

aurifrons (Townsend), 1892e: 67 (*Plagia*).—Pa.; Mich. to Mass., s. to Ill. and Va.

ruralis (Fallén), 1810b: 265 (*Tachina*).—Sweden; B.C., U.S. se. to Miss. and D.C., Mexico, Europe.

americana Wulp, 1890: 102 (*Plagia*).—Mexico.

Genus WAGNERIA Robineau-Desvoidy

Wagneria Robineau-Desvoidy, 1830: 126. Type-species, *gagatea* Robineau-Desvoidy (mon.)=*costata* (Fallén).

REFERENCE: Reinhard, 1955c (rev.).

cornuta Curran, 1928b: 49.—Mass.; N.B. to Ohio and Conn.
major Curran, 1928b: 48.—Ark.; Mass., Miss., Ga., Fla.
ocellaris Reinhard, 1955c: 55.—Oreg.; Wash., Idaho, Calif.
pacata Reinhard, 1955c: 56.—Ohio; Alta., N. Dak., Calif., Colo.
vernata West, 1925: 127.—N.Y.; Minn. to Que., s. to Colo. and Conn., also B.C., Idaho, Calif.

Subfamily DEXIINAE

The food habits of the subfamily Dexiinae are varied. In addition to many lepidopterous larvae, some coleopterous larvae are parasitized, and a scattering of parasitism on other insects has been reported.

The divergent usage of the name Dexiinae is discussed under the genus *Dexia* below.

Tribe THELAIRINI

The hosts of the tribe Thelairini are lepidopterous larvae, notably in the families Noctuidae and Arctiidae.

Genus SPATHIDEXIA Townsend

Spathidexia Townsend, 1912a: 110. Type-species, *clemonsi* Townsend (orig. des.).

REFERENCE: Arnaud, 1960 (rev.).

cerussata Reinhard, 1934a: 152.—Ohio; Mich., Conn.
clemonsi Townsend, 1912a: 110.—Mass.; Mich. to Maine, s. to Ky. and N.C., also N. Mex., Tex.
creolensis Reinhard, 1955a: 131.—Fla.; La.
dunningii (Coquillett), 1895j: 54 (*Thryptocera*).—Ill.; Alta. to N.B., entire U.S., also P.R., Jamaica.
    rasilis Reinhard, 1934a: 153.—Wis.
nexa Reinhard, 1953d: 94.—Mexico (D.F.); Ariz.
reinhardi Arnaud, 1960: 26.—Tex.; Sask. to N.B., s. to Calif. and Miss.

Genus THELAIRA Robineau-Desvoidy

Thelaira Robineau-Desvoidy, 1830: 214. Type-species, *abdominalis* Robineau-Desvoidy (Townsend, 1916d: 9)=*nigripes* (Fabricius).

americana Brooks, 1945a: 88.—Que.; B.C. to N.S., entire U.S., Mexico.
    leucozona, authors, not Panzer.
    nigripes, authors, not Fabricius.
bryanti Curran, 1925f: 281.—Alta.; Alaska, Alta. to Calif. and Colo.

Tribe DEXIINI

Genus DEXIA Meigen

Dexia Meigen, 1826: 33. Type-species, *Musca volvulus* Fabricius (Westwood, 1840: 139).
Phyllomya Robineau-Desvoidy, 1830: 213. Type-species, *Musca volvulus* Fabricius (mon.).
Neadmontia Townsend, 1912d: 164. Type-species, *Admontia limata* Coquillett (orig. des.).
Ocypterosoma Townsend, 1915a: 19. Type-species, *Admontia polita* Coquillett (orig. des.).

The generic name *Dexia*, and the higher categories Dexiini and Dexiinae founded upon it, have been subject to very diverse usage. The nomenclaturally correct use, based on the oldest valid type designation for *Dexia*, is that adopted by Townsend and followed here. European and older American literature use the name *Phyllomya* for the present genus, and *Dexia* instead of *Dexilla* of this catalog (see tribe Dexillini, subfamily Proseninae).

REFERENCE: Curran, 1927m (key, as *Phyllomya*).

fuscicosta (Curran), 1927m: 147 (*Phyllomya*).—B.C.
limata (Coquillett), 1902g: 105 (*Admontia*).—Idaho; Calif., Colo., Man.
polita (Coquillett), 1898c: 234 (*Admontia*).—N.Y.; Mich. and Ohio to Maine, also Fla.

Tribe MACQUARTIINI

Species of the tribe Macquartiini parasitize a variety of hosts, including larvae of chrysomelid beetles, of various lepidopterous families, and of sawflies.

Genus ALASKOPHYTO Townsend

Alaskophyto Townsend, 1915c: 285. Type-species, *Muscopteryx obscura* Coquillett (orig. des.).

obscura (Coquillett), 1902g: 116 (*Muscopteryx*).—Alaska.

Genus ANTHOMYIOPSIS Townsend

Anthomyiopsis Townsend, 1916e: 20. Type-species, *cypseloides* Townsend (orig. des.).

cypseloides Townsend, 1916e: 21.—N.H.; Alaska.

Genus ATHANATUS Reinhard

Phytopsis Townsend, 1915a: 20 (preocc. Hall, 1947). Type-species, *Amobia californica* Coquillett (orig. des.).
Athanatus Reinhard, 1947a: 15. Type-species, *knowltoni* Reinhard (orig. des.)=*californicus* (Coquillett).

Phytopsis Hall, though proposed for "?Plantae," is based on a type-species now considered to represent the work of an animal.

californicus (Coquillett), 1895k: 100 (*Amobia*).—Calif.; Oreg., Idaho, Utah, Colo. **N. comb.**
   knowltoni Reinhard, 1947a: 16.—Utah.

Genus DYSCOLOMYIA Reinhard

Dyscolomyia Reinhard, 1959b: 226. Type-species, *lucina* Reinhard (orig. des.).

lucina Reinhard, 1959b: 227.—Calif.

Genus EUBRACHYMERA Townsend

Eubrachymera Townsend, 1919c: 162. Type-species, *debilis* Townsend (orig. des.).

debilis Townsend, 1919c: 162.—S. Dak.; N. Dak. to Maine, s. to Tenn.

Genus EULASIONA Townsend

Eulasiona Townsend, 1892a: 119. Type-species, *comstocki* Townsend (orig. des.).

REFERENCE: Curran, 1927m (key).

Subgenus EULASIONA Townsend

comstocki Townsend, 1892a: 120.—N.Y.; Minn. to Que., s. to Kans. and Va., also B.C., Calif., Ariz., Ga.

nigra Curran, 1924g: 194.—B.C.; B.C. to Calif. and Ariz., also ?Mich.

Subgenus PARAMUSCOPTERYX Townsend

Paramuscopteryx Townsend, 1915f: 218 (as genus). Type-species, *genalis* Townsend (orig. des.). **N. status.**

Townsendina Curran, 1934a: 427, 464. Type-species, *fasciata* Curran (orig. des.). **N. syn.** (genus and subg.).

fasciata (Curran), 1934a: 464 (*Townsendina*).—Colo. **N. comb.**

genalis (Townsend), 1915f: 219 (*Paramuscopteryx*).—Idaho; Wash. to Calif. and Colo., also Vt.

Genus GIBSONOMYIA Curran

Gibsonomyia Curran, 1925f: 281. Type-species, *nigricosta* Curran (orig. des.).=*washingtoniana* (Bigot).

washingtoniana (Bigot), 1888a: 269 (*Morinia*).—Wash.; B.C. to Mont., s. to Calif. and Colo.

nigricosta Curran, 1925f: 282.—Alta.

Genus GRISDALEMYIA Curran

Steinia Brauer and Bergenstamm, 1893: 136 (1893: 48) (preocc. Diesing, 1866). Type-species, *Tachina protuberans* Zetterstedt (mon.) =*callida* (Meigen).
Steiniella Berg, 1898: 17 (n. name for *Steinia* Brauer and Bergenstamm, but preocc. Schuett, 1895). Type-species, *Tachina protuberans* Zetterstedt (aut.) =*callida* (Meigen).
Grisdalemyia Curran, 1926c: 133. Type-species, *bigelowi* Curran (orig. des.).
Psiloneura Aldrich, 1926d: 23. Type-species, *flavisquama* Aldrich (orig. des.) =*bigelowi* Curran.

aldrichi Curran, 1926c: 135.—Calif.; B.C.
bigelowi Curran, 1926c: 134.—Ont.; Calif., Mich., Ohio.
    flavisquama Aldrich, 1926d: 23 (*Psiloneura*).—Mich.
setosa Reinhard, 1937: 72.—S. Dak.; Nebr.

Genus ICTERICOPHYTO Townsend

Ictericophyto Townsend, 1916h: 626. Type-species, *Eulasiona spinosa* Coquillett (orig. des.).
Celotrophus Reinhard, 1958d: 280. Type-species, *soporis* Reinhard (orig. des.) =*catskillensis* (West). **N. syn.**, H. J. Reinhard *in litt.*

catskillensis (West), 1925: 125 (*Eulasiona*).—N.Y.; Mich., Ohio.
    soporis Reinhard, 1958d: 281 (*Celotrophus*).—Ohio. **N. syn.**, H. J. Reinhard *in litt.*
spinosa (Coquillett), 1897: 53 (*Eulasiona*).—Alaska; B.C., Ont., Que.
tibialis (Curran), 1927m: 150 (*Eulasiona*).—Que.; B.C.

Genus LASIONALIA Curran

Lasionalia Curran, 1934a: 437, 467. Type-species, *cinerea* Curran (orig. des.).

cinerea Curran, 1934a: 467.—Minn.; Wash., Calif., Ohio, N.C.

Genus MELEDONUS Aldrich

Meledonus Aldrich, 1926d: 15. Type-species, *latipennis* Aldrich (orig. des.).

albiceps Reinhard, 1956a: 126.—Calif.; Utah.
latipennis Aldrich, 1926d: 16.—Calif.
lindensis Reinhard, 1953a: 57.—Wash.

Genus METOPOMUSCOPTERYX Townsend

Metopomuscopteryx Townsend, 1915f: 219. Type-species, *Muscopteryx tibialis* Coquillett (orig. des.).

fatigantis Reinhard, 1958d: 284.—Calif.
incurata Reinhard, 1958d: 283.—Ariz.
tibialis (Coquillett), 1902g: 115 (*Muscopteryx*).—Idaho; B.C. to Calif., Ariz., and Mo.

Genus MINELLA Robineau-Desvoidy

Minella Robineau-Desvoidy, 1830: 209. Type-species, *nitida* Robineau-Desvoidy (mon.) =*nigrita* (Fallén).

americana Reinhard, 1943c: 14.—Ohio.

Genus MUSCOPTERYX Townsend

Muscopteryx Townsend, 1892e: 170. Type-species, *chaetosula* Townsend (orig. des.).
Psammoppia Townsend, 1915a: 20. Type-species, *Brachycoma pulverea* Coquillett (orig. des.) =*chaetosula* Townsend.

REFERENCE: Reinhard, 1944a (rev.).

chaetosula Townsend, 1892e: 171.—Mexico (Chih.); B.C., Wash., Calif., Ariz., Kans., Tex.
   pulverea Coquillett, 1897: 132 (*Brachycoma*).—Tex.
hiemalis Reinhard, 1944a: 355.—Wis.
hilaris Reinhard, 1944a: 355.—Tex.; Miss., Ga.
hinei Reinhard, 1944a: 354.—Fla.
longiseta Reinhard, 1944a: 355.—Calif.
nitida Reinhard, 1944a: 357.—Calif.
parilis Reinhard, 1944a: 356.—Tex.; Calif., Ariz., Wyo.
petentis Reinhard, 1958d: 280.—Ariz.; Mexico (Baja Calif.).

Genus MYIOCLONIA Reinhard

Myioclonia Reinhard, 1945a: 28. Type-species, *erythrocera* Reinhard (orig. des.).

albertana Reinhard, 1945a: 29.—Alta.
erythrocera Reinhard, 1945a: 29.—Tenn.; Ont.
nigricornis Reinhard, 1945a: 30.—Wash.

Genus OCHROCERA Townsend

Ochrocera Townsend, 1916e: 17. Type-species, *vaginalis* Townsend (orig. des.).

vaginalis Townsend, 1916e: 18.—N.H.; Maine.

Genus PELATACHINA Meade

Hyria Robineau-Desvoidy, 1863a: 1100 (preocc. Lamarck, 1819). Type-species, *Tachina tibialis* Fallén (as Meigen; orig. des.).

Tachina, subg. **Pelatachina** Meade, 1894: 109 (n. name for *Hyria* Robineau-Desvoidy). Type-species, *tibialis* Fallén (aut.).

Eohyria Townsend, 1915a: 23. Type-species, *Pelatachina pellucida* Coquillett (orig. des.). **N. syn.**, L. P. Mesnil *in litt.*

REFERENCE: Curran, 1927d (key).

limata Coquillett, 1902g: 107.—Idaho; B.C. to N. Dak., s. to Calif.
orillia Curran, 1927d: 23.—Ont.; Mich.
pellucida Coquillett, 1897: 65.—U.S.; B.C., Wash., Oreg., Colo., Ont.

Genus PSEUDOMORINIA Wulp

Pseudomorinia Wulp, 1891b: 213, 259. Type-species, *pictipennis* Wulp (sub. mon., Wulp, 1891b: 260).

pictipennis Wulp, 1891b: 260.—Mexico (Guerrero); Ariz.

Genus XANTHOCERA Townsend

Xanthocera Townsend, 1915a: 22. Type-species, *clistoides* Townsend (orig. des.).

atra Townsend, 1919b: 569.—Ill.; N.Y., N.J.
clistoides Townsend, 1915a: 22.—Ill.; Kans.
    johnsoni Townsend of Coquillett, 1897: 64 (in *Hyalurgus*).
lucentis Reinhard, 1962a: 174.—N.Y.

Tribe HYALURGINI

The species of *Hyalurgus* parasitize Tenthredinidae.

Genus Hyalurgus Brauer and Bergenstamm

Hyalurgus Brauer and Bergenstamm, 1893: 136 (1893: 48). Type-species, *Tachina crucigera* Zetterstedt (mon.; misident.) = *lucidus* (Meigen).

lucidus (Meigen), 1824: 268 (*Tachina*; n. name for *diaphana* Fallén).—Sweden; Europe, introd. Man., ?estab.
    diaphana Fallén, 1820g: 33 (*Tachina*; preocc. Fabricius, 1805).—Sweden.

Tribe ICELIINI
Genus ERVIA Robineau-Desvoidy

Ervia Robineau-Desvoidy, 1830: 225. Type-species, *Ocyptera triquetra* Olivier (mon.).
Paranaphora Townsend, 1908: 72. Type-species, *diademoides* Townsend (orig. des.)=*triquetra* (Olivier).

triquetra (Olivier), 1812: 423 (*Ocyptera*).—Carolina; Kans. to N.Y., s. to Cent. Amer. and Ga.
    mestor Walker, 1849: 741 (*Tachina*).—N. Amer. (as New Zealand, error).
    diademoides Townsend, 1908: 73 (*Paranaphora*).—Miss.

Tribe ZELIINI

The hosts of the species of Zeliini are coleopterous larvae.

Genus METADEXIA Coquillett

Metadexia Coquillett, 1899d: 220. Type-species, *tricolor* Coquillett (orig. des., as gen. n., sp. n.).

tricolor Coquillett, 1899d: 220.—La.; Nev. to Pa., s. to Mexico, La., and Md.

Genus MINTHOZELIA Townsend

Minthozelia Townsend, 1919b: 556. Type-species, *montana* Townsend (orig. des.).
    REFERENCE: Reinhard, 1946b (rev.).

argentosa Reinhard, 1946b: 55.—Tex.; Ariz.
gracilis Reinhard, 1946b: 56.—N. Mex.; Calif. to Tex.
metalis Reinhard, 1946b: 57.—Ohio; Pa., N.C.
mira Reinhard, 1946b: 56.—Tex.
montana Townsend, 1919b: 557.—Ariz.
nitens Reinhard, 1946b: 58.—Ariz.; Calif., Nev., N. Mex.
ruficauda Reinhard, 1946b: 58.—N.Y.

Genus ZELIA Robineau-Desvoidy

Zelia Robineau-Desvoidy, 1830: 314. Type-species, *rostrata* Robineau-Desvoidy (Coquillett, 1910b: 621)=*vertebrata* (Say).
Leptoda Wulp, 1885: 196. Type-species, *Dexia gracilis* Wiedemann (Wulp, 1891b: 250)=*vertebrata* (Say).

Euzelia Townsend, 1915a: 23. Type-species, *Zelia wildermuthii* Walton (orig. des.).

rufina (Bigot), 1885c: xxvi (*Homodexia*).—Calif.
vertebrata (Say), 1829: 176 (1859b: 366) (*Dexia*).—Ind.; Wash. to to Colo. to Que. and Maine, s. to Mexico and Fla.
  gracilis Wiedemann, 1830: 373 (*Dexia*).—Unknown.
  rostrata Robineau-Desvoidy, 1830: 315.—N. Amer.
wildermuthii Walton, 1914a: 177.—N. Mex.; Calif., Colo.
zonata (Coquillett), 1895a: 315 (*Gymnodexia*).—Fla.; Ind. to N.J., s. to Fla.

Tribe URAMYINI

Species of the tribe Uramyini parasitize various lepidopterous larvae.

Genus ANAPORIA Townsend

Anaporia Townsend, 1919b: 560. Type-species, *Aporia limacodis* Townsend (orig. des.).
  REFERENCE: Aldrich, 1921a (key, in *Pseudeuantha*).

limacodis (Townsend), 1892m: 275 (*Aporia*).—N.Y.; Mich. to Maine, s. to Ga.
  coquilletti Aldrich, 1921a: 90 (*Pseudeuantha*).—Mass.
pristis (Walker), 1849: 841 (*Dexia*).—Mass.; Mich. to N.H., s. to Ga.
  basalis Walker, 1852: 281 (*Tachina*; preocc. Walker, 1836).—Unknown.
  johnsoni Townsend, 1892e: 81 (*Macquartia*).—Pa. **N. syn.**
  isae Coquillett, 1897: 96 (*Exorista*).—D.C. **N. syn.**
rubripes (Aldrich), 1921a: 91 (*Pseudeuantha*).—Fla.; Tex.

Genus PARAPORIA Townsend

Aporia Macquart, 1846: 296 (1846: 168) (preocc. Hübner, 1819). Type-species, *quadrimaculata* Macquart (mon.).
Neaporia Townsend, 1908: 67 (n. name for *Aporia* Macquart, but preocc. Gorham, 1897). Type-species, *Aporia quadrimaculata* Macquart (aut.).
Paraporia Townsend, 1912c: 48 (n. name for *Neaporia* Townsend). Type-species, *Aporia quadrimaculata* Macquart (aut.).

quadrimaculata (Macquart), 1846: 297 (1846: 169) (*Aporia*).—Colombia; Ariz., Mexico, Neotropical.

Genus PSEUDEUANTHA Townsend

Pseudeuantha Townsend, 1915g: 416. Type-species, *linellii* Townsend (orig. des.)=*indita* (Walker).

REFERENCE: Aldrich, 1921a (key).

indita (Walker), 1861: 306 (?*Lydella*).—Mexico; Ariz., El Salvador.
 caloptera Bigot, 1888a: 263 (*Tricoliga*).—Mexico.
umbratilis Reinhard, 1935b: 164.—Tex.; Ariz., Mexico.

Genus URAMYA Robineau-Desvoidy

Uramya Robineau-Desvoidy, 1830: 215. Type-species, *producta* Robineau-Desvoidy (mon.).
Uromacquartia Townsend, 1916h: 626. Type-species, *halisidotae* Townsend (orig. des.).

aldrichi Reinhard.—Not n. of Mexico.
halisidotae (Townsend), 1916h: 626 (*Uromacquartia*).—Oreg.; B.C. to Calif. and Ariz.

Tribe URODEXIINI

The species of Urodexiini are chiefly parasites of beetle larvae, and a few attack roaches.

Genus CHAETONODEXODES Townsend

Chaetonodexodes Townsend, 1916i: 321. Type-species, *rafaeli* Townsend (orig. des.).

The species of *Chaetonodexodes* parasitize chrysomelid larvae.

vanderwulpi (Townsend), 1892h: 131 (*Myothyria*).—Fla.; Calif. and Colo. to Fla. **N. comb.**

Genus CHOLOMYIA Bigot

Cholomyia Bigot, 1884d: xxxvii. Type-species, *inaequipes* Bigot (mon.).

inaequipes Bigot, 1884d: xxxvii.—Mexico; Wis. and Ont. to Mass., s. to Ariz., Cent. Amer., and Fla.
 nigriceps Williston, 1908: 353, fig. 146.—Unknown.
 longipes, authors, not Fabricius.
 flavipes Coquillett *in* Johnson, 1905b: 78 (*Metadexia*). Nomen nudum.

Genus PHYLLOPHILOPSIS Townsend

Phyllophila Townsend, 1915a: 21 (preocc. Guenée, 1852). Type-species, *Chaetona nitens* Coquillett (orig. des.).
Phyllophilopsis Townsend, 1915b: 78 (n. name for *Phyllophila* Townsend). Type-species, *Chaetona nitens* Coquillett (aut.).

evanida Reinhard, 1958b: 233.—Va.; N.Y. and Mass. to Va.
nitens (Coquillett), 1899d: 221 (*Chaetona*).—N.H.; Ohio to Maine, s. to Ga.
    americana, authors (as *Chaetona*), not Bigot.

Tribe EBENIINI

Genus CHAETONOPSIS Townsend

Chaetonopsis Townsend, 1915a: 21. Type-species, *Chaetona spinosa* Coquillett (orig. des.).
Neonyctia Townsend, 1919c: 163. Type-species, *ciliata* Townsend (orig. des.)=*spinosa* (Coquillett). **N. syn.**

spinosa (Coquillett), 1899d: 222 (*Chaetona*).—Pa.; Kans. to N.Y. and R.I., s. to Okla. and Ga.
    ciliata Townsend, 1919c: 163 (*Neonyctia*).—Kans. **N. syn.**

Tribe SOPHIINI

Genus EUANTHA Wulp

Euantha Wulp, 1885: 198. Type-species, *Dexia dives* Wiedemann (mon.)=*litturata* (Olivier).

litturata (Olivier), 1812: 423 (*Ocyptera*).—Carolina; Colo., Pa., Ky., Ga., Fla., Mexico.
    liturata, emend. or error.
    dives Wiedemann, 1830: 377 (*Dexia*).—Ky.
    pictipennis Macquart, 1843a: 224 (1843: 67) (*Sericocera*).—Pa.

Genus EUCORDYLIGASTER Townsend

Eucordyligaster Townsend, 1917d: 123. Type-species, *Cordyligaster septentrionalis* Townsend (orig. des.).
Cordyligaster, authors, not Macquart.

septentrionalis (Townsend), 1909: 250 (*Cordyligaster*).—Md.; Kans. to Pa., s. to La. and Fla.
    minuscula, authors, not Wulp.

Tribe ERIOTHRIXINI

The species of Eriothrixini are parasites of lepidopterous larvae.

Genus ATROPHOPALPUS Townsend

Atrophopalpus Townsend, 1892h: 130. Type-species, *angusticornis* Townsend (orig. des.).

REFERENCE: Reinhard, 1934d (rev., in *Ceratomyiella*).

angusticornis Townsend, 1892h: 131.—Fla.; Ga.
orbitalis (Reinhard), 1934d: 16 (*Ceratomyiella*).—N.J.; Miss. **N. comb.** The species is not renamed because it has escaped from the secondary homonymy created in *Ceratomyiella* when Townsend (1939b: 181) transferred *Paradidyma orbitalis* Coquillett, 1904 (Neotropical), to that genus. If *Atrophopalpus* and *Ceratomyiella* finally prove to be synonymous, as once suggested, *orbitalis* Reinhard will require a new name.

Genus CATEMOPHRYS Townsend

Wulpia Brauer and Bergenstamm, 1893: 128, 188 (1893: 40, 100) (preocc. Bigot, 1886). Type-species, *aperta* Brauer and Bergenstamm (orig. des., as gen. n., sp. n.)=*sequens* (Townsend).
Catemophrys Townsend, 1908: 65. Type-species, *Vanderwulpia sequens* Townsend (orig. des.).
Brauerimyia Townsend, 1908: 65 (n. name for *Wulpia* Brauer and Bergenstamm). Type-species, *Wulpia aperta* Brauer and Bergenstamm (aut.)=*sequens* (Townsend).

sequens (Townsend), 1892e: 172 (*Vanderwulpia*).—N. Mex.; Ariz. to Ala., s. to Mexico.
  aperta Brauer and Bergenstamm, 1893: 128, 188 (1893: 40, 100) (*Wulpia*).—Mexico (V.C.).

Genus CERATOMYIELLA Townsend

Ceratomyiella Townsend, 1891c: 379. Type-species, *conica* Townsend (orig. des.).

REFERENCE: Reinhard, 1934d (rev.).

bicincta Reinhard, 1934d: 13.—Tex.; Calif. to Ark., s. to Mexico.
conica Townsend, 1891c: 380.—Ill.; Kans. to N.Y., s. to Tex. and Ga.

Genus EPIDEXIA Townsend

Epidexia Townsend, 1912a: 112. Type-species, *filamentosa* Townsend (orig. des.) =*pulverea* (Coquillett).

mimela Reinhard, 1953c: 96.—Tex.; Mexico.
pulverea (Coquillett), 1897: 115 (*Masicera*).—Ga., Fla.; Tex., La., Md. to Fla.
  filamentosa Townsend, 1912a: 112.—Fla.

Genus MICROMINTHO Townsend

Micromintho Townsend, 1919b: 554. Type-species, *melania* Townsend (orig. des.).
Metallicomintho Townsend, 1919b: 555. Type-species, *abdominalis* Townsend (orig. des.) =*melania* Townsend.

REFERENCE: Reinhard, 1955c (rev.).

melania Townsend, 1919b: 555.—Ariz.; Alta., Calif. to New England and Fla.
  abdominalis Townsend, 1919b: 555 (*Metallicomintho*).—Ariz.
  cinerosa Reinhard, 1923: 268 (*Metachaeta*; preocc. Coquillett, 1902, when in *Wagneria*).—Tex.
  distincta Curran, 1928b: 49 (*Wagneria*; n. name for *cinerosa* Reinhard).—Tex.

Genus PARADIDYMA Brauer and Bergenstamm

Paradidyma Brauer and Bergenstamm, 1891: 382 (1891: 78). Type-species, *Didyma validinervis* Wulp (mon.).
Atrophopoda Townsend, 1891c: 373. Type-species, *singularis* Townsend (orig. des.).
Lachnomma Townsend, 1892a: 103. Type-species, *magnicornis* Townsend (orig. des.) =*singularis* (Townsend).
Phytoadmontia Townsend, 1916h: 626. Type-species, *Admontia setigera* Coquillett (orig. des.).

REFERENCE: Reinhard, 1934d (rev.).

affinis Reinhard, 1934d: 35.—Tex; B.C., Calif. to Mass., s. to Mexico and Fla.
apicalis Reinhard, 1934d: 33.—Tex.; Ariz., N.C., Fla., Mexico, Guatemala.
aristalis Reinhard, 1934d: 28.—N. Mex.; Ariz.
cinerescens Reinhard, 1934d: 26.—Utah.
crassiseta Reinhard, 1934d: 27.—Mexico (Chih.); Ariz., N. Mex.

neglecta (West), 1925: 125 (*Eulasiona*).—N.Y.; Iowa.
neomexicana Reinhard, 1934d: 23.—N. Mex.; Ariz., ?Idaho.
obliqua Reinhard, 1934d: 22.—Idaho; Wash. and Idaho, s. to Calif. and Tex.
    retracta Reinhard, 1934d: 26.—Utah. **N. syn.**
petiolata Reinhard, 1934d: 39.—Ind.; Iowa to Mich. to N.J. and Va., also Ga.
setigera (Coquillett) *in* Baker, 1904: 36 (*Admontia*).—Calif.; Ariz.
singularis (Townsend), 1891c: 374 (*Atrophopoda*).—Ill.; Mont. to Mass., s. to Calif., Mexico, and Fla.
    magnicornis Townsend, 1892a: 104 (*Lachnomma*).—N. Mex.

Genus PERISCEPSIA Gistel

Scopolia Robineau-Desvoidy, 1830: 268 (preocc. Hübner, 1825 or earlier). Type-species, *Musca carbonaria* Panzer (Zetterstedt, 1844: 1239).
Periscepsia Gistel, 1848: x (n. name for *Scopolia* Robineau-Desvoidy). Type-species, *Musca carbonaria* Panzer (aut.).
Phoricheta Rondani, 1861a: 8 (n. name for *Scopolia* Robineau-Desvoidy). Type-species, *Musca carbonaria* Panzer (aut.).
Metachaeta Coquillett, 1895k: 98. Type-species, *atra* Coquillett (orig. des.) =*laevigata* (Wulp).
Eutricogena Townsend, 1915a: 23. Type-species, *Tricogena setipennis* Coquillett (orig. des.) =*clesides* (Walker).
Neophorichaeta Smith, 1915b: 100. Type-species, *johnsoni* Smith (orig. des.) =*clesides* (Walker).
Polideosoma Townsend, 1915f: 226. Type-species, *rohweri* Townsend (orig. des.).
Wagneria, authors, in part, not Robineau-Desvoidy.

    REFERENCE: Reinhard, 1955c (rev., *Wagneria* sens. lat.).

cinerosa (Coquillett), 1902g: 116 (*Phorichaeta*).—Ariz.; B.C. to Mont., s. to Calif., Ariz., and Colo. **N. comb.**
clesides (Walker), 1849: 757 (*Tachina*).—N. Amer.; Alaska to Sask., s. to Oreg., also Minn. to N.B. and Maine, s. to Ohio. **N. comb.**
    setipennis Coquillett, 1897: 130 (*Tricogena*).—N.H. **N. comb.**
    johnsoni Smith, 1915b: 100 (*Neophorichaeta*).—Maine. **N. comb.**
    ontario Curran, 1926b: 87 (*Peteina*).—Ont. **N. comb.**

helymus (Walker), 1849: 795 (*Tachina*).—Maine; Alaska, B.C., and Que., U.S., s. to Calif., N. Mex., Kans., and Conn. **N. comb.**
    sequax Williston *in* Cook, 1884: 424 (1884: 5; 1885: 83) (*Scopolia*).—Mich. **N. comb.**
labradorensis (Brooks), 1945a: 92 (*Petinops*).—Labr. **N. comb.**
laevigata (Wulp), 1890: 205 (*Rhinophora*).—Mexico; B.C. and Que., U.S., s. to Calif., Tex., and Va., Mexico, Guatemala. **N. comb.**
    atra Coquillett, 1895k: 99 (*Metachaeta*).—Ill. **N. comb.**
    carbonaria, authors, not Panzer.
    helymus, authors, not Walker.
polita (Brooks), 1945a: 92 (*Eutricogena*).—B.C.; Oreg. **N. comb.**
rohweri (Townsend), 1915f: 227 (*Polideosoma*).—Colo.; Wash. to Mont. and Colo. **N. comb.**

Genus VANDERWULPIA Townsend

Vanderwulpia Townsend, 1891c: 381. Type-species, *atrophopodoides* Townsend (orig. des.).

atrophopodoides Townsend, 1891c: 381.—N. Mex.; Calif. to Tex., s. to Mexico (Chih.).

Tribe LESKIINI

Species of the tribe Leskiini are parasites of lepidopterous larvae, especially of concealed kinds such as borers, leaf rollers, and case-makers.

The generic classification of the tribe is subject to wide difference of opinion. It has been suggested that *Leskiomima*, *Dejeaniopalpus*, and *Leskiella* are synonyms of *Genea*, and perhaps this will ultimately be the arrangement. However, it seems better to recognize them as distinct taxa until the classification of the entire tribe can be thoroughly reviewed, rather than to propose synonymy that might soon have to be revised. The four nominal genera are covered by James (1947) in his revision of the Leskiini with setulose first vein.

Dexia analis Say, a name once used commonly in this and other tribes, has been applied to several species and is best left unrecognized.

REFERENCE: James, 1947 (partial rev.).

Genus DEJEANIOPALPUS Townsend

Dejeaniopalpus Townsend, 1916i: 312. Type-species, *texensis* Townsend (orig. des.).

texensis Townsend, 1916i: 312.—Tex.; Tex., Mich., Que. to Va.

Genus EUMYOBIA Townsend

Eumyobia Townsend, 1911a: 146 (also 1912b: 312). Type-species, *flava* Townsend (orig. des., as gen. n., sp. n.).

Unnamed sp.—Ga.

Genus GENEA Rondani

Genea Rondani, 1850a: 172. Type-species, *maculiventris* Rondani (mon.)=*trifaria* (Wiedemann).

aurea James, 1947: 112.—Va.; Kans. to Mich. and N.Y., s. to Fla.

Genus JAYNESLESKIA Townsend

Jaynesleskia Townsend, 1934c: 395. Type-species, *Leskiomima jaynesi* Aldrich (orig. des.).

Unnamed sp.—Kans., Fla.

Genus LESKIELLA James

Leskiella James, 1947: 96. Type-species, *brevirostris* James (orig. des.).

brevirostris James, 1947: 97.—Ga.; Ill. to Md., s. to Fla.

Genus LESKIOMIMA Brauer and Bergenstamm

Leskiomima Brauer and Bergenstamm, 1891: 372, 406 (1891: 68, 102) (as *Lesciomima*, p. 372 (68) in part). Type-species, *Stomoxys tenera* Wiedemann (mon.).

Leskiomera (error) Banks, 1912: 110.

cinerea James, 1947: 101.—Fla.

tenera (Wiedemann), 1830: 251 (*Stomoxys*).—N. Amer. (as Unknown; Townsend, 1939b: 224); Wis. to Que. and N.S., s. to Tex. and Fla.

Genus MYOBIOPSIS Townsend

Myobiopsis Townsend, 1916h: 628. Type-species, *similis* Townsend (orig. des.).

Leskiopalpus Townsend, 1916h: 629. Type-species, *calidus* Townsend (orig. des.)=*depilis* (Coquillett). **N. syn.**

depilis (Coquillett), 1895a: 313 (*Myiobia*).—Fla.; Mich. to Mass., s. to Miss. and Fla. **N. comb.**

calidus Townsend, 1916h: 629 (*Leskiopalpus*).—N.Y. **N. comb.**

similis Townsend, 1916h: 628.—Mass.; Nebr., Iowa, N.Y., Maine, Conn., N.J.

pallida Harris, 1835: 599 (*Stomoxys*). Nomen nudum.

Genus SIPHOCLYTIA Townsend

Siphoclytia Townsend, 1892a: 116. Type-species, *robertsonii* Townsend (orig. des.).

robertsonii Townsend, 1892a: 117.—Fla.; Tex., La., Ga.

Genus SIPHOLESKIA Townsend

Sipholeskia Townsend, 1916h: 628. Type-species, *Drepanoglossa occidentalis* Coquillett (orig. des.).

loriola Reinhard, 1955d: 126.—Tex.; Mexico.
occidentalis (Coquillett), 1895c: 126 (*Drepanoglossa*).—Calif.; Calif. to Colo., s. to Mexico (Chih.).
    gilensis Townsend, 1897e: 40 (*Myobia*).—N. Mex.
    eucerata, authors, not Bigot.

Tribe TELOTHYRIINI

Genus LESKIOPSIS Townsend

Leskiopsis Townsend, 1916h: 627. Type-species, *Myiobia thecata* Coquillett (orig. des.).

thecata (Coquillett), 1895k: 105 (*Myiobia*).—Pa.; Ohio to Mass., s. to Fla.

Subfamily GONIINAE
(Exoristinae)

Tribe ACEMYINI

The hosts of the tribe Acemyini are members of the order Orthoptera.

Genus ACEMYA Robineau-Desvoidy

Acemya Robineau-Desvoidy, 1830: 202. Type-species, *subrotunda* Robineau-Desvoidy (Rondani, 1856: 75)=*acuticornis* (Meigen).
Acemyia, emend. or error.
Euacemyia Townsend, 1912d: 163. Type-species, *Acemyia tibialis* Coquillett (orig. des.).

masurius (Walker), 1849: 753 (*Tachina*).—N. Amer. Unrecognized.
tibialis Coquillett, 1897: 116.—Calif.; B. C. to Que., s. to Calif. and Tex.

Genus CERACIA Rondani

Ceracia Rondani, 1865: 221 (1865: 49). Type-species, *mucronifera* Rondani (mon.).

dentata (Coquillett), 1895a: 311 (*Acemyia*).—Calif., Ala., Fla.; B.C. to Ont., all U.S., Mexico.

Genus EUHALIDAYA Walton

Euhalidaya Walton, 1914d: 130 (as *Euhallidaya*, error). Type-species, *severinii* Walton (orig. des.)=*genalis* (Coquillett).
Orphanotrophus Reinhard, 1943b: 82. Type-species, *orbitalis* Reinhard (orig. des.)=*genalis* (Coquillett).

The sole species of this genus has been reared from *Diapheromera femorata*, a common walking stick.

genalis (Coquillett), 1897: 83 (*Biomyia*).—Ga.; Calif., Ariz., Kans., Wis., Ohio, Pa., Md., Va.
   severinii Walton, 1914d: 130.—Wis.
   orbitalis Reinhard, 1943b: 83 (*Orphanotrophus*).—Ohio.

Genus HEMITHRIXION Brauer and Bergenstamm

Hemithrixion Brauer and Bergenstamm, 1891: 357 (1891: 53). Type-species, *oestriforme* Brauer and Bergenstamm (orig. des., as gen. n., sp. n.).
Coquillettina Walton, 1915a: 104. Type-species, *plankii* Walton (orig. des.).

oestriforme Brauer and Bergenstamm, 1891: 357 (1891: 53).—Colo.; B.C. to Man., s. to Tex.
plankii (Walton), 1915a: 105 (*Coquillettina*).—N.J.; Wis., Ill.

Tribe BLONDELIINI

(Compsilurini, Trypherini in part)

The tribe Blondeliini corresponds to that recognized by Mesnil (1960). It includes most of the Compsilurini of Townsend and a large share of his tribe Trypherini. A variety of hosts is parasitized including larvae of Lepidoptera, Coleoptera, and tenthredinid Hymenoptera.

Genus ADORYPHOROPHAGA Townsend

Adoryphorophaga Townsend, 1931b: 469. Type-species, *Doryphorophaga aberrans* Townsend (orig. des.).

aberrans (Townsend), 1916c: 217 (*Doryphorophaga*).—Va.; S. Dak. to Ont. and Mass., s. to Mo. and Va.
sedula (Reinhard), 1935c: 392 (*Doryphorophaga*).—Ohio; Colo. to N.Y., s. to Tex. and Fla.
    patrita Reinhard, 1935c: 394 (*Doryphorophaga*).—Ohio. **N. syn.**, H. J. Reinhard *in litt.*

Genus ANOXYNOPS Townsend

Anoxynops Townsend, 1927: 274, 286. Type-species, *conicus* Townsend (orig. des.).

aldrichi (Curran), 1926d: 217 (*Paralispe*).—Que.; Mich. to Que., s. to Va., also Ariz., Tex. **N. comb.**

Genus APACHEPROSPHERYSA Townsend

Apacheprospherysa Townsend, 1926: 27. Type-species, *orbitalis* Townsend (orig. des.).

orbitalis Townsend, 1926: 28.—N. Mex.

Genus APLOMYIOPSIS Villeneuve

Aplomyiopsis Villeneuve, 1933: 125. Type-species, *galerucellae* Villeneuve (mon.).
Synaplomyia Villeneuve, 1934: 181 (unjustified n. name for *Aplomyiopsis* Villeneuve). Type-species, *Aplomyiopsis galerucellae* Villeneuve (aut.).

The hosts are larvae of Chrysomelidae and of the leaf-feeding coccinellid, *Epilachna*.

epilachnae (Aldrich), 1923e: 95 (*Paradexodes*).—Mexico; introd. 19 States, not estab. **N. comb.**
galerucellae Villeneuve, 1933: 125.—Oreg.; Calif.
lathami (Curran), 1925f: 284 (*Lydella*).—N.Y.; Ill. and Mich. to Mass., s. to Tenn. **N. comb.**
nana (Curran), 1929f: 506 (*Dexodes*).—B.C.; Calif.
opaca (Reinhard), 1945a: 30 (*Lixophaga*).—Wash.; Alta., Wyo., Mich., Que. **N. comb.**
seticauda (Reinhard), 1930b: 200 (*Masicera*).—Tex. **N. comb.**
unicolor (Smith), 1917b: 137 (*Pilatea*).—Maine, N.H., Mass., Conn.; Mont., N. Mex., N.Y., Labr. **N. comb.**
vexans (Curran), 1925c: 151 (*Hypostena*).—N.S.
xylota (Curran), 1927d: 22 (*Dexodes*).—Ont.; Wis. to Ont. and R.I.

Genus APOROTACHINA Meade

Tachina, subg. **Aporotachina** Meade, 1894: 109. Type-species, *angelicae* Meigen (Coquillett, 1910b: 509).
Neothelaira Townsend, 1912a: 109. Type-species, *dexina* Townsend (orig. des.).
Aulicomyia Reinhard, 1943b: 80. Type-species, *invulnerata* Reinhard (orig. des.)=*dexina* (Townsend). **N. syn.**

chaetoneura (Coquillett), 1897: 115 (*Masicera*).—N.H.; B.C. and Alta., s. to Calif. and Colo., also Mich. to Maine and Conn. **N. comb.**
dexina (Townsend), 1912a: 109 (*Neothelaira*).—Mass.; Ohio to Maine and R.I. **N. comb.**
    aurifrons Coquillett, 1897: 115 (*Masicera*; preocc. Doleschall, 1858).—N.H. **N. comb.**
    invulnerata Reinhard, 1943b: 81 (*Aulicomyia*).—Ohio. **N. comb.**
pusilla (Reinhard), 1953b: 244 (*Neothelaira*).—Mo.; N.C. **N. comb.**

Genus BLONDELIA Robineau-Desvoidy

Blondelia Robineau-Desvoidy, 1830: 122. Type-species, *Tachina nigripes* Fallén (Coquillett, 1910b: 515; Suspension of the Rules required).
Phrynolydella Townsend, 1919b: 572. Type-species, *polita* Townsend (orig. des.).
Pseudoeribea Townsend, 1926: 26. Type-species, *paradexoides* Townsend (orig. des.).
Anetia, authors, not Robineau-Desvoidy.
Lydella, authors, not Robineau-Desvoidy.

Acceptance of *nigripes* as the type-species of *Blondelia* is desirable for reasons of stability and universality. The earliest valid designation, by Townsend (1916d: 6), named *B. pallidipalpis* Robineau-Desvoidy, but this is a nomen dubium. Coquillett (1910b), although in agreement with long-established usage, technically failed to designate as type one of the originally included species, and Suspension of the Rules is required to validate his designation of *nigripes*.

connecta (Curran), 1925f: 286 (*Lydella*).—B.C. **N. comb.**
eufitchiae (Townsend), 1892d: 286 (*Masicera*).—Colo. **N. comb.**
hyphantriae (Tothill), 1922: 43 (*Lydella*).—B.C.; Minn. to Ont. and Mass., s. to D.C., also B.C., Wash., Colo. **N. comb.** ?=*obconica*.
    hyphantriae Townsend, 1893h: 467 (*Nemoraea*). Nomen nudum.

inclusa (Hartig), 1838: 285 (*Tachina*).—Germany; Europe, introd. Que., not estab.

nigripes nigripes (Fallén), 1810b: 270 (*Tachina*).—Sweden; Europe, introd. N.H., Mass., Conn., not estab.

   ssp. *piniariae* (Hartig), 1838: 283 (*Tachina*; as sp.).—Germany; Europe, introd. Maine, N.H., R.I., not estab.

obconica (Walker), 1852: 296 (*Tachina*).—U.S. **N. comb.**

paradexoides (Townsend), 1926: 27 (*Pseudoeribea*).—Mass. **N. comb.** ?=*eufitchiae*.

polita (Townsend), 1919b: 572 (*Phrynolydella*).—Ariz.; N. Mex., Mexico (Chih.). **N. comb.**

   strigata, authors, not Wulp.

Genus CELATORIA Coquillett

Celatoria Coquillett, 1890: 235. Type-species, *crawii* Coquillett (orig. des.) = *diabroticae* (Shimer).

Chaetophleps Coquillett, 1895j: 51. Type-species, *setosa* Coquillett (orig. des.).

Neocelatoria Walton, 1914b: 13. Type-species, *ferox* Walton (mon.) = *setosa* (Coquillett).

diabroticae (Shimer), 1871: 219 (*Tachina*).—Ill.; Wash. to Mass., s. to Calif., Mexico, and Ga.

   crawii Coquillett, 1890: 235.—Calif.

   galerucae Harris, 1835: 599 (*Ocyptera*). Nomen nudum.

setosa (Coquillett), 1895j: 51 (*Chaetophleps*).—Md.; Oreg. to Mass., s. to Calif. and Ga.

   ferox Walton, 1914b: 13 (*Neocelatoria*).—Md.

Genus CHARASOMA Reinhard

Charasoma Reinhard, 1952b: 10. Type-species, *subole* Reinhard (orig. des.).

pammelan Reinhard, 1952b: 12.—Oreg.; Calif.
reside Reinhard, 1952b: 12.—Ohio; Mich.
subole Reinhard, 1952b: 11.—Tex.; N.Y.

Genus COMPSILURA Bouché

Compsilura Bouché, 1834: 58. Type-species, *Tachina concinnata* Meigen (Coquillett, 1910b: 526).

The common *Compsilura concinnata* is probably the most polyphagous of the tachinids, with an estimated 200 known hosts among larvae of Lepidoptera. It is an especially valuable parasite of the gypsy moth.

concinnata (Meigen), 1824: 412 *(Tachina).*—Europe; introd. and estab., Minn. to P.E.I., s. to N.J., also B.C. to Calif.

Genus CRYPTOMEIGENIA Brauer and Bergenstamm

Cryptomeigenia Brauer and Bergenstamm, 1891: 311 (1891: 7). Type-species, *setifacies* Brauer and Bergenstamm (mon.).
Emphanopteryx Townsend, 1892a: 120. Type-species, *eumyothyroides* Townsend (orig. des.).
Eumyothyria Townsend, 1892a: 121. Type-species, *illinoiensis* Townsend (orig. des.).
Meigeniella Coquillett, 1902g: 104. Type-species, *hinei* Coquillett (orig. des.).

The species of *Cryptomeigenia* parasitize *Phyllophaga* larvae. The genus itself is easily recognized, but specific identification is virtually impossible at present.

REFERENCE: Curran, 1926k (rev., Canadian spp.)

brimleyi Reinhard, 1947a: 17.—N.C.; Ga.
crassipalpis Reinhard, 1947a: 18.—N.Y.
demylus (Walker), 1849: 779 *(Tachina).*—?N. Amer.
dubia Curran, 1926k: 164.—Ont.
eumyothyroides (Townsend), 1892a: 121 *(Emphanopteryx).*—N.Y.
flavibasis Curran, 1927m: 145.—Ill.
hinei (Coquillett), 1902g: 104 *(Meigeniella).*—Ohio.
illinoiensis (Townsend), 1892a: 122 *(Eumyothyria).*—Ill.
  illinoisensis, error.
menapis (Walker), 1849: 769 *(Tachina).*—N. Amer. (as Sweden, error).
  conica Harris, 1835: 599 *(Tachina).* Nomen nudum.
muscoides Curran, 1926k: 157.—Man.
nigripes Curran, 1926k: 162.—Ont.
nigripilosa Curran, 1926k: 161.—B.C.
ochreigaster Curran, 1926k: 165.—Man.
ontario Curran, 1926k: 159.—Ont. ?=*menapis*.
pedestris (Walker), 1852: 313 *(Dexia).*—U.S.
prisca (Walker), 1849: 780 *(Tachina).*—N.S.
simplex Curran, 1926k: 163.—Que.
theutis (Walker), 1849: 778 *(Tachina).*—N.S.
triangularis Curran, 1926k: 160.—Que.

Genus DOLICHOTARSUS Brooks

Dolichotarsus Brooks, 1945a: 94. Type-species, *kingi* Brooks (orig. des.).

griseus Brooks, 1945a: 95.—B.C.
kingi Brooks, 1945a: 95.—Sask.; Wash., N. Mex.
livescens Reinhard, 1958c: 226.—N. Mex.

Genus DORYPHOROPHAGA Townsend

Doryphorophaga Townsend, 1912d: 164. Type-species, *Lydella doryphorae* Riley (orig. des.).
REFERENCE: Reinhard, 1935c (rev.).
australis Reinhard, 1935c: 389.—Tex.; Ariz., Mexico, Fla.
doryphorae (Riley), 1869: 111 (*Lydella*).—Mo.; B.C. to N.S., s. to Ariz., Mexico, and S.C.
 aerata Coquillett, 1897: 100 (*Exorista*).—Ill. **N. syn.**
macella Reinhard, 1935c: 390.—Tex.; Man. to Mass., s. to Calif., Mexico, and Fla.

Genus ELEPHANTOCERA Townsend

Elephantocera Townsend, 1915l: 98. Type-species, *greenei* Townsend (orig. des.).
greenei Townsend, 1915l: 99 .—N.J.; Ont.

Genus ERIBELLA Mesnil

Eribella Mesnil, 1960: 654. Type-species, *Masicera polita* Coquillett (mon.).
The known hosts for the genus *Eribella* are larvae of Pterophoridae.
exilis (Coquillett), 1897: 156 (*Masicera*).—Mass.; Mich. to N.B., s. to R.I. **N. comb.**
 tothilli Curran, 1925c: 152 (*Hypostena*).—N.B. **N. syn., n. comb.**
 tenthredinidarum Townsend of Coquillett, 1897: 114 (*Masicera*).
polita (Coquillett), 1902g: 114 (*Masicera*).—N. Mex.; B.C., Sask., Colo., Iowa, Mich., Ohio, Vt.

Genus ERYNNIOPSIS Townsend

Erynniopsis Townsend, 1926: 30. Type-species, *rondanii* Townsend (orig. des.).
extricata (West), 1925: 127 (*Hyperecteina*).—N.Y. **N. comb.**
 ?=*rondanii.*
rondanii Townsend, 1926: 30.—Europe; introd. 10 States, Idaho and Calif. to Mass., estab. only Calif.
 nitida, authors (in *Erynnia*), not Robineau-Desvoidy.

Genus EUCELATORIA Townsend

Eucelatoria Townsend, 1909: 249. Type-species, *Tachina armigera* Coquillett (orig. des.).

armigera (Coquillett), 1889: 332 (*Tachina*).—Calif.; Calif. to Tex. and Mexico, also West Indies, introd. and estab. Hawaii.
rubentis (Coquillett), 1895a: 310 (*Achaetoneura*).—Fla.; Tenn., S.C., Ga.

Genus EUCELATORIOPSIS Townsend

Eucelatoriopsis Townsend, 1927: 276. Type-species, *teffeensis* Townsend (orig. des.).
Tachinophytopsis Townsend, 1927: 277. Type-species, *carinata* Townsend (orig. des.). **N. syn.**

Our species has been reared from the larvae of several cassidine Chrysomelidae.

dimmocki (Aldrich), 1932a: 5 (*Anetia*).—Mass.; N. Dak. to N.H., s. to Ariz., Mexico, and Fla. **N. comb.**

Genus LIXOPHAGA Townsend

Lixophaga Townsend, 1908: 86. Type-species, *parva* Townsend (orig. des.).
Euzenillia Townsend, 1911a: 148. Type-species, *aurea* Townsend (orig. des., as gen.n., sp.n.)=*variabilis* (Coquillett).
Euzenilliopsis Townsend, 1916g: 76. Type-species, *diatraeae* Townsend (orig. des.).
Erycioides Curran, 1930j: 103. Type-species, *thoracica* Curran (orig. des.). **N. syn.**
Prolixophaga Townsend, 1934c: 404. Type-species, *Lixophaga plumbea* Aldrich (orig. des.).

Variability, relationships, and synonymy in the commonly occurring genus *Lixophaga* are poorly understood.

REFERENCES: Aldrich, 1925d (rev.); Curran, 1935b (key, males).

alberta (Curran), 1925c: 154 (*Hypostena*).—Alta.
diatraeae (Townsend), 1916g: 76 (*Euzenilliopsis*).—Cuba (as La., error); Neotropical, introd., La., Fla., estab. Fla.
fasciata Curran, 1930j: 100.—N.Y.; Mich., Ohio.
impatiens (Curran), 1925c: 154 (*Hypostena*).—Man.
jennei Aldrich, 1926d: 18.—Ark.; La., Fla.
mediocris Aldrich, 1925d: 136.—Va.; Ill. to N.J., s. to Miss. and Fla.
nigribasis Curran, 1930j: 100.—N.Y.; Mich.
orbitalis Aldrich, 1926d: 17.—Calif.

parva Townsend, 1908: 86.—Tex.
plumbea Aldrich, 1925d: 134.—Va.; Utah to Ont. and Conn., s. to Ariz. and Fla.
thoracica (Curran), 1930j: 103 (*Erycioides*).—N.Y. **N. comb.**
variabilis (Coquillett), 1895j: 57 (*Hypostena*).—Ill.; Man. to Maine, s. to Tex. and Fla., also B.C., Calif.
    aurea Townsend, 1911a: 148 (*Euzenillia*).—Mass.

Genus LYDELLOHOUGHIA Townsend

Lydellohoughia Townsend, 1927: 280. Type-species, *nana* Townsend (orig. des.).
tenella (Reinhard), 1937: 65 (*Dexodes*).—Calif.; Ariz. **N. comb.**

Genus Lydinolydella Townsend

Lydinolydella Townsend, 1927: 278. Type-species, *metallica* Townsend (orig. des.).
metallica Townsend, 1927: 325.—Brazil; introd. N.J., Md., not estab.

Genus MEIGENIA Robineau-Desvoidy

Meigenia Robineau-Desvoidy, 1830: 198. Type-species, *Tachina floralis* Fallén (as Robineau-Desvoidy; Robineau-Desvoidy, 1863a: 1065; misident.)=*mutabilis* (Fallén).
Collatia Curran, 1934a: 423, 464. Type-species, *Zenillia submissa* Aldrich and Webber (mon.). **N. syn.**
albicincta (Reinhard), 1924b: 272 (*Pilatea*).—Tex. **N. comb.**
mutabilis (Fallén), 1810b: 273 (*Tachina*).—Sweden; Europe, introd. N.J., not estab.
    floralis, authors, not Fallén.
submissa (Aldrich and Webber), 1924a: 30 (*Zenillia*).—N. Mex.; Alta. to S. Dak., s. to Ariz. and N. Mex., also Ohio. **N. comb.**
    ornata Reinhard, 1937: 71 (*Collatia*).—Colo. **N. syn., n. comb.**

Genus MEIGENIELLOIDES Townsend

Meigenielloides Townsend, 1919b: 573. Type-species, *cinerea* Townsend (orig. des.).
Synoris Aldrich, 1926d: 12. Type-species, *coquilletti* Aldrich (orig. des.)=*cinerea* Townsend.

cinerea Townsend, 1919b: 574.—N. Mex.; Wash. to Wyo., s. to Mexico.
   coquilletti Aldrich, 1926d: 13 (*Synoris*).—Calif.
   pedestris, authors (in *Hypostena*), not Walker.

Genus METADORIA Brauer and Bergenstamm

Metadoria Brauer and Bergenstamm, 1893: 117 (1893: 29). Type-species, *mexicana* Brauer and Bergenstamm (orig. des., as gen. n., sp. n.)=*barbata* (Bigot).
levis (Aldrich and Webber), 1924a: 86 (*Phorocera*).—N.J.; Nebr. and Kans. to N.Y. and N.J. **N. comb.**

Genus MYIOPHARUS Brauer and Bergenstamm

Myiopharus Brauer and Bergenstamm, 1889: 161 (1889: 93). Type-species, *metopia* Brauer and Bergenstamm (orig. des., as gen. n., sp. n.).
Parkeriellus Smith, 1916: 96. Type-species, *flavipalpis* Smith (orig. des.)=*dorsalis* (Coquillett).
canadensis Reinhard, 1945a: 31.—Que.; Ont.
dorsalis (Coquillett), 1898c: 236 (*Exorista*).—Pa.; Idaho to Ont. and N.Y., s. to Calif., Tex., Iowa, and N.J.
   flavipalpis Smith, 1916: 97 (*Parkeriellus*).—Mont.
securis Reinhard, 1945a: 30.—Tex.

Genus OPSOMEIGENIA Townsend

Opsomeigenia Townsend, 1919b: 577. Type-species, *Hypostena pusilla* Coquillett (orig. des.).
flavipalpis (Reinhard), 1934b: 187 (*Anetia*).—Tex. **N. comb.**
pusilla (Coquillett), 1895j: 58 (*Hypostena*).—Ill.; Mo. to Md., also Tex.
   parvula Reinhard, 1934b: 186 (*Anetia*).—Tex. **N. syn., n. comb.**

Genus OSWALDIA Robineau-Desvoidy

Oswaldia Robineau-Desvoidy, 1863a: 840. Type-species, *muscaria* Robineau-Desvoidy (orig. des.)=*muscaria* (Fallén).
Aubaeanetia Townsend, 1919b: 569. Type-species, *assimilis* Townsend (orig. des.).
Parameigenia Townsend, 1919b: 576. Type-species, *Paradexodes albifacies* Townsend (orig. des.).

Subgenus DEXODES Brauer and Bergenstamm

Dexodes Brauer and Bergenstamm, 1889: 87, 128 (1889: 19, 60) (as genus). Type-species, *Tachina spectabilis* Meigen (Brauer, 1893: 476; misident.)=*albisquama* (Zetterstedt).
Paradexodes Townsend, 1908: 101. Type-species, *aurifrons* Townsend (orig. des.).

aurifrons (Townsend), 1908: 101 (*Paradexodes*).—Mass.; Utah, Mich., Ohio.
minor (Curran), 1925f: 283 (*Lydella*).—Que.; Que. to Mich. and N.J., also Kans., N.C. **N. comb.**
sartura (Reinhard), 1959b: 234 (*Dexodes*).—N.Y.; Ohio. **N. comb.**

Subgenus OSWALDIA Robineau-Desvoidy

albifacies (Townsend), 1908: 102 (*Paradexodes*).—N.H.; Ont. to N.B., s. to Md.
anorbitalis (Brooks), 1945a: 96 (*Aubaeanetia*).—Que.; Ont., Md. **N. comb.**
assimilis (Townsend), 1919b: 570 (*Aubaeanetia*).—N.H.; N. Dak. to N.B., s. to Miss. and Ga.
conica (Reinhard), 1934b: 189 (*Dexodes*).—Wis. **N. comb.**
valida (Curran), 1927d: 21 (*Dexodes*).—Ont.; Wis. to Ont., s. to Ohio and Conn. **N. comb.**

Genus OXYNOPS Townsend

Oxynops Townsend, 1912a: 110. Type-species, *serratus* Townsend (orig. des.)=*anthracinus* (Bigot).
Euchaetophleps Townsend, 1916h: 625. Type-species, *Chaetophleps polita* Coquillett (orig. des.)=*anthracinus* (Bigot).

anthracinus (Bigot), 1888a: 259 (*Degeeria*).—Mexico; N. Dak. to Mass., s. to N. Mex., Mexico, and Fla.
  nitens Coquillett, 1897: 63 (*Hypostena*).—Fla.
  polita Coquillett, 1902g: 107 (*Chaetophleps*).—S. Dak.
  serratus Townsend, 1912a: 110.—Fla.

Genus PANACEMYIA Townsend

Panacemyia Townsend, 1919c: 164. Type-species, *panamensis* Townsend (orig. des.).
Nimiocauda Reinhard, 1943b: 78. Type-species, *erilis* Reinhard (orig. des.).

erilis (Reinhard), 1943b: 79 (*Nimiocauda*).—N.Y.; Mass.
pallipes Reinhard, 1953b: 246.—Tex.; Ga.
verticalis Reinhard, 1953b: 247.—Ohio; Mass.

Genus PARALISPE Brauer and Bergenstamm

Paralispe Brauer and Bergenstamm, 1891: 337 (1891: 33). Type-species, *brasiliana* Brauer and Bergenstamm (orig. des., as gen. n., sp. n.).

Stomatolydella Townsend, 1919b: 570. Type-species, *infernalis* Townsend (orig. des.).

infernalis (Townsend), 1919b: 570 (*Stomatolydella*).—N. Mex.; Ariz. to Ill. and Ohio, s. to Mexico and Fla.
    barbata Bigot of Allen, 1929: 683 (as *barbarta*).
    inconspicua, authors (in *Didyma*), not Wulp.

Genus PHOENICIOMYIA Townsend

Phoeniciomyia Townsend, 1915f: 231. Type-species, *arizonica* Townsend (orig. des.).

arizonica Townsend, 1915f: 231.—Ariz.

Genus PHRYNOFRONTINA Townsend

Phrynofrontina Townsend, 1919b: 579. Type-species, *convexa* Townsend (orig. des.) = *discalis* (Coquillett).

discalis (Coquillett), 1902g: 114 (*Sturmia*).—Wis.; Minn. to Maine, s. to Ga. also Alta., Utah.
    convexa Townsend, 1919b: 580.—Md.

Genus PSEUDOMYOTHYRIA Townsend

Pseudomyothyria Townsend, 1892a: 131. Type-species, *indecisa* Townsend (orig. des.) = *ancilla* (Walker).

Epidexiopsis Townsend, 1916i: 308. Type-species, *orbitalis* Townsend (orig. des.) = *ancilla* (Walker). **N. syn.**

ancilla (Walker), 1852: 299 (*Tachina*).—U.S.; Utah to Mich. and N.Y., s. to Mexico and Fla.
    indecisa Townsend, 1892a: 132.—Ill.
    orbitalis Townsend, 1916i: 308 (*Epidexiopsis*).—Fla. **N. syn., n. comb.**

Genus SCHIZOCEROPHAGA Townsend

Schizocerophaga Townsend, 1916g: 77. Type-species, *leibyi* Townsend (orig. des.).

leibyi Townsend, 1916g: 77.—N.C.; Kans. to Va., s. to Ariz., La., and S.C.

Genus SITOPHAGA Gistel

Fabricia Meigen, 1838: 250 (preocc. de Blainville, 1828). Type-species, *Tachina pacta* Meigen (mon.)=*cinerea* (Fallén).
Sitophaga Gistel, 1848: ix (n. name for *Fabricia* Meigen). Type-species, *Tachina pacta* Meigen (aut.)=*cinerea* (Fallén).
Biomya Rondani, 1856: 72. Type-species, *Tachina pacta* Meigen (present des.)=*cinerea* (Fallén). **N. syn.**
Biomyia, emend.
Viviania Rondani, 1861a: 48, 53 (as *Viviana*, p. 48) (n. name for *Fabricia* Meigen). Type-species, *Tachina pacta* Meigen (aut.)=*cinerea* (Fallén). **N. syn.**
Pseudatractocera Townsend, 1892a: 107. Type-species, *neomexicana* Townsend (orig. des.). **N. syn.**
Eubiomyia Townsend, 1916g: 74. Type-species, *calosomae* Townsend (orig. des.). **N. syn.**

The difficult genus *Sitophaga* is in need of revision. The hosts are adult Coleoptera, including *Phyllophaga*, *Carabus*, and *Eleodes*.

One of the commonly used names for the genus, *Biomya*, is easily confused with *Byomya* Robineau-Desvoidy (1830), a muscid. Both names have been emended to *Biomyia*.

angustifrons (Reinhard), 1930a: 104 (*Viviania*).—Tex.; Fla., Cuba. **N. comb.**
arrisor (Reinhard), 1959a: 157 (*Viviania*).—Calif. **N. comb.**
aurigera (Coquillett), 1895a: 309 (*Masiphya*).—Fla.; Miss. **N. comb.**
    pedita Reinhard, 1959a: 158 (*Viviania*).—Fla. **N. syn., n. comb.**
calosomae (Townsend), 1916g: 74 (*Eubiomyia*).—Mass.; N.B., N.S., Pa. **N. comb.**
    obscura Curran, 1925f: 285 (*Lydella*).—N.S. **N. syn.,** J. F. McAlpine *in litt.*, **n. comb.**
    calosomae Coquillett *in* Burgess, 1897: 431 (1897: 83) (*Pseudatractocera*). Nomen nudum.
eleodivora (Walton), 1918: 25 (*Biomyia*).—Nebr. **N. comb.**
georgiae (Brauer and Bergenstamm), 1891: 312 (1891: 8) (*Viviania*).—Ga.; Kans. to N.Y. and Conn., s. to Tex. and Fla. **N. comb.**
    lachnosternae Townsend, 1908: 106 (*Viviania*).—Ill. **N. comb.**
lateralis (Curran), 1925f: 284 (*Lydella*).—B.C. **N. comb.**
leechi (Curran), 1932c: 12 (*Erycia*).—B.C. **N. comb.**
mutabilis (Coquillett) *in* Baker, 1904: 37 (*Biomya*).—Nev. **N. comb.**

neomexicana (Townsend), 1892a: 108 (*Pseudatractocera*).—N. Mex.; Wash., Idaho, Kans., Calif. to Tex. **N. comb.**
nocturnalis (Reinhard), 1930a: 104 (*Viviania*).—Tex.; Ark. **N. comb.**
sordicolor (Townsend), 1891c: 359 (*Masicera*).—Ill.; Nebr. to N.Y., s. to Ark. and S.C. **N. comb.**

Genus SPATHIMEIGENIA Townsend

Spathimeigenia Townsend, 1915a: 19. Type-species, *spinigera* Townsend (orig. des.).
Hylotomomyia Townsend, 1916e: 31. Type-species, *Admontia hylotomae* Coquillett (orig. des.).
Neoswaldia Mesnil, 1960: 655. Type-species, *Hylotomomyia buckelli* Curran (mon.). **N. syn.**

The species of *Spathimeigenia* are parasites of tenthredinid larvae. The status and correct identification of some of the commonest forms are still in doubt.

REFERENCES: Aldrich, 1931a (rev.); Reinhard, 1958e (notes).

aurifrons Curran, 1930k: 246.—Que.; Wis. to Que. and Pa.
bridwelli Aldrich, 1931a: 9.—Kans.
buckelli (Curran), 1926d: 216 (*Hylotomomyia*).—B.C.
dolopis Reinhard, 1958e: 208.—Oreg.
dolosa Reinhard, 1958e: 209.—Tex.
erecta Aldrich, 1931a: 8.—Mich.; Ont.
erronis Reinhard, 1958e: 212.—Tex.
fivoris Reinhard, 1958e: 211.—Calif.
hylotomae (Coquillett), 1898c: 233 (*Admontia*).—Mass.; Mich. to Mass., s. to Tex. and Fla.
    consternata West, 1925: 126 (*Hyperecteina*).—N.Y. **N. syn.**
mexicana Aldrich, 1931a: 5.—Mexico (Michoacan); Ariz.
nigriventris Smith, 1917b: 139.—Mass.
spinigera Townsend, 1915a: 19.—Md.; Pa. to Tex. and Fla.
    demylus Walker of Coquillett, 1897: 54 (*Admontia*).
texensis Aldrich, 1931a: 9.—Tex.; Ariz., Mexico.

Genus TACHINOPHYTO Townsend

Tachinophyto Townsend, 1892a: 130. Type-species, *floridensis* Townsend (orig. des.).

floridensis Townsend, 1892a: 131.—Fla.; Md., Mexico.

Genus THELAIRODORIA Townsend

Thelairodoria Townsend, 1927: 266. Type-species, *thrix* Townsend (orig. des.).

setinervis (Coquillett), 1910a: 129 (*Exorista*).—Tenn.; Tex. to Ohio and Fla. **N. comb.**
  floscula Reinhard, 1958b: 241.—Fla. **N. syn.**

Genus THEMATHECA Reinhard

Thematheca Reinhard, 1961: 207. Type-species, *medeola* Reinhard (orig. des.).

medeola Reinhard, 1961: 208.—Mexico (Oaxaca); Ariz.

Genus TINALYDELLA Townsend

Tinalydella Townsend, 1927: 265. Type-species, *tinensis* Townsend (orig. des.).

procincta (Reinhard), 1935b: 170 (*Anetia*).—Tex.; Fla.

Genus TORYNOTACHINA Townsend

Torynotachina Townsend, 1915l: 102. Type-species, *quinteri* Townsend (orig. des.).

quinteri Townsend 1915l: 102.—Md.; N.Y., Pa.

Genus TRYPHEROMYIA Reinhard

Trypheromyia Reinhard, 1945a: 32. Type-species, *pallens* Reinhard (orig. des.).

pallens Reinhard, 1945a: 32.—Ohio; Iowa, Va., Mexico.

Genus UROPHYLLOPSIS Townsend

Urophyllopsis Townsend, 1916h: 625. Type-species, *Admontia retiniae* Coquillett (orig. des.).

retiniae (Coquillett), 1897: 54 (*Admontia*).—Calif.; Oreg.

Genus VELOCIA Robineau-Desvoidy

Velocia Robineau-Desvoidy, 1863a: 950. Type-species, *cursoria* Robineau-Desvoidy (Townsend, 1916d: 9)=*luctuosa* (Meigen).
Amedoria Brauer and Bergenstamm, 1889: 106 (1889: 38). Type-species, *Hypostena medorina* Schiner (mon.)=*luctuosa* (Meigen).

Methypostena Townsend, 1908: 67. Type-species, *Hypostena barbata* Coquillett (orig. des.).
Odontosoma Townsend, 1916h: 633. Type-species, *Celatoria spinosa* Coquillett (orig. des.)=*barbata* (Coquillett).
Arrhinomyia, authors, not Brauer and Bergenstamm.
Degeeria, authors, not Meigen.
Medina, authors, not Robineau-Desvoidy.

In the latest European monograph of the family (Mesnil, 1962), *Velocia* is placed as a subgenus of *Medina*. The two are unquestionably closely related, but we prefer to regard them as distinct genera.

barbata (Coquillett), 1895j: 57 (*Hypostena*).—N.H.; Alaska, B.C. and Idaho to N.B., thence s. to Va., also Calif., Kans., Tex., La., Ga.
    spinosa Coquillett, 1897: 60 (*Celatoria*).—N.H.
    ouelleti Curran, 1925c: 150 (*Hypostena*).—Que.
    luctuosa, authors, not Meigen.

Genus Vibrissina Rondani

Vibrissina Rondani, 1861a: 35. Type-species, *Tachina demissa* Meigen (orig. des.; misident.)=*turrita* (Meigen).

turrita (Meigen), 1824: 401 (*Tachina*).—Europe; Asia, Japan, introd. Man., ?estab.

Genus XIPHOMYIA Townsend

Xiphomyia Townsend, 1917d: 125. Type-species, *gladiatrix* Townsend (orig. des.).
auriceps Aldrich, 1926d: 11.—Va.; Ga.
texana Reinhard, 1923: 267.—Tex.; Ariz., Miss., Ohio, Va.
    insignis Reinhard, 1934b: 188 (*Dexodes*).—Tex. **N. syn., n. comb.**

Tribe NEOMINTHOINI
Genus EUPELECOTHECA Townsend

Eupelecotheca Townsend, 1919c: 169. Type-species, *celer* Townsend (orig. des.).
Pantagathus Reinhard, 1935b: 168. Type-species, *alogus* Reinhard (orig. des.)=*celer* Townsend.

celer Townsend, 1919c: 169.—D.C.; S. Dak. and Iowa to Conn., s. to Fla.
    alogus Reinhard, 1935b: 169 (*Pantagathus*).—Iowa.
    rufilabris Wulp of Coquillett, 1897: 103 (*Phorocera*).

Genus PELECOTHECA Townsend

Pelecotheca Townsend, 1919c: 168. Type-species, *panamensis* Townsend (orig. des.).

curulis (Reinhard), 1943c: 18 (*Pantagathus*).—La.; N.Y., Tex., Miss., Ga.
    panamensis Townsend of Allen, 1929: 687.

Tribe EXORISTINI
(including Phorocerini and Phoriniini of Townsend)

The tribe Exoristini contains numerous parasites of lepidopterous larvae and some of tenthredinid larvae, plus one that attacks a weevil, *Listroderes*. Some species of the tribe are notably polyphagous, and some are of great importance in biological control.

In the latest European revision (Mesnil, 1960), *Exorista* includes *Guerinia*, *Podotachina*, and *Tachinomyia* as subgenera, but we prefer to recognize them as distinct genera. Another major difference is that Mesnil considers *Stomatomyia* a subgenus of *Spoggosia*, and *Euphorocera* a synonym of *Spoggosia* in the strict sense. There is clearly a close relationship, but we prefer to retain them as separate genera.

It is noteworthy that true *Phorocera*, one of the most widely used genera in the Nearctic literature, does not occur in North America. *Phorocera* species in the sense of the well-known revision by Aldrich and Webber (1924a) are referred in this catalog to 13 genera in 5 different tribes, in addition to a number left for the present in the "Unplaced species of Erycini."

    REFERENCES: Mesnil, 1956, 1960 (keys).

Genus BESSA Robineau-Desvoidy

Bessa Robineau-Desvoidy, 1863b: 164. Type-species, *secutrix* Robineau-Desvoidy (orig. des.) = *selecta* (Meigen).
Ptychomyia Brauer and Bergenstamm, 1889: 89 (1889: 21). Type-species, *Tachina selecta* Meigen (mon.).
Daeochaeta Townsend, 1892a: 97. Type-species, *harveyi* Townsend (orig. des.).

Our species of *Bessa* are parasites of tenthredinid larvae, especially of the subfamily Nematinae.

    REFERENCE: Mesnil, 1960 (key).

harveyi (Townsend), 1892a: 98 (*Daeochaeta*).—Maine; s. N.W.T. to Nfld., s. to Wash. and Ohio, also Calif., Ariz.
    harvei (error) Mesnil, 1960: 632.

tenthredinidarum Townsend, 1892d: 285 (*Masicera*).—Ont. *selecta*, authors, not Meigen.

selecta (Meigen), 1824: 377 (*Tachina*).—Europe; introd. N. Amer., distribution uncertain, probably confused with *harveyi*.

Genus CHAETEXORISTA Brauer and Bergenstamm

Chaetexorista Brauer and Bergenstamm, 1894: 80 (1895: 616). Type-species, *javana* Brauer and Bergenstamm (mon.).

javana Brauer and Bergenstamm, 1894: 80 (1895: 616).—Java; Oriental Region, introd. Mass., estab. Host: The Oriental moth, *Cnidocampa flavescens* (Walker).

Genus DIPLOSTICHUS Brauer and Bergenstamm

Diplostichus Brauer and Bergenstamm, 1889: 93 (1889: 25). Type-species, *tenthredinum* Brauer and Bergenstamm (orig. des., as gen. n., sp. n.)=*janitrix* (Hartig).

The species are parasites of tenthredinid larvae.

lophyri (Townsend), 1892d: 289 (*Phorocera*).—Ont.; B.C. to N.S., s. to Calif., Ohio, and Fla. **N. comb.**

petiolata Coquillett, 1897: 98 (*Exorista*).—Va. **N. syn., n. comb.**

hamata Aldrich and Webber, 1924a: 62 (*Phorocera*).—Conn. **N. syn.**

sellersi Hall, 1939: 243.—Oreg.; B.C., Calif.

Genus Epiplagiops Blanchard

Epiplagiops Blanchard, 1943: 450. Type-species, *littoralis* Blanchard (orig. des.).

The sole included species is a parasite of larvae of the vegetable weevil, *Listroderes*. The genus is in the *Euphorocera-Stomatomyia* complex, and may prove to be a synonym.

littoralis Blanchard, 1943: 451.—Argentina; Uruguay, introd. Calif., not estab.

Genus EUPHOROCERA Townsend

Euphorocera Townsend, 1892a: 112. Type-species, *tachinomoides* Townsend (orig. des.).

Neophorocera Townsend, 1912d: 163. Type-species, *Phorocera edwardsii* Williston (orig. des.).

Plagiotachina Townsend, 1927: 261.   Type-species, *peruviana* Townsend (orig. des.).

arnaudi (Reinhard), 1956c: 106 (*Phorocera*).—Calif.; Nev., Utah, Ariz., Nebr. **N. comb.**
claripennis (Macquart), 1848a: 209 (1848: 49) (*Phorocera*).—N. Amer.; recorded Alaska to Ont., all U.S., Mexico, but probably a mixture with *edwardsii*.
clunalis (Reinhard), 1956c: 107 (*Phorocera*).—Tex. **N. comb.**
edwardsii (Williston), 1889a: 1921 (*Phorocera*).—Unknown; distribution confused with *claripennis* (see above). **N. comb.**
flaviceps (Bigot), 1887b: cxli (*Cryptopalpus*).—Rocky Mts. Unrecognized. **N. comb.**
floridensis Townsend, 1916c: 217.—Fla.; Ariz. to Md., s. to Mexico and Fla.
indivisa (Aldrich and Webber), 1924a: 64 (*Phorocera*).—Tex. **N. comb.**
omissa (Reinhard), 1934b: 193 (*Phorocera*).—Tex.; N. Mex., Kans., Ark., S.C.
subnitens (Aldrich and Webber), 1924a: 65 (*Phorocera*).—Va. **N. comb.**
tachinomoides Townsend, 1892a: 112.—N. Mex.; Calif. to S. Dak., s. to Mexico and Tex.
vibrissata (Brauer and Bergenstamm), 1891: 351 (1891: 47) (*Podotachina*).—N.Y. ?=*edwardsii* or *claripennis*. **N. comb.**

Genus EXORISTA Meigen

Exorista Meigen, 1803: 280.   Type-species, *Musca larvarum* Linnaeus (mon.).
Eutachina Brauer and Bergenstamm, 1889: 98 (1889: 30).   Type-species, *Musca larvarum* Linnaeus (mon.).
Cyclotaphrys Townsend, 1909: 246.   Type-species, *anser* Townsend (orig. des.) =*fallax* (Meigen).
Tachina, authors, not Meigen.

REFERENCE: Mesnil, 1960 (key).

fallax (Meigen), 1824: 321 (*Tachina*).—Germany; Palaearctic, ?introd. New England.
  anser Townsend, 1909: 246 (*Cyclotaphrys*).—U.S.S.R.
fasciata fasciata (Fallén), 1820g: 5 (*Tachina*).—Sweden; Greenland, n. Europe.
  ssp. *moreti* (Robineau-Desvoidy), 1853b: 534 (*Tachina*; as sp.).—France; s. Europe, n. Africa, introd. New England, not estab.
    segregata Rondani, 1859: 181 (*Chetogena*).—Italy.

japonica (Townsend), 1909: 247 (*Tachina*).—Japan; China, introd. New England, not estab.
larvarum (Linnaeus), 1758: 596 (*Musca*).—Europe; introd. New England, estab.
mella (Walker), 1849: 767 (*Tachina*).—N.S.; B.C. to N.S., all U.S.
melba, error.
orgyiae LeBaron, 1871: 16 (*Tachina*).—Ill.
clisiocampae Townsend, 1891f: 83 (*Tachina*).—Maine.
orgyiae Townsend, 1892d: 284 (*Tachina*; preocc. LeBaron, 1871).—W. Va.
fernaldi Williston *in* Forbush and Fernald, 1896: 387 (*Achaetoneura*).—Mass.
orgyiarum Townsend, 1908: 107 (*Tachina*; n. name for *orgyiae* Townsend).—W. Va.
larvarum, authors, not Linnaeus.

Genus GUERINIA Robineau-Desvoidy

Guerinia Robineau-Desvoidy, 1830: 196. Type-species, *festiva* Robineau-Desvoidy (Coquillett, 1910b: 548)=*rustica* (Fallén).
dydas (Walker), 1849: 748 (*Tachina*).—Ont.; s. Canada, all U.S.
spinosula Townsend, 1891c: 353 (*Tachina*).—Ill.
tenthredinivora Townsend, 1892d: 285 (*Tachina*).—Ont.
rustica, authors, not Fallén.
simulans, authors, not Meigen.
rufostomata (Bigot), 1888a: 260 (*Tachina*).—Rocky Mts. Unrecognized.
rustica (Fallén), 1810b: 264 (*Tachina*).—Sweden; ?N. Amer., Europe.
simulans Meigen, 1824: 306 (*Tachina*).—Europe.
trudis Reinhard, 1951: 8.—Utah; B.C. to N. Dak., s. to Calif. and Ariz.

Genus GUERINIOPSIS Reinhard

Gueriniopsis Reinhard, 1943a: 166. Type-species, *plausilis* Reinhard (orig. des.)=*setipes* (Coquillett).
setipes (Coquillett), 1902g: 112 (*Frontina*).—S. Dak.; Man. to Pa., s. to S. Dak. and Iowa. **N. comb.**
plausilis Reinhard, 1943a: 167.—Man. **N. syn.**

Genus PALPEXORISTA Townsend

Palpexorista Townsend, 1926: 28. Type-species, *phoroceroides* Townsend (orig. des.)=*imitator* (Aldrich and Webber).

Yahuartachina Townsend, 1927: 261. Type-species, *yahuarphrynoides* Townsend (orig. des.). **N. syn.**

REFERENCE: Aldrich and Webber, 1924a (key, in *Phorocera*).

cocciphila (Aldrich and Webber), 1924a: 53 (*Phorocera*).—D.C.; Ohio. **N. comb.**

coccyx (Aldrich and Webber), 1924a: 64 (*Phorocera*).—Va.; Mich. to N.Y., s. to Ga., also Tex. **N. comb.**
  heros, authors, not Schiner.
  longiuscula, authors, not Walker.

einaris (Smith), 1912: 119 (*Phorocera*).—Mass.; Mich. to Mass., s. to Fla., also Tex. **N. comb.**

imitator (Aldrich and Webber), 1924a: 63 (*Phorocera*).—Conn.; Ohio to Mass., s. to Fla., also Tex.
  phoroceroides Townsend, 1926: 29.—Mass.

pellecta (Reinhard), 1957: 106 (*Phorocera*).—Ariz. **N. comb.**

stolida (Reinhard), 1957: 107 (*Phorocera*).—Ohio; Ga. **N. comb.**

sulcata (Aldrich and Webber), 1924a: 66 (*Phorocera*).—Ga.; R.I. to Ga., also Tex. **N. comb.**

tuxedo (Curran), 1930j: 110 (*Phorocera*).—N.Y. **N. comb.**

virilis (Aldrich and Webber), 1924a: 52 (*Phorocera*).—D.C.; Ohio, N.Y. **N. comb.**

Genus PARASETIGENA Brauer and Bergenstamm

Parasetigena Brauer and Bergenstamm, 1891: 339, 401 (1891: 35, 97). Type-species, *Chetogena segregata* Rondani (mon.; misident.) = *agilis* (Robineau-Desvoidy).

Under various names, one species was introduced as a parasite of the gypsy moth, *Porthetria dispar* (Linnaeus).

agilis (Robineau-Desvoidy), 1830: 132 (*Phorocera*).—France; Europe, introd. and estab., New England, N.Y.
  silvestris Robineau-Desvoidy, 1863a: 531 (*Duponchelia*).—France.
  sylvestris, emend.
  segregata, authors, not Rondani.

Genus Phorocera Robineau-Desvoidy

Phorocera Robineau-Desvoidy, 1830: 131. Type-species, *agilis* Robineau-Desvoidy (Robineau-Desvoidy, 1863a: 509; misident.) = *assimilis* (Fallén).

Chetogena Rondani, 1856: 68. Type-species, *Tachina gramma* Meigen (orig. des.) = *assimilis* (Fallén).

Chaetogena, emend. or error.

TACHINIDAE 1057

Robineau-Desvoidy (1863a) designated *assimilis*, and showed *agilis*, an originally included species, in its synonymy. However, this *agilis* was a misidentification, or possibily based on a nontypical original specimen. The true *agilis* is a *Parasetigena* (q.v.).

Phorocera in the sense of the well-known revision by Aldrich and Webber (1924a) is a mixture of at least 13 genera distributed among 5 tribes in this catalog.

obscura (Fallén), 1810b: 283 (*Tachina*).—Sweden; introd. N.S., not estab.

Genus Podotachina Brauer and Bergenstamm

americana   Brauer and Bergenstamm, 1891: 351 (1891: 47).— Unknown (as N.Y., error) =*sorbillans* (Wiedemann). Not Nearctic.

Genus PSEUDOTACHINOMYIA Smith

Pseudotachinomyia Smith, 1917a: 54. Type-species, *webberi* Smith (orig. des.)=*slossonae* (Townsend).

REFERENCE: Reinhard, 1935a (rev.).

aequalis Reinhard, 1935a: 133.—Tex.; Colo. to N.Y., s. to Tex.
compascua Reinhard, 1935a: 135.—Wash.; Alta., Utah.
longiforceps Brooks, 1945a: 89.—Ont.; Que., N.Y., Conn.
slossonae (Townsend), 1908: 108 (*Euphorocera*).—N.H.; Ill. to Ont. and N.H., s. to S.C., also Tex.
webberi Smith, 1917a: 54.—Mass.
cinerea, authors, not Wulp.

Genus SPOGGOSIA Rondani

Spoggosia Rondani, 1859: 182. Type-species, *occlusa* Rondan[i] (mon).=*echinura* (Robineau-Desvoidy).
Murdockiana Townsend, 1916h: 622. Type-species, *Euphorocera gelida* Coquillett (orig. des.). **N. syn.**, L. P. Mesnil *in litt.*

gelida (Coquillett), 1897: 101 (*Euphorocera*).—Alaska; Y.T., N.W.T. (Southampton I.). **N. comb.**

Genus STOMATOMYIA Brauer and Bergenstamm

Stomatomyia Brauer and Bergenstamm, 1889: 98 (1889: 30). Type-species, *Chetogena filipalpis* Rondani (mon.).
Plagiprospherysa Townsend, 1892a: 113. Type-species, *valida* Townsend (orig. des.)=*parvipalpis* (Wulp).

Tachinopsis Coquillett, 1897: 120. Type-species, *mentalis* Coquillett (orig. des.) = *parvipalpis* (Wulp).

Plagiops Townsend, 1911a: 141. Type-species, *littoralis* Townsend (orig. des., as gen. n., sp. n.) = *floridensis* (Townsend).

floridensis (Townsend), 1892a: 114 (*Plagiprospherysa*).—Fla.; Nebr., Tex., Ga.
> *littoralis* Townsend, 1911a: 141 (also 1912a: 107) (*Plagiops*).—Fla.
> *litoralis*, error.

parvipalpis (Wulp), 1890: 124 (*Prospherysa*).—Mexico; B.C. to Calif., Mexico, and Tex.
> *valida* Townsend, 1892a: 113 (*Plagiprospherysa*).—N. Mex.
> *mentalis* Coquillett, 1897: 120 (*Tachinopsis*).—Wash.

Genus TACHINOMYIA Townsend

Tachinomyia Townsend, 1892a: 96. Type-species, *robusta* Townsend (orig. des.) = *panaetius* (Walker).

REFERENCE: Webber, 1941 (rev.).

acosta Webber, 1941: 299.—N.J.; Mich. to Mass. and N.J., also B.C. to Calif., Ga.

apicata Curran, 1926k: 171.—Que.; Man. to N.S., s. to Mo. and Va., also B.C., Alta.

cana Webber, 1941: 298.—Mass.; Man. to Mass., s. to Tex.

dakotensis Webber, 1941: 302.—S. Dak.; B.C. to Man. and Mich., s. to Calif.

floridensis Townsend, 1892a: 97.—Fla.

montana (Smith), 1917b: 140 (*Allophorocera*).—Mont.; B.C. to Mont., s. to Calif. and Ariz.

nigricans Webber, 1941: 301.—Mass.; N. Dak. to Que., s. to S. Dak. and N.J., also B.C., Ga.

panaetius (Walker), 1849: 767 (*Tachina*).—N.S.; S. Dak. to N.S., s. to Kans. and N.J.
> *pansa* Walker, 1849: 787 (*Tachina*).—N. Amer., N.S.
> *violenta* Walker, 1849: 788 (*Tachina*).—N.S.
> *irrequieta* Walker, 1849: 789 (*Tachina*).—N.S.
> *robusta* Townsend, 1892a: 96.—Mich. (Curran, 1926k: 169, lectotype).

similis (Williston), 1893d: 256 (*Prospherysa*).—Calif.; B.C. to Man., s. to Calif. and Ariz.
> *occidentalis* Curran, 1926k: 170.—B.C.

variata Curran, 1926k: 169.—Ont.; B.C. to N.B., s. to Calif., Mexico, Kans., and N.J.

Tribe SIPHONINI

(Crocutini, Actiini)

With few exceptions, the species of the tribe Siphonini are small and parasitize the larvae of various Microlepidoptera such as *Laspeyresia, Rhyacionia, Tortrix,* and *Tinea.*
There is wide disagreement on the makeup of the tribe. As here constituted, it is a combination of Townsend's two tribes, the Actiini and Siphonini, minus a scattering of genera referred in this catalog to eight other tribes. Some of the remaining genera (e.g., *Clausicella* and *Elfia*) have also been referred elsewhere, some even into the Tachininae, by Mesnil in his work on the Palaearctic fauna, but for the present we leave them in the Siphonini, with which they are customarily associated by general habitus and biology. We have grouped the genera in three subtribes to assist in future analysis of the relationships.

Subtribe CORONIMYIINA

The genera *Crocinosoma, Drepanoglossa,* and *Philocalia* are atypical.

Genus CLAUSICELLA Rondani

Clausicella Rondani, 1856: 61. Type-species, *suturata* Rondani (orig. des., as gen. n., sp. n.).
Siphophyto Townsend, 1892a: 127. Type-species, *floridensis* Townsend (orig. des.). **N. syn.**
Epigrimyia, authors, in part, not Townsend.
    REFERENCE: Reinhard, 1946c (rev.).

floridensis (Townsend), 1892a: 128 (*Siphophyto*).—Fla.; B.C., Ont., all U.S. **N. comb.**
neomexicana (Townsend), 1892a: 128 (*Siphophyto*).—N. Mex.; Calif. to Kans. and Tex. **N. comb.**
opaca (Coquillett), 1895c: 128 (*Siphophyto*).—Calif.; Calif. to Iowa and Tex. **N. comb.**
politura (Reinhard), 1946c: 85 (*Siphophyto*).—Que.; S. Dak., Mich., N.H., N.B. **N. comb.**
    americana Brauer and Bergenstamm *in* Coquillett, 1897: 75 (*Gymnopareia*). Nomen nudum.
setigera (Coquillett), 1895c: 127 (*Siphophyto*).—Calif.; B.C. to Calif. and Wyo., also Wis. to Ont. and N.Y., s. to La. and Fla. **N. comb.**
    trinidadensis Townsend of Fattig, 1949: 31 (*Siphocrocuta*).
turmalis (Reinhard), 1946c: 86 (*Siphophyto*).—N.Y.; Wis. to Que., s. to Tex. and Ga., also ?Calif. **N. comb.**

Genus CORONIMYIA Townsend

Coronimyia Townsend, 1892a: 128. Type-species, *geniculata* Townsend (orig. des.).
Epigrimyia, authors, in part, not Townsend.

REFERENCE: Reinhard, 1946c (rev.).

geniculata Townsend, 1892a: 129.—Ill.; Ill. to Ont. and N.Y., s. to La. and Ga.
melitarae Reinhard, 1946c: 91.—Calif.

Genus CROCINOSOMA Reinhard

Crocinosoma Reinhard, 1947a: 20. Type-species, *cornuale* Reinhard (orig. des.).
cornuale Reinhard, 1947a: 21.—Tex.; Va. to Ga. to Mexico.

Genus DREPANOGLOSSA Townsend

Drepanoglossa Townsend, 1891c: 377. Type-species, *lucens* Townsend (orig. des.).
lucens Townsend, 1891c: 378.—N. Mex.; N. Dak., Colo., Calif., Ariz.
pavonacea (Reinhard), 1939a: 72 (*Siphoclytia*).—Ohio; Ill. to N.Y., s. to Ga., also Tex. **N. comb.**

Genus GINGLYMIA Townsend

Ginglymia Townsend, 1892a: 118. Type-species, *acrirostris* Townsend (orig. des.).
Ginglymyia, error or emend.

acrirostris Townsend, 1892a: 119.—Mich.; N.Y., Ohio, Tex.

Genus LASIONEURA Coquillett

Lasioneura Coquillett, 1895j: 50. Type-species, *johnsoni* Coquillett (orig. des.).
bicolor (Curran), 1923l: 246 (*Ginglymia*).—B.C. ?=*johnsoni*.
johnsoni Coquillett, 1895j: 50.—Wash.; B.C. to Wyo., s. to Calif. and Mexico.

Genus PHILOCALIA Reinhard

Philocalia Reinhard, 1939a: 70. Type-species, *tenuirostris* Reinhard (orig. des.).

Evanalia Strickland, 1941: 64. Type-species, *medicinensis* Strickland (orig. des.)=*tenuirostris* Reinhard.

tenuirostris Reinhard, 1939a: 71.—Mont.; Alta., Ill., Ind.
  medicinensis Strickland, 1941: 64 (*Evanalia*).—Alta.

Subtribe SIPHONINA

Several atypical genera have been left here for the present, until their relationships can be studied: *Eucoronimyia, Evidomyia, Lispideosoma, Paralispidea, Phantasiomyia, Phasiostoma,* and *Slossonaemyia.*

Genus ACTIA Robineau-Desvoidy

Actia Robineau-Desvoidy, 1830: 85. Type-species, *pilipennis* Robineau-Desvoidy (Emden, 1954: 63; Suspension of the Rules required)=*lamia* (Meigen).
Gymnophtalma Lioy, 1864a: 1341. Type-species, *Tachina crassicornis* Meigen (mon.; misident.)=*pilipennis* (Fallén).
Polychaetoneura Walton, 1914c: 90. Type-species, *elyii* Walton (orig. des.)=*americana* (Townsend).
Actiopsis Townsend, 1917d: 121. Type-species, *autumnalis* Townsend (orig. des.).
Xanthoactia Townsend, 1919b: 585. Type-species, *Lasioneura valloris* Coquillett (orig. des.).
Thryptocera, authors, not Macquart.

The genus *Actia* was proposed for two new species, *cingulata* and *pilipennis*. The first reviser, Robineau-Desvoidy (1850, 1851b), restricted *Actia* to *pilipennis* and relatives and referred *cingulata* to a new genus *Elfia*. Unfortunately, Coquillett (1910b), the first to validly designate a type for *Actia*, cited *cingulata*. Although usage is somewhat divided, most authors have followed Robineau-Desvoidy's clear intent, as we have done above. However, Suspension of the Rules is required to suppress the designation by Coquillett.

Many undescribed species are known, and the available key is of limited use.

REFERENCE: Curran, 1933a (rev.).

americana (Townsend), 1892e: 69 (*Thryptocera*).—D.C.; Wis. to Maine, s. to Ariz., Mexico, and Fla.
  elyii Walton, 1914c: 91 (*Polychaetoneura*).—Conn.
autumnalis (Townsend), 1917d: 122 (*Actiopsis*).—Md.; N.Y., N.H., Va.
diffidens Curran, 1933a: 5.—N.S.; B.C. to N.S., s. to Mich. and N.Y.
flavipes (Coquillett), 1897: 58 (*Thryptocera*).—N.H.; N.Y.

interrupta Curran, 1933a: 6.—N.Y.; B.C. to N.B., s. to Calif., Mexico, and D.C.
   pilipennis, authors, not Fallén.
nudibasis Stein, 1924: 135.—Germany; introd. Ont., Mass., and Conn., not estab.
ontario Curran, 1933a: 4.—Ont.; Mich., Ohio, N.Y.
palloris (Coquillett), 1895j: 50 (*Lasioneura*).—N.H.; Ont. and Que., s. to N.Y. and Maine.
panamensis Curran, 1933a: 3.—Panama Canal Zone; Calif., Tex.
rufescens (Greene), 1934a: 34 (*Actiopsis*).—S. Dak.; Ill., Ohio, Ga.

Genus APHANTORHAPHA Townsend

Aphantorhapha Townsend, 1919b: 586. Type-species, *arizonica* Townsend (orig. des.).

arizonica Townsend, 1919b: 586.—Ariz.
atoma Reinhard, 1947a: 19.—Tex.
hurdi Reinhard, 1959a: 161.—Calif.

Genus CHAETOSTIGMOPTERA Townsend

Chaetostigmoptera Townsend, 1916h: 624. Type-species, *Chaetophleps crassinervis* Walton (orig. des.).

crassinervis (Walton), 1913: 51 (*Chaetophleps*).—Md.; Md. to Fla. to La., also Ohio.

Genus EUCORONIMYIA Townsend

Isoglossa Coquillett, 1895c: 125 (preocc. Casey, 1893). Type-species, *hastata* Coquillett (orig. des.).
Eucoronimyia Townsend, 1908: 84 (n. name for *Isoglossa* Coquillett). Type-species, *Isoglossa hastata* Coquillett (aut.).

Eucoronimyia is an aberrant genus that may not belong here.

hastata (Coquillett), 1895c: 126 (*Isoglossa*).—Calif.; Colo., Tex.

Genus EVIDOMYIA Reinhard

Evidomyia Reinhard, 1958c: 226. Type-species, *infida* Reinhard (orig. des.).

Evidomyia is close to *Eucoronimyia*, and like the latter it may belong elsewhere.

infida Reinhard, 1958c: 226.—Calif.

Genus LISPIDEOSOMA Reinhard

Lispideosoma Reinhard, 1943a: 164. Type-species, *flavipes* Reinhard (orig. des.).

flavipes Reinhard, 1943a: 165.—Wis.; Wash.

Genus MESSIOMYIA Reinhard

Messiomyia Reinhard, 1955d: 124. Type-species, *triconis* Reinhard (orig. des.).

triconis Reinhard, 1955d: 125.—Tex.

Genus PARALISPIDEA Townsend

Paralispidea Townsend, 1915a: 20. Type-species, *Admontia unispinosa* Coquillett (orig. des.).

aperta Reinhard, 1958b: 237.—Mass.; N.Y., Ohio.
unispinosa (Coquillett), 1898c: 234 (*Admontia*).—La.; Md.

Genus PHANTASIOMYIA Townsend

Phantasiomyia Townsend, 1915f: 225. Type-species, *gracilis* Townsend (orig. des.).

atripes (Coquillett), 1897: 58 (*Thryptocera*).—Mass. (?error for Calif.), Ariz.
gracilis Townsend, 1915f: 226.—N. Mex.; Calif., Ariz.

Genus PHASIOSTOMA Townsend

Phasiostoma Townsend, 1915f: 224. Type-species, *aristale* Townsend (orig. des.).

aristale Townsend, 1915f: 225.—N. Mex.

Genus PSEUDOSIPHONA Townsend

Pseudosiphona Townsend, 1916h: 622. Type-species, *Siphona brevirostris* Coquillett (orig. des.).

brevirostris (Coquillett), 1897: 76 (*Siphona*).—?R.I. (as Mo., error); N.H., Conn.

Genus SIPHONA Meigen

Crocuta Meigen, 1800: 39. Type-species, *Musca geniculata* De Geer (Coquillett, 1910b: 528). Suppressed by I.C.Z.N., 1963b: 339.

Siphona Meigen, 1803: 281. Type-species, *Conops irritans* Linnaeus (as *Stomoxys irritans* Fabricius; mon.; misident.)= *geniculata* (De Geer).
Bucentes Latreille, 1809: 339. Type-species, *cinereus* Latreille (mon.) =*geniculata* (De Geer).
Siphonopsis Townsend, 1916h: 622. Type-species, *Siphona plusiae* Coquillett (orig. des.).

REFERENCE: Curran, 1932d (key).

conata (Reinhard), 1959a: 162 (*Siphonopsis*).—Calif. **N. comb.**
geniculata (De Geer), 1776: 38 (*Musca*).—Europe; Man. to N.H., s. to Kans. and Ga.
illinoiensis Townsend, 1891c: 368.—Ill.; Kans. to Ont., s. to Tex. and Fla.
intrudens (Curran), 1932d: 14 (*Bucentes*).—Pa.; Ont., N.Y.
lurida Reinhard, 1943c: 20.—Oreg.; Calif.
lutea (Townsend), 1919b: 584 (*Crocuta*).—N.H.; Minn., Mich.
plusiae Coquillett, 1895c: 125.—Calif.; Ariz., Tex., Mexico.
tenuis Curran, 1933b: 10.—Ont.; Mich., Ohio, Que., Vt.

Genus SLOSSONAEMYIA Townsend

Slossonaemyia Townsend, 1916h: 624. Type-species, *Chaetophleps rostrata* Coquillett (orig. des.).

angulicornis (Curran), 1930j: 97 (*Elephantocera*).—N.Y. **N. comb.**
rostrata (Coquillett), 1898c: 235 (*Chaetophleps*).—Fla.

Subtribe NEAERINA
Genus ACRONARISTA Townsend

Acronarista Townsend, 1908: 85. Type-species, *mirabilis* Townsend (orig. des.).

cornuta Reinhard, 1931a: 26.—La.; Ga.
mirabilis Townsend, 1908: 86.—Fla.

Genus ACRONARISTOPSIS Townsend

Acronaristopsis Townsend, 1919c: 178. Type-species, *bahamensis* Townsend (orig. des.).

Unnamed spp.—Ark., Fla.

Genus APHELOGLUTUS Greene

Apheloglutus Greene, 1934a: 28, 32 (also as *Apeloglutus*, p. 32). Type-species, *latifrons* Greene (orig. des.).

latifrons Greene, 1934a: 32.—Colo.; B.C.

Genus DICHAETONEURA Johnson

Dichaetoneura Johnson, 1907a: 9. Type-species, *leucoptera* Johnson (mon.).

REFERENCE: Thompson, 1953a (biol., immatures).

leucoptera Johnson, 1907a: 9.—Mass.; Minn. to P.E.I., s. to Md.

Genus ELFIA Robineau-Desvoidy

Elfia Robineau-Desvoidy, 1850: 190. Type-species, *Actia cingulata* Robineau-Desvoidy (Robineau-Desvoidy, 1863a: 672).
Craspedothrix Brauer and Bergenstamm, 1893: 150 (1893: 62). Type-species, *vivipara* Brauer and Bergenstamm (mon.).
Lispidea Coquillett, 1895j: 51. Type-species, *palpigera* Coquillett (orig. des.). **N. syn.**
Plectops Coquillett, 1897: 57. Type-species, *melissopodis* Coquillett (orig. des.). **N. syn.**
Goliathocera Townsend, 1915a: 21. Type-species, *Clausicella antennalis* Coquillett (orig. des.)=*setigera* (Thomson). **N. syn.**
Lophosiocera Townsend, 1916h: 623. Type-species, *curriei* Townsend (orig. des.). **N. syn.**
Nephopteropsis Townsend, 1916h: 623. Type-species, *Clausicella johnsoni* Coquillett (orig. des.). **N. syn.**
Phylacteropoda Townsend, 1916h: 623. Type-species, *Clausicella tarsalis* Coquillett (orig. des.). **N. syn.**

aenea (Coquillett), 1895j: 57 (*Hypostena*).—Calif.; N. Mex., Mexico (Chih.). **N. comb.**
amplicornis (James), 1955d: 83 (*Plectops*).—Wash.; Oreg., Colo. **N. comb.**
curriei (Townsend), 1916h: 623 (*Lophosiocera*).—Wyo.; Utah. **N. comb.**
erisma (Reinhard), 1962b: 219 (*Plectops*).—Ariz. **N. comb.**
erotema (Reinhard), 1958b: 238 (*Nephopteropsis*).—N.Y.; Va. **N. comb.**
johnsoni (Coquillett), 1897: 56 (*Clausicella*).—Pa.; Minn. to Mass., thence to Fla. **N. comb.**
manca (Greene), 1934a: 31 (*Plectops*).—Md.; Tenn., Va., Ga. **N. comb.**
melissopodis (Coquillett), 1897: 57 (*Plectops*).—D.C.; Mich. to Mass., s. to Tex. and Fla. **N. comb.**
nigra (Brooks), 1945a: 93 (*Phylacteropoda*).—Man.; N.Y., N.H., Md. **N. comb.**
palpigera (Coquillett), 1895j: 52 (*Lispidea*).—Ill.; Ill. to Mass., s. to Miss. **N. comb.**

pruinosa (Malloch), 1927b: 91 (*Plectops*).—Md.; Calif., Va., Ga. **N. comb.**

setigera (Thomson), 1869: 527 (*Lophosia*).—Calif.; Oreg., Utah. **N. comb.**

  antennalis Coquillett, 1895j: 56 (*Clausicella*).—Calif. **N. comb.**

tarsalis (Coquillett), 1895j: 56 (*Clausicella*).—Ill.; Ont. and Que., s. to Ill. and Md. **N. comb.**

usitata (Coquillett), 1897: 56 (*Clausicella*).—N.H.; Alaska, Mo., Maine. **N. comb.**

Genus EURYCEROMYIA Townsend

Euryceromyia Townsend, 1892a: 115. Type-species, *robertsonii* Townsend (orig. des.).

robertsonii Townsend, 1892a: 116.—Ill.; Ill. and Mich. to Mass., also Va.

Genus PHYTOMYPTERA Rondani

Phytomyptera Rondani, 1845b: 32, 33. Type-species, *nitidiventris* Rondani (mon.)=*nigrina* (Meigen).

walleyi Brooks, 1945a: 91.—Que.

Genus PSEUDAPINOPS Coquillett

Pseudapinops Coquillett, 1902g: 108. Type-species, *niger* Coquillett (orig. des.).

niger Coquillett, 1902g: 108.—Idaho; B.C. to Calif., also Pa. to Va.

rogalis Reinhard, 1955a: 132.—Calif.

Genus SCHIZACTIA Townsend

Schizactia Townsend, 1926: 31. Type-species, *Schizotachina vitinervis* Thompson (orig. des.).

ruficornis (Greene), 1934a: 33 (*Schizotachina*).—Fla. **N. comb.**

vitinervis (Thompson), 1911: 268 (*Schizotachina*).—Mass.; Minn. to N.H., thence s. to Fla.

Genus SCHIZOTACHINA Walker

Tachina, subg. **Schizotachina** Walker, 1852: 264. Type-species, *convecta* Walker (Coquillett, 1910b: 261).

convecta (Walker), 1852: 276 (*Tachina*).—U.S.; Calif. to Mich., Mass., and Fla.
exul (Walker), 1852: 277 (*Tachina*).—U.S. Unrecognized.
longicornis (Coquillett), 1902g: 106 (*Neaera*).—N.Y.; S. Dak. to Ont. and Mass., s. to Calif. and Fla.

Tribe GRAPHOGASTRINI

The flies of the tribe Graphogastrini parasitize larvae of Microlepidoptera.

REFERENCE: Brooks, 1942 (rev.).

Genus CLASTONEURIOPSIS Reinhard

Clastoneuriopsis Reinhard, 1939a: 68. Type-species, *meralis* Reinhard (orig. des.).
Neopsalidopteryx Brooks, 1942: 142. Type-species, *Clistomorpha alberta* Curran (orig. des.). **N. syn.**, H. J. Reinhard *in litt*.

alberta (Curran), 1927i: 298 (*Clistomorpha*).—Alta.; Wash., Utah. **N. comb.**
deceptor (Curran), 1927i: 298 (*Clistomorpha*).—B.C.; Wash., Utah, Colo. **N. comb.**
meralis Reinhard, 1939a: 69.—Wash.

Genus PSALIDOPTERYX Townsend

Psalidopteryx Townsend, 1916e: 21. Type-species, *slossonae* Townsend (orig. des.).

alaskensis Brooks, 1942: 148.—Alaska.
brunnea Brooks, 1942: 147.—Alta.; B.C., Idaho, Utah, Colo., Wis.
dorsalis (Coquillett), 1902g: 108 (*Hyalomyodes*).—Idaho; B.C., Nev., Utah.
fuscisquamis Brooks, 1942: 149.—Alta.
 fuscipennis (error) Strickland, 1946: 173.
grandis Brooks, 1942: 149.—Wyo.; Wash., N. Mex., Colo.
macdunnoughi Brooks, 1942: 146.—P.E.I.
nuda Brooks, 1942: 147.—B.C.
orientalis Brooks, 1942: 146.—Va.
pollinosa Brooks, 1942: 148.—B.C.
pseudonuda Brooks, 1942: 147.—B.C.; Idaho.
psilocorsiphaga Brooks, 1942: 149.—Que.; Ohio, N.S.
slossonae Townsend, 1916e: 22.—N.H.; Que.

Tribe ADMONTIINI
(Trichopareiini)

The members of the tribe Admontiini are parasites of tipulid larvae.

Genus ADMONTIA Brauer and Bergenstamm

Admontia Brauer and Bergenstamm, 1889: 104 (1889: 36). Type-species, *podomyia* Brauer and Bergenstamm (orig. des., as gen. n., sp. n.).
Trichopareia Brauer and Bergenstamm, 1889: 103 (1889: 35). Type-species, *Tachina seria* Meigen (mon.).
Trichoparia, error or emend.
Admontiopsis Townsend, 1915a: 19. Type-species, *Admontia tarsalis* Coquillett (orig. des.).
Euadmontia Townsend, 1915a: 19. Type-species, *Admontia pergandei* Coquillett (orig. des.).
Euhyperecteina Townsend, 1915a: 19. Type-species, *Admontia nasoni* Coquillett (orig. des.).
Xenadmontia Townsend, 1915a: 22. Type-species, *Hypostena degeerioides* Coquillett (orig. des.).
Iconomedina Townsend, 1916h: 626. Type-species, *Degeeria washingtonae* Coquillett (orig. des.).

REFERENCE: Curran, 1927i (partial key).

badiceps Reinhard, 1958d: 279.—Wash.
degeerioides (Coquillett), 1895j: 58 (*Hypostena*).—D.C.; Mich. to N.H., s. to Va., also Utah.
    aestivalis West, 1925: 126 (*Hyperecteina*).—N.Y. **N. syn.**
    bishopi West, 1925: 126 (*Hyperecteina*).—N.Y. **N. syn.**
dubia Curran, 1927i: 297.—Sask. ?=*pollinosa*.
duospinosa (West), 1925: 127 (*Hyperecteina*).—N.Y.; Pa., D.C.
nasoni Coquillett, 1895j: 55.—Ill.; Wis., Mich., Ind., N.J.
    rufochaeta Curran, 1927i: 296.—Wis.
    ruficeps (error) Aldrich, 1929d: 35.
offella Reinhard, 1962b: 223.—Ariz.
pergandei Coquillett, 1895j: 54.—D.C.; Tenn. to Md., s. to Miss. and Ga., also Mass.
    americana Brauer and Bergenstamm, 1891: 410 (1891: 106). Nomen nudum.
pollinosa Curran, 1927i: 296.—Man.; Sask., S. Dak.
tarsalis Coquillett, 1898c: 234.—La.; Oreg., Fla.
washingtonae (Coquillett), 1895k: 104 (*Degeeria*).—N.H.

Tribe HYPERECTEININI

The species of Hyperecteinini parasitize Orthoptera, adult beetles, and a few lepidopterous larvae such as those of *Diatraea*.

Genus CARTOCOMETES Aldrich

Cartocometes Aldrich, 1929c: 9. Type-species, *io* Aldrich (orig. des.).
io Aldrich, 1929c: 10.—N.Y.; Ont., Vt., Mass.

Genus EUTHYPROSOPA Townsend

Euthyprosopa Townsend, 1892a: 106. Type-species, *petiolata* Townsend (orig. des.).
petiolata Townsend, 1892a: 107.—N. Mex.

Genus GILVELLA Mesnil

Gilvella Mesnil, 1960: 654. Type-species, *Hypostena gilvipes* Coquillett (mon.).
  REFERENCE: Reinhard, 1942c (key, as *Oedematocera*, part).
gilvipes (Coquillett), 1897: 61 (*Hypostena*).—Ga. (unpublished); Mass. to Fla. to Tex., also Mo.
optata (Reinhard), 1942c: 107 (*Oedematocera*).—N.Y.; Ohio and N.Y. to Fla., also Tex., La. **N. comb.**

Genus GREMLINOTROPHUS Reinhard

Gremlinotrophus Reinhard, 1943a: 163. Type-species, *derisus* Reinhard (orig. des.).
derisus Reinhard, 1943a: 163.—Ohio; Alta., Mont., Calif., Colo.

Genus HYPERECTEINA Schiner

Hyperecteina Schiner, 1861b: 143. Type-species, *metopina* Schiner (orig. des., as gen. n., sp. n.)=*cinerea* (Perris).
Centeter Aldrich, 1923d: 3. Type-species, *cinerea* Aldrich (orig. des.)= *aldrichi* Mesnil.

The species are parasites of adult Scarabaeidae.

aldrichi Mesnil, 1953a: 50 (n. name for *cinerea* Aldrich).—Japan; introd., estab., N.Y., Pa., N.J., Mass., Conn., D.C.
  cinerea Aldrich, 1923d: 4 (*Centeter*; preocc. Perris, 1852).—Japan.

cinerea (Perris), 1852: 206 (*Metopia*).—France; Europe, n. Africa, introd. N.Y., ?estab.
    metopina Schiner, 1861b: 144 (also 1862a: 537).—Austria.
longicornis (Fallén), 1810b: 282 (*Tachina*).—Sweden; Europe, introd. N.Y., ?estab.

Genus MIAMIMYIA Townsend

Miamimyia Townsend, 1916i: 308. Type-species, *cincta* Townsend (orig. des.).

cincta Townsend, 1916i: 309.—Fla.

Genus OEDEMATOCERA Townsend

Oedematocera Townsend, 1916h: 621. Type-species, *Hypostena flaveola* Coquillett (orig. des.).

The hosts are various Orthoptera.

REFERENCE: Reinhard, 1942c (key).

flaveola (Coquillett), 1897: 61 (*Hypostena*).—N.H.; Kans. to Ont. and N.H., s. to Ga.

Genus ORAPHASMOPHAGA Reinhard

Oraphasmophaga Reinhard, 1958d: 284. Type-species, *Paraphasmophaga pictipennis* Reinhard (orig. des.).

pictipennis (Reinhard), 1935b: 167 (*Paraphasmophaga*).—Tex.

Genus PHASMOPHAGA Townsend

Phasmophaga Townsend, 1909: 243. Type-species, *antennalis* Townsend (orig. des.).

antennalis Townsend, 1909: 244.—Wis.; Kans., N.Y., Tex.
meridionalis Townsend, 1909: 244.—Tex.

Genus ROESELIOPSIS Townsend

Roeseliopsis Townsend, 1915a: 23. Type-species, *Racodineura americana* Coquillett (orig. des.).

americana (Coquillett), 1897: 66 (*Racodineura*).—Ga.; Ill., Fla.
floridensis Greene, 1934a: 30.—Fla.

Genus STENONEURA Reinhard

Stenoneura Reinhard, 1945a: 33. Type-species, *serotina* Reinhard (orig. des.).

serotina Reinhard, 1945a: 33.—Tex.; Fla.

Tribe ERYTHROCERINI

The hosts of the Erythrocerini are adult scarabaeid beetles, larvae of Microlepidoptera, and earwigs (*Forficula*).

Genus CANELOMYIA Reinhard

Canelomyia Reinhard, 1958d: 282. Type-species, *fumator* Reinhard (orig. des.).

fumator Reinhard, 1958d: 282.—Ariz.

Genus Elodia Robineau-Desvoidy

Elodia Robineau-Desvoidy, 1863a: 936. Type-species, *gagatea* Robineau-Desvoidy (orig. des.)=*tragica* (Meigen).

flavipalpis Aldrich, 1933a: 21.—Japan; Korea, introd. ne. U.S., not estab.

subfasciata Aldrich, 1933a: 22.—Japan; introd. ne. U.S., not estab.

tragica (Meigen), 1824: 408 (*Tachina*).—Germany; Europe, introd. Ont., not estab.

Genus ERYNNIA Robineau-Desvoidy

Erynnia Robineau-Desvoidy, 1830: 125. Type-species, *nitida* Robineau-Desvoidy (mon.)=*ocypterata* (Fallén).

Tortriciophaga Townsend, 1916h: 625. Type-species, *Pseudomyothyria tortricis* Coquillett (orig. des.).

Anachaetopsis, authors, not Brauer and Bergenstamm.

tortricis (Coquillett), 1895j: 55 (*Pseudomyothyria*).—Calif.; B.C. to Nfld., s. to Calif., Mexico, and Va.

vagans Aldrich, 1923a: 54 (*Anachaetopsis*).—Oreg. **N. comb.**

columbia Curran, 1925c: 153 (*Hypostena*).—B.C. **N. comb.**

imitator Curran, 1925c: 154 (*Hypostena*).—B.C. **N. comb.**

Genus OCYTATA Gistel

Roeselia Robineau-Desvoidy, 1830: 145 (preocc. Hübner, 1825). Type-species, *arvensis* Robineau-Desvoidy (Townsend, 1916d: 8)=*pallipes* (Fallén).

Ocytata Gistel, 1848: x (n. name for *Roeselia* Robineau-Desvoidy). Type-species, *Roeselia arvensis* Robineau-Desvoidy (aut.) =*pallipes* (Fallén).
Racodineura Rondani, 1861a: 31 (n. name for *Roeselia* Robineau-Desvoidy). Type-species, *Roeselia arvensis* Robineau-Desvoidy (aut.) =*pallipes* (Fallén).
Rhacodineura Bezzi *in* Bezzi and Stein, 1907: 378, emend.

The only known species of *Ocytata* is a well-known parasite of earwigs.

pallipes (Fallén), 1820g: 22 (*Tachina*).—Sweden; Europe, introd. Oreg., ?estab. **N. comb.**
 antiqua Meigen, 1824: 412 (*Tachina*).—Europe. **N. comb.**

Tribe FRONTININI

The known hosts of the tribe Frontinini are lepidopterous larvae and adult beetles of the genus *Eleodes*.

Genus CLOACINA Reinhard

Cloacina Reinhard, 1945a: 34. Type-species, *filialis* Reinhard (orig. des.).

filialis Reinhard, 1945a: 35.—Tex.; Utah, Ariz., N. Mex.

Genus ELEODIPHAGA Walton

Eleodiphaga Walton, 1918: 23. Type-species, *caffreyi* Walton (orig. des.).

The hosts of *Eleodiphaga* are beetles of the genus *Eleodes*.

caffreyi Walton, 1918: 24.—N. Mex.; Idaho, Ariz.
martini Reinhard, 1937: 69.—Idaho.
pollinosa Walton, 1918: 24.—N. Mex.; Calif.

Genus HYPERTROPHOCERA Townsend

Hypertrophocera Townsend, 1891c: 360. Type-species, *parvipes* Townsend (orig. des.).
Neotractocera Townsend, 1892a: 105. Type-species, *anomala* Townsend (orig. des.) =*parvipes* Townsend.

parvipes Townsend, 1891c: 361.—N. Mex.; Mexico (Chih.).
 anomala Townsend, 1892a: 106 (*Neotractocera*).—N. Mex.

Genus ISTOCHETA Rondani

Istocheta Rondani, 1859: 171. Type-species, *frontosa* Rondani (mon.) =*marmorata* (Fabricius).
Histochaeta, emend.

claripennis Reinhard, 1943c: 15.—Utah; Wash., Wyo.

Genus ORGANOMYIA Townsend

Organomyia Townsend, 1915f: 232. Type-species, *frontalis* Townsend (orig. des.).

frontalis Townsend, 1915f: 232.—N. Mex.; Tex.
    owenii Reinhard, 1930c: 263 (*Macromeigenia*).—Tex. **N. syn.**

Genus PARAPHASMOPHAGA Townsend

Paraphasmophaga Townsend, 1915f: 223. Type-species, *clavis* Townsend (orig. des.).

clavis Townsend, 1915f: 223.—N. Mex.; Calif., Ariz.
dissita Reinhard, 1962b: 221.—Ariz.; Calif.

Tribe GONIINI

The species of the tribe Goniini parasitize a great variety of lepidopterous larvae, especially cutworms and army worms.

We have divided the genera between two subtribes, the Goniina and the Chaetogaediina. Mesnil (1949b) has referred some genera of the latter to the Sturmiini, but we are inclined to believe that they are closer to the typical Goniini.

Subtribe GONIINA

Genus ACROGLOSSA Williston

Acroglossa Williston, 1889a: 1916. Type-species, *hesperidarum* Williston (mon.).
Cnephalomyia Townsend, 1911a: 144. Type-species, *floridana* Townsend (orig. des., as gen. n., sp. n.) ?=*hesperidarum* Williston.
Cnephaliops Townsend, 1915a: 23. Type-species, *Pseudogonia ruficauda* Townsend (orig. des.)=*hesperidarum* Williston.

finitima (Snow), 1895a: 184 (*Cnephalia*).—N. Mex. ?=*hesperidarum*.
floridana (Townsend), 1911a: 144 (*Cnephalomyia*).—Fla. ?=*hesperidarum*.

hesperidarum Williston, 1889a: 1917.—N.H.; Calif. to S. Dak. to N.B., s. to Mexico and Fla.
   obsoleta Townsend, 1892e: 66 (*Pseudogonia*).—N.Y.
   ruficauda Townsend, 1892e: 66 (*Pseudogonia*).—S. Dak.

Genus ARAVAIPA Townsend

Aravaipa Townsend, 1919b: 589. Type-species, *atrophopoda* Townsend (orig. des.).

atrophopoda Townsend, 1919b: 589.—Ariz.

Genus CHAETOCRANIA Townsend

Chaetocrania Townsend, 1915a: 23. Type-species, *Spallanzania antennalis* Coquillett (orig. des.).

antennalis (Coquillett), 1897: 136 (*Spallanzania*).—Calif.; Nev., Utah, Ariz.

Genus DISTICHONA Wulp

Distichona Wulp, 1890: 44. Type-species, *varia* Wulp (mon.).
Pseudogermaria Brauer and Bergenstamm, 1891: 352 (1891: 48). Type-species, *georgiae* Brauer and Bergenstamm (mon.).
Paragermaria Townsend, 1909: 247. Type-species, *autumnalis* Townsend (orig. des.) =*georgiae* (Brauer and Bergenstamm).

georgiae (Brauer and Bergenstamm), 1891: 352 (1891: 48) (*Pseudogermaria*).—Ga.; Ill. to Mass., s. to Ala. and Ga.
   facialis Coquillett, 1902g: 117 (*Gaediopsis*).—Ga. **N. syn., n. comb.**
   autumnalis Townsend, 1909: 247 (*Paragermaria*).—Mass.
   auriceps, authors, not Coquillett.
   varia, authors, in part, not Wulp.

Genus EUCNEPHALIA Townsend

Eucnephalia Townsend, 1892e: 166. Type-species, *gonoides* Townsend (orig. des.).

gonoides Townsend, 1892e: 167.—N. Mex.; Ariz.

Genus GONIA Meigen

Salmacia Meigen, 1800: 38. Type-species, *Musca capitata* De Geer (Coquillett, 1910b: 602). Suppressed by I.C.Z.N., 1936b: 339.

Gonia Meigen, 1803: 280. Type-species, *bimaculata* Wiedemann (present des.). The long-recognized type-species, *Musca capitata* De Geer, designated by Curtis, 1835a: pl. 533, was first referred to *Gonia* in 1826. However, Wiedemann (1819) had described two species in *Gonia*, and these are the first included species in the sense of the Code.

Reaumuria Robineau-Desvoidy, 1830: 79. Type-species, *Musca capitata* De Geer (Robineau-Desvoidy, 1863a: 733; misident.)=*ornata* Meigen.

Cystogonia Townsend, 1915a: 21. Type-species, *Gonia turgida* Coquillett (orig. des.).

Knabia Townsend, 1915c: 286. Type-species, *hirsuta* Townsend (orig. des.)=*frontosa* Say.

Cnephalogonia Townsend, 1916b: 178. Type-species, *Gonia distincta* Smith (orig. des.).

Phosococephalops Townsend, 1927: 237. Type-species, *fulvus* Townsend (orig. des.)=*pallens* Wiedemann.

Setigonia Brooks, 1944a: 221. Type-species, *Gonia setigera* Tothill (orig. des.). **N. syn.**

Fuscigonia Brooks, 1944a: 223. Type-species, *Gonia fuscicollis* Tothill (orig. des.). **N. syn.**

Rhedia Robineau-Desvoidy of Brooks, 1944a: 223.

Numerous genera have been proposed for the species included here in *Gonia*, but we do not regard them as even of subgeneric significance.

REFERENCES: Morrison, 1940 (rev., males); Brooks, 1944a (rev.).

albagenae Morrison, 1940: 349.—B.C.; B.C. to Calif. and Colo.
aldrichi Tothill, 1924a: 209.—Alta.; B.C. to N.B., s. to Ariz. and Va.
atra Cockerell, 1889b: 3 (also 1889c: 106) (as *frontosa* var.; preocc. Meigen, 1826).—Colo. Unrecognized; ?=*frontosa*.
aturgida Brooks, 1944a: 233.—Idaho; Wash. to Sask., s. to Calif. and Ariz.
breviforceps Tothill, 1924a: 210.—Alta.; B.C. to Mont., s. to Calif. and Ariz., also Mich.
brevipulvilli Tothill, 1924a: 211.—N.B.; B.C. to N.B., s. to Calif. and Mass., also D.C.
capitata (De Geer).—Not Nearctic.
carinata Tothill, 1924a: 208.—Utah; Calif., Wyo.
contumax Brooks, 1944a: 233.—Wash.; Calif., Utah.
crassicornis (Fabricius), 1794: 328 (*Musca*).—West Indies; s. Fla., Neotropical.
pallens, authors, not Wiedemann.

distincta (Smith, 1915b: 99.—Mass.; Alta. to N.B., s. to Ohio and Conn., also Colo., N.C.
frontosa Say, 1829: 175 (1859b: 365).—Upper Missouri River; Alaska, B.C. to Que., s. to Calif. and N.C.
    philadelphica Macquart, 1843a: 208 (1843: 51).—Pa.
    albifrons Walker, 1849: 798.—Ont.
    hirsuta Townsend, 1915c: 287 (*Knabia*).—Sask.
    basalis Harris, 1835: 599. Nomen nudum.
    tarda Harris, 1835: 599. Nomen nudum.
fuscicollis Tothill, 1924a: 207.—Ind.; Sask. and Man., s. to Nebr. and Tenn., also Md.
grandipulvilli Morrison, 1940: 345.—Alta.; B.C. to Que., s. to Calif.
longiforceps Tothill, 1924a: 208.—Alta.; B.C. to Ont., s. to Calif. and Okla.
    discalis Morrison, 1940: 347.—B.C.
longipulvilli Tothill, 1924a: 211.—B.C.; B.C. to N. Dak., s. to Calif., Mexico, and Tex., also Hawaii (immigrant).
nigra (Brooks), 1944a: 225 (*Rhedia*).—Sask.; B.C.
occidentalis Brooks, 1944a: 234.—B.C.; B.C. to Mont., s. to Calif. and Ariz.
pilosa Brooks, 1944a: 231.—Man.; Wash. to Man., s. to Calif. and Utah.
porca Williston, 1887a: 10.—Oreg.; B.C. to Man., s. to Calif. and Ariz.
reinhardi Brooks, 1944a: 232.—Colo.; Wash. to Wyo., s. to Calif. and Ariz.
robusta Brooks, 1944a: 234.—Alta.; Calif.
sagax Townsend, 1892e: 65.—Iowa; Colo. and Okla. to Wis. and Conn.
senilis Williston, 1887a: 10.—Kans.; Colo. and Ariz. to Fla., also Mich., N.Y.
sequax Williston, 1887a: 12.—Calif.; B.C. to Mass., s. to Calif., Mexico, Mich., and Conn.
    exul Williston, 1887a: 11.—N.Y., Mass., Conn.
    vertebrata Harris, 1835: 599. Nomen nudum.
setifacies (Brooks), 1944a: 225 (*Rhedia*).—Sask.; Idaho, Man. **N. comb.**
setigera Tothill, 1924a: 199.—Mass.; B.C. to Ont. and Mass., s. to Calif., Ariz., and N.J.
smithi Brooks, 1944a: 235.—Ont.; Man., P.E.I.
texensis Reinhard, 1924c: 357.—Tex.; Miss., Mexico. ?=*crassicornis*.
turgida Coquillett, 1897: 134.—Calif.; B.C., Oreg., Idaho, Mont.

TACHINIDAE 1077

Genus IMAGUNCULA Reinhard

Imaguncula Reinhard, 1958c: 230. Type-species, *tabida* Reinhard (orig. des.).

tabida Reinhard, 1958c: 230.—Ariz.; N. Mex., Mexico.

Genus OLENOCHAETA Townsend

Olenochaeta Townsend, 1892a: 114. Type-species, *kansensis* Townsend (orig. des.).

kansensis Townsend, 1892a: 115.—Kans.; Kans. to Ohio and Md., s. to Calif. and Fla.
  varia, authors, in part, not Wulp.

Genus ONYCHOGONIA Brauer and Bergenstamm

Onychogonia Brauer and Bergenstamm, 1889: 100 (1889: 32). Type-species, *Gonia interrupta* Rondani (mon.).
Goniocnephalia Townsend, 1915f: 222. Type-species, *melanica* Townsend (orig. des.). **N. syn.**

REFERENCE: Brooks, 1944a (key).

fissiforceps (Tothill), 1924a: 207 (*Gonia*).—B.C.; B.C. and Alta., s. to Calif. and Colo.
magna Brooks, 1944a: 227.—Utah; Wash. to Mont., s. to Calif. and Colo.
melanica (Townsend), 1915f: 222 (*Goniocnephalia*).—N. Mex.; Wyo. **N. comb.**
tenuiforceps (Morrison), 1940: 356 (*Gonia*).—Labr.; Alaska, Alta., Idaho.
yukonensis (Tothill), 1924a: 210 (*Gonia*).—Y.T.; Y.T., s. to Oreg. and Colo., also Ont.

Genus SPALLANZANIA Robineau-Desvoidy

Spallanzania Robineau-Desvoidy, 1830: 78. Type-species, *gallica* Robineau-Desvoidy (Coquillett, 1910b: 606) =*hebes* (Fallén).
Cnephalia Rondani, 1856: 62. Type-species, *Tachina hebes* Fallén (orig. des.).

colludens Reinhard, 1958c: 231.—Tex.; Nev., Ariz., S.C.
hebes (Fallén), 1820g: 11 (*Tachina*).—Sweden; B.C. to Mass., s. to Calif., Mexico, and Fla., Palearctic.
  bucephala Meigen, 1824: 252 (*Tachina*).—Austria.
  pansa Snow, 1895a: 182 (*Cnephalia*).—N. Mex.

Subtribe CHAETOGAEDIINA, N. subtribe

Certain genera referred to the Sturmiini by Mesnil appear to us to be more closely related to the typical Goniini. However, they differ notably from the Goniina in having the ocellar bristles proclinate, and they are segregated here as a distinct group.

Genus CHAETOGAEDIA Brauer and Bergenstamm

Chaetogaedia Brauer and Bergenstamm, 1891: 336 (1891: 32). Type-species, *Prospherysa vilis* Wulp (Townsend, 1908: 94).
Phrissopolia Townsend, 1908: 93. Type-species, *Prospherysa crebra* Wulp (mon.).
Eophrissopolia Townsend, 1926: 36. Type-species, *acroglossoides* Townsend (orig. des.) = *townsendi* Sabrosky and Arnaud.
Frontinogaedia Townsend, 1926: 37. Type-species, *Baumhaueria analis* Wulp (orig. des.).

The name *crebra* has been widely misapplied in the past to most of the species here recognized as distinct.

analis (Wulp), 1867: 148 (*Baumhaueria*).=Wis.; Wis. to Vt., s. to Kans. and Va., also ?N. Mex., ?Tex., ?Fla.
    acroglossoides Townsend, 1891c: 367 (*Frontina*).—Ill.
crebra (Wulp), 1890: 120 (*Prospherysa*).—Mexico; Calif., Ariz., N. Mex.
desertorum (Townsend), 1908: 94 (*Phrissopolia*).—N. Mex.; Ariz.
monticola (Bigot), 1887b: cxl (also 1888a: 91) (*Blepharipeza*).—Rocky Mts.; Calif. to Tex., introd. and estab. Hawaii.
rufifrons (Wulp), 1890: 121 (*Prospherysa*).—Mexico; N. Mex.
townsendi Sabrosky and Arnaud (for *acroglossoides* Townsend).— Md.; N. Dak., Ohio, N.C. **N. name.**
    acroglossoides Townsend, 1926: 37 (*Eophrissopolia*; preocc. Townsend, 1891).—Md.

Genus CHAETOGLOSSA Townsend

Chaetoglossa Townsend, 1892a: 125. Type-species, *picticornis* Townsend (orig. des.).

nigripalpis Townsend, 1892a: 126.—Fla. ?=*picticornis*.
picticornis Townsend, 1892a: 126.—Fla.; S. Dak., Kans., Ill., Ariz., Tex., Mexico.
    violae Townsend, 1892a: 126.—Fla.

Genus GAEDIOPHANA Brauer and Bergenstamm

Gaediophana Brauer and Bergenstamm, 1893: 123, 201 (1893: 35, 113). Type-species, *atra* Brauer and Bergenstamm (mon.) =*lugubris* (Wulp).
Chaetogaediopsis Townsend, 1916h: 620. Type-species, *Gaediopsis cockerellii* Coquillett (orig. des.) =*lugubris* (Wulp).

The genus *Gaediophana* was long considered to consist of one species, as shown below, but Reinhard (1951: 5) has described a second species from Mexico, *G. monnula*, recognized chiefly by distinctive male genitalia. It is different from *cockerellii*, of which the type is before us, but the relationship of these two to earlier names may now be questioned, pending restudy of the types of the latter.

lugubris (Wulp), 1890: 53 (*Mystacella*).—Mexico; Ariz. to Tex.
    tenebricosa Wulp, 1890: 77 (*Phorocera*).—Mexico. **N. syn., n. comb.**, J. M. Aldrich notes.
    atra Brauer and Bergenstamm, 1893: 123, 201 (1893: 35, 113).—Mexico.
    cockerellii Coquillett, 1902g: 117 (*Gaediopsis*).—N. Mex.

Genus GAEDIOPSIS Brauer and Bergenstamm

Gaediopsis Brauer and Bergenstamm, 1891: 336, 401 (1891: 32, 97) (also 1893: 190; 1893: 102). Type-species, *mexicana* Brauer and Bergenstamm (orig. des., as gen. n., sp. n.).
Poliophrys Townsend, 1908: 90. Type-species, *sierricola* Townsend (orig. des.).

Subgenus GAEDIOPSIS Brauer and Bergenstamm

mexicana Brauer and Bergenstamm, 1891: 336 (1891: 32).—Mexico; Ariz., Tex.
organensis (Townsend), 1908: 93 (*Poliophrys*).—N. Mex.
sierricola (Townsend), 1908: 93 (*Poliophrys*).—Mexico (Chih.); Calif. to Tex.
    serricolor, error.

Subgenus EUGAEDIOPSIS Townsend

Eugaediopsis Townsend, 1916h: 620 (as genus). Type-species, *Gaediopsis ocellaris* Coquillett (orig. des.).
Eugaedia Townsend, 1916h: 621. Type-species, *Gaediopsis setosa* Coquillett (orig. des.).

flavipes Coquillett, 1895k: 100.—Ala.; Ark. to S.C., s. to Fla., also Va.
ocellaris Coquillett, 1902g: 118.—Ohio; Mich. to N.H., s. to Ill. and Va.
rubentis (Reinhard), 1961: 209 (*Eugaediopsis*).—Ariz.; Mexico. **N. comb.**
setosa Coquillett, 1897: 136.—Calif.; B.C. to Calif., Mexico, and N. Mex.
> *monticola* Townsend, 1898c: 269.—N. Mex. **N. syn.**

vinnula (Reinhard), 1961: 210 (*Eugaediopsis*).—Ariz. **N. comb.**

Genus HESPEROMYIA Brauer and Bergenstamm

Hesperomyia Brauer and Bergenstamm, 1889: 114 (1889: 46). Type-species, *erythrocera* Brauer and Bergenstamm (orig. des., as gen. n., sp. n.).
Oestroplagia Townsend, 1919b: 566. Type-species, *petiolata* Townsend (orig. des.).

erythrocera Brauer and Bergenstamm, 1889: 114 (1889: 46).—Tex.; N. Mex., Okla.
petiolata (Townsend), 1919b: 567 (*Oestroplagia*).—Ariz.

Genus TOROSOMYIA Reinhard

Torosomyia Reinhard, 1935b: 171. Type-species, *parallela* Reinhard (orig. des.).

parallela Reinhard, 1935b: 172.—Kans.

Tribe BELVOSIINI

The generally bulky flies of the Belvosiini parasitize large heterocerous larvae.

Genus ATACTA Schiner

Atacta Schiner, 1868: 328. Type-species, *brasiliensis* Schiner (orig. des.).
Atactomima Townsend, 1916j: 15. Type-species, *crescentis* Townsend (orig. des.). **N. syn.**

> REFERENCE: Aldrich, 1925c (rev.).

brasiliensis Schiner, 1868: 328.—Brazil; Tex. to Fla., thence n. to Va., Neotropical.
> *apicalis* Coquillett, 1897: 83.—Ga.

crassiceps Aldrich, 1925c: 30.—Va.; Okla. and Tex. to Ga., also Md., Va.

Genus ATACTOPSIS Townsend

Atactopsis Townsend, 1917c: 229. Type-species, *facialis* Townsend (orig. des.).
Paratacta Reinhard, 1923: 266. Type-species, *facialis* Reinhard (orig. des.)=*reinhardi* Sabrosky and Arnaud.

reinhardi Sabrosky and Arnaud (for *facialis* Reinhard).—Tex. **N. name.**
    facialis Reinhard, 1923: 266 (*Paratacta*; preocc. Townsend, 1917).—Tex.

Genus BELVOSIA Robineau-Desvoidy

Belvosia Robineau-Desvoidy, 1830: 103. Type-species, *bicincta* Robineau-Desvoidy (mon.).
Belvoisia, error or emend.
Latreillia Robineau-Desvoidy, 1830: 104 (preocc. Roux, 1830). Type-species, *Musca bifasciata* Fabricius (Coquillett, 1910b: 558).
Willistonia Brauer and Bergenstamm, 1889: 97 (1889: 29). Type-species, *Musca esuriens* Fabricius (mon.; misident.)= *aldrichi* (Townsend).
Latreillimyia Townsend, 1908: 104 (n. name for *Latreillia* Robineau-Desvoidy). Type-species, *Musca bifasciata* Fabricius (aut.).
Belvosiopsis Townsend, 1927: 248. Type-species, *brasiliensis* Townsend (orig. des.)=*weyenberghiana* Wulp.

    REFERENCE: Aldrich, 1928 (rev.).

argentifrons Aldrich, 1928: 32.—Va.; N.Y., Ga.
auratilis Reinhard, 1951: 1.—Mexico (Michoacan); Tex.
bicincta Robineau-Desvoidy, 1830: 103.—Carolina, Antilles; Kans., Carolina, Ariz. to Tex., s. through Cent. Amer. and West Indies to Brazil.
bifasciata (Fabricius), 1775: 777 (*Musca*).—West Indies (as Amer.); Calif. to Mass., s. to Mexico and West Indies.
    bifuscata, error.
borealis Aldrich, 1928: 28.—Pa.; Mich. to Vt., s. to N. Mex. and Fla. also Calif.
    orion Brimley, 1928: 205.—N. C. **N. syn.**
canadensis Curran, 1927m: 152.—Sask.; Alta. to Mich. and Ohio, s. to Colo. and Ark.
ciliata Aldrich, 1928: 22.—Mexico; Mo., Panama, Brazil.
pollinosa Rowe, 1933: 123.—Ill.

semiflava Aldrich, 1928: 11.—N. Mex.; Calif. to Ohio, s. to Mexico and Ga.
splendens Curran, 1927m: 153.—Sask.; Sask. to Mich., s. to Kans.
townsendi Aldrich, 1928: 33.—Va.; Kans. to N.J., s. to Tex. and Fla.

Genus TRIACHORA Townsend

Triachora Townsend, 1908: 105. Type-species, *Latreillia unifasciata* Robineau-Desvoidy (mon.).
Goniomima Townsend, 1908: 105. Type-species, *Belvosia luteola* Coquillett (mon.). **N. syn.**

REFERENCE: Aldrich, 1928 (rev., in *Belvosia*).

luteola (Coquillett), 1900g: 253 (*Belvosia*).—P.R.; Tex., Fla. **N. comb.**
omissa (Aldrich), 1928: 21 (*Belvosia*).—Va.; Ohio, Md., D.C., Ark., Ga. **N. comb.**
slossonae (Coquillett), 1895a: 312 (*Belvosia*).—Fla.; La.
unifasciata (Robineau-Desvoidy), 1830: 105 (*Latreillia*).—Pa.; Wis. to Maine, s. to Tex. and Ga., also Cuba.
 flavicauda Riley, 1870c: 51 (*Exorista*).—Mo.

Tribe HARRISIINI

The species of Harrisiini parasitize lepidopterous larvae.
REFERENCE: Brooks, 1947 (rev.).

Genus LESCHENAULTIA Robineau-Desvoidy

Leschenaultia Robineau-Desvoidy, 1830: 324. Type-species, *cilipes* Robineau-Desvoidy (Townsend, 1916d: 7) = *leucophrys* (Wiedemann).
Blepharipeza Macquart, 1843a: 211 (1843: 54). Type-species, *rufipalpis* Macquart (mon.) = *leucophrys* (Wiedemann).
Rileya Brauer and Bergenstamm, 1893: 121 (1893: 33) (preocc. Ashmead, 1888). Type-species, *americana* Brauer and Bergenstamm (orig. des., as gen. n., sp. n.).
Rileymyia Townsend, 1893a: 277 (n. name for *Rileya* Brauer and Bergenstamm). Type-species, *Rileya americana* Brauer and Bergenstamm (aut.).

adusta (Loew), 1872a: 92 (Cent. 10, no. 67) (*Blepharipeza*).—Calif.; Ariz., N. Mex.
americana (Brauer and Bergenstamm), 1893: 121 (1893: 33) (*Rileya*).—Calif.; B.C. to Mont., s. to Calif.
 triseta Brooks, 1947: 181 (*Rileymyia*).—B.C.

exul (Townsend), 1892e: 64 (*Blepharipeza*).—N.H.; Ont. to N.B., s. to Ohio and Va.
    incauta Harris, 1835: 599 (*Tachina*). Nomen nudum.
fulvipes (Bigot), 1887b: cxl (*Blepharipeza*).—Wash.; B.C. to Sask., s. to Ariz. and Tex., also ?Md.
    rufescens Townsend, 1892a: 90 (*Blepharipeza*).—?Md.
grossa Brooks, 1947: 176.—Ariz.; Nev., N. Mex.
halisidotae Brooks, 1947: 176.—Ont.; Wis. to Que., s. to Ohio and N.J.
hospita Reinhard, 1952b: 7.—Mexico (Michoacan); N. Mex.
leucophrys (Wiedemann), 1830: 308 (*Tachina*).—Brazil; B.C. to Ont., all U.S., Neotropical.
    rufipalpis Macquart, 1843a: 212 (1843: 55) (*Blepharipeza*).—Mexico, Cuba.

Genus PARACHAETA Coquillett

Parachaeta Coquillett, 1897: 123. Type-species, *Blepharipeza bicolor* Macquart (orig. des.; misident.) = *fusca* Townsend.

bicolor (Macquart), 1846: 286 (1846: 158) (*Blepharipeza*).—Tex.; Mexico.
fusca Townsend, 1916d: 11.—N.Y.; Kans. to Maine and Md.
    bicolor, authors, in part, not Macquart.
    inermis, authors, not Bigot.

Tribe STURMIINI

The tribe Sturmiini as here recognized is basically that of Townsend, minus *Winthemia* and relatives (see next tribe). The hosts are lepidopterous caterpillars, with a few records from tenthredinid larvae.

Genus BLEPHARIPA Rondani

Blepharipa Rondani, 1856: 71. Type-species, *Senometopia ciliata* Macquart (orig. des.) = *scutellata* (Robineau-Desvoidy).
Ugimyia Rondani, 1870: 137. Type-species, *sericariae* Rondani (mon.) ? = *zebina* (Walker). ?Syn.
Blepharipoda Brauer and Bergenstamm, 1889: 96 (1889: 28) (preocc. Randall, 1840). Type-species, *Nemoraea scutellata* Robineau-Desvoidy (mon.).
Crossocosmia Mik, 1890: 313. Type-species, *Ugimyia sericariae* Rondani (as Cornalia; orig. des.) ? = *zebina* (Walker). ?Syn.

The status of *Ugimyia* and *Crossocosmia* is complicated by the questionable identity of Rondani's *sericariae*, which was based on larvae and pupae that might conceivably have belonged to a different genus than *Blepharipa*. The adult of *"sericariae,"* described later in 1870 by Cornalia, and accepted by some as probably conspecific with Rondani's species, is a synonym of *B. zebina* (Walker).

Blepharipa scutellata is one of the most successful parasites of the gypsy moth.

schineri (Mesnil), 1939: 32 (*Blepharipoda*).—Austria; Europe, China, Japan, introd. New England, not estab.

> *flavoscutellata* Zetterstedt of Schiner, 1862a: 482.
> *sericariae*, authors, not Rondani or Cornalia.

scutellata (Robineau-Desvoidy), 1830: 73 (*Nemoraea*).—Europe; introd. and estab. New England, N.Y.

Genus BOLOMYIA Brauer and Bergenstamm

Bolomyia Brauer and Bergenstamm, 1891: 347 (1891: 43). Type-species, *Mystacella violacea* Wulp (mon.) = *rufata* (Bigot).

rufata (Bigot), 1888a: 257 (*Exorista*).—Mexico; Ariz., Cent. Amer.

Genus DRINO Robineau-Desvoidy

Drino Robineau-Desvoidy, 1863a: 250. Type-species, *volucris* Robineau-Desvoidy (orig. des.) = *lota* (Meigen).

Zygosturmia Townsend, 1911a: 142. Type-species, *inca* Townsend (orig. des., as gen. n., sp. n.).

Laximasicera Curran, 1927d: 14. Type-species, *sexualis* Curran (orig. des.) = *bakeri* (Coquillett). **N. syn.**, L. P. Mesnil *in litt*.

Cubaemyia Townsend, 1931a: 179. Type-species, *Tachina cubaecola* Jaennicke (orig. des.). **N. syn.**

Sturmia, authors, not Robineau-Desvoidy.

antennalis (Reinhard), 1922: 330 (*Pseudochaeta*).—Tex.; Ariz. to S.C. **N. comb.**

bakeri (Coquillett), 1897: 112 (*Sturmia*).—Colo.; B.C. to Sask., s. to Colo., also Mich., N.B. **N. comb.**

> *sexualis* Curran, 1927d: 14 (*Laximasicera*).—Alta. **N. syn., n. comb.**

bohemica Mesnil, 1949a: 23.—Czechoslovakia; Europe, introd. and estab. Ont. to Nfld. and N.S., also introd. but not estab. B.C., Sask., Man.

> *inconspicua*, authors, in part, not Meigen.

crescentis (Reinhard), 1944b: 69 (*Sturmia*).—Tex.; Colo. and Kans. to Mexico, also N.Y., Ohio, N.C. **N. comb.**
cubaecola (Jaennicke), 1867: 382 (1868: 74) (*Tachina*).—Cuba; s. Tex., P.R. **N. comb.**
gilva (Hartig), 1838: 288 (*Tachina*).—Germany; Palaearctic.
    var. *aurora* Mesnil, 1949a: 15 (as sp.).—Japan; introd. N.B., not estab.
inca (Townsend), 1911a: 142 (also 1912b: 323) (*Zygosturmia*).—Peru; s. Calif. to Fla., also Va., N.C.
incompta (Wulp), 1890: 99 (*Brachycoma*).—Mexico (Guerrero); Calif. to Maine, s. to Mexico and Fla.
    inquinata Wulp, 1890: 107 (*Masicera*).—Mexico (Morelos). **N. comb.**
inconspicua (Meigen), 1830: 369 (*Tachina*).—Germany; Europe, introd. Ont. to P.E.I. and New England. Introduced as *inconspicua*, but now considered probably a complex of up to 5 species, only *bohemica* being recovered.
    gilva, authors, not Hartig.
pilatei (Coquillett), 1897: 111 (*Sturmia*).—Ga.; Calif. to Ga., also Tenn., N.C. **N. comb.**
rhoeo (Walker), 1849: 778 (*Tachina*).—Unknown; N. Dak. to N.Y., s. to Calif., Mexico, and Fla., also Jamaica. **N. comb.**
    protoparcis Townsend, 1892i: 70 (*Masicera*).—Jamaica. **N. syn.**
    distincta, authors, not Wiedemann.

Genus GYMNOERYCIA Townsend

Gymnoerycia Townsend, 1916i: 312. Type-species, *rubra* Townsend (orig. des.).

rubra Townsend, 1916i: 313.—Fla.

Genus MASIPHYOMYIA Reinhard

Masiphyomyia Reinhard, 1944b: 65. Type-species, *alearis* Reinhard (orig. des.).

    REFERENCE: Reinhard, 1944b (rev.).

alearis Reinhard, 1944b: 66.—Tex.
comosa Reinhard, 1944b: 67.—Colo.; B.C. and Alta. to Calif. and Colo.
longicornis Reinhard, 1944b: 67.—Tex.; Oreg., Calif., Ariz.
paralis Reinhard, 1944b: 67.—N. Mex.; Calif., Ariz.

Genus MICROSILLUS Aldrich

Microsillus Aldrich, 1926d: 20. Type-species, *Houghia baccharis* Reinhard (orig. des.).

baccharis (Reinhard), 1922: 332 (*Houghia*).—Tex.

Genus MILONIUS Reinhard

Milonius Reinhard, 1955d: 127. Type-species, *scordalus* Reinhard (orig. des.).

scordalus Reinhard, 1955d: 127.—Tex.; Mexico.

Genus MIMOLOGUS Reinhard

Mimologus Reinhard, 1955d: 129. Type-species, *effetus* Reinhard (orig. des.).

effetus Reinhard, 1955d: 130.—Calif.

Genus MYOTHYRIOPSIS Townsend

Myothyriopsis Townsend, 1919b: 575. Type-species, *bivittata* Townsend (orig. des.) ?=*picta* (Wulp).

picta (Wulp), 1890: 108 (*Masicera*).—Mexico; Tex.

Genus MYSTACELLA Wulp

Mystacella Wulp, 1890: 51. Type-species, *solita* Wulp (Coquillett 1910b: 573).
Macromeigenia Brauer and Bergenstamm, 1891: 311 (1891: 7). Type-species, *Tachina chrysoprocta* Wiedemann (mon.). **N. syn.**

REFERENCE: Reinhard, 1930c (rev.).

chrysoprocta (Wiedemann), 1830: 309 (*Tachina*).—Unknown; Mich. to Mass., s. to Tex. and Ga., also Mexico to Costa Rica. **N. comb.**
  interrupta Walker, 1852: 295 (*Tachina*).—Ga. **N. syn.**
frioensis (Reinhard), 1922: 329 (*Ernestia*).—Tex.; B.C., Idaho, Ariz., N. Mex. **N. comb.**

Genus Pales Robineau-Desvoidy

Pales Robineau-Desvoidy, 1830: 154. Type-species, *florea* Robineau-Desvoidy (Coquillett, 1910b: 582) = *pavida* (Meigen). Not *Pales* Meigen, 1800, suppressed by I.C.Z.N., 1963b.

Ctenophorocera Brauer and Bergenstamm, 1891: 342 (1891: 38). Type-species, *experta* Brauer and Bergenstamm (Townsend, 1916d: 6) = *sarcophagaeformis* (Jaennicke).

Neopales Coquillett, 1910b: 575 (n. name for *Pales* Robineau-Desvoidy). Type-species, *florea* Robineau-Desvoidy (aut.) = *pavida* (Meigen).

pavida (Meigen), 1824: 398 (*Tachina*).—Germany; introd. New England, not estab.

Genus SIPHOSTURMIA Coquillett

Siphosturmia Coquillett, 1897: 83. Type-species, *Argyrophylax rostrata* Coquillett (orig. des.).

Siphosturmiopsis Townsend, 1915k: 91. Type-species, *rafaeli* Townsend (orig. des.).

Mesnil (1949a, b, 1950) considers this group to be a subgenus of *Drino*.

REFERENCES: Reinhard, 1931b (rev., partial key), 1934c (partial key).

confusa Reinhard, 1931b: 6.—Tex.; Calif. to Tex.
  normula authors, not Wulp.
maltana Reinhard, 1951: 7.—Mont.; Calif., N. Mex.
melampyga (Reinhard), 1931b: 9 (*Siphosturmiopsis*).—Tex.; Wash., N. Mex., Ill. to Vt., Fla. **N. comb.**
melitaeae (Coquillett), 1897: 121 (*Demoticus*).—Calif. **N. comb.**
  rufiventris Reinhard, 1934c: 18 (*Siphosturmiopsis*).—Calif. **N. syn., n. comb.**
oteroensis (Reinhard), 1934c: 16 (*Siphosturmiopsis*).—N. Mex. **N. comb.**
phyciodis (Coquillett), 1897: 109 (*Sturmia*).—Md., Mass.; Kans. to Mass., s. to Tex. and N.C. **N. comb.**
rostrata (Coquillett), 1895k: 106 (*Argyrophylax*).—Ala.; Ohio to Md., s. to Tex. and Fla.

Genus STURMIA Robineau-Desvoidy

Sturmia Robineau-Desvoidy, 1830: 171. Type-species, *vanessae* Robineau-Desvoidy (Robineau-Desvoidy, 1863a: 888) = *bella* (Meigen).

Although our species are atypical *Sturmia*, they are left here until their proper generic position can be established.

chrysoprocta Reinhard, 1924b: 273.—Kans.; Colo., N. Mex., Tex.

harrisinae Coquillett, 1897: 111.—U.S.; Ohio, N.Y., Calif. to Tex., Fla., Mexico.
  tuxedo Curran, 1930j: 102 (*Erycia*).—N.Y.
  unispinosa Reinhard, 1930b: 199 (*Masicera*).—Tex.

Genus TOWNSENDIELLOMYIA Baranov

Sturmia, subg. **Townsendiellomyia** Baranov, 1932: 73. Type-species *Zygobothria nidicola* Townsend (orig. des.).

nidicola (Townsend), 1908: 99 (*Zygobothria*).—Austria; Europe, introd. and estab. New England.

Genus ZIZYPHOMYIA Townsend

Zizyphomyia Townsend, 1916i: 317. Type-species, *celer* Townsend (orig. des.) =*limata* (Coquillett). ?=*Drino*.

limata (Coquillett), 1902g: 113 (*Sturmia*).—La.; Kans., Ohio, Tex. to Miss.
  celer Townsend, 1916i: 318.—Tex.

Tribe WINTHEMIINI

Genus HEMISTURMIA Townsend

Hemisturmia Townsend, 1927: 262, 316 (as *Humisturmia*, p. 262). Type-species, *carcelioides* Townsend (orig. des.).
Ceratochaeta Brauer and Bergenstamm of Coppel, 1960: 97.

tortricis (Coquillett), 1897: 103 (*Phorocera*).—Mo.; B.C. to Vt., s. to Calif., Mexico, Miss., and N.J. **N. comb.**

Genus NEMORILLA Rondani

Nemorilla Rondani, 1856: 66. Type-species, *Tachina maculosa* Meigen (orig. des.).

The principal hosts are Microlepidoptera.

  REFERENCE: Aldrich and Webber, 1924a (rev.).

floralis (Fallén), 1810b: 287 (*Tachina*).—Sweden; Palaearctic, introd. Ont., Ohio, Mass., Conn. ?estab.
insolens Aldrich and Webber, 1924a: 6.—Mass.; Ohio, Conn., R.I., N.C.
maculosa (Meigen), 1824: 265 (*Tachina*).—France, Germany; Palaearctic, introd. Ont., ?estab.
parva (Coquillett), 1897: 100 (*Exorista*).—Colo.

pyste (Walker), 1849: 754 (*Tachina*).—N.S.; B.C. to N.S., all U.S., Mexico.
   phycitae LeBaron, 1872: 123 (*Tachina*).—Ill.
   scudderi Williston, 1889a: 1921 (*Exorista*).—Tex.
   arvicola Meigen of Parrott and Schoene, 1912: 30 (*Exorista*).
   floralis, N. Amer. authors, not Fallén.
   maculosa, N. Amer. authors, not Meigen.

Genus OMOTOMA Lioy

Omotoma Lioy, 1864a: 1338. Type-species, *Tachina amoena* Meigen (Townsend, 1916d: 8).
Homotoma, emend. (preocc. Guérin-Méneville, 1844).
Nemosturmia Townsend, 1926: 34. Type-species, *pilosa* Townsend (orig. des.) =*fumiferanae* (Tothill).
fumiferanae (Tothill), 1912: 2 (*Winthemia*).—B.C.; B.C. to Nfld., s. to Ariz. and Ga.
   pilosa Townsend, 1926: 35 (*Nemosturmia*).—N.H.
   amoena, authors, not Meigen.

Genus ORASTURMIA Reinhard

Orasturmia Reinhard, 1947a: 21. Type-species, *vallicola* Reinhard (orig. des.).
vallicola Reinhard, 1947a: 22.—Tex.

Genus PSEUDOLOMYIA Reinhard

Pseudolomyia Reinhard, 1962a: 175. Type-species, *scissilis* Reinhard (orig. des.).
scissilis Reinhard, 1962a: 176.—Ohio; Ariz., Tex.

Genus WINTHEMIA Robineau-Desvoidy

Winthemia Robineau-Desvoidy, 1830: 173. Type-species, *Tachina variegata* Meigen (as Fabricius; Robineau-Desvoidy, 1863a: 207).
Trisisyropa Townsend, 1916e: 28. Type-species, *vesiculata* Townsend (orig. des.).
Okea Townsend, 1916g: 74. Type-species, *Winthemia okefenokeensis* Smith (orig. des.).
Neowinthemia Townsend, 1919b: 583. Type-species, *abdominalis* Townsend (orig. des.).

Winthemia quadripustulata and *W. rufopicta*, here listed in the usual sense, may be either extremely variable or complexes of easily confused species. Mesnil (1949b: 82) has expressed the opinion that *quadripustulata* of Reinhard (1931c) is actually *variegata* Meigen, but we have seen no North American specimens that agree with the latter and many that agree with European *quadripustulata*.

REFERENCE: Reinhard, 1931c (rev.).

abdominalis (Townsend), 1919b: 583 (*Neowinthemia*).—Va.; Ohio, Pa., Conn., Ga.
antennalis Coquillett, 1902g: 115.—Calif.; N. Mex.
borealis Reinhard, 1931c: 27.—Man.; B.C., Oreg., Man. to N.S.
cecropia (Riley), 1870a: 101 (*Exorista*).—Ill.; Kans., Mich., Ont., N.Y., Conn.
    platysamiae Townsend, 1892d: 288 (*Exorista*).—N.Y.
    cecropiae Reinhard, 1931c: 34.—Ill.
citheroniae Sabrosky, 1948c: 65.—Fla.; Ill., Md.
datanae (Townsend), 1892d: 288 (*Exorista*).—N.Y.; B.C. to Ont., s. to Oreg., Ill., and Fla., also N. Mex., Mexico.
deilephilae (Osten Sacken), 1887b: 164 (*Tachina*).—Unknown (presumed Mo.); Oreg. to Mich., s. to Calif., Mexico, and Fla.
duplicata Reinhard, 1931c: 17.—N. Mex.
imitator Reinhard, 1931c: 39.—Tex.
infesta (Williston) *in* Forbes, 1885: 65 (*Exorista*).—Ill., Pa. Unrecognized, ?=*quadripustulata*.
intermedia Reinhard, 1931c: 41.—Tex.; Utah to Mass., s. to Tex. and Fla., also Mexico to Brazil.
intonsa Reinhard, 1931c: 28.—B.C.
leucanae (Kirkpatrick), 1861: 358 (*Exorista*).—Ohio. Unrecognized.
    leucaniae, error.
militaris (Walsh), 1861: 367 (*Senometopia*).—Ill. Unrecognized.
montana Reinhard, 1931c: 36.—Ariz.; N.Y., Mexico.
occidentis Reinhard, 1931c: 22.—B.C.; Wash., Mont., Man., Ariz., N. Mex., Mexico, Mich., also introd. Ont. and Nfld., ?estab.
okefenokeensis Smith, 1916: 95.—Ga.; Fla., Cuba.
ostensackenii (Kirkpatrick), 1861: 358 (*Exorista*).—Ohio. Unrecognized.
polita Reinhard, 1931c: 21.—Mass.; Ohio.
quadripustulata (Fabricius), 1794: 324 (*Musca*).—Germany; Wash. to Vt., s. to Calif. and Fla.
    illinoensis Robertson, 1901: 286.—Ill.
rufopicta (Bigot), 1888a: 259 (*Chetolyga*).—Rocky Mts.; Man. to Vt. and Mass., s. to Calif. and Fla., also B.C.

rufonotata Bigot, 1888a: 257 (*Chetolyga*).—Rocky Mts. **N. syn.**
    ciliata Townsend, 1891c: 363 (*Exorista*).—Ill.
sinuata Reinhard, 1931c: 25.—Md.; N. Dak. to Maine, s. to Mexico and Ga.
texana Reinhard, 1931c: 19.—Tex.
vesiculata (Townsend), 1916e: 28 (*Trisisyropa*).—Maine.

Tribe MASIPHYINI

Genus IGNOTOMYIA Reinhard

Ignotomyia Reinhard, 1961: 210. Type-species, *cunina* Reinhard (orig. des.).
cunina Reinhard, 1961: 211.—Ariz.

Genus PHASIOPSIS Townsend

Phasiopsis Townsend, 1912a: 108. Type-species, *floridana* Townsend (orig. des.).
floridana Townsend, 1912a: 108.—Fla.

Genus PROMASIPHYA Townsend

Promasiphya Townsend, 1927: 379. Type-species, *Masiphya confusa* Aldrich (orig. des.).
confusa (Aldrich), 1925b: 109 (*Masiphya*).—Fla.; Ariz. to Md., s. to Mexico and Fla.
    brasiliana Brauer and Bergenstamm of Coquillett, 1897: 82.
    var. **irrisor** Reinhard, 1962b: 222.—Calif.; Nev., Ariz., Tex., Ill.
townsendi (Aldrich), 1925b: 110 (*Masiphya*; as *confusa* var.)—Ariz.; Tex., Mexico.

Tribe CARCELIINI

Most species of the tribe Carceliini parasitize lepidopterous larvae; a few attack sawflies.

Genus ANGUSTIOPSIS Reinhard

Angustiopsis Reinhard, 1959b: 231. Type-species, *saginata* Reinhard (orig. des.).
saginata Reinhard, 1959b: 231.—Tex.

Genus CARCELIA Robineau-Desvoidy

Carcelia Robineau-Desvoidy, 1830: 176. Type-species, *bombylans* Robineau-Desvoidy (Townsend, 1916d: 6).

Parexorista Brauer and Bergenstamm, 1889: 87, 161 (as *Paraexorista*, p. 87) (1889: 19, 93). Type-species, *Exorista cheloniae* Rondani (mon.) = *lucorum* (Meigen).

REFERENCE: Sellers, 1943 (rev.).

amplexa (Coquillett), 1897: 98 (*Exorista*).—N.H.; Minn. to Que. and Maine, s. to Kans. and Fla., also Ariz., P.R.

hemerocampae Townsend, 1909: 248 (*Sisyropa*).—Mass.

griseomicans, authors, not Wulp.

diacrisiae Sellers, 1943: 61.—Ind.; Ind. to Conn., s. to Ala. and Fla.

formosa (Aldrich and Webber), 1924a: 23 (*Zenillia*).—Mass.; Mich. to Mass., s. to Mo. and Fla., also Ariz., N. Mex.

gnava (Meigen), 1824: 330 (*Tachina*).—Europe; introd. Wash., Maine, N.H., Mass., Conn., not estab.

bombylans, authors, not Robineau-Desvoidy.

inflatipalpis (Aldrich and Webber), 1924a: 24 (*Zenillia*).—Va.; B.C., Ind., N.Y., Tex., Fla.

lagoae (Townsend), 1891d: 159 (*Exorista*).—Mexico; Ohio, D.C. to Fla., also Tex., Miss., Mexico, Jamaica.

flavirostris, authors, not Wulp.

laxifrons Villeneuve, 1912: 90.—Europe; introd. and estab. New England. Host: *Nygmia phaeorrhoea* (Donovan).

cheloniae, authors, in part, not Rondani.

malacosomae Sellers, 1943: 50.—Mass.; B.C. to N.B. and N.J., also Calif., Colo., Tex., Tenn., Ga.

cheloniae, authors, in part, not Rondani.

olenensis Sellers, 1943: 67.—Mass.; B.C., Que.

perplexa Sellers, 1943: 65.—N.J.; N.Y., Maine, Mass., Conn.

protuberans (Aldrich and Webber), 1924a: 15 (*Zenillia*).—Vt.

reclinata (Aldrich and Webber), 1924a: 32 (*Zenillia*).—N. Mex.; B.C., Ont., all U.S.

separata (Rondani), 1859: 134 (*Exorista*; as sp.).—Italy; Palaearctic, introd. ne. U.S., not estab. = *excisa* (Fallén) var.

tibialis (Reinhard), 1931c: 11 (*Winthemia*).—France (as Mass., error). = *processioneae* (Ratzeburg). Not Nearctic.

yalensis Sellers, 1943: 69.—Idaho; B.C., Calif., Ont. to Maine and Mass.

Genus EUSISYROPA Townsend

Eusisyropa Townsend, 1908: 97. Type-species, *Tachina blanda* Osten Sacken (mon.).

REFERENCES: Sellers, 1943 (key, in *Zenillia*); Thompson, 1953a (biol., immature stages).

blanda (Osten Sacken), 1887b: 162 (*Tachina*).—Mass. (as Unknown); B.C., Wash., Ont. to P.E.I., s. to Ariz. and Fla.
    proserpina Williston, 1889a: 1919 (*Exorista*; as var.).— Unknown.
boarmiae (Coquillett), 1897: 95 (*Exorista*).—Miss. (Townsend, 1908: 98; Sellers, 1943: 22); Ark. to Md., s. to Tex. and Fla.
    hypenae Coquillett *in* Howard, 1897: 47 (*Exorista*). Nomen nudum.
collina (Reinhard), 1944b: 68 (*Zenillia*).—Ariz.; Calif., Okla. **N. comb.**
virilis (Aldrich and Webber), 1924a: 40 (*Zenillia*; as *blanda* ssp.).—Ill. (as N.Y., error); B.C., Wash., Calif. to Maine, s. to Mexico and Ga.

Genus GYMNOCARCELIA Townsend

Gymnocarcelia Townsend, 1919b: 582. Type-species, *ricinorum* Townsend (orig. des.).

ricinorum Townsend, 1919b: 582.—Fla.; Calif. to Ont. and N.H., s. to Mexico and Fla.
    albifrons Walker, 1852: 283 (*Tachina*; preocc. Walker, 1836).—U.S.
    noctuae Harris, 1835: 599 (*Tachina*). Nomen nudum.

Genus HUBNERIA Robineau-Desvoidy

Hubneria Robineau-Desvoidy, 1847: 601. Type-species, *Carcelia nigripes* Robineau-Desvoidy (Robineau-Desvoidy, 1863a: 279) = *affinis* (Fallén).
Huebneria, emend.

estigmenensis (Sellers), 1943: 79 (*Aplomya*).—Vt.; Idaho to Ont. and Maine, s. to Calif., Ariz., and R.I. **N. comb.**
    affinis, authors, not Fallén.

Genus HYPHANTROPHAGA Townsend

Hyphantrophaga Townsend, 1892l: 247. Type-species, *Meigenia hyphantriae* Townsend (orig. des.).

desmiae (Sellers), 1943: 16 (*Zenillia*).—Calif.
euchaetiae (Sellers), 1943: 13 (*Zenillia*).—N.Y.; Ohio to Mass., s. to Md.
hyphantriae (Townsend), 1891h: 176 (*Meigenia*).—N. Mex.; B.C., Mich., Ohio, Calif. to N.C. and Fla.
  ceratomiae Coquillett, 1897: 101 (*Exorista*).—Kans., Mo., Tex.

Genus SISYROPA Brauer and Bergenstamm

Sisyropa Brauer and Bergenstamm, 1889: 163 (1889: 95). Type-species, *Tachina thermophila* Wiedemann (mon.).
Oxexorista Townsend, 1912d: 165. Type-species, *Exorista eudryae* Townsend (orig. des.). Opinion differs on whether the type-species was misidentified (see Townsend, 1936b: 210, 211; Sellers, 1943: 98).

REFERENCE: Sellers, 1943 (rev.).

alypiae Sellers, 1943: 101.—Mass.; B.C., Mont., Vt., Conn., Miss., Mexico.
eudryae (Townsend), 1892d: 287 (*Exorista*).—N.Y.; Wis. to Que. and Maine, s. to Kans., Miss., and N.C.
  thompsoni Townsend, 1915a: 21 (*Oxexorista*).—Ohio.

Tribe EUMASICERINI
(Trypherini, in part)

Townsend's tribe Trypherini is not recognized in this catalog. Many of its well-known genera (*Lixophaga, Doryphorophaga*, etc.) are referred to the Blondeliini, and a few to other tribes. There remains a characteristic group of genera, with strong Neotropical affinities, which appears to merit tribal status near the Carceliini, though possibly a subtribe of the latter. As far as known, the flies deposit microtype eggs on plant surfaces.

Genus EUMASICERA Townsend

Eumasicera Townsend, 1909: 249. Type-species, *coccidella* Townsend (orig. des.) = *sternalis* (Coquillett).

sternalis (Coquillett), 1897: 109 (*Sturmia*).—Mo.; N.Y. and Mass. to Va., also Mo., Tex.
  coccidella Townsend, 1909: 249.—Mass.

Genus HOUGHIA Coquillett

Houghia Coquillett, 1897: 118. Type-species, *setipennis* Coquillett (orig. des.).

setipennis Coquillett, 1897: 118.—Ga.; N.C. to Tex. and Fla.

Genus HYPERTROPHOMMA Townsend

Hypertrophomma Townsend, 1915l: 99. Type-species, *opacum* Townsend (orig. des.).

opacum Townsend, 1915l: 100.—N.J.; N.Y., Ohio, Tex.

Genus METOPIOPS Townsend

Metopiops Townsend, 1912b: 338. Type-species, *mirabilis* Townsend (orig. des.). ?=*Pseudochaeta*.

pyralidis (Coquillett), 1897: 117 (*Pseudochaeta*).—D.C.; Kans. to Ont. and Mass., s. to Tex. and Fla.

Genus PACIDIANUS Reinhard

Pacidianus Reinhard, 1943b: 88. Type-species, *hirsutus* Reinhard (orig. des.).

hirsutus Reinhard, 1943b: 89.—Mich.; Ohio.
persimilis Reinhard, 1943b: 90.—N.Y.; Mich., Ont., Va.

Genus PHAENOPSIS Townsend

Phaenopsis Townsend, 1912b: 362. Type-species, *arabella* Townsend (orig. des.).

REFERENCE: Reinhard, 1946a (rev.).

venusta Reinhard, 1946a: 120.—Tex.

Genus PSEUDOCHAETA Coquillett

Pseudochaeta Coquillett, 1895a: 309. Type-species, *argentifrons* Coquillett (orig. des.).
Cylindromasicera Townsend of Townsend, 1941a: 260, in part (La. sp.).

REFERENCE: Reinhard, 1946a (rev.).

argentifrons Coquillett, 1895a: 310.—Calif. (Coquillett, 1897: 116); Ont., Que., N.Y., Calif. to Fla.
canadensis Brooks, 1945a: 93.—Que. **N. syn.**

brooksi Sabrosky and Arnaud, 1963: 155.—Fla.; Tex., Miss., Va., s. to Fla.
  argentifrons Coquillett of Reinhard, 1946a: 116.
clurina Reinhard, 1946a: 119.—Tex.
finalis Reinhard, 1946a: 114.—Tex.
frontalis Reinhard, 1946a: 114.—Tex.; Kans., Ariz., La.
marginalis Reinhard, 1946a: 118.—Tex.
perdecora Reinhard, 1946a: 115.—Tex.; Fla.
robusta (Reinhard), 1924b: 271 (*Oxynops*).—Tex.; Md. **N. comb.**
siminina Reinhard, 1946a: 117.—Ohio; Mich., Ont., Mass., Ark., Tex.

Tribe ERYCIINI

(Masicerini, Lydellini, Phrynoini)

The species of Eryciini parasitize a great variety of lepidopterous larvae, with a few parasitic on sawfly larvae.

Genus ANGUSTIA Sellers

Angustia Sellers, 1943: 107. Type-species, *Zenillia angustivitta* Aldrich and Webber (orig. des.).

angustivitta (Aldrich and Webber), 1924a: 18 (*Zenillia*).—N.Y.; Pa.

Genus APLOMYA Robineau-Desvoidy

Aplomya Robineau-Desvoidy, 1830: 184. Type-species, *zonata* Robineau-Desvoidy (Robineau-Desvoidy, 1863a: 459, 460) = *confinis* (Fallén).
Aplomyia, emend.
Platymya Robineau-Desvoidy, 1830: 116. Type-species, *aestivalis* Robineau-Desvoidy (Robineau-Desvoidy, 1863a: 191) = *nemestrina* (Meigen).
Platymyia, emend.
Phebellia Robineau-Desvoidy, 1846: 37. Type-species, *aestivalis* Robineau-Desvoidy (mon.). In *Aplomya*, the type-species is a secondary homonym, but it will not be renamed pending study of the classification.
Eumea Robineau-Desvoidy, 1863a: 302. Type-species, *locuples* Robineau-Desvoidy (orig. des.) ?=*mitis* (Meigen).
Myxexoristops Townsend, 1911b: 170. Type-species, *Myxexorista pexops* Brauer and Bergenstamm (mon.)=*blondeli* (Robineau-Desvoidy).
Epimasicera Townsend, 1912c: 51. Type-species, *Tachina westermanni* Zetterstedt (orig. des.)=*spernenda* (Zetterstedt)

Exorista, authors, not Meigen.
Zenillia, authors, not Meigen.

Aplomya is retained here in its broad sense, chiefly because there is still much difference of opinion on the restricted genera or subgenera to be recognized, and even on the generic or subgeneric assignment of some of the species. Certain Nearctic forms appear to be intermediates, and cast doubt on some of the segregates. Until the classification has stabilized, and with the advantage of a modern revision of the Nearctic species, we prefer to maintain *Aplomya* in its wider application.

Certain of the restricted segregates adopted in the current European literature, and occasionally used in North America, are listed here as synonyms, in order that the references and type fixations will be available for future reference. The species *polita* and *theclarum* are typical *Aplomya*, but *caesar* and *mitis* would be referred to *Eumea*, *fronto*, *neurotomae*, and *nox* to *Myxexoristops*, *epicydes*, *imitator*, and *trichiosomae* to *Phebellia* (a group of parasites on cimbicid sawflies), *trisetosa* to *Phonomyia*, and *confusionis* and *hortulana* to *Platymya*. At least some of these groupings may prove to be morphologically and biologically justified, at some hierarchial level.

REFERENCES: Sellers, 1943 (rev.); Mesnil, 1954 (key to genera of "Masicerina").

blondeli (Robineau-Desvoidy), 1830: 161 (*Phryxe*).—Europe; introd. Man., ?estab.

caesar (Aldrich), 1916b: 20 (*Exorista*).—Ont.; B.C. to N.B., s. to Calif., Tex., and Va.

    nigripalpis Townsend, 1896: 330 (*Exorista*; preocc. Macquart, 1849).—?Ill.

cerurae Sellers, 1943: 94.—Mass.

confusionis Sellers, 1943: 86.—N.H.; B.C. to N.H., s. to Calif. and Mass., also D.C.

crassiseta (Aldrich and Webber), 1924a: 29 (*Zenillia*).—Ind.; Man. to Mass., s. to Ark. and Ga.

doloma Reinhard, 1958b: 239.—W. Va.; Mont., Wyo.

epicydes (Walker), 1849: 786 (*Tachina*).—Ont.; B.C. to Ont. and Maine, s. to Calif., Wyo., and R.I.

    coerulea Aldrich and Webber, 1924a: 23 (*Zenillia*).—Mass.

fronto (Coquillett), 1897: 96 (*Exorista*).—N.H.; B.C., Idaho, Calif., Wyo., Ohio.

helvina (Coquillett), 1897: 96 (*Exorista*).—N.H.; B.C., Wash., Minn. to Maine and N.J.

hortulana (Meigen), 1824: 330 (*Tachina*).—Germany; B.C. to Ont., Europe.

imitator Sellers, 1943: 93.—B.C.; Alta., Wash.
mitis (Meigen), 1824: 335 (*Tachina*).—Europe; introd. e. Canada, Mich. to Mass., s. to Ind. and Pa., ?estab.
neurotomae Sellers, 1943: 85.—N.J.; Oreg., Nev., N.H. to Pa. and N.J.
nox (Hall), 1937a: 203 (*Zenillia*).—Japan; introd. B.C. and N.B., not estab. =*blondeli* ssp. *stolida* (Stein). **N. syn.**, L. P. Mesnil *in litt.*, **n. comb.**
pheosiae Sellers, 1943: 95.—Maine; Mass.
polita (Coquillett), 1897: 99 (*Exorista*).—Ga.
theclarum (Scudder) *in* Osten Sacken, 1887b: 166 (*Tachina*).—Unknown; B.C. to Que. and Maine, s. to Calif., Miss., and Va.
  chrysophani Townsend, 1891e: 197 (*Exorista*).—Iowa.
  confinis, authors, not Fallén.
trichiosomae Sellers, 1943: 94.—Maine; ?B.C.
trisetosa (Coquillett), 1902g: 110 (*Exorista*).—Idaho; Wash. to Pa., s. to Calif. and N. Mex.

Genus BUQUETIA Robineau-Desvoidy

Buquetia Robineau-Desvoidy, 1847: 286. Type-species, *musca* Robineau-Desvoidy (mon.).
Eipogona Rondani, 1868c: 588. Type-species, *Masicera setifacies* Rondani (mon.) =*musca* Robineau-Desvoidy.
Eupogona, emend.

Both European and North American species are parasites of *Papilio* larvae.

obscura (Coquillett), 1897: 124 (*Winthemia*).—Md.; Iowa to Mass., also Tex., Md.
  pollinosa Reinhard, 1922: 333 (*Blepharipeza*).—Tex.
  americana Curran, 1927m: 148 (*Eipogona*).—Iowa.

Genus CATAGONIOPSIS Townsend

Catagoniopsis Townsend, 1926: 29. Type-species, *infernalis* Townsend (orig. des.) =*specularis* (Aldrich and Webber).

facialis (Coquillett), 1897: 105 (*Phorocera*).—Tex.; Calif., Ariz., N. Mex. **N. comb.**
meracanthae (Greene), 1921c: 126 (*Phorocera*).—Md.; Va.
specularis (Aldrich and Webber), 1924a: 70 (*Phorocera*).—N. Mex.; B.C. to Calif. and N. Mex.
  infernalis Townsend, 1926: 30.—N. Mex.

TACHINIDAE

Genus CEROMASIA Rondani

Masicera, subg. **Ceromasia** Rondani, 1856: 71. Type-species, *florum* Macquart (Brauer, 1893: 476; misident.)=*rubrifrons* (Macquart).

auricaudata Townsend, 1908: 102.—Idaho; B.C. to Mont., s. to Calif. and N. Mex., also introd. Man. to Nfld. and N.B., ?estab.
  rutila Meigen of Curran, 1927d: 16 (*Erycia*), in part.
aurifrons Townsend, 1908: 102.—N.H.; Mich., Ohio, Mass., N.S.
  festinans, authors, not Meigen.
  rutila Meigen of Curran, 1927d: 16 (*Erycia*), in part.
borealis Brooks, 1945a: 89.—N.W.T.; Alta., ?Alaska.

Genus CHRYSOEXORISTA Townsend

Chrysoexorista Townsend, 1915g: 435. Type-species, *viridis* Townsend (orig. des.).
Chrysomasicera Townsend, 1915f: 230. Type-species, *borealis* Townsend (orig. des.)=*ochracea* (Wulp). **N. syn.**

dawsoni (Sellers), 1943: 29 (*Zenillia*).—Ariz.
fulgoris (Sellers), 1943: 24 (*Zenillia*).—N. Mex. **N. comb.**
lineata (Wulp), 1890: 54 (*Mystacella*).—Mexico (Tabasco); Colo., Ariz., N. Mex., Tex., Cent. Amer.
marginata (Aldrich and Webber), 1924a: 17 (*Zenillia*).—Colo.
ochracea (Wulp), 1890: 63 (*Exorista*).—Mexico; N. Mex., Costa Rica. **N. comb.**
  borealis Townsend, 1915f: 230 (*Chrysomasicera*).—N. Mex. **N. comb.**

Genus CLEMELIS Robineau-Desvoidy

Clemelis Robineau-Desvoidy, 1863a: 481. Type-species, *Zenillia ciligera* Robineau-Desvoidy (orig. des.)=*pullata* (Meigen).

unipilum (Aldrich and Webber), 1924a: 83 (*Phorocera*).—Oreg. **N. comb.**

Genus CYZENIS Robineau-Desvoidy

Cyzenis Robineau-Desvoidy, 1863a: 544. Type-species, *Phryno hemisphaerica* Robineau-Desvoidy (orig. des.)=*albicans* (Fallén).

Pseudodidyma Townsend, 1915c: 287. Type-species, *pullula* Townsend (orig. des.).

albicans (Fallén), 1810b: 286 (*Tachina*).—Sweden; Europe, introd. and estab. N.S.
pullula (Townsend), 1915c: 288 (*Pseudodidyma*).—Alaska; B.C., Idaho.
pullula, authors (*Didyma*), not Wulp.

Genus EUCEROMASIA Townsend

Euceromasia Townsend, 1912a: 112. Type-species, *spinosa* Townsend (orig. des.).

floridensis Reinhard, 1957: 109.—Fla.; Tex., N.C.
neptis Reinhard, 1947a: 23.—Tex.; Mo.
solata Reinhard, 1947a: 24.—Tex.
spinosa Townsend, 1912a: 112.—Mass.

Genus EUEXORISTA Townsend

Euexorista Townsend, 1912d: 166. Type-species, *Tachina futilis* Osten Sacken (orig. des.).

futilis (Osten Sacken), 1887b: 161 (*Tachina*).—Unknown (?Mass.); B.C. to Que. and Maine, s. to Calif. and N.C.
vanessae Harris, 1835: 599 (*Tachina*). Nomen nudum.

Genus EUFRONTINA Brooks

Eufrontina Brooks, 1945a: 90. Type-species, *Frontina spectabilis* Aldrich (orig. des.).

ethniae Brooks, 1945a: 91.—Ont.
spectabilis (Aldrich), 1916b: 21 (*Frontina*).—Ont.; B.C., Man. to Que. and Maine, also W. Va.

Genus FRONTINIELLA Townsend

Frontiniella Townsend, 1918d: 21. Type-species, *parancilla* Townsend (orig. des., as gen. n., sp. n.).

parancilla Townsend, 1918d: 21 (as *pararcilla*, error).—Mich.; Mich. to Conn., s. to Calif., Mexico, and Fla.
stilla Reinhard, 1943c: 17 (*Achaetoneura*).—Tex.
ancilla, authors, not Walker.

Genus LESPESIA Robineau-Desvoidy

Lespesia Robineau-Desvoidy, 1863a: 567. Type-species, *Erycia ciliata* Macquart (mon.).
Achaetoneura Brauer and Bergenstamm, 1891: 334 (1891: 30). Type-species, *hesperus* Brauer and Bergenstamm (Coquillett, 1910b: 502) =*frenchii* (Williston).
Parafrontina Brauer and Bergenstamm, 1893: 115 (1893: 27). Type-species, *apicalis* Brauer and Bergenstamm (orig. des., as gen. n., sp. n.) =*archippivora* (Riley).
Rileyella Townsend, 1909: 249. Type-species, *Tachina aletiae* Riley (orig. des.).
Masiceropsis Townsend, 1916b: 178. Type-species, *Masicera pauciseta* Coquillett (orig. des.) =*archippivora* (Riley).
Ypophaemyia Townsend, 1916g: 75. Type-species, *malacosomae* Townsend (orig. des.) =*archippivora* (Riley).
Frontina, authors, not Meigen.

REFERENCES: Webber, 1930 (rev., as *Achaetoneura*); Beneway, 1963 (rev.).

aletiae (Riley), 1879: 162 (*Tachina*).—Fla. (as Unknown; Townsend, 1941a: 224); Ill. to Ont. and N.J., s. to Calif. and Fla.
  fraterna Comstock, 1880b: 303 (*Tachina*).—Ala.
anonyma (Riley), 1872: 129 (*Tachina*).—Mo. Unrecognized, **N. comb.**
archippivora (Riley), 1871: 150 (*Tachina*).—Mo.; B.C. to Ont., all U.S., Mexico. This name is commonly credited to Williston (1889a).
  websteri Townsend, 1891k: 206 (*Meigenia*).—Ind.
  promiscua Townsend, 1891f: 84 (*Phorocera*).—Maine.
  apicalis Brauer and Bergenstamm, 1893: 115 (1893: 27) (*Parafrontina*).—Tex. (as N. Amer.; Townsend, 1941a: 218).
  pauciseta Coquillett, 1897: 114 (*Masicera*).—Calif.
  malacosomae Townsend, 1916g: 75 (*Ypophaemyia*).—N.C.
ciliata (Macquart), 1834b: 294 (1834: 158) (*Erycia*).—France (but apparently accidentally introd. from N. Amer.); B.C. to N.S., s. to Calif. and Ariz. in West, and to D.C. in East.
  samiae Webber, 1930: 15 (*Achaetoneura*).—Maine.
cuculliae (Webber), 1930: 18 (*Achaetoneura*).—Va.; Man. to Maine, s. to Tex. and Fla.
datanarum (Townsend), 1892d: 287 (*Masicera*).—Minn., Ont., N.Y.; Que., R.I., Conn., N.J.
  anisotae Webber, 1930: 13 (*Achaetoneura*).—N.J.

dimmocki (Webber), 1930: 20 (*Achaetoneura*).—Mass.; Man., Ohio, Pa., Md.
dubia (Williston), 1889a: 1924 (*Masicera*).—Unknown. Unrecognized, **N. comb.**
euchaetiae (Webber), 1930: 11 (*Achaetoneura*).—N.Y.; S. Dak. to Ont. and N.H., s. to Kans. and N.J.
ferruginea (Reinhard), 1924b: 269 (*Frontina*).—Tex.
frenchii (Williston), 1889a: 1923 (*Masicera*).—Maine, B.C. to N.S., all U.S.
  longicornis Wiedemann, 1830: 325 (*Tachina*; preocc. Fallén, 1810).—Unknown. **N. comb.**
  hesperus Brauer and Bergenstamm, 1891: 334 (1891: 30) (*Achaetoneura*).—N. Amer.
  sphingivora Townsend, 1892d: 286 (*Masicera*).—W. Va.
  malacosomae Curran, 1925c: 155 (*Frontina*; preocc. Townsend, 1916).—N.S.
  sordida Curran, 1926k: 171 (*Frontina*; n. name for *malacosomae* Curran).—N.S.
fulvipalpis (Bigot), 1888a: 263 (*Masicera*).—Rocky Mts. Unrecognized, **N. comb.**
laniiferae (Webber), 1930: 33 (*Achaetoneura*).—Mexico; Fla.
melalophae (Allen), 1926a: 192 (*Achaetoneura*).—Miss.; B.C. to N.B., s. to Calif., Tex., and Va.
pholi (Webber), 1930: 27 (*Achaetoneura*).—Mass.; Mo.
rileyi (Williston), 1889a: 1924 (*Masicera*).—Unknown; Tex. to Fla.
schizurae (Townsend), 1891i: 187 (*Masicera*).—Kans.; B.C. to Ariz. and N. Mex. in the West, Kans. to Ont. and Mass., s. to Tex. and Fla. in the East.
  schizurae Coquillett, 1897: 113 (*Sturmia*; preocc. Townsend, 1891).—Wash.
  piperi Townsend, 1908: 98 (*Argyrophylax*; n. name for *schizurae* Coquillett).—Wash.
testacea (Webber), 1930: 25 (*Achaetoneura*).—D.C.; N.Y. to N.C., also Kans., Tex.
  violenta, authors, not Walker.
texana (Webber), 1930: 24 (*Achaetoneura*).—Tex.; Mexico.
westonia (Webber), 1930: 24 (*Achaetoneura*).—Calif.

Genus LYDELLA Robineau-Desvoidy

Lydella Robineau-Desvoidy, 1830: 112. Type-species, *grisescens* Robineau-Desvoidy (Robineau-Desvoidy, 1863a: 855).

Lydella, subg. *Anetia* Robineau-Desvoidy, 1863a: 868 (preocc. Hübner, 1823). Type-species, *occlusa* Robineau-Desvoidy (orig. des.)=*grisescens* Robineau-Desvoidy.
REFERENCE: Herting, 1959 (rev.).

deckeri (Curran), 1929a: 12 (*Erycia*).—Iowa. Unrecognized. ?=*radicis*.
grisescens Robineau-Desvoidy.—Not Nearctic, not introd.
radicis (Townsend), 1916a: 19 (*Andrina*).—D.C.; B.C. to Que. and Maine, s. to Calif., Mexico, Miss., and Va.
   nigrita Townsend, 1891c: 358 (*Masicera*; preocc. Robineau-Desvoidy, 1863).—Ill.
   intermedia Villeneuve, 1932: 272 (as *stabulans* var.).—Mich., France.
   myoidaea, N. Amer. authors, not Robineau-Desvoidy.
   senilis, N. Amer. authors, in part, not Meigen.
stabulans (Meigen).—Not Nearctic.
thompsoni Herting, 1959: 428.—Germany; Palaearctic, introd. and estab. Man. to Que., s. to Mo., Ala., and N.C. Host: *Ostrinia* [formerly *Pyrausta*] *nubilalis* (Hübner).
   grisescens, authors, not Robineau-Desvoidy.
   senilis, authors, in part, not Meigen.

Genus MADREMYIA Townsend

Madremyia Townsend, 1916i: 305. Type-species, *parva* Townsend (orig. des.)=*saundersii* (Williston).

saundersii (Williston), 1889a: 1922 (*Phorocera*).—New England; B.C. to Nfld., s. to Calif., Mexico, Ill., and Mass.
   parva Townsend, 1916i: 306.—Mexico (Chih.).

Genus Masicera Macquart

Masicera Macquart, 1834b: 285 (1834: 149). Type-species, *Tachina silvatica* Fallén (Robineau-Desvoidy, 1863a: 872, 880).

silvatica (Fallén), 1810b: 268 (*Tachina*).—Sweden; Europe, introd. New England, not estab.
   sylvatica, emend.

Genus Metagonistylum Townsend

Metagonistylum Townsend, 1927: 379. Type-species, *minense* Townsend (orig. des.).

minense Townsend, 1927: 381.—Brazil; introd. La., Fla., not estab.

Genus NILEA Robineau-Desvoidy

Nilea Robineau-Desvoidy, 1863a: 275. Type-species, *innoxia* Robineau-Desvoidy (orig. des.).
Phyllophorocera Townsend, 1916h: 621. Type-species, *Phorocera sternalis* Coquillett (orig. des.). **N. syn.**
Eutritochaeta Townsend, 1919b: 580. Type-species, *carpocapsae* Townsend (orig. des.) ?=*noctuiformis* (Smith). **N. syn.**

carpocapsae (Townsend), 1919b: 580 (*Eutritochaeta*).—Ark. ?=*noctuiformis*. **N. comb.**
noctuiformis (Smith), 1915b: 101 (*Neopales*).—Mass.; Mich., Maine. **N. comb.**
sternalis (Coquillett), 1902g: 111 (*Phorocera*).—N.H.; B.C. to Que. and Maine, s. to Wyo. and Ohio. **N. comb.**

Genus OBOLOCERA Townsend

Obolocera Townsend, 1919c: 180. Type-species, *Homoeonychia rapae* Smith (orig. des.).

rapae (Smith), 1917b: 139 (*Homoeonychia*).—Mass.

Genus PATELLOA Townsend

Patelloa Townsend, 1916h: 619. Type-species, *Phorocera leucaniae* Coquillett (orig. des.).
Pateloa (error) Curran, 1934a: 444.
Patelloapsis Townsend, 1927: 263, 345. Type-species, *similis* Townsend (orig. des.). **N. syn.**

fuscimacula (Aldrich and Webber), 1924a: 48, 73 (*Phorocera*; as *fusicimacula*, p. 48).—Calif.; B.C.
leucaniae (Coquillett), 1897: 104 (*Phorocera*).—D.C.; Mich. to N.B., s. to Tex. and Fla.
pachypyga (Aldrich and Webber), 1924a: 70 (*Phorocera*).—Mass.; B.C., Idaho, Calif., Mich. to Mass., s. to Mo. and Md.
pluriseriata (Aldrich and Webber), 1924a: 73 (*Phorocera*).—Calif.; Utah, Colo., Mexico. **N. comb.**
reinhardi (Aldrich and Webber), 1924a: 74 (*Phorocera*).—Mich.; B.C., Kans., Ohio, N.Y.
setifrons (Aldrich and Webber), 1924a: 71 (*Phorocera*).—Sask.; ?Calif. **N. comb.**
silvatica (Aldrich and Webber), 1924a: 72 (*Phorocera*).—B.C.; Oreg., Calif., Minn. to Mass. **N. comb.**

Genus PHRYXE Robineau-Desvoidy

Phryxe Robineau-Desvoidy, 1830: 158. Type-species, *athaliae* Robineau-Desvoidy (Robineau-Desvoidy, 1863a: 329, 358) = *vulgaris* (Fallén).
Plagiophryxe Townsend, 1926: 32. Type-species, *pecosensis* Townsend (orig. des.).

REFERENCE: Sellers, 1943 (rev.).

pecosensis (Townsend), 1926: 33 (*Plagiophryxe*).—N. Mex.; Alaska, B.C. to Nfld., s. to Calif., N. Mex., and Mass.
vulgaris (Fallén), 1810b: 282 (*Tachina*).—Sweden; B.C. and Idaho to Calif., also Wis. to N.B. and N.J.
    hirsuta Osten Sacken, 1887b: 163 (*Tachina*).—Unknown.

Genus PILATEA Townsend

Pilatea Townsend, 1916b: 178. Type-species, *Masicera celer* Coquillett (orig. des.).
Sisyrosturmia Townsend, 1926: 35. Type-species, *chaetosa* Townsend (orig. des.). **N. syn.**
Erycia, authors, not Robineau-Desvoidy.
Masicera, N. Amer. authors, in part, not Macquart.

REFERENCE: Curran, 1927d (key, as *Erycia*).

aldrichi (Curran), 1927d: 17 (*Erycia*).—B.C.; Alta., Idaho, Calif. **N. comb.**
arator (Aldrich), 1925c: 32 (*Masicera*).—Pa.; Mich. to Que., s. to Va., also Tex. **N. comb.**
australis (Coquillett), 1897: 110 (*Sturmia*).—Fla. **N. comb.**
celer (Coquillett), 1897: 114 (*Masicera*).—La.; Calif. to Mich. and Fla.
chaetosa (Townsend), 1926: 36 (*Sisyrosturmia*).—N.H.; ?N.Y. **N. comb.**
delecta (Curran), 1927d: 16 (*Erycia*).—Man.; B.C., Vt. **N. comb.**
flavitarsa (Reinhard), 1934b: 191 (*Erycia*).—Calif.; Wash., Oreg. **N. comb.**
occidentalis (Coquillett), 1897: 110 (*Sturmia*).—Wash.; B.C. to Calif. and Colo. **N. comb.**
    levata Reinhard, 1934b: 193 (*Erycia*).—Calif. **N. comb.**
picata (Reinhard), 1953a: 58 (*Erycia*).—Calif. **N. comb.**
ruficornis Smith, 1917b: 138.—N.H.
sectilis (Reinhard), 1953a: 59 (*Erycia*).—Wyo. **N. comb.**
varifrons (Curran) 1927d: 18 (*Erycia*).—B.C. **N. comb.**

Genus PROPHRYNO Townsend

Prophryno Townsend, 1927: 262, 353. Type-species, *aurulans* Townsend (orig. des.).

marginalis (Aldrich and Webber), 1924a: 84 (*Phorocera*).—Tex. N. comb.

parviteres (Aldrich and Webber), 1924a: 80 (*Phorocera*).—Tex.; Ariz. to Md. and Ga. N. comb.
flavicauda, authors, in part, not Wulp.

Genus PSEUDOPERICHAETA Brauer and Bergenstamm

Pseudoperichaeta Brauer and Bergenstamm, 1889: 92 (1889: 24). Type-species, *major* Brauer and Bergenstamm (orig. des., as gen. n., sp. n.) =*palesioidea* (Robineau-Desvoidy).
Anoxycampta, authors, in part, not Bigot.

erecta (Coquillett), 1902g: 112 (*Phorocera*).—Ark.; B.C. to Que., all U.S.
loxostegeae Reinhard, 1922: 331 (*Exorista*).—Tex.

insidiosa (Robineau-Desvoidy), 1863a: 338 (*Phryxe*).—France; Europe, Japan, introd. Mich. to N.S., s. to Ind. and Pa., ?estab.
roseanae Brauer and Bergenstamm, 1891: 332 (1891: 28) (*Myxexorista*).—Unknown.
major, authors, not Brauer and Bergenstamm.

palesioidea (Robineau-Desvoidy), 1830: 160 (*Phryxe*).—France; Oreg., Europe.
trizonata Zetterstedt, 1844: 1166 (*Tachina*).—Sweden.

Genus THELYMYIA Brauer and Bergenstamm

Thelymyia Brauer and Bergenstamm, 1891: 330 (1891: 26). Type-species, *loewii* Brauer and Bergenstamm (mon.) = *saltuum* (Meigen).

REFERENCE: Sellers, 1943 (rev.).

curriei (Coquillett), 1897: 94 (*Exorista*).—N. Dak.; Alta. to N.Y., s. to S. Dak. and Ohio, also Tex.

disparis Reinhard, 1959b: 230.—Ariz.

erecta Sellers, 1943: 105.—Idaho; B.C., Wash., Utah.

mathesoni (Reinhard), 1937: 68 (*Zenillia*).—N.Y.; Man., Ohio. N. comb., H. J. Reinhard *in litt*.

Genus TSUGAEA Hall

Tsugaea Hall, 1939: 240. Type-species, *nox* Hall (orig. des.).

nox Hall, 1939: 242.—Oreg.; B.C., Calif., Nev.

Genus ZENILLIA Robineau-Desvoidy

Zenillia Robineau-Desvoidy, 1830: 152. Type-species, *Musca libatrix* Panzer (Robineau-Desvoidy, 1863a: 471).

Our native species may not be true *Zenillia*, but it has not been possible to place them at this time.

REFERENCE: Sellers, 1943 (rev.).

angustata (Wulp), 1890: 70 (*Exorista*).—Mexico (Guerrero); Tex.
  coquilletti Aldrich and Webber, 1924a: 18.—Tex.
blandita (Coquillett), 1897: 96 (*Exorista*).—N.H.; Wis. to Maine, s. to Ohio and R.I., also D.C.
browni Curran, 1933b: 11.—Ont.
libatrix (Panzer), 1798: 12 (*Musca*).—Germany; Europe, introd. New England, not estab.
scolex Reinhard, 1953a: 56.—Calif.

Unplaced Species of Eryciini

The following species may be found to fall in genera known at present only from the Palaearctic Region. All except the later described species may be identified in keys to *Phorocera* and *Zenillia* in Aldrich and Webber (1924a).

anassa Reinhard, 1959a: 160 (*Phorocera*).—Idaho.
comstocki Williston, 1889a: 1922 (*Phorocera*).—S.C. (Aldrich and Webber, 1924a: 78); Ill., N.C., Ga., Tex.
festinans Aldrich and Webber, 1924a: 85 (*Phorocera*).—N. Mex.; B.C., Calif.
halisidotae Aldrich and Webber, 1924a: 84 (*Phorocera*).—Idaho; B.C., Calif.
incrassata Smith, 1912: 121 (*Phorocera*).—Idaho; B.C., Oreg., also introd. Ont., N.B., Nfld., ?estab.
lobeliae Coquillett, 1897: 97 (*Exorista*).—Md., D.C., Va.; Mich. to Conn., s. to Tex. and Fla.
mitis Curran, 1930j: 108 (*Phorocera*).—N.Y.
regilla Reinhard, 1959a: 158 (*Phorocera*).—Calif.; Ariz.
signata Aldrich and Webber, 1924a: 86 (*Phorocera*).—S.C.; Tenn., Miss.
  xanthura Aldrich and Webber, 1924a: 82 (*Phorocera;* preocc. Wulp, 1890).—Tenn. **N. syn.**
  nitelae Aldrich and Webber, 1924b: 195 (*Phorocera;* n. name for *xanthura* Aldrich and Webber).—Tenn. **N. syn.**
  flavicauda, authors, in part, not Wulp.
texana Aldrich and Webber, 1924a: 79 (*Phorocera*).—Tex.; Ariz., Mexico.

ustulata Reinhard, 1959a: 159 (*Phorocera*).—Calif.; Utah, N.Y., N.J.
valens Aldrich and Webber, 1924a: 20 (*Zenillia*).—N.Y.; Minn. and Nebr. to Mass.
victoria Aldrich and Webber, 1924b: 195 (*Phorocera*; n. name for *tenuiseta* Aldrich and Webber).—B.C.
  tenuiseta Aldrich and Webber, 1924a: 82 (*Phorocera*; preocc. Macquart, 1846).—B.C.

Unplaced Genus and Species of Tachinidae

Genus OTOMASICERA Townsend

Otomasicera Townsend, 1912a: 113. Type-species, *patella* Townsend (orig. des.).

patella Townsend, 1912a: 113.—Mass. Egg and 1st stage maggot; adult lost.

Unplaced Species of Tachinidae

abzoe Walker, 1849: 846 (*Dexia*).—Ga.
americana Bigot, 1888a: 256 (*Evibrissa*).—Wash. Territory.
analis Robineau-Desvoidy, 1830: 315 (*Zelia*).—Carolina.
analis Say, 1829: 177 (1859b: 367) (*Dexia*).—Ind. ?Leskiini.
antennata Walker, 1852: 298 (*Tachina*).—U.S.
apicalis Robineau-Desvoidy, 1830: 316 (*Zelia*).—Carolina.
atra Robineau-Desvoidy, 1830: 288 (*Clytia*).—Carolina. ?*Myiophasia*.
candens Walker, 1849: 720 (*Tachina*).—N.S. (?error).
cremides Walker, 1849: 842 (*Dexia*).—N. Amer.
hybreas Walker, 1849: 785 (*Tachina*).—Ont. ?*Ceromasia*.
inermis Bigot, 1887b: cxl (*Blepharipeza*).—N. Amer.
maculosa Coquillett, 1895a: 313 (*Hypostena*).—Fla.
melanocera Robineau-Desvoidy, 1830: 312 (*Dexia*).—Carolina.
nigrifacies Bigot, 1888a: 258 (*Chetolyga*).—Rocky Mts.
ogoa Walker, 1849: 841 (*Dexia*).—N.S.
parva Bigot, 1888a: 260 (*Phorocera*).—Rocky Mts.
postica Walker, 1852: 310 (*Dexia*).—Ga.
pumila Robineau-Desvoidy, 1863a: 622 (*Peleteria*).—Amer.
rufonotata Bigot, 1888a: 269 (*Morphomyia*).—Calif.
septentrionalis Walker *in* Lord, 1866: 339 (*Eurigaster*).—B.C.
serva Walker, 1852: 349 (*Musca*).—U.S.
setinervis Coquillett, 1898c: 236 (*Hypostena*).—Fla.
triangularis Wulp, 1867: 149 (*Dexia*).—Wis.
velox Robineau-Desvoidy, 1830: 316 (*Zelia*).—Carolina.

Family CUTEREBRIDAE

By J. G. Chillcott

Larvae of the Nearctic species develop in the skin of various rodents and lagomorphs. The adults are known as rodent bot flies and rabbit bot flies. The published keys are not entirely satisfactory.

REFERENCES: Swenk, 1905 (synopsis); Bau, 1929c (key).

Genus CUTEREBRA Clark

Cuterebra Clark, 1815: 64, 70. Type-species, *Oestrus cuniculi* Clark (Desmarest, 1859: 247).
Cutiterebra, emend.
Trypoderma Wiedemann, 1830: 256. Type-species, *Musca americana* Fabricius (Coquillett, 1910b: 618).
Bogeria Austen, 1895: 391. Type-species, *princeps* Austen (orig. des.).
Atrypoderma Townsend, 1919b: 592. Type-species, *Musca americana* Fabricius (orig. des.).
Orthocuterebra Bau, 1929b: 543. Type-species, *Cuterebra lepusculi* Townsend (Aldrich, 1931b: 117)=*princeps* (Austen).
Paracuterebra Bau, 1929b: 543. Type-species, *Oestrus cuniculi* Clark (Aldrich, 1931b: 117).
Protocuterebra Bau, 1929b: 543. Type-species, *Musca americana* Fabricius (Aldrich, 1931b: 117).
Pseudobogeria Bau, 1931b: 206. Type-species, *Musca buccata* Fabricius (orig. des.).

aldrichi Austen, 1933: 704.—Calif.
americana (Fabricius), 1775: 774 (*Musca*).—Amer.; Wyo., Colo., Calif. to N. Mex., also Ga.
    cauterium Clark, 1815: 70.—Ga.
    polita Coquillett, 1898a: 10.—Wyo.
angustifrons Dalmat, 1942a: 418.—Iowa.
approximata Walker *in* Lord, 1866: 338.—B.C.; ?Ariz.
beameri Hall, 1943: 25.—Kans.
buccata (Fabricius), 1776: 305 (*Oestrus*).—Carolina; Minn. to N.S., s. to Tex. and Fla.
    baccata (error) Lugger, 1897: 229.
    purivora Clark, 1815: 70.—Amer.
    scudderi Townsend, 1917f: 27 (*Bogeria*).—Tex., D.C., Md.
cuniculi (Clark), 1797: 299 (*Oestrus*).—Ga.; Fla.
cyanella P. R. Jones, 1906: 391.—Nebr.

emasculator Fitch, 1856b: 478 (1859: 160).—N.Y.; Mich. to Que. and Mass., s. to Ala.
    scutellaris Brauer, 1863: 230.—N. Amer.
    buccator (error) Williston, 1884a: 426.
enderleini Bau, 1929c: 5(key) (1931a: 20, descr.)—Tex., Ga.
fasciata Swenk, 1905: 184.—Nebr.; B.C., N.B., N. Mex.
fontinella Clark, 1827: 410.—Ill.; Minn. to Ont. and Mass., s. to Ill. and Ga.
    fontanella, error or emend.
grisea Coquillett, 1904a: 11.—B.C.; N.W.T. to N.S., s. to Utah and N.J.
horripilum Clark, 1815: 70.—Ga.; Minn. to N.S., s. to Tex. and Fla.
    abdominalis Swenk, 1905: 182.—Nebr.
    cuniculi, authors, not Clark.
jellisoni Curran, 1942a: 78.—Oreg.; Nev., Utah, Ariz.
latifrons Coquillett, 1898a: 10.—Calif.
lepivora Coquillett, 1898a: 9.—Calif., Wyo.
    leporivora Aldrich, 1905: 419, emend.
nitida Coquillett, 1898a: 10.—Calif.
peromysci Dalmat, 1942b: 45.—Iowa.
princeps (Austen), 1895: 393 (*Bogeria*).—Mexico (Son.); Wash. to S. Dak., s. to Ariz., Mexico, and Tex.
    lepusculi Townsend, 1897d: 8.—N. Mex.
    albifrons Swenk, 1905: 183.—Wyo.
ruficrus (Austen), 1933: 711 (*Bogeria*).—Colo.; Oreg. to Colo., s. to Calif. and Tex.
similis Johnson, 1903a: 101.—N. Mex.
sterilator Lugger, 1897: 229.—Minn.
subbuccata Bau, 1929a: 305.—Nev.
tenebrosa Coquillett, 1898a: 11.—Oreg., Calif., Colo.; B.C. to S. Dak., s. to Calif. and Colo.
thomomuris Jellison, 1949: 483.—Mont.

Family OESTRIDAE
(including Hypodermatidae)
By J. G. Chillcott

The larvae are endoparasitic in mammals.

REFERENCES: James, 1948 (synopsis, figs.); Zumpt, 1957 (classif.).

Subfamily OESTRINAE

The larvae develop in the nasal cavities and pharynges of various Artiodactyla and Perrisodactyla. Our common species is the sheep bot fly.

Genus OESTRUS Linnaeus

Oestrus Linnaeus, 1758: 584. Type-species, *ovis* Linnaeus (Curtis, 1826: pl. 106).
Cephalemyia Latreille *in* S.N.A., 1818: 273. Type-species, *Oestrus ovis* Linnaeus (mon.).

ovis Linnaeus, 1758: 585.—Sweden; Alta., Ont., throughout U.S., cosmopolitan.

Subfamily CEPHENEMYIINAE

Genus CEPHENEMYIA Latreille

Cephenemyia Latreille *in* S.N.A., 1818: 271. Type-species, *Oestrus trompe* Modeer (mon.).
Cephenomyia, emend.
Cephalemya (error) Curran, 1934a: 414.

The larvae develop in the nasal cavities and pharynges of Cervidae. These nose bots occur throughout the U.S. and much of Canada, but the distribution of the species is not well known.

REFERENCE: Bennett and Sabrosky, 1962 (rev., adults and larvae).

apicata Bennett and Sabrosky, 1962: 438.—Calif.; B.C., Alta., Mont.
jellisoni Townsend, 1941b: 161.—Mont.; B.C. to w. Ont.; s. to Calif., cent. Tex., and Colo.
macrotis Brauer, 1863: 211, 279.—N. Amer. Unrecognized.
   macrotis Bau, 1928: 458.—N. Amer.
phobifer (Clark), 1815: 69 (*Oestrus*).—Ga.; N. Dak. to Ont. and Maine, s. to Ga., also Tex. and La.
   phobifera, emend.
   abdominalis Aldrich, 1915b: 149.—N.Y.
pratti Hunter, 1916: 170.—Tex.; Oreg. to Mont., s. to Ariz., n. Mexico, and Tex.

trompe (Modeer), 1786: 134 (*Oestrus*).—Lapland; n. Alaska to Labr. and Nfld., n. Holarctic.
  nasalis, authors, not Linnaeus. See *Gasterophilus*.
ulrichii Brauer.—Not Nearctic.

Subfamily HYPODERMATINAE
Genus HYPODERMA Latreille

Hypoderma Latreille *in* S.N.A., 1818: 272. Type-species, *Oestrus bovis* Linnaeus (mon.).
Lithohypoderma Townsend, 1917e: 129. Type-species, *Musca ascarides* Scudder (orig. des.).

Mature larvae are found in the skin of Artiodactyla. North American species of the subgenus *Hypoderma* are cattle grubs; *H. (Oedemagena) tarandi* is a reindeer grub.

REFERENCES: Bishopp et al., 1926 (biol.); Patton, 1936 (rev.); James, 1948 (synopsis).

Subgenus HYPODERMA Latreille

bovis (Linnaeus), 1758: 584 (*Oestrus*).—Europe; Alaska, Canada, n. U.S., s. to Calif., n. Tex., and N.C., Holarctic. Far-south records in the U.S. and West Indies are from imported cattle.
lineatum (Villers), 1789: 349 (*Oestrus*).—Europe; B.C. to Que., throughout U.S., cosmopolitan.
  supplens Walker, 1849: 685 (*Oestrus*).—N.S.
  bonassi Brauer, 1875: 75.—Colo.

Subgenus OEDEMAGENA Latreille

Oedemagena Latreille *in* S.N.A., 1818: 272 (as genus). Type-species, *Oestrus tarandi* Linnaeus (mon.).

tarandi (Linnaeus), 1758: 584 (*Oestrus*).—Sweden; Arctic to Hudsonian zones, s. to B.C., Ont., Nfld., ?Labr.
  terraenovae Knab, 1913: 155.—Nfld.

Unplaced Genus and Species of Oestridae
Genus SUIOESTRUS Townsend

Suioestrus Townsend, 1921: 134. Type-species, *cookii* Townsend (orig. des.).

The larvae were found in a domestic pig.

cookii Townsend, 1921: 134.—W. Va.

Unplaced Genus of Diptera

Scatophora Robineau-Desvoidy, 1830: 811. Type-species, *carolinensis* Robineau-Desvoidy (Spuler, 1924d: 378). The type-species is unrecognized.

Unplaced Species of Diptera

americana Walker, 1848: 28 (*Asthenia*).—Ont. Unrecognized.
atra Fabricius, 1780: 203 (*Tipula*).—Greenland. Unrecognized.
carolinensis Robineau-Desvoidy, 1830: 811 (*Scatophora*).—Carolina. Unrecognized.
hematophila Laboulbène, 1882: 223 (*Simulia*).—Nfld. Unrecognized.
qualis Say, 1829: 176 (1859b: 366) (*Cordylura*).—Ind. Unrecognized.
tenuis Walker, 1861: 331 (?*Asteia*).—U.S. Unrecognized.

Nomina Nuda

The following nomina nuda have not been identified.

abdominalis Harris, 1835: 597 (*Stratyomys*).
achaeta James, 1938d: 73 (*Asilus*).
acrididarum Townsend, 1893h: 468 (*Sarcophaga*).
acuminatus Van Duzee *in* Parent, 1933: 166 (*Condylostylus*).
aenobarbus Loew *in* Osten Sacken, 1878c: 82 (*Neoitamus*).
    aeneobarba, error.
albionense Rothfels, 1956: 120 (*Prosimulium*).
albitarsis Harris, 1835: 595 (*Chironomus*).
aliata Hopkins, 1895: 159 (*Sciara*).
alternata Harris, 1835: 595 (*Tipula*).
amoenifrons Harris, 1835: 597 (*Stratyomys*).
analis Harris, 1835: 598 (*Milesia*).
angulatus Harris, 1835: 598 (*Syrphus*).
annulipes Harris *in* Johnson, 1925c: 85 (*Myopa*).
ansatus Harris, 1835: 596 (*Asilus*).
antennata Coquillett *in* Hine, 1904b: 91 (*Gonia*).
appendiculata Harris, 1835: 597 (*Platypeza*).
appensus Harris, 1835: 598 (*Helophilus*).
aquilae Brodie and White, 1883: 57 (*Hippobosca*). ?=*Lynchia americana*.
articulosa Harris, 1835: 596 (*Bibio*).
atribarbis Say *in* Harris, 1835: 596 (*Asilus*).
atritarsata Harris, 1835: 596 (*Empis*).
aurata Harris, 1835: 596 (*Thereva*).

auratilis Sherman, 1920: 14 (*Allodia*).
auricincta Harris, 1835: 597 (*Leptis*).
auricoma Sherman, 1920: 13 (*Boletina*).
aurinota Harris *in* Johnson, 1925c: 88 (*Tachina*). ?=*Archytas apicifer* or *californiae*.
basalis Harris, 1835: 598 (*Volucella*).
bidentata Fisher *in* Shaw, 1941a: 170 (*Platyura*).
binotatus Harris *in* Johnson, 1925c: 79 (*Syrphus*).
biplagiata Say *in* Johnson, 1925c: 64 (*Leia*).
biplagiatus Harris, 1835: 595 (*Tanypus*).
borealis Stein *in* Baker, 1904: 33 (*Limnophora*).
calceatum Harris, 1835: 595 (*Simulium*).
calceola Harris, 1835: 600 (*Anthomyia*).
calceolus Say *in* Johnson, 1925c: 76 (*Asilus*).
cauta Harris, 1835: 600 (*Tetanocera*).
cernua Sherman, 1920: 13 (*Leia*).
cinctipes Harris, 1835: 600 (*Trypeta*).
claracollis Harris, 1835: 595 (*Chironomus*).
colon Harris, 1835: 600 (*Ortalis*).
consors Loew *in* Hagen, 1881a: 44 (*Aricia*).
crataegiplica Osten Sacken, 1878c: 6 (*Cecidomyia*; orig. as *crataegi plica* Walsh, 1869: 80).
cumberlandensis Alexander *in* Rogers, 1930: 49 (*Molophilus*).
debilipes Harris, 1835: 595 (*Ceratopogon*).
dubius Say *in* Johnson, 1925c: 72 (*Mulio*)
elevatus Harris, 1835: 596 (*Hybos*).
encaustus Harris, 1835: 598 (*Syrphus*).
facialis Harris *in* Johnson, 1925c: 85 (*Conops*).
fascicollis Harris, 1835: 598 (*Eristalis*).
fascipennis Harris, 1835: 597 (*Xylophagus*).
flavipes Coquillett *in* Cockerell, 1898b: 155 (*Siphonella*).
floridensis Bromley, 1950a: 234 (*Bombomima*).
fulviana Harris *in* Johnson, 1925c: 76 (*Laphria*).
glauconotatus Harris, 1835: 596 (*Asilus*).
globulus Osten Sacken, 1878c: 6 (*Cecidomyia*; orig. as *crataegi globulus* Walsh, 1869: 80).
gnava Harris, 1835: 598 (*Milesia*).
goniphora Harris, 1835: 600 (*Anthomyia*).
    goniophora, error.
gophori Brodie *in* Brodie and White, 1883: 54 (*Cuterebra*).
guttulatus Harris, 1835: 595 (*Anopheles*; as *quadrimaculatus* var.).
heros Harris, 1835: 597 (*Xylophagus*).
hirtidorsum Tothill, 1924b: 269 (*Fabriciella*).
humeralis Harris, 1835: 597 (*Leptis*).

incisuralis Harris, 1835: 597 (*Xylophagus*).
interrupta Harris, 1835: 599 (*Conops*).
isabella Fisher *in* Shaw and Fisher, 1952: 194 (*Azana*).
ischiaca Harris, 1835: 600 (*Anthomyia*).
ischiaca Harris, 1835: 598 (*Milesia*).
ischiacus Harris, 1835: 597 (*Chrysops*).
johannseni Fisher *in* Strickland, 1946: 161 (*Dynastosoma*).
johannseni Sherman, 1920: 13 (*Neuratelia*).
lapsans Harris, 1835: 600 (*Lauxania*). =*Minettia* sp.
lateralis Harris, 1835: 595 (*Chironomus*).
lateralis Harris, 1835: 599 (*Dexia*). =*Ptilodexia* sp.
lenis Harris, 1835: 600 (*Anthomyia*).
limbipennis Coquillett *in* Cockerell, 1898b: 155 (*Pachycerina*).
limbus Osten Sacken, 1878c: 6 (*Cecidomyia;* orig. as *crataegi limbus* Walsh, 1869: 80).
longiseta Fisher *in* Wray, 1950: 28 (*Platyura;* as *discoloria* ssp.).
lunatifrons Harris, 1835: 599 (*Anthomyia*).
maculipennis Harris *in* Johnson, 1925c: 63 (*Chironomus*).
majuscula Sherman, 1920: 13 (*Boletina*).
maxima Sherman, 1920: 14 (*Allodia*).
neobscura Shannon *in* Johannsen, 1928b: 801 (*Xylota*).
nigrina Sherman, 1920: 14 (*Trichonta*).
noctivaga Sherman, 1920: 13 (*Boletina*).
notatifrons Harris, 1835: 600 (*Anthomyia*).
nudipennis Loew *in* Packard, 1869a: 408 (*Sarcophaga*).
nuphera Harris, 1835: 600 (*Ortalis*). ?=*Eurosta asteris* or *solidaginis*.
obesa Harris, 1835: 599 (*Echinomyia*). ?=*Peleteria iterans*.
obliqua Harris, 1835: 598 (*Milesia*).
obscura Fisher *in* Shaw, 1940: 50 (*Mycetophila;* as *fungorum* var.).
oedipodinis Townsend, 1893h: 468 (*Sarcophaga*). ?=*Blaesoxipha* sp.
ornata Harris *in* Johnson, 1925c: 68 (?*Oxycera*).
partiarius Say *in* Johnson, 1925c: 76 (*Asilus*).
parviceps Loew *in* Hagen, 1881a: 44 (*Aricia*).
pecten Fisher *in* Strickland, 1946: 161 (*Coelosia*).
pectenipes Say *in* Johnson, 1925c: 78 (*Rhamphomyia*).
placida Harris, 1835: 600 (*Cordylura*).
plagiata Harris, 1835: 596 (*Thereva*).
politus Harris, 1835: 597 (*Xylophagus*).
protenta Sherman, 1920: 14 (*Allodia*).
pterelas Harris, 1835: 598 (*Eristalis*).
pulchella Bromley, 1950a: 234 (*Bombomima*).
pulsator Harris *in* Johnson, 1925c: 97 (*Seioptera*).
putricola Cole *in* Barrett, 1932: 295 (*Euxesta*).
quadrangula Sherman, 1920: 14 (*Allodia*).

quadrifasciata Harris, 1835: 600 (*Trypeta*).
quadripunctata Harris, 1835: 597 (*Stratyomys*).
reces Harris, 1835: 600 (*Anthomyia*).
rubidapex Harris, 1835: 599 (*Tachina*). ?=*Peleteria haemorrhoa*.
ruficornis Harris, 1835: 598 (*Chrysogaster*).
rufipes Harris *in* Johnson, 1925c: 83 (*Xylota*).
schinophora Loew *in* Hagen, 1881a: 44 (*Aricia*).
scutellaris Loew *in* Parent, 1929a: 173 (*Sciopus*).
scutellata Harris, 1835: 595 (*Trichocera*).
scutellatus Harris, 1835: 597 (*Sargus*).
seorsa Sherman, 1920: 13 (*Phthinia*).
septentrionalis Alexander *in* Johannsen, 1928b: 691 (*Antocha*).
serrosa Pettey, 1918a: 331 (*Sciara*).
simulatus Harris, 1835: 598 (*Syrphus*).
spuria Curran *in* Dobroscky, 1925: 280 (*Sarcophaga*).
squamosella Kieffer, 1913a: 23 (*Neolasioptera;* n. name for nonexistent "*squamosa* Felt, 1909: 483").
    squamosa Felt *in* Kieffer, 1913a: 23.
stigmata Harris *in* Johnson, 1925c: 71 (*Leptis*).
stricklandi Fisher *in* Strickland, 1946: 161 (*Exechia*).
subnigra Fisher, 1938b: 222 (*Sceptonia*).
subsimplex Loew *in* Hagen, 1881a: 50 (*Coenosia*).
tarsata Loew *in* Hagen, 1881a: 49 (*Hylemyia*).
tenebrosa Coquillett *in* Slosson, 1895: 6 (*Chaetona*).
tenebrosa Sherman, 1920: 13 (*Boletina*).
teretus Harris, 1835: 598 (*Syrphus*).
timida Harris, 1835: 600 (*Anthomyia*).
trichosa Fisher *in* Strickland, 1946: 160 (*Macrocera*).
trifasciata Harris, 1835: 597 (*Leptis*).
triplagiata Harris, 1835: 595 (*Leia*).
tuberculata Harris, 1835: 598 (*Sericomyia*).
tudicornis Harris, 1835: 598 (*Merodon*).
univittata Sherman, 1920: 13 (*Coelosia*).
vancouverensis Sherman, 1920: 13 (*Paratinia*).
velox Harris, 1835: 600 (*Phora*).
viridicincta Harris, 1835: 597 (*Stratyomys*).
vittata Sherman, 1920: 13 (*Phthinia*).
vittatifrons Harris, 1835: 598 (*Syrphus*).
vittatum Harris *in* Johnson, 1925c: 63 (*Asindulum*).
vittipennis Harris, 1835: 595 (*Tipula*).
vorax Harris, 1835: 596 (*Asilus*).
weidhausii Crafts *in* Hamilton, 1957: 92 (also *in* Porter, Spilker, and Walker, 1959: 59) (*Phytomyza*).
ziczac Harris, 1835: 595 (*Leia*).

SELECTED BIBLIOGRAPHY OF NORTH AMERICAN DIPTERA

By Jack R. Coulson, Curtis W. Sabrosky, and Irmgard Muller

This bibliography includes those references, nearly 4,800 in number, employed in the present catalog or in annotations within the bibliography itself. All references except one (Schrank, 1795) have been examined by us, or in a very few cases by others for us. In a few cases of duplicate publication (e.g., see Loew, 1850c), we have seen either separate or serial, but not both.

Under each author, or authors, references are arranged chronologically by year, but within each year the references are grouped by journal and the subletters do not show chronology except for each journal. A single reference is used for the two or more parts of a paper or a numbered series of papers, that appeared within one continuously paged volume within a single year, or over a number of years if no other reference intervenes (exception: Loew's Centuriae). No reference includes parts from two or more volumes even within a single year unless the volumes are paged continuously.

Where two or more parts are included in a single reference, the parts are separated in the title by semicolons, and the corresponding pages likewise. In continuously paged entries covering two or more years, semicolons are used to separate the pages of different dates of publication, each date appearing in italics at the end of its respective series of pages. Commas are used in the same manner within single-year references to indicate publication of a series of pages at different times within the year, e.g., in a later number of a journal, or in a separate fascicle.

If it was necessary to divide references to continued articles or works because of their appearance in different volumes and/or at different dates, they are marked continued. The abbreviation "cont.", when used after the title, indicates that the work is a continuation, and when used after the pages, that it is continued. The abbreviation may appear in both places, although preceding and following parts do not necessarily appear in this bibliography. The abbreviation "concl." marks the concluding portion.

Double citations. Many published works are reissued in some way, not counting author's "reprints" and "separates." Where both date and pagination of a reissued work are like the original, a double citation is usually not given in the catalog, although the information may appear in the bibliography. For other reissues, double citations such as the following are used in the catalog:

(1) Macquart, 1838b: 5 (1839: 121).
(2) Townsend, 1911a: 146 (also 1912b: 312).
(3) Loew, 1863a: 35 (Cent. 3, no. 67).
(4) Say, 1830: 188 (1859b: 371).

The first example signifies that the work was republished in its original form, in this instance with a different pagination. Information on this reissue is contained in an annotation under the original (i.e., first cited) reference, not in a separate bibliographic entry. In the second example, "also" indicates that the taxon or designation referred to can also be found in the second bibliographic entry. This type of double citation is usually employed when a genus or species is described as new in two different papers, or is redescribed in more detail. For Loew's Centuriae (example 3), the second or parenthetical reference consists of the Centuria number and the species number, which are the same for both original publication and reissue (although the actual pagination differs), and which have often been used in the literature as the sole references to Loew's species, as in Aldrich, 1905. In the fourth example, the second citation refers to the LeConte edition (1859a, b) of the "Complete Writings of Thomas Say," which is cited here because it is more widely available than many of Say's original papers.

Authors. Names of authors are boldfaced except where the reference is merely an explanation of one of the "series" referred to in the bibliography (see below). Where two or more spellings appear in the literature, the most commonly used is adopted here, but others are listed, and where necessary, cross-referenced. Minutes of meetings, etc., are attributed to the recording secretary, except in the case of the "Bulletin de la Société entomologique de France," where individual authorship is noted for portions of the minutes.

Reference to a small section by one author that appears as an integral part of a larger work by (or edited by) a different author is treated in one of two ways. Where an individual title is given to the smaller portion, we have cited that part with its author, as for example:

Adams, C. F.
1907 New species of *Mycetophila*. P. 37. *In* Banta, A. M., The fauna of Mayfield's Cave. Carnegie Inst. Wash., Pub. 67: 1–114.

This is referred to in the body of the catalog as "Adams, 1907: 37." Had the smaller portion been untitled, only the main reference would have been listed in the bibliography, and the catalog citation would read "Adams *in* Banta, 1907: 37." Cross-references are used in both methods.

Dates of Publication. Every effort has been made to determine the correct date of publication. The actual date, at the left hand margin, is enclosed in parentheses if given only in a later volume of the same work or journal, and in brackets if discovered in some other way. In the latter case, the reference is annotated to show the reason for adoption of the date. Where the work itself shows a date that is not the actual date of publication, the former is given in parentheses between the journal name and volume number, or, for books and "series," after the place of publication.

Titles. Titles are given as they appeared in the original, with errors indicated by "sic." Misnumbered parts are sometimes corrected in brackets following the "sic." Only German nouns and proper names in English are capitalized. Generic and specific names are italicized, whether or not they were so used in the original. Titles in Oriental and Cyrillic alphabets have been translated and placed in brackets, unless the original work also contained a translated title, in which case the translated title is given in parentheses.

Journals. Names of journals are abbreviated according to the system used by the U.S. Department of Agriculture Library, as shown in Whitlock (1939). For others, abbreviations using the same general system have been devised. At the end of the bibliography is a list of the abbreviations used herein, showing the full name of the journal, place and dates of publication, name changes, series, and other information that may be helpful in locating the periodical in libraries. Journal series numbers have been given, in front of the volume numbers. Where both the whole series number and a separate series number appear on the journal, both are given in the entry, e.g., "ser. 2, 6 [=whole ser., 28]: 15–32." Numbers of parts within the journal volume are given only when they are separately paged, and appear in parentheses between volume number and pages, as **6 (3): 2–16.** In a very few cases, two subdivisions are needed within the parentheses, but their meaning is then indicated. In some cases (e.g., Kansas University Science Bulletin), a journal forms part of another named journal, both with a separate system of volume numbers. For completeness and for assistance in case of confused references in the literature, the other name and volume number are given in parentheses at the end of the reference, and the alternate name is explained in the appended list of periodicals.

"Series." Several important series of publications, such as Wytsman's "Genera Insectorum" and Lindner's "Die Fliegen der palaearktischen Region," appeared under an overall title and editorship, but at irregular intervals. References within the "series" appear under the separate authors, and are shown to be *"in"* the particular "series" involved. The notation "q.v." refers the reader to an information entry for the whole "series," inserted in the bibliography under its own editorship, which name is not boldfaced.

Books. Single-volume books are cited simply by title, pages, and place of publication. Multiple-volume works, if the volumes are separately paged, are divided into a number of references, one for each volume used. Complete information on the entire work is included with the first reference to it.

Illustrations. The number of figures and plates, if any, is placed at the end of a reference. Figures are listed first, unless only part of a plate refers to the particular article cited, in which case the plate number, followed by the relevant figure numbers, is given.

Aczél, M. L.
1939   Das System der Familie Dorylaidae. Dorylaiden-Studien I. Zool. Anz. **125**: 15–23, 3 figs.
1940   Vorarbeiten zu einer Monographie der Dorylaiden (Dipt.). Dorylaiden-Studien V. Zool. Anz. **132**: 149–169, 6 figs.
1948   Grundlagen einer Monographie der Dorilaiden (Diptera). Dorilaiden-Studien VI. Acta Zool. Lilloana **6**: 5–168, 13 figs.
1949   Notes on Tylidae. II. Argentine species of the subfamily Tylinae in the entomological collection of the Miguel Lillo Foundation. Acta Zool. Lilloana **8**: 219–280, 13 figs., 2 pls.
1950   Géneros y especies de la tribu "Trypetini." I. Dos géneros y tres especies nuevos de la Argentina (Tephritidae, Diptera). Acta Zool. Lilloana **9**: 307–323, 4 figs., 1 pl.
1951   Morfología externa y división sistemática de las "Tanypezidiformes" con sinópsis de las especies argentinas de "Tylidae" ("Micropezidae") y "Neriidae" (Dipt.). Acta Zool. Lilloana **11**: 483–589, 24 figs., 4 pls.
1955   Fruit flies of the genus *Tomoplagia* Coquillett (Diptera, Tephritidae). U.S. Natl. Mus. Proc. **104**: 321–411, 13 figs., 8 pls.
1956a   Revisión parcial de las Pyrgotidae neotropicales y antárticas, con sinópsis de los géneros y especies (Diptera, Acalyptratae). Parte I. Rev. Brasil. de Ent. **4**: 161–184, fig. A.
1956b   Revisión parcial de las Pyrgotidae neotropicales y antárticas, con sinópsis de los géneros y especies (Diptera, Acalyptratae). Parte II. Rev. Brasil. de Ent. **5**: 1–70, figs. 1–60, pls. 1–3.

Aczél, M. L.—Continued
1956c Revisión parcial de las Pyrgotidae neotropicales y antárticas, con sinópsis de los géneros y especies (Diptera, Acalyptratae). Parte III (Conclusão). Rev. Brasil. de Ent. **6:** 1–38, figs. 61–96, pls. 4–5.

Adachi, J. (See Matsumura and Adachi)

Adams, C. F.
1903a Dipterological contributions. Kans. Univ. Sci. Bul. **2** [=whole ser., 12]: 21–47. (=Kans. Univ. Bul. 4 (6).)
1903b Descriptions of six new species. Pp. 221–223. *In* Snow, F. H., A preliminary list of the Diptera of Kansas. Kans. Univ. Sci. Bul. **2** [=whole ser., 12]: 211–223. (=Kans. Univ. Bul. 4 (6).)
1904a Descriptions of new Oscinidae. Ent. News 15: 303–304.
1904b Notes on and descriptions of North American Diptera. Kans. Univ. Sci. Bul. **2** [=whole ser., 12]: 433-455. (=Kans. Univ. Bul. 4 (6).)
1904c On the North American species of *Siphonella*. Psyche **11:** 103–104.
1905a On the North American species of *Oscinis*. Ent. News **16:** 108–111.
1905b Diptera africana. I. Kans. Univ. Sci. Bul. **3** [=whole ser., 13]: 149–208. (=Kans. Univ. Bul. 6 (2).)
1907 New species of *Mycetophila*. P. 37. *In* Banta, A. M. The fauna of Mayfield's Cave. Carnegie Inst. Wash., Pub. **67:** 1–114.
1908 Notes on North American species of *Crassiseta* v. Ros. N.Y. Ent. Soc. Jour. **16:** 151–152.

Addis, C. J.
1945 *Phlebotomus* (*Dampfomyia*) *anthophorus*, n. sp., and *Phlebotomus diabolicus* Hall from Texas (Diptera: Psychodidae). Jour. Parasitol. **31:** 119–127, 2 pls.

Agassiz, L.
1842–
1847
Nomenclator zoologicus, nomina systematica generum animalium tam viventium quam fossilium, secundem ordinem alphabeticum disposita, adjectis auctoribus, libris in quibus reperiuntur, anno editionis, etymologia et familiis, ad quas pertinent in variis classibus. 12 Fascs. Soloduri [=Solothurn, Switzerland].

The entire work consists of 12 fascicles, Soloduri, 1842–1847. Some fascicles are in several parts, and/or involve several authors. Fasc. 9/10, which includes the Diptera by Agassiz and H. Loew,

Agassiz, L.—Continued
    consists of xlii pp. of introductory material, and 10 unnumbered, separately paged parts by different authors, a part for each taxon listed in its title, some of which are only 1 or 2 pages in length.
    See Agassiz and Loew.

———, **and Loew, H.**
1846   Nomina systematica generum dipterorum. [Pt. 4], 42 pp. *In* Agassiz, L., Nomenclator zoologicus [q.v.]. Fasc. 9/10: Titulum et praefationem operis, Mollusca, Lepidoptera, Strepsiptera, Diptera, Myriapoda, Thysanura, Thysanoptera, Suctoria, Epizoa et Arachnidas. Soloduri [=Solothurn, Switzerland].

Ahlberg, O.
1939   *Diarthronomyia chrysanthemi* nom. nov. (=*hypogaea* Felt nec Löw). Ent. Tidskr. **60**: 274.

Aitken, T. H. G.
1939   The *Anopheles maculipennis* complex of western America (Diptera, Culicidae). Pan-Pacific Ent. **15**: 191–192.
1941   A new American subgenus and species of *Aedes* (Diptera, Culicidae). Pan-Pacific Ent. **17**: 81–84.
1945   Studies on the anopheline complex of western America. Calif. Univ., Pubs., Ent. **7**: 273–364, 39 figs.

Aldrich, J. M.   (See also Brues, 1903; Van Duzee, Cole, and Aldrich)
1892a   New species of *Phora*. Canad. Ent. **24**: 142–146, 5 figs.
1892b   A new genus and species of Tabanidae. Psyche **6**: 236–237, 3 figs.
1893a   Revision of the genera *Dolichopus* and *Hygroceleuthus*. Kans. Univ. Quart. **2**: 1–26, 1 pl.
1893b   New genera and species of Psilopinae. Kans. Univ. Quart. **2**: 47–50.
1893c   The dolichopodid genus *Liancalus* Loew. Psyche **6**: 569–571.
1894   New genera and species of Dolichopodidae. Kans. Univ. Quart. **2**: 151–157.
1895   The tipulid genera *Bittacomorpha* and *Pedicia*. Psyche **7**: 200–202, 1 fig.
1896a   The dipterous genera *Tachytrechus* and *Macellocerus*. Amer. Ent. Soc. Trans. **23**: 81–84.
1896b   Dolichopodidae; Phoridae. Pp. 309–345, pl. 12, figs. 108–119; pp. 435–438. *In* Williston, S. W., On the Diptera of St. Vincent (West Indies). Ent. Soc. London, Trans. **1896**: 253–446, pls. 8–14.

Aldrich, J. M.—Continued

1897   A collection of Diptera from Indiana caves. Pp. 186–190. *In* Blatchley, W. S., The fauna of Indiana caves. Ind. Dept. Geol. and Nat. Resources, Ann. Rpt. (1896) **21**: 175–212.

1901   Supplement. Dolichopodidae. Pp. 333–366, pl. 6, figs. 7–24. *In* Godman, F.D., and Salvin, O., eds., Biologia Centrali-Americana [q.v.]. Zoologia-Insecta-Diptera, Vol. 1, 378 pp., 6 pls. London.

1902   Dolichopodidae of Grenada, W. I. Kans. Univ. Sci. Bul. **1**: [=whole ser., 11]: 75–95, pl. 4. (=Kans. Univ. Bul. 2 (8).)

1904   A contribution to the study of American Dolichopodidae. Amer. Ent. Soc. Trans. **30**: 269–286.

1905   A catalogue of North American Diptera. Smithsn. Inst., Smithsn. Misc. Collect. **46** (2 [=pub. 1444]): 1–680. For addendum, see Aldrich, 1907b.

1906   The dipterous genus *Calotarsa*, with one new species. Ent. News **17**: 123–127, 1 pl.

1907a   The dipterous genus *Scellus*, with one new species. Ent. News **18**: 133–136.

1907b   Additions to my catalogue of North American Diptera. N.Y. Ent. Soc. Jour. **15**: 2–9.

1909   The fruit-infesting forms of the dipterous genus *Rhagoletis*, with one new species. Canad. Ent. **41**: 69–73, 1 pl.

1911a   A revision of the North American species of the dipterous genus *Hydrophorus*. Psyche **18**: 45–70, 1 pl.

1911b   The dipterous genus *Diostracus* Loew. Psyche **18**: 70–73, 1 pl.

1912a   The biology of some western species of the dipterous genus *Ephydra*. N.Y. Ent. Soc. Jour. **20**: 77–99, 3 pls.

1912b   Two western species of *Ephydra*. N.Y. Ent. Soc. Jour. **20**: 99–102.

1913   The North American species of *Lispa* (Diptera: Anthomyidae). N.Y. Ent. Soc. Jour. **21**: 126–146.

1914a   A new *Leucopis* with yellow antennae. Jour. Econ. Ent. **7**: 404–405.

1914b   Description of *Sarcophaga kellyi*. Pp. 443–445, pl. 40. *In* Kelly, E. O. G., A new sarcophagid parasite of grasshoppers. Jour. Agr. Res. **2**: 435–445, pl. 40.

1915a   A new *Sarcophaga* parasitic on *Allorhina nitida*. Jour. Econ. Ent. **8**: 151–152, 1 fig.

1915b   The deer bot-flies (genus *Cephenomyia* Latr.). N.Y. Ent. Soc. Jour. **23**: 145–150, 1 pl.

Aldrich, J. M.—Continued

1915c   New American species of *Asteia* and *Sigalsoësa* [sic]. Psyche **22**: 94–98, 2 figs.

>   *Sigalsoësa* is correct as *Sigaloëssa* in the Table of Contents and Index of Vol. 22 of the journal, and in the reprint.

1915d   The dipterous genus *Symphoromyia* in North America. U.S. Natl. Mus. Proc. **49**: 113–142, 10 figs.

1916a   *Sarcophaga* and allies in North America. [Vol. 1], 302 pp., 16 pls. *In* Entomological Society of America, Thomas Say Foundation [q.v.]. Lafayette, Ind.

1916b   Two new Canadian Diptera. Canad. Ent. **48**: 20–22.

1918a   The kelp flies of North America (genus *Fucellia*, family Anthomyidae). Calif. Acad Sci. Proc. **ser. 4, 8**: 157–179, 10 figs.

1918b   Two new *Hydrotaeas* (Diptera, Anthomyidae). Canad. Ent. **50**: 311–314.

1918c   New and little-known Canadian Oscinidae. Canad. Ent. **50**: 336–343, 5 figs.

1918d   The anthomyid genus *Pogonomyia* (Dip.). Ent. News **29**: 179–185, 1 fig.

1919a   *Leiomyza* in North America (Dipt., Drosophilidae). Ent. News **30**: 137–141, 1 fig.

1919b   Two new genera of Anthomyidae (Dipt.). Ent. Soc. Wash. Proc. **21**: 106–109, 1 fig.

1919c   Description of a new species of *Hylemyia*. Pp. 380–381. *In* Plath, O. E., The prevalence of *Phormia azurea* Fallén (larva parasitic on nestling birds) in the Puget Sound region and data on two undescribed flies of similar habit. Ent. Soc. Amer. Ann. **12**: 373–381.

1921a   The muscoid genera *Pseudeuantha* and *Uramyia* (Diptera). Insecutor Inscitiae Menstruus **9**: 83–92.

1921b   The anthomyiid genus *Atherigona* in America (Diptera). Insecutor Inscitiae Menstruus **9**: 93–98, fig. 2.

1922a   A new genus of Helomyzidae. Brooklyn Ent. Soc. Bul. **17**: 108–109.

1922b   Two-winged flies of the genera *Dolichopus* and *Hydrophorus* collected in Alaska in 1921, with new species of *Dolichopus* from North America and Hawaii. U.S. Natl. Mus. Proc. **61** (25): 1–18.

1922c   A new genus of two-winged fly with mandible-like labella. Ent. Soc. Wash. Proc. **24**: 145–152, 1 pl.

Aldrich, J. M.—Continued

1923a   A new tachinid parasite of the codling moth (Dip.). Ent. News **34:** 53–54.
1923b   Notes on the dipterous family Hippoboscidae. Insecutor Inscitiae Menstruus **11:** 75–79.
1923c   New genera of two-winged flies of the subfamily Leptogasterinae of the family Asilidae. U.S. Natl. Mus. Proc. **62 (20):** 1–6, 3 figs.
1923d   Two Asiatic muscoid flies parasitic on the so-called Japanese beetle. U.S. Natl. Mus. Proc. **63 (6):** 1–4.
1923e   A new parasitic fly reared from the bean beetle. Ent. Soc. Wash. Proc. **25:** 95–96.
1925a   A new *Leucopis* from San Francisco. Pan-Pacific Ent. **1:** 152.
1925b   Notes on some types of American muscoid Diptera in the collection of the Vienna Natural History Museum [cont.]. Ent. Soc. Amer. Ann. **18:** 107–130, 2 figs. [cont.].
1925c   New Diptera or two-winged flies in the United States National Museum. U.S. Natl. Mus. Proc. **66 (18):** 1–36, 1 fig.
1925d   Two new species of the tachinid genus *Lixophaga*, with notes and key. Ent. Soc. Wash. Proc. **27:** 132–136.
1926a   Notes on muscoid flies with retracted hind crossvein, with key and several new genera and species. Amer. Ent. Soc. Trans. **52:** 7–28.
1926b   North American two-winged flies of the genus *Cylindromyia* Meigen (*Ocyptera* of authors). U.S. Natl. Mus. Proc. **68 (23):** 1–27, 1 pl.
1926c   American two-winged flies of the genus *Microphthalma* Macquart, with notes on related forms. U.S. Natl. Mus. Proc. **69 (13):** 1–8.
1926d   Descriptions of new and little known Diptera or two-winged flies. U.S. Natl. Mus. Proc. **69 (22):** 1–26.
    This was published in December.
1926e   On the status of the generic name *Anthrax* Scopoli. Insecutor Inscitiae Menstruus **14:** 12–15.
1926f   A new genus of Helomyzidae from Chile with key to genera. Insecutor Inscitiae Menstruus **14:** 96–102.
1927   Notes on muscoid synonymy. Brooklyn Ent. Soc. Bul. **22:** 18–25.
1928   A revision of the American parasitic flies belonging to the genus *Belvosia*. U.S. Natl. Mus. Proc. **73 (8):** 1–45.

Aldrich, J. M.—Continued

1929a A revision of the two-winged flies of the genus *Procecidochares* in North America, with an allied new genus. U.S. Natl. Mus. Proc. **76** (2): 1–13.

1929b Revision of the two-winged flies of the genus *Coelopa* Meigen in North America. U.S. Natl. Mus. Proc. **76** (11): 1–6.

1929c New genera and species of muscoid flies. U.S. Natl. Mus. Proc. **76** (15): 1–13.

1929d Notes on synonymy of Diptera, No. 3. Ent. Soc. Wash. Proc. **31**: 32–36.

1929e Three new acalyptrate Diptera. Ent. Soc. Wash. Proc. **31**: 89–91.

1930a New two-winged flies of the family Calliphoridae from China. U.S. Natl. Mus. Proc. **78** (1): 1–5, 3 figs.

1930b American two-winged flies of the genus *Stylogaster* Macquart. U.S. Natl. Mus. Proc. **78** (9): 1–27.

1930c Notes on the types of American two-winged flies of the genus *Sarcophaga* and a few related forms, described by the early authors. U.S. Natl. Mus. Proc. **78** (12): 1–39, 3 pls.

1931a North American two-winged flies of the genus *Spathimeigenia*, with descriptions of five new species. U.S. Natl. Mus. Proc. **80** (11): 1–10.

1931b Notes on Diptera, No. 5. Ent. Soc. Wash. Proc. **33**: 116–121.

1932a Records of dipterous insects of the family Tachinidae reared by the late George Dimmock, with description of one new species and notes on the genus *Anetia* Robineau-Desvoidy. U.S. Natl. Mus. Proc. **80** (20): 1–8.

1932b New Diptera, or two-winged flies, from America, Asia, and Java, with additional notes. U.S. Natl. Mus. Proc. **81** (9): 1–28, 2 figs., 1 pl.

1933a Notes on the tachinid genus *Elodia* R. D., with three new species of *Elodia* and *Phorocera* (Diptera) from Japan. Ent. Soc. Wash. Proc. **35**: 19–23.

1933b Notes on Diptera, No. 6. Ent. Soc. Wash. Proc. **35**: 165–170.

————, **and Darlington, P. S.**

1908 The dipterous family Helomyzidae. Amer. Ent. Soc. Trans. **34**: 67–100, 2 pls.

————, **and Shannon, R. C.** (See Shannon, 1923b)

Aldrich, J. M., and Webber, R. T.
1924a The North American species of parasitic two-winged flies belonging to the genus *Phorocera* and related genera. U.S. Natl. Mus. Proc. **63** (17): 1–90.
1924b Change of preoccupied names. Ent. Soc. Wash. Proc. **26**: 195.

Alexander, C. P. (see also Curran, 1934a)
1911a New Tipulidae (Diptera). Canad. Ent. **48**: 286–288.
1911b Notes on two Tipulidae (Dipt.). Ent. News **22**: 349–354, 4 figs.
1911c Synonymical, and other notes on the Tipulidae (Diptera). Psyche **18**: 192–203, 1 pl.
1912a On the tropical American Rhipidiae (Tipulidae, Dipt.). Brooklyn Ent. Soc. Bul. **8**: 6–17, 1 pl.
1912b New species of *Furcomyia* (Tipulidae). Canad. Ent. **44**: 333–342, 361–364, 1 pl.
1912c The American species of *Adelphomyia* Bergroth (Tipulidae, Dipt.). Pomona Col. Jour. Ent. **4**: 829–831, 1 fig.
1912d New nearctic Tipulidae (Diptera). Psyche **19**: 163–171, 1 pl.
1913a A synopsis of part of the neotropical crane-flies of the subfamily Limnobinae. U.S. Natl. Mus. Proc. **44**: 481–549, 4 pls.
1913b Report on a collection of Japanese crane-flies (Tipulidae), with a key to the species of *Ptychoptera*. Canad. Ent. **45**: 197–210, pls. 3–4.
1914a New or little-known craneflies from the United States and Canada. Tipulidae, Diptera. Acad. Nat. Sci. Phila. Proc. **66**: 579–606, 3 pls.
1914b On a collection of crane-flies from British Guiana (Tipulidae, Diptera). Amer. Ent. Soc. Trans. **40**: 223–255, pls. 3–4.
1915a New or little-known crane-flies from the United States and Canada: Tipulidae, Diptera. Part 2. Acad. Nat. Sci. Phila. Proc. **67**: 458–514, 6 pls.
1915b A new nearctic *Gonomyia* (Tipulidae, Diptera). Ent. News **26**: 170–172, 3 figs.
1915c New nearctic crane-flies in the United States National Museum (Diptera, Tipulidae). Insecutor Inscitiae Menstruus **3**: 127–142.
1915d A biological reconnaissance of the Okefenokee Swamp in Georgia. The Tipulidae (Diptera). Wash. Univ. [St. Louis] Studies **2** (pt. 1): 97–98.

Alexander, C. P.—Continued

1916a New or little-known crane-flies from the United States and Canada: Tipulidae, Ptychopteridae, Diptera. Part 3. Acad. Nat. Sci. Phila. Proc. **68:** 486–549, 7 pls.

1916b New nearctic crane-flies (Tipulidae, Diptera). Canad. Ent. **48:** 42–53.

1916c New North American species of the genus *Gonomyia* Meigen (Tipulidae, Diptera). Canad. Ent. **48:** 316–325.

1916d New limnophiline crane-flies from the United States and Canada (Tipulidae, Diptera). N.Y. Ent. Soc. Jour. **24:** 118–125, 1 pl.

1916e New species of crane-flies from the West Indies (Tipulidae, Dip.). Ent. News **27:** 343–347, 6 figs.

1917 New nearctic crane-flies (Tipulidae). Part II; (Tipulidae, Diptera). Part III. Canad. Ent. **49:** 22–31, 61–64; 199–211, pl. 12.

1918a New nearctic crane-flies (Tipulidae, Diptera). Part IV; Part V; Part VI; Canad. Ent. **50:** 60–71; 158–165, 242–246, 1 fig.; 381–386, 411–416.

1918b New species of crane-flies from California (Dip.). Ent. News **29:** 285–288.

1919a New nearctic species of the genus *Erioptera* Meigen (Tipulidae, Diptera). Brooklyn Ent. Soc. Bul. **14:** 104–108.

1919b The crane-flies collected by the Canadian Arctic Expedition, 1913–18. Pp. 3–30, pls. 1–6. *In* Anderson, R. M., ed., Report of the Canadian Arctic Expedition 1913–18 [q.v.]. Vol. 3: Insects, Pt. C: Diptera, 90 pp., 2 figs., 10 pls. Ottawa, Ont.

1919c New nearctic crane-flies (Rhyphidae and Tipulidae, Diptera). Part VII; (Tipulidae, Diptera). Part VIII. Canad. Ent. **51:** 162–172; 191–199.

1919d New or little-known crane-flies from Japan. Part I. (Tipulidae, Diptera). Ent. Mag. [Japan] **3:** 122–127.

1919e Notes on the genus *Dicranoptycha* Osten Sacken (Tipulidae, Diptera). Ent. News **30:** 19–22.

1919f Two new crane-flies from California (Tipulidae, Diptera). Ent. News **30:** 214–215.

1919g New species of eriopterine crane-flies from the United States (Tipulidae, Diptera). Insecutor Inscitiae Menstruus **7:** 143–148.

1919h The crane-flies of New York. Part I. Distribution and taxonomy of the adult flies. N.Y. (Cornell) Agr. Expt. Sta. Mem. **25:** 763–993, 12 figs., 31 pls.

    Also bound in N.Y. (Cornell) State Col. Agr., and Agr. Expt. Sta., Ann. Rpt. 32: 763–993.

Alexander, C. P.—Continued

1919i   Undescribed species of Japanese crane-flies (Tipulidae, Diptera). Ent. Soc. Amer. Ann. **12:** 327–348.

1920a   The crane-flies collected by the Swedish expedition (1895–1896) to southern Chile and Tierra del Fuego. Arkiv för Zool. **13 (6):** 1–32, 2 pls.

1920b   Undescribed Tipulidae (Diptera) from western North America. Calif. Acad. Sci. Proc. **ser. 4, 10:** 35–46.

1920c   Two undescribed pediciine crane-flies from the United States (Tipulidae, Diptera). Canad. Ent. **52:** 78–80.

1920d   New nearctic crane-flies (Tipulidae, Diptera). Part IX; Part X. Canad. Ent. **52:** 109–112; 224–229.

1920e   New or little-known crane-flies in the Queensland Museum (Tipulidae, Diptera). Queensland Mus. Mem. **7:** 52–63.

1920f   An undescribed species of *Ptychoptera* from the western United States (Ptychopteridae, Diptera). Ent. News **31:** 3–4.

1920g   New species of crane-flies from the United States and Canada (Tipulidae, Diptera). Jour. Ent. and Zool. **12:** 85–92.

1920h   The crane-flies of New York. Part II. Biology and phylogeny. N.Y. (Cornell) Agr. Expt. Sta. Mem. **38:** 699–1133, 87 pls.

      Also bound in N.Y. (Cornell) State Col. Agr., and Agr. Expt. Sta., Ann. Rpt. 33: 699–1133.

1920i   Scientific results of the Katmai Expedition of the National Geographic Society. The crane-flies (Tipulidae, Diptera). Ohio Jour. Sci. **20:** 193–203.

1921a   Undescribed Tipulidae (Diptera) from western North America, Part II. Calif. Acad. Sci. Proc. **ser. 4, 11:** 103–107.

1921b   Dipterous insects of the family Tipulidae from the Pribilof Islands, Alaska. Calif. Acad. Sci. Proc. **ser. 4, 11:** 183–184.

1921c   New nearctic crane-flies (Tipulidae, Diptera). Part XI. Canad. Ent. **53:** 132–137.

1921d   A new species of *Tipula* injurious to pasture lands (Tipulidae, Diptera). Insecutor Inscitiae Menstruus **9:** 135–137.

1922a   The crane-flies of New York: First supplementary list. Brooklyn Ent. Soc. Bul. **17:** 58–62.

1922b   New species of crane-flies from southern Indiana. Mich. Univ. Mus. Zool. Occas. Papers **127:** 1–7.

Alexander, C. P.—Continued

1922c Undescribed crane-flies in the Paris National Museum (Tipulidae, Diptera): Part IV [sic] [=Part III] [concl.]. Paris, Mus. Natl. d'Hist. Nat. Bul. **28**: 73–75.

1923 Insects, arachnids, and chilopods of the Pribilof Islands, Alaska. Diptera. Suborder Orthorrhapha. Division Nematocera. Families Tipulidae and Rhyphidae. North Amer. Fauna **46**: 159–169, pls. 10–11.

1924a New or little-known crane-flies from New England. Boston Soc. Nat. Hist. Occas. Papers **5**: 115–118.

1924b The crane-flies of New York: Second supplementary list. Brooklyn Ent. Soc. Bul. **19**: 57–64.

1924c Undescribed species of nematocerous Diptera from North America and Japan. Insecutor Inscitiae Menstruus **12**: 81–84.

1924d New species of two-winged flies from western North America belonging to the family Tipulidae. U.S. Natl. Mus. Proc. **64** (10): 1–16.

1924e New or little-known crane-flies from northern Japan (Tipulidae, Diptera). Philippine Jour. Sci. **24**: 531–611, pls. 1–2.

1925a The crane-flies (Tipulidae) of New England: First supplementary list. Boston Soc. Nat. Hist. Occas. Papers **5**: 169–174.

1925b Undescribed species of crane-flies from the eastern United States and Canada (Dipt.: Tipulidae). Part I. Ent. News **36**: 200–204, 229–230.

1926a Undescribed species of the genus *Limnophila* from eastern North America. Part I. (Tipulidae, Diptera). Brooklyn Ent. Soc. Bul. **21**: 109–115.

1926b Records of crane-flies (Tipulidae) from Ontario (Diptera). Canad. Ent. **58**: 236–240.

1926c Undescribed species of crane-flies from the eastern United States and Canada (Dipt.: Tipulidae). Part II; Part III. Ent. News **37**: 44–51; 291–297.

1926d Undescribed species of crane-flies from Cuba and Jamaica (Tipulidae, Diptera). N.Y. Ent. Soc. Jour. **34**: 223–230.

1926e Undescribed species of crane-flies from the United States and Canada. [Part I]. (Diptera, Tipulidae); Part II. (Diptera, Tipulidae). Insecutor Inscitiae Menstruus **14**: 19–24; 114–122.

1926f Undescribed species of *Dicranoptycha* from eastern North America (Tipulidae, Diptera). Psyche **33**: 54–59.

Alexander, C. P.—Continued

1926g Three undescribed eriopterine crane-flies from California (Tipulidae, Diptera). Pan-Pacific Ent. **3**: 77–79.

1927a The crane-flies (Tipulidae) of New England: Second supplementary list. Boston Soc. Nat. Hist. Occas. Papers **5**: 223–231.

1927b Undescribed species of the genus *Limnophila* from eastern North America (Tipulidae, Diptera). Part II; Part III. Brooklyn Ent. Soc. Bul. **22**: 56–64, 6 figs.; 110–115.

1927c Diptera. Fam. Tipulidae. Subfam. Cylindrotominae. Fasc. 187, 16 pp., 2 pls. *In* Wytsman, P., ed., Genera insectorum [q.v.]. Bruxelles.

1927d New or little-known nearctic species of Trichoceridae (Diptera). Part I. Canad. Ent. **59**: 66–73, 2 figs.

1927e New nearctic crane-flies (Tipulidae, Diptera). Part XII. Canad. Ent. **59**: 184–193.

1927f Records and descriptions of crane-flies from Alberta (Tipulidae, Diptera). I. Canad. Ent. **59**: 214–225, 2 figs.

1927g Undescribed species of crane-flies from the eastern United States and Canada (Dipt.: Tipulidae). Part IV. Ent. News **38**: 181–184.

1927h Records and descriptions of crane-flies from the eastern United States (Tipulidae, Diptera). N.Y. Ent. Soc. Jour. **35**: 55–63.

1927i Undescribed species of Tipulidae from Utah (Diptera). Pan-Pacific Ent. **3**: 143–145.

1927j Undescribed crane flies from the Holarctic Region in the United States National Museum. U.S. Natl. Mus. Proc. **72** (2): 1–17, 1 pl.

1928a Records of crane-flies (Tipulidae) from Ontario (Diptera). Part II. Canad. Ent. **60**: 54–60.

1928b New or little known species of the genus *Tipula* from Labrador (Tipulidae, Dipt.). Canad. Ent. **60**: 95–101.

1928c Studies on the crane-flies of Mexico. Part IV. (Order Diptera, superfamily Tipuloidea). Ent. Soc. Amer. Ann. **21**: 101–119.

1929a The crane-flies of New York: Third supplementary list; Fourth supplementary list. Brooklyn Ent. Soc. Bul. **24**: 22–29; 295–302.

1929b Undescribed species of the genus *Limnophila* from eastern North America (Tipulidae, Diptera). Part IV. Brooklyn Ent. Soc. Bul. **24**: 187–191.

Alexander, C. P.—Continued

1929c Crane-flies (Tipulidae, Trichoceridae, Tanyderidae). Pt. 1, 240 pp., 3 figs., 12 pls. *In* British Museum (Natural History), Diptera of Patagonia and south Chile [q.v.]. London.

1929d New nearctic crane-flies (Tipulidae, Diptera). Part XIII. Canad. Ent. **61**: 15–22, 4 figs.

1929e A list of the crane-flies of Quebec (Diptera). I. Canad. Ent. **61**: 231–236, 247–251.

1929f Undescribed species of crane-flies from the eastern United States and Canada (Dipt.: Tipulidae). Part V. Ent. News **40**: 44–49.

1929g Undescribed species of eriopterine crane-flies from the United States and Canada (Tipulidae, Diptera). Part I. N.Y. Ent. Soc. Jour. **37**: 49–58.

1929h New or little known Tipulidae from the Philippines (Diptera). Philippine Jour. Sci. **40**: 239–273, 1 pl.

1930a The crane-flies (Tipulidae) of New England: Third supplementary list. Boston Soc. Nat. Hist. Occas. Papers **5**: 267–278.

1930b New or insufficiently-known crane-flies from the Nearctic Region (Tipulidae, Diptera). Part I; Part II. Brooklyn Ent. Soc. Bul. **25**: 71–77; 276–282.

1931a A new genus and species of bibionid Diptera. Brooklyn Ent. Soc. Bul. **26**: 7–11, 2 figs.

1931b New or insufficiently-known crane-flies from the Nearctic Region (Tipulidae, Diptera). Part III. Brooklyn Ent. Soc. Bul. **26**: 177–184.

1931c A list of the crane-flies of Quebec (Diptera). II. Canad. Ent. **63**: 135–147.

1932 Appendix. A new species of *Tipula* from Wisconsin. Pp. 240–242, fig. 197. *In* Dickinson, W. E., The crane-flies of Wisconsin. Milwaukee Publ. Mus. Bul. **8**: 139–260, figs. 29–197, pls. 22–24.

1933a Undescribed species of eriopterine crane-flies from the United States and Canada (Tipulidae, Diptera), Part II. N.Y. Ent. Soc. Jour. **41**: 91–100.

1933b New or little-known Tipulidae from eastern Asia (Diptera). XVII. Philippine Jour. Sci. **52**: 395–442, 5 pls.

1934 The exploration of Southampton Island, Hudson Bay by George Miksch Sutton, sponsored by Mr. John Bonner Semple 1929–1930. Part II. Zoology. Section 4. Spiders and insects (in part) of Southampton Island. II.

Alexander, C. P.—Continued

Diptera collected on Southampton Island by George Miksch Sutton. Trichoceridae and Tipulidae. Pittsburgh, Carnegie Inst., Carnegie Mus. Mem. **12 (pt. 2, sect. 4** [=pub. 162]): 3–10, 1 pl.

1936a The crane-flies (Tipulidae) of New England: Fourth supplementary list. Boston Soc. Nat. Hist. Occas. Papers **8**: 273–292, 4 figs.

1936b New or little-known Tipulidae from eastern Asia (Diptera). XXX. Philippine Jour. Sci. **60**: 165–204, 3 pls.

1937 New species of Ptychopteridae (Diptera). Brooklyn Ent. Soc. Bul. **32**: 140–143.

1938a New or little-known Tipulidae from eastern Asia (Diptera). XXXVI. Philippine Jour. Sci. **66**: 93–134, 3 pls.

1938b New or insufficiently-known crane-flies from the Nearctic Region (Tipulidae, Diptera). Part IV. Brooklyn Ent. Soc. Bul. **33**: 71–78.

1939a New or insufficiently-known crane-flies from the Nearctic Region (Tipulidae, Diptera). Part V. Brooklyn Ent. Soc. Bul. **34**: 92–100.

1939b *Tipula (Lunatipula) reesi* new species. Pp. 143–144. *In* Rees, B. E., and Ferris, G. F., The morphology of *Tipula reesi* Alexander (Diptera: Tipulidae). Microentomology **4**: 143–176, 20 figs.

1940a Records and descriptions of North American crane-flies (Diptera). Part I. Tipuloidea of the Great Smoky Mountains National Park, Tennessee. Amer. Midland Nat. **24**: 602–644, 48 figs.

1940b New or insufficiently-known crane-flies from the Nearctic Region (Tipulidae, Diptera). Part VI. Brooklyn Ent. Soc. Bul. **35**: 84–89.

1940c New nearctic crane-flies (Tipulidae, Diptera). Part XIV. Canad. Ent. **72**: 151–155.

1940d Undescribed species of crane-flies from the eastern United States and Canada (Dipt.: Tipulidae). Part VI. Ent. News **51**: 83–85, 99–103.

1940e Studies on the crane-flies of Mexico. Part VII. (Order Diptera, superfamily Tipuloidea). Ent. Soc. Amer. Ann. **33**: 140–161.

1941a Records and descriptions of North American crane-flies (Diptera). Part II. Tipuloidea of mountainous western North Carolina. Amer. Midland Nat. **26**: 281–319, 26 figs.

Alexander, C. P.—Continued

- 1941b New or insufficiently-known crane-flies from the Nearctic Region (Tipulidae, Diptera). Part VII. Brooklyn Ent. Soc. Bul. **36**: 12–17.
- 1941c New nearctic crane-flies (Tipulidae, Diptera). Part XV; Part XVI. Canad. Ent. **73**: 85–90; 206–213.
- 1941d Undescribed species of crane-flies from the eastern United States and Canada (Dipt.: Tipulidae). Part VII. Ent. News **52**: 192–196.
- 1942a New nearctic crane-flies (Tipulidae, Diptera). Part XVII. Canad. Ent. **74**: 206–212.
- 1942b Guide to the insects of Connecticut. Part VI. The Diptera or true flies of Connecticut. First fascicle [part]. Family Tanyderidae; Family Ptychopteridae; Family Trichoceridae; Family Anisopodidae; Family Tipulidae. Conn. State Geol. and Nat. Hist. Survey, Bul. **64**: 183–184, fig. 18; 184–187, fig. 19; 188–192, fig. 20; 192–196, fig. 21; 196–486, figs. 22–55.
- 1942c *Tipula (Lunatipula) triplex colei* Alexander subsp. nov. P. 67. *In* Porter, J. B., *Massospora tipulae* sp. nov. and *Tipula triplex colei* Alexander subsp. nov. Elisha Mitchell Sci. Soc. Jour. **58**: 65–68.
- 1943a Records and descriptions of North American crane-flies (Diptera). Part III. Tipuloidea of the Upper Gunnison Valley, Colorado. Amer. Midland Nat. **29**: 147–179, 38 figs.
- 1943b Records and descriptions of North American crane-flies (Diptera). Part IV. Tipuloidea of the Yellowstone National Park. Amer. Midland Nat. **30**: 718–764, 50 figs.
- 1943c New species of Ptychopteridae (Diptera). Part II. Brooklyn Ent. Soc. Bul. **38**: 37–42.
- 1943d New nearctic crane-flies (Tipulidae, Diptera). Part XVIII; Part XIX. Canad. Ent. **75**: 13–20; 139–145.
- 1943e Undescribed species of western nearctic Tipulidae (Diptera). I. Great Basin Nat. **4**: 89–100.
- 1943f Undescribed species of crane-flies from the western United States and Canada (Dipt.: Tipulidae). Part I; Part II. Ent. News **54**: 45–51; 253–258.
- 1944a New nearctic crane-flies (Tipulidae, Diptera). Part XX; Part XXI; Part XXII. Canad. Ent. **76**: 57–62; 166–172; 217–222.
- 1944b Undescribed species of Tipulidae from the western United States (Diptera). Part I. Pan-Pacific Ent. **20**: 91–97.

Alexander, C. P.—Continued

1944c New species of crane-flies from the United States and Canada (Tipulidae, Diptera). Part II. Jour. Ent. and Zool. **36:** 89–94.

1944d Undescribed species of crane-flies from the eastern United States and Canada (Dipt.: Tipulidae). Part VIII; Part IX. Ent. News **55:** 125–129; 241–247.

1945a Records and descriptions of North American crane-flies (Diptera). Part V. Tipuloidea of the Grand Teton National Park and Teton National Forest, Wyoming. Amer. Midland Nat. **33:** 391–439, 46 figs.

1945b Undescribed species of *Tipula* from western North America (Diptera, Tipulidae). Part I. Brooklyn Ent. Soc. Bul. **40:** 33–37.

1945c New nearctic crane-flies (Tipulidae, Diptera). Part XXIII. Canad. Ent. **77:** 1–6.

1945d Undescribed species of crane-flies from the western United States and Canada (Dipt.: Tipulidae). Part III; Part IV. Ent. News **56:** 126–132; 155–161.

1945e New or little-known crane-flies from California (Tipulidae, Diptera). I. South. Calif. Acad. Sci. Bul. **44:** 33–45, pls. 13–16.

1945f Undescribed species of Tipulidae from the western United States (Diptera). Part II. Pan-Pacific Ent. **21:** 91–97.

1946a Records and descriptions of North American crane-flies (Diptera). Part VI. Tipuloidea of Arizona, New Mexico and Trans-Pecos Texas, 1. Amer. Midland Nat. **35:** 484–531, 27 figs.

1946b Undescribed species of *Tipula* from western North America (Diptera, Tipulidae). Part II; Part III. Brooklyn Ent. Soc. Bul. **41:** 45–51; 65–71.

1946c New nearctic crane-flies (Tipulidae, Diptera). Part XXIV; Part XXV; Part XXVI. Canad. Ent. (1945) **77:** 140–144; 186–191; 204–208.

1946d New nearctic crane-flies (Tipulidae, Diptera). Part XXVII. Canad. Ent. **78:** 155–159.

1946e Undescribed species of crane-flies from the western United States and Canada (Dipt.: Tipulidae). Part V; Part VI. Ent. News **57:** 65–71; 173–179.

1946f New or little-known crane-flies from California (Tipulidae, Diptera). II. South. Calif. Acad. Sci. Bul. **45:** 1–16, 2 pls.

1946g Undescribed species of western nearctic Tipulidae (Diptera). II. Great Basin Nat. (1944) **5:** 93–103.
    Dated at end of issue.

Alexander C. P.—Continued

1947a New species of Ptychopteridae (Diptera). Part III. Brooklyn Ent. Soc. Bul. **42**: 19–24.

1947b New nearctic crane-flies (Tipulidae, Diptera). Part XXVIII. Canad. Ent. **79**: 68–75.

(1947c) Undescribed species of crane-flies from the eastern United States and Canada (Dipt.: Tipulidae). Part X. Ent. News (1946) **57**: 245–252.

1947d Undescribed species of crane-flies from the western United States and Canada (Dipt.: Tipulidae). Part VII; Part VIII. Ent. News **58**: 61–67; 205–209.

1947e Undescribed species of Tipulidae from the western United States (Diptera). Part III. Pan-Pacific Ent. **23**: 91–96.

1947f Notes on the tropical American species of Tipulidae (Diptera). II. The primitive Eriopterini: *Sigmatomera, Trentepohlia, Gnophomyia, Neognophomyia, Gonomyia* and allies. Rev. de Ent. **18**: 65–100, 31 figs.

1947g New or little-known crane-flies from California (Tipulidae, Diptera). III. South. Calif. Acad. Sci. Bul. **46**: 35–50, 2 pls.

1948a Records and descriptions of North American crane-flies (Diptera). Part VII. The Tipuloidea of Utah, I. Amer. Midland Nat. **39**: 1–82, 62 figs.

1948b Undescribed species of crane-flies from the western United States and Canada. (Dipt.: Tipulidae). Part IX; Part X. Ent. News **59**: 121–128; 207–214.

1948c New or insufficiently-known crane-flies from the Nearctic Region (Diptera, Tipulidae). Part VIII. Brooklyn Ent. Soc. Bul. (1947) **42**: 131–135.

1949a Records and descriptions of North American crane-flies (Diptera). Part VIII. The Tipuloidea of Washington, I. Amer. Midland Nat. **42**: 257–333, 65 figs.

1949b New or insufficiently-known crane-flies from the Nearctic Region (Diptera, Tipulidae). Part IX; Part X; Part XI. Brooklyn Ent. Soc. Bul. **44**: 15–20; 98–104; 152–157.

1949c Undescribed species of crane-flies from the western United States and Canada (Dipt.: Tipulidae). Part XI. Ent. News **60**: 39–45.

1949d New nearctic crane-flies (Tipulidae, Diptera). Part XXIX. Canad. Ent. (1948) **80**: 166–171.

1950a New or insufficiently-known crane-flies from the Nearctic Region (Diptera, Tipulidae). Part XII; Part XIII. Brooklyn Ent. Soc. Bul. **45**: 41–47; 156–160.

Alexander C. P.—Continued

1950b Notes on the tropical American species of Tipulidae (Diptera). VI. The tribe Limoniini, genus *Limonia:* Subgenera *Limonia, Neolimnobia, Discobola* and *Rhipidia.* Rev. de Ent. **21:** 161–221, 42 figs.

1950c Undescribed species of crane-flies from the western United States and Canada (Dipt.: Tipulidae). Part XII. Ent. News **61:** 29–35.

1950d Undescribed species of crane-flies from the eastern United States and Canada (Dipt.: Tipulidae). Part XI. Ent. News **61:** 163–171.

1950e Undescribed species of Tipulidae from the western United States (Diptera). Part IV. Pan-Pacific Ent. **26:** 81–85.

1951 New or insufficiently-known crane-flies from the Nearctic Region (Diptera, Tipulidae). Part XIV. Brooklyn Ent. Soc. Bul. **46:** 85–91.

1952a Undescribed species of nematocerous Diptera. Part I. Brooklyn Ent. Soc. Bul. **47:** 88–94.

1952b Undescribed species of crane-flies from the western United States and Canada (Dipt.: Tipulidae). Part XIII. Ent. News **63:** 233–237, 267–271.

1953a Undescribed species of nematocerous Diptera. Part II. Brooklyn Ent. Soc. Bul. **48:** 41–49.

1953b The oriental Tipulidae in the collection of the Indian Museum. Part III. Indian Mus. Rec. **50:** 321–357, 15 figs.

1954 Records and descriptions of North American crane-flies (Diptera). Part IX. The Tipuloidea of Oregon, 1. Amer. Midland Nat. **51:** 1–86, 77 figs.

1955a Undescribed species of crane-flies from the western United States and Canada (Dipt.: Tipulidae). Part XIV; Part XV. Ent. News **66:** 15–21; 125–132.

1955b The crane flies of Alaska and the Canadian Northwest (Tipulidae, Diptera). The genus *Erioptera* Meigen. Mich. Univ. Mus. Zool. Misc. Pub. **90:** 1–33, 38 figs.

1956a Undescribed species of crane-flies from the eastern United States and Canada (Dipt.: Tipulidae). Part XII. Ent. News **67:** 177–185.

1956b Undescribed species of crane-flies from the western United States and Canada (Dipt.: Tipulidae). Part XVI. Ent. News **67:** 210–216.

1956c Two new crane-flies from Point Barrow, Alaska (Tipulidae: Diptera). Pan-Pacific Ent. **32:** 123–125.

Alexander C. P.—Continued
- 1956d The crane-flies of South Africa in the Natal Museum (Diptera: Tipulidae). Natal Mus. Ann. **13**: 395–433, 40 figs.
- 1958a Undescribed species of nematocerous Diptera. Part V. Brooklyn Ent. Soc. Bul. **53**: 48–52.
- 1958b Undescribed species of crane-flies from the western United States and Canada (Dipt.: Tipulidae). Part XVII; Part XVIII. Ent. News **69**: 129–136; 215–221.
- 1958c Geographical distribution of the net-winged midges (Blepharoceridae, Diptera). Pp. 813–828, 23 figs. *In* Becker, E.C., ed., Proceedings of the Tenth International Congress of Entomology [q.v.]. Vol. 1, 941 pp., 597 figs. Ottawa, Ont.
- 1958d Undescribed species of western nearctic Tipulidae (Diptera). III. Great Basin Nat. **18**: 31–36.
- 1959a Undescribed species of crane-flies from the western United States and Canada (Dipt.: Tipulidae). Part XIX; Part XX. Ent. News **70**: 47–54; 69–78.
- 1959b Undescribed species of nematocerous Diptera. Part VI; Part VII. Brooklyn Ent. Soc. Bul. **54**: 37–43; 53–60.
- 1959c Undescribed species of Tipulidae from the western United States (Diptera). Part V. Pan-Pacific Ent. **35**: 129–134.
- 1961a Undescribed species of western nearctic Tipulidae (Diptera). IV. Great Basin Nat. **21**: 10–16.
- 1961b Undescribed species of nearctic Tipulidae (Diptera). I. Great Basin Nat. **21**: 79–86.
- 1962 Undescribed species of nearctic Tipulidae (Diptera). II. Great Basin Nat. **22**: 1–7.
- 1963 Guide to the insects of Connecticut. Part VI. The Diptera or true flies of Connecticut. Eighth fascicle [part]. Family Blepharoceridae. Conn. State Geol. and Nat. Hist. Survey, Bul. **93**: 39–71, pls. 8–13.

Alexander, C. P., and Leonard, M.D. (See Alexander, 1914a)

Alexander, M. L. (See Patterson and Alexander)

Allen, H. W.
- 1924 Notes on Miltogramminae with descriptions of two new species. Boston Soc. Nat. Hist. Occas. Papers **5**: 89–92.
- 1926a Notes on some North American species of *Achaetoneura* with a description of one new species (Diptera, Tachinidae). Amer. Ent. Soc. Trans. **52**: 187–198, 1 pl.
- 1926b North American species of two-winged flies belonging to the tribe Miltogrammini. U.S. Natl. Mus. Proc. **68 (9)**: 1–106, 2 figs., 5 pls.

Allen H. W.—Continued
1929   An annotated list of the Tachinidae of Mississippi. Ent. Soc. Amer. Ann. **22**: 676–690, 1 fig.

[Alluaud, C. A., and Jeannel, R., eds.]
1912— Voyage de Ch. Alluaud et R. Jeannel en Afrique orientale
1929   (1911–1912). Résultats scientifiques. [11 "vols."] i.e., 20 separately paged divisions in 57 nos. Paris.

> The work was issued in 57 numbers on various zoological subjects by a number of authors. Libraries may bind this variously; the U.S. Department of Agriculture Library binds the work in 11 volumes. Each taxonomic group is separately paged, there being 20 such divisions, including some general material included in "Vol. 1." The Coleoptera are in 2 separately paged volumes. The single volume on "Insectes Diptères" (Vol. 5 in the U.S. Dept. Agr. set) consists of nos. 6, 25, 30, 33, 37, and 53, and was issued 1913–1923. A "no. 58", on Lepidoptera, was issued in 1932, in Soc. Zool. de France, Mém., and is bound with nos. 1–57 in the U.S. Dept. Agr. Library.
> See Kieffer, 1913d.

Andersen, F. S.
1937   Ueber die Metamorphose der Ceratopogoniden und Chironomiden Nordost-Grönlands. Meddel. om Grønland (Copenhagen) **116** (1): 3–95, 56 figs.
1946   East Greenland lakes as habitats for chironomid larvae. Studies on the systematics and biology of Chironomidae. II. Meddel. om Grønland (Copenhagen) **100** (10): 1–65, 1 fig., 2 pls.

[Anderson, R. M., ed.]
1919— Report of the Canadian Arctic Expedition, 1913–18. 16
1946   vols. Ottawa, Ont.

> This is to consist of 16 volumes upon completion, issued in parts by various authors. The latest published part found by us was dated 1946, at which time the series was still incomplete. Volume 3, on the insects, includes an introduction, and separately paged Parts A–L, issued 1919–1922. The pages of the various parts are marked with the part number also, eg., p. "19c." Part C, on the Diptera, was issued in 1919.
> See Alexander, 1919b; Dyar, 1919a; Malloch, 1919f.

d'Andretta, C., Jr. (See Lane and d'Andretta)

d'Andretta, M. A. V. (See also Guimarães and d'Andretta)

————, **and Carrera, M.**
1950   Sobre as espécies brasileiras de Toxophorinae (Diptera, Bombylliidae [sic]). Dusenia **1**: 351–374, 45 figs.

Andrews, H. (See Cockerell and Andrews)

Annandale, N.
- 1910   A new genus of psychodid Diptera from the Himalayas and Travancore. Indian Mus. Rec. **5**: 141–144, 1 fig., 1 pl.

[Araoz Alfaro, G., sec.]
- 1904   Segundo Congreso Médico Latino-Americano, Buenos Aires, Abril 4–11 de 1904. 2 vols. Buenos Aires.
  See Lahille, 1904.

Argo, N. G.   (See Melander and Argo)

Arnaud, P. H. (Jr.)   (See also Sabrosky and Arnaud)
- 1951   A study of the genus *Paradejeania* Brauer and Bergenstamm (Diptera: Tachinidae or Larvaevoridae). Canad. Ent. **83**: 317–329, 9 figs.
- 1952   *Reinhardiana* new name for *Hypenomyia* Townsend (Diptera: Tachinidae or Larvaevoridae). Pan-Pacific Ent. **28**: 58.
- 1956   The heleid genus *Culicoides* in Japan, Korea and Ryukyu Islands (Insecta: Diptera). Microentomology **21**: 84–207, figs. 1, 3, 5, 44–94.
- 1958a   *Cressonomyia* new name for *Plagiopsis* Cresson preoccupied (Diptera: Ephydridae). Ent. News **69**: 24.
- 1958b   A synopsis of the genus *Melanderia* Aldrich (Diptera, Dolichopodidae). Ent. Soc. Wash. Proc. **60**: 179–186, 8 figs.
- 1958c   The entomological publications of Charles Henry Tyler Townsend [1863–1944], with lists of his new generic and specific names. Microentomology **23**: 1–63, 1 fig.
- 1960   A review of the genus *Spathidexia* Townsend (Diptera: Tachinidae). Wasmann Jour. Biol. **18**: 1–36, 33 figs.

——, **and Hoyt, C. P.**
- 1956   Description of a new species of *Diadocidia* from California (Diptera: Mycetophilidae). Pan-Pacific Ent. **32**: 87–90, 1 fig.

Arriaga, J. J., Urbina, M., and Rebollar, R. [eds.]   (See Herrera, 1892)

Arribálzaga, E. Lynch   (See Lynch Arribálzaga, E.)

Arribálzaga, F. Lynch   (See Lynch Arribálzaga, F.)

Ashmead, W. H.
- 1880   Orange insects. A treatise on the injurious and beneficial insects found on the orange trees of Florida. 78 pp., 23 figs., 4 pls. Jacksonville, Fla.

Astaurov, B. L.   (See Frolova and Astaurov)

Aubertin, D.
1933   Notes on certain species of the genus *Orthellia*, with a description of one new species. Ann. and Mag. Nat. Hist. ser. **10, 11**: 139–144.

Audouin, J. V., Blanchard, E., Doyère, L., and Edwards, H. Milne
[1836–   [Pt. 2]. Orthoptères, Hémiptères, Névroptères, Hy-
1849]   ménoptères, Lépidoptères, Rhipiptères et Diptères. Pls. 76–182. *In* [Cuvier, G. C. L. D.], Le règne animal [The so-called "disciples edition"] [q.v.]. Les Insectes, Atlas, 182 pls. Paris.

> This Atlas was issued in a number of livraisons, and bound in 2 (parts) in 3. The Diptera, pls. 160–182, are by Blanchard, and their dates are uncertain. We have followed Horn and Schenkling, 1928: 87, and dated them all 1849.

Augustson, G. F.
1943   A new parasitic fly from bats (Diptera: Pupipara). South. Calif. Acad. Sci. Bul. **42**: 52–53, 1 pl.

Aurivillius, C.   (See Holmgren and Aurivillius)

Austen, E. E.
1895   On the specimens of the genus *Cutiterebra* and its allies (family Oestridae) in the collection of the British Museum, with descriptions of a new genus and three new species. Ann. and Mag. Nat. Hist. ser. **6, 15**: 377–396, 1 pl.

1901   An addition to the British Stratiomyidae, with the description of a new genus. Ent. Monthly Mag. **37** [=ser. 2, 12]: 241–246.

1907   The synonymy and generic position of certain species of Muscidae (sens. lat.) in the collection of the British Museum, described by the late Francis Walker. Ann. and Mag. Nat. Hist. ser. **7, 19**: 326–347.

1933   New and little-known species of *Cuterebra* Clark and *Bogeria* Austen (Diptera: Family Oestridae). Zool. Soc. London, Proc. Gen. Mtg. **1933**: 699–713, 1pl.

Avery, J. G.   (See Wood et al.)

Ayers, E. L.   (See Steiner et al.)

Back, E. A.
1904   New species of North American Asilidae. Canad. Ent. **36**: 289–293.

1909   The robber-flies of America north of Mexico, belonging to the subfamilies Leptogastrinae and Dasypogoninae. Amer. Ent. Soc. Trans. **35**: 137–400, 11 pls.

Baird, S. F.   (See Verrill, 1873)

Baker, A. C., Stone, W. E., Plummer, C. C., and McPhail, M.
1944 A review of studies on the Mexican fruitfly and related Mexican species. U.S. Dept. Agr. Misc. Pub. **531:** 1–155, 82 figs., 10 pls.

Baker, C. F.
1904 Diptera. Reports on Californian and Nevadan Diptera, I. Invertebrata Pacifica **1:** 17–39.

Baker, F. C.
1936 A new species of *Orthopodomyia*, *O. alba* sp. n. (Diptera, Culicidae). Ent. Soc. Wash. Proc. **38:** 1–7, 1 pl.

Balduf, W. V.
1959 Obligatory and facultative insects in rose hips: Their recognition and bionomics. Ill. Biol. Monog. **26:** 1–194, 12 pls.

Banks, N. (See also Cresson, 1910–1911; Hine, 1919; McAtee and Banks)
1894 Some Psychodidae from Long Island, N.Y. Canad. Ent. **26:** 329–333.
1895 Notes on *Psychoda*. Canad. Ent. **27:** 324.
1901 The eastern species of *Psychoda*. Canad. Ent. **33:** 273–275.
1907 The Psychodidae of the vicinity of Washington. Ent. Soc. Wash. Proc. **8:** 148–151.
1909 A new species of *Systropus* (Bombylidae). Ent. News **20:** 18.
1911a Four new species of Asilidae. Canad. Ent. **43:** 128–130, 1 fig.
1911b A curious habit of one of our phorid flies. Ent. Soc. Wash. Proc. **13:** 212–213, 1 fig.
1912 At the *Ceanothus* in Virginia. Ent. News **23:** 102–110.
1913 Notes on Diptera. Ent. Soc. Wash. Proc. **15:** 52.
1914a Two new species of *Psychoda* (Dipt.). Ent. News **25:** 127–128.
1914b Notes on Asilidae, with two new species. Psyche **21:** 131–133.
1914c A new ortalid fly. Ent. Soc. Wash. Proc. **16:** 138.
1915a Notes on some Virginian species of *Platypeza* (Platypezidae, Dipt.). N.Y. Ent. Soc. Jour. **23:** 213–216, 1 pl.
1915b Notes and descriptions of Pipunculidae. Psyche **22:** 166–170, 1 pl.
1915c A new species of *Mycetaulus* (Diptera, Sepsidae). Ent. Soc. Wash. Proc. **17:** 145.
1916 Synopses of *Zodion* and *Myopa* with notes on other Conopidae. Ent. Soc. Amer. Ann. **9:** 191–200.

Banks, N.—Continued
1917a Synopsis of the genus *Dasyllis* (Asilidae). Brooklyn Ent. Soc. Bul. **12**: 52–55.
1917b Notes on some new species of the genus *Dioctria* (Asilidae). Psyche **24**: 117–119.
1920 Descriptions of a few new Diptera. Canad. Ent. **52**: 65–67.
1926 Descriptions of a few new American Diptera. Psyche **33**: 42–44.
1931 Some Psychodidae from the Carolina mountains. Brooklyn Ent. Soc. Bul. **26**: 227–228.
———, **Greene, C. T., McAtee, W. L., and Shannon, R. C.** (See Shannon, 1916a)

Banta, A. M. (See also Adams, 1907)
1907 The fauna of Mayfield's cave. Carnegie Inst. Wash., Pub. **67**: 1–114.

Baranov, N. (or Baranoff)
1932 Zur Kenntnis der formosanischen Sturmien (Dipt. Larvaevor.). Neue Beitr. z. System. Insektenkunde **5**: 70–82, pl. 2.

Barber, H. S.
1908 Note on *Omomyia hirsuta* Coquillett (Diptera, Phycodromidae). Ent. Soc. Wash. Proc. **9**: 28–29.

Barnes, H. F.
1926 An undescribed mushroom-feeding gall-midge. Ent. Monthly Mag. **62** [=ser. 3, 12]: 89–92, 6 figs.
1927 British gall midges. I. Ent. Monthly Mag. **63** [=ser. 3, 13]: 164–172, 211–221, 9 figs.
1946a Gall midges of economic importance. Vol. 1: Gall midges of root and vegetable crops, 104 pp., 10 pls. London.
   The entire work consists of 7 volumes, London, 1946–1956.
1946b Gall midges of economic importance. Vol. 2: Gall midges of fodder crops, 160 pp., 4 pls. London.
1948a Gall midges of economic importance. Vol. 3: Gall midges of fruit, 184 pp., 9 pls. London.
1948b Gall midges of economic importance. Vol. 4: Gall midges of ornamental plants and shrubs, 165 pp., 2 figs., 10 pls. London.
1949 Gall midges of economic importance. Vol. 6: Gall midges of miscellaneous crops, 229 pp., 14 pls. London.
1951 Gall midges of economic importance. Vol. 5: Gall midges of trees, 270 pp., 8 pls. London.
1956 Gall midges of economic importance. Vol. 7: Gall midges of cereal crops, 261 pp., 7 figs., 16 pls. London.

Barnes, H. F.—Continued
1958 Experimental interbreeding of hessian fly from Kansas, U.S.A., Germany, and England. Ztschr. f. Pflanzenkrank. (Pflanzenpath.) u. Pflanzenschutz **65**: 333–343, 2 figs.

Barr, A. R.
1957 A new species of *Culiseta* (Diptera: Culicidae) from North America. Ent. Soc. Wash. Proc. **59**: 163–167, 1 fig.
1958 The mosquitoes of Minnesota (Diptera: Culicidae: Culicinae). Minn. Agr. Expt. Sta. Tech. Bul. **228**: 1–154, 132 figs.

Barrett, R. E.
1932 An annotated list of the insects and arachnids affecting the various species of walnuts or members of the genus *Juglans* Linn. Calif. Univ., Pubs., Ent. **5**: 275–309.

Basden, E. B.
1961 Type collections of Drosophilidae (Diptera). 1. The Strobl Collection. Beitr. z. Ent. (Berlin) **11**: 160–224, 14 figs., 2 pls.

Basham, E. H.
1948 *Culex* (*Melanoconion*) *mulrennani*, a new species from Florida (Diptera: Culicidae). Ent. Soc. Amer. Ann. **41**: 1–7, 2 pls.

Bastin, F. (See Goetghebuer and Bastin)

Bates, M.
1933 Notes on American Trypetidae (Diptera) II. Psyche **40**: 48–56, 4 figs.
1935 Notes on American Trypetidae (Diptera). III. The genus *Tephrella*. Pan-Pacific Ent. **11**: 103–114, 3 figs.
1949 The natural history of mosquitoes. 379 pp., 8 figs., 16 pls. New York.

Bau, A.
1928 Die Gattung *Cephenomyia*, Latreille (Diptera, olim Oestridae). Centbl. f. Bakt., Parasitenk. u. Infektionskrank., Abt. II **75**: 458–459.
1929a *Cuterebra conflans* und *subbuccata*, spec. novae, sowie Bemerkung über *C. schroederi* Enderlein (Diptera, olim Oestridae). Stettin. Ent. Ztg. **90**: 303–307, 1 pl.
1929b Versuch einer Teilung der Gattung *Cuterebra* (Diptera, olim Oestridae) in vier Untergattungen. Centbl. f. Bakt., Parasitenk. u. Infektionskrank., Abt. II **77**: 542–544, 1 fig.

Bau, A.—Continued
1929c   *Cuterebra semilutea*, sp. n. (Ins. Dipt.), sowie Schlüssel zur Bestimmung der *Cuterebra*-Arten. Senckenbergiana **11**: 1–7, 2 figs.
1931a   *Cuterebra* (*Atryposoma*) *enderleini* spec. nov. (Dipt.). Deut. Ent. Gesell. Mitt. **2**: 20–21, 2 figs.
1931b   Ueber das Genus *Cuterebra*, Clark, (Diptera, Oestridae). Einteilungen desselben in sechs Untergattungen. Beschreibung neuer Species und Aufstellung einer Bestimmungstabelle der mittel- und südamerikanischen Arten. Konowia **10**: 197–240, 2 figs., 2 pls.

Bause, E.
1913   Die Metamorphose der Gattung *Tanytarsus* und einiger verwandter Tendipedidenarten. Ein Beitrag zur Systematik der Tendipediden. Inaugural-Dissertation, Westfälische Wilhelms-Universität zu Münster. 2 + 126 pp., 12 pls. Münster, Germany.

> Also published, pp. 1–128, pls. 1–12, In Thienemann, A., Vorarbeiten für eine Monographie der Chironomiden-Metamorphose. Lieferung 1, in Arch. f. Hydrobiol. u. Planktonkunde, Sup.-Band 2: 1–241, 92 figs., 16 pls., *1914*.

Bean, J. L.
1949   A study of male hypopygia of the species of *Tubifera* (Syrphidae, Diptera) that occur north of Mexico. Canad. Ent. **81**: 140–152, 21 figs.

Becher, E.
[1886]   F. Insecten von Jan Mayen. Gesammelt von Dr. F. Fischer, Arzt der österreichischen Expedition auf Jan Mayen. Pp. 59–66, pl. 5. *In* K. Akademie der Wissenschaften, Wien, Die Internationale Polarforschung 1882–1883 [q.v.]. Vol. 3, Pt. 6: Zoology, 132 pp., 9 pls. Wien.

> Dated from information in Sharp, 1887: 5.

Beck, E. C.   (See also Beck and Beck)
1951   A new *Culicoides* from Florida (Diptera, Ceratopogonidae). Fla. Ent. **34**: 135–136, 1 fig.
1956   A new species of *Culicoides* from Florida with additional distribution data for the genus (Diptera: Heleidae). Fla. Ent. **39**: 133–138, 2 figs.
1957   Two new species of *Culicoides* from Florida (Diptera: Heleidae). Fla. Ent. **40**: 103–105, 2 figs.
1961   Two new Chironomidae (Diptera) and additional state records from Florida. Fla. Ent. **44**: 125–128, 2 figs.
1962   Five new Chironomidae (Diptera) from Florida. Fla. Ent. **45**: 89–92, 6 figs.

Beck, W. M., and Beck, E. C.
1958   A new species of *Xenochironomus* from Florida (Diptera: Chironomidae). Fla. Ent. **41**: 27–28, 1 fig.

Becker, E. C., ed.
1958   Proceedings of the Tenth International Congress of Entomology, Montreal, August 17–25, 1956. 4 vols. Ottawa, Ont.

> Each individually paged volume consists of a number of sections, each edited by a Section Editor; there are a total of 16 sections. The Section on Geographical Distribution, Vol. 1, pp. 613–846, 153 figs., is edited by W. R. M. Mason, W. R. Richards, and P. F. Bruggemann. The Section on Medical and Veterinary Entomology, Vol. 3, pp. 495–892, 74 figs., 4 pls., is edited by C. R. Twinn.
> See Alexander, 1958c; Downes, 1958.

Becker, T.   (See also Strobl, 1894b)
1889a Altes und Neues aus der Schweiz. Ein dipterologischer Beitrag. Wien. Ent. Ztg. **8**: 73–84, 1 pl.
1889b Berichtigung. Wien. Ent. Ztg. **8**: 285.
1891   Neues aus Süd-Tyrol und Steiermark. Ein dipterologischer Beitrag. Wien. Ent. Ztg. **10**: 281–288, 1 pl.
1894   Dipterologische Studien. I. Scatomyzidae. Berlin. Ent. Ztschr. **39**: 77–196, 1 pl.
1895   Dipterologische Studien. II. Sapromyzidae. Berlin. Ent. Ztschr. **40**: 171–264, 1 pl.
1896   Dipterologische Studien. IV. Ephydridae. Berlin. Ent. Ztschr. **41**: 91–276, 4 pls.
1897   Beitrag zur Dipteren-Fauna von Nowaja-Semlja. [Akad. Nauk S.S.S.R.] Imp. Akad. Nauk St. Petersburg, Zool. Muz. Ezheg. (Acad. Imp. des Sci. de St. Pétersbourg, Zool. Mus. Ann.) **2**: 396–404.
1900   Beiträge zur Dipteren-Fauna Sibiriens. Nordwest-Sibirische Diptera. Finska Vetensk. Soc., Acta Soc. Sci. Fenn. **26** (9): 1–66, 2 pls.
1901   Die Phoriden. K.-k. Zool.-Bot. Gesell. Wien, Abhandl. **1** (1): 1–100, 1 fig., 5 pls.
1902– Aegyptische Dipteren. Berlin Zool. Mus. Mitt. **2** (h. 2,
1903   art. 1): 1–66, 1 pl., *1902;* **2** (h. 3): 67–195, 4 pls., *1903.*
1904a Die paläarktischen Formen der Dipterengattung *Lispa* Latr. Ztschr. f. Ent. Breslau **n. ser., 29**: 1–70.
1904b Die Dipterengattung *Peletophila* Hagenbach. Ztschr. f. System. Hymenopt. u. Dipt. **4**: 129–133.
1906   Die Ergebnisse meiner dipterologischen Frühjahrsreise nach Algier und Tunis, 1906. Ztschr. f. System. Hymenopt. u. Dipt. **6**: 273–287, 353–367, 1 fig. [cont.].

Becker, T.—Continued
1907a   Die Dipteren-Gruppe Milichinae. Budapest Magyar Nemzeti Muz., Ann. Hist. Nat. **5**: 507–550, 2 figs., 1 pl.
1907b   Ein Beitrag zur Kenntnis der Dipterenfauna Nordsibiriens. [Akad. Nauk S.S.S.R.] Imp. Akad. Nauk St. Petersburg, Zap., Fiz.-Mat. Otd. (Acad. Imp. des Sci. de St. Pétersbourg, Mém., Cl. Phys.-Math.) **ser. 8, 18 (10)**: 1–6.
1907c   Appendix. Beschreibung von 3 neuen Dipteren aus Ost-Grönland. Pp. 411–414. *In* Nielsen, J. C., The insects of East Greenland. Meddel. om Grønland (Copenhagen) **29**: 363–414.
1907d   Die Ergebnisse meiner dipterologischen Frühjahrsreise nach Algier und Tunis, 1906 [concl.]. Ztschr. f. System. Hymenopt. u. Dipt. **7**: 33–61, 97–128, 225–256, 369–407, 454, 12 figs.
1908   Dipteren der Kanarischen Inseln. Berlin Zool. Mus. Mitt. **4**: 1–180, 4 pls.
1909a   *Microphorus* Macq. und seine nächsten Verwandten (Diptera). Wien. Ent. Ztg. **28**: 25–28, 3 figs.
1909b   Collections recueillies par M. Maurice de Rothschild dans l'Afrique orientale anglaise. Insectes: Diptères nouveaux. Paris, Mus. Natl. d'Hist. Nat. Bul. **15**: 113–121.
1910a   Chloropidae. Eine monographische Studie. Arch. Zool. Budapest **1**: 23–174, 2 pls. [cont.].
1910b   Orthorrhapha Brachycera; Cyclorrhapha; Holometopa. Pp. 636–652; 652–657; 658–665. *In* Becker, T., Kuntze, A., Schnabl, J., and Villeneuve, J., Dipterologische Sammelreise nach Korsika (Dipt.). Deut. Ent. Ztschr. **1910**: 635–665.
1912a   Chloropidae. Eine monographische Studie [concl.]. Budapest Magyar Nemzeti Muz., Ann. Hist. Nat. **10**: 21–256, 2 figs., 1 pl.
1912b   Berichtigungen zur Monographie der Chloropiden. Budapest Magyar Nemzeti Muz., Ann. Hist. Nat. **10**: 645–646.
1913   Genera Bombyliidarum. [Akad. Nauk S.S.S.R.] Imp. Akad. Nauk St. Pétersburg, Zool. Muz. Ezheg. (Acad. Imp. des Sci. de St. Pétersbourg, Zool. Mus. Ann.) **17**: 421–502, 36 figs.
1915   Orthorrhapha Brachycera; Cyclorrhapha Aschiza; Cyclorrhapha Schizophora. Pp. 53–60, figs. 1–3; 60–62, fig. 4; 63–67, figs. 5–7. *In* Becker, T., Dziedzicki, H., Schnabl, J., and Villeneuve, J., Résultats scientifiques de l'expédition des frères Kuznecov (Kouznetzov) à l'Oural arctique en 1909, sous la direction de H. Backlund.

Becker, T.—Continued
- Livr. 7. (Diptera). [Akad. Nauk S.S.S.R.] Imp. Akad. Nauk Petrograd, Zap., Fiz.-Mat. Otd. (Acad. Imp. des Sci. de Petrograd, Mém., Cl. Phys.-Math.) **ser. 8, 28** (7): 1–67, 7 figs., 3 pls.
- 1922 Dipterologische Studien. Dolichopodidae. B. Nearktische und neotropische Region. Zool.-Bot. Gesell. Wien, Abhandl. (1921) **13** (1): 1–394, 147 figs.
- 1926 Ephydridae. [Fam.] 56, pp. 1–48, figs. 1–65 (=lfg. 10); pp. 49–115, figs. 66–134 (=lfg. 11). *In* Lindner, E., ed., Die Fliegen der palaearktischen Region [q.v.]. Vol. 6, Pt. 1. Stuttgart.

————, **Bezzi, M., Bischof, J., Kertész, K., and Stein, P.**, eds.
- 1903a Katalog der paläarktischen Dipteren. Vol. 1, 383 pp. Budapest.

  Bischof actually did not participate; see Becker et al., 1905: foreword.

  The entire work consists of 4 volumes, Budapest, 1903–1907. Only the first 2 volumes show Bischof as co-editor (see above). Authorship within the volumes was divided between the editors.

- 1903b Katalog der paläarktischen Dipteren. Vol. 2, 396 pp. Budapest.

  Bischof actually did not participate; see Becker et al., 1905: foreword.

————, **Bezzi, M., Kertész, K., and Stein, P.**, eds. (See also Bezzi and Stein, 1907)
- 1905 Katalog der paläarktischen Dipteren. Vol. 4, 328 pp. Budapest.

  See Becker et al., 1903a, for volumes 1 and 2.

- 1907 Katalog der paläarktischen Dipteren. Vol. 3, 828 pp. Budapest.

————, **Dziedzicki, H., Schnabl, J., and Villeneuve, J.** (See Becker, 1915; Schnabl, 1915)

————, **Kuntze, A., Schnabl, J., and Villeneuve, J.** (See Becker, 1910b; Schnabl, 1911)

Beekey, C. E.
- 1938 The immature and adult stages of a new species of *Scatopse* from Maine (Diptera: Scatopsidae). Ent. News **49**: 151–154, 1 pl.

Beier, M. (See Strouhal and Beier)

Belkin, J. N., and Hogue, C. L.
- 1959 A review of the crabhole mosquitoes of the genus *Deinocerites* (Diptera, Culicidae). Calif. Univ., Pubs., Ent. **14**: 411–458, 41 figs.

Belkin, J. N., and McDonald, W. A.
1957   A new species of *Aedes* (*Ochlerotatus*) from tree holes in southern Arizona and a discussion of the *varipalpus* complex (Diptera: Culicidae). Ent. Soc. Amer. Ann. **50**: 179–191, 3 figs.

Bellardi, L.
1859   Saggio di ditterologia messicana. Parte I. Pt. 1, 80 pp., 2 pls. Torino.
> Also published in R. Accad. delle Sci. Torino, Mem. **19**: 201–277, 2 pls. *1861*.
> The entire work consists of 2 parts and an appendix, Torino, 1859–1862.

1861   Saggio di ditterologia messicana. Parte II. Pt. 2, 99 pp., 2 pls. Torino.
> Also published in R. Accad. delle Sci. Torino, Mem. **21**: 103–199, 2 pls., *1864*.

1862   Saggio di ditterologia messicana. Appendice. App., 28 pp., 1 pl. Torino.
> Also published in R. Accad. delle Sci. Torino, Mem. **21**: 200–225, 1 pl. *1864*.

Beneway, D. F.
1963   A revision of the flies of the genus *Lespesia* (=*Achaetoneura*) in North America (Diptera: Tachinidae). Kans. Univ. Sci. Bul. **44**: 627–686, 34 figs.

Benjamin, D. M.   (See Kearby and Benjamin)

Benjamin, F. H.
1934   Descriptions of some native trypetid flies with notes on their habits. U.S. Dept. Agr. Tech. Bul. **401**: 1–95, 44 figs.

Bennett, G. F., and Sabrosky, C. W.
1962   The nearctic species of the genus *Cephenemyia* (Diptera, Oestridae). Canad. Jour. Zool. **40**: 431–448, figs. 1–7.

Bequaert, J.
1920   A new nemestrinid fly from central Texas. N.Y. Ent. Soc. Jour. (1919) **27**: 301–307, 1 fig.

1924   Notes upon Surcouf's treatment of the Tabanidae in the Genera Insectorum and upon Enderlein's proposed new classification of this family. Psyche **31**: 24–40.

[1926a]   Notes on Hippoboscidae. I. *Lynchia* Weyenbergh and *Lynchia* Speiser are not congeneric. Psyche (1925) **32**: 265–277.
> This December issue was marked as being received by the U.S. Dept. Agr. Bur. Ent. on Jan. 25, 1926. Bequaert himself (1953: 366) dates it as 1926.

Bequaert J.—Continued
- 1926b Medical report of the Hamilton Rice Seventh Expedition to the Amazon, in conjunction with the Department of Tropical Medicine of Harvard University, 1924–1925. Part II. Medical and economic entomology. Harvard Univ., Inst. Trop. Biol. and Med. Contrib. **4:** 155–257, 9 figs., pls. 61–67.
- 1926c The date of publication of the Hymenoptera and Diptera described by Guérin in Duperrey's "Voyage de La Coquille." Ent. Mitt. **15:** 186–195.
- 1930 Notes on American Nemestrinidae. Psyche **37:** 286–297.
- 1931 The genus *Lasia* (Diptera, Cyrtidae) in North America, with descriptions of two new species. Amer. Mus. Nat. Hist., Amer. Mus. Novitates **455:** 1–11, 1 fig.
- 1933a Description of a new North American species of *Lasia* (Diptera, Cyrtidae). Amer. Mus. Nat. Hist., Amer. Mus. Novitates **617:** 1–2, 1 fig.
- 1933b Notes on the Tabanidae described by the late C. P. Whitney. Boston Soc. Nat. Hist. Occas. Papers **8:** 81–88.
- 1934 Notes on American Nemestrinidae. Second paper. N.Y. Ent. Soc. Jour. **42:** 163–184, 1 fig.
- 1935 The American species of *Lipoptena* (Diptera, Hippoboscidae). Brooklyn Ent. Soc. Bul. **30:** 170.
- 1942a A monograph of the Melophaginae, or kedflies, of sheep, goats, deer and antelopes (Diptera, Hippoboscidae). Ent. Amer. **22:** 1–210, 19 figs.
- 1942b *Carnus hemapterus* Nitzsch, an ectoparasitic fly of birds, new to America (Diptera). Brooklyn Ent. Soc. Bul. **37:** 140–149, 1 fig.
- 1947 Catalogue of recent and fossil Nemestrinidae of America north of Mexico. Psyche **54:** 194–207.
- 1953 The Hippoboscidae or louse-flies (Diptera) of mammals and birds. Part I. Structure, physiology and natural history. Ent. Amer. (1952) **32:** 1–209, 19 figs.; (1953) **33:** 211–442, figs. 20–21.
- 1954– The Hippoboscidae or louse-flies (Diptera) of mammals and
- 1957 birds. Part II. Taxonomy, evolution and revision of American genera and species. Ent. Amer. **34:** 1–232, figs. 22–44, *1954;* **35:** 233–416, figs. 45–82, *1955;* (1956) **36:** 417–611, figs. 83–104, *1957.*

Bequaert, M.
- 1961 Contribution à la connaissance morphologique et à la classification des Mydaidae (Diptera). Brussels Inst. Roy. des Sci. Nat. de Belg. Bul. **37 (19):** 1–18, 17 figs.

Berdén, S.
1952 Taxonomical notes on Psychodidae (Dipt. Nem.). 1. *Psychoda lativentris* n. sp., a species hitherto confused with *alternata* Say. Opusc. Ent. (Soc. Ent. Lund.) **17**: 110–112, 6 figs.

Berg, C.
1898 Substitución de nombres genéricos. [I]; II.; III. Buenos Aires Mus. Nac. Comun. **1**: 16–19, *1898;* 41–43, *1898;* 77–80, *1899.*

Berg, C. O. (See also Neff and Berg)
1950a *Hydrellia* (Ephydridae) and some other acalyptrate Diptera reared from *Potamogeton.* Ent. Soc. Amer. Ann. **43**: 374–398, 4 pls.
1950b Biology of certain Chironomidae reared from *Potamogeton.* Ecol. Monog. **20**: 83–101, 2 pls.
1953 Sciomyzid larvae (Diptera) that feed on snails. Jour. Parasitol. **39**: 630–636.
1961 Biology of snail-killing Sciomyzidae (Diptera) of North America and Europe. Pp. 197–202. *In* Strouhal, H., and Beier, M., eds., XI. Internationaler Kongress für Entomologie [q.v.]. Vol. 1, 803 pp., 305 figs., 20 pls. Wien.

See Glumac, 1961, for note on date.

Berg, V. L.
1940 The external morphology of the immature stages of the bee fly, *Systoechus vulgaris* Loew, (Diptera, Bombyliidae), a predator of grasshopper egg pods. Canad. Ent. **72**: 169–178, 6 figs.

Bergenstamm, J. E. von (See Brauer and Bergenstamm)

Bergroth, E.
1888a Ueber einige nordamerikanische Tipuliden. I.; II. Wien. Ent. Ztg. **7**: 193–201; 239–240.
1888b On some South African Tipulidae. Ent. Tidskr. **9**: 127–141, 1 pl.
1889 Zwei neue Dipteren. Wien. Ent. Ztg. **8**: 295–298.
1901 Über eine auf Eulen schmarotzende Hippoboscide. Soc. pro Fauna et Flora Fenn., Meddel. **27**: 146–150.
1913 A new genus of Tipulidae from Turkestan, with notes on other forms. Ann. and Mag. Nat. Hist. **ser. 8, 11**: 575–584, 2 figs.
1915 Some tipulid synonymy. Psyche **22**: 54–59.

Bernice P. Bishop Museum
1954–1962 Insects of Micronesia. 20 vols. Honolulu.

> This series is to consist of 20 volumes upon completion. Most volumes are issued in consecutively paged numbers as completed. Only volumes 1 and 2 are complete to date, with an additional 15 volumes begun. Volume 12, which includes the Diptera Nematocera, is not yet complete, consisting of 4 numbers to date, issued 1956–1959.
> See Tokunaga and Murachi, 1959.

Berté, E. (See Rondani and Berté)

Berthelot, S. (See Webb and Berthelot)

Berthold, A. A.
1827 Latreille's Natürliche Familien der Thierreichs. Aus dem Französischen. Mit Anmerkungen und Zusätzen. 606 pp. Weimar.

> German translation of Latreille, 1825.

Betten, C. (See Needham and Betten)

Beutenmüller, W.
1892 Catalogue of gall-producing insects found within fifty miles of New York City, with descriptions of their galls, and some new species. Amer. Mus. Nat. Hist. Bul. **4**: 245–278, 8 pls.

1907a New species of gall-producing Cecidomyiidae. Amer. Mus. Nat. Hist. Bul. **23**: 385–400, 5 pls.

1907b Descriptions of new species of Cecidomyidae. Canad. Ent. **39**: 305–307.

1908 On some apparently new Cecidomyiidae. Canad. Ent. **40**: 73–75.

1913a Notes on some species of Cecidomyiidae. Canad. Ent. **45**: 413–416.

1913b A new *Empis* from the Black Mountains, North Carolina (Diptera, Empididae). Insecutor Inscitiae Menstruus **1**: 130.

Bezzi, M. (See also Becker et al., 1903a, 1903b; Becker et al., 1905, 1907)

1895 Contribuzioni alla fauna ditterologica italiana. Soc. Ent. Ital. Bul. **27**: 39–78.

1902 Neue Namen für einige Dipteren-Gattungen. Ztschr. f. System. Hymenopt. u. Dipt. **2**: 190–192.

1906 Noch einige neue Namen für Dipterengattungen. Ztschr. f. System. Hymenopt. u. Dipt. **6**: 49–55.

1907 Nomenklatorisches über Dipteren. Wien. Ent. Ztg. **26**: 51–56.

Bezzi, M.—Continued
1908 Nomenklatorisches über Dipteren. III. Wien. Ent. Ztg. **27**: 74–84.
1909 Einige neue paläarktische *Empis*-Arten. Deut. Ent. Ztschr. **1909** (**Beih.**): 85–103 [cont.].
1910 Revisio systematica generis dipterorum *Stichopogon*. Budapest Magyar Nemzeti Muz., Ann. Hist. Nat. **8**: 129–159.
1913 Taumaleidi (Orfnefilidi) italiani con descrizione di nuove specie. Portici, R. Scuola Super. d'Agr., Lab. Zool. Gen. e Agr. Bol. **7**: 227–266, 8 figs.
1920 Further notes on the Lonchaeidae (Dipt.), with description of new species from Africa and Asia. Bul. Ent. Res. **11**: 199–210.
1921 On the bombyliid fauna of South Africa (Diptera) as represented in the South African Museum. So. African Mus. Ann. **18**: 1–180, pls. 1–2.
1923 The genus *Urophora* Robineau-Desvoidy in America (Diptera: Trypaneidae). Amer. Ent. Soc. Trans. **49**: 1–6, 1 pl.
1924a Una nuova *Tipula* delle Alpi, con ali ridotte anche nel maschio (Dipt.). Genoa Mus. Civico di Storia Nat. Ann. **51**: 228–233.
1924b The Bombyliidae of the Ethiopian Region. 390 pp., 46 figs. British Museum (Natural History), London.

———, **and Stein, P.**
1907 Cyclorrapha Aschiza. Cyclorrapha Schizophora: Schizometopa. Pp. 1–189; 190–749. *In* Becker, T., Bezzi, M., Kertész, K., and Stein, P., eds., Katalog der paläarktischen Dipteren, Vol. 3, 828 pp. Budapest.
    See Becker et al., 1903a, for details of this work

Bigot, J. M. F.
1854 Essai d'une classification générale et synoptique de l'ordre des Insectes Diptères. IIIe. Soc. Ent. de France, Ann. ser. **3, 2**: 447–482.
1856 Essai d'une classification générale et synoptique de l'ordre des Insectes Diptères. IVe. Soc. Ent. de France, Ann. ser. **3, 4**: 51–91.
1857a Dipteros. Pp. 328–349, [pl. 20 of insects and crustaceans, bound in vol. 8]. *In* Sagra, R. de la, Historia fisica, politica y natural de la Isla de Cuba [q.v.]. Vol. 7, 371 pp. Paris, "1856".
    The corrections and addenda on p. 350 are dated Oct., 1857; Woodward, 1913: 1781, calls the date 1856 on the title page of the volume an error. This portion on Diptera was also published as pp. 783–829 of the French edition, in Paris, 829 pp., *1857*.

Bigot, J. M. F.—Continued

1857b Essai d'une classification générale et synoptique de l'ordre des Insectes Diptères. V^e. Soc. Ent. de France, Ann. **ser. 3, 5:** 517–564.

1859a Dipterorum aliquot nova genera. Rev. et Mag. de Zool. **ser. 2, 11:** 305–315, pl. 11.

1859b Essai d'une classification générale et synoptique de l'ordre des Insectes Diptères. VIIe. Soc. Ent. de France, Ann. **ser. 3, 7:** 201–231.

1875a Diptères nouveaux ou peu connus. 4e partie, V: Asilides exotiques nouveaux; 5e partie, VI: Espèces exotiques nouvelles des genres *Sphixea* (Rondani) et *Volucella* (auctorum); 5e partie, VII: Espèces nouvelles du genre *Cyphomyia*. Soc. Ent. de France, Ann. **ser. 5, 5:** 237–248; 469–482; 483–488.

1875b (Description d'une nouvelle espèce de Diptère). Soc. Ent. de France, Ann. **ser. 5, 5 (Bul.):** clxxiv–clxxvi.

1877 Diptères nouveaux ou peu connus. 7e partie, IX: Genre *Somomyia* (Rondani) *Lucilia* (Rob.-Desv.) *Calliphora, Phormia, Chrysomyia* (id.); 8e partie, X: Genre *Somomya* (Rondani) *Calliphora, Melinda, Mufetia, Lucilia, Chrysomyia* (alias *Microchrysa* Rond.) Robineau-Desvoidy. Soc. Ent. de France, Ann. **ser. 5, 7:** 35–48; 243–259.

1878 Diptères nouveaux ou peu connus. 9e partie, XII: Genus *Phumosia, Pyrellia, Cosmina, Ochromyia* et *Curtonevra;* 9e partie, XIII: Genres *Ocyptera* (Latr.), *Ocypterula, Exogaster* (Rond.); 9e partie, XIV: Notes et mélanges; 10e partie, XV: Tribu des Asilidi curies des Laphridae et Dasypogonidae. Soc. Ent. de France, Ann. **ser. 5, 8:** 31–40; 40–47; 48; 213–240, 401–446.

1879a Diptères nouveaux ou peu connus. 11e partie, XVI: Curiae Xylophagidarum et Stratiomydarum (Bigot). Soc. Ent. de France, Ann. **ser. 5, 9:** 183–234.

1879b [Note]. Soc. Ent. de France, Ann. **ser. 5, 9 (Bul.):** l-li.

1879c (Diagnoses de trois genres nouveaux de Diptères). Soc. Ent. de France, Ann. **ser. 5, 9 (Bul.):** lxvii–lxviii.

1880a Diptères nouveaux ou peu connus. 14e partie, XXI: Syrphidi (mihi).—Genre *Eristalis* (Fabr.). Soc. Ent. de France, Ann. **ser. 5, 10:** 213–230.

1880b (Diagnoses de trois genres nouveaux de Diptères). Soc. Ent. de France, Ann. **ser. 5, 10 (Bul.):** xlvi–xlviii.

1882a (Descriptions de quatre genres nouveaux de la tribu des Syrphides (Syrphidae auctorum), ainsi que celles de deux nouvelles espèces). Soc. Ent. de France, Ann. **ser. 6, 2 (Bul.):** lxvii–lxviii.

Bigot, J. M. F.—Continued
1882b (Description de deux nouvelles espèces de Diptères propres à la Californie, dont l'une est le type d'un genre nouveau). Soc. Ent. de France, Ann. **ser. 6, 2 (Bul.)**: xci-xcii.
1882c (Diagnoses de genres et espèces inédits de Syrphides. 1re partie; 2e partie; 3e partie). Soc. Ent. de France, Ann. **ser. 6, 2 (Bul.)**: cxiv-cxv; cxx-cxxi; cxxviii-cxxix.
1883a Diptères nouveaux ou peu connus. 20e partie, XXXI: Genres *Volucella* (Geoffr., Hist. des Insectes, 1764) et *Phalachromyia* (Rondani, Esame d. var. spec. d'Insett. Ditteri Brasiliani, Torino. 1848); 21e partie, XXXII: Syrphidi (1re partie); 22e partie, XXXII: Syrphidi (2e partie). Espèces nouvelles, no. I; 23e partie, XXXII: Syrphidi (2e partie). Espèces nouvelles, no. II. Soc. Ent. de France, Ann. **ser. 6, 3**: 61-88; 221-258; 315-356; 535-560.
1883b (Description d'un nouveau genre de Diptères de la tribu des Syrphides). Soc. Ent. de France, Ann. **ser. 6, 3 (Bul.)**: xx-xxi.
1883c (Diagnose d'un nouveau genre et d'une nouvelle espèce de Diptères de la tribu des Antomyzides [sic]). Soc. Ent. de France, Ann. **ser. 6, 3 (Bul.)**: xxx.
1884a Diptères nouveaux ou peu connus. 24e partie, XXXII: Syrphidi (2e partie). Espèces nouvelles, no. III. Soc. Ent. de France, Ann. **ser. 6, 4**: 73-116.
1884b (Diagnoses d'un genre et d'une espèce de Diptères). Soc. Ent. de France, Ann. (1883) **ser. 6, 3 (Bul.)**: cviii-cix.
1884c (Description d'un nouveau genre et d'une nouvelle espèce de Diptères de la famille des Ortalidae). Soc. Ent. de France, Ann. **ser. 6, 4 (Bul.)**: xxix.
1884d (Description d'un nouveau genre et d'une nouvelle espèce de Diptères de la famille des Dexidae). Soc. Ent. de France, Ann. **ser. 6, 4 (Bul.)**: xxxvii.
1885a Diptères nouveaux ou peu connus. 25e partie, XXXIII: Anthomyzides nouvelles. Soc. Ent. de France, Ann. (1884) **ser. 6, 4**: 263-304.
1885b Diptères nouveaux ou peu connus. 27e partie, XXXV: Famille des Anomalocerati (mihi) (Coriacae, Pupipara, Nycteribidae, auctor). Soc. Ent. de France, Ann. **ser. 6, 5**: 225-246.
1885c (Diagnoses de deux genres nouveaux de Diptères du groupe des Déxiaires). Soc. Ent. de France, Ann. **ser. 6, 5 (Bul.)**: xxv-xxvi.

Bigot, J. M. F.—Continued

1885d (Les diagnoses de deux genres nouveaux de Diptères appartenant à la famille des Ortalidae). Soc. Ent. de France, Ann. **ser. 6,** 5 (**Bul.**): clxv-clxvi.

1886a Diptères nouveaux ou peu connus. 29e partie (suite), XXXVII: 2e. Essai d'une classification synoptique du groupe des Tanypezidi (mihi) et description de genres et d'espèces inédits. Soc. Ent. de France, Ann. **ser. 6,** 6: 369–392.

1886b (Diagnoses nouvelles d'un genre et d'une espèce de l'ordre des Diptères). Soc. Ent. de France, Ann. **ser. 6, 6** (**Bul.**): ciii-civ.

1887a Diptères nouveaux ou peu connus. 31e partie, XXXIX: Descriptions de nouvelles espèces de Stratiomydi et de Conopsidi; 32e partie, XL: Descriptions de nouvelles espèces de Myopidi. Soc. Ent. de France, Ann. **ser. 6, 7:** 20–46; 203–208.

1887b (Diagnoses de quelques espèces nouvelles de Diptères). Soc. Ent. de France, Ann. **ser. 6, 7** (**Bul.**): cxxxix-cxlii.

1887c (Diagnoses abrégées de quelques Diptères nouveaux, provenant de l'Amérique du Nord). Soc. Ent. de France, Ann. **ser. 6, 7** (**Bul.**): clxxx-clxxxii.

1887d Diptères nouveaux ou peu connus. Leptidi; Muscidi. Soc. Zool. de France, Bul. **12:** 97–118; 581–617.

1888a Diptères nouveaux ou peu connus. 33e partie, XLI: Tachinidae; 34e partie, XLII: Diagnoses de nouvelles espèces. Soc. Ent. de France, Ann. **ser. 6, 8:** 77–101; 253–270.

1888b (Notes critiques sur les Diptères). Soc. Ent. de France, Ann. **ser. 6, 8** (**Bul.**): xxiv.

1888c (Diagnoses d'espèces nouvelles de Dolichopodi). Soc. Ent. de France, Ann. **ser. 6, 8** (**Bul.**): xxix-xxx.

1889 Diptères nouveaux ou peu connus. 34e partie, XLII: Empidi; 35e partie, XLIII: Cyrtidi; 35e partie, XLIV: Therevidi. Soc. Ent. de France, Ann. **ser. 6, 9:** 111–134; 313–320; 321–328.

1890 Diptères nouveaux ou peu connus. 36e partie, XLV: Dolichopodi. Essai d'une classification générale. Soc. Ent. de France, Ann. **ser. 6, 10:** 261–296.

1892a Diptères nouveaux ou peu connus. 37e partie, XLVI: Bombylidi (mihi). Soc. Ent. de France, Ann. **61:** 321–376.

1892b Descriptions de Diptères nouveaux. Tabanidi. Soc. Zool. de France, Mém. **5:** 602–691.

1892c Nova genera dipterorum. Wien. Ent. Ztg. **11:** 161–162.

Bilimek, D.
1867   Fauna der grotte Cacahuamilpa in Mexico. K.-k. Zool.-
Bot. Gesell. Wien, Verhandl. **17** (**Abhandl.**): 901–908.

Billberg, G. J.
1820   Enumeratio insectorum in Museo Gust. Joh. Billberg. 138
pp. [Stockholm].

Bischof, J. (See Becker et al.)

Bishop, S. C.
1911   A new root gall midge from *Smilacina* (Dipt.). Ent. News
22: 346.

Bishopp, F. C., Laake, E. W., Brundrett, H. M., and Wells, R. W.
1926   The cattle grubs or ox warbles, their biologies and suggestions for control. U.S. Dept. Agr., Dept. Bul. **1369**:
1–119, 38 figs.

Blackwelder, R. E.
1947   The dates and editions of Curtis' British Entomology.
Smithsn. Inst., Smithsn. Misc. Collect. **107** (**5** [=pub.
3894]): 1–27, 4 pls.
1949   Studies on the dates of books on Coleoptera. I; II; Studies
on the dates of works on Coleoptera, III. Coleopt.
Bul. **3**: 42–46; 76; 92–94.

Blanc, F. L. (See also Foote and Blanc)
1959   A new species of *Chaetostoma* from California. Pan-Pacific Ent. **35**: 201–203, 1 fig.

——, **and Foote, R. H.**
1961   A new genus and five new species of California Tephritidae
(Diptera). Pan-Pacific Ent. **37**: 73–83, 4 figs.

Blanchard, E. (See also Audouin et al.)
1840   Histoire naturelle des insectes. Orthoptères, Névroptères,
Hémiptères, Hyménoptères, Lépidoptères et Diptères.
Vol. 3, 672 pp., 67 pls. *In* Laporte, F. L. de, Histoire
naturelle des animaux articulés [q.v.]. Paris.
1852   Orden IX. Dipteros. Pp. 327–468. *In* Gay, C., ed.,
Historia fisica y politica de Chile [q.v.]. Zoologia, Vol.
7, 471 pp. Paris.

Blanchard, E. E.
1942   Nuevos dipteros e himenopteros parasitos, de la Republica
Argentina. Soc. Ent. Argentina, Rev. **11**: 340–379, 17
figs.
1943   Un nuevo exoristido, importante parasito del gorgojo de las
hortalizas. Soc. Ent. Argentina, Rev. **11**: 450–454, 1 pl.

Blanchard, R.
1901 Observations sur quelques moustiques. Soc. de Biol. [Paris] Compt. Rend. **53**: 1045–1046.
1902 Nouvelle note sur les moustiques. Soc. de Biol. [Paris] Compt. Rend. **54**: 793–795.
1905 Les moustiques. Histoire naturelle et médicale. 673 pp., 316 figs. Paris.

[**Blanford, W. T., ed.**]
1888– Fauna of British India. 77 vols. London.
1950
> 77 volumes, including first and second editions, have been issued to 1950, the date of the last volume of which we have a record, and more volumes are proposed. The volumes are serially numbered only within taxonomic groups; two editions are sometimes involved. Editorship has been changed a number of times, Blanford being the first; A. E. Shipley edited the series 1908–1927. The name of the series changed in the 1940's to "Fauna of India, including Pakistan, Ceylon, Burma and Malaya". The 6 volumes on Diptera, the first two unnumbered, were issued 1912–1940, and more are proposed.
> See Brunetti, 1912.

Blanton, F. S. (See also Wirth and Blanton)
1937 A new North American fly belonging to the genus *Rivellia* (Otitidae). Canad. Ent. **69**: 139–141, 1 fig.

Blatchley, W. S. (See Aldrich, 1897)

Boesel, M. W.
1948 *Holoconops* in the western Lake Erie region (Diptera: Heleidae). Ohio Jour. Sci. **48**: 69–72, 1 pl.

Bohart, G. E.
1938 Synopsis of the genus *Dalmannia* in North America (Diptera, Conopidae). Pan-Pacific Ent. **14**: 132–136.

——, **Stephen, W. P., and Eppley, R. K.**
1960 The biology of *Heterostylum robustum* (Diptera: Bombyliidae), a parasite of the alkali bee. Ent. Soc. Amer. Ann. **53**: 425–435, 15 figs.

Bohart, R. M. (See also Freeborn and Bohart)
(1949) The subgenus *Neoculex* in America north of Mexico (Diptera, Culicidae). Ent. Soc. Amer. Ann. (1948) **41**: 330–345, 3 pls.
1950 A new species of *Orthopodomyia* from California (Diptera, Culicidae). Ent. Soc. Amer. Ann. **43**: 399–404, 1 pl.

Boheman, C. H.
1858　Bidrag till Lapplands dipter-fauna. K. Svenska Vetensk. Akad. Öfversigt af . . . Förhandl. **15**: 55–57.
1865　Spetsbergen insekt-fauna. K. Svenska Vetensk. Akad. Öfversigt af . . . Förhandl. **22**: 563–577, 1 pl.

Bollow, H.
1954　Die landwirtschaftlich wichtigen Haarmücken. Ztschr. f. Pflanzenbau u. Pflanzenschutz **49** [=n. ser., 5]: 197–232, 20 figs.

Bolwig, N.
1940　The description of *Scatophila unicornis* Czerny, 1900 (Ephydridae, Diptera). Roy. Ent. Soc., London, Proc. Ser. B: Taxonomy **9**: 129–137, 8 figs.

Bonhag, P. F.
1951　The skeleto-muscular mechanism of the head and abdomen of the adult horsefly (Diptera: Tabanidae). Amer. Ent. Soc. Trans. **77**: 131–202, 1 text fig. + figs. 1–29.

Borgmeier, T.
1923　Contribuição para o conhecimento dos phorideos do Brasil (Phoridae-Diptera). [Rio de Janeiro] Mus. Nac. do Rio de Janeiro, Arch. **24**: 323–346, 12 figs.
1924　Novos generos e especies de phorideos do Brasil. [Rio de Janeiro] Mus. Nac. de Rio de Janeiro, Bol. **1**: 167–202 23 figs.
1928　Nota previa sobre alguns phorideos que parasitam formigas cortadeiras dos generos *Atta* e *Acromyrmex*. Bol. Biol. **1928**: 119–126, 2 figs.
1931　Sobre alguns phorideos que parasitam a Saúva e outras formigas cortadeiras (Diptera-Phoridae). Inst. Biol. de Defesa Agr. e Anim. [São Paulo] Arch. **4**: 209–228, pls. 21–25.
1932　Revisão do genero ecitophilo *Xanionotum* Brues, com a descripção de duas especies novas (Dipt. Phoridae). Rev. de Ent. **2**: 369–380, 2 figs.
1937　Uma nova especie de *Apocephalus* (Dipt. Phoridae), endoparasita de *Chauliognathus fallax* Germ. (Col. Cantharidae). Rev. de Ent. **7**: 207–216, 12 figs.
1958　Neue Beitraege zur Kenntnis der neotropischen Phoriden (Diptera, Phoridae). Studia Ent. (Rio de Janeiro) n. ser., **1**: 305–406, figs. I–III, 1–106.
1960　Gefluegelte und ungefluegelte Phoriden aus der neotropischen Region, nebst Beschreibung von sieben neuen Gattungen (Diptera, Phoridae). Studia Ent. (Rio de Janeiro) n. ser., **3**: 257–374, 119 figs.

Borgmeier, T.—Continued
1961   Weitere Beitraege zur Kenntnis der neotropischen Phoriden, nebst Beschreibung einiger *Dohrniphora*-Arten aus der indo-australischen Region (Diptera, Phoridae). Studia Ent. (Rio de Janeiro) **n. ser., 4**: 1–112, 150 figs.
1962a Some new North American species of the dipterous family Phoridae. Rev. Brasil. de Biol. **22**: 65–82, 17 figs.
1962b Versuch einer Uebersicht ueber die neotropischen *Megaselia*-Arten, sowie neue oder wenig bekannte Phoriden verschiedener Gattungen (Dipt. Phoridae). Studia Ent. (Rio de Janeiro) **n. ser., 5**: 289–488, 132 figs.
1963   A new species of *Megaselia* from Arizona (Diptera, Phoridae). Chicago Nat. Hist. Mus., Fieldiana, Zool. **44**: 133–135, fig. 24.

Bory, J. B. G. M. (Baron de Saint-Vincent)
1829   Dictionnaire classique d'histoire naturelle. Vol. 15: Rua–S, 754 pp. Paris.
> The entire work consists of 17 volumes, Paris, 1822–1831. Many authors are involved.

Bosc, L. A. G. (See also S.N.A., 1803)
1792   *Keroplatus*. Soc. d'Hist. Nat. de Paris, Actes **1**: 42–43, 1 fig.

Bottimer, L. J. (See Wirth and Bottimer)

Bouché, P. F.
1833   Naturgeschichte der schädlichen und nützlichen Garten-Insecten und die bewährtesten Mittel zur Vertilgung der ersteren. 176 pp. Berlin.
1834   Naturgeschichte der Insekten, besonders in Hinsicht ihrer ersten Zustände als Larven und Puppen. Erste Lieferung. 216 pp., 10 pls. Berlin.
> This was the only Lieferung published.

1847   Beiträge zur Kenntniss der Insekten-Larven. Stettin Ent. Ztg. **8**: 142–146.

Boyce, A. M.
1935   Bionomics of the walnut husk fly, *Rhagoletis completa*. Hilgardia (1934) **8**: 363–579, 77 figs.

Boyd, M. F., ed.
1949   Malariology. A comprehensive survey of all aspects of this group of diseases from a global standpoint. Vol. 1, 787 pp., 237 figs., 1 pl. Philadelphia and London.
> This work consists of 2 volumes, continuously paged 1–1643, 436 figs., 1 pl., published at Philadelphia and London, 1949.

Bradley, G. H. (See King et al.)

Braschnikov, W. C.
1897   Zur Biologie und Systematik einiger Arten minierender Diptera. Moskva Selsk. Khoz. Inst. Izv. **3 (pt. 2):** 19–43, 7 figs., pl. 2.

Brauer, F.
1858   Neue Beiträge zur Kenntniss der europäischen Oestriden. K.-k. Zool.-Bot. Gesell. Wien, Verhandl. **8 (Abhandl.):** 449–470, 1 pl.

1863   Monographie der Oestriden. 292 pp., 10 pls. Wien.

1875   Beschreibung neuer und ungenügend bekannter Phryganiden und Oestriden. K.-k. Zool.-Bot. Gesell. Wien Verhandl. **25 (Abhandl.):** 69–78, 1 pl.

1882   Die Zweiflügler des Kaiserlichen Museums zu Wien. II. K. Akad. der Wiss. Wien, Math.-Nat. Cl. Denkschr. **44 (1):** 59–110, pls. 1–2.

    Also published separately, 54 pp., 2 pls., Wien, *1882*.

1883   Die Zweiflügler des Kaiserlichen Museums zu Wien. III. K. Akad. der Wiss. Wien, Math.-Nat. Cl. Denkschr. **47:** 1–100, 5 pls.

    Also published separately, 100 pp., 5 pls., Wien, *1883*.

1885   Systematisch-zoologische Studien. K. Akad. der Wiss. Wien, Math.-Nat. Cl. Sitzber. Abt. 1 **91:** 237–413, 1 pl.

1893   Vorarbeiten zu einer Monographie der Muscaria Schizometopa (exclusive Anthomyidae) von Prof. Dr. Fr. Brauer und Julius Edl. v. Bergenstamm. K.-k. Zool.-Bot. Gesell. Wien, Verhandl. **43 (Abhandl.):** 447–525.

1898   Beiträge zur Kenntniss der Muscaria Schizometopa. K. Akad. der Wiss. Wien, Math.-Nat. Cl. Sitzber. Abt. 1 **107:** 493–546.

———, **and Bergenstamm, J. E. von**
1889   Die Zweiflügler des Kaiserlichen Museums zu Wien. IV. Vorarbeiten zu einer Monographie der Muscaria Schizometopa (exclusive Anthomyidae). Pars I. K. Akad. der Wiss. Wien, Math.-Nat. Cl. Denkschr. **56 (1):** 69–180, 11 pls.

    Also published separately in Wien, *1889*, 112 pp. It is possible that the separate takes precedence. Some of the included genera are dated in Neave (1939–1950) from the Denkschrift as 1890.

1891   Die Zweiflügler des Kaiserlichen Museums zu Wien. V. Vorarbeiten zu einer Monographie der Muscaria Schizometopa (exclusive Anthomyidae). Pars II. K. Akad. der Wiss. Wien, Math.-Nat. Cl. Denkschr. **58:** 305–446.

    Also published separately in Wien, *1891*, 142 pp.

Brauer, F., and Bergenstamm, J. E. von—Continued
- 1893 Die Zweiflügler des Kaiserlichen Museums zu Wien. VI. Vorarbeiten zu einer Monographie der Muscaria Schizometopa (exclusive Anthomyidae). Pars III. K. Akad. der Wiss. Wien, Math.-Nat. Cl. Denkschr. **60:** 89–240.

    Also published separately in Wien, *1893*, 152 pp.

- 1894 Die Zweiflügler des Kaiserlichen Museums zu Wien. VII. Vorarbeiten zu einer Monographie der Muscaria Schizometopa (exclusive Anthomyidae). Pars IV. Pt. 7, 88 pp. Wien.

    Also published in K. Akad. der Wiss. Wien, Math.-Nat. Cl. Denkschr. *61:* 537–624, *1895*.

    The entire work consists of 7 parts, Wien, 1889–1894, the first 3 being by Brauer alone, and only the last 4 being subtitled "Vorarbeiten zu einer Monographie . . . [etc.]". These 7 individually paged parts were also published in volumes 42, 44, 47, 56, 58, 60, and 61 of K. Akad. der Wiss. Wien, Math.-Nat. Cl. Denkschr., from 1880–1895.

Brauns, A.
- 1954a Untersuchungen zur angewandten Bodenbiologie. Vol. 1: Terricole Dipterenlarven, 179 pp., 96 figs., 6 pls. Göttingen, Frankfurt, Berlin.

    The entire work consists of 2 volumes to date, Göttingen, [etc.], 1954.

- 1954b Untersuchungen zur angewandten Bodenbiologie. Vol. 2: Puppen terricole Dipterenlarven, 156 pp., 75 figs. Göttingen, Frankfurt, Berlin.

Brennan, J. M. (See also Philip, 1941d)
- 1935 The Pangoniinae of nearctic America, Diptera: Tabanidae. Kans. Univ. Sci. Bul. **22** [=whole ser., 32]: 249–401, 9 pls. (=Kans. Univ. Bul. 36 (14).)

Brèthes, J.
- 1906 *Sarcophaga caridei* una nueva mosca langosticida. Buenos Aires Mus. Nac. An. **13** [=ser. 3, 6]: 297–301, 3 figs.

Brewster, D., ed.
- [1817– The Edinburgh encyclopaedia. Vol. 12, pp. 1–376, pls.
- 1818] 330–340, *1817*; pp. 377–749, pls. 341–346, *1818*. Edinburgh.

    Pp. 1–376 of this volume were published in November, 1817. The entire work consists of 18 volumes, published in Edinburgh, 18-?–1830. A "Third edition" was reprinted in 1830. For dating of the first edition, see Sherborn, 1937: 112.

Brimley, C. S.

1923   Additional Syrphidae (Diptera) from North Carolina, with descriptions of two supposed new species. Ent. News **34**: 277–279.

1924   Three supposed new species of *Ceraturgus* (Diptera, Asilidae) from North Carolina. Ent. News **35**: 8–12.

1925   New species of Diptera from North Carolina. Ent. News **36**: 73–77.

1927   Two new species of Diptera from North Carolina (Tachinidae, Conopidae). Ent. News **38**: 235–236.

1928   Some new wasps (Hymenoptera) and two new Diptera from North Carolina. Elisha Mitchell Sci. Soc. Jour. **43**: 199–206.

1938   The insects of North Carolina. 560 pp. N.C. Dept. of Agriculture, Raleigh.

> Two supplements have been published, 1942 and 1950, the last by D. L. Wray.

Brinck, P. (See Hanström, Brinck, and Rudebeck)

Brischke, C. G. A.

1881   Die Blattminirer in Danzig's Umgebung. Naturf. Gesell. in Danzig, Schr. n. ser., **5**: 233–290, 5 figs.

British Museum (Natural History)

1927–   Insects of Samoa and other Samoan terrestrial Arthropoda.
1935   5 vols. London.

> This consists of 9 parts bound in 5 volumes, and was issued in a number of fascicles by different authors. Volume 4 consists of Part 6, the Diptera, and was issued in 9 fascicles, 1927–1935. See Edwards, 1928b.

1929–   Diptera of Patagonia and south Chile. 7 pts. London.
1951

> Some of the parts were issued in several fascicles, each by one or more authors. Part 5 was issued in 3 fascicles, 1930–1932; Part 6, in 6 fascicles, 1929–1948; and Part 7 in 3 fascicles, 1934–1937.
> See Alexander, 1929c; Hall, 1937b; Malloch, 1933c, 1934a.

1939–   Ruwenzori Expedition 1934–5. 3 vols. London.
1957

> Volume 2 consists of 9 consecutively paged numbers on Diptera, issued 1939–1957. Additional material from this Expedition is included in the series "Ruwenzori Expedition 1952".
> See Richards, 1951.

Brock, E. M.

1960   Mutualism between the midge *Cricotopus* and the alga *Nostoc*. Ecology **41**: 474–483, 6 figs.

Brodie, W.

1894   Canadian galls and their occupants. Biol. Rev. Ontario **1**: 13–15, 44–46, 73–75, 109–111, 1 fig.

1909   Galls found in the vicinity of Toronto. No. 3. Canad. Ent. **41**: 157–160.

Brodie, W., and White, J. E.

1883   Check list of insects of the Dominion of Canada. 67 pp. Toronto, Ont.

Brohmer, P. (See Enderlein, 1914a, 1920)

Brohmer, P., Ehrmann, P., and Ulmer, G., eds.

(1927–1937)   Die Tierwelt Mitteleuropas. 7 vols. Leipzig.

> This was to be composed of 7 volumes when completed, and was issued in "Lieferungen" without dates. Dates of completion were given on the title pages of completed volumes. The last portion known to us was issued in 1937, at which time the work was still incomplete. An "Ergänzungsband" was begun in 1932, with a supplement to Vol. 6. It is not known whether this series continues. Each volume of the series concerns a major taxonomic group, in which each order is assigned an "Abtheilung" number and is separately paged within the volume. Volumes 4–6, which included the insects, are complete. Vol. 6, including the Diptera, was issued in 3 parts, 1927–1936, and includes 5 orders, nos. 14–18.
>
> See Enderlein, 1936a.

Bromley, S. W.

1924   New robber-flies (Asilidae, Diptera). Boston Soc. Nat. Hist. Occas. Papers **5**: 125–127, 4 figs.

1925   The *Bremus* resembling *Mallophorae* of the southeastern United States (Diptera, Asilidae). Psyche **32**: 190–194.

1926   The external anatomy of the black horsefly, *Tabanus atratus* Fab. (Diptera: Tabanidae). Ent. Soc. Amer. Ann. **19**: 440–460, 12 figs.

1928   Notes on the genus *Proctacanthus* with the descriptions of two new species. Psyche **35**: 12–15.

1929   Notes on the asilid genera *Bombomima* and *Laphria* with descriptions of three new species and two new varieties (Diptera). Canad. Ent. **61**: 157–161, 1 fig.

1931   New Asilidae, with a revised key to the genus *Stenopogon* Loew: (Diptera). Ent. Soc. Amer. Ann. **24**: 427–435.

1934a   The robber flies of Texas (Diptera, Asilidae). Ent. Soc. Amer. Ann. **27**: 74–113, 2 pls.

1934b   Two new dasypogonine robber flies from the Southwest (Asilidae: Diptera). N.Y. Ent. Soc. Jour. **42**: 225–226.

Bromley, S. W.—Continued
- 1935a Notes on Texas robber flies with the description of a new species of *Proctacanthella* (Asilidae: Diptera). Mich. Univ. Mus. Zool. Occas. Papers **304**: 1–7.
- 1935b The laphriine robber flies of North America (Diptera: Asilidae). Ohio State Univ., Abs. Doctors' Diss. (1933–1934) **14**: 125–134.
- 1936 The genus *Diogmites* in the United States of America with descriptions of new species (Diptera: Asilidae). N.Y. Ent. Soc. Jour. **44**: 225–237.
- 1937a The genus *Stenopogon* Loew in the United States of America (Asilidae: Diptera). N.Y. Ent. Soc. Jour. **45**: 291–309, 4 figs.
- 1937b New and little-known Utah Diptera with notes on the taxonomy of the Diptera. Utah Acad. Sci., Arts and Letters, Proc. **14**: 99–109, 6 figs.
- 1938 A new *Neoitamus* from Utah (Diptera: Asilidae). Utah Acad. Sci., Arts and Letters, Proc. **15**: 61.
- 1940 New U.S.A. robber flies (Diptera: Asilidae). Brooklyn Ent. Soc. Bul. **35**: 13–21, 5 figs.
- 1946 Guide to the insects of Connecticut. Part VI. The Diptera or true flies of Connecticut. Third fascicle. Asilidae. Conn. State Geol. and Nat. Hist. Survey, Bul. **69**: 1–48, 3 pls.
- 1950a Florida Asilidae (Diptera) with description of one new species. Ent. Soc. Amer. Ann. **43**: 227–239, 1 pl.
- 1950b Records and descriptions of Asilidae in the collection of the University of Michigan Museum of Zoology. Mich. Univ. Mus. Zool. Occas. Papers **527**: 1–5.
- 1951 Asilid notes (Diptera) with descriptions of thirty-two new species. Amer. Mus. Nat. Hist., Amer. Mus. Novitates **1532**: 1–36, 7 figs.

Brookman, B.
- 1941 A new Californian *Stenopogon* (Diptera, Asilidae). Pan-Pacific Ent. **17**: 78–80.

——, **and Reeves, W. C.**
- 1950 A new name for a California mosquito. Pan-Pacific Ent. **26**: 159–160.

Brooks, A. R.
- 1942 *Clistomorpha, Psalidopteryx* and allies (Diptera, Tachinidae). Canad. Ent. **74**: 140–150, 4 figs.
- 1943 A review of the Canadian species of *Ernestia* sens. lat. (Tachinidae, Diptera). Canad. Ent. **75**: 66–78, 2 pls.

Brooks, A. R.— Continued
- 1944a A review of the North American species of *Gonia* sens. lat. (Diptera, Tachinidae). Canad. Ent. (1943) **75**: 219–236, 2 pls.
- 1944b A review of the North American species of *Linnaemya* sens. lat. (Diptera, Tachinidae). Canad. Ent. **76**: 193–206, 1 pl.
- 1945a New Canadian Diptera (Tachinidae). Canad. Ent. **77**: 78–96.
- 1945b A revision of the North American species of the *Phasia* complex (Diptera, Tachinidae). Sci. Agr. **25**: 647–679, 4 pls.
- 1946a A revision of the North American species of the *Rhodogyne* complex (Diptera, Larvaevoridae). Canad. Ent. (1945) **77**: 218–230, 2 pls.
- 1946b A new Canadian *Tabanus* (Diptera, Tabanidae). Canad. Ent. (1945) **77**: 234.
- [1947] A revision of the North American species of *Leschenaultia* sens. lat. (Diptera, Larvaevoridae). Canad. Ent. (1946) **78**: 169–182, 3 pls.

    This issue was first received by the Smithsonian Institution library in Mar., 1947. Thompson, 1949: 7, gives a hint as to the correctness of 1947 as the date of publication.

- 1949 New North American larvaevorine flies. Canad. Ent. **81**: 21–24.
- 1951 Identification of the root maggots (Diptera: Anthomyiidae) attacking cruciferous garden crops in Canada, with notes on biology and control. Canad. Ent. **83**: 109–120, 32 figs.
- 1952 Identification of bombyliid parasites and hyperparasites of Phalaenidae of the prairie provinces of Canada, with descriptions of six other bombyliid pupae (Diptera). Canad. Ent. **84**: 357–373, 46 figs.

Brown, B.
- 1897 Two new species of asilids from New Mexico. Kans. Univ. Quart. **ser. A, 6**: 103.

Brown, N. R., and Clark, R. C.
- 1956 Studies of predators of the balsam woolly aphid *Adelges piceae* (Ratz.) (Homoptera: Adelgidae). I. Field identification of *Neoleucopis obscura* (Hal.), *Leucopina americana* (Mall.) and *Cremifania nigrocellulata* Cz. (Diptera: Chamaemyiidae). Canad. Ent. **88**: 272–279, figs. 1–23.

Brues, C. T. (See also Melander and Brues)

1901a A new species of *Dolichopus* from Texas. Ent. News **12**: 44–45, 1 fig.
1901b Two new myrmecophilous genera of aberrant Phoridae from Texas. Amer. Nat. **35**: 337–356, 11 figs.
1902a New and little known guests of the Texan legionary ants. Amer. Nat. **36**: 365–378, 7 figs.
1902b Notes on the larvae of some Texan Diptera. Psyche **9**: 351–354, 3 figs.
1903 A monograph of the North American Phoridae. Amer. Ent. Soc. Trans. **29**: 331–403, 5 pls.
1905 Phoridae from the Indo-Australian Region. Budapest Magyar Nemzeti Muz., Ann. Hist. Nat. **3**: 541–555, 1 fig.
1906a Diptera. Fam. Phoridae. Fasc. 44, 21 pp., 2 pls. *In* Wytsman, P., ed., Genera insectorum [q.v.]. Bruxelles.
1906b Two new species of Phoridae. Wis. Nat. Hist. Soc. Bul. **4**: 100–102.
1908 Some new North American Phoridae. N.Y. Ent. Soc. Jour. **16**: 199–201.
1909 Some further remarks on the systematic affinities of the Phoridae, with descriptions of two new North American species. Wis. Nat. Hist. Soc. Bul. **7**: 103–108, 1 fig.
1913 A new species of Phoridae from New England. Psyche **20**: 90–91.
1914 The phorid genus *Platyphora* in America. Psyche **21**: 76–79, 5 figs.
1915 A synonymic catalog of the dipterous family Phoridae. Wis. Nat. Hist. Soc. Bul. (1914) **12**: 85–152.
1916a A remarkable new species of *Phora* (*Trineura*). Canad. Ent. **48**: 394–395.
1916b A new species of *Aphiochaeta* (Diptera, Phoridae) from New England. Psyche **23**: 175–176.
1919a New North American Phoridae of the genus *Aphiochaeta* (Diptera). Insecutor Inscitiae Menstruus (1918) **6**: 183–194.
1919b The Phoridae of Grenada. Harvard Univ., Mus. Compar. Zool. Bul. **62**: 499–506, 2 figs.
1924a Notes on some New England Phoridae (Diptera). Psyche **31**: 41–44, 1 fig.
1924b New and unrecorded American species of the family Phoridae (Diptera). Psyche **31**: 155–161.

Brues, C. T.—Continued.

1936   A new ecitophilous North American phorid fly. Brooklyn Ent. Soc. Bul. **31**: 68–70.

1943   Some North American species of *Chaetopleurophora* (Diptera, Phoridae). Psyche **50**: 50–52.

1950   Guide to the insects of Connecticut. Part VI. The Diptera or true flies of Connecticut. Fourth fascicle [part]. Family Phoridae. Conn. State Geol. and Nat. Hist. Survey, Bul. **75**: 33–85, 8 figs.

———, **and Melander, A. L.**

1932   Classification of insects. Harvard Univ., Mus. Compar. Zool. Bul. **73**: 1–672, 1121 figs.

———, **Melander, A. L., and Carpenter, F. M.**

1954   Classification of insects. (Revised ed.). Harvard Univ., Mus. Compar. Zool. Bul. **108**: 1–917, 1219 figs.

Brullé, A.

1832   Mémoire sur un genre nouveau de Diptères, de la famille des Tipulaires. Soc. Ent. de France, Ann. [ser. 1], **1**: 205–209.

Brundin, L.

1949   Chironomiden und andere Bodentiere der südschwedischen Urgebirgsseen. Ein Beitrag zur Kenntnis der bodenfaunistischen Charakterzüge schwedischer oligotropher Seen. Sweden, Fishery Bd., Inst. Freshwater Res. (Drottningholm), Rpt. **30**: 1–915, 241 figs.

1956   Zur Systematik der Orthocladiinae (Dipt. Chironomidae). Sweden, Fishery Bd., Inst. Freshwater Res. (Drottningholm), Rpt. **37**: 5–185, 137 figs.

Brundrett, H. M.   (See Bishopp et al.)

Brunetti, E.

1911a   New oriental Nemocera. Indian Mus. Rec. **4**: 259–316.

1911b   Revision of the oriental Tipulidae with descriptions of new species. Indian Mus. Rec. **6**: 231–314.

1912   Diptera Nematocera (excluding Chironomidae and Culicidae). Diptera, [Vol. 1], 581 pp., 44 figs., 12 pls. *In* [Blanford, W. T., ed.], Fauna of British India [q.v.]. London.

1920   Catalogue of oriental and south Asiatic Nemocera. Indian Mus. Rec. **17**: 1–300.

1923   Second revision of the oriental Stratiomyidae. Indian Mus. Rec. **25**: 45–180.

Bryce, D.

1960 Studies on the larvae of the British Chironomidae (Diptera), with keys to the Chironominae and Tanypodinae. Soc. Brit. Ent. Trans. **14**: 19–62, 15 figs.

Buren, W. F.

1947 A new *Aedes* from the Florida Keys (Diptera, Culicidae). Ent. Soc. Wash. Proc. **49**: 228–229.

Burgess, A. F.

1897 Notes on certain Coleoptera known to attack the gypsy moth. Pp. 412–431, pls. 3–5 of app. *In* Wood, E. W., Stetson, S. S., Avery, J. G., Pratt, A., Sargent, F. W., and Sessions, W. R., Report of the State Board of Agriculture on the work of extermination of the gypsy moth. Mass. State Bd. Agr. Ann. Rpt. Sec. (1896) **44**: 349–434, pl. 1, 3 photos, pls. 1–5 of app.

Also published separately (pp. 64–83), Boston, 85 pp., *1897*.

Burgess, E. (See also Comstock, 1880)

1878 Two interesting American Diptera, *Glutops singularis* and *Epibates osten-sackenii*. Boston Soc. Nat. Hist. Proc. (1876–1878) **19**: 320–324, 1 pl.

Burke, H. E.

1905 Black check in western hemlock. U.S. Dept. Agr., Bur. Ent., Cir. [ser. 2], **61**: 1–10, 5 figs.

Buxton, P. A.

1954 British Diptera associated with fungi. 2.–Diptera bred from Myxomycetes. Roy. Ent. Soc., London, Proc. Ser. A: Gen. Ent. **29**: 163–171.

1960 British Diptera associated with fungi. III. Flies of all families reared from about 150 species of fungi. Ent. Monthly Mag. **96** [=ser. 4, 21]: 61–94.

Byers, G. W. (See also Rogers and Byers)

1961 The crane fly genus *Dolichopeza* in North America. Kans. Univ. Sci. Bul. **42**: 665–924, 244 figs.

Call, R. E.

1897 Some notes on the flora and fauna of Mammoth Cave, Ky. Amer. Nat. **31**: 377–392, pls. 10–11.

Cameron, A. E.

1926 Bionomics of the Tabanidae (Diptera) of the Canadian prairie. Bul. Ent. Res. **17**: 1–42, 18 figs., 5 pls.

Campbell, R. E., and Davidson, W. M.
1924   Notes on aphidophagous Syrphidae of southern California. South. Calif. Acad. Sci. Bul. **23:** 1–9, 59–71, pls. A, B. E, N, O.
    Reprint is paged continuously, 1–20.

Camras, S.
1943   Notes on the North American species of the *Zodion obliquefasciatum* group (Diptera: Conopidae). Ent. News **54:** 187–191.
1944   Notes on the North American species of the *Zodion fulvifrons* group (Diptera: Conopidae). Pan-Pacific Ent. **20:** 121–128.
1945   A study of the genus *Occemyia* in North America (Diptera: Conopidae). Ent. Soc. Amer. Ann. **38:** 216–222.
1953   A review of the genus *Myopa* in North America (Diptera: Conopidae). Wasmann Jour. Biol. **11:** 97–114.
1954   A new species of *Zodion* from California (Diptera: Conopidae). Pan-Pacific Ent. **30:** 165–166.
1955   A review of the New World flies of the genus *Conops* and allies (Diptera: Conopidae). U.S. Natl. Mus. Proc. **105:** 155–187.
1957   A review of the New World *Physocephala* (Diptera: Conopidae). Ent. Soc. Amer. Ann. **50:** 213–218.

———, **and Hurd, P. D., Jr.**
1957   The conopid flies of California. Calif. Insect Survey, Bul. **6:** 19–49, 4 figs.

Capelle, K. J.
1953   A revision of the genus *Loxocera* in North America with a study of geographical variation in *L. cylindrica* (Diptera: Psilidae). Ent. Soc. Amer. Ann. **46:** 99–114, 4 pls.
1956   The genus *Rhopalosyrphus*, with a description of a new species from Arizona (Diptera, Syrphidae). Kans. Ent. Soc. Jour. **29:** 170–175, 5 figs.

Carpenter, F. M.   (See Brues, Melander, and Carpenter)

Carpenter, G. D. H.   (See also Collin, 1937a)
1938   Notes on insects collected in west Greenland by the Oxford University Greenland Expedition, 1936, with descriptions of a new species of *Angitia* (Hymenoptera, Ichneumonidae) by A. Roman, and of *Fannia* (Diptera, Anthomyidae) by J. E. Collin. Ann. and Mag. Nat. Hist. **ser. 11, 1:** 529–553.

Carpenter, S. J., and LaCasse, W. J.
1955   Mosquitoes of North America (north of Mexico). 360 pp., 288 figs., 127 pls. Berkeley and Los Angeles.

Carrera, M. (See also d'Andretta and Carrera)
———, **Lopes, H. de S., and Lane, J.**
1947 Contribuição ao conhecimento dos "Microdontinae" neotrópicos e descrição [sic] de duas novas espécies de *"Nausigaster"* Williston (Diptera, Syrphidae). Rev. Brasil. de Biol. **7**: 471–486, 49 figs.

Carson, H. L.
1958 The population genetics of *Drosophila robusta*. Adv. in Genet. **9**: 1–40, 4 figs.

Carter, H. F.
1919 New West African Ceratopogoninae. Ann. Trop. Med. and Parasitol. **12**: 289–302, 4 figs., 1 pl.
1921 A revision of the genus *Leptoconops*, Skuse. Bul. Ent. Res. **12**: 1–28, 10 figs.

Castelnau, Count de (See Laporte, F. L. de)

Cazier, M. A.
1941 A generic review of the family Apioceratidae with a revision of the North American species (Diptera-Brachycera). Amer. Midland Nat. **25**: 589–631, 2 figs., 4 pls.
1954 New species and notes on flies belonging to the genera *Rhaphiomidas* and *Apiocera* (Diptera, Apioceratidae). Amer. Mus. Nat. Hist., Amer. Mus. Novitates **1696**: 1–10.

Chamberlin, T. R. (See Rockwood, Zimmerman, and Chamberlin)

Chandler, H. P. (See Usinger et al.)

Chen, S. H.
1939 Étude sur les Diptères Conopides de la Chine. [Shanghai] Mus. Heude, Notes d'Ent. Chin. **6**: 161–231, 45 figs.

Chenu, J. C., ed.
1850–
1861 Encyclopédie d'histoire naturelle, ou traité complet de cette science. 22 vols. + 9 "Tables alphabetiques". Paris.

> None of the 31 volumes are numbered.
> See Desmarest, 1859.

Chillcott, J. G.
1958a Notes on a rare fly, *Megagrapha pubescens* (Loew) (Diptera: Empididae). Canad. Ent. **90**: 608–611, 6 figs.
1958b A new nearctic species of *Symballophthalmus* Becker (Diptera: Empididae). Canad. Ent. **90**: 647–649, 6 figs.
1959 Studies on the genus *Rhamphomyia* Meigen: A revision of the nearctic species of the *basalis* group of the subgenus *Pararhamphomyia* Frey (Diptera: Empididae). Canad. Ent. **91**: 257–275, 28 figs.

Chillcott, J. G.—Continued
1961a A revision of the nearctic species of Fanniinae (Diptera: Muscidae). Canad. Ent. (1960) **92** (**Sup. 14**): 1–295, 289 figs., 61 maps.
1961b A revision of the genus *Roederioides* Coquillett (Diptera: Empididae). Canad. Ent. **93**: 419–428, 29 figs.
1961c The genus *Bolbomyia* Loew (Diptera: Rhagionidae). Canad. Ent. **93**: 632–636, 8 figs.
1962 A revision of the *Platypalpus juvenis* complex in North America (Diptera: Empididae). Canad. Ent. **94**: 113–143, 50 figs.

Chittenden, F. H.
1927 *Tritoxa flexa* Wied., the black onion fly (Ortalidae, Dipt.). Canad. Ent. **59**: 1–4, 1 fig.

Christenson, L. D. (See also Steiner et al.)
——, and **Foote, R. H.**
1960 Biology of fruit flies. Ann. Rev. Ent. **5**: 171–192.

Christophers, S. R.
1960 *Aëdes aegypti* (L.) the yellow fever mosquito. Its life history, bionomics and structure. 739 pp., 86 figs. Cambridge, England.

Chu, H. F.
1949 How to know the immature insects. 234 pp., 631 figs. Dubuque, Iowa.

Chun, C. (See Leuckart and Chun)

Claassen, P. W.
1922 The larva of a chironomid (*Trissocladius equitans* n. sp.) which is parasitic upon a may-fly nymph (*Rithrogena* sp.). Kans. Univ. Sci. Bul. **14** [=whole ser., 24]: 395–399, 3 pls. (=Kans. Univ. Bul. 23 (18).)

Clark, B.
1797 Observations on the genus *Oestrus*. Linn. Soc. (London), Trans. **3**: 289–329, 1 pl.
1815 An essay on the bots of horses and other animals. 72 pp., 2 pls. London.
1827 Of the insect called Oistros by the ancients, and of the true species intended by them under this appellation: In reply to the observations of W. S. MacLeay, Esq., and the French naturalists. To which is added, a description of a new species of *Cuterebra*. Linn. Soc. London, Trans. **15**: 402–411.

Clark, R. C. (See Brown and Clark)

Clausen, C. P.
- 1940 Entomophagous insects. 688 pp., 257 figs. New York.
- 1956 Biological control of insect pests in the continental United States. U.S. Dept. Agr. Tech. Bul. **1139**: 1–151, 1 fig.

Cockerell, T. D. A.
- [1889a] Thistle insects. Colo. Biol. Assoc. Rpt. **6**: 1–2.
    > Although itself undated, this unpaged report bears a printed statement at the end to the effect that it was also published in the Jan. 16, 1889, issue of "Custer County Courant", a newspaper.
- [1889b] Notes. Colo. Biol. Assoc. Rpt. **10**: 2–3.
    > From dates mentioned in the text of this unpaged report, its date can be established as early 1889, probably February. Unlike the earlier reports, this does not bear a date of publication in the "Custer County Courant", although it may have appeared there.
- 1889c Contributions towards a list of the fauna and flora of Wet Mountain Valley, Colorado. I. West Amer. Sci. **6**: 103–106.
- 1889d Entomological notes from Colorado. Ent. Monthly Mag. **25**: 324.
- 1890a Notes on some species of gall-gnats (Cecidomyiae). Entomologist **23**: 278–282, 3 figs.
- 1890b The *Bigelovia* cecid. Ent. Monthly Mag. **26** [=ser. 2, 1]: 109.
- 1890c *Trypeta bigeloviae*, n. sp. Ent. Monthly Mag. **26** [=ser. 2, 1]: 324.
- 1895 *Cecidomyia atriplicis* n. sp. Amer. Nat. **29**: 766–767.
- 1896 New species of insects taken on a trip from the Mesilla Valley to the Sacramento Mts., New Mexico. N.Y. Ent. Soc. Jour. **4**: 201–207.
- 1898a New North American insects. Ann. and Mag. Nat. Hist. ser. 7, **2**: 321–331.
- 1898b Contributions to the entomology of New Mexico. II. Some records of Diptera. Davenport Acad. Nat. Sci. Proc. **7**: 149–156.
- 1900a *Diplosis partheniicola*, n. sp. Entomologist **33**: 201.
- 1900b *Asphondylia mentzeliae*, n. sp. Entomologist **33**: 302.
- 1901 A new cecidomyiid on *Gutierrezia*. Canad. Ent. **33**: 23.
- 1902 Some gall-insects. Canad. Ent. **34**: 183–184.
- 1904 Three new cecidomyiid flies. Canad. Ent. **36**: 155–156.
- 1905 Miscellaneous notes. Canad. Ent. **37**: 361–362.
- 1907 A gall-gnat of the prickly-pear cactus. Canad. Ent. **39**: 324.

Cockerell, T. D. A.—Continued
1908a   The dipterous family Nemestrinidae. Amer. Ent. Soc. Trans. **34**: 247–253, 1 pl.
1908b   A new gall on *Aster*. Canad. Ent. **40**: 89.
1908c   A remarkable cecidomyiid fly. Canad. Ent. **40**: 421–422, 1 fig.
1909a   A new gall-gnat on *Artemisia*. Canad. Ent. **41**: 150–151.
1909b   Fossil insects from Colorado. Entomologist **42**: 170–174, 3 figs.
1910a   A new dipterous parasite of bats. Canad. Ent. **42**: 59–60.
1910b   Fossil insects and a crustacean from Florissant, Colorado. Amer. Mus. Nat. Hist. Bul. **28**: 275–288, 4 figs.
1913a   A new gall on *Peritoma serrulatum*. Jour. Econ. Ent. **6**: 279–280, 1 fig.
1913b   A fossil asilid fly from Colorado. Entomologist **46**: 213–214.
1914a   A new cecidomyiid fly. Jour. Econ. Ent. **7**: 460, 1 fig.
1914b   A new dipterous gall on *Stanleya*. Jour. Ent. and Zool. **6**: 240–241, 2 figs.
1914c   The fossil and recent Bombyliidae compared. Amer. Mus. Nat. Hist. Bul. **33**: 229–236, 20 figs.
1915   A new fly of the family Phoridae from California. Canad. Ent. **47**: 351–352.
1916   Two new Diptera of the genus *Rhamphomyia* from Colorado. Canad. Ent. **48**: 123–124, 1 fig.
1917   New Tertiary insects. U.S. Natl. Mus. Proc. **52**: 373–384, pl. 31.
1918   The mosquitoes of Colorado. Jour. Econ. Ent. **11**: 195–200.
1926   The genus *Dixa* in Colorado (Diptera: Dixidae). Ent. Soc. Wash. Proc. **28**: 166.

———, and **Andrews, H.**
1916   Some Diptera (*Microdon*) from nests of ants. U.S. Natl. Mus. Proc. **51**: 53–56, 2 figs.

Coe, R. L.
1938   Rediscovery of *Callicera yerburyi*, Verrall (Diptera: Syrphidae): Its breeding habits, with a description of the larva. Entomologist **71**: 97–102, pl. 1.
1942   The British species of the genus *Chamaemyia* (Dipt., Chamaemyiidae). Ent. Monthly Mag. **78** [=ser. 4, 3]: 173–180, 5 figs.
1943   *Chamaemyia juncorum* Fall. and *C. herbarum* R.-D. (Dipt., Chamaemyiidae): A correction to my recent paper on British species of the genus. Ent. Monthly Mag. **79** [=ser. 4, 4]: 128–129, 1 fig.

Coe, R. L.—Continued
1953 Diptera: Syrphidae. Pt. 1, 98 pp., 46 figs. *In* Royal Entomological Society of London, Handbooks for the identification of British insects [q.v.]. Vol. 10: Cyclorrhapha. London.

Coher, E. I.
1950 Correction: A new genus of fungus gnats (Diptera: Mycetophilidae). Rev. de Ent. **21:** 114.
1959 A synopsis of American Mycomyiini with descriptions of new species (Diptera: Mycetophilidae). Ent. Amer. **37:** 1–155, 1 fig., 17 pls.

Cole, F.R. (See also Ferris and Cole; Latta and Cole; Van Duzee, Cole, and Aldrich)
1912a Notes on the Diptera of Laguna Beach. Pomona Col. Jour. Ent. **4:** 837–840, 3 figs.
1912b Some Diptera of Laguna Beach. Pomona Col., Laguna Mar. Lab., Ann. Rpt. **1:** 150–162, 10 figs.
1916a A new species of *Exoprosopa* (Dip.). Ent. News **27:** 463, 1 fig.
1916b New species of Asilidae from southern California. Psyche **23:** 63–69, 3 pls.
1917 Notes on Osten Sacken's group "*Poecilanthrax*", with descriptions of new species. N.Y. Ent. Soc. Jour. **25:** 67–80, 5 pls.
1919 The dipterous family Cyrtidae in North America. Amer. Ent. Soc. Trans. **45:** 1–79, 15 pls.
1921 Insects of the Pribilof Islands, Alaska. Diptera from the Pribilof Islands, Alaska. Calif. Acad. Sci. Proc. **ser. 4, 11:** 169–177, 4 figs.
1923a Expedition of the California Academy of Sciences to the Gulf of California in 1921. The Bombyliidae (bee flies); Diptera from the islands and adjacent shores of the Gulf of California. II. General report. Calif. Acad. Sci. Proc. **ser. 4, 12:** 289–314, 38 figs.; 457–481, 16 figs.
1923b Notes on California Bombyliidae with descriptions of new species. Jour. Ent. and Zool. **15:** 21–26, 11 figs.
1923c Notes on the dipterous family Cyrtidae. Psyche **30:** 46–48.
1923d A revision of the North American two-winged flies of the family Therevidae. U.S. Natl. Mus. Proc. **62 (4):** 1–140, 3 figs., 13 pls.
1924a Notes on Diptera of the syrphid genus *Sphegina*. Ent. News **35:** 39–44, 1 pl.
1924b Notes on the dipterous family Asilidae, with descriptions of new species. Pan-Pacific Ent. **1:** 7–13, 6 figs.

Cole, F. R.—Continued
- 1925 Notes on the dipterous family Therevidae. Canad. Ent. **57**: 84–88, 2 figs.
- 1927 A study of the terminal abdominal structures of male Diptera (two-winged flies). Calif. Acad. Sci. Proc. **ser. 4, 16**: 397–499, 287 figs.
- 1952 New bombyliid flies reared from anthophorid bees (Diptera: Brachycera). Pan-Pacific Ent. **28**: 126–130, 2 figs.
- 1957 New bombyliid flies from Chiapas, Mexico (Diptera). Pan-Pacific Ent. **33**: 200–202, 1 fig.
- 1959 A new name proposed in the genus *Thereva* (Diptera: Therevidae). Pan-Pacific Ent. **35**: 148.
- 1960a Stiletto-flies of the genus *Furcifera* Krober (Diptera: Therevidae). Ent. Soc. Amer. Ann. **53**: 160–169, 2 pls.
- 1960b New names in Therevidae and Bombyliidae (Diptera). Pan-Pacific Ent. **36**: 118.

——, and **Lovett, A. L.**
- 1919 New Oregon Diptera. Calif. Acad. Sci. Proc. **ser. 4, 9**: 221–255, 6 pls.
- 1921 An annotated list of the Diptera (flies) of Oregon. Calif. Acad. Sci. Proc. **ser. 4, 11**: 197–344, 54 figs.

——, and **Wilcox, J.**
- 1938 The genera *Lasiopogon* Loew and *Alexiopogon* Curran in North America (Diptera-Asilidae). Ent. Amer. **18**: 1–90, 3 pls.

Coleman, R. W.
- 1953 A new blackfly species from California (Diptera, Simuliidae). Ent. Soc. Wash. Proc. **55**: 45–46, 4 figs.

Collart, A.
- 1933 Description d'un Hélomyzide troglophile nouveau de Belgique (Diptera). Soc. Ent. de Belg. Bul. et Ann. **73**: 402–405, 2 figs.

Collin, J. E. (See also Carpenter, 1938)
- 1902 Four new species of Diptera (fam. Borboridae) found in Britain. Ent. Monthly Mag. **38** [=ser. 2, 13]: 55–60, 6 figs.
- 1920 *Eumerus strigatus* Fallén and *tuberculatus* Rondani (Diptera, Syrphidae). Ent. Monthly Mag. **56** [=ser. 3, 6]: 102–106, pl. 3.
- 1923 Diptera (Orthorrhapha Brachycera and Cyclorrhapha) from Spitzbergen and Bear Island. Results of the Oxford University Expedition to Spitzbergen, 1921. Ann. and Mag. Nat. Hist. **ser. 9, 11**: 116–123.

Collin, J. E.—Continued
- 1927 On some characters of possible generic importance in the *Hylemyia-Chortophila* group of the Anthomyidae (Diptera). Ent. Monthly Mag. **63** [=ser. 3, 13]: 129–135.
- 1930a A revision of the Greenland species of the anthomyid genus *Limmophora* sens. lat. (Diptera), with figures of the male genitalia of these and many other palaearctic species. Ent. Soc. London, Trans. **78**: 255–281, 13 pls.
- 1930b Some species of the genus *Meoneura* (Diptera). Ent. Monthly Mag. **66** [=ser. 3, 16]: 82–89, 1 pl.
- 1930c Some new species of the dipterous genus *Scatella* Dsv. and the differentiation of *Stictoscatella* gen. nov. (Ephydridae). Ent. Monthly Mag. **66** [=ser. 3, 16]: 133–139, 7 figs.
- 1931 The Oxford University Expedition to Greenland, 1928.— Diptera (Orthorrhapha Brachycera and Cyclorrhapha) from Greenland. Ann. and Mag. Nat. Hist. **ser. 10, 7**: 67–91, 27 figs.
- 1935 Diptera (Cyclorrhapha) from Akpatok Island, Ungava Bay, Canada. Results of the Oxford University Hudson Strait Expedition, 1931. Ann. and Mag. Nat. Hist. **ser. 10, 15**: 369–381, 11 figs.
- 1937a Description of *Rhamphomyia hirticula* sp. n. (Empididae). Pp. 407–409, figs. a–b. *In* Carpenter, G. D. H. Notes on insects collected in north-west Greenland by the Oxford University Ellesmere Land Expedition, 1934. Ann. and Mag. Nat. Hist. **ser. 10, 20**: 401–409, 2 figs.
- 1937b Two new species of the genus *Meoneura* (Diptera, Carnidae). Ent. Monthly Mag. **73** [=ser. 3, 23]: 250–252, 2 figs.
- 1939 On various new or little known British Diptera, including several species bred from the nests of birds and mammals. Ent. Monthly Mag. **75** [=ser. 3, 25]: 134–154, 7 figs.
- 1943 The British species of *Prosalpia* Pok. (Dipt., Anthomyidae). Ent. Monthly Mag. **79** [=ser. 4, 4]: 83–86.
- 1946 The British genera and species of Oscinellinae (Diptera, Chloropidae). Roy. Ent. Soc., London, Trans. **97**: 117–148, 6 figs.
- 1948 A short synopsis of the British Sapromyzidae (Diptera). Roy. Ent. Soc., London, Trans. **99**: 225–242, 3 figs.
- 1951 A new arctic species of *Fucellia* (Diptera Anthomyidae) with maculated wings. Ent. Meddel. **26**: 187–190, 2 figs.
- 1952a On the European species of the genus *Odinia* R.–D. (Diptera Odiniidae). Roy. Ent. Soc., London, Proc. Ser. B: Taxonomy **21**: 110–116.

Collin, J. E.—Continued
- 1952b On the subdivisions of the genus *Pipizella* Rnd., and an additional British species (Diptera, Syrphidae). Soc. Brit. Ent. Jour. **4**: 85–88, 3 figs.
- 1953a On the British species of *Scaptomyza* Hardy and *Parascaptomyza* Duda (Dipt., Drosophilidae). Entomologist **86**: 148–151.
- 1953b A revision of the British (and notes on other) species of Lonchaeidae (Diptera). Soc. Brit. Ent. Trans. **11**: 181–207, 24 figs.
- 1953c Some additional British Anthomyiidae (Diptera). Soc. Brit. Ent. Jour. **4**: 169–177.
- 1954 The genus *Chiastochaeta* Pokorny (Diptera: Anthomyiidae). Roy. Ent. Soc., London, Proc. Ser. B: Taxonomy **23**: 95–102, 3 figs.
- 1956 On the identity of Fallén's *Drosophila glabra* (Dipt., Camillidae). Ent. Monthly Mag. **92** [=ser. 4, 17]: 225–226.
- 1958 Notes on some British species of *Fannia* (Dipt., Muscidae), with the description of a new species. Ent. Monthly Mag. **94** [=ser. 4, 19]: 86–92, 1 fig.
- 1961 Empididae. Vol. 6, 782 pp., 317 figs. *In* [Verrall, G. H., ed.], British flies [q.v.]. Cambridge, England.

This was first published in 3 separate parts in the same year.

Collinge, W. E.
- 1909 Observations on the life-history and habits of *Thereva nobilitata*, Fabr., and other species. Jour. Econ. Biol. **4**: 14–18, pl. 4.

Colyer, C. N.
- 1957 A new species of *Plastophora* (Dipt., Phoridae) from England. A short discussion of the evolution of the present concept of the genus and a key for the identification of the world species. Brotéria **sér. trimest.: Cien. Nat., 26**: 75–89, 21 figs.

Comstock, A. B. (See Comstock and Comstock)

Comstock, J. H.
- [1880a] Report upon cotton insects. 511 pp., 77 figs., 3 pls. Washington, D.C., *1879*.

Dated May 18, 1880, from Comstock, 1881: 275.

- 1880b Report of the entomologist. [U.S. Dept. Agr.] Comnr. Agr. Rpt. **1879**: 185–348, 16 pls.

Published Oct. 18, 1880, according to Comstock, 1881: 275.

Comstock, J. H.—Continued
- 1881 Report of the entomologist. [U.S. Dept. Agr.] Comnr. Agr. Rpt. **1880:** 253–373, 24 pls.
- 1882 Report on miscellaneous insects. Pp. 195–214, pls. 14–20. *In* Riley, C. V., Report of the entomologist. [U.S. Dept. Agr. Comnr. Agr. Rpt. **1881/1882:** 61–214, pls. 1–20.
- 1918 The wings of insects. 430 pp., 427 figs., 10 pls. Ithaca, N.Y.
- 1940 An introduction to entomology. Ed. 9, 1064 pp., 1228 figs. Ithaca, N.Y.

———, **and Comstock, A. B.**
- 1895 A manual for the study of insects. 701 pp., 797 figs., 6 pls. Ithaca, N.Y.

Condrashoff, S. F.
- 1961 Three new species of *Contarinia* Rond. (Diptera: Cecidomyiidae) in Douglas-fir needles. Canad. Ent. **93:** 123–130, 24 figs.

Cook, A. J.
- 1884 Practical entomology. Notes from the entomological laboratory of the Michigan Agricultural College. Mich. State Bd. Agr. Ann. Rpt. Sec. (1882–1883) **22:** 422–450, 14 figs.

  Also published separately, with a title-page bearing the heading "Notes on injurious insects," Mich. Agr. Col., without a date, 32 pp., 14 figs., [?*1884*]. Also published in Mich. State Hort. Soc. Ann. Rpt. Sec. (1884) **14:** 81–109, 14 figs., *1885*.

Cook, E. F. (See also Kraft and Cook; Meade and Cook)
- 1955 A contribution toward a monograph of the family Scatopsidae (Diptera). Part I. A revision of the genus *Rhegmoclema* Enderlein (=*Aldrovandiella* Enderlein) with particular reference to the North American species; Part II. The genera *Rhegmoclemina* Enderlein, *Parascatopse* n. g., and a new species of *Rhegmoclema*. Ent. Soc. Amer. Ann. **48:** 240–251, figs. 2–3, pls. 1–5; 351–364, pls. 1–4.
- 1956a A contribution toward a monograph of the Scatopsidae (Diptera). Part III. The genus *Rhexoza* Enderlein; Part IV. The genus *Swammerdamella* Enderlein; Part V. The genus *Colobostema* Enderlein. Ent. Soc. Amer. Ann. **49:** 1–12, figs. 1–5; 15–29, figs. 1–6; 325–332, figs. 1–3.
- 1956b The nearctic Chaoborinae (Diptera: Culicidae). Minn. Agr. Expt. Sta. Tech. Bul. **218:** 1–102, 34 figs.

Cook, E. F.—Continued
1957   A contribution toward a monograph of the Scatopsidae (Diptera). Part IV. The genera *Scatopse* Geoffroy and *Hoploplagia* Enderlein. Ent. Soc. Amer. Ann. (1956) **49:** 593–611, 7 pls.
1958   A contribution toward a monograph of the Scatopsidae (Diptera). Part VII. The genus *Psectrosciara* Kieffer. Ent. Soc. Amer. Ann. **51:** 587–595, 5 figs.

Cooper, J. L., and Rapp, W. F., Jr.
1944   Check list of the Dixidae of the world. Canad. Ent. **76:** 247–252.

Coppel, H. C.   (See also McLeod, McGugan, and Coppel)
1960   Key to adults of dipterous parasites of spruce budworm, *Choristoneura fumiferana* (Clem.) (Lepidoptera: Tortricidae). Ent. Soc. Amer. Ann. **53:** 94–97, 2 pls.

Coquerel, C.
1858   Note sur les larves appartenant à une espèce nouvelle de Diptère (*Lucilia hominivorax*) developpées dans les sinus frontaux de l'homme à Cayenne. Soc. Ent. de France, Ann. ser. 3, **6:** 171–176, 1 pl.

Coquillett, D. W.   (See also Davidson, 1894, 1896; Call, 1897; Slingerland, 1897; Pergande, 1901; Johnson, 1903a; Baker, 1904)
1886a Monograph of the Lomatina of North America. Canad. Ent. **18:** 81–87.
1886b The North American genera of Anthracina. Canad. Ent. **18:** 157–159.
1886c The North American species of *Toxophora*. Ent. Amer. **1:** 221–222.
1887a Monograph of the genus *Anthrax* north of Mexico. Amer. Ent. Soc. Trans. and Acad. Nat. Sci. Phila., Ent. Sect. Proc. **14:** 159–182.
1887b Notes on the genus *Exoprosopa*. Canad. Ent. **19:** 12–14.
1887c Synopsis of the North American species of *Lordotus*. Ent. Amer. **3:** 115–116.
1889   The corn worm or boll worm in California. Insect Life (U.S. Dept. Agr., Div. Ent. Period. Bul.) **1:** 331–332.
1890   The dipterous parasite of *Diabrotica soror*. Insect Life (U.S. Dept. Agr., Div. Ent. Period. Bul.) **2:** 233–236, 1 fig.
1891a A new *Rhaphiomidas*. West Amer. Sci. **7:** 84–86.
1891b New Bombylidae from California. West Amer. Sci. **7:** 197–200.

Coquillett, D. W.—Continued

1891c   New Bombylidae of the group *Paracosmus*. West Amer. Sci. **7**: 219–222.

1891d   Revision of the bombylid genus *Aphoebantus*. Pp. 6–16. First published, Mar., 1891, as a preprint from West Amer. Sci.

         Also published in West Amer. Sci. **7**: 254–264, Oct., *1891*.

1892a   Revision of the species of *Anthrax* from America, north of Mexico. Amer. Ent. Soc. Trans. **19**: 168–187.

1892b   Revision of the bombylid genus *Epacmus* (*Leptochilus*). Canad. Ent. **24**: 9–11.

1892c   Notes and descriptions of Bombylidae. Canad. Ent. **24**: 123–126.

1892d   A new genus of Diptera allied to *Rhaphiomidas*. Canad. Ent. **24**: 314–315.

1892e   A new *Dalmannia* from California. Ent. News and Acad. Nat. Sci. Phila., Ent. Sect. Proc. **3**: 150–151.

1892f   The dipterous parasite of *Melanoplus devastator* in California. Insect Life (U.S. Dept. Agr., Div. Ent. Period. Bul.) **5**: 22–24.

1893a   Synopsis of the asilid genus *Anisopogon*. Canad. Ent. **25**: 20–22.

1893b   Synopsis of the asilid genus *Blacodes*. Canad. Ent. **25**: 33–34.

1893c   Synopsis of the asilid genus *Dioctria*. Canad. Ent. **25**: 80.

1893d   Synopsis of the asilid genera *Mallophora* and *Nicocles*. Canad. Ent. **25**: 118–120.

1893e   A new asilid genus related to *Erax*. Canad. Ent. **25**: 175–177.

1893f   Synopsis of the dipterous genus *Thereva*. Canad. Ent. **25**: 197–201.

1893g   Synopsis of the dipterous genus *Psilocephala*. Canad. Ent. **25**: 222–229.

1893h   An anomalous empid. Ent. News and Acad. Nat. Sci. Phila., Ent. Sect. Proc. **4**: 208–210, 1 fig.

1894a   Notes and descriptions of North American Bombylidae. Amer. Ent. Soc. Trans. **21**: 89–112.

1894b   New North American Trypetidae. Canad. Ent. **26**: 71–75.

1894c   Two interesting new Diptera from Washington. Ent. News and Acad. Nat. Sci. Phila., Ent. Sect. Proc. **5**: 125–126.

1894d   Synopsis of the dipterous genus *Symphoromyia*. N.Y. Ent. Soc. Jour. **2**: 53–56.

Coquillett, D. W.—Continued
- 1894e Revision of the dipterous family Therevidae. N.Y. Ent. Soc. Jour. **2**: 97–101.
- 1894f A new *Anthrax* from California. N.Y. Ent. Soc. Jour. **2**: 101–102.
- 1895a Descriptions of new genera and species. Pp. 307–319. *In* Johnson, C. W., Diptera of Florida. Acad. Nat. Sci. Phila. Proc. **1895**: 303–340.
- 1895b A synopsis of the dipterous genus *Phora*. Canad. Ent. **27**: 103–107.
- 1895c New Tachinidae with a slender proboscis. Canad. Ent. **27**: 125–128.
- 1895d New North American Mycetophilidae. Canad. Ent. **27**: 199–201.
- 1895e A new *Volucella* from Washington. Ent. News **6**: 131–132.
- 1895f On the occurrence of the tachinid genus *Heteropterina* Macq. in North America. Ent. News **6**: 207–208.
- 1895g A cecidomyiid that lives on poison oak. Insect Life (U.S. Dept. Agr., Div. Ent. Period. Bul.) **7**: 348.
- 1895h Two dipterous insects injurious to flowers. Insect Life (U.S. Dept. Agr., Div. Ent. Period. Bul.) **7**: 399–402, 1 fig.
- 1895i A new wheat pest. Insect Life (U.S. Dept. Agr., Div. Ent. Period. Bul.) **7**: 406–408, 1 fig.
- 1895j Notes and descriptions of Tachinidae. N.Y. Ent. Soc. Jour. **3**: 49–58.
- 1895k New genera and species of Tachinidae. N.Y. Ent. Soc. Jour. **3**: 97–107.
- 1895l The bombylid genus *Acreotrichus* in America. Psyche **7**: 273.
- [1895m] Revision of the North American Empididae.—A family of two-winged insects. U.S. Natl. Mus. Proc. (1896) **18**: 387–440.

    Dated from information by Aldrich, 1905: 19.
- 1896a New Culicidae from North America. Canad. Ent. **28**: 43–44.
- 1896b A new subfamily of Ephydridae. Ent. News **7**: 220–221, 1 fig.
- 1896c A new dipterous genus related to *Gnoriste*. Ent. Soc. Wash. Proc. **3**: 321–322, 1 fig.
- 1897 Revision of the Tachinidae of America north of Mexico. A family of parasitic two-winged insects. U.S. Dept. Agr., Div. Ent., Tech. Ser. **7**: 1–156.

    Some copies lack the complete index and the errata, and consist of only 154 pp.

Coquillett, D. W.—Continued

1898a On *Cuterebra emasculator*, with descriptions of several allied species. Canad. Ent. **30:** 9–11.
1898b On the dipterous genus *Eusiphona*. Canad. Ent. **30:** 53.
1898c Additions to my synopsis of the Tachinidae. Canad. Ent. **30:** 233–237.
1898d New species of Sapromyzidae. Canad. Ent. **30:** 277–280.
1898e Synopsis of the asilid genus *Ospriocerus*. Ent. News **9:** 37.
1898f Notes and descriptions of Oscinidae. N.Y. Ent. Soc. Jour. **6:** 44–49.
1898g On the dipterous family Scatophagidae. N.Y. Ent. Soc. Jour. **6:** 160–165.
1898h A new dipterous genus belonging to the Therevidae. N.Y. Ent. Soc. Jour. **6:** 187–188.
1898i The buffalo-gnats, or black-flies, of the United States. (A synopsis of the dipterous family Simuliidae). U.S. Dept. Agr., Div. Ent., Bul. **n. ser.,** 10: 66–69, 2 figs.
1898j Report on a collection of Japanese Diptera, presented to the U.S. National Museum by the Imperial University of Tokyo. U.S. Natl. Mus. Proc. **21:** 301–340.
1899a Description of a new *Psilopa*. Canad. Ent. **31:** 8.
1899b New genera and species of Nycteribidae and Hippoboscidae. Canad. Ent. **31:** 333–336.
1899c A new dipterous family related to the Chironomidae. Ent. News **10:** 60–61, 1 fig.
1899d New genera and species of Dexidae. N.Y. Ent. Soc. Jour. **7:** 218–222.
1899e Notes and descriptions of Trypetidae. N.Y. Ent. Soc. Jour. **7:** 259–268.
[1899f] Order Diptera. Pp. 341–346. *In* Jordan, D.S., The fur seals and fur-seal islands of the North Pacific Ocean [q.v.]. Vol. 4, 384 pp., 86 pls., pl. A. Washington, D.C., "1898."

> Dated by means of Coquillett's own dating of his included new genus, *Eutanypus*, in 1910b: 543, and from many other authors.

[1899g] A cecidomyiid injurious to seeds of sorghum. U.S. Dept. Agr., Div. Ent., Bul. (1898) **n. ser.,** 18: 81–82.

> The copy of the tear-sheet of this article in the files at the at the U.S. National Museum bears a notation in Coquillett's handwriting that Bul. 18 was issued Jan. 7, 1899.

1900a New genera and species of Ephydridae. Canad. Ent. **32:** 33–36.

Coquillett, D. W.—Continued
- 1900b Two new genera of Diptera. Ent. News **11:** 429–430, 1 fig.
- 1900c New Scenopinidae from the United States. Ent. News **11:** 500–501.
- 1900d Notes and descriptions of Ortalidae. N.Y. Ent. Soc. Jour. **8:** 21–25.
- 1900e Two new cecidomyians destructive to buds of roses. U.S. Dept. Agr., Div. Ent., Bul. **n. ser., 22:** 44–48, 1 fig.
- 1900f A new violet pest (*Diplosis violicola* n. sp.). U.S. Dept. Agr., Div. Ent., Bul. **n. ser., 22:** 48–51, 1 fig.
- 1900g Report on a collection of dipterous insects from Puerto Rico. U.S. Natl. Mus. Proc. **22:** 249–270.
- 1900h Papers from the Harriman Alaska Expedition. IX. Entomological results (3): Diptera. Wash. Acad. Sci. Proc. **2:** 389–464.

  Also published separately, as Harriman Alaska Expedition, Vol. 9, Pt. 2, 78 pp., in New York, *1904*.

- 1901a Three new species of Culicidae. Canad. Ent. **33:** 258–260.
- 1901b A new genus of Ortalidae. Ent. News **12:** 15.
- 1901c Three new species of Diptera. Ent. News **12:** 16–18.
- 1901d A new anthomyiid injurious to lupines. Ent. News **12:** 206–207.
- 1901e Types of anthomyid genera. N.Y. Ent. Soc. Jour. **9:** 134–146.
- 1901f Original descriptions of new Diptera. Pp. 585–586, pl. 15, figs. 5–8. *In* Needham, J. G., and Betten, C., Aquatic insects in the Adirondacks. N.Y. State Mus. Bul. **47:** 383–612, 42 figs., 36 pls. (=[Vol. 9, part: Ent. 13].)

  Also published in N.Y. State Mus. Ann. Rpt. (1900) 54 (vol. 4): 585-586, *1902*.

- 1901g Some insects of the Hudsonian Zone in New Mexico. Diptera. Psyche **9:** 149–150.
- 1901h New Diptera in the U.S. National Museum. U.S. Natl. Mus. Proc. **23:** 593–618.
- 1901i Papers from the Hopkins Stanford Galapagos Expedition, 1898–1899. II. Entomological results (2): Diptera. Wash. Acad. Sci. Proc. **3:** 371–379.
- 1902a New cyclorhaphous Diptera from Mexico and New Mexico. Canad. Ent. **34:** 195–202.
- 1902b Three new species of *Culex*. Canad. Ent. **34:** 292–293.
- 1902c Three new species of nemoatcerous [sic] Diptera. Ent. News **13:** 84–85.

Coquillett, D. W.—Continued
- 1902d New orthorraphous Diptera from Mexico and Texas. N.Y. Ent. Soc. Jour. **10**: 136–141.
- 1902e New acalyptrate Diptera from North America. N.Y. Ent. Soc. Jour. **10**: 177–191.
- 1902f New forms of Culicidae from North America. N.Y. Ent. Soc. Jour. **10**: 191–194.
- 1902g New Diptera from North America. U.S. Natl. Mus. Proc. (1903) **25**: 83–126.
- 1903a The genera of the dipterous family Empididae, with notes and new species. Ent. Soc. Wash. Proc. **5**: 245–272.
- 1903b The occurrence of the phorid genus *Aenigmatias* in America. Canad. Ent. **35**: 20–22.
- 1903c A new culicid genus related to *Corethra*. Canad. Ent. **35**: 189–190.
- 1903d Four new species of *Culex*. Canad. Ent. **35**: 255–257.
- 1903e A new *Anopheles* with unspotted wings. Canad. Ent. **35**: 310.
- 1904a Several new Diptera from North America. Canad. Ent. **36**: 10–12.
- 1904b Notes on *Culex nigritulus*. Ent. News **15**: 73–74.
- 1904c Diptera from southern Texas with descriptions of new species. N.Y. Ent. Soc. Jour. **12**: 31–35.
- 1904d The genera of the dipterous family Empididae. (Addenda.) Ent. Soc. Wash. Proc. **6**: 51–52.
- 1904e New Diptera from Central America. Ent. Soc. Wash. Proc. **6**: 90–98.
- 1904f New North American Diptera. Ent. Soc. Wash. Proc. **6**: 166–192.
- 1905a A new cecidomyiid on cotton. Canad. Ent. **37**: 200.
- 1905b A new subapterous tipulid from New Mexico. Canad. Ent. **37**: 347.
- 1905c A new dexiid parasite of a Cuban beetle. Canad. Ent. **37**: 362.
- 1905d New nematocerous Diptera from North America. N.Y. Ent. Soc. Jour. **13**: 56–69.
- 1906a A new *Tabanus* related to *punctifer*. Ent. News **17**: 48.
- 1906b A new *Culex* near *curriei*. Ent. News **17**: 109.
- 1906c On the breaking-up of the old genus *Culex*. Science n. ser., **23**: 312–314.
- 1907a New genera and species of Diptera. Canad. Ent. **39**: 75–76.
- 1907b A new phorid genus with horny ovipositor. Canad. Ent. **39**: 207–208, 2 figs.

Coquillett, D. W.—Continued
- 1907c Discovery of blood-sucking Psychodidae in America. Ent. News **18**: 101–102.
- 1907d Notes and descriptions of Hippoboscidae and Streblidae. Ent. News **18**: 290–292, 1 fig.
- 1908 New genera and species of Diptera. Ent. Soc. Wash. Proc. (1907) **9**: 144–148.
- 1909 A new stratiomyid from Texas. Canad. Ent. **41**: 212.
- 1910a New genera and species of North American Diptera. Ent. Soc. Wash. Proc. **12**: 124–131.
- 1910b The type-species of the North American genera of Diptera. U.S. Natl. Mus. Proc. **37**: 499–647.
- 1910c New species of North American Diptera. Canad. Ent. **42**: 41–47.
- 1910d Corrections to my paper on the type-species of the North American genera of Diptera. Canad. Ent. **42**: 375–378.

Corti, E.
- 1909 Contributo alla conoscenza del gruppo delle "Crassisete" in Italia (Ditteri). Soc. Ent. Ital. Bul. (1908) **40**: 121–162.

Costa, A.
- 1856 De quibusdam novis insectorum generibus, ossia caratteri di nuova generi e specie in quella descritte, spettanti tutti alla fauna Napoletano. R. Accad. delle Sci. (Soc. R. Borbonica) Napoli, Rend. n. ser., **5**: 17–20, 1 pl.
    > A summary of a paper published later in R. Accad. delle Sci. Napoli, Mem. **2**: 219–233, 1 pl., *1856*.
- 1857 Contribuzione alla fauna ditterologica Italiana. Giambattista Vico, Gior. Sci. (Naples) **2**: 438–460.
- 1863 Nuovi studii sulla entomologia della Calabria Ulteriore. Accad. delle Sci. Fis. e Mat. Napoli, Atti **1** (**2**): 1–80, 4 pls.

Costa Lima, A. da (See Lima, A. da Costa)

Couden, E. D.
- 1908 A gall-maker of the family Agromyzidae (*Agromyza tiliae* n. sp.). Ent. Soc. Wash. Proc. **9**: 34–36, 1 fig.

Crampton, G. C.
- 1942 Guide to the insects of Connecticut. Part VI. The Diptera or true flies of Connecticut. First fascicle [part]. The external morphology of the Diptera. Conn. State Geol. and Nat. Hist. Survey, Bul. **64**: 1–509, 55 figs., 4 pls.

Crane, A. E.
- 1961 A study of the habits of *Rhamphomyia scutellaris* Coquillett (Diptera: Empididae). Wasmann Jour. Biol. **19**: 247–263.

Crawford, J. C. (See Hyslop, 1910)

Creager, D. B., and Spruijt, F. J.
1935   The relation of certain fungi to larval development of *Eumerus tuberculatus* Rond. (Syrphidae, Diptera). Ent. Soc. Amer. Ann. **28**: 425–436, 5 figs., pl. 1.

Cresson, E. T., Jr.
1906   Some North American Diptera from the South West. Paper I. Amer. Ent. Soc. Trans. **32**: 279–288, 1 pl.

1907a Some North American Diptera from the South West. Paper II. Amer. Ent. Soc. Trans. **33**: 99–108, 1 pl.

1907b The North American species of the dipterous family Scenopinidae. Amer. Ent. Soc. Trans. **33**: 109–114, 1 pl.

1908   Dipterological notes. I. Micropezidae. Amer. Ent. Soc. Trans. **34**: 1–12, 2 pls.

1910– Studies in North American dipterology: Pipunculidae.
1911   Amer. Ent. Soc. Trans. **36**: 267–290, *1910*; 291–329, pls. 5–9, *1911*.

1912   Studies of some Pipunculidae from the eastern United States (Diptera). Ent. News **23**: 452–456.

1913   Descriptions of two new species of the dipterous genera *Chaetopsis* and *Stenomyia*, with notes on other species. Ent. News **24**: 317–321, 7 figs.

1914a Descriptions of new genera and species of the dipterous family Ephydridae. –I. Ent. News **25**: 241–250, 1 pl.

1914b Some nomenclatorial notes on the dipterous family Trypetidae. Ent. News **25**: 275–279.

1914c Descriptions of new North American acalyptrate Diptera. –I. Ent. News **25**: 457–460.

1915a Descriptions of new genera and species of the dipterous family Ephydridae. –II. Ent. News **26**: 68–72.

1915b A new genus and some new species belonging to the dipterous family Bombyliidae. Ent. News **26**: 200–207, 3 figs.

1915c Some North American Diptera from the Southwest. Paper III. A revision of the species of the genus *Mythicomyia*. Ent. News **26**: 448–456, 1 fig.

1916a Studies in the American Ephydridae (Diptera). [I]. Amer. Ent. Soc. Trans. **42**: 101–124, 1 pl.

1916b Descriptions of new genera and species of the dipterous family Ephydridae. –III. Ent. News **27**: 147–152.

1916c Dipterological notes. II. A study of the *lateralis*-group of the bombyliid genus *Villa* (*Anthrax* of authors, in part). Ent. News **27**: 439–444.

Cresson, E. T., Jr.—Continued
- 1917a Studies in the American Ephydridae (Diptera). II. A revision of the species of the genera *Notiphila* and *Dichaeta*. Amer. Ent. Soc. Trans. **43**: 27–66, 2 pls. .
- 1917b Descriptions of new genera and species of the dipterous family Ephydridae. IV. Ent. News **28**: 340–341.
- 1918a Costa Rican Diptera collected by Philip P. Calvert, Ph. D., 1909–1910. Paper III. A report on the Ephydridae. Amer. Ent. Soc. Trans. **44**: 39–68, 1 pl.
- 1918b New North American Diptera (Scatophagidae). Ent. News **29**: 133–137.
- 1919 Dipterological notes and descriptions. Acad. Nat. Sci. Phila. Proc. **71**:. 171–194.
- 1920a A revision of the nearctic Sciomyzidae (Diptera, Acalyptratae). Amer. Ent. Soc. Trans. **46**: 27–89, 3 pls.
- 1920b Descriptions of new North American acalyptrate Diptera. –II. (Trypetidae, Sapromyzidae). Ent. News **31**: 65–67.
- 1920c Description of a new species of the asilid genus *Pogosoma* (Diptera). Ent. News **31**: 211–215.
- 1922a Studies in American Ephydridae (Diptera). III. A revision of the species of *Gymnopa* and allied genera constituting the subfamily Gymnopinae. Amer. Ent. Soc. Trans. (1921) **47**: 325–343, 1 pl.
- 1922b Descriptions of new genera and species of the dipterous family Ephydridae. –V. Ent. News **33**: 135–137.
- 1923 Records of some western Diptera, with descriptions of two new species of the family Bombyliidae. Acad. Nat. Sci. Phila. Proc. **75**: 365–367.
- 1924a Studies in the dipterous family Ortalidae, with descriptions of new species, mostly from North America. Amer. Ent. Soc. Trans. **50**: 225–241.
- 1924b Descriptions of new genera and species of the dipterous family Ephydridae. Paper VI. Ent. News **35**: 159–164.
- 1925 Descriptions of new genera and species of the dipterous family Ephydridae. –VII. Ent. News **36**: 165–167.
- 1926 Descriptions of new genera and species of Diptera (Ephydridae and Micropezidae). Amer. Ent. Soc. Trans. **52**: 249–274.
- 1929 A revision of the North American species of fruit flies of the genus *Rhagoletis* (Diptera: Trypetidae).—Amer. Ent. Soc. Trans. **55**: 401–414, 1 pl.
- 1930a Studies in the dipterous family Ephydridae.—Paper III. Amer. Ent. Soc. Trans. **56**: 93–131.

Cresson, E. T., Jr.—Continued
- 1930b Descriptions of new genera and species of the dipterous family Ephydridae. Paper VIII. Ent. News **41**: 76–81.
- 1930c Notes on and descriptions of some neotropical Neriidae and Micropezidae (Diptera). Amer. Ent. Soc. Trans. **56**: 307–362.
- 1931a Notes on the *abstersa*-group of the genus *Tephritis*, and a description of a new species from California (Diptera: Trypetidae). Ent. News **42**: 3–5.
- 1931b Descriptions of new genera and species of the dipterous family Ephydridae. Paper IX; Paper X. Ent. News **42**: 104–108; 168–170.
- 1933a Descriptions of new species of the dipterous family Ephydridae. Ent. News **44**: 65–70.
- 1933b A new genus and species of the dipterous family Ephydridae reared from duck weed. Ent. News **44**: 229–231.
- 1934ı Descriptions of new genera and species of the dipterous family Ephydridae. –XI. Amer. Ent. Soc. Trans. **60**: 199–222, 1 pl.
- 1934b Two new species of the genus *Hydrellia* from Mount Desert, Maine (Diptera: Ephydridae). Ent. News **45**: 234–236.
- 1935a Descriptions of genera and species of the dipterous family Ephydridae. Amer. Ent. Soc. Trans. **61**: 345–372.
- 1935b A new species of *Micropeza* from Colorado (Diptera: Micropezidae). Ent. News **46**: 229–230.
- 1936 Descriptions and notes on genera and species of the dipterous family Ephydridae. II. Amer. Ent. Soc. Trans. **62**: 257–270.
- 1938a The Neriidae and Micropezidae of America north of Mexico (Diptera). Amer. Ent. Soc. Trans. **64**: 293–366, 1 fig., 3 pls.
- 1938b Descriptions of some North American Micropezidae (Diptera). Ent. News **49**: 72–76.
- 1938c Notes on, and descriptions of, some neotropical Ephydridae (Dipt.). Rev. de Ent. **8**: 24–40.
- 1939 Description of a new genus and ten new species of Ephydridae, with a discussion of the species of the genus *Discomyza* (Diptera). Acad. Nat. Sci. Phila. Notulae Nat. **21**: 1–12.
- 1940 Descriptions of new genera and species of the dipterous family Ephydridae. Paper XII. Acad. Nat. Sci. Phila. Notulae Nat. **38**: 1–10.
- 1941a New genera and species of North American Ephydridae (Diptera). Ent. News **52**: 35–38.

Cresson, E. T., Jr.—Continued
- 1941b The species of the neotropical genus *Nostima* (Diptera: Ephydridae). Acad. Nat. Sci. Phila. Notulae Nat. **78**: 1–9, 5 figs.
- 1942a Synopses of North American Ephydridae (Diptera). [I]. Amer. Ent. Soc. Trans. **68**: 101–128.
- 1942b Descriptions of two new nearctic species of the genus *Hydrellia* reared from pond-weed (Diptera: Ephydridae). Ent. News **53**: 78–79.
- 1943 The species of the tribe Ilytheini (Diptera: Ephydridae: Notiphilinae). Amer. Ent. Soc. Trans. **69**: 1–16, 2 pls.
- 1944a Descriptions of new genera and species of the dipterous family Ephydridae. Paper XIV. Acad. Nat. Sci. Phila. Notulae Nat. **135**: 1–9.
- 1944b Synopses of North American Ephydridae (Diptera). Parts IA and II. Amer. Ent. Soc. Trans. **70**: 159–180.
- 1946 Synopses of North American Ephydridae (Diptera). III. The tribe Notiphilini of the subfamily Notiphilinae. Amer. Ent. Soc. Trans. **72**: 227–240.
- 1949 A systematic annotated arrangement of the genera and species of the North American Ephydridae. IV. The subfamily Napaeinae. Amer. Ent. Soc. Trans. (1948) **74**: 225–260.

Criddle, N. (See also Curran, 1925j)
- 1921 The entomological record, 1920. Ent. Soc. Ontario, Ann. Rpt. (1920) **51**: 72–90.

Crosby, C. R. (See Johannsen and Crosby)

Crow, J. F. (See Patterson and Crow)

Curran, C. H. (See also Van Duzee and Curran)
- 1921a Revision of the *Pipiza* group of the family Syrphidae (flower-flies) from north of Mexico. Calif. Acad. Sci. Proc. ser. 4, **11**: 345–393, 108 figs.
- 1921b A revision of *Syrphus* species belonging to the *ribesii* group (Dipt.). Canad. Ent. **53**: 152–160.
- 1921c New species of Syrphidae (Diptera). Canad. Ent. **53**: 171–176.
- 1922a A new western syrphid (Diptera). Canad. Ent. (1921) **53**: 258–260.
- 1922b A genus and species of Syrphidae new to Canada. Canad. Ent. (1921) **53**: 260.
- 1922c New species of Canadian Syrphidae (Diptera). Pt. I. Canad. Ent. (1921) **53**: 275–276.

Curran, C. H.—Continued
- 1922d New species of Canadian Syrphidae (Diptera). Pt. II. Canad. Ent. **54**: 14–19.
- 1922e New species of the syrphid genus *Chilosia* from Canada (Diptera). Canad. Ent. **54**: 19–20, 67–72.
- 1922f New and little-known Canadian Syrphidae (Diptera). Canad. Ent. **54**: 94–96, 117–119.
- 1922g Correction. Canad. Ent. **54**: 96.
- 1922h Notes and corrections (Syrphidae, Diptera). Canad. Ent. **54**: 191.
- 1922i New Diptera in the Canadian National Collection. Canad. Ent. **54**: 277–287.
- 1922j The syrphid genera *Hammerschmidtia* and *Brachyopa* in Canada. Ent. Soc. Amer. Ann. **15**: 239–255.
- 1923a Two new North American Diptera. Boston Soc. Nat. Hist. Occas. Papers **5**: 59–61.
- 1923b Two undescribed syrphid flies from New England. Boston Soc. Nat. Hist. Occas. Papers **5**: 65–67.
- 1923c Our North American *Leucozona*, a variety of *lucorum* (Syrphidae, Diptera). Canad. Ent. **55**: 38.
- 1923d The *Stenosyrphus sodalis* group (Syrphidae, Diptera). Canad. Ent. **55**: 59–64.
- 1923e An apparently undescribed species of *Scellus* (Dolichopodidae, Diptera). Canad. Ent. **55**: 73–74.
- 1923f Changes of names. Canad. Ent. **55**: 74.
- 1923g Studies in Canadian Diptera. I. Revision of the asilid genus *Cyrtopogon* and allied genera; II. The genera of the family Blepharoceridae. Canad. Ent. **55**: 92–95, 116–125, 132–142, 169–174, 185–190, pl. 6; 266–269.
- 1923h Undescribed Canadian Dolichopodidae, with key to *Chrysotimus* (Diptera). Canad. Ent. **55**: 190–192.
- 1923i Change of name. Canad. Ent. **55**: 198.
- 1923j Apparently undescribed Canadian Asilidae and Dolichopodidae (Diptera). Canad. Ent. **55**: 207–211.
- 1923k A new dolichopid from Ontario (Diptera). Canad. Ent. **55**: 236–237.
- 1923l New North American Diptera. Canad. Ent. **55**: 245–246.
- 1923m A new syrphid from Ontario (Diptera). Canad. Ent. **55**: 269.
- 1923n New cyclorrhaphous Diptera from Canada. Canad. Ent. **55**: 271–279.
- 1923o Two examples of sexual dimorphism in the genus *Sericomyia* (Diptera, Syrphidae). Insecutor Inscitiae Menstruus **11**: 136–141.

Curran, C. H.—Continued

1923p Two varieties of *Eurosta solidaginis* Fitch (Trypetidae, Dipt.). Ent. News **34:** 302.

1924a Notes on the genus *Pipizella* Rondani, with descriptions of new species (Diptera, Syrphidae). Amer. Ent. Soc. Trans. **49:** 339–345.

1924b The generic position of *Beris viridis* Say (Stratiomyidae, Diptera). Canad. Ent. **56:** 24.

1924c Synopsis of the genus *Chrysotoxum* with notes and descriptions of new species (Syrphidae, Diptera). Canad. Ent. **56:** 34–40.

1924d *Rhagoletis symphoricarpi*, a new trypaneid from British Columbia (Dipt.). Canad. Ent. **56:** 62–63.

1924e A new species of *Nothosympycnus* (Dolichopodidae) with synopsis of Canadian species. Canad. Ent. **56:** 108–110.

1924f Seven new species of *Rhaphium* (Dolichopodidae, Diptera). Canad. Ent. **56:** 133–141.

1924g New Canadian Diptera, with a synopsis of the genus *Cynorhina*. Canad. Ent. **56:** 193–196.

1924h Some apparently new Canadian Psychodidae (Dipt.). Canad. Ent. **56:** 215–220.

1924i Four apparently undescribed Diptera from Canada. Canad. Ent. **56:** 250–253.

1924j Two undescribed species of *Cyrtopogon*, with notes (Diptera). Canad. Ent. **56:** 277–280.

1924k A new Canadian syrphid (Diptera). Canad. Ent. **56:** 288.

1924l On the genus *Arctophyto* Townsend in North America (Tachinidae, Diptera). Canad. Ent. **56:** 302–303.

1924m A new *Dolichopus* from British Columbia (Dolichopodidae, Dipt.). Canad. Ent. **56:** 304–305.

1924n New species of *Ernestia* and *Mericia* (Dipt.: Tachinidae). Ent. News **35:** 245–250.

1924o Brief diagnoses of some Diptera occurring in New England. Psyche **31:** 226–228.

1924p New species of Syrphidae. Boston Soc. Nat. Hist. Occas. Papers **5:** 79–82, 2 figs.

1925a Three new Diptera from Labrador. Canad. Ent. **57:** 24–26.

1925b New species of *Xylota* (Syrphidae, Dipt.). Canad. Ent. **57:** 44–45.

1925c New Tachinidae in the Canadian National Collection (Diptera). Canad. Ent. **57:** 150–156.

Curran, C. H.—Continued
1925d *Buckellia*, a new genus of Asilidae (Diptera). Canad. Ent. **57:** 156.
1925e Four new nearctic Diptera. Canad. Ent. **57:** 254–257.
1925f Some apparently new nearctic Tachinidae (Diptera). Canad. Ent. **57:** 281–286.
1925g Three new nearctic Tachinidae (Dipt.). Ent. News **36:** 13–18.
1925h The American species of the tachinid genus *Peleteria* Desv. (Diptera). Roy. Soc. Canada, Proc. and Trans. **ser. 3, 19** [=whole ser., 31](5): 225–257, 2 pls.
1925i Revision of the genus *Neoascia* Williston (Diptera: Syrphidae). Ent. Soc. Wash. Proc. **27:** 51–62.
1925j Diptera. Pp. 99–101. *In* Criddle, N., The entomological record, 1924. Ent. Soc. Ontario, Ann. Rpt. (1924) **55:** 89–106.
1925k Contribution to a monograph of the American Syrphidae from north of Mexico. Kans. Univ. Sci. Bul. (1924) **15** [=whole ser., 25]: 7–216, 12 pls. (=Kans. Univ. Bul. 26 (7).)
1926a New Diptera from the West Indies. Amer. Mus. Nat. Hist., Amer. Mus. Novitates **220:** 1–14.
1926b New nearctic Diptera mostly from Canada. Canad. Ent. **58:** 81–89.
1926c *Grisdalemyia*, a new genus of Tachinidae (Diptera). Canad. Ent. **58:** 133–135, 2 figs.
Published in June.
1926d Descriptions of new Canadian Diptera. Canad. Ent. **58:** 170–175, 211–218.
1926e Two new Canadian Psychodidae (Dipt.). Canad. Ent. **58:** 228–229.
1926f The species of *Hilara* occurring in Banff and vicinity (Empididae, Diptera). Canad. Ent. **58:** 245–249.
1926g A new species of *Comantella* (Asilidae, Diptera). Canad. Ent. **58:** 310–312.
1926h Partial synopsis of American species of *Volucella* with notes on Wiedemann's types. Ent. Soc. Amer. Ann. **19:** 50–66.
1926i Appendix. New Diptera from Jamaica. Pp. 102–114. *In* Gowdey, C. C., Catalogus insectorum jamaicensis. Jamaica Dept. Agr. Ent. Bul. **4 (1/2):** 1–114.
1926j The nearctic species of the genus *Rhaphium* Meigen (Dolichopodidae, Dipt.). Roy. Canad. Inst. Trans. **15:** 249–260 [cont.].

Curran, C. H.—Continued

1926k The Canadian species of the tachinid genera *Cryptomeigenia* B. B. and *Tachinomyia* Tns. (Dipt.). Roy. Soc. Canada, Proc. and Trans. **ser. 3, 20** [=whole ser., 32] **(5):** 155–171, 2 pls.

1927a New neotropical and oriental Diptera in the American Museum of Natural History. Amer. Mus. Nat. Hist., Amer. Mus. Novitates **245:** 1–9, 1 fig.

1927b Synopsis of males of the genus *Platycheirus* St. Fargeau and Serville with descriptions of new Syrphinae (Diptera). Amer. Mus. Nat. Hist., Amer. Mus. Novitates **247:** 1–13.

1927c Four new American Diptera. Amer. Mus. Nat. Hist., Amer. Mus. Novitates **275:** 1–4, 3 figs.

1927d Studies in Canadian Diptera. III. The species of the tachinid genera related to *Lydella*, as represented in the Canadian National Collection. Canad. Ent. **59:** 11–24.

1927e A new species of *Mallochiella*. Canad. Ent. **59:** 49–50.

1927f Descriptions of nearctic Diptera. Canad. Ent. **59:** 79–92, 5 figs.

1927g Notes on Syrphidae (Diptera). Canad. Ent. **59:** 205–207.

1927h Some new Canadian Scatophagidae (Diptera). Canad. Ent. **59:** 253–261.

1927i Some new North American Diptera. Canad. Ent. **59:** 290–303, 1 pl.

1927j Synopsis of the syrphid genus *Copestylum* Macq. (Diptera). Ent. News **38:** 43–46.

1927k The nearctic species of the genus *Rhaphium* Meigen (Dolichopodidae, Dipt.) [concl.]. Roy. Canad. Inst. Trans. **16:** 99–179, 4 pls.

1927l Synopsis of the Canadian Stratiomyidae. Roy. Soc. Canada, Proc. and Trans. **ser. 3, 21** [=whole ser., 33]**(5):** 191–228.

1927m Some new American Tachinidae (Diptera). Brooklyn Ent. Soc. Bul. **22:** 144–154.

1928a New species of *Ommatius* from America, with Key. Amer. Mus. Nat. Hist., Amer. Mus. Novitates **327:** 1–6.

1928b Two new species of *Wagneria* (Tachinidae, Dipt.). Canad. Ent. **60:** 48–49.

1928c Revision of the American species of *Archytas* (Tachinidae, Diptera). Canad. Ent. **60:** 201–208, 218–226, 249–256, 275–282, 3 pls.

1928d Insects of Porto Rico and the Virgin Islands. Diptera or two-winged flies. Pt. 1, pp. 1–118, 39 figs. *In* New York Academy of Sciences, Scientific Survey of Puerto Rico and The Virgin Islands [q.v.]. Vol. 11. New York.

Curran, C. H.—Continued
- 1928e New eastern species of *Medeterus* (Dolichopodidae, Diptera). N.Y. State Mus. Bul. **274**: 199–204.
- 1929a New Diptera in the American Museum of Natural History. Amer. Mus. Nat. Hist., Amer. Mus. Novitates **339**: 1–13, 2 figs.
- 1929b The genus *Myxosargus* Brauer (Stratiomyidae, Diptera). Amer. Mus. Nat. Hist., Amer. Mus. Novitates **378**: 1–4.
- 1929c Some new nearctic Diptera. Canad. Ent. **61**: 30–34, 1 fig.
- 1929d A new syrphid from Canada. Canad. Ent. **61**: 45–46.
- 1929e New species of Scatophagidae (Diptera). Canad. Ent. **61**: 130–134.
- 1929f New Syrphidae and Tachinidae. Ent. Soc. Amer. Ann. **22**: 489–510.
- 1930a New species of Diptera belonging to the genus *Baccha* Fabricius (Syrphidae). Amer. Mus. Nat. Hist., Amer. Mus. Novitates **403**: 1–16.
- 1930b New species of *Lepidanthrax* and *Parabombylius* (Bombyliidae, Diptera). Amer. Mus. Nat. Hist., Amer. Mus. Novitates **404**: 1–7.
- 1930c New Diptera belonging to the genus *Mesogramma* Loew (Syrphidae). Amer. Mus. Nat. Hist., Amer. Mus. Novitates **405**: 1–14, 3 figs.
- 1930d New species of Eristalinae with notes (Syrphidae, Diptera). Amer. Mus. Nat. Hist., Amer. Mus. Novitates **411**: 1–27.
- 1930e New species of Volucellinae from America (Syrphidae, Diptera). Amer. Mus. Nat. Hist., Amer. Mus. Novitates **413**: 1–23, 1 fig.
- 1930f New Diptera from North and Central America. Amer. Mus. Nat. Hist., Amer. Mus. Novitates **415**: 1–16, 1 fig.
- 1930g New Syrphidae from Central America and the West Indies. Amer. Mus. Nat. Hist., Amer. Mus. Novitates **416**: 1–11.
- 1930h New American Asilidae (Diptera). [I.] Amer. Mus. Nat. Hist., Amer. Mus. Novitates **425**: 1–21, 3 figs.
- 1930i Three new Diptera from Canada. N.Y. Ent. Soc. Jour. **38**: 73–76.
- (1930j) Report on the Diptera collected at the Station for the Study of Insects, Harriman Interstate Park, N.Y. Amer. Mus. Nat. Hist. Bul. (1931) **61**: 21–115.
- 1930k A new tachinid parasitic on a sawfly. Canad. Ent. **62**: 246–247.
- 1930l A new *Gymnopternus* from Oregon (Dolichopidae, Diptera). Canad. Ent. **62**: 287.

Curran, C. H.—Continued
1931a First supplement to the 'Diptera of Porto Rico and the Virgin Islands'. Amer. Mus. Nat. Hist., Amer. Mus. Novitates **456**: 1–23, 4 figs.
1931b New American Asilidae (Diptera). II. Amer. Mus. Nat. Hist., Amer. Mus. Novitates **487**: 1–25.
1931c Twelve new Diptera. Amer. Mus. Nat. Hist., Amer. Mus. Novitates **492**: 1–13, 7 figs.
1931d The nearctic species of the nemestrinid genus *Rhynchocephalus* Fischer (Diptera). Canad. Ent. **63**: 68–72.
1931e Four new Diptera in the Canadian National Collection. Canad. Ent. **63**: 93–98.
1931f Some new North American Diptera. Canad. Ent. **63**: 249–254.
1932a The genus *Dictya* Meigen (Tetanoceridae, Diptera). Amer. Mus. Nat. Hist., Amer. Mus. Novitates **517**: 1–7, 13 figs.
1932b New American Syrphidae (Diptera), with notes. Amer. Mus. Nat. Hist., Amer. Mus. Novitates **519**: 1–9.
1932c New North American Diptera, with notes on others. Amer. Mus. Nat. Hist., Amer. Mus. Novitates **526**: 1–13, 9 figs.
1932d New American Diptera. Amer. Mus. Nat. Hist., Amer. Mus. Novitates **534**: 1–15.
1932e New species of Trypaneidae, with a key to the North American genera. Amer. Mus. Nat. Hist., Amer. Mus. Novitates **556**: 1–19, 9 figs.
1933a The North American species of *Actia* in the American Museum of Natural History. Amer. Mus. Nat. Hist., Amer. Mus. Novitates **614**: 1–7.
1933b New North American Diptera. Amer. Mus. Nat. Hist., Amer. Mus. Novitates **673**: 1–11, 1 fig.
1933c The North American species of *Anorostoma* Loew (Helomyzidae: Diptera). Amer. Mus. Nat. Hist., Amer. Mus. Novitates **676**: 1–7.
1933d Some North American Diptera. Amer. Mus. Nat. Hist., Amer. Mus. Novitates **682**: 1–11.
1934a The families and genera of North American Diptera 512 pp., 235 figs., 2 pls. New York.
1934b The North American Lonchopteridae (Diptera). Amer. Mus. Nat. Hist., Amer. Mus. Novitates **696**: 1–7, 3 figs.
1934c Notes on the Syrphidae in the Slosson Collection of Diptera. Amer. Mus. Nat. Hist., Amer. Mus. Novitates **724**: 1–7.

Curran, C. H.—Continued
- 1934d New American Asilidae (Diptera). III. Amer. Mus. Nat. Hist., Amer. Mus. Novitates **752**: 1–18, 1 fig.
- 1934e The Diptera of Kartabo, Bartica District, British Guiana, with descriptions of new species from other British Guiana localities. Amer. Mus. Nat. Hist. Bul. **66**: 287–532, 54 figs.
- 1935a New American Asilidae (Diptera). IV. Amer. Mus. Nat. Hist., Amer. Mus. Novitates **806**: 1–12.
- 1935b New American Diptera. Amer. Mus. Nat. Hist., Amer. Mus. Novitates **812**: 1–24, 4 figs.
- 1938 New American Diptera. Amer. Mus. Nat. Hist., Amer. Mus. Novitates **975**: 1–7.
- 1939a The dipterous genus *Chrysotachina* Brauer and Bergenstamm (Tachinidae). Amer. Mus. Nat. Hist., Amer. Mus. Novitates **1021**: 1–3.
- 1939b Synopsis of the American species of *Volucella* (Syrphidae: Diptera). Part I. Table of species. Amer. Mus. Nat. Hist., Amer. Mus. Novitates **1027**: 1–7.
- 1939c Two new American Diptera with notes on *Asemosyrphus* Bigot. Amer. Mus. Nat. Hist., Amer. Mus. Novitates **1031**: 1–3.
- 1939d The species of *Temnostoma* related to *bombylans* Linné (Syrphidae, Diptera). Amer. Mus. Nat. Hist., Amer. Mus. Novitates **1040**: 1–3.
- 1940 Some new neotropical Syrphidae (Diptera). Amer. Mus. Nat. Hist., Amer. Mus. Novitates **1086**: 1–14.
- 1941 New American Syrphidae. Amer. Mus. Nat. Hist. Bul. **78**: 243–304.
- 1942a American Diptera. Amer. Mus. Nat. Hist. Bul. **80**: 51–84.
- 1942b Guide to the insects of Connecticut. Part VI. The Diptera or true flies of Connecticut. First fascicle [part]. Key to families. Conn. State Geol. and Nat. Hist. Survey, Bul. **64**: 175–182.
- 1947a New and little known American Tachinidae. Amer. Mus. Nat. Hist. Bul. **89**: 45–122, 134 figs.
- 1947b New species of *Volucella* from Hawaii and United States. Amer. Mus. Nat. Hist., Amer. Mus. Novitates **1361**: 1–6.
- 1951 Synopsis of the North American species of *Spilomyia* (Syrphidae, Diptera). Amer. Mus. Nat. Hist., Amer. Mus. Novitates **1492**: 1–11.

Curran, C. H.—Continued

1953a The Asilidae and Mydaidae of the Bimini Islands, Bahamas, British West Indies (Diptera). Amer. Mus. Nat. Hist., Amer. Mus. Novitates **1644:** 1–6.

1953b Notes and descriptions of some Mydaidae and Syrphidae. Amer. Mus. Nat. Hist., Amer. Mus. Novitates **1645:** 1–15.

1960 Review of the tachinid genus *Juriniopsis* Townsend (Diptera). Amer. Mus. Nat. Hist., Amer. Mus. Novitates **2014:** 1–7.

———, and Fluke, C. L.

1926 Revision of the nearctic species of *Helophilus* and allied genera. Wis. Acad. Sci., Arts, Letters, Trans. **22:** 207–281, 3 pls.

Curry, L. L.

1958 Larvae and pupae of the species of *Cryptochironomus* (Diptera) in Michigan. Limnol. and Oceanog. **3:** 427–442, 26 figs.

Curtis, J.

1824– British entomology: Being illustrations and descriptions of
1834   the genera of insects found in Great Britain and Ireland. Vol. 1, pls. 1–50, *1824*; Vol. 2, pls. 51–98, *1825*; Vol. 3, pls. 99–146, *1826*; Vol. 4, pls. 147–194, *1827*; Vol. 5, pls. 195–241, *1828*; Vol. 6, pls. 242–289, *1829*; Vol. 7, pls. 290–337, *1830*; Vol. 8, pls. 338–383, *1831*; Vol. 9, pls. 384–433, *1832*; Vol. 10, pls. 434–481, *1833*; Vol. 11, pls. 482–529, *1834*. London, "1823–1840."

> This was published originally in 192 parts, London, 1824–1839, each part containing at first five, and later four, plates, each plate with two unnumbered pages of text. One part was issued each month beginning January, 1824. These were then to be bound in 16 volumes, with the 769 plates arranged serially. A second printing was bound in 8 volumes, with plates systematically arranged. The plates on Diptera are scattered in the first printing, but in the second, are all bound in Volume 8: Diptera-Omaloptera. See Blackwelder, 1947: 1, for a discussion of the dates and editions of this work.

1835a British entomology: Being illustrations and descriptions of the genera of insects found in Great Britain and Ireland. Vol. 12, pls. 530–577. London.

> Each plate is accompanied by two unnumbered pages of text.

1835b Descriptions, &c. of the insects brought home by Commander James Clark Ross, R.N., F.R.S., &c. Pp. lix-lxxx, pl. A. *In* Ross, J., Narrative of a second voyage

Curtis, J.—Continued
in search of a North-West Passage [q.v.]. Appendix, Natural History, cxliv pp., 5 pls. London.

    Plate A was published in 1834, but contains no names.

1836   British entomology: Being illustrations and descriptions of the genera of insects found in Great Britain and Ireland. Vol. 13, pls. 578–625. London.

    Each plate is accompanied by two unnumbered pages of text.

1837a   British entomology: Being illustrations and descriptions of the genera of insects found in Great Britain and Ireland. Vol. 13, pls. 626–673. London.

    Each plate is accompanied by two unnumbered pages of text.

1837b   A guide to an arrangement of British insects. Ed. 2, 294 pp. London.

1838–1839   British entomology: Being illustrations and descriptions of the genera of insects found in Great Britain and Ireland. Vol. 15, pls. 674–721, *1838;* Vol. 16, pls. 722–769, *1839.* London.

    Each plate is accompanied by two unnumbered pages of text.

1845   The potato flies. Gard. Chron. and Agr. Gaz. **1845:** 816–817, 5 figs.

    Article under the pseudonym "Ruricola."

1846   The holly leaf fly (*Phytomyza ilicis*). Gard. Chron. and Agr. Gaz. **1846:** 444, 1 fig.

    Article under the pseudonym "Ruricola."

1847   Observations on the natural history and economy of a weevil affecting the pea-crops, and various insects which injure or destroy the mangold-wurzel and beet. Roy. Agr. Soc. England, Jour. **8:** 399–416, pl. S.

Cushing, E. C., and Patton, W. S.
1933   Studies on the higher Diptera of medical and veterinary importance. *Cochliomyia americana* sp. nov., the screw-worm fly of the New World. Ann. Trop. Med. and Parasitol. **27:** 539–551, 7 figs.

[**Cuvier, G. C. L. D.**]
1829–1830   Le règne animal distribué d'après son organisation, pour servir de base à l'histoire naturelle des animaux, et d'introduction a l'anatomie comparée. Ed. 2, 5 vols. Paris.

    Volumes 4 and 5 are by P. A. Latreille, and are titled "Les Crustacés, les Arachnides et les Insectes," Volumes 1 and 2. See Latreille, 1829.

[Cuvier, G. C. L. D.]—Continued
[1836– Le règne animal distribué d'après son organisation, pour
1849] servir de base à l'histoire naturelle des animaux, et d'introduction à l'anatomie comparée, par une réunion de disciples de Cuvier [i.e. by 11 students of Cuvier].

> [The so-called "disciples edition"], 17 vols. (in 20). Paris. Issued in 262 livraisons. The unnumbered volumes are arranged into 10 taxonomic groups. That of "Les Insectes," by Audouin, Blanchard, Doyère, and Edwards, includes 2 volumes of text, and an Atlas of 2 volumes (in 3).
> See Audouin et al., 1836–1849.

Czerny, L.

1900 Eine neue *Scatophila* (Dipt.) aus Oesterreich. Wien. Ent. Ztg. **19**: 205–206.

1902 Bemerkungen zu den Arten der Gattungen *Anthomyza* Fll. und *Ischnomyia* Lw. Wien. Ent. Ztg. **21**: 249–256.

1903a Revision der Heteroneuriden. Wien. Ent. Ztg. **22**: 61–107, 3 pls.

1903b Ueber *Drosophila costata* und *fuscimana* Ztt. (Dipt.). Ztschr. f. System. Hymenopt. u. Dipt. **3**: 198–201.

1904a *Cremifania nigrocellulata*, eine neue Ochthiphiline. Systematische Stellung und Gattungen-Diagnose der Ochthiphilinen. Wien. Ent. Ztg. **23**: 167–170.

1904b Revision der Helomyziden. I. Teil. Wien. Ent. Ztg. **23**: 199–244, 263–286, 1 pl.

1906 Zwei neue *Chortophila*-Arten aus Oberösterreich (Dipt.). Wien. Ent. Ztg. **25**: 251–254.

1924 Monographie der Helomyziden (Dipteren). Zool.-Bot. Gesell. Wien, Abhandl. **15** (1): 1–166, 1 pl.

1926 Ergänzungen und Berichtigungen zu meiner Monographie der Helomyziden. Konowia **5**: 53–56.

1928a Ergänzungen zu meiner Monographie der Helomyziden. III. Konowia **7**: 52–55.

1928b Clusiidae. [Fam.] 54a, pp. 1–12, 18 figs. (=lfg. 28). *In* Lindner, E., ed., Die Fliegen der palaearktischen Region [q.v.]. Vol. 6, Pt. 1. Stuttgart.

1930a Clythiidae (Platypezidae). [Fam.] 34, pp. 1–29, 26 figs. (=lfg. 47, part). *In* Lindner, E., ed., Die Fliegen der palaearktischen Region [q.v.]. Vol. 4, Pt. 7. Stuttgart.

1930b Tylidae und Neriidae. [Fams.] 42a–42b, pp. 1–18, 17 figs. (=lfg. 47, part). *In* Lindner, E., ed., Die Fliegen der palaearktischen Region [q.v.]. Vol. 5. Stuttgart.

1932 Tyliden und Neriiden des zoologischen Museums in Hamburg (Dipt.). Stettin. Ent. Ztg. **93**: 267–302.

Czerny, L.—Continued
1933 Ergänzungen zu meiner Monographie der Helomyziden. VIII. Konowia **12**: 236–238.
1934a Musidoridae (Lonchopteridae). [Fam.] 30, pp. 1–16, 30 figs. (=lfg. 83, part). *In* Lindner, E., ed., Die Fliegen der palaearktischen Region [q.v.]. Vol. 4, Pt. 5. Stuttgart.
1934b Lonchaeidae. [Fam.] 43, pp. 1–40, 19 figs., 3 pls. (=lfg. 83, part). *In* Lindner, E., ed., Die Fliegen der palaearktischen Region [q.v.]. Vol. 5. Stuttgart.

Daecke, E.
1905 Two new species of Diptera from New Jersey. Ent. News **16**: 249–251, 2 figs.
1907 Annotated list of the species of *Chrysops* occurring in New Jersey and descriptions of two new species. Ent. News **18**: 139–146, 2 figs., 1 pl.
1910 Trypetid galls and *Eurosta elsa* n. sp. Ent. News **21**: 341–343, 1 pl.

Dahl, F.
1897 *Puliciphora*, eine neue, flohähnliche Fliegengattung. Zool. Anz. **20**: 409–412.
1898 Über den Floh und seine Stellung im System. Gesell. Naturf. Freunde, Berlin, Sitzber. **1898**: 185–199, 15 figs.
1909 Die Gattung *Limosina* und die biocönotische Forschung. Gesell. Naturf. Freunde, Berlin, Sitzber. **1909**: 360–377.

Dahl, F., ed.
1925– Die Tierwelt Deutschlands und der angrenzenden Meeres-
1961 teile nach ihren Merkmalen und nach ihrer Lebensweise. 48 pts. Jena.

> This series is issued irregularly in separately paged "Teile," each covering a particular taxon which is published at the time of its completion. Through 1961, there have been 48 such parts published.
> See Hendel, 1928b; Karl, 1928.

Dahl, R. G.
1959 Studies on Scandinavian Ephydridae (Diptera, Brachycera). Opusc. Ent. (Soc. Ent. Lund.), Sup. **25**: 1–225, 84 figs.

Dale, J. C.
1842 Descriptions, &c. of a few rare or undescribed species of British Diptera, principally from the collection of J. C. Dale, Esq., M. A., F. L. S., &c. Ann. and Mag. Nat. Hist. **8**: 430–433.

Dallas, W. S.

1866   Insecta. Pp. 381–710. *In* Günther, A. C. L. G., ed., The record of zoological literature [q.v.]. Vol. 2 (1865), 798 pp. London.

Dalman, J. W.

1816   Försök till systematisk uppställning af Sveriges fjärilar. *Chionea araneoides*, ett nytt inländskt insekt af Tvavingarnas Ordning. K. Vetensk. Acad. Handl. [ser. 3], **1816**: 48–101, 102–105.

Dalmat, H. T.

1942a   A new *Cuterebra* (Diptera: Cuterebridae) from Iowa with notes on certain facial structures. Amer. Midland Nat. **27**: 418–421, 1 fig.

1942b   A new parasitic fly (Cuterebridae) from the northern white-footed mouse. N.Y. Ent. Soc. Jour. **50**: 45–58, 4 pls.

Dampf, A.

1938   Un nuevo *Phlebotomus* (Insecta, Diptera, fam. Psychodidae) procedente de Texas, E.U.A. [Mex.] Esc. Nac. de Cien. Biol., An. **1**: 119–131, 4 pls.

1944   Notas sobre flebotomidos mexicanos. Observaciones generales y descripcion de dos especies nuevas (*Phlebotomus oppidanus* y *Phl. vindicator*), encontradas en la ciudad de México (Ins. Diptera, fam. Phlebotomidae). Soc. Mex. de Hist. Nat. Rev. **5**: 237–254, pls. 5–8.

Darlington, P. S.   (See Aldrich and Darlington)

Davidson, A.

1894   On the parasites of wild bees in California. Ent. News and Acad. Nat. Sci. Phila., Ent. Sect. Proc. **5**: 170–172.

1896   Parasites of spider eggs. Ent. News **7**: 319–320.

Davidson, W. M.   (See also Campbell and Davidson)

1917   Early spring Syrphidae in California and a new *Pipiza* (Dip.). Ent. News **28**: 414–419, 1 fig.

1922   Notes on certain species of *Melanostoma* (Diptera: Syrphidae). Amer. Ent. Soc. Trans. **48**: 35–47, 3 figs., 1 pl.

1926   A new Californian syrphid (Diptera). Ent. News **37**: 40–42, 1 fig.

Davies, D. M.   (See also Syme and Davies)

1949   Description of *Simulium euryadminiculum*, a new species of blackfly (Simuliidae: Diptera). Canad. Ent. **81**: 45–49, 6 figs.

———, Peterson, B. V., and Wood, D. M.
1962 The black flies (Diptera: Simuliidae) of Ontario. Part I. Adult identification and distribution with descriptions of six new species. Ent. Soc. Ontario, Proc. (1961) **92**: 69–154, 90 figs.

Davis, J. J.
1919 Contributions to a knowledge of the natural enemies of *Phyllophaga*. Ill. State Nat. Hist. Survey, Bul. **13**: i–iv, 53–138, 46 figs., pls. 3–15.

Davis, K. C. (See Needham et al.)

Day, L. T.
1881 Notes on Sciomyzidae, with descriptions of new species. Canad. Ent. **13**: 85–89.
1882 The species of *Odontomyia* found in the United States. Acad. Nat. Sci. Phila. Proc. **1882**: 74–88.

DeFoliart, G. R. (See also Peterson and DeFoliart; Stone and DeFoliart)
———, and Peterson, B. V.
1960 New North American Simuliidae of the genus *Cnephia* Enderlein (Diptera). Ent. Soc. Amer. Ann. **53**: 213–219, 25 figs.

De Geer, C.
1776 Mémoires pour servir à l'histoire des Insectes. Vol. 6, 523 pp., 30 pls. Stockholm.
  The entire work consists of 7 volumes, Stockholm, 1752–1778.

De Leon, D.
1935 A study of *Medetera aldrichii* Wh. (Diptera-Dolichopodidae), a predator of the mountain pine beetle (*Dendroctonus monticolae* Hopk., Coleo.-Scolytidae). Ent. Amer. **15**: 59–91, pls. 4–7.

Del Ponte, E.
1938 Las especies argentinas del género *Cochliomyia* T. T. (Dipt. Musc.). Rev. de Ent. **8**: 273–281, 2 figs.

Del Rosario, F.
1936 The American species of *Psychoda* (Diptera: Psychodidae). Philippine Jour. Sci. **59**: 85–148, 1 fig., 6 pls.

Delucchi, V., and Pschorn-Walcher, H.
1955 Les espèces du genre *Cnemodon* Egger (Diptera, Syrphidae) prédatrices de *Dreyfusia* (*Adelges*) *piceae* Ratzeburg (Hemiptera, Adelgidae). I. Révision systématique et répartition géographique des espèces du genre *Cnemodon* Egger. Ztschr. f. Angew. Ent. **37**: 492–506, 7 figs.

Demerec, M. (See Spencer, 1938)

Dendy, J. S., and Sublette, J. E.
1959 The Chironomidae (=Tendipedidae: Diptera) of Alabama with descriptions of six new species. Ent. Soc. Amer. Ann. **52:** 506–519, 24 figs.

Desmarest, E.
1859 Annelés. 312 pp., 279 figs., 40 pls. *In* Chenu, J. C., ed., Encyclopédie d'histoire naturelle [q.v.]. Paris.

De Stefani Perez, T.
1908 Una nuova interessante cecidomia. Marcellia (1907) **6:** 174–176.

Deutsche Akademie der Wissenschaften zu Berlin
1957– Abhandlungen zur Larvalsystematik der Insekten. 6 vols.
1962 Berlin.
> Six volumes on various orders have been published to date. See Fittkau, 1962.

Díaz Nájera, A. (See Vargas and Díaz Nájera; Vargas, Martínez Palacios, and Díaz Nájera)

Dietz, W. G.
1913 A synopsis of the described North American species of the dipterous genus *Tipula* L. Ent. Soc. Amer. Ann. **6:** 461–484.
1914 The *hebes* group of the dipterous genus *Tipula* Linnaeus. Amer. Ent. Soc. Trans. **40:** 345–363, 2 pls.
1915a Two new Tipulidae from northern Alberta. Canad. Ent. **47:** 329–332, 2 figs.
1915b A preoccupied specific name in *Tipula* (Dipt.). Ent. News **26:** 125.
1916 Synoptical table of the North American species of *Ormosia* Rondani (*Rhypholophus* Kolenati), with descriptions of new species (Diptera). Amer. Ent. Soc. Trans. **42:** 135–146, 1 pl.
1917 Key to the North American species of the *tricolor* group of the dipterous genus *Tipula* Linnaeus, with descriptions of four new species. Ent. News **28:** 145–151, 1 pl.
1918 A revision of the North American species of the tipulid genus *Pachyrhina* Macquart, with descriptions of new species (Diptera). Amer. Ent. Soc. Trans. **44:** 105–140, 4 pls.
1919 The *streptocera* group of the dipterous genus *Tipula* Linnaeus. Ent. Soc. Amer. Ann. **12:** 85–93, 1 pl.
1920 Three new crane-flies from eastern Canada. Canad. Ent. **52:** 5–8.

Dietz, W. G.—Continued
1921a A list of the crane-flies taken in the vicinity of Hazleton, Pennsylvania (Diptera). Amer. Ent. Soc. Trans. **47:** 233–268.
1921b The *impudica* group of the dipterous genus *Tipula* Linnaeus. Ent. Soc. Amer. Ann. **14:** 1–15, 1 pl.
1921c Description of two new species of the *angustipennis* group of the dipterous genus *Tipula* Linnaeus with table of species. Ent. News **32:** 299–302.

Disciples de Cuvier (See Cuvier, 1836–1849)

Dixon, T. J.
1960 Key to and descriptions of the third instar larvae of some species of Syrphidae (Diptera) occurring in Britain. Roy. Ent. Soc., London, Trans. **112:** 345–379, 8 figs.

Doane, R. W.
1898 A new trypetid of economic importance. Ent. News **9:** 69–72.
1899 Notes on Trypetidae with descriptions of new species. N.Y. Ent. Soc. Jour. **7:** 177–193, pls. 3–4.
1900 New North American Tipulidae. N.Y. Ent. Soc. Jour. **8:** 182–198, 2 pls.
1901 Descriptions of new Tipulidae. N.Y. Ent. Soc. Jour. **9:** 97–127.
1908a Notes on the tipulid genus *Dicranomyia*. Ent. News **19:** 5–9, 4 figs.
1908b New North American *Pachyrhina*, with a table for determining the species. Ent. News **19:** 173–179.
1908c New species of the tipulid genus *Rhypholophus*, with a table for determining the North American species. Ent. News **19:** 200–202.
1908d A new species of *Tipula* with vestigial wings. Psyche **15:** 47–49, 1 fig.
1909 More *Tipula* with vestigial wings. Psyche **16:** 17–19.
1911 *Tipula fallax* and others. Psyche **18:** 160–166, 2 figs.
1912 New western *Tipula*. Ent. Soc. Amer. Ann. **5:** 41–61.

Dobroscky, I. D. (See also Shannon and Dobroscky)
1925 External parasites of birds and the fauna of birds' nests. Biol. Bul. **48:** 274–281.

Dobzhansky, T. (See also Sturtevant and Dobzhansky)
1935 *Drosophila miranda*, a new species. Genetics **20:** 377–391, 33 figs.

Dobzhansky, T., and Epling, C.
1944 Contributions to the genetics, taxonomy, and ecology of *Drosophila pseudoobscura* and its relatives. Carnegie Inst. Wash., Pub. **554**: 1–183, 24 figs., 4 pls.

Dodge, H. R.
1947 A new species of *Wyeomyia* from the pitcher plant (Diptera, Culicidae). Ent. Soc. Wash. Proc. **49**: 117–122, 1 pl.
1955a New muscid flies from Florida and the West Indies. Fla. Ent. **38**: 147–151.
1955b Sarcophagid flies parasitic on reptiles (Diptera, Sarcophagidae). Ent. Soc. Wash. Proc. **57**: 183–187, 11 figs.
1956a New North American Sarcophagidae, with some new synonymy (Diptera). Ent. Soc. Amer. Ann. **49**: 182–190, 2 pls.
1956b A new sarcophagid genus with descriptions of 15 new species (Diptera). Ent. Soc. Amer. Ann. **49**: 242–263, 3 pls.

Doleschall, C. L.
[1858] Derde bijdrage tot de kennis der dipteren fauna van Nederlandsch Indië. Natuurk. Tijdschr. v. Nederland. Indië (1858–1859) **17** [=ser. 4, 3]: 73–128.

Dated by Horn and Schenkling, 1928: 268.

Dollfus, R. P. (See Stoll et al.)

Dorogostajskij, V., Rubzov, I. A., and Vlasenko, N.
1935 (Notes on taxonomy, biology and geographical distribution of black flies in east Siberia.) Akad. Nauk S.S.S.R., Zool. Inst., Parazitol. Sborn. (Acad. des Sci. de l'U.R.S.S., Inst. Zool., Mag. de Parasitol.) **5**: 107–204, 21 figs.

In Russian; English summary, pp. 199–204.

Doucette, C. F., Latta, R., Martin, C. H., Schopp, R., and Eide, P. M.
1942 Biology of the narcissus bulb fly in the Pacific Northwest. U.S. Dept. Agr. Tech. Bul. **809**: 1–66, 28 figs.

Douwe, C. van   (See Kieffer and Douwe)

Downes, J. A.
1958 The genus *Culicoides* (Diptera: Ceratopogonidae) in Canada: An introductory review. Pp. 801–808. *In* Becker, E. C., ed., Proceedings of the Tenth International Congress of Entomology [q.v.]. Vol. 3, 895 pp., 134 figs., 4 pls. Ottawa, Ont.

Downs, W. G.   (See Lopes and Downs)

Doyère, L.   (See Audouin et al.)

Drew, W. A.

1963   Nearctic genera of Anthomyiini (Muscidae, Diptera) with a list of nearctic species. Mich. State Univ., Mus., Pubs., Biol. Ser. **2**: 193–272.

Drury, D.

1770   Illustrations of natural history. Wherein are exhibited upwards of 240 figures of exotic insects, according to their different genera. [Vol. 1], 130 pp., 4 figs., 50 pls. London.

> The entire work consists of 3 volumes, London, 1770–1782. The index to this volume, in which the plates and descriptions are named, was published with Volume 2 in 1773, from which year the names must date.

1773   Illustrations of natural history. Wherein are exhibited upwards of 240 figures of exotic insects, according to their different genera. Vol. 2, 90 pp., 50 pls., and index to Vols. 1–2 (4 unnumbered pages). London.

Duda, O.

1918   Revision der europäischen Arten der Gattung *Limosina* Macquart (Dipteren). Zool.-Bot. Gesell. Wien, Abhandl. **10 (1)**: 1–240, 8 pls.

1923a   Revision der altweltlichen Arten der Gattung *Borborus* (*Cypsela*) Meigen. Arch. f. Naturgesch. **Abt. A, 89 (4)**: 35–112. 14 figs.

1923b   Die orientalischen und australischen Drosophiliden-Arten (Dipteren) des Ungarischen National-Museums zu Budapest. Budapest Magyar Nemzeti Muz., Ann. Hist. Nat. **20**: 24–59.

1924a   Beitrag zur Systematik der Drosophiliden unter besonderer Berücksichtigung der paläarktischen u. orientalischen Arten (Dipteren). Arch. f. Naturgesch. **Abt. A, 90 (3)**: 172–234, 7 pls.

1924b   Berichtigung zur Revision der europäischen Arten der Gattung *Limosina* Macq. (Dipteren) nebst Beschreibung von sechs neuen Arten. Zool.-Bot. Gesell. Wien, Verhandl. (1923) **73**: 163–180, 7 figs.

1924c   Beitrag zur Systematik der Limosinen-Untergattungen *Trachyopella* und *Elachisoma* und Beschreibung von *Elachisoma pilosa* n. sp. ♀ (Dipteren). Konowia **3**: 5–9, 3 figs.

1924d   Revision der europäischen u. grönlandischen sowie einiger südostasiat. Arten der Gattung *Piophila* Fallén (Dipteren). Konowia **3**: 97–113, 153–203.

Duda, O.—Continued
- 1925a Die aussereuropäischen Arten der Gattung *Leptocera* Olivier=*Limosina* Macquart mit Berücksichtigung der europäischen Arten. Arch. f. Naturgesch. (1924) Abt. A, **90** (11): 5–215, 4 pls.
- 1925b Die costaricanischen Drosophiliden des Ungarischen National-Museums zu Budapest. Budapest Magyar Nemzeti Muz., Ann. Hist. Nat. **22**: 149–229, 14 figs.
- [1926] Monographie der Sepsiden (Dipt.). I. Wien, Naturhist. Mus. Ann. (1925) **39**: 1–153, 2 figs., 7 pls.
  Dated by Malloch, 1928b: 611.
- 1927 Revision der altweltlichen Astiidae (Dipt.). Deut. Ent. Ztschr. **1927**: 113–147, 3 figs., 2 pls.
- 1928a Scatopsidae. [Fam.] 5, pp. 1–62, 56 figs., 2 pls. (=lfg. 26). *In* Lindner, E., ed., Die Fliegen der palaearktischen Region [q.v.]. Vol. 2, Pt. 1. Stuttgart.
- 1928b Beitrag zur Kenntnis der aussereuropäischen Scatopsiden (Dipt.). Konowia **7**: 259–297, 25 figs., 1 pl.
- 1929 Die Ausbeute der Deutschen Chaco-Expedition 1925/26 (Diptera). VI. Sepsidae, VII. Piophilidae, VIII. Cypselidae, IX. Drosophilidae und X. Chloropidae; X. Chloropidae (Fortsetzung). Konowia **8**: 33–50, 6 figs.; 165–169, 1 fig.
- 1930 Die neotropischen Chloropiden (Dipt.). Folia Zool. et Hydrobiol. **2**: 46–128.
- 1931 Die neotropischen Chloropiden (Dipt.). 1. Fortsetzung: Nachtrag, Ergänzungen, Berichtigungen und Index. Folia Zool. et Hydrobiol. **3**: 159–172.
- 1932– Chloropidae. [Fam.] 61, pp. 1–48, figs. 1–17, pls. 1–3 1933 (=lfg. 64), *1932*; pp. 49–112 (=lfg. 68), *1933*; pp. 113–176, figs. 18–20 (=lfg. 70), *1933*; pp. 177–248, fig. 21 (=lfg. 72), *1933*. *In* Lindner, E., ed., Die Fliegen der palaearktischen Region [q.v.]. Vol. 6, Pt. 1. Stuttgart
- 1934a Camillidae. [Fam.] 58f, pp. 1–7, 8 figs. (=lfg. 81). *In* Lindner, E., ed., Die Fliegen der palaearktischen Region [q.v.]. Vol. 6, Pt. 1. Stuttgart.
- 1934b Drosophilidae. [Fam.] 58g, pp. 1–64, figs. 1–20, pls. 1–2 (=lfg. 84) [cont.]. *In* Lindner, E., ed., Die Fliegen der palaearktischen Region [q.v.]. Vol. 6, Pt. 1. Stuttgart.
- 1935 Drosophilidae [concl.]. [Fam.] 58g, pp. 65–118, figs. 21–30, pls. 3–5 (=lfg. 86). *In* Lindner, E., ed., Die Fliegen der palaearktischen Region [q.v.]. Vol. 6, Pt. 1. Stuttgart.
- 1940 Revision der afrikanischen Drosophiliden (Diptera). II. Budapest Magyar Nemzeti Muz., Ann. Hist. Nat. **33 (Zool.)**: 19–53, 6 figs.

Duméril, C.
1800   Exposition d'une méthode naturelle pour la classification et l'étude des Insectes, présentée à la société philomatique le brumaire an 9. Jour. de Phys., de Chim., d'Hist. Nat. et des Arts **51**: 427–439, 5 pls.

Duncan, J.
1837   Characters and descriptions of the dipterous insects indigenous to Britain. Mag. Zool. and Bot. **1**: 145–167, 7 figs.

[**Duperrey, L. I., ed.**]
1826–   Voyage autour du monde sur la corvette de sa majesté La
[1839?]   Coquille, pendant les années 1822, 1823, 1824, et 1825. 6 vols., and 5 vols. of Atlases. Paris, "1826–1830."

> This work was issued in livraisons and is grouped into 4 sections: "Histoire du Voyage," "Hydrographie," "Botanique," and "Zoologie." The latter, the only section actually completed, is by R. P. Lesson and P. Garnot, was issued in 28 livraisons, 1826–1839? (it is possible that it was completed by 1838), and consists of 2 volumes and an atlas. Volume 2 of "Zoologie" is bound into two parts, Part 2 of which has two separately paged divisions. The Atlas, issued in 1831, consists of a total of 157 plates, numbered serially within each taxonomic group. The plates on insects are numbered 1–21.
> See Guérin-Méneville, 1831, 1838.

Dupuis, C.
1963   Essai monographique sur les Phasiinae (Diptères Tachinaires parasites d'Hétéroptères). Paris, Mus. Natl. d'Hist. Nat. Mém. **n. ser.** [=ser. 3], ser. A: Zool., **26**: 1–461, 73 figs.

Dyar, H. G.   (See also Howard, Dyar, and Knab)
1904a   The larva of *Culex punctor* Kirby, with notes on an allied form. N.Y. Ent. Soc. Jour. **12**: 169–171, 1 pl.
1904b   Brief notes on mosquito larvae. N.Y. Ent. Soc. Jour. **12**: 243–246.
1905a   A new mosquito. N.Y. Ent. Soc. Jour. **13**: 74.
1905b   Our present knowledge of North American corethrid larvae. Ent. Soc. Wash. Proc. **7**: 13–16, 1 fig.
1905c   Remarks on genitalic genera in the Culicidae. Ent. Soc. Wash. Proc. **7**: 42–49, 1 fig.
1907   Report on the mosquitoes of the coast region of California, with descriptions of new species. U.S. Natl. Mus. Proc. **32**: 121–129.
1916a   Mosquitoes at San Diego, California. Insecutor Inscitiae Menstruus **4**: 46–51.
1916b   New *Aedes* from the mountains of California (Diptera, Culicidae). Insecutor Inscitiae Menstruus **4**: 80–90.

Dyar, H. G.—Continued

1917a The mosquitoes of the mountains of California (Diptera, Culicidae). Insecutor Inscitiae Menstruus **5**: 11–21.

1917b The mosquitoes of the Pacific Northwest (Diptera, Culicidae). Insecutor Inscitiae Menstruus **5**: 97–102, 1 pl.

1917c Notes on the *Aedes* of Montana (Diptera, Culicidae). Insecutor Inscitiae Menstruus **5**: 104–121.

1917d A new *Aedes* from the Rocky Mountain region (Diptera, Culicidae). Insecutor Inscitiae Menstruus **5**: 127–128.

1918a A second note on the species of *Culex* of the Bahamas (Diptera, Culicidae). Insecutor Inscitiae Menstruus (1917) **5**: 183–187.

1918b A revision of the American species of *Culex* on the male genitalia. Insecutor Inscitiae Menstruus **6**: 86–111, 2 pls.

1918c New American mosquitoes (Diptera, Culicidae). Insecutor Inscitiae Menstruus **6**: 120–129.

1919a The mosquitoes collected by the Canadian Arctic Expedition, 1913–18 (Diptera, Culicidae). Pp. 31–33, 2 figs. *In* Anderson, R. M., ed., Report of the Canadian Arctic Expedition 1913–18 [q.v.]. Vol. 3: Insects, Pt. C: Diptera, 90 pp., 2 figs., 10 pls. Ottawa, Ont.

1919b Westward extension of the Canadian mosquito fauna (Diptera, Culicidae). Insecutor Inscitiae Menstruus **7**: 11–39.

1920a The mosquitoes of British Columbia and Yukon Territory, Canada (Diptera, Culicidae). Insecutor Inscitiae Menstruus **8**: 1–27, 1 pl.

1920b A new mosquito from Mexico (Diptera, Culicidae). Insecutor Inscitiae Menstruus **8**: 81–82.

1920c The American *Aedes* of the *stimulans* group (Diptera, Culicidae). Insecutor Inscitiae Menstruus **8**: 106–120.

1920d The *Aedes* of the mountains of California and Oregon (Diptera, Culicidae). Insecutor Inscitiae Menstruus **8**: 165–173.

1921a Ring-legged *Culex* in Texas (Diptera, Culicidae). Insecutor Inscitiae Menstruus **9**: 32–34.

1921b Notes on the North American species of *Choeroporpa* (Diptera, Culicidae). Insecutor Inscitiae Menstruus **9**: 37–39.

1921c The American *Aedes* of the *punctor* group (Diptera, Culicidae). Insecutor Inscitiae Menstruus **9**: 69–80, 1 pl.

1921d Illustrations of certain mosquitoes (Diptera, Culicidae). Insecutor Inscitiae Menstruus **9**: 114–118, 1 pl.

Dyar, H. G.—Continued
1922a New mosquitoes from Alaska (Diptera, Culicidae). Insecutor Inscitiae Menstruus **10**: 1–3.
1922b Mosquito notes (Diptera, Culicidae). Insecutor Inscitiae Menstruus **10**: 92–99.
1923   The mosquitoes of the Yellowstone National Park (Diptera, Culicidae). Insecutor Inscitiae Menstruus **11**: 36–46.
1924a *Phoniomyia* and *Dendromyia* Theobald (Diptera, Culicidae). Insecutor Inscitiae Menstruus **12**: 107–113.
1924b Two new mosquitoes from California (Diptera, Culicidae). Insecutor Inscitiae Menstruus **12**: 125–127.
1924c The American forms of *Aedes cinereus* Meigen (Diptera, Culicidae). Insecutor Inscitiae Menstruus **12**: 179–180.
1925   A new North American *Dixa* and note (Diptera, Culicidae). Insecutor Inscitiae Menstruus **13**: 217–218.
1926a Three psychodids from the Glacier National Park (Diptera, Psychodidae). Insecutor Inscitiae Menstruus **14**: 103–106, 1 pl.
1926b Some apparently new American psychodids (Diptera, Psychodidae). Insecutor Inscitiae Menstruus **14**: 107–111.
1927   American Psychodidae –I (Diptera). Ent. Soc. Wash. Proc. **29**: 162–164, 6 figs.
1928a American Psychodidae –II (Diptera). Ent. Soc. Wash. Proc. **30**: 87–89, 1 pl.
1928b The mosquitoes of the Americas. Carnegie Inst. Wash., Pub. **387**: 1–616, 123 pls.
1929a The present knowledge of the American species of *Phlebotomus* Rondani (Diptera, Psychodidae). Amer. Jour. Hyg. **10**: 112–124, 4 figs.
1929b A new species of mosquito from Montana with annotated list of the species known from the state. U.S. Natl. Mus. Proc. **75 (23)**: 1–8.
1929c American Psychodidae –III (Diptera). Ent. Soc. Wash. Proc. **31**: 63–64.

———, and Knab, F.
1906a Diagnoses of new species of mosquitoes. Biol. Soc. Wash. Proc. **19**: 133–141.
1906b Notes on some American mosquitoes with descriptions of new species. Biol. Soc. Wash. Proc. **19**: 159–172.
1906c The larvae of Culicidae classified as independent organisms. N.Y. Ent. Soc. Jour. **14**: 169–230, 13 pls.

Dyar, H. G., and Knab, F.—Continued
- 1906d The species of mosquitoes in the genus *Megarhinus*. Smithsn. Inst., Smithsn. Misc. Collect. (1907) **48**: 241–258, 1 fig. (=pub. 1657). (=Quart. Issue 3(3) [=pub. 1656].)
- 1907a New American mosquitoes. N.Y. Ent. Soc. Jour. **15**: 100–101.
- 1907b Descriptions of three new North American mosquitoes. N.Y. Ent. Soc. Jour. **15**: 213–214.
- 1908 Descriptions of some new mosquitoes from tropical America. U.S. Natl. Mus. Proc. **35**: 53–70.
- 1909a Mosquito comment. Canad. Ent. **41**: 101–102.
- 1909b Descriptions of some new species and a new genus of American mosquitoes. Smithsn. Inst., Smithsn. Misc. Collect. **52**: 253–266, 1 fig. (=pub. 1822). (=Quart. Issue 5(2) [=pub. 1813].)
- 1909c On the identity of *Culex pipiens* Linnaeus (Diptera, Culicidae). Ent. Soc. Wash. Proc. **11**: 30–39, 3 pls.
- 1910 Description of three new American mosquitoes (Diptera, Culicidae). Ent. Soc. Wash. Proc. (1909) **11**: 173–174.
- 1918a New American mosquitoes (Diptera, Culicidae). Insecutor Inscitiae Menstruus (1917) **5**: 165–169.
- 1918b The genus *Culex* in the United States (Diptera, Culicidae). Insecutor Inscitiae Menstruus (1917) **5**: 170–183.

———, and Ludlow, C. S.
- 1921 Two new American mosquitoes (Diptera, Culicidae). Insecutor Inscitiae Menstruus **9**: 46–50.

———, and Shannon, R. C.
- 1924a The American species of *Uranotaenia* (Diptera, Culicidae). Insecutor Inscitiae Menstruus **12**: 187–192.
- 1924b Some new species of American *Dixa* Meigen (Diptera, Culicidae). Insecutor Inscitiae Menstruus **12**: 193–201.
- 1924c The American Chaoborinae (Diptera, Culicidae). Insecutor Inscitiae Menstruus **12**: 201–216.
- 1924d The American species of Thaumalidae (Orphnephilidae) (Diptera). Wash. Acad. Sci. Jour. **14**: 432–434.
- 1925a New mosquitoes from Brazil (Diptera, Culicidae). Wash. Acad. Sci. Jour. **15**: 39–41.
- 1925b The mosquitoes of Peary's North Pole expedition of 1908 (Diptera, Culicidae). Wash. Acad. Sci. Jour. **15**: 77–78.
- 1927 The North American two-winged flies of the family Simuliidae. U.S. Natl. Mus. Proc. **69** (10): 1–54, 7 pls.

Dziedzicki, H. (See also Becker et al.; Schnabl and Dziedzicki)
1884   Przyczynek do fauny owadów dwuskrzydłych. Gatunki rodzajów: *Mycothera, Mycetophila, Staegeria*. Pamiet. Fizyiograficzny **4**: 298–324, pls. 5–9.

    A German translation, "Beitrag zur Fauna der zweiflügeligen Insecten Arten der Gattungen *Mycothera, Mycetophila*, and *Staegeria*", was published with some minor additions and without the plates, in 1886–1887, in Wien. Ent. Ztg. **5**: 153–156, 189–190, 229–231, 251–253, 265–266, 326–327, 346–347, *1886;* and **6**: 37–43, *1887*.

1885   Przyczynek do fauny owadów dwuskrzydłych. Rodzaje nowe: *Hertwigia*, nov. gen., *Eurycera*, nov. gen. i gatunki rodzajów: *Boletina, Sciophila*. Pamiet. Fizyiograficzny **5 (3)**: 164–194, pls. 4–9.

1889   Revue des espèces européennes du genre *Phronia* Winnertz, avec la description des deux genres nouveaux: *Macrobrachius* et *Megophthalmidia*. Russ. Ent. Obshch. Trudy (Soc. Ent. Ross. Horae) **23**: 404–532, 17 figs., pls. 12–21.

1923   Revue des espèces européennes du genre *Anatella* Winnertz avec la description des deux genres nouveaux: *Heteropygium* et *Allophallus*. Towar. Nauk Warszawsk., Arch. Nauk Biol. (1922) **1 (15)**: 1–8, 2 pls.

Eaton, A. E.
1875   Breves dipterarum uniusque lepidopterarum Insulae Kerguelensi indigenarum diagnoses. Ent. Monthly Mag. **12**: 58–61.

1904   New genera of European Psychodidae. Ent. Monthly Mag. **40** [=ser. 2, 15]: 55–59.

Edwards, F. W. (See also Thienemann, 1941; Tonnoir and Edwards)
1913   Notes on British Mycetophilidae. Ent. Soc. London, Trans. **1913**: 334–382, pls. 12–18.

1915a   On the British species of *Simulium*. 1. The adults. Bul. Ent. Res. **6**: 23–42, 6 figs.

1915b   Three new species of the dipterous genus *Olbiogaster*, O.-S., in the British Museum Collection. Ann. and Mag. Nat. Hist. ser. 8, **16**: 502–505.

1920   On the use of the generic name *Ceratopogon*, Meigen (Diptera, Chironomidae). Ann. and Mag. Nat. Hist. ser. 9, **6**: 127–130, 1 fig.

1921a   British Limnobiidae. Some records and corrections. Ent. Soc. London, Trans. **1921**: 196–230, pls. 1–2.

1921b   Diptera Nematocera from Arran and Loch Etive. Scot. Nat. **1921**: 59–61, 89–92, 121–125.

Edwards, F. W.—Continued
1922a *Deuterophlebia mirabilis*, gen. et sp. n., a remarkable dipterous insect from Kashmir. Ann. and Mag. Nat. Hist. ser. 9, **9**: 379–387, 2 figs., 1 pl.
1922b Results of the Oxford University Expedition to Spitsbergen, 1921. No. 14. Diptera Nematocera. Ann. and Mag. Nat. Hist. ser. 9, **10**: 193–215, 17 figs.
1922c On some Malayan and other species of *Culicoides*, with a note on the genus *Lasiohelea*. Bul. Ent. Res. **13**: 161–167, 1 pl.
1923 Notes on the dipterous family Anisopodidae. Ann. and Mag. Nat. Hist. ser. 9, **12**: 475–493, 3 figs., 1 pl.
1924a A note on the genus *Protanypus* Kieffer (Diptera: Chironomidae). Ann. de Biol. Lacustre **13**: 119–122, 2 figs.
1924b Results of the Merton College Expedition to Spitsbergen, 1923. No. 4. Diptera Nematocera. Ann. and Mag. Nat. Hist. ser. 9, **14**: 162–174, 1 fig.
1925a British fungus-gnats (Diptera, Mycetophilidae) with a revised generic classification of the family. Ent. Soc. London, Trans. **1924**: 505–670, pls. 49–61.
1925b Notes on the types of Mycetophilidae (Diptera) described by Staeger and Zetterstedt. Ent. Tidskr. (1924) **45**: 160–168.
1926 On the British biting midges (Diptera, Ceratopogonidae). Ent. Soc. London, Trans. **74**: 389–426, 3 figs., pls. 91–92.
1928a Diptera. Fam. Protorhyphidae, Anisopodidae, Pachyneuridae, Trichoceridae. Fasc. 190, 40 pp., 2 pls. *In* Wytsman, P., ed., Genera insectorum [q.v.]. Bruxelles.
1928b Nematocera. Pp. 23–102, figs. 1–20 (=fasc. 2, part). *In* British Museum (Natural History), Insects of Samoa [q.v.]. Vol. 4: Pt. 6 (Diptera), 366 pp., 118 figs. London.
1929a A revision of the Thaumaleidae (Dipt.). Zool. Anz. **82**: 121–142, 46 figs.
1929b British non-biting midges (Diptera, Chironomidae). Ent. Soc. London, Trans. **77**: 279–430, 15 figs., pls. 17–19.
1929c Notes on the Ceroplatinae, with descriptions of new Australian species (Diptera, Mycetophilidae). Linn. Soc. N. S. Wales, Proc. **54**: 162–175.
1931 Some suggestions on the classification of the genus *Tipula* (Diptera, Tipulidae). Ann. and Mag. Nat. Hist. ser. **10**, **8**: 73–82.
1932a Diptera. Fam. Culicidae. Fasc. 194, 258 pp., 5 pls. *In* Wytsman, P., ed., Genera insectorum [q.v.]. Bruxelles.

Edwards, F. W.—Continued
1932b Notes on highland Diptera, with descriptions of six new species. Scot. Nat. **1932:** 43–52, 19 figs.
1933 Oxford University Expedition to Hudson's Strait, 1931: Diptera Nematocera. With notes on some other species of the genus *Diamesa*. Ann. and Mag. Nat. Hist. **ser. 10, 12:** 611–620, 2 figs.
1935a Diptera Nematocera from East Greenland. Ann. and Mag. Nat. Hist. **ser. 10, 15:** 467–473, 2 figs.
1935b Diptera from Bear Island. Ann. and Mag. Nat. Hist. **ser. 10, 15:** 531–543, 2 figs.
1936 Bombyliidae from Chile and western Argentina. Rev. Chilena de Hist. Nat. Pura y Apl. **40:** 31–41, pl. 5.
1937 Chironomidae (Diptera) collected by Prof. Aug. Thienemann in Swedish Lappland. Ann. and Mag. Nat. Hist. **ser. 10, 20:** 140–148, 2 figs.
1938a On the British Lestremiinae, with notes on exotic species. (Diptera, Cecidomyiidae). 1; 2; 3; 4; 5; 6; 7. Roy. Ent. Soc., London, Proc. Ser. B: Taxonomy **7:** 18–24; 25–32, figs. 1–3; 102–108, figs. 4–7; 173–182, figs. 8–10; 199–210, figs. 11–16; 229–243, figs. 17–22; 253–265, figs. 23–28.
1938b British short-palped craneflies. Taxonomy of adults. Soc. Brit. Ent. Trans. **5:** 1–168, 31 figs., 5 pls.
1938c I. A new genus of the subfamily Podonominae. Pp. 152–154, figs. 1–2. *In* Edwards, F. W., and Thienemann, A., Neuer Beitrag zur Kenntnis der Podonominae (Dipt. Chironomidae). Chironomidae aus Lappland IV. Zool. Anz. **122:** 152–158, 6 figs.
1939 A new species of *Orthopodomyia* (Diptera, Culicidae). Roy. Ent. Soc., London, Proc. Ser. B: Taxonomy **8:** 121–123, 1 fig.
1941 Notes on British fungus-gnats (Dipt., Mycetophilidae). Ent. Monthly Mag. **77** [=ser. 4, 2]: 21–32, 67–82, 9 figs.
———, **and Thienemann, A.** (See Edwards, 1938c)

Edwards, H. Milne (See Audouin et al.)

Egger, J.
1860 [Dipterologische Beiträge.] Beschreibung neuer Zweiflügler [cont.]. K.-k. Zool.-Bot. Gesell. Wien, Verhandl. **10 (Abhandl.):** 795–802 [cont.].
1862 Dipterologische Beiträge. Fortsetzung der Beschreibung neuer Dipteren [cont.]. K.-k. Zool.-Bot. Gesell. Wien, Verhandl. **12 (Abhandl.):** 777–784, 1233–1236 [cont.].

Egger, J.—Continued
1863 Dipterologische Beiträge. Fortsetzung der Beschreibung neuer Dipteren [cont.]. K.-k. Zool.-Bot. Gesell. Wien, Verhandl. **13 (Abhandl.)**: 1101–1110 [cont.].
1865 Dipterologische Beiträge. Fortsetzung der Beschreibung neuer Zweiflügler [concl.]. K.-k. Zool.-Bot. Gesell. Wien, Verhandl. **15 (Abhandl.)**: 573–574.

Egglishaw, H. J.
1960 Studies on the family Coelopidae (Diptera). Roy. Ent. Soc., London, Trans. **112**: 109–140, 24 figs.

Ehrmann, P. (See Brohmer, Ehrmann, and Ulmer)

Eide, P. M. (See Doucette et al.)

Eldridge, B. F., and James, M. T.
1957 The typical muscid flies of California (Diptera: Muscidae, Muscinae). Calif. Insect Survey, Bul. **6**: 1–17, 3 pls.

Emden, F. I. van
1949 The scientific name of the common tachinid parasite of *Diatraea* spp. (Lep. Pyral.) in Central and South America, with notes on related species (Dipt.). Rev. de Ent. **20**: 499–508, 6 figs.
1954 Tachinidae and Calliphoridae. Sect. a, 133 pp., 42 figs. *In* Royal Entomological Society of London, Handbooks for the identification of British insects [q.v.]. Vol. 10: Cyclorrhapha, Pt. 4: Calyptrata (I). London.

Emmons, E.
1854 Insects of New York. Vol. 5, 272 pp., 47 pls. *In* New York Natural History Survey, Natural History of New York [q.v.]. [Div. 5]: Agriculture of New York. Albany, N.Y.

Enderlein, G. (See also Kieffer, 1912a)
1910a The Percy Sladen Trust Expedition to the Indian Ocean in 1905 under the leadership of Mr. J. Stanley Gardiner, M. A. Volume III. No. V. Diptera, Mycetophilidae. Linn. Soc. London, Trans. **ser. 2, Zool., 14**: 59–81, 6 figs.
1910b Neue Gattungen und Arten aussereuropäischer Fliegen. Stettin. Ent. Ztg. (1911) **72**: 135–209, 4 figs.
1911a Die phyletischen Beziehungen der Lycoriiden (Sciariden) zu den Fungivoriden (Mycetophiliden) und Itonididen (Cecidomyiiden) und ihre systematische Gliederung. Arch. f. Naturgesch. **77 (h. 1, sup. 3)**: 116–201, 2 figs., 2 pls.

Enderlein, G.—Continued
1911b Klassifikation der Oscinosominen. Gesell. Naturf. Freunde, Berlin, Sitzber. **1911**: 185–244, 11 figs.
1911c Einige neue Gattungen und Arten aussereuropäischer Chloropinen. Zool. Anz. **38**: 122–126, 1 fig.
1912a Zur Kenntnis der Zygophthalmen. Über die Gruppierung der Sciariden und Scatopsiden. Zool. Anz. **40**: 261–282, 15 figs.
1912b Studien über die Tipuliden, Limoniiden, Cylindrotomiden und Ptychopteriden. Zool. Jahrb., Abt. f. System., Geog. u. Biol. Tiere **32**: 1–88, 51 figs.
1912c Die Phoridenfauna Süd-Brasiliens. Stettin. Ent. Ztg. **73**: 16–45, 2 figs.
1912d *Paryphoconus*, eine neue Chironomiden-Gattung aus Brasilien. Stettin. Ent. Ztg. **73**: 57–60, 4 figs.
1914a Ord. Diptera, Fliegen (Zweiflügler). Pp. 272–334, figs. 212–248. *In* Brohmer, P., Fauna von Deutschland. 587 pp., 516 figs., 20 pls. Leipzig.
1914b Dipterologische Studien. IX. Zur Kenntnis der Stratiomyiiden mit 3-ästiger Media und ihre Gruppierung. A. Formen, bei denen der 1. Cubitalast mit der Discoidalzelle durch Querader verbunden ist oder sie nur in einem Punkte berührt (Subfamilien: Geosarginae, Analcocerinae, Stratiomyiinae). Zool. Anz. **43**: 577–615, 8 figs.
1914c Dipterologische Studien. XI. Zur Kenntnis tropischer Asiliden. Zool. Anz. **44**: 241–263, 8 figs.
1920 Ord. Diptera, Fliegen Zweiflügler. Pp. 265–315, figs. 316–352, pl. 12. *In* Brohmer, P., Fauna von Deutschland. Ed. 2, 472 pp., 569 figs., 19 pls. Leipzig.
1921a Das System der Kriebelmuecken (Simuliidae). Deut. Tierärztl. Wchnschr. **29**: 197–200.
1921b Dipterologische Studien. XVII. Zool. Anz. **52**: 219–232.
1921c Über die phyletisch älteren Stratiomyiidensubfamilien (Xylophaginae, Chiromyzinae, Solvinae, Beridinae, und Coenomyiinae). Berlin Zool. Mus. Mitt. **10**: 151–214, 16 figs.
1922a Klassifikation der Micropeziden. Arch. f. Naturgesch. **Abt. A, 88 (5)**: 140–229, 1 fig.
1922b Weitere Beiträge zur Kenntnis der Simuliiden. Konowia **1**: 67–76.
1922c Ein neues Tabanidensystem. Berlin Zool. Mus. Mitt. **10**: 335–351.
1923 Vorläufige Diagnosen neuer Tabanidengenera (Dipt.). Deut. Ent. Ztschr. **1923**: 544–545.

Enderlein, G.—Continued
- 1924   Zur Klassifikation der Phoriden und über vernichtende Kritik. Ent. Mitt. **13:** 270–281, 1 fig.
- 1925a   Studien an blutsaugenden Insekten. I. Grundlagen eines neuen Systems der Tabaniden. Berlin Zool. Mus. Mitt. **11:** 255–409, 5 figs.
- 1925b   Zur Klassifikation der Tabaniden, eine Berichtigung. Zool. Anz. **62:** 180–181.
- 1925c   Weitere Beiträge zur Kenntnis der Simuliiden und ihrer Verbreitung. Zool. Anz. **62:** 201–211.
- 1926a   Zur Kenntnis der Scatopsiden. Zool. Anz. **68:** 137–142.
- 1926b   Zur Kenntnis der Bombyliiden-Subfamilie Systropodinae (Dipt.). Wien. Ent. Ztg. **43:** 69–92, 1 fig.
- 1927   Dipterologische Studien. XIX. Stettin. Ent. Ztg. **88:** 102–109.
- 1928   Klassifikation der Sarcophagiden. Arch. f. Klassifikatorische u. Phylogenetische Ent. **1:** 1–56, 7 figs.
- 1930a   Der heutige Stand der Klassifikation der Simuliiden. Arch. f. Klassifikatorische u. Phylogenetische Ent. **1:** 77–97, 7 figs.
- 1930b   Dipterologische Studien. XX. Deut. Ent. Ztschr. **1930:** 65–71.
- 1934a   Weiterer Ausbau des Systems der Simuliiden (Dipt.). Deut. Ent. Ztschr. **1933:** 273–292.
- 1934b   Dipterologica. I. Gesell. Naturf. Freunde, Berlin, Sitzber. **1933:** 416–429.
- 1934c   Aussereuropäische Simuliiden aus dem Wiener Museum. Gesell. Naturf. Freunde, Berlin, Sitzber. **1934:** 190–195.
- 1935a   Neue Simuliiden, besonders aus Afrika. Gesell. Naturf. Freunde, Berlin, Sitzber. **1934:** 358–364.
- 1935b   Zur Klassifikation der Psychodinen. Gesell. Naturf. Freunde, Berlin, Sitzber. **1935:** 246–249.
- 1936a   22. Ordnung: Zweiflügler, Diptera. Abt. 16, 259 pp., 317 figs. (=lfg. 2, part). *In* Brohmer, P., Ehrmann, P., and Ulmer, G., eds., Die Tierwelt Mitteleuropas [q.v.]. Vol. 6: Insekten III Teil. Leipzig.

    This lieferung was issued with the title page for Vol. 6, which was dated 1936.
- 1936b   Notizen zur Klassifikation der Blepharoceriden (Dipt.). Blepharoceridae, Blepharocerinae. Deut. Ent. Gesell. Mitt. **7:** 42–43.
- 1936c   Simuliologica. I. Gesell. Naturf. Freunde, Berlin, Sitzber. **1936:** 113–130.
- 1937a   Klassifikation der Psychodiden (Dipt.). Deut. Ent. Ztschr. **1936:** 81–112, 1 fig.

Enderlein, G.—Continued
1937b Dipterologica. IV. Gesell. Naturf. Freunde, Berlin, Sitzber. **1936**: 431–443.
1938   Beiträge zur Kenntnis der Syrphiden. Gesell. Naturf. Freunde, Berlin, Sitzber. **1937**: 192–237, 1 fig.
1939   Zur Kenntnis der Klassifikation der Tetanoceriden. Bremen, Deut. Kolon.—u. Uebersee-Mus., Veröffentl. **2**: 201–210.

Engel, E. O.
1918   Das Dipterengenus *Atalanta* Mg. (*Clinocera* ol.). Deut. Ent. Ztschr. **1918**: 1–80, 197–268, 72 figs.
1925–   Asilidae. [Fam.] 24, pp. 1–8, figs. 1–7 (=lfg. 5), *1925;* pp.
1930   9–64, figs. 8–29 (=lfg. 9), *1926;* pp. 65–128, figs. 30–89 (=lfg. 20), *1927;* pp. 129–192, figs. 90–151 (=lfg. 27), *1928;* pp. 193–256, figs. 152–200 (=lfg. 29), *1928;* pp. 257–320, figs. 201–230 (=lfg. 36), *1929;* pp. 321–384, figs. 231–247 (=lfg. 37), *1930;* pp. 385–448, figs. 248–275 (=lfg. 42), *1930;* pp. 449–491, figs. 276–284 (=lfg. 44), *1930*. *In* Lindner, E., ed., Die Fliegen der palaearktischen Region [q.v.]. Vol. 4, Pt. 2. Stuttgart.
1932–   Bombyliidae. (Fam.] 25, pp. 1–48, figs. 1–16 (=lfg. 65),
1937   *1932;* pp. 49–96, figs. 17–41, pls. 1–2 (=lfg. 67), *1932;* pp. 97–144, figs. 42–71, pl. 3 (=lfg. 69), *1933;* pp. 145–192, figs. 72–103, pl. 4 (=lfg. 76), *1933;* pp. 193–256, figs. 104–113 (=lfg. 80), *1934;* pp. 257–304, figs. 114–127, pls. 5–6 (=lfg. 87), *1935;* pp. 305–352, figs. 128–135 (=lfg. 89), *1935;* pp. 353–400, figs. 136–160, pl. 7 (=lfg. 91), *1935;* pp. 401–448, figs. 161–185, pls. 8–10 (=lfg. 99), *1936;* pp. 449–512, figs. 186–200, pls. 11–12 (=lfg. 101), *1936;* pp. 513–560, figs. 201–224, pl. 13 (=lfg. 105), *1936;* pp. 561–619, figs. 225–239, pls. 14–15 (=lfg. 111), *1937*. *In* Lindner, E., ed., Die Fliegen der palaearktischen Region [q.v.]. Vol. 4, Pt. 3. Stuttgart.
1938–   Empididae. [Fam.] 28, pp. 1–40, figs. 1–30, pl. 1 (=lfg.
1954   120), *1938;* pp. 41–104, figs. 31–54 (=lfg. 124), *1939;* pp. 105–152, figs. 55–97, pls. 2–6 (=lfg. 130), *1939;* pp. 153–192, figs. 98–103, pls. 7–13 (=lfg. 132), *1940;* pp. 193–272, figs. 104–162, pls. 14–20 (=lfg. 142), *1941;* pp. 273–320, figs. 163–187, pls. 21–25 (=lfg. 150), *1943;* pp. 321–384, figs. 188–207, pls. 26–32 (=lfg. 154), *1945;* pp. 385–399, figs. 208–211, pl. 33 (=lfg. 177, part), *1954* [cont. by R. Frey]. *In* Lindner, E., ed., Die Fliegen der palaearktischen Region [q.v.]. Vol. 4, Pt. 4. Stuttgart.

English, K. M. I.
1947 Notes on the morphology and biology of *Apiocera maritima* Hardy (Diptera, Apioceridae). Linn. Soc. N. S. Wales, Proc. (1946) **71**: 296–302, 13 figs.

Entomological Society of America
1916– Thomas Say Foundation. 6 vols. Lafayette, Ind., Wash-
1959   ington, D.C.

> This is a series of unrelated entomological works published by the Society's Thomas Say Foundation, first at Lafayette, and later at Washington. To date, 6 volumes have been published; serial numbering began with Volume 3.
> See Aldrich, 1916a; Hall, 1948; Stone, Knight, and Starcke, 1959.

Entomologiska Foreningen i Stockholm
1903– Svensk Insektfauna. 13 vols. Stockholm.
1961

> This series will consist of 13 volumes upon completion, each volume devoted to a particular numbered order of insects. The first portions, 1903–1915, were published in the periodical "Entomologisk Tidskrift," with one exception; separate publication of this series actually began in 1917. It is issued in "rekvisions" chronologically numbered (except that some earlier material was republished for inclusion in the series and assigned "rekvision" numbers out of order), and published whenever a portion is complete. There are 48 rekvisions published through 1962. The volumes [=Orders] are broken into subgroups which are continuously paged if not complete within one rekvision. Pagination begins in all volumes in consideration of the portions previously published in Ent. Tidskr. Diptera is Order 11, and is divided, and was issued, as shown below:

[Order] 11: Tvåvingar. Diptera.—Incomplete. 1905–1959.

  Suborder I: Orthorapha.—Complete, all by E. Wahlgren. 1905–1922.

   Group 1: Nemocera. 1905–1922.

    Fams. 1–9. In Ent. Tidskr. 26: 91–154 [rekv. 5], 1905.

    Fams. 10–11. Pp. 69–140 [rekv. 22], 1919.

    Fams. 12–13. Pp. 141–204, 205–273 [rekvs. 24, 25], 1921–1922.

   Group 2: Brachycera. 1907–1912.

    Fams. 14–23. In Ent. Tidskr. 28: 129–192 [rekv. 9], 1907.

    Fam. 24. In Ent. Tidskr. 31: 41–95 [rekv. 14]. 1910.

    Fams. 25–26. In Ent. Tidskr. 33: 1–56 [rekv. 17], 1912.

Entomologiska Foreningen i Stockholm—Continued
    Suborder II: Cyclorapha.—Incomplete. 1909-1959.
    Group 1: Aschiza.—Complete, by E. Wahlgren. 1909-1910.
        Fam. 1. In Ent. Tidskr. 30: 1-86 [rekv. 12], 1909.
        Fams. 2-4. In Ent. Tidskr. 31: 209-235 [rekv. 15], 1910.
    Group 2: Schizophora.—Incomplete. 1917-1959.
        [Subgroup] Holometopa.—Complete, by E. Wahlgren. 1917-1927.
        Fams. 5-12. Pp. 113-224 [rekv. 20], 1917.
        Fams. 13-20. Pp. 225-322 [rekv. 21], 1919.
        Fams. 21-26. Pp. 323-416 (rekv. 30), 1927.
        With index to Aschiza and Holometopa, pp. i-xii.
        [Subgroup] Schizometopa.—Incomplete. 1954-1959.
        Fam. 1. Pp. 1-92, 93-196, 197-334 (rekvs. 44, 45, 47), by O. Ringdahl, 1954-1959.
        [Subgroup] Pupipara.—Not begun.
    See Ringdahl, 1954-1959; Wahlgren, 1917.

Epling, C. (See Dobzhansky and Epling)

Eppley, R. K. (See Bohart, Stephen, and Eppley)

Erichson, W. F. (See also Ménétries, 1851)
1840 IV. Die Henopier, eine Familie aus der Ordnung der Dipteren. Pp. 135-175, 1 pl. *In his* Entomographien, Untersuchungen in dem Gebiete der Entomologie mit besonderer Benutzung der Königl. Sammlung zu Berlin. Heft 1. 180 pp., 2 unnumbered pls. Berlin.
    Only 1 heft was ever issued.

Eschscholz, J. F.
1823 Entomographien. Naturw. Abhandl. aus Dorpat **1**: 57-186, 2 pls.

Esselbaugh, C. O. (See Sommerman, Sailer, and Esselbaugh)

Essig, E. O.
1958 Insects and mites of western North America. Ed. 2, 1050 pp., 766 figs. New York.

Ewen, A. B., and Saunders, L. G.
1958 Contributions toward a revision of the genus *Atrichopogon* based on characters of all stages (Diptera, Heleidae). Canad. Jour. Zool. **36**: 671-724, 21 figs.

Fabricius, J. C.
1775 Systema entomologiae, sistens insectorum classes, ordines, genera, species adiectis synonymis, locis, descriptionibus, observationibus. 832 pp. Flensburgi et Lipsiae [= Flensburg and Leipzig].

Fabricius, J. C.—Continued
1776 Genera insectorum eorumque characteres naturales secundum numerum, figuram, situm et proportionem omnium partium oris adiecta mantissa specierum nuper detectarum. 310 pp. Chilonii [=Kiel].
1781 Species insectorum exhibentes eorum differentias specificas, synonyma, auctorum, loca natalia, metamorphosin. Vol. 2, 517 pp. Hamburgi et Kilonii [=Hamburg and Kiel].
> The entire work consists of 2 volumes published at Hamburg and Kiel, 1781.

1787 Mantissa insectorum sistens species nuper detectas. Vol. 2, 382 pp. Hafniae [=Copenhagen].
> The entire work consists of 2 volumes published at Copenhagen, 1787.

1794 Entomologia systematica emendata et aucta. Vol. 4, 472 pp. Hafniae [=Copenhagen].
> The entire work consists of 4 volumes published at Copenhagen, 1792-1794.

1798 Supplementum entomologiae systematicae. 572 pp. Hafniae [=Copenhagen].
1805 Systema antliatorum secundum ordines, genera, species. 373 pp., + 30 pp. Brunsvigae [=Brunswick].

Fabricius, O.
1780 Fauna Groenlandica. 452 pp., 1 pl. Hafniae [=Copenhagen].

Fairchild, G. B. (See also Philip and Fairchild)
1934 Notes on Tabanidae. Boston Soc. Nat. Hist. Occas. Papers **8**: 139-144, 1 pl.
1935 A new *Tabanus* (Diptera) from Florida. Fla. Ent. **18**: 53-54.
1937a A preliminary list of the Tabanidae (Diptera) of Florida. Fla. Ent. **19**: 58-63 [cont.].
1937b A preliminary list of the Tabanidae (Diptera) of Florida [concl.]. Fla. Ent. **20**: 10-11.
1942 Notes on Tabanidae (Dipt.) from Panama. VII. The subgenus *Neotabanus* Ad. Lutz. Ent. Soc. Amer. Ann. **35**: 153-182, 2 pls.
1950 The generic names for Tabanidae (Diptera) proposed by Adolfo Lutz. Psyche **57**: 117-127.

———, **and Harwood, R. F.**
1962 *Phlebotomus* sandflies from animal burrows in eastern Washington (Diptera: Psychodidae). Ent. Soc. Wash. Proc. (1961) **63**: 239-245, 12 figs.

Fairchild, G. B., and Hertig, M.
1957   Notes on the *Phlebotomus* of Panama. XIII. The *vexator* group, with descriptions of new species from Panama and California. Ent. Soc. Amer. Ann. **50**: 325–334, 3 pls.

Fallén, C. F.
1810a   Specim. entomolog. novam Diptera disponendi methodum exhibens. 26 pp., 1 pl. Lund.
1810b   Försök att bestämma de i Sverige funne flugarter, som kunna föras till slägtet *Tachina*. K. Vetensk. Acad. Nya Handl. [ser. 2], **31**: 253–287.
1813   Beskrifning öfver några i Sverige funna vattenflugor (Hydromyzides). K. Vetensk. Acad. Handl. [ser. 3], **1813**: 240–257.
1815   Empidiae Sveciae. Pp. 1–16 [cont.]. Lundae [=Lund].

    The entire work consists of 34 pp., Lund, 1815–1816.

1816a   Empidiae Sveciae [concl.]. Pp. 17–24, 25–34. Lundae [=Lund].
1816b   Syrphici Sveciae. Pp. 1–14, 15–22 [cont.]. Lundae [=Lund].

    The entire work consists of 62 pp., Lund, 1816–1817.

1817a   Syrphici Sveciae [concl.]. Pp. 23–30, 31–42, 43–62. Lundae [=Lund].
1817b   Stratiomydae Sveciae. 14 pp. Lundae [=Lund].
1817c   Scenopinii et Conopsariae Sveciae. 14 pp. Lundae [=Lund].
1817d   Beskrifning öfver de i Sverigefunna fluge arter, som kunna föras till slägtet *Musca*. Första afdelningen. K. Vetensk. Acad. Handl. [ser. 3], **1816**: 226–254.
1819   Scatomyzides Sveciae. 10 pp. Lundae [=Lund].
1820a   Oscinides Sveciae. 10 pp. Lundae [=Lund].
1820b   Sciomyzides Sveciae. 16 pp. Lundae [=Lund].
1820c   Ortalides Sveciae. 34 pp. Lundae [=Lund].
1820d   Opomyzides Sveciae. 12 pp. Lundae [=Lund].
1820e   Heteromyzides Sveciae. 10 pp. Lundae [=Lund].
1820f   Rhizomyzides Sveciae. 10 pp. Lundae [=Lund].
1820g   Monographia Muscidum Sveciae. Pp. 1–12, 13–24, 25–40 [cont.]. Lundae [=Lund].

    The entire work consists of 94 pp., Lund, 1820–1825.

1821   Monographia Muscidum Sveciae [cont.]. Pp. 41–48 [cont.]. Lundae [=Lund].
1823a   Agromyzides Sveciae. 10 pp. Lundae [=Lund].
1823b   Phytomyzides et Ochtidiae Sveciae. 10 pp. Lundae [=Lund].

Fallén, C. F.—Continued.
- 1823c Monographia Dolichopodum Sveciae. Pp. 1–16, 17–24. Lundae [=Lund].
- 1823d Hydromyzides Sveciae. 12 pp. Lundae [=Lund].
- 1823e Geomyzides Sveciae. 8 pp. Lundae [=Lund].
- 1823f Monographia Muscidum Sveciae [cont.]. Pp. 49–56, 57–64 [cont.]. Lundae [=Lund].
- 1824–1825 Monographia Muscidum Sveciae [concl.]. Pp. 65–72, *1824;* 73–80, 81–94, *1825.* Lundae [=Lund].
- 1826 Supplementum Dipterorum Sveciae. Pp. 1–8, 9–16. Lundae [=Lund].

Fattig, P. W.
- 1949 The Larvaevoridae (Tachinidae) or parasitic flies of Georgia. Emory Univ. (Ga.) Mus. Bul. **8:** 1–40.

Felt, E. P. (See also Johannsen, 1908; Needham, 1908, 1925; Pritchard and Felt; Weiss and West, 1921; Woods, 1916)
- [1898a] Additional notes on *Sciara.* The fungus gnats (Ord. Diptera, Fam. Mycetophilidae). Pp. 223–228, pl. 6. *In* Lintner, J. A., Reports on the injurious and other insects of the State of New York. Rpt. 12: for 1896, pp. 161–399, 10 figs., 15 pls. Albany, N.Y., "1897."

    Also published in N.Y. State Mus. Ann. Rpt. (1896) **50 (1):** 223–228, *1898.*
    Dated by Felt, 1899: 398.
    See Lintner, 1883, for details of Lintner's Reports.

- [1898b] *Phora albidihalteris* n. sp. A mushroom *Phora* (Ord. Diptera: Fam. Phoridae). Pp. 228–229. *In* Lintner, J. A., Reports on the injurious and other insects of the State of New York. Rpt. 12: for 1896, pp. 161–399, 10 figs., 15 pls. Albany, N.Y., "1897."

    Also published in N.Y. State Mus. Ann. Rpt. (1896) **50 (1):** 228–229, *1898.*
    Dated by Felt, 1899: 398.

- 1899 Supplement to the 14th report of the State Entomologist, 1898. Memorial of life and entomologic [sic] work of Joseph Albert Lintner, Ph.D., State Entomologist, 1874–98. N.Y. State Mus., Bul. **5:** 300–611 (=Bul. 24: [Ent. 6]).

    Also published in N.Y. State Mus. Ann. Rpt. (1898) **52 (vol. 1, app.):** 300–611, *1900.*

Felt, E. P.—Continued
1904   Mosquitoes or Culicidae of New York State. N.Y. State Mus. Bul. **79** [=Ent. 22]: 239–400, 113 figs., 57 pls. (=N.Y. State Univ. Bul. 323.)

> The Appendix is paged 391a–f.
> Also published in N.Y. State Mus. Ann. Rpt. (1903) **57** (vol. 1, pt. 2, app. 5): 239–400, *1905*.

1905a   *Culex brittoni* n. sp. Ent. News **16**: 79–80.

1905b   Studies in Culicidae. Pp. 442–497, figs. 1–21, pls. 1–19. *In his* 20th report of the State Entomologist on injurious and other insects of the State of New York, 1904. N.Y. State Mus. Bul. **97** [=Ent. 24]: 359–564, 24 figs., 19 pls. (=N.Y. State Educ. Dept. Bul. 357.)

> Also published in N.Y. State Mus. Ann. Rpt. (1904) **58** (vol. 5, app. 7): 442–497, *1906*.

1906   Studies in Cecidomyiidae. Pp. 116–132, figs. 15–48. *In his* 21st report of the State Entomologist on injurious and other insects of the State of New York, 1905. N.Y. State Mus. Bul. **104** [=Ent. 26]: 49–186, 48 figs., 9 pls. (=N.Y. State Educ. Dept. Bul. 382.)

> Also published in N.Y. State Mus. Ann. Rpt. (1905) **59** (vol. 2, app. 5): 116–132, *1907*.

1907a   *Cecidomyia acarivora* n. sp. Ent News **18**: 242.

1907b   Appendix: New species of Cecidomyiidae. Pp. 97–165. *In his* 22nd report of the State Entomologist on injurious and other insects of the State of New York, 1906. N.Y. State Mus. Bul. **110** [=Ent. 28]: 39–186, 2 figs., 3 pls. (=N.Y. State Educ. Dept. Bul. 403.)

> Also published in N.Y. State Mus. Ann. Rpt. (1906) **60** (vol. 1, app. 3): 97–165, *1908*.

1908a   *Contarinia gossypii* n. sp. Ent. News **19**: 210–211.

1908b   Appendix D. Pp. 286–422, figs. 29–49, pls. 33–44. *In his* 23d report of the State Entomologist on injurious and other insects of the State of New York, 1907. N.Y. State Mus., Mus. Bul. **124**: 5–541, 49 figs., 44 pls. (=N.Y. State Univ. Educ. Dept. Bul. 433.)

> Also published in N.Y. State Mus. Ann. Rpt. (1907) **61** (vol. 2, app. 2): 286–422, *1908*.

1909a   Additional rearings in Cecidomyiidae. Jour. Econ. Ent. **2**: 286–293.

Felt, E. P.—Continued

1909b Injurious insects. Pp. 13–40, figs. 1–12. *In his* 24th report of the State Entomologist on injurious and other insects of the State of New York, 1908. N.Y. State Mus., Mus. Bul. **134:** 1–206, 22 figs., 17 pls. (=N.Y. State Univ. Educ. Dept. Bul. 455.)

    Also published in N.Y. State Mus. Ann. Rpt. (1908) **62** (vol. **2, app. 4, bul. 134**): 13–40, *1909*.

1910a Two new Cecidomyiidae. Ent. News **21:** 10–12.

1910b Gall midges of *Aster, Carya, Quercus* and *Salix*. Jour. Econ. Ent. **3:** 347–356.

1911a *Endaphis* Kieff. in the Americas (Dipt.). Ent. News **22:** 128–129.

1911b Four new gall midges (Dipt.). Ent. News **22:** 301–305.

1911c *Rhopalomyia grossulariae* n. sp. Jour. Econ. Ent. **4:** 347.

1911d New species of gall midges. Jour. Econ. Ent. **4:** 476–484, 546–559, 1 fig.

1911e A generic synopsis of the Itonidae. N.Y. Ent. Soc. Jour. **19:** 31–62.

1911f Appendix: *Miastor americana* Felt, an account of pedogenesis. Pp. 82–104, figs. 7–10, pls. 22–35. *In his* 26th report of the State Entomologist on injurious and other insects of the State of New York, 1910. N.Y. State Mus., Mus. Bul. **147:** 5–180, 10 figs., 35 pls. (=N.Y. State Univ. Educ. Dept. Bul. 490.)

    Also published in N.Y. State Mus. Ann. Rpt. (1910) **64** (vol. **2, app. 3**): 82–104, *1912*.

1912a *Diarthronomyia californica* n. sp. (Diptera, Itonidae). Pomona Col. Jour. Ent. **4:** 752.

1912b Notes for the year. Pp. 98–123, text-figs. 3–6, pl. 8, figs. 3–4, pls. 25–27. *In his* 27th report of the State Entomoligist on injurious and other insects of the State of New York, 1911. N.Y. State Mus., Mus. Bul. **155:** 1–198, 6 figs., 27 pls. (=N.Y. State Univ. Educ. Dept. Bul. 510.)

    Also published in N.Y. State Mus. Ann. Rpt. (1911) **65** (vol. **2, app. 3, bul. 155**): 98–123, *1913*.

1912c New Itonididae (Dipt.). N.Y. Ent. Soc. Jour. **20:** 102–107.

1912d New gall midges or Itonidae (Dipt.). N.Y. Ent. Soc. Jour. **20:** 146–156.

Felt, E. P.—Continued
1912e Studies in Itonididae. N.Y. Ent. Soc. Jour. **20:** 236–248.
1912f Observations on the identity of the wheat midge. Jour. Econ. Ent. **5:** 286–289.
1912g *Itonida inopis* O. S. Jour. Econ. Ent. **5:** 368–369.
1912h *Arthrocnodax occidentalis* n. sp. (Dipt.). Jour. Econ. Ent. **5:** 402.
1912i The gall midge fauna of western North America. Pomona Col. Jour. Ent. **4:** 753–757.
1913a Three new gall midges (Diptera). Canad. Ent. **45:** 304–308.
1913b Two new Canadian gall midges. Canad. Ent. **45:** 417–419.
1913c *Cystodiplosis eugeniae* n. sp. (Dipt.). Ent. News **24:** 175–176.
1913d *Itonida anthici* n. sp. (Dipt.). Jour. Econ. Ent. **6:** 278–279.
1913e *Arthrocnodax carolina* n. sp. Jour. Econ. Ent. **6:** 488–489.
1913f Descriptions of gall midges (Diptera). N.Y. Ent. Soc. Jour. **21:** 213–219.
1913g Appendix: A study of gall midges. Pp. 127–226, figs. 16–79, pls. 10–14. *In his* 28th report of the State Entomologist on injurious and other insects of the State of New York, 1912. N.Y. State Mus., Mus. Bul. **165:** 5–265, 79 figs., 14 pls. (=N.Y. State Univ. Bul. 547.)

Also published in N.Y. State Mus. Ann. Rpt. (1912) **66** (**vol. 2, app. 3**): 127–226, *1914*.

1913h The gall midge fauna of New England. Psyche **20:** 133–147.
1913i *Didactylomyia capitata* sp. nov. Psyche **20:** 174.
1914a *Hormomyia bulla*, n. sp. Canad. Ent. **46:** 286–287.
1914b New gall midges (Itonididae). Insecutor Inscitiae Menstruus **2:** 117–123.
1914c *Aplonyx sarcobati* n. sp. Jour. Ent. and Zool. **6:** 93–94.
1914d Descriptions of gall midges. N.Y. Ent. Soc. Jour. **22:** 124–134.
1914e Gall midges as forest insects. Ottawa Nat. **28:** 76–79.
1914f Additions to the gall midge fauna of New England. Psyche **21:** 109–114.
1915a New North American gall midges. Canad. Ent. **47:** 226–232.
1915b *Mycodiplosis macgregori* n. sp. Jour. Econ. Ent. **8:** 149.
1915c New gall midges. Jour. Econ. Ent. **8:** 405–409.

Felt, E. P.—Continued

1915d   Appendix: A study of gall midges. II. Pp. 79–213, figs. 14–36, pls. 1–15. *In his* 29th report of the State Entomologist on injurious and other insects of the State of New York, 1913. N.Y. State Mus., Mus. Bul. **175**: 5–257, 36 figs., 16 pls. (=N.Y. State Univ. Bul. 589.)

> Also published in N.Y. State Mus. Ann. Rpt. (1913) **67** (app. 3): 79–213, *1915*.

1915e   Appendix: A study of gall midges. III. Pp. 127–288, 101 figs., pls. 2, 4–19. *In his* 30th report of the State Entomologist on injurious and other insects of the State of New York, 1914. N.Y. State Mus., Mus. Bul. (1916) **180**: 5–336, 101 figs., 19 pls. (=N.Y. State Univ. Bul. 606.)

> Also published in N.Y. State Mus. Ann. Rpt. (1914) **68** (app. 2): 127–288, *1916*.

1915f   New genera and species of gall midges. U.S. Natl. Mus. Proc. **48**: 195–211, 15 figs.

1916a   New gall midges. Canad. Ent. **48**: 29–33.

1916b   Gall midges of certain Chenopodiaceae (Dip.). Ent. News **27**: 201–203.

1916c   New North American gall midges (Dipt.). Ent. News **27**: 412–417.

1916d   31st report of the State Entomologist on injurious and other insects of the State of New York, 1915. N.Y. State Mus. Bul. **186**: 7–215, 39 figs., 18 pls.

> Includes "A study of gall midges. IV", pp. 101–172, 39 figs., pls. 14–18.
> Also published in N.Y. State Mus. Ann. Rpt. (1915) **69** (vol. 2, app. 3): 7–215, *1918*.

1916e   New western gall midges. N.Y. Ent. Soc. Jour. **24**: 175–196.

1917a   *Asphondylia websteri* n. sp. Jour. Econ. Ent. **10**: 562.

1917b   New gall midges. N.Y. Ent. Soc. Jour. **25**: 193–196.

1918a   Appendix: A study of gall midges. V. Pp. 101–252, figs. 2–53, pls. 3–8. *In his* 32d report of the State Entomologist on injurious and other insects of the State of New York, 1916. N.Y. State Mus. Bul. (1917) **198**: 7–276, 54 figs., 8 pls.

> Also published in N.Y. State Mus. Ann. Rpt. (1916) **70** (vol. 2, app. 3): 101–252, *1919*.

1918b   Appendix: A study of gall midges. VI. Pp. 76–205, figs. 10–73, pls. 4–12. *In his* 33rd report of the State Ento-

Felt, E. P.—Continued

    mologist on injurious and other insects of the State of New York, 1917. N.Y. State Mus. Bul. (1917) **202:** 7–240, 82 figs., 12 pls.
        Also published in N.Y. State Mus. Ann. Rpt. (1917) **71** (vol. 2, app. 2, bul. 202): 76–205, *1919*.

1918c New gall midges (Dipt.). Jour. Econ. Ent. **11:** 380–384.

1919 Five non-gall-making midges (Dip., Cecidomyidae). Ent. News **30:** 219–223.

1920 New gall midges or Itonididae from the Adirondacks. N.Y. Ent. Soc. Jour. (1919) **27:** 277–292.

1921a The number of antennal segments in gall midges and a new species. Brooklyn Ent. Soc. Bul. **16:** 93–95.

1921b Observations on *Johnsonomyia* Felt with the description of a new species. Canad. Ent. **53:** 96.

1921c Appendix: A study of gall midges. VII. Pp. 81–240, figs. 3–56, pls. 8–20. *In his* 34th report of the State Entomologist on injurious and other insects of the State of New York, 1918. N.Y. State Mus. Bul. (1920) **231/232:** 7–288, 56 figs., 20 pls.

1921d Three new sub-tropical gall midges (Itonididae, Dipt.). Ent. News **32:** 141–143.

1921e New species of reared gall midges (Itonididae). N.Y. Ent. Soc. Jour. **29:** 115–118.

1922a A new gall midge on rushes (Dipt., Cecidomyiidae). Ent. News **33:** 166–168.

1922b A new and remarkable fig midge. Fla. Ent. **6:** 5–6.

1922c A new cecidomyiid parasite of the white fly. U.S. Natl. Mus. Proc. **61 (23):** 1–2.

1925a New gall midges. Jour. Ent. and Zool. **17:** 15.

1925b Key to gall midges. (A resumé of Studies I–VII. Itonididae). N.Y. State Mus. Bul **257:** 1–239, 57 figs., 8 pls.

1926a New gall midges from New England. Boston Soc. Nat. Hist. Occas. Papers **5:** 207–208.

1926b A new spruce gall midge. Canad. Ent. **58:** 229–230; [Editor's note, p. 240].

1926c New non-gall making Itonididae (Diptera). Canad. Ent. **58:** 265–268.

1926d A new predaceous midge on roses (Dipt.: Cecidomyiidae). Ent. News **37:** 141.

1926e Three western gall midges. Jour. Ent. and Zool. **18:** 79–81.

1928 A new western gall midge. Jour. Ent. and Zool. **20:** 58.

1932 A new predaceous gall midge for California. Pan-Pacific Ent. **8:** 167–168.

Felt, E. P.—Continued
- 1933 A new *Lestodiplosis*. Psyche **40**: 113–114.
- 1934a New gall midges. Brooklyn Ent. Soc. Bul. **29**: 77–78.
- 1934b A new gall midge on fig (Diptera: Itonididae). Ent. News **45**: 131–133.
- 1934c Two western species of *Asphondylia*. Jour. Ent. and Zool. **26**: 34.
- 1935a A new melon gall midge. Brooklyn Ent. Soc. Bul. **30**: 79–80.
- 1935b *Trisopsis* in the United States (Dipt., Itonididae or Cecidomyiidae). Ent. News **46**: 75–77.
- 1935c New species of gall midges from Texas. Kans. Ent. Soc. Jour. **8**: 1–8.
- 1935d A new gall midge. N.Y. Ent. Soc. Jour. **43**: 47.
- 1935e A gall midge on pine cones. N.Y. Ent. Soc. Jour. **43**: 48–49.
- 1936a Two new cockle burr midges (Diptera: Cecidomyiidae). Ent. News **47**: 231–233.
- 1936b New midges on pine and grass. N.Y. Ent. Soc. Jour. **44**: 7–9.
- 1939a A new juniper midge (Diptera: Cecidomyiidae). Ent. News **50**: 159–160.
- 1939b A new gall midge on *Rhododendron*. N.Y. Ent. Soc. Jour. **47**: 41–42.
- 1940 Plant galls and gall makers. 364 pp., 344 figs., 41 pls. Ithaca, N.Y.
- 1958 Introduction; II. Subfamily Heteropezinae; III. Subfamily Itonidinae (=Cecidomyiinae). Pp. 47–49; 89–92; 93–206. *In* Pritchard, A. E., and Felt, E. P., Guide to the insects of Connecticut. Part VI. The Diptera or true flies of Connecticut. Sixth fascicle [part]. Itonididae (Cecidomyiidae). Conn. State Geol. and Nat. Hist. Survey, Bul. **87**: 47–206, pls. 10–15.

———, and **Young, D. B.**
- 1904 Importance of isolated rearings from culicid larvae. Science n. ser., **20**: 312–313.

Fernald, C. H. (See Forbush and Fernald)

Ferris, G. F. (See also Rees and Ferris)
- 1916 Some ectoparasites of bats (Dipt.). Ent. News **27**: 433–438, 1 fig., 2 pls.
- 1923 Observations on the larvae of some Diptera Pupipara, with description of a new species of Hippoboscidae. Parasitology **15**: 54–58, 4 figs.

Ferris, G. F.—Continued
1924 The New World Nycteribiidae (Diptera Pupipara). Ent. News **35**: 191–199, 1 fig., 1 pl.
1927 Some American Hippoboscidae (Diptera Pupipara). Canad. Ent. **59**: 246–251, 4 figs.

———, **and Cole, F. R.**
1922 A contribution to the knowledge of the Hippoboscidae (Diptera Pupipara). Parasitology **14**: 178–205, 20 figs.

Fischer von Waldheim, G.
1819 Programme d'invitation à la séance publique de la Société Impérial des Naturalistes, qui aura lieu le 15 Décembre, contenant une notice d'une mouche carnivore. 11 pp., 1 pl. Moscou.

Fisher, E. G.   (See also Shaw and Fisher)
1934 Four new species of Mycetophilidae (Diptera). Canad. Ent. **66**: 276–278, 4 figs.
1937 New North American fungus gnats (Mycetophilidae). N.Y. Ent. Soc. Jour. **45**: 387–401, 2 pls.
1938a North American fungus gnats. II. (Diptera: Mycetophilidae). Amer. Ent. Soc. Trans. **64**: 195–200, pl. 9.
1938b A comparative study of the male terminalia of the Mycetophilidae of nearctic America. (Abstract). Cornell Univ., Abs. Theses **1937**: 219–222.
1940 New Mycetophilidae from North Carolina (Diptera). Ent. News **51**: 243–247.
1941 Distributional notes and keys to American Ditomyiinae, Diadocidiinae and Ceroplatinae with descriptions of new species (Diptera: Mycetophilidae). Amer. Ent. Soc. Trans. **67**: 275–301, pls. 23–24.
1946 The genus *Monoclona* Mik (Diptera: Mycetophilidae). Acad. Nat. Sci. Phila. Notulae Nat. **175**: 1–4, 5 figs.

Fitch, A.
1845a Insects injurious to vegetation. No. 2. Insects of the genus *Cecidomyia*, including the Hessian and wheat-fly. Amer. Quart. Jour. Agr. and Sci. **1**: 255–269, 1 pl.
1845b Insects injurious to vegetation. No. 3. The wheat fly. Amer. Quart. Jour. Agr. and Sci. **2**: 233–264, 1 pl.
1847 Winter insects of eastern New York. Amer. Jour. Agr. and Sci. **5**: 274–284.
1855 Report on the noxious, beneficial and other insects of the State of New York. [I]. N.Y. State Agr. Soc. Trans. (1854) **14**: 705–880, 28 figs.

  Also published separately in his "Reports on noxious, beneficial and other insects of the State of New York," Rpts. 1/2, pp. 1–176, Albany, *1856*.

Fitch, A.—Continued
1856a Report on the noxious, beneficial and other insects of the State of New York. [II]. N.Y. State Agr. Soc. Trans. (1855) **15**: 409–559, 10 figs., 4 pls.

> Also published separately in his "Reports on noxious, beneficial and other insects of the State of New York," Rpts. 1/2, pp. 177–336, Albany, *1856*.

1856b Third report on the noxious and other insects of the State of New York. N.Y. State Agr. Soc. Trans. **16**: 315–490, 1 fig., 4 pls.

> Also published separately in his "Reports on the noxious, beneficial and other insects of the State of New York," Rpts. 3/5, [Pt. I (=3rd rpt.)], pp. 1–172, Albany, *1859*.

1859 Fifth report on the noxious and other insects of the State of New York. N.Y. State Agr. Soc. Trans. (1858) **18**: 781–854, 9 figs.

> Also published separately in his "Reports on the noxious, beneficial and other insects of the State of New York", Rpts. 3/5, [Pt. 3 (=5th rpt.)], pp. 1–74, Albany, *1859*.

1861 Sixth report on the noxious and other insects of the State of New York. N.Y. State Agr. Soc. Trans. (1860) **20**: 745–868, 2 pls.

> Also published separately in his "Reports on the noxious, beneficial and other insects of the State of New York," [Rpts. 6/9], pp. 1–126, Albany, *1865*.

1864 The Nebraska bee-killer. Country Gent. **24**: 63.
1872 Fourteenth report on the noxious, beneficial and other insects of the State of New York. N.Y. State Agr. Soc. Trans. (1870) **30**: 355–381.

Fittkau, E. J.
1957 *Thienemannimyia* und *Conchapelopia*, zwei neue Gattungen innerhalb der *Ablabesmyia-costalis*-Gruppe (Diptera, Chironomidae). (Chironomidenstudien VII). Arch. f. Hydrobiol. n. ser., **53**: 313–322, 11 figs.
1962 Die Tanypodinae (Diptera: Chironomidae). (Die Tribus Anatopyniini, Macropelopiini und Pentaneurini). Vol. 6, 453 pp., 409 figs. *In* Deutsche Akademie der Wissenschaften zu Berlin, Abhandlungen zur Larvalsystematik der Insekten [q.v.]. Berlin.

Fluke, C. L., Jr. (See also Curran and Fluke; Hull and Fluke)
1922 Syrphidae of Wisconsin. Wis. Acad. Sci., Arts, Letters, Trans. **20**: 215–253, 2 pls.
1929 The known predacious and parasitic enemies of the pea aphid in North America. Wis. Agr. Expt. Sta. Res. Bul. **93**: 1–47, 2 figs., 6 pls.

Fluke, C. L., Jr.—Continued
- 1930 High-altitude Syrphidae with descriptions of new species (Diptera). Ent. Soc. Amer. Ann. **23**: 133–144, 5 figs.
- 1931 Notes on certain *Syrphus* flies related to *Xanthogramma* (Diptera Syrphidae) with descriptions of two new species. Wis. Acad. Sci., Arts, Letters, Trans. **26**: 289–309, 2 pls.
- 1933 Revision of the *Syrphus* flies of America north of Mexico (Diptera, Syrphidae, *Syrphus* s. l.). Part I. Wis. Acad. Sci., Arts, Letters, Trans. **28**: 63–127, 3 pls.
- 1935 Revision of the *Epistrophe* flies of America north of Mexico (Diptera, Syrphidae). Ent. Amer. **15**: 1–56, 3 pls.
- 1939 New Syrphidae (Diptera) from Central and North America. Ent. Soc. Amer. Ann. **32**: 365–375, 1 pl.
- 1942 Revision of the neotropical Syrphini related to *Syrphus* (Diptera, Syrphidae). Amer. Mus. Nat. Hist., Amer. Mus. Novitates **1201**: 1–24, 51 figs.
- 1949 Some Alaskan syrphid flies, with descriptions of new species. U.S. Natl. Mus. Proc. **100**: 39–54, 2 figs.
- 1950 The male genitalia of *Syrphus*, *Epistrophe* and related genera (Diptera, Syrphidae). Wis. Acad. Sci., Arts, Letters, Trans. **40**: 115–148, 10 pls.
- 1952 The *Metasyrphus* species of North America (Syrphidae). Amer. Mus. Nat. Hist., Amer. Mus. Novitates **1590**: 1–27, 54 figs.
- 1953a New Syrphidae from North America. Kans. Ent. Soc. Jour. **26**: 125–129.
- 1953b Some syrphid fly synonomy [sic]. Ent. News **64**: 208–209.
- 1954 Two new North American species of Syrphidae, with notes on *Syrphus* (Diptera). Amer. Mus. Nat. Hist., Amer. Mus. Novitates **1690**: 1–10, 20 figs.
- 1958 A study of the male genitalia of the Melanostomini (Diptera-Syrphidae). Wis. Acad. Sci., Arts, Letters, Trans. (1957) **46**: 261–279, 6 pls.

———, **and Hull, F. M.**
- (1946) Syrphid flies of the genus *Cheilosia*, subgenus *Chilomyia* in North America (Part II). Wis. Acad. Sci., Arts, Letters, Trans. (1944) **36**: 327–347, 17 figs.
- 1947 The *Cartosyrphus* flies of North America (Syrphidae). Wis. Acad. Sci., Arts, Letters, Trans. (1945) **37**: 221–263, 3 pls.

———, **and Weems, H. V., Jr.**
- 1956 The Myoleptini of the Americas (Diptera, Syrphidae). Amer. Mus. Nat. Hist., Amer. Mus. Novitates **1758**: 1–23, 33 figs.

Foote, B. A.
1959a   A new species of *Pteromicra* reared from land snails, with a key to the nearctic species of the genus (Diptera, Sciomyzidae). Ent. Soc. Wash. Proc. **61**: 14–16, 3 figs
1959b   Biology and life history of the snail-killing flies belonging to the genus *Sciomyza* Fallén (Diptera, Sciomyzidae). Ent. Soc. Amer. Ann. **52**: 31–43, 19 figs.
1961a   The marsh flies of Idaho and adjoining areas (Diptera: Sciomyzidae). Amer. Midland Nat. **65**: 144–167, 30 figs.
1961b   A new species of *Antichaeta* Haliday, with notes on other species of the genus (Diptera: Sciomyzidae). Ent. Soc. Wash. Proc. **63**: 161–164, 3 figs.

Foote, R. H. (See also Blanc and Foote; Christenson and Foote; McFadden and Foote; Steyskal and Foote)
1953   A new gall midge infesting holly (Diptera: Itonididae). Ent. News **64**: 197–201, 11 figs.
1956   Gall midges associated with cones of western forest trees (Diptera: Itonididae). Wash. Acad. Sci. Jour. **46**: 48–57, 9 figs.
1958   The genus *Euarestoides* in the United States and Mexico (Diptera, Tephritidae). Ent. Soc. Amer. Ann. **51**: 288–293, 9 figs.
1959a   A new North American species of *Tephritis*, with some observations on its generic position (Diptera, Tephritidae). Brooklyn Ent. Soc. Bul. **54**: 13–17, 3 figs.
1959b   Notes on the genus *Euleia* Walker in North America (Diptera: Tephritidae). Kans. Ent. Soc. Jour. **32**: 145–150, 8 figs.
1959c   A new synonymy in the Tephritidae (Diptera). Ent. Soc. Wash. Proc. **61**: 59.
1960a   *Paraterellia*, a new genus of Tephritidae from the western United States (Diptera). Ent. Soc. Amer. Ann. **53**: 121–125, 7 figs.
1960b   The genus *Trypeta* Meigen in America north of Mexico (Diptera, Tephritidae). Ent. Soc. Amer. Ann. **53**: 253–260, 14 figs.
1960c   A new North American fruit fly genus, *Procecidocharoides* (Diptera: Tephritidae). Ent. Soc. Amer. Ann. **53**: 671–675, 4 figs.
1960d   A new tephritid genus, *Rhagoletoides*, with notes on its distribution and systematic position (Diptera, Tephritidae). Ent. News **71**: 145–149.

Foote, R. H.—Continued
1960e The genus *Tephritis* Latreille in the Nearctic Region north of Mexico: Descriptions of four new species and notes on others (Diptera: Tephritidae). Kans. Ent. Soc. Jour. **33**: 71–85, 4 figs.
1960f The species of the genus *Neotephritis* Hendel in America north of Mexico (Diptera, Tephritidae). N.Y. Ent. Soc. Jour. **68**: 145–151, 3 figs.
1960g A revision of the genus *Trupanea* in America north of Mexico (Diptera, Tephritidae). U.S. Dept. Agr. Tech. Bul. **1214**: 1–29, 25 figs.
1960h Notes on some North American Tephritidae, with descriptions of two new genera and two new species (Diptera). Biol. Soc. Wash. Proc. **73**: 107–118, 4 figs.
1962 The types of North American Tephritidae in the Snow Museum, the University of Kansas (Diptera). Kans. Ent. Soc. Jour. **35**: 170–179.

———, and **Blanc, F. L.**
1959 A new genus of North American fruit flies. Pan-Pacific Ent. **35**: 149–156, 3 figs.

———, and **Pratt, H. D.**
1954 The *Culicoides* of the eastern United States (Diptera, Heleidae). U.S. Dept. Health, Educ., Welfare, U.S. Publ. Health Serv., Publ. Health Monog. **18**: 1–53, 11 pls. (=Publ. Health Serv. Pub. 12.)

Forbes, S. A.
1885 Notes on insects injurious to wheat. Pp. 34–69, 1 text-fig., pl. 1, figs. 5–6, pls. 2–5. *In his* Third annual report of S. A. Forbes for the year 1884. Noxious and Beneficial Insects of State of Ill., State Ent. Rpt. (1884) **14**: 1–136, 1 fig., 12 pls.
1890 The meadow maggots or leather-jackets (*Tipula bicornis* Loew, MSS., et al.). Pp. 78–83, pl. 6, fig. 4. *In his* Fifth report of S. A. Forbes for the years 1887 and 1888. Noxious and Beneficial Insects of State of Ill., State Ent. Rpt. **16**: 1–104, 6 pls.
1912 On the black-flies and buffalo-gnats (*Simulium*) as possible carriers of Pellagra in Illinois. Noxious and Beneficial Insects of State of Ill., State Ent. Rpt. **27**: 21–55, figs. 1–25.
  In the sixteenth report of S. A. Forbes.

Forbush, E. H., and Fernald, C. H.
1896 The gypsy moth. *Porthetria dispar* (Linn.). A report of the work of destroying the insect in the Commonwealth of Massachusetts, together with an account of its history and habits both in Massachusetts and Europe. 495 pp. + c pp. Massachusetts State Board of Agriculture, Boston.
> Part I, pp. 1–252, is by Forbush; Part II, pp. 253–495, is by Fernald. Pages i-c of appendices are included at the end of the volume.

Forest, J. (See Stoll et al.)

Forster, J. R.
1771 Novae species insectorum. Centuria I. 100 pp. London.
> This is only Centuria ever published.

Fox, I.
1946 Two new biting midges or *Culicoides* from western United States (Diptera, Ceratopogonidae). Ent. Soc. Wash. Proc. **48**: 244–246, 2 figs.
1948 *Hoffmania*, a new subgenus in *Culicoides* (Diptera: Ceratopogonidae). Biol. Soc. Wash. Proc. **61**: 21–28, 1 pl.
1955 A catalogue of the bloodsucking midges of the Americas (*Culicoides*, *Leptoconops*, and *Lasiohelea*) with keys to the subgenera and nearctic species, a geographic index, and bibliography. Puerto Rico Univ. Jour. Agr. **39**: 214–285.

———, **and Hoffman, W. A.**
1944 New neotropical biting sandflies of the genus *Culicoides* (Diptera: Ceratopogonidae). Puerto Rico Jour. Publ. Health and Trop. Med. **20**: 108–111, 5 figs.

Fox, R. M., and Stabler, R. M.
1953 *Basilia calverti* n. sp. (Diptera: Nycteribiidae) from the interior long-legged bat. Jour. Parasitol. **39**: 22–27, 2 figs.

Fox, W. J., sec.
1905 (Minutes of the meeting of the Feldman Collecting Social, Philadelphia, Dec. 21, 1904.) Pp. 61–62. *In* Skinner, H., ed., Doings of societies. Ent. News **16**: 55–64.

França, C., and Parrot, L.
1921 Essai de classification des Phlébotomes. Inst. Pasteur de l'Afrique du Nord, Arch. **1**: 279–284, 6 figs.

Frauenfeld, G. R. von
1867 Zoologische Miscellen. XI. K.-k. Zool.-Bot. Gesell. Wien, Verhandl. **17** (**Abhandl.**): 425-502, pl. 12.

Fredeen, F. J. H. (See Shewell and Fredeen)
Freeborn, S. B. (See also Boyd, 1949)
1926   A new chaoborid gnat, *Chaoborus lacustris* sp. nov. (Chaoboridae, Diptera). Pan-Pacific Ent. **2**: 161–163.
——, **and Bohart, R. M.**
1951   The mosquitoes of California. Calif. Insect Survey, Bul. **1**: 25–78, 49 figs.
Freeman, C. C.
1958   *Liriomyza guytona*: A new species of agromyzid leaf miner. Ent. Soc. Amer. Ann. **51**: 344–345, 1 fig.
Freeman, P.
1955–   A study of the Chironomidae (Diptera) of Africa south of
1956   the Sahara. Part I; Part II. Brit. Mus. (Nat. Hist.) Bul., Ent. **4**: 1–67, 15 figs., 1 pl., *1955;* 285–366, 17 figs., *1956*.
1957   A study of the Chironomidae (Diptera) of Africa south of the Sahara. Part III. Brit. Mus. (Nat. Hist.) Bul., Ent. **5**: 321–426, 18 figs., 1 pl.
1958   A study of the Chironomidae (Diptera) of Africa south of the Sahara. Part IV. Brit. Mus. (Nat. Hist.) Bul., Ent. **6**: 261–363, 15 figs., 2 pls.
French, G. H.
1900   A parasite the supposed cause of some cases of epilepsy. Canad. Ent. **32**: 263–264, 1 fig.
Frey, R. (See also Lundström and Frey)
1908   Über die in Finnland gefundenen Arten des Formenkreises der Gattung *Sepsis* Fall. (Dipt.). Deut. Ent. Ztschr. **1908**: 577–588.
1915   Résultats scientifiques de l'expédition polaire Russe en 1900–1903, sous la direction du Baron E. Toll. Section E: Zoologie. Volume II, Livr. 10. Diptera Brachycera aus den arktischen Kunstengegenden Sibiriens. [Akad. Nauk S.S.S.R.] Ross. Akad. Nauk Petrograd, Zap., Fiz.-Mat. Otd. (Acad. des Sci. de Russie, Mém., Cl. Phys.-Math.) ser. **8, 29 (10)**: 1–35, 2 pls.
1918a   Mitteilungen über südamerikanische Dipteren. Finska Vetensk. Soc., Öfversigt af . . . Förhandl., Afd. A: Mat. och Naturvetensk. (1919) **60 (14)**: 1–35, 1 pl.
      Separate is dated 1918.
1918b   Beitrag zur Kenntnis der Dipterenfauna des nördl. europäischen Russlands. II. Dipteren aus Archangelsk. Soc. pro Fauna et Flora Fenn. Acta **46 (2)**: 1–32.

Frey, R.—Continued
- 1921 Studien über den Bau des Mundes der niederen Diptera Schizophora nebst Bemerkungen über die Systematik dieser Dipterengruppe. Soc. pro Fauna et Flora Fenn. Acta **48 (3)**: 1–245, 10 pls.
- 1922 Vorarbeiten zu einer Monographie der Gattung *Rhamphomyia* Meig. (Dipt., Empididae). Notulae Ent. **2**: 1–10, 33–45, 65–77.
- 1924 Die nordpaläarktischen *Tetanocera*-Arten (Dipt., Sciomyzidae). Notulae Ent. **4**: 47–53, 12 figs.
- 1925 Zur Systematik der paläarktischen Psiliden (Dipt.). Notulae Ent. **5**: 47–50.
- 1927 Zur Systematik der Diptera Haplostomata. III. Fam. Micropezidae. Notulae Ent. **7**: 65–76.
- 1942 Entwurf einer neuen Klassifikation der Mückenfamilie Sciaridae (Lycoriidae). Notulae Ent. **22**: 5–44, 12 figs.
- 1945 Tiergeographische Studien über die Dipterenfauna der Azoren. 1. Verzeichnis der bisher von den Azoren bekannten Dipteren. Finska Vetensk. Soc., Comm. Biol. **8 (10)**: 1–114, 33 figs.
- 1946a Übersicht der Gattungen der Syrphiden-Unterfamilie Syrphinae (Syrphinae+Bacchinae). Notulae Ent (1945) **25**: 152–172.
- 1946b Anteckningar om Finlands Agromyzider. Notulae Ent. **26**: 13–55.
- 1946c Neue Diptera Brachycera aus Finnland und angrenzenden Ländern. IV. Notulae Ent. **26**: 65–69, 2 figs.
- 1948 Entwurf einer neuen Klassifikation der Mückenfamilie Sciaridae (Lycoriidae). II. Die nordeuropäischen Arten. Notulae Ent. (1947) **27**: 33–112, 10 figs., 20 pls.
- 1950 Neue paläarktische *Rhamphomyia*-Arten nebst Bestimmungstabelle der *Rhamphomyia*-Subgenera. Notulae Ent. (1949) **29**: 91–119.
- 1954–1956 Empididae [cont. from E. O. Engel]. [Fam.] 28, pp. 400–432, figs. 212–217, pls. 34–36 (=lfg. 177, part), *1954*; pp. 433–480, figs. 218–223, pls. 37–42 (=lfg. 181), *1955*; pp. 481–528, figs. 224–229, pls. 43–48 (=lfg. 183), *1955*; pp. 529–576, fig. 230, pls. 49–54 (=lfg. 184), *1955*; pp. 577–639, figs. 231–246, pls. 55–57 (=lfg. 188), *1956*. *In* Lindner, E., ed., Die Fliegen der palaearktischen Region [q.v.]. Vol. 4, Pt. 4. Stuttgart.
- 1960 Studien über indoaustralische Clusiiden (Dipt.) nebst Katalog der Clusiiden. Finska Vetensk. Soc., Comm. Biol. **22 (1)**: 1–31, 6 figs.

Frick, K. E.

1951 *Liriomyza langei*, a new species of leaf miner of economic importance in California (Diptera: Agromyzidae). Pan-Pacific Ent. **27**: 81–88, 2 figs.

1952 A generic revision of the family Agromyzidae (Diptera) with a catalogue of New World species. Calif. Univ., Pubs., Ent. **8**: 339–452, 34 figs.

1953 Some additions and corrections to the species list of North American Agromyzidae (Diptera). Canad. Ent. **85**: 68–76.

1954 Three North American *Phytomyza* species closely related to *P. nigritella* Zetterstedt (Agromyzidae: Diptera). Ent. Soc. Amer. Ann. **47**: 367–374, 1 fig.

1955 Nearctic species in the *Liriomyza pusilla* complex. No. 3. *L. alliovora*, new name for the Iowa onion miner (Diptera: Agromyzidae). Kans. Ent. Soc. Jour. **28**: 88–92, 1 fig.

1956 Revision of the North American *Calycomyza* species north of Mexico (*Phytobia*: Agromyzidae: Diptera). Ent. Soc. Amer. Ann. **49**: 284–300, 1 fig., 2 pls.

1957a Nearctic species in the *Liriomyza pusilla* complex. No. 2. *L. munda* and two other species attacking crops in California (Diptera: Agromyzidae). Pan-Pacific Ent. **33**: 59–70.

1957b Nomenclatural changes and type designations of some New World Agromyzidae (Diptera). Ent. Soc. Amer. Ann. **50**: 198–205.

1958 *Liriomyza dianthi* n. sp., a new pest of carnations in California (Diptera: Agromyzidae). Ent. Soc. Wash. Proc. **60**: 1–5, 4 figs.

1959 Synopsis of the species of agromyzid leaf miners described from North America (Diptera). U.S. Natl. Mus. Proc. **108**: 347–465, 170 figs.

Friend, R. B.

1942 Guide to the insects of Connecticut. Part VI. The Diptera or true flies of Connecticut. First fascicle [part]. Taxonomy. Wing venation. Conn. State Geol. and Nat. Hist. Survey, Bul. **64**: 166–174, figs. 15–17.

Fries, B. F.

1824 Observationes entomologicae. Pars I. 20 pp., 1 pl. Lundae [=Lund].

This is the only part ever published.

Froggatt, W. W.
1919   The lantana fly (*Agromyza lantanae*). Agr. Gaz. N.S. Wales **30**: 665–668, 1 pl.

Frohne, W. C.
1952   Mosquito news from Alaska. Mosquito News **12**: 263.
1959   Predation of dance flies (Diptera: Empididae) upon mosquitoes in Alaska, with especial reference to swarming. Mosquito News **19**: 7–11.

Frolova, S. L., and Astaurov, B. L.
1929   Die Chromosomengarnitur als systematisches Merkmal. (Eine vergleichende Untersuchung der russischen und amerikanischen *Drosophila obscura* Fall.). Ztschr. f. Zellforsch. u. Mikros. Anat. (1930) **10**: 201–213, 9 figs.

Frost, S. W.   (See also Needham, Frost, and Tothill)
1919   Two species of *Pegomyia* mining the leaves of dock. Jour. Agr. Res. **16**: 229–243, 1 fig., 2 pls.
1924   A study of the leaf mining Diptera of North America. N.Y. (Cornell) Agr. Expt. Sta. Mem. **78**: 1–228, 14 pls.
1927   Three new species of *Phytomyza* (Agromyzidae, Diptera). Ent. Soc. Amer. Ann. **20**: 217–220.
1928   Notes on *Phytomyza* with a description of a new species (Diptera). Canad. Ent. **60**: 77–78.
1930   The leaf-miners of *Aquilegia*, with a description of a new species. Ent. Soc. Amer. Ann. **23**: 457–460, 6 figs.
1931   New North American Agromyzidae (Dipt.). Canad. Ent. **63**: 275–277.
1934   A new species related to *Agromyza virens* Loew (Dipt.: Agromyzidae). Ent. News **45**: 40–41.
1943   Three new species of Diptera related to *Agromyza pusilla* Meig. N.Y. Ent. Soc. Jour. **51**: 253–263, 2 pls.
1954   A new name for *Phytomyza subpusilla* Frost (Diptera). Ent. News **65**: 73.
1962   *Liriomyza archboldi*, a new species (Dipt., Agromyzidae). Ent. News **73**: 51–53, 3 figs.

Frota-Pessoa, O.
1945   Sobre o subgênero "*Hirtodrosophila*", com descrição de um nova espécie (Diptera, Drosophilidae, *Drosophila*). Rev. Brasil. de Biol. **5**: 469–483, 4 figs.
1946   *Drosophila mallochi* nom. nov. Ent. News **57**: 155.

Fulton, B. B.
1941   A luminous fly larva with spider traits (Diptera, Mycetophilidae). Ent. Soc. Amer. Ann. **34**: 289–302, 2 pls.

Fyles, T. W.
1883 Description of a dipterous parasite of *Phylloxera vastatrix*. Canad. Ent. (1882) **14**: 237–239, 1 fig.

Gabritschevsky, E.
1924 Farbenpolymorphismus und Vererbung mimetischer Varietäten der Fliege *Volucella bombylans* und anderer "hummelähnlicher" Zweiflügler. Ztschr. f. Induktive Abstam. u. Vererbungslehre **32**: 321–353, 4 figs.
1926 Convergence of coloration between American pilose flies and bumblebees (*Bombus*). Biol. Bul. **51**: 269–287, 4 pls.

Galindo, P. (See Mangabeira and Galindo)

Gambel, W., sec.
1849 [Minutes of the meeting of the Academy of Natural Sciences of Philadelphia for Aug. 21, 1849.] Acad. Nat. Sci. Phila. Proc. (1850) **4**: 194–195.

Garrett, C. B. D. (See also Gill, 1962)
1921 Notes on Helomyzidae and descriptions of new species (Diptera). Insecutor Inscitiae Menstruus **9**: 119–132.
1922a Two new Blepharoceridae (Diptera). Insecutor Inscitiae Menstruus **10**: 89–91.
1922b New species of Helomyzidae (Diptera). Insecutor Inscitiae Menstruus **10**: 175–177.
1922c New Tipulidae from British Columbia (Diptera). Ent. Soc. Wash. Proc. **24**: 58–64, 1 fig.
1923 Two new Diptera in the Canadian National Collection. Canad. Ent. **55**: 244.
1924a New American Dixidae. 7 pp. Cranbrook, B. C.
1924b Some new American Helomyzidae (Diptera). Insecutor Inscitiae Menstruus **12**: 26–34.
1924c On British Columbian Mycetophilidae (Diptera). I; II. Insecutor Inscitiae Menstruus **12**: 60–67; 159–169.
1925a Sixty-one new Diptera. 12 pp. Cranbrook, B. C.
1925b Seventy new Diptera. 16 pp. Cranbrook, B. C.

Gay, C., ed.
1844–1871 Historia fisica y politica de Chile. 28 vols. Paris and Santiago.

> The volumes of this series are divided into sections according to content. The section on "Zoologia" consists of 8 numbered volumes, by various authors, published 1847–1854. Diptera forms a part of Volume 7, 1852.
> See Blanchard, 1852.

Géhin, J. J. B.

1857   Notes pour servir à l'histoire des Insectes nuisibles à l'agriculture dans le Départment de la Moselle. No. 2: Insectes qui attaquent les blés, 38 pp. Metz.

> The entire work consists of 5 numbers, published at Metz, 1857–1860. The third and fifth were also published in Soc. d'Hist. Nat. du Dépt. de la Moselle, Bul., Vols. 8 and 9, 1857 and 1860, respectively.

Genung, W. G.

1959   Biological and ecological observations on *Mydas maculiventris* Westwood (Diptera: Mydaidae) as a predator of white grubs. Fla. Ent. **42**: 35–37.

Geoffroy, E. L.

1762   Histoire abrégée des Insectes qui se trouvent aux environs de Paris. Vol. 2, 690 pp., 11 pls. Paris.

> The entire work, which was issued anonymously, consists of 2 volumes, Paris, 1762. It was reissued in 1764, with Geoffroy given as author.

Gerhardt, R. W. (See Owen and Gerhardt)

Gerstäcker, A.

1856   Beitrag zur Kenntniss der Henopier. Stettin Ent. Ztg. **17**: 339–361.

1857   Beitrag zur Kenntniss exotischer Stratiomyiden. Linnaea Ent. **11**: 261–350, pl. 3.

1860   Beschreibung einiger ausgezeichneten neuen Dipteren aus der Familie Muscariae. Stettin Ent. Ztg. **21**: 163–208, 1 pl.

1864   Uebersicht der in der Umgegend Berlins bis jetzt beobachteten Dolichopoden. Stettin Ent. Ztg. **25**: 20–48.

1868   Systematische Uebersicht der bis jetzt bekannt gewordenen Mydaiden (Mydasii Latr.). Stettin Ent. Ztg. **29**: 65–103, 1 pl.

Gervais, [F. L.] P.

1844   Atlas de zoologie, ou collection de 100 planches comprenant 257 figures d'animaux nouveaux ou peu connus classés d'après la méthode de M. de Blainville. 32 pp., 100 pls. Paris.

Gibson, A.

1917   The entomological record, 1916. Ent. Soc. Ontario, Ann. Rpt. (1916) **47**: 137–171.

Giglio-Tos, E.
1891a Nuove specie di Ditteri del Museo Zoologico di Torino. V. [Turin Univ.] Mus. Zool. ed Anat. Comp., Bol. **6 (102):** 1–4.
> The pages are unnumbered.

1891b Diagnosi di quattro nuovi generi di Ditteri. [Turin Univ.] Mus. Zool. ed Anat. Comp., Bol. **6 (108):** 1–6, 3 figs.

1892a Ditteri del Messico. Pt. 1, 72 pp., 1 pl. Torino.
> Also published in R. Accad. delle Sci. Torino, Mem. **43** (Cl. di Sci. Fis., Mat. e Nat.): 99–168, 1 pl., *1893*.
> The entire work consists of 4 separately paged parts, Torino, 1892–1895, each of which was also published in R. Accad. delle Sci. Torino, Mem.

1892b Sui due generi Sirfidi *Rhopalosyrphus* ed *Omegasyrphus*. [Turin Univ.] Mus. Zool. ed Anat. Comp., Bol. **7 (118):** 1–3.

1892c Diagnosi di nuove specie di Ditteri. VI. Sirfidi del Messico. [Turin Univ.] Mus. Zool. ed Anat. Comp., Bol. **7 (123):** 1–7.

1893a Diagnosi di nuovi generi e di nuove specie di Ditteri. VIII. [Turin Univ.] Mus. Zool. ed Anat. Comp., Bol. **8 (147):** 1–11.

1893b Diagnosi di nuovi generi e di nuove specie di Ditteri. IX. [Turin Univ.] Mus. Zool. ed Anat. Comp., Bol. **8 (158):** 1–14.

1893c Ditteri del Messico. Pt. 2, 80 pp., 1 pl. Torino.
> Also published in R. Accad. delle Sci. Torino, Mem. **43** (Cl. di Sci. Fis., Mat. e Nat.): 321–398, 1 pl., *1893*.

1894   Ditteri del Messico. Pt. 3, 76 pp., 1 pl. Torino.
> Also published in R. Accad. delle Sci. Torino, Mem. **44** (Cl. di Sci. Fis., Mat. e Nat.): 473–546, 1 pl., *1894*.

1895a Ditteri del Messico. Pt. 4, 74 pp., 1 pl. Torino.
> Also published in R. Accad. delle Sci. Torino, Mem. **45** (Cl. di Sci. Fis., Mat. e Nat.): 1–74, 1 pl., *1896*.

1895b Mission scientifique de M. Ch. Alluaud aux Îles Séchelles (Mars–Avril–Mai 1892). 5e mémoire. Diptères. Soc. Ent. de France, Ann. **64:** 353–368.

Giles, G. M.
1900   A handbook of the gnats or mosquitoes giving the anatomy and life history of the Culicidae. 374 pp., 16 figs., 7 pls. London.

Gill, G. D.
1962   The heleomyzid flies of America north of Mexico (Diptera: Heleomyzidae). U.S. Natl. Mus. Proc. **113**: 495–603, 96 figs.

Gillette, C. P.
1890   A new cecidomyiid infesting box elder (*Negundo aceroides*). Psyche **5**: 392–393, 1 fig.
1904   Some of the more important insects of 1903. Colo., Agr. Expt. Sta. Bul. **94**: 3–15, pl. 1. (=Tech. Ser. 6.)
    Bulletin 94 is the Report of the Entomologist, 1903.

Giraud, J.
1861   Fragments entomologiques. K.-k. Zool.-Bot. Gesell. Wien, Verhandl. **11** (**Abhandl**:): 446–494, 1 pl.

Girschner, E.
1893   Beitrag zur Systematik der Musciden. Berlin. Ent. Ztschr. **38**: 297–312, 3 figs.
1897   Über die Postalar-Membran (Schüppchen, Squamulae) der Dipteren. Illus. Wchnschr. f. Ent. **2**: 534–539, 553–559, 567–571, 586–589, 603–607, 641–645, 666–670, 6 pls.
1901   Ueber eine neue Tachinide und die Scutellarbeborstung der Musciden. Wien. Ent. Ztg. **20**: 69–72, pl. 1.

Gistel, J.
1848   Naturgeschichte des Thierreichs. Für höhere Schulen. xvi+216 pp., 32 pls+4 pp. of explanation. Stuttgart.

Gjullin, C. M.   (See also Stage, Gjullin, and Yates)

———, **Sailer, R. I., Stone, A., and Travis, B. V.**
1961   The mosquitoes of Alaska. U.S. Dept. Agr., Agr. Handb. **182**: 1–98, 80 figs.

Glover, T.
1874   Manuscript notes from my journal, or illustrations of insects, native and foreign. Diptera, or two-winged flies, 120 pp., pls. 1–12+A. Washington.
    The entire work consists of 2 volumes, Washington, 1874–1876, both lithographed and printed on 1 side of paper only. The second volume is on Hemiptera. Only 45 of the first and 53 of the second volume were published. There is another work titled "Manuscript notes from my journal," by Glover, published in Washington, 1877, but with a different subtitle and subject.

Glumac, S.
(1961) Phylogenetical system of the syrphid-flies (Syrphidae, Diptera) based upon the male genitalia structure and the type of larvae. Pp. 202–206, 1 fig. *In* Strouhal, H., and Beier, M., eds., XI. Internationaler Kongress für Entomologie [q.v.]. Vol. 1, 803 pp., 305 figs., 20 pls. Wien.

> According to Prof. Max Beier *in litt.*, this volume was published in December, 1961. The date 1961 was later published in the Table of Contents of Vol. 2, 1962.

Gmelin, J. F.
[1790] Caroli a Linné, Systema naturae per regna tria naturae, . . . Ed. 13. Vol. 1: Regnum Animale, Pt. 5, pp. 2225–3020. Lipsiae [=Leipzig].

> Dated by Woodward, 1910: 1128.
> The entire work consists of 3 volumes in 10 parts, Lipsiae, 1788–1793. Volume 1, on the Animal Kingdom, consists of 7 continuously paged parts, 1788–1792.

Goddard, W. H.
1938 The description of the puparia of fourteen British species of Sphaeroceridae (Borboridae, Diptera). Soc. Brit. Ent. Trans. **5**: 235–258, 13 figs.

Godman, F. D., and Salvin, O., eds.
1879– Biologia Centrali-Americana, or, contributions to the knowl-
1915 edge of the fauna and flora of Mexico and Central America. 57 vols. London.

> This was issued in 240 parts and bound in single volumes, or in a series of separately numbered volumes, for major taxonomic groups. The Diptera comprise 3 volumes, 1886–1903: Vol. 1, 1886–1901; Vol. 2, 1888–1903; Vol. 3, 1891–1903. For a gazetteer of place names used in this work, see Selander and Vaurie, 1962.
> See Aldrich, 1901; Osten Sacken, 1886a, 1887a; Williston, 1891, 1892b, 1900–1901; Wulp, 1888–1890, 1891b, 1895–1900, 1903.

Goeldi, E. A.
1905 Os mosquitos no Pará. Pará, Mus. Goeldi (Mus. Paraense) de Hist. Nat. e Ethnog., Mem. **4**: 1–154, 144 figs., 5 pls.

Goetghebuer, M. (See also Remy, 1928)
1921 Chironomides de Belgique et spécialement de la zone des Flandres. Brussels Mus. Roy. d'Hist. Nat. de Belg. Mém. **31** [=vol. 8, no. 4]: 1–211, 233 figs., 1 pl.

Goetghebuer, M.—Continued
1922   Nouveaux matériaux pour l'étude de la faune des Chironomides de Belgique. Ann. de Biol. Lacustre **11:** 38–62, 19 figs.
1928   Diptères (Nématocères): Chironomidae. III. Chironomariae. Vol. 18, 174 pp., 275 figs. *In* Office Central de Faunistique de la Federation Française des Sociétés de Sciences Naturelles, Faune de France [q.v.]. Paris.
1932a   Diptères Chironomidae. IV. (Orthocladiinae, Corynoneurinae, Clunioninae, Diamesinae). Vol. 23, 204 pp., 315 figs. *In* Office Central de Faunistique de la Federation Française des Sociétés de Sciences Naturelles, Faune de France [q.v.]. Paris.
1932b   Ceratopogonidae et Chironomidae nouveaux ou peu connus d'Europe. Soc. Ent. de Belg. Bul. et Ann. **72:** 125–130, 9 figs.
1933   Chironomides du Groenland oriental, du Svalbard et de la Terre de François Joseph. Norges Svalbard- og Ishavs-Undersøk., Skr. om Svalbard og Ishavet **53:** 19–31, 16 figs.
1934   Catalogue des Chironomides de Belgique. V.–Chironominae. Soc. Ent. de Belg. Bul. et Ann. **74:** 391–405.

———, **and Bastin, F.**
1925   Contribution à l'étude des Sepsidae de Belgique. Soc. Ent. de Belg. Bul. et Ann. **65:** 123–137, 1 fig., 2 pls.

———, **and Lenz, F.**
1933–   Heleidae (Ceratopogonidae). [Fam.] 13a, pp. 1–48, figs.
1934   1–42, pls. 1–6 (=lfg. 77), *1933;* pp. 49–133, figs. 43–177, pls. 7–12 (=lfg. 78), *1934.* *In* Lindner, E., ed., Die Fliegen der palaearktischen Region [q.v.]. Vol. 3. Stuttgart.
1936   Tendipedidae-Pelopiinae (Chironomidae-Tanypodinae). [Fam.] 13b, pp. 1–48, figs. 1–16, pls. 1–3 (=lfg. 97); pp. 49–81, figs. 17–100, pls. 4–6 (=lfg. 100). *In* Lindner, E., ed., Die Fliegen der palaearktischen Region [q.v.]. Vol. 3. Stuttgart.
1937–   Tendipedidae-Tendipedinae (Chironomidae-Chironominae).
1938   [Fam.] 13c, pp. 1–48, figs. 1–11, pls. 1–7 (=lfg. 107), *1937;* pp. 49–72, figs. 12–24, pls. 8–14 (=lfg. 109), *1937;* pp. 73–128, figs. 25–36, pls. 15–19 (=lfg. 118), *1938* [cont.]. *In* Lindner, E., ed., Die Fliegen der palaearktischen Region [q.v.]. Vol. 3. Stuttgart.

Goetghebuer, M., and Lenz, F. (Continued)

1939a Tendipedidae-Diamesinae (Chironomidae-Diamesinae). [Fam.] 13d, pp. 1–30, figs. 1–19, pls. 1–4 (=lfg. 127). *In* Lindner, E., ed., Die Fliegen der palaearktischen Region [q.v.]. Vol. 3. Stuttgart.

1939b Tendipedidae-Podonominae (Chironomidae-Podonominae). [Fam.] 13e, pp. 1–16, figs. 1–30 (=lfg. 131, part). *In* Lindner, E., ed., Die Fliegen der palaearktischen Region [q.v.]. Vol. 3. Stuttgart.

1939c Tendipedidae-Corynoneurinae (Chironomidae-Corynoneurinae). [Fam.] 13f, pp. 1–19, figs. 1–39 (=lfg. 131, part). *In* Lindner, E., ed., Die Fliegen der palaearktischen Region [q.v.]. Vol. 3. Stuggart.

1940–1944 Tendipedidae-Orthocladiinae (Chironomidae-Orthocladiinae). [Fam.] 13g, pp. 1–24, figs. 1–9, pls. 1–3 (=lfg. 137), *1940;* pp. 25–64, figs. 10–24, pls. 4–8 (=lfg. 144), *1942;* pp. 65–112, figs. 25–64, pls. 9–14 (=lfg. 148), *1943;* pp. 113–144, figs. 65–91, pls. 15–18 (=lfg. 152), *1944* [cont.]. *In* Lindner, E., ed., Die Fliegen der palaearktischen Region [q.v.]. Vol. 3. Stuttgart.

1950a Tendipedidae-Orthocladiinae (Chironomidae-Orthocladiinae) [cont.]. [Fam.] 13g, pp. 145–208, figs. 92–104, pls. 19–23 (=lfg. 162) [cont.]. *In* Lindner, E., ed., Die Fliegen der palaearktischen Region [q.v.]. Vol. 3. Stuttgart.

1950b Tendipedidae-Clunioninae (Chironomidae-Clunioninae). [Fam.] 13 h, pp. 1–23, figs. 1–45 (=lfg. 163). *In* Lindner, E., ed., Die Fliegen der palaearktischen Region [q.v.]. Vol. 3. Stuttgart.

1954–1960 Tendipedidae-Tendipedinae (Chironomidae-Chironominae) [cont.]. [Fam.] 13c, pp. 129–168, figs. 37–121 (=lfg. 176), *1954;* pls. 20–23 (=lfg. 178), *1954;* pp. 169–200, figs. 122–135 (=lfg. 195), *1957;* pp. 201–232, figs. 236–346 (=lfg. 208), *1960* [cont.]. *In* Lindner, E., ed., Die Fliegen der palaearktischen Region [q.v.]. Vol. 3. Stuttgart.

————, **and Lindroth, C. H.**

1931 Fam. Ceratopogonidae und Chironomidae. Pp. 274–285, figs. 3–6. *In* Lindroth, C. H., Die Insektenfauna Islands und ihre Probleme. Uppsala Univ. Zool. Bidr. **13**: 105–599, 49 figs. + figs. 11–50.

Goetghebuer, M., and Tonnoir, A. L.
1920 Catalogue raisonné des Tipulidae de Belgique. Soc. Ent. de Belg. Bul. **2**: 104–112, 131–147, 5 figs., 2 pls. [cont.].
1921 Catalogue raisonné des Tipulidae de Belgique [concl.]. Soc. Ent. de Belg. Bul. **3**: 47–58, 105–125, figs. 33–43.

Goffe, E. R.
1933 Synonymic notes on the dipterous family Syrphidae. Ent. Soc. So. England, Trans. (1932) **8**: 77–83.
1944a The generic name *Leiota* (*Liota*) Rondani (Dipt., Syrphidae). Ent. Monthly Mag. **80** [=ser. 4, 5]: 28–29.
1944b The synonymy of *Zelima* Meigen, 1800, *Tubifera* Meigen, 1800, and allied genera (Dipt., Syrphidae). Ent. Monthly Mag. **80** [=ser. 4, 5]: 109–117.
1944c Some changes in generic nomenclature in Syrphidae (Diptera). Ent. Monthly Mag. **80** [=ser. 4, 5]: 128–132.
1944d The genera *Cheilosia* (*Chilosia*) Panzer, 1809, Meigen, 1822, Hoffmannsegg, *Chilomyia* Shannon, *Cartosyrphus* Bigot (Dipt., Syrphidae). Ent. Monthly Mag. **80** [=ser. 4, 5]: 238–248.
1944e On subdividing the genus *Epistrophe* Walker, 1852 (Dipt., Syrphidae), as used by Sack (in Lindner, 1930). Entomologist **77**: 135–140.
1945a The genus *Ceria* Fabricius, 1794, nec Scopoli, 1763, and the several names proposed or used in its stead (Diptera, Syrphidae). Entomologist **78**: 120–122.
1945b Note on the type-species of some genera of Syrphidae (Diptera). Soc. Brit. Ent. Jour. **2**: 276–279.
1952 An outline of a revised classification of the Syrphidae (Diptera) on phylogenetic lines. Soc. Brit. Ent. Trans. **11**: 98–124.

Gojmerac, W. L.
1956 Description of the sugar beet root maggot, *Tetanops myopaeformis* (von Roder), with observations on reproductive capacity. Ent. News **67**: 203–210, 13 figs.

Goode, G. Brown
1889 Report upon the condition and progress of the U.S. National Museum during the year ending June 30, 1887. Smithsn. Inst. Ann. Rpt. **1887 (2)**: 1–62.

Goureau, C. C.
1851 Mémoire pour servir à l'histoire des Diptères dont les larves minent les feuilles des plantes. Soc. Ent. de France, Ann. **ser, 2, 9**: 131–176, 3 pls.

Gowdey, C. C. (See Curran, 1926i)
Grabham, M.
1905   Notes on some Jamaican Culicidae. Canad. Ent. **37:** 401–411, 7 figs.
1906   A new *Corethrella* from Jamaica. Ent. News **17:** 343–345, 1 fig.
Graenicher, S.
1910a   Some new and rare Diptera from Wisconsin. Canad. Ent. **42:** 26–29.
1910b   A preliminary list of the flies of Wisconsin belonging to the families Bombylidae, Syrphidae and Conopidae. Wis. Nat. Hist. Soc. Bul. **8:** 32–44.
1910c   The bee-flies (Bombyliidae) in their relations to flowers. Wis. Nat. Hist. Soc. Bul. **8:** 91–101.
1913   Records of Wisconsin Diptera. Wis. Nat. Hist. Soc. Bul. (1912) **10:** 171–185.
Grassé, P.-P., ed.
1949–   Traité de zoologie, anatomie, systématique, biologie. 17
1960   vols. Paris.
> This is to consist of 17 volumes upon completion. As of 1960, the work was incomplete. Volume 10, which includes the Diptera, consists of 2 separately bound but continuously paged fascicles, a total of 1948 pp., 1648 figs., 6 pls., published in 1951. See Séguy, 1951.

Gravenhorst, G. H.
1834   VI. Bericht der entomologischen Abtheilung. Pp. 718–747. *In* Versammlung der Naturforscher und Aertze zu Breslau, Bericht über die eilfte [sic] Versammlung der deutschen Naturforscher und Aertze in Breslau, 1833. Isis (Oken's) **1834:** 545–759, pl. 14.
1837   Bericht der entomologischen Section vom Jahre 1836. Schles. Gesell. f. Vaterländ. Cult., Uebers. d. Arb. u. Veränderungen **1836:** 82–88.
Gray, G. (See Griffith and Pidgeon, 1832)
Greathead, D. J.
1958   Notes on the larva and life history of *Cyrtonotum cuthbertsoni* Duda (Dipt., Drosophilidae), a fly associated with the desert locust *Schistocerca gregaria* (Forskal). Ent. Monthly Mag. **94** [=ser. 4, 19]: 36–37, pl. 3.
Greene, C. T. (See also Banks et al.; Malloch, Greene, and McAtee)
1917   Two new cambium miners (Diptera). Jour. Agr. Res. **10:** 313–318, 1 pl.

Greene, C. T.—Continued
- 1918   Three new species of Diptera. Ent. Soc. Wash. Proc. **20**: 69–71.
- 1919   A new genus in Scatophagidae (Diptera). Ent. Soc. Wash. Proc. **21**: 126–129, 1 fig.
- 1921a   An illustrated synopsis of the puparia of 100 muscoid flies (Diptera). U.S. Natl. Mus. Proc. **60** (10): 1–39, 20 pls.
- 1921b   A new genus of Bombyliidae (Diptera). Ent. Soc. Wash. Proc. **23**: 23–24, 1 fig.
- 1921c   Two new species of Diptera. Ent. Soc. Wash. Proc. **23**: 125–127, 1 fig.
- 1922   Synopsis of the North American flies of the genus *Tachytrechus*. U.S. Natl. Mus. Proc. **60** (17): 1–21, 1 pl.
- 1923   A new species of *Volucella* (Diptera). Ent. Soc. Wash. Proc. **25**: 165–168, 1 pl.
- 1924a   Synopsis of the North American flies of the genus *Scellus*. U.S. Natl. Mus. Proc. **65** (16): 1–18, 3 pls.
- 1924b   New species of *Mythicomyia* and its relationship, with a new genus (Diptera). Ent. Soc. Wash. Proc. **26**: 60–64, 3 figs.
- 1925.   The puparia and larvae of sarcophagid flies. U.S. Natl. Mus. Proc. **66** (29): 1–26, 9 pls.
- 1929   Characters of the larvae and pupae of certain fruit flies. Jour. Agr. Res. **38**: 489–504, 4 figs.
- 1934a   Tachinid flies with an evanescent fourth vein, including a new genus and five new species. Ent. Soc. Wash. Proc. **36**: 27–40, 4 pls.
- 1934b   A revision of the genus *Anastrepha* based on a study of the wings and on the length of the ovipositor sheath (Diptera: Trypetidae). Ent. Soc. Wash. Proc. **36**: 127–179, 4 figs., 5 pls.
- 1938   A new genus and two new species of the dipterous family Phoridae. U.S. Natl. Mus. Proc. **85**:. 181–185, 1 fig.
- 1941a   Two new species of cecidomyiid flies from phlox. U.S. Natl. Mus. Proc. **90**: 547–551, 1 fig.
- 1941b   A remarkable new species of the genus *Pseudacteon* (Diptera: Phoridae). Ent. Soc. Wash. Proc. **43**: 183–184, 1 fig.
- 1955   Larvae and pupae of the genera *Microdon* and *Mixogaster* (Diptera, Syrphidae). Amer. Ent. Soc. Trans. **81**: 1–20, 19 figs.

Griffen, A. B. (See Patterson, Stone, and Griffen; Stone, Griffen, and Patterson)

Griffin, F. J.
1932   On the contents of the parts and dates of publication of 'Schiner (J. R.), Fauna Austriaca.—Die Fliegen (Diptera)', 1860–1864. Ann. and Mag. Nat. Hist. ser. 10, **10**: 570.
1937   A further note on "Palisot de Beauvois, Insectes Rec. Afr. Amér." 1805–1821. Soc. Bibliog. Nat. Hist. Jour. **1**: 121–122.
1938   On the date of publication of Latreille (in Sonnini's Buffon), An X (sic), Hist. Nat. Gén. Partic. Crust. Ins. 3. Soc. Bibliog. Nat. Hist. Jour. **1**: 157.

Griffith, E., and Pidgeon, E.
1832   The class Insecta arranged by the Baron Cuvier, with supplementary additions to each order, and notices of new genera and species by George Gray, Esq. Volume the second. Vol. 15, 793 pp., pls. 2, 3, 6, 14, 21, 38, 43, 47, 49, 50, 53, 54, 58–60, 62–95, 97, 99–123, 125–134, 136–138, 140. *In* Griffith, E., and others, The animal kingdom arranged in conformity with its organization by the Baron Cuvier [q.v.]. London.

Griffith, E., and others
1827–   The animal kingdom arranged in conformity with its
1835   organization by the Baron Cuvier. 15 vols., and Index. London.

> The 2 volumes on insects (marked "First" and "Second") are by Griffith and E. Pidgeon, and make up Volumes 14 and 15 of the whole work. These volumes, both published in 1832, contain new genera and species by G. Gray.
> See Griffith and Pidgeon, 1832.

Grigarick, A. A.
1959   Bionomics of the rice leaf miner, *Hydrellia griseola* (Fallén), in California (Diptera: Ephydridae). Hilgardia **29**: 1–80, 50 figs.

Grimshaw, P. H.
1901   Part I. Diptera. Pp. 1–77, pls. 1–3. *In* Sharp, D., ed., Fauna Hawaiiensis [q.v.]. Vol. 3, 704 pp., 19 pls. Cambridge, England.

Grossbeck, J. A.
1904a   Description of a new *Culex*. Canad. Ent. **36**: 332.
1904b   Description of two new species of *Culex*. Ent. News **15**: 332–333.
1905   New species of Culicidae. Canad. Ent. **37**: 359–360.
1906   Notes on *Culex squamiger*, Coq., with description of a closely-allied species. Canad. Ent. **38**: 129–131, 1 fig.

Grossbeck, J. A.—Continued
1912   Types of insects, except Lepidoptera and Formicidae, in the American Museum of Natural History additional to those previously listed. Amer. Mus. Nat. Hist. Bul. **31**: 353–379.

Grossenbacher, J. G.
1915   Medullary spots and their cause. Torrey Bot. Club, Bul. **42**: 227–239, 2 pls.

Grote, A. R.
1867   Description of two new species of North American brachycerous Diptera. Ent. Soc. Phila. Proc. **6**: 445.

Günther, A. C. L. G., ed.
1865–   The record of zoological literature. 6 vols. London.
1870
> The continuation is now published by the Zoological Society of London, as "The Zoological Record" [q.v.].
> See Dallas, 1866.

Guérin-Méneville, F. E.   (See also Latreille et al.)
[1831]   Pls. 20–21 (=livrs. 23–24, part). *In* Duperrey, L. I., ed., Voyage autour du monde sur la corvette de sa majesté La Coquille [q.v.]. Zoologie, Atlas, Insectes, 21 pls. Paris.
> Dated by Sherborn and Woodward, 1901: 391, and Bequaert, 1926c: 186.

1835   Pls. 92–104 (=livrs.?). *In his* Iconographie du règne animal de G. Cuvier [q.v.]. Vol. 2: Planches des animaux invertébrés, Insectes, 104 pls. Paris.
> This date is taken from that appearing on some of the plates. However, the plates may have been issued later, and/or at separate times, since the plates were published in mixed fascicles of 10.

[1838]   Première division. Crustacés, arachnides et insectes. Div. 1, pp. 1–216, 217–319 (=livrs. 25, 26). *In* Duperrey, L. I., ed., Voyage autour du monde sur la corvette de sa majesté La Coquille [q.v.]. Zoologie, Vol. 2, Pt. 2. Paris.
> Dated by Sherborn and Woodward, 1901: 391, and Bequaert, 1926c: 186.

[1844]   Douzième ordre. Les Diptères. Pp. 531–559 (=livr. 50). *In his* Iconographie du règne animal de G. Cuvier [q.v.]. Vol. 3: Texte explicatif, Insectes, 576 pp. (=livrs. 45–50). Paris.
> The date on the title-page of the section "Insectes" is 1829–1838. Actual dates are from Sherborn, 1922: lxiii.

Guérin-Méneville, F. E.
1829– Iconographie du règne animal de G. Cuvier, ou représenta-
1844   tion d'après nature de l'une des espèces les plus re-
       marquables et souvent non encore figurées, de chaque
       genre d'animaux. 3 vols. Paris.
> Each of the 10 taxonomic sections of this work has an independent numeration of both text and plates. The 3 volumes (2 of plates and 1 of text) were issued in 50 livraisons, each containing portions of different sections. The work is bound variously by libraries, but most commonly in 7 parts. Both colored and black-and-white plates are known.
> See Guérin-Méneville, 1835, 1844.

Guettard, J. E.
1762   Observations qui peuvent servir à former quelques caractères de coquillages. [Paris] Acad. Roy. des Sci., Hist. avec Mém. Math. et Phys. **1756** (**Mém.**): 145–183.

Guimarães, L. R., and d'Andretta, M. A. V.
1956   Sinopse dos Nycteribiidae (Diptera) do novo mundo. Arq. de Zool. de Estado de São Paulo **10**: 1–184, 232 figs.

Guthrie, E.
1917   New Mycetophilidae from California. Ent. Soc. Amer. Ann. **10**: 314–319, 3 pls.

Hackman, W.
1955   On the genera *Scaptomyza* Hardy and *Parascaptomyza* Duda (Dipt., Drosophilidae). Notulae Ent. **35**: 74–91, 32 figs.
1959   On the genus *Scaptomyza* Hardy (Dipt., Drosophilidae), with descriptions of new species from various parts of the world. Acta Zool. Fenn. **97**: 1–73, 74 figs.

Hagen, H. A.
1880   A new species of *Simulium*, with a remarkable nympha [sic] case. Boston Soc. Nat. Hist. Proc. (1878–1880) **20**: 305–307.
1881a List of N. American Anthomyidae, examined by R. H. Meade, Esq., Bradford, England. Canad. Ent. **13**: 43–51.
1881b List of N. American Sarcophagidae, examined by R. H. Meade, Esq., Bradford, England. Canad. Ent. **13**: 146–150.

Haldeman, S. S.
1847   Descriptions of several new and interesting animals. Amer. Jour. Agr. and Sci. **6**: 191–194, 4 figs.

Haliday, A. H. (See also Curtis, 1837b; Walker, 1851a; Westwood, 1840)

1832 The characters of two new dipterous genera, with indications of some generic subdivisions and several undescribed species of Dolichopidae. Zool. Jour. (London) (1830–1831) **5**: 350–367, pl. 15.

1833 Catalogue of Diptera occurring about Holywood in Downshire. Ent. Mag (London) **1**: 147–180.

1836 British species of the dipterous tribe Sphaeroceridae. Ent. Mag. (London) **3**: 315–336.

1838 New British insects indicated in Mr. Curtis's Guide [concl.]. Ann. Nat. Hist. (1839) **2**: 183–190.

1839 Remarks on the generic distribution of the British Hydromyzidae (Diptera). Ann. Nat. Hist. **3**: 217–224, 401–411.

1851 Family XXI. Dolichopidae. Pp. 144–221. *In* (Walker, F., Stainton, H. T., and Wilkinson, S. J.), Insecta Britannica [q.v.]. [Vol. 1]: Diptera Vol. 1 (by Walker), 314 pp., 10 pls. London.

1855 Descriptions of insects figured, and references to plates illustrating the notes on Kerry insects. Nat. Hist. Rev. **2** (**Proc.**): 59–64, pls. 2, 3 (part).

1856 Addenda and corrigenda. Pp. xi–xv. *In* (Walker, F., Stainton, H. T., and Wilkinson, S. J.), Insecta Britannica [q.v.]. [Vol. 4]: Diptera Vol. 3 (by Walker), xxiv+352 pp., pls. 21–30. London.

Hall, D. G.

1927 A new species of *Sarcophaga* (Diptera) from Ohio. Insecutor Inscitiae Menstruus (1926) **14**: 176–178, 1 pl.

1928 *Sarcophaga pallinervis* and related species in the Americas. Ent. Soc. Amer. Ann. **21**: 331–352, 4 pls.

1929 The North American species of *Sarcophaga* belonging to the "A" group (Dip.: Sarcophagidae). Ent. News **40**: 319–324, 1 fig.

1930 Three new West Indian Sarcophaginae (Diptera). Amer. Mus. Nat. Hist., Amer. Mus. Novitates **423**: 1–4, 3 figs.

1931a A new *Sarcophaga* from South Carolina (Diptera: Sarcophagidae). Ent. News **42**: 217–219, 1 fig.

1931b New Texas Sarcophaginae (Diptera: Calliphoridae). Ent. News **42**: 280–286, 4 figs.

1931c New North American Sarcophagidae. Ent. Soc. Amer. Ann. **24**: 181–182, 1 fig.

1931d Two new species of Sarcophaginae from California. Pan-Pacific Ent. **8**: 52–54, 2 figs.

Hall, D. G.—Continued
1932a Canadian Sarcophagidae. Canad. Ent. **64:** 102–103, 1 fig.
1932b A new biting *Culicoides* from saltmarshes in the south eastern states. Ent. Soc. Wash. Proc. **34:** 88–89, 1 fig.
1933a A new species of *Sarcophaga* inhabiting nests of paper wasps. Ent. Soc. Wash. Proc. **35:** 110–111, 1 fig.
1933b The Sarcophaginae of Panama (Diptera: Calliphoridae). Amer. Mus. Nat. Hist. Bul. **66:** 251–285, 26 figs.
1936 *Phlebotomus* (*Brumptomyia*) *diabolicus*, a new species of biting gnat from Texas (Diptera: Psychodidae). Ent. Soc. Wash. Proc. **38:** 27–29, 1 fig.
1937a New muscoid flies (Diptera) in the United States National Museum. U.S. Natl. Mus. Proc. **84:** 201–216, 8 figs.
1937b Sarcophaginae. Pp. 347–375, figs. 61–71, 1 pl. (=fasc. 3, part). *In* British Museum (Natural History), Diptera of Patagonia and south Chile [q.v.]. Pt. 7, 384 pp., 74 figs., 1 pl. London.
1938 New genera and species of South American Sarcophagidae (Diptera). Arb. über Morph. u. Taxonom. Ent. **5:** 253–259, 6 figs.
1939 Two new species of Tachinidae parasitic upon hemlock sawfly larvae in North America (Diptera: Tachinidae). Ent. Soc. Wash. Proc. **41:** 239–243.
1943 A new species of *Cuterebra* from Kansas (Diptera: Cuterebridae). Ent. Soc. Wash. Proc. **45:** 25–26.
1948 The blowflies of North America. [Vol. 4], 477 pp., 51 pls. *In* Entomological Society of America, Thomas Say Foundation [q.v.]. Washington, D. C.

Hall, J. C. (See also Painter and Hall)
1952 A new species of *Lordotus* from southern California (Bombyliidae: Diptera). Pan-Pacific Ent. **28:** 49–50.
1954a A revision of the genus *Lordotus* Loew in North America (Diptera: Bombyliidae). Calif. Univ., Pubs., Ent. **10:** 1–33, 24 figs.
1954b Notes on the biologies of three species of Bombyliidae, with a description of one new species. Ent. News **65:** 145–149.
1956 A new species of *Anastoechus* Osten Sacken with notes on the congeners. Ent. News **67:** 199–203.
1957 Notes and descriptions of new California Bombyliidae (Diptera). Pan-Pacific Ent. **33:** 141–148.
1958 A change of name in the bombyliid genus *Anastoechus* (Diptera). Ent. News **69:** 195.

Hallock, H. C.
1938   New Sarcophaginae (Diptera). Ent. Soc. Wash. Proc. **40**: 95–99, 2 figs.

Hamilton, C. C.
1957   Holly pests in the East. Natl. Hort. Mag. **36**: 91–102, figs. 31–34.

Hammer, O.
1941   Biological and ecological investigations on flies associated with pasturing cattle and their excrement. Dansk Naturhist. For. København, Vidensk. Meddel. **105**: 141–393, 50 figs.

Handlirsch, A.
1909   Zur Phylogenie und Flügelmorphologie der Ptychopteriden (Dipteren). Wien, K. Naturhist. Hof-Mus. Ann. **23**: 263–272, 1 pl.

Hanson, W. J.
1958   A revision of the subgenus *Melanonemotelus* of America north of Mexico (Diptera, Stratiomyidae). Kans. Univ. Sci. Bul. **38**: 1351–1391, 32 figs.
1961   Notes on *Nemotelus* (Diptera: Stratiomyidae). Kans. Ent. Soc. Jour. **34**: 214–215.

Hanström, B., Brinck, P., and Rudebeck, G., eds.
1955–   South African animal life. (Results of the Lund University
1961   Expedition in 1950–1951). 8 vols. Lund University, Uppsala.

> The volumes are issued in continuously paged parts, and later as complete volumes. The work consists of 8 volumes to date.
> See McAlpine, 1960c.

Hardy, D. E.
1936   A new Bibionidae (Diptera) from Utah. Utah Acad. Sci., Arts and Letters, Proc. **13**: 195.
1937   New Bibionidae (Diptera) from nearctic America. Utah Acad. Sci., Arts and Letters, Proc. **14**: 199–213, 1 pl.
1938a New Bibionidae from British Columbia. Canad. Ent. **70**: 207–210, 10 figs.
1938b New Therevidae (Diptera) from Utah. Ent. Soc. Amer. Ann. **31**: 144–146, 3 figs.
1939   New nearctic Pipunculidae (Diptera). Kans. Ent. Soc. Jour. **12**: 16–25, 1 pl.
1940a Studies in New World *Plecia* (Bibionidae-Diptera). Part I. Kans. Ent. Soc. Jour. **13**: 15–27, 1 pl.

Hardy, D. E.—Continued
- 1940b Dorylaidae notes and descriptions (Pipunculidae-Diptera). Kans. Ent. Soc. Jour. **13:** 101–114, 2 pls.
- 1942a New western Asilidae. Kans. Ent. Soc. Jour. **15:** 57–61, 6 figs.
- 1942b Studies in New World *Philia* (Bibionidae). Part I. Kans. Ent. Soc. Jour. **15:** 127–134, 1 pl.
- 1943a New Therevidae and Asilidae in the Snow Entomological Collection. Kans. Ent. Soc. Jour. **16:** 24–29, 4 figs.
- 1943b Studies in *Phyllomydas* (Mydaidae-Diptera). Kans. Ent. Soc. Jour. **16:** 50–52.
- 1943c A revision of nearctic Dorilaidae (Pipunculidae). Kans. Univ. Sci. Bul. **29:** 3–231, 18 pls.
- 1944a New Asilidae and Mydaidae (Diptera) in the Snow Collection. Canad. Ent. **76:** 226–230, 1 pl.
- 1944b A revision of North American Omphralidae (Scenopinidae). Kans. Ent. Soc. Jour. **17:** 31–40, 42–51, 1 pl.
- 1945 Revision of nearctic Bibionidae including neotropical *Plecia* and *Penthetria* (Diptera). Kans. Univ. Sci. Bul. **30:** 367–547, 13 pls.
- 1947a The genus *Leptopteromyia* (Asilidae-Diptera). Kans. Ent. Soc. Jour. **20:** 72–75, 2 figs.
- 1947b Notes and descriptions of Dorilaidae (Pipunculidae-Diptera). Kans. Ent. Soc. Jour. **20:** 146–153, 1 pl.
- 1948a Notes and descriptions of Dorilaidae (Pipunculidae-Diptera). Part II. Kans. Ent. Soc. Jour. **21:** 88–91, 1 pl.
- 1948b New and little known Diptera from the California Academy of Sciences collection (Rhagionidae and Dorilaidae). Wasmann Collect. **7:** 129–137, 1 pl.
- 1949 The North American *Chrysopilus* (Rhagionidae-Diptera). Amer. Midland Nat. **41:** 143–167, 15 figs.
- 1950 The nearctic *Nomoneura* and *Nemomydas* (Diptera: Mydaidae). Wasmann Jour. Biol. **8:** 9–37, 2 pls.
- 1952 Additions and corrections to Bryan's check list of the Hawaiian Diptera. Hawaii. Ent. Soc. Proc. (1951) **14:** 443–484d.
- 1954 Studies in New World Dorilaidae (Pipunculidae: Diptera). Kans. Ent. Soc. Jour. **27:** 121–127, 5 figs.
- 1959a A new Bibionidae from California (Diptera). Pan-Pacific Ent. **35:** 209–211, 2 figs.
- 1959b A review of the genus *Pseudiastata* Coquillett (Drosophilidae, Diptera). Hawaii. Ent. Soc. Proc. (1958) **17:** 76–82, 2 figs.

Hardy, D. E.—Continued
1959c The Walker types of fruit flies (Tephritidae-Diptera) in the British Museum collection. Brit. Mus. (Nat. Hist.) Bul., Ent. **8**: 159–242, pls. 11–16.
1960a A new *Bibio* from the Sierra Mountains, California, and a new *Plecia* from Malaya (Diptera: Bibionidae). Hawaii. Ent. Soc. Proc. (1959) **17**: 255–259, 2 figs.
1960b Diptera: Nematocera-Brachycera (except Dolichopodidae). Vol. 10, 368 pp., 120 figs. *In* Zimmerman, E. C., ed., Insects of Hawaii [q.v.]. Honolulu.

———, **and Knowlton, G. F.**
1939a New and little known Utah Pipunculidae (Diptera). Canad. Ent. **71**: 87–91, 15 figs.
1939b New and little known western Pipunculidae (Diptera). Ent. Soc. Amer. Ann. **32**: 113–124, 1 pl.

———, **and McGuire, J. U.**
1947 The nearctic *Ptiolina* (Rhagionidae-Diptera). Kans. Ent. Soc. Jour. **20**: 1–15, 1 pl.

———, **and Nagatomi, A.**
1960 An unusual new Nematocera from Japan (Diptera), and a new family name. Pacific Insects **2**: 263–267, 1 fig.

———, **and Wheeler, M. R.**
1960 *Paracacoxenus*, new genus, with notes on *Cacoxenus indagator* Loew (Diptera: Drosophilidae). Ent. Soc. Amer. Ann. **53**: 356–359, 2 figs.

Hardy, J.
1849 Notes on the remedies for the turnip-fly amongst the ancients, and on the turnip-fly of New Holland, with a notice of a new genus and species of Diptera. Berwickshire Nat. Club, Hist. (1842–1849) **2**: 359–362.
1850 On the effects produced by some insects, &c. upon plants. Ann. and Mag. Nat. Hist. ser. 2, **6**: 182–188.

Harmston, F. C.
1939 A new *Scellus* (Dolichopodidae: Diptera) with key to males. Utah Acad. Sci., Arts and Letters, Proc. **16**: 71–73, 6 figs.
1951a New species of Dolichopodidae from California and Utah (Diptera). Great Basin Nat. **11**: 11–17.
1951b New species of Dolichopodidae in the University of Kansas collection (Diptera). Kans. Ent. Soc. Jour. **24**: 103–109.
1952 New species of Dolichopodidae in the U.S. National Museum (Diptera). Ent. Soc. Wash. Proc. **54**: 281–294.

———, **and James, M. T.** (See Harmston and Knowlton, 1942b)

Harmston, F. C., and Knowlton, G. F.
1939a New Utah Dolichopodidae (Diptera). Ent. News **50**: 256–259, 5 figs.
1939b New Dolichopodidae (Diptera). Ent. Soc. Amer. Ann. **32**: 349–352, 10 figs.
1939c Three new Dolichopodidae. Kans. Ent. Soc. Jour. **12**: 83–86, 1 pl.
1939d A new *Dolichopus* from Iowa (Diptera). Ent. Soc. Wash. Proc. **41**: 87–88, 1 fig.
1940a *Tachytrechus* studies (Dolichopodidae, Diptera). Canad. Ent. **72**: 111–115, 5 figs.
1940b New and little-known Utah Dolichopodidae (Diptera). Ent. News **51**: 129–134, 2 figs.
1940c The genus *Sympycnus* in Utah (Dolichopodidae: Diptera). Ent. Soc. Amer. Ann. **33**: 395–403, 28 figs.
1904d New mid-western Dolichopodidae (Diptera). Kans. Ent. Soc. Jour. **13**: 58–61, 9 figs.
1940e Two new California Dolichopodidae (Diptera). Pan-Pacific Ent. **16**: 108–110, 3 figs.
1940f Four new *Hercostomus* from Utah (Dolichopodidae: Diptera). Ent. Soc. Wash. Proc. **42**: 125–128, 9 figs.
1941a New species of *Hercostomus* from western North America (Diptera, Dolichopodidae). Canad. Ent. **73**: 127–132. 15 figs.
1941b New western Dolichopodidae (Diptera). Kans. Ent. Soc. Jour. **14**: 92–97, 8 figs.
1942a The dipterous genus *Campsicnemus* in North America. Brooklyn Ent. Soc. Bul. **37**: 10–17, 9 figs.
1942b New Dolichopodidae from Utah and Colorado (Diptera). Canad. Ent. **74**: 80–85, 10 figs.
1942c New Dolichopodidae of western North America. Ent. Soc. Amer. Ann. **35**: 17–22, 13 figs.
1942d Three new *Syntormon* (Diptera: Dolichopodidae) from western United States. Ent. Soc. Wash. Proc. **44**: 22–26, 11 figs.
1943a Five new western Dolichopodidae (Diptera). Brooklyn Ent. Soc. Bul. **38**: 101–107.
1943b New species of *Parasyntormon* from the United States (Diptera, Dolichopodidae). Canad. Ent. **75**: 63–65.
1945 New Dolichopodidae from Michigan (Diptera). Kans. Ent. Soc. Jour. **18**: 77–81.
1946a New and little known Dolichopodidae from Indiana (Diptera). Amer. Midland Nat. **36**: 671–674.

Harmston, F. C., and Knowlton, G. F.—Continued
1946b Two new *Paraclius* (Diptera: Dolichopodidae). Kans. Ent. Soc. Jour. **19**: 23–25.
1946c Three new western Dolichopodidae. Canad. Ent. (1945) **77**: 137–139.

Harriot, S. C.
1942a Notes on the genus *Seioptera* Kirby (Otitidae, Diptera). N.Y. Ent. Soc. Jour. **50**: 195–197.
1942b A new genus and a new species of Otitidae from North America (Diptera). N.Y. Ent. Soc. Jour. **50**: 249–250.
1942c New species of Otitidae from California (Diptera). Pan-Pacific Ent. **18**: 23–26, 1 fig.

Harris, M.
1776– An exposition of English insects with curious observations
[1780] and remarks wherein each insect is particularly described, its parts and properties considered, the different sexes distinguished, and the natural history faithfully related. Decad I, pp. 1–40, 2 pls. + pls. 1–10, *1776*; Decad II, pp. 41–72, pls. 11–20, [?*1776*]; Decads III, IV, V, pp. 73–99, 100–138, 139–166, pls. 21–30, 31–40, 41–50 + 1 pl., [?*1780*]. London, "1776".

> The dates of the latter portions of this work are in doubt. Decad II was issued sometime between 1776 and 1780; we have arbitrarily used 1776. See Lisney, 1960: 170, for a discussion of this work and its dating.

Harris, T. W.
1833 VIII. Insects. Pp. 566–595. *In* Hitchcock, E., Report on the geology, mineralogy, botany, and zoology of Massachusetts. [Ed. 1], 700 pp., 104 figs., + Atlas, 17 pls. Amherst, Mass.
1835 VIII. Insects. Pp. 553–602. *In* Hitchcock, E., Report on the geology, mineralogy, botany, and zoology of Massachusetts. Ed. 2, 702 pp., 104 figs., + Atlas, 17 pls. Amherst, Mass.
1841 A report on the insects of Massachusetts, injurious to vegetation. 459 pp. Cambridge, Mass.

> Reissued under the title "A treatise on some of the insects of New England, which are injurious to vegetation", with the same pagination, Cambridge, 1842.

1862 A treatise on some of the insects injurious to vegetation. Ed. 3, 640 pp., 278 figs., 8 pls. Boston.

> A "New edition", edited by C. L. Flint, was published in 1863.

Harris, T. W.—Continued
1869 Descriptions of insects selected from the manuscripts of Dr. Harris. Pp. 325–336. *In* Scudder, S. H., ed., Entomological correspondence of Thaddeus William Harris, M.D. Boston Soc. Nat. Hist. Occas. Papers **1:** 1–375, 4 pls.

Hart, C. A. (See also Forbes, 1912)
1896 On the entomology of the Illinois River and adjacent waters. First paper. Ill. State Lab. Nat. Hist. Bul. (1895–1896) **4:** 149–273, pls. 1–2, 5–15.

Hartig, T.
1838 Ueber die parasitischen Zweiflügler des Waldes. Jahresber. über die Fortschr. der Forstw. u. Forstl. Naturk. **1:** 275–306.

Hartley, J. C.
1961 A taxonomic account of the larvae of some British Syrphidae. Zool. Soc. London, Proc. **136:** 505–573, 117 figs.

Harwood, R. F. (See Fairchild and Harwood)

Haseman, L.
1907 A monograph of the North American Psychodidae including ten new species and an aquatic psychodid from Florida. Amer. Ent. Soc. Trans. **33:** 299–333, pls. 5–8.
1908 Notes on the Psychodidae. Ent. News **19:** 274–285, 2 figs.

Hasselquist, F. (See Linnaeus, 1762)

Hauber, U. A.
1945 Tanypodinae of Iowa (Diptera). I. The genus *Pentaneura* Philippi (*Tanypus*). Amer. Midland Nat. **34:** 496–503, 12 figs.
1947 The Tendipedinae of Iowa (Diptera). Amer. Midland Nat. **38:** 456–465, 20 figs.

[Hausmann, J. F. L.]
1799 Entomologische Bemerkungen. 64 pp. Braunschweig.
> This work consists of four individually titled articles. It was issued anonymously, but was credited to Hausmann by Horn and Schenkling, 1928: 525.

Hays, K. L.
1961 *Tabanus aranti* sp. nov. (Diptera: Tabanidae) from Alabama. Ent. News **72:** 127–129, 1 fig.

Hearle, E.
1923 A new mosquito from British Columbia (Culicidae, Diptera). Canad. Ent. **55:** 4–5.

Hearle, E.—Continued
1927 A new Canadian mosquito (Culicidae). Canad. Ent. **59**: 101–103.
1932 The blackflies of British Columbia (Simuliidae, Diptera). Ent. Soc. Brit. Columbia, Proc. **29**: 5–19, 7 figs.

Hedlin, A. F.
1959 Description and habits of a new species of *Phytophaga* (Diptera: Cecidomyiidae) from western red cedar cones. Canad. Ent. **91**: 719–723, 15 figs.

Heiss, E. M.
1938 A classification of the larvae and puparia of the Syrphidae of Illinois exclusive of aquatic forms. Ill. Biol. Monog. **16** (4): 1–142, 17 pls. (=Ill. Univ. Bul. 36(1).)

Heller, K. M.
1902 *Strongylophthalmyia* nom. nov. für *Strongylophthalmus* Hendel. Wien. Ent. Ztg. **21**: 226.

(Hellwig, J. C. L., and Illiger, J. C. W.)
1795– Fauna Etrusca. Sistens Insecta quae in provinciis Floren-
1807 tina et Pisana praesertim collegit Petrus Rossius. (Mantissae priore parte adjecta), iterum edita et annotatis perpetuis aucta. 2 vols. Helmstadii [=Helmstedt, Germany].

> This is a German translation and expansion of Rossi's "Fauna Etrusca" [q.v.] and "Mantissa insectorum", the latter appearing only in Volume 1. Volume 1 is by Hellwig, Volume 2 by Illiger. A "Nachricht", dated 1808, is sometimes bound with Volume 2. See Illiger, 1807.

Hemming, F.
1945 On the importance of facsimile reproductions of rare works of importance in systematic zoology, with special reference to Meigen (J. G.), 1800, Nouvelle classification des mouches à deux ailes. Internatl. Comn. Zool. Nomencl., Bul. Zool. Nomencl. **1**: 119–160.

Hemming, F., ed.
1939– Opinions and Declarations rendered by the International
1959 Commission on Zoological Nomenclature. 20 vols. London.

> The volumes were issued in continuously paged parts, the first 11 parts being titled "Opinions rendered . . . [etc.]." Volume 2, 1939–1955, began publication first, with Opinion 134, and contains all matters relating to the Lisbon Congress of 1936. It is divided into 2 continuously paged sections, A and B. Volume 1, 1943–1958, containing Declarations 1–9, a facsimile of Opinions 1–133 (originally published by the Smithsonian Institution), and Directions of the Commission concerning matters prior to 1936,

Hemming, F., ed.—Continued
>is divided into 5 separately paged sections, A–F. Volume 3 began with Opinion 182, and the work ends with Opinion 568 in Volume 20. The dates of the volumes used in the Bibliography are: 3, 1944–1954; 5, 1954; 8, 1954–1955; 10, 1955–1956; 11, 1955–1956; 15, 1957; 19, 1958–1959; 20, 1959. Subsequent Opinions and Declarations are being published in Internatl. Comn. Zool. Nomencl., Bul. Zool. Nomencl., commencing in Volume 7, 1959.
>>See I.C.Z.N., 1954a–c, 1955a, b, 1957a, b, 1958a, b, 1959a–c, 1961a, b, 1963a, b.

Hendel, F. (See also Meijere, 1911a)

1900 Untersuchungen über die europäischen Arten der Gattung *Tetanocera* im Sinne Schiner's. Eine dipterologische Studie. K.–k. Zool.-Bot. Gesell. Wien, Verhandl. **50**: 319–358.

1901a Zur Kenntnis der Tetanocerinen (Dipt.). Természet. Füzetek **24**: 138–142.

1901b Dipterologische Anmerkungen. Wien. Ent. Ztg. **20**: 197–199.

1902a *Strongylophthalmus*, eine neue Gattung der Psiliden (Dipt.). Wien. Ent. Ztg. **21**: 179–181, 2 figs.

1902b Dipterologische Anmerkungen. Wien. Ent. Ztg. **21**: 265.

1902c Revision der paläarktischen Sciomyziden (Dipteren-Subfamilie). K.-k. Zool.-Bot. Gesell. Wien., Abhandl. **2** (**1**): 1–94, 1 pl.

1903 Kritische Bemerkungen zur Systematik der Muscidae Acalypteratae. Wien. Ent. Ztg. **22**: 249–252.

1907a Nomina nova für mehrere Gattungen der acalyptraten Musciden. Wien. Ent. Ztg. **26**: 98.

1907b Neue und interessante Dipteren aus dem Kaiserl. Museum in Wien. (Ein Beitrag zur Kenntnis der acalyptraten Musciden.) Wien. Ent. Ztg. **26**: 223–245, 1 pl.

1908a Diptera. Fam. Muscaridae, Subfam. Lauxaninae. Fasc. 68, 66 pp., 3 pls. *In* Wytsman, P., ed., Genera insectorum [q.v.]. Bruxelles.

1908b Nouvelle classification des mouches à deux ailes (Diptera L.), d'après un plan tout nouveau par J. G. Meigen, Paris, an VIII (1800 v. s.). Mit einem Kommentar. K.-k. Zool.-Bot. Gesell. Wien, Verhandl. **58**: 43–69.

1909 Beitrag zur Kenntnis der Ulidiinen (Dipt.). Wien. Ent. Ztg. **28**: 247–270.

1910a Diptera. Fam. Muscaridae, Subfam. Ulidiinae. Fasc. 106, 76 pp., 4 pls. *In* Wytsman, P., ed., Genera insectorum [q.v.]. Bruxelles.

Hendel, F.—Continued
1910b Ueber die Nomenklatur der Acalyptratengattungen nach Th. Beckers Katalog der paläarktischen Dipteren, Bd. 4. Wien. Ent. Ztg. **29**: 307–313.
1911 Über von Professor J. M. Aldrich erhaltene und einige andere amerikanische Dipteren. Wien. Ent. Ztg. **30**: 19–46.
1913 Neue amerikanische Dipteren. 1. Beitrag. Deut. Ent. Ztschr. **1913**: 617–636, 5 figs.
1914a Diptera. Fam. Muscaridae, Subfam. Platystominae. Fasc. 157, 179 pp., 15 pls. *In* Wytsman, P., ed., Genera insectorum [q.v.]. Bruxelles.
1914b Neue amerikanische Dipteren. 2. Beitrag. Deut. Ent. Ztschr. **1914**: 151–176, 16 figs.
1914c Die Bohrfliegen Südamerikas. [Dresden] K. Zool. u. Anthrop.-Ethnog. Mus. Abhandl. u. Ber. (1912) **14** (**3**): 1–84, 4 pls.
1914d Namensänderung (Dipt.). Ent. Mitt. **3**: 73.
1914e Die Gattungen der Bohrfliegen. (Analytische Übersicht aller bisher bekannten Gattungen der Tephritinae). Wien. Ent. Ztg. **33**: 73–98.
1914f Die Arten der Platystominen. K.-k. Zool.-Bot. Gesell. Wien, Abhandl. **8** (**1**): 1–409, 4 pls.
1914g H. Sauter's Formosa-Ausbeute. Acalyptrate Musciden (Dipt.). III. Sup. Ent. **3**: 90–117, 7 figs.
1917 Beiträge zur Kenntnis der acalyptraten Musciden. Deut. Ent. Ztschr. **1917**: 33–47, 3 figs.
1920a Die paläarktischen Agromyziden (Dipt.). (Prodromus einer Monographie). Arch. f. Naturgesch. (1918) **Abt. A, 84** (**7**): 109–174, 4 figs.
1920b Zwei neue europäische Dipterengattungen. Wien. Ent. Ztg. **38**: 53–56, 2 figs.
1922 Blattminierende Fliegen (Musciden). Wien. Ent. Ztg. **39**: 65–72.
1923 Neue europäische *Melanagromyza*-Arten (Dipt.). Konowia **2**: 142–145.
1924 Ueber das Genus *Parallelomma* Beck. und seine Verwandten in Europa (Dipt., Cordyl.). Ent. Mitt. **13**: 82–84.
1925–1926 Neue Übersicht über die bisher bekannt gewordenen Gattungen der Lauxaniiden, nebst Beschreibung neuer Gattungen und Arten. Encyclopédie Ent., Sér. B, [Div.] II: Diptera (1925) **2**: 103–112, *1925;* 113–142, 4 figs., *1926.*

Hendel, F.—Continued
1927a   Trypetidae. [Fam.] 49, pp. 1–64, figs. 1–23, pls. 1–4 (=lfg. 16); pp. 65–128, figs. 24–59, pls. 5–8 (=lfg. 17); pp. 129–192, figs. 60–73, pls. 9–12 (=lfg. 18); pp. 193–221, figs. 74–79, pls. 13–17 (=lfg. 19). *In* Lindner, E., ed., Die Fliegen der palaearktischen Region [q.v.] Vol. 5. Stuttgart.
1927b   Beiträge zur Systematik der Agromyziden. 10. Beitrag zur Blattminenkunde Europas. Zool. Anz. **69**: 248–271, 3 figs.
1928a   Neue oder weniger bekannte Bohrfliegen (Trypetidae) meist aus dem Deutschen Entomologischen Institut Berlin-Dahlem. Ent. Mitt. **17**: 341–370.
1928b   Zweiflügler oder Diptera. II. Allgemeiner Teil. T. 11, 135 pp., 224 figs. *In* Dahl, F., ed., Die Tierwelt Deutschlands [q.v.]. Jena.
1930a   Die Ausbeute der deutschen Chaco-Expedition 1925/26. Diptera. XIX. Ephydridae. Konowia **9**: 127–155.
1930b   Entomologische Ergebnisse der schwedischen Kamtchatka-Expedition 1920–1922. 28. Diptera Brachycera. 2. Fam. Cordyluridae und Dryomyzidae. Arkiv för Zool. **21** ([sect.] **A, no. 18**): 1–12 (=h. 3, part).
1931a   Agromyzidae. [Fam.] 59, pp. 1–64, figs. 1–80 (=lfg. 52); pp. 64–128, figs. 81–148 (=lfg. 54); pp. 129–192, figs. 149–218 (=lfg. 56); pp. 193–256, figs. 219–259 (=lfg. 58) [cont.]. *In* Lindner, E., ed., Die Fliegen der palaearktischen Region [q.v.]. Vol. 6, Pt. 2. Stuttgart.
1931b   Neue aegyptische Dipteren aus der Gruppe der acalyptraten Musciden gesammelt von Prof. Efflatoun Bey. Soc. Roy. Ent. d'Egypte, Bul. **15**: 59–73, 1 fig.
1932a   Agromyzidae [cont.]. [Fam.] 59, pp. 257–320, figs. 260–330 (=lfg. 66) [cont.]. *In* Lindner, E., ed., Die Fliegen der palaearktischen Region [q.v.]. Vol. 6, Pt. 2. Stuttgart.
1932b   Die Ausbeute der deutschen Chaco-Expedition 1925/26. Diptera (Fortsetzung). XXX–XXXVI. Sciomyzidae, Lauxaniidae, Tanypezidae, Lonchaeidae, Tylidae, Drosophilidae, Milichiidae. Konowia **11**: 98–110, 115–145, figs. 1–6 + 1–9.
1933a   Pyrgotidae. [Fam.] 36, 15 pp., 2 figs., 1 pl. (=lfg. 73). *In* Lindner, E., ed., Die Fliegen der palaearktischen Region [q.v.]. Vol. 5. Stuttgart.

Hendel, F.—Continued
1933b Ueber einige Typen Wiedemann's und Schiner's von acalyptraten Musciden aus Südamerika, nebst einigen verwandten Arten (Dipt.). Rev. de Ent. **3:** 58–83, 1 fig.
1933c Neue acalyptrate Musciden aus der paläarktischen Region (Dipt.). Deut. Ent. Ztschr. **1933:** 39–56, 1 fig.
1934a Agromyzidae [cont.]. [Fam.] 59, pp. 321–368, figs. 331–373, pls. 1–4 (=lfg. 85) [cont.]. *In* Lindner, E., ed., Die Fliegen der palaearktischen Region [q.v.].—Vol. 6, Pt. 2. Stuttgart.
1934b Revision der Tethiniden (Dipt. Muscid. Acal.). Tijdschr. v. Ent. **77:** 37–54, 7 figs.
1935a Agromyzidae [cont.]. [Fam.] 59, pp. 369–416, figs. 374–422, pls. 5–9 (=lfg. 90); pp. 417–464, figs. 423–467, pls. 10–12 (=lfg. 92); pp. 465–512, figs. 468–496, pls. 13–14 (=lfg. 94) [cont.]. *In* Lindner, E., ed., Die Fliegen der palaearktischen Region [q.v.]. Vol. 6, Pt. 2. Stuttgart.
1935b Bemerkungen zu "The families and genera of North American Diptera" by C. H. Curran, New-York 1934. Konowia **14:** 51–57.
1936a Agromyzidae [concl.]. [Fam.] 59, pp. 513–570, figs. 497–498, pls. 15–16 (=lfg. 96). *In* Lindner, E., ed., Die Fliegen der palaearktischen Region [q.v.]. Vol. 6, Pt. 2. Stuttgart.
1936b Ergebnisse einer zoologischen Sammelreise nach Brasilien, insbesondere in das Amazonasgebiet, ausgeführt von Dr. H. Zerny. X. Teil. Diptera: Muscidae Acalyptratae (excl. Chloropidae). Wien, Naturhist. Mus. Ann. **47:** 61–106, 5 figs.

Hendrickson, J. A.
1961 Notes on two species of *Condylostylus* and description of a new species of *Neurigona* from California (Diptera: Dolichopodidae). Wasmann Jour. Biol. **19:** 277–281.

Hennig, W.
1934 Revision der Tyliden (Dipt., Acalypt.). I. Teil: Die Taeniapterinae Amerikas. Stettin. Ent. Ztg. **95:** 65–108, 294–330, 4 pls. [cont.].
1935a Revision der Tyliden (Dipt., Acalypt.). I. Teil: Die Taeniapterinae Amerikas [concl.]. Stettin. Ent. Ztg. **96:** 27–67.
1935b Revision der Tyliden (Dipt., Acalypt.). II. Teil: Die ausseramerikanischen Taeniapterinae, die Trepidariinae und Tylinae. Konowia **14:** 68–92, 192–216, 289–310, 9 figs. [cont.].

Hennig, W.—Continued
1936   Revision der Tyliden (Dipt., Acalypt.). II. Teil: Die ausseramerikanischen Taeniapterinae, die Trepidariinae und Tylinae [concl.]. Konowia **15**: 129–144, 201–239, figs. 10–15.
1937   Übersicht über die Arten der Neriiden und über die Zoogeographie dieser Acalyptraten-Gruppe (Diptera). Stettin. Ent. Ztg. **98**: 240–280.
1938a Tyliden aus Japan. Insecta Matsumurana **13**: 1–14, 13 figs.
1938b Beiträge zur Kenntnis der Clusiiden und ihres Kopulationsapparates (Dipt. Acalypt.). Encyclopédie Ent., Sér. B, [Div.] II: Diptera (1937–1938) **9**: 121–138, 4 figs.
1939   Otitidae (46. Pterocallidae und 47. Ortalidae). [Fams.] 46–47, pp. 1–48, figs. 1–13, pls. 1–6 (=lfg. 126); pp. 49–78, figs. 14–26, pls. 7–12 (=lfg. 128). *In* Lindner, E., ed., Die Fliegen der palaearktischen Region [q.v.]. Vol. 5. Stuttgart.
1940   Ulidiidae. [Fam.] 45, 34 pp., 12 figs., 6 pls. (=lfg. 133). *In* Lindner, E., ed., Die Fliegen der palaearktischen Region [q.v.]. Vol. 5. Stuttgart.
1943   Piophilidae. [Fam.] 40, 52 pp., 32 figs., 2 pls. (=lfg. 151). *In* Lindner, E., ed., Die Fliegen der palaearktischen Region [q.v.]. Vol. 5. Stuttgart.
1948   Die Larvenformen der Dipteren. Pt. 1, 185 pp., 63 figs., 3 pls. Berlin.
   The entire work consists of 3 parts, Berlin, 1948–1952.
1949   Sepsidae. [Fam.] 39a, pp. 1–48, figs. 1–72, pls. 1–5 (=lfg. 157); pp. 49–91, figs. 73–82, pls. 6–10 (=lfg. 159). *In* Lindner, E., ed., Die Fliegen der palaearktischen Region [q.v.]. Vol. 5. Stuttgart.
1950   Die Larvenformen der Dipteren. Pt. 2, 458 pp., 236 figs., 10 pls. Berlin.
1952   Die Larvenformen der Dipteren. Pt. 3, 628 pp., 338 figs., 21 pls. Berlin.
1953   Übersicht über die europäischen Arten der Gattung *Chiastochaeta* (Diptera: Muscidae). Beitr. z. Ent. (Berlin) **3**: 655–668, 27 figs.
1954   Flügelgeäder und System der Dipteren, unter Berücksichtigung der aus dem Mesozoikum beschriebenen Fossilien. Beitr. z. Ent. (Berlin) **4**: 245–388, 272 figs.

Hennig, W.—Continued
- 1955–1957 Muscidae. [Fam.] 63b, pp. 1–48, figs. 1–16, pls. 1–3 (=lfg. 182), *1955;* pp. 49–96, figs. 17–19, pls. 4–6 (=lfg. 185), *1955;* pp. 97–144, figs. 20–42 (=lfg. 194), *1956;* pp. 145–192, figs. 43–44, pls. 7–9 (=lfg. 197), *1957* [cont.]. *In* Lindner, E., ed., Die Fliegen der palaearktischen Region [q.v.]. Vol. 7. Stuttgart.
- 1958a Die Familien der Diptera Schizophora und ihre phylogenetischen Verwandtschaftsbeziehungen. Beitr. z. Ent. (Berlin) **8**: 505–688, 365 figs.
- 1958b Muscidae [cont.]. [Fam.] 63b, pp. 193–232, figs. 45–48 (=lfg. 199) [cont.]. *In* Lindner, E., ed., Die Fliegen der palaearktischen Region [q.v.]. Vol. 7. Stuttgart.
- 1959–1963 Muscidae [cont.]. [Fam.] 63b, pp. 233–288, figs. 49–58, pls. 10–12 (=lfg. 204), *1959;* pp. 289–336, figs. 59–63, pls. 13–15 (=lfg. 205), *1959;* pp. 337–384, figs. 64–70, pls. 16–18 (=lfg. 207), *1959;* pp. 385–432, figs. 71–152, pls. 19–20 (=lfg. 209), *1960;* pp. 433–480, figs. 153–184, pls. 21–22 (=lfg. 213), *1960;* pp. 481–528, figs. 185–212, pls. 23–26 (=lfg. 215), *1961;* pp. 529–576, figs. 213–218, pls. 27–30 (=lfg. 217), *1961;* pp. 577–624, figs. 219–239, pl. 31 (=lfg. 223), *1962;* pp. 625–672, figs. 240–262 (=lfg. 225), *1962;* pp. 673–720, figs. 263–305, pls. 32–33 (=lfg. 227), *1962;* pp. 721–768, figs. 306–313 (=lfg. 229), *1962;* pp. 769–816, figs. 314–334 (=lfg. 233), *1963;* pp. 817–864, figs. 335–347 (=lfg. 234), *1963* [cont.]. *In* Lindner, E., ed., Die Fliegen der palaearktischen Region [q.v.]. Vol. 7. Stuttgart.

Henriksen, K. L., and Lundbeck, W.
- 1917 Conspectus faunae groenlandicae, pars secunda. II. Landarthropoder (Insecta et Arachnida). Meddel. om Grønland (Copenhagen) **22**: 481–823, 4 figs.

Henshaw, S.
- 1889 Part III. The more important writings of Charles Valentine Riley. Pt. 3, pp. 97–454. *In* U.S. Department of Agriculture Division of Entomology, Bibliography of the more important contributions to American economic entomology [q.v.]. Pts. 1/3, 454 pp. Washington, "1890".

Hering, E. M.
- 1935 Drei neue Bohrfliegen-Arten aus der Mark Brandenburg (Dipt. Trypetidae). (6. Beitrag zur Kenntnis der Trypetidae). Märkische Tierwelt **1**: 169–174, 5 figs.

Hering, E. M.—Continued
1940a   Neue alt- und neuweltliche Bohrfliegen (Diptera: Trypetidae). Arb. über Morph. u. Taxonom. Ent. **7**: 50–57, 6 figs.
1940b   I. Neue Arten und Gattungen. Siruna Seva **1**: 1–16, 6 figs.
1940c   II. Kleinere Bemerkungen und Namensänderung. Siruna Seva **1**: 16.
1940d   Neue Arten und Gattungen. Siruna Seva **2**: 1–16, 5 figs.
1942   Neue Gattungen und Arten palaearktischer und exotischer Fruchtfliegen. Siruna Seva **4**: 1–31, 25 figs.
1944   Neue Gattungen und Arten von Fruchtfliegen der Erde. Siruna Seva **5**: 1–17, 8 figs.
1947   Neue Gattungen und Arten der Fruchtfliegen. Siruna Seva **6**: 1–16, 8 figs.
1951   Neue paläarktische und nearktische Agromyziden (Dipt.). Notulae Ent. **31**: 31–45, 3 figs.

Hermann, F.
1905   Beitrag zur Kenntnis der Asiliden (Dipt.). Berlin. Ent. Ztschr. **50**: 14–42, 29 figs.
1912   Beiträge zur Kenntnis der südamerikanischen Dipterenfauna auf Grund der Sammelergebnisse einer Reise in Chile, Peru und Bolivia, ausgeführt in den Jahren 1902–1904 von W. Schnuse. Fam. Asilidae. K. Leopoldinisch-Carolinisch. Deut. Akad. d. Naturf., Abhandl. (Acad. C. Leopoldino-Carolinae Germ. Nat. Curio., Nova Acta) **96**: 1–275, 87 figs., pls. 1–5.
1914   H. Sauter's Formosa-Ausbeute. Mydaidae et Asilidae (Dasypogoninae, Laphrinae et Leptogastrinae) (Dipt.). Ent. Mitt. **3**: 33–44, 83–95, 102–112, 129–136, figs. 1–11.

Herrera, A. L.
1892   Fauna cavernicola de Cacahuamilpa. Pp. 268–281, pls. 14–15. In Arriaga, J. J., Urbina, M., and Rebollar, R., [eds.], Expedición a la gruta de Cacahuamilpa. Estudio, Semenario de Cien. Méd. (Inst. Méd. Nac. de Mex.) (1893) **4**: 268–281, 339–344, 455–461, pls. 14–15, 17.

Herrick, A.
1884   A final report on the Crustacea of Minnesota included in the orders Cladocera and Copepoda. Minn. Geol. and Nat. Hist. Survey, Ann. Rpt. **12** (5): 1–191, 30 pls.

    Parts 1 through 4 of the 12th report are paged continuously, parts 5 and 6 separately.

Hertig, M.   (See Fairchild and Hertig)

Herting, B.
1959 Revision einiger europäischer Raupenfliegen (Dipt., Tachinidae). Wien, Naturhist. Mus. Ann. **63:** 423–429.
1960 Biologie der westpaläarktischen Raupenfliegen. Dipt., Tachinidae. Monog. z. Angew. Ent. **16:** 1–188, 12 figs.

Hesse, A. J.
1938 A revision of the Bombyliidae (Diptera) of southern Africa. [Part I.] So. African Mus. Ann. **34:** 1–1053, 332 figs.
1956 A revision of the Bombyliidae (Diptera) of southern Africa. Part II; Part III. So. African Mus. Ann. **35:** 1–464, figs. 1–170, pl. 1; 465–972, figs. 171–286, pls. 2–3.

Hincks, W. D. (See Kloet and Hincks)

Hine, J. S. (See also Webb and Wells, 1924)
1900 Description of two new species of Tabanidae. Canad. Ent. **32:** 247–248.
1901a Change of name. Canad. Ent. **33:** 28.
1901b Description of new species of Stratiomyidae with notes on others. Ohio Nat. **1:** 112–114, 1 fig.
1902 New or little known Diptera. Ohio Nat. **2:** 228–230.
1903a Some Diptera from Arizona. Canad. Ent. **35:** 244–246.
1903b The genus *Peditia* with one new species. Ohio Nat. **3:** 416–417.
1903c Tabanidae of Ohio with a catalogue and bibliography of the species from America north of Mexico. Ohio Acad. Sci. Spec. Paper **5:** 1–63, 2 pls. (=Contributions from the Ohio State University Department of Zoology and Entomology No. 10.)
1904a New species of North American Tabanidae. Canad. Ent. **36:** 55–56.
1904b The Diptera of British Columbia. (First part.) Canad. Ent. **36:** 85–92.
1904c On Diptera of the family Ephydridae. Ohio Nat. **4:** 63–65.
1904d The Tabanidae of western United States and Canada. Ohio Nat. **5:** 217–249. (=Contributions from the Ohio State University Department of Zoology and Entomology No. 21.)
1904e Insects injurious to stock in the vicinity of the Gulf Biologic Station. Pp. 57–60. *In* Howard, L. O., ed., Some miscellaneous results of the work of the Division of Entomology. VII. U.S. Dept. Agr., Div. Ent., Bul. [n. ser.], **44:** 1–99, 19 figs., 1 pl.

Hine, J. S.—Continued
- 1905   New species of North American *Chrysops*. Ohio Nat. **6**: 391–393.
- 1906a   The North American species of *Tabanus* with a uniform mid-dorsal stripe. Ohio Nat. **7**: 19–28.
- 1906b   Two new species of Diptera belonging to Asilinae. Ohio Nat. **7**: 29–30.
- 1907a   Robber flies of the genus *Philonicus*. Ohio Nat. **7**: 115–118, 1 fig.
- 1907b   Descriptions of new North American Tabanidae. Ohio Nat. **8**: 221–230.
- 1907c   Records of Diptera from Lake Temagami, Ont. Canad. Ent. **39**: 98–99.
- 1908   Two new species of Asilidae from British Columbia. Canad. Ent. **40**: 202–204, 2 figs.
- 1909   Robberflies of the genus *Asilus*. Ent. Soc. Amer. Ann. **2**: 136–172, 2 pls. (=Contributions from the Ohio State University Department of Zoology and Entomology No. 32.)
- 1911a   Robberflies of the genera *Promachus* and *Proctacanthus*. Ent. Soc. Amer. Ann. **4**: 153–172. (=Contributions from the Ohio State University Department of Zoology and Entomology No. 33.)
- 1911b   A new species of *Nothomyia*. Ohio Nat. **11**: 301–302, 1 fig.
- 1911c   New species of Diptera of the genus *Erax*. Ohio Nat. **11**: 307–311.
- 1912   Five new species of North American Tabanidae. Ohio Nat. **12**: 513–516, 4 figs.
- 1916   Descriptions of robber flies of the genus *Erax*. Ohio Jour. Sci. **17**: 21–22.
- 1917   Descriptions of North American Tabanidae. Ohio Jour. Sci. **17**: 269–271.
- 1918a   Notes on robber flies from southwest Texas, collected by the Bryant Walker Expedition, with a description of a new species of *Erax*. Mich. Univ. Mus. Zool. Occas. Papers **61**: 1–7.
- 1918b   Descriptions of seven species of *Asilus* (family Asilidae). Ohio Jour. Sci. **18**: 319–322.
- 1919   Robberflies of the genus *Erax*. Ent. Soc. Amer. Ann. **12**: 103–154, 3 pls. (=Contributions from the Ohio State University Department of Zoology and Entomology No. 57.)

Hine, J. S.—Continued
- 1920 Descriptions of horseflies from Middle America. I; II. Ohio Jour. Sci. **20**: 185–192; 311–319, 4 figs. (=Contributions from the Ohio State University Department of Zoology and Entomology No. 60.)
- 1922a Some robberflies in the University of Michigan Museum of Zoology, and the description of a new species. Mich. Univ. Mus. Zool. Occas. Papers **121**: 1–7.
- 1922b Descriptions of Alaskan Diptera of the family Syrphidae. Ohio Jour. Sci. **22**: 143–147.
- 1923a Horseflies collected by Dr. J. M. Aldrich in Alaska in 1921. Canad. Ent. **55**: 143–146.
- 1923b Some notes on American Tabanidae with the description of a new species from Africa. Ohio Jour. Sci. **23**: 204–206.
- 1925 Tabanidae of Mexico, Central America and the West Indies. Mich. Univ. Mus. Zool. Occas. Papers **162**: 1–35.

Hirvenoja, M.
- 1961 Description of the larva of *Corynocera ambigua* Zett. (Dipt. Chironomidae) and its relation to the subfossil species *Dryadotanytarsus edentulus* Anders. and *D. duffi* Deevey. Ann. Ent. Fenn. **27**: 105–110, 1 fig.

Hitchcock, E. (See Harris, 1833, 1835)

Hodson, W. E. H.
- 1927 The bionomics of the lesser bulb flies, *Eumerus strigatus*, Flyn., and *Eumerus tuberculatus*, Rond., in south-west England. Bul. Ent. Res. **17**: 373–384, pls. 21–22.
- 1932 The large narcissus fly, *Merodon equestris*, Fab. (Syrphidae). Bul. Ent. Res. **23**: 429–448, 1 fig., pl. 39.

Hoffman, W. A. (See also Fox and Hoffman; Root and Hoffman)
- 1924 *Stilobezzia mallochi* and *Atrichopogon gilva* (Dipt.: Chironomidae). Ent. News **35**: 282–284.
- 1925 A review of the species of *Culicoides* of North and Central America and the West Indies. Amer. Jour. Hyg. **5**: 274–301, 1 fig., 2 pls.
- 1926a Two new species of American *Leptoconops* (Diptera, Chironomidae). Bul. Ent. Res. **17**: 133–136, 1 fig.
- 1926b Notes on Ceratopogoninae (Diptera). Ent. Soc. Wash. Proc. **28**: 156–159, 4 figs.

Hogue, C. L. (See Belkin and Hogue)

Holmgren, A. E.
1869 Bidrag till kännedomen om Beeren Eilands och Spetsbergens insekt-fauna. K. Svenska Vetensk. Akad. Handl. **n. ser.** [=ser. 4], **8 (5)**: 3–55.
1872 Insekter från Nordgrönland, samlade af Prof. A. E. Nordenskiöld år 1870. K. Svenska Vetensk. Akad. Öfversigt af . . . Förhandl. **29 (6)**: 97–105.
1880 Novas species insectorum cura et labore A. E. Nordenskiöldii e Novaia Semlia coactorum. 24 pp. Holmiae [=Stockholm].
1883 Diptera. Pp. 162–190. *In* Holmgren, A. E., and Aurivillius, C., Insecta a viris doctissimis Nordenskiöld illun ducem sequentibus in insulis Waigatsch et Novaja Semlia anno 1875 collecta. Ent. Tidskr. **4**: 139–194, 8 pls.

Holmgren, A. E., and Aurivillius, C. (See Holmgren, 1883)

Holtedahl, O., ed.
1922– Report of the scientific results of the Norwegian Expedition
1930 to Novaya Zemlya 1921. 45 rpts. Oslo.
   These separately paged reports are variously bound by libraries in 1 or more volumes.
   See Kieffer, 1922b; Sack, 1923.

Hooker, C. W. (See Russell and Hooker)

Hopkins, A. D.
1895 Notes on the habits of certain mycetophilids, with description of *Epidapus scabiei* n. sp. Ent. Soc. Wash. Proc. **3**: 149–159, 20 figs.

Horn, W., and Schenkling, S.
1928– Index litteraturae entomologicae. Serie I: Die Welt-
1929 Literatur über die gesamte Entomologie bis inklusiv 1863. Vol. 1, pp. 1–352, pl. 1, *1928;* Vol. 2 pp. 353–704, pl. 2, *1928;* Vol. 3, pp. 705–1056, pl. 3, *1928;* Vol. 4, pp. 1057–1426, pl. 4, *1929*. Berlin-Dahlem.
   "Series II: 1864–1900" is in process of publication.

Horsfall, W. R.
1955 Mosquitoes. Their bionomics and relation to disease. 723 pp. New York.

Hough, G. de N. (See also Hunter, 1898)
1898a Two new American species of *Cynomyia*, a study in chaetotaxy. Ent. News **9**: 105–111, 2 figs.
1898b A third American species of *Cynomyia*. Ent. News **9**: 165–166, 1 fig.

Hough, G. de N.—Continued
1899a Synopsis of the Calliphorinae of the United States. Zool. Bul. **2**: 283–290, 11 figs.
1899b Some Muscinae of North America. Biol. Bul. (1900) **1**: 19–33, 19 figs.
1899c Some North American genera of the dipterous group, Calliphorinae Girschner. Ent. News **10**: 62–66.
1899d Studies in Diptera Cyclorhapha. 1. The Pipunculidae of the United States. Boston Soc. Nat. Hist. Proc. **29**: 77–86.

Howard, L. O. (See also Hine, 1904e; Riley and Howard)
1897 Some insects affecting the hop plant. Pp. 40–51, figs. 35–38. *In his* Some miscellaneous results of the work of the Division of Entomology. U.S. Dept. Agr., Div. Ent., Bul. **n. ser.**, **7**: 1–87, 44 figs.
1901 Mosquitoes: How they live, how they carry disease, how they are classified, how they may be destroyed. 241 pp., 50 figs. New York.

———, **Dyar, H. G., and Knab, F.**
1913 The mosquitoes of North and Central America and the West Indies. Volume 2, Plates. Carnegie Inst. Wash., Pub. (1912) **159** (**2**): i–x, 150 pls.

> Stamped date of issue is Feb. 24, 1913.

1915–
1917 The mosquitoes of North and Central America and the West Indies. Volume 3: Systematic description (in two parts). Part I; Volume 4: Systematic description (in two parts). Part II. Carnegie Inst. Wash., Pub. **159** (**3**): 1–523, 1 fig., *1915;* (**4**): 523–1064, *1917*.

> Volume 1 of Pub. 159 (1912) is separately paged; Volume 2 contains the plates.

Hoyt, C. P. (See Arnaud and Hoyt)

Hubert, A. A. (See also Wirth and Hubert)
1953 Another species of *Rhamphomyia* predaceous on mosquitoes (Diptera: Empididae). Pan-Pacific Ent. **29**: 190.

Huckett, H. C. (See also James and Huckett)
1924 A systematic study of the Anthomyiinae of New York, with especial reference to the male and female genitalia. N.Y. (Cornell) Agr. Expt. Sta. Mem. **77**: 1–91, 1 fig., 18 pls.
1927 A new kelp fly from Long Island (*Fucellia*, Diptera). Brooklyn Ent. Soc. Bul. **22**: 163–165, 1 fig.

Huckett, H. C.—Continued
- 1928 Little known anthomyid flies that commonly occur on the catkins of willow (Muscidae, Diptera). Brooklyn Ent. Soc. Bul. **23**: 70–82, 1 fig., 1 pl.
- 1929 New Canadian anthomyids belonging to the genus *Hylemyia* Rob.-Desv. (Muscidae, Diptera). Canad. Ent. **61**: 93–96, 110–119, 136–144, 161–168, 180–190, 4 pls.
- 1932 The North American species of the genus *Limnophora* Robineau-Desvoidy, with descriptions of new species (Muscidae, Diptera). N.Y. Ent. Soc. Jour. **40**: 25–76, 105–158, 279–338, 7 pls.
- 1934 A revision of the North American species belonging to the genus *Coenosia* Meigen and related genera (Diptera: Muscidae). Part I. The subgenera *Neodexiopsis, Coenosia, Hoplogaster* and related genera *Allognota, Bithoracochaeta* and *Schoenomyza*; Part II. The subgenus *Limosia* (*Coenosia* of authors). Amer. Ent. Soc. Trans. **60**: 57–119, 4 pls.; 133–198, 6 pls.
- 1936 A revision of the connectant forms between coenosian and limnophorine genera occurring in North America (Diptera, Muscidae). N.Y. Ent. Soc. Jour. **44**: 187–222, 2 pls.
- 1939 Descriptions of new North American Anthomyiidae belonging to the genus *Pegomyia* Rob.-Desv. (Diptera). Amer. Ent. Soc. Trans. **65**: 1–36, 3 pls.
- 1940 The North American species of the genera *Leucophora* Robineau-Desvoidy and *Proboscimyia* Bigot (Muscidae, Diptera). N.Y. Ent. Soc. Jour. **48**: 335–364, 4 pls.
- 1941 A revision of the North American species belonging to the genus *Pegomyia* (Diptera: Muscidae). Amer. Ent. Soc. Mem. **10**: 1–131, 9 pls.
- 1944a A revision of the North American species belonging to the genus *Hydrophoria* Robineau-Desvoidy (Diptera: Muscidae). Ent. Soc. Amer. Ann. **37**: 261–297, 6 pls.
- 1944b A revision of the North American genus *Eremomyioides* Malloch (Diptera, Muscidae). N.Y. Ent. Soc. Jour. **52**: 361–368.
- 1946 The subgenera *Craspedochaeta* and *Acrostilpna* in North America, genus *Hylemyia* sens. lat. (Diptera, Muscidae). Brooklyn Ent. Soc. Bul. **41**: 110–125, 1 pl.
- 1947 The North American species of the subgenus *Botanophila* Lioy, genus *Hylemyia* sens. lat. (Diptera, Muscidae). N.Y. Ent. Soc. Jour. **55**: 1–33, 2 pls.

Huckett, H. C.—Continued
- 1948 The subgenus *Phorbia* Robineau-Desvoidy in North America, genus *Hylemyia* sens. lat. (Diptera, Muscidae). Brooklyn Ent. Soc. Bul. (1947) **42**: 109–125, 2 pls.
- 1949 The subgenus *Pycnoglossa* Coquillett in North America, genus *Hylemyia* sens. lat. (Muscidae, Diptera). N.Y. Ent. Soc. Jour. **57**: 51–65, 1 pl.
- 1950 The genus *Paraprosalpia* (Villeneuve) in North America, (=*Prosalpia* Pokorny preoc.) Muscidae. Brooklyn Ent. Soc. Bul. **45**: 121–132, 133–143, 1 pl.
- 1951a The genus *Eremomyia* Stein in North America, with descriptions of new species (Muscidae: Diptera). N.Y. Ent. Soc. Jour. **59**: 75–91, 1 pl.
- 1951b The *setiventris*-complex in the genus *Hylemya* Rob. Desv., with descriptions of new species and subspecies from North America (Diptera, Muscidae). Ent. Soc. Wash. Proc. **53**: 251–260, 1 pl.
- 1952 Males of the genus *Hylemya* sens. lat. from North America, having dorsal bristles on mid metatarsus, with descriptions of new species (Muscidae, Diptera). Brooklyn Ent. Soc. Bul. **47**: 113–122 [cont.].
- 1953a Males of the genus *Hylemya* sens. lat. from North America, having dorsal bristles on mid metatarsus, with descriptions of new species (Muscidae, Diptera) [concl.]. Brooklyn Ent. Soc. Bul. **48**: 10–19, pl. 3.
- 1953b A new species of the anthomyiid genus *Hylemya* Rob.-Desv. from Oregon, reared from fir cones (Muscidae, Diptera). Brooklyn Ent. Soc. Bul. **48**: 107–110, pl. 6.
- 1954 A review of the North American species belonging to the genus *Hydrotaea* Robineau-Desvoidy (Diptera, Muscidae). Ent. Soc. Amer. Ann. **47**: 316–342.

Hull, F. M. (See also Fluke and Hull)
- 1922 New Syrphidae (Diptera) from Mississippi. Ent. Soc. Amer. Ann. **15**: 370–373.
- 1923 Notes on the family Syrphidae (Diptera) with descriptions of new species. Ohio Jour. Sci. **23**: 295–298.
- 1924 *Milesia* in North America (Dipt.: Syrphidae). Ent. News **35**: 280–282.
- 1925a Notes on the North American species of the genus *Didea* with the description of a new species. Ent. Soc. Amer. Ann. **18**: 277–280, 1 pl.
- 1925b A review of the genus *Eristalis* Latreille in North America. [Part I]; Part II. Ohio Jour. Sci. **25**: 11–45, 2 pls.; 285–312, 2 pls.

Hull, F. M.—Continued
- 1930a Some new species of Syrphidae (Diptera) from North and South America. Amer. Ent. Soc. Trans. **56:** 139–148, 1 pl.
- 1930b Notes on several species of North American Pachygasterinae (Diptera: Stratiomyidae) with the description of a new species. Ent. News **41:** 103–106.
- 1930c Some notes and descriptions of cerioidine wasp-waisted flies (Syrphidae, Diptera). Psyche **37:** 178–181.
- 1935a Descriptions of new species of the genus *Sphegina* with a key to those known from North America (Syrphidae: Diptera). Amer. Ent. Soc. Trans. **61:** 373–382, 1 pl.
- 1935b Some mimetic flies, with description of two new species from North America (Syrphidae: Diptera). Psyche **42:** 99–102, 1 pl.
- 1941a Some new species of Syrphidae. Brooklyn Ent. Soc. Bul. **36:** 166–168.
- 1941b Some new syrphid flies from North and South America (Diptera). Ent. News **52:** 157–163.
- 1941c Some new species of Syrphidae from Florida, Cuba and Brazil (Diptera). Ent. News **52:** 278–283.
- 1942a The flies of the genus *Meromacrus* (Syrphidae). Amer. Mus. Nat. Hist., Amer. Mus. Novitates **1200:** 1–10, 1 pl.
- 1942b Notes and descriptions of North American Stratiomyidae. Brooklyn Ent. Soc. Bul. **37:** 70–72.
- 1942c Some new species of Syrphidae. Psyche **49:** 19–24.
- 1942d The genus *Ferdinandea* Rondani. Wash. Acad. Sci. Jour. **32:** 239–241.
- 1942e Some new species of *Baccha* and *Mesogramma* (Dipt. Syrphidae). Rev. de Ent. **13:** 44–49.
- 1943a The genus *Mesogramma*. Ent. Amer. **23:** 1–41, 7 pls.
- 1943b The New World species of the genus *Baccha*. Ent. Amer. **23:** 42–99, 2 figs., 10 pls.
  See Hull, 1949a, for continuation.
- 1943c Some new American syrphid flies (Diptera). Ent. News **54:** 29–37.
- 1943d New species of syrphid flies in the National Museum. Wash. Acad. Sci. Jour. **33:** 39–43.
- 1943e New species of American syrphid flies. Brooklyn Ent. Soc. Bul. **38:** 48–53.
- 1944a New syrphid flies from North and South America. Brooklyn Ent. Soc. Bul. **39:** 35–40.
- 1944b Additional species of the genus *Baccha* from the New World. Brooklyn Ent. Soc. Bul. **39:** 56–64.

Hull, F. M.—Continued

1944c Some new forms of *Platycheirus* of the family Syrphidae. Brooklyn Ent. Soc. Bul. **39**: 75–79.

1944d Notes upon flies of the genus *Solva* Walker. Ent. News **55**: 263–265.

1944e Studies on syrphid flies in the Museum of Comparative Zoölogy. Psyche **51**: 22–45.

1945a New syrphid flies. Ent. News **56**: 182–187, 210–217.

1945b Some undescribed syrphid flies. New England Zool. Club, Proc. **23**: 71–78.

[1946] New syrphid flies from Mississippi. Ent. News (1945) **56**: 268–272.

> This is in the December issue, stamped as received at the Smithsonian Institution library, Feb. 11, 1946. The November issue was mailed Dec. 14, 1945, and there commonly were delays of 1 to 3 months in the publication of the numbers of this journal during 1944–1946.

1949a The genus *Baccha* from the New World. Ent. Amer. (1947) **27**: 89–285, 47 pls.

> A continuation of Hull, 1943b.

1949b The morphology and inter-relationship of the genera of syrphid flies, recent and fossil. Zool. Soc. London, Trans. **26**: 257–408, 25 figs.

1954 The genus *Mixogaster* Macquart (Diptera, Syrphidae). Amer. Mus. Nat. Hist., Amer. Mus. Novitates **1652**: 1–28, 32 figs.

1956 A new southwestern species of *Mallota* Meigen (Diptera: Syrphidae). Psyche **63**: 24–26.

1957a Some new species of robber flies (Diptera: Asilidae). Psyche **64**: 70–75.

1957b Some flies of the family Asilidae (Diptera). Psyche **64**: 90–96.

1958a New genera of robber flies (Diptera, Asilidae). Rev. Brasil. de Biol. **18**: 317–324.

1958b Some robber flies (Diptera: Asilidae). Ent. News **69**: 99–108.

1961 The genus *Psilocurus* Loew. Ent. News **72**: 101–104.

1962 Robberflies of the World. The genera of the family Asilidae. Part 1; Part 2. U.S. Natl. Mus. Bul. **224**: 1–430, figs. 1–29, 1 pl.; 431–907, figs. 30–35+1–2536.

1963 A new species of *Laphystia* Loew (Diptera: Asilidae). Ent. News **74**: 202–203.

Hull, F. M., and Fluke, C. L., Jr.
1950   The genus *Cheilosia* Meigen (Diptera, Syrphidae). The subgenera *Cheilosia* and *Hiatomyia*. Amer. Mus. Nat. Hist. Bul. **94**: 299–402, 173 figs.

Humboldt, F. H. A. von
1819   Voyage aux régions équinoxiales du Nouveau Continent, fait en 1799, 1800, 1801, 1802, 1803 et 1804 par Al. de Humboldt et A. Bonpland. Vol. 2, 722 pp. (=livr. 5–8). *In* [Humboldt, F. H. A., ed.], (Voyage de MM Alexandre de Humboldt et Aimé Bonpland) [q.v.]. Pt. I: Relation historique. Paris.

[Humboldt, F. H. A. von, ed.]
1805–1837   (Voyage de MM Alexandre de Humboldt et Aimé Bonpland.) 6 pts., in 23 vols. Paris.

> This work was issued in livraisons, many authors being involved. The arrangement and collation of the volumes vary. They are divided among six parts, each part with its own title, though the half-title as given above appears on most of the volumes. The 6 parts are: I. Relation historique; II. [Zoologique]; III. Essai politique; IV. Astronomique; V. [Physique générale]; and VI. Botanique. Part I is authored by Humboldt and consists of 3 numbered volumes (a fourth planned volume was never issued), and 2 Atlases, Paris, 1814–1837.
> See Humboldt, 1819.

Hungerford, H. B.
1915   A parasite of the cottonwood borer beetle (Col., Dip.). Ent. News **26**: 135.

Hunter, S. J.
1898   Parasitic influences on *Melanoplus*. Kans. Univ. Quart. ser. A, **7**: 207–210.

Hunter, W. D.
1896a   A contribution to the knowledge of North American Syrphidae. [I]. Canad. Ent. **28**: 87–101.
1896b   A summary of the members of the genus *Chilosia*, Meig., in North America, with descriptions of new species. Canad. Ent. **28**: 227–233.
1896c   A new species of *Tropidia* (Syrphidae) and note on the generic position of *Melanostoma rufipes* Williston. Ent. News **7**: 215–216.
1896d   [Name change]. Ent. News **7**: 305.

Hunter, W. D.—Continued
1897   A contribution to the knowledge of North American Syrphidae. II. Canad. Ent. **29:** 121–144, 1 pl.
1916   A new species of *Cephenomyia* from the United States (Diptera, Oestridae). Ent. Soc. Wash. Proc. **17:** 169–173, 1 pl.
———, **Pratt, F. C., and Mitchell, J. D.**
1913   The principal cactus insects of the United States. U.S. Dept. Agr., Bur. Ent., Bul. [n. ser.], **113:** 1–71, 7 pls.

Hurd, P. D., Jr.   (See Camras and Hurd)

Hutton, F. W.
1901   Synopsis of the Diptera Brachycera of New Zealand. New Zeal. Inst. Trans. and Proc. (1900) **33** [=n. ser., 16]: 1–95.

Hyslop, J.
1910   (Biological notes on *Thereva egressa* Coquillett). P. 98. *In* [Crawford, J. C., sec.], (Short notes and exhibition of specimens). Ent. Soc. Wash. Proc. **12:** 92–98.

I.C.Z.N.   (See International Commission on Zoological Nomenclature)

Ide, F. P.
1926   Descriptions of two new species of *Platycheirus* (Syrphidae, Diptera). Canad. Ent. **58:** 155–156.

Illiger, J. C. W.   (See also Hellwig and Illiger)
1801   Neue Insekten. Mag. f. Insektenkunde (1802) **1:** 163–208.
1807   Vol. 2, 511 pp., 9 pls. *In* (Hellwig, J. C. L., and Illiger, J. C. W.), Fauna Etrusca [q.v.]. Helmstadii [=Helmstedt, Germany].

Illingworth, J. F.
1912   Cherry fruit-flies and how to control them. N.Y. (Cornell) Agr. Expt. Sta. Bul. **325:** 190–204, figs. 45–66.
    Also bound in N.Y. (Cornell) State Col. Agr., and Agr. Expt. Sta., Ann. Rpt. **26:** 190–204, figs. 45–66, *1913*.

Imms, A. D.
1957   A general textbook of entomology including the anatomy, physiology, development and classification of insects. Revised by O. W. Richards and R. G. Davies. Ed. 9, 886 pp., 609 figs. London.

Ingram, A., and Macfie, J. W. S.
1921   West African Ceratopogoninae. Ann. Trop. Med. and Parasitol. **15:** 313–374, 23 figs., 1 pl.

Instituto Oswaldo Cruz   (See Lutz, 1909)

International Commission on Zoological Nomenclature

1914 Opinion 57. Names dating from Hasselquist's "Iter Palaestinum," 1757, and the translation, 1762, are untenable. Pp. 131–134. *In their* Opinions rendered by the International Commission on Zoological Nomenclature, Opinions 57 to 65. Smithsn. Inst. Pub. **2256:** 131–169.

> Also published in Hemming, F., Opinions and declarations rendered by the International Commission on Zoological Nomenclature, Vol. 1, Sect. B, pp. 131–134, *1958*.

1925 Opinion 82. Suspension of rules for *Musca* Linnaeus, 1758A, type *Musca domestica.* Pp. 1–7. *In their* Opinions rendered by the International Commission on Zoological Nomenclature, Opinions 82–90. Smithsn. Inst., Smithsn. Misc. Collect. **73** (3 [=pub. 2830]): 1–40.

> Also published in Hemming, F., Opinions and declarations rendered by the International Commission on Zoological Nomenclature, Vol. 1, Sect. B, pp. 295–301, *1958*.

1928 Opinion 98. Brauer and Bergenstamm. Pp. 1–3. *In their* Opinions rendered by the International Commission on Zoological Nomenclature, Opinions 98 to 104. Smithsn. Inst., Smithsn. Misc. Collect. **73** (5 [=pub. 2973]): 1–28.

> Also published in Hemming, F., Opinions and declarations rendered by the International Commission on Zoological Nomenclature, Vol. 1, Sect. B, pp. 369–371, *1958*.

1954a Opinion 205. Rejection of the generic name *Phoranthella* Townsend (Class Insecta, Order Diptera), as published in 1915, as a nomen nudum. Pp. 309–318 (=pt. 24). *In* Hemming, F., ed., Opinions and declarations rendered by the International Commission on Zoological Nomenclature [q.v.]. Vol. 3, 448 pp., 2 pls. London.

1954b Opinion 256. Emendation to *Phlebotomus* of the generic name *Flebotomus* Rondani, 1840 (Class Insecta, Order Diptera) under the Plenary Powers. Pp. 199–230 (=pt. 24). *In* Hemming, F., ed., Opinions and declarations rendered by the International Commission on Zoological Nomenclature [q.v.]. Vol. 5, 426 pp., 4 pls. London.

1954c Opinion 290. Validation, under the Plenary Powers, of the generic names *Acantholyda* Costa, 1894 (Class Insecta, Order Hymenoptera) and *Acanthocnema* Becker, 1894 (Class Insecta, Order Diptera). Pp. 89–98 (=pt. 7). *In* Hemming, F., ed., Opinions and declarations rendered by the International Commission on Zoological Nomenclature [q.v.]. Vol. 8, 404 pp. London.

International Commission on Zoological Nomenclature—Continued

1955a Opinion 348. Suppression, under the Plenary Powers, of the generic name *Titania* Meigen, 1800, for the purpose of validating the generic name *Chlorops* Meigen, 1803 (Class Insecta, Order Diptera). Pp. 421–436 (=pt. 15). *In* Hemming, F., ed., Opinions and declarations rendered by the International Commission on Zoological Nomenclature [q.v.]. Vol. 10, 562 pp., 1 pl. London.

1955b Opinion 369. Suppression under the Plenary Powers of the generic names *Tylos* Meigen, 1800 (Class Insecta, Order Diptera) and *Tylos* Heyden, 1826 (Class Arachnida) and validation thereby of the generic names *Tylos* Audouin, [1826] (Class Crustacea, Order Isopoda) and *Micropeza* Meigen, 1803 (Class Insecta, Order Diptera). Pp. 265–300 (=pt. 19). *In* Hemming, F., ed., Opinions and declarations rendered by the International Commission on Zoological Nomenclature [q.v.]. Vol. 11, 480 pp., 1 pl. London.

1957a Opinion 441. Validation under the Plenary Powers of the names for five genera in the Order Diptera (Class Insecta) published in 1762 by Geoffroy (E. L.) in the work entitled "Histoire abrégée des Insectes qui se trouvent aux environs de Paris" ("Opinion" supplementary to "Opinion" 228). Pp. 83–120 (=pt. 6). *In* Hemming, F., ed., Opinions and declarations rendered by the International Commission on Zoological Nomenclature [q.v.]. Vol. 15, 530 pp. London.

1957b Opinion 442. Validation under the Plenary Powers of the generic name "*Stratiomys*" Geoffroy, 1762 (Class Insecta, Order Diptera). Pp. 121–162 (=pt. 7). *In* Hemming, F., ed., Opinions and declarations rendered by the International Commission on Zoological Nomenclature [q.v.]. Vol. 15, 530 pp. London.

1958a Opinion 526. Direction that *Lestodiplosis* be treated as the valid original spelling for the generic name published by Kieffer in 1894 both with the above spelling and with the spelling *Leptodiplosis* (Class Insecta, Order Diptera). Pp. 291–300 (=pt. 13). *In* Hemming, F., ed., Opinions and declarations rendered by the International Commission on Zoological Nomenclature [q.v.]. Vol. 19, 436 pp. London.

International Commission on Zoological Nomenclature—Continued

1958b   Opinion 531. Validation under the Plenary Powers of the generic name *"Campsicnemus"* Haliday, 1851 (Class Insecta, Order Diptera). Pp. 349–360 (=pt. 20). *In* Hemming, F., ed., Opinions and declarations rendered by the International Commission on Zoological Nomenclature [q.v.]. Vol. 19, 436 pp. London.

1959a   Opinion 547. Designation under the Plenary Powers of a type species in harmony with accustomed usage for the nominal genus *Anopheles* Meigen, 1818 (Class Insecta, Order Diptera). Pp. 153–164 (=pt. 14). *In* Hemming, F., ed., Opinions and declarations rendered by the International Commission on Zoological Nomenclature [q.v.]. Vol. 20, 448 pp., 2 pls. London.

1959b   Opinion 548. Validation under the Plenary Powers of the generic name *Toxorhynchites* Theobald (Class Insecta, Order Diptera), as published in 1901 in the Journal of Tropical Medicine. Pp. 165–174 (=pt. 15). *In* Hemming, F., ed., Opinions and declarations rendered by the International Commission on Zoological Nomenclature [q.v.]. Vol. 20, 448 pp., 2 pls. London.

1959c   Opinion 550. Suppression under the Plenary Powers of the generic name *Taeniorhynchus* Lynch-Arribálzaga, 1891, for the purpose of protecting the generic name *Mansonia* Blanchard, 1901 (Class Insecta, Order Diptera) and matters relating to other names in the same class and in the Class Aves incidental thereto. Pp. 185–198 (=pt. 17). *In* Hemming, F., ed., Opinions and declarations rendered by the International Commission on Zoological Nomenclature [q.v.]. Vol. 20, 448 pp., 2 pls. London.

1961a   Opinion 597. *Prothechus* Rondani, 1856, and *Alloneura* Rondani, 1856 (Insecta, Diptera), suppressed under the Plenary Powers. Internatl. Comn. Zool. Nomencl., Bul. Zool. Nomencl. **18**: 230–235.

1961b   Opinion 616. *Tanytarsus* van der Wulp, 1874 (Insecta, Diptera). Designation of a type-species under the Plenary Powers. Internatl. Comn. Zool. Nomencl., Bul. Zool. Nomencl. **18**: 361–362.

1963a   Opinion 652. *Dasiops* Rondani, 1856 (Insecta, Diptera): Designation of a type-species under the Plenary Powers. Internatl. Comn. Zool. Nomencl., Bul. Zool. Nomencl. **20**: 114–116.

International Commission on Zoological Nomenclature—Continued
 1963b Opinion 678. The suppression under the Plenary Powers of the pamphlet published by Meigen, 1800. Internatl. Comn. Zool. Nomencl., Bul. Zool. Nomencl. **20**: 339–342.

Ishida, H.
 1959 The catalogue of the Japanese Tipulidae, with keys to the genera and subgenera (Diptera). V. Limoniinae, tribe Hexatomini. Hyogo Univ. Agr. (Sasayama), Sci. Rpt., Ser. Nat. Sci. **4**: 1–11.

Izquierdo, J. J.
 1916 Investigaciones sobre paludismo en Puebla. Tesis, Colegio del Estado de Puebla. 104 pp., 18 figs. [Puebla, Mexico].

Jacobson, G.
 1898 Compte rendu de l'expédition, envoyée par l'Académie Impériale des Sciences à Novaia Zemlia en été 1896. IV. [Zoological explorations on Novaya Zemlya. Insects of Novaya Zemlya]. [Akad. Nauk S.S.S.R.] Imp. Akad. Nauk St. Petersburg, Zap., Fiz.-Mat. Otd. (Acad. Imp. de St. Pétersbourg, Mém., Cl. Phys.-Math.) **ser. 8, 8 (1)**: 171–244.
> In Russian. Reference has been seen to a separately-paged work of 74 pp., also 1898, which may be either a reprint or a separate publication.

Jaennicke, F.
 1867 Neue exotische Dipteren. Senckenb. Naturf. Gesell. Abhandl. **6**: 311–408, pls. 43–44.
> Also published separately as "Neue exotische Dipteren aus den Museen zu Frankfurt a. M. und Darmstadt", 100 pp., 2 pls., Frankfurt, *1868*.

James, M. T. (See also Eldridge and James; Harmston and James)
 1932a New Stratiomyidae in the American Museum of Natural History. Amer. Mus. Nat. Hist., Amer. Mus. Novitates **571**: 1–7, 5 figs.
 1932b New and little-known Colorado Diptera. N.Y. Ent. Soc. Jour. **40**: 435–438.
 1933a New Stratiomyidae in the Snow Entomological Collection. Kans. Ent. Soc. Jour. **6**: 66–71.
 1933b New Asilidae from Colorado. Amer. Mus. Nat. Hist., Amer. Mus. Novitates **596**: 1–3.
 1934a *Hoplitimyia*, a new genus of Stratiomyidae. Ent. Soc. Amer. Ann. **27**: 443–444.
 1934b Taxonomic notes on some Colorado Asilidae. Pan-Pacific Ent. **10**: 83–86.

James, M. T.—Continued
1935a The nearctic species of *Adoxomyia* (Diptera, Stratiomyidae). Pan-Pacific Ent. **11**: 62–64.
1935b The genus *Hermetia* in the United States (Diptera, Stratiomyidae). Brooklyn Ent. Soc. Bul. **30**: 165–170.
1936a A proposed classification of the nearctic Stratiomyinae (Diptera: Stratiomyidae). Amer. Ent. Soc. Trans. **62**: 31–36, 2 figs.
1936b Some new western Bibionidae (Diptera). Amer. Mus. Nat. Hist., Amer. Mus. Novitates **832**: 1–6, 1 fig.
1936c Notes on *Nemotelus* (Dipt., Stratiomyidae). Brooklyn Ent. Soc. Bul. **31**: 86–91.
1936d The genus *Odontomyia* in America north of Mexico (Diptera, Stratiomyidae). Ent. Soc. Amer. Ann. **29**: 517–550, 1 pl.
1936e New species and records of Colorado Diptera. N.Y. Ent. Soc. Jour. **44**: 341–344.
1936f New Stratiomyidae in the collection of the California Academy of Sciences. Pan-Pacific Ent. **12**: 86–90.
1936g A review of the nearctic Geosarginae (Diptera, Stratiomyidae). Canad. Ent. (1935) **67**: 267–275.
1937a Some new and little-known neotropical and subtropical Stratiomyidae. Brooklyn Ent. Soc. Bul. **32**: 149–155.
1937b New Colorado Asilidae (Diptera). Ent. News **48**: 12–15.
1937c The genus *Comantella* Curran (Diptera, Asilidae). Pan-Pacific Ent. **13**: 61–63.
1938a Notes on some North American Mydaidae (Diptera). Ent. News **49**: 63–64.
1938b The dipterous families Nemestrinidae, Cyrtidae, and Scenopinidae in Colorado. Kans. Ent. Soc. Jour. **11**: 21–23.
1938c A second species of *Scoliopelta* (Diptera, Stratiomyidae). Pan-Pacific Ent. **14**: 156–157.
1938d A systematic and ecological study of the robber flies (Asilidae) of Colorado. Pp. 70–74. *In* [Lester, O. C., ed.], Abstracts of theses and reports for higher degrees, 1938. Colo. Univ., Studies (1938–1941) **ser. A: Gen., 26** (**1**): 21–160. (=Colo. Univ. Bul. 38 (19) [=Gen. Ser. 423].)
1939a The genus *Dolichopus* in Colorado (Diptera, Dolichopodidae). Amer. Ent. Soc. Trans. **65**: 209–226.
1939b Notes on my monograph of *Odontomyia* (Diptera, Stratiomyidae). Brooklyn Ent. Soc. Bul. **34**: 220.

James, M. T.
- 1939c   A review of the nearctic Beridinae (Diptera, Stratiomyidae). Ent. Soc. Amer. Ann. **32**: 543–548.
- 1939d   Studies in neotropical Stratiomyidae (Diptera). Kans. Ent. Soc. Jour. **12**: 32–36, 37–46.
- 1939e   The species of *Euparyphus* related to *crotchii* O. S. (Diptera, Stratiomyidae). Pan-Pacific Ent. **15**: 49–56.
- 1940   Studies in neotropical Stratiomyidae (Diptera). Rev. de Ent. **11**: 119–149, 3 figs.
- 1941a   Notes on the nearctic Geosarginae (Diptera: Stratiomyidae). Ent. News **52**: 105–108.
- 1941b   The robber flies of Colorado (Diptera, Asilidae). Kans. Ent. Soc. Jour. **14**: 27–36, 37–53.
- 1942a   Additions to the "Robber flies of Colorado". Kans. Ent. Soc. Jour. **15**: 124–126, 1 fig.
- 1942b   A review of the Myxosargini (Diptera, Stratiomyidae). Pan-Pacific Ent. **18**: 49–60, 7 figs.
- 1942c   A new *Empis* of the subgenus *Pachymeria* (Diptera, Empididae). Pan-Pacific Ent. **18**: 163–164.
- 1943a   The genus *Culicoides* in northern Colorado (Diptera, Ceratopogonidae). Pan-Pacific Ent. **19**: 148–153, 3 figs.
- 1943b   A revision of the nearctic species of *Adoxomyia* (Diptera, Stratiomyidae). Ent. Soc. Wash. Proc. **45**: 163–171, 3 figs.
- 1946   The dipterous family Tylidae (Micropezidae) in Colorado. Ent. News **57**: 128–131, 1 fig.
- 1947   A review of the larvaevorid flies of the tribe Leskiini with the setulose first vein (R_1). U.S. Natl. Mus. Proc. **97**: 91–115, 1 fig.
- [1948] The flies that cause myiasis in man. U.S. Dept. Agr. Misc. Pub. (1947) **631**: 1–175, 98 figs.
    Dated by Sabrosky, 1950f: 315.
- 1949   Some new and poorly known Therevidae (Diptera) from Colorado. Ent. Soc. Amer. Ann. **42**: 10–13, 3 figs.
- 1950a   The genus *Scopeuma* in the western United States and southwestern Canada (Diptera, Scopeumatidae). Ent. Soc. Amer. Ann. **43**: 343–353.
- 1950b   Some new and poorly-known Adoxomyiinae (Diptera, Stratiomyidae) from the Southwest. Kans. Ent. Soc. Jour. **23**: 71–73.
- 1951   The Stratiomyidae of Alaska (Diptera). Ent. Soc. Wash. Proc. **53**: 342–343.
- 1952   The ethiopian genera of Sarginae, with descriptions of new species. Wash. Acad. Sci. Jour. **42**: 220–226, 5 figs.

James, M. T.
1953   Notes on the distribution, systematic position, and variation of some Calliphorinae, with particular reference to the species of western North America (Diptera, Calliphoridae). Ent. Soc. Wash. Proc. **55**: 143–148.
1955a   The blowflies of California (Diptera: Calliphoridae). Calif. Insect Survey, Bul. **4**: 1–34, 1 fig., 2 pls.
1955b   The genus *Cordilura* in America north of Mexico. Ent. Soc. Amer. Ann. **48**: 84–100, 2 pls.
1955c   Two new Diptera from the Pacific Coast States. Kans. Ent. Soc. Jour. **28**: 47–48, 1 fig.
1955d   A new tachina fly of economic importance. Pan-Pacific Ent. **31**: 83–85, 1 fig.
1955e   A new sarcophagid parasite of *Nomia* bees (Diptera). Ent. Soc. Wash. Proc. **57**: 283–285.
1957a   The genus *Eulalia* in Florida and the West Indies. Fla. Ent. **40**: 15–18.
1957b   A new *Stratiomys* from California (Diptera: Stratiomyidae). Pan-Pacific Ent. **33**: 43–44.

———, **and Huckett, H. C.**
1952   The Diptera collected by I. O. Buss in southwestern Yukon Territory during the summer of 1950. Canad. Ent. **84**: 265–269.

———, **and Steyskal, G. C.**
1952   A review of the nearctic Stratiomyini (Diptera, Stratiomyidae). Ent. Soc. Amer. Ann. **45**: 385–412, 18 figs.

Jamnback, H. A. (See also Stone and Jamnback)
———, **and Wirth, W. W.**
1963   The species of *Culicoides* related to *obsoletus* in eastern North America (Diptera: Ceratopogonidae). Ent. Soc. Amer. Ann. **56**: 185–198, 3 pls.

Jaques, H. E.
[1939]   A preliminary list of the Tetanoceridae of Iowa. Iowa Acad. Sci. Proc. (1938) **45**: 283.

> This is stamped as having been received by the Smithsonian Library Aug. 22, 1939, a year's delay in publication as was ordinary during that period.

Jarvis, T. D.
1907   Insect galls of Ontario. Ent. Soc. Ontario, Ann. Rpt. (1906) **37**: 56–72, pls. A–F.

Jeannel, R. (See Alluaud and Jeannel)

Jellison, W. L.
1949   *Cuterebra thomomuris* sp. nov., a warble from the pocket gopher, *Thomomys talpoides* (Rodentia: Geomyidae). Jour. Parasitol. **35**: 482–486, 3 figs.

Jenks, G. E.
1938   Marvels of metamorphosis. Natl. Geog. Mag. **74**: 807–828, 23 figs.

Jobling, B.
1938   A revision of the species of the genus *Trichobius* (Diptera Acalypterae, Streblidae). Parasitology **30**: 358–387, 14 figs.
1939   On some American genera of the Streblidae and their species, with the description of a new species of *Trichobius* (Diptera, Acalypterae). Parasitology **31**: 486–497, 4 figs.
1949   Host-parasite relationship between the American Streblidae and the bats, with a new key to the American genera and a record of the Streblidae from Trinidad, British West Indies (Diptera). Parasitology **39**: 315–329, 3 figs.

Jörgensen, P.   (See Kieffer and Jörgensen)

Johannsen, O. A.   (See also Tilbury, 1913)
1903a   Notes on some Adirondack Diptera collected by Messrs. MacGillivray and Houghton. Ent. News **14**: 14–17.
1903b   Part 6. Aquatic nematocerous Diptera. Pp. 328–441, pls. 32–50. *In* Needham, J. G., MacGillivray, A. D., Johannsen, O. A., and Davis, K. C., Aquatic insects of New York State. N.Y. State Mus. Bul. **68** [=Ent. 18]: 197–517, 26 figs., 52 pls. (=N.Y. State Univ. Bul. 295.)

   Also published in N.Y. State Mus. Ann. Rpt. (1902) **57 (vol. 3, app. 6)**: 328–441, *1904*.

1905   Aquatic nematocerous Diptera II. Pp. 76–315, figs. 16–18, pls. 16–37. *In* Needham, J. G., Morton, K. J., and Johannsen, O. A., May flies and midges of New York. Third report on aquatic insects. N.Y. State Mus. Bul. **86** [=Ent. 23]: 7–352, 18 figs., 37 pls. (=N.Y. State Educ. Dept. Bul. 343.)

   Also published in N.Y. State Mus. Ann. Rpt. (1904) **58 (vol. 5** [=app. 7]): 76–315, *1906*.

1907a   Appendix. Some new species of Kansas Chironomidae. Pp. 109–112. *In* Tucker, E. S., Some results of desultory collecting of insects in Kansas and Colorado. Kans. Univ. Sci. Bul. **4** [=whole ser., 14]: 51–112. [=Kans. Univ. Bul. 7 (5).)

Johannsen, O. A.—Continued
- 1907b Notes on the Chironomidae. Pp. 400–401. *In* [Skinner, H., ed.], Notes and news. Ent. News **18**: 400–402.
- 1908 New North American Chironomidae. Pp. 264–285. *In* Felt, E. P., 23d report of the State Entomologist on injurious and other insects of the State of New York, 1907. N.Y. State Mus., Mus. Bul. **124**: 5–541, 49 figs., 44 pls. (=N.Y. State Univ., Educ. Dept. Bul. 433.)

  Also published in N.Y. State Mus. Ann. Rpt. (1907) **61** (**vol. 2, app. 2**): 264–285, *1908*.

- 1909 Diptera. Fam. Mycetophilidae. Fasc. 93, 141 pp., 7 pls. *In* Wytsman, P., ed., Genera insectorum [q.v.]. Bruxelles.
- 1910a The fungus gnats of North America. The Mycetophilidae of North America. Part I. Maine Agr. Expt. Sta. Bul. (1909) [**ser. 2**], **172**: 209–276, 3 pls. [i.e., figs. 46–107].

  Also bound as Maine Agr. Expt. Sta. Ann. Rpt. (1909) 35 [sic] [=**25**]: 209–276.

- 1910b The fungus gnats of North America. The Mycetophilidae of North America. Part II. Maine Agr. Expt. Sta. Bul. [**ser. 2**], **180**: 125–192, 4 pls. [i.e., figs. 83–147].

  Also bound as Maine Agr. Expt. Sta. Ann. Rpt. **26**: 125–192.

- [1912a] The fungus gnats of North America. The Mycetophilidae of North America. Part III. Maine Agr. Expt. Sta. Bul. (1911) [**ser. 2**], **196**: 249–328, 5 pls. [i.e., figs. 98–245].

  Dated from some copies which are stamped "First copies mailed Mar. 8, 1912".

  Also bound as Maine Agr. Expt. Sta. Ann. Rpt. (1911) **27**: 249–328.

- 1912b The fungus gnats of North America. The Mycetophilidae of North America. Part IV. Maine Agr. Expt. Sta. Bul. [**ser. 2**], **200**: 57–146, 7 pls. [i.e., figs. 24–267].

  Also bound as Maine Agr. Expt. Sta. Ann. Rpt. **28**: 57–146.

- 1914 *Sciara congregata* sp. nov. (Diptera). Psyche **21**: 93.
- 1916 New eastern Anthomyiidae (Diptera). Amer. Ent. Soc. Trans. **42**: 385–398, 1 pl.
- 1917 Some North American Anthomyiidae (Dipt.). Ent. News **28**: 323–327.
- 1921 The genus *Diamesa* Meigen (Diptera, Chironomidae). Ent. News **32**: 229–232.
- 1923a North American Dixidae. Psyche **30**: 52–58, 1 fig.
- 1923b Stratiomyiid larvae and puparia of the North Eastern States. N.Y. Ent. Soc. Jour. (1922) **30**: 141–153, pls. 9–10.

Johannsen, O. A.—Continued
- 1924a A new chloropid subgenus and species from New York. Canad. Ent. **56**: 89.
- 1924b A new species of *Dixa* from California. Psyche **31**: 45–46.
- 1925 A new sciarid from the eastern United States (Dipt.: Mycetophilidae). Ent. News **36**: 266–267.
- 1926 The genus *Trichotanypus* Kieffer (Chironomidae, Dipt.). Canad. Ent. **58**: 99–100.
- 1928a Two new species of western chironomids (Diptera). Jour. Ent. and Zool. **20**: 33–35.
- 1928b Order Diptera. Pp. 687–868. *In* Leonard, M. D., List of the insects of New York with a list of the spiders and certain other allied groups. N.Y. (Cornell) Agr. Expt. Sta. Mem. (1926) **101**: 1–1121.
- 1929a A new species of *Sciara* from Canada (Diptera). Canad. Ent. **61**: 223–224.
- 1929b A new species of *Blepharocera* from Massachusetts (Diptera). Psyche **36**: 123–124.
- 1929c A new sciarid from Luray Cavern, Virginia (Diptera: Mycetophilidae). Ent. Soc. Wash. Proc. **31**: 88.
- 1934a New species of North American Ceratopogonidae and Chironomidae. N.Y. Ent. Soc. Jour. **42**: 343–352.
- 1934b Aquatic Diptera. Part I. Nemocera, exclusive of Chironomidae and Ceratopogonidae. N.Y. (Cornell) Agr. Expt. Sta. Mem. **164**: 1–71, 24 pls.
- 1935 Aquatic Diptera. Part II. Orthorrhapha-Brachycera and Cyclorrhapha. N.Y. (Cornell) Agr. Expt. Sta. Mem. **177**: 1–62, 12 pls.
- 1937a Aquatic Diptera. Part III. Chironomidae: Subfamilies Tanypodinae, Diamesinae, and Orthocladiinae. N.Y. (Cornell) Agr. Expt. Sta. Mem. **205**: 1–84, 18 pls.
- 1937b Part IV. Chironomidae: Subfamily Chironominae. Pp. 3–52, pls. 1–9. *In* Johannsen, O. A., and Thomsen, L. C., Aquatic Diptera. Parts IV and V. N.Y. (Cornell) Agr. Expt. Sta. Mem. **210**: 1–80, 18 pls.
- 1942 Immature and adult stages of new species of Chironomidae (Diptera). Ent. News **53**: 70–75, 12 figs.
- 1943a Two new species of American Ceratopogonidae (Diptera). Ent. Soc. Amer. Ann. **36**: 761–762.
- 1943b A generic synopsis of the Ceratopogonidae (Heleidae) of the Americas, a bibliography, and a list of the North American species. Ent. Soc. Amer. Ann. **36**: 763–791, 3 pls.

Johannsen, O. A.—Continued
- 1943c Adult and immature stages of *Cricotopus elegans* n. sp. (Chironomidae, Diptera). Ent. News **54**: 77–79.
- 1945 Two new species of Cecidomyiidae from Florida. Fla. Ent. **28**: 8–10.
- 1946 Revision of the North American species of the genus *Pentaneura* (Tendipedidae: Chironomidae, Diptera). N.Y. Ent. Soc. Jour. **54**: 267–289, 1 pl.
- 1947 A new species of *Hydrobaenus* (*Chaetocladius*) from Connecticut with notes on related forms (Diptera, Chironomidae). Ent. News **58**: 171–174, 3 figs.
- 1950 A new *Pterobosca* from Florida with a key to American species. Fla. Ent. **33**: 141–144.
- 1952a Family Tendipedidae (=Chironomidae) except Tendipedini. Pp. 3–26, 3 figs., pls. 1–2. *In* Johannsen, O. A., and Townes, H. K., Jr., Guide to the insects of Connecticut. Part VI. The Diptera or true flies of Connecticut. Fifth fascicle: Midges and gnats [part]. Tendipedidae (Chironomidae). Conn. State Geol. and Nat. Hist. Survey, Bul. **80**: 3–147, 232–250 (bibliography and index of entire 5th fascicle), 3 figs., 22 pls.
- 1952b Guide to the insects of Connecticut. Part VI. The Diptera or true flies of Connecticut. Fifth fascicle: Midges and gnats [part]. Family Heleidae (=Ceratopogonidae). Conn. State Geol. and Nat. Hist. Survey, Bul. **80**: 149–175, 232–250 (bibliography and index of entire 5th fascicle), pls. 1–3.

———, **and Crosby, C. R.**
- 1913 The life history of *Thrypticus muhlenbergiae* sp. nov. (Diptera). Psyche **20**: 164–166, 1 fig.

———, **and Thomsen, L. C.** (See also Johannsen, 1937b; Thomsen, 1937)
- 1937 Aquatic Diptera. Parts IV and V. N.Y. (Cornell) Agr. Expt. Sta. Mem. **210**: 1–80, 18 pls.

———, **and Townes, H. K., Jr.** (See Johannsen, 1952a; Townes, 1952)

Johansen, F., and Nielsen, J. C. (See Nielsen, 1910)

Johnson, C. W. (See also Coquillett, 1895a; Skinner, 1899)
- 1894 List of the Diptera of Jamaica with descriptions of new species. Acad. Nat. Sci. Phila. Proc. **1894**: 271–281.
- 1895a Diptera of Florida. Acad. Nat. Sci. Phila. Proc. **1895**: 303–340.

Johnson, C. W.—Continued

1895b A review of the *Stratiomyia* and *Odontomyia* of North America. Amer. Ent. Soc. Trans. **22**: 227–278, 2 pls.

1897 Some notes and descriptions of new Leptidae. Ent. News **8**: 117–120, 3 figs.

1898 Notes and descriptions of new Syrphidae from Mt. St. Elias, Alaska. Ent. News **9**: 17–18.

1900a New North American Ortalidae. Canad. Ent. **32**: 246–247.

1900b Some notes and descriptions of seven new species and one new genus of Diptera. Ent. News **11**: 323–328, 3 figs.

1900c Order Diptera. Pp. 617–700, figs. 291–328. *In* Smith, J. B., Insects of New Jersey. A list of the species occurring in New Jersey with notes on those of economic importance. N.J., State Bd. Agr. Ann. Rpt. (1899) **27** (**Sup.**): 1–755, 328 figs.

1902a New North American Diptera. Canad. Ent. **34**: 240–242.

1902b Remarks on *Tephronota ruficeps* and description of a new species. Ent. News **13**: 143–144, 1 fig.

1903a Diptera of Beulah, New Mexico. Pp. 101–104, 105–106, 1 fig. *In* Skinner, H., ed., A list of the insects of Beulah, New Mexico. Amer. Ent. Soc. Trans. **29**: 35–56, *1902*; 57–88, 89–104, 105–117, 1 fig., *1903*.

1903b Some notes and descriptions of three new Leptidae. Ent. News **14**: 22–26, 2 figs.

1903c Two new species of the family Pipunculidae. Ent. News **14**: 107–108.

1903d A new genus and four new species of Asilidae. Psyche **10**: 111–114, 4 figs.

1903e Descriptions of three new Diptera of the genus *Phthiria*. Psyche **10**: 184–185.

1904 Some notes, and descriptions of four new Diptera. Psyche **11**: 15–20.

1905a Synopsis of the tipulid genus *Bittacomorpha*. Psyche **12**: 75–76.

1905b Recent entomological literature. Psyche **12**: 77–78.

1906 Descriptions of two new Diptera of the family Dolichopodidae. Psyche **13**: 59–60.

1907a A new genus and species of the family Tachinidae, parasitic on *Archips cerasivorana*. Psyche **14**: 9–10.

1907b Some North American Syrphidae. Psyche **14**: 75–80.

1907c A review of the species of the genus *Bombylius* of the eastern United States. Psyche **14**: 95–100.

1908a Notes on New England Bombyliidae, with a description of a new species of *Anthrax*. Psyche **15**: 14–15.

Johnson, C. W.—Continued

1908b   A note on *Calotarsa*, and descriptions of two new species of *Callimyia*. Psyche **15:** 58–59.

1908c   The Diptera of the Bahamas, with notes and description of one new species. Psyche **15:** 69–80.

1909a   New and little known Tipulidae. Boston Soc. Nat. Hist. Proc. **34:** 115–133, 2 pls.

1909b   Notes on the distribution of some Trypetidae with description of a new species. Psyche **16:** 113–114.

1910a   Some additions to the dipteran fauna of New England. Psyche **17:** 228–235.

> Although this has been cited as 1911 in the literature (Melander, 1932: 92, states it was mailed January, 1911), the early date of receipt by the Smithsonian Institution library (Jan. 3, 1911) indicates a mailing date sometime in 1910.

1910b   Order Diptera. Pp. 703–814, figs. 293–340. *In* Smith, J. B., The insects of New Jersey. N.J. State Mus. Ann. Rpt. **1909:** 15–888, 340 figs.

1911   Notes on the dipterous genera proposed by Billberg in his Enumeratio Insectorum. Psyche **18:** 73–74.

1912a   New North American Diptera. Psyche **19:** 1–5, 6 figs.

1912b   The velutinous species of the genus *Chrysopilus*. Psyche **19:** 108–110.

1912c   New and interesting Diptera. Psyche **19:** 151–153, 1 fig.

1912d   The North American species of the genus *Haematopota*. Psyche **19:** 181–183.

1913a   Insects of Florida. I. Diptera. Amer. Mus. Nat. Hist. Bul. **32:** 37–90.

1913b   The North American species of the genera *Arthropeas* and *Arthroceras*. Canad. Ent. **45:** 9–12.

1913c   On the *Criorhina intersistens* Walker and an allied species (Dipt.). Ent. News **24:** 293–295.

1913d   The dipteran fauna of Bermuda. Ent. Soc. Amer. Ann. **6:** 443–452, 2 figs.

1913e   Species of the genus *Gaurax* of the eastern United States. Psyche **20:** 34–35.

1913f   A study of the Clusiodidae, (Heteroneuridae) of the eastern United States. Psyche **20:** 97–101.

1914a   Some new and interesting species of *Sapromyza*. Psyche **21:** 20–23, 1 pl.

1914b   A new stratiomyid. Psyche **21:** 158–159.

1915a   A new species of the genus *Nephrocerus*. Canad. Ent. **47:** 54–56.

1915b   Two new species of Borboridae. Psyche **22:** 21–22.

Johnson, C. W.—Continued
- 1915c A new species of *Pseudotephritis*. Psyche **22**: 49.
- 1916a Further studies on the Platypezidae. Psyche **23**: 27–33, 3 figs.
- 1916b Some New England Syrphidae. Psyche **23**: 75–80.
- 1916c The *Volucella bombylans* group in America. Psyche **23**: 160–163.
- 1917a A new maritime anthomyid (Diptera). Canad. Ent. **49**: 148.
- 1917b Species of the genus *Brachyopa* of the eastern United States (Diptera). Canad. Ent. **49**: 360–362.
- 1917c A new species of *Criorhina* from New England. Psyche **24**: 153–154.
- 1918   Notes on the species of the genus *Dioctria*. Psyche **25**: 102–103.
- 1919a The North American Diptera described by Nils S. Swederus. Canad. Ent. **51**: 32.
- 1919b New species of the genus *Villa* (*Anthrax*). Psyche **26**: 11–13.
- 1919c On the variation of *Tabanus atratus* Fabricius. Psyche **26**: 163–165.
- 1919d A new species of the genus *Ulidia*. Psyche **26**: 165–166.
- 1920   A revision of the species of the genus *Loxocera*, with a description of a new allied genus and a new species. Psyche **27**: 15–19.
- 1921a New species of Diptera. Boston Soc. Nat. Hist. Occas. Papers **5**: 11–17, 2 figs.
- 1921b A review of the American species of the genus *Palloptera*. Psyche **28**: 20–23, 1 fig.
- 1921c New Diptera from Texas and Mexico. Psyche **28**: 56–59.
- 1922   New genera and species of Diptera. Boston Soc. Nat. Hist. Occas. Papers **5**: 21–26, 11 figs.
- 1923a A review of the Platypezidae of eastern North America. Boston Soc. Nat. Hist. Occas. Papers **5**: 51–58, 1 pl.
- 1923b New and interesting species of Diptera. Boston Soc. Nat. Hist. Occas. Papers **5**: 69–72, 2 figs.
- 1923c New species of North American Cyrtidae. Psyche **30**: 49–51.
- 1924   A review of the New England species of *Chrysotoxum*. Boston Soc. Nat. Hist. Occas. Papers **5**: 97–100.
- 1925a A new species of the genus *Gaurax*. Psyche **32**: 47.
- 1925b Fauna of New England. 15. List of the Diptera or two-winged flies. Boston Soc. Nat. Hist. Occas. Papers **7 (15)**: 1–326, 1 fig.

Johnson, C. W.—Continued
- 1925c Diptera of the Harris Collection. Boston Soc. Nat. Hist. Proc. (1925-1928) **38**: 57-99.
- 1926a A revision of some of the North American species of Mydaidae. Boston Soc. Nat. Hist. Proc. (1925-1928) **38**: 131-145, 1 pl.
- [1926b] New species of Diptera from North Carolina and Florida. Psyche (1925) **32**: 299-302.
    > This December issue was marked as received by the U.S. Dept. Agr. Bur. Ent. on Jan. 25, 1926, and included Bequaert, 1926a, dated 1926 by Bequaert, 1953: 366.
- 1926c The synonymy of *Actina viridis* (Say). Psyche **33**: 88-91.
- 1926d A note on *Beris annulifera* (Bigot). Psyche **33**: 108-109.
- 1927 New species of Scatophagidae. Psyche **34**: 100-103.
- 1929a Diptera of Labrador. Psyche **36**: 129-146.
- 1929b Notes on the Syrphidae collected at Jaffrey and Mount Monadnock, N.H., with a description of a new species. Psyche **36**: 370-375.
- 1931 Two new species of fungus gnats of the genus *Apemon*. Psyche **38**: 22-24, 2 figs.

Johnson, D. E.
- 1942 A new *Cyrtopogon* (Asilidae, Diptera) from Utah. Great Basin Nat. **3**: 1-4, 1 fig.
- 1958 A new species of *Mallophora* from the Great Salt Lake Desert (Diptera: Asilidae). Great Basin Nat. **18**: 41-42.

——, **and Johnson, L. M.**
- 1957 New *Poecilanthrax*, with notes on described species (Diptera: Bombyliidae). Great Basin Nat. **17**: 1-26, 2 pls.
- 1959a New and insufficiently known *Exoprosopa* from the Far West (Diptera: Bombyliidae). Great Basin Nat. (1958) **18**: 69-84, 28 figs.
- 1959b Notes on the genus *Lordotus* Loew, with descriptions of new species (Diptera: Bombyliidae). Great Basin Nat. **19**: 9-26.
- 1960 Taxonomic notes on North American beeflies, with descriptions of new species (Diptera: Bombyliidae). Great Basin Nat. (1959) **19**: 67-74, 8 figs.

——, **and Maughan, L.**
- 1953 Studies in Great Basin Bombyliidae. Great Basin Nat. **13**: 17-27, 9 figs.

Johnson, F. (See Slingerland and Johnson)

Johnson, L. M. (See Johnson and Johnson)

Johnson, N. E.
1963   *Contarinia washingtonensis* (Diptera: Cecidomyiidae), new species infesting the cones of Douglas-fir. Ent. Soc. Amer. Ann. **56**: 94–103, 6 figs.

Jones, B. J.
1906   Catalogue of the Ephydridae, with bibliography and description of new species. Calif. Univ., Pubs., Ent. **1**: 153–198, 4 figs., 1 pl.

Jones, C. M.   (See Philip and Jones)

Jones, C. R.
1917   New species of Colorado Syrphidae. Ent. Soc. Amer. Ann. **10**: 219–231.
1922   A contribution to our knowledge of the Syrphidae of Colorado. Colo., Agr. Expt. Sta. Bul. **269**: 1–72, 8 pls.

Jones, F. M.
1916   Two insect associates of the California pitcher plant, *Darlingtonia californica* (Dipt.). Ent. News **27**: 385–392, 2 pls.
1920   Another pitcher-plant insect (Diptera, Sciarinae). Ent. News **31**: 91–94, 1 pl.

Jones, P. R.
1906   A new *Cuterebra* from Nebraska. Ent. News **17**: 391–392.
1907a   A preliminary list of the Asilidae of Nebraska with description of new species. Amer. Ent. Soc. Trans. **33**: 273–286.
1907b   A preliminary list of Nebraska Syrphidae with descriptions of new species. N.Y. Ent. Soc. Jour. **15**: 87–100.

Jones, R. H.   (See also Wirth and Jones)
1956   New species of *Culicoides* from Wisconsin (Diptera, Heleidae). Ent. Soc. Wash. Proc. **58**: 25–33, 16 figs.
1961a   Observations on the larval habitats of some North American species of *Culicoides* (Diptera: Ceratopogonidae). Ent. Soc. Amer. Ann. **54**: 702–710.
1961b   Descriptions of pupae of thirteen North American species of *Culicoides* (Diptera: Ceratopogonidae). Ent. Soc. Amer. Ann. **54**: 729–746, 6 pls.

———, **and Wirth, W. W.**
1958   New records, synonymy, and species of Texas *Culicoides* (Diptera, Heleidae). Kans. Ent. Soc. Jour. **31**: 81–91, 6 figs.

Jordan, D. S.
1898– The fur seals and fur-seal islands of the North Pacific
1899   Ocean. 4 vols. Washington, D.C.

> The portion on Diptera is in Part 4, C: Appendices, I.
> Reports upon the insects, spiders, mites, and myriapods collected by Dr. L. Stejneger and Mr. G. E. H. Barrett-Hamilton on the Commander Islands, pp. 328–352, pl. A, edited by W. H. Ashmead.
> See Coquillett, 1899f.

K. Akademie der Wissenschaften, Wien
1886   Die Internationale Polarforschung 1882–1883. Die Oesterreichische Polarstation Jan Mayen, ausgerüstet durch seine Excellenz Graf Hanns Wilczek. Geleitet vom K. K. Corvetten-Capitän Emil Edlen von Wohlgemuth. Beobachtungs-Ergebnisse. 3 vols. Wien.

> Volume 3 includes separately paged parts 6, 7, and 8.
> See Becher, 1886.

K. Svenska Vetenskaps-Akademien
1857– Kongliga svenska fregatten Eugenies resa omkring
1910   jorden under befäl af C. A. Virgin, åren 1851–1853. 3 pts. Stockholm, Uppsala.

> The work is in 3 parts as follows: I. Botanik; II. Zoologi; and III. Fysik. Pt. 2, 1857–1910, is in 3 separately paged sections: I. Insecta; II. Arachnider; and III. Annulater. The whole work was issued in 16 hefts, some including portions of more than one section. The section on Insecta includes portions from hefts 4, 6, 7, 10, and 12, and was published 1858–1869.
> See Thomson, 1869.

Kahl, P. H. I.
1897   New species of the syrphid genera *Mixogaster* Macq. and *Ceria* Fabr., with notes. Kans. Univ. Quart. **ser. A, 6:** 137–146.

Kaltenbach, J. H.
1856   Die deutschen Phytophagen aus den Klasse der Insekten, oder Versuch einer Zusammenstellung der auf Deutschlands Pflanzen beobachteten Bewohner und deren Feinde. Naturhist. Ver. der Preuss. Rheinlande u. Westfalens, Verhandl. **13** [=n. ser., 3]: 165–265 [cont.].

1867   Die deutschen Phytophagen aus der Klasse der Insekten [cont.]. Naturhist. Ver. der Preuss. Rheinlande u. Westfalens, Verhandl. **24** [=ser. 3, 4]: 21–117 [cont.].

Kaltenbach, J. H.—Continued
1869 Die deutschen Phytophagen aus der Klasse der Insekten [concl.]. Naturhist. Ver. der Preuss. Rheinlande u. Westfalens, Verhandl. **26** [=ser. 3, 6]: 106–224.
1874 Die Pflanzenfeinde aus der Klasse der Insekten. 848 pp., 401 figs. Stuttgart.

Karabinos, J. V. (See Kessel and Karabinos)

Karl, O.
1928 Zweiflügler oder Diptera. III. Muscidae. T. 13, 232 pp., 114 figs. *In* Dahl, F., ed., Die Tierwelt Deutschlands [q.v.]. Jena.
1930 I. Wissenschaftliche Mitteilungen. 1. Ergänzungen und Berichtigungen zu meiner Arbeit über die Musciden (Prof. Dr. Fr. Dahl, Die Tierwelt Deutschlands, Teil 13). Teil II. Zool. Anz. **86**: 161–174, 7 figs.
1934 Ergänzungen und Berichtigungen zur meiner Arbeit über die Musciden. (Prof. Dr. Fr. Dahl, Die Tierwelt Deutschlands, Teil 13). Teil IV. Zool. Anz. **107**: 90–93, 3 figs.

Karsch, F.
1880 Die Spaltung der Dipterengattung *Systropus* Wiedemann. Ztschr. f. die Gesam. Naturw. **53**: 654–658.

Kaston, B. J.
1937 Notes on dipterous parasites of spiders. N.Y. Ent. Soc. Jour. **45**: 415–420, 5 figs.

Kaufmann, B. P. (See Stone, Griffen, and Patterson, 1941)

Kearby, W. H., and Benjamin, D. M.
1963 A new species of *Thecodiplosis* (Diptera: Cecidomyiidae) on red pine in Wisconsin. Canad. Ent. **95**: 414–417, 2 figs.

Keating, W. H.
1824 Major Long's second expedition. Narrative of an expedition to the source of St. Peter's River, Lake Winnepeek, Lake of the Woods, &c. &c., performed in the year 1823, by order of the Hon. J. C. Calhoun, Secretary of War, under the command of Stephen Long, Major, U.S.T.E. 2 vols. Philadelphia.

> There is another edition of this work published in London, 1825.
> See Say, 1824a.

Kellogg, V. L.
1893   Insect notes. Kans. Acad. Sci. Trans. Ann. Mtgs. **13**: 112–115, 1 fig.
1900a   An extraordinary new maritime fly. Biol. Bul. **1**: 81–87, 3 figs.
1900b   A new blepharocerid. Psyche **9**: 39–41, 2 figs.
1901   An aquatic psychodid. Ent. News **12**: 46–49, 2 figs.
1903   The net-winged midges (Blepharoceridae) of North America. Calif. Acad. Sci. Proc. ser. **3**, Zool., **3**: 187–232, 1 fig., 5 pls.

Kelly, E. O. G.   (See Aldrich, 1914b)

Kennedy, H. D.
1958   Biology and life history of a new species of mountain midge, *Deuterophlebia nielsoni*, from eastern California (Diptera: Deuterophlebiidae). Amer. Micros. Soc. Trans. **77**: 201–228, 7 pls.
1960   *Deuterophlebia inyoensis*, a new species of mountain midge from the alpine zone of the Sierra Nevada Range, California (Diptera: Deuterophlebiidae). Amer. Micros. Soc. Trans. **79**: 191–210, 7 pls.

Kertész, K.   (See also Becker et al., 1903a, 1903b; Becker et al., 1905, 1907)
1900a   Bemerkungen über Pipunculiden (Dipt.). Wien. Ent. Ztg. **19**: 244–245.
1900b   Nachtrag zu meinen Bemerkungen über Pipunculiden. Wien. Ent. Ztg. **19**: 270.
1902a   Catalogus dipterorum hucusque descriptorum. Vol. 1, 339 pp. Lipsiae, Budapestini [=Leipzig, Budapest].
    The entire work consists of seven volumes published in Leipzig and Budapest, 1902–1910.
1902b   Catalogus dipterorum hucusque descriptorum. Vol. 2, 359 pp. Lipsiae, Budapestini [=Leipzig, Budapest].
1907a   Ein neuer Dipteren-Gattungsname. Budapest Magyar Nemzeti Muz., Ann. Hist. Nat. **5**: 499.
1907b   Vier neue *Pipunculus*-Arten. Budapest Magyar Nemzeti Muz., Ann. Hist. Nat. **5**: 579–583.
1908   Catalogus dipterorum hucusque descriptorum. Vol. 3, 367 pp. Lipsiae, Budapestini [=Leipzig, Budapest].
1909a   Catalogus dipterorum hucusque descriptorum. Vol. 4, 349 pp. Lipsiae, Budapestini [=Leipzig, Budapest].
1909b   Catalogus dipterorum hucusque descriptorum. Vol. 5, 199 pp. Lipsiae, Budapestini [=Leipzig, Budapest].

Kertész, K.—Continued
- 1909c Catalogus dipterorum hucusque descriptorum. Vol. 6, 362 pp. Lipsiae, Budapestini [=Leipzig, Budapest].
- 1910 Catalogus dipterorum hucusque descriptorum. Vol. 7, 470 pp. Lipsiae, Budapestini [=Leipzig, Budapest].
- 1911 Ueber die generische Hinzugehörigkeit der bis jetzt beschriebenen *Pachygaster*-Arten. Pp. 29–32. *In* Severin, G., sec., Ier Congrès international d'entomologie [q.v.]. Vol. 2: Mém., 520 pp., 14 figs., 27 pls. Bruxelles.
- 1912 The Percy Sladen Trust Expedition to the Indian Ocean in 1905, under the leadership of Mr. J. Stanley Gardiner, M.A. Volume IV. No. VI. Diptera, Stratiomyiidae. Linn. Soc. London, Trans. **ser. 2, Zool., 15**: 95–99, 4 figs.
- 1915 Contributions to the knowledge of the Dorylaidae. Budapest Magyar Nemzeti Muz., Ann. Hist. Nat. **13**: 386–392, 5 figs.
- 1923 Vorarbeiten zur einer Monographie der Notacanthen. XLV–L. Budapest Magyar Nemzeti Muz., Ann. Hist. Nat. **20**: 85–129, 18 figs.

Kessel, B. B. (See Kessel and Kessel)

Kessel, E. L.
- 1947 American smoke flies (*Microsania:* Clythiidae). Wasmann Collect. **7**: 23–30.
- 1948 A review of the genus *Platypezina* Wahlgren, announcement of its presence in the New World, and the description of a new species (Diptera: Clythiidae). Wasmann Collect. **7**: 47–64, 5 figs.
- 1949a New species of *Callomyia* from California (Diptera: Clythiidae). Wasmann Collect. (1948) **7**: 139–148.
    Date of issue is stamped on cover.
- 1949b Two new species of *Agathomyia* from the Pacific coast of North America (Diptera: Clythiidae). Wasmann Collect. **7**: 215–219.
- 1950a *Protoclythia*, a new genus of flat-footed flies, and the description of two new species (Diptera: Clythiidae). Wasmann Collect. (1949) **7**: 257–275.
    Date of issue is stamped on cover.
- 1950b A new species of *Clythia* from California (Diptera: Clythiidae). Wasmann Jour. Biol. **8**: 77–80.
- (1952a) *Metaclythia*, a new genus of flat-footed flies, and the description of a new species (Diptera: Clythiidae). Wasmann Jour. Biol. (1951) **9**: 347–350.

Kessel, E. L.—Continued
- 1952b A key to the genera of Clythiidae (Diptera). Wasmann Jour. Biol. **10:** 201–204.
- 1952c Another American fly attracted to smoke (Diptera: Empididae). Pan-Pacific Ent. **28:** 56–58.
- 1955 The mating activities of balloon flies. System. Zool. **4:** 97–104, 3 figs.
- 1957 Distribution and variation in *Agathomyia notata* (Loew) (Diptera: Platypezidae). Wasmann Jour. Biol. **15:** 69–80.
- 1958 The smoke fly, *Hormopeza copulifera* Melander (Diptera: Empididae). Pan-Pacific Ent. **34:** 86.
- 1959a A new species of flat-footed fly from Alaska (Diptera: Platypezidae). Wasmann Jour. Biol. **17:** 19–22.
- 1959b Introducing *Hilara wheeleri* Melander as a balloon maker, and notes on other North American balloon flies (Diptera: Empididae). Wasmann Jour. Biol. **17:** 221–230.
- 1960a The systematic positions of *Platycnema* Zetterstedt and *Melanderomyia*, new genus, together with the description of the genotype of the latter (Diptera: Platypezidae). Wasmann Jour. Biol. **18:** 87–101.
- 1960b The response of *Microsania* and *Hormopeza* to smoke (Diptera: Platypezidae and Empididae). Pan-Pacific Ent. **36:** 67–68.
- (1961a) The life cycle of *Clythia agarici* (Willard) (Diptera: Platypezidae). Wasmann Jour. Biol. (1960) **18:** 263–270, 3 figs.
- 1961b A new species of *Platypezina* from North America (Diptera: Platypezidae). Wasmann Jour. Biol. **19:** 187–190.
- 1961c New species of flat-footed flies from North America (Diptera: Platypezidae). Wasmann Jour. Biol. **19:** 191–227.
- 1961d The immature stages of *Callomyia*, with the description of a new species of this genus (Diptera: Platypezidae). Calif. Acad. Sci. Occas. Papers **30:** 1–10, 2 figs., 1 pl.
- 1963 The genus *Calotarsa*, with special reference to *C. insignis* Aldrich (Diptera: Platypezidae). Calif. Acad. Sci. Occas. Papers **39:** 1–24, 9 figs., 1 pl.

———, and **Karabinos, J. V.**
- 1947 *Empimorpha geneatis* Melander, a balloon fly from California, with a chemical examination of its balloons (Diptera: Empididae). Pan-Pacific Ent. **23:** 181–192.

Kessel, E. L., and Kessel, B. B.
1939   Diptera associated with fungi. Wasmann Collect. **3:** 73–92.
1951   A new species of balloon-bearing *Empis* and an account of its mating activities (Diptera: Empididae). Wasmann Jour. Biol. **9:** 137–146.

Kessel, Q. C.
1925   A synopsis of the Streblidae of the World. N.Y. Ent. Soc. Jour. **33:** 11–33, 4 pls.

Khalaf, K. T.
1952a   The male of *Culicoides weesei* Khalaf (Heleidae, Diptera). Kans. Ent. Soc. Jour. **25:** 65.
1952b   The *Culicoides* of the Wichita Refuge, Oklahoma. Taxonomy and seasonal incidence (Diptera, Heleidae). Ent. Soc. Amer. Ann. **45:** 348–358, 6 figs.
1954   The speciation of the genus *Culicoides* (Diptera, Heleidae). Ent. Soc. Amer. Ann. **47:** 34–51.
1957   Light trap survey of the *Culicoides* of Oklahoma (Diptera, Heleidae). Amer. Midland Nat. **58:** 182–221, 22 figs.

Kieffer, J. J. (See also Bause, 1914; Enderlein, 1911a; Thienemann, 1916)
1889   Neue Beiträge zur Kenntniss der Gallmücken. Ent. Nachr. **15:** 183–194.
1892   Beobachtungen über Gallmücken mit Beschreibung einiger neuen Arten. Wien. Ent. Ztg. **11:** 212–224, 1 pl.
1894a   Description de quelques larves de Cécidomyes. Feuille des Jeunes Nat. **24:** 83–88, 7 figs., 119–121, 5 figs., 147–152, 8 figs., 185–189, 6 figs.

        Separates were published with the following paginations: 1–6, 7 figs.; and 1–3, 5 figs., 3–7, 8 figs., 8–11, 6 figs.

1894b   Sur le groupe *Epidosis* de la famille des Cecidomyidae. Soc. Ent. de France, Ann. **63:** 311–350, 8 figs., 2 pls.
1894c   (Trois genres nouveaux du groupe des *Diplosis* (Dipt.). Soc. Ent. de France, Ann. **63 (Bul.):** xxviii–xxix.
1894d   Ueber die Heteropezinae. Wien. Ent. Ztg. **13:** 200–212, 1 pl.
1895a   Essai sur le groupe *Campylomyza*. Misc. Ent. **3:** 46–47, 57–63, 73–79, 91–97, 109–113, 10 figs. [cont.].
1895b   Ueber moosbewohnende Gallmückenlarven. Ent. Nachr. **21:** 113–123.
1895c   (Note préliminaire sur le genre *Campylomyza* (Dipt.).) Soc. Ent. de France, Ann. (1894) **63 (Bul.):** clxxv–clxxvi.

Kieffer, J. J.—Continued
- 1895d (Genres nouveaux dans le groupe des *Diplosis* (Dipt.).) Soc. Ent. de France, Ann. (1894) **63 (Bul.)**: cclxxx.
- 1895e Nouvelles observations sur le groupe des *Diplosis* et description de cinq genres nouveaux (Dipt.). Soc. Ent. de France, Ann. **64 (Bul.)**: cxcii–cxciv, 1 fig.
- 1896a Essai sur le groupe *Campylomyza* [concl.]. Misc. Ent. (1895) **3**: 129–133, 2 pls.
- 1896b Quatre nouveaux genres du groupe *Diplosis*. Misc. Ent. **4**: 4–5.
- 1896c Neue Mittheilungen über Gallmücken. Wien. Ent. Ztg. **15**: 85–105.
- 1896d Observations sur les *Diplosis*, et diagnoses de cinq espèces nouvelles (Dipt.). Soc. Ent. de France, Bul. **1896**: 382–384.
- 1898a Synopse des Cécidomyies d'Europe et d'Algérie décrites jusqu'à ce jour. Soc. d'Hist. Nat. de Metz, Bul. **20** [=ser. 2, 8]: 1–64.
- 1898b Description d'un nouveau genre et d'une nouvelle espèce de Sciaride (Dipt.). Soc. Ent. de France, Bul. **1898**: 194–196.
- 1899 Description d'un nouveau genre et tableau des genres européens de la famille des Chironomides (Dipt.). Soc. Ent. de France, Bul. **1899**: 66–70, 1 fig.
- 1901 Synopsis des zoocécides d'Europe. Soc. Ent. de France, Ann. **70**: 233–579.
- 1903 Description de trois genres nouveaux et de cinq espèces nouvelles de la famille des Sciaridae (Diptères). Sci. de Bruxelles, Ann. **27 (Mém.)**: 196–204, 3 figs., 1 pl.
- 1904a Étude sur les Cécidomyies gallicoles. Soc. Sci. de Bruxelles, Ann. **28 (Mém.)**: 329–350.
- 1904b Nouvelles Cécidomyies xylophiles. Soc. Sci. de Bruxelles, Ann. **28 (Mém.)**: 367–410, 1 pl.
- 1906a Diptera. Fam. Chironomidae. Fasc. 42, 78 pp., 4 pls. *In* Wytsman, P., ed., Genera insectorum [q.v.]. Bruxelles.
- 1906b Description de nouveaux Diptères Nématocères d'Europe. Soc. Sci. de Bruxelles, Ann. **30 (Mém.)**: 311–348, 21 figs.
- 1906c Description d'un genre nouveau et de quelques espèces nouvelles de Diptères de l'Amérique du Sud. Soc. Sci. de Bruxelles, Ann. **30 (Mém.)**: 349–358, 3 figs.
- 1908a Description d'une espece nouvelle de Chironomides d'Egypte. Budapest Magyar Nemzeti Muz., Ann. Hist. Nat. **6**: 576–577.

Kieffer, J. J.—Continued
- 1908b I. Neue und bekannte Chironomiden. Pp. 1–10, 33–39, 78–84, figs. 1–16. *In* Kieffer, J. J., and Thienemann, A., Neue und bekannte Chironomiden und ihre Metamorphose. Ztschr. f. Wiss. Insektenbiol. **n. ser., 4** [=whole ser., 13]: 1–10, 33–39, 78–84, 124–128, 184–190, 214–219, 256–259, 277–286, 57 figs.
- 1909 Diagnoses de nouveaux Chironomides d'Allemagne. Soc. d'Hist. Nat. de Metz, Bul. **26**: [=ser. 3, 2]: 37–56.
- 1910 Description de quelques Diptères exotiques. Portici, R. Scuola Super. d'Agr., Lab. Zool. Gen. e Agr. Bol. **4**: 327–328, 1 fig.
- 1911a Description de nouveaux Chironomides de l'Indian Museum de Calcutta. Indian Mus. Rec. **6**: 113–177, 2 pls.
- 1911b Nouvelles descriptions de Chironomides obtenus d'éclosion. Soc. d'Hist. Nat. de Metz, Bul. **27** [=ser. 3, 3]: 1–60.
- 1911c Description d'un Chironomide d'Amérique formant un genre nouveau. Soc. d'Hist. Nat. de Metz, Bul. **27** [=ser. 3, 3]: 103–105.
- 1911d Nouveaux Tendipédides du groupe *Orthocladius* (Dipt.). (1re note); (2me note). Soc. Ent. de France, Bul. **1911**: 181–187; 199–202.
- 1912a Anhang. Beschreibung neuer Sciariden von den Seychellen-Inseln. Pp. 192–194, pl. 9, figs. 11–14. *In* Enderlein, G., The Percy Sladen Trust Expedition to the Indian Ocean in 1905, under the leadership of Mr. J. Stanley Gardiner, M. A. Volume IV. No. XIII. Diptera, Sciaridae. Linn. Soc. London, Trans. **ser. 2, Zool., 15**: 181–194, pl. 9.
- 1912b Neue Gallmücken-Gattungen. 2 pp. Bitsch, France.
    This is quoted in full, pp. x–xi, *in* Trotter, A., Bibliografia e recensioni, in Marcellia **11**: i–xxxv, **1912**.
- 1912c Nouveaux Chironomides (Tendipedidae) de Ceylan. Spolia Zeylanica **8**: 1–24, 9 figs.
- 1912d Description de quatre nouveaux Insectes exotiques. Portici, R. Scuola Super. d'Agr., Lab. Zool. Gen. e Agr. Bol. **6**: 171–175, 1 fig.
- 1913a Diptera. Fam. Cecidomyidae. Fasc. 152, 346 pp., 15 pls. *In* Wytsman, P., ed., Genera insectorum [q.v.]. Bruxelles.
- 1913b Nouveaux Chironomides (Tendipédides) d'Allemagne. Soc. d'Hist. Nat. de Metz, Bul. **28** [=ser. 3, 4]: 7–35.
- 1913c Glanures diptérologiques. Soc. d'Hist. Nat. de Metz, Bul. **28** [=ser. 3, 4]: 45–55.

Kieffer, J. J.—Continued

1913d I. Chironomidae et Cecidomyidae. Pp. 1–43, 1 fig. (=no. [6]). *In* [Alluaud, C. A., and Jeannel, R., eds.], Voyage de Ch. Alluaud et R. Jeannel en Afrique orientale [q.v.]. Insectes Diptères [="Vol. 5"], 351 pp., 35 figs., 1 pl. Paris.

1913e Nouvelle étude sur les Chironomides de l'Indian Museum de Calcutta. Indian Mus. Rec. **9**: 119–197, 2 pls.

1913f 1. *Dasyhelea halophila* n. sp., eine neue halophile Zuckmücke. Pp. 255–256. *In* [Kieffer, J. J., and Douwe, C. van], Zur Flora und Fauna der Strandtümpel von Rovigno (in Istria). Biol. Centbl. **33**: 254–258, 3 figs.

1914 South African Chironomidae (Diptera). So. African Mus. Ann. **10**: 259–270.

1915 Neue Chironomiden aus Mitteleuropa. Brotéria sér. Zool., **13**: 65–87.

1917 Chironomides d'Amérique conservés au Musée National Hongrois de Budapest. Budapest Magyar Nemzeti Muz., Ann. Hist. Nat. **15**: 292–364, 48 figs.

1918a Beschreibung neuer, auf Lazarettschiffen des östlichen Kriegsschauplatzes und bei Ignalino in Litauen von Dr. W. Horn gesammelter Chironomiden, mit Uebersichtstabellen einiger Gruppen von paläarktischen Arten (Dipt.). Ent. Mitt. **7**: 35–53, 94–110, 163–170, 177–188, 14 figs.

1918b Chironomides d'Afrique et d'Asie conservés au Museum National Hongrois de Budapest. Budapest Magyar Nemzeti Muz., Ann. Hist. Nat. **16**: 31–136, 48 figs.

1919a Chironomiden der nördlichen Polarregion. Pp. 40–48, 110–120, 18 figs. *In* Kieffer, J. J., and Thienemann, A., Chironomiden, gesammelt von Dr. A. Koch (Münster i. W.) aus den Lofoten, der Bäreninsel und Spitzbergen (Dipt.). Ent. Mitt. **8**: 38–48, 110–124, 18 figs.

1919b Chironomides d'Europe conservés au Musée National Hongrois de Budapest. Budapest Magyar Nemzeti Muz., Ann. Hist. Nat. **17**: 1–160, 60 figs.

1919c Observations sur les Chironomides (Dipt.) décrits par J. R. Malloch. Soc. Ent. de France, Bul. **1919**: 191–194.

1920 Tableau synoptique des Chironomides paléarctiques appartenant aux genres *Polypedilum* et *Limnochironomus*. Soc. Sci. de Bruxelles, Ann. **39** (**Doc. et Compt. Rend.**): 159–167.

Kieffer, J. J.—Continued
1921a Chironomides nouveaux ou peu connus de la région paléarctique. Soc. d'Hist. Nat. de la Moselle, Bul. **29** [=ser. 3, 5]: 51–109.
1921b Chironomides de l'Afrique équatoriale (l^{re} partie). Soc. Ent. de France, Ann. **90**: 1–56, 2 pls.
1921c Sur quelques Diptères piqueurs de la tribu des Ceratopogoninae. Inst. Pasteur de l'Afrique du Nord, Arch. **1**: 107–115, 6 figs.
1921d Synopse de la tribu des Chironomariae (Diptères). Soc. Sci. de Bruxelles, Ann. **40 (Doc. et Compt. Rend.)**: 269–277.
1922a Diagnose de nouveaux genres et espèces de Chironomides (Dipt.). Soc. Ent. de France, Bul. **1921**: 287–289.
1922b Chironomides de la Nouvelle-Zemble. Rpt. 2, 24 pp., 15 figs. *In* Holtedahl, O., ed., Report of the scientific results of the Norwegian Expedition to Novaya Zemlya 1921 [q.v.]. Oslo.
1922c Nouveaux Chironomides à larves aquatiques. Soc. Sci. de Bruxelles, Ann. **41 (Doc. et Compt. Rend.)**: 355–367.
1922d Chironomides nouveaux ou peu connus de la région paléarctique. Soc. Sci. de Bruxelles, Ann. **42 (Mém.)**: 71–128.
1922e Notice sur quelques Chironomides d'Amérique et de Nouvelle-Zélande. Soc. Linn. de Lyon, Ann. (1921) **ser. 2, 68**: 145–148.
1922f Chironomides de l'Afrique équatoriale (2e partie). Soc. Ent. de France, Ann. **91**: 1–72, 4 pls.
1923a Diagnose de quelques nouveaux Tanypodines (Dipt.). Soc. Ent. de France, Bul. **1922**: 296–297.
1923b Chironomides de l'Afrique équatoriale (3e partie). Soc. Ent. de France, Ann. **92**: 149–204, pls. 1–2.
1924a Quelques Chironomides nouveaux et remarquables du nord de l'Europe. Soc. Sci. de Bruxelles, Ann. **43 (Doc. et Compt. Rend.)**: 390–397.
1924b Quelques nouveaux Chironomides de Scandinavie. Soc. Sci. de Bruxelles, Ann. **44 (Doc. et Compt. Rend.)**: 80–86.
1924c Chironomides nouveaux ou rares de l'Europe centrale. Soc. d'Hist. Nat. de la Moselle, Bul. **30** [=ser. 3, 6]: 11–110.
1925a Diptères (Nématocères piqueurs): Chironomidae Ceratopogoninae. Vol. 11, 139 pp., 83 figs. *In* Office Central de Faunistique de la Fédération Française des Sociétés de Sciences Naturelles, Faune de France [q.v.]. Paris.

Kieffer, J. J.—Continued
- 1925b Nouveaux genres et nouvelles espèces de Chironomides piqueurs. Inst. Pasteur d'Algérie, Arch. **3**: 405–430, 7 figs.
- 1925c Deux nouveaux genres des groupes *Corynoneura* et *Bezzia* (Dipt. Chironomidae). Soc. Ent. de France, Ann. **94**: 54.
- 1925d Deux genres nouveaux et plusieurs espèces nouvelles du groupe des Orthocladiariae (Diptères, Chironomidae). Soc. Sci. de Bruxelles, Ann. 44 (**Doc. et Compt. Rend.**): 555–566.
- 1926 Chironomiden der 2. Fram-Expedition (1898–1902). Norsk Ent. Tidsskr. (1925) **2**: 78–89, 11 figs.

———, **and Douwe, C. van** (See Kieffer, 1913f)

———, **and Jörgensen, P.**
- 1910 Gallen und Gallentiere aus Argentinien. Centbl. f. Bakt., Parasitenk. u. Infektionskrank., Abt. II **27**: 362–444, 61 figs.

———, **and Lundbeck, W.**
- 1911 Diptera. Pp. 272–275, figs. 1–5. *In* Koenig, A., Avifauna spitzbergensis. 294 pp., 100 figs., 34 pls. Bonn.
  The Chironomidae are by Kieffer.

———, **and Thienemann, A.** (See Kieffer, 1908b, 1919a)

Kincaid, T.
- 1897 The Psychodidae of Washington. Ent. News **8**: 143–146.
- 1899 The Psychodidae of the Pacific Coast. Ent. News **10**: 30–36, 13 figs.
- 1901 Notes on American Psychodidae. Ent. News **12**: 193–196, 1 pl.

King, J. L.
- 1916 Observations on the life history of *Pterodontia flavipes* Gray (Diptera). Ent. Soc. Amer. Ann. **9**: 309–321, pls. 15–16.

King, W. V.
- 1939 Varieties of *Anopheles crucians* Wied. Amer. Jour. Trop. Med. **19**: 461–471, 2 figs.

———, **Bradley, G. H., Smith, C. N., and McDuffie, W. C.**
- 1960 A handbook of the mosquitoes of the southeastern United States. U.S. Dept. Agr., Agr. Handb. **173**: 1–188, 25 figs., 10 pls.

Kingsley, J. S., ed.
- 1883– The standard natural history. 6 vols. Boston.
- 1885   Volume 2, which contains the Diptera, was issued 1883–1884, as follows: pp. 1–48, 49–96, 1883; pp. 97–144, 145–192, 193–240, 241–336, 337–432, 433–555, 1884.
  See Williston, 1884a.

Kirby, W.
1824 (Land invertebrate animals). Pp. ccxiv–ccxix. *In* [Parry, W. E.], A supplement to the appendix of Captain Parry's voyage for the discovery of a North-West Passage, in the years 1819–20. Containing an account of the subjects of natural history. [240 pp.] [i.e., pp. clxxi–cccx], 6 pls. London.

> This Supplement was issued entirely separate from the main volume issued in 1821, but with continuing page numbers.

1837 The insects. Pt. 4, 325 pp., 8 pls. *In* Richardson, J., Fauna Boreali-Americana [q.v.]. Norwich, London.

———, and Spence, W.
1817 An introduction to entomology, or elements of the natural history of insects. [Ed. 1], Vol. 2, 529 pp., pls. 1–2. London.

> The entire work consists of 4 volumes, London, 1815–1826. Later editions appeared up to the seventh.

1826 An introduction to entomology, or elements of the natural history of insects. [Ed. 1], Vol. 4, 634 pp., pls. 21–30. London.

Kirkpatrick, J.
1861 The army worm. Ohio State Bd. Agr., Ann. Rpt. (1860) **15**: 350–358.

Kloet, G. S., and Hincks, W. D.
1945 A check list of British insects. 483 pp. Arbroath, Scotland.

> The Orthorrhapha are by R. L. Coe.

Knab, F. (See also Dyar and Knab; Howard, Dyar, and Knab)
1913 A new bot-fly from reindeer (Diptera: Muscoidea). Biol. Soc. Wash. Proc. **26**: 155–156.
1914a A new mesembrine fly. Canad. Ent. **46**: 325–326.
1914b Two North American Syrphidae. Insecutor Inscitiae Menstruus **2**: 151–153.
1914c A review of our species of *Trigonometopus* (Diptera: Lauxaniidae). Psyche **21**: 123–126.
1914d Ceratopogoninae sucking the blood of caterpillars. Ent. Soc. Wash. Proc. **16**: 63–66.
1915a Two new species of *Pipunculus* (Diptera: Pipunculidae). Biol. Soc. Wash. Proc. **28**: 83–85, 1 pl.
1915b A new *Simulium* from Texas (Diptera, Simuliidae). Insecutor Inscitiae Menstruus **3**: 77–78.
1916a A new mosquito from the eastern United States. Biol. Soc. Wash. Proc. **29**: 161–163.

Knab, F.—Continued
1916b   Dispersal of some Ortalidae. Brooklyn Ent. Soc. Bul. **11**: 40–46, 3 figs.
1917   On some North American species of *Microdon* (Diptera: Syrphidae). Biol. Soc. Wash. Proc. **30**: 133–143.
———, and **Shannon, R. C.**
1916   Tanypezidae in the United States (Diptera Acalyptrata). Insecutor Inscitiae Menstruus **4**: 33–36.
———, and **Yothers, W. W.**
1914   Papaya fruit fly. Jour. Agr. Res. **2**: 447–453, pls. 41–42.

Knight, K. L. (See Mattingly, Stone, and Knight; Stone, Knight, and Starcke)

Knipling, E. F. (See also Wells and Knipling)
1937   The biology of *Sarcophaga cistudinis* Aldrich (Diptera), a species of Sarcophagidae parasitic on turtles and tortoises. Ent. Soc. Wash. Proc. **39**: 91–101, 2 pls.

Knowlton, G. F. (See also Hardy and Knowlton; Harmston and Knowlton; Rowe and Knowlton; Stains and Knowlton)
———, and **Rowe, J. A.**
1934   New blood-sucking flies from Utah (Simuliidae, Diptera). Ent. Soc. Amer. Ann. **27**: 580–584, 10 figs.

Koebele, A.
1893   Studies of parasitic and predaceous insects in New Zealand, Australia, and adjacent islands. U.S. Dept. Agr. [Rpt.] [**51**]: 1–39.

Koenig, A. (See Kieffer and Lundbeck, 1911)

Kolenati, F. A.
1860   Einige neue Insekten-Arten vom Altvater. Wien. Ent. Monatschr. **4**: 381–394.

Komp, W. H. W.
1926   A new *Culex* from Honduras. Insecutor Inscitiae Menstruus **14**: 44–45.

Kowarz, F. (See also Lintner, 1892)
1873   Beitrag zur Dipteren-Fauna Ungarns. K.-k. Zool.-Bot. Gesell. Wien, Verhandl. **23** (**Abhandl.**): 453–464.
1874   Die Dipteren-Gattung *Chrysotus* Meig. K.-k. Zool.-Bot. Gesell. Wien, Verhandl. **24** (**Abhandl.**): 453–478, pl. 13.

Kraft, K. J., and Cook, E. F.
1961   A revision of the Pachygasterinae (Diptera, Stratiomyidae) of America north of Mexico. Ent. Soc. Amer. Misc. Pub. **3**: 1–24, 6 figs.

Kramer, H.
1908a *Sarcophaga*-Arten der Oberlausitz. Ent. Wchnbl. **25**: 152–153, 3 figs.
1908b *Sarcophaga affinis* Fll. und Verwandte. Ent. Wchnbl. **25**: 200–201, 2 figs.

Krivosheina, P., and Mamajev, B. M.
1962 (Larvae of the European species of the genus *Temnostoma* (Diptera, Syrphidae).) Ent. Obozr. **41**: 921–930, 29 figs.
In Russian; English summary, p. 930.

Kröber, O.
1911 Die Thereviden Süd- und Mittelamerikas. Budapest Magyar Nemzeti Muz., Ann. Hist. Nat. **9**: 475–529, 2 figs.
1912 Die Thereviden Nordamerikas. Stettin. Ent. Ztg. **73**: 209–272.
1913 Diptera. Fam. Therevidae. Fasc. 148, 69 pp., 3 pls. *In* Wytsman, P., ed., Genera insectorum [q.v.]. Bruxelles.
1914 Beiträge zur Kenntnis der Thereviden und Omphraliden. Jahrb. der Hamburg. Wiss. Anst. (1913) **31** (**Beih. 2**): 29–74, 3 figs. (=Naturhist. Mus. in Hamburg, Mitt. 31.)
1915a Die Gattung *Zodion* Latr. Arch. f. Naturgesch. **Abt. A, 81** (4): 84–117.
1915b Die nord- und südamerikanischen Arten der Gattung *Conops*. Arch. f. Naturgesch. **Abt. A, 81** (5): 121–160.
1916 Die *Myopa*-Arten der nicht-palaearktischen Regionen. Arch. f. Naturgesch. (1915) **Abt. A, 81** (7): 23–39.
1919a Katalog der Conopiden, nebst Beschreibung der Gattungen und Bestimmungstabellen der Gattungen und Arten. Arch. f. Naturgesch. (1917) **Abt. A, 83** (8): 1–91 [cont.].
1919b Katalog der Conopiden, nebst Beschreibung der Gattungen und Bestimmungstabellen der Gattungen und Arten [concl.]. Arch. f. Naturgesch. (1917) **Abt. A, 83** (9): 1–52, 10 pls.
1920 Die *Chrysops*-Arten der paläarktischen Region. Zool. Jahrb., Abt. f. System., Geol. u. Biol. Tiere **43**: 41–160, 12 figs.
1926 Die *Chrysops*-Arten Nordamerikas einschl. Mexicos. Stettin. Ent. Ztg. **87**: 209–353, 2 pls.
1928 Die amerikanischen Arten der Tabaniden-Subfamilie Diachlorinae End. Arch. f. Schiffs.- u. Tropen-Hyg. **32** (2): 1–55, 26 figs.
1929 Neue Beiträge zur Kenntnis der Thereviden und Tabaniden (Dipt.). Deut. Ent. Ztschr. **1928**: 417–434, 19 figs.
1931a Dreizehn neue neotropische *Tabanus*arten. Konowia **10**: 291–300.

Kröber, O.—Continued
1931b Die *Tabanus*-Gruppen *Straba* End. und *Poecilosoma* Lutz (=*Hybostraba* End. und *Hybopelma* End.) der neotropischen Region. Zool. Anz. **94**: 67–89, 20 figs.
1933 Das Subgenus *Neotabanus* der Tabanidengattung *Tabanus* s. lat. Rev. de Ent. **3**: 337–367.
1937 Ein Beitrag zur Kenntnis der Omphraliden (Scenopiniden), Diptera. Stettin. Ent. Ztg. **98**: 211–231, 2 figs.

Krogstad, B. O.
1959 Some aspects of the ecology of *Axymyia furcata* McAtee (Diptera: Sylvicolidae). Minn. Acad. Sci. Proc. **27**: 175–177, 2 figs.

Kruseman G., Jr.
1933 Tendipedidae Neerlandicae. Pars I. Genus *Tendipes* cum generibus finitimis. Tijdschr. v. Ent. **76**: 119–216, 67 figs.

Kumm, H. W.
1936 The Jamaican species of *Hippelates* and *Oscinella* (Diptera, Chloropidae). Bul. Ent. Res. **27**: 307–329, 6 figs., 1 pl.

Kuntze, A. (See Becker et al.)

Kuster, K. C.
1934 A study of the general biology, morphology of the respiratory system, and respiration of certain aquatic *Stratiomyia* and *Odontomyia* larvae (Diptera). Mich. Acad. Sci., Arts, and Letters, Papers **19**: 605–658, figs. 20–21, pls. 80–83.

Laake, E. W. (See Bishopp et al.)

Laboulbène, A.
1873 Métamorphoses de la Cécidomyie du buis, *Cecidomyia* (*Diplosis*) *buxi*. Soc. Ent. de France, Ann. ser. **5, 3**: 313–326, 1 pl.
1882 Note sur l'Insecte Diptère nuisible de Terre-Neuve signalé par M. le docteur Treille. Arch. de Méd. Nav. **38**: 222–224.

LaCasse, W. J. (See Carpenter and LaCasse)

Lackschewitz, P.
1934 Ueber einige hochnordische *Trichocera*-Arten (Diptera Nematocera). Tromsø Mus. Årsh. (1931) **54 (1** [=Naturhist. Avd. 9]): 3–8, 4 figs.
1935 Zur Kenntnis der polyneuren Nematoceren (Dipt.) des nördlichen Norwegens. Tromsø Mus. Årsh. (1930) **53 (4** [=Naturhist. Avd. 8]): 3–27, 2 figs.

Laffoon, J. L.
1957 A revision of the nearctic species of *Fungivora* (Diptera, Mycetophilidae). Iowa State Col. Jour. Sci. (1956) **31**: 141–340, 12 pls.

Lahille, F.
1904 Notes sur la classification des moustiques. Pp. 71–95, 4 figs., 1 pl. *In* Araoz Alfaro, G., sec., Segundo Congreso Médico Latino-Americano [q.v.]. Vol. 2, 270 pp., 27 figs., 7 pls. Buenos Aires.

Lamb, C. G.
1914 The Percy Sladen Trust Expedition to the Indian Ocean in 1905, under the leadership of Mr. J. Stanley Gardiner, M.A. Vol. V. No. XV. Diptera: Heteroneuridae, Ortalidae, Trypetidae, Sepsidae, Micropezidae, Drosophilidae, Geomyzidae, Milichiidae. Linn. Soc. London, Trans. ser. 2, Zool., **16**: 307–372, 48 figs., 3 pls.

Lamore, D. H.
1960 Cases of parasitism of the basilica spider, *Allepeira lemniscata* (Walckenaer), by the dipteran endoparasite, *Ogcodes dispar* (Macquart) (Araneida: Argiopidae and Diptera: Acroceridae). Ent. Soc. Wash. Proc. **62**: 65–85, 13 figs.

Landrock, K.
1918 Die Pilzmückengattung *Dynatosoma* Winn. Arch. f. Naturgesch., (*1916*) Abt. A, **82 (12)**: 38–51, 17 figs.
1923 Die Pilzmücken Mährens. 1. Nachtrag. Wien. Ent. Ztg. **40**: 163–171.
1925a Dipterologische Miszellen. Mycetophilidae. Wien. Ent. Ztg. **42**: 179–182.
1925b Neue Mycetophiliden. Natuurhist. Maandbl. **14**: 37–40, 15 figs.
1926–1927 Fungivoridae (Mycetophilidae). [Fam.] 8, pp. 1–48, figs. 1–14, pls. 1–3 (=lfg. 12), *1926*; pp. 49–96, figs. 15–21, pls. 4–6 (=lfg. 13), *1926*; pp. 97–144, figs. 22–24, pls. 7–9 (=lfg. 14), *1927*; pp. 145–196, figs. 25–26, pls. 10–13 (=lfg. 15), *1927*. *In* Lindner, E., ed., Die Fliegen der palaearktischen Region [q.v.]. Vol. 2, Pt. 1. Stuttgart.

Lane, J. (See also Carrera, Lopes, and Lane)
1947 Espécies brasileiras de *Stilobezzia* (Dipt. Ceratopogonidae) e *Zygoneura stonei* nov. nom. (Dipt. Mycetophilidae). Rev. de Ent. **18**: 197–214, 9 figs.

Lane, J., and d'Andretta, C., Jr.
1958   Neotropical Anisopodidae (Diptera, Nematocera). Studia Ent. (Rio de Janeiro) **n. ser.**, **1**: 497–528, 29 figs.

Laporte, F. L. de (or La Porte) (Count de Castelnau)
1840   Histoire naturelle des animaux articulés. Annelides, Crustacés, Arachnides, Myriapodes et Insectes. 4 vols. Paris.

> The first 2 volumes are by Laporte, the third by E. Blanchard, and the fourth by P. H. Lucas.
> See Blanchard, 1840.

La Rivers, I.   (See Usinger et al.)

Latreille, P. A.   (See also Bory, 1829; Olivier, 1812; S.N.A., 1803, 1818)

[1796]   Précis des caractères génériques des Insectes, disposés dans un ordre naturel. 179 pp. Paris, "An V".

> An V=Sept. 22, 1796 to Sept. 21, 1797. This work was reviewed in the Soc. Philomath. de Paris, Bul. 1: 118–119, the issue for Dec. 21, 1796–Feb. 19, 1797. This fact indicates publication of Latreille's work in late 1796 or possibly very early 1797. Established practice dates it 1796. See Sherborn, 1902: xxxiii.

[1802]   Histoire naturelle, générale et particulière, des Crustacés et des Insectes, Tome troisième [Vol. 3]. [Vol. 95], 468 pp. *In* Sonnini, C. S., ed., Histoire naturelle par Buffon [q.v.]. Paris, "An X".

> Evidence points to the publication of this volume in An XI (=Sept. 22, 1802, to Sept. 21, 1803) (see Griffin, 1938: 157). The fact that the date on the title page was not corrected indicates publication early in An XI. Established usage dates this volume 1802.

1804   Tableau méthodique des Insectes. Pp. 129–200. *In* Société de Naturalistes et d'Agriculteurs, Nouveau dictionnaire d'histoire naturelle, appliqué aux arts, principalement à l'agriculture et à l'économie rurale et domestique. Vol. 24, [sect. 3]: Tableaux méthodiques d'histoire naturelle, 238 pp., 5 pls. Paris.

> See S.N.A., 1803, for details of this work.

[1805]   Histoire naturelle, générale et particulière, des Crustacés et des Insectes, Tome quatorzième [Vol. 14]. [Vol. 106], 432 pp., pls. 104–112. *In* Sonnini, C. S., ed., Histoire naturelle par Buffon [q.v.]. Paris, "An XIII".

> An XIII=Sept. 22, 1804, to Sept. 21, 1805. Established usage dates this 1805. The use of this date also gives the clear indication of its publication later than Latreille, 1804. See Sabrosky, 1941b: 744, footnote 4.

Latreille, P. A.—Continued
- 1809 Genera crustaceorum et insectorum secundum ordinem naturalem in familias disposita, iconibus exemplisque plurimis explicata. Vol. 4, 399 pp., 4 pls. Parisiis et Argentorat [=Paris and Strasbourg].

    The entire work consists of 4 volumes, Paris and Strasbourg, 1806–1809.

- 1810 Considérations générales sur l'ordre naturel des animaux. 444 pp. Paris.
- 1825 Familles naturelles du règne animal. 570 pp. Paris.

    See Berthold, 1827, the German translation in which the names are first latinized.

- 1829 Les Crustacés, les Arachnides et les Insectes, Tome second [Vol. 2]. Vol. 5, 556 pp., 5 pls. *In* [Cuvier, G. C. L. D.], Le règne animal . . . Ed. 2 [q.v.]. Paris.

———, **Lepeletier, A. L. M., Serville, J. G. A., and Guérin-Méneville, F. E.**
- 1825– Entomologie, ou histoire naturelle des Crustacés, des
- [1828] Arachnides et des Insectes [pt. 2 (=Insectes [i.e., Arthropoda] pt. 7)]. Vol. 10, pp. 1–344 (=livr. 96), *1825;* pp. 345–833 (=livr. 100), [*1828*]. *In* Société de Gens de Lettres, de Savans et d'Artistes, Encyclopédie méthodique [q.v.]. Histoire naturelle. Paris.

    For dating see Sherborn and Woodward, 1899: 595.

Latta, R. (See also Doucette et al.)
———, **and Cole, F. R.**
- 1933 A comparative study of the species of *Eumerus* known as the lesser bulb flies. Calif. Dept. Agr., Monthly Bul. **22:** 142–152, 16 figs.

Laurence, B. R.
- 1955 The ecology of some British Sphaeroceridae (Borboridae, Diptera). Jour. Anim. Ecol. **24:** 187–199, 5 figs.

Lavigne, R.
- 1962 Immature stages of the stalk-eyed fly, *Sphyracephala brevicornis* (Say) (Diptera: Diopsidae) with observations on its biology. Brooklyn Ent. Soc. Bul. **57:** 5–14, 1 fig., 1 pl.

Leach, W. E. (See also Brewster, 1817–1818)
- 1817 On the genera and species of eproboscideous insects. 20 pp., 3 pls. Edinburgh, Sept., 1817.

    Also published in Wernerian Nat. Hist. Soc. Mem. **2:** 547–566, *1818.*

Leathers, A. L.
1922   *Chironomus braseniae*, new species (Dip., Chironomidae). Ent. News **33**: 8.

LeBaron, W.
1871   Insects injurious to the apple tree. Noxious [and Beneficial] Insects of State of Ill., [State Ent.] Rpt. [**2**]: 13–46, 4 figs.

In the first annual report of W. LeBaron.

1872   Insects injurious to the apple. Noxious [and Beneficial] Insects of State of Ill., [State Ent.] Rpt. [**3**]: 99–133, 3 figs.

In the second annual report of W. LeBaron.

LeConte, J. L., ed.
1859a   The complete writings of Thomas Say on the entomology of North America. Vol. 1, 412 pp., 54 pls. New York.

The entire work consists of 2 volumes, New York, 1859. They were reprinted under the imprint Philadelphia, 1891, and also published under the title "American entomology: A description of the insects of North America, by Thomas Say, with illustrations drawn and colored after nature," Boston, undated.

1859b   The complete writings of Thomas Say on the entomology of North America. Vol. 2, 814 pp. New York.

Lehmann, J. G. C.
1822   Observationes zoologicae praesertim in faunam hamburgensem. Pugillus primus. Indicem scholarum publice privatimque in Hamburgensium Gymnasio Academico. 55 pp. Hamburgi [=Hamburg].

The section on Diptera (pp. 38–46) was republished in K. Leopoldinisch-Carolinische Akad. d. Naturf., Verhandl. (Acad. C. Leopoldino-Carolinae Nat. Curio., Nova Acta Phys.-Med.) **12**: 239–248, *1824*, with the addition of a colored plate.

Lengersdorf, F.
1931   Neue *Sciara-(Lycoria)* Arten aus der Sammlung des Zoologischen Instituts der Universität Halle. Zool. Anz. **96**: 251–255, 3 figs.

Lenz, F.   (See also Goetghebuer and Lenz)
1941   Die Jugenstadien der Sectio Chironomariae (Tendipedini) connectentes (Subf. Chironominae=Tendipedinae). Zusammenfassung und Revision. Arch. f. Hydrobiol. (1942) **n. ser., 38**: 1–69, 95 figs.

Leonard, M. D. (See also Alexander and Leonard; Curran, 1931f; Johannsen, 1928b)

1913 Additions to the New Jersey Tipulidae (Diptera), with the description of a new species. Ent. News **24:** 247–249, 1 fig.

1928 List of the insects of New York with a list of the spiders and certain other allied groups. N.Y. (Cornell) Agr. Expt. Sta. Mem. (1926) **101:** 1–1121.

1930 A revision of the dipterous family Rhagionidae (Leptidae) in the United States and Canada. Amer. Ent. Soc. Mem. **7:** 1–181, 3 figs., 3 pls.

1931 Two new species of *Symphoromyia* (Rhagionidae, Diptera) from the eastern United States. Amer. Mus. Nat. Hist., Amer. Mus. Novitates **497:** 1–2.

Lepeletier, A. L. M. (or Le Peletier) (Count de Saint-Fargeau) (See Latreille et al.)

——, **and Serville, J. G. A.** (See Latreille et al., 1828)

Lester, O. C. (See James, 1938d)

Leuckart, R., and Chun, C., eds.

1887– Bibliotheca zoologica. Original-Abhandlungen aus dem
1921   Gesammtgebiete der Zoologie. 71 hefts. Cassel, Stuttgart.

> This is a series of separately paged memoirs. The last issued, as far as known, was Heft 71, in 1921. Some are in 2 or more parts, sometimes separately paged, and some have a "Nachtrag". Heft 20 was issued in 4 continuously paged Lieferungen, Stuttgart, 1895–1898. The name of the series was changed to "Zoologica", etc., with the Nachtrag to Heft 21, in 1898 (Heft 11, 1892, also had this latter title). The early hefts were published in Cassel.
> See Rübsaamen, 1898.

Levi-Castillo, R.

1951 Los mosquitos del género *Haemagogus*-Williston, 1896 en América del Sur. 77 pp., 28 figs. Cuenca, Ecuador.

Lewis, F. B.

1956 Two new species of Ceratopogonidae (Diptera). Psyche **63:** 46–49, 2 figs.

Lichtenstein, A. A. H.

1800 Beschreibung eines neu entdeckten Wasserinsekts. Arch. f. Zool. u. Zootomie **1 (1):** 168–175, 1 pl.

Lichtwardt, B.

1909 Beitrag zur Kenntnis der Nemestriniden (Dipt.). Deut. Ent. Ztschr. **1909:** 113–123, 507–514, 643–651, 8 figs.

Lima, A. da Costa   (See also Lutz, Neiva, and Lima)
1921   Sobre os streblideos americanos (Diptera-Pupipara). Arch. da Esc. Super. Agr. e Med. Vet. [Nictheroy, Rio de Janeiro] **5:** 17–34, 2 pls.

Lindner, E.
1925–   Handbuch. Pp. 1–32, figs. 1–17, pl. 1 (=lfg. 4), *1925;*
1929   pp. 33–48, figs. 18–48, pls. 2–3 (=lfg. 21), *1927;* pp. 49–64, figs. 49–66, pls. 4–5 (=lfg. 23), *1927;* pp. 65–80, figs. 67–80, pls. 6, 8 (=lfg. 25), *1928;* pp. 81–96, figs. 81–105, pls. 7, 9 (=lfg. 31), *1929* [cont.]. *In* Lindner, E., ed., Die Fliegen der palaearktischen Region [q.v.]. Vol. 1, 422 pp., 481 figs., 28 pls. Stuttgart.

1930   Revision der amerikanischen Dipteren-Familie der Rhopalomeridae. Deut. Ent. Ztschr. **1930:** 122–137, 1 fig.

1931–   Handbuch [cont.]. Pp. 97–144, figs. 106–220, pl. 10 (=lfg.
1935   57), *1931;* pp. 145–160, figs. 221–256, pls. 11–12 (=lfg. 60), *1932;* pp. 161–208, figs. 257–357, pl. 13 (=lfg. 74), *1933;* pp. 209–240, figs. 358–389, pls. 14–15 (=lfg. 95), 1935 [cont.]. *In* Lindner, E., ed., Die Fliegen der palaearktischen Region [q.v.]. Vol. 1, 422 pp., 481 figs., 28 pls. Stuttgart.

1936a   Handbuch [cont.]. Pp. 241–280, figs. 390–429, pls. 16–17 (=lfg. 102) [cont.]. *In* Lindner, E., ed., Die Fliegen der palaearktischen Region [q.v.]. Vol. 1, 422 pp., 481 figs., 28 pls. Stuttgart.

1936b   Stratiomyidae. [Fam.] 18, pp. 1–48, figs. 1–38 (=lfg. 104) [cont.]. *In* Lindner, E., ed., Die Fliegen der palaearktischen Region [q.v.]. Vol. 4, Pt. 1. Stuttgart.

1937–   Stratiomyidae [concl.]. [Fam.] 18, pp. 49–96, figs. 39–55,
1938   pls. 1–4 (=lfg. 108), *1937;* pp. 97–144, figs. 56–81, pls. 5–6 (=lfg. 110), *1937;* pp. 145–176, figs. 82–103, pl. 7 (=lfg. 114), *1937;* pp. 177–218, figs. 104–136 (=lfg. 116), *1938.* *In* Lindner, E., ed., Die Fliegen der palaearktischen Region [q.v.]. Vol. 4, Pt. 1. Stuttgart.

1940–   Handbuch [concl.]. Pp. 281–312, figs. 430–444, pls. 18–19
1949   (=lfg. 134), *1940;* pp. 313–336, figs. 445–447, pls. 20–21 (=lfg. 136), *1940;* pp. 337–368, figs. 448–472, pls. 22–23 (=lfg. 146), *1943;* pp. 369–416, figs. 473–481, pls. 24–25 (=lfg. 156), *1948;* pp. 417–422, pls. 26–28 (=lfg. 158), *1949.* *In* Lindner, E., ed., Die Fliegen der palaearktischen Region [q.v.]. Vol. 1, 422 pp., 481 figs., 28 pls. Stuttgart.

Lindner, E., ed.
1924– Die Fliegen der palaearktischen Region. 8 vols. Stutt-
1963 gart.

    This series is issued in "Lieferungen", numbered chronologically as they appear, and continues to date. Each separately paged family is given a number and assigned to a volume, but a family is not necessarily composed of consecutive Lieferungen. Some of the volumes were later divided into parts. The arrangement of the work, with the dates for each volume and part, is given below:

    Vol. 1. Complete, introductory material, no families. 1925–1949.
    Vol. 2. Incomplete. 1926–1930.
        Pt. 1. Complete, families 1–5, 7–8. 1926–1930.
        Pt. 2. Not yet begun, family 6.
    Vol. 3. Incomplete, families 9–17. 1929–1962.
    Vol. 4. Incomplete. 1924–1958.
        Pt. 1. Complete, families 18–23. 1924–1938.
        Pt. 2. Complete, family 24. 1925–1930.
        Pt. 3. Complete, families 25–27. 1925–1937.
        Pt. 4. Complete, family 28. 1938–1956.
        Pt. 5. Incomplete, families 29–30. 1930–1941.
        Pt. 6. Complete, families 31–32. 1928–1935.
        Pt. 7. Incomplete, families 33–35. 1925–1958.
    Vol. 5. Complete, families 36–53. 1927–1949.
    Vol. 6. Complete. 1926–1938.
        Pt. 1. Complete, families 54–58, 60–61. 1926–1938.
        Pt. 2. Complete, family 59. 1931–1936.
    Vol. 7. Incomplete, families 62–63. 1937–1963.
    Vol. 8. Incomplete, families 64–66. 1930–1963.

    See Becker, 1926; Czerny, 1928b, 1930a, 1930b, 1934a, 1934b; Duda, 1928a, 1932–1933, 1934a, 1934b, 1935; Engel, 1925–1930, 1932–1937, 1938–1954; Frey, 1954–1956; Goetghebuer and Lenz, 1933–1934, 1936, 1937–1938, 1939a, 1939b, 1939c, 1940–1944, 1950a, 1950b, 1954–1960; Hendel, 1927a, 1931a, 1932a, 1933a, 1934a, 1935a, 1936a; Hennig, 1939, 1940, 1943, 1949, 1955–1957, 1958b, 1959–1963; Landrock, 1926–1927; Martini, 1929–1931; Mesnil, 1944, 1949b, 1950–1952, 1953b, 1954–1963; Sack, 1928–1932, 1937; Stackelberg, 1930–1941.

Lindroth, C. H. (See Goetghebuer and Lindroth)

Linnaeus, C. (or Linné, C. von)
1758 Systema naturae per regna tria naturae. Ed. 10, Vol. 1, 824 pp. Holmiae [=Stockholm].
    The entire work consists of 2 volumes, Holmiae, 1758–1759.

1761 Fauna svecica sistens animalia Sveciae regni. Ed. 2, 578 pp., 2 pl. Stockholmiae [=Stockholm].
    The first edition, published in 1746, is outside the scope of zoological nomenclature.

Linnaeus, C.—Continued
1762   Zweyter Theil, enthalt Beschreibungen verschiedener wichtiger Naturalien. Pp. 267–606. *In* Hasselquist, F., Reise nach Palästina in den Jahren von 1749 bis 1752. 606 pp. Rostock, Germany.

> This is the German translation by T. H. Gadebusch of Hasselquist's "Iter Palaestinum, eller resa til Heliga Landet, foerrättad ifran år 1749 til 1752", 619 pp.., Holmiae, 1757.
> Suppressed by I.C.Z.N., 1914: 131.

1763   Amoenitates Academicae, seu dissertationes variae physicae, medicae, botanicae, antehac seorsim editae, nunc collectae et auctae cum tabulis aeneis. Vol. 6, 486 pp., 5 pls. Holmiae [=Stockholm].

> The entire work consists of 7 volumes, Holmiae and Lipsia. [=Leipzig], 1749–1769.

1767   Systema naturae per regna tria naturae. Ed. 12 (rev.), Vol. 1, Pt. 2, pp. 533–1327. Holmiae [=Stockholm].

> The entire work consists of 3 volumes, Holmiae, 1766–1768e
> The first volume is in 2 continuously paged parts, 1766–1767, and consists of 1327 pp. and 36 unnumbered pages of index.

Linné, C. von   (See Linnaeus, C.)

Lintner, J. A.   (See also Felt, 1898a, 1898b)
1879a   The clover-seed fly—a new insect pest. Canad. Ent. **11**: 44.
1879b   On *Cecidomyia leguminicola*, n. sp. Canad. Ent. **11**: 121–124.
1883   Reports on the injurious and other insects of the State of New York. Rpt. 1: First annual report [for 1882], 381 pp., 84 figs. Albany, N.Y.

> There were 13 such reports by Lintner, published in Albany, 1883–1899, only the first being marked "Annual report". The first 2 reports, for 1882 and 1885, were published only as separates, 1883 and 1886, respectively. There was another entirely different report by Lintner for 1885 published in the N.Y. State Mus. Ann. Rpt. in 1887. This was not counted in the numbering. Lintner's report for 1886 (published in 1888), later designated the "3rd report", and his succeeding numbered reports, were issued both in the N.Y. State Mus. Ann. Rept. and as separates with identical pagination (except for the 4th report). The 14th report and succeeding ones are by E. P. Felt, and were published only in publications of the N.Y. State Museum.

1888a   Reports on the injurious and other insects of the State of New York. Rpt. 4: [for 1887], 237 pp., 68 figs. Albany, N.Y.

> Also published as "Report of the State Entomologist", in N.Y. State Mus. Ann. Rpt. (1887) **41**: 123–358, 68 figs., *1888*.

Lintner, J. A.—Continued
- 1888b The melon plant louse. Cult. and Country Gent. **53:** 725.
- [1892] Reports on the injurious and other insects of the State of New York. Rpt. 7: [for 1890], pp. 197–405, 40 figs. Albany, N.Y., "1891."

    Also published as "Report of the State Entomologist", in N.Y. State Mus. Ann. Rpt. (1890) **44:** 197–405, 40 figs., *1892*.
    Dated by Felt, 1899: 374.

- 1895 Report of the State Entomologist for the year 1894. N.Y. State Mus. Ann. Rpt. (1894) **48(1):** 339–636, 24 figs., 4 pls.

    Also published separately as his "Reports on the injurious and other insects of the State of New York", Rpt. 10, pp. 339–636, 24 figs., 4 pls., Albany, "1895" [=*1896*].
    Separate dated by Felt, 1899: 392.

- [1897] Reports on the injurious and other insects of the State of New York. Rpt. 11: for 1895, pp. 85–326, 25 figs., 16 pls. Albany, N.Y., "1896."

    Also published as "Report of the State Entomologist", in N.Y. State Mus. Ann. Rpt. (1895) **49(1):** 85–326, 25 figs., 16 pls., *1897*.
    Dated by Felt, 1899: 395.

- [1898] Reports on the injurious and other insects of the State of New York. Rpt. 12: for 1896, pp. 161–399, 10 figs., 15 pls. Albany, N.Y., "1897."

    Also published as "Report of the State Entomologist", in N.Y. State Mus. Ann. Rpt. (1896) **50(1):** 161–399, 10 figs., 15 pls., *1898*.
    Dated by Felt, 1899: 398.

Lioy, P.
- 1863 I ditteri distribuiti secondo un nuovo metodo di classificazione naturale. I.R. Ist. Veneto di Sci., Let. ed Arti, Atti **ser. 3, 9:** 187–236 [cont.].
- 1864a I ditteri distribuiti secondo un nuovo metodo di classificazione naturale [cont.]. I.R. Ist. Veneto di Sci., Let. ed Arti, Atti **ser. 3, 9:** 499–518, 569–604, 719–771, 879–910, 989–1027, 1087–1126, 1311–1352 [cont.].
- 1864b I ditteri distribuiti secondo un nuovo metodo di classificazione naturale [concl.]. I.R. Ist. Veneto di Sci., Let. ed Arti, Atti **ser. 3, 10:** 59–84.

Lisney, A. A.
- 1960 A bibliography of British Lepidoptera 1608–1799. 315 pp., 39 pls. London.

Löw, F.
1874 Beiträge zur Kenntniss der Gallmücken. K.-k. Zool.-Bot. Gesell. Wien, Verhandl. **24 (Abhandl.)**: 143–162, 11 pls.

Loew, H. (or Löw, H.) (See also Agassiz and Loew; Osten Sacken, 1862b)
1840 Ueber die im Grosherzogthum Posen aufgefundenen Zweyflügler. Isis (Oken's) **1840**: 512–583, pl. 1 (part).
1844a Beschreibung einiger neuen Gattungen der europäischen Dipterenfauna. Stettin Ent. Ztg. **5**: 114–130, 154–173, 2 pls.
1844b Kritische Untersuchung der europäischen Arten des Genus *Trypeta*, Meig. Ztschr. f. die Ent. (Leipzig) **5**: 312–437, 2 pls.
[1845] (Dipterologischer Beitrag) [I]. K. Friedrich-Wilhelms-Gymnasiums zu Posen, Öffentl. Prüf. d. Schüler **1845**: 1–52, 1 pl.

> Since this publication included a program of the examinations given Mar. 17, 1845, established practice assumes publication close to that date. It was also issued separately, titled "Dipterologische Beiträge," with the same pagination, under the imprint Posen, *1845*. This same information applies to the "Zweiter Theil", the program being for Mar. 26, 1847, and the separate dated Posen, 1847. The two articles which later were issued as the "Dritter Theil," Posen, 1847, were first published in Naturw. Ver. zu Posen, Jahresber. 1846: 1–24, 25–44, in 1847.

1846a *Helophilus*. Stettin Ent. Ztg. **7**: 116–127, 141–150, 164–169.
1846b Fragmente zur Kenntniss der europäischen Arten einiger Dipterengattungen. Linnaea Ent. **1**: 319–530, pl. 3.
1847a Ueber *Tetanocera stictica* und ihre nächsten Verwandten, nebst der Beschreibung zweier anderen neuen *Tetanocera*-Arten. Stettin Ent. Ztg. **8**: 114–124.
1847b Ueber *Tetanocera ferruginea* und die ihr verwandten Arten. Stettin Ent. Ztg. **8**: 194–202.
1847c Ueber die europäischen Raubfliegen (Diptera-Asilica). Linnaea Ent. **2**: 384–568, 587–591 [cont.].
1848 Ueber die europäischen Raubfliegen (Diptera-Asilica) [cont.]. Linnaea Ent. **3**: 386–495 [cont.].
1849 Ueber die europäischen Raubfliegen (Diptera-Asilica) [concl.]. Linnaea Ent. **4**: 1–155.
1850a *Meghyperus* und *Arthropeas*, zwei neue Dipterengattungen. Stettin Ent. Ztg. **11**: 302–308, 1 pl.

Loew, H.—Continued

[1850b] Dipterologische Beiträge. Vierter Theil. K. Friedrich-Wilhelms-Gymnasiums zu Posen, Öffentl. Prüf. d. Schüler **1850**: 1–40, 1 pl.

> Since this publication included a program of the examinations given early in 1850 (actual dates not known), established practice assumes publication close to that date. This last Beitrag was also issued separately, with the same pagination, under the imprint Posen, *1850*. The "program" has not been seen by us.

1850c   Ueber den Bernstein und die Bernsteinfauna. K. Realschule zu Meseritz, Programm **1850**: 1–44.

> Also issued separately, with the same pagination, under the imprint Berlin, *1850*. The Programm has not been seen.

1851   Beschreibung einiger neuen Tipularia terricola. Linnaea Ent. **5**: 385–406, pl. 2.

1853   Neue Beiträge zur Kenntniss der Dipteren. Erster Beitrag. K. Realschule zu Meseritz, Programm **1853**: 1–38.

> Also issued separately, with the same pagination, under the imprint Berlin, *1853*. The Programm has not been seen.

1854   Neue Beiträge zur Kenntniss der Dipteren. Zweiter Beitrag. K. Realschule zu Meseritz, Programm **1854**: 1–24.

> Also issued separately, with the same pagination, under the imprint Berlin, *1854*. The Programm has not been seen.

1855a   Einige Bemerkungen über die Gattung *Sargus*. Zool.-Bot. Ver. Wien, Verhandl. **5 (Abhandl.)**: 131–148, 4 figs.

1855b   Neue Beiträge zur Kenntniss der Dipteren. Dritter Beitrag. K. Realschule zu Meseritz, Programm **1855**: 1–52.

> Also issued separately, with the same pagination, under the imprint Berlin, *1855*. The Programm has not been seen.

1857a   Neue Beiträge zur Kenntniss der Dipteren. Fünfter Beitrag. K. Realschule zu Meseritz, Programm **1857**: 1–56.

> Also issued separately, with the same pagination, under the imprint Berlin, *1857*. The Programm has not been seen.

1857b   Bidrag till kännedomen om Afrikas Diptera [cont.]. K. Svenska Vetensk. Akad. Öfversigt af . . . Förhandl. **14**: 337–383 [cont.].

1857c   Dipterologische Mittheilungen. Wien. Ent. Monatschr. **1**: 33–56, 1 pl.

Loew, H.—Continued
1857d Die europäischen Arten der Gattung *Cheilosia*. Zool.-Bot. Ver. Wien, Verhandl. **7 (Abhandl.):** 579–616.

> Although "Schluss folgt" appears at the end of the article, a continuation never appeared.

1858a Ueber einige neue Fliegengattungen. Berlin. Ent. Ztschr. **2:** 101–122, 1 pl.

1858b Drei neue *Ortalis*-Arten. Berlin. Ent. Ztschr. **2:** 374–376.

1858c Zur Kenntniss der europäischen *Tabanus*-Arten. K.-k. Zool.-Bot. Gesell. Wien, Verhandl. **8 (Abhandl.):** 573–612.

1858d Bidrag till kännedomen om Afrikas Diptera [cont.]. K. Svenska Vetensk. Akad. Öfversigt af . . . Förhandl. **15:** 335–341 [cont.].

1858e Zehn neue Diptern. Wien. Ent. Monatschr. **2:** 7–15.

1859a Neue Beiträge zur Kenntniss der Dipteren. Sechster Beitrag. K. Realschule zu Meseritz, Programm **1859:** 1–50.

> Also issued separately, with the same pagination, under the imprint Berlin, *1859*.

1859b Die nordamerikanische Arten der Gattungen *Tetanocera* und *Sepedon*. Wien. Ent. Monatschr. **3:** 289–300.

1860a Neue Beiträge zur Kenntniss der Dipteren. Siebenter Beitrag. K. Realschule zu Meseritz, Programm **1860:** 1–46.

> Also issued separately, with the same pagination, under the imprint Berlin, *1860*.

1860b Diptera Americana ab Osten-Sackenio collecta. Decas prima. Wien. Ent. Monatschr. **4:** 79–84.

> This was the only such "Decas" published. Probably the forerunner of his Centuriae (see Loew, 1861b, etc.).

1860c Bidrag till kännedomen om Afrikas Diptera [cont.]. K. Svenska Vetensk. Akad. Öfversigt af . . . Förhandl. (1861) **17:** 81–97 [cont.].

> The date on the title page of this volume for 1860 is 1861, but the part concerned here was issued in 1860.

1860d Die Dipteren-Fauna Südafrika's. Erste Abtheilung. Naturw. Ver. f. Sachsen u. Thüringen, Abhandl. (1858–1861) **2:** 57–402, 2 pls.

> Also published separately, xi + 330 pp., 2 pls., Berlin, *1860*.

Loew, H.—Continued
1861a   Neue Beiträge zur Kenntniss der Dipteren. Achter Beitrag [part]. K. Realschule zu Meseritz, Programm **1861**: 1–60 [cont.].

> Also issued separately, with the same pagination, under the imprint Berlin, without a date, and later with its continuation under the imprint Berlin, *1861* (see Loew, 1861e).

1861b   Diptera Americae septentrionalis indigena. Centuria prima. Berlin. Ent. Ztschr. **5**: 307–359.

> This was apparently also issued separately, pp. 1–53, with a title page dated 1861, and in Loew, 1864d: 1–53.

1861c   Diptera aliquot in insula Cuba collecta. Wien. Ent. Monatschr. **5**: 33–43.

1861d   *Blaesoxipha grylloctona*, nov. genus et species. Wien. Ent. Monatschr. **5**: 384–387.

1861e   Neue Beiträge zur Kenntniss der Dipteren. Beitrag 8, 100 pp. Berlin.

> Also published in K. Realschule zu Meseritz, Programm **1861**: 1–60, *1861* (see Loew, 1861a), and **1862**: 61–100, *1862*. There were a total of 8 of these "Neue Beiträge" published in Berlin, 1853–1861, and also in K. Realschule zu Meseritz, Programm, 1853–1862.

1862a   Die europäischen Bohrfliegen (Trypetidae). 128 pp., 26 pls. Wien.

1862b   Diptera Americae septentrionalis indigena. Centuria secunda. Berlin. Ent. Ztschr. **6**: 185–232.

> Also published in Loew, 1864d: 55–102.

1862c   Monographs of the Diptera of North America. Part I. Smithsn. Inst., Smithsn. Misc. Collect. **6**(1[=pub. 141]): 1–221, figs. 1–3+1–12, 2 pls.

1862d   Ueber einige bei Varna gefangene Dipteren. Wien. Ent. Monatschr. **6**: 161–175.

(1862e) Ueber die europäischen Helomyzidae und die in Schlesien vorkommenden Arten derselben. Ztschr. f. Ent. Breslau (1859) **13**: 1–80.

> For date see Ztschr. f. Ent. Breslau n. ser., 30 (Vereinsnachr.): iii, footnote, 1905.

1863a   Diptera Americae septentrionalis indigena. Centuria tertia. Berlin. Ent. Ztschr. **7**: 1–55.

> Also published in Loew, 1864d: 103–157.

1863b   Diptera Americae septentrionalis indigena. Centuria quarta. Berlin. Ent. Ztschr. **7**: 275–326.

> Also published in Loew, 1864d: 159–210.

Loew, H.—Continued

1864a Diptera Americae septentrionalis indigena. Centuria quinta. Berlin. Ent. Ztschr. **8:** 49–104.

    Also published in Loew, 1864d: 211–261.

1864b Die Arten der Gattung *Balioptera*. Berlin. Ent. Ztschr. **8:** 347–356.

1864c Ueber die europäischen Arten der Gattung *Diastata*. Berlin. Ent. Ztschr. **8:** 357–368.

1864d Diptera Americae septentrionalis indigena. Vol. 1: [Centuriae 1–5], 266 pp. Berolini [=Berlin], "1861."

    A reissue of Loew, 1861b, 1862b, 1863a, 1863b, and 1864a. Centuria 1 was apparently also issued separately, its title page, date 1861, being used as the title page for this volume. We have no knowledge of separate publications of Centuriae 2–5 other than in the above volume, the date of which, therefore, was probably not before 1864.
    The entire work consists of 2 volumes, Berlin, [1864?]–1872.

1864e Monographs of the Diptera of North America. Part II. Smithsn. Inst., Smithsn. Misc. Collect. **6** (2 [=pub. 171]): 1–360, 5 pls.

1864f Acht neue *Cordylura*-Arten. Wien. Ent. Monatschr. **8:** 17–26.

(1864g) Ueber die schlesischen Arten der Gattung *Tachypeza* Meig. Ztschr. f. Ent. Breslau (1860) **14:** 3–32.

    For date see Ztschr. f. Ent. Breslau n. ser., 30 (Vereinsnachr.): iii, footnote, 1905.

1866a Diptera Americae septentrionalis indigena. Centuria sexta. Berlin. Ent. Ztschr. (1865) **9:** 127–186.

    Also published in Loew, 1872b: 1–60.

1866b Diptera Americae septentrionalis indigena. Centuria septima. Berlin. Ent. Ztschr. **10:** 1–54.

    Also published in Loew, 1872b: 61–114.

(1866c) Ueber die bisher in Schlesien aufgefundenen Arten der Gattung *Chlorops* Macq. Ztschr. f. Ent. Breslau (1861) **15:** 3–96.

    For date see Ztschr. f. Ent. Breslau n. ser., 30 (Vereinsnachr.): iii, footnote, 1905.

1868a Die amerikanischen Ulidina. Berlin. Ent. Ztschr. (1867) **11:** 283–326.

    The pertinent plate is Pl. 2 of Volume 12.

1868b Die europäischen Ortalidae. Ztschr. f. die Gesam. Naturw. **32:** 1–11.

Loew, H.—Continued
- 1869a Beschreibung europäischer Dipteren. Systematische Beschreibung der bekannten europäischen zweiflügeligen Insecten, von Johann Wilhelm Meigen. Vol. 1: Achter Theil oder zweiter Supplementband, 310 pp. Halle.

The entire work, a continuation of Meigen, 1818–1838, consists of 3 volumes, Halle, 1869–1873.
- 1869b Diptera Americae septentrionalis indigena. Centuria octava. Berlin. Ent. Ztschr. **13**: 1–52.

Also published in Loew,, 1872b: 115–166.
- 1869c Diptera Americae septentrionalis indigena. Centuria nona. Berlin. Ent. Ztschr. **13**: 129–186.

Also published in Loew, 1872b: 167–224.
- 1869d Ueber Dypteren [sic] der Augsburger Umgegend. Naturhist. Ver. in Augsburg, Ber. **20**: 39–59.
- 1871 Beschreibung europäischer Dipteren. Systematische Beschreibung der bekannten europäischen zweiflügeligen Insecten, von Johann Wilhelm Meigen. Vol. 2: Neunter Theil oder dritter Supplementband, 319 pp. Halle.
- 1872a Diptera Americae septentrionalis indigena. Centuria decima. Berlin. Ent. Ztschr. **16**: 49–115.

Also published in Loew, 1872b: 225–291.
- 1872b Diptera Americae septentrionalis indigena. Vol. 2: [Centuriae 6–10], 300 pp. Berolini [=Berlin].

A reissue of Loew, 1866a, 1866b, 1869b, 1869c, and 1872a, apparently as a complete volume (see Vol. 1, Loew, 1864d).
- 1873a Diptera nova, in Pannoniâ inferiori et in confinibus Daciae regionibus a Ferd. Kowarzio capta. Berlin. Ent. Ztschr. **17**: 33–52.
- 1873b Monographs of the Diptera of North America. Part III. Smithsn. Inst., Smithsn. Misc. Collect. **11**: (3 [=pub. 256]): 1–351, 4 pls.
- 1873c Beschreibung europäischer Dipteren. Systematische Beschreibung der bekannten europäischen zweiflügeligen Insecten, von Johann Wilhelm Meigen. Vol. 3: Zehnter Theil oder vierter Supplementband, 320 pp. Halle.
- 1874a Neue nordamerikanische Dasypogonina. Berlin. Ent. Ztschr. **18**: 353–377.
- 1874b Neue nordamerikanische Diptera. Berlin. Ent. Ztschr. **18**: 378–384.
- 1876 Beschreibung neuer amerikanischen Dipteren. Ztschr. f. die Gesam. Naturw. **48**: 317–340.
- 1878 Neue nordamerikanischen Ephydriden. Ztschr. f. die Gesam. Naturw. **51**: 192–203.

Long, W. H., Jr.
1902 New species of *Ceratopogon*. Biol. Bul. **3:** 3-14, 7 figs.

Lopes, H. de S. (See also Carrera, Lopes, and Lane)
1946 Sarcophagidae do México, capturados pelo professor A. Dampf (Diptera). Inst. Oswaldo Cruz, Mem. **44:** 119-146, 62 figs.

―――, **and Downs, W. G.**
1951 Contribuição ao conhecimento das espécies do genero "*Acanthodotheca*" Townsend (Diptera-Sarcophagidae). Inst. Oswaldo Cruz, Mem. (1949) **47:** 571-603, 59 figs.

Lord, J. K.
1866 The naturalist in Vancouver Island and British Columbia. Vol. 2, 375 pp., 4 figs. London.

> The entire work consists of 2 volumes, London, 1866.

Lovett, A. L. (See also Cole and Lovett)
1920 Two new species of Syrphidae (Diptera). Calif. Acad. Sci. Proc. ser. **4, 10:** 51-52.

Lowe, H. (See Smith and Lowe)

Lucas, P. H. (See d'Orbigny, 1848)

Ludlow, C. S. (See also Dyar and Ludlow)
1904 Mosquito notes. [-No. 1]. Canad. Ent. **36:** 233-236.
1905a A new North American *Taeniorhynchus*. Canad. Ent. **37:** 231-232.
1905b Mosquito notes. -No. 4. Canad. Ent. **37:** 385-388 [cont.].
1906a Mosquito notes. -No. 4 [concl.]. Canad. Ent. **38:** 132-134.
1906b An Alaskan mosquito. Canad. Ent. **38:** 326-328.
1906c A new American mosquito. George Wash. Univ. Bul. **5** (4): 83-84.

> Also published in George Wash. Univ. Pubs., Nat. and Phys. Sci. Ser. **1:** 85-86, *1907*.

1907 Mosquito notes. -No. 5 [cont.]. Canad. Ent. **39:** 129-131, 266-268 [cont.].
1911 A new Alaskan mosquito. Canad. Ent. **43:** 178-179.

Lugger, O.
[1897] Parasites of man and domesticated animals. Minn. State Expt. Sta., Ent. Ann. Rpt. (1896) **2:** 44-231, figs. 24-187 (includes parts of pls. 6-7, and all of pls. 9-13, and 16).

> Appeared in 1897 as evidenced by the date of the letter of transmittal, i.e., Jan. 1, 1897.

Lundbeck, W. (See also Henriksen and Lundbeck; Kieffer and Lundbeck)

1898   Diptera groenlandica. Naturhist. For. Kjøbenhavn, Vidensk. Meddel. **1898** [=ser. 5, 10]: 236–314, 2 pls. [cont.].

1901   Diptera groenlandica [concl.]. Naturhist. For. Kjøbenhavn, Vidensk. Meddel. **1900** [=ser. 6, 2]: 281–316, 5 figs.

1910   Diptera Danica. Genera and species of flies hitherto found in Denmark. Vol. 3: Empididae, 329 pp., 141 figs. Copenhagen.

        The entire work consists of 7 volumes, Copenhagen, 1907–1927, and was never completed.

1912   Diptera Danica. Genera and species of flies hitherto found in Denmark. Vol. 4: Dolichopodidae, 416 pp., 130 figs. Copenhagen.

1916   Diptera Danica. Genera and species of flies hitherto found in Denmark. Vol. 5: Lonchopteridae-Syrphidae, 603 pp., 202 figs. Copenhagen.

1922   Diptera Danica. Genera and species of flies hitherto found in Denmark. Vol. 6: Pipunculidae-Phoridae, 455 pp., 132 figs. Copenhagen.

1927   Diptera Danica. Genera and species of flies hitherto found in Denmark. Vol. 7: Platypezidae-Tachinidae, 571 pp., 116 figs. Copenhagen.

Lundström, C.

1906   Beiträge zur Kenntnis der Dipteren Finlands. I. Mycetophilidae. Soc. pro Fauna et Flora Fenn. Acta **29 (1)**: 1–50, 4 pls., 1 map.

1909   Beiträge zur Kenntnis der Dipteren Finlands. IV. Supplement Mycetophilidae. Soc. pro Fauna et Flora Fenn. Acta **32 (2)**: 1–67, 14 pls.

1910   Beiträge zur Kenntnis der Dipteren Finlands. V. Bibionidae. Soc. pro Fauna et Flora Fenn. Acta **33 (1)**: 1–16, 1 pl.

1911a   Beiträge zur Kenntnis der Dipteren Finlands. VII. Melusinidae (Simuliidae). Soc. pro Fauna et Flora Fenn. Acta **34 (12)**: 1–23, 1 pl.

1911b   Neue oder wenig bekannte europäische Mycetophiliden. Budapest Magyar Nemzeti Muz., Ann. Hist. Nat. **9**: 390–419, pls. 11–15.

1912   Beiträge zur Kenntnis der Dipteren Finlands. VIII. Supplement 2. Mycetophilidae, Tipulidae, Cylindrotomidae und Limnobiidae. Soc. pro Fauna et Flora Fenn. Acta **36 (1)**: 1–70, 7 pls.

Lundström, C.—Continued
1913   Neue oder wenig bekannte europäische Mycetophiliden.
III. Budapest Magyar Nemzeti Muz., Ann. Hist. Nat.
11: 305–322, pls. 15–16.
1915   Résultats scientifiques de l'Expédition Polaire Russe en 1900–1903, sous la direction du Baron E. Toll. Section E: Zoologie. Volume II, livr. 8. Diptera Nematocera aus den arctischen Gegenden Sibiriens. [Akad. Nauk S.S.S.R.] Imp. Akad. Nauk Petrograd, Zap., Fiz.-Mat. Otd. (Acad. Imp. des Sci. de Petrograd, Mém., Cl. Phys.-Math.) **ser. 8, 29** (8): 1–33, 2 pls.

    Reprinted in 1918, and commonly cited as such.

———, **and Frey, R.**
1913   Beitrag zur Kenntnis der Dipterenfauna des nördl. europäischen Russlands. Soc. pro Fauna et Flora Fenn. Acta **37** (10): 3–20.

Lutz, A.
1909   Collecção de tabánidas. Pp. 28–30, 2 figs. *In* Instituto Oswaldo Cruz, Instituto Oswaldo Cruz, em Manguinos, Rio de Janeiro. 47 pp., 50 figs. Rio de Janeiro.

    This publication, including the article by Lutz, was issued anonymously. See Fairchild, 1950: 118, for discussion of authorship.

1913a   Sobre a systematica dos tabanideos, sub-familia Tabaninae. Brazil Med. **27**: 486–487.
1913b   Contribuição para o estudo das Ceratopogoninas hematofagas encontradas no Brazil.—Beiträge zur Kenntniss der blutsaugenden Ceratopogoninen Brasiliens. Inst. Oswaldo Cruz, Mem. **5**: 45–73, 3 pls.

———, **and Neiva, A.**
1914   As Tabanidae do Estado do Rio de Janeiro.—Ueber die Tabaniden des Staates Rio de Janeiro. Inst. Oswaldo Cruz, Mem. **6**: 69–80.

———, **Neiva, A., and Lima, A. da Costa**
1915   Sobre "Pupipara" ou "Hippoboscidae" de aves brasileiras. Inst. Oswaldo Cruz, Mem. **7**: 173–199, 2 pls.

Lynch Arribálzaga, E.
1879   Asílides argentinos. Soc. Cient. Argentina, An. **8**: 145–153 [cont.].
1880   Asílides argentinos [cont.]. Soc. Cient. Argentina, An. **9**: 26–33, 49–57, 224–230, 252–265 [cont.].

Lynch Arribálzaga, F.
[1891a] Dipterología argentina. La Plata Mus. Rev. (1890–1891) **1:** 345–377 [cont.].

> As evidenced by many 1891 dates in Vol. 1, this could not have appeared in 1890. The separate appeared in 1891, and most authorities give this as the correct date.

1891b Dipterología argentina [concl.]. La Plata Mus. Rev. **2:** 131–174, 5 pls.

Macfie, J. W. S. (See also Ingram and Macfie)
1932 Ceratopogonidae from the wings of dragonflies. Tijdschr. v. Ent. **75:** 266–283, 7 figs.
1934 Report on a collection of Ceratopogonidae from Malaya. Ann. Trop. Med. and Parasitol. **28:** 177–194, 279–293, 9 figs.
1937 Ceratopogonidae from Trinidad. Ann. and Mag. Nat. Hist. ser. 10, **20:** 1–18, 6 figs.
1938 Notes on Ceratopogonidae (Diptera). Roy. Ent. Soc., London, Proc. Ser. B: Taxonomy **7:** 157–166, 9 figs.
1939 A report on a collection of Brazilian Ceratopogonidae (Dipt.). Rev. de Ent. **10:** 137–219, 24 figs.
1940a The genera of Ceratopogonidae. Ann. Trop. Med. and Parasitol. **34:** 13–30.
1940b Ceratopogonidae (Diptera) from British Guiana and Trinidad. Roy. Ent. Soc., London, Proc. Ser. B: Taxonomy **9:** 179–195, 5 figs.
1948 Some species of *Culicoides* (Diptera, Ceratopogonidae) from the State of Chiapas, Mexico. Ann. Trop. Med. and Parasitol. **42:** 67–87, 10 figs.

MacGillivray, A. D. (See Needham et al.)

Mackerras, I. M.
1954 The classification and distribution of Tabanidae (Diptera). I. General review. Austral. Jour. Zool. **2:** 431–454, 10 figs.
1955 The classification and distribution of Tabanidae (Diptera). II. History. Morphology. Classification. Subfamily Pangoniinae; III. Subfamilies Scepsidinae and Chrysopinae. Austral. Jour. Zool. **3:** 439–511, 39 figs.; 583–633, 26 figs.

Macleay, W. S.
1829 Notice of *Ceratitis citriperda*, an insect very destructive to oranges. Zool. Jour. (London) **4:** 475–482.

Macquart, J.

1823   Monographie des Insectes Diptères de la famille des Empides, observés dans le nord-ouest de la France. Soc. d'Amateurs des Sci., de l'Agr. et des Arts, Lille, Rec. des Trav. **1819/1822:** 137–165.

        See Macquart, 1827b.

1826a   Insectes Diptères du nord de la France. Tipulaires. Soc. d'Amateurs des Sci., de l'Agr. et des Arts, Lille, Rec. des Trav. **1823/1824:** 59–224, 4 pls.

        Also published separately as his "Insectes Diptères du nord de la France", [Vol. 1], 175 pp., 4 pls., Lille, *1826*. Pages 167–175, not present in the journal article, contain index and explanation to figures.

1826b   Insectes Diptères du nord de la France. Asiliques, Bombyliers, Xylotomes, Leptides, Stratiomydes, Xylophagites et Tabaniens. Soc. des Sci., de l'Agr. et des Arts, Lille, Rec. des Trav. **1825:** 324–499, 3 pls.

        Also published separately as his "Insectes Diptères du nord de la France", [Vol. 2], 178 pp., 3 pls., Lille, *1826*.

1827a   Insectes Diptères du nord de la France. Platypézines, Dolichopodes, Empides, Hybotides. Soc. des Sci., de l'Agr. et des Arts, Lille, Rec. des Trav. **1826/1827:** 213–291, 1 pl.

        This includes only Platypezidae and Dolichopodidae. Also published in Macquart, 1827b, pp. 1–76.

1827b   Insectes Diptères du nord de la France. [Vol. 3]: Platypézines, Dolichopodes, Empides, Hybotides, 159 pp., 4 pls. Lille.

        Platypezidae and Dolichopodidae are reprinted from Macquart, 1827a. Empididae and Hybotidae are revised from Macquart, 1823.

        The entire work consists of 5 volumes published at Lille, 1826–1834, and for the most part also in Soc. Roy. des Sci., de l'Agr. et des Arts, Lille, Mém.

1829   Insectes Diptères du nord de la France. Syrphies. Soc. Roy. des Sci., de l'Agr. et des Arts, Lille, Mém. **1827/1828:** 149–371, 4 pls.

        Also published separately as his "Insectes Diptères du nord de la France", [Vol. 4], 223 pp., 4 pls., Lille, *1829*.

1834a   Histoire naturelle des Insectes.—Diptères, Tome premier. Diptera, Vol. 1, 578 pp., 12 pls. *In* [Roret, N. E., ed.], (Collection des suites à Buffon) [q.v.]. Paris.

Macquart, J.—Continued

1834b Insectes Diptères du nord de la France. Athéricères: Créophiles, Oestrides, Myopaires, Conopsaires, Scénopiniens, Céphalopsides. Soc. Roy. des Sci., de l'Agr. et des Arts, Lille, Mém. **1833**: 137–368, 6 pls.

> Also published separately as his "Insectes Diptères du nord de la France", [Vol. 5], 232 pp., 6 pls., Lille, *1834*.

1835 Histoire naturelle des Insectes. Diptères, Tome deuxième. Diptera, Vol. 2, 703 pp., 12 pls. *In* [Roret, N. E., ed.], (Collection des suites à Buffon) [q.v.]. Paris.

1836 Description d'un nouveau genre d'Insectes Diptères de la famille des Tanystomes. Soc. Ent. de France, Ann. **[ser. 1], 5**: 517–520, 1 pl.

1838a Diptères exotiques nouveaux ou peu connus. Soc. Roy. des Sci., de l'Agr. et des Arts, Lille, Mém. **1838 (2)**: 9–225, 25 pls.

> Also published separately as his "Diptères exotiques nouveaux ou peu connus", Vol. 1, Pt. 1, pp. 5–221, 25 pls., Paris, *1838*.

1838b **Diptères exotiques nouveaux ou peu connus.** Vol. 1, Pt. 2, pp. 5–207, 14 pls. Paris.

> Also published in Soc. Roy. des Sci., de l'Agr. et des Arts, Lille, Mém. **1838 (3)**: 121–323, 14 pls., *1839*.
>
> The entire work consists of 2 volumes (in 5) and 5 supplements, Paris, 1838–1855. Vol. 1 has 2 separately paged parts, Vol. 2 has 3. The work was also published in the periodical Soc. des Sci., de l'Agr. et des Arts, Lille, Mém. from 1838 through 1855.

[1839] Diptères. Pp. 97–119, pl. 4, figs. 2–11 (=livr. 44, part). *In* Webb, P. B., and Berthelot, S., eds., Histoire naturelle des Îles Canaries [q.v.]. Vol. 2, Pt. 2: Zoologie, [Sect. 6]: Entomologie, 119 pp., 7 pls. Paris, "1836–1844".

> For dating see Stearn, 1937: 55.

1840 Diptères exotiques nouveaux ou peu connus. Vol. 2, Pt. 1, pp. 5–135, 21 pls. Paris.

> Also published in Soc. Roy. des Sci., de l'Agr. et des Arts, Lille, Mém. **1840**: 283–413, 21 pls., *1841*.

1842 Diptères exotiques nouveaux ou peu connus. Soc. Roy. des Sci., de l'Agr. et des Arts, Lille, Mém. **1841 (1)**: 65–200, 22 pls.

> Also published separately as his "Diptères exotiques nouveaux ou peu connus", Vol. 2, Pt. 2, pp. 5–140, 22 pls., Paris, *1842*.

1843a Diptères exotiques nouveaux ou peu connus. Soc. Roy. des Sci., de l'Agr. et des Arts, Lille, Mém. **1842**: 162–460, 36 pls.

> Also published separately as his "Diptères exotiques nouveaux ou peu connus", Vol. 2, Pt. 3, pp. 5–304, 36 pls., Paris, *1843*.

Macquart, J.—Continued
- 1843b Description d'un nouveau genre d'Insectes Diptères. Soc. Ent. de France, Ann. **ser. 2, 1:** 59–63, 1 pl.
- 1846 Diptères exotiques nouveaux ou peu connus. [Ier] Supplément. Soc. Roy. des Sci. de l'Agr. et des Arts, Lille, Mém. (1845) **1844:** 133–364, 20 pls.

    Also published separately as his "Diptères exotiques nouveaux ou peu connus. Supplément" [I], pp. 5–238, 20 pls., Paris, *1846*.

- 1847 Diptères exotiques nouveaux ou peu connus. 2e supplément. Soc. Roy. des Sci., de l'Agr. et des Arts, Lille, Mém. **1846:** 21–120, 6 pls.

    Also published separately as his "Diptères exotiques nouveaux ou peu connus. Supplément II", pp. 5–104, 6 pls., Paris, ?*1847*. (The separate may have appeared in 1848.)

- 1848a Diptères exotiques nouveaux ou peu connus. Suite de 2me supplément [i.e., 3e supplément]. Soc. Roy. des Sci., de l'Agr. et des Arts, Lille, Mém. **1847 (2):** 161–237, 7 pls.

    Also published separately as his "Diptères exotiques nouveaux ou peu connus. Supplément III", pp. 1–77, 7 pls., Paris, *1848*.

- 1848b Nouvelles observations sur les Diptères d'Europe de la tribu des Tachinaires [cont.]. Soc. Ent. de France, Ann. **ser. 2, 6:** 85–138, pls. 3–6 [cont.].
- 1849 Nouvelles observations sur les Diptères d'Europe de la tribu des Tachinaires [cont.]. Soc. Ent. de France, Ann. **ser. 2, 7:** 353–418, pls. 10–12 [cont.].
- 1850 Diptères exotiques nouveaux ou peu connus. 4e supplément [part]. Soc. des Sci., de l'Agr. et des Arts, Lille, Mém. **1849:** 309–465 (text), 466–479 (explanation of figs., index), pls. 1–14.

    Also published separately as his "Diptères exotiques nouveaux ou peu connus. Supplément IV" [part], pp. 5–161 (text), Paris, ?*1850*. (The separate of this part of Supplement 4 may have appeared in 1851 with its continuation.) Pages 311–317 (explanation of figs.), 324–336 (combined index of the 2 parts of this supplement), and pls. 1–14, were issued with the second part of Supplement 4.

- 1851 Diptères exotiques nouveaux ou peu connus. Suite du 4e supplément. Soc. Natl. des Sci., de l'Agr. et des Arts, Lille, Mém. **1850:** 134–282 (text), 283–294 (explanation of figs., index), pls. 15–28.

    Also published separately as his "Diptères exotiques nouveaux ou peu connus. Supplément IV" [part], pp. 161–309 (text), 317–323 (explanation of figs.), 324–336 (combined index of the 2 parts of this supplement), pls. 15–28, Paris, *1851*.

Macquart, J.—Continued
- 1854 Nouvelles observations sur les Diptères d'Europe de la tribu des Tachinaires [cont.]. Soc. Ent. de France, Ann. ser. 3, **2:** 373–446, pls. 13–15 [cont.].
- 1855 Diptères exotiques nouveaux ou peu connus. 5ᵉ supplément. Soc. Imp. des Sci., de l'Agr. et des Arts, Lille, Mém. **1854:** 25–156, 7 pls.
    > Also published separately as his "Diptères exotiques nouveaux ou peu connus. Supplément V", pp. 5–136, 7 pls., Paris, *1855*.

Madwar, S.
- 1937 Biology and morphology of the immature stages of Mycetophilidae (Diptera, Nematocera). Roy. Soc. London, Phil. Trans., Ser. B **227:** 1–110, 392 figs.

Mainland, G. B. (See also Patterson and Mainland)
- 1941 Studies in *Drosophila* speciation. III. The *Drosophila macrospina* group. Genetics **26:** 160–161.

Malloch, J. R. (See also Cole and Lovett, 1919; Melander, 1913c)
- 1909 A division of the dipterous genus *Phora*, Latr., into subgenera. Glasgow Nat. **1:** 24–28.
- 1910 Scottish Phoridae, with tables of all the British species, and notes of localities. Ann. Scot. Nat. Hist. **1910:** 15–21, 87–92.
- 1912a New American dipterous insects of the family Pipunculidae. U.S. Natl. Mus. Proc. **43:** 291–299, 9 figs.
- 1912b The insects of the dipterous family Phoridae in the United States National Museum. U.S. Natl. Mus. Proc. **43:** 411–529, 7 pls.
- 1912c One new genus and eight new species of dipterous insects in the United States National Museum collection. U.S. Natl. Mus. Proc. (1913) **43:** 649–658, 1 pl.
- 1913a Notes on the synonymy of some genera and species in the Chloropidae (Diptera). Canad. Ent. **45:** 175–178.
- 1913b Three new North American Diptera. Canad. Ent. **45:** 273–275.
- 1913c New North American Diptera. Canad. Ent. **45:** 282–284.
- 1913d The genus *Parodinia* Coquillett (Geomyzidae, Dipt.). Ent. News **24:** 274–276.
- 1913e A revision of the species in *Agromyza* Fallén, and *Cerodontha* Rondani (Diptera). Ent. Soc. Amer. Ann. **6:** 269–336, 4 pls.
- 1913f A new genus and two new species of Chloropidae (Diptera). Insecutor Inscitiae Menstruus **1:** 46–48.
- 1913g Four new species of North American Chloropidae (Diptera). Insecutor Inscitiae Menstruus **1:** 60–64.

Malloch, J. R.—Continued
- 1913h   A new species of Agromyzidae (Diptera). Insecutor Inscitiae Menstruus **1:** 109–110.
- 1913i   A new genus and three new species of Phoridae from North America, with notes on two recently erected genera (*Crepidopachys* and *Pronomiophora* Enderlein). Psyche **20:** 23–26, 1 fig.
- 1913j   Descriptions of new species of American flies of the family Borboridae. U.S. Natl. Mus. Proc. **44:** 361–372.
- 1913k   Two new species of Diptera in the United States National Museum collection. U.S. Natl. Mus. Proc. **44:** 461–463.
- 1913l   Notes on some American Diptera of the genus *Fannia*, with descriptions of new species. U.S. Natl. Mus. Proc. **44:** 621–631, 1 pl.
- 1913m   Three new species of Anthomyidae (Diptera) in the United States National Museum collection. U.S. Natl. Mus. Proc. **45:** 603–607.
- 1913n   A synopsis of the genera of Agromyzidae, with descriptions of new genera and species. U.S. Natl. Mus. Proc. **46:** 127–154, 3 pls.
- 1913o   The genera of flies in the subfamily Botanobiinae with hind tibial spur. U.S. Natl. Mus. Proc. **46:** 239–266, 2 pls.
- 1913p   A new species of *Simulium* from Texas. Ent. Soc. Wash. Proc. **15:** 133–134.
- 1913q   Two new species of Borboridae from Texas. Ent. Soc. Wash. Proc. **15:** 135–137, 1 fig.
- 1914a   Some undescribed North American Sapromyzidae. Biol. Soc. Wash. Proc. **27:** 29–41, 1 pl.
- 1914b   Synopsis of the genus *Probezzia*, with description of a new species (Diptera). Biol. Soc. Wash. Proc. **27:** 137–139.
- 1914c   Notes on Illinois Phoridae (Diptera) with descriptions of three new species. Brooklyn Ent. Soc. Bul. **9:** 56–60.
- 1914d   A synopsis of the genera in Chloropidae, for North America. Canad. Ent. **46:** 113–120.
- 1914e   New American Diptera. Ent. News **25:** 172–178.
- 1914f   Notes on North American Agromyzidae (Dipt.). Ent. News **25:** 308–314.
- 1914g   Notes on North American Diptera, with descriptions of new species in the collection of the Illinois State Laboratory of Natural History. Ill. State Lab. Nat. Hist. Bul. (1915) **10:** 213–243, 3 pls.
- 1914h   Synopsis of North American species of the genus *Bezzia* (Chironomidae). N.Y. Ent. Soc. Jour. **22:** 281–285.

Malloch, J. R.—Continued

1914i   Four new North American Chloropidae (Diptera). Psyche **21**: 24–26.

1914j   American black flies or buffalo gnats. U.S. Dept. Agr., Bur. Ent., Tech. Ser. **26**: 1–72, 6 pls.

1914k   Notes on the dipterous genus *Chyromya* R. D. Ent. Soc. Wash. Proc. **16**: 179–181.

1914l   Costa Rican Diptera collected by Philip P. Calvert, Ph. D., 1909–10. Amer. Ent. Soc. Trans. **40**: 1–36, 1 pl.

1915a   Four new North American Diptera. Biol. Soc. Wash. Proc. **28**: 45–48, 2 figs.

1915b   Two new North American Diptera. Brooklyn Ent. Soc. Bul. **10**: 64–66.

1915c   An undescribed sapromyzid (Diptera). Brooklyn Ent. Soc. Bul. **10**: 86–88.

1915d   North American Diptera. Canad. Ent. **47**: 12–16.

1915e   A new species of *Neogaurax* (Chloropidae, Dipt.). Ent. News **26**: 108.

1915f   A revision of the North American Pachygasterinae with unspined scutellum (Diptera). Ent. Soc. Amer. Ann. **8**: 305–320, 1 pl.

1915g   The Chironomidae, or midges, of Illinois, with particular reference to the species occurring in the Illinois River. Ill. State Lab. Nat. Hist. Bul. **10**: 275–543, 24 pls.

1915h   Some additional records of Chironomidae for Illinois and notes on other Illinois Diptera. Ill. State Lab. Nat. Hist. Bul. (1918) **11**: 305–363, 5 pls.

1915i   Flies of the genus *Agromyza*, related to *Agromyza virens*. U.S. Natl. Mus. Proc. **49**: 103–108, 1 fig., 1 pl.

1915j   Notes on the flies of the genus *Pseudodinia*, with description of a new species. U.S. Natl. Mus. Proc. **49**: 151–152.

1915k   Notes on North American Chloropidae (Diptera). Ent. Soc. Wash. Proc. **17**: 158–162.

1916a   A new genus and species of Helomyzidae (Diptera). Brooklyn Ent. Soc. Bul. **11**: 14–16.

1916b   A new genus and species of North American Chloropidae (Diptera). Brooklyn Ent. Soc. Bul. **11**: 86–87.

1916c   A key to the males of the anthomyid genus *Hydrotaea* recorded from North America (Diptera). Brooklyn Ent. Soc. Bul. **11**: 109–111.

1916d   Three new North American species of the genus *Agromyza* (Diptera). Psyche **23**: 50–54.

Malloch, J. R.—Continued
- 1917a A key to the North American genera of Coenosiinae (Diptera, Anthomyiidae). Brooklyn Ent. Soc. Bul. **12:** 35–37.
- 1917b A new genus and species of Anthomyiidae (Diptera). Brooklyn Ent. Soc. Bul. **12:** 37–38.
- 1917c A new genus of Anthomyiidae (Diptera). Brooklyn Ent. Soc. Bul. **12:** 113–115.
- 1917d A new North American species of the genus *Tetramerinx* (Diptera, Anthomyiidae). Canad. Ent. **49:** 225–226.
- 1917e The anthomyiid genus *Phyllogaster* (Diptera). Canad. Ent. **49:** 227–228.
- 1917f Key to the subfamilies of Anthomyiidae. Canad. Ent. **49:** 406–408.
- 1917g A preliminary classification of Diptera, exclusive of Pupipara, based upon larval and pupal characters, with keys to imagines in certain families. Part 1. Ill. State Lab. Nat. Hist. Bul. (1918) **12:** 161–409, pls. 28–57.
- 1918a Diptera from the southwestern United States. Paper IV. Anthomyiidae. Amer. Ent. Soc. Trans. **44:** 263–319, 1 pl.
- 1918b Notes and descriptions of some anthomyiid genera. Biol. Soc. Wash. Proc. **31:** 65–68.
- 1918c A new species of *Hartomyia* from Illinois (Ceratopogonidae, Diptera). Brooklyn Ent. Soc. Bul. **13:** 18.
- 1918d Notes on Chloropidae, with descriptions (Diptera). Brooklyn Ent. Soc. Bul. **13:** 19–21.
- 1918e A new species of *Orthocladius* (Chironomidae, Diptera). Brooklyn Ent. Soc. Bul. **13:** 42.
- 1918f An undescribed North American species of *Hydrotaea* (Diptera, Anthomyiidae). Brooklyn Ent. Soc. Bul. **13:** 93–94.
- 1918g Three new North American Chloropidae (Diptera). Brooklyn Ent. Soc. Bul. **13:** 108–111.
- 1918h A partial key to species of the genus *Agromyza* (Diptera). [First paper]; second paper. Canad. Ent. **50:** 76–80; 130–132.
- 1918i Key to the North American species of *Agromyza* related to *simplex* Loew. Canad. Ent. **50:** 178–179.
- 1918j A new North American species of Anthomyiidae (Diptera). Canad. Ent. **50:** 310–311.
- 1918k *Anthracophaga distichliae* sp. n. Jour. Econ. Ent. **11:** 386–387.

Malloch, J. R.—Continued
- 1918l Two new North American Phoridae (Diptera). Ent. News **29**: 146–147.
- 1918m A new species of *Johannsenomyia* (Ceratopogonidae, Diptera). Ent. News **29**: 229–230.
- 1918n A revision of the dipterous family Clusiodidae (Heteroneuridae). Ent. Soc. Wash. Proc. **20**: 2–8.
- 1918o The genus *Cnemedon* Egger in North America (Diptera Syrphidae). Ent. Soc. Wash. Proc. **20**: 127–128.
- 1919a One new genus and two new species of Anthomyiidae from the vicinity of Washington, D.C. (Diptera). Biol. Soc. Wash. Proc. **32**: 1–4, 1 fig.
- 1919b Two new North American Anthomyiidae (Diptera). Biol. Soc. Wash. Proc. **32**: 133–134.
- 1919c Some new eastern Anthomyiidae (Diptera). Biol. Soc. Wash. Proc. **32**: 207–210.
- 1919d A new phorid from Illinois (Diptera, Phoridae). Brooklyn Ent. Soc. Bul. **14**: 47–48.
- 1919e New species of flies (Diptera) from California. Calif. Acad. Sci. Proc. ser. **4, 9**: 297–312.
- 1919f The Diptera collected by the Canadian Expedition, 1913–1918 (excluding the Tipulidae and Culicidae). Pp. 34–90, pls. 7–10. *In* Anderson, R. M., ed., Report of the Canadian Arctic Expedition 1913–18 [q.v.]. Vol. 3: Insects, Pt. C: Diptera, 90 pp., 2 figs., 10 pls. Ottawa, Ont.
- 1919g A new species of *Hylemyia* from Canada (Diptera, Anthomyiidae). Canad. Ent. **51**: 95–96.
- 1919h A new species of *Coenosia* from Canada (Diptera, Anthomyiidae). Canad. Ent. **51**: 96.
- 1919i A new species of the genus *Tachydromia* from Illinois (Diptera, Empididae). Canad. Ent. **51**: 248.
- 1919j A new species of Phoridae from Illinois (Diptera). Canad. Ent. **51**: 256–257.
- 1919k Three new Canadian Anthomyiidae (Diptera). Canad. Ent. **51**: 274–276.
- 1919l A new anthomyiid from Labrador (Diptera). Canad. Ent. **51**: 277–278.
- 1919m On an undescribed species of *Medeterus* (Diptera, Dolichopodidae). Ent. News **30**: 7–9.
- 1919n The generic status of *Zodion palpalis* Robertson (Diptera, Conopidae), with generic key to the family. Ent. Soc. Wash. Proc. **21**: 204–205.

Malloch, J. R.—Continued
- 1920a Descriptions of new North American Anthomyiidae (Diptera). Amer. Ent. Soc. Trans. **46**: 133–196, 3 pls.
- 1920b A synopsis of the North American species of the genus *Pegomyia* Robineau-Desvoidy (Diptera, Anthomyiidae). Brooklyn Ent. Soc. Bul. **15**: 121–127.
- 1920c Descriptions of new genera and species of Scatophagidae (Diptera). Ent. Soc. Wash. Proc. **22**: 34–38.
- 1920d Some new North American Sapromyzidae (Diptera). Canad. Ent. **52**: 126–128.
- 1920e Some new species of Lonchaeidae from America (Diptera). Canad. Ent. **52**: 129–132.
- 1920f Some new species of the genus *Lonchaea* (Diptera, Lonchaeidae). Canad. Ent. **52**: 246–247.
- 1920g A synoptic revision of the anthomyiid genus *Hydrophoria* Robineau-Desvoidy (Diptera). Canad. Ent. **52**: 253–257.
- 1920h A new species of *Coenosia* from the western United States (Diptera, Anthomyiidae). Ent. News **31**: 103–104.
- 1920i The genus *Aspistes* Meigen in North America (Diptera, Scatopsidae). Ent. News **31**: 275–276.
- 1920j Scientific results of the Katmai Expedition of the National Geographic Society. XII. Descriptions of Diptera of the families Anthomyiidae and Scatophagidae. Ohio Jour. Sci. **20**: 267–288, 3 pls.
- 1921a A synopsis of the anthomyiid genus *Trichopticus* Rondani (Diptera). Canad. Ent. (1920) **52**: 271–274.
- 1921b Synopses of the anthomyiid genera *Mydaea*, *Ophyra*, *Phyllogaster*, *Tetramerinx*, and *Eulimnophora* (Diptera). Canad. Ent. **53**: 9–13.
- 1921c A synopsis of the North American species of the genera *Melanochelia* Rondani and *Limnophora* R.-D. (Diptera, Anthomyiidae). Canad. Ent. **53**: 61–64.
- 1921d A synopsis of the North American species of the genus *Helina* R.-D., sens. lat. (Diptera, Anthomyiidae). Canad. Ent. **53**: 103–109.
- 1921e The North American species of the anthomyiid genus *Hebecnema* Schnabl (Diptera). Canad. Ent. **53**: 214–215.
- 1921f A new genus of Agromyzidae (Diptera). Brooklyn Ent. Soc. Bul. (1920) **15**: 147–148.
- 1921g A new genus of Anthomyiidae (Diptera). Brooklyn Ent. Soc. Bul. **16**: 53.

Malloch, J. R.—Continued
- 1921h Exotic Muscaridae (Diptera). –I. Ann. and Mag. Nat. Hist. **ser. 9, 7:** 161–173.
- 1921i Exotic Muscaridae (Diptera). –IV. Ann. and Mag. Nat. Hist. **ser. 9, 8:** 414–425.
- 1921j Forest insects in Illinois. I. The subfamily Ochthiphilinae (Diptera, family Agromyzidae). Ill. State Nat. Hist. Survey, Bul. (1922) **13:** 345–361, 2 pls.
- 1921k Notes on some of van der Wulp's species of North American Anthomyiidae (Diptera). Ent. News **32:** 40–45.
- 1921l A synopsis of the genera of the anthomyiid subfamily Coenosiinae (Diptera). Ent. News **32:** 106–107.
- 1921m Two new species of the genus *Coenosia* (Anthomyiidae, Diptera). Ent. News **32:** 134.
- 1921n A key to the species of the genus *Coenosia* Meigen (Dipt., Anthomyiidae). Ent. News **32:** 201–205.
- 1921o Some notes on Drosophilidae (Diptera). Ent. News **32:** 311–312.
- 1921p Dipterous insects of the family Anthomyiidae from the Pribilof Islands, Alaska. Calif. Acad. Sci. Proc. **ser. 4, 11:** 178–182.
- 1922a Seven new species of the syrphid genus *Sphegina* Meigen (Diptera). Biol. Soc. Wash. Proc. **35:** 141–144.
- 1922b Notes on Clusiodidae (Diptera). Boston Soc. Nat. Hist. Occas. Papers **5:** 47–50.
- 1922c A synopsis of the North American species of the dipterous genus *Amaurosoma* Becker, with descriptions of new species. Brooklyn Ent. Soc. Bul. **17:** 77–78.
- 1922d A new borborid from Maryland (Diptera, Borboridae). Brooklyn Ent. Soc. Bul. **17:** 87.
- 1922e Two new species of the genus *Helina* R.-D. (Diptera, Anthomyiidae). Brooklyn Ent. Soc. Bul. **17:** 95–96.
- 1922f Keys to the syrphid genus *Sphegina* Meigen (Dip.). Ent. News **33:** 266–270.
- 1923a Flies of the anthomyiid genus *Phaonia* Robineau-Desvoidy and related genera, known to occur in North America. Amer. Ent. Soc. Trans. **48:** 227–282, 3 pls.
- 1923b A new North American species of the genus *Beckerina* (Phoridae, Diptera). Brooklyn Ent. Soc. Bul. **18:** 32–33.
- 1923c A new genus of Phoridae (Diptera). Brooklyn Ent. Soc. Bul. **18:** 143–144.

Malloch, J. R.—Continued
- 1923d Expedition of the California Academy of Sciences to the Gulf of California in 1921. Anthomyidae and Lonchaeidae (kelp flies and their allies). Calif. Acad. Sci. Proc. ser. **4, 12:** 425–428.
- 1923e An amended synopsis of the genus *Mydaea* (Diptera, Anthomyiidae). Canad. Ent. **55:** 220–221.
- 1923f A new species of *Forcipomyia* from the eastern United States (Diptera, Ceratopogonidae). Ent. News **34:** 4–5.
- 1923g A new empid from the eastern United States (Diptera). Ent. News **34:** 5.
- 1923h The cordylurid genus *Paralleloma* and its nearest allies (Dipt.). Ent. News **34:** 139–140, 175–180.
- 1923i Insects, arachnids, and chilopods of the Pribilof Islands, Alaska. Diptera (except Tipulidae, Rhyphidae and Calliphoridae). North Amer. Fauna **46:** 170–227, pls. 12–15.
- 1923j Some new genera and species of Lonchaeidae and Sapromyzidae (Diptera). Ent. Soc. Wash. Proc. **25:** 45–53.
- 1924a Exotic Muscaridae (Diptera). –XII. Ann. and Mag. Nat. Hist. ser. **9, 13:** 409–424.
- 1924b Exotic Muscaridae (Diptera). –XIV. Ann. and Mag. Nat. Hist. ser. **9, 14:** 513–522, 1 fig.
- 1924c A new North American species of *Amiota* Loew (Diptera). Brooklyn Ent. Soc. Bul. **19:** 51–52.
- 1924d Three new species of *Agromyza* and synonymical notes (Diptera, Agromyzidae). Canad. Ent. **56:** 191–192.
- 1924e The North American species of the genus *Hoplogaster* (Diptera: Anthomyiidae). Ent. News **35:** 171–172.
- 1924f Two new Phoridae from the eastern United States (Diptera). Ent. News **35:** 355–357.
- 1924g Two new cordylurid flies from the Pacific Coast. Pan-Pacific Ent. **1:** 14–15.
- 1924h New and little-known calyptrate Diptera from New England. Psyche **31:** 193–205, 1 pl.
- 1924i A new species of *Canacea* from the United States (Diptera: Ephydridae). Ent. Soc. Wash. Proc. **26:** 52–53.
- 1924j A new genus and species of Muscidae (Diptera). Ent. Soc. Wash. Proc. **26:** 74.
- 1924k Descriptions of neotropical two-winged flies of the family Drosophilidae. U.S. Natl. Mus. Proc. **66 (3):** 1–11.
- 1924l Notes on Australian Diptera. No. IV. Linn. Soc. N.S. Wales, Proc. **49:** 348–359, 5 figs.

Malloch, J. R.—Continued
- 1925a A new borborid from Maryland (Diptera, Borboridae). Brooklyn Ent. Soc. Bul. **20**: 97.
- 1925b A new North American species of *Hydrotaea* (Diptera). Brooklyn Ent. Soc. Bul. **20**: 184–185.
- 1925c Notes on Australian Diptera. No. VI; No. VII. Linn. Soc. N. S. Wales, Proc. **50**: 80–97, 12 figs.; 311–340, 23 figs.
- 1925d A synopsis of New World flies of the genus *Sphaerocera* (Diptera: Borboridae). Ent. Soc. Wash. Proc. **27**: 117–123.
- 1925e Exotic Muscaridae (Diptera). –XVII. Ann. and Mag. Nat. Hist. ser. **9, 16**: 361–377.
- 1926 New genera and species of acalyptrate flies in the United States National Museum. U.S. Natl. Mus. Proc. **68 (21)**: 1–35, 2 pls.
- 1927a A new species of the genus *Fannia* R.—D. from North America (Diptera, Anthomyiidae). Ent. News **38**: 176, 2 figs.
- 1927b Descriptions of a new genus and three new species of Diptera. Ent. Soc. Wash. Proc. **29**: 90–93, 1 fig.
- 1927c A new agromyzid fly of economic importance from Africa. Ann. and Mag. Nat. Hist. ser. **9, 19**: 575–577.
- 1928a Notes on American two-winged flies of the family Sapromyzidae. U.S. Natl. Mus. Proc. **73 (23)**: 1–18, 5 figs.
- 1928b Notes on Australian Diptera. No. XVII. Linn. Soc. N.S. Wales, Proc. **53**: 598–617.
- 1929 Exotic Muscaridae (Diptera). –XXVI. Ann. and Mag. Nat. Hist. ser. **10, 4**: 97–120, 7 figs.
- 1930 Notes on Australian Diptera. No. XXVI. Linn. Soc. N.S. Wales, Proc. **55**: 488–492, 3 figs.
- 1931a Exotic Muscaridae (Diptera). –XXXV. Ann. and Mag. Nat. Hist. ser. **10, 8**: 425–446, 6 figs.
- 1931b Notes on some acalpytrate flies in the United States National Museum. U.S. Natl. Mus. Proc. **78 (15)**: 1–32, 10 figs.
- 1931c Flies of the genus *Pseudotephritis* Johnson (Diptera: Ortalidae). U.S. Natl. Mus. Proc. **79 (34)**: 1–6.
- 1932a Notes on exotic Diptera. (1); (2). Stylops **1**: 112–120, 3 figs.; 121–126, 6 figs.
- 1932b Exotic Muscaridae (Diptera). –XXXVII. Ann. and Mag. Nat. Hist. ser. **10, 10**: 297–330, 11 figs.
- 1933a The genus *Coelopa* Meigen (Diptera, Coelopidae). Ann. and Mag. Nat. Hist. ser. **10, 11**: 339–350, 1 fig.

Malloch, J. R.—Continued
- 1933b Some acalyptrate Diptera from the Marquesas Islands. Bernice P. Bishop Mus. Bul. **114:** 1–31, 9 figs.
- 1933c Acalyptrata. Pp. 177–391, figs. 36–68, pls. 2–7 (=fasc. 4) [cont.]. *In* British Museum (Natural History), Diptera of Patagonia and south Chile [q.v.]. Pt. 6, 499 pp., 84 figs., 8 pls. London.
- 1934a Acalyptrata [concl.]. Pp. 393–489, figs. 69–84, pl. 8 (=fasc. 5). *In* British Museum (Natural History), Diptera of Patagonia and south Chile [q.v.]. Pt. 6, 499 pp., 84 figs., 8 pls. London.
- 1934b The exploration of Southampton Island, Hudson Bay by George Miksch Sutton, sponsored by Mr. John Bonner Semple 1929–1930. Part II. Zoology. Section 4. Spiders and insects (in part) of Southampton Island. III. Chironomidae, Sciaridae, Phoridae, Syrphidae, Piophilidae, Helomyzidae, Calliphoridae, Oestridae, and Tachinidae. Pittsburgh, Carnegie Inst., Carnegie Mus. Mem. **12 (pt. 2, sect. 4** [=pub. 162]): 13–32, 14 figs.
- 1935a Exotic Muscaridae (Diptera). –XXXVIII. Ann. and Mag. Nat. Hist. **ser. 10, 15:** 242–266, 22 figs.
- 1935b Exotic Muscaridae (Diptera). –XL. Ann. and Mag. Nat. Hist. **ser. 10, 16:** 562–572, 5 figs.
- 1935c The North American species of the dipterous genus *Microsania*, Zett. Stylops **4:** 65–66, 1 fig.
- 1940 The North American genera of the dipterous subfamily Chamaemyiinae. Ann. and Mag. Nat. Hist. **ser. 11, 6:** 265–274.
- 1941a The American genus *Paracantha* Coquillett (Diptera, Trypetidae). Rev. de Ent. **12:** 32–42, 10 figs.
- 1941b Florida Diptera. Fla. Ent. **24:** 49–51.
- 1942 Notes on two genera of American flies of the family Trypetidae. U.S. Natl. Mus. Proc. **92:** 1–20, 1 fig.

————, **Greene, C. T., and McAtee, W. L.**
- 1931 District of Columbia Diptera: Rhagionidae. Ent. Soc. Wash. Proc. **33:** 213–220.

————, **and McAtee, W. L.**
- 1924a Flies of the family Drosophilidae of the District of Columbia region, with keys to genera, and other notes, of broader application. Biol. Soc. Wash. Proc. **37:** 25–41, pls. 8–9.
- 1924b Keys to flies of the families Lonchaeidae, Pallopteridae, and Sapromyzidae of the eastern United States, with a list of the species of the District of Columbia region. U.S. Natl. Mus. Proc. **65 (12):** 1–26, 2 pls.

Maltais, J. B. (See Petch and Maltais)

Mamajev, B. M. (See Krivosheina and Mamajev)

Mangabeira, O., Filho, and Galindo, P.
1944   The genus *Flebotomus* in California.   Amer. Jour. Hyg. **40**: 182–198, 3 pls.

Mani, M. S.
1946   Studies on Indian Itonididae (Cecidomyiidae: Diptera). VIII.–Keys to the genera from the Oriental Region. Indian Jour. Ent. (1945) **7**: 189–235, 117 figs.

Manis, H. C.
1941   Bionomics and morphology of the black onion fly, *Tritoxa flexa* (Wied.) (Diptera, Ortalidae). Iowa State Col. Jour. Sci **16**: 96–98.

Mansbridge, G. H.
1933   On the biology of some Ceroplatinae and Macrocerinae (Diptera, Mycetophilidae) with an appendix on the chemical nature of the web fluid in larvae of Ceroplatinae, by Harold W. Buston.   Roy. Ent. Soc., London, Trans. **81**: 75–92, 15 figs.

Marchand, W.
1920   The early stages of Tabanidae (horse-flies).   Rockefeller Inst. Med. Res. Monog. **13**: 1–203, 15 pls.

Marcovitch, S.
1915   The biology of the juniper berry insects, with descriptions of new species.   Ent. Soc. Amer. Ann. **8**: 163–181, 7 pls.

Marten, J.
1882   New Tabanidae.   Canad. Ent. **14**: 210–212.
1883   New Tabanidae.   Canad. Ent. **15**: 110–112.
1893   Description of a new species of gall-making Diptera.   Ohio Agr. Expt Sta. Tech. Ser. Bul. **1**: 155–156, 1 fig. (=bul. 3, part).

Martin, C. H. (See also Doucette et al.; Wilcox and Martin)
1934   Notes on the larval feeding habits and life history of *Eumerus tuberculatus* Rondani.   Brooklyn Ent. Soc. Bul. **29**: 27–36, pls. 2–3.
1953   Intraspecific variation of taxonomic characters in *Coleomyia* and two new species (Diptera: Asilidae).   Pan-Pacific Ent. **29**: 25–34.
1955a New species in the genus *Parataracticus* Cole from southern California (Diptera: Asilidae).   Kans. Ent. Soc. Jour. **28**: 116–120.
1955b Notes on the genus *Haplopogon* and a new species (Diptera: Asilidae).   Ohio Jour. Sci. **55**: 315–316.

Martin, C. H.—Continued
- 1957   A revision of the Leptogastrinae in the United States (Diptera, Asilidae). Amer. Mus. Nat. Hist. Bul. **111**: 345–385, 55 figs.
- 1959a   The *Holopogon* complex of North America, excluding Mexico, with the descriptions of a new genus and a new subgenus (Diptera, Asilidae). Amer. Mus. Nat. Hist., Amer. Mus. Novitates **1980**: 1–40, 21 figs.
- 1959b   New species of *Cerotainiops* Curran (Diptera: Asilidae). Kans. Ent. Soc. Jour. **32**: 49–53.
- 1961   The misidentification of *Erax* Scopoli in the Americas (Diptera: Asilidae). Kans. Ent. Soc. Jour. **34**: 1–4.
- 1962a   The mistaken identity of *Proctacanthus arno* Townsend and a new species (Diptera: Asilidae). Kans. Ent. Soc. Jour. **35**: 185–188.
- 1962b   Changes in the status of *Efferia barbata* (Fab.) and *Efferia albibarbis* (Macq.) (Diptera: Asilidae). Kans. Ent. Soc. Jour. **35**: 247–253.

Martínez Palacios, A. (See Vargas, Martínez Palacios, and Díaz Nájera)

Martini, E.
- 1929–1931   Culicidae. (11. Dixidae–12. Culicidae). [Fams.] 11–12, pp. 1–48, figs. 1–67 (=lfg. 33), *1929;* pp. 49–96, figs. 68–127 (=lfg. 35), *1929;* pp. 97–144, figs. 128–168, pl. 1 (=lfg. 38), *1929;* pp. 145–192, figs. 169–250 (=lfg. 40), *1930;* pp. 193–256, figs. 251–306 (=lfg. 46), *1930;* pp. 257–320, figs. 307–360 (=lfg. 48), *1930;* pp. 321–398, figs. 361–431 (=lfg. 52), *1931*. *In* Lindner, E., ed., Die Fliegen der palaearktischen Region [q.v.], Vol 3. Stuttgart.

Matheson, R.
- 1933   A new species of mosquito from Colorado (Diptera, Culicidae). Ent. Soc. Wash. Proc. **35**: 69–71, 1 fig.

Matsumura, S.
- 1911   Erster Beitrag zur Insekten-Fauna von Sachalin. Tohôku Imp. Univ., Col. Agr., Jour. **4** (**1**): 1–145, pls. 1–2.
- 1916   (Thousand insects of Japan. Additamenta.) Vol. 2, pp. 185–474, pls. 16–25. Tokyo.

  In Japanese, with some English descriptions.

  The entire work consists of 4 continuously paged volumes, a total of 962 pp.+43 pp., 71 pls., published in Tokyo, 1913–1921. The work is an "Additamenta" to Matsumura's 4 volumes entitled ("Thousand insects of Japan"), Tokyo, 1904–1907, and his 4 volumes of ("Thousand insects of Japan, Supplement"), Tokyo, 1909–1912, of both of which there are 3 editions.

Matsumura, S., and Adachi, J.
1916 Synopsis of the economic Syrphidae of Japan. Pl. [sic] [=Pt.] I. Ent. Mag. [Japan] **2:** 1–36, pl. 1.
1917a Synopsis of the economic Syrphidae of Japan. Pt. II. Ent. Mag. [Japan] **2:** 133–156, pl. 6.
1917b Synopsis of the economic Syrphidae of Japan. Pt. III. Ent. Mag. [Japan] **3:** 14–46.
1919 Synopsis of the economic Syrphidae of Japan. Pt. III [sic] [=IV]. Ent. Mag. [Japan] **3:** 128–144, pl. 3.

Mattingly, P. F., Stone, A., and Knight, K. L.
1962 *Culex aegypti* Linnaeus, 1762 (Insecta, Diptera): Proposed validation and interpretation under the Plenary Powers of the species so named. Z.N.(S.) 1216. Internatl. Comn. Zool. Nomencl., Bul. Zool. Nomencl. **19:** 208–219, 5 figs., pl. 5.

Maughan, L. (See also Johnson, L. M.; Johnson and Maughan)
1935 A systematical and morphological study of Utah Bombyliidae, with notes on species from intermountain states. Kans. Ent. Soc. Jour. **8:** 27–36, 37–80, 4 pls.

McAlpine, J. F.
1951 A review of the North American species of *Herina* Robineau-Desvoidy (=*Tephronota* Loew) (Diptera: Otitidae). Canad. Ent. **83:** 308–314, 1 pl.
1956 Cone-infesting lonchaeids of the genus *Earomyia* Zett., with descriptions of five new species from western North America (Diptera: Lonchaeidae). Canad. Ent. **88:** 178–196, 21 figs.
1960a A new species of *Leucopis* (*Leucopella*) from Chile and a key to the world genera and subgenera of Chamaemyiidae (Diptera). Canad. Ent. **92:** 51–58, 5 figs.
1960b First record of the family Camillidae in the New World (Diptera). Canad. Ent. **92:** 954–956, 6 figs.
1960c Chapter XVI. Diptera (Brachycera) Lonchaeidae. Pp. 327–376, 115 figs. *In* Hanström, B., Brinck, P., and Rudebeck, G., eds., South African animal life [q.v.]. Vol. 7, 488 pp., 496 figs. Uppsala.
1961a A new species of *Dasiops* (Diptera: Lonchaeidae) injurious to apricots. Canad. Ent. **93:** 539–544, 15 figs.
1961b Variation, distribution and evolution of the *Tabanus* (*Hybomitra*) *frontalis* complex of horse flies (Diptera: Tabanidae). Canad. Ent. **93:** 894–924, 31 figs.
1962 A revision of the genus *Campichoeta* Macquart (Diptera: Diastatidae). Canad. Ent. **94:** 1–10, 18 figs.

McAlpine, J. F.—Continued
- 1963 Relationships of *Cremifania* Czerny (Diptera: Chamaemyiidae) and description of a new species. Canad. Ent. **95**: 239–253, 12 figs.

McAtee, W. L.   (See also Banks et al.; Malloch, Greene, and McAtee; Malloch and McAtee)
- 1919a Key to the nearctic species of the genus *Laphria* (Diptera, Asilidae). Ohio Jour. Sci. (1918) **19**: 143–172, 2 pls.
- 1919b Notes on the nearctic *Nusa* (Diptera, Asilidae). Ohio Jour. Sci. **19**: 244–248, 5 figs.
- 1921a Description of a new genus of Nemocera (Dipt.). Ent. Soc. Wash. Proc. **23**: 49.
- 1921b District of Columbia Diptera: Scatopsidae. Ent. Soc. Wash. Proc. **23**: 120–124.
- 1922 Notes on nearctic bibionid flies. U.S. Natl. Mus. Proc. (1921) **60** (**11**): 1–27.
- 1923 Descriptions of *Bibio* (Diptera) from the Carolinas. Ent. Soc. Wash. Proc. **25**: 62–64.
- 1927 Bird nests as insect and arachnid hibernacula. Ent. Soc. Wash. Proc. **29**: 180–184.

———, **and Banks, N.**
- 1920 District of Columbia Diptera: Asilidae. Ent. Soc. Wash. Proc. **22**: 13–20, 21–33, 2 figs.

———, **and Walton, W. R.**
- 1918 District of Columbia Diptera: Tabanidae. Ent. Soc. Wash. Proc. **20**: 188–206, 1 fig., pl. 10.

McBride, S. J.
- 1870 The so-called web-worm of young trout. Amer. Ent. and Bot. **2**: 365–367.

McCracken, I.
- 1904 *Anopheles* in California, with description of a new species. Ent. News **15**: 9–14, 1 pl.

McDonald, W. A.   (See Belkin and McDonald)

McDuffie, W. C.   (See King et al.)

McDunnough, J.
- 1921 A revision of the Canadian species of the *affinis* group of the genus *Tabanus* (Diptera). Canad. Ent. **53**: 139–144.
- 1922 Two new Canadian Tabanidae (Diptera). Canad. Ent. **54**: 238–240.

McFadden, M. W., and Foote, R. H.
- (1961) The genus *Orellia* R.–D. in America north of Mexico (Diptera: Tephritidae). Ent. Soc. Wash. Proc. (1960) **62**: 253–261, 4 figs.

McGugan, B. M. (See McLeod, McGugan, and Coppel)

McGuire, J. U. (See Hardy and McGuire)

McLeod, J. H., McGugan, B. M., and Coppel, H. C.
1962   A review of the biological control attempts against insects and weeds in Canada. [Canada] Commonwealth Inst. Biol. Control, Tech. Commun. **2**: 1–216, 16 maps.

Meade, A. B., and Cook, E. F.
1961   Notes on the biology of *Scatopse fuscipes* (Meigen) (Diptera: Scatopsidae). Ent. News **72**: 13–18.

Meade, R. H.
1883   Annotated list of British Anthomyiidae [cont.]. Ent. Monthly Mag. **19**: 213–220 [cont.].
1885   Description of a new maritime fly belonging to the family Scatomyzides, Fallén. Ent. Monthly Mag. **22**: 152–154.
1891   Additions to the list of British Anthomyiidae. Ent. Monthly Mag. **27** [=ser. 2, 2]: 42–43.
1894   Supplement to annotated list of British Tachiniidae [sic]. Ent. Monthly Mag. **30** [=ser. 2, 5]: 69–73, 107–110, 156–160.

Meigen, J. W.
1800   Nouvelle classification des mouches à deux ailes (Diptera L.) d'après un plan tout nouveau. 40 pp. Paris.

   Facsimile reproduction appears in Hemming, 1945: 121–160. This publication has recently been suppressed by the International Commission on Zoological Nomenclature (I.C.Z.N., 1963b). See Melville, 1961, for discussion.

1803   Versuch einer neuen Gattungseintheilung der europäischen zweiflügeligen Insekten. Mag. f. Insektenkunde **2**: 259–281.
1804   Klassifikazion und Beschreibung der europäischen zweiflügeligen Insecten (Diptera Linn.). Erster Band. Abt. I, pp. i–xxviii + 1–152, pls. 1–8; Abt. II, pp. i–vi + 153–314, pls. 9–15. Braunschweig.

   This was the only volume ever published.

1818   Systematische Beschreibung der bekannten europäischen zweiflügeligen Insekten. Vol. 1, xxxvi + 333 pp., pls. 1–11. Aachen.

   A second edition, with minor changes, was published with xxiv + 259 pp., pls. 1–11, at Halle, *1851*.
   The entire work consists of 7 volumes, issued 1818–1838, the first two at Aachen, the last five at Hamm. Volume 7 is also called a "Supplementband". A second edition of volumes 1 and 2 was published at Halle, 1851. The work was continued by Loew, 1869a, 1871, and 1873c.

Meigen, J. W.—Continued
1820 Systematische Beschreibung der bekannten europäischen zweiflügeligen Insekten. Vol. 2, x + 365 pp., pls. 12–21. Aachen.
       A second edition, with minor changes, was published with vi + 276 pp., pls. 12–21, at Halle, *1851*.
1822 Systematische Beschreibung der bekannten europäischen zweiflügeligen Insekten. Vol. 3, x + 416 pp., pls. 22–32. Hamm.
1824 Systematische Beschreibung der bekannten europäischen zweiflügeligen Insekten. Vol. 4, xii + 428 pp., pls. 33–41. Hamm.
1826 Systematische Beschreibung der bekannten europäischen zweiflügeligen Insekten. Vol. 5, xii + 412 pp., pls. 42–54. Hamm.
1830 Systematische Beschreibung der bekannten europäischen zweiflügeligen Insekten. Vol. 6, iv + 401 pp., pls. 55–66. Hamm.
1838 Systematische Beschreibung der bekannten europäischen zweiflügeligen Insekten. Vol. 7: "oder Supplementband", xii + 434 pp., pls. 67–74. Hamm.

Meijere, J. C. H. de
1901 Ueber die Metamorphose von *Callomyia amoena* Meig. Tijdschr. v. Ent. (1900) **43**: 223–231, pl. 13.
1909 Zur Kenntnis der Metamorphose der Lauxaninae. Ztschr. f. Wiss. Insektenbiol. **n. ser.**, 5 [=whole ser., 14]: 152–155, 2 figs.
1910 Studien über südostasiatische Dipteren. IV. Die neue Dipterenfauna von Krakatau. Tijdschr. v. Ent. **53**: 58–194, pls. 4–8.
1911a Zur Kenntnis der Metamorphose von *Platypeza* und der verwandtschaftlichen Beziehungen der Platypezinen. Tijdschr. v. Ent. **54**: 241–254, pl. 17.
1911b Studien über südostasiatische Dipteren. VI. Tijdschr. v. Ent. **54**: 258–432, pls. 18–22.
1913a H. Sauter's Formosa Ausbeute. Sepsinae (Dipt.). Budapest Magyar Nemzeti Muz., Ann. Hist. Nat. **11**: 114–124, 2 figs.
1913b Studien über südostasiatische Dipteren. VII. Tijdschr. v. Ent. **56**: 317–354, pls. 15–17.
1918 Neue holländische Dipteren. Tijdschr. v. Ent. **61**: 128–141, pl. 8.
1928 Vierde supplement op de nieuwe naamlijst van nederlandsche Diptera. Tijdschr. v. Ent. **71**: 11–83.

Meijere, J. C. H. de—Continued
1944 Over de metamorphose van *Metopia leucocephala* Rossi, *Cacoxenus indagator* Löw, *Palloptera saltuum* L., *Paranthomyza nitida* Mg. en *Hydrellia nigripes* Zett. (Dipt.). Tijdschr. v. Ent. (1943) **86**: 57–61, 23 figs.
1946 *Oxydiscus* (Diptera) changed in [sic] *Oxyrhiza*. Ent. Ber. **12**: 68.

Meinert, F.
1864 *Miastor metraloas:* Yderligere oplysning om den af Prof. Nic. Wagner nyligt beskrevne insektlarve, som formerer sig ved spiredannelse. Naturhist. Tidsskr. **ser. 3, 3**: 37–43.
1865 Endnu et par ord om *Miastor:* Tilligemed bemaerkninger om spiredannelsen hos en anden *Cecidomyia*-larve og om aeggets dannelse og udvikling i dyreriget overhovedet. Naturhist. Tidsskr. **ser. 3, 3**: 225–238.
1890a *Philornis molesta*, en paa fugle snyltende tachinarie. Naturhist. For. Kjøbenhavn, Vidensk. Meddel. **1889** [=ser. 5, 1]: 304–317, 1 pl.
1890b *Aenigmatias blattoides*, dipteron novum apterum. Ent. Meddel. (1889–1890) **2**: 212–226, pl. 4.

Melander, A. L. (See also Brues and Melander; Brues, Melander, and Carpenter)
1900 A decade of Dolichopodidae. Canad. Ent. **32**: 134–144, 2 figs.
1901 Gynandromorphism in a new species of *Hilara*. Psyche **9**: 213–215, 2 figs.
1902a A monograph of the North American Empididae. Part I. Amer. Ent. Soc. Trans. **28**: 195–367, 5 pls.
   This was the only "part" ever published.
1902b Notes on the Acroceridae. Ent. News **13**: 178–182, 1 fig.
1903a An interesting new *Chrysotus*. Ent. News **14**: 72–75, 1 fig.
1903b A review of the North American species of *Nemotelus*. Psyche **10**: 171–183, pl. 4.
1904 Notes on North American Stratiomyidae. Canad. Ent. **36**: 14–24, 53–54.
1906 Some new or little-known genera of Empididae. Ent. News **17**: 370–379, 5 figs.
1910 The genus *Tachydromia*. Psyche **17**: 41–62, 1 pl.
1912 The dipterous genus *Bibiodes*. Amer. Mus. Nat. Hist. Bul. **31**: 337–341, 4 figs.
1913a A synopsis of the dipterous groups Agromyzinae, Milichiinae, Ochthiphilinae and Geomyzinae. N.Y. Ent. Soc. Jour. **21**: 219–273, 283–300, 1 pl.

Melander, A. L.—Continued
1913b A synopsis of the Sapromyzidae. Psyche **20:** 57–82, 1 pl.
1913c Some acalyptrate Muscidae. Psyche **20:** 166–169, 1 fig.
1913d Note on two preoccupied muscid names. Psyche **20:** 205.
1916 The dipterous family Scatopsidae. Wash. Agr. Expt. Sta. Bul. **130:** 1–21, 2 pls.
1918 The dipterous genus *Drapetis* Meigen (family Empididae). Ent. Soc. Amer. Ann. **11:** 183–221.
1920a Review of the nearctic Tetanoceridae. Ent. Soc. Amer. Ann. **13:** 305–332, 1 pl.
1920b Synopsis of the dipterous family Psilidae. Psyche **27:** 91–101.
1923a The genus *Cyrtopogon* (Diptera: Asilidae). Psyche **30:** 102–119.
1923b The genus *Lasiopogon* (Diptera: Asilidae). Psyche **30:** 135–145.
1923c Studies in Asilidae (Diptera). Psyche **30:** 207–219.
1924a New species of *Platypalpus* occurring in New England. Boston Soc. Nat. Hist. Occas. Papers **5:** 83–87.
1924b Review of the dipterous family Piophilidae. Psyche **31:** 78–86.
1927 Diptera. Fam. Empididae. Fasc. 185, 434 pp., 8 pls. *In* Wytsman, P., ed., Genera insectorum [q.v.]. Bruxelles.
1932 The entomological publications of C. W. Johnson. Psyche **39:** 87–99.
1940a The dipterous genus *Microphorus* (Diptera, Empididae). I. Phylogeny; II. Taxonomy. Pan-Pacific Ent. **16:** 5–11, 11 figs., 1 pl.; 59–69.
1940b *Hilara granditarsis*, a balloon-maker. Psyche **47:** 55–56.
1945 Ten new species of Empididae (Diptera). Psyche **52:** 79–87.
1946a The nearctic species of *Iteaphila* and *Apalocnemis*. Brooklyn Ent. Soc. Bul. **41:** 29–40, 1 fig.
1946b *Apolysis, Oligodranes* and *Empidideicus* in America (Diptera, Bombyliidae). Ent. Soc. Amer. Ann. **39:** 451–495, 1 pl.
1946c Synopsis of *Coptophlebia*, with descriptions of new American and Oriental species (Diptera, Empididae). Pan-Pacific Ent. **22:** 105–117, 6 figs.
1947 Synopsis of the Hemerodromiinae (Diptera, Empididae). N.Y. Ent. Soc. Jour. **55:** 237–273, 2 pls.
1950a *Aphoebantus* and its relatives *Epacmus* and *Eucessia* (Diptera: Bombyliidae). Ent. Soc. Amer. Ann. **43:** 1–45, 2 pls.

Melander, A. L.—Continued
1950b Taxonomic notes on some smaller Bombyliidae (Diptera). Pan-Pacific Ent. **26**: 139–144, 145–156.
(1952a) The North American species of Tethinidae (Diptera). N.Y. Ent. Soc. Jour. (1951) **59**: 187–212, 6 figs.
1952b The American species of Trixoscelidae. N.Y. Ent. Soc. Jour. **60**: 37–52.
[1958] A new *Tachyempis* (Diptera: Empididae). Ent. Soc. Wash. Proc. (1957) **59**: 296.

> Issue was published Jan., 1958, according to R. H. Foote, editor of the Proceedings.

1960 A new species of *Charadromia* from California (Diptera: Empididae). Wasmann Jour. Biol. **18**: 129–130.
(1961) The genus *Mythicomyia* (Diptera: Bombyliidae). Wasmann Jour. Biol. (1960) **18**: 161–261.

———, **and Argo, N. G.**
1924 Revision of the two-winged flies of the family Clusiidae. U.S. Natl. Mus. Proc. **64** (11): 1–54, 4 pls.

———, **and Brues, C. T.**
1900 New species of *Hygroceleuthus* and *Dolichopus*, with remarks on *Hygroceleuthus*. Biol. Bul. **1**: 123–148, 22 figs.
1903 Guests and parasites of the burrowing bee *Halictus*. Biol. Bul. **5**: 1–27, 7 figs.

———, **and Spuler, A.**
1917 The dipterous families Sepsidae and Piophilidae. Wash. Agr. Expt. Sta. Bul. **143**: 1–103, 1 pl.

Melin, D.
1923 Contributions to the knowledge of the biology, metamorphosis and distribution of the Swedish asilids in relation to the whole family of asilids. Uppsala Univ. Zool. Bidr. **8**: 1–317, 305 figs.

Melville, R. V. (See also Stoll et al.)
1961 Report on Mr. C. W. Sabrosky's proposal for the suppression under the Plenary Powers of the pamphlet entitled "Nouvelle classification des mouches à deux ailes" by J. W. Meigen, 1800. Internatl. Comn. Zool. Nomencl., Bul. Zool. Nomencl. **18**: 9–64.

Ménétries, E.
1851 Insecten. Pp. 43–76, pl. 3. *In* Middendorff, A. T. von, Reise in den äussersten Norden und Osten Sibiriens [q.v.]. Vol. 2, Pt. 1: Wirbellose Thiere, 516 pp., 32 pls. St. Petersburg.

Mesnil, L. P.
1934   A propos de deux Diptères nouveaux de la famille des Opomyzidae. Rev. Franç. d'Ent. **1**: 191–207, 27 figs.
1939   Essai sur les Tachinaires (Larvaevoridae). [France] Min. de l'Agr., [Cent. Natl. de Rech. Agron., Versailles], Monog. **[7]**: 1–67, 2 pls.
1944   Larvaevorinae (Tachininae). [Fam.] 64g, pp. 1–48, figs. 1–28, pls. 1–2 (=lfg. 153) [cont.]. *In* Lindner, E. ed., Die Fliegen der palaearktischen Region [q.v.]. Vol. 8. Stuttgart.
1949a   Essai de révision des espèces du genre *Drino* Robineau-Desvoidy Sturmiinae à oeufs macrotypes. Brussels Inst. Roy. des Sci. Nat. de Belg. Bul. **25 (42)**: 1–38.
1949b   Larvaevorinae (Tachininae) [cont.]. [Fam.] 64g, pp. 49–104, figs. 29–30 (=lfg. 161) [cont.]. *In* Lindner, E., ed., Die Fliegen der palaearktischen Region [q.v.]. Vol. 8. Stuttgart.
1950–1952   Larvaevorinae (Tachininae) [cont.]. [Fam.] 64g, pp. 105–160, pls. 6–7 (=lfg. 164), *1950*; pp. 161–208, pls. 3–5 (=lfg. 166), *1951*; pp. 209–256, pls. 8–9 (=lfg. 168), *1952* [cont.]. *In* Lindner, E., ed., Die Fliegen der palaearktischen Region [q.v.]. Vol. 8. Stuttgart.
1953a   Note synonymique. Soc. Ent. de France, Bul. **58**: 50.
1953b   Larvaevorinae (Tachininae) [cont.]. [Fam.] 64g, pp. 257–304, figs. 31–34 (=lfg. 172) [cont.]. *In* Lindner, E., ed., Die Fliegen der palaearktischen Region [q.v.]. Vol. 8. Stuttgart.
1954–1956   Larvaevorinae (Tachininae) [cont.]. [Fam.] 64g, pp. 305–368, figs. 35–36 (=lfg. 175), *1954*; pp. 369–416, figs. 37–40 (=lfg. 179), *1954*; pp. 417–464 (=lfg. 186), *1955*; pp. 465–512, pls. 10–14 (=lfg. 189), *1956*; pp. 513–560, pls. 15–17 (=lfg. 192), *1956* [cont.]. *In* Lindner, E., ed., Die Fliegen der palaearktischen Region [q.v.]. Vol. 8. Stuttgart.
1960–1963   Larvaevorinae (Tachininae) [cont.]. [Fam.] 64g, pp. 561–608 (=lfg. 210), *1960*; pp. 609–656 (=lfg. 212), *1960*; pp. 657–704 (=lfg. 219), *1961*; pp. 705–752 (=lfg. 221), *1962*; pp. 753–800 (=lfg. 224), *1962*; pp. 801–848 (=lfg. 235), *1963* [cont.]. *In* Lindner, E., ed., Die Fliegen der palaearktischen Region [q.v.]. Vol. 8. Stuttgart.

Metcalf, C. L.
1911   Life histories of Syrphidae. II. Ohio Nat. **12**: 397–405, pl. 19.

Metcalf, C. L.—Continued
- 1913   The Syrphidae of Ohio. Ohio Biol. Survey, Bul. **1:** 7–122, 3 figs., 11 pls. (=Bul. 1). (=Ohio State Univ. Bul. 17 (31).)
- 1916a   A list of Syrphidae of North Carolina. Elisha Mitchell Sci. Soc. Jour. **32:** 95–112.
- 1916b   Syrphidae of Maine. Maine Agr. Expt. Sta. Bul. [**ser. 2**], **253:** 193–264, figs. 28–37.
  > Also bound in Maine Agr. Expt. Sta. Ann. Rpt. (1916) **32:** 193–264, figs. 28–37, 1917.
- 1917a   Two new Syrphidae (Diptera) from eastern North America. Ent. News **28:** 209–212, 1 pl.
- 1917b   Syrphidae of Maine. Second report. Maine Agr. Expt. Sta. Bul. [**ser. 2**], **263:** 153–176, figs. 8–12.
  > Also bound in Maine Agr. Expt. Sta. Ann. Rpt. **33:** 153–176 figs. 8–12.
- 1921   The genitalia of male Syrphidae: Their morphology, with especial reference to its taxonomic significance. Ent. Soc. Amer. Ann. **14:** 169–226, pls. 9–19. (=Contributions from the Ohio State University Department of Zoology and Entomology No. 67.)

Metz, C. W.
- 1938   *Sciara reynoldsi*, a new species which hybridizes with *Sciara ocellaris* Comst. Jour. Hered. **29:** 176–178, fig. 7.

Meunier, F.
- 1899   Révision des Diptères fossiles types de Loew conservés au Musée Provincial de Koenigsberg. Misc. Ent. **7:** 161–165, 169–182, pls. 1–4.

Mickel, C. E.   (See Nicholson and Mickel)

Middendorff, A. T. von
- 1847–1875   Reise in den äussersten Norden und Osten Sibiriens während der Jahre 1843 und 1844. 4 vols.+Atlas. St. Petersburg.
  > Each volume consists of 2 separately paged parts, which in most cases were issued in lieferungen.
  > See Ménétries, 1851.

Middlekauff, W. W.
- 1944   A new species of *Aedes* from Florida (Diptera: Culicidae). Ent. Soc. Wash. Proc. **46:** 42–44, 1 fig.

Mik, J.
- 1864   Dipterologische Beiträge. K.-k. Zool.-Bot. Gesell. Wien, Verhandl. **14 (Abhandl.):** 785–798.
- 1866   Beitrag zur Dipterenfauna des österreichischen Küstenlandes. K.-k. Zool.-Bot. Gesell. Wien, Verhandl. **16 (Abhandl.):** 301–310, pl. 1, A.

Mik, J.—Continued
1867   Dipterologische Beiträge zur "Fauna austriaca". K.-k. Zool.-Bot. Gesell. Wien, Verhandl. **17 (Abhandl.)**: 412–423, 1 pl.
1874   Beitrag zur Dipteren-Fauna Oesterreich's. K.-k. Zool.-Bot. Gesell. Wien, Verhandl. **24 (Abhandl.)**: 329–354, pl. 7.
> Also published separately, 26 pp. Wien, *1874*. Separate has not been seen; information is from Mik, 1891b: 69.

1878a   Dipterologische Beiträge. K.-k. Zool.-Bot. Gesell. Wien, Verhandl. **28 (Abhandl.)**: 617–632, pl. 10.
1878b   Dipterologische Untersuchungen. (Wien) K.k. Akad. Gymnasium, Jahresber. **1878**: 1–24, 1 pl.
> Also published separately, 26 pp., 1 pl., Wien, *1878*. The journal has not been seen; information is from Mik, 1891b: 70.

1881   Dipterologische Mittheilungen, II. K.-k. Zool.-Bot. Gesell. Wien, Verhandl. **31 (Abhandl.)**: 315–330, pl. 16.
1884   Fünf neue österreichische Dipteren. K.-k. Zool.-Bot. Gesell. Wien, Verhandl. (1883) **33 (Abhandl.)**: 251–262, 4 figs.
1886   Dipterologische Miscellen. I; II; III. Wien. Ent. Ztg. **5**: 101–102; 276–279; 317–318.
1890   *Ugimyia sericariae* Rond., der Parasit des japanischen Seidenspinners. Ein dipterologischer Beitrag. Wien. Ent. Ztg. **9**: 309–316, pl. 3, figs. 7–8.
1891a   Dipterologischen Miscellen. XVII. Wien. Ent. Ztg. **10**: 1–5.
1891b   Repertorium meiner entomologischen Publicationen bis zum Schlusse des Jahres 1890. Wien. Ent. Ztg. **10**: 67–96.
1891c   Vorläufige Notiz über *Parathalassius blasigii*, ein neues Dipteron aus Venedig. Wien. Ent. Ztg. **10**: 216–217.
1894a   Ueber *Echinomyia popelii* Portsch. Wien. Ent. Ztg. **13**: 100.
1894b   Über eine neue *Agromyza*, deren Larven in den Blüthenknospen von *Lilium martagon* leben. Wien. Ent. Ztg. **13**: 284–290, 1 pl.
1895   Bemerkungen zu den Dipteren-Gattungen *Pelecocera* Meig. und *Rhopalomera* Wied. Wien. Ent. Ztg. **14**: 133–136.
1897   Einige Bemerkungen zur Dipteren-Familie der Syrphiden. Wien. Ent. Ztg. **16**: 61–66, 113–119.
1899   *Verrallia* nov. gen. Pipunculidarum (Dipt.). Wien. Ent. Ztg. **18**: 133–137.

Miller, D.
 1945   Generic name changes in Diptera. Roy. Ent. Soc., London, Proc. Ser. B: Taxonomy **14**: 72.

Miller, D. D.
 1944   *Drosophila melanura*, a new species of the *melanica* group. N.Y. Ent. Soc. Jour. **52**: 85–97, 2 figs.
 1958   Geographical distributions of the American *Drosophila affinis* subgroup species. Amer. Midland Nat. **60**: 52–70, 1 fig.

Miller, R. B.
 1941   A contribution to the ecology of the Chironomidae of Costello Lake, Algonquin Park, Ontario. Toronto Univ. Studies, Biol. Ser. **49**: 1–63, 17 figs. (=Ontario Fisheries Res. Lab. Pub. 60.)

Mills, H.   (See Snow and Mills)

Mitchell, E. G.
 1908   Descriptions of nine new species of gnats. N.Y. Ent. Soc. Jour. **16**: 7–14, 12 figs.

Mitchell, J. D.   (See Hunter, Pratt, and Mitchell)

Modeer, A.
 1786   Styng-flug-slägtet (*Oestrus*). K. Vetensk. Acad. Nya Handl. [ser. 2], **7**: 125–158.

Möhn, E.
 1954   Eine neue zoophage Gallmücken-Art an Tannenläusen. Ztschr. f. Angew. Ent. **36**: 462–468, 7 figs.

Molliard, M.
 1903   La galle du *Cecidomyia cattleyae* n. sp. Marcellia (1902) **1**: 165–170, pl. 2.

Moodie, R. L.
 1905   A new *Milesia* from Arizona with notes on some Wyoming Syrphidae. Ent. News **16**: 138–143.

Morge, G.
 1956   Über Morphologie und Lebensweise der bisher unbekannten Larven von *Palloptera usta* Meigen, *Palloptera ustulata* Fallén und *Stegana coleoptrata* Scopoli (Diptera). Beitr. z. Ent. (Berlin) **6**: 124–137, 25 figs.
 1959   Monographie der palaearktischen Lonchaeidae (Diptera). Beitr. z. Ent. (Berlin) **9**: 1–92, 323–371, 909–945, 171 figs. [cont.].

Morris, H. M.
1921   The larval and pupal stages of the Bibionidae.   Bul. Ent. Res. **12**: 221–232, 17 figs.
1922   The larval and pupal stages of the Bibionidae.   –Part II. Bul. Ent. Res. **13**: 189–195, 10 figs., pl. 9.

Morris, M. H.   (See Gambel, 1849)

Morrison, F. O.
1940   A revision of the American species of *Gonia* Meigen (Diptera: Tachinidae).   Canad. Jour. Res., Sect. D, Zool. Sci. **18**: 336–362, 3 pls.

Morrissey, T.
1950   Tanypodinae of Iowa (Diptera).   III. Amer. Midland Nat. **43**: 88–91, 4 figs.

Morton, K. J.   (See Needham, Morton, and Johannsen)

Müller, F.
1895   Contribution towards the history of a new form of larvae of Psychodidae (Diptera) from Brazil.   Ent. Soc. London, Trans. **1895**: 479–482, 2 pls.

Müller, O. F.
1764   Fauna insectorum Fridrichsdalina, sive methodica descriptio insectorum agri Fridrichsdalinensis, cum characteribus genericis et specificis, nominibus trivialibus, locis natalibus, iconibus allegatis, novisque pluribus speciebus additis. xxiv + 96 pp.   Hafniae et Lipsiae [=Copenhagen and Leipzig].
1776   Zoologiae Danicae prodromus, seu animalium Daniae et Norvegiae indigenarum characteres, nomina, et synonyma imprimis popularium. xxxii + 282 pp.   Hafniae [= Copenhagen].

Murachi, E. K.   (See Tokunaga and Murachi)

Muttkowski, R. A.
1915   New insect life histories.   I.   Wis. Nat. Hist. Soc. Bul. **13**: 109–122, 5 figs.

Nagatomi, A.   (See Hardy and Nagatomi)

Nájera, A. Díaz   (See Díaz Nájera, A.)

Namba, R.
1956   A revision of the flies of the genus *Rivellia* (Otitidae, Diptera) of America north of México.   U.S. Natl. Mus. Proc. **106**: 21–84, 10 figs.

Natural History Society of Wisconsin
1885– Proceedings of the Natural History Society of Wisconsin.
1889   231 pp. Milwaukee, Wis.
> These Proceedings were issued in continuously paged signatures as follows: Proc. for Oct., 1884–Mar., 1885, pp. 3–44, 1885; Proc. for Apr.–Dec., 1885, pp. 45–86, [1886]; Proc. for Jan.–Dec., 1886, pp. 89–140, [1887]; Proc. for Mar.–Dec., 1887, pp. 141–190, [1888]; Proc. for Jan.–Dec., 1888, pp. 191–231 [1889]. (These continue the Proceedings which appeared in the society's "Naturhist. Ver. von Wis., Jahresber.," and succeeding Proceedings were published in the society's "Occas. Papers" and later its "Bul.".)
> See Wheeler, 1889.

Neave, S. A., ed.
1939– Nomenclator zoologicus. 5 vols. London.
1950

Needham, J. G.
1908   Appendix C. Report of the entomological field station conducted at Old Forge, N.Y., in the summer of 1905. Pp. 156–248, figs. 2–16, pls. 4–32. *In* Felt, E. P., 23d report of the State Entomologist on injurious and other insects of the State of New York, 1907. N.Y. State Mus., Mus. Bul. **124**: 5–541, 49 figs., 44 pls.
> Also published in N.Y. State Mus. Ann. Rpt. (1907) 61 (vol. 2, app. 2): 156–248, figs. 2–16, pls. 4–32, *1908*.

1925   Observations on a flower gall of the chaparral. Jour. Ent. and Zool. **17**: 17–20, 1 fig.

Needham, J. G., and Betten, C. (See Coquillett, 1901f)
———, **Frost, S. W., and Tothill, B. H.**
1928   Leaf-mining insects. 351 pp., 91 figs., 3 pls. Baltimore, Md.

———, **MacGillivray, A. D., Johannsen, O. A., and Davis, K. C.** (See Johannsen, 1903b)

———, **Morton, K. J., and Johannsen, O. A.** (See Johannsen, 1905)

Neff, S. E., and Berg, C. O.
1962   Biology and immature stages of *Hoplodictya spinicornis* and *H. setosa* (Diptera: Sciomyzidae). Amer. Ent. Soc. Trans. **88**: 77–93, pls. 3–4.

Neiva, A. (See Lutz and Neiva; Lutz, Neiva, and Lima)

Neveu-Lemaire, M.
1902a Sur la classification des Culicides. Soc. de Biol. [Paris] Compt. Rend. **54**: 1329–1332.
1902b Classification de la famille des Culicidae. Soc. Zool. de France, Mém. **15**: 195–227, 12 figs.

New York Academy of Sciences
1919– Scientific survey of Puerto Rico and the Virgin Islands.
1960   19 vols. New York.
> This series will consist of 19 volumes upon completion, only 7 being finished by 1960. The volumes are issued in continuously paged parts whenever a portion is completed. The Diptera are included in Vol. 11, only 1 part of which has been published through 1960.
> See Curran, 1928d.

New York Natural History Survey
1842– Natural History of New York. 30 vols. Albany, N.Y.
1894
> This work is divided into 6 divisions: Zoology, Botany, Mineralogy, Geology, Agriculture, and Paleontology, each division by a different author and consisting of various numbers of Parts and Volumes. Division 5, which includes the insects, is entitled "Agriculture of New-York", is by E. Emmons, and was published in Albany, 1846–1854; it consists of 5 volumes, each with a subtitle.
> See Emmons, 1854.

Newell, W.
1930   The Mediterranean fruit fly situation. Jour. Econ. Ent. **23**: 512–535.

Newman, E.
1838   Entomological notes [cont.]. Ent. Mag. (London) **5**: 372–402, 4 figs. [cont.].
1841   Entomological notes [cont.]. Entomologist (Newman's) (1840–1842) [**1**]: 220–223 [cont.].

Nicholson, H. P., and Mickel, C. E.
1950   The black flies of Minnesota (Simuliidae). Minn. Agr. Expt. Sta. Tech. Bul. **192**: 1–64, 32 figs.

Nielsen, J. C. (See also Becker, 1907c)
1907   The insects of East-Greenland. Meddel. om Grønland (Copenhagen) **29**: 363–414.
1910   II. A catalogue of the insects of north-east Greenland with descriptions of some larvae. Pp. 55–68, 2 pls. *In* Johansen, F., and Nielsen, J. C., The insects of the "Danmark" Expedition. Meddel. om Grønland (Copenhagen) (1911) **43**: 35–68, 2 pls.

Nielsen, P.
1951   *Limonia* (*Dicranomyia*) *vibei* n. sp. from Grønland (Dipt. Tipul.). Ent. Meddel. **26**: 185–186, 2 figs.

Nitzsch, C. L.
1818   Die Familien und Gattungen der Thierinsekten (Insecta epizoica) als Prodromus einer Naturgeschichte derselben. Mag. d. Ent. (Halle) **3**: 261–316.

Novitski, E.   (See Sturtevant and Novitski)

Nowell, W. R.
1951   The dipterous family Dixidae in western North America (Insecta: Diptera). Microentomology **16**: 187–270, 15 figs.

Nowicki, M. S.
1873   Beiträge zur Kenntniss der Dipterenfauna Galiziens. 35 pp. Krakau, Poland.

Nutting, W. L.
1953   The biology of *Euphasiopteryx brevicornis* (Townsend) (Diptera, Tachinidae), parasitic in the cone-headed grasshoppers (Orthoptera, Copiphorinae). Psyche **60**: 69–81, 1 fig., pl. 4.

Office Central de Faunistique de la Fédération Française des Sociétés de Sciences Naturelles
1921–   Faune de France. 63 vols. Paris.
1959

> This series of separately paged volumes on various zoological subjects by a number of authors consists of 63 volumes to date and continues.
> See Goetghebuer, 1928, 1932; Kieffer, 1925a; Parent, 1938.

Okada, I.
1939   Studien über die Pilzmücken (Fungivoridae) aus Hokkaido (Diptera, Nematocera). Hokkaido Imp. Univ., Faculty Agr., Jour. **42**: 267–336, pls. 15–18.

Oldenberg, L.
1910   Vier neue paläarktische Akalypteren (Dipt.). Deut. Ent. Ztschr. **1910**: 284–287.

1914   Beitrag zur Kenntnis der europäischen Drosophiliden. Arch. f. Naturgesch. **Abt. A, 80 (2)**: 1–42, 3 figs.

1915   Berichtigung zu meiner Drosophilidenarbeit. Arch. f. Naturgesch. (1914) **Abt. A, 80 (9)**: 93.

1923   Neue Acalyptraten (Dipt.) meiner Ausbeute. Deut. Ent. Ztschr. **1923**: 307–319.

Oldroyd, H.
1949   Diptera: Introduction and key to families. Pt. 1, 49 pp., 97 figs., 1 pl. *In* Royal Entomological Society of London, Handbooks for the identification of British insects [q.v.]. Vol. 9: Nematocera and Brachycera. London.

Olfers, I. F. M. von
1816   De vegetativis et animatis corporibus in corporibus animatis reperiundus. Pars I. 112 pp., 1 pl. Berlin.

> This, the only part issued, is cited by Horn and Schenkling, 1928: 899, as a "Dissert. inaug.", Goettingae, 1815. This latter has not been located, even in the University Library at Göttingen, and it may not have been an actual formal publication.

Oliver, D. R.
1959   Some Diamesini (Chironomidae) from the Nearctic and Palearctic. Ent. Tidskr. **80**: 48–64, 19 figs.

Olivier, G. A.
1811–   Insectes [(i.e., Arthropoda) Pt. 5]. Vol. 8, pp. 1–360
[1812] (=livr.?), *1811*; pp. 361–722 (=livr. 77), [*1812*]. *In* Société de Gens de Lettres, de Savans et d'Artistes, Encyclopédie méthodique [q.v.]. Histoire naturelle. Paris.

> For dating, see Sherborn and Woodward, 1899: 595.

1813   Premier mémoire sur quelques Insectes qui attaquent les céréales. Soc. d'Agr. du Dépt. de la Seine, Mém. d'Agr., d'Écon. Rurale et Dom. **16** [=whole ser., (44)]: 477–495, 2 pls.

O'Neill, K.   (See Randolph and O'Neill)

d'Orbigny, C., ed.
[1848] Dictionnaire universel d'histoire naturelle. Vol. 11, 816 pp. Paris.

> For dating, see Sherborn and Palmer, 1898: 350.
> The entire work consists of 13 volumes of text, and 3 of plates, published at Paris, 1838–1849.

Ortiz, I.
1951   Estudios en *Culicoides* (Diptera, Ceratopogonidae). IX. Sobre los caracteres diferenciales entre *Culicoides paraensis* (Goeldi, 1905), *C. stellifer* (Coquillett, 1901) y *C. lanei* (Ortiz, 1950). Descripción de cuatro nuevas especies con la redescripción de algunas otras poco conocidas. Rev. de Sanid. y Asistencia Social (Caracas) **16**: 573–591, 8 pls.

Osburn, R. C.
1908   British Columbia Syrphidae, new species and additions to the list. Canad. Ent. **40**: 1–14, 1 fig.
1910   Studies on Syrphidae. I. *Syrphus arcuatus* Fallén and a related new species; II. The invalidity of *Scaeva* (=*Catabomba*) as a genus; III. An interesting meristic variation in *Syrphus perplexus*. N. Y. Ent. Soc. Jour. **18**: 53–57, pl. 1; 58–62, pl. 2; 62–66, pl. 3.

Osburn, R. C.—Continued
1926   A new species of the genus *Condidea* (Diptera, Syrphidae). Ent. News **37**: 51–53.

Oshanin, B.
1910   Tables générales des publications de la Société entomologique de Russie ainsi que des articles, des synopsis et des formes nouvelles y contenues. 1859–1908. Russ. Ent. Obshch. Trudy (Soc. Ent. Ross. Horae) **38 (Sup.)**: 1–282.

Osten Sacken, C. R.   (See also Lintner, 1888a; Packard, 1877)
1858   Catalogue of the described Diptera of North America. Smithsn. Inst., Smithsn. Misc. Collect. **3 (1** [=pub. 102])**: vii-xx, 1–92.

1859   New genera and species of North American Tipulidae with short palpi, with an attempt at a new classification of the tribe. Acad. Nat. Sci. Phila. Proc. **1859**: 197–256, 1 fig., 2 pls.

1860   Appendix to the paper entitled New Genera and species of North American Tipulidae with short palpi, &c. Acad. Nat. Sci. Phila. Proc. **1860**: 15–17.

1861   Description of nine new North American Limnobiaceae. Acad. Nat. Sci. Phila. Proc. **1861**: 287–292.

1862a   Characters of the larvae of Mycetophilidae. Ent. Soc. Phila. Proc. (1861–1863) **1**: 151–172, 1 pl.

1862b   V. On the North American Cecidomyidae. Pp. 173–205, figs. 1–12, 2 pls. *In* Loew, H., Monographs of the Diptera of North America. Part I. Smithsn. Inst., Smithsn. Misc. Collect. **6 (1** [=pub. 141])**: 1–221, 13 figs., 2 pls.

1862c   Entomologische Notizen [concl.]. Stettin Ent. Ztg. **23**: 408–415.

1863   *Lasioptera* reared from a gall of the golden-rod. Ent. Soc. Phila. Proc. (1861–1863) **1**: 368–370.

1864   Descriptions of several new North American Ctenophorae. Ent. Soc. Phila. Proc. **3**: 45–49.

1865   Description of some new genera and species of North American Limnobina. Part I. Ent. Soc. Phila. Proc. **4**: 224–242.

1866   Two new North American Cecidomyidae. Ent. Soc. Phila. Proc. **6**: 219–220.

1868   Description of a new species of Culicidae. Amer. Ent. Soc. Trans. **2**: 47–48.

1869a   Biological notes on Diptera (galls on *Solidago*). Amer. Ent. Soc. Trans. **2**: 299–303.

Osten Sacken, C. R.—Continued

1869b Monographs of the Diptera of North America. Part IV. Smithsn. Inst., Smithsn. Misc. Collect. 8 (1 [=pub. 219]): 1–345, 7 figs., 4 pls.

1870–
1871 Biological notes on Diptera. (Article 2nd.); (Article 3d.). Amer. Ent. Soc. Trans. 3: 51–54, *1870*; 345–347, *1871*.

1874a Report on the Diptera collected by Lieut. W. L. Carpenter in Colorado during the summer of 1873. [U.S. Dept. Int.] U.S. Geol. and Geog. Survey of the Ter., Ann. Rpt. [7]: 561–566.

1874b A list of the Leptidae, Mydaidae and Dasypogonina of North America. Buffalo Soc. Nat. Sci. Bul. (1874–1875) **2**: 169–187.

1875a Chapter X. Report upon the collection of Diptera made in portions of Colorado and Arizona during the year 1873. Pp. 803–807, 3 figs. *In* Wheeler, G. M., ed., Report upon geographical and geological explorations and surveys west of the one hundredth meridian [q.v.]. Vol. 5, 1021 pp., 3 figs., 45 pls. Washington.

1875b Prodrome of a monograph of the Tabanidae of the United States. Part I. The genera *Pangonia*, *Chrysops*, *Silvius*, *Haematopota*, *Diabasis*. Boston Soc. Nat. Hist. Mem. **2**: 365–397.

1875c On the North American species of the genus *Syrphus* (in the narrowest sense). Boston Soc. Nat. Hist. Proc. (1875–1876) **18**: 135–153.

1875d A list of the North American Syrphidae. Buffalo Soc. Nat. Sci. Bul. **3**: 38–71.

1875e Three new galls of Cecidomyiae. Canad. Ent. **7**: 201–202.

1876a Prodrome of a monograph of the Tabanidae of the United States. Part II. The genus *Tabanus*. Boston Soc. Nat. Hist. Mem. **2**: 421–479.

1876b Report on the Diptera brought home by Dr. Bessels from the Arctic Voyage of the "Polaris", in 1872. Boston Soc. Nat. Hist. Proc. (1876–1877) **19**: 41–43.

1877 Western Diptera: Descriptions of new genera and species of Diptera from the region west of the Mississippi and especially from California. [U.S. Dept. Int.] U.S. Geol. and Geog. Survey of the Ter., Bul. **3**: 189–354.

1878a Prodrome of a monograph of the Tabanidae of the United States. Supplement. Boston Soc. Nat. Hist. Mem. **2**: 555–560.

1878b Bemerkungen über Blepharoceriden. Deut. Ent. Ztschr. [=Berlin. Ent. Ztschr.] **22**: 405–416.

Osten Sacken, C. R.—Continued
- 1878c Catalogue of the described Diptera of North America. [Ed. 2]. Smithsn. Inst., Smithsn. Misc. Collect. **16** (2 [=pub. 270]): 1–276.
- 1883a Synonymica concerning exotic dipterology. No. II. Berlin. Ent. Ztschr. **27**: 295–298.
- 1883b A singular North American fly (*Opsebius pterodontinus* n. sp.). Berlin. Ent. Ztschr. **27**: 299–300.
- 1885 Elenco delle pubblicazioni entomologiche del Professor Camillo Rondani. Soc. Ent. Ital. Bul. **17**: 149–162.
- 1886a Diptera, Vol. I [part]. Pp. 1–24, 25–48, 49–72, 73–104, 105–128, pls. 1–2 [cont.]. *In* Godman, F. D., and Salvin, O., eds., Biologia Centrali-Americana [q.v.]. Zoologia-Insecta-Diptera, Vol. 1, 378 pp., 6 pls. London.
- 1886b Studies on Tipulidae. Part I. Review of the published genera of the Tipulidae longipalpi. Berlin. Ent. Ztschr. **30**: 153–188.
- 1887a Diptera, Vol. I [part, concl.]. Pp. 129–160, 161–176, 177–208, 209–216, pl. 3. *In* Godman, F. D., and Salvin, O., eds., Biologia Centrali-Americana [q.v.]. Zoologia-Insecta-Diptera, Vol. 1, 378 pp., 6 pls. London.
- 1887b Some North American Tachinae. Canad. Ent. **19**: 161–166.

Ouellet, J.
- 1934 Description d'un nouveau Diptère du genre *Chloropisca*. Nat. Canad. **61** [=ser. 3, 5]: 320–323.
- 1940 Un nouveau Diptère du genre *Enicita* (Sepsides). Nat. Canad. **67** [=ser. 3, 11]: 225–228, 1 fig.
- 1942 Deux nouveaux Diptères (Empididae, Tachinidae). Nat. Canad. **69** [=ser. 3, 13]: 78–85, 4 figs.

Owen, W. B., and Gerhardt, R. W.
- 1957 The mosquitoes of Wyoming. Wyo. Univ. Pubs. **21**: 74–141, 11 figs.

Packard, A. S. (See also Verrill, 1873)
- 1869a Guide to the study of insects and a treatise on those injurious and beneficial to crops. 702 pp., 651 figs., 10 pls. Salem, Mass.
- 1869b On insects inhabiting salt water. Essex Inst. Proc. **6** (**Commun.**): 41–51, 6 figs.
- 1871 On insects inhabiting salt water. No. 2. Amer. Jour. Sci. and Arts ser. **3, 1** [=whole ser., 101]: 100–110, 5 figs.
- 1877 On a new cave fauna in Utah. [U.S. Dept. Int.] U.S. Geol. and Geog. Survey of the Ter., Bul. **3**: 157–169, figs. 5–10.
- 1878 Insects affecting the cranberry, with remarks on other injurious insects. [U.S. Dept. Int.] U.S. Geol. and Geog. Survey of the Ter., Ann. Rpt. (1876) **10**: 521–531, 9 figs.

Pagast, F.
1947   Systematik und Verbreitung der um die Gattung *Diamesa* gruppierten Chironomiden. Arch. f. Hydrobiol. n. ser., **41**: 435–596, 84 figs.

Painter, E. M.   (See Painter and Painter)

Painter, R. H.
1925   A review of the genus *Lepidophora* (Diptera, Bombyliidae). Amer. Ent. Soc. Trans. **51**: 119–127, 1 fig.

1926a   Notes on the genus *Parabombylius* (Diptera). Ent. News **37**: 73–78.

1926b   The *lateralis* group of the bombylid genus *Villa*. Ohio Jour. Sci. **26**: 205–212.

1930a   A review of the bombyliid genus *Heterostylum* (Diptera). Kans. Ent. Soc. Jour. **3**: 1–7.

1930b   Notes on some Bombyliidae (Diptera) from the Republic of Honduras. Ent. Soc. Amer. Ann. **23**: 793–807, 1 pl.

1932a   A monographic study of the genus *Geron* Meigen as it occurs in the United States (Diptera: Bombyliidae). Amer. Ent. Soc. Trans. **58**: 139–167, pls. 10–11.

1932b   A review of the genus *Apiocera* Westwood from North America, (Apioceridae, Diptera). Ent. Soc. Amer. Ann. **25**: 350–356, 1 pl.

1933   New subgenera and species of Bombyliidae (Diptera). Kans. Ent. Soc. Jour. **6**: 5–18.

1934   Two new species of North American *Exoprosopa* (Bombyliidae, Diptera). Kans. Ent. Soc. Jour. **7**: 68–70.

(1938) The family Apioceratidae (Diptera) in North America. Kans. Univ. Sci. Bul. (1936) **24** [=whole ser., 34]: 187–302, 1 fig., pl. 8 (=Kans. Univ. Bul. 37 (14).)

   Dated from list of publications on inside of cover of Vol. 29, Pt. 2, and subsequent volumes of the bulletin.

[1940] Notes on type specimens and descriptions of new North American Bombyliidae. Kans. Acad. Sci. Trans. (1939) **42**: 267–301, 2 pls.

   The date of issue stamped on the separates is Feb. 21, 1940.

1962   The taxonomy and biology of *Systoechus* and *Anastoechus* bombyliid (Diptera) predators in grasshopper egg pods. Kans. Ent. Soc. Jour. **35**: 255–269, 9 figs.

———, **and Hall, J. C.**
1960   A monograph of the genus *Poecilanthrax* (Diptera: Bombyliidae). Kans. Agr. Expt. Sta. Tech. Bul. **106**: 1–132, 31 maps, 34 photos, 8 pls.

Painter, R. H., and Painter, E. M.
  1962 Notes on and redescriptions of types of North American Bombyliidae (Diptera) in European museums. Kans. Ent. Soc. Jour. **35**: 2–164, 20 figs.

Palacios, A. Martinez (See Martinez Palacios, A.)

Palisot de Beauvois, A. M. F. J.
  1805– Insectes recueillis en Afrique et en Amérique dans les royaumes d'Oware et de Benin, à Saint-Domingue et dans les États-Unis, pendant les années 1786–1797. Pp. i–xvi+1–24, 6 pls. (=livr. 1), *1805;* pp. 25–40, 6 pls. (=livr. 2), [*1805*]; pp. 41–56, 6 pls. (=livr. 3), [*1806*]; pp. 57–72, 6 pls. (=livr. 4), [*1807*]; pp. 73–88, 6 pls. (=livr. 5), [*1807*]; pp. 89–100, 6 pls. (=livr. 6), [*1809*]; pp. 101–120, 6 pls. (=livr. 7), [*1811*]; pp. 121–136, 6 pls. (=livr. 8), [*1811*]; pp. 137–156, 6 pls. (=livr. 9), [*1817*]; pp. 157–172, 6 pls. (=livr. 10), [*1817*]; pp. 173–190, 6 pls. (=livr. 11), [*1818*]; pp. 191–208, 6 pls. (=livr. 12), [*1818*]; pp. 208–224, 6 pls. (=livr. 13), [*1819*]; pp. 225–240, 6 pls. (=livr. 14), [*1820*]; pp. 241–276, 6 pls. (=livr. 15, by J. G. A. Serville), [*1821*]. Paris, "1805."
  [1821]

  For dating, see Griffin, 1937: 121. The plates are not numbered consecutively. The 3 plates for Diptera are in livraisons 3, 6, and 13.

Palmer, T. S. (See Sherborn and Palmer)

Pandellé, L.
  1896 Études sur les Muscides de France. IIe partie [concl.]. Rev. d'Ent. (Caen) **15**: 1–219, 221–230.
  1898 Études sur les Muscides de France. IIIe partie. Rev. d'Ent. (Caen) **17 (spec. pagination):** 1–80 [cont.].
  1899 (Études sur les Muscides de France (IIIe partie)) [cont.]. Rev. d'Ent. (Caen) **18 (spec. pagination):** 81–120, 121–208, 209–220 [cont.].
  1900 (Études sur les Muscides de France (IIIe partie)) [cont.]. Rev. d'Ent. (Caen) **19 (spec. pagination):** 221–244, 245–260, 261–276, 277–292, 293–308 [cont.].

Pantel, J.
  1910 Recherches sur les Diptères à larves entomobies. I. Caractères parasitiques aux points de vue biologique, éthologique et histologique. Cellule **26**: 25–216, 26 figs., 5 pls.
  1913 Recherches sur les Diptères à larves entomobies. II. Les enveloppes de l'oeuf avec les formations qui en dépendent, les dégâts indirects du parasitisme. Cellule **29**: 5–289, 26 figs., 7 pls.

Panzer, G. W. F.
[1798] Faunae insectorum germanicae initiae oder Deutschlands Insecten. H. 54, 24 pp., 24 pls. Nürnberg.

> The entire work consists of 190 hefts, the first 109 by Panzer, published in Nürnberg, 1792–?1813. Heft 110 was by M. Geyer, published in ?, in Augsburg, and hefts 111–190 by G. A. W. Herrich-Schaeffer, in Regensburg, 1829–1844. For dating of the hefts, see Sherborn, 1923: 566. The date of heft 109 is given by Horn and Schenkling, 1928: 911, as "wohl 1810".

[1804] Faunae insectorum germanicae initiae oder Deutschlands Insecten. H. 86, 24 pp., 24 pls. Nürnberg.

> For dating, see Sherborn, 1923: 566.

[1809a] Faunae insectorum germanicae initiae oder Deutschlands Insecten. H. 104, 24 pp., 24 pls. Nürnberg.

> The date is uncertain. Hefts 101–107 were issued sometime between 1806 and 1809, and we have arbitrarily dated Hefts 104 and 105 as 1809. See Sherborn, 1923: 566.

[1809b] Faunae insectorum germanicae initiae oder Deutschlands Insecten. H. 105, 24 pp., 24 pls. Nürnberg. See Panzer, 1809a for note on date.

[1809c] Faunae insectorum germanicae initiae oder Deutschlands Insecten. H. 108, 24 pp., 24 pls. Nürnberg.

> For dating, see Sherborn, 1923: 566.

[1813] Faunae insectorum germanicae initiae oder Deutschlands Insecten. H. 109, 24 pp., 24 pls. Nürnberg.

> The date is uncertain. Sherborn, 1923: 566, gives 1813, but Horn and Schenkling, 1928: 911, say "wohl 1810".

Paramonov, S. J. (See also Waterhouse and Paramonov)
1930 Beiträge zur Monographie der Gattungen *Cytherea*, *Anastoechus* etc. Bombyliidae (Diptera). [Akad. Nauk U. R. S. R.] Vseukrains'ka Akad. Nauk, Fiz.-Mat. Vidd., Trudy (Acad. des Sci. de l'Ukraïne, Cl. des Sci. Phys. et Math., Mém.) **15**: 355–481. (=Zool. Muz., Z'irnik Prats' (Mus. Zool., Trav.) 9.)

1947 Zur Kenntnis der amerikanischen Bombyliiden-Gattung *Triploechus* Edw. (Diptera). Rev. de Ent. **18**: 183–192.

1949 Revision of the species of *Lepidophora* Westw. (Bombyliidae, Diptera). Rev. de Ent. **20**: 631–643.

Parent, O.
1928 Étude sur les Diptères Dolichopodides exotiques conservés au Zoologisches Staatsinstitut und Zoologisches Museum de Hambourg. Zool. Staatsinst. u. Zool. Mus. Hamburg. Mitt. **43**: 155–198, 85 figs.

Parent, O.—Continued
- 1929a Étude sur les Dolichopodides exotiques de la collection von Roder. Soc. Sci. de Bruxelles, Ann. **ser. B, 49:** 169–246, 124 figs.
- 1929b Études sur les Dolichopodides. Encyclopédie Ent., Sér. B, [Div.] II: Diptera **5:** 1–18, 22 figs.
- 1933 Contribution à la faune diptérologique de l'Amérique centrale et méridionale (Diptères Dolichopodides). Soc. Ent. de Belg. Bul. et Ann. **73:** 163–186, 36 figs.
- 1934a Étude sur les types de Dolichopodides exotiques de Francis Walker, conservés au British Museum. Ann. and Mag. Nat. Hist. **ser. 10, 13:** 1–38, 70 figs.
- [1934b] Diptères Dolichopodides exotiques. Soc. Natl. des Sci. Nat. et Math. de Cherbourg, Mém. (1929–1933) **41:** 257–308, pls. 67–79.

    Parent dates this 1934 in his own bibliography (see Parent 1935: 4).

- [1935] Vingt ans d'activité diptérologique. 4 pp. Aire, France.

    The last publications listed in this bibliography are those published early in 1935. It is therefore assumed the bibliography was issued sometime in 1935.

- 1938 Diptères Dolichopodidae. Vol. 35, 720 pp., 1002 figs. *In* Office Central de Faunistique de la Federation Française des Sociétés de Sciences Naturelles, Faune de France [q.v.]. Paris.

Parker, J. R., and Wakeland, C.
- 1957 Grasshopper egg pods destroyed by larvae of bee flies, blister beetles and ground beetles. U.S. Dept. Agr. Tech. Bul. **1165:** 1–29, 3 figs.

Parker, R. R.
- 1914a Sarcophagidae of New England: Males of the genera *Ravinia* and *Boettcheria*. Boston Soc. Nat. Hist. Proc. **35:** 1–77, 8 pls.
- 1914b A new sarcophagid scavenger from Montana. Canad. Ent. **46:** 417–423, 1 pl.
- 1916a New species of New England Sarcophagidae. Canad. Ent. **48:** 359–364, 422–427, 2 figs.
- 1916b Sarcophagidae of New England: Genus *Sarcophaga*. Jour. Econ. Ent. **9:** 438–441, 1 fig.
- 1916c Sarcophagidae of New England: Genus *Sarcophaga*. N.Y. Ent. Soc. Jour. **24:** 171–175.
- 1916d Sarcophagidae of New England. III. *Sarcofahrtia ravinia*, new genus and new species. Psyche **23:** 131–139, 1 fig.

Parker, R. R.—Continued
- 1917 A new *Sarcophaga* from New York. Canad. Ent. **49:** 157–161, 1 fig.
- 1918a A new species of *Sarcophaga* from British Columbia. Canad. Ent. **50:** 122–124, 1 fig.
- 1918b A new species of *Sarcophaga* from California. Jour. Ent. and Zool. **10:** 32–33, 1 fig.
- 1918c A new species of *Sarcophaga* from Niagara Falls. N.Y. Ent. Soc. Jour. **26:** 28–30, 1 fig.
- 1919a North American Sarcophagidae: Flies of the genus *Metoposarcophaga* Townsend. Canad. Ent. **51:** 154–158, 1 fig.
- 1919b North American Sarcophagidae: New species of the genus *Sarcofahrtia* R. Parker (Dip.). Ent. News **30:** 201–203, 6 figs.
- 1920a Another new species of *Sarcophaga* from Niagara Falls. N.Y. Ent. Soc. Jour. (1919) **27:** 265–267, 1 fig.
- 1920b North American Sarcophagidae: New species from British Columbia and Alaska. Brooklyn Ent. Soc. Bul. **15:** 105–110, 3 figs.
- 1921 North American Sarcophagidae: A new genus and several new species from the south-west United States. Brooklyn Ent. Soc. Bul. **16:** 112–115, 14 figs.
- 1923 New Sarcophagidae from Asia, with data relating to the *dux* group. Ann. and Mag. Nat. Hist. **ser. 9, 11:** 123–129, 4 figs.

Parrot, L. (See França and Parrot)

Parrott, P. J., and Schoene, W. J.
- 1912 The apple and cherry ermine moths. N.Y. (Geneva) Agr. Expt. Sta. Tech. Bul. **24:** 1–40, 10 figs., 9 pls.

Parry, W. E. (See Kirby, 1824)

Parsons, C. T.
- 1940 The Conopidae of the West Indies and Bermuda (Diptera). Psyche **47:** 27–37, 1 pl.
- 1948 A classification of North American Conopidae. Ent. Soc. Amer. Ann. **41:** 223–246, 1 fig.

Paterson, H. E. (See Zumpt and Paterson)

Patterson, J. T. (See also Stone, Griffen, and Patterson; Sturtevant and Novitski, 1941; Wheeler, 1949a, 1949b, 1952b, 1954a, 1957)
- 1919 Polyembryony and sex. Jour. Hered. **10:** 344–352, figs. 4–5.
- 1941 The *virilis* group of *Drosophila* in Texas. Amer. Nat. **75:** 523–539.

Patterson, J. T.—Continued
1942 I. Interspecific hybridization in the genus *Drosophila*. Pp. 7–15. *In his* Studies in the genetics of *Drosophila*. II. Gene variation and evolution. Tex. Univ. Pub. **4228:** 1–200, 31 figs., 8 pls.
1943 I. The Drosophilidae of the Southwest. Pp. 7–216, 66 figs., pls. 1–10. *In his* Studies in the genetics of *Drosophila*. III. The Drosophilidae of the Southwest. Tex. Univ. Pub. **4313:** 1–327, 66 figs., 25 pls.
1944 A new member of the *virilis* group. Pp. 102–103. *In his* Studies in the genetics of *Drosophila*. IV. Papers dealing with the taxonomy, nutrition, cytology and interspecific hybridization in *Drosophila*. Tex. Univ. Pub. **4445:** 1–223, 8 figs., 23 pls.
1952 Revision of the *montana* complex of the *virilis* species group. Pp. 20–34, 1 fig. *In his* Studies in the genetics of *Drosophila*. VII. Further articles on genetics, cytology and taxonomy. Tex. Univ. Pub. **5204:** 1–251, 64 figs.

———, **and Alexander, M. L.**
1952 *Drosophila wheeleri*, a new member of the *mulleri* subgroup. Pp. 129–136, 1 fig. *In* Patterson, J. T., Studies in the genetics of *Drosophila*. VII. Further articles on genetics, cytology and taxonomy. Tex. Univ. Pub. **5204:** 1–251, 64 figs.

———, **and Crow, J. F.**
1940 XII. Hybridization in the *mulleri* group of *Drosophila*. Pp. 251–256. *In* Patterson, J. T., Studies in the genetics of *Drosophila*. [I]. Tex. Univ. Pub. **4032:** 1–256, 18 figs., 4 pls.

———, **and Mainland, G. B.**
1944 The Drosophilidae of Mexico. Pp. 9–101, 16 pls. *In* Patterson, J. T., Studies in the genetics of *Drosophila*. IV. Papers dealing with the taxonomy, nutrition, cytology and interspecific hybridization in *Drosophila*. Tex. Univ. Pub. **4445:** 1–223, 8 figs., 23 pls.

———, **and Stone, W. S.**
1952 Evolution in the genus *Drosophila*. 610 pp., 68 figs. New York.

———, **Stone, W. S., and Griffen, A. B.**
1940 XI. Evolution of the *virilis* group in *Drosophila*. Pp. 218–250, 1 fig., 1 pl. *In* Patterson, J. T., Studies in the genetics of *Drosophila*. [I]. Tex. Univ. Pub. **4032:** 1–256, 18 figs., 4 pls.

Patterson, J. T., and Ward, C. L.
1952   *Drosophila euronotus*, a new member of the *melanica* species group. Pp. 158–161, 1 fig. *In* Patterson, J. T., Studies in the genetics of *Drosophila*. VII. Further articles on genetics, cytology and taxonomy. Tex. Univ. Pub. **5204**: 1–251, 64 figs.

———, **and Wheeler, M. R.**
1942   Description of new species of the subgenera *Hirtodrosophila* and *Drosophila*. Tex. Univ. Pub. **4213**: 67–109.

Patton, W. H.
1897   A principle to observe in naming galls: Two new gall-making Diptera. Canad. Ent. **29**: 247–248.

Patton, W. S.   (See also Cushing and Patton)
1936   Studies on the higher Diptera of medical and veterinary importance. The warble flies of the genus *Hypoderma*. Ann. Trop. Med. and Parasitol. **30**: 453–468, 12 figs.

Pechuman, L. L.
1938a   A synopsis of the New World species of *Vermileo* (Diptera-Rhagionidae). Brooklyn Ent. Soc. Bul. **33**: 84–89.
1938b   Two new nearctic *Silvius* (Diptera). Canad. Ent. **70**: 165–171, 1 fig.
1945   A new species of *Glutops* (Diptera, Coenomyiidae). Canad. Ent. **77**: 134–135.
1949   Some notes on Tabanidae (Diptera) and the description of two new *Chrysops*. Canad. Ent. **81**: 77–84.
1956   An unusual new *Tabanus* from Arizona (Diptera: Tabanidae). Pan-Pacific Ent. **32**: 39–42, 8 figs.
1960   Some new and little-known North American Tabanidae (Diptera). Canad. Ent. **92**: 793–799, 11 figs.
1962   A new nearctic *Tabanus* of the *fulvulus* group. Brooklyn Ent. Soc. Bul. **57**: 66–70.

Pennak, R. W.
1945   Notes on mountain midges (Deuterophlebiidae) with a description of the immature stages of a new species from Colorado. Amer. Mus. Nat. Hist., Amer. Mus. Novitates **1276**: 1–10, 3 figs.

Perez, T. De Stefani   (See De Stefani Perez, T.)

Pergande, T.
1901   The ant-decapitating fly. Ent. Soc. Wash. Proc. **4**: 497–501, figs. 20–21.

Perkins, R. C. L.
1905 Leafhoppers and their natural enemies. (Pt. IV. Pipunculidae). Hawaii. Sugar Planters' Assoc. Expt. Sta., Div. Ent. Bul. **1**: 123–157, pls. 5–7.

Perris, É.
1847 Lettre de M. Édouard Perris à M. M sur une excursion dans les Grandes Landes. Acad. Roy. des Sci., Belles-Let. et Arts de Lyon, Mém., Sect. des Sci. **2**: 433–506.

1849 Notes pour servir à l'histoire des métamorphoses de diverses espèces de Diptères (Première partie). 1. Notice sur une larve de *Mycetophila* qui se couvre de ses excréments. Soc. Ent. de France, Ann. ser. **2, 7**: 51–61, pl. 3 (I).

1852 Seconde excursion dans les Grandes-Landes. Soc. Linn. de Lyon, Ann. **1850/1852**: 145–216.

1870 Histoire des Insectes du pin maritime. Diptères. Soc. Ent. de France, Ann. ser. **4, 10**: 135–232, 321–366, pls. 1–5.

Perty, M.
[1830–1833] Insecta brasiliensia. Pp. 1–60, pls. 1–12, [*1830*]; pp. 61–124, pls. 13–24, [*1832*]; pp. 125–224, pls. 25–40, [*1833*]. *In his* Delectus animalium articulatorum quae in itinere per Brasiliam annis MDCCCXVII–MDCCCXX jussu et auspiciis Maximiliani Josephi I. Bavariae regis augustissimi peracto collegerunt Dr. J. B. de Sphix et Dr. C. F. Ph. de Martius. 44 + 224 pp., 40 pls. Monachii [=Munich], "1830–1834".

   For dating, see Horn and Schenkling, 1928: 932.

Petch, C. E., and Maltais, J. B.
1932 A preliminary list of the insects of the Province of Quebec. Part II, Diptera, by A. F. Winn and G. Beaulieu. Revised and supplemented. Quebec Soc. Protect. Plants, Ann. Rpt. **23/24 (Sup. to 24)**: 1–100.

Peterson, A.
1916 The head-capsule and mouth-parts of Diptera. Ill. Biol. Monog. **3 (2)**: 3–112, 25 pls.

1951 Larvae of insects. An introduction to nearctic species. Vol. 2, 416 pp., 104 figs. Columbus, Ohio.

   Diptera appear on pp. 219–349, figs. D1–D38.
   The entire work consists of 2 volumes, Columbus, 1948–1951.

Peterson, B. V. (See also Davies, Peterson, and Wood; DeFoliart and Peterson; Stone and Peterson)

Peterson, B. V.—Continued
1960a New distribution and host records for bat flies, and a key to the North American species of *Basilia* Ribeiro (Diptera: Nycteribiidae). Ent. Soc. Ontario, Proc. [=Ann. Rpt.] (1959) **90**: 30–37, 16 figs.
1960b The Simuliidae (Diptera) of Utah, Part I. Keys, original citations, types and distribution. Great Basin Nat. **20**: 81–104.
1962 *Cnephia abdita*, a new black fly (Diptera: Simuliidae) from eastern North America. Canad. Ent. **94**: 96–102, 23 figs.

———, and **DeFoliart, G. R.**
1960 Four new species of *Prosimulium* from western United States. Canad. Ent. **92**: 85–102, 34 figs.

Pettey, F. W.
1918a A revision of the genus *Sciara* of the family Mycetophilidae (Diptera). Ent. Soc. Amer. Ann. **11**: 319–343, 2 pls.
1918b A new species of *Sciara* from red clover crowns. Jour. Econ. Ent. **11**: 420, 1 fig., 1 pl.

Peus, F.
1934 Dixiden und Culiciden aus Lettland. Notulae Ent. **14**: 69–78, 3 figs.

Philip, C. B.
1931 The Tabanidae (horseflies) of Minnesota. With special reference to their biologies and taxonomy. Minn. Agr. Expt. Sta. Tech. Bul. **80**: 1–132, 3 figs., 4 pls.
1936a *Tabanus rhombicus* and related western horseflies. Canad. Ent. **68**: 148–160.
1936b An interesting new horsefly from North Carolina (Diptera: Tabanidae). Ent. News **47**: 229–231, 1 fig.
1936c A new horsefly from the southeastern United States. Kans. Ent. Soc. Jour. **9**: 100–101.
1936d New Tabanidae (horseflies) with notes on certain species of the *longus* group of *Tabanus*. Ohio Jour. Sci. **36**: 149–156.
1936e The *furcatus* group of western North American flies of the genus *Chrysops* (Diptera: Tabanidae). Ent. Soc. Wash. Proc. (1935) **37**: 153–161.
1937a Notes on certain males of North American horseflies (Tabanidae). II. The *affinis* or "red-sided" group of *Tabanus* sens. lat., with a key to the females. Canad. Ent. **69**: 35–40, 49–58.

Philip, C. B.—Continued
- 1937b New horseflies (Tabanidae, Diptera) from the southwestern United States. Pan-Pacific Ent. **13:** 64–67.
- 1941a Notes on three western genera of flies (Diptera, Tabanidae). Brooklyn Ent. Soc. Bul. **36:** 185–199.
- 1941b Comments on the supra-specific categories of nearctic Tabanidae (Diptera). Canad. Ent. **73:** 2–14.
- 1941c Notes on nearctic Tabaninae (Diptera). Part I. *Stenotabanus, Atylotus,* and *Tabanus* s. str.; Part II. *Tabanus* s. lat. and *Hybomitra.* Canad. Ent. **73:** 105–110; 142–153.
- 1941d Notes on nearctic Pangoniinae (Diptera, Tabanidae). Ent. Soc. Wash. Proc. **43:** 113–130, 1 pl.
- 1942a Further notes on nearctic Tabanidae (Diptera). New England Zool. Club, Proc. **21:** 55–68.
- 1942b Notes on nearctic Tabaninae. Part III. The *Tabanus lineola* complex. Psyche **49:** 25–40.
- 1947 A catalog of the blood-sucking fly family Tabanidae (horseflies and deerflies) of the Nearctic Region north of Mexico. Amer. Midland Nat. **37:** 257–324.
- 1950a New North American Tabanidae (Diptera). Part I. Pangoniinae. Ent. Soc. Amer. Ann. (1949) **42:** 451–460, 2 figs.
- 1950b New North American Tabanidae (Diptera). Part II. Tabanidae [sic] [=Tabaninae]; III. Notes on *Tabanus molestus* and related horseflies with a prominent single row of triangles on the abdomen. Ent. Soc. Amer. Ann. **43:** 115–122, 2 figs.; 240–248.
- 1950c Corrections and addenda to a catalog of nearctic Tabanidae. Amer. Midland Nat. **43:** 430–437.
- 1953 The genus *Chrysozona* Meigen in North America (Diptera, Tabanidae). Ent. Soc. Wash. Proc. **55:** 247–251.
- 1954a New North American Tabanidae (Diptera). VI. Descriptions of Tabaninae and new distributional data. Ent. Soc. Amer. Ann. **47:** 25–33.
- 1954b New North American Tabanidae. VIII. Notes on and keys to the genera and species of Pangoniinae exclusive of *Chrysops.* Rev. Brasil. de Ent. **2:** 13–60, 10 figs.
- 1955 New North American Tabanidae. IX. Notes on and keys to the genus *Chrysops* Meigen. Rev. Brasil. de Ent. **3:** 47–128, 7 figs.
- 1957 New records of Tabanidae (Diptera) in the Antilles. Amer. Mus. Nat. Hist., Amer. Mus. Novitates **1858:** 1–16, 4 figs.

Philip, C. B.—Continued
1959   New North American Tabanidae. X. Notes on synonymy, and description of a new species of *Chrysops*. Amer. Ent. Soc. Trans. **85**: 193–217.
1960a New North American Tabanidae. XI. Supplemental notes pertinent to a catalog of nearctic species. Ent. Soc. Amer. Ann. **53**: 364–369.
1960b New North American Tabanidae. XII. A new variety of *Tabanus imitans* Walker. Fla. Ent. **43**: 171–174, 1 fig.
1961a New North American Tabanidae. XIII. Change of name for a well-known species of *Chrysops*. Ent. News **72**: 160–162.
1961b Notes on palaeartic [sic] *Nemorius* (Diptera: Tabanidae) with description of one new species. Soc. Roy. d'Ent. de Belg. Bul. et Ann. **97**: 225–236, 9 figs.
1962   New North American Tabanidae. XVI. A new species from the South Texas Gulf Coast (Diptera). Ent. Soc. Wash. Proc. **64**: 171–174, 3 figs.

————, **and Fairchild, G. B.**
1956   American biting flies of the genera *Chlorotabanus* Lutz and *Cryptotylus* Lutz (Diptera, Tabanidae). Ent. Soc. Amer. Ann. **49**: 313–324, 1 fig., 1 pl.

————, **and Jones, C. M.**
1962   New North American Tabanidae. XV. Additions to records of *Chrysops* in Florida. Fla. Ent. **45**: 67–69, 1 fig.

————, **and Steffan, W. A.**
1962   New North American Tabanidae. XIV. An undescribed *Apatolestes* from the California coast (Diptera). Pan-Pacific Ent. **38**: 41–43, 1 fig.

Philippi, R. A.
1865   Aufzählung der chilenischen Dipteren. K.-k. Zool.-Bot. Gesell. Wien, Verhandl. **15** (**Abhandl.**): 595–782, 7 pls.

Phillips, E. F.
1925   The bee-louse, *Braula coeca*, in the United States. U.S. Dept. Agr., Dept. Cir. **334**: 1–11.

Phillips, V. T.
1923   A revision of the Trypetidae of northeastern America. N.Y. Ent. Soc. Jour. **31**: 119–155, 2 pls.
1946   The biology and identification of trypetid larvae (Diptera: Trypetidae). Amer. Ent. Soc. Mem. **12**: 1–161, 3 figs., 16 pls.

Pickett, A. D.
1937   Studies on the genus *Rhagoletis* (Trypetidae) with special reference to *Rhagoletis pomonella* (Walsh). Canad. Jour. Res., Sect. D, Zool. Sci. **15**: 53–75, 25 figs.

Pidgeon, E.   (See Griffith and Pidgeon)

Pipkin, S. B.
1961   Taxonomic relationships within the *Drosophila victoria* species group, subgenus *Pholadoris* (Diptera: Drosophilidae). Ent. Soc. Wash. Proc. **63**: 145–161, 2 figs.

Plath, O. E.   (See Aldrich, 1919c; Townsend, 1919a)

Pleske, T.
1922   Revue critique des genres, espèces, et sousespèces paléarctiques des sousfamilles des Stratiomyiinae et des Pachygastrinae (Diptères). [Akad. Nauk S.S.S.R.] Ross. Akad. Nauk Petrograd, Zool. Muz. Ezheg. (Acad. des Sci. de Russie, Zool. Mus. Ann.) **23**: 325–338.

Poey, F.
1851   Memorias sobre la historia natural de la Isla de Cuba, acompañadas de sumarios latinos y extractos en frances. Vol. 1, 463 pp., 34 pls. Habana.

> The entire work consists of 2 volumes, Habana, 1851–[1861], the second volume being dated "1856–1858" (but information within the volume shows that at least portions of it were published through 1861).

Pokorny, E.
1887a   (III.) Beitrag zur Dipterenfauna Tirols. K.-k. Zool.-Bot. Gesell. Wien, Verhandl. **37 (Abhandl.)**: 381–420, pl. 7.
1887b   Neue Tipuliden aus den österreichischen Hochalpen. Wien. Ent. Ztg. **6**: 50–60, pl. 1.
1889   (IV.) Beitrag zur Dipterenfauna Tirols. K.-k. Zool.-Bot. Gesell. Wien, Verhandl. **39 (Abhandl.)**: 543–574.
1893a   Eine alte und einige neue Gattungen der Anthomyiden. Wien. Ent. Ztg. **12**: 53–64.
1893b   Bemerkungen und Zusätze zu Prof. G. Strobl's "Die Anthomyinen Steiermarks". K.-k. Zool.-Bot. Gesell. Wien, Verhandl. **43 (Abhandl.)**: 526–544.

Porter, H. L., Spilker, O. W., and Walker, J. T.
1959   The control of insects and plant diseases in the nursery. Ed. 6, 158 pp., 53 figs. Ohio Dept. Agr., Reynoldsburg, Ohio.

Porter, J. B.   (See Alexander, 1942c)

Pratt, A.   (See Wood et al.)

Pratt, F. C.   (See Hunter, Pratt, and Mitchell)

Pratt, H. D. (See Foote and Pratt)

Prescott, H. W.
1960 Suppression of grasshoppers by nemestrinid parasites (Diptera). Ent. Soc. Amer. Ann. **53:** 513–521, 6 figs.

Priddy, R. B.
1954 Three new species of nearctic *Conophorus* (Diptera, Bombyliidae). Kans. Ent. Soc. Jour. **27:** 53–56.
1958 The genus *Conophorus* in North America (Diptera, Bombyliidae). Kans. Ent. Soc. Jour. **31:** 1–33, 6 figs.

Pritchard, A. E.
1935 New Asilidae from the southwestern United States (Diptera). Amer. Mus. Nat. Hist., Amer. Mus. Novitates **813:** 1–13.
1938a Synopsis of North and Central American *Holcocephala* with a description of a new species (Diptera: Asilidae). N.Y. Ent. Soc. Jour. **46:** 11–21.
1938b Revision of the robberfly genus *Taracticus* Loew with descriptions of three new species (Diptera: Asilidae). N.Y. Ent. Soc. Jour. **46:** 179–190.
1938c The genus *Hodophylax* James, with a description of *basingeri*, new species (Diptera, Asilidae). Pan-Pacific Ent. **14:** 129–131, 1 fig.
1941 The genus *Haplopogon* in the New World, with the description of *erinus* n. sp. (Diptera: Asilidae). Ent. Soc. Amer. Ann. **34:** 350–354, 1 pl.
1942 A revision of the genus *Cerotainiops* Curran (Diptera, Asilidae). Kans. Ent. Soc. Jour. **15:** 19–24.
1943 Revision of the genus *Cophura* Osten Sacken (Diptera: Asilidae). Ent. Soc. Amer. Ann. **36:** 281–309, 1 pl.
1947 The North American gall midges of the tribe Micromyini, Itonididae (Cecidomyiidae) Diptera. Ent. Amer. **27:** 1–44, 45–87, 2 pls.
1948a The North American gall midges of the tribe Catotrichini and Catochini (Diptera: Itonididae (Cecidomyiidae)). Ent. Soc. Amer. Ann. (1947) **40:** 662–671, 1 fig.
1948b *Clinodiplosis pucciniae*, a new gall midge feeding on a rust (Diptera: Itonididae). Pan-Pacific Ent. **24:** 29–30, 1 fig.
1951 The North American gall midges of the tribe Lestremiini, Itonididae (Cecidomyiidae) Diptera. Calif. Univ., Pubs., Ent. **8:** 239–275, 3 pls.
1952 A new gall midge pest of toyon berries (Diptera: Itonididae). Pan-Pacific Ent. **28:** 16.

Pritchard, A. E.—Continued
- 1953a The gall midges of California. Diptera: Itonididae (Cecidomyiidae). Calif. Insect Survey, Bul. **2**: 125–150, 1 pl.
- 1953b The white clover flower midge as differentiated from the red clover flower midge (Diptera: Itonididae). Pan-Pacific Ent. **29**: 128–132, 2 figs.
- 1958 I. Subfamily Lestremiinae. Pp. 50–87, pls. 10–15. *In* Pritchard, A. E., and Felt, E. P., Guide to the insects of Connecticut. Part VI. The Diptera or true flies of Connecticut. Sixth fascicle [part]. Itonididae (Cecidomyiidae). Conn. State. Geol. and Nat. Hist. Survey, Bul. **87**: 47–206, pls. 10–15.
- 1960a A new classification of the paedogenic gall midges formerly assigned to the subfamily Heteropezinae (Diptera: Cecidomyiidae). Ent. Soc. Amer. Ann. **53**: 305–316, 8 figs.
- 1960b Two new species of catochine gall midges, with a key to genera of the Catochini (Diptera: Cecidomyiidae). Pan-Pacific Ent. **36**: 195–197, 2 figs.
- 1961a *Lasioptera allioides*, a new gall midge on grass (Diptera: Cecidomyiidae). Ent. Soc. Wash. Proc. **63**: 55–57, 2 figs.
- 1961b *Phaenobremia doutti*, a new gall midge predator of aphids in California (Diptera: Cecidomyiidae). Ent. Soc. Wash. Proc. **63**: 100–101, 2 figs.

———, **and Felt, E. P.** (See Felt, 1958; Pritchard, 1958)

Pschorn-Walcher, H. (See Delucchi and Pschorn-Walcher)

Quate, L. W.
- 1954 A revision of the Psychodidae of the Hawaiian Islands (Diptera). Hawaii. Ent. Soc. Proc. (1953) **15**: 335–356, 45 figs.
- 1955 A revision of the Psychodidae (Diptera) in America north of Mexico. Calif. Univ., Pubs., Ent. **10**: 103–273, 105 figs.
- 1959 Classification of the Psychodini (Psychodidae: Diptera). Ent. Soc. Amer. Ann. **52**: 444–451, 5 figs.
- 1960a New species and records of nearctic Psychodidae (Diptera). Pan-Pacific Ent. **36**: 143–149, 2 figs.
- 1960b Guide to the insects of Connecticut. Part VI. The Diptera or true flies of Connecticut. Seventh fascicle. Psychodidae. Conn. State Geol. and Nat. Hist. Survey, Bul. **92**: 1–48, 7 pls.

———, **and Wirth, W. W.**
- 1951 A taxonomic revision of the genus *Maruina* (Diptera: Psychodidae). Wasmann Jour. Biol. **9**: 151–166, 2 pls.

Quisenberry, B. F.
1949a   A new genus of Tephritidae, near *Xanthomyia* (Diptera). Brooklyn Ent. Soc. Bul. **44:** 49–52, 1 fig.
1949b   Notes and descriptions of North American Tephritidae (Diptera). Kans. Ent. Soc. Jour. **22:** 81–88, 6 figs.
1949c   The genus *Oxyna* in the Nearctic Region north of Mexico (Diptera: Tephritidae). Pan-Pacific Ent. **25:** 71–76.
1950   The genus *Euaresta* in the United States (Diptera: Tephritidae). N.Y. Ent. Soc. Jour. **58:** 9–38, 3 figs.
1951   A study of the genus *Tephritis* Latreille in the Nearctic Region north of Mexico (Diptera: Tephritidae). Kans. Ent. Soc. Jour. **24:** 56–72, 1 fig., 1 pl.

Rafinesque, C. S.
1815   Analyse de la Nature ou tableau de l'Univers et des corps organisés. 224 pp. Palerme.

Randolph, N. M., and O'Neill, K.   (See T. S. H. D., 1944)

Rapp, W. F., Jr.   (See also Cooper and Rapp)
1943a   A new Dorilaidae (Diptera). Ent. News **54:** 118.
1943b   Some new North American Pipunculidae (Diptera). Ent. News **54:** 222–224.
1944   A new species of *Psychoda* from New York (Psychodidae, Diptera). Ent. News **55:** 232–233, 1 fig.
1945   *Pseudolutzomyia*, new name for *Lutzomyia* Curran, 1934 (Diptera). Ent. Soc. Wash. Proc. **47:** 278.
1946a   Two new Nemocera Diptera (Sciaridae and Cecidomyidae). Amer. Ent. Soc. Trans. (1945) **71:** 125–128, 2 figs.
1946b   The generic name *Pandora*. Ann. and Mag. Nat. Hist. (1945) **ser. 11, 12:** 499–500.

Rathvon, S. S.   (See Reigert, 1855)

Rau, P.
1940   Some mud-daubing wasps of Mexico and their parasites. Ent. Soc. Amer. Ann. **33:** 590–595.

Rebollar, R.   (See Arriaga, Urbina, and Rebollar)

(Redtenbacher, L., and Schiner, I. R.)
1847–   Fauna Austriaca. Nach der analytischen Methode be-
1864   arbeitet. 3 vols. Wien.

> This "series" consists, as far as known, only of 2 works by the 2 authors shown above, under the same title and format without any other association, such as overall editorship or series volume designation. The volumes (Redtenbacher's volume on Coleoptera, 1847–1849, of which there are 2 later editions, and Schiner's 2 volumes on Diptera, 1860–1864) were issued in continuously paged hefts. The Diptera volumes were issued in 14 hefts, Vol. 1, 1860–1862; Vol. 2, 1862–1864. The date on the title pages of all 3 volumes is that of the final heft of the volume.
> See Schiner, 1860a, 1861a, 1862a.

Rees, B. E., and Ferris, G. F. (See Alexander, 1939b)
Reeves, W. C. (See Brookman and Reeves)
Reiche, L.
 1857–   (Description sommaire de cinq espèces nouvelles d'Insectes,
 1864   provenant de l'expédition aux mers arctiques, effectuée en
         1856, sous la direction de S. A. I. le prince Napoléon.)
         Soc. Ent. de France, Ann. ser. 3, 5 (Bul.): viii–x.
Reigert, J. F.
 1855   A treatise on the cause of cholera, an interesting discovery.
         15 pp. Lancaster, Pa.
Reinhard, E. G.
 1938   The egg-laying and early stages of the robber fly, *Erax aestuans* (Diptera: Asilidae). Ent. News **49**: 281–283, 1 fig.
Reinhard, H. J.
 [1922] Some new species of Texas Tachinidae (Diptera). Ent. Soc. Amer. Ann. (1921) **14**: 329–336, pl. 28.
> The stated date of issue in the volume is Dec. 31, 1921, but dates of receipt for the Smithsonian Library (Jan. 28, 1922) and U.S. Department of Agriculture Library (Jan. 27, 1922) indicate publication in Jan., 1922. This is also dated 1922 in J. M. Aldrich's card catalog at the U.S. National Museum.

 1923   New Tachinidae from Texas (Diptera). Ent. News **34**: 266–269.
 1924a   A new southern tachinid fly (Diptera). Ent. News **35**: 54–56.
 1924b   New muscoid Diptera. Ent. News **35**: 269–274.
 1924c   A new species of *Gonia* from Texas (Diptera). Ent. News **35**: 357–358.
 1929   Notes on the muscoid flies of the genera *Opelousia* and *Opsodexia* with the description of three new species. U.S. Natl. Mus. Proc. **76 (20)**: 1–9.
 1930a   On the genus *Viviania* with the description of two new species from Texas (Tachinidae, Diptera). Brooklyn Ent. Soc. Bul. **25**: 102–107.
 1930b   Two new North American species of muscoid flies (Tachinidae, Diptera). Brooklyn Ent. Soc. Bul. **25**: 199–202.
 1930c   A synopsis of the genus *Macromeigenia* including the description of one new species (Diptera: Tachinidae). Ent. News **41**: 261–264.
 1931a   A new species of two-winged fly belonging to the genus *Acronarista* (Diptera: Tachinidae). Ent. News **42**: 26–27.

Reinhard, H. J.—Continued
- 1931b The two-winged flies belonging to *Siphosturmia* and allied genera, with descriptions of two new species. U.S. Natl. Mus. Proc. **79 (11)**: 1–11.
- 1931c Revision of the American parasitic flies belonging to the genus *Winthemia*. U.S. Natl. Mus. Proc. **79 (20)**: 1–54, 1 pl.
- 1934a North American parasitic flies of the genus *Spathidexia* with descriptions of two new species. Brooklyn Ent. Soc. Bul. **29**: 150–154.
- 1934b New North American Tachinidae. Brooklyn Ent. Soc. Bul. **29**: 186–195.
- 1934c Two new species of the tachinid genus *Siphosturmiopsis* with key and notes (Diptera). Ent. News **45**: 15–19.
- 1934d American muscoid flies of the genera *Ceratomyiella* and *Paradidyma*. U.S. Natl. Mus. Proc. **83**: 9–43.
- 1934e Revision of the American two-winged flies belonging to the genus *Cuphocera*. U.S. Natl. Mus. Proc. **83**: 45–70.
- 1935a Notes on the tachinid genus *Pseudotachinomyia* with descriptions of two new species (Diptera). Ent. News **46**: 132–135.
- 1935b New genera and species of American muscoid flies (Tachinidae: Diptera). Ent. Soc. Amer. Ann. **28**: 160–173.
- 1935c North American two-winged flies of the genus *Doryphorophaga* (Tachinidae, Diptera). N.Y. Ent. Soc. Jour. **43**: 387–394.
- 1937   New North American muscoid Diptera. Brooklyn Ent. Soc. Bul. **32**: 62–74.
- 1938   Four new nearctic species of *Fabriciella* (Tachinidae, Diptera). Canad. Ent. **70**: 8–11.
- 1939a New genera and species of muscoid Diptera. Brooklyn Ent. Soc. Bul. **34**: 61–74.
- 1939b A review of the muscoid genus *Eumacronychia* with key and descriptions of new species (Diptera). N.Y. Ent. Soc. Jour. **47**: 57–68.
- 1941   A new nearctic species of *Exopalpus* (Tachinidae, Diptera). Kans. Ent. Soc. Jour. **14**: 58–60.
- 1942a Notes on *Fabriciella* with descriptions of five new species (Tachinidae, Diptera). Brooklyn Ent. Soc. Bul. **37**: 24–30.
- 1942b New North American Tachinidae belonging to the genera *Microchaetina* and *Hypenomyia* with key to the known species (Diptera). Canad. Ent. **74**: 88–91.

Reinhard, H. J.—Continued
- 1942c A new species of *Oedematocera* with notes and key (Tachinidae: Diptera).   Ent. News **53**: 106–108.
- 1942d A new parasitic muscoid fly from Texas.   Ent. Soc. Wash. Proc. **44**: 17–18.
- 1943a New genera of North American muscoid Diptera.   Canad. Ent. **75**: 163–169.
- 1943b New Tachinidae from northeastern United States (Diptera). Brooklyn Ent. Soc. Bul. **38**: 78–90.
- 1943c New North American Muscoidea (Tachinidae, Diptera). Kans. Ent. Soc. Jour. **16**: 14–23.
- 1944a New North American Tachinidae belonging to the genus *Muscopteryx* (Diptera).   Ent. Soc. Amer. Ann. **37**: 352–358.
- 1944b New muscoid Diptera from the United States.   Kans. Ent. Soc. Jour. **17**: 57–72.
- 1944c Change of name in Diptera (Tachinidae).   Kans. Ent. Soc. Jour. **17**: 159.
- 1945a New genera and species of North American Tachinidae (Diptera).   Canad. Ent. **77**: 28–36.
- 1945b New genera and species of muscoid flies.   Kans. Ent. Soc. Jour. **18**: 67–77.
- 1945c A new muscoid parasite reared from beetles in California (Diptera).   Pan-Pacific Ent. **21**: 11–13.
- 1946a The tachinid genera *Pseudochaeta* and *Phaenopsis* in North America (Diptera).   Canad. Ent. **78**: 111–121.
- 1946b The genus *Minthozelia* in the United States (Diptera, Tachinidae).   Kans. Ent. Soc. Jour. **19**: 52–59.
- 1946c A review of the tachinid genera *Siphophyto* and *Coronimyia* (Diptera).   Ent. Soc. Wash. Proc. **48**: 79–92.
- 1947a New genera and species of muscoid Diptera.   Kans. Ent. Soc. Jour. **20**: 15–24.
- 1947b New North American muscoid Diptera.   Kans. Ent. Soc. Jour. **20**: 95–116, 117–126, 3 pls.
- 1951 New American muscoid Diptera.   Brooklyn Ent. Soc. Bul. **46**: 1–9.
- 1952a New North American muscoid Diptera.   Canad. Ent. **84**: 140–147.
- 1952b New genera and species of muscoid Diptera.   Brooklyn Ent. Soc. Bul. **47**: 1–12.
- 1952c Muscoid flies of the genus *Chaetophlepsis* (Diptera).   Kans. Ent. Soc. Jour. **25**: 13–21.
- 1953a New muscoid Diptera from the western United States. Pan-Pacific Ent. **29**: 49–59.

Reinhard, H. J.—Continued
- 1953b Notes on muscoid synonymy with descriptions of three new species (Diptera). Ent. Soc. Wash. Proc. **55:** 243–247.
- 1953c New Mexican Tachinidae (Diptera). Kans. Ent. Soc. Jour. **26:** 95–102.
- 1953d New species of Tachinidae from Mexico (Diptera). Brooklyn Ent. Soc. Bul. **48:** 89–96.
- 1954 Parasitic flies of the genus *Prosenoides* (Tachinidae, Diptera). Canad. Ent. **86:** 408–413.
- 1955a New nearctic Sarcophagidae and Tachinidae (Diptera). Brooklyn Ent. Soc. Bul. **50:** 128–133.
- 1955b North American Muscoidea (Diptera: Tachinidae). Ent. News **66:** 233–238.
- 1955c North American tachinid flies of the genus *Wagneria* (Diptera). Kans. Ent. Soc. Jour. **28:** 49–59.
- 1955d New genera and species of North American Tachinidae (Diptera). Kans. Ent. Soc. Jour. **28:** 123–130.
- 1956a New Tachinidae (Diptera). Ent. News **67:** 121–129.
- 1956b A synopsis of the tachinid genus *Leucostoma* (Diptera). Kans. Ent. Soc. Jour. **29:** 155–168.
- 1956c New muscoid Diptera mainly from California. Pan-Pacific Ent. **32:** 103–110.
- 1957 New American muscoid Diptera (Sarcophagidae, Tachinidae). Ent. News **68:** 99–111.
- 1958a Parasitic flies of the genus *Mochlosoma* (Tachinidae, Diptera). Canad. Ent. **90:** 98–110.
- 1958b New American Tachinidae (Diptera). Ent. News **69:** 233–242.
- 1958c New genera and species of North American Tachinidae (Diptera). Kans. Ent. Soc. Jour. **31:** 225–232.
- 1958d North American Tachinidae (Diptera). Kans. Ent. Soc. Jour. **31:** 277–284.
- 1958e Notes on *Spathimeigenia* with descriptions of four new species (Diptera, Tachinidae). Ent. Soc. Wash. Proc. **60:** 207–212.
- 1959a New nearctic Tachinidae (Diptera). Pan-Pacific Ent. **35:** 157–163.
- 1959b New North American Tachinidae (Diptera). Ent. News **70:** 225–234.
- 1960 Change of generic name in Tachinidae (Diptera). Ent. News **71:** 103.
- 1961 New American Tachinidae and Sarcophagidae (Diptera). Kans. Ent. Soc. Jour. **34:** 204–213.

Reinhard, H. J.—Continued
- 1962a North American muscoid Diptera. Ent. News **73:** 169–178.
- 1962b New North American Tachinidae (Diptera). Pan-Pacific Ent. **38:** 215–224.
- 1963a New American Sarcophagidae (Diptera). Ent. News **74:** 75–83.
- 1963b Correction (*Golcondamyia*, Tachinidae). Ent. News **74:** 152.

Remmert, H.
- 1959 Untersuchungen an zwei nahe verwandten *Coelopa*-Formen (Diptera, Coelopidae). Ztschr. f. Wiss. Zool. (Abt. A) **162:** 128–143, 10 figs.

Rempel, J. G.
- 1937a Notes on the genus *Chasmatonotus* with descriptions of three new species (Diptera, Chironomidae). Canad. Ent. **69:** 250–255, 1 pl.
- 1937b A new species of the subgenus *Kribioxenus* (Diptera, Chironomidae). Canad. Ent. **69:** 274–275.
- 1939 Neue Chironomiden aus Nordostbrasilien. Zool. Anz. **127:** 209–216, 10 figs.

Remy, P.
- 1928 Arthropodes terrestres récoltés au Groenland au cours de la croisière du "Pourquoi-Pas?" en 1926 (l^{re} liste). Soc. Linn. de Lyon, Bul. Bi-Mens. **7:** 51–53.

Reuter, E.
- 1895 Zwei neue Cecidomyinen. Soc. pro Fauna et Flora Fenn. Acta **11** (8): 1–15, 2 pls.

Ribeiro, A. de M.
- 1903 *Basilia ferruginea* genero novo e especie nova da familia das Nycteribias. [Rio de Janeiro] Mus. Nac. do Rio de Janeiro, Arch. **12:** 175–179, pl. 1.

Richards, O. W.
- 1930 The British species of Sphaeroceridae (Borboridae, Diptera). Zool. Soc. London, Proc. Gen. Mtg. **1930:** 261–345, 23 figs., 1 pl.
- 1944 *Leptocera downesi* sp. n. (Diptera, Sphaeroceridae) breeding in sprouting wheat. Roy. Ent. Soc., London, Proc. Ser. B: Taxonomy **13:** 137–139, 7 figs.
- 1951 Brachypterous Sphaeroceridae. Pp. 829–851, 10 figs. (=no. 8). *In* British Museum (Natural History), Ruwenzori Expedition 1934–5 [q.v.]. Vol. 2, 1054 pp., 490 figs., 10 pls. London.

Richards, O. W.—Continued
1960   On two N. American species of *Leptocera* Oliv., subgenus *Coproica* Rdi., with a review of the subgenus (Dipt., Sphaeroceridae). Ann. and Mag. Nat. Hist. (1959) **ser. 13, 2:** 199–208, pls. 3–4.
1961   Notes on the names of some Diptera Sphaeroceridae. Ann. and Mag. Nat. Hist. (1960) **ser. 13, 3:** 561–564.

Richardson, J.
1829–   Fauna Boreali-Americana, or the zoology of the northern
1837   parts of British America. 4 pts. Norwich, London.
> Sometimes bound in 3 volumes.
> See Kirby, 1837.

Riley, C. V.   (See also Comstock, 1882; McBride, 1870; Walsh and Riley; Williston, 1893d)
1867   Queries answered. Insects affecting apple tree roots. Prairie Farmer **n. ser., 19** [=whole ser., 35]: 397.
1869   First annual report on the noxious, beneficial and other insects of the State of Missouri. Mo., State Bd. Agr. Ann. Rpt. (1868) **4 (spec. pagination):** 1–181, 98 figs.
> Also appeared separately, with the same pagination, Jefferson City, Mo., *1869*.
> There were 9 annual reports by Riley, from 1869–1877, each bound in the Mo. State Bd. Agr. Ann. Rpt., and as a separate.

1870a   The *Cecropia* moth. Amer. Ent. **2:** 97–102, figs. 59–67.
1870b   Cypress-gall. Amer. Ent. and Bot. **2:** 244, fig. 153.
1870c   Second annual report on the noxious, beneficial and other insects of the State of Missouri. Mo., State Bd. Agr. Ann. Rpt. (1869) **5 (spec. pagination):** 1–135, 99 figs.
> Also appeared separately, with the same pagination, Jefferson City, Mo., *1870*.

1871   Third annual report on the noxious, beneficial and other insects of the State of Missouri. Mo., State Bd. Agr. Ann. Rpt. (1870) **6 (spec. pagination):** 1–175, 73 figs.
> Also appeared separately, with the same pagination, Jefferson City, Mo., *1871*.

1872   Fourth annual report on the noxious, beneficial and other insects of the State of Missouri. Mo., State Bd. Agr. Ann. Rpt. (1871) **7 (spec. pagination):** 1–145, 66 figs.
> Also appeared separately, with the same pagination, Jefferson City, Mo., *1872*.

1873   Fifth annual report on the noxious, beneficial and other insects of the State of Missouri. Mo., State Bd. Agr. Ann. Rpt. (1872) **8 (spec. pagination):** 1–160, 75 figs.
> Also appeared separately, with the same pagination, Jefferson City, Mo., *1873*.

Riley, C. V.—Continued
- 1874 Descriptions and natural history of two new species which brave the dangers of *Sarracenia variolaris*. Acad. Sci. St. Louis, Trans. (1868–1877) **3**: 235–240, 2 figs.
- 1875 Seventh annual report on the noxious, beneficial and other insects of the State of Missouri. Mo. State Bd. Agr. Ann. Rpt. (1874) **10 (spec. pagination)**: 1–196, 40 figs.
  > Also appeared separately, with the same pagination, Jefferson City, Mo., *1875*.
- 1877 Ninth annual report on the noxious, beneficial and other insects of the State of Missouri. Mo. State Bd. Agr. Ann. Rpt. (1876) **12 (spec. pagination)**: 1–129, 33 figs.
  > Also appeared separately, with the same pagination, Jefferson City, Mo., *1877*.
- 1878 Chapter XI. Invertebrate enemies. Pp. 284–334, figs. 21–66. *In* U.S. Entomological Commission, Reports of the U.S. Entomological Commission [q.v.]. Vol. 1: First annual report for 1877, relating to the Rocky Mountain locust, 477 pp., 111 figs., 5 pls. + App. of 294 pp. Washington.
- 1879 Parasites of the cotton worm. Canad. Ent. **11**: 161–162.
- 1883 Dipterous enemies of *Phylloxera vastatrix*. Canad. Ent. **15**: 39.
- [1885] Report of the entomologist. [U.S. Dept. Agr.] Comnr. Agr. Rpt. **1884**: 285–418, 10 pls.
  > Dated by Henshaw, 1889: 326.
- [1886a] Report of the entomologist. [U.S. Dept. Agr.] Comnr. Agr. Rpt. **1885**: 207–343, 9 pls.
  > Dated by Henshaw, 1889: 353.
- [1886b] Fourth report of the United States Entomological Commission, being a revised edition of Bulletin No. 3, and the final report on the cottonworm, together with a chapter on the bollworm. Vol. 4, 399 pp., 45 figs. *In* U.S. Entomological Commission, Reports of the U.S. Entomological Commission [q.v.]. Washington, "1885".
  > Dated by Henshaw, 1889: 336.
- 1887 Report of the entomologist. [U.S. Dept. Agr.] Comnr. Agr. Rpt. **1886**: 459–592, 11 pls.
- 1889 Notes on the cochineal insect. Insect Life (U.S. Dept. Agr., Div Ent., Period. Bul.) **1**: 258–259.

———, **and Howard, L. O.**
- 1889 *Hermetia mucens* infesting bee hives. Insect Life (U.S. Dept. Agr., Div. Ent., Period. Bul.) **1**: 353–354.

Riley, N. D. (See Stoll et al.)

Ringdahl, O.

1916 Einige neue Anthomyiden aus Schweden. Ent. Tidskr. **37**: 233–239.

1918 Neue nordische Anthomyiden. Ent. Tidskr. **39**: 148–194.

1920 Neue skandinavische Dipteren. Ent. Tidskr. **41**: 24–40.

1922 Två nya anthomyidsläkten. Ent. Tidskr. **43**: 1–4.

1924 Översikt av de hittill i vårt land funna arterna tillhörande släktena *Mydaea* R. D. och *Helina* R. D. (Muscidae). Ent. Tidskr. **45**: 39–66.

1926 Neue nordische Musciden nebst Berichtigung und Namensänderungen. Ent. Tidskr. **47**: 101–118.

1928a Neue skandinavische Dipteren. Ent. Tidskr. **49**: 18–21, 2 figs.

1928b *Pseudochirosia*, eine neue Anthomyiinengattung (Type: *Chirosia fractiseta* Stein). Ent. Tidskr. **49**: 22–23.

1928c Beiträge zur Kenntnis der Anthomyidenfauna des nördlichen Norwegens. Tromsø Mus. Årsh. (1926) **49** (3): 1–60, 1 fig.

1929a Bestämningstabeller till svenska muscidsläkten. 1 avd. Muscinae. Ent. Tidskr. **50**: 8–13.

1929b Übersicht der in Schweden gefundenen *Hylemyia*-Arten mit posteroventraler Apikalborste an den Hinterschienen. Ent. Tidskr. **50**: 269–273, 4 figs.

1930 Entomologische Ergebnisse der schwedischen Kamchatka-Expedition 1920–1922. 30. Diptera Brachycera. 3. Fam. Muscidae. Arkiv för Zool. **21** ([sect.] **A, no. 20**): 1–16, 6 figs. (=h.3, part).

1931 Einige Mitteilungen über lappländische Dipteren. Ent. Tidskr. **52**: 171–174.

1932 Einige neue Musciden aus Schweden. Ent. Tidskr. **53**: 156–160.

1933a Översikt av i Sverige funna *Hylemyia*-arter. Ent. Tidskr. **54**: 1–35, 54 figs.

1933b Tachiniden und Musciden aus Nordost-Grönland. Norges Svalbard- og Ishavs-Undersøk., Skr. om Svalbard og Ishavet **53**: 15–18.

1935 Neue fennoskandische Musciden. Notulae Ent. **15**: 26–31, 3 figs.

1936 Anteckningar till svenska arter av familjen Scopeumatidae (Diptera). Ent. Tidskr. **57**: 158–179, 6 figs.

Ringdahl, O.—Continued

1937   Neue schwedische Hylemyiinen. Opusc. Ent. (Soc. Ent. Lund.) **2:** 124–129, 21 figs.

1938   Översikt av svenska *Pegomyia*-arter (Diptera: Muscidae). Ent. Tidskr. **59:** 190–213, 4 pls.

1939   Diptera der Fam. Muscidae, (die Gattungen *Aricia* und *Anthomyza*) von Zetterstedt in "Insecta Lapponica" und "Diptera Scandinaviae" beschrieben. Opusc. Ent. (Soc. Ent. Lund.) **4:** 137–159.

1941   Bidrag till kännedomen om flugfaunan (Diptera Brachycera) på Hallands Väderö. Ent. Tidskr. **62:** 1–23.

1947   Bestimmungstabelle der mir bekannten europäischen und nordamerikanischen Arten von *Alloeostylus* und *Lasiops* (Dipt.: Musc.). Opusc. Ent. (Soc. Ent. Lund.) **12:** 90–95, 22 figs.

1951   Dipterologische Notizen 8. Neue Musciden aus schwedisch Lappland und Jämtland. Opusc. Ent. (Soc. Ent. Lund.) **16:** 33–36.

1954–
1959   1. Fam. Muscidae. H. 1, pp. 1–92, 14 figs.+figs. 1–121 (=rekv. 44), *1954;* h. 2, pp. 93–196, 11 figs.+figs. 1–222 (=rekv. 45), *1956;* h. 3, pp. 197–334, 19 figs.+figs. 1–322 (=rekv. 47), *1959*. *In* Entomologiska Foreningen i Stockholm, Svensk Insektfauna [q.v.]. [Order] 11: Diptera, [Suborder II]: Cyclorapha, [Group 2]: Schizophora, [Subgroup] Schizometopa. Stockholm.

Ritcher, P. O.

1940   Kentucky white grubs. Ky. Agr. Expt. Sta. Bul. **401:** 73–157, 40 figs., 6 pls.

    Also bound in Ky. Agr. Expt. Sta. Ann. Rpt. **53:** 73–157.

Roback, S. S.

1951   A classification of the muscoid calyptrate Diptera. Ent. Soc. Amer. Ann. **44:** 327–361, 2 figs., 7 pls.

1952   New species of Sarcophagini (Diptera: Sarcophagidae). Wash. Acad. Sci. Jour. **42:** 45–49, 6 figs.

1953   New records of *Symbiocladius equitans* (Claassen) with some notes on the genus (Diptera: Tendipedidae). Acad. Nat. Sci. Phila. Notulae Nat. **251:** 1–2.

1954   The evolution and taxonomy of the Sarcophaginae (Diptera, Sarcophagidae). Ill. Biol. Monog. **23 (3/4):** 1–181, 1 fig., 34 pls.

1955   The tendipedid fauna of a Massachusetts cold spring (Diptera: Tendipedidae). Acad. Nat. Sci. Phila. Notulae Nat. **270:** 1–8, 2 pls.

Roback, S. S.—Continued
1957a   The immature tendipedids of the Philadelphia area (Diptera: Tendipedidae). Acad. Nat. Sci. Phila. Monog. **9:** 1–152, 28 pls.
1957b   Some Tendipedidae from Utah. Acad. Nat. Sci. Phila. Proc. **109:** 1–24, 1 fig., 5 pls.
1959a   The subgenus *Ablabesmyia* of *Pentaneura* (Diptera: Tendipedidae: Pelopiinae). Amer. Ent. Soc. Trans. **85:**. 113–135, 17 pls.
1959b   Some Tendipedidae from Montana. Acad. Nat. Sci. Phila. Notulae Nat. **315:** 1–4, 10 figs.
1960   A new *Calopsectra* from Kansas (Diptera: Tendipedidae: Calopsectrini). Acad. Nat. Sci. Phila. Notulae Nat. **328:** 1–4, 17 figs.
1963   The genus *Xenochironomus* (Diptera: Tendipedidae) Kieffer, taxonomy and immature stages. Amer. Ent. Soc. Trans. (1962) **88:** 235–245, 1 fig., pls. 18–21.

Robertson, C.
1901   Some new Diptera. Canad. Ent. **33:** 284–286.

Robineau-Desvoidy, J. B.
1827   Essai sur la tribu des Culicides. Soc. d'Hist. Nat. de Paris, Mém. **3:** 390–413, pl. 10, part.
1830   Essai sur les Myodaires. [Paris] Inst. de France, [Cl. des] Sci. Math. et Phys., Acad. Roy. des Sci., Mém. présentés par divers Savans [**ser. 2**], **2:** 1–813.

> Considered as published subsequent to Wiedemann, 1830. See Aubertin, 1933: 141.

1842   Notice sur le genre Fucellie, *Fucellia*, R. D., et en particulier sur le *Fucellia arenaria*. Soc. Ent. de France, Ann. (1841) [**ser. 1**], **10:** 269–272.
1846   Myodaires des environs de Paris [cont.]. Soc. Ent. de France, Ann. **ser. 2, 4:** 17–38 [cont.].
1847   Myodaires des environs de Paris [cont.]. Soc. Ent. de France, Ann. **ser. 2, 5:** 255–287, 591–617 [cont.].
1848   Myodaires des environs de Paris [cont.]. Soc. Ent. de France, Ann. **ser. 2, 6:** 429–477 [cont.].
1849   [Note on *Lucilia dispar* and *Phormia regina*.] Soc. Ent. de France, Ann. **ser. 2, 7 (Bul.):** iv–v.
1850   Myodaires des environs de Paris [cont.]. Soc. Ent. de France, Ann. **ser. 2, 8:** 183–209 [cont.].

Robineau-Desvoidy, J. B.—Continued
- 1851a Descriptions d'Agromyzes et de Phytomyzes écloses chez M. le colonel Goureau. Rev. et Mag. de Zool. **ser. 2, 3:** 391–405.
- 1851b Myodaires des environs de Paris [cont.]. Soc. Ent. de France, Ann. **ser. 2, 9:** 177–190 [cont.].
- 1853a Diptères des environs de Paris (1). Famille des Myopaires. Soc. des Sci. Hist. et Nat. de l'Yonne, Bul. **7:** 83–160.
- 1853b Sur les éclosions de plusieurs espèces de Diptères, obtenues par le docteur Moret, Médicin a Auxerre. Soc. des Sci. Hist. et Nat. de l'Yonne, Bul. **7:** 531–536.
- 1863a Histoire naturelle des Diptères des environs de Paris. Vol. 1, xvi + 1143 pp. Paris.
    The entire work consists of 2 volumes, Paris, 1863.
- 1863b Histoire naturelle des Diptères des environs de Paris. Vol. 2, 920 pp. Paris.

Robinson, H.
- 1960 Four new Dolichopodidae from the eastern United States (Diptera). Ohio Jour. Sci. **60:** 270–273.

Robinson, I.
- 1953 The postembryonic stages in the life cycle of *Aulacigaster leucopeza* (Meigen) (Diptera Cyclorrhapha: Aulacigasteridae). Roy. Ent. Soc., London, Proc. Ser. A: Gen. Ent. **28:** 77–84, 10 figs.

Rockwood, L. P., Zimmerman, S. K., and Chamberlin, T. R.
- 1947 The wheat stem maggots of the genus *Meromyza* in the Pacific Northwest. U.S. Dept. Agr. Tech. Bul. **928:** 1–18, 1 fig.

Röder, V. von
- 1881 Dipterologische Notizen. Berlin. Ent. Ztschr. **25:** 210–216.
- 1885 Dipteren von der Insel Portorico. Stettin Ent. Ztg. **46:** 337–349.
- 1886 Ueber die nordamerikanischen Lomatiina von Mr. Coquillett in dem "Canadian Entomologist". Wien. Ent. Ztg. **5:** 263–265.
- 1887 *Asyndulum montanum* n. spec. (Ein dipterologischer Beitrag). Wien. Ent. Ztg. **6:** 116.
- 1889 Ueber *Myopa clausa* Lw. Wien. Ent. Ztg. **8:** 5.
- 1890 Zwei neue nordamerikanischen Dipteren. Wien. Ent. Ztg. **9:** 230–232.

Rogers, J. S.
1926a   A new *Dicranomyia* allied to *Dicranomyia immodesta* Osten Sacken-Tipulidae, Diptera. Fla. Ent. **9:** 49–52, 4 figs.
1926b   Some notes on the feeding habits of adult crane-flies. Fla. Ent. **10:** 5–8.
1927a   Notes on the life history, distribution and ecology of *Diotrepha mirabilis* Osten Sacken. Ent. Soc. Amer. Ann. **20:** 23–36, pl. 3.
1927b   Notes on the biology of *Atarba picticornis* Osten Sacken. Tipulidae-Diptera. Fla. Ent. **10:** 49–54, 2 figs.
1930   The summer crane-fly fauna of the Cumberland Plateau in Tennessee. Mich. Univ. Mus. Zool. Occas. Papers **215:** 1–50, 5 pls.
1931   Notes on a small collection of crane-flies from Oklahoma, with description of new species: Tipulidae-Diptera. Okla. Univ. Biol. Survey, Pubs. **3:** 331–338, 3 figs.
1933a   The ecological distribution of the crane-flies of northern Florida. Ecol. Monog. **3:** 1–74, 25 figs.
1933b   Contributions toward a knowledge of the natural history and immature stages of the crane-flies. 1. The genus *Polymera* Wiedemann. Mich. Univ. Mus. Zool. Occas. Papers **268:** 1–14, 2 pls.
1942   The crane flies (Tipulidae) of the George Reserve, Michigan. Mich. Univ. Mus. Zool. Misc. Pub. **53:** 1–128, 8 pls.
1949   The life history of *Megistocera longipennis* (Macquart) (Tipulidae, Diptera), a member of the neuston fauna. Mich. Univ. Mus. Zool. Occas. Papers **521:** 1–14, 2 pls.

———, **and Byers, G. W.**
1956   The ecological distribution, life history, and immature stages of *Lipsothrix sylvia* (Diptera: Tipulidae). Mich. Univ. Mus. Zool. Occas. Papers **572:** 1–14, 7 figs.

Rohdendorf, B. B. (or Rodendorf)
1928   [Flies of the family Sarcophagidae, parasitic on grasshoppers.] Tashkent, Uzbek. Opytn. Sta. Zashch. Rast. [Trudy] **14:** 1–64+i–ii, 5 pls.
    In Russian.
1931   Calliphorinen-Studien IV (Dipt.). Eine neue Calliphorinen-Gattung aus Ostsibirien. Zool. Anz. **95:** 175–177, 1 fig.

Rohwer, G. G. (See Steiner et al.)

Rondani, C.

1840a Memoria per servir alla ditterologia italiana. No. 1: Sopra una specie di insetto dittero, 16 pp., 1 pl. Parma.

> There were 20 numbered (with the exception of a No. 15) and individually titled papers in this series by Rondani, 1840-1862. The first 2 were issued only as separates, Parma, 1840. Of the remainder, 13 appeared in one of 3 Italian journals and also as separates; 5 were published in one of 2 French journals without separates as far as is known. A No. 15 was never published as such, but one of several revisions of earlier numbers is usually considered to be No. 15 of the series. The title of the series was given in Latin, "Fragmentum ad inserviendum dipterologiae italicae," in the French journals, and in the last 4 papers as "Commentarium pro dipterologia italica." In some other cases it was abbreviated.

1840b Memoria per servir alla ditterologia italiana. No. 2: Sopra alcuni nuovi generi di insetti ditteri, 28 pp., 1 pl. Parma.

1843 Quattro specie di insetti ditteri proposti come tipi di genere nuovi. Memoria sesta per servir alla ditterologia italiana. [Accad. delle Sci. dell' Ist. Bologna], Nuovi Ann. delle Sci. Nat. **10**: 32-46, 1 pl.

> A separate of 15 pp. was issued according to Osten Sacken, 1885: 151; it has not been seen and its place of publication and date is unknown to us.

1844a Proposta della formazione di un genere nuovo per due specie di insetti ditteri. Memoria nona per servir alla ditterologia italiana. Soc. Agr. e Accad. delle Sci. dell' Ist. Bologna, Nuovi Ann. delle Sci. Nat. e Rend. **ser. 2,** 1 [sic] [=2]: 193-202, pl. 2.

> A separate of 12 pp. was issued according to Osten Sacken, 1885: 151; it has not been seen and its place of publication and date is unknown to us.

1844b Ordinis dipterorum. Stirps II. Muscidae Rndn. Athericera Lat. Mac. Zett. West. Soc. Agr. e Accad. delle Sci. dell' Ist. Bologna, Nuovi Ann. delle Sci. Nat. e Rend. **ser. 2,** 1 [sic] [=2]: 447-459 [cont.].

1845a Sulle differenze sessuali delle Conopinae e Myopinae negli insetti ditteri. Memoria undecima per servir alla ditterologia italiana. Soc. Agr. e Accad. delle Sci. dell' Ist. Bologna, Nuovi Ann. delle Sci. Nat. e Rend. **ser. 2, 3**: 5-16.

> A separate of 16 pp. was issued according to Osten Sacken, 1885: 152; it has not been seen and its place of publication and date is unknown to us.

Rondani, C.—Continued

1845b Descrizione di due generi nuovi di insetti ditteri. Memoria duodecima per servire alla ditterologia italiana. R. Accad. delle Sci. dell' Ist. Bologna, Nuovi Ann. delle Sci. Nat. e Rend. **ser. 2, 3:** 25–36, pl. 1.

 A separate of 16 pp. was issued according to Osten Sacken, 1885: 152; it has not been seen and its place of publication and date is unknown to us.

1846 Compendio della seconda memoria ditterologia (publicata 1840) con algune aggiunte et correzioni. Soc. Agr. e Accad. delle Sci. dell' Ist. Bologna, Nuovi Ann. delle Sci. Nat. e Rend. **ser. 2, 6:** 363–376, 1 pl.

 A separate of 14 pp. was issued according to Osten Sacken, 1885: 153; it has not been seen and its place of publication and date is unknown to us. This is one of the revisions of earlier numbers and considered by some to be No. 15 of Rondani's series "Memoria per servir alla ditterologia italiana". This article would fit chronologically as No. 15.

1848 Esame di varie specie d'insetti ditteri brasiliani. Studi Ent. (Turin) **1:** 63–112.

1850a Osservazioni sopra alquante specie di esapodi ditteri del Museo Torinese. Accad. delle Sci. dell' Ist. e Soc. Agr. Bologna, Nuovi Ann. delle Sci. Nat. e Rend. **ser. 3, 2:** 165–197, pl. 4.

1850b Nota sexta pro dipterologia italica de nova specie generis *Ceriae* Fabricii. Soc. Ent. de France, Ann. **ser. 2, 8:** 211–214, pl. 7, fig. 1.

1856* Dipterologiae Italicae prodromus. Vol. 1: Genera Italica ordinis dipterorum ordinatim disposita et distincta et in familias et stirpes aggregata, 228 pp. Parmae [= Parma].

 The entire work was to have consisted of 84 stirps distributed among 32 families. Only stirps 1–25 (families 1–5, and part of 6) were completed in 8 volumes, published at Parma, Milano, and Firenze, 1856–1880. The first 5 volumes were published only as separate works, each with continuous pagination. Volumes 6–8 were published in portions of various size both as separates and in one of 3 Italian journals. The composition, pagination, dates, etc. of the completed volumes are given below. No mention of the journals is made here, as information on them is given later under the individual references used in this bibliography. A Facsimile Edition of Rondani's work was published by W. Junk in Berlin, 1914. However, it is incomplete and also is taken partly from the journal articles and partly from the separate works. Those references in this bibliography which do *not* appear in the facsimile edition are so noted; all others, marked with an asterisk (*), can be found there. The

Rondani, C.—Continued

    composition of the facsimile edition, the dates of the various parts of the entire original work and its publication in the various journals, are discussed in detail by Sabrosky, 1961c.

  Vol. 1: Genera Italica . . . 228 pp.
      Parmae, 1856.
      [A general outline]
  Vol. 2: Species Italicae . . . Pars 1. 264 pp.
      Parmae, 1857.
      [Families 1–3=Stirps 1–10]
  Vol. 3: Species Italicae . . . Pars 2. 243 pp.
      Parmae, 1859.
      [Family 4 (part)=Stirps 11–12 (part)]
  Vol. 4: Species Italicae . . . Pars 3. 174 pp.
      Parmae, 1861.
      [Family 4 (part)=Stirps 12 (end)]
  Vol. 5: Species Italicae . . . Pars 4. 239 pp.
      Parmae, 1862.
      [Family 4 (end)=Stirps 13–16]
  Vol. 6: Species Italicae . . . Pars 5. 151 pp. + Sup., 94 pp. + 24 pp.
      Milano, [?Firenze], "1865"[?1866]–[?1871].
      [Family 5 (part)=Stirps 17; and supplement to families 1–4]
        (Vol. is called "Pars 6" in some of its portions.)
    [Stirps 17]: Anthomyinae. 151 pp. Milano, "1865"[1866].
      (Revised and enlarged, 304 pp., Parma, 1877.)
    Supplemento. 4 fascs., 94 pp. + 24 pp. Milano, "1865" [?1866]–[?1871].
    Fasc. 1. pp. 1–20. Milano, "1865"[?1866].
    Fasc. 2. pp. 21–59. Milano, "1865"[?1866].
    Fasc. 3. pp. 61–94. Milano, [?1868].
    Fasc. 4. 24 pp. [?Firenze], [?1870 or 1871].
      (Material in fasc. 4 later included in above revision of Anthomyinae, 1877.)
  Vol. 7: Species Italicae . . . Pars 6. 4 fascs. [in 5 parts].
      "Mediolani, Modenae, Florentiae" [Milano, Firenze], 1866–1871.
      [Family 5 (part)=Stirps 18–20]
        (Vol. is called "Pars 7" in some of its portions.)
    Fasc. 1 [Stirps 18]. 51 pp. Milano, 1866.
    Fasc. 2 [Stirps 19]. 60 pp. Milano, 1868.
      (Revised and enlarged, 78 pp., Modenae, 1877.)
    Fasc. 3 [Stirps 20 (part)]. 37 pp. [?Firenze], 1869.
    Fasc. 4 [Stirps 20 (end)]. [2 pts.] Firenze, 1870–1871.
      [Pt. 1]. 59 pp. Firenze, 1870.
      [Pt. 2]. 53 pp. Firenze, 1871.

Rondani, C.—Continued

    Vol. 8: Species Italicae . . . Pars 7. [5 pts.].
    "Florentiae" [=Firenze], 1874-1880.
    [Family 5 (end) and Family 6 (part)=Stirps 21-25]
    [Pt. 1] Stirps 21. 16 pp. Firenze, 1874.
    [Pt. 2] Stirps 22. 32 pp. Firenze, 1874.
    [Pt. 3] Stirps 23. 23 pp. Firenze, 1875.
    [Pt. 4] Stirps 24. 12 pp. Firenze, 1876.
    [Pt. 5] Stirps 25. 43 pp. Firenze, 1880.
    See Rondani, 1856, 1857, 1859, 1861a, 1862 (Vols. 1-5); 1866b, 1868b (Vol. 7, part). Also journal articles for Vol. 6, Rondani, 1865, 1866a, 1868a, 1871a; for rest of Vol. 7, Rondani, 1869, 1870a, 1871b; and for part of Vol. 8, Rondani, 1874, 1875a, 1880. Also revision, Rondani, 1877.

1857*   Dipterologiae Italicae prodromus. Vol. 2: Species Italicae ordinis dipterorum in genera characteribus definita, ordinatim collectae, methodo analitica distinctae, et novis vel minus cognitis descriptis, Pars prima: Oestridae, Syrpfhidae [sic], Conopidae, 264 pp., 1 fig. Parmae [=Parma].

    *See discussion under Rondani, 1856.

1859*   Dipterologiae Italicae prodromus. Vol. 3: Species Italicae ordinis dipterorum in genera characteribus definita, ordinatim collectae, methodo analitica distinctae, et novis vel minus cognitis descriptis, Pars secunda: Muscidae, Siphoninae et (partim) Tachininae, 243 pp., 1 pl. Parmae [=Parma].

    *See discussion under Rondani, 1856.

1860   Stirpis cecidomyarum. Genera revisa. Nota undecima, pro dipterologia italica. Soc. Ital. di Sci. Nat. Atti (1859-1860) **2**: 286-294, 1 pl.

1861a*   Dipterologiae Italicae prodromus. Vol. 4: Species Italicae ordinis dipterorum in genera characteribus definita, ordinatim collectae, methodo analatica [sic] distinctae, et novis vel minus cognitis descriptis, Pars tertia: Muscidae, Tachininarum complementum, 174 pp. Parmae [=Parma].

    *See discussion under Rondani, 1856.

1861b   Species europeae generis *Phasiae* Latreillei observatae et distinctae. Soc. Ital. di Sci. Nat. Atti **3**: 205-220, 1 pl.

Rondani, C.—Continued

1862* Dipterologiae Italicae prodromus. Vol. 5: Species Italicae ordinis dipterorum in genera characteribus definita, ordinatim collectae, methodo analitica distinctae, et novis vel minus cognitis descriptis, Pars quarta: Muscidae, Phasiinae-Dexinae-Muscinae-Stomoxidinae, 239 pp. Parmae [=Parma].

*See discussion under Rondani, 1856.

1863 Diptera exotica revisa et annotata. 99 pp., 1 pl. Modena.

Also published under the title "Dipterorum species et genera aliqua exotica", in Arch. per Zool. l'Anat. e Fis. (Modena (1863) **3 (1):** 1–99, pl. 5, **1864.**

1865 Diptera Italica non vel minus cognita descripta vel annotata observationibus nonnullis additis. Fasc. I. Oestridae-Syrphidae-Conopidae; Fasc. II. Muscidae. Soc. Ital. di Sci. Nat. Atti **8:** 127–146; 193–231.

These fascicles were also published separately under the same titles, pp. 1–20, and 21–59, Milano, "*1865*" [?1866]. They are the first parts of the supplement to Vols. 1–5, and form part of Vol. 6, of "Dipterologiae Italicae prodromus". Although both have the imprint date 1865, it is not known whether they were issued separately or together with the separate of Rondani, 1866a [q.v.]. Neither appear in the Facsimile Edition.

1866a Anthomyinae Italicae collectae distinctae et in ordinem dispositae. Soc. Ital. di Sci. Nat. Atti **9:** 68–217.

Also published separately under the same title, 151 pp., Milano, *1866*. A title page issued with it is titled "Species Italicae ordinis dipterorum a Prof. Camillo Rondani collectae, distinctae, et in ordinem dispositae novis, vel minus cognitis descriptis. Pars quinta. Anthomyinae. Dipter: Ital: Prodromi cum supplemento. Pars [=Vol.] 6'', and is dated 1865. It is not known whether the 2 parts of Rondani, 1865, were issued together with this as the "supplemento", or separately. This portion on Anthomyinae was revised, expanded, and incorporated with Rondani, 1871a, to form Rondani, 1877 [q.v.]. It is not in the Facsimile Edition.

1866b* Scatophaginae Italicae collectae distinctae et in ordinem dispositae. Dipterol. Ital. prodomi Pars [=Vol.] VII, Fasc. I. Fasc. 1, 51 pp. *In his* Dipterologiae Italicae prodromus. Vol. 7: Species Italicae ordinis dipterorum a Prof. Camillo Rondani collectae, distinctae, et in ordinem dispositae novis, vel minus cognitis descriptis, Pars sexta: Scatophaginae Sciomyzinae Ortalidinae. Milano.

Also published under the same title in Soc. Ital. di Sci. Nat. Atti **10:** 85–135, *1867*.

*See discussion under Rondani, 1856.

Rondani, C.—Continued

1868a Diptera Italica non vel minus cognita descripta vel annotata observationibus nonnullis additis. Fasc. III. Soc. Ital. di Sci. Nat. Atti **11**: 21–54.

> Also published separately under the same title, pp. 61–94, Milano, [?*1868*]. It is part of the supplement to Vols. 1–5, and forms part of Vol. 6, of "Dipterologiae Italicae prodromus". The date of the separate is not known but is presumably not before 1868. This does not appear in the Facsimile Edition.

1868b Sciomizinae [sic] Italicae collectae, distinctae et in ordinem dispositae. Dipterol. Ital. prodromi Pars [=Vol.] VII, Fasc. II. Fasc. 2, 60 pp. *In his* Dipterologiae Italicae prodromus. Vol. 7: Species Italicae ordinis dipterorum a Prof. Camillo Rondani collectae, distinctae, et in ordinem dispositae novis, vel minus cognitis descriptis, Pars sexta: Scatophaginae Sciomyzinae Ortalidinae. Milano.

> Also published under the same title, corrected, in Soc. Ital. di Sci. Nat. Atti **11**: 199–256, *1868*. This is not in the Facsimile Edition. A revised edition of this fascicle was published separately, 78 pp., Modenae, 1877, and reprinted in Soc. Nat. di Modena Ann. **11**: 7–79, 1877, which edition does appear in the Facsimile.

1868c Specierum italicarum ordinis dipterorum catalogus notis geographicus. Soc. Ital. di Sci. Nat. Atti **11**: 559–603.

1869* Ortalidinae Italicae collectae, distinctae et in ordinem dispositae. Dipterologiae Italicae prodromi Pars [=Vol.] VII—Fasc. 3. Linea A. Ortaloidi. Soc. Ent. Ital. Bul. **1**: 5–37.

> Also published separately under the same title, 37 pp. [?Firenze], *1869*, as fascicle 3 of his "Dipterologiae Italicae prodromus," Vol. 7. Our copy of the separate lacks an imprint with place of publication.

*See discussion under Rondani, 1856.

1870 Sull' insetto Ugi. Soc. Ent. Ital. Bul. **2**: 134–137.

1871a Diptera Italica non vel minus cognita descripta aut annotata. Fasc. IV. Addenda Anthomyinis Prodr. Vol. VI. Soc. Ent. Ital. Bul. (1870) **2**: 317–338.

> A separate of 24 pp. was published according to Osten Sacken, 1885: 159; it has not been seen and its place of publication [?Firenze] and date [?1870 or 1871] is unknown to us. This work concludes the supplement in Vol. 6 of "Dipterologiae Italicae prodromus." It does not appear in the Facsimile Edition. The work was incorporated with Rondani, 1866a, to form Rondani, 1877 [q.v.].

Rondani, C.—Continued

1871b* Ortalidinae Italicae collectae, distinctae et in ordinem dispositae. Dipterologiae Italicae prodromi Pars [=Vol.] VII. –Fasc. 4 [concl.]. Soc. Ent. Ital. Bul. **3:** 3–24, 161–188.

> Also published separately under the same title, 53 pp., Firenze, *1871*, as fascicle 4, [pt. 2] of his "Dipterologiae Italicae prodromus," Vol. 7.
> *See discussion under Rondani, 1856.

1874* Species Italicae ordinis dipterorum (Muscaria Rndn.). Stirps XXI. –Tanipezinae Rndn. collectae et observatae. Soc. Ent. Ital. Bul. **6:** 167–182.

> Also published separately as "Dipterologiae Italicae prodromus," Vol. 8, Stirps 21 [=Pt. 1], 16 pp., Firenze, *1874*.
> *See discussion under Rondani, 1856.

1875a* Species Italicae ordinis dipterorum (Muscaria Rndn.) collectae et observatae. Stirps XXIII. Agromyzinae. Soc. Ent. Ital. Bul. **7:** 166–191.

> Also published separately as "Dipterologiae Italicae prodromus," Vol. 8, Stirps 23 [=Pt. 3], 26 pp., Firenze, *1875*.
> *See discussion under Rondani, 1856.

1875b Muscaria exotica musei civici januensis. Fragmentum III. Species in insula Bonae Fortunae (Borneo), Provincia Sarawak, annis 1865–1868 lectae a March. J. Doria et Doct. O. Beccari. Genoa Mus. Civico di Storia Nat. Ann. **7:** 421–464, 5 figs.

1877* Species Italicae ordinis dipterorum ordinatim dispositae, methodo analitica distinctae, et novis vel minus cognitis descriptis. Pars quinta. Stirps XVII—Anthomyinae. 304 pp. Parmae [=Parma].

> A revision and expansion of Rondani, 1866a, and including his 1871a.
> *See discussion under Rondani, 1856.

1878 Muscaria exotica musei civici januensis observata et distincta. Fragmentum IV. Hippoboscita exotica non vel minus cognita. Genoa Mus. Civico di Storia Nat. Ann. **12:** 150–170, 1 fig.

1880* Species Italicae ordinis dipterorum (Muscaria Rndn.) collectae et observatae. Stirps XXV. Copromyzinae Zett. Soc. Ent. Ital. Bul. **12:** 3–45.

> Also published separately as "Dipterologiae Italicae prodromus," Vol. 8, Stirps 25 [=Pt. 5], 43 pp., Firenze, *1880*.
> *See discussion under Rondani, 1856.

———, **and Berté, E.** (See Rondani, 1840a)

Root, F. M., and Hoffman, W. A.
1937   The North American species of *Culicoides*. Amer. Jour. Hyg. **25**: 150–176, 8 pls.

[Roret, N. E. ed.]
1834–   (Collection des suites à Buffon, formant avec les oeuvres de
1890   cet auteur un cours complet d'histoire naturelle.) 82 vols. +11 atlases. Paris.

> A collection of works on natural history by many authors. The work is sometimes called only "Suites à Buffon," and the volumes are known collectively as "Roret's Suites à Buffon." The volumes of each taxonomic group are independently numbered, there being no overall volume numbering, nor in most cases, an overall title. There are fifteen such groups, and many other miscellaneous single volumes. The Diptera were published in 2 volumes by Macquart, Paris, 1834–1835.
> See Macquart, 1834a, 1835.

Roser, C. von
1840   Erster Nachtrag zu dem in Jahre 1834 bekannt gemachten Verzeichnisse in Württemberg vorkommender zweiflügliger Insekten. K. Württemb. Landw. Ver., Stuttgart, Correspondenzbl. **37** [=n. ser., 17] (**1**): 49–64.

Ross, A.
1961   Biological studies on bat ectoparasites of the genus *Trichobius* (Diptera: Streblidae) in North America, north of Mexico. Wasmann Jour. Biol. **19**: 229–246.

Ross, E. S.
1943   The identity of *Aedes bimaculatus* (Coquillett) and a new subspecies of *Aedes fulvus* (Wiedemann) from the United States (Diptera, Culicidae). Ent. Soc. Wash. Proc. **45**: 143–151, 4 figs.

Ross, H. H.
1947   The mosquitoes of Illinois (Diptera, Culicidae). Ill. Nat. Hist. Survey, Bul. **24**: 1–96, 184 figs.

Ross, J.
1835   Narrative of a second voyage in search of a North-West Passage and of a residence in the Arctic Regions during the years 1829, 1830, 1831, 1832, 1833. 1 vol., and separate appendix. London.

> The appendix is divided into separately paged parts. That on Natural History is by J. C. Ross, and is entitled "Account of the objects in the several departments of natural history, seen and discovered during the present expedition." It consists of cxliv pp., and 5 pls., the latter published in 1834.
> See Curtis, 1835b.

Rossi, F. W.

1848   Systematisches Verzeichniss der zweiflügelichten Insecten (Diptera) des Erzherzogthumes Österreich. iii–x + 86 pp. Wien.

Rossi, P. (or Rossius)

1790a   Fauna Etrusca. Sistens insecta quae in provinciis Florentina et Pisana praesertim collegit. Vol. 1, 272 pp. Liburni [=Livorno].

> The entire work consists of 2 volumes, Liburni, 1790. For a German translation and expansion of this work and of Rossi's "Mantissa insectorum," see Hellwig and Illiger, 1795–1807 whose work is sometimes also listed as another edition of "Fauna Etrusca."

1790b   Fauna Etrusca. Sistens insecta quae in provinciis Florentina et Pisana praesertim collegit. Vol. 2, 348 pp., 10 pls. Liburni [=Livorno].

Roth, L. M.

1945   The male and larva of *Psorophora* (*Janthinosoma*) *horrida* (Dyar and Knab) and a new species of *Psorophora* from the United States (Diptera: Culicidae). Ent. Soc. Wash. Proc. **47**: 1–23, 19 figs.

Rothfels, K. H.

1956   Black flies: Siblings, sex, and species grouping. Jour. Hered. **47**: 113–121, figs. 4–9.

Roubaud, E.

1906   Aperçus nouveaux, morphologiques et biologiques, sur les Diptères piqueurs du groupe des Simulies. [Paris] Acad. des Sci. Compt. Rend. **143**: 519–521.

Rowe, J. A. (See also Knowlton and Rowe)

1931   A revision of the males of the nearctic species in the genus *Fabriciella* (Tachinidae). Ent. Soc. Amer. Ann. **24**: 643–678, 4 pls.

1933   Records of Tachinidae from Illinois with description of one new species (Diptera). Ent. News **44**: 122–126, 1 fig.

———, and **Knowlton, G. F.**

1935   The genus *Tabanus* in Utah. Canad. Ent. **67**: 238–244, 10 figs.

1936   Pangoniinae of Utah (Tabanidae: Diptera). Ohio Jour. Sci. **36**: 253–259, 1 pl.

Royal Entomological Society of London
1949– Handbooks for the identification of British insects. 10
1963 vols. London.

> This series is to consist of 10 volumes upon completion. The volumes are issued in separately paged parts as completed. All have been begun except Vol. 3. The Diptera are included in Volumes 9 and 10, neither of which are complete. Volume 9 consists of Parts 1 and 2, issued 1949–1950; Volume 10 consists of Parts 1 and 4a, issued 1953–1954.
> See Coe, 1953; Emden, 1954; Oldroyd, 1949.

Rozeboom, L. E.
1934 A new nycteribiid from Florida. Jour. Parasitol. **20**: 315–316, 1 fig.

Rubzov, I. A. (or Rubtsov, Rubtzov) (See also Dorogostajskij, Rubzov, and Vlasenko)
1940a [Fam. Simuliidae.] No. 6, 532 pp., 93 figs. (=n. ser., 23). *In* [Zoologicheskii Institut Akademii Nauk SSSR], [Fauna SSSR, Novaia Seriia] [q.v.]. Diptera, Vol. 6. Moscow, Leningrad.

> In Russian.

1940b (Notes on the black-flies of Transbaikalia.) Akad. Nauk S.S.S.R., Zool. Inst., Parazitol. Sborn. (Acad. des Sci. de l'U.R.S.S., Inst. Zool., Mag. de Parasitol.) (1939) **7**: 193–201, 2 figs.

> In Russian; English summary, pp. 199–201.

Rudebeck, G. (See Hanström, Brinck, and Rudebeck)

Rübsaamen, E. H.
1892 Die Gallmücken des Königl. Museums für Naturkunde zu Berlin. Berlin. Ent. Ztschr. **37**: 321–411, pls. 7–18.
1893 Vorläufige Beschreibung neuer Cecidomyiden. Ent. Nachr. **19**: 161–166.
1894 Die aussereuropäischen Trauermücken des Königl. Museums für Naturkunde zu Berlin. Berlin. Ent. Ztschr. **39**: 17–42, 3 figs., 3 pls.
1895a Cecidomyidenstudien. Ent. Nachr. **21**: 177–194.
1895b Ueber Cecidomyiden. Wien. Ent. Ztg. **14**: 181–193, 1 pl.
1898 Zoologische Ergebnisse der von der Gesellschaft für Erdkunde zu Berlin unter Leitung Dr. von Drygalski's ausgesandten Grönlandsexpedition. VIII. Grönlandische Mycetophiliden, Sciariden, Cecidomyiden, Psylliden, Aphiden und Gallen. Pp. 103–119, 11 figs., pls. 5–6 (=lfg. 4, part). *In* Leuckart, R., and Chun, C., eds., Bibliotheca zoologica [q.v.]. H. 20, 132 pp., 26 figs., 6 pls. Stuttgart.

Rübsaamen, E. H.—Continued
- 1910 Ueber deutsche Gallmücken und Gallen. Ztschr. f. Wiss. Insektenbiol. **6:** [=whole ser., 15]: 125–132, 199–204, 283–289, 336–342, 415–425, figs. 1–23 [cont.].
- 1911 Ueber deutsche Gallmücken und Gallen [cont.]. Ztschr. f. Wiss. Insektenbiol. **7** [=whole ser., 16]: 120–125, 168–172, 278–282, 350–353, 390–394, figs. 24–54 [cont.].

Rueger, M. E.
- 1958 Aedes (Ochlerotatus) barri, a new species of mosquito from Minnesota (Diptera, Culicidae). Kans. Ent. Soc. Jour. **31:** 34–46, 2 pls.

Russell, H. M., and Hooker, C. W.
- 1908 A new cecidomyiid on oak. Ent. News **19:** 349–352, pl. 14.

Ruthe, J. F. von
- 1831 Einige Bemerkungen und Nachträge zu Meigen's "Systematischer Beschreibung der europäischen zweiflügeligen Insecten." Isis (Oken's) **1831:** 1203–1222.

S. N. A. (See Société de Naturalistes et d'Agriculteurs)

Sabrosky, C. W. (See also Bennett and Sabrosky; Stoll et al.)
- 1935a The Chloropidae of Kansas (Diptera). Amer. Ent. Soc. Trans. **61:** 207–268, 1 fig.
- 1935b Notes on the taxonomic status of certain species of the genus *Chlorops* (Diptera, Chloropidae). Ent. News **46:** 77–84.
- 1935c The vittate species of the genus *Madiza* (Diptera, Chloropidae). Kans. Ent. Soc. Jour. **8:** 105–116, 1 pl.
- 1936 A synopsis of the nearctic species of *Oscinella* and *Madiza*, based on a study of the types (Diptera, Chloropidae). Ent. Soc. Amer. Ann. **29:** 707–728.
- 1938 Taxonomic notes on the dipterous family Chloropidae. I. N.Y. Ent. Soc. Jour. **46:** 417–434, 4 figs.
- 1939a The European frit fly and its forms in North America. Ent. Soc. Amer. Ann. **32:** 321–324.
- 1939b A new North American species of *Asteia* (Diptera, Asteiidae). Pan-Pacific Ent. **15:** 165–167, 1 fig.
- 1940 Twelve new North American species of *Oscinella* (Diptera, Chloropidae). Canad. Ent. **72:** 214–230, 1 pl.
- 1941a The *Hippelates* flies or eye gnats: Preliminary notes. Canad. Ent. **73:** 23–27.
- 1941b An annotated list of the genotypes of the Chloropidae of the world (Diptera). Ent. Soc. Amer. Ann. **34:** 735–765.
- 1941c The genus *Ectecephala* in North America (Diptera, Chloropidae). Ent. Soc. Wash. Proc. **43:** 75–80.

Sabrosky, C. W.—Continued
- 1943a A revised synopsis of nearctic *Thaumatomyia* (=*Chloropisca*) (Diptera, Chloropidae). Canad. Ent. **75:** 109–117.
- 1943b A new species of and notes on Acroceridae (Diptera). Ent. News **54:** 176–182.
- 1943c New genera and species of Asteiidae (Diptera), with a review of the family in the Americas. Ent. Soc. Amer. Ann. **36:** 501–514, 5 figs.
- 1944 A revision of the American spider parasites of the genera *Ogcodes* and *Acrocera* (Diptera, Acroceridae). Amer. Midland Nat. **31:** 385–413, 1 pl.
- 1947 *Rhodesiella:* A genus new to the Western Hemisphere (Diptera: Chloropidae). Ent. News **58:** 174–177.
- 1948a A further contribution to the classification of the North American spider parasites of the family Acroceratidae (Diptera). Amer. Midland Nat. **39:** 382–430, 2 pls.
- 1948b A synopsis of the nearctic species of *Elachiptera* and related genera (Diptera, Chloropidae). Wash. Acad. Sci. Jour. **38:** 365–382, 14 figs.
- 1948c *Winthemia citheroniae*, new species, with notes on the correct name of *W. cecropia* (Diptera, Larvaevoridae). Ent. Soc. Wash. Proc. **50:** 63–67.
- 1949a The North American heleomyzid genus *Lutomyia*, with description of a new species. Mich. Univ. Mus. Zool. Occas. Papers **517:** 1–6, 5 figs.
- 1949b "*Leptocera lutosa*": A complex of nearctic species (Diptera, Sphaeroceratidae). Ent. Soc. Wash. Proc. **51:** 1–24, 3 pls.
- 1949c The muscid genus *Ophyra* in the Pacific Region (Diptera). Hawaii. Ent. Soc. Proc. (1948) **13:** 423–432, 12 figs.
- 1950a A new species of *Pseudogaurax* from Florida (Diptera, Chloropidae). Brooklyn Ent. Soc. Bul. **45:** 33–34, 1 fig.
- 1950b A new species of *Eribolus* from California (Diptera, Chloropidae). Pan-Pacific Ent. **26:** 91–92.
- 1950c A synopsis of the chloropid genera *Chaetochlorops* and *Eugaurax* (Diptera). Wash. Acad. Sci. Jour. **40:** 183–188.
- 1950d Notes on Trichopodini (Diptera, Larvaevoridae), with description of a new parasite of cotton stainers in Puerto Rico. Wash. Acad. Sci. Jour. **40:** 361–371, 11 figs.
- 1950e The genus *Dicraeus* in North America (Diptera, Chloropidae). Ent. Soc. Wash. Proc. **52:** 53–62, 1 pl.

Sabrosky, C. W.—Continued
- 1950f Date of publication of James' "The flies that cause myiasis in man". Ent. Soc. Wash. Proc. **52**: 315.
- 1951a A revision of the nearctic species of the genus *Gaurax* (Diptera, Chloropidae). Amer. Midland Nat. **45**: 407–431, 8 figs.
- 1951b A review of the nearctic species of *Lasiopleura* (Diptera, Chloropidae). Canad. Ent. **83**: 336–343, 4 figs.
- 1951c Two new species of *Pseudiastata* (Dipt., Drosophilidae) predaceous on the pineapple mealybug. Bul. Ent. Res. **41**: 623–627, 8 figs.
- 1953a A new North American species of *Metopia* (Diptera, Sarcophagidae). Brooklyn Ent. Soc. Bul. **48**: 50–53.
- 1953b Two new species of Milichiidae, with miscellaneous notes on the family (Diptera). Ent. News **64**: 38–42.
- 1953c Taxonomy and host relations of the tribe Ormiini in the Western Hemisphere (Diptera, Larvaevoridae). I; II. Ent. Soc. Wash. Proc. **55**: 167-183; 289–305.
- 1955a A third species of *Eusiphona*, with remarks on the systematic position of the genus (Diptera, Milichiidae). Ent. News **66**: 169–173.
- 1955b The taxonomic status of the armyworm parasite known as *Archytas piliventris* (van der Wulp) (Diptera: Larvaevoridae). Fla. Ent. **38**: 77–83, 6 figs.
- 1956 A new species of the *Leptocera lutosa* complex (Diptera: Sphaeroceridae). Ent. News. **67**: 74–76.
- 1957 Synopsis of the New World species of the dipterous family Asteiidae. Ent. Soc. Amer. Ann. **50**: 43–61, 15 figs.
- 1958 New species and notes on North American acalyptrate Diptera. Ent. News **69**: 169–176.
- 1959a The nearctic species of the filth fly genus *Meoneura* (Diptera, Milichiidae). Ent. Soc. Amer. Ann. **52**: 17–26, 1 pl.
- 1959b A revision of the genus *Pholeomyia* in North America (Diptera, Milichiidae). Ent. Soc. Amer. Ann. **52**: 316–331, 1 pl.
- 1959c Flies of the genus *Odinia* in the Western Hemisphere (Diptera: Odiniidae). U.S. Natl. Mus. Proc. **109**: 223–236, 1 pl.
- 1961a A new nearctic species of *Stenoscinis*, with key to the species of the Western Hemisphere (Diptera, Chloropidae). Ent. News **72**: 19–23, 1 fig.
- 1961b Three new nearctic acalypterate Diptera. Ent. News **72**: 229–234, 1 fig.

Sabrosky, C. W.—Continued
1961c   Rondani's "Dipterologiae Italicae Prodromus". Ent. Soc. Amer. Ann. **54:** 827–831.

———, **and Arnaud, P. H., Jr.**
1963   A holotype problem and a new specific name in *Pseudochaeta* (Diptera: Tachinidae). Ent. News **74:** 155–156.

Sack, P.
1923   Dipteren aus Nowaja Semlja. Rpt. 15, 10 pp., 3 figs. *In* Holtedahl, O., ed., Report of the scientific results of the Norwegian Expedition to Novaya Zemlya 1921 [q.v.]. Oslo.

1928–   Syrphidae. [Fam.] 31, pp. 1–48, figs. 1–66, pl. 1 (=lfg. 30),
1932   *1928;* pp. 49–96, figs. 67–145 (=lfg. 32), *1929;* pp. 97–144, figs. 146–210 (=lfg. 34), *1929;* pp. 145–176, figs. 211–250, pls. 2–8 (=lfg. 41), *1930;* pp. 177–250, figs. 251–283, pls. 9–10 (=lfg. 49), *1930;* pp. 241–288, figs. 284–320, pl. 11 (=lfg. 55), *1931;* pp. 289–336, figs. 321–341, pls. 12–14 (=lfg. 59), *1931;* pp. 337–384, figs. 342–369, pls. 15–16 (=lfg. 61), *1932;* pp. 385–451, figs. 370–389, pls. 17–18 (=lfg. 63), *1932*. *In* Lindner, E., ed., Die Fliegen der palaearktischen Region [q.v.]. Vol. 4, Pt. 6. Stuttgart.

1937   Cordyluridae. [Fam.] 62a, pp. 1–48, figs. 1–16, pls. 1–3 (=lfg. 112); pp. 49–103, figs. 17–21, pls. 4–6 (=lfg. 113). *In* Lindner, E., ed., Die Fliegen der palaearktischen Region [q.v.]. Vol. 7. Stuttgart.

Sagra, R. de la
1839–   Historia fisica, politica y natural de la Isla de Cuba. 13 vols.
1861   Paris.

> Part I: Historia fisica y politica, consists of Volumes 1 and 2, and Volumes 3–12 comprise Part II: Historia natural. Volume 13 is a supplement to Part I. Much of this series was issued at the same time in a French edition in Paris.
> See Bigot, 1857a.

Sahlberg, J.
1886   *Lynchia fumipennis* n. sp. en på *Pandion haliaëtus* lefvande hippoboscid. Soc. pro Fauna et Flora Fenn., Meddel. **13:** 149–152, 1 fig.

Sailer, R. I.   (See Gjullin et al.; Sommerman, Sailer, and Esselbaugh)
Saint-Fargeau, Count de   (See Lepeletier, A. L. M.)
Saint-Vincent, Baron de   (See Bory, J. B. G. M.)
Salvin, O.   (See Godman and Salvin)

Sanjean, J.
1957 Taxonomic studies of *Sarcophaga* larvae of New York, with notes on the adults. N.Y. (Cornell) Agr. Expt. Sta. Mem. **349**: 1–115, 149 figs.

Sargent, F. W. (See Wood et al.)

Sasakawa, M.
1958 The female terminalia of the Agromyzidae, with description of a new genus (I). Saikyo Univ. (Kyoto), Sci. Rpt., Agr. **10**: 133–150, 5 pls.

Satchell, G. H.
1947 The larvae of the British species of *Psychoda* (Diptera: Psychodidae). Parasitology **38**: 51–69, 84 figs.
1953 New and little known Samoan Psychodidae and a new species from Rarotonga. Roy. Ent. Soc., London, Proc. Ser. B: Taxonomy **22**: 181–188, 5 figs.

Saunders, L. G. (See also Ewen and Saunders)
1925 On the life history, morphology and systematic position of *Apelma* Kieff. and *Thyridomyia* n.g. (Diptera, Nemat. Ceratopogoninae). Parasitology **17**: 252–277, 9 figs., 1 pl.
1928 Some marine insects of the Pacific coast of Canada. Ent. Soc. Amer. Ann. **21**: 521–545, 9 figs.
1956 Revision of the genus *Forcipomyia* based on characters of all stages (Diptera, Ceratopogonidae). Canad. Jour. Zool. **34**: 657–705, 19 figs.

[**Saunders, W. W.**, ed.]
1850–1869 Insecta Saundersiana: Or characters of undescribed insects in the collection of William Wilson Saunders, Esq., F.R.S., F.L.S., &c. 4 vols. London.

> Only volumes 1 (on Diptera, by F. Walker, the title page of which is dated 1856) and 3 (on Buprestidae, by Saunders) have volume designations. (Volumes 2 and 4 are on Curculionidae, by H. Jekel, and Homoptera, by Walker, respectively.) The volumes were usually issued in continuously paged parts. Volume 3 was never completed.
> See Walker, 1850, 1851b, 1852, 1856b.

Savtshenko, E. N. (or Savchenko)
1956 [Survey of the palaearctic species of crane-flies (Diptera, Tipulidae) of the *Tipula juncea* Meig. group.] Kiïv. Derzhavnii Univ. im. T. G. Shevchenka, Nauk. Zap. **15** (3): 129–148, 6 figs. (=Z'irnik [Sborn.] Zoomuz. [or Zool. Muz. Trudy] 5.)
> In Ukrainian.

Savtshenko, E. N.—Continued
1961 [Fam. Tipulidae, Subfam. Tipulinae, Genus *Tipula* L. (Part 1).] No. 3, 487 pp., 295 figs. (=n. ser., 79). *In* [Zoologicheskii Institut Akademii Nauk SSSR], [Fauna SSSR, Novaia Seriia] [q.v.]. Diptera, Vol. 2. Moscow, Leningrad.
   In Russian.

Say, T.
1817a Description of several new species of North American insects. Acad. Nat. Sci. Phila. Jour. **1**: 19–23.
   Also *in* LeConte, 1859b: 1–4.

1817b Some account of the insect known by the name of Hessian fly, and of a parasitic insect that feeds on it. Acad. Nat. Sci. Phila. Jour. **1**: 45–48.
   Also *in* LeConte, 1859b: 4–7.

1823 Descriptions of dipterous insects of the United States. Acad. Nat. Sci. Phila. Jour. **3**: 9–54, 73–104.
   Also *in* LeConte, 1859b: 38–66, 67–89.

1824a Appendix. Part I.—Natural History. 1. Zoology. E. Class Insecta. Pp. 268–378. *In* Keating, W. H., Major Long's second expedition [q.v.]. Vol. 2, 459 pp., pls. 6–15. Philadelphia.
   Also *in* LeConte, 1859a: 176–258.

1824b American entomology, or descriptions of the insects of North America. [Vol. 1], [101 pp.], pls. 1–18. Philadelphia.
   Also *in* LeConte, 1859a: 1–35, pls. 1–18.
   The entire work consists of 3 volumes, with a total of 54 plates, Philadelphia, 1824–1828. Neither the pages nor the volumes are numbered. This work is not the same as Say's "American entomology" published in 1817.

1825 American entomology, or descriptions of the insects of North America. [Vol. 2], [121 pp.], pls. 19–36. Philadelphia.
   Also *in* LeConte, 1859a: 35–81, pls. 19–36.

1828 American entomology, or descriptions of the insects of North America. [Vol. 3]. [136 pp.], pls. 37–54. Philadelphia.
   Also *in* LeConte, 1859a: 81–121, pls. 37–54.

1829– Descriptions of North American dipterous insects. Acad.
1830 Nat. Sci. Phila. Jour. (1829–1830) **6**: 149–178, *1829*; 183–188, *1830*.
   Also *in* LeConte, 1859b: 348–368, 368–371.

Say, T.—Continued
1831 Descriptions of new species of North American insects, found in Louisiana by Joseph Barabino. 19 pp. New Harmony, Ind.

> This does not appear in LeConte's work (See Scudder, 1899).

1832 New species of North American insects, found by Joseph Barabino, chiefly in Louisiana. 16 pp. New Harmony, Ind.

> Also *in* LeConte, 1859a: 300–309.

Schaeffer, C.
1916 New Diptera of the family Asilidae with notes on known species. N.Y. Ent. Soc. Jour. 24: 65–69.

Schaffner, J. V., Jr.
1959 Microlepidoptera and their parasites reared from field collections in the northeastern United States. U.S. Dept. Agr. Misc. Pub. 767: 1–97.

Schellenberg, J. R.
1803 Genres des mouches Diptères, représentés en XLII planches projettées et dessinées et expliquées par deux amateurs de l'entomologie. 95 pp., 42 pls. Zurich.

Schenkling, S. (See Horn and Schenkling)

Schiner, I. R. (Schiner's first name was Ignaz, but his works are signed "J. R. Schiner.")
1856 Anmerkungen zu dem Bande V, Pag. 13 dieser Verhandlungen abgedruckten Aufsatze Frauenfeld's: Beitrag zur Insekten-Geschichte. Zool.-Bot. Ver. Wien, Verhandl. 6 (**Abhandl.**): 215–224.
1860a H. 1, pp. 1–72; h. 2, pp. 73–184, pl. 1. *In* (Redtenbacher, L., and Schiner, I. R.), Fauna Austriaca [q.v.]. Die Fliegen (Diptera) (by Schiner), Vol. 1, lxxx+674 pp., 2 pls. Wien, "1862."

> See Griffin, 1932: 570, for pagination and dating of hefts of Schiner's work.

1860b Vorläufiger Commentar zum dipterologischen Theile der "Fauna Austriaca", mit einer näheren Begründung der in derselben aufgenommenen neuen Dipteren-Gattungen. I. Wien. Ent. Monatschr. 4: 47–55.
1860c Vorläufiger Commentar zum dipterologischen Theile der "Fauna Austriaca." II. Wien. Ent. Monatschr. 4: 208–216.

Schiner, I. R.—Continued
- 1861a H. 3/4, pp. 185–368; h. 5, 369–440. *In* (Redtenbacher, L., and Schiner, I. R.), Fauna Austriaca [q.v.]. Die Fliegen (Diptera) (by Schiner), Vol. 1, lxxx+674 pp., 2 pls. Wien, "1862."
  <small>See Griffin, 1932: 570, for pagination and dating of hefts of Schiner's work.</small>
- 1861b Vorläufiger Commentar zum dipterologischen Theile der "Fauna Austriaca". III. Wien. Ent. Monatschr. **5**: 137–144 [cont.].
- 1862a H. 6/7, pp. 441–656; h. 8 (part), pp. 657–674+i-lxxx, pl. 2. *In* (Redtenbacher, L., and Schiner, I. R.), Fauna Austriaca [q.v.]. Die Fliegen (Diptera) (by Schiner), Vol. 1, lxxx+674 pp., 2 pls. Wien, "1862".
  <small>See Griffin, 1932: 570, for pagination and dating of hefts of Schiner's work.</small>
- 1862b Vorläufiger Commentar zum dipterologischen Theile der "Fauna Austriaca". V. Wien. Ent. Monatschr. **6**: 428–436 [cont.].
- 1863 Vorläufiger Commentar zum dipterologischen Theile der "Fauna Austriaca". V. [concl.]. Wien. Ent. Monatschr. **7**: 217–226.
- 1864 Catalogus systematicus dipterorum Europae. 115 pp. Vindobonae [=Vienna].
- 1866a Die Wiedemann'schen Asiliden, interpretirt und in die seither errichteten neuen Gattungen eingereiht. K.-k. Zool.-Bot. Gesell. Wien, Verhandl. **16** (**Abhandl.**): 649–722.
- 1866b Bericht über die von der Weltumseglungsreise [sic] der k. Fregatte Novara mitgebrachten Dipteren. K.-k. Zool.-Bot. Gesell. Wien, Verhandl. **16** (**Abhandl.**): 927–934.
- 1867a Zweiter Bericht über die von der Weltumsgelungsreise der k. Fregatte Novara mitgebrachten Dipteren. K.-k. Zool.-Bot. Gesell. Wien, Verhandl. **17** (**Abhandl.**): 303–314.
- 1867b Neue oder weniger bekannte Asiliden der k. zoologischen Hofcabinetes in Wien. K.-k. Zool.-Bot. Gesell. Wien, Verhandl. **17** (**Abhandl.**): 355–412.
- 1868 Diptera. [Art. 1], 388 pp., 4 pls. *In* [Wüllerstorf-Urbair, B. von, In charge], Reise der österreichischen Fregatte Novara [q.v.]. Zool., Vol. 2, Abt. 1, [Sect.] B. Wien.

Schiødte, J. C. (See also Staeger, 1843)
- 1844 Forhandlinger i det skandinaviske entomologiske selskab. Naturhist. Tidsskr. (1844–1845) ser. 2, **1**: 16–70.

Schirmer, K.

1913   Zwei neue Dipteren aus dem Norden und Süden Europas. Wien. Ent. Ztg. **32**: 221-222.

Schlinger, E. I.

1952   The emergence, feeding habits, and host of *Opsebius diligens* Osten Sacken (Diptera Acroceridae). Pan-Pacific Ent. **28**: 7-12, 1 fig.

1960a   A revision of the genus *Ogcodes* Latreille with particular reference to species of the Western Hemisphere. U.S. Natl. Mus. Proc. **111**: 227-336, 9 figs., pls. 1-13.

1960b   A review of the genus *Eulonchus* Gerstaecker. Part I. The species of the *smaragdinus* group (Diptera: Acroceridae). Ent. Soc. Amer. Ann. **53**: 416-422, 2 figs., pl. 1.

1961   New species of *Acrocera* from Arizona and *Ocnaea* from California, with synonymical notes on the genus *Ocnaea* (Diptera: Acroceridae). Ent. News **72**: 7-12, 3 figs.

Schmitz, H.

[1913]   Biologisch-anatomische Untersuchungen an einer höhlenbewohnenden Mycetophilidenlarve (*Polylepta leptogaster* Winn.). Erste Mitteilung. Natuurhist. Genootsch. in Limburg, Jaarb. [=Meded.] **1912**: 65-96, 3 pls.

> The volume for 1912 is not dated, but includes an article titled "Wintervergadering, gehouden te Roermond op 30 December 1912," and so the date of publication must have been 1913.

1915   Neue Beiträge zur Kenntnis der myrmecophilen und termitophilen Phoriden. Nummer 2-15. Deut. Ent. Ztschr. **1915**: 465-507, pls. 7-11.

1919a   Neue europäische *Aphiochaeta*-Arten III (Phoridae, Dipt.). Ent. Ber. **5**: 139-146.

1919b   Die Phoriden von holländisch Limburg. Mit Bestimmungstabellen aller bisher kenntlich beschriebenen europäischen Phoriden. Dritter Teil. Natuurhist. Genootsch. in Limburg, Jaarb. **1918**: 147-164.

1920a   Die Phoriden von holländisch Limburg. Mit Bestimmungstabellen aller bisher kenntlich beschriebenen europäischen Phoriden. Vierter Teil. Natuurhist. Genootsch. in Limburg, Jaarb. **1919**: 91-152, 1 pl.

1920b   Eine neue nordamerikanische *Phora*-Art (Phoridae, Dipt.). Ent. Ber. **5**: 223-226.

1921   Neue europäische *Aphiochaeta*-Arten IV (Phoridae, Dipt.). Ent. Ber. **5**: 319-327.

Schmitz, H.—Continued
1922 Über das Vorkommen von Kreuzborstenreihen bei Phoriden. Phys. Ökonom. Gesell. Königsberg, Schr. **63**: 130–131.
1924a Een nieuwe phoride, *Cremersia zikani* n. g. n. sp. Natuurhist. Maandbl. **13**: 32–34, 1 fig.
1924b Mitteilungen über allerlei Phoriden (Phoridae orb. terr., Diptera). Natuurhist. Maandbl. **13**: 148–150.
1925a *Trophithauma*, eine neue Phoridengattung aus Costa Rica und den Philippinen. Natuurhist. Maandbl. **14**: 40, 1 fig.
1925b Neue Gattungen und Arten europaeische Phoriden. Encyclopédie Ent., Sér. B, [Div.] II: Diptera **2**: 73–85, 2 figs.
1927a Die paläarktischen Arten der Gattung *Phora* Latr. Bestimmungsschlüssel und neue Arten. Konowia **6**: 144–160, 3 figs.
1927b Klassifikation der Phoriden und Gattungsschluessel [concl.]. Natuurhist. Maandbl. **16**: 9–12, 19–28.
1927c Revision der Phoridengattungen mit Beschreibung neuer Gattungen und Arten. Natuurhist. Maandbl. **16**: 30–39, 45–50, 59–68, 72–79, 92–100, 110–116, 128–132, 142–148, 164, 176, 28 figs. [cont.].
1929a Ergebnisse der Groenlandreise der "Pourquoi pas?" 1926. Eine neue *Megaselia*-Art (Untergattung *Aphiochaeta*) von Ost-Groenland. Natuurhist. Maandbl. **18**: 85–86, 2 figs.
1929b Revision der Phoriden nach forschungsgeschichtlichen und nomenklatorischen, systematischen und anatomischen biologischen und faunistischen Gesichtspunkten. 212 pp., 49 figs., 2 pls. Berlin, Bonn.
1930 Zwei neue nordamerikanische *Phora*-Arten. Natuurhist. Maandbl. **19**: 59–60, 3 figs.
1932 Neue *Stichillus*- und *Phalacrotophora*-Arten, mit einer Aufteilung von *Phalacrotophora* in drei Untergattungen. Tijdschr. v. Ent. **75** (**Sup.**): 115–127, 1 fig.
1937 Over eene nieuwe europeesche *Borophaga* (Diptera, Phoridae). *Borophaga* s. str. *o'kellyi* n. sp. Natuurhist. Maandbl. **26**: 91–92, 1 fig.
1952a Neue Arten von *Gymnophora* und *Megaselia* aus U.S.A. (Diptera, Phoridae). Brotéria **sér. trimest.**: Cien. Nat., **21**: 177–188, 2 figs.
1952b Neue Arten von *Stichillus* und *Borophaga* aus U.S.A. (Phoridae, Diptera). Natuurhist. Maandbl. **41**: 102–104.

Schmitz, H., and Wirth, W. W.
1954 A review of the North American species of the genus *Phora* Latreille (Diptera: Phoridae). Wasmann Jour. Biol. **12**: 113–127, 4 figs.

Schnabl, J. (See also Becker, Dziedzicki et al.; Becker, Kuntze et al.)
[1887] Contributions à la faune diptérologique. Genre *Aricia*. Russ. Ent. Obshch. Trudy (Soc. Ent. Ross. Horae) (1885–1887) **20**: 271–440, 3 figs., 21 pls.

> Dated by Oshanin, 1910: 7.

1888a Contributions à la faune diptérologique. Additions aux descriptions précedentes des *Aricia* et descriptions des espèces nouvelles. Russ. Ent. Obshch. Trudy (Soc. Ent. Ross. Horae) **22**: 378–486.

1888b *Alloeostylus* nov. gen. Anthomyidarum. Ent. Nachr. **14**: 49–50.

1889 Contributions à la faune diptérologique. Russ. Ent. Obshch. Trudy (Soc. Ent. Ross. Horae) **23**: 313–347.

1902 *Limnospila*, nov. gen. Anthomyidarum. Wien. Ent. Ztg. **21**: 111–114, 4 figs.

1911 Anthomyidae. Pp. 62–100. *In* Becker, T., Kuntze, A., Schnabl, J., and Villeneuve, J., Dipterologische Sammelreise nach Korsika (Dipt.) [concl.]. Deut. Ent. Ztschr. **1911**: 62–100, 117–130, 3 figs.

1915 Anthomyidae. Pp. 2–51, 3 pls. *In* Becker, T., Dziedzicki, H., Schnabl, J., and Villeneuve, J., Résultats scientifiques de l'expédition des frères Kuznecov (Kouznetzov) à l'Oural arctique en 1909, sous la direction de H. Backlund. Livr. 7. (Diptera). [Akad. Nauk S. S. S. R.] Imp. Akad. Nauk Petrograd, Zap., Fiz.-Mat. Otd. (Acad. Imp. des Sci. de Petrograd, Mém., Cl. Phys.-Math.) ser. 8, **28** (7): 1–67, 7 figs., 3 pls.

———, **and Dziedzicki, H.**
1911 Die Anthomyiden. K. Leopoldinisch-Carolinisch. Deut. Akad. d. Naturf., Abhandl. (Acad. C. Leopoldino-Carolinae Germ. Nat. Curio., Nova Acta) **95**: 53–358, pls. 3–37.

> Also paged separately, 1–306.

Schoene, W. J. (See Parrott and Schoene)

Schopp, R. (See Doucette et al.)

Schrank, F. von P.
1795   Naturhistorische und ökonomische Briefe über das Donaumoor. 211 pp., 1 pl. Mannheim.
> This reference has not been seen. The date is sometimes given as 1796.

1803   Fauna Boica. Durchgedachte Geschichte der in Baiern einheimischen und zahmen Thiere. Vol. 3, Pt. 1, 272 pp. Landshut.
> The entire work consists of 3 volumes, each in 2 parts, Nürnberg, Ingolstadt, Landshut, 1798-1804. The parts are separately paged except those of Volume 1. The 2 parts of Volume 3 were issued 1803-1804.

Schummel, T. E.   (See also Gravenhorst, 1834, 1837)
1829   Beschreibung der, in Schlesien einheimischen Arten einiger Dipteren-Gattungen. I. *Limnobia*. Beitr. z. Ent., Besonders in Bezug auf Schles. Fauna (Breslau) **1**: 97-201, 5 pls.

Schweinitz, L. D. de
1822   IV. Synopsis fungorum Carolinae superioris, secundum observationes. Naturf. Gesell. zu Leipzig, Schr. **1**: 28-131, pls. 1-2.

Scopoli, J. A.
1763   Entomologia carniolica exhibens insecta carnioliae indigene et distributa in ordines, genera, species, varietates methodo Linnaeana. 421 pp. Vindobonae [=Vienna].

Scotland, M. B.
1934   The animals of the *Lemna* association. Ecology **15**: 290-294, 2 figs.
1940   Review and summary of studies of insects associated with *Lemna minor*. N.Y. Ent. Soc. Jour. **48**: 319-332, 4 pls.

Scudder, S. H.
1888- Butterflies of the eastern United States and Canada, with
1889   special reference to New England. 3 vols., 1958 pp., 89 pls. Cambridge, Mass.
> The volumes are continuously paged and were issued in 12 monthly parts.
> See Williston, 1889a.

Scudder, S. H.   (See also Harris, 1869; Osten Sacken, 1887b)
1899   An unknown tract on American insects by Thomas Say. Psyche **8**: 306-308.

Seago, J. M.
1953 A new species of *Fannia* from North America (Diptera, Muscidae). Kans. Ent. Soc. Jour. **26**: 141–142.
1954 The *pusio* group of the genus *Fannia* Robineau-Desvoidy with descriptions of new species (Diptera, Muscidae). Amer. Mus. Nat. Hist., Amer. Mus. Novitates **1699**: 1–13, 20 figs.

Seamans, H. L.
1917 A new species of *Tropidia* (Syrphidae) from Montana (Dipt.). Ent. News **28**: 342.
1923 An undescribed anthomyid in the Canadian National Collection (Diptera). Canad. Ent. **55**: 221–222.
1926 A new species of muscid from Alberta (Diptera). Canad. Ent. **58**: 175–176.

Sedman, Y. S.
1959 Male genitalia in the subfamily Cheilosiinae. Genus *Chrysogaster* s. l. (Diptera, Syrphidae). Ent. Soc. Wash. Proc. **61**: 49–58, 2 pls.
1961 Male genitalia in the sub-family Cheilosiinae. Genus *Brachyopa* (Diptera: Syrphidae). Ent. Soc. Wash. Proc. **63**: 53–55, 5 figs.

Séguy, E.
1923 Description d'Anthomyiaires nouveaux (Diptères). Soc. Ent. de France, Ann. (1922) **91**: 360–368, 4 figs.
1928 Étude sur quelques Mydaidae nouveaux ou peu connus. Encyclopédie Ent., Sér. B, [Div.] II: Diptera **4**: 129–156, 44 figs.
1937 Diptera. Fam. Muscidae. Fasc. 205, 604 pp., 9 pls. *In* Wytsman, P., ed., Genera insectorum [q.v.]. Bruxelles.
1938 Notes sur les Anthomyiides (Muscidae). 12ᵉ note. Encyclopédie Ent., Sér. B, [Div.] II: Diptera (1937–1938) **9**: 109–120.
1941 Diptères recueillis par M. L. Chopard d'Alger à la Côte d'Ivoire. Soc. Ent. de France, Ann. (1940) **109**: 109–130, 6 figs.
1950 La biologie des Diptères. Encyclopédie Ent., Sér. A **26**: 1–609, 225 figs., 10 pls.
1951 Ordre des Diptères (Diptera Linné, 1758). Pp. 449–744, figs. 438–713, pl. 4. *In* Grassé, P.-P., ed., Traité de zoologie, anatomie, systématique, biologie [q.v.]. Vol. 10: Insectes supérieurs et hémiptéroïdes, Fasc. 1: Névroptéroïdes-mécoptéroïdes hyménoptéroïdes, 975 pp., 905 figs., 5 pls. Paris.

Séguy, E.—Continued
1952   Diptera. Fam. Scatophagidae. Fasc. 209, 107 pp., 44 figs. *In* Wytsman, P., ed., Genera insectorum [q.v.]. Bruxelles.
1953   Un Lycoriide nouveau du Groenland. Encyclopédie Ent., Sér. B, [Div.] II: Diptera (1947–1953) **11**: 118.

Seín, F., Jr.
1933   *Anastrepha* fruit flies in Puerto Rico. Puerto Rico Dept. Agr. Jour. **17**: 183–196, 5 pls.

Selander, R. B., and Vaurie, P.
1962   A gazetteer to accompany the "Insecta" volumes of the "Biologia Centrali-Americana." Amer. Mus. Nat. Hist., Amer. Mus. Novitates **2099**: 1–70, 8 figs.

Sellers, W. F.
1943   The nearctic species of parasitic flies belonging to *Zenillia* and allied genera. U.S. Natl. Mus. Proc. **93**: 1–108.

Serville, J. G. A.   (See Latreille et al.; Lepeletier and Serville)

Sessions, W. R.   (See Wood et al.)

Severin, G., sec.
1911–   Ier Congrès international d'entomologie, Bruxelles, Août,
1912   1910. 2 vols. Bruxelles.
> Volume 2, the "Mémoires," was issued in 1911, Volume 1, the "Historique et procès-verbaux," in 1912.
> See Kertész, 1911.

Shannon, R. C.   (See also Aldrich and Shannon; Dyar and Shannon; Knab and Shannon)
1915a   A new eastern *Brachyopa* (Diptera, Syrphidae). Insecutor Inscitiae Menstruus **3**: 144–145.
1915b   An eastern *Chilosia* with hairy eyes (Diptera, Syrphidae). Ent. Soc. Wash. Proc. **17**: 168.
1916a   Appendix. Systematic and synonymic notes. Pp. 195–203. *In* Banks, N., Greene, C. T., McAtee, W. L., and Shannon, R. C., District of Columbia Diptera: Syrphidae. Biol. Soc. Wash. Proc. **29**: 173–203.
1916b   Two new North American Diptera. Insecutor Inscitiae Menstruus **4**: 69–72, 1 fig.
1916c   Notes on some genera of Syrphidae with descriptions of new species. Ent. Soc. Wash. Proc. **18**: 101–113.
1921   A reclassification of the subfamilies and genera of the North American Syrphidae. Brooklyn Ent. Soc. Bul. **16**: 65–72 [cont.].
1922a   A reclassification of the subfamilies and genera of the North American Syrphidae [cont.]. Brooklyn Ent. Soc. Bul. (1921) **16**: 120–128 [cont.].

Shannon, R. C.—Continued
- 1922b A reclassification of the subfamilies and genera of the North American Syrphidae [concl.]. Brooklyn Ent. Soc. Bul. **17**: 30–43, pl. 2.
- 1922c Revision of the Chilosini (Diptera, Syrphidae). Insecutor Inscitiae Menstruus **10**: 117–145.
- 1923a A reclassification of the subfamilies and genera of North American Syrphidae (Diptera). Appendix. Brooklyn Ent. Soc. Bul. **18**: 17–21.
- 1923b Genera of nearctic Calliphoridae, blowflies, with revision of the Calliphorini (Diptera). Insecutor Inscitiae Menstruus **11**: 101–118, 3 pls.
- 1924a A new *Cynorhinella* (Syrphidae, Diptera). Boston Soc. Nat. Hist. Occas. Papers **5**: 123–124.
- 1924b Nearctic Calliphoridae, Luciliini (Diptera). Insecutor Inscitiae Menstruus **12**: 67–81.
- 1924c Change of preoccupied name (Dip.). Ent. Soc. Wash. Proc. **26**: 178.
- 1924d North American species of *Ferdinandea* (Diptera: Syrphidae). Ent. Soc. Wash. Proc. **26**: 214–215.
- 1925a The genus *Chalcomyia* (Diptera: Syrphidae). Boston Soc. Nat. Hist. Occas. Papers **5**: 151–153.
- 1925b The syrphid-flies of the subfamily Ceriodinae in the U.S. National Museum Collection. Insecutor Inscitiae Menstruus **13**: 48–52, 53–65.
- 1925c North American *Sphecomyia* (Diptera, Syrphidae). Pan-Pacific Ent. **2**: 43–44.
- 1925d Some American Syrphidae (Diptera). Ent. Soc. Wash. Proc. **27**: 107–112.
- 1926a Review of the American xylotine syrphid-flies. U.S. Natl. Mus. Proc. **69** (9): 1–52.
- 1926b The chrysotoxine syrphid-flies. U.S. Natl. Mus. Proc. **69** (11): 1–20.
- 1926c The occurrence of an American genus in Europe and a European genus in America (Diptera: Syrphidae, Sepsidae). Ent. Soc. Wash. Proc. **28**: 112–114.
- 1926d Synopsis of the American Calliphoridae (Diptera). Ent. Soc. Wash. Proc. **28**: 115–139.
- 1939   *Temnostoma bombylans* and related species (Syrphidae, Diptera). Ent. Soc. Wash. Proc. **41**: 215–224.
- 1940   Highland Syrphidae (Diptera) of North Carolina. Ent. Soc. Wash. Proc. **42**: 117–120.

Shannon, R. C., and Dobroscky, I. D.
1924   The North American bird parasites of the genus *Protocalliphora* (Calliphoridae, Diptera). Wash. Acad. Sci. Jour. **14**: 247–253.

Sharp, D.
1887   Insecta. [Sect.] Insecta, 330 pp. *In* Zoological Society of London, The zoological record [q.v.]. Vol. 23 (1886). London.

Sharp, D., ed.
1899–   Fauna Hawaiiensis, being the land-fauna of the Hawaiian
1913   Islands. 3 vols. Cambridge, England.

     The volumes were issued in continuously paged parts at irregular intervals, with the subtitle "or the zoology of the Sandwich (Hawaiian) Isles." Volume 3, which includes the Diptera, was issued in 6 parts, 1901–1910, with the title page dated 1913.
     See Grimshaw, 1901.

Shaw, F. R.
1934   A new species of *Sciara* (Diptera). Canad. Ent. **66**: 233, 2 figs.
1935a   A new species of Sciarinae. Brooklyn Ent. Soc. Bul. **30**: 160, 1 pl.
1935b   Some new Mycetophilidae. Canad. Ent. **67**: 227–230, 1 pl.
1935c   Notes on the Mycetophilidae with descriptions of new species. Psyche **42**: 84–91, 1 pl.
1940   Some new Mycetophilidae. Canad. Ent. **72**: 48–51, 1 pl.
1941a   Fungus gnats from the southern Appalachians. –I. Amer. Midland Nat. **26**: 168–173, 11 figs.
1941b   New Sciarinae from Oklahoma and New Mexico. Amer. Midland Nat. **26**: 320–324, 6 figs.
1941c   Some new species of the genus *Sciara* from Canada. Canad. Ent. **73**: 174–175, 2 figs.
1941d   A new species of *Ceroplatus* (Diptera, Mycetophilidae). Psyche **48**: 20–21.
1947   Some observations on the genus *Leptomorphus* with a description of a new subspecies. Brooklyn Ent. Soc. Bul. (1946) **41**: 155–157.
1948   A new genus and species of fungus-gnats (Mycetophilidae). Brooklyn Ent. Soc. Bul. **43**: 94–96, 1 pl.
1951a   Some new Mycetophilidae from the western United States. Brooklyn Ent. Soc. Bul. **46**: 65–70, 1 pl.
1951b   Some new species of western Mycetophilidae (Diptera). Ent. Soc. Wash. Proc. **53**: 275–280, 1 pl.

Shaw, F. R.—Continued
(1952a) Some notes on synonymy of the Mycetophilidae (Diptera). Psyche (1951) **58:** 148.
1952b The external anatomy of *Palaeoplatyura johnsoni* Joh. (Diptera—Mycetophilidae). Amer. Ent. Soc. Trans. **78:** 21–31, pl. 3.
1953a Some new Diptera with remarks on the affinities of the genus *Pnyxia* Joh. Psyche **60:** 62–68, pl. 3.
1953b A review of some of the more important contributions to our knowledge of the systematic relationships of the Sciaridae (Diptera). Hawaii. Ent. Soc. Proc. (1952) **15:** 25–32.
1962 A key to the North American species of *Bolitophila* (Diptera: Mycetophilidae), with some observations on those described by C. B. D. Garrett. Ent. Soc. Amer. Ann. **55:** 99–101, 1 pl.

————, **and Fisher, E. G.**
1952 Guide to the insects of Connecticut. Part VI. The Diptera or true flies of Connecticut. Fifth fascicle: Midges and gnats [part]. Family Fungivoridae (= Mycetophilidae). Conn. State Geol. and Nat. Hist. Survey, Bul. **80:** 177–231, 232–250 (bibliography and index of entire 5th fascicle), pls. 4–9.

————, **and Shaw, M. M.**
1951 Relationships of certain genera of fungus gnats of the family Mycetophilidae. Smithsn. Inst., Smithsn. Misc. Collect. **117 (3** [=pub. 4053]): 1–23, 45 figs.

Shaw, M. M. (See Shaw and Shaw)

Sherborn, C. D.
1902 Index animalium sive index nominum quae ab A. D. MDCCLVIII generibus et speciebus animalium imposita sunt. Sect. 1, lix + 1195 pp. Cantabrigiae [=Cambridge].

> The entire work is divided into 2 sections: Sect. 1: 1758-1800; and Sect. 2: 1801–1850. Section 1 is complete in 1 volume, 1195 pp., Cambridge, 1902. Section 2 was issued in 33 parts, London, 1922–1933: Pts. 1–28, 1922–1931, continuously paged i-cxxxii, 1–7056; Pts. 29–33 (Addenda), 1932–1933, paged i-vii, cxxxiii-cxlviii, 1–1098.

1922 Introduction, bibliography and Index A-Aff. Pt. 1, pp. i-cxxxi + 1–28. *In his* Index animalium sine index nominum quae ab MDCCLVIII generibus et speciebus animalium imposita sunt. Sect. 2. London.

Sherborn, C. D.—Continued
1923 On the dates of G. W. F. Panzer's 'Fauna Insect. German.', 1792–1844. Ann. and Mag. Nat. Hist. ser. **9, 11:** 566–567.
1937 Brewster's Edinburgh Encyclopaedia. Issued in 18 vols. from 18— to 1830. Soc. Bibliog. Nat. Hist. Jour. **1:** 112.

———, **and Palmer, T. S.**
1898 Dates of Charles d'Orbigny's Dictionnaire universel d'histoire naturelle. Ann. and Mag. Nat. Hist. ser. **7, 3:** 350–352.

———, **and Woodward, B. B.**
1899 On the dates of the 'Encyclopédie Méthodique': Additional note. Zool. Soc. London, Proc. Gen. Mtg. **1899:** 595.
1901 Notes on the dates of publication of the natural history portions of some French voyages. Part I. Ann. and Mag. Nat. Hist. ser. **7, 7:** 388–392.

Sherman, R. S.
1920 Notes on the Mycetophilidae of British Columbia. Ent. Soc. Brit. Columbia, Proc. (1919) **14:** 12–15.
1921 New species of Mycetophilidae. Ent. Soc. Brit. Columbia, Proc. (1920) **16:** 16–21.

Shewell, G. E.
1938 The Lauxaniidae (Diptera) of southern Quebec and adjacent regions. Canad. Ent. **70:** 102–110, 111–118, 133–142, 6 pls.
1939a A revision of the genus *Camptoprosopella* Hendel (Diptera, Lauxaniidae). Canad. Ent. **71:** 130–144, 145–153, 3 pls.
1939b New North American species of *Homoneura* Wulp (Diptera, Lauxaniidae). Canad. Ent. **71:** 264–266.
1940 Preoccupied names in the genus *Homoneura* (Diptera, Lauxaniidae). Canad. Ent. **72:** 86.
1950 A new species of *Sarcophaga* reared from the Columbian ground squirrel (Diptera: Sarcophagidae). Canad. Ent. **82:** 245–246, 2 figs.
1952 New Canadian black flies (Diptera: Simuliidae). I. Canad. Ent. **84:** 33–42, 4 figs.
1955 A new species of *Hilara* from arctic Canada (Diptera: Empididae). Canad. Ent. **87:** 45–46.
1959 New Canadian black flies (Diptera: Simuliidae). II; III. Canad. Ent. **91:** 83–87, 9 figs.; 686–697, 42 figs.

Shewell, G. E.—Continued
- 1960 Notes on the family Odiniidae with a key to the genera and descriptions of new species (Diptera). Canad. Ent. **92**: 625–633, 13 figs.
- 1962 A new Canadian species of *Stenopterina* Macq. with notes on the species allied to *brevipes* (Fab.) (Diptera: Otitidae). Canad. Ent. **94**: 194–200, 18 figs.

——, **and Fredeen, F. J. H.**
- 1958 Two new black flies from Saskatchewan (Diptera: Simuliidae). Canad. Ent. **90**: 733–738, 20 figs.

Shimer, H.
- 1868 Description of a new species of *Cecidomyia*. Amer. Ent. Soc. Trans. **1**: 281–283.
- 1869 A summers study of hickory galls, with descriptions of supposed new insects bred therefrom. Amer. Ent. Soc. Trans. **2**: 386–398.
- 1871 Additional notes on the striped squash beetle. Amer. Nat. **5**: 217–220, 2 figs.

Shiraki, T.
- 1949 Studies on the Syrphidae. 1. The classification of the subfamilies. Mushi **20**: 59–73.

Siebke, H.
- 1863 Beretning om en i sommeren 1861 foretagen entomologisk reise. Nyt Mag. for Naturvidensk. **12**: 105–192.

Sintenis, F.
- 1889 Über *Limnophila pilicornis* Zett. [Tartu, Ülikooli (Dorpat Univ.)] Naturf. Gesell. bei der Univ. Dorpat, Sitzber. (1888) **8**: 396–398.

Sjöstedt, Y., ed.
- 1907– Wissenschaftliche Ergebnisse der schwedischen zoologischen
- 1910 Expedition nach dem Kilimandjaro, dem Meru und den umgebenden Massai Steppen Deutsch-Ostafrikas 1905–1906. 3 vols. Stockholm.

  The entire work is by a number of authors and consists of 22 separately paged parts on the various zoological taxa. The work was issued in hefts and was bound into 3 volumes. Volume 2, 1907–1910, consists of Pts. 8–14, the Diptera being contained in Pt. 10, 1907–1910.
  See Speiser, 1909.

Skinner, F. E.
- 1962 Pacific Coast Entomological Society, Proceedings. Pan-Pacific Ent. **38**: 65–71.

Skinner, H. (See also Fox, 1905; Johannsen, 1907b; Johnson, 1903a)
- 1899 Doings of societies. Ent. News **10**: 79–80.

Skuse, F. A. A.
1888   Diptera of Australia. Part III. The Mycetophilidae. Linn. Soc. N.S. Wales, Proc. **ser. 2, 3** [=whole ser., 12]: 1123–1222, pls. 31–32.

1889   Diptera of Australia. Part VI. The Chironomidae. Linn. Soc. N.S. Wales, Proc. **ser. 2, 4** [=whole ser., 13]: 215–311, pls. 11–14 + 14 bis.

1890a   Diptera of Australia. Part VII. The Tipulidae brevipalpi. Linn. Soc. N.S. Wales, Proc. (1889) **ser. 2, 4** [=whole ser., 13]: 757–892, pls. 21–24.

1890b   Diptera of Australia. Nematocera. Supplement II. Linn. Soc. N.S. Wales, Proc. **ser. 2, 5** [=whole ser., 14]: 595–640, pl. 19.

Slingerland, M. V.
1897   The raspberry-cane maggot (*Phorbia rubivora* Coquillett). Canad. Ent. **29**: 162–163.

——, **and Johnson, F.**
1904   Two grape pests. N.Y. (Cornell) Agr. Expt. Sta. Bul. **224**: 65–73, figs. 26–29.

> Also bound in N.Y. (Cornell) State Col. Agr., and Agr. Expt. Sta., Ann. Rpt. **18**: 65–73, figs. 26–29, *1905*.

Slosson, A. T.
1895   Additional list of insects taken in alpine region of Mt. Washington. [I]; [II]. Ent. News **6**: 4–7; 316–321.

1897   Additional list of insects taken in alpine region of Mt. Washington. [IV]. Ent. News **8**: 237–240.

Smart, J.
1944   Notes on Simuliidae (Diptera). II. Roy. Ent. Soc., London, Proc. Ser. B: Taxonomy **13**: 131–136.

1945   The classification of the Simuliidae (Diptera). Roy. Ent. Soc., London, Trans. **95**: 463–528.

1946   A new name in Simuliidae (Diptera). Entomologist **79**: 22.

Smith, C. N. (See King et al.)

Smith, H. E.
1912   A contribution on North American dipterology. Ent. Soc. Wash. Proc. **14**: 118–127.

1915a   A new genus of Tachinidae from the Canadian Northwest. Canad. Ent. **47**: 153–155.

1915b   New species of Tachinidae (Diptera) from New England. Psyche **22**: 98–102.

1916   New Tachinidae from North America. Ent. Soc. Wash. Proc. **18**: 94–98.

Smith, H. E.—Continued
1917a Notes on New England Tachinidae, with the description of one new genus and two new species. Psyche **24**: 54–58.
1917b Five new species of North American Tachinidae. Psyche **24**: 137–141.
1917c Notes on North American Tachinidae, including the description of one new genus. Ent. Soc. Wash. Proc. **19**: 122–126.

Smith, J. B. (See also Johnson, 1900c, 1910d)
1890 The insects injuriously affecting cranberries. N.J. Agr. Expt. Sta. Spec. Bul. **K**: 1–43, 26 figs.

Smith, L. M.
1928 Distinction between three species of *Eumerus* (Syrphidae, Diptera), with description of a new species. Pan-Pacific Ent. **4**: 137–139, 9 figs.

———, and **Lowe, H.**
1948 The black gnats of California. Hilgardia **18**: 157–183, 16 figs.

Smith, M. E.
1952a A new northern *Aedes* mosquito, with notes on its close ally, *Aedes diantaeus* H., D., & K. (Diptera, Culicidae). Brooklyn Ent. Soc. Bul. **47**: 19–28, 29–40, 3 pls.
1952b Immature stages of the marine fly, *Hypocharassus pruinosus* Wh., with a review of the biology of immature Dolichopodidae. Amer. Midland Nat. **48**: 421–432, 9 figs.

Smith, R. C.
1934 Notes on the Neuroptera and Mecoptera of Kansas, with keys for the identification of species. Kans. Ent. Soc. Jour. **7**: 120–145, 1 pl.

Snodgrass, R. E.
1904 The hypopygium of the Tipulidae. Amer. Ent. Soc. Trans. **30**: 179–236, pls. 8–18.
1924 Anatomy and metamorphosis of the apple maggot, *Rhagoletis pomonella* Walsh. Jour. Agr. Res. **28**: 1–36, 8 figs., 6 pls.
1935 Principles of insect morphology. 667 pp., 319 figs. New York, London.

Snow, F. H. (See also Adams, 1903b)
1904 Lists of Coleoptera, Lepidoptera, Diptera and Hemiptera collected in Arizona by the entomological expeditions of the University of Kansas in 1902 and 1903. Kans. Univ. Sci. Bul. **2** [=whole ser., 12]: 323–350. (=Kans. Univ. Bul. 4 (9).)

Snow, W. A.

1891   The moose fly—a new *Haematobia.* Canad. Ent. **23**: 87–89.

1892   Notes and descriptions of Syrphidae. Kans. Univ. Quart. **1**: 33–38, 1 pl.

1894a   Descriptions of North American Trypetidae, with notes. Kans. Univ. Quart. **2**: 159–174, 2 pls.

1894b   American Platypezidae. Kans. Univ. Quart. **3**: 143–152, 1 pl.

1895a   *Cnephalia* and its allies. Kans. Univ. Quart. **3**: 177–186.

1895b   A new species of *Pelecocera.* Kans. Univ. Quart. **3**: 187.

1895c   American Platypezidae. Paper II. Kans. Univ. Quart. **3**: 205–207.

1895d   Diptera of Colorado and New Mexico. Syrphidae. Kans. Univ. Quart. **3**: 225–247.

1896   List of Asilidae, supplementary to Osten Sacken's Catalogue of North American Diptera. 1878–1895. Kans. Univ. Quart. **4**: 173–190.

———, **and Mills, H.**

1900   The destructive *Diplosis* of the Monterey pine. Ent. News **11**: 489–494, 1 pl.

Snyder, F. M.

1940   The genus *Acridomyia* Stackelberg in North America (Diptera: Muscidae). Amer. Mus. Nat. Hist., Amer. Mus. Novitates **1076**: 1–2, 3 figs.

1941   Contribution to a revision of neotropical Mydaeini (Diptera: Muscidae.) Amer. Mus. Nat. Hist., Amer. Mus. Novitates **1134**: 1–22.

1949a   Nearctic *Helina* Robineau-Desvoidy (Diptera, Muscidae). Amer. Mus. Nat. Hist. Bul. **94**: 111–160.

1949b   Review of nearctic Mydaea, sensu stricto, and *Xenomydaea* (Diptera, Muscidae). Amer. Mus. Nat. Hist., Amer. Mus. Novitates **1401**: 1–38.

1954   A review of nearctic *Lispe* Latreille (Diptera, Muscidae). Amer. Mus. Nat. Hist., Amer. Mus. Novitates **1675**: 1–40, 77 figs.

1956   Notes and descriptions of *Muscina* and *Dendrophaonia* (Diptera: Muscidae). Ent. Soc. Amer. Ann. (1955) **48**: 445–452, 7 figs., 1 pl.

1958   A review of New World *Neodexiopsis* (Diptera, Muscidae). Amer. Mus. Nat. Hist., Amer. Mus. Novitates **1892**: 1–27, 16 figs.

Société de Gens de Lettres, de Savans et d'Artistes
1782- Encyclopédie méthodique, ou par ordre de matières. 196
1832 vols. Paris, Liège.

> This large work, sometimes titled "Dictionnaire encyclopédie méthodique," is by many authors, and is divided into 26 major sections, with many additional miscellaneous volumes. The work covers all branches of arts and sciences; it was never completed. There are 14 volumes of text and 13 volumes of plates covering zoology. The text volumes consist of 10 volumes numbered in series and subtitled "Histoire naturelle", 1782–1828, 3 volumes subtitled "Histoire naturelle des Vers", 1789–1832, and 1 volume subtitled "Histoire naturelle des Zoophytes", 1824–1827. The volumes of plates are entitled "Tableau encyclopédique et méthodique des trois règnes de nature", as are those on botany. The 7 parts on Arthropoda are contained in text volumes 4–10 of the above section "Histoire naturelle", which were published in Paris, 1789–1828. The 397 plates were issued in 2 parts: Pls. 1–267 in 1797, under the supervision of J. P. Bonnaterre, and with their 142 pp. of explanation by F. E. Guérin-Méneville published in 1818; Pls. 268–397 and 38 pp. of explanation by P. A. Latreille, in 1818. The text was issued in livraisons and several authors were involved, both in authorship of whole volumes and in articles within the volumes. Volumes 4–8 were mainly by G. A. Olivier, and were further subtitled "Insectes"; Volumes 9–10 were by Latreille and others, and were further subtitled "Entomologie, ou histoire naturelle des Crustacés, des Arachnides et des Insectes".
> See Latreille et al., 1825–1828; Olivier, 1811–1812.

Société de Naturalistes et d'Agriculteurs (See also Latreille, 1804)
1803 Nouveau dictionnaire d'histoire naturelle, appliqué aux arts, principalement à l'agriculture et à l'économie rurale et domestique. Vol. 4, 575 pp., pls. B1–B12, B14–B15. Paris.

> The entire work, by many authors, consists of 24 volumes, Paris, 1803–1804, of which the 24th consists of 5 separately paged sections.

1818 Nouveau dictionnaire d'histoire naturelle, appliqué aux arts, principalement à l'agriculture et à l'économie rurale et domestique. Nouvelle edition. [Ed. 2], Vol. 23, 612 pp., 6 pls. Paris.

> This "new edition" [=2nd] of the preceding consists of 36 volumes, published at Paris, 1816–1819.

Sommerman, K. M.
1958 Two new species of Alaskan *Prosimulium*, with notes on closely related species (Diptera, Simuliidae). Ent. Soc. Wash. Proc. **60**: 193–202, 32 figs.
1962a *Prosimulium doveri*, a new species from Alaska, with keys to related species (Diptera: Simuliidae). Ent. Soc. Wash. Proc. (1961) **63**: 225–235, 1 pl.

Sommerman, K. M.—Continued
1962b   Alaskan snipe fly immatures and their habitat (Rhagionidae: *Symphoromyia*). Mosquito News **22**: 116–123, 16 figs.

———, **Sailer, R. I., and Esselbaugh, C. O.**
1955   Biology of Alaskan black flies (Simuliidae, Diptera). Ecol. Monog. **25**: 345–385, 8 figs.

Sonnini, C. S., ed.
[1799]–   Histoire naturelle par Buffon, nouvelle édition accom-
1808   pagnée de notes. 127 vols. Paris, "An VII"–1808.

> Many authors are involved in this work. Only the first 64 volumes were consecutively numbered; the last 63 appeared separately, later being called "Suites à Buffon" with volume numbers being assigned. The 14 volumes by Latreille called "Histoire naturelle, générale et particulière, des Crustacés et des Insectes", form volumes [93–106] of the "Suites". They were published in Paris, "An X–XIII" [=1802–1805].
> See Latreille, 1802, 1805.

Souza Lopes, H. de   (See Lopes, H. de S.)

Spärck, R.
1922–   Beiträge zur Kenntnis der Chironomidenmetamorphose
1923   I–IV. Ent. Meddel. **14** [=ser. 2, 9]: 32–48, *1922*; 49–110, figs. 1–4 + 1–8, *1923*.

Speiser, P.
1899   Eine neue, auf Halbaffen lebende Hippobosciden-Art (Dipt.). Wien. Ent. Ztg. **18**: 197–202, 1 fig.
1902a   Besprechung einiger Gattungen und Arten der Diptera Pupipara. Természet. Füzetek **25**: 327–338.
1902b   Studien über Diptera Pupipara. Ztschr. f. System. Hymenopt. u. Dipt. **2**: 145–180.
1904a   Zur Nomenclatur blutsaugender Dipteren Amerikas. Insekten-Börse **21**: 148.
1904b   Besprechung einiger Gattungen und Arten der Diptera Pupipara. II. Budapest Magyar Nemzeti Muz., Ann. Hist. Nat. **2**: 386–395.
1905   Beiträge zur Kenntnis der Hippobosciden (Dipt.). Ztschr. f. System. Hymenopt. u. Dipt. **5**: 347–360.
1909   4. Orthorhapha. Pp. 31–65, figs. 1–15 [cont.]. *In* Sjöstedt, Y., ed., Wissenschaftliche Ergebnisse der schwedischen zoologischen Expedition nach dem Kilimandjaro [q.v.]. Vol. 2, Pt. 10: Diptera, 202 pp., 19 figs., 2 pls. Stockholm.
1914   Ein neues Beispiel vicariierender Dipterenarten in Nordamerika und Europa. Zool. Anz. **44**: 91–94.

Spence, W. (See Kirby and Spence)

Spencer, W. P. (See also Patterson, 1943; Stalker and Spencer)
- 1938   *Drosophila virilis americana*, a new sub-species. [Abstract]. Pp. 169–170. *In* Demerec, M., sec., Abstracts of papers presented at the 1937 meetings of the Genetics Society of America, Woods Hole, Massachusetts, August 30–September 1, 1937, Indianapolis, Indiana, December 28–30, 1937. Genetics **23**: 139–177.
- 1940   Levels of divergence in *Drosophila* speciation. Amer. Nat. **74**: 299–311.
- 1942   New species in the *quinaria* group of the subgenus *Drosophila*. Tex. Univ. Pub. **4213**: 53–66.

Spilker, O. W. (See Porter, Spilker, and Walker)

Spruijt, F. J. (See Creager and Spruijt)

Spuler, A. (See also Melander and Spuler)
- 1923   North American genera and subgenera of the dipterous family borboridae. Acad. Nat. Sci. Phila. Proc. **75**: 369–378.
- 1924a   Species of subgenera *Collinella* and *Leptocera* of North America. Ent. Soc. Amer. Ann. **17**: 106–117, 1 pl.
- 1924b   North American species of the genera *Sphaerocera* and *Aptilotus* (Diptera-Borboridae). Pan-Pacific Ent. **1**: 66–71, 4 figs.
- 1924c   North American species of the subgenera *Opacifrons* Duda and *Pteremis* Rondani of the genus *Leptocera* Olivier (Diptera, Borboridae). Psyche **31**: 121–135, 10 figs.
- 1925a   North American species of *Borborus* Meigen, and *Scatophora* Robineau-Desvoidy. Brooklyn Ent. Soc. Bul. **20**: 1–16, 13 figs.
- 1925b   Studies in North American Borboridae (Diptera). Canad. Ent. **57**: 99–104, 116–124, 1 pl.
- 1925c   North American species of the subgenus *Scotophilella* Duda (Diptera, Borboridae). N.Y. Ent. Soc. Jour. **33**: 70–84, 147–162, 1 pl.

Stabler, R. M. (See Fox and Stabler)

Stackelberg, A. A. (or Stakelberg, Shtakel'berg, Schtakel'berg)
- 1928   Dolichopodiden-Studien. I. Die paläarktischen *Dolichopus*-Arten mit gelben Schenkeln und schwarzen Postocularcilien. Zool. Anz. **79**: 260–269, 6 figs.
- 1929   (Über eine neue Muscide, die als Parasit in *Locusta migratoria* L. auftritt.) [Leningrad] Gosud. Inst. Opytn. Agron. (State Inst. Expt. Agron.) Otd. Prikl. Ent. (Bur. Appl. Ent.) Isv. Prikl. Ent. **4**: 121–129, 7 figs.
  In Russian; German summary, pp. 126–128.

Stackelberg, A. A.—Continued
- 1930– Dolichopodidae. [Fam.] 29, pp. 1–64, figs. 1–31, pls. 1–2
- 1941 (=lfg. 51), *1930;* pp. 65–128, figs. 32–61, pls. 3–4 (=lfg. 71), *1933;* pp. 129–176, figs. 62–143 (=lfg. 82), *1934;* pp. 177–224, figs. 144–169, pls. 5–12 (=lfg. 138), *1941* [cont.]. *In* Lindner, E., ed., Die Fliegen der palaearktischen Region [q.v.]. Vol. 4, Pt. 5. Stuttgart.
- 1952 [New species of Cordyluridae (Diptera) from northern U.S.S.R.] Akad. Nauk S.S.S.R., Zool. Inst., Trudy **12:** 405–407, 4 figs.
  In Russian.

Staeger, C. (See also Zetterstedt, 1849)
- 1839 Systematisk fortegnelse over de i Danmark hidtil fundne Diptera. Naturhist. Tidsskr. (1838–1839) **2:** 549–600 [cont.].
- 1840 Systematisk fortegnelse over de i Danmark hidtil fundne Diptera [concl.]. Naturhist. Tidsskr. (1840–1841) **3:** 1–58, 228–288.
- 1843 (Udsigt over den danske faunas arter af antliatslaegten *Platycheirus.*) Pp. 320–327. *In* Schiødte, J. C., Forhandlinger i det Skandinaviske Entomologiske Selskab. Naturhist. Tidsskr. (1842–1843) **4:** 315–360.
- 1845a Bemaerkninger til synonymien af *Sciomyza glabricula* Fall. Naturhist. Tidsskr. (1844–1845) **ser. 2, 1:** 38–40.
- 1845b Grønlands antliater. Naturhist. Tidsskr. (1844–1845) **ser. 2, 1:** 346–369.

Stage, H. H., Gjullin, C. M., and Yates, W. W.
- 1952 Mosquitoes of the northwestern states. U.S. Dept. Agr., Agr. Handb. **46:** 1–95, 39 figs.

Stains, G. S., and Knowlton, G. F.
- 1939 A new *Myopa* from Utah (Diptera). Utah Acad. Sci., Arts and Letters, Proc. **16:** 51.
- 1940 Three new western Simuliidae (Diptera). Ent. Soc. Amer. Ann. **33:** 77–80, 1 fig.
- 1943 A taxonomic and distributional study of Simuliidae of western United States. Ent. Soc. Amer. Ann. **36:** 259–280, 3 pls.

Stainton, H. T. (See Walker, Stainton, and Wilkinson)

Stalker, H. D.
- 1945 On the biology and genetics of *Scaptomyza graminum* Fallén (Diptera, Drosophilidae). Genetics **30:** 266–279, 1 fig.
- 1953 Taxonomy and hybridization in the *cardini* group of *Drosophila.* Ent. Soc. Amer. Ann. **46:** 343–358, 3 figs.

Stalker, H. D.—Continued
- 1956   On the evolution of parthenogenesis in *Lonchoptera* (Diptera). Evolution (Lancaster, Pa.) **10**: 345–359, 2 figs.

Stalker, H. D., and Spencer, W. P.
- 1939   Four new species of *Drosophila*, with notes on the *funebris* group. Ent. Soc. Amer. Ann. **32**: 105–112, 6 figs.

Stannius, F. H.
- 1831a   Die europäischen Arten der Zweyflüglergattung *Dolichopus*. Isis (Oken's) **1831**: 28–68, 122–144, 248–271, pl. 1 (part).
- 1831b   Observationes de speciebus nonnullis generis *Mycetophila* vel novis vel minus cognitus. 30 pp., 1 pl. Breslau.

Starcke, H.   (See Stone, Knight, and Starcke)

Stearn, W. T.
- 1937   On the dates of publication of Webb and Berthelot's Histoire naturelle des Iles Canaries. Soc. Bibliog. Nat. Hist. Jour. **1**: 49–63.

Stebbins, F. A.
- 1910   Insect galls of Springfield, Massachusetts, and vicinity. Springfield [Mass.] Mus. Nat. Hist. Bul. **2**: 1–64, 32 pls.

Steenberg, C. M.
- 1924   Étude sur deux espèces de *Phronia* dont les larves se forment de leurs excréments une couche protectrice: La *Phronia strenua* Winn. et la *Phronia johannae* n. sp. (Diptera Nematocera); Errata. Dansk Naturhist. For. København, Vidensk. Meddel. **78** [=ser. 8, 5]: 1–51, 13 figs., 8 pls.; 211.
- 1938   Recherches sur la métamorphose d'un Mycétophile *Delopsis aterrima* (Zett.) (Diptera Nematocera). K. Danske Vidensk. Selsk., Biol. Meddel. **14** (**1**): 1–29, 8 pls.
- 1943   Études sur les larves du genre *Phronia* (Fungivoridae, Nematocera). Ent. Meddel. **23**: 337–351, 2 pls.

Stefani Perez, T. De   (See De Stefani Perez, T.)

Steffan, W. A.   (See Philip and Steffan)

Stein, P.   (See also Becker et al., 1903a, 1903b; Becker et al., 1905, 1907; Bezzi and Stein)
- 1892   Drei neue merkwürdige *Homalomyia*-Arten. Ein dipterologischer Beitrag. Wien. Ent. Ztg. **11**: 69–77.
- 1895   Die Anthomyidengruppe *Homalomyia* nebst ihren Gattungen und Arten. Berlin. Ent. Ztschr. **40**: 1–141.
- 1897   Anthomyiden mit *Lispa*-ähnlich erweiterten Tastern. Ent. Nachr. **23**: 317–323.

Stein, P.—Continued
1898   Nordamerikanische Anthomyiden. Beitrag zur Dipterenfauna der Vereinigten Staaten. Berlin. Ent. Ztschr. (1897) **42**: 161–288.
1899   *Euryomma*, eine neue Gattung der Anthomyidengruppe. Ent. Nachr. **25**: 19–22.
1900   Einige neue Anthomyiden. Ent. Nachr. **26**: 305–324.
1901   Die Walker'schen aussereuropäischen Anthomyiden in der Sammlung des British Museum zu London. Ztschr. f. System. Hymenopt. u. Dipt. **1**: 185–221.
1903   Die europäischen Arten der Gattung *Hydrotaea* Rob.-Desv. K.-k. Zool.-Bot. Gesell. Wien, Verhandl. **53**: 285–337.
1904   Die amerikanischen Anthomyiden des Königlichen Museums für Naturkunde zu Berlin und des Ungarischen National-Museums zu Budapest. Budapest Magyar Nemzeti Muz., Ann. Hist. Nat. **2**: 414–495.
1906   Die mir bekannten europäischen *Pegomyia*-Arten. Wien. Ent. Ztg. **25**: 47–107.
1907   Zur Kenntniss der Dipteren von Central-Asien. II. [Akad. Nauk S.S.S.R.] Imp. Akad. Nauk St. Petersburg, Zool. Muz. Ezheg. (Acad. Imp. des Sci. de St. Pétersbourg, Zool. Mus. Ann.) **12**: 318–372.
1908   Analytische Übersicht aller mir bekannten breitstirnigen Anthomyiden-Männchen mit Ausschluss der Gattungen *Lispa* und *Fucellia* (Dipt.). Wien. Ent. Ztg. **27**: 1–15.
1910   Zur Kenntniss der Gattung *Fucellia* Rob. Desv. Wien. Ent. Ztg. **29**: 11–27.
1911   Die von Schnuse in Südamerika gefangenen Anthomyiden. Arch. f. Naturgesch. **77** (**vol. 1, h. 1**): 61–189.
1914   Versuch, die Gattungen und Arten unserer Anthomyiden nur nach dem weiblichen Geschlecht zu bestimmen, nebst Beschreibung einiger neuen Arten. Arch. f. Naturgesch. (1913) **Abt. A, 79** (**8**): 4–55.
1916   Die Anthomyiden Europas. Tabellen zur Bestimmung der Gattungen und aller mir bekannten Arten, nebst mehr oder weniger ausführlichen Beschreibungen. Arch. f. Naturgesch. (1915) **Abt. A, 81** (**10**): 1–224.
1918   Zur weitern Kenntnis aussereuropaeischer Anthomyiden. Budapest Magyar Nemzeti Muz., Ann. Hist. Nat. **16**: 147–244.
1919   Die Anthomyidengattungen der Welt, analytisch bearbeitet, nebst einem kritisch-systematischen Verzeichniss aller aussereuropäischen Arten. Arch. f. Naturgesch. (1917) **Abt. A, 83** (**1**): 85–178.

Stein, P.—Continued
1920 Nordamerikanische Anthomyiden. 2. Beitrag. Arch. f. Naturgesch. (1918) **Abt. A, 84 (9):** 1–106.
1924 Die verbreitetsten Tachiniden Mitteleuropas nach ihren Gattungen und Arten. Arch. f. Naturgesch. **Abt. A, 90 (6):** 1–271.

Steiner, L. F., Rohwer, G. G., Ayers, E. L., and Christenson, L. D
1961 The role of attractants in the recent Mediterranean fruit fly eradication program in Florida. Jour. Econ. Ent. **54:** 30–35.

Stenhammar, C.
1844 Försök till gruppering och revision af de svenska Ephydrinae. K. Vetensk. Akad. Handl. [ser. 3], 1843: 75–272, 1 pl.
[1854] Skandinaviens Copromyzinae. K. Vetensk. Akad. Handl. [ser. 3], **1853:** 257–442.
> This portion of the 1853 volume could not have appeared until at least 1854, as evidenced by the many 1854 dates mentioned in the text, including the statement "Inlemnad den 19 December 1854" on the title page of the article by Stenhammar. This was also published separately under the title "Copromyzinae Scandinaviae," 184 pp., Holmiae [=Stockholm], *1855*. The separate dates the journal article as 1854.

Stephen, W. P. (See Bohart, Stephen, and Eppley)

Stephens, J. F.
1829a The nomenclature of British insects. 68 pp. London.
> Published June 1, 1829. See Blackwelder, 1949: 92.

1829b A systematic catalogue of British insects: Being an attempt to arrange all the hitherto discovered indigenous insects in accordance with their natural affinities. Vol. 2, 388 pp. London.
> Published July 15, 1829. See Blackwelder, 1949: 92. The entire work consists of 2 volumes, London, 1829.

1841 A list of insects found near Harrietsham, in Kent. Together with the description of a new genus and species of Yponomeutidae. Entomologist (Newman's) (1840–1842) [1]: 199–202.

Stetson, S. S. (See Wood et al.)

Steyskal, G. C. (See also James and Steyskal)
1938a New Stratiomyidae and Tetanoceridae (Diptera) from North America. Mich. Univ. Mus. Zool. Occas. Papers **386:** 1–10, 2 pls.
1938b The pre-copulatory behavior of the male of *Dolichopus omnivagus* Van Duzee (Diptera, Dolichopidae). Brooklyn Ent. Soc. Bul. **33:** 193–194.

Steyskal, G. C.—Continued
- 1939   A new species of *Dictya* (Sciomyzidae, Diptera). Canad. Ent. **71**: 78, 1 fig.
- 1941a   A curious habit of an empidid fly. Brooklyn Ent. Soc. Bul. **36**: 117, 1 fig.
- 1941b   A new species of *Euparyphus* from Michigan (Diptera, Stratiomyidae). Brooklyn Ent. Soc. Bul. **36**: 123–124.
- 1941c   A new species of *Pterodontia* (Diptera, Acroceridae), Brooklyn Ent. Soc. Bul. **36**: 140.
- 1942a   A new species of *Phyllomyza* from Virginia (Diptera: Milichiidae). Ent. News **53**: 84–85, 1 fig.
- 1942b   Notes on the genus *Dolichopus* (Diptera, Dolichopodidae). Paper 2. Brooklyn Ent. Soc. Bul. **37**: 62–67.
- 1942c   A curious habit of an empidid fly: Further notes. Brooklyn Ent. Soc. Bul. **37**: 67.
- 1943a   A new species of *Pholeomyia*, with a key to the North American species (Diptera, Milichiidae). Ent. News **54**: 99–102.
- 1943b   Old World Sepsidae in North America, with a key to the American genera (Diptera). Pan-Pacific Ent. **19**: 93–95, 1 fig.
- 1944a   A key to the North American species of the genus *Suillia* R.-D. (Diptera, Helomyzidae). Brooklyn Ent. Soc. Bul. **39**: 173–176.
- 1944b   A new ant-attacking fly of the genus *Pseudacteon*, with a key to the females of the North American species (Diptera, Phoridae). Mich. Univ. Mus. Zool. Occas. Papers **489**: 1–4.
- 1946   *Themira nigricornis* Meigen in North America, with a revised key to the nearctic species of *Themira* (Diptera: Sepsidae). Ent. News **57**: 93–95.
- 1947a   A revision of the nearctic species of *Xylomyia* and *Solva* (Diptera, Erinnidae). Mich. Acad. Sci., Arts, and Letters, Papers (1945) **31**: 181–190, 8 figs.
- 1947b   The genus *Diacrita* Gerstaecker (Diptera, Otitidae). Brooklyn Ent. Soc. Bul. (1946) **41**: 149–154, 2 figs.
- 1947c   Notes on the genus *Dolichopus* (Diptera, Dolichopodidae). Paper 3. Brooklyn Ent. Soc. Bul. **42**: 34–38.
- 1949a   New Diptera from Michigan (Stratiomyidae, Sarcophagidae, Sciomyzidae). Mich. Acad. Sci., Arts, and Letters, Papers (1947) **33**: 173–180, 1 fig.
- 1949b   A new anomalous acalyptrate fly (Diptera). Brooklyn Ent. Soc. Bul. **44**: 134–137, 1 pl.

Steyskal, G. C.—Continued
- 1950   A curious habit of an empidid fly: Third note. Brooklyn Ent. Soc. Bul. **45**: 155.
- 1951a   The dipterous fauna of tree trunks. Mich. Acad. Sci., Arts, and Letters, Papers (1949) **35**: 121–134.
- 1951b   A new species of *Euparyphus* from Ontario (Diptera, Stratiomyidae). Ent. Soc. Wash. Proc. **53**: 273–274.
- (1951c)   The genus *Sepedon* Latreille in the Americas (Diptera: Sciomyzidae). Wasmann Jour. Biol. (1950) **8**: 271–297, 5 pls.
- 1953a   A suggested classification of the lower brachycerous Diptera. Ent. Soc. Amer. Ann. **46**: 237–242.
- 1953b   Further notes on Diptera of tree trunks with descriptions of two new species of *Drapetis* (Diptera: Empididae) and an abstract of a Finnish paper on Diptera by Tuomikoski. Mich. Acad. Sci., Arts, and Letters, Papers (1952) **38**: 255–260, 6 figs.
- 1954a   *Colobaea* and *Hedria*, two genera of Sciomyzidae new to America (Diptera: Acalyptratae). Canad. Ent. **86**: 60–65, 2 figs.
- 1954b   The American species of the genus *Dictya* Meigen (Diptera, Sciomyzidae). Ent. Soc. Amer. Ann. **47**: 511–539, 2 figs., 8 pls.
- 1954c   The Sciomyzidae of Alaska (Diptera). Ent. Soc. Wash. Proc. **56**: 54–71, 12 figs.
- 1954d   The genus *Pteromicra* Lioy (Diptera, Sciomyzidae) with especial reference to the North American species. Mich. Acad. Sci., Arts, and Letters, Papers (1953) **39**: 257–269, 9 figs.
- 1956a   The genus *Seioptera* Kirby (Diptera, Otitidae). Ent. Soc. Amer. Ann. **49**: 30–32, 2 figs.
- 1956b   The eastern species of *Nemomydas* Curran (Diptera: Mydaidae). Mich. Univ. Mus. Zool. Occas. Papers **573**: 1–5, 1 pl.
- 1956c   New species and taxonomic notes in the family Sciomyzidae (Diptera, Acalyptratae). Mich. Acad. Sci., Arts, and Letters, Papers (1955) **41**: 73–87, 17 figs.
- 1957   A revision of the family Dryomyzidae (Diptera, Acalyptratae). Mich. Acad. Sci., Arts, and Letters, Papers (1956) **42**: 55–68, 14 figs.
- 1958a   Notes on nearctic Helcomyzidae and Dryomyzidae (Diptera, Acalyptratae). Mich. Acad. Sci., Arts, and Letters, Papers (1957) **43**: 133–143, 14 figs.

Steyskal, G. C.—Continued
[1958b] A new species of the genus *Pteromicra* associated with snails (Diptera, Sciomyzidae). Ent. Soc. Wash. Proc. (1957) **59**: 271–272, 3 figs.
> This issue was published Jan. 1958, according to R. H. Foote, editor of the Proceedings.

1958c Notes on the Richardiidae, with a review of the species known to occur in the United States (Diptera, Acalyptratae). Ent. Soc. Amer. Ann. **51**: 302–310, 3 pls.

1959a The American species of the genus *Tetanocera* Duméril (Diptera). Mich. Acad. Sci., Arts, and Letters, Papers (1958) **44**: 53–91, 40 figs.

1959b *Dolichopus correus*, new species, and notes on other Dolichopodidae (Diptera, Brachycera). Mich. Univ. Mus. Zool. Occas. Papers **604**: 1–6, 9 figs.

1960a The genus *Antichaeta* Haliday, with special reference to the American species (Diptera, Sciomyzidae). Mich. Acad. Sci., Arts, and Letters, Papers (1959) **45**: 17–26, 11 figs.

1960b New North and Central American species of Sciomyzidae (Diptera: Acalyptratae). Ent. Soc. Wash. Proc. **62**: 33–43, 18 figs.

1961a Two new species of *Sepsisoma* from Kansas (Diptera: Richardiidae). Kans. Ent. Soc. Jour. **34**: 83–85, 2 figs.

1961b The genera of Platystomatidae and Otitidae known to occur in America north of Mexico (Diptera, Acalyptratae). Ent. Soc. Amer. Ann. **54**: 401–410, 16 figs.

1961c The North American Sciomyzidae related to *Pherbellia fuscipes* (Macquart) (Diptera Acalyptratae). Mich. Acad. Sci., Arts, and Letters, Papers (1960) **46**: 405–415, 7 figs.

1962a The American species of the genera *Melieria* and *Pseudotephritis* (Diptera: Otitidae). Mich. Acad. Sci., Arts, and Letters, Papers (1961) **47**: 247–262, 8 figs.

1962b The genus *Curranops* Harriot (Diptera: Otitidae). Ent. Soc. Wash. Proc. **64**: 117–118, 1 fig.

1963 A second North American species of *Traginops* Coquillett (Diptera, Odiniidae). Ent. Soc. Wash. Proc. **65**: 51–54, 6 figs.

———, **and Foote, R. H.**
1962 Notes on the genus *Icterica* Loew in North America (Diptera: Tephritidae). Ent. Soc. Wash. Proc. **64**: 166, 2 figs.

Stoll, N. R., Dollfus, R. P., Forest, J., Riley, N. D., Sabrosky, C. W., Wright, C. W., and Melville, R. V.
1961 International code of zoological nomenclature adopted by the XV International Congress of Zoology. 176 pp. London.

Stone, A. (See also Gjullin et al.; Mattingly, Stone, and Knight; Wirth and Stone)
1933 Two new species of *Tabanus* from North America (Diptera). Ent. Soc. Wash. Proc. **35**: 75–77.
1935 Notes on Tabanidae (Diptera). Ent. Soc. Wash. Proc. **37**: 11–21, 1 pl.
1938 The horseflies of the subfamily Tabaninae of the Nearctic Region. U.S. Dept. Agr. Misc. Pub. **305**: 1–171, 79 figs.
1939a A revision of the genus *Pseudodacus* Hendel (Dipt., Trypetidae). Rev. de Ent. **10**: 282–289, 4 figs., 1 pl.
1939b A new genus of Trypetidae near *Anastrepha* (Diptera). Wash. Acad. Sci. Jour. **29**: 340–350, 16 figs.
1940 Two new nearctic Tabanidae and some new records and corrections (Diptera). Ent. Soc. Wash. Proc. **42**: 59–63, 6 figs.
1941 The generic names of Meigen 1800 and their proper application (Diptera). Ent. Soc. Amer. Ann. **34**: 404–418.
1942 The fruitflies of the genus *Anastrepha*. U.S. Dept. Agr. Misc. Pub. **439**: 1–112, 22 figs., 23 pls.
1948 *Simulium virgatum* Coquillett and a new related species (Diptera: Simuliidae). Wash. Acad. Sci. Jour. **38**: 399–404, 10 figs.
1949 A new genus of Simuliidae from Alaska (Diptera). Ent. Soc. Wash. Proc. **51**: 260–267, 2 pls.
1952 The Simuliidae of Alaska (Diptera). Ent. Soc. Wash. Proc. **54**: 69–96, 4 figs.
1953 New tabanid flies of the tribe Merycomyiini. Wash. Acad. Sci. Jour. **43**: 255–258, 3 figs.
1954 The genus *Bolbodimyia* Bigot (Diptera, Tabanidae). Ent. Soc. Amer. Ann. **47**: 248–254.
1962 A correction in mosquito nomenclature (Diptera: Culicidae). Ent. Soc. Wash. Proc. (1961) **63**: 246.

———, **and DeFoliart, G. R.**
1959 Two new black flies from the western United States (Diptera, Simuliidae). Ent. Soc. Amer. Ann. **52**: 394–400, 4 pls.

———, **and Jamnback, H. A.**
1955 The black flies of New York State (Diptera: Simuliidae). N.Y. State Mus. Bul. **349**: 1–144, 23 pls.

Stone, A., Knight, K. L., and Starcke, H.
1959   A synoptic catalog of mosquitoes of the world (Diptera, Culicidae). Vol. 6, 358 pp. *In* Entomological Society of America, Thomas Say Foundation [q.v.]. Washington.

———, **and Peterson, B. V.**
1958   *Simulium defoliarti*, a new black fly from the western United States (Diptera, Simuliidae). Brooklyn Ent. Soc. Bul. **53**: 1–6, 2 pls.

———, **and Wirth, W. W.**
1947   On the marine midges of the genus *Clunio* Haliday (Diptera, Tendipedidae). Ent. Soc. Wash. Proc. **49**: 201–224, 3 pls.

Stone, W. S.   (See also Patterson and Stone; Patterson, Stone, and Griffen)

———, **Griffen, A. B., and Patterson, J. T.**
1941   *Drosophila montana*, a new species of the *virilis* group. [Abstract.] P. 172. *In* Kaufmann, B. P., sec., Abstracts of papers presented at the 1941 meetings of the Genetics Society of America. Cold Spring Harbor, New York, August 27–29, 1941, Dallas, Texas, December 29–31, 1941. Genet. Soc. Amer. Rec. **10** (**spec. pagination**): 129–176.

    A preprint without change of pagination from Genetics 27: 129–176, *1942*.

Storå, R.
1939   Mitteilungen über die Nematoceren Finnlands II. Notulae Ent. **19**: 16–30, 23 figs.

Strand, E.
1927   Animaux divers (à l'exclusion des Arachnides, Lépidoptères et Hyménoptères) nommés jusqu'en 1926 dans les travaux de M. le professeur Embrik Strand. Arch. f. Naturgesch. (1925) **Abt. A, 91** (8): 62–66.

1928   Miscellanea nomenclatorica zoologica et paleontologica. I–II. Arch. f. Naturgesch. (1926) **Abt. A, 92** (8): 30–75.

Streisinger, G.
1946   The *cardini* species group of the genus *Drosophila*. N.Y. Ent. Soc. Jour. **54**: 105–113, 1 fig.

Strickland, E. H.
1938   An annotated list of the Diptera (flies) of Alberta. Canad. Jour. Res., Sect. D, Zool. Sci. **16**: 175–219, 1 fig.

1941   A new genus of the family Tachinidae from Alberta. Canad. Ent. **73**: 64–66, 2 figs.

Strickland, E. H.—Continued
1946 An annotated list of the Diptera (flies) of Alberta. Additions and corrections. Canad. Jour. Res., Sect. D, Zool. Sci. **24**: 157–173.

Strobl, G.
1892 Zur Kenntniss und Verbreitung der Phoriden Oesterreichs. Wien. Ent. Ztg. **11**: 193–204.
1893a Neue österreichische Muscidae Acalypterae. I. Theil; II. Theil; III. Theil; IV. Theil. Wien. Ent. Ztg. **12**: 225–231; 250–256; 280–285; 306–308.
1893b Die Anthomyinen Steiermarks. (Mit Berücksichtigung der Nachbarländer.) K.-k. Zool.-Bot. Gesell. Wien, Verhandl. **43** (**Abhandl.**): 213–276.
1894a Anmerkungen zu Herrn Em. Pokorny's Aufsatz in den Verhandlungen der k. k. Zoologisch-Botanischen Gesellschaft in Wien, Jahrg. 1893, pag. 526–544. Wien. Ent. Ztg. **13**: 65–76.
1894b Die Dipteren von Steiermark. II. Theil. Naturw. Ver. f. Steiermark, Mitt. (1893) **30**: 1–152.
1898a Die Dipteren von Steiermark. IV. Theil. Naturw. Ver. f. Steiermark, Mitt. (1897) **34**: 192–298.
1898b Fauna Ditera Bosne, Hercegovine i Dalmatije. (Sarajevo), Zemaljsk. Muz. u Bosni i Hercegovini, Glasnik **10**: 387–466, 561–616.
1900 Dipterenfauna von Bosnien, Herzegovina und Dalmatia. Wiss. Mitt. aus Bosnien u. Herzegowina **7**: 552–670.

> A German translation of Strobl, 1898b, with some additional material.

1909 Die Dipteren von Steiermark. [V. Theil]. Naturw. Ver. f. Steiermark, Mitt. **46**: 45–293.

Strouhal, H., and Beier, M., eds.
1961– XI. Internationaler Kongress für Entomologie, Wien, 17.
1962 bis 25. August 1960, Verhandlungen. 3 vols. Wien.

> The third volume was issued in 2 separately paged parts. See Berg, 1961; Glumac, 1961.

Stuckenberg, B. R.
1954 Studies on *Paragus*, with descriptions of new species (Diptera Syrphidae). Rev. de Zool. et Bot. Africaines **49**: 97–139, figs. 1–43.

Sturtevant, A. H.
1916 Notes on North American Drosophilidae with descriptions of twenty-three new species. Ent. Soc. Amer. Ann. **9**: 323–343.

Sturtevant, A. H.—Continued
1918a A synopsis of the nearctic species of the genus *Drosophila* (sensu lato). Amer. Mus. Nat. Hist. Bul. **38**: 441–446.
1918b Acalypterae (Diptera) collected in Mobile County, Alabama. N.Y. Ent. Soc. Jour. **26**: 34–40.
1919 A new species closely resembling *Drosophila melanogaster*. Psyche **26**: 153–155, 1 fig.
1921 The North American species of *Drosophila*. Carnegie Inst. Wash., Pub. **301**: 1–150, 49 figs., 3 pls.
1923 New species and notes on synonymy and distribution of Muscidae Acalypteratae (Diptera). Amer. Mus. Nat. Hist., Amer. Mus. Novitates **76**: 1–12, 2 figs.
1939 On the subdivision of the genus *Drosophila*. Natl. Acad. Sci. Proc. **25**: 137–141.
1942 The classification of the genus *Drosophila*, with descriptions of nine new species. Tex. Univ. Pub. **4213**: 5–51.
1954 Nearctic flies of the family Periscelidae (Diptera) and certain Anthomyzidae referred to the family. U.S. Natl. Mus. Proc. **103**: 551–561.

———, **and Dobzhansky, T.**
1936 Observations on the species related to new forms of *Drosophila affinis*, with descriptions of seven. Amer. Nat. **70**: 574–584, 4 figs.

> Sometimes cited as ". . . related to *Drosophila affinis*, with descriptions of seven new forms," apparently the intended title.

Sturtevant, A. H., and Novitski, E.
1941 Sterility in crosses of geographical races of *Drosophila micromelanica*. Natl. Acad. Sci. Proc. **27**: 392–394.

———, **and Wheeler, M. R.**
1954 Synopses of nearctic Ephydridae (Diptera). Amer. Ent. Soc. Trans. (1953) **79**: 151–261.

Sublette, J. E. (See also Dendy and Sublette)
1960 Chironomid midges of California. I. Chironominae, exclusive of Tanytarsini (=Calopsectrini). U.S. Natl. Mus. Proc. **112**: 197–226, 2 figs.

Swederus, N. S.
1787 Et nytt genus och femtio nya species af insecter. K. Vetensk. Acad. Nya Handl. [ser. 2], **8**: 181–201, 276–290.

Swenk, M. H.
1905 The North American species of *Cuterebra*. N.Y. Ent. Soc. Jour. **13**: 181–185.
1916 Descriptions and records of North American Hippoboscidae. N.Y. Ent. Soc. Jour. **24**: 126–136.

Syme, P. D., and Davies, D. M.
1958 Three new Ontario black flies of the genus *Prosimulium* (Diptera: Simuliidae). 1. Descriptions, morphological comparisons with related species, and distribution. Canad. Ent. **90**: 697–719, 28 figs.

Szilády, Z.
1923 New or little known horseflies (Tabanidae). Biol. Hungarica **1** (1): 1–39, 29 figs., 1 pl.
1926a New and Old World horseflies. Biol. Hungarica **1** (7): 1–30, 7 figs., 1 pl.
1926b Dipterenstudien. Budapest Magyar Nemzeti Muz., Ann. Hist. Nat. **24**: 586–611, 18 figs.

T.S.H.D. (See Texas State Health Department)

Takada, H. (See Wheeler and Takada)

Tarwid, K.
1936 Note sur l'identité systématique des genres: *Zelmira* Meig. (=*Platyura* Meig.) et *Asindulum* Latr. (Diptera, Fungivoridae). Towar. Nauk. Warszawsk., Sprawozd., Wyzd. 4, **29**: 1–9, 10 figs.

Taylor, R. L.
1928 A new species of *Lonchaea* Fallén (Lonchaeidae, Diptera). Brooklyn Ent. Soc. Bul. **23**: 191–194, 1 fig.

Telford, H. S.
1939 The Syrphidae of Minnesota. Minn. Agr. Expt. Sta. Tech. Bul. **140**: 1–76, 2 pls.
1949 A monograph of the genus *Eristalis* Latreille (Syrphidae, Diptera) from North America north of Mexico. [Abstract]. Minn. Univ., Grad Sch., Sums. Ph.D. Theses (1940–1941) **4**: 35–40.

Texas State Health Department
1944 The mosquitoes of Texas. Compiled by the Division of Medical Entomology, Bureau of Laboratories. 100 pp., 32 figs. Austin, Tex.

> Authorship of an included new species is being attributed to Randolph, N. M., and O'Neill, K., in accordance with information in the Preface, p. 3.

Theobald, F. V. (See also Howard, 1901)
1900 Report on the collections of mosquitoes (Culicidae) received at the British Museum (Natural History) from various parts of the world in connection with the investigations into the causes of malaria conducted by the Colonial Office and the Royal Society. 12 pp. British Museum (Natural History), London.

Theobald, F. V.—Continued
- 1901a A monograph of the Culicidae or mosquitoes. Mainly compiled from the collections received at the British Museum from various parts of the world. Vol. 1, 424 pp., 151 figs. London.
    > The entire work consists of 5 volumes and an Atlas, published at London, 1901-1910.
- 1901b A monograph of the Culicidae or mosquitoes. Mainly compiled from the collections received at the British Museum from various parts of the world. Vol. 2, 391 pp., figs. 152-318. London.
- 1901c The classification of mosquitoes. Jour. Trop. Med. **4**: 229-235, 5 figs., 1 pl.
- 1902   The classification of the Anophelina. Jour. Trop. Med. **5**: 181-183, 8 figs.
- 1903a A monograph of the Culicidae or mosquitoes. Mainly compiled from the collections received at the British Museum from various parts of the world. Vol. 3, 359 pp., 193 figs., 17 pls. London.
- 1903b Description of a new North American *Culex*. Canad. Ent. **35**: 211-213.
- 1903c Notes on Culicidae and their larvae from Pecos, New Mexico, and description of a new *Grabhamia*. Canad. Ent. **35**: 311-316, 2 figs.
- 1903d Two new Jamaican Culicidae. Entomologist **36**: 281-283.
- 1904   New Culicidae from the Federated Malay States. Entomologist **37**: 12-15, 36-39, 77-78, 111-113, 163-165, 211-213, 236-239.
- 1905   The mosquitoes or Culicidae of Jamaica. 40 pp., 9 pls. Institute of Jamaica, Kingston.
- 1906   A new *Megarhinus*. Entomologist **39**: 241.
- 1907   A monograph of the Culicidae or mosquitoes. Mainly compiled from collections received at the British Museum. Vol. 4, 639 pp., 297 figs., 16 pls. London.

Thienemann, A. (See also Bause, 1914; Edwards and Thienemann; Kieffer and Thienemann; Zavřel and Thienemann)
- 1916   Schwedische Chironomiden. Mit Beschreibungen neuer Arten von Prof. Dr. J. J. Kieffer (Bitsch). Pp. 483-554, 35 figs., pls. 17-18. *In his* Vorarbeiten für eine Monographie der Chironomiden-Metamorphose [cont.]. Arch. f. Hydrobiol. u. Planktonkunde, Sup.-Band **2**: 483-654, 131 figs., pls. 17-18 [cont.].

Thienemann, A.—Continued
- 1926a Hydrobiologische Untersuchungen an den kalten Quellen und Bäcken der Halbinsel Jasmund auf Rügen. Arch. f. Hydrobiol. **n. ser.**, **17**: 221–336, 3 figs., pl. 17.
- 1926b Hydrobiologische Untersuchungen an Quellen. VII. Insekten aus norddeutschen Quellen mit besonderer Berücksichtigung der Dipteren. Deut. Ent. Ztschr. **1926**: 1–50.
- 1936 Alpine Chironomiden. Ergebnisse von Untersuchungen in der Gegend von Garmisch-Partenkirchen, Oberbayern. Arch. f. Hydrobiol. **n. ser.**, **30**: 167–262, figs. A–E + 1–15, pls. 6–11.
- 1937 Podonominae, eine neue Unterfamilie der Chironomiden. (Chironomiden aus Lappland. I.) Internatl. Rev. Gesam. Hydrobiol. u. Hydrog. **35**: 65–112, 21 + 6 figs.
- 1941 Lappländische Chironomiden und ihre Wohngewässer. Ergebnisse von Untersuchungen im Abiskogebiet in Schwedisch-Lappland. Arch. f. Hydrobiol., Sup.-Band **17**: 1–253, 47 figs., 13 pls.
- 1944 Bestimmungstabellen für die bis jetzt bekannten Larven und Puppen der Orthocladiinen (Diptera Chironomiden). Arch. f. Hydrobiol. **n. ser.**, **39**: 551–664, pls. 4–30.
- 1954 Die Binnengewässer. Einzeldarstellungen aus der Limnologie und ihren Nachbargebieten. Vol. 20: *Chironomus*: Leben, Verbreitung und wirtschaftliche Bedeutung der Chironomiden, 834 pp., 300 figs., 31 pls. Stuttgart.

> "Die Binnengewässer", edited by Thienemann, consists of 21 volumes published in Stuttgart, 1925–1955. It is not known whether the series continues. Thienemann died in 1960.

Thomas, F. L.
- 1914 Three new species of Trypetidae from Colorado. Canad. Ent. **46**: 425–429, 3 figs.

Thompson, M. T.
- 1907 Three galls made by cyclorrhaphous flies. Psyche **14**: 71–74, 3 figs.
- 1915 An illustrated catalogue of American insect galls. 116 pp., 21 pls. Nassau, N.Y.

Thompson, W. R.
- 1910 A new species of the genus *Leucopis*. Canad. Ent. **42**: 238–242.

Thompson, W. R.—Continued
1911   Tachinidae, new and old. Canad. Ent. **43**: 265–272, 313–317, 1 pl.
1943a   Parasites of the Arachnida and Coleoptera. Pt. 1, 151 pp. *In his* A catalogue of the parasites and predators of insect pests [q.v.]. Sect. 1. Belleville, Ont.
1943b   Parasites of the Dermaptera and Diptera. Pt. 2, 99 pp. *In his* A catalogue of the parasites and predators of insect pests [q.v.]. Sect. 1. Belleville, Ont.
1944a   Parasites of the Hemiptera. Pt. 3, 149 pp. *In his* A catalogue of the parasites and predators of insect pests [q.v.]. Sect. 1. Belleville, Ont.
1944b   Parasites of the Hymenoptera, Isopoda and Isoptera. Pt. 4, 130 pp. *In his* A catalogue of the parasites and predators of insect pests [q.v.]. Sect. 1. Belleville, Ont.
1944c   Parasites of the Lepidoptera. Pp. 1–130 (=pt. 5: A–Ch) [cont.]. *In his* A catalogue of the parasites and predators of insect pests [q.v.]. Sect. 1. Belleville, Ont.
1945–1947   Parasites of the Lepidoptera [concl.]. Pp. 131–258 (=pt. 6: Ci–F), *1945;* pp. 259–385 (=pt. 7: G–M), *1946;* pp. 386–523 (=pt. 8: N–P), *1946;* pp. 524–627 (=pt. 9: Q–Z), *1947*. *In his* A catalogue of the parasites and predators of insect pests [q.v.]. Sect. 1. Belleville, Ont.
1949   [Editorial.] Canad. Ent. (1948) **80**: 7–8.
1950a   Index of parasites of the Lepidoptera. Pt. 10, 107 pp. *In his* A catalogue of the parasites and predators of insect pests [q.v.]. Sect. 1. Belleville, Ont.
1950b   Neuroptera, Odonata, Orthoptera, Psocoptera, Siphonaptera, Thysanoptera. Pt. 11, 35 pp. *In his* A catalogue of the parasites and predators of insect pests [q.v.]. Sect. 1. Belleville, Ont.
1951   Hosts of the Coleoptera and Diptera. Pt. 1, 147 pp. *In his* A catalogue of the parasites and predators of insect pests [q.v.]. Sect. 2. Ottawa, Ont.
1953a   The tachinid parasites of *Archips cerasivorana* Fitch. (1) *Dichaetoneura leucoptera* Johns. (Diptera); (2) *Eusiyropa blanda* O. S. (Diptera). Canad. Ent. **85**: 19–30, 3 pls.; 393–404, figs. 16–19, 2 pls.
1953b   Hosts of the Hymenoptera. Pp. 1–190 (=pt. 2) [cont.]. *In his* A catalogue of the parasites and predators of insect pests [q.v.]. Sect. 2. Ottawa, Ont.
1954   *Hyalomyodes triangulifera* Loew (Diptera, Tachinidae). Canad. Ent. **86**: 137–144, 2 pls.

Thompson, W. R.—Continued

1955– Hosts of the Hymenoptera [concl.]. Pp. 191–332 (=pt. 3),
1958 *1955;* pp. 333–561 (=pt. 4), *1957;* pp. 562–698 (=pt. 5, with Lepidoptera and Strepsiptera), *1958. In his* A catalogue of the parasites and predators of insect pests [q.v.]. Sect. 2. Ottawa, Ont.

Thompson, W. R.

1943– A catalogue of the parasites and predators of insect pests.
1958 2 sects. Belleville, Ottawa, Ont.

> The entire work is divided into 2 sections: 1. Parasite host catalogue; 2. Host parasite catalogue. The sections were issued in parts, each taxonomic grouping being separately paged. Section 1 consists of 11 parts (7 separately paged groups), Belleville, 1943–1950. Section 2 consists of 5 parts (2 separately paged groups), Ottawa, 1951–1958. The work as originally proposed was to consist of 4 sections. It is not known whether this will be continued.
>
> See Thompson, 1943a, 1943b, 1944a, 1944b, 1944c, 1945–1947, 1950a, 1950b, 1951, 1953b, 1955–1958.

Thomsen, L. C. (See also Johannsen and Thomsen)

1935 New species of New York State Ceratopogonidae. N.Y. Ent. Soc. Jour. **43**: 283–296, 2 pls.

1937 Part V. Ceratopogonidae. Pp. 57–80, pls. 10–18. *In* Johannsen, O. A., and Thomsen, L. C., Aquatic Diptera. Parts IV and V. N.Y. (Cornell) Agr. Expt. Sta. Mem. **210**: 1–80, 18 pls.

Thomson, C. G.

[1869] 6. Diptera. Species nova descripsit. Pp. 443–614, pl. 9 (=h. 12, no. 2). *In* K. Svenska Vetenskaps-Akademien, Kongliga svenska fregatten Eugenies resa omkring jorden [q.v.]. Pt. 2: Zoologie, [Sec.] 1: Insekter, 617 pp., 9 pls. Stockholm, "1868."

> Although the title page for Diptera is dated 1868, contemporary evidence, such as reviews and receipts by societies, indicates that it was published early in 1869. For example, it was listed as received in the library of the Kgl. Danske Videnskaberne Selskabs at the June 18th, 1869, meeting of that society (see K. Danske Vidensk. Selsk., Oversigt over . . . Forhandl. 1869: 14). Woodward, 1915: 2057, dates the work 1869, and it does not appear in "Zoological Record" until the volume for 1870, although "1868" is quoted there as its date.

Thorpe, W. H.

1930 The biology of the petroleum fly (*Psilopa petrolii* [sic] Coq.). Ent. Soc. London, Trans. **78**: 331–344, 4 figs., pls. 28–29.

> The name *petrolii* in the title is corrected to *petrolei* in a correction slip attached to the issue and reprints.

Thorpe, W. H.—Continued
1931a   The biology, post-embryonic development, and economic importance of *Cryptochaetum iceryae* (Diptera, Agromyzidae) parasitic on *Icerya purchasi* (Coccidae, Monophlebini). Zool. Soc. London, Proc. Gen. Mtg. **1930**: 929–971, 23 figs., 5 pls.
1931b   The biology of the petroleum fly. Science n. ser. **73**: 101–103.

Throckmorton, L. H.   (See Wheeler and Throckmorton)

Thurman, E. B., and Winkler, E. C.
1950   A new species of mosquito in California, *Aedes* (*Ochlerotatus*) *bicristatus* (Diptera, Culicidae). Ent. Soc. Wash. Proc. **52**: 237–250, 1 fig., 5 pls.

Tiensuu, L.
1935   Die bisher aus Finnland bekannten Musciden. Soc. pro Fauna et Flora Fenn. Acta **58** (4): 1–56, 16 figs.

Tilbury, M. R.
1913   Notes on the feeding and rearing of the midge, *Chironomus cayugae* Johannsen. N.Y. Ent. Soc. Jour. **21**: 305–308, 1 fig.

Tilden, J. W.   (See Skinner, 1962)

Tokunaga, M.
1933   Chironomidae from Japan (Diptera). I. Clunioninae. Philippine Jour. Sci. **51**: 87–99, pls. 1–2.

———, **and Murachi, E. K.**
1959   Diptera: Ceratopogonidae. Pp. 103–434, 98 figs. *In* Bernice P. Bishop Museum, Insects of Micronesia [q.v.]. Vol. 12, 434 pp., 119 figs. [cont.]. Honolulu, Hawaii.

Tollet, R.
1948   Un Mycetophilidae (Diptera) nouveau des États-Unis d'Amérique. Brussels Mus. Roy. d'Hist. Nat. de Belg. Bul. **24** (40): 1–4, 6 figs.
1959   Note systématique sur les Corynoscelidae fam. nov. (Diptera) du globe et description d'un Corynoscelidaè nouveau de l'hémisphère austral. Soc. Roy. d'Ent. de Belg. Bul. et Ann. **95**: 132–153, 6 figs.

Tonnoir, A. L.   (See also Goetghebuer and Tonnoir)
1920   Notes sur quelques Psychodidae africains. Rev. Zool. Africaine **8**: 127–147, 3 figs.
1922   Synopsis des espèces européennes du genre *Psychoda* (Diptères). Soc. Ent. de Belg. Ann. **62**: 49–88, 16 figs.
1924   New Zealand Dixidae (Dipt.). (Canterbury Col., Christchurch) Canterbury Mus. Rec. **2**: 221–233, 12 figs.

Tonnoir, A. L.—Continued
  1929 Australian Mycetophilidae. Synopsis of the genera. Linn. Soc. N. S. Wales, Proc. **54**: 584–614, 7 figs., pls. 22–23.

Tonnoir, A. L., and Edwards, F. W.
  1927 New Zealand fungus gnats (Diptera, Mycetophilidae). New Zeal. Inst. Trans. and Proc. **57**: 747–878, fig. A, pls. 58–80.

Tothill, B. H. (See Needham, Frost, and Tothill)

Tothill, J. D.
  1912 Systematic notes on North American Tachinidae. Canad. Ent. **44**: 1–5.
  1921 A revision of the nearctic species of the tachinid genus *Ernestia* R. D. (Diptera). Canad. Ent. **53**: 199–205, 226–230, 247–252, 270–274, 1 fig.
  1922 The natural control of the fall webworm (*Hyphantria cunea* Drury) in Canada together with an account of its several parasites. Canada Dept. Agr. Bul. n. ser., **3**: 1–107, 99 figs., 6 pls. (=Ent. Branch Bul. 19.)
  1924a A revision of the nearctic species of the genus *Gonia* (Diptera, Tachinidae). Canad. Ent. **56**: 196–200, 206–212.
  1924b A revision of the nearctic species in the genus *Fabriciella* (Tachinidae). Canad. Ent. **56**: 257–269.

Townes, H. K., Jr. (See also Johannsen and Townes)
  1945 The nearctic species of Tendipedini (Diptera, Tendipedidae (=Chironomidae)). Amer. Midland Nat. **34**: 1–206, 261 figs.
  1952 Tribe Tendipedini (=Chironomini). Pp. 27–103, pls. 3–22. *In* Johannsen, O. A., and Townes, H. K., Jr., Guide to the insects of Connecticut. Part VI. The Diptera or true flies of Connecticut. Fifth fascicle: Midges and gnats [part]. Tendipedidae (Chironomidae). Conn. State Geol. and Nat. Hist. Survey, Bul. **80**: 3–147, 232–250 (bibliography and index of entire 5th fascicle), 3 figs., 22 pls.

Townsend, C. H. T.
  1891a The North American genera of calyptrate Muscidae. Paper I. Ent. Soc. Wash. Proc. **2**: 89–100.
  1891b Notes on North American Tachinidae sens. lat., with descriptions of new species. Paper I. Ent. Soc. Wash. Proc. **2**: 134–146.

Townsend, C. H. T.—Continued
- 1891c Notes on North American Tachinidae sens. str. with descriptions of new genera and species. Paper II. Amer. Ent. Soc. Trans. **18**: 349–382.

  This was published in December.
- 1891d An *Exorista* parasitic on *Lagoa opercularis*. Ent. News and Acad. Nat. Sci. Phila., Ent. Sect. Proc. **2**: 159–160.
- 1891e A tachinid parasite of *Chrysophanus dione*. *Exorista chrysophani* n. sp. Ent. News and Acad. Nat. Sci. Phila., Ent. Sect. Proc. **2**: 197–199.
- 1891f Two new tachinids. Psyche **6**: 83–85.
- 1891g A new *Simulium* from southern New Mexico. Psyche **6**: 106–107.
- 1891h A parasite of the fall web-worm. Psyche **6**: 176–177.
- 1891i A tachinid parasite of the oak unicorn prominent. Psyche **6**: 187–188.
- 1891j Description of a muscid bred from swine dung, with notes on two muscid genera. Canad. Ent. **23**: 152–155.
- 1891k A tachinid bred from a chrysalis. Canad. Ent. **23**: 206–207.
- 1892a Notes on North American Tachinidae sens. str., with descriptions of new genera and species. Paper III. Amer. Ent. Soc. Trans. **19**: 88–132.

  No "Paper IV" under this title was ever published. For Papers V–VII, see Townsend, 1892d and 1892e.
- 1892b The North American genera of calyptrate Muscidae. Paper II; Paper III; Paper IV; Paper V. Amer. Ent. Soc. Trans. **19**: 133–144; 273–278; 279–284; 290–294.
- 1892c The North American genera of nemocerous Diptera. Amer. Ent. Soc. Trans. **19**: 144–160.
- 1892d Notes on North American Tachinidae, with descriptions of new species. Paper VII. Amer. Ent. Soc. Trans. **19**: 284–289.
- 1892e Notes on North American Tachinidae, with descriptions of new genera and species. Paper V; Paper VI. Canad. Ent. **24**: 64–70, 77–82; 165–172.
- 1892f A sarcophagid parasite of *Cimbex americana*. Canad. Ent. **24**: 126–127.
- 1892g A preliminary grouping of the described species of *Sapromyza* of North America, with one new species. Canad. Ent. **24**: 301–304.
- 1892h New North American Tachinidae. Ent. News and Acad. Nat. Sci. Phila., Ent. Sect. Proc. **3**: 80–81, 129–131.

Townsend, C. H. T.—Continued
- 1892i A tachinid bred from larva of *Protoparce jamaicensis* (Butl.) in Jamaica. Inst. Jamaica, Jour. **1**: 70–71.
- 1892j A dexiid parasite of a longicorn beetle. Inst. Jamaica, Jour. **1**: 105–106.
- 1892k Description of a *Sarcophaga* bred from *Helix*. Psyche **6**: 220–221.
- 1892l A new genus of Tachinidae. Psyche **6**: 247.
- 1892m An *Aporia* bred from *Limacodes* sp. Psyche **6**: 275–276.
- 1893a Review.—Part III of Brauer and Bergenstamm's Monograph of the Muscaria Schizometopa. Ent. News and Acad. Nat. Sci. Phila., Ent. Sect. Proc. **4**: 276–277.
- 1893b On the horse-flies of New Mexico and Arizona. Kans. Acad. Sci. Trans. Ann. Mtgs. **13**: 133–135.
- 1893c On a peculiar acalyptrate muscid found near Turkey Tanks, Arizona. Kans. Acad. Sci. Trans. Ann. Mtgs. **13**: 135–136.
- 1893d A nycteribiid from a New Mexico bat. N.Y. Ent. Soc. Jour. **1**: 79–80.
- 1893e A tachinid reared from cells of a mud-dauber wasp. Ohio Agr. Expt. Sta. Tech. Ser. Bul. **1**: 165–166 (=bul. 3, part).
- 1893f An interesting blood-sucking gnat of the family Chironomidae. Psyche **6**: 369–371, 1 pl.
- 1893g Description of a new and interesting phasiid-like genus of Tachinidae s. str. Psyche **6**: 429–430.
- 1893h Hosts of North American Tachinidae, etc. I. Psyche **6**: 466–468.
- 1893i A cabbage-like cecidomyidous gall on *Bigelovia*. Psyche **6**: 491–492, 1 fig.
- 1893j Fleshy cecidomyiid twig gall on *Atriplex canescens*. Amer. Nat. **27**: 1021.
- 1893k A trypetid bred from galls on *Bigelovia*. Canad. Ent. **25**: 48–52.
- 1893l Comments on Mr. van der Wulp's recent diagnoses of new species of Mexican Phasiidae, Gymnosomatidae, Ocypteridae, and Phaniidae. Canad. Ent. **25**: 165–168.
- 1894a A very remarkable and anomalous syrphid, with peculiarly developed hind tarsi. Canad. Ent. **26**: 50–52, 2 figs.
- 1894b A cone-like cecidomyid gall on *Bigelovia*. Psyche **7**: 176.
- 1895a Contributions to the dipterology of North America. I. Syrphidae; II. Tabanidae, Conopidae, Tachinidae, etc. Amer. Ent. Soc. Trans. **22**: 33–55; 55–80.

Townsend, C. H. T.—Continued
- 1895b Notes on the Diptera of Baja California. Calif. Acad. Sci. Proc. **ser. 2, 4:** 593–620.
- 1896 Notes on the species of *Exorista* of temperate North America. Psyche **7:** 329–331.
- 1897a Contributions from the New Mexico Biological Station. No. IV. Diptera from the Sacramento and the White Mountains, in southern New Mexico. I. Ann. and Mag. Nat. Hist. ser. **6, 19:** 138–149.
- 1897b Contributions from the New Mexico Biological Station. No. II (continued). On a collection of Diptera from the lowlands of the Rio Nautla, in the state of Vera Cruz. II. Ann. and Mag. Nat. Hist. **ser. 6, 20:** 19–33, 272–291.
- 1897c Diptera from the lower Rio Grande or Tamaulipan region of Texas. I. N.Y. Ent. Soc. Jour. **5:** 171–178.
- 1897d Description of the bot-fly of the cotton-tail rabbit in New Mexico, *Cuterebra lepusculi* n. sp. Psyche **8:** 8–9.
- 1897e Diptera from the headwaters of the Gila River. I; II. Psyche **8:** 38–41; 92–94.
- 1898a Some characteristic maritime Diptera from the south end of Padre Island and the adjacent Texas coast. I. Ent. News **9:** 167–169.
- 1898b Diptera from the lower Rio Grande or Tamaulipan region of Texas. II. N.Y. Ent. Soc. Jour. **6:** 50–52.
- 1898c Diptera of the Organ Mountains in Southern New Mexico. II. Psyche **8:** 267–269.
- 1901 New and little-known Diptera from the Organ Mountains and vicinity in New Mexico. Amer. Ent. Soc. Trans. **27:** 159–164.
- 1908 The taxonomy of the muscoidean flies, including descriptions of new genera and species. Smithsn. Inst., Smithsn. Misc. Collect. **51 (2** [=pub. 1803]): 1–138.
- 1909 Descriptions of some new Tachinidae. Ent. Soc. Amer. Ann. **2:** 243–250.
- 1911a Announcement of further results secured in the study of muscoid flies. Ent. Soc. Amer. Ann. **4:** 127–152.
- 1911b Review of work by Pantel and Portchinski on reproductive and early stage characters of muscoid flies. Ent. Soc. Wash. Proc. **13:** 151–170.
- 1912a Foundation of some new genera and species of muscoid flies mainly on reproductive and early-stage characters. N.Y. Ent. Soc. Jour. **20:** 107–119.

Townsend, C. H. T.—Continued
- 1912b Descriptions of new genera and species of muscoid flies from the Andean and Pacific Coast regions of South America. U.S. Natl. Mus. Proc. **43**: 301–367.
- 1912c A readjustment of muscoid names. Ent. Soc. Wash. Proc. **14**: 45–53.
- 1912d Six new genera of nearctic Muscoidea. Ent. Soc. Wash. Proc. **14**: 163–166.
- 1913a A jumping maggot which lives in cactus blooms (*Acucula saltans*, gen. et sp. nov.). Canad. Ent. **45**: 262–265.
- 1913b On *Trichiopoda* Latreille, *Polistomyia* Townsend and *Trichopodopsis* new genus. N.Y. Ent. Soc. Jour. **21**: 147–148, (313, a correction).
- 1913c On the tribe Dejeaniini of the muscoid family Hystriciidae, with five new genera. Psyche **20**:. 102–106.
- 1914 New muscoid flies mainly Hystriciidae and Pyrrhosiinae from the Andean Montanya [cont.]. Insecutor Inscitiae Menstruus **2**: 10–16, 29–32, 42–48, 81–96, 123–128, 133–144, 153–160, 169–176 [cont.].
- 1915a Proposal of new muscoid genera for old species. Biol. Soc. Wash. Proc. **28**: 19–23.
- 1915b *Phyllophilopsis*, new name. Canad. Ent. **47**: 78.
- 1915c New Canadian and Alaskan Muscoidea. Canad. Ent. **47**: 285–292.
- 1915d Revision of *Myiophasia*. Ent. Soc. Wash. Proc. **17**: 107–114.
- 1915e New Masiceratidae and Dexiidae from South America. N.Y. Ent. Soc. Jour. **23**: 61–68.
- 1915f New western and southwestern Muscoidea. N.Y. Ent. Soc. Jour. **23**: 216–234.
- 1915g New neotropical muscoid flies. U.S. Natl. Mus. Proc. **49**: 405–440.
- 1915h Correction of the misuse of the generic name *Musca*, with description of two new genera. Wash. Acad. Sci. Jour. **5**: 433–436.
- 1915i A new generic name for the screw-worm fly. Wash. Acad. Sci. Jour. **5**: 644–646.
- 1915j New muscoid flies, mainly Hystriciidae and Pyrrhosiinae from the Andean Montanya [concl.]. Insecutor Inscitiae Menstruus (1914) **2**: 183–187.
- 1915k Nine new tropical American genera of Muscoidea. Insecutor Inscitiae Menstruus **3**: 91–97.
- 1915l New genera of muscoid flies from the Middle Atlantic States. Insecutor Inscitiae Menstruus **3**: 97–104.

Townsend, C. H. T.—Continued
1915m Synonymical notes on Muscoidea. Insecutor Inscitiae Menstruus **3:** 115–122.
1916a *Andrina radicis* Townsend, new name. Canad. Ent. **48:** 19.
1916b New muscoid genera (Dip.). Ent. News **27:** 178.
1916c Description of two new tachinids (Dip.). Ent. News **27:** 217.
1916d Designations of muscoid genotypes, with new genera and species. Insecutor Inscitiae Menstruus **4:** 4–12.
1916e Elucidations of New England Muscoidea. Insecutor Inscitiae Menstruus **4:** 17–33.
1916f Muscoid flies from the southern United States. Insecutor Inscitiae Menstruus **4:** 41–59.
1916g Some new North American muscoid forms. Insecutor Inscitiae Menstruus **4:** 73–78.
1916h Diagnoses of new genera of muscoid flies founded on old species. U.S. Natl. Mus. Proc. **49:** 617–633.
1916i New genera and species of muscoid flies. U.S. Natl. Mus. Proc. **51:** 299–323.
1916j New and noteworthy Brazilian Muscoidea collected by Herbert H. Smith. Amer. Mus. Nat. Hist. Bul. **35:** 15–22.
1917a New genera and species of American muscoid Diptera. Biol. Soc. Wash. Proc. **30:** 43–50.
1917b Genera of the dipterous tribe Sarcophagini. Biol. Soc. Wash. Proc. **30:** 189–197.
1917c Second paper on Brazilian Muscoidea collected by Herbert H. Smith. Amer. Mus. Nat. Hist. Bul. **37:** 221–233.
1917d Miscellaneous muscoid notes and descriptions. Insecutor Inscitiae Menstruus (1916) **4:** 121–128.
1917e *Lithohypoderma*, a new fossil genus of oestrids. Insecutor Inscitiae Menstruus (1916) **4:** 128–130.
1917f A synoptic revision of the Cuterebridae, with synonymic notes and the description of one new species. Insecutor Inscitiae Menstruus **5:** 23–28.
1918a New muscoid genus from the Chiricahua Mountains, Arizona (Dip.). Ent. News **29:** 177–178.
1918b New genera of Amobiinae (Diptera). Insecutor Inscitiae Menstruus (1917) **5:** 157–165.
1918c New muscoid genera, species and synonymy (Diptera). Insecutor Inscitiae Menstruus **6:** 151–156 [cont.].
1918d Some muscoid synonymy, with one new genus. Ent. Soc. Wash. Proc. **20:** 19–21.

Townsend, C. H. T.—Continued

1919a Description of the new species of *Phormia*. Pp. 379–380. *In* Plath, O. E., The prevalence of *Phormia azurea* Fallén (larva parasitic on nestling birds) in the Puget Sound region and data on two undescribed flies of similar habit. Ent. Soc. Amer. Ann. **12**: 373–381.

1919b New genera and species of muscoid flies. U.S. Natl. Mus. Proc. **56**: 541–592.

1919c New muscoid genera, species and synonymy (Diptera) [concl.]. Insecutor Inscitiae Menstruus (1918) **6**: 157–182.

1921 Some new muscoid genera ancient and recent. Insecutor Inscitiae Menstruus **9**: 132–134.

1926 New holarctic Muscoidea (Diptera). Insecutor Inscitiae Menstruus **14**: 24–41.

1927 Synopse dos generos muscideos da região humida tropical da America, con generos e especies novas. [São Paulo] Mus. Paulista, Rev. **15**: 203–385, 7 figs.

> Townsend consistently cited 1926 for this paper, but the journal was issued Jan. 20, 1927. See Arnaud, 1958c: 4, footnote 3.

1931a Notes on American oestromuscoid types. Rev. de Ent. **1**: 65–104, 157–183.

1931b New genera and species of American oestromuscoid flies. Rev. de Ent. **1**: 313–354, 437–479.

1934a Manual of myiology in twelve parts. Pt. 1: Development and structure, 280 pp. São Paulo.

> The entire work consists of 12 separately bound parts, São Paulo, 1934–1942.

1934b Muscoid notes and descriptions. Rev. de Ent. **4**: 110–112.

1934c New neotropical oestromuscoid flies. Rev. de Ent. **4**: 201–212, 390–406.

1935a Manual of myiology in twelve parts. Pt. 2: Muscoid classification and habits, 296 pp., 9 pls. São Paulo.

1935b New muscoid genera, mainly from the Neotropical Region. Rev. de Ent. **5**: 68–74.

1936a Manual of myiology in twelve parts. Pt. 3: Oestroid classification and habits (Gymnosomatidae to Tachinidae), 255 pp. São Paulo.

1936b Manual of myiology in twelve parts. Pt. 4: Oestroid classification and habits (Dexiidae and Exoristidae), 303 pp. São Paulo.

1936c Notes on Aldrich's 1926 species of *Cylindromyia* (Dipt.). Rev. de Ent. **6**: 488.

Townsend, C. H. T.—Continued
1937   Manual of myiology in twelve parts. Pt. 5: Muscoid generic diagnoses and data (Glossinini to Agriini), 234 pp. São Paulo.
1938a   Manual of myiology in twelve parts. Pt. 6: Muscoid generic diagnoses and data (Stephanostomatini to Moriniini), 309 pp. São Paulo.
1938b   Manual of myiology in twelve parts. Pt. 7: Oestroid generic diagnoses and data (Gymnosomatini to Senostomatini), 434 pp. São Paulo.
1939a   Manual of myiology in twelve parts. Pt. 8: Oestroid generic diagnoses and data (Microtropesini to Voriini), 408 pp. São Paulo.
1939b   Manual of myiology in twelve parts. Pt. 9: Oestroid generic diagnoses and data (Thelairini to Clythoini), 270 pp. São Paulo.
1940   Manual of myiology in twelve parts. Pt. 10: Oestroid generic diagnoses and data (Anacamptomyiini to Frontinini), 335 pp. São Paulo.
1941a   Manual of myiology in twelve parts. Pt. 11: Oestroid generic diagnoses and data (Goniini to Trypherini), 342 pp. São Paulo.
1941b   An undescribed American *Cephenemyia*. N.Y. Ent. Soc. Jour. **49**: 161–163.
1942   Manual of myiology in twelve parts. Pt. 12: General consideration of the Oestromuscaria, 365 pp., 82 pls. São Paulo.

Travis, B. V. (See Gjullin et al.)

Tripp, H. A.
1955   Descriptions and habits of Cecidomyiidae (Diptera) from white spruce cones. Canad. Ent. **87**: 253–263, 30 figs.

Tucker, E. S. (See also Johannsen, 1907a)
1907   Some results of desultory collecting of insects in Kansas and Colorado. Kans. Univ. Sci. Bul. **4** [=whole ser., 14]: 51–112. (=Kans. Univ. Bul. 7 (5).)
1908   Incidental studies of new species of *Oscinis*. Ent. News **19**: 272–274.
1909   Additional results of collecting insects in Kansas and Colorado. Kans. Acad. Sci. Trans. **22**: 276–304.
1911   Description of a new fly of the family Dolichopodidae, with remarks and corrections of preceding papers. Kans. Acad. Sci. Trans. **23/24**: 105–107.

Tuomikoski, R.
1958 Mitteilungen über die Empididen (Dipt.) Finnlands. V. Die Gattung *Iteaphila* Zett. s. str. Ann. Ent. Fenn. **24:** 125-131.
1960 Zur Kenntnis der Sciariden (Dipt.) Finnlands. Suom. Eläin-ja Kasvitiet. Seura Vanamo, Eläintiet. Julkaisu. (Soc. Zool. Bot. Fenn. Vanamo, Ann. Zool.) **21 (4):** 1-164, 33 figs.

Turner, R. L.
1924 A new mosquito from Texas (Diptera, Culicidae). Insecutor Inscitiae Menstruus **12:** 84.

Twinn, C. R.
1936 The blackflies of eastern Canada (Simuliidae, Diptera). Canad. Jour. Res., Sect. D, Zool. Sci. **14:** 97-150, 15 figs.
1938 Blackflies from Utah and Idaho, with descriptions of new species (Simuliidae, Diptera). Canad. Ent. **70:** 48-55.

U.S. Department of Agriculture Division of Entomology
1889–
1905 Bibliography of the more important contributions to American economic entomology. 8 pts. Washington.

> Parts 1-5 are by S. Henshaw and parts 6-8 by N. Banks. The first 3 parts are continuously paged and were published 1889-1890, with an index and a title page dated 1890. This bibliography is continued by the "Index to the literature of American economic entomology", which began in 1917 with Volume 1, by Banks, and which continued under various authorship.
> See Henshaw, 1889.

U.S. Entomological Commission
1878–
1890 Reports of the United States Entomological Commission. 5 vols. Washington.

> Authorship of the reports varies. The first report, by C. V. Riley, A. S. Packard, and C. Thomas, is called an "Annual Report".
> See Riley, 1878, 1886b.

Ulmer, G. (See Brohmer, Ehrmann, and Ulmer)

Underwood, W. L.
1903 A new mosquito. Science n. ser., **18:** 182-184.

Urbina, M. (See Arriaga, Urbina, and Rebollar)

Usinger, R. L. (See also Wirth and Stone, 1956)

———, **La Rivers, I., Chandler, H. P., and Wirth, W. W.**
1948 Entomology 133. Biology of aquatic and littoral insects. Calif. Univ. Syllabus Ser. **SS:** 1-244.

> The Diptera, by Wirth, are on pp. 86-136.

Vaillant, F.
1959   The Thaumaleidae (Diptera) of the Appalachian Mountains. N.Y. Ent. Soc. Jour. **67**: 31–37, 2 pls.
1960   Quelques Empididae Atalantinae des monts des Appalaches (Dipt.). Soc. Ent. de France, Bul. **65**: 117–123, 2 pls.

van der Wulp, F. M. (See Wulp, F. M. van der)

Van Duzee, M. C. (See also Curran, 1926j)
1913a   Synoptical table of the North American species of the dipterous genus *Sympycnus*, with the description of a new species. Ent. News **24**: 269–272, 2 figs.
1913b   A revision of the North American species of the dipterous genus *Neurigona* (Dolichopodidae). Ent. Soc. Amer. Ann. **6**: 22–61, 2 pls.
1914a   Notes on *Sciapus*, with descriptions of three new species. Canad. Ent. **46**: 389–393.
1914b   New species of North American Dolichopodidae (Diptera). Ent. News **25**: 404–407.
1914c   New species of North American Dolichopodidae (Dip.). Ent. News **25**: 433–443, pl. 18.
1915a   A revision of the North American species of the dipterous genus *Diaphorus*. Buffalo Soc. Nat. Sci. Bul. **11** (**2**): 161–194, 1 fig.
1915b   Descriptions of three new species of the dipterous genus *Sciapus* with a key to the North American species. Ent. News **26**: 17–26, 3 figs.
1915c   Table of North American species of the dipterous genus *Thrypticus*, with descriptions of four new species. Psyche **22**: 84–88, 1 pl.
1915d   A biological reconnaissance of the Okefenokee Swamp region in Georgia. The Dolichopodidae (Diptera). Wash. Univ. [St. Louis] Studies **2** (**pt. 1**): 87–95, 8 figs.
1916a   Notes on *Chrysotimus* with the description of a new species (Diptera). Canad. Ent. **48**: 23–24.
1916b   Table of males of the North American species of the genus *Asyndetus* with descriptions of six new species. Psyche **23**: 88–94, 1 fig.
1917a   New North American species of Dolichopodidae (Dip.). Ent. News **28**: 123–128, 4 figs.
1917b   New North American species of Dolichopodidae (Diptera). Canad. Ent. **49**: 337–342, 1 fig.
1917c   Descriptions of a few new *Diaphorus* from the western states (Diptera). Psyche **24**: 33–39.
1918   New North American species of Dolichopodidae (Diptera). Ent. News **29**: 45–51, 1 fig.

Van Duzee, M. C.—Continued
- 1919a Key to the North American species of the dipterous genus *Medeterus* with descriptions of new species. Calif. Acad. Sci. Proc. ser. **4, 8**: 257–270.
- 1919b Two new *Asyndetus* with a table of the North American species (Dolichopodidae, Diptera). Ent. News **30**: 248–250, 4 figs.
- 1920 Three new species of Dolichopodidae (Diptera) from California and Nevada. Calif. Acad. Sci. Proc. ser. **4, 10**: 47–49.
- 1921a A new species of the dipterous family Dolichopodidae from the Pribilof Islands, Alaska. Calif. Acad. Sci. Proc. ser. **4, 11**: 167–168.
- 1921b Notes and descriptions of a few North American Dolichopodidae (Diptera). Psyche **28**: 120–129, 9 figs.
- 1921c Classification. Pp. 9–296, 1 fig., 16 pls. *In* Van Duzee, M. C., Cole, F. R., and Aldrich, J. M., The dipterous genus *Dolichopus* Latreille in North America. U.S. Natl. Mus. Bul. **116**: 1–304, 1 fig., 16 pls.
- 1922a The genus *Xiphandrium* Loew in North America (Diptera Dolichopodidae). Amer. Ent. Soc. Trans. **48**: 79–87, 10 figs.
- 1922b Three new species of *Parasyntormon* with a table of species (Dolichopodidae, Diptera). Canad. Ent. **54**: 88–90.
- 1923a New and known species of *Porphyrops* from North America (Diptera, Dolichopodidae). Ent. News **34**: 239–243.
- 1923b The *Pelastoneurus* of North America (Dolichopodidae, Diptera). Ent. Soc. Amer. Ann. **16**: 30–48, 1 pl.
- 1923c Scientific results of the Katmai Expedition of the National Geographic Society. Diptera of the family Dolichopodidae. Ohio Jour. Sci. **23**: 241–263, 1 pl.
- 1923d New species of North American Dolichopodidae. Psyche **30**: 63–73, 1 fig.
- 1924a New species of the dipterous family Dolichopodidae. Boston Soc. Nat. Hist. Occas. Papers **5**: 101–106.
- 1924b A revision of the North American species of the dipterous genus *Chrysotus*. Buffalo Soc. Nat. Sci. Bul. **13 (3)**: 3–53, 1 pl.
- 1924c New Canadian Dolichopodidae (Diptera). Canad. Ent. **56**: 244–249.
- 1924d A new western dolichopodid. Pan-Pacific Ent. **1**: 43–44.
- 1924e North American species of *Paraphrosylus* Becker, a subgenus of *Aphrosylus* Walker. Pan-Pacific Ent. **1**: 73–78.

Van Duzee, M. C.—Continued
- 1924f   Notes and descriptions of two-winged flies of the family Dolichopodidae from Alaska. U.S. Natl. Mus. Proc. (1923) **63 (21):** 1–16, 1 pl.
- 1925a   The dipterous genus *Syntormon* in North America (Dolichopodidae). Amer. Ent. Soc. Trans. **50:** 275–287, 16 figs.
- 1925b   *Scellus virago* Aldrich (a two-winged fly) and two forms closely related to it. Calif. Acad. Sci. Proc. ser. **4, 14:** 175–183, 23 figs.
- 1925c   Dolichopodids, new or imperfectly known. Pan-Pacific Ent. **1:** 153–155.
- 1925d   A revision of the North American species of the genus *Argyra* Macquart, two-winged flies of the family Dolichopodidae. U. S. Natl. Mus. Proc. **66 (23):** 1–43, 1 pl.
- 1925e   New species of North American Dolichopodidae (Diptera). Psyche **32:** 178–189.
- 1926a   A new dolichopodid genus, with descriptions of five new species (Diptera). Amer. Ent. Soc. Trans. **52:** 39–46, 15 figs.
- 1926b   The genus *Thinophilus* in North America (Dolichopodidae, Diptera). Ent. Soc. Amer. Ann. **19:** 35–49.
- 1926c   New species of North American Dolichopodidae. Psyche **33:** 45–52.
- 1926d   A new species of Scenopinidae from California (Diptera). Pan-Pacific Ent. **2:** 164.
- 1926e   The genus *Micropeza* in North America. Pan-Pacific Ent. **3:** 1–4.
- 1926f   A table of the North American species of *Hydrophorus* with the description of a new form (Diptera). Pan-Pacific Ent. **3:** 4–11.
- 1926g   Further new Dolichopodidae in the Canadian National Collection (Diptera). Canad. Ent. **58:** 56–59.
- 1926h   New *Dolichopus* in the Canadian National Collection. Canad. Ent. **58:** 230–232.
- 1927a   A contribution to our knowledge of the North American Conopidae (Diptera). Calif. Acad. Sci. Proc. ser. **4, 16:** 573–604.
- 1927b   The North American *Nematoproctus* (Dipt.: Dolichopodidae). Ent. News **38:** 53–54.
- 1927c   Three new species of *Psilopus* from North America, and notes on *caudatus* Wied. (Dipt.: Dolichopodidae). Ent. News **38:** 72–76, 4 figs.
- 1927d   North American species of *Polymedon* (Diptera Dolichopodidae). Ent. Soc. Amer. Ann. **20:** 123–126.

Van Duzee, M. C.—Continued

- 1927e   Four new dolichopids in the collection of the California Academy of Sciences. Pan-Pacific Ent. **3**: 146–148.
- 1928a   New Mycetophilidae taken in California and Alaska. Calif. Acad. Sci. Proc. **ser. 4, 17**: 31–65, 34 figs.
- 1928b   Three new dolichopodids from western Canada (Dipt.). Canad. Ent. **60**: 40–42.
- 1928c   Three new species of *Rhaphium* (Diptera). Pan-Pacific Ent. **4**: 166–168.
- 1928d   New North American species of Dolichopodidae (Diptera). Pan-Pacific Ent. **5**: 87–89.
- 1928e   Table of the North American species of *Medeterus*, with descriptions of three new forms. Psyche **35**: 36–43.
- 1929   Tropical American Diptera or two-winged flies of the family Dolichopodidae from Central and South America. U.S. Nat. Mus. Proc. **74 (10)**: 1–64, 2 pls.
- 1930a   New Dolichopidae from Connecticut. Amer. Mus. Nat. Hist., Amer. Mus. Novitates **439**: 1–5.
- 1930b   New species of Dolichopodidae from North America and the West Indies. Canad. Ent. **62**: 84–87, 1 fig.
- 1930c   New species of Dolichopodidae from North America (Diptera). Ent. News **41**: 53–55, 70–73, 1 fig.
- 1930d   Three new dolichopids from California and Colorado (Diptera). Pan-Pacific Ent. **6**: 123–126.
- 1930e   The dipterous genus *Sympycnus* Loew in North America and the West Indies. Pan-Pacific Ent. **7**: 35–47, 49–63, 2 pls.
- 1930f   The dolichopodid genus *Nematoproctus* Loew in North America. Psyche **37**: 167–172.
- 1931   A new species of *Physocephala* from Ontario, Canada (Conopidae, Diptera). Canad. Ent. **63**: 284.
- 1932a   New species of Dolichopidae from North America and Cuba, with notes on known species. Amer. Mus. Nat. Hist., Amer. Mus. Novitates **521**: 1–14, 13 figs.
- 1932b   New North and South American Dolichopidae, with notes on previously described species. Amer. Mus. Nat. Hist., Amer. Mus. Novitates **569**: 1–22, 31 figs.
- 1932c   Three new species of Dolichopodidae from North America and Cuba, with notes on *Diaphorus leucostola* Loew and its allies (Diptera). Ent. News **43**: 183–187, 3 figs.

Van Duzee, M. C.—Continued
1933a New Dolichopidae from North America with notes on several described species. Amer. Mus. Nat. Hist., Amer. Mus. Novitates **599**: 1–27, 52 figs.
1933b New American Dolichopidae. Amer. Mus. Nat. Hist., Amer. Mus. Novitates **655**: 1–20, 33 figs.
1933c Preoccupied names of dolichopodid flies and the new names proposed for the species (Diptera). Ent. News **44**: 151–152.
1933d On five species of Diptera, new and old. Pan-Pacific Ent. **9**: 63–67, 6 figs.
1934 Conopidae from North Dakota and the Rocky Mountain region. Ent. Soc. Amer. Ann. **27**: 315–323, 4 figs.
———, **Cole, F. R., and Aldrich, J. M.** (See also Van Duzee, 1921c)
1921 The dipterous genus *Dolichopus* Latreille in North America. U.S. Natl. Mus. Bul. **116**: 1–304, 1 fig., 16 pls.
———, **and Curran, C. H.**
1934a Key to the males of nearctic *Dolichopus* Latreille (Diptera). Amer. Mus. Nat. Hist., Amer. Mus. Novitates **683**: 1–26.
1934b Key to the females of nearctic *Dolichopus* Latreille (Diptera). Amer. Mus. Nat. Hist., Amer. Mus. Novitates **684**: 1–17.
van Emden, F. I. (See Emden, F. I. van)
Vargas, L.
1939 Datos acerca del *A. pseudopunctipennis* y de un *Anopheles* nuevo de California. Medecina, Rev. Mex. **19**: 356–362.
1943 *Anopheles earlei* Vargas, 1942, n. sp. norteamericana del grupo *maculipennis*. Pan Amer. Sanit. Bur., Bol. de la Ofic. Sanit. Panamer. **22**: 8–12, 6 figs.
1946 *Corethrella (Corethrella) laneana* n. sp. (Diptera, Culicidae), procente de Monterrey, N. L. [Mex.] Inst. de Salubridad y Enferm. Trop. Rev. **7**: 63–67, 2 pls.
1949 *Culicoides travisi* Vargas n. n. [Mex.] Inst. de Salubridad y Enferm. Trop. Rev. **10**: 233–234.
1953 *Beltranmyia* n. subg. de *Culicoides* (Insecta: Heleidae). [Mex.] Inst. de Salubridad y Enferm. Trop. Rev. **13**: 33–36.
1960 The subgenera of *Culicoides* of the Americas (Diptera, Ceratopogonidae). Rev. de Biol. Trop. **8**: 35–47, fig. 1.

Vargas, L., and Díaz Nájera, A.
1949 Claves para identificar las pupas de los simulidos de Mexico. Descripción de *Simulium* (*Dyarella*) *freemani* n. sp. de *Simulium* (*Neosimulium*) *encisoi* n. sp. y referencias adicionales sobre *S. anduzei* y *S. ruizi.* [Mex.] Inst. de Salubridad y Enferm. Trop. Rev. **10**: 283–319, 10 pls.

———, **Martínez Palacios, A., and Díaz Nájera, A.**
1946 Simulidos de Mexico. Datos sobre sistemática y morfologia, descripción de nuevos subgéneros y especies. [Mex.] Inst. de Salubridad y Enferm. Trop. Rev. **7**: 97–192, pls. 1–25.

———, **and Wirth, W. W.**
1955 *Culicoides blantoni* n. sp. (Diptera, Heleidae). [Mex.] Inst. de Salubridad y Enferm. Trop. Rev. **15**: 33–35, 1 pl.

Vaurie, P. (See Selander and Vaurie)

Verrall, G. H.
1873 Additions and corrections to the list of British Syrphidae. Ent. Monthly Mag. **9**: 251–256.
1877 Description of a new genus and species of Phoridae parasitic on ants. Linn. Soc. (London), Jour., Zool. (1878) **13**: 258–260.
1892 Two new English species of *Homalomyia.* Ent. Monthly Mag. **28** [=ser. 2, 3]: 149.
1901 Platypezidae, Pipunculidae, and Syrphidae of Great Britain. Vol. 8, 691 pp., 457 figs. *In* [Verrall, G. H., ed.], British flies [q.v.]. London.
   Catalogues of the 3 families involved, dated 1900, and separately paged 1–121, are bound with this volume.
1909 Stratiomyidae and succeeding families of the Diptera Brachycera of Great Britain. Vol. 5, 780 pp., 407 figs. *In* [Verrall, G. H., ed.], British flies [q.v.]. London.
1912 Another hundred new British species of Diptera [concl.]. Ent. Monthly Mag. **48** [=ser. 2, 23]: 20–27, 56–59, 144–147, 190–197.

[Verrall, G. H., ed.]
1901– British flies. 3 vols. London, Cambridge.
1961
   This series, planned to consist of 14 volumes, was begun in 1901. The first 2 volumes (8 and 5) were issued in London (1901 and 1909, respectively) by Verrall, with a note that Collin was to author a third volume. This third volume (Vol. 6) was not published until 1961, in Cambridge, long after Verrall's death, but his original prospectus was followed.
   See Collin, 1961; Verrall, 1901, 1909.

Verrill, A. E.
1873 Report upon the invertebrate animals of Vineyard Sound and the adjacent waters, with an account of the physical characters of the region. Pp. 295–778, 4 figs., pls. 1–38. *In* Baird, S. F., Report on the condition of the sea fisheries of the south coast of New England in 1871 and 1872. U.S. Comn. Fish and Fisheries, [Rpt. of Comnr.] (1871–1872) [**1**]: 1–852, 23 figs., 40 pls.

Versammlung der Naturforscher und Aertze zu Breslau (See Gravenhorst, 1834)

Vigé, L.
1939 A new syrphid fly from Louisiana (Diptera). Ent. News **50**: 66–68.

Villeneuve, J. (See also Becker, 1908; Becker, Dziedzicki et al.; Becker, Kuntze et al.)
1912 Des espèces européennes du genre *Carcelia* R. D. (Diptères). Feuille des Jeunes Nat. **42** [=ser. 5, 2]: 89–92.
1922a Descriptions d'Anthomyides nouveaux. Paris, Mus. Natl. d'Hist. Nat. Bul. **28**: 509–513.
1922b Descriptions de Tachinides nouveaux (Dipt. Musc.). Paris, Mus. Natl. d'Hist. Nat. Bul. **28**: 514–516.
1924 Contribution à la classification des "Tachinidae" paléarctiques. Ann. des Sci. Nat., Zool. ser. **10, 7**: 5–39, 7 figs.
1928 Sur *Trixa alpina* Meig. Konowia **7**: 303–306.
1932 Notes diptérologiques. Soc. Ent. de France, Bul. **37**: 271–272.
1933 Description de *Aplomyiopsis galerucellae* n. gen., n. sp. (Tachinidae), parasite de *Galerucella luteola* (F. Müll.) en Amérique du Nord. Portici, R. Ist. Super. Agr., Lab. Zool. Gen. e Agr. Bol. **27**: 125–126.
1934 Notes diptérologiques. Rev. Franç. d'Ent. **1**: 180–183.

Villers, C. de
1789 Caroli Linnaei entomologia, faunae suecicae descriptionibus aucta. Vol. 3, 657 pp., 4 pls. Lugduni [=Lyon, France].
The entire work consists of 4 volumes, Lugduni, 1789.

Vlasenko, N. (See Dorogostajskij, Rubzov, and Vlasenko)

Vockeroth, J. R.
1952 A new nearctic species of *Rhaphium*, with notes on other species (Diptera: Dolichopodidae). Canad. Ent. **84**: 276–280, 2 figs.
1954 Notes on northern species of *Aedes*, with descriptions of two new species (Diptera: Culicidae). Canad. Ent. **86**: 109–116, 6 figs.

Vockeroth, J. R.—Continued
1958   Two new nearctic species of *Spilomyia* (Diptera: Syrphidae), with a note on the taxonomic value of wing microtrichia in the Syrphidae. Canad. Ent. **90**: 284–291, 12 figs.
1960   Taxonomy of the genus *Cecidomyia* (Diptera: Cecidomyiidae) with special reference to the species occurring on *Pinus banksiana* Lamb. Canad. Ent. **92**: 65–79, 13 figs.
1961a   The North American species of the family Opomyzidae (Diptera: Acalypterae). Canad. Ent. **93**: 503–522, 42 figs.
1961b   [Note on *Scathophora* Meigen, 1803.] P. 296. *In* [China, W. E., ed.], Comments on the report on C. W. Sabrosky's proposed suppression of Meigen's "Nouvelle Classification", 1800. Internatl. Comn. Zool. Nomencl., Bul. Zool. Nomencl. **18**: 296.
1962   A new species of *Dolichopus* from North Carolina (Diptera: Dolichopodidae). Canad. Ent. **94**: 502–505, 2 figs.

Wachtl, F. A.
1884   Eine neue und eine verkannte Cecidomyide. Wien. Ent. Ztg. **3**: 161–166, 1 pl.
1894   Analytische Uebersicht der europäischen Gattungen aus dem Verwandtschaftskreise von *Echinomyia* Duméril, nebst Beschreibung einer neuen *Eudora*. Wien. Ent. Ztg. **13**: 140–144.

Wadley, F. M.
1931   Ecology of *Toxoptera graminum*, especially as to factors affecting importance in the northern United States. Ent. Soc. Amer. Ann. **24**: 325–395, 22 figs.

Wahlberg, P. F.
1839   Bidrag till svenska dipternas kännedom. K. Vetensk. Acad. Handl. [ser. 3], **1838**: 1–23.
1844a   Om *Rhaphium flavipalpe* Zett. K. Svenska Vetensk. Akad. Öfversigt af . . . Förhandl. **1**: 37–38.
1844b   Nya Diptera från Norrbotten och Luleå Lappmark. K. Svenska Vetensk. Akad. Öfversigt af . . . Förhandl. **1**: 64–68.
1845   Nytt dipter slägte från Luleå Lappmark. K. Svenska Vetensk. Akad. Öfversigt af . . . Förhandl. (1844) **1**: 217–219, 1 pl.
1847   Tvänne nya dipter-genera af agromyzidernas familj. K. Svenska Vetensk. Akad. Öfversigt af . . . Förhandl. **4**: 259–263, 1 pl.
1851   Nya Diptera. K. Svenska Vetensk. Akad. Öfversigt af . . . Förhandl. (1850) **7**: 215–225.

Wahlgren, E.

1904 Über einige Zetterstedt'sche Nemocerentypen. Arkiv för Zool. **2** (7): 1–19, 8 figs.

1910 Zur Kenntnis schwedischer Dipteren. II. Ent. Tidskr. **31**: 28–34, 2 figs.

1917 Fam. 5–12. Conopidae, Cordyluridae, Coelopidae, Cypselidae, Dryomyzidae, Clusiidae, Helomyzidae, Sciomyzidae. Pp. 113–224, figs. 63–100 [=rekv. 20]. *In* Entomologiska Foreningen i Stockholm, Svensk Insektfauna [q.v.]. [Order] 11: Diptera, Suborder II: Cyclorapha, Group 2: Schizophora, [Subgroup Holometopa]. Stockholm.

1918 Zur Kenntnis schwedischer Dipteren. III. Ent. Tidskr. **39**: 1–9.

Wakeland, C. (See Parker and Wakeland)

Walker, F. (See also Haliday, 1851, 1856; Lord, 1866)

1833 Observations on the British species of Sepsidae. Ent. Mag. (London) **1**: 244–256.

1834 Observations on the British species of Pipunculidae. Ent. Mag. (London) **2**: 262–270.

1836 Descriptions of the British Tephritites. Ent. Mag. (London) **3**: 57–85, 1 pl.

1837 Notes on Diptera. Ent. Mag. (London) **4**: 226–230.

1848– List of the specimens of dipterous insects in the collection
1849 of the British Museum. Vol. 1, pp. 1–229, *1848;* Vol. 2, pp. 231–484, *1849;* Vol. 3, pp. 485–687, *1849;* Vol. 4, pp. 689–1172, *1849*. London.

> The entire series consists of 7 volumes, London, 1848–1855, 1172 pp., and 774 pp. (vols. 5–7), the last 3 volumes being supplements.

1850 Diptera. Vol. 1, pp. 1–76, 2 pls. [cont.]. *In* [Saunders, W. W., ed.], Insecta Saundersiana [q.v.]. London, "1856."

1851a Diptera. Vol. 1. [Vol. 1], 314 pp., pls. 1–10. *In* (Walker, F., Stainton, H. T., and Wilkinson, S. J.), Insecta Britannica [q.v.]. London.

1851b Diptera [cont.]. Vol. 1, pp. 77–156, 2 pls. [cont.]. *In* [Saunders, W. W., ed.], Insecta Saundersiana [q.v.]. London, "1856."

1852 Diptera [cont.]. Vol. 1, pp. 157–252, 253–414, 4 pls. [cont.]. *In* [Saunders, W. W., ed.], Insecta Saundersiana [q.v.]. London, "1856."

Walker, F.—Continued
- 1853   Diptera Vol. 2. [Vol. 2], 297 pp., pls. 11–20. *In* (Walker, F., Stainton, H. T., and Wilkinson, S. J.), Insecta Britannica [q.v.]. London.
- 1854   List of the specimens of dipterous insects in the collection of the British Museum. Vol. 5: Supplement I, pp. 1–330, 2 figs.; Vol. 6: Supplement II, pp. 331–506, 8 figs. London.
- 1856a   Diptera. Vol. 3. [Vol. 4], xxiv + 352 pp., pls. 21–30. *In* (Walker, F., Stainton, H. T., and Wilkinson, S. J.), Insecta Britannica [q.v.]. London.
- 1856b   Diptera [concl.]. Vol. 1, pp. 415–474. *In* [Saunders, W. W., ed.], Insecta Saundersiana [q.v.]. London, "1856."
- 1857–1858   Characters of undescribed Diptera in the collection of W. W. Saunders. Ent. Soc. London, Trans. **n. ser.** [=ser. 2], **4**: 119–158, *1857*; 190–235, *1858*.
- 1859   Catalogue of dipterous insects collected at Makessar in Celebes, by Mr. A. R. Wallace, with descriptions of new species. Linn. Soc. (London), Jour. of Proc., Zool. **4**: 90–144 [cont.].
- 1860a   Characters of undescribed Diptera in the collection of W. W. Saunders. Ent. Soc. London, Trans. (1858–1861) **n. ser.** [=ser. 2], **5**: 268–296 [cont.].
- 1860b   Catalogue of the dipterous insects collected in Amboyna by Mr. A. R. Wallace, with descriptions of new species. Linn. Soc. (London), Jour. of Proc., Zool. **5**: 144–168.
- 1861   Characters of undescribed Diptera in the collection of W. W. Saunders [concl.]. Ent. Soc. London, Trans. (1858–1861) **n. ser.** [=ser. 2], **5**: 297–334.

(Walker, F., Stainton, H. T., and Wilkinson, S. J.)
- 1851–1859   Insecta Britannica. 5 vols. London.

> This series, initiated by Stainton, was to consist of a number of volumes on British insects by different authors. As far as known, only 5 volumes ever appeared, without overall editorship: 2 on Lepidoptera (1 by Stainton, 1 by Wilkinson), and the 3 volumes on Diptera by Walker.
> See Haliday, 1851, 1856; Walker, 1851a, 1853, 1856a.

Walker, J. T.   (See Porter, Spilker, and Walker)

Wallengren, H. D. J.
- 1881   Revision af skandinaviens Tipulidae. Ent. Tidskr. **2**: 177–208.

Walley, G. S.
- 1925   New Canadian Chironomidae of the genus *Tanypus* (Dipt.). Canad. Ent. **57**: 271–278, 1 pl.
- 1926a   New Canadian Chironomidae. Canad. Ent. **58**: 64–65.
- 1926b   Four new Canadian Chironomidae. Canad. Ent. **58**: 205–207.
- 1927a   Two new species of Empididae from Ontario (Empididae, Dipt.). Canad. Ent. **59**: 96–98, 1 fig.
- 1927b   Review of the Canadian species of the dipterous family Blephariceridae. Canad. Ent. **59**: 112–116, pl. 2.
- 1928a   A new species of *Cricotopus* with a key to the genus (Dipt., Chironomidae). Canad. Ent. **60**: 21–22.
- 1928b   Genus *Tanypus* in Canada, with a key to the North American species (Diptera, Chironomidae). Ent. Soc. Amer. Ann. **21**: 581–593, 1 pl.
- 1932   A new species of *Forcipomyia* (Dipt., Chironomidae). Canad. Ent. **64**: 165–166, 1 fig.
- 1933   A note on some *Fabriciella* types (Dipt., Tachinidae). Canad. Ent. **65**: 168.

Walsh, B. D.   (See also Felt, 1925b)
- 1861   Insects injurious to vegetation in Illinois. Ill. State Agr. Soc. Trans. (1859–1860) **4**: 335–372, 3 figs., 1 pl.
- 1864a   On dimorphism in the hymenopterous genus *Cynips*, with an appendix, containing hints for a new classification of Cynipidae, including descriptions of several new species, inhabiting the oak-galls of Illinois. Ent. Soc. Phila. Proc. (1863–1864) **2**: 443–500, 6 figs.
- 1864b   On the insects, coleopterous, hymenopterous and dipterous, inhabiting the galls of certain species of willow. Part 1st. Diptera. Ent. Soc. Phila. Proc. **3**: 543–644.
- 1864c   On certain remarkable or exceptional larvae, coleopterous, lepidopterous and dipterous, with descriptions of several new genera and species, and of several species injurious to vegetation, which have been already published in agricultural journals. Boston Soc. Nat. Hist. Proc. (1862–1863) **9**: 286–308.
- 1866   On the insects, coleopterous, hymenopterous and dipterous, inhabiting galls of certain species of willow. Part 2nd and last. Ent. Soc. Phila. Proc. (1866–1867) **6**: 223–288.
- 1867   The apple worm and the apple maggot. *Carpocapsa pomonella* (Linnaeus), *Trypeta pomonella* (new species). Amer. Jour. Hort. and Florist's Companion **2**: 338–343, 2 figs.

Walsh, B. D.—Continued
- 1869   Mr. Couper's thorn leaf gall. Canad. Ent. **1:** 79–80.
- 1870   Larvae in the human bowels. Amer. Ent. **2:** 137–141, 1 fig.

———, **and Riley, C. V.**
- 1869a   The apple-root plant-louse (*Eriosoma* [*pemphigus*] *pyri*, Fitch). Amer. Ent. **1:** 81–84, figs. 70–72.
- 1869b   Galls and their architects. Amer. Ent. **1:** 101–110, figs. 78–90.
- 1870   Answers to correspondents. Amer. Ent. **2:** 24–32, figs. 17–27.

Waltl, J.
- 1837   Neue Gattungen von Mucken bey München. Isis (Oken's) **1837:** 283–287.

Walton, W. R.   (See also McAtee and Walton)
- 1910   A new species of *Dasyllis* from Pennsylvania. Ent. News **21:** 243–244, 1 pl.
- 1911   Notes on Pennsylvanian Diptera, with two new species of Syrphidae. Ent. News **22:** 318–322, 1 pl.
- 1912   New North American Diptera. Ent. News **23:** 463–464.
- 1913   New North American Tachinidae (Dipt.). Ent. News **24:** 49–52, 1 pl.
- 1914a   Report on some parasitic and predaceous Diptera from northeastern New Mexico. U.S. Natl. Mus. Proc. **48:** 171–186, 2 pls.
- 1914b   A new tachinid parasite of *Diabrotica vittata*. Ent. Soc. Wash. Proc. **16:** 11–14, 1 pl.
- 1914c   Four new species of Tachinidae from North America. Ent. Soc. Wash. Proc. **16:** 90–95, 7 figs.
- 1914d   A new tachinid parasite of *Diapheromera femorata* Say. Ent. Soc. Wash. Proc. **16:** 129–132, 1 pl.
- 1915a   A new and interesting genus of North American Tachinidae. Ent. Soc. Wash. Proc. **17:** 104–107, 1 pl.
- 1915b   A new nocturnal species of Tachinidae. Ent. Soc. Wash. Proc. **17:** 162–164, 3 figs.
- 1918   Three new tachinid parasites of *Eleodes*. Ent. Soc. Wash. Proc. (1917) **19:** 22–26, pl. 5.

Ward, C. L.   (See Patterson and Ward)

Washburn, F. L.
- 1905   The Diptera of Minnesota. Two-winged flies affecting the farm, garden, stock and household. Minn. Agr. Expt. Sta. Bul. **93:** 19–168, 160 figs., 1 pl.

    Also bound in Minn. Agr. Expt. Sta. Ann. Rpt. (1904–1905) **13:** 19–168.

Waterhouse, C. D.
1887   Note on a new parasitic dipterous insect of the family Hippoboscidae. Zool. Soc. London, Proc. Sci. Mtg. 1887: 163–164, 1 fig.

Waterhouse, D. F., and Paramonov, S. J.
1950   The status of the two species of *Lucilia* (Diptera, Calliphoridae) attacking sheep in Australia. Austral. Jour. Sci. Res., Ser. B: Biol. Sci. **3**: 310–336, 13 figs., 1 pl.

Webb, J. L., and Wells, R. W.
1924   Horse-flies: Biologies and relation to western agriculture. U.S. Dept. Agr., Dept. Bul. **1218**: 1–36, 18 figs., 4 pls.

Webb, P. B., and Berthelot, S., eds.
1835–   Histoire naturelle des Îles Canaries. 3 vols., and Atlas.
1850   Paris.

> The volumes are in 2 separately paged parts each, some with a further breakdown, and were issued in livraisons. Volume 2, part 2, on Zoology, has 6 separately paged sections and was issued in Paris, 1836–1844. Volume 1, part 1, was issued in 1839. See Macquart, 1839.

Webber, R. T.   (See also Aldrich and Webber)
1930   A revision of the North American tachinid flies of the genus *Achaetoneura*. U.S. Natl. Mus. Proc. **78** (10): 1–37, 14 figs.
1931   A new parasitic fly of the genus *Chaetophlepsis*. U.S. Natl. Mus. Proc. **78** (20): 1–4.
1941   Synopsis of the tachinid flies of the genus *Tachinomyia*, with descriptions of new species. U.S. Natl. Mus. Proc. **90**: 287–304.

Weems, H. V., Jr.   (See Fluke and Weems)

Wehr, E. E.
1922a   A synopsis of the Tabanidae of Nebraska, with a description of a new species from Colorado. Nebr. Univ., Univ. Studies **22**: 107–118.

> Also paged separately 1–12.

1922b   A synopsis of the Syrphidae of Nebraska, with descriptions of new species from Nebraska and Colorado. Nebr. Univ., Univ. Studies **22**: 119–162.

> Also paged separately 13–56.

Weiss, H. B., and West, E.
1921   Notes on the insects of the spreading dogbane, *Apocynum androsaemifolium* L., with a description of a new dogbane midge, by Dr. E. P. Felt. Canad. Ent. **53**: 146–152.

Welch, P. S.

1912 Observations on the life history of a new species of *Psychoda* Ent. Soc. Amer. Ann. **5**: 411–418, 2 pls.

Wells, R. W. (See also Bishopp et al.; Webb and Wells)

――――, **and Knipling, E. F.**

1938 A report of some recent studies on species of *Gasterophilus* occurring in horses in the United States. Iowa State Col. Jour. Sci. **12**: 181–203, 2 pls.

West, E. (See Weiss and West)

West, L. S.

1924 New northeastern Dexiinae (Dipterae [sic], Tachinidae). Psyche **31**: 184–192.

1925 New Phasiidae and Tachinidae from New York State. N.Y. Ent. Soc. Jour. **33**: 121–135.

1928 Family Tachinidae. Pp. 807–821. *In* Leonard, M.D., A list of the insects of New York with a list of the spiders and certain other allied groups. N.Y. (Cornell) Agr. Expt. Sta. Mem. **101**: 1–1121.

> This is in the portion on Diptera edited by O. A. Johannsen.

1951 The housefly, its natural history, medical importance and control. 584 pp., 176 figs. Ithaca, N.Y.

Westwood, J. O.

1835a Insectorum novorum exoticorum (ex ordine dipterorum) descriptiones. London and Edinb. Phil. Mag. and Jour. Sci. **ser. 3, 6**: 280–281, 447–449.

1835b Insectorum nonnullorum novorum (ex ordine dipterorum) descriptiones. Soc. Ent. de France, Ann. **[ser. 1], 4**: 681–685.

(1839) Description of a new genus of dipterous insects from New South Wales. Ent. Soc. London, Trans. (1837–1840) **2**: 151–152, 1 pl.

1840 Order XIII. Diptera Aristotle (Antliata Fabricius. Halteriptera Clairv.). Pp. 125–128 (=signature I, part), 129–144 (=signature K), 145–158 (=signature L). *In his* An introduction to the modern classification of insects. Synopsis of the genera of British insects, 158 pp. London.

> The entire work consists of 2 volumes and the separately paged "Synopsis . . . ," which is usually bound at the end of Volume 2. The 2 volumes were issued in a number of signatures, as was the "Synopsis," London, 1838–1840. See Blackwelder, 1949: 46, for details in dating this work.

Westwood, J. O.—Continued
[1841] Plates XIII and XIV. Synopsis of the dipterous family Midasidae, with descriptions of numerous species. Pp. 49–56, pls. 13–14. *In his* Arcana entomologica: Or illustrations of new, rare, and interesting insects. Vol. 1, 192 pp., 48 pls. London, "1845."

> The entire work consists of 2 volumes, both dated 1845, but which were issued in 24 bimonthly numbers, London, 1841–1845. Volume 1 contains numbers 1–13, and was issued 1841–1843. Although the first 4 numbers are not dated, they were evidently published in 1841, since the 5th number is dated Jan. 1, 1842. The above reference on the "Midasidae" was in the 4th number.

1847 The willow-twig midge. Gard. Chron. and Agr. Gaz. **1847**: 588, 1 fig.

> This article is signed simply "W".

(1848) Descriptions of some new exotic species of Acroceridae (Vesiculosa, Latr.), a family of dipterous insects. Ent. Soc. London, Trans. (1847–1849) **5**: 91–98.

(1852) Observations on the destructive species of dipterous insects known in Africa under the names of the tsetse, zimb and tsaltsalya, and on their supposed connexion with the fourth plague of Egypt. Zool. Soc. London, Proc. (1850) **18**: 258–270, 1 pl.

1876 Notae dipterologicae. No. 3. Descriptions of new genera and species of the family Acroceridae. Ent. Soc. London, Trans. **1876**: 507–518, 2 pls.

Weyenbergh, H.
1881 Dos nuevas especies del grupo de los dipteros pupiparos. Soc. Cient. Argentina, An. **11**: 193–200.

Wheeler, G. M., ed.
1875– Report upon geographical and geological explorations and
1889 surveys west of the one hundredth meridian. 7 vols., and 1 sup. Washington.

> See Osten Sacken, 1875a.

Wheeler, M. R. (See also Hardy and Wheeler; Patterson and Wheeler; Sturtevant and Wheeler)
1949a XII. The subgenus *Pholadoris* (*Drosophila*) with descriptions of two new species. Pp. 143–156. *In* Patterson, J. T., Studies in the genetics of *Drosophila*. VI. Articles on genetics, cytology and taxonomy. Tex. Univ. Pub. **4920**: 1–223, 3 figs., 24 pls.

Wheeler, M. R.—Continued
- 1949b XIII. Taxonomic studies on the Drosophilidae. Pp. 157–195, 2 figs. *In* Patterson, J. T., Studies in the genetics of *Drosophila*. VI. Articles on genetics, cytology and taxonomy. Tex. Univ. Pub. **4920**: 1–223, 3 figs., 24 pls.
- 1952a The dipterous family Canaceidae in the United States. Ent. News **63**: 89–94.
- 1952b XI. The Drosophilidae of the Nearctic Region, exclusive of the genus *Drosophila*. Pp. 162–218, 1 fig. *In* Patterson, J. T., Studies in the genetics of *Drosophila*. VII. Further articles on genetics, cytology and taxonomy. Tex. Univ. Pub. **5204**: 1–251, 64 figs.
- 1954a V. Taxonomic studies on American Drosophilidae. Pp. 47–64, 2 figs. *In* Patterson, J. T., Studies in the genetics of *Drosophila*. VIII. Articles on genetics, taxonomy, cytology, and radiation. Tex. Univ. Pub. **5422**: 1–307, 82 figs., 10 pls.
- 1954b *Rhinotora diversa* Giglio Tos from the southwestern United States (Diptera: Rhinotoridae). Wasmann Jour. Biol. **12**: 35–39, 2 figs.
- 1955 *Paraneossos*, a new genus of Trixoscelidae (Diptera: Acalyptrata). Wasmann Jour. Biol. **13**: 107–112, 2 figs.
- 1956 *Latheticomyia*, a new genus of acalyptrate flies of uncertain family position. U.S. Natl. Mus. Proc. **106**: 305–314, 2 figs.
- 1957 VII. Taxonomic and distributional studies of nearctic and neotropical Drosophilidae. Pp. 79–114, 25 figs. *In* Patterson, J. T., Studies in the genetics of *Drosophila*. IX. Articles on genetics, taxonomy, cytology, and radiation. Tex. Univ. Pub. **5721**: 1–316, 120 figs., 1 pl.
- 1959 A nomenclatural study of the genus *Drosophila*. Pp. 180–205, 1 fig. *In* Wheeler, M. R., ed., Biological contributions. A collection of essays and research articles dedicated to John Thomas Patterson on the occasion of his eightieth birthday. Tex. Univ. Pub. **5914**: 1–271, 50 figs.
- 1960a A new subgenus and species of *Stegana* Meigen (Diptera: Drosophilidae). Ent. Soc. Wash. Proc. **62**: 109–111, 1 fig.
- 1960b A new subgenus and two new species of *Pseudiastata* Coquillett (Diptera, Drosophilidae). Brooklyn Ent. Soc. Bul. **55**: 67–70, 3 figs.
- 1960c Three new North American Drosophilidae. Southwest. Nat. **5**: 89–91, 2 figs.

Wheeler, M. R.—Continued
1960d   New species of the *quinaria* group of *Drosophila* (Diptera, Drosophilidae). Southwest. Nat. **5:** 160–164, 12 figs.
1961   New species of southwestern acalyptrate Diptera. Southwest. Nat. **6:** 86–91, 1 fig.

Wheeler, M. R., and Takada, H.
1963   A revision of the American species of *Mycodrosophila* (Diptera: Drosophilidae). Ent. Soc. Amer. Ann. **56:** 392–399, 35 figs.

————, **and Throckmorton, L. H.**
1961   Notes on Alaskan Drosophilidae (Diptera), with the description of a new species. Brooklyn Ent. Soc. Bul. (1960) **45:** 134–143, 2 pls.

Wheeler, W. M.
[1889]   On two new species of cecidomyid flies producing galls on *Antennaria plantaginifolia*. Pp. 209–216. *In* Natural History Society of Wisconsin, Proceedings of the Natural History Society of Wisconsin [q.v.]. Milwaukee, Wis.

> The signature in which this article appeared contains the Proceedings for 1888, through the Dec. 17th meeting of the society, and is otherwise undated. Aldrich, 1905: 69, dates the signature Apr. 1889.

1890   Descriptions of some new North American Dolichopodidae. Psyche **5:** 337–343, 355–362, 373–379.
1896a   The genus *Ochthera*. Ent. News **7:** 121–123, 2 figs.
1896b   Two dolichopodid genera new to America. Ent. News **7:** 152–156.
1896c   A new genus and species of Dolichopodidae. Ent. News **7:** 185–189, 1 fig.
1896d   A new empid with remarkable middle tarsi. Ent. News **7:** 189–192, 3 figs.
1897   A genus of maritime Dolichopodidae new to America. Calif. Acad. Sci. Proc. ser. 3, **1 (Zool.):** 145–152, 1 pl.
1898   A new genus of Dolichopodidae from Florida. Zool. Bul. **1:** 217–220, 1 fig.
1899   New species of Dolichopodidae from the United States. Calif. Acad. Sci. Proc. ser. 3, **2 (Zool.):** 1–84, 4 pls.
1906   A new wingless fly (*Puliciphora borinquenensis*) from Porto Rico. Amer. Mus. Nat. Hist. Bul. **22:** 267–271, pl. 34.
1918   *Vermileo comstocki*, sp. nov., an interesting leptid fly from California. New England Zool. Club, Proc. **6:** 83–84.

White, J. E. (See Brodie and White)

White, M. J. D.

1950   Cytological studies on gall midges (Cecidomyiidae). Tex. Univ. Pub. **5007**: 1–80, 51 figs.

Whitlock, C.

1939   Abbreviations used in the Department of Agriculture for titles of publications. U.S. Dept. Agr. Misc. Pub. **337**: 1–278.

Whitney, C. P.

1879   Descriptions of some new species of Tabanidae. Canad. Ent. **11**: 35–38.

1904   Descriptions of some new species of Tabanidae. Canad. Ent. **36**: 205–207.

1914   Descriptions of four new Tabanidae, with remarks upon *Chrysops cursim*. Canad. Ent. **46**: 343–346.

1915   A new *Tabanus*. Canad. Ent. **47**: 380–381.

Wiedemann, C. R. W.

1817   Neue Zweiflügler (Diptera Linn.) aus der Gegend um Kiel. Zool. Mag. (Wiedemann's) **1** (1): 61–86.

1818   Neue Insecten vom Vorgebirge der Guten Hoffnung. Zool. Mag. (Wiedemann's) **1** (2): 40–48.

1819   Brasilianische Zweiflügler. Zool. Mag. (Wiedemann's) **1** (3): 40–56.

1820a   Diptera exotica. [Ed. 1], Pt. I, pp. i–xix + 1–42, 1 fig. Kiliae [=Kiel].

> Also published in the enlarged edition, Wiedemann, 1821c, pp. 1–42.
> This first edition consists of 2 parts, Kiliae, 1820–1821. Part I is as above; Part II [="Sectio II"], consists of pp. i–iv + 43–50, and pp. 1–101 (see Wiedemann, 1821a and 1821b).

1820b   Munus rectoris in Academia Christiano-Albertina iterum aditurus nova dipterorum genera offert iconibusque illustrat. 23 pp., 1 pl. Kiliae Holsatorum [=Kiel].

> This work is generally known as "Nova dipterorum".

1821a   Diptera exotica. [Ed. 1], [Pt.] II, pp. i–iv + 43–50 [cont.]. Kiliae [=Kiel].

> Also published in the enlarged edition, Wiedemann, 1821c, pp. 43–50.

1821b   Diptera exotica. [Ed. 1], [Pt.] II [concl.], pp. 1–101, 2 pls. Kiliae [=Kiel].

> Also published in the enlarged edition, Wiedemann, 1821c, pp. 51–151.

Wiedemann, C. R. W.—Continued
1821c Diptera exotica. [Ed. 2], 244 pp., 1 fig., 2 pls. Kiliae [=Kiel].

> The pages 152-244 contain material not in the first edition, pp. 1-151 being the same as the first edition's pp. 1-50 and 1-101.

1824 Munus rectoris in Academia Christiana Albertina aditurus analecta entomologica ex Museo Regio Havniensi. 60 pp., 1 pl. Kiliae [=Kiel].

> This work is generally known as "Analecta entomologica".

1828 Aussereuropäische zweiflügelige Insekten. Vol. 1, xxxii + 608 pp., 7 pls. Hamm.

> The entire work consists of 2 volumes, Hamm, 1828-1830.

1830 Aussereuropäische zweiflügelige Insekten. Vol. 2, xii + 684 pp., 5 pls. Hamm.

> Considered as published prior to Robineau-Desvoidy, 1830. See Aubertin, 1933: 141.

1831 Monographia generis Midarum. K. Leopoldinisch-Carolinisch. Akad. d. Naturf., Verhandl. (Acad. C. Leopoldino-Carolinae Nat. Curio., Nova Acta Phys.-Med.) **15** [=ser. 2, 5 (=Verhandl. 7)] (**2**): 19-56, pls. 52-54.

Wilcox, J. (See also Bromley, 1937a; Cole and Wilcox)
1935a New asilid flies of the genus *Ablautus* with a key to the species. Canad. Ent. **67**: 222-227.
1935b Description of the male of *Willistonina bilineata* (Williston) together with a new form (Diptera-Asilidae). Pan-Pacific Ent. **11**: 31-34.
1936a New *Ommatius* with a key to the species (Diptera, Asilidae). Brooklyn Ent. Soc. Bul. **31**: 172-176, 1 pl.
1936b *Laphria vultur* Osten Sacken and two related species (Diptera-Asilidae). Canad. Ent. **68**: 7-11, 1 fig.
1936c A new robber fly, with a key to the species of *Callinicus* and *Chrysoceria* (Diptera: Asilidae). Ent. News **47**: 208-210.
1936d Asilidae, new and otherwise, from the Southwest, with a key to the genus *Stichopogan* [sic]. Pan-Pacific Ent. **12**: 201-212, 1 fig. [cont.].
1937 Asilidae, new and otherwise, from the Southwest, with a key to the genus *Stichopogon* (concl.). Pan-Pacific Ent. **13**: 37-45.

Wilcox, J.—Continued
1941   New *Heteropogon* with a key to the species (Diptera, Asilidae). Brooklyn Ent. Soc. Bul. **36**: 50–56.
1946   New *Nicocles* with a key to the species (Diptera, Asilidae). Brooklyn Ent. Soc. Bul. (1945) **40**: 161–165.
1949   The genus *Itolia* Wilcox (Diptera: Asilidae). Pan-Pacific Ent. (1948) **24**: 191–193.
1959   The *clausa* group of *Cophura* Osten Saken [sic] (Diptera: Asilidae). Brooklyn Ent. Soc. Bul. **54**: 121–127, 1 fig.
1960   *Laphystia* Loew in North America (Diptera: Asilidae). Ent. Soc. Amer. Ann. **53**: 328–346, pls. 1–2.
1961   The genus *Hodophylax* James (Diptera: Asilidae). Brooklyn Ent. Soc. Bul. **56**: 112–116.

———, **and Martin, C. H.**
1935   The genus *Coleomyia* (Diptera–Asilidae). Brooklyn Ent. Soc. Bul. **30**: 204–213, 1 fig.
1936a A review of the genus *Cyrtopogon* Loew in North America (Diptera—Asilidae). Ent. Amer. **16**: 1–94, 5 pls.
1936b The genus *Nannocyrtopogon* (Diptera, Asilidae). Ent. Soc. Amer. Ann. **29**: 449–459, 1 pl.
1941   The genus *Dioctria* Meigen in North America (Diptera-Asilidae). Ent. Amer. **21**: 1–22, 1 pl.
1942   Change in name in Diptera. Brooklyn Ent. Soc. Bul. **37**: 35.
1945   Contributions from the Los Angeles Museum Channel Islands Biological Survey. 29. Robber flies (Diptera, Asilidae). South. Calif. Acad. Sci. Bul. **44**: 10–17, pl. 7.
1957a *Nannocyrtopogon* (Diptera-Asilidae). Ent. Soc. Amer. Ann. **50**: 376–392, 2 figs.
1957b *Backomyia* (Diptera-Asilidae), a new genus. Kans. Ent. Soc. Jour. **30**: 1–5.

Wilkinson, S. J.   (See Walker, Stainton, and Wilkinson)

Willard, F.
1914   Two new species of *Platypeza* found at Stanford University. Psyche **21**: 166–168, 7 figs.

Williams, F. X.
1909   The Monterey pine resin midge—*Cecidomyia resinicoloides* n. sp. Ent. News **20**: 1–8, 1 pl.
1939   Biological studies in Hawaiian water-loving insects. Part III. Diptera or true flies. A, Ephydridae and Anthomyiidae. Hawaii. Ent. Soc. Proc. (1938) **10**: 85–119, 1 fig., 9 pls.

Williams, R. W. (See also Wirth and Williams)
- 1951a Observations on the bionomics of *Culicoides tristriatulus* Hoffman with notes on *C. alaskensis* Wirth and other species at Valdez, Alaska, summer 1949 (Diptera, Heleidae). Ent. Soc. Amer. Ann. **44:** 173–183.
- 1951b The immature stages of *Culicoides tristriatulus* Hoffman (Diptera, Heleidae). Ent. Soc. Amer. Ann. **44:** 430–440, 3 pls.
- 1953 Notes on the bionomics of the *Alluaudomyia* of Baker County, Georgia. I. Observations on breeding habitats of *bella* and *needhami* (Diptera, Heleidae). Ent. Soc. Wash. Proc. **55:** 283–285, 7 figs.
- 1955 Two new species of *Culicoides* from Cheboygan County, Michigan (Diptera, Heleidae). Ent. Soc. Wash. Proc. **57:** 269–274, 12 figs.
- 1956 The biting midges of the genus *Culicoides* found in the Bermuda Islands (Diptera, Heleidae). Jour. Parasitol. **42:** 297–305, 4 figs.
- 1957 Two new species of *Alluaudomyia* from Cheboygan County, Michigan, with a note on the synonymy of *para* [sic] and *downesi* (Diptera, Heleidae). Ent. Soc. Wash. Proc. (1956) **58:** 327–331, 4 figs.

Williston, S. W. (See also Adams, 1903a; Aldrich, 1896b; Cook, 1884; Forbes, 1885; Forbush and Fernald, 1896; Hunter, 1897; Kellogg, 1893; Riley, 1889)
- 1880 Some interesting new Diptera. Conn. Acad. Arts and Sci., Trans. **4:** 243–246, 1 fig.
- 1882a Contribution to a monograph of the North American Syrphidae. Amer. Phil. Soc. Proc. **20:** 299–332.
- 1882b New or little known genera of North American Syrphidae. Canad. Ent. **14:** 77–80.
- 1882c The North American species of *Conops*. Conn. Acad. Arts and Sci., Trans. **4:** 325–342, 1 pl.
- 1883a On the North American Asilidae (Dasypogoninae, Laphrinae), with a new genus of Syrphidae. Amer. Ent. Soc. Trans. and Acad. Nat. Sci. Phila., Ent. Sect. Proc. (1884) **11:** 1–35, 2 pls.
- 1883b The North American species of Nemestrinidae. Canad. Ent. **15:** 69–72, 1 fig.
- 1883c North American Conopidae: *Stylogaster, Dalmannia, Oncomyia*. Conn. Acad. Arts and Sci., Trans. (1882–1885) **6:** 91–98.

Williston, S. W.—Continued
- 1884a Order VII. Diptera. Pp. 403–432, 433, figs. 508–549. *In* Kingsley, J. S., ed., The standard natural history [q.v.]. Vol. 2, 555 pp., 666 figs., 20 pls. Boston.
- 1884b Eine merkwürdige neue Syrphiden-Gattung. Wien. Ent. Ztg. **3**: 185–186, 1 fig.
- 1885a On the North American Asilidae (Part II). Amer. Ent. Soc. Trans. and Acad. Nat. Sci. Phila., Ent. Sect. Proc. **12**: 53–76.
- 1885b On the classification of North American Diptera. (First paper). Brooklyn Ent. Soc. Bul. **7**: 129–139, 3 figs.
- 1885c On the classification of North American Diptera. (Second paper); (Third paper). Ent. Amer. **1**: 10–13; 114–116, 152–155.
- 1885d Notes and descriptions of North American Xylophagidae and Stratiomyidae. Canad. Ent. **17**: 121–128.
- 1885e North American Conopidae: Conclusion. Conn. Acad. Arts and Sci., Trans. (1882–1885) **6**: 377–394, 1 pl.
- 1886a Dipterological notes and descriptions. Amer. Ent. Soc. Trans. and Acad. Nat. Sci. Phila., Ent. Sect. Proc. **13**: 287–307.
- 1886b On two interesting new genera of Leptidae. Ent. Amer. **2**: 105–108.
- 1887a North American Tachinidae. *Gonia*. Canad. Ent. **19**: 6–12.
- 1887b Notes and descriptions of North American Tabanidae. Kans. Acad. Sci. Trans. Ann. Mtgs. **10**: 129–142.
- [1887c] Synopsis of the North American Syrphidae. U.S. Natl. Mus. Bul. (1886) **31**: i–xxx, 1–335, 12 pls.
  Dated by Goode, 1889: 29.
- 1888a Synopsis of the families and genera of North American Diptera, exclusive of the genera of the Nematocera and Muscidae, with bibliography and new species, 1878–88. 84 pp., 1 fig. New Haven, Conn.

  This is the first edition of Williston's "Manual," see Williston, 1896a, 1908.
- 1888b *Hilarimorpha* and *Apiocera*. Psyche **5**: 99–102.
- 1888c An Australian parasite of *Icerya purchasi*. Insect Life (U.S. Dept. Agr., Div. Ent., Period. Bul.) **1**: 21–22, 1 fig.
- 1889a The dipterous parasites of North American butterflies. Pp. 1912–1924, pl. 89 (=pt. 12, part). *In* Scudder, S. H., Butterflies of the eastern United States and Canada [q.v.]. Vol. 3, pp. 1775–1958, pls. 1–89. Cambridge, Mass.

Williston, S. W.—Continued
- 1889b A new species of *Haematobia*. Ent. Amer. **5:** 180–181, 1 fig.
- 1889c Notes on Asilidae. Psyche **5:** 255–259.
- 1891 Fam. Syrphidae. Pp. 1–56 [cont.]. *In* Godman, F. D., and Salvin, O., eds., Biologia Centrali-Americana [q.v.]. Zoologia-Insecta-Diptera, Vol. 3, 127 pp., 2 pls. London.
- 1892a A new species of *Criorhinia* [sic], with notes on synonymy. Ent. News and Acad. Nat. Sci. Phila., Ent. Sect. Proc. **3:** 145–146.
- 1892b Fam. Syrphidae [concl.]; Fam. Conopidae; Fam. Pipunculidae. Pp. 57–72, 73–79, pl. 1+pl. 2, figs. 1–12; pp. 79–86, pl. 2, figs. 13–14; pp. 86–88. *In* Godman, F. D., and Salvin, O., eds., Biologia Centrali-Americana [q.v.]. Zoologia-Insecta-Diptera, Vol. 3, 127 pp., 2 pls. London.
- 1892c Diptera Brasiliana. Part II. Kans. Univ. Quart. **1:** 43–46.
- 1893a The North American Psychodidae. Ent. News and Acad. Nat. Sci. Phila., Ent. Sect. Proc. **4:** 113–114.
- 1893b Description of a species of *Chlorops* reared from galls on *Muhlenbergia mexicana*, by F. M. Webster. Ohio Agr. Expt. Sta. Tech. Ser. Bul. **1:** 156–157 (=bul. 3, part).
- 1893c New or little-known Diptera. Kans. Univ. Quart. **2:** 59–78.
- 1893d List of Diptera of the Death Valley Expedition. Pp. 253–259. *In* Riley, C. V., The Death Valley Expedition. A biological survey of parts of California, Nevada, Arizona, and Utah. Part II. 4. Report on a small collection of insects made during the Death Valley Expedition. North Amer. Fauna **7:** 235–268.
- 1894 On the genus *Dolichomyia*, with the description of a new species from Colorado. Kans. Univ. Quart. **3:** 41–43.
- 1895a A new tachinid with remarkable antennae. Ent. News **6:** 29–32, 2 figs.
- 1895b *Dialysis* and *Triptotricha*. Kans. Univ. Quart. **3:** 263–266, 1 fig.
- 1895c New Bombyliidae. Kans. Univ. Quart. **3:** 267–269.
- 1895d Two remarkable new genera of Diptera. Kans. Univ. Quart. **4:** 107–109.
- 1896a Manual of the families and genera of North American Diptera. Ed. 2, 167 pp. New Haven, Conn.
- 1896b A new genus of Hippoboscidae. Ent. News **7:** 184–185.
- 1896c On the Diptera of St. Vincent (West Indies). Ent. Soc. London, Trans. **1896:** 253–446, pls. 8–14.

Willston, S. W.—Continued
- 1897 Diptera Brasiliana. Part IV. Kans. Univ. Quart. **ser. A, 6:** 1–12.
- 1898 Notes and descriptions of Mydaidae. Kans. Acad. Sci. Trans. Ann. Mtgs. **15:** 53–58.
- 1899 On the genus *Thlipsogaster* Rond. Psyche **8:** 331–332.
- 1900– Supplement [part]. Pp. 217–248, *1900;* pp. 249–264,
- 1901   265–272, 273–296, 297–328, 329–332, pls. 4–5+pl. 6, figs. 1–6, *1901. In* Godman, F. D., and Salvin, O., eds., Biologia Centrali-Americana [q.v.]. Zoologia-Insecta-Diptera, Vol. 1, 378 pp., 6 pls. London.
- 1907 Dipterological notes. N.Y. Ent. Soc. Jour. **15:** 1–2.
- 1908 Manual of North American Diptera. Ed. 3, 405 pp., 163 figs. New Haven, Conn.
- 1917 *Camptopelta,* a new genus of Stratiomyidae. Ent. Soc. Amer. Ann. **10:** 23–26.

Wilson, J. W.
- 1932 Coleoptera and Diptera collected from a New Jersey sheep pasture. N.Y. Ent. Soc. Jour. **40:** 77–93.

Winkler, E. C. (See Thurman and Winkler)

Winnertz, J.
- 1846 Beschreibung einiger neuen Gattungen aus der Ordnung der Zweiflügler. Stettin Ent. Ztg. **7:** 11–20, 2 pls.
- 1852a Beitrag zur Kenntniss der Gattung *Ceratopogon* Meigen. Linnaea Ent. **6:** 1–80, 8 pls.
- 1852b Dipterologisches. Stettin Ent. Ztg. **13:** 49–58, 1 pl.
- 1853 Beitrag zu einer Monographie der Gallmücken. Linnaea Ent. **8:** 154–322, 4 pls.
- 1863 Beitrag zu einer Monographie der Pilzmücken. K.-k. Zool.-Bot. Gesell. Wien, Verhandl. **13 (Abhandl.):** 637–964, 4 pls.
- 1867 Beitrag zu einer Monographie der Sciarinen. 187 pp., 1 pl. Wien.
- 1870 Die Gruppe der Lestreminae. K.-k. Zool.-Bot. Gesell. Wien, Verhandl. **20 (Abhandl.):** 9–36, 2 pls.

Wirth, W. W. (See also Jamnback and Wirth; Jones and Wirth; Quate and Wirth; Schmitz and Wirth; Stone and Wirth; Usinger et al.; Vargas and Wirth)
- 1949 A revision of the clunionine midges with descriptions of a new genus and four new species (Diptera: Tendipedidae). Calif. Univ., Pubs., Ent. **8:** 151–182, 7 figs.
- 1951a A revision of the dipterous family Canaceidae. Bernice P. Bishop Mus. Occas. Papers **20:** 245–275, 6 figs.

Wirth, W. W.—Continued
1951b   The genus *Culicoides* in Alaska (Diptera, Heleidae). Ent. Soc. Amer. Ann. **44**: 75–86, 2 pls.
1951c   A new mountain midge from California (Diptera: Deuterophlebiidae). Pan-Pacific Ent. **27**: 49–57, 2 figs.
1951d   The genus *Probezzia* in North America (Diptera, Heleidae). Ent. Soc. Wash. Proc. **53**: 25–34, 1 fig.
1951e   A new biting midge of the genus *Leptoconops* from Florida, with new records of other American species (Diptera, Heleidae). Ent. Soc. Wash. Proc. **53**: 281–284, 7 figs.
1951f   New species and records of Virginia Heleidae (Diptera). Ent. Soc. Wash. Proc. **53**: 313–326, 1 pl.
1952a   The Heleidae of California. Calif. Univ., Pubs., Ent. **9**: 95–266, 33 figs.
1952b   The genus *Alluaudomyia* Kieffer in North America (Diptera, Heleidae). Ent. Soc. Amer. Ann. **45**: 423–434, 11 figs.
1952c   The status of the genus *Parabezzia* Malloch (Diptera, Heleidae). Ent. Soc. Wash. Proc. **54**: 22–26, 1 fig.
1952d   Three new nearctic species of *Systenus* with a description of the immature stages from tree cavities (Diptera, Dolichopodidae). Ent. Soc. Wash. Proc. **54**: 236–244, 8 figs.
1952e   The immature stages of two species of Florida salt marsh sand flies (Diptera, Heleidae). Fla. Ent. **35**: 91–100, 2 pls.
1953a   Biting midges of the heleid genus *Stilobezzia* in North America. U.S. Natl. Mus. Proc. **103**: 57–85, 2 figs.
1953b   American biting midges of the heleid genus *Monohelea*. U.S. Natl. Mus. Proc. **103**: 135–154, 2 figs.
1954a   A new intertidal fly from California, with notes on the genus *Nocticanace* Malloch (Diptera: Canaceidae). Pan-Pacific Ent. **30**: 59–62, 1 fig.
1954b   A new species of *Glutops* and other new records of California Tabanoidea (Diptera). Pan-Pacific Ent. **30**: 137–142.
1954c   A new genus and species of Ephydridae (Diptera) from a California sulphur spring. Wasmann Jour. Biol. **12**: 195–202, 2 figs.
1955   Three new species of *Culicoides* from Texas (Diptera: Heleidae). Wash. Acad. Sci. Jour. **45**: 355–359, 3 figs.
1956a   New species and records of biting midges ectoparasitic on insects (Diptera, Heleidae). Ent. Soc. Amer. Ann. **49**: 356–364, 5 figs.
1956b   The biting midges ectoparasitic on blister beetles (Diptera, Heleidae). Ent. Soc. Wash. Proc. **58**: 15–23, 5 figs.

Wirth, W. W.—Continued
- 1956c   The heleid midges involved in the pollination of rubber trees in America (Diptera, Heleidae). Ent. Soc. Wash. Proc. **58**: 241–250, 1 pl.
- 1956d   The Ephydridae (Diptera) of the Bahama Islands. Amer. Mus. Nat. Hist., Amer. Mus. Novitates **1817**: 1–20, 2 figs.
- 1957   The species of *Cricotopus* midges living in the blue-green alga *Nostoc* in California (Diptera: Tendipedidae). Pan-Pacific Ent. **33**: 121–126, 3 figs.
- 1958   A review of the genus *Gastrops* Williston, with descriptions of two new species (Diptera, Ephydridae). Ent. Soc. Wash. Proc. **60**:. 247–250.
- 1959   *Pachyhelea*, a new genus of American Ceratopogonidae related to *Palpomyia* (Diptera). Brooklyn Ent. Soc. Bul. **54**: 50–52.
- 1962a   The North American species of the biting midge genus *Jenkinshelea* Macfie (Diptera: Ceratopogonidae). Brooklyn Ent. Soc. Bul. **57**: 1–4, 6 figs.
- 1962b   A reclassification of the *Palpomyia-Bezzia-Macropeza* groups, and a revision of the North American Sphaeromiini (Diptera, Ceratopogonidae). Ent. Soc. Amer. Ann. **55**: 272–287, 14 figs.

———, **and Blanton, F. S.**
- 1956   A new species of salt-marsh sand fly from Florida, the Bahamas, Panama and Ecuador: Its distribution, and taxonomic differentiation from *Culicoides furens* (Poey) (Diptera, Heleidae). Fla. Ent. **39**: 157–162, 2 figs.

———, **and Bottimer, L. J.**
- 1956   A population study of the *Culicoides* midges of the Edwards Plateau region of Texas. Mosquito News **16**: 256–266, 10 figs.

———, **and Hubert, A. A.**
- 1960a   Ceratopogonidae (Diptera) reared from cacti, with a review of the *copiosus* group of *Culicoides*. Ent. Soc. Amer. Ann. **53**: 639–658, pls. 1–4.
- 1960b   Philippine Zoological Expedition 1946–1947. *Camptopterohelea*, a new genus of Ceratopogonidae from the Philippines (Diptera). Chicago Nat. Hist. Mus., Fieldiana, Zool. **42**: 89–91, fig. 25.
- 1962   The species of *Culicoides* related to *piliferus* Root and Hoffman in eastern North America (Diptera, Ceratopogonidae). Ent. Soc. Amer. Ann. **55**: 182–195, 10 figs., 1 pl.

Wirth, W. W., and Jones, R. H.—Continued
 1956   Three new North American species of tree-hole *Culicoides* (Diptera, Heleidae). Ent. Soc. Wash. Proc. **58:** 161–168, 3 figs.
 1957   The North American subspecies of *Culicoides variipennis* (Diptera, Heleidae). U.S. Dept. Agr. Tech. Bul. **1170:** 1–35, 9 figs.
——, **and Stone, A.**
 1956   Chapter 14. Aquatic Diptera. Pp. 372–482, 64 figs. *In* Usinger, R. L., ed., Aquatic insects of California, with keys to North American genera and California species. 508 pp., 501 figs. Berkeley, Calif.
——, **and Williams, R. W.**
 1957   The biting midges of the Bermuda Islands, with descriptions of five new species (Diptera, Heleidae). Ent. Soc. Wash. Proc. **59:** 5–14, 2 figs.

Wollaston, T. V.
 1858   Brief diagnostic characters of undescribed Madeiran insects. Ann. and Mag. Nat. Hist. **ser. 3, 1:** 18–28, 113–125, 2 pls.

Wood, D. M.   (See Davies, Peterson, and Wood)
Wood, E. W., Stetson, S. S., Avery, J. G., Pratt, A., Sargent, F. W., and Sessions, W. R.   (See Burgess, 1897)
Wood, J. H.
 1909   On the British species of *Phora* (Part II) [cont.]. Ent. Monthly Mag. **45** [=ser. 2, 20]: 24–29, 59–63, 113–120, 143–149, 191–195, 240–244, 6 figs. [cont.].
 1910   On the British species of *Phora* (Part II) [concl.]. Ent. Monthly Mag. **46** [=ser. 2, 21]: 149–154, 195–202, 243–249.

Woods, W. C.
 1916   Blueberry insects in Maine. Maine Agr. Expt. Sta. Bul. (1915) [**ser. 2**], **244:** 249–288, 62 figs.
   Also published in Maine Agr. Expt. Sta. Ann. Rpt. (1915) **31:** 249–288, *1916*.

Woodward, B. B.   (See also Sherborn and Woodward)
 1903– Catalogue of the books, manuscripts, maps and drawings in
 1915   the British Museum (Natural History). Vol. 1, pp. 1–500, **1903**; Vol. 2, pp. 501–1038, **1904**; Vol. 3, pp. 1039–1494, **1910**; Vol. 4, pp. 1495–1956, **1913**; Vol. 5, pp. 1957–2403, **1915**. London.
   The entire work consists of 8 volumes, London, 1903–1940. Volumes 1–5 are continuously paged as above; volumes 6–8, the supplement, London, 1922–1940, are continuously paged 1–1480.

Wray, D. L.
1950   Insects of North Carolina. Second supplement. 59 pp. N.C. Dept. Agr., Raleigh, N.C.

> The main work and first supplement are by C. S. Brimley, 1938 and 1942, respectively.

Wright, C. W. (See Stoll et al.)

[Wüllerstorf-Urbair, B. von, In Charge]

1861–   Reise der österreichischen Fregatte Novara um die Erde in
1875   den Jahren 1857, 1858, 1859, unter den Befehlen des Commodore B. von Wüllerstorf-Urbair. 9 pts. Wien.

> This work, by many authors, consists of 15 volumes, some of which are further subdivided into separately paged Abtheilungen, and which are grouped into 9 Theile according to subject matter. Some of the Theile have an individual overall editorship; the Zoologischer Theil does not. This latter part consists of 2 volumes (in 6), 1864–1875. Volume 2, 1864–1875, consists of 3 Abtheilungen and an Atlas. Abtheilung 1 is further divided into separately bound sections A and B. The articles within the Zoologischer Theile were issued separately and with separate pagination. [Section] B, Wien, "1868" [1866–1868], contains 2 articles, including that on Diptera.
> See Schiner, 1868.

Wulp, F. M. van der

1858   Beschrijving van eenige nieuwe of twijfelachtige soorten van Diptera uit de familie der Nemocera. Tijdschr. v. Ent. **2**: 159–185, pls. 1–12.

1867   Eenige Noord-Americaansche Diptera. Tijdschr. v. Ent. **10** [=ser. 2, 2]: 125–164, pls. 3–5.

1869   Nog iets over Noord-Americaansche Diptera. Tijdschr. v. Ent. **12** [=ser. 2, 4]: 80–86.

1870   Opmerkingen omtrent uitlandsche asiliden. Tijdschr. v. Ent. **13** [=ser. 2, 5]: 207–217, 1 pl.

1874   Dipterologische aanteekeningen. Tijdschr. v. Ent. **17**: 109–148, 1 pl.

1876   Verslag van de dertigste zomervergadering der Nederlandsche Entomologische Vereeniging gehouden te Amsterdam op Zaturdag 24 Julij 1875. Tijdschr. v. Ent. **19** (**spec. pagination**): i–liv, 2 figs.

> Wulp was the Secretary of the Vereeniging.

1881   Amerikaansche Diptera. Tijdschr. v. Ent. **24**: 141–168, pl. 15 [cont.].

Wulp, F. M. van der—Continued
1882a Remarks on certain American Diptera in the Leyden Museum and descriptions of nine new species. Leyden Mus. Notes **4**: 73–92.
1882b Amerikaansche Diptera [cont.]. Tijdschr. v. Ent. (1881–1882) **25**: 77–136, pls. 9–10 [cont.].
Aldrich, 1905: 65, incorrectly dated this 1883.
1883 Amerikaansche Diptera [concl.]. Tijdschr. v. Ent. **26**: 1–60, pls. 1–2.
1884 Quelques Diptères exotiques. Soc. Ent. de Belg. Ann. **28** (**Bul.**): cclxxxviii–ccxcvii, 3 figs.
1885 Langwerpige dexinen-vormen. Tijdschr. v. Ent. **28**: 189–200, 1 pl.
1887 *Sarcophagula*, een nieuw geslacht der Sarcophaginae. Tijdschr. v. Ent. **30**: 173–174.
1888– Fam. Muscidae. Pp. 2–40, pls. 1–2, *1888*; pp. 41–56, 57–
1890 88, 89–112, 113–144, 145–176, 177–200, 201–208, pls. 3–4, *1890* [cont.]. *In* Godman, F. D., and Salvin, O., eds., Biologia Centrali-Americana [q.v.]. Zoologia-Insecta-Diptera, Vol. 2, 489 pp., 11 figs., 13 pls. London.
1891a Eenige uitlandsche Diptera. Tijdschr. v. Ent. **34**: 193–218, 1 pl.
1891b Fam. Muscidae [cont.]. Pp. 209–224, 225–248, 249–264, pls. 5–6 [cont.]. *In* Godman, F. D., and Salvin, O., eds., Biologia Centrali-Americana [q.v.]. Zoologia-Insecta-Diptera, Vol. 2, 489 pp., 11 figs., 13 pls. London.
1892 Diagnoses of new Mexican Muscidae. Tijdschr. v. Ent. **35**: 183–195.
1895– Fam. Muscidae [concl.]. Pp. 265–272, *1895*; pp. 273–280,
1900 281–288, 289–304, 305–312, 313–320, 321–344, pls. 7–8, *1896*; pp. 345–360, 361–368, 369–376, pl. 9, *1897*; pp. 377–384, *1898*; pp. 385–392, 393–408, 409–416, pls. 10–11, *1899*; pp. 417–428, 11 figs., pl. 12, *1900*. *In* Godman, F. D., and Salvin, O., eds., Biologia Centrali-Americana [q.v.]. Zoologia-Insecta-Diptera, Vol. 2, 489 pp., 11 figs., 13 pls. London.
1903 Fam. Hippoboscidae. Pp. 429–432, pl. 13, figs. 1–6. *In* Godman, F. D., and Salvin, O., eds., Biologia Centrali-Americana [q.v.]. Zoologia-Insecta-Diptera, Vol. 2, 489 pp., 11 figs., 13 pls. London.

Wyatt, I. J.
1959 A new genus and species of Cecidomyiidae (Diptera) infesting mushrooms. Roy. Ent. Soc., London, Proc. Ser. B: Taxonomy **28**: 175–179, 11 figs.

Wytsman, P., ed.
1902–
1959 Genera insectorum. 213 vols. Bruxelles.

> Each volume [=fascicle] involves a major group of insects, and is by a separate author. To date, 213 fascicles have been published.
>
> See Alexander, 1927c; Brues, 1906a; Edwards, 1928a, 1932a; Hendel, 1908a, 1910a, 1914a; Johannsen, 1909; Kieffer, 1906a, 1913a; Kröber, 1913b; Melander, 1927; Séguy, 1937, 1952.

Yates, W. W. (See Stage, Gjullin, and Yates)

Yothers, W. W. (See Knab and Yothers)

Young, D. B. (See Felt and Young)

Zavřel, J., and Thienemann, A.
1921 Die Metamorphose der Tanypinen (II. Teil). Pp. 655–784, 76 figs. *In* Thienemann, A., Vorarbeiten für eine Monographie der Chironomiden-Metamorphose [concl.]. Arch. f. Hydrobiol., Sup.-Band **2**: 655–850, 127 figs.

Zeller, P. C.
1842 Dipterologische Beyträge. Zweyte Abtheilung. Isis (Oken's) **1842**: 807–847, 1 pl.

Zenker, J. C.
1833 Miscellen. Not. aus dem Geb. der Natur- u. Heilk. [Froriep's] **35**: 344.

Zetterstedt, J. W.
1833 Resa genom Umeå Lappmarker i Vesterbottens Län, förrättad år 1832. 398 pp., 3 pls. Örebro, Sweden.
1837 Conspectus familiarum, generum et specierum dipterorum, in fauna insectorum Lapponica descriptorum. Isis (Oken's) **1837**: 28–67.
1838 Dipterologis Scandinaviae. Sect. 3: Diptera, pp. 477–868. *In his* Insecta Lapponica. vi + 1,140 (pp.) Lipsiae [=Leipzig].

> The pages are double-columned, each column numbered; a total of 1,140 columns. The work was issued in 6 hefts, Lipsiae, 1838–1840, as follows: 1–256, 257–476, 477–868, 1838; 869–1014, 1839; 1015–1036, 1037–1140 + i-vi + title page dated 1840, 1840.

Zetterstedt, J. W.—Continued

1842–
1860
Diptera Scandinaviae. Disposita et descripta. Vol. 1, pp. iii–xvi+1–440, *1842;* vol. 2, pp. 441–894, *1843;* vol. 3, pp. 895–1280, *1844;* vol. 4, pp. 1281–1738, *1845;* vol. 5, pp. 1739–2162, *1846;* vol. 6, pp. 2163–2580, *1847;* vol. 7, pp. 2581–2934, *1848;* vol. 8, pp. 2935–3366, *1849;* vol. 9, pp. 3367–3710, *1850;* vol. 10, pp. 3711–4090, *1851;* vol. 11, pp. v–xii+4091–4546, *1852;* vol. 12: Sup. 3, pp. v–xx+4547–4942, *1855;* vol. 13: Sup. 4, pp. v–xvi+4943–6190, *1859;* vol. 14, pp. 6191–6609, *1860.* Lundae [=Lund.].

> This work is complete here. It is sometimes bound in 7 volumes.

Zimin, L. S.

1951 [Fam. Muscidae (Tribes Muscini, Stomoxydini).] No. 4, 285 pp., 472 figs. (=n. ser., 45). *In* [Zoologicheskii Institut Akademii Nauk SSSR], [Fauna SSSR, Novaia Seriia] [q.v.]. Diptera, Vol. 18. Moscow, Leningrad.

> In Russian.

Zimmerman, E. C., ed.

1948–
1960
Insects of Hawaii. 9 vols. Honolulu, Hawaii.

> This series of volumes by different authors continues to date; it consists of Volumes 1–8, and 10, as of 1960.
> See Hardy, 1960b.

Zimmerman, S. K. (See Rockwood, Zimmerman, and Chamberlin)

Zoological Society of London

1865–
1961
The zoological record. 96 vols. London.

> Volumes 1–6 were edited by A. C. L. G. Günther as "The record of zoological literature" [q.v.]. Volumes 7–22 were published by the Zoological Record Association, 1871–1886. The Zoological Society of London began publishing the series with Volume 23, 1887, and publication continues. (Volumes 43–52, 1908–1917, were published by the Royal Society of London as Volumes 6–15 of their "International Catalogue of Scientific Literature, Section K".) Volumes 1–22 were continuously paged, but beginning with Volume 13, the volumes have been divided into separately paged taxonomic sections, issued separately (e.g., Vol. 23 has 19 such sections).
> See Sharp, 1887.

[Zoologicheskii Institut Akademii Nauk SSSR]

1935– [Fauna SSSR. Novaia Seriia.] 79 nos. Moscow, Lenin-
1961 grad.

>This work in Russian is a continuation of ["Fauna Rossii"] which began publication in St. Petersburg [=Petrograd, Leningrad] in 1911. Each major taxonomic group is assigned an inpendent system of volume numbering, and the various parts of that taxon are issued, upon completion by different authors, in separately paged numbers. There was no overall system of numbering until the beginning of the "new series" in 1935, in which the numbers are chronologically numbered, and which continues to date. We have no complete record of the number of issues in the old series. Early volumes had an additional French title. Upon completion, the portion covering Diptera is to consist of 20 volumes, of which the following 11 numbers (all in the present series) have been issued through 1961:
>
>Vol. 2, no. 3 (Tipulidae, part), 1961 (n. ser., 79).
>
>Vol. 3, no. 2 (Psychodidae), 1937 (n. ser., 10).
> no. 4 (Culicidae, part), 1937 (n. ser., 11).
>
>Vol. 6, no. 6 (Simuliidae), 1940 (n. ser., 23). [Revised, 1956 (n. ser., 64)]
>
>Vol. 7, no. 2 (Tabanidae), 1937 (n. ser., 9).
>
>Vol. 9, no. 2 (Bombyliidae), 1940 (n. ser., 25).
>
>Vol. 17, no. 1 (Gastrophilidae), 1955 (n. ser., 60).
>
>Vol. 18, no. 4 (Muscidae, part), 1951 (n. ser., 45).
>
>Vol. 19, no. 1 (Sarcophagidae, part), 1937 (n. ser., 12).
> no. 3 (Oestridae), 1957 (n. ser., 68).
>
>See Rubzov, 1940a; Savtshenko, 1961; Zimin, 1951.

Zumpt, F.

1957 Some remarks on the classification of the Oestridae s. lat. (Diptera). Ent. Soc. South. Africa, Jour. **20**: 154–161, 1 pl.

――――, **and Paterson, H. E.**

1953 Studies on the family Gasterophilidae, with keys to the adults and maggots. Ent. Soc. South. Africa, Jour. **16**: 59–72, 17 figs.

Periodicals Cited, with Explanations

This list is arranged alphabetically by the abbreviations used in the bibliography. The abbreviations follow the systems of the U.S. Department of Agriculture Library and the "Union list of serials in libraries of the United States and Canada". As far as possible, complete information is given on changes of name (which are cross referenced), series numbers, and other details helpful in locating a desired periodical in libraries. A plus sign (+) following a year indicates continuation to date.

Acad. Nat. Sci. Phila. Jour. = Journal of the Academy of Natural Sciences of Philadelphia. Philadelphia. 1–8, 1817–1842; Ser. 2, 1–16, 1847–1918.

Acad. Nat. Sci. Phila. Monog. = Monographs of the Academy of Natural Sciences of Philadelphia. Philadelphia. 1, 1935+

Acad. Nat. Sci. Phila. Notulae Nat. = Notulae Naturae. The Academy of Natural Sciences, Philadelphia. 1, 1939+

Acad. Nat. Sci. Phila. Proc. = Proceedings of the Academy of Natural Sciences of Philadelphia. Philadelphia. 1, 1841+

> Numbering of the volumes ceased with vol. 8, 1856, and was resumed with vol. 53, 1901.

Acad. Roy. des Sci., Belles-Let. et Arts de Lyon, Mém., Sect. des Sci. = Mémoires de l'Académie Royale des Sciences, Belles-Lettres et Arts de Lyon. Section des Sciences. Lyon. 1–2, 1845–1849. (N. ser., 1, 1851, as "Mémoires de l'Académie Nationale des Sciences, . . . de Lyon. Classe des Sciences"; N. ser., 2–17, 1852–1870, as "Mémoires de l'Académie Impériale des Sciences, . . . de Lyon. Classe des Sciences"; N. ser., 18–31, 1870–1892, as "Mémoires de l'Académie des Sciences, . . . de Lyon. Classe des Sciences"; United with "Classe des Lettres" to form Ser. 3, 1, 1893+ as Mémoires de l'Académie des Sciences, . . . de Lyon. Classe des Sciences et Lettres.")

Acad. Sci. St. Louis, Trans. = Transactions of the Academy of Science of St. Louis. St. Louis, Mo. 1, 1856+

[Accad. delle Sci. dell'Ist. Bologna], Nuovi Ann. delle Sci. Nat. = Nuovi Annali delle Scienze Naturali. See **Accad. delle Sci. dell'Ist. e Soc. Agr. Bologna, Nuovi Ann. delle Sci. Nat. e Rend.**

Accad. delle Sci. dell'Ist. e Soc. Agr. Bologna, Nuovi Ann. delle Sci. Nat. e Rend. = Nuovi Annali delle Scienze Naturali e Rendiconto

1485

dei Lavori dell'Accademia delle Scienze dell'Istituto e della Società Agragria di Bologna. Bologna. Ser. 3, 1–2, 1850. (1–10, 1838–1843, as [**Accad. delle Sci. dell'Ist. Bologna**], Nuovi Ann. delle Sci. Nat. [q.v.]; Ser. 2, 1–10, 1844–1848, as **Soc. Agr. e Accad. delle Sci. dell'Ist. Bologna, Nuovi Ann. delle Sci. Nat. e Rend.** [q.v.]; Ser. 3, 3–10, 1851–1854, as "Nuovi Annali delle Scienze Naturali e Rendiconto dei Lavori dell'Accademia delle Scienze dell'Istituto di Bologna con Appendica Agraria.")

Supersedes "Annali di Storia Naturale."

Accad. delle Sci. Fis. e Mat., Atti=Atti dell'Accademia delle Scienze Fisiche e Mathematiche. Società Reale di Napoli, Naples. 1–5, 1861–1873. (6–9, 1875–1882, as "Atti della R. Accademia . . . Matematiche"; Ser. 2, 1–20, 1888–1935, as "Atti della Reale Accademia . . . Matematiche.")

Supersedes "Atti della Reale Accademia delle Scienze sezione della Reale Società Borbonica." Suspended between 1917–1926. Issued in separately paged numbers.

Acta Zool. Fenn.=Acta Zoologica Fennica. Helsinki. 1, 1926+

Acta Zool. Lilloana=Acta Zoologica Lilloana. Universidad Nacional de Tucumán, Instituto "Miguel Lillo", Tucumán, Argentina. 1, 1943+

Adv. in Genet.=Advances in Genetics. New York. 1, 1947+

Agr. Gaz. N.S. Wales=Agricultural Gazette of New South Wales. Sydney. 1, 1890+

[**Akad. Nauk S.S.S.R.**] **Imp. Akad. Nauk St. Petersburg (or Petrograd), Zap., Fiz.-Mat. Otd. (Acad. Imp. des Sci. de St.-Petérsbourg, Mém., Cl. Phys.-Math.)**=Zapiski Imperatorskoi Akademii Nauk (Mémoires de l'Académie Impériale des Sciences de St.-Pétersbourg [or Petrograd]). Fiziko-Matematicheskomu Otd'leniiu (Classe Physico-Mathématique). St. Petersburg [=Petrograd, Leningrad]. Ser. 8, 1–32, 1894–1914 (St. Petersburg) and 33–34, 1914–1916 (Petrograd). (Ser. 8, 35–36, 1918–1923, (Petrograd) and 37 (1) (Leningrad) as [**Akad. Nauk S.S.S.R.**] **Ross. Akad. Nauk Petrograd, Zap., Fiz.-Mat. Otd.** [q.v.]; Ser. 8, 37 (end), 1926–1930, as "Zapiski Akademii Nauk Soyuza S.S.R. (Mémoires de l'Académie des Sciences de l'U.R.S.S.). Fiziko- . . . (Classe . . .)".)

The numbers of vols. 28–34 of ser. 8 were issued irregularly. Up to 1914 their titles were the same as that on the title pages of vols. 1–32; during 1914–1915 their titles became like that of the title pages of vols. 33–34; during 1915–1916, their titles varied from that of the title pages of vols. 33–34 and that of vols. 35–36.

[Ser. 1], 1–14, 1726–1746, as "Commentarii Academiae Scientiarum Imperialis Petropolitanae"; [Ser. 2], 1–120, 1747–1775, as "Novi Commentarii

Academiae . . . "; [Ser. 3], 1–6, 1777–1782, as "Acta Academia . . . ";
[Ser. 4], 1–15, 1788–1802, as "Nova Acta Academiae . . . "; [Ser. 5], 1–11,
1803–1822, as "Mémoires de la Académie. Avec l'Histoire de l'Académie";
Ser. 6, as "Mémoires de l'Académie Impériale des Sciences de St.-Pétersbourg," in 2 sections: "Sciences Politique, Histoire et Philologie," 1–9,
1830–1859, and "Sciences Mathématiques, Physiques et Naturelles," 1–10,
1831–1859 (vols. 3–10 of latter further divided into "Sciences Mathématiques
et Physiques" and "Sciences Naturelles"); Ser. 7, 1–42, 1859–1897, as
"Mémoires de l'Académie . . . de St.-Pétersbourg"; Ser. 8, in 2 classes:
"Fiziko-Matematicheskomu Otd'liniiu" as outlined above, and "Istoricheskomu Nauk i Filologii Otd'leniiu (Classe Historico-Philologique),"
1–13, 1895–1922.

Russian spelling and French titles vary slightly. The volumes of ser.
7 and 8 were usually issued in separately paged numbers.

**[Akad. Nauk S.S.S.R.] Imp. Akad. Nauk St. Petersburg, Zool. Muz.
Ezheg. (Acad. Imp. des Sci. de St.-Pétersbourg, Zool. Mus. Ann.)**
=Ezhegodnik Zoologicheskogo Muzeia Imperatorskoi Akademii
Nauk (Annuaire du Musée Zoologique de l'Académie Impériale des
Sciences de St.-Pétersbourg). See **[Akad. Nauk S.S.S.R.] Ross.
Akad. Nauk Petrograd, Zool. Muz. Ezheg.**

**[Akad. Nauk S.S.S.R.] Ross. Akad. Nauk Petrograd, Zap., Fiz.-Mat.
Otd. (Acad. des Sci. de Russie, Mém., Cl. Phys.-Math.]**=Zapiski
Rossiiskoi Akademii Nauk (Mémoires de l'Académie des Sciences de
Russie). Fiziko-Matematicheskomu Otd'leniiu (Classe PhysicoMathématique). See **[Akad. Nauk S.S.S.R.] Imp. Akad. Nauk St.
Petersburg, Zap., Fiz.-Mat. Otd.**

**[Akad. Nauk S.S.S.R.] Ross. Akad. Nauk Petrograd, Zool. Muz.
Ezheg. (Acad. des Sci. de Russie, Zool. Mus. Ann.)**=Ezhegodnik
Zoologicheskogo Muzeia Rossiiskoi Akademii Nauk (Annuaire du
Musée Zoologique de l'Académie des Sciences de Russie). Petrograd [=St. Petersburg, Leningrad]. 22–24, 1917–1923. (1–18,
1896–1914 (St. Petersburg) and 19–20, 1914–1915 (Petrograd) as
**[Akad. Nauk S.S.R.] Imp. Akad. Nauk St. Petersburg, Zool. Muz.
Ezheg.** [q.v.]; 21, 1916, as "Ezhegodnik Zoologicheskogo Muzeia
Akademii Nauk (Annuaire . . . l'Académie des Sciences)";
25–32, 1924–1931, as "Ezhegodnik Zoologicheskogo Muzeia Akademii Nauk S.S.S.R. (Annuaire . . . l'Académie des Sciences de
l'U.R.S.S.).")

Superseded by **Akad. Nauk S. S. S. R., Zool. Inst., Trudy** [q.v.].

**Akad. Nauk S.S.S.R., Zool. Inst., Parazitol. Sborn. (Acad. des Sci. de
l'U.R.S.S., Inst. Zool., Mag. de Parasitol.)**=Parazitologicheskii
Sbornik Zoologicheskogo Instituta Akademii Nauk S.S.S.R. (Magasin de Parasitologie de l'Institut Zoologique de l'Académie des
Sciences de l'U.R.S.S.). Leningrad. 1, 1930+

The French title was included up to vol. 8, 1940.

Akad. Nauk S.S.S.R., Zool. Inst., Trudy=Trudy Zoologicheskogo Instituta Akademii Nauk S.S.S.R. Leningrad. 1, 1932+

> Supersedes [**Akad. Nauk S.S.S.R.**] Ross. Akad. Nauk Petrograd, Zool. Muz. Ezheg. [q.v.].

[**Akad. Nauk U.R.S.R.**] **Vseukrains'ka Akad. Nauk, Fiz.-Mat. Vidd., Trudy (Acad. des Sci. de l'Ukraïne, Cl. des Sci. Phys. et Math., Mém.**)=Trudy Fizichno Matematichnii Viddil. Vseukrains'ka Akademiia Nauk (Mémoires de la Classe des Sciences Physiques et Mathématiques. Académie des Sciences de l'Ukraïne). Kiev. 1–15, 1923–1930.

> Superseded by "Prirodnicho-takhnichnii Viddil." Spelling of this periodical varies slightly. Some of its volumes also form part of other series, separately numbered, e.g., the "Z'irnik Prats' Zoologicheskii Muzeia (Travaux du Musée Zoologique)."

Amer. Ent.=The American Entomologist: An Illustrated Magazine of Popular and Practical Entomology. New York. 1–3, 1868–1880.

> The April-December numbers of vol. 2 were called **Amer. Ent. and Bot.** [q.v.]. Vol. 3, 1880, also called ser. 2, vol. 1.

Amer. Ent. and Bot.=The American Entomologist and Botanist. See **Amer. Ent.**

Amer. Ent. Soc. Mem.=Memoirs of the American Entomological Society. Philadelphia. 1, 1916+

Amer. Ent. Soc. Trans.=Transactions of the American Entomological Society. Philadelphia. 1–6, 1867–1877, and 17, 1890+ (7–16, 1878–1889, as **Amer. Ent. Soc. Trans. and Acad. Nat. Sci. Phila., Ent. Sect. Proc.** [q.v.].

Amer. Ent. Soc. Trans. and Acad. Nat. Sci. Phila., Ent. Sect. Proc. =Transactions of the American Entomological Society and Proceedings of the Entomological Section of the Academy of Natural Sciences. See **Amer. Ent. Soc. Trans.**

Amer. Jour. Agr. and Sci.=American Journal of Agriculture and Science. Albany, N.Y. 5–7, 1847–1848. (1–4, 1845–1846, as **Amer. Quart. Jour. Agr. and Sci.** [q.v.].

Amer. Jour. Hort. and Florist's Companion=The American Journal of Horticulture and Florist's Companion. Boston. 1–5, 1867–1869. (6–8, 1869–1870, as "Tilton's Journal of Horticulture and Floral Magazine"; 9, 1871, as "Tilton's Journal of Horticulture and Florist's Companion.")

Amer. Jour. Hyg.=American Journal of Hygiene. Baltimore, Md. 1, 1921+

Amer. Jour. Sci. and Arts=American Journal of Science and Arts. New Haven, Conn.; New York. 2–50, 1820–1845; Ser. 2, 1–50, 1846–1870; Ser. 3, 1–19, 1871–1879. (1, 1818–1819, and Ser. 3,

20–50, 1880–1895, Ser. 4, 1–50, 1896–1920, and Ser. 5, 1–36, 1921–1938, and 237, 1939+ as "American Journal of Science.")

> Volumes are also numbered in whole series, 1–236, 1818–1938. This periodical also known as "Silliman's Journal."

Amer. Jour. Trop. Med.=The American Journal of Tropical Medicine. Baltimore, Md. 1–31, 1921–1951.

> Superseded by "The American Journal of Tropical Medicine and Hygiene" with the absorption of "The Journal of the National Malaria Society."

Amer. Micros. Soc. Trans.=Transactions of the American Microscopical Society. Indianapolis, Ind.; Menasha, Wis. 17, 1895+ ([1], 1878, as "Proceedings of the National Microscopical Congress"; [2]–13, 1879–1891, as "Proceedings of the American Society of Microscopists"; 14–16, 1892–1894, as "Proceedings of the American Microscopical Society.")

> Volume numbering began with vol. 10, 1888.

Amer. Midland Nat.=American Midland Naturalist. Notre Dame, Ind. 1, 1909+ Vol. 1, nos. 1–4, as "Midland Naturalist."

Amer. Mus. Nat. Hist., Amer. Mus. Novitates=American Museum Novitates. American Museum of Natural History, New York. 1, 1921+

Amer. Mus. Nat. Hist. Bul.=Bulletin of the American Museum of Natural History. New York. 1, 1881+

Amer. Nat.=The American Naturalist. Philadelphia, Boston, New York. 1, 1867+

Amer. Phil. Soc. Proc.=Proceedings of the American Philosophical Society. Philadelphia. 1, 1838+

Amer. Quart. Jour. Agr. and Sci.=American Quarterly Journal of Agriculture and Science. See **Amer. Jour. Agr. and Sci.**

Ann. and Mag. Nat. Hist.=Annals and Magazine of Natural History. London. 6–20, 1841–1847; Ser. 2, 1–20, 1848–1857; Ser. 3, 1–20, 1858–1867; Ser. 4, 1–20, 1868–1877; Ser. 5, 1–20, 1878–1887; Ser. 6, 1–20, 1888–1897; Ser. 7, 1–20, 1898–1907; Ser. 8, 1–20, 1908–1917; Ser. 9, 1–20, 1918–1927; Ser. 10, 1–20, 1928–1937; Ser. 11, 1–14, 1938–1947; Ser. 12, 1–10, 1948–1957; Ser. 13, 1, 1958+ (1–5, 1838–1840, as **Ann. Hist. Nat.** [q.v.].)

Ann. de Biol. Lacustre=Annales de Biologie Lacustre. Brussels. 1–15, 1906–1926.

Ann. des Sci. Nat., Zool.=Annales des Sciences Naturelles. Série Zoologie. Paris. Ser. 2, 1–20, 1834–1843; Ser. 3, 1–20, 1844–1853; Ser. 4, 1–20, 1854–1863; Ser. 5, 1–20, 1864–1874; Ser. 9, 1–20, 1905–1915; Ser. 10, 1–20, 1916–1937. (Ser. 6, 1–20, 1875–1885, Ser. 7, 1–20, 1886–1895, and Ser. 8, 1–20, 1896–1904, as "Annales . . . Série Zoologie et Paléontologie," with incorporation of

"Annales des Sciences Géologiques"; Ser. 11, 1, 1938+ as "Annales des Sciences Naturelles. Zoologie et Biologie Animale.")

[Ser. 1], 1–30, 1824–1833, as "Annales des Sciences Naturelles," [combined Zoologie and Botanique]; then split into 2 series: Zoologie, as above, and Série Botanique, Ser. 2–10, vols. 1–20 each, 1834–1938, and Ser. 11, 1, 1939+

Ann. Ent. Fenn.=Annales Entomologici Fennicae. Suomen Hyönteistieteellinem Aikakauskirja (Entomological Society of Finland), Helsinki. 1, 1935+

Ann. Nat. Hist.=Annals of Natural History. See **Ann. and Mag. Nat. Hist.**

Ann. Rev. Ent.=Annual Review of Entomology. Stanford, Calif. 1, 1956+

Ann. Scot. Nat. Hist.=Annals of Scottish Natural History. See **Scot. Nat.**

Ann. Trop. Med. and Parasitol.=Annals of Tropical Medicine and Parasitology. Liverpool School of Tropical Medicine, Liverpool, England. 1, 1907+

Supersedes "Memoirs of the Liverpool School of Tropical Medicine."

Arb. über Morph. u. Taxonom. Ent.=Arbeiten über Morphologische und Taxonomische Entomologie aus Berlin-Dahlem. Berlin-Dahlem. 1–11, 1934–1944.

Arch. da Esc. Super. Agr. e Med. Vet. [Nichtheroy, Rio de Janeiro]=Archivos da Escola Superior de Agricultura e Medicina Veterinaria. Nichtheroy, Brazil. 1–10, 1917–1933.

Arch. de Méd. Nav.=Archives de Médicine Navale. Paris. 1–53, 1864–?, and 67–94, ?–?. (54–66, ?–1910, as "Archives de Médicine Navale et Coloniale"; 95–130, 1911–1942, as "Archives de Médicine et de Pharmacie Navale".)

Superseded by "Revue de Médicine Navale Métropole et Outremer."

Arch. f. Hydrobiol.=Archiv für Hydrobiologie. Berlin, Stuttgart. N. ser., 12, 1918+ (1–12, 1893–1905, as "Forschungsberichte aus der Biologische Station zu Plön"; N. ser., 1–11, 1905–1917, as "Archiv für Hydrobiologie und Planktonkunde.")

A supplement volume began in 1911, see **Arch. f. Hydrobiol., Sup.-Band.**

Arch. f. Hydrobiol., Sup.-Band=Archiv für Hydrobiologie. Supplement-Band. Berlin, Stuttgart. 2 (4), 1921+ (1–2 (3), 1911–1916, as **Arch. f. Hydrobiol. u. Planktonkunde, Sup.-Band.** [q.v.].)

Arch. f. Hydrobiol. u. Planktonkunde, Sup.-Band=Archiv für Hydrobiologie und Planktonkunde. Supplement-Band. See **Arch. f. Hydrobiol., Sup.-Band.**

Arch. f. Klassifikatorische u. Phylogenetische Ent.=Archiv für Klassifikatorische und Phylogenetische Entomologie. Vienna. 1, 1928-1930.

Arch. f. Naturgesch.=Archiv für Naturgeschichte. Berlin. 1-77, 1835-1911; continued in 2 sections: "Abteilung A. Original-Arbeiten," 78-92, 1912-1926, and "Abteilung B. Jahres-Berichte," 78-89, 1912-1926. (N. ser., 1 1932+ as "Zeitschrift für Wissenschaftliche Zoologie. Abteilung B. Archiv für Naturgeschichte."

>The volumes were issued in separately paged hefts.

Arch. f. Schiffs- u. Tropen-Hyg.=Archiv für Schiffs- und Tropenhygiene, Leipzig. 1-44, 1897-1940. (45-48, 1941-1944, as "Deutsche Tropenmedizinische Zeitschrift.")

Arch. f. Zool. u. Zootomie=Archiv für Zoologie und Zootomie. Berlin, Brunswick. 1-5, 1800-1806.

>This was issued in separately paged numbers. Superseded by "Neues Archiv für Zoologie und Zootomie." The periodical is sometimes known as "Wiedemann's Archiv."

Arch. Zool. Budapest=Archivum Zoologicum. Budapest. 1, 1909-1910.

Arkiv för Zool.=Arkiv för Zoologi. K. Svenska Vetenskapsakademien, Stockholm, Uppsala. 1, 1903+

>This is issued in separately paged numbers, and beginning with Vol. 17, 1925, in 2 sections, A and B, each with their own separately paged numbers (although still bound as a single volume).

Arq. de Zool. de Estado de São Paulo=Arquivos de Zoologia do Estado de São Paulo. São Paulo. 1, 1940+

>Supersedes [São Paulo] Mus. Paulista, Rev. [q.v.], vols. 1-3, 1940-1942, also being numbered 24-26, in continuation of that periodical.

Austral. Jour. Sci. Res., Ser. B: Biol. Sci.=Australian Journal of Scientific Research. Series B: Biological Sciences. Melbourne. 1-5, 1948-1952. (6, 1953+ as "Australian Journal of Biological Sciences.")

Austral. Jour. Zool.=Australian Journal of Zoology. Melbourne. 1, 1953+

Beitr. z. Ent. (Berlin)=Beiträge zur Entomologie. Berlin. 1, 1951+

Beitr. z. Ent., Besonders in Bezug auf Schles. Fauna (Breslau)=Beiträge zur Entomologie, Besonders in Bezug auf Schlesische Fauna. Breslau. 1, 1829.

>A periodical with the title "Beiträge zur Entomologie, Besonders in Bezug auf Schlesien," was also issued at Breslau, 1-3, 1832-1833.

Berlin Zool. Mus. Mitt. = Mitteilungen aus dem Zoologischen Museum in Berlin. Berlin. 1, 1898+

> This was sometimes issued in separately paged numbers. It was suspended 1942–1948.

Berlin. Ent. Ztschr. = Berliner Entomologische Zeitschrift. Berlin. 1–18, 1857–1874, and 25–28, 1881–1914. (19–24, 1875–1880, as **Deut. Ent. Ztschr.** [=**Berlin, Ent. Ztschr.**] [q.v.].)

Bernice P. Bishop Mus. Bul. = Bernice P. Bishop Museum. Bulletin. Honolulu, Hawaii. 1, 1922+

Bernice P. Bishop Mus. Occas. Papers = Occasional Papers of Bernice P. Bishop Museum. Honolulu, Hawaii. 1, 1898+

> Suspended 1924–1929.

Berwickshire Nat. Club. Hist. = History of the Berwickshire Naturalist's Club. Edinburgh; Alnwick; Berwick; London. 1, 1831+

> Running title is "Proceedings of the . . . Club."

Biol. Bul. = Biological Bulletin of the Marine Biological Laboratory Woods Hole, Mass. Boston; Woods Hole. 1, 1899+

> Supersedes **Zool. Bull.** [q.v.].

Biol. Centbl. = Biologisches Centralblatt. Leipzig. 1–36, 1881–1916. (37, 1917+ as "Biologisches Zentralblatt.")

Biol. Hungarica = Biologica Hungarica. Budapest. 1, 1923–1926.

> This has 7 separately paged numbers.

Biol. Rev. Ontario = Biological Review of Ontario. Toronto. 1, 1894.

Biol. Soc. Wash. Proc. = Proceedings of the Biological Society of Washington. Washington, D.C. 1, 1880+

Bol. Biol. = Boletim Biologica. São Paulo. 1–21, 1926–1932; N. ser., 1, 1933+

Boston Soc. Nat. Hist. Mem. = Memoirs of the Boston Society of Natural History. Boston. 1–9, 1862–1936.

> Supersedes "Boston Journal of Natural History."

Boston Soc. Nat. Hist. Occas. Papers = Occasional Papers of the Boston Society of Natural History. Boston. 1–8, 1869–1941.

> Vol. 7 is separately paged.

Boston Soc. Nat. Hist. Proc. = Proceedings of the Boston Society of Natural History. Boston. 1–42, 1841–1942.

> Prior to 1841, the Proceedings of the Society were published in "American Journal of Science."

Brazil Med. = Brazil Medico. Rio de Janeiro. 1–46, 1887–1926. (47, 1927+ as "Brasil Medico.")

Bremen, Deut. Kolon.- u. Uebersee-Mus., Veröffentl.=Veröffentlichungen aus dem Deutschen Kolonial- und Uebersee-Museum in Bremen. Bremen. 1-3, 1935-1942.

Superseded by "Veröffentlichungen aus dem Museum für Natur-, Völker- und Handelskunde," N. ser., 1, 1949+

Brit. Mus. (Nat. Hist.) Bul., Ent.=Bulletin of the British Museum (Natural History), Entomology. London. 1, 1950+

There are also series for Botany, Geology and Palaeontology, History, Mineralogy, and Zoology.

Brooklyn Ent. Soc. Bul.=Bulletin of the Brooklyn Entomological Society. Brooklyn, N.Y. 1, 1878+

Vols. 1-7 were published 1878-1885. A "new series" was begun in 1912, with vol. 8, and volume numbering is continuous.

Brotéria=Brotéria. Revista de Sciéncias Naturae. Lisboa. 1-5, 1902-1906; continued in 3 "series": "Série Botanico," 6-25, 1907-1931, "Série Zoológica", 6-27, 1907-1931, and "Série de Vulgar Sciencias," 6-22, 1907-1924; the first two "series" then united to form "Série Trimestral: Ciências Naturaes," 1, 1932+, and the third then became "Série Mensal: Fé Sciencias-letras," 1, 1925+

Brussels Inst. Roy. des Sci. Nat. de Belg. Bul.=Institut Royal des Sciences Naturelles de Belgique. Bulletin. Brussels. 25, 1949+ (1-24, 1882-1948, as **Brussels Mus. Roy. d'Hist. de Belg. Bul.** [q.v.].)

Suspended 1889-1930. Separately paged numbers were issued from vol. 6, 1930+. Title also given in Dutch.

Brussels Mus. Roy. d'Hist. Nat. de Belg. Bul.=Bulletin du Musée Royal d'Histoire Naturelle de Belgique. See **Brussels Inst. Roy. des Sci. Nat. de Belg. Bul.**

Brussels Mus. Roy. d'Hist. Nat. de Belg. Mém.=Mémoires du Musée Royal d'Histoire Naturelle de Belgique. Brussels. [1-30], 1900-1920; 31-110, 1921-1948. (111, 1949+ as "Institut Royal des Sciences Naturelles de Belgique. Mémoires."

Nos. 1-31, 1900-1921, are assigned to "Volumes" 1-8, without their own independent numbering (except No. 31). The title of the journal is also given in Dutch. A "Ser. 2," 1, 1935+, runs concurrently with the above periodical.

Budapest Magyar Nemzeti Muz., Ann. Hist. Nat.=Annales Historico-Naturales Musei Nationales Hungarici. Magyar Nemzeti Muzeum, Budapest. 1, 1903+

Vols. 29-41, 1935-1948, are in 3 separately paged parts: "Pars Zoologica," "Pars Botanica," and "Pars Mineralogica, Geologica, Palaeontologica." Vols. [42-49], 1951-1957, are called N. ser., 1-8; vol. 50, 1958, also as N. ser., 9. The periodical supersedes **Természet. Füzetek** [q.v.]. The title is also given in Hungarian: "Természetrajzi Osztályainak Folyóirata" (1-31, 1903-1938), "Országos Magyar Természettudmányi Muzeum Folyóirata" (32-41, 1939-1948), and "Országos Természettudományi Múzeum Evikönve ([42], 1951+).

Buenos Aires Mus. Nac. An.=Anales del Museo Nacional de Buenos Aires. Buenos Aires. 4-20, 1895-1911. (1-3, 1864-1894, as "Anales del Museo Público de Buenos Aires"; 21-31, 1912-1923, as "Anales del Museo Nacional de Historia Natural de Buenos Aires"; 32-36, 1923-1931, as "Anales del Museo Nacional de Historia Natural Bernardino Rivadavia Buenos Aires"; 37-42, 1931-1947, as "Anales del Museo Argentino de Ciencias Naturales 'Bernardino Rivadavia'.")

Also numbered in series: Vols. 4-7, 1895-1902, as Ser. 2, 1-4; vols. 8-22, 1902-1913, as Ser. 3, 1-15; vols. 23-42, with no additional series numbers. Beginning in 1948, all publications of the Museum were issued by the Instituto Nacional de Investigación de las Ciencias Naturales.

Buenos Aires Mus. Nac. Comun.=Comunicaciones del Museo Nacional de Buenos Aires. Buenos Aires. 1, 1898-1901. (2, 1921-1925, as "Comunicaciones del Museo Nacional de Historia Natural Bernardino Rivadavia Buenos Aires.")

Suspended 1902-1920. A new series was begun in 1947-1948, as "Comunicaciones del Instituto Nacional de Investigación de las Ciencias Naturales," in 3 sections: "Ciencias Zoológicas," 1 (9), 1948+ (nos. 1-8, 1947-1948, as "Comunicaciones del Museo Argentino de Ciencias Naturales 'Bernardino Rivadavia'," and were without a volume designation); "Ciencias Botánicas," 1, 1948+; and "Ciencias Geológicas," 1, 1948+. The numbers of these series are separately paged, and up to 1955, also have the Museum's name in their titles.

Buffalo Soc. Nat. Sci. Bul.=Bulletin of the Buffalo Society of Natural Sciences. Buffalo, N.Y. 1, 1873+

This is issued in separately paged numbers from vol. 11, 1914+

Bul. Ent. Res.=Bulletin of Entomological Research. London. 1, 1910+

Calif. Acad. Sci. Occas. Papers=Occasional Papers of the California Academy of Sciences. San Francisco. 1, 1890+

Suspended 1906-1921, 1932-1939.

Calif. Acad. Sci. Proc.=Proceedings of the California Academy of Sciences. San Francisco. 1-7, 1854-1876; Ser. 2, 1-6, 1888-1896; Ser. 3, in 4 parts: Zoology, 1-4, 1897-1906, Botany, 1-2, 1897-1904, Geology, 1-2, 1897-1905, and Mathematics-Physics, 1, 1897-1903; Ser. 4, 1, 1907+

Calif. Dept. Agr., Monthly Bul.=Monthly Bulletin of the Department of Agriculture. State of California. Sacramento. 9-24, 1919-1934. (1-8, 1911-1919, as "The Monthly Bulletin of the State

Commission of Horticulture, State of California"; 25–29, 1935–1940, as "Bulletin of the Department of Agriculture, State of California"; 30, 1941+ as "State of California, Department of Agriculture, Bulletin.")

> Supersedes "Bulletin of the State Board of Horticulture." Some individual numbers have the title "Annual Report."

Calif. Insect Survey, Bul.=Bulletin of the California Insect Survey. Berkeley, Los Angeles. 1, 1950+

Calif. Univ., Pubs., Ent.=University of California. Publications in Entomology. Berkeley. 1, 1906+

Calif. Univ. Syllabus Ser.=University of California Syllabus Series. Berkeley, Los Angeles. 1, 1905+

> These are irregularly published study manuals, etc., usually classed as separates by libraries. Some issues are given arbitrary letter designations in place of regular numbers in sequence.

Canad. Ent.=The Canadian Entomologist. London, Ont. 1, 1868+ A consecutively numbered supplement began with vol. 88, 1, 1956+ (Supplement 4, 1957, is called 1.)

Canad. Jour. Res., Sec. D, Zool. Sci.=Canadian Journal of Research. Section D, Zoological Sciences. See **Canad. Jour. Zool.**

Canad. Jour. Zool.=Canadian Journal of Zoology. Ottawa, Ont. 29, 1951+ (13–28, 1935–1950, as **Canad. Jour. Res., Sec. D, Zool. Sci.** [q.v.].)

> The latter journal was continuously paged without sections, vols. 1–12, 1929–1934; vols. 13–21, 1935–1943, were issued in 2 parts, Secs. A–B, and C–D, each section separately paged; vols. 22–28, 1944–1950, were issued in 6 parts, Sections A–F, each of which became a separate journal with vol. 29, 1951. The other sections were: A. Physical Sciences; B. Chemical Sciences; C. Botanical Sciences; E. Medical Sciences; and F. Technology.

[Canada] Commonwealth Inst. Biol. Control, Tech. Commun.= Technical Communication. Commonwealth Institute of Biological Control. Commonwealth Agricultural Bureaux, Ottawa, Canada; Trinidad. 1, 1960+

Canada Dept. Agr. Bul.=Dominion of Canada. Department of Agriculture. Bulletin. New Series. Ottawa, Ont. 1–181, 1922–1935.

> This combines the bulletins of various branches of the Department of Agriculture previously published separately into one serially numbered series, the "New Series." The bulletin number in the series of the particular Branch involved is often included. The title of the Bulletin is sometimes given in French. This was superseded by "Farmer's Bulletin," a subseries of the Department of Agriculture, Publications, beginning with Pub. 475, 1935.
>
> One of the Branches whose bulletins were combined in the above "New Series" was the Entomological Branch, beginning with its Bulletin 19, 1922. Nos. 10–18, 1915–1920, were published separately as "Dominion of Canada,

Department of Agriculture, Entomological Branch, Bulletin"; previously its bulletins were published in "Dominion of Canada, Department of Agriculture, Experimental Farms, Bulletin" (1–98, 1887–1922; Ser. 2, 1–49, 1898–1921), which was also later included in the above "New Series." The early Entomological Branch Bulletins were published in the Experimental Farms Bulletins as follows: Nos. 1–4, 1911–1912, as "Division of Entomology, Bulletin," in Expt. Farms Bul. Ser. 2, 7; [old ser.], 69, 70; Ser. 2, 9, respectively. Nos. 5–8, 1912–1914, as "Entomological Bulletin," in Expt. Farms Bul. Ser. 2, 9–10, 12, 17–18, respectively; No. 9, 1915, was issued separately as "Entomological Bulletin."

(Canterbury Col., Christchurch) Canterbury Mus. Rec. = Records of the Canterbury Museum. Canterbury College, Christchurch, New Zealand. 1, 1907+

Carnegie Inst. Wash., Pub. = Carnegie Institution of Washington. Publications. Washington, D.C. 1, 1902+

Cellule = La Cellule. Louvain, France. 1, 1884+

Centbl. f. Bakt., Parasitenk. u. Infektionskrank., Abt. II = Centralblatt für Bakteriologie, Parasitenkunde und Infektionskrankheiten. Zweite Abteilung. Jena. 2–77, 1896–1929. (1, 1895, as "Centralblatt für Bakteriologie und Parasitenkunde. Zweite Abteilung"; 78–106, 1929–1945, as "Zentralblatt für Bakteriologie, Parasitenkunde und Infektionskrankheiten. Zweite Abteilung"; 107, 1952+ as "Zentralblatt für Bakteriologie, Parasitenkunde, Infektionskrankheiten und Hygiene. Zweite Abteilung.")

Abt. I, 17, 1895+, continues the volume numbering of "Centralblatt für Bakteriologie und Parasitenkunde" (1–16, 1887–1894) upon its split into the 2 Abteilungen.

Chicago Nat. Hist. Mus., Fieldiana, Zool. = Fieldiana: Zoology. Chicago Natural History Museum, Chicago. 31, 1945+ (1–8, 1895–1907, as "Publications of Field Columbian Museum. Zoological Series"; 9–30, 1908–1944, as "Publications of Field Museum of Natural History. Zoological Series.")

There were also series for Botany, Anthropology, History, Technique, Ornithology (which joined with Zoology) and a Report Series. Early volumes are sometimes divided into separately paged parts.

Coleopt. Bul. = The Coleopterist's Bulletin. Washington, D.C. 1, 1947+

Colo., Agr. Expt. Sta. Bul. = Bulletin. Agricultural Experiment Station. State Agricultural College of Colorado (1886–1887); State Agricultural College (1888–1899); Agricultural College of Colorado (1900–1905); Colorado Agricultural College (1905–1934); Colorado State College (1935–1944); Colorado Agricultural and Mechanical College (1944–1957); Colorado State University (1958+). Fort Collins, Colo. 6–52, 53–96, 97–282, 1889–1923; 496S, 1957. (1–3, 1886–1887, as "Bulletin. State Agricultural College of Colorado";

4–5, 1888, and 497S, 1958+, as "Bulletin. Experiment Station"; 283–461, 1923–1940, as "Bulletin. Colorado Experiment Station"; 462–493, 1940–1946, and 494S–495S, 1956, as "Bulletin. Colorado Agricultural Experiment Station.")

> Suspended 1947–1955; upon resumption of publication the letter "S" was added to bulletin numbers.

Colo. Biol. Assoc. Rpt.=Report of the Colorado Biological Association. ?Place. 1–12, 1888–1889.

> No. 12, 1889, was the last seen by us. Some of these reports also appeared in the "Custer County Courant," a newspaper.

Colo. Univ., Studies=The University of Colorado Studies. Boulder, Colo. 1–29, 1902–1957.

> (Vols. 1–25, 1902–1937, are not marked in any series; vols. 26–29, 1938–1957, have separately paged numbers, and are marked "General series (A)" (Ser. B [humanities] and Ser. C [social studies] began in 1939, and Ser. D [physical and biological sciences] began in 1940); there was a reorganization into 10 separate series in 1948–1950, each beginning with a new volume numbering. Beginning with vol. 11, 1925, of [Ser. A], the numbers are also marked as "University of Colorado Bulletins," with its series number.

Conn. Acad. Arts and Sci., Trans.=Transactions of the Connecticut Academy of Arts and Sciences. New Haven, Conn. 1, 1866+

> Volume numbering is irregular.

Conn. State Geol. and Nat. Hist. Survey, Bul.=State of Connecticut. State Geological and Natural History Survey. Bulletin. Hartford, Conn. 1, 1904+

Cornell Univ., Abs. Theses=Abstracts of Theses accepted in partial satisfaction of the requirements for the Doctor's Degree. Cornell University. Ithaca, N.Y. 1937+

> No volume numbers used. The last volume we have seen is for 1947.

Country Gent.=The Country Gentleman. Albany, N.Y.; Philadelphia, Pa. 1–26, 1853–1865; 63, 1898+   (27–62, 1866–1897, as **Cult. and Country Gent.** [q.v.].)

Cult. and Country Gent.=Cultivator and Country Gentleman. See **Country Gent.**

Dansk Naturhist. For. København, Vidensk. Meddel.=Viddenskabelige Meddelelser fra Dansk Naturhistoriske Forening i København. 70, 1919+   (1–10, 1849–1859, Ser. 2, 1–10, 1860–1868, and Ser. 3, 1–7, 1869–1874, as "Videnskabelige Meddelelser fra den Naturhistoriske Forening i Köbenhavn"; Ser. 3, 8–10, 1875–1878, Ser. 4, 1–10, 1879–1888, Ser. 5, 1–10, 1890–1898, Ser. 6, 1–10, 1899–1908, and Ser. 7, 1–2, 1909–1910, and 63, 1911, as "Videnskabelige Meddelelser fra den Naturhistoriske Forening i Kjøbenhavn"; 64–

69, 1912–1918, as "Videnskabelige Meddelelser fra Dansk Naturhistoriske i Kjøbenhavn.")

> Vols. 63–70, 1911–1919, are also called Ser. 7, 3–10; vols. 71–86, 1920–1928, also as Ser. 8, 1–9 (some of the volumes of Ser. 8 are in 2 separately paged parts, although each part is assigned a consecutive whole series number); 87, 1929+ without additional series numbers.

Davenport Acad. Nat. Sci. Proc.=Proceedings of the Davenport Academy of Natural Sciences. Davenport, Iowa. 1–7, 1867–1898. (8–13, 1899–1914, as "Proceedings of the Davenport Academy of Sciences.")

Deut. Ent. Gesell. Mitt.=Mitteilungen der Deutschen Entomologischen Gesellschaft. Berlin. 1, 1930+

Deut. Ent. Ztschr.=Deutsche Entomologische Zeitschrift. Berlin. 25–31, 1881–1887; 1888–?; N. ser., 1, 1954+

> Vols. 19–24, 1875–1880, of the **Berlin. Ent. Ztschr.** were called **Deut. Ent. Ztschr.** [=**Berlin. Ent. Ztschr.**] [q.v.]. In 1881, **Deut. Ent. Ztschr.** became a separate periodical but the volume numbering of the **Berlin. Ent. Ztschr.** was continued up through 1887, after which no volume numbers were given (until the N. ser.). At times, separately paged "Beihefte" have been published with individual volumes.

Deut. Ent. Ztschr. [=**Berlin. Ent. Ztschr.**]=Deutsche Entomologische Zeitschrift [=Berliner Entomologische Zietschrift]. See **Berlin. Ent. Ztschr.**

Deut. Tierärztl. Wchnschr.=Deutsche Tierärztliche Wochenschrift. Karlsruhe, Hannover. 1–50, 1893–1942, and 53?, 1946+ (51–52, 1943–1944, absorbed "Tierärztliche Rundschau", both titles being given, and the volume numbers of the latter also given, i.e., its vols. 49–50; 1944–1945, combined with 2 other journals to form "Tierärztliche Zeitschrift," which then was discontinued, splitting again into the 4 journals.)

> Vols. 53–54, ?1946–1947, have not been seen by us.

Dresden K. Zool. u. Anthrop.-Ethnog. Mus. Abhandl. u. Ber.=Abhandlungen und Berichte des Königlichen Zoologischen und Anthropologische-Ethnographischen Museums zu Dresden. Berlin. 1–15, 1886–1922. (15–20, 1924–1939, as "Abhandlungen und Berichte der Staatliches Museen für Tierkunde und Völkerkunde zu Dresden"; two Museums were formed in 1952, 1 for zoology and 1 for ethnology, and the above periodical continued in part 21, 1953, as "Abhandlungen und Berichte aus dem Staatlichen Museum für Tierkunde—Forschungsinstitut-Dresden"; 22–23 (1), 1954–1956, as "Abhandlungen und Berichte aus dem Staatlichen Museum für Tierkunde—Forschungsstelle-Dresden"; 23 (2) 1957+, as

"Abhandlungen und Berichte aus dem Staatlichen Museum für Tierkunde in Dresden.")

> Vols. 1-20, 1886-1939, have separately paged numbers, and sometimes separate "Beihefte."

Dusenia=Dusenia. Publicatio Periodica de Scientia Naturali. Curitiba, Brazil. 1, 1950+

Ecol. Monog.=Ecological Monographs. Ecological Society of America, Durham, N.C. 1, 1931+

Ecology=Ecology. Brooklyn Botanical Gardens, Brooklyn, N.Y. 1, 1920+

> Supersedes "Plant World."

Elisha Mitchell Sci. Soc. Jour.=Journal of the Elisha Mitchell Scientific Society. Chapel Hill, N.C. 1, 1883+

> Vols. 12-18, 1895-1903, have some separately paged numbers [="parts"]. Vol. 18, 1902-1903, is also included in the series "The University of North Carolina, The University's Bulletins."

Emory Univ. (Ga.) Mus. Bul.=Emory University Museum Bulletin. Atlanta, Ga. 1, 1943+

Encyclopédie Ent., Sér. A.=Encyclopédie Entomologique. Série A. Travaux Généraux. Paris. 1, 1924+

Encyclopédie Ent., Sér. B, [Div.] II: Diptera=Encyclopédie Entomologique. Série B. Mémoires et Notes. II. Diptera. Paris. 1, 1924+

> [Division] I covers Coleoptera, and III, Lepidoptera.

Ent. Amer.=Entomologica Americana. Lancaster, Pa. 1, 1885+

> Suspended 1891-1926; vol. 7, 1926+, are marked N. ser.

Ent. Ber.=Entomologische Berichten. Den Haag. 1, 1901+

Ent. Mag. [Japan]=Entomological Magazine. Kyoto, Japan. 1-3, 1915-1919.

Ent. Mag. (London)=The Entomological Magazine. London. 1-5, 1832-1838.

> Superseded by **Entomologist (Newman's)** [q.v.]

Ent. Meddel.=Entomologiske Meddelelser. Copenhagen. 1, 1887+

> Vols. 6-14, 1897-1925, also called Ser. 2, 1-9. Vol. 15 began publication in 1921.

Ent. Mitt.=Entomologische Mitteilungen. Berlin-Dahlem. 1-17, 1912-1928.

> Supersedes "Deutsche Entomologische National-Bibliothek."

Ent. Monthly Mag.=The Entomologist's Monthly Magazine. London. 1, 1864+

> Vols. 26–50, 1890–1914, also called Ser. 2, 1–25; vols. 51–75, 1915–1939, also as Ser. 3, 1–25; vols. 76, 1940+, also as Ser. 4, 1+

Ent. Nachr.=Entomologisches Nachrichten. Berlin. 1–26, 1875–1900.

Ent. News=Entomological News. Philadelphia. 6, 1896+ (1–5, 1890–1895, as **Ent. News and Acad. Nat. Sci. Phila., Ent. Sect. Proc.** [q.v.].)

Ent. News and Acad. Nat. Sci. Phila., Ent. Sect. Proc.=Entomological News and Proceedings of the Entomological Section of the Academy of Natural Sciences of Philadelphia. See **Ent. News.**

Ent. Obozr.=Entomologicheskoe Obozrenie. Leningrad. 25, 1931+ (1–24, 1901–1930, as "Russkoe Entomologicheskoe Obozrenie".)

> Suspended 1939–1944.

Ent. Soc. Amer. Ann.=Annals of the Entomological Society of America. Columbus, Ohio; Washington, D.C. 1, 1908+

Ent. Soc. Amer. Misc. Pub.=Miscellaneous Publications of the Entomological Society of America. Washington, D.C. 1, 1959+

Ent. Soc. Brit. Columbia, Proc.=Proceedings of the Entomological Society of British Columbia. Victoria, Vancouver, Vernon, B. C. 3, 1913+ (1–2, 1911–1912, as "Proceedings of the British Columbia Entomological Society.")

> Vols. 8–20, 1916–1922, alternate as an Economic Series, and a Systematic Series. Vols. 1–7, 1911–1915, called New Series (superseding "Bulletin of the British Columbia Entomological Society").

Ent. Soc. London, Trans.=The Transactions of the Entomological Society of London. See **Roy. Ent. Soc., London, Trans.**

Ent. Soc. Ontario, Ann. Rpt.=Annual Report of the Entomological Society of Ontario. See **Ent. Soc. Ontario, Proc.**

Ent. Soc. Ontario, Proc.=Proceedings of the Entomological Society of Ontario. Toronto, Ont. 90, 1959+ (1, 1869–1870, as "First Annual Report of the Noxious Insects of the Province of Ontario"; [2–6], 1871–1875, and [8–12], 1877–1881, as "Annual Report of the Entomological Society of the Province of Ontario"; [7], 1876, and 15–89, 1884–1958, as **Ent. Soc. Ontario, Ann. Rpt.** [q.v.]; [13–14], 1882–1883, as "Report of the Entomological Society of Ontario."

> Vols. [11–14] are unnumbered, although [11–14] are numbered on the first page of the reports. Vol. 90, 1959, is also called "Annual Report." Some of the earlier reports were also published with different pagination in the various reports of the Ontario Department of Agriculture.

Ent. Soc. Phila. Proc.=Proceedings of the Entomological Society of Philadelphia. Philadelphia. 1-6, 1861-1867.

Superseded by **Amer. Ent. Soc. Trans.** [q.v.].

Ent. Soc. So. England, Trans.=Transactions of the Entomological Society of the South of England. Southampton, England. 6-8, 1930-1933. (1-4, 1924-1928, as "Transactions of the Hampshire Entomological Society"; 5, 1929, as "Transactions of the Entomological Society of Hampshire and the South of England.")

Superseded by **Soc. Brit. Ent. Trans.** [q.v.].

Ent. Soc. South. Africa, Jour.=The Journal of the Entomological Society of Southern Africa. Pretoria. 1, 1939+

Ent. Soc. Wash. Proc.=Proceedings of the Entomological Society of Washington. Washington, D.C. 1, 1884+

Ent. Tidskr.=Entomologisk Tidskrift. Uppsala, Sweden. 1, 1880+

Ent. Wchnbl.=Entomologisches Wochenblatt. Leipzig, Stuttgart. 24-25, 1907-1908. (1-23, 1884-1906, as **Insekten-Börse** [q.v.]; 26-27, 1909-1910, as "Entomologische Rundschau," which included a separately paged "Insektenbörse"; then split into a textual portion called "Entomologische Rundschau," 28-56, 1911-1939, and an advertising sheet (Anzeigenblatt) called "Insektenbörse," a Beilage, 28, 1911+).

"Entomologische Rundschau" absorbed "Societas Entomologica" in 1931, and vols. 48-49, 1931-1932, carry in addition, volume numbers 46-47 of the latter journal. "Entomologische Rundschau" merged into "Entomologische Zeitschrift" in 1939, and vol. 53, 1939-1940, of the latter is also called vol. 56 of "Entomologische Rundschau" (although merger did not take place until the 36th number of the volume).

Volumes 43-52, 1926-1935, of "Insektenbörse" were issued as an advertising sheet to both "Entomologische Rundschau" (vols. 43-52) and "Entomologische Zeitschrift" (vols. 40-49), and thereafter continued (with vol. 53, 1936) as a Beilage to the latter (vol. 50).

Entomologist=The Entomologist. London. 10, 1876+ ([1]-9 1840-1875, as **Entomologist (Newman's)** [q.v.], with a half-title, page reading "Newman's Entomologist.")

Vol. 1, 1840-1842, is unnumbered. Suspended 1843-1863. The subtitle of vols. 10-85, 1876-1952, varies, and from vol. 86, 1953+, it is absent.

Entomologist (Newman's)=The Entomologist. (Newman's Entomologist). See **Entomologist**.

Essex Inst. Proc.=Proceedings of the Essex Institute. Salem, Mass. 1-6,1848/1856-1867/1871.

Vols. 4-6, 1864-1871, include separately paged "Communications." The covers of the numbers of vol. 6, 1867-1871, read "Proceedings and Communications of the Essex Institute," but the title page of the volume reads only as Proceedings.

Estudio, Semenario de Cien. Méd. (Inst. Méd. Nac. de Mex.)=El Estudio. Semenario de Ciencias Médicas. Instituto Médico Nacional de Mexico, Mexico City. 1–4, 1889–1893.

> Superseded by "Anales del Instituto Médico Nacional."

Evolution (Lancaster, Pa.)=Evolution. Society for the Study of Evolution, Lancaster, Pa. 1, 1947+

Feuille des Jeunes Nat.=Feuille des Jeunes Naturalistes. Paris. 1–44, 1870–1914. (45–47, 1924–26, and N. ser., 1–7, 1946–1952, as "Feuille des Naturalistes"; N. ser., 8, 1953+, as "Cahiers des Naturalistes.")

> Vols. 24–30, 1894–1900, also called Ser. 3, 24–30; vols. 31–40, 1900–1910, also as Ser. 4, 1–10; 41–44, 1911–1914, also as Ser. 5, 1–4.
> The issues are numbered serially, 1–518, 1870–1914, and N. ser., 1–34, 1924–1926.

Finska Vetensk. Soc., Acta Soc. Sci. Fenn.=Acta Societatis Scientiarum Fennicae. Finska Vetenskaps-Societeten, Helsingfors. 3–50, 1846–1926. (1–2, 1842–1845, as "Commentationes Societatis Scientiarum Fennicae"; continues as "Acta Societatis Scientiarum," "N. ser. A: Opera Physico-Mathematica," 1, 1926+ and "N. ser. B: Opera Biologica," 1, 1938+)

Finska Vetensk. Soc., Comm. Biol.=Commentationes Biologicae. Finska Vetenskaps-Societeten, Helsingfors. 1, 1924+

Finska Vetensk. Soc., Öfversigt af . . . Förhandl., Afd. A: Mat. och Naturvetensk.=Öfversigt af Finska Vetenskaps-Societetens Förhandlingar. Afd. A. Matematik och Naturvetenskaper. Helsingfors. 51–64, 1909–1922.

> The "Öfversigt . . . Förhandlingar" began as a single journal, 1–50, 1838–1908; with vol. 51, 1909, it split into 3 separate series: A, as above; B, Humanistika Vetenskaper; and C, Redogörelser och Förhandlingar. It was superseded by the various Commentationes of the Society (e.g., see **Finska Vetensk. Soc., Comm. Biol.**). The 3 sections may be bound by libraries together as volumes of the "Öfversigt" or separately, and each was issued in separately paged numbers.

Fla. Ent.=The Florida Entomologist. Gainesville, Fla. 4, 1920+ (1–3, 1917–1920, as "The Florida Buggist.").

Folia Zool. et Hydrobiol.=Folia Zoologica et Hydrobiologica. Latvijas (Lettländischen) Universität, Riga. 1, 1929+

[France] Min. de l'Agr., [Cent. Natl. de Rech. Agron., Versailles], Monog.=Monographies publiées par les Stations et Laboratoires de Recherches Agronomiques. Ministère de l'Agriculture, Direction de l'Agriculture [Centre National de Recherches Agronomiques, Versailles], Paris. [5], 1938–? [there are at least 8, through 1939]. ([1–2], 1933–1935, as "Collection de Monographies publiées par les Stations et Laboratoires de Recherches Agronomiques de

Versailles"; [3], 1935, not titled in series; [4], 1936, as "Monographies publiées par le Centre National de Recherches Agronomiques No. 3 [sic].")

> These are usually classed as separates by libraries. No serial numbers appear on most; they are assigned from lists at the end of the later monographs.

Gard. Chron. and Agr. Gaz.=Gardeners' Chronicle and Agricultural Gazette. London. 1844–1873. (1841–1843, and Ser. 2, 1–26, 1874–1886, and Ser. 3, 1, 1887+, as "Gardeners' Chronicle.")

> No volume numbers were used in the first series.

Genet. Soc. Amer. Rec.=Records of the Genetics Society of America. Lancaster, Pa.; Menasha, Wis.; Cold Spring Harbor, N.Y.; Columbus, Ohio; Madison, Wis., etc. 2, 1933+ ([1], 1932, as "The Genetics Society of America. Program of the first Annual Meeting" [preprinted from **Amer. Nat.,** vol. 67].)

Genetics=Genetics. Princeton, N.J.; Menasha, Wis. 1, 1916+

George Wash. Univ. Bul.=George Washington University Bulletin. Washington, D.C., 1, 1902+

> Early volumes issued in separately paged numbers.

Genoa Mus. Civico di Storia Nat. Ann.=Annali del Museo Civico di Storia Naturale di Genova. Genoa, Italy. 1–46, 1870–1915. (47, 1916+, as "Annali del Museo Civico di Storia Naturale Giacomo Doria.")

> Subtitle varies. Vols. 21–40, 1884–1901, also called Ser. 2, 1–20; vols. 41–50, 1904–1926, also called Ser. 3, 1–10.

Gessell. Naturf. Freunde, Berlin, Sitzber.=Sitzungsberichte der Gesellschaft Naturforschender Freunde zu Berlin. Berlin. 1839/1859+

> Volume numbers are not used. Material for the years 1839–1859 was issued in a single volume, in 1912.

Giambattista Vico, Gior. Sci. (Naples)=Il Giambattista Vico. Giornale Scientifico. Naples. 1–4, 1857.

Glasgow Nat.=The Glasgow Naturalist. Glasgow. 1, 1908+

> Suspended 1922–1925, 1927–1929.

Great Basin Nat.=The Great Basin Naturalist. Provo, Utah. 1, 1939+

Harvard Univ., Inst. Trop. Biol. and Med. Contrib.=Contributions from the Institute for Tropical Biology and Medicine. Harvard University, Cambridge, Mass. 1–4, 1925–1926. (5–6, 1930–1934, as "Contributions from the Department of Tropical Medicine and the Institute for Tropical Biology and Medicine.")

Harvard Univ., Mus. Compar. Zool. Bul.=Bulletin of the Museum of Comparative Zoology at Harvard College. Harvard University, Cambridge, Mass. 1, 1863+

Hawaii. Ent. Soc. Proc.=Proceedings of the Hawaiian Entomological Society. Honolulu. 1, 1905+

Hawaii. Sugar Planters' Assoc. Expt. Sta., Div. Ent. Bul.=Bulletin. Division of Entomology. Report of the Work of the Experiment Station of the Hawaiian Sugar Planters' Association. Honolulu. 1-8, 1905-1909.   (9-14, 1910-1919, as "Bulletin. Entomological Series. Report . . . Association"; 15-21, 1924-1936, and 22, 1942, as "Bulletin of the Experiment Station of the Hawaiian Sugar Planters' Association. Entomological Series.")

> Bul. 1, 1905, has 10 continuously paged parts. Buls. 2-5, 1906-1907, are in a "Volume 2," separately paged. There are several other series of Bulletins.

Hilgardia=Hilgardia. Agricultural Experiment Station, Berkeley. Calif. 1, 1925+

Hokkaido Imp. Univ., Faculty Agr., Jour.=Journal of the Faculty of Agriculture. Hokkaido Imperial University. Sapporo, Sendai, Japan. 20-48, 1929-1949.   (1-2, 1876-1907, as "Journal of the Sapporo Agricultural College"; 3-7, 1908-1917, as **Tohôku Imp. Univ., Col. Agr., Jour.** [q.v.]; 8-19, 1918-1928, as "Journal of the College of Agriculture. Hokkaido Imperial University"; 47, 1950+, as "Journal of the Faculty of Agriculture. Hokkaido University.")

> The word "Imperial" was dropped after 1950, with the latter numbers of vols. 47 and 48. Volumes were occasionally issued in separately paged numbers.

Hyogo Univ. Agr. (Sasayama), Sci. Rpts., Ser. Nat. Sci.=The Science Reports of the Hyogo University of Agriculture. Series Natural Science. Sasayama, Japan. 1, 1953+

I.R. Ist. Veneto di Sci., Let. ed Arti, Atti=Atti dell' I.R. Istituto Veneto di Scienze, Lettere ed Arti. Venice. Ser. 3, 1-11, 1856-1865.   (1-7, 1840-1848, and Ser. 2, 1-6, 1850-1855, as "Atti delle Adunanze dell' I.R. Istituto Veneto di . . . Arti"; Ser. 3, 12-16, 1866-1871, and Ser. 4, 1-3, 1871-1874, as "Atti dell' Regio Istituto Veneto di . . . Arti"; Ser. 5, 1-8, 1874-1882, Ser. 6, 1-7, 1883-1889, Ser. 7, 1-10, 1890-1898, Ser. 8, 1-18, 1898-1916, Ser. 9, 1-10, 1917-1926, and 86-102, 1926-1943, as "Atti del Reale Istituto Veneto di . . . Arti"; 103, 1943+, as "Atti dell' Istituto Veneto di . . . Arti.")

> Volume and series numbers appeared only on the signatures and not volume title pages of Series 1 and 2, and Ser. 3, vol. 1, 1840-1856. (Ser. 2, vol. 6, 1854/1855, is erroneously called "Ser. 3, vol. 1.") Volumes of series 7-9, 1890-1926, are also called vols. 48-85 (of the whole series). Volumes of series 8-9, 1898-1926, and volumes 86-93, 1926-1934, are in 2 separately paged sections: "Parte I," and "Parte II"; 94, 1934+, are in 3 separately paged Sections: "Parte I," "Parte II: Scienze Morali e Lettere" ("Classe

di Scienze . . . Lettere," 97, 1937+), and "Parte II: Scienze Matematiche e Naturali" ("Classe di Scienze . . . Naturali," 97, 1937+).

Ill. Biol. Monog.=Illinois Biological Monographs. Urbana, Ill. 1, 1914+

> Volumes 1–24, 1914–1954, have separately paged parts, and thereafter each monograph is a volume in itself. Volumes 1–17 (1), 1914–1938, also form part of the "University of Illinois Bulletin" series.

Ill. Nat. Hist. Survey, Bul.=Bulletin of the Illinois Natural History Survey. Bloomington, Peoria, Urbana, Ill. 22–26, 1941–1955. ([1] (1), 1876, as "Illinois Museum of Natural History. Bulletin"; [1] (2–6), 1878–1883, as "Illinois State Laboratory of Natural History. Bulletin"; 2–12, 1884–1917, as **Ill. State Lab. Nat. Hist. Bul.** [q.v.]; 13–19, 1918–1932, as **Ill. State Nat. Hist. Survey, Bul.** [q.v.]; 20–21, 1932–1941, and 27, 1957+, as "Illinois Natural History Survey Bulletin.")

> Volume 1, 1876–1883, is unnumbered, and consists of 6 separately paged numbers. Volumes 13–21, 1918–1941, are as shown above only on the title page of the volumes; title on individual numbers reads "State of Illinois. Department of Registration and Education. Division of the Natural History Survey. Bulletin."

Ill. State Agr. Soc. Trans.=Transactions of the Illinois State Agricultural Society. Springfield, Ill. 1–8, 1853–1870. (9–56, 1871–1918, as "Transactions of the Department of Agriculture of the State of Illinois.")

> Superseded by "Annual Report of the Department of Agriculture. State of Illinois." Volumes 9–56, 1871–1918, are also called N. ser., 1–48. Volume 19, 1881, is called "Annual Report."

Ill. State Lab. Nat. Hist. Bul.=Bulletin of the Illinois State Laboratory of Natural History. See **Ill. Nat. Hist. Survey, Bul.**

Ill. State Nat. Hist. Survey, Bul.=Bulletin of the Illinois State Natural History Survey. See **Ill. Nat. Hist. Survey, Bul.**

Illus. Wchnschr. f. Ent.=Illustrierte Wochenschrift für Entomologie. See **Ztschr. f. Wiss. Insektenbiol.**

Ind. Dept. Geol. and Nat. Resources, Ann. Rpt.=Annual Report. Department of Geology and Natural Resources, Indiana. Indianapolis. 17–35, 1891–1911. (1–8/10, 1869–1879, as "Annual Report of the Geological Survey of Indiana"; 11–16, 1881–1889, as "Annual Report. Department of Geology and Natural History. Indiana"; 36–41, 1911–1919, as "Annual Report of Department of Geology and Natural History. Indiana.")

> Superseded by the Annual Report of the Indiana Department of Conservation. No reports were issued for 1887, 1890, and 1892.

Indian Jour. Ent.=The Indian Journal of Entomology. New Delhi. 1, 1939+

Indian Mus. Rec.=Records of the Indian Museum. Calcutta. 1, 1907+

Insect Life (U.S. Dept. Agr., Div. Ent., Period. Bul.)=Insect Life. U.S. Department of Agriculture. Division of Entomology. Periodical Bulletin. Washington, D.C. 1–7, 1888–1895.

Insecta Matsumurana=Insecta Matsumurana. Sapporo, Japan. 1, 1926+

Insecutor Inscitiae Menstruus=Insecutor Inscitiae Menstruus. Washington, D.C. 1–14, 1913–1926.

Insekten-Börse=Insekten-Börse. See **Ent. Wchnbl.**

Inst. Biol. de Defesa Agr. e Anim. [São Paulo] Arch.=Archivos do Instituto Biológico de Defesa Agricola e Animal. São Paulo. 1–4, 1928–1931. (5–8, 1934–1937, as "Archivos do Instituto Biológico"; 9–12, 1939–1941, as "Arquivos do Instituto Biológico"; 13, 1942+, as "Arquivos do Instituto Biológico, Departamento da Defesa, Sanitaria da Agricultura.")

Inst. Jamaica, Jour.=Journal of the Institute of Jamaica. Kingston. 1–2, 1891–1899.

Inst. Oswaldo Cruz, Mem.=Memorias do Instituto Oswaldo Cruz. Rio de Janeiro. 1, 1909+

Inst. Pasteur d'Algérie, Arch.=Archives des Institut Pasteur d'Algérie. Algiers. 1, 1923+

> Vol. 1, no. 1, 1923, is also called vol. 3, no. 2, of **Inst. Pasteur de l' Afrique du Nord, Arch.** [q.v.].

Inst. Pasteur de l'Afrique du Nord, Arch.=Archives des Instituts Pasteur de l'Afrique du Nord. Algiers. 1–2, 1921–1922.

> Replaces "Archives des Instituts Pasteur de Tunis" during 1921–1922, which resumed with vol. 12, no. 1, in 1923 (also called vol. 3, no. 1, of the above series). Superseded by **Inst. Pasteur d'Algérie, Arch.** [q.v.], vol. 1, no. 1, 1923, of which is also called vol. 3, no. 2, of the above series.

Internatl. Comn. Zool. Nomencl., Bul. Zool. Nomencl.=The Bulletin of Zoological Nomenclature. International Commission on Zoological Nomenclature, London. 1, 1943+

Internatl. Rev. Gesam. Hydrobiol. u. Hydrog.=International Revue der Gesamten Hydrobiologie und Hydrographie. Leipzig. 1–43, 1908–1943.

Invertebrata Pacifica=Invertebrata Pacifica. Claremont, Calif. 1, 1903–1907.

Iowa Acad. Sci. Proc.=Proceedings of the Iowa Academy of Science. Des Moines. [1], 1887/1893+

> Volumes 1–12, 1887–1904, read "Proceedings . . . Sciences"; vol. 1 is unnumbered. There was also a "Proceedings of the Iowa Academy of Sciences" published at Iowa City, 1875–1880.

Iowa State Col. Jour. Sci.=Iowa State College Journal of Science. Ames, Iowa. 1–33, 1926–1959. (34, 1959+ as "Iowa State Journal of Science.")

Isis (Oken's)=Isis. Jena, Leipzig. 1–41, 1817–1848.

> Sometimes cited as "Isis von Oken," which appears on half-title page to the volumes.

Jahrb. der Hamburg. Wiss. Anst.=Jahrbuch der Hamburgischen Wissenschaftlichen Anstalten. Hamburg. 1–34, 1883–1917.

> One or more Beihefts accompany many of the volumes. Contains [vols. 1–10] of **Zool. Staatsinst. u. Zool. Mus. Hamburg, Mitt.** [q.v.], 1884–1892, and vols. 11–38 of the same journal appear as Beihefts, 1893–1922.

Jahresber. über die Fortschr. der Forstw. u. Forstl. Naturk.=Jahresberichtes über die Fortschritte der Forstwissenschafte und der Forstlichen Naturkunden. Berlin. 1, 1836–1839.

Jamaica Dept. Agr. Ent. Bul.=Department of Agriculture, Jamaica. Entomological Bulletin. Kingston. 1–5, 1921–1930. (6, 1932, as "Department of Science and Agriculture, Jamaica. Entomological Bulletin.")

> Bul. 4 is in 2 separately paged parts, Pt. 1/2, 1926, as above; Pt. 3, 1928, as in Bul. 6.

Jour. Agr. Res.=Journal of Agricultural Research. Washington, D.C. 1–78, 1913–1949.

> Suspended 1921–1922.

Jour. Anim. Ecol.=The Journal of Animal Ecology. London. 1, 1932+

Jour. de Phys., de Chim., d'Hist. Nat. et des Arts=Journal de Physique, de Chimie, d'Histoire Naturelle et des Arts. Paris. 44–96, 1794–1823. (1, 1773, as "Observations et Mémoires sur la Physique, sur l'Histoire Naturelle et sur les Arts et Metiers"; 2–43, 1773–1793, as "Observations sur la Physique, sur l'Histoire Naturelle et sur les Arts.")

> Supersedes "Introduction aux Observations sur la Physique."

Jour. Econ. Biol.=The Journal of Economic Biology. London. 1–10, 1905–1915.

> Superseded by "The Journal of Zoological Research."

Jour. Econ. Ent.=Journal of Economic Entomology. Concord, N.H.; Geneva, N.Y.; Menasha, Wis.; Baltimore, Md. 1, 1908+

Jour. Ent. and Zool.=Journal of Entomology and Zoology. Claremont, Calif. 5–42, 1913–1951. (1, 1909, as "Pomona Journal of Entomology"; 2–4, 1910–1912, as **Pomona Col. Jour. Ent.** [q.v.].)

Jour. Hered.=The Journal of Heredity. Washington, D.C. 5, 1914+ (1–4, 1910–1913, as "American Breeder's Magazine".)

Jour. Parasitol.=The Journal of Parasitology. Lancaster, Pa. 1, 1914+

Jour. Trop. Med.=The Journal of Tropical Medicine. London. 1–9, 1898–1906. (10, 1907+, as "The Journal of Tropical Medicine and Hygiene.")

K. Akad. der Wiss. Wien, Math.-Nat. Cl. Denkschr.=Denkschriften der Kaiserlichen Akademie der Wissenschaften. Mathematisch-Naturwissenschaftliche Classe. Wien. 1–70, 1850–1901. (71–92, 1907–1916, as "Denkschriften der Kaiserlichen Akademie der Wissenschaften. Mathematisch-Naturwissenschaftliche Klasse"; 93–95, 1917–1918, as "Kaiserliche Akademie der Wissenschaften in Wien. Mathematisch-Naturwissenschaftliche Klasse. Denkschriften"; 96–105, 1919–1943, as "Akademie der Wissenschaften in Wien. Mathematisch-Naturwissenschaftliche Klasse. Denkschriften" [vol. 105 is the last seen by us].)

Vols. 1–56, 1850–1889, are in 2 separately paged parts (except 35, 40, and 52); vols. 57, 1890+, are not in parts (except 66 which is in 3 continuously paged parts).

The other Class of the Denkschriften is "Philosophisch-Historische Klasse"; there is also a "Denkschriften der Gesamtakademie," 1, 1947+

K. Akad. der Wiss. Wien, Math.-Nat. Cl. Sitzber. Abt. I=Sitzungsberichte der Mathematisch-Naturwissenschaftlichen Classe der Kaiserlichen Akademie der Wissenschaften. Abteilung I. Wien. 43–111, 1861–1902. (112–123, 1903–1914, as "Sitzungsberichte der Mathematisch-Naturwissenschaftlichen Klasse der . . . Wissenschaften. Abteilung I."; 124–126, 1915–1917, as "Kaiserliche Akademie der Wissenschaften. Mathematisch-Naturwissenschaftliche Klasse. Sitzungsberichte. Abteilung I."; 127–155, 1918–1946, as "Akademie der Wissenschaften. Mathematisch-Naturwissenschaftliche Klasse. Sitzungsberichte. Abteilung I"; 156, 1947+, as "Österreichische Akademie der Wissenschaften. Mathematisch-Naturwissenschaftliche Klasse. Sitzungsberichte. Abteilung I.")

Volumes 1–42, 1848–1860, of the Sitzungsberichte were without a breakdown into Abteilungen; 43–64, 1861–1871, and 163, 1954+, were with the separate Abteilungen I and II; 65–96, 1872–1887, and 132–162, 1923–1953, were with 3 separate Abteilungen, I, II, and III; 97–131, 1888–1922, were with 4, Abteilungen I, IIA, IIB, and III.

Volumes 1–123, 1848–1914, also as "Sitzungsberichte der Kaiserliche Akademie der Wissenschaften. Mathematisch-Naturwissenschaftliche Classe [=Klasse]."

K. Danske Vidensk. Selsk., Biol. Meddel.=Det Kgl. Danske Videnskabernes Selskab. Biologiske Meddelelser. Copenhagen. 1, 1917+

This is issued in separately paged numbers.

K. Friedrich-Wilhelms-Gymnasiums zu Posen, Öffentl. Prüf. d. Schüler=Zu der Öffentlichen Prüfung der Schüler des Königlichen Friedrich-Wilhelms-Gymnasiums zu Posen. Posen. Dates unknown, include 1845 and 1850.

An annual "school program."

K.-k. Zool.-Bot. Gesell. Wien, Abhandl.=Abhandlungen der K.-k. Zool. Botan. Gesellschaft in Wien. See **Zool.-Bot. Gesell. Wien, Abhandl.**

K.-k. Zool.-Bot. Gesell. Wien, Verhandl.=Verhandlungen der Kaiserlich-königlichen Zoologisch-Botanischen Gesellschaft in Wien. See **Zool.-Bot. Gesell. Wien, Verhandl.**

K. Leopoldinisch-Carolinisch. Akad. d. Naturf., Verhandl. (Acad. C. Leopoldino-Carolinae Nat. Curio., Nova Acta Phys.-Med.)=Verhandlungen der Kaiserlichen Leopoldinisch-Carolinischen Akademie der Naturforscher (Nova Acta Physico-Medica Academiae Caesareae Leopoldino-Carolinae Naturae Curiosorum). See **K. Leopoldinisch-Carolinisch. Deut. Akad. d. Naturf., Abhandl.**

K. Leopoldinisch-Carolinisch. Deut. Akad. d. Naturf., Abhandl. (Acad. C. Leopoldino-Carolinae Germ. Nat. Curio., Nova Acta)= Abhandlungen der Kaiserlichen Leopoldinisch-Carolinischen Deutschen Akademie der Naturforscher (Nova Acta. Academiae Caesareae Leopoldino-Carolinae Germanicae Naturae Curiosorum). Nürnberg, Erlangen, Bonn, Breslau, Jena, Dresden, Halle. 64–103, 1895–1918. (1–8, 1757–1791, as "Nova Acta Physico-Medica Academiae Caesareae Leopoldino-Carolinae Naturae Curiosorum"; 9, 1818, as "Verhandlungen der Leopoldinisch-Carolinischen Academie der Naturforscher (Nova Acta Physico-Medica Academiae Caesareae Leopoldino-Carolinae Naturae Curiosorum)"; 10–19, 1821–1842, as **K. Leopoldinisch-Carolinisch. Akad. d. Naturf., Verhandl.** [q.v.]; 20–26, 1843–1858, as "Verhandlungen der Kaiserlichen Leopoldinisch-Carolinischen Akademie der Naturforscher (Novorum Actorum Academiae Caesareae Leopoldino-Carolinae Naturae Curiosorum)"; 27–35, 1860–1869, as "Verhandlungen der Kaiserlichen Leopoldinisch-Carolinischen Deutschen Akademie der Naturforscher (Novorum Actorum Academiae Caesareae Leopoldino-Carolinae Germanicae Naturae Curiosorum)"; 36–63, 1873–1895, as "Verhandlungen . . . Naturforscher (Nova Acta. Academiae Caesareae Leopoldino-Carolinae Germanicae Naturae Curiosorum)"; 104–110, 1919–1928, as "Abhandlungen der Leopoldinisch-Carolinischen Deutschen Akademie der Naturforscher (Nova Acta. Academiae Leopoldino-Carolinae Germanicae Naturae Curiosorum)"; N. ser., 1–14, 1932–1945, as "Nova Acta Leopoldina (Abhandlungen der Kaiserlich Leopoldinisch-Carolinisch Deutschen Akademie der Naturforscher)"; N. ser., 15, 1952+, as "Nova Acta

Leopoldina (Abhandlungen der Deutschen Akademie der Naturforscher [Leopoldina] zu Halle/Saale".)

> Supersedes "Acta Physico-Medica Academiae Caesareae Leopoldino-Carolinae Naturae Curiosorum." The "Verhandlungen" 9–26, 1818–1858, are also numbered 1–18, and in addition vols. 11–19, 1823–1842, are also called Ser. 2, 1–10, vols. 20–29, 1843–1861, also as Ser. 3, 1–10, and vols. 30–35, 1862–1869, also as Ser. 4, 1–6. Some early volumes are accompanied by 1 or more supplements. Some volumes are in parts, sometimes separately paged, and later the individual numbers were both continuously and separately paged. The volumes of the new series are continuously paged.

K. Realschule zu Meseritz, Programm=Programm der Königlichen Realschule zu Meseritz. Meseritz [=Miedzyrzecz, Poland]. 1840–1868.

> The 29 separately paged, unnumbered, issues were annual "school programs."

K. Svenska Vetensk. Akad. Handl.=Kongliga Svenska Vetenskaps-Akademiens Handlingar. Stockholm. N. ser., 1–63, 1855–1923. (1813–1842, as **K. Vetensk. Acad. Handl.** [q.v.]; 1843–1854, as K. **Vetensk. Akad. Handl.** [q.v.]; Ser. 3, 1–25, 1924–1948, and Ser. 4, 1, 1949+, as "Kungliga Svenska Vetenskapsakademiens Handlingar."

> Supersedes **K. Vetensk. Acad. Nya Handl.** [q.v.], which in turn supersedes an earlier "Handlingar." Libraries often count the first series, 1813–1854 (in which no volume numbers were used) as "Ser. 3," and the N. ser., 1855–1923, as "Ser. 4," Ser. 3, 1924–1948, as "Ser. 5," and Ser. 4, 1949+ as "Ser. 6." The Handlingar is issued in separately paged numbers from N. ser., vol. 2, 1856+

K. Svenska Vetensk. Akad. Öfversigt af . . . Förhandl.=Öfversigt af Kongl. Vetenskaps-Akademiens Förhandlingar. Stockholm. 1–63, 1844–1921.

> Volumes 29–63, 1872–1921, consist of separately paged numbers.

K. Vetensk. Acad. Handl.=Kongl. Vetenskaps-Academiens Handlingar. See **K. Svenska Vetensk. Akad. Handl.**

K. Vetensk. Acad. Nya Handl.=Kongl. Vetenskaps Academiens Nya Handlingar. Stockholm. 1–33, 1780–1812.

> Supersedes "Kongl. Svenska Vetenskaps Academiens Handlingar" (1–40, 1739–1779 [1–7, 1739–1746, as "Kongl. Swenska Wetenskaps Academiens Handlingar"]). Superseded by **K. Vetensk. Acad. Handl.** [q.v.]. Libraries often count these "Nya Handlingar" as Ser. 2 of the whole series.

K. Vetensk. Akad. Handl.=Kongl. Vetenskaps-Akademiens Handlingar. See **K. Svenska Vetensk. Akad. Handl.**

K. Württemb. Landw. Ver., Stuttgart, Correspondenzbl.=Correspondenzblatt des Königlich Württembergischen Landwirthschaftlichen Vereins. Stuttgart, Tübingen. 1-54, 1822-1848.

> Vols. 21-54, ?-1848, also called N. ser., 1-34. It is not known if this was issued in separately paged numbers.

Kans. Acad. Sci. Trans.=Transactions of the Kansas Academy of Science. Topeka. 18, 1901+ (1-17, 1868-1900, as **Kans. Acad. Sci. Trans. Ann. Mtgs.** [q.v.].)

Kans. Acad. Sci. Trans. Ann. Mtgs.=Transactions of the Annual Meetings of the Kansas Academy of Science. See **Kans. Acad. Sci. Trans.**

Kans. Agr. Expt. Sta. Tech. Bul.=Technical Bulletin. Agricultural Experiment Station. Kansas State Agricultural College (1916-1930); Kansas State College of Agriculture and Applied Science (1931-1958); Kansas State University of Agriculture and Applied Science (1959+). Topeka; Manhattan. 1, 1916+

Kans. Ent. Soc. Jour.=Journal of the Kansas Entomological Society. McPherson; Manhattan. 1, 1928+

Kans. Univ. Quart.=The Kansas University Quarterly. Lawrence, Kans. 1-10, 1892-1901.

> Superseded by **Kans. Univ. Sci. Bul.** [q.v.]. Beginning with vol. 6, 2 separate series appeared: Series A. Science and Mathematics, 6-10, 1897-1901, and Series B. Philology and History, 6-8, 1897-1899.

Kans. Univ. Sci. Bul.=The Kansas University Science Bulletin, Lawrence, Kans. 1, 1902+

> Supersedes **Kans. Univ. Quart.** [q.v.]; vols. 1-26, 1902-1940, are also numbered in continuation of that periodical, 11-36. Those volumes are also included in the series "Bulletin of the University of Kansas."

Kiïv. Derzhavnii Univ. im. T. G. Shevchenka, Nauk. Zap.=Kiïvskii Derzhavnii Universitet imeni T. G. Shevchenka. Naukovi Zapiski. Kiev. 4, 1938+ (1-3, 1935-1937, as "Kiïvskii Derzhavnii Universitet (Universitet d'État de Kiev), Naukovi Zapiski (Bulletin Scientifique) N. Ser. "[of the University's "Universitetskiia Izvyestiia"].)

> Each separately paged number ("Vid") is also numbered serially as one of many separate series, e.g., "Biologichnii Z'irnik," "Khemichinii Z'irnik", "Matematichnii Z'irnik," or "Z'irnik Zoomuzeiu" (=Sbornik Zoomuzeia; =Zool. Muz. Trudy).

Konowia=Konowia. Zeitschrift für Systematische Insektenkunde. Wien. 1-17, 1922-1939.

Ky. Agr. Expt. Sta. Bul.=Bulletin. Kentucky Agricultural Experiment Station. State College of Kentucky (1885-1908); State University (1908+). Lexington. 1-7, 1885-1886 and 9-15, 16-

134, 135–180, 181–673, 1887–1961. (8, 1886, as "Bulletin. Kentucky Experiment Station"; 674, 1961+, as "Bulletin. Agricultural Experiment Station.")

> Nos. 48–419, 1894–1941, are continuously paged within each year and later bound and indexed in the Station's Annual Report. Many Bulletins are also called "Research Bulletins".

La Plata Mus. Rev.=Revista del Museo de La Plata. Universidad Nacional de La Plata [after 1905], Buenos Aires. 1–34, 1890–1934; N. ser., in 6 secciones: Antropología, 1, 1936+; Botánica, 1, 1936+; Geología, 1, 1936+; Oficial, 1, 1935+; Paleontología, 1, 1936+; and Zoología, 1, 1937+.

> Vols. 14–24, 1907–1919, of old series, are also called Ser. 2, 1–11 [last as Vol. "12"]; vols. 25–34, 1922–1934, also as Ser. 3, 1–10. Vols. 23–24, 1916–1919, are in 2 separately paged parts.

[Leningrad] Gosud. Inst. Opytn. Agron. (State Inst. Expt. Agron.) Otd. Prikl. Ent. (Bur. Appl. Ent.) Isv. Prikl. Ent.=Izvestia Otdela Prikladnoi Entomologii. Gosudarstvennyi Institut Opytnoi Agronomii (State Institute of Experimental Agronomy). Otdel Prikladoi Entomologii (Bureau of Applied Entomology). Petrograd (1921–1922) (=Leningrad [1927–1930]). 3–4 (1), 1927–1929. (1, 1921, as "Izvestia Otdela Pricladoi Entomologii. Sel'sko-Khoziaistvennia Uchen'l Komitet (Agricultural Scientific Committee)"; 2, 1922, as "Izvestia . . . Entomologii. Sel'sko- . . . Komitet. Gosudarstvennyi Institut Opytnoi Agronomii"; 4(2), 1930, as "Izvestia . . . Entomologii. Vsesioznaia Akademiia C.-Kh. Nauk im Lenina (Lenin Academy of Agricultural Sciences in U.S.S.R.). Institut Zashchity Rastenii (Institute for Plant Protection). Otdel Prikladnoi Entomologii i Zoologii (Bureau of Applied Entomology and Zoology)".)

> The "Izvestia" have the additional English title as follows: vols. 1–2, "Reports of the Bureau of Applied Entomology of the Agricultural Scientific Committee"; vol. 3, "Reports of the Bureau of Applied Entomology"; vol. 4, "Reports on Applied Entomology."
> Superseded by the entomological series of "Trudy po Zashchity Rastenii."

Leyden Mus. Notes=Notes from the Leyden Museum. Rijks Museum van Natuurlijke Historie, Leiden. 1–36, 1879–1914.

> Superseded by "Zoologische Mededeelingen uitg. vanwege's Rijks Museum van Natuurlijke Historie te Leiden."

Limnol. and Oceanog.=Limnology and Oceanography. Baltimore Md. 1, 1956+

Linn. Soc. (London), Jour., Zool.=The Journal of the Linnean Society. Zoology. London. 8–35, 1865–1924 (and on individual numbers through Vol. 38, 1934.) (1–7, 1857–1864, as **Linn.**

Soc. (London), Jour. of Proc., Zool. [q.v.]; 36, 1924+, as "The Journal of the Linnean Society of London. Zoology.")

There is also a series for Botany.

Linn. Soc. (London), Jour. of Proc., Zool.=Journal of the Proceedings of the Linnean Society. Zoology. See **Linn. Soc. (London), Jour.**

Linn. Soc. London, Trans.=The Transactions of the Linnean Society of London. London. 7–30, 1804–1875; Ser. 2, in 2 sections: Zoology, 1–19, 1875–1936 and Botany, 1–9, 1875–1922; Ser. 3, 1, 1939+ (1–6, 1791–1802, as **Linn. Soc. (London), Trans.** [q.v.]).

Linn. Soc. (London), Trans.=Transactions of the Linnean Society. See **Linn. Soc. London, Trans.**

Linn. Soc. N. S. Wales, Proc.=Proceedings of the Linnean Society of New South Wales. Sydney. 1, 1875+

Vols. 11–20, 1886–1895, are called Ser. 2, 1–10, with the addition of the whole series volume number. Supplements accompany some volumes.

Linnaea Ent.=Linnaea Entomologica. Berlin. 1–16, 1846–1858.

Supersedes Ztschr. f. die Ent. (Leipzig) [q.v.].

London and Edinb. Phil. Mag. and Jour. Sci.=The London and Edinburgh Philosophical Magazine and Journal of Science. London. Ser. 3, 1–16, 1832–1840. (1–42, 1798–1813, and N. ser., 1–11, 1827–1832, Ser. 7, 36–40, 1945–1955, and Ser. 8, 1, 1956+, as "The Philosophical Magazine"; 43–68, 1814–1826, as "The Philosophical Magazine and Journal"; Ser. 3, 17–37, 1841–1850, Ser. 4, 1–50, 1851–1875, Ser. 5, 1–50, 1876–1900, Ser. 6, 1–50, 1901–1925, and Ser. 7, 1–35, 1926–1944, as "The London, Edinburgh and Dublin Philosophical Magazine and Journal of Science" [and as volume title pages of vols. 36–39, 1945–1948].)

Supplements accompany some volumes.

Märkische Tierwelt=Märkische Tierwelt. Berlin. 1–4, 1934–1941.

Mag. d. Ent. (Halle)=Magazin der Entomologie, Herausgegeben von Dr. Ernst Friedrich Germar. Halle. 1–4, 1813–1821.

Also known as "Germar's Magazin."

Mag. f. Insektenkunde=Magazin für Insektenkunde, herausgegeben von Karl Illiger. Braunschweig. 1–6, 1801–1807.

Also known as "Illiger's Magazin."

Mag. Zool. and Bot.=Magazine of Zoology and Botany. Edinburgh; London. 1–2, 1837–1838.

United with "Companion to the Botanical Magazine" to form **Ann. Nat. Hist.** [q.v.].

Maine Agr. Expt. Sta. Bul.=Bulletin. Maine Agricultural Experiment Station. Maine State College (1885–1897); University of Maine (1897+). Augusta; Orono. Ser. 2, 1, 1889+ ([Ser. 1], 1–20, 1885–1886, as "Bulletin. Maine Experiment Station"; [Ser. 1], 21–26, 1887–1888, as "Bulletin. Agricultural Experiment Station.")

> Buls. 1–20 of the old series were published in the newspaper "Whig and Courier". Nos. 1–5, 1885, were later reprinted. Of the second series, Buls. 1–52, 1889–1899, and 459, 1948+, are separately paged (6–52, 1894–1899, were later also published as continuously paged appendices to the annual Part 2 [="Report of the Director of the Agricultural Experiment Station"] of the "Annual Report of the Maine State College" (later University of Maine). Buls. 53–449, 1899–1947, were published continuously paged within each year and later bound and indexed as part of the annual reports of the Maine Agricultural Experiment Station; Buls. 450–458, 1947–1948, were also continuously paged but not indexed in the annual reports. The words "second series" appear only up through Bul. 25, 1896.

Marcellia=Marcellia; Rivista Internazionale de Cecidologia. Padova, Avellino, Portici, Italy. 1, 1902+

Mass. State Bd. Agr. Ann. Rpt. Sec.=Annual Report of the Secretary of the Massachusetts State Board of Agriculture. Boston. 2–62, 1854–1914. (1, 1853, as "Annual Report of the Secretary of the Board of Agriculture"; 63–65, 1915–1917 (1918), as "Annual Report of the Massachusetts Board of Agriculture.")

> Superseded by "Annual Report of the Massachusetts State Department of Agriculture." Vols. 63–65, 1915–1917, are in 2 separately paged parts each.

Meddel. om Grønland (Copenhagen)=Meddelelser om Grønland. Copenhagen. 1, 1879+

> Vol. 72, 1934+, are issued in separately paged numbers, although many volumes are issued as single volumes.

Medecina, Rev. Mex.=Medecina; Revista Mexicana. Escuela Nacional de Medecino, Mexico City. 1, 1920+

[Mex.] Esc. Nac. de Cien. Biol., An.=Anales de la Escuela Nacional de Ciencias Biológicas. Mexico City. 1, 1938+

[Mex.] Inst. de Salubridad y Enferm. Trop. Rev.=Revista del Instituto de Salubridad y Enfermedades Tropicales. Mexico City. 1, 1939+

Mich. Acad. Sci., Arts, and Letters, Papers=Papers of the Academy of Science, Arts, and Letters. Ann Arbor, Mich. 1, 1921+

> Vol. 24, 1938, is in 4 separately paged parts.

Mich. State Bd. Agr. Ann. Rpt. Sec.=Annual Report of the Secretary of the State Board of Agriculture of the State of Michigan. Lansing, Mich. 1–87, 1862–1949. (88, 1950,+ as "Annual Report. Secretary of the State Board of Agriculture. State of Michigan.")

Mich. State Univ., Mus., Pubs., Biol. Ser.=Publications of the Museum. Michigan State University. Biological Series. East Lansing, Mich. 1, 1957+

Mich. Univ. Mus. Zool. Misc. Pub.=Miscellaneous Publications. Museum of Zoology, University of Michigan. Ann Arbor, Mich. 1, 1916+

Mich. Univ. Mus. Zool. Occas. Papers=Occasional Papers of the Museum of Zoology. University of Michigan, Ann Arbor. 1, 1913+

Microentomology=Microentomology. Stanford University, Palo Alto, Calif. 1, 1936+

Milwaukee Publ. Mus. Bul.=Bulletin of the Public Museum of the City of Milwaukee. Milwaukee, Wis. 1, 1910+

Minn. Acad. Sci. Proc.=Proceedings of the Minnesota Academy of Science. Northfield; Minneapolis. 4, 1936+ (1–4, 1932–1936, as "Program of the Annual Meeting of the Minnesota Academy of Science.")

Two "volumes" were issued in 1936.

Minn. Agr. Expt. Sta. Bul.=Bulletin. Agricultural Experiment Station. University of Minnesota. St. Anthony Park; St. Paul; Minneapolis. 8–415, 1889–1952. (1–7, 1888–1889, as "Bulletin. Experiment Station of the College of Agriculture"; 416, 1953+ as "Station Bulletin. Agricultural Experiment Station.")

Buls. 1–7, 1888–1889, and 129, 1913+, are separately paged. Buls. 8–128, 1889–1912, are continuously paged within each year; nos. 26–128, 1893–1912, are then bound as part of the annual reports of the Agricultural Experiment Station.

Minn. Agr. Expt. Sta. Tech. Bul.=Technical Bulletin. Agricultural Experiment Station. University of Minnesota. St. Paul; Minneapolis. 1, 1921+

Minn. Geol. and Nat. Hist. Survey, Ann. Rpt.=The Geological and Natural History Survey of Minnesota. Annual Report. Minneapolis. 1–24, 1872–1899.

Some reports have special separately paged portions.

Minn. State Expt. Sta., Ent. Ann. Rpt.=Annual Report of the Entomologist of the State Experiment Station of the University of Minnesota to the Governor. Minneapolis; St. Paul; St. Anthony Park. 1–7, 1895–1902. (8–11, 1903–1906, as "Annual Report of the State Entomologist of Minnesota to the Governor"; 12–19,

1907–1922, as "Report of the State Entomologist of Minnesota to the Governor.")

> Nos. 3 (1897) and 7 (1902) were published also as **Minn. Agr. Expt. Sta. Bul.** 55 and 77, respectively.

Minn. Univ., Grad. Sch., Sums. Ph. D. Theses=Summaries of Ph. D. Theses. University of Minnesota. Minneapolis. 1, 1939+

Misc. Ent.=Miscellanea Entomologica. Narbonne; Séte; Toulouse; Paris. 1, 1892+

> Suspended 1940–1943.

Mo. State Bd. Agr. Ann. Rpt.=Annual Report of the State Board of Agriculture of the State of Missouri. Jefferson City, Mo. 10–25, 1874–1892. (1–2, 1865–1866, and 26–46, 1893–1913, as "Annual Report of the Missouri State Board of Agriculture"; 3–8, 1867–1872, as **Mo., State Bd. Agr. Ann. Rpt.** [q.v.]; 9, 1873, as "Proceedings of the State Board of Agriculture and of the State Horticultural Society"; [47–54, 1914–1922, as "The Missouri Yearbook of Agriculture. Annual Report. Missouri State Board of Agriculture"; 55, 1923, as "Yearbook Bulletin. Annual Report. Missouri State Board of Agriculture"; 1929/1930, as "The Biennial Agricultural Report. Missouri State Board of Agriculture"]; 1931–1932, as "Report of the Missouri State Board of Agriculture.")

> Superseded and continued by "Biennial Report of the Missouri State Department of Agriculture."
> Nos. 55–56, 1923–1924, were published in "The Monthly Bulletin. Missouri State Board of Agriculture," vols. 21 and 22. No reports issued between 1924 and 1929/1930. The report for 1929/1930 was published in "The Bulletin of the Missouri State Board of Agriculture," vol. 28.

Mo., State Bd. Agr. Ann. Rpt.=Annual Report of the State Board of Agriculture. Missouri. See **Mo. State Bd. Agr. Ann. Rpt.**

Monog. Z. Angew. Ent.=Monographien zur Angewandten Entomologie. Berlin. 1, 1917+

> A supplement to **Ztschr. f. Angew. Ent.** [q.v.].

Moskva Selsk. Khoz. Inst. Izv.=Izvestiia Moskovskago Sel'skokhoziaistvennago Instituta (Annales de l'Institut Agronomique de Moscou). Moscow. 1–22, 1895–1917.

> Supersedes and is superseded by "Izvestiia Petrovskoi Sel'skokhoziaistvennago Akademii" (later "Izvestia Sel'skokhoziaistvennoi Akademii imeni K. A. Tmiriazeva").

Mosquito News=Mosquito News. New Brunswick, N.J.; Albany, N.Y. 1, 1941+

Mushi=Mushi. Fukuoka, Japan. 1, 1928+

N.J. Agr. Expt. Sta. Spec. Bul.=Special Bulletin. New Jersey.

Agricultural Experiment Station. New Brunswick, N.J. A–T, 1882–1902.

N.J., State Bd. Agr. Ann. Rpt.=Annual Report of the State Board of Agriculture. Trenton, N.J. 15–44, 1887–1916 (1917). (1–14, 1874–1886, as "Annual Report of the New Jersey State Board of Agriculture.")

> Some volumes are accompanied by separate supplements.

N.J. State Mus. Ann. Rpt.=Annual Report of the New Jersey State Museum. Trenton, N.J. 1902–1914 (13 vols.). (1901, as "Report of the New Jersey State Museum.")

> No volume numbers used, the volume for each year being published the following year. Superseded by "New Jersey. Department of Conservation and Development. Annual Report."

N.Y. (Cornell) Agr. Expt. Sta. Bul.=Bulletin. Agricultural Experiment Station of the College of Agriculture. Cornell University. Albany; Ithaca, N.Y. 216–361, 1904–1915. (1–25, 1888–1890, as "Bulletin of the Agricultural Experiment Station"; 26–211, 1891–1903, and 378, 1916+, as "Bulletin. Cornell University Agricultural Experiment Station"; 212–215, 1903–1904, as "Bulletin. Experiment Station of the College of Agriculture"; 362–377, 1915–1916, as "Bulletin. Experiment Station of the New York State College of Agriculture.")

> Nos. 1–410, 1888–1922, are continuously paged within each year and later bound in the annual reports of the Agricultural Experiment Station. Nos. 411, 1922+, are separately paged.

N.Y. (Cornell) Agr. Expt. Sta. Mem.=Memoir. Agricultural Experiment Station. Cornell University. Ithaca, N.Y. 1, 1913+

> Nos. 1–60, 1913–1922, are paged continuously within each year and later bound with the annual reports of the Agricultural Experiment Station. 61, 1922+, are separately paged.

N.Y. Ent. Soc. Jour.=Journal of the New York Entomological Society. New York. 1, 1893+

N.Y. (Geneva) Agr. Expt. Sta. Tech. Bul.=Technical Bulletin. New York Agricultural Experiment Station. Geneva, N.Y. 1–93, 1906–1924. (94–288, 1923–1949, as "Technical Bulletin. New York State Agricultural Experiment Station.")

N.Y. State Agr. Soc. Trans.=Transactions of the New York State Agricultural Society. Albany. 1–34, 1842–1886 (1889).

> The Transactions for 1890–1900 appeared in the "Annual Report of the New York State Agricultural Society," 50–59. This was continued in 1910, as "Proceedings of the Annual Meetings of the New York State Agricultural Society," 70–76, 1910–1916, and 85–88, 1917–1920, which were published in the "Department of Agriculture Bulletin" (later "Agricultural Bulletin").

Prior to 1890, the "Annual Report" was somewhat irregular in publication, and at times was included in the early "Transactions." There was no printed report 60–69, 1900–1909, or 77–84.

N.Y. State Mus. Ann. Rpt.=New York State Museum. Annual Report. Albany. 43–72, 1889–1918 (1920). (1–5, 1847–1851, as "Annual Report of the Regents of the University, on the Condition of the State Cabinet of Natural History"; 6–23, 1852–1870, as "Annual Report of the Regents of the University of the State of New York, on the Condition . . . History"; 24–38, 1872–1884, as "Annual Report on the New York State Museum of Natural History"; 39–42, 1885–1888, as "Annual Report of the Trustees of the State Museum of Natural History.")

Rpt. 1, 1847 (1848) has separately paged sections. Rpts. 48, 1894 (1895) +, usually consist of 2 or more separately paged volumes, sometimes with a further subdivision, and include the "Memoirs" and the **N.Y. State Mus. Bul.** [q.v.] (nos. 12–208) as a number of variously paged appendices. From Rpt. 61, 1907 (1908) +, the "Annual Report" is made up entirely of these Memoirs and Bulletins. The University of the State of New York published all reports except nos. 58–66, 1906–1912 which were published by the New York State Education Department.

N.Y. State Mus. Bul.=New York State Museum Bulletin. Albany. 47–118, 1901–1908, and 181–351, 1916–1956. (1, 1892, and 8–46, 1889–1901, as **N.Y. State Mus., Bul.** [q.v.]; 2–7, 1887–1889, as "Bulletin of the New York State Museum of Natural History"; 119–180, 1908–1916, as **N.Y. State Mus., Mus. Bul.** [q.v.]; 352, 1956+, as "New York State Museum and Science Service Bulletin.")

Publication began with No. 2 in 1887; No. 1 was published in 1892. Nos. 1–54, 1887–1902, were assigned to "volumes" numbered 1–10, the bulletins within each volume being continuously paged (on nos. 3–7 and 47–54, the volume number is not shown). Nos. 55–57, 1912, are separately paged. Beginning with No. 52, 1902, each bulletin is also numbered in a series according to subject matter, e.g., Botany, Paleontology, Zoology, Entomology, etc. Previous bulletins were retroactively included in this numbering, which system continued through No. 266, 1925 (although only nos. 52–118, 1902–1908, were so marked on covers).

Nos. 12–208, 1895–1918, were published for later inclusion in volumes of **N.Y. State Mus. Ann. Rpt.** [q.v.], as one of the appendices, and were variously paged accordingly. (Generally, nos. 58–208, 1912–1918, were grouped according to subject, each subject being a separately paged appendix, the bulletins of that appendix usually being continuously paged.) Some Bulletins were published as double issues, e.g., 207/208, 1918.

Nos. 52–180, 1902–1916, were also included in the separate series "University of the State of New York Bulletin" (nos. 52–79, 1902–1904, and 165–180, 1913–1916), and "New York State Education Department Bulletin" (nos. 80–164, 1905–1913 [as "New York State University, Education Department Bulletin" for nos. 119-164, 1908-1913]), and were given additional bulletin numbers within those series.

N.Y. State Mus., Bul.=Bulletin of the New York State Museum. See **N.Y. State Mus. Bul.**

N.Y. State Mus., Mus. Bul.=New York State Museum. Museum Bulletin. See **N.Y. State Mus. Bul.**

Nat. Canad.=Le Naturaliste Canadien. Quebec. 1, 1868+ Vols. 21-56, 1894-1929, also called Ser. 2, 1-36; vol. 57, 1930+, also as Ser. 3, 1+. Vol. 6, nos. 4 and 6, 1874, were omitted in numbering.

Nat. Hist. Rev.=Natural History Review. Dublin; London. 1-7, 1854-1860; [N. ser.], 1-5, 1861-1865.

> Vols. 2-7, 1856-1860, have separately paged "Proceedings" and other separately paged sections. Subtitle of the first series varies; that of the new series is "A Quarterly Journal of Biological Science."

Natal Mus. Ann.=Annals of the Natal Museum. London; Dorking, England. 1, 1906+

Natl. Acad. Sci. Proc.=Proceedings of the National Academy of Sciences of the United States of America. Washington, D.C. 1, 1915+

> An earlier Proceedings was published in 1 volume, in 3 parts, 1877-1895.

Natl. Geog. Mag.=The National Geographic Magazine. Washington, D.C. 1, 1889+

Natl. Hort. Mag.=National Horticultural Magazine. Baltimore, Md. 1, 1922+

Naturf. Gesell. in Danzig, Schr.=Schriften der Naturforschenden Gesellschaft in Danzig. Danzig. N. ser., 1-20, 1863-1938. (1-6, 1820-1862, as "Neueste Schriften der Naturforschenden Gesellschaft zu Danzig.")

Naturf. Gesell. zu Leipzig, Schr.=Schriften der Naturforschenden Gesellschaft zu Leipzig. Leipzig. 1, 1822.

Naturhist. For. Kjøbenhavn, Vidensk. Meddel.=Videnskabelige Meddelelser fra den Naturhistorisk Forening i Kjøbenhavn. See **Dansk Naturhist. For. København, Vidensk. Meddel.**

Naturhist. Tidsskr.=Naturhistorisk Tidsskrift. Copenhagen. 1-4, 1837-1843; Ser. 2, 1-2, 1844-1849; Ser. 3, 1-14, 1861-1884.

Naturhist. Ver. der Preuss. Rheinlande u. Westfalen, Verhandl.= Verhandlungen des Naturhistorischen Vereines der Preussischen Rheinlande und Westfalens. Bonn. 6-41, 1849-1884, and 63-90, 1906-1933. (1-5, 1844-1848, as "Verhandlungen des Naturhistorischen Vereines der Preussischen Rheinlande"; 42-62, 1885-1905, as "Verhandlungen des Naturhistorischen Vereins der Preussischen Rheinlande, Westfalens und des Reg.-Bezirks Osnabrück"; 91, 1935+, as "Decheniana. Verhandlungen des Naturhistorischen Vereins der Rheinlande und Westfalens.")

> Vols. 11-20, 1854-1863, also called N. ser., 1-10; vols. 21-30, 1864-1873, also as Ser. 3, 1-10; vols. 31-40, 1874-1883, also as Ser. 4, 1-10; vols. 41-50, 1884-1893, also as Ser. 5, 1-10.
>
> Vols. 95-100, 1937-1941, were issued in 2 separately paged sections: A. Geologische Abteilung; and B. Biologische Abteilung.

Naturhist. Ver. in Augsburg, Ber.=Bericht des Naturhistorischen Vereins in Augsburg. Augsburg. 1-29, 1848-1885. (29-50, 1887-1933, as "Bericht des Naturwissenschaftlichen Vereins für Schwaben und Neuburg.")

> Superseded by "Die Schwäbische Naturkunde."

Naturw. Abhandl. aus Dorpat=Naturwissenschaftliche Abhandlungen aus Dorpat. Dorpat [=Tartu, Estonia]. 1, 1823.

Naturw. Ver. f. Sachsen u. Thüringen, Abhandl.=Abhandlungen des Naturwissenschaftlichen Vereins für Sachsen und Thüringen. Berlin. 1-2, 1856-1861.

Naturw. Ver. f. Steiermark, Mitt.=Mitteilungen des Naturwissenschaftlichen Vereins für Steiermark. Graz. 1, 1862+

Natuurhist. Genootsch. in Limburg, Jaarb.=Jaarboek van het Natuurhistorisch Genootschap in Limburg. See **Natuurhist. Maandbl.**

Natuurhist. Maandbl.=Natuurhistorisch Maandblad. Limburg. 13, 1924+ (1911, as "Mededeelingen van het Natuurhistorisch Genootschap in Limburg"; 1912-1923, as **Natuurhist. Genootschap in Limburg, Jaarb.** [q.v.].)

> No volume numbers used 1911-1923. The volume for 1912 is subtitled "Mededeelingen . . ."

Natuurk. Tijdschr. v. Nederland. Indië=Natuurkundig Tijdschrift voor Nederlandsch-Indië. Bogor, Bandung; Jakarta; The Hague, etc. 1-100, 1850-1940. (101-102, 1941-1946, as "Natuurwetenschappelijk Tijdschrift voor Nederlandsch-Indië"; 103-106, 1947-1950, as "Chronica Naturae" [with an additional Indonesian title]; 107, 1951+ as "Madjalak ilmu alam untuk Indonesia (Indonesian Journal for Natural Science)".)

> Vols. 4-10, 1853-1856, also called N. ser. 1-7; 11-14, 1856-1857, also as Ser. 3, 1-4; 15-20, 1858-1860, also as Ser. 4, 1-6; 21-25, 1860-1863, also as Ser. 5, 1-5; 26-30, 1864-1868, also as Ser. 6, 1-5; 31-39, 1870-1880, also as Ser. 7, 1-9; 40-51, 1881-1892, also as Ser. 8, 1-12; 52-56, 1893-1897, also as Ser. 9, 1-5; 57-66, 1898-1907, also as Ser. 10, 1-10; 67, 1908+, with no additional series numbers.

Nebr. Univ., Univ. Studies=The University Studies of the University of Nebraska. Lincoln. 1–40, 1888–1940. (41, 1941–1945, and N. ser., 1, 1946+, as "University of Nebraska Studies.")

> Old series was superseded by 3 separate series: "Studies in the Humanities," "Studies in Science and Technology," and "Studies in Social Science." These 3 then united with "Studies in language, literature, and criticism" to form the New Series above in 1946.
> Vols. 1–40 were issued in separately paged numbers, but with continuous pagination also given [except in vol. 40].

Neue Beitr. z. System. Insektenkunde=Neue Beiträge zur Systematische Insektenkunde. Berlin. 1–5, 1915–1932.

> A supplement to Ztschr. f. Wiss. Insektenbiol. [q.v.].

New England Zool. Club, Proc.=Proceedings of the New England Zoölogical Club. Cambridge, Mass. 1–24, 1899–1947.

New Zealand Inst. Trans. and Proc.=Transactions and Proceedings of the New Zealand Institute. Wellington. 1–63, 1868–1933. (64–79, 1935–1952, as "Transactions and Proceedings of the Royal Society of New Zealand"; 80–84, 1952–1958, as "Transactions of the Royal Society of New Zealand" [which includes the Proceedings]; 85, 1958+, split into "Transactions . . . New Zealand," and "Royal Society of New Zealand. Proceedings" [85–88, 1958–1960; "Proceedings of the Royal Society of New Zealand," 89, 1961+].)

Norges Svalbard- og Ishavs-Undersøk., Skr. om Svalbard og Ishavet= Skrifter om Svalbard og Ishavet. Norges Svalbard- og Ishavs-Undersøkelser, Oslo. 13–81, 1928–1940. (1–11, 1922–1929, as "Resultater av de Norske Statsunderstottede Spitsbergenekspeditioner [Skrifter om Svalbard og Ishavet]"; 12, 1927, as "Skrifter om Svalbard og Nordishavet"; 82–89, 1941–1947, as "Norges Svalbard- og Ishavs-Undersøkelser. Skrifter"; 90, 1948+, as "Norsk Polarinstitut. Skrifter.")

> Nos. 1–11 were published nonchronologically to form 1 volume with title indicated above.

Norsk Ent. Tidsskr.=Norsk Entomologisk Tidsskrift. Kristiania (later Oslo). 1, 1920+

North Amer. Fauna=North American Fauna. Washington, D.C. 1, 1889+

> Nos. 1–56, 1889–1938, were issued by the U.S. Department of Agriculture, Bureau of Biological Survey (early issues by Division of Ornithology and Mammalogy); 57, 1941+, are issued by U.S. Department of the Interior, Fish and Wildlife Service. Nos. 6 and 9 were never published.

Not. aus dem Geb. der Natur- u. Heilk. [Froriep's]=Notizen aus dem Gebiete der Natur- und Heilkunde. Erfurt; Weimar. 1–50, 1821–

1836; Ser. 3, 1–11, 1847–1849.   (Ser. 2, 1–40, 1837–1846, as "Neue Notizen . . . Heilkunde.")

> Superseded by "Tagsberichte über die Fortschritte der Natur- und Heilkunde." Sometimes known as "Froriep's Notizen."

Notulae Ent.=Notulae Entomologicae. Helsingfors. 1, 1921+

Noxious and Beneficial Insects of State of Ill., State Ent. Rpt.= Report of the State Entomologist on the Noxious and Beneficial Insects of the State of Illinois. Springfield; Champaign; Chicago. 6–29, 1876/1877–1916.   ([1], 1867/1868, as "First Annual Report on the Noxious Insects of the State of Illinois"; [2–4], 1871–1873, as "First (-Third) Annual Report on the Noxious Insects of the State of Illinois"; [5], 1874, as "Fourth Annual Report on the Noxious and Beneficial Insects of the State of Illinois.")

> The first Report was by Walsh; Reports 2–5, were by LeBaron (his Reports 1–4); Reports 6–11, by Thomas; and Reports 12–29, were by Forbes.

Nyt Mag. for Naturvidensk.=Nyt Magazin for Naturvidenskaberne. Oslo. 1–74, 1836–1934.   (75, 1936+, as "Nytt Magazin for Naturvidenskapens.")

> Vols. 21–26, 1875–1881, also called Ser. 2, 1–6; Vols. 27–32, 1882–1892, also as Ser. 3, 1–6.

Ohio Acad. Sci. Spec. Paper=Ohio Academy of Science. Special Paper. Columbus. 1–7, 1899–1903.

> Superseded by and continued in "Proceedings of the Ohio Academy of Science".

Ohio Agr. Expt. Sta. Tech. Ser. Bul.=Bulletin of the Ohio Agricultural Experiment Station. Technical Series. Wooster; Norwalk. 1–10, 1889–1932.

> Buls. 1–4, 1889–1896, are continuously paged and bound into "Volume 1." The rest of the numbers are separately paged.

Ohio Biol. Survey, Bul.=Ohio Biological Survey. Bulletin. Columbus. 1–8, 1913–1954 (=Buls. 1–43).   (N. ser., 1, 1959+, as "Bulletin of the Ohio Biological Survey.")

> The old series has continuously paged volumes, made up of a number of Bulletins, numbered consecutively 1–43, which are also assigned numbers within each volume. Bulletins 1–31 (=vol. 6, no. 2), 1913–1935, are also included and assigned numbers in the series "The Ohio State University Bulletin"; Buls. 32–43, 1936–1954, are included in the series "Ohio State University Studies Series." The volumes of the New Series are issued in separately paged numbers, and the numbers are only numbered consecutively within each volume.

Ohio Jour. Sci.=The Ohio Journal of Science. Columbus. 16, 1916+   (1 (1–2), 1900, as "O. S. U. Naturalist"; 1 (3)–15, 1900–1915, as **Ohio Nat.** [q.v.].)

Ohio Nat.=The Ohio Naturalist. Columbus. See **Ohio Jour. Sci.**
Ohio State Bd. Agr., Ann. Rpt.=Annual Report of the Ohio State Board of Agriculture. Columbus. 1–68, 1846–1913.
> Superseded and continued in various Ohio Department of Agriculture publications.

Ohio State Univ., Abs. Doctors' Diss.=Abstracts of Doctors' Dissertations. Ohio State University, Columbus. 1–25, 1929–1937. (26, 1938+, as "Abstracts of Doctoral Dissertations.")
Okla. Univ. Biol. Survey, Pubs.=Publications of the University of Oklahoma, Biological Survey. Norman, Okla. 1–5, 1929–1933.
> Vol. 1, nos. 1–3, 1929, are also numbered in the series "University of Oklahoma Bulletin."

Opusc. Ent. (Soc. Ent. Lund.)=Opuscula Entomologica. Societas Entomologica Lundensis, Lund. 1, 1936+
Opusc. Ent. (Soc. Ent. Lund.), Sup.=Opuscula Entomologica. Supplementum. Societas Entomologica Lundensis, Lund. 1, 1938+
Ottawa Nat.=The Ottawa Naturalist. Ottawa, Ont. 1–32, 1887–1919. (33, 1919/1920+, as "The Canadian Field-Naturalist.")
> Supersedes "Ottawa Field-Naturalists' Club, Transactions," and Vols. 1–33, 1887–1919/1920, are also subtitled "Transactions of the Ottawa Field-Naturalists' Club," vols. 3–35.

Pacific Insects=Pacific Insects. Honolulu. 1, 1959+
Pamiet. Fizyiograficzny=Pamietnik Fizyiograficzny. Warsaw. 1–27, 1881–1922.
> The numbers are separately paged from Vol. 5+

Pan Amer. Sanit. Bur., Bol. de la Ofic. Sanit. Panamer.=Boletín de la Oficina Sanitaria Panamericana. Pan American Sanitary Bureau, Washington, D.C. 3, 1923+ (1–2, 1922–1923, as "Boletín Pan-americano de Sanidad de la Oficina Sanitaria Internacional.")
Pan-Pacific Ent.=The Pan-Pacific Entomologist. San Francisco. 1, 1924+
Pará, Mus. Goeldi (Mus. Paraense) de Hist. Nat. e Ethnog., Mem.=Memorias do Museu Goeldi (Museu Paraense) de Historia Natural e Ethnographie. Pará, Brazil. 3–4, 1902–1905. (1–2, 1900, as "Memorias do Museu Paraense de Historia Natural e Ethnographie.")
Parasitology=Parasitology. London. 1, 1908+
[Paris] Acad. des Sci. Compt. Rend.=Comptes Rendus Hebdomadaires des Séances de l'Académie des Sciences. Paris. 1, 1835+
> Earlier the Comptes Rendus were in the Academy's "Procès-Verbaux des Séances de l'Académie tenus depuis la fondation de l'Institut jusqu'à mois d'Août 1835."

[Paris] Acad. Roy. des Sci., Hist. avec Mém. Math. et Phys.=
Histoire de l'Académie Royale des Sciences. Avec les Mémoires
de Mathématique et Physique. Paris. 1699–1788 (1732–1791).
(1789 [1793], as "Histoire de l'Académie des Sciences. Avec les
Mémoires . . ."; 1790 [1797], as "Mémoires de l'Académie des
Sciences.")

>The Histoire and the Mémoires are separately paged. No volume numbers were used.
>
>Supersedes "Mémoires de l'Académie Royale des Sciences, depuis 1666 jusqu'à 1699" (earlier "Histoire de l'Académie Royale des Sciences"). Superseded by "Mémoires de l'Institut National des Sciences et Arts. Sciences Mathématiques et Physiques" (later "Mémoires de l'Institut des Sciences, Lettres et Arts").

[Paris] Inst. de France, [Cl. des] Sci. Math. et Phys., Acad. Roy. des Sci., Mém. présentés par divers Savans=Mémoires présentés par divers Savans à l'Académie Royale des Sciences de l'Institut de France. Sciences Mathématiques et Physiques. Paris. [Ser. 2], 1–9, 1827–1846. ([Ser. 1], 1–2, 1806–1811, as "Mémoires présentés à l'Institut des Sciences, Lettres et Arts, par divers Savans et lus dans ses Assemblés. Sciences Mathématiques et Physiques"; [Ser. 2], 10–11, 1848–1851, as "Mémoires présentés par divers Savants à l'Académie des Sciences de l'Institut National de France. Sciences . . ."; [Ser. 2], 12–19, 1854–1865 (1868), as "Mémoires présentés . . . de l'Institut Impérial de France. Sciences . . ."; [Ser. 2], 20–35, 1872–1914, as "Mémoires présentés . . . de l'Institut de France. Sciences . . .")

>"Series 2" does not appear on the volumes until Vol. 22, 1876+, at which time the volumes began to be issued in separately paged numbers.

Paris, Mus. Natl. d'Hist. Nat. Bul.=Bulletin du Muséum National d'Histoire Naturelle. Paris. 13–34, 1907–1928; Ser. 2, 1, 1929+
 (1–12, 1895–1906, as "Bulletin du Muséum d'Histoire Naturelle.")

Paris, Mus. Natl. d'Hist. Nat. Mém.=Mémoires du Muséum National d'Histoire Naturelle. Paris. N. sér., 1–30, 1935–1951; N. ser. [=Ser. 3], in 3 series: A. Zoologie, 1, 1950+, B. Botanique, 1, 1950+, and C. Sciences de la Terre, 1, 1950+ ([Ser. 1], 1–20, 1815–1832, as "Mémoires du Muséum d'Histoire Naturelle.")

>The first series supersedes "Annales du Muséum d'Histoire Naturelle," and is superseded by "Nouvelles Annales . . . "

Philippine Jour. Sci.=The Philippine Journal of Science. Manila. 1, 1906+

>Vols. 2–4, 1907–1909, were published in 3 separately paged sections; vols. 5–13, 1910–1918, issued in 4 separately paged sections.

Phys. Ökonom. Gesell. Königsberg,Schr.=Schriften der Physikalisch-Ökonomischen Gesellschaft zu Königsberg in Pr. Königsberg. 1–72, 1860–1941.

Pittsburgh, Carnegie Inst., Carnegie Mus. Mem.=Memoirs of the Carnegie Museum. Carnegie Institute, Pittsburgh, Pa. 1–12, 1901–1936.

> Vol. 12, 1932–1936, was issued in separately paged parts, with a further breakdown into sections. Each number of the Memoirs is also assigned a Carnegie Institute publication number.

Pomona Col. Jour. Ent.=Pomona College Journal of Entomology. See Jour. Ent. and Zool.

Pomona Col., Laguna Mar. Lab., Ann. Rpt.=Annual Report of the Laguna Marine Laboratory at Laguna Beach, Orange County, California. Pomona College, Claremont, Calif. 1–6, 1912–1918.

Portici, R. Ist. Super. Agr., Lab. Zool. Gen. e Agr. Bol.=Bollettino del Laboratorio di Zoologia Generale e Agraria della R. Istituto Superiore Agrario in Portici. Portici. 21–30, 1928–1938. (1–20, 1907–1927, as Portici, R. Scuola Super. d'Agr., Lab. Zool. Gen. e Agr. Bol. [q.v.]; 31, 1939+, as "Bollettino de Laboratorio di Zoologia Generale e Agraria della Facoltà Agraria in Portici.")

> Suspended in 1944–1955.

Portici, R. Scuola Super. d'Agr., Lab. Zool. Gen. e Agr. Bol.=Bollettino del Laboratorio di Zoologia Generale e Agraria della R. Scuola Superiore d'Agricoltura in Portici. See Portici, R. Ist. Super. Agr., Lab. Zool. Gen. e Agr. Bol.

Prairie Farmer=(The) Prairie Farmer. Chicago; Indianapolis; New York. 3–17, 1843–1857, and 19, 1859+ (1–2, 1841–1842, as "The Union Agriculturist and Western Prairie Farmer"; 18, 1858, as "Emery's Journal of Agriculture and the Prairie Farmer.")

> Vols. 18–37, 1858–1868, also called N. ser., 2–21, continuing the volume numbers of "Emery's Journal of Agriculture" absorbed in 1858. Vol. 38 is omitted in numbering. Nos. 1–6 of vol. 54, 1882, are called "Peoples' Illustrated Weekly and the Prairie Farmer." Beginning with Vol. 98, 1926, an Indiana edition was also published; with Vol. 103, 1931, an Illinois edition; and with Vol. 117, 1945, a Michigan edition began publication; the main edition is called "General edition."

Psyche=Psyche. A Journal of Entomology. Cambridge, Mass. 1, 1874+

Puerto Rico Dept. Agr. Jour.=The Journal of the Department of Agriculture of Puerto Rico. See Puerto Rico Univ. Jour. Agr.

Puerto Rico Jour. Publ. Health and Trop. Med.=The Puerto Rico Journal of Public Health and Tropical Medicine. San Juan. 8–26, 1933–1950. (1–2, 1925–1927, as "The Porto Rico Health Review"; 3–4, 1927–1929, as "The Porto Rico Review of Public Health and Tropical Medicine"; 5–7, 1929–1932, as "The Porto Rico Journal of Public Health and Tropical Medicine.")

Puerto Rico Univ. Jour. Agr.=The Journal of Agriculture of the University of Puerto Rico. Rio Piedras. 18, 1934+ (1 (1), 1917, as "The Journal of the Board of Commissioners of Agriculture. Porto Rico"; 1 (2)–17, 1917–1933, as **Puerto Rico Dept. Agr. Jour.** [q.v.].)

Quebec Soc. Protect. Plants, Ann. Rpt.=Annual Report of the Quebec Society for the Protection of Plants. Quebec, Que. 13–27, 1920/1921–1934/1935. (1–12, 1908/1909–1919/1920, as "Annual Report of the Quebec Society for the Protection of Plants from Insects and Fungous Diseases"; 28, 1936/1943+, as "Report of the Quebec Society for the Protection of Plants.")

> Nos. 7–17, 1914/1915–1924/1925, were issued as supplements of the "Report of the Department of Agriculture, Province of Quebec." The numbers are sometimes accompanied by separately paged supplements.

Queensland Mus. Mem.=Members of the Queensland Museum. Brisbane. 1, 1912+

R. Accad. delle Sci. (Soc. R. Borbonica) Napoli, Rend.=Rendiconto della Reale Accademia delle Scienze. Società Reale Borbonica, Naples. N. Ser., 5–6, 1856–1857. (1, 1842, as "Rendiconto delle Adunanze e de' Lavori dell' Accademia delle Scienze sezione della Società Reale Borbonica di Napoli"; 2–5, 1843–1846, as "Rendiconto delle Adunanze e de' Lavori dell' Accademia . . . di Napoli"; 6–9, 1847–1849, as "Rendiconto delle Adunanze e de' Lavori dell' Accademia Napolitana della Scienze sezione della Società Reale Borbonica"; N. ser., 1, 1852, as "Rendiconto della Reale Accademia delle Scienze sezione della Società Reale Borbonica"; N. ser., 2–4, 1853–1855, as "Rendiconto della Società Reale Borbonica.")

> Superseded by "Rendiconto della Reale Accademia delle Scienze Fisiche e Matematische."

Rev. Brasil. de Biol.=Revista Brasileira de Biologia. Rio de Janeiro. 1, 1941+

Rev. Brasil. de Ent.=Revista Brasileira de Entomologia. São Paulo. 1, 1954+

Rev. Chilena de Hist. Nat. Pura y Apl.=Revista Chilena de Historia Natural Pura y Aplicada. Valparaiso. 28, 1924+ (1–27, 1897–1923, as "Revista Chilena de Historia Natural.")

> Vols. 16, 18, and 26 were never published.

Rev. d'Ent. (Caen)=Revue d'Entomologie. Caen. 1–28, 1882–1910.

Rev. de Biol. Trop.=Revista de Biologia Tropical. San José, Costa Rica. 1, 1953+

Rev. de Ent.=Revista de Entomologia. São Paulo; Rio de Janeiro. 1–22, 1931–1951.

Rev. de Sanid. y Asistencia Social (Caracas)=Revista de Sanidad y Asistencia Social. Caracas. 5, 1940+ (1–4, 1936–1939, as "Boletín. Ministerio de Sanidad y Asistencia Social, Caracas, Venezuela.")

Rev. de Zool. et Bot. Africaines=Revue de Zoologie et de Botanique Africaines. Brussels. 16, 1929+ (1–15, 1911–1928, as **Rev. Zool. Africaine** [q.v.].)

Rev. et Mag. de Zool.=Revue et Magasin de Zoologie Pure et Appliquée. Paris. Ser. 2, 1–23, 1849–1872; Ser. 3, 1–7, 1873–1879. (1–11, 1838–1848, as "Revue Zoologique, par la Société Cuvierienne".)

> Supersedes "Bulletin Zoologique." "Revue Zoologique" united with "Magasin de Zoologie" to form Ser. 2 above. Ser. 3 is also numbered as whole series vols. 36–[42] (except last 2 volumes).

Rev. Franç. d'Ent.=Revue Française d'Entomologie. Paris. 1, 1934+

Rev. Zool. Africaine=Revue Zoologique Africaine. See **Rev. de Zool. et Bot. Africaines.**

[Rio de Janeiro] Mus. Nac. do Rio de Janeiro, Arch.=Archivos do Museu Nacional do Rio de Janeiro. Rio de Janeiro. 1–28, 1876–1926. (29–36, 1927–1934, as "Archivos do Museu Nacional"; 37, 1943+, as "Arquivos do Museu Nacional.")

> Vol. 9, 1896, as "Revista do Museu Nacional do Rio de Janeiro."

[Rio de Janeiro] Mus. Nac. do Rio de Janeiro, Bol.=Boletim do Museu Nacional do Rio de Janeiro. Rio de Janeiro. 1–3, 1923–1927. (4–14/17, 1928–1938/1941, as "Boletim do Museu Nacional"; N. ser., as "Boletim do Museu Nacional" in 4 sections: Antropologia, 1, 1942+; Botanica, 1, 1942+; Geologia, 1, 1943+; and Zoologia, 1, 1942+).

> Vols. 2–5, 1925–1929, and the N. ser. are issued in separately paged numbers.

Rockefeller Inst. Med. Res. Monog.=Monographs of the Rockefeller Institute for Medical Research. New York. 1–23, 1910–1930.

Roy. Agr. Soc. England, Jour.=(The) Journal of the Royal Agricultural Society of England. London. 1–25, 1839–1864; Ser. 2, 1–25, 1865–1889; Ser. 3, 1–11, 1890–1900; 62, 1901+

Roy. Canad. Inst. Trans.=Transactions of the Royal Canadian Institute. Toronto. 11, 1915+ (1–10, 1889–1915, as "Transactions of the Canadian Institute.")

> Supersedes "Proceedings of the Canadian Institute."

Roy. Ent. Soc., London, Proc. Ser. A: Gen. Ent.=Proceedings of the Royal Entomological Society of London. Series A. General Entomology. London. 11, 1936+

> The Proceedings were not broken into series until 1936. Vols. 1–7, 1926–1932, as "Proceedings of the Entomological Society of London"; vols. 8–10, 1933–1936, as "Proceedings of the Royal . . .," which then continued in the 3 series, A (as above), B, and C. Previous to 1926, the Proceedings appeared in the Society's Transactions.

Roy. Ent. Soc., London, Proc. Ser. B: Taxonomy=Proceedings of the Royal Entomological Society of London. Series B. Taxonomy. London. 5, 1936+   (1–4, 1932–1935, as **Stylops** [q.v.].)

Roy. Ent. Soc. London, Trans.=Transactions of the Royal Entomological Society of London. London. 81, 1933+   (1–5, 1833–1849, and N. ser., 1–5, 1850–1861, and Ser. 3, 1–5, 1862–1867, and 1868–1925 [without volume numbers], and 74–80, 1926–1932, as **Ent. Soc. London, Trans.** [q.v.].)

Roy. Soc. Canada, Proc. and Trans.=Proceedings and Transactions of the Royal Society of Canada. Ottawa; Montreal. 1–12, 1882–1894; Ser. 2, 1–12, 1895–1906; Ser. 3, 1, 1907+

> From Ser. 2+, the journal is issued in 5 separately paged sections: 1. Littérature Française; 2. English Literature; 3. Mathematical, Physical and Chemical Sciences; 4. Geological Sciences; and 5. Biological Sciences.

Roy. Soc. London, Phil. Trans., Ser. B=Philosophical Transactions of the Royal Society of London. B: Containing papers of a biological character. London. 178, 1887+

> The "Philosophical Transactions" did not split into sections A and B until 1887. Vols. 1–65, 1665–1775, as "Philosophical Transactions" (with subtitle); 66–81, 1776–1791, and 1792–1852 (without volume numbers), and 143, 1853+, as "Philosophical Transactions of the Royal Society of London." From vol. 178, 1887+, it appeared in 2 sections: A. Containing papers of a mathematical or physical character; and B. as above.
>
> No volume numbers were used 1792–1852. Vols. 1–17, 21–25, 29–30, and 34–35, were issued as continuously paged volumes. Vols. 41–186, 1739–1895, were issued in 2 or more parts. There was an interruption in publication between vol. 12, 1679, and 13, 1682, during which time a periodical, "Philosophical Collections" (nos. 1–7, 1679–1682) was published.

Russ. Ent. Obshch. Trudy (Soc. Ent. Ross. Horae)=Trudy Russkago Entomologicheskago Obshchestva v' S.-Peterburg'. Horae Societatis Entomologicae Rossicae Variis Sermonibus in Rossia Usitatis Editae. St. Petersburg [=Petrograd, Leningrad]. 1–2, 1861–1863 (in 2 editions) and 17–42, 1882–1932.   (3–16, 1865–1881, as "Horae Societatis Entomologicae Rossicae.")

> The "Trudy" appeared as a separate publication during 1865–1882 (vols. 3–13), the 2 periodicals recombining in 1882 as above. Russian spelling varies slightly. Suspended 1917–1931. A supplement appeared irregularly with some of the volumes.

Saikyo Univ. (Kyoto), Sci. Rpt., Agr.=The Scientific Reports of the Saikyo University. Agriculture. Kyoto, Japan. 1, 1951+

> There is also a series on "Natural Science."

[São Paulo] Mus. Paulista, Rev.=Revista do Museu Paulista. São Paulo. 1–23, 1895–1938; N. ser., 1, 1947+

> The old series was superseded by **Arq. de Zool. de Estado de São Paulo** [q.v.].

(Sarajevo), Zemaljsk. Muz. u Bosni i Hercegovini, Glasnik=Glasnik Zemaljskog Muzeja u Bosni i Hercegovini. Sarajevo, Yugoslavia. 1–49, 1889–1937. [50–52], as 1–3, 1938–1940, as "Glasnik Zemalskog Muzeja Kraljevine Jugoslavije"; N. ser., 1, 1946, as "Glasnik Drzhavnog Muzeja u Sarajevu"; N. ser., 2, 1947+, as "Glasnik Zemaljskog Muzeja u Sarajevu.")

> Vols. 39–52, 1927–1940, in 2 sections: "I. Sveska za Prirodne Nauk," and "II. Sveska za Historiju i Etnografiju," separately paged from vol. 40, 1928+; N. ser. 9–12, 1954–1957, in 2 separate sections: "Arheologija" and "Istorija," and "Ethnografija." N. ser., 13, 1958+, in 3 separate sections: "Arheologija," "Istorija," and "Ethnografija." Titles also given in Russian. Spellings vary slightly.

Schles. Gesell. f. Vaterländ. Cult., Uebers. d. Arb. u. Veränderungen= Uebersicht der Arbeiten und Veränderungen der Schlesischen Gesellschaft für Vaterländische Cultur. Breslau. 1824–1949. (28–116, 1850–1943, as "Jahresbericht der Schlesischen Gesellschaft für Vaterländische Cultur.")

> No volume numbers used 1824–1849. Vols. 109–116, 1937–1943, are in 2 separate sections: "Naturwissenschaftliche-Medizinische Reihe," 1–8, 1937–1939, and "Geistwissenschaftliche Reihe," 1–6, 1937–1940.
> Supersedes "Correspondenzblatt der Schlesischen Gesellschaft . . ." The words "Cultur" and "Jahresbericht" are sometimes spelled "Kultur" and "Jahres-Bericht."

Sci. Agr.=Scientific Agriculture. La Revue Agronomique Canadienne. Ottawa. 1, 1921+

Science=Science. New York. 1–23, 1883–1894; N. ser., 1, 1895+

Scot. Nat.=The Scottish Naturalist. Edinburgh. 1912–1938, and 60, 1948+   (1892–1911, as **Ann. Scot. Nat. Hist.** [q.v.].)

> Volume numbering began with Vol. 60, 1948; suspended 1940–1947. Supersedes "The Scottish Naturalist and Journal of the Perthshire Society of Natural Science."

Senckenb. Naturf. Gesell. Abhandl.=Abhandlungen herausgegeben von der Senckenbergischen Naturforschenden Gesellschaft. Frankfurt a. M. 1–43, 1854–1933 (=Nos. 1–427).   (428, 1934+

as "Abhandlungen der Senckenbergischen Naturforschenden Gesellschaft.")

> Nos. 1–427, 1854–1933, are continuously paged within each volume, although consecutively numbered. Nos. 428, 1934+ are each separately paged and are not assigned to a volume.

Senckenbergiana=Senckenbergiana. Wissenschaftliche Mitteilungen herausgegeben von der Senckenbergischen Naturforschenden Gesellschaft in Frankfurt a. M. Frankfurt. 1–34, 1918–1953. (Continues in 2 separate series: "Senckenbergiana Biologica," 35, 1954+, and "Senckenbergiana Lethaea," 35, 1954+).

[Shanghai] Mus. Heude, Notes d'Ent. Chin.=Notes d'Entomologie Chinoise. Musée Heude, Université l'Aurore, Shanghai. 1, 1929+

> One of a series of "Notes." Vol. 12, 1948, is the last volume seen by us.

Siruna Seva=Siruna Seva. Blätter für Fruchtfliegen-Kunde. Berlin. 1, 1940+

Smithsn. Inst. Ann. Rpt.=Annual Report of the Board of Regents of the Smithsonian Institution. Washington, D.C. 3–10, 1849–1855, and 1856+ (1847, as "Report of the Board of Regents"; 1848, as "Report of the Board of Regents of the Smithsonian Institution.")

> No volume numbers used in first 2 reports or from the 11th on. From 1884–1950, the report was in 2 annual volumes, the second volume being subtitled "Report of the United States National Museum," 1884–1906 (called "Part 2" 1884–1887), and "Report on the Progress and Condition of the United States National Museum," 1907–1950; this continues separately as "The United States National Museum, Annual Report," 1951+

Smithsn. Inst. Pub.=Smithsonian Institution Publication. Washington, D.C. 1, 1848+

> Most publications are included in one of the other series of the Smithsonian Institution, but a few were issued irregularly as separates and not included in any other series.

Smithsn. Inst., Smithsn. Misc. Collect.=Smithsonian Miscellaneous Collections. Smithsonian Institution, Washington. D.C. 1, 1862 (1852)+

> Most volumes consist of a number of separately paged (except as shown below) miscellaneous publications which are issued previously as separates, each assigned an overall **Smithsn. Inst. Pub.** [q.v.] number. Many volumes, however, are issued as whole volumes, and a few in two parts. All volumes are also assigned publication numbers when completed. Vols. 1–4, all dated 1862, consist of previously published separates dating as far back as 1852. Publication numbering began with vol. 5, 1864, in which previous Miscellaneous Collections were numbered retroactively. There is no indication on the separate numbers of vols. 1–43, 1862–1901, as to which volume they belong. The separate numbers of vols. 44–55, 1904–1910, have the indication "Part of

vol.—"; thereafter (vol. 56, 1912+) each separate upon its publication is assigned a specific number within a volume.

Vols. 45 (1903), 47-48 (1905), 50 and 52 (both 1908) make up vols. 1-5 of a "Quarterly Issue," which contain continuously paged material, each article of which is issued first as a separate, and then combined into quarterly numbers, and finally into volumes; all 3 forms (i.e., separates, numbers, and volumes) are also assigned Publication numbers. Vol. 67 (1924) and 75 (1928) also consist of continuously paged numbers, but are not called "Quarterly Issue."

So. African Mus. Ann.=Annals of the South African Museum. London; Edinburgh; Cape Town. 1, 1898+

Soc. Agr. e Accad. delle Sci. dell' Ist. Bologna, Nuovi Ann. delle Sci. Nat. e Rend.=Nuovi Annali delle Scienze Naturali e Rendiconto delle Sessioni della Società Agragria, e dell' Accademia delle Scienze dell' Istituto di Bologna. See **Accad. delle Sci. dell' Ist. e Soc. Agr. Bologna, Nuovi Ann. delle Sci. Nat. e Rend.**

Soc. Bibliog. Nat. Hist. Jour.=The Journal of the Society for the Bibliography of Natural History. London. 1, 1936+

Soc. Brit. Ent. Jour.=The Journal of the Society for British Entomology. Southampton. 1, 1934+

Supersedes "Journal of the Entomological Society of the South of England."

Soc. Brit. Ent. Trans.=Transactions of the Society for British Entomology. Southampton. 1, 1934+

Supersedes **Ent. Soc. So. England, Trans.** [q.v.].

Soc. Cient. Argentina, An.=Anales de la Sociedad Cientifica Argentina. Buenos Aires. 1, 1876+

Soc. d'Agr. du Dépt. de la Seine, Mém. d'Agr. d'Écon. Rurale et Dom.=Mémoires d'Agriculture, d'Économie Rurale et Domestique, publiés par la Société d'Agriculture du Départment de la Seine. Paris. 1-16 [=whole ser., 29-44], 1801-1813. [1], 1761, as "Recueil contenant les déliberations de la Société Royale d'Agriculture de la généralité de Paris"; [2], 1762-1784, as [Renfermant divers Mémoires publiés par la Société Royale d'Agriculture de Paris de 1762 à 1784]; [3-27], 1785-1791, as "Mémoires d'Agriculture, d'Économie Rurale et Domestique publiés par la Société Royal d'Agriculture de Paris"; [28], 1799, as "Compte Rendu à la Société Royal d'Agriculture de Paris"; [45-87], 1814-1847, as "Mémoires d'Agriculture publiés par la Société Royale et Centrale d'Agriculture du Départment de la Seine"; [88-93], 1848-1852, as "Mémoires d'Agriculture publiés par la Société Nationale et Centrale d'Agriculture du Départment de la Seine"; [94-115], 1853-1869, as "Mémoires d'Agriculture publiés par la Société Imperiale et Centrale d'Agriculture de France"; [116-117], 1870-1872, as "Mémoires d'Agriculture publiés par la Société Centrale d'Agriculture de France"; [118–124], 1873–1876, as "Mémoires publiés par la Société Centrale d'Agriculture de France"; [125]–144, 1877–1916, as "Mémoires publiés par la Société Nationale d'Agriculture de France.")

Volume numbers were not given until Vol. 126, 1881 (except 1801–1813).
Superseded by "Travaux et Notices publiés par l'Académie d'Agriculture de France."

Soc. d'Amateurs des Sci., de l'Agr. et des Arts, Lille, Rec. des Trav.=Recueil des Travaux de la Société d'Amateurs des Sciences, de l'Agriculture et des Arts à Lille. See **Soc. des Sci., de l'Agr. et des Arts, Lille, Mém.**

Soc. d'Hist. Nat. de la Moselle, Bul.=Bulletin de la Société d' Histoire Naturelle de la Moselle. Metz. 29, 1921+ (1, 1843, as "Mémoires de la Société d'Histoire Naturelle de la Départmente de la Moselle"; 2–12, 1844–1870, as "Bulletin de la Société d' Histoire Naturelle du Départmente de la Moselle"; 13–28, 1871–1913, as **Soc. d'Hist. Nat. de Metz, Bul.** [q.v.].)

Vols. 13–24, 1874–1905, also called Ser. 2, 1–12; 25–35, 1908–1938, also as Ser. 3, 1–11. Suspended 1914–1920 and 1939–1949.

Soc. d'Hist. Nat. de Metz, Bul.=Bulletin de la Société d'Histoire Naturelle de Metz. See **Soc. d'Hist. Nat. de la Moselle, Bul.**

Soc. d'Hist. Nat. de Paris, Actes=Actes de la Société d'Histoire Naturelle de Paris. Paris. 1, 1792.

Soc. d'Hist. Nat. de Paris, Mém.=Mémoires de la Société d'Histoire Naturelle de Paris. Paris. 1–5, 1823–1834.

There was an earlier journal of the same name, 1, 1799.

Soc. de Biol. [Paris] Compt. Rend.=Comptes Rendus Hebdomadaires des Séances et Mémoires de la Société de Biologie. Paris. 36–135, 1888–1941. (Ser. 1, 1–5, 1849–1853, Ser. 2, 1–5, 1854–1858, and Ser. 3, 1–5, 1859–1863, and 16–35, 1864–1884, as "Comptes Rendus des Séances et Mémoires de la Société de Biologie"; 136, 1942+, as "Comptes Rendus Hebdomadaires des Séances de la Société de Biologie.")

Vols. 16–20, 1864–1868, also called Ser. 4, 1–5; vols 21–25, 1869–1873, also as Ser. 5, 1–5; vols. 26–30, 1874–1878, also as Ser. 6, 1–5; vols. 31–35, 1879–1884, also as Ser. 7, 1–5; 36–40, 1884–1888, also as Ser. 8, 1–5; vols. 41–45, 1889–1893, also as Ser. 9, 1–5; 46–50, 1894–1898, also as Ser. 10, 1–5; vol. 51, 1899, also as Ser. 11, 1; vols. 52, 1900+, without additional series numbers.

The Comptes Rendus and the Mémoires formed 2 separately paged sections in vols. 1–46, 1849–1894, and the volumes thereafter were continuously paged.

The volumes for 1904–1940 (vols. 56–134) were also assigned Année numbers (An. 56–92), sometimes more than 1 volume (with same Année number) being issued per year. Thereafter (Vol. 135, 1941+) the Année numbers were dropped and there was but 1 volume issued per year, continuously paged, although sometimes in 2 parts.

Soc. des Sci., de l'Agr. et des Arts, Lille, Mém.=Mémoires de la Société des Sciences, de l'Agriculture et des Arts de Lille. Lille. 1848–1849, and Ser. 3, 8–14, 1870–1874, Ser. 4, 1–22, 1876–1895, and Ser. 5, 1895+ (1806–1819, as "Séances publiques de la Société d'Amateurs des Sciences et Arts de la Ville de Lille"; 1819–1824, as **Soc. d'Amateurs des Sci., de l'Agr. et des Arts, Lille, Rec. des Trav.** [q.v.]; 1825–1827, as **Soc. des Sci., de l'Agr. et des Arts, Lille, Rec. des Trav.** [q.v.]; 1827–1847, as **Soc. Roy. des Sci., de l'Agr. et des Arts, Lille, Mém.** [q.v.]; 1850–1851, as **Soc. Natl. des Sci., de l'Agr. et des Arts, Lille, Mém.** [q.v.]; 1852–1853, and Ser. 2, 1–10, 1854–1863, and Ser. 3, 1–7, 1864–1869, as **Soc. Imp. des Sci., de l'Agr. et des Arts, Lille, Mém.** [q. v.].)

> The first series consists of 34 volumes, the 5 separately paged parts of the "Séances . . . ," 1806–1819, being considered as Vol. 1, with 4 volumes of the "Recueil des Travaux," 1819–1827, and 29 volumes of the "Mémoires," 1827–1853. (The volumes [=Années] of the old series were without volume numbers, and often covered more than a single year.) Some of the volumes were issued in separately paged parts. Beginning with Ser. 5, the Mémoires were issued in the form of monographs, with special title page and a half title for the series name.

Soc. des Sci., de l'Agr. et des Arts, Lille, Rec. des Trav.=Recueil des Travaux de la Société des Sciences, de l'Agriculture et des Arts de Lille. See **Soc. des Sci., de l'Agr. et des Arts, Lille, Mém.**

Soc. des Sci. Hist. et Nat. de l'Yonne, Bul.=Bulletin de la Société des Sciences Historiques et Naturelles de l'Yonne. Auxerre. 1, 1847+

> Vols. 21–32, 1868–1878, also called Ser. 2, 1–12; vols. 33–50, 1879–1896, also as Ser. 3, 1–20 (irregularly numbered); vols. 51–70, 1897–1916, also as Ser. 4, 1–20; vols. 71–90, 1917–1936, also as Ser. 5, 1–20; Vols. 91, 1937+, without additional series numbers. Vols. 89 (1935) and 92 (1938) consist of 2 separately paged volumes each (the second one of vol. 89 is erroneously called vol. 90).

Soc. Ent. Argentina, Rev.=Revista de la Sociedad Entomológica Argentina. Buenos Aires. 1, 1926+

Soc. Ent. de Belg. Ann.=Annales de la Société Entomologique de Belgique. See **Soc. Roy. d'Ent. de Belg. Bul. et Ann.**

Soc. Ent. de Belg. Bul.=Bulletin de la Société Entomologique de Belgique. Bruxelles. 1–6, 1919–1924.

> Previously in **Soc. Ent. de Belg. Ann.** [q.v.] and recombined with it to form **Soc. Ent. de Belg. Bul. et Ann.** [q. v.].

Soc. Ent. de Belg. et Ann.=Bulletin & Annales de la Société Entomologique de Belgique. See **Soc. Roy. d'Ent. de Belg. Bul. et Ann.**

Soc. Ent. de France, Ann.=Annales de la Société Entomologique de France. Paris. 1–11, 1832–1842; Ser. 2, 1–10, 1843–1852; Ser. 3,

1-8, 1853-1860; Ser. 4, 1-10, 1861-1870; Ser. 5, 1-10, 1871-1880; Ser. 6, 1-10, 1881-1890; 60, 1891+

> Volumes 1-64, 1832-1895, include a separately paged "Bulletin Entomologique" (vol. 1—Ser. 2, vol. 7, 1832-1849, and Ser. 4, vol. 1—Ser. 5, vol. 2, 1861-1872), "Bulletins Trimestriels" (Ser. 2, vol. 8—Ser. 4, vol. 10, 1850-1860), and "Bulletin des Séances (Ser. 5, vol. 3—vol. 64, 1873-1895). This portion began separate publications in 1896 as **Soc. Ent. de France, Bul.** [q.v.].

Soc. Ent. de France, Bul.=Bulletin de la Société Entomologique de France. Paris. 1896-1931 (without volume numbers) and 37, 1932+

> Previous to 1896, the Bulletin was a separately paged section of **Soc. Ent. de France, Ann.** [q.v.].

Soc. Ent. Ital. Bul.=Bullettino della Società Entomologica Italiana. Firenze. 1-54, 1869-1922. (55, 1923+, as "Bollettino della Società Entomologica Italiana.")

Soc. Fouad 1er d'Ent. Bul.=Bulletin de la Société Fouad 1er d'Entomologie. Cairo. 22-38, 1939-1954. (1-6, 1907-1922, and 39, 1955+, as "Bulletin de la Société Entomologique d'Egypte"; 7-21, 1923-1937, as **Soc. Roy. Ent. d'Egypte, Bul.** [q. v.].)

> A "new series" was begun with vol. 11, 1927, but volume numbers are continuous.

Soc. Imp. des Sci., de l'Agr. et des Arts, Lille, Mém.=Mémoires de la Société Impériale des Sciences, de l'Agriculture et des Arts de Lille. See **Soc. des Sci., de l'Agr. et des Arts, Lille, Mém.**

Soc. Ital. di Sci. Nat. Atti=Atti della Società Italiana di Scienze Naturali. Milano. 1, 1855+

Soc. Linn. de Lyon, Ann.=Annales de la Société Linnéenne de Lyon. Lyon. [1-4], 1826-1852; Ser. 2, 1-80, 1852-1936.

> No volume numbers were used in the first series.

Soc. Linn. de Lyon, Bul. Bi-Mens.=Bulletin Bi-Mensuel de la Société Linnéenne de Lyon. Lyon. 1-10, 1922-1931.

> Superseded by "Bulletin Mensuel de la Société . . ."

Soc. Mex. de Hist. Nat. Rev.=Revista de la Sociedad Mexicana de Historia Natural. Mexico City. 1, 1939+

Soc. Natl. des Sci., de l'Agr. et des Arts, Lille, Mém.=Mémoires de la Société Nationale des Sciences, de l'Agriculture et des Arts de Lille. See **Soc. des Sci., de l'Agr. et des Arts, Lille, Mém.**

Soc. Natl. des Sci. Nat. et Math. de Cherbourg, Mém.=Mémoires de la Société Nationale des Sciences Naturelles et Mathématiques de Cherbourg. Paris; Cherbourg; St. Lo. 22, 1879+ (1-2, 1852-1854, as "Mémoires de la Société de Sciences Naturelles de Cherbourg"; 3-15, 1855-1870, as "Mémoires de la Société Impériale des Sciences Naturelles de Cherbourg"; 16–21, 1871–1878, as "Mémoires de la Société Nationale des Sciences Naturelles de Cherbourg.")

Vols. 11–20, 1865–1876, also called Ser. 2, 1–10; vols. 21–30, 1877–1897, also as Ser. 3, 1–10; vols. 31–40, 1897–1928, also as Ser. 4, 1–10; 41, 1929+, also as Ser. 5, 1+   Vol. 42, 1933/1936, is the last seen by us. Continues?

Soc. pro Fauna et Flora Fenn. Acta=Acta Societas pro Fauna et Flora Fennica. Helsingfors. 1, 1875+

Issued in separately paged numbers.

Soc. pro Fauna et Flora Fenn., Meddel.=Meddelanden af Societas pro Fauna et Flora Fennica. Helsingfors. 1–50, 1873–1924.

Superseded by "Memoranda Societatis pro Fauna . . . "

Soc. Roy. d'Ent. de Belg. Bul. et Ann.=Bulletin & Annales de la Société Royale d'Entomologique de Belgique. Bruxelles. 91, 1955+   (1–7, 1857–1863, as "Annales de la Société Entomologique Belge"; 8–64, 1864–1924, as **Soc. Ent. de Belg. Ann.** [q.v.]; 65–90, 1925–1954, as **Soc. Ent. de Belg. Bul. et Ann.** [q.v.].)

Vols. 10–22, 1866–1879, have a separately paged "Comptes-Rendus de Séances", and vols. 28–35, 1884–1891, a separately paged "Bulletin"; the material involved in these sections is continuously paged in the other volumes and later was issued in a separate publication. **Soc. Ent. de Belg. Bul.** [q.v.], which later recombined with the present periodical.

Soc. Roy. des Sci., de l'Agr. et des Arts, Lille, Mém.=Mémoires de la Société Royale des Sciences, de l'Agriculture et des Arts de Lille. See **Soc. des Sci., de l'Agr. et des Arts, Lille, Mém.**

Soc. Roy. Ent. d'Egypte, Bul.=Bulletin de la Société Royale Entomologique d'Egypte. Cairo. 7–21, 1923–1937.   (1–6, 1907–1922, and 39, 1955+, as "Bulletin de la Société Entomologique d'Egypte"; 22–38, 1939–1954, as "Bulletin de la Société Fouad 1er d'Entomologie.")

A "new series" was begun with vol. 11, 1927, but volume numbers are continuous.

Soc. Sci. de Bruxelles, Ann.=Annales de la Société Scientifique de Bruxelles. Bruxelles. 1, 1875+

Vols. 1–46, 1875–1926, without series; vols. 47–56, 1926–1936, in 4 series: A. Sciences Mathématiques; B. Sciences Physiques et Naturelles; C. Sciences Médicales; and D. Sciences Agronomiques et Techniques; vols. 57, 1937+, in 3 series: 1. Sciences Mathématiques et Physiques; 2. Sciences Naturelles et Médicales; and 3. Sciences Économiques. Vols. 1–46 consist of separately paged "Mémoires" and "Documents et Comptes Rendus."

Soc. Zool. de France, Bul.=Bulletin de la Société Zoologique de France. Paris. 1, 1876+

Soc. Zool. de France, Mém.=Mémoires de la Société Zoologique de France. Paris. 1–29, 1888–1932.

South. Calif. Acad. Sci. Bul.=Bulletin of the Southern California Academy of Sciences. Los Angeles. 1, 1902+

Southwest. Nat.=The Southwestern Naturalist. Dallas, Tex. 1, 1956+

Spolia Zeylanica=Spolia Zeylanica. Colombo, Ceylon. 1–12, 1903–1924, and 23, 1944+ (13–22, 1924–1940, as "Spolia Zeylanica. The Ceylon Journal of Science. Section B. Zoology and Geology.")

> The "Ceylon Journal of Science," Sections A–G, was formed of a number of independent journals in 1924. All sections continue. However, "Spolia Zeylanica" which joined in 1924, as Section B, broke away in 1944.

Springfield [Mass.] Mus. Nat. Hist. Bul.=Springfield Museum of Natural History. Bulletin. Springfield, Mass. 1–2, 1904–1910. (3, 1924+ as "Museum of Natural History. Springfield, Massachusetts. Bulletin.")

> Issued irregularly, No. 3 being the last known to us.

Stettin Ent. Ztg.=Entomologische Zeitung. See **Stettin. Ent. Ztg.**

Stettin. Ent. Ztg.=Stettiner Entomologische Zeitung. Stettin. 55, 1894+ (1–54, 1840–1893, as **Stettin Ent. Ztg.** [q.v.] [vols. 55–75, 1894–1914, under both titles].)

Studi Ent. (Turin)=Studi Entomologici. Turin. 1, 1848.

> Published by Baudi and Truqui, whose names are sometimes used in referring to this periodical.

Studia Ent. (Rio de Janeiro)=Studia Entomologica. Rio de Janeiro. 1–3, 1952–1955; N. ser., 1, 1958+

Stylops=Stylops. A Journal of Taxonomic Entomology. See **Roy. Ent. Soc., London, Proc. Ser. B: Taxonomy.**

Suom. Eläin-ja Kasvitiet. Seura Vanamo, Eläintiet. Julkaisu. (Soc. Zool. Bot. Fenn. Vanamo, Ann. Zool.)=Suomalaisen Eläin-ja Kasvitieteellisen Seuran Vanamon Eläintieteellisiä Julkaisuja (Annales Zoologici Societatis Zoolog.-Botanicae Fennicae Vanamo). Helsinki. 1, 1932+

> Supersedes "Suomalaisen . . . Seuran Vanamon Julkaisuja (Annales Societatis . . . Vanamo)," and runs concurrently with "Suomalaisen . . . Seuran Vanamon Kasvitieteellisiä Julkaisuja (Annales Botanici Societatis . . . Vanamo)." These journals are issued in separately paged numbers.

Sup. Ent.=Supplementa Entomologica. Berlin. 1–17, 1912–1929.

Sweden, Fishery Bd., Inst. Freshwater Res. (Drottningholm), Rpt.=Institute of Freshwater Research. Drottningholm. Report. Fishery Board of Sweden. Stockholm; Lund. 29, 1949+ (1–28, 1933–1947, as "Meddelanden från Statens Undersöknings-och Försöksanstalt för Sötvattensfisket (Mitteilungen der Anstalt für Binnenfischerei bei Drottningholm)".)

System. Zool.=Systematic Zoology. Washington D.C. 1, 1952+

[Tartu, Ülikooli (Dorpat Univ.)] Naturf. Gesell. bei der Univ. Dorpat, Sitzber.=Sitzungsberichte der Naturforscher-Gesellschaft bei der Universität Dorpat. Tartu, Estonia. 5–10 (1), 1881–1893. ([1]–4, 1853/1860–1878, as "Sitzungsberichte der Naturforscher-Gesellschaft zu Dorpat"; 10 (2)–12, 1894–1900, as "Sitzungsberichte der Naturforscher-Gesellschaft bei der Universität Jurjeff (Dorpat) (Protokoly Obshchestvo Estestvoispytatelei pri Imperatorskomu Yur'evskomu Universitet)"; 13–23, 1901–1914, as "Sitzungsberichte . . . Universität Jurjew (Dorpat) (Protokoly . . .)"; 24–40, 1915–1933, as "Tartu Ülikooli juures oleva Loodusuurijate Seltsi Aruanded (Sitzungsberichte der Naturforscher-Gesellschaft bei der Universität Dorpat)"; 41–46, 1934–1940, as "Tartu . . . Aruanded (Annales Societatis rebus Naturae Investigandis in Universitate Dorpatiensi Constitutae)"; [Vol. 47 not seen]; 48, 1955+, as "Loodusuurijate Seltsi Aastaraamat.")

> Vol. 1, 1853–1860, is not numbered. The numbers of Vols. 1–4 are called "Sitzungsberichte der Dorpater Naturforscher-Gesellschaft." The spelling of the Universität Dorpat of the German subtitle of vols. 24–40, changes to "Tartu" for vols. 32–40, 1925–1933.

Tashkent, Uzbek. Opytn. Sta. Zashch. Rast. [Trudy]=Uzbekstanskai Opytnaia Stantsia Zashchity Rastenii [Publications]. Tashkent. [2]–20, 1925–1930. ([1], 1925, as "Sredne-Aziatskaia Opytnaia Stantsia Zashchity Rastenii [Trudy]"; 21–24, 1931, as "Sredne-Aziatskii Institut Zashchity Rastenii [Trudy]"; 25, 1931, as "Sredne-Aziatskii Issledovatel'ski Institut Zashchity Rastenii [Trudy]".)

> Nos. 1–9, 1925–1927, were not numbered; numbers were assigned them on cover of No. 11. Nos. 1–11 are usually classed as separates by libraries.

Természet. Füzetek=Természetrajzi Füzetek. Budapest. 1–25, 1877–1902.

> Superseded by **Budapest Magyar Nemzeti Muz., Ann. Hist. Nat.** [q.v.].

Tex. Univ. Pub.=The University of Texas Publications. Austin, Tex. 3801, 1938+ (1–379, 1901–1914, and 1–72, 1915, as "Bulletin of the University of Texas"; 1–72, 1916, and 1701–3772, 1917–1937, as "University of Texas Bulletin.")

> Many Bulletins and Publications also form part of other series of journals of the University of Texas, but many were also issued as separates. For 1901–1914, the Bulletins were numbered consecutively; 1915–1916, they were numbered consecutively within each year; for 1917+, the first 2 digits of the Bulletin or Publication numbers indicate the year of issue and the last 2 digits show the position in the yearly series. (Since the numbers are assigned to various branches and departments of the University at the beginning of each year, they are not chronological and some numbers may be missing in the yearly series.)

Tijdschr. v. Ent.=Tijdschrift voor Entomologie. Amsterdam; The Hague. 1 (2), 1857+ (1 (1), 1857, as "Mémoires d'Entomologie.")

> Vols. 9-16, 1865-1873, also called Ser. 2, 1-8. A separately paged supplement sometimes accompanies a volume.

Tohôku Imp. Univ., Col. Agr., Jour.=Journal of the College of Agriculture. Tohôku Imperial University. See **Hokkaido Imp. Univ., Faculty Agr., Jour.**

Toronto Univ. Studies, Biol. Ser.=University of Toronto Studies. Biological Series. Toronto. 1-58, 1898-1948. (59, 1951+, as "University of Toronto. Biological Series.")

> This is one of many series of the "Studies;" it supersedes the "University of Toronto, Department of Biology, Publications from the Biological Laboratory." The series "Publications of the Ontario Fisheries Research Laboratory," 1, 1922+, is contained in the "Biological Series."

Torrey Bot. Club, Bul.=Bulletin of the Torrey Botanical Club. New York; Menasha, Wis. 1, 1870+

Towar. Nauk. Warszawsk., Arch. Nauk Biol.=Archiwum Nauk Biologicznych Towarzystiva Naukowego Warszawskiego. Lwów; Warsaw. 1, 1921+

> The volumes consist of separately paged numbers. Vol. 1 has an additional Latin title "Disciplinarum Biologicarum Archivum Societatis Scientiarum Varsaviensis"; in Vols. 2, 1929+, the additional title is in French as "Archives de Biologie de la Société des Sciences et des Lettres de Varsovie." Vol. 8, no. 2, 1938, is the last seen by us. Continues?

Towar. Nauk. Warszawsk., Sprawozd., Wyzd. 4=Sprawozdania z Posiedzeń Towarzystwa Naukowego Warszawskiego. Wyzdiał IV. Nauk Biologiczynch. Lwów; Warsaw. 22, 1929+

> Beginning with the 1950 volume, Wyzdiał 4 also includes the separately issued "Nauk Technicznych," 1, 1950+
> Vols. 1-5, 1908-1912, of the "Sprawozdania" were without class breakdown; vols. 6-11, 1913-1918, and 19-21, 1926-1928, appeared in 3 classes: Wyzdiał I, II, and III; vol. 12/18, 1918-1925, was a synoptic issue for the years 1918-1925 published in 1927; vols. 22, 1929+ are issued in 4 classes: Wyzdiał I, II, III, and IV (the latter, as above, previously combined with Wyzd. 3). The volumes also have the French title "Comptes Rendus des Séances de la Société et des Lettres de Varsovie" (Wyzdiał="Classe").

Tromsø Mus. Årsh.=Tromsø Museums Årshefter. Tromsø, Norway. 41-70, 1918-1947 (1950). (1-40, 1878-1917, as "Tromsø Museums Aarshefter.")

> Superseded by "Acta Borealia" in 2 series. Vols. 40, 1917+, were issued in separately paged numbers; from Vol. 51, 1928+, the numbers were also numbered consecutively in 2 series: Humanistisk Avd. (No. 4 called Kulturhistorisk Avd.) and Naturhistorisk Avd.

[Turin Univ.] Mus. Zool. ed Anat. Comp., Bol.=Bollettino dei Musei

di Zoologia ed Anatomia Comparata della R. Università di Torino.
Turin University, Turin. 1–49, 1886–1942.

> Superseded by "Bolletino dell'Istituto e Museo di Zoologia della Università di Torino." Vols. 1–27, 1886–1922, were issued in separately paged numbers, numbered serially 1–746; vols. 38–40, 1923–1925, were issued in separately paged numbers, numbered N. ser., 1–38; vols. 41–48, 1926–1940, were issued in continuously paged numbers, numbered Ser. 3, 1–115; and vol. 49, 1941/1942, was issued in continuously paged numbers, Ser. 4, 116–130.

U.S. Comn. Fish and Fisheries, Rpt. of Comnr.=Report of the Commissioner. United States Commission of Fish and Fisheries (1871–1903); Dept. of Commerce and Labor (1904–1921); Dept. of Commerce (1922–1927); U.S. Dept. of Commerce (1928–1939); U.S. Dept. of the Interior (1940). Washington, D.C. 2–29, 1872–1903. ([1], 1871/1872, as "Report on the conditions of the sea fisheries of the south coast of New England for 1871 and 1872"; 1904, as "Report of the Bureau of Fisheries"; 1905–1912, as "Report of the Commissioner of Fisheries. Bureau of Fisheries"; 1913–1940 (1950), as "Report of the United States Commissioner of Fisheries. Bureau of Fisheries.")

> Reports for 1872–1903 are called "Parts" 2–29; other reports are not numbered. Reports for 1905–1923 contain separately paged "special papers" and appendices. Ceases with the 1940 report?

U.S. Dept. Agr., Agr. Handb.=Agriculture Handbook. United States Department of Agriculture, Washington, D.C. 1, 1950+

U.S. Dept. Agr., Bur. Ent., Bul.=Bulletin. United States Department of Agriculture. Bureau of Entomology. Washington, D.C. N. ser., 49–127, 1904–1915. (1–33, 1883–1895, and N. ser. 1–48, 1895–1904, as **U.S. Dept. Agr., Div. Ent., Bul.** [q.v.].)

> From No. 41, 1903+, the term "new series" no longer appears on the numbers.

U.S. Dept. Agr., Bur. Ent., Cir.=Circular. United States Department of Agriculture. Bureau of Entomology. Washington, D.C. [Ser. 2], 56–173, 1904–1913. ([Ser. 1], 1–8, 1885, as "Silk Culture Circular. U.S. Department of Agriculture"; [Ser. 1], 9–25, 1885–1886, as "Circular. U.S. Department of Agriculture. Division of Entomology"; [Ser. 1], 26–49, 1886–1894, as "Div. Ent. No. (or Ent. No.) U.S. Department of Agriculture"; Ser. 2, 1–55, 1891–1903, as "Circular. U.S. Department of Agriculture. Division of Entomology.")

> Series 1 consists of miscellaneous sheets printed by the Division of Entomology. The words "Second series" appear only up through No. 52, 1903.

U.S. Dept. Agr., Bur. Ent., Tech. Ser.=Technical Series. United States Department of Agriculture. Bureau of Entomology.

Washington, D.C. 10–27, 1905–1915. (1–9, 1895–1901, as **U.S. Dept. Agr., Div. Ent., Tech. Ser.** [q.v.].)

[**U.S. Dept. Agr.**] **Comnr. Agr. Rpt.**=Report of the Commissioner of Agriculture. U.S. Department of Agriculture, Washington, D.C. 1862–1888 (26 vols.). (1849–1861 [13 vols.] as "Report of the Commissioner of Patents. Agriculture"; 1889–1896, and 1924+, as "Report of the Secretary of Agriculture"; 1897–1923, as "Annual Reports of the Department of Agriculture" [which includes the "Report of the Secretary of Agriculture" and continuously paged Departmental reports].)

> Prior to 1849, the reports were embodied in the annual reports of the Commissioner of Patents. From 1894–1937, the reports of the Secretary were also published in "Yearbook of Agriculture." The reports for 1924+ are usually bound with the separately paged reports of the various Bureau Chiefs of the Department of Agriculture.

U.S. Dept. Agr., Dept. Bul.=Department Bulletin. United States Department of Agriculture. Washington, D.C. 1126–1500, 1923–1927. (1–242, 1913–1915, as "Bulletin of the U.S. Department of Agriculture"; 243–1125, 1915–1923, as "Bulletin. U.S. Department of Agriculture.")

> Superseded by **U.S. Dept. Agr. Tech. Bul.** [q.v.].

U.S. Dept. Agr., Dept. Cir.=Department Circular. United States Department of Agriculture. Washington, D.C. 1–425, 1919–1927.

> Nos. 10–12, and 21–24, were never published. Superseded by "Circular. U.S. Department of Agriculture."

U.S. Dept. Agr., Div. Ent., Bul.=Bulletin. United States Department of Agriculture. Division of Entomology. See **U.S. Dept. Agr., Bur. Ent., Bul.**

U.S. Dept. Agr., Div. Ent., Tech. Ser.=Technical Series. United States Department of Agriculture. Division of Entomology. See **U.S. Dept. Agr., Bur. Ent., Tech. Ser.**

U.S. Dept. Agr. Misc. Pub.=United States Department of Agriculture. Miscellaneous Publications. Washington, D.C. 1, 1929+

U.S. Dept. Agr. Rpt.=Report. United States Department of Agriculture. Washington, D.C. 59–117, 1899–1918. ([1–58], 1862–1898, were miscellaneous papers published by the Department; a list of them and their assigned numbers is printed on the cover pages of No. 59.)

U.S. Dept. Agr. Tech. Bul.=Technical Bulletin. United States Department of Agriculture, Washington, D.C. 1, 1927+

> Supersedes **U.S. Dept. Agr., Dept. Bul.** [q.v.].

U.S. Dept. Health, Educ., Welfare, U.S. Publ. Health Serv., Publ. Health Monog.=Public Health Monograph. United States Department of Health, Education, and Welfare. Public Health Service. Washington, D.C. 8, 1953+ (1–2, 1950, as "Public Health Technical Monograph. Federal Security Agency. Public Health Service"; 3–7, 1951–1952, as "Public Health Monograph. Federal Security Agency. Public Health Service.")

> Published concurrently with and indexed in "Public Health Reports." Usually classed as separates by libraries.

[U.S. Dept. Int.], U.S. Geol. and Geog. Survey of the Ter., Ann. Rpt.= Annual Report of the United States Geological and Geographical Survey of the Territories. [U.S. Department of the Interior], Washington, D.C. [7]–12, 1874–1883. ([1–2], 1867–1868, were included in the reports of the Commissioner of the General Land-Office for those years; [3], 1869, as "Preliminary Field Report of the United States Geological Survey of Colorado and New Mexico"; [4], 1871, as "Preliminary Report of the United States Geological Survey of Wyoming and portions of Contiguous Territories (being a fifth Annual Report of Progress)"; 6, 1873, as "Annual Report of the United States Geological Survey of the Territories.")

> Superseded by "Annual Report of the United States Geological Survey." Vols. 1–4 and 7–8 (1874–1875) are unnumbered. A reprint of the first 3 reports titled "First, second and third annual reports of the United States Geological Survey of the Territories for the years 1867, 1868, and 1869, under the Department of the Interior," was published in Washington, in 1873.

[U.S. Dept. Int.] U.S. Geol. and Geog. Survey of the Ter., Bul.= Bulletin of the United States Geological and Geographical Survey of the Territories. [U.S.] Department of the Interior, Washington, D.C. 1–6, 1875–1882.

> Superseded by "Bulletin of the United States Geological Survey." Vol. 1 has 2 series of Bulletins, the first consisting of 2 separately paged numbers, and the second series consisting of continuously paged numbers, as in Vols. 2–6.

U.S. Natl. Mus. Bul.=United States National Museum. Bulletin. Washington, D.C. 1, 1875+

> Nos. 1–16, 1875-1882, are also bound in **Smithsn. Inst., Smithsn. Misc. Collect.** [q.v.] vols. 13 (1878) and 23–24 (1882–1883), as "Bulletins of the United States National Museum," Vols. 1–3, respectively. Some of the later volumes are in parts which are sometimes separately paged.

U.S. Natl. Mus. Proc.=Proceedings of the United States National Museum. Washington, D.C. 1, 1878+

> Vols. 1–4, 1878–1881, are also bound in **Smithsn. Inst., Smithsn. Misc. Collect.** [q.v.] vols. 19 and 22. Each number of the Proceedings is also assigned a **Smithsn. Inst. Pub.** [q.v.] number. Vols. 60–82, 1921–1933, were issued in separately paged numbers.

Uppsala Univ. Zool. Bidr.=Zoologiska Bidrag från Uppsala. Uppsala University, Uppsala; Stockholm; Göteborg. 1, 1911+

Utah Acad. Sci., Arts and Letters, Proc.=Proceedings of the Utah Academy of Sciences, Arts and Letters. Provo; Salt Lake City. 11–27/28, 1934–1949/1951. (1–2, 1918–1921, as "Transactions of the Utah Academy of Sciences"; [3], 1926, as "Utah Academy of Sciences. Nineteenth Annual Convention"; [4–5], 1927–1928, as "Utah Academy of Sciences. Abstracts of Papers"; 6–8, 1929–1931, as "Proceedings. Utah Academy of Sciences"; 9–10, 1932–1933, as "Proceedings of the Utah Academy of Sciences"; 29, 1951/1952+, as "Utah Academy of Sciences, Arts and Letters. Proceedings.")

Some volumes are continuously paged double volumes. Vols. 3–9 were later bound together with a cover title "Proceedings. Utah Academy of Sciences, Arts and Letters. Vols. 3–9, inclusive." Vols. 3–5 are unnumbered.

Wash. Acad. Sci. Jour.=Journal of the Washington Academy of Sciences. Washington, D.C. 1, 1911+

Supersedes **Wash. Acad. Sci. Proc.** [q.v.].

Wash. Acad. Sci. Proc.=Proceedings of the Washington Academy of Sciences. Washington, D.C. 1–13, 1899–1911.

Superseded by **Wash. Acad. Sci. Jour.** [q.v.].

Wash. Agr. Expt. Sta. Bul.=Bulletin. Agricultural Experiment Station. Washington Agricultural College and School of Science (1891–1892); State Agricultural College and School of Science (1892–1894); Washington State Agricultural College and School of Science (1894–1905); State College of Washington (1906–1959); Washington State University (1959+). Pullman, Wash. 73–470, 1906–1945. (1–72, 1891–1905, as "Bulletin. Experiment Station"; 471–519, 1946–1950, as "Bulletin. Agricultural Experiment Stations. Institute of Agricultural Sciences"; 520, 1950+, as "Bulletin. Washington Agricultural Experimental Stations. Institute of Agricultural Sciences.")

Nos. 1–8, 1891–1893, were continuously paged; nos. 9–11, 1894, were also continuously paged. From no. 12, 1894+, the bulletins are separately paged.

Wash. Univ. [St. Louis] Studies=Washington University Studies. St. Louis, Mo. 1–13, 1913–1926; N. ser., in 3 separate series: Language and Literature, 1, 1927+; Sciences and Technology, 1, 1928+; and Social and Philosophical Sciences, 1, 1927+ (and for a while an "Annual Bibliography," 1923/1925+).

Vols. 1–4, 1913–1916, of the old series were issued in 2 separately paged "Parts" corresponding to the 2 later "Series," beginning with vol. 5: Humanistic Series, 5–13, 1917–1926; and Scientific Series, 5–13, 1917–1926. (Part 1 of vols. 1–4 became "Scientific Series," Part 2 the Humanistic Series.)

Wasmann Collect.=The Wasmann Collector. See **Wasmann Jour. Biol.**

Wasmann Jour. Biol.=Wasmann Journal of Biology. San Francisco. 8, 1950+ (1-2, 1937-1938, as "Wasmann Club Collector"; 3-7, 1938-1949, as **Wasmann Collect.** [q.v.].)

West Amer. Sci.=The West American Scientist. San Diego, Calif. 1-21, 1884-1919.

Wien, K. Naturhist. Hof-Mus. Ann.=Annalen des K. k. Naturhistorischen Hof-Museums. See **Wien, Naturhist. Mus. Ann.**

Wien. Ent. Montaschr.=Wiener Entomologische Monatschrift. Wien. 1-8, 1857-1864.

Wien. Ent. Ztg.=Wiener Entomologische Zeitung. Wien. 1-50, 1882-1933.

> Merged into "Koleopterologische Rundschau."

(Wien) K.k. Akad. Gymnasium, Jahresber.=Jahresberichte des Kais. kön. Akademische Gymnasium. Wien. 1858-?

> An annual school "program." Dates not known, but includes one for 1894 (as "Jahres-Bericht über das K. k. Akademische Gymnasium in Wien").

Wien, Naturhist. Mus. Ann.=Annalen des Naturhistorischen Museums in Wien. Wien. 33, 1920+ (1-31, 1886-1917, as **Wien, K. Naturhist. Hof-Mus. Ann.** [q.v.]; 32, 1918, as "Annalen des Naturhistorischen Hofmuseums.")

Wis. Acad. Sci., Arts, Letters, Trans.=Transactions of the Wisconsin Academy of Sciences, Arts, and Letters. Madison. 1, 1870+

> Some volumes are in 2 continuously paged parts.

Wis. Agr. Expt. Sta. Res. Bul.=Research Bulletin. Agricultural Experiment Station. University of Wisconsin. Madison. 1, 1909+

> Nos. 1-35, 1909-1915, were continuously paged within each year and later bound as part of the "Annual Report of the Agricultural Experiment Station of the University of Wisconsin"; from No. 36, 1915+, the Research Bulletins are separately paged.

Wis. Nat. Hist. Soc. Bul.=Bulletin of the Wisconsin Natural History Society. Milwaukee. 1-13, 1900-1915.

> Called "N. ser." throughout.

Wiss. Mitt. aus Bosnien u. Herzegowina=Wissenschaftliche Mitteilungen aus Bosnien und der Herzegowina. Wien. 1-13, 1893-1916.

Wyo. Univ. Pubs.=University of Wyoming Publications. Laramie, Wyo. 2, 1935+ (1, 1922-1934, as "University of Wyoming Publications in Science" in 2 separately paged series: Botany, and Geology.)

Zool. Anz.=Zoologischer Anzeiger. Leipzig. 1, 1878+

    Suspended 1944–1950.

Zool.-Bot. Gesell. Wien, Abhandl.=Abhandlungen der Zoologisch-Botanischen Gesellschaft in Wien. Wien. 10, 1918+ (1–9, 1901–1917, as **K.-k. Zool.-Bot. Gesell. Wien, Abhandl.** [q.v.].)

    Issued in separately paged numbers through vol. 16, 1936.

Zool.-Bot. Gesell. Wien, Verhandl.=Verhandlungen der Zoologisch-Botanischen Gesellschaft in Wien. Wien. 68, 1918+ (1–7, 1851–1857, as **Zool.-Bot. Ver., Wien, Verhandl.** [q.v.]; 8–67, 1858–1917, as **K.-k. Zool.-Bot. Gesell. Wien, Verhandl.** [q.v.].)

    Vols. 1–44, 1851–1894, have the separately paged sections "Abhandlungen" and "Sitzungsberichte." The "Kaiserlich-königlichen" appears as "K.-k." on the individual numbers.

Zool.-Bot. Ver. Wien, Verhandl.=Verhandlungen des Zoologisch-Botanischen Vereins in Wien. See **Zool.-Bot. Gesell. Wien, Verhandl.**

Zool. Bul.=Zoölogical Bulletin. Boston. 1–2, 1897–1899.

    Superseded by **Biol. Bul.** [q.v.].

Zool. Jahrb., Abt. f. System., Geog. u. Biol. Tiere=Zoologische Jahrbücher. Abteilung für Systematik, Geographie und Biologie der Tiere. Jena. 3–51, 1888–1925. (52, 1926+, as "Zoologische Jahrbücher. Abteilung für Systematik, Ökologie und Geographie der Tiere.")

    The first 2 volumes of "Zoologische Jahrbücher" (1886–1887) were not divided into separate Abteilungen. In 1888, it was published in 3 separate sections: one, as above, and the other 2 as "Abteilung für Allgemeine Zoologie und Physiologie der Tiere," and "Abteilung für Anatomie und Ontogenie der Tiere," and also a Supplement. The old spellings of "Abteilung" and "Tiere" were "Abtheilung" and "Thiere."

Zool. Jour. (London)=The Zoological Journal. London. 1–5, 1824–1834.

Zool. Mag. (Wiedemann's)=Zoologisches Magazin. Kiel; Altona. 1–2, 1817–1823.

    Published by C. R. W. Wiedemann, and sometimes known as "Wiedemann's Magazin." Issued in separately paged numbers; Vol. 2 consists of only No. 1.

Zool. Soc. London, Proc.=Proceedings of the Zoological Society of London. London. 1–28, 1833–1860, and 107, 1937+ (1861–1890, as **Zool. Soc. London, Proc. Sci. Mtg.** [q.v.]; 1891–1936, as **Zool. Soc. London, Proc. Gen. Mtg.** [q.v.].)

    Supersedes "Proceedings of the Committee of Science and Correspondence of the Zoological Society of London"; however, upon resumption of volume numbering in 1937, the 2 volumes of this journal (1831–1832) were included in the reckoning.

Vols. 107–108, 1937–1939, were issued in 3 separately paged series: A. General and Experimental; B. Systematic and Morphological; and C. Abstracts of Papers to be communicated. Vols. 109–113, 1939–1945, were issued only in the 2 series, A and B. Volumes for 1900–1905 were issued in 2 separately paged "Volumes"; those volumes for 1906-1936, though bound in 2 parts, are continuously paged.

Zool. Soc. London, Proc. Gen. Mtg.=Proceedings of the General Meetings for Scientific Business of the Zoological Society of London. See **Zool. Soc. London, Proc.**

Zool. Soc. London, Proc. Sci. Mtg.=Proceedings of the Scientific Meetings of the Zoological Society of London. See **Zool. Soc. London, Proc.**

Zool. Soc. London, Trans.=Transactions of the Zoological Society of London. London. 1, 1835+

Suspended 1917–1925.

Zool. Staatsinst. u. Zool. Mus. Hamburg, Mitt.=Mitteilungen aus dem Zoologischen Staatsinstitut und Zoologischen Museum in Hamburg. Hamburg. 39–46, 1922–1936. ([1–10], 1884–1892, are counted as material included in **Jahrb. der Hamburg. Wiss. Anst.** [q.v.] only [10] being a special section titled "Mitteilung aus dem Naturhistorischen Museum"; 11–38, 1893–1922, were published as Beihefts of **Jahrb. der Hamburg. Wiss. Anst.** [q.v.] [11–31, 1893–1914, as "Mitteilungen aus dem Naturhistorischen Museum in Hamburg"; 32, 1915, as "Mitteilungen aus dem Naturhistorischen (Zoologischen) Museum in Hamburg"; 33–37, 1916–1919, as "Mitteilungen aus dem Zoologischen Museum in Hamburg"; and 38, 1922, as "Mitteilungen aus dem Zoologischen Staatsinstitut und Zoologischen Museum in Hamburg"]; 47, 1938+ as "Mitteilungen aus dem Hamburgischen Zoologischen Museum und Institut.")

Ztschr. f. Angew. Ent.=Zeitschrift für Angewandte Entomologie. Berlin. 1, 1914+

Suspended 1944–1949. For supplement, see **Monog. z. Angew. Ent.**

Ztschr. f. die Ent. (Leipzig)=Zeitschrift für Entomologie. Leipzig. 1–5, 1839–1844.

Published by E. F. Germar, and sometimes known as "Germar's Magazin". Superseded by **Linn. Ent.** (q.v.).

Ztschr. f. die Gesam. Naturw.=Zeitschrift für die Gesammten Naturwissenschaften. Berlin; Leipzig; Stuttgart; Halle. 1–54, 1853–1881. (55, 1882+, as "Zeitschrift für Naturwissenschaften.")

Supersedes "Jahresberichte des Naturwissenschaftlichen Vereines in Halle." Vol. 92, 1938, is the last volume seen by us. Continues?
Vols. 35–48, 1869–1876, also called N. ser., 1–14; vols. 49–54, 1877–1881, also as Ser. 3, 1–6; vols. 55–62, 1882–1889, also as Ser. 4, 1–8; vols. 63–86,

1890–1918, also as Ser. 5, 1–24 (last vol. erroneously called "Ser. 6"); 87, 1925+, without additional series numbers.

Another separate journal titled "Zeitschrift für die Gesamte Naturwissenschaft" began publication at Brunswick and Berlin in 1935.

Ztschr. f. Breslau=Zeitschrift für Entomologie. Breslau. 1–15, 1847–1861; N. ser. 1–32, 1870–1907; N. ser. [=ser. 3], 15–18, 1927–1937. ([Ser. 3], 1–14, 1908–1924, as "Jahresheft des Vereins für Schlesische Insektenkunde zu Breslau.")

   [Ser. 1], vol. 7, was never published. [Ser. 3], 1, 1908, is also called N. ser. [=ser. 2], vol. 33. Some of the volumes (=hefts) are double volumes.

Ztschr. f. Induktiv Abstam. u. Vererbungslehre=Zeitschrift für Induktiv Abstammungs- und Vererbungslehre. Berlin. 1–88, 1908–1957. (89, 1958+, as "Zeitschrift für Vererbungslehre.")

   Suspended 1944–1947.

Ztschr. f. Pflanzenbau u. Pflanzenschutz=Zeitschrift für Pflanzenbau und Pflanzenschutz. Stuttgart; Munich. 45–50, 1950–1955. (1–5, 1898–1902, as "Praktische Blätter für Pflanzenschutz"; 6–22, 1903–1919, and 24–44, 1925–1945, and 51, 1956+, as "Praktische Blätter für Pflanzenbau und Pflanzenschutz"; 23, 1924, as "Praktische Blätter der Bayer.Landesanstalt für Pflanzenbau und Pflanzenschutz.")

   Vols. 6–22, 1903–1919, are also called [n. ser.], vols. 1–27; vols. 23–44, 1924–1945, also as [n. ser.], 1–22; vols. 45–50, 1950–1955, also as n. ser., 1–6.

Ztschr. f. Pflanzenkrank. (Pflanzenpath.) u. Pflanzenschutz=Zeitschrift für Pflanzenkrankheiten (Pflanzenpathologie) und Pflanzenschutz. Stuttgart. 39, 1929+   (1–31, 1891–1921, as "Zeitschrift für Pflanzenkrankheiten"; 32–35, 1922–1925, as "Zeitschrift für Pflanzenkrankheiten und Gallenkunde"; 36–38, 1926–1928, as "Zeitschrift für Pflanzenkrankheiten und Pflanzenschutz.")

   Suspended 1944–1948.

Ztschr. f. System. Hymenopt. u. Dipt.=Zeitschrift für Systematische Hymenopterologie und Dipterologie. Mecklenberg. 1–8, 1901–1908.

   Merged with **Deut. Ent. Ztschr.** [q.v.]

Ztschr. f. Wiss. Insektenbiol.=Zeitschrift für Wissenschaftliche Insektenbiologie. Neudamm;Husum;Berlin,etc. (10–36), 1905–1937, =N. ser., 1–27.   (1–2, 1896–1897, as **Illus. Wchnschr. f. Ent.** [q.v.]; 3–5, 1898–1900, as "Illustrierte Zeitschrift für Entomologie"; 6–9, 1901–1904, as "Allgemeine Zeitschrift für Entomologie.")

   "Illustrierte" is sometimes spelled with an initial 'J'.

Ztschr. f. Wiss. Zool. (Abt. A)=Zeitschrift für Wissenschaftliche Zoologie. Abteilung A. Leipzig. 141, 1932+

> Vols. 1–140, 1848–1932, were without division into Abteilungen, simply as "Zeitschrift für Wissenschaftliche Zoologie". For Abteilung B of this series, beginning in 1932, see **Arch. f. Naturgesch.**

Ztschr. f. Zellforsch. u. Mikros. Anat.=Zeitschrift für Zellforschung und Mikroskopische Anatomie. Berlin. 2–28, 1925–1938, and 34, 1947+ (1, 1924, as "Zeitschrift für Zellen- und Gewebelehre"; 29–33, 1939–1944, as "Zeitschrift . . . Anatomie. Abteilung A: Allgemeine Zellforschung und Mikroskopische Anatomie" [Abt. B: "Chromosoma," 1–2, 1939–1944, continued with vol. 3, 1947+, as an independent journal].

> Vols. 1–20, 1924–1934, form Abteilung B of "Zeitschrift für Wissenschaftliche Biologie."

INDEX

Names are indexed as follows:
CAPITALS: All names, whether valid or synonymous, for taxa above the generic level;
Boldface: Valid generic and subgeneric names;
Roman: Valid specific, subspecific, and varietal names;
Italics: All synonyms, nomina nuda, and names of extra-limital taxa even though valid.
Parentheses around a generic name indicate the original genus for a species of uncertain position. A † before a page number refers to a misidentification. Synonyms of species-group names are listed in their original spelling, and thus do not necessarily agree in gender with the genus to which they are presently referred. For the purposes of the index, emendations and errors are assigned the authorship of the correct names, as they are usually cited in the literature.

	Page		Page
aar Philip, Tabanus	332	abdominalis Johnson & Johnson, Lordotus	412
aasa Philip, Hybomitra	337	abdominalis Loew, Euxesta	651
aatos Philip, Hybomitra	337	*abdominalis* Loew, Xylophagus	297
abacta (Giglio-Tos), Quadrularia	889	abdominalis Philip, Silvius	323
abactor Philip, Tabanus	332	abdominalis (Reinhard), Melanomya	932
abaestuans Philip, Chrysops	324	abdominalis Robineau-Desvoidy, (Estheria)	989
abana Curran, Euxesta	651	abdominalis (Say), Argyra	524
abanus Curran, Cricotopus	157	abdominalis (Say), Holcocephala	374
Abaristophora Schmitz	535	abdominalis (Say), Ospriocerus	381
abatus Philip, Chrysops	324	abdominalis (Say), Thrypticus	511
abaureus (Philip), Pilimas	321	abdominalis (Say), Tipula	26
abbreviata (Loew), Nephrotoma	20	*abdominalis* Say, Zodion	629
abbreviata Loew, Parydra	749	abdominalis (Stenhammar), Lytogaster	752
abbreviata (Loew), Rondaniella	228	*abdominalis* Swenk, Cuterebra	1110
abbreviata (Loew), Tenthredomyia	616	*abdominalis* (Townsend), Micromintho	1032
abbreviata (Loew), Thecophora	631	abdominalis (Townsend), Winthemia	1090
abbreviata Malloch, Agromyza	795	abdominalis Wiedemann, Volucella	600
abbreviata Van Duzee, Mycomya	220	abdominalis Williston, Nicocles	380
abbreviata Walker, (Sciara)	235	abdominalis (Williston), Pseudohecamede	737
abbreviatus Johnson, Saropogon	383	*abdominoflavatus* Picado, Metriocnemus	160
abbreviatus Loew, Bibio	192	aberrans Borgmeier, Diplonevra	534
abbreviatus Loew, Pelastoneurus	501	*aberrans* (Johannsen), Microtendipes	173
abbreviatus Malloch, Chironomus	167	aberrans (Malloch), Earomyia	716
abbreviatus (Van Duzee), Calyxochaetus	528	aberrans Malloch, Phaonia	905
abbreviatus Van Duzee, Dolichopus	487	aberrans Philip, Chrysops	324
abbreviatus (Zetterstedt), Metasyrphus	560	aberrans Reinhard, Amobia	934
abcirus Walker, Empis	458	aberrans Sabrosky, Dicraeus	783
abdena Hall, Hylemya	848	aberrans (Townsend), Adoryphorophaga	1038
abdita Coquillett, Rhamphomyia	465	aberrans Wheeler, Medetera	509
abdita Johannsen, Sciara	230	aberrantis (Curran), Metasyrphus	560
abdita Peterson, Cnephia	185	aberrata Felt, Dasineura	258
abditus Philip, Tabanus	332	aberrata Lundbeck, Diamesa	151
abdominalis Adams, Boletina	216	aberratus Hardy & Knowlton, Pipunculus	553
abdominalis Adams, Mydas	359	*abfitchii* (Felt), Aedes	112
abdominalis Adams, Nemotelus	309	abhamata Felt, Lasioptera	270
abdominalis Aldrich, Cephenemyia	1111	abiens (Stein), Helina	886
abdominalis Aldrich, Hydrotaea	900	abiesemia Foote, Dasineura	258
abdominalis Aldrich, Plagiomima	1019	abietis Huckett, Hylemya	855
abdominalis Back, Stichopogon	385	abietum McAlpine, Earomyia	716
abdominalis Becker, Siphonella	786	abina Hall, Acronesia	928
abdominalis (Brown), Cerotainiops	388, †388	**Abiskomyia** Edwards	154
abdominalis Cazier, Rhaphiomidas	357	abitibiensis Huckett, Hylemya	852
abdominalis Coquillett, Chlorops	789	abitus Adams, Zodion	628
abdominalis (Curran), Ceriana	615	**Ablabesmyia** Johannsen	148
abdominalis Fabricius, Tabanus	332	*Ablautatus* Loew	364
abdominalis Felt, (Itonida)	289	ablautoides Melander, Cyrtopogon	367
abdominalis Felt, Lobopteromyia	276	**Ablautus** Loew	364
abdominalis Harris, (Stratiomys)	1113	abluta Doane, Tipula	26

1549

744–243—65——98

	Page		Page
abnormalis Garrett, Acantholeria	814	*Acidia* Robineau-Desvoidy	†677
abnormalis (Malloch), Phytobia	798	**Acidogona** Loew	663
abnormis Coquillett, Aphoebantus	429	*acidusa* (Walker), Anastrepha	†673
abnormis Cresson, Micropeza	634	acifer Townes, Polypedilum	174
abnormis (Dietz), Paradelphomyia	59	**Acinia** Robineau-Desvoidy	664
abnormis (Felt), Asteromyia	274	acirostre Reinhard, Leucostoma	976
abnormis Felt, Rhopalomyia	265	*Aciura* Robineau-Desvoidy	†669, †670, †671
abnormis Stein, Pegomya	862	**Aciurina** Curran	670
abominator Dyar & Knab, Culex	119	aclista Reinhard, Peleteria	996
aboriginis Dyar, Aedes	111	**Acnemia** Winnertz	222
aboriginis Harmston & Knowlton, Dolichopus	487	**Acoenonia** Pritchard	247
abortivus Malloch, Chironomus	166	ACOENONIINI	247
abrasus Van Duzee, Dolichopus	487	**Acontistoptera** Brues	545
abrevena Garrett, Neuratelia	224	acosta Webber, Tachinomyia	1058
abrupta Alexander, Limnophila	64	acra (Curran), Mallophorina	397
abrupta (Curran), Melangyna	565	*acra* (Walker), Fannia	895
abrupta Johannsen, Exechia	206	acricola Van Duzee, Dolichopus	487
abrupta Malloch, Fannia	893	*acrididarum* Townsend, (Sarcophaga)	1113
abrupta (Wiedemann), Bombyliopsis	992	**Acridiophaga** Townsend	944
abruptus Aldrich, Dolichopus	487	acridiophagoides (Lopes & Downs), Blaesoxipha	943
abruptus (Garrett), Procladius	149	**Acridomyia** Stackelberg	915
abruptus (Townes), Chironomus	169	*Acridophaga* Townsend	944
absaroka Alexander, Ormosia	87	acrirostris Townsend, Ginglymia	1060
absaroka Alexander, Tipula	30	**Acritochaeta** Grimshaw	875
abscondita Snow, Platypeza	549	**Acrocera** Meigen	405
absconus Van Duzee, Dolichopus	487	ACROCERIDAE	403
abserratus (Felt & Young), Aedes	111	ACROCERINAE	405
absobrina Felt, Rhabdophaga	256	**Acrochordonodes** Bigot	609
absobrina (Felt), Rhizomyia	262	*Acrodrosophila* Duda	763
absobrinus (Felt), Culiseta	117	**Acroglossa** Williston	1073
absoluta Johannsen, Exechia	206	acroglossoides (Townsend), Chaetogaedia 1078,	1078
absonus Van Duzee, Dolichopus	487	**Acrometopa** Schiner	707
abstersus (Loew), Euarestoides	669	**Acrometopia** Schiner	707
absurda Johannsen, Exechia	206	**Acronarista** Townsend	1064
absurdus (Johannsen), Cricotopus	157	**Acronaristopsis** Townsend	1064
abundans Spuler, Leptocera	723	**Acronesia** Hall	928
abutilon Felt, Asphondylia	267	acropennis Martin, Holopogon	375
abyssa Alexander, Gonomyia	75	acrophila (Dodge), Sarcophaga	957
abzoe Walker, (Dexia)	1108	*acrophilus* Dyar, Aedes	113
acadiana Reinhard, Sarcophaga	957	**Acroptena** Pokorny	863
Acallomyia Melander	458	acroptera Melander, Microphorella	552
ACALYPTRATAE	633	acrostichalis Malloch, Hylemya	853
Acandotheca Townsend	943	acrostichalis Melander, Oligodranes	421
Acanthocnema Becker	835	**Acrosticta** Loew	650
Acanthodotheca Townsend	943	**Acrostilpna** Ringdahl	856
Acantholeria Garrett	814	**Acrotaenia** Loew	664
acanthophallus Alexander, Dicranoptycha	52	acrum Curran, Temnostoma	614
acarivora (Felt), Mycodiplosis	282	actaeon Alexander, Pedicia	55
ACARTOPHTHALMIDAE	808	*actaeon* Osten Sacken, Tabanus	335
Acartophthalmus Czerny	808	**Actenoptera** Czerny	714
accidentalis Harmston & Knowlton, Dolichopus	487	**Actia** Robineau-Desvoidy	1061
accola Vockeroth, Cecidomyia	287	ACTIINI	1059
accurata Alexander, Tipula	31	**Actina** Meigen	300
acela (Walker), Lispe	878	actinobola (Loew), Trupanea	667
Acemya Robineau-Desvoidy	1036	actinosa (Reinhard), Nowickia	993
Acemyia Robineau-Desvoidy	1036	**Actiopsis** Townsend	1061
ACEMYINI	1036	actites Melander, Mythicomyia	416
acerba Alexander, Limonia	45	actites Philip & Steffan, Apatolestes	320
acerba (Walker), Ravinia	954	actius Melander, Lasiopogon	377
acerifolia (Felt), Anaretella	245	acton Coquillett, Rhaphiomidas	357
acerifolia Felt, Dasineura	258	**Actora** Meigen	678
acerifolia (Felt), Feltiella	281	actuaria Johannsen, Allodia	205
acerifolia (Felt), Janetiella	264	actuosa (Johannsen), Bradysia	232
acerifolia Felt, Rhabdophaga	256	*Acucula* Townsend	716
acerifolia (Felt), Sackenomyia	265	aculeata (Aldrich), Blaesoxipha	944
acerina (Felt), Lobodiplosis	281	*aculeata* (Pandellé), Onesia	†930
acerinus (Felt), Arthrocnodax	283	aculeata (Schmitz), Megaselia	540
aceris (Felt), Mayetiola	263	acuminata (Tothill), Nowickia	994
aceris Felt, Winnertzia	252	acuminata (Townsend), Leucostoma	976
aceris (Greene), Phytobia	798	acuminatus Cresson, Pipunculus	554
aceris (Shimer), Rhabdophaga	257	acuminatus Loew, Dolichopus	487
acernea (Felt), Parallelodiplosis	286	*acuminatus* Van Duzee (Condylostylus)	1113
aceus Garrett, Docosia	226	acuta Adams, Psilocephala	350
achaeta James, (Asilus)	1113	acuta Doane, Tipula	34
Achaetella Malloch	829	acuta Garrett, Bolitophila	197
Achaetina Malloch	899	acuta Garrett, Sciophila	225
Achaetomelina Cresson	685	acuta (Johannsen), Bradysia	232
Achaetomus Coquillett	814	acuta Melander, Mythicomyia	416
Achaetoneura Brauer & Bergenstamm	1101	acuta Stein, Hydrotaea	899
achilleae Johnson, Neaspilota	672	acutangula (Zetterstedt), Leptocera	725
Achlyomyia Pleske	316	acutangulus (Thomson), Euarestoides	669
achlytarsis Chillcott, Platypalpus	478	acuticauda Huckett, Hylemya	852
Acicephala Coquillett	828	*acuticornis* (Loew), Nanna	831
aciculata (Loew), Cressonomyia	742	acuticornis (Malloch), Spilogona	879
aciculifera Alexander, Gonomyia	75	acuticornis Shewell, Camptoprosopella	696

INDEX 1551

	Page		Page
acuticornis (Van Duzee), Peloropeodes	517	aenea (Coquillett), Villa	433
acuticornis (Wulp), Hoplodictya	691	aenea (Fabricius), Physiphora	653
acuticornis Wiedemann, Dolichopus	†492	*aenea* (Jones), Epistrophe	564
acutilabella Stalker, Drosophila	763	aenea Shannon & Dobroscky, Protocalliphora	925
acutipennis Malloch, Pegomya	862	aenea (Staeger), Peleteria	997
acutipleura Doane, Tipula	34	aenea (Wiedemann), Chaetopsis	650
acutus (Bigot), Tabanus	332	*aenea* (Wiedemann), Myiophasia	983
acutus (Fabricius), Meromacrus	622	*aenea* (Wulp), Chaetopsis	650
acutus Van Duzee, Dolichopus	487	aeneipes (Meijere), Decachaetophora	681
Acyphona Osten Sacken	83	aeneiventris Van Duzee, Medetera	509
adachiae Sabrosky, Stenoscinis	778	*aeneobarba* Loew, (Neoitamus)	1113
adaequatus Van Duzee, Dolichopus	487	aeneonigra (Loew), Cressonomyia	742
adaleonora (Steyskal), Caloparyphus	308	aeneoventris (Williston), Alophorella	967
adamsi Augustson, Trichobius	921	aenescens (Stenhammar), Pelina	752
adamsi (Brues), Stichillus	536	aenescens (Wiedemann), Ophyra	901
adamsi Cresson, Scatophila	758	*aeneus* (De Geer), Dolichopus	497
adamsi Laffoon, Exechia	206	aeneus (Scopoli), Eristalis	625
adamsi Sabrosky, Chlorops	789	*aeneus* Van Duzee, Medetera	509
adaptatus Schlinger, Ogcodes	406	*aeneus* (Walker), Chrysogaster	591
addenda (West), Melanomya	932	aenicolor Shannon, Ferdinandea	588
addicta (Huckett), Spilogona	880	aenigma Cresson, Notiphila	747
addita (Walker), Lypha	1011	*aenigma* (Reinhard), Blaesoxipha	945
Adejeania Townsend	999	**Aenigmatias** Meinert	537
adela Pritchard, Polyardis	249	AENIGMATIINAE	537
adelpha Steyskal, Stratiomys	311	*aenobarbus* Loew (Neoitamus)	1113
adenostoma Felt, Asphondylia	267	aepalia (Walker), Caliprobola	608
adfinis (Cresson), Trimerinoides	743	aequa (Becker), Conioscinella	783
adhesa (Felt), Camptoneuromyia	275	aequa Stein, Azelia	899
Adiamesa Kieffer	151	*aequale* Loew, Temnostoma	614
Adiplosis Felt	289	aequale Van Duzee, Rhaphium	512
adirondacensis Alexander, Elephantomyia	71	aequalis Doane, Tipula	34
adirondacensis (Alexander), Limonia	45	*aequalis* Fabricius, Bombylius	409
adirondacensis Alexander, Ormosia	85	aequalis Johannsen, Metriocnemus	161
adjacens Walker, (Dolichopus)	530	aequalis Loew, Doros	569
adjecta (Doane), Limonia	50	aequalis (Loew), Euaresta	665
adjuncta Cresson, Ceropsilopa	741	aequalis Loew, Platypalpus	478
adjuncta Dietz, Limnophila	66	aequalis (Malloch), Homoneura	698
Admontia Brauer & Bergenstamm	1068	aequalis (Malloch), Jenkinshelea	137
ADMONTIINI	1068	aequalis Painter, Geron	423
Admontiopsis Townsend	1068	aequalis Reinhard, Pseudotachinomyia	1057
adolescens (Walker), Melangyna	568	*aequalis* Stein, Lispoides	879
adonis Osten Sacken, Astrophanes	441	aequalis Van Duzee, Dolichopus	487
Adoryphorophaga Townsend	1038	aequalis Van Duzee, Exechia	206
Adoxomyia Bezzi	305	aequalis Van Duzee, Medetera	509
Adoxomyia Kertész	305	*aequalis* (Van Duzee), Thecophora	631
adrogans Cresson, Cordilura	827	aequalis (Wood), Megaselia	540
adsimilis (Goetghebuer), Chaetocladius	160	aequetincta (Becker), Hybomitra	338
adspersa Coquillett, Euaresta	665	aequiatra Alexander, Limnophila	65
adultus Van Duzee, Dolichopus	487	aequicornis Melander, Platypalpus	479
adumbrata (Coquillett), Villa	437	aequifasciata (Dendy & Sublette), Ablabesmyia	148
adumbratus Johannsen, Procladius	149		
adunca Borgmeier, Cremersia	544	aequifrons (Stein), Lispoides	879
adusta (Loew), Americina	841	aequus Cresson, Pipunculus	553
adusta (Loew), Leschenaultia	1082	*aerata* (Coquillett), Doryphorophaga	1042
adusta (Loew), Scaptomyza	769	aeratus Coquillett, Platycheirus	576
adusta (Loew), Villa	433	aeratus Van Duzee, Dolichopus	487
adusta Osten Sacken, Limnophila	65	*aerea* Bigot, Caliprobola	608
adusta (Wulp), Juriniopsis	1001	aerea (Fallén), Spilogona	879, †896, ††897
adustoides Alexander, Limnophila	65	aerea (Loew), Lejota	590
adustus Loew, Saropogon	383	aerea Schmitz, Phora	536
adustus Van Duzee, Diaphorus	519	aeripes Melander, Empis	458
advenae Cresson, Hydrellia	743	aerobatica Melander, Empis	458
adventitia Melander, Dactylolabis	61	*Aeschnasoma* Johnson	18
adversa Coquillett, Rhamphomyia	461	*Aeshnasoma* Johnson	18
aea Walker (Cordylura)	915	*aescytes* (Walker), Cheilosia	586
aeacides Loew, Ospriocerus	381	aestiva Felt, Camptomyia	256
aeacidinus (Williston), Ospriocerus	381	aestiva Felt, Mycodiplosis	282
aeacus (Wiedemann), Ospriocerus	381	aestiva Kessel, Agathomyia	547
aeata Walker, Laphria	389	aestiva (Meigen), Hylemya	854
Aecothea Haliday	811	aestiva Van Duzee, Neurigona	518
Aedes Meigen	110	*aestivalis* (Dyar), Aedes	114
Aedimorphus Theobald	116	*aestivalis* (Harris), Hybomitra	341
aedon Townsend, Ogcodes	407	aestivalis Townes, Stenochironomus	177
aegeriformis (Gray), Lepidophora	425	*aestivalis* (West), Admontia	1068
aegeriiformis (Gray), Lepidophora	425	aestivum Davies, Peterson, & Wood, Simulium	185
aegiale (Walker), Eclimus	427		
Aegialomyia Philip	329	aestivus Melander, Bibiodes	195
aegra Martin, Leptogaster	362	aestuans (Linnaeus), Efferia	393, †395
aegra Walker, Sarcophaga	958	*aestuans* Loew, Hydrophorus	506
aegrotus Osten Sacken, Tabanus	332	aestuans Wulp, Chrysops	324
aegypti (Linnaeus), Aedes	116	aestuum Aldrich, Fucellia	842
aelops (Walker), Beskia	972	*aestuum* Loew, Hydrophorus	506
aemene (Walker), Hylemya	847	aesyctes (Walker), Cheilosia	586
aemula Loew, Diacrita	644	*aeta* (Walker), Holcocephala	375
aemulator (Loew), Nicocles	380	aethiops Alexander, Pedicia	54
aenea (Coquillett), Elfia	1065	aethiops Malloch, Fannia	893

1552 A CATALOG OF DIPTERA OF NORTH AMERICA

	Page		Page
aethiops Malloch, Milichia	732	amiatis McAtee, Laphria	389
aethiops (Townes), Chironomus	169	ainsliei (Curran), Oscinella	781
aethiops Van Duzee, Dolichopus	487	ainsliei Van Duzee, Dolichopus	487
aethiops (Walker), Physocephala	628	aitkeni Cazier, Rhaphiomidas	357
a*fer* McAtee, Bibio	193	aitkeni Philip, Apatolestes	320
affinalis Frost, Phytomyza	804	aitkeniana Alexander, Tipula	40
affine (Wheeler), Syntormon	515	aix Townes, Pseudochironomus	176
affinis Adams, Culex	118	akeleyi Johannsen, Rymosia	207
affinis (Bellardi), Efferia	393	akpatokensis Edwards, Boletina	217
affinis Cresson, Pipunculus	553	akrina (Roback), Tanytarsus	180
affinis (Fallén), Hubneria	†1093	**Akronia** Hine	310
affinis Fallén, Phytomyza	†804	aktis McAtee, Laphria	389
affinis (Fallén), Pseudosarcophaga	†941	alaba (Walker), Hylemya	847
affinis Fisher, Anatella	205	alabamensis Sturtevant, Drosophila	767
affinis Garrett, Docosia	226	alacer Van Duzee, Dolichopus	488
affinis Johnson, Chaetoclusia	806	alachua Jamnback & Wirth, Culicoides	128
affinis (Kirby), Hybomitra	338	alamedensis Sublette, Paralauterborniella	173
affinis Loew, Chrysotus	521	alascaensis (Alexander), Limonia	45
affinis (Lundbeck), Ormosia	85	alascaensis Alexander, Tipula	29
affinis Macquart, Laphria	389	*alascensis* (Townsend), Mesembrina	910
affinis Malloch, Hebecnema	890	alaska Alexander, Tipula	34
affinis Malloch, Lonchaea	717, †717	alaskaensis Hardy, Chrysopilus	345
affinis (Malloch), Ophiomyia	797	alaskaensis (Ludlow), Culiseta	117
affinis Meigen, Piophila	712, †712	alaskana (Alexander), Neolimnophila	73
affinis Quisenberry, Stenopa	671	alaskense Cresson, Philotelma	756
affinis Reinhard, Paradidyma	1032	*alaskensis* (Alexander), Arctoconopa	80
affinis Say, Scaeva	562	alaskensis (Brooks), Alophorella	967
affinis (Snow), Neotephritis	669	alaskensis Brooks, Psalidopteryx	1067
affinis Stein, Pegomya	858, †862	alaskensis Chillcott, Fannia	893
affinis Sturtevant, Drosophila	768	alaskensis (Coquillett), Paraclunio	163
affinis Walker, Dolichopus	487	alaskensis Fluke, Helophilus	617
affinis (Williston), Neoitamus	397	alaskensis Hunter, Cheilosia	583
affinis (Williston), Physocephala	628	alaskensis James, Sargus	302
afflicta Dietz, Tipula	31	alaskensis Kessel, Agathomyia	547
afflictus (Osten Sacken), Dolichopus	487	alaskensis (Malloch), Anatopynia	145
affluens Van Duzee, Dolichopus	487	alaskensis (Malloch), Hydrophoria	863
agalma Wheeler, Hydrophorus	504	alaskensis (Malloch), Megaselia	540
agapenor Walker, Microdon	598	alaskensis Malloch, Phaonia	908
agarici (Lintner), Megaselia	538	alaskensis (Shannon), Acronesia	928
agarici Willard, Platypeza	549	alaskensis (Shannon), Chrysosyrphus	592
agasicles Walker, Rhamphomyia	461	*alaskensis* Shannon, Francilia	926
agassis Garrett, Sciophila	225	alaskensis Wirth, Culicoides	131
agassizii Loew, Exoprosopa	443	**Alaskophyto** Townsend	1022
agasthus Walker, Empis	458	**Alassomyia** Felt	262
Agatachys Meigen	476	alata (Aldrich), Sarcodexia	956
Agathomyia Reinhard	1016	alata Guthrie, Mycetophila	210
Agathomyia Verrall	547	alatus Beck, Chironomus	166
Agathon Röder	99	alba Baker, Orthopodomyia	108
agelenae Melander, Opsebius	405	alba (Loew), Neaspilota	672
agens Cresson, Polytrichophora	740	*alba* Roback, Pentaneura	147
agens (Curran), Carposcalis	575	albagenae Morrison, Gonia	1075
agens (Melander), Tachyempis	475	albaria (Coquillett), Jenkinshelea	137
ageratae Benjamin, Trupanea	667	albarius Painter, Geron	422
agilis Aldrich, Dolichopus	489	albata Coquillett, Rhamphomyia	461
agilis (Harris), Compsobata	633	albatus Johnson, Chaoborus	103
agilis (Meigen), Onesia	930	*albavictoria* Patterson & Mainland, Amiota	761
agilis Melander, Mythicomyia	416	albeola Huckett, Fucellia	842
agilis Reinhard, Ptilodexia	988	*alberta* (Alexander), Hexatoma	69
agilis (Robineau-Desvoidy), Parasetigena	1056	alberta (Curran), Clastoneuriopsis	1067
Agkistrocerus Philip	341	alberta (Curran), Cordilura	830
agnella Reinhard, Eumacronychia	935	*alberta* Curran, Cordilura	828
agnesea Hardy, Tomosvaryella	555	alberta (Curran), Gonarcticus	832
agnon (Walker), Metasyrphus	560	*alberta* (Curran), Hedriodiscus	315
Agnotemyia Williston	343	*alberta* (Curran), Lixophaga	1043
Agnotomyia Williston	342	alberta Curran, Mericia	1007
Agonosoma Guérin-Méneville	†486	alberta Curran, Mycetophila	210
AGONOSOMATINAE	482	alberta Curran, Peleteria	996
agraria Felt, (Cecidomyia)	289	*alberta* (Curran), Pericoma	92
agraria (Felt), Lycoriella	231	alberta Fisher, Bolitophila	197
agrayloides (Kieffer), Lauterborniella	172	alberta (Huckett), Spilogona	879
agrestis (Coquillett), Lepidanthrax	440	alberta Leonard, Ptiolina	347
agrestis (Coquillett), Lepidanthrax	440	*albertae* Dyar, Aedes	114
Agria Robineau-Desvoidy	†941	albertana Reinhard, Myioclonia	1025
agrimoniae Felt, Contarinia	277	albertensis Alexander, Chionea	72
agrion (Jaennicke), Proctacanthus	399	albertensis Alexander, Ormosia	85
agrippina (Osten Sacken), Villa	433	albertensis Alexander, Tipula	31
Agrolimna Cresson	747	albertensis Cole, Psilocephala	350
Agromyza Fallén	795	albertensis Curran, Dolichopus	488
AGROMYZIDAE	794	albertensis Johnson, Palloptera	715
agromyzina (Hendel), Aldrichiomyza	730	albertensis Wirth & Jones, Culicoides	132
agromyzina Meigen, Phytomyza	804	albescens (Townes), Phaenopsectra	174
AGROMYZINAE	795	*albiapicatus* Parent, Condylostylus	484
agronomus Melander & Brues, Dolichopus	487	albibarba (Loew), Acanthocnema	835
agrorum (Fabricius), Eristalis	623	albibarbis (Bigot), Rhagio	345
agrostidis Kieffer, Neolasioptera	272	albibarbis (Macquart), Efferia	393
agrostis Felt, Neolasioptera	272	albibarbum (Van Duzee), Rhaphium	512
agrostis (Osten Sacken), Asteromyia	274	albibarbus Curran, Eucyrtopogon	373

INDEX 1553

	Page		Page
albibasalis (Zetterstedt), Lasiops	903	albipectus (Macquart), Villa	433
albibasis Bigot, Stenopogon	383	albipennis (Brooks), Phasia	969
albibasis (Malloch), Mallochohelea	139	albipennis Felt, Rhopalomyia	265
albibasis (Malloch), Spilochroa	817	albipennis (Loew), Neaspilota	672
albibasis Stein, Coenosia	870	albipennis Meigen, Chironomus	†170
albicalceata (Cresson), Pteromicra	687	albipennis Ringdahl, Pseudochirosia	844
albicans (Fallén), Cyzenis	1100	albipennis (Rondani), Leptocera	726
albicans Loew, Argyra	524	albipennis Say, Bibio	192
albicans (Meigen), Hecamede	737	albipennis (Say), Lynchia	918
albicapillus Loew, Bombylius	408	albipennis (Williston), Telmatoscopus	93
albiceps Loew, Gymnopternus	498	albipes (Bigot), Neoascia	594
albiceps (Loew), Paroxyna	665	albipes Cresson, Seioptera	648
albiceps Loew, Thereva	352	albipes Felt, Neolasioptera	273
albiceps Macquart, Eristalis	623	albipes (Johannsen), Chaoborus	103
albiceps (Macquart), Laphystia	376	albipes (Johnson), Dolichopeza	23
albiceps (Macquart), Polydontomyia	621	albipes Leonard, Limnophila	67
albiceps (Malloch), Dasiops	716	albipes Meigen, Phytomyza	805
albiceps Meigen, Phytomyza	804	albipes Walker, (Drosophila)	772
albiceps Reinhard, Meledonus	1024	albipes (Walker), Metachela	471
albiceps (Wulp), Compsobata	633	albipes (Zetterstedt), Micromorphus	530
albiciliata (Aldrich), Mesorhaga	482	albipila Kraft & Cook, Zabrachia	318
albiciliata Van Duzee, Medetera	509	albipilis Curran, Microdon	597
albiciliatus Loew, Dolichopus	488	albipilis Snow, Mallota	621
albicincta Cole, Villa	439	albipilosa Curran, Cerotainia	387
albicincta (Fallén), Calythea	865	albipilosa Hardy, Belosta	355
albicincta (Reinhard), Meigenia	1044	albipilosa Williston, Pipiza	579
albicinctum Van Duzee, Keirosoma	†515	albipilosus Adams, Euparyphus	307
albicinctus Cole, Ogcodes	406	albipilosus Brennan, Apatolestes	320
albicollaris Painter, Exoprosopa	444	albipilosus Curran, Holopogon	375
albicoma Reinhard, Emblemasoma	949	albipleura (Curran), Neocnemodon	581
albicomus Hine, Asilus	392	albipodus Harmston & Knowlton, Hercostomus	497
albicornis Meigen, (Coenosia)	877	albipunctata (Curran), Melangyna	565
albicornis (Say), Rhagio	344	albipunctatus (Williston), Telmatoscopus	93
albicornis Wilcox & Martin, Dioctria	371	albirostris Macquart, Nemotelus	309
albicosta (Walker), Ischnomyia	819	albiscutellata (Harris), Rhagoletis	674
albicoxa Aldrich, Dolichopus	488	albiscutellatus (Macquart), Leucotabanus	†331
albicoxa James, Cordilura	829	albiseta Coquillett, Lauxania	699
albicoxa Van Duzee, Argyra	524	albiseta (Cresson), Tomosvaryella	556
albicoxa (Walker), Condylostylus	483	albiseta (Roser), Leucophora	867
albicrus (Townes), Stictochironomus	177	albiseta (Zetterstedt), Chelipoda	472
albidihalteris (Felt), Megaselia	539	albisquama (Ringdahl), Spilogona	879
albidipennis Loew, Geron	422	albistria Walker, Chironomus	166
albidipennis (Loew), Neaspilota	672	albistylum Johnson, Berkshiria	318
albidohalteralis Malloch, Orthocladius	155	albistylum Macquart, Ectecephala	791
albidohalterata (Malloch), Phytobia	798	albistylum Macquart, Tropidia	609
albidorsata Malloch, Bezzia	141	albitarsis (Banks), Pericoma	92
albidosa Huckett, Eremomyia	857	albitarsis Curran, Cyrtopogon	368
albidus Cole & Wilcox, Lasiopogon	377	albitarsis (Felt), Cecidomyia	287
albidus Hall, Lordotus	412	albitarsis (Felt), Neolasioptera	273
albifacies (Johnson), Coenosia	873	albitarsis Harris, (Chironomus)	1113
albifacies Johnson, Cyrtopogon	367	albitarsis Meigen, Agromyza	795
albifacies Parent, Diaphorus	519	albitarsis (Osten Sacken), Hexatoma	69
albifacies (Townsend), Oswaldia	1046	albitarsis Stein, Lispe	877
albifacies Van Duzee, Zodion	629	albitarsis Zetterstedt, Homalocephala	652
albifacies Williston, Promachus	400	albitarsis Zetterstedt, Lonchaea	717
albifasciatum Hardy, Omphralosoma	355	albius Walker, Dioctria	370
albifasciatus (Back), Dizonias	372	albiventris Johnson, Ogcodes	407
albifascies Adams, Chlorops	789	albiventris Loew, Argyra	524
albifrons Back, Saropogon	383	albiventris (Loew), Probezzia	138
albifrons Curran, Exoprosopa	444	albivitta Walker, Pedicia	53
albifrons Say, Thereva	352	albocalyptrata Malloch, Phaonia	905
albifrons Spuler, Leptocera	723	albocaudata Doane, Tipula	27
albifrons Swenk, Cuterebra	1110	albocincta Doane, Tipula	35
albifrons Walker, Gonia	1076	albocincta Van Duzee, Pseudatrichia	355
albifrons (Walker), Gymnocarcelia	1093	albocostata (Fallén), Pherbellia	686
albifrons (Walker), Ptilodexia	988	albofascia Doane, Tipula	35
albifrons Wiedemann, Eristalis	623	albofasciatus Hough, Pipunculus	554
albifrons Wilcox & Martin, Cyrtopogon	367	albofasciatus Macquart, Anthrax	431
albifrons (Zetterstedt), Limnospila	876	alboflorens (Walker), Hydrophorus	504
albihalter Wirth, Mallochohelea	139	alboguttata (Wahlberg), Amiota	761
albihalteralis Cole, Pherocera	349	albohirta Felt, Dasineura	258
albihirta (Alexander), Hexatoma	69	albohirtus Van Duzee, Chrysotus	521
albilata Walker, Tipula	26	albolineata Felt, Neolasioptera	273
albimacula Doane, Tipula	26	albomacula Stone, Stonemyia	321
albimaculata Welch, Psychoda	95	albomaculata (Felt), Asteromyia	274
albimana (Macquart), Taeniaptera	636	albomaculatus (Jaennicke), Epalpus	1002
albimanus (Fabricius), Platycheirus	576	albomaculatus Van Duzee, Hydrophorus	504
albimanus (Meigen), Clusiodes	807	albomanicata (Alexander), Limnophila	67
albimanus (Meigen), Paratendipes	173	albonotata Doane, Tipula	34
albimanus Wiedemann, Anopheles	106	albonotatum Townsend, Zodion	628
albimanus Wirth, Systenus	517	albonotatus Loew, Syneches	448
albimarginatus James, Nemotelus	309	albonotatus (Loew), Tachytrechus	502
albimargo Pandellé, Pegomya	†859	albopenicillatus (Bigot), Parabombylius	410
albinodus Townes, Polypedilum	174	albopilosa Coquillett, Rhamphomyia	461
albionense Rothfels, (Prosimulium)	1113		
albipectus Macquart, Bombylius	409		

1554   A CATALOG OF DIPTERA OF NORTH AMERICA

	Page		Page
albopilosa (Cresson), Adoxomyia	305	*Aldrichiella* Rohdendorf	929
albopilosa Kröber, Thereva	352	aldrichii Brues, Conicera	536
albosetosa Hine, Cophura	366	aldrichii Coquillett, Psilocephala	350
albosetosa Van Duzee, Medetera	509	*aldrichii* (Doane), Tephritis	668
albosparsus (Bigot), Anthrax	431	*aldrichii* (Felt), Anarete	246
albospinosa Van Duzee, Neurigona	518	aldrichii Hine, Promachus	400
albotarsus Felt, (Cecidomyia)	289	aldrichii Johannsen, Palaeoplatyura	203
albovaria (Coquillett), Pherbellia	686	aldrichii Melander, Empis	458
albovarians Curran, Cyrtopogon	367	aldrichii Melander, Lasiopogon	377
albovenosa Coquillett, Hyadina	751	*aldrichii* (Snow), Agathomyia	548
alboviridis Malloch, Chironomus	166	aldrichii Stein, Pentacricia	877
albovitta (Walsh), Dasineura	258	aldrichii Sturtevant, Chymomyza	770
albovittata Doane, Tipula	40	aldrichii Townsend, Hyalomya	968
albovittata (Macquart), Villa	439	aldrichii Van Duzee, Neurigona	518
albovittata Malloch, Oxycera	307	aldrichii (Wheeler), Dolichopus	488
albovittata (Walsh), Dasineura	258	aldrichii Wheeler, Medetera	509
albrighti Alexander, Ormosia	85	**Aldrichina** Townsend	929
albula (Fallén), Hylemya	847	**Aldrichiomyza** Hendel	730
albula (Loew), Tethina	727	*Aldrichocyptera* Townsend	972
albulum Townes, Polypedilum	176	*Aldrovandiella* Enderlein	238
albulus Melander, Metapogon	378	alea Laffoon, Mycetophila	210
alcanor Walker, Laphria	391	alearis Reinhard, Masiphyomyia	1085
alcathoe (Walker), Hylemya	845	aleator Alexander, Limnophila	64
alcedo (Aldrich), Blaesoxipha	943	*Aleomyia* Phillips	659
alcedo (Loew), Chrysotachina	1005	alesia (Walker), Compsobata	634
alcestis Alexander, Tipula	31	*alethes* (Walker), Tolmerus	401
alcidice (Walker), Metasyrphus	560	aletiae (Comstock), Megaselia	538
alcis (Snow), Lyperosiops	914	aletiae (Riley), Lespesia	1101
alcyon (Say), Poecilanthrax	441	aleutica Alexander, Limnophila	65
aldrichanus (Dyar), Telmatoscopus	94	aleutica Alexander, Tipula	29
aldrichi (Alexander), Arctoconopa	79	alexanderi Felt, Hormomyia	287
aldrichi Alexander, Limnophila	64	*alexanderi* (Felt), Xylopriona	249
aldrichi Allen, Phrosinella	938	alexanderi James, Bibio	192
aldrichi Austen, Cuterebra	1109	alexanderi (Johnson), Gonomyia	75
aldrichi Cresson, Canace	733	alexanderi (Laffoon), Mycetophila	210
aldrichi Cresson, Setacera	754	alexanderi Shaw, Exechia	206
aldrichi (Curran), Anoxynops	1038	alexanderi (Shaw), Orfelia	202
aldrichi Curran, Grisdalemyia	1024	alexanderi Wirth & Hubert, Culicoides	128
aldrichi (Curran), Hesperophasiopsis	978	alexandriana Dietz, Tipula	27
aldrichi Curran, Peleteria	997	alexandriana Garrett, Chionea	72
aldrichi (Curran), Pilatea	1105	**Alexandriaria** Garrett	49
aldrichi Czerny, Eccoptomera	811	*Alexiopogon* Curran	377
aldrichi Del Ponte, Cochliomyia	923	alexippus Walker, Platypalpus	478
aldrichi Dyar & Knab, Aedes	114	algens (Coquillett), Anatopynia	145
aldrichi (Garrett), Anypotacta	815	algens Curran, Arctophyto	985
aldrichi Hendel, Tetanops	648	algens Leonard, Symphoromyia	343
aldrichi Johnson, Argyra	525	algens Wheeler, Hydrophorus	504
aldrichi Johnson, Odontomyia	316	algens (Wiedemann), Nowickia	995
aldrichi Johnson, Phthiria	420	algonquin Alexander, Tipula	24
aldrichi (Hull), Ceriana	615	algonquin Sturtevant & Dobzhansky, Drosophila	768
aldrichi Hunter, Cheilosia	583	algonquina Malloch, Helina	886
aldrichi James, Chrysopilus	346	alhambra Hull, Eristalis	623
aldrichi (Malloch), Berkshiria	318	alia Doane, Tipula	31
aldrichi Malloch, Oxycera	307	*aliata* Hopkins, (Sciara)	1113
aldrichi (Malloch), Palpomyia	140	alicia (Alexander), Cheilotrichia	78
aldrichi Malloch, Pogonomyia	901	aliciae Johannsen, Dixa	102
aldrichi (Malloch), Scatophaga	838	*aliena* (Cresson), Ditrichophora	739
aldrichi Melander, Parathalassius	452	aliena Hardy, Tomosvaryella	555
aldrichi Mesnil, Hyperecteina	1069	aliena Malloch, Beckerina	537
aldrichi Painter, Apiocera	356	aliena Malloch, Coenosia	870
aldrichi (Parker), Metoposarcophaga	950	aliena Malloch, Hylemya	853
aldrichi Parker, Sarcophaga	957	alienum (Becker), Oscinisoma	778
aldrichi Patterson & Crow, Drosophila	763	alienus McAtee, Bibio	192
aldrichi Reinhard, Uramya	1029	alienus Van Duzee, Diaphorus	519
aldrichi Ringdahl, Lasiops	904	aliternigra Melander, Drapetis	477
aldrichi Sabrosky, Lutomyia	811	alkalinella (Cresson), Ptilomyia	736
aldrichi Shannon, Caliprobola	608	*Allactina* Curran	300
aldrichi Snyder, Helina	886	**Allanthalia** Melander	450
aldrichi (Sturtevant & Wheeler), Parydra	750	**Allarete** Pritchard	245
aldrichi Tothill, Gonia	1075	allecta (Melander), Phytobia	799
aldrichi (Townsend), Mericia	1007	alleghani Melander, Neoplasta	472
aldrichi Van Duzee, Chrysotus	521	*Allenanicia* Townsend	937
aldrichi Van Duzee, Diaphorus	519	alleni Back, Cyrtopogon	367
aldrichi Van Duzee, Pelastoneurus	501	*alleni* (Brues), Diplonevra	†535
aldrichi (Van Duzee), Rhaphium	512	alleni Cazier, Apiocera	356
aldrichi Van Duzee, Sympycnus	527	alleni Johnson, Limnophila	63
aldrichi Wheeler, Aphaniosoma	822	*alleni* Turner, Aedes	115
aldrichi Wilcox & Martin, Cyrtopogon	367	allia (Frost), Liriomyza	801
aldrichi Williston, Dialysis	343	allioides Pritchard, Lasioptera	270
aldrichi (Williston), Leptocera	723	allioniae Felt, Lasioptera	270
aldrichi (Williston), Nemotelus	310	*Alliops* Malloch	857
Aldrichia Coquillett	414		
aldrichia (Shannon), Acronesia	928	**Alliopsis** Schnabl & Dziedzicki	857
aldrichiana Alexander, Tipula	30	alliovora Frick, Liriomyza	801
aldrichiana (Enderlein), Simulium	188	*alliterata* (Huckett), Spilogona	882
Aldrichiella Hendel	730	*Allobezzia* Kieffer	142

INDEX 1555

	Page		Page
Allocotocera Mik	222	alticola James, Odontomyia	315
Allocotus Loew	427	alticola Malloch, Coenosia	870
Allodia Winnertz	204	alticola Malloch, Phaonia	905
Alloeostylus Schnabl	904	alticola (Malloch), Spilogona	879
Allognosta Osten Sacken	300	alticrista Alexander, Limnophila	64
Allognota Pokorny	874	altifila (Felt), Schizomyia	267
Allograpta Osten Sacken	568	altifilus (Felt), Holoneurus	254
Allomethus Hardy	555	altilega Huckett, Hydrophoria	863
Allomyella Malloch	834	altissima (Osten Sacken), Nephrotoma	20
Allomyia Felt	262	altitudinum Bromley, Laphria	389
Allomyia Malloch	834	*altiusculus* Dyar, Aedes	112
Allophora Robineau-Desvoidy	967	altivagus Aldrich, Hydrophorus	504
Allophyla Loew	809	alumnus Melander, Platypalpus	478
Allopiophila Hendel	711	*alvea* (Walker), Eurosta	663
Allosphaerocera Hendel	719	alveofrons McAlpine, Dasiops	716
Allotrichoma Becker	736	alypiae Sellers, Sisyropa	1094
Alluaudomyia Kieffer	133	amabilis Alexander, Limnophila	67
allynii Marten, Tabanus	335	amabilis (Osten Sacken), Conophorus	413
alma (Meigen), Lispocephala	876	amabilis Parent, Sciapus	486
almquistii (Holmgren), Spilogona	879	amachaerus (Townes), Chironomus	167
alneti (Fallén), Didea	562	*Amalopis* Haliday	54
Aloceuxesta Hendel	651	amalopis (Osten Sacken), Dasysyrphus	563
alogus (Reinhard), Eupelecotheca	1051	amanda Cresson, Renocera	692
alone (Walker), Macrorchis	875	amans (Cresson), Bucephalina	832
alopecis (Reinhard), Blaesoxipha	946	amaranthi Felt, Asphondylia	267
alopecuri (Reuter), Dasineura	258	*amastris* (Walker), Nicocles	380
alopex (Osten Sacken), Crioprora	610	*amatus* (Walker), Condylostylus	485
Alophora Robineau-Desvoidy	967	**Amauromyza** Hendel	798
Alophorella Townsend	967	*Amaurosoma* Becker	830
Alophorellopsis Townsend	968	amazon Daecke, Chrysops	324
Alophoropsis Townsend	967	ambigua Cresson, Micropeza	634
aloponotum Dyar, Aedes	111	ambigua Fallén, Agromyza	795
alpestre Dorogostajskij, Rubzov, & Vlasenko, Prosimulium	182	ambigua (Fallén), Hydrophoria	863, †864
alpestris Cook, Scatopse	238	*ambigua* (Macquart), Jurinella	†1001
alpha Hull, Eristalis	624	ambigua Zetterstedt, Corynocera	181
alpha (Osten Sacken), Poecilanthrax	442	ambiguum (Fallén), Melanostoma	574, †575
alpha (Phillips), Rhynencina	659	*ambiguum* (Loew), Dictyacium	690
alphaeus (Sublette), Chironomus	166	*ambiguus* (Macquart), Efferia	394
alpica (Zetterstedt), Spilogona	879	ambiguus Stone, Leucotabanus	331
alpicola Rondani, Pogonomyia	902	amblycoryphae (Coquillett), Blaesoxipha	946
alpicola (Zetterstedt), Cricotopus	157	*Amblycoryphenes* Townsend	946
alpicola (Zetterstedt), Phaonia	905	ambocnema Chillcott, Rhamphomyia	461
alpina (Cresson), Parydra	749	*Ambopogon* Greene	711
alpina Harmston & Knowlton, Medetera	509	ambrosiae Felt, Neolasioptera	273
alpina Huckett, Hydrophoria	863	ambrosiae Frick, Phytobia	800
alpina (Ringdahl), Pegomya	862	*Amedoria* Brauer & Bergenstamm	1050
alpina (Zetterstedt), Calliphora	929	amelanchieris (Greene), Phytobia	797
alpinensis Fluke & Hull, Cheilosia	586	ameles Pritchard, Cophura	366
alpinus Cresson, Pipunculus	552	americana (Alexander), Neocladura	72
alpinus (Linnaeus), Aedes	†113	americana (Alexander), Paradelphomyia	59
alta Doane, Tipula	31	americana Bezzi, Thaumalea	120
alta Felt, Colpodia	253	americana Bigot, (Evibrissa)	†1030, 1108
alta (Tucker), Villa	437	*americana* (Bigot), Pterodontia	405
altacola Martin, Leptogaster	362	americana Borgmeier, Hypocera	534
alter Parent, Hydrophorus	504	*americana* Brauer & Bergenstamm, Admontia	1068
alterans Williston, Phthiria	420	*americana* (Brauer & Bergenstamm), Clausicella	1059
altercinctus Melander, Aphoebantus	429	*americana* (Brauer & Bergenstamm), Gymnoclytia	965
Altermetoponia Miller	300	americana (Brauer & Bergenstamm), Leschenaultia	1082
alterna (Walker), Nephrotoma	20	*americana* Brauer & Bergenstamm, Podotachina	1057
alternans Garrett, Anorostoma	812	americana Brooks, Thelaira	1021
alternans Loew, Euxesta	651	americana Carter, Leptoconops	122
alternans (Loew), Paracleius	500	americana (Coquillett), Phthiria	419
alternans Loew, Temnostoma	614	americana (Coquillett), Roeseliopsis	1070
alternatherae Dendy & Sublette, Nanocladius	155	americana Cresson, Hydrellia	743
alternata Aldrich, Plagiomima	1019	americana (Curran), Bezzimyia	963
alternata (Dietz), Nephrotoma	22	*americana* (Curran), Buquetia	1098
alternata Felt, Mycodiplosis	282	americana Curran, Leucozona	563
alternata Fisher, Mycomya	219	*americana* Cushing & Patton, Cochliomyia	924
alternata Harris, (Tipula)	1113	*americana* Darlington, Eccoptomera	811
alternata Huckett, Hylemya	851	americana Day, Odontomyia	316
alternata (Loew), Diplotoxa	792	americana (Fabricius), Cuterebra	1109
alternata Reinhard, Eumacronychia	935	americana Felt, Asynapta	254
alternata Rübsaamen, (Sciara)	236	americana Felt, Brachineura	252
alternata Say, Psychoda	95	americana (Felt), Catotricha	243
alternata (Say), Villa	433	americana (Felt), (Cecidomyia)	289
alternatus Cresson, Pipunculus	553	americana Felt, Cincticornia	269
alternatus (Say), Sylvicola	190	americana Felt, Colpodia	253
alternicula Quate, Psychoda	96	*americana* Felt, Cordylomyia	249
althaea Hull, Xylota	606	americana Felt, Dasineura	258
alticola Aldrich, Cylindromyia	973	americana Felt, Endaphis	276
alticola Alexander, Tipula	31	americana Felt, Feltiella	281
alticola (Cockerell), Rhopalomyia	265	*americana* (Felt), Henria	252
alticola Huckett, Pegomya	858		
alticola James, Coleomyia	366		

	Page		Page
americana Felt, Hormomyia	287	amoeba (Stein), Helina	889
americana Felt, Janetiella	264	**Amoebaleria** Garrett	815
americana Felt, Mayetiola	263	amoena Cresson, Orthacheta	831
americana Felt, Miastor	247	*amoena* (Harris), Orthellia	913
americana Felt, Odontodiplosis	288	amoena (Loew), Chymomyza	770
americana Felt, Toxomyia	276	amoena (Meigen), Omotoma	†1089
americana Fitch, Meromyza	793	*amoenifrons* Harris, (Stratiomys)	1113
americana Fittkau, Ablabesmyia	148	ampelophila Felt, Contarinia	277
americana (Fittkau), Pentaneura	147	ampelophila Felt, Dasineura	258
americana Garrett, Heleomyza	816	ampelophila (Felt), Winnertzia	253
americana Guérin-Méneville, Toxophora	425	*ampelophila* Loew, Drosophila	768
americana Hardy, Leptopteromyia	363	ampelus (Walker), Mericia	1008
americana Hardy, Plecia	192	amphericus Melander & Brues, Dolichopus	488
americana Hering, Paroxyna	665	**Amphicnephes** Loew	655
americana (Hough), Cynomyopsis	930	**Amphicosmus** Coquillett	427
americana (Johannsen), Eucorethra	104	*Amphinome* Meigen	42
americana Johnson, Spania	348	amphinome (Walker), Eccritosia	393
americana Johnson, Volucella	601	**Amphipogon** Wahlberg	711
americana (Kieffer), Polypedilum	176	amphitea Walker, Toxophora	425
americana Kincaid, Pericoma	93	amphitrite (Townes), Chironomus	167
americana (Leach), Lynchia	918	ampla (Doane), Pedicia	54
americana Loew, Arthropeas	298	ampla Garrett, Mycomya	219
americana Malloch, Fannia	893	amplectens Aldrich, Hydrophorus	504
americana Malloch, Leucopis	709	amplectens Melander, Proclinopyga	467
americana Malloch, Megaphthalma	833	amplexa (Coquillett), Carcelia	1092
americana Malloch, Minettia	701	amplicella Coquillett, Phthiria	420
americana Melander, Iteaphila	454	amplicella Coquillett, Rhamphomyia	462
americana Melander, Phyllodromia	473	amplicornis (James), Elfia	1065
americana Needham, Dolichopeza	23	*ampliforceps* (Rowe), Nowickia	995
americana Osten Sacken, Cylindrotoma	41	amplifrons (Brooks), Cistogaster	964
americana Osten Sacken, Haematopota	330	amplifrons Cole, Psilocephala	350
americana Patterson, Gitona	762	*amplifrons* Kröber, Tabanus	333
americana Reinhard, Minella	1025	amplipedis Coquillett, Rhamphomyia	462
americana Robineau-Desvoidy, Graphomya	910	amplipennis Van Duzee, Dolichopus	488
americana Schmitz & Wirth, Phora	536	amplus (Coquillett), Caloparyphus	308
americana (Shannon), Xylota	607	amplus Curran, Platycheirus	576
americana Spencer, Drosophila	763	amplus Curran, Scellus	506
americana (Stein), Proboscimyia	868	amplus Townes, Glyptotendipes	171
americana Steyskal, Colobaea	685	ampulla (Aldrich), Sarcodexia	955
americana Steyskal, Sepedon	693	ampullaceus Van Duzee, Hydrophorus	504
americana Swederus, (Musca)	625	ampullaria Felt, Rhopalomyia	265
americana (Townsend), Actia	1061	*amyotii* (Fitch), Sitodiplosis	278
americana (Townsend), Lydina	1014	amystis Walker, Empis	458
americana (Townsend), Lydina	1014	amytis Walker, Empis	458
americana (Townsend), Myiophasia	984	anabnormis Huckett, Pegomya	863
americana Walker, (Asthenia)	1113	*Anacampta* Loew	643
americana (Wiedemann), Bradysia	†234	anacapai (Wilcox & Martin), Efferia	393
americana (Wiedemann), Homoneura	698	anaces Walker, (Sarcophaga)	961
americana Wiedemann, Rhamphomyia	461	*Anachaetopsis* Brauer & Bergenstamm	†1071
americana (Wiedemann), Xylomya	299	**Anacimas** Enderlein	329
americana (Wulp), Exoprosopa	444	*Anaclinia* Winnertz	224
americana (Wulp), Voria	1020	*Anacrostichus* Bezzi	457
americanella Shewell, Minettia	701	anale Kröber, Zodion	629
americanum Chillcott, Euryomma	898	*analis* (Adams), Exechia	206
americanum Cresson, Clanoneurum	740	analis Borgmeier, Apocephalus	543
americanum (Wheeler), Chrysotus	521	analis (Coquillett), Mycetophila	210
americanum Wiedemann, Zodion	629	*analis* (Fabricius), Archytas	††999
americanus (Brauer Bergenstamm), Senotainia	939	analis (Fabricius), Physoconops	627, †627
americanus (Curran), Orrhodops	381	*analis* Harris, (Milesia)	1113
americanus Drury, Tabanus	332	analis Kirby, Aspistes	241
americanus Forster, Tabanus	332	*analis* Macquart, Anthrax	431
americanus (Johannsen), Chaoborus	103	analis (Macquart), Blera	610
americanus Kieffer, Glyptotendipes	172	*analis* Macquart, Laphria	390
americanus (Kieffer), Tanypus	150	*analis* (Macquart), Laphria	390
americanus (Malloch), Clusiodes	807	analis (Meigen), Antichaeta	688
americanus Melander, Platypygus	415	analis Melander, Oligodranes	421
americanus (Palisot de Beauvois), Diachlorus	328	*analis* (Robineau-Desvoidy), Bonnetia	1004
americanus Van Duzee, Campsicnemus	526	analis Robineau-Desvoidy (Zelia)	1108
americanus (Van Duzee), Rhaphium	513	analis Say, Anthrax	431
americanus (Wiedemann), Metasyrphus	560	analis Say, (Dexia)	1108
Americaptilotus Richards	723	analis Schnabl, Pegomya	862
Americina Malloch	841	analis Westwood, Pterodontia	405
amethystinus (Macquart), Archytas	999, †999	analis Williston, Ogcodocera	428
amicabilis (West), Carinosillus	989	analis Williston, Xylota	605
amicta Reinhard, Clairvillia	975	analis (Wulp), Chaetogaedia	1078
amida, Walker, (Sapromyza)	706	anana Hall, Acronesia	928
Amiota Loew	760	*ananassa* (Osten Sacken), Thecodiplosis	278
amissas (Walker), Baccha	573	ananassae Doleschall, Drosophila	768
amithaon (Walker), Crioprora	610	anandra (Dodge), Ravinia	954
Ammobates Stannius	502	anane (Walker), Hylemya	852
Ammobia Bezzi & Stein	934	**Anapausis** Enderlein	241
ammophiloides Townsend, Systropus	424	**Anaporia** Townsend	1028
ammophilus Loew, Asyndetus	524	**Anarete** Haliday	246
amnicola (Melander & Brues), Dolichopus	488	**Anaretella** Enderlein	244
Amobia Robineau-Desvoidy	934	anarmostus (Melander), Gymnopternus	498
Amobiopsis Townsend	936	*Anarostomoides* Malloch	810

INDEX 1557

Entry	Page
anas Townes, Pseudochironomus	176
Anasimyia Schiner	618
anassa Reinhard, (Phorocera)	1107
anastasia Hull, Volucella	600
Anastoechus Osten Sacken	411
Anastrepha Schiner	673
Anatella Winnertz	205
anatolicus Chillcott, Platypalpus	478
Anatopynia Johannsen	144
anausis Walker, Helophilus	620
anaxias (Walker), Peleteria	997
anaxo Walker, Rhamphomyia	462
Anaxylophagus Malloch	297
anceps (Zetterstedt), Helina	†888
Anceus Roback	171
anchineuria Speiser, Ornithomyia	917
anchista Steyskal, Sepedon	693
anchora (Loew), Pseudogaurax	782
ancilla (Osten Sacken), Philorus	100
ancilla (Walker), Pseudomyothyria	1047, †1100
ancora (Coquillett), Dasyhelea	126
ancoralis (Coquillett), Tenthredomyia	616
ancyla Quate, Pericoma	92
ancysta Roback, Diamesa	151
andersoni Bromley, Stenopogon	383
andersoni Curran, Dolichopus	488
andersoni Leonard, Chrysopilus	346
Andrenosoma Rondani	386
androclus (Walker), Helophilus	618, †623
androgynes Felt, Dicrodiplosis	279
anemone Felt, Dasineura	258
anepsia Pritchard, Anarete	246
Anetia Robineau-Desvoidy	†1039, 1103
Aneurina Lioy	532
Anevrina Lioy	532
Angarotipula Savtshenko	30
angelica Telford, Cheilosia	583
angelicae Beutenmüller, (Cecidomyia)	293
angelicae Frost, Melanagromyza	796
angelicella Frost, Phytomyza	804
Angioneura Brauer & Bergenstamm	932
anglofennica Edwards, Allodia	205
angularis (Alexander), Erioptera	84
angularis Loew, Stratiomys	312
angulata (Adams), Mycomya	220
angulata Curran, Peleteria	997
angulata (Felt), Mycodiplosis	282
angulata (Loew), Phytobia	798
angulata Loew, Tipula	32
angulata (Malloch), Spilogona	879
angulata Shewell, Camptoprosopella	696
angulata (Thomson), Leptocera	723
angulatus Harris, (Syrphus)	1113
angulatus (Karsch), Systropus	424
angulatus (Van Duzee), Tachytrechus	502
angulicincta James, Stratiomys	312
angulicornis (Curran), Slossonaemyia	1064
angulicornis (Malloch), Liriomyza	801
angulineura (Reinhard), Euscopolia	974
angulus Osten Sacken, Lepidanthrax	440
angus Cresson, Pipunculus	552
angusta Banks, Eutreta	661
angusta Melander, Mythicomyia	416
angusta Osten Sacken, Baccha	572
angusta Sabrosky, Elachiptera	777
angusta Stein, Hylemya	850
angustafona Rapp, Psychoda	96
angustata Coquillett, Stenomicra	820
angustata Cresson, Lytogaster	752
angustata Cresson, Psila	639
angustata Van Duzee, Argyra	525
angustata Van Duzee, Exechia	206
angustata (Van Duzee), Orfelia	202
angustata (Wulp), Zenillia	1107
angustatum Williston, Melanostoma	575
angustatus Aldrich, Dolichopus	488
angustatus (Zetterstedt), Platycheirus	576
Angustia Sellers	1096
angusticornis Townsend, Atrophopalpus	1031
angusticornis Van Duzee, Dolichopus	488
angusticornis (Van Duzee), Thecophora	632
angustifacies Hardy, Chrysopilus	346
angustifrons (Aldrich), Blaesoxipha	944
angustifrons (Banks), Amoebaleria	815
angustifrons Dalmat, Cuterebra	1109
angustifrons Loew, Cordilura	827
angustifrons Loew, Paragus	578
angustifrons (Meigen), Hylemya	847
angustifrons Melander, Tethina	727
angustifrons (Reinhard), Sitophaga	1048
angustifrons Sabrosky, Elachiptera	777
angustifrons Williston, Asilus	393
angustifrons (Williston), Physoconops	627
angustifrons (Wulp), Lynchia	918
angustifurca Melander, Bicellaria	451
angustigena Foote, Trypeta	676
angustigena Sabrosky, Phlebosotera	824
Angustiopsis Reinhard	1091
angustior Alexander, Limnophila	64
angustipennis (Alexander), Arctoconopa	80
angustipennis Hine, Asilus	392
angustipennis Loew, Ctenophora	20
angustipennis Loew, Diogmites	371
angustipennis Loew, Rhamphomyia	462
angustipennis Loew, Tachytrechus	503
angustipennis (Loew), Tephritis	668
angustipennis Loew, Tipula	36
angustipennis Melander, Strongylophthalmyia	641
angustipennis (Melander), Tethina	727
angustipennis (Staeger), Pteromicra	687
angustistylum Sabrosky, Elachiptera	777
angustitarsis Malloch, Hylemya	847
angustitarsis Malloch, Lonchaea	717
angustitarsis (Malloch), Paraprosalpia	866
angustiventralis Huckett, Hylemya	847
angustiventris Loew, Xylota	605
angustiventris Malloch, Hylemya	847
angustivitta (Aldrich & Webber), Angustia	1096
angustula Alexander, Limnophila	65
angustum Townes, Polypedilum	175
anhydor Dyar, Uranotaenia	108
anicula (Zetterstedt), Mydaea	†891
anilis Fallén, Dryomyza	680
aninotata Huckett, Pegomya	858
anips Dyar, Culex	119
aniseta Stein, Hylemya	850
Anisomera Meigen	69
ANISOPODIDAE	190
ANISOPODINAE	190
ANISOPODOIDEA	190
Anisopogon Loew	373
Anisopus Meigen	190
anisotae (Webber), Lespesia	1101
Anisotamia Macquart	†428
anna (Coquillett), Villa	439
anna Cresson, Limnellia	758
anna Harriot, Euxesta	651
anna Williston, Volucella	600
annae Steyskal, Tetanocera	693
annamariae (Brimley), Labostigmina	313
annexa Huckett, Leucophora	867
anniae Brimley, Helophilus	619
annonae (Fabricius), Euxesta	651
annosa (Zetterstedt), Quadrularia	889
annularis (De Geer), Chironomus	166
annularis (Hine), Hamatabanus	341
annularis (Meigen), Paratendipes	173
annularis (Melander), Lyciella	700
annularis Melander, Tachypeza	473
annulata Adams, Chlorops	789
annulata (Fallén), Periscelis	710
annulata (Felt), Dicrodiplosis	279
annulata Hull, Laphystia	376
annulata (Linnaeus), Limonia	44
annulata Meigen, Trichocera	15
annulata (Melander), Lyciella	¹ 700
annulata Melander, Mythicomyia	416
annulata (Say), Ablabesmyia	148
annulata (Say), Dolichopeza	23
annulata (Walker), Thaumatomyia	787
annulata Westwood, Gynoplistia	69
annulatus Cook, Chaoborus	103
annulatus (Meigen), Symmerus	199, †199
annulatus (Say), Leucotabanus	331
annulatus (Say), Psilonyx	363
annulatus Van Duzee, Chrysotus	521
annulatus Van Duzee, Gymnopternus	498
annulatus (Williston), Tolmerus	401
annulicornis Malloch, Johannsenomyia	137
annulicornis Malloch, Sphaerocera	718
annulicornis Say, Tipula	25
annulicrus (Townes), Stictochironomus	177

	Page		Page
annulifera (Bigot), Beris	301	antennata Felt, Dasineura	258
annulifera Bigot, Xylota	605	antennata Felt, Dicrodiplosis	279
annulimanus Wulp, Anopheles	106	antennata Felt, Porricondyla	255
annulipes Harris, (Myopa)	1113	antennata Stein, Fucellia	842
annulipes Johnson, Heteromeringia	806	antennata Walker, (Tachina)	1108
annulipes (Johnson), Telmatoscopus	94	antennata (Winnertz), Monardia	250
annulipes Lundström, Synneuron	237	antennatus Malloch, Apocephalus	543
annulipes Macquart, Asilus	392	antennatus Wilcox & Martin, Nannocyrtopogon	379
annulipes (Macquart), Callopistromyia	643		
annulipes Macquart, Chyliza	640	*antennipes* (Say), Rainieria	636
annulipes (Meigen), Themira	684	**Anthalia** Zetterstedt	450
annulipes (Schmitz), Megaselia	540	*anthemon* McAtee, Laphria	391
annulipes Walsh, (Cecidomyia)	291	**Anthepiscopus** Becker	455
annulipes Zetterstedt, Dolichopus	496	anthici Felt, (Itonida)	289
annulipes (Zetterstedt), Pherbellia	686	*Anthoica* Rondani	916
annuliventris (Malloch), Brillia	154	*anthomydea* (Bigot), Myospila	890
annulus (Meigen), Limonia	44	**Anthomyia** Meigen	865
annulus (Walker), Copromyza	720	**Anthomyiella** Malloch	865
Anocha Pritchard	244	ANTHOMYIIDAE	826
Anolisimyia Dodge	942	ANTHOMYIINAE	843
anomala Adams, Thereva	353	ANTHOMYIINI	844
anomala (Bellardi), Efferia	393	**Anthomyiopsis** Townsend	1022
anomala (Cole), Coelopina	680	**Anthomyza** Fallén	819
anomala (Johannsen), Epicypta	209	*Anthomyza* Meigen	865
anomala Johnson, Cylindrotoma	41	ANTHOMYZIDAE	819
anomala (Malloch), Megaselia	540	anthonomus Melander, Oligodranes	421
anomala Malloch, Piophila	711	anthophila Melander, Empis	458
anomala Melander, Mythicomyia	416	anthophila (Osten Sacken), Rhopalomyia	265
anomala (Osten Sacken), Erioptera	80	*Anthophilina* Zetterstedt	819
anomala Painter, Exoprosopa	444	anthophilus Loew, Apostrophus	972
anomala (Townsend), Hypertrophocera	1072	anthophorinus (Fallén), Eristalis	623
anomala Townsend, Neophyto	952	anthophorus Addis, Phlebotomus	91
anomala (Townsend), Plagiomima	1019	*anthorinus* (Fallén), Eristalis	623
anomala Wheeler, Paramycodrosophila	771	anthracina Bigot (Anthomyia)	869
anomala Wilcox & Martin, Backomyia	364	*anthracina* Brennan, Bequaertomyia	319
anomala Williston, Desmatomyia	420	anthracina (Czerny), Hylemya	855
Anomalempis Melander	447	anthracina (Doane), Procecidochares	660
Anomalochaeta Frey	821	anthracina Malloch, Coenosia	871
anomalus Cole, Cyrtopogon	368	*anthracina* Malloch, Hylemya	850
anomalus Malloch, Chrysotus	521	anthracina (Thompson), Thompsonomyia	1007
anomalus Painter, Sparnopolius	413	ANTHRACINAE	431
anomalus (Shannon), Chalcosyrphus	589	anthracinus Bigot, Chrysopilus	346
anonyma (Riley), Lespesia	1101	anthracinus (Bigot), Oxynops	1046
anonyma (Williston), Monochaetoscinella	778	antbracinus Zetterstedt, Chironomus	165
Anopheles Meigen	105	antbracodes Coquillett, Rhamphomyia	462
ANOPHELINAE	105	*anthracodes* Malloch, Hylemya	850
anopla Steyskal, Pteromicra	687	**Anthracophaga** Loew	788
Anoplodonta James	314	anthrax Loew, Platypeza	549
Anoplomerus Rondani	507	*Anthrax* Scopoli	431, †433
anorbitalis (Brooks), Oswaldia	1046	*anthrax* Williston, Laphria	389
anormostus (Melander), Gymnopternus	498	anthreas Walker, Xylcta	606
Anorostoma Loew	812	antica Garrett, Boletina	217
Anorostomoides Malloch	810	*antica* (Walker), Bithoracochaeta	874
anorufa Stein, Pegomya	858	antica Walker, (Sciomyza)	695
Anorycampta Bigot	†1106	**Antichaeta** Haliday	688
Anoxynops Townsend	1038	anticus (Walker), Microtendipes	172
ansatus Harris (Asilus)	1113	antigua Wheeler, Stegana	760
anser (Townsend), Exorista	1054	antimachus (Walker), Tolmerus	401
antaea (Walker), Lampria	388	*Antineura* Melander	802
anteapicalis Alexander, Limonia	45	*antiopa* Dietz, Tipula	27
antecedens Walker, Anthrax	431	**Antiopa** Meigen	582
antecessor Melander, Mythicomyia	416	*antiqua* (Meigen), Hylemya	847
anteilis (Roback), Orthocladius	155	*antiqua* (Meigen), Ocytata	1072
antennaepes (Say), Rainieria	636	*antitheus* Walker, Chrysogaster	591
antennalis (Coquillett), Chaetocrania	1074	*Antocha* Osten Sacken	51
antennalis (Coquillett), Elfia	1066	antoinetta Dyar & Knab, Wyeomyia	107
antennalis (Coquillett), Stilobezzia	134	antoma Garrett, Boletina	217
antennalis Coquillett, Winthemia	1090	antrozoi (Townsend), Basilia	922
antennalis (Fitch), Camptoprosopella	696	*anxia* (Walker), Ravinia	954
antennalis Malloch, Pseudodinia	708	**Anypotacta** Czerny	815
antennalis Melander, Leptopeza	451	**Anzamyia** Reinhard	1016
antennalis (Reinhard), Drino	1084	**Aochletus** Osten Sacken	307
antennalis Stein, Coenosia	870	apache Alexander, Atarba	71
antennalis Townsend, Phasmophaga	1070	apache Alexander, Gnophomyia	73
antennalis Townsend, Xanthophyto	1009	apache Alexander, Hexatoma	70
antennaria (Doane), Hexatoma	70	apache Alexander, Tipula	24
antennariae (Wheeler), Asphondylia	267	apache Painter & Hall, Poecilanthrax	442
antennariae (Wheeler), Rhopalomyia	265	apache Sabrosky, Thaumatomyia	787
antennata Aldrich, Lispe	877	**Apachekolos** Martin	361
antennata Alexander, Pedicia	55	**Apachemyia** Townsend	1010
antennata (Banks), Atomosiella	387	**Apacheprospherysa** Townsend	1038
antennata Coquillett (Gonia)	1113	**Apalocnemis** Philippi	454
antennata Coquillett, Limnophila	67	**Apatolestes** Williston	320
antennata Felt, Camptomyia	256	*Apaulina* Hall	925
antennata Felt, Caryomyia	270	*Apedilum* Townes	173
antennata Felt (Cecidomyia)	289		

INDEX 1559

	Page
Apeloglutus Greene	1064
Apemon Johannsen	204
aperta Alexander, Tipula	35
aperta Brauer & Bergenstamm, Catemophrys	1031
aperta (Coquillett), Hesperoconopa	78
aperta (Coquillett), Pedicia	54
aperta Loew, Rhamphomyia	462
aperta Melander, Apolysis	421
aperta Melander, Euthyneura	449
aperta Reinhard, Paralispidea	1063
aperta Röder, Myopa	630
aperta (Stein), Helina	886
apertella (Parker), Blaesoxipha	943
Aphaniosoma Becker	822
Aphanotrigonum Duda	785
Aphantorhapha Townsend	1062
Aphantotimus Wheeler	511
apheles Melander & Brues, Dolichopus	488
Apheloglutus Greene	1064
Aphelomyia Roback	956
aphidivora Felt (Itonida)	289
Aphidoletes Kieffer	280
Aphiochaeta Brues	540
Aphoebantus Loew	429
Aphria Robineau-Desvoidy	1010
APHRIINI	1010
Aphriosphyria Townsend	997
Aphritis Latreille	597
APHROSYLINAE	509
Aphrosylus Haliday	509
apicales Harmston & Knowlton, Peloropeodes	517
apicalis Adams, Culex	118, †118
apicalis Adams, Keroplatus	201
apicalis Alexander, Ormosia	85
apicalis (Banks), Telmatoscopus	94
apicalis (Brauer & Bergenstamm), Lespesia	1101
apicalis (Cole), Curranops	644
apicalis Coquillett, Atacta	1080
apicalis (Coquillett), Erythrandra	941
apicalis Coquillett, Euparyphus	307
apicalis (Coquillett), Geomyza	821
apicalis Curtis, Scatophaga	838
apicalis Felt, Asynapta	254
apicalis Felt (Cecidomyia)	289
apicalis Felt, Lobopteromyia	276
apicalis Hardy & Knowlton, Pipunculus	553
apicalis Johnson, Chaetopsis	650
apicalis Loew, Chyliza	640
apicalis Loew, Platypalpus	478
apicalis (Loew), Suillia	809
apicalis Loew, Tipula	35
apicalis Loew, Volucella	600
apicalis Malloch, Emmesomyia	858
apicalis Malloch, Gaurax	781
apicalis (Meigen), Geomyza	821
apicalis Meigen, Gnoriste	218
apicalis Reinhard, Paradidyma	1032
apicalis Robineau-Desvoidy (Zelia)	†990, 1108
apicalis (Shaw), Orfelia	202
apicalis (Stein), Pegomya	858
apicalis Stein, Phaonia	905
apicalis (Wahlberg), Homalocephala	652
apicalis Walker (Discocephala)	401
apicalis (Walker), Microphthalma	982
apicalis Walker, Myopa	630
apicalis (Walker), Peleteria	997
apicalis (Walley), Pentaneura	147
apicalis (Wiedemann), Efferia	394
apicalis (Williston), Acrosticta	650
apicalis Wirth, Systenus	517
apicalis (Zetterstedt), Clusiodes	807
apicarinus Hardy & Knowlton, Pipunculus	552
apicata (Alexander), Limonia	49
apicata Bennett & Sabrosky, Cephenemyia	1111
apicata Curran, Tachinomyia	1058
apicata Felt, Dasineura	258
apicata Felt, Rhopalomyia	265
apicata Johannsen, Phaonia	905
apicata (Loew), Pteromicra	687
apicata Malloch, Bezzia	141
apicata Osten Sacken, Ctenophora	19
apicata (Thomas), Mylogymnocarena	669
apicata (Thomson), Scaptomyza	769
apicatum Townes, Polypedilum	175
apicatus Malloch, Hippelates	775
apicauda Curran, Pyrophaena	578

	Page
apicaudus (Curran), Xylota	607
apichaetus Curran, Chamaesyrphus	595
apicifer (Walker), Archytas	999, †999
apicifera Townsend, Volucella	600
apicinebula (Malloch), Physoptera	541
apicis Kieffer, (Cecidomyia)	289
apiciseta Ringdahl, Hylemya	853
apicispina Alexander, Gonomyia	74
apicola Cole, Villa	438
Apicomyia Shannon	588
apicula Curran, Pyrophaena	578
apicula Garrett, Docosia	226
apicula Loew, Stratiomys	312
apiculata Alexander, Limnophila	64
apiculatus Malloch, Clusiodes	807
apiculus Coquillett, Lordotus	412
apiforme (Fabricius), Temnostoma	†614
apila (Bromley), Laphria	389
apina (Walker), Paraprosalpia	867
Apinocyptera Townsend	973
Apinops Coquillett	971
Apiocera Westwood	356
APIOCERIDAE	356
APIOCERINAE	356
apiphilus Felt, Arthrocnodax	283
apisaon (Walker), Parapenium	580
apivora (Aldrich), Neohylemyia	867
apivora (Fitch), Promachus	400
Aplomya Robineau-Desvoidy	1096
Aplomyia Robineau-Desvoidy	1096
Aplomyiopsis Villeneuve	1038
Aplonyx De Stefani Perez	275
aplopappi (Coquillett), Aciurina	670
Apocephalus Coquillett	543
apocyni Felt (Cecidomyia)	289
apocyni Felt, Lasioptera	270
apocynifiorae Felt, Lestodiplosis	284
Apolysis Loew	421
Apomidas Coquillett	357
Aporia Macquart	1028
aporia Pritchard, Polyardis	250
Aporotachina Meade	1039
Apostrophus Loew	972
appalachicola Alexander, Neolimnophila	73
apparens (Becker), Conioscinella	783
Appendicia Stein	1007
appendiculata Grabham, Corethrella	105
appendiculata Harris (Platypeza)	1113
appendiculata (Herrick), Chaoborus	103
appendiculata (Loew), Mycomya	220
appendiculata Loew, Paralimna	748
appendiculata Loew, Parydra	749
appendiculata Loew, Tipula	32
appendiculata (Macquart), Lepidophora	425
appendiculata Malloch, Hylemya	853
appendiculata (Zetterstedt), Clinocera	468
appendiculatum Harmston & Knowlton, Parasyntormon	516
appendiculatus Bigot (Holopogon)	401
appendiculatus Cresson, Pipunculus	554
appendiculatus Loew, Asyndetus	524
appendiculatus Van Duzee, Dolichopus	488
appendiculatus Wheeler, Medetera	511
appendipes (Cresson), Tomosvaryella	555
appensus Harris (Helophilus)	1113
appoximata Banks, Pseudotephritis	647
appressa James, Adoxomyia	305
appropinqua (Adams), Thaumatomyia	787
approximans (Walker), Diachlorus	328
approximata Banks, Pseudotephritis	647
approximata (Dietz), Nephrotoma	20
approximata Malloch, Leptocera	725
approximata (Malloch), Megaselia	541
approximata (Malloch), Neophyllomyza	730
approximata Malloch, Oxycera	307
approximata Sturtevant & Wheeler, Nostima	745
approximata Walker, Cuterebra	1109
approximatonervis (Zetterstedt), Lasiosina	791
apricata Melander, Mythicomyia	416
aprilina Alexander, Tipula	27
aprilina Malloch, Agromyza	796
aprilina (Meigen), Bradysia	234
aprilina Osten Sacken, Limnophila	64
aprilis Felt, (Itonida)	289
Apriona Kieffer	251
Aprionus Kieffer	251
apsectra Edwards, Bryomyia	250

	Page
Apsectrotanypus Fittkau	145
apta Stein, Phaonia	908
Aptanogyna Börner	235
aptena Wirth, Tethymyia	163
Aptilotus Mik	†723
Aptorthus Aldrich	482
Apystomyia Melander	424
aquatilis Aldrich, Hydrophorus	504
aquavicinus (Hardy), Pipunculus	553
aquilae Brodie & White, (Hippobosca)	1113
aquilegiana Frost, Phytomyza	804
aquilonia Kessel, Agathomyia	547
aquilonia McAlpine, Earomyia	716
aquilonius Fairchild & Harwood, Phlebotomus	91
aquita (Dietz), Limonia	45
aquitima Huckett, Hylemya	847
Arabiopsis Townsend	939
Arachnidomyia Townsend	957
araneae (Coquillett), Pseudoganrax	782
araneosa (Coquillett), Tephritis	668
araneosa Felt, Clinodiplosis	282
aranti Hays, Tabanus	332
arapaho Alexander, Ormosia	84
arapahoensis Alexander, Molophilus	88
arator (Aldrich), Pilatea	1105
arator Melander, Nemotelus	310
aratrix Pandellé, Sarcophaga	957
aratus Roback, Cricotopus	157
Aravaipa Townsend	1074
arborcola Martin, Leptogaster	362
arborealis Stone, Tabanus	337
arboreum Curran, Rhaphium	512
arboreus Van Duzee, Pelastoneurus	502
arboricola Root & Hoffman, Culicoides	129
arbustorum (Linnaeus), Eristalis	623
archboldi Frost, Liriomyza	801
Archilestes Schiner	364
Archilestris Loew	364
Archilimnophila Alexander	61
Archimimus Reinhard	942
Archimyia Enderlein	297
archippivora (Riley), Lespesia	1101
archon Melander, Anomalempis	447
Archytas Jaennicke	999
arctica (Becker), Themira	684
arctica (Boheman), Diamesa	151
arctica (Curran), Xylota	606
arctica Curtis, Tipula	30
arctica Holmgren, Boletina	217
arctica Holmgren, Piophila	711
arctica Johnson, Volucella	601
arctica (Lundbeck), Phytoliriomyza	803
arctica Lundström, Trichocera	15
arctica (Malloch), Copromyza	720
arctica Malloch, Hydrophoria	865
arctica Malloch, Microprosopa	835
arctica Malloch, Peleteria	997
arctica (Malloch), Pseudodiamesa	153
arctica Malloch, Ptiolina	347
arctica Malloch, Smittia	162
arctica (Ringdahl), Pegomya	860
arctica (Sack), Tenuirostra	1006
arctica Schirmer, Sericomyia	603
arctica Van Duzee, Medetera	510
arctica (Zetterstedt), Melangyna	566
arctica (Zetterstedt), Helophilus	879
arcticola Alexander, Nephrotoma	21
arcticola Huckett, Pegomya	860
arcticum Malloch, Simulium	188
arcticus (Becker), Gonarcticus	832
arcticus (Coquillett), Atrichopogon	122
arcticus Melander, Platypalpus	478
arcticus Zetterstedt, Helophilus	†618
arctiventris James, Chrysopilus	346
Arctobiella Coquillett	716
Arctoconopa Alexander	79
Arctopelopia Fittkau	146
Arctophila Schiner	604
Arctophyto Townsend	985
Arctopiophila Duda	712
Arctosyrphus Frey	620
arctotibia Chillcott, Rhamphomyia	462
Arctotipula Alexander	29
arcuaria (Felt), Caryomyia	270
arcuata Coquillett, Rhamphomyia	462
arcuata (Doane), Ormosia	85

	Page
arcuata (Fabricius), Palloptera	715
arcuata Felt, Rhopalomyia	265
arcuata Garrett, Sciara	230
arcuata (Loew), Euthycera	690
arcuata (Loew), Milichiella	733
arcuata Loew, Odontomyia	315
arcuata (Malloch), Plastophora	542
arcuata (Meigen), Gymnophora	†542
arcuata Sherman, Tetragoneura	228
arcuata (Stein), Gymnodia	884
arcuata (Tothill), Mericia	1008
arcuata Van Duzee, Neurigona	518
arcuata (Walker), Tritoxa	649
arcuatus (Fallén), Dasysyrphus	†560, †561, 563
arcuatus (Garrett), Procladius	149
arcuatus (Say), Xanthomelanodes	967
arcuatus Van Duzee, Campsicnemus	526
arcuatus Van Duzee, Chrysotus	521
arcucinctus (Walker), Metasyrphus	560
arculata (Loew), Strauzia	676
ardeae (Macquart), Lynchia	918
ardens Macquart, Mallophora	396
Ardoptera Macquart	470
arelate (Walker), Paraprosalpia	866
arenaria Cresson, Scatophila	758
arenicola Cole, Villa	440
arenicola James, Leptogaster	362
arenicola (Johnson), Villa	434
arenicola Johnson & Johnson, Exoprosopa	444
arenicola Melander, Aphoebantus	429
arenicola (Osten Sacken), Lasiopogon	377
arenicola Painter, Geron	423
arenicola Reinhard, Senotainia	938
arenicola Wilcox, Stichopogon	385
arenosa (Coquillett), Villa	437
arenosa (Ringdahl), Spilogona	880
arenosus Pritchard, Omniablautus	380
areolata (Osten Sacken), Prolimnophila	63
areolatus Walker, Chrysops	327
areos (Walker), Lydina	1014
arethusa (Hull), Mesograpta	571
arethusa (Osten Sacken), Poecilanthrax	442
arge Dyar & Shannon, Dixa	101
argentata Coquillett, Coenosia	871
argentata (Loew), Johannsenomyia	137
argentata (Walker), Brachydeutera	753
argentata (Williston), Adoxomyia	305
argentatus (Cole), Anthrax	431
argentatus Coquillett, Nicocles	380
argentatus Van Duzee, Chrysotus	521
argentatus Van Duzee, Hydrophorus	504
argentea (Bigot), Rhamphomyia	464
argentea Curran, Rhamphomyia	465
argentea Doane, Dicranota	57
argentea (Rowe), Nowickia	995
argentea Snyder, Lispe	877
argentea (Townsend), Cylindromyia	973, †973
argenteceps (Alexander), Limonia	49
argenteus (Say), Stichopogon	385
argenteus (Townes), Chironomus	167
argenteus (Wiedemann), Leucotabanus	331
argenti Felt, Trotteria	276
argenticeps (Curran), Myopina	843
argenticeps Malloch, Coenosia	871
argenticeps Malloch, Spilogona	880
argenticolor Stein, Coenosia	871
argentifacies Van Duzee, Diaphorus	519
argentifacies Van Duzee, Hydrophorus	504
argentifacies (Van Duzee), Physoconops	627
argentifacies (Williston), Perasis	382
argentifer (Loew), Nicocles	380
argentifera (Kröber), Ozodiceromyia	349
argentifrons Aldrich, Belvosia	1081
argentifrons (Brooks), Hyalomya	968
argentifrons Cole, Psilocephala	350
argentifrons Coquillett, Pseudochaeta	1095, †1096
argentifrons (Hine), Efferia	394
argentifrons Townsend, Gymnoprosopa	936
argentifrons (Townsend), Senotainia	939
argentifrons Williston, Desmatoneura	430
argentina Bigot, Ophyra	†901
argentipes Van Duzee, Dolichopus	488
argentipila Fluke & Hull, Cheilosia	586
argentisquama Felt, Lasioptera	270
argentiventris (Malloch), Spilogona	880

INDEX 1561

	Page		Page
argentiventris Van Duzee, Argyra	525	arizonensis Patterson & Wheeler, Drosophila	763
Argentoepalpus Townsend	1002	*arizonensis* Quisenberry, Tephritis	668
argentosa Reinhard, Minthozelia	1027	arizouensis (Schaeffer), Cophura	366
argoi Shannon, Xylota	605	arizonensis Wilcox, Heteropogon	374
Argoravinia Townsend	943	arizonensis (Williston), Philonicus	398, †398
argryofrons Hardy & Knowlton, Pipunculus	553	arizonensis Wirth & Hubert, Culicoides	130
argus Melander, Dolichocephala	470	arizonica Alexander, Dioptopsis	99
argus (Roback), Chironomus	167	*arizonica* Alexander, Tipula	38
argus (Say), Limonia	44	arizonica (Parker), Metoposarcophaga	951
argus Williston, Simulium	187	arizonica Townsend, Aphantorhapha	1062
argus (Zetterstedt), Ernoneura	837	arizonica Townsend, Phoeniciomyia	1047
arguta Alexander, Pilaria	68	arizonica (Townsend), Sarcophaga	957
argutus Painter, Geron	423	arizonicus Alexander, Molophilus	88
Argyra Macquart	524	*arizonicus* (Banks), Physoconops	627
Argyramoeba Schiner	431	arizonicus Banks, Systropus	424
Argyrandrus Bezzi	458	arizonicus Harmston, Dolichopus	488
argyrata Curran, Hilara	456	arizonicus Wheeler, Paraneossos	817
argyria Melander, Euthyneura	449	arizoniensis Felt, Winnertzia	253
argyriceps (Curran), Nanna	831	arkansensis (Malloch), Dasiops	716
argyrocephala Macquart, (Sarcophaga)	961	arkansensis Van Duzee, Chrysotus	521
argyrocephala (Meigen), Metopia	937	armata Aldrich, Cylindromyia	973
Argyromoeba Schiner	431	armata Cresson, Mythicomyia	416
argyropus Becker, Tachytrechus	503	armata Doane, Tipula	35
argyropya Wiedemann, Anthrax	431	armata Garrett, Mycomya	220
argyropygus Wiedemann, Anthrax	431	armata Hardy, Tomosvaryella	555
argyrosoma (Hine) Efferia	394	armata (Hine), Efferia	394
argyrostoma (Cresson) Ditrichophora	739	*armata* (Malloch), Xenomydaea	892
argyrostoma (Robineau-Desvoidy), Sarcophaga	957	armata (Meigen), Fannia	893
Arhipidia Alexander	45	armata Melander, Drapetis	477
Aricia Robineau-Desvoidy	886	armata Osten Sacken, Erioptera	82
Ariciella Malloch	890	*armata* Robineau-Desvoidy, Strauzia	676
ariciiformis (Holmgren), Fucellia	842	*armata* Shewell, Homoneura	698
arida Cole, Leptogaster	362	armata (Stein), Hylemya	847
arida (Curran), Ptilodexia	988	armatipes Malloch, Sphegina	593
arida Reinhard, Psilopleura	1019	armatipes (Malloch), Xenomydaea	892
arida (Williston), Efferia	394	armatipes Wirth, Palpomyia	140
aridela Alexander, Ulomorpha	68	armatum Curran, Rhaphium	512
aridella (Fallén), Chamaemyia	707	*armatus* Andersen, Psectrocladius	159
aridum Sabrosky, Astiosoma	824	armiger Huckett, Coenosia	871
aridus Cole & Wilcox, Lasiopogon	377	armiger Van Duzee, Thinophilus	507
aridus Curran, Saropogon	383	armigera Alexander, Gonomyia	75
aridus James, Hodophylax	374	armigera (Coquillett), Eucelatoria	1043
aridus Malloch, Apocephalus	543	armillaris Osten Sacken, Erioptera	83
aridus Painter, Geron	423	armillata (Osten Sacken), Blera	610
aridus (Williston), Tomosvaryella	556	armillatus Melander, Platypalpus	478
ariel (Sublette), Chironomus	167	armipes Cresson, Mythicomyia	416
arietinus (Coquillett), Podonomus	150	armipes (Fallén), Hydrotaea	899
aristale Townsend, Phasiostoma	1063	armipes (Loew), Acantholeria	814
aristalis (Coquillett), Parectecephala	792	armipes Loew, Empis	458
aristalis (Coquillett), Ptychoneura	938	armipes Loew, Sepedon	693
aristalis (Coquillett), Sciomyza	688	*armipes* (Malloch), Xenomydaea	892
aristalis (Curran), Ptychoneura	938	armipes Melander, Limnia	691
aristalis (Malloch), Megaselia	538	armipes Melander, Microphorus	452
aristalis Reinhard, Paradidyma	1032	*armipes* (Stein), Spilogona	†883
aristalis Townsend, Ostracophyto	1009	arnaudi Foote, Euarestoides	669
aristata (Aldrich), Microcerella	951	arnaudi Harmston, Medetera	510
aristata Malloch, Aecothea	811	arnaudi (Laffoon), Mycetophila	210
aristata Malloch, Agromyza	795	arnaudi Melander, Charadrodromia	475
aristatus Aldrich & Shannon, Boreellus	925	arnaudi Quate, Telmatoscopus	93
aristatus James, Nannocyrtopogon	379	arnaudi (Reinhard), Euphorocera	1054
aristatus (Johnson), Chalcosyrphus	589	arnaudi Wilcox & Martin, Nannocyrtopogon	379
arizona (Snyder), Coenosia	874	arnaudi Wirth, Nocticanace	734
arizonae Kieffer, Lasioptera	271	*arno* Townsend, Proctacanthus	†399
arizonaensis (Hardy), Dilophus	195	arnolitra Huckett, Hylemya	847
arizonaensis Quisenberry, Tephritis	668	aromaticae Felt, Dasineura	258
arizonense Cook, Colobostoma	238	arossi Kessel, Agathomyia	548
arizonensis Alexander, Epiphragma	61	arpadi (Szilády), Hybomitra	338
arizonensis Bequaert, Hirmoneura	402	arpidia Malloch, Oscinoides	781
arizonensis Bohart, Culex	118	*Arrhenica* Osten Sacken	69
arizonensis Borgmeier, Coniceromyia	536	*Arrhinomyia* Brauer & Bergenstamm	1051
arizonensis (Bromley), Osprioceros	381	arrisor (Reinhard), Sitophaga	1048
arizonensis Chillcott, Fannia	893	artata (Reinhard), Melanomya	933
arizonensis Cole, Acrocera	405	artemisiae Felt, Asphondylia	268
arizonensis Cole, Psilocephala	350	artemisiae Felt, Diarthronomyia	262
arizonensis Coquillett, Orimarga	52	artemisiae (Kaltenbach), Phytobia	800
arizonensis (Coquillett), Villa	437	artemita Hull, Xylota	605
arizonensis Cresson, Scatella	757	**Arthria** Kirby	241
arizonensis Felt, Asphondylia	268	arthritica Melander, Empis	458
arizonensis Felt, Lasioptera	271	**Arthroceras** Williston	298
arizonensis (Felt), Phaenolauthia	261	ARTHROCERATINAE	298
arizonensis Johnson & Johnson, Amphicosmus	427	**Arthrocnodax** Rübsaamen	283
arizonensis Johnson & Johnson, Lordotus	412	**Arthropeas** Loew	298
arizonensis (Malloch), Megaselia	540	articulatus Say, Bibio	192
arizonensis Malloch, Trupanea	667	articulosa (Felt), Xylopriona	249
		articulosa Harris, (Bibio)	1113
		articus (Malloch), Copromyza	720

	Page		Page
artifer (Curran), Polypedilum	175	astuta Stein, Mydaea	892
artisia (Roback), Pagastia	152	astutus Fluke, Metasyrphus	560
arubae Fox & Hoffman, Culicoides	129	astutus Williston, Asilus	392
arundani (Hardy), Pipunculus	553	astylata Melander, Oedalea	450
arvicola (Meigen), Nemorilla	†1089	**Asynapta** Loew	254
asackeni Wilcox, Laphria	389	**Asyndetus** Loew	523
asakakae Smart, Simulium	187	*Asyndulum* Latreille	199
Asaphomyia Stone	321	ata Garrett, Mycomya	220
asbestos Philip, Chrysops	324	**Atacta** Schiner	1080
ASCHIZA	530	**Atactomima** Townsend	1080
Ascia Meigen	594	**Atactopsis** Townsend	1081
asciaeformis Becker, Pelastoneurus	501	*ataenia* (Macquart), Diachlorus	328
ascita Cresson, Hydrellia	743	**Ataenogera** Kröber	349
asclepiae Felt, Lestodiplosis	284	*Atalanta* Meigen	468
asclepiae Felt, Neolasioptera	273	**Atarba** Osten Sacken	70
ascriptiva Hendel, Oedopa	653	*atelestes* (Hardy), Dilophus	196
asellus Wheeler, Parasyntormon	516	**Ateloglossa** Coquillett	986
asema Melander, Empis	458	*Atelogossa* Coquillett	986
Asemosyrphus Bigot	620	*Atemnocera* Bigot	600
asemus Pritchard, Aprionus	251	ater Brennan, Apatolestes	320
asiatica Alexander, Erioptera	82	ater Coquillett, Apinops	971
Asiconops Chen	626	*ater* (Coquillett), Parabombylius	410
Asicya Lynch Arribálzaga	376	ater (Cresson), Oligodranes	421
ASILIDAE	360	ater Macquart, (Chlorops)	793
ASILINAE	391	*ater* Macquart, Chrysops	324
ASILOIDEA	348	ater Malloch, Oscinoides	781
asiloides (De Geer), Mydas	358	ater Meigen, Pipunculus	552
Asilus Linnaeus	391	ater Melander, Syneches	448
Asindulum Latreille	199	ater Melander & Argo, Clusiodes	807
asiovora Shannon & Dobroscky, Protocalliphora	926	*ater* (Palisot de Beauvois), Whitneyomyia	331, †337
		ater (Townes), Chironomus	170
Asmeringa Becker	751	*aterrima* (Banks), Pericoma	92
Aspathia Enderlein	185	*aterrima* Becker, Piophila	711
aspera (Roback), Ablabesmyia	148	aterrima Bigot, Nemopoda	682
aspera Walker, Chionea	72	*aterrima* Coquillett, Chaetoplagia	1017
aspera (Walker), Ravinia	955	aterrima (Doane), Oxyna	665
aspersa Doane, Tipula	26	aterrima (Fabricius), Phora	537
aspersulus Alexander, Molophilus	88	aterrima Johnson, Xylomya	299
Asphondylia Loew	267	aterrima (Malloch), Earomyia	716
ASPHONDYLIINI	267	aterrima (Meigen), Smittia	162
aspidoptera Alexander, Tipula	27	aterrima (Melander), Rhexoza	240
aspidoptera (Coquillett), Pedicia	54	aterrima Robineau-Desvoidy, (Peckia)	961
Aspilomyia Hendel	672	aterrima Wulp, Pogonomyia	902
Aspilota Loew	672	aterrimum (Villers), Leucostoma	976
aspilota Melander, Mythicomyia	417	aterrimus (Bigot), Anthrax	431
aspinosa Saunders, Forcipomyia	124	aterrimus (Robineau-Desvoidy), Archytas	1000
aspinosa (Shewell), Lyciella	700	athabasca Alexander, Tipula	32
Aspistes Meigen	241	athabasca Sturtevant & Dobzhansky, Drosophila	768
ASPISTINAE	241		
asplenifolia (Felt), Janetiella	264	athabascae (Alexander), Limonia	46
asquamatus Andersen, Limnophyes	160	**Athanatus** Reinhard	1022
Asseclamyia Reinhard	1012	*Atherigona* Rondani	875
assidua Johannsen, Exechia	206	**Atherix** Meigen	344
assidua (Walker), Ravinia	†953, 954	**Athrycia** Robineau-Desvoidy	1016
assimilis (Fallén), Drapetis	477	**Athyroglossa** Loew	735
assimilis (Fallén), Muscina	909	**Atissa** Haliday	735
assimilis Huckett, Pegomya	858	*Atissiella* Cresson	736
assimilis (Loew), Suillia	809	ATISSINI	735
assimilis (Macquart), Thaumatomyia	787	atkinsoni Curran, Rhaphium	512
assimilis (Macquart), Villa	436	atlanis (Aldrich), Blaesoxipha	945
assimilis Malloch, Fucellia	842	atlanis Huckett, Pegomya	858
assimilis (Malloch), Liriomyza	801	atlanis Malloch, Phaonia	905
assimilis Reinhard, Prosenoides	988	**atlantica** (Brues), Megaselia	539
assimilis (Townsend), Oswaldia	1046	*atlantica* Cresson, Atissa	735
Assipala Philip	323	atlantica Felt, Hormomyia	287
astacus Garrett, Boletina	217	atlantica Fisher, Bolitophila	197
astarte Hull, Xylota	606	*atlantica* (Parker), Erythrandra	941
Asteia Meigen	824	atlantica Shannon, Xylota	605
ASTEIIDAE	823	atlantica Steyskal, Dictya	689
ASTEIINAE	824	atlantica Wirth & Williams, Bezzia	141
astericaulis Felt, Rhopalomyia	265	atlanticus Dyar & Knab, Aedes	111
asteriflorae Felt, Rhopalomyia	265	atlanticus Hough, Pipunculus	553
asterifoliae (Beutenmüller), Asteromyia	274	atlanticus (Johnson), Stenotabanus	329
asteris (Felt), Lestodiplosis	284	atlanticus Pechuman, Chrysops	324
asteris (Harris), Eurosta	663	atlantis Wirth & Williams, Dasyhelea	126
Asteromyia Felt	274	atoma Reinhard, Aphantorhapha	1062
asterspinosae White, Lasioptera	271	atomaria (Zetterstedt), Hylemya	854
astictopus Dyar & Shannon, Chaoborus	103	atomarius (Zetterstedt), Limnophyes	160
astigma Alexander, Dicranota	57	atomella (Malloch), Megaselia	540
astigmatica Alexander, Elliptera	51	*Atomogaster* Macquart	898
Astiosoma Duda	824	**Atomosia** Macquart	386
astis (Roback), Orthocladius	156	**Atomosiella** Wilcox	387
Astrolabis Osten Sacken	57	*Atonia* Williston	387
Astrophanes Osten Sacken	441	*Atoniomyia* Hermann	387
astur Osten Sacken, Laphria	389, †389	atra (Adams), Leptocera	721
asturina (Bromley), Laphria	389	atra Borgmeier, Chaetopleurophora	533
astuta (Osten Sacken), Hybomitra	338	*atra* Brauer & Bergenstamm, Gaediophana	1079

INDEX 1563

	Page		Page
atra Cockerell, Gonia	1075	atripennis Cole & Wilcox, Lasiopogon	377
atra Cole, Boletina	217	atripennis Coquillett, Chaetoplagia	1017
atra Cole, Stratiomys	311	atripennis James, Cordilura	827
atra (Coquillett), Periscepsia	1034	atripennis (Say), Xanthomelanodes	967
atra Cresson, Micropeza	634	atripennis Stone, Tabanus	335
atra Cresson, Mythicomyia	416	*atripennis* (Townsend), Xanthomelanodes	967
atra (Curran), Carposcalis	575	atripes Bigot, Symphoromyia	343
atra (Curran), Chalcosyrphus	590	atripes (Coquillett), Phantasiomyia	1063
atra (Curran), Trigonomma	788	atripes Cresson, Notiphila	747
atra Fabricius, (Tipula)	1113	atripes (Malloch), Nanna	831
atra Loew, Hilara	456	*atripes* (McAtee), Cerotainiops	388
atra (Loew), Procecidochares	660	*atripes* McAtee, Laphria	389
atra (Loew), Typopsilopa	743	atripes (Melander), Trichina	449
atra Macquart, Penthetria	192	atripes Rempel, Chasmatonotus	159
atra Malloch, Forbesomyia	246	atripes Robineau-Desvoidy, (Neria)	636
atra Malloch, Macateeia	868	atripes Stein, Fannia	893
atra Meigen, Conicera	536	atripes Wilcox, Itolia	376
atra (Meigen), Copromyza	719	atripes Wilcox, Lestomyia	378
atra (Meigen), Pseudonapomyza	803	*atripes* Wilcox & Martin, Dioctria	371
atra (Meigen), Xylopriona	249	atripes Wilcox & Martin, Nannocyrtopogon	379
atra Painter, Villa	434	atripes Wirth, Mallochohelea	139
atra Robineau-Desvoidy, (Clytia)	1108	atripilosus James, Bibio	192
atra (Röder), Cylindromyia	973	atriplicicola (Cockerell), Asphondylia	268
atra Sherman, Tetragoneura	228	atriplicis (Townsend), Asphondylia	268
atra (Stein), Fannia	893	atrisetis Cresson, Notiphila	747
atra Townsend, Leucostoma	976	*atrisquama* (Ringdahl), Spilogona	880
atra Townsend, Xanthocera	1026	atrisquamula Hennig, Spilogona	880
atra Van Duzee, Boletina	217	atrisumma Doane, Tipula	35
atra (Walker), Archytas	1000	atrita Melander, Mythicomyia	416
Atractocclesia Townsend	1020	*atritarsata* Harris, (Empis)	1113
atramontensis (Banks), Dorylomorpha	556	atritarsis Malloch, Lonchaea	717
atrata Coquillett, Rhamphomyia	462	atritibia Malloch, Chironomus	165
atrata (Coquillett), Trichoclytia	970	atritibia Ringdahl, Coenosia	871
atrata (Coquillett), Villa	437	atriventris (Coquillett), Caloparyphus	308
atrata Cresson, Ditrichophora	739	at.'iventris Cresson, Dichaeta	748
atrata Curran, Pipiza	579	*atriventris* (Graenicher), Cephalochrysa	302
atrata (Hine), Bolbodimyia	330	atriventris Hendel, Rivellia	655
atrata Jones, Cerotainia	387	atriventris Wirth, Probezzia	138
atrata Melander, Psila	639	atrobasis (McDunnough), Hybomitra	338
atrata Say, (Scatopse)	†238, 241	atrocapilla Hull & Fluke, Cheilosia	583
atrata Say, (Sciara)	235	atrocera (Dietz), Nephrotoma	21
atrata Walker, Coenosia	871	atrocitrea Malloch, Phaonia	905
atrata Wirth, Dasyhelea	126	atrocularis Walsh, (Cecidomyia)	291
atrata Zetterstedt, Cordilura	827	atrocyanea Ringdahl, Phaonia	905
atratulus (Loew), Conophorus	413	atroglauca Coquillett, Hydrellia	744
atratulus (Zetterstedt), Metriocnemus	161	atropalpus (Coquillett), Aedes	115
atratus Coquillett, Microphorus	452	**Atropharista** Townsend	1009
atratus Fabricius, Tabanus	332	**Atrophopalpus** Townsend	1031
atratus Malloch, Pogonomyioides	902	*Atrophopoda* Townsend	1032
atratus Theobald, Culex	119	atrophopoda Townsend, Aravaipa	1074
atratus Van Duzee, Chrysotus	521	atrophopodoides Townsend, Vanderwulpia	1034
atrella (Townes), Chironomus	165	atropos Dyar & Knab, Anopheles	105
atribarbis Say, Asilus	1113	*atropos* Osten Sacken, Chrysops	325
atricapillus Loew, Bombylius	408	atrota (Reinhard), Gymnoclytia	964
atricauda (Zetterstedt), Hylemya	856	*atrovirens* Harris, Dolichopus	488
atriceps Dietz, Ormosia	85	atrovirens (Loew), Setacera	755
atriceps Loew, Bombylius	408	atroviridis (Townes), Chironomus	164
atriceps Loew, Eristalis	623	atrovittata Malloch, Hylemya	847
atriceps (Loew), Stenoscinis	778	atrox Bromley, Promachus	400
atriceps Reinhard, Chaetophlepsis	1015	atrox (Williston), Dasylechia	388
atriceps (Zetterstedt), Boreellus	925	*Atrypoderma* Townsend	1109
Atrichia Schrank	†355	*Attamyia* Greene	544
Atrichomelina Cresson	685	attenuata Alexander, Pedicia	54
Atrichopogon Kieffer	122	attenuata Felt, Dasineura	259
atricornis Alexander, Tipula	40	attenuata Malloch, Hylemya	847
atricornis Bigot, Chrysops	327	attenuata Rübsaamen, (Sciara)	236
atricornis Harris, Dolichopus	491	attenuatus Hine, Syrphus	559
atricornis Harris, Loxocera	638	attenuatus Walker, Chironomus	165
atricornis (Malloch), Gonarcticus	832	attonsa (Laffoon), Mycetophila	210
atricornis (Meigen), Allophyla	809	attrita Johannsen, Exechia	206
atricornis Meigen, Phytomyza	804	aturgida Brooks, Gonia	1075
atricornis Walsh, (Cecidomyia)	291	**Atylotus** Osten Sacken	330
atricorpus Philip, Whitneyomyia	331	*Aubaeanetia* Townsend	1045
atridorsalis Back, Leptogaster	362	*Aubaeina* Enderlein	974
atrifacies Aldrich, Leucopis	709	auceps (Wulp), Neoitamus	397
atrifrons Cole, Holopogon	375	aucta (Fallén), Verrallia	551
atrifrons (Coquillett), Neochirosia	842	*aucta* (Zetterstedt), Clinocera	468
atrifrons Malloch, Gimnomera	840	*audax* McAtee, Laphria	391
atrifrons Melander & Spuler, Piophila	711	audax Osten Sacken, Mydas	358
atrilabre Cresson, Allotrichoma	736	audibertiae Felt, Rhopalomyia	265
atrilinea Sabrosky, Gaurax	781	augur Osten Sacken, Apiocera	356
atrimana (Loew), Leptopsilopa	741	*augusta* (Curran), Pericoma	93
atrimanus Coquillett, Chironomus	169	augusta Curran, Ptiolina	347
atripalpis Aldrich, Phytomyza	804	augusta Felt, Dasineura	259
atripalpus Sabrosky, Gaurax	781	AULACEPHALINI	981
atripennis Back, Holopogon	375	**Aulacigaster** Macquart	823
		AULACIGASTRIDAE	823

Entry	Page
Aulacogaster Macquart	823
Aulicomyia Reinhard	1039
aurantiaca Coquillett, Psilocephala	350
aurantiaca (Doane), Xenochaeta	664
aurantiaca Hough, Muscina	909
aurantiaca Townsend, Trichopoda	966
aurantionota Alexander, Tipula	39
aurata Bellardi, Hermetia	304
aurata Coquillett, Hilara	456
aurata (Coquillett), Macronychia	936
aurata Curran, Criorhina	612
aurata (Doane), Hexatoma	69
aurata Felt, Mycodiplosis	282
aurata Harris, (Thereva)	1113
aurata Jones, Parydra	749
aurata Robineau-Desvoidy, Hemyda	974
auratilis Reinhard, Belvosia	1081
auratilis Sherman, (Allodia)	1114
auratus (Aldrich), Tachytrechus	503
auratus Cole, Cyrtopogon	368
auratus Johnson, Asilus	392
auratus Loew, Chrysotus	521
auratus Priddy, Conophorus	413
auratus (Townsend), Physoconops	626
auratus Williston, Eclimus	426
aurea (Aldrich), Megaselia	538
aurea James, Genea	1035
aurea (Johannsen), Pentaneura	147
aurea Lovett, Criorhina	611
aurea Malloch, Forcipomyia	125
aurea Malloch, Phaonia	905
aurea (Townsend), Lixophaga	1044
aurea (Townsend), Metoposarcophaga	951
aureiventris (Brues), Anevrina	532
aureola Melander, Mythicomyia	416
aureopilis Townsend, Helophilus	619
aureum Fries, Simulium	185
aureum (Guthrie), Dynatosoma	209
aureus Johannsen, Psectrocladius	159
auriannulatus (Hine), Asilus	392
auricaudata Brooks, Paraphorantha	969
auricaudata Townsend, Ceromasia	1099
auricaudata Williston, Myolepta	589
auricaudatus Bigot, Paragus	578
auriceps Aldrich, Plagiomima	1019
auriceps Aldrich, Xiphomyia	1051
auriceps Coquillett, Distichona	†1074
auriceps (Melander), Phytobia	798
auricincta Harris, (Leptis)	1114
auricoma Sherman, (Boletina)	1114
auricomus Alexander, Molophilus	89
auricomus Hine, Asilus	392
auricornis Frost, Phytomyza	804
auriensis (Roback), Albabesmyia	148
aurifacies Aldrich, Dolichopus	488
aurifacies Van Duzee, Pelastoneurus	501
aurifacies Van Duzee, Physocephala	628
aurifacies Van Duzee, Sympycnus	527
aurifacies Wilcox & Martin, Dioctria	370
aurifer (Coquillett), Aedes	111
aurifer Osten Sacken, Bombylius	408
aurifer (Thomson), Hercostomus	497
aurifera Melander, Mythicomyia	416
aurifera (Thomson), Dioxyna	666
aurifex Osten Sacken, Cyrtopogon	368
aurifex Van Duzee, Dolichopus	488
aurifex Wiedemann, Microdon	†599
aurifrons Brooks, Juriniopsis	1001
aurifrons (Coquillett), Aporotachina	1039
aurifrons Curran, Spathimeigenia	1049
aurifrons (Reinhard), Deopalpus	1004
aurifrons (Stein), Macrorchis	875
aurifrons (Townsend), Amobia	934
aurifrons Townsend, Ceromasia	1099
aurifrons (Townsend), Oswaldia	1046
aurifrons (Townsend), Voria	1020
aurigera (Coquillett), Sitophaga	1048
aurihirta Felt, Dasineura	259
aurilimbus (Stone), Hybomitra	338
aurimystacea (Hine), Efferia	394
aurinota Harris, (Tachina)	1114
aurinota Hine, Myolepta	589
aurinota Walker, Baccha	573
aurinotatus Van Duzee, Thrypticus	511
auripennis Alexander, Limnophila	65
auripennis (Osten Sacken), Pedicia	54
auripes Aldrich, Ephydra	754
auripila Curran, Hilara	456
auripila Felt, Asphondylia	268
auripila (Hine), Efferia	394
auripila Metcalf, Callicera	595
auripilosa Johnson, Ocnaea	403
auripilosus Wilcox & Martin, Cyrtopogon	368
auripilus (Bigot), Eclimus	426
auripleura (Curran), Neocnemodon	581
auripuncta Painter, Aldrichia	414
aurivittata Wheeler, Medetera	510
aurofasciata Kröber, Thereva	353
auroides (Felt), Aedes	114
aurora Mesnil, Drino	1085
aurulans (Robineau-Desvoidy), Cynomyopsis	930
aurulentus (Fabricius), Ceraturgus	365
aurulentus (Fabricius), Microdon	597
ausoba (Walker), Macrorchis	875
austera (Doane), Hexatoma	69
austinana (Reinhard), Blaesoxipha	947
austini Painter, Bombylius	409
australe (Johnson), Parapenium	580
australina Alexander, Pseudolimnophila	62
australis Aldrich, Procecidochares	660
australis Alexander, Dicranoptycha	52
australis (Blanchard), Blaesoxipha	945
australis Byers, Dolichopeza	23
australis (Coquillett), Pilatea	1105
australis Doane, Tipula	35
australis Namba, Rivellia	655
australis Reinhard, Doryphorophaga	1042
australis Townsend, Myiophasia	983
australis (Townsend), Phaenicia	927
australis Van Duzee, Diaphorus	519
australis Van Duzee, Neurigona	518
australis Wirth & Jones, Culicoides	132
austrina (Coquillett), Dimecoenia	755
Austrolimnophila Alexander	61
Automola Loew	642
autumnalis (Alexander), Pedicia	54
autumnalis Banks, Asilus	392
autumnalis (Banks), Telmatoscopus	94
autumnalis Beutenmüller, Asphondylia	268
autumnalis (Cole), Poecilanthrax	442
autumnalis De Geer, Musca	913
autumnalis Fluke, Syrphus	559
autumnalis Garrett, Mycomya	220
autumnalis Garrett, Sceptonia	215
autumnalis (Townsend), Actia	1061
autumnalis (Townsend), Distichona	1074
autumnals Garrett, Sceptonia	215
Auxanommatidia Borgmeier	544
auxiliaria Johannsen, Exechia	206
Avaritia Fox	128
avia Loew, Notiphila	746
aviceps Townes, Polypedilum	175
avicularia (Linnaeus), Ornithomyia	†917
aviculata Shaw, Exechia	206
avida Coquillett, Empis	459
avida Coquillett, Rhamphomyia	462
avida Osten Sacken, Volucella	600
avida (Walker), Ravinia	955
avidus Loew, Scellus	506
avidus (Wulp), Machimus	396
avis (Alexander), Dicranota	58
avium (Curran), Brachicoma	940
avium Shannon & Dobroscky, Protocalliphora	926
awanichi Alexander, Tipula	35
Axymyia McAtee	196
Axysta Haliday	751
aylmeri (Garrett), Dioptopsis	99
azaleae Felt, Asphondylia	268
azaleae (Felt), Mayetiola	263
azaleae Shannon, Bombylius	409
Azana Walker	223
Azelia Robineau-Desvoidy	898
azrael Alexander, Hexatoma	69
azteca Sturtevant & Dobzhansky, Drosophila	768
azurea (Fallén), Protocalliphora	926
azurea (Fluke), Xylota	606
azygos Malloch, Phaonia	905
baal (Townsend), Tabanus	336
babista Walker, Baccha	573

INDEX

Entry	Page
babiyi (Rempel), Nilothauma	173
baboquivari Wilcox, Ommatius	398
baccata Coquillett, Psilocephala	350
baccata (Fabricius), Cuterebra	1109
Baccha Fabricius	572
baccharis (Coquillett), Eutreta	661
baccharis (Felt), Rhopalomyia	265
baccharis (Reinhard), Microsillus	1086
bacchides (Walker), Allograpta	569
Bacchina Williston	572
BACCHINI	572
bacchoides (Bigot), Baccha	573
bacilliger Kieffer, Trichocladius	158
Backomyia Wilcox & Martin	364
baculifer Melander, Hilara	456
bacuntius (Walker), Teuchocnemis	608
badgeri Dyar, Culex	118
badia Coquillett, Phthiria	420
badia (Doane), Austrolimnophila	61
badia (Loew), Tipulogaster	363
badia (Walker), Blera	610
badia (Walker), Limonia	43, †47
badia (Walker), Paraprosalpia	866
badia Walker, Stratiomys	311
badiceps Reinhard, Admontia	1068
badius Kröber, Diachlorus	328
badius Van Duzee, Chrysotus	521
baeomyia Wheeler, Drosophila	767
baffinense Twinn, Simulium	185
bahamensis Dyar & Knab, Culex	118
bahamensis (Johnson), Dasyhelea	126
baia Cresson, Ochthera	753
bakeri (Aldrich), Oxysarcodexia	952
bakeri Alexander, Tipula	32
bakeri Cole, Dolichopus	488
bakeri Cole, Thereva	353
bakeri Cook, Psectrosciara	240
bakeri Coquillett, Acrocera	405
bakeri (Coquillett), Drino	1084
bakeriana Alexander, Tipula	29
balachowskyi Mesnil, Geomyza	821
balanus (Walker), Mallota	621
Baldratia Kieffer	†274
Balioptera Loew	821
balioptera (Loew), Neoempheria	222
balioptera Loew, Tipula	30
baliopteroides Alexander, Tipula	30
baliopterus Loew, Microdon	599
ballistrarius Melander, Platypalpus	478
ballucatus Melander, Platypalpus	478
balsamicola (Felt), Campylomyza	248
balsamicola (Lintner), Dasineura	259
balsamifera (Felt), Contarinia	277
balsamifera Felt, Mayetiola	263
balteata (Holmgren), Pegomya	858
balteatus Melander, Aphoebantus	429
baltica (Ringdahl), Spilogona	880
baltimorensis Macquart, Tabanus	336
baltimoreus Macquart, (Tanypus)	181
baltimoricus Macquart, Bibio	192
balyras (Walker), Temnostoma	614
banana Garrett, Diamesa	151
banffi Huckett, Pegomya	858
banffi (Seamans), Spilogona	881
banffiana Alexander, Tipula	32
banksi (Aczél), Pipunculus	552
banksi (Curran), Parapenium	580
banksi (Hull), Blera	610
banksi Johnson, Dioctria	371
banksi Johnson, Dipalta	441
banski Johnson, Platypeza	549
banski (Malloch), Cordilura	829
banksi Melander, Chelifera	471
banksi Townes, Pseudochironomus	176
banksi Van Duzee, Rhaphium	512
banksi Wilcox & Martin, Cyrtopogon	368
banksianae Vockeroth, Cecidomyia	287
banksii (Van Duzee), Condylostylus	483
baptisiae (Frost), Liriomyza	801
baracoa (Cresson), Grallipeza	635
barbara McAlpine, Earomyia	716
barbaricus Van Duzee, Dolichopus	488
barbarta Bigot, Paralispe	1047
Barbastoma Garrett	810
barbata Aldrich, Symphoromyia	343
barbata Bigot, Paralispe	1047
barbata (Coquillett), Velocia	1051
barbata Doane, Tipula	32
barbata (Garrett), Orbellia	810
barbata Loew, Empis	458
barbata Loew, Stratiomys	311
barbata Loew, Xylota	605
barbata Sabrosky, Leptocera	721
barbata Thomson, Sarcophaga	957
barbata (Zetterstedt), Pogonota	833
barbatimanus Kieffer, Psectrocladius	159
barbatoides Melander, Empis	458
barbatulus Loew, Gymnopternus	498
barbatus (Fabricius), Efferia	†393, 395
barbatus (Loew), Chrysotus	521
barbatus Melander, Aphoebantus	430
barbatus Osten Sacken, Anastoechus	411
barberi Alexander, Limnophila	63
barberi Borgmeier, Trophodeinus	546
barberi Coquillett, Anopheles	105
barberi (Coquillett), Bezzia	141
barberi (Coquillett), Pentaneura	147
barberi (Darlington), Suillia	809
barberi (Felt), Allarete	245
barberi (Felt), Eucatocha	244
barberi Felt, Porricondyla	255
barberi Frick, Agromyza	795
barberi (Malloch), Conicera	536
barberi (Malloch), Megaselia	538
barberi Sabrosky, Lasiopleura	774
barberi Shannon, Temnostoma	614
barberi (Townsend), Phaenicia	928
barberi Wirth, Serromyia	136
Barberiella Shannon	592
barbicauda Van Duzee, Dolichopus	488
barbicornis (Linnaeus), Orthocladius	156
barbicula Huckett, Hylemya	846
barbimanus (Edwards), Psectrocladius	159
barbipes (Staeger), Glyptotendipes	172
barbipes Van Duzee, Argyra	525
barbipes Van Duzee, Chrysotus	521
barbipes Van Duzee, Dolichopus	488
barbipes (Van Duzee), Rhaphium	512
barbosai Wirth & Blanton, Culicoides	129
barda Alexander, Tipula	32
bardus (Say), Eristalis	†621, 623
barei Curran, Volucella	600
barlowi Felt, Campylomyza	248
barnesi Edwards, Mycophila	250
Barnesina Pritchard	251
Baromyia Reinhard	975
baroni Felt, Asphondylia	268
baroni Williston, Cheilosia	583
baroni (Williston), Thecophora	631
baropeodes Melander, Euhybus	448
barpana (Walker), Helina	886
barri Rueger, Aedes	111
barrus (McAtee), Scatopse	238
barycnemus Coquillett, Dolichopus	488
barypoda Coquillett, Rhamphomyia	462
basalis (Banks), Telmatoscopus	94
basalis Felt, Aphidoletes	280
basalis Felt, Lestodiplosis	284
basalis (Felt), Neolasioptera	273
basalis Harris, Gonia	1076
basalis Harris, (Volucella)	1114
basalis Loew, Bibio	194
basalis Loew, Hilara	456
basalis Loew, Rhamphomyia	462
basalis (Malloch), Paratendipes	174
basalis (Malloch), Trichocladius	158
basalis Melander, Trichina	449
basalis (Staeger), Cricotopus	157
basalis (Stein), Coenosia	874
basalis Van Duzee, Argyra	525
basalis Van Duzee, Diaphorus	519
basalis Van Duzee, Phronia	215
basalis (Van Duzee), Thecophora	631
basalis (Walker), Anaporia	1028
basalis (Walker), Blaesoxipha	947
basalis (Walker), Diogmites	371
basalis (Walker), Euxesta	651
basalis (Walker), Palpomyia	140
basalis (Walley), Ablabesmyia	148
basalis (Zetterstedt), Helina	886
basdeni Collin, Hydrotaea	899
basdeni Wheeler, Drosophila	767

744–243—65——99

Name	Page
basicornis Curran, Microdon	597
basidens Townes, Paratendipes	173
basiflava Felt, Lasioptera	271
basifulvus (Walker), Archytas	†999
Basila Cresson	738
basilaris Coquillett, Rivellia	656
basilaris Curran, Brachyopa	592
basilaris Macquart, Eristalis	623
basilaris (Say), Chrysopilus	346
basilaris (Wiedemann), Mesograpta	571
Basilia Ribeiro	922
basingeri (Hall), Protocalliphora	926
basingeri Pritchard, Hodophylax	374
basingeri Wilcox & Martin, Cyrtopogon	368
basiola (Osten Sacken), Rhagoletis	674
basiseta Malloch, Phaonia	906
basistylatus Parent, Sympycnus	527
basitarsalis Sabrosky, Gaurax	781
bastardi (Macquart), Efferia	393
bastardi (Macquart), Villa	433
bastardii Macquart, Eristalis	623
bastardii (Macquart), Promachus	400
batatas (Osten Sacken), Rhabdophaga	257
bathyphila Kieffer, Prodiamesa	152
batillifer Loew, Dolichopus	489
baton Walker, Xylota	607
Batula Cresson	741
baueri Hoffman, Culicoides	131
baueri Wirth, Forcipomyia	124
baumhaueri Meigen, Dioctria	370
bausei Kieffer, Tanytarsus	180
bautias (Walker), Mallota	621
bea Felt, Asphondylia	268
beameri Alexander, Hexatoma	69
beameri Brennan, Chrysops	324
beameri (Dodge), Sarcophaga	957
beameri (Hall), Blaesoxipha	943
beameri Hall, Cuterebra	1109
beameri Hardy, Bibio	192
beameri Hardy, Chrysopilus	346
beameri (Hardy), Scenopinus	356
beameri Hardy, Tomosvaryella	555
beameri Harmston & Knowlton, Dolichopus	488
beameri James, Merosargus	303
beameri James, Nemotelus	310
beameri James, Stratiomys	311
beameri Kraft & Cook, Zabrachia	318
beameri Painter, Apiocera	357
beameri Philip, Pilimas	321
beameri (Reinhard), Deopalpus	1004
beameri (Reinhard), Metoposarcophaga	951
beameri Sabrosky, Oscinella	779
beameri Stains & Knowlton, Simulium	187
beameri Wilcox, Ommatius	398
beameri Wilcox & Martin, Cyrtopogon	368
beameri Wilcox & Martin, Dioctria	370
Beameromyia Martin	361
beata Aldrich, Asteia	824
beata Johannsen, Allodia	205
beatifica (Whitney), Whitneyomyia	331
beatricis Steyskal, Pherbellia	686
beatula Osten Sacken, Tipula	24
beatus Van Duzee, Dolichopus	488
beaulieui Dietz, Tipula	36
beauvoisii Robineau-Desvoidy, Prionella	654
Bebryx Gistel	903
beckae Wirth, Stilobezzia	134
beckeri Shaw, Rymosia	207
beckeri (Wood), Megaselia	540
Beckerina Malloch	537
BECKERININI	537
becquaerti (Kieffer), Leptoconops	122
bedeguar Osten Sacken, (Cecidomyia)	293
belfragei (Hine), Efferia	394
belkini Dendy & Sublette, Cricotopus	157
bella Cole, Chromolepida	350
bella (Cole), Furcifera	350
bella (Coquillett), Alluaudomyia	133
bella Johannsen, Allodia	205
bella Johannsen, Exechia	206
bella Kröber, Thereva	353
bella (Loew), Cophura	367
bella (Loew), Euaresta	665
bella Loew, Leucopis	709
bella Loew, Notiphila	746
bella Loew, Tipula	28
bella Melander, Hilara	456
bella Melander, Mythicomyia	416
bella Townsend, Milesia	614
bella (Williston), Cynorhinella	588
bellamyi Alexander, Atarba	71
bellardii Jaennicke, Eristalis	623
bellardii Shannon, Polybiomyia	616
bellardii Szilády, Tabanus	337
Bellardina Edwards	26
belli Curran, Negasilus	397
bellingeri Shaw, Bradysia	232
bellula Johannsen, Exechia	206
bellula Johannsen, Trichonta	216
bellula Snow, Euaresta	665
bellula Williston, Chrysogaster	591
bellula Williston, Leucopis	709
bellulus (James), Hedriodiscus	314
bellulus Melander, Nemotelus	309
bellulus Van Duzee, Chrysotus	521
bellulus Williston, Keroplatus	†200
bellus Adams, Chrysopilus	347
bellus (Loew), Caloparyphus	308
bellus (Loew), Procladius	149
bellus Van Duzee, Chrysotus	521
Belosta Hardy	355
beltrami Telford, Eristalis	623
Beltranmyia Vargas	131
Belvoisia Robineau-Desvoidy	1081
Belvosia Robineau-Desvoidy	1081
BELVOSIINI	1080
Belvosiopsis Townsend	1081
belzebul Schiner, Mallophora	396
benanderi (Ringdahl), Paraprosalpia	866
benedicti (Bromley), Efferia	394
benedictus Whitney, Tabanus	335
benefica Malloch, Pseudoleucopis	708
benjamini Malloch, Fannia	893
bentincki (Laffoon), Mycetophila	210
bequaerti Hull, Mallota	621
bequaerti (Kieffer), Leptoconops	122
Bequaertomyia Brennan	319
BEQUAERTOMYIINAE	319
berberis Curran, Rhagoletis	674
Bercaeopsis Townsend	957
Bergenstammia Mik	468
bergenstammii (Mik), Spiniphora	533
bergi Cresson, Hydrellia	744
bergi Steyskal, Dryomyza	681
bergi Steyskal, Renocera	692
bergi Steyskal, Tetanocera	694
bergrothi Becker, Cosmetopus	834
bergrothi (Williston), Prionocera	19
bergrothiana Alexander, Tipula	30
BERIDINAE	300
beringensis Malloch, Cordilura	827
beringensis Malloch, Macrocera	201
Beris Latreille	301
Berkshiria Johnson	318
bermudae Wirth & Williams, Dasyhelea	126
bermudaensis (Melander), Tethina	727
bermudensis Johnson, Odontomyia	315
bermudensis Parsons, Conops	626
bermudensis Williams, Culicoides	131
bernardinensis Alexander, Tipula	35
bertae Kessel, Callomyia	547
bertrami Edwards, Diamesa	151
Beskia Brauer & Bergenstamm	972
Bessa Robineau-Desvoidy	1052
besselsi Osten Sacken, Tipula	29
besselsoides Alexander, Tipula	29
Besseria Robineau-Desvoidy	972
bessophila Quate, Pericoma	92
bestigma (Coquillett), Limonia	43
beta Hull, Eristalis	624
betae (Curtis), Pegomya	860
betarum (Lintner), Hylemya	853
betheli Cockerell, Asphondylia	268
betheli Felt, Oligotrophus	264
betheliana Cockerell, Rhopalomyia	265
betsyae Pritchard, Eucatocha	244
betulae Sabrosky, Odinia	794
betulae (Winnertz), Oligotrophus	265
beutenmuelleri (Dietz), Nephrotoma	21
bexarensis (Bromley), Efferia	394
Bezzia Kieffer	141
Bezzimyia Townsend	963

INDEX 1567

	Page
biacus Alexander, Gonomyia	74
biangulata Curran, Peleteria	997
biannulata Malloch, Sphegina	593
biannulata (Say), Stylogaster	632
biannulata Wirth, Bezzia	141
biappendiculata Goetghebuer, Diamesa	151
biarmata Alexander, Antocha	51
biarmata Doane, Tipula	35
Bibio Geoffroy	192
Bibiocephala Osten Sacken	99
Bibiodes Coquillett	195
BIBIONIDAE	191
BIBIONINAE	192
BIBIONOIDEA	191
Bibionus Curran	99
bibosa Melander, Mythicomyia	416
bicarina (Tothill), Mericia	1008
bicarinata Williston, Cyphomyia	304
bicaudata (Hine), Efferia	394
bicaudata Malloch, Hylemya	854
Bicellaria Macquart	451
biciliata (Coquillett), Hylemya	847
bicincta Loew, Clinocera	468
bicincta Reinhard, Ceratomyiella	1031
bicincta Robineau-Desvoidy, Belvosia	1081
bicincta (Staeger), Exechia	206
bicinctus Loew, Dizonias	372
bicinctus (Meigen), Cricotopus	157
bickleyi Wirth & Hubert, Culicoides	129
biclavata Shewell, Twinnia	182
bicolor Adams, Zodion	628
bicolor (Banks), Threticus	95
bicolor (Bellardi), Efferia	394
bicolor (Coquillett), Ectrepesthoneura	227
bicolor Coquillett, Euthera	978
bicolor Coquillett, Hippelates	775
bicolor (Coquillett), Melanomya	933
bicolor Coquillett, Phthiria	420
bicolor Coquillett, Phytomyza	804
bicolor (Cresson), Euxesta	651
bicolor Cresson, Notiphila	746
bicolor (Curran), Lasioneura	1060
bicolor (Day), Hedriodiscus	315
bicolor (Dodge), Sarcophaga	957
bicolor (Fabricius), Paragus	578
bicolor Foote, Gymnocarena	676
bicolor Garrett, Macrocera	201
bicolor Garrett, Sciophila	225
bicolor Hardy, Heteromydas	358
bicolor Johnson, Saropogon	383
bicolor Johnson, Senopterina	657
bicolor Lane, Stilobezzia	134
bicolor (Loew), Boreodromia	466
bicolor (Loew), Sciapus	486
bicolor Loew, Sigaloessa	824
bicolor Loew, Xylota	605
bicolor (Macquart), Parachaeta	1083, †1083
bicolor (Macquart), Polydontomyia	621
bicolor Macquart, Tabanus	332
bicolor (Malloch), Trichocladius	158
bicolor Meigen, Psila	639
bicolor Melander, Oligodranes	421
bicolor Oldenberg, Acartophthalmus	808
bicolor Rempel, Chasmatonotus	159
bicolor Shaw and Fisher, Synapha	219
bicolor Stone, Pseudodacus	674
bicolor Van Duzee, Medetera	510
bicolor (Van Duzee), Peloropeodes	517
bicolor Walker, Scatophaga	838
bicolor (Walker), Tephrochlamys	810
bicolor (Wiedemann), Atylotus	330
bicolor (Wiedemann), Brachycara	†306
bicolor (Wiedemann), Lampria	388
bicolor (Wiedemann), Pegomya	858
bicolor (Williston), Xanthoepalpus	1003
bicolor (Zetterstedt), Suillia	†809
bicolorata (Malloch), Helina	886
bicolorata (Malloch), Trophithauma	543
bicolorata Shannon, Cheilosia	583
bicomata Alexander, Pedicia	54
bicorne Dorogostajskij, Rubzov, & Vlasenko, Simulium	185
bicorne (Townes), Nilothauma	173
bicornis Forbes, Tipula	35
bicornis Melander, Platypalpus	478
bicornis Say, Sphyracephala	638
bicristatus Thurman & Winkler, Aedes	111

	Page
bicruciata (Malloch), Paraprosalpia	866
bicruciata Stein, Fucellia	843
bicruciatum Bigot, (Melanostoma)	625
bicuspidata Alexander, Ormosia	84
bidactylus (Hardy), Pipunculus	553
bidens (Cresson), Tomosvaryella	555
bidens Johannsen, Asphondylia	268
bidens Loew, Tipula	39
bidens Ringdahl, Hylemya	853
bidentata Alexander, Gonomyia	75
bidentata Felt, Dasineura	259
bidentata Felt, Porricondyla	255
bidentata Fisher, (Platyura)	1114
bidentata (Malloch), Calythea	865
bidentata Ringdahl, Phaonia	906
bidentatus (Cresson), Mosillus	734
bidenticulata Alexander, Cryptolabis	78
bidenticulata Alexander, Gonomyia	75
bidentifera Alexander, Pedicia	54
bidwelli (Hine), Phasia	969
bifalcata Doane, Tipula	35
bifarius Melander, Oligodranes	421
bifasciata (Coquillett), Pentaneura	147
bifasciata (Fabricius), Belvosia	1081
bifasciata Garrett, Zygomyia	216
bifasciata Williston, Sericomyia	603
bifasciatum (Walker), Dynatosoma	209
bifasciatus Malloch, Orthocladius	156
bifasciatus (Say), Leptomorphus	223
bifenestrata (Bigot), Villa	437
bifida Alexander, Limnophila	64
bifida Dietz, Tipula	32
bifida Garrett, Sciophila	225
bifida (Hardy), Beameromyia	361
bifida Steyskal, Sepedon	693
bifidaria Alexander, Ormosia	84
bifidus (Garrett), Procladius	149
bifidus Hardy, Nemomydas	359
bifila Alexander, Tipula	39
bifilata Coquillett, Rhamphomyia	462
bifimbriata Collin, Fannia	893
bifimbriata Huckett, Spilogona	880
bifimbriatus Kieffer, Chironomus	165
biflexuosa Strobl, Sepsis	683
bifolia Stebbins, (Cecidomyia)	293
biformis (Lundbeck), Bradysia	232
bifractus Loew, Dolichopus	488
bifrons (Walker), Pelastoneurus	501
bifurca (Becker), Siphonella	786
bifurca Loew, Exoprosopa	444
bifurca Patterson & Wheeler, Drosophila	763
bifurcata Fisher, Exechia	206
bifurcata Wirth, Dasyhelea	126
bifuscata (Fabricius), Belvosia	1081
bigeloviae (Cockerell), Aciurina	670
bigeloviae Cockerell, (Cecidomyia)	291
bigeloviae (Cockerell), Rhopalomyia	265
bigeloviae (Townsend), Aciurina	670
bigeloviaebrassicoides Townsend, (Cecidomyia)	293
bigeloviaestrobiloides Townsend, (Cecidomyia)	293
bigelovioides Felt, Rhopalomyia	265
bigelowi Chillcott, Fannia	893
bigelowi Curran, Cheilosia	584
bigelowi (Curran), Chrysosyrphus	592
bigelowi Curran, Cyrtopogon	368
bigelowi Curran, Grisdalemyia	1024
bigelowi (Curran), Lejota	590
bigelowi (Curran), Opsotheresia	990
bigelowi Curran, Platycheirus	576
bigelowi Curran, Syrphus	559
bigelowi (Curran), Xylota	605
bigelowi Walley, Rhamphomyia	462
bigeminata Alexander, Tipula	35
bigladia Alexander, Limnophila	67
Bigonichaeta Rondani	1012
Bigonicheta Rondani	1012
Bigonochaeta Rondani	1012
bigoti Brunetti, Sargus	302
bigoti Melander, Empis	458
Bigotomyia Malloch	909
bigradata (Loew), Villa	435
biguttata Sabrosky, Odinia	794
biguttatus (Coquillett), Culicoides	129
bihamata Alexander, Gonomyia	75

1568　A CATALOG OF DIPTERA OF NORTH AMERICA

Name	Page
bilimbatum (Bigot), Callinicus	365
bilimekii (Brauer & Bergenstamm), Ormia	980
bilinearis Williston, Helophilus	618
bilineata (Adams), Diplotoxa	792
bilineata Adams, Elachiptera	777
bilineata Dietz, Ormosia	85
bilineata Painter, Apiocera	357
bilineata Van Duzee, Zodion	629
bilineata Walker, Laphria	390
bilineata (Williston), Willistonina	386
bilineata Wirth, Bezzia	142
bilineatus Curtis, Helophilus	618
bilineatus Loew, Diogmites	371
bilineatus (Melander), Microphorus	452
bilineatus Melander, Oligodranes	421
bilobata Garrett, Bolitophila	197
bilobata Shaw, Exechia	206
bilobatus Brundin, Diplocladius	154
bilobifera Cresson, Hydrellia	744
bilobus (Hardy), Pipunculus	553
bilychnis Melander, Mythicomyia	416
bimacula (Curran), Thecophora	631
bimacula (Walker), Cyrtopogon	368
bimacula Walker, Trichocera	15
bimaculata Cole, Thereva	353
bimaculata (Fabricius), Mycetophila	213
bimaculata (Hendel), Otites	646
bimaculata Loew, Acrocera	405
bimaculata (Loew), Clinohelea	136
bimaculata Loew, Cordilura	829
bimaculata (Loew), Neoempheria	222
bimaculata (Loew), Tachydromia	474, †474
bimaculata (Lovett), Melangyna	566
bimaculata (Meigen), Hydrotaea	899
bimaculata (Melander), Rhegmoclemina	239
bimaculata Van Duzee, Argyra	525
bimaculata (Walker), Dilophus	195
bimaculata Williston, Sphaerocera	718
bimaculatus (Coquillett), Aedes	111
bimaculatus Johnson, Thinophilus	507
bimaculatus Osten Sacken, Chasmatonotus	159
bimaculus Walker (Chironomus)	181
binodatus Harmston & Knowlton, Sympycnus	527
binodatus Loew, Tachytrechus	503
binotata (Bigot), Cylindromyia	973
binotata Chillcott, Fannia	893
binotata (Cresson), Hyadina	751
binotata Loew, Clinocera	468
binotata Melander, Tachypeza	473
binotatus Harris, (Syrphus)	1114
binotatus (Loew), Hedriodiscus	315
binummus Loew, Brachystoma	447
Biomya Rondani	1048
Biomyia Robineau-Desvoidy	913
Biomyia Rondani	1048
bipars (Walker), Rivellia	655
bipartita Osten Sacken, Erioptera	82
bipartita (Walker), Mallota	621
bipartitus Painter, Lordotus	412
bipiliaris Sturtevant & Wheeler, Scatophila	758
biplagiata Say, (Leia)	1114
biplagiatus Harris (Tanypus)	1114
biproducta Alexander, Tipula	35
bipunctata Curran, Rhamphomyia	463
bipunctata (Fallén), Piophila	712
bipunctata Kincaid, Pericoma	92
bipunctata (Linnaeus), Forcipomyia	125
bipunctata (Loew), Homalocephala	653
bipunctata Loew, Mycetophila	210
bipunctatus Greene, Tachytrechus	503
bipunctatus (Scopoli), Sargus	301
bipunctipennis Wheeler, Scaptomyza	769
biquadrata (Walker), Spilogona	881
biramus Quate, Pericoma	92
birdi Curran, Psilocurus	382
birdi Curran, Rhamphomyia	462
birdi Curran, Saropogon	383
birdi Felt, Ficiomyia	263
birdiei Whitney, Tabanus	332
birulai Lundström, Boletina	217
bisaccata Cook, Rhegmoclemina	239
biscaynei Cresson, Pipunculus	554
Bischofia Hendel	688
bisepta Hardy, Bibio	193
biseriata Cresson, Notiphila	747
biseriata (Loew), Mycomya	220
biseriata (Stein), Helina	888
biseta (Townes), Chironomus	165
bisetata (Garrett), Heleomyza	816
bisetosa Coquillett, Micropeza	635
bisetosa (Coquillett), Paracoenia	755
bisetosa (Coquillett), Trupanea	667
bisetosa Doane, Tipula	35
bisetosa Hall, Onesia	930
bisetosa (Huckett), Spilogona	880
bisetosa Parker, Boettcheria	948
bisetosus Van Duzee, Dolichopus	488
bisetulata (Malloch), Megaselia	538
bishopi (West), Admontia	1068
bishoppi Aldrich, Sarcophaga	957
bishoppi Brennan, Chrysops	324
bishoppi (Del Rosario), Philosepedon	95
bishoppi Sabrosky, Hippelates	775
bishoppi Stone, Tabanus	332
bisignata Cresson, Scatophila	758
bisignata Melander, Milichiella	733
bisinuatis Doane, Cryptolabis	78
bisinuatum Van Duzee, Syntormon	515
bison Aldrich, Sarcophaga	957
bispina (Fisher), Bradysia	232
bispina (Loew), Homoneura	698
bispina (Malloch), Oscinella	779
bispina Melander, Drapetis	477
bispina Van Duzee, Mycetophila	213
bispinigera Alexander, Erioptera	83
bispinosa Cook, Swammerdamella	239
bispinosa Cresson, Notiphila	747
bispinosa (Loew), Homoneura	698
bispinosa Malloch, Helina	886
bispinosa (Malloch), Nanna	831
bispinosa (Malloch), Triphleba	534
bispinosa (Melander & Spuler), Themira	684
bispinosa (Zetterstedt), Hydrotaea	899
bispinosus Alexander, Molophilus	88
bistellatus Daecke, Chrysops	324
bistigma (Coquillett), Limonia	43
bistria (Walker), Zodion	629
bistriata Parent, Medetera	510
bistriata (Walker), Thaumatomyia	787
bisulca Alexander, Erioptera	83
bisulcata Macquart, Olfersia	919
bisulcus Osten Sacken, Aphoebantus	429
biterminata (Walker), Limnophila	65
Bithoracochaeta Stein	874
Bittacomorpha Westwood	98
Bittacomorphella Alexander	98
BITTACOMORPHINAE	98
bituberculata Alexander, Trichocera	15
bituberculata Doane, Tipula	24
bituberculata Loew, Parydra	749
biuncus Doane, Tipula	35
bivisualis Patterson, Gitona	762
bivittata (Coquillett), Bezzia	141
bivittata (Coquillett), Dolichocodia	986
bivittata Curran, Johnsonia	950
bivittata Loew, Psila	639
bivittata Lovett, Xylota	606
bivittata Malloch, Leucopis	709
bivittata Say, Leia	227
bivittata Stein, Pegomya	859
bivittata Van Duzee, Neurigona	518
bivittatum Malloch, Simulium	186
bivittatus (Cresson), Oligodranes	421
bivittatus Loew, Lasiopogon	377
bivulneris Melander, Mythicomyia	416
Blacodes Loew	366
Blaesoxipha Loew	943
Blaesoriphotheca Townsend	946
blaisdelli (Cresson), Calliopum	696
blaisdelli Cresson, Dalmannia	632
blakeae Dodge, Anolisimyia	942
blanchardiana (Jaennicke), Villa	436
blanda Curran, Peleteria	997
blanda (Osten Sacken), Eusisyropa	1093
blanda Osten Sacken, Gonomyia	74
blandita (Coquillett), Zenillia	1107
blandus Van Duzee, Dolichopus	488
blantoni Bromley, Asilus	392
blantoni Bromley, Regasilus	400
blantoni Hering, Procecidochares	660
blantoni Vargas & Wirth, Culicoides	131
blarina (Townes), Chironomus	167
blax Dyar & Shannon, Dixa	101

	Page		Page
Blax Loew	366	*borealis* (Coquillett), Diamesa	152
Blepharepium Rondani	364	*borealis* Cresson, Euthycera	690
Blepharicera Macquart	99	borealis Cresson, Hydrellia	744
BLEPHARICERIDAE	98	borealis (Cresson), Parydra	749
Blepharigena Rondani	1016	borealis Curran, Allomyella	834
Blepharipa Rondani	1083	borealis Curran, Dictya	689
Blepharipeza Macquart	1082	borealis Curran, Lonchoptera	531
Blepharipoda Brauer & Bergenstamm	1083	borealis Curtis, (Chironomus)	181
Blephariptera Macquart	816	borealis Czerny, Trichochlamys	816
Blepharocera Macquart	99	borealis (Doane), Limonia	43
Blepharoprocta Loew	447	borealis (Fabricius), Rhamphomyia	462
Blepharoptera Macquart	99	*borealis* (Felt), Anarete	246
Blepharoptera Macquart	816	borealis Felt, Aphidoletes	280
Blera Billberg	610	borealis Felt, Asynapta	254
blondeli (Robineau-Desvoidy), Aplomya	1097	borealis Felt, Bremia	280
Blondelia Robineau-Desvoidy	1039	borealis Felt, Dasineura	259
BLONDELIINI	1037	borealis (Felt), Dicrodiplosis	279
boardmani Rozeboom, Basilia	922	borealis Felt, Lobodiplosis	281
boarmiae (Coquillett), Eusisyropa	1093	*borealis* (Felt), Neocatocha	243
boehmeriae Beutenmüller, (Cecidomyia)	293	borealis (Felt), Peromyia	251
Boettcheria Parker	948	borealis Felt, Porricondyla	255
Bogeria Austen	1109	*borealis* Felt, Tritozyga	243
boharti Alexander, Limnophila	63	borealis Foote, Antichaeta	688
boharti Bromley, Stenopogon	383	*borealis* Frey, Tetanocera	694
boharti Brookman & Reeves, Culex	118	*borealis* Garrett, Diamesa	151
boharti James, Stratiomys	311	borealis Goetghebuer, Limnophyes	160
boharti Philip, Tabanus	332	borealis James, Odontomyia	316
boharti Schlinger, Ogcodes	406	borealis Johnson, Phthiria	420
Bohartia Hull	370	borealis (Kieffer), Ceratopogon	132
bohartorum Camras, Myopa	630	borealis Lackschewitz, Trichocera	15
bohartorum (Laffoon), Mycetophila	210	*borealis* (Ludlow), Aedes	111
bohemica Becker, Lispe	878	borealis (Malloch), Cnephia	184
bohemica Mesnil, Drino	1084	*borealis* (Malloch), Eupogonomyia	902
bohemani Goetghebuer, Diamesa	151	borealis Malloch, Hydrophoria	864
bohemani (Ringdahl), Helina	886	borealis (Malloch), Leptocera	723
Bolbodimyia Bigot	330	borealis (Malloch), Liriomyza	801
Bolbomyia Loew	298	borealis (Malloch), Megaselia	538
Boletina Staeger	216	*borealis* Malloch, Piophila	711
boletina (Zetterstedt), Odinia	794	borealis Patterson, Drosophila	763
Bolitophila Meigen	197	borealis Reinhard, Johnsonia	950
Bolitophilella Landrock	197	borealis Reinhard, Winthemia	1090
BOLITOPHILINAE	197	borealis Rübsaamen, (Sciara)	236
Bolomyia Brauer & Bergenstamm	1084	borealis Shewell, Camptoprosopella	696
bolsteri Van Duzee, Dolichopus	488	borealis Staeger, Helophilus	617
bombiformis Townsend, Microdon	598	*borealis* Stein, (Limnophora)	1114
Bombobrachycoma Townsend	940	borealis Steyskal, Sepedon	693
bomboides Hunter, Pocota	610	*borealis* (Townsend), Chrysoexorista	1099
bomboides (Wiedemann), Mallophora	396	borealis Van Duzee, Exechia	206
Bombomima Enderlein	388	borealis Walker, Tipula	32
bombylans (Fabricius), Temnostoma	††614	*borealis* Wheeler, Scaptomyza	769
bombylans (Linnaeus), Volucella	600	borealis Zetterstedt, Leptopeza	451
bombylans Robineau-Desvoidy, Carcelia	†1092	**Boreellus** Aldrich & Shannon	925
BOMBYLIIDAE	407	boregoensis Alexander, Tipula	35
BOMBYLIINAE	408	**Boreochlus** Edwards	150
Bombyliomyia Brauer & Bergenstamm	992	**Boreodromia** Coquillett	466
Bombyliopsis Townsend	992	*Boreomyia* Coquillett	466
Bombylius Linnaeus	408	*boreus* (Stone), Hybomitra	340
bonassi Brauer, Hypoderma	1112	*boreus* Van Duzee, Dolichopus	492
Bonellia Robineau-Desvoidy	1006	borinquenensis Wheeler, Puliciphora	545
Bonellimyia Townsend	1006	borinqueni Fox & Hoffman, Culicoides	†131
bonita Huckett, Coenosia	871	**Borophaga** Enderlein	535
bonnarius Johnson, Nemotelus	309	boscii (Macquart), Mesograpta	571
Bonnetia Robineau-Desvoidy	1004	boscii (Macquart), Rhagio	344
boonei Curran, Mydas	358	*boscii* (Robineau-Desvoidy), Archytas	1000
BORBORIDAE	718	boscii (Robineau-Desvoidy), Limnia	691
Borborillus Duda	720	*boscii* (Robineau-Desvoidy), Psorophora	109
Borboropsis Czerny	809	boscii Robineau-Desvoidy, Rivellia	655
Borborus Meigen	†719	*Botanobia* Lioy	†779, 781
borea (Snyder), Coenosia	874	**Botanophila** Lioy	851
boreale Cook, Rhegmoclema	238	*botaurinorum* (Swenk), Lynchia	918
boreale (Van Duzee), Raphium	512	botaurus (Townes), Chironomus	169
borealis Aldrich, Belvosia	1081	bottimeri Wirth, Culicoides	129
borealis (Aldrich), Mesorhaga	482	*boulderensis* (Felt), Campylomyza	†248, 248
borealis Alexander, Rhabdomastix	77	boulderi Felt, Campylomyza	248
borealis Brooks, Ceromasia	1099	bournei (Shaw), Bradysia	232
borealis Brues, Apocephalus	543	BOUVIEROMYIINI	322
borealis Byers, Dolichopeza	23	bovis (Linnaeus), Hypoderma	1112
borealis Chillcott, Fannia	893	boycei Cresson, Rhagoletis	674
borealis Cole, Aphoebantus	429	boydi Beck, Chironomus	167
borealis (Cole), Helophilus	619	*boydi* Vargas, Anopheles	106
borealis Cole, Ogcodes	406	brachialis Coquillett, Chironomus	169
borealis Cole, Tabuda	352	brachialis (Melander), Tachypeza	473
borealis Cole, Thereva	353	**Brachicoma** Rondani	940
borealis Cook, Chaoborus	103	**Brachineura** Rondani	252
borealis (Coquillett), Arctophyto	985	BRACHINEURINI	252
borealis Coquillett, Cheilosia	584	*Brachiosoma* Theobald	120
borealis Coquillett, Diadocidia	198	*Brachycampta* Winnertz	205

1570 A CATALOG OF DIPTERA OF NORTH AMERICA

	Page
Brachycara Thomson	306
BRACHYCERA	296
brachycera (Osten Sacken), Hexatoma	69
brachychaeta Cresson, Notiphila	747
Brachycoma Rondani	940
Brachycrotaphus Czerny	637
Brachydeutera Loew	753
brachygaster Hull, Sphegina	593
Brachymyia Williston	611
brachyneura Alexander, Rhabdomastix	77
brachyneura Malloch, Metriocnemus	161
Brachyneura Rondani	252
brachynteroides Osten Sacken, (Cecidomyia)	279
Brachyopa Meigen	592
Brachypalpus Macquart	608
Brachypeza Winnertz	205
brachyphallus Alexander, Cryptolabis	78
Brachypogon Kieffer	133
Brachypremna Osten Sacken	18
brachypterna (Loew), Heleomyza	816
Brachypteromyia Williston	917
brachyrhabda Alexander, Ormosia	85
brachyrhynchus (Macquart), Physoconops	626, †627
brachyrhynchus (Osten Sacken), Nemomydas	359
brachysoma Coquillett, Empis	459
brachysoma Coquillett, Sapromyza	704
brachystigmaticus Hardy & Knowlton, Pipunculus	554
brachystoma (Coquillett), Hilara	457
brachystoma Coquillett, Sapromyza	704
Brachystoma Meigen	447
brachystoma (Stenhammar), Leptocera	722
BRACHYSTOMATINAE	447
bracteatum Coquillett, Simulium	185
bracteatus Van Duzee, Chrysotus	521
bradleii Van Duzee, Sciapus	486
bradleyi (Alexander), Cladura	72
bradleyi Bequaert, Hirmoneura	402
bradleyi Bromley, Stenopogon	383
bradleyi Cresson, Axysta	751
bradleyi King, Anopheles	105
bradorei Chillcott, Fannia	893
Bradysia Winnertz	232
braggii Jones, Eupeodes	562
brakeleyi (Coquillett), Corethrella	105
branickii (Nowicki), Pseudodiamesa	153
braseniae (Leathers), Polypedilum	175
brasiliana (Brauer & Bergenstamm), Promasiphya	†1091
brasiliana Brauer & Bergenstamm, Synthesiomyia	911
brasiliensis Schiner, Atacta	1080
brassicae (Bouché), Hylemya	847
brassicae (Riley), Liriomyza	801
brassicoides Packard, (Cecidomyia)	292
brassicoides Townsend, (Cecidomyia)	293
Brauerimyia Townsend	1031
Braula Nitzsch	726
BRAULIDAE	726
braycnemus Coquillett, Dolichopus	488
brehmei Knab, Culex	119
Bremia Rondani	280
brennani Philip, Chrysops	324
brennani (Stone), Hybomitra	338
Brennania Philip	321
breviaria Felt, Janetiella	264
brevicalcarata Alexander, Ormosia	85
brevicauda Chillcott, Fannia	893
brevicauda Felt, Asphondylia	268
brevicauda (Felt), Camptoneuromyia	275
brevicauda Felt, Janetiella	264
brevicauda Huckett, Pseudocoenosia	875
brevicauda Loew, Dichaeta	748
brevicauda Van Duzee, Dolichopus	488
brevicauda Van Duzee, Hydrophorus	505
brevicauda Van Duzee, Sympycnus	527
breviceps Loew, Dilophus	196
breviceps Loew, Parydra	750
brevicercum Knowlton & Rowe, Simulium	188
brevichaeta (Shannon), Cheilosia	586
breviciliatus Van Duzee, Dolichopus	488
breviciliatus Van Duzee, Dolichopus	489
breviclava Alexander, Pedicia	54
brevicollis Alexander, Tipula	28

	Page
brevicomus (Hine), Neoitamus	397
brevicorne (Van Duzee), Rhaphium	512
brevicornis Alexander, Trichocera	16
brevicornis (Brooks), Metavoria	1018
brevicornis (Felt), Corinthomyia	249
brevicornis (Felt), Janetiella	264
brevicornis Johnson, Allognosta	300
brevicornis Loew, Euparyphus	308
brevicornis Loew, Hormopeza	453
brevicornis Loew, Leptogaster	362
brevicornis Loew, Sparnopolius	413
brevicornis (Malloch), Spilogona	880
brevicornis (Meigen), Swammerdamella	239
brevicornis (Melander), Piophila	713
brevicornis Osten Sacken, Sphecomyia	612
brevicornis (Robinson), Lamprochromus	530
brevicornis (Say), Sphyracephala	638
brevicornis Sherman, Rymosia	207
brevicornis (Townsend), Euphasiopteryx	980
brevicornis (Williston), Cophura	367
brevicornis (Wirth), Nilobezzia	138
brevicornis (Zetterstedt), Hylemya	854
brevicosta Kieffer, Psectrocladius	159
brevicostalis (Malloch), Phytagromyza	803
brevicostalis Melander, Limnia	691
brevifasciata Johnson, Rivellia	655
brevifilosa Alexander, Limnophila	65
breviforceps Tothill, Gonia	1075
brevifrons Melander, Hemerodromia	470
brevifrons Walker, Hesperinus	191
brevifurca Melander, Bicellaria	451
brevifurca Osten Sacken, Limnophila	67
brevifurcata (Alexander), Pedicia	54
brevifurcata Alexander, Tipula	27
brevihirta Steyskal, Tetanocera	694
brevijuncta Hardy, Tomosvaryella	555
brevilamellatum Van Duzee, Rhaphium	512
breviligula Alexander, Tipula	31
brevimanus Loew, Dolichopus	488
brevinervis (Malloch), Nanocladius	155
brevinervis Malloch, Xanthaciura	671
brevinervis Van Duzee, Diaphorus	519
brevineura (West), Alophorella	967
breviorcornis (Doane), Nephrotoma	20
brevioricornis Alexander, Hexatoma	69
brevipalpis Chillcott, Fannia	893
brevipalpis Cook, Scatopse	238
brevipalpis Huckett, Hylemya	852
brevipalpis (Kieffer), Tanytarsus	179
brevipennis Cook, Psectrosciara	240
brevipennis (Loew), Besseria	972
brevipennis (Macquart), Forcipomyia	125
brevipennis Malloch, Allomyella	834
brevipennis Meigen, Dolichopus	489
brevipennis (Olivier), Labostigmina	314
brevipennis (Wiedemann), Proctacanthus	399
brevipes Aldrich, Lispe	878
brevipes (Fabricius), Senopterina	657
brevipes Van Duzee, Argyra	525
brevipes (Van Duzee), Gymnopternus	498
brevipes Van Duzee, Sympycnus	527
brevipes Van Duzee, Thinophilus	507
brevipetiolata (Shaw), Lycoriella	231
brevipetiolata Van Duzee, Exechia	206
brevipila (Alexander), Hexatoma	70
brevipila Loew, Hilara	456
brevipilosa Malloch, Mydaea	891
brevipilosa (Shannon), Xylota	607
brevipilosus Van Duzee, Dolichopus	489
brevipulvilli Tothill, Gonia	1075
brevirostris Alexander, Ornithodes	56
brevirostris Coquillett, Eugnoriste	231
brevirostris (Coquillett), Pseudosiphona	1063
brevirostris James, Leskiella	1035
brevirostris (Macquart), Sparnopolius	413
brevirostris (Osten Sacken), Helius	51
brevirostris Reinhard, Plagiomima	1019
brevirostris (Tothill), Nowickia	993
brevirostris Van Duzee, Physocephala	627
brevirostris (Van Duzee), Thecophora	631
brevis Banks, Dioctria	370
brevis Coquillett, Paradmontia	1015
brevis Cresson, Pipunculus	554
brevis (Cresson), Renocera	692
brevis Garrett, Dixa	101
brevis Huckett, Proboscimyia	868

INDEX 1571

Entry	Page
brevis (Loew), Empis	459
brevis Loew, Rhamphomyia	462
brevis (Rondani), Fannia	†895, 895
brevis (Stein), Helina	886, †888
brevis (Van Duzee), Peloropeodes	517
brevis (Walker), Allognosta	300
brevis Walker, (Drosophila)	772
brevis Walker, (Ephydra)	759
Breviscapus Quate	93
breviseta (Thomson), Hydrophorus	505
brevisetus (Malloch), Copromyza	720
brevispina Malloch, Phaonia	906
brevispinosa Alexander, Dicranota	58
brevissima Alexander, Neolimnophila	73
brevistylata McAlpine, Earomyia	716
brevistylata (Williston), Atoniomyia	387
brevistylata Williston, Exoprosopa	444
brevistylus Coquillett, Aphoebantus	429
brevitarsata Malloch, Hylemya	847
brevitarsis (Malloch), Lasiops	904
brevitarsis Malloch, Lispocephala	876
brevitarsis (Malloch), Paraprosalpia	866
brevitibia Melander, Clinocera	469
brevitibia Van Duzee, Brachypeza	205
brevitibialis Zetterstedt, Chironomus	169
Brevitrichia Hardy	355
breviusculoides Bromley, Stenopogon	383
breviusculus Loew, Stenopogon	383
brevivena (Osten Sacken), Limonia	46
breviventris Kahl, Mixogaster	599
breviventris Van Duzee, Sympycnus	527
brevivenula Alexander, Limonia	46
brevivitta (Coquillett), Mycomya	220
Bricinnia Walker	657
Bricinniella Giglio-Tos	657
bridwelli Aldrich, Spathimeigenia	1049
bridwelli Cole, Sphegina	593
bridwelli (Hine), Phasia	969
Brillia Kieffer	154
brimleyi Curran, Homalactia	1013
brimleyi Hardy, Allomethus	555
brimleyi Hine, Chrysops	324
brimleyi Reinhard, Cryptomeigenia	1041
brimleyi Sabrosky, Rhodesiella	781
brimleyi Shannon, Sphegina	593
brimleyi Steyskal, Dictya	689
brittoni Alexander, Rhabdomastix	77
brittoni (Felt), Culiseta	117
brizia (Walker), Eustalomyia	†868, 868
Brochella Melander	455
bromeliae Sturtevant, Drosophila	763
Bromeloecia Spuler	723
bromleyi Curran, Mallophora	396
bromleyi Hardy, Nicocles	380
bromleyi (Hull), Dioctria	370
bromleyi Pritchard, Ommatius	398
bromleyi Wilcox, Laphystia	376
Bromleyus Hardy	365
brookmani Alexander, Gonomyia	74
brookmani Harriot, Euxesta	651
brookmani Wilcox, Laphystia	376
brookmani Wirth, Culicoides	132
brookmani Wirth, Dasyhelea	126
brookmani Wirth, Forcipomyia	125
brooksi Chillcott, Fannia	893
brooksi (Curran), Helophilus	620
brooksi Johnson, Agathomyia	548
brooksi (Johnson), Stenomyia	654
brooksi Pipkin, Drosophila	767
brooksi Sabrosky & Arnaud, Pseudochaeta	1096
brousii Williston, Eristalis	623
broweri Alexander, Ormosia	85
broweri Alexander, Tipula	32
broweriana Alexander, Limonia	47
broweriana Alexander, Prionocera	19
browni Curran, Cheilosia	584
browni Curran, Cordilura	829
browni Curran, Empis	459
browni Curran, Hydrophorus	505
browni (Curran), Lyciella	700
browni Curran, Rhaphium	512
browni Curran, Stratiomys	311
browni Curran, Zenillia	1107
browni Kellogg, Eretmoptera	163
browni (Shaw), Bradysia	232
browni Townes, Stenochironomus	177
browni (Twinn), Prosimulium	183
browni (Van Duzee), Gymnopternus	500
browningi (Laffoon), Mycetophila	210
brucensis Steyskal, Euparyphus	308
bruesi Cresson, Ephydra	754
bruesi Graenicher, Helophilus	618
bruesi Van Duzee, Dolichopus	489
bruesii Johnson, Phyllomydas	360
bruesii Melander, Nemotelus	309
brumalis Czerny, Oldenbergiella	810
brumalis Fitch, Trichocera	16
brumalis (Long), Forcipomyia	126
Brumptomyia França & Parrot	91
bruneifacies Van Duzee, Dolichopus	489
bruneri Johnson, Stratiomys	311
bruneri Wilcox & Martin, Nannocyrtopogon	379
brunetta (Huckett), Paraprosalpia	866
Brunettia Annandale	94
brunicosa (Robineau-Desvoidy), Bufolucilia	927
brunnea (Bromley), Laphria	390
brunnea Brooks, Psalidopteryx	1067
brunnea Cole, Thereva	353
brunnea Coquillett, Empis	459
brunnea (Doane), Limonia	46
brunnea Kröber, Psilocephala	351
brunnea (Latreille), Pseudolynchia	919
brunnea Melander, Hemerodromia	470
brunnea Roback, Anatopynia	145
brunnea Stone, Merycomyia	322
brunneicosta (Johnson), Nanna	831
brunneifacies (Robinson), Gymnopternus	498
brunneifrons (Zetterstedt), Hydrophoria	864
brunneinervis (Stein), Phaonia	907
brunneipennis Johannsen, Chironomus	165
brunneipennis Johnson, Orthacheta	831
brunneipennis Leonard, Rhagio	344
brunneipes (Cresson), Rainieria	636
brunneisquama (Malloch), Spilogona	882
brunneistigma Doane, Neaspilota	672
brunneonitens Cresson, Discocerina	738
brunneostigmata Doane, Neaspilota	672
brunnescens Malloch, Coenosia	872
brunnescens (Zetterstedt), Hylemya	847
brunnesquama (Malloch), Spilogona	882
brunneus Aldrich, Dolichopus	489
brunneus Hine, Chrysops	324
brunneus Johnson, Mydas	358
brunneus Van Duzee, Dolichopus	489
brunneus (Walker), Micropsectra	178
brunnicans Walley, Cricotopus	157
brunnicollis (Becker), Tricimba	785
brunnipennis Becker, Chlorops	789
brunnipennis (Malloch), Lamproscatella	756
brunnipennis Melander, Clinocera	468
brunnipes (Fabricius), Bibio	192
brunnipes Johnson, Beris	301
brunnipes (Malloch), Megaselia	538
brunnipes (Meigen), Pherbellia	686
brunnipes (Melander & Spuler), Sepsis	683
brunnipes (Zetterstedt), Micropsectra	178
bryanti Alexander, Limnophila	67
bryanti Curran, Peleteria	997
bryanti Curran, Thelaira	1021
bryanti Felt, Coquillettomyia	281
bryanti (Felt), Corinthomyia	249
bryanti Felt, Hyperdiplosis	284
bryanti Johnson, Bibio	193
bryanti (Johnson), Limonia	45
bryanti Malloch, Campsicnemus	526
bryanti Malloch, Cosmetopus	834
bryanti Van Duzee, Dolichopus	489
bryantiana Alexander, Erioptera	81
bryantiana Alexander, Phyllolabis	60
bryantii (Johnson), Melangyna	568
Bryomyia Kieffer	250
bubonis (Packard), Lynchia	918
buccalis Van Duzee, Physocephala	628
buccata (Fabricius), Cuterebra	1109
buccata (Fallén), Heterocheila	679
buccata (Loew), Ferdinandea	588
buccata (Macquart), Psilota	593
buccata Malloch, Xenomydaea	892
buccata Melander, Trixoscelis	817
buccata Wheeler, Amiota	761
buccator Fitch, Cuterebra	1110
buccatum (Cresson), Hydrochasma	738

1572   A CATALOG OF DIPTERA OF NORTH AMERICA

	Page
bucculenta (Coquillett), Hylemya	847
Bucentes Latreille	1064
bucephala (Meigen), Spallanzania	1077
Bucephalina Malloch	832
Bucephalomyia Malloch	885
bucera Alexander, Ormosia	87
bucera Alexander, Tipula	35
bucera Shaw, Bolitophila	197
bucerus Coquillett, Lordotus	412
buchanani Malloch, Minettia	701
bucinator Melander, Euthyneura	449
bucinator Melander, Mythicomyia	416
buckelli (Curran), Spathimeigenia	1049
Buckellia Curran	366
buddleia Felt, Asphondylia	268
buenoi Alexander, Tipula	36
Bufolucilia Townsend	926
bulbirostris (Loew), Physoconops	626
bulbosa (Fluke), Melangyna	566
bulbosa Johannsen, Allodia	205
bulbosa (Melander), Anthalia	450
bulbosa (Osten Sacken), Metatrichia	354
bulbula Felt, Rhopalomyia	266
bulla (Felt), Trishormomyia	288
bulla Thomsen, Stilobezzia	134
bulla Walsh, (Cecidomyia)	294
bulla Westwood, Acrocera	405
bullans (Wiedemann), Euaresta	665
bullata Melander, Hormopeza	453
bullata Parker, Sarcophaga	957
bullatus (Bromley), Haplopogon	373
bullifera Kessel & Kessel, Empis	459
bumeliae Felt, Asphondylia	268
Buquetia Robineau-Desvoidy	1098
bureni Wirth, Thalassomya	163
burgessi Alexander, Gonomyia	74
burgessi (Malloch), Melanagromyza	796
burgessi (Williston), Physocephala	627
burkei Shannon, Cheilosia	584
burnesi Van Duzee, Dolichopus	489
burneyensis Alexander, Ormosia	85
burra (Alexander), Cheilotrichia	79
buscki (Felt), Anarete	246
busckii Coquillett, Drosophila	766
busckii Townsend, Bezzimyia	963
buskii Coquillett, Drosophila	766
bussi James, Psilocephala	351
buteonis (Swenk), Ornithoctona	917
butleri Johnson & Johnson, Exoprosopa	444
butleri Johnson & Johnson, Poecilanthrax	442
butleri Wirth & Hubert, Culicoides	130
buxi (Laboulbène), Monarthropalpus	289
byersi (Laffoon), Mycetophila	210
Byomya Robineau-Desvoidy	913
bysia (Walker), Phaonia	906
byssina (Schrank), Smittia	162
Byssodon Enderlein	186
caca Pritchard, Cophura	367
caccabata (Tucker), Heleomyza	816
cacheae Harmston & Knowlton, Hercostomus	497
cachinnans Osten Sacken, Bombylius	409
cacopiloga (Hine), Proctacanthella	398
cacothius Dyar, Aedes	111
Cacotrophus Reinhard	950
Cacozelus Reinhard	1012
cacti Richards, Leptocera	725
cacticola Wirth & Hubert, Culicoides	131
cactifera Shewell, Homoneura	698
cactorum Wirth & Hubert, Dasyhelea	126
cacuminifer Melander, Empis	459
cadaverina (Linnaeus), Pyrellia	912
cadaverina (Robineau-Desvoidy), Cynomyopsis	930
cadaverum Kirby, (Musca)	915
Cadrema Walker	776
caduca Huckett, Pegomya	858
caducus Townes, Microtendipes	172
caeligena Melander, Empis	459
caelobata, Spuler, Leptocera	722
caelum Townes, Microtendipes	172
caena Roback, Diamesa	151
Caenia Robineau-Desvoidy	755
caenosa (Rondani), Leptocera	720
Caenotus Cole	424
caerulea (Bigot), Sargus	302

	Page
caerulea (Malloch), Boreellus	925
caerulea Van Duzee, Mesorhaga	482
caerulea Walker, Lonchaea	717
caeruleiformis Macquart, Nemopoda	682
caeruleiviridis (Macquart), Phaenicia	927
caerulescens Loew, Senopterina	657
caerulescens Malloch, Medetera	510
caerulescens (Robineau-Desvoidy), Cochliomyia	924
caerulescens Stein, Hydrotaea	900
caerulescens (Stein), Phaonia	906
caerulifrons (Johnson), Merosargus	303
caerulus Van Duzee, Chrysotus	521
caesar (Aldrich), Aplomya	1097
caesar (Johannsen), Lycoriella	231
caesar (Linneaus), Lucilia	†927
caesariatus Martin, Holopogon	375
caesarion (Meigen), Orthellia	913
caesia (Coquillett), Minettia	701
caesia (Meigen), Amoebaleria	815
caesia (Reinhard), Euphyto	935
caesia Stein, Pegomya	858
caesia Stein, Phaonia	907
caesius Melander, Cyrtopogon	368
caffreyi Walton, Eleodiphaga	1072
cahuilla Alexander, Tipula	24
cala Melander, Mythicomyia	416
calainus Melander & Brues, Dolichopus	489
Calamoncosis Enderlein	786
calaveras Alexander, Tipula	26
calcaneus Loew, Callinicus	365
calcar (Osten Sacken), Pedicia	54
calcarata (Coquillett), Forcipomyia	124
calcarata (Coquillett), Mycomya	220
calcarata Curran, Sericomyia	603
calcarata Doane, Tipula	35
calcarata Williston, Tropidia	609
calcaratum Van Duzee, Rhaphium	512
calcaratus Aldrich, Dolichopus	489
calcaratus Curran, Eucyrtopogon	373
calcaratus (Loew), Condylostylus	483
calcaratus (Loew), Neocnemodon	581
calcaratus Van Duzee, Campsicnemus	526
calcaratus Van Duzee Sympycnus	527
calcaroides Alexander, Pedicia	55
calceata Loew, Argyra	525
calceata (Snow), Calotarsa	550
calceatum Harris (Simulium)	1114
calceola Harris (Anthomyia)	1114
calceolus Say (Asilus)	1114
calciequina Felt, Winnertzia	253
calcitrans (Curran), Lejota	590
calcitrans (Linnaeus), Stomoxys	914
calcitrans Loew, Argyra	525
calcitrans (Spuler), Copromyza	719
caldaria (Lintner), Bradysia	232
calens Linnaeus, Tabanus	332
calidus (Townsend), Myobiopsis	1035
californiae (Walker), Archytas	999
californiae (Walker), Exoprosopa	444
californiae (Walker), Stenopogon	383
californica Alexander, Gonomyia	74
californica (Banks), Laphria	389
californica (Bigot), Cylindromyia	973
californica Bohart, Orthopodomyia	108
californica Capelle, Loxocera	638
californica (Cole), Psectrosciara	240
californica Coquillett, Rhamphomyia	462
californica (Coquillett), Valentibulla	670
californica (Cresson), Leptopsilopa	741
californica (Davidson), Heringia	580
californica Doane, Tephritis	668
californica (Doane), Tipula	40
californica Doane, Tipula	36
californica Felt, Dasineura	259
californica Felt, Diarthronomyia	262
californica Felt, Dicrodiplosis	279
californica Felt, Mayetiola	263
californica Felt, Rhabdophaga	257
californica Felt, Rhopalomyia	266
californica (Felt), Thomasiniana	281
californica Fisher, Phronia	215
californica Greene, Mythicomyia	416
californica Hall, Melanodexia	931
californica (Hardy), Pseudonomoneura	360
californica James, Beris	301

INDEX 1573

	Page		Page
californica Johannsen, Dixa	102	*Calliope* Westwood	696
californica Kessel, Protoclythia	549	**Calliopum** Strand	696
californica Kincaid, Pericoma	92	callipedilus Loew, Cyrtopogon	368
californica Kröber, Thereva	353	**Calliphora** Robineau-Desvoidy	929
californica (Malloch), Coenosia	873	CALLIPHORIDAE	922
californica Malloch, Sphegina	594	CALLIPHORINAE	926
californica Malloch, Trupanea	667	CALLIPHORINI	928
californica (Marten), Hybomitra	338	*Calliplatyura* Malloch	202
californica Martin, Leptogaster	362	*Calliprobola* Rondani	608
californica (Osten Sacken), Hexatoma	70	callipus Garrett, Eccoptomera	811
californica (Osten Sacken), Limonia	44	*callithotrys* Dyar, Aedes	111
californica (Packard), Hydropyrus	754	callithrix Melander, Platypalpus	478
californica (Parker), Blaesoxipha	946	*Callitroga* Brauer	923
californica Sabrosky, Elachiptera	777	*Callitroga* Hall	923
californica Sabrosky, Meoneura	729	**Callomyia** Meigen	547
californica (Schaeffer), Efferia	394	*Callopistria* Loew	643
californica Shannon, Myolepta	589	**Callopistromyia** Hendel	643
californica Steyskal, Pseudotephritis	647	*Calloplatyura* Malloch	202
californica Sturtevant, Drosophila	763	*callosa* James, Dieuryneura	306
californica Townsend, Hesperophasiopsis	978	CALOBATIDAE	633
californica (Van Duzee), Anevrina	532	**Calobatina** Enderlein	635
californica Van Duzee, Argyra	525	CALOBATINAE	633
californica Van Duzee, Micropeza	635	caloceps Bigot (Exochostoma)	319
californica Van Duzee, Mycomya	220	**Caloforcipomyia** Saunders	125
californicum Blanc, Chetostoma	677	*Calomyia* Meigen	547
californicum Camras, Zodion	629	**Caloparyphus** James	308
californicum Harmston, Syntormon	515	*Calopelta* Greene	413
californicum Smith, Spilochaetosoma	1006	*calophaga* Loew, Coenosia	874
californicum Sublette, Polypedilum	175	**Calopsectra** Kieffer	179
californicus (Bigot), Eclimus	426	*calopteni* (Riley), Hylemya	850
californicus (Bigot), Pilimas	321	*caloptera* (Bigot), Pseudeuantha	1029
californicus (Cole), Poecilanthrax	442	*Caloptera* Guérin-Méneville	69
californicus Cole & Wilcox, Lasiopogon	377	caloptera Loew, Tipula	27
californicus (Coquillett), Athanatus	102	*caloptera* Say, Erioptera	82
californicus Fairchild & Hertig, Phlebotomus	92	*Calopterella* Coquillett	772
californicus Harmston, Aphrosylus	509	calopteroides Alexander, Tipula	27
californicus Harmston & Knowlton, Polymedon	503	*calopterus* (Mitchell), Polypedilum	175
californicus Hine, Asilus	392	*calopus* Bigot, Stratiomys	311
californicus Johannsen, Chironomus	169	calopus Loew, Nothomyia	317
californicus Leonard, Rhagio	344	calopyga Loew, Coenosia	874
californicus Melander, Neossos	817	*calorhina* (Bigot), Sphecomyia	613
californicus Sabrosky, Eribolus	777	caloris Painter, Apiocera	357
californicus Townsend, Hyalomyodes	977	calosomae (Coquillett), Sitophaga	1048
californicus Townsend, Xanthomelanodes	967	calosomae (Townsend), Sitophaga	1048
californicus Van Duzee, Chrysotus	521	**Calotarsa** Townsend	550
californicus Van Duzee, Diaphorus	519	caltha (Shannon), Cheilosia	586
californicus Van Duzee, Dolichopus	489	calva Coquillett, Oestrophasia	979
californicus Van Duzee, Gymnopternus	499	calva (Coquillett), Procatharosia	971
californicus Wilcox, Ablautus	364	*calva* Doane, Tipula	40
californicus Wilcox & Martin, Cyrtopogon	369	calva (Loew), Holcocephala	375
californiensis Alexander, Rhabdomastix	77	calva Melander, Mythicomyia	416
californiensis (Kellogg), Maruina	97	calva (Melander), Tachyempis	475
californiensis (Macquart), Deopalpus	1004	calverti (Cresson), Zeros	746
californiensis Malloch, Bigotomyia	909	*calverti* Fox & Stabler, Basilia	922
californiensis Malloch, Coenosia	873	calvicrura (Coquillett), Mydaea	892
californiensis (Malloch), Megaselia	540	calvimontis Cockerell, Rhamphomyia	462
californiensis (Malloch), Phyllogaster	876	calvimontis James, Dolichopus	489
californiensis Wheeler, Medetera	510	calvus Loew, Geron	423
californioides Bromley, Stenopogon	383	**Calycomyza** Hendel	799
Califortalis Curran	649	calypso Hull, Baccha	572
caliginosa Cresson, Hydrellia	744	*calyptera* (Say), Eutreta	661
caliginosus Foote, Procecidocharoides	661	calyptrata (Coquillett), Hyalomya	969
caliginosus (Johannsen), Clinotanypus	145	calyptrata (Hendel), Phytobia	800
caliginosus (Johannsen), Glyptotendipes	172	calyptrata West, Ateloglossa	986
caligula Melander, Mythicomyia	416	calyptrata (Zetterstedt), Pegomya	861, 1862
calinota (Dietz), Nephrotoma	21	CALYPTRATAE	826
Caliprobola Rondani	608	**Calyptrosomus** Reinhard	975
caliptera Fitch, (Cecidomyia)	291	**Calythea** Schnabl & Dziedzicki	865
caliptera Say, Erioptera	82	**Calyxochaetus** Bigot	528
caliptera (Say), Eutreta	661	*Camaromyia* Hendel	665
caliptera (Say), Exoprosopa	444	camatus Reinhard, Archimimus	942
Calirrhoe Meigen	987	cambrica Edwards, Bryomyia	251
calla Kessel, Callomyia	547	*Camerania* Giglio-Tos	600
Callachna Aldrich	660	*cameroni* Curran, Chrysopilus	346
Callicera Panzer	595	**Camilla** Haliday	773
CALLICERINI	595	camillae Weems, Myolepta	589
callichroma Hull & Fluke, Cheilosia	584	CAMILLIDAE	773
callida Johannsen, Allodia	265	campanulata Robertson, Sphegina	594
callidula Philip, Chrysops	324	*campestre* Curran, Peleteria	998
callidus Osten Sacken, Chrysops	324	campestre Curran, Rhaphium	512
callidus Williston, Tolmerus	401	campestris (Coquillett), Lepidanthrax	440
callima Melander, Mythicomyia	416	campestris Curran, Mericia	1008
Callimyia Meigen	547	campestris Dyar & Knab, Aedes	111
Callinapaea Sturtevant & Wheeler	750	campestris (Fallén), Metopia	937
Callinicus Loew	365	*campestris* (Huckett), Hylemya	845
		campestris (Robineau-Desvoidy), Hylemya	855

1574 A CATALOG OF DIPTERA OF NORTH AMERICA

	Page		Page
Campichaeta Macquart	772	canadensis Snyder, Helina	886
Campichoeta Macquart	772	canadensis Snyder, Lispe	878
Campogaster Rondani	976	canadensis (Theobald), Aedes	111
camprestris (Coquillett), Lepidanthrax	440	canadensis (Tothill), Nowickia	993
CAMPSICNEMINAE	526	canadensis Van Duzee, Chrysotus	521
Campsicnemus Haliday	526	canadensis Van Duzee, Dolichopus	489
Camptocladius Wulp	162	canadensis (Van Duzee), Lamprochromus	530
Camptomyia Kieffer	256	*canadensis* Walker, Scatophaga	838
Camptoneura Macquart	644	canadensis Walker, (Stratiomys)	319
Camptoneuromyia Felt	275	canadensis (Westwood), Limonia	49, †71
Camptopelta Williston	309	canalicula Johannsen, Exechia	206
Camptoprosopella Hendel	696	canaliculatus Thomson, Dolichopus	489
Camptops Aldrich	948	canalis (Coquillett), Geminaria	413
Camptopsis Townsend	950	canariensis (Macquart), Pseudolynchia	919
Camptopterohelea Wirth & Hubert	133	canaster Melander, Empis	459
Camptopyga Aldrich	951	cancer Theobald, Deinocerites	120
Camptosceles Haliday	526	candens Walker, (Tachina)	1108
CAMPYLOCHETINI	1015	candicans Loew, Rhamphomyia	462
Campylomyza Meigen	248	candida Coquillett, Efferia	394
cana Coquillett, Hilara	456	candida Huckett, Coenosia	871
cana Cresson, Ditrichophora	739	candidata Loew, Thereva	353
cana (Hine), Efferia	394	candidatus Melander, Parathalassius	452
cana (Huckett), Spilogona	880	candidipennis Foote, Tephritis	668
cana (Loew), Melieria	645	candidipes Foote, Cecidomyia	287
cana (Macquart), Hylemya	850	candidulus Loew, Systoechus	410
cana Melander, Chersodromia	475	canella (Bromley), Efferia	394
cana Melander, Iteaphila	454	**Canelomyia** Reinhard	1071
cana Melander, Minettia	701	*canescens* (Meigen), Tephrochlamys	810
cana (Walker), Erioptera	81	*canescens* Stein, Coenosia	871
cana Webber, Tachinomyia	1058	canescens (Walker), Ptilodexia	988
Canace Haliday	733	canescens (Wheeler), Hydrophorus	505
Canacea Haliday	733	caniceps Cresson, Ilythea	746
CANACEIDAE	733	canicularis (Linnaeus), Fannia	894
Canaceoides Cresson	733	canifrons Cresson, Ditrichophora	739
canada Hull & Fluke, Cheilosia	584	*canifrons* Walker, Chrysops	325
canadense (Brooks), Gymnosoma	965	canis Williston, Laphria	389
canadense Curran, Rhaphium	512	*canithorax* Hoffman, Culicoides	131
canadense Hearle, Simulium	188	canities Van Duzee, Hydrophorus	505
canadensis Aldrich, Lasiosina	791	canities Van Duzee, Thinophilus	508
canadensis Brooks, Pseudochaeta	1095	canonicola (Dyar & Shannon), Simulium	185
canadensis Cole, Psilocephala	351	canora Reinhard, Chaetophlepsis	1015
canadensis (Cresson), Beris	301	cantator (Coquillett), Aedes	111
canadensis Cresson, Pelina	752	cantenea (Roback), Metoposarcophaga	951
canadensis (Curran), Antichaeta	688	cantities Van Duzee, Thinophilus	508
canadensis (Curran), Arctosyrphus	620	canus Cole & Wilcox, Lasiopogon	377
canadensis Curran, Belvosia	1081	canus Melander, Caenotus	424
canadensis Curran, Bombylius	409	canus Melander, Platypalpus	478
canadensis (Curran), Cephalochrysa	302	canuta (Wulp), Tricharaea	960
canadensis (Curran), Chrysosyrphus	592	capax Coquillett, Aphoebantus	429
canadensis Curran, Coenosia	871	capax (Coquillett), Lasiopleura	774
canadensis Curran, Cynorhinella	589	capax (Coquillett), Oligodranes	421
canadensis Curran, Heringia	580	capax Cresson, Hydrochasma	738
canadensis (Curran), Hybomitra	339	capensis (Wiedemann), Ophyra	901
canadensis Curran, Laphystia	376	*capillaris* (Brues), Megaselia	541
canadensis (Curran), Metasyrphus	560	*capillata* (Coquillett), Ectecephala	†791
canadensis Curran, Nicoletes	380	capillata Johannsen, Exechia	206
canadensis (Felt), Asteromyia	274	capillata (Loew), Acanthocnema	835
canadensis Felt, Asynapta	254	capillata Loew, Cheilosia	586
canadensis Felt, Cincticornia	269	capistrano Alexander, Tipula	24
canadensis Felt, Contarinia	277	*capitata* (De Geer), Gonia	1075
canadensis Felt, Cystiphora	261	capitata Felt, Colpodia	253
canadensis Felt, Dasineura	259	capitata Felt, Didactylomyia	254
canadensis Felt, Dirhiza	255	capitata Felt, Rhopalomyia	266
canadensis Felt, (Itonida)	289	capitata Loew, Blepharicera	99
canadensis Felt, Monardia	250	capitata Townsend, Cockerelliana	1012
canadensis Felt, Porricondyla	255	capitata (Wiedemann), Ceratitis	672
canadensis (Felt), Trishormomyia	288	capitata (Zetterstedt), Phytobia	799
canadensis Garrett, Aecothea	811	capitella Alexander, Antocha	51
canadensis (Garrett), Agathon	99	capitis (Curran), Polybiomyia	616
canadensis Hall, Sarcophaga	958	capito (Coquillett), Hylemya	848
canadensis Hardy, Dorylomorpha	556	capito Loew, Oedopa	653
canadensis Huckett, Hylemya	847	capito Osten Sacken, Pantarbes	414
canadensis Johnson, Agathomyia	548	capnioptera Alexander, Neolimnophila	73
canadensis (Johnson), Herina	645	capnopterus Melander, Oreogeton	453
canadensis Kröber, Chrysops	325	caprea (Coquillett), Villa	439
canadensis (Loew), Epochra	677	capreolata (Laffoon), Mycetophila	210
canadensis Loew, Nemotelus	310	*capsularis* Patterson, (Cecidomyiaceltis)	293
canadensis Loew, Tipula	30	capsularis Patton, (Cecidomyiaceltis)	293
canadensis Macquart, Bibio	194	*captiosa* Johannsen, Rymosia	207
canadensis (Macquart), Trypetoptera	695	captiva Felt, Mycodiplosis	282
canadensis Malloch, Agromyza	795	captiva Johannsen, Exechia	206
canadensis Malloch, Fannia	893	captonis (Marten), Hybomitra	338
canadensis Reinhard, Myiopharus	1045	captus Coquillett, Empis	459
canadensis Schlinger, Ogcodes	406	captus Coquillett, Hemerodromia	470
canadensis (Shannon), Cheilosia	587	capucina (Fabricius), Exoprosopa	†432, 444
canadensis (Smith), Imitomyia	975	*cara* (West), Hyalomya	968
canadensis Snyder, Acridomyia	915	**carata** Roback, Boettcheria	948

INDEX 1575

	Page		Page
carbicolor (Lovett), Cynorhinella	588	carolinensis (Metcalf), Sericomyia	603
carbona (Cresson), Otites	646	carolinensis (Robineau-Desvoidy), Heleomyza	816
carbona Curran, Anorostoma	812		
carbonaria Borgmeier, Triphleba	534	carolinensis Robineau-Desvoidy, (Neria)	636
carbonaria Felt, Dasineura	259	*carolinensis* (Robineau-Desvoidy), Orthellia	913
carbonaria (Hendel), Nanna	831	carolinensis Robineau-Desvoidy, (Scatophora)	†719, 1113
carbonaria Kessel, Protoclythia	549		
carbonaria (Loew), Oscinella	779	carolinensis Schiner, Laphria	389
carbonaria Macquart, Limnophila	63	carolinensis Schiner, Leptogaster	362
carbonaria (Meigen), Fannia	894, †894	carolinensis Van Duzee, Dolichopus	489
carbonaria Melander, Hilara	456	caroliniana Alexander, Tipula	30
carbonaria (Panzer), Periscepsia	†1034	carolinus Hardy, Bibio	193
carbonaria Patterson & Wheeler, Drosophila	763	carolus Alexander, Dolichopeza	23
carbonaria (Rondani), Fannia	897	*Carphotricha* Loew	†662
carbonaria Snow, Laphria	389	carpini (Felt), Parallelodiplosis	286
carbonaria (Walker), Cordilura	827	carpini (Felt), Phaenolauthia	261
carbonarius Bosc, Keroplatus	200	carpini (Felt), Polyardis	250
carbonarius Giglio-Tos, Euparyphus	307	carpini Felt, Winnertzia	253
carbonarius Loew, Nemotelus	310	carpinicola (Kieffer), Parallelodiplosis	286
carbonarius Osten Sacken, Aphoebantus	429	carpocapsae (Townsend), Nilea	1104
carbonarius Walker, Chrysops	324	*Carpolonchaea* Bezzi	718
carbonella (Zetterstedt), Spilogona	†882	carpophaga (Tripp), Mayetiola	263
carbonifer (Loew), Neurigona	518	**Carposcalis** Enderlein	575
carbonifer Osten Sacken, Mydas	358	carptura Melander, Mythicomyia	416
carbonifera (Felt), Asteromyia	274	carri Curran, Bibio	193
carbonifera (Osten Sacken), Asteromyia	274	carri (Curran), Xylota	607
carbonifera (Walker), Archytas	1000	carruthi Shaw, Mycetophila	210
carbonipes (Alexander), Arctoconopa	79	carsoni Alexander, Erioptera	83
carbonitens Cockerell, Lasioptera	271	carsoni Wheeler, Drosophila	764
Carcelia Robineau-Desvoidy	1092	*carteri* Hoffman, Leptoconops	122
CARCELIINI	1091	**Cartocometes** Aldrich	1069
Carcinomyia Townsend	931	**Cartosyrphus** Bigot	586
Cardiacephala Macquart	633	carunculata Alexander, Tipula	35
cardini Sturtevant, Drosophila	763	*carus* Cresson, Lordotus	412
Cardiocladius Kieffer	154	carus (Townes), Chironomus	165
carduorum Huckett, Pegomya	858	caryae Felt, Cincticornia	269
carectorum Chillcott, Platypalpus	478	caryae (Felt), Dentifibula	276
caribea Sturtevant, Drosophila	768	caryae Felt, (Dirhiza)	291
Caricea Robineau-Desvoidy	870, †886	caryae Felt, Lasioptera	271
caricicola (Kieffer), Dasineura	259	caryae Felt, (Oligotrophus)	289
Cariciella Malloch	874	caryae (Felt), Parallelodiplosis	286
caricis (Felt), Bremia	280	caryae (Felt), Trotteria	276
caricis Felt, Dasineura	259	caryae Osten Sacken, (Diplosis)	291
caricis Felt, Lobopteromyia	276	caryaecola Felt, Schizomyia	267
caridei (Brèthes), Blaesoxipha	†944, 944	caryaecola Osten Sacken, (Cecidomyia)	270
carinata (Curran), Carposcalis	575	*caryae nucicola* Osten Sacken, (Cecidomyia)	270
carinata Doane, Tipula	27	**Caryomyia** Felt	270
carinata (Dodge), Sarcophaga	958	cascadensis Alexander, Pedicia	54
carinata Loew, Notiphila	746	cascadensis Weems, Xylota	607
carinata Spuler, Leptocera	723	cascadica Alexander, Dicranota	57
carinata Sturtevant & Wheeler, Scatophila	758	casei (Linnaeus), Piophila	711
carinata Tothill, Gonia	1075	cassiae Felt, Lasioptera	271
carinatus (Curran), Neocnemodon	581	*casta* Johannsen, Exechia	206
carinatus (Townes), Chironomus	167	*casta* Loew, Tipula	27
Carinosillus Reinhard	989	castanea (Bigot), Myopa	630
carlatus (Roback), Orthocladius	156	castanea (Jaennicke), Villa	438
carlina Schmitz, Phora	537	castaneae Felt, Rhopalomyia	266
carlynensis (Malloch), Megaselia	540	castaneae Stebbins, (Cecidomyia)	293
carnea (Fabricius), Pentaneura	†147	castaneicoxa Borgmeier, Dohrniphora	535
carneosa (Fittkau), Pentaneura	147	castanipes (Bigot), Villa	435
carneus Walker, Nemotelus	310	castanipes Jaennicke, Bibio	193
CARNIDAE	728	*castanoptera* (Loew), Physocephala	628
CARNINAE	728	casteeli Pipkin, Drosophila	767
Carnus Nitzsch	729	*castor* Wheeler, Polymedon	503
carolae Capelle, Rhopalosyrphus	599	castus Wheeler, Polymedon	503
caroli (Malloch), Spilogona	880	casuarius (Townes), Chironomus	167
carolina Banks, Pericoma	92	cata (Melander & Brues), Megaselia	538
carolina Felt, Arthrocnodax	283	*Catabomba* Osten Sacken	562
carolina Felt, Epimyia	252	**Cataclinusa** Schmitz	545
carolina Felt, Mycodiplosis	282	**Catagoniopsis** Townsend	1098
carolina Felt, Porricondyla	255	catalina (Curran), Didea	562
carolina Felt, Rhopalomyia	266	*catalina* (Parker), Metoposarcophaga	951
carolina Fisher, Phthinia	225	catalina Shannon, Cheilosia	584
carolinae (Felt), Colpodia	253	catalinae Alexander, Limonia	46
carolinae (Felt), Lestodiplosis	284	catalinensis Alexander, Tipula	26
carolinae (Felt), Peromyia	251	**Catalinovoria** Townsend	10¹⁷
carolinae Martin & Wilcox, Asilus	392	catalpae Comstock, (Diplosis)	289
carolinae (Robineau-Desvoidy), Cylindromyia	974	catalpae (Malloch), Goniopsita	786
		catana (Curran), Megaselia	538
carolinensis Alexander, Ormosia	85	cataphylla Dyar, Aedes	111
carolinensis Alexander, Teucholabis	76	**Catatasina** Enderlein	316
carolinensis (Bigot), Condylostylus	485	catawba Alexander, Pedicia	56
carolinensis Brauer & Bergenstamm, Ptilodexia	988	catawba Alexander, Tipula	35
		catawbae (Boesel), Leptoconops	122
carolinensis Bromley, Echthopoda	372	catawbiana Alexander, Tipula	27
carolinensis Cole & Wilcox, Lasiopogon	377	catawbiensis Alexander, Dicranota	56
carolinensis (Macquart), Hamatabanus	†338, 341		

	Page		Page
catawbiensis Shaw, Phthinia	225	cedrica Huckett, Pegomya	858
Catemophrys Townsend	1031	celarata (Aldrich), Blaesoxipha	945
catenarius Melander, Aphoebantus	429	celastri Felt, Neolasioptera	273
catenatus Walker, Tabanus	333	celastri Stebbins, (Cecidomyia)	293
Catharosia Rondani	971	celator (Laffoon), Mycetophila	210
CATHAROSIINI	971	**Celatoria** Coquillett	1040
cathistes Pritchard, Heteropezina	247	celatus Pechuman, Chrysops	325
Catocha Haliday	243	celer Cole, Eclimus	426
CATOCHINI	243	celer Cole, Exoprosopa	444
Catotricha Edwards	243	celer (Coquillett), Pilatea	1105
CATOTRICHINI	243	*celer* Osten Sacken, Chrysops	324
catskillensis (West), Ictericophyto	1024	celer Townsend, Eupelecotheca	1051
Cattasoma Reinhard	941	celer Townsend, Hyalomya	968
cattleyae (Felt), Parallelodiplosis	286	celer Townsend, Zizyphomyia	1088
cattleyae (Molliard), Parallelodiplosis	286	celer (Wiedemann), Villa	435
catulina (Coquillett), Villa	435	celeripes Van Duzee, Dolichopus	489
catulus Coquillett, Aphoebantus	429	celeripes Winnertz, Corynoneura	164
catulus Osten Sacken, Stichopogon	385	celesteana Pritchard, Anocha	244
caudata Curran, Criorhina	611	cellarius Melander, Platypalpus	478
caudata (Curran), Volucella	602	cellularis Spuler, Leptocera	723
caudata (Fallén), Dichaeta	748	*Celotrophus* Reinhard	1024
caudata Felt, Asynapta	254	celtiphyllia Felt, Mayetiola	263
caudata Felt, Hormomyia	287	*cena* Melander, Minettia	701
caudata (Felt), Mayetiola	263	**Cenosoma** Wulp	979
caudata Felt, Porricondyla	255	centerensis Felt, Lasioptera	271
caudata Felt, Trotteria	276	*Centeter* Aldrich	1069
caudata (Lundbeck), Rhabdomastix	77	*Centor* Loew	788
caudata Staeger, Mycetophila	210	centralis (Brues), Megaselia	538
caudata (Townsend), Blaesoxipha	947	centralis Hanson, Nemotelus	310
caudata Van Duzee, Mesorhaga	482	*centralis* Loew, Dixa	102
caudatula Oldenberg, Chymomyza	770	centralis Loew, Oxycera	307
caudatulus (Loew), Condylostylus	483	centralis Loew, Tipula	30
caudatum Shewell, Prosimulium	182	centralis Malloch, Anatopynia	145
caudatum Van Duzee, Raphium	512	centralis Wheeler, Paramycodrosophila	771
caudatus Cresson, Pipunculus	553	*centron* (Marten), Hybomitra	340
caudatus Edwards, Protanypus	153	*centrotus* Howard, Dyar, & Knab, Aedes	111
caudatus Van Duzee, Asyndetus	524	*ceparum* (Meigen), Hylemya	847
caudatus Van Duzee, Chrysotus	521	*cepetorum* (Meade), Hylemya	847
caudatus Van Duzee, Sympycnus	527	cephala Garrett, Leia	227
caudatus (Wiedemann), Condylostylus	483	cephalanthi Felt, Rhabdophaga	257
caudelli (Malloch), Dorylomorpha	556	*Cephalemya* Latreille	1111
caudellii (Coquillett), Mallochohelea	139	*Cephalemyia* Latreille	1111
caudifera Alexander, Limnophila	65	**Cephalia** Meigen	643
caulfieldi Garrett, Mycomya	220	**Cephalochaeta** Kertész	302
caulicola (Coquillett), Clinodiplosis	282	**Cephalomyza** Hendel	798
caulicola Felt, Lasioptera	271	*Cephalops* Fallén	551
caulicola Felt, Mayetiola	263	**Cephalosphaera** Enderlein	554
caulicola Felt, Rhabdophaga	257	cephalotes Cresson, Lamproscatella	756
caurina (Doane), Urophora	659	**Cephenemyia** Latreille	1111
cuarina (Laffoon), Mycetophila	210	CEPHENEMYIINAE	1111
cuarinus McAtee, Dilophus	195	*Cephenomyia* Latreille	1111
cauta Harris, (Tetanocera)	1114	*Cephenus* Berthold	424
cauta Townsend, Catalinovoria	1017	*cephus* Fabricius, Anthrax	431
cauterium Clark, Cuterebra	1109	**Ceracia** Rondani	1037
cautor (Coquillett), Villa	439	*Cerantichir* Enderlein	637
cavaticus Wirth & Jones, Culicoides	129	ceras (Townsend), Silvius	323
cavatus Van Duzee, Dolichopus	489	*cerasi* (Felt), Asynapta	254
cavernicola (Brues), Megaselia	538	*cerasi* (Felt), Bryomyia	251
cavernicola Melander, Hilara	456	cerasi Felt, Lestodiplosis	284
cavicola Townsend, Chiricahuia	978	cerasi (Felt), Rhizomyia	262
cavicola Townsend, Neomuscina	911	cerasifolia (Felt), Mycodiplosis	282
cavifrons (Kröber), Scenopinus	356	cerasiphila Felt, (Cecidomyia)	291
cavillator (Laffoon), Mycetophila	210	cerasiserotinae Osten Sacken, (Cecidomyia)	293
cayensis Fairchild, Tabanus	333	*cerastes* (Osten Sacken), Hamatabanus	342
caymanicus Fairchild, Tabanus	337	cerata (Walker), Ptilodexia	988
cayuga (Alexander), Dicranota	58	**Ceratempis** Melander	467
cayuga (Alexander), Paradelphomyia	59	**Ceratinostoma** Meade	840
cayuga Alexander, Tipula	27	**Ceratitis** Macleay	672
cayuga (Malloch), Megaselia	538	**Ceratobarys** Coquillett	776
cayugae Johannsen, Chironomus	165	**Ceratochaeta** Brauer & Bergenstamm	†1088
cayugae Johannsen, Phaonia	907	**Ceratocystia** Dyar and Knab	110
cazieri Alexander, Tipula	24	*Ceratolophus* Kieffer	136
cazieri Brookman, Stenopogon	383	ceratomiae (Coquillett), Hyphantrophaga	1094
cazieri Harriot, Tetanops	648	*Ceratomyia* Felt	250
cazieri Martin, Hadrokolos	373	**Ceratomyiella** Townsend	1031
cazieri Wilcox, Laphystia	376	**Ceratopogon** Meigen	132
cazieriana Alexander, Dicranota	58	CERATOPOGONIDAE	121
ceanothi Felt, Asphondylia	268	CERATOPOGONINAE	127
Cecidomyia Meigen	287	CERATOPOGONINI	132
Cecidomyiaceltis Patton	291	*Ceratoxys* Rondani	643
CECIDOMYIIDAE	241	*Ceraturgopsis* Johnson	365
CECIDOMYIINAE	251	**Ceraturgus** Wiedemann	365
CECIDOMYIINI	279	cerea (Coquillett), Gimnomera	840
cecropia (Riley), Winthemia	1090	*cerealis* Fitch, (Cecidomyia)	292
cecropiae Reinhard, Winthemia	1090	cerealis (Gillette), Hylemya	848
cedens Walker, Anthrax	431	**Ceria** Fabricius	615

INDEX 1577

Entry	Page
Ceriana Rafinesque	615
ceringogaster Chillcott, Fannia	894
Ceriodes Rondani	615
Cerioides Rondani	615
CERIOIDINI	615
ceris Roback, Cricotopus	157
cernua Sherman, (Leia)	1114
cerocarpi Felt, Dasineura	259
Cerodontha Rondani	801
Ceromasia Rondani	1099
Ceroplatus Bosc	200
Ceropsilopa Cresson	741
Cerotainia Schiner	387
Cerotainiops Curran	387
Cerotelion Rondani	200
Ceroxys Macquart	643
cerrita Alexander, Ormosia	85
certima (Walker), Cochliomyia	924
certimus Adams, Chlorops	789
cerurae Sellers, Aplomya	1097
cerussata Reinhard, Spathidexia	1021
cerussatus (Osten Sacken), Nannocyrtopogon	379
cerutias Loew, Hydrophorus	505
cervi (Linnaeus), Lipoptena	920
cervicula Doane, Tipula	27
cervina (Alexander), Pedicia	54
cervina (Doane), Limonia	46
cervina Loew, Rhamphomyia	462
cervinus Loew, Aphoebantus	429
cessator (Aldrich), Blaesoxipha	946
cesta (Haliday), Axysta	751
Cetema Hendel	788
cevelatus (Curran), Neocnemodon	581
ceyx (Loew), Poecilanthrax	442
Chactochlorops Malloch	785
Chaetexorista Brauer & Bergenstamm	1053
chaetifrons Steyskal, Texasa	654
chaetilamellus Harmston & Knowlton, Hercostomus	497
chaetoala (Sublette), Chironomus	167
Chaetoapnaea Hendel	749
Chaetochlorops Malloch	785
Chaetocladius Kieffer	160
Chaetoclusia Coquillett	806
Chaetocrania Townsend	1074
Chaetogaedia Brauer & Bergenstamm	1078
CHAETOGAEDIINA	1078
Chaetogaediopsis Townsend	1079
Chaetogena Rondani	1056
Chaetoglossa Townsend	1078
Chaetolabis Townes	164
Chaetolonchaea Czerny	716
Chaetomacera Cresson	693
chaetomera Czerny, Suillia	809
Chaetomus Czerny	815
chaetoneura (Coquillett), Aporotachina	1039
Chaetoneura Malloch	532
chaetoneura (Malloch), Megaselia	538
Chaetoneurophora Malloch	532
Chaetonodexodes Townsend	1029
Chaetonopsis Townsend	1030
Chaetopeleteria Mik	996, †996
Chaetophleps Coquillett	1040
Chaetophlepsis Townsend	1015
Chaetoplagia Coquillett	1017
Chaetopleurophora Schmitz	533
chaetopoda (Davidson), Carposcalis	575
chaetopodus Williston, Platycheirus	577
Chaetopsis Loew	650
Chaetoravinia Townsend	954
Chaetosa Coquillett	837
chaetosa (Townsend), Pilatea	1105
Chaetostigmoptera Townsend	1062
Chaetostoma Rondani	677
chaetosula Townsend, Muscopteryx	1025
chaetosus Cole & Wilcox, Lasiopogon	377
chagnoni Curran, Cyrtopogon	368
chagnoni (Johnson), Pyrgotella	658
chagnoni Philip, Chrysops	325
chagrinensis Stalker & Spencer, Drosophila	767
CHALARINAE	550
Chalarus Walker	550
chalcochrus Loew, Gymnopternus	500
chalcogaster (Wiedemann), Ophyra	901
Chalcomyia Williston	590
chalcopyga Loew, Sericomyia	603
Chalcosyrphus Curran	589
chalepus (Walker), Helophilus	618
chalonensis (Nowell), Dixa	102
chalybaea (Coquillett), Melinocera	1007
chalybea (Hendel), Cirrula	753
chalybea Wiedemann, Xylota	606
chalybescens Williston, Cheilosia	584
chalybeus Macquart, Eristalis	623
chalybeus Melander, Oligodranes	421
chalybeus (Van Duzee), Condylostylus	483
Chamaedipsia Mik	469
Chamaemyia Meigen	707
CHAMAEMYIIDAE	706
CHAMAEMYIINAE	706
CHAMAEMYIINI	706
Chamaepsila Hendel	639
Chamaesyrphus Mik	595
chamberlini (Laffoon), Mycetophila	210
champlaini Curran, Microdon	597
champlaini (Philip), Haematopota	330
champlainii (Walton), Laphria	389
CHAOBORIDAE	102
CHAOBORINAE	103
CHAOBORINI	103
Chaoborus Lichtenstein	103
chaoi Shaw, Trichonta	216
chapmani (Edwards), Trichocladius	158
characta Kraft & Cook, Pachygaster	318
Charadrodromia Melander	475
Charasoma Reinhard	1040
Chasmatonotus Loew	159
cheaini (Garrett), Dioptopsis	100
Cheilosia Meigen	583
Cheilosia Panzer	578
CHEILOSIINI	583
Cheilotrichia Rossi	78
chelana Melander, Tachydromia	474
Chelifera Macquart	471
Cheligaster Macquart	684
cheliopterus Rondani, Tabanus	333
Chelipoda Macquart	472
Chelisia Rondani	844
CHELISIINI	844
chelonia (Townes), Chironomus	165
cheloniae (Rondani), Carcelia	††1092
chen Townes, Pseudochironomus	176
cherokeana Alexander, Tipula	32
cherokeensis Alexander, Limnophila	67
cherokeensis (Jones), Melangyna	566
Chersodromia Walker	475
Chetogena Rondani	1056
Chetostoma Rondani	677
Chiastochaeta Pokorny	844
Chiastocheta Pokorny	844
chiclayae Greene, Anastrepha	673
chidesteri Dyar, Culex	118
chillcotti Cook, Swammerdamella	239
chillcotti Fluke, Metasyrphus	560
Chilomyia Shannon	583
chilosia (Curran), Carposcalis	576
Chilosia Meigen	583
chilosioides (Shannon), Chrysosyrphus	592
chimaera (Osten Sacken), Villa	435, †435
chinook Shannon, Chrysotoxum	582
chinooki Priddy, Conophorus	413
chinquapin Beutenmüller, (Cecidomyia)	293
chintimini Lovett, Cheilosia	584
Chionea Dalman	72
chionthrix Hull & Fluke, Cheilosia	587
chiopterus (Meigen), Culicoides	128
chiragra Melander, Microphorella	452
chiricahuensis Alexander, Tipula	34
Chiricahuiini Townsend	978
CHIRICAHUIINI	978
Chiromantis Rondani	472
Chiromyia Robineau-Desvoidy	822
CHIROMYZINAE	300
CHIRONOMIDAE	142
CHIRONOMINAE	164
CHIRONOMINI	164
Chironomus Meigen	164
Chirosia Rondani	844
chirosphena Hull, Platycheirus	576
chitona Melander, Mythicomyia	419
chittendeni (Coquillett), Wohlfahrtia	942
chlamydata (Melander), Liriomyza	801

1578 A CATALOG OF DIPTERA OF NORTH AMERICA

	Page
chlanoflavus Harmston & Knowlton, Chrysotus	521
chloratica Johannsen, Mycomya	221
chloricus Wheeler, Chrysotus	521
Chlorohystricia Townsend	992
Chlorometaphyto Townsend	1005
Chloromyia Duncan	303
chlorophylla Osten Sacken, Erioptera	81
chlorophylloides Alexander, Erioptera	81
CHLOROPIDAE	773
CHLOROPINAE	787
Chloropisca Loew	787
CHLOROPOIDEA	777
Chloroprocta Wulp	923
Chlorops Meigen	789
chloropus Bergroth, Ornithomyia	917
Chlorotabanus Lutz	329
Choeroporpa Dyar	119
Cholomyia Bigot	1029
chorea (Fabricius), Lonchaea	717
chorea Lundbeck, Diamesa	151
chorea (Meigen), Limonia	46
choreus (Meigen), Procladius	149
choricus Wheeler, Chrysotus	521
Choristomma Stein	808
Choristoneura Rübsaamen	275
Chortophila Macquart	846
christata Garrett, Zygomyia	216
christiansoni Wirth & Hubert, Forcipomyia	125
christulata Garrett, Zygomyia	216
Chromatocera Townsend	1012
Chromolepida Cole	350
chromolepida Cole, Villa	434
chrysanthemi Ahlberg, Diarthronomyia	262
chrysanthemi Kowarz, Phytomyza	804
Chrysanthrax Osten Sacken	437
chrysida Huckett, Pegomya	858
chrysites Osten Sacken, Mydas	358
Chrysoceria Williston	365
Chrysochlamys Walker	588
chrysochlamys Williston, Cheilosia	584
Chrysochlora Latreille	†305
Chrysochlorina James	305
CHRYSOCHLORINAE	305
Chrysoclamis Walker	588
chrysocoma Osten Sacken, Erioptera	81
chrysocoma (Osten Sacken), Goniops	322
chrysocomoides Alexander, Erioptera	81
Chrysoexorista Townsend	1099
Chrysogaster Meigen	590
CHRYSOGASTRINI	590
chrysologus (Walker), Hydrophorus	505
chrysolygus Walker, Hydrophorus	505
Chrysomasicera Townsend	1099
chrysomela Bromley, Mallophora	396
Chrysomya Robineau-Desvoidy	†923
Chrysomyia Robineau-Desvoidy	†923
CHRYSOMYINAE	923
CHRYSOMYINI	923
Chrysomyza Fallén	653
chrysophani (Townsend), Aplomya	1098
chrysopila Loew, Hermetia	304
Chrysopila Macquart	345
Chrysopilus Macquart	345
CHRYSOPINAE	322
CHRYSOPINI	322
chrysopogon Loew, Cyrtopogon	368
chrysoprasi (Walker), Condylostylus	483
chrysoprasius (Walker), Condylostylus	483
chrysoprocta Reinhard, Sturmia	1087
chrysoprocta (Wiedemann), Mystacella	1086
chrysops Martin, Beameromyia	361
Chrysops Meigen	323
chrysopsidis (Loew), Rhopalomyia	266
chrysorrhoea (Meigen), Protocalliphora	926
CHRYSOSOMATINAE	482
Chrysosomidia Curran	608
chrysostoma Loew, Schoenomyza	870
chrysostoma Osten Sacken, Mydas	358
chrysostomus Loew, Dolichopus	489
chrysostomus (Wiedemann), Helophilus	618
Chrysosyrphus Sedman	592
Chrysotachina Brauer & Bergenstamm	1005
chrysothamni Felt, Asphondylia	268
chrysothamni Felt, Asteromyia	274
chrysothamni Felt, Rhopalomyia	266

	Page
chrysothrix Hull & Fluke, Cheilosia	587
Chrysotimus Loew	529
CHRYSOTOXINI	582
chrysotoxoides Macquart, Sericomyia	603
Chrysotoxum Meigen	582
Chrysotus Meigen	521
Chrysozona Meigen	330
chumash Alexander, Tipula	40
churchillensis Alexander, Erioptera	83
churchillensis Alexander, Tipula	30
churchillensis Chillcott, Platypalpus	478
churchilli Malloch, Chaetosa	837
churchilliana Alexander, Prionocera	19
Chyliza Fallén	640
CHYLIZINAE	640
Chylizosoma Hendel	840
Chymomyza Czerny	770
Chymophila Macquart	599
Chyromantis Rondani	472
Chyromya Robineau-Desvoidy	822
Chyromyia Robineau-Desvoidy	822
CHYROMYIDAE	821
cibolae James, Adoxomyia	305
ciliata Aldrich, Belvosia	1081
ciliata Coquillett, Rhamphomyia	462
ciliata (Fabricius), Hydrotaea	899
ciliata (Fabricius), Psorophora	109
ciliata (Macquart), Lespesia	1101
ciliata (Townsend), Chaetonopsis	1030
ciliata (Townsend), Winthemia	1091
ciliata Van Duzee, Argyra	525
ciliata Van Duzee, Medetera	510
ciliata Van Duzee, Neurigona	518
ciliata (Walker), Hylemya	854
ciliata Winnertz, Anatella	205
ciliata (Winnertz), Forcipomyia	125
ciliata (Zetterstedt), Megaselia	†540
ciliatipes James, Cordilura	827
ciliatissima Chillcott, Fannia	894
ciliatum Curran, Rhaphium	512
ciliatus (Aldrich), Dolichopus	494
ciliatus Bigot, Platycheirus	576
ciliatus (Loew), Condylostylus	484
ciliatus Malloch, Chrysotus	521
ciliatus (Van Duzee), Microprosopa	835
ciliatus Walker, Dolichopus	494
cilicauda Malloch, Coenosia	871
cilicrura (Rondani), Hylemya	850
cilifemoratus (Van Duzee), Calyxochaetus	528
cilifera (Malloch), Gymnodia	884
cilifera (Malloch), Homoneura	698
cilifera Malloch, Hylemya	848
ciliipes (Aldrich), Condylostylus	483
cilimanus Van Duzee, Neurigona	518
cilioraca (Rondani), Hylemya	850
cilipes (Coquillett), Forcipomyia	125
cilipes (Haliday), Azelia	899
cilipes (Say), Rhamphomyia	462
cilipes Wiedemann, Trichopoda	966
cimarronensis Rogers, Tipula	24
cimbiciformis (Fallén), Mallota	†621
cimbicis (Aldrich), Diplonevra	534
cimbicis (Townsend), Boettcheria	948
cimiciforme Loew, Notogramma	646
cimmeria Speiser, Tipula	27
Cina Fabricius	615
cincinna Felt, Corinthomyia	249
cincinnata Johannsen, Exechia	206
cincta (Coquillett), Dasyhelea	127
cincta (Coquillett), Leia	227
cincta (Fabricius), Hybomitra	338
cincta (Fallén), Melangyna	566
cincta Felt, Hormomyia	287
cincta Felt, (Itonida)	290
cincta Felt, Lestodiplosis	285
cincta Felt, Lobodiplosis	281
cincta Felt, Mycodiplosis	282
cincta Felt, Rhizomyia	262
cincta Johannsen, Boletina	217
cincta Johannsen, Trichonta	216
cincta (Loew), Camptoprosopella	696
cincta (Loew), Spaziphora	836
cincta Loew, Tipula	28
cincta (Meigen), Tricimba	785
cincta Olivier, Odontomyia	315
cincta Townsend, Miamimyia	1070

Entry	Page
cincta Van Duzee, Allodia	205
cinctella Kieffer, (Cecidomyia)	290
cinctella (Zetterstedt), Meliscaeva	568
Cincticornia Felt	269
cincticornis Walker, Chrysops	324
cinctipennis (Alexander), Arctoconopa	79
cinctipes (Coquillett), Forcipomyia	125
cinctipes (Coquillett), Mochlonyx	104
cinctipes Felt, Parallelodiplosis	286
cinctipes Harris, (Trypeta)	1114
cinctipes (Johannsen), Ablabesmyia	148
cinctipes (Say), Limonia	44
cinctocornis Doane, Tipula	36
Cinctotipula Alexander	24
cinctum Townes, Polypedilum	175
cincturus (Coquillett), Oligodranes	421
cinctus Banks, Pipunculus	553
cinctus (Felt), Arthrocnodax	283
cinctus (Osten Sacken), Euparyphus	308
cinctus Townes, Stenochironomus	177
Cinderella Steyskal	825
cinefacta Coquillett, Rhamphomyia	462
cinefacta (Coquillett), Villa	437
cineracea (Coquillett), Neodeceia	702
cineracea Coquillett, Rhamphomyia	462
cineracea Coquillett, Tipula	32
cineralis Hull & Fluke, Cheilosia	584
cinerapennis Adams, Chlorops	789
cineraria (Loew), Acantholeria	814
cinerascens Back, Stenopogon	383
cinerascens Cole, Thereva	353
cinerascens (Townsend), Senotainia	939
cinerea (Aldrich), Hyperecteina	1069
cinerea (Alexander), Hexatoma	70
cinerea (Back), Laphria	389
cinerea Banks, Psychoda	96
cinerea (Bigot), Empis	458
cinerea Cole, Psilocephala	351
cinerea Cole, Villa	438
cinerea Coquillett, Atelogiossa	986
cinerea (Coquillett), Athrycia	1016
cinerea (Coquillett), Milichiella	†733
cinerea (Coquillett), Neodeceia	702
cinerea (Coquillett), Trixoscelis	817
cinerea Curran, Lasionalia	1024
cinerea (Doane), Gonomyia	75
cinerea Felt, Lasioptera	271
cinerea (Felt), Neolasioptera	273
cinerea James, Leskiomima	1035
cinerea James, Wilcoxia	385
cinerea Johnson, Symphoromyia	343
cinerea Jones, Ephydra	754
cinerea (Loew), Olcella	784
cinerea Macquart, Lestremia	244
cinerea (Meigen), Bebryx	†903
cinerea Meigen, Bolitophila	197
cinerea Melander, Tachyempis	475
cinerea Patterson & Wheeler, Drosophila	767
cinerea (Perris), Hydrophorus	506
cinerea (Perris), Hyperecteina	1070
cinerea Snow, Platypeza	549
cinerea Townsend, Hesperodinera	987
cinerea Townsend, Meigenielloides	1045
cinerea Wulp, Microchaetina	984
cinerea (Wulp), Pseudotachinomyia	†1057
cinereibarbis Bigot, Coenomyia	296
cinereicolor Alexander, Pedicia	54
cinereifrons (Frost), Phytobia	798
cinereipennis Lundström, Limonia	46
cinereipleura (Alexander), Cheilotrichia	79
cinereiventre (Loew), Syntormon	515
cinereiventre Van Duzee, Zodion	629
cinereiventris (Zetterstedt), Phaonia	908
cinerella (Fallén), Hylemya	855
cinerella (Wulp), Helina	886
cinereoborealis (Felt & Young), Aedes	115
cinereovittata Bigot, Brachyopa	592
cinerescens Reinhard, Paradidyma	1032
cinereum Bellardi, Simulium	187
cinereum Curran, Anorostoma	812
cinereus Bigot, Bombylius	409
cinereus Cole, Lasiopogon	377
cinereus Meigen, Aedes	116
cinereus Melander, Oligodranes	421
cinerivus Painter, Bombylius	409
cinerosa (Coquillett), Periscepsia	1033
cinerosa (Reinhard), Micromintho	1032
cinerosa (Zetterstedt), Hylemya	845
cingarus (Aldrich), Oxysarcodexia	952
cingulata (Dietz), Nephrotoma	21
cingulata Kröber, Thereva	353
cingulata Johnson & Johnson, Exoprosopa	445
cingulata (Loew), Argyra	525
cingulata Loew, Phthiria	419
cingulata (Loew), Rhagoletis	674
cingulata Rübsaamen, Sciara	230
cingulatulus (Macquart), Mesograpta	571
cingulatus Johnson & Johnson, Lordotus	412
cingulatus Loew, Pipunculus	552
cingulatus (Macquart), Mesograpta	571
cingulatus Macquart, Tabanus	333
cingulum Meigen, Mycetophila	210
cinifera Becker, Lispe	878
Cinochira Zetterstedt	975
CINOCHIRINI	975
cintalapa Cole, Anthrax	431
Cinxia Meigen	603
cio Quate, Telmatoscopus	94
Circia Malloch	843
circinata (Loew), Icterica	664
circularis Alexander, Tipula	40
cirrata Melander, Chelifera	471
cirratus Melander, Epacmus	428
cirrhatus Osten Sacken, Pycnopogon	382
cirripes Melander, Microphorus	452
cirriventris Schmitz, Megaselia	540
Cirrula Cresson	753
cirsioni Felt, Dasineura	259
cismarina McAtee, Anapausis	241
Cistogaster Latreille	964
cistudinis (Aldrich), Cistudinomyia	948
Cistudinomyia Townsend	948
Citabaena Walker	†596
citellivora Shewell, Sarcophaga	958
citheroniae Sabrosky, Winthemia	1090
citima Vockeroth, Spilomyia	613
citreibasis Malloch, Phaonia	905
citreifrons (Malloch), Homoneura	698
citreifrons (Malloch), Phytobia	800
citricola Osten Sacken, (Cecidomyia)	293
citrina (Doane), Limonia	46
citrina Osten Sacken, (Cecidomyia)	293
citrulli Felt, (Itonida)	290
citus Hine, Asilus	392
claasseni Alexander, Tipula	39
cladacantha Alexander, Tipula	35
Cladochaeta Coquillett	771
Cladopelma Kieffer	166
Cladotanytarsus Kieffer	180
Cladura Osten Sacken	72
claggi Alexander, Limnophila	66
claggi (Alexander), Pedicia	54
Clairvillia Robineau-Desvoidy	975
clandestina (Dietz), Nephrotoma	21
clandicans Loew, Campsicnemus	526
Clanoneurum Becker	740
clara Curran, Nausigaster	596
clara Curran, Peleteria	997
clara Doane, Tipula	38
clara Kessel, Callomyia	547
clara Loew, Macrocera	201
clara Loew, Tetanocera	694
clara (Melander), Phytobia	800
clara (Schmitz), Megaselia	540
claracollis Harris, (Chironomus)	1114
clarans (Huckett), Spilogona	880
clarifrons Townsend, Gymnoprosopa	936
claripalpis (Wulp), Paratheresia	991
claripennis (Coquillett), Camptoprosopella	697
claripennis James, Adoxomyia	305
claripennis (Kröber), Chrysops	328
claripennis (Lundbeck), Orthocladius	156
claripennis (Macquart), Euphorocera	1054
claripennis Malloch, Chironomus	168
claripennis Malloch, Palloptera	715
claripennis (Malloch), Procladius	149
claripennis (Malloch), Trixoscelis	817
claripennis Melander, Prorates	424
claripennis Reinhard, Istochaeta	1073
claripennis Van Duzee, Hydrophorus	505
clarkeae Felt, Parallelodiplosis	286
clarkei Felt, Lasioptera	271
clarkei Felt, Rhopalomyia	266
clarkei (Felt), Trishormomyia	288

1580 A CATALOG OF DIPTERA OF NORTH AMERICA

	Page		Page
clarki Curran, Exoprosopa	444	Clinorhyncha Loew	275
clarki Curran, Volucella	600	Clinotanypus Kieffer	145
clarum (Dyar & Shannon), Simulium	186	*Cliochloria* Enderlein	653
Clasiopa Stenhammar	738	*Cliopeza* Enderlein	635
Clasiopella Hendel	742	*clisiocampae* (Townsend), Exorista	1055
classicum Harmston & Knowlton, Parasyntormon	516	clistoides (Townsend), Myiophasia	983
		clistoides Townsend, Xanthocera	1026
classicus Dyar, Aedes	114	Clistomorpha Townsend	977
Clastoneuriopsis Reinhard	1067	*Clitellaria* Meigen	†305
Clastopteromyia Malloch	771	CLITELLARIINAE	305
claterna Reinhard, Impeccantia	1013	CLITELLARIINI	305
clathrata Dietz, Tipula	32	*clivicola* (Malloch), Spilogona	882
clathrata (Loew), Paroxyna	666	cloacaris Fabricius, (Musca)	842
clauda Coquillett, Empis	459	Cloacina Reinhard	1072
clauda Coquillett, Rhamphomyia	462	*Clogmia* Enderlein	93
clauda Pritchard, Wasmanniella	245	clunalis Melander, Epacmus	428
claudicans Loew, Campsicnemus	526	clunalis (Reinhard), Euphorocera	1054
clausa Brauer & Bergenstamm, Oestrophasia	979	Clunio Haliday	163
clausa (Coquillett), Cophura	367	clurina Reinhard, Pseudochaeta	1096
clausa Coquillett, Empis	459	Clusia Haliday	806
clausa Loew, Myopa	630	*Clusiaria* Malloch	807
clausa Osten Sacken, Elliptera	51	CLUSIIDAE	805
clausa (Osten Sacken), Trichopsidea	403	CLUSIINAE	806
clausicella (Macquart), Mallophorina	397	Clusiodes Coquillett	807
Clausicella Rondani	1059	CLUSIODIDAE	805
clausus Coquillett, Keroplatus	200	CLUSIODINAE	807
clavata Edwards, Diamesa	151	cluvia (Walker), Phaenicia	927
clavata (Fabricius), Baccha	572	*clypeata* (Malloch), Thaumatomyia	787
clavata Garrett, Bolitophila	197	clypeatus (Meigen), Platycheirus	576
clavata Garrett, Sciara	230	*Clythia* Meigen	549
clavata (Loew), Borophaga	535, †535	CLYTHIIDAE	546
clavata Loew, Dixa	102	*Clytocerus* Eaton	93
clavata Van Duzee, Mycetophila	210	*Cnemodon* Egger	581
clavator Coquillett, Rhamphomyia	462	*Cnephalia* Rondani	1077
Clavator Philippi	†378	Cnephaliops Townsend	1073
clavatum Van Duzee, Syntormon	515	Cnephalogonia Townsend	1075
clavatus Beck, Tanypus	150	Cnephalomyia Townsend	1073
clavatus (Drury), Mydas	358	Cnephia Enderlein	184
clavatus (Van Duzee), Condylostylus	483	Cnodacophora Czerny	633
clavatus Van Duzee, Sympycnus	527	coachellensis (Hall), Ravinia	954
clavicauda Van Duzee, Mesorhaga	482	*coactus* (Wiedemann), Mallota	621
clavicornis Brennan, Chrysops	324	coaequalis (Schmitz), Megaselia	540
clavicornis (Saunders), Smittia	162	coalescens (Walker), Mesograpta	571
clavicornis (Van Duzee), Calyxochaetus	528	coangustata Schmitz, Phora	537
claviculatus Loew, Paracleius	500	*coarctata* (Fallén), Hylemya	†849
claviger (Townes), Chironomus	167	coarctatus Loew, Microdon	599
clavigera Loew, Rhamphomyia	462	coarctatus Stone, Tabanus	333
clavigera Osten Sacken, Phyllolabis	60	cobala Reinhard, Mericia	1008
clavigerellus Wheeler, Teuchophorus	529	*cobaltinus* Van Duzee, Chrysotus	522
clavinervis Van Duzee, Macrocera	201	*Coboldia* Melander	237
clavipennis (Coquillett), Camptoprosopella	697	cocci Felt, Dentifibula	276
clavipes (Fabricius), Bittacomorpha	98	*coccidella* Townsend, Eumasicera	1094
clavipes (Loew), Ectaetia	241	Coccidomyia Felt	262
clavipes Loew, Tachypeza	473	cocciphila (Aldrich & Webber), Palpexorista	1056
clavipes Meigen, Trichina	449	cocciphila (Coquillett), Syneura	542
clavis Townsend, Paraphasmophaga	1073	coccyx (Aldrich & Webber), Palpexorista	1056
clavis (Williston), Sargus	302	cochleata (Rübsaamen), Lycoriella	232
clavula (Beutenmüller), Lasioptera	271	Cochliomyia Townsend	923
claytonae Wheeler & Takada, Mycodrosophila	771	cockerellae Aldrich, Sarcophaga	958
		cockerelli Alexander, Gnophomyia	73
claytoniae Felt, (Cecidomyia)	290	cockerelli Cole, Thereva	353
clelia Osten Sacken, Stonyx	440	cockerelli (Coquillett), Ormosia	85
clematidis Felt, Asphondylia	268	*cockerelli* (Felt), Anarete	246
clematidis Felt, Contarinia	277	cockerelli Felt, Rhopalomyia	266
clematidis Felt, Dasineura	259	cockerelli Felt, Thecodiplosis	278
clematidis (Felt), Neolasioptera	273	*cockerelli* Jones, Microdon	598
clematidis Loew, Phytomyza	804	cockerelli Malloch, Bezzia	141
clematiflorae Felt, Lestodiplosis	285	cockerelli Melander, Oligodranes	421
clemativora Coquillett, Phytomyza	804	*cockerelli* (Van Duzee), Condylostylus	485
Clemelis Robineau-Desvoidy	1099	Cockerelliana Townsend	1012
clementi Alexander, Dicranota	57	*cockerelli* (Coquillett), Culicoides	128
clementi (Wilcox & Martin), Efferia	394	*cockerellii* (Coquillett), Gaediophana	1079
clemonsi Townsend, Spathidexia	1021	cocklei (Townsend), Sphenometopa	939
cleoae Metcalf, Sphaerophoria	570	coconino Alexander, Tipula	35
Cleodiplosis Felt	280	coeca (Greene), Cremersia	544
Cleona Meigen	547	coeca Nitzsch, Braula	726
clepsydra Fisher, Exechia	206	Coelodiazesis Dyar & Knab	105
clepsydrus Coquillett, Orthocladius	156	Coelomyia Haliday	897
cleptes Osten Sacken, Mydas	358	Coelopa Meigen	679
clesides (Walker), Periscepsia	1033	COELOPIDAE	679
cleta Kessel, Callomyia	547	Coelopina Malloch	680
Climacura Howard, Dyar, & Knab	117	Coelosia Winnertz	218
Clinocera Meigen	468	Coelotanypus Kieffer	146
CLINOCERINAE	466	Coenia Robineau-Desvoidy	755
Clinoceroides Hendel	835	Coenomyia Latreille	296
Clinodiplosis Kieffer	282	COENOMYIIDAE	296
Clinohelea Kieffer	136		

INDEX 1581

	Page		Page
Coenosia Meigen	870	coloradensis Alexander, Rhabdomastix	77
coenosiaeformis Stein, Hylemya	848	coloradensis Bigot, Chrysops	324
COENOSIINAE	869	coloradensis (Bigot), Tropidia	609
Coenosopsia Malloch	905	coloradensis (Brues), Aenigmatias	537
coercens Walker, Dolichopus	489	coloradensis (Cockerell), Neolasioptera	273
coerula Fluke & Hull, Cheilosia	584	*coloradensis* Cockerell & Andrews, Microdon	598
coerula Williston, Baccha	573	*coloradensis* Cole & Wilcox, Lasiopogon	377
coerulea (Aldrich & Webber), Aplomya	1097	coloradensis Doane, Tipula	32
coerulea Cole, Ocnaea	403	coloradensis Felt, Contarinia	277
coeruleifrons Macquart, Nemopoda	682	*coloradensis* Felt, Cordylomyia	249
coeruleiviridis (Macquart), Phaenicia	927	coloradensis Felt, Janetiella	264
coerulescens (Robineau-Desvoidy), Cochliomyia	924	*coloradensis* Felt, Neolasioptera	273
		coloradensis Felt, Procystiphora	258
coerulescens (Williston), Carposcalis	576	coloradensis Grote, Sparnopolius	413
coesiofasciatus Macquart, Tabanus	332	coloradensis Harmston, Hercostomus	497
coffeatus Macquart, Tabanus	335	coloradensis Harmston & James, Scellus	506
cognata Aldrich, Plagiomima	1019	coloradensis Hough, Calliphora	929
cognata Cresson, Notiphila	747	coloradensis (James), Diogmites	371
cognata Cresson, Rivellia	655	coloradensis James, Leptogaster	362
cognata Doane, Tipula	28	coloradensis James, Psilocephala	351
cognata Loew, Baccha	572	*coloradensis* Malloch, Hydrophoria	863
cognata Stein, Pegomya	858	coloradensis Malloch, Lonchaea	717
cognatella Osten Sacken, Gonomyia	75	*coloradensis* (Malloch), Phytobia	799
cognatus Hardy, Bibio	193	coloradensis Pennak, Deuterophlebia	100
cognatus Loew, Pelastoneurus	501	coloradensis Quisenberry, Paroxyna	666
cognatus (Melander & Brues), Dolichopus	492	coloradensis Van Duzee, Chrysotus	521
coheri Shaw, Rymosia	207	coloradensis Van Duzee, Condylostylus	483
Colcondamyia Reinhard	948	coloradensis Wirth, Bezzia	141
coleana Alexander, Tipula	32	coloradica Alexander, Gonomyia	74
colei Alexander, Tipula	39	colorata Coquillett, Rhamphomyia	462
colei Alexander, Trichocera	16	*colorata* Melander, Psila	640
colei Bromley, Stichopogon	385	colorata Walker, Drosophila	764
colei Felt, Colpodia	253	colorati Felt, Lasioptera	271
colei James, Odontomyia	316	**Colpodia** Winnertz	253
colei Johnson & Johnson, Poecilanthrax	442	colteri Alexander, Tipula	28
colei Kessel, Agathomyia	548	*columbia* (Curran), Erynnia	1071
colei (Malloch), Stenochironomus	177	columbiae Curran, Cheilosia	584
colei Melander, Oligodranes	421	*columbiae* Curran, Mallota	621
colei Namba, Rivellia	655	columbiae (Curran), Melangyna	566
colei Philip, Apatolestes	320	*columbiae* (Dyar & Knab), Psorophora	110
colei Sabrosky, Ogcodes	406	columbiaensis Hardy, Bibio	193
colei Wirth, Helophilus	619	columbiana Alexander, Limnophila	66
colemani Wirth, Forcipomyia	124	columbiana Alexander, Trichocera	16
Coleomyia Wilcox & Martin	366	columbiana Chillcott, Fannia	894
Coleophasia Townsend	977	columbiana (Kertész), Tomosvaryella	555
coleophora Melander, Hemerodromia	470	columbiana Sherman, Dziedzickia	218
coleoptrata (Scopoli), Stegana	760	*columbianum* Curran, Chrysotoxum	582
collaris Loew, Loxocera	638	columbica Walker, Laphria	389
collaris Loew, Psila	640	**Columbiella** Malloch	807
collaris (Melander), Scatopse	238	columbiensis Priddy, Conophorus	414
collaris Say, Tipula	25	colute Harmston & James, Rhaphium	512
collata Alexander, Limnophila	64	comanche Alexander, Tipula	28
collateralis Melander, Platypalpus	480	comanche Painter, Bombylius	409
Collatia Curran	1044	comanche Painter, Villa	435
collator (Townes), Chironomus	167	comans Sabrosky, Pholeomyia	732
collina Melander, Mythicomyia	416	**Comantella** Curran	366
collina (Reinhard), Eusisyropa	1093	comantis (Coquillett), Empis	459
Collinella Duda	721	*comantis* Coquillett. Empis	459
Collinellula Aldrich	529	comantis Curran, Eucyrtopogon	373
Collinellula Strand	721	**Comasarcophaga** Hall	949
collini Hall, Acronesia	928	*Comastes* Osten Sacken	411
collini Priddy, Conophorus	413	comastes Williston, Apatolestes	320
collini Ringdahl, Hylemya	856	*comastes* (Williston), Hybomitra	338
collinsonioides Beutenmüller, (Cecidomyia)	293	comata Aldrich, Hydrotaea	900
collinsonifolia Beutenmüller, (Cecidomyia)	293	*comata* Bigot, Symphoromyia	343
colludens Reinhard, Spallanzania	1077	*comata* (Coquillett), Empis	459
collusor (Melander), Metachela	471	comata (Huckett), Spilogona	880
collusor (Townsend), Hippelates	775	comata (Laffoon), Mycetophila	211
Colobaea Zetterstedt	685	comata Loew, Thereva	353
Coloboneura Melander	475	comata Melander, Wiedemannia	↓469
Colobostema Enderlein	238	*comatus* (Doane), Molophilus	88
colombensis Macquart, Tabanus	333	comatus (Loew), Condylostylus	484
colon (Harris), Aedes	114	comatus Loew, Dolichopus	489
colon Harris, (Ortalis)	1114	*comatus* (Schiner), Condylostylus	†484
colon Loew, Seioptera	648	combinata (Linnaeus), Geomyza	821
colonica Walker, Empis	459	*combinata* (Loew), Limnia	691
colonus Bergroth, Molophilus	88	combustus Loew, Saropogon	383
Coloradalia Curran	1015	comes (Walker), Hybomitra	338
coloradella Cockerell, (Diplosis)	291	*comes* (Walker), Ravinia	954
coloradense Curran, Rhaphium	512	cometes Steyskal, Empis	459
coloradense Garrett, Anorostoma	812	comita (Huckett), Coenosia	871
coloradense Greene, Chrysotoxum	582	*comitatus* Dyar & Knab, Culex	118
coloradensis (Aldrich), Blaesoxipha	947	comma Melander, Mythicomyia	416
coloradensis Aldrich, Dolichopus	489	comma (Wiedemann), Eurosta	663
coloradensis Alexander, Elliptera	51	commiscibilis Doane, Tipula	26
coloradensis Alexander, Ptychoptera	97	commixtus Walker, Tabanus	333
		Commoptera Brues	546

744-243—65——100

1582   A CATALOG OF DIPTERA OF NORTH AMERICA

	Page		Page
communis (De Geer), Aedes	111	concinna (Williston), Limonia	50
communis Felt, Dasineura	259	concinna (Williston), Suragina	344
communis Hanson, Nemotelus	310	concinnata (Meigen), Compsilura	1041
communis James, Odontomyia	315	concinnum Snow, Melanostoma	575
communis (Osten Sacken), Limonia	49	concinnus (Coquillett), Aphoebantus	429
communis Parker, Ravinia	955	concinnus (Coquillett), Coelotanypus	146
communis Van Duzee, Diaphorus	520	*concinnus* (Coquillett), Pseudonomoneura	360
communis (Walker), Euryomma	898	concisa (Macquart), Villa	436
communis Walker, Xylota	605	concolor Malloch, Forcipomyia	125
Comops Brennan	321	concolor (Malloch), Gymnochiromyia	822
comosa Loew, Cheilosia	586	concolor (Stein), Spilogona	880
comosa Reinhard, Masiphyomyia	1085	concolor (Thomson), Anthomyza	819
comosus Hine, Asilus	392	concrescens Melander, Mythicomyia	416
comosus Melander, Oligodranes	421	*Condidea* Coquillett	603
comosus Van Duzee, Thrypticus	511	condomina Harmston & Knowlton, Argyra	525
compacta (West), Opsotheresia	990	**Condylostylus** Bigot	483
compactus Van Duzee, Dolichopus	489	condylus Harmston & Knowlton, Teuchophorus	529
compactus Walker, Eristalis	623	confera Garrett, Cordyla	208
compar Cresson, Micropeza	634	confertus Fluke, Metasyrphus	560
compar Cresson, Pelina	752	*confinis* (Fallén), Aplomya	†1098
compar (Robineau-Desvoidy), Orthellia	913	confinis Walker, Dolichopus	489
comparata Melander, Mythicomyia	416	*confinis* (Walker), Micropsectra	178
comparata (Tucker), Villa	438	confinnis (Lynch Arribálzaga), Psorophora	110
comparilis (Reinhard), Oxysarcodexia	952	conflictus Curran, Microdon	598
compascua Reinhard, Pseudotachinomyia	1057	*confluens* (Loew), Diplotoxa	792
compascua (Wulp), Peleteria	997	confluens Loew, Hydromyza	840
compedita (Loew), Homoneura	698	confluens (Loew), Strauzia	676
compes (Coquillett), Stictochironomus	177	*confluens* Say, Ornithoica	916, †916
completa Cresson, Rhagoletis	674	*confluenta* (Say), Ornithoica	916, †916
completa Malloch, Phaonia	906	conforma Huckett, Coenosia	871
completus (Macquart), Efferia	395	*conformis* (Fallén), Pegomya	†862
completus Van Duzee, Dolichopus	489	conformis (Holmgren), Trissocladius	155
complexa Osten Sacken, Teucholabis	76	conformis Loew, Hydrellia	744
complexa Quate, Pericoma	92	*conformis* Malloch, Chironomus	165
complosa (Reinhard), Blaesoxipha	943	*conformis* (Malloch), Lasiops	903
compositarum (Verrall), Melangyna	566	conformis (Reinhard), Deopalpus	1004
compositus Hine, Asilus	392	*confraternus* Banks, Pipunculus	554
compressa Aldrich, Cylindromyia	974	confundens (Townsend), Macronychia	936
compressa (Dodge), Erythrandra	941	confusa (Aldrich), Promasiphya	1091
compressa Painter, Villa	434	confusa (Alexander), Dicranota	57
compressa (Reinhard), Blaesoxipha	944	confusa Cook, Swammerdamella	240
compressa (Robineau-Desvoidy), Cynomyopsis	930	confusa (Curran), Carposcalis	576
compressa Stein, Coenosia	871	*confusa* (Curran), Eurosta	663
compressa (Tothill), Nowickia	993	confusa Curran, Laphystia	376
compressiceps Borgmeier, Lecanocerus	546	confusa (Curran), Madiza	731
Compsilura Bouché	1040	*confusa* Curran, Peleteria	997
COMPSILURINI	1037	confusa James, Odontomyia	316
Compsobata Czerny	633	confusa Johnson, Blera	610
Compsomyia Brauer & Bergenstamm	†923	confusa Loew, Cordilura	828
Compsomyiops Townsend	924	*confusa* Malloch, Dictya	689
compta (Cole), Acrosticta	650	confusa Reinhard, Siphosturmia	1087
compta Coquillett, Empis	459	confusa Shannon, Xylota	605
compta Coquillett, Leptopeza	451	confusa Shewell, Camptoprosopella	696
compta Coquillett, Rhamphomyia	462	confusa (Wahlgren), Amoebaleria	815
compta (Meigen), Psilopa	740	confusa (West), Ptilodexia	988
compta Melander, Mythicomyia	416	*confusa* (Westwood), Stylogaster	632
comptus Van Duzee, Dolichopus	489	confusio Martin, Apachekolos	361
comstocki (Aldrich), Spiniphora	533	confusionis Sellers, Aplomya	1097
comstocki Alexander, Gnophomyia	73	confusus (Garrett), Podonomus	150
comstocki (Kellogg), Agathon	99	*confusus* Harris, Chrysops	325
comstocki Townsend, Eulasiona	1023	confusus Kröber, Chrysops	324
comstocki Wheeler, Vermileo	342	confusus Malloch, Tanytarsus	179
comstocki (Williston), Acrochordonodes	609	*confusus* Walker, Tabanus	334
comstocki Williston, Hermetia	304	congareenarum (Dyar & Shannon), Simulium	185
comstocki Williston (Phorocera)	1107	congenita (Dietz), Shannonomyia	67
comstocki Williston, Volucella	601	conglomerata (Malloch), Megaselia	540
comstockiana Alexander, Tipula	32	conglomerata (Pettey), Bradysia	232
comta (Fallén), Bonnetia	1004	congregaria Melander, Hilara	456
comutata Curran, Heringia	580	congregata Johannsen, Sciara	230
conabilis (Reinhard), Metoposarcophaga	951	congregata (Malloch), Ophiomyia	797
Conarete Pritchard	245	*congrua* Malloch, Hydrophoria	864
conata (Reinhard), Siphona	1064	conica Curran, Neoascia	594
concava Alexander, Tipula	28	*conica* (Fabricius), Delphinia	644
concava Spuler, Leptocera	723	*conica* (Harris), Cryptomeigenia	1041
concavifrons Kröber, Thereva	353	conica (Malloch), Rhyncophoromyia	543
concavus Leonard, Rhagio	345	conica (Reinhard), Oswaldia	1046
concavus (Say), Syrphus	559	conica Townsend, Ceratomyiella	1031
Conchapelopia Fittkau	146	conica (Wiedemann), Hydrophoria	864
Conchyliastes Theobald	109	*conicans* Huckett, Hylemya	846
conciliata Cresson, Polytrichophora	740	**Coniceps** Loew	642
concinna Doane, Tipula	37	*coniceps* Macquart, Exoprosopa	444
concinna (Laffoon), Mycetophila	211	coniceps (Malloch), Ophiomyia	797
concinna (Meigen), Diplonevra	†535	**Conicera** Meigen	536
concinna (Meigen), Probezzia	138	**Coniceromyia** Borgmeier	536
concinna Melander, Mythicomyia	416	conicola Foote, Mycodiplosis	282
concinna Williston, Hermetia	304		

INDEX 1583

conicola (Greene), Hapleginella — 778
conifacies (Macquart), Villa — 437
conifera Sturtevant & Wheeler, Scatophila — 758
conifrons (Zetterstedt), Paraprosalpia — 866
Conioscinella Duda — 783
Coniosternum Becker — 837
Conisternum Becker — 837
conjuncta (Adams), Trupanea — 667
conjuncta (Coquillett), Iteaphila — 454
conjuncta Curran, Peleteria — 997
conjuncta (Johnson), Homoneura — 698
conjuncta Johnson, Pseudotephritis — 647
conjuncta Loew, Clinocera — 468
conjuncta Loew, Rhamphomyia — 462
conjuncta Loew, Rivellia — 655
conjuncta (Osburn), Melangyna — 566
conjuncta Thomson, Myopa — 630
conjunctimontis Frick, Xyraeomyia — 803
conjunctivus Hardy, Bibio — 192
conjungens Walker, Epistrophe — 564
connecta (Curran), Blondelia — 1039
connecta Felt, Cincticornia — 269
connectans (Curran), Condylostylus — 484
connectans Garrett, Bolitophila — 197
connectens Melander, Epacmus — 428
connexa (Kieffer), Micropsectra — 178
connexa (Macquart), Villa — 434
connexa Say, (Sapromyza) — 706
connexa Stein, Pegomyia — 858
connexa (Steyskal), Elgiva — 690
connexionis Benjamin, Xanthaciura — 671
connexus Johnson, Chrysopilus — 346
Conophorus Meigen — 413
CONOPIDAE — 625
CONOPINAE — 626
CONOPINI — 626
Conops Linnaeus — 626
conostomus Williston, Helophilus — 620
consanguinea Loew, Lispe — 878, †878
consanguinea (Macquart), Villa — 433
consanguineus (Harmston), Gymnopternus — 499
consanguineus (Loew), Ospriocerus — 381
consanguineus (Wheeler), Dolichopus — 489
conscriptus Huckett, Eremomyioides — 857
consentiens Curran, Cheilosia — 584
conservativa Malloch, Rhamphomyia — 462
consessor (Coquillett), Villa — 434
consimilata Malloch, Helina — 886
consimilis Dietz, Limnophila — 66
consimilis (Holmgren), Lycoriella — 232
consobrina Felt, Caryomyza — 270
consobrina Felt, Dasineura — 259
consobrina Felt, Lasioptera — 271
consobrina (Felt), Lobopteromyia — 276
consobrina Felt, Porricondyla — 255
consobrina Felt, Rhabdophaga — 257
consobrina (Felt), Trishormomyia — 288
consobrina Huckett, Hylemya — 856
consobrina Macquart, Lucilia — 927
consobrina Robineau-Desvoidy (Myophora) — 961
consobrina (Zetterstedt), Phaonia — 906, ††908
consobrinum Zetterstedt, Rhaphium — 512
consobrinus (Holmgren), Orthocladius — 156
consobrinus Robineau-Desvoidy, Culex — 118
consonans (Laffoon), Mycetophila — 211
consors Loew (Aricia) — 1114
consors (Walker), Paracleius — 500
consors (Zetterstedt), Mydaea — 891
conspectus Van Duzee, Dolichopus — 489
conspicua Dietz, Tipula — 28
conspicua Malloch, Fannia — 894
conspicua Osten Sacken, Asphondylia — 269
conspicua Sabrosky, Odinia — 794
conspicua (Walker), Psilocephala — 351
conspicualis (Malloch), Megaselia — 538
conspicuus (Del Rosario), Telmatoscopus — 94
conspicuus Van Duzee, Dolichopus — 489
conspurcata Doane, Eurosta — 663
constans (Doane), Pedicia — 54
constans (Loew), Hoplitimyia — 313
consternata (West), Spathimeigenia — 1049
constricta Condrashoff, Contarinia — 277
constrictor (Malloch), Paraprosalpia — 866
constrictus Banks, Pipunculus — 554
constrictus Becker, Chlorops — 789
consul (Osten Sacken), Villa — 438

contaminata Doane, Tipula — 26
contaminata Winnertz, Exechia — 206
Contarinia Rondani — 277
CONTARINIINI — 276
contempta(Osten Sacken),Pseudolimnophila — 62
contermina Walker, Pedicia — 53
conterminus Walker, Dolichopus — 489
conterminus Walker, Tabanus — 335
conterrens (Walker), Psorophora — 109
contigua (Loew), Anthrax — 431
contigua Macquart, Sphaerophoria — 570
contigua Walker, Musca — 913
contigua Walker, Mycetophila — 211
contiguus Melander, Aphoebantus — 429
contiguus Melander, Platypalpus — 478
contiguus (Reinhard), Deopalpus — 1004
contiguus Walker, Dolichopus — 490
continentalis Alexander, Tipula — 28
continentalis Dyar & Knab, Uranotaenia — 108
contingens Walker (Dolichopus) — 504
contorta Curran, Euxesta — 651
contorta (Hardy), Tomosvaryella — 555
contortrix Alexander, Tipula — 40
contortus Bromley, Diogmites — 371
contracta Felt, Mycodiplosis — 282
contracta Melander, Chelipoda — 472
contractifrons (Zetterstedt), Spilogona — 880
contumax Brooks, Gonia — 1075
contumax (Osten Sacken), Melangyna — 568
conurus Osten Sacken, Aphoebantus — 429
conus Hardy, Bibio — 194
convecta (Walker), Schizotachina — 1067
convena (Reinhard), Blaesoxipha — 947
convergens Aldrich, Dolichopus — 490
convergens Huckett, Pegomya — 859
convergens (Malloch), Sobarocephala — 806
convergens Painter, Apiocera — 357
convergens Van Duzee, Chrysotus — 522
convergens (Van Duzee), Hercostomus — 498
convergens (Walker), Suillia — 809
convexa Alexander, Pedicia — 54
convexa Cole, Acrocera — 405
convexa Curran, Peleteria — 997
convexa Spuler, Leptocera — 722
convexa Townsend, Phrynofrontina — 1047
convexiforceps Brooks, Archytas — 1000
convexifrons Malloch, Schoenomyza — 870
convexus Loew, Hippelates — 775
convictum (Walker), Polypedilum — 175
convoluta (Felt), Asteromyia — 274
convolvuli Felt, Lasioptera — 271
cookii Townsend, Suioestrus — 1112
cooleyi Parker, Sarcophaga — 958
cooleyi (Seamans), Helophilus — 620
cooperi Sabrosky, Eusiphona — 732
Copecrypta Townsend — 1003
Copestylum Macquart — 602
cophas Walker, Rhamphomyia — 462
Cophura Osten Sacken — 366
copiosa (Thomsen), Bezzia — 141
copiosa (Wulp), Helina — †888
copiosus Root & Hoffman, Culicoides — 131
Coproica Rondani — 725
Copromyza Fallén — 719
Coprophila Duda — 725
coprophila Felt, Cordylomyia — 249
coprophila (Lintner), Bradysia — 232
Coptophlebia Bezzi — 458
copulifera Melander, Hormopeza — 453
coquilletti (Aldrich), Anaporia — 1028
coquilletti Aldrich, Cyrtophleba — 1017
coquilletti Aldrich, Dolichopus — 490
coquilletti (Aldrich), Meigenielloides — 1045
coquilletti Aldrich & Webber, Zenillia — 1107
coquilletti Back, Saropogon — 383
coquilletti (Bezzi), Stichopogon — 385
coquilletti Cresson, Ceropsilopa — 742
coquilletti Dyar & Knab, Uranotaenia — 108
coquilletti (Hendel), Geomyza — 821
coquilletti (Hendel), Physoclypeus — 702
coquilletti Hendel, Rivellia — 656
coquilletti Johnson, Phthiria — 420
coquilletti (Kertész), Tomosvaryella — 555
coquilletti (Kieffer), Forcipomyia — 125
coquilletti Kieffer, Stilobezzia — 134
coquilletti Landrock, Dynatosoma — 209

	Page		Page
coquilletti Malloch, Apocephalus	543	corpulenta (Cresson), Paroxyna	666
coquilletti (Malloch), Lasiops	903	corpulentus Ewen, Atrichopogon	122
coquilletti (Malloch), Phytobia	799	correus Steyskal, Dolichopus	490
coquilletti Melander, Euhybus	448	corrupta Huckett, Pegomya	859
coquilletti Melander, Platypalpus	479	corticalis (Loew), Pseudotephritis	647
coquilletti Painter, Villa	438	corticalis (Melander), Tachypeza	473
coquilletti Sedman, Melangyna	566	corticis Felt, Dasineura	259
coquilletti Wilcox, Ablautus	364	corticis Felt, Parallelodiplosis	286
coquilletti Williston, Criorhina	611	corticis Taylor, Lonchaea	717
coquilletti (Williston), Parabombylius	410	*corusca* (Wiedemann), Psilocephala	352
Coquillettia Williston	432	corvallis (Curran), Neocnemodon	581
Coquillettidia Dyar	108	corvina Fabricius, Musca	†913
coquillettii Hine, Chrysops	324	corvina Kessel, Callomyia	547
coquillettii (Hine), Efferia	394	*corvina* Loew, Rhamphomyia	462
coquillettii (Hine), Neoitamus	397	corvina Malloch, Paraleucopis	708
coquillettii Hine, Proctacanthus	399	corvina (Verrall), Fannia	894
coquillettii McAtee, Laphria	389	coryli Felt, Mycodiplosis	282
Coquillettina Walton	1037	coryli (Felt), Parallelodiplosis	286
Coquillettomyia Felt	281	coryli (Felt), Phaenolauthia	261
coqulus Garrett, Symmerus	199	corylifolia Felt, Mycodiplosis	282
coracina Alexander, Tipula	41	coryloides Foote, Mycodiplosis	282
coracina (Loew), Fannia	894	*coryloides* Osten Sacken, (Cecidomyia)	295
coracina (Zetterstedt), Phaenopsectra	174	*coryloides* (Packard), Schizomyia	267
corallogaster (Bigot), Lampria	388	*Coryneta* Meigen	474
corax Osten Sacken, Dolichopus	490	**Corynocera** Zetterstedt	181
coraxa (Kessel), Platypeza	549	**Corynoneura** Winnertz	164
corbis Twinn, Simulium	188	**CORYNONEURINI**	164
corbis (Walker), Mesograpta	571	corynorhini Cockerell, Trichobius	921
cordata (Reinhard), Anzamyia	1016	corynorhini (Ferris), Basilia	922
cordiforceps (Rowe), Nowickia	994	**CORYNOSCELIDIDAE**	237
Cordilura Fallén	827	*corythus* (Walker), Xanthomelanodes	967
CORDILURIDAE	826	*cosmeta* Reinhard, Sarcophaga	960
Cordilurina James	829	**Cosmetopus** Becker	834
Cordyla Meigen	208	cossae Shimer, (Cecidomyia)	291
Cordyligaster Macquart	†1030	costalis (Aldrich), Sciapus	486
Cordylomyia Felt	248	costalis (Coquillett), Webersteriana	985
Cordylura Fallén	827	*costalis* Curran, Cordylurella	836
Cordylurella Malloch	836	costalis Curran, Xenopterella	705
cordyluroides Stein, Phyllogaster	876	costalis Gerstäcker, Diacrita	644
Corethra Meigen	103	costalis Loew, Chrysotus	522
Corethrella Coquillett	104	*costalis* (Loew), Limnia	691
CORETHRELLINAE	104	costalis Loew, Psilocephala	351
coriacea Hull & Fluke, Cheilosia	587	costalis (Melander), Piophila	712
Corinthomyia Felt	249	*costalis* Say, Tipula	29
cormus (Walker), Iteaphila	454	costalis Stein, Fucellia	842
cornellia (Shannon), Myolepta	589	costalis Stein, Pegomya	859, †862
corni (Felt), Anarete	246	costalis Van Duzee, Hercostomus	498
corni Felt, Lasioptera	271	costalis (Walker), Seioptera	648
cornicina (Fabricius), Orthellia	†913	*costalis* Wiedemann, Baccha	572
cornicola (Beutenmüller), Neolasioptera	273	*costalis* Wiedemann, Tabanus	335
cornicola Williston, Haematobia	914	costalis (Williston), Efferia	394
cornifera (Dietz), Nephrotoma	21	costalis Wirth, Ceropsilopa	742
cornifera (Walker), Strauzia	676	costaloides Alexander, Tipula	35
cornifolia (Felt), Cincticornia	269	costata Coquillett, Limnophila	66
cornigera Curran, Peleteria	996	costata (Loew), Elachiptera	777
cornigera (Walker), Strauzia	676	costata Say, Baccha	572
cornu (Osten Sacken), Mayetiola	264	costata (Say), Villa	439
cornuale Reinhard, Crocinosoma	1060	*costatus* Loew Chrysotus	522
cornuta (Bigot), Dohrniphora	535	*costatus* Loew, Ogcodes	407
cornuta Byers, Dolichopeza	23	costatus (Loew), Rhagio	345
cornuta Curran, Peleteria	996	costomaculata (Dietz), Limonia	49
cornuta Curran, Wagneria	1020	*costomarginata* (Dietz), Nephrotoma	21
cornuta (Doane), Ormosia	85	costopunctatus Dietz, Molophilus	88
cornuta Felt, Colpodia	253	cothurnata (Rondani), Helina	886
cornuta Johannsen, Dixa	102	cothurnatus Bigot, Microdon	598
cornuta Kraft & Cook, Zabrachia	318	cotidiana Snyder, Lispe	878
cornuta (Loew), Orthacheta	831	cottrelli Telford, Cheilosia	584
cornuta Reinhard, Acronarista	1064	coxale Loew, Asindulum	199
cornuta (Walker), Delina	841	coxalis (Coquillett), Neuratelia	224
cornuta (Walsh), Rhabdophaga	257	coxalis (Curran), Neocnemodon	581
cornuticaudata Curran, Peleteria	996	coxalis Garrett, Zygomyia	216
cornuticaudata (Walley), Pentaneura	147	coxalis Loew, Gymnopternus	499
cornutus Loew, Chrysotus	522	coxata Ferris, Nycterophilia	921
cornutus Loew, Dolichopus	489	coxata (Stenhammar), Leptocera	722
cornutus Van Duzee, Asyndetus	524	coxata Wheeler, Chymomyza	770
cornutus (Van Duzee), Peloropeodes	517	coxendix (Fitch), Oscinella	779
cornutus (Wiedemann), Ceraturgus	365	coyote Bromley, Stenopogon	383
corona (Cresson), Hyadina	751	craigheadii Walton, Microdon	598
coronado Alexander, Tipula	40	cramptonella (Alexander), Erioptera	83
coronata (Guérin-Méneville), Labostigmina	313	*cramptoni* (Alexander), Limonia	46
coronata (Loew), Pelomyia	726	cramptoni Alexander, Molophilus	88
coronata (Loew), Phytobia	800	cramptoniana Alexander, Limonia	46
coronata (Ringdahl), Hylemya	853	*cramptoniana* Alexander, Ormosia	85
coronata Sabrosky, Odinia	794	**Cramptonomyia** Alexander	196
coronator Dyar & Knab, Culex	118	cranbrookensis Curran, Sphaerophoria	570
Coronimyia Townsend	1060	*cranbrooki* (Felt), Tritozyga	243
CORONIMYIINA	1059	cranbrooki Garrett, Mycomya	220

INDEX

	Page		Page
crandalli Curran, Spilomyia	613	crevecoeuri (Coquillett), Pseudogriphoneura	703
Craspedochaeta Macquart	855	cribellum (Loew), Pseudotephritis	647
Craspedochoeta Macquart	855	*cribellum* (Osten Sacken), Stenotabanus	†329
Craspedothrix Brauer & Bergenstamm	1065	*cribraria* (Harris), Euthycera	690
crassata Garrett, Pseudoleria	812	cribrata (Stenhammar), Scatophila	758
crassicauda Loew, Gymnopternus	499	*cribrata* (Wulp), Eurosta	663
crassicauda Van Duzee, Boletina	217	*cribripennis* (Harris), Eutreta	662
crassicaudatus Malloch, Chironomus	165	*cribrum* (Loew), Pseudotephritis	647
crassiceps Aldrich, Atacta	1080	crickmayi Curran, Hilara	456
crassicollis (Walker), Smittia	162	crickmeri (Alexander), Gonomyodes	79
crassicornis Aldrich, Dolichopus	490	**Cricotopus** Wulp	157
crassicornis (Fabricius), Gonia	1075	criddlei (Aldrich), Oscinella	779
crassicornis (Greene), Glabellula	415	criddlei (Brooks), Hybomitra	338
crassicornis (Stannius), Allodia	205	criddlei Curran, Cordilura	828
crassicornis (Williston), Dialineura	352	*criddlei* (Curran), Pericoma	93
crassicosta (Becker), Neoscatella	756	criddlei Dietz, Tipula	32
crassifemorata Malloch, Serromyia	136	criddlei Van Duzee, Hydrophorus	505
crassifemoris (Fitch), Platypalpus	479	*Criddleria* Curran	811
crassimana (Haliday), Leptocera	723	crinipes (Ringdahl), Allomyella	834
crassimana (Loew), Hydropyrus	754	crinipes Van Duzee, Scellus	506
crassinervis Loew, Rhamphomyia	466	crinita Coquillett, Scatophaga	838
crassinervis (Walton), Chaetostigmoptera	1062	crinita Martin, Apachekolos	361
crassinervis (Zetterstedt), Procladius	149	crinitus (Aldrich), Condylostylus	484
crassipalpis Macquart, Sarcophaga	958	crinitus Ewen, Atrichopogon	122
crassipalpis Reinhard, Cryptomeigenia	1041	crinitus Martin, Holopogon	375
crassipalpis Reinhard, Eumacronychia	935	*Crinura* Schnabl	846
crassipalpis Ringdahl, Myopina	843	*Crinurina* Karl	856
crassipes Bigot, Pipiza	579	**Crioprora** Osten Sacken	610
crassipes Cresson, Hydrellia	744	**Criorhina** Meigen	611
crassipes (Loew), Neoleria	†813	*Criorrhina* Meigen	611
crassipes (Meigen), Rhaphium	512	crispata (Felt), Phaenolauthia	261
crassipes (Wood), Megaselia	540	cristasta (Coquillett), Comantella	366
crassiseta (Aldrich & Webber), Aplomya	1097	cristata Malloch, Hydrotaea	899
crassiseta (Laffoon), Mycetophila	211	cristata Melander, Mythicomyia	417
crassiseta Reinhard, Paradidyma	1032	*cristatus* (Curran), Hybomitra	338
Crassiseta Roser	776	cristatus Fabricius, Chironomus	166
crassiseta Zetterstedt, Phytomyza	804	cristatus Painter, Conophorus	414
crassitibia (Van Duzee), Peloropeodes	517	*cristatus* Van Duzee, Pelastoneurus	501
crassivena Alexander, Erioptera	83	cristipes Schmitz & Wirth, Phora	537
crassivenis Curran, Medetera	510	crocata (Coquillett), Euthyneura	449
crassiventris (Huckett), Spilogona	884	crocea McAtee, Laphria	389
crassulina (Cockerell), Rhopalomyia	266	croceum Painter, Heterostylum	411
crassus Loew, Nemotelus	310	crocina (Coquillett), Villa	437
crassus Townes, Pseudochironomus	176	crocina Melander, Mythicomyia	417
crataegibedeguar Osten Sacken, (Cecidomyia)	293	**Crocinosoma** Reinhard	1060
		crocota (Coquillett), Euthyneura	449
crataegifolia Felt, Lestodiplosis	285	crocotus Loew, Chlorops	789
crataegifolia (Felt), Trishormomyia	288	*Crocuta* Meigen	1063
crataegi globulus Walsh, (Lasioptera)	1114	CROCUTINI	1059
crataegi limbus Walsh, (Lasioptera)	1115	croesus (Osten Sacken), Ferdinandea	588
crataegiplica Osten Sacken, (Cecidomyia)	1114	crosbyi Van Duzee, Chrysotus	522
cratorhina Hull & Fluke, Cheilosia	584	*Crossocosmia* Mik	1083
crawfordi Coquillett, Pseudacteon	544	**Crossopalpus** Bigot	477
crawfordi Shannon, Caliprobola	608	crotalariae Stebbins, (Cecidomyia)	293
crawii Coquillett, Celatoria	1040	crotchi (Osten Sacken), Caloparyphus	308
crebra Pritchard, Conarete	245	croxtoni Nicholson & Mickel, Simulium	186
crebra Pritchard, Xylopriona	249	crucians Wiedemann, Anopheles	106
crebra (Wulp), Chaetogaedia	1078	cruciata Hendel, Pelomyia	726
Cremersia Schmitz	544	cruciata Snyder, Helina	886
cremides Walker, (Dexia)	†986, 1108	cruciator (Laffoon), Mycetophila	211
Cremifania Czerny	706	cruciatus (Say), Ceraturgus	365
CREMIFANIINAE	706	*crucifera* Huckett, Hylemya	848
crenatus (Osten Sacken), Dolichopus	490	*crucigera* (Wiedemann), Meromacrus	622
creolensis Reinhard, Spathidexia	1021	crucigerus (Coquillett), Caloparyphus	308
creper (Snow), Dasysyrphus	563	crudelis Bromley, Diogmites	371
crepidarius Melander, Platypalpus	479	cruenta Coquillett, Symphoromyia	343
crepuscula Arnaud, Melanderia	507	cruentum (McAtee), Andrenosoma	386
crepuscula Hull, Solva	299	crumbi Wilcox & Martin, Nannocyrtopogon	379
crepuscularis (Bequaert), Chlorotabanus	330	crumborum Martin, Coleomyia	366
crepuscularis Malloch, Culicoides	131	**Crumomyia** Macquart	720
crepuscularis (Stein), Helina	888	cruralis Coquillett, Hydrellia	744
crepusculenta (Huckett), Spilogona	880	cruziana Felt, Rhopalomyia	266
cresca Huckett, Pegomya	859	*Crymobia* Loew	810
crescentis (Reinhard), Drino	1085	crypta Gill, Eccoptomera	811
cressoni Aczél, Tomoplagia	678	cryptica Cook, Rhexoza	240
cressoni Alexander, Limnophila	63	cryptica Sabrosky, Leptocera	721
cressoni Fisher, Keroplatus	200	*Cryptineura* Bigot	591
cressoni Johnson, Pipunculus	†554	*Cryptochaetum* Rondani	825
cressoni (Hine), Efferia	394	CRYPTOCHETIDAE	825
cressoni Hull, Sphegina	594	**Cryptochetum** Rondani	825
cressoni Malloch, Hydrotaea	899	**Cryptochironomus** Kieffer	166
cressoni (Malloch), Palpomyia	140	*Cryptochoetum* Rondani	825
cressoni Melander, Mythicomyia	417	**Cryptolabis** Osten Sacken	78
cressoni Van Duzee, Chrysotus	522	*Cryptolucilia* Brauer & Bergenstamm	912
Cressonomyia Arnaud	742	**Cryptomeigenia** Brauer & Bergenstamm	1041
cretaceus Osten Sacken, Cyrtopogon	369	**Cryptotreta** Blanc & Foote	662
cretans (Huckett), Spilogona	880	cryptus Harmston & Knowlton, Hercostomus	498

	Page		Page
csikii Aczél, Verrallia	551	currani Hardy, Phyllomydas	360
ctenistes Melander, Wiedemannia	469	currani Harriot, Seioptera	648
Ctenoceria Rondani	97	currani (James), Caloparyphus	308
ctenocnema Melander, Empis	459	currani (James), Hedriodiscus	314
Ctenodactylomyia Felt	262	currani James, Stratiomys	311
Ctenophora Meigen	19	currani Kessel, Metaclythia	550
Ctenophorocera Brauer & Bergenstamm	1087	currani Leonard, Symphoromyia	343
ctenopyga Alexander, Limonia	46	currani Martin, Holopogon	375
ctites Dyar, Psorophora	109	*currani* (Ouellet), Microtrichomma	1001
cubaecola (Jaennicke), Drino	1085	currani Pritchard, Heteropogon	374
Cubaemyia Townsend	1084	currani (Shewell), Lyciella	700
cubana Curran, Allograpta	569	*currani* Steyskal, Dictya	689
cubana Hull, Baccha	572	currani Steyskal, Rhamphomyia	463
cubensis Macquart, Eristalis	623	currani Van Duzee, Argyra	525
cubita Garrett, Corynoneura	164	currani Van Duzee, Chrysotus	522
cubita Garrett, Prodiamesa	153	currani (Van Duzee), Gymnopternus	499
cubitalis (Osten Sacken), Dactylolabis	61	currani (Walley), Pentaneura	147
cuclux Whitney, Chrysops	324	*currani* Wehr, Chrysotoxum	582
cuculliae (Webber), Lespesia	1101	**Curranops** Harriot	644
Cucullomyia Roback	956	*curreyi* (Felt), Corinthomyia	249
cucumeris (Johannsen), Bradysia	232	*curriei* (Coquillett), Aedes	112
cucumeris (Lintner), Aphidoletes	280	curriei (Coquillett), Clinohelea	136
cucurbitae Felt (Itonida)	290	curriei (Coquillett), Thelymyia	1106
cucurbitae Felt, Mycodiplosis	282	curriei (Malloch), Pseudacteon	544
cuervana (Hardy), Efferia	394	curriei (Townsend), Elfia	1065
Culex Linnaeus	117	curriei (Townsend), Peleteria	996
Culicada Felt	110	cursim Whitney, Chrysops	324
Culicella Felt	117	curta Johannsen, Phthinia	225
Culicelsa Felt	111	curta (Loew), Villa	435
CULICIDAE	105	curticornis Kröber, Myopa	630
culiciformis (De Geer), Mochlonyx	104	curtilamellatus Malloch, Chironomus	167
culiciformis (Linnaeus), Procladius	149	curtipennis Spuler, Leptocera	724
CULICINAE	107	curtipennis Wilcox & Martin, Cyrtopogon	368
CULICINI	107	curtipes James, Bibio	193
CULICOIDEA	98	curtirostris Alexander, Elephantomyia	71
Culicoides Latreille	127	curtistylus Curran, Cyrtopogon	368
CULICOIDINI	127	*curtistylus* Goetghebuer, Orthocladius	156
culicoidithorax Hoffman, Ceratopogon	132	*curtiventris* Stenhammar, Leptocera	724
Culiseta Felt	116	CURTONOTIDAE	759
culmicola Morris, (Cecidomyia)	291	**Curtonotum** Macquart	759
culta (Wiedemann), Paracantha	662	curtus (Hardy), Pipunculus	553
cultaris (Coquillett), Paracantha	662	curtus (Hine), Metasyrphus	560
cultaventris Martin, Leptogaster	362	curulis (Reinhard), Pelecotheca	1052
cultriger Kieffer, Diplocladius	154	*curva* Curran, Iteaphila	454
cumatilis Grote, Sparnopolius	413	curvaria (Curran), Xylota	606
cumberlandensis Alexander, (Molophilus)	1114	curvata Alexander, Ormosia	85
cunctans (Meigen), Lasiops	†903	curvata Fisher, Mycomya	220
cunctans Say, Tipula	27	curvicauda Gerstäcker, Toxotrypana	659
cuneata Loew, Tritoxa	649	curvicauda (Meigen), Coenia	755
cuneatum Wehr, Chrysotoxum	582	curvicauda (Zetterstedt), Hylemya	846
cuneatus (Townes), Chironomus	167	curvinervis (Becker), Anevrina	532
cuneicornis (Zetterstedt), Hylemya	856	curvinervis Borgmeier, Aenigmatias	537
cuneiformis Van Duzee, Medetera	510	curvinervis Curran, Nausigaster	596
cuneipennis Melander, Platypalpus	479	curvinervis Malloch, Phaonia	906
cuneola (Adams), Leia	227	curvinervis (Zetterstedt), Leiomyza	823
cunicula (Osten Sacken), Villa	439	curvipennis (Fallén), Stegana	760
cuniculator Condrashoff, Contarinia	277	curvipes Coquillett, Rhamphomyia	463
cuniculi (Clark), Cuterebra	1109, †1110	curvipes Latreille, Sphaerocera	718
cuniculus Van Duzee, Dolichopus	495	*curvipes* Loew, Xylota	†606
cunina Reinhard, Ignotomyia	1091	curvipes Malloch, Fannia	894
Cuphocera Macquart	†1003	curvipes Malloch, Hylemya	848
CUPHOCERINI	1003	curvipes Melander, Stilpon	476
cuprarius (Linnaeus), Sargus	301	curvipes (Stein), Phaonia	906
cuprea Shannon & Dobroscky, Protocalliphora	926	curvipes (Van Duzee), Melanderia	507
cupreifrons (Walker), Hylemya	850	curvipes (Wiedemann), Polydontomyia	621
cupressi Schweinitz, (Merulius)	290	*curvirostris* (Bigot), Prosenoides	†988
cupressiananassa (Osten Sacken), Thecodiplosis	278	curvispina Van Duzee, Campsicnemus	526
cupreus Say, Dolichopus	490	*curvitibia* Melander & Spuler, Sepsis	683
cupricrus Walker, (Cordylura)	869	curvitibiae Hardy, Pipunculus	552
cuprilineata Wheeler, Ochthera	753	*curvivena* (Coquillett), Gonomyia	75
cuprilineata Williston, Ochthera	753	cushmani Johnson, Agathomyia	548
cuprina (Hall), Protocalliphora	926	*cuspidata* Doane, Tipula	39
cuprina (Wiedemann), Phaenicia	928	cuspidatus Melander, Euhybus	448
cuprinus (Linnaeus), Sargus	301	**Cuterebra** Clark	1109
cuprinus Wheeler, Sympycnus	527	CUTEREBRIDAE	1109
cuprinus Wiedemann, Dolichopus	490	cuticornis Huckett, Pegomya	863
cuprovittatus Wiedemann, Eristalis	625	*Cutiterebra* Clark	1109
currani Alexander, Dicranota	57	*cutta* (Wiedemann), Paracantha	662
currani Alexander, Gonomyia	75	**Cyamops** Melander	820
currani Cole & Wilcox, Lasiopogon	377	cyanea Hunter, Cheilosia	587
currani (Fluke), Helophilus	619	cyanea (Smith), Lejota	590
currani (Fluke), Melangyna	566	cyaneiventris (Wulp), Chlorohystricia	992
currani Fluke, Syrphus	559	cyanella Jones, Cuterebra	1109
currani Garrett, Anorostoma	812	cyanella (Osten Sacken), Crioprora	610
currani Hardy, Bibio	194	cyanescens Camras, Zodion	629
		cyanescens (Coquillett), Psorophora	109

INDEX 1587

cyanescens Loew, Cheilosia... 587
cyaneus Wheeler, Pelastoneurus... 501
cyanicolor Zetterstedt, Pyrellia... 912
cyanoceps Johnson, Phthiria... 420
cyanococci Felt, Dasineura... 259
cyanococci Felt, Mycodiplosis... 253
cyanogaster (Loew), Crioprora... 6.6
cyanogaster Wheeler, Medetera... 5 6
Cyanomyia Robineau-Desvoidy... 930
Cyanus Hall... 930
cyathiformis Melander, Renocera... 692
cybele (Coquillett), Anthrax... 431
cybira Walker, Stomoxys... 914
cyclocerculus Dyar, Aedes... 113
Cyclodionaea Townsend... 976
cyclops Melander, Sapromyza... 704
cyclops Osten Sacken, Aphoebantus... 429
CYCLORRHAPHA... 530
Cyclotaphrys Townsend... 1054
cylindrata Doane, Tipula... 32
cylindrica (Curran), Ceriana... 615
cylindrica (Fabricius), Baccha... †573
cylindrica (Fabricius), Nemopoda... 682
cylindrica Loew, Argyra... 525
cylindrica Pettey, Sciara... 230
cylindrica Say, Loxocera... 638
cylindrica (Say), Sphaerophoria... 570
cylindricornis (Fabricius), Lauxania... 699
cylindricus (Stein), Eremomyioides... 857
cylindricus (Van Duzee), Nematoproctus... 515
cylindrigallae Felt, Lasioptera... 271
Cylindromasicera Townsend... 1095
Cylindromyia Meigen... 972
CYLINDROMYIINI... 971
Cylindrotoma Macquart... 41
CYLINDROTOMINAE... 41
cylla Melander, Mythicomyia... 417
CYLLENIINAE... 426
cymatophorus Osten Sacken, Tabanus... 333
cymbalista Osten Sacken, Cyrtopogon... 368
cymballista Melander, Oreogeton... 453
cynipsea (Linnaeus), Sepsis... 683
cynipsea Osten Sacken, (Cecidomyia)... 270
cynocephala Hine, Sericomyia... 603
cynoglossi Frick, Phytobia... 800
Cynomya Robineau-Desvoidy... 930
Cynomyia Robineau-Desvoidy... 930
Cynomyopsis Townsend... 930
cynoprosopa Hull & Fluke, Cheilosia... 584
cynops Snow, Brachyopa... 592
Cynorhina Williston... 610
Cynorhinella Curran... 588
Cyphocera Macquart... 1004
Cyphomyia Wiedemann... 304
CYPHOMYIINAE... 303
cypris (Meigen), Villa... 437
Cypsela Meigen... †719
cypseloides Townsend, Anthomyiopsis... 1022
CYRTIDAE... 403
Cyrtoma Meigen... 451
cyrtoneurina Stein, Limnophora... 885
Cyrtonotum Macquart... 759
Cyrtophleba Rondani... 1017
Cyrtophlebia Rondani... 1017
Cyrtophloeba Rondani... 1017
Cyrtopogon Loew... 367
cyrtopogona Cole, Cophura... 366
CYRTOSIINAE... 415
Cystiphora Kieffer... 261
Cystodiplosis Kieffer & Jörgensen... 289
Cystogonia Townsend... 1075
CYTHEREINAE... 414
Cyzenis Robineau-Desvoidy... 1099
czekanowskii Stackelberg, Dolichopus... 490
czerni (Garrett), Neoleria... 813
czernyi Collart, Heleomyza... 816
czernyi Garrett, Anypotacta... 815
czernyi (Garrett), Neoleria... 813
czernyi Johnson, Clusia... 806

dacetoptera Phillips, Trupanea... 667
dacne Philip, Chrysops... 325
dacotensis (Dyar & Shannon), Cnephia... 184
dactylica Melander, Empis... 459
Dactylocladius Kieffer... 155
Dactylolabis Osten Sacken... 61

Dactylomyia Aldrich... 518
daeckei (Cresson), Calobatina... 635
daeckei (Hine), Hybomitra... 338
daeckei Johnson, Brachyopa... 592
daeckei Johnson, Nephrocerus... 551
daedalus Macfie, Culicoides... 129
daedalus Stone, Stenotabanus... 329
Daeochaeta Townsend... 1052
dahlgrueni (Enderlein), Simulium... 188
dakota Alexander, Dolichopeza... 23
dakota Hull & Fluke, Cheilosia... 584
dakotana Steyskal, Ulidiotites... 649
dakotensis Aldrich, Dolichopus... 490
dakotensis Dyar, Psychoda... 95
dakotensis Harmston, Chrysotus... 522
dakotensis Harmston & Knowlton, Argyra... 525
dakotensis (Townsend), Erythrandra... 941
dakotensis Townsend, Euscopolia... 974
dakotensis (Townsend), Nowickia... 995
dakotensis Van Duzee, Physocephala... 628
dakotensis Webber, Tachinomyia... 1058
Dalmannia Robineau-Desvoidy... 632
DALMANNIINAE... 632
damnosus (Say), Aedes... 115
dampfi (Duda), Rhexoza... 240
Dampfomyia Addis... 91
damula (Osten Sacken), Dactylolabis... 62
dana Walker, Rhamphomyia... 463
danthoniae Felt, Lasioptera... 271
daphne (Hull), Metasyrphus... 560
daphne (Osten Sacken), Anthrax... 431
daphne Pritchard, Cophura... 367
daphne (Wiedemann), Trupanea... †667
dapsilis Reinhard, Calyptrosomus... 975
darbyi (Sublette), Chironomus... 167
daria Walker, Rhamphomyia... 463
darlingtoniae (Jones), Aphanotrigonum... 785
dascyllus (Walker), Efferia... 395
Dasineura Rondani... 258
Dasiops Rondani... 716
Dasyhelea Kieffer... 126
DASYHELEINAE... 126
Dasyholopogon Martin... 375
Dasylechia Williston... 388
Dasyllis Loew... 388
dasyllis Williston, Cyrtopogon... 368
dasylloides Williston, Cyrtopogon... 368
Dasymolophilus Goetghebuer... 88
Dasyneura Rondani... 258
Dasyommia Enderlein... 337
Dasyopa Malloch... 784
dasyops Malloch, Dolichopus... 490
Dasypleuron Malloch... 840
dasypodus Coquillett, Dolichopus... 490
DASYPOGONINAE... 364
dasyprocta (Loew), Okeniella... 834
Dasyrhamphomyia Frey... 461
Dasysyrphus Enderlein... 563
dasythrix Becker, Scatophaga... 838
datanae (Townsend), Winthemia... 1090
datanarum (Townsend), Lespesia... 1101
dauci (Meigen), Conicera... 536
Daulopogon Loew... 377
davidsoni Coquillett, Brachicoma... 940
davidsoni Curran, Pipiza... 579
davidsonii Coquillett, Sarcophaga... 958
daviesi Peterson & DeFoliart, Prosimulium... 182
davisensis Steyskal, Pseudotephritis... 647
davisi Alexander, Ormosia... 85
davisi Edwards, Diamesa... 151
davisi Felt, Dyodiplosis... 287
davisi Felt, Feltiella... 281
davisi Johnson, Chrysopilus... 346
davisi Johnson, Psilocephala... 351
davisii (Walton), Napomyza... 804
dawsoni Philip, Chrysops... 325
dawsoni Philip, Tabanus... 336
dawsoni (Sellers), Chrysoexorista... 1099
dayana Alexander, Hexatoma... 70
dayi Cresson, Dryomyza... 681
deani Painter, Heterostylum... 411
debilipalpis Lutz, Culicoides... 129
debilipennis (Lundbeck), Metriocnemus... 161
debilipes Harris (Ceratopogon)... 1114
debilis Becker, Condylostylus... 484
debilis Coquillett, Syneches... 448

	Page		Page
debilis Loew, Chaetopsis	650	degenera Walker, Psychoda	96
debilis Loew, Elliponeura	793	degenerata (Alexander), Pedicia	54
debilis Loew, Gymnopternus	499	degeneri Alexander, Tipula	35
debilis Loew, Philygria	745	*degustator* Dyar, Culex	119
debilis Loew, Platypalpus	479	deilephilae (Osten Sacken), Winthemia	1090
debilis Loew, Rhamphomyia	463	**Deinocerites** Theobald	120
debilis (Osten Sacken), Sphecomyiella	658	**Dejeania** Robineau-Desvoidy	†999
debilis Stein, Pegomya	859	*dejeanii* Robineau-Desvoidy, Delina	841
debilis Townsend, Eubrachymera	1023	dejeanii Robineau-Desvoidy, (Nerina)	915
debilis Walker, Sargus	302	DEJEANIINI	999
debilis (Williston), Gymnodia	884	**Dejeaniopalpus** Townsend	1034
debilis (Williston), Pedicia	54	dejecta Walker, Tipula	28
Decachaetophora Duda	681	delecta (Curran), Pilatea	1105
decedens (Walker), Anatopynia	145	delessei Séguy, (Lycoria)	236
Deceia Malloch	703	deleta (Stein), Phaonia	906
decemarticulatum (Twinn), Prosimulium	182	**Delia** Robineau-Desvoidy	846
decemguttata (Walker), Diastata	772	delicata Huckett, Hylemya	845
decemmaculata Walsh (Cecidomyia)	291	delicata Johannsen, Boletina	217
decens Cresson, Hydrellia	744	delicatula (Alexander), Neocladura	72
decens Townsend, Eumacronychia	935	delicatulus Hine, Asilus	392
deceptiva (Aldrich), Blaesoxipha	947	delicatulus Melander, Lasiopogon	377
deceptiva Fitch, Hylemya	850	delicatulus Osten Sacken, Chrysops	325
deceptiva Malloch, Cordilura	829	delicatus Aldrich, Dolichopus	490
deceptiva (Malloch), Liriomyza	801	delicatus Loew, Chrysotimus	529
deceptor (Curran), Clastoneuriopsis	1067	delicatus Van Duzee, Thinophilus	508
deceptor Curran, Drapetis	477	delicatus (Walker), Condylostylus	484
deceptor Dyar & Knab, Culex	120	delila (Loew), Anthrax	432
deceptor (Malloch), Homoneura	698	**Delina** Robineau-Desvoidy	841
decipiens (Loew), Elachiptera	777	DELININI	840
decipiens Loew, Paralimna	748	delita Johannsen, Allodia	205
decisa (Townsend), Senotainia	939	delongi Johnson, Mixogaster	599
decisus (Walker), Helophilus	618	*Delopsis* Skuse	209
decisus (Walker), Pararchytas	1002	**Delphinia** Robineau-Desvoidy	644
deckeri (Curran), Lydella	1103	delphiniae Frost, Phytomyza	804
declarator Dyar & Knab, Culex	118	delta (Hine), Esenbeckia	322
declinata Becker, Anthracophaga	788	delumbis Melander, Empis	·459
decolor (Fallén), Pogonomyioides	†902	delusus Tucker, Tolmerus	401
decolor Kessel, Agathomyia	548	demandata (Fabricius), Physiphora	653
decolor Melander, Platypalpus	479	*Demeijerea* Kruseman	169
decolor Shewell, Camptoprosopella	696	demissus Van Duzee, Dolichopus	490
decolorata (Malloch), Anatopynia	145	demogorgon (Walker), Poecilanthrax	442
decoloratus (Malloch), Polypedilum	175	*demorgon* (Walker), Poecilanthrax	442
decora Aldrich, Cylindromyia	973	demylus (Walker), Cryptomeigenia	1041, †1049
decora Doane, Tipula	32	dena Roback, Anatopynia	145
decora Loew, Exoprosopa	444	denali Alexander, Erioptera	83
decora (Loew), Leia	227	denaria Davies, Peterson, & Wood, Cnephia	185
decora (Loew), Poecilographa	692	**Dendrolimnophila** Alexander	67
decora Macquart, Somula	612	*Dendromyza* Hendel	797
decora (Staeger), Limonia	48	**Dendrophaonia** Malloch	905
decoratus (Holmgren), Orthocladius	156	*Dendrophila* Lioy	868
decorior Steyskal, Pholeomyia	732	*deniedmannii* (Ludlow), Aedes	114
decoris Williston, Notiphila	747	denningi Foote & Pratt, Culicoides	132
decorum Walker, Simulium	189	denningi Pritchard, Cordylomyia	248
decorus Johannsen, Chironomus	165	*denningi* Shaw, Mycetophila	210
decorus Say, Sargus	302	densursi Alexander, Tipula	35
decorus Van Duzee, Dolichopus	490	*dentata* (Bigot), Limnophora	885
decorus Williston, Xylophagus	297	dentata Bigot, Stratiomys	312
decticus Howard, Dyar, & Knab, Aedes	112	dentata (Coquillett), Ceracia	1037
decumbens Hull & Fluke, Cheilosia	584	dentata Felt, Camptomyia	256
decumbens Malloch, Chironomus	165	dentata Felt, Coquillettomyia	281
decurvata Alexander, Antocha	51	dentata Fisher, Mycomya	220
decussata Alexander, Ormosia	85	dentata Lundström, Mycetophila	211
dedita Alexander, Ormosia	85	dentata Malloch, Bezzia	141
defecta James, Labostigmina	313	dentata Stone, Lucumaphila	673
defecta (Loew), Thanategia	472	dentica (Guthrie), Brachypeza	205
defecta Van Duzee, Docosia	226	*denticauda* Malloch, Hylemya	853
defecta (Winnertz), Anaretella	245	denticauda (Zetterstedt), Paraprosalpia	866
defectiva Felt, Campylomyza	248	denticornis Malloch, Coenosia	872
defectus Van Duzee, Dolichopus	490	denticornis (Panzer), Cerodontha	801
defessa (Osten Sacken), Amoebaleria	815	denticulata Felt, Dasineura	259
definita Cresson, Hydrellia	744	denticulatus Wirth & Hubert, Culicoides	129
deflecta (Johannsen), Micropsectra	178	*dentiens* (Pandellé), Pegomya	861
deflecta (Loew), Thanategia	472	dentifera Alexander, Ormosia	85
deflecta Malloch, Drosophila	764	**Dentifibula** Felt	276
deflectus Van Duzee, Dolichopus	490	dentipes (Fabricius), Hydrotaea	900
deflexa (Walley), Microtendipes	173	*dentipes* Williston, Mallota	621
deflorata (Holmgren), Spilogona	880	dentiventris (Ringdahl), Paraprosalpia	866
defoliarti Stone & Peterson, Simulium	188	denuda Wilcox & Martin, Dioctria	371
deformis Hardy, Tomosvaryella	555	denudata (Holmgren), Spilogona	880
deformis Van Duzee, Neurigona	518	denudata (Melander), Tethina	727
defrenata Alexander, Ormosia	85	denudata Melander, Aphoebantus	429
defuncta (Osten Sacken), Limonia	50	**Deopalpus** Townsend	1003
Degeeria Meigen	†1051	depilis (Coquillett), Myobiopsis	1035
degeerioides (Coquillett), Admontia	1068	depressa Huckett, Hydrotaea	900
degener Wheeler, Campsicnemus	526	*depressa* (Malloch), Leucophora	868
degenera (Walker), Mericia	1008	depressa (Say), Lipoptena	920

INDEX 1589

	Page		Page
depressa (Stein), Fannia	894	DIAMESINAE	151
depressa Stein, Hylemya	848	diana (Osten Sacken), Eutreta	661
depressus Fluke, Metasyrphus	560	diantaeus Howard, Dyar, & Knab, Aedes	112
depressus (Shannon), Chalcosyrphus	590	diantherae (Malloch), Melanagromyza	796
depressus Van Duzee, Thinophilus	508	di ntherec e (Malloch), Melanagromyza	796
deprivata Alexander, Paradelphomyia	60	dianthi Frick, Liriomyza	801
depuncta (Fallén), Helina	887	diaphana (Doane), Pedicia	54
derbyi Doane, Tipula	40	*diaphana* (Fallén), Hyalurgus	1026
derelicta Dietz, Tipula	32	diaphanus (Wiedemann), Lasiops	904
derelicta (Walker), Ravinia	954	DIAPHORINAE	519
deremptus Walker, (Orthochile)	530	**Diaphorus** Meigen	519
derisus Reinhard, Gremlinotrophus	1069	diara Roback, Corynoneura	164
derivatum Walker, Chrysotoxum	582	Diarthronomyia Felt	262
derivatus Walker, Tabanus	333	**Diastata** Meigen	772
Deromyia Philippi	†371	DIASTATIDAE	772
dertona (Walker), Eurosta	663	diatraeae (Townsend), Lixophaga	1043
deserta Melander, Trixoscelis	818	*Diazoma* Wallengren	15
deserta Patterson, (Cecidomyiaceltis)	293	**Diazosma** Bergroth	15
deserta Patton, (Cecidomyiaceltis)	293	*dicentum* Dyar and Shannon, Prosimulium	183
desertensis Wirth & Hubert, Forcipomyia	125	dichaeta Fisher, Mycomya	220
deserti Wilcox & Martin, Nannocyrtopogon	379	dichaeta (Loew), Lamproscatella	756
deserticola Hall, Anastoechus	411	*dichaeta* Malloch, Coenosia	872
desertorum Alexander, Tipula	24	**Dichaeta** Meigen	748
desertorum (Townsend), Chaetogaedia	1078	dichaeta (Shaw), Bradysia	232
desertus Coquillett, Aphoebantus	429	**Dichaetoneura** Johnson	1065
desiderata Melander, Empis	459	**Dichocera** Williston	1013
desideratus (Johnson), Nemomydas	359	*Dichoceropsis* Townsend	1014
desidiosa Johannsen, Neuratelia	224	dichopticus Stone, Gymnopais	182
Desmatomyia Williston	420	dichroa Loew, Acrosticta	650
Desmatoneura Williston	430	dichroa Loew, Docosia	226
desmiae (Sellers), Hyphantrophaga	1094	*Dichrochira* Hendel	687
desmodii Felt, Lasioptera	271	*Dichrodiplosis* Kieffer	279
Desmometopa Loew	731	dichromata Snow, Xenochaeta	664
Desmometopina Curran	731	dickei Jones, Culicoides	129
Desmomyza Curran	730	dickinsoni Alexander, Tipula	30
despecta (Haliday), Scatophila	758	**Diclasiopa** Hendel	739
despecta (Walker), Bithoracochaeta	874	**Dicolonus** Loew	370
despecta (Walker), Phronia	215	**Dicraeus** Loew	783
despectus (Kieffer), Paraphaenocladius	161	*Dicranoclista* Bezzi	432
despicatus Loew, Gymnopternus	499	**Dicranomyia** Stephens	45
destructor (Say), Mayetiola	263	**Dicranophragma** Osten Sacken	65
determinata (Walker), Hylemya	854	**Dicranoptycha** Osten Sacken	52
deterra Walley, Empis	459	**Dicranota** Zetterstedt	56
detersus Loew, Dolichopus	490	*Dicrobezzia* Kieffer	138
Dettopsomyia Lamb	771	*Dicrodiplosis* Kieffer	279
Deuterophlebia Edwards	100	*Dicrohelea* Kieffer	137
DEUTEROPHLEBIIDAE	100	**Dicrotendipes** Kieffer	169
Deutominettia Hendel	697	**Dictya** Meigen	689
deutschi Zetterstedt, Lonchaea	717	**Dictyacium** Steyskal	690
devia Dietz, Tipula	39	*Dictyomyia* Cresson	690
devia (Fallén), Brachicoma	940	dicum Dyar & Shannon, Prosimulium	182
devia (Laffoon), Mycetophila	211	**Dicyphoma** James	303
deviata Dietz, Ormosia	86	**Didactylomyia** Felt	254
deviatus (Malloch), Tanytarsus	179	**Didea** Macquart	562
devinctus (Say), Stictochironomus	177	diderma Garrett, (Sciara)	236
devineyae Beck, Chironomus	170	dido Alexander, Tipula	35
devulsa (Reinhard), Blaesoxipha	945	didyma (Loew), Clistomorpha	977
Dexia Meigen	†981, 1021	didyma (Loew), Neoempheria	222
DEXIINAE	†979, 1020	diegoensis Painter, Bombylius	408
DEXIINI	†981, 1021	diervillae (Felt), Anarete	246
Dexilla Westwood	981	diervillae Felt, Asphondylia	268
DEXILLINI	981	diervillae (Felt), Colpodia	253
dexina (Townsend), Aporotachina	1039	dietrichi Pechuman, Tabanus	333
Dexodes Brauer & Bergenstamm	1046	dietrichi Shaw, Rymosia	207
Dexosarcophaga Townsend	953	dietziana Alexander, Limonia	44
Diabasis Macquart	328	dietziana Alexander, Tipula	35
diabolica Alexander, Tipula	35	dietziella Alexander, Nephrotoma	21
diabolicus Hall, Phlebotomus	92	dietzii Felt, Porricondyla	255
diabolus Becker, Siphonella	786	**Dieuryneura** James	306
diabroticae (Shimer), Celatoria	1040	differens Garrett, Boletina	217
diacanthophora Alexander, Tipula	35	difficilis (Becker), Conioscinella	⟩ 783
DIACHLORINI	328	difficilis Garrett, Anatella	205
Diachlorus Osten Sacken	328	difficilis Garrett, Mycomya	220
diacrisinae Sellers, Carcelia	1092	difficilis Gill, Heleomyza	816
Diacrita Gerstäcker	644	difficilis Johannsen, Phronia	215
diadela Melander, Mythicomyia	417	difficilis Loew, Gymnopternus	499
diademata Bigot, (Stratiomys)	319	difficilis (Lundbeck), Orthocladius	156
diademoides (Townsend), Ervia	1027	difficilis (Malloch), Megaselia	540
Diadocidia Ruthe	198	difficilis (Stein), Fannia	894
DIADOCIDIINAE	198	difficilis (Wiedemann), Hybomitra	338
diagonalis (Loew), Villa	439	diffidens Curran, Actia	1061
dialata Van Duzee, Docosia	226	*diffinis* (Malloch), Lasiops	904
Dialineura Rondani	352	diffissa Johannsen, Rymosia	207
Dialysis Walker	342	diffissa Johannsen, Trichonta	216
Dialyta Meigen	904	diffusa (Snow), Gymnocarena	676
Diamesa Waltl	151	diffusus (Wiedemann), Condylostylus	484

	Page		Page
diflava Alexander, Tipula	32	*Diplosis* Loew	287
digitalis Fisher, Neoempheria	222	**Diplostichus** Brauer & Bergenstamm	1053
digitalis Fluke, Pyrophaena	578	**Diplotoxa** Loew	792
digitarius Cresson, Geron	423	diplotoxoides Becker, Elliponeura	793
digitata Sabrosky, Meoneura	729	*dipseticus* Dyar & Knab, Culex	118
digitatus Malloch, Chironomus	167	dipura Melander, Mythicomyia	417
digitifer Townes, Polypedilum	175	*dira* Robineau-Desvoidy, Stomoxys	914
digitus Van Duzee, Dolichopus	490	directus (Dendy & Sublette), Chironomus	167
Digonichaeta Rondani	1012	direptor Melander, Platypalpus	479
Digonicheta Rondani	1012	direptor Wheeler, Aphrosylus	509
Digonochaeta Rondani	1012	dirhaphis Alexander, Molophilus	88
dilatata Alexander, Erioptera	82	**Dirhiza** Loew	255
dilatata (Brues), Megaselia	540	diropeda Melander, Mythicomyia	417
dilatata Felt, Campylomyza	248	diruta (Stein), Phaonia	906
dilatata Felt, Porricondyla	255	dis Alexander, Tipula	24
dilatata Felt, Trishormomyia	288	discalis Brooks, Catalinovoria	1017
dilaticosta Van Duzee, Polymedon	503	discalis (Coquillett), Phrynofrontina	1047
dilatus Cresson, Chrysopilus	346	discalis Loew, Stratiomys	311, †1311
dilatus Rowe & Knowlton, Chrysops	327	discalis (Malloch), Liriomyza	801
diligens Osten Sacken, Opsebius	405	discalis Melander, Drapetis	477
diloga Melander, Mythicomyia	418	discalis Melander, Microphorus	453
Dilophus Meigen	195	*discalis* Morrison, Gonia	1076
diluta Adams, Macrocera	201	discalis Williston, Chrysops	325
diluta Doane, Tipula	28	discalis (Williston), Physoconops	627
diluta (Johannsen), Bradysia	232	discaloides Curran, Stratiomys	311
diluta (Loew), Orfelia	202	discata Cook, Psectrosciara	240
dilutior Melander, Platypalpus	479	discessus Walker, Dolichopus	490
dilutus Fisher, Symmerus	199	discifer Loew, Platypalpus	479
Dimecoenia Cresson	755	*discifer* Stannius, Dolichopus	493
dimensus (Walker), Allograpta	569	discifera Melander, Tachypeza	473
Dimeraspis Newman	597	discimana Malloch, Mydaea	891
dimicki Cole & Wilcox, Lasiopogon	377	discimanoides Snyder, Mydaea	891
dimidiata (Adams), Clinohelea	136	discimanus Loew, Platycheirus	577
dimidiata (Cresson), Cordilura	829	**Discobola** Osten Sacken	44
dimidiata Cresson, Psilopa	741	*Discocephala* Macquart	374
dimidiata Dietz, Tipula	35	**Discocerina** Macquart	737
dimidiata Loew, Brachydeutera	753	DISCOCERININI	737
dimidiata (Loew), Mycodrosophila	771	*discoida* Say, Mycetophila	214
dimidiata (Loew), Neurigona	518	*discolor* Banks, Pipunculus	553
dimidiata (Loew), Prionocera	19	discolor (Coquillett), Anatopynia	145
dimidiata Loew, Psila	639	discolor (Coquillett), Psorophora	110
dimidiata Loew, Rhamphomyia	463	discolor (Cresson), Lyciella	700
dimidiata Macquart, Ogcodocera	428	discolor Loew, Chrysotus	522
dimidiata Melander, Trixoscelis	818	discolor (Loew), Dialysis	343
dimidiata Say, (Sciara)	236	discolor Loew, Diogmites	371
dimidiata (Sturtevant & Wheeler), Philygria	745	discolor (Loew), Spanoparea	813
dimidiata (Walker), Physocephala	628	*discolor* Loew, Tipula	34
dimidiata Walker, Phytomyza	804	discolor Robinson, Peloropeodes	517
dimidiatus Curran, Promachus	400	discolor Van Duzee, Dolichopus	490
dimidiatus (Loew), Condylostylus	484	discolor Zetterstedt, Rhaphium	512
dimidiatus Loew, Dilophus	195	discolorata James, Odontomyia	315
dimidiatus Loew, Paragus	578	discoloria (Meigen), Orfelia	202
dimidiatus (Loew), Rhagio	345	**Discomyza** Meigen	740
dimidiatus Macquart, Syrphus	559	DISCOMYZINI	740
dimidiatus (Macquart), Taracticus	385	disconcerta Curran, Rhamphomyia	463
dimidiatus Wiedemann, Eristalis	623	disconvenita Melander, Empis	459
diminucosta Harmston & Knowlton, Teuchophorus	529	discors (Laffoon), Mycetophila	211
diminuta Walker, Phytomyza	804	discrepans (Pandellé), Styloneuria	964
dimmocki (Aldrich), Eucelatoriopsis	1043	discreta (Meigen), Hylemya	853
dimmocki Hine, Chrysops	325	discreta Stein, Limnophora	885
dimmocki (Webber), Lespesia	1102	disfascia Martin, Beameromyia	361
dimorphus Malloch, Chironomus	170	disgregus (Snow), Dasysyrphus	563
Dinera Robineau-Desvoidy	986	*disiuncta* (Wiedemann), Lepidanthrax	440
Dioctria Meigen	370	*disiuncta* (Wiedemann), Microphthalma	982
Dioctrodes Coquillett	385	*disjectus* (Williston), Epistrophe	564
Diogmites Loew	371	disjuncta Cresson, Scatophila	758
Diomonus Walker	223	disjuncta (Garrett), Synapha	219
Dioprosopa Hull	572	disjuncta (Johnson), Homoneura	698
DIOPSIDAE	637	*disjuncta* Loew, Bolitophila	197
diopsides (Walker), Homalocephala	653	disjuncta Melander, Apolysis	421
Dioptopsis Enderlein	99	disjuncta Van Duzee, Neurigona	518
Diostracus Loew	508	disjuncta Walker, Tipula	35
diota Garrett (Sciara)	236	disjuncta (Wiedemann), Microphthalma	982
Diotrepha Osten Sacken	53	disjunctus Van Duzee, Chrysotus	522
Dioxyna Frey	666	disjunctus (Wiedemann), Lepidanthrax	440
Dipalta Osten Sacken	441	*disjunctus* (Williston), Epistrophe	564
Diphaomyia Vargas	131	*dispar* Banks, Laphria	389
Diphuia Cresson	737	dispar (Becker), Pholeomyia	732
diplaci Felt, Asphondylia	268	dispar Bigot, Dialysis	343
diplaci Felt, Lasioptera	271	dispar Bromley, Stenopogon	383
diplasus Hall, Lordotus	412	dispar Coquillett, Saropogon	383
Diplocentra Loew	759	dispar (Coquillett), Villa	437
Diplocladius Kieffer	154	dispar Cresson, Ceropsilopa	742
Diploneura Lioy	534	dispar (Macquart), Ogcodes	406
Diplonevra Lioy	534	*dispar* (Williston), Pseudogaurax	782
		dispar Wulp, Chrysopilus	347

INDEX 1591

	Page		Page
disparella Banks, Laphria	389	diversa (Coquillett), Stilobezzia	134
disparicauda Borgmeier, Apocephalus	543	diversa Coquillett, Thereva	353
disparilis Bergroth, Dialysis	343	diversa Curran, Camptoprosopella	697
disparilis Coquillett, Rhamphomyia	463	diversa Dietz, Tipula	35
disparilis Melander, Leptopeza	451	diversa (Garrett), Neoleria	813
disparis Reinhard, Thelymyia	1106	diversa Giglio-Tos, Rhinotora	819
dispellens (Walker), Brachypremna	18	diversa Johnson, Agromyza	795
disphana (Doane), Pedicia	54	diversa Johnson, Brachyopa	592
disrupta (Cockerell), Aciurina	670	diversa (Johnson), Platypezina	548
dissecta (Meigen), Pegomya	863	diversa Melander, Drapetis	477
dissidens (Tucker), Hippelates	775	diversa (Osten Sacken), Limonia	49
dissidia Parker, Sarcophaga	960	*diversa* (Osten Sacken), Limonia	46
dissimilicosta Spuler, Leptocera	724	diversifasciata (Knab), Meliscaeva	568
dissimilipes Melander, Platypalpus	479	diversipennis Borgmeier, Abaristophora	535
dissimilipes Van Duzee, Syntormon	515	diversipennis Curran, Dolichopus	490
dissimilipes Wheeler, Pelastoneurus	501	*diversipennis* Curran, Mallota	621
dissimilis Aldrich, Hydrotaea	899	diversipes Coquillett, Platypalpus	479
dissimilis Brennan, Chrysops	325	diversipes Curran, Microprosopa	835
dissimilis Hardy, Tomosvaryella	555	diversipes (Fitch), Sylvicola	190
dissimilis Johannsen, Tanytarsus	179	*diversipes* (Macquart), Meliscaeva	568
dissimilis Malloch, Orthachaeta	831	diversipes Melander, Drapetis	477
dissimilis Malloch, Parecetecephala	792	diversipes Melander, Tachydromia	474
dissimilis Malloch, Phaonia	906	diversipilosis Curran, Eucyrtopogon	373
dissimilis (Malloch), Pleurochaetella	834	diversipilosus Curran, Microdon	598
dissimilis Melander, Oligodranes	421	diversipunctata (Curran), Melangyna	566
dissimilis Walker, Dialysis	343	diversoides (Dietz), Limonia	46
dissita Reinhard, Paraphasmophaga	1073	diversus Coquillett, Lordotus	412
distans Loew, Empis	459	dives (Hardy), Pipunculus	553
distans Melander, Tachypeza	473	dives (Johannsen), Micropsectra	178
distans (Osten Sacken), Limonia	46	dives Johannsen, Sciara	230
distichliae (Malloch), Chlorops	789	dives (Loew), Nicocles	380
Distichona Wulp	1074	dives (Osten Sacken), Ferdinandea	588
distilobatus Alexander, Molophilus	89	*dives* (Wiedemann), Euantha	1030
distincta Alexander, Erioptera	82	*dives* (Williston), Pilimas	321
distincta (Curran), Micromintho	1032	*divexus* (Doane), Ormosia	85
distincta Curran, Pipiza	579	dividua Melander, Drapetis	477
distincta (Doane), Limonia	49	dividuus Van Duzee, Chrysotus	522
distincta Garrett, Dixa	101	divigatus Harmston, Dolichopus	490
distincta Garrett, Lutomyia	811	divisa Alexander, Limonia	46
distincta Garrett, Macrocera	201	*divisa* (Coquillett), Alophorella	968
distincta (Garrett), Neuratelia	224	divisa (Coquillett), Exoprosopa	444
distincta Garrett, Sciophila	225	*divisa* (Loew), Gymnoclytia	965
distincta Greene, Anastrepha	673	divisa (Meigen), Hydrophoria	864
distincta Quisenberry, Paroxyna	666	divisa (Williston), Epistrophe	564
distincta (Robineau-Desvoidy), Bonnetia	1004	divisor (Banks), Laphria	389
distincta Smith, Gonia	1076	divisus Cresson, Lordotus	412
distincta (Townsend), Erythrandra	941	divisus Hardy, Chrysopilus	346
distincta (Wiedemann), Drino	†1085	*divisus* Harris, Tabanus	335
distincta (Wiedemann), Microphthalma	982	divisus Loew, Helophilus	619
distincta Williston, Neoascia	594	divisus Melander, Oligodranes	421
distinctum Malloch, Simulium	187	divisus Walker, Chrysops	325
distinctus Chillcott, Roederioides	468	**Dixa** Meigen	101
distinctus (Malloch), Trichocladius	158	**Dixapuella** Dyar & Shannon	101
distinctus Melander, Oligodranes	421	**Dixella** Dyar & Shannon	102
distinctus Van Duzee, Chrysotus	522	DIXIDAE	100
distinctus Van Duzee, Dolichopus	491	**Dizonias** Loew	372
distinctus Van Duzee, Medetera	510	**Dizygomyza** Hendel	799
distinctus (Wiedemann), Proctacanthus	399	doanei Dietz, Tipula	32
distinctus (Williston), Helophilus	619	doanei (Kellogg), Agathon	99
distinctus (Williston), Neoitamus	397	doanei Melander, Dioctria	370
distinguendus Kieffer, Chironomus	165	doaneiana Alexander, Tipula	36
distorta (Allen), Amobia	934	doclea (Walker), Pogonortalis	655
distortus (Van Duzee), Calyxochaetus	528	**Docosia** Winnertz	226
distractus Walker, Dolichopus	490	dodgei (Whitney), Anacimas	329
distus Fisher, Bolitophila	197	dodrans Osten Sacken, Exoprosopa	444
Ditanytarsus Kieffer	180	dodrina Curran, Exoprosopa	444
Ditomyia Winnertz	198	*dohaniani* Johnson, Automola	642
DITOMYIINAE	198	**Dohrniphora** Dahl	535
Ditrichophora Cresson	739	dolabraria Melander, Empis	459
divaricata (Aldrich), Dohrniphora	535	dolens Felt, Asynapta	254
divaricata Alexander, Dicranota	57	dolens (Johannsen), Bradysia	232
divaricata Cresson, Taeniaptera	636	dolicheretma Melander, Clinocera	468
divaricata (Felt), Asteromyia	274	**Dolichocephala** Macquart	470
divaricata Felt, Contarinia	277	**Dolichocodia** Townsend	986
divaricata (Loew), Orfelia	202	*Dolichogaster* Stein	868
divergens (Coquillett), Ormosia	85	*Dolichoglossa* Stein	868
divergens Dietz, Ormosia	86	**Dolichomyia** Wiedemann	425
divergens Johannsen, Brachypeza	206	**Dolichopeza** Curtis	22
divergens (Loew), Agathomyia	548	dolichophallus (Alexander), Hesperoconopa	78
divergens Loew, Drapetis	477	DOLICHOPIDAE	482
divergens (Malloch), Megaselia	540	DOLICHOPODIDAE	482
divergens Van Duzee, Sciapus	486	DOLICHOPODINAE	487
divergens Walker, Mycetobia	191	**Dolichopus** Latreille	487
diversa (Coquillett), Alophorella	968	**Dolichotarsus** Brooks	1041
diversa Coquillett, Phthiria	419	*Doliomyia* Johannsen	776
diversa Coquillett, Rhamphomyia	463	doloma Reinhard, Aplomya	1097

	Page		Page
dolomata Vockeroth, Geomyza	821	drapetoides Walker, Microphorus	453
dolopis Reinhard, Spathimeigenia	1049	dreisbachi (Hardy), Pipunculus	553
dolorosa Melander, Tachypeza	473	*dreisbachi* Harmston & Knowlton, Hercotomus	498
dolorosa (Williston), Camptoprosopella	697		
dolorosus Melander, Oligodranes	421	dreisbachi Townes, Glyptotendipes	171
dolorosus (Williston), Parabombylius	410	**Drepanoglossa** Townsend	1060
dolosa Benjamin, Neaspilota	672	*Drepanomyia* Wheeler	508
dolosa Reinhard, Spathimeigenia	1049	**Drino** Robineau-Desvoidy	1084
dolosus Parent, Dolichopus	490	**Drosophila** Fallén	763
domestica Haseman, Psychoda	96	DROSOPHILIDAE	760
domestica Linnaeus, Musca	913	DROSOPHILINAE	763
domestica (Meigen), Rymosia	207	DROSOPHILOIDEA	734
domestica (Osten Sacken), Limonia	45	druias Melander, Apolysis	421
domesticus Van Duzee, Dolichopus	490	*dryadis* (Holmgren), Melangyna	568
dominicana (Townsend), Euphasiopteryx	980	dryas Johannsen, Leia	227
dondiae Felt, Asphondylia	268	**Drymodesmyia** Vargas	130
donuca (Walker), Paraprosalpia	867	**Dryomyia** Kieffer	261
donysa (Walker), Eurosta	663	**Dryomyza** Fallén	680
dora Pritchard, Cophura	367	DRYOMYZIDAE	680
dorcadion Osten Sacken, Exoprosopa	444	dryomyzina Zetterstedt, Sciomyza	688
dorenus (Roback), Orthocladius	156	*Dryope* Robineau-Desvoidy	680
DORILAIDAE	550	dubia Bigot, Sphaerophoria	570
Dorilas Meigen	551	dubia Curran, Admontia	1068
doris Osten Sacken, Exoprosopa	444	dubia Curran, Cryptomeigenia	1041
doris (Townes), Chironomus	167	*dubia* Curran, Lonchoptera	531
dorneri Malloch, Chironomus	171	dubia (Curran), Ptilodexia	988
Doros Meigen	569	dubia Curtis, (Anthomyia)	915
dorothea Alexander, Erioptera	83	dubia (Fallén), Lypha	1011
dorothea Alexander, Hexatoma	70	dubia Garrett, Pseudoleria	812
dorothea Alexander, Tipula	32	dubia Johnson, Agathomyia	548
dorotheae Brimley, Odontomera	642	*dubia* (Johnson), Trypeta	677
dorri (Johnson), Oscinoides	781	dubia (Malloch), Micropsectra	178
dorsalis (Coquillett), Myiopharus	1045	*dubia* (Meigen), Pentaneura	†148
dorsalis (Coquillett), Psalidopteryx	1067	dubia (Shannon), Xylota	606
dorsalis (Fabricius), Hedriodiscus	314	dubia Siebke, Bolitophila	197
dorsalis Garrett, Dixa	102	dubia (West), Gymnoclytia	964
dorsalis (Harris), Psila	639	dubia (Williston), Efferia	394
dorsalis (Johnson), Dolichopeza	23	dubia (Williston), Lespesia	1102
dorsalis (Loew), Cerodontha	801	dubiosa Johnson, Seioptera	648
dorsalis (Loew), Gaurax	781	dubiosa Van Duzee, Bolitophila	198
dorsalis Loew, Hippelates	775	*dubiosus* Van Duzee, Sciapus	486
dorsalis (Loew), Oscinella	779	dubitata Malloch, Agromyza	795
dorsalis Loew, Schoenomyza	870	*dubitata* (Malloch), Megaselia	540
dorsalis (Loew), Sciapus	486	dubitatus Johannsen, Orthocladius	156
dorsalis Loew, Syndyas	448	dubium (Van Duzee), Rhaphium	513
dorsalis (Malloch), Cordilura	829	dubium (Zetterstedt), Melanostoma	575
dorsalis (Meigen), Aedes	112	*dubius* Cresson, Pipunculus	554
dorsalis Meigen, Chironomus	165	dubius (Macquart), Gaurax	782
dorsalis (Say), Allognosta	300	dubius Say (Mulio)	1114
dorsalis Van Duzee, Chrysotus	522	dubius Van Duzee, Chrysotus	522
dorsalis Van Duzee, Dolichopus	490	dubius Williston, Cyrtopogon	368
dorsalis (Van Duzee), Hercostomus	498	dulcis Osten Sacken, Erioptera	82
dorsalis Walker, Ctenophora	20	*dulcis* Stein, Phaonia	907
dorsata Felt, Porricondyla	255	dulichii Felt, Thecodiplosis	278
dorsata (Loew), Oscinella	779	dumosae (Felt), Asteromyia	274
dorsata (Zetterstedt), Spilogona	880, †881	duncani Alexander, Teucholabis	76
dorsatum (Say), Pogonosoma	391	duncani Bromley, Stenopogon	383
dorsifer Walker, Tabanus	333	duncani Curran, Callicera	595
dorsilinea (Stein), Hylemya	848	duncani Curran, Eristalis	624
dorsilinea (Wulp), Muscina	909	duncani Painter, Bombylius	409
Dorsilopha Sturtevant	766	duncani Sturtevant, Drosophila	767
dorsimacula Walker, Tipula	36	duncani (Wilcox), Atoniomiyia	387
dorsimaculata Felt, Lasioptera	272	duncani Wilcox, Heteropogon	374
dorsolineata Doane, Tipula	24	duncani Wilcox, Laphystia	376
dorsonotatus Macquart, Tabanus	333	dunningii (Coquillett), Spathidexia	1021
dorsopunctus Fairchild, Chrysops	325	duospinosa (West), Admontia	1068
dorsovittata Malloch, Coenosia	871	dupla (Cresson), Paroxyna	666
dorsovittata (Malloch), Gymnodia	884	dupla Cresson, Psilopa	741
dorsovittatus Hine, Chrysops	325	dupla Garrett, Bolitophila	198
dorycerus Loew, Dolichopus	490	duplare Reinhard, Mochlosoma	987
Dorylas Meigen	551	duplex Shewell & Fredeen, Simulium	186
Dorylomorpha Aczél	556	duplex (Walker), Atylotus	330
doryphorae (Riley), Doryphorophaga	1042	duplex (Walker), Euhybus	448
Doryphorophaga Townsend	1042	duplex Walker, Tipula	36
dosiades (Walker), Cylindromyia	974	*duplicata* (Doane), Limonia	50
doutti (Pritchard), Aphidoletes	280	duplicata Fluke, Pyrophaena	578
dovei Hall, Culicoides	129	*duplicata* (Hall), Ravinia	955
doveri Sommerman, Prosimulium	182	duplicata Johnson, Chaetopsis	650
downesi Alexander, Tipula	36	*duplicata* Malloch, Clusiodes	807
downesi Richards, Leptocera	721	duplicata (Malloch), Pegomya	859
downesi Wirth, Alluaudomyia	134	duplicata (Meigen), Helina	887
downesi Wirth & Hubert, Culicoides	129	duplicata Reinhard, Eumacronychia	935
drabicola Cole, Lasiopogon	377	duplicata Reinhard, Winthemia	1090
draco Hull, Meromacrus	622	*duplicata* (Wiedemann), Mesograpta	†571
drakei (Pritchard), Cophura	366	duplicatus Aldrich, Dolichopus	490
Drapetis Meigen	477	duplicatus (Johannsen), Paratendipes	174

INDEX 1593

	Page		Page
dupliciformis Alexander, Tipula	36	effilatum (Wheeler), Rhaphium	513
duplicis Coquillett, Rhamphomyia	463	effrenata (Walker), Ravinia	954
duplicis Coquillett, Thereva	353	effrenatum Reinhard, Leucostoma	976
dupreei (Coquillett), Aedes	112	effrenus (Coquillett), Poecilanthrax	442
dura Curran, Phytomyza	804	effulgens Huckett, Coenosia	871
dura Garrett, Mycomya	220	*egberti* Dyar & Knab, Culex	119
durani Cresson, Setacera	755	egerminans Loew, Phthiria	419
durani (Davidson), Ceriana	615	EGINIINAE	826
durani (Shannon), Ceriana	615	**Egle** Robineau-Desvoidy	854
duryi Hine, Proctacanthus	399	egleformis Huckett, Hylemya	848
dux (Johannsen), Bradysia	232	egregia Snow, Platypeza	549
dux Johannsen, Chironomus	170	egregius Becker, Chlorops	789
dyari Alexander, Erioptera	81	egressa Coquillett, Thereva	353
dyari (Coquillett), Anatopynia	145	egula (Reinhard), Nowickia	993
dyari (Coquillett), Culiseta	117	ehrmanii Coquillett, Aldrichia	414
dyari Felt, Lestremia	244	ehrmanni Aldrich, Hypocera	534
dyari Garrett, Dixa	102	*ehrmanni* Coquillett, Aldrichia	414
dyari (Malloch), Megaselia	539	einaris (Smith), Palpexorista	1056
dyari Shannon, Sphecomyia	612	*Finfeldia* Kieffer	164
dyari (Townes), Phaenopsectra	174	*Eipogona* Rondani	1098
dychei Williston, Helophilus	618	eiseni Alexander, Erioptera	82
dydas (Walker), Guerinia	1055	eiseni Townsend, Hermetia	304
dymka (Kessel), Platypeza	549	ejuncida Say, Xylota	605, †606
Dynatosoma Winnertz	209	**Elachiptera** Macquart	776
Dyodiplosis Rübsaamen	287	**Elachisoma** Rondani	725
dysanor Dyar, Aedes	111	elachista (Townes), Paralauterborniella	173
Dyscolomyia Reinhard	1023	*Flaeophila* Rondani	64
Dyscrasis Aldrich	644	*Flaeotoma* Costa	386
Dyseuaresta Hendel	668	elana Roback, Corynoneura	164
dysmicus Quate, Telmatoscopus	94	elanis Reinhard, Sarcophaga	958
Dziedzickia Johannsen	218	**Elaphropeza** Macquart	478
		elata (Fabricius), Tetanocera	694
eadsi Philip, Tabanus	333	elata Johannsen, Allodia	205
earlei Vargas, Anopheles	106	elatus Roback, Psectrocladius	159
Earomyia Zetterstedt	716	*eldoradensis* Wheeler, Hydrophorus	506
EBENIINI	1030	electa Alexander, Prionocera	19
ebenina Alexander, Erioptera	81	electa (Say), Zonosemata	675
eboreus Painter, Bombylius	409	electa (Zetterstedt), Mydaea	891
ebyi (Bromley), Ospriocerus	381	electra Felt, Mayetiola	263
ecalcar Alexander, Erioptera	83	electus Adams, Scenopinus	356
ecalcarata Thomson, Sepsis	683	elegans Aldrich, Dolichopus	490
Eccemyia Robineau-Desvoidy	631	elegans Coquillett, Amphicosmus	427
Eccoptomera Loew	811	elegans Coquillett, Johnsonia	950
Eccritosia Schiner	393	elegans (Coquillett), Orfelia	202
Ecculex Felt	116	*elegans* (Coquillett), Probezzia	138
ecetra Walker, Rhamphomyia	463	elegans Curran, Plunomia	706
echemon (Walker), Atomosia	386	*elegans* (Doane), Limonia	45
echinaspis Spuler, Leptocera	721	elegans Johannsen, Cricotopus	157
echinata Garrett, Mycomya	220	elegans Kincaid, Psychoda	96
echinata (Séguy), Hylemya	848	elegans Loew, Sargus	302
echinata (Thomson), Protodejeania	1002	elegans Malloch, Oscinoides	781
echinochloa Felt, Lasioptera	271	elegans Osten Sacken, Ula	53
Echinogaster Lioy	996	*elegans* Panzer, Chamaemyia	†707, 707
Echinohelea Macfie	135	elegans Spuler. Leptocera	724
Echinomya Latreille	998	elegans Stein, Chelisia	844
ECHINOMYINI	993	*elegantipes* (Ouellet), Themira	684
Echinomyodes Townsend	993	elegantula Johannsen, Monoclona	224
Echthopoda Loew	372	elegantula (Johannsen), Stilobezzia	135
Ecitomyia Brues	545	elegantula Malloch, Forcipomyia	125
Eclimus Loew	426	elegantula (Williston), Orfelia	202
eclipta Benjamin, Trupanea	667	elegantulum (Meigen), Rhaphium	513
Ectaetia Enderlein	241	elegantulus Röder, Agathon	99
ECTAETIINAE	241	*Fleodiomyia* Townsend	943
Ectecephala Macquart	791	**Eleodiphaga** Walton	1072
Ectemnia Enderlein	185	eleodis (Aldrich), Blaesoxipha	944
Ectrepesthoneura Enderlein	227	eleodivora (Walton), Sitophaga	1048
edactylus Loew, Dolichopus	491	**Elephantocera** Townsend	1042
edentata Alexander, Limnophila	64	**Elephantomyia** Osten Sacken	71
edentata Stone, Anastrepha	673	elevata (West), Ptilodexia	988
edentula Johannsen, Mycetophila	212	*elevatus* Harris (Hybos)	1114
edeta (Walker), Ptiolina	347	**Elfia** Robineau-Desvoidy	1065
Edioamoeba Garrett	814	**Elgiva** Meigen	690
edititia (Say), Villa	437	elinguis Melander, Apystomyia	424
edititoides Painter, Villa	439	elisae (Meigen), Calliopum	696
edmundsi Alexander, Tipula	28	elita Townsend, Eumacronychia	935
edura Johannsen, Mycetophila	211	elizabethae Brimley, Ceraturgus	365
edwardi (Alexander), Pilaria	68	*elizabethae* Quisenberry, Chetostoma	678
edwardsi Jones, Metriocnemus	161	**Elliponeura** Loew	792
edwardsi (Kruseman), Chironomus	167	**Elliptera** Schiner	51
edwardsi Pritchard, Anarete	246	elnora Dyar & Shannon, Thaumalea	120
edwardsii (Coquillett), Villa	435	elodeae (Townes), Chironomus	167
edwardsii (Loew), Paracosmus	427	**Elodia** Robineau-Desvoidy	1071
edwardsii (Williston), Euphorocera	1054	**Eloeophila** Rondani	64
effera Coquillett, Rhamphomyia	463	elongata Aldrich, Sarcophaga	958
Efferia Coquillett	393	elongata Allen, Eumacronychia	935
effetus Reinhard, Mimologus	1086	*elongata* Banks, Platypeza	549

1594 A CATALOG OF DIPTERA OF NORTH AMERICA

	Page		Page
elongata Chillcott, Fannia	894	**Enicita** Westwood	684
elongata Cook, Psectrosciara	240	**Enicomira** Duda	685
elongata (Fabricius), Baccha	572	*Enicopus* Walker	684
elongata (Felt), Pararete	245	enigma Coquillett, Ptilomyia	736
elongata Malloch, Hydrophoria	865	enigma Melander & Brues, Dolichopus	491
elongata Malloch, Paracantha	662	enigmata Chillcott, Fannia	894
elongata (Melander), Chelipoda	472	*enigmatica* Dietz, Ormosia	87
elongata (Say), Dialysis	343	**Enlinia** Aldrich	529
elongata (Shannon), Bufolucilia	927	*Ennyomma* Townsend	983
elongata Williston, Xylota	605	*Ennyommopsis* Townsend	983
elongatum Van Duzee, Rhaphium	513	enodis Melander, Empis	459
elongatus (Curran), Neocnemodon	581	*Enoplempis* Bigot	457
elongatus Felt, Holoneurus	254	enoria Melander, Mythicomyia	417
elongatus (Hough), Cyanus	930	*enotahensis* Seago, Fannia	893
elongatus Sabrosky, Dicraeus	783	ensifera Melander, Chelifera	471
Elophila Meigen	622	*Ensina* Robineau-Desvoidy	†665
Elophilus Meigen	622	*Enteromyia* Enderlein	916
elsa Alexander, Dicranoptycha	52	*Enteromyza* Rondani	915
elsa Daecke, Eurosta	663	entomophthorae Alexander, Tipula	32
eltoni (Edwards), Limnophyes	160	*Eohyric* Townsend	1026
eluta Loew, Diastata	772	*Eophrissopolia* Townsend	1078
eluta Loew, Euxesta	651	**Epacmus** Osten Sacken	428
eluta Loew, Tipula	28	**Epalpus** Rondani	1002
eluta Pritchard, Conarete	245	epeirae (Brues), Phalacrotophora	542
eluthera Dyar & Shannon, Chaoborus	103	epheba (Osten Sacken), Villa	439
elyii (Walton), Actia	1061	ephedrae Cockerell, Lasioptera	271
elymi Felt, Rhabdophaga	257	ephedricola Cockerell, Lasioptera	271
emarginata Felt (Cecidomyia)	290	*Ephelia* Schiner	64
emarginata (Felt), Feltiella	282	*ephemerae* Kieffer, Smittia	162
emarginata Felt, Giardomyia	284	*ephippium* (Zetterstedt), Gaurax	782
emarginata Macquart, Exoprosopa	445	**Ephydra** Fallén	754
emarginata (Malloch), Cordilura	829	EPHYDRIDAE	734
emarginata (Say), Epistrophe	564	EPHYDRINAE	753
emarginata (Tothill), Nowickia	995	EPHYDRINI	753
emarginata Van Duzee, Medetera	510	*Epibates* Osten Sacken	426
emarginatum Davies, Peterson, & Wood, Simulium	186	epicautae Wirth, Atrichopogon	123
		Epichlorops Becker	791
emarginatum Wheeler, Parasyntormon	516	epicyles (Walker), Aplomya	1097
emarginatus McAtee, Dilophus	195	**Epicypta** Winnertz	209, †215
emarginatus Van Duzee, Chrysotus	522	*Epidapus* Haliday	235
emarginicorne Curran, Parasyntormon	516	**Epidexia** Townsend	1032
errasculator Fitch, Cuterebra	1110	*Epidexiopsis* Townsend	1047
Emblemasoma Aldrich	949	**Epidiplosis** Felt	285
emergens Stone, Cnephia	184	**Epigrimyia** Townsend	974, †1059, †1060
eminens Johannsen, Neuratelia	224	epilachnae (Aldrich), Aplomyiopsis	1038
emmelina Alexander, Limnophila	64	epilepsalis French, (Gastrophilus)	961
emmesia Malloch, Pegomya	858	*Epimasicera* Townsend	1096
Emmesomyia Malloch	858	epimicta Alexander, Limnophila	66
emorsus (Townes), Chironomus	167	**Epimyia** Felt	252
Empalia Winnertz	219	**Epiphragma** Osten Sacken	60
Empeda Osten Sacken	79	**Epiplagiops** Blanchard	1053
empedoides (Alexander), Erioptera	80	*Epiplatea* Loew	641
Empedomorpha Alexander	80	epista Pritchard, Gongromastix	245
Emphanopteryx Townsend	1041	epistates (Osten Sacken), Hybomitra	338
Empheria Winnertz	222	**Epistrophe** Walker	564
EMPIDAE	446	*Episyrphus* Matsumura	568
EMPIDIDAE	446	epitheca Reinhard, Sarcophaga	958
Empidideicus Becker	415	**Epochra** Loew	677
empidiformis (Say), Hemerodromia	470	*Epochroa* Loew	677
Empidigeron Painter	423	*Epomyia* Cole	350
EMPIDINAE	452	*Epourenia* Rapp	256
EMPIDOIDEA	446	epytus (Walker), Cylindromyia	974
empiformis (Say), Hemerodromia	470	equalis Hine, Tabanus	333
Empimorpha Coquillett	457	equalis (Van Duzee), Orfelia	202
empirica (Hutton), Leptocera	724	equatorialis (Townsend), Chrysotachina	1005
Empis Linnaeus	457	eques (Johannsen), Forcipomyia	124
emulata Painter, Villa	439	equestris (Fabricius), Merodon	617
Emusca Malloch	913	*equi* (Clark), Gasterophilus	916
Enarmostus Walker	367	equifrons Huckett, Hylemya	848
encausta Osten Sacken, Phyllolabis	60	equina Fallén, Copromyza	719
encaustus Harris, (Syrphus)	1114	equinus Theobald, Haemagogus	116
enceliae Felt, Asphondylia	268	equitans (Claassen), Symbiocladius	157
enceliae Felt, Rhopalomyia	266	*Erax* Scopoli	†393
encisoi Vargas & Díaz Nájera, Simulium	187	erecta Aldrich, Spathimeigenia	1049
Endaphis Kieffer	276	erecta (Coquillett), Pseudoperichaeta	1106
enderleini Bau, Cuterebra	1110	erecta Sellers, Thelymyia	1106
Endochironomus Kieffer	169	erecta (Wood), Megaselia	538
endymion Osten Sacken, Tabanus	333	erectus Becker, Condylostylus	484
enecator Melander, Tachydromia	474	*eregeroni* Brodie (Diplosis)	291
enervatus Melander, Platypalpus	479	erema Pritchard, Moehnia	246
engelhardti (Bromley), Laphria	389	*eremi* (Felt), Campylomyza	248
engelhardti Bromley, Stenopogon	384	eremica Melander, Mythicomyia	417
engelhardti Painter, Heterostylum	411	eremicola Melander, Aphoebantus	429
engelhardti Shannon, Polybiomyia	616	eremicus Painter & Hall, Poecilanthrax	442
engelhardti Wilcox, Nicocles	380	eremita Osten Sacken, Exoprosopa	444
engelmannia Alexander, Dicranota	57	eremites Shewell, Cnephia	184

INDEX 1595

eremitis Melander, Oligodranes................ 421
Eremomyia Stein............................ 857
Eremomyioides Malloch..................... 857
Eretmoptera Kellogg....................... 163
Eribella Mesnil........................... 1042
Eribolus Becker........................... 777
ericameriae Felt, Rhopalomyia................. 266
ericia (Pettey), Bradysia..................... 232
eriensis Alexander, Tipula.................... 36
erigeroni Brodie (Diplosis)................... 291
erigerontis (Felt), Neolasioptera............. 273
erigerontis Felt, Rhopalomyia................. 266
erii Felt, Coccidomyia........................ 262
erilis (Reinhard), Panacemyia................. 1046
erinacioides Malloch, Rhamphomyia............. 463
Erinna Meigen............................... 297
ERINNIDAE..................................... 296
erinus Pritchard, Haplopogon.................. 373
Eriocera Macquart......................... 69
Eriogaster Macquart......................... 457
eriophora (Williston), Hexatoma............... 70
eriophthalma (Zetterstedt), Hylemya......... 855
Erioptera Meigen.......................... 80
ERIOPTERINI................................... 71
ERIOTHRIXINI.................................. 1031
Eriphia Meigen.............................. 903
eriphides Walker, Heteromyza................ 810
erisma (Reinhard), Elfia...................... 1065
ERISTALINI.................................... 617
Eristalis Latreille....................... 622
Eristaloides Rondani........................ 622
Eristalomya Rondani......................... 622
Eristalomyia Rondani........................ 622
ermae Hall, Lordotus.......................... 412
Ernestia Robineau-Desvoidy.................. †1007
ERNESTIINI.................................... 1007
Ernoneura Becker.......................... 837
eronis Curran, Peleteria.................... 997
eronis Curran, Pipunculus................... 554
erostrata Alexander, Limonia.................. 48
erotema (Reinhard), Elfia..................... 1065
errabunda (Harris), Archytas................ 1000
errabunda (Wulp), Ravinia..................... 954
errans Garrett, Dynatosoma.................... 209
errans Malloch, Coenosia...................... 871
errans (Meigen), Phaonia...................... 906
erratica (Felt), Pectinodiplosis.............. 278
erraticus Curran, Platycheirus................ 577
erraticus (Dyar & Knab), Culex................ 119
erro Aldrich, Emblemasoma..................... 949
erronis Reinhard, Spathimeigenia.............. 1049
erubescens Osten Sacken (Cecidomyia).......... 293
erucicida Knab, Forcipomyia................. 125
erucicola (Coquillett), Euptilopareia......... 1018
Erucophaga Reinhard....................... 949
erudita Melander, Chyliza..................... 640
Ervia Robineau-Desvoidy................... 1027
Erycia Robineau-Desvoidy.................... †1105
ERYCIINI...................................... 1096
Erycioides Curran........................... 1043
Erynnia Robineau-Desvoidy................. 1071
Erynniopsis Townsend...................... 1042
erythocnemius Hine, Asilus.................... 392
erythraeus (Bigot), Tabanus................... 333
Erythrandra Brauer & Bergenstamm.......... 941
erythrocephala (Hendel), Otites............... 646
erythrocephala (Leach), Ornithoctona.......... 917
erythrocephala (Meigen), Calliphora......... 929
erythroceora Brauer & Bergenstann, Hesperomyia.. 1080
erythrocera Loew, Notiphila................... 746
erythrocera Reinhard, Myioclonia.............. 1025
erythrocera (Robineau-Desvoidy), Lispocephala.................................... 877
erythrocera (Thomson), Arctophyto............. 986
ERYTHROCERINI................................. 1071
erythronota (Strobl), Chaetopleurophora....... 533
Erythrophasia Townsend...................... 967
erythrophrys (Williston), Nephrotoma........ 20
erythropleura Sabrosky, Elachiptera........... 777
erythrotelus Walker, Tabanus................ 336
erythrothorax Dyar, Culex..................... 118
erythrura (Loew), Furcifera................. 350
erythrura (Wulp), Amobia...................... 934
eschata Pritchard, Conarete................... 245

Esenbeckia Rondani........................ 322
Esenbekia Rondani........................... 322
essigi Felt, Rhabdophaga...................... 257
essigi Wirth, Palpomyia....................... 140
estella Alexander, Erioptera.................. 83
estigmenensis (Sellers), Hubneria............. 1093
estotilandica Rondani (Scatina)............... 842
esuriens Bromley, Diogmites................... 372
esuriens Fabricius, Volucella............... ††601
ethellia Curran, Rhamphomyia................ 462
ethniae Brooks, Eufrontina.................... 1100
Euacemyia Townsend.......................... 1036
Euadmontia Townsend......................... 1068
Euantha Wulp.............................. 1030
Euaraba Townsend............................ 939
Euaresta Loew............................. 665
euarestoides (Bates), Valentibulla.......... 670
Euarestoides Benjamin..................... 669
Euarmostus Walker........................... 367
Eubiomyia Townsend.......................... 1048
Eubrachycoma Townsend....................... 941
Eubrachymera Townsend..................... 1023
Eucalliphora Townsend..................... 928
Eucatocha Edwards......................... 244
Eucelatoria Townsend...................... 1043
Eucelatoriopsis Townsend.................. 1043
eucephala (Loew), Pseudocalliope.............. 703
eucera (Loew), Nephrotoma..................... 21
eucera (Loew), Parectecephala................. 792
eucera Melander, Niphogenia................... 467
eucera Osten Sacken, Dicranota................ 57
eucerata (Bigot), Parafischeria.......... 1011, †1036
Euceratomyia Williston...................... 595
euceroides Alexander, Nephrotoma.............. 21
Euceromasia Townsend...................... 1100
Euceroplatus Edwards...................... 200
Eucessia Coquillett....................... 431
euchaetiae (Sellers), Hyphantrophaga.......... 1094
euchaetiae (Webber), Lespesia................. 1102
Euchaetogyne Townsend..................... 990
Euchaetophleps Townsend..................... 1046
euchenor (Walker), Cylindromyia............... 973
Euchlorops Malloch.......................... 791
Euclesia Girschner........................ 1019
Euclitellaria Kertész..................... 305
Euclytia Townsend......................... 968
Eucnephalia Townsend...................... 1074
Eucordyligaster Townsend.................. 1030
Eucorethra Underwood...................... 104
EUCORETHRINAE................................. 104
Eucoronimyia Townsend..................... 1062
Eucosmoptera Phillips....................... 671
Eucyrtophloeba Townsend................... 1017
Eucyrtopogon Curran....................... 373
Eudactylolabis Alexander.................. 62
eudactylus Loew, Dolichopus................... 491
eudamides Walker, Empis....................... 459
Eudicrana Loew............................ 223
eudicrana Loew, Leptogaster................... 362
Eudicranota Alexander..................... 56
Eudioctria Wilcox & Martin................ 370
eudora (Coquillett), Villa.................... 438
Eudorylas Aczél........................... 552
Eudrapetis Melander......................... 477
eudryae (Smith), Chaetophlepsis............... 1016
eudryae (Townsend), Sisyropa.................. 1094
euedes Howard, Dyar, & Knab, Aedes.......... 112
euethes Reinhard, Plagiomima.................. 1019
Euexorista Townsend....................... 1100
eufitchiae (Townsend), Blondelia.............. 1039
Euforcipomyia Malloch..................... 124
Eufrontina Brooks......................... 1100
Eugaedia Townsend........................... 1079
Eugaediopsis Townsend..................... 1079
Eugauraex Malloch......................... 783
Eugenacephala Johnson....................... 682
eugenia Williston, Volucella.................. 601
eugeniae Felt, Cystodiplosis.................. 289
eugeniae Felt, Dasineura...................... 259
Eugeniamyia Williston....................... 593
Eugnophomyia Alexander.................... 73
Eugnoriste Coquillett..................... 231
eugonatus Loew, Ogcodes....................... 406
Eugymnochaeta Townsend...................... 1005
Eugymnogaster Townsend.................... 970

1596 A CATALOG OF DIPTERA OF NORTH AMERICA

	Page		Page
Euhalidaya Walton	1037	**Euryneura** Schiner	306
Euhallidaya Walton	1037	*Euryneurasoma* Johnson	306
Euhilarella Townsend	938	eurynotus (Brues), Aenigmatias	537
Euhybos Coquillett	448	**Euryomma** Stein	898
Euhybus Coquillett	448	*Euryparia* Ringdahl	852
Euhyperecteina Townsend	1068	euryptera Alexander, Pedicia	54
Eukiefferiella Thienemann	154	*euryptera* (Loew), Paroxyna	665
Eukraiohelea Ingram & Macfie	135	**Euscopolia** Townsend	974
Eulalia Meigen	315	*Eusenotainia* Townsend	938
Eularraevora Townsend	993	**Eusimulium** Roubaud	185
Eulasiona Townsend	1023	**Eusiphona** Coquillett	732
Euleia Walker	677	**Eusisyropa** Townsend	1093
Eulimnophila Alexander	68	**Eustalomyia** Kowarz	868
Eulimnophora Malloch	884	*Eustigoptera* Cresson	757
Euloewia Townsend	983	*Eutachina* Brauer & Bergenstamm	1054
Eulonchus Gerstäcker	404	*Eutanypus* Coquillett	151
eulophus (Loew), Ceratobarys	776	euthamiae Stebbins, (Cecidomyia)	293
Eumacronychia Townsend	935	**Euthera** Loew	978
Eumasicera Townsend	1094	**Eutheresia** Townsend	990
EUMASICERINI	1094	EUTHERINI	978
Eumea Robineau-Desvoidy	1096	**Euthycera** Latreille	690
Eumegaparia Townsend	981	**Euthyneura** Macquart	449
eumenes (Osten Sacken), Villa	435	**Euthyprosopa** Townsend	1069
Eumenos Meigen	604	**Eutonia** Wulp	63
EUMERINI	596	**Eutreta** Loew	661
Eumeros Meigen	604	**Eutrichopoda** Townsend	965
Eumerus Meigen	596	*Eutricogena* Townsend	1033
Eumerus Meigen	604	*Eutritochaeta* Townsend	1104
Eumesembrina Townsend	910	**Eutrixa** Coquillett	981
Eumetopia Macquart	651	**Eutrixopsis** Townsend	978
Eumetopiella Hendel	651	eutrophus Loew, Osprioceras	381
Eumicrodon Curran	599	euxesta Alexander, Limnophila	67
Eumicrophthalma Townsend	982	**Euxesta** Loew	651
Eumicrotipula Alexander	34	*Euxestina* Enderlein	651
Eumusca Townsend	913	*Euzelia* Townsend	1028
Eumyiolepta Shannon	589	**Euzenillia** Townsend	1043
Eumyobia Townsend	1035	*Euzenilliopsis* Townsend	1043
Eumyothyria Townsend	1041	euzona Loew, Ditomyia	198
eumyothyroides (Townsend), Cryptomeigenia	1041	*Evanalia* Strickland	1061
eunicea (Hull & Fluke), Hiatomyia	587	*evanescens* (Enderlein), Melangyna	566
eunota (Loew), Elachiptera	776	*evanescens* (Tucker), Leptocera	724
euochrus Howard, Dyar, & Knab, Aedes	116	evanida (Reinhard), Nowickia	993
Eupachygaster Kertész	318	evanida Reinhard, Phyllophilopsis	1030
Euparyphus Gerstäcker	307	evansi (James), Odontomyia	315
eupatoriflorae Beutenmüller, (Cecidomyia)	293	evarthae (Malloch), Megaselia	538
eupatoriflorae (Felt), Clinorhyncha	275	evasa (Dietz), Nephrotoma	21
eupatorii Felt, Asphondylia	268	evecta Walker, Volucella	601
eupatorii Felt, Brachineura	252	everes Walker, Eristalis	624
eupatorii (Felt), Lestodiplosis	285	evergladea Alexander, Erioptera	82
eupatorii (Felt), Neolasioptera	273	*evermanni* Aldrich, Fucellia	843
eupatorii (Kaltenbach), Liriomyza	801	evidens Alexander, Tipula	36
Eupeitenus Macquart	192	evidens Osten Sacken, Cyrtopogon	368
Eupelecotheca Townsend	1051	**Evidomyia** Reinhard	1062
Eupeodes Osten Sacken	562	evisceratus Melander, Microphorus	453
Euphasiopteryx Townsend	980	evolvens Parent, Dolichopus	491
Euphoranta Townsend	967	exagitans Johannsen, Metriocnemus	161
Euphoria Robineau-Desvoidy	912	examinis Felt, Clinodiplosis	282
Euphormia Townsend	924	excavata Felt, (Cecidomyia)	290
Euphorocera Townsend	1053	excavata Felt, Lasioptera	271
Euphrosyne Meigen	201	excavata (Sturtevant & Wheeler), Lytogaster	752
Euphyto Townsend	935	*excavationis* Felt, (Cecidomyia)	290
Euphytomima James	935	excavatus Kieffer, (Chironomus)	181
Fupodermodera Townsend	995	excelsior (Bergroth), Nephrotoma	21
Eupogona Rondani	1098	excelsius Townes, Polypedilum	176
Eupogonomyia Malloch	902	excentricum (Harris), Temnostoma	614
Euptilopareia Townsend	1018	excessus Stone, Tabanus	334
Euptilostena Alexander	74	excipiens Becker, Medetera	510
Euravinia Townsend	953	excipiens Becker, Siphonella	786
eurekana (Reinhard), Nowickia	995	excisa (Aldrich), Blaesoxipha	944
Eurhimyia Bigot	619	excisa (Becker), Spiniphora	533
eurhinata (Bigot), Villa	†439	excisa Melander, Tachypeza	473
Euribia Meigen	659	*excisa* (Thomson), Atherigona	875
Eurimyia Bigot	619	excisum Davies, Peterson, & Wood, Simulium	186
euronotus Patterson & Ward, Drosophila	764	excisus Aldrich, Chrysotus	522
europaea Egger, Microphthalma	982	excisus Kieffer (Chironomus)	181
Eurosta Loew	662	excisus (Wiedemann), Physoconops	626
Eurostina Curran	662	*excitans* (Walker), Aedes	116
euryadminiculum Davies, Simulium	186	excitans Walker, Chrysops	325
Eurycephala Röder	648	exclusus Walker, Dolichopus	491
Eurycephalomyia Hendel	648	excrucians (Walker), Aedes	112
Eurycera Dziedzicki	222	*ezculpta* Osten Sacken, Triogma	42
Euryceromyia Townsend	1066	**Exechia** Winnertz	206
eurycerus Philip, Tabanus	333	EXECHIINI	204
Eurycnemus Wulp	†154	**Exepacmus** Coquillett	430
Eurygarka Quate	95	exhibitor Melander, Hemerodromia	470

	Page		Page
exigens Dyar & Shannon, Prosimulium	183	**Fabriciella** Bezzi	††993, 994
exigua (Adams), Leptocera	725	fabricii (Holmgren), Hylemya	848
exigua Cresson, Ditrichophora	739	*Fabriciodes* Townsend	993
exigua Loew, Rhamphomyia	463	faceta (Laffoon), Mycetophila	211
exigua (Meade), Hylemya	848	faceta Reinhard, Plagiomima	1019
exigua (Meigen), Campogaster	977	faceta Sherman, Rymosia	207
exigua Say (Sciara)	236	faciale Aldrich, Emblemasoma	949
exigua (Zetterstedt), Phronia	215	facialis (Coquillett), Catagoniopsis	1098
exiguella Spuler, Leptocera	725	*facialis* (Coquillett), Distichona	1074
exiguus Johannsen, Tanytarsus	180	*facialis* (Coquillett), Lauxaniella	700
exiguus Loew, Gymnopternus	499	*facialis* Coquillett, Xylota	606
exiguus Van Duzee, Chrysotus	523	facialis Cresson, Bombylius	409
exile Curran, Rhaphium	513	facialis Curran, Eutreta	661
exilidens Hardy, Tomosvaryella	555	*facialis* Harris, (Conops)	1114
exilipalpis Stone, Tabanus	333	facialis Hendel, Pseudohecamede	737
exilis (Coquillett), Atrichopogon	123	*facialis* Hunter, Mallota	621
exilis Coquillett, Empis	459	facialis Malloch, Hylemya	855
exilis (Coquillett), Epichlorops	791	facialis Melander, Drapetis	477
exilis (Coquillett), Eribella	1042	*facialis* (Reinhard), Atactopsis	1081
exilis (Coquillett), Eutrixa	981	facialis Reinhard, Metavoria	1019
exilis Coquillett, Metacosmus	427	facialis Townsend, Chrysops	325
exilis Cresson, Scatophila	758	*facialis* Williston, Volucella	601
exilis Johnnsen, Cricotopus	157	facirecedens Harmston & Knowlton, Dolichopus	491
exilis Loew, Gymnopternus	499		
exilis (Malloch), Dorylomorpha	557	facula Reinhard, Homalactia	1013
exilis Say (Sciara)	236	fairchildi Stone, Tabanus	333
exilis (Stein), Helina	887	fairfaxensis Wirth, Forcipomyia	124
exilis Van Duzee, Chrysotus	522	falcata Chillcott, Fannia	894
exilistyla (Alexander), Cheilotrichia	79	*falcata* Johannsen, Allodia	205
eximia (Macquart), Ogcodocera	428	falcata Johannsen, Mycetophila	211
eximia Stenhammar, Coelopa	680	falcata Melander, Empis	459
eximia (Wiedemann), Phaenicia	927	falcata (Pettey), Bradysia	232
Exodontha Rondani	301	falcata Van Duzee, Medetera	510
exoloma (Doane), Pedicia	55	*falcatus* Aldrich, Pelastoneurus	501
Exopalpus Macquart	†1001	falcatus Bergroth, Molophilus	88
Exoprosopa Macquart	443	falcifera Alexander, Pedicia	53
EXOPROSOPINAE	433	falcifera Reinhard, Colcondamyia	949
Exoptata Coquillett	443	falciformis (Aldrich), Blaesoxipha	947
Exorista Meigen	1054, †1097	falcipedia Chillcott, Rhamphomyia	463
EXORISTINAE	1036	*falconis* (Harris), Lynchia	918
EXORISTINI	1052	*falculata* Pandellé, Sarcophaga	957
Exoristoides Coquillett	1013	fallax Curran, Melanostoma	575
exotica (Wiedemann), Allografta	569	fallax (Greene), Conophorus	414
exotica Wiedemann (Scatophaga)	842	fallax (Johannsen), Polypedilum	175
expansa Aldrich, Pholeomyia	732	*fallax* (Johnson), Labostigmina	313
expansa Sabrosky, Asteia	824	fallax (Johnson), Limonia	44
expansa Steyskal, Dictya	689	*fallax* Loew, Mycetophila	211
explicata Felt, (Cecidomyia)	290	fallax (Loew), Psila	640
expolita (Coquillett), Bezzia	142	fallax Loew, Tipula	32
expolita (Coquillett), Bradysia	232	*fallax* (Meigen), Exorista	1054
exporrecta Melander, Proclinopyga	467	*fallax* Osten Sacken, Chrysops	326
expulsa Walker, Rhamphomyia	463	fallax Sherman, Tetragoneura	228
exquisita (Malloch), Megagrapha	476	fallax Van Duzee, Dolichopus	491
exquisita (Osten Sacken), Proctacanthella	398	fallei (Back), Comantella	366
erquisitus (Mitchell), Stenochironomus	177	falleni Wheeler, Drosophila	764
exsculpta Loew, Ochthera	753	*falsisetae* (Cresson), Ptilomyia	736
exsculpta Osten Sacken, Triogma	42	falto (Walker), Cyrtopogon	368
exstincta Loew, Mycetophila	211	*familiaris* (Harris), Pollenia	931
exta Cazier, Apiocera	357	**Fannia** Robineau-Desvoidy	892
extatus Roback, Trichocladius	158	FANNIINAE	892
extensa (Felt), Parallelodiplosis	286	*farcus* Williston, Xylomya	299
extensa Huckett, Hylemya	851	*farinalis* Walley, Chironomus	168
extensa (Malloch), Spilogona	881	farinosa (Osten Sacken), Lasioptera	271
extensivena Alexander, Gonomyia	75	farinosus Johnson & Maughan, Oestranthrax	446
extenta Johannsen, Mycetophila	212	farri Shaw, Bradysia	232
extenuata Huckett, Hylemya	848	farri Wirth, Atrichopogon	123
extera (Cresson), Axysta	752	*fasciapennis* Cresson, Stenomyia	654
extrarius Aldrich, Hydrophorus	505	fasciapennis (Say), Epiphragma	61
extrema Becker, Siphonella	786	*fasciata* (Adams), Eutreta	661
extrema (Enderlein), Melangyna	567	*fasciata* (Coquillett), Eumacronychia	935
extrema (Holmgren), Smittia	162	fasciata (Curran), Eulasiona	1023
extremis Day, Odontomyia	315	fasciata Curran, Lixophaga	1043
extremitata Malloch, Hylemya	848	*fasciata* (Curran), Pseudomeriania	1009
extremitis (Coquillett), Villa	439	fasciata (Fabricius), Hoplitimyia	313
extricata (West), Erynniopsis	1042	fasciata (Fallén), Exorista	1054
exuberans Pandelle, Sarcophaga	†959	*fasciata* (Fallén), Phalacrotophora	†542
exul Osten Sacken, Tabanus	336	fasciata (Hardy), Jassidophaga	551
exul (Townsend), Leschenaultia	1083	fasciata (Johannsen), Dziedzickia	218
exul (Walker), Schizotachina	1067	fasciata Johnson & Johnson, Dicranoclista	433
erul Williston, Gonia	1076	fasciata Loew, Ptiolina	347
erusta Johannsen, Mycetophila	212	*fasciata* Loew, Tipula	25
exustus (Walker), Scellus	506	fasciata Macquart, Didea	563
		fasciata Macquart, Exoprosopa	444
Fabricia Latreille	994	fasciata Macquart, Volucella	601
Fabricia Meigen	1048	fasciata Melander, Glabellula	415
Fabricia Robineau-Desvoidy	994	fasciata Say, Heteromyia	137

744–243—65——101

Entry	Page
fasciata Say, Sciophila	225
fasciata (Say), Xylomya	299
fasciata Swenk, Cuterebra	1110
fasciata Walker (Helomyza)	816
fasciata Wiedemann, Acrocera	405
fasciatum (Fabricius), Hoplocheiloma	636
fasciatus (Garrett), Keroplatus	200
fasciatus (Harris), Ogcodes	407
fasciatus Johnson & Johnson, Poecilanthrax	442
fasciatus Loew, Pipunculus	553
fasciatus (Say), Chrysopilus	346
fasciatus Walker, Ceraturgus	365
fasciatus Walker, Helophilus	618
fasciatus Walker, Xylophagus	297
fascicollis Harris (Eristalis)	1114
fasciculata (Loew), Fannia	†894
fasciculata (Schnabl), Hydrophoria	864
fasciger (Curran), Procladius	149
fascigera Alexander, Rhabdomastix	77
fascinator (Laffoon), Mycetophila	211
fasciola Coquillett, Amphicnephes	655
fasciolata Melander, Mythicomyia	417
fasciolata Osten Sacken, Limnophila	65
fasciolatum (De Geer), Chrysotoxum	582
fasciolus Coquillett, Keroplatus	200
fasciolus (Coquillett), Oligodranes	421
fascipennis Coquillett, Chasmatonotus	159
fascipennis Curran, Eurosta	663
fascipennis Harris (Xylophagus)	1114
fascipennis (Macquart), Baccha	573
fascipennis Macquart, Chrysops	325
fascipennis (Say), Epiphragma	61
fascipennis (Say), Exoprosopa	444
fascipennis (Say), Orfelia	202
fascipennis Sturtevant & Wheeler, Glenanthe	737
fascipennis Wiedemann, Baccha	573
fascipennis (Zetterstedt), Ormosia	85
fascipes (Becker), Spaziophora	†836
fascipes (Coquillett), Microtendipes	172
fascipes (Meigen), Rhaphium	513
fasciventris Curran, Cordilura	829
fasciventris (Curran), Promericia	1009
fasciventris (Loew), Dialysis	343
fasciventris Malloch, Chironomus	166
fasciventris Malloch, Gimnomera	840
fasciventris Melander, Platypalpus	479
fasciventris Ringdahl, Hylemya	848
fasciventris Van Duzee, Argyra	525
fasciventris Van Duzee, Sympycnus	527
fastigatus (Townes), Chironomus	167
fastigiorum Schmitz, Gymnophora	541
fastosa Johannsen, Mycetophila	213
fastuosa (Johannsen), Natarsia	146
fatigans (Johannsen), Bradysia	233
fatigans Johannsen, Tanytarsus	179
fatigantis Reinhard, Metopomuscopteryx	1025
fatima (Huckett), Spilogona	881
fattigi Bromley, Asilus	392
fattigi (Bromley), Laphria	389
fattigi Cook, Scatopse	238
fattigi (Dodge), Erythrandra	941
fattigi (Dodge), Sarcophaga	958
fattigiana Alexander, Tipula	36
fattigiana Alexander, Trichocera	16
fatua Johannsen, Mycetophila	211
fauna (Fabricius), Villa	†434
fausta (Osten Sacken), Rhagoletis	674
faustina Alexander, Tipula	26
faustina (Osten Sacken), Villa	434
fautricoides Curran, Mallophora	396
favillacea Loew, Scatella	757
favillaceus Loew, Leptogaster	362
fax Townsend, Volucella	602
felis Bromley, Stenopogon	384
felis (Osten Sacken), Laphria	389
felix Curran, Platycheirus	577
felix (Osten Sacken), Epistrophe	564
felti (Malloch), Liriomyza	801
felti (Pettey), Bradysia	233
felti Pritchard, Anarete	246
Feltiella Rübsaamen	281
femineum (Van Duzee), Rhaphium	513
femineus Curran, Platycheirus	577
femoraatra Alexander, Erioptera	82
femoralis (Loew), Lauxaniella	700
femoralis Loew, Pipiza	579
femoralis Macquart, Asilus	392
femoralis (Meigen), Cerodontha	†801
femoralis Meigen, Scatopse	238
femoralis Reinhard, Sarcofahrtia	942
femoralis (Schiner), Tricharaea	†960
femoralis (Stein), Fannia	894
femoralis (Thomson), Trupanea	667
femoralis Van Duzee, Argyra	525
femorata (Loew), Fannia	896
femorata Loew, Hilara	456
femorata (Loew), Hoplocyrtoma	452
femorata (Macquart), Efferia	394
femorata (Malloch), Bucephalomyia	885
femorata (Meigen), Borophaga	535
femorata (Meigen), Serromyia	136
femorata Melander, Anthalia	450
femorata Say (Sciara)	236
femorata Williston, Crioprora	610
femoratα (Williston), Myiophthiria	917
femoratur (Van Duzee), Rhaphium	513
femoratum Van Duzee, Syntormon	515
femoratus Cresson, Pipunculus	552
femoratus Melander, Bibiodes	195
femoratus (Say), Condylostulus	484, †484
femoratus Van Duzee, Diaphorus	519
femoratus Wiedemann, Bibio	193
femorella (Fallén), Chyromya	822
fenderi Alexander, Dicranota	58
fenderi Alexander, Lipsothrix	77
fenderi Alexander, Molophilus	88
fenderi Alexander, Tipula	36
fenderi Kessel, Agathomyia	548
fenderiana Alexander, Bittacomorphella	98
fenderiana Alexander, Pedicia	55
fenderiana Alexander, Phyllolabis	60
Fenderomyia Shaw	200
fenebris Alexander, Tipula	31
fenestra Felt, Arthrocnodax	283
fenestra (Felt), Trishormomyia	288
fenestra Felt, Tritozyga	243
fenestralis (Cresson), Zeros	746
fenestralis (Fallén), Aecothea	811
fenestralis Fisher, Keroplatus	200
fenestralis (Linnaeus), Scenopinus	356
fenestralis (Scopoli), Sylvicola	190
fenestrata (Bigot), Phasia	969
fenestrata Coquillett, Mycetophila	212
fenestrata (Coquillett), Myiomyrmica	646
fenestrata Foote, Metatephritis	669
fenestrata Hull, Baccha	573
fenestrata (Macquart), Pyrgota	657
fenestrata (Malloch), Lestremia	244
fenestrata (Malloch), Megaselia	540
fenestrata (Say), Tachypeza	473
fenestrata Snow, Eurosta	663
fenestratoides (Coquillett), Villa	437
fenestratus (Kröber), Physoconops	626
fenestratus (Osten Sacken), Conophorus	414
fennica (Karl), Hylemya	848
fennicus Storå, Psectrocladius	159
fenyesi Malloch, Metopina	544
fera Coquillett, Symphoromyia	344
fera (Williston), Stonemyia	321
Ferdinandea Rondani	588
fernaldi Alexander, Ormosia	85
fernaldi Alexander, Trichocera	16
fernaldi (Back), Laphria	389
fernoldi Parker, Boettcheria	948
fernaldi Shaw, Keroplatus	200
fernaldi (Williston), Exorista	1055
Feronia Leach	919
ferox (Humboldt), Psorophora	109
ferox (Walton), Celatoria	1040
ferox Williston, Laphria	389
ferrisi (Bequaert), Neolipoptena	920
ferrugatus (Fabricius), Diachlorus	328
ferruginata (Stenhammar), Leptocera	725
ferruginea (Doane), Aciurina	670
ferruginea (Fabricius), Nephrotoma	21
ferruginea (Fallén), Brachyopa	593
ferruginea Fallén, Tetanocera	694
ferruginea Lovett, Cheilosia	584
ferruginea Macquart, Odontomera	642
ferruginea Melander, Dryomyza	681
ferruginea (Reinhard), Lespesia	1102
ferruginea (Scopoli), Coenomyia	296

INDEX 1599

Entry	Page
erruginea (Townsend), Gymnoclytia	964
ferruginea (Walker), Scoliocentra	814
ferrugineovittatus Zetterstedt, Chironomus	166
ferrugineus Cresson, Oidematops	685
ferrugineus Macquart, Chironomus	165
ferrugineus Palisot de Beauvois, Tabanus	333
ferruginosa (Meigen), Diadocidia	198
ferruginosus (Wiedemann), Culex	119
fervida Curran, Euxesta	651
festina Coquillett, Psilocephala	351
festina (Dietz), Nephrotoma	21
festinans Aldrich & Webber, (Phorocera)	1107
festinans (Meigen), Ceromasia	†1099
festinans Reinhard, Cattasoma	941
festinans Zetterstedt, Dolichopus	495
festiva Hull, Polybiomyia	616
festiva (Loew), Euaresta	665
festiva (Loew), Heteromyia	137
festiva Meigen, Pipiza	†579
festiva Wirth, Dasyhelea	127
festiva (Zetterstedt), Eustalomyia	868
festivus (Holmgren), Chaetocladius	160
festivus Loew, Gaurax	782
festivus Say, Chironomus	171
fibrinflatum Twinn, Simulium	189
fibulans Huckett, Hylemya	852
fibulata Felt, Mycodiplosis	283
fibulata (Wulp), Paroxyna	666
ficana Walker, Rhamphomyia	463
Ficiomyia Felt	263
fidelis (Osten Sacken), Limonia	45
filamentosa Alexander, Tipula	36
filamentosa Townsend, Epidexia	1032
filata (Fabricius), Mydas	358
filia Walker, Sericomyia	603
filialis Reinhard, Cloacina	1072
filicauda Alexander, Gonomyia	75
filicauda Henriksen & Lundbeck, Rhamphomyia	463
filicis (Felt), Arthrocnodax	283
filicis Felt, Bremia	280
filicis Felt, Clinorhyncha	275
filicis Felt, Dasineura	259
filicis (Felt), Lobopteromyia	276
filicis Huckett, Hylemya	845
filicornis (Aldrich), Condylostylus	485
filifer Aldrich, Paracleius	500
filifer Loew, Scellus	506
filiformis Alexander, Gonomyia	75
filiola Loew, Gymnosoma	965
filiola (Loew), Sphecomyiella	658
filipalpis (Thomson), Peleteria	998
filipalpus Allen, Gymnoprosopa	936
filipes Loew, Rymosia	207
filipes (Loew), Sciapus	486
filipes Walker, Tipula	†28, 38
filius (Walker), Rhagio	345
fimbriata (Coquillett), Forcipomyia	125
fimbriata Coquillett, Rhamphomyia	463
fimbriata (Huckett), Spilogona	880
fimbriata (Macquart), Paracantha	662
fimbriata (Schnabl), Spilogona	881
fimbriata (Walker), Pentaneura	147
fimbriata (Waterhouse), Myiophthiria	917
fimbriatus (Coquillett), Lasiops	903
fimbriatus Loew, Gymnopternus	499
finalis (Becker), Conioscinella	783
finalis (Loew), Neotephritis	669
finalis Reinhard, Pseudochaeta	1096
finitima (Snow), Acroglossa	1073
finitima Stein, Pegomya	859
finitima (Walker), Bombyliopsis	992
finitimus (Stone), Agkistrocerus	341
finitus Walker, Dolichopus	491
finlandica Edwards, Mycetophila	211
Finlaya Theobald	115
firmum Steyskal, Dictyacium	690
fisherae (Laffoon), Mycetophila	211
fisherae Shaw, Macrocera	201
fisheri Dyar, Aedes	115
fisheri (Malloch), Megaselia	538
fisheri Shaw, Macrocera	201
fisherii (Walton), Melangyna	566
fissa (Bigot), Villa	440
fissicarina (Tothill), Mericia	1008
fissiforceps (Tothill), Onychogona	1077

Entry	Page
fistulator Melander, Proclinopyga	467
fitchii Felt, Prodiplosis	284
fitchii (Felt & Young), Aedes	112
fitchii Osten Sacken, Promachus	400
fitchii Osten Sacken, Protoplasa	90, †196
fivoris Reinhard, Spathimeigenia	1049
flabellata Van Duzee, Phronia	215
flabellatus (Meigen), Tanytarsus	180
flabellifer Osten Sacken, Polymedon	503
Flabellifera Meigen	19
flabellum Hull, Platycheirus	577
flagellaria Garrett, Hesperinus	191
flagellatus (Harmston), Paracleius	500
flagellitenens Wheeler, Dolichopus	491
flagrans Osten Sacken, Arctophila	604
flammifer Melander, Platypalpus	479
flava (Alexander), Rhabdomastix	77
flava Coquillett, Anthalia	450
flava Coquillett, Zagonia	818
flava Day, Odontomyia	315
flava (Fallén), Megaselia	538
flava Felt, Didactylomyia	254
flava Felt, Porricondyla	255
flava (Garrett), Pilaria	68
flava (Linnaeus), Chyromya	822
flava (Meigen), Clusia	806
flava Melander & Argo, Sobarocephala	806
flava Sabrosky, Eusiphona	732
flava Sabrosky, Stomosis	730
flava (Staeger), Coelosia	218
flava Stein, Phaonia	906
flava (Townsend), Euclytia	968
flava Townsend, Neofischeria	1010
flava Van Duzee, Neurigona	518
flava (Winnertz), Asindulum	199
flavapila Doane, Limnophila	66
flavellus (Zetterstedt), Tanytarsus	180
flavens (Malloch), Smittia	162
flavens Melander, Trixoscelis	818
flaveola Alexander, Limonia	42
flaveola Coquillett, Acnemia	222
flaveola (Coquillett), Eumegaparia	981
flaveola (Coquillett), Minettia	701
flaveola (Coquillett), Oedematocera	1070
flaveola (Coquillett), Ormosia	87
flaveola Coquillett, Phthiria	420
flaveola Coquillett, Trypeta	677
flaveola (Fabricius), Dryomyza	681
flaveola (Fallén), Liriomyza	801
flaveola (Meigen), Scaptomyza	769
flaveola (Osten Sacken), Dicranota	56
flaveola Sabrosky, Acrocera	405
flaveolum (Coquillett), Astiosoma	824
flavescens Dietz, Limonia	46
flavescens Fabricius, Tipula	27
flavescens (Felt), Camptoneuromyia	275
flavescens Felt, Dasineura	259
flavescens Johannsen, Allocotocera	223
flavescens Johnson, Sepsisoma	642
flavescens (Loew), Euthycera	690
flavescens Macquart, Laphria	389
flavescens (Müller), Aedes	112
flavescens (Reinhard), Melanomya	932
flavescens Shannon, Brachyopa	592
flavescens (Tucker), Conioscinella	783
flaviantenna (Hardy & Knowlton), Tomosvaryella	555
flaviantennum (Stains & Knowlton), Prosimulium	183
flavibarbis Adams, Chrysopilus	346
flavibarbis Harris, Laphria	390
flavibasis Alexander, Gonomyia	75
flavibasis Alexander, Tipula	36
flavibasis Curran, Cryptomeigenia	1041
flavibasis Huckett, Coenosia	873
flavibasis Macquart, Bombylius	409
flavibasis Malloch, Cricotopus	157
flavibasis Malloch, Phaonia	906
flavibasis (Malloch), Smittia	162
flavibasis (Stein), Fannia	894
flavicans Fabricius, Tipula	27
flavicans (Meigen), Chaoborus	103
flavicans (Stein), Pegomya	859
flavicauda Coquillett, Thereva	353
flavicauda (Malloch), Phaenopsectra	174
flavicauda (Riley), Triachora	1082

744–243—65——102

	Page		Page
flavicauda Van Duzee, Chrysotus	522	flavifrons (Johannsen), Pentaneura	147
flavicauda Winnertz, Coelosia	218	flavifrons Macquart, Chrysotoxum	582
flavicauda Wulp (Phorocera)	†1106, †1107	flavifrons Melander, Empidideicus	415
flavicaudata Bigot, Hylemya	845	flavifrons Sabrosky, Sigaloessa	824
flaviceps (Bigot), Euphorocera	1054	flavifrons Spuler, Leptocera	722
flaviceps (Coquillett), Belosta	355	flavifrons Stein, Coenosia	871
flaviceps Cresson, Ilythea	746	flavifrons (Walker), Pegomya	859
flaviceps (Johannsen), Palpomyia	140	flavifrons Walker, Xylota	605
flaviceps (Loew), Hippelates	775	flavihirtus Van Duzee, Hydrophorus	505
flaviceps (Loew), Poecilanthrax	442	flavilacertus Van Duzee, Dolichopus	491
flaviceps (Loew), Pseudocalliope	703	flavimana Loew, Rivellia	656
flaviceps Macquart (Conops)	632	flavimana Malloch, Sphegina	594
flaviceps (Macquart), Labostigmina	313	flavimana Meigen, Sepsis	683
flaviceps Sabrosky, Elachiptera	777	flavinervis Frost, Phytomyza	804
flaviceps Stein, Schoenomyza	870	flavinervis (Malloch), Megaselia	538
flaviciliatus Van Duzee, Dolichopus	491	flavinervis Malloch, Microprosopa	835
flaviciliatus Van Duzee, Gymnopternus	499	flavinervis Malloch, Pogonomyia	902
flavicincta Loew, Lispe	878	flavinotum Frick, Phytobia	800
flavicincta Loew, Tnereva	353	*flavipalpis* Adams, Symphoromyia	343
flavicinctus (Loew), Clinotanypus	146	flavipalpis (Aldrich), Blaesoxipha	947
flavicingulus (Walker), Stictochironomus	177	flavipalpis Aldrich, Elodia	1071
flaviocollis Say, Laphria	389	flavipalpis (Haliday), Chamaemyia	707
flavicoma Doane, Tipula	36	*flavipalpis* Macquart, Bombyliomyia	†992
flavicornis Aldrich, Leucopis	709	flavipalpis Macquart (Cynomyia)	931
flavicornis (Brauer), Melinocera	1007	*flavipalpis* Malloch, Megaselia	541
flavicornis Coquillett, Mydaea	891	*flavipalpis* Malloch, Meromyza	793
flavicornis Fallén, Phytomyza	804	flavipalpis (Reinhard), Opsomeigenia	1045
flavicornis Felt, Dasineura	259	*flavipalpis* Smith, Myiopharus	1045
flaticornis James, Nemotelus	310	flavipalpis Stein, Fannia	894
flavicornis Johnson, Nemotelus	310	flavipalpis (Zetterstedt), Pegomya	859
flavicornis Loew, Platypeza	549	flavipalpus Van Duzee, Asyndetus	524
flavicornis Malloch, Ariciella	890	*flavipennis* Bigot, Musca	913
flavicornis Malloch, Plunomia	707	flavipennis Cole, Psilocephala	351
flavicornis (Meigen), Microchrysa	303	*flavipennis* Coquillett, Hylemya	845
flavicornis Melander, Zagonia	818	flavipennis (Coquillett), Melanomya	933
flavicornis (Olivier), Labostigmina	313	flavipennis (Fallén), Hylemya	845
flavicornis (Reinhard), Deopalpus	1004	flavipennis Hull & Fluke, Cheilosia	584
flavicornis Robineau-Desvoidy, Trichopoda	966	flavipennis Morge, Lonchaea	717
flavicornis (Townsend), Senotainia	939	*flavipennis* Stein, Pogonomyia	902
flavicornis Van Duzee, Argyra	525	flavipennis Van Duzee, Hydrophorus	505
flavicornis Van Duzee, Chrysotimus	529	flavipennis Walker (Cordylura)	869
flavicornis (Van Duzee), Hercostomus	498	flavipennis Williston, Melanophrys	1010
flaticornis Wulp, Leptogaster	362	flavipes (Aldrich), Blaesoxipha	947
flavicosta Van Duzee, Medetera	510	flavipes (Aldrich), Condylostylus	484
flavicoxa Johannsen, Zygoneura	231	flavipes Aldrich, Procecidochares	660
flavicoxa Melander, Anthepiscopus	455	*flavipes* Banks, Dioctria	370
flavicoxa Melander, Proclinopyga	467	flavipes Cole, Pherocera	349
flavicoxa Melander & Spuler, Themira	684	flavipes Coquillett, Ablautus	364
flavicoza Stein, Coenosia	871	flavipes (Coquillett), Actia	1061
flavicoxa Van Duzee, Argyra	525	*flavipes* (Coquillett), Cholomyia	1029
flavicoxa Van Duzee, Dolichopus	491	flavipes Coquillett, Gaediopsis	1080
flavicoxa Van Duzee, Nematoproctus	515	flavipes Coquillett, Laphystia	376
flavicoxa Van Duzee, Parasyntormon	516	flavipes Coquillett, Prosenoides	988
flavicoxa (Van Duzee), Rhaphium	513	*flavipes* Coquillett, (Siphonella)	1114
flavicoxalis Cresson, Hydrellia	744	*flavipes* (Coquillett), Taracticus	385
flavida Alexander, Phyllolabis	60	flavipes (Coquillett), Xanthomelanodes	967
flavida Cook, Parascatopse	239	flavipes Cresson, Discocerina	738
flavida Coquillett, Hexamitocera	841	flavipes Cresson, Mythicomyia	417
flavida Coquillett, Psilopa	741	*flavipes* (Fallén), Pegomya	†861
flavida Felt, Aphidoletes	280	flavipes Felt, Lasioptera	271
flavida Felt, Asynapta	254	flavipes Gray, Pterodontia	404
flavida (Felt), Bryomyia	251	*flavipes* (Hardy), Furcifera	350
flavida Melander, Mythicomyia	416	flavipes Johannsen, Cricotopus	157
flavida Melander, Trixoscelis	818	flavipes Loew, Cordilura	828
flavida (Wiedemann), Pachycerina	702	*flavipes* (Loew), Euthycera	690
flavida Williston, Elachiptera	†777	*flavipes* Loew, Hippelates	†775
flavidipalpis Huckett, Coenosia	873	flavipes Loew, Leptogaster	362
flavidipalpis Malloch, Mydaea	891	flavipes (Loew), Xanthogramma	569
flavidipennis Macquart, Psilota	593	flavipes (Macquart), Helius	51
flavidipennis Zetterstedt, Lonchaea	†717	flavipes Meigen, Campylomyza	248
flavidisquama Huckett, Hylemya	853	*flavipes* (Meigen), Leptopeza	451
flavidorsus Hardy, Bromleyus	365	flavipes (Meigen), Palpomyia	140
flavidula (Malloch), Mallochohelea	139	flavipes (Meigen), Phaenopsectra	174
flavidulus Malloch, Gaurax	782	flavipes (Meigen), Phytosciara	230
flavidus Bigot, Chrysopilus	347	flavipes Meigen, Trichina	449
flavidus Coquillett, Parepalpus	1003	*flavipes* Melander & Spuler, Piophila	711
flavidus (Hine), Stenotabanus	329	flavipes Reinhard, Lispideosoma	1063
flavidus Kieffer, Procladius	149	*flavipes* Stein, Coenosia	871
flavidus Wiedemann, Chrysops	325	flavipes (Sturtevant & Wheeler), Lytogaster	752
flavifacies Bigot, Helophilus	619	flavipes Van Duzee, Argyra	525
flavifacies Collin, Neoneura	729	flavipes Van Duzee, Mesorhaga	482
flavifacies (Enderlein), Melangyna	567	*flavipes* Van Duzee, Peloropeodes	517
flavifacies Van Duzee, Dolichopus	491	*flavipes* Walker, Eristalis	623
flavifascies (Coquillett), Clusiodes	807	*flavipes* Walker, Heteromyza	810
flavifrons (Bigot), Oncopsia	637	*flavipes* (Wiedemann), Hybomitra	338
flavifrons (Johannsen), Brillia	154	flavipes Williston, Coenosia	871

INDEX

Entry	Page
flavipes (Williston), Heteromeringia	806
flavipes Williston, Hirmoneura	402
flavipes (Williston), Neoitamus	397
flavipes (Williston), Zeros	746
flavipila Macquart, Laphria	389
flavipilis (Jones), Zabrops	386
flavipilosa Cole, Psilocephala	352
flavipilosa Cole, Thereva	353
flavipilosa Cole, Villa	439
flavipilosus Cole, Bombylius	409
flavipilosus (Coquillett), Anthepiscopus	455
flavirostris Loew, Platypalpus	479
flavirostris Walker, Rhamphomyia	463
flavirostris (Wulp), Carcelia	†1092
flaviseta (Johnson), Sobarocephala	806
flavisetus Malloch, Chrysotus	522
flavisquama (Aldrich), Grisdalemyia	1024
flavisquama Malloch, Hylemya	853
flavisquama (Zetterstedt), Helina	886
flavissimus Foote, Proecidocharoides	661
flavitarsa (Reinhard), Pilatea	1105
flavitarsis Cresson, Typopsilopa	743
flavitarsis Darlington, Tephrochlamys	810
flavitarsis Malloch, Bezzia	142
flavitarsis (Van Duzee), Gymnopternus	499
flavitibia Bigot, Xylota	605
flavitibia Johannsen, Dialyta	904
flavitibia Stein, Fannia	894
flavitibialis Van Duzee, Polymedon	503
flavvaria (Coquillett), Coelomyia	898
flaviventris Cole, Micropeza	635
flaviventris Hackman, Scaptomyza	769
flaviventris (James), Caloparyphus	308
flaviventris (Johannsen), Polypedilum	175
flaviventris (Johnson), Liriomyza	802
flaviventris Melander, Mythicomyia	417
flaviventris (Wulp), Peleteria	997
flavoabdominalis Felt, Dasineura	259
flavoanulata (Felt), Asteromyia	274
flavocalyptrata (Stein), Helina	886
flavocauda Doane, Tipula	36
flavocinctus (Bellardi), Hybomitra	341
flavocostalis Painter, Villa	434
flavofemorata Malloch, Phaonia	908
flavofemoratus (Hine), Neoitamus	397
flavofemoratus (Malloch), Aenigmatias	537
flavoferruginea Osten Sacken, Cladura	72
flavohalterata Malloch, Hydrophoria	864
flavohirta (Coquillett), Mycomya	220
flavohirta Kröber, Thereva	353
flavolinea Felt, Contarinia	277
flavolunata (Felt), Asteromyia	274
flavomaculata (Felt), Asteromyia	274
flavomaculata Felt, Neolasioptera	273
flavomaculata (Hough), Dorylomorpha	557
flavomaculata Malloch, Sphegina	594
flavomarginata Doane, Tipula	36
flavomarginata (Felt), Lestodiplosis	285
flavomontana Patterson, Drosophila	764
flavoniger (Coquillett), Rhagio	345
flavonigra (Coquillett), Liriomyza	801
flavonigra (Coquillett), Probezzia	138
flavonotata (Macquart), Zonosemata	†674, 675
flavopedalis (Felt), Peromyia	251
flavopilosa Kröber, Myopa	630
flavopinicola Wheeler, Drosophila	764
flavoscuta (Felt), Asteromyia	274
flavoscuta Felt, (Cecidomyia)	290
flavoscuta Felt, Dasineura	259
flavoscuta (Felt), Peromyia	251
flavoscuta Felt, Trichopteromyia	250
flavoscutellata Becker, Cephalia	643
flavoscutellata Fallén, Phytomyza	805
flavoscutellata Steyskal, Pterodontia	404
flavoscutellata (Zetterstedt), Blepharipa	†1084
flavoscutellatus Malloch, Orthocladius	156
flavosericea Hull & Fluke, Cheilosia	584
flavosignatus (Hull), Melangyna	566
flavoterminata Jones, Mallota	621
flavotessellata (Walton), Ptilodexia	988
flavotestacea (Zetterstedt), Amoebaleria	815
flavotibialis (Felt), Phaenolauthia	261
flavoumbrosa Alexander, Tipula	36
flavovaria (Coquillett), Bucephalina	832
flavoventris Felt, Neolasioptera	273
flavus (Adams), Euarestoides	669
flavus Coquillett, Phasiops	981
flavus (Curran), Neorhynchocephalus	402
flavus Ewen, Atrichopogon	122
flavus (Jaennicke), Lordotus	412
flavus (Johannsen), Polypedilum	175
flavus (Johannsen), Psectrocladius	159
flavus Kieffer, Coelotanypus	146
flavus Loew, Gymnopternus	499
flavus (Macquart), Chlorotabanus	330
Flebotomus Rondani	91
fletcheri (Aldrich), Blaesoxipha	945
fletcheri Bromley, Saropogon	383
fletcheri (Coquillett), Aedes	112
fletcheri (Malloch), Pseudocoenosia	875
Fletcherimyia Townsend	945
flexa (Curran), Xylota	607
flexa Van Duzee, Neuratelia	224
flexa (Wiedemann), Tritoxa	649
flexuosa Coquillett, Rhamphomyia	463
flinti (Malloch), Periscelis	710
flintiana Alexander, Gonomyia	74
floccosa Felt, Diarthronomyia	262
floccosa (Macquart), Hylemya	847
floralis Coquillett, Phthiria	419
floralis (Fabricius), Mesograpta	571
floralis (Fallén), Hylemya	848
floralis (Fallén), Nemorilla	†1044, 1088, †1089
floralis (McAtee), Swammerdamella	240
florella Shannon, Cheilosia	584
florens Alexander, Gonomyia	75
florens (Johannsen), Anatopynia	145
florescentiae (Linnaeus), Orellia	†671
floricola (Felt), Prodiplosis	284
floricola Sturtevant, Drosophila	767
florida Felt, Asphondylia	268
florida Felt, Dasineura	259
florida (Felt), Lestodiplosis	285
florida (Felt), Parallelodiplosis	286
florida Hull, Volucella	601
floridana (Aldrich), Rhytidops	679
floridana Camras, Physocephala	627
floridana Felt, Kalodiplosis	281
floridana Felt, Lestremia	244
floridana (Johannsen), Forcipomyia	123
floridana Johannsen, Lestodiplosis	285
floridana Johnson, Macrocera	201
floridana Johnson, Rivellia	656
floridana (Macquart), Villa	436
floridana Malloch, Clastopteromyia	771
floridana (Osten Sacken), Limonia	46
floridana (Townsend), Acroglossa	1073
floridana Townsend, Phasiopsis	1091
floridanus Camras, Physoconops	626
floridanus (Dyar & Knab), Culex	120
floridanus Johnson, Chrysops	327
floridanus Szilády, Tabanus	335
floridanus Wheeler, Pelastoneurus	501
floridense (Dyar & Knab), Psorophora	110
floridense Townes, Polypedilum	175
floridensis (Aldrich), Ravinia	954
floridensis (Aldrich), Rhytidops	679
floridensis Aldrich, Tachytrechus	503
floridensis Alexander, Molophilus	88
floridensis Alexander, Tipula	28
floridensis Beck, Culicoides	129
floridensis Bromley, Asilus	392
floridensis Bromley, (Bombomima)	1114
floridensis Bromley, Stenopogon	384
floridensis Cresson, Notiphila	747
floridensis Curran, Pseudogriphoneura	703
floridensis (Felt), Mayetiola	263
floridensis Fisher, Monoclona	224
floridensis Greene, Roeseliopsis	1070
floridensis Hardy, Tomosvaryella	555
floridensis (Hine), Stenotabanus	329
floridensis (Johnson), Beameromyia	361
floridensis (Malloch), Coenosia	874
floridensis Malloch, Eugaurax	783
floridensis Reinhard, Euceromasia	1100
floridensis Reinhard, Nicephorus	990
floridensis Sabrosky, Ogcodes	407
floridensis Sabrosky, Pseudogaurax	782
floridensis (Shannon), Monoceromyia	616
floridensis Snyder, Helina	887
floridensis Steyskal, Dictya	689
floridensis Steyskal, Sepedon	693
floridensis Steyskal, Stratiomys	311
floridensis (Townsend), Amobia	934

1602 A CATALOG OF DIPTERA OF NORTH AMERICA

	Page
floridensis (Townsend), Clausicella	1059
floridensis Townsend, Euphorocera	1054
floridensis Townsend, Juriniopsis	1001
floridensis (Townsend), Stomatomyia	1058
floridensis Townsend, Tachinomyia	1058
floridensis Townsend, Tachinophyto	1049
floridensis Wirth, Leptoconops	122
floridica Haseman, Psychoda	95
floridula Wheeler, Neurigona	518
florilega (Zetterstedt), Hylemya	†848, 848
florum (Fabricius), Xylota	†605
florum (Walker), Nowickia	994
floscula Reinhard, Thelairodoria	1050
fluginatus Hardy, Biblo	193
flukei (Curran), Blera	611
flukei (Curran), Xylota	605
flukei Hardy, Bibio	193
flukei Jones, Culicoides	129
flukei (Jones), Metasyrphus	560
flukei Snyder, Muscina	909
flutatus Harmston & Knowlton, Hercostomus	498
fluviatilis Dyar, Pericoma	92
fochi (Pettey), Bradysia	233
foecunda Johannsen, Mycetophila	211
foeda Loew, Trichonta	216
foedus Loew, Chrysopilus	346
foetedi Felt, Lobopteromyia	276
foliata (Felt), Cordylomyia	249
foliatum Curran, Rhaphium	513
foliora Russell & Hooker (Cecidomyia)	299
folliculi (Felt), Dryomyia	261
fontanella Clark, Cuterebra	1110
fontanum Syme & Davies, Prosimulium	183
fontinalis (Fallén), Leptocera	721
fontinella Clark, Cuterebra	1110
footeana Alexander, Tipula	28
footei Steyskal, Pherbellia	686
footei Wirth, Alluaudomyia	133
footei Wirth & Jones, Culicoides	131
forbesi Malloch, Simulium	186
Forbesomyia Malloch	246
FORBESOMYIINI	246
forceps (Alexander), Dicranota	58
forceps Pettey, Sciara	230
forceps Sabrosky, Leptocera	721
forceps (Townes), Chironomus	167
forcipata Borgmeier, Triphleba	534
forcipata Cook, Psectrosciara	240
forcipata Ferris, Basilia	922
forcipata (Lundström), Arctoconopa	80
forcipata Sabrosky, Meoneura	729
Forcipomyia Meigen	123
FORCIPOMYIINAE	122
forcipula Meijere, Limonia	48
forcipulata (Lundbeck), Bradysia	233
forcipulus (Osten Sacken), Molophilus	88
forficula Benjamin, Paracantha	662
formicarum (Melander), Scatopse	238
formiceps Huckett, Hylemya	852
formosa (Aldrich & Webber), Carcelia	1092
formosa (Coquillett), Urophora	659
formosa (Loew), Echthopoda	372
formosa (Loew), Elachiptera	777
formosa Loew, Hydrellia	744
formosa Loew, Macrocera	201
formosa Melander, Mythicomyia	417
formosa (Scopoli), Chloromyia	303
formosa Wiedemann, Trichopoda	966
formosula Melander, Rhamphomyia	463
formosum Shewell, Prosimulium	183
formosus (Cresson), Oliogodranes	422
formosus Hine, Asilus	392
formosus Van Duzee, Dolichopus	491
fornax Townsend, Volucella	603
fornicata Martin, Leptogaster	362
fortis Aldrich, Dolichopus	491
fortisa Reinhard, Sarcophaga	958
fortunatus (Wheeler), Calyxochaetus	528
fossae Becker, Chlorops	789
fossulatum (Loew), Trigonomma	788
foveolata Meigen, Piophila	713
foxi Cole, Thereva	353
foxi (Van Duzee), Physoconops	627
foxleei Shewell, Stenopterina	657
foxleei Vockeroth, Spilomyia	613

	Page
fracida Reinhard, Mactomyia	1014
fracta Malloch, Hylemya	849
fracta Osten Sacken, Allograpta	569
fractiseta (Stein), Pseudochirosia	844
fractura (Coquillett), Trypeta	677
fragariae Felt, (Cecidomyia)	290
fragariae Malloch, Agromyza	795
fragila (Fluke), Epistrophe	564
fragilina Alexander, Tipula	32
fragilis Back, Stichopogon	385
fragilis Bromley, Diogmites	372
fragilis (Loew), Mycomya	220
fragilis Loew, Tipula	32
fragilis (Walley), Pentaneura	147
fragmentata Dietz, Tipula	40
Francilia Shannon	926
franciscana Bigot, Laphria	390
franciscana Hering, Paroxyna	665
franciscanus McCracken, Anopheles	106
franconiae Felt, Lestremia	244
franconiensis (Malloch), Megaselia	540
frankliniana (Townsend), Blaesoxipha	945
fratellus Wiedemann, Bombylius	409
fratellus Williston, Tabanus	337
frater Cockerell, (Cecidomyia)	291
frater Robinson, Peloropeodes	517
frater (Walker), Helophilus	618
fratercula (Brues), Triphleba	534
fraterculus (Malloch), Homoneura	698
fraterculus Van Duzee, Parasyntormon	516
fraterculus (Van Duzee), Physoconops	627
fraterculus (Wheeler), Thrypticus	511
fraterculus (Wiedemann), Anastrepha	673
fraterna (Comstock), Lespesia	1101
fraterna Garrett, Dixa	101
fraterna (Loew), Homoneura	698
fraterna Loew, Scoliocentra	814
fraterna Loew, Tipula	28
fraterna Macquart, Lucilia	927
fraterna Malloch, Coenosia	871
fraterna Malloch, Phaonia	906
fraterna Say, (Sciara)	236
fraternum Twinn, Simulium	188
fraternus Harris, Dilophus	196
fraternus Kröber, Chrysops	327
fraternus Loew, Bibio	193
fraternus Van Duzee, Paracleius	500
fratria (Loew), Euleia	677
fratria Osten Sacken, Limnophila	66
fraudigera Williston, Lestomyia	378
frauditor Loew, Dolichopus	491
fraudulenta (Loew), Parapenium	580
fraudulenta Williston, Volucella	601
fraudulentus Johnson, Bombylius	409
fraudulosa Loew, Xylota	607
fraxini (Felt), Arthrocnodax	283
fraxini (Felt), Mayetiola	263
fraxinicola Felt, Dasineura	259
fraxinifolia Felt, Lasioptera	271
fraxinifolia Felt, Lestodiplosis	285
fraxinifolia (Felt), Rhizomyia	262
freeborni Aitken, Anopheles	106
freeborni Alexander, Limnophila	63
freeborni Wirth, Leptoconops	122
frenata (Holmgren), Phaonia	908
frenchii (Marten), Hybomitra	339
frenchii (Williston), Lespesia	1102
frequens Johannsen, Chironomus	167
frequens Johannsen, Mycomya	220
frequens Loew, Gymnopternus	499
FRERAEINI	970
fretus (Stone), Hybomitra	339
freyii (Ringdahl), Spilogona	882
freytagi DeFoliart & Peterson, Cnephia	184
frickii Ludlow, Culex	118
friendi Alexander, Tipula	25
frigida (Alexander), Pedicia	55
frigida (Boheman), Exechia	206
frigida Coquillett, Scatophaga	838
frigida Erichson, Lispe	878
frigida (Fabricius), Coelopa	680
frigida (Holmgren), Allomyella	835
frigida Walker, Tipula	41
frigida (Zetterstedt), Pegomya	859
frigidus Osten Sacken, Chrysops	325

	Page		Page
frigidus (Zetterstedt), Orthocladius	156	fuliginosa (Say), Tipula	36
frigipennis (Spuler), Copromyza	720	fuliginosum Robineau-Desvoidy, Gymnosoma	965
fringilla Malloch, Pegomya	859		
fringillina Curtis, Ornithomyia	917	fuliginosus (Felt), Mochlonyx	104
frioensis (Reinhard), Mystacella	1086	fuliginosus Fitch, (Molobrus)	236
frisoni Alexander, Ormosia	85	fuliginosus (Loew), Poecilanthrax	442
frisoni (Dyar & Shannon), Prosimulium	183	fuliginosus Wiedemann, Chrysops	325
frisoni Malloch, Coenosia	871	fultonensis (Alexander), Hexatoma	70
frit (Linnaeus), Oscinella	780	fultonensis Alexander, Molophilus	88
frohnei Sommerman, Prosimulium	183	fultonensis Alexander, Tipula	30
frontalia Melander, Mythicomyia	417	fultonensis Felt, Porricondyla	255
frontalis Aldrich, Johnsonia	950	fultoni (Fisher), Orfelia	202
frontalis Brooks, Gymnocheta	1005	fulva Banks, Mallophora	396
frontalis Brooks, Lypha	1011	fulva Beutenmüller, (Cecidomyia)	293
frontalis Cole, Psilocephala	351	fulva (Bigot), Hebecnema	890
frontalis Coquillett, Empis	459	fulva Coquillett, Zacompsia	654
frontalis Coquillett, Notiphila	747	fulva (Doane), Limonia	46
frontalis Coquillett, Psila	640	fulva Felt, Aphidoletes	280
frontalis Curran, (Diclasiopa)	759	fulva Felt, Camptoneuromyia	275
frontalis Curran, Eutreta	661	fulva Felt, Cordylomyia	248
frontalis (Fallén), Trixoscelis	818	fulva Felt, Dasineura	259
frontalis (Loew), Calyxochaetus	528	fulva Felt, Dicrodiplosis	279
frontalis Loew, Rhamphomyia	463	fulva Felt, Hormomyia	287
frontalis Osten Sacken, Ctenophora	20	*fulva* Gray, Toxophora	425
frontalis Reinhard, Pseudochaeta	1096	fulva Johannsen, Diamesa	151
frontalis Say, Thereva	353	fulva (Johnson), Agathomyia	548
frontalis Schmitz, Phora	537	*fulva* (Kieffer), Odontomesa	153
frontalis (Smith), Procatharosia	971	fulva Steyskal, Antichaeta	688
frontalis (Thomson), Orthellia	913	fulvacrura Snyder, Muscina	909
frontalis (Tothill), Appendicia	1007	fulvalineata Patterson & Wheeler, Drosophila	764
frontalis Townsend, Organomyia	1073	fulvaster Osten Sacken, Chrysops	325
frontalis (Tucker), Coniocinella	784	*fulvescens* (Johannsen), Polypedilum	176
frontalis Van Duzee, Chrysotus	522	*fulvescens* (Walker), Atylotus	330
frontalis Van Duzee, Dolichopus	491	*fulviana* Harris, (Laphria)	1114
frontalis Van Duzee, Medetera	510	fulviana (Say), Villa	434
frontalis (Van Duzee), Thecophora	631	*fulvibarba* (Loew), Pogonota	833
frontalis Van Duzee, Thinophilus	508	*fulvibasis* Macquart, Bombylius	†408, 409
frontalis (Walker), Hybomitra	339	fulvibasoides Painter, Bombylius	408
frontata (Zetterstedt), Hydrophoria	864	fulvicallus Philip, Tabanus	333
frontatum Wheeler, Aphaniosoma	822	fulvicauda Alexander, Prionocera	19
Frontina Meigen	†1101	fulvicauda (Felt), Bradysia	233
Frontiniella Townsend	1100	fulvicauda (Say), Andrenosoma	386
FRONTININI	1072	fulviceps Holmgren, Piophila	711
Frontinogaedia Townsend	1078	fulviceps (Strobl), Borboropsis	809
fronto (Coquillett), Aplomya	1097	fulvicollis Walker, Rachicerus	296
fronto Osten Sacken, Tabanus	333	fulvicolor Alexander, Pedicia	55
fronto Walker, (Drosophila)	772	fulvicoma (Coquillett), Villa	439
fronto (Williston), Physoconops	627	fulvicornis Bigot, Cephalia	643
frontoorbitalis Sabrosky, Oscinella	780	fulvicornis (Coquillett), Phrosinella	938
frontosa Hine, Akronia	310	fulvicornis (Curran), Labostigmina	313
frontosa Say, Gonia	1076	*fulvicornis* James, Nemotelus	310
frontosus Loew, Brachypalpus	608	fulvicoxa Walker, Dilophus	195
frontosus Lovett, Platycheirus	577	fulvicoxalis (Bigot), Sepsis	683
frontulenta Huckett, Hylemya	849	*fulvida* (Bigot), Bombyliopsis	992
frosti Alexander, Limnophila	66	fulvida Coquillett, Phithiria	420
frosti Bromley, Asilus	392	fulvidorsum Van Duzee, Hydrophorus	505
frosti Felt, Asynapta	254	fulvidum Coquillett, Dynatosoma	209
frosti Johnson, Leptocera	721	*fulvifrons* (Bigot), Milesia	614
frosti Pechuman, Hybomitra	339	fulvifrons Illiger, Mydas	358
fructuaria Felt, Lasioptera	271	fulvifrons (Macquart), Chaetopsis	650
frustra Pritchard, Mallophorina	397	fulvifrons Say, Zodion	629
frustrator (Laffoon), Mycetophila	211	fulvifrons (Walker), Stilbometopa	918
fucata (Fabricius), Acinia	†664	fulvilateralis (Macquart), Hybomitra	339
fucata Loew, Thereva	353	fulvilineata Doane, Tipula	28
fucata (Wulp), Deopalpus	†1004	fulvilineis Philip, Tabanus	335
fucatoides Bromley, Thereva	353	fulvinodus Doane, Tipula	36
fucatus Coquillett, Aphoebantus	429	fulvipalpis (Bigot), Lespesia	1102
fucatus Van Duzee, Dolichopus	491	*fulvipennis* (Hine), Cressonomyia	742
Fucellia Robineau-Desvoidy	842	*fulvipennis* (Macquart), Physocephala	628
FUCELLIINAE	842	*fulvipennis* Tucker, Anastoechus	411
Fucellina Schnabl & Dziedzicki	842	fulvipes (Bigot), Leschenaultia	1083
Fucomyia Haliday	680	fulvipes Bigot, Symphoromyia	343
fucorum (Fallén), Fucellia	842	fulvipes Coquillett, Acrosticta	650
fugax Alexander, Ormosia	84	*fulvipes* (Coquillett), Villa	434
fugax (Johannsen), Cricotopus	157	fulvipes Cresson, Hiatus	645
fugax (Meigen), Hylemya	853	fulvipes Loew, Dolichopus	491
fugax Osten Sacken, Chrysops	324	*fulvipes* Macquart, Sarcophaga	†960
fulcifrons Macquart, Tabanus	336	fulvipes Philip, Apatolestes	320
fulgens Wiedemann, Microdon	599	fulvipes Walker, (Sarcophaga)	961
fulgida Melander, Mythicomyia	417	fulvipes Walker, Tabuda	352
fulgidus Parent, Sciapus	486	*fulvipes* Walsh, Mydas	359
fulgoris (Sellers), Chrysoexorista	1099	fulvipilus Rempel, Chironomus	165
fuliginea Patterson & Wheeler, Drosophila	763	fulvipluma (Kieffer), Smittia	162
fuliginosa (Meigen), Forcipomyia	125	fulvisquama (Zetterstedt), Helina	887
fuliginosa Melander, Iteaphila	454	fulvistigma Hine, Chrysops	325
fuliginosa (Osten Sacken), Hexatoma	70	*fulvithorax* Curran, Cordilura	829

1604 A CATALOG OF DIPTERA OF NORTH AMERICA

	Page		Page
fulvithorax Fabricius, Laphria	391	fungicola (Coquillett), Megaselia	538
fulvithorax (Malloch), Bezzia	141	fungicola Felt, Hyperdiplosis	284
fulvithorax Shewell, Prosimulium	183	fungicola Felt, Mycophila	250
fulvitibia Van Duzee, Mycomya	220	fungicola Felt, Winnertzia	253
fulviventris (Bigot), Helina	889	fungiperda Felt, Mycodiplosis	283
fulviventris (Johannsen), Pseudochironomus	176	*Fungivora* Meigen	209
fulviventris Macquart, Mallophora	396	FUNGIVORIDAE	196
fulviventris Macquart, Proctacanthus	399	**Fungobia** Lioy	720
fulviventris Schaeffer, Asilus	392	fungorum (De Geer), Mycetophila	211
fulviventris Wiedemann, (Sciara)	236	*fungorum* (Malloch), Megaselia	538
fulvofrater Walker, Tabanus	333	*fur* (Johnson), Forcipomyia	123
fulvohirta (Wiedemann), Villa	437	fur (Osten Sacken), Anthrax	432
fulvohirtus Van Duzee, Chrysotus	522	fur (Williston), Cophura	367
fulvoides Alexander, Limonia	46	*fur* Williston, Tabanus	334
fulvomedia Alexander, Hexatoma	70	furacis Reinhard, Dolichocodia	986
fulvopedalis Felt, Asphondylia	268	*furax* (Williston), Efferia	393
fulvopilosus Johnson, Tabanus	332	furca Walker, Tipula	28
fulvosetosus Parent, Micromorphus	530	furcata (Coquillett), Notiphila	747
fulvulus Wiedemann, Tabanus	333	furcata Curran, Medetera	510
fulvum (Coquillett), Prosimulium	183	*furcata* Edwards, Diamesa	151
fulvus (Johannsen), Cardiocladius	154	furcata (Fallén), Lonchoptera	531
fulvus Johannsen, Chironomus	167	furcata (Felt), Asynapta	254
fulvus (Wiedemann), Aedes	112	furcata Johannsen, Monoclona	224
fulvus (Wiedemann), Sparnopolius	413	furcata McAtee, Axymyia	196
fuma Fluke & Hull, Cheilosia	584	furcata (Say), Scatophaga	838
fumata (Doane), Ormosia	84	*furcata* (Wulp), Deopalpus	1004
fumatella Lundbeck, (Sciara)	236	furcatum Malloch, Parasimulium	182
fumator Reinhard, Canelomyia	1071	furcatus (Kieffer), Chaetocladius	160
fumator Reinhard, Chromatocera	1012	furcatus (Kincaid), Telmatoscopus	94
fumicosta Malloch, Chamaemyia	707	furcatus (Stein), Lasiops	903
fumicosta (Malloch), Liriomyza	801	furcatus (Van Duzee), Condylostylus	484
fumicosta Painter, Villa	434	furcatus Walker, Chrysops	325
fumida (Coquillett), Dryomyza	681	furcifer Curran, Rhaphium	513
fumida Coquillett, Empis	459	furcifer Loew, Pelastoneurus	501
fumida (Coquillett), Villa	439	furcifer Melander, Bicellaria	451
fumida (Johannsen), Bradysia	233	furcifera Alexander, Erioptera	81
fumidicosta Alexander, Limnophila	66	*furcifera* (Bigot), Toxotrypana	659
fumidus Coquillett, Aphoebantus	429	**Furcifera** Kröber	350
fumidus Johannsen, Chironomus	169	furcillata (Williston), Physocephala	628
fumiferanae (Tothill), Omotoma	1089	*Furcomyia* Meigen	45
fumipenne Gill, Anorostoma	812	furculatum (Shewell), Simulium	186
fumipennis Alexander, Dicranota	57	furens (Poey), Culicoides	129
fumipennis (Bigot), Cylindromyia	973	furensoides Williams, Culicoides	129
fumipennis Brooks, Lypha	1011	furialis Alexander, Tipula	24
fumipennis Coquillett, Loxocera	639	furibunda Alexander, Ormosia	85
fumipennis Huckett, Coenosia	873	furtiva Huckett, Coenosia	871
fumipennis Huckett, Eremomyia	857	furtiva Stein, Mydaea	891
fumipennis Huckett, Pegomya	859	furva Cresson, Lytogaster	752
fumipennis (Loew), Rhaphium	514	*furva* (West), Alophorella	967
fumipennis (Malloch), Gaurax	782	fusca Bromley, Holcocephala	375
fumipennis Melander, Lasiopogon	377	fusca (Cresson), Pherbellia	686
fumipennis Melander, Minettia	701	fusca Felt, Johnsonomyia	252
fumipennis Melander, Trixoscelis	818	*fusca* (Felt), Neocatocha	243
fumipennis Osten Sacken, Ctenophora	20	fusca Garrett, Sciophila	225
fumipennis Painter, Systoechus	410	fusca (Garrett), Thaumalea	120
fumipennis (Sahlberg), Olfersia	919	fusca (Hardy), Scenopinus	356
fumipennis (Say), Chrysopilus	347	fusca Huckett, Leucophora	867
fumipennis Spuler, Leptocera	721	*fusca* (Johannsen), Cardiocladius	154
fumipennis (Thomson), Metasyrphus	560	fusca Kraft & Cook, Eupachygaster	318
fumipennis Van Duzee, Hydrophorus	505	fusca Loew, Dixa	101
fumipennis Walker, Bibio	193	fusca Loew, Spilomyia	613
fumipennis Westwood, Acrocera	405	fusca (Macquart), Heleomyza	816
fumipennis Wiedemann, Tabanus	333	*fusca* (Macquart), Lynchia	918
fumipennis Wilcox & Martin, Cyrtopogon	368	*fusca* Meigen, Bolitophila	198
fumipennis (Zetterstedt), Spilogona	879, †880	fusca Meigen, Limonia	50
fumisquama Snyder, Acridomyia	915	fusca Philip, Brennania	321
fumosa Allen, Phrosinella	938	*fusca* Shewell, Sapromyza	704
fumosa (Coquillett), Alophorella	968	*fusca* (Stein), Phaonia	906
fumosa Cresson, Exoprosopa	444	fusca Townsend, Parachaeta	1083
fumosa Doane, Tipula	28	fusca Van Duzee, Mycetophila	212
fumosa Loew, Rhamphomyia	463, †464	fusca Winnertz, Campylomyza	248
fumosa (Stein), Spilogona	881	fuscana Huckett, Phaonia	906
fumosa Vaillant, Wiedemannia	469	fuscanipennis (Macquart), Chloroprocta	923
fumosalis Cresson, Paracoenia	755	*fuscanipennis* (Macquart), Platycheirus	577
fumosina (Curran), Smittia	162	fuscatus (Fabricius), Sylvicola	190
fumosus (Coquillett), Aphoebantus	429	fuscatus (Hine), Philonicus	398
fumosus Hardy, Nemomydas	359	fuscatus Wirth, Cricotopus	157
fumosus (Johannsen), Limnophyes	160	fuscibasis Aldrich, Goniochaeta	1018
fumosus Van Duzee, Dolichopus	491	fuscibasis (Malloch), Homoneura	698
fundata Alexander, Tipula	32	fuscicauda Huckett, Pegomya	859
funditor Loew, Dolichopus	491	fuscicauda Malloch, Phaonia	906
funebris (Fabricius), Drosophila	764	*fusciceps* (Zetterstedt), Hylemya	850, †850
funebris (Macquart), Baccha	573	fuscicollis Tothill, Gonia	1076
funebris (Meigen), Diplonevra	534	*fuscicornis* Loew, Philygria	745
funeralis Parent, Diaphorus	519	fuscicornis Malloch, Chironomus	170
funestus (Osten Sacken), Eclimus	426	fuscicosta (Curran), Dexia	1022

INDEX

Entry	Page
fuscicosta Van Duzee, Rhaphium	513
fuscicostatus Hine, Tabanus	334
fuscifer Harris, Tipula	28
fuscifrons Malloch, Coenosia	871
Fuscigonia Brooks	1075
fuscimacula (Aldrich & Webber), Patelloa	1104
fuscinervis Stein, Pegomya	859
fuscinervis Stein, Phaonia	906
fuscinervis (Van Duzee), Sciapus	486
fuscipalpis Becker, Oscinella	780
fuscipalpis (Bigot), Hybomitra	340
fuscipalpis Schmitz, Borophaga	535
fuscipalpis Van Duzee, Mycomya	220
fuscipalpis (Zetterstedt), Francilia	926
fuscipenne (Meigen), Polypedilum	175
fuscipennis Brooks, Psalidopteryx	1067
fuscipennis Cole, Psilocephala	351
fuscipennis (Coquillett), Dziedzickia	218
fuscipennis (Haliday), Leptocera	721
fuscipennis Hendel, Orthacheta	831
fuscipennis (Jaennicke), Archytas	1000
fuscipennis Loew, Clinocera	468
fuscipennis Loew, Sepedon	693
fuscipennis (Loew), Tipula	30
fuscipennis Macquart, Bibio	193
fuscipennis (Macquart), Microdon	598
fuscipennis (Macquart), Poecilanthrax	443
fuscipennis (Macquart), Pyrgota	657
fuscipennis Say, Baccha	573
fuscipes (Bigot), Sparnopolius	413
fuscipes Loew, Didea	563
fuscipes (Macquart), Pherbellia	686
fuscipes Malloch, Eremomyioides	857
fuscipes (Meigen), Metriocnemus	161
fuscipes Meigen, Scatopse	238
fuscipes Melander, Gloma	454
fuscipes (Van Duzee), Peloropeodes	517
fuscipes Zetterstedt, Cordilura	829
fuscisquamis Brooks, Psalidopteryx	1067
fuscitarsis Adams, Pipunculus	553
fuscitarsis (Say), Allognosta	300
fuscitibia Stein, Fannia	894
fuscitibia Sublette, Paratendipes	174
fuscitibialis (Harmston & Knowlton), Calyxochaetus	528
fusciventris Malloch, Chironomus	170
fusciventris (Malloch), Lyciella	700
fusciventris Van Duzee, Phronia	215
fusciventris Van Duzee, Trichonta	216
fusciventris (Wiedemann), Ornithoctona	918
fuscoantennata Dietz, Erioptera	81
fuscoanulata Felt (Baldratia)	291
fuscocapitata (Malloch), Trivialia	705
fuscofasciata Malloch, Pegomya	862
fuscohalterata Malloch, Hylemya	854
fuscohalterata Melander, Platypalpus	479
fuscolinea (Garrett), Neoleria	813
fuscomarginata (Huckett), Spilogona	884
fuscomarginata (Malloch), Xenomydaea	892
fusconervosus Macquart, Tabanus	334
fuscopedunculata (Malloch), Megaselia	539
fuscopicea Gill, Schroederella	814
fuscopunctata Macquart, (Coenosia)	877
fuscopunctatus Macquart, Tabanus	334
fuscopyga Alexander, Ormosia	86
fuscostigmosa Alexander, Nephrotoma	22
fuscovaria Osten Sacken, Limnophila	65
fuscovenosa Alexander, Limnophila	66
fuscula (Fallén), Fannia	895
fuscula Wirth, Stilobezzia	134
fusculus (Coquillett), Atrichopogon	122
fuscum Syme & Davies, Prosimulium	183
fuscus Loew, Pipunculus	†552, 552
fuscus Osten Sacken, Aedes	116
fuscus Van Duzee, Diaphorus	519
fusicimacula (Aldrich & Webber), Patelloa	1104
fusicornis (Coquillett), Forcipomyia	123
fusiformis (Doane), Ormosia	85
fusiformis Felt, Rhopalomyia	266
fusinervis (Malloch), Atrichopogon	122
fusistylus (Goetghebuer), Trissocladius	155
fusitarsis Van Duzee, Condylostylus	484
futilis Johannsen, Sciara	230
futilis Malloch, Scatophaga	839
futilis (Osten Sacken), Euexorista	1100
futilis West, Dinera	986
futilis (Wulp), Pentaneura	147
gabrieli Cole & Wilcox, Lasiopogon	377
Gaediophana Brauer & Bergenstamm	1079
Gaediopsis Brauer & Bergenstamm	1079
gagatigaster Steyskal, Labostigmina	313
gagatina Loew, Cordilura	828
gagatinus Loew, Opsebius	405
gaigei Rogers, Gonomyia	74
gaigei Steyskal, Dictya	689
galactodes Loew, Microstylum	379
Galactomyia Townsend	966
galactopoda Alexander, Limnophila	67
galactoptera (Bergroth), Rhabdomastix	77
galapterus (Townes), Chironomus	168
galbana (Johannsen), Megalopelma	223
galbea Melander, Mythicomyia	417
galeata (Aldrich), Oxysarcodexia	952
galeata Malloch, Hydrophoria	864
galeator (Townes), Chironomus	168
galeopsidis Felt, Lasioptera	271
galerucae (Harris), Celatoria	1040
galerucellae Villeneuve, Aplomyiopsis	1038
galii Felt, Dasineura	258
galiorum (Kieffer), Dasineura	258
Ganperdea Aldrich	867
garretti Alexander, Ormosia	84
garretti Alexander, Trichocera	16
garretti Chillcott, Fannia	895
garretti Cooper & Rapp, Dixa	101
garretti (Curran), Blera	611
garretti Curran, Hilara	456
garretti (Curran), Melangyna	566
garretti Felt, Lestremia	244
garretti Gill, Eccoptomera	811
garretti Huckett, Hylemya	849
garretti Snyder, Helina	887
garretti Sublette & Sublette, Diamesa	151
garretti (Walley), Pentaneura	147
garryae Felt, Asphondylia	268
gaspeana Alexander, Erioptera	81
gaspeana Alexander, Shannonomyia	67
gaspensis (Alexander), Hexatoma	70
gaspensis Alexander, Ormosia	85
gaspensis Alexander, Tipula	32
gaspicola (Alexander), Arctoconopa	80
gaspicola Alexander, Limonia	45
GASTEROPHILIDAE	915
GASTEROPHILINAE	915
Gasterophilus Leach	915
Gastrophilus Leach	915
Gastrops Williston	752
gastrostacta (Wiedemann), Baccha	573
Gastrus Leach	915
gaudialis (Cockerell), Diplonevra	534
gaudens Cresson, Psilopa	740
Gaurax Loew	781
gausa Melander, Mythicomyia	417
gavia (Aldrich), Blaesoxipha	944
gaylussacii Felt, (Cecidomyia)	295
gazophylax (Loew), Ligyra	445
gelascens (Walker), Stichopogon	385
geldrius (Walker), Lasiops	904
gelida (Coquillett), Spoggosia	1057
gelida Coquillett, Tipula	32
gelidus Kieffer, Orthocladius	156
gemella (Coquillett), Euaresta	665
gemella (Coquillett), Villa	440
gemina Huckett, Hylemya	851
Geminaria Coquillett	413
geminata Hine, Meryocomyia	322
geminata Johannsen, Macrocera	201
geminata Kieffer, Diamesa	151
geminata (Loew), Jamesomyia	663
geminata Sabrosky, Spilochroa	817
geminata Townsend, Nausigaster	596
geminatus (Reinhard), Deopalpus	1004
geminatus (Say), Cricotopus	157
geminatus (Say), Toxomerus	571
geminatus Wiedemann, Chrysops	326
gemini (Shannon), Cheilosia	587
gemma Brimley, Ommatius	398
gemmae Felt, Dasineura	259
gemmae Felt, Rhabdophaga	257
gemmaria Stebbins, (Cecidomyia)	294

1606 A CATALOG OF DIPTERA OF NORTH AMERICA

	Page		Page
gemmifer (Walker), Condylostylus	485	gibbosa (Doane), Hexatoma	70
genalis (Coquillett), Euhalidaya	1037	gibbosa (Felt), Bryomyia	251
genalis (Coquillett), Heleomyza	816	gibbosus Van Duzee, Diaphorus	520
genalis Coquillett, Metaphyto	1008	*gibbus* Loew, Heteropogon	374
genalis Huckett, Paraprosalpia	866	gibbus Loew, Lordotus	412
genalis Malloch, Paracantha	662	*gibsoni* (Alexander), Limonia	46
genalis Melander, Phytomyza	804	gibsoni Curran, Rhaphium	513
genalis Thomson, Hippelates	775	gibsoni Felt, Dasineura	259
genalis (Thomson), Paroxyna	666	gibsoni (Malloch), Melanagromyza	796
genalis (Townsend), Eulasiona	1023	gibsoni (Malloch), Spilogona	881
genarum Becker, Chlorops	789	gibsoni (Twinn), Prosimulium	183
Genea Rondani	1035	**Gibsonomyia** Curran	1023
geneatis (Melander), Empis	459	*gideon* (Fabricius), Villa	†436
geniculata Bigot, Rhamphomyia	465	gigantea Alexander, Pedicia	55
geniculata (Bouché), Pegomya	859	gigantea Cresson, Cirrula	753
geniculata (De Geer), Siphona	1064	gigantea Frick, Phytobia	800
geniculata (Fallén), Coenosia	†871	*giganteus* De Geer, Tabanus	332
geniculata (Kirby), Iteaphila	454	giganteus Hine, Promachus	400
geniculata (Macquart), Melangyna	566	gigantulus (Loew), Silvius	323
geniculata Townsend, Coronimyia	1060	*gigas* (Garrett), Scoliocentra	815
geniculata (Zetterstedt), Chamaemyia	707	gigas Lovett, Brachyopa	592
geniculati Reuter, Stenodiplosis	278	gigas Root & Hoffman, Culicoides	131
geniculatus (Macquart), Copromyza	719	gigas Sabrosky, Oscinella	780
geniseta (Malloch), Deutominettia	697	gilanus Townsend, Tabanus	334
genitalis Schnabl, Hylemya	846	*gilensis* (Townsend), Sipholeskia	1036
genitivus Melander, Euhybus	448	gillettei Felt, Dicrodiplosis	279
gentilis Hering, Paracantha	662	gillettei (Townsend), Arctophyto	986, †986
gentilis Huckett, Paraprosalpia	866	*gilletti* (Felt), Xylopriona	249
gentneri Fluke, Metasyrphus	560	*gilletti* (Townsend), Arctophyto	986
gentneri Pritchard, Dasineura	259	gilloglyorum Kessel, Callomyia	547
genualis Coquillett, Clinocera	468	*gilsoni* (Schmitz), Triphleba	534
genualis (Johannsen), Orfelia	202	gilva (Hartig), Drino	1085, †1085
genualis (Loew), Forcipomyia	125	gilva (Linnaeus), Laphria	390
genualis Loew, Liancalus	507	gilva (Zetterstedt), Pegomya	859
genualis (Loew), Physocephala	628	**Gilvella** Mesnil	1069
genualis Loew, Phytomyza	805	gilvihirta (Coquillett), Anthalia	450
genualis (Melander), Phytobia	799	gilvipes (Coquillett), Gilvella	1069
genualis (Stein), Fannia	895	gilvipes Coquillett, Metapogon	378
genualis Van Duzee, Dolichopus	491	gilvipes (Coquillett), Nostima	745
genualis (Williston), Melangyna	566	*gilvipes* Coquillett, Platypalpus	481
geometra (Robineau-Desvoidy), Calobatina	635	gilvipes Hine, Asilus	392
geometroides (Cresson), Calobatina	635	gilvipes Loew, Drapetis	477
Geomyza Fallén	821	gilvipes (Loew), Pogonota	833
GEOMYZIDAE	820	gilvipes Loew, Rhamphomyia	463
geomyzinus (Fallen), Clusiodes	807	*gilvipes* Loew, Thereva	353
georgei Alexander, Erioptera	81	gilvipes Van Duzee, Chrysotus	522
georgiae Alexander, Polymera	60	gilvipilosa Coquillett, Rhamphomyia	463
georgiae (Brauer & Bergenstamm), Distichona	1074	gilvus (Coquillett), Atrichopogon	122
georgiae (Brauer & Bergenstamm), Sitophaga	1048	*gilvus* Hardy, Bibio	193
georgiae Melander, Limnia	691	gilvus Philip, Chrysops	324
georgiana Alexander, Tipula	36	**Gimnomera** Rondani	840
georgiana Townsend, Aphria	1010	**Ginglychaeta** Aldrich	1018
georgianus Hardy, Chrysopilus	346	**Ginglymia** Townsend	1060
georgianus King, Anopheles	106	*Ginglymyia* Townsend	1060
georgica (Macquart), Archytas	1000	giraudii (Egger), Megaselia	539
georgicus Macquart, Anthrax	432	*giraulti* (Townsend), Phaenicia	928
georgina Wiedemann, Laphria	390	**Gitona** Meigen	762
georgina Wiedemann, Sarcophaga	958	glabanum (Johannsen), Megalopelma	223
Geosargus Bezzi	301	**Glabellula** Bezzi	415
gera (Roback), Ablabesmyia	148	GLABELLULINAE	415
Geranomyia Haliday	49	glaber Loew, Nemotelus	310
germana Osten Sacken, Dicranoptycha	52	glaber (Walker), Hydrophorus	†504, 505
germana (Walker), Psilocephala	351	glaberrima (Wiedemann), Silba	718
germanum (Johannsen), Megalopelma	223	glabra (Coquillett), Bezzia	141
germanus Wheeler, Dolichopus	496	glabra (Fallén), Camilla	773
GERMARIINI	1012	glabra Fallén, Madiza	731
germinationis (Linnaeus), Opomyza	821	glabra Loew, Cordilura	829
Geron Meigen	422	glabra, Loew, Rhamphomyia	463
geronimo Alexander, Limonia	46	glabra (Meigen), Athyroglossa	735
geronimo Alexander, Tipula	24	glabra (Meigen), Thaumatomyia	787
geropogon Philip, Anacimas	329	glabra (Stein), Pegomya	859
gerronis Alexander, Ormosia	85	glabra West, Ateloglossa	986
gesta Roback, Pentaneura	147	glabrata (Say), Atomosia	386
gesticulor Melander, Platypalpus	479	glabrata (Wiedemann), Tipulogaster	363
getzendaneri Wilcox, Cophura	367	*glabrescens* Edwards, Tanytarsus	179
geyserensis Alexander, Limonia	46	glabricula (Bigot), Melanodexia	931
ghanii Shaw, Mycetophila	211	glabricula (Fallén), Ocydromia	451
giardi (Kieffer), Zeuxidiplosis	278	glabricula (Fallén), Pteromicra	687
Giardomyia Felt	284	glabrifrons Meigen, Scenopinus	355
gibba (Loew), Callachna	660	glacialis (Alexander), Pedicia	55
gibba Melander, Mythicomyia	417	glacialis Edwards, Cricotopus	157
gibber (Williston), Metapogon	378	*glacialis* (Johnson), Melangyna	566
gibbera (Coquillett), Bezzia	141	*glacialis* (Kieffer), Orthocladius	156
gibbera (Meigen), Azelia	899	*glacialis* Loew, Helophilus	617
gibbera Melander, Mythicomyia	417	*glacialis* (Lundbeck), Lycoriella	232
		glacialis Malloch, Puliciphora	545

INDEX 1607

	Page		Page
glacialis Melander, Platypalpus	479	*Goniocnephalia* Townsend	1077
glacialis Rübsaamen, (Sciara)	236	gonioides Coquillett, Opsidia	937
glacialis (Zetterstedt), Alliopsis	857	*Goniomima* Townsend	1082
gladiator Melander, Empis	459	*Goniomyia* Meigen	73
gladiator Mik, Hypocharassus	508	*goniophora* Harris, (Anthomyia)	1114
gladiator (Osten Sacken), Limonia	46	**Goniops** Aldrich	322
gladiator Stone, Tabanus	334	**Goniopsita** Duda	786
gladius Van Duzee, Dolichopus	491	*Gonioscelis* Schiner	383
glandis Felt, Dasineura	259	goniphora Harris, (Anthomyia)	1114
Glaphyroptera Winnertz	227	*goniphora* Harris, Pedicia	53
glaphyropus Loew, Athyroglossa	735	*Gonirhyncus* Rondani	630
glarealis Melander, Cyrtopogon	368	gonoides Townsend, Eucnephalia	1074
glauca (Aldrich), Amoebaleria	815	*Gonomyella* Alexander	73
glauca (Brooks), Linnaemyia	1006	**Gonomyia** Meigen	73
glauca (Coquillett), Chiastochaeta	844	**Gonomyodes** Alexander	79
glauca (Coquillett), Minettia	701	*Gonypes* Latreille	362
glauca (Coquillett), Oedoparena	681	gopheri Johnson, Pegomya	859
glauca Coquillett, Rhamphomyia	463	*gophori* Brodie, (Cuterebra)	1114
glauca Macfie, Stilobezzia	134	gossypii Felt, Contarinia	277
glauca Melander, Apolysis	421	gothicana Alexander, Gonomyia	74
glauca (Stein), Spilogona	881	gothicana Alexander, Tipula	26
glaucellus (Stenhammar), Hecamedoides	738	gouldingi Stone, Simulium	186
glaucescens (Zetterstedt), Fannia	895	**Grabhamia** Theobald	110
glauconotatus Harris, (Asilus)	1114	**Graceus** Goetghebuer	172
Glaucops Szilády	337	graciae Alexander, Tipula	32
glaucum Coquillett, Simulium	187	gracile Curran, Rhaphium	513
glaucurus Wiedemann, Chironomus	166	gracilentus Holmgren, Tanytarsus	179
GLAUROCHARINI	979	gracilicornis (Loew), Nephrotoma	21
gleditchiae (Osten Sacken), Dasineura	259	*gracilipalpis* (Hine), Hybomitra	338
gleditschiae (Osten Sacken), Dasineura	259	gracilipennis Spuler, Leptocera	724
Glenanthe Haliday	737	*gracilipes* (Aldrich), Neurigona	518
glendenningi Alexander, Tipula	28	gracilipes Alexander, Molophilus	88
gleniensis (Huckett), Hylemya	845	gracilipes Loew, Cordilura	830
globifer (Lundström), Limnophyes	160	gracilipes (Loew), Pseudogriphoneura	703
globistylus Roback, Cricotopus	157	gracilipes Malloch, Hylemya	849
globithorax (Osten Sacken), Limonia	43	gracilis Aldrich, Dolichopus	497
globosa (Aldrich), Johnsonia	950	gracilis Aldrich, Sarcophaga	958
globosa Felt, Cincticornia	269	*gracilis* Alexander, Chionea	72
globosa Felt, Lestodiplosis	285	gracilis Bromley, Proctacanthus	399
globosa Felt, Rhadbophaga	257	*gracilis* (Coquillett), Melangyna	566
globosa Pettey, Sciara	230	*gracilis* (Dietz), Limonia	44
globosa (Townsend), Myiophasia	983	*gracilis* (Doane), Limonia	46
globosa (Walker), Neoascia	595	gracilis Fallén, Anthomyza	819
globosus Cresson, Pipunculus	553	*gracilis* Felt, Corinthomyia	249
globosus (Fabricius), Microdon	598	gracilis Fisher, Cordyla	208
globosus Walton, Neochrysops	328	gracilis Hunter, Cheilosia	587
globulicauda Van Duzee, Sympycnus	527	gracilis Johannsen, Boletina	217
globulifera Alexander, Limnophila	67	gracilis Johannsen, Coelosia	218
globulus Osten Sacken, Asphondylia	268	gracilis (Johnson), Rhagio	345
globulus Osten Sacken, (Cecidomyia)	1114	gracilis Loew, Hilara	456
Gloma Meigen	454	gracilis Loew, Rhamphomyia	463
glomerata Walker, Tipula	30	*gracilis* (Macquart), Villa	434, †434
gloriosa (Sabrosky), Ocnaea	403	*gracilis* (McAtee), Ectaetia	241
gloriosus Hull, Meromacrus	622	gracilis Melander, Mythicomyia	417
gloydae (Alexander), Cheilotrichia	79	*gracilis* Packard, Ephydra	754
gluteatus Melander, Aphoebantus	429	gracilis Reinhard, Minthozella	1027
glutinosa Felt, Rhopalomyia	266	gracilis Shewell, Camptoprosopella	697
glutinosa Osten Sacken, (Cecidomyia)	270	gracilis Townsend, Phantasiomyia	1063
Glutops Burgess	298	*gracilis* Walker, Bibio	194
Glyptotendipes Kieffer	171	*gracilis* Walker, Trichocera	16
gmundensis (Egger), Micropsectra	178	gracilis Wiedemann, Asilus	392
Gnamptopsilopus Aldrich	485	*gracilis* Wiedemann, (Limnobia)	90
gnaphalodis Felt, Rhopalomyia	266	gracilis Wiedemann, Tabanus	334
gnaphaloides Osten Sacken, (Cecidomyia)	292	*gracilis* (Wiedemann), Zelia	1028
gnata (Dietz), Nephrotoma	21	gracilis Williston, Dolichomyia	425
gnava Harris, (Milesia)	1114	*gracilis* Williston, Physoconops	626
gnava (Meigen), Carcelia	1092	*gracilis* Williston, Platyura	204
Gnophomyia Osten Sacken	73	gracilis Williston, Xylophagus	297
Gnoriste Meigen	218	gracilistyla Alexander, Prionocera	19
GNORISTINI	216	graenicheri (Van Duzee), Condylostylus	484
Gnus Rubzov	188	grahami (Aldrich), Aldrichina	929
Golcondamyia Reinhard	949	**Grallipeza** Rondani	635
Goliathocera Townsend	1065	*Grallomya* Rondani	636
gomphus Townes, Polypedilum	175	*Grallopoda* Rondani	636
Gonarcticoides Curran	843	*graminea* Felt, Campylomyza	248
Gonarcticus Becker	832	gramineus Coquillett, Chlorops	789
Gonatherus Rondani	830	graminicola (Lundbeck), Orthocladius	156
gonea Garrett, Amoebaleria	815	graminicola (Zetterstedt), Piezura	898
Gonempeda Alexander	79	graminis (Felt), Colpodia	253
Gongromastix Enderlein	245	graminis Felt, Dasineura	259
Gonia Meigen	1074	graminis Fitch, (Cecidomyia)	292
GONIINA	1073	graminivora Alexander, Tipula	32
GONIINAE	1036	graminum (Fallén), Scaptomyza	769, †770
GONIINI	1073	grande (Curran), Rhaphium	513
gonimus Dyar & Knab, Aedes	114	grande Darlington, Anorostoma	812
Goniochaeta Townsend	1018	grande Garrett, Dynatosoma	209

	Page		Page
grandicornis Borgmeier,'Myrmosicarius	544	griseomicans (Wulp), Carcelia	†1092
grandifemoralis Curran, Pipiza	579	griseopunctatum (Malloch), Polypedium	175
grandilobus Brundin, Chaetocladius	160	**Griseosilvius** Philip	323
grandipulvilli Morrison, Gonia	1076	grisescens (Fallén), Dinera	986
grandis Aldrich, Dolichopus	491	grisescens (Hull), Helophilus	619
grandis (Bergroth), Holorusia	18	grisescens (Meigen), Pherbellia	686
grandis Brooks, Psalidopteryx	1067	*grisescens* Robineau-Desvoidy, Lydella 1103, †1103	
grandis (Coquillett), Paraphorantha	969	grisescens (Sabrosky), Conioscinella	784
grandis Garrett, Neuratelia	224	griseum Coquillett, Simulium	187
grandis Greene, Pseudacteon	544	griseus Brooks, Dolichotarsus	1042
grandis (Hardy), Pipunculus	553	griseus Hine, Machimus	396
grandis (Hine), Efferia	394	griseus Malloch, Chironomus	168
grandis (James), Exodontha	301	groenlandensis Goetghebuer, Orthocladius	156
grandis Johnson, Psilocephala	351	groenlandica Andersen, Micropsectra	178
grandis Lovett, Criorhina	611	*groenlandica* (Enderlein), Simulium	187
grandis Osten Sacken, Bibiocephala	99	groenlandica (Holmgren), Bradysia	233
grandis Painter, Geron	423	groenlandica (Lundbeck), Eupogonomyia	902
grandis (Pettey), Bradysia	233	groenlandica Lundbeck, Gnoriste	219
grandis Reinhard, Prosenoides	988	groenlandica (Lundbeck), Megaselia	540
grandis Shannon, Melanodexia	931	groenlandica Malloch, Limnophora	885
grandis Spuler, Leptocera	722	groenlandica (Nielsen), Melangyna	566
grandis Williston, Hadromyia	609	groenlandica Staeger, Boletina	217
grandissima Sabrosky, Oscinella	780	*groenlandica* (Zetterstedt), Protophormia	925
granditarsa (Forster), Pyrophaena	578	*groenlandicum* Enderlein, Simulium	189
granditarsis Curran, Hilara	456	*groenlandicus* Andersen, Limnophyes	160
granditarsis Greene, Tachytrechus	503	groenlandicus Curran, Platycheirus	577
grandivillosa Huckett, Hylemya	849	groenlandicus (Fabricius), Helophilus	618
granulosa (Cresson), Athyroglossa	735	*groenlandicus* Goetghebuer, Orthocladius	156
granulosa Wirth, Bezzia	141	groenlandicus Zetterstedt, Dolichopus	491
graphica Doane, Tipula	28	groenlandiensis Andersen, Limnophyes	160
graphica Osten Sacken, Erioptera	83	grossa Brooks, Leschenaultia	1083
GRAPHOGASTRINI	1067	grossa Cook, Rhexoza	240
Graphomya Robineau-Desvoidy	910	grossa (Fabricius), Laphria	390
Graphomyia Robineau-Desvoidy	910	grossbecki Dyar & Knab, Aedes	112
grassator (Fyles), Lestodiplosis	285	grossbecki (Johnson), Anthrax	432
grassator Wheeler, Aphrosylus	509	grossulariae Felt, Rhopalomyia	266
grata Coquillett, Exoprosopa	444	grossulariae Fitch, (Cecidomyia)	292
grata (Loew), Thaumatomyia	787	grossulariae (Meigen), Epistrophe	564
grata Loew, Tipula	33	grossus Bromley, Diogmites	372
gratiosus Aldrich, Hydrophorus	505	gruiformis Alexander, Limnophila	63
gratus Loew, Dolichopus	491	guasa Alexander, Tipula	24
gratus Loew, Stenopogon	384	guatemalanus Hine, Tabanus	337
gravida (Loew), Lytogaster	752	gubernans Melander, Wiedemannia	469
gravidus Melander, Platypalpus	479	**Guerinia** Robineau-Desvoidy	1055
gravipes Wulp, Leucostoma	976	**Gueriniopsis** Reinhard	1055
gravis Haliday, Coelopa	680	guerlus (Roback), Tanytarsus	179
greenei Foote, Coulson, & Robinson, Tachytrechus	503	guildiana (Williston), Mallophorina	397
greenei Shannon, Chrysogaster	591	gulosa Coquillett, Empis	459
greenei Shannon, Temnostoma	614	*gurges* (Walker), Mesograpta	571
greenei Townsend, Elephantocera	1042	gurneyi Shaw, Leptomorphus	223
greenei Van Duzee, Dolichopus	493	gutierreziae (Cockerell), Rhopalomyia	266
gregaria Frick, Phytomyza	804	gutierreziae Felt, Asteromyia	274
gregarius Kieffer, Tanytarsus	179	guttata (Coquillett), Pherbellia	686
gregsoni Edwards, Diamesa	151	*guttata* Dziedzicki, Mycetophila	210
Gremlinotrophus Reinhard	1069	guttata (Fallén), Melangyna	567
grenfelli Alexander, Tipula	28	guttata Wheeler, Leucophenga	762
greylockensis (Johnson), Caloparyphus	308	guttatulus (Townsend), Stenotabanus	329
griffithi Johnson, Chrysopilus	346	guttatus Hardy & Wheeler, Paracacoxenus	762
grindeliae Aldrich, Procecidochares	660	gutatus Wiedemann, Tabanus	334
grindeliae (Coquillett), Urophora	659	guttifera Walker, Drosophila	764
grindeliae Felt, Asteromyia	274	*guttiferus* (Harris), Hybomitra	339
grindeliae Felt, Rhopalomyia	266	**Guttipelopia** Fittkau	148
Grisdalemyia Curran	1024	guttipennis (Coquillett), Culicoides	129
grisea Coquillett, Cuterebra	1110	guttipennis (Wulp), Pentaneura	148
grisea (Coquillett), Dasyhelea	127	guttipennis (Zetterstedt), Anomalochaeta	821
grisea (Coquillett), Melanomya	933	guttularis (Coquillett), Anatopynia	145
grisea (Curran), Bibiocephala	99	guttulata Hull, Sphaerophoria	570
grisea Curran, Ptiolina	347	*guttulatus* Harris, (Anopheles)	1114
grisea (Ludlow), Aedes	113	guttulus (Wiedemann), Condylostylus	484
grisea Malloch, Lasiopleura	774	guttulus (Wiedemann), Holopogon	375
grisea Malloch, Scatophaga	838	guytona Freeman, Liriomyza	802
grisea Patterson & Wheeler, Drosophila	767	**Gymnocarcelia** Townsend	1093
grisea Robineau-Desvoidy, (Araba)	961	**Gymnocarena** Hering	675
grisea (Walker), Paraprosalpia	867	**Gymnochaeta** Robineau-Desvoidy	1005
grisea Walker, (Sciophila)	229	**Gymnocheta** Robineau-Desvoidy	1005
griseata Curran, Stratiomys	311	**Gymnochiromyia** Hendel	822
griseipleura Alexander, Limnophila	66	*Gymnoclasiopa* Hendel	739
griseocaerulea Malloch, Pseudophaonia	909	**Gymnoclytia** Brauer & Bergenstamm	964
griseogaster Snyder, Helina	887	**Gymnodia** Robineau-Desvoidy	884
griseola (Coquillett), Brevitrichia	355	**Gymnoerycia** Townsend	1085
griseola (Fallén), Hydrellia	744	*Gymnogaster* Townsend	970
griseola (Fallén), Leucopis	709	*Gymnomera* Rondani	840
griseola (Fallén), Pherbellia	686	*Gymnomyza* Strobl	714
griseola (Zetterstedt), Campichoeta	773	*Gymnopa* Fallén	734
griseolus (Wulp), Rhagio	345	GYMNOPAIDINI	182
		Gymnopais Stone	182

INDEX 1609

	Page
Gymnophania Brauer & Bergenstamm	†971
Gymnophora Macquart	541
Gymnophtalma Lioy	1061
GYMNOPINI	734
Gymnoprosopa Townsend	936
Gymnopsoa Townsend	951
Gymnopternus Loew	498
Gymnosoma Meigen	965
GYMNOSOMATINAE	964
GYMNOSOMATINI	964
Gymnosphen Frey	636
Gyneuryparia Enderlein	304
Gynoplistia Westwood	69
Gyroconops Camras	627
haagii Jaennicke, Volucella	601
habilis Johannsen, Sciara	230
habilis Johannsen, Sciophila	225
habilis (Snow), Melangyna	567
habra Melander, Mythicomyia	417
Hadrokolos Martin	373
Hadromyia Williston	609
Hadroneura Lundström	219
Haemagogus Williston	116
haemaphorus (Marten), Hybomitra	339
Haematobia Lepeletier & Serville	914
haematodes (Fabricius), Xylota	607
Haematomyidium Goeldi	128
Haematopota Meigen	330
haematopotides (Bigot), Tabanus	337
HAEMATOPOTINI	330
haematopotus Malloch, Culicoides	131
haemorrhoa (Wulp), Peleteria	997
haemorrhoa (Zetterstedt), Pegomya	859
Haemorrhoestrus Townsend	916
haemorrhoidalis (Fallén), Linnaemya	1006
haemorrhoidalis (Fallén), Sarcophaga	958
haemorrhoidalis (Linnaeus), Gasterophilus	916
haemorrhoidalis (Lundbeck), Lycoriella	232
haemorrhoidalis (Macquart), Psilocephala	351
haemorrhoidalis (Meigen), Microprosopa	835
haeretica (Osten Sacken), Limonia	46, †47
hagani Felt, Protaplonyx	274
hageni Aldrich, (Cecidomyia)	294
Hagenomyia Shewell	188
Haigia Steyskal	644
halcyon (Say), Poecilanthrax	441
halictorum (Melander & Brues), Phalacrotophora	542
Halidayella Hendel	696
Halidayina Duda	725
halisidotae Aldrich & Webber, (Phorocera)	1107
halisidotae Brooks, Leschenaultia	1083
halisidotae (Townsend), Uramya	1029
Halithea Haliday	842
hallahani Alexander, Ormosia	85
halli Schlinger, Eulonchus	404
halli Wilcox, Hodophylax	374
halone (Walker), Ptilodexia	988
halophila Packard, Ephydra	754
halophilus Hanson, Nemotelus	310
halophilus Packard, (Chironomus)	181
halterale (Coquillett), Polypedilum	175
halteralis (Adams), Thaumatomyia	787
halteralis Coquillett, Bibiodes	195
halteralis (Coquillett), Leptometopa	731
halteralis (Cresson), Parydra	750
halteralis (Loew), Bicellaria	451
halteralis Malloch, Johannsenomyia	137
halteralis Van Duzee, Chrysotus	522
halteralis Van Duzee, Medetera	510
halterata (Malloch), Oscinella	780
halterata Melander & Spuler, Piophila	712
halterata (Osten Sacken), Limonia	46
halteratum (Meigen), Rhegmoclema	238
halteratus Melander, Aphoebantus	429
halteratus Sabrosky, Cyamops	820
halterella (Edwards), Limonia	46
hamamelidis (Felt), Aphidoletes	280
hamamelidis (Felt), Camptoneuromyia	275
hamamelidis (Felt), Neolasioptera	273
hamata (Aldrich), Blaesoxipha	944
hamata (Aldrich & Webber), Diplosichus	1053
hamata Alexander, Tipula	39
hamata Borgmeier, Diplonevra	535
hamata (Dietz), Nephrotoma	22

	Page
hamata Felt, Dirhiza	255
hamata Felt, Lasioptera	271
hamata (Felt), Neolasioptera	273
hamata Felt, Porricondyla	255
hamata Garrett, Mycomya	220
hamata Melander, Mythicomyia	417
hamata (Pettey), Bradysia	233
Hamatabanus Philip	341
hamatofila Patterson & Wheeler, Drosophila	764
hamatus Johannsen, Metriocnemus	161
hamatus Loew, Helophilus	619
Hamaxia Walker	979
hamifera Lowe, Spilomyia	613
hamifera Melander, Wiedemannia	469
hamifera Wheeler, Notiphila	747
hamilla Reinhard, Mericia	1008
Hammaspis Duda	785
Hammerschmidtia Schummel	593
Hammobates Stannius	502
Hammomyia Rondani	867
hammondi Brooks, Pararchytas	1002
handfordi Curran, Villa	434
hannai (Alexander), Pedicia	55
hannai (Cole), Heterocheila	679
hannai (Malloch), Mydaea	892
hannai Wilcox & Martin, Backomyia	364
hansoni Alexander, Rhabdomastix	77
hansoni Shaw, Trichonta	216
Hapleginella Duda	778
Haplobasis Osten Sacken	82
Haplomyza Hendel	802
Haplopogon Engel	373
harbecki Van Duzee, Dolichopus	491
Harbeckia Aldrich	952
harbeckii Van Duzee, Asyndetus	524
hardyi Alexander, Erioptera	83
hardyi Alexander, Tipula	27
hardyi Bromley, Neoitamus	397
hardyi Cole, Furcifera	350
hardyi Harmston, Dolichopus	491
hardyi Harmston & Knowlton, Sympycnus	527
hardyi (Stains & Knowlton), Prosimulium	183
harmstoni Alexander, Gonomyia	75
harmstoni Hardy & Knowlton, Pipunculus	553
Harnischia Kieffer	166
harpasa (Walker), Ptilodexia	988
harpax Pandellé, Sarcophaga	958
harperi (Alexander), Austrolimnophila	61
harpestylis Chillcott, Platypalpus	479
harpiger Melander, Platypalpus	479
harpyia Harris, Musca	913
harrimani Coquillett, Ornithodes	56
harringtoni (Coquillett), Homalactia	1013
harrisi (Osten Sacken), Eclimus	426
harrisi Reinhard, Exoristoides	1013
HARRISIINI	1082
harrisinae Coquillett, Sturmia	1088
harrisoni Alexander, Molophilus	88
harrisoni Alexander, Pilaria	68
harrisoniana Alexander, Ormosia	85
harti Cresson, Hydrellia	744
harti Malloch, Aspistes	241
harti Malloch, (Chironomus)	181
harti (Malloch), Homoneura	698
harti Malloch, Phaonia	906
harti Malloch, Tachydromia	474
hartii (Johannsen), Bradysia	233
hartmaniae Felt, (Itonida)	290
Hartomyia Malloch	134
haruspex Melander, Hemerodromia	470
haruspex Osten Sacken, Apiocera	357
harvei (Townsend), Bessa	1052
harveyi (Hine), Efferia	394
harveyi (Hine), Villa	434
harveyi Osburn, Arctophila	604
harveyi (Townsend), Bessa	1052
hastata (Coquillett), Eucoronimyia	1062
hastata (Johannsen), Bradysia	233
hastatus Loew, Dolichopus	491
hastatus Melander, Platypalpus	479
hastatus (Van Duzee), Calyxochaetus	528
hastatus Van Duzee, Chrysotus	522
hastingsae Alexander, Tipula	36
haustellata Reinhard, Plagiomima	1019
haynei Dodge, Wyeomyia	107
hazeltonensis (Dietz), Paradelphomyia	59

1610   A CATALOG OF DIPTERA OF NORTH AMERICA

	Page
Hearlea Rubzov	188
Hearlea Vargas, Martínez, and Díaz	188
hearlei (Philip), Hybomitra	339
hearlei Twinn, Simulium	187
Hebecnema Schnabl	890
hebes Curran, Trupanea	667
hebes (Fallén), Spallanzania	1077
hebes Johannsen, Sciophila	225
hebes Loew, Tipula	32
hebes Loew, Trichosia	229
hebes Melander, Neoplasta	472
hebes (Walker), Pelastoneurus	501
hebicornuta Kraft & Cook, Zabrachia	318
Hecamede Haliday	737
Hecamedoides Hendel	738
hecate Felt, Lasioptera	271
hecate (Hull & Fluke), Hiatomyia	587
hecate Melander & Spuler, Sepsis	683
hecticus (Jaennicke), Mesograpta	571
hedgesi Alexander, Tipula	24
hedleya Huckett, Hylemya	852
Hedria Steyskal	691
Hedriodiscus Enderlein	314
Hedroneura Hendel	690
hegemonica Dyar & Shannon, Dixa	101
Helaeomyia Cresson	741
Helcomyza Curtis	678
HELCOMYZIDAE	678
helderbergensis Alexander, Tipula	32
Helea Meigen	132
HELEIDAE	121
helena Felt, (Choristoneura)	292
helena Felt, Dicrodiplosis	279
helenae (Bromley), Efferia	394
helenae (James), Brevitrichia	355
helenae James, Dolichopus	491
Heleodromia Haliday	467
Heleomyza Fallén	†808, 815
HELEOMYZIDAE	808
HELEOMYZINAE	809
helianthi (Brodie), Trishormomyia	288
helianthi (Felt), Neolasioptera	273
helianthibulla Walsh, (Cecidomyia)	294
helianthiflorae Felt, Asphondylia	268
helianthiglobulus Osten Sacken, Asphondylia	268
helicis (Dyar), Eurygarka	95
helicis (Townsend), Helicobia	949
helicivora (Dodge), Sarcophaga	958
Helicobia Coquillett	949
Helina Robineau-Desvoidy	885
Heliophilus Meigen	604
Helioplagia Townsend	1013
helis Roback, Trichocladius	158
Helius Lepeletier & Serville	50
helluo Osten Sacken, Ocnaea	403
Helobia Lepeletier & Serville	80
Helodon Enderlein	184
Helomyza Fallén	815
HELOMYZIDAE	808
helophila Alexander, Gonomyia	75
Helophila Fabricius	622
Helophilus Meigen	617
Helophilus Meigen	622
helva (Doane), Limonia	47
helvina (Coquillett), Aplomya	1097
helvinus Loew, Xanthochlorus	529
helvocincta Doane, Tipula	33
helvola (Loew), Amoebaleria	815
helvolus (Loew), Stenopogon	384
helvum (Loew), Curtonotum	759
helymus (Walker), Periscepsia	1034, †1034
hemapterus Nitzsch, Carnus	729
hematophila Laboulbène, (Simulia)	1113
hemerocampae (Townsend), Carcelia	1092
Hemerodromia Meigen	470
HEMERODROMIINAE	470
Hemeromyia Coquillett	728
hemiata Garrett, Leia	227
Hemiberis Enderlein	300
Hemicnetha Enderlein	187
Hemineurina Frey	231
hemingfordensis Shewell, Sapromyza	704
Hemipenthes Loew	435
hemiptera (Curran), Lutomyia	811
Hemiscaptomyza Hackman	769
Hemisturmia Townsend	1088
	Page
hemiteleus Dyar, Aedes	116
Hemithrixion Brauer & Bergenstamm	1037
Hemyda Robineau-Desvoidy	974
hendeli Aldrich, Dyscrasis	644
hendeli (Johnson), Stenomyia	654
hendersoni Cockerell, Aedes	115
hendersoni Harmston & Knowlton, Parasyntormon	516
Henicomyia Coquillett	349
hennei Wilcox & Martin, Cophura	367
hennigi Schlinger, Ogcodes	407
Henops Meigen	406
Henria Wyatt	251
henshawi Johnson, Dioctria	370
henshawi Malloch, Eupachygaster	318
henshawi Wheeler, Dolichopus	495
hepatica Alexander, Limnophila	63
heptacantha Alexander, Ormosia	85
Heptagyia Philippi	152
hera (Osten Sacken), Brennania	321
heraea (Walker), Orthellia	913
herbarum (Robineau-Desvoidy), Chamaemyia	707
Hercostomus Loew	497
herennius (Walker), Diogmites	372
Herina Robineau-Desvoidy	645
Heringia Rondani	579
hermannia Alexander, Tipula	25
Hermetia Latreille	304
HERMETIINAE	304
herminius Walker, Asilus	393
hermiona Hull & Fluke, Cheilosia	584
Hermione Meigen	306
Hermoneura Meigen	402
hermsi Sabrosky, Hippelates	775
heros Harris, (Xylophagus)	1114
heros (Schiner), Palpexorista	†1056
heros (Wiedemann), Proctacanthus	399
herrickii (Theobald), Toxorhynchites	107
herscheli Malloch, Rhamphomyia	463
herschelli Malloch, Rhamphomyia	463
Hertwigia Dziedzicki	218
hertzogi Rapp, Pipunculus	552
Heryngia Rondani	580
hespera Banks, Eutreta	661
hespera Steyskal, Tetanocera	694
Hesperempis Melander	455
hesperia Alexander, Gonomyia	74
hesperia Felt, Campylomyza	248
hesperia (Pritchard), Cophura	367
hesperia Sabrosky, Oscinella	780
hesperia (Shannon), Cheilosia	584
hesperia Shannon, Xylota	605
hesperia Shannon & Dobroscky, Protocalliphora	926
hesperia Sturtevant & Wheeler, Scatophila	758
hesperidarum Williston, Acroglossa	1074
HESPERININAE	191
Hesperinus Walker	191
hesperis Martin, Leptogaster	362
hesperis (Sublette), Chironomus	170
Hesperoconopa Alexander	78
Hesperodes Coquillett	200
Hesperodinera Townsend	987
Hesperomyia Brauer & Bergenstamm	1080
Hesperophasia Townsend	977
HESPEROPHASIINI	977
Hesperophasiopsis Townsend	978
Hesperotipula Alexander	40
hesperus (Brauer & Bergenstamm), Lespesia	1102
hessei Hall, Anastoechus	411
hesterna Reinhard, Sarcophaga	958
hetaera (Reinhard), Blaesoxipha	948
heterocerus Dietz, Molophilus	88
Heterocheila Rondani	678
Heterochrysops Kröber	323
heterogamus Melander, Oreogeton	453
Heteromeringia Czerny	806
Heteromydas Hardy	358
Heteromyia Say	136
HETEROMYIINI	136
Heteromyza Fallén	810
heteromyzina (Zetterstedt), Microprosopa	835
heteroneura Brauer & Bergenstamm, Parahypochaeta	1016
Heteroneura Fallén	807

… INDEX 1611

Entry	Page	Entry	Page
heteroneura (Haliday), Leptocera	724	Hineomyia Townsend	1005
heteroneura (Meigen), Taxigramma	940	hinkleyi (Malloch), Conioscinella	784
heteroneurus (Macquart), Paracleius	500	hinmani Khalaf, Culicoides	131
Heteropeza Winnertz	247	hinnulus Wheeler, Parasyntormon	516
Heteropezina Pritchard	247	**Hippelates** Loew	774
HETEROPEZINI	247	HIPPOBOSCIDAE	916
Heteropogon Loew	373	*hippoboscoides* Aldrich, Goniops	322
Heteroptera Macquart	725	hircina (Coquillett), Villa	438
heteroptera Macquart, Mallophora	396	hirculus (Coquillett), Neophyto	952
heteroptera (Say), Penthetria	192	*hircus* (Zetterstedt), Pogonota	833
Heteropterina Macquart	940	**Hirmoneura** Meigen	402
Heteropterna Skuse	200	HIRMONEURINAE	402
heteropterus (Macquart), Paracleius	500	*Hirmophloeba* Rondani	402
heteropus (Coquillett), Diamesa	152	hirsuta Coquillett, Omomyia	714
Heterostoma Rondani	678	hirsuta Doane, Tipula	36
Heterostylum Macquart	411	*hirsuta* (Felt), Corinthomyia	249
heterotricha Bohart, Dalmannia	632	hirsuta Felt, Neolasioptera	273
Heterotrissocladius Spärck	155	hirsuta Ferris, Lynchia	918
HETEROTROPINAE	423	hirsuta Hull & Fluke, Cheilosia	584
Heterotropus Loew	423	hirsuta Loew, Macrocera	201
hewitti Alexander, Tipula	41	hirsuta Melander, Apalocnemis	454
hewitti Hearle, Aedes	113	*hirsuta* (Osten Sacken), Phryxe	1105
Hexamitocera Becker	841	hirsuta Sturtevant & Wheeler, Scatophila	758
Hexatoma Latreille	69	*hirsuta* (Townsend), Gonia	1076
HEXATOMINI	59	*hirsuta* Van Duzee, Dalmannia	632
Hexodonta Rondani	301	hirsuta Wheeler, Scaptomyza	770
hexodontus Dyar, Aedes	113	*hirsuteron* (Theobald), Aedes	114
hians Melander, Aphoebantus	429	hirsuticallus Philip, Chrysops	326
hians Melander, Platypalpus	479	*hirsutifrons* (Curran), Tetropismenus	649
hians (Say), Hydropyrus	754	hirsutissima James, Stratiomys	311
hians (Zetterstedt), Phaonia	908	hirsutitarsis Harmston, Dolichopus	491
hiantha Hull & Fluke, Cheilosia	584	*hirsutula* (Dietz), Nephrotoma	21
hiata Melander, Mythicomyia	417	*hirsutulus* (Zetterstedt), Lasiops	†903
Hiatomyia Shannon	587	hirsutus Becker, Condylostylus	484
Hiatus Cresson	645	hirsutus Coquillett, Aphoebantus	429
hiawatha Shannon, Cheilosia	584	hirsutus Melander, Anthepiscopus	455
hibisci (Felt), Neolasioptera	273	hirsutus Reinhard, Pacidianus	1095
hibisci Felt, Trisopsis	288	hirsutus Townsend, Deopalpus	1004
hickmanae Alexander, Dicranota	58	hirsutus Van Duzee, Diaphorus	520
hicoriae (Felt), Lestodiplosis	285	hirta (Coquillett), Pseudonomoneura	360
hiemalis (De Geer), Trichocera	16	hirta (Curtis), Peleteria	998
hiemalis (James), Scatophaga	838	hirta Felt, Rhizomyia	262
hiemalis (Loew), Orbellia	810	hirta Hough, Cynomya	931
hiemalis Reinhard, Muscopteryx	1025	hirta Johnson, Symphoromyia	343
hiera Quate, Pericoma	92	hirta (Loew), Lasiopleura	774
hieroglyphica (Bigot), Chrysogaster	591	hirta (Malloch), Hylemya	855
hieroglyphica (Olivier), Labostigmina	313	hirta Malloch, Lonchaea	717
hieroglyphicus Malloch, Culicoides	132	hirta (Malloch), Palpomyia	140
hieroglyphicus (Meigen), Sphaerophoria	570	hirta Meigen, Sciophila	225
highlandica Cole, Cophura	367	hirta (Townsend), Metaphyto	1008
Hilara Meigen	456	hirta Van Duzee, Argyra	525
Hilarella Rondani	936	*hirticauda* (Malloch), Spilogona	882
hilarella (Zetterstedt), Actenoptera	714	hirticauda Van Duzee, Allodia	205
hilarella (Zetterstedt), Hilarella	936	hirticauda Van Duzee, Mycomya	220
hilariformis (Stein), Dendrophaonia	905	hirticeps Loew, Thereva	353
Hilarimorpha Schiner	348	hirticeps (Stein), Fannia	895
HILARIMORPHIDAE	348	*hirticollis* (Harris), Pollenia	931
hilaris Osten Sacken, Chrysops	326	hirticollis (Say), Mycomya	220
hilaris Reinhard, Muscopteryx	1025	hirticornis Felt, Rhabdophaga	257
hilaris (Walker), Stenochironomus	177	hirticrus Melander, Empis	459
Hilaromorpha Schiner	348	hirticula Collin, Rhamphomyia	463
hildebrandi Felt, Rhabdophaga	257	*hirtidorsum* Tothill, (Fabriciella)	1114
hillifera (Aldrich), Peckia	953	*hirtifrons* (Johnson), Rhinoleucophenga	763
hiltoni Priddy, Conophorus	414	hirtiloba Alexander, Phyllolabis	60
Himantostoma Loew	975	hirtimanum Van Duzee, Rhaphium	513
Hinea Townsend	1005	*hirtioculatus* (Macquart), Hamatabanus	342
hinei Aldrich, Fucellia	842	hirtipalpis Malloch, Phyllomyza	729
hinei Aldrich, Sarcophaga	958	hirtipennis (Loew), Anatopynia	145
hinei Alexander, Tipula	30	hirtipennis (Malloch), Forcipomyia	124
hinei (Allen), Amobia	934	hirtipennis (Osten Sacken), Molophilus	88
hinei Brennan, Apatolestes	320	hirtipennis Van Duzee, Macrocera	201
hinei Bromley, Proctacanthus	399	*hirtipes* Coquillett, Empis	459
hinei Bromley, Promachus	400	hirtipes Coquillett, Leptogaster	363
hinei Chillcott, Fannia	895	*hirtipes* (Fabricius), Trichopoda	966
hinei Cole & Wilcox, Lasiopogon	377	*hirtipes* (Fries), Prosimulium	183
hinei (Coquillett), Cryptomeigenia	1041	hirtipes Johnson, Orthacheta	832
hinei (Cresson), Cressonomyia	742	hirtipes Loew, Campsicnemus	526
hinei (Curran), Xylota	605	hirtipes Loew, Rhamphomyia	464
hinei Daecke, Chrysops	326	hirtipes Malloch, Neohydrotaea	901
hinei Fluke, Syrphus	559	hirtipes Melander, Tachydromia	474
hinei Garrett, Anorostoma	812	*hirtipes* (Mitchell), Polypedilum	175
hinei (Johnson), Hybomitra	339	hirtipes (Osten Sacken), Rhopalomyia	268
hinei Malloch, Hylemya	849	hirtipes Stein, Chirosia	844
hinei Reinhard, Muscopteryx	1025	hirtipes Van Duzee, Chrysotus	522
hinei Wilcox & Martin, Coleomyia	366	hirtipes Van Duzee, Hydrophorus	505
hinellus Philip, Tabanus	334	*hirtipes* Van Duzee, Zodion	629

	Page		Page
hirtirostris Sturtevant & Wheeler, Scatophila.	758	**Hoplolabis** Osten Sacken	82
hirtithorax (Aldrich), Earomyia	716	*hordoides* (Osten Sacken), Rhabdophaga	258
hirtibibia (Stein), Hylemya	849	*horizontala* Haseman, Psychoda	96
hirtiventris (Malloch), Xenomydaea	†891, 892	hormatha Melander, Mythicomyia	417
hirtocculata James, Odontomyia	315	**Hormomyia** Loew	287
Hirtodrosophila Duda	767	**Hormopeza** Zetterstedt	453
hirtoides Sabrosky, Lasiopleura	774	**Hormosomyia** Felt	253
hirtula (Bigot), Hybomitra	341	*horneae* Foote & Pratt, Culicoides	130
hirtula (Rondani), Leptocera	725	*hornigi* (Cresson), Piophila	713
hirtula Zetterstedt, Rhamphomyia	464	horrida Becker, Chlorops	790
hirtulus (Coquillett), Culicoides	128	horrida Coquillett, Cyrtophleba	1017
hirtulus (Zetterstedt), Lasiops	903	horrida (Dyar & Knab), Psorophora	109
hirtum (Aldrich), Astiosoma	824	horrida (Giglio-Tos), Eucyrtophloeba	1017
hirtus Loew, Bibio	192	horripilans (Melander), Tethina	727
hirtus Loew, Eristalis	624	horripilum Clark, Cuterebra	1110
hirtus (Loew), Rhagio	345	hortensia (Alexander), Dactylolabis	61
hirtus Loew, Tetropismenus	649	*hortensis* (Fitch), Thaumatomyia	787
hirudo Shannon & Dobroscky, Protocalliphora	926	hortulana (Meigen), Aplomya	1097
		horvathi Kertész, Pipunculus	552
hirundo Shannon & Dobroscky, Protocalliphora	926	hospes Melander, Caenotus	424
		hospita Reinhard, Leschenaultia	1083
hispa Alexander, Ormosia	86	houghi Aldrich, Sarcophaga	958
hispida Cazier, Apiocera	357	houghi Kertész, Pipunculus	552
hispida (Curran), Bombyliomyia	992	houghi Malloch, Hydrotaea	900
hispida (Felt), Rhizomyia	262	houghi Malloch, Macrophorbia	856
hispida (Tothill), Nowickia	995	**Houghia** Coquillett	1095
hispida Walker, Lispe	878	houghii (Coquillett), Homoneura	698
hispida (Wulp), Trafoia	1007	houghii (Melander), Chersodromia	475
Histochaeta Rondani	1073	houghii (Stein), Bigotomyia	909
histrio Walker, Trichopoda	966	hovgaardii Holmgren, Rhamphomyia	464
histrio (Wiedemann), Psilonyx	363	howardi Felt, Lasioptera	271
histrio (Zetterstedt), Eustalomyia	868	howardi Malloch, Fannia	895
hitchcocki Shaw, Phronia	215	howardi Coquillett, Psorophora	109
hiulca (Laffoon), Mycetophila	211	howlandi Wilcox, Laphystia	376
Hodophylax James	374	howlandi Wilcox & Martin, Nannocyrtopogon	379
hoeli Frey, Rhamphomyia	464		
hoelsi Frey, Rhamphomyia	464	hsui Hackman, Scaptomyza	769
hoemorrhoidalis (Macquart), Psilocephala	351	huachucae Wheeler, Amiota	761
Hoffmania Fox	128	*huachucanus* (Hardy), Pipunculus	553
hohensis Alexander, Erioptera	81	huachucanus Hardy, Stenopogon	384
Holcocephala Jaennicke	374	huachucensis Alexander, Limonia	45
hollandi Alexander, Tipula	33	hubachecki Cook, Rhegmoclema	238
hollensis (Melander & Brues), Culicoides	131	*hubbardi* Cole, Acrocera	406
holmgreni (Boheman), Pegomya	859	hubbardii Coquillett, Henicomyia	349
holmgreni (Jacobson), Chaetocladius	160	hubbardii (Coquillett), Minettia	701
Holoclera Schiner	†452	hubbelli Alexander, Ormosia	86
Holoconops Kieffer	122	hubbelli Bromley, Asilus	392
Holoneura Kieffer	254	hubbelli Philip, Chrysops	324
Holoneurus Kieffer	254	*hubbelli* Rogers, Erioptera	82
Holopogon Loew	375	*hubelli* Rogers, Erioptera	82
holoptera (Lutz, Neiva, & Lima), Lynchia	918	**Hubneria** Robineau-Desvoidy	1093
holoptica Dietz, Erioptera	81	hucketti Ringdahl, Hylemya	853
holopticus Stone, Gymnopais	182	hucketti West, Ptilodexia	988
Holorusia Loew	18	hudsoni Felt, (Cecidomyia)	290
holosericea Schmitz, Phora	537	*hudsoni* (Felt), Chaoborus	103
holosericeus Walker, Geron	423	hudsoni Felt, Mycodiplosis	283
holosericus Melander, Platypalpus	479	hudsonica Alexander, Dactylolabis	61
holotricha Felt, Mycodiplosis	283	hudsonica Alexander, Rhabdomastix	77
holotricha (Osten Sacken), Caryomyia	270	hudsonica Felt, Giardomyia	284
holotricha (Osten Sacken), Ormosia	86	hudsonica (Osten Sacken), Limonia	44
Holtedahlia Kieffer	179	hudsonica Steyskal, Dictya	689
holti (Malloch), Liriomyza	802	hudsonici (Felt), Thecodiplosis	278
holtii McAtee, Bibio	193	hudsonici Felt, Winnertzia	253
Homalactia Townsend	1013	*Huebneria* Robineau-Desvoidy	1093
Homalocephala Zetterstedt	652	huffae Alexander, Pedicia	55
Homalomyia Bouché	893	huliphilum Garrett, Dynatosoma	209
homichlophila Alexander, Limonia	47	hulli (Cresson), Parydra	750
hominivorax (Coquerel), Cochliomyia	924	hulli Painter, Exoprosopa	445
homoepas Dyar & Ludlow, Culex	119	hulli Steyskal, Stratiomys	311
Homoneura Wulp	697	*humeralis* Harris, (Leptis)	1114
Homotoma Lioy	1089	humeralis Loew, Amiota	761
honestus Osten Sacken, Rachicerus	296	*humeralis* Loew, Gymnopternus	499
hoodiana Alexander, Ormosia	84	humeralis Malloch, Oscinoides	781
hoodiana (Bigot), Cheilosia	584	humeralis Melander, Empidideicus	415
hoodiana Bigot, Odontomyia	316	humeralis Melander, Tachypeza	473
hoodianus (Bigot), Rhagio	345	*humeralis* Osten Sacken, Ogcodes	407
hopkinsi Felt, Asynapta	254	*humeralis* (Robertson), Hyalomya	959
hopkinsi Felt, (Cecidomyia)	290	humeralis (Roser), Phytobia	800
hopkinsii (Coquillett), Boletina	217	humeralis Say, (Limnobia)	90
hoplites Melander, Mythicomyia	417	humeralis Stein, Eremomyia	857, †857
hoplites Spuler, Leptocera	721	*humeralis* Van Duzee, Physocephala	628
Hoplitimyia James	313	*humeralis* Walker, Bibio	194
Hoplocheiloma Cresson	636	humeralis (Williston), Blera	611
Hoplocyrtoma Melander	452	*humeralis* (Zetterstedt), Mydaea	†891
Hoplodictya Cresson	691	humicola Ewen, Atrichopogon	122
Hoplogaster Rondani	873	humicola Lundbeck, (Sciara)	236

INDEX

	Page		Page
humida Garrett, Mycomya	220	HYDRELLIINI	743
humidicola (Osten Sacken), Limonia	†43, 47	*Hydrina* Robineau-Desvoidy	†745, 751
humilis (Coquillett), Atissa	735	*Hydrobaenus* Fries	†155
humilis Coquillett, Empis	460	**Hydrochasma** Hendel	738
humilis (Felt), Holoneurus	255	**Hydrodromia** Macquart	468
humilis Loew, Chrysopilus	346	hydroleonoides Johnson, Odontomyia	316
humilis (Loew), Dioxyna	666	**Hydromyza** Fallén	840
humilis Loew, Gymnopternus	499	hydrophilus Aldrich, Liancalus	507
humilis (Loew), Herina	645	**Hydrophoria** Robineau-Desvoidy	863
humilis (Loew), Pherbellia	686	HYDROPHORINAE	504
humilis Meigen, Coenosia	870	**Hydrophorus** Fallén	504
humilis Osten Sacken, Phthiria	420	**Hydropyrus** Cresson	754
humilis Parent, Chrysotus	522	**Hydrotaea** Robineau-Desvoidy	899
humilis (Stein), Helina	887	*hyemalis* (Fitch), Anopheles	106
humilis Van Duzee, Dolichopus	491	*Hyetodesia* Rondani	886
humilis Williston, Parydra	750	*Hygroceleuthus* Loew	487
humilis (Zetterstedt), Gymnodia	885	hygropetrica Alexander, Erioptera	83
Humisturmia Townsend	1088	hygropetricus (Kieffer), Metriocnemus	161
humulicaulis Felt, Lasioptera	271	hylecoites (Pritchard), Henria	252
hungerfordi Brennan, Chrysops	327	**Hylemya** Robineau-Desvoidy	845
hungerfordi Sabrosky, Acrocera	†405, 406	*Hylemyia* Robineau-Desvoidy	845
hunteri Coquillett, Hermetia	304	*hylemyioides* Malloch, Helina	886
hunteri (Curran), Blera	610	**Hylemyza** Schnabl & Dziedzicki	845
hunteri Curran, Cheilosia	584	*Hylephila* Rondani	867
hunteri (Curran), Epistrophe	564	hylotomae (Coquillett), Spathimeigenia	1049
hunteri (Hough), Blaesoxipha	945	*Hylotomomyia* Townsend	1049
hunteri (Kessel), Platypeza	549	hynesi Alexander, Ormosia	86
hunteri Malloch, Simulium	189	hynesiana Alexander, Pedicia	55
huntsmaniana Dietz, Tipula	33	hyoscyami (Panzer), Pegomya	859, †862
hurdi Hall, Lordotus	412	*Hyparctius* Martini	111
hurdi Reinhard, Aphantorhapha	1062	*Hypaspistomyia* Hendel	731
hurdi Wirth, Forcipomyia	125	*hypenae* (Coquillett), Eusisyropa	1093
huron Alexander, Molophilus	89	*Hypenomyia* Townsend	985
huron Alexander, Tipula	33	*Hyperalonia* Rondani	†445
huron (Bromley), Laphria	390	hyperborea (Boheman), Spilogona	881
huronensis Steyskal, Tetanocera	694	hyperborea (Greene), Piophila	711
huronis Alexander, Ormosia	85	hyperborea (Osten Sacken), Nasiternella	56
hutchingsi Curran, Microdon	598	hyperboreus Staeger, Chironomus	†165, 165
hyacintha (Hull & Fluke), Hiatomyia	587	hyperboreus (Staeger), Platycheirus	†577, 577
Hyadina Haliday	751	**Hyperdiplosis** Felt	284
HYADININI	751	**Hyperecteina** Schiner	1069
hyaenoides Melander, Platypalpus	479	HYPERECTEININI	1069
Hyalanthrax Osten Sacken	433	HYPEROSCELIDIDAE	237
hyalina (Coquillett), Leia	227	**Hypertrophocera** Townsend	1072
hyalina (Garrett), Scoliocentra	814	**Hypertrophomma** Townsend	1095
hyalinata (Meigen), Sapromyza	704	hyphantriae (Tothill), Blondelia	1039
hyalinipennis Cole, Lepidanthrax	440	*hyphantriae* (Townsend), Blondelia	1039
hyalinipennis Hine, Tabanus	333	hyphantriae (Townsend), Hyphantrophaga	1094
hyalinipennis Painter & Hall, Poecilanthrax	442	**Hyphantrophaga** Townsend	1094
hyalinipennis Zetterstedt, Lonchaea	717	*hypocera* Becker, Cetema	788
hyalinus Cole, Opsebius	405	**Hypocera** Lioy	534
hyalinus Coquillett, Chasmatonotus	159	**Hypocerina** Malloch	536
hyalinus Coquillett, Leptomorphus	223	**Hypocharassus** Mik	508
hyalinus Coquillett, Saropogon	383	**Hypoderma** Latreille	1112
hyalinus Coquillett, Synechus	448	HYPODERMATIDAE	1111
hyalinus Shannon, Chrysops	328	HYPODERMATINAE	1112
hyalipennis Cole, Exoprosopa	445	**Hypodermodes** Townsend	910
hyalipennis Shaw, Macrocera	201	*hypogaea* Felt, Diarthronomyia	262
hyalipennis Zetterstedt, Lonchaea	717	*hypoleuca* Loew, Hydrellia	744
Hyalistata Wheeler	762	hypomelas James, Labostigmina	314
hyalomoides Townsend, Clistomorpha	977	hypomelas (Loew), Saropogon	383
Hyalomya Robineau-Desvoidy	968	hypomelas (Macquart), Villa	434
Hyalomyia Robineau-Desvoidy	968	*Hypopelta* Aldrich	951
Hyalomyiopsis Brooks	968	hypopygialis Ringdahl, Fucellia	842
Hyalomyodes Townsend	977	hypopygialis Schaeffer, Asilus	392
hyaloptera Alexander, Trichocera	16	*hypopygialis* Stein, Coenosia	872
HYALURGINI	1026	hypopygialis (Townsend), Microcerella	951
Hyalurgus Brauer & Bergenstamm	1026	*Hypovoria* Villeneuve	1017
Hybomitra Enderlein	337	*Hyria* Robineau-Desvoidy	1026
Hybos Meigen	447	**Hyrmophlaeba** Rondani	402
hybos Melander, Mythicomyia	417	*Hyrmophloeba* Rondani	402
HYBOTIDAE	446	*Hystricia* Macquart	†992
HYBOTINAE	447	*Hyetricocnema* Townsend	945
hybreas Walker (Tachina)	1108	*hystricoides* (Williston), Archytas	1000
hybrida (Dietz), Prionocera	19	hystricosa (Williston), Protodejeania	1002
hybrida (Meigen), Bolitophila	198	hystrix Brues, Xanionotum	545
hybrida (Meigen), Erioptera	81, †81	*hystrix* (Fabricius), Archytas	†1000
hybrida (Schnabl), Phaonia	906		
hybridus Loew, Helophilus	618	ibis (Alexander), Limonia	49
hybridus Melander, Paracleius	500	*Icaria* Schiner	661
hybus Coquillett, Geron	423	*icasta* Alexander, Gonomyia	74
hydationis Dyar & Shannon, Simulium	189	ICELIINI	1027
hydei Sturtevant, Drosophila	764	iceryae (Williston), Cryptochetum	825
hydrangeae Felt, Asphondylia	268	ichneumonea Say, Mycetophila	212
Hydrellia Robineau-Desvoidy	743	**Ichneumonops** Townsend	974
		Iconomedina Townsend	1068

1614 A CATALOG OF DIPTERA OF NORTH AMERICA

	Page		Page
icterica (Holmgren), Pegomya	860	*imitans* (Walker), Atylotus	330
Icterica Loew	664	imitans Walker, Tabanus	334
Ictericophyto Townsend	1024	imitata Alexander, Dactylolabis	62
Icteromyza Hendel	799	imitator (Aldrich & Webber), Palpexorista	1056
idahensis Stein, Chirosia	844	*imitator* (Curran), Erynnia	1071
idahoa Shannon, Cheilosia	587	imitator Johannsen, Boletina	217
idahoana Steyskal, Pseudotephritis	647	*imitator* Johannsen, Mycetophila	213
idahoense Twinn, Simulium	186	imitator Johannsen, Rymosia	207
idahoensis (Aldrich), Dolichopus	492	imitator Reinhard, Winthemia	1090
idahoensis Alexander, Tipula	33	imitator Sellers, Aplomya	1098
idahoensis Fisher, Monoclona	224	imitatrix (Malloch), Homoneura	698
idahoensis (Hall), Melanodexia	931	imitatrix Malloch, Phaonia	907
idahoensis Harmston & Knowlton, Medetera	510	imitatrix (Malloch), Spilogona	881
idahoensis James, Odontomyia	315	imitatrix Shewell, Camptoprosopella	697
idahoensis Steyskal, Pherbellia	686	imitatus Sturtevant, Cyamops	820
idahoensis (Theobald), Aedes	114	**Imitomyia** Townsend	975
idahoensis Van Duzee, Chrysotus	522	**IMITOMYIINI**	975
idahoensis Wilcox & Martin, Cyrtopogon	368	immaculata Alexander, Erioptera	82
idahona Harmston & Knowlton, Argyra	525	immaculata Alexander, Teucholabis	76
Idana Loew	645	immaculata Cole, Dziedzickia	218
idei Alexander, Tipula	25	immaculata (Coquillett), Phytoliriomyza	803
idei (Walley), Ablabesmyia	148	immaculata Johnson, Macrocera	201
idessa (Walker), Graphomya	911	immaculata (Macquart), Gymnoclytia	964
Idia Wiedemann	923	*immaculata* Malloch, Fannia	894
Idiocera Dale	74	immaculatus Johnson, Nemotelus	309
Idiognophomyia Alexander	73	immanis Alexander, Limonia	47
Idiolimnophila Alexander	64	immarginatus (Zetterstedt), Platycheirus	577
Idioplasta Osten Sacken	90	immatura (Osten Sacken), Limonia	44
Idioptera Macquart	65	*immemor* (Osten Sacken), Limonia	47
idonea Aldrich, Sarcophaga	958	immensa (Spuler), Copromyza	720
Idoneamima Dodge	957	immigrans Sturtevant, Drosophila	764
idoneus Van Duzee, Dolichopus	492	immodesta (Osten Sacken), Limonia	47
idyla (Walker), Hydrotaea	900	immodestoides Alexander, Limonia	47
ignipes (Reinhard), Blaesoxipha	947	immunda (Zetterstedt), Pogonota	833
ignobilis Loew, Tipula	33	immutica Collin, Fannia	895
ignobilis Loew, Zygomyia	216	impar (Aldrich), Blaesoxipha	946
ignobilis Walker, (Limnobia)	90	impar Davies, Peterson, & Wood, Simulium	186
ignobilis (Zetterstedt), Hydrophoria	864	impar Johannsen, Sciophila	225
ignota Alexander, Tipula	34	*Imparia* Roback	945
Ignotomyia Reinhard	1091	imparilis Hardy, Bibio	193
ilerda (Walker), Eucalliphora	929	impatiens (Curran), Lixophaga	1043
ilicicola Foote, Asphondylia	268	impatiens (Johannsen), Bradysia	233
ilicicola Loew, Phytomyza	804	impatiens Johannsen, Neoempheria	222
ilicis Curtis, Phytomyza	804	impatiens (Walker), Culiseta	117
ilicis Loew, Phytomyza	804	impatientifolia Felt, Lasioptera	271
ilicoides Felt, Asphondylia	268	impatientis Felt, Mycodiplosis	283
Ilione Haliday	690	impatientis (Osten Sacken), Schizomyia	267
Ilione Hendel	690	**Impeccantia** Reinhard	1013
Ilisia Rondani	83	impedita Loew, Rhamphomyia	464
illini Alexander, Elliptera	51	impedita Stein, Mydaea	891
illinoense (Malloch), Polypedilum	175	impellans (Johannsen), Mycetophila	212
illinoensis (Malloch), Ablabesmyia	148	impensus (Walker), Metriocnemus	161
illinoensis (Malloch), Cerodontha	801	imperator Walker, (Cordylura)	877
illinoensis Robertson, Epigrimyia	974	*imperator* Walley, Chironomus	166
illinoensis Robertson, Mallota	621	*imperfecta* Alexander, Tipula	35
illinoensis Robertson, Winthemia	1090	imperfecta (Coquillett), Trupanea	667
illinoiensis (Alexander), Pilaria	68	imperfecta (Loew), Microsania	547
illinoiensis Alexander, Tipula	33	imperfectus Van Duzee, Dolichopus	492
illinoiensis (Felt), Xylopriona	249	imperialis (Curran), Melangyna	567
illinoiensis (Townsend), Cryptomeigenia	1041	impersonata Huckett, Hylemya	853
illinoiensis Townsend, Siphona	1064	impetuum Curran, Rhaphium	513
illinoisensis Malloch, Palpomyia	140	*impexa* Loew, Hydrotaea	900
illinoisensis Townsend, Cryptomeigenia	1041	impexus Melander, Platypalpus	479
illocale Reinhard, Mochlosoma	987	*impiger* (Coquillett), Villa	437
illota (Osten Sacken), Hybomitra	339	impiger (Walker), Aedes	†112, 113
illotus (Williston), Leptocera	725	*implacabilis* (Walker), Aedes	114
illucens (Linnaeus), Hermetia	304	implicata Huckett, Hydrophoria	864
illudens (Laffoon), Mycetophila	212	implicatus Vockeroth, Aedes	113
illustris Alexander, Limonia	47	impolita Huckett, Eremomyia	857
illustris Doane, Tipula	30	importuna (Walker), Metoposarcophaga	951
illustris Johannsen, Neoempheria	222	impressa (Bigot), Stilbometopa	918
illustris (Meigen), Lucilia	927	impressus Becker, Hippelates	775
illustris Melander, Mythicomyia	417	impudica Doane, Tipula	36
Ilythea Haliday	746	*impudica* (Reiche), Scatophaga	†838, 839
ILYTHEINI	746	impudicus Wheeler, Hercostomus	498
Imaguncula Reinhard	1077	impula Huckett, Hylemya	846
Imatisma Macquart	620	impunctata Malloch, Coenosia	871
imbecilla (Osten Sacken), Pilaria	68	impunctus Kröber, Chrysops	326
imbecillus (Karsch), Systropus	424	inadusta Alexander, Tipula	36
imbellis Alexander, Tipula	33	inaequalis (Coquillett), Trypeta	677
imbellis Melander, Mythicomyia	417	*inaequalis* Loew, Bibio	193
imfurcatus Philip, Chrysops	327	inaequalis Loew, Odontomyia	†315, 316
imitabilis Namba, Rivellia	656	inaequalis (Malloch), Homoneura	698
imitans (Johannsen), Bradysia	233	inaequalis Malloch, Hylemya	849
imitans Johannsen, Mycomya	220	*inaequalis* (Malloch), Megaselia	540
imitans Loew, Parydra	750	inaequalis Malloch, Pseudotephritis	647

INDEX 1615

Name	Page
inaequalis Shewell, Camptoprosopella	697
inaequalis Stebbins, (Cecidomyia)	294
inaequalis Van Duzee, Argyra	525
inaequalis Van Duzee, Sympycnus	527
inaequata Namba, Rivellia	656
inaequipes Bigot, Cholomyia	1029
inamollae Fox & Hoffman, Culicoides	128
inanis Felt, Caryomyia	270
inarmata (Hunter), Xylota	607
inauratus (Coquillett), Lepidanthrax	440
inca (Townsend), Drino	1085
incallida Johannsen, Sciophila	225
incallida (Walker), Diamesa	152
incana Townsend, Tropidia	609
incanus Johnson, Bombylius	409
incanus Reinhard, Isidotus	981
incauta (Harris), Leschenaultia	1083
incauta (Huckett), Spilogona	881
incerta (Adams), Phronia	215
incerta Becker, Oscinella	780
incerta (Malloch), Bezzia	141
incerta (Malloch), Homoneura	698
incerta Malloch, Phaonia	908
incerta (Walker), Quadrularia	889
incerta Walker, (Sarcophaga)	961
incerta West, Ptilodexia	989
incertus (Macquart), Archytas	1000, †1000
incertus Walker, Chrysotus	522
inchoata (Melander), Chersodromia	475
incidatus (Townes), Chironomus	168
incidens (Thomson), Culiseta	117
incisa Cook, Rhexoza	240
incisa Curran, Dictya	689
incisa Doane, Tipula	36
incisa Felt, Trishormomyia	288
incisa (Meigen), Phytobia	798
incisa (Ringdahl), Paraprosalpia	866
incisa (Tothill), Mericia	1008
incisa (Walker), Villa	436
incisa (Wiedemann), Esenbeckia	322
incisiva Painter, Villa	436
incisiva Stein, Pegomya	860
incisum (Coquillett), Hydrochasma	738
incisuralis Harris, (Xylophagus)	1115
incisuralis (Loew), Dohrniphora	535
incisuralis Loew, Dolichopus	492
incisuralis Loew, Leptogaster	363
incisuralis (Macquart), Atylotus	331
incisuralis (Macquart), Efferia	393
incisuralis Macquart, Eristalis	623
incisuralis (Say), Esenbeckia	322
incisurata Edwards, Anatella	205
incisurata Malloch, Gimnomera	840
incisurata Melander & Spuler, Themira	684
incisurata Wulp, Coenosia	871
incisurata (Zetterstedt), Fannia	895
incisus (Felt), Arthrocnodax	283
incisus (Loew), Rhagio	345
incisus Macquart, Mydas	359
incisus (Walker), Hybomitra	339
inclinata Becker, Diplotoxa	792
inclusa Dietz, Tipula	33
inclusa (Hartig), Blondelia	1040
incognitus Malloch, Chironomus	169
incola Becker, Scatophaga	838
incommoda Cresson, Parydra	749
incompleta Loew, Rhamphomyia	464
incompleta (Stein), Pegomya	860
incompletus (Curran), Apostrophus	972
incompta Johannsen, Mycomya	220
incompta (Wulp), Drino	1085
incompta (Zetterstedt), Phaenopsectra	174
inconcinna Loew, Macrocera	201
incongrua (Malloch), Pseudogriphoneura	704
incongrua (Reinhard), Peleteria	998
incongrua Walker, Hamaxia	979
incongruus Aldrich, Dicraeus	783
incongruus Wheeler, Dolichopus	492
inconspicua Huckett, Hylemya	849
inconspicua (Malloch), Pentaneura	147
inconspicua (Malloch), Phytobia	798
inconspicua (Malloch), Tomosvaryella	555
inconspicua (Meigen), Drino	†1084, 1085
inconspicua (Wulp), Paralispe	†1047
inconspicuus Ewen, Atrichopogon	123
inconspicuus (Grossbeck), Aedes	115
inconstans Fitch, (Molobrus)	236
inconstans (Osten Sacken), Pedicia	55
incontesta Curran, Peleteria	998
incrassata Malloch, Limnophora	885
incrassata (Schmitz), Megaselia	539
incrassata Smith (Phorocera)	1107
incrassata (Stein), Hylemya	849
increpitus Dyar, Aedes	113
incriminator Dyar & Knab, Culex	119
inculta (Coquillett), Drapetis	477
inculta (Coquillett), Villa	440
inculta Loew, Mycetophila	214
inculta (Stein), Bigotomyia	909
incultus Osten Sacken, Ogcodes	407
incurata Reinhard, Metopomuscopteryx	1025
incursa Malloch, Hylemya	853
incurva (Aldrich), Metoposarcophaga	951
incurva Doane, Tipula	33
incurva (Loew), Nephrotoma	20
incurva Loew, Tritoxa	649
incurvus Melander, Platypalpus	479
indecisa (Malloch), Phytobia	798
indecisa Townsend, Pseudomyothyria	1047
indecora Kieffer, Forcipomyia	125
indecora (Loew), Pholeomyia	732
indecora (Malloch), Phytobia	797
indentus Aldrich, Hydrophorus	506
index McAtee, Laphria	390
indiana Dyar, Dixa	102
indianensis Alexander, Erioptera	83
indianus Harmston, Hercostomus	498
indianus Harmston & Knowlton, Dolichopus	492
indianus (Harmston & Knowlton), Tachytrechus	503
indicata Melander, Mythicomyia	417
indicta Huckett, Pegomya	860
indifferens Curran, Rhagoletis	674
indigena Johannsen, Macrocera	201
indigena (Osten Sacken), Limonia	43
indigenoides (Alexander), Limonia	43
indigenus Van Duzee, Dolichopus	492
indistincta Doane, Limnophila	63
indistinctus Bromley, Stenopogon	384
indistinctus Malloch, Chironomus	169
indita (Walker), Pseudeuantha	1029
indivisa (Aldrich & Webber), Euphorocera	1054
indivisa Osten Sacken, Cladura	72
indivisa Townsend, Trichopoda	966
indivisus (Aldrich), Sarcofahrtia	942
indubitans Dyar & Shannon, Mansonia	108
indulgens Johannsen, Neoempheria	222
indus Osten Sacken, Chrysops	326
industrius Knab, Pipunculus	554
indutile Reinhard, Mochlosoma	987
ineptifrons Huckett, Hylemya	849
ineptus (Stein), Lasiops	904
ineptus (Walker), Pelastoneurus	501
inermis Allen, Metopia	937
inermis Bigot, (Blepharipeza)	†1083, 1108
inermis Brauer, Gasterophilus	916
inermis Coquillett, Aphoebantus	429
inermis (Coquillett), Parabezzia	136
inermis Doane, Tipula	39
inermis (Loew), Condylostylus	484
inermis (Loew), Parallelomma	840
inermis Robineau-Desvoidy, Strauzia	676
inermis Stein, Phyllogaster	876
inermis Steyskal, Pteromicra	687
iners (Meigen), Schroederella	814
inextricata Dyar & Shannon, Dixa	101
inferialis Melander, Platypalpus	479
infernalis Townsend, Catagoniopsis	1098
infernalis (Townsend), Paralispe	1047
infernalis (Townsend), Peleteria	996
infesta Becker, Oscinella	780
infesta (Williston), Winthemia	1090
infestus (Stein), Lasiops	904
infida Reinhard, Evidomyia	1062
infinita Cazier, Apiocera	357
infirma Felt, (Cecidomyia)	290
infirmatus Dyar & Knab, Aedes	113
inflata (Hine), Efferia	394
inflata Johannsen, Rymosia	208
inflaticornis Allen, Gymnoprosopa	936
inflatifrons Shannon, Chrysogaster	591
inflatipalpis (Aldrich & Webber), Carcelia	1092

744-243—65——103

	Page		Page
inflatus Aldrich, Dolichopus	492	insignis Lutz, Culicoides	128
inflatus Felt, Holoneurus	255	insignis Melander, Mythicomyia	417
inflatus Macquart, Eristalis	624	*insignis* (Reinhard), Xiphomyia	1051
inflexus Walker, Eristalis	623	*insignis* (Stein), Bithoracochaeta	874
infumata Coquillett, Empis	460	insignita Melander, Chersodromia	475
infumata Haliday, Platypeza	†549	insolens Aldrich & Webber, Nemorilla	1088
infumata (Malloch), Megaselia	539	insolens Coquillett, Paracosmus	427
infumata (Malloch), Phytobia	801	insolens Felt, Dicrodiplosis	279
infumata Melander, Drapetis	477	insolens Felt, Hyperdiplosis	284
infumata (Thomson), Sphaerophoria	570	insolita (Osburn), Melangyna	567
infumatus (Aldrich), Sciapus	486	insolita (Walker), Atropharista	1009
infumatus (Bigot), Chrysosyrphus	592	insolitum Curran, Rhaphium	513
infuscata (Doane), Limonia	50	insolitus (Van Duzee), Calyxochaetus	528
infuscata Gill, Amoebaleria	815	instabilis Curran, Archytas	1000
infuscata Loew, Sphegina	594	instabilis Melander, Oligodranes	422
infuscata Loew, Tipula	27	instans (Huckett), Spilogona	881
infuscata Malloch, Probezzia	138	*insuetus* (Osten Sacken), Atylotus	331
infuscata (Townsend), Phaenicia	928	insulanus Van Duzee, Thinophilus	508
infuscata Van Duzee, Neurigona	518	*insularis* (Cresson), Tomosvaryella	556
infuscatus Cole, Cyrtopogon	368	insularis Johnson, Limnophila	66
infuscatus (Fluke), Epistrophe	564	insularis (Malloch), Oscinella	780
infuscatus (Karsch), Systropus	424	*insularis* (Williston), Palaeosepsis	682
infuscatus Leonard, Chrysopilus	346	insulicola (Quate), Trichopsychoda	95
infuscatus (Malloch), Cricotopus	157	insulsa Johannsen, Phronia	215
infuscatus Van Duzee, Diaphorus	520	*insurgens* (Aldrich), Metoposarcophaga	951
infuscipes (Van Duzee), Thecophora	631	intacta Walker, Coenosia	873
ingens Cresson, Exoprosopa	445	*intacta* (Walker), Lispocephala	877
ingens Johnson & Johnson, Poecilanthrax	442	integer Loew, Helophilus	619
ingloria Alexander, Ormosia	86	integer Loew, Tetanops	648
ingrami Carter, Forcipomyia	125	integra Alexander, Tipula	39
ingrata Dietz, Tipula	33	*integra* (Becker), Thaumatomyia	788
ingratus (Loew), Dicraeus	783	integre Williston, Chrysotoxum	582
ingratus Williston, Chlorops	789	integrifoliae Felt, Asphondylia	268
inhabilis Alexander, Limonia	47	integriloba Alexander, Dicranota	58
inimica Fitch, (Cecidomyia)	292	integripes Parent, Dolichopus	492
inimica Robineau-Desvoidy, Stomoxys	914	intensica Curran, Heringia	580
innocens (Osten Sacken), Ormosia	84	*intensivus* Townsend, Tabanus	333
innocens (Shewell), Simulium	186	intentus Aldrich, Hydrophorus	506
innocua Malloch, Hylemya	847	intentus Curran & Fluke, Helophilus	618
innocuus Curran, Metriocnemus	161	intentus Melander & Brues, Dolichopus	492
innocuus (Zetterstedt), Lasiops	903	interdicta (Dyar), Philosepedon	95
innominata (Williston), Stomosis	730	interfrontalis (Melander), Metopomyza	802
innotatus (Curran), Hedriodiscus	315	interjectus Van Duzee, Dolichopus	492
innotatus Loew, Hydrophorus	505	*intermedia* (Becker), Eribolus	778
innoxium Comstock & Comstock, Simulium	188	intermedia Brooks, Lypha	1011
innubila Spencer, Drosophila	764	intermedia Chillcott, Fannia	895
innuitus Dyar & Knab, Aedes	113	*intermedia* (Cresson), Neoscatella	756
inopa (Adams), Limnia	691	intermedia (Curran), Cordilura	828
inopis Osten Sacken, Cecidomyia	287	intermedia Felt, Mycodiplosis	283
inops Coquillett, Boletina	217	intermedia Fisher, Mycomya	220
inops (Coquillett), Orfelia	202	*intermedia* (Garrett), Limonia	49
inops (Coquillett), Villa	436	intermedia Garrett, Pseudoleria	812
inops Melander, Platypalpus	479	intermedia Johnson, Criorhina	612
inops Townsend, Volucella	602	intermedia (Loew), Strauzia	676
inornata Alexander, Tipula	36	intermedia Lundbeck, Fucellia	842
inornata (Coquillett), Neotephritis	669	intermedia (Malloch), Pseudogriphoneura	704
inornata Coquillett, Phthiria	420	*intermedia* (Meigen), Cylindromyia	973
inornata (Malloch), Megaselia	539	intermedia Melander, Mythicomyia	417
inornata Melander, Anthalia	450	intermedia (Reinhard), Metopotachina	993
inornata (Osten Sacken), Pseudolimnophila	62	intermedia Reinhard, Winthemia	1090
inornata Stein, Hylemya	852	intermedia (Sherman), Orfelia	202
inornata (Williston), Culiseta	117	*intermedia* (Townsend), Sarcophaga	959
inornatus (Aldrich), Condylostylus	484	*intermedia* (Tucker), Mallophorina	397
inornatus Cole, Caenotus	424	intermedia Van Duzee, Medetera	510
inornatus Loew, Eristalis	624	*intermedia* Villeneuve, Lydella	1103
inornatus (Van Duzee), Calyxochaetus	528	intermedia Walker, Scatophaga	838
inornatus Van Duzee, Diaphorus	520	*intermedia* (Wiedemann), Odontomyia	316
inquilinus (Coquillett), Chaetochlorops	785	intermedium Banks, Zodion	629
inquilinus Felt, Oligotrophus	265	intermedius (Curran), Neocnemodon	581
inquinata (Wulp), Drino	1085	*intermedius* (Walker), Atylotus	331
inquinatus Loew, Stenopogon	384	intermedius Walker, Rhagio	345
inquisitor Aldrich, Symphoromyia	343	*interrogans* (Walker), Toxomerus	571
inquisitor Felt, Rhopalomyia	266	interrogator Dyar & Knab, Culex	118
inscitus (Walker), Hybomitra	338	interrupta (Banks), Pericoma	93
inscripta (Meigen), Neoleria	813	*interrupta* (Becker), Chlorops	789
insecta Coquillett, Rhamphomyia	464	interrupta Curran, Actia	1062
insecta (Loew), Xanthaclura	671	interrupta Curran, Eutheresia	990
insidiosa (Robineau-Desvoidy), Pseudoperichaeta	1106	*interrupta* Harris, (Conops)	1115
insignifica Alexander, Tipula	33	*interrupta* Hull, Laphystia	376
insignifica Shaw, Neuratelia	224	interrupta Jones, Sphaerophoria	570
insignilobus Kieffer, Micropsectra	178	interrupta (Macquart), Efferia	394
insignis Aldrich, Calotarsa	550	interrupta (Macquart), Rivellia	656
insignis (Banks), Laphria	390	interrupta Malloch, Zygomyia	216
insignis Felt, Walshomyia	265	interrupta Melander, Anthalia	450
insignis Loew, Stratiomys	312	interrupta Olivier, Odontomyia	316
		interrupta Painter, Apiocera	357

	Page		Page
interrupta Stone, Anastrepha	673	isabellina (Doane), Limonia	47
interrupta (Walker), Mystacella	1086	isabellina (Wiedemann), Stonemyia	321
interrupta Williston, Spilomyia	613	isabellina Williston, Volucella	601
interrupta (Zetterstedt), Exechia	206	*isae* (Coquillett), Anaporia	1028
interruptus Coquillett, Aphoebantus	429	*ischiaca* Harris, (Anthomyia)	1115
interruptus (Loew), Asyndetus	524	*ischiaca* Harris, (Milesia)	1115
interruptus Malloch, Gaurax	782	ischiaca Loew, Hydrellia	744
interruptus (Malloch), Melangyna	567	*ischiaca* Walker, Stratiomys	311
interruptus Painter & Hall, Poecilanthrax	442	*ischiacus* Harris, (Chrysops)	1115
intersecta (Meigen), Pegomya	860	**Ischnomyia** Loew	819
intersistens (Walker), Blera	610	**Ischyrosyrphus** Bigot	565
intertropica (Walker), Lynchia	919	**Isidotus** Reinhard	981
intestinalis (De Geer), Gasterophilus	916	islandica Becker, Scatophaga	838
intonsa Cazier, Apiocera	357	islandicus Malloch, Tanytarsus	179
intonsa Reinhard, Winthemia	1090	isoaristus Harmston & Knowlton, Sympycnus	527
intonsus Hardy, Nemomydas	359	isocerus Townes, Polypedilum	175
intricata (Alexander), Limonia	47	*Isoecacta* Garrett	133
intrita Alexander, Hexatoma	70	*Isoglossa* Coquillett	1062
introrsa Melander, Mythicomyia	417	**Isohelea** Kieffer	133
intrudens Aldrich, Rhagoletis	674	isolata Alexander, Gonomyia	76
intrudens (Curran), Siphona	1064	isolata Malloch, Agromyza	795
intrudens Dyar, Aedes	113	isolatus (West), Carinosillus	989
intrudens Malloch, Leptocera	723	isommatus Melander, Microphorus	453
intrudens (Osten Sacken), Dasysyrphus	563	*Isopenthes* Osten Sacken	435
intrudus Harmston, Chrysotus	522	*Isosargus* James	302
inurbana Aldrich, Symphoromyia	343	**Istocheta** Rondani	1073
inusitata Alexander, Tipula	36	*isura* (Walker), Fannia	894
inusitata (Melander), Chersodromia	475	*Itamus* Loew	397
inusta (Melander), Tachypeza	473	itasca (Philip), Hybomitra	339
inustorum Felt, Lasioptera	271	itascae (Laffoon), Mycetophila	212
invaria Walker, Agromyza	795	itascae Sabrosky, Lasiopleura	774
invasor Reinhard, Plagiosippus	1015	**Iteaphila** Zetterstedt	454
invelata (Reinhard), Nowickia	994	iterans (Walker), Peleteria	996
invenusta (Walker), Cnephia	185	ithaca (Shannon), Chrysosyrphus	592
inversa Curran, Cordilura	828	ithacae Felt, Feltiella	282
inversa (Walker), Clastopteromyia	771	ithacana Alexander, Ormosia	86
inversus Curran, Cyrtopogon	368	*ithacanensis* (Johannsen), Glyptotendipes	172
inversus Ide, Platycheirus	577	*ithacensis* Huckett, Hylemya	849
inversus Melander, Aphoebantus	429	ithypyga McAtee, Laphria	390
invigora (Curran), Epistrophe	564	**Itolia** Wilcox	376
involuta Van Duzee, Argyra	525	*Itonida* Meigen	287
invulnerata (Reinhard), Aporotachina	1039	ITONIDIDAE	241
inyoensis Alexander, Tipula	33		
inyoensis Kennedy, Deuterophlebia	100	jaceana (Hering), Urophora	659
inyoensis Reinhard, Senotainia	939	jacintoensis Alexander, Tipula	28
inyoi Wilcox & Martin, Nannocyrtopogon	379	jacksonensis Alexander, Tipula	29
io Aldrich, Cartocometes	1069	jacksoni (Alexander), Limonia	43
io Williston, Bombylius	†409	*jacobi* Hardy, Bibio	194
iola Pritchard, Anaretella	245	jacobi Malloch, Pegomya	860
iolambdis Dyar, Culex	119	jacobus Alexander, Tipula	28
ioogoon Alexander, Rhabdomastix	77	jactator (Loew), Mercurymyia	568
iota Osten Sacken, Exoprosopa	445	jacumbae Dyar & Shannon, Simulium	189
iowa Alexander, Dicranota	57	jaennickeana (Osten Sacken), Villa	436
iowaensis Harmston & Knowlton, Dolichopus	492	jamaicensis Edwards, Culicoides	131
iowana Wheeler, Scatophila	758	*jamaicensis* (Theobald), Wyeomyia	107
iowensis Alexander, Limnophila	66	*jamesi* Cresson, Micropeza	634
iowensis (Rogers), Limonia	47	jamesi Fox, Culicoides	132
iowensis Steyskal, Tetanocera	694	jamesi Gill, Anorostoma	812
iraeos (Robineau-Desvoidy), Phytobia	799	jamesi Hanson, Nemotelus	310
irata Alexander, Erioptera	83	*jamesi* (Hardy), Dilophus	195
irene Alexander, Limnophila	64	*jamesi* (Harmston & Knowlton), Sympycnus	527
iridipennis Dyar, Aedes	113	jamesi Pritchard, Proctacanthella	398
iridipennis (Zetterstedt), Bradysia	233	jamesi Snyder, Lispe	878
iridis (Cockerell), Anarete	246	jamesi Steyskal, Stratiomys	312
iridis (Hendel), Phytobia	799	jamesi Wilcox, Laphystia	376
iron Steyskal, Dictya	689	**Jamesomyia** Quisenberry	663
iroquoiana (Malloch), Megaselia	539	jamnbacki Wirth &Hubert, Culicoides	129
iroquois Alexander, Tipula	28	**Janetiella** Kieffer	264
iroquois Sturtevant & Dobzhansky, Drosophila	768	janta (Roback), Ablabesmyia	148
irrasus (Walker), Pelastoneurus	501	**Janthinosoma** Lynch Arribálzaga	109
irregularis Loew, Rhamphomyia	464	janus McAtee, Laphria	390
irregularis Stebbins, (Cecidomyia)	294	**Japanagromyza** Sasakawa	796
irrequieta (Walker), Tachinomyia	1058	japonica (Townsend), Exorista	1055
irrisor Reinhard, Promasiphya	1091	japonicus Tokunaga, Telmatogeton	163
irrisoris (Reinhard), Boettcheria	948	jaquesi Harmston & Knowlton, Dolichopus	492
irritans (Fallén), Hydrotaea	900	**Jassidophaga** Aczél	551
irritans (Linnaeus), Haematobia	914	jauva Rapp, Niadina	235
irritatum Lugger, Simulium	189	javana Brauer & Bergenstamm, Chaetoxorista	1053
irrorata (Fallén), Dolichocephala	470	javana Townsend, Eutrixopsis	978
irrorata Johnson, Limnophila	67	**Jaynesleskia** Townsend	1035
irroratus Coquillett, Traginops	794	j-beameri Wilcox & Martin, Nannocyrtopogon	379
irroratus Say, Anthrax	432	jeanae DeFoliart & Peterson, Cnephia	184
irrupta Melander, Mythicomyia	417	jeanae Pechuman, Silvius	323
irvinei Wilcox & Martin, Nannocyrtopogon	379	jeanneae Quate, Telmatoscopus	94
isabella Fisher, (Azana)	1115		

1618 A CATALOG OF DIPTERA OF NORTH AMERICA

	Page		Page
jejuna Johnson, Tipula	25	johnsoni Painter, Geron	423
jellisoni Curran, Cuterebra	1110	*johnsoni* Painter, Villa	434
jellisoni Philip, Tabanus	336	*johnsoni* Slingerland & Johnson, Contarinia	277
jellisoni Townsend, Cephenemyia	1111	*johnsoni* (Smith), Periscepsia	1033
jemezi Wilcox & Martin, Cyrtopogon	368	johnsoni Spuler, Leptocera	722
Jenkinshelea Macfie	137	johnsoni (Stein), Leucophora	867
Jenkinsia Kieffer	137	johnsoni (Thomas), Aciurina	670
jennei Aldrich, Lixophaga	1043	johnsoni (Tothill), Mericia	1008
jenningsi Malloch, Simulium	189	*johnsoni* (Townsend), Anaporia	†1026, 1028
jepsoni Alexander, Tipula	26	johnsoni Van Duzee, Argyra	525
jersei Garrett, Anorostoma	812	johnsoni Van Duzee, Asyndetus	524
Joannisia Kieffer	251	johnsoni Van Duzee, Chrysotus	522
joculator Laffoon, Leia	227	johnsoni (Van Duzee), Rhaphium	513
joffrei (Pettey), Bradysia	233	johnsoni West, Arctophyto	986
johannis Dyar & Shannon, Thaumalea	120	**Johnsonia** Coquillett	949
johannseni (Enderlein), Bradysia	233	johnsoniana Alexander, Tipula	36
johannseni (Felt), Micromya	250	*johnsonii* (Wheeler), Hypocharassus	508
johannseni Fisher, (Dynatosoma)	1115	**Johnsonomyia** Felt	252
johannseni Fisher, Keroplatus	200	*Johnsonomyia* Malloch	318
johannseni Garrett, Dixa	101	johnsonorum Painter & Hall, Poecilanthrax	442
johannseni Garrett, Sceptonia	215	johnstoni Shaw, Peyerimhoffia	235
Johannseni Guthrie	209	johnstonii (Grabham), Psorophora	109
johannseni Hart, Simulium	186	jonesi (Aldrich), Blaesoxipha	945
johannseni (Malloch), Dasyhelea	127	jonesi Cresson, Arthropeas	298
johannseni (Malloch), Megaselia	539	jonesi (Cresson), Bequaertomyia	319
johannseni (Roback), Ablabesmyia	148	jonesi Cresson, Exoprosopa	445
johannseni Shaw, Epidapus	235	*jonesi* (Cresson), Pilimas	321
johannseni Shaw, Zygoneura	231	jonesi Curran, Euaresta	665
johannseni Sherman, Dziedzickia	218	jonesi Curran, Eutreta	661
johannseni Sherman, (Neuratelia)	1115	jonesi Curran, Trupanea	667
johannseni Thomsen, Forcipomyia	124	jonesi Fluke, Syrphus	559
johannseni Wirth, Monohelea	135	*jonesi* James, Stratiomys	312
Johannseniella Williston	136	jonesi (Johnson), Bittacomorphella	98
johannsenii (Bause), Stempellina	179	jonesi (Quate), Threticus	95
Johannsenomyia Malloch	137	jonesi Wirth & Hubert, Culicoides	131
johannsoni Garrett	215	jonesii (Johnson), Nemomydas	359
johannus Alexander, Tipula	31	jordani Kellogg, Blepharicera	99
johanseni Alexander, Tipula	41	josephus Alexander, Tipula	26
johanseni (Townsend), Boreellus	925	jubata Chillcott, Rhamphomyia	464
johnsonella (Alexander), Dolichopeza	23	jubata (Williston), Efferia	304
johnsoni Aldrich, Dolichopus	492	jubatoides Bromley, Stenopogon	384
johnsoni Aldrich, Lispe	878	jubatus Coquillett, Stenopogon	384
johnsoni Aldrich, Sarcophaga	958	jucunda Becker, Mesorhaga	483
johnsoni Alexander, Limnophila	64	jucunda Garrett, Boletina	217
johnsoni (Alexander), Pedicia	55	jucunda (Johannsen), Bradysia	233
johnsoni (Back), Pycnopogon	382	jucunda Johannsen, Mycetophila	212
johnsoni (Brues), Megaselia	540	jucunda Loew, Palloptera	715
johnsoni (Brues), Stichillus	536	jucunda (Wulp), Phytobia	800
johnsoni Cole, Pterodontia	404	jucundus (Loew), Condylostylus	484
johnsoni (Coquillett), Anatopynia	145	jucundus Van Duzee, Nematoproctus	515
johnsoni (Coquillett), Bezzia	142	jucundus Walker, Chironomus	170
johnsoni (Coquillett), Blera	611	jugalis Tucker, Dolichopus	492
johnsoni (Coquillett), Elfia	1065	jugata Johannsen, Mycetophila	212
johnsoni Coquillett, Exepacmus	430	*jugatoria* (Say), Trichopoda	966
johnsoni Coquillett, Exoristoides	1013	juglandis Cresson, Rhagoletis	674
johnsoni Coquillett, Hesperodes	200	jugulator Melander, Hemerodromia	470
johnsoni Coquillett, Hilara	456	juli Brues, Plastophora	542
johnsoni Coquillett, Lasioneura	1060	julia (Curran), Psectrocladius	159
johnsoni Coquillett, Psilocephala	351	juliaetta Aldrich, Sarcophaga	958
johnsoni Coquillett, Stenoxenus	139	julietta Shannon, Cheilosia	584
johnsoni Coquillett, Symphoromyia	343	junceus Coquillett, Lordotus	412
johnsoni Coquillett, Thereva	353	junci Felt, Procystiphora	258
johnsoni (Cresson), Antichaeta	688	*junci* (Meigen), Tanytarsus	180
johnsoni Cresson, Hydrellia	744	juncorum (Fallén), Chamaemyia	707
johnsoni Cresson, Renocera	692	juncta Coquillett, Cylindrotoma	41
johnsoni (Curran), Labostigmina	314	juncta Coquillett, Euxesta	651
johnsoni (Curran), Melangyna	567	junctura (Coquillett), Villa	438
johnsoni Darlington, Porsenus	808	junctus Coquillett, Roederiodes	468
johnsoni (Felt), Anarete	246	*junctus* Coquillett, Tachytrechus	503
johnsoni Felt, Asphondylia	268	junctus Van Duzee, Chrysotus	522
johnsoni Felt, Contarinia	277	junctus Van Duzee, Diaphorus	520
johnsoni Felt, Porricondyla	255	*junctus* Van Duzee, Nematoproctus	515
johnsoni (Felt), Trishormomyia	288	juniperi (Felt), Alassomyia	262
johnsoni Hine, Tabanus	334	juniperina Felt, Contarinia	277
johnsoni (Hine), Tolmerus	401	juniperina (Felt), Lestodiplosis	285
johnsoni Hull, Mixogaster	599	juniperina Felt, Walshomyia	265
johnsoni Hunter, Callicera	595	*juniperinus* Marcovitch, Rhagoletis	675
johnsoni Johannsen, Palaeoplatyura	203	juno Curran, Hilara	456
johnsoni (Jones), Carposcalis	576	*juno* (Curran), Telmatoscopus	94
johnsoni Kessel, Platypezina	548	junta Curran, Exoprosopa	445
johnsoni Kieffer, Chironomus	170	junus Roback, Cricotopus	157
johnsoni Malloch, Clusiodes	807	**Jurinella** Brauer & Bergenstamm	1001
johnsoni Malloch, Coenosia	872	JURINIINI	1002
johnsoni Malloch, Helina	887	**Juriniopsis** Townsend	1001
johnsoni (Malloch), Lasiops	904	*jurinoides* Townsend. Atropharista	1009
johnsoni Melander, Empis	460	jussiaeae Vigé, Toxomerus	571

INDEX 1619

	Page		Page
juvenalis Felt, Lasioptera	271	kincaidia (Shannon), Cheilosia	586
juvenalis Felt, Porricondyla	256	kincaidii (Coquillett), Pyritis	604
juvenilis (Stein), Pegomya	860	kingi Brooks, Dolichotarsus	1042
juvenis Melander, Platypalpus	479	*Kingia* Malloch	856
		Kingiella Malloch	856
kahli Kessel, Melanderomyia	547	*Kingiella* Séguy	856
kahli Snow, Spilomyia	613	kirbyana Alexander, Tipula	33
kaibabensis Alexander, Tipula	36	kirkwoodi Alexander, Tipula	37
kaiseri (Shaw), Bradysia	233	kisliuki Stone, Tabanus	334
kallstroemia Felt, Lasioptera	271	kleini Curran, Dolichopus	492
Kalodiplosis Felt	281	klettii Osten Sacken, Lasia	404
kamloopsi Hearle, Simulium	187	*klotsi* Matheson, Aedes	113
kamtchatica Ringdahl, Fucellia	843	kluane (Alexander), Arctoconopa	80
kamtchotica (Ringdahl), Hydrophoria	865	kluaneana Alexander, Erioptera	81
kamtschatkense (Hendel), Nanna	831	knabeni Goetghebuer, Orthocladius	156
kansensis Adams, Nemotelus	309	knabi Alexander, Erioptera	82
kansensis Aldrich, Dolichopus	492	knabi Coquillett, Metriocnemus	161
kansensis (Aldrich), Pelastoneurus	501	knabi (Cresson), Oligodranes	422
kansensis Alexander, Gonomyia	76	*knabi* Cresson, Setacera	755
kansensis Alexander, Tipula	36	*knabi* (Dyar), Chaoborus	104
kansensis Felt, Lestremia	244	knabi Shannon, Syrphus	559
kansensis (Hardy), Pipunculus	553	*Knabia* Townsend	1075
kansensis Harmston, Chrysotus	522	knowltoni Alexander, Dactylolabis	62
kansensis (Hine), Efferia	394	knowltoni Beck, Culicoides	131
kansensis James, Bibio	193	knowltoni (Bromley), Efferia	395
kansensis Townsend, Olenochaeta	1077	*knowltoni* Curran, Euxesta	651
kansensis (Townsend), Senotainia	939	knowltoni Curran, Myxosargus	317
karnerensis Felt, Campylomyza	248	knowltoni Hardy, Bibio	193
karnerensis Felt, Clinorhyncha	275	*knowltoni* (Hardy), Tomosvaryella	556
karnerensis Felt, Dasineura	259	knowltoni Harmston, Scellus	506
karnerensis (Felt), Mochlonyx	104	knowltoni James, Nemotelus	309
karnerensis (Felt), Odontodiplosis	288	knowltoni Kraft & Cook, Zabrachia	318
karnerensis Felt, Porricondyla	256	*knowltoni* Reinhard, Athanatus	1022
karnerensis Felt, Trotteria	276	knowltoni Sabrosky, Elachiptera	777
karnerensis Felt, Winnertzia	253	knowltoni Twinn, Simulium	189
Karschomyia Felt	282	knowltonia (Alexander), Gonomyodes	79
Karshomyia Felt	282	knowltoniana Alexander, Gonomyia	74
kasloensis (Felt), Polyardis	250	knulli Bromley, Asilus	392
katahdin (Alexander), Pedicia	55	koebelei (Felt), Cleodiplosis	280
katmai (Alexander), Arctoconopa	80	*Kompia* Aitken	111
katmai Dyar & Shannon, Simulium	189	*Koniosternum* Becker	837
katmaiensis Alexander, Tipula	41	*Konisomyia* Felt	243
katmaiensis Malloch, Hydrophoria	865	*kooteniensis* (Garrett), Limonia	49
katmaiensis (Malloch), Nanna	831	*Kophosoma* Van Duzee	517
kaw (Sabrosky), Elachiptera	776	kowarzi (Verrall), Fannia	895
kawiensis Martin, Beameromyia	361	*Kribiozenus* Kieffer	†173
keeni Foote, Rubsaamenia	255	krombeini Sabrosky, Metopia	937
keeniana Williston, Sphegina	594	**Kronomyia** Felt	252
keenii (Coquillett), Stonyx	440	kuiterti (Hardy), Scenopinus	356
Keirosoma Van Duzee	515	kulshanensis Alexander, Dicranota	57
kelloggi Garrett, Bibiocephala	99	kulshanicus Alexander, Molophilus	89
kelloggi (Snow), Carposcalis	576	kummi Edwards, Orthopodomyia	108
kelloggi Wilcox, Eucyrtopogon	373	*Kurtomyia* Roback	946
kelloggi Wilcox, Stenopogon	384	kuschei Alexander, Limonia	50
kelloggii Theobald, Culex	119		
kellyi (Aldrich), Blaesoxipha	946	laakei (Hall), Ravinia	954
Kellymyia Townsend	945	labecula Foote, Tephritis	668
kennedyi Harmston & Knowlton, Syntormon	515	labeculosum (Mitchell), Polypedilum	175
Kennesawmyia Dodge	940	labiata Loew, Empis	460
kennicotti Alexander, Tipula	28	labida Borgmeier, Triphleba	534
kennicotti (Banks), Compsobata	634	*labiosa* (Boheman), Spilogona	881
keremeosa Snyder, Helina	887	labiosus Melander, Epacmus	428
kermodei (Townsend), Nowickia	995	labis (Coquillett), Xanthophyto	1009
kernae Martin, Cerotainiops	388	**Labostigmina** Enderlein	313
kernensis Wirth, Palpomyia	140	labradorensis (Brooks), Periscepsia	1034
KEROPLATINAE	199	*labradorensis* Dyar & Shannon, Aedes	113
Keroplatus Bosc	200	*labradorensis* (Enderlein), Hybomitra	339
kerteszi Kieffer, Leptoconops	122	labradorensis Johnson, Bibio	193
kesseli Hardy, Dialysis	343	labradorensis Malloch, Pegomya	860
kesseli Hendrickson, Neurigona	518	*labradorica* Alexander, Tipula	31
kesseli Philip, Tabanus	334	*Labrundinia* Fittkau	146
keyensis Buren, Aedes	115	*Lacchoprosopa* Curran	940
khalafi Beck, Culicoides	129	*Laccoprosopa* Townsend	940
kiamichii Shaw, Mycomya	220	lacerata Bigot, Stratiomys	312
kiefferi (Garrett), Podonomus	150	lacertosus Melander, Platypalpus	480
Kiefferulus Goetghebuer	170	laceyi Curran, Xylophagus	297
kincaidi (Aldrich), Spilogona	881	*Lachnomma* Townsend	1032
kincaidi Aldrich, Symphoromyia	343	*lachnosternae* (Townsend), Sitophaga	1048
kincaidi Alexander, Tipula	30	lachnosternum Melander Argo, Sobarocephala	806
kincaidi Coquillett, Criorhina	611	lacicola Patterson, Drosophila	764
kincaidi (Coquillett), Hadroneura	219	lacinariae Felt, Asphondylia	268
kincaidi (Coquillett), Okeniella	834	lacinia Martin, Beameromyia	362
kincaidi Hardy, Chrysopilus	346	laciniatus Coquillett, Dolichopus	492
kincaidi (Johnson), Hoplodictya	691	lacki Edwards, Rymosia	208
kincaidi Malloch, Agromyza	795	lacroixi (Alexander), Limonia	47
kincaidi Quate, Pericoma	92	*lacteipennis* Curran ,Bibio	193

	Page		Page
lacteipennis (Fallén), Meoneura	†729	lanceolatus Adams, Microdon	598
lacteipennis Johannsen, Trichocladius	158	lancifer Osten Sacken, Bombylius	409
lacteipennis Kieffer, Anarete	246	laneana Vargas, Corethrella	105
lacteipennis (Kröber), Furcifera	350	lanei Alexander, Tipula	29
lacteipennis (Loew), Diclasiopa	739	lanei Wirth, Echinohelea	135
lacteipennis (Loew), Milichiella	733	lanei Wirth, Monohelea	135
lacteipennis Malloch, Orthocladius	156	langei Frick, Liriomyza	802
lacteipennis (Malloch), Pseudonapomyza	803	languidus Becker, Chlorops	789
lacteipennis Melander, Anthalia	450	lanhami James, Laphystia	376
lacteipennis Zetterstedt, Ceratopogon	132	*lani̇̆ ria* (Wiedemann), Cochliomyia	†923
lacteipennis (Zetterstedt), Lispocephala	877	lanifera (McDunnough), Hybomitra	339
lacteipes Alexander, Tipula	29	laniger Cresson, Eclimus	426
lacteum Cresson, Allotrichoma	736	laniiferae (Webber), Lespesia	1102
lactuca Frost, Phytomyza	804	lanipes (Fabricius), Trichopoda	966
lactucae Felt, Lasioptera	271	laniventris Eschscholtz, Empis	460
lacunaris (Coquillett), Villa	439	lantanae Frick, Phytobia	800
lacustris (Alexander), Cryptolabis	78	lantanae (Froggatt), Ophiomyia	797
lacustris Freeborn, Chaoborus	103	lanuginosa (Doane), Ormosia	86
lacustris Stone, Tabanus	334	laparoides Bromley, Saropogon	383
laeta (Fallén), Phaonia	907	**Laphria** Meigen	388
laeta Huckett, Coenosia	873	laphriformis Curran, Cyrtopogon	368
laeta Walker, Mycetophila	212	LAPHRIINAE	386
laetabilis Melander, Platypalpus	480	laphroides (Wiedemann), Mallophorina	397
laeticornis Borgmeier, Stichillus	536	*Laphyctis* Loew	376
laetifica Alexander, Tipula	34	**Laphystia** Loew	376
laetifica (Robineau-Desvoidy), Quadrularia	889	*lapilcei* (Robineau-Desvoidy), Nowickia	995
laetoides Curran, Microdon	598	lappa Stebbins, (Cecidomyia)	294
laetum (Meigen), Polypedilum	175	*lappona* (Linnaeus), Sericomyia	603
laetus Loew, Helophilus	619	lapponica Duda, Scatopse	238
laetus Loew, Microdon	†598, 598	lapponica (Ringdahl), Scatophaga	838
laetus Loew, Pelastoneurus	501	*lapponicus* Loew, Chrysops	†328
laetus Loew, Platypalpus	480	lapponicus (Zetterstedt), Metasyrphus	560, †563
laeviana (Felt), Asteromyia	274	*Lapria* Meigen	388
laevifrons Becker, Ectecephala	†791	*lapsans* Harris, (Lauxania)	1115
laevifrons Jones, Cheilosia	586	lara Steyskal, Nemomydas	359
laevigata Alexander, Tipula	37	*larga* (Aldrich), Metoposarcophaga	951
laevigata Loew, Empis	460	laricata Malloch, Coenosia	872
laevigata Loew, Rhamphomyia	464	laricicola Alexander, Limnophila	67
laevigata (Meigen), Leiomyza	823	laricicola Alexander, Molophilus	89
laevigata (Wulp), Periscepsia	1034	*Larvaevora* Meigen	998
laevigatus Loew, Gymnopternus	499	LARVAEVORIDAE	961
laevigatus Van Duzee, Sympycnus	527	LARVAEVORINAE	992
laevinus (Walker), Promachus	400	LARVAEVORINI	993
laevis Alexander, Dicranoptycha	52	*Larvaevoropsis* Townsend	994
laevis Alexander, Erioptera	83	larvarum (Linnaeus), Exorista	1055, †1055
laevis Becker, Chlorops	789	lasciva (Fabricius), Taeniaptera	636
laevis (Bigot), Cheilosia	586	lasciva (Zetterstedt), Hylemya	845
laevis (Cresson), Athyroglossa	735	*lascivum* (Twinn), Cnephia	184
laevis Loew, Allophyla	809	lascivus Adams, Chlorops	789
laevis (Stein), Fannia	895	lasia Reinhard, Myiophasia	983
laevis (Stein), Hylemya	849	**Lasia** Wiedemann	404
laffooni Gill, Spanoparea	813	*Lasiobezzia* Kieffer	141
laffooni Kessel, Agathomyia	548	lasiocercum Cresson, Allotrichoma	736
lagganensis Alexander, Phyllolabis	60	**Lasiodiamesa** Kieffer	150
lagoae (Townsend), Carcelia	1092	lasiofemoratus Hardy & Knowlton, Pipunculus	553
lagotis Wheeler, Parasyntormon	516	**Lasiohelea** Kieffer	123
laguna Wilcox, Laphystia	376	**Lasiomastix** Osten Sacken	63
lagunae Cole, Euparyphus	307	*lasiomerus* Walker, Chironomus	171
Laiomyia Izquierdo	117	**Lasiomma** Stein	855
lambda James, Nemotelus	310	**Lasionalia** Curran	1024
lambens (Wiedemann), Sarcodexia	†955	**Lasioneura** Coquillett	1060
lamellata (Becker), Homoneura	698	*lasiophthalma* (Loew), Labostigmina	313
lamellata Collin, Meoneura	729	lasiophthalma (Macquart), Hybomitra	339
lamellata Doane, Tipula	37	lasiophthalma Malloch, Hydrotaea	900
lamellatus Loew, Diaphorus	520	lasiophthalma (Malloch), Smittia	162
lamellatus Loew, Pelastoneurus	501	lasiophthalma Williston, Cheilosia	584
lamellicauda Huckett, Hylemya	849	*Lasiophthicus* Rondani	562
lamellicornis Thomson, (Dolichopus)	530	*Lasiopticus* Rondani	562
lamellipes Walker, Dolichopus	492	**Lasiopiophila** Duda	712
lamnia Walker, (Eriphia)	869	**Lasiopleura** Becker	774
Lampetia Meigen	617	**Lasiopogon** Loew	377
lampra Pritchard, Mycophila	250	lasiops Malloch, Fannia	895
lampra Steyskal, Cinderella	825	lasiops (Malloch), Smittia	162
Lampria Macquart	388	**Lasiops** Meigen	903
Lamprochromus Mik	530	**Lasioptera** Meigen	270
Lamproclasiopa Hendel	738	LASIOPTERINI	270
Lampromydas Séguy	358	**Lasiopteryx** Stephens	†261, 291
Lamproscatella Hendel	756	*Lasiopthricus* Rondani	562
lamprurus (Bigot), Cheilosia	586	*Lasiopticus* Rondani	562
lampuris Reinhard, Juriniopsis	1001	*lasiopus* Walker, Chironomus	165
lanata (Enderlein), Melangyna	566	**Lasioscelus** Becker	833
lanata Lundbeck, Scatophaga	838	**Lasiosina** Becker	791
lanata Martin, Leptogaster	363	*Lasiosoma* Winnertz	225
lanceolata Felt, Rhopalomyia	266	lasiosterna Snyder, Helina	887
lanceolata (Kincaid), Maruina	97	lasipes Wiedemann, Laphria	390
lanceolata Melander, Oedalea	450		

INDEX 1621

Name	Page
lasius Melander, Oligodranes	422
lassata Johannsen, Mycetophila	211
lassenensis Johnson & Johnson, Bombylius	409
lassenica Quate, Pericoma	92
lata (Coquillett), Drapetis	477
lata (Coquillett), Sericomyia	603
lata Cresson, Hydrellia	744
lata (Loew), Adoxomyia	305
lata Macquart, Laphria	390
lata (Walker), Allognosta	300
lata Walker, Coenosia	872
lata Walker, (Ephydra)	759
lata (Walker), Hydrotaea	900
lata Wiedemann, Volucella	†601
latebrosa Felt, Rhabdophaga	257
latelimbata Curran, Notiphila	747
latelimbatus (Bigot), Anthrax	432
latelimbatus Curran, Euparyphus	307
latens (Aldrich), Heleomyza	816
laterale (Loew), Allotrichoma	736
laterale Loew, Chrysotoxum	582
lateralis Adams, Psilocephala	351
lateralis (Curran), Sitophaga	1048
lateralis (Fallén), Napomyza	804
lateralis Felt, (Monardia)	292
lateralis Harris, (Chironomus)	1115
lateralis Harris, Cylindromyia	973
lateralis Harris, (Dexia)	1115
lateralis Johnson, Volucella	601
lateralis (Linnaeus), Fannia	894
lateralis Loew, Platypalpus	487
lateralis Loew, Psila	640
lateralis (Macquart), Archytas	1000
lateralis (Macquart), Efferia	394
lateralis (Macquart), Metopia	937
lateralis (Macquart), Phytobia	799
lateralis (Meigen), Aedes	†114
lateralis (Say), Neurigona	518
lateralis (Say), Villa	434
lateralis Stein, Hylemya	847
lateralis (Walker), Clusia	806
lateralis Walker, Pipunculus	552
lateralis Wiedemann, Chrysops	326
lateralis (Williston), Phytobia	800
lateralis (Wulp), Leia	227
laterella (Zetterstedt), Phytobia	†799
latericaudus Curran, Cyrtopogon	369
latericia (Loew), Suillia	809
lateriflorae Felt, Rhopalomyia	266
lateritius Rondani, Tabanus	336
lateropili Allen, Metopia	937
latevittata (Dietz), Nephrotoma	21
lathami (Curran), Aplomyiopsis	1038
lathana Stone, Anastrepha	673
Latheticomyia Wheeler	825
Lathyrophthalmus Mik	625
latianulum (Tothill), Nowickia	995
laticallus Brennan, Silvius	323
laticallus (Philip), Hybomitra	339
laticauda Loew, Steneretma	653
laticaudatus (Curran), Dasysyrphus	563
laticaudus (Curran), Dasysyrphus	563
laticeps Alexander, Erioptera	83
laticeps (Bigot), Heterostylum	411
laticeps Hine, Tabanus	334
laticeps Loew, Stratiomys	312
laticeps (Wulp), Holcocephala	375
laticlavia Melander, Mythicomyia	417
laticornis Coquillett, Ectecephala	791
laticornis (Enderlein), Hybomitra	341
laticornis (Hine), Hybomitra	339
laticornis Loew, Dolichopus	492
laticornis Malloch, Phaonia	907
laticornis Meigen, Lonchaea	717
laticornis (Ringdahl), Hydrophoria	864
laticosta Borgmeier, Triphleba	534
laticrus Van Duzee, Tachytrechus	503
latifacies (Tothill), Nowickia	994
latifacies Van Duzee, Diaphorus	520
latifacies Van Duzee, Pelastoneurus	502
latifacies Van Duzee, Rhaphium	513
latifasciaeformis Duda, Drosophila	767
latifasciata Reinhard, Gymnoprosopa	936
latifasciatus (Macquart), Metasyrphus	560
latifibula Frey, Tetanocera	694
latiflagrum (Enderlein), Esenbeckia	322

Name	Page
latifolia Alexander, Phyllolabis	60
latiforceps Sabrosky, Leptocera	721
latiforceps (Tothill), Nowickia	995
latifrons (Aldrich), Catalinovoria	1017
latifrons Brennan, Chrysops	326
latifrons Cole, Psilocephala	350
latifrons Coquillett, Cuterebra	1110
latifrons (Curran), Bezzimyia	963
latifrons Curran, Pipiza	579
latifrons Greene, Apheloglutus	1064
latifrons Hardy, Chalarus	550
latifrons (Hough), Eucalliphora	929
latifrons Loew, Cordilura	828
latifrons (Loew), Dasyopa	784
latifrons Loew, Eristalis	624
latifrons (Loew), Eurosta	663
latifrons Loew, Helophilus	618
latifrons (Loew), Sobarocephala	806
latifrons Malloch, Fannia	895
latifrons Malloch, Pogonomyia	902
latifrons Reinhard, Metavoria	1019
latifrons (Tothill), Nowickia	995
latifrons (Wulp), Rhagoletoides	675
latifrontalis Huckett, Hylemya	853
latifrontata (Malloch), Helina	889
latigena McAlpine, Campichoeta	773
latigena (Tothill), Nowickia	994
latilamina (Collin), Spilogona	881
latimanus Van Duzee, Thinophilus	508
latipalpis Bigot, Symphoromyia	343
latipalpis (Stein), Pegomya	863
latipenis Quate, Telmatoscopus	94
latipennis Aldrich, Meledonus	1024
latipennis Banks, Pipunculus	553
latipennis (Coquillett), Steganopsis	705
latipennis Curran, Hemyda	974
lotipennis Felt, Campylomyza	248
latipennis (Felt), Mayetiola	263
latipennis Hine, Asilus	392
latipennis (Loew), Ospriocerus	381
latipennis Loew, Tipula	33
latipennis Lundbeck, (Sciara)	236
latipennis (Macquart), Eutreta	662
latipennis (Malloch), Lasiops	904
latipennis Melander, Drapetis	477
latipennis (Zetterstedt), Hylemya	856, †856
latipes Felt, Mayetiola	263
latipes (Loew), Dolichopus	492
latipes (Meigen), Leptometopa	731
latipes Meigen, Piophila	713
latipes (Meigen), Simulium	186
latipilosa Curran, Criorhina	611
latisetosa Parker, Ravinia	954
latisterna Parker, Boettcheria	948
latistyla Alexander, Dicranota	57
latitarsata Melander & Spuler, Themira	684
latitarsis (Egger), Neocnemodon	581
latitarsis Hunter, Helophilus	618
latitarsis Van Duzee, Sympycnus	527
latitarsis (Zetterstedt), Pegomya	862
latiterebrus (Czerny), Dasiops	716
latiuscula (Loew), Baccha	573
latiusculus (Loew), Ceroxys	643
lativentre Graenicher, Zodion	629
lativentris Bellardi, Hermetia	305
lativentris Berdén, Psychoda	96
lativentris Loew, Stratiomys	312, †313
lativittata Malloch, Drosophila	765
l.tivittata Malloch, Pegomya	859
lativittotus (Coquillett), Aedes	112
latrans (Walker), Cheilosia	586
latrappensis Ouellet, Empis	460
Latreillia Robineau-Desvoidy	1081
latreillii Robineau-Desvoidy, Gymnosoma	965
latreillii Robineau-Desvoidy, Mesembrina	910
Latreillimyia Townsend	1081
latro Walker, Helophilus	618
latronis Van Duzee, Dolichopus	492
latruncula (Williston), Efferia	395
latum Curran, Melanosoma	575
latus (Coquillett), Haplopogon	373
latus (Loew), Chrysosyrphus	592
latus Van Duzee, Asyndetus	524
laurentiana Steyskal, Dictya	689
lauta (Loew), Dialysis	343
lauta Wheeler, Ochthera	753

Name	Page
Lauterborniella Bause	172
lautus (Coquillett), Haplopogon	373
lautus (Coquillett), Lepidanthrax	440
lautus (Hardy), Pipunculus	553
lautus Loew, Heteropogon	374
lautus (Loew), Symmerus	199
Lauxania Latreille	699
Lauxaniella Malloch	700
LAUXANIIDAE	695
LAUXANIOIDEA	695
lavendula (Melander), Tethina	727
Laverania Theobald	106
laxa Collin, Lonchaea	717
laxa Cresson, Scatella	757
laxa Osten Sacken, Didea	563
laxifrons Reinhard, Eutrixa	981
laxifrons Villeneuve, Carcelia	1092
laxifrons (Zetterstedt), Helina	887
Laximasicera Curran	1084
Laxina Curran	483
lazarensis (Felt & Young), Aedes	111
lebanoensis (Fluke), Metasyrphus	561
lebanonensis Wheeler, Drosophila	767
Lecanocerus Borgmeier	546
lecontei Alexander, Limonia	45
lecta Melander, Clinocera	469
lectus Becker, Sciapus	486
lecythus Walker, Asilus	392
leechi Alexander, Tipula	37
leechi (Curran), Sitophaga	1048
leechi Hall, Eclimus	426
leechi Kessel, Agathomyia	548
legata Alexander, Ormosia	86
leguminicola Felt, Campylomyza	248
leguminicola (Lintner), Dasineura	259
leguminicola Melander, Chyliza	640
Leia Meigen	227
leibyi Townsend, Schizocerophaga	1047
leidyi (Walsh), Fannia	895
LEIINI	226
leiocantha Alexander, Tipula	37
Leiomyza Macquart	823
Leiota Rondani	590
Leiponeura Skuse	74
Leja Meigen	227
Lejops Rondani	†618
Lejota Rondani	590
Lemnaphila Cresson	745
Lemtopeza Macquart	451
lemur Osten Sacken, Baccha	573
lenis Cresson, Ditrichophora	739
lenis Felt, Mycodiplosis	283
lenis Harris, (Anthomyia)	1115
lenis Johannsen, Mycetophila	212
lenis Osten Sacken, Ptychoptera	97
lenta Johannsen, Mycetophila	212
lenta (Osten Sacken), Shannonomyia	67
lenta (Williston), Volucella	602
lenticularis Melander, Mythicomyia	417
leona Roback, Diamesa	152
leonardi Alexander, Rhabdomastix	77
leonardi (Van Duzee), Condylostylus	484
leonardi (West), Trochilodes	1015
lepida Johannsen, Coelosia	218
lepida (Melander), Wiedemannia	469
Lepidanthrax Osten Sacken	440
lepidii Felt, Dasineura	259
lepidipes Hardy, Tomosvaryella	555
lepidocera (Wiedemann), Lepidophora	425, †426
Lepidomyia Loew	590
Lepidophora Westwood	425
Lepidosia Coquillett	109
Lepidostola Mik	590
lepidota (Osten Sacken), Villa	436, †438
lepidotoides Johnson, Villa	438
lepidus Hine, Asilus	392
lepivora Coquillett, Cuterebra	1110
leporivora Coquillett, Cuterebra	1110
Lepromyia Williston	590
LEPTIDAE	342
Leptis Fabricius	344
leptis (Osten Sacken), Arthroceras	298
Leptocera Olivier	720
Leptochilus Loew	428
LEPTOCONOPINAE	121
Leptoconops Skuse	121
Leptoda Wulp	1027
Leptodiplosis Kieffer	284
leptodoma Alexander, Rhabdomastix	77
Leptogaster Meigen	362
leptogaster (Winnertz), Speolepta	219
leptogastra Loew, Empis	460
LEPTOGASTRINAE	361
Leptometopa Becker	731
Leptomorphus Curtis	223
Leptopeza Macquart	451
Leptopsilopa Cresson	741
Leptopteromyia Williston	363
leptorhabda Alexander, Ormosia	87
leptostyla Alexander, Erioptera	81
Leptosyna Kieffer	†252
LEPTOSYNINI	251
leptotarsus Curran, Cyrtopogon	368
lepus Van Duzee, Parasyntormon	516
lepusculi Townsend, Cuterebra	1110
Lepyria Roback	956
Leria Robineau-Desvoidy	816
lerneri Curran, Leptogaster	363
Leschenaultia Robineau-Desvoidy	1082
Lesciomima Brauer & Bergenstamm	1035
Leskiella James	1035
LESKIINI	1034
Leskiomera Brauer & Bergenstamm	1035
Leskiomima Brauer & Bergenstamm	1035
Leskiopalpus Townsend	1035
Leskiopsis Townsend	1036
Lespesia Robineau-Desvoidy	1101
Lestodiplosis Kieffer	284
Lestomyia Williston	378
lestomyiformis Wilcox & Martin, Nannocyrtopogon	379
Lestophonus Williston	825
Lestremia Macquart	244
LESTREMIINAE	243
LESTREMIINI	244
lesueurii (Macquart), Epistrophe	564
leucacra James, Dolichopus	492
leucanae (Kirkpatrick), Winthemia	1090
leucaniae (Coquillett), Patelloa	1104
leucaniae (Kirkpatrick), Winthemia	1090
leucaniae Townsend, (Sarcophaga)	961
leucocephala (Rossi), Metopia	937
leucocoma (Williston), Efferia	395
leucogaster Chillcott, Fannia	895
leucogaster (James), Hedriodiscus	314
leucogaster (Zetterstedt), Spilogona	881
leucogastra (Loew), Pholeomyia	732
leucomelas Walker, Tabanus	334
leuconotips Dyar, Aedes	113
leucoparea Loew, Cheilosia	586
leucopeza (Meigen), Aulacigaster	823
leucopeza (Meigen), Monohelea	135
leucopeza (Meigen), Pteromicra	687
leucophaea Doane, Tipula	30
leucophaea (Meigen), Lestremia	244
Leucophenga Mik	761
Leucophora Robineau-Desvoidy	867
leucophorus (Bigot), Hybomitra	340
leucophrys (Wiedemann), Leschenaultia	1083
Leucopina Malloch	708
LEUCOPINI	708
Leucopis Meigen	708
leucopogon (Williston), Proctacanthella	399
Leucopomyia Malloch	709
leucoprocta (Wiedemann), Bithoracochaeta	874
leucoprocta (Wiedemann), Ogcodocera	428
leucoproctum (Loew), Hydrochasma	739
leucoptera Johnson, Dichaetoneura	1065
leucoptera Loew, Hilara	456
leucoptera Loew, Rhamphomyia	464
leucoptera West, Ptilodexia	989
leucopyga Wiedemann, Toxophora	425
leucoscelis (Townes), Chironomus	169
leucosticta (Meigen), Fannia	895
leucostola Loew	524
leucostoma Loew, Amiota	761
leucostoma Loew, Diaphorus	520
leucostoma (Loew), Neoleria	813
leucostoma Loew, Schoenomyza	870
Leucostoma Meigen	976
leucostoma (Meigen) Psilopa	741
leucostoma (Robineau-Desvoidy), Archytas	1000

	Page		Page
leucostoma (Wiedemann), Ophyra	901	*Limnochironomus* Kieffer	169
leucostoma Williston, Zodion	629	**Limnophila** Macquart	63
leucostoma (Zetterstedt), Nanna	831	**Limnophora** Robineau-Desvoidy	885
LEUCOSTOMATINI	975	limnophorina (Stein), Spilogona	881
Leucotabanus Lutz	331	**LIMNOPHORINAE**	879
leucotelus (Walker), Ogcodocera	428	**Limnophyes** Eaton	160
leucothrix Melander, Pteromicra	687	**Limnospila** Schnabl	876
leucozona Loew, Cyrtopogon	368	**Limonia** Meigen	42
leucozona (Panzer), Thelaira	†1021	**LIMONIINAE**	42
Leucozona Schiner	563	**LIMONIINI**	42
levata (Reinhard), Pilatea	1105	limosa (Fallén), Leptocera	721
levata (West), Ptilodexia	988	**Limosia** Robineau-Desvoidy	870
levicula Coquillett, Empis	460	**Limosina** Macquart	723
levicula (Coquillett), Villa	440	limpidipennis (Hine), Philonicus	398
leviculus Coquillett, Aphoebantus	429	*limpidipennis* Loew, Parydra	750
levifrons Spuler, Leptocera	724	limpidipennis (Wilcox), Backomyia	364
levigata Loew, Psilocephala	351	limpiensis Mainland, Drosophila	764
levigata Melander, Mythicomyia	417	*limuva* (Brimley), Physoconops	627
levigena Spuler, Leptocera	724	*Linacerus* Garrett	150
levis (Aldrich & Webber), Metadoria	1045	lindensis Reinhard, Meledonus	1024
levis (Coquillett), Atrichopogon	123	linderae Beutenmüller, Lasioptera	271
levis Loew, Psila	639	lindrothi Goetghebuer, Diamesa	152
lewisiana Alexander, Pedicia	53	lindseyi Alexander, Gonomyia	74
lherminieri (Robineau-Desvoidy), Cochliomyia	924	lindsleyi Sturtevant & Wheeler, Asmeringa	751
		lineafrons Spuler, Leptocera	725
lherminieri (Robineau-Desvoidy), Ravinia	954	linearis Alexander, Tipula	39
lherminierii Macquart, Eristalis	623	linearis Malloch, Helina	887
lherminierii (Macquart), Sparnopolius	413	linearis Stein, Hylemya	849
Liancalus Loew	507	linearis Walker, (Drosophila)	772
libatrix (Panzer), Zenillia	1107	lineariventris (Zetterstedt), Hylemya	849
libella Alexander, Ormosia	84	*lineata* (Day), Elgiva	690
libera Aldrich, Sarcophaga	958	lineata Loew, Clinocera	468
liberta (Osten Sacken), Limonia	47	lineata Maequart, Baccha	572
libertoides (Alexander), Limonia	47	lineata (Meigen), Palpomyia	140
libo Walker, Xylota	607	*lineata* (Scopoli), Nephrotoma	21
lienosa Wheeler, Neurigona	518	lineata Van Duzee, Micropeza	634
lignator Steyskal, Sepedon	693	lineata Wheeler, Latheticomyia	825
ligni Felt, Janetiella	264	lineata (Wulp), Chrysoexorista	1099
lignivora (Felt), Monardia	250	lineata (Zetterstedt), Microprosopa	835
ligulata Shaw, Exechia	207	lineatum (Aldrich), Astiosoma	824
Ligyra Newman	445	lineatum (Villers), Hypoderma	1112
lilaea (Walker), Eucalliphora	929	*lineatus* Fabricius, Helophilus	†620
lilliana Alexander, Ormosia	86	*lineatus* Fabricius, Tabanus	332
lima (Melander), Liriomyza	802	*lineatus* Jaennicke, Chrysops	327
limacodis (Townsend), Anaporia	1028	lineatus Loew, Sympycnus	527
limae Stone, Anastrepha	673	*lineatus* Say, Chironomus	171
limata (Coquillett), Dexia	1022	lineella (Fallén), Tricimba	785
limata (Coquillett), Myoleja	669	lineifrons Sabrosky, Ormia	980
limata Coquillett, Pelatachina	1026	lineola (Adams), Leia	227
limata Coquillett, Psilocephala	351	lineola Curran, Meromyza	793
limata Coquillett, Rhamphomyia	464	lineola Fabricius, Tabanus	334
limata Coquillett, Symphoromyia	343	*lineola* (Say), Xenochironomus	171
limata (Coquillett), Zizyphomyia	1088	lineola (Zetterstedt), Melangyna	567
limata (Laffoon), Mycetophila	212	lineolata Macquart, (Stratiomys)	319
limatula Coquillett, Laphystia	376	lineotarsus Curran, Cyrtopogon	369
limatulus Say, Anthrax	432	**Linnaemya** Robineau-Desvoidy	1005
limatus (Hine), Dasysyrphus	563	**Linnaemyia** Robineau-Desvoidy	1005
limbata Aldrich, Cylindromyia	973	**LINNAEMYINI**	1004
limbata Loew, Rhamphomyia	464	linsdalei Alexander, Limonia	47
limbata Steyskal, Odontomera	642	linsdalei Alexander, Tipula	40
limbata (Thomson), Suillia	809	linsleyi Wirth, Palpomyia	140
limbata Williston, Drosophila	768	lintea Fluke & Hull, Cheilosia	585
limbatellus (Holmgren), Psectrocladius	159	*lintneri* (Felt), Mochlonyx	104
limbatinevris Macquart, Tabanus	332	**Liochrysops** Philip	328
limbatus Leonard, Rhagio	345	*liogaster* (Thomson), Euleia	677
limbatus (Loew), Conophorus	414	**Liogma** Osten Sacken	42
limbatus Palisot de Beauvois, Tabanus	332	*Liohippelates* Duda	774
limbatus Van Duzee, Liancalus	507	*Liomyza* Macquart	823
limbatus (Williston), Opomydas	360	lionota Holmgren, Tipula	41
limbellatus Enderlein, Anacimas	329	**Liopiophila** Duda	712
limbinervis Stein, Phaonia	907	liorhina (Philip), Hybomitra	339
limbipennis (Bigot), Chrysopilus	347	*Lioya* Enderlein	688
limbipennis Coquillett, (Pachycerina)	1115	**Lipara** Meigen	786
limbipennis Macquart, Sericomyia	603	**Lipochaeta** Coquillett	750
limbipennis (Williston), Volucella	602	**LIPOCHAETINI**	750
limbiventris Curran, Euparyphus	307	*Lipoleucopis* Meijere	708
limbiventris (Thomson), Mesograpta	571	lipophleps Alexander, Rhabdomastix	77
limbrocutris Adams, Euparyphus	307	*Lipophleps* Bergroth	74
limbus Osten Sacken, (Cecidomyia)	1115	**Lipoptena** Nitzsch	920
limbus Williston, Microdon	598	*Lipoptenella* Bequaert	920
limenitis Steyskal, Pherbecta	692	**LIPOPTENINI**	920
limitatum (Becker), Rhopalopterum	779	lipsia (Walker), Pegomya	860
Limnellia Malloch	757	**Lipsothrix** Loew	77
Limnia Robineau-Desvoidy	691	liriodendri (Felt), Neolasioptera	273
Limnoagromyza Malloch	796	liriodendri (Osten Sacken), Thecodiplosis	278
Limnobia Meigen	43	**Liriomyza** Mik	801

Entry	Page
Liriope Meigen	97
LIRIOPEIDAE	97
Lispa Latreille	877
Lispe Latreille	877
Lispidea Coquillett	1065
Lispideosoma Reinhard	1063
lispina (Thomson), Schoenomyza	870
LISPINAE	877
Lispocephala Pokorny	876
Lispoides Malloch	879
Litanomyia Melander	472
Lithohypoderma Townsend	1112
liticen Melander, Mythicomyia	417
litoralis Allen, Senotainia	939
litoralis (Cole), Atissa	735
litoralis Curran, Laphystia	376
litoralis Curran, Spaziphora	836
litoralis Painter, Geron	423
litoralis Painter & Hall, Poecilanthrax	442
litoralis Van Duzee, Dolichopus	492
litoralis Townsend, Stomatomyia	1058
litorea (Aldrich), Trixoscelis	818
litorea Cresson, Glenanthe	737
litorea (Fallén), Scatophaga	†838, 839
litorea (Fallén), Spilogona	†880
litorella (Fallén), Schoenomyza	870
Litorella Rondani	869
Litorhynchus Macquart	†443
litorosa (Reinhard), Boettcheria	948
littoralis Blanchard, Epiplagiops	1053
littoralis (Dyar), Telmatoscopus	94
littoralis (Malloch), Homoneura	698
littoralis (Malloch), Paraprosalpia	866
littoralis Malloch, Phyllogaster	876
littoralis (Say), Mycomya	221
littoralis (Townsend), Stomatomyia	1058
littoris Cole, Lasiopogon	377
litturata (Olivier), Euantha	1030
liturata Loew, Rhamphomyia	464
liturata (Meigen), Hylemya	848
liturata Melander & Spuler, Piophila	711
liturata Williston, Acrocera	406
liturata Williston, Spilomyia	613
liturata (Olivier), Euantha	1030
lituratus Adams, Chlorops	789
lituratus Becker, Hippelates	775
lituratus (Loew), Teuchocnemis	608
litus (Coquillett), Epacmus	428
lituus Osten Sacken, (Cecidomyia)	295
livescens Reinhard, Dolichotarsus	1042
livia (Osten Sacken), Villa	434
livida Hall, Calliphora	929
livida Wehr, Cheilosia	585
lividifrons Van Duzee, Gymnopternus	499
lividipes Van Duzee, Hydrophorus	505
lividiventris Huckett, Pegomya	860
lividiventris (Zetterstedt), Lasiops	904
lividus Curran, Mydas	358
livingstoni (Coquillett), Calliopum	696
Lixophaga Townsend	1043
lobata (Felt), Coquillettomyia	281
lobata (Felt), Hyperdiplosis	284
lobata Felt, Rhopalomyia	266
lobata Garrett, Dixa	101
lobata Huckett, Hylemya	846
lobata Loew, Sphegina	594
lobata Shewell, Minettia	701
lobatus Hardy, Bibio	193
lobatus Loew, Dolichopus	492
lobatus Van Duzee, Medetera	509
lobeliae Coquillett, (Exorista)	1107
lobicornis (Osten Sacken), Myelaphus	379
lobifera Alexander, Limnophila	67
lobifera Curran, Dictya	689
lobiferus (Say), Glyptotendipes	172
lobiger (Kieffer), Chironomus	169
lobiochaeta Sturtevant & Wheeler, Ptilomyia	736
Lobioptera Wahlberg	732
lobipennis Van Duzee, Dolichopus	497
Lobodiplosis Felt	281
Lobopteromyia Felt	276
lobosa (Pettey), Bradysia	233
lobulatus Curran, Archytas	1000
lobulifera Malloch, Sphegina	594
lobus Beck, Chironomus	169
loewi Banks, Leptogaster	362
loewi (Becker), Nanna	831
loewi Cole, Ocnaea	403
loewi Cresson, Notiphila	747
loewi Garrett, Suillia	809
loewi James, Cordilura	828
loewi Painter, Phthiria	420
loewi Sedman, Baccha	573
loewi Steyskal, Limnia	691
loewi Steyskal, Tetanocera	694
loewiana Alexander, Tipula	37
Loewiella Williston	366
loewii Hendel, Phytomyza	804
loewii Kertész, Pipunculus	553
loewii (Williston), Ceriana	615
Loewimyia Sabrosky	825
loganensis Alexander, Tipula	30
logani (Johannsen), Micropsectra	178
loisae Stone & Jamnback, Cnephia	185
loloensis Alexander, Limonia	43
lomae Cole, Nicocles	380
LOMATIINAE	428
Lonchaea Fallén	717
LONCHAEIDAE	715
Lonchoptera Meigen	531
LONCHOPTERIDAE	530
LONCHOPTEROIDEA	530
longala Patterson & Wheeler, Drosophila	767
longeoblita Steyskal, Empis	460
longibara Van Duzee, Rhaphium	513
longicarina (Tothill), Mericia	1008
longicauda (Hendel), Otites	646
longicauda Loew, Pelastoneurus	502
longicauda Loew, Rhamphomyia	464
longicauda (Strobl), Hylemya	849
longicauda Van Duzee, Tetragoneura	228
longicauda Van Duzee, Thrypticus	511
longicauda (Zetterstedt), Pseudocoenosia	875
longicella Macquart, (Asilus)	401
longiceps Loew, Dilophus	195
longicorne Van Duzee, Parasyntormon	516
longicorne (Van Duzee), Rhaphium	512
longicornis Banks, Dioctria	371
longicornis Bigot, Stictomyia	654
longicornis (Coquillett), Dziedzickia	218
longicornis (Coquillett), Odontoloxozus	637
longicornis (Coquillett), Pseudocalliope	703
longicornis (Coquillett), Schizotachina	1067
longicornis Dietz, Ula	53
longicornis (Doane), Ormosia	86
longicornis (Fallén), Hyperectenia	1070
longicornis Felt, Dicrodiplosis	279
longicornis Huckett, Pegomya	860
longicornis Johannsen, Boletina	217
longicornis Loew, Rhamphomyia	464
longicornis Loew, Spilomyia	613
longicornis Loew, Xylophagus	297
longicornis Melander, Ceratempis	467
longicornis Patterson & Wheeler, Drosophila	764
longicornis Reinhard, Masiphyomyia	1085
longicornis (Say), Thecophora	631
longicornis Snow, Eutreta	661
longicornis Stannius, Dolichopus	492
longicornis Sturtevant & Wheeler, Hyadina	751
longicornis (Walker), Hexatoma	70
longicornis (Wiedemann), Lespesia	1102
longicornis Williston, Lonchaea	717
longicornis (Zetterstedt), Helina	887
longicornus (Doane), Ormosia	86
longicosta Spuler, Leptocera	724
longiforceps Brooks, Pseudotachinomyia	1057
longiforceps Tothill, Gonia	1076
longifringa Haseman, Psychoda	96
longifrons (Brues), Phalacrotophora	542
longifurca Kieffer, Brillia	154
longifurca Melander, Clinocera	468
longigena Garrett, Pseudoleria	812
longigenoidea Gill, Pseudoleria	812
longiglossa (Philip), Hybomitra	339
longihirtus Van Duzee, Chrysotus	522
longilamellatus Parent, Pelastoneurus	502
longilamellus Harmston & Knowlton, Hercostomus	498
longilobum Peterson & DeFoliart, Prosimulium	183
longimaculata Stein, Coenosia	872
longimana Fallén, Tanypeza	641

INDEX 1625

	Page		Page
longimana (Felt), Didactylomyia	254	longus Osten Sacken, Tabanus	334
longimana Melander, Mythicomyia	417	longus (Walker), Cosmetopus	834
longimana (Pokorny), Pegomya	860	longus (Wiedemann), Proctacanthus	399
longimana Van Duzee, Medetera	510	lonicera Felt, Rhopalomyia	266
longimanus (Kieffer), Diamesa	152	lonicerae (Robineau-Desvoidy), Phytagromyza	803
longimanus Loew, Chrysotus	522		
longimanus Loew, Cyrtopogon	369	lopha Kraft & Cook, Zabrachia	318
longimanus Loew, Dolichopus	492	**Lophosceles** Ringdahl	908
longinasus Shannon, Cynorhinella	589	*Lophosiocera* Townsend	1065
longinervis Van Duzee, Medetera	510	lophyri (Townsend), Diplostichus	1053
longinervis Van Duzee, Sympycnus	527	*loraria* (Loew), Thecophora	631
longinqua Van Duzee, Medetera	510	**Lordiphosa** Basden	767
longipalpe Curran, Rhaphium	513	**Lordotus** Loew	412
Longipalpifer Levi-Castillo	116	*lorena* (Roback), Blaesoxipha	946
longipalpis Hardy, Chrysopilus	346	loreta Cresson, Ochthera	753
longipalpis Malloch, Hylemya	854	loricatus Melander, Oligodranes	422
longipalpis Melander, Anthepiscopus	455	loriola Reinhard, Sipholeskia	1036
longipalpis Roth, Psorophora	110	loripedis Coquillett, Empis	460
longipalpis (Van Duzee), Thecophora	632	loripedis Coquillett, Rhamphomyia	464
longipalpus Randolph & O'Neill, Psorophora	110	**Lotobia** Lioy	719
longipennis (Fabricius), Minettia	701	lotus (Williston), Dasysyrphus	564
longipennis (Felt), Aprionus	251	lotus Williston, Eclimus	426
longipennis Loew, Dolichopus	492	loughnani Edwards, Culicoides	131
longipennis (Loew), Phytobia	799	louisianae Melander, Limnia	691
longipennis (Loew), Piophila	712	louisianensis (Dodge), Sarcophaga	958
longipennis Loew, Rhamphomyia	464	lovetti (Curran), Neocnemodon	581
longipennis (Loew), Sphaeromias	138	lovetti Curran, Xylota	606
longipennis (Loew), Suillia	809	lovetti Melander, Chelifera	471
longipennis (Macquart), Megistocera	18	*lovetti* (Van Duzee), Robertsonomyia	630
longipennis (Malloch), Megaselia	539	lowii Theobald, Uranotaenia	108
longipennis Melander, Syneches	448	**Loxocera** Meigen	638
longipennis Melander, Tachyempis	475	loxocerata (Fallén), Hexamitocera	841
longipennis (Ringdahl), Paraprosalpia	866	*loxostegeae* (Reinhard), Pseudoperichaeta	1106
longipennis (Schummel), Limonia	47	*lucasi* (Bellardi), Dizonias	372
longipennis (Wiedemann), Strauzia	676	*lucens* Hardy, Bibio	193
longipennis (Wirth), Ceratopogon	133	lucens Loew, Sargus	302
longipes (Fabricius), Cholomyia	†1029	lucens Meigen, Lipara	786
longipes Hardy & Knowlton, Pipunculus	553	lucens, Melander, Mythicomyia	418
longipes Johnson, Tipula	38	lucens Townsend, Drepanoglossa	1060
longipes Loew, Bibio	193	lucentis Reinhard, Xanthocera	1026
longipes (Loew), Bicellaria	451	lucescens Huckett, Eremomyia	857
longipes Loew, Empis	460	lucia Alexander, Erioptera	83
longipes (Loew), Penthetria	192	lucida Alexander, Teucholabis	76
longipes (Loew), Renocera	692	lucida Chillcott, Fannia	895
longipes (Loew), Rhaphium	513	lucida Doane, Tipula	37
longipes (Loew), Stenoscinis	779	*lucida* Van Duzee, Physocephala	628
longipes (Walker), Clinocera	468	lucidiventris (Zetterstedt), Hydrophoria	864
longipile Gill, Anorostoma	812	lucidus (Meigen), Hyalurgus	1026
longipilis Banks, Myopa	630	lucifer Adams, Chrysopilus	346
longipilosus (Wehr), Cheilosia	587	lucifer (Fabricius), Poecilanthrax	442
longiplata (Haseman), Telmatoscopus	94	*lucifer* Johannsen, Chironomus	169
longipulvilli Tothill, Gonia	1076	*lucifer* (Osten Sacken), Eclimus	426
longirostris Johnson, Rhynencina	659	lucifuga Kessel, Agathomyia	548
longirostris Macquart, Exoprosopa	444	lucifuga Spuler, Leptocera	724
longirostris Melander, Oligodranes	422	**Lucilia** Robineau-Desvoidy	927
longiseta (Curran), Neocnemodon	581	LUCILIINI	926
longiseta Fisher (Platyura)	1115	lucina Reinhard, Dyscolomyia	1023
longiseta Reinhard, Muscopteryx	1025	lucipeda Reinhard, Metopia	937
longisetosa Alexander, Trichocera	16	*lucorum* (Fallén), Quadrularia	889
longispina (Pettey), Bradysia	233	lucorum (Linnaeus), Leucozona	563
longispina Van Duzee, Mycomya	220	lucta Snow, Cheilosia	586
longispinosa Malloch, Coenosia	872	luctator Melander, Platypalpus	480
longispinosa (Malloch), Liriomyza	802	luctifer (Osten Sacken), Eclimus	426
longistylata Alexander, Limonia	50	luctifera (Loew), Pherbellia	686
longistylata McAlpine, Earomyia	716	luctifera Loew, Rhamphomyia	464
longistylatum Shewell, Simulium	188	luctuosa Bigot, (Alophora)	970
longitalus (Van Duzee), Condylostylus	484	luctuosa Cresson, Hydrellia	744
longitarsis (Malloch), Forcipomyia	126	luctuosa (Kirby), Iteaphila	454
longitarsis Melander, Microphorella	452	luctuosa Loew, Rhamphomyia	464
longitarsus Goetghebuer, Metriocnemus	161	luctuosa Meigen, Mycetophila	212
longitibia Goetghebuer, Smittia	162	luctuosa (Meigen), Phytobia	799
longitudinalis (Loew), Strauzia	676	*luctuosa* (Meigen), Velocia	†1051
longiunguis (Tothill), Nowickia	994	luctuosa Melander, Gloma	454
longiuscula (Walker), Palpexorista	†1056	luctuosa Osten Sacken, Gnophomyia	73
longiusculus Hine, Tabanus	334	luctuosa Spuler, Leptocera	724
longiventris (Johannsen), Elachiptera	777	luctuosum Meigen, Orygma	682
longiventris (Loew), Baccha	573	**Lucumaphila** Stone	673
longiventris Loew, Tipula	31	lucyae Martin, Cerotainiops	388
longua Garrett, Sciophila	225	ludens (Loew), Anastrepha	673
longula (Becker), Lasiopleura	774	ludificata Quate, Pericoma	92
longula (Fallén), Hylemya	849	ludis (Coquillett), Heteropogon	374
longula (Johnson), Strongylophthalmyia	641	ludoviciana Alexander, Tipula	28
longulus (Loew), Eribolus	778	ludoviciana Wirth, Probezzia	138
longulus (Loew), Ospriocerus	381	lugens (Johannsen), Bradysia	233
Longurio Loew	18	lugens Loew, Baccha	573
longus Aldrich, Dolichopus	492	*lugens* Loew, Bibio	194

	Page		Page
lugens (Loew), Nephrotoma	21	luteipennis Loew, Mydas	358
lugens Loew, Rhamphomyia	464	luteipennis (Osten Sacken), Pseudolimnophila	62
lugens Loew, Scatella	757		
lugens Loew, Xylophagus	297	luteipes (Camras), Thecophora	631
lugens Melander, Oligodranes	422	luteipes Johnson, Beris	301
lugens (Wiedemann), Allodia	205	luteipes Melander & Spuler, Sepsis	683
lugens Wiedemann, Chrysops	326	luteipes (Van Duzee), Calyxochaetus	528
luggeri (Aldrich), Anevrina	532	luteipes (Williston), Exodontha	301
luggeri Nicholson & Mickel, Simulium	189	luteipyga Alexander, Dactylolabis	62
luggeri Townsend, Metopia	937	luteisquama (Zetterstedt), Helina	887
luglani Jones & Wirth, Culicoides	129	luteiventer Curran, Rhamphomyia	464
lugubre Loew, Rhaphium	513	luteiventris Loew, Rhamphomyia	464
lugubria Sabrosky, Oscinella	780	luteiventris Schmitz, Gymnophora	542
lugubrina (Zetterstedt), Fannia	895	*luteoala* (Garrett), Schroederella	814
lugubris Cresson, Ditrichophora	739	luteogrisea Wirth & Williams, Dasyhelea	127
lugubris Harris, Penthetria	192	luteola Alexander, Limnophila	66
lugubris Loew, Pelastoneurus	502	*luteola* (Coquillett), Stomosis	730
lugubris (Macquart), Physocephala	628	luteola (Coquillett), Triachora	1082
lugubris (Macquart), Whitneyomyia	331	luteola Dietz, Ormosia	86
lugubris (Meigen), Triphleba	534	luteola Malloch, Beckerina	537
lugubris Melander, Bicellaria	451	luteola Malloch, Cordilura	830
lugubris Williston, Cheilosia	588	*luteola* Malloch, Pegomya	859
lugubris (Wulp), Gaediophana	1079	luteola (Pettey), Bradysia	233
lugubris (Wulp), Nowickia	995	luteolus Hall, Lordotus	412
lumbalis Harris, Chrysops	325	luteopalpus Curran, Chrysotimus	529
luna (Felt), Corinthomyia	249	luteopennis Philip, Chrysops	326
luna Hull & Fluke, Cheilosia	585	*luteopilosum* Curran, Chrysotoxum	582
luna Lovett, Criorhina	611	luteovenus Root & Hoffman, Culicoides	128
lunata (Fabricius), Stomorhina	923	lutescens Curran, Anorostoma	812
lunatifrons Harris, (Anthomyia)	1115	lutescens Johnson & Johnson, Lordotus	412
lunatifrons (Zetterstedt), Pegomya	860	luteus Coquillett, Saropogon	383
Lunatipula Edwards	34	luteus Curran, Chrysotimus	529
lundbecki Collin, Melanostoma	575	luteva (Walker), Phaonia	906
lundbecki Curran, Dolichopus	492	**Lutomyia** Aldrich	811
lundbecki (Nielsen), Nephrotoma	21	lutosa (Stenhammar), Leptocera	721
lundbecki (Soot-Ryen), Metasyrphus	†561	lutosopra Garrett, Prodiamesa	153
lundbeckii Johannsen, Metriocnemus	161	lutosus (Townes), Stictochironomus	178
lundbeckii Ringdahl, Pegomya	860	*lutzi* Curran, Cophura	367
lundstroemi Landrock, Exechia	206	lutzi Curran, Euxesta	651
Lundstroemia Kieffer	180	lutzi Curran, Exoprosopa	445
lundstromi Goetghebuer, Procladius	149	lutzi Curran, Jurinella	1001
lunifer Loew, Gymnopternus	499	lutzi Curran, Lepidanthrax	441
lunifera Hering, Epochra	677	lutzi (Curran), Nowickia	994
luniger (Meigen), Metasyrphus	561	lutzi Curran, Volucella	601
Lunomyia Curran & Fluke	620	lutzii Sturtevant, Drosophila	767
lunula Martin, Beameromyia	362	*Lutzomyia* Curran	963
lunulata Bigot, Myolepta	589	*Lycia* Robineau-Desvoidy	700
lunulatus (Meigen), Dasysyrphus	564	**Lyciella** Collin	700
lunulatus Meigen, Helophilus	619	lycopi Felt, Lasioptera	272
lupatus Melander, Platypalpus	480	*Lycoria* Meigen	230
lupina (Williston), Criorhina	611	**Lycoriella** Frey	231
lupini (Coquillett), Hylemya	849	LYCORIIDAE	229
lupini Felt, Dasineura	260	**Lydella** Robineau-Desvoidy	†1039, 1102
lupini Felt, Lasioptera	272	LYDELLINI	1096
lupulina (Fabricius), Minettia	701	**Lydellohoughia** Townsend	1044
lupus Whitney, Chrysops	325	**Lydina** Robineau-Desvoidy	1014
luravi (Johannsen), Bradysia	233	**Lydinolydella** Townsend	1044
lurida (Coquillett), Orfelia	203	**Lygistorrhina** Skuse	204
lurida (Garrett), Heptagyia	152	LYGISTORRHININAE	204
lurida (Loew), Geomyza	821	lygropis Alexander, Tipula	37
lurida Reinhard, Siphona	1064	**Lynchia** Weyenbergh	918
lurida Walker, (Sciara)	236	**Lynchiella** Lahille	107
luridipennis Loew, Tetanops	648	*Lyperosia* Rondani	914
lustrans (Reinhard), Catharosia	971	**Lyperosiops** Townsend	914
lutaria (Fabricius), Scatophaga	839	**Lypha** Robineau-Desvoidy	1011
lutatius (Walker), Cyrtopogon	369	LYPHINI	1011
lutea Becher, Trichocera	16	lyraformis Shewell, Minettia	701
lutea Cole, Agathomyia	548	lyrata Williston, Dichocera	1013
lutea (Coquillett), Aciurina	670	lyratus Osten Sacken, Cyrtopogon	369
lutea (Coquillett), Physoclypeus	702	lyrifera Dietz, Tipula	37
lutea Doane, Limnophila	66	lyristes Melander, Platypalpus	480
lutea Loew, Cordilura	829	lysimachiae (Beutenmüller), Dasineura	260
lutea Loew, Hilara	456	*lysinoe* (Walker), Helina	889
lutea (Loew), Neoleria	813	**Lytogaster** Becker	752
lutea (Malloch), Stilobezzia	135		
lutea (McAtee), Andrenosoma	386	mabelae Brimley, Ceraturgus	365
lutea (Meigen), Liriomyza	802	mabelae (Cresson), Ptilomyia	736
lutea (Meigen), Megaselia	539	mabelae (Melander), Hesperempis	455
lutea Painter, Lepidophora	426	mabelana Alexander, Erioptera	83
lutea Tollet, Neoempheria	222	macalpinei Chillcott, Fannia	895
lutea (Townsend), Siphona	1064	macateei (Alexander), Limonia	43
luteiceps Sabrosky, Oscinella	780	macateei (Alexander), Pedicia	55
luteicornis Cresson, Pipunculus	552	*macateei* James, Bibio	194
luteipennis (Curran), Carpocalis	576	macateei (Malloch), Anevrina	532
luteipennis Knab & Shannon, Tanypeza	641	*macateei* Malloch, Canace	733
luteipennis Loew, Dolichopus	493	macateei (Malloch), Stenochironomus	177

	Page		Page
Macateeia Malloch	868	macula Martin, Beameromyia	362
macclurei (Shaw), Bradysia	233	macularis Johannsen, Neoempheria	222
maccrackenae Dyar & Knab, Culiseta	117	macularis (Zetterstedt), Melangyna	567
maccus Loew, (Diplosis)	292	maculata (Cole), Aciurina	670
macdonaldi Quate, Telmatoscopus	94	*maculata* (Coquillett), Olfersia	919
macdunnoughi Brooks, Psalidopteryx	1067	maculata (Cresson), Pherbellia	686
macdunnoughi Cole, Thereva	353	*maculata* (Curran), Peronyma	660
macella Reinhard, Doryphorophaga	1042	maculata (Doane), Dicranota	57
macellaria Egger, Ephydra	754	maculata Felt, Colpodia	253
macellaria (Fabricius), Cochliomyia	924	maculata Felt, Didactylomyia	254
Macellocerus Mik	502	maculata Loew, Clinocera	468
macer Loew, Systropus	424	*maculata* (Meigen), Limonia	45
macerinus (Walker), Heteropogon	374	maculata Olivier, Oxycera	307
macfarlanei (Jones), Bradysia	233	maculata (Scopoli), Graphomya	910
macfiei Wirth, Monohelea	135	maculata (Stein), Leucophora	867
macgillisi Chillcott, Bolbomyia	298	maculata Thompson, Leucopis	709
macgregori Felt, Mycodiplosis	283	maculata Van Duzee, Neurigona	518
Machimus Loew	396	maculata Wilcox, Itolia	376
macilenta Loew, Rhamphomyia	464	*maculata* (Williston), Epalpus	1002
mackenziei Steyskal, Tujunga	649	maculatipennis Say, Tipula	41
macnabeana Alexander, Chionea	72	maculatum Darlington, Anorostoma	812
macnabi Alexander, Cladura	72	maculatus Banks, Ommatius	398
macnabi Alexander, Tipula	37	maculatus Cazier, Rhaphiomidas	357
macneilli Kessel, Agathomyia	548	*maculatus* (Curran), Neorhynchocephalus	402
macneilli Quate, Telmatoscopus	94	maculatus Melander, Aphoebantus	429
Macorcoenosia Malloch	870	maculatus Melander, Oligodranes	422
macquarti (Banks), Laphria	390	maculiceps Becker, Parectecephala	792
macquarti Cresson, Hyadina	751	maculicornis (Hine), Neopachygaster	317
macquarti Philip, Chrysops	326	maculicornis Sabrosky, Gaurax	782
macquarti Shannon, Polyblomyia	616	maculicosta (Coquillett), Limonia	43
macquarti Zetterstedt, Iteaphila	454	maculifemorata Hering, Paroxyna	666
MACQUARTIINI	1022	*maculifer* (Bigot), Hybomitra	340
macra Johannsen, Gnoriste	219	maculifer (Bigot), Rhagio	345
macracantha Alexander, Tipula	37	maculifer Walker, Bombylius	409
Macrobrachius Dziedzicki	214	maculifrons (Bigot), Melangyna	567
macrocephala (Giglio-Tos), Volucella	601	*maculifrons* (Bigot), Myopa	630
macrocephala (Hendel), Tetanops	649	maculigera Foote, Trupanea	667
Macrocera Meigen	201	maculinervis James, Heteropogon	374
macrocera (Say), Cerotainia	387	maculipennis Aldrich, Lypha	1011
macrocera (Say), Limnophila	63	maculipennis Bigot, Cephalia	643
macrocera (Say), Nephrotoma	21	maculipennis (Cole), Phthiria	420
macrochaeta Loew, Notiphila	747	maculipennis (Coquillett), Monohelea	135
Macrochira Zetterstedt	711	maculipennis Greene, Lestodiplosis	285
Macrocoenia Malloch	870	*maculipennis* Harris, (Chironomus)	1115
Macrocoenosia Malloch	870	maculipennis (Macquart), Cyrtopogon	369
macrodon Frey, Bradysia	233	maculipennis (Macquart), Sphecomyiella	658
macrofemoralis Curran, Neoascia	595	maculipennis Malloch, Camptoprosopella	697
macrofemoralis Curran, Pipiza	579	maculipennis Meigen, Trichocera	16
macrofila (Felt), Schizomyia	267	*maculipennis* Melander, Dryomyza	681
macrofilus (Felt), Arthrocnodax	283	maculipennis Rempel, Chasmatonotus	159
macroglossus Westwood, Pangonius	320	*maculipennis* (Say), Leia	228
macroides Curran, Gnoriste	219	*maculipennis* Say, Tipula	41
macrolabis Loew, Tipula	37	maculipennis (Spuler), Copromyza	720
macrolabis (Wiedemann), Efferia	395	maculipennis Spuler, Leptocera	724
macrolaboides Alexander, Tipula	37	*maculipennis* Townsend, Eristalis	624
Macromeigenia Brauer & Bergenstamm	1086	maculipennis Van Duzee, Hydrophorus	505
Macromelanderia Curran	678	maculipennis Walker, Tachydromia	†474, 474
Macromya Robineau-Desvoidy	992	maculipennis (Wiedemann), Discomyza	740
MACROMYINI	992	*maculipennis* Wulp, Cordilura	829
macroneura (Malloch), Neurohelea	137	maculipennis (Zetterstedt), Helina	887
Macronevra Macquart	198	*maculipes* (Bigot), Neoplasta	472
Macronichia Rondani	936	maculipes Sabrosky, Gaurax	782
Macronychia Rondani	936	maculipes (Walker), Pelastoneurus	502
Macropelopia Thienemann	145	maculipleura Alexander, Tipula	28
macrophallus Alexander, Pedicia	55	maculitarsis Curran, Themira	684
macrophallus (Dietz), Tipula	24	maculitarsis Van Duzee, Dolichopus	493
Macrophorbia Malloch	856	maculiventris Huckett, Coenosia	872
macropogon (Bigot), Menetus	1018	*maculiventris* (Malloch), Azelia	899
Macroptera Becker	476	maculiventris (Van Duzee), Gymnopternus	499
macroptera Loew, Hilara	456	maculiventris Westwood, Mydas	359
macroptera Patterson & Wheeler, Drosophila	764	maculosa Coquillett, (Hypostena)	1108
macroptera (Pettey), Bradysia	233	maculosa (Coquillett), Leucophenga	761
macropterus Loew, Geron	423	maculosa (Coquillett), Trypeta	677
macropus Walker, Sepedon	693	maculosa Felt, Contarinia	277
macropyga (Frey), Spilogona	884	maculosa Felt, Dasineura	260
Macrorchis Rondani	875	*maculosa* Guthrie, Mycetophila	212
Macrosargus Bigot	303	maculosa (Loew), Parectecephala	792
macrospina Stalker & Spencer, Drosophila	764	maculosa Loew, Stratiomys	312
macrotis Bau, Cephenemyia	1111	maculosa (Malloch), Phytobia	798
macrotis Brauer, Cephenemyia	1111	*maculosa* (Meigen), Nemorilla	1088, †1089
macrura Coquillett, Rhampomyia	462	maculosa Namba, Rivellia	656
macswaini Wirth, Forcipomyia	126	maculosus (Coquillett), Eucyrtopogon	373
macswaini Wirth, Telmatogeton	163	maculosus Ewen, Atrichopogon	123
Mactomyia Reinhard	1014	maculosus Painter, Parabombylius	410
macula (Cole), Villa	437	maddocki (Alexander), Paradelphomyia	60
macula (Loew), Poecilominettia	703	madera Doane, Tipula	33

	Page
madina Dietz, Tipula	37
madisoni Parker, Sarcofahrtia	942
Madiza Fallén	730, †786
madizans (Fallén), Trimerina	742
madizina (Hendel), Tylomyza	797
MADIZINAE	729
Madremyia Townsend	1103
maehleri Cazier, Rhaphiomidas	357
Mackistocera Wiedemann	18
magdalena Dietz, Limnophila	63
magdalenae Cresson, Euxesta	651
magdalenae Cresson, Tetanops	648
magdalenae Wheeler, Hydrophorus	505
magistri (Aldrich), Enlinia	529
magna (Aldrich), Blaesoxipha	944
magna Brooks, Onychogonia	1077
magna (Coquillett), Minettia	701
magna (Coquillett), Rhysophora	743
magna Cresson, Chaetopsis	650
magna Garrett, Boletina	217
magna Garrett, Mycomya	221
magna Johnson, Arthropeas	298
magna Osten Sacken, Toxorhina	90
magna (Walker), Ocnaea	403
magnafumosa Stalker & Spencer, Drosophila	764
magnantenna James, Dolichopus	493
magnapennis (Johnson), Phasia	969
magnaquinaria Wheeler, Drosophila	764
magnicallus (Stone), Stenotabanus	329
magnicauda (Lundström), Limonia	47
magnicornis Cresson, (Zabrachia)	319
magnicornis (Loew), Phytobia	799, †799
magnicornis (Townsend), Paradidyma	1033
magnicornis Van Duzee, Chrysotus	522
magnicornis Van Duzee, Paracleius	501
magnicornis (Zetterstedt), Phaonia	908
magnicornis (Zetterstedt), Tachina	998
magnifica Hine, Pedicia	53
magnifica (Walker), Archilestris	364
magnificus (Johannsen), Leptomorphus	223
magnifolia Alexander, Tipula	25
magnipalpis (Becker), Goniopsita	786
magnipalpus Van Duzee, Thinophilus	508
magnipennis (Felt), Culiseta	117
magnipennis (Johannsen), Jenkinshelea	137
magnipunctata (Malloch), Spilogona	881
magnistyla Alexander, Cryptolabis	78
magnum Dyar & Shannon, Prosimulium	183
magnus Bellardi, Promachus	400
magnus (Coquillett), Pachyhelea	140
magnus (Osten Sacken), Eclimus	426
mahican Sturtevant & Dobzhansky, Drosophila	768
mainensis Alexander, Gonomyia	76
mainensis (Alexander), Helius	51
mainensis Alexander, Tipula	37
mainensis Cresson, Pipunculus	552
mainlandi Patterson, Drosophila	764
majalis Osten Sacken (Cecidomyia)	294
major (Alexander), Dicranota	58
major Becker, Diplotoxa	792
major (Bigot), Ptilodexia	989
major Brauer & Bergenstamm, Pseudoperichaeta	†1106
major Coquillett, Trichobius	921
major Curran, Wagneria	1020
major Felt, Neolasioptera	273
major Felt, Rhopalomyia	266
major (Hine), Caloparyphus	308
major Linnaeus, Bombylius	409
major (Malloch), Coenosia	874
major (Malloch), Dasyhelea	127
major Malloch, Leucopis	709
major (Malloch), Pegomya	860
major Malloch, Phytomyza	805
major (Strobl), Ophiomyia	797
major Van Duzee, Chrysotus	522
majus Cook, Rhegmoclema	239
majuscula Alexander, Pedicia	55
majuscula (Coquillett), Macrorchis	875
majuscula Frick, Phytobia	800
majuscula Loew, Ptiolina	347
majuscula Sherman (Boletina)	1115
malacosomae (Curran), Lespesia	1102
malacosomae Sellers, Carcelia	1092
malacosomae (Townsend), Lespesia	1101

	Page
malaisei (Ringdahl), Spilogona	881
malformans Melander & Spuler, Themira	684
malheurensis Alexander, Tipula	35
mali Fitch (Molobrus)	236
mali (Kieffer), Dasineura	260
malifero Garrett, Procladius	149
malkini Alexander, Tipula	24
malleola (Bigot), Peleteria	998
mallitosa Huckett, Phyllogaster	876
mallochi Alexander, Tipula	37
mallochi (Cole), Ceratopogon	132
mallochi (Curran), Melangyna	567
mallochi (Duda), Pseudogaurax	782
mallochi Frota-Pessoa, Drosophila	765
mallochi Hardy & McGuire, Ptiolina	347
mallochi Hoffman, Stilobezzia	135
mallochi Huckett, Pegomya	860
mallochi Kieffer, Chironomus	167
mallochi Kieffer, Orthocladius	156
mallochi Sabrosky, Oscinella	780
mallochi Shewell, Camptoprosopella	697
mallochi Steyskal, Tetanocera	694
mallochi (Sturtevant), Pelomyiella	727
mallochi (Walley), Ablabesmyia	148
mallochi Wirth, Bezzia	142
mallochi Wirth, Nilobezzia	138
Mallochiella Melander	731
mallochii Huckett, Neohylemyia	867
Mallochohelea Wirth	139
Mallochomyia Townsend	925
Mallochomyza Hendel	697
Mallophora Macquart	396
Mallophorina Curran	397
mallos Walker, Rhamphomyia	464
Mallota Meigen	620
Malpighia Enderlein	20
maltana Reinhard, Siphosturmia	1087
malvae (Burgess), Phytobia	800
malyschevi Dorogostajskij, Rubzov, & Vlasenko, Simulium	188
mamillata Loew, Tropidia	609
mana Pritchard, Micromyia	250
manahatta Alexander, Tipula	28
manca Coquillett, Empis	460
manca Coquillett, Rhamphomyia	464
manca (Greene), Elfia	1065
manca Johannsen, Cordyla	208
manca Osten Sacken, Gonomyia	75
Mancia Coquillett	441
mancipennis Coquillett, Metacosmus	427
mandalota Melander, Anthalia	450
mandan Alexander, Tipula	33
mandibulata Aldrich, Melanderia	507
maneei (Hine), Tolmerus	401
manicata (Doane), Ormosia	86
manicata (Meigen), Fannia	895
manicatus Collin, Hydrophorus	505
manicula Van Duzee, Dolichopus	493
manifestus Walker, Tabanus	336
manitoba Brooks, Mericia	1008
manitobensis (Alexander), Arctoconopa	80
manitobensis Curran, Microdon	598
mannerheimi Zetterstedt, Dolichopus	493
manni Chilcott, Platypalpus	480
Manota Williston	204
MANOTINAE	204
Mansonia Blanchard	107
manteri (Johnson), Platyura	204
Mantidophaga Townsend	946
Mantipeza Rondani	471
mantis (De Geer), Ochthera	753
mantivora Riley, (Sarcophaga)	961
manuleata (Loew) Lauxaniella	700
marceda Sherman, Tetragoneura	228
marchandi Alexander, Limnophila	63
marcida Coquillett, Psilocephala	351
marcidus Coquillett, Aphoebantus	429
marcidus Wheeler, Sympycnus	527
margarita Alexander, Erioptera	83
margarita Alexander, Pedicia	53
margarita Alexander, Protanyderus	90
margarita Alexander, Rhabdomastix	77
margarita Alexander, Tipula	33
margarita Hull & Fluke, Cheilosia	585
margaritus Hull, Brachypalpus	608
margella (Reinhard), Metopotachina	993

INDEX 1629

	Page		Page
marginalis Adams, Mycetobia	191	*Marshamia* Robineau-Desvoidy	1004
marginalis (Aldrich & Webber), Prophryno	1106	martinensis Cole & Wilcox, Lasiopogon	378
marginalis (Banks), Pericoma	92	martini Bromley, Stenopogon	384
marginalis (Cresson), Oligodranes	422	martini Reinhard, Eleodiphaga	1072
marginalis Curran, Arctophyto	986	martini Wilcox, Laphystia	376
marginalis Fabricius, Tabanus	334	martinorum Painter, Apiocera	357
marginalis Frost, Phytomyza	805	**Maruina** Müller	96
marginalis Harris, Tipula	35	MARUININI	96
marginalis Johannsen, Mycomya	221	*marylandana* Del Rosario, Psychoda	96
marginalis Loew, Cyrtopogon	369	marylandica (Malloch), Leucophora	867
marginalis Loew, Stratiomys	312	marylandica Malloch, Lonchaea	717
marginalis (Malloch), Liriomyza	802	marylandica Malloch, Phaonia	907
marginalis (Malloch), Megaselia	541	marylandicus Malloch, Neossos	817
marginalis (Malloch), Oscinella	780	*masamae* Dyar, Aedes	112
marginalis Reinhard, Pseudochaeta	1096	*masconina* Curran, Cordilura	828
marginalis Williston, Xylota	607	masculans Huckett, Hylemya	846
marginata (Aldrich), Blaesoxipha	947	masculina (Aldrich), Blaesoxipha	944
marginata (Aldrich & Webber), Chrysoexorista	1099	**Masicera** Macquart	1103, †1105
marginata Becker, Meromyza	793	MASICERINI	1196
marginata Cook, Swammerdamella	240	*Masiceropsis* Townsend	1101
marginata Coquillett, Phthiria	420	MASIPHYINI	1091
marginata Felt, Aphidoletes	280	**Masiphyomyia** Reinhard	1085
marginata Felt, Rhabdophaga	257	masoni Chillcott, Platypalpus	480
marginata Huckett, Pegomya	860	masoni Chillcott, Symballophthalmus	476
marginata Loew, Cyphomyia	304	*masoni* Coquillett, Rhamphomyia	464
marginata Loew, Dixa	101	massyla (Walker), Chaetopsis	650
marginata (Loew), Phytobia	799	masurius (Walker), Acemya	1036
marginata (Malloch), Cordilura	830	mathesoni Alexander, Gonomyia	74
marginata Melander, Mythicomyia	417	mathesoni Belkin & Hogue, Deinocerites	120
marginata Rübsaamen (Sciara)	236	mathesoni (Curran), Ptilodexia	989
marginata (Say), Idana	645	mathesoni Middlekauff, Aedes	111
marginata (Say), Mesograpta	571	mathesoni (Reinhard), Thelymyia	1106
marginata (Say), Physocephala	628	matura Melander, Euthyneura	449
marginata Say, Volucella	602	*maturus* Johannsen, Chironomus	165
marginata Stein, Hylemya	†845, 852	matutinus Melander, Oligodranes	421
marginata (Walker), Hylemya	850	maudae Coquillett, Platyura	204
marginatum Loew, Anorostoma	812	*maura* (Bigot), Pseudolynchia	919
marginatus Aldrich, Dolichopus	493	maura (Meigen), Ophiomyia	797
marginatus Cole, Aphoebantus	430	maura Wheeler, Medetera	510
marginatus Cole, Ogcodes	406	**Mauromyia** Coquillett	1014
marginatus Johnson & Johnson, Poecilanthrax	443	*maverna* Walker, Trupanea	667
marginatus (Jones), Metasyrphus	561	maxima Coquillett, Toxophora	425
marginatus (Osten Sacken), Eclimus	426	maxima Felt, Hormomyia	288
marginatus Osten Sacken, Eulonchus	404	maxima Johannsen, Mycomya	221
marginatus (Say), Sylvicola	190	*maxima* Sherman, (Allodia)	1115
marginatus Wulp, Sargus	302	maximus (Hardy), Pipunculus	554
marginella (Malloch), Anatopynia	145	*maximus* Stein, Phyllogaster	876
marginella Malloch, Hylemya	852	mayedai Alexander, Tipula	24
marginepunctata (Macquart), Paracantha	662	*Mayetia* Kieffer	263
marguerita Snyder, Dendrophaonia	905	**Mayetiola** Kieffer	263
marguerita Snyder, Helina	887	mazamae Rondani, Lipoptena	920
maria Alexander, Erioptera	82	mcateei Wirth, Forcipomyia	124
mariae Steyskal, Drapetis	477	mcclayi Martin, Cerotainiops	388
marialiciae Brimley, Eulonchus	404	mcclureana Alexander, Limnophila	65
marianapolitana (Ouellet), Thaumatomyia	787	mcdunnoughi Alexander, Limnophila	67
mariannae Alexander, Tipula	37	*mcgregori* (Bromley), Hodophylax	374
marilandica Felt, Asynapta	254	mckinleyana Alexander, Erioptera	83
marilandica Felt, Neocatocha	243	*meadei* Pokorny, Pogonomyia	902
marina Doane, Tipula	33	meadii (Jones), Metasyrphus	561
marina Felt, Aphidoletes	280	*Meckelia* Robineau-Desvoidy	643
marina (Macquart), Fucellia	†842	**Mecynocorpus** Roback	950
marina (Saunders), Smittia	162	medeola Reinhard, Themaatheca	1050
marinensis Cresson, Parascatella	757	**Medetera** Fischer von Waldheim	509
marionae Shaw, Rymosia	207	medetera Melander, Drapetis	477, †478
mariposa Alexander, Hexatoma	70	MEDETERINAE	509
mariposa Alexander, Tipula	37	*Medeterus* Fischer von Waldheim	509
mariposa (James), Calopargrphus	308	media Banks, Dioctria	370
maritima Alexander, Tipula	27	media (Coquillett), Bezzia	141
maritima Curran, Criorhina	612	media Garrett, Pseudoleria	812
maritima Felt, Dasineura	260	*media* James, Stratiomys	311
maritima Haliday, Fucellia	†842, †843	media Shewell, Camptoprosopella	697
maritima (Melander), Pelomyiella	727	media Williston, Brachyopa	592
maritima (Zetterstedt), Chamaemyia	707	mediaconstricta (Fluke), Melangyna	567
maritimus (Townsend), Stenotabanus	329	*mediana* Banks, Platypeza	549
marivirginiae Brimley, Somula	612	mediana Felt, Asynapta	254
markii (Garrett), Dioptopsis	100	mediana Reinhard, Peleteria	998
marmorata Bigot, Baccha	574	medicaginis Huckett, Eremomyia	857
marmorata (Osten Sacken), Limonia	47	*medicinensis* (Strickland), Philocalia	1061
marmorata (Zetterstedt), Helina	†888	*Medina* Robineau-Desvoidy	†1051
marmoratus (Becker), Copromyza	720	mediocre Reinhard, Cattasoma	941
marmoratus Bigot, Microdon	598	mediocris Aldrich, Lixophaga	1043
marmoratus (Townsend), Archytas	1000	*mediolineata* (Ludlow), Aedes	112
marmoreus Johnson & Johnson, Poecilanthrax	442	mediovena Alexander, Rhabdomastix	77
marmoreus (Townes), Stictochironomus	178	mediovittatum Knab, Simulium	187
marshalli Stone & Wirth, Clunio	163	meditabunda (Fabricius), Myospila	890
		medius (Jones), Metasyrphus	561

1630 A CATALOG OF DIPTERA OF NORTH AMERICA

Name	Page
medius (Stein), Lasiops	904
medorae Painter, Bombylius	409
megacantha Alexander, Rhabdomastix	77
megacephala (Loew), Bucephalina	832
megacephala (Loew), Hedriodiscus	315
megacephala (Loew), Volucella	†601, 601
megacephala Malloch, Hylemya	849
megacera Alexander, Ormosia	86
megacera (Macquart), Lampria	388
megacera (Osten Sacken), Hexatoma	69
megacetes Melander, Chersodromia	475
megachile (Coquillett), Mallophorina	397
Megacyttarus Bigot	461
Megaeuloewia Townsend	983
Megagrapha Melander	476
megalabiata Alexander, Tipula	37
megalodonta Alexander, Tipula	25
megalogaster Snow, Microdon	598
Megalopelma Enderlein	223
megalops Alexander, Dicranota	57
Meganeria Cresson	635
meganosa Hull & Fluke, Cheilosia	585
megantica (Curran), Labostigmina	314
megantica (Curran), Pericoma	93
megaparamera Williams, Alluaudomyia	133
Megaparia Wulp	982
Megapariopsis Townsend	982
megaphallus Alexander, Dicranoptycha	52
Megaphorus Bigot	396
Megaphthalma Becker	832
Megaphthalmoides Ringdahl	833
Megapollion Walker	379
Megapollyon Walker	379
megarhabda (Alexander), Erioptera	83
Megarhina Lepeletier & Serville	50
MEGARHININAE	106
Megarhinus Robineau-Desvoidy	107
megarrhina Osten Sacken, Gnoriste	219
Megaselia Rondani	538
Megaselida Rondani	538
megastoma (Boheman), Spilogona	881
megatarsa Fluke & Hull, Cheilosia	586
megatergata Alexander, Tipula	39
megatricha Kertész, Hylemya	†848
megaura Doane, Tipula	37
megerlei (Wiedemann), Agkistrocerus	341
Meghyperus Loew	449
Megistocera Wiedemann	18
Megophtalmidia Dziedzicki	228
megophthalma Alexander, Erioptera	81
Megophthalma Becker	832
Megophthalmidia Dziedzicki	228
Megophthalmum Becker	832
megorchis Melander, Neoplasta	472
meibomiae Beutenmüller (Cecidomyia)	292
meibomiifoliae (Beutenmüller), Hyperdiplosis	284
Meigenia Robineau-Desvoidy	1044
Meigeniella Coquillett	1041
Meigenielloides Townsend	1044
meigenii (Osten Sacken), Ormosia	86
meigenii (Schiner), Wohlfahrtia	942
meigenii Wiedemann, Eristalis	†623
meigenii (Wiedemann), Exoprosopa	445
meigenii Wiedemann, Stratiomys	312
meijerei Collin, Odinia	794
melaena (Loew), Leia	227
melalophae (Allen), Lespesia	1102
mclampodia Loew, Psilocephala	351
melampus (Becker), Elachiptera	776
melampus (Loew), Condylostylus	484
melampus Loew, Ogcodes	407
melampus (Loew), Rhaphium	513
melampyga (Aldrich), Sarcodexia	956
melampyga (Loew), Liriomyza	802
melampyga (Reinhard), Siphosturmia	1087
melampygia Alexander, Dicranoptycha	52
Melanagromyza Hendel	796
melancholica (Becker), Coniosicinella	784
melancholica Johannsen, Boletina	217
melanderi Aldrich, Leiomyza	823
melanderi (Alexander), Hesperoconopa	78
melanderi Alexander, Limonia	47
melanderi Banks, Myopa	630
melanderi (Banks), Nemotelus	310
melanderi Becker, Dolichopus	494
melanderi Brues, Acontistoptera	545
melanderi Cole, Acrocera	406
melanderi Cole, Parathalassius	452
melanderi Cole, Sphegina	594
melanderi Cook, Rhegmoclemina	239
melanderi (Cresson), Athyroglossa	735
melanderi Cresson, Ochthera	753
melanderi Cresson, Parascatella	757
melanderi (Curran), Melangyna	566
melanderi Curran, Melanostoma	575
melanderi Duda, Sepsis	683
melanderi Fisher, Palaeoplatyura	203
melanderi (Garrett), Eccoptomera	811
melanderi (Johnson), Homoneura	699
melanderi Kieffer (Chironomus)	181
melanderi (Malloch), Lasiops	903
melanderi Quate, Pericoma	92
melanderi Sabrosky, Leptocera	721
melanderi Steyskal, Dryomyza	681
melanderi Steyskal, Sepedon	693
melanderi Sturtevant, Drosophila	764
melanderi (Sturtevant), Pelomyiella	727
melanderi Van Duzee, Dolichopus	493
melanderi Vockeroth, Trixoscelis	818
Melanderia Aldrich	507
melanderiana Alexander, Erioptera	82
Melanderomyia Kessel	547
melandri Becker, Dolichopus	494
melanella Frost, Phytomyza	805
melaneura Garrett, Morpholeria	814
Melangyna Verrall	565
melania Townsend, Micromintho	1032
melanica Sturtevant, Drosophila	764
melanica (Townsend), Onychogonia	1077
melanimon Dyar, Aedes	113
melanis (Curran), Epistrophe	564
melanis Hardy & Knowlton, Pipunculus	554
melanissima Sturtevant, Drosophila	764
melanocera Robineau-Desvoidy (Dexia)	1108
melanoceratus Bigot, Conophorus	414
melanocerus Loew, Chlorops	790
melanocerus Loew, Dolichopus	493
melanocerus Melander, Platypalpus	479
melanocerus Wiedemann, Tabanus	334
Melanochaeta Bezzi	776
melanochaeta Melander, Cophura	367
Melanochelia Rondani	†879
Melanoconion Theobald	119
Melanodexia Williston	931
MELANODEXIINI	931
Melanodexiopsis Hall	931
melanogaster (Bigot), Theresia	991
melanogaster Meigen, Drosophila	768
melanogaster Melander, Platypalpus	480
melanogaster Schlinger, Acrocera	406
melanogaster Wiedemann, Laphria	390
melanohalteralis Tucker, Anastoechus	411
Melanomya Rondani	932
MELANOMYINI	932
Melanomyza Malloch	703
Melanonemotelus Hanson	309
melanoneura Loew, Thereva	353
melanopalpa Patterson & Wheeler, Drosophila	765
melanophleba (Loew), Dialineura	352
Melanophora Meigen	963
MELANOPHORINAE	963
MELANOPHRYCTINI	1009
MELANOPHRYINI	1009
Melanophrys Williston	1010
melanopilosus Hardy, Bibio	193
melanopleurus Loew, Cyrtopogon	368
melanopoda (Williston), Thecophora	631
melanopogon (Bigot), Anthrax	432
melanopogon (Bigot), Eclimus	426
melanopogon Hermann, Atomosia	387
melanopogon Steyskal, Nemomydas	359
melanopogon Wiedemann, Laphria	389
melanoprocta Loew, Psilocephala	351
melanops (Meigen), Pentaneura	147
melanoptera (Wiedemann), Pogonosoma	391
melanopus James, Ptecticus	303
melanopygatus (Bigot), Xanthoepalpus	1003
melanorhina (Bigot), Hybomitra	339
melanosa Williston, Sphaerophoria	570
melanoscuta Coquillett, Phthiria	420

	Page
melanosoma (Goetghebuer), Pentaneura	147
melanosoma (Huckett), Spilogona	881
melanosoma Melander, Antichaeta	688
melanosoma Melander, Hemerodromia	471
melanostigma Steyskal, Tetanocera	694
melanostoma (Loew), Clusiodes	807
melanostoma Loew, Stratiomys	312
Melanostoma Schiner	574
MELANOSTOMATINI	574
melanostomus Loew, Eristalis	623
melanosus Johnson & Johnson, Lordotus	412
melanosus Williston, Eclimus	426
melanothrix Melander, Pteromicra	687
melanotum Sabrosky, Gaurax	782
melantera James, Odontomyia	316
melanura Bigot Exoprosopa	444
melanura Cazier, Apiocera	357
melanura Chillcott, Fannia	895
melanura (Coquillett), Culiseta	117
melanura (Loew), Acidogona	663
melanura Meigen, Sarcophaga	958
melanura Miller, Drosophila	765
melanus Harmston & Knowlton, Campsicnemus	526
melas Bigot, Myelaphus	379
melasoma (Loew), Orfelia	203
melasoma (Wulp), Villa	437
melastoma Loew, Stratiomys	312
melba (Walker), Exorista	1055
Meledonus Aldrich	1024
Meleterus Aldrich	1018
meliceris Reinhard, Phasiomyia	970
Melieria Robineau-Desvoidy	645
Meligramma Frey	565
meliloti Felt, Dasineura	260
Melina Robineau-Desvoidy	686
Melinia Ringdahl	855
Melinocera Townsend	1007
Meliscaeva Frey	568
MELISONEURINI	984
melissopodis (Coquillett), Elfia	1065
melitaeae (Coquillett), Siphosturmia	1087
melitarae Reinhard, Coronimyia	1060
Melithreptus Loew	570
mella (Walker), Exorista	1055
mellea (Johannsen), Bradysia	233
melleicauda (Alexander), Limonia	47
melleus (Coquillett), Culicoides	129
melleus Loew, Chlorops	790
melleus Melander, Platypalpus	480
mellifrons Townsend, Metatachina	1014
melliginis (Fitch), Rivellia	656
mellinum (Linnaeus), Melanostoma	575, †575
mellipes (Coquillett), Ceropsilopa	742
mellissoides Hull, Eristalis	624
melobosis (Walker), Lypha	1011
Meloehelea Wirth	123
MELOPHAGINAE	920
MELOPHAGINI	920
Melophagus Latreille	920
Melusina Meigen	15
MELUSINIDAE	15
memorandus Melander, Philetus	455
menapis (Walker), Cryptomeigenia	1041
mendax Curran, Rhagoletis	674
mendax Johannsen, Mycomya	221
mendax Reinhard, Sarcophaga	959
mendica (Loew), Orfelia	203
mendosa (Loew), Orfelia	203
mendotae Muttkowski, Diamesa	152
Menetus Aldrich	1018
meniscus Loew, Gymnopternus	499
mentalis (Coquillett), Stomatomyia	1059
mentalis (Williston), Melangyna	567
menthae Felt, Giardomyia	28
menthae Felt, Neolasioptera	273
menthastri (Linnaeus), Sphaerophoria	570
mentzeliae Cockerell, Asphondylia	268
Meoneura Rondani	728
mercanthae (Greene), Catagoniopsis	1098
meralis Reinhard, Clastoneuriopsis	1067
Merapioidus Bigot	612
mercatorum Patterson & Wheeler, Drosophila	765
mercedis (Coquillett), Villa	439

	Page
mercieri Parent, Dolichopus	491
mercurator Dyar, Aedes	114
Mercurymyia Fluke	568
merdaria (Fabricius), Scatophaga	839
Mericia Robineau-Desvoidy	1007
meridiana Doane, Tipula	28
meridiana Patterson & Wheeler, Drosophila	765
meridiana (Staeger), Pilaria	68
meridionale Riley, Simulium	186
meridionalis Alexander, Phyllolabis	60
meridionalis Chillcott, Fannia	895
meridionalis Cole, Villa	440
meridionalis Dendy & Sublette, Glyptotendipes	172
meridionalis Felt, Aphidoletes	280
meridionalis Johannsen, Chironomus	165
meridionalis Townsend, Phasmophaga	1070
Meringodixa Nowell	102
Merodon Meigen	617
MERODONTINI	617
Meromacrus Rondani	622
Meromyza Meigen	793
Meroplius Rondani	681
Merosargus Loew	303
Merycomyia Hine	322
mesae (Tucker), Asilus	392
Mesamyia Malloch	841
Mesembrina Meigen	910
mesensis (Townsend), Myiophasia	983
mesocera Alexander, Ormosia	86
mesochra (Shaw), Bradysia	233
Mesocyphona Osten Sacken	82
Mesogramma Loew	570
mesogramma (Loew), Scatophila	758
mesogrammus Loew, Platypalpus	480
Mesograpta Loew	570
mesopedalis Sabrosky, Gaurax	782
Mesophila Walker	597
Mesophora Borgmeier	543
mesopora Steyskal, Tetanocera	694
Mesorhaga Schiner	482
Mesoscaptomyza Hackman	769
mesotergata Alexander, Tipula	37
mesquite (Bromley), Efferia	395
Messiomyia Reinhard	1063
messoria (Fallén), Diplotoxa	792
mestor (Walker), Ervia	1027
metabola (McDunnough), Hybomitra	340
Metachaeta Coquillett	1033
Metachela Coquillett	471
Metaclythia Kessel	550
Metacosmus Coquillett	427
Metadexia Coquillett	1027
Metadioctria Wilcox & Martin	371
Metadiplosis Felt	285
Metadoria Brauer & Bergenstamm	1045
Metaforcipomyia Saunders	125
Metagonistylum Townsend	1103
Metalimnobia Matsumura	44
metalis Reinhard, Minthozelia	1027
metallica Cole, Parydra	749
metallica Felt, Trotteria	276
metallica Kröber, Thereva	353
metallica Townsend, Lydinolydella	1044
metallica (Townsend), Myiophasia	983
metallica (Townsend), Protocalliphora	926
metallica Walker, (Chyliza)	641
metallica Walker, Ptychoptera	97
metallica Wiedemann, Xylota	607
metallica (Williston), Neoascia	595
metallica (Wulp), Rivellia	656
Metallicomintho Townsend	1032
metallicus (Robineau-Desvoidy), Archytas	1000, ††1001
metallicus Van Duzee, Nematoproctus	515
metallifera Bigot, Xylota	607
metallifera (Walker), Archytas	1000
Metangela Rübsaamen	229
Metapelastoneurus Aldrich	501
Metaphragma Coquillett	352
Metaphyto Coquillett	1008
Metaplagia Coquillett	1018
Metapogon Coquillett	378
Metasyrphus Matsumura	560
Metatachina Townsend	1014

1632 A CATALOG OF DIPTERA OF NORTH AMERICA

	Page		Page
metatarsalis Melander, Euhybus	448	microcera Melander, Psila	639
metatarsalis Melander, Glabellula	415	*Microcerata* Felt	246
metatarsalis (Thomson), Gymnopternus	499	**Microcerella** Macquart	951
metatarsata Stein, Hydrotaea	900	microcerus Melander, Platypalpus	479
metatarsata (Stein), Pogonomyia	902	**Microchaetina** Wulp	984
Metatephritis Foote	669	**Microchrysa** Loew	303
Metatrichia Coquillett	354	**Microdon** Meigen	597
Metavoria Townsend	1018	MICRODONTINI	597
metcalfi (Curran), Blera	611	**Microdrosophila** Malloch	771
metcalfi (Fluke), Epistrophe	564	microfulcrum (James), Compsobata	634
meteorica (Linnaeus), Hydrotaea	900	**Microlynchia** Lutz, Neiva, & Lima	919
Methypostena Townsend	1051	micromelanica Patterson, Drosophila	765
Metopia Meigen	936	**Micromintho** Townsend	1032
METOPIIDAE	933	**Micromorphus** Mik	530
Metopina Macquart	544	**Micromya** Rondani	250
metopina Schiner, Hyperecteina	1070	*Micromyia* Rondani	250
METOPININAE	537	MICROMYINI	248
METOPININI	538	*Microperiscelis* Oldenberg	710
Metopiops Townsend	1095	**Micropeza** Meigen	634
metopium Osten Sacken, Bombylius	409	MICROPEZIDAE	633
Metopobrachia Enderlein	634	MICROPEZINAE	634
Metopomuscopteryx Townsend	1025	MICROPEZOIDEA	633
Metopomyza Enderlein	802	microphallus Alexander, Dicranoptycha	52
Metoponia Macquart	300	microphallus Alexander, Limnophila	66
Metoposarcophaga Townsend	950	microphona Melander, Charadromia	475
Metopotachina Townsend	993	*Microphor* Macquart	452
metraloas Meinert, Miastor	247	**Microphorella** Becker	452
METRIOCNEMINI	160	**Microphorus** Macquart	452
Metriocnemus Wulp	160	**Microphthalma** Macquart	982
metzi Johnson, Pseudotephritis	647	*Microphthalmia* Macquart	982
mevarna (Walker), Trupanea	667	**Microprosopa** Becker	835
mexicana Aldrich, Spathimeigenia	1049	microps Melander, Loxocera	639
mexicana (Bellardi), Psorophora	110	**Micropsectra** Kieffer	178
mexicana Brauer & Bergenstamm, Gaediopsis	1079	micropteryx (Thomson), Calythea	865
mexicana Cazier, Apiocera	357	micropyga Melander, Aphoebantus	430
mexicana Cole, Chromolepida	350	micropyga Melander, Drapetis	477
mexicana Enderlein, Oncopsia	637	micropygus Wahlberg, Dolichopus	493
mexicana (Hine), Efferia	395	**Microsania** Zetterstedt	547
mexicana (Macquart), Phaenicia	928	*Microsciasma* Townsend	971
mexicana (Macquart), Stenopterina	657	**Microsenotainia** Townsend	938
mexicana Macquart, Volucella	601	*microsetulosa* Malloch, Trupanea	667
mexicana (Townsend), Microchaetina	984	**Microsillus** Aldrich	1086
mexicana (Wiedemann), Dyseuaresta	668	*microstigma* Curran, Trupanea	667
mexicanum Borgmeier, Xanionotum	545	microstoma Loew, Odontomyia	316
mexicanus Cole, Stenopogon	384	**Microstylum** Macquart	379
mexicanus Curran, Xanthandrus	574	**Microtabanus** Fairchild	329
mexicanus (Linnaeus), Chlorotabanus	†330	**Microtendipes** Kieffer	172
mexicanus (Macquart), Asemosyrphus	†620	**Microtrichomma** Giglio-Tos	1001
mexicanus Wiedemann, Bombylius	409	*Microxylota* Jones	596
miamensis (Townsend), Johnsonia	950	micrura (Curran), Xylota	606
Miamimyia Townsend	1070	micrura (Osten Sacken), Allograpta	569
Miastor Meinert	247	mida Moodie, Milesia	614
MIASTORINI	247	midas Brauer, Asilus	392
micans (Hendel), Campichoeta	773	*Midas* Fabricius	358
micans Loew, Rivellia	656	midas Hull, Ferdinandea	588
micans (Macquart), Morellia	912	middlekauffi Townes, Pseudochironomus	176
micans (Ringdahl), Spilogona	881	migrata Walker, Hilara	456
micans Schiner, Proctacanthus	399	mikii Strobl, Platycoenosia	898
micheneri Alexander, Blephariceira	99	mikii Williston, Hilarimorpha	348
micheneri Alexander, Tipula	40	*Mikimyia* Bigot	659
micheneri (Fluke), Lepidomyia	590	milanoensis Reinhard, Gymnoprosopa	936
micheneri James, Adoxomyia	306	milbertii Macquart, Proctacanthus	399
micheneri James, Labostigmina	314	**Milesia** Latreille	614
micheneri (James), Pseudonomoneura	360	MILESIINAE	583
micheneri Philip, Esenbeckia	322	MILESIINI	604
michigana Alexander, Limonia	47	**Milichia** Meigen	732
michigana Sabrosky, Leptocera	721	**Milichiella** Giglio-Tos	733
michiganensis (Hardy), Pipunculus	553	MILICHIIDAE	728
michiganensis Namba, Rivellia	656	MILICHIINAE	732
michiganensis Quisenberry, Tephritis	668	MILICHIOIDEA	718
michiganensis (Townsend), Microphthalma	982	milichioides (Melander), Phycomyza	727
michiganus Harmston & Knowlton, Dolichopus	495	militaris Gerstäcker, Mydas	359
		militaris Johannsen, Chironomus	166
micidus Reinhard, Opsidiotrophus	938	militaris Johannsen, Keroplatus	200
mickeli Hardy, Bibio	193	militaris (Meigen), Hydrotaea	900
Micrempis Melander	475	militaris Walker, Sericomyia	603
microcellula Alexander, Erioptera	83	militaris (Walsh), Winthemia	1090
microcentrus Coquillett, Hippelates	775	millardi Alexander, Molophilus	89
microcephala Kraft & Cook, Zabrachia	318	*Millardia* Curran	504
microcephala (Loew), Triphleba	534	*millbrae* Jones, Ephydra	754
microcephala (Osten Sacken), Hybomitra	340	millefolii Wachtl, Clinorhyncha	275
microcephalus Wehr, Silvius	323	*millepunctata* (Loew), Sphecomyiella	658
microcera Alexander, Hexatoma	69	milleri (Townes), Chironomus	169
microcera Alexander, Pilaria	68	milleri Whitney, Tabanus	336
microcera (Loew), Lasiosina	791	milnei Steyskal, Phyllomyza	729
		Milonius Reinhard	1086

INDEX

	Page		Page
MILTOGRAMMINAE	933	minuta (Townsend), Catharosia	971
MILTOGRAMMINI	933	minuta Walker (Drosophila)	772
Miltoravinia Townsend	953	*minuta* (Wiedemann), Meroplius	682
milvina Bromley, Laphria	390	minuta Williston, Townsendia	385
mima (Hennig), Compsobata	634	minuticornis Van Duzee, Sympycnus	527
mime Hull, Tenthredomyia	616	minutipalpis (Stein), Fannia	896
mimela Reinhard, Epidexia	1032	minutissima Hall, Sarcophaga	958
mimesis Dyar, Aedes	112	minutissima Melander, Apolysis	421
mimetica Malloch, Helina	888	minutissimus (Zetterstedt), Culicoides	129
mimetica Malloch, Hylemya	855	minutula Alexander, Cryptolabis	78
mimetica Malloch, Paracantha	662	*minutum* Lugger, Simulium	189
Mimocalla Hull	574	minutus Cole, Caenotus	424
Mimologus Reinhard	1086	*minutus* (Johnson), Copromyza	720
mimoris Reinhard, Sarcophaga	958	minutus Loew, Gymnopternus	499
mimosae Felt, Asphondylia	268	minutus (Meigen), Atrichopogon	123
mimula (Johannsen), Orfelia	203	minutus Parent, Thrypticus	517
mimuli Felt, Neolasioptera	273	minutus Van Duzee, Paracleius	501
mimus Melander, Platypalpus	480	*minutus* Van Duzee, Pelastoneurus	501
mimus Osten Sacken, Ablautus	364	minutus (Van Duzee), Systenus	517
minacis Hardy, Tomosvaryella	555	*minutus* Walker, Rhamphomyia	464
Minella Robineau-Desvoidy	1025	minutus Wilcox & Martin, Nannocyrtopogon	379
minense Townsend, Metagonistylum	1103	minutus (Zetterstedt), Orthocladius	156
Minettia Robineau-Desvoidy	701	minytus Walker, Rhamphomyia	464
mingusae Martin, Holopogon	375	mira (Bigot), Empis	460
mingusi Wilcox & Martin, Nannocyrtopogon	375	mira Coquillett, Eusiphona	732
mingwe Alexander, Tipula	36	mira (Coquillett), Villa	438
minima Alexander, Dicranoptycha	52	mira Melander, Mythicomyia	418
minima (Becker), Gymnochiromyia	822	mira Reinhard, Minthozelia	1027
minima Cresson, Notiphila	747	mirabile (Townes), Nilothauma	173
minima (Malloch), Melanagromyza	796	mirabilis Adams, Scenopinus	356
minima Malloch, Phaonia	908	mirabilis (Collin), Leptocera	724
minima Melander, Mythicomyia	418	mirabilis Melander, Helcomyza	678
minima Shannon, Cochliomyia	924	mirabilis (Osten Sacken), Orimarga	53
minima Van Duzee, Medetera	510	mirabilis Townsend, Acronarista	1064
minimum Steyskal, Sepsisoma	642	mirabilis Townsend, Ichneumonops	974
minimus Alexander, Longurio	18	miranda Dobzhansky, Drosophila	768
minimus (Meigen), Limnophyes	160	miriamae (Shaw), Orfelia	203
minimus Van Duzee, Hydrophorus	505	mirifica Melander, Mythicomyia	418
minimus (Van Duzee), Micromorphus	530	mirificus Melander, Gymnopternus	499
miniscula Cook, Rhexoza	240	miripes (Coquillett), Anatopynia	145
ministra Melander, Mythicomyia	418	miscella (Coquillett), Villa	440
minnesotae Barr, Culiseta	117	miscelli (Coquillett), Deopalpus	1004
minnesotensis Telford, Cheilosia	585	miscellus Coquillett, Lordotus	412
minor (Adams), Oscinella	780	misella Osten Sacken, Pterodontia	405
minor Alexander, Ptychoptera	97	misellus Loew, Diogmites	372
minor Cresson, Euxesta	651	*misellus* Melander, Dolichopus	494
minor Cresson, Pipunculus	554	miseranda Alexander, Tasiocera	88
minor Curran, Chrysotoxum	582	*Misgomyia* Coquillett	298
minor (Curran), Oswaldia	1046	mississippensis Hull, Mallota	621
minor (Felt), Neuromyia	261	mississippiensis Hoffman, Culicoides	131
minor Felt, Porricondyla	256	mississippiensis Hull, Somula	612
minor (Haliday), Themira	685	*mississippii* Dyar, Aedes	114
minor (Malloch), Amiota	761	missouriensis Bromley, Diogmites	372
minor Malloch, Leucopis	709	*missouriensis* (Riley), Proctacanthus	399
minor Malloch, Pogonomyia	902	mitchellae (Dyar), Aedes	113
minor Melander, Mythicomyia	418	mitchellae Felt, Lasioptera	272
minor Melander, Wiedemannia	469	mitchellensis Alexander, Ormosia	86
minor West, Ptilodexia	989	mitchellensis (Shaw), Orfelia	203
minor (Wirth), Nilobezzia	138	mitchelli Brimley, Ceraturgus	365
minor (Zetterstedt), Megaselia	†539	mitchelli Malloch, Leptocera	725
minorata Banks, Platypeza	549	mitchellii (Theobald), Wyeomyia	107
minos Osten Sacken, Ospriocerus	381	mitis (Cresson), Oligodranes	422
Minthozelia Townsend	1027	mitis (Curran), Bolbomyia	298
minus Cook, Rhegmoclema	239	*mitis* (Curran), Liriomyza	801
minus (Dyar & Shannon), Cnephia	184	mitis Curran, Metriocnemus	161
minuscula Goureau, Phytomyza	805	mitis Curran, (Phorocera)	1107
minuscula (Hine), Hybomitra	340	mitis (Johannsen), Mycetophila	212
minuscula Melander, Mythicomyia	418	mitis (Meigen), Aplomya	1098
minuscula (Wulp), Eucordyligaster	†1030	mitis Osten Sacken, Chrysops	326
minusculus Hine, Promachus	400	mitis (Reinhard), Cinochira	975
minuta (Aldrich), Megaselia	539	mitis (Reinhard), Melanomya	932
minuta (Alexander), Paradelophmyia	60	mitrata Dietz, Tipula	37
minuta Banks, Psychoda	96	mitrata Melander, Mythicomyia	418
minuta Brooks, Gymnoclytia	964	mitrephorus Melander, Oreogeton	453
minuta Curran, Neoascia	595	miwok Alexander, Tipula	37
minuta (Felt), Feltiella	282	mixis Philip, Tabanus	334
minuta (Frost), Haplomyza	802	**Mixiogaster** Macquart	599
minuta Greene, Mythicomyia	418	mixopolia Melander, Empis	460
minuta Hull, Chrysogaster	591	mixta Alexander, Cryptolabis	78
minuta Loew, Argyra	525	mixta (Curran), Xylota	606
minuta Macquart, Mallophora	396	mixta Hine, Merycomyia	322
minuta Malloch, Pegomya	860	mixta Steyskal, Hedria	691
minuta Malloch, Scatophaga	839	*Mixtemyia* Macquart	613
minuta (Melander), Micrempis	476	mixtum Syme & Davies, Prosimulium	183
minuta (Snow), Procecidochares	660	mixtus Coquillett, Aphoebantus	430
minuta Stein, Fannia	894	mixtus (Holmgren), Orthocladius	156

744-243—65——104

1634 A CATALOG OF DIPTERA OF NORTH AMERICA

	Page		Page
m-nigrum (Zetterstedt), Desmometopa	731	monki (Bromley), Efferia	395
mobile (Coquillett), Villa	436	*monki* Bromley, Ospriocerus	381
MOCHLONYCHINI	104	mono Alexander, Tipula	37
Mochlonyx Loew	104	**Monoceromyia** Shannon	616
Mochlosoma Brauer & Bergenstamm	987	*monochaeta* Loew, Mycetophila	212
Mochlostyrax Dyar & Knab	119	*Monochaetophora* Hendel	689
moderator Stone, Tabanus	334	**Monochaetoscinella** Duda	778
modesta (Brues), Megaselia	539	*monochroma* Dietz, Tipula	38
modesta (Coquillett), Euphyto	935	*monochroma* (Harris), Trichocera	15
modesta Coquillett, Symphoromyia	344	monochromus Wulp, Chironomus	168
modesta (Felt), Asteromyia	274	**Monoclona** Mik	224
modesta Felt, Dasineura	260	**Monoculicoides** Khalaf	131
modesta Felt, Mycodiplosis	283	monoculus Swederus (Musca)	625
modesta (Felt), Peromyia	251	monoensis Alexander, Ptychoptera	97
modesta (Felt), Trishormomyia	288	monoensis Hine, Tabanus	334
modesta (Felt), Xylopriona	249	monoensis Wirth, Culicoides	129
modesta Johannsen, Coelosia	218	monogramma Melander, Proclinopyga	467
modesta Johannsen, Dixa	101	monohammi Townsend, Eutheresia	990
modesta Loew, Coenosia	872	**Monohelea** Kieffer	135
modesta Meigen, Heleomyza	816	*monophlebi* Skuse, Cryptochetum	†825
modesta (Meigen), Limonia	47	*Monorhipidia* Alexander	45
modesta Melander, Diastata	772	monostigma Melander, Geomyza	821
modesta Melander, Mythicomyia	418	monotheca (Edwards), Polyardis	250
modesta (Osten Sacken), Dicranota	57	monrovia Wilcox & Martin, Dioctria	370
modesta (Reinhard), Opsidia	937	monrovia Wilcox & Martin, Nannocyrtopogon	379
modesta Sturtevant, Drosophila	766	monstri James, Bibio	193
modesta Van Duzee, Medetera	510	monstrosus Osten Sacken, Scellus	506
modesta Van Duzee, Polylepta	225	montana Aldrich, Symphoromyia	343
modesta (Wiedemann), Argoravinia	943	*montana* Alexander, Dicranota	57
modesta (Williston), Thecophora	631	montana Allen, Eumacronychia	935
modesta Williston, Trichopteromyia	250	*montana* (Brues), Phora	537
modestus Ide, Platycheirus	577	montana Cook, Scatopse	238
modestus Knab, Microdon	598	montana Coquillett, Bolitophila	198
modestus Loew, Chrysopilus	346	montana (Coquillett), Eugymnogaster	971
modestus (Loew), Epacmus	428	montana Cresson, Ditrichophora	739
modestus (Loew), Parapenium	580	montana Day, Tetanocera	694
modestus Loew, Stenopogon	384	montana (Dietz), Nephrotoma	21
modestus Say, Chironomus	169	montana Felt, Bremia	280
modestus Williston, Helophilus	620	montana Felt, Camptomyia	256
modestus (Williston), Psilocurus	382	montana (Felt), Campylomyza	248
modica Dietz, Ormosia	86	montana Felt, Dirhiza	255
modoc Alexander, Limnophila	64	montana Felt, Giardomyia	284
modoc Alexander, Tipula	37	montana Felt, Hormomyia	288
modocensis (Sublette), Chironomus	171	montana Felt, Odontodiplosis	288
modocensis Wirth, Bezzia	141	montana (Felt), Parallelodiplosis	286
moechus Loew, Tachytrechus	503	montana Garrett, Boletina	217
moechus Osten Sacken, Chrysops	326	montana Hunter, Tropidia	609
Moehnia Pritchard	246	montana Kraft & Cook, Pachygaster	318
MOEHNIINI	246	montana Malloch, Hylemya	849
moerens (Johannsen), Orfelia	203	montana (Osten Sacken), Dactylolabis	62
moerens Walker, Chrysops	324	*montana* Patterson & Wheeler, Drosophila	765
moerens (Zetterstedt), Paraprosalpia	866	montana (Reinhard), Orthosimyia	984
moesta (Holmgren), Hylemya	856	montana Reinhard, Winthemia	1090
moesta (Johannsen), Orfelia	203	montana (Shannon), Acronesia	928
moffitti Painter & Hall, Poecilanthrax	442	montana (Smith), Tachinomyia	1058
mohave Wirth, Culicoides	129	montana (Snow), Procecidochares	660
mohavea Melander, Apolysis	421	montana Steyskal, Dictya	689
mohavea Melander, Trixoscelis	818	montana Stone, Griffen & Patterson, Drosophila	765
mohavensis Alexander, Tipula	37	*montana* (Townsend), Euphasiopteryx	980
mojavensis Patterson & Crow, Drosophila	765	montana Townsend, Minthozelia	1027
molestum Harris, Simulium	189	montana (Townsend), Nowickia	995
molestus Say, Tabanus	334	montana West, Eutheresia	990
molestus (Wiedemann), Psorophora	109	montana Wheeler, Leucophenga	762
molitor (Loew), Villa	434	montana Wheeler, Scaptomyza	769
mollicula (Fallén), Coenosia	873	*montanensis* Hallock, Sarcophaga	958
mollis Becker, Scatophaga	839	montanensis Parker, Sarcofahrtia	942
mollis Melander, Platypalpus	479	montanipes Hull & Fluke, Cheilosia	585
mollis (Walker), Phormia	924	montanum Garrett, Dynatosoma	209
mollissima Haliday, Coelomyia	898	montanum (Van Duzee), Rhaphium	513
molophiloides Alexander, Cryptolabis	78	montanum Röder, Asindulum	199
Molophilus Curtis	88	montanus Aldrich, Meleterus	1018
mombinpraeoptans Sefn, Anastrepha	673	montanus Coquillett, Gaurax	782
monacantha (Collin), Spilogona	881	montanus (Curran), Metasyrphus	561
monacha Melander, Mythicomyia	419	*montanus* Ferris & Cole, Melophagus	921
monacha Melander, Tachydromia	474	*montanus* Garrett, Dixa	101
monacha Osten Sacken, Asphondylia	268	montanus Harmston & Knowlton, Campsicnemus	526
monachus (Hull), Melangyna	566	montanus Hine, Asilus	392
monardi Brodie (Diplosis)	292	montanus James, Nemotelus	309
Monardia Kieffer	250	montanus Loew, Cyrtopogon	369
Monarthropalpus Rübsaamen	289	montanus Melander, Oligodranes	422
monela (Reinhard), Microcerella	952	*montanus* (Melander), Platypalpus	481
moneta (Osten Sacken), Villa	434	montanus Osten Sacken, Chrysops	326
monilicornis (Coquillett), Forcipomyia	124	montanus Painter & Hall, Poecilanthrax	442
moniliformis (Doane), Limonia	47		
monilis (Felt), Campylomyza	248, †248		
monilis (Linnaeus), Ablabesmyia	†148		

Entry	Page
montanus Sabrosky, Hippelates	775
montanus Van Duzee, Sympycnus	527
montanus Williston, Eristalis	623
montcalmi (Blanchard), Aedes	116
montensis Snow, Callicera	595
monticola Alexander, Antocha	51
monticola (Alexander), Limonia	48
monticola Alexander, Rhabdomastix	77
monticola Alexander, Tipula	37
monticola (Becker), Thaumatomyia	788
monticola Belkin & McDonald, Aedes	113
monticola (Bigot), Calythea	865
monticola (Bigot), Chaetogaedia	1078
monticola (Dodge), Sarcophaga	958
monticola (Felt), Aprionus	251
monticola Johnson, Agathomyia	548
monticola Johnson & Johnson, Poecilanthrax	443
monticola (Jones), Carposcalis	576
monticola (Malloch), Megaselia	541
monticola Malloch, Phaonia	907
monticola Malloch, Scatophaga	839
monticola Malloch, Sphegina	594
monticola (Malloch), Spilogona	882
monticola Martin, Beameromyia	362
monticola Melander, Brochella	455
monticola Melander, Lasiopogon	378
monticola Melander, Platypalpus	480
monticola Melander, Sapromyza	704
monticola (Osten Sacken), Ormosia	86
monticola Townsend, Gaediopsis	1080
monticola (Van Duzee), Calyxochaetus	528
monticola Van Duzee, Dolichopus	493
monticola Vockeroth, Geomyza	821
montigena Hunter, Pyritis	604
montiradicis James, Empis	460
montiradicis James, Psilocephala	351
montis Cole, Lestomyia	378
montivaga Coquillett, Psilocephala	351
montivagans Huckett, Hylemya	849
montivagum Johnson, Melanostoma	575
montivagum Wheeler, Parasyntormon	516
montivagus (Hardy), Pipunculus	554
montivagus (Snow), Metasyrphus	561
morata Coquillett, Psilocephala	351
morator Melander, Piophila	712
moravica Landrock, Mycetophila	212
morbosus Stone, Tabanus	334
mordellaria (Fallén), Hypocera	534
Morellia Robineau-Desvoidy	912
moreti (Robineau-Desvoidy), Exorista	1054
moriens (Zetterstedt), Hylemya	853
morinioides (Townsend), Myiophasia	983
morio (Brischke), Phytobia	800
morio (Linnaeus), Villa	436
morio (Zetterstedt), Phaonia	†907, 907
morioides (Osten Sacken), Limonia	47
morioides (Say), Villa	436
morionella (Aldrich), Helicobia	949
morionella (Zetterstedt), Hylemya	855
morissoni (Bigot), Hadromyia	610
morissoni Bigot, Rhamphomyia	465
mormon (Alexander), Hesperoconopa	78
mormon Alexander, Tipula	37
mormon Melander, Aphoebantus	430
morna Curran, Lepidanthrax	441
morosa Aldrich, Sarcophaga	958
morosa (Meigen), Phytobia	799
morosum Loew, Microstylum	379
morosus Loew, Stenopogon	384
morosus Osten Sacken, Chrysops	326
morosus (Walker), Polydontomyia	621
Morpholeria Garrett	813
morrilli (Townsend), Orthellia	913
morrisoni Alexander, Tipula	37
morrisoni (Bigot), Hadromyia	610
morrisoni Brooks, Phoranthella	970
morrisoni Cresson, Hydrellia	744
morrisoni (Malloch), Coenosia	873
morrisoni Malloch, Fannia	896
morrisoni Malloch, Phaonia	908
morrisoni Osten Sacken, Paracosmus	427
morrisoni (Townsend), Metopia	937
morrisoni Townsend, Phoranthella	970
morsicans Melander, Epacmus	428
morsitans (Theobald), Culiseta	117
morticia Shannon, Calliphora	929
mortisequa (Kirby), Cynomyopsis	930
mortuorum (Linnaeus), Cynomya	931
moscowa Garrett, Acantholeria	814
mosellana (Géhin), Sitodiplosis	278
Mosillus Latreille	734
mucens (Linnaeus), Hermetia	304
mucida Osten Sacken, Atomosia	387
mucidoides Bromley, Atomosia	387
mucorea (Loew), Villa	435
muelleri Dyar, Aedes	113
muhlenbergiae Johannsen & Crosby, Thrypticus	511
muhlenbergiae (Marten), Asteromyia	274
mulaiki Alexander, Tipula	25
mularis Stone, Tabanus	335
mulcata (Giglio-Tos), Helina	887
muliebris Osten Sacken, Toxorhina	90
mulinum Van Duzee, Parasyntormon	516
mulleri Sturtevant, Drosophila	765
mulrennani Basham, Culex	119
mulrennani Beck, Culicoides	129
mulsea Melander, Mythicomyia	418
multianulata Felt, Dasineura	260
multiarticulata Felt, Dirhiza	255
multiarticulata Felt, Monardia	250
multidentatum (Twinn), Prosimulium	183
multifila (Felt), Cincticornia	269
multineurus Felt, Holoneurus	255
multinoda (Felt), Camptomyia	256
multinodus Felt, Holoneurus	255
multipunctata (Curran), Pentaneura	148
multipunctata Loew, Drosophila	764
multipunctata Sabrosky, Asteia	824
multipunctata Williston, Paralimna	748
multipunctatus Malloch, Culicoides	132
multiseriata (Aldrich), Chaetopleurophora	533
multiseriata Malloch, Helina	887
multiseta (Felt), Lycoriella	231
multisetifera Pettey, Sciara	230
multisetosa Chillcott, Fannia	896
multisetosa (Holmgren), Scatophaga	839
multisetosa (Strobl), Quadrularia	889
multisetosus Van Duzee, Dolichopus	493
multistylata Alexander, Gonomyia	74
Mumetopia Melander	820
munda (Coquillett), Valentibulla	670
munda Frick, Liriomyza	802
munda (Johannsen), Bradysia	234
munda Loew, Cordilura	830
munda Loew, Psilocephala	351
munda Namba, Rivellia	656
munda Osten Sacken, Limnophila	63
munda Spencer, Drosophila	765
mundoides Alexander, Limnophila	63
mundula (Coquillett), Valentibulla	670
mundus (Coquillett), Probezzia	138
mundus Loew, Diaphorus	520
mundus (Wiedemann), Condylostylus	484
muralis (Walker), Lucilia	927
Murdockiana Townsend	1057
muricatus (Osten Sacken), Eclimus	427
murina (Doane), Paroxyna	666
murina Loew, Leptogaster	363
murina Melander, Mythicomyia	418
murina (Zetterstedt), Limonia	48
murphyi Chillcott, Platypalpus	480
murtfeldtiana Felt, Lasioptera	272
mus (Bigot), Oligodranes	422
mus (Osten Sacken), Aphoebantus	430
Musca Linnaeus	913
muscaria (Coquillett), Villa	435
muscaria (Fabricius), Hylemya	854
muscaria (Loew), Xenochaetina	705
MUSCIDAE	†826, 869
muscina (Meigen), Phytobia	799
Muscina Robineau-Desvoidy	909
MUSCINAE	909
MUSCOIDEA	826
muscoides Curran, Cryptomeigenia	1041
Muscopteryx Townsend	1025
muscosa Stebbins, (Cecidomyia)	294
musicus (Say), Psorophora	109
Musidora Meigen	531
MUSIDORIDAE	530
mutabilis Adams, Euparyphus	307

	Page		Page
mutabilis (Coquillett), Dasyhelea	127	mystacea (Linnaeus), Hypodermodes	910
mutabilis (Coquillett), Sitophaga	1048	mystaceae Curran, Criorhina	611
mutabilis (Fabricius), Hoplitimyia	313	**Mystacella** Wulp	1086
mutabilis (Fallén), Meigenia	1044	mystaceus (Macquart), Rhagio	345
mutabilis Loew, Hilara	456	mystica Dyar & Shannon, Dixa	101
mutabilis Loew, Rhamphomyia	464	mysticus (Becker), Tachytrechus	503
mutabilis Melander, Mythicomyia	418	mysticus Rapp, Allomethus	555
mutabilis Sherman, Mycomya	221	**Mythicomyia** Coquillett	416
mutans Huckett, Hylemya	849	**MYTHICOMYIINAE**	415
mutata (Malloch), Cnephia	184	*Myxeoristops* Townsend	1096
mutatus Dyar, Aedes	113	**MYXOSARGINI**	317
mutatus Walker, Tabanus	333	**Myxosargus** Brauer	317
mutica Dietz, Tipula	40		
mutica Loew, Mycetophila	212	nacta Johannsen, Boletina	217
mutica (Zetterstedt), Fannia	896	nacta (Johannsen), Scatopsciara	235
muticus Johannsen, Tanytarsus	179	nadineae (Cresson), Discocerina	738
Mutiloptera Coquillett	821	naevia Peus, Dixa	102
mutor (Adams), Docosia	226	naevus (Mitchell), Stictochironomus	178
mutua (Johannsen), Bradysia	234	naica Melander, Drapetis	477
Mycetaulus Loew	712	nais (Townes), Chironomus	168
Mycetobia Meigen	191	naknek Shannon, Xylota	606
MYCETOBIINAE	191	*nana* Cole, Thereva	353
Mycetophila Meigen	209	nana (Coquillett), Chersodromia	475
MYCETOPHILIDAE	196	nana Coquillett, Mancia	441
MYCETOPHILINAE	204	nana Coquillett, Senotainia	939
MYCETOPHILINI	208	nana (Curran), Aplomyiopsis	1038
MYCETOPHILOIDEA	196	nana (Fallén), Pherbellia	686
Mycodiplosis Rübsaamen	282	nana Loew, Bolbomyia	298
Mycodrosophila Oldenberg	770	nana (Loew), Cordylurella	836
Mycomya Rondani	219	nana Loew, Rhamphomyia	464
Mycomyia Rondani	219	nana Melander, Micrempis	476
MYCOMYINI	219	nana (Townsend), Cylindromia	973
Mycophila Felt	250	nana Walker (Ephydra)	759
Mycothera Winnertz	209	*nana* (Zetterstedt), Coenosia	870
Mydaea Robineau-Desvoidy	891	nanciae Brimley, Tetanocera	694
Mydaeina Malloch	884	nanella Cole, Thereva	353
MYDAEINAE	885	nanella Melander, Glabellula	415
MYDAIDAE	357	*nanellum* Loew, Zodion	629
mydas Brauer, Asilus	392	**Nanna** Becker	830
Mydas Fabricius	358	**Nannocyrtopogon** Wilcox & Martin	379
MYDASIDAE	357	**Nannodioctria** Wilcox & Martin	371
MYDIDAE	357	nannomorpha Hull & Fluke, Cheilosia	585
Myelaphus Bigot	379	**Nanocladius** Kieffer	154
myermus (Curran), Neocnemodon	581	nanseni Kieffer, Orthocladius	156
myersi Alexander, Teucholabis	76	nanseni (Kieffer), Prosmittia	162
Myiocera Robineau-Desvoidy	989	nanseni Kieffer, Psectrocladius	159
Myioclonia Reinhard	1025	nantuckensis Hine, Tabanus	3..
Myiolepta Newman	589	*nanus* (Coquillett), Psorophora	110
Myiolia Rondani	669	*nanus* (Macquart), Stenotabanus	329
Myiomyrmica Steyskal	646	*nanus* (Meigen), Metriocnemus	†161
Myiopharus Brauer & Bergenstamm	1045	nanus Root & Hoffman, Culicoides	129
Myiophasia Brauer & Bergenstamm	983	nanus (Zetterstedt), Eribolus	778
MYIOPHASIINI	983	napaea (Laffoon), Mycetophila	212
Myiophthiria Rondani	917	napaea Melander, Iteaphila	454
Myiospila Robineau-Desvoidy	890	napaea Melander, Mythicomyia	418
Mylogymnocarena Foote	669	*Napaea* Robineau-Desvoidy	749
Myobiopsis Townsend	1035	**Napomyza** Westwood	803
Myocera Robineau-Desvoidy	†989	narcissi (Fabricius), Merodon	617
Myoceropsis Townsend	988	narcissi Smith, Eumerus	597
Myodina Robineau-Desvoidy	648	narina (Walker), Spilogona	882
myoidea (Robineau-Desvoidy), Cynomyopsis	930	narona Snyder, Mydaea	891
myoidea (Robineau-Desvoidy), Lydella	†1103	narona (Walker), Limnophora	885
Myoleja Rondani	669	narragansett Sturtevant & Dobzhansky, Drosophila	768
Myolepta Newman	589	narytia (Walker), Herina	645
MYOLEPTINI	589	nasalis (Linnaeus), Gasterophilus	916, †1112
Myopa Fabricius	630	*nasalis* Melander, Exepacmus	430
myopa Melander, Pholeomyia	732	nasalis (Thomson), Physogenua	702
myopaeformis (Röder), Tetanops	648	nasellensis Reinhard, Chaetophlepsis	1016
Myophthiria Rondani	917	nasica Hull & Fluke, Cheilosia	585
myopina (Fallén), Myopina	843	nasica Osburn, Sphecomyia	612
Myopina Robineau-Desvoidy	843	nasica Say, Rhingia	588
MYOPINAE	628	*Nasiterna* Wallengren	56
MYOPININI	843	**Nasiternella** Wahlgren	56
myosotus Osten Sacken, Dolichopus	493	naso (Walker), Platycheirus	577
Myospila Rondani	890	nasoni Coquillett, Admontia	1068
Myothyriopsis Townsend	1086	nasoni Coquillett, Rhamphomyia	464
myricae Beutenmüller (Cecidomyia)	290	*nasoni* (Coquillett), Taxigramma	940
myriosticta Alexander, Phyllolabis	60	nasoni (Cresson), Cnodacophora	633
myrmecoides (Loew), Myrmecothea	646	nasoni Cresson, Stenomyia	654
Myrmecomya Robineau-Desvoidy	643	*nasoni* Malloch, Helina	888
Myrmecomyia Robineau-Desvoidy	643	nasoni (Malloch), Megaselia	541
Myrmecothea Hendel	646	nasoni Stein, Lispe	878
Myrmosicarius Borgmeier	544	*Nasonimyia* Townsend	940
myrrhea (Brauer & Bergenstamm), Juriniopsis	1001	nassauensis Felt, Lasioptera	272
mystacea (Coquillett), Hylemya	855	nasuta (Bigot), Neoascia	595

INDEX

	Page		Page
nasuta Cresson, Ceropsilopa	742	NEMATOCERA	15
nasuta (Melander), Tylomyza	797	*nematolinus* Enderlein	309
nasuta Spuler, Leptocera	724	**Nematoproctus** Loew	515
Natarsia Fittkau	146	*Nematotelus* Geoffroy	309
nativa Johannsen, Exechia	207	NEMESTRINIDAE	401
natvigi Goetghebuer, Micropsectra	178	NEMESTRININAE	402
natvigi (Goetghebuer), Trissocladius	155	*Nemestrinopsis* Cockerell	402
Nausigaster Williston	596	**Nemochaeta** Wulp	1000
NAUSIGASTRINI	596	**Nemomydas** Curran	359
navajo Alexander, Nephrotoma	21	nemophila Kessel, Agathomyia	548
Neacreotrichus Cockerell	419	**Nemopoda** Robineau-Desvoidy	682
Neadmontia Townsend	1021	*nemoptera* (Alexander), Dicranota	58
neadusta Alexander, Limnophila	66	nemoralis (Hine), Efferia	395
NEAERINA	1064	nemoralis (Meigen), Bradysia	234
neali Steyskal, Sarcophaga	958	nemoralis (Meigen), Neuratelia	224
Neaporia Townsend	1028	**Nemorilla** Rondani	1088
nearctica Chillcott, Piezura	898	**Nemorimyza** Frey	798
nearctica Hardy, Plecia	192	nemorum (Fabricius), Xylota	607
nearctica McAlpine, Cremifania	706	nemorum (Linnaeus), Eristalis	624
nearctica Parker, Sarcophaga	959	nemorum (Meigen), Suillia	809
nearctica Shewell, Hilara	456	*Nemosturmia* Townsend	1089
nearcticus Dyar, Aedes	113	NEMOTELINI	309
nearcticus Sabrosky, Eribolus	778	*Nemotelinus* Enderlein	309
nearno Martin, Proctacanthus	399	**Nemotelus** Geoffroy	309
Neaspilota Osten Sacken	672	neoaffinis Fluke, Helophilus	618
Neatonia Bromley	387	neoaliciae Garrett, Dixa	102
nebraskensis Quate, Telmatoscopus	94	**Neoalticomerus** Hendel	794
Nebritus Coquillett	349	**Neoascia** Williston	594
nebritus Coquillett, Epacmus	428	neoborealis (Snyder), Eupogonomyia	902
nebularum Aldrich, Coelopa	680	*neobscura* Shannon, (Xylota)	1115
nebulinervis Alexander, Tipula	27	neobscura Snyder, Mydaea	891
nebulipennis Alexander, Tipula	33	**Neocatocha** Felt	243
nebulo (Coquillett). Villa	435	*Neocelatoria* Walton	1040
nebulo (Osten Sacken), Eucyrtopogon	373	*Neocerata* Coquillett	258
nebulosa (Alexander), Limonia	50	*Neocerotopogon* Malloch	133
nebulosa (Coquillett), Catharosia	971	**Neochirosia** Malloch	841
nebulosa Coquillett, Cladochaeta	771	**Neochrysops** Walton	328
nebulosa (Coquillett), Cordylurella	836	**Neocladura** Alexander	72
nebulosa (Coquillett), Heleomyza	816	**Neocnemodon** Goffe	581
nebulosa Coquillett, Macrocera	201	**Neocoelopa** Malloch	680
nebulosa (Coquillett), Monohelea	135	*Neocota* Coquillett	461
nebulosa Coquillett, Pseudiastata	762	neocoxendix Sabrosky, Oscinella	780
nebulosa (Coquillett), Sphenometopa	939	neocryptus Harmston & Knowlton, Hercostomus	498
nebulosa (Coquillett), Villa	438		
nebulosa (Fallén), Diastata	772	neoculatus Wilcox & Martin, Nannocyrtopogon	379
nebulosa Garrett, Docosia	226		
nebulosa (Johannsen), Phronia	215	**Neoculex** Dyar	117
nebulosa Johnson, Xylota	606	neocynipsea Melander & Spuler, Sepsis	683
nebulosa Kröber, Thereva	353	**Neocyptera** Townsend	973
nebulosa (Loew), Grallipeza	635	**Neodeceia** Malloch	702
nebulosa (Malloch), Clinohelea	136	**Neodexiopsis** Malloch	873
nebulosa (Say), Ornithoctona	917	**Neodichocera** Walton	1014
nebulosa Sturtevant, Drosophila	768	*Neodioctria* Wilcox & Martin	371
nebulosus Coquillett, Gastrops	752	*Neodionaea* Townsend	975
nebulosus Melander, Cyamops	820	**Neodiplocampta** Curran	441
nebulosus Palisot de Beauvois, Tabanus	335	**Neodiplotoxa** Malloch	793
nebulosus Walker, Eristalis	623	neoelegans Alexander, Limonia	45
nebulosus (Walker), Leptomorphus	223	**Neoempheria** Osten Sacken	222
necotus Hardy, Bibio	193	*Neoepicypta* Coher	215
needhami Alexander, Erioptera	82	*Neoeriphia* Schnabl & Dziedzicki	901
needhami Cresson, Setacera	755	neofalcatus Alexander, Molophilus	89
needhami Felt, Hormomyia	288	**Neofischeria** Townsend	1010
needhami Johannsen, Setacera	755	neoflavellus Malloch, Tanytarsus	179
needhami Thomsen, Alluaudomyia	133	**Neoforcipomyia** Tokunaga	124
needhamii (Johannsen), Polypedilum	176	neofusca Felt, Lasioptera	272
nefarius Hine, Tabanus	335	*Neogaurax* Malloch	781
Negasilus Curran	397	**Neogcodes** Schlinger	407
neglecta Becker, Siphonella	786	**Neogimnomera** Malloch	832
neglecta Curran, Peleteria	998	*Neoglaphyroptera* Osten Sacken	227
neglecta (Johannsen), Bradysia	234	**Neogriphoneura** Malloch	702
neglecta Johannsen, Cordyla	208	neoheves Garrett, Sciophila	225
neglecta (Malloch), Copromyza	719	**Neohirmoneura** Bequaert	402
neglecta Malloch, Mydaea	891	**Neohydrotaea** Malloch	901
neglecta Sturtevant & Wheeler, Hyadina	751	**Neohylemyia** Malloch	867
neglecta (Townsend), Peleteria	997	**Neoitamus** Osten Sacken	397
neglecta (West), Paradidyma	1035	*neojacobi* Hardy, Bibio	194
neglecta Williston, Stylogaster	632	neojubatus Wilcox & Martin, Stenopogon	384
neglectus Bromley, Stenopogon	384	**Neolasioptera** Felt	272
neglectus Wheeler, Pelastoneurus	502	**Neoleria** Malloch	813
neglectus Wheeler, Thinophilus	508	**Neoleucophenga** Oldenberg	762
negligens (Alexander), Limonia	50	**Neoleucopis** Malloch	709
negundifolia Felt, Contarinia	277	**Neolimnophila** Alexander	72
negundinis Gillette, (Cecidomyia)	292	**Neolipophleps** Alexander	75
neili Steyskal, Sepedon	693	**Neolipoptena** Bequaert	920
nelliana (Alexander), Limonia	50	neolurida Alexander, Rhabdomastix	77
nemakagonensis (Graenicher), Villa	439	neomexicana (Alexander), Dicranota	58

	Page		Page
neomexicana Alexander, Erioptera	83	nervosa Walker, (Thereva)	354
neomexicana Chillcott, Fannia	896	*nervosus* Loew, Bibio	194
neomexicana (Cockerell), Asphondylia	269	nervosus Staeger, Chironomus	169
neomexicana Cole, Thereva	353	**Nesotipula** Alexander	31
neomexicana (Felt), Peromyia	251	netta Townes, Pseudochironomus	176
neomexicana Malloch, Hylemya	849	neuquenensis (Blanchard), Blaesoxipha	945
neomexicana Reinhard, Paradidyma	1033	**Neuratelia** Rondani	224
neomexicana (Townsend), Clausicella	1059	**Neurigona** Rondani	517
neomexicana Townsend, Leucostoma	976	NEURIGONINAE	517
neomexicana (Townsend), Myiophasia	983	*Neuroctena* Rondani	680
neomexicana (Townsend), Sitophaga	1049	*Neurogona* Rondani	517
neomexicanus Harmston, Dolichopus	493	*Neurogonia* Rondani	517
neomexicanus (Melander), Toreus	457	**Neurohelea** Kieffer	137
NEOMINTHOINI	1051	**Neuromyia** Felt	261
neomodestus Malloch, Chironomus	169	*Neurota* Curran	306
neomorio (Alexander), Limonia	48	neurotomae Sellers, Aplomya	1098
neomorum (Linnaeus), Eristalis	624	neuter Melander, Oligodranes	422
Neomusca Malloch	911	nevadae Bigot, Stratiomys	312
Neomuscina Townsend	911	nevadaensis (Hardy), Pipunculus	554
neonebulosa Alexander, Limonia	50	nevadana Steyskal, Haigia	644
Neonyctia Townsend	1030	nevadensis Alexander, Limnophila	66
Neopachygaster Austen	317	nevadensis Cook, Swammerdamella	240
Neopales Coquillett	1087	nevadensis Wilcox & Martin, Nannocyrtopogon	379
Neopelomyia Hendel	727	newcomeri Doane, Tipula	33
neoperplexus (Curran), Metasyrphus	561	nexa Johannsen, Exechia	207
Neophorichaeta Smith	1033	nexa Reinhard, Spathidexia	1021
Neophorocera Townsend	1053	nexilis (Dietz), Nephrotoma	20
Neophyllomyza Melander	729	nexilis (Johannsen), Orfelia	203
Neophyto Townsend	952	nexilis (Reinhard), Comasarcophaga	949
Neoplasta Coquillett	472	**Niadina** Rapp	235
Neoplatyura Malloch	202	niagarana (Parker), Udamopyga	960
neopoeciloptera Malloch, Helina	887	**Nicephorus** Reinhard	990
Neopogon Bezzi	384	*nicholsoni* Benjamin, Eurosta	663
neopolychaeta Chillcott, Fannia	896	**Nicocles** Jeannicke	380
Neopsalida Townsend	976	nicolensis (Curran), Caloparyphus	308
Neopsalidopteryx Brooks	1067	nictans (Dietz), Nephrotoma	22
neopulicaris Wirth, Culicoides	128	nidicola Aldrich, Hylemya	855
Neorhegmoclemina Cook	239	nidicola Cole, Anthrax	432
Neorhynchocephalus Lichtwardt	402	nidicola Malloch, Fannia	896
Neorondania Osten Sacken	304	nidicola Malloch, Hydrotaea	900
Neoscatella Malloch	756	nidicola (Townsend), Townsendiellomyia	1088
Neosciara Pettey	232	nielseniana Alexander, Limonia	48
Neosimulium Rubzov	187	nielsoni Kennedy, Deuterophlebia	100
Neosimulium Vargas, Martínez, & Díaz	187	nigella (Bromley), Laphria	390
Neosolieria Townsend	1010	nigella Johannsen, Docosia	226
Neossos Malloch	817	nigella Johannsen, Polylepta	225
Neossus Malloch	817	nigella (Reinhard), Nowickia	994
Neostilobezzia Goetghebuer	134	niger Andersen, Tanytarsus	179
Neoswaldia Mesnil	1049	niger (Banks), Telmatoscopus	94
Neosyntormon Curran	516	niger Cole, Ogcodes	407
Neotabanus Lutz	332	niger Coquillett, Pseudapinops	1066
Neotephritis Hendel	668	*niger* Cresson, Lordotus	412
neoternatus (Bromley), Diogmites	372	niger (Hardy), Dilophus	196
neotexensis Brooks, Peleteria	998	niger Leonard, Rachicerus	296
Neothelaira Townsend	1039	niger Loew, Coniceps	642
neotibialis West, Ptilodexia	989	niger Macquart, (Ceraturgus)	401
neotomaria Chillcott, Fannia	896	niger Macquart, Chrysops	†324, 326
Neotractocera Townsend	1072	niger Macquart, Eclimus	427
NEOTTIOPHILIDAE	714	*niger* Macquart, Eristalis	623
Neotylus Hendel	635	niger Martin, Paratacticus	381
neouliginosa Snyder, Lispe	878	niger Melander & Argo, Clusiodes	807
neovaria Wheeler, Leucophenga	761	*niger* Palisot de Beauvois, Tabanus	332
Neowinthemia Townsend	1089	niger Roback, Psectrocladius	159
neoxena Alexander, Phalacrocera	41	niger Root and Hoffman, Culicoides	129
nepenthe Hull, Baccha	574	*niger* (Wiedemann), Efferia	393
nephophila Alexander, Pedicia	54	niger Williston, Gastrops	752
nephophila Alexander, Tipula	28	nigerrima Loew, Odontomyia	†316, 316
Nephopteropsis Townsend	1065	nigerrima Lundbeck, Piophila	712
nephopterus (Mitchell), Stenochironomus	177	nigerrima Malloch, Hydrophoria	864
NEPHROCERINI	551	*nigerrima* Rondani, Saltella	683
Nephrocerus Zetterstedt	551	nigerrimum (Melander), Calliopum	696
Nephrotoma Meigen	20	*nigra* Aldrich, Cylindromyia	973
nepticula (Loew), Neoempheria	222	nigra Back, Townsendia	385
neptis (Loew), Phytobia	798	*nigra* (Bigot), Adoxomyia	305
neptis Reinhard, Euceromasia	1100	nigra (Brooks), Elfia	1065
neptun Dietz, Tipula	33	nigra (Brooks), Gonia	1076
Neptunimyia Felt	244	nigra (Brooks), Phasia	969
Nerax Hull	393	nigra Brundin, Parakiefferiella	161
nereis (Townes), Chironomus	168	nigra (Cole), Orfelia	203
neria (Curran), Smittia	162	nigra Cresson, Villa	434
NERIIDAE	637	nigra Curran, Eulasiona	1023
Neriocephalus Enderlein	634	*nigra* Day, Odontomyia	317
nervosa Loew, Phytomyza	805	nigra Fisher, Mycomya	221
nervosa Meigen, (Sciara)	236	nigra Greene, Volucella	601
nervosa Melander, Tachyempis	475	*nigra* Johannsen, Leia	227
nervosa Stein, Phaonia	907		

INDEX 1639

Name	Page
nigra Loew, Myolepta	589
nigra (Ludlow), Aedes	115
nigra Macquart, Microphthalma	982
nigra (Meigen), Copromyza	720
nigra Meigen, Drapetis	†477, 477
nigra (Meigen), Megaselia	539
nigra Meigen, Phytomyza	805
nigra (Meigen), Sceptonia	215
nigra Meigen, Spania	348
nigra (Perty), Lynchia	918
nigra Say, (Thereva)	354
nigra Stein, Fannia	893
nigra (Tucker), Oscinella	780
nigra (Van Duzee), Thecophora	631
nigra (Walker), Muscina	909
nigra (Wiedemann), Phorodonta	230
nigra (Williston), Blera	611
nigra (Williston), Helaeomyia	741
nigra Williston, Mallophora	396
nigra Wirth, Neurohelea	137
nigrans (Melander), Sapromyza	704
Nigrasilus Hine	398
nigratarsi Fluke, Sphaerophoria	570
nigratus (Fabricius) Exoprosopa	444
nigrella Cole, Milichiella	733
nigrens (Wulp), Phasia	†969
nigrescens Arnaud, Paradejeania	1002
nigrescens (Cresson), Philygria	745
nigrescens (Curran), Nigrobonellia	1006
nigrescens Hull & Fluke, Cheilosia	585
nigrescens Palisot de Beauvois, Tabanus	335
nigrescens Stein, Coenosia	872
nigrescens Wheeler, Amiota	761
nigrescens Wheeler, Pelastoneurus	501
nigresceum Knowlton & Rowe, Simulium	188
nigriapicalis Van Duzee, Dolichopus	493
nigribarba James, Adoxomyia	306
nigribarba (Loew), Parapenium	580
nigribarba Shannon, Calliphora	929
nigribarba (Van Duzee), Orfelia	203
nigribarbis Bigot, Sargus	302
nigribarbus Loew, Gymnopternus	499
nigribarbus Van Duzee, Hydrophorus	505
nigribasis Curran, Lixophaga	1043
nigribasis Curran, Trichopalpus	836
nigribasis Malloch, Helina	888
nigribasis McAlpine, Herina	645
nigribasis Stein, Hylemya	††852
nigribimbo Whitney, Chrysops	326
nigribucca Hough, Calliphora	929
nigricana (Garrett), Heleomyza	816
nigricans (Enderlein), Bradysia	233
nigricans (Johannsen), Bradysia	233
nigricans Johannsen, Chironomus	170
nigricans Johannsen, Phaonia	907
nigricans Loew, Hormopeza	453
nigricans (Loew), Neodiplotoxa	793
nigricans Loew, Rhamphomyia	465
nigricans Meigen, Diaphorus	520
nigricans Melander, Mythicomyia	418
nigricans (Stein), Helina	887
nigricans Webber, Tachinomyia	1058
nigricans (Wiedemann), Hybomitra	340
nigricauda (Adams), Mycomya	221
nigricauda Alexander, Cladura	72
nigricauda Bigot, (Hydrophoria)	869
nigricauda Curran, Brachyopa	592
nigricauda (Loew), Villa	434
nigricauda Malloch, Phaonia	907
nigricauda Metcalf, Sphaerophoria	570
nigricauda Van Duzee, Dolichopus	493
nigricaudata Huckett, Hylemya	850
nigriceps Bigot, (Curtonevra)	915
nigriceps (Bigot), Ursophyto	991
nigriceps Curran, Euxesta	651
nigriceps Huckett, Hylemya	855
nigriceps Loew, Dalmannia	632
nigriceps (Loew), Elachiptera	777
nigriceps (Loew), Megaselia	541
nigriceps Macquart, Piophila	713
nigriceps Meigen, Piophila	713
nigriceps Williston, Cholomyia	1029
nigricera Van Duzee, Gymnopternus	499
nigriciliatus Van Duzee, Chrysotus	522
nigricola (Walker), Gnophomyia	73
nigricolor (Coquillett), Nannocyrtopogon	374
nigricolor (Fallén), Hebecnema	890
nigricormis Greene, Myxosargus	317
nigricornis Adams, Pipunculus	553
nigricornis Allen, Eumacronychia	935
nigricornis (Coquillett), Trupanea	667
nigricornis (Curran), Neocnemodon	581
nigricornis (Doane), Myoleja	669
nigricornis Egger, Leucopis	709
nigricornis Greene, Myxosargus	317
nigricornis Hunter, Tropidia	609
nigricornis (Loew), Cephalochrysa	302
nigricornis (Loew), Monochaetoscinella	778
nigricornis Meigen, Dolichopus	493
nigricornis Meigen, Psila	639
nigricornis (Meigen), Themira	684
nigricornis Reinhard, Myioclonia	1025
nigricornis Reinhard, Opsotheresia	990
nigricornis (Townsend), Leucostoma	976
nigricornis (Townsend), Mericia	1008
nigricornis Van Duzee, Leia	227
nigricornis Van Duzee, Neurigona	518
nigricornis Van Duzee, Pelastoneurus	502
nigricornis (Wiedemann), Physocephala	628
nigricornis (Williston), Mericia	1008
nigricosta Curran, Gibsonomyia	1023
nigricosta Rondani, Tetanocera	694
nigricoxa (Loew), Rhaphium	513
nigricoxa Melander & Spuler, Piophila	712
nigricoxa Stein, Coenosia	872
nigricoxa Van Duzee, Argyra	525
nigricoxa Van Duzee, Dolichopus	493
nigricoxa Van Duzee, Gymnopternus	499
nigricoxum Stone, Simulium	188
nigrifacies Bigot, (Chetolyga)	1108
nigrifacies (Curran), Melangyna	567
nigrifacies Van Duzee, Hercostomus	498
nigrifascia Stone, Anastrepha	673
nigrifematus Hardy, Bibio	193
nigrifemoratus (Macquart), Copromyza	719
nigrifemoratus Van Duzee, Polymedon	504
nigrifemoratus (Walker), Condylostylus	484
nigrifemur Malloch, Sphaerocera	719
nigrifrons Kröber, Zodion	629
nigrifrons Malloch, Meoneura	729
nigrifrons Parent, Chrysotus	522
nigrifrons Spuler, Leptocera	724
nigrifrons (Stein), Spilogona	882
nigrifrons (Townsend), Myiophasia	984
nigrifrons (Walker), Lasiops	904
nigrifrons Walker, (Stratiomys)	319
nigrihirta Van Duzee, Mycomya	221
nigrilinea (Doane), Lipsothrix	77
nigrilineatus Van Duzee, Dolichopus	493
nigrimana Cole, Sphegina	594
nigrimana Coquillett, Lauxania	699
nigrimana (Coquillett), Mumetopia	820
nigrimana Kröber, Psilocephala	351
nigrimana Kröber, Thereva	353
nigrimana (Malloch), Gymnochiromyia	822
nigrimana Meigen, Piophila	712
nigrimana (Meigen), Pteromicra	687
nigrimana (Williston), Leptopsilopa	741
nigrimana (Zetterstedt), Acanthocnema	836
nigrimanum Loew, Brachystoma	447
nigrimanus (Bigot), Physoconops	627
nigrimanus Van Duzee, Dolichopus	493
nigrimanus Van Duzee, Neurigona	518
nigrina (Bigot), Phormia	924
nigrina Felt, Asteromyia	274
nigrina Johannsen, Dynatosoma	209
nigrina Kröber, Psilocephala	351
nigrina Sherman, (Trichonta)	1115
nigrina Westwood, Acrocera	406
nigrinervis Frost, Phytomyza	805
nigrinervis Van Duzee, Hydrophorus	505
nigrinotata Hennig, Grallipeza	633
nigrinus Fallén, Nemotelus	310
nigrinus (Zetterstedt), Acartophthalmus	808
nigripalpis Becker, Scatophaga	839
nigripalpis Malloch, Clusiodes	807
nigripalpis (Malloch), Siphonella	787
nigripalpis (Townsend), Aplomya	1097
nigripalpis (Townsend), Bufolucilia	927
nigripalpis Townsend, Chaetoglossa	1078
nigripalpis Van Duzee, Chrysotus	522
nigripalpus Theobald, Culex	118

	Page		Page
nigripalpus (Walker), Sapromyza	704	nigriventris Macquart, Meromyza	793
nigripennis (Cole), Poecilanthrax	442	nigriventris Macquart, Proctacanthus	399
nigripennis (Curran), Hesperophasiopsis	978	nigriventris Smith, Spathimeigenia	1049
nigripennis Fallén, Phytomyza	805	nigriventris Van Duzee, Argyra	525
nigripennis (Gray), Pyrgota	657	nigriventris Walton, Criorhina	612
nigripennis (Loew), Conophorus	414	*nigriventris* Wulp, Sepedon	693
nigripennis (Schnabl), Helina	887	nigriventris (Zetterstedt), Spilogona	882
nigripennis Van Duzee, Aphrosylus	509	nigriverticellus Bromley, Stenopogon	384
nigripennis (Walker), Helina	888	nigroaenea Walker, (Gymnopa)	759
nigripennis (Williston), Chrysosyrphus	592	*nigroantennata* (Dietz), Nephrotoma	22
nigripes (Aldrich), Mesorhaga	483	nigroapicalis Van Duzee, Dolichopus	493
nigripes (Camras), Thecophora	631	nigroapicata Curran, Cheilosia	585
nigripes (Coquillett), Diplotoxa	792	nigrobarba Hull & Fluke, Cheilosia	585
nigripes (Curran), Blera	611	**Nigrobonellia** Brooks	1006
nigripes Curran, Cryptomeigenia	1041	nigrocaudatus Van Duzee, Paracleius	501
nigripes (Fabricius), Thelaira	1021	*nigrocella* Wheeler, Scaptomyza	769
nigripes (Fallén), Blondelia	1040	nigrocellulata Czerny, Cremifania	706
nigripes Harmston & Knowlton, Parasyntormon	516	nigrociliata Malloch, Lonchaea	717
nigripes Hine, Promachus	400	nigrociliatum Curran, Rhaphium	513
nigripes (Jones), Eucyrtopogon	373	nigroclavata Alexander, Limonia	45
nigripes Loew, Argyra	525	nigrocoerulea Lovett, Cheilosia	587
nigripes Loew, Chrysogaster	591	nigrocomus Hull, Metasyrphus	561
nigripes Loew, Medetera	510	nigrocornea (Tothill), Pseudomeriania	1009
nigripes Loew, Pipunculus	554	nigrocorporis Doane, Tipula	31
nigripes Meigen, Agromyza	795	nigrocyanea Hull & Fluke, Cheilosia	587
nigripes (Meigen), Bradysia	234	nigrodorsalis Alexander, Ulomorpha	68
nigripes (Meigen), Palpomyia	140	nigrofasciata Curran, Cheilosia	585
nigripes (Melander), Drapetis	477	nigrofemorata Alexander, Limnophila	67
nigripes Melander, Euhybus	448	nigrofemorata Wilcox, Willistonina	386
nigripes Melander & Argo, Heteromeringia	806	nigrofemoratus Hine, Proctacanthus	399
nigripes Osten Sacken, Dicranoptycha	52	nigrofemoratus Painter, Parabombylius	410
nigripes (Osten Sacken), Ferdinandea	588	nigrofemoratus (Walker), Condylostylus	484
nigripes Painter, Geron	423	nigrogeniculata Alexander, Limnophila	66
nigripes (Robineau-Desvoidy), Bonnetia	1004	nigrogenualis Alexander, Dicranoptycha	52
nigripes (Robineau-Desvoidy), Pseudolimnophora	885	nigrohalteralis (Malloch), Paralauterborniella	173
nigripes Van Duzee, Asyndetus	524	nigrohalterata Duda, Drosophila	767
nigripes Van Duzee, Campsicnemus	526	nigrohydei Patterson & Wheeler, Drosophila	765
nigripes Van Duzee, Thrypticus	511	*nigrolanata* Cresson, Scatophaga	838
nigripes Wiedemann, Tabanus	335	nigrolimbata Cresson, Scatophaga	839
nigripes Williston, Ceraturgus	365	nigromaculata Jones, Xylota	607
nigripes (Williston), Criorhina	612	*nigromaculata* Stein, Lispe	878
nigripes (Zetterstedt), Aedes	113	nigromaculis (Ludlow), Aedes	113
nigripes Zetterstedt, Chrysops	326	nigromelanica Patterson & Wheeler, Drosophila	765
nigripila (Johannsen), Micropsectra	178	nigronasica Painter, Villa	439
nigripila (Osten Sacken), Ormosia	86	nigronitida Alexander, Ulomorpha	68
nigripilosa Cole, Thereva	353	nigronotum Wilcox, Ablautus	364
nigripilosa Curran, Cryptomeigenia	1041	nigropalpis (Tothill), Mericia	1008
nigripilosa Hardy & McGuire, Ptiolina	347	*nigroparvum* Twinn, Simulium	189
nigripilosa Wilcox & Martin, Dioctria	371	nigropecta Cresson, Villa	434
nigripilosa Williston, Pipiza	†579, 579	*nigropilosa* (Cresson), Adoxomyia	305
nigripilosus Van Duzee, Thinophilus	508	*nigropilosa* (Curran), Melangyna	566
nigripilus Loew, Bibio	193	*nigropilosa* (Wulp), Parepalpus	†1003
nigripleura (Alexander & Leonard), Pseudolimnophila	62	nigropilosus Schaeffer, Promachus	400
nigrirostris (Loew), Anoplodonta	314	nigropolita Malloch, Dialyta	904
nigrisquama (Malloch), Phytobia	800	nigropterus Fairchild, Chrysops	324
nigrita Bigot, Rhamphomyia	466	*nigroscutellata* Becker, Elachiptera	777
nigrita (Fabricius), Exoprosopa	444	nigroscutellata (Stein), Pegomya	863
nigritarse (Curran), Parapenium	580	nigroseta Hull & Fluke, Cheilosia	584
nigritarsis (Stein), Coenosia	873	nigrospiracula Patterson & Wheeler, Drosophila	765
nigritarsis (Zetterstedt), Pegomya	860	*nigrostigma* Curran, Euparyphus	307
nigritarsus Wilcox & Martin, Cyrtopogon	368	nigrotibiata Curran, Pipiza	579
nigritella Huckett, Coenosia	873	*nigrotuberculatus* (Fairchild), Hybomitra	338
nigritella (Melander), Piophila	712	nigroventris Fluke, Metasyrphus	561
nigrithorax Stein, Hylemya	850	nigroviridis Walker, Chyliza	640
nigritibia Walker, (Chironomus)	181	nigrovittata (Loew), Chrysogaster	591
nigritulus Coquillett, Stenopogon	384	nigrovittata Lovett, Cheilosia	585
nigritum Townes, Polypedilum	175	nigrovittata (Malloch), Dettopsomyia	771
nigritus Curran, Bibio	193	nigrovittatum Curran, Rhaphium	513
nigritus Malloch, Orthocladius	156	nigrovittatus Macquart, Tabanus	335
nigriventris Bigot, Chrysops	326	nigrovittatus Malloch, Chironomus	168
nigriventris Cresson, Discocerina	738	nigrum (Van Duzee), Rhaphium	513
nigriventris Huckett, Pseudocoenosia	875	**Nilea** Robineau-Desvoidy	1104
nigriventris (Johannsen), Platyura	204	**Nilobezzia** Kieffer	138
nigriventris Johnson & Johnson, Lordotus	412	**Nilodorum** Kieffer	170
nigriventris Loew, Hilara	456	**Nilohelea** Kieffer	133
nigriventris Loew, Stratiomys	312	**Nilotanypus** Kieffer	148
nigriventris (Macquart), Delphinia	644	**Nilothauma** Kieffer	173
		nimbipennis Alexander, Ormosia	86
		Nimiocauda Reinhard	1046
		Nimioglossa Reinhard	987
		nimius Aldrich, Polymedon	504
		niphadias (Alexander), Tasiocera	88
		niphadopsis Dyar & Knab, Aedes	113

INDEX 1641

	Page
Niphogenia Melander	467
nipigonense Curran, Temnostoma	614
nipigonensis (Curran), Eutheresia	990
nippontucki Philip, Tabanus	336
Nippotipula Matsumura	26
nitelae Aldrich & Webber, (Phorocera)	1107
nitens (Coquillett), Ospriocerus	381
nitens (Coquillett), Oxynops	1046
nitens (Coquillett), Phyllophilopsis	1030
nitens (Loew), Paramyia	730
nitens Malloch, Hemeromyia	728
nitens Melander, Mumetopia	820
nitens Melander, Neophyllomyza	730
nitens Melander, Platypalpus	480
nitens (Melander & Spuler), Pseudodinia	708
nitens Reinhard, Minthozelia	1027
nitens (Stein), Pogonomyia	902
nitens Walker, Scatopse	238
nitens (Wiedemann), Copecrypta	†1003
nitens (Zetterstedt), Metasyrphus	†560
nitida Adams, Tetragoneura	228
nitida (Banks), Brunettia	94
nitida Banks, Platypeza	549
nitida (Coquillett), Alophorella	968
nitida Coquillett, Cuterebra	1110
nitida Cresson, Parydra	749
nitida Curran, Cyrtophleba	1017
nitida (Felt), Asteromyia	274
nitida Harris, Scatopse	238
nitida Hendel, Milichiella	733
nitida Johannsen, Docosia	226
nitida Johnson, Heteromeringia	806
nitida (Macquart), Leptogaster	362
nitida Malloch, Hemeromyia	728
nitida (Malloch), Phytagromyza	803
nitida Melander, Drapetis	478
nitida Melander, Mythicomyia	418
nitida Melander, Phytomyza	805
nitida Melander, Pseudodinia	708
nitida Melander, Trichina	449
nitida Reinhard, Muscopteryx	1025
nitida (Robineau-Desvoidy), Erynniopsis	†1042
nitida (Stein), Helina	888
nitida Sturtevant & Wheeler, Diphuia	737
nitida Van Duzee, Neurigona	518
nitida Van Duzee, Sciophila	226
nitida Wiedemann, Chrysogaster	591
nitida Williston, Dioctria	370
nitida (Wulp), Nowickia	996
nitida (Wulp), Pseudodinia	708
nitidellus (Malloch), Trichocladius	158
nitidicollis Lundström, Exechia	207
nitidicollis (Meigen), Epistrophe	565
nitidifacies Hine, Nigrasilus	398
nitidifrons (Brues), Diplonevra	535
nitidifrons Cresson, Scatella	757
nitidifrons Hardy & McGuire, Ptiolina	347
nitidifrons Malloch, Hippelates	775
nitidifrons (Stein), Spilogona	881
nitidipleura Melander, Platypalpus	480
nitidissima (Meigen), Oscinella	780
nitidissima Melander & Spuler, Piophila	711
nitidithorax (Alexander), Dactylolabis	62
nitidiuscula Alexander, Limonia	50
nitidiventris Bigot, (Holopogon)	401
nitidiventris (Curran), Carposcalis	576
nitidiventris Loew, Euxesta	652
nitidiventris Loew, Pipunculus	552
nitidiventris Melander, Trixoscelis	818
nitidiventris Van Duzee, Medetera	510
nitidivittata Macquart, Rhamphomyia	465
nitidor Cresson, Micropeza	634
nitidoscutellatus Lundström, Orthocladius	156
Nitidotipula Alexander	24
nitidula (Coquillett), Hylemya	846
nitidula Curran, Chrysogaster	591
nitidula (Fallén), Nemopoda	682
nitidula (Malloch), Pseudonapomyza	803
nitidula (Meigen), Diplonevra	535
nitidula Melander, Mythicomyia	418
nitidula Sabrosky, Pholeomyia	732
nitidula Zetterstedt, Coelopa	680
nitidula Zetterstedt, Iteaphila	455
nitidulus Alexander, Molophilus	89
nitidulus (Coquillett), Paratendipes	174

	Page
nitidulus (Fabricius), Anastoechus	†411
nitidus Adams, Xylophagus	297
nitidus Cole, Cyrtopogon	369
nitidus Cole, Epacmus	428
nitidus Cole, Metacosmus	427
nitidus Coquillett, Molophilus	89
nitidus (Hennig), Micropeza	634
nitidus Johnson, Rachicerus	296
nitidus (Malloch), Trichocladius	158
nitidus Melander, Meghyperus	449
nitidus Melander & Argo, Clusiodes	807
nitidus Van Duzee, Polymedon	504
nitidus Wehr, Eristalis	624
nitidus Wilcox & Martin, Nannocyrtopogon	379
nitoris (Coquillett), Clairvillia	976
nitoris Coquillett, Thereva	354
nitoris Martin, Leptogaster	363
nivalis Curran, Archytas	1000
nivalis (Tothill), Nowickia	995
nivalis (Zetterstedt), Quadrularia	890
nivea Kröber, Thereva	354
nivea Loew, Coenosia	872
niveiceps Cresson, Ephydra	754
niveifrons Hull & Fluke, Cheilosia	587
niveipennis Kröber, Thereva	354
niveipennis Zetterstedt, Agromyza	795
niveipila Osten Sacken, (Cecidomyia)	294
niveitarsis Osten Sacken, Limnophila	67
niveivenosa Cresson, Nostima	745
niveus Cresson, Geron	423
nivicola Doane, Chionea	72
nivitarsis (Coquillett), Aedes	111
nivoriunda (Fitch), Diamesa	152
nivosa Cresson, Lamproscatella	756
nivosus Osten Sacken, Tabanus	334
nixe (Townes), Chironomus	168
nixoni Felt, (Cecidomyia)	290
nobilis Felt, Asynapta	254
nobilis Johnson, Macrocera	201
nobilis Loew, Hippelates	775
nobilis (Loew), Hydrellia	744
nobilis (Loew), Tipula	26
nobilis (Stein), Spilogona	882
Nobilotipula Alexander	25
nocheles Dyar & Shannon, Dixa	102
nocivum (Harris), Culicoides	128
Nocticanace Malloch	734
noctifer Osten Sacken, Chrysops	326
Noctivaga Sherman, (Boletina)	1115
noctivaga Van Duzee, Exechia	207
noctivagans (Alexander), Cheilotrichia	79
noctuae (Harris), Gymnocarcelia	1093
noctuiformis (Smith), Nilea	1104
noctula (Wiedemann), Exoprosopa	444
nocturnala Haseman, Psychoda	95
nocturnalis (Reinhard), Sitophaga	1049
nocturnalis Walton, Neophyto	952
nodatus (Loew), Calyxochaetus	528
nodicornis (Osten Sacken), Liogma	42
nodipennis Van Duzee, Dolichopus	493
nodipes Melander, Empis	460
nodipoplitea Steyskal, Empis	460
nodosa (Felt), Janetiella	264
nodosus Curran, Platycheirus	577
nodulicornis Zetterstedt, Tipula	30
nodulosa Beutenmüller, Lasioptera	272
nodulus (Osten Sacken), Rhabdophaga	257
Noeza Meigen	447
nokomis Hull & Fluke, Cheilosia	585
nomadus Harmston & Knowlton, Dolichopus	493
nomiivora (James), Euphyto	935
Nomoneura Bezzi	†360
nonacantha Alexander, Ormosia	86
nooksackensis Alexander, Dicranota	58
nooksackiae Alexander, Dicranota	57
nora (Doane), Xanthomyia	663
Norellia Robineau-Desvoidy	827
Norelliosoma Hendel	827
Norellisoma Hendel	827
Norellisoma Wahlgren	827
norena Roback, Pentaneura	147
norma Wiedemann, Stratiomys	312
normae Fluke, Platycheirus	577
normalis Malloch, Hylemya	850
normaniana Felt, Rhabdophaga	257
normula Loew, Stratiomys	312

Name	Page
normula (Wulp), Siphosturmia	†1087
norvegica (Ringdahl), Spilogona	882
Nostima Coquillett	745
nostocicola Wirth, Cricotopus	158
notabilis Alexander, Dicranota	56
notabilis (Walker), Hybomitra	339
notabilis (Williston), Orfelia	203
notata (Bigot), Laphystia	376
notata (Coquillett), Aciurina	670
notata (Fallén), Homoneura	699
notata Hull, Sphegina	594
notata (Linnaeus), Scatopse	238
notata (Loew), Agathomyia	548
notata Loew, Baccha	572
notata (Loew), Chelifera	471
notata Loew, Chyliza	640
notata Loew, Dixa	101
notata Loew, Phthiria	420
notata Loew, Stratiomys	312
notata Osten Sacken, Brachyopa	592
notata Painter, Apiocera	357
notata (Wiedemann), Blera	611
notata (Wiedemann), Euxesta	652
notata (Wiedemann), Psilocephala	351
notatifrons Harris, (Anthomyia)	1115
notatum Adams, Simulium	187
notatus (Bigot), Silvius	323
notatus (Loew), Calyxochaetus	528
notatus Macquart, Toxomerus	571
notatus (Wiedemann), Tolmerus	401
notescens Johannsen, Boletina	217
notha Williston, Xylota	606
Nothomyia Loew	317
Nothosympycnus Wheeler	528
Nothra Westwood	404
NOTHYBOIDEA	637
Notiphila Fallén	746
NOTIPHILINAE	743
NOTIPHILINI	746
notiphiloides Cresson, Hydrellia	744
notmani Alexander, Dicranota	57
notmani Alexander, Ormosia	86
notmani Curran, Themira	684
notmani Felt, Parwinnertzia	254
Notochaetopsis Townsend	943
Notogramma Loew	646
notomaculata Sabrosky, Pterodontia	405
Notonaulax Becker	785
Notosympycnus Wheeler	528
nova Curran, Nausigaster	596
nova (Dyar & Shannon), Twinnia	182
nova Van Duzee, Medetera	510
nova Walker, Dixa	102
novacaesariensis Alexander, Molophilus	89
novaeangliae Alexander, Limnophila	66
novaeangliae Alexander, Limonia	44
novaeangliae Felt, Porricondyla	256
novaeangliae Johnson, Sphaerophoria	570
novaeangliae (Malloch), Spilogona	884
novaeangliae (West), Carinosillus	989
novaeboracensis (Fitch), Eutreta	662
novaescoeiae Macquart, Tabanus	335
novaescotiae Macquart, Helophilus	618
novaescotiae Macquart, Tabanus	335
novaescotiae (Macquart), Tolmerus	401
novaescotiae (Shewell), Lyciella	700
novamexicana Patterson, Drosophila	765
novangliae Felt, Lestodiplosis	285
novata Johannsen, Sciophila	225
noveboracensis Alexander, Chionea	72
noveboracensis Alexander, Dicranota	57
noveboracensis Alexander, Gonomyia	76
noveboracensis (Alexander), Pseudolimnophila	62
noveboracensis Alexander, Tipula	29
noveboracensis Felt, Giardomyia	284
noveboracensis (Felt), Heteropeza	247
noveboracensis (Van Duzee), Condylostylus	485
novecarolina Beutenmüller, Rhamphomyia	465
novella Coquillett, Thereva	354
novella Steyskal, Labostigmina	314
novemmaculata (Zetterstedt), Spilogona	882
novum (Williston), Heterostylum	411
Nowickia Wachtl	993
nox (Hall), Aplomya	1098
nox (Hall), Melanodexia	931
nox (Hall), Microcerella	952
nox Hall, Tsugaea	1106
nubecula (Johnson), Homalocephala	652
nubecula Osten Sacken, Ctenophora	19
nubeculosa Coquillett, Phthiria	420
nubeculosa (Meigen), Limonia	43
nubeculosum (Meigen), Polypedilum	175
nubeculosus Zetterstedt, Sargus	†301
nubiapex Philip, Chrysops	324
nubifer (Coquillett), Procladius	149
nubifer Van Duzee, Dolichopus	493
nubifera (Coquillett), Clinohelea	136
nubifera Coquillett, Scatophaga	839
nubila (Melander), Homoneura	699
nubila Melander, Pelomyia	726
nubila (Osten Sacken), Ormosia	87
nubila Say, (Mycetophila)	229
nubila Stein, Mydaea	891
nubilifera (Malloch), Homoneura	699
nubilipennis Schmitz, Megaselia	539
nubilipes Say, Scenopinus	356
nubilis Harris, Tipula	36
nubilosa Alexander, Erioptera	84
nubilum Van Duzee, Syntormon	515
nubilus Say, (Chrysotus)	530
nucicola Osten Sacken, (Cecidomyia)	270
nucis James, Hermetia	305
nuda (Adams), Conioscinella	784
nuda Brooks, Psalidopteryx	1067
nuda (Coquillett), Trixoscelis	818
nuda (Dyar), Trichomyia	91
nuda Hull & Fluke, Cheilosia	584
nuda (James), Anoplodonta	314
nuda Loew, Empis	460
nuda (Malloch), Neochirosia	842
nuda (McDunnough), Hybomitra	340
nuda (Schnabl), Hydrophoria	864
nuda (Townsend), Besseria	972
Nudaria Karl	862
nudatus (Cresson), Canaceoides	733
nudibasis Snyder, Helina	888
nudibasis Stein, Actia	1062
nudicornis Cresson, Parallelomma	840
nudifacies Snyder, Lispe	878
nudifemorata Malloch, Lonchaea	717
nudifemur (Malloch), Homoneura	699
nudifrons (Curran), Melangyna	567
nudifrons (Curran), Neocnemodon	581
nudipalpis Malloch, Pulicophora	545
nudipennis Loew, (Sarcophaga)	1115
nudipes Cresson, Notiphila	747
nudipes Stein, Coenosia	872
nudipes Van Duzee, Chrysotus	523
nudiseta Curran, Heterocheila	679
nudiseta Stein, Coenosia	872
nudiseta (Stein), Xenomydaea	892
nudiseta (Wulp), Synthesiomyia	911
nudiuscula (Loew), Oscinella	780
nudiusculus Loew, Psilocurus	382
nudum (Van Duzee), Rhaphium	514
nudus (Coquillett), Oedenops	748
nudus Cresson, Chrysopilus	346
nudus Loew, Dolichopus	493
nudus Painter, Geron	423
nudus (Rapp), Tomosvaryella	555
nudus Van Duzee, Diaphorus	520
nugator (Coquillett), Villa	437
nugator Osten Sacken, Cyrtopogon	369
nugatoria Johannsen, Exechia	207
nugatoria Johannsen, Mycomya	221
nugax Johannsen, Exechia	207
nugax Johannsen, Sciophila	225
nugax Melander, Hilara	456
nuntia Alexander, Tipula	29
Nupedia Karl	862
nuperus Reinhard, Meleterus	1018
nuphera Harris, (Ortalis)	1115
nupta Alexander, Limnophila	65
nupta (Zetterstedt), Spilogona	884
nuptialis Alexander, Dicranota	58
nuptus Melander, Anthepiscopus	455
nura (Melander), Trichina	449
Nusa Walker	†386, †387
nuttingi Alexander, Rhabdomastix	76
nuttingi Sabrosky, Euphasiopteryx	980

INDEX 1643

Name	Page
nyblaei (Zetterstedt), Chaoborus	104
NYCTERIBIIDAE	922
nycteris Alexander, Limnophila	67
nycteris (Alexander), Limonia	48
Nycterophilia Ferris	921
nycthemerus (Macquart), Villa	436
nyctichroma Hull & Fluke, Cheilosia	587
nyctops (Alexander), Cheilotrichia	79
nylanderi (Felt), Anocha	244
nymphis Walker, Stratiomys	312
NYMPHOMYIIDAE	98
nyssaecola Beutenmüller, (Cecidomyia)	292
Nyssorhynchus Blanchard	106
oarus (Walker), Brachypalpus	608
obconica (Walker), Blondelia	1040
obcordatus Aldrich, Dolichopus	493
obediens Johannsen, Exechia	207
obediens (Johannsen), Phaenopsectra	174
obediens Johannsen, Polylepta	225
obesa (Fabricius), Ornidia	600
obesa Felt, Contarinia	277
obesa (Felt), Paradiplosis	286
obesa (Fitch), Thaumatomyia	787
obesa Harris, (Echinomyia)	1115
obesa Hull & Fluke, Cheilosia	585
obesa Loew, Empis	460
obesa (Loew), Rhinoleucophenga	763
obesa Loew, Stratiomys	313
obesa Malloch, Alliopsis	857
obesa Townsend, Opsotheresia	990
obesa (Wiedemann), Jurinella	†1001
obesa Winnertz, Trichonta	216
obesula Johannsen, Boletina	217
obesulus (Loew), Conophorus	414
obesulus Loew, Dilophus	195
obesus Coquillett, Trixodes	980
obesus (Harris), Pterodontia	404
obesus Van Duzee, Medetera	511
obfuscata (Tucker), Leptocera	724
obispoensis Alexander, Tipula	40
oblata (Townsend), Opsidia	937
oblectabilis (Loew), Leia	227
oblidens (Walker), Orthocladius	156
obliqua Harris, (Milesia)	1115
obliqua Melander, Micrempis	476
obliqua Reinhard, Paradidyma	1033
obliqua (Say), Allograpta	569
obliqua (Say), Mycomya	221
obliqua (Say), Tomoplagia	678
obliquefasciatum (Macquart), Zodion	629
obliquus (Hine), Caloparyphus	308
obliquus Loew, Chrysotus	523
obliterata (Dietz), Nephrotoma	22
oblitus Becker, Chlorops	790
Obolocera Townsend	1104
Obolodiplosis Felt	286
obreptus (Townes), Chironomus	170
obscoena (Eschscholz), Calliphora	930
obscura Bigot, Hilarimorpha	348
obscura Bigot, Pollenia	931
obscura (Coquillett), Alaskophyto	1022
obscura (Coquillett), Bolbomyia	298
obscura (Coquillett), Buquetia	1098
obscura (Coquillett), Dieuryneura	306
obscura Coquillett, Docosia	226
obscura Coquillett, Hemeromyia	728
obscura Coquillett, Psilocephala	351
obscura (Coquillett), Villa	438
obscura (Curran), Sitophaga	1048
obscura (Dyar and Shannon), Mochlonyx	104
obscura Fallén, Drosophila	768
obscura (Fallén), Phorocera	1057
obscura (Fallén), Scatophaga	839
obscura Felt, Myocodiplosis	283
obscura Fisher, (Mycetophila)	1115
obscura Garrett, Neuratelia	224
obscura Haliday, Leucopis	709
obscura Harris, Sciara	230
obscura Johannsen, Boletina	217
obscura Johnson, Agathomyia	548
obscura (Johnson), Dolichopeza	23
obscura (Loew), Minettia	702
obscura Loew, Platypeza	549
obscura Loew, Xylota	†605, 605
obscura (Malloch), Bezzia	141
obscura Malloch, Mydaeina	884
obscura (Olivier), Labostigmina	314
obscura Sabrosky, Pholeomyia	732
obscura Sabrosky, Sigaloessa	824
obscura (Say), Carposcalis	576
obscura (Stein), Mydaea	†891, 891
obscura (Townsend), Melanomya	932
obscura Van Duzee, Argyra	525
obscura (Walker), Forcipomyia	126
obscura Walker, (Mycetophila)	229
obscura Walker, (Scatopse)	241
obscura West, Ptilodexia	989
obscura Williston, Discocerina	738
obscura (Williston), Hexatoma	69
obscura Williston, Paralimna	748
obscura Williston, Scatella	757
obscurata (Meigen), Helina	888
obscuratoides (Schnabl), Helina	888
obscuratus Malloch, Chironomus	170
obscurella (Fallén), Discocerina	738
obscurella (Fallén), Meoneura	729
obscurella Malloch, Mydaea	891
obscurella (Zetterstedt), Trichopalpus	837
obscuriceps (Cresson), Neoscatella	756
obscuriceps Loew, Hydrellia	744
obscuricornis Loew, Baccha	572
obscuricornis Loew, Chlorops	790
obscuricornis (Loew), Melieria	645
obscurinervis Becker, Scatophaga	839
obscurinervis (Stein), Helina	888
obscurior Collin, Pegomya	862
obscuripennis (Bigot), Chrysopilus	347
obscuripennis Bigot, Nemopoda	682
obscuripennis Johnson, Gaurax	782
obscuripennis Johnson, Leptogaster	363
obscuripennis Loew, Chlorops	790
obscuripennis Loew, Rachicerus	296
obscuripennis (Snow), Paroxyna	666
obscuripennis (Stein), Spilogona	882
obscuripennis Van Duzee, Medetera	510
obscuripennis (Williston), Physoconops	627
obscuripes (Holmgren), Metriocnemus	161
obscuripes Johnson, Actina	300
obscuripes Loew, Ephydra	754
obscuripes Loew, Leptogaster	363
obscuripes (Zetterstedt), Helina	†886
obscuris (Doane), Gonomyia	76
obscuriventris (Loew), Allognosta	300
obscuriventris Loew, Stenopogon	384
obscurum Banks, Zodion	629
obscurum Loew, Temnostoma	614
obscurus (Coquillett), Dasiops	716
obscurus Coquillett, Microphorous	453
obscurus (Cresson), Oligodranes	422
obscurus (Cresson), Zerox	746
obscurus Felt, Arthrocnodax	283
obscurus Hine, Philonicus	398
obscurus (Johannsen), Cardiocladius	154
obscurus Loew, Bibio	194
obscurus Loew, Eristalis	624
obscurus Loew, Helophilus	618
obscurus (Loew), Oreogeton	453
obscurus (Say), Gymnopternus	499
obscurus (Wulp), Philornis	911
obsoleta Curran, Peleteria	998
obsoleta Johnson, Loxocera	638
obsoleta Leonard, Ptiolina	347
obsoleta (Loew), Anthrax	432
obsoleta (Loew), Chelifera	471
obsoleta Loew, Scatella	757
obseleta (Malloch), Spilogona	882
obsoleta (Townsend), Acroglossa	1074
obsoleta Wulp, Acrocera	406
obsoleta (Wulp), Microphthalma	982
obsoletus Loew, Helophilus	619
obsoletus (Meigen), Culicoides	128
obsoletus Van Duzee, Dolichopus	493
obsoletus Wiedemann, Chrysops	326
obsoletus Wiedemann, Eristalis	†625
obtectus Melander, Aphoebantus	430
obtruncata (Loew) Mycomya	221
obtusa Alexander, Antocha	51
obtusa Cook, Swammerdamella	240
obtusa (Fallén), Pherbellia	686
obtusa Malloch, Plunomia	707

	Page		Page
obtusa (Malloch), Thaumatomyia	788	occidentalis Spencer, Drosophila	765
obtusa Osten Sacken, Pedicia	53	occidentalis Sturtevant, Periscelis	710
obtusa (Zetterstedt), Leucophora	868	occidentalis Sturtevant & Wheeler, Nostima	745
obtusicauda Van Duzee, Gymnopternus	499	occidentalis Sturtevant & Wheeler, Ptilomyia	736
obtusifibula Melander, Tetanocera	694	occidentalis Townsend, Aphria	1010
obtusilamellata (Malloch), Lyciella	700	occidentalis Townsend, Gymnoclytia	965
obtusilobus Malloch, Chironomus	168	occidentalis Van Duzee, Asyndetus	524
obtusum (Dyar & Shannon), Simulium	187	occidentalis Van Duzee, Diaphorus	520
obtusum Townes, Polypedilum	175	occidentalis Wheeler, Pelastoneurus	502
obtusum Van Duzee, Rhaphium	514	occidentalis Williston, Cheilosia	585
obumbrata Loew, Eudicrana	223	occidentalis Williston, Eristalis	624
obumbratus Johannsen, Orthocladius	156	occidentalis Williston, Pelomyia	726
obversa Huckett, Eremomyia	857	occidentalis Wirth, Atrichopogon	123
obvius (Walker), Psectrocladius	159	occidentalis Wirth, Forcipomyia	126
Occemyia Robineau-Desvoidy	631	occidentalis Wirth & Jones, Culicoides	132
Occemyia Robineau-Desvoidy	631	occidentata Malloch, Phora	537
occidens Alexander, Limnophila	67	*occidentis* Banks, Stratiomys	312
occidens Coquillett, Meghyperus	449	occidentis Fairchild & Hertig, Phlebotomus	92
occidens (Hardy), Dorylomorpha	557	occidentis (Hardy), Beameromyia	362
occidensis (Walker), Thecophora	631	occidentis Reinhard, Winthemia	1090
occidentale (Aldrich), Parasyntormon	516	occidentis Walker, (Hyalomya)	970
occidentale Banks, Zodion	629	*occidentis* (Walker), Pyrellia	912
occidentale Curran, Chrysotoxum	582	*occidentis* (Walker), Stomoxys	914
occidentale Curran, Gymnosoma	965	occidua (Fabricius), Tricharaea	960
occidentale Melander, Brachystoma	447	occidua (Walker), Gymnoclytia	965
occidentale (Townsend), Parapenium	580	*occimons* Snyder, Mydaea	892
occidentale Townsend, Simulium	186	occipitale Curran, Rhaphium	514
occidentalis (Adams), Leptocera	724	occipitalis Adams, Psilocephala	351
occidentalis (Adams), Trupanea	667	occipitalis Coquillett, Dilophus	195
occidentalis Aldrich, Bittacomorpha	98	occipitalis (Johnson), Labostigmina	314
occidentalis Aldrich, Dolichopus	493	occipitalis (Loew), Nephrotoma	22
occidentalis Aldrich, Sarcophaga	959	occipitalis Melander, Mumetopia	820
occidentalis Alexander, Dicranoptycha	52	occipitalis Melander & Spuler, Piophila	712
occidentalis Alexander, Phalacrocera	41	occulta (Meigen), Hydrotaea	900
occidentalis Banks, Stratiomys	312	occulta Wulp, Rivellia	656
occidentalis (Bigot), Condylostylus	485	oceanicus (Packard), Cricotopus	158
occidentalis (Brooks), Alophorella	968	ocellaris Coquillett, Gaediopsis	1080
occidentalis Brooks, Gonia	1076	*ocellaris* (Curran), Chaetophlepsis	1016
occidentalis Brooks, Mericia	1008	ocellaris Curran, Microdon	598
occidentalis (Brues), Triphleba	534	ocellaris (Linnaeus), Epiphragma	61
occidentalis Cole, Hercostomus	498	ocellaris Malloch, Leucopis	709
occidentalis Cole, Psilocephala	352	ocellaris Osten Sacken, (Cecidomyia)	†229, 294
occidentalis Coquillett, Eugnoriste	231	ocellaris Reinhard, Wagneria	1020
occidentalis Coquillett, Melieria	645	ocellaris Sabrosky, Gaurax	782
occidentalis Coquillett, Metaplagia	1018	ocellaris (Townsend), Camptoprosopella	697
occidentalis (Coquillett), Pilatea	1105	*ocellata* (Johannsen), Mycetophila	210
occidentalis (Coquillett), Prodiamesa	153	*ocellata* Osten Sacken, (Cecidomyia)	294
occidentalis (Coquillett), Sipholeskia	1036	ocellus Walker, Mycetophila	212
occidentalis Cresson, Ditrichophora	739	**Ochlerotatus** Lynch Arribálzaga	110
occidentalis Cresson, Notiphila	747	*Ochlerothatus* Lynch Arribálzaga	110
occidentalis Curran, Lonchoptera	531	ochracea (Aldrich), Ravinia	954
occidentalis Curran, Platycheirus	577	ochracea (Bigot), Euphasiopteryx	980
occidentalis Curran, Tachinomyia	1058	*ochracea* (Doane), Limonia	48
occidentalis Curran, Toxomerus	572	ochracea (Wulp), Chrysoexorista	1099
occidentalis (Doane), Nephrotoma	22	ochraceus (Loew), Rhagio	345
occidentalis Doane, Tipula	37	*ochraceus* Schiner, Leptogaster	363
occidentalis Dyar & Knab, Anopheles	106	ochraceus Wulp, (Stenopogon)	401
occidentalis Felt, Arthrocnodax	283	*ochrapesus* (Rathvon), Chyromya	822
occidentalis Felt, Diathronomyia	262	ochreatus (Townes), Chironomus	164
occidentalis Felt, Lestremia	244	*ochreiceps* (Bigot), Physocephala	628
occidentalis (Felt), Mayetiola	263	ochreicornis (Townsend), Arctophyto	986
occidentalis Felt, Rhabdophaga	257	ochreifrons Curran, Laphystia	376
occidentalis Garrett, Dixa	102	ochreigaster Curran, Cryptomeigenia	1041
occidentalis Harmston, Chrysotimus	529	ochricollis Melander, Platypalpus	480
occidentalis (Hine), Machimus	396	ochricornis (Loew), Melieria	645
occidentalis Hine, Proctacanthus	399	*ochricornis* (Townsend), Arctophyto	986
occidentalis (Huckett), Spilogona	880	ochrifacies Van Duzee, Thinophilus	508
occidentalis James, Odontomyia	316	*ochripes* (Bigot), Melanostoma	575
occidentalis Johannsen, Megophthalmidia	228	ochripes (Meigen), Leptocera	724
occidentalis (Johnson), Anthrax	432	ochripes Sabrosky, Oscinella	780
occidentalis Kraft & Cook, Neopachygaster	317	ochripes (Thomson), Hylemya	855
occidentalis Malloch, Clusia	806	ochripyga (Wulp), Oxysarcodexia	952
occidentalis (Malloch), Homoneura	699	**Ochrocera** Townsend	1025
occidentalis Malloch, Hydrophoria	864	ochrogaster (Thomson), Fannia	896
occidentalis Malloch, Hylemya	847	ochrolabis Loew, Sciara	230
occidentalis Malloch, Lonchaea	717	*Ochromeigenia* Townsend	979
occidentalis Malloch, Microsania	547	*ochropus* (Dyar & Knab), Mansonia	108
occidentalis Malloch, Mydaea	891	ochrostomus (Zetterstedt), Metasyrphus	561
occidentalis Malloch, Norellia	827	**Ochthera** Latreille	753
occidentalis Malloch, Pipunculus	554	**Ochtheroidea** Williston	735
occidentalis Malloch, Sphegina	594	*Ochthiphila* Fallén	707
occidentalis (Melander & Brues), Puliciphora	545	*Ochtiphila* Fallén	707
occidentalis Osburn, Sphecomyia	612	OCHTIPHILIDAE	706
occidentalis Sabrosky, Tricimba	785	**Ocnaea** Erichson	403
occidentalis Sherman, Dziedzickia	218	ocreata Melander, Mythicomyia	418
occidentalis (Snow), Orellia	671	ocresia (Walker), Anastrepha	673

INDEX 1645

	Page		Page
octoguttata (Zetterstedt), Hylemya	855	oklahomensis (Bromley), Ceraturgus	365
octomaculata (Zetterstedt), Coenosia	873	oklahomensis Cole & Wilcox, Lasiopogon	378
octomaculatus (Curran), Caloparyphus	308	oklahomensis (Hardy), Dilophus	195
octonotata (Walker), Hyadina	751	oklahomensis Khalaf, Culicoides	129
octopunctata (Zetterstedt), Coenosia	873	oklahomensis Pritchard, Promachus	400
octopunctatus (Say), Tacticus	385	*oklahomensis* (Stone), Hybomitra	340
oculata Fallén, Heteromyza	810	okoboji (Walley), Pentaneura	147
oculata (Townsend), Phaenicia	927	**Olbiogaster** Osten Sacken	190
oculatus Wilcox & Martin, Nannocyrtopogon	380	**Olcella** Enderlein	784
oculeus Melander, Platypalpus	480	oldenbergi (Goetghebuer), Chironomus	170
oculifera Bigot, Anthomyia	865	**Oldenbergiella** Czerny	810
Ocydromia Meigen	451	olenensis Sellers, Carcelia	1092
OCYDROMIINAE	449	**Olenochaeta** Townsend	1077
ocymi (Fabricius), Pyrophaena	578	*oleous* (Rapp), Pipunculus	553
Ocyptamus Macquart	573	**Olfersia** Leach	919
Ocyptera Latreille	†972, 972	**OLFERSIINI**	919
ocypterata Townsend, Aphria	1010	olga Aldrich, Diostracus	508
Ocypterodes Townsend	972	olga Cresson, Psilopa	741
Ocypterosoma Townsend	1021	*olia* Doane, Tipula	31
Ocytata Gistel	1071	oligacanthus Alexander, Molophilus	89
Odinia Robineau-Desvoidy	794	*Oligarces* Meinert	247
ODINIIDAE	793	**Oligodranes** Loew	421
Odonatisca Savtshenko	31	**OLIGOTROPHINI**	256
Odontocera Macquart	801	**Oligotrophus** Latreille	264
Odontocyptera Townsend	973	*Olina* Robineau-Desvoidy	†719
Odontodiplosis Felt	288	**Olinea** Richards	719
Odontoloxozus Enderlein	637	olivacea Cresson, Notiphila	747
Odontomera Macquart	642	olivacea (Meigen), Prodiamesa	153
Odontomesa Pagast	153	olivacea Melander, Clinocera	468
Odontomyia Meigen	315	olivia Hull & Fluke, Cheilosia	587
Odontomyiina Enderlein	316	ollius Walker, Empis	460
ODONTOMYIINI	314	olmaba Brimley, Neophyto	952
Odontonyx Rübsaamen	230	**olympia** Cole & Wilcox, Lasiopogon	377
Odontopoda Aldrich	224	olympia Doane, Tipula	37
Odontosoma Townsend	1051	olympia (Kincaid), Telmatoscopus	94
Odontotipula Alexander	24	olympia Wheeler, Chymomyza	770
Oecacta Poey	128	olympiae (Aldrich), Anevrina	532
Oecothea Haliday	811	olympiae (Aldrich), Tachytrechus	503
Oedalea Meigen	450	olympica Alexander, Limnophila	66
Oedaspis Loew	†660	omani (Hall), Blaesoxipha	944
Oedaspissolidago Patton	660	**Omapanta** Schmitz	542
Oedemagena Latreille	1112	omega Sabrosky, Leptocera	721
Oedemasoma Townsend	972	**Omegasyrphus** Giglio-Tos	599
Oedematocera Townsend	1070	ominosa (Alexander), Prionocera	19
Oedematopteryx Townsend	967	omissa (Aldrich), Triachora	1082
oedemerarum Storå, Atrichopogon	123	omissa (Reinhard), Euphorocera	1054
Oedenops Becker	748	**Omisus** Townes	173
Oedicarena Loew	675	**Ommatius** Wiedemann	398
oedicnema Garrett, Acantholeria	814	**Omniablaútus** Pritchard	380
oediemus Garrett, Acantholeria	814	omnivagus Van Duzee, Dolichopus	493
oedipodinis Townsend, (Sarcophaga)	1115	*omole* (Walker), Muscina	909
oedipus Fabricius, Anthrax	†432	**Omomyia** Coquillett	714
oedipus Wheeler, Campsicnemus	526	**Omotoma** Lioy	1089
Oedopa Loew	653	*Omphrale* Meigen	355
Oedoparea Loew	678	**OMPHRALIDAE**	354
Oedoparena Curran	681	**Omphralosoma** Kröber	355
oestraceus (Linnaeus), Eristalis	624	omus Pritchard, Cerotainiops	388
Oestranthrax Bezzi	446	*Oncodes* Latreille	406
OESTRIDAE	1111	**Oncodocera** Macquart	428
oestriforme Brauer & Bergenstamm, Hemithrixion	1037	*Oncomyia* Robineau-Desvoidy	631
oestriformis (Walker), Eristalis	624	**Oncopsia** Enderlein	637
OESTRINAE	1111	onerosa Alexander, Ormosia	86
Oestrohilarella Townsend	938	**Onesia** Robineau-Desvoidy	930
OESTROIDEA	922	**Onodiplosis** Felt	289
onondagensis (Felt), Aedes	112
Oestrophasia Brauer & Bergenstamm	979	ontariensis Hall, Sarcophaga	959
Oestroplagia Townsend	1080	ontario Curran, Actia	1062
Oestrus Linnaeus	1111	ontario Curran, Cheilosia	586
offella Reinhard, Admontia	1068	ontario Curran, Chrysogaster	591
Ogcodes Latreille	406	ontario Curran, Cordilura	828
Ogcodocera Macquart	428	ontario Curran, Cryptomeigenia	1041
ogilvii (Malloch), Condylostylus	485	*ontario* (Curran), Persicepsia	1033
ogoa Walker, (Dexia)	1108	ontario (Curran), Xylota	607
ohioensis (Hall), Blaesoxipha	947	ontario (Davidson), Carposcalis	576
ohioensis (Hine), Atylotus	331	ontario (Walley), Polypedilum	175
ohioensis Melander, Oedalea	450	ontarioensis (Curran), Ceriana	615
ohioensis Spencer, Drosophila	764	ontarioensis (Curran), Neocnemodon	581
ohioensis Steyskal, Stratiomys	313	onteona Roback, Diamesa	152
Oidematops Cresson	685	onusta (Loew), Mycomya	221
Okanagania Townsend	1008	onychodactylum Dyar & Shannon, Prosimulium	184
Okea Townsend	1089
okefenoke (Alexander), Nephrotoma	22	**Onychogonia** Brauer & Bergenstamm	1077
okefenokeensis Smith, Winthemia	1090	onyx Steyskal, Pseudacteon	544
okellyi Schmitz, Borophaga	535	*opaca* Aldrich, Leptocera	724
Okenia Zetterstedt	834	opaca Allen, Metopia	937
Okeniella Hendel	834	opaca (Coquillett), Aciurina	670
Okenina Malloch	833

	Page		Page
opaca (Coquillett), Alophorella	968	orbitalis Aldrich, Lixophaga	1043
opaca (Coquillett), Clausicella	1059	*orbitalis* (Hall), Ravinia	954
opaca (Coquillett), Hybomitra	340	orbitalis (Loew), Polytrichophora	740
opaca Coquillett, Laphystia	376	orbitalis Malloch, Clusiodes	807
opaca (Coquillett), Megapariopsis	982	orbitalis Malloch, Leucopis	709
opaca (Coquillett), Wohlfahrtia	942	orbitalis (Melander) Phytagromyza	803
opaca (Loew), Bezzia	141	orbitalis (Reinhard), Atrophopalpus	1031
opaca (Loew), Lauxaniella	700	*orbitalis* (Reinhard), Euhalidaya	1037
opaca (Reinhard), Aplomyiopsis	1038	orbitalis (Reinhard), Metopotachina	993
opaca (Schnabl), Spilogona	882	orbitalis Townsend, Apacheprospherysa	1038
opaca Shannon, Caliprobola	608	*orbitalis* (Townsend), Pseudomyothyria	1047
opaca (Williston), Clastopteromyia	771	orbitalis Walsh, (Cecidomyia)	292
opacella Richards, Leptocera	724	orbitalis Webber, Chaetophlepsis	1016
opacifrons Aldrich, Oscinella	780	*orbitalis* (Williston), Villa	436
Opacifrons Duda	722	orbitaseta (Stein), Helina	888
opacipennis Foote, Tephritis	668	orbospiracula Patterson & Wheeler, Drosophila	767
opacithorax (Malloch), Palpomyia	140		
opacithorax Malloch, Rhamphomyia	465	orcasae Malloch, Scatophaga	839
opacivittata (Dietz), Nephrotoma	22	orchestes Melander, Mythicomyia	418
opaculus (Loew), Lasiopogon	378	orchestris Melander, Iteaphila	454
opacum Coquillett, Anorostoma	813	orcina (Wiedemann), Mallophora	396
opacum Townsend, Hypertrophomma	1095	ordinaria Coquillett, Drosophila	765
opacus (Coquillett), Vermilio	342	ordinaria (Melander), Poecilominettia	703
opacus Loew, Diaphorus	520	ordinaria Spuler, Leptocera	724
opacus Loew, Gymnopternus	499	ordinaria Sturtevant & Wheeler, Scatophila	758
opacus (Williston), Verrallia	551	ordinaria (West), Melanomya	932
opalescens Townsend, Volucella	601	ordinata Becker, Athyroglossa	735
opalia (Walker), Ophyra	901	oreas Melander, Apalocnemis	454
opalizans Osten Sacken, Antocha	51	oreas Osten Sacken, Systoechus	410
Opelodexia Reinhard	933	oreas (Wheeler), Calyxochaetus	528
Opelousia Townsend	932	*oregona* (Aldrich), Zagonia	818
operta Chillcott, Fannia	896	oregona Alexander, Cladura	72
Opetiophora Loew	783	oregona Cole, Dziedzickia	218
ophioides Townes, Polypedilum	175	oregona Curran, Eutreta	661
Ophiomyia Braschnikov	797	*oregona* Curran, Xylota	606
Ophthalmomyia Williston	733	oregona Lovett, Pipiza	579
Ophyra Robineau-Desvoidy	901	oregona Steyskal, Pherbellia	686
opifera (Coquillett), Blaesoxipha	945	oregonense Harmston & Knowlton, Syntormon	515
opifex Alexander, Ormosia	86		
opima (Loew), Leia	227	oregonensis Alexander, Erioptera	81
opinator Osten Sacken, Syrphus	559	oregonensis Alexander, Limnophila	64
opipare Reinhard, Mochlosoma	987	oregonensis Chillcott, Fannia	896
opiparis Reinhard, Senotainia	939	oregonensis Cook, Psectrosciara	240
opisthocera Dietz, Tipula	40	oregonensis Felt, Hormosomyia	253
Opistholoba Mik	210	oregonensis Foote, Contarinia	277
opisthopus Komp, Culex	119	oregonensis Malloch, Coenosia	872
Oplogaster Rondani	873	oregonensis (Malloch), Helina	888
Opomydas Curran	360	oregonensis Townsend, Myiophasia	984
Opomyza Fallén	820	oregonensis Van Duzee, Dolichopus	494
OPOMYZIDAE	820	oregonensis Van Duzee, Hydrophorus	505
oporina Melander, Mythicomyia	418	oregonensis Van Duzee, Medetera	510
oppidana (Scopoli), Chyromya	822	oregonica Alexander, Ormosia	84
oppidans Huckett, Hylemya	850	oregonica Alexander, Prionocera	19
oppidanus Dampf, Phlebotomus	92	oregonicola Alexander, Molophilus	89
Oppiopsis Townsend	952	**Orellia** Robineau-Desvoidy	671
opportunus Van Duzee, Dolichopus	494	**Oreogeton** Schiner	453
opposita (Banks), Philosepedon	95	**Oreomyza** Pokorny	31
opposita Loew, Philygria	745	**Oreophila** Lackschewitz	87
oppressa Thomsen, Dasyhelea	127	*Oreophyto* Townsend	985
Opsebius Costa	405	**Oreothalia** Melander	467
Opsidia Coquillett	937	**Orestilia** Reinhard	984
Opsidiopsis Townsend	937	**Orfelia** Costa	202
Opsidiotrophus Reinhard	938	organensis (Townsend), Gaediopsis	1079
Opsiomyia Coquillett	836	**Organomyia** Townsend	1073
Opsodexia Townsend	932	orgyiae (LeBaron), Exorista	1055
Opsodexiopsis Townsend	932	*orgyiae* (Townsend), Exorista	1055
Opsolasia Coquillett	891	orgyiarum (Townsend), Exorista	1055
Opsomeigenia Townsend	1045	orichalcea (Stein), Pseudophaonia	909
Opsophyto Townsend	943	orichalcea (Zetterstedt), Mydaea	891
Opsotheresia Townsend	990	orichalceoides Huckett, Pseudophaonia	909
optata Melander, Mythicomyia	418	oriens Martin, Holopogon	375
optata (Reinhard), Gilvella	1069	*oriens* Melander & Spuler, Piophila	711
optiva Alexander, Tipula	31	orientale Curran, Rhaphium	514
opuntiae Felt, Asphondylia	269	orientalis Brooks, Psalidopteryx	1067
opuntiae Felt, (Cecidomyia)	290	orientalis (Coquillett), Neodichocera	1014
Oraphasmophaga Reinhard	1070	*orientalis* Huckett, Pegomya	861
Orasturmia Reinhard	1089	orientalis Schiner, Atherigona	875
orbatus (Say), Dilophus	195	orientalis Townsend, Metavoria	1019
Orbellia Robineau-Desvoidy	810	orientalis (Townsend), Nowickia	995
orbicallus Philip, Tabanus	335	*orillia* (Curran), Leptometopa	731
orbicula (Fabricius), Acrocera	406	orillia Curran, Pelatachina	1026
orbicularis Harmston, Hercostomus	498	*orillia* (Curran), Telmatoscopus	94
orbiculata (Felt), Obolodiplosis	286	orilliaensis Curran, Cheilosia	585
orbiculata Van Duzee, Medetera	510	**Orimarga** Osten Sacken	52
orbicus (Townes), Chironomus	168	orion Brimley, Belvosia	1081
orbitalis Aldrich, Hydrotaea	900	orion (Hull), Helophilus	619

INDEX 1647

	Page		Page
orion Osten Sacken, Tabanus	333	osceola Alexander, Tipula	29
Ormia Robineau-Desvoidy	980	**Oscinella** Becker	779
ORMIINI	979	**OSCINELLINAE**	774
ormioides (Townsend), Hamaxia	979	oscinina (Fallén), Siphonella	787
Ormosia Rondani	84	*Oscinis* Latreille	†779, 789
ornata Fabricius, Milesia	614	**Oscinisoma** Lioy	778
ornata Hardy, Dorylomorpha	557	**Oscinoides** Malloch	781
ornata Harris, (Oxycera)	1115	*Oscinosoma* Lioy	778
ornata (Johnson), Spilochroa	817	oscitans (Walker), Limnellia	758
ornata Loew, Zygomyia	216	oslari (Alexander), Shannonomyia	68
ornata (Meigen), Fannia	896	oslari (Dietz), Prionocera	19
ornata (Meigen), Mycomya	221	**Ospriocerus** Loew	381
ornata (Meigen), Pentaneura	147	ostensackeni Kellogg, Blepharicera	99
ornata Melander, Mythicomyia	418	ostensackenii (Burgess), Eclimus	427
ornata (Reinhard), Meigenia	1044	ostensackenii (Kirkpatrick), Winthemia	1090
ornata Say, (Cecidomyia)	292	ostenta Melander, Mythicomyia	418
ornata Van Duzee, Neurigona	518	ostiorum (Curtis), Ceratinostoma	840
ornata Wirth, Monohelea	135	**Ostracophyto** Townsend	1009
ornaticauda Van Duzee, Hercostomus	498	**Oswaldia** Robineau-Desvoidy	1045
ornaticollis (Meigen), Allodia	205	oteroensis (Reinhard), Siphosturmia	1087
ornatifrons Frey, Tetanocera	694	otiosa Coquillett, Empis	460
ornatipennis Van Duzee, Dolichopus	494	otiosa Coquillett, Rhamphomyia	465
ornatipes (Johnson), Homoneura	699	otiosa Coquillett, Thereva	354
ornatipes (Kieffer), Stictochironomus	177	*otiosa* (Coquillett), Villa	437
ornatipes Melander, Microphorella	452	otiosa (Stein), Xenomydaea	892
ornatipes (Townsend), Calotarsa	550	**Otites** Latreille	646
ornatipes (Van Duzee), Hercostomus	498	**OTITIDAE**	642
ornatipes Van Duzee, Syntormon	516	**OTITINAE**	643
ornatithorax (Enderlein), Grallipeza	633	**Otomasicera** Townsend	1108
ornatula Melander, Mythicomyia	418	otroeda (Walker), Rivellia	656
ornatum (Van Duzee), Rhaphium	514	*ottawaense* Twinn, Simulium	189
ornatus Kröber, Chrysops	327	*ottawaensis* Dietz, Tipula	33
ornatus (Say), Chrysopilus	346	ottawensis Melander, Limnia	691
ornatus (Van Duzee), Hercostomus	498	*ouelleti* (Curran), Velocia	1051
ornatus (Wiedemann), Sphecomyia	613	ouelleti (Curran), Xylota	606
ornatus Williston, Euparyphus	307	*ouelleti* (Shewell), Lyciella	700
Ornidia Lepeletier & Serville	600	*Ouelletia* Curran	938
Ornithoctona Speiser	917	ousairani Khalaf, Culicoides	129
Ornithodes Coquillett	56	ovalis (Adams), Oscinella	780
Ornithoeca Rondani	916	ovalis Alexander, Tipula	40
Ornithoica Rondani	916	ovalis (Edwards), Peromyia	251
ORNITHOICINAE	916	ovata Felt, Colpodia	253
Ornithomya Rondani	917	ovata Fisher, Exechia	207
Ornithomyia Rondani	917	ovata (Pettey), Bradysia	234
ORNITHOMYIINAE	917	ovata Stein, Coenosia	874
ORNITHOMYIINI	917	*ovata* Van Duzee, Mycetophila	213
ornithophilia Davies, Peterson, & Wood, Cnephia	184	ovaticornis (Van Duzee), Gymnopternus	499
Ornithoponus Aldrich	918	ovatipennis Foote, Tephritis	668
oronoensis (Metcalf), Melangyna	568	ovatus Loew, Dolichopus	494
Oropeza Needham	23	ovatus Van Duzee, Paracleius	501
oropezoides Johnson, Tipula	25	oviducta (Garrett), Boletina	217
Orphanotrophus Reinhard	1037	*oviformis* Patterson, (Cecidomyiaceltis)	294
Orphnabaccha Hull	573	oviformis Patton, (Cecidomyiaceltis)	294
orphne (Walker), Neoitamus	397	ovinus (Linnaeus), Melophagus	920
ORPHNEPHILIDAE	120	ovis Linnaeus, Oestrus	1111
orphnephiloides Malloch, Beckerina	537	*owenii* (Reinhard), Organomyia	1073
Orrhodops Hull	381	*Ozexorista* Townsend	1094
ORTALIDAE	642	oxia Steyskal, Tetanocera	694
Ortalimyia Curran	646	oxoniana Edwards, Smittia	162
Ortalis Fallén	†643, 647	oxybeles Steyskal, Dictya	689
Orthacheta Becker	831	*Ozycephala* Macquart	657
Orthellia Robineau-Desvoidy	912	**Oxycera** Meigen	306
Orthochaeta Becker	831	**OXYCERINI**	306
ORTHOCLADIINAE	153	oxycoccana (Johnson), Dasineura	260
ORTHOCLADIINI	154	*Oxydiscus* Meijere	59
Orthocladius Wulp	155	**Oxydosphyria** Townsend	996
Orthocuterebra Bau	1109	**Oxyna** Robineau-Desvoidy	664
orthogonia Oliver, Pagastia	152	**Oxynops** Townsend	1046
Orthogonis Hermann	391	*Oxyrhiza* Meijere	59
orthomera Alexander, Erioptera	81	**Oxysarcodexia** Townsend	952
Orthoneura Macquart	591	oxytona Alexander, Tipula	37
Orthoneuromyia Williston	382	**Ozodiceromyia** Bigot	349
Orthonevra Macquart	591		
Orthopodomyia Theobald	108	pabulorum (Fallén), Pararicia	910
Orthosia Reinhard	984	pacalis Alexander, Limnophila	66
Orthosimyia Reinhard	984	pacata Reinhard, Wagneria	1020
oruca Walsh, (Cecidomyia)	294	pachyceras Williston, Symphoromyia	343
Orygma Meigen	682	**Pachycerina** Macquart	702
osborni (Alexander), Pilaria	68	pachycerus Williston, Chrysops	327
osborni (Stains & Knowlton), Cnephia	184	*Pachychoeta* Bezzi	776
osburni (Curran), Dasysyrphus	564	pachycnemus Loew, Dolichopus	494
osburni (Hine), Hybomitra	340	pachycnemus Loew, Platypalpus	480
osceola Alexander, Erioptera	81	pachycondylae (Brues), Cataclinusa	545
osceola Alexander, Limnophila	66	**Pachyconops** Camras	626
osceola Alexander, Ptychoptera	97	**Pachydiamesa** Oliver	153
		Pachygaster Meigen	318

Entry	Page
pachygaster Westwood, Mydas	359
PACHYGASTRINAE	317
Pachygraphia Brauer & Bergenstamm	953
Pachyhelea Wirth	139
Pachyleptus Walker	137
pachymera Bigot, Rhamphomyia	465
pachymera (Williston), Pachyhelea	140
Pachymeria Stephens	457
Pachyneres Greene	415
pachyneura (Loew), Triphleba	534
Pachyneurella Brues	545
PACHYNEURIDAE	196
Pachyophthalmus Brauer & Bergenstamm	934
pachyphallus Alexander, Cryptolabis	78
pachyprocta (Hagen), Metoposarcophaga	951
pachyprocta (Parker), Metoposarcophaga	951
pachyproctosa Parker, Metoposarcophaga	951
pachypus Bigot, Atherix	344
pachypyga (Aldrich & Webber), Patelloa	1104
Pachyrhina Macquart	20
pachyrhinoides Alexander, Tipula	25
Pachyrhyna Macquart	20
Pachyrina Macquart	20
Pachyrrhina Macquart	20
Pachysphyria Enderlein	574
pachystylum Williston, Microdon	598
pachytarse Bigot, Melanostoma	575
Pacidianus Reinhard	1095
pacifica Alexander, Bittacomorphella	98
pacifica Alexander, Hexatoma	70
pacifica (Alexander), Paradelphomyia	60
pacifica Banks, Dalmannia	632
pacifica Cole, Thereva	354
pacifica Cresson, Sepedon	693
pacifica (Cresson), Setacera	755
pacifica Curran, Aciurina	670
pacifica (Curran), Blera	610
pacifica (Curran), Coleophasia	977
pacifica Curran, Comantella	366
pacifica Curran, Eutreta	661
pacifica Curran, Odontomyia	315
pacifica Curran, Renocera	692
pacifica (Doane), Tephritis	668
pacifica Doane, Tipula	26
pacifica Fisher, Mycetophila	210
pacifica (Hall), Melanodexia	931
pacifica Hunter, Cheilosia	585
pacifica Kessel, Platypezina	548
pacifica Kincaid, Psychoda	96
pacifica Lovett, Cheilosia	587
pacifica Lovett, Didea	563
pacifica Malloch, Fucellia	843
pacifica (Melander), Liriomyza	802
pacifica (Saunders), Smittia	162
pacifica Shannon, Chrysogaster	591
pacificensis Hearle, Aedes	111
pacificum (Cole), Calliopum	696
pacificus Cole & Wilcox, Lasiopogon	378
pacificus (Lovett), Dasysyrphus	564
pacilus (Walker), Platycheirus	577
packardi Alexander, Tipula	33
packardi Felt, Mycodiplosis	283
packardi Felt, Sackenomyia	265
packardi Malloch, Hydrophoria	864
packardi Van Duzee, Dolichopus	494
paeneadusta Alexander, Limnophila	66
pagana (Fabricius), Mydaea	†891
paganus Meigen, Chironomus	165
Pagastia Oliver	152
pagella (Reinhard), Blaesoxipha	947
pagetonotum Dyar & Knab, Aedes	115
pahasapa (Alexander), Arctoconopa	80
pahasapa Alexander, Pedicia	55
painteri (Alexander), Arctoconopa	80
painteri Bromley, Promachus	400
painteri Cazier, Rhaphiomidas	357
painteri (Felt), Mayetiola	263
painteri Hull, Microdon	599
painteri James, Bibio	194
painteri James, Odontomyia	316
painteri Maughan, Poecilanthrax	442
painteri Priddy, Conophorus	414
painteri Pritchard, Cophura	367
painteri Sabrosky, Oscinella	780
painterorum Johnson & Johnson, Exoprosopa	445
paiuta Alexander, Gonomyia	76
paiuta Alexander, Tipula	33
Palaeosepsis Duda	682
palaestrica (Meigen), Hydrotaea	900
palaestricus Loew, Dolichopus	494
palaga (Reinhard), Orthosimyia	984
Paleoplatyura Meunier	203
Pales Meigen	20
Pales Robineau-Desvoidy	1086
palesioidea (Robineau-Desvoidy), Pseudoperichaeta	1106
Pallasia Robineau-Desvoidy	964
pallasii (Townsend), Gymnoclytia	964
pallens Alexander, Trichocera	16
pallens Bigot, Exoprosopa	444
pallens Coquillett, Culex	118
pallens (Coquillett), Pentaneura	147
pallens (Coquillett), Pseudodacus	674
pallens Reinhard, Trypheromyia	1050
pallens, Ross, Aedes	112
pallens Twinn, Simulium	185
pallens (Wiedemann), Gonia	†1075
pallens (Wiedemann), Sciapus	486
pallens Wiedemann, Volucella	602
pallens Wirth, Dasyhelea	127
pallescens Alexander, Cylindrotoma	41
pallescens (Bigot), Condylostylus	484
pallescens Johnson & Maughan, Bombylius	409
pallescens (Shannon), Phaenicia	928
palliata (Coquillett), Haplomyza	802
palliata (Loew), Villa	439
palliatus (Coquillett), Stictochironomus	178
palliatus McAtee, Bibio	194
pallicauda (Zetterstedt), Microprosopa	835
palliceps Johnson, Leptocera	723
pallicornis Stein, Phaonia	906
pallicornis Van Duzee, Mesorhaga	483
pallida Alexander, Dicranoptycha	52
pallida Alexander, Dicranota	56
pallida Allen, Gymnoprosopa	936
pallida (Cole), Cryptotreta	662
pallida (Coquillett), Apachemyia	1010
pallida (Coquillett), Villa	437
pallida Day, Dryomyza	680
pallida (Fallén), Megaphthalma	833
pallida (Harris), Clusia	806
pallida (Harris), Myobiopsis	1035
pallida (Johnson), Mallophorina	397
pallida Kröber, Psilocephala	351
pallida (Loew), Cadrema	776
pallida Loew, Empis	460
pallida Loew, Rivellia	656
pallida Loew, Tipula	34
pallida (Loew), Trypetoptera	695
pallida Macquart, Chrysops	325
pallida Malloch, Pogonota	833
pallida Malloch, Probezzia	138
pallida (Malloch), Scatophaga	838
pallida Say, Coenomyia	296
pallida Say, (Mesembrina)	915
pallida Say, Ornithomyia	917
pallida Stein, Coenosia	872
pallida Stein, Pegomya	859
pallida (Stein), Phaonia	907
pallida (Walker), Cordilura	829, †839
pallida Williston, Drosophila	769
pallida (Winnertz), Forcipomyia	126
pallida (Zetterstedt), Allanthalia	450
pallida (Zetterstedt), Anthomyza	819
pallida (Zetterstedt), Scaptomyza	770
pallidescens Philip, Tabanus	335
pallidicauda (Zetterstedt), Microprosopa	835
pallidiciliatus (Van Duzee), Gymnopternus	499
pallidicornis Van Duzee, Mesorhaga	483
pallidifemur Malloch, Xylomya	299
pallidilabris (Rondani), Lynchia	919
pallidilabris (Rondani), Lynchia	799
pallidipalpis Cresson, Notiphila	747
pallidipennis (Cresson), Paroxyna	666
pallidipennis Loew, Ogcodes	407
pallidipes Malloch, Gaurax	782
pallidipes (Malloch), Nanna	831
pallidiseta Malloch, Agromyza	795
pallidisquama (Zetterstedt), Phaonia	907
pallidiventris (Holmgren), Bradysia	233
pallidiventris Malloch, Fannia	896

INDEX

Entry	Page
pallidiventris (Malloch), Megaselia	541
pallidiventris (Malloch), Stilobezzia	134
pallidivittatus Malloch, Chironomus	166
pallidocera Dietz, Tipula	40
pallidohirta (Grossbeck), Aedes	116
pallidosa Huckett, Phaonia	907
pallidula Coquillett, Phaonia	907
pallidula (Coquillett), Villa	438
pallidula (Hine), Efferia	395
pallidus Bellardi, Chrysops	325
pallidus Cresson, Epacmus	428
pallidus (Johannsen), Microtendipes	172
pallidus Palisot de Beauvois, Tabanus	332
pallifrons (Curran), Metasyrphus	560
pallinervis (Thomson), Ravinia	954
pallipalpis (Zetterstedt), Lispocephala	876
pallipennis Curran, Microdon	599
pallipennis (Zetterstedt), Chaetolonchaea	716
pallipes Bigot, Sargus	302
pallipes Cresson, Lytogaster	752
pallipes Fabricius, (Stratiomys)	319
pallipes (Fallén), Ocytata	1072
pallipes Johnson, Pipunculus	552
pallipes (Loew), Calotarsa	550
pallipes Loew, Cheilosia	586
pallipes Loew, Chrysotus	523
pallipes (Loew), Hippelates	775
pallipes (Loew), Solva	299
pallipes Malloch, Hebecnema	890
pallipes Reinhard, Panacemyia	1046
pallipes Say, Bibio	194
pallipes (Say), Compsobata	634
pallipes Say, Nemotelus	309
pallipes Say, Scenopinus	356
pallipes Say, Sciophila	225
pallipes Stein, Coenosia	872
pallipes Walker, (Sarcophaga)	961
pallipes (Wiedemann), Allognosta	300
pallitarse Curran, Melanostoma	575
palliventris (Curran), Metasyrphus	561
Palloptera Fallén	715
PALLOPTERIDAE	714
PALLOPTEROIDEA	710
palloris (Coquillett), Actia	1062
palloris (Coquillett), Chelifera	471
pallustris Hull, Psilocurus	382
palmare (Loew), Syntormon	516
palmarum Alexander, Tipula	38
palmata Johannsen, Exechia	207
palmerae James, Culicoides	129
palmerae Jones, Mallota	621
palmeri Felt, (Cecidomyia)	294
palmi (Shaw), Orfelia	203
palmulosus Snow, Platycheirus	577
palomarensis Alexander, Hexatoma	70
palomaricus Alexander, Molophilus	89
palpale Curran, Rhaphium	514
palpalis Adams, Chlorops	790
palpalis (Adams), Rhagio	345
palpalis (Coquillett), Metopotachina	993
palpalis (Coquillett), Oxyna	665
palpalis (Coquillett), Trichopalpus	837
palpalis Dietz, Ormosia	86
palpalis Malloch, Scatophaga	838
palpalis Melander, Oligodranes	422
palpalis (Robertson), Robertsonomyia	629
palpalis Stein, Mydaea	892
palpata (Brues), Megaselia	541
palpata Stein, Pegomya	860, †861
Palpexorista Townsend	1055
palpiger (Wheeler), Chrysotus	523
palpigera (Coquillett), Elfia	1065
palpinus Palisot de Beauvois, Tabanus	335
Palpomyia Meigen	140
PALPOMYIINI	139
palposa (Loew), Orellia	671
palposa Malloch, Puliciphora	545
palposa (Stein), Pegomya	861
palposa (Walker), Lispe	878
PALPOSTOMATINI	978
paltidus Hardy, Bibio	193
paludicola (Alexander), Molophilus	89
paludicola (Alexander), Pedicia	55
paludis (Johannsen), Leucophora	868
paludosa Meigen, Tipula	26
paludum (Meigen), Scatella	757
paluster Chillcott, Platypalpus	480
paluster Melander & Brues, Dolichopus	494
palustris Dyar, Aedes	112
palustris (Felt), Cecidomyia	287
palustris Felt, Hormomyia	288
palustris Felt, Lasioptera	272
palustris Felt, Rhopalomyia	266
palustris Felt, Winnertzia	253
palustris (Melander), Pherbellia	687
palustris (Saunders), Forcipomyia	124
palustris Spencer, Drosophila	765
pammelan Reinhard, Charasoma	1040
Panacemyia Townsend	1046
panaetius (Walker), Tachinomyia	1058
panamensis Curran, Actia	1062
panamensis Townsend, Pelecotheca	†1052
pancerastes Dyar & Shannon, Prosimulium	183
panda Martin, Leptogaster	363
Pandasyopthalmus Stuckenberg	578
Pandora Haliday	682
PANGONIINAE	320
PANGONIINI	320
Pangonius Latreille	320
panici Felt, Lasioptera	272
paniculata (Felt), Asteromyia	274
panneus Melander, Oligodranes	422
PANOPINAE	403
Panoplites Theobald	107
pansa (Snow), Spallanzania	1077
pansa (Walker), Tachinomyia	1058
Pantagathus Reinhard	1051
Pantarbes Osten Sacken	414
pantherinus (Gerstäcker), Nemomydas	359
pantomimus Melander & Brues, Dolichopus	494
Panzeriopsis Townsend	996
papillata Felt, Porricondyla	256
papillifera Melander, Tetanocera	694
papulata Cresson, Parydra	749
par (Coquillett), Brillia	154
par (Johannsen), Brillia	154
par (Walker), Chrysopilus	346
par Walker, Gymnosoma	965
Parabezzia Malloch	136
Parabombylius Williston	410
Paracacoxenus Hardy	762
Paracalobata Hendel	†633
Paracantha Coquillett	662
Parachaeta Coquillett	1083
Paracladura Brunetti	16
Paracleius Bigot	500
Paraclius Bigot	500
Paraclunio Kieffer	163
Paracoenia Cresson	755
Paracollinella Duda	720
Paracosmus Osten Sacken	427
Paracrocera Mik	405
Paracuterebra Bau	1109
Paradejeania Brauer & Bergenstamm	1001
Paradelphomyia Alexander	59
Parademoticus Townsend	1011
Paradexodes Townsend	1046
paradexoides (Townsend), Blondelia	1040
paradichaeta (Shaw), Bradysia	234
paradichroa Fisher, Docosia	226
Paradicranota Alexander	57
Paradidyma Brauer & Bergenstamm	1032
Paradionaea Townsend	976
Paradiplosis Felt	286
paradisea Alexander, Ormosia	84
Paradixa Tonnoir	102
Paradmontia Coquillett	1015
paradorenus (Roback), Orthocladius	156
paradoxa (Jaennicke), Neodiplocampta	441
paradoxa (Johannsen), Mycetophila	213
paradoxa Osten Sacken, Cryptolabis	78
paradusta Wheeler, Scaptomyza	770
paraensis (Goeldi), Culicoides	130
Paraezorista Brauer & Bergenstamm	1092
parafacialis Huckett, Eremomyia	857
Parafischeria Townsend	1011
Parafrontina Brauer & Bergenstamm	1101
Paragermaria Townsend	1074
PARAGINI	578
Paragopsis Matsumura	596
Paragus Latreille	578
Parahypochaeta Brauer & Bergenstamm	1016

744–243—65——105

	Page		Page
Parakiefferiella Thienemann	161	paravittata Wheeler, Scaptomyza	770
Paralauterborniella Lenz	173	*Parcipromus* Reinhard	1010
Paraleucophenga Oldenberg	762	paradalinus James, Euparyphus	307
Paraleucopis Malloch	708	pardus Osten Sacken, Exoprosopa	445
Paralimna Loew	748	pardus Townes, Polypedilum	176
Paralimnophora Malloch	879	**Parectecephala** Becker	792
Paralipophleps Alexander	75	**Paregle** Schnabl	854
paralis Reinhard, Masiphyomyia	1085	parens Bigot, Eristalis	624
Paralispe Brauer & Bergenstamm	1047	parens (Williston), Xylomya	299
Paralispidea Townsend	1063	**Parepalpus** Coquillett	1003
parallela Aldrich, Sarcophaga	959	**Parephydra** Coquillett	735
parallela (Doane), Ormosia	86	*Parexorista* Brauer & Bergenstamm	1092
parellela Johannsen, Fannia	897	**Parhelophilus** Girschner	619
parallela Malloch, Leucopis	709	*Parhydrophorus* Wheeler	504
parallela (Melander), Chersodromia	475	parietina (Osten Sacken), Limonia	44
parallela Reinhard, Torosomyia	1080	parilis Cresson, Ditrichophora	739
parallela Walker, (Sciomyza)	695	parilis (Johannsen), Bradysia	234
Parallelodiplosis Rübsaamen	286	parilis Reinhard, Muscopteryx	1025
Parallelomma Becker	840	paripes (Edwards), Glyptotendipes	172
Parallelomma Becker	829	parkeri Cazier, Apiocera	357
Paralucilia Brauer & Bergenstamm	924	parkeri Cazier, Rhaphiomidas	357
PARAMACRONYCHIINI	940	parkeri Malloch, Eremomyioides	857
Paramadiza Malloch	731	parkeri Melander, Aphoebantus	430
Paramadiza Melander	728	parkeri Melander, Oliogodranes	422
Parameigenia Townsend	1045	parkeri Philip, Apatolestes	320
paramelanica Patterson, Drosophila	765	*Parkeriellus* Smith	1045
Parametopia Townsend	937	parksi Bromley, Ospriocerus	381
Paramicroprosopa Ringdahl	836	parksi (Reinhard), Deopalpus	1004
paramunda Alexander, Limnophila	64	*Parlimnophora* Malloch	879
Paramuscopteryx Townsend	1023	parma Felt, Janetiella	264
Paramycodrosophila Duda	771	parma Melander, Mythicomyia	418
Paramyia Williston	730	parmatus Van Duzee, Diaphorus	520
Paranaphora Townsend	1027	*parmensis* Rondani, Saltella	683
parancilla Townsend, Frontiniella	1100	parnassum Malloch, Simulium	189
Paraneossos Wheeler	817	*Parodinia* Coquillett	817
Parantichaeta Enderlein	688	parodites (Dyar), Culiseta	117
Parapenium Collin	580	**Paroedopa** Coquillett	653
parapentheres Alexander, Limonia	49	**Paromphrale** Kröber	355
Paraphaenocladius Thienemann	161	*paron* (Walker), Odontomyia	317
Paraphasia Townsend	969	paropus (Walker), Tolmerus	401
Paraphasmophaga Townsend	1073	**Paroxyna** Hendel	665
Paraphorantha Townsend	969	parrii (Kirby), Prionocera	19
Paraphrosylus Becker	509	parrioides (Alexander), Tipula	30
Paraphyto Coquillett	942	parshleyi Alexander, Tipula	38
parapicata Hackman, Scaptomyza	769	partheniicola (Cockerell), Paradiplosis	286
Paraplagia Brauer & Bergenstamm	1016	parthenocissi (Stebbins), Dasineura	260
Paraporia Townsend	1028	*partiarius* Say, (Asilus)	1115
Paraprosalpia Villeneuve	866	partica (Roback), Pagastia	152
Paraprosena Brauer & Bergenstamm	†990	particeps (Adams), Culiseta	117
Parapsalida Townsend	976	particeps (Becker), Hippelates	775
Pararchytas Brauer & Bergenstamm	1002	particeps (Doane), Limonia	48
pararcilla Townsend, Frontiniella	1100	partita Huckett, Pegomya	861
Pararete Pritchard	245	*partita* Malloch, Schoenomyza	870
Pararhamphomyia Frey	461	partitor (Banks), Laphria	390
Pararicia Brauer & Bergenstamm	910	*partitus* Becker, Hippelates	775
pararoralis Duda, Leptocera	721	partitus Melander & Brues, Dolichopus	494
parascalaenum Beck, Polypedilum	176	*parva* (Adams), Lasiosina	791
Parascaptomyza Duda	770	parva (Adams), Olcella	784
Parascatella Cresson	757	parva (Adams), Robertsonomyia	630
Parascatopse Cook	239	parva Bigot, (Phorocera)	1108
parascopula Fisher, Mycomya	221	parva Brooks, Lypha	1011
Parasetigena Brauer & Bergenstamm	1056	parva Cook, Rhegmoclemina	239
PARASIMULIINI	182	parva (Coquillett), Nemorilla	1088
Parasimulium Malloch	182	parva Coquillett, Rhamphomyia	465
parasita Fabricius, Stomoxys	914	parva Cresson, Parydra	750
Parasphaerocera Spuler	718	*parva* (Doane), Hexatoma	69
paraspina Wirth, Alluaudomyia	134	parva Edwards, Diamesa	152
Paraspiniphora Malloch	533	parva Garrett, Cordyla	208
Parasteinia Cockerell	876	parva Garrett, Dixa	101
Parastenophora Malloch	534	parva Garrett, Sciophila	226
Parasymmictus Bigot	403	parva Hardy, Pseudatrichia	355
Parasymmyctus Bigot	403	parva (Holmgren), Bradysia	234
Parasyntormon Wheeler	516	parva Johannsen, Brillia	154
Paratacta Reinhard	1081	parva Kraft & Cook, Zabrachia	318
Paratanypus Garrett	150	*parva* Loew, Discocerina	738
Parataracticus Cole	381	parva (Lundbeck), Smittia	162
Paratendipes Kieffer	173	parva (Malloch), Leptocera	724
Paraterellia Foote	671	parva Osten Sacken, Erioptera	82
Parathalassius Mik	452	*parva* (Robineau-Desvoidy), Hylemya	††854
Paratheresia Townsend	991	parva Shannon, Chrysogaster	591
Parathyroglossa Hendel	735	*parva* Shannon & Dobrosoky, Protocalliphora	926
Paratidia Malloch	827	parva Townsend, Lixophaga	1044
Paratinia Mik	224	*parva* Townsend, Madremyia	1103
Paratissa Coquillett	740	parva Van Duzee, Medetera	510
Paratrichobius Lima	921	parva (Walker), Forcipomyia	126
Paravilla Painter	438	parva Walker, (Lasioptera)	292

INDEX

Entry	Page
parva Walker, Mycetophila	212
parva (Williston), Xylota	608
parva Wirth, Alluaudomyia	134
parvaeformis (Schnabl), Hylemya	854
parvaicornis Malloch, Hylemys	853
parvella Alexander, Dicranota	57
parvemarginata Alexander, Tipula	28
parvicauda (Van Duzee), Condylostylus	485
parvicauda Van Duzee, Rymosia	208
parvicella (Coquillett), Napomyza	804
parvicellula Alexander, Pedicia	54
parviceps Loew, (Aricia)	1115
parviceps Malloch, Phaonia	907
parviceps (Malloch), Thaumatomyia	788
parvicornis Loew, Agromyza	795
parvicornis Loew, Baccha	572
parvicornis Loew, Gymnopternus	499
parvicornis (Loew), Villa	438
parvicornis Malloch, Hylemya	853
parvicornis (Meigen), Dasiops	716
parvicornis Melander, Drapetis	478
parvicornis Van Duzee, Asyndetus	524
parvicornis Van Duzee, Chrysotus	523
parvicornis Van Duzee, Dolichopus	494
parvidus Painter, Geron	423
parvilamellatus Malloch, Chironomus	167
parviloba Alexander, Dactylolabis	62
parvimaculata (Stein), Phyllogaster	876
parvimaculata (Stein), Spilogona	882
parvimaculata Van Duzee, Mycetophila	212
parvimanus Van Duzee, Dolichopus	494
parvipalpis (Wulp), Stomatomyia	1058
parvipennis Spuler, Leptocera	722
parvipes Townsend, Hypertrophocera	1072
parvipyga Malloch, Tetanocera	694
parvis (Adams), Robertsonomyia	630
parvisquama Malloch, Coenosia	873
parvistigma Vockeroth, Geomyza	821
parvitarsus Garrett, Pseudoleria	812
parviteres (Aldrich & Webber), Prophryno	1106
parvoclava Martin, Leptogaster	363
parvula (Coquillett), Allocotocera	223
parvula Coquillett, Dioctria	371
parvula Haliday, Coelopa	680
parvula (Loew), Mesograpta	571
parvula (Loew), Tethina	727
parvula (Reinhard), Opsomeigenia	1045
parvulus Daecke, Chrysops	327
parvulus Schaeffer, Ommatius	398
parvulus Van Duzee, Chrysotus	523
parvulus Westwood, Mydas	359
parvum Townes, Polypedilum	176
parvum (Williston), Melanostoma	575
parvus Aldrich, Pelastoneurus	502
parvus Loew, Hydrophorus	505
parvus (Van Duzee), Telmaturgus	529
Parwinnertzia Felt	254
Parydra Stenhammar	749
PARYDRINAE	749
PARYDRINI	749
Paryphoconus Enderlein	139
pascuorum (Meigen), Pararicia	910
passiva Curran, Cordilura	828
patella Townsend, Otomasicera	1108
patellans (Pandellé), Pegomya	863
patellata Aldrich, Lispe	878
Patelloa Townsend	1104
Patelloapsis Townsend	1104
Pateloa Townsend	1104
patens Beutenmüller, Asphondylia	268
patens Johannsen, Trichonta	216
paterifera Alexander, Tipula	27
patibulatus (Say), Condylostylus	485
patibulus Quate, Telmatoscopus	94
patrita (Reinhard), Adoryphorophaga	1038
patruelis (Coquillett), Heteropogon	374
pattersoni White, Oligotrophus	265
pattonii Williston, Sphecomyia	613
patula Martin, Leptogaster	363
patulosa Garrett, Bolitophila	198
patulus (Walker), Hybomitra	340
pauca Hardy, Tomosvaryella	555
paucifili Felt, (Cecidomyia)	290
pauciseta (Coquillett), Lespesia	1101
pauciseta (Felt), Lycoriella	231
pauciseta Melander, Stilpon	476
paucus Osten Sacken, Opsebius	405
paula (Loew), Mycetophila	212
paulina Hall, Sarcophaga	959
paullula Loew, Parydra	750
paulus Bergroth, Molophilus	89
paulus Pritchard, Taracticus	385
pauper Hull, Platycheirus	577
pauper (Loew), Anthrax	432
paupera Osten Sacken, Ula	53
paurosoma (Sturtevant & Wheeler), Paracoenia	756
paurosomus Pritchard, Heteropogon	374
pauxilla (Holmgren), Spilogona	881
pauxillus (Williston), Dasysyrphus	564
pavida Coquillett, Psilocephala	351
pavida (Meigen), Pales	1087
pavidus Coquillett, Aphoebantus	430
pavidus (Holmgren), Cricotopus	158
pavonacea (Reinhard), Drepanoglossa	1060
pavonina (Osten Sacken), Epiphragma	61
paxillata (Laffoon), Mycetophila	212
pearcei Cazier, Apiocera	357
pearyi Dyar & Shannon, Aedes	113
pearyi (Malloch), Spilogona	881
peas (Walker), Baccha	573
peayi Alexander, Erioptera	83
peccator Dyar & Knab, Culex	119
pechumani Curran, Euxesta	652
pechumani Philip, Chrysops	327
pechumani Philip, Tabanus	334
pechumani Reinhard, Vibrissotheresia	991
pechumani Sabrosky, Elachiptera	777
Peckia Robineau-Desvoidy	953
pecosensis (Townsend), Phryxe	1105
pecosensis Wirth, Culicoides	130
pecten Fisher, (Coelosia)	1115
pectenipes Say, (Rhamphomyia)	1115
pectinata (Aldrich), Ravinia	954
pectinata Felt, Camptomyia	256
pectinata (Felt), Colpodia	253
pectinata Felt, Winnertzia	253
pectinata (Johannsen), Helina	888
pectinata (Loew), Pseudoleria	812
pectinata Loew, Rhamphomyia	465
pectinata Melander, Bicellaria	451
pectinata Sherman, Rymosia	208
pectinatellae (Dendy & Sublette), Chironomus	168
pectinator Melander, Platypalpus	480
pectinerata Garrett, Pseudoleria	812
pectinifer Wheeler, Thinophilus	508
pectiniger Melander, Stilpon	476
pectinipes Melander, Gloma	454
pectinipes Van Duzee, Hydrophorus	505
Pectinodiplosis Felt	278
pectinulata Cresson, Ephydra	754
pectita Johannsen, Mycetophila	213
pectoralis (Coquillett), Oscinella	780
pectoralis Coquillett, Platyura	204
pectoralis Huckett, Coenosia	874
pectoralis Loew, Loxocera	639
pectoralis Macquart, Sepsis	†683
pectoralis Van Duzee, Chrysotus	523
pectoralis Van Duzee, Mycetophila	212
pectoralis Van Duzee, Neurigona	518
pectoralis Van Duzee, Sympycnus	527
pectoris Coquillett, Rhamphomyia	465
pectorosa (Hendel), Pteromicra	687
pecuarum (Riley), Cnephia	184
pedalis Felt, Dasineura	260
pedatum Townes, Polypedilum	176
pedella (Fallén), Coenosia	872
pedellus (De Geer), Microtendipes	172
pedestris (Malloch), Pegomya	861
pedestris (Walker), Cryptomeigenia	1041
pedestris (Walker), Meigenielloides	†1045
Pedicella Bigot	303
pedicellata Felt, Rhopalomyia	266
pedicellata (Williston), Polybiomyia	616
Pedicellina James	302
Pedicia Latreille	53
PEDICIINI	53
pedicillaris (Huckett), Paraprosalpia	866
pediformis Shaw, Rymosia	208
pediontis (McAlpine), Hybomitra	340

	Page		Page
pedita (Reinhard), Sitophaga	1048	pentalineata Wirth & Hubert, Dasyhelea	127
Peditia Latreille	53	**Pentaneura** Philippi	146
pedunculata (Hall), Sarcodexia	956	**Pentapedilum** Kieffer	176
pedunculata (Loew), Nephrotoma	22	pentastigma (Thomson), Scatella	757
Pegohylemyia Schnabl	852	*pentatoma* Kieffer, Zavrelia	180
Pegomya Robineau-Desvoidy	858	*Penthesilea* Meigen	611
Pegomyia Robineau-Desvoidy	858	**Penthetria** Meigen	192
Pelastoneurus Loew	501	*Penthoptera* Schiner	69
Pelatachina Meade	1026	penumbra Alexander, Nephrotoma	22
Pelecocera Meigen	595	peodes Osten Sacken, Aphoebantus	430
PELECOCERINI	595	perangusta Alexander, Pedicia	54
PELECORHYNCHIDAE	319	perarticulata Felt, Ficiomyia	263
Pelecotheca Townsend	1052	**Perasis** Hermann	382
peleensis (Walley), Ablabesmyia	148	perbrevis Van Duzee, Neurigona	518
Peleteria Robineau-Desvoidy	996	percita Alexander, Tipula	35
Peleteriopsis Townsend	996	percomplexa Alexander, Gonomyia	76
Peletieria Robineau-Desvoidy	996	percursa (Laffoon), Mycetophila	213
Pelignellus Sturtevant & Wheeler	737	perdecora Reinhard, Pseudochaeta	1096
Pelignus Cresson	737	*perdita* (Dietz), Nephrotoma	20
Pelina Haliday	752	perdita (Malloch), Megaselia	541
pellecta (Reinhard), Palpexorista	1056	peregrina (Malloch), Megaselia	539
pellex Osten Sacken, (Cecidomyia)	294	peregrina (Melander), Empis	460
pellita (Fisher), Orfelia	203	*peregrinata* Walker, Ocydromia	451
pellucida Coquillett, Empis	460	peregrinum (Meigen), Euryomma	898
pellucida Coquillett, Geminaria	413	*peregrinus* Edwards, Podonomus	150
pellucida Coquillett, Pelatachina	1026	peregrinus (Johannsen), Atrichopogon	123
pellucida Coquillett, Toxophora	425	perennis (Meigen), Chaetocladius	160
pellucida Doane, Tipula	38	*perexigua* Alexander, Tipula	33
pellucida Spuler, Leptocera	722	perfecta (Alexander), Limonia	49
pellucida (Stein), Fannia	896	perfecta (Loew), Strauzia	676
pellucidiguttata (Dietz), Limonia	50	perfecta (Pettey), Bradysia	234
pellucidus (Coquillett), Aphoebantus	430	*perfida* (Dietz), Nephrotoma	21
pellucidus Coquillett, Nebritus	349	perfida Stein, Phaonia	907
pellucidus (Walker), Tanytarsus	180	perfidiosa Alexander, Tipula	38
Pelnia Enderlein	100	perfidiosus (Hunter), Helophilus	619
Pelomyia Williston	726	perflavens (Alexander), Cheilotrichia	79
Pelomyiella Hendel	727	perflaveolus Alexander, Molophilus	89
Pelopia Meigen	149	perflavus Walker, Chlorops	790
pelops Melander, Oreothalia	467	perfoliata Felt, Contarinia	277
Pelorempis Johannsen	104	perfoliata (Felt), Neolasioptera	273
Peloropeodes Wheeler	516	perfuscus Malloch, Metriocnemus	161
peltastes (Spuler), Copromyza	720	*Pergandea* Aldrich	867
peltata (Aldrich), Oxysarcodexia	952	pergandei Coquillett, Admontia	1068
peltatoides Curran, Platycheirus	577	pergandei Coquillett, Apocephalus	543
peltatus (Meigen), Platycheirus	577	pergandei (Coquillett), Forcipomyia	126
Peltopyga Townsend	945	pergandei Felt, Dasineura	260
pemetica Alexander, Dactylolabis	62	pergandei (Williston), Pelecocera	595
pemetica Alexander, Limonia	50	*Pergandia* Aldrich	867
pemeticus (Johnson), Atylotus	331	*peribleptus* Dyar & Knab, Culex	119
pemphigae Malloch, Leucopis	709	periclymeni Hendel, Phytomyza	805
pendens Cole, Bombylius	408	**Pericoma** Walker	92
pendula Alexander, Ptychoptera	97	PERICOMINI	92
pendulifera Alexander, Tipula	27	**Peridiplosis** Felt	281
pendulus (Linnaeus), Helophilus	†618	*perigrina* (Melander), Empis	460
penelope (Osten Sacken), Procecidocharoides	661	perimerus Melander, Platypalpus	481
penepretiosa Chillcott, Fannia	896	perincisa Alexander, Nephrotoma	22
peneprorsus Chillcott, Platypalpus	480	PERISCELIDIDAE	710
penicillaris (Stein), Hylemya	846	**Periscelis** Loew	710
penicillata (Alexander), Limonia	47	*Periscepsia* Gistel	1033
penicillata Alexander, Tipula	38	perissa Pritchard, Acoenonia	247
penicillatus Parent, Pelastoneurus	502	perissa Reinhard, Sarcophaga	959
penicillatus Van Duzee, Dolichopus	494	perissa Steyskal, Pteromicra	687
penicilli Cresson, Hydrellia	744	*perissum* Dyar & Shannon, Simulium	189
penicula Melander, Mythicomyia	418	perita Johannsen, Mycetophila	213
peniculata Parker, Ravinia	954	peritomatis Cockerell, (Cecidomyia)	294
peninsula (Snyder), Coenosia	874	perlata Garrett, Bolitophila	198
peninsularis Patterson & Wheeler, Drosophila	765	*perlonga* Johannsen, Mycetophila	213
penita (Adams), Elachiptera	777	perlongipes Johnson, Tipula	38
penitalis (Coquillett), Pyraustomyia	1011	perlongum Coquillett, Zodion	629
penna (Pettey), Bradysia	234	*permata* Guthrie, Mycetophila	211
pennarista Harmston & Knowlton, Sympycnus	527	permutata (Lundbeck), Lycoriella	232
pennipes (Fabricius), Trichopoda	966	*permutatum* (Dyar & Shannon), Cnephia	184
pennipes (Fabricius), Trichopoda	966	pernicis Coquillett, Efferia	395
pennsylvanica (Dietz), Limonia	48	pernix Melander & Brues, Dolichopus	494
pennsylvanica Felt, Coccidomyia	262	pernodosa Alexander, Ormosia	86
pennsylvanica (Felt), Peromyia	251	pernotata (Malloch), Homoneura	699
pennsylvanica Felt, Youngomyia	279	perocculta (Cockerell), Mayetiola	263
pennyslvanica (Malloch),Chaetopleurophora	533	**Peromyia** Kieffer	251
pennulae Felt, Porricondyla	256	peromysci Dalmat, Cuterebra	1110
penobscot Alexander, Tipula	33	**Peronyma** Loew	660
pensus Aldrich, Dolichopus	494	*perpallidus* (Bigot), Eupeodes	562
pensus Aldrich, Hydrophorus	505	perpallidus Verrall, Platycheirus	577
Pentacricia Stein	877	perparvula Alexander, Tipula	33
pentaformis Huckett, Hylemya	851	*perparvus* (Johannsen), Telmaturgus	529
		perplexa (Brues), Dohrniphora	535
		perplexa Camras, Myopa	630

	Page		Page
perplexa (Coquillett), Villa	439	petrei Mesnil, Opomyza	821
perplexa Curran, Brachyopa	593	petrolei (Coquillett), Helaeomyia	741
perplexa Dietz, Ormosia	86	Petrosarcophaga Townsend	957
perplexa (Felt), Anarete	246	petulca Wheeler, Medetera	510
perplexa Felt, Mycodiplosis	283	*petulca* Williston, Cheilosia	584
perplexa Garrett, Amoebaleria	815	peus Speiser, Culex	118
perplexa Hull, Sphegina	594	**Peyerimhoffia** Kieffer	235
perplexa Johnson, Agathomyia	548	phacodes Melander, Mythicomyia	418
perplexa (Malloch), Megaselia	539	**Phaenicia** Robineau-Desvoidy	927
perplexa Sellers, Carcelia	1092	Phaenobremia Kieffer	280
perplexa Van Duzee, Neurigona	518	**Phaenolauthia** Kieffer	261
perplexens Ludlow, Anopheles	106	phaenops (Osten Sacken), Hybomitra	340
perplexum Johnson, Chrysotoxum	582	**Phaenopsectra** Kieffer	174
perplexus (Back), Diogmites	372	**Phaenopsis** Townsend	1095
perplexus Hull, Eristalis	623	**Phaeobalia** Mik	469
perplexus Johnson & Johnson, Lordotus	412	phaeonota Alexander, Pilaria	†67, 68
perplexus (Osburn), Metasyrphus	561	phaeonotus Loew, Holopogon	375
perplexus Philip, Chrysops	326	phaethus Hardy & Knowlton, Pipunculus	552
perplexus Van Duzee, Dolichopus	494	**Phaiosterna** Cresson	748
perpolita (Johnson), Psila	640	*phairi* Curran, Peleteria	996
perpulcher (James), Sargus	301	**Phalacrocera** Schiner	41
perpulchra (Mitchell), Lauterborniella	172	*Phalacrodexia* Townsend	932
perpusilla (Meigen), Phytoliriomyza	803	*Phalacrodira* Enderlein	565
perpusilla (Walker), Mallophora	397	**Phalacrophyto** Townsend	984
perpusilla Walker, (Sciara)	236	**Phalacrotophora** Enderlein	542
perrarum Reinhard, Leucostoma	976	phalaenoides (Linnaeus), Psychoda	96
perretti Alexander, Tipula	31	*Phalaenula* Meigen	93
perrima (Walker), Hylemya	850	phalangioides Alexander, Limonia	49
Perrisia Rondani	258	phalerata Meigen, Drosophila	765
persequus Malker, Xylophagus	297	phalerata Melander, Mythicomyia	418
persica (Riley), Mycetobia	191	**Phantasiomyia** Townsend	1063
persicae Frick, Phytomyza	805	**Phantolabis** Alexander	78
persicoides (Osten Sacken), Caryomyia	270	**Phaonia** Robineau-Desvoidy	905
persimilis (Alexander), Dicranota	58	PHAONIINAE	899
persimilis Alexander, Limnophila	66	pharetra Melander, Mythicomyia	418
persimilis (Banks), Philonicus	398	phaseolunata (Frost), Liriomyza	802
persimilis Dobzhansky & Epling, Drosophila	768	**Phasia** Latreille	969
persimilis Felt, Rhabdophaga	257	*phasiatrata* Smith, Phasia	969
persimilis (Johannsen), Boreochlus	150	PHASIINAE	964
persimilis Malloch, Mydaea	891	PHASIINI	967
persimilis Reinhard, Pacidianus	1095	*Phasioclista* Townsend	983
perspectabilis Alexander, Ormosia	86	phasioides (Coquillett), Alophorella	968
perspicax Aldrich, Sarcophaga	959	**Phasiomyia** Townsend	969
perspicax Cole, Cyrtopogon	369	**Phasiops** Coquillett	981
perspicua Huckett, Coenosia	872	**Phasiopsis** Townsend	1091
perspicua Johannsen, Exechia	207	*Phasiopteryx* Brauer & Bergenstamm	980
perspicua Wulp, Trichonta	216	**Phasiostoma** Townsend	1063
perspicuum Sommerman, Prosimulium	183	**Phasmophaga** Townsend	1070
persuasa (Osten Sacken), Oedicarena	675	*Phebellia* Robineau-Desvoidy	1096
pertenuis Johnson, Mydas	359	phemius Walker, Rhamphomyia	465
pertinax (Garrett), Pseudodiamesa	153	phengites Melander, Tachydromia	474
pertinax Williston, Chrysops	326	pheosiae Sellers, Aplomya	1098
perturbans (Walker), Mansonia	108	**Pherbecta** Steyskal	692
pertusa Huckett, Pegomya	861	**Pherbellia** Robineau-Desvoidy	685
pertusa (Loew), Villa	437	**Pherocera** Cole	349
pertusus Loew, Amphicnephes	655	*philadelphica* Macquart, Exoprosopa	444
peruanus Becker, Hippelates	775	*philadelphica* Macquart, Gonia	1076
pervagus Walker, Eristalis	624	philadelphica (Macquart), Homoneura	699
petaini (Pettey), Bradysia	234	philadelphica Robineau-Desvoidy, (Meckelia)	654
Petaurista Meigen	15	*philadelphica* Robineau-Desvoidy, Phormia	924
PETAURISTIDAE	15	*philadelphicus* Macquart, Bombylius	†409, 409
petchi Curran, Rhaphium	514	*philadelphicus* Macquart, Eristalis	625
petena Melander, Mythicomyia	418	philadelphicus Macquart, Proctacanthus	399
petentis Reinhard, Muscopteryx	1025	philadelphicus Macquart, Syrphus	559
Petersina Enderlein	565	*philadelphicus* Schiner, Holopogon	375
petersoni (Malloch), Orbellia	810	*philadelphicus* Schiner, Promachus	400
petersoni Stone & DeFoliart, Simulium	189	**Philetus** Melander	455
petes Melander, Mythicomyia	418	*Philhelius* Coquillett	569
Petia Coquillett	971	**Philia** Meigen	195
Petinarctia Villeneuve	1015	philipi Pechuman, Silvius	323
petiolata (Alexander), Dicranota	57	philipi (Stone), Hybomitra	340
petiolata (Coquillett), Diplostichus	1053	**Philocalia** Reinhard	1060
petiolata Coquillett, Sphegina	594	philoctetes Wheeler, Campsicnemus	526
petiolata Malloch, Parabezzia	136	**Philolutra** Mik	469
petiolata Melander, Anthalia	450	philombrius Wheeler, Hydrophorus	505
petiolata Melander, Apolysis	421	**Philonicus** Loew	398
petiolata Melander, Mythicomyia	418	**Philornis** Meinert	911
petiolata Reinhard, Paradidyma	1033	*Philorus* Kellogg	100
petiolata Townsend, Euthyprosopa	1069	**Philosepedon** Eaton	95
petiolata (Townsend), Hesperomyia	1080	**Philotelma** Becker	756
petiolata (Townsend), Reinhardiana	985	philtrum Melander, Chrysotus	523
petiolatum Van Duzee, Parasyntormon	516	**Philygria** Stenhammar	745
petiolatus Hine, Tabanus	335	PHILYGRIINI	745
petiolicola (Felt), Asteromyia	274	*Philygriola* Hendel	745
petiolicola Felt, Schizomyia	267		

	Page		Page
Phlebosotera Duda	824	PHYSOCEPHALINI	627
PHLEBOTOMINAE	91	**Physoclypeus** Hendel	702
Phlebotomus Rondani	91	**Physoconops** Szilády	626
Phloophila Hull	299	*Physogenia* Macquart	702
Phloridosa Sturtevant	767	**Physoptera** Borgmeier	541
phlox Greene, Hyperdiplosis	284	**Phytagromyza** Hendel	803
phobifer (Clark), Cephenemyia	1111	*Phytoadmontia* Townsend	1032
phobifera (Clark), Cephenemyia	1111	*Phytobia* Lioy	797
phoca Aldrich, Hydrophorus	505	*Phytodes* Coquillett	952
Phoeniciomyia Townsend	1047	**Phytoliriomyza** Hendel	803
phoenicurus Loew, Heteropogon	374	**Phytomyptera** Rondani	1066
Pholadoris Sturtevant	767	**Phytomyza** Fallén	804
Pholeomyia Bilimek	732	PHYTOMYZINAE	797
pholi (Webber), Lespesia	1102	*Phytophaga* Rondani	256, †1263
Phoneutisca Loew	474	*Phytopsis* Townsend	1022
Phora Latreille	536	**Phytosciara** Frey	230
Phorantha Rondani	969	**Phytotendipes** Goetghebuer	171
Phoranthella Brooks	970	*Pialeoidea* Westwood	403
Phoranthella Townsend	970	*Pialoidea* Westwood	403
Phorbia Robineau-Desvoidy	846	pica Townes, Omisus	173
Phoricheta Rondani	1033	*picaea* (Felt), Mayetiola	263
PHORIDAE	531	picata (Reinhard), Pilatea	1105
PHORINAE	532	picciola (Bigot), Dioxyna	666
PHORINIINI	1052	picea (Robineau-Desvoidy), Nowickia	994
Phormia Robineau-Desvoidy	924	picea (Rübsaamen), Bradysia	234
PHORMIINI	924	picea (Walker), Scatella	757
Phorocera Robineau-Desvoidy	1056	*picea* (Winnertz), Forcipomyia	†126
PHOROCERINI	1052	piceae (Felt), Mayetiola	263
phoroceroides Townsend, Palpexorista	1056	piceifrons (Townsend), Nowickia	994
phoroctenia Alexander, Tipula	33	piceipes (Melander & Spuler), Sepsis	683
Phoroctenia Coquillett	20	piceus Hine, Asilus	392
Phorodonta Coquillett	230	*piceus* (Macquart), Bonnetia	1004
PHOROIDEA	531	piceus Philip, Chrysops	327
phorophragma Alexander, Limnophila	63	picinus Hanson, Nemotelus	310
Phortica Schiner	761	picipes Brauer & Bergenstamm, Erythrandra	941
Phorticoides Malloch	710	*picipes* Loew, Stratiomys	311
Phosococephalops Townsend	1075	*picipes* Malloch, Scatophaga	839
photiniae Pritchard, Asphondylia	269	picipes (Meigen), Metriocnemus	161
photophila Felt, Dasineura	260	picta (Coquillett), Amiota	761
photophila Felt, Dirhiza	255	picta (Coquillett), Phytobia	798
photophila (Felt) Giardomyia	284	picta (Fabricius), Delphinia	644
photophila (Felt), Peromyia	251	picta (Fallén), Nostima	745
photophilus (Felt), Holoneurus	255	picta (Lehmann), Megaselia	539
phragmites Felt, Asteromyia	275	picta (Loew), Odinia	794
Phrissopolia Townsend	1078	*picta* (Meigen), Linnaemya	1006
Phronia Winnertz	214	picta (Meigen), Melieria	645
Phrosina Robineau-Desvoidy	938	picta Melander, Mythicomyia	418
Phrosinella Robineau-Desvoidy	938	picta Sturtevant & Wheeler, Scatophaga	758
Phryne Meigen	190	picta Williston, Dalmannia	632
PHRYNEIDAE	190	picta (Wulp), Myothyriopsis	1086
Phrynofrontina Townsend	1047	picta Wulp, Oxycera	307
PHRYNOINI	1096	*picta* (Zetterstedt), Graphomya	911
Phrynolydella Townsend	1039	pictella (Thomson), Liriomyza	802
Phryxe Robineau-Desvoidy	1105	*picticauda* (Bigot), Allograpta	569
phthia (Walker), Syneches	449	picticornis Bigot, Sargus	302
Phthinia Winnertz	225	*picticornis* Bigot, Symphoromyia	343
Phthiria Meigen	419	picticornis (Coquillett), Poecilominettia	703
PHTHIRIINAE	419	picticornis Knab & Shannon, Tanypeza	641
phyciodis (Coquillett), Siphosturmia	1087	picticornis Loew, Chrysotus	523
phycitae (LeBaron), Nemorilla	1089	picticornis Loew, Cordilura	828
PHYCODROMIDAE	679	picticornis (Loew), Stenopogon	384
Phycomyza Melander	727	picticornis Osten Sacken, Atarba	71
Phylacteropoda Townsend	1065	picticornis Townsend, Chaetoglossa	1078
Phylidorea Bigot	65	pictilis (Reinhard), Metopotachina	993
phyllocerus Bigot, Phyllomydas	360	pictipennis Becker, Fucellia	843
phyllocerus Vockeroth, Dolichopus	494	pictipennis (Loew), Chrysogaster	591
Phyllodromia Zetterstedt	†472, 473	pictipennis Loew, Cordilura	828
PHYLLODROMIINAE	470	pictipennis (Macquart), Conophorus	414
Phyllogaster Stein	876	*pictipennis* (Macquart), Euantha	1030
Phyllolabis Osten Sacken	60	pictipennis Oldenberg, Scatophaga	839
Phyllomya Robineau-Desvoidy	1021	*pictipennis* (Reinhard), Oraphasmophaga	1070
Phyllomydas Bigot	360	pictipennis Wheeler, Pelastoneurus	502
Phyllomyza Fallén	729	pictipennis (Wiedemann), Furcifera	350
PHYLLOMYZIDAE	728	*pictipennis* Williston, Myopa	631
phyllophagae Curran, Microphthalma	982	pictipennis Wulp, Pseudomorinia	1026
Phyllophila Townsend	1030	pictipennis (Zetterstedt), Anatopynia	145
Phyllophilopsis Townsend	1030	pictipes (Bigot), Blera	611
phyllophora Melander, Tetanocera	694	pictipes Bigot, Melanostoma	575
Phyllophorocera Townsend	1104	pictipes (Bigot), Xanthomelanodes	967
phyllophorus Loew, Gymnopternus	499	pictipes Coquillett, Mythicomyia	418
phylloxerae Riley, Leucopis	709	pictipes Hagen, Simulium	188
Phyllozomyia Dyar	107	pictipes Harmston & Knowlton, Sympycnus	527
Physegenua Macquart	702	pictipes (Loew), Beameromyia	362
Physiphora Fallén	653	pictipes (Loew), Dictya	689
Physocephala Schiner	627	pictipes (Zetterstedt), Clusiodes	807

INDEX 1655

	Page		Page
pictitarsis (Bigot), Callinicus	365	pilosa (Zettersteat), Orthacheta	832
pictiventris Duda, Drosophila	767	pilosella (Becker), Gaurax	782
pictiventris (Malloch), Lyciella	700	pilosella (Coquillett), Cordilura	830
pictiventris Wheeler, Pseudiastata	762	pilosella (Loew), Pentaneura	147
pictula (Loew), Tenthredomyia	616	pilosella (Osten Sacken), Ulomorpha	68
pictula Meigen, Mycetophila	213	pilosellus Brundin, Cricotopus	158
pictulum Williston, Temnostoma	614	piloseta Malloch, Hylemya	852
pictulum Williston, Zodion	629	pilosicornis Walker, (Porphyrops)	530
picturata Alexander, Tipula	36	pilosifrons Allen, Phrosinella	938
picturata Coquillect, Phthiria	420	pilosipes Hull & Fluke, Cheilosia	585
picturata (Snow), Acinia	664	pilosissima Stein, Coenosia	872
pictus (Bigot), Chyliza	640	pilosula (Dietz), Nephrotoma	21
pictus Cole, Metapogon	378	pilosulus Becker, Chlorops	790
pictus (Coquillett), Stilobezzia	134	pilosulus (Becker), Gaurax	782
pictus (Loew), Nicocles	380	pilosum (Knowlton & Rowe), Simulium	185
pictus Malloch, Apocephalus	543	pilosus (Cole), Poecilanthrax	442
pieta Roback, Diamesa	152	pilosus (Coquillett), Scoliocentra	815
Piezura Rondani	898	pilosus (Dyar & Knab), Culex	120
pifanoi Ortiz, Culicoides	130	pilosus Hunter, Helophilus	618
pigra (Fabricius), Xylota	607	pilosus James, Myxosargus	317
pigra (Lovett), Xylota	607	pilosus Leonard, Chrysopilus	346
pigra (Osten Sacken), Stonemyia	321	pilosus Loew, Eristalis	624
pikei Shannon, Cheilosia	585	pilosus Roback, Psectrocladius	159
pikei Whitney, Chrysops	327	pilosus Schlinger, Eulonchus	404
Pilaria Sintenis	68	pilulae (Beutenmüller), Cincticornia	269
pilata (Pettey), Bradysia	234	pilumnus Kröber, Chrysops	326
Pilatea Townsend	1105	pima Painter, Villa	436
pilatei (Coquillett), Drino	1085	pimpla Coquillett, Tetragoneura	228
pilatei (Hough), Phaenicia	927	pinea (Felt), Colpodia	253
pilatei Johnson, Dizonias	372	pingreensis (Fluke), Metasyrphus	561
pilatei (Macquart), Ligyra	445	pingreensis James, Bibio	194
pilatus Melander, Platypalpus	480	pingreensis James, Dolichopus	494
pilatus Van Duzee, Dolichopus	494	pinguis (Loew), Clinotanypus	146
Pilica Curran	386	pinguis Loew, Mycetophila	213
pilicauda Huckett, Hylemya	849	pinguis (Walker), Culiseta	117
pilicaudata (Walley), Pentaneura	147	pinguis (Walker), Parydra	750
piliceps Alexander, Tipula	30	pini (Felt), Anaretella	245
pilicornis (Aldrich), Condylostylus	485	pini (Felt), Feltiella	282
pilicornis (Coquillett), Setacera	755	pini (Felt), Janetiella	264
pilicornis (Fabricius), Chironomus	165	pini Felt, Porricondyla	256
pilifemur Ringdahl, Hylemya	850	pini Felt, Rhopalomyia	266
piliferus Root & Hoffman, Culicoides	130	piniariae (Hartig), Blondelia	1040
piligeronis Coquillett, Rhamphomyia	465	pinicola Malloch, Leucopis	709
pilimana Loew, Odontomyia	316	pinicola Sturtevant, Drosophila	765
pilimana (Ringdahl), Eremomyia	857	pinicorticis (Felt), Aprionus	251
pilimana Stein, Hylemya	850	pinicorticis Felt, Winnertzia	253
Pilimas Brennan	321	pinifoliae Felt, (Itonida)	290
pilipennis (Alexander), Hesperoconopa	78	pininopis Osten Sacken, Cecidomyia	287
pilipennis (Fallén), Actia	†1062	piniperda Malloch, Leucopis	709
pilipes Borgmeier, Cremersia	544	piniradiatae (Snow & Mills), Thecodiplosis	278
pilipes (Fabricius), Erioptera	80	piniresinosae Kearby & Benjamin, Thecodiplosis	278
pilipes (Loew), Bicellaria	451	pinirigidae Packard, (Diplosis)	279
pilipes Malloch, Orthocladius	156	pionips Dyar, Aedes	113
pilipes Stein, Hydrotaea	900	**Piophila** Fallén	711
pilipleura Borgmeier, Coniceromyia	536	PIOPHILIDAE	710
pilirostris (Ringdahl), Trichopalpus	836	piperi (Coquillett), Parademoticus	1011
pilitarsis Malloch, Hydrophorus	505	piperi Dyar & Shannon, Simulium	189
pilitarsis Stein, Hylemya	850	piperi Knab, Microdon	598
pilitarsis (Stein), Paraprosalpia	866	piperi (Townsend), Lespesia	1102
pilitibia (Ringdahl), Hylemya	854	piperitae Felt, (Cecidomyia)	290
pilitibia Stein, Hydrotaea	900	piperitae Felt, Dasineura	260
pilivenstris (Wulp), Archytas	†1000	pipiens Linnaeus, Culex	118
piloala (Garrett), Lasiodiamesa	150	pipiens (Linnaeus), Syritta	609
pilosa Aldrich, Symphoromyia	343	**Pipiza** Fallén	579
pilosa Brooks, Gonia	1076	*Pipizella* Rondani	†580
pilosa (Coquillett), Belosta	355	PIPIZINI	579
pilosa (Coquillett), Forcipomyia	126	PIPUNCULIDAE	550
pilosa Cresson, Ochthera	753	PIPUNCULINAE	551
pilosa Day, Odontomyia	316	PIPUNCULINI	551
pilosa Dietz, Ormosia	87	**Pipunculus** Latreille	551
pilosa (Drury), Archytas	†1000	pirata Loew, Hydrophorus	505
pilosa Felt, Rhopalomyia	266	piscataquis Alexander, Limonia	48
pilosa Garrett, Macrocera	201	piscicidium Riley, Simulium	189
pilosa (Garrett), Trichothaumalea	120	pistica (Williston), Heringia	580
pilosa Garret., Zygomyia	216	pisticoides (Williston), Neocnemodon	581
pilosa (Hine), Efferia	395	pistillata (Lundström), Allodia	205
pilosa Kröber, Psilocephala	351	placida (Aldrich), Sarcodexia	†956
pilosa Melander, Drapetis	477	placida Coquillett, Psilocephala	351
pilosa Staeger, Piophila	712	placida Harris, (Cordilura)	1115
pilosa Stein, Pegomya	861	placida (Huckett), Spilogona	882
pilosa (Tothill), Nowickia	995	placidum Johannsen, Dynatosoma	209
pilosa (Townsend), Omotoma	1089	placidus (Curran), Neocnemodon	581
pilosa Wilcox, Promachina	399	plagens Williston, Symphoromyia	343
pilosa Williston, Myopa	630	*Plagia* Meigen	1020
pilosa (Zetterstedt), Jassidophaga	551		

	Page		Page
plagiata (Harris), Epalpus	1002	plebeia (Malloch), Megaselia	541
plagiata Harris, (Thereva)	1115	plebeia (Walker), Exechia	207
plagiata (Melander), Phytagromyza	803	plebeia Walker, Hilara	456
plagiata (Walker), Stichopogon	385	plebeia Williston, Myopa	630
Plagiocera Macquart	622	plebeja Johannsen, Leia	227
plagioides Townsend, Goniochaeta	1018	plebeja Loew, Odontomyia	317
Plagiomima Brauer & Bergenstamm	1019	plebeja (Loew), Palpomyia	140
PLAGIONEURINAE	504	plebeja Loew, Tetanocera	694
Plagioneurus Loew	504	plebeja Lopes, Oxysarcodexia	952
Plagiophryze Townsend	1105	plebejus Loew, Hippelates	775
Plagiops Cresson	742	*plebia* Loew, Odontomyia	317
Plagiops Townsend	1058	plebs Melander, Trixoscelis	818
Plagiopsis Cresson	742	**Plecia** Wiedemann	192
Plagiosippus Reinhard	1015	PLECIINAE	192
Plagiotachina Townsend	1054	*Plecticus* Loew	302
Plagiotoma Loew	678	*Plectops* Coquillett	1065
Plagiprospherysa Townsend	1057	**Plectromyia** Osten Sacken	57
plagosa (Coquillett), Villa	438	plectrum Melander, Empis	460
plana Steyskal, Diacrita	644	plena (Hine), Efferia	395
planensis Johannsen, Pentaneura	147	pleomenda Reinhard, Sarcophaga	959
Planetolestes Lynch Arribálzaga	365	plesia Curran, Xylota	607
plangens Curran, Archytas	1000	*Plesiastina* Winnertz	199
plangens Wiedemann, Chrysops	325	**Plesiomma** Macquart	382
planiceps (Fallén), Gonatherus	830	plesius Curran, Anthrax	432
planiceps (Loew), Metaphragma	352	*plesius* Curran, Nemotelus	309
planicollis (Becker), Eribolus	778	**Plethochaeta** Coquillett	841
planicornia Doane, Tipula	38	pleuracantha Alexander, Ormosia	86
planicosta Reinhard, Nimioglossa	987	pleuracicula Alexander, Tipula	38
planiforceps (Tothill), Nowickia	995	pleurale Malloch, Prosimulium	183
planifrons (Aldrich), Ravinia	955	*pleuralis* (Adams), Rhagio	345
planipalpis (Stein), Hylemya	850	pleuralis Aldrich, Procecidochares	660
planipes Borgmeier, Stichillus	536	pleuralis Banks, Dioctria	370
planipes (Harris), Sphaerocera	718	pleuralis (Becker), Heleomyza	816
planipes Van Duzee, Neurigona	518	*pleuralis* (Coquillett), Heleomyza	816
planiscuta Thomson, (Lauxania)	706	pleuralis (Coquillett), Palaeosepsis	682
planitarsis Reinhard, Sphenometopa	939	*pleuralis* (Cresson), Pteromicra	687
planitarsus Wilcox & Martin, Cyrtopogon	369	pleuralis Curran, Symphoromyia	344
planiventris (Loew), Mesograpta	571	pleuralis (Dietz), Paradelphomyia	60
plankii (Walton), Hemithrixion	1037	*pleuralis* Malloch, Dasyopa	784
plantaginis Robineau-Desvoidy, Phytomyza	805	pleuralis Malloch, Orthocladius	156
planus Osten Sacken, Lordotus	412	pleuralis (Malloch), Phytobia	800
Plastophora Brues	542	pleuralis (Williston), Gonomyia	75
Plastosciara Berg	235	pleuralis (Wood), Megaselia	541
platancala Loew, Psilocephala	351	*pleuricellus* Williston, Anthrax	432
plantanifolia Felt, Lestodiplosis	285	pleuriseta (Cresson), Ptilomyia	736
platura (Meigen), Hylemya	850	pleuriseta Malloch, Lonchaea	717
Platurocypta Enderlein	215	pleuritica Loew, Cordilura	830
platycarina (Tothill), Mericia	1008	*pleuritica* Loew, Loxocera	638
platycauda Curran, Cyrtopogon	369	pleuriticus Melander, Stilpon	476
platycera Hine, Cheilosia	586	*Pleurochaeta* Becker	834
platyceras Hine, Asilus	392	**Pleurochaetella** Vockeroth	834
platycheira Melander, Mythicomyia	418	plexipus Garrett, Dixa	101
Platycheirus Lepeletier & Serville	576	plexipus Garrett, Eudicrana	223
platychira (Malloch), Megaselia	539	plicata Felt, Rhabdophaga	257
Platychirus Lepeletier & Serville	576	plicata Kraft & Cook, Zabrachia	318
Platycoenosia Strobl	898	plinthopyga (Wiedemann), Blaesoxipha	946
Platygaster Zetterstedt	415	*Ploas* Latreille	413
platygastra Cresson, Hydrellia	744	plumasana (Reinhard), Nowickia	994
platymera Walker, Tipula	31	plumata (De Geer), Volucella	601
Platymya Robineau-Desvoidy	1096	*plumata* Loew, Cheilosia	587
Platymyia Robineau-Desvoidy	1096	plumata (Loew), Suillia	809
Platypalpus Macquart	478	plumata (Wulp), Camptoprosopella	697
platypelta Cresson, Paracoenia	756	plumbea Aldrich, Lixophaga	1044
Platypeza Meigen	549	plumbea Aldrich, Symphoromyia	344
PLATYPEZIDAE	546	plumbea Alexander, Gonomyia	74
Platypezina Wahlgren	548	*plumbea* (Zetterstedt), Phaonia	†907
Platypezoides Johnson	548	plumbeus Aldrich, Hydrophorus	506
platyphallus Alexander, Limnophila	66	*plumbeus* Drury, Tabanus	332
Platyphora Verrall	537	plumbeus (Say), Rhagio	345
platyprosopus Loew, Dolichopus	492	plumeum Johnson, Chrysotoxum	582
platyptera (Loew), Xanthomyia	664	plumipes (Fabricius), Trichopoda	966
Platyptera Meigen	457	plumipes Melander, Drapetis	478
platyptera (Thomson), Phytobia	800	plumipes (Scopoli), Dolichopus	494
platypterus Loew, Diogmites	372	plumiseta Frost, Phytomyza	805
platypus (Coquillett), Cardiocladius	154	plumitarsis Fallén, Dolichopus	494
PLATYPYGINAE	415	plumosa Coquillett, Cheilosia	587
Platypygus Loew	415	plumosa Loew, Tetanocera	694
platyrostra (Alexander), Limonia	48	plumosa Van Duzee, Rymosia	208
platysamiae (Townsend), Winthemia	1090	*plumosa* (Van Duzee), Sciapus	486
PLATYSTOMATIDAE	655	plumosus Aldrich, Dolichopus	494
PLATYSTOMIDAE	655	plumosus (Linnaeus), Chironomus	166
Platytipula Matsumura	27	**Plunomia** Curran	706
Platyura Meigen	†202, 204	pluricellus Williston, Anthrax	432
plausilis Reinhard, Gueriniopsis	1055	pluriseriata (Aldrich & Webber), Patelloa	1104
plausor Osten Sacken, Cyrtopogon	369	plusiae Coquillett, Siphona	1064
plebeia Malloch, Fannia	896	pluto Hull, Sphegina	594

INDEX

	Page
pluto Hull & Fluke, Cheilosia	585
pluto Melander, Platypalpus	480
pluto Wiedemann, Anthrax	432
plutonia Hunter, Cheilosia	587
plutonis Alexander, Tipula	30
pluvialis (Dyar & Shannon), Trichothaumalea	120
pluvialis (Huckett), Spilogona	883
pluvialis (Linnaeus), Anthomyia	865
pluvialis Malloch, Forcipomyia	125
pluvialis Malloch, Hylemya	850
Pneumaculex Dyar	108
Pnyxia Johannsen	235
Pocota Lepeletier & Serville	610
poculum Osten Sacken, (Cecidomyia)	294
podagra Melander, Empis	460
podagra (Newman), Microdon	598
podagrae (Beutenmüller), Neolasioptera	273
podagrae Felt, Cincticornia	269
podagrae Felt, Rhabdophaga	257
podagratus (Zetterstedt), Platycheirus	577
podagrica (Loew), Morellia	912
PODONOMINAE	150
Podonomus Phillippi	150
podophylli Felt, Bremia	280
podophylli (Felt), Youngomyia	279
podopostyla Speiser, Stilbometopa	918
Podotachina Brauer & Bergenstamm	1057
Poecilanthrax Osten Sacken	441
poecilogaster (Osten Sacken), Poecilanthrax	442
Poecilognathus Jaennicke	420
Poecilographa Melander	692
Poecilominettia Hendel	703
Poecilomyia Melander	692
poeciloptera Loew, Empis	460
poeciloptera (Malloch), Helina	887
poecilopterus (Mitchell), Stenochironomus	177
Poecilosomella Duda	722
Poemyza Hendel	798
poenitens (Wheeler), Hercostomus	498
poetica Osten Sacken, Limnophila	67
poeyi (Garrett), Alluaudomyia	133
Pogonephydra Hendel	753
pogonias (Wiedemann), Efferia	395
pogonoides Snyder, Xenomydaea	892
Pogonomyia Rondani	901
Pogonomyioides Malloch	902
Pogonomyza Schnabl	845
Pogonortalis Hendel	655
Pogonosoma Rondani	391
Pogonota Zetterstedt	833
pokornyi (Stein), Platycoenosia	898
polaris (Frey), Hybomitra	340
polaris Kieffer, Cricotopus	158
polaris Kieffer, Diamesa	152
polaris (Kieffer), Diamesa	152
polaris Kieffer, Metriocnemus	161
polaris Kieffer, Psectrocladius	159
polaris (Kieffer), Smittia	162
polaris Kirby, Chironomus	165
Polidaria Curran	1014
Polidea Macquart	1014
Polideosoma Townsend	1033
polidoides (Townsend), Lydina	1014
Polietella Ringdahl	909
polingi Alexander, Gonomyia	74
polingi Alexander, Tipula	38
poliocephala Alexander, Gonomyia	76
poliochros Dyar, Aedes	115
Potiophrys Townsend	1079
polistensis Hall, Sarcophaga	959
Polistomyia Townsend	966
polita Brooks, Alophorella	968
polita Brooks, Chaetophlepsis	1016
polita (Brooks), Periscepsia	1034
polita (Coquillett), Aplomya	1098
polita (Coquillett), Cordilura	828
polita Coquillett, Cuterebra	1109
polita (Coquillett), Dexia	1022
polita (Coquillett), Eribella	1042
polita Coquillett, Lispe	878
polita (Coquillett), Oxynops	1046
polita Coquillett, Tetanops	648
polita Coquillett, Zabrachia	318
polita (Harris), Cordilura	829
polita (Johnson), Dolichopeza	23
polita (Linnaeus), Microchrysa	303
polita Loew, Mycetophila	210
polita (Loew), Procecidochares	661
polita Loew, Rhamphomyia	465
polita Loew, Syndyas	448
polita Malloch, Hydrophoria	864
polita (Malloch), Micropsectra	179
polita Malloch, Pseudodinia	708
polita (Malloch), Spilochroa	817
polita Reinhard, Winthemia	1090
polita Sabrosky, Meoneura	729
polita Say, Lonchaea	717
polita (Say), Mesograpta	571
polita Say, Sciara	230
polita (Townsend), Blondelia	1040
polita Townsend, Epigrimyia	974
polita Townsend, Gymnoprosopa	936
polita (Williston), Leptocera	724
politellus Melander, Platypalpus	480
politissima Alexander, Limnophila	64
Politomyia Reinhard	974
politura (Reinhard), Clausicella	1059
politus (Coquillett), Cricotopus	158
politus (Coquillett), Monohelea	135
politus Hanson, Nemotelus	310
politus Harris, (Xylophagus)	1115
politus (Johnson), Hybomitra	339
politus Loew, Gymnopternus	499
politus Malloch, Xylophagus	297
politus Melander, Platypalpus	480
politus (Say), Nicocles	380
politus (Williston), Leptocera	†723
polius Melander, Oligodranes	422
polleni Garrett, Mycomya	221
Pollenia Robineau-Desvoidy	931
POLLENIINAE	931
POLLENIINI	931
pollenius (Cole), Callinicus	365
pollens Osten Sacken, Microstylum	379
pollex Alexander, Molophilus	89
pollex Osten Sacken, Dolichopus	494
pollex Shaw, Exechia	207
pollex (Van Duzee), Rhaphium	514
pollicis (Pettey), Bradysia	234
pollinaria (Fluke), Helophilus	620
pollinaris Reinhard, Euphyto	935
pollinosa Brooks, Paraphorantha	969
pollinosa Brooks, Psalidopteryx	1067
pollinosa Cole, Psilocephala	351
pollinosa Cole, Tritoxa	649
pollinosa Curran, Admontia	1068
pollinosa Curran, Cophura	367
pollinosa Curran, Eutreta	661
pollinosa (Melander), Phytobia	799
pollinosa (Reinhard), Buquetia	1098
pollinosa Ringdahl, Pegomya	861
pollinosa Rowe, Belvosia	1081
pollinosa Walton, Eleodiphaga	1072
pollinosa (Williston), Paratissa	740
pollinosa Wirth, Dasyhelea	127
pollinosum Williston, Arthroceras	298
pollinosus Leonard, Rhagio	345
pollinosus Van Duzee, Sciapus	486
pollinosus Wilcox, Nicocles	380
pollinosus Williston, Silvius	323
Polyangaeus Doane	56
Polyardis Pritchard	249
Polybiomyia Shannon	616
Polyblepharis Bezzi	458
polycantha Alexander, Erioptera	83
polycantha Alexander, Tipula	38
polychaeta Patterson & Wheeler, Drosophila	765
polychaeta (Pettey), Bradysia	234
polychaeta (Stein), Fannia	896
Polychaetoneura Walton	1061
Polydonta Macquart	621
Polydontomyia Williston	621
polygena Melander, Mythicomyia	418
polygoni Seamans, Pegomya	861
polygrammus (Loew), Asemosyrphus	620
polygynus Melander, Anthepiscopus	455
Polylepta Winnertz	225
Polymeda Meigen	80
Polymedon Osten Sacken	503
polymera (Loew), Nephrotoma	22
Polymera Wiedemann	60

1658 A CATALOG OF DIPTERA OF NORTH AMERICA

	Page
polymeroides (Alexander), Dicranota	58
polymorpha Andersen, Smittia	162
Polypedilum Kieffer	174
polypori (Melander), Piophila	712
polypori Willard, Platypeza	549
polystigma (Meigen), Chamaemyia	707
Polytrichophora Cresson	739
Polyxena Meigen	208
polyzona (Loew), Dziedzickia	218
pomeroyi Parent, Chrysotus	523
pomiflorae Felt, Campylomyza	248
pomifolia Felt, Campylomyza	248
pomonella (Walsh), Rhagoletis	674
pompale (Reinhard), Microtrichomma	1001
pomum (Osten Sacken), Schizomyia	267
pomus (Curran), Metasyrphus	561
ponderosa (Curran), Ptilodexia	989
ponderosus Melander, Epacmus	428
ponecta Felt, Rhabdophaga	257
pontiaca Shannon, Cheilosia	585
poplitea Loew, Empis	460
popoffana (Townsend), Acronesia	928
populi Felt, Campylomyza	248
populi Felt, Dicrodiplosis	279
populi Felt, Kronomyia	252
populi Felt, Rhabdophaga	257
populi Steyskal, Drapetis	477
populi Steyskal, Milichiella	733
populi Steyskal, Sobarocephala	806
populi Wheeler & Throckmorton, Drosophila	768
populicola (Walker), Phytagromyza	803
populifolia Felt, Lestodiplosis	285
populifolia Felt, Mycodiplosis	283
porca Williston, Gonia	1076
porcina Hull & Fluke, Cheilosia	585
porcus (Walker), Helophilus	619
Porphyrops Meigen	†512, 524
porphyrops Van Duzee, Dolichopus	495
porrecta Felt, Colpodia	253
porrecta Felt, Porricondyla	256
porrecta Felt, Rhabdophaga	257
porrectus Melander, Platypalpus	480
Porricondyla Rondani	255
PORRICONDYLINI	252
Porsenus Darlington	808
portaecola (Walker), Tachypeza	473
portenkoi (Stackelberg), Allomyella	835
portensis Huckett, Hylemya	846
porterae (Cockerell), Sackenomyia	265
porteri Dodge, Philornis	911
portoricensis (Macquart), Condylostylus	485
pose Dyar & Knab, Culex	119
positivus Osten Sacken, Cyrtopogon	369
Posthosyrphus Enderlein	560
postiana Alexander, Dactylolabis	62
postica Say, Volucella	602
postica (Stein), Fannia	896
postica Walker, (Dexia)	1108
postica (Walker), Tachypeza	473
posticalis (Lundbeck), Trichotanypus	†150, 151
posticata Curran, Peleteria	997
posticata (Fabricius), Mallota	621
posticata (Meigen), Phytobia	798
posticata Say, Laphria	390
postilena Harris, Scatophaga	838
postilla (Reinhard), Blaesoxipha	946
Postleria Garrett	813
postpositus Melander, Platypalpus	480
potamogeti (Townes), Chironomus	168
potanini Rohdendorf, Blaesoxipha	944
potentillaecaulis Stebbins, (Cecidomyia)	294
potomaca Fisher, Ditomyia	198
potrix Melander, Mythicomyia	418
Potthastia Kieffer	151
praeapicalis Shewell, Homoneura	699
praecisa Loew, Tipula	38
praecox (Lehmann), Hydrophorus	506
praecox Meigen, Phytomyza	805
praecox Meijere, Lipoleucopis	708
praedator Wheeler, Aphrosylus	509
praelauta Alexander, Tipula	26
praelonga Felt, Cordylomyia	248
praemiosa Giglio-Tos, Sepedon	693
praenubila (Johannsen), Mycetophila	212
praepes Williston, Cyrtopogon	369
praestans Melander, Chelipoda	472

	Page
praeusta Loew, Cordilura	828
praeusta (Loew), Palpomyia	140
praeustus Loew, Dolichopus	495
praevolans (Wulp), Boettcheria	948
prairiensis (Bromley), Efferia	395
prairiensis Martin, Beameromyia	362
prairiensis Tucker, Tolmerus	401
prasinata Melander, Clinocera	468
prasinus Johnson, Thinophilus	508
prasinus Loew, Diostracus	508
prasinus Meigen, Chironomus	166
Praspedomyza Hendel	800
pratensis Felt, Colpodia	253
pratensis (Felt), Parallelodiplosis	286
pratensis Felt, Rhabdophaga	257
pratensis (Meigen), Hylemya	851
pratincola (Panzer), Anthomyiella	865
pratincola Wheeler, Chrysotus	523
pratorum Kirby, (Tipula)	90
pratorum Meigen, Meromyza	793
pratti Alexander, Dolichopeza	23
pratti (Dodge), Sarcophaga	959
pratti Hunter, Cephenemyia	1111
pratti Shaw, Exechia	207
prattii Coquillett, Heteromyia	137
prattii (Hine), Efferia	395
prava Chillcott, Rhamphomyia	465
pravipes Melander, Mythicomyia	418
pravus Reinhard, Carinosillus	989
precatoria (Fallén), Chelifera	471
predator Curran, Cyrtopogon	369
presignis Chillcott, Fannia	896
pressipes Parent, Sciapus	486
pretans (Grossbeck), Aedes	114
pretiosa (Coquillett), Villa	435
pretiosa Hull, Myolepta	589
pretiosa (Schiner), Fannia	†896
pretiosa (Walker), Paraprosalpia	866
pretiosina (Curran), Azelia	899
pretiosus (Banks), Caloparyphus	308
pretiosus Banks, Ommatius	398
prexaspes (Walker), Ptilodexia	989
priapulus Loew, Rhamphomyia	465
pribilofensis Alexander, Tipula	31
pribilofensis Malloch, Eupogonomyia	902
pribilovia Alexander, Tipula	31
pricei Dyar, Aedes	112
prima Hendel, Trixoscelis	818
prima Hunter, Cheilosia	586
prima Malloch, Coenosopsia	905
primavera Hull, Xylota	607
primigenia Melander, Ragas	455
primitiva Alexander, Chionea	72
Primochrysotoxum Shannon	582
primoris Reinhard, Orestilla	984
primoveris Alexander, Prionocera	19
primoveris Shannon, Cheilosia	585
princeps (Austen), Cuterebra	1110
princeps Osten Sacken, Cyrtopogon	369
princeps Wheeler, Medetera	510
princeps Williston, Promachus	400
Prionella Robineau-Desvoidy	654
Prionocera Loew	18
Prionolabis Osten Sacken	63
prisca Stein, Phaonia	907
prisca (Walker), Cryptomeigenia	1041
pristis (Walker), Anaporia	1028
pritchardi Bromley, Diogmites	372
pritchardi Bromley, Saropogon	383
pritchardi Bromley, Stichopogon	385
pritchardi Martin, Hadrokolos	373
pritchardi Wirth, Dasyhelea	127
privernus (Walker), Toxomerus	571
privigna Melander, Piophila	712
Probezzia Kieffer	138
problematica (Johnson), Phaenicia	928
probohemica Speiser, Lispe	878
proboscidalis (Malloch), Hylemya	845
proboscidalis Malloch, Neohylemya	867
proboscidea Lengersdorf, Rhynchosciara	231
proboscidea (Malloch), Rhyncophoromyia	543
proboscidea (Strobl), Ophiomyia	797
proboscidea Zetterstedt, Cordilura	828
proboscideus (Loew), Lepidanthrax	441
proboscideus Williston, Hippelates	775
Proboscidomyia Bigot	868

INDEX

	Page		Page
Proboscimyia Bigot	868	propinqua Walker, Mycetophila	213
Procatharosia Villeneuve	971	propinquina Huckett, Hylemya	851
Proccecidochares Hendel	660	propinquus Bromley, Stenopogon	384
Proccecidocharoides Foote	661	*propinquus* (Melander & Brues), Dolichopus	489
procedens (Walker), Helina	888	propinquus Van Duzee, Hydrophorus	506
procera (Loew), Cetema	788	*propinquus* (Walker), Chrysopilus	346
procera (Loew), Hoplocyrtoma	452	propinquus Wheeler, Paracleius	501
procera Loew, Mycetophila	213	propleuralis Melander, Empidideicus	416
Proceroplatus Edwards	202	propusilla Sabrosky & Arnaud, Cylindromyia	973
procerus Van Duzee, Dolichopus	492		
procerus (Wheeler), Gymnopternus	499	**Porates** Melander	424
Prochyliza Walker	713	prorsus Melander, Platypalpus	480
procincta (Reinhard), Tinalydella	1050	*Prosalpia* Pokorny	866
Procistogaster Townsend	964	**Prosena** Lepeletier & Serville	987
Procladius Skuse	149	PROSENINAE	979
proclinata Cresson, Hydrellia	744	PROSENINI	985
Proclinopyga Melander	467	**Prosenoides** Brauer & Bergenstamm	988
proclivis Osten Sacken, Chrysops	327	proserpina Alexander, Gonomyia	74
procnemis (Williston), Chymomyza	770	*proserpina* (Williston), Eusisyropa	1093
procnemoides Wheeler, Chymomyza	770	PROSIMULIINAE	182
Proctacantha Macquart	399	PROSIMULIINI	182
Proctacanthella Bromley	398	**Prosimulium** Roubaud	182
Proctacanthus Macquart	399	**Prosmittia** Brundin	162
procteri Cresson, Hydrellia	744	prosopidis Cockerell, Asphondylia	269
procteriana Alexander, Pedicia	54	*prospectus* Tucker, Tolmerus	401
proctori Cresson, Hydrellia	744	**Protanyderus** Handlirsch	90
procyon (Osten Sacken), Hybomitra	340	PROTANYPODINI	153
Procystiphora Felt	258	**Protanypus** Kieffer	†144, 153
Prodiamesa Kieffer	152	**Protaplonyx** Felt	274
Prodiplosis Felt	284	protea (Alexander), Pedicia	55
prodotes Dyar, Aedes	111	proteana (Felt), Trishormomyia	288
producta Cook, Scatopse	238	*protenta* Sherman, (Allodia)	1115
producta (Felt), Bryomyia	251	*Protenthes* Johannsen	150
producta Felt, Youngomyia	279	protervus Melander, Tachytrechus	503
producta (Hine), Efferia	395	protextus (Townes), Chironomus	170
producta Johannsen, Phronia	215	*Prothecus* Rondani	†551
producta (Walker), Hyadina	751	**Protocalliphora** Hough	925
producta Walker, Micropeza	635	**Protoclythia** Kessel	549
productella Alexander, Tipula	33	*Protocuterebra* Bau	1109
productus Hine, Tabanus	335	**Protodejeania** Townsend	1002
productus Loew, Chlorops	790	*Protodexia* Townsend	943
profectus Fisher, Boletina	217	*Protomacleaya* Theobald	115
Proforcipomyia Saunders	124	protoparcis (Townsend), Drino	1085
profuga Stein, Hylemya	853	**Protophormia** Townsend	925
profunda (Alexander), Limonia	48	**Protopiophila** Duda	713
profunda Alexander, Ormosia	86	**Protoplasa** Osten Sacken	90
profusa (Townes), Phaenopsectra	174	*Protoplasta* Osten Sacken	90
profuscata Steyskal, Odontomyia	316	*protritus* (Osten Sacken), Epistrophe	565
profusus Osten Sacken, Cyrtopogon	369	*protrudens* (West), Opsotheresia	990
profusus (Walker), Blera	611	protuberans (Aldrich & Webber), Carcelia	1092
prognatha (Melander), Tethina	728	protuberans Malloch, Macateeia	868
Progonomyia Alexander	73	protuberans Malloch, Phaonia	907
prohibita (Aldrich), Blaesoxipha	944	*Protylos* Aczél	635
Prohippelates Malloch	776	provocans (Becker), Olcella	784
projecta (Becker), Megaselia	541	*provocans* (Walker), Aedes	114
projecta (Malloch), Olcella	784	provocans Walker, Chrysops	327
prolepsis (Reinhard), Blaesoxipha	947	proxima Alexander, Ormosia	87
prolifica (Felt), Bradysia	234	*proxima* (Curran), Tenthredomyia	616
prolifica (Osten Sacken), Thaumatomyia	787	proxima Johnson, Callomyia	547
Prolimnophila Alexander	63	proxima Malloch, Hydrophoria	864
prolixa Alexander, Tipula	25	*proxima* (Malloch), Oscinella	780
prolixa Reinhard, Eumacronychia	935	*proxima* (Say), Syritta	609
prolixa Sherman, Rymosia	208	*proxima* (Walker), Phormia	924
Prolirophaga Townsend	1043	proxima West, Ptilodexia	989
prolixus Dyar, Aedes	112	proxima (Wulp), Phaonia	907
Promachina Bromley	399	proximus Aldrich, Pelastoneurus	502
Promachus Loew	400	*proximus* (Cresson), Tomosvaryella	555
Promasiphya Townsend	1091	proximus McAtee, Dilophus	195
Promericia Brooks	1009	proximus Say, Chlorops	790
promethea Hull & Fluke, Cheilosia	585	proximus (Walker), Chrysopilus	346
prominens Alexander, Prinocera	19	proximus Walker, Tabanus	335
prominens (Becker), Neoleria	813	prudens Curran, Hydrellia	744
prominens (Walker), Helius	51	*prudens* Curran, Psychoda	96
promiscua Ferris & Cole, Ornithoica	916	prudens Pritchard, Mallophorina	397
promiscua (Townsend), Lespesia	1101	prudens (Walley), Ablabesmyia	148
promissa Frick, Phytobia	800	pruinella (Huckett), Spilogona	882
Promusca Townsend	913	pruinescens Thomsen, Palpomyia	140
propepusilla Frost, Liriomyza	802	pruinosa Bigot, Melanostoma	575
properans Bromley, Diogmites	372	pruinosa (Coquillett), Bezzia	141
Prophryno Townsend	1106	*pruinosa* Coquillett, Microphthalma	982
propinqua (Adams), Thecophora	632	pruinosa Coquillett, Oedalea	450
propinqua Bromley, Dioctria	371	*pruinosa* (Coquillett), Phytobia	798
propinqua Hardy, Tomosvaryella	555	pruinosa Coquillett, Tachypeza	473
propinqua (Macquart), Fannia	897	pruinosa (Cresson), Hyadina	751
propinqua Schiner, Euryneura	306	pruinosa (Macquart), Phaonia	907
		pruinosa (Malloch), Elfia	1066

	Page		Page
pruinosa Melander, Mythicomyia	418	*Pseudopandora* Rapp	683
pruinosa Melander, Pseudodinia	708	**Pseudoperichaeta** Brauer & Bergenstamm	1106
pruinosa (Robertson), Hyalomya	968	*Pseudopetina* Ringdahl	1015
pruinosa Wirth, Stilobezzia	134	**Pseudophaonia** Malloch	909
pruinosus (Bigot), Arthroceras	298	*Pseudophortica* Sturtevant	763
pruinosus Bigot, Tabanus	†332	pseudopiliferus Wirth & Hubert, Culicoides	130
pruinosus Coquillett, Sciapus	486	*Pseudopogonota* Malloch	838
pruinosus (Hine), Atylotus	331	**Pseudopsila** Johnson	639
pruinosus (Johnson), Longurio	18	pseudopunctipennis Theobald, Anopheles	106
pruinosus Van Duzee, Thinophilus	508	*Pseudopyrellia* Girschner	912
pruinosus (Wheeler), Hypocharassus	508	**Pseudosarcophaga** Kramer	941
pruinosus (Wirth), Ceratopogon	133	*Pseudosciara* Kieffer	235
pruni (Grossenbacher), Phytobia	798	*Pseudosciomyza* Malloch	678
prunivora (Walsh), Fannia	894	**Pseudosiphona** Townsend	1063
Psaeroptera Wahlberg	652	*Pseudostenophora* Malloch	532
Psaeropterella Hendel	649	pseudostigma Johnson, Gaurax	782
Psairoptera Wahlberg	652	**Pseudotachinomyia** Smith	1057
Psalidopteryx Townsend	1067	pseudotener (Goetghebuer), Chironomus	168
psammina Cole, Villa	440	**Pseudotephritina** Malloch	647
psammophilus (Osten Sacken), Stenotabanus	329	**Pseudotephritis** Johnson	647
Psammoppia Townsend	1025	pseudotruncorum Alexander, Tipula	33
Psectrocladius Kieffer	159	pseudotsugae Condrashoff, Contarinia	277
Psectrosciara Kieffer	240	pseudovicina Hering, Trupanea	667
PSECTROSCIARINAE	240	pseudoviridis (Malloch), Pseudochironomus	177
Psectrotanypus Kieffer	145	psi Cresson, Phthiria	420
psederae Felt, Lasioptera	272	**Psila** Meigen	639
Pselaphephila Becker	831	PSILIDAE	638
pseliophora Melander, Tachydromia	474	PSILINAE	638
pseudacaciae (Fitch), Dasineura	260	*Psilocephala* Zetterstedt	350
Pseudacicephala Malloch	830	**Psiloconopa** Zetterstedt	83
Pseudacteon Coquillett	544	psilocorsiphaga Brooks, Psalidopteryx	1067
Pseudapinops Coquillett	1066	**Psilocurus** Loew	382
Pseudatractocera Townsend	1048	*Psilodiamesa* Kieffer	151
Pseudatrichia Osten Sacken	355	*Psilomyia* Latreille	639
Pseudeuantha Townsend	1029	*Psiloneura* Aldrich	1024
Pseudeuleia Hering	669	**Psilonyx** Aldrich	363
Pseudiastata Coquillett	762	**Psilopa** Fallén	740
Pseudoarchytas Townsend	999	**Psilopelmia** Enderlein	186
Pseudobezzia Malloch	141	**Psilopiella** Van Duzee	486
Pseudobogeria Bau	1109	PSILOPINAE	734
pseudobscura Wirth, Bezzia	141	PSILOPINI	740
Pseudocalliope Malloch	703	*Psilopleura* Reinhard	1019
Pseudochaeta Coquillett	1095	PSILOPODINAE	482
Pseudochironomus Malloch	176	*Psilopodinus* Bigot	†483, 485
Pseudochirosia Ringdahl	844	*Psilopus* Meigen	†483, 485
Pseudochlorops Malloch	787	**Psilota** Meigen	593
Pseudoclista Brauer & Bergenstamm	983	**Psilotanypus** Kieffer	149
Pseudocoenosia Stein	875	**Psilozia** Enderlein	187
pseudoconfusa Philip, Chrysops	324	psittacinus (Loew), Sciapus	486
pseudoculata Cole, Thereva	354	psittacinus Meigen, Chironomus	168
Pseudoculex Dyar	111	psittacus Pettey, Sciara	230
Pseudoculicoides Malloch	126	**Psorophora** Robineau-Desvoidy	109
Pseudodacus Hendel	674	**Psychoda** Latreille	95
pseudodecora (Becker), Pholeomyia	732	PSYCHODIDAE	91
Pseudodiamesa Goetghebuer	153	PSYCHODINAE	92
pseudodiantaeus Smith, Aedes	112	PSYCHODINI	95
Pseudodidyma Townsend	1100	PSYCHODOIDEA	90
Pseudodinia Coquillett	708	psylla Loew, Acnemia	222
pseudodispar (Frey), Spilogona	882	**Ptecticus** Loew	302
pseudodissecta (Ringdahl), Pegomya	863	**Pterallastes** Loew	609
Pseudoeribea Townsend	1039	*pterelas* Harris, (Eristalis)	1115
Pseudogaurax Malloch	782	**Pteremis** Rondani	722
Pseudogermaria Brauer & Bergenstamm	1074	*Pterempis* Bezzi	457
Pseudogeron Cresson	421	**Pterobosca** Macfie	123
pseudoglobosus Curran, Microdon	598	**Pterocalla** Rondani	647
Pseudogriphoneura Hendel	703	PTEROCALLIDAE	642
Pseudohecamede Hendel	736	*Pterocanthus* Malloch	904
Pseudohippelates Malloch	774	*Pterochionea* Alexander	72
Pseudohystricia Brauer & Bergenstamm	1001	**Pterodontia** Gray	404
Pseudoleria Garrett	811	*pterodontinus* Osten Sacken, Opsebius	405
Pseudoleucopis Malloch	708	**Pterogramma** Spuler	723
Pseudolfersia Coquillett	919	**Pteromicra** Lioy	687
Pseudolimnophila Alexander	62	*pterophorina* Gerstäcker, Pyrgota	657
Pseudolimnophora Strobl	885	pteropleuralis Sabrosky, Meoneura	729
Pseudolomyia Reinhard	1089	pteropoecila (Alexander), Dactylolabis	62
Pseudolutzomyia Rapp	963	pterospilus Townes, Polypedilum	176
Pseudolynchia Bequaert	919	**Ptilodexia** Brauer & Bergenstamm	988
pseudomaxima Fisher, Mycoma	221	**Ptilomyia** Coquillett	736
pseudomelanica Sturtevant, Drosophila	765	**Ptiolina** Zetterstedt	347
Pseudomeriania Brooks	1009	ptyarion Steyskal, Dictya	689
Pseudomorinia Wulp	1026	*Ptychomyia* Brauer & Bergenstamm	1052
Pseudomyothyria Townsend	1047	**Ptychoneura** Brauer & Bergenstamm	938
Pseudonapomyza Hendel	803	**Ptychoptera** Meigen	97
Pseudonomoneura Bequaert	360	PTYCHOPTERIDAE	97
pseudonuda Brooks, Psalidopteryx	1067	PTYCHOPTERINAE	97
pseudoobscura Frolova, Drosophila	768	pubera (Loew), Atrichomelina	685

INDEX 1661

	Page
pubera Loew, Echthopoda	372
pubera Loew, Tipula	38
puberiseta Parent, Dolichopus	495
puberula (Ringdahl), Spilogona	882
pubescens (Day), Limnia	692
pubescens Day, Odontomyia	316
pubescens (Delucchi & Pschorn-Walcher), Neocnemodon	581
pubescens Loew, Chlorops	790
pubescens Loew, Chrysotoxum	582, †583
pubescens (Loew), Megagrapha	476
pubescens (Loew), Parapenium	580
pubescens (Loew), Scolicoentra	815
pubescens Melander, Platypalpus	480
pubescens Patterson (Cecidomyiaceltis)	294
pubescens Patton (Cecidomyiaceltis)	294
pubescens Stein, Fannia	†896
pubescens Walker, Scatophaga	838
pubescens Williston, Laphria	390
pubiceps (Stein), Helina	889
pubicornis (Coquillett), Sarcophaga	959
pubipennis (Osten Sacken), Limonia	50
pubipennis (Osten Sacken), Molophilus	89
pubitarsis (Zetterstedt), Trichocladius	158
pucciniae Pritchard, Clinodiplosis	282
pudens Melander, Platypalpus	480
pudibunda Osten Sacken (Cecidomyia)	294
pudica Felt, Hormomyia	288
pudica (Loew), Empis	460
pudica Meigen, Cordilura	828
pudica (Osten Sacken), Limonia	48
pudicoides Alexander, Limonia	48
pudicus Osten Sacken, Chrysops	327
pudoa Hall, Phaonia	907
pudorosa Felt, Dasineura	260
pueblensis Jaennicke, Exoprosopa	445
puella (Wiedemann), Atomosia	387
puella Williston, Pipiza	579
puellus Bromley, Psilocurus	382
puellus Williston, Lordotus	412
puer Alexander, Gonomyia	75
puerinus Melander. Platypalpus	480
pugetense (Dyar & Shannon), Simulium	186
pugetensis Alexander, Ormosia	87
pugeti Cole & Wilcox, Lasiopogon	495
pugil Loew, Dolichopus	378
pugil Wheeler, Sympycnus	528
pugilis (Alexander), Hesperoconopa	78
pugionis Felt (Itonida)	290
pulchella Bromley (Bombomima)	1115
pulchella Cresson, Euxesta	652
pulchella (Johnson), Agathomyia	548
pulchella (Macquart), Labostigmina	313
pulchella (Meigen), Ditrichophora	739
pulchella Williston, Chrysogaster	591
pulchella Williston, Cophura	367
pulchella (Williston), Parapenium	580
pulchellus (Banks), Piophila	712
pulchellus (Kröber), Physoconops	627
pulchellus Loew, Bombylius	409
pulcher Back, Cyrtopogon	369
pulcher (Back), Diogmites	372
pulcher (Johannsen), Leptomorphus	223
pulcher (Johannsen), Procladius	149
pulcher Melander, Oligodranes	422
pulcher Painter, Parabombylius	410
pulcher Walker, Dolichopus	495
pulcherrima Back, Townsendia	385
pulcherrima Patterson & Mainland, Leucophenga	762
pulcherrimus Williston, Lordotus	412
pulchra Curran, Morellia	912
pulchra (Huckett), Spilogona	882
pulchra Loew, Diastata	772
pulchra Loew, Pachygaster	318
pulchra Loew, Rhamphomyia	463
pulchra Pritchard, Mallophorina	397
pulchra (Snow), Protoclythia	549
pulchra Sturtevant & Wheeler, Scatophila	758
pulchra (Williston), Caliprobola	608
pulchra (Williston), Nostima	745
pulchra Williston, Platyura	204
pulchrifrons Loew, Notiphila	747
pulchrimanus (Bigot), Dolichopus	495
pulchripennis (Coquillett), Stenochironomus	177
pulchripennis (Lundbeck), Ablabesmyia	148

	Page
pulchripes Loew, Cheilosia	586
pulchripes (Loew), Neodiplotoxa	793
pulchripes Loew, Psilopa	741
pulchrissimus Williston, Lordotus	412
pulicans (Linnaeus), Culicoides	128
pulicaria (Fallén), Megaselia	539
pulicaria Loew, Epicypta	209
pulicaria Loew, Scatopse	†238
pulicaria Meigen, (Sciara)	236
pulicaria (Zetterstedt), Coenosia	872
pulicaris (Linnaeus), Culicoides	128
Puliciphora Dahl	545
pulla (Adams), Thaumatomyia	788
pulla Aldrich, Sarcophaga	959
pulla (Brues), Megaselia	541
pulla Coquillett, Mauromyia	1014
pulla Cresson, Hydrellia	744
pulla Loew, Rhamphomyia	465
pulla Melander, Mythicomyia	418
pulla Snow, Platypeza	549
pulla (Stein), Xenomydaea	892
pullata (Coquillett), Dziedzickia	218
pullata (Coquillett), Orfelia	203
pullata Coquillett, Symphoromyia	344
pullata (Coquillett), Villa	436
pullata (Melander), Heleodromia	467
pullata Melander, Trichina	449
pullata (Wirth), Mallochohelea	139
pullatus (Coquillett), Aedes	113
pullatus Foote, Procecidocharoides	661
pullatus Melander, Oligodranes	422
pullicornis Sabrosky, Oscinella	779
pullipes (Coquillett), Thaumatomyia	788
pullula (Snow), Melangyna	567
pullula (Townsend), Cyzenis	1100
pullula (Wulp), Cyzenis	†1100
pullula (Zetterstedt), Hylemya	855
pullus (Wiedemann), Amphicnephes	655
pulmitarsis Fallén, Dolichopus	494
pulsator Harris, (Seioptera)	1115
pulvera (Harbeck), Anthracophaga	789
pulverea (Coquillett), Alophorella	968
pulverea (Coquillett), Bezzia	141
pulverea (Coquillett), Epidexia	1032
pulverea (Coquillett), Muscopteryx	1025
pulvereus Melander, Epacmus	428
pulverulentus Melander, Platypalpus	481
pulvicrura (Huckett), Spilogona	882
pulvillata (Stein), Phaonia	907
pulvillus Van Duzee, Sympycnus	528
pulvinariae Malloch, Leucopis	709
pulvinata (Grimshaw), Atherigona	875
pumila Alexander, Pedicia	55
pumila (Fallén), Coenosia	872
pumila Melander, Glabellula	415
pumila Robineau-Desvoidy, (Peleteria)	1108
pumilio (Holmgren), Limnophyes	160
pumilio Loew, Paracleius	501
Pumilio Schembi	596
pumilus Coquillett, Stenopogon	384
pumilus Macquart, Eristalis	625
pumilus Macquart, Tabanus	335
punctata Cole, Sphegina	594
punctata Coquillett, Acrometopia	707
punctata (Coquillett), Ormia	980
punctata Coquillett, Stictomyia	654
punctata Meigen, Mycetophila	211
punctata Robineau-Desvoidy, Dinera	986
punctata (Robineau-Desvoidy), Helina	888
punctata Robineau-Desvoidy, Ormia	980
punctata (Stein), Quadrularia	890
punctata Walker, (Sciara)	236
punctatus (Fabricius), Sylvicola	†190
punctatus Wirth, Glutops	298
puncticeps (Coquillett), Poecilominettia	703
puncticollis (Dietz), Tipula	25
puncticollis (Zetterstedt), Epichlorops	791
puncticornis Macquart, Tipula	41
punctifer Becker, Meromyza	793
punctifer Bigot, Sargus	302
punctifer (Malloch), Poecilominettia	703
punctifer Osten Sacken, Tabanus	335
punctifera Malloch, Eupachygaster	318
punctifera (Walker), Peleteria	996
punctifrons (Becker), Olcella	784

	Page		Page
punctifrons Curran, Salpingogaster	574	pusio Loew, Rhamphomyia	465
punctifrons (Hendel), Tetanops	649	pusio Osten Sacken, Dioctria	370
punctigera Coquillett, Paroedopa	653	pusio Osten Sacken, Pantarbes	414
punctigera (Townsend), Phoranthella	970	pusio (Schiner), Palaeosepsis	682
punctipennis Becker, Fucellia	843	pusio (Wiedemann), Fannia	896
punctipennis Coquillett, Metapogon	378	pusiola (Wulp), Ravinia	955
punctipennis Coquillett, Trigonometopus	705	pustulata (Felt), Asteromyia	275
punctipennis Curran, Brachyopa	593	pustulata Felt, Cincticornia	269
punctipennis (Macquart), Hybomitra	339	pustuloides Beutenmüller, (Cecidomyia)	294
punctipennis (Meigen), Erioptera	†81	puta Laffoon, Sciophila	226
punctipennis Meigen, Tanypus	150	putilla (Reinhard), Blaesoxipha	947
punctipennis (Melander), Eucyrtopogon	373	*putricola* Cole, (Euxesta)	1115
punctipennis Melander, Nicocles	380	putrida Felt, (Itonida)	290
punctipennis Melander, Spilochroa	817	putrida Sturtevant, Drosophila	765
punctipennis (Say), Anopheles	106	putris (Linnaeus), Themira	684
punctipennis (Say), Chaoborus	103	**Pycnoglossa** Coquillett	845
punctipennis (Say), Pelastoneurus	502	**Pycnopogon** Loew	382
punctipennis (Say), Rhagio	345	pygmaea (Alexander), Ormosia	87
punctipennis Walker, Phthria	420	pygmaea (Becker), Olcella	784
punctipennis (Wiedemann), Paralimna	748	*pygmaea* Cole, Thereva	353
punctipennis (Williston), Bezzia	141	pygmaea (Haliday), Atissa	735
punctipennis Wulp, Syneches	449	*pygmaea* Hull, Solva	299
punctipes (Meigen), Chaetosa	837	pygmaea (Loew), Swammerdamella	240
punctipes (Wiedemann), Phaenopsectra	174	pygmaea Shannon, Tropidia	609
punctipes Wirth, Stilobezzia	134	pygmaea (Theobald), Psorophora	110
punctistigma Benjamin, Neaspilota	672	pygmaea Winnertz, Heteropeza	247
punctitarse Curran, Rhaphium	514	*pygmaea* (Zetterstedt), Coenosia	†872
punctodes Dyar, Aedes	114	*pygmaeum* Williston, Zodion	629
punctor (Kirby), Aedes	114	pygmaeus Borgmeier, Trophodeinus	546
punctulata Becker, Elachiptera	777	pygmaeus Fabricius, Bombylius	409
punctulata Hunter, Cheilosia	585	pygmaeus Hull, Psilocurus	382
punctulata Macquart, Haematopota	330	pygmaeus (Williston), Microtabanus	329
punctulata Williston, Nausigaster	596	pygochroa Melander, Trixoscelis	818
punctulatus Enderlein, Odontoloxozus	637	pygophora Coquillett, Coelosia	218
punctum (Becker), Thaumatomyia	788	*Pygostolus* Loew	380
punctum (Fabricius), Sepsis	683	*pylone* (Walker), Quadrularia	889
punctum (Loew), Nephrotoma	22	pyralidis (Coquillett), Metopiops	1095
punctum (Stannius), Platurocypta	215	*pyramis* Doane, Tipula	38
punctus Garrett, Boletina	217	pyrastri (Linnaeus), Scaeva	562
pungens (Townes), Chironomus	166	**Pyraustomyia** Townsend	1011
pungens Wiedemann, Culex	119	**Pyrellia** Robineau-Desvoidy	912
punicea Martin, Beameromyia	362	**Pyrgota** Wiedemann	657
punicei Brodie, (Diplosis)	294	**Pyrgotella** Curran	658
punifer Bigot, Sargus	302	PYRGOTIDAE	657
pura Alexander, Tipula	26	pyri (Bouché), Dasineura	260
pura (Loew), Tephritis	668	**Pyritis** Hunter	604
purivora Clark, Cuterebra	1109	pyrivora (Riley), Contarinia	277
purpurascens Townsend, Hyalomya	969	*Pyropa* Illiger	837
purpurata Bequaert, Lasia	404	**Pyrophaena** Schiner	578
purpuratum Cole, Notogramma	646	*Pyrrempis* Melander	458
purpuratus (Van Duzee), Gymnopternus	499	pyrrhina Bigot, Sphaerophoria	570
purpurea Felt, Dasineura	260	pyrrhocephala (Loew), Otites	646
purpurea Townsend, Lucilia	927	*pyrrhosoma* Melander & Spuler, Sepsis	684
purpurea Walker, Xylota	606	*pyrrhostoma* Bigot, Odontomyia	316
purpureipes Aitken, Aedes	114	pyste (Walker), Nemorilla	1089
purpureus (Walker), Euhybus	448		
purpurops Steyskal, Traginops	794	q-oruca Walsh, (Cecidomyia)	294
purus Curran, Hydrophorus	506	quadradentis Hardy, Tomosvaryella	556
purus Curran, Saropogon	383	*quadrangula* Sherman, (Allodia)	1115
pusilla Aldrich, Cylindromyia	973	quadrata (Fallén), Limnellia	758
pusilla (Coquillett), Opsomeigenia	1045	quadrata Johannsen, Exechia	207
pusilla (Fallén), Sphaerocera	719	quadrata (Osten Sacken), Pilaria	68
pusilla (Huckett), Spilogona	882	quadrata (Say), Tropidia	609
pusilla Johnson, Hilarimorpha	348	quadrata Steyskal, Pherbellia	686
pusilla Loew, Sepedon	693	quadrata (Sturtevant), Microdrosophila	771
pusilla Loew, Tachydromia	474	quadrata Williston, Volucella	602
pusilla Macquart, Atomosia	387	quadratula (Loew), Sciophila	226
pusilla Macquart, Volucella	602, †602	quadratus (Say), Chrysopilus	347
pusilla (Meigen), Megaselia	541	quadratus (Say), Platycheirus	577
pusilla Meigen, Piophila	711, †712	*quadratus* (Stains & Knowlton), Simulium	185
pusilla Melander, Mythicomyia	418	quadratus Van Duzee, Diaphorus	520
pusilla (Reinhard), Aporotachina	1039	*quadratus* (Van Duzee), Thinophilus	508
pusilla (Speiser), Microlynchia	919	quadratus (Wiedemann), Promachus	400
pusilla Tonnoir, Psychoda	96	quadriboscis Lovett, Criorhina	612
pusilla (Walker), Swammerdamella	240	*quadricincta* Van Duzee, Pelastoneurus	502
pusilla (Zetterstedt), Themira	684	quadricornis Melander, Neophyllomyza	730
pusillans Huckett, Hylemya	845	*quadridentata* (Walker), Actina	300
pusillus Loew, Gymnopternus	499	quadrifasciata Curran, Chaetopsis	650
pusillus (Loew), Procladius	149	*quadrifasciata* Harris, (Trypeta)	1116
pusillus Loew, Syneches	449	quadrifasciata (Macquart), Rivellia	656
pusillus Lutz, Culicoides	128	quadrifasciata Say, Ptychoptera	97
pusillus Macquart, Eristalis	624	quadrifasciata (Say), Spilomyia	613
pusillus Wiedemann, Dilophus	196	*quadrigemina* Loew, Stratiomys	312
pusio Loew, Chrysotimus	529	quadrilamellatus Loew, Dolichopus	495
pusio Loew, Euxesta	652	quadrilinea Walker, (Loxocera)	641
pusio Loew, Hippelates	775	quadrilineata (Bigot), Chrysochlorina	305

INDEX 1663

	Page		Page
quadrilineata (Loew), Lyciella	701	quinquelineata (Adams), Olcella	785
quadrilineata Melander, Renocera	692	quinquelineata Ringdahl, Hylemya	856
quadrimaculata (Ashmead), Baccha	573	quinquelineata (Say), Rhamphomyia	465
quadrimaculata Loew, Xylota	606	quinquelineata (Zetterstedt), Spilogona	883
quadrimaculata (Macquart), Paraporia	1028	*quinquelineatus* Macquart, Tabanus	337
quadrimaculata (Panzer), Pipiza	579	*quinquemaculatus* Hine, Tabanus	334
quadrimaculata (Walker), Leucophenga	761	quinquenotata Cresson, Nostima	745
quadrimaculatus Becker, Chlorops	790	quinquenotatus (Johnson), Oligodranes	422
quadrimaculatus Cresson, Euparyphus	308	quinquepunctata (Say), Suillia	809
quadrimaculatus Say, Anopheles	106	quinquepunctatus Loew, Chlorops	790
quadrinotata (Bigot), Neoascia	595	quinquevittatus Wiedemann, Tabanus	335
quadrinotata Cresson, Scatella	757	quintana Cole, Tetragoneura	228
quadrinotatus Aldrich, Paracleius	501	quinteri Townsend, Torynotachina	1050
quadrinotatus Bigot, (Dasypogon)	401	*quintilis* Malloch, Hylemya	856
quadrinotatus Johnson, Nemotelus	309	*quintius* (Walker), Mesograpta	571
quadriplagiatus (Harris), Argyra	524	quirinus Philip, Tabanus	336
quadripunctata Harris, (Stratiomys)	1116	quivera Painter, Villa	438
quadripunctata (Malloch), Megaselia	541		
quadripunctatus (Banks), Telmatoscopus	94	ra Harriot, Tritoxa	649
quadripunctatus Malloch, Chironomus	170	*Rabdophaga* Westwood	256
quadripustulata (Fabricius), Winthemia	1090	rabelloi Lane, Stilobezzia	134
quadrisetosa (Becker), Lamproscatella	756	*rabida* (Walker), Ravinia	955
quadrisetosa (Coquillett), Ravinia	954	rabiosa Alexander, Tipula	38
quadrisetosa (Malloch), Liriomyza	802	rabunensis (Dodge), Sarcophaga	959
quadrisetosa Malloch, Pogonomyia	902	*Rabunmyia* Dodge	941
quadrisetosa (Thomson), Myospila	890	raca Garrett, Anorostoma	812
quadrisetosa Thomson, Notiphila	747	raca Garrett, Bolitophila	198
quadrisetosum (Thomson), Calliopum	696	racemi Felt, Rhabdophaga	257
quadrisetosus Kessel, Trichobius	921	racemi Stebbins, (Cecidomyia)	294
quadrisetosus (Malloch), Dasiops	716	*racemicola* Felt, Rhopalomyia	266
quadrisignata Coquillett, Myospila	890	racemicola (Osten Sacken), Rhopalomyia	266
quadrispinosa Malloch, Pegomyia	861	**Rachicerus** Walker	296
quadrispinosa (Pettey), Bradysia	234	rachiphaga Tripp, Dasineura	260
quadrispinosus Alexander, Molophilus	89	**Rachispoda** Lioy	721
quadrituberculata Loew, Parydra	749	*Racodineura* Rondani	1072
quadrivittata Alexander, Dicranoptycha	52	*racua* Osten Sacken, Brachyopa	593
quadrivittata (Macquart), Euxesta	651	*radaca* (Walker), Baccha	573
quadrivittata Meigen, Hilara	456	radialis Melander, Neoplasta	472
quadrivittata (Sabrosky), Olcella	784	radialis (Shaw), Bradysia	234
quadrivittatum Malloch, Aphaniosoma	822	radians (Macquart), Condylostylus	485
quadrivittatus Jones, Lasiopogon	378	*radiata* Loew, Trichopoda	966
quadrivittatus (Say), Silvius	323	*radiatae* (Snow & Mills), Thecodiplosis	278
Quadrula Pandellé	889	radicis Felt, Mycodiplosis	283
Quadrularia Huckett	889	radicis (Townsend), Lydella	1103
quaesitus Stone, Tabanus	335	*radicum* (Fabricius), Mericia	1008
quagga (Townes), Stictochironomus	178	radicum (Linnaeus), Hylemya	855
qualis Say, (Cordylura)	1113	*radicum* (Walsh & Riley), Heringia	580
quartomollis Schmitz, Gymnophora	†542	radifera (Coquillett), Trupanea	667
quatei Cook, Rhexoza	240	radifolii Felt, Dasineura	260
quatei Wirth, Forcipomyia	126	radiosum Shannon, Chrysotoxum	583
quatei Wirth, Thiomyia	755	**Rafaelia** Townsend	955
quaternaria Loew, Euxesta	652	**Ragas** Walker	455
quaternaria Loew, Stratiomys	†312, 312	rainierensis Alexander, Molophilus	89
quatuornotata Loew, Mycetophila	214	rainieri Huckett, Hylemya	851
quaylei Dyar & Knab, Aedes	112	rainieri Shannon, Xylota	606
quaylii Doane, Tipula	38	rainieri Wilcox & Martin, Coleomyia	366
quebecense Twinn, Simulium	186	rainieri Wilcox & Martin, Cyrtopogon	369
querceti (Bouché), Dendrophaonia	905	rainieria Alexander, Erioptera	83
quercicola (Kieffer), Henria	252	rainieria (Alexander), Pedicia	55
querciflorae Felt, Lasioptera	272	**Rainieria** Rondani	636
quercifolia Felt, Cincticornia	269	rainiericola Alexander, Tipula	38
quercifolia (Felt), Thecodiplosis	278	*raleighi* Macfie, Forcipomyia	125
quercina Felt, (Cecidomyia)	290	ramaleyi James, Scenopinus	356
quercina Felt, Dasineura	260	ramifer Loew, Dolichopus	495
quercina Felt, Lasioptera	272	ramona Alexander, Tipula	38
quercina (Felt), Lobodiplosis	281	ramosa (Reinhard), Oxysarcodexia	953
quercina (Felt), Peridiplosis	281	*ramosus* Van Duzee, Pelastoneurus	501
quercina (Felt), Peridiplosis	281	ramuli Felt, (Cecidomyia)	290
quercina Felt, Porricondyla	256	ramuscula (Beutenmüller), Neolasioptera	273
quercina Felt, Youngomyia	279	ramuscula Felt, Rhabdophaga	257
querciperda Felt, Lasioptera	272	*rangiferina* Alexander, Tipula	37
quercirami Felt, Lasioptera	272	ranunculi (Holmgren), Spilogona	880
quercivora (Felt), Henria	252	*ranunculi* (Schrank), Phytomyza	805
quercus (Felt), Henria	252	rapae (Smith), Obolocera	1104
quercus majulis Osten Sacken, (Cecidomyia)	294	*rapax* Loew, Ochthera	753
quercus pilulae (Walsh), Cincticornia	269	rapax Loew, Tachypeza	473
queres Alexander, Tipula	36	rapax (Osten Sacken), Efferia	395
querula Alexander, Dicranota	58	rapax Osten Sacken, Laphria	390
querula (Walker), Ravinia	955	rapax (Walker), Helicobia	949
querulus Osten Sacken, Liancalus	507	*raphani* (Harris), Hylemya	847
quieta Stein, Phaonia	907	rara Johnson, Haematopota	330
quinaria Loew, Drosophila	766	rara (Osten Sacken), Limonia	50
quinquecellula Alexander, Ufomorpha	68	rara Spuler, Leptocera	724
quinquefasciatus Say, Culex	118	*rarotongensis* Satchell, Psychoda	96
quinquelimbata (Bigot), Melangyna	567	rasa (Loew), Stonemyia	321
		rasilis Reinhard, Spathidexia	1021

Name	Page
rastrifera Melander, Chelifera	471
rastristyla Alexander, Tipula	26
ratibidae Felt, Asphondylia	269
rattus Osten Sacken, Aphoebantus	430
rattus Osten Sacken, Cyrtopogon	369
rauterbergi (Wheeler), Chrysotus	523
rava Coquillett, Efferia	395
rava Foote, Neotephritis	669
rava Loew, Rhamphomyia	465
rava (Wulp), Hesperodinera	†987
ravida Coquillett, Empis	460
ravida Coquillett, Rhamphomyia	465
ravida Reinhard, Nimioglossa	987
ravidus Coquillett, Microphorus	453
ravinia Parker, Sarcofahrtia	942
Ravinia Robineau-Desvoidy	953
ravus Loew, Bombylius	409
ravus Melander, Microphorus	453
Reaumuria Robineau-Desvoidy	1075
recedens (Walker), Hybomitra	339
recedens (Walker), Parapenium	580
recedens Walker, Tabanus	334
recens Johannsen, Cordyla	208
recens Walker, Dixa	101
recens Wheeler, Drosophila	766
reces Harris, (Anthomyia)	1116
reciproca Walker, Empis	460
reclinata (Aldrich & Webber), Carcelia	1092
reclusum Banks, Zodion	629
recondita Osten Sacken, (Asphondylia)	295
recondita (Osten Sacken), Pilaria	68
recta (Johannsen), Mycetophila	213
recticosta Aldrich, Dolichopus	495
rectifrons Huckett, Pegomya	863
rectilineatus Schmitz, Stichillus	536
rectinervis Becker, Chlorops	790
rectinervis Bigot, (Somomyia)	928
rectispina (Alexander), Cheilotrichia	79
rectoides (Curran), Melangyna	567
rectus Osten Sacken, Syrphus	559
recula (Laffoon), Mycetophila	213
recurva (Adams), Diplotoxa	792
recurva Alexander, Erioptera	84
recurva Garrett, Bolitophila	198
recurva Johannsen, Mycomya	221
recurva Johannsen, Paratinia	224
recurva Malloch, Hylemya	851
recurva (Thomson), Morellia	912
recurvata Felt, Aphidoletes	280
recurvata Felt, (Cecidomyia)	290
recurvata Goetghebuer, Micropsectra	179
recurvata Melander, Thanatella	472
recurvatus Chillcott, Roederiodes	468
recurvus (Coquillett), Heterostylum	411
recurvus Melander, Platypalpus	481
redeuns Walker, Chironomus	165
rediviva (Walker), Ravinia	954
redlandae Cole, Lestomyia	378
reducta (Alexander), Dicranota	58
reducta Cook, Swammerdamella	240
reducta Felt, Asteromyia	275
reducta Felt, Mycodiplosis	283
reeksi Vockeroth, Cecidomyia	287
reesi Alexander, Tipula	34
reevesi Bohart, Culex	118
reevesi Wirth, Culex	118
reevesi Wirth, Culicoides	130
referens Walker, Sepsis	683
referta Alexander, Tipula	37
reflecta (Huckett), Spilogona	883
reflectens Walker, Xylophagus	297
reflectipennis (Curran), Dasysyrphus	564
reflectus Aldrich, Dolichopus	495
reflexa Alexander, Gonomyia	76
reflexa Felt, (Itonida)	290
reflexa Robineau-Desvoidy, Myopina	843
reflexus (Walker), Chrysopilus	347
refulgens Harmston & Knowlton, Dolichopus	494
refusa (Giglio-Tos) Helina	888
regalis Curran, Peleteria	997
Regasilus Curran	400
regelationis (Linnaeus), Trichocera	16
regilla Reinhard, (Phorocera)	1107
regina (Meigen), Phormia	924
regina West, Arctophyto	986
reginae Felt, (Itonida)	290
regularis Curran, Omomyia	714
Reichertella Enderlein	237
reicherti Fairchild, Chrysops	325
reinhardi (Aldrich & Webber), Patelloa	1104
reinhardi Arnaud, Spathidexia	1021
reinhardi Bromley, Nicocles	380
reinhardi Bromley, Psilocurus	382
reinhardi Brooks, Gonia	1076
reinhardi Hull, Polybiomyia	616
reinhardi James, Hermetia	305
reinhardi Sabrosky, Euphasiopteryx	980
reinhardi Sabrosky & Arnaud, Atactopsis	1081
Reinhardiana Arnaud	985
reinhardii (Hall), Ravinia	954
reinwardtii Wiedemann, Tabanus	336
reipublicae Walker, Pipunculus	554
rejecta Aldrich, Fucellia	843
rejecta (Johannsen), Dziedzickia	218
rejectus Osten Sacken, Cyrtopogon	369
rejectus Williston, Heteropogon	374
relata Stein, Hylemya	847
relictus (Curran & Fluke), Helophilus	619
reliquens Huckett, Hylemya	851
relima Wheeler, Drosophila	766
remingtoni Alexander, Limonia	49
remipes Wahlberg, Dolichopus	495
remotus (Curran), Melangyna	566
remotus Walker, Dolichopus	495
rempeli Vockeroth, Aedes	114
rempelii Thienemann, Chironomus	165
remulus Van Duzee, Diaphorus	520
remus Van Duzee, Dolichopus	495
remuscula (Beutenmüller), Neolasioptera	273
renidescens Melander & Brues, Dolichopus	495
reniformis Hull, Neopachygaster	317
reniformis Stebbins, (Cecidomyia)	295
Renocera Hendel	692
repanda Johannsen, Exechia	207
repandus Van Duzee, Diaphorus	520
reperta (Reinhard), Blaesoxipha	944
repleta Huckett, Hylemya	851
repleta (Walker), Diastata	772
repleta Wollaston, Drosophila	766
replicata Huckett, Hylemya	856
reptans Fallén, Agromyza	795
repulsa Alexander, Tipula	25
reses Giglio-Tos, Scatophaga	839
reside Reinhard, Charasoma	1040
resinicola (Osten Sacken), Cecidomyia	287
resinicoloides Williams, Cecidomyia	287
resinosa (Wiedemann), Camptoprosopella	697
resplendens Loew, Dioctria	371
resplendens Wahlberg, Mesembrina	910
restorata Huckett, Hylemya	856
restricta Alexander, Erioptera	83
restuans Theobald, Culex	119
resurgens Walker, Tipula	33
retardata (Malloch), Megaselia	541
reticulata (Alexander), Limonia	50
reticulata Harris, Tetanocera	695
reticulata (Johnson), Acrometopia	707
reticulata Snow, Eurosta	663
reticulatum Cook, Rhegmoclema	239
reticulus Van Duzee, Dolichopus	495
retinens Alexander, Tipula	25
retinens Van Duzee, Dolichopus	495
retiniae (Coquillett), Urophyllopsis	1050
Retinodiplosis Kieffer	287
retorta Wulp, (Tipula)	90
retracta Aldrich, Uclesia	1020
retracta Reinhard, Paradidyma	1033
retrorsa Alexander, Cryptolabis	78
retrorsus Melander, Oligodranes	422
retroversus Chillcott, Roederiodes	468
retusa Doane, Tipula	38
reverberata (Malloch), Liriomyza	802
reversa (Aldrich), Blaesoxipha	944
reversus Walker, Hybos	447
rex Bromley, Mallophora	397
rex (Curran & Fluke), Helophilus	619
rex Hauber, Pseudochironomus	177
rex (Osten Sacken), Villa	438
reynoldsi (Metz), Bradysia	234
Rhabdomastix Skuse	76

INDEX

	Page		Page
Rhabdophaga Westend	256	riggsi Khalaf, Culicoides	130
Rhabdopselaphus Bigot	421	rigidae (Osten Sacken), Mayetiola	263
Rhachicerus Walker	296	rigidae Packard, (Diplosis)	279
Rhachoepalpus Townsend	1003	*rigidirostris* (Wulp), Catalinovoria	1017
Rhachogaster Townsend	995	*Rileya* Brauer & Bergenstamm	1082
Rhacodineura Rondani	1072	*rileyana* (Enderlein), Simulium	189
rhadamanthus Loew, Ospriocerus	381	rileyana Felt, Rhabdophaga	257
rhaeba Melander, Mythicomyia	418	*Rileyella* Townsend	1101
Rhaeboza Enderlein	237	rileyi (Aldrich), Blaesoxipha	945
Rhagio Fabricius	344	rileyi Coquillett, Mythicomyia	†417, 419
RHAGIONIDAE	342	*rileyi* (Williston), Eristalis	624
RHAGIONINAE	342	rileyi (Williston), Lespesia	1102
Rhagoletis Loew	674	*rileyi* (Williston), Xylota	607
Rhagoletoides Foote	675	*Rileymyia* Townsend	1082
Rhagoletotrypeta Aczél	675	rindgei Reinhard, Chaetophlepsis	1016
Rhamphidia Meigen	51	*ringdahl* Drew, Hylemya	853
Rhamphinina Bigot	†988	ringdahli Drew, Hylemya	853
Rhamphomyia Meigen	461	ringdahli Stein, Hydrotaea	900
Rhaphidolabina Alexander	56	rioensis Patterson, Drosophila	765
Rhaphidolabis Osten Sacken	58	riparella (Hendel), Melanagromyza	796
RHAPHIINAE	512	riparia Fallén, Ephydra	754
Rhaphiomidas Osten Sacken	357	riparia Felt, Lasioptera	272
Rhaphium Meigen	512	*riparia* (Malloch), Melanagromyza	796
rhathyme Dyar & Shannon, Dixa	101	riparia Meigen, Notiphila	747
rhea Osten Sacken, Exoprosopa	445	riparius Dyar & Knab, Aedes	114
Rhedia Robineau-Desvoidy	1075	riparius (Malloch), Procladius	149
Rhegmoclema Enderlein	238	riparius Meigen, Chironomus	166
RHEGMOCLEMATINI	238	ripicola Melander, Lasiopogon	378
Rhegmoclemina Enderlein	239	*rita* Curran, Cheilosia	586, 586
Rheotanytarsus Bause	180	rita (Curran), Neocnemodon	581
Rhexoza Enderlein	240	ritae Patterson & Wheeler, Drosophila	766
Rhicnoessa Loew	727	*rivalis* Melander, Dioctria	371
rhicnoptiloides (Alexander), Dactylolabis	62	**Rivellia** Robineau-Desvoidy	655
Rhingia Scopoli	588	rivertonensis (Johnson), Longurio	18
RHINIINAE	923	rivinae Felt, Schizomyia	267
Rhinogastrophilus Townsend	916	rivosa (Meigen), Minettia	702
Rhinoleucophenga Hendel	762	rivularis Osten Sacken, Dicranota	57
RHINOPHORINAE	963	rivuli Twinn, Simulium	186
rhinoprosopa Hull & Fluke, Cheilosia	585	*rivulorum* Kieffer, Orthocladius	156
Rhinotora Schiner	819	rixosa Reinhard, Euphyto	935
RHINOTORIDAE	818	robertsoni (Coquillett), Pholeomyia	732
Rhipidia Meigen	45	robertsoni Curran, Dolichopus	488
rhipidioides (Alexander), Limonia	47	robertsoni Sabrosky, Hippelates	776
Rhizomyia Kieffer	262	robertsonii Coquillett, Brachystoma	447
rhizophorum Stone & Jamnback, Prosimulium	183	*robertsonii* (Townsend), Alophorella	967
Rhodesiella Adams	781	robertsonii Townsend, Euryceromyia	1066
rhododendri Felt, Giardomyia	284	robertsonii Townsend, Siphoclytia	1036
Rhodogyne Meigen	965	**Robertsonomyia** Malloch	629
rhodoides (Osten Sacken), Rhabdophaga	257	robiginosa Melander, Antichaeta	688
rhodophaga (Coquillett), Dasineura	260	robiginosa Melander, Mythicomyia	419
rhoeo (Walker), Drino	1085	*robii* (Jones), Eumerus	597
rhoina (Felt), Cincticornia	269	robiniae (Haldeman), Obolodiplosis	286
rhoinus Felt, Arthrocnodax	283	robinsoni (Townsend), Nowickia	996
rhois (Coquillett), Dasineura	260	robur Garrett, Tetragoneura	228
rhombica (Osten Sacken), Hybomitra	340	robusta (Aldrich), Blaesoxipha	946
RHOPALOMERIDAE	679	robusta Bromley, Proctacanthella	399
Rhopalomyia Rübsaamen	265	robusta Brooks, Dichocera	1013
Rhopalopternum Duda	779	robusta Brooks, Gonia	1076
Rhopalopterum Duda	779	robusta (Brooks), Hyalomya	969
Rhopalosyrphus Giglio-Tos	599	rubusta Chillcott, Fannia	896
Rhymosia Winnertz	207	robusta Coquillett, Chyliza	640
Rhynchanthrax Painter	438	robusta Coquillett, Myiophasia	984
Rhynchodexia Bigot	†988	robusta Cresson, Rhysophora	743
Rhynchoderia Bigot	†988	robusta (Curran), Blera	611
Rhynchomilichia Hendel	732	robusta (Curran), Ptilodexia	989
Rhynchopeteina Townsend	1015	robusta (Curran), Scatophaga	839
Rhynchosciara Rübsaamen	231	robusta Curran, Sphaerophoria	570
Rhyncophoromyia Malloch	543	robusta Felt, Didactylomyia	254
Rhynencina Johnson	659	robusta Felt, Mycodiplosis	283
RHYPHIDAE	190	robusta Garrett, Pseudoleria	812
Rhypholophus Kolenati	84	robusta Hine, Cheilosia	585
Rhyphus Latreille	190	robusta Johnson, Argyra	525
Rhysophora Cresson	743	robusta Johnson, Phyllogaster	876
Rhytidops Lindner	679	robusta Loew, Tetanocera	694
ribesii (Linnaeus), Syrphus	559	robusta (Reinhard), Pseudochaeta	1096
ribicola Doane, Rhagoletis	674	robusta Shannon, Chrysogaster	591
RICHARDIIDAE	641	robusta Stein, Hylemya	852
richardsi Chillcott, Platypalpus	481	robusta Sturtevant, Drosophila	766
richardsi Sabrosky, Leptocera	721	*robusta* Townsend, Tachinomyia	1058
richardsoni Malloch, Pseudochironomus	177	robusta Walker, (Sciara)	236
richardsoni Wilcox & Martin, Nannocyrtopogon	380	*robusta* Walker, Stratiomys	312
ricinorum Townsend, Gymnocarcelia	1093	*robusta* (Wiedemann), Peletaria	†998
ridingsi Cresson, Pogonosoma	391	robustum Curran, Raphium	514
riederi Reinhard, Cacozelus	1012	robustum (Osten Sacken), Heterostylum	411
		robustus Brennan, Chrysops	324
		robustus Cresson, Geron	423

	Page		Page
robustus Johnson & Johnson, Poecilanthrax	442	rubella (Osten Sacken), Ormosia	87
robustus Melander, Microphorus	453	rubens Coquillett, Ablautus	364
robustus Telford, Microdon	598	rubens Coquillett, Eucessia	431
robustus Townsend, Hyalomyodes	977	rubentis (Coquillett), Eucelatoria	1043
robustus (Van Duzee), Gymnopternus	499	rubentis (Reinhard), Gaediopsis	1080
rodecki James, Proctacanthus	399	rubescens (Alexander), Dicranota	58
Roederella Engel	469	rubescens Alexander, Teucholabis	76
roederi (Williston), Euchaetogyne	990	rubi Brischke, Agromyza	795
Roederia Mik	469	rubi (Felt), Schizomyia	267
Roederiodes Coquillett	467	rubia Alexander, Erioptera	82
Roederioides Coquillett	467	*rubicunda* Malloch, Scatophaga	838
roena Roback, Smittia	162	*rubicundulus* (VanDuzee), Physoconops	627
Roeselia Robineau-Desvoidy	1071	rubicundus Adams, Chlorops	790
Roeseliopsis Townsend	1070	rubicundus Hine, Asilus	392
rogalis Reinhard, Pseudapinops	1066	rubicundus Melander, Paracosmus	427
rogatoris Coquillett, Hemerodromia	471	rubida Alexander, Limnophila	67
rogersella Alexander, Ulomorpha	68	rubida (Bigot), Myopa	630
rogersi Alexander, Dicranoptycha	52	rubida Coquillett, Dioctria	371
rogersi Alexander, Dicranota	57	rubida (Coquillett), Thaumatomyia	788
rogersi (Alexander), Dolichopeza	23	rubida Curran, Euxesta	652
rogersi (Beck & Beck), Chironomus	171	rubida Felt, Dicrodiplosis	279
rogersiana (Alexander), Dicranota	58	rubida Felt, Winnertzia	253
rogersiana (Alexander), Limonia	50	rubida Felt, Youngomyia	279
rogersiana Alexander, Polymera	60	rubida (Loew), Acrosticta	650
rohweri Allen, Eumacronychia	935	rubida Martin, Coleomyia	366
rohweri Doane, Tipula	33	*rubida* Van Duzee, Physocephala	628
rohweri (Townsend), Periscepsia	1034	*rubidapex* Harris, (Tachina)	1116
romanovichiana Alexander, Ormosia	87	rubidiapex (Reinhard), Steveniopsis	985
Rondania Jaennicke	304	rubidum (Coquillett), Chetostoma	678
Rondaniella Johannsen	228	rubidum (Williston), Andrenosoma	386
rondanii Guimarães & d'Andretta, Basilia	922	rubidus Cole, Parataracticus	382
rondanii (Ringdahl), Hylemya	851	rubidus (Coquillett), Heteropogon	374
rondanii (Strobl), Fannia	896	*rubidus* (Robineau-Desvoidy), Psorophora	109
rondanii Townsend, Erynniopsis	1042	rubiflorae Felt, Dasineura	260
Ropalocera Meigen	388	rubifolia Felt, Camptoneuromyia	275
ROPALOMERIDAE	679	*rubiginigaster* Bigot, Xylota	607
roralis (Linnaeus), Melanophora	963	*rubiginis* (Walker), Promachus	400
rosacea Felt, Rhabdophaga	257	rubiginosa (Alexander), Pedicia	55
rosae (Fabricius), Psila	639	rubiginosa Loew, Holorusia	18
rosarum (Fabricius), Pyrophaena	578	*rubiginosa* Macquart, Exoprosopa	444
rosarum (Hardy), Dasineura	260	rubiginosus Bigot, (Seilopogon)	401
rosea (Felt), Asteromyia	275	rubisolita (Felt), Parallelodiplosis	286
roseanae (Brauer & Bergenstamm), Pseudoperichaeta	1106	rubivora (Coquillett), Pegomya	861
rosenscholdi (Zetterstedt), Stictochironomus	178	rubra (Felt), Asteromyia	275
rosewalli Wirth, Probezzia	138	rubra Felt, Johnsonomyia	252
rosivora (Coquillett), Aphidoletes	280	rubra Hull, Laphystia	376
rossi Harmston & Knowlton, Rhaphium	514	*rubra* James, Stratiomys	312
rossi Pechuman, Glutops	298	rubra Kieffer, Anarete	246
rossi Philip, Apatolestes	320	rubra Townsend, Gymnoerycia	1085
rostellata (Doane), Prionocera	19	rubribrunnea Sturtevant & Wheeler, Scatophila	759
rostrata (Bigot), Carposcalis	576	rubriceps Huckett, Pegomya	861
rostrata (Coquillett), Siphosturmia	1087	rubriceps (Macquart), Altermetoponia	300
rostrata (Coquillett), Slossonaemyia	1064	rubricornis Borgmeier, Chaetopleurophora	533
rostrata (Hendel), Neopelomyia	727	*rubricornis* (Zetterstedt), Lispocephala	877
rostrata Loew, Tachypeza	473	rubrifasciatus Bromley, Heteropogon	374
rostrata (Ringdahl), Spilogona	883	rubrifrons (Bigot), Peleteria	998
rostrata Robineau-Desvoidy, Zelia	1028	rubrifrons Patterson & Wheeler, Drosophila	766
rostrata (Say), Limonia	49	*rubrifrons* (Robineau-Desvoidy), Zodion	629
rostrata (Tothill), Nowickia	996	*rubrifrons* Townsend, Calliphora	930
rostratum (Melander & Brues), Trophithauma	543	rubrilata (Philip), Hybomitra	341
rostrifer Alexander, Molophilus	89	rubrina Huckett, Coenosia	872
rostrifera Jaennicke, Exoprosopa	445	rubrinota (Alexander), Hexatoma	70
rostrifera (Osten Sacken), Limonia	48	rubripalpis (Wulp), Ariciella	890
rotgeri James, Comantella	366	rubripes (Aldrich), Anaporia	1028
rothi Ringdahl, Helina	888	*rubripleuralis* (Curran), Meliscaeva	568
rotundata Collin, Pseudolimnophora	885	rubriventris (Macquart), Lampria	388
rotundata Felt, Mycodiplosis	283	rubriventris Macquart, Senotainia	939
rotundiceps (Aldrich), Sciapus	486	rubrivittata Sabrosky, Thaumatomyia	788
rotundiceps (Loew), Rhaphium	514	rubrivittatus Adams, Chlorops	790
rotundicorne Van Duzee, Parasyntormon	516	rubroflava (Hull & Fluke), Hiatomyia	587
rotundicornis Loew, Sapromyza	704	rubroscuta (Felt), Parallelodiplosis	286
rotundicornis Loew, Tetanocera	694	**Rubsaamenia** Kieffer	255
rotundicornis Malloch, Neoleria	813	rubtzovi Smart, Simulium	189
rotundifolia (Felt), Chaoborus	103	rudbeckiae (Beutenmüller), Lestodiplosis	285
rotundiloba Alexander, Tipula	38	rudbeckiae Felt, Lasioptera	272
rotundipennis Greene, Tachytrechus	503	rudbeckiaeconspicua Osten Sacken, Asphondylia	269
rotundipennis Loew, Chrysopilus	347	rudimentis Alexander, Limnophila	64
rotundipennis (Loew), Eutreta	661	rudis (Aldrich), Blaesoxipha	944
rotundipennis Macquart, (Sciara)	236	rudis Cresson, Pelina	752
rotundipennis Melander, Glabellula	415	rudis (Fabricius), Pollenia	931
royalensis (Bromley), Laphria	390	rudis Garrett, Dixa	101
rubbiginigaster Bigot, Xylota	607	*Ruebsaamenia* Kieffer	255
rubefactus Melander, Platypalpus	481	rufa Cresson, Automola	642
rubella (Loew), Neurigona	519	rufa (Loew), Palpomyia	140

INDEX 1667

	Page		Page
rufa Malloch, Sphegina	593	rufipes (Fabricius), Bibio	192
rufa (Panzer), Elgiva	690	rufipes (Fabricius), Promachus	†400, 400
rufa Spuler, Leptocera	722	rufipes (Fallén), Pegomya	861
rufa (Van Duzee), Platyura	204	rufipes Harris, (Xylota)	1116
rufa (Williston), Brachyopa	593	rufipes (Jones), Taracticus	385
rufalipes Hardy, Bibio	194	rufipes Loew, Odontomyia	316
rufata (Bigot), Bolomyia	1084	rufipes Loew, Xylophagus	297
rufescens Coquillett, Chlorops	790	rufipes Macquart, Atomosia	387
rufescens (Greene), Actia	1062	rufipes (Macquart), Eumetopiella	651
rufescens Loew, Empis	461	rufipes (Macquart), Microdon	598
rufescens (Macquart), Diogmites	372	rufipes Meigen, Cephalia	643
rufescens Stein, Pegomya	861	rufipes (Meigen), Megaselia	539
rufescens (Townsend), Leschenaultia	1083	rufipes Meigen, Phytomyza	805
rufiabdominalis Jones, Brachyopa	593	rufipes (Williston), Melanostoma	575
rufibarbis Bromley, Stenopogon	384	rufipilis Hull, Eristalis	623
rufibarbis (Macquart), Efferia	395	rufipunctatus (Curran), Metasyrphus	561
rufibarbis Van Duzee, Thinophilus	508	rufirostra Say, Rhamphomyia	465
rufibarboides Bromley, Stenopogon	384	rufirostris Say, Rhamphomyia	465
rufibasis Malloch, Phaonia	907	rufisquama (Schnabl), Lasiops	904
rufibasis Osten Sacken, Limnophila	64	rufitarsis Macquart, Lonchaea	717
rufibasis Stein, Coenosia	872	rufitarsis (Macquart), Rivellia	656
ruficauda (Brauer), Mericia	1008	rufitarsis (Stein), Spilogona	883
ruficauda Curran, Acanthocnema	836	rufithoracicum (Curran), Parapenium	580
ruficauda (Fabricius), Orellia	671	rufithorax Brues, Chaetopleurophora	533
ruficauda Reinhard, Minthozelia	1027	rufithorax (Say), Dialysis	343
ruficauda (Snow), Metasyrphus	561	rufithorax Wiedemann, Bibio	194
ruficauda (Townsend), Acroglossa	1074	rufitibia Stein, Coenosia	874
ruficauda (Wulp), Copecrypta	1003	rufitibia Stein, Dialyta	904
ruficauda (Zetterstedt), Neoleria	813	rufitibia Stein, Fucellia	843
ruficaudus Curran, Taracticus	385	rufitibia (Stein), Helina	889
ruficeps Aldrich, Microphthalma	982	rufitibia (Stein), Phyllogaster	876
ruficeps Curran, Admontia	1068	rufitibia (Stein), Spilogona	881
ruficeps (Curran), Orygma	682	rufitibia (Wulp), Johnsonia	950
ruficeps (Macquart), Atylotus	330	rufitibialis Hardy, Bibio	194
ruficeps (Meigen), Dicraeus	†783	rufiventris (Coquillett), Senotainia	939
ruficeps (Meigen), Hylemya	†850, 855	rufiventris Curran, Archytas	1000
ruficeps Reinhard, Euphyto	935	rufiventris Curran, Cerotainiops	388
ruficeps Stein, Pegomya	861	rufiventris Curran, Laphystia	376
ruficeps Wulp, Herina	645	rufiventris Hendel, Acrosticta	650
ruficeps Wulp, Micropeza	635	rufiventris (Loew), Furcifera	350
ruficollis Haliday, Rachicerus	296	rufiventris Loew, Mydas	359
ruficollis (Meigen), Clusiodes	807	rufiventris Loew, Sphegina	594
ruficollis (Meigen), Leptopeza	451	rufiventris Macquart, Eristalis	624
ruficollis Meigen, Mycetophila	213	rufiventris (Macquart), Holcocephala	374
ruficornis (Bigot), Pilimas	321	rufiventris Macquart, Proctacanthus	†399
ruficornis (Bigot), Theresia	991	rufiventris (Meigen), Tephrochlamys	810
ruficornis Fabricus, Tabanus	332	rufiventris Melander, Platypalpus	481
ruficornis (Greene), Schizactia	1066	rufiventris (Reinhard), Siphosturmia	1087
ruficornis Harris, (Chrysogaster)	1116	rufiventris (Townsend), Sarcodexia	955
ruficornis Loew, Dolichopus	495	rufoabdominalis Cole, Nemotelus	309
ruficornis Macquart, (Thereva)	354	rufoabdominalis Cole, Ogcodes	407
ruficornis Malloch, Lonchaea	717	rufocaudata (Bigot), Tephromyiopsis	960
ruficornis (Meigen), Megaselia	539	rufochaeta Curran, Admontia	1068
ruficornis Smith, Pilatea	1105	rufocincta Osten Sacken, Ptychoptera	97
ruficornis (Townsend), Myiophasia	984	rufofasciata Curran, Laphystia	376
ruficornis Van Duzee, Physocephala	628	rufofrater Walker, Tabanus	336
ruficornis Williston, Gymnocheta	1005	rufofrator Walker, Tabanus	336
ruficoxa (Macquart), Saltella	683	rufolimbatus (Bigot), Lepidanthrax	441
ruficrus (Austen), Cuterebra	1110	rufomaculata Jones, Volucella	601
ruficrus (Wiedemann), Meromacrus	622	rufonotata Bigot, (Morphomyia)	1108
ruficrus Williston, Microdon	598	rufonotata (Bigot), Winthemia	1091
rufifacies Macquart, Calliphora	929	rufopicta (Bigot), Winthemia	1090
rufifrons (Wulp), Chaetogaedia	1078	rufopuncta Curran, Hilara	456
rufifrons Wulp, Tetanops	648	rufoscutella Dodge, Neomuscina	911
rufigena Aldrich, Ursophyto	991	rufostomata (Bigot), Guerinia	1055
rufigena (Bigot), Phormia	925	rufotarsus Back, Cyrtopogon	369
rufilabris (Wulp), Eupelecotheca	†1051	rufotibialis Back, Ablautus	364
rufimana Meigen, Cordilura	828	rufotibialis Johnson & Johnson, Lordotus	412
rufina (Bigot), Zelia	1028	rufula Curran, Cordylurella	836
rufina (Fallén), Pegomya	861	rufulus (Osten Sacken), Conophorus	414
rufinervis (Pokorny), Xenomydaea	892	rufus Back, Saropogon	383
rufipalpis (Bigot), Trafoia	1007	rufus Felt, Arthrocnodax	283
rufipalpis Brooks, Gymnocheta	1005	rufus (Loew), Oreogeton	454
rufipalpis (Jaennicke), Phormia	924	rufus Loew, Syneches	449
rufipalpis (Macquart), Adejeania	†999	rufus Palisot de Beauvois, Tabanus	333
rufipalpis Macquart, Calliphora	929	rufus Williston, Myelaphus	379
rufipalpis (Macquart), Leschenaultia	1083	rufus Williston, Nicocles	380
rufipedalis Felt, Dasineura	260	rufus Williston, Proctacanthus	†399, 399
rufipennis Doane, Tephritis	668	rugglesi Nicholson & Mickel, Simulium	186
rufipennis Hine, Philonicus	398	rugia (Walker), Phaonia	†892, 907
rufipennis James, Labostigmina	314	ruginosus Becker, Chlorops	790
rufipennis (Macquart), Ptilodexia	989	rugosa (Felt), Lestodiplosis	285
rufipes (Bigot), Melanostoma	575	rugosa (Felt), Trichopteromyia	250
rufipes (Brooks), Epalpus	1002	ruidoso Alexander, Tipula	38
rufipes (Curran), Rhamphomyia	463	rumicifoliae Huckett, Pegomya	861
rufipes (Curran), Urophora	659	rumicis Felt, Lestodiplosis	285

	Page		Page
rumicis (Loew), Contarinia	277	sagana Shaw, Trichonta	216
rupecula (Bigot), Hylemya	850	sagata (Loew), Villa	436
rupestris Haliday, Dolichopus	495	sagax (Johannsen), Orfelia	202
rupestris (McDunnough), Hybomitra	340	sagax Osten Sacken, Tabanus	336
rupestris Vaillant, Oreothalia	467	sagax Townsend, Gonia	1076
rupicola (Bigot), Phormia	925	saginata Reinhard, Angustiopsis	1091
rupicola Doane, Tipula	26	sagittaria (Say), Physocephala	628
rupium Fabricius, Eristalis	624	sagittarius Van Duzee, Chrysotus	523
ruralis (Fallén), Voria	1020	sagittata Cook, Swammerdamella	240
ruralis (Meigen), Hydrophoria	†864	sagittata Stone, Lucumaphila	673
ruricola Felt, (Cecidomyia)	290	sagittifera Alexander, Tipula	38
rurika Roback, Pentaneura	147	sagulata Melander, Trixoscelis	818
russelli Curran, Thinophilus	508	sahlbergi (Becker), Pogonota	833
rustica (Fabricius), Dexilla	982	sahlbergii Loew, Chlorops	790
rustica (Fallén), Guerinia	†1055, 1055	saileri Stone, Cnephia	184
rustica Loew, Rhamphomyia	465	salax (Loew), Heringia	580
rustica (Osten Sacken), Adoxomyia	306	salax Wheeler, Peloropeodes	517
rustica Winnertz, Phronia	215	salebrosa Painter, Villa	435
rusticata (Malloch), Megaselia	539	salicifolia Felt, Dasineura	260
rusticola Doane, Tipula	25	salicifolia Felt, Rhabdophaga	257
ruthae (Brimley), Dryomyza	681	salicifoliae Osten Sacken (Cecidomyia)	295
rutila (Meigen), Ceromasia	††1099	salicifolius Felt, Oligotrophus	265
rutila Sherman, Dziedzickia	218	saliciperda Felt, Asynapta	254
rutila Van Duzee, Psilopiella	486	*salicis* (Fitch), Mayetiola	263
rutilans (Fabricius), Theresia	991	salicis (Malloch), Melanagromyza	796
rutiliceps Melander, Agromyza	795	salicis (Schrank), Rhabdophaga	257
rutilioides (Jaennicke), Paradejeania	1002	salicisbatatas (Osten Sacken), Rhabdophaga	257
rutilus (Coquillett), Toxorhynchites	107	salicisbrassicoides Packard, (Cecidomyia)	292
Rutylapa Edwards	202	*saliciscornu* (Osten Sacken), Mayetiola	264
ryckmani Wirth & Hubert, Culicoides	131	saliciscoryloides Osten Sacken (Cecidomyia)	295
ryckmani Wirth & Hubert, Dasyhelea	127	salicisgnaphaloides Osten Sacken (Cecidomyia)	292
Rymosia Winnertz	207		
Rypholophus Kolenati	84	*salicis hordeoides* (Osten Sacken), Rhabdophaga	258
		salicishordoides (Osten Sacken), Rhabdophaga	258
SABETHINI	107	salicisnodulus (Osten Sacken), Rhabdophaga	257
sabina (Osten Sacken), Villa	435	salicisrhodoides (Osten Sacken) Rhabdophaga	257
sabinae (Felt), Mayetiola	264		
sabinae Felt, Rhopalomyia	266	*salicis siliqua* (Walsh), Mayetiola	264
sabinae Patterson, Rhopalomyia	266	salicisstrobiliscus Osten Sacken (Cecidomyia)	292
Sabinata Parker	950	*salicisstrobiloides* (Osten Sacken) Rhabdophaga	258
sabrina Alexander, Limnophila	65		
sabroskyi (Dodge), Sarcophaga	959	salicistriticoides (Osten Sacken), Rhabdophaga	257
sabroskyi Gill, Amoebaleria	815		
sabroskyi Hanson, Nemotelus	310	salicisverruca (Osten Sacken), Trishormomyia	288
sabroskyi (Hardy), Pipunculus	554		
sabroskyi James, Euparyphus	307	salicola Huckett, Hylemya	854
sabroskyi Schlinger, Ogcodes	407	salicornia Quate, Psychoda	96
sabroskyi Steyskal, Dictya	689	salictaria Felt, Asphondylia	269
sabroskyi Steyskal, Sepsisoma	642	salihi Khalaf, Culicoides	130
sabroskyi Wirth, Probezzia	138	salina Aldrich, Lispe	878
sabuleti Steyskal, Melieria	645	*salina* (Curran), Hydropyrus	754
sabulona (Osten Sacken), Lestomyia	378	*salinaria* (Curran), Oscinella	779
sabulonus Becker, Chlorops	790	salinaria (Sturtevant & Wheeler), Lamproscatella	756
sabulosa (Coquillett), Villa	438		
sacajawea Alexander, Tipula	26	salinarius Coquillett, Culex	119
sacajaweae (Shannon), Xylota	608	salinus Cresson, Pelignus	737
Sacandaga Alexander	77	salinus (Malloch), Stichopogon	385
sacandaga Alexander, Gonomyia	75	sallei (Bellardi), Diogmites	372
sachtlebeni (Aczél), Tomosvaryella	556	*salmacia* Meigen	1074
sackeni Aldrich, Symphoromyia	344	salmani Alexander, Trichocera	16
sackeni Banks, Asilus	392	**Salpingogaster** Schiner	574
sackeni (Banks), Laphria	390	salpinx Melander, Mythicomyia	419
sackeni Edwards, Olbiogaster	190	*salsa* (Johnson), Orygma	682
sackeni Fairchild, Tabanus	336	saltans Vaillant, Wiedemannia	469
sackeni Felt, Tritozyga	243	*saltator* (Harris), Trichocera	16
sackeni Garrett, Amoebaleria	815	saltatrix (Linnaeus), Meromyza	793
sackeni Hine, Chrysops	327	**Saltella** Robineau-Desvoidy	682
sackeni Hine, Promachus	400	saltonensis Wirth, Culicoides	128
sackeni Wilcox, Laphria	389	saltus Stone & Jamnback, Prosimulium	183
sackeni Williston, Dioctria	371	salutans Bromley, Diogmites	372
sackeni (Williston), Heterostylum	411	salvia Martin, Leptogaster	363
sackeni Williston, Mallota	621	salviae Felt, Rhopalomyia	266
sackeniana Alexander, Tipula	29	salvum (Aldrich), Mecynocorpus	950
sackenianus Alexander, Molophilus	89	sambuci Felt, Asphondylia	269
sackenii (Coquillett), Poecilanthrax	443	*sambuci* (Felt), Lestremia	244
sackenii Johnson & Maughan, Conophorus	414	sambuci (Felt), Neolasioptera	273
sackenii (Röder), Bittacomorphella	98	sambucifolia Felt, Contarinia	277
sackenii (Williston), Neorhynchocephalus	402	sambucifolius Felt, Arthrocnodax	283
sackenii Williston, Ptecticus	303	*sambuci umbellicola* (Osten Sacken), Youngomyia	279
Sackenomyia Felt	265		
sacra Alexander, Tipula	30	*samiae* (Webber), Lespesia	1101
sacrator Walker, Laphria	390	sanctaeluciae Alexander, Tipula	40
sadales Walker, Laphria	390	sanctaemariae Wirth, Dasyhelea	127
sadyates (Walker), Tolmerus	401	sanctaeritae Alexander, Orimarga	52
sadytes (Walker), Tolmerus	401	sanctaeritae Alexander, Tipula	38
saffrana Fabricius, Laphria	390		

INDEX

	Page		Page
sanctipauli (Malloch), Spilogona	883	sayi Brennan, Silvius	323
sanduca Melander, Hesperempis	455	sayi (Dyar & Knab), Psorophora	109
sanguinea Felt, Colpodia	253	sayi Dyar & Shannon, Simulium	189
sanguinea Felt, Janetiella	264	sayi Felt, Epidiplosis	285
sanguinea Hendel, Euxesta	652	sayi (Jaennicke), Phaenicia	928
sanguinea Williston, Volucella	601	sayi (Johnson), Dolichopeza	23
sanguinia Felt, (Cecidomyia)	290	sayi Shannon, Polybiomyia	616
sanguinolenta (Loew), Parectecephala	792	sayi (Theobald), Psorophora	109
sanguinolenta (Osten Sacken), Caryomyia	270	sayi (Wiedemann), Sciapus	486
sanguisuga (Coquillett), Culicoides	128	sayii Johnson, Atomosia	387
saniosa (Say), Lampria	388	saylori Alexander, Tipula	38
sansoni Curran, Cyrtopogon	369	sayloriana Alexander, Tipula	25
sansoni Dyar & Knab, Aedes	112	**Sayomyia** Coquillett	103
sanus Osten Sacken, Tachytrechus	503	scaber Coquillett, Chlorops	790
sapphira (Hall), Protocalliphora	926	*scaber* (Loew), Condylostylus	485
sapphirina (Osten Sacken), Uranotaenia	108	scabiei (Hopkins), Pnyxia	235
sapphirinus Osten Sacken, Eulonchus	404	scabra (Giglio-Tos), Dendrophaonia	905
Sapromyia Roback	957	scabra Spuler, Sphaerocera	719
Sapromyza Fallén	704	scabrum (Aldrich), Aphanotrigonum	785
Sapromyzama Malloch	697	**Scaeva** Fabricius	562
SAPROMYZIDAE	695	scalaenum (Schrank), Polypedilum	176
Sapromyzosoma Malloch	697	scalaris (Fabricius), Fannia	897
saratogensis (Fitch), Limnia	692	scalaris (Loew), Megaselia	539
sarcinata (Loew), Peronyma	660	*scalaris* Loew, Mycetophila	213
sarcobati Felt, Aplonyx	275	scalaris Loew, Notiphila	747
sarcobati Felt, Onodiplosis	289	scalaris Melander, Aphoebantus	430
Sarcoclista Townsend	941	*scalaris* (Wiedemann), Olbiogaster	190
Sarcodexia Townsend	955	scalus (Haseman), Telmatoscopus	94
Sarcodexiopsis Townsend	955	scamboides Curran, Platycheirus	577
Sarcofahrtia Parker	941	scambus (Staeger), Platycheirus	577
Sarcofahrtiamyia Hall	956	scambus (Zetterstedt), Hydrotaea	900
Sarcofahrtiopsis Hall	956	scanloni Wirth & Hubert, Culicoides	130
Sarcomarcronychia Townsend	934	*scaphula* Alexander, Tipula	38
Sarcomyia Roback	953	**Scaptomyza** Hardy	769
Sarcophaga Meigen	956	*Scaptosciara* Edwards	235
SARCOPHAGIDAE	933	scapularis (Adams), Robertsonomyia	629
sarcophagina (Coquillett), Phalacrophyto	985	scapularis (Bigot), Apachekolos	361
sarcophagina (Townsend), Brachicoma	940	scapularis (Johannsen), Orfelia	203
SARCOPHAGINAE	942	scapularis (Loew), Coelotanypus	146
sarcophagoides (Townsend), Amobia	934	scapularis Loew, Cordilura	830
Sarcophagula Wulp	960	scapularis Loew, Dolichopus	495
Sarcophodexia Townsend	943	*scapularis* Loew, Hydrellia	744
Sarcotachinella Townsend	956	scapularis (Loew), Neoplasta	472
Sardiocera Brauer & Bergenstamm	991	*scapularis* (Loew), Rhagio	345
SARGINAE	301	scapularis Melander, Oliogodranes	422
Sargus Fabricius	301	scapularis (Rondani), Aedes	114
sarnia (Walker), Paracantha	662	scapulata (Bigot), Morellia	912
Saropogon Loew	382	scapulata Melander, Mythicomyia	419
saroria Williston, Sapromyza	704	scapulatus Melander, Oliogodranes	422
sarotes Loew, Dolichopus	495	*Scarabaeophaga* Townsend	957
Sarothromyia Brauer & Bergenstamm	960	scaria Alexander, Limnophila	64
Sarpedia Roback	946	**Scatella** Robineau-Desvoidy	757
sarraceniae Riley, Sarcophaga	959	SCATELLINI	755
sarrecenioides Aldrich, Sarcophaga	959	*Scathophaga* Meigen	837
Sarraceniomyia Townsend	957	*Scatomyza* Fallén	837
sartura (Reinhard), Oswaldia	1046	SCATOMYZIDAE	826
saskatchewana Shewell & Fredeen, Cnephia	184	**Scatophaga** Meigen	837
Saskatchewania Smith	975	scatophagina (Fallén), Leiomyza	823
sassafras Felt, Dasineura	260	scatophagina Melander, Empis	461
satanica Bigot, Xylota	607	*scatophagina* (Zetterstedt), Hylemya	847
satanica Shannon, Melanodexia	931	SCATOPHAGINAE	826
satchelli Quate, Psychoda	96	SCATOPHAGINI	827
satellitia Dyar, Pericoma	93	**Scatophila** Becker	758
satiata Felt, Lestodiplosis	285	scatophora (Perris), Epicypta	209
satiata Johannsen, Exechia	207	**Scatophora** Robineau-Desvoidy	†719,1113
satisfacta West, Eutheresia	990	*Scatopsciara* Edwards	235
sativae (Johannsen), Lycoriella	231	**Scatopse** Geoffroy	237
satrapa (Wheeler), Lamprochromus	530	SCATOPSIDAE	237
satura Osten Sacken, Volucella	602	SCATOPSINAE	237
saturata (Alexander), Hexatoma	70	SCATOPSINI	237
saturni (Felt), Trishormomyia	288	scaurissima Wheeler, Rhamphomyia	465
satyr Alexander, Tipula	39	*scelesta* (Hall), Oxysarcodexia	953
satyriacus Melander, Platypalpus	481	**Scellus** Loew	506
Saucropus Loew	518	SCENOPINIDAE	354
saundersi Kumm, Hippelates	776	**Scenopinus** Latreille	355
saundersii (Williston), Madremyia	1103	**Sceptonia** Winnertz	215
savaiiensis Edwards, Psychoda	96	**Schadonophasma** Dyar & Shannon	103
savonoskii Malloch, Phaonia	908	schaefferi Back, Leptogaster	363
savoryi (Parker), Blaesoxipha	944	schaefferi (Coquillett), Dicyphona	304
saxatilis Grossbeck, Culex	118	schalis Pritchard, Gongromastix	245
saxemontana Alexander, Tipula	38	schineri (Giraud), Melanagromyza	796
saxicola Osten Sacken, Antocha	51	schineri (Mesnil), Blepharipa	1084
saxorum Wiedemann, Eristalis	624	*schinophora* Loew, (Aricia)	1116
sayi (Aldrich), Neuratelia	224	**Schistostoma** Becker	453
sayi Alexander, Tipula	29	**Schizactia** Townsend	1066
sayi Banks, Zodion	629	**Schizocerophaga** Townsend	1047

	Page		Page
Schizohelea Kieffer	135	scopifer James, Dolichopus	495
schizomera Alexander, Tipula	26	scopiventris Harmston & Knowlton, Thinophilus	508
Schizomyia Kieffer	267		
SCHIZOPHORA	633	*Scopolia* Robineau-Desvoidy	1033
schizophorus Melander, Philetus	455	scopula Fisher, Mycomya	221
schizopinax Dyar, Aedes	114	scopula (Townes), Chironomus	171
Schizotachina Walker	1066	scordalus Reinhard, Milonius	1086
schizura Kincaid, Psychoda	95	scoriacea Loew, Euxesta	652
schizurae (Coquillett), Lespesia	1102	*scoriacea* (Loew), Typopsilopa	743
schizurae (Townsend), Lespesia	1102	scorpio McAtee, Laphria	390
schlingeri Hall, Aphoebantus	430	scotiae (Curran), Pericoma	92
schlingeri Wilcox & Martin, Backomyia	364	scotias Loew, Gymnopternus	500
schmitzi (Duda), Leptocera	724	scotica Edwards, Mycetophila	213
schnablei (Williston), Ceriana	615	scotlandae Cresson, Lemnaphila	745
schoenherri (Fallén), Pherbellia	686	*Scotophilella* Duda	723
schoenherri Zetterstedt, Anthalia	450	*scripta* (Linnaeus), Sphaerophoria	†720
Schoenomyza Haliday	869	scripta (Malloch), Copromyza	720
Schroederella Enderlein	814	*Scriptotricha* Loew	662
Schroederia Enderlein	814	scriptus Coquillett, Aphoebantus	430
schuhi Bromley, Asilus	392	*scriptus* Say, Anthrax	432
Schummelia Edwards	25	scrobicollis (Melander), Rhegmoclemina	239
schwardti Philip, Tabanus	336	scrobiculata (Loew), Villa	435
schwarzi (Alexander), Limonia	45	scrobiculata Melander, Chyliza	640
schwarzi (Malloch), Megaselia	539	*scrobinator* (Loew), Condylostylus	485
schwarzii Coquillett, Aenigmatias	537	scrofa (Aldrich), Microcerella	951
schwarzii (Coquillett), Nilobezzia	139	scrophulariae (Felt), Lestodiplosis	285
schwarzii Coquillett, Tachydromia	474	scrotifera Melander, Chelifera	471
SCIAPODINAE	482	*scudderi* (Townsend), Cuterebra	1109
Sciapus Zeller	†483, 485	*scudderi* (Williston), Nemorilla	1089
Sciara Meigen	230	sculleni Alexander, Hexatoma	70
SCIARIDAE	229	sculleni Alexander, Ptychoptera	97
sciarina Staeger, Boletina	217	sculleni Fluke, Metasyrphus	561
Sciasma Coquillett	971	sculleni Wilcox, Cophura	367
sciaspidis Spuler, Leptocera	722	sculleni Wilcox & Martin, Coleomyia	366
scilla Hull & Fluke, Cheilosia	585	scutata Felt, Dasineura	260
scimitarus (Townes), Chironomus	168	scutatus Aldrich, Pelastoneurus	502
scintillans (Loew), Sciapus	486	scutatus (Meigen), Platycheirus	577
Sciodromia Westwood	467	scutellaris Adams, Nausigaster	596
Sciomyza Fallén	688	*scutellaris* Brauer, Cuterebra	1110
SCIOMYZIDAE	685	*scutellaris* (Bromley), Laphria	390
SCIOMYZINAE	685	scutellaris (Coquillett), Curranops	644
SCIOMYZOIDEA	678	scutellaris Coquillett, Rhamphomyia	465
SCIONINI	322	scutellaris Cresson, Nostima	745
sciophila Loew, Sciara	230	scutellaris Curran, Euxesta	652
Sciophila Meigen	†219, 225	scutellaris (Fabricius), Eristalis	624
sciophila (Osten Sacken), Limonia	43	*scutellaris* (Fallén), Saltella	683
SCIOPHILINAE	216	scutellaris (Loew), Furcifera	350
SCIOPHILINI	222	*scutellaris* Loew, (Sciapus)	1116
Sciopus Zeller	485	scutellaris Melander, Anthalia	450
scissa Melander, Drapetis	478	scutellaris Melander, Empidideicus	416
scissilis Reinhard, Pseudolomyia	1089	scutellaris (Reinhard), Deopalpus	1004
scissurae Macfie, Dasyhelea	127	*scutellaris* (Swenk), Lynchia	918
scita (Johannsen), Bradysia	234	scutellaris Van Duzee, Argyra	525
scita Johannsen, Cordyla	208	*scutellaris* Walker, Tabanus	336
scita Walker, Bibio	194	scutellarmata Lovett, Xylota	606
scita Walker, Chionea	72	scutellata Coquillett, Mythicomyia	419
scitula (Coquillett), Villa	438	*scutellata* Felt, Cordylomyia	248
scitula Johannsen, Neuratelia	224	scutellata Garrett, Amoebaleria	815
scitula (Laffoon), Mycetophila	213	*scutellata* Garrett, Cordyla	208
scitula Reinhard, Eumacronychia	935	*scutellata* Harris, (Trichocera)	1116
scitula (Williston), Blera	611	scutellata Hull, Milesia	614
scitula (Williston), Cophura	367	*scutellata* (Loew), Baccha	573
scituloides Shaw, Neuratelia	224	scutellata (Malloch), Pseudogriphoneura	704
scitulus (Coquillett), Brillia	154	scutellata (Meigen), Dasyhelea	127
scitulus Williston, Microdon	598	scutellata (Robineau-Desvoidy), Blepharipa	1084
scitulus (Williston), Phyllomydas	360	scutellata Say, (Campylomyza)	292
scitus (Walker), Hamatabanus	341	scutellata Say, Trichocera	16
Scleropogon Loew	383	*scutellata* (Williston), Polybiomyia	616
Scleroprocta Edwards	84	scutellata Winnertz, Corynoneura	164
scobinator (Loew), Condylostylus	485	*scutellatus* (Harris), Condylostylus	484
scolex Reinhard, Zenillia	1107	*scutellatus* Harris, (Sargus)	1116
Scoliaphleps Becker	830	scutellatus Melander & Argo, Clusiodes	807
Scoliocentra Loew	814	scutellatus Say, (Ceratopogon)	142
Scoliopelta Williston	301	scutifer Knab, Microdon	598
scolopacea (Say), Rhamphomyia	465	scylla (Osten Sacken), Villa	436
scolopax Osten Sacken, Phthiria	420	scyphocerca Chillcott, Fannia	897
scopalis (Brues), Megaselia	541	seagoi (Dodge), Sarcophaga	959
scoparia Coquillett, Empis	461	*seamansi* Huckett, Hylemya	849
scoparia Pandellé, Sarcophaga	959	seamansi Shewell, Neoalticomerus	794
scoparia (Zetterstedt), Myopina	844	secabile (Walker), Blepharepium	365
scoparius (Cresson), Tomosvaryella	556	seclusa (Laffoon), Mycetophila	213
scoparius Loew, Dolichopus	491	sectata (Huckett), Spilogona	883
Scopeuma Meigen	837	sectilis (Reinhard), Pilatea	1105
SCOPEUMATIDAE	826	sectus McAtee, Dilophus	195
scopifer (Coquillett), Oreogeton	454	*secunda* (Brues), Megaselia	541
scopifer Harmston, Asyndetus	524		

INDEX 1671

Name	Page
secunda (Melander & Spuler), Sepsis	684
securicornis Fallén, Phyllomyza	729
securifera Coquillett, Symphoromyia	344
securifera Villeneuve, Sarcophaga	958
securiferus (Macquart), Allograpta	569
securis Reinhard, Myiopharus	1045
seducta Collin, Meoneura	729
sedula Alexander, Limnophila	64
sedula Huckett, Hylemya	856
sedula Johannsen, Boletina	217
sedula (Reinhard), Adoryphorophaga	1038
sedulus Van Duzee, Dolichopus	495
segnis (Holmgren), Pogonomyioides	902
segnis (Linnaeus), Xylota	606
segregata (Rondani), Exorista	1054, †1056
Seioptera Kirby	647
sejuncta (Loew), Limnellia	758
selecta (Meigen), Bessa	†1053, 1053
selene (Osten Sacken), Villa	437
selenitica (Meigen), Scaeva	562
Selfia Khalaf	132
sellata Loew, Rhamphomyia	465
sellersi Hall, Diplostichus	1053
semenivora (Beutenmüller), Dasineura	260
semenrumicis Patterson, (Cecidomyiaceltis)	295
semenrumicis Patton, (Cecidomyiaceltis)	295
semicomatus (Van Duzee), Condylostylus	485
semidea Alexander, Tipula	30
semifacta Alexander, Limnophila	66
semifasciatus Cresson, Pipunculus	553
semiflava Aldrich, Belvosia	1082
semifulvipes Painter, Villa	434
semifuscus (Banks), Physoconops	627
semiglabra Sabrosky, Sigaloessa	824
semiglobosa (Ringdahl), Spilogona	883
semiinterrupta (Fluke), Melangyna	567
semilucida (Bates), Aciurina	670
semilutea (Loew), Paratissa	740
semimarginalis Hall, Sarcophaga	959
semimetallicus Macquart, Eristalis	623
seminiger Becker, Chlorops	790
seminigra (Loew), Villa	436
seminigra Sherman, Rymosia	208
seminole Alexander, Erioptera	81
seminole Alexander, Tipula	38
seminole Bromley, Dioctria	370
seminole Sturtevant & Dobzhansky, Drosophila	768
seminole Townes, Glyptotendipes	171
seminuda Banks, Myopa	630
semiothisae Brooks, Chaetophlepsis	1016
semiplumatus Becker, Pelastoneurus	502
semirufa (Meigen), Orfelia	203
semitaria Coquillett, Thereva	354
semitarius Melander, Cyrtopogon	369
semitecta (Coquillett), Laphria	390
semiustus Coquillett, Saropogon	383
semivirens Kieffer, Orthocladius	156
semiviridis Wulp, Chloroprocta	923
semivitta Malloch, Allognota	874
semivittatus Sabrosky, Gaurax	782
Semnomyia Frey	234
senaria Loew, Stratiomys	312
senata (Walley), Pentaneura	147
senator Melander, Hormopeza	453
senega Alexander, Tipula	34
senescens Melander & Spuler, Piophila	713
Senetainia Macquart	938
senex (Johannsen), Trichocladius	158
senex Melander, Heterotropus	423
senex Osten Sacken, Saropogon	383
senex Walker, Thereva	354
seniculus Loew, Holopogon	375
senilis (Bigot), Heteropogon	374
senilis (Fabricius), Psilocephala	351
senilis (Haliday), Paradelphomyia	†59
senilis (Johannsen), Glyptotendipes	171
senilis Knab, Microdon	598
senilis (Meigen), Lydella	††1103
senilis (Townsend), Leucostoma	976
senilis Williston, Gonia	1076
senilis Wulp, Bibio	193
Senopterina Macquart	657
Senotaina Macquart	938
Senotainia Macquart	938
sensua Curran, Cheilosia	586
sentis Alexander, Ormosia	87
Seoptera Kirby	647
seersa Sherman, (Phthinia)	1116
separata Alexander, Blepharicera	99
separata Hull, Mallota	621
separata (Johannsen), Dziedzickia	218
separata Malloch, Calythea	866
separata (Rondani), Carcelia	1092
separata Stein, Fucellia	843
separata (Walker), Villa	439
separatus Hine, Chrysops	327
separatus Melander, Aphoebantus	429
separatus Walker, Dolichopus	495
Sepedon Latreille	693
sepentaria (Harris), Strauzia	676
sepia (Meigen), Hylemya	846, †846
sepiella (Zetterstedt), Paraprosalpia	866
SEPSIDAE	681
Sepsidimorpha Frey	683
Sepsis Fallén	683
Sepsisoma Johnson	642
septemmaculata (Walsh), Lestodiplosis	285
septemmaculatus (Adams), Caloparyphus	308
septemtrionalis Rübsaamen, (Sciara)	236
septemtrionis Alexander, Dicranoptycha	52
septemtrionis Osten Sacken, Erioptera	81
septenaria (Harris), Strauzia	676
septentrionale Curran, Rhaphium	514
septentrionalis Alexander, (Antocha)	1116
septentrionalis (Bergroth), Pedicia	55
septentrionalis (Collin), Neoleria	813
septentrionalis (Curran), Pseudomeriania	1009
septentrionalis (Dyar & Knab), Toxorhynchites	107
septentrionalis (Hendel), Norellia	827
septentrionalis (Loew), Hybomitra	339
septentrionalis Loew, Tipula	31
septentrionalis Melander, Drapetis	478
septentrionalis Melander, Limnia	692
septentrionalis Rübsaamen, (Sciara)	236
septentrionalis (Stein), Lasiops	904
septentrionalis (Townsend), Eucordyligaster	1030
septentrionalis Walker, (Eurigaster)	1108
septris Roback, Trichocladius	158
sepulta (Laffoon), Mycetophila	213
sequax (Garrett), Diamesa	152
sequax Johannsen, Mycomya	221
sequax Walker, Dolichopus	494
sequax Williston, Chrysops	327
sequax Williston, Gonia	1076
sequax (Williston), Hybomitra	340
sequax (Williston), Periscepsia	1034
sequens (Townsend), Catemophrys	1031
sequoia Sabrosky, Ocnaea	404
sequoia Wilcox & Martin, Nannocyrtopogon	380
sequoiae Alexander, Molophilus	89
sequoiae Chillcott, Fannia	897
sequoiae Quate, Trichomyia	91
sequoiarum (Alexander), Dioptopsis	100
sequoiarum Alexander, Limnophila	64
sequoiarum Alexander, Ormosia	87
sequoiarum Alexander, Tipula	38
sequoicola Alexander, Tipula	34
sequoiensis Alexander, Phyllolabis	60
sera Roback, Brillia	154
sera (Walker), Limonia	48
serena Cresson, Hydrellia	744
serena (Fallén), Fannia	897
serena (Meigen), Pyrellia	†912
Sergentia Kieffer	174
seriata Aldrich, Ginglychaeta	1018
seriata Loew, Hilara	456
seriata (Loew), Icterica	664
seriata (Melander), Tethina	728
seriata Stein, Hylemya	851
sericariae (Rondani), Blepharipa	†1084
sericata (Loew), Icterica	664
sericata (Meigen), Phaenicia	928
sericata Van Duzee, Argyra	525
sericata Hardy, Bibio	194
sericatus Melander, Platypalpus	481
sericea Malloch, Hylemya	853
sericea Say, Laphria	390
sericea (Say), Rymosia	208
sericeus Say, Asilus	393
Serichlamys Curran	597

1672 A CATALOG OF DIPTERA OF NORTH AMERICA

	Page
Sericomyia Meigen	603
SERICOMYIINI	603
seriepunctatus (Osten Sacken), Anthrax	432
seripila Hull & Fluke, Cheilosia	585
serotina Felt, Lasioptera	272
serotina Reinhard, Stenoneura	1071
serotinae Osten Sacken, (Cecidomyia)	293
serotinella Alexander, Limnophila	65
serotinus Loew, Dilophus	195
serpentaria (Harris), Strauzia	676
serpentina Alexander, Erioptera	82
serpentina Osten Sacken, Dipalta	441
serpentina (Wiedemann), Anastrepha	673
serperastrorum Melander, Empis	460
serrata Chillcott, Fannia	897
serrata Cook, Psectrosciara	240
serrata Felt, Cincticornia	269
serrata Garrett, Dixa	102
serrata (Linnaeus), Heleomyza	816
serrata (Malloch), Lyciella	701
serrata Robineau-Desvoidy, Haematobia	914
serrataria (Garrett), Heleomyza	816
serraticollis Walker, Dilophus	195
serratula Loew, Brachystoma	447
serratus (Coquillett), Conophorus	414
serratus (Lewis), Ceratopogon	133
serratus Townsend, Oxynops	1046
serratus Van Duzee, Dolichopus	495
serricolor (Townsend), Gaediopsis	1079
serridens Alexander, Ormosia	87
serripes Johannsen, Rymosia	208
Serromyia Meigen	136
serrosa Pettey, (Sciara)	1116
serrulata Loew, Tipula	31
serrulatae (Osten Sacken), Dasineura	260
serrulatum Loew, Brachystoma	447
serta Loew, Tipula	34
sertata (Laffoon), Mycetophila	213
serus Malloch, Chironomus	166
serva (Meigen), Phaonia	†906, 908
serva Walker, (Musca)	1108
Servaisia Robineau-Desvoidy	946
servilis (Aldrich), Sarcodexia	956
servillei (Guérin-Méneville), Chrysopilus	346
sessilis (Alexander), Dicranota	58
sessilis Alexander, Dolichopeza	23
setacea (Aldrich), Megaselia	†538, 539
Setacera Cresson	754
setaria (Meigen), Pegomya	861
setariae Felt, (Cecidomyia)	290
setaventris Cresson, Micropeza	634
seticauda (Malloch), Homoneura	699
seticauda Malloch, Hydrophoria	864
seticauda (Reinhard), Aplomyiopsis	1038
seticauda Van Duzee, Pelastoneurus	502
seticellula Alexander, Limnophila	65
seticosta (Ringdahl), Spilogona	883
seticosta Spuler, Leptocera	722
seticoxa Steyskal, Pherbellia	686
setifacies Brooks, Euthera	978
setifacies (Brooks), Gonia	1076
setifacies Reinhard, Microchaetina	984
setifacies (West), Lypha	1011
setifer Chillcott, Fannia	897
setifer Loew, Dolichopus	495
setifer Malloch, Hylemya	851
setifirma Huckett, Hylemya	851
setiformis Huckett, Pegomya	861
setifrons (Aldrich & Webber), Patelloa	1104
setifrons (Brooks), Metaphyto	1008
setifrons (Melander), Melanagromyza	796
setigena (Coquillett), Chromatocera	1012
setiger (Johannsen), Orfelia	203
setiger Malloch, Neochirosia	842
setigera (Adams), Leptocera	724
setigera (Aldrich), Blaesoxipha	947
setigera Alexander, Rhabdomastix	77
setigera (Cole), Coleomyia	366
setigera (Coquillett), Clausicella	1059
setigera (Coquillett), Hineomyia	1005
setigera (Coquillett), Paradidyma	1033
setigera (Coquillett), Procecidochares	660
setigera (Cresson), Polytrichophora	740
setigera (Johannsen), Hylemya	852
setigera (Lintner), Contarinia	277
setigera Malloch, Amiota	761

	Page
setigera Malloch, Coenosia	872
setigera (Osten Sacken), Volucella	601
setigera (Stein), Hylemya	848
setigera (Thomson), Elfia	1066
setigera Tonnoir, Psychoda	96
setigera Tothill, Gonia	1076
setigera Townsend, Myiophasia	984
Setigeresta Benjamin	665
Setigonia Brooks	1075
setilamellata (Huckett), Spilogona	883
setinervis Coquillett (Hypostena)	1108
setinervis (Coquillett), Thelairodoria	1050
setinervis (Huckett), Spilogona	883
setipalpis Sabrosky, Phlebosotera	824
setipalpis Shewell, Camptoprosopella	697
setipennis Coquillett, Houghia	1095
setipennis (Coquillett), Periscepsia	1033
setipennis (Fallén), Bigonicheta	†1012
setipes (Coquillett), Bucephalina	832
setipes (Coquillett), Gueriniopsis	1055
setipes (Coquillett), Nilobezzia	139
setipes Malloch, Lispocephala	877
setipes Melander & Argo, Sobarocephala	806
setipes Van Duzee, Argyra	525
setiseriata Huckett, Hylemya	851
setisissima Huckett, Hylemya	851
setitarsata Huckett, Hylemya	851
setitibia Shewell, Homoneura	699
setitibialis (Spuler), Copromyza	720
setiventris Stein, Hylemya	851
setosa (Bigot), Dryomyza	681
setosa Coquillett, Brachicoma	940
setosa (Coquillett), Celatoria	1040
setosa Coquillett, Gaediopsis	1080
setosa (Coquillett), Hoplodictya	691
setosa (Coquillett), Neophyto	952
setosa (Coquillett), Neoscatella	756
setosa Coquillett, Oestrophasia	979
setosa Coquillett, Rhamphomyia	465
setosa Curran, Peleteria	998
setosa (Doane), Rhagoletis	674
setosa Felt, Dasineura	260
setosa Felt, Lestremia	244
setosa Felt, Porricondyla	256
setosa Garrett, Docosia	226
setosa Garrett, Sciophila	226
setosa Loew, Cordilura	828
setosa (Loew), Phytobia	798
setosa Loew, Pyrellia	912
setosa Melander, Palloptera	715
setosa Melander & Spuler, Piophila	711
setosa Reinhard, Grisdalemyia	1024
setosa Townsend, Hesperophasia	977
setosimanus (Goetghebuer), Tanytarsus	180
setosivena Alexander, Trichocera	16
setosus (Cresson), Oligodranes	422
setosus Loew, Dolichopus	495
setosus (Stein), Eremomyioides	857
setosus (Van Duzee), Hercostomus	498
setosus Van Duzee, Sympycnus	528
setulicosta Allen, Senotainia	939
setulosa (Loew), Bezzia	141
setulosa (Malloch), Siphonella	787
setulosa Melander, Drapetis	477
severa Cresson, Limnia	692
severa Johannsen, Sciophila	226
severini Blanton, Rivellia	656
severini Harmston & Knowlton, Asyndetus	524
severini Shewell, Homoneura	699
severini Tonnoir, Psychoda	†96
severinii Walton, Euhalidaya	1037
severnensis Curran, Pipiza	579
sevierensis Alexander, Gonomyia	76
sexarticulatus Loew, Dolichopus	496
sexdentata (Pettey), Bradysia	234
sexfasciata (Hine), Hybomitra	340
sexfasciata (Say), Laphystia	376
sexfasciata Walker, Sericomyia	603, †603
sexfasciatus (Stone), Hamatabanus	342
sexmaculata Felt, Neolasioptera	273
sexmaculata Macquart, Hermetia	305
sexmaculata Walker, Coenosia	870
sexmaculatus Palisot de Beauvois (Syrphus)	625
sexnotata Meigen, Coenosia	872
sexpunctata Loew, Volucella	602
sexquadrata (Walker), Melangyna	567

INDEX 1673

Name	Page	Name	Page
sexualis (Curran), Drino	1084	signatipennis (Wulp), Cylindromyia	973
shannoni Alexander, Gonomyia	74	signatiseta Hunter, Cheilosia	587
shannoni Alexander, Limnophila	65	signatoides Dziedzicki, Mycetophila	213
shannoni (Alexander), Limonia	45	*signatus* Hardy, Bibio	194
shannoni Cresson, Limnia	692	signatus (Loew), Pseudogaurax	782
shannoni (Curran), Cheilosia	586	*signatus* (Wulp), Allograpta	569
shannoni Curran, Rhaphium	514	*signicosta* (Walker), Leucophenga	761
shannoni Dyar, Phlebotomus	92	signifer Coquillett, Hydrophorus	506
shannoni Sabrosky, Gaurax	782	signifer (Walker), Epalpus	1002
shannoni (Townsend), Microphthalma	982	signifera Coquillett, Neaspilota	672
shannoni Wirth, Systenus	517	signifera (Coquillett), Orthopodomyia	109
Shannonomyia Alexander	67	signifera (Loew), Ceriana	616
sharonae Johnson & Johnson, Exoprosopa	445	signifera Melander, Trixoscelis	818
shasta Alexander, Lipsothrix	77	*signifera* Melander & Spuler, Sepsis	683
shasta Alexander, Tipula	36	*signifera* Townsend, Paratheresia	991
shasta Wirth, Deuterophlebia	100	signiferum (Osten Sacken), Rhaphium	514
shastensis Alexander, Blepharicera	99	signiferum Wulp, Cenosoma	979
shastensis Alexander, Pedicia	55	*signipennis* (Coquillett), Limonia	47
shastensis Alexander, Tipula	26	signipennis (Coquillett), Psorophora	110
shawi (Felt), Trishormomyia	288	*signius* (Walker), Lasiops	904
shawi Fisher, Exechia	207	sila (Reinhard), Neosolieria	1010
shawi (Laffoon), Mycetophila	213	**Silba** Macquart	718
shawii (Johnson), Villa	435	*Siligo* Aldrich	818
sheldoni Alexander, Erioptera	81	*siliqua* (Osten Sacken), Mayetiola	264
sheldoni (Coquillett), Homoneura	699	sillersi Hull, Laphystia	377
sheldoni (Coquillett), Neophyto	952	silvacola Martin, Beameromyia	362
shelfordi Alexander, Limonia	50	silvana Felt, Campylomyza	248
shelfordi Curran, Dolichopus	496	silvana Felt, Mycodiplosis	283
shepherdiae Felt, Asphondylia	269	silvarum (Meigen), Bufolucilia	927
shermani Cole & Wilcox, Lasiopogon	378	silvatica (Aldrich & Webber), Patelloa	1104
shermani Garrett, Boletina	217	silvatica (Fallén), Masicera	1103
shermani Garrett, Leia	227	silvatica Johannsen, Neuratelia	224
shermani Garrett, Mycomya	221	silvatica (Meigen), Leptocera	724
shermani Hine, Chrysops	327	silvatica Meigen, Tetanocera	694
shermani Parker, Sarcophaga	959	silvestra Doane, Tipula	38
shewelli Peterson & De Foliart, Prosimulium	183	silvestrii (Kieffer), Bradysia	234
shewelli Sabrosky, Lasiopleura	774	silvestris (Fallén), Paraprosalpia	867
shewelli Sabrosky, Ogcodes	407	silvestris Johannsen, Anatella	205
shewelli Sabrosky, Phlebosotera	824	*silvestris* (Robineau-Desvoidy), Parasetigena	1056
shoshone Alexander, Erioptera	84	silvicola Harmston, Chrysotus	523
shoshone Alexander, Tipula	34	silvicola Harmston, Dolichopus	496
sialia Shannon, Cheilosia	586	**Silvius** Meigen	322
sialia Shannon & Dobroscky, Protocalliphora	926	silvus Cole, Bombylius	409
siberita (Fabricius), Prosena	987	sima Aldrich, Sarcophaga	959
sibilans (Haliday), Lamproscatella	756	sima Osten Sacken, Exoprosopa	445
sibirita (Fabricius), Prosena	987	simiaceps Cresson, Ditrichophora	739
sibleyi West, Eutheresia	990	similans Johannsen, Blepharicera	99
sica Alexander, Cryptolabis	78	similans (Johannsen), Lycoriella	231
sicaria Melander, Chelipoda	472	*similas* Garrett, Tetragoneura	228
sicarium (McAtee), Andrenosoma	386	similata Malloch, Beckerina	537
sicarius Van Duzee, Dolichopus	496	*similata* (Malloch), Cordilura	829
sicca Cresson, Notiphila	748	*similata* (Malloch), Homoneura	698
siccana (Reinhard), Boettcheria	948	similata (Malloch), Melanagromyza	796
Sicodus Rafinesque	474	similata (Malloch), Micropsectra	179
sicula McAtee, Laphria	390	similata Malloch, Rhamphomyia	465
sicula Quate, Pericoma	92	simile (Curran), Parapenium	580
Sicus Latreille	474	simil!s (Aldrich), Condylostylus	485
sierra Quate, Telmatoscopus	94	similis Aldrich, Liancalus	507
sierrae Hardy, Bibio	194	similis Alexander, Limnophila	66
sierrae (Laffoon), Mycetophila	213	*similis* Becker, Ectecephala	791
sierrensis Alexander, Paradelphomyia	60	similis (Beekey), Rhexoza	240
sierrensis (Ludlow), Aedes	114	*similis* Brennan, Apatolestes	320
sierricola Alexander, Ulomorpha	68	*similis* (Brown), Cerotainiops	388
sierricola (Townsend), Gaediopsis	1079	similis (Coquillett), Leptopsilopa	741
Sigaloessa Loew	824	similis (Coquillett), Nanna	831
SIGALOESSINAE	823	similis Coquillett, Phthiria	420
sigilla Reinhard, Myiophasia	984	similis (Cresson), Homalocephala	653
sigilla Reinhard, Sarcophaga	959	similis (Cresson), Pherbellia	687
sigillata Dziedzicki, Mycetophila	213	similis Felt, Caryomyia	270
sigma (Coquillett), Oligodranes	422	*similis* Felt, Dasineura	260
sigma Johannsen, Mycomya	221	*similis* Fitch, Hylemya	850
sigma Kincaid, Psychoda	96	similis Garrett, Docosia	226
sigma (Phillips), Trypeta	677	similis Garrett, Macrocera	201
sigmoides Loew, Drosophila	768	similis Garrett, Pseudoleria	812
sigmoides Loew, Mycetophila	213	similis Garrett, Tetragoneura	228
sigmoideus Alexander, Molophilus	89	similis (Giglio-Tos), Volucella	602
signata Aldrich & Webber (Phorocera)	1107	similis Hall, Paracosmus	427
signata (Banks), Pericoma	93	similis Harmston & Knowlton, Argyra	525
signata Cresson, Notiphila	748	similis (Hough), Tomosvaryella	556
signata Foote, Trupanea	667	similis James, Bibio	194
signata (Meigen), Amobia	†934	similis (James), Cephalochrysa	302
signatifrons Cole, Pherocera	349	*similis* Johannsen, Chironomus	165
signatifrons (Coquillett), Lauxaniella	700	similis Johannsen Dixa	101
signatipennis (Cole), Poecilanthrax	443	similis Johannsen, Phronia	215
signatipennis Cole, Psilocephala	352	similis Johannsen, Sciophila	226
signatipennis Foote, Tephritis	668	*similis* (Johnson), Acrosticta	650

	Page		Page
similis Johnson, Ceraturgus	365	simulans Sturtevant, Drosophila	768
similis Johnson, Chyliza	640	simulans Townes, Polypedilum	176
similis Johnson, Cuterebra	1110	simulans Van Duzee, Chrysotus	523
similis (Johnson), Dolichopeza	23	simulans Van Duzee, Dolichopus	496
similis (Johnson), Labostigmina	314	*simulans* Van Duzee, Physocephala	628
similis Johnson, Palloptera	715	simulans (Walker), Limonia	†50, 50
similis (Johnson), Platyura	204	*simulans* Walker, Tabanus	335
similis Jones, Microdon	598	simulata Huckett, Hylemya	851
similis (Jones), Stenopogon	384	*simulata* Walker, Tipula	34
similis Jones, Syrphus	559	simulata Walley, Forcipomyia	126
similis Kieffer, Metriocnemus	161	simulator Felt, Campylomyza	248
similis (Loew), Allognosta	300	simulator Felt, Dasineura	261
similis (Loew), Melieria	645	simulatus Greene, Tachytrechus	503
similis Macquart, Helophilus	618	*simulatus* Harris, (Syrphus)	1116
similis Macquart, Sepsis	683	Simulia Latreille	185
similis Macquart, Tabanus	336	SIMULIIDAE	181
similis Malloch, Apocephalus	543	SIMULIINAE	184
similis Malloch, Corynoneura	164	**Simulium** Latreille	185
similis Malloch, Eremomyioides	857	sinawava Alexander, Erioptera	84
similis (Malloch), Lasiosina	792	sincerus Melander, Dolichopus	496
similis Malloch, Pogonomyia	902	*sincerus* Walker, Eristalis	625
similis Malloch, Simulium	188	singularis (Aldrich), Blaesoxipha	947
similis Rubzov, Simulium	189	singularis Burgess, Glutops	298
similis Shannon, Cheilosia	586	singularis (Spuler), Copromyza	720
similis Stein, Hylemya	850	singularis (Townsend), Paradidyma	1033
similis Steyskal, Pteromicra	687	singularis Van Duzee, Gymnopternus	500
similis Townsend, Myobiopsis	1035	*singularis* Van Duzee, Mycetophila	210
similis (Townsend), Plagiomima	1019	sini Cole, Villa	440
similis (Townsend), Senotainia	939	sinipalpis Allen, Metopia	937
similis Van Duzee, Diaphorus	520	sinistra Dietz, Tipula	38
similis VanDuzee, Medetera	510	**Sinophthalmus** Coquillett	761
similis (Walker), Muscina	909	sinopsis Reinhard, Senotainia	939
similis (Walker), Tachypeza	473	sinuata Collin, Limnophora	885
similis Williston, Ctenophora	20	sinuata Malloch, Hylemya	846
similis (Williston), Efferia	395	sinuata Meigen, Sarcophaga	959
similis Williston, Ptecticus	303	sinuata Reinhard, Winthemia	1091
similis (Williston), Tachinomyia	1058	sinuata Townsend, Steveniopsis	985
similissima Dietz, Chrysopila	32	sinuosa (Bigot), Chrysogaster	591
simillima (Kieffer), Dasineura	260	sinuosa (Coquillett), Pentaneura	147
simillima Steyskal, Xylomya	299	sinuosa (Wiedemann), Villa	436
simillima Walker, Lispe	878	sinuosus (Curran), Neocnemodon	581
simillimus (Walker), Chrysopilus	346	siouana Alexander, Limnophila	66
siminina Reinhard, Pseudochaeta	1096	**Siphlodora** Patterson & Mainland	768
simpatica Dziedzicki, Anatella	205	sipho Melander, Oligodranes	422
simpla Coquillett, Hylemya	851	sipho (Say), Condylostylus	†484, 485
simpla Felt, Cincticornia	269	**Siphoclytia** Townsend	1036
simplex (Aldrich), Tricharaea	960	**Siphoeleskia** Townsend	1036
simplex Alexander, Limnophila	64	*Siphomyia* Williston	774
simplex Alexander, Tipula	37	**Siphona** Meigen	†914, 1063
simplex (Bigot), Ptilodexia	989	*siphonalis* (Grossbeck), Aedes	112
simplex Bigot, (Stratiomys)	319	**Siphonella** Macquart	786
simplex Coquillett. Eccontomera	811	SIPHONINA	1061
simplex (Coquillett), Mycomya	221	siphonina Bigot, Proboscimyia	868
simplex Curran, Cryptomeigenia	1041	*siphonina* (Zetterstedt), Hilarella	†936
simplex Doane, Tipula	38	SIPHONINI	1059
simplex (Fallén), Leucostoma	976	*Siphonopsis* Townsend	1064
simplex Fallén, Sciomyza	688	*Siphopallasia* Brooks	964
simplex Garrett, Bolitophila	198	*Siphophyto* Townsend	1059
simplex Garrett, Dixa	102	*Siphoplagia* Townsend	1019
simplex Garrett, Monoclona	224	**Siphoplagiopsis** Townsend	1019
simplex Goetghebuer & Bastin, Sepsis	683	**Siphosturmia** Coquillett	1087
simplex Kieffer, Diamesa	152	*Siphosturmiopsis* Townsend	1087
simplex (Loew), Allotrichoma	736	siskiyou Felt, Janetiella	264
simplex Loew, Dicolonus	370	siskiyouensis Alexander, Tipula	38
simplex Loew, Dryomyza	681	**Sisyropa** Brauer & Bergenstamm	1094
simplex Loew, Leucopis	709	*Sisyrosturmia* Townsend	1105
simplex (Loew), Melanagromyza	796	sitchana Kincaid, Pericoma	93
simplex Loew, Mydas	359	sitiens (Collin), Paraprosalpia	867
simplex (Loew), Wiedemannia	469	sitiens Huckett, Pegomya	861
simplex Melander, Euhybus	448	sitiens Wirth & Hubert, Culicoides	131
simplex Snow, Nemotelus	310	**Sitodiplosis** Kieffer	278
simplex Stein, Coenosia	874	**Sitophaga** Gistel	1048
simplex Thomas, Eutreta	661	skinneri Coquillett, Trochilodes	1015
simplex Walker, Syneches	449	skinneri (Cresson), Cressonomyia	742
simplicipes Aldrich, Dolichopus	496	skinneri Johnson, Cheilosia	585
simplicipes (Becker), Pleurochaetella	834	*slossonae* Aldrich, Leiomyza	823
simplicipes Curran, Medetera	510	slossonae Alexander, Gonomyia	74
simplicipes Curran, Rhaphium	514	slossonae Cockerell, Bibio	194
simplicipes Melander, Platypalpus	481	slossonae Cole & Wilcox, Lasiopogon	378
simplicis Curran, Bibio	193	*slossonae* Coquillett, Cordilura	830
simplicistyla (Alexander), Pedicia	55	slossonae (Coquillett), Helioplagia	1013
simplicitarse Van Duzee, Syntormon	516	slossonae Coquillett, Hybos	447
simson Fabricius, Anthrax	432	slossonae Coquillett, Lipochaeta	751
simulans (Johannsen), Psectrocladius	159	slossonae Coquillett, Nostima	745
simulans (Meigen), Guerinia	†1055, 1055	slossonae (Coquillett), Palpomyia	140
simulans Root & Hoffman, Culicoides	130	slossonae Coquillett, Psilocephala	352

Entry	Page
slossonae Coquillett, Sapromyza	704
slossonae (Coquillett), Triachora	1082
slossonae (Cresson), Pseudohecamede	737
slossonae (Curran), Mesograpta	571
slossonae Curran, Sericomyia	603
slossonae Dyar & Shannon, Simulium	186
slossonae Felt, Catocha	243
slossonae (Hine), Efferia	395
slossonae (Johnson), Anthrax	431
slossonae (Johnson), Brachycera	306
slossonae Johnson, Nemotelus	310
slossonae Johnson, Nephrocerus	551
slossonae (Johnson), Rhaphium	514
slossonae Malloch, Cricotopus	158
slossonae Malloch, Pegomya	863
slossonae (Malloch), Spiniphora	533
slossonae (Shannon), Cheilosia	586
slossonae Shewell, Camptoprosopella	697
slossonae Townsend, Psalidopteryx	1067
slossonae (Townsend), Pseudotachinomyia	1057
slossonae Van Duzee, Diaphorus	520
slossonae Van Duzee, Dolichopus	496
slossonae Van Duzee, Keirosoma	515
slossonae (Williston), Pericoma	93
Slossonaemyia Townsend	1064
smaragdinus Gerstäcker, Eulonchus	404
smilacifolia Felt, Dasineura	261
smilacinae Felt, Asphondylia	269
smilacinae Bishop, Dasineura	261
smithae Alexander, Pedicia	55
smithi Brooks, Gonia	1076
smithi (Dodge), Sarcophaga	959
smithi Jenks, Ocnaea	404
smithi Lewis, Mallochohelea	139
smithi Sabrosky, Ocnaea	404
smithi Shaw, Fenderomyia	200
smithi (Wulp), Microtrichomma	1001
smithii (Brues), Megaselia	539
smithii Brues, Xanionotum	545
smithii (Coquillett), Probezzia	138
smithii (Coquillett), Wyeomyia	107
Smittia Holmgren	162
snodgrassii Coquillett, Canace	733
snoqualmiensis Alexander, Limnophila	66
snoqualmiensis Alexander, Tipula	38
snowhilli (Del Rosario), Telmatoscopus	94
snowi (Adams), Tenthredomyia	616
snowi Back, Holopogon	375
snowi (Cresson), Otites	646
snowi (Curran), Metasyrphus	561
snowi (Hart), Labostigmina	314
snowi Hering, Paroxyna	666
snowi (Hine), Efferia	395
snowi Painter, Geron	423
snowi (Wehr), Metasyrphus	561
snowi Wilcox, Laphystia	377
snowi Wirth & Jones, Culicoides	130
Snowiellus Hine	330
snowii Bezzi, Stichopogon	385
snowii (Doane), Nephrotoma	22
snowii (Haseman), Telmatoscopus	93
snowii (Hine), Tolmerus	401
snowii Van Duzee, Diaphorus	520
snyderi (Dodge), Sarcophaga	959
snyderi Seago, Fannia	897
Snyderia James	827
Sobarocephala Czerny	806
sobria Johannsen, Boletina	217
sobrians Huckett, Hylemya	851
sobrina (Collin), Hylemya	853
sobrina (Felt), Cincticornia	269
sobrina Osten Sacken, Dicranoptycha	52
sobrinus Quate, Telmatoscopus	94
sobrinus (Wheeler), Calyxochaetus	528
soccata Loew, Rhamphomyia	465
soccata (Walker), Phaonia	908
soccatus Melander, Platypalpus	481
soccatus Walker, Dolichopus	496
socculata (Zetterstedt), Pegomya	861
socia (Cresson), Parydra	750
socia Johannsen, Mycetophila	214
socia Osten Sacken, Exoprosopa	445
sociabilis Loew, Lispe	878
sociabilis (Osten Sacken), Limonia	44
sociabilis (Williston), Rhamphomyia	465
socialis (Felt), **Asteromyia**	275
socialis Felt, Mayetiola	264
socialis Namba, Rivellia	656
socialis (Stein), Emmesomyia	858
socialis (Zetterstedt), Fannia	897
sociata (Meigen), Leucophora	868
sociella (Zetterstedt), Fannia	897
socium (Johannsen), Megalopelma	224
socius Loew, Dolichopus	496
socius (Osten Sacken), Hybomitra	338
sodalis Loew, Diaphorus	520
sodalis (Loew), Nephrotoma	22
sodalis Osten Sacken, Cophura	367
sodalis Wheeler, Hydrophorus	506
sodalis Williston, Eclimus	427
sodalis (Williston), Hybomitra	341
sodalis (Williston), Melangyna	568
sodalitatis Felt, Rhabdophaga	258
sodom (Williston), Villa	440
solani (Felt), Neolasioptera	273
solarii Stone, Simulium	187
solaris (Loew), Trupanea	667
solata Reinhard, Euceromasia	1100
solatrix (Osten Sacken), Epiphragma	61
soleata Melander, Rhamphomyia	465
solenopsidis Brues, Commoptera	546
solidaginis Beutenmüller, (Asphondylia)	292
solidaginis Felt, Lestodiplosis	285
solidaginis (Felt), Lestremia	244
solidaginis Felt, Trotteria	276
solidaginis Felt, Winnertzia	253
solidaginis (Fitch), Eurosta	663
solidaginis (Kaltenbach), Phytobia	800
solidaginis (Loew), Rhopalomyia	266
solidaginis Osten Sacken, Lasioptera	272
solidus Van Duzee, Dolichopus	496
solita (Reinhard), Appendicia	1007
solita Walker, (Coenosia)	877
solita Walker, Notiphila	748
solita Walker, Phytomyza	805
solita (Walker), Quadrularia	889
solitaria (Osten Sacken), Limonia	44
solitaria Stein, Pegomya	861
solitaria Stein, Phaonia	905
solitaria (Zetterstedt), Pseudocoenosia	875
solitarius (Johnson), Nemomydas	359
solitarius Knab, Hypodermodes	910
solitus (Walker), Systoechus	410
solivaga Melander, Proclinopyga	467
sollicitans (Walker), Aedes	114
solonensis Wirth, Forcipomyia	126
solor Alexander, Hexatoma	70
solox Enderlein, Hybomitra	340
solstitialis Alexander, Limnophila	65
solus Bromley, Saropogon	383
Solva Walker	299
sombrea Harmston & Knowlton, Neurigona	519
sommermanae Stone, Cnephia	184
somnolenta Dyar & Shannon, Dixa	102
Somula Macquart	612
sonoita Wheeler, Gitona	762
sonomensis (Osten Sacken), Hybomitra	340
sonora Wirth, Forcipomyia	125
sonorensis Cook, Parascatopse	239
sonorensis Wirth & Jones, Culicoides	132
sonoriana Shannon, Cheilosia	585
sootryeni Ringdahl, Mydaea	892
SOPHIINI	1030
Sophophora Sturtevant	768
soporis (Reinhard), Icterichophyto	1024
sorbens Melander, Mythicomyia	419
sordens (Johannsen), Nanocladius	155
sordens (Wulp), Polypedilum	176
sordicolor (Townsend), Sitophaga	1049
sordida Aldrich, Lispe	878
sordida Bigot, Olfersia	920
sordida (Curran), Lespesia	1102
sordida (Fallén), Desmometopa	731
sordida Giglio-Tos, Microphthalma	982
sordida Loew, Empis	461
sordida Loew, Exoprosopa	445
sordida (Loew), Prionocera	19
sordida Loew, Rhamphomyia	466
sordida Packard, Mycetobia	191
sordida (Wiedemann), Neogriphoneura	702
sordida Wirth, Bezzia	141
sordida Wulp, Mycetophila	214

	Page		Page
sordida Zetterstedt, Copromyza	720	Speolepta Edwards	219
sordidatus Van Duzee, Dolichopus	496	spernata Dietz, Tipula	39
sordidellus Becker, Chlorops	790	spernax Osten Sacken, Tipula	29
sordidellus (Zetterstedt), Culicoides	128	sperryana Alexander, Gonomyia	74
sordidus Osten Sacken, Chrysops	327	sperryana Alexander, Tipula	39
sordipes (Adams), Leptocera	724	sperryorum Melander, Aphoebantus	430
sordipes Melander, Euhybus	448	speyeri (Barnes), Mycophila	250
sorex (Townes), Chironomus	168	sphaeristes Brues, Dolichopus	496
sorghicola (Coquillett), Contarinia	277	**Sphaerocera** Latreille	718
soror Alexander, Molophilus	89	SPHAEROCERIDAE	718
soror Bigot, Atomosia	387	**Sohaeromias** Curtis	138
soror (Macquart), Oscinella	780	SPHAEROMIINI	137
soror Melander, Tetanocera	694	sphaeronotus Jobling, Trichobius	921
soror (Williston), Bombyliomyia	992	sphaerophoria Curran, Neoascia	595
sororcula Czerny, Suillia	809	**Sphaerophoria** Lepeletier & Serville	569
sororcula (Loew), Rondaniella	228	sphagnicola (Alexander), Limonia	48
sororcula Osten Sacken, Dicranoptycha	52	sphagnicola Alexander, Nephrotoma	22
sororcula (Wiedemann), Dioxyna	†666	sphagnicola Shaw, Mycomya	221
sororcula Williston, Cheilosia	587	sphagnumensis Williams, Culicoides	131
sororculus Williston, Lordotus	412	**Sphecomyia** Latreille	612
sororia Williston, Sapromyza	704	**Sphecomyiella** Hendel	658
sororius Cresson, Pipunculus	552	**Sphegina** Meigen	593
sorosis (Williston), Caliprobola	608	sphenofrons Reinhard, Asseclamyia	1012
sorosis (Williston), Liriomyza	802	**Sphenometopa** Townsend	939
sorotes Loew, Dolichopus	495	*Sphenomyia* Aldrich	879
sospita (Huckett), Spilogona	883	sphenura Steyskal, Pteromicra	687
spadix (Cresson), Pherbellia	686	*sphingivora* (Townsend), Lespesia	1102
spaldingi Dietz, Tipula	38	*Sphixea* Rondani	614
spaldingi Painter, Villa	439	*Sphiximorpha* Rondani	615
Spallanzania Robineau-Desvoidy	1077	*sphondylii* Robineau-Desvoidy, Phytomyza	805
Spania Meigen	348	sphondylii (Schrank), Saltella	683
Spanipalpus Townsend	1003	**Sphyracephala** Say	638
Spanoparea Czerny	813	**Sphyromyia** Bigot	997
Sparnopolius Loew	413	*Sphyroperiscelis* Sturtevant	710
sparsa Alexander, Erioptera	84	*Sphyxea* Rondani	614
sparsa (Loew), Limnia	691	*Sphyximorpha* Rondani	615
sparsa (Wiedemann), Eutreta	661	spiculatus Alexander, Molophilus	89
sparsimacula Alexander, Dactylolabis	62	*Spilaria* Schnabl	886, †889
sparsipilosum Back, Dicolonus	370	**Spilochaetosoma** Smith	1006
sparsipuncta (Alexander), Pedicia	55	**Spilochroa** Williston	817
sparsus Van Duzee, Diaphorus	520	*Spilogaster* Macquart	886
sparus Alexander, Molophilus	89	**Spilogona** Schnabl	879
sparus Whitney, Tabanus	336	*Spilographa* Loew	676
spatha Doane, Tipula	39	**Spilomyia** Meigen	613
Spathichira Bigot	487	**Spilopteromyia** Malloch	911
Spathidexia Townsend	1021	spilota (Curtis), Ilythea	746
Spathimeigenia Townsend	1049	spina Van Duzee, Argyra	525
spathiophora Malloch, Fannia	897	spinata (Pettey), Bradysia	234
Spathiophora Rondani	836	spineiventer (Tothill), Nowickia	996
Spathiphora Rondani	836	spinerecta Alexander, Tipula	39
spathulata (Zetterstedt), Coelomyia	898	spinicauda Van Duzee, Rymosia	208
Spatichira Bigot	487	spinicornis (Loew), Hoplodictya	691
spatulata (Aldrich), Blaesoxipha	945	spinicosta (Malloch), Cremersia	544
spatulata (Shewell), Lyciella	701	*spinicosta* (Malloch), Spilogona	883
spatulata Stone, Anastrepha	673	spinicosta (Zetterstedt), Helina	889
spatulatus (Harmston & Knowlton), Calyxochaetus	528	spinicostalis Huckett, Spilogona	883
spatulatus (Malloch), Pseudacteon	544	spinidens Malloch, Hylemya	852
spatulatus Pritchard, Heteropogon	374	spinifemorata Huckett, Hydrotaea	900
Spaziphora Rondani	836	spinifer Alexander, Gonomyia	76
speciosa Felt, Lobodiplosis	281	spinifer (Johannsen), Psectrocladius	159
speciosa Felt, Schizomyia	267	*spinifer* Malloch, Chrysotus	523
speciosa Loew, Tipula	36	spinifera (Alexander), Limonia	48
speciosa (Lopes), Blaesoxipha	947	spinifera (Leach), Olfersia	920
Speciosia Roback	947	spinifex Alexander, Ormosia	87
speciosus Van Duzee, Dolichopus	496	*spi-iformis* Patterson, (Cecidomyiaceltis)	295
spectabilis (Aldrich), Eufrontina	1100	spiniformis Patton, (Cecidomyiaceltis)	295
spectabilis (Aldrich), Peckia	953	spiniger Curran, Eucyrtopogon	373
spectabilis Doane, Tipula	37	*spiniger* (Malloch), Oscinella	779
spectabilis (Loew), Amoebaleria	815	spiniger (Stein), Lasiops	904
spectabilis (Loew), Clusia	806	spinigera Felt, Porricondyla	256
spectabilis Loew, Diaphorus	520	spinigera Malloch, Tricimba	785
spectabilis Loew, Empis	461	spinigera Townsend, Spathimeigenia	1049
spectabilis Loew, Gymnopternus	500	*spinigera* Van Duzee, Mycetophila	210
spectabilis Melander, Drapetis	478	spinigerella Malloch, Pegomya	861
spectabilis (Townes), Chironomus	168	spinilamellata Malloch, Helina	889
specularis(Aldrich & Webber), Catagoniopsis	1098	*spinilamellata* Malloch, Hylemya	852
specularis (Coquillett), Forcipomyia	125	spinilamellata Stein, Hylemya	852
speculifer Melander, Oligodranes	422	spinimana (Fallén), Norellia	827
speculifera (Walker), Linnaemya	1006	spinimanus (Zetterstedt), Scellus	506
speculum Hull & Fluke, Cheilosia	585	*spinipennis* (Coquillett), Helioplagia	1013
specus (Aldrich), Aecothea	811	spinipennis (Haliday), Leptocera	725
Spelobia Spuler	723	spinipennis (Meigen), Bigonicheta	1012
spenceri Alexander, Cramptonomyia	196	*spinipes* Aldrich, Lispe	878
spenceriana Alexander, Tipula	27	spinipes (Bigot), Hylemya	851
spencerii (Theobald), Aedes	114	spinipes (Coquillett), Anevrina	532
		spinipes (Fallén), Hydrotaea	899

INDEX 1677

	Page		Page
spinipes Melander, Euthyneura	449	spuleri Sabrosky, Leptocera	721
spinipes Melander, Stilpon	476	spurca Aldrich, Lutomyia	811
spinipes Say, Dilophus	195	*spuria* Curran, (Sarcophaga)	1116
spinipes Scopoli, Sepedon	693	spuria (Fallén), Bicellaria	451
spinipes Van Duzee, Thinophilus	508	spuria Malloch, Helina	889
spinipes Wirth, Mallochohelea	139	spurius (Fallén), Chalarus	550
Spiniphora Malloch	533	squalens (Zetterstedt), Helina	889
spinitalus Van Duzee, Diaphorus	520	*squalida* Meigen, Scatophaga	838
spinitarse Curran, Rhaphium	514	squamatus Andersen, Limnophyes	160
spinitarsis Aldrich, Pogonomyia	902	*squamifera* Stein, Pegomya	861
spinitarsis Harmston, Asyndetus	524	squamiger (Coquillett), Aedes	114
spinitarsis (Van Duzee), Tachytrechus	503	squamigera (Coquillett), Villa	435
spinitibia (Ringdahl), Spilogona	882	*squamipes* Cole, Ablautus	364
spiniventris Coquillett, Hylemya	852	squamipes (Coquillett), Forcipomyia	126
spiniventris (Hine), Efferia	395	squamosa Felt, Neolasioptera	273
spinosa (Coquillett), Chaetonopsis	1030	*squamosa* Felt, (Neolasioptera)	1116
spinosa (Coquillett), Ictericophyto	1024	squamosa Felt, Trotteria	276
spinosa (Coquillett), Velocia	1051	squamosa Hardy, Psilocephala	352
spinosa (Felt), Anarete	246	*squamosa* (Johnson), Brunettia	94
spinosa (Felt), Anocha	244	*squamosa* (Stein), Spilogona	879
spinosa (Felt), Feltiella	282	*squamosella* Kieffer, (Neolasioptera)	1116
spinosa Felt, Metadiplosis	285	squamosus Alexander, Molophllus	89
spinosa (Garrett), Eccoptomera	811	squamosus Coquillett, Aphoebantus	†429, 430
spinosa Hendel, Ischnomyia	819	squamosus Van Duzee, Dolichopus	496
spinosa (Loew), Dimecoenia	755	squamulae (Curran), Carposcalis	576
spinosa (Osten Sacken), Hexatoma	70	squamulae (Curran), Neocnemodon	581
spinosa (Tothill), Nowickia	994	squamulicola Stebbins, (Cecidomyia)	295
spinosa Townsend, Euceromasia	1100	squarrosae (Felt), Asteromyia	275
spinosa Walker, (Coenosia)	877	squiresi Alexander, Tasiocera	88
spinosa (Walker), Helina	889	stabulans (Fallén), Muscina	909
spinosissima Alexander, Dicranoptycha	52	*stabulans* (Meigen), Lydella	1103
spinosissima Malloch, Hylemya	846	staegeri Lundbeck, Chironomus	166
spinosissima Stein, Pegomya	861	*staegeri* (Zetterstedt), Azelia	899
spinosissima (Strobl), Chaetopleurophora	†533, 533	**Staegeria** Rondani	836
spinosulâ (Bigot), Plagiomima	†1016, 1019	Staegeria Wulp	224
spinosula (Townsend), Guerinia	1055	staffordi Cresson, Ceropsilopa	742
spinosus Melander, Platypalpus	481	stagmomantidis (Townsend), Blaesoxipha	947
spinosus Root & Hoffman, Culicoides	130	stagnalis (Fallén), Scatella	757
spinosus Van Duzee, Hydrophorus	506	stagnalis (Haliday), Clinocera	469
spinulae Felt, Lasioptera	272	stainsi Alexander, Dicranota	57
spinulosa Cole, Tethina	728	stainsi (Hardy), Pipunculus	554
spinulosa (Malloch), Spiniphora	533	*stalactoides* Doane, Tipula	32
spirae (Felt), Parallelodiplosis	286	stalagmites Alexander, Tipula	39
spiraeae Kaltenbach, Agromyza	795	stalkeri Wheeler, Drosophila	766
spiraeaflorae Felt, (Cecidomyia)	290	stalkeri Wheeler & Takada, Mycodrosophila	771
spiraeafolia Felt, Lasioptera	272	stamfordi (Johannsen), Chaetocladius	160
spiraeafolia Felt, Lestodiplosis	285	staminea (Williston), Efferia	395
spiraeina (Felt), Anaretella	245	stanfordensis Arnaud & Hoyt, Diadocidia	198
spiraeina Felt, Contarinia	277	stanfordi Harmston & Knowlton, Hercostomus	498
spiraeina Felt, Dasineura	261	stanleyae (Cockerell), Dasineura	261
spiraeina Felt, (Itonida)	290	stansburyi Johnson, Acrocera	406
spiralis Garrett, Dixa	102	stanwoodae (Alexander), Pilaria	68
spirifera Melander, Tetanocera	695	statuta Reinhard, Sarcophaga	959
Spirobolomyia Townsend	947	**Stearibia** Lioy	713
spitzbergensis (Kieffer), Orthocladius	156	steensensis Alexander, Pedicia	55
spizellae Huckett, Hylemya	855	**Stegana** Meigen	760
splendens Adams, Hippelates	775	**Steganina** Wheeler	760
splendens Curran, Belvosia	1082	STEGANINAE	760
splendens Doane, Cylindrotoma	41	**Steganolauxania** Frey	705
splendens Doane, Tipula	39	**Steganopsis** Meijere	704
splendens (Macquart), Microdon	599	steganoptera Malloch, Amiota	761
splendens Melander, Platypalpus	481	stegna (Say), Carposcalis	576
splendens Van Duzee, Argyra	526	**Stegomyia** Theobald	115
splendida Alexander, Erioptera	82	**Stegopterna** Enderlein	184
splendida (Coquillett), Phasiomyia	970	*steini* Johnson, Coenosia	871
splendida Macquart, (Calliphora)	933	*steini* (Schnabl), Hylemya	854
splendida (Macquart), Protocalliphora	†926	*steini* Verrall, Coenosia	873
splendida (Stein), Fannia	896	**Steinia** Brauer & Bergenstamm	1024
splendida Van Duzee, Argyra	526	**Steiniella** Berg	1024
splendida (Winnertz), Alluaudomyia	†133	*Steinomyia* Malloch	893
splendida Wirth, Forcipomyia	125	stejnegeri Aldrich, Coelopa	679
splendidulus Loew, Dolichopus	487	stellans (Loew), Anthrax	432
splendidus Loew, Dolichopus	490	stellaris Melander, Hemerodromia	471
splendidus Malloch, Gaurax	782	stellatus Coquillett, Tanypus	150
spleniata (Laffoon), Mycetophila	214	stellatus Martin, Holopogon	375
Spodius Loew	191	stellifer (Coquillett), Culicoides	130
Spoggosia Rondani	1057	*stelligera* (Coquillett), Euaresta	665
Spogostylum Macquart	431	**Stempellina** Bause	179
spoliata Loew, Euxesta	652	stenammatis (Long), Forcipomyia	126
spoliata Williston, Euxesta	652	*Stenaulacotheca* Townsend	943
spondylii Robineau-Desvoidy, Phytomyza	805	**Steneretma** Loew	653
spongivora Walker, (Cecidomyia)	292	stenhammari Zetterstedt, Dolichopus	496
Spongostylum Macquart	431	stenhammari (Zetterstedt), Limnellia	758
sporadicus Harmston & Knowlton, Dolichopus	496	*Stenhammaria* Duda	722
spretor (Reinhard), Blaesoxipha	944	**Stenochironomus** Kieffer	177

	Page		Page
Stenodiplosis Reuter	278	stimulans (Walker), Aedes	114
stenofrons Wilcox & Martin, Cyrtopogon	369	stimulans Walker (Sarcophaga)	961
Stenolaucotheca Townsend	943	*stipator* Osten Sacken, Eristalis	624
Stenomicra Coquillett	820	stipatus Walker, Helophilus	620
Stenomyia Loew	653	stolida (Reinhard), Palpexorista	1056
Stenoneura Reinhard	1071	stolida Walker, Mycetophila	214
Stenopa Loew	671	stolidus Alexander, Molophilus	89
stenophallus Alexander, Dicranoptycha	52	*Stomatolydella* Townsend	1047
Stenophora Malloch	532	**Stomatomyia** Brauer & Bergenstamm	1057
Stenopogon Loew	383	*Stomatorrhina* Rondani	923
stenoptera (Alexander), Dicranota	58	**Stomorhina** Rondani	923
stenoptera Loew, Empis	461	*Stomorrhina* Rondani	923
Stenopterina Macquart	657	**Stomosis** Melander	730
stenorhabda Alexander, Tipula	25	STOMOXYINAE	913
Stenoscinis Malloch	778	**Stomoxys** Geoffroy	914
Stenosyrphus Matsumura & Adachi	565	stoneana Alexander, Chionea	72
Stenotabanus Lutz	329	stonei Alexander, Tipula	25
STENOXENINI	139	stonei Blanc & Foote, Procecidochares	661
Stenoxenus Coquillett	139	stonei Cook, Psectrosciara	240
stenozona (Loew), Villa	435	stonei James, Culicoides	130
stentor (Melander), Anthepiscopus	455	stonei Kessel, Agathomyia	548
STEPHANOSTOMATIDAE	933	stonei Philip, Tabanus	336
stercoraria (De Geer), Smittia	162	*stonei* Stains & Knowlton, Simulium	189
stercoraria (Linnaeus), Scatophaga	839	stonei Wilcox & Martin, Nannocyrtopogon	380
stercoraria (Meigen), Copromyza	719	stonei Wirth, Monohelea	135
stercorarius (Robineau-Desvoidy), Meroplius	682	stonei Wirth, Palpomyia	140
sterilator Lugger, Cuterebra	1110	stonei Wirth, Stilobezzia	135
Steringomyia Pokorny	928	**Stonemyia** Brennan	321
sternalis Allen, Eumacronychia	935	**Stonyx** Osten Sacken	440
sternalis (Coquillett), Eumasicera	1094	*straminea* (Doane), Orellia	671
sternalis (Coquillett), Nilea	1104	straminea Loew, Opetiophora	783
sternalis Curran, Volucella	602	straminea (Malloch), Megaselia	539
sternalis Loew, Psila	639	straminea Osten Sacken, Erioptera	81
sternalis (Reinhard), Archimimus	942	*stramineus* (Williston), Efferia	395
sternata Doane, Tipula	39	straminipes (Malloch), Megaselia	539
sternodontis Townsend, Sarcodexia	955	strataegum (Wheeler), Syntormon	516
Steveniopsis Townsend	985	stratifrons Huckett, Hylemya	846
stewarti Coleman, Cnephia	184	*Stratiohorborus* Duda	720
stewarti Mangabeira & Galindo, Phlebotomus	92	*Stratiomyia* Geoffroy	311
		STRATIOMYIDAE	299
steyskali Foote, Pteromicra	687	STRATIOMYINAE	310
steyskali Namba, Rivellia	656	STRATIOMYINI	311
steyskali Sabrosky, Acrocera	406	**Stratiomys** Geoffroy	311
Sthenopleura Aldrich	1017	*Straussia* Robineau-Desvoidy	676
Sthenopyga Aldrich	950	**Strauzia** Robineau-Desvoidy	676
Stichillus Enderlein	535	STREBLIDAE	921
Stichopogon Loew	384	*strepens* Alexander, Limnophila	66
stictica (Meigen), Erioptera	84	strepens Loew, Tipula	29
stictica Meigen, Phora	537	streptocera Doane, Tipula	40
sticticum (Loew), Trypetisoma	705	striata Malloch, Sphaerocera	719
sticticus (Meigen), Aedes	114	striata Schiner, Anastrepha	673
stictipennis Wirth, Alluaudomyia	134	striata Stein, Pegomya	861
Stictocephala Loew	647	striata (Stein), Phaonia	908
Stictochironomus Kieffer	177	*striata* (Walker), Scatophila	758
Stictomyia Bigot	654	striata (Williston), Leia	227
Stictoscatella Collin	757	striatella Wulp, Rhagoletis	674
stigma (Alexander), Dicranota	58	*striatifrons* (Kröber), Physoconops	627
stigma (Fabricius), Notogramma	646	striatifrons Malloch, Lonchaea	717
stigma (Hendel), Otites	646	striatipes Walker, Bibio	194
stigmalis Coquillett, Anthalia	450	striatus (Malloch), Trichocladius	158
stigmalis Coquillett, Ceratopogon	133	striatus Osten Sacken, Chrysops	327
stigmata (Doane), Limonia	48	striatus Painter, Lordotus	412
stigmata Harris (Leptis)	1116	*stricklandi* (Curran), Deopalpus	1004
stigmata Williston, Chrysogaster	591	stricklandi Curran, Helophilus	618
stigmatella (Zetterstedt), Oedalea	†450	stricklandi Curran, Pipunculus	554
stigmaterus Say, Chironomus	166	*stricklandi* Fisher (Exechia)	1116
stigmaterus Say, Dilophus	196	stricklandi Harmston & Knowlton, Dolichopus	496
stigmatias Loew, Euxesta	652		
stigmatica Coquillett, Euaresta	665	stricklandi (Laffoon), Mycetophila	214
stigmatica (Coquillett), Tephritis	668	stricklandi Steyskal, Tetanocera	695
stigmatica (Dietz), Nephrotoma	21	stricta Steyskal, Dictya	689
stigmatica (Osten Sacken), Cheilotrichia	79	strigata Coquillett, Tipula	27
stigmatica Wulp, Micropeza	635	strigata Staeger, Mycetophila	214
stigmaticalis Becker, Chlorops	790	strigata Staeger, Sphaerophoria	570
stigmaticalis Loew, Euparyphus	307	*strigata* Stein, Hylemya	845
stigmaticalis (Thomson), Phormia	924	*strigata* (Wulp), Blondelia	†1040
stigmaticus Malloch, Pipunculus	554	strigatus (Fallén), Eumerus	597
stigmatipennis Lovett, Xylota	606	strigifemur Stein, Coenosia	872
stigmatosoma Dyar, Culex	118	strigilata Loew, Myolepta	589
stigmatus Becker, Chlorops	790	strigilatus McAtee, Dilophus	196
Stilbometopa Coquillett	918	*strigilecula* (Ferris), Ornithoctona	918
stilla (Reinhard), Frontiniella	1100	strigilifer Melander, Microphorus	453
Stilobezzia Kieffer	134	stripipes Curran, Lestomyia	378
STILOBEZZIINI	134	stripipes Johnson, Orthacheta	832
stilobezzioides Foote & Pratt, Culicoides	130	stripipes Loew, Thereva	354
Stilpon Loew	476	strigosa (Reinhard), Johnsonia	950

INDEX

Entry	Page
strigula Loew, Pterocalla	647
striolata (Zetterstedt), Hylemya	853
strobiligemma Stebbins, (Cecidomyia)	295
strobiliscus Osten Sacken, (Cecidomyia)	292
strobiloides (Osten Sacken), Rhabdophaga	258
strobiloides Townsend, (Cecidomyia)	293
strobilophilus Foote, Holoneurus	255
STRONGYGASTRINI	976
Strongylophthalmus Hendel	640
Strongylophthalmyia Heller	640
STRONGYLOPHTHALMYIINAE	640
strumaticus Melander, Euhybus	448
struthio Walker, Tetanocera	695
stulta (Osten Sacken), Limonia	48
stupkai Alexander, Limnophila	66
stuprator Melander, Thanategia	472
Sturmia Robineau-Desvoidy	†1084, 1087
STURMIINI	1083
Stygeropis Loew	18
stygia (Bromley), Orthogonis	391
stygia (Coquillett), Leptocera	724
stygia (Fabricius), Calliphora	929
stygia Harris, Melanophora	963
stygia Meigen, (Anthomyia)	869
stygius (Enderlein), Whitneyomyia	331
stygius Say, Dilophus	196
stygius Say, Tabanus	336
stygius Townes, Microtendipes	173
stylata (Brauer & Bergenstamm), Petinarctia	1015
stylata Coquillett, Rhamphomyia	466
stylata (Fabricius), Stylogaster	632
stylifer Alexander, Ormosia	86
stylifer Van Duzee, Physocephala	628
stylifera Dietz, Tipula	34
stylifera Johannsen, Chironomus	168
Styloconops Kieffer	122
Stylogaster Macquart	632
STYLOGASTRINAE	632
Stylomyia Westwood	632
Styloneuria Brauer & Bergenstamm	964
stylosa Curran, Cophura	367
suavis (Loew), Rhagoletis	675
subaenescens Aldrich, Sarcophaga	959
subaequalis (Malloch), Paratendipes	174
subalbipes (Johnson), Dolichopeza	23
subangulata (Malloch), Phytobia	799
subannulata (Loew), Mesograpta	571
subapache Alexander, Tipula	25
subapicata Alexander, Limonia	49
subaptera (Alexander), Pedicia	55
subaptera (Malloch), Copromyza	719
subapterogyne Alexander, Dicranota	57
subarctica Alexander, Rhabdomastix	77
subarctica Alexander, Tipula	31
subarctica Ringdahl, Chirosia	844
subarcuata Johnson, Palloptera	715
subarcuata Schmitz, Gymnophora	542
subarida (Bromley), Efferia	395
subarmatum Curran, Rhaphium	514
subasper (Coquillett), Palpomyia	140
subaterrima (Malloch), Smittia	162
subatomella (Malloch), Megaselia	541
subatra Fisher, Boletina	217
subauratus Loew, Geron	423
subbarbata Alexander, Tipula	34
subbifasciata Fitch, Sphyracephala	638
subborealis (Alexander), Cheilotrichia	79
subbuccata Bau, Cuterebra	1110
subcaerulea Thomsen, Dasyhelea	127
subcaeruleus (Coquillett), Leptomorphus	223
subcantans (Felt), Aedes	114
subcaudata Alexander, Rhabdomastix	77
subchalybea Curran, Cheilosia	585
subchalybea Curran, Neoascia	595
subchlorophylla Alexander, Erioptera	81
subciliata (Malloch), Megaselia	540
subciliatus Loew, Dolichopus	496
subcincta (Townes), Paralauterborniella	173
subcinerea Doane, Tipula	26
subcinerea Osten Sacken, Gonomyia	76
subcornuta Alexander, Ormosia	87
subcostata (Alexander), Limnophila	66
subcostata Dietz, Ormosia	87
subcostatus Loew, Chrysotus	523
subcostatus Van Duzee, Dolichopus	496
subcultellatum Sublette, Polypedilum	176
subcuprea (Schaeffer), Efferia	395
subdentifera Alexander, Ormosia	87
subdilatatus Loew, Gymnopternus	500
subdirectus Van Duzee, Dolichopus	496
subdiscalis Aldrich, Sarcophaga	959
subdivisa (Townsend), Trichopoda	966
subdola (Johnson), Piophila	713
subelata Malloch, Allodia	205
subeluta Johnson, Tipula	29
suberecta Sabrosky, Leptocera	721
subexcisum Edwards, Simulium	186
subfasciata Aldrich, Elodia	1071
subfasciata Curran, Eurosta	663
subfasciata (Curran), Melangyna	568
subfasciata (Harris), Chymomyza	770
subfasciata Loew, Tipula	34
subfasciata Loew, Xylota	606
subfasciata (Say), Laphystia	376
subfasciata Westwood, Acrocera	406
subfascigera Alexander, Rhabdomastix	77
subflavus Painter, Parabombylius	410
subflavus Van Duzee, Dolichopus	496
subfronto Philip, Tabanus	333
subfunebris Stalker & Spencer, Drosophila	766
subfurcatum Van Duzee, Rhaphium	514
subfurcifera Alexander, Erioptera	81
subfusca Malloch, Phaonia	908
subfusca (Malloch), Triphleba	534
subfuscata Felt, Trotteria	276
subfuscinervis (Zetterstedt), Phaonia	908
subgrandis (Shaw), Bradysia	234
subgrisea Malloch, Pegomya	858
subinfumata Malloch, Empis	461
subinfumata (Malloch), Phytobia	800
subita (Laffoon), Mycetophila	214
subjacens Walker, Gasterophilus	916
subjectus Van Duzee, Chrysotus	523
subjectus (Walker), Euhybus	448
sublettei Beck, Chironomus	168
subligulata Shaw, Exechia	207
sublittoralis Shaw, Mycomya	221
sublongus Stone, Tabanus	336
sublugubrina (Malloch), Leptocera	†723
sublunata (Loew), Leia	228
sublutea (Malloch), Megaselia	541
submacula Banks, Platypeza	549
submaculata Loew, Tipula	39
submanicata (Malloch), Megaselia	541
submarginalis (Curran), Epistrophe	565
submarginalis (Malloch), Megaselia	541
submarginalis (Sabrosky), Olcella	785
submissa (Aldrich & Webber), Meigenia	1044
subniger Coquillett, Chlorops	790
subniger Coquillett, Tabanus	336
subnigra Fisher, (Sceptonia)	1116
subnigripes Malloch, Agromyza	796
subnitens (Aldrich & Webber), Euphorocera	1054
subnitens Alexander, Molophilus	89
subnitens (Cockerell), Neorhynchocephalus	402
subnitens Cresson, Hydrellia	744
subnitens (Cresson), Tomosvaryella	556
subnitens (Reinhard), Steveniopsis	985
subnitida Malloch, Hylemya	854
subnitida Sturtevant & Wheeler, Hyadina	751
subnotata Johnson, Psilocephala	352
subnubila Alexander, Ormosia	87
subnubila Cresson, Ditrichophora	739
subnuda (Alexander), Tasiocera	88
subnudipennis Goetghebuer, Smittia	163
subnudus Sturtevant & Wheeler, Pelignellus	737
subobscurata (Malloch), Megaselia	541
subobsoleta (Alexander), Catotricha	243
subobtusa Alexander, Pedicia	54
suboccidentalis Spencer, Drosophila	766
subole Reinhard, Charasoma	1040
subopaca (Coquillett), Alophorella	968
subopaca (Coquillett), Euphyto	935
subopaca Loew, Ephydra	754
subopacus Loew, Pipunculus	554
subpalustris Spencer, Drosophila	766
subparallelus Malloch, Orthocladius	157
subpellucens Malloch, Hydrophoria	864
subpellucens (Zetterstedt), Coelomyia	898
subpellucida Malloch, Hydrophoria	864
subpicta (Malloch), Megaselia	539

1680 A CATALOG OF DIPTERA OF NORTH AMERICA

Name	Page
subpiligera (Malloch), Leptocera	†721
subpilosa (Schaeffer), Efferia	395
subpilosus (Kieffer), Heterotrissocladius	155
subpleuralis (Wood), Megaselia	541
subpolaris Alexander, Tipula	41
subpolita (Brooks), Linnaemya	1006
subpolita Malloch, Lonchaea	717
subpolita Malloch, Scatophaga	839
subpubescens Collin, Fannia	896
subpura (Johnson), Tephritis	668
subpusilla (Frost), Liriomyza	802
subquatuornotata Shaw, Mycetophila	214
subquinaria Spencer, Drosophila	766
subrobusta Gill, Pseudoleria	812
subrostrata (Stein), Spilogona	883
subrostratus (Zetterstedt), Lasiops	903
subrufa Cole, Psilocephala	352
subseptemtrionis Alexander, Erioptera	81
subserrata (Shewell), Lyciella	701
subserta Alexander, Tipula	34
subsessilis (Alexander), Dicranota	58
subsimilis Alexander, Limnophila	66
subsimplex Loew, (Coenosia)	1116
subsimus Fluke, Metasyrphus	561
subsinuatum (Alexander), Diazosma	15
subspina Van Duzee, Dolichopus	496
subspinata Huckett, Hylemya	852
subsquamosa Kieffer, Neolasioptera	273
substituta Richards, Leptocera	†723
substituta (Walker), Pegomya	860
substriata (Stein), Hylemya	853
substriatella (Malloch), Pegomya	861
subsultans (Linnaeus), Sphaerocera	†534, †718
subtendens (Townes), Chironomus	170
subtenella Frost, Phytomyza	805
subtenuicornis (Alexander), Limnophila	63
subtenuicornis Doane, Tipula	25
subteresa Garrett, Bolitophila	198
subterminalis (Say), Orfelia	203
subtilis Doane, Tipula	39
subtilis (Hardy), Pipunculus	553
subtilis Quate, Telmatoscopus	94
subtrivialis (Pettey), Bradysia	234
subtropica Curran, Xylota	607
subtruncata Alexander, Dicranota	58
subtruncata (Felt), Parallelodiplosis	[286
subtruncifer Alexander, Dicranota	57
subtusradiata Duda, Amiota	761
Subula Meigen	299
Subulaomyia Williston	299
subulata Coquillett, Lipoptena	†920, 920
subulata (Loew), Adoxomyia	306
subulatus Loew, Gymnopternus	500
subulatus (Wiedemann), Stenopogon	[384
subunica (Alexander), Austrolimnophila	61
subusta Malloch, Palloptera	715
subvarius Johnson, Bombylius	410
subvenosa Alexander, Dolichopeza	23
subvirens (Malloch), Melanagromyza	796
subvirescens (Loew), Tomosvaryella	556
subvittata (Loew), Cetema	788
subvittata (Malloch), Gimnomera	840
subvittatus Malloch, Hippelates	775
succedens Stein, Hydrotaea	900
succedens Walker, Ctenophora	20
succinata (Wiedemann), Rivellia	656
succincta Alexander, Tipula	29
succincta (Coquillett), Villa	436
succinctum (Townes), Paralauterborniella	173
sudator Osten Sacken, Cyrtopogon	369
sudeticus Becker, Eribolus	778
sudiai (Dodge), Sarcophaga	959
sudigeronis Coquillett, Rhamphomyia	466
sueta (Wulp), Ravinia	955
suffalcatus Alexander, Molophilus	89
suffasciatus Melander, Platypalpus	480
sufflavus Van Duzee, Dolichopus	496
sufflexa Melander, Hemerodromia	471
suffumata Alexander, Ormosia	84
suffusa Alexander, Limnophila	63
suffusa Melander, Trixoscelis	818
suffusca (Garrett), Limonia	49
suffusca Spencer, Drosophila	766
sugens (Loew), Imitomyia	975
sugens Melander, Mythicomyia	419
sugens (Wiedemann), Physoconops	626
suilla (Fabricius), Scatophaga	839
Suillia Robineau-Desvoidy	808
SUILLIINAE	808
Suioestrus Townsend	1112
suisterci (Townsend), Scatophaga	838
sulaceps Townes, Polypedilum	176
sulcata (Aldrich & Webber), Palpexorista	1056
sulcata Sabrosky, Ectecephala	791
sulcata Sturtevant, Drosophila	764
sulcatifemur Borgmeier, Anevrina	532
sulcifrons Coquillett, Ectecephala	791
sulcifrons Macquart, Tabanus	336
sulcocarina (Tothill), Mericia	1008
sulculata (Aldrich), Metoposarcophaga	951
sulfuriceps Malloch, Schoenomyza	870
sulfuriceps Strobl, Agromyza	795
sulfurifrons (Duda), Thaumatomyia	†787
sulphurea Doane, Tipula	29
sulphurea Loew, Phthiria	420
sulphurella Osten Sacken, Gonomyia	75
sulphureus Loew, Chlorops	790
sulphureus (Palisot de Beauvois), Chlorotabanus	†330
sulphuripes Loew, Opsebius	405
sulphuripes (Thomson), Sphaerophoria	570
sundewalli Kloet & Hincks, Elgiva	690
sunwapta Alexander, Erioptera	81
superba Foote, Paraterellia	671
superba Loew, Palloptera	715
superbiens (Loew), Neurigona	518
superbus (Banks), Telmatoscopus	94
superbus Van Duzee, Dolichopus	496
superbus (Wiedemann), Condylostylus	485
superjumentarius Whitney, Tabanus	336
superlineata Doane, Limnophila	65
supernumeraria Alexander, Dactylolabis	62
superstitiosa Say, Hemerodromia	471
supina (Coquillett), Villa	435
supplens (Walker), Hypoderma	1112
supplicata Alexander, Tipula	40
Suragina Walker	344
surda (Zetterstedt), Spilogona	883
surdus Curran, Chlorops	790
surdus Osten Sacken, Chrysops	327
suspecta Cresson, Hydrellia	744
suspecta Dietz, Tipula	31
suspecta Loew, Tipula	32
suspecta (Malloch), Spilogona	883
suspecta (Malloch), Triphleba	534
suspensa (Loew), Anastrepha	673
sussurans Jaennicke, Helophilus	618
susurra (Marten), Hybomitra	340
sutilis Reinhard, Sarcophaga	959
sutor Melander, Platypalpus	481
suttoni Alexander, Tipula	30
suturalis (Loew), Nephrotoma	21
swainei Felt, Rhabdophaga	258
Swammerdamella Enderlein	239
SWAMMERDAMELLINI	239
swannanoa Brimley, Cheilosia	585
sweetae Alexander, Tipula	41
sweetmani Alexander, Erioptera	84
swezeyi Wilcox & Martin, Cyrtopogon	369
sybleae (Wirth), Mallochohelea	139
sybleae Wirth, Stilobezzia	134
sycophanta Quate, Brunettia	94
sycophantor (Melander), Microphorus	453
syletor Melander, Charadromia	475
sylphida (Bigot), Phaenicia	927
sylvana (Felt), Catocha	243
sylvania Kessel, Agathomyia	548
sylvarum (Meigen), Bufolucilia	927
sylvatica Brues, Puliciphora	545
sylvatica (Fallén), Masicera	1103
silvatica (Meigen), Tomosvaryella	556
sylvestrii (Kieffer), Bradysia	234
sylvestris (Fabricius), Cricotopus	158
sylvestris (Felt), Arthrocnodax	283
sylvestris Felt, Asteromyia	275
sylvestris Felt, Bremia	280
sylvestris Felt, Colpodia	253
sylvestris (Felt), Cordylomyia	248
sylvestris (Felt), Dirhiza	255
sylvestris Felt, Lestremia	244
sylvestris Felt, Porricondyla	256
sylvestris (Kieffer), Bradysia	234

INDEX 1681

	Page		Page
sylvestris (Robineau-Desvoidy), Parasetigena	1056	tacoma Cresson, Ditrichophora	739
sylvestris (Theobald), Aedes	116	tacoma Melander, Tachydromia	474
sylvia (Alexander), Lipsothrix	77	tacomae Melander, Anomalempis	447
sylvicola Doane, Tipula	39	*tacomae* Melander, Cyrtopogon	368
sylvicola (Grossbeck), Aedes	112	tacomae Melander, Microphorus	453
Sylvicola Harris	190	tacomicola Alexander, Tipula	31
SYLVICOLIDAE	190	tacta Huckett, Pegomya	861
sylvosus (Williston), Physoconops	627	taediosa (Aldrich), Blaesoxipha	944
Symballophthalmus Becker	476	taeniapennis (Coquillett), Stenochironomus	177
Symbiocladius Kieffer	157	**Taeniaptera** Macquart	636
symmachus Loew, Diogmites	372	**TAENIAPTERINAE**	635
Symmerus Walker	199	taeniata Alexander, Gonomyia	76
symmetrica Osten Sacken, (Cecidomyia)	295	taeniata (Bellardi), Olbiógaster	191
symphoricarpi Curran, Rhagoletis	675	taeniata Snow, Platypeza	549
Symphoromyia Frauenfeld	343	*taeniata* (Winnertz), Orfelia	203
Symplecta Meigen	80	taeniatifrons (Enderlein), Cnephia	185
symplocarpi Felt, Lobopteromyia	276	*taeniatus* (Wiedemann), Aedes	116
SYMPYCNINAE	526	taenigaster Alexander, Tipula	31
Sympycnus Loew	527	taeniocera Dietz, Ormosia	87
Synamphotera Loew	†466	*taeniola* (Coquillett), Phytobia	798
Synapha Meigen	219	taenionotus Say, Chironomus	171
Synaplomyia Villeneuve	1038	*Taeniorhynchus* Lynch Arribálzaga	110
Synarthrus Loew	515	taeniorhynchus (Wiedemann), Aedes	115
synchroa Alexander, Tipula	25	*Taeniotabanus* Kröber	332
syndesmus (Coquillett), Parabombylius	410	tagax Melander, Mythicomyia	419
Syndiamesa Kieffer	151	tagax (Williston), Efferia	395
Syndyas Loew	448	tagax (Williston), Zabrops	386
Syneches Walker	448	tahoensis Alexander, Ormosia	87
Syneura Brues	542	tahoensis (Coquillett), Caloparyphus	308
Synneuron Lundström	237	*tahoensis* Dyar, Aedes	111
Synoris Aldrich	1044	tahoensis Harmston & Knowlton, Tachytrechus	503
Syntemna Winnertz	†218		
Synthesiomyia Brauer & Bergenstamm	911	tahoensis (Reinhard), Nowickia	994
syntheta Dyar & Shannon, Uranotaenia	108	talaris Melander, Platypalpus	481
synthetica (Townsend), Blaesoxipha	945	talpina Reinhard, Phrosinella	938
Synthyridomyia Saunders	124	talpula (Loew), Agathomyia	548
Syntomogaster Schiner	977	talus Fluke, Metasyrphus	561
syntormoides Wheeler, Asyndetus	524	talus Van Duzee, Dolichopus	496
Syntormon Loew	515	tamaulipana Townsend, Volucella	602
Syritta Lepeletier & Serville	608	tamia (Melander), Melanagromyza	796
SYRPHIDAE	557	*tanbarkenis* Painter & Hall, Poecilanthrax	443
Syrphidis Goffe	559	tanbarkensis Painter & Hall, Poecilanthrax	443
SYRPHINAE	558	*tandrec* Robineau-Desvoidy, Theresia	991
SYRPHINI	558	*tangomus* (Rapp), Tomosvaryella	556
SYRPHOIDEA	550	*Tanipoda* Rondani	636
syrphoides Hull, Didea	563	tanneri Alexander, Tipula	39
Syrphus Fabricius	558	tanneri (Bromley), Efferia	396
syrtis (Coquillett), Villa	439	tanneri Hardy, Zionea	349
Systellapha Enderlein	635	*tantalisa* Huckett, Hylemya	854
Systenus Loew	517	tantalus Hull & Fluke, Cheilosia	585
Systoechus Loew	410	tantilla Alexander, Erioptera	82
SYSTROPODINAE	424	tantilla (Coquillett), Villa	438
Systropus Wiedemann	424	tantilla (Loew), Mycomya	221
Sziladynus Enderlein	337	**TANYDERIDAE**	90
		Tańypeza Fallén	641
TABANIDAE	319	**TANYPEZIDAE**	641
TABANINAE	328	*Tanypoda* Rondani	636
TABANINI	330	**TANYPODINAE**	144
tabanivorus (Hall), Carinosillus	989	**Tanyptera** Latreille	19
TABANOIDEA	296	*tanypus* Loew, Dolichopus	493
Tabanus Linnaeus	332	tanypus Loew, Phthinia	225
tabellaria (Fitch), Rhagoletis	675	**Tanypus** Meigen	149
tabescens (Banks), Efferia	395	**TANYTARSINI**	178
tabida Reinhard, Imanguncula	1077	**Tanytarsus** Wulp	179
Tabuda Walker	352	taos Melander, Clinocera	469
Tachina Meigen	998, †1054	tapetis (Coquillett), Euaresta	665
Tachinalia Curran	992	*Taphrophila* Rondani	43
TACHINIDAE	961	**Taracticus** Loew	385
TACHININAE	992	tarandi (Linnaeus), Hypoderma	1112
TACHININI	993	*tarandi* (Walker), Hybomitra	341
tachinomoides Townsend, Euphorocera	1054	*tarchetius* Walker, Baccha	572
Tachinomyia Townsend	1058	*tarda* Harris, Gonia	1076
Tachinophyto Townsend	1049	tarda Snow, Cheilosia	587
Tachinophytopsis Townsend	1043	tardus Coquillett, Aphoebantus	430
Tachinopsis Coquillett	1058	taris Roback, Corynoneura	164
Tachista Loew	474	tarsalis Banks, Pipunculus	554
tachistiformis Melander, Platypalpus	481	tarsalis Becker, Chlorops	790
Tachydromia Meigen	474, †478	tarsalis Coquillett, Admontia	1068
TACHYDROMIINAE	473	tarsalis Coquillett, Culex	119
Tachyempis Melander	475	tarsalis (Coquillett), Elfia	1066
Tachypeza Meigen	473	*tarsalis* (Curran), Calyxochaetus	528
Tachysta Loew	474	tarsalis Felt, Holoneurus	255
Tachytrechus Haliday	502	tarsalis Johnson, Cylindrotoma	41
Tachytrechus Stannius	502	tarsalis Loew, Desmometopa	731
tacoma (Alexander), Gonomyodes	79	tarsalis (Malloch), Cordilura	830
tacoma Alexander, Pedicia	55		

744–243—65——107

	Page		Page
tarsalis (Schummel), Platycheirus	577	tenera (Loew), Agathomyia	548
tarsalis Townsend, Chaetophlepsis	†1016	tenera (Wiedemann), Lesklomima	1035
tarsalis Van Duzee, Chrysotus	523	tenera (Zetterstedt), Pegomya	862
tarsalis Van Duzee, Neurigona	519	tenessensis (Bigot), Tabanus	334
tarsalis Walker, (Gymnopa)	759	teneus Hardy, Bibio	194
tarsalis (Walker), Lasiops	904	tennessa Alexander, Dicranoptycha	52
tarsata Aldrich, Sarcophaga	959	tennessa Alexander, Elliptera	51
tarsata Fallén, Lonchaea	717	tennessa Alexander, Tipula	27
tarsata Felt, Trotteria	276	tennesseensis Alexander, Ormosia	87
tarsata Loew, (Hylemya)	1116	tenta Hall, Sarcofahrtiamyia	956
tarsata (Ringdahl), Hylemya	851	tentaculata (De Geer), Lispe	878
tarsata (Williston), Hydropyrus	754	tentans Fabricius, Chironomus	166
tarsata (Wulp), Pegomya	862	tentatrix Loew, Euthera	978
tarsata (Zetterstedt), Melangyna	568	tenthes Melander, Mythicomyia	419
Tarsohomoneura Hendel	697	tenthredinidarum (Townsend), Bessa	†1042, 1053
Tasiocera Skuse	87	tenthredinivora (Townsend), Guerinia	1055
tau Bigot, Volucella	602	tenthredinoides (Wulp), Xylomya	299
tau Philip, Chrysops	327	**Tenthredomyia** Shannon	616
tau Sabrosky, Elachiptera	777	*tenuicalx* Enderlein, Prosimulium	†183, 183
taughannock Alexander, Tipula	35	tenuicauda Van Duzee, Gymnopternus	500
tausa Huckett, Coenosia	873	tenuicaudatus Malloch, Chironomus	168
taxiconis Foote, Lestodiplosis	285	tenuicornis Cresson, Sepedon	693
Taxigramma Perris	939	tenuicornis (Malloch), Palpomyia	140
taxodii Felt, (Itonida)	290	tenuicornis Osten Sacken, Limnophila	63
taxodii Felt, (Retinodiplosis)	290	tenuiforceps (Morrison), Onychogonia	1077
taylori Philip, Chrysops	326	tenuilinea Alexander, Tipula	29
teapanus Aldrich, Chrysotus	523	tenuimanus Van Duzee, Dolichopus	496
teate (Walker), Hydrophoria	865	tenuior Walker, (Cordylura)	877
tecomae Felt, (Bremia)	290	tenuipes Aldrich, Dolichopus	496
tectura (Adams), Zodion	629	*tenuipes* Coquillett, Bibio	192
tectus Osten Sacken, Tabanus	336	tenuipes (Loew), Apachekolos	361
tectus Van Duzee, Thrypticus	511	tenuipes (Loew), Nemomydas	359
tegminipennis (Say), Poecilanthrax	443	tenuipes (Loew), Pherbellia	687
tehama Alexander, Dicranota	58	tenuipes (Osten Sacken), Dicranota	59
tehamicola Alexander, Dicranota	58	tenuipes (Say), Pilaria	68
teinoptera Hackman, Scaptomyza	769	tenuipes (Walker), Blepharicera	99
Tejasomyia Reinhard	949	tenuipes (Walker), Drosophila	766
teleta Reinhard, Microchaetina	984	*tenuipes* (Zetterstedt), Limonia	48
telfordi Alexander, Erioptera	84	**Tenuirostra** Ringdahl	1006
teligera Fluke, Mesograpta	571	tenuirostris Reinhard, Philocalia	1061
teliosis Reinhard, Chaetophlepsis	1016	tenuis Bromley, Cyrtopogon	369
telluris (Coquillett), Villa	440	tenuis Curran, Siphona	1064
Telmaphilus Becker	214	*tenuis* Drew, Hylemya	850
Telmatogeton Schiner	163	tenuis (Loew), Anthomyza	819
TELMATOGETONINI	163	tenuis (Loew), Nephrotoma	22
Telmatoscopus Eaton	93	tenuis (Loew), Neurigona	519
Telmaturgus Mik	529	tenuis Loew, Stenomyia	654
TELOTHYRIINI	1036	*tenuis* (Meigen), Tanytarsus	180
temeraria (Dietz), Nephrotoma	21	tenuis Melander, Platypalpus	481
temerarium (Becker), Rhaphium	514	tenuis (Osburn), Melangyna	568
temeritatis Felt, Colpodia	253	tenuis Walker, (Asteia)	1113
Temnocera Lepeletier & Serville	600	tenuis (Walker), Mycomya	221
Temnostoma Lepeletier & Serville	613	tenuiseta Aldrich & Webber, (Phorocera)	1108
temporalis Thomson, Eristalis	624	tenuiseta Greene, Tachytrechus	503
tenaculata Sabrosky, Leptocera	721	tenuispina (Loew), Homoneura	699
tenax Johannsen, Hylemya	845	tenuissima (Hendel), Stenomyia	654
tenax (Linnaeus), Eristalis	624	tenuistylus Hanson, Nemotelus	310
tenax Melander, Platypalpus	481	tenuistylus Wirth, Culicoides	132
tenaya Alexander, Tipula	39	tenuitas Felt, Mycodiplosis	283
TENDIPEDIDAE	142	tenuitas Felt, Neolasioptera	273
tendipes (Malloch), Spilogona	883	*Tephomyiella* Townsend	943
Tendipes Meigen	164	*Tephrella* Bezzi	†670
tenebrarum (Aldrich), Leptocera	724	**TEPHRITIDAE**	658
tenebricosa (Wulp), Gaediophana	1079	**Tephritis** Latreille	667
tenebrosa Coquillett, (Chaetona)	1116	**TEPHRITOIDEA**	641
tenebrosa Coquillett, Cuterebra	1110	*Tephritoides* Benjamin	668
tenebrosa (Coquillett), Dasyhelea	127	tephrocephala Loew, Tipula	29
tenebrosa Coquillett, Empis	461	**Tephrochlamys** Loew	810
tenebrosa Coquillett, Empis	460	*Tephromyiella* Townsend	943
tenebrosa (Coquillett), Paroxyna	666	**Tephromyiopsis** Townsend	960
tenebrosa Coquillett, Phronia	215	*Tephronota* Loew	645
tenebrosa Coquillett, Tipula	29	tepida Alexander, Limnophila	66
tenebrosa Sherman, (Boletina)	1116	terebrans Alexander, Limnophila	64
tenebrosa Spencer, Drosophila	766	*Terelliosoma* Rondani	648
tenebrosus (Coquillett), Ospriocerus	381	teres Melander, Empis	461
tenebrosus Coquillett, Platycheirus	577	*terelus* Harris, (Syrphus)	1116
tenebrosus (Coquillett), Podonomus	150	tergata (Coquillett), Sphenometopa	939
tenebrosus (Williston), Machimus	396	tergata Doane, Tipula	39
tenella (Reinhard), Lydellohoughia	1044	tergata Fitch, (Cecidomyia)	292
tenella (Zetterstedt), Coelosia	218	tergisa (Say), Psilocephala	352
tenellus (Hardy & Knowlton), Tomosvaryella	556	*tergissa* Say, Laphria	390
tenellus Melander, Platypalpus	481	*tergissa* (Say), Psilocephala	352
tener Coquillett, Hippelates	776	terminale (Van Duzee), Rhaphium	514
tener Loew, Dolichopus	496	*terminalis* Banks, Pipunculus	552
tener (Loew), Sciapus	486	terminalis Coquillett, Keroplatus	200
tener Osten Sacken, Tabanus	336	*terminalis* (Coquillett), Phytobia	798

INDEX

	Page
terminalis (Curran), Epistrophe	565
terminalis (Hine), Neoitamus	397
terminalis Loew, Cordilura	827
terminalis Loew, Dolichopus	496
terminalis (Loew), Mumetopia	820
terminalis Loew, Palloptera	715
terminalis (Loew), Rhagio	345
terminalis (Loew), Scaptomyza	769
terminalis Melander & Argo, Clusiodes	807
terminalis Van Duzee, Chrysotus	523
terminalis (Van Duzee), Nematoproctus	515
terminalis Van Duzee, Neurigona	519
terminalis (Van Duzee), Thecophora	631
terminalis Walker, (Anthomyia)	915
terminata Garrett, Mycomya	221
terminatus Cazier, Rhaphiomidas	357
terminatus Walker, Dolichopus	496
terna Loew, Dixa	101
ternaria Loew, Tipula	34
ternatus Loew, Diogmites	372
terraenovae Alexander, Limnophila	66
terraenovae (Alexander), Limonia	48
t erraenovae (Knab), Hypoderma	1112
terraenovae Macquart, Calliphora	929
terraenovae (Macquart), Hybomitra	341
terraenovae Macquart, Laphria	390
terraenovae (Macquart), Protophormia	925
terraenovae (Robineau-Desvoidy), Protophormia	925
terrena (Coquillett), Villa	440
terrena Felt, Colpodia	253
terrestris Felt, (Cecidomyia)	291
terricola (Johnson), Lasiopogon	378
territans Walker, Culex	118, †119
tersa Coquillett, Empis	461
tersa Coquillett, Rhamphomyia	466
tersa Namba, Rivellia	656
tersesthes Townsend	121
tersus Coquillett, Platypalpus	481
tertia (Brues), Megaselia	540
tertia Cockerell, Lasioptera	272
tertianus Loew, Sympycnus	528
tesca (Quate), Philosepedon	95
tescorum Chillcott, Fannia	897
tessa (Hull & Fluke), Hiatomyia	587
tessellata (Aldrich), Neophyto	952
tessellata Allen, Metopia	937
tessellata (Brooks), Linnaemya	1006
tessellata (Fabricius), Peleteria	997
tessellata Loew, Tipula	31
tessellata (Zetterstedt), Monohelea	†135
testacea (Edwards), Platurocypta	215
testacea Loew, Hilara	456
testacea Loew, Rhamphomyia	466
testacea (Malloch), Plethochaeta	841
testacea Melander, Antichaeta	688
testacea Melander, Iteaphila	455
testacea Melander, Micrempis	476
testacea Melander, Oedalea	450
testacea (Melander), Piophila	713
testacea Roser, Drosophila	766
testacea Ruthe, Thaumalea	†120
testacea Stein, Hylemya	847
testacea (Webber), Lespesia	1102
testaceicornis Macquart, Eristalis	625
testaceipes Bigot, Chrysopilus	347
testaceus Cole & Wilcox, Lasiopogon	378
testaceus Loew, Longurio	18
testaceus (Loew), Tipulogaster	363
testaceus Macquart, Chlorops	790
testaceus Melander, Syneches	449
testaceus Townes, Glyptotendipes	172
testaceus (Wulp), Mansonia	108
testata (Edwards), Platurocypta	215
testudinalis Wirth & Hubert, Culicoides	130
testudinea (Loew), Acrotaenia	664
Tetanocera Duméril	693
TETANOCERIDAE	685
TETANOCERINAE	688
Tetanops Fallén	648
Tethina Haliday	727
TETHINIDAE	726
Tethymyia Wirth	163
tethys (Townes), Chironomus	168
tetona Reinhard, Sarcophaga	960
tetonensis Alexander, Gonomyia	76
tetonensis Alexander, Trichocera	16
tetonensis Wheeler, Chymomyza	770
tetonica Alexander, Ormosia	84
tetonicola Alexander, Dicranota	57
tetonicola Alexander, Limnophila	67
tetra Aldrich, Sarcophaga	960
tetra Alexander, Tipula	28
tetracantha (Loew), Fannia	895
tetrachaeta (Malloch), Spilogona	883
Tetrachaeta Stein	876
Tetraciura Hendel	670
Tetradiscus Bigot	640
tetradymia (Felt), Mayetiola	264
Tetragoneura Winnertz	228
TETRAGONEURINI	226
tetragrammus (Loew), Lasiopogon	378
Tetramerinx Berg	876
Tetraneurella Dahl	476
Tetraspila Frey	640
tetraspilus (Loew), Caloparyphus	308
tetraspina (Phillips), Xanthaciura	671
tetrasticta Kieffer, Pentaneura	147
Tetraxyphus Kieffer	249
Tetrechus Haliday	502
tetrica (Marten), Hybomitra	341
tetricus Loew, Dolichopus	497
Tetropismenus Loew	649
tetropsis Bigot, Tabanus	337
Tetropsis Coquillett	935
Teuchocnemis Osten Sacken	608
Teucholabis Osten Sacken	76
Teuchophorus Loew	529
texana (Aldrich), Sarcodexia	955
texana Aldrich & Webber, (Phorocera)	1107
texana (Banks), Efferia	396
texana Bromley, Cophura	367
texana Bromley, Leptogaster	363
texana Cockerell, Hirmoneura	402
texana Cresson, Micropeza	635
texana Cresson, Paralimna	748
texana Curran, Exoprosopa	445
texana Curran, Nausigaster	596
texana Curran, Salpingogaster	574
texana Enderlein, Calobatina	635
texana (Felt), Conarete	245
texana Felt, Coquillettomyia	281
texana (Felt), Cordylomyia	249
texana Felt, (Itonida)	291
texana (Felt), Mayetiola	264
texana Felt, Walshomyia	265
texana Hall, Comasarcophaga	949
texana Hering, Trupanea	667
texana (Hough), Neomuscina	911
texana James, Adoxomyia	306
texana Johnson, Spilomyia	613
texana Lane & d'Andretta, Olbiogaster	191
texana (Long), Forcipomyia	126
texana (Malloch), Ophiomyia	797
texana (Malloch), Tethina	728
texana Malloch, Trupanea	667
texana (Melander), Cephalochrysa	302
texana Namba, Rivellia	656
texana Painter, Villa	438
texana Patterson, Drosophila	763
texana Reinhard, Winthemia	1091
texana Reinhard, Xiphomyia	1051
texana Shannon, Chrysogaster	591
texana Shewell, Camptoprosopella	697
texana (Townsend), Microcerella	952
texana (Webber), Lespesia	1102
texana (Williston), Physocephala	628
texana Wirth, Monohelea	135
texanum (Dyar & Knab), Psorophora	110
texanus Bromley, Diogmites	372
texanus (Bromley), Hadrokolos	373
texanus Bromley, Promachus	400
texanus Bromley, Stenopogon	384
texanus Dampf, Phlebotomus	92
texanus (Greene), Myrmosicarius	544
texanus Hine, Tabanus	337
texanus Hull, Eristalis	625
texanus Malloch, Hippelates	775
texanus Painter, Bombylius	410
texanus Pechuman, Silvius	323
texanus Van Duzee, Asyndetus	524
texanus Van Duzee, Diaphorus	520

1684 A CATALOG OF DIPTERA OF NORTH AMERICA

	Page
Texasa Steyskal	654
texasiana (Johnson), Labostigmina	314
texensis Aldrich, Spathimeigenia	1049
texensis Alexander, Tipula	39
texensis Curran, Dictya	689
texensis Curran, Laphystia	377
texensis Curran, Myxosargus	317
texensis Curran, Peleteria	998
texensis Malloch, Phaonia	908
texensis Reinhard, Gonia	1076
texensis Sabrosky, Pholeomyia	732
texensis Stone, Asaphomyia	321
texensis (Townsend), Cynomyopsis	930
texensis Townsend, Dejeaniopalpus	1034
texensis Townsend, Lipochaeta	751
texensis (Wheeler), Nocticanace	734
texensis Wirth, Mallochohelea	139
thalactri (Felt), Mayetiola	264
Thalassomya Schiner	163
Thalassomyia Schiner	163
thalictri Felt, Asphondylia	269
thallasinus Van Duzee, Thinophilus	508
Thambeta Williston	53
Thanategia Melander	472
tharpiana Alexander, Limonia	47
thatuna Aldrich, Sarcophaga	960
thatuna (Shannon), Bufolucila	927
thatuna Shannon, Psilota	593
Thaumalea Ruthe	120
THAUMALEIDAE	120
Thaumastoptera Mik	51
Thaumatomyia Zenker	787
thecata (Coquillett), Leskiopsis	1036
theclarum (Scudder), Aplomya	1098
Thecodiplosis Kieffer	278
Thecomyia Rondani	631
Thecophora Rondani	631
Thelaira Robineau-Desvoidy	1021
THELAIRINI	1021
Thelairodoria Townsend	1050
thelcter Dyar, Aedes	115
Thelodiscus Aldrich	941
Thelylepticocnema Townsend	950
Thelymyia Brauer & Bergenstamm	1106
Thematheca Reinhard	1050
Themira Robineau-Desvoidy	684
Theobaldia Neveu-Lemaire	116
Theresia Robineau-Desvoidy	991
THERESIINI	989
Thereva Latreille	352
THEREVIDAE	348
Therioplectes Zeller	†337
thermarum Collin, Scatella	757
thermophila Cresson, Ephydra	754
thermophilus Townes, Paratendipes	174
thersites Wheeler, Campsicnemus	527
theutis (Walker), Cryptomeigenia	1041
Thevenemyia Bigot	426
Thevenetimyia Bigot	426
thibaulti Dyar & Knab, Aedes	115
Thienemanniella Kieffer	164
Thienemannimyia Fittkau	146
thinobia (Thomson), Fucellia	843
Thinodromia Melander	475
Thinophilus Wahlberg	507
Thiomyia Wirth	755
thioptera Shaw, Mycetophila	214
thlipsomyzoldes (Jaennicke), Phthiria	420
thomae (Curran), Dioxyna	666
thomae Loew, Euxesta	652
Thomasia Rübsaamen	280
Thomasiniana Strand	280
thomi Felt, Rhopalomyia	266
thomomuris Jellison, Cuterebra	1110
thompsoni Cole, Cyrtopogon	369
thompsoni (Felt), Caryomyia	270
thompsoni Felt, Rhopalomyia	266
thompsoni Frick, Phytobia	799
thompsoni Herting, Lydella	1103
thompsoni Möhn, Aphidoletes	280
thompsoni (Townsend), Sisyropa	1094
Thompsonomyia Brooks	1006
thomsenae Wirth, Dasyhelea	127
thomsenae Wirth, Stilobezzia	134
thomsoni Gill, Suillia	809
thomsoni (Williston), Peleteria	998

	Page
thones Dyar & Shannon, Dixa	102
thoracica Collin, Scoliocentra	814
thoracica Coquillett, Dynatosoma	209
thoracica Curran, Aciurina	670
thoracica (Curran), Lixophaga	1044
thoracica Fabricius, Laphria	391
thoracica Fitch (Cecidomyia)	292
thoracica (Meigen), Anevrina	532
thoracica Robineau-Desvoidy, Delphinia	644
thoracica Van Duzee, Argyra	526
thoracicus (Fabricius), Chrysopilus	347
thoracicus (Hine), Atylotus	331
thoracicus (Loew), Clinotanypus	146
thoracicus Loew, Pterallastes	609
thoracicus Say, Bibio	194
thoracicus Say, Dilophus	195
thoracicus (Say), Syneches	449
thoracis Williston, Drosophila	767
Thoracochaeta Duda	722
thornburghae Vaillant, Thaumalea	120
thornburghi Vaillant, Thaumalea	120
Threticus Eaton	95
thriambus Dyar, Culex	119
Thricops Rondani	903
thrinax Quate, Psychoda	96
thrixia Huckett, Pegomya	862
Thrypticus Gerstäcker	511
Thryptocera Macquart	†1061
Thryptochaeta Rondani	772
Thryptocheta Rondani	772
thujae (Hedlin), Mayetiola	264
thulensis Alexander, Tipula	30
thurmanae Foote, Valentibulla	670
thurstoni Brodie (Diplosis)	292
thyceae (Reinhard), Blaesoxipha	944
thylax Hull, Platycheirus	577
THYREOPHORIDAE	713
Thyridanthrax Osten Sacken	437
Thyridomyia Saunders	124
tibblesi Stone & Jamnback, Twinnia	182
tibialis Banks, Dioctria	371
tibialis Coquillett, Acemya	1036
tibialis Coquillett, Cyrtopogon	369
tibialis (Coquillett), Metopomuscopteryx	1025
tibialis Coquillett, Mythicomyia	419
tibialis (Coquillett), Synapha	219
tibialis Cresson, Hydrellia	744
tibialis Cresson, Mosillus	734
tibialis Cresson, Parydra	749
tibialis Curran, Holopogon	375
tibialis (Curran), Ictericophyto	1024
tibialis (Fallén), Paragus	578
tibialis Felt, Lasioptera	272
tibialis (Fitch), Oscinella	780
tibialis (Hardy), Pipunculus	554
tibialis Hull, Psilocurus	382
tibialis Loew, Dilophus	196
tibialis (Macquart), Dicraeus	783
tibialis (Macquart), Efferia	393
tibialis (Macquart), Tolmerus	401
tibialis Malloch, Fannia	897
tibialis (Malloch), Gimnomera	840
tibialis Malloch, Plunomia	707
tibialis (McAtee), Colobostema	238
tibialis (Meigen), Palpomyia	140
tibialis (Reinhard), Carcelia	1092
tibialis Robineau-Desvoidy (Estheria)	989
tibialis Say, Ommatius	398
tibialis (Say), Physocephala	628
tibialis Say, (Tanypus)	181
tibialis Staeger (Tanypus)	181
tibialis Stein, Coenosia	874
tibialis Van Duzee, Chrysotus	523
tibialis Van Duzee, Gymnopternus	500
tibialis (Van Duzee), Hercostomus	498
tibialis Van Duzee, Pelastoneurus	502
tibialis Wiedemann, Mydas	359
tibialis (Zetterstedt), Neoleria	†813, 813
tiburonensis Cole, Exoprosopa	445
tidwelli Philip & Jones, Chrysops	327
tigrina Alexander, Dicranoptycha	52
tigrina (Fabricius), Coenosia	870
tigrinum (Osten Sacken), Carposcalis	576
tigrinus (De Geer), Anthrax	432
tiliacea (Felt), Janetiella	264
tiliae (Couden), Melanagromyza	796

INDEX

Entry	Page
tiliae (Felt), Lobopteromyia	276
tiliae citrina Osten Sacken, (Cecidomyia)	293
tiliaginea Felt, Neolasioptera	273
timberlakei Alexander, Tipula	39
timberlakei Blanc & Foote, Urophora	659
timberlakei Bromley, Stenopogon	384
timberlakei Curran, Volucella	602
timberlakei (Felt), Mayetiola	264
timberlakei Melander, Aphoebantus	430
timberlakei Melander, Apolysis	421
timberlakei (Walton), Clairvillia	976
timberlakei Wilcox, Itolia	376
timberlakei Wilcox & Martin, Nannocyrtopogon	380
timida Harris, (Anthomyia)	1116
Tinalydella Townsend	1050
tincta (Walker), Scoliocentra	814
tincta (Zetterstedt), Mydaea	†891
tinctinervis Malloch, Lispocephala	877
tinctipennis Malloch, Circia	843
tinctipennis (Stein), Helina	887
tinctisquama Huckett, Pegomya	862
tingaureus (Philip), Atylotus	331
tingi Alexander, Tipula	38
tingi (Hardy), Dilophus	196
tinia (Walker), Hylemya	850
tinkhami Bromley, Stenopogon	384
tinkhami (Hardy), Pseudonomoneura	360
tinkhami Philip, Esenbeckia	322
Tipula Linnaeus	24
TIPULIDAE	16
tipulina Osten Sacken, Phalacrocera	42
TIPULINAE	17
TIPULINI	17
tipulivora Malloch, Phaonia	908
Tipulodina Enderlein	29
Tipulogaster Cockerell	363
TIPULOIDEA	15
Titania Meigen	789
Titanogrypa Townsend	956
titillans (Walker), Mansonia	108
titillans (Winnertz), Forcipomyia	125
titubans Osten Sacken, Exoprosopa	445
toga Snyder, Helina	889
togata Fitch, (Cecidomyia)	292
togata (Melander), Haplomyza	803
togatus Melander, Oligodranes	422
toklat (Alexander), Cheilotrichia	79
tolandi Wilcox, Cophura	367
tolandi Wilcox, Hodophylax	374
tolandi Wilcox, Laphystia	377
tolandi Wilcox & Martin, Nannocyrtopogon	380
tolhurstae Felt ,(Cecidomyia)	291
Tolmerus Loew	401
tomentosa Osten Sacken, (Musca)	625
tomentosus Bigot, Chrysopilus	347
tomentosus Melander, Epacmus	428
Tomoplagia Coquillett	678
Tomoplagina Curran	660
Tomosvaryella Aczél	555
tonnoiri Dyar, Psychoda	96
tonsus (Aldrich), Condylostylus	485
tonsus Loew, Dolichopus	497
topazina Osten Sacken, Ctenophora	20
Toreus Melander	457
tormentor Dyar & Knab, Aedes	115
tornensis (Ringdahl), Spilogona	883
tornetraskensis (Edwards), Trissocladius	155
torontoensis Felt, Dasineura	261
Torosomyia Reinhard	1080
torosus (Reinhard), Deopalpus	1004
torpida Hull, Laphystia	377
*torrens (Townsend), Leptoconops	122
torreyae (Johannsen), Spilogona	883
torreyi Felt, (Cecidomyia)	295
torrida Harmston, Neurigona	519
torridus Harmston & Knowlton, Hercostomus	498
torta Reinhard, Peleteria	998
tortilis Coquillett, Trypeta	677
tortilis Reinhard, Eumacronychia	935
tortilis (Theobald), Aedes	115
Tortriciophaga Townsend	1071
tortricis (Coquillett), Erynnia	1071
tortricis (Coquillett), Hemisturmia	1088
torva Williston, Baccha	573
torvus Osten Sacken, Syrphus	559
Torynotachina Townsend	1050
toshua Huckett, Coenosia	873
tothastica Quate, Psychoda	96
tothilli (Curran), Eribella	1042
tothilli Parker, Metoposarcophaga	951
totifuscus Sublette, Stenochironomus	177
toweri Felt, Dasineura	261
townesi Alexander, Dicranota	58
townesi Alexander, Ormosia	87
townesi Alexander, Ptychoptera	97
townesi Hardy, Bibio	194
townesi Shaw, Sciara	231
townesi Wirth, Forcipomyia	126
townesiana Alexander, Pedicia	55
townsendi Aldrich, Belvosia	1082
townsendi Aldrich, Dolichopus	497
townsendi (Aldrich), Promasiphya	1091
townsendi Camras, Physoconops	626
townsendi Curran, Peleteria	998
townsendi Hall, Onesia	930
townsendi Hunter, Cheilosia	587
townsendi Malloch, Pipunculus	552
townsendi Sabrosky & Arnaud, Chaetogaedia	1078
townsendi (Smith), Chaetophlepsis	1016
townsendi (Snow), Polybiomyia	616
townsendi (Williston), Opomydas	360
Townsendia Williston	385
Townsendiellomyia Baranov	1088
townsendii (Aldrich), Mesorhaga	483
Townsendina Curran	1023
toxicodendri (Felt), Adiplosis	289
toxicodendri (Felt), Xylopriona	249
toxicodendron (Felt), Xylopriona	249
toxodentis (Hardy & Knowlton), Tomosvaryella	556
Toxomerus Macquart	571
Toxomyia Felt	276
toxoneura (Osten Sacken), Austrolimnophila	61
toxoneura (Osten Sacken), Metangela	229
Toxophora Meigen	425
TOXOPHORINAE	425
Toxorhina Loew	89
Toxorhynchites Theobald	106
TOXORHYNCHITINAE	106
Tozorrhina Loew	90
Toxotrypana Gerstäcker	659
trabeculata (Loew), Pherbellia	687
Trachyleucophenga Hendel	763
Trachyopella Duda	725
traenis (Roback), Orthocladius	157
Trafoia Brauer & Bergenstamm	1007
tragica (Meigen), Elodia	1071
Traginops Coquillett	794
tranquilla (Osten Sacken), Stonemyia	321
transatlanticus Walker, Paragus	578
Transculicia Dyar	117
transducta (Walker), Pherbellia	686
transfuga Walker, Hilara	456
transiens Rubzov, Simulium	186
transiens (Walker), Atrichopogon	123
transita (Townsend), Oxysarcodexia	953
transitus (Coquillett), Aphoebantus	430
translata (Walker), Tomosvaryella	556
translucida Doane, Tipula	39
translucipennis (West), Ptilodexia	989
transversa Cresson, Parydra	750
transversa Fallén, Drosophila	766
transversa (Felt), Cincticornia	269
transversa (Hull), Allograpta	569
transversa Malloch, Plunomia	707
transversa (Osburn), Sericomyia	603
transversa Sturtevant & Wheeler, Athyroglossa	735
transversa Van Duzee, Neurigona	519
transversa Walker, Notiphila	748
transversalis Curran, Syrphus	559
transversalis (Malloch), Lasiosina	791
transversalis Meigen, Merodon	617
transversa Wiedemann, Eristalis	625
transversus Wirth, Atrichopogon	123
trapezoidalis (Bellardi), Promachina	399
traverae Thomsen, Dasyhelea	127
travisi Stone, Prosimulium	183
travisi Vargas, Culicoides	130
tremulus (Linnaeus), Cricotopus	158

Name	Page
trepida (McDunnough), Hybomitra	341
Tressinus Schmitz	536
triacantha Shaw, Mycomya	221
Triachora Townsend	1082
triadenii Beutenmüller (Cecidomyia)	295
trialbawhorla (Haseman), Telmatoscopus	94
triangula (Coquillett), Iteaphila	455
triangula (Fallén), Pseudolimnophora	885
triangularis Alexander, Ormosia	87
triangularis Collin, Meoneura	729
triangularis Curran, Cryptomeigenia	1041
triangularis Curran, Mericia	1008
triangularis (Felt), Lestodiplosis	285
triangularis Felt, Lobodiplosis	281
triangularis (Felt), Parallelodiplosis	286
triangularis Giglio-Tos, Eristalis	625
triangularis Johannsen, Trichonta	216
triangularis Loew, Tetanocera	694
triangularis Say, Xylophagus	297
triangularis Shaw, Rymosia	208
triangularis Van Duzee, Chrysotus	523
triangularis Wulp, (Dexia)	1108
triangulata (Becker), Conioscinella	784
triangulata Garrett, Amoebaleria	815
triangulatum (Van Duzee), Rhaphium	514
triangulatus Martin, Haplopogon	373
triangulatus Van Duzee, Diaphorus	520
triangulifer (Loew), Hyalomyodes	977
triangulifera (Zetterstedt), Melangyna	568
triangulifera (Zetterstedt), Spilogona	884
trianguligera Malloch, Fannia	897
trianguligera (Zetterstedt) Spilogona	†884
Tribelos Townes	170
tribulator Alexander, Tipula	30
tribulatum Lugger, Simulium	187
tribulis (Harris), Tephritis	668
tributaria Alexander, Limonia	43
tricaudatum (Van Duzee), Rhaphium	514
tricella (Bromley), Efferia	396
tricellula Cole, Villa	439
Trichacrostylia Enderlein	315
trichaeta Cresson, Hydrellia	744
trichaetus Malloch, Pipunculus	552
Tricharaea Thomson	960
Tychephelia Alexander	64
Trichiaspis Duda	719
Trichina Meigen	449
Trichiopoda Berthold	966
trichiosomae Sellers, Aplomya	1098
TRICHOBIINAE	921
Trichobius Gervais	921
Trichocera Meigen	15
TRICHOCERIDAE	15
Trichochlamys Czerny	816
Trichocladius Kieffer	158
Trichoclytia Townsend	970
trichodactyla (Rondani), Hylemya	849
Trichodejeania Townsend	999
trichomerus Walker, (Chironomus)	181
Trichomyia Curtis	91
TRICHOMYIINAE	91
trichonota Loew, Mycetophila	214
Trichonta Winnertz	215
Trichopalpus Rondani	836
Trichopareia Brauer & Bergenstamm	1068
TRICHOPAREIINI	1068
trichoparia Brauer & Bergenstamm	1068
trichophora Alexander, Rhabdomastix	77
trichophora Alexander, Tipula	25
Trichophticus Rondani	903
Trichopoda Berthold	966
TRICHOPODINI	965
Trichopodopsis Townsend	966
Trichops Rondani	903
Trichopsidea Westwood	403
TRICHOPSIDEINAE	403
Trichopsychoda Tonnoir	95
Trichoptera Lioy	772
Trichoptera Meigen	95
trichoptera (Osten Sacken), Paracladura	16
Trichopteromyia Williston	250
Trichopticoides Ringdahl	901
Trichopticus Rondani	903
trichopus (Thomson), Carposcalis	576
trichosa Fisher, (Macrocera)	1116
Trichoscelis Rondani	817
Trichosia Winnertz	229
Trichotanypus Kieffer	151
Trichothaumalea Edwards	120
Trichotipula Alexander	24
trichura Melander, Drapetis	477
trichurus (Dyar), Aedes	115
Tricimba Lioy	785
tricincta Bigot, Baccha	572
tricincta Bigot, Volucella	602
tricincta (Loew), Anastrepha	673
tricincta Loew, Boletina	217
tricincta (Loew), Hexamitocera	841
tricinctus (Meigen), Cricotopus	158
Triclis Loew	†382
tricolor Bigot, Ischyrosyrphus	565
tricolor Coquillett, Criorhina	612
tricolor Coquillett, Metadexia	1027
tricolor Curran, Pipiza	579
tricolor (Doane), Gymnocarena	676
tricolor Fabricius, Tipula	29
tricolor Jaennicke, Eristalis	624
tricolor (Loew), Coelotanypus	146
tricolor Osten Sacken, Euparyphus	307
tricolor Snow, Eutreta	661
tricolor Wheeler, Latheticomyia	825
tricoloripes Curran, Syntormon	516
triconis Reinhard, Messiomyia	1063
Tricopalpus Rondani	836
Tricophthicus Rondani	903
tricornis Alexander, Ormosia	87
tricosa Reinhard, Eumacronychia	935
Tricyphona Zetterstedt	54
tridactila (Kincaid), Philosepedon	95
tridens (Felt), Anaretella	245
tridens (Loew), Tenthredomyia	616
tridens Malloch, Hylemya	853
tridens Van Duzee, Neurigona	519
tridens Walton, Neodichocera	1014
tridentata Coquillett, Empis	461
tridentata Hardy, Dorylomorpha	557
tridentata Rübsaamen, (Sciara)	236
tridentatae Rübsaamen, Rhopalomyia	266
tridenticulata Alexander, Dolichopeza	23
trifaria Melander, Mythicomyia	419
trifarius Loew, Ablautus	364
trifasciata Coquillett, Mycetophila	212
trifasciata Curran, Peleteria	997
trifasciata Harris, (Leptis)	1116
trifasciata Loew, Trichopoda	966
trifasciata (Say), Chaetopsis	650, 650
trifasciata (Say), Microphthalma	982
trifasciata Walker, Lela	228
trifasciata Wirth, Palpomyia	140
trifasciata (Zetterstedt), Thaumatomyia	788
trifasciatum Robertson, Temnostoma	614
trifasciatus (Hausmann), Milesia	614
trifasciatus Hull, Brachypalpus	608
trifasciatus (Panzer), Cricotopus	158
trifasciatus (Say), Stichopogon	385
trifascipennis (Zetterstedt), Anatopynla	145
trifida Alexander, Limnophila	64
trifidus Melander, Oligodranes	422
trifolii (Burgess), Liriomyza	802
trifolii (Felt), Colpodia	253
trifolii Felt, Contarinia	277
trifolii (Lintner), Dasineura	259
trifolii (Loew), Dasineura	261
trifolii (Pettey), Lycoriella	231
trifolium Osten Sacken, Silvius	323
trifolius (Garrett), Procladius	149
triformis Alexander, Gonomyia	76
triformis Melander, Mythicomyia	419
trifurca (Pettey), Bradysia	234
trifurcata Huckett, Hylemya	852
Triglyphus Loew	†580
trigonata (Zetterstedt), Spilogona	884
trigonifera Chillcott, Fannia	897
trigonifera (Zetterstedt), Spilogona	884
trigonolabis Edwards, Orthocladius	157
Trigonometopus Macquart	705
Trigonomma Enderlein	788
trigonus Townes, Polypedilum	176
trigramma (Loew), Olcella	785
trijunctus Walker, Tabanus	337
triligatus (Walker), Hybomitra	338
trilineata (Huckett), Spilogona	884

Entry	Page
trilineata (Stein), Hylemya	854
trilineata (Wulp), Senotainia	939
trilitura Blanc & Foote, Aciurina	670
trilobatus Kieffer, Paraclunio	163
Trilobomyza Hendel	800
trilobus (Kincaid), Breviscapus	93
Trilophyrobata Hennig	634
triloris Reinhard, Erucophaga	949
triluminata Cresson, Cardiacephala	633
trimaculata (Emmons), Tipula	41
trimaculata (Macquart), Strauzia	676
trimaculata Painter, Apiocera	357
trimaculata (Stein), Fannia	†895
trimaculatus Palisot de Beauvois, Tabanus	337
trimera Felt, Neolasioptera	273
Trimerina Macquart	742
Trimerinoides Cresson	743
Trimicra Osten Sacken	80
trinervis (Becker), Triphleba	534
Trineura Meigen	536
trinidadensis (Townsend), Clausicella	†1059
trinodulosa Tonnoir, Psychoda	96
trinotata Macquart, Chrysops	326
trinotata Staeger, Mycetophila	214
trinotatus Melander, Nemotelus	309
triocellata (Osten Sacken), Limonia	44
Triodites Osten Sacken	429
Triodonta Williston	621
Triogma Schiner	42
triorbiculata (Sabrosky), Conioscinella	784
tripartitum (Frey), Rhaphium	514
triphaea Alexander, Limonia	44
Triphleba Rondani	533
tripilus Van Duzee, Sympycnus	528
triplagiata Harris, (Leia)	1116
triplasia Wulp, Sarcophaga	960
Triplasius Loew	†411
triplex Melander, Trixoscelis	818
triplex (Walker), Euhybus	448
triplex Walker, Tipula	39
triplicseta Schmitz & Wirth, Phora	537
Triploechus Edwards	174
Tripodura Townes	272
tripsaci Felt, Lasioptera	342
Triptotricha Loew	44
tripunctata (Fabricius), Limonia	44
tripunctata Felt, Neolasioptera	273
tripunctata Loew, Drosophila	766
tripunctata (Wulp), Neomuscina	911
tripus Sabrosky, Gaurax	782
triquetra (Olivier), Ervia	1027
triquetra (Wiedemann), Azelia	899
triradiata (Hull), Mesograpta	571
triseriata Malloch, Hylemya	851
triseriatus (Say), Aedes	115
triseta (Brooks), Leschenaultia	1082
triseta (Coquillett), Parascatella	757
triseta Malloch, Microprosopa	835
triseta Malloch, Pegomya	862
triseta Stein, Coenosia	873
trisetosa (Coquillett), Aplomya	1098
trisetosa Van Duzee, Medetera	510
trisetosus Van Duzee, Dolichopus	497
trisetosus Van Duzee, Sciapus	486
Trishormomya Kieffer	288
Trishormomyia Kieffer	288
Trisisyropa Townsend	1089
Trisopsis Kieffer	288
trisphenata Wheeler, Leucophenga	762
trispila (Wiedemann), Hybomitra	341
trispina Wheeler, Drosophila	766
trispinosa Lundström, Tipula	39
trispinosa (Malloch), Spiniphora	533
trispinum Becker, Allotrichoma	736
Trissocladius Kieffer	155, †155
triste Bigot, Zodion	629
tristellus Edwards, Metriocnemus	161
tristicula (Holmgren), Hylemya	855
tristigma (Osten Sacken), Limonia	44
tristimonia (Alexander), Cheilotrichia	79
tristina (Hall), Melanodexia	932
tristiola (Zetterstedt), Spilogona	†884
tristis (Alexander), Hexatoma	70
tristis (Bigot), Nemotelus	310
tristis Doane, Tipula	34
tristis Felt, Bremia	280
tristis Loew, Cheilosia	587
tristis Loew, Eulonchus	404
tristis Loew, Gymnopternus	500
tristis Loew, Hilara	457
tristis Loew, Microdon	†598, 598
tristis (Loew), Morpholeria	814
tristis (Loew), Symmerus	199
tristis Melander, Mythicomyia	419
tristis (Ringdahl), Hydrophoria	865
tristis Schiner, Mesorhaga	483
tristis (Walker), Dizonias	372
tristis Walker, Rhamphomyia	466
tristis Williston, Bibio	194
tristis Williston, Melanodexia	932
tristissima (Garrett), Heleomyza	816
tristissima Osten Sacken, Gnophomyia	73
tristriatulus Hoffman, Culicoides	128
tristyla Wirth, Dasyhelea	127
trisulcata (Adams), Tricimba	785
tritici (Coquillett), Bradysia	234
tritici Felt, (Itonida)	291
tritici (Fitch), Meoneura	729
tritici (Kirby), Contarinia	277
triticicola Kieffer, (Cecidomyia)	291
triticoides (Osten Sacken), Rhabdophaga	257
triton Alexander, Tipula	39
Tritonia Meigen	613
Tritoxa Loew	649
Tritozyga Loew	243
trituberculata (Sturtevant & Wheeler), Parydra	750
tritum (Walker), Polypedilum	176
Trivialia Malloch	705
trivialis (Johannsen), Bradysia	234
trivialis (Loew), Palpomyia	140
trivialis Loew, Platypalpus	†479, 481
trivialis (Loew), Thaumatomyia	787
trivialis Malloch, Phaonia	908
trivialis (Wulp), Oxysarcodexia	953
trivittata Bigot, Symphoromyia	344
trivittata (Cresson), Pherbellia	687
trivittata Curran, Ateloglossa	986
trivittata Curran, Eutheresia	990
trivittata Frost, Phytomyza	805
trivittata Johnson, Macrocera	201
trivittata (Loew), Chaoborus	104
trivittata Loew, Hilara	457
trivittata (Loew), Lauxaniella	700
trivittata Macquart, Taeniaptera	636
trivittata (Sabrosky), Olcella	784
trivittata Say, Tipula	34
trivittata (Stein), Hylemya	854
trivittatum Malloch, Simulium	187
trivittatus (Coquillett), Aedes	115
trivittatus (Fabricius), Helophilus	†618
trivittatus Melander, Lasiopogon	378
trivittatus (Say), Hedriodiscus	314
trivittatus (Say), Ptecticus	303
trivittatus Van Duzee, Diaphorus	520
trixa Curran, Aciurina	670
TRIXINI	980
Trixoclista Townsend	941
Trixodes Coquillett	980
TRIXODINI	980
trixcides (Walker), Microphthalma	982
TRIXOSCELIDIDAE	817
Trixoscelis Rondani	817
trizonata (Zetterstedt), Pseudoperichaeta	1106
trochanterata Collin, Scaptomyza	769
trochanterata Cresson, Discocerina	738
trochanterata (Malloch), Leptocera	722
trochanteratus (Malloch), Neocnemodon	581
trochanteratus (Malloch), Tomosvaryella	555
Trochilodes Coquillett	1015
trochilus (Coquillett), Oligodranes	422
trochilus (Coquillett), Rhaphiomidas	357
Trochobola Osten Sacken	44
troene (Walker), Helina	889
troi Cresson, Scatella	757
trompe (Modeer), Cephenemyia	1112
Trophithauma Schmitz	543
Trophodeinus Borgmeier	546
tropicalis Townsend, Baccha	572
Tropidia Meigen	609
Trotteria Kieffer	275
trucis Coquillett, Symphoromyia	344

	Page		Page
truculenta Alexander, Tipula	39	turbata (Coquillett), Villa	438
trudis Reinhard, Guerinia	1055	turbida (Curran), Paracoenia	756
trunca (Coquillett), Cophura	367	turbida Huckett, Eremomyia	857
trunca Melander, Clinocera	468	turbidus Wiedemann, Tabanus	337
truncata Alexander, Pedicia	54	turcana Townsend, Micropeza	634
truncata Edwards, Allodia	205	turgida Coquillett, Gonia	1076
truncata Felt, Contarinia	277	turgida Hardy, Tomosvaryella	556
truncata (Felt), Cordylomyia	249	turgidula Cresson, Discocerina	738
truncata (Felt), Rhopalomyia	267	turitella Fisher, Mycomya	221
truncata Felt, Rhopalomyia	267	*turmale* Twinn, Simulium	189
truncata (Hine), Efferia	396	turmalis (Reinhard), Clausicella	1059
truncata Hull & Fluke, Cheilosia	584	*turpis* (Walker), Limonia	50
truncata Kincaid, Pericoma	93	*turpis* (Zetterstedt), Prodiamesa	153
truncata Lundström, Coelosia	218	turrita (Meigen), Vibrissina	1051
truncatipennis (Dodge), Brachicoma	940	tuscarora Alexander, Tipula	39
truncatula Loew, Pelina	752	tuxedo (Curran), Palpexorista	1056
truncatum Cook, Rhegmoclema	239	*tuxedo* (Curran), Sturmia	1088
truncatus Chillcott, Platypalpus	481	tuxis Curran, Chironomus	166
truncatus Harmston & Knowlton, Hercostomus	498	twightae Alexander, Tipula	39
		twinni Stains & Knowlton, Simulium	189
truncorum Meigen, Tipula	†33	**Twinnia** Stone & Jamnback	182
truncula Foote, Rhopalomyia	267	twogwoteeana Alexander, Tipula	30
Trupanea Guettard	666	TYLIDAE	633
Trupanea Macquart	400	*Tylomyia* Roback	955
Trupanea Schrank	666	**Tylomyza** Hendel	797
TRUPANEIDAE	658	*Tylos* Meigen	634
Trupheoneura Malloch	533	*Tylostypia* Enderlein	337
truquii (Bellardi), Hedriodiscus	315	tylus (Townes), Chironomus	168
truquii (Bellardi), Philonicus	398	typhus (Whitney), Hybomitra	341
truquii Bellardi, Promachus	400	typica (Curran), Strauzia	676
trux McAtee, Laphria	391	**Typopsilopa** Cresson	743
Trypanea Guettard	666	TYPOPSILOPINI	743
Trypeta Meigen	676	*Tyrophaga* Kirby	711
TRYPETIDAE	658		
Trypetisoma Malloch	705	*ucayali* Townsend, Chaetophlepsis	†1016
trypetophora Dietz, Tipula	41	**Uclesia** Girschner	1019
Trypetoptera Hendel	695	*Uclesiopsis* Townsend	1020
TRYPHERINI	1037, 1094	**Udamopyga** Hall	960
Trypheromyia Reinhard	1050	*Ugimyia* Rondani	1083
Trypoderma Wiedemann	1109	uhleri Banks, Stenopogon	384
tryposylonis (Townsend), Amobia	934	*uinta* Alexander, Limonia	48
Tryptochaeta Rondani	772	uinta Harmston & Knowlton, Neurigona	519
tsugae Felt, Camptomyia	256	uinta Peterson & DeFoliart, Prosimulium	183
tsugae Felt, Campylomyza	248	uintaense Harmston & Knowlton, Syntormon	516
tsugae (Felt), Lestodiplosis	285	uinticola Alexander, Prionocera	19
tsugae (Felt), Mayetiola	264	**Ula** Haliday	53
tsugae Felt, Mycodiplosis	283	ULIDIIDAE	642
Tsugaea Hall	1106	ULIDIINAE	650
tuba Stebbins (Cecidomyia)	295	**Ulidiotites** Steyskal	649
tuberans (Williston), Myolepta	589	uliginosa Alexander, Erioptera	81
tuberculata (Coquillett), Efferia	396	uliginosa Alexander, Limonia	48
tuberculata (Curran), Xylota	606	*uliginosa* (Fallén), Helina	888
tuberculata Felt (Lasioptera)	292	uliginosa Fallén, Lispe	878
tuberculata Harris (Sericomyia)	1116	uliginosa Felt (Hormomyia)	291
tuberculata Loew, Ochthera	753	uliginosa Kramer, Sarcophaga	960
tuberculata Malloch, Helina	887	uliginosus Van Duzee, Dolichopus	497
tuberculata Rondani, Hydrotaea	900	ulmea Felt, Dasineura	261
tuberculata (Zetterstedt), Fannia	897	ulmi (Beutenmüller), Mayetiola	264
tuberculatus Rondani, Eumerus	597	ulmi (Felt), Heteropeza	247
tuberculatus (Townes), Chironomus	166	*ulmi* Frost, Agromyza	795
tuberculosa Cresson, Scatophila	759	**Ulomorpha** Osten Sacken	68
tuberosa (Curtis), Fannia	897	*ulrichii* Brauer, Cephenemyia	1112
tuberosum (Lundström), Simulium	189	ultima Alexander, Tipula	27
tubicen Melander, Mythicomyia	419	ultima (Osten Sacken), Neolimnophila	73
tubicula (Osten Sacken), Caryomyia	270	*ultimus* (Walker), Promachus	400
Tubifera Meigen	622	*ultimus* Whitney, Chrysops	326
tubifera Melander, Microphorella	452	umbellatarum (Fabricius), Melangyna	568
tuckeri Felt, Campylomyza	248	umbellicola (Osten Sacken), Youngomyia	279
tuckeri Felt, Porricondyla	256	umbilicata Loew, Rhamphomyia	466
tudicornis Harris (Merodon)	1116	umbra Alexander, Tipula	26
Tujunga Steyskal	649	umbra Felt, Asynapta	254
tularensis Gill, Amoebaleria	815	umbracula Quate, Psychoda	96
tularensis James, Stratiomys	313	umbrarum (Linnaeus), Dictya	689
tulipiferae Osten Sacken, (Cecidomyia)	295	umbratica (Aldrich), Exechia	207
tumescens Melander, Mythicomyia	419	umbratica (Meigen), Hebecnema	890
tumida Banks, Odontomyia	316	umbratilis Reinhard, Pseudeuantha	1029
tumida (Felt), Corinthomyia	249	umbratilis (Williston), Blera	611
tumida Hardy, Tomosvaryella	556	*umbrina* (Stein), Spilogona	883
tumida Melander, Trixoscelis	818	umbrinus Back, Holopogon	375
tumidosa Felt, Porricondyla	256	*umbrinus* Loew, Diogmites	371
tumidosae Felt, Dasineura	261	umbripictus Becker, Pelastoneurus	502
tumidosae Felt, Mayetiola	264	umbroides Curran, Dictya	689
tumifica Beutenmüller, Lasioptera	272	umbrosa Loew, Hilara	457
tundrae (Schnabl), Spilogona	884	umbrosa (Loew), Oscinella	781
tundrarum Chillcott, Fannia	897	umbrosa Loew, Rhamphomyia	466
tunicata (Zetterstedt), Pegomya	862	umbrosa Loew, Sapromyza	704

INDEX 1689

	Page		Page
umbrosa Loew, Tipula	39	unifasciatus (Say), Sciapus	486
umbrosa (Snow), Protoclythia	549	uniformata Haseman, Psychoda	96
umbrosa Van Duzee, Exechia	207	uniformis Aldrich, Cylindromyia	973
umbrosus Van Duzee, Dolichopus	497	*uniformis* Cresson, Euthycera	690
umiat (Alexander), Cheilotrichia	79	*uniformis* Del Rosario, Psychoda	96
unacus Townes, Glyptotendipes	171	*uniformis* Felt, Rhopalomyia	267
uncata Melander, Mythicomyia	419	*uniformis* Hine, Tabanus	333
uncata (Wulp), Blaesoxipha	947	uniformis Malloch, Hydrophoria	865
uncinata (Coquillett), Euleia	677	unigera Alexander, Pedicia	55
uncinata Hardy, Dorylomorpha	557	unilimbata Loew, Stratiomys	312
uncinata Hendel, Clasiopella	742	unilineata (Zetterstedt), Leucophora	868
uncinata Johannsen, Stilobezzia	136	unilineatus (Zetterstedt), Megaphthalmoides	833
uncinata (Laffoon), Mycetophila	214	unimaculata Coquillett, Phthiria	420
uncinata (Melander), Scatopse	238	unimaculata Loew, Rhamphomyia	466
undata Wiedemann, Pyrgota	657	unimaculata (Loew), Tipula	25
underwoodi Underwood, Eucorethra	104	unimaculata Townsend, Nausigaster	596
undine (Townes), Chironomus	168	unimaculatus Loew, Chasmatonotus	159
undosa (Coquillett), Orellia	671	unimicra (Alexander), Prionocera	19
undi lata (Cresson), Parydra	750	unipila Loew, Drapetis	478
undulata Melander, Clinocera	469	unipilum (Aldrich & Webber), Clemelis	1099
undulata (Say), Copecrypta	1003	uniplagia Alexander, Dicranota	59
unduligera Alexander, Erioptera	82	unipunctata Curran, Volucella	602
unguicauda (Malloch), Bradysia	234	*unipunctata* (Haseman), Maruina	97
unguicula Beutenmüller, (Cecidomyia)	292	unipunctata Meigen, Mycetophila	214
unguicula Felt, Dasineura	261	unipunctum (Zetterstedt), Scaptomyza	769
unguiculata (Malloch), Allomyella	835	uniqua Garrett, Macrocera	201
unguiculata Malloch, Pegomya	862	uniseriata Malloch, Phaonia	908
unguiculata Westwood, Acrocera	406	*uniseriata* (Stein), Hylemya	849
unguiculatus (Cresson), Tomosvaryella	556	uniseta Curran, Lonchoptera	531
unguiculatus (Malloch), Stictochironomus	178	*uniseta* (Stein), Helina	888
ungulata Chillcott, Fannia	897	uniseta Stein, Limnophora	885
ungulata Doane, Tipula	39	unispiculatus Alexander, Molophilus	89
ungulata (Felt), Rhizomyia	262	unispinosa (Coquillett), Paralispidea	1063
ungulata Loew, Rhamphomyia	466	unispinosa (Malloch), Nanna	831
ungulatus (Linnaeus), Dolichopus	497	*unispinosa* (Reinhard), Sturmia	1088
ungulivena (Walker), Sciapus	486	unispinosa Stein, Hydrotaea	900
unica (Johannsen), Probezzia	138	unistriata (Zetterstedt), Leucophora	868
unica Malloch, Emmesomyia	858	unistylatum Vockeroth, Rhaphium	514
unica (Osten Sacken), Austrolimnophila	61	unituberculata Loew, Parydra	750
unica Spuler, Leptocera	722	universalis (Melander), Tachyempis	475
unica (Stein), Phyllogaster	876	universitatis Cockerell, Dixa	102
unica (Townsend), Amobia	934	univitta (Walker), Composobata	633
unicincta Doane, Tipula	39	univittata (Coquillett), Minettia	702
unicincta Van Duzee, Exechia	207	univittata (Roser), Pegomya	862
unicolor Aldrich, Camptops	948	*univittata* Sherman, (Coelosia)	1116
unicolor Becker, Diplotoxa	792	univittata Van Duzee, Medetera	511
unicolor (Brooks), Gymnoclytia	965	*univittata* (Walker), Compsobata	633
unicolor (Coquillett), Culicoides	†129, 130	univittata Coquillett, Chasmatonotus	159
unicolor Coquillett, Pseudatrichia	355	univittatus Loew, Plagioneurus	504
unicolor Curran, Lestomyia	378	*univittatus* Loew, Stenopogon	384
unicolor (Curran), Neocnemedon	581	univittatus Macquart, Chrysops	†326, 327
unicolor Curtis, Scaeva	562	unusa Garrett, Boletina	217
unicolor Hull, Diogmites	372	*Upodemocera* Townsend	995
unicolor (Loew), Anastrepha	673	upsilon Alexander, Ormosia	87
unicolor (Loew), Ectecephala	791	upsilon Philip, Chrysops	327
unicolor Loew, Hercostomus	498	uralensis Villeneuve, Calliphora	929
unicolor Loew, Hilara	457	*uralica* Stein, Pseudocoenosia	875
unicolor Loew, Nemotelus	310	**Uramya** Robineau-Desvoidy	1029
unicolor Loew, Notiphila	747	URAMYINI	1028
unicolor Loew, Plesiomma	382	**Uranotaenia** Lynch Arribálzaga	108
unicolor Loew, Tetanocera	695	urbana Malloch, Milichiella	733
unicolor (Lundbeck), Allodia	205	urbana (Meigen), Anevrina	532
unicolor Macquart, Tabanus	336	urbana (Meigen), Mydaea	892
unicolor Shannon, Chrysogaster	591	urbana Richards, Leptocera	725
unicolor (Smith), Aplomyiopsis	1038	*Urellia* Robineau-Desvoidy	666
unicolor Snow, Platypeza	550	urnicola Osten Sacken, (Cecidomyia)	295
unicolor Stein, Pegomya	862	urocera (Dietz), Nephrotoma	22
unicolor Stein, Pogonomyia	901	urodela Sabrosky, Leptocera	722
unicolor (Townsend), Phaenicia	928	URODEXIINI	1029
unicolor Van Duzee, Exechia	207	*Uromacquartia* Townsend	1029
unicolor (Walker), Pentaneura	147	**Urophora** Robineau-Desvoidy	659
unicolor (Walker), Mycomya	221	**Urophyllopsis** Townsend	1050
unicolor (Williston), Laphria	391	*ursa* (Kieffer), Diamesa	152
unicolor Wulp, Chironomus	168	ursamajor Alexander, Trichocera	16
unicorn Garrett, (Sciara)	236	ursina Malloch, Lonchaea	717
unicornis Alexander, Ormosia	87	ursina (Malloch), Megaselia	540
unicornis Czerny, Scatophila	759	*ursina* Malloch, Rhamphomyia	466
unictus Townes, Stenochironomus	177	ursina (Osten Sacken), Tasiocera	88
unicum (Twinn), Prosimulium	183	ursinella Melander, Rhamphomyia	466
unidorsalis Huckett, Hylemya	851	ursinum (Edwards), Prosimulium	183
unifasciata Blanc & Foote, Myoleja	669	ursinus (Holmgren), Metriocnemus	161
unifasciata Curran, Neoascia	595	**Ursophyto** Aldrich	991
unifasciata Loew, Oxycera	307	ursula Melander, Aphoebantus	430
unifasciata (Loew), Tipula	25	ursus Alexander, Molophilus	89
unifasciata (Robineau-Desvoidy), Triachora	1082	urticae (Felt), Mycodiplosis	282

	Page
urtifolia Felt, (Cecidomyia)	291
u-signata Cresson, Discomyza	740
usingeri Quate, Pericoma	93
usingeri Wirth, Clinohelea	136
usingeri Wirth, Culicoides	130
usitata (Coquillett), Elfia	1066
usitata Doane, Tipula	39
usitatus Van Duzee, Diaphorus	520
ustulata Fallén, Palloptera	715
ustulata Kröber, Thereva	354
ustulata (Loew), Chrysogaster	591
ustulata Reinhard, (Phorocera)	1108
ustulata Zetterstedt, Cordilura	830
uta Alexander, Ptychoptera	97
utahense Harmston & Knowlton, Syntormon	516
utahense (Knowlton & Rowe), Simulium	185
utahensis Banks, Nicocles	380
utahensis (Bromley), Efferia	396
utahensis Bromley, Stenopogon	384
utahensis Felt, Rhopalomyia	267
utahensis Fox, Culicoides	130
utahensis Hardy, Bibio	194
utahensis Hardy, Thereva	354
utahensis (Hardy & Knowlton), Tomosvaryella	556
utahensis Harmston & Knowlton, Asyndetus	524
utahensis Harmston & Knowlton, Campsicnemus	527
utahensis Harmston & Knowlton, Dolichopus	497
utahensis Harmston & Knowlton, Hercostomus	498
utahensis Harmston & Knowlton, Paracleius	501
utahensis Harmston & Knowlton, Sympycnus	528
utahensis Harmston & Knowlton, Tachytrechus	503
utahensis Harmston & Knowlton, Teuchophorus	529
utahensis Johnson & Johnson, Excprosopa	445
utahensis Malloch, Chironomus	166
utahensis Maughan, Villa	437
utahensis Quisenberry, Oxyna	665
utahensis (Rowe & Knowlton), Atylotus	331
utahensis (Smith), Macronychia	936
utahensis Stains & Knowlton, Myopa	630
utahensis Wilcox, Laphystia	377
utahicola Alexander, Tipula	39
utahna Harmston, Argyra	526
utahnum Van Duzee, Parasyntormon	516
ute Snyder, Helina	889
utilis Aldrich, Sarcophaga	960
uvens Melander, Bicellaria	451
uxama (Walker), Fannia	895
uxorcula Van Duzee, Dolichopus	497
vacans (Coquillett), Villa	435
vaccinii Osten Sacken, (Cecidomyia)	295
vaccinii (Smith), Dasineura	260
vacua (Fabricius), Volucella	†602
vacua (Fallén), Dexilla	982
vacua Osten Sacken, Brachyopa	593
vafra Alexander, Gonomyia	76
vaga Namba, Rivellia	656
vagabunda (Knab), Tomosvaryella	556
vagabunda (Meigen), Pollenia	931
vagans (Aldrich), Erynnia	1071
vagans (Fallén), Meoneura	729
vagans (Haliday), Leptocera	726
vagans Loew, Diastata	772
vagans Loew, Pelastoneurus	502
vaginalis Fallén, Lonchaea	717
vaginalis Townsend, Ochrocera	1025
vaginifer Melander, Empis	461
vagrans Loew, Pelastoneurus	502
valens Aldrich & Webber, (Zenillia)	1108
valens Cresson, Ditrichophora	739
Valentibulla Foote & Blanc	670
valentis Coquillett, Empis	461
valga Coquillett, Rhamphomyia	466
valga Harris, Chionea	72
valgata (Reinhard), Microcerella	952
valgus (Hardy), Scenopinus	356
valgus Melander, Platypalpus	481
valida (Brauer & Bergenstamm) Theresia	991

	Page
valida (Curran), Oswaldia	1046
valida Curran, Peleteria	998
valida (Harris), Sphecomyiella	658
valida (Loew), Chelifera	471
valida Loew, Hydrellia	744
valida Loew, Tetanocera	695
valida Loew, Tipula	40
valida (Townsend), Microchaetina	984
valida (Townsend), Stomatomyia	1058
valida (Walker), Poecilominettia	703
valida (Walker), Taeniaptera	636
valida (Wiedemann), Ogcodocera	428
valida (Wiedemann), Theresia	991
validicornis Lundbeck, (Sciara)	236
validum Brauer & Bergenstamm, Mochlosoma	987
validus Loew, Bombylius	410
validus Loew, Chrysotus	521
validus Meigen, Merodon	617
validus Wiedemann, Tabanus	332
vallicola Reinhard, Orasturmia	1089
valverdensis Alexander, Limonia	49
vana (Coquillett), Villa	438
vana (Zetterstedt), Spilogona	884
vancouverensis Alexander, Limnophila	64
vancouverensis Alexander, Phalacrocera	42
vancouverensis Parker, Sarcophaga	960
vancouverensis Sherman, (Paratinia)	1116
vandalicum Dyar & Shannon, Simulium	189
vanderwulpi (Schnabl), Helina	886
vanderwulpi (Townsend), Chaetonodexodes	1029
Vanderwulpia Townsend	1034
vanduzeei (Alexander), Dicranota	59
vanduzeei (Alexander), Limonia	49
vanduzeei (Alexander), Protanyderus	90
vanduzeei Alexander, Ulomorpha	68
vanduzeei Cole, Amphicosmus	427
vanduzeei Cole, Thereva	354
vanduzeei Cole, Villa	440
vanduzeei (Cresson), Calliopum	696
vanduzeei Cresson, Coelopa	680
vanduzeei (Cresson), Parydra	750
vanduzeei Curran, Campsicnemus	527
vanduzeei Curran, Dolichopus	497
vanduzeei Curran, Gymnopternus	500
vanduzeei Curran, Medetera	511
vanduzeei Curran, Pipiza	579
vanduzeei Curran, Rhaphium	514
vanduzeei Curran, Syntormon	516
vanduzeei Dyar & Knab, Wyeomyia	107
vanduzeei James, Euparyphus	307
vanduzeei Johnson, Agathomyia	548
vanduzeei Malloch, Pegomya	862
vanduzeei Robinson, Diaphorus	520
vanduzeei Wilcox, Ablautus	364
vanduzeei Wilcox & Martin, Cyrtopogon	369
vanduzeei Wilcox & Martin, Nannocyrtopogon	380
vandykei (Coquillett), Dicranoclista	433
vandykei Van Duzee, Hydrophorus	506
vandykei Wilcox & Martin, Cyrtopogon	369
vandykei Wilcox & Martin, Nannocyrtopogon	380
vanessae (Harris), Euexorista	1100
vaporariorum Haliday, Sphaerocera	719
vapulare Reinhard, Leucostoma	976
vara Loew, Rhamphomyia	466
vara (Staeger), Zygomyia	216
varans Curran, Cyrtopogon	369
varelus (Roback), Tanytarsus	179
varia (Coquillett), Sciomyza	688
varia Coquillett, Senopterina	657
varia (Curran), Nigrobonellia	1006
varia Doane, Tipula	35
varia Jones, Notiphila	746
varia Loew, Parydra	750
varia Meigen, Atherigona	†875
varia (Melander), Phytobia	800
varia (Stein), Helina	889
varia Walker, Leia	228
varia (Walker), Leucophenga	761
varia Walker, (Thereva)	354
varia Wulp. Distichona	†1074, †1077
variabilis (Brues), Anevrina	532
variabilis Chillcott, Fannia	897
variabilis (Coquillett), Lixophaga	1044

INDEX 1691

	Page
variabilis Cresson, Scatophila	758
variabilis (Doane), Paroxyna	666
variabilis Felt, Mycodiplosis	283
variabilis Hanson, Nemotelus	310
variabilis Loew, Bibio	194
variabilis Loew, Cordilura	828
variabilis Loew, Dolichopus	497
variabilis (Loew), Oscinella	781
variabilis Loew, Rivellia	656
variabilis (Staeger), Cricotopus	158
variabilis (Van Duzee), Campsicnemus	527
variabilis Van Duzee, Diaphorus	520
varians Banks, Myopa	630
varians (Johannsen), Bradysia	234
varians Malloch, Lispocephala	877
variata Alexander, Tipula	34
variata (Coquillett), Villa	438
variata Curran, Tachinomyia	1058
variata (Fallén), Hylemya	845
variata (Malloch), Liriomyza	802
variatum Cook, Colobostema	238
variceps (Coquillett), Pseudocalliope	703
variceps (Loew), Thaumatomyia	787
varicolor (Bigot), Anthrax	432
varicolor (Coquillett), Bezzia	141
varicolor Coquillett, Plethochaeta	841
varicolor (Meigen), Hylemya	852
varicorne (Coquillett), Colobostema	238
varicornis Curran, Cordilura	828
varicornis Curran, Microprosopa	835
varicornis Curran, Uclesia	1020
varicornis (Loew), Symphoromyia	344
varicosta (Malloch), Leptocera	724
varicoxa Van Duzee, Nematoproctus	515
variegata Huckett, Pegomya	862
variegata Kincaid, Pericoma	93
variegata (Loew), Anthomyza	819
variegata Loew, Psilocephala	352
variegata Olivier, Oxycera	307
variegata Walker, Atherix	344
variegata Wirth, Mallochohelea	139
variegatum Harmston, Syntormon	516
variegatus Fabricius, Tabanus	336
variegatus (Loew), Sciapus	486
varifrons Coquillett, Agromyza	796
varifrons (Curran), Pilatea	1105
varihalteratus (Malloch), Gaurax	782
variipennis (Coquillett), Culicoides	132
variofacialis Sturtevant & Wheeler, Scatophila	759
variola Garrett, Macrocera	201
varipalpus (Coquillett), Aedes	†114, 115
varipennis Coquillett, Acnemia	222
varipennis (Coquillett), Eucyrtopogon	373
varipennis (Coquillett), Lauterborniella	172
varipennis (Coquillett), Paraterellia	671
varipennis Coquillett, Tachydromia	474
varipennis (Curran), Hilara	457
varipennis Van Duzee, Scellus	506
varipennis Wirth & Williams, Forcipomyia	126
varipes Coquillett, Cricotopus	158
varipes Coquillett, Dolichopus	497
varipes (Coquillett), Leptopsilopa	741
varipes (Coquillett), Phaonia	906
varipes Coquillett, Pseudodinia	708
varipes (Coquillett), Psorophora	110
varipes (Curran), Hylemya	848
varipes Curran, Laphystia	377
varipes Curran, Platycheirus	577
varipes (Dietz), Limonia	43
varipes (Holmgren), Scatophaga	839
varipes (Johnson), Calobatina	635
varipes Kröber, Henicomyia	349
varipes (Loew), Beameromyia	362
varipes Loew, Empis	461
varipes (Loew), Eumetopiella	651
varipes (Loew), Hedriodiscus	315
varipes Loew, Myolepta	589
varipes Loew, Stilpon	476
varipes Malloch, Sphaerocera	718
varipes (Malloch), Triphleba	534
varipes McAtee, Laphria	389
varipes Meigen, Piophila	712
varipes Van Duzee, Chrysotus	523
varipes Van Duzee, Mesorhaga	483
varipes (Walker), Cordilura	829
varipes (Williston), Efferia	396

	Page
varipila Fluke & Hull, Cheilosia	585
variseta Fluke & Hull, Cheilosia	585
variseta (Melander), Tethina	728
varitarsis (Curran), Telmatoscopus	94
varius Coquillett, Aphoebantus	430
varius Cresson, Pipunculus	552
varius Fabricius, Anthrax	†432
varius Fabricius, Bombylius	410
varius Painter & Hall, Poecilanthrax	443
varius (Townes), Stictochironomus	178
varius (Walker), Pelastoneurus	502
varix Melander, Chelifera	471
varus Goetghebuer, Chironomus	168
vasatus Melander, Aphoebantus	430
vasta Coquillett, Toxophora	425
vasta (Coquillett), Villa	439
vates Melander, Hemerodromia	471
vau (Say), Psuedotephritis	647
vecors Osten Sacken, Xylota	607
vegeta (Laffoon), Mycetophila	214
vegetus Harmston, Dolichopus	497
vegetus (Wheeler), Calyxochaetus	528
velata Vockeroth, Geomyza	821
velcidus Hardy, Bibio	194
veles Loew, Medetera	511
Velocia Coquillett	445
Velocia Robineau-Desvoidy	1050
velorum McAtee, Bibio	194
velox Harris, (Phora)	1116
velox Melander, Platypalpus	481
velox Robineau-Desvoidy, (Zelia)	1108
velutina (Bigot), Stonemyia	321
velutina Curran, Stratiomys	311
velutina Johnson, Callomyia	547
velutina Loew, Hilara	457
velutina Loew, Platypeza	550
velutina (Lundbeck), Smittia	163
velutina (Malloch), Spilogona	882
velutina Meigen, Phora	537
velutina Van Duzee, Argyra	526
velutinipes (Brues), Woodiphora	538
velutinus Cresson, Pipunculus	552
velutinus Loew, Chrysopilus	347
velutinus (Ruthe), Mochlonyx	104
velutinus (Williston), Ischyrosyrphus	565
velveta (Doane), Hexatoma	70
venablesi (Curran), Metasyrphus	561
venae Felt, Lobopteromyia	276
venae Stebbins, (Cecidomyia)	295
venalicia (Osten Sacken), Leptocera	723
venata Curran, Medetera	511
venaticus Melander, Platypalpus	481
venator Dyar & Shannon, Simulium	187
venatoria Felt, Feltiella	282
venatoris (Coquillett), Parafischeria	1011
venitalis Felt, Dicrodiplosis	279
venitalis Felt, Lobopteromyia	276
venosa Dietz, Trichocera	16
venosa (Johnson), Dolichopeza	23
venosa Loew, Dixa	102
venosa (Staeger), Lycoriella	232
venosa Wulp, Megaparia	982
venosus (Loew), Nemomydas	359
venteris (Curran), Neocnemodon	581
ventralis (Aldrich), Dexilla	982
ventralis Borgmeier, Megaselia	541
ventralis Coquillett, Ospriocerus	381
ventralis Cresson, Micropeza	634
ventralis (Fallén), Pherbellia	687
ventralis Gerstäcker, Mydas	359
ventralis Loew, Gymnopternus	500
ventralis Say, Lasioptera	272
ventralis Say, Leia	228
ventralis Stein, Hylemya	848
ventralis Williston, Laphria	391
ventricosa (Wulp), Oxysarcodexia	953
ventricosum Loew, Chrysotoxum	583
Ventrops Williston	752
ventrovittis Dyar, Aedes	115
venus Philip, Chrysops	327
venusta (Bergroth), Limonia	50
venusta (Coquillett), Anatopynia	145
venusta (Coquillett), Dohrniphora	535
venusta Curran, Allograpta	569
venusta Johannsen, Phronia	215
venusta (Laffoon), Mycetophila	214

	Page		Page
venusta (Meigen), Geomyza	821	vesiculosa (Felt), Asteromyia	275
venusta Osten Sacken, Erioptera	84	vesiculosa Say, Myopa	630
venusta Reinhard, Phaenopsis	1095	vesparia (Meade), Fannia	897
venusta Snow, Callomyia	547	*vespertilio* Gerstäcker, Pyrgota	657
venustoides Hart, Simulium	187	vespertina (Fallén), Hebecnema	890
venustum Say, Simulium	189	vespertina Osten Sacken, Erioptera	81
venustum Williston, Temnostoma	614	vespertina Vockeroth, Geomyza	821
venustus Hoffman, Culicoides	128	vespiforme (Linnaeus), Temnostoma	614
venustus (Malloch), Nematoproctus	515	*vespiformis* (Gorski), Sphecomyia	†612
venustus (Meigen), Dasysyrphus	563	*vespillo* (Fabricius), Pollenia	931
venustus Melander, Nematoproctus	515	vespoides (Bigot), Heteropogon	374
venustus Osten Sacken, Tabanus	337	*vespoides* (James), Hoplitimyia	313
vera Back, Dioctria	370	vestigipennis Alexander, Dactylolabis	62
vera (Pritchard), Cophura	367	vestigipennis Doane, Tipula	40
verbenae Beutenmüller, (Cecidomyia)	291	**Vestiplex** Bezzi	30
verbenae Felt, Asphondylia	269	vestis Melander, Mythicomyia	419
verbenae Felt, Lasioptera	272	vestita (Walker), Villa	435
verbenae (Hering), Phytobia	800	vestitipennis (Kieffer), Procladius	149
verbenicola Hering, Liriomyza	802	*vestitus* (Melander), Gymnopternus	500
verbenifolia Felt, Lestodiplosis	285	vestitus Walker, Bibio	194
verbesinae Beutenmüller, (Cecidomyia)	295	*veterinus* (Clark), Gasterophilus	916
verbosa (Walker), Criorhina	612	vetitus (Melander), Gymnopternus	500
verecunda (Laffoon), Mycetophila	214	*vetius* (Melander), Gymnopternus	500
verecundum Stone & Jamnback, Simulium	189	vetula Melander, Iteaphila	455
vericauda (Coquillett), Blaesoxipha	946	vetusta Walker, Lepidophora	426
verio Garrett, Cordyla	208	vexans (Curran), Aplomyiopsis	1038
Vermileo Macquart	342	vexans (Meigen), Aedes	116
VERMILEONINAE	342	vexativus Painter & Hall, Poecilanthrax	443
vermontana Alexander, Pilaria	68	vexator Coquillett, Phlebotomus	92
verna (Fabricius), Lispocephala	877	vexatrix (Osten Sacken), Adejeania	999
vernaae Harmston & Knowlton, Dolichopus	497	vialis Osten Sacken, Thereva	354
vernaculus Van Duzee, Gymnopternus	500	*viana* (Walker), Hylemya	850
vernale Shewell, Prosimulium	183	viatica Felt, Contarinia	277
vernalis (Felt), Allarete	245	*Viatica* Garrett	811
vernalis Felt, Dasineura	261	vibei Collin, Fucellia	843
vernalis Felt, Oligotrophus	265	*vibei* Nielsen, Limonia	48
vernalis Felt, Porricondyla	256	vibex Townes, Polypedilum	176
vernalis Huckett, Eremomyia	857	vibrans (Linnaeus), Seioptera	648
vernalis (Malloch), Psectrocladius	159	vibrissata (Brauer & Bergenstamm), Euphorocera	1054
vernalis (Osten Sacken), Pedicia	56		
vernal's (Robineau-Desvoidy), Tachina	998	*vibrissata* (Malloch), Ophiomyia	797
vernalis Sherman, Dziedzickia	218	vibrissatus (Malloch), Dasiops	716
vernata Alexander, Limnophila	65	**Vibrissina** Rondani	1051
vernata West, Wagneria	1020	**Vibrissotheresia** Reinhard	991
vernilis (Reinhard), Sarcodexia	955	viburni (Felt), Dentifibula	276
vernoniae (Beutenmüller), Lasioptera	272	viburni (Felt), Karschomyia	282
vernoniae Felt, Asphondylia	269	viburni Felt, Lasioptera	272
vernoniae Felt, Youngomyia	279	viburni Felt, Schizomyia	267
vernoniae (Loew), Neaspilota	672	viburnicola (Beutenmüller), Neolasioptera	274
vernoniflorae Felt, Lasioptera	272	viburnifolia Felt, Cystiphora	261
verpus Melander, Platypalpus	481	viburnifolia Felt, Sackenomyia	265
verralli Collin, Coenosia	873	vicaria Huckett, Pegomya	862
Verrallia Mik	551	vicaria Walker, Myopa	630
verruca (Osten Sacken), Trishormomyia	288	vicaria Walker, Sepsis	684
verrucicola Osten Sacken, (Cecidomyia)	295	vicarius Walker, Platypalpus	481
versatilis (Curran), Paraterellia	671	*vicarius* (Walker), Tabanus	335
versicolor Chillcott, Rhamphomyia	466	*vicina* (Cresson), Cordilura	830
versicolor (Felt), Polyardis	250	vicina Cresson, Parydra	750
versicolor (Loew), Diplotoxa	792	vicina Dietz, Tipula	29
versicolor Loew, Tipula	33	vicina Hardy & McGuire, Ptiolina	347
versicolor Painter, Geron	423	vicina Johannsen, Sciara	231
versicolor Stein, Phaonia	908	vicina Lintner, Pegomya	†1858, 860
versicolor Van Duzee, Diaphorus	520	*vicina* (Macquart), Labostigmina	313
versipellis (Williston), Chrysosyrphus	592	*vicina* Macquart, Musca	913
versipes Melander, Platypalpus	481	vicina Macquart. Tetanocera	695
versutus Melander, Platypalpus	481	vicina Robineau-Desvoidy, Calliphora	929
versutus Van Duzee, Dolichopus	497	vicina (Walker), Ornithoica	916
vertebrata (Bromley), Efferia	396	vicina (Walker), Psilocephala	352
vertebrata Cole, Dioctria	370	vicina (Wulp), Trupanea	667
vertebrata Harris, Gonia	1076	vicinalis Reinhard, Euptilopareia	1018
vertebrata (Say), Zelia	1028	*vicinus* Aldrich, Paracleius	500
vertebratus (Say), Hedriodiscus	315	vicinus Cresson, Zeros	746
veretebratus (Say), Promachus	400	*vicinus* Macquart, Bombylius	409
vertebratus (Say), Rhagio	345	*vicinus* Macquart), Efferia	394
verticalis Borgmeier, Borophaga	535	vicinus Macquart (Tabanus)	342
verticalis (Loew), Camptoprosopella	697	victoria Aldrich & Webber (Phorocera)	1108
verticalis Malloch, Leucopis	709	victoria Sturtevant, Drosophila	767
verticalis Reinhard, Panacemyia	1046	victoria Townsend (Echinomyia)	999
verticina (Zetterstedt), Hydrophoria	865	victoria Williston, Volucella	602
vesca (Laffoon), Mycetophila	2¹4	vidua Alexander, Pseudolimnophila	62
vescus Hine, Asilus	393	vidua Walker (Atherix)	348
vesicata Huckett, Hylemya	851	vidua Wheeler, Medetera	511
vesiculana Curran, Volucella	602	viduata Alexander, Ormosia	85
vesicularia Curran, Volucella	602	*viduus* Cresson, Pipunculus	552
vesiculata (Townsend), Winthemia	1091	vierecki (Cresson), Anthrax	432
vesiculosa (Fabricius), Volucella	†602	*vierecki* Cresson, Cordilura	828

INDEX 1693

Entry	Page
vierecki Cresson, Heterostylum	411
vierecki Garrett, Docosia	226
vierecki Malloch, Pipunculus	554
vierecki Melander, Platypalpus	481
vietus Van Duzee, Thrypticus	511
vigil Osten Sacken, Scellus	506
vigil (Walker), Wohlfahrtia	942
vigilans Aldrich, Dolichopus	497
vigilans Allen, Senotainia	939
vigilans (Coquillett), Villa	439
vilis Melander, Mythicomyia	419
vilis (Stein), Hylemya	850
Villa Lioy	433
villicrura (Coquillett), Lasiops	904
villihumilis Snyder, Helina	889
villipes Coquillett, Rhamphomyia	466
villipes (Wulp), Metoposarcophaga	951
villosa DeFoliart & Peterson, Cnephia	184
villosa Garrett, Macrocera	201
villosa Osten Sacken, Erioptera	82
villosa Robineau-Desvoidy, Prionella	654
villosipennis Root & Hoffman, Culicoides	130
villosulum Bigot, Chrysotoxum	†582, 583
villosulus (Bigot), Apatolestes	320
villosus Bigot, Merapioidus	612
vinelandi (Curran), Metasyrphus	562
vinetorum (Fabricius), Eristalis	625
vinnipegensis Dyar, Aedes	114
vinnula (Reinhard), Gaediopsis	1080
vinus (Garrett), Heleomyza	816
violacea Meigen, Sepsis	†683, 684
violacea Say, Volucella	601
violaceiventris (Enderlein), Melangyna	567
violaceus (Van Duzee), Gymnopternus	500
violae Reinhard, Sphenometopa	939
violae Townsend, Chaetoglossa	1078
violenta (Walker), Tachinomyia	1058, †1102
violicola (Coquillett), Mayetiola	264
vipio (Osten Sacken), Protanyderus	90
virago Aldrich, Scellus	506
virens Harmston & Knowlton, Parasyntormon	516
virens (Loew), Melanagromyza	796
virescens (Loew), Limonia	49
virescens (Loew), Nephrotoma	22
virga Coquillett, Dolichopus	497
virgata Coquillett, Empis	461
virgata Coquillett, Leptogaster	363
virgata (Coquillett), Oscinella	781
virgata Coquillett, Rhamphomyia	466
virgata (Dietz), Nephrotoma	21
virgata Doane, Gonomyia	76
virgata Melander, Mythicomyia	419
virgata Osten Sacken, Toxophora	425
virgator Melander, Hormopeza	453
virgatum Coquillett, Simulium	187
virgatus (Townes), Stictochironomus	178
virginiana Felt, Mayetiola	264
virginianiae (Felt), Contarinia	277
virginianus (Wirth), Ceratopogon	133
virginianus (Wulp), Efferia	395
virginica (Banks), Laphria	391
virginica Banks, Myopa	630
virginica Banks, Verrallia	551
virginica (Felt), Camptoneuromyia	275
virginica Felt, Lasioptera	272
virginicus Banks, Asilus	393
virginiensis (Drury), Milesia	614
virginiensis Malloch, Puliciphora	545
virginiensis (Macquart), Archytas	1000
virginiensis Van Duzee, Dolichopus	497
virgo Edwards, Abiskomyia	154
virgo Melander, Mythicomyia	419
virgo (Wiedemann), Condylostylus	483
virgo (Wiedemann), Odontomyia	317
virgulatus Bellardi, Chrysops	327
virgultus Theobald, Culex	†118
virida (Hine), Dimecoenia	755
viridaenea Shannon, Xylota	605
viridana (Meigen), Earomyia	716
viridella Sturtevant & Wheeler, Scatophila	759
viridescens Quisenberry, Neaspilota	672
viridescens Robineau-Desvoidy, Calliphora	†929, 929
viridescens Robineau-Desvoidy (Myophora)	961
viridicans (Doane), Limonia	47
viridicincta Harris (Stratiomys)	1116
viridicollis Wulp, Chironomus	166
viridicoxa (Aldrich), Condylostylus	485
viridifacies Van Duzee, Hydrophorus	506
viridifacies Van Duzee, Medetera	511
viridifacies Van Duzee, Thinophilus	508
viridifemora Macquart (Chrysotus)	530
viridiflava Felt, Contarinia	277
viridiflos (Walker), Hydrophorus	506
viridinota Brues, Phora	537
viridis (Bellardi), Labostigmina	314
viridis (Coquillett), Stilobezzia	134
viridis (Fallén), Gymnocheta	†1005
viridis (Frost), Melanagromyza	796
viridis Hine, Nothomyia	317
viridis Parent, Condylostylus	485
viridis (Say), Actina	300
viridis Say, Sargus	302
viridis Townsend, Microdon	599
viridis Van Duzee, Dolichopus	497
viridis Van Duzee, Neurigona	519
viridis (Wiedemann), Orthellia	913
viridiventris Malloch, Tanytarsus	180
viridivittatus (Robinson), Sciapus	486
viridula Alexander, Erioptera	82
viridula (Coquillett), Japanagromyza	796
viridulans Robineau-Desvoidy, Rivellia	656
viridulus (Linnaeus), Chironomus	169
virilis (Aldrich & Webber), Eusisyropa	1093
virilis (Aldrich & Webber), Palpexorista	1056
virilis Sturtevant, Drosophila	766
vitabilis Alexander, Tipula	40
vitalis (Cresson), Pherbellia	687
vitellina Kieffer, Pentaneura	147
viticola Osten Sacken, (Cecidomyia)	295
vitinea (Felt), Neolasioptera	274
vitinea (Felt), Polyardis	250
vitinervis (Thompson), Schizactia	1066
vitiosa Coquillett, Dalmannia	632
vitis Felt, Brachineura	252
vitis Felt, Dasineura	261
vitis (Felt), Rhizomyia	262
vitis Osten Sacken, Lasioptera	272
vitiscoryloides (Packard), Schizomyia	267
vitislituus Osten Sacken, (Cecidomyia)	295
vitispomum (Osten Sacken), Schizomyia	267
vitrea (Coquillett), Epicypta	209
vitrea Hull, Neopachygaster	317
vitrea Wulp, Tipula	29
vitrinervis (Malloch), Triphleba	534
vitripennis (Curran), Cophura	367
vitripennis (Doane), Pedicia	55
vitripennis Loew, Geron	423
vitripennis (Meigen), Cricotopus	158
vitripennis Meigen, Syrphus	559
vitripennis Shannon, Chrysops	328
vittafrons Shannon, Syrphus	559
vittata Cole, Sphegina	594
vittata (Coquillett), Dziedzickia	218
vittata Coquillett, Hexamitocera	841
vittata (Coquillett), Scaptomyza	770
vittata (Coquillett), Stegana	760
vittata Loew, Hemerodromia	471
vittata Loew, Ischnomyia	819
vittata Loew, Notiphila	747
vittata Loew, Rhamphomyia	466
vittata (Macquart), Cochliomyia	924
vittata Meigen, Ctenophora	20
vittata Melander, Drapetis	478
vittata Melander, Limnia	692
vittata Sabrosky, Elachiptera	777
vittata Sherman, (Phthinia)	1116
vittata (Theobald), Aedes	113
vittata (Townes), Phaenopsectra	174
vittata Van Duzee, Medetera	511
vittata (Wiedemann), Specomyia	613
vittatas Van Duzee, Diaphorus	520
vittatifrons Harris, (Syrphus)	1116
vittatipennis Doane, Tipula	40
vittatum Harris, (Asindulum)	1116
vittatum (Meigen), Parallelomma	840
vittatum Zetterstedt, Simulium	187
vittatus Coquillett, Aphoebantus	430
vittatus Curran, Parabombylius	410
vittatus Johnson, Ogcodes	406
vittatus Loew, Dolichopus	497

Name	Page
vittatus Loew, Trigonometopus	705
vittatus Macquart, Eristalis	625
vittatus Macquart, Mydas	359
vittatus (Malloch), Lasiosina	792
vittatus Melander, Platypalpus	481
vittatus Painter, Parabombylius	410
vittatus Sabrosky, Eugaurax	783
vittatus Wiedemann, Chrysops	327
vittatus Wilcox, Callinicus	365
vittifacies (Curran), Melangyna	568
vittiger Melander, Platypalpus	479
vittiger Thomson, Tabanus	337
vittigera Coquillett, Sapromyza	704
vittigera (Coquillett), Zonosemata	675
vittigera (Loew), Strauzia	676
vittigera (Zetterstedt), Pegomya	862
vittipennis Harris, (Tipula)	1116
vittipennis (Walker), Tachypeza	473
vittipes Loew, Cordilura	829
vittipes Sabrosky, Gaurax	782
vittipes (Zetterstedt), Eustalomyia	869
vittisternum Sabrosky, Ogcodes	407
vittiventris Coquillett, Phthiria	420
vittula Loew, Ischnomyia	819
vittula (Loew), Nephrotoma	22
viva (Rapp), Rhabdophaga	258
vivax Fabricius, (Musca)	625
vivax Osten Sacken, Tabanus	337
vivax Williston, Laphria	391
Viviana Rondani	1048
Viviania Rondani	1048
vivida (Harris), Bombyliopsis	992
vivida Williston, Gymnocheta	1005
vivida (Winnertz), Scatopsciara	235
vividus Loew, Chrysotus	523
vix Townsend, Pterodontia	405
vocatoria (Fallén), Chelipoda	472
vockerothi Chillcott, Roederiodes	468
vockerothi Fluke, Metasyrphus	562
vockerothi Martin, Holopogon	375
vockerothi Sabrosky, Pholeomyia	732
volaticus (Williston), Neorhynchocephalus	402
volitans Melander, Tachytrechus	503
Volucella Geoffroy	600
VOLUCELLINI	600
Volucellosia Curran	603
volucricaput (Walker), Microprosopa	835
volucris Johannsen, Cordyla	208
volucris Osten Sacken, Eupeodes	562
volucris (Wulp), Peckia	953
vomitoria (Linnaeus), Calliphora	930
vorax (Bromley), Laphria	391
vorax Harris, (Asilus)	1116
vorax Loew, Tachytrechus	503
Voria Robineau-Desvoidy	1020
VORIINI	1016
vulgaris Aldrich, Cylindromyia	973
vulgaris (Cresson), Parydra	750
vulgaris Curran, Archytas	999
vulgaris (Fallén), Phryxe	1105
vulgaris Fallén, Piophila	711
vulgaris (Fitch), Camptoprosopella	697
vulgaris Fitch, (Molobrus)	236
vulgaris Garrett, Mycomya	221
vulgaris Garrett, Pseudoleria	812
vulgaris Loew, Systoechus	410
vulgaris Loew, Trichonta	216
vulgaris Martin, Beameromyia	362
vulgaris Van Duzee, Chrysotus	523
vulgata (Bergroth), Limonia	48
vulgata (Malloch), Megaselia	541
vulnerata (Loew), Stenopa	671
vulnerata Melander, Mythicomyia	419
vulneratus Melander, Cyrtopogon	369
vulnificus Melander, Platypalpus	481
vulpecula Coquillett, Aphoebantus	430
vulpina Coquillett, Scatophaga	839
vulpina (Coquillett), Villa	438
vulsus Van Duzee, Diaphorus	520
vultur Osten Sacken, Laphria	391
Wagneria Robineau-Desvoidy	1020, †1033
Wahlgrenia Ringdahl	908
waldorfi (Felt), Asteromyia	275
walkeri Curtis, Leptomorphus	223
walkeri Garrett, Spanoparea	813
walkeri Theobald, Anopheles	106
walkeri Van Duzee, Dolichopus	497
walleyi (Alexander), Dolichopeza	23
walleyi Alexander, Limnophila	64
walleyi Alexander, Limonia	49
walleyi Brooks, Phytomyptera	1066
walleyi Townes, Polypedilum	176
walshii Felt, Mayetiola	264
Walshomyia Felt	265
waltoni (Malloch), Phytobia	798
warmkei Wirth, Atrichopogon	123
warneri Alexander, Tipula	26
wasatchensis Alexander, Ormosia	84
wasatchensis Harmston & Knowlton, Hercostomus	498
washingtona (Melander), Hemeromyia	728
washingtona Melander, Psila	639
washingtonae (Coquillett), Admontia	1068
washingtonensis Johnson, Contarinia	277
washingtoniana (Bigot), Gibsonomyia	1023
Wasmanniella Kieffer	245
watsoni Curran, Lonchaea	717
watsoni Felt, Ctenodactylomyia	262
waughi Curran, Chionea	72
webberi Johnson, Villa	436
webberi Smith, Pseudotachinomyia	1057
webbii Doane, Tephritis	668
weborgi (Fluke), Epistrophe	565
websteri (Aldrich), Blaesoxipha	947
websteri (Coquillett), Atrichopogon	123
websteri Felt, Asphondylia	269
websteri (Townsend), Lespesia	1101
Websteriana Walton	985
weedii (Coquillett), Rhamphomyia	466
weedii Townsend, Hyalomyodes	977
weemsi Sabrosky, Leptocera	722
weesei Khalaf, Culicoides	130
weidhausii Crafts, (Phytomyza)	1116
welchi (Hall), Sarcodexia	956
weldi Felt, Lasioptera	272
weldi Felt, Rhopalomyia	267
weldoni Jones, Eupeodes	562
wellsi (Felt), Mayetiola	264
wellsi Felt, Porricondyla	256
werneri Alexander, Atarba	71
werneri Alexander, Tipula	34
weslacensis (Bromley), Apachekolos	361
westonia (Webber), Lespesia	1102
westwoodi Osten Sacken, Elephantomyia	71
westwoodi Sabrosky, Pterodontia	405
wetmorei Alexander, Orimarga	52
wetmorei (Malloch), Pseudogriphoneura	704
whartoni (Needham), Limonia	49
wheeleri Brues, Apocephalus	543
wheeleri Brues, Ecitomyia	545
wheeleri Curran, Trupanea	667
wheeleri Foote, Coulson, and Robinson, Medetera	511
wheeleri Hackman, Scaptomyza	770
wheeleri (Hough), Paralucilia	924
wheeleri (Long), Forcipomyia	126
wheeleri Melander, Hilara	457
wheeleri Melander, Nemotelus	309
wheeleri Melander, Pelastoneurus	502
wheeleri (Melander & Brues), Dolichopus	497
wheeleri Patterson & Alexander, Drosophila	766
wheeleri Sabrosky, Leiomyza	823
wheeleri Spuler, Leptocera	722
wheeleri (Sturtevant), Periscelis	710
wheeleri Van Duzee, Campsicnemus	527
wheeleri Van Duzee, Rhaphium	514
wheeleri West, Ateloglossa	986
whitmani (Melander), Tethina	727
whitneyi Alexander, Tipula	34
whitneyi (Johnson), Merycomyia	322
Whitneyomyia Bequaert	331
whittakeri (James), Scenopinus	356
wickhami Townsend, Arctophyto	986
wiedemanni (Johnson), Metasyrphus	560
wiedemanni Kröber, Chrysops	327
wiedemanni Osten Sacken, Tabanus	337
wiedemanni Townsend, Lonchaea	717
Wiedemannia Zetterstedt	469
wierzejskii (Mik), Hydrophoria	865
wilburi Cresson, Hydrellia	744
wilburi (Hardy), Tomosvaryella	556

	Page		Page
wilburi Sabrosky, Dicraeus	783	wirthi Wheeler, Chymomyza	770
wilcoxi (Bromley), Efferia	396	wirthi Williams, Alluaudomyia	134
wilcoxi Bromley, Proctacanthella	399	**Wirthia** Arnaud	507
wilcoxi Bromley, Stenopogon	384	**Wirthiella** Sublette	171
wilcoxi Curran, Anorostoma	813	*wisconensis* Wheeler, Chrysotus	523
wilcoxi (Curran), Gimnomera	840	wisconsinensis Fluke & Hull, Cheilosia	587
wilcoxi James, Cyrtopogon	369	wisconsinensis Jones, Culicoides	131
wilcoxi James, Heteropogon	374	wisconsinensis Wheeler, Chrysotus	523
wilcoxi Martin, Holopogon	375	**Wohlfahrtia** Brauer & Bergenstamm	942
wilcoxi Painter, Villa	436	*Wohlfahrtiopsis* Townsend	957
wilcoxi Pritchard, Cerotainiops	388	wolcotti (Swenk), Lynchia	919
wilcoxi (Pritchard), Cophura	367	*wolffi* (Cresson), Euaresta	665
Wilcoxia James	385	woodgatei Alexander, Limnophila	65
wildermuthii Walton, Zelia	1028	woodi Alexander, Tipula	40
wileyae Philip, Chrysops	327	*Woodia* Malloch	533
willamettensis Alexander, Limonia	49	**Woodiphora** Schmitz	538
willametti Cole & Wilcox, Lasiopogon	378	wrangeliensis Parker, Sarcophaga	960
williami (Shaw), Orfelia	203	wrighti (Whitney), Hybomitra	339
williamsae Alexander, Blepharicera	99	*Wulpia* Brauer & Bergenstamm	1031
williamsiana Alexander, Tipula	30	wulpiana (Bergroth), Nephrotoma	22
williamsii Doane, Tipula	40	wyalusingensis (Dietz), Nephrotoma	22
willingii (Smith), Arctosyrphus	620	**Wyeomyia** Theobald	107
willissmithi Alexander, Tipula	40	wyliei Martin, Parataracticus	382
willistoni Banks, Myopa	631	wyomingense Stone & DeFoliart, Simulium	186
willistoni Brennan, Apatolestes	320	wyomingensis James, Stratiomys	312
willistoni Cockerell, Lasioptera	272		
willistoni Cole, Dialineura	352	**Xanionotum** Brues	545
willistoni Coquillett, Euxesta	652	xantha (Roback), Micropsectra	179
willistoni (Cresson), Leptopsilopa	741	**Xanthaciura** Hendel	670
willistoni Cresson, Lytogaster	752	**Xanthandrus** Verrall	574
willistoni Curran, Chrysotoxum	583	xanthas Philip, Chrysops	325
willistoni Curran, Cyrtopogon	370	*Xanthempis* Bezzi	457
willistoni Curran, Protodejeania	1002	xanthii Felt, Asphondylia	269
willistoni (Day), Hedriodiscus	315	xanthina Painter, Exoprosopa	445
willistoni Giles, Culex	119	xanthina Painter, Villa	439
willistoni Hine, Asilus	393	*xanthippe* Williston, Laphria	389
willistoni (Hine), Efferia	396	*Xanthoactia* Townsend	1061
willistoni (Kahl), Ceriana	616	xanthobasis James, Thereva	354
willistoni Laffoon, Platyura	204	xanthocal Harmston & Knowlton, Chrysotus	523
willistoni Osten Sacken, Pantarbes	414	*xanthocauda* Curran, Spilomyia	613
willistoni (Philip), Haematopota	330	xanthocera Collin, Odinia	794
willistoni Sabrosky, Elachiptera	777	*xanthocera* (Loew), Diclasiopa	739
willistoni Sabrosky, Lasiopleura	774	**Xanthocera** Townsend	1026
willistoni Sedman, Melanostoma	575	xanthochiton Melander, Platypalpus	481
willistoni Snow, Cheilosia	588	XANTHOCHLORINAE	526
willistoni Sturtevant, Drosophila	769	**Xanthochlorus** Loew	529
willistoni (Wheeler), Thrypticus	511	xanthocnemus Loew, Dolichopus	497
Willistonia Brauer & Bergenstamm	1081	**Xanthoepalpus** Townsend	1003
willistonii Aldrich, Dolichopus	495	*Xanthoernestia* Townsend	1009
willistonii (Coquillett), Poecilanthrax	443	xanthogaster (Kieffer), Probezzia	138
willistonii (Snow), Chamaesyrphus	595	**Xanthogramma** Schiner	569
Willistonina Back	386	*xanthomela* (Dietz), Tipula	40
wilsoni Blanc & Foote, Neaspilota	672	*Xanthomelana* Wulp	966
wilsoni Pechuman, Tabanus	337	**Xanthomelanodes** Townsend	966
wilsoni (Walsh), Fannia	897	**Xanthomyia** Phillips	663
wilsonii (Osten Sacken), Hexatoma	70	*zanthopareus* (Williston), Physoconops	626
winburni Painter, Geron	423	*zanthopennis* Wilcox & Martin, Dioctria	370
winifredae Namba, Rivellia	656	**Xanthophyto** Townsend	1009
winnemana Alexander, Dicranoptycha	52	xanthopilis Townsend, Microdon	599
winnemana Alexander, Tipula	36	xanthopoda Melander & Spuler, Piophila	711
winnemana Malloch, Hylemya	851	xanthopoda Williston, Drapetis	478
winnemana Malloch, Leptocera	723	xanthopodus Melander, Platypalpus	481
winnemana Malloch, Mydaea	892	xanthopterus Low, Mydas	359
winnemana Malloch, Plastophora	542	xanthopus Hardy, Chrysopilus	347
winnemana McAtee, Laphria	391	*zanthopus* Melander, Platypalpus	481
winnemanae Malloch, Lonchaea	717	xanthopus Wiedemann, Bibio	194
winnemanae (Malloch), Melanagromyza	796	*zanthopus* Wiedemann, Sargus	302
winnemanae Malloch, Phaonia	908	xanthosoma Alexander, Dicranota	59
winnemannae Malloch, Pipunculus	554	xanthostigma Dietz, Tipula	29
winnertzi (Tarwid), Asindulum	199	xanthostigma (Loew), Nephrotoma	22
Winnertzia Rondani	252	*zanthostoma* Melander & Spuler, Piophila	711
winthemi Kröber, Tabanus	335	xanthostoma (Walker), Piophila	713
winthemi (Meigen), Pegomya	†860, 862	xanthostoma (Williston), Epistrophe	565
winthemi Zetterstedt, Tachypeza	473	*xanthostomus* (Williston), Physoconops	626
Winthemia Robineau-Desvoidy	1089	xanthura Aldrich & Webber (Phorocera)	1107
winthemii Lehmann, Leia	228	xanthus Alexander, Molophilus	89
WINTHEMIINI	1088	xanthus Melander, Oreogeton	454
wirthi Chillcott, Roederiodes	468	*zantippe* (Banks), Plastophora	542
wirthi Cook, Parascatopse	230	xavia Dyar & Shannon, Dixa	101
wirthi Foote & Pratt, Culicoides	130	xena Roback, Corynoneura	164
wirthi Harmston, Aphrosylus	509	*Xenadmontia* Townsend	1068
wirthi (Laffoon), Mycetophila	214	*Xenaricia* Malloch	890
wirthi Quate, Pericoma	93	**Xenochaeta** Snow	664
wirthi Quate, Trichomyia	91	**Xenochaetina** Malloch	705
wirthi Sabrosky, Meoneura	729	**Xenochironomus** Kieffer	171
wirthi Saunders, Forcipomyia	125	*Xenocoenosia* Malloch	873

Name	Page
xenolabis Kieffer, Chironomus	171
Xenoleucopis Malloch	709
Xenomydaea Malloch	892
Xenophorbia Malloch	854
Xenophytomyza Frey	801
Xenoppia Townsend	951
Xenopterella Malloch	705
xeres Curran, Euxesta	652
xerophila Hardy, Tomosvaryella	556
xerophila Wheeler, Medetera	511
Xestoprosopa Hull	574
xethis (Roback), Orthocladius	157
Xiphandrium Loew	512
xipheres (Wheeler), Rhaphium	514
Xiphomyia Townsend	1051
Xiphura Brullé	19
xuthogaster Schlinger, Ocnaea	404
Xylomya Rondani	299
XYLOMYIDAE	298
XYLOPHAGIDAE	296
XYLOPHAGINAE	296
Xylophagus Meigen	297
xylophila Edwards, Cordylomyia	249
Xylopriona Kieffer	249
xylota (Curran), Aplomyiopsis	1038
Xylota Meigen	604
Xylotodes Shannon	607
xylotoides (Johnson), Ischyrosyrphus	565
Xylotomima Shannon	606
Xyraeomyia Frick	803
Yahuartachina Townsend	1056
yakimensis Melander, Microphorus	453
yalensis Sellers, Carcelia	1092
Yamatotipula Matsumura	27
yaqui Painter, Villa	436
yellowstonei (Cole), Poecilanthrax	443
yellowstonensis (Alexander), Cheilotrichia	79
yellowstonensis Alexander, Limonia	50
yellowstonensis Alexander, Tipula	34
Yetodesia Rondani	886
yohoensis (Alexander), Gonomyodes	79
yonahlossee Alexander, Dicranota	56
yorkii Parker, Sarcophaga	960
yosemite Alexander, Tipula	40
yosemite Cresson, Allotrichoma	736
yosemite Cresson, Eclimus	427
yosemite (Osten Sacken), Philorus	100
youngi Alexander, Tipula	40
youngi (Malloch), Tylomyza	797
Youngomyia Felt	279
Ypophaemyia Townsend	1101
ypsilon Alexander, Limonia	49
ypsilon Foote, Paraterellia	671
ypsilon Johannsen, Leptomorphus	223
ypsilon Williston, Chrysotoxum	583
yuccae Felt, Dasineura	261
yuccae Felt, Lestodiplosis	285
yukonensis Alexander, Erioptera	82
yukonensis Cole & Wilcox, Lasiopogon	378
yukonensis (Cresson), Parydra	750
yukonensis Hoffman, Culicoides	128
yukonensis Shannon, Cheilosia	585
yukonensis (Tothill), Onychogonia	1077
yulenus Philip, Tabanus	335
Zabrachia Coquillett	318
Zabrops Hull	386
Zacompsia Coquillett	654
Zagonia Coquillett	818
zamon (Townsend), Eccritosia	393
zauschneriae Felt, Thecodiplosis	278
Zavrelia Kieffer	180
Zavrelimyia Fittkau	146
zeae (Riley), Hylemya	850
zeas (Riley), Hylemya	850
Zelia Robineau-Desvoidy	1027
ZELIINI	1027
Zelima Meigen	604
Zelmira Meigen	202
zelotypa Alexander, Tipula	40
Zenillia Robineau-Desvoidy	†1097, 1107
zephyria Snow, Rhagoletis	675
Zeros Cresson	746
zetterstedti (Loew), Suillia	†809
zetterstedtii (Ringdahl), Hydrophoria	865
zetterstedtii (Ringdahl), Spilogona	884
zetterstedtii Rondani, Azelia	899
Zeuxidiplosis Kieffer	278
Zeuximyia Philip	323
ziczac Harris, (Leia)	1116
zigiae Felt, Lasioptera	272
zinzalus Philip, Chrysops	328
zionana Alexander, Limonia	49
Zionea Hardy	349
zionensis Alexander, Blepharicera	99
zionensis Alexander, Orimarga	52
zionensis Alexander, Phyllolabis	60
zionensis Harmston & Knowlton, Neurigona	519
zionensis Johnson & Johnson, Poecilanthrax	443
zionicola Alexander, Gonomyia	74
ziziae Felt, Lasioptera	272
Zizyphomyia Townsend	1088
Zodion Latreille	628
zonalis Curran, Uclesia	1020
zonalis (Kirby), Hybomitra	341
zonata (Coquillett), Zelia	1028
zonata Hardy & McGuire, Ptiolina	347
zonata (Hine), Efferia	396
zonatus Bigot, Eristalis	625
zonatus Cole & Wilcox, Lasiopogon	378
zonopterus (Mitchell), Stenochironomus	177
Zonosema Loew	674
Zonosemata Benjamin	675
zonus Coquillett, Lordotus	412
zoosophus Dyar & Knab, Aedes	115
zuelaniae Stone, Anastrepha	673
zukeli Alexander, Erioptera	84
Zygastropyga Townsend	950
Zygomyia Winnertz	216
Zygoneura Meigen	231
Zygosturmia Townsend	1084
zygota (Philip), Hybomitra	341
zythicolor Philip, Tabanus	337